UNIVERSE

ELEVENTH
EDITION

UNIVERSE

ELEVENTH EDITION

Robert M. Geller
University of California, Santa Barbara

Roger A. Freedman
University of California, Santa Barbara

William J. Kaufmann III
San Diego State University

w.h.freeman
Macmillan Learning

New York

Vice President, STEM: Daryl Fox

Program Director: Brooke Suchomel

Program Manager: Lori Stover

Developmental Editor: Meg Rosenburg

Associate Editor: Kevin Davidson

Marketing Manager: Shannon Howard

Marketing Assistant: Savannah DiMarco

Director of Digital Production: Keri deManigold

Senior Media and Supplements Editor: Victoria Garvey

Director, Content Management Enhancement: Tracey Kuehn

Senior Managing Editor: Lisa Kinne

Senior Content Project Manager: Vivien Weiss

Permissions Manager: Jennifer MacMillan

Photo Researcher: Krystyna Borgen, Lumina Datamatics, Inc.

Director of Design, Content Management: Diana Blume

Design Services Manager: Natasha A. S. Wolfe

Cover Design Manager: John Callahan

Text Designer: Lumina Datamatics, Inc.

Senior Workflow Supervisor: Paul W. Rohloff

Production Supervisor: José Olivera

Art Manager: Matthew McAdams

Composition: Lumina Datamatics, Inc.

Printing and Binding: LSC Communications

Cover image: Andrew Studer Photography

Library of Congress Control Number: 2018957815

ISBN-13: 978-1-319-03944-8
ISBN-10: 1-319-03944-8

Printed in the United States of America
1 2 3 4 5 6   23 22 21 20 19 18

W. H. Freeman and Company
One New York Plaza
Suite 4500
New York, NY
10004-1562
www.macmillanlearning.com

About the Authors

Max Lurie

Robert M. Geller teaches and conducts research in astrophysics at the University of California, Santa Barbara (UCSB), where he obtained his PhD. His doctoral research was in observational cosmology under Professor Robert Antonucci, and they are still involved in a long-term effort to detect "cosmological halos." He also leads a collaboration of scientists and physicians exploring the use of magnetotactic bacteria to enhance the effectiveness of both radiation and chemotherapy in treating cancer. Dr. Geller places a strong emphasis on education, and he received the Distinguished Teaching Award at UCSB in 2003. His favorite outdoor activity is rock climbing, even developing new routes in the Santa Barbara mountains; in his garage, he built an unusual telescope with lenses made of water and a system to control the 3D motion of magnetotactic bacteria with either a joystick or pre-programmed trajectories.

Roger A. Freedman is on the faculty of the Department of Physics at the University of California, Santa Barbara. He grew up in San Diego, California, and was an undergraduate at the University of California campuses in San Diego and Los Angeles. He did his doctoral research in nuclear theory and its astrophysical applications at Stanford University under the direction of Professor J. Dirk Walecka. Dr. Freedman joined the faculty at UCSB in 1981 after three years of teaching and doing research at the University of Washington. Dr. Freedman holds a commercial pilot's license, and when not teaching or writing he can frequently be found flying with his wife, Caroline. He has flown across the United States and Canada.

Caroline J. Robillard

William J. Kaufmann III was the author of the first four editions of *Universe*. Dr. Kaufmann earned his bachelor's degree magna cum laude in physics from Adelphi University in 1963, a master's degree in physics from Rutgers in 1965, and a PhD in astrophysics from Indiana University in 1968. At 27, he became the youngest director of any major planetarium in the United States when he took the helm of the Griffith Observatory in Los Angeles. During his career, he also held positions at San Diego State University, UCLA, Caltech, and the University of Illinois. A prolific author, his many books include *Black Holes and Warped Spacetime*, *Relativity and Cosmology*, *Planets and Moons*, *Stars and Nebulas*, and *Galaxies and Quasars*. Dr. Kaufmann died in 1994.

Contents Overview

Contents

Explore Scientific Reasoning by Exploring the Universe

One of the oldest fields of scientific discovery, the discipline of astronomy encompasses endless mysteries to unravel. Recent missions and advancements that extend beyond the edge of our solar system have raised more questions than they have answered. The eleventh edition of *Universe* invites students into the study of astronomy, encouraging them to ask their own questions.

To accomplish this, *Universe* makes a point of introducing transparency to scientific discoveries by detailing disproven theories and explaining the research that excluded them. Rather than simply presenting facts, *Universe* allows students to explore scientific reasoning, while also guiding them toward an understanding of the key foundational concepts of astronomy. Available with the Sapling Plus homework system, this edition provides the latest science updates, new pedagogical support, and continued refinement of the content from previous editions.

NEW! Visual Key Ideas

The updated **Key Ideas** section at the end of each chapter now includes the most important figures from the chapter, to serve as visual guideposts for the chapters and to provide students with the opportunity to review and assess what they have learned.

FIGURE 12-4 R I V U X G

Juno Spacecraft and a Whole New View of Jupiter Juno produces the highest resolution images of Jupiter ever taken, and the spacecraft orbits the poles to give views of Jupiter never seen before. This is an enhanced-color view, looking directly down at Jupiter's south pole (appearing blue). Cyclones (or hurricanes) were not expected at Jupiter's poles, but are seen here as circular or oval in shape and appear white, or a whiter shade within blue. The cyclones range in size, but can reach widths of 1000 km or so. Only longer-term observations will tell us if these cyclones are a persistent feature, or if they come and go. (NASA/JPL-Caltech/SwRI/MSSS/Betsy Asher Hall/Gervasio Robles)

Updated Science

Chapter 3: New photos of the 2017 total solar eclipse

Chapter 8: Includes a revised explanation of the formation of the solar system to reflect our current understanding; new techniques for detecting exoplanets and determining habitability

Chapter 9: Enhanced emphasis on climate change

Chapter 11: Updated to reflect Mars rover discoveries and discovery of water on Mars; includes information on upcoming 2020 Mars rover

Chapter 12: New photos from the Juno mission

Chapter 13: Updated information on proposed missions to outer planets

Chapter 14: New images of Pluto and new understanding of its surface

Chapter 15: Rosetta and Dawn missions, new Ceres (asteroid) and 67P comet photos

Chapter 21: Consolidated and clarified information on neutron stars and stellar remnants; added most recent LIGO updates for gravitational waves; updated to our current understanding of heavy element production

Chapter 27: Enhanced discussion of exoplanets and habitability; added new targets for astrobiology research

Cosmic Connections figures summarize key ideas visually

The *Cosmic Connections* feature presents particularly challenging or central concepts as full-page figures, to accentuate their importance and help students synthesize information and draw connections between varying applications of the same principles.

COSMIC CONNECTIONS
Universal Gravitation

Gravity is one of the fundamental forces of nature. We can see its effects here on Earth as well as in the farthest regions of the observable universe.

The weight of an ordinary object is just the gravitational force exerted on that object by Earth.

Earth is held together by the mutual gravitational attraction of its parts.

Milky Way Galaxy

Sun

26,000 light-years

Sun

The mutual gravitational attraction of all the matter in the Milky Way Galaxy holds it together. The gravitational force of the Galaxy on our Sun and solar system holds us in an immense orbit around the galactic center.

Gravitational forces exerted by the Sun keep the planets in their orbits. The farther a planet is from the Sun, the weaker the gravitational force that acts on the planet.

100

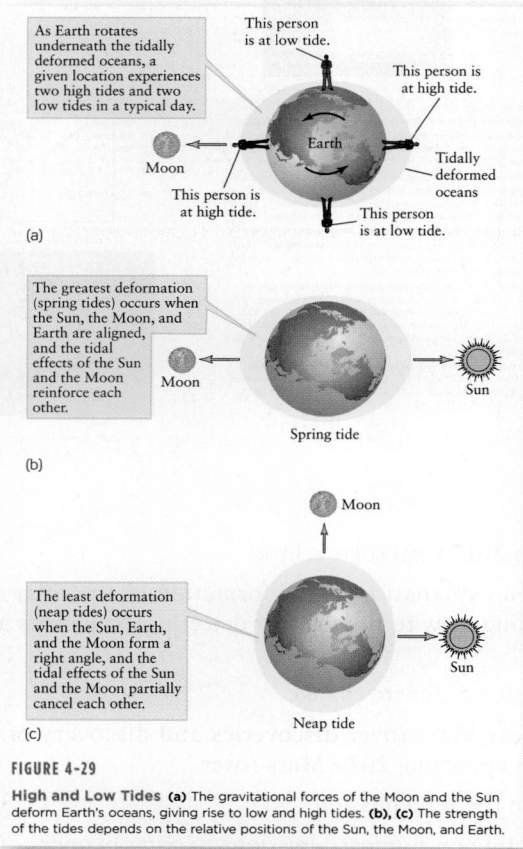

As Earth rotates underneath the tidally deformed oceans, a given location experiences two high tides and two low tides in a typical day.

This person is at low tide.

This person is at high tide.

Earth

Moon

Tidally deformed oceans

This person is at high tide.

This person is at low tide.

(a)

The greatest deformation (spring tides) occurs when the Sun, the Moon, and Earth are aligned, and the tidal effects of the Sun and the Moon reinforce each other.

Moon

Sun

Spring tide

(b)

Moon

The least deformation (neap tides) occurs when the Sun, Earth, and the Moon form a right angle, and the tidal effects of the Sun and the Moon partially cancel each other.

Sun

Neap tide

(c)

FIGURE 4-29

High and Low Tides **(a)** The gravitational forces of the Moon and the Sun deform Earth's oceans, giving rise to low and high tides. **(b), (c)** The strength of the tides depends on the relative positions of the Sun, the Moon, and Earth.

Explanatory Art helps students identify key concepts in figures

Many figures in the text utilize balloon captions to help students interpret complex figures. Select photos also contain brief labels that point out the most important concepts.

Fold-Out Photos highlight some of astronomy's most majestic sights

Throughout the text are special gatefold photo spreads, which highlight awe-inspiring astronomy photography in full, poster-worthy detail.

NASA SDO

Guest Essays and News Articles showcase new perspectives on astronomy

Featured at the end of 13 chapters are guest essays from celebrated scientists or articles from prominent science news organizations that present unique perspectives on modern astronomy research. *Universe*, Eleventh Edition, includes updated essays, as well as five brand new essays on topics ranging from supernovae and gravitational waves to climate change.

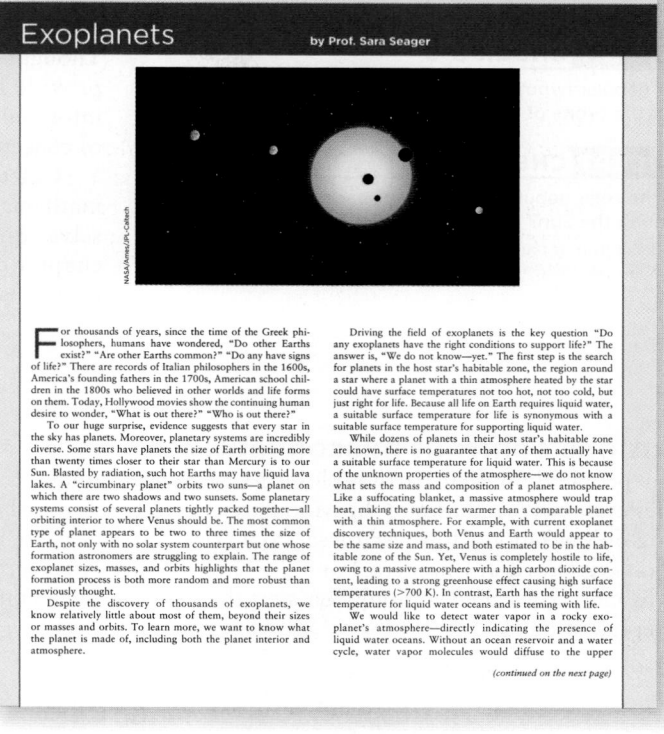

Exoplanets
by Prof. Sara Seager

For thousands of years, since the time of the Greek philosophers, humans have wondered, "Do other Earths exist?" "Are other Earths common?" "Do any have signs of life?" There are records of Italian philosophers in the 1600s, America's founding fathers in the 1700s, American school children in the 1800s who believed in other worlds and life forms on them. Today, Hollywood movies show the continuing human desire to wonder, "What is out there?" "Who is out there?"

To our huge surprise, evidence suggests that every star in the sky has planets. Moreover, planetary systems are incredibly diverse. Some stars have planets the size of Earth orbiting more than twenty times closer to their star than Mercury is to our Sun. Blasted by radiation, such hot Earths may have liquid lava lakes. A "circumbinary planet" orbits two suns—a planet on which there are two shadows and two sunsets. Some planetary systems consist of several planets tightly packed together—all orbiting interior to where Venus should be. The most common type of planet appears to be two to three times the size of Earth, not only with no solar system counterpart but one whose formation astronomers are struggling to explain. The range of exoplanet sizes, masses, and orbits highlights that the planet formation process is both more random and more robust than previously thought.

Despite the discovery of thousands of exoplanets, we know relatively little about most of them, beyond their sizes or masses and orbits. To learn more, we want to know what the planet is made of, including both the planet interior and atmosphere.

Driving the field of exoplanets is the key question "Do any exoplanets have the right conditions to support life?" The answer is, "We do not know—yet." The first step is the search for planets in the host star's habitable zone, the region around a star where a planet with a thin atmosphere heated by the star could have surface temperatures not too hot, not too cold, but just right for life. Because all life on Earth requires liquid water, a suitable surface temperature for life is synonymous with a suitable surface temperature for supporting liquid water.

While dozens of planets in their host star's habitable zone are known, there is no guarantee that any of them actually have a suitable surface temperature for liquid water. This is because of the unknown properties of the atmosphere—we do not know what sets the mass and composition of a planet atmosphere. Like a suffocating blanket, a massive atmosphere would trap heat, making the surface far warmer than a comparable planet with a thin atmosphere. For example, with current exoplanet discovery techniques, both Venus and Earth would appear to be the same size and mass, and both estimated to be in the habitable zone of the Sun. Yet, Venus is completely hostile to life, owing to a massive atmosphere with a high carbon dioxide content, leading to a strong greenhouse effect causing high surface temperatures (>700 K). In contrast, Earth has the right surface temperature for liquid water oceans and is teeming with life.

We would like to detect water vapor in a rocky exoplanet's atmosphere—directly indicating the presence of liquid water oceans. Without an ocean reservoir and a water cycle, water vapor molecules would diffuse to the upper

(continued on the next page)

LEARNING GOALS

By reading the sections of this chapter, you will learn

8-1 The key characteristics of the solar system that must be explained by any theory of its origins

8-2 How the abundances of chemical elements in the solar system and beyond explain the sizes of the planets

8-3 How we can determine the age of the solar system by measuring abundances of radioactive elements

8-4 Why scientists think the Sun and planets all formed from a cloud called the solar nebula

8-5 How the solar nebula model explains the formation of the terrestrial planets

8-6 How the jovian planets formed and migrated in the early solar system

8-7 How astronomers search for planets around other stars

Learning Goals keep students on track

Learning Goals for each section, explicitly laid out at the beginning of each chapter, make it easier to plan lectures and provide students with a checklist of what they should expect to learn.

Motivation Statements connect the details to the broader picture

Free-standing motivation statements introduce the main theme of paragraphs or sections within the text and help students connect to the big-picture ideas of astronomy.

4-3 Tycho Brahe's astronomical observations disproved ancient ideas about the heavens

On November 11, 1572, a bright star suddenly appeared in the constellation Cassiopeia. At first, it was even brighter than Venus, but then it began to grow dim. After 18 months, it had faded from view. Modern astronomers recognize this event as a supernova explosion, the violent death of a massive star.

In the sixteenth century, however, the vast majority of scholars held with the ancient teachings of Aristotle and Plato, who had argued that the heavens are permanent and unalterable. Consequently, the "new star" of 1572 could not really be a star at all, because the heavens do not change; it must instead be some sort of bright object quite near Earth, perhaps not much farther away than the clouds overhead.

A supernova explosion and a comet revealed to Tycho that our universe is more dynamic than had been imagined.

Concept Check and Calculation Check Questions make reading active

Thought-provoking **Concept Check** questions at the end of each section provide immediate assessment, leading students to draw conclusions informed by the text and to demonstrate applied thinking and synthesis of concepts.

Calculation Check questions appear only in sections that present mathematical reasoning, offering students the opportunity to test themselves quantitatively by solving mathematical problems related to the chapter concepts. Answers to Concept Check and Calculation Check questions are included at the end of each chapter.

CONCEPT CHECK 8-5

If the nebular hypothesis is correct, what must it explain about the orbits of the planets?

CONCEPT CHECK 8-6

Was the solar nebula cold until nuclear reactions began powering the Sun?

Answers appear at the end of the chapter.

Cautions confront misconceptions

Many people think that Earth is closer to the Sun in summer than in winter and that the phases of the Moon are caused by Earth's shadow falling on the Moon. However, these "common-sense" ideas are incorrect (as explained in Chapters 2 and 3). Throughout *Universe*, paragraphs marked by the **Caution!** icon alert the reader to such common conceptual pitfalls.

CAUTION!

Figure 8-4 shows that there is about 10 times more hydrogen than helium in our galaxy when comparing the *number of atoms*. Figure 8-1 shows that our solar system is made of 71% hydrogen versus 27% helium when comparing *mass*. The explanation of this seeming inconsistency lies in the difference between comparing the mass of atoms versus their numbers: Each helium atom has about 4 times the mass of a hydrogen atom, which makes helium's contribution to total mass larger than its contribution to the total number of atoms.

ANALOGY

Figure skaters make use of the conservation of angular momentum. When a spinning skater pulls her arms and legs in close to her body, the rate at which she spins automatically increases (**Figure 8-7**). Even if you are not a figure skater, you can demonstrate this by sitting on a rotating office chair. Sit with your arms outstretched and hold a weight, like a brick or a full water bottle, in either hand. Now use your feet to start your body and the chair rotating, lift your feet off the ground, and then pull your arms inward. Your rotation will speed up quite noticeably.

Analogies relate astronomy concepts to everyday experiences

Throughout *Universe*, **Analogy** paragraphs connect astronomical ideas with more familiar experiences on Earth to help students relate and develop a physical intuition for the relevant phenomena.

Tools of the Astronomer's Trade and STAR guide student problem solving

Tools of the Astronomer's Trade boxes present in-depth and more quantitative discussions of the tools that researchers use to acquire and interpret data. All the worked examples in *Universe* are contained within Tools of the Astronomer's Trade boxes and follow a logical and consistent sequence of steps called **STAR**: assess the Situation, select the Tools, find the Answer, and Review the answer. This framework helps students divide a big problem into smaller steps. Feedback from instructors indicates that when students follow these steps in their own work, they more rapidly develop important problem-solving skills.

BOX 10-1 TOOLS OF THE ASTRONOMER'S TRADE

Calculating Tidal Forces

The gravitational force between any two objects decreases with increasing distance between the objects. This principle leads to a simple formula for estimating the tidal force that Earth exerts on parts of the Moon.

Consider two small objects, each of mass m, on opposite sides of the Moon. Because one object is farther from Earth's center than the other, Earth exerts different forces F_{near} and F_{far} on them (see the accompanying illustration). The closer object is pulled harder toward Earth than the more distant one; as a result, they are pulled away from each other and away from the center of the Moon (see Section 4-8, especially Figure 4-27). The *tidal force* on these two objects, which is the force that pulls them apart, is the difference between the forces acting on the individual objects:

$$F_{tidal} = F_{near} - F_{far}$$

Both F_{near} and F_{far} are inversely proportional to the square of the distance from Earth (see Section 4-7). If this distance were doubled, both forces would decrease to $\frac{1}{2^2} = \frac{1}{4}$ of their original values. But the tidal force F_{tidal}, which is the *difference* between these two forces, decreases even more rapidly with distance: It is inversely proportional to the *cube* of the distance from Earth to the Moon. That is, doubling the Earth–Moon distance decreases the tidal force to $\frac{1}{2^3} = \frac{1}{8}$ of its original value. If we use the symbols M_{Earth} for Earth's mass (5.974×10^{24} kg), d for the Moon's diameter, r for the center-to-center distance from Earth to the Moon, and G for the universal constant of gravitation (6.67×10^{-11} newton m^2/kg^2), the tidal force is given approximately by the formula

$$F_{tidal} = \frac{2GM_{Earth}md}{r^3}$$

Objects, each of mass m

r_{near} r_{far}

Moon, diameter d

Earth

Earth–Moon distance = r

rocks ($m = 1$ kg). The Earth–Moon distance (r) and the diameter of the Moon (d, equal to the separation between the two rocks) are given in Table 10-1.

Tools: We use the equation given for the tidal force F_{tidal}.

Answer: To use the given values in the equation, we must first convert the values for the average Earth–Moon distance r and the Moon's diameter d from kilometers to meters:

$$d = 3476 \text{ km} \times \frac{10^3 \text{ m}}{1 \text{ km}} = 3.476 \times 10^6 \text{ m}$$

$$r = 384{,}400 \text{ km} \times \frac{10^3 \text{ m}}{1 \text{ km}} = 3.844 \times 10^8 \text{ m}$$

The tidal force is then

$$F_{tidal} = \frac{2(6.67 \times 10^{-11})(5.974 \times 10^{24})(1)(3.476 \times 10^6)}{(3.844 \times 10^8)^3}$$

BOX 4-3 ASTRONOMY DOWN TO EARTH

Newton's Laws in Everyday Life

In our study of astronomy, we use Newton's three laws of motion to help us understand the motions of objects in the heavens. But you can see applications of Newton's laws every day in the world around you. By considering these everyday applications, we can gain insight into how Newton's laws apply to celestial events that are far removed from ordinary human experience.

Newton's *first* law, or principle of inertia, says that an object at rest naturally tends to remain at rest and that an object in motion naturally tends to remain in motion. This law explains the sensations that you feel when riding in an automobile. When you are waiting at a red light, your car and your body are both at rest. When the light turns green and you press on the gas pedal, the car accelerates forward but your body attempts to stay where it was. Hence, the seat of the accelerating car pushes forward into your body, and it feels as though you are being pushed back in your seat.

Once the car is up to cruising speed, your body wants to keep moving in a straight line at this cruising speed. If the car makes a sharp turn to the left, the right side of the car will move toward you. Thus, you will feel as though you are being thrown to the car's right side (the side on the outside of the turn). If you bring the car to a sudden stop by pressing on the brakes, your body will continue moving forward until the seat belt stops you. In this case, it feels as though you are being thrown toward the front of the car.

Newton's *second* law states that the net outside force on an object equals the product of the object's mass and its acceleration. You can accelerate a crumpled-up piece of paper to a pretty good speed by throwing it with a moderate force. But if you try to throw a heavy rock by using the same force, the acceleration

will be much less because the rock has much more mass than the crumpled paper. Because of the smaller acceleration, the rock will leave your hand moving at only a slow speed.

Automobile airbags are based on the relationship between force and acceleration. It takes a large force to bring a fast-moving object suddenly to rest because this requires a large acceleration. In a collision, the driver of a car not equipped with airbags is jerked to a sudden stop and the large forces that act can cause major injuries. But if the car has airbags that deploy in an accident, the driver's body will slow down more gradually as it contacts the airbag, and the driver's acceleration will be less. (Remember that *acceleration* can refer to slowing down as well as to speeding up.) Hence, the force on the driver and the chance of injury will both be greatly reduced.

Newton's *third* law, the principle of action and reaction, explains how a car can accelerate at all. It is not correct to say that the engine pushes the car forward, because Newton's second law tells us that it takes a force acting from outside the car to make the car accelerate. Rather, the engine makes the wheels and tires turn, and the tires push backward on the ground. From Newton's third law, the ground must exert an equally large forward force on the car, and this is the force that pushes the car forward. You can see this backward force in action when a car drives through wet ground and sprays mud backward from the tires.

You use the same principles when you walk: You push backward on the ground with your foot, and the ground pushes forward on you. Icy pavement or a freshly waxed floor have greatly reduced friction. In these situations, your feet and the surface under you can exert only weak forces on each other, and it is much harder to walk.

Astronomy Down to Earth reveals the universal applicability of the laws of nature

Astronomy Down to Earth boxes illustrate how the same principles astronomers use to explain celestial phenomena can also explain everyday behavior here on Earth, from the color of the sky to why diet soft drinks can float in water.

FIGURE 8-12 R I **V** U X G

Primitive Meteorite This is a cross-section from a fragment of the Allende meteorite that landed in Chihuahua, Mexico in 1969. A fireball was observed at 1:05 A.M. local time, and over the next 25 years, about 3 tons of meteorite fragments were collected over an 8 km by 50 km (5 mi by 31 mi) area. Radioactive dating yields an age of 4.57 billion years, corresponding to the very early solar system. The meteorite contains rocks and metals that are consistent with the inner solar system and is thought to come from a smashed-up planetesimal. In 2009, astronomers where shocked to discover that the meteorite is also magnetized in a manner that suggests the original planetesimal had a chemically differentiated iron core with a magnetic dynamo (similar to the source of Earth's magnetic field). (Matteo Chinellato/Shutterstock)

Wavelength Tabs are included with photographic images

These icons show whether an image was made with radio waves (R), infrared radiation (I), visible light (V), ultraviolet light (U), x rays (X), or gamma rays (G).

The Solar System and Stars and Galaxies tailor the text to your course

In addition to the complete 27-chapter version of *Universe*, two shorter versions of the book are also available. *Universe: The Solar System* includes Chapters 1–16 and 27; it omits the chapters on stars, stellar evolution, galaxies, and cosmology. *Universe: Stars and Galaxies* includes Chapters 1–8 and 16–27; it omits the detailed chapters on the solar system but retains the overview of the solar system in Chapters 7 and 8.

SaplingPlus for *Universe* combines the rigor of Sapling homework with the formative power of LearningCurve adaptive quizzing. In the Sapling homework, every question includes hints; targeted, answer-specific feedback; and fully worked-out solutions to guide students through their progress toward understanding.

Assignment Score: 12.2% Resources Give Up? Hint Check Answer

Question 10 of 24

Determine which statements support or do not support the current theory of Solar System formation.

Support theory of Solar System formation	Do not support theory of Solar System formation

Answer Bank

- The Sun uses nuclear fusion to produce heat.
- Outer planets are made of ices and gases, whereas the inner ones are rocky.
- Planets orbit around the Sun in the same direction.
- Flattening dust clouds exist around young stars.
- The planets in our Solar System lie nearly in the plane of the earlier protoplanetary disk.
- Many moons rotate in the same directions as the planets.
- Some exoplanets have random orbits.

Image credit: C.R. O'Dell/Rice University/NASA/ESA

Assignment Score: 15.7% Resources Give Up? Solution Next Question

Question 10 of 24   ⊘ Attempt 4

Solution

The Solar System is believed to have formed from the collapse of a rotating cloud of gas and dust, called a solar nebula. This contraction formed a protostar that later developed into the Sun. The planets accreted from material in a protoplanetary disk rotating around the protostar.

There is strong observational evidence for this formation theory. The planets in our Solar System revolve in the same plane with the Sun at its center. A common orbital plane would exist if the planets formed from one solar nebula and protoplanetary disk. The planets also orbit in the same direction around the Sun and many of their moons orbit similarly around their planets. The planets and moons, for the most part, rotate about their axes in the same direction as their orbits. None of this motion is random, as it can be explained by formation from the same protoplanetary disk.

Protoplanetary Disks in the Orion Nebula
A Hubble Space Telescope view of a small portion of the Orion Nebula revealing five young stars. Four of the stars are surrounded by gas and dust trapped as the stars formed, but were left in orbit about the star. These are possibly protoplanetary disks that might evolve on to form planets.

Supporting evidence also includes the compositions of the terrestrial and jovian planets. It was too warm in the inner Solar System region for hydrogen to condense. In the cooler regions of the outer Solar System hydrogen was able to condense. As a result, the inner terrestrial planets are small and solid and the outer jovian planets are much larger and gaseous.

Additional evidence includes what astronomers can observe in other star systems. New, young stars are being discovered with flattening dust clouds around them, as depicted in the image of the Orion Nebula. These debris disks resemble the type of protoplanetary disk formation expected around new stars with potential planets. This evidence supports the idea that planets in other solar systems can form like to ours.

The Sun does use nuclear fusion to produce heat, but this alone does not provide support for the Solar System to have formed from a solar nebula. The random orbits of some exoplanets do not match the current Solar System formation theory. Multiple theories have been proposed to explain these exoplanet anomalies.

LearningCurve adaptive quizzing provides material at different difficulty levels and topics based on student performance. Students work to achieve a target score, allowing highly prepared students to complete the activity promptly, and providing students who are less familiar with the material the necessary time to practice. The student and instructor dashboards provide the accuracy of student responses for each topic, giving insight into important topics to review in class or study.

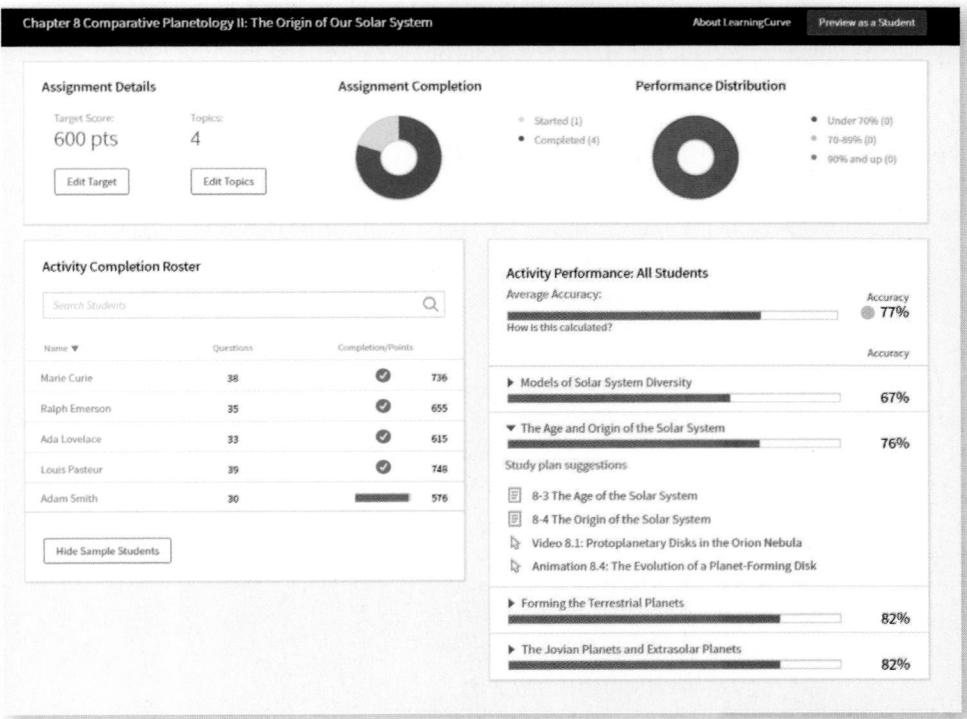

The mobile-accessible e-book included with every course contains embedded animations to augment the striking visuals found in the text, as well as many tools for studying, such as shareable highlighting, note-taking, and flashcards.

Starry Night™

Starry Night™ is a brilliantly realistic planetarium software package. It is designed for easy use by anyone with an interest in the night sky. See the sky from anywhere on Earth or lift off and visit any solar system body or any location up to 20,000 light-years away. A download code for *Starry Night*™ is available with the text upon request.

Instructor's Resources

The instructor's resources for *Universe* are all available for download from SaplingPlus, as well as from the Macmillan Learning catalog site.

The **Test Bank** offers more than 3500 multiple-choice questions referenced by section, available in editable Microsoft Word documents or Diploma files for quick exam creation.

The **Instructor's Manual** accompanying the text provides instructors with a summary and table of contents of each chapter, pedagogical advice on how to teach the chapter, additional questions for in-class discussion, and the answers to book problems all in one place.

Lecture Slides and Images in PowerPoint allow instructors to tailor their lectures to suit their own needs using images and notes from the textbook, pre-organized into slides.

Clicker Questions can be used as lecture launchers with or without a classroom response system such as iClicker. Each chapter includes questions that relate to figures in the text and address common misconceptions, as well as writing questions for instructors who would like to add a writing or class discussion element to their lectures.

Acknowledgments

We would like to thank our supplement authors and accuracy reviewers for their special work on this edition.

Peter A. Becker, *George Mason University*
Katie Berryhill, *Solano Community College*
James Dickinson, *Clackamas Community College*
Michael Dunham, *State University of New York at Fredonia*
Davin Flateau, *University of Cincinnati*
Elizabeth Holden, *University of Wisconsin at Platteville*

Jack T. Howard, *Rowan-Cabarrus Community College*
Susannah E. Lomant, *Perimeter College, Georgia State University*
Norman L. Markworth, *Stephen F. Austin State University*
James D. Martin, Jr., *Lake Sumter State College*
Barton J. Pritzl, *University of Wisconsin at Oshkosh*
Takahashi Sato, *Kwantlen Polytechnic University*

We would also like to thank our colleagues who have carefully scrutinized the manuscript of this edition. *Universe* is a stronger and better textbook because of their conscientious efforts.

Viviana Acquaviva, *New York City College of Technology*
Peter A. Becker, *George Mason University*
Katie Berryhill, *Solano Community College*
Erin Wells Bonning, *Emory University*
Bram Boroson, *Clayton State University*
Bruce W. Bridenbecker, *Copper Mountain College*
Peter J. Brown, *Texas A&M University*
Tatyana Brusentsova, *Valencia College*
Tabitha Buehler, *University of Utah*
Marianne Caldwell, *Hillsborough Community College*
John Carzoli, *Oakton Community College*
Emmanouil Chatzopoulos, *Louisiana State University*
Erik N. Christensen, *South Florida Community College*
Greg Clements, *Metropolitan Community College*
James Dickinson, *Clackamas Community College*
Gabriel Dinca, *Eastern Florida State College*
James Dove, *Metropolitan State University of Denver*
Christian Draper, *Utah Valley University*
Bryan Dunne, *University of Illinois at Urbana–Champaign*
Adriana Durbala, *University of Wisconsin–Stevens Point*
Robert A. Egler, *North Carolina State University*
Gregory J. Falabella, *Wagner College*
Tim Farris, *Volunteer State Community College*
Gregory Feiden, *University of North Georgia*
Davin Flateau, *University of Cincinnati*
Steven Furlanetto, *University of California, Los Angeles*
Matthew Ganis, *Pace University*
Brian Geislinger, *Gadsden State Community College*
Doug Gobeille, *University of Rhode Island*
Dr. Albert Gottlieb, *Union County College*
Susan Hartley, *University of Minnesota Duluth*
Javier Hasbun, *University of West Georgia*
David Hedin, *Northern Illinois University*

Inge Heyer, *Loyola University Maryland*
Sarah Higdon, *Georgia Southern University*
Michael Hood, *Mt. San Antonio College*
Jack T. Howard, *Rowan-Cabarrus Community College*
Adam Jensen, *University of Nebraska at Kearney*
Kishor Kapale, *Western Illinois University*
Mario Klaric, *Midlands Technical College*
Todd M. Krueger, *College of The Albemarle*
Geoff Lenters, *Grand Valley State University*
Susannah Lomant, *Perimeter College, Georgia State University*
Naibi Marinas, *University of Florida*
N. L. Markworth, *Stephen F. Austin State University*
James D. Martin, Jr., *Lake Sumter State College*
Paul Martin, *Eastern Florida State College*
Audeliz Matias, *SUNY Empire State College*
David Mayeux, *San Juan College*
Steven Merriman, *Moraine Valley Community College*
Peter A. Milne, *University of Arizona*
Vesna Milosevic-Zdjelar, *University of Winnipeg*
Anatoly Miroshnichenko, *University of Southern Mississippi*
Majid Noori, *Cumberland County College*
Kerry A. Ozmelek, *Dr. Robert Ketterer Charter School*
Christopher Palma, *Pennsylvania State University*
Marina S. Papenkova, *East Los Angeles College*
David R. Patton, *Trent University*
Jon Pedicino, *College of the Redwoods*
Dr. Elisha Polomski, *St. Cloud State University*
Barton J. Pritzl, *University of Wisconsin at Oshkosh*
Joshua Ridley, *Murray State University*
Takahashi Sato, *Kwantlen Polytechnic University*
Eric M. Schlegel, *University of Texas at San Antonio*
Paul Schmidtke, *Arizona State University*
Ann Schmiedekamp, *Penn State Abington*

Bruce A. Sherman, *Southeastern Louisiana University*
Christopher Sirola, *University of Southern Mississippi*
G. Roger Stanley, *San Antonio College*
Chris Stockdale, *Marquette University*
Dave Syndergaard, *American Public University System*
Christopher Taylor, *California State University, Sacramento*
B. L. Team, *University of West Georgia*
T. E. Tessier, *Central New Mexico Community College*
Gregory Tucker, *Brown University*
Gajendra Tulsian, *Daytona State College*

Toshiya Ueta, *University of Denver*
Dirk Valk, *Polk State College*
Tiffany Watkins, *Boise State University*
James R. Webb, *Florida International University*
Stephanie Welch, *Southeastern Louisiana University*
Samanthi Wickramarachchi, *Virginia Commonwealth University*
Christopher J. Wood, *Schoolcraft College*
David A. Wood, Jr., *San Antonio College*
Brian Woodahl, *Indiana University–Purdue University Indianapolis*
Zheng Zheng, *University of Utah*

We would also like to renew our thanks to those who have reviewed past editions
of *Universe:*

Robert Antonucci, *University of California, Santa Barbara*
Nahum Arav, *Virginia Tech University*
Ann Bragg, *Marietta College*
Richard Bowman, *Bridgewater College*
Robert Braunstein, *Northern Virginia Community College, Loudoun*
Thomas Campbell, *Allen Community College*
Brian Carter, *Grossmont College*
Stephen Case, *Olivet Nazarene University*
Demian H. J. Cho, *Kenyon College*
Richard A. Christie, *Okanagan College*
Eric Collins, *California State University, Northridge*
Anca Constantin, *James Madison University*
George J. Corso, *DePaul University*
John Cottle, *University of California, Santa Barbara*
Tamara Davis, *University of Queensland*
Jamie Day, *Transylvania University*
Steven Desch, *Arizona State University*
Peter Detterline, *Kutztown University*
James Dickinson, *Clackamas Community College*
Jeanne Digel, *Canada College*
Edward Dingler, *Southwest Virginia Community College*
David Duluk, *Los Angeles City College/Glendale Community College*
Jacqueline Dunn, *Midwestern State University*
Jean W. Dupon, *Menlo College*
Stephanie Fawcett, *McKendree University*
Kent Fisher, *Columbus State Community College*
Martin Fisk, *Oregon State University*
Juhan Frank, *Louisiana State University*
Tony George, *Columbia Basin College*
Erika Gibb, *University of Missouri, St. Louis*
Daniel Greenberger, *City College of New York*
Wayne K. Guinn, *Lon Morris College*
Hunt Guitar, *New Mexico State University*
Joshua Gundersen, *University of Miami*
Kathleen A. Harper, *Denison University*
Lynn Higgs, *University of Utah*
Andy Howell, *University of California, Santa Barbara*
Melinda Hutson, *Portland Community College*
Douglas R. Ingram, *Texas Christian University*
Adam G. Jensen, *University of Colorado, Boulder*
Darell Johnson, *Missouri Western State University*
Michael Joner, *Brigham Young University*
Lauren Jones, *Denison University*
Steve Kawaler, *Iowa State University*
William Keel, *University of Alabama*
Charles Kerton, *Iowa State University*
Kevin Kimberlin, *Bradley University*

H. S. Krawczynski, *Washington University, St. Louis*
Erik Kubik, *Providence College*
Lauren Likkel, *University of Wisconsin, Eau Claire*
Dennis Machnik, *Plymouth State University*
Franck Marchis, *University of California, Berkeley*
Peter Meinhold, *University of California, Santa Barbara*
Anatoly Miroshnichenko, *University of North Carolina, Greensboro*
Michele M. Montgomery, *University of Central Florida*
Patrick Motl, *Indiana University, Kokomo*
Krishna Mukherjee, *Slippery Rock University*
Edward M. Murphy, *University of Virginia*
Donna Naples, *University of Pittsburgh*
Gerald H. Newsom, *The Ohio State University*
Brian Oetiker, *Sam Houston State University*
Ronald P. Olowin, *Saint Mary's College*
Fritz Osell, *Northern Oklahoma College*
Michael J. O'Shea, *Kansas State University*
J. Douglas Patterson, *Johnson County Community College*
Stan Peale, *University of California, Santa Barbara*
Nicolas Pereyra, *University of Texas, Pan American*
Charles Peterson, *University of Missouri*
Ylva Pihlstrom, *University of New Mexico*
Kevin Plaxco, *University of California, Santa Barbara*
Richard Rand, *University of New Mexico*
Raghavan Rangarajan, *Physical Research Laboratory, Ahmedabad, India*
Sean Raymond, *Laboratoire d'Astrophysique de Bordeaux*
Ian Redmount, *Saint Louis University*
Adam Rengstorf, *Purdue University, Calumet*
Mike Reynolds, *Florida Community College*
Todd Rimkus, *Marymount University*
Frederick Ringwald, *California State University, Fresno*
Andrew Rivers, *Northwestern University*
Paul Robinson, *Westchester Community College*
Robert Rosner, *University of Chicago*
Louis Rubbo, *Coastal Carolina University*
Alan Rubin, *University of California, Los Angeles*
Kyla Scarborough, *Pittsburg State University*
Paul Schmidtke, *Arizona State University*
Doug Showell, *University of Nebraska, Omaha*
Caroline Simpson, *Florida International University*
Larry K. Smith, *Snow College*
Frank Spera, *University of California, Santa Barbara*
Glenn Spiczak, *University of Wisconsin, River Falls*
James D. Stickler, *Allegany College of Maryland*
Jack Sweeney, *Salem State University*
Chris Taylor, *California State University, Sacramento*

Charles M. Telesco, *University of Florida*
Dale Trapp, *Concordia University, St. Paul*
Nicole Vogt, *New Mexico State University*

Lauren White, *NASA/JPL*
Mark Whittle, *University of Virginia*
Michael E. Wysession, *Washington University, St. Louis*

Many others have participated in the preparation of this book, and we thank them for their efforts. We are grateful for the guidance we have received from Lori Stover, our program manager. We're also deeply grateful to our development editor on *Universe,* Meg Rosenburg—whose own extensive background in astronomy (PhD, Caltech) is a gift to authors and readers alike—and for the contributions of associate editor, Kevin Davidson. Vicky Garvey, the media and supplements editor, supervised the creation of the superb multimedia. Special thanks go to senior content project manager Vivien Weiss, who skillfully kept all three versions of this book on track and on target. Design manager Natasha Wolfe deserves credit for the design of the book. We also thank Lumina Datamatics for coordinating and producing the excellent artwork, and extend our gratitude to photo editor Krystyna Borgen for finding just the right photographs. Linda DeBruyn deserves a medal for wading through the prose and copyediting it into proper English. Finally, thank you to Marcel W. Bergman, T. Alan Clark, and William J. F. Wilson for their outstanding work keeping the *Starry Night*™ exercises up to date.

On a personal note, Robert Geller would like to extend special thanks to Robert Antonucci (aka Ski). Ski has helped with numerous topics throughout this textbook, was Robert's PhD thesis adviser, and—very conveniently—is his next door neighbor. While some discussions take the usual form of pointing to data plots on the computer, many discussions are literally over the fence. In addition to benefiting from Ski's seemingly unlimited expertise, and ceaseless desire to talk science, he has made astronomy FUN from the first day Robert was a teaching assistant for Ski's astronomy class.

Robert Geller would also like to thank his parents for nurturing a love of science. Although some home appliances that were dismantled in the name of science were never put back together, he always had their support. Most of all, Robert would like to thank his family for their support through years of long hours researching and writing for this textbook. His wife, Susanne, has even contributed scientific results to this book as part of the team that discovered the Higgs particle in 2012; if her next project is successful, she can add dark matter to her list. With the completion of this edition, Robert and his daughter, Zoe, will have some time to climb together, and maybe Robert will pick up that guitar. . . .

Roger Freedman would like to thank his father, Richard Freedman, for first cultivating his interest in space many years ago, and for his father's personal contributions to the exploration of the universe as an engineer for the Atlas and Centaur launch vehicle programs. Most of all, Roger thanks his charming wife, Caroline, for putting up with his long nights slaving over the computer!

Although we have made a concerted effort to produce an error-free edition, some mistakes may have crept in unbidden. We would appreciate hearing from anyone who finds an error or wishes to comment on the text.

How to Get the Most from *Universe*

If you're like most students just opening this textbook, you're enrolled in one of the few science courses you'll take in college. As you study astronomy, you'll probably do relatively little reading compared to a literature or history course—at least in terms of the number of pages. But your readings will be packed with information, much of it new to you and (we hope) exciting. You can't read this textbook like a novel and expect to learn much from it. Don't worry, though. We wrote this book with you in mind. We'll take you on a guided tour of the book and media, with suggestions about how *Universe* can help you succeed in your astronomy course.

Apply these techniques to studying astronomy

- **Read before each lecture.** You'll get the most out of your astronomy course if you read each chapter before hearing a lecture about its subject matter. That way, many of the topics will already be clear in your mind, and you'll understand the lecture better. You'll be able to spend more of your listening and note-taking time on the more challenging ideas presented in the lecture.

- **Take notes as you read and make use of office hours.** Keep a notebook handy as you read, and write down the key points of each section so that you can review them later. If any parts of the section don't seem clear on first reading, make a note of them, too, including the page numbers. Once you've gone through the chapter, reread it with special emphasis on the ideas that gave you trouble the first time. If you're still unsure after the lecture, consult your instructor, either during office hours or after class. Bring your notes with you so your instructor can see which concepts are posing difficulty. Once your instructor has helped clarify things for you, revise your notes so you'll remember your newfound insights. You'll end up with a chapter summary in your own words—a tremendous help when studying for exams!

- **Make use of your fellow students.** Many students find it useful to form study groups for astronomy. You can hash out challenging topics together and have a good time while doing it. But make sure you write up your homework by yourself, because the penalties for copying or plagiarizing other students' work can be severe in the extreme. Some students find individual assistance useful, too. If you think a tutor will be helpful, link up with one early. Getting a tutor late in the course, in the belief that you'll be able to catch up with what you missed earlier on, is almost always a lost cause.

- **Take advantage of the website and e-book.** Take some time to explore the *Universe* website (www.whfreeman.com/universe11e). You'll find review materials, animations, videos, interactive exercises, flashcards, and many other features keyed to chapters in *Universe*. All of these features are designed to help you learn and enjoy astronomy, so make sure to take full advantage of them. You can also use the e-book version of this textbook, which combines the complete content of the book and the website in a convenient online format.

- **Try astronomy for yourself with your star charts.** At the back of this book is a set of star charts for each month of the year in the northern hemisphere. (For a set of southern hemisphere star charts, see the *Universe* website.) Star charts can get you started with your own observations of the universe. Hold the chart overhead in the same orientation as the compass points, with southern horizon toward the south and western horizon toward the west. (To save strain on your arms, you may want to cut these pages out of the book.) Depending on the version of this textbook that your instructor requested, it may also include access to the easy-to-use *Starry Night*™ planetarium program, which you can use to view the sky on any date and time as seen from anywhere on Earth.

Here's the most important advice of all

We haven't mentioned the most important thing you should do when studying astronomy: Have fun! Of all the different kinds of scientists, astronomers are among the most excited about what they do and what they study. Let some of that excitement about the universe rub off on you, and you'll have a great time with this course and with this textbook.

In preparing this edition of *Universe,* we've tried very hard to make it the kind of textbook that a student like you will find useful. We're very interested in your comments and opinions! Please feel free to e-mail or write to us and we will respond personally.

Best wishes for success in your studies!

Robert M. Geller
Department of Physics
University of California, Santa Barbara
Santa Barbara, CA 93106
rhmg@physics.ucsb.edu

Roger A. Freedman
Department of Physics
University of California, Santa Barbara
Santa Barbara, CA 93106
airboy@physics.ucsb.edu

This magnificent object is called, quite fittingly, the Butterfly Nebula—it's just one of many "planetary nebulae." Its colorful wings are made of gas, and hiding at the center within its body is a single star. Near the end of a star's life, it expands hundreds of times larger than its normal size, becoming a red giant. Shortly after that, gases from the star's outer layers are expelled into space. However, these gases do not emit any visible light on their own, and the light we see isn't due to the gas's temperature. The light we see results from two steps: First, the hot core of the dying central star (called a white dwarf) emits copious amounts of invisible ultraviolet light absorbed by the gas; then the gas "glows" by emitting visible light. Each atom (hydrogen, oxygen, etc.) glows with a different color, giving rise to this astral mosaic. Many planetary nebulae are spherical in shape, but the Butterfly Nebula contains a thick doughnut-shaped torus that funnels the gas and ultraviolet light along an axis. The wings of this nebula span about 3 light-years from tip to tip. (NASA/ESA/Hubble)

I Introducing Astronomy

Many details are revealed by looking at ultraviolet light emitted by the Sun. The bright arches result from magnetic fields poking out of the solar surface. Electrons accelerated by these magnetic fields then emit ultraviolet light. Parts of these features can also be seen in visible light, where they are called solar flares. More than just emitting light, solar flares can hurl mountain-sized quantities of matter from the Sun. Sometimes the solar flares are directed at Earth, where they can damage communication satellites and pose risks to astronauts. (NASA SDO)

R I V **U** X G

1 | Astronomy and the Universe

LEARNING GOALS

By reading the sections of this chapter, you will learn

Imagine yourself looking skyward on a clear, dark, moonless night, far from the glare of city lights. As you gaze upward, you see a panorama that no poet's words can truly describe and that no artist's brush could truly capture. Literally thousands of stars are scattered from horizon to horizon, many of them grouped into a luminous band called the Milky Way (which extends up and down across the middle of the above photograph). As you watch, the entire spectacle swings slowly overhead from east to west as the night progresses.

For thousands of years, people have looked up at the heavens and contemplated the universe. Like our ancestors, we find our thoughts turning to profound questions as we gaze at the stars. How was the universe created? Where did Earth, the Moon, and the Sun come from? What are the planets and stars made of? And how do we fit in? What is our place in the cosmic scope of space and time?

Wondering about the universe is a key part of what makes us human. Our curiosity, our desire to explore and discover, and, most important, our ability to reason about what we have discovered are qualities that distinguish us from other animals. The study of the stars transcends all boundaries of culture, geography, and politics. In a literal sense, astronomy is a universal subject—its subject is the entire universe.

The night sky seen from Mauna Kea in Hawaii. The feature extending across the sky is the Milky Way, consisting of hundreds of billions of stars. The Milky Way is the galaxy in which we reside. (John Hook/Moment/Getty Images)

R I V U X G

1-1 To understand the universe, astronomers use the laws of physics to construct testable theories and models

Astronomy has a rich heritage that dates back to the myths and legends of antiquity. Centuries ago, the heavens were thought to be populated with demons and heroes, gods and goddesses. Astronomical phenomena were explained as the result of supernatural forces and divine intervention.

The course of civilization was greatly affected by a profound realization: *The universe is comprehensible.* For example, ancient Greek astronomers discovered that by observing the heavens and carefully reasoning about what they saw, they could learn something about how the universe operates. As we shall see in Chapter 3, ancient Greek astronomers measured the size of Earth and were able to understand and predict eclipses without appealing to supernatural forces. Modern science is a direct descendant of astronomy, which had many contributors from the Middle East, Africa, Asia, Central America, and, eventually, Greece.

The Scientific Method

TUTORIAL 1.1 Like art, music, or any other human creative activity, science makes use of intuition and experience. But the approach used by scientists to explore physical reality differs from other forms of intellectual endeavor in that it is based fundamentally on *observation, logic,* and *skepticism.* This approach, called the **scientific method,** requires that our ideas about the world around us be consistent with what we actually observe.

The scientific method goes something like this: A scientist trying to understand some observed phenomenon proposes a **hypothesis,** which is a collection of ideas that seems to explain what is observed. It is in developing hypotheses that scientists are at their most creative, imaginative, and intuitive. But their hypotheses must always agree with existing observations and experiments, because a discrepancy with what is observed implies that the hypothesis is wrong. The scientist then uses logic to work out the implications of the hypothesis and to make predictions that can be tested. A hypothesis is on firm ground only after it has accurately forecast the results of new experiments or observations.

> Hypotheses, models, theories, and laws are essential parts of the scientific way of knowing.

Scientists describe reality in terms of **models,** which are hypotheses that have withstood observational or experimental tests. A model tells us about the properties and behavior of some object or phenomenon. A familiar example is a model of the atom, which scientists picture as electrons orbiting a central nucleus. Another example, which we will encounter in Chapter 18, is a model that tells us about physical conditions (for example, temperature, pressure, and density) in the interior of the Sun (**Figure 1-1**). A well-developed model uses mathematics—one of the most powerful tools for logical thinking—to make detailed predictions. For instance, a successful model of the Sun's interior should describe the values of temperature, pressure,

FIGURE 1-1 R I V U X G

Our Star, the Sun The Sun is a typical star. Its diameter is about 1.39 million kilometers (roughly a million miles), and its surface temperature is about 5500°C (10,000°F). A detailed scientific model of the Sun tells us that it draws its energy from nuclear reactions occurring at its center, where the temperature is about 15 million °C (27 million °F). (Trifonov_Evgeniy/Getty Images)

and density at each depth within the Sun, as well as the relations between these quantities. For this reason, mathematics is one of the most important tools used by scientists.

A body of related hypotheses can be pieced together into a self-consistent description of nature called a **theory.** An example from Chapter 4 is the theory that the planets are held in their orbits around the Sun by the Sun's gravitational force (**Figure 1-2**). Without models and theories there is no understanding and no science, only collections of facts.

CAUTION!

In everyday language, the word *theory* is often used to mean an idea that looks good on paper, but has little to do with reality. In science, however, a good theory is one that explains reality very well and that can be applied to explain new observations. An excellent example is the theory of gravitation (Chapter 4), which was devised by the English scientist Isaac Newton in the late 1600s to explain the orbits of the six planets known at that time. When astronomers of later centuries discovered the planets Uranus and Neptune and the dwarf planet Pluto, they found that these planets also moved in accordance with Newton's theory. The same theory describes the motions of satellites around Earth as well as the orbits of planets around other stars.

FIGURE 1-2

Planets Orbiting the Sun An example of a scientific theory is the idea that Earth and the planets orbit the Sun due to the Sun's gravitational attraction. This theory is universally accepted because it makes predictions that have been tested and confirmed by observation. (The Sun and planets are actually much smaller than this illustration would suggest.) (adventtr/Getty Images)

An important part of a scientific theory is its ability to make predictions that can be tested by other scientists. Predictions that are verified by observation lend support to the theory and suggest that it might be correct. If the predictions are *not* verified, the theory must be modified or completely replaced. For example, an old theory held that the Sun and planets orbit around a stationary Earth. This theory led to certain predictions that could be checked by observation, as we will see in Chapter 4. In the early 1600s, the Italian scientist Galileo Galilei used one of the first telescopes to show that these predictions were incorrect. As a result, the theory of a stationary Earth was rejected, eventually to be replaced by the modern model shown in Figure 1-2 in which Earth and other planets orbit the Sun.

An idea that *cannot* be tested by observation or experiment does not qualify as a scientific theory. An example is the idea that there is a little man living in your refrigerator who turns the inside light on or off when you open and close the door. The little man is invisible, weightless, and makes no sound, so you cannot detect his presence. While this is an amusing idea, it cannot be tested and thus cannot be considered science.

Skepticism is an essential part of the scientific method. New hypotheses must be able to withstand the close scrutiny of other scientists. The more radical the hypothesis, the more skepticism and critical evaluation it will receive from the scientific community, because the general rule in science is that *extraordinary claims require extraordinary evidence.* That is why scientists as a rule do not accept claims that people have been abducted by aliens

and taken aboard UFOs. The evidence presented for these claims is flimsy, secondhand, and unverifiable.

At the same time, scientists must be open-minded. They must be willing to discard long-held ideas if these ideas fail to agree with new observations and experiments, as long as the new data have survived critical review. If an alien spacecraft really did land on Earth, scientists would be the first to accept that aliens existed—provided they could take a careful look at the spacecraft and its occupants. That is why the validity of scientific knowledge can be temporary. With new evidence, some ideas only need modification, while occasionally, some ideas may need to be entirely replaced. Throughout this book, you will encounter many instances wherein new observations have transformed our understanding of Earth, the planets, the Sun and stars, and indeed the very structure of the universe.

Theories that accurately describe the workings of physical reality have a significant effect on civilization. For example, basing his conclusions in part on observations of how the planets orbit the Sun, Isaac Newton deduced a set of fundamental principles that describe how *all* objects move. These theoretical principles, which we will encounter in Chapter 4, work equally well on Earth as they do in the most distant corner of the universe. They represent our first complete, coherent description of how objects move in the physical universe. **Newtonian mechanics** had an immediate practical application in the construction of machines, buildings, and bridges. It is no coincidence that the Industrial Revolution followed hard on the heels of these theoretical and mathematical advances inspired by astronomy.

Newtonian mechanics and other physical theories have stood the test of time and been shown to have great and general validity. Proven theories of this kind are collectively referred to as the **laws of physics.** Thus, the most reliable theories with the broadest applicability can eventually be considered laws of physics. Astronomers use these laws to interpret and understand their observations of the universe. The laws governing light and its relationship to matter are of particular importance, because the only information we can gather about distant stars and galaxies is in the light that we receive from them. Using the physical laws that describe how objects absorb and emit light, astronomers have measured the temperature of the Sun and even learned what the Sun is made of. By analyzing starlight in the same way, they have discovered that our own Sun is a rather ordinary star and that the observable universe may contain 10 billion trillion stars just like it.

Technology in Science

An important part of science is the development of new tools for research and new techniques of observation. As an example, until fairly recently, everything we knew about the distant universe was based on visible light. Astronomers would peer through telescopes to observe and analyze visible starlight. By the end of the nineteenth century, however, scientists had begun to discover forms of light invisible to the human eye: x rays, gamma rays, radio waves, microwaves, and ultraviolet and infrared radiation.

As we will see in Chapter 6, in recent years astronomers have constructed telescopes that can detect such nonvisible forms of light (**Figure 1-3**). These instruments give us views of the universe vastly different from anything our eyes can see. These new views have allowed us to see through the atmospheres of distant planets, to study the thin but incredibly violent gas that surrounds our Sun, and even to observe new solar systems being formed around distant stars. Aided by high-technology telescopes, today's astronomers

FIGURE 1-3 R I **V** U X G

A Telescope in Space Because it orbits outside Earth's atmosphere in the near-vacuum of space, the Hubble Space Telescope (HST) can detect not only visible light but also ultraviolet and near-infrared light coming from distant stars and galaxies. These forms of nonvisible light are absorbed by our atmosphere and hence are difficult or impossible to detect with a telescope on Earth's surface. This photo of HST was taken by the crew of the space shuttle *Columbia* after a servicing mission in 2002. (Courtesy of Scientific American/NASA/AAT)

carry on the program of careful observation and logical analysis begun thousands of years ago by their ancient Greek predecessors.

CONCEPT CHECK 1-1

Which is held in higher regard as an approximation of the truth by professional astronomers: a hypothesis or a theory? Explain your answer.

Answer appears at the end of the chapter.

1-2 By exploring the planets, astronomers uncover clues about the formation of the solar system

The science of astronomy allows our intellects to voyage across the cosmos. We can think of three stages in this voyage: from Earth to other parts of the solar system, from the solar system to the stars, and from stars to galaxies and the grand scheme of the universe.

> Studying planetary science gives us a better perspective on our own unique Earth.

The star we call the Sun and all the celestial bodies that orbit the Sun—including Earth, the other planets, all their various moons, and smaller bodies such as asteroids and comets—make up the **solar system.** Since the 1960s, a series of robotic spacecraft has been sent to explore each of the planets (**Figure 1-4**). Using the remote "eyes" of such spacecraft, we have flown over Mercury's cratered surface, peered beneath Venus's poisonous cloud cover, and discovered enormous canyons and extinct volcanoes on Mars. We have found active volcanoes on a moon of Jupiter, probed the atmosphere of Saturn's moon Titan, seen the rings of Uranus up close, and looked down on the active atmosphere of Neptune.

Along with rocks brought back by the Apollo astronauts from the Moon (the only world beyond Earth visited by humans), new information from spacecraft has revolutionized our understanding of the origin and evolution of the solar system. We have come to realize that many of the planets and their satellites were shaped by collisions with other objects. Craters on the Moon and on many other worlds are the relics of innumerable impacts by bits of interplanetary rock. The Moon may itself be the result of a catastrophic collision between Earth and a planet-sized object shortly after the solar system was formed. Such a collision could have torn sufficient material from the primordial Earth to create the Moon.

The oldest objects found on Earth are **meteorites,** chemically distinct bits of interplanetary debris that sometimes fall to our planet's surface. By using radioactive age-dating techniques, scientists have found that the oldest meteorites are 4.56 billion years old—older than any other rocks found on Earth or the Moon. The conclusion is that our entire solar system, including the Sun and planets, formed 4.56 billion years ago. The few thousand years of recorded human history is no more than the twinkling of an eye compared to the long history of our solar system.

The discoveries that we have made in our journeys across the solar system are directly relevant to the quality of human life on our own planet. Until recently, our understanding of geology, weather, and climate was based solely on data from Earth.

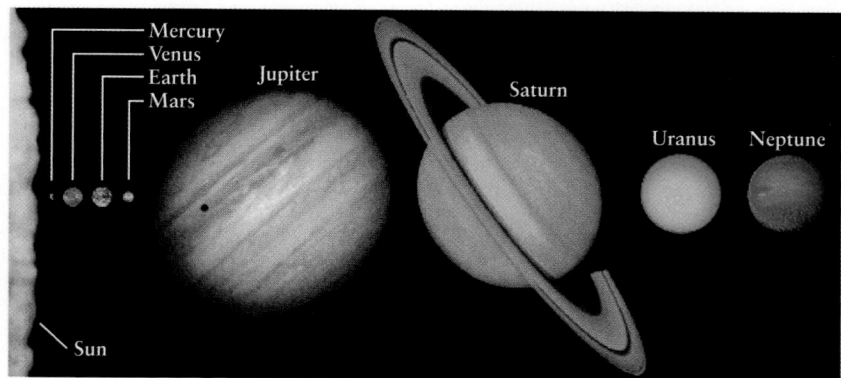

FIGURE 1-4

The Sun and Planets to Scale This montage of images from various spacecraft and ground-based telescopes shows accurate relative sizes of the planets and the Sun. The Sun is so large compared to the planets that only a portion of it fits into this illustration. The distances from the Sun to each planet are not shown to scale; the actual distance from the Sun to Earth, for instance, is 12,000 times greater than Earth's diameter.
(Calvin J. Hamilton and NASA/JPL)

Since the advent of space exploration, however, we have been able to compare and contrast other worlds with our own. This new knowledge gives us valuable insight into our origins, the nature of our planetary home, and the limits of our natural resources.

1-3 By studying stars and nebulae, astronomers discover how stars are born, grow old, and die

The nearest of all stars to Earth is the Sun. Although humans have used the Sun's warmth since the dawn of our species, it was only in the 1920s and 1930s that physicists figured out how the Sun shines. At the center of the Sun, thermonuclear reactions—so called because they require extremely high temperatures—convert hydrogen (the Sun's primary constituent) into helium. This violent process releases a vast amount of energy, which eventually makes its way to the Sun's surface and escapes as light (see Figure 1-1). Thus, through nuclear reactions, hydrogen acts as a "fuel" for stars. All the stars you can see in the nighttime sky also shine by nuclear reactions (Figure 1-5). By 1950, physicists could reproduce such nuclear reactions here on Earth in the form of a hydrogen bomb (Figure 1-6). In the future, converting hydrogen into helium might someday offer a cleaner method for the production of nuclear energy.

Because nuclear reactions consume the original material of which stars are made, the way an engine consumes its fuel, stars cannot last forever. Rather, they must form, evolve, and eventually die.

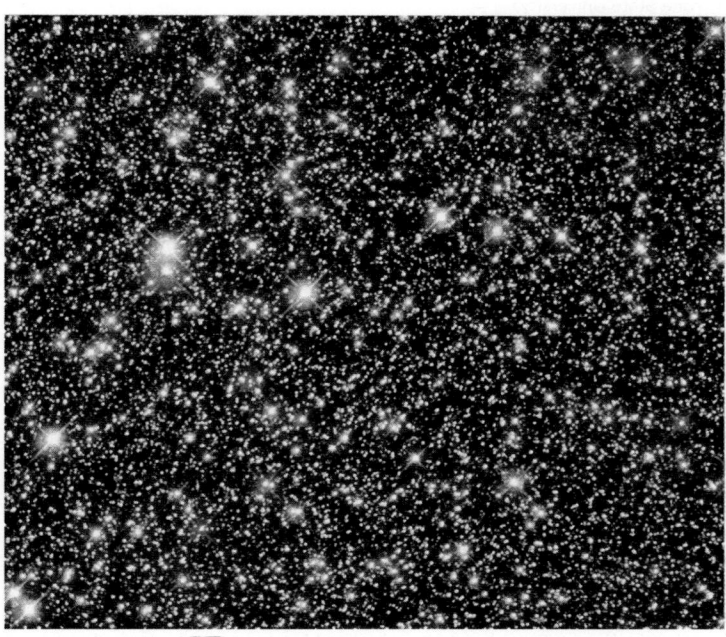

FIGURE 1-5 R I **V** U X G

Stars like Grains of Sand This Hubble Space Telescope image shows thousands of stars in the constellation Sagittarius. Each star shines because of thermonuclear reactions that release energy in its interior. Different colors indicate stars with different surface temperatures: Stars with the hottest surfaces appear blue, while those with the coolest surfaces appear red.
(The Hubble Heritage Team, AURA/STScI/NASA)

FIGURE 1-6 R I **V** U X G

A Thermonuclear Explosion A hydrogen bomb uses the same physical principle as the thermonuclear reactions at the Sun's center: the conversion of matter into energy by nuclear reactions. This thermonuclear detonation on October 31, 1952, had an energy output equivalent to 10.4 million tons of TNT (almost 1000 times greater than the nuclear bomb detonated over Hiroshima in World War II). This is a mere 10-billionth of the amount of energy released by the Sun in 1 second. (The Nuclear Weapon Archive)

> **CAUTION!**
>
> Astronomers often use biological terms such as "birth" and "death" to describe stages in the evolution of inanimate objects like stars. Keep in mind that such terms are used only as *analogies,* which help us visualize these stages. They are not to be taken literally!

The Life Stories of Stars

The rate at which stars emit energy in the form of light tells us how rapidly they are consuming their nuclear "fuel" (hydrogen), and hence how long they can continue to shine before reaching the end of their life spans. More massive stars have more hydrogen—and thus more nuclear "fuel"—but consume it at such a prodigious rate that they live out their lives in just a few million years. Less massive stars have less material to consume, but their nuclear reactions proceed so slowly that their life spans are measured in billions of years. (Our own star, the Sun, is in early middle age: It is 4.56 billion years old, with a projected lifetime of 10 billion years.)

While no astronomer can watch a single star go through all of its life stages, we have been able to piece together the life stories of stars by observing many different stars at different points in their life cycles. Important pieces of the puzzle have been discovered by studying huge clouds of interstellar gas, called **nebulae**

FIGURE 1-8 R I V U X G

The Crab Nebula—Wreckage of an Exploded Star When a dying star exploded in a supernova, it left behind this elegant funeral shroud of glowing gases blasted violently into space. A thousand years after the explosion these gases are still moving outward at about 1800 kilometers per second (roughly 4 million miles per hour). The Crab Nebula is 6500 light-years from Earth and about 13 light-years across. (NASA, ESA, J. Hester, and A. Loll/Arizona State University)

FIGURE 1-7 R I V U X G

The Orion Nebula—Birthplace of Stars This beautiful nebula is a stellar "nursery" where stars are formed out of the nebula's gas. Intense ultraviolet light from newborn stars excites the surrounding gas and causes it to glow. Many of the stars embedded in this nebula are less than a million years old, a brief interval in the lifetime of a typical star. The Orion Nebula is some 1500 light-years from Earth and is about 30 light-years across. (NASA, ESA, M. Robberto [Space Telescope Science Institute/ESA] and the Hubble Space Telescope Orion Treasury Project Team)

(singular **nebula**), which are found scattered across the sky. Within some nebulae, such as the Orion Nebula shown in Figure 1-7, stars are born from the material of the nebula itself. Other nebulae reveal what happens when nuclear reactions stop and a star dies. Some stars that are far more massive than the Sun end their lives with a spectacular detonation called a **supernova** (plural **supernovae**) that blows the star apart. The Crab Nebula (Figure 1-8) is a striking example of a remnant left behind by a supernova.

> Studying the life cycles of stars is crucial for understanding our own origins.

Dying stars can produce some of the strangest objects in the sky. Some dead stars become **pulsars,** which spin rapidly at rates of tens or hundreds of rotations per second. And some stars end their lives as almost inconceivably dense objects called **black holes,** whose gravity is so powerful that nothing—not even light—can escape. Even though a black hole itself emits essentially no radiation, a number of black holes have been discovered beyond our solar system. This is typically done by detecting the x rays emitted by gas falling toward a black hole. However, other methods are still being explored, and in 2015, two merging large black holes were detected by their unique emission of gravitational waves.

During their death throes, stars return the gas of which they are made to interstellar space. (Figure 1-8 shows these expelled gases expanding away from the site of a supernova explosion.) This gas contains heavy elements—that is, elements heavier than hydrogen

FIGURE 1-9   **R I V U X G**

A Galaxy This spectacular galaxy, called M63, contains about 100 billion stars. M63 has a diameter of about 60,000 light-years and is located about 35 million light-years from Earth. Along this galaxy's spiral arms you can see a number of glowing clumps. Like the Orion Nebula in our own Milky Way Galaxy (see Figure 1-7), these are sites of active star formation. (2004–2013 R. Jay GaBany, Cosmotography.com)

and helium—that were created during the star's lifetime by nuclear reactions in its interior. Interstellar space thus becomes enriched with newly manufactured atoms and molecules. The Sun and its planets were formed from interstellar material that was enriched in this way. This means that the atoms of iron and nickel that make up Earth, as well as the carbon in our bodies and the oxygen we breathe, were created deep inside ancient stars. By studying stars and their evolution, we are really studying our own origins.

CONCEPT CHECK 1-2

If a star were twice as massive as our Sun, would it shine for a longer or shorter span of time? Why?

Answer appears at the end of the chapter.

1-4 By observing galaxies, astronomers learn about the origin and fate of the universe

Stars are not spread uniformly across the universe but are grouped together in huge assemblages called **galaxies.** Galaxies come in a wide range of shapes and sizes. Our Sun is just one star in a galaxy we call the Milky Way (see the figure that opens this chapter). A typical galaxy, like our Milky Way, contains several hundred billion stars. Some galaxies are much smaller, containing only a few million stars. Others are monstrosities that devour neighboring galaxies in a process called "galactic cannibalism."

Our Milky Way Galaxy has arching spiral arms like those of the galaxy shown in **Figure 1-9**. These arms are particularly active sites of star formation. In recent years, astronomers have discovered a mysterious object at the center of the Milky Way with a mass millions of times greater than that of our Sun. It now seems certain that this curious object is an enormous black hole.

Some of the most intriguing galaxies appear to be in the throes of violent convulsions and are rapidly expelling matter. The centers of these strange galaxies, which may harbor even more massive black holes, are often powerful sources of x rays and radio waves.

Even more awesome sources of energy are found still deeper in space. Often located at distances so great that their light takes billions of years to reach Earth, we find the mysterious **quasars.** Although in some ways quasars look like nearby stars (**Figure 1-10**), they are among the most distant and most luminous objects in the sky. A typical quasar shines with the brilliance of a

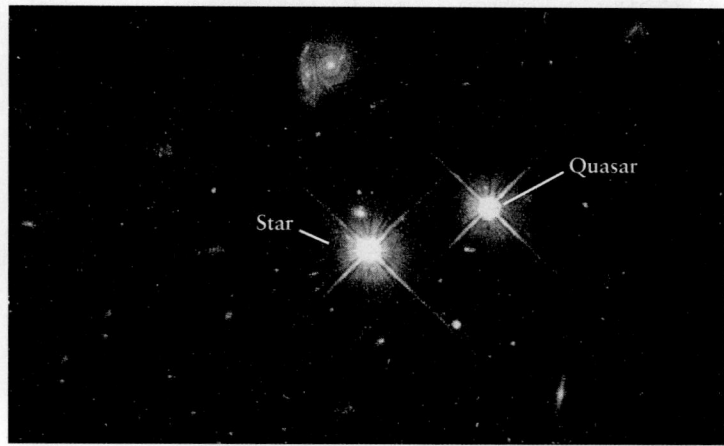

FIGURE 1-10   **R I V U X G**

A Quasar The two bright starlike objects in this image look almost identical, but they are dramatically different. The object on the left is indeed a star that lies a few hundred light-years from Earth. But the "star" on the right is actually a quasar about 9 billion light-years away. To appear so bright even though they are so distant, quasars like this one must be some of the most luminous objects in the universe. The other objects in this image are galaxies like that in Figure 1-9. (Charles Steidel, California Institute of Technology/NASA) hundred galaxies. Detailed observations of quasars imply that they draw their energy from material falling into enormous black holes.

Galaxies and the Expanding Universe

The motions of distant galaxies reveal that they are moving away from us and from each other. In other words, the universe is *expanding*. Extrapolating into the past, we learn that the universe must have been born from an incredibly dense state some 13.7 billion years ago. A variety of evidence indicates that at that moment—the beginning of time— the universe began with a cosmic explosion, known as the **Big Bang,** which occurred throughout all space.

> The motions of distant galaxies motivate the ideas of the expanding universe and the Big Bang.

Thanks to the combined efforts of astronomers and physicists, we are making steady advances in understanding the nature and history of the universe. This understanding may reveal the origin of some of the most basic properties of physical reality. Studying the most remote galaxies is also helping to answer questions about the ultimate fate of the universe. Such studies suggest that the expansion of the universe will continue forever and that it is actually gaining speed.

LOOKING DEEPER 1.1 The work of unraveling the deepest mysteries of the universe requires specialized tools, including telescopes, spacecraft, and computers. But for many purposes, the most useful device for studying the universe is the human brain itself. Our goal in this book is to help you use *your* brain to share in the excitement of scientific discovery.

In the remainder of this chapter, we introduce some of the key concepts and mathematics that we will use in subsequent chapters. Study these carefully, for you will use them over and over again throughout your own study of astronomy.

1-5 Astronomers use angles to denote the positions and apparent sizes of objects in the sky

Whether they study planets, stars, galaxies, or the very origins of the universe, astronomers must know where to point their telescopes. For this reason, an important part of astronomy is keeping track of the positions of objects in the sky. A system for measuring angles is an essential part of this aspect of astronomy (**Figure 1-11**).

An **angle** measures the opening between two lines that meet at a point. A basic unit to express angles is the **degree,** designated by the symbol °. A full circle is divided into 360°, and a right angle measures 90° (Figure 1-11a). As Figure 1-11b shows, if you draw

> Angles are a tool that we will use throughout our study of astronomy.

lines from your eye to each of the two "pointer stars" in the Big Dipper, the angle between these lines—that is, the **angular distance** between these two stars—is about 5°. (In Chapter 2, we will see that these two stars "point" to Polaris, the North Star.) The angular distance between the stars that make up the top and bottom of the Southern Cross, which is visible from south of the equator, is about 6° (Figure 1-11c).

Astronomers also use angles to describe the apparent size of a celestial object—that is, how wide the object appears in the sky. For example, the angle covered by the diameter of the full moon is about ½° (Figure 1-11a). We therefore say that the **angular diameter** (or **angular size**) of the Moon is ½°. Alternatively, astronomers say that the Moon **subtends,** or extends over, an angle of ½°. Ten full moons could fit side by side between the two pointer stars in the Big Dipper.

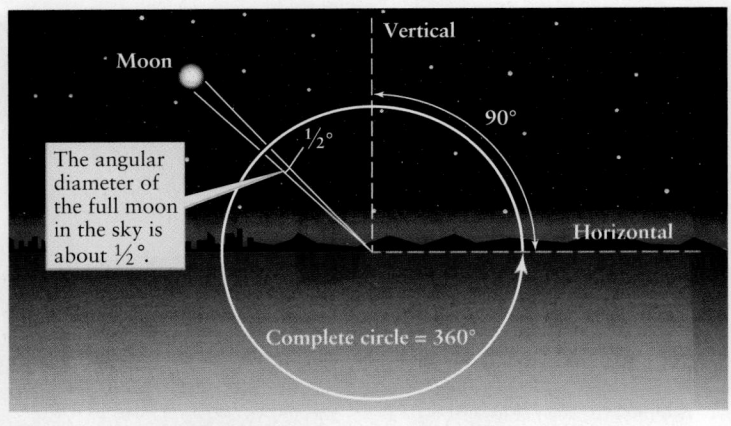

(a) Measuring angles in the sky

(b) Angular distances in the northern hemisphere

(c) Angular distances in the southern hemisphere

FIGURE 1-11

Measuring Angles (a) Angles are measured in degrees (°). There are 360° in a complete circle and 90° in a right angle. For example, the angle between the vertical direction (directly above you) and the horizontal direction (toward the horizon) is 90°. The angular diameter of the full moon in the sky is about 1/2°. **(b)** The seven bright stars that make up the Big Dipper can be seen from anywhere in the northern hemisphere. The angular distance between the two "pointer stars" at the front of the Big Dipper is about 5°. **(c)** The four bright stars that make up the Southern Cross can be seen from anywhere in the southern hemisphere. The angular distance between the stars at the top and bottom of the cross is about 6°.

FIGURE 1-12

Estimating Angles with Your Hand The adult human hand extended to arm's length can be used to estimate angular distances and angular sizes in the sky.

The average adult human hand held at arm's length provides a means of estimating angles, as Figure 1-12 shows. For example, a fist covers an angle of about 10°, whereas a fingertip is about 1° wide. You can use various segments of your index finger extended to arm's length to estimate angles a few degrees across.

To talk about smaller angles, we subdivide the degree into 60 **arcminutes** (also called minutes of arc), which is commonly abbreviated as 60 arcmin or 60′. An arcminute is further subdivided into 60 **arcseconds** (or seconds of arc), usually written as 60 arcsec or 60″. Thus,

$$1° = 60 \text{ arcmin} = 60'$$
$$1' = 60 \text{ arcsec} = 60''$$

For instance, on January 1, 2007, the planet Saturn had an angular diameter of 19.6 arcsec (19.6″) as viewed from Earth. That is a convenient, precise statement of how big the planet appeared in Earth's sky on that date. (Because this angular diameter is so small, to the naked eye Saturn appears simply as a point of light. To see any detail on Saturn, such as the planet's rings, requires a telescope.)

If we know the angular size of an object as well as the distance to that object, we can determine the actual linear size of the object (measured in kilometers or miles, for example). Box 1-1 describes how this is done.

BOX 1-1   TOOLS OF THE ASTRONOMER'S TRADE

The Small-Angle Formula

ANIMATION 1.2 ➤ You can estimate the angular sizes of objects in the sky with your hand and fingers (see Figure 1-12). Using rather more sophisticated equipment, astronomers can measure angular sizes to a fraction of an arcsecond. Keep in mind, however, that *angular* size is not the same as *actual* size. As an example, if you extend your arm while looking at a full moon, you can completely cover the Moon with your thumb. That's because from your perspective, your thumb has a larger angular size (that is, it subtends a larger angle) than the Moon. But the actual size of your thumb (about 2 centimeters) is much less than the actual diameter of the Moon (more than 3000 kilometers).

The accompanying figure shows how the angular size of an object is related to its linear size. Part (**a**) of the figure shows that for a given angular size, the more distant the object, the larger its actual size. For example, your fingertip held at arm's length covers the full moon, but the Moon is much farther away and is far larger in linear size. Part (**b**) shows that for a given linear size, the angular size decreases the farther away the object. This is why a car looks smaller and smaller as it drives away from you.

We can put these relationships together into a single mathematical expression called the **small-angle formula**. Suppose that an object subtends an angle α (the Greek letter alpha) and is at a distance *d* from the observer, as in part (**c**) of the figure. If the angle α is small, as is almost always the case for objects in the sky, the width or linear size (*D*) of the object is given by the following expression:

(a) For a given angular size α, the more distant the object, the greater its actual (linear) size.

(b) For a given linear size, the more distant the object, the smaller its angular size.

(c) Relating an object's linear size *D*, angular size α, and distance *d*

(a) Two objects that have the same angular size may have different linear sizes if they are at different distances from the observer. **(b)** For an object of a given linear size, the angular size is smaller the farther the object is from the observer. **(c)** The small-angle formula relates the linear size *D* of an object to its angular size α and its distance *d* from the observer.

The small-angle formula

$$D = \frac{\alpha d}{206,265}$$

(continued on the next page)

BOX 1-1 *(continued)*

D = width or linear size of an object

α = angular size of the object, in arcsec

d = distance to the object

The number 206,265 is required in the formula so that the units of α are in arcseconds. (206,265 is the number of arcseconds in a complete 360° circle divided by the number 2π.) As long as the same units are used for D and d, any units for linear distance can be used (km, light-years, etc.).

The following examples show two different ways to use the small-angle formula. In both examples we follow a four-step process: Evaluate the *situation* given in the example, decide which *tools* are needed to solve the problem, use those tools to find the *answer* to the problem, and *review* the result to see what it tells you. Throughout this book, we'll use these same four steps in *all* examples that require the use of formulas. We encourage you to follow this four-step process when solving problems for homework or exams. You can remember these steps by their acronym: *S.T.A.R.*

EXAMPLE: On December 11, 2006, Jupiter was 944 million kilometers from Earth and had an angular diameter of 31.2 arcsec. From this information, calculate the actual diameter of Jupiter in kilometers.

Situation: The astronomical object in this example is Jupiter, and we are given its distance d and its angular size α (the same as angular diameter). Our goal is to find Jupiter's diameter D.

Tools: The equation to use is the small-angle formula, which relates the quantities d, α, and D. Note that when using this formula, the angular size α must be expressed in arcseconds.

Answer: The small-angle formula as given is an equation for D. Plugging in the given values $\alpha = 31.2$ arcsec and $d = 944$ million kilometers,

$$D = \frac{31.2 \times 944,000,000 \text{ km}}{206,265} = 143,000 \text{ km}$$

Because the distance d to Jupiter is given in kilometers, the diameter D is also in kilometers.

Review: Does our answer make sense? From Appendix 2 at the back of this book, the equatorial diameter of Jupiter measured by spacecraft flybys is 142,984 kilometers, so our calculated answer is very close.

EXAMPLE: Under excellent conditions, a telescope on Earth can see details with an angular size as small as 1 arcsec. What is the greatest distance at which you could see details as small as 1.7 meters (the height of a typical person) under these conditions?

Situation: Now the object in question is a person, whose linear size D we are given. Our goal is to find the distance d at which the person has an angular size α equal to 1 arcsec.

Tools: Again we use the small-angle formula to relate d, α, and D.

Answer: We first rewrite the formula to solve for the distance d, then plug in the given values $D = 1.7$ m and $\alpha = 1$ arcsec:

$$d = \frac{206,265 D}{\alpha} = \frac{206,265 \times 1.7 \text{ m}}{1} = 350,000 \text{ m} = 350 \text{ km}$$

Review: This value is much less than the distance to the Moon, which is 384,000 kilometers. Thus, even the best telescope on Earth could not be used to see an astronaut walking on the surface of the Moon.

CONCEPT CHECK 1-3

Is it possible for a basketball to look bigger than the Moon? Smaller?

Answer appears at the end of the chapter.

1-6 Powers-of-ten notation is a useful shorthand system for writing numbers

Astronomy is a subject of extremes. Astronomers investigate the largest structures in the universe, including galaxies and clusters of galaxies. But they must also study atoms and atomic nuclei, among the smallest objects in the universe, in order to explain how and why stars shine. They also study conditions ranging from the incredibly hot and dense centers of stars to the frigid near-vacuum of interstellar space. To describe such a wide range of phenomena, we need an equally wide range of both large and small numbers.

> Learning powers-of-ten notation will help you deal with very large and very small numbers.

Powers-of-Ten Notation: Large Numbers

Astronomers avoid such confusing terms as "a million billion billion" by using a standard shorthand system called **powers-of-ten notation**. All the cumbersome zeros that accompany a large number are consolidated into one term consisting of the number 10 followed by an **exponent**, which is written as a superscript.

The exponent indicates how many zeros you would need to write out the long form of the number. Thus,

$$10^0 = 1 \text{ (one)}$$

$$10^1 = 10 \text{ (ten)}$$

$$10^2 = 100 \text{ (one hundred)}$$

$$10^3 = 1000 \text{ (one thousand)}$$

$$10^4 = 10,000 \text{ (ten thousand)}$$

$$10^6 = 1,000,000 \text{ (one million)}$$

$$10^9 = 1,000,000,000 \text{ (one billion)}$$

$$10^{12} = 1,000,000,000,000 \text{ (one trillion)}$$

and so forth. The exponent also tells you how many tens must be multiplied together to give the desired number, which is why the exponent is also called the **power of ten.** For example, ten thousand can be written as 10^4 ("ten to the fourth" or "ten to the fourth power") because $10^4 = 10 \times 10 \times 10 \times 10 = 10,000$.

In powers-of-ten notation, numbers are written as a figure between 1 and 10 multiplied by the appropriate power of ten. The approximate distance between Earth and the Sun, for example, can be written as 1.5×10^8 kilometers (or 1.5×10^8 km, for short). Once you get used to it, this is more convenient than writing "150,000,000 kilometers" or "one hundred and fifty million kilometers." (The same number could also be written as 15×10^7 or 0.15×10^9, but the preferred form is *always* to have the first figure be between 1 and 10.)

Calculators and Powers-of-Ten Notation

Most electronic calculators use a shorthand for powers-of-ten notation. To enter the number 1.5×10^8, you first enter 1.5, then press a key labeled "EXP" or "EE," then enter the exponent 8. (The EXP or EE key takes care of the " $\times 10$" part of the expression.) The number will then appear on your calculator's display as "1.5 E 8," "1.5 8," or some variation of this; typically the " $\times 10$" is not displayed as such. There are some variations from one kind of calculator to another, so you should spend a few minutes reading your calculator's instruction manual to make sure you know the correct procedure for working with numbers in powers-of-ten notation. You will be using this notation continually in your study of astronomy, so this is time well spent.

CAUTION!

Confusion can result from the way that calculators display powers-of-ten notation. Since 1.5×10^8 is displayed as "1.5 8" or "1.5 E 8," it is not uncommon to think that 1.5×10^8 is the same as 1.5^8. That is not correct, however; 1.5^8 is equal to 1.5 multiplied by itself 8 times, or 25.63, which is not even close to $150,000,000 = 1.5 \times 10^8$. Another, not uncommon, mistake is to write 1.5×10^8 as 15^8. If you are inclined to do this, perhaps you are thinking that you can multiply 1.5 by 10, then tack on the exponent later—another process that does not work: 15^8 is equal to 15 multiplied by itself 8 times, or 2,562,890,625, which again is nowhere near 1.5×10^8. Reading the manual for your calculator will help you to avoid these common errors.

Powers-of-Ten Notation: Small Numbers

You can use powers-of-ten notation for numbers that are less than 1 by using a minus sign in front of the exponent. A negative exponent tells you to *divide* by the appropriate number of tens. For example, 10^{-2} ("ten to the minus two") means to divide by 10 twice, so $10^{-2} = 1/10 \times 1/10 = 1/100 = 0.01$. This same idea tells us how to interpret other negative powers of ten:

$10^0 = 1$ (one)

$10^{-1} = 1/10 = 0.1$ (one-tenth)

$10^{-2} = 1/10 \times 1/10 = 1/10^2 = 0.01$ (one-hundredth)

$10^{-3} = 1/10 \times 1/10 \times 1/10 = 1/10^3 = 0.001$ (one-thousandth)

$10^{-4} = 1/10 \times 1/10 \times 1/10 \times 1/10 = 1/10^4 = 0.0001$ (one-ten-thousandth)

$10^{-6} = 1/10 \times 1/10 \times 1/10 \times 1/10 \times 1/10 \times 1/10 = 1/10^6 = 0.000001$ (one-millionth)

$10^{-12} = 1/10 \times 1/10 \times 1/10 \times 1/10 \times 1/10 \times 1/10 \times 1/10 \times 1/10 \times 1/10 \times 1/10 \times 1/10 \times 1/10 = 1/10^{12} = 0.000000000001$ (one-trillionth)

and so forth.

As these examples show, negative exponents tell you how many tenths must be multiplied together to give the desired number. For example, one-ten-thousandth, or 0.0001, can be written as 10^{-4} ("ten to the minus four") because $10^{-4} = 1/10 \times 1/10 \times 1/10 \times 1/10 = 0.0001$.

A useful shortcut in converting a decimal to powers-of-ten notation is to notice where the decimal point is. For example, the decimal point in 0.0001 is four places to the left of the "1," so the exponent is -4, that is, $0.0001 = 10^{-4}$.

You can also use powers-of-ten notation to express a number like 0.00245, which is not a multiple of $1/10$. For example, $0.00245 = 2.45 \times 0.001 = 2.45 \times 10^{-3}$. (Again, the standard for powers-of-ten notation is that the first figure is a number between 1 and 10.) This notation is particularly useful when dealing with very small numbers. A good example is the diameter of a hydrogen atom, which is much more convenient to state in powers-of-ten notation (1.1×10^{-10} meter, or 1.1×10^{-10} m) than as a decimal (0.00000000011 m) or a fraction (110 trillionths of a meter.)

Because it bypasses all the awkward zeros, powers-of-ten notation is ideal for describing the size of objects as small as atoms or as big as galaxies (**Figure 1-13**). **Box 1-2** explains how powers-of-ten notation also makes it easy to multiply and divide numbers that are very large or very small.

1-7 Astronomical distances are often measured in astronomical units, light-years, or parsecs

Astronomers use many of the same units of measurement as do other scientists. They often measure lengths in meters (abbreviated m), masses in kilograms (kg), and time in seconds (s). (You can read more about these units of measurement, as well as techniques

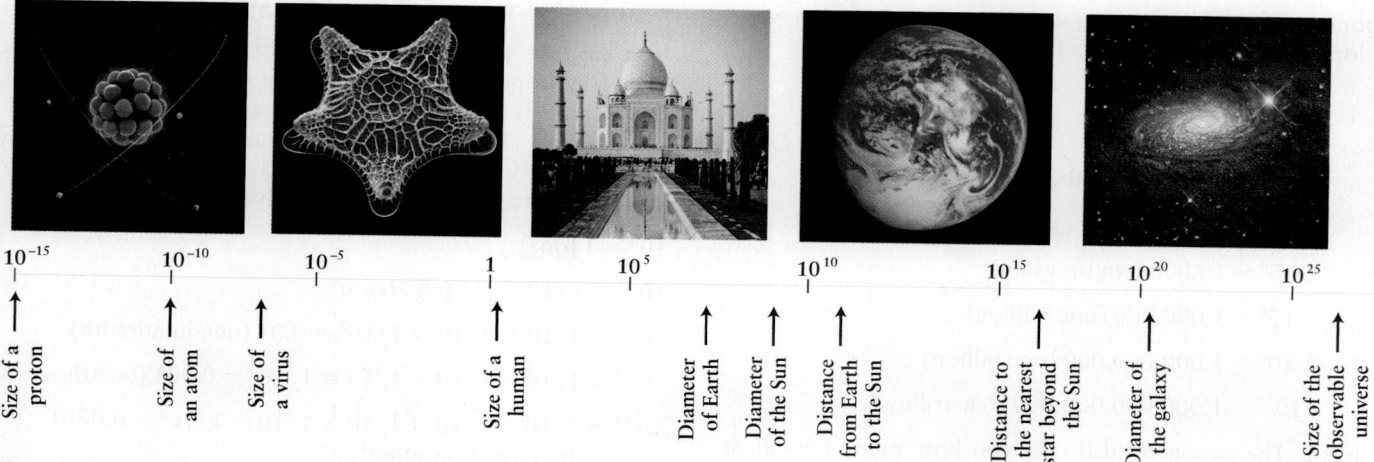

FIGURE 1-13

Examples of Powers-of-Ten Notation This scale gives the sizes of objects in meters, ranging from subatomic particles at the left to the entire observable universe at the right. The protons and neutrons in the illustration on the left form the nucleus of an atom; the electrons orbit about 100,000 times farther out. Second from left is the cell wall of a single-celled aquatic organism called a diatom, 10^{-4} meter (0.1 millimeter) in size. At the center is the Taj Mahal, about 60 meters tall and within reach of our unaided senses. Second from right is the planet Earth, about 10^7 meters in diameter. At the far right is a galaxy, 10^{21} meters (100,000 light-years) in diameter. (Science Photo Library – Andrzej Wojcicki/Getty Images; Steve Gschmeissner/Getty Images; iStockphoto/Thinkstock/Getty Images; NASA/JPL/Johns Hopkins University; 2004–2013 R. Jay GaBany, Cosmotography.com)

BOX 1-2   TOOLS OF THE ASTRONOMER'S TRADE

Arithmetic with Powers-of-Ten Notation

Using powers-of-ten notation makes it easy to multiply numbers. For example, suppose you want to multiply 100 by 1000. If you use ordinary notation, you have to write a lot of zeros:

$$100 \times 1000 = 100{,}000 \text{ (one hundred thousand)}$$

By converting these numbers to powers-of-ten notation, we can write this same multiplication more compactly as

$$10^2 \times 10^3 = 10^5$$

Because $2 + 3 = 5$, we are led to the following general rule for *multiplying* numbers expressed in terms of powers of ten: Simply *add* the exponents.

EXAMPLE: $10^4 \times 10^3 = 10^{4+3} = 10^7$.

To *divide* numbers expressed in terms of powers of ten, remember that $10^{-1} = 1/10$, $10^{-2} = 1/100$, and so on. The general rule for any exponent n is

$$10^{-n} = \frac{1}{10^n}$$

In other words, dividing by 10^n is the same as multiplying by 10^{-n}. To carry out a division, you first transform it into multiplication by changing the sign of the exponent, and then carry out the multiplication by adding the exponents.

EXAMPLE: $\dfrac{10^4}{10^6} = 10^4 \times 10^{-6} = 10^{4+(-6)} = 10^{4-6} = 10^{-2}$

Usually a computation involves numbers like 3.0×10^{10}, that is, an ordinary number multiplied by a factor of 10 with an exponent. In such cases, to perform multiplication or division, you can treat the numbers separately from the factors of 10^n.

EXAMPLE: We can redo the first numerical example from Box 1-1 in a straightforward manner by using exponents:

$$D = \frac{31.2 \times 944{,}000{,}000 \text{ km}}{206{,}265}$$

$$= \frac{3.12 \times 10 \times 9.44 \times 10^8}{2.06265 \times 10^5} \text{ km}$$

$$= \frac{3.12 \times 9.44 \times 10^{1+8-5}}{2.06265} \text{ km} = 14.3 \times 10^4 \text{ km}$$

$$= 1.43 \times 10 \times 10^4 \text{ km} = 1.43 \times 10^5 \text{ km}$$

for converting between different sets of units, in Box 1-3.)

Like other scientists, astronomers often find it useful to combine these units with powers of ten and create new units using prefixes. As an example, the number $1000 (= 10^3)$ is represented by the prefix "kilo," so a distance of 1000 meters is the same as 1 kilometer (1 km). Here are some of the most common prefixes, with examples of how they are used:

Specialized units make it easier to comprehend immense cosmic distances.

$$\text{one-billionth meter} = 10^{-9}\ \text{m} = 1\ \text{nanometer}$$
$$\text{one-millionth second} = 10^{-6}\ \text{s} = 1\ \text{microsecond}$$
$$\text{one-thousandth arcsecond} = 10^{-3}\ \text{arcsec} = 1\ \text{milliarcsecond}$$
$$\text{one-hundredth meter} = 10^{-2}\ \text{m} = 1\ \text{centimeter}$$
$$\text{one thousand meters} = 10^3\ \text{m} = 1\ \text{kilometer}$$
$$\text{one million tons} = 10^6\ \text{tons} = 1\ \text{megaton}$$

In principle, we could express all sizes and distances in astronomy using units based on the meter. Indeed, we will use kilometers to give the diameters of Earth and the Moon, as well as the Earth–Moon distance. But, while a kilometer (roughly equal to three-fifths of a mile) is an easy distance for humans to visualize, a megameter (10^6 m) is not. For this reason, astronomers have devised units of measurement that are more appropriate for the tremendous distances between the planets and the far greater distances between the stars.

When discussing distances across the solar system, astronomers use a unit of length called the **astronomical unit** (abbreviated **au**). This is the average distance between Earth and the Sun:

$$1\ \text{au} = 1.496 \times 10^8\ \text{km} = 92.96\ \text{million miles}$$

Thus, the average distance between the Sun and Jupiter can be conveniently stated as 5.2 au.

ANIMATION 1.1 To talk about distances to the stars, astronomers use two different units of length. The **light-year** (abbreviated **ly**) is the distance that light travels in 1 year. This is a useful concept because the speed of light in empty space always has the same value, 3.00×10^5 km/s (kilometers per second) or 1.86×10^5 mi/s (miles per second). In terms of kilometers or astronomical units, 1 light-year is given by

$$1\ \text{ly} = 9.46 \times 10^{12}\ \text{km} = 63{,}240\ \text{au}$$

This distance is roughly equal to 6 trillion miles.

CAUTION!

Keep in mind that despite its name, the light-year is a unit of distance and *not* a unit of time. As an example, Proxima Centauri, the nearest star other than the Sun, is a distance of 4.2 light-years from Earth. This means that light takes 4.2 years to travel to us from Proxima Centauri.

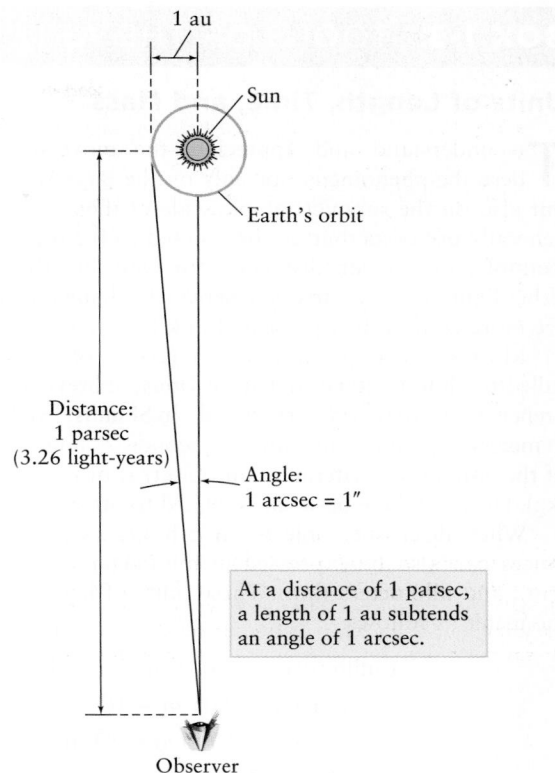

FIGURE 1-14

A Parsec The parsec, a unit of length commonly used by astronomers, is equal to 3.26 light-years. The parsec is defined as the distance at which 1 au perpendicular to the observer's line of sight subtends an angle of 1 arcsec.

Physicists often measure interstellar distances in light-years because the speed of light is one of nature's most important numbers. But many astronomers prefer to use another unit of length, the *parsec,* because its definition is closely related to a method of measuring distances to the stars.

Imagine taking a journey far into space, beyond the orbits of the outer planets. As you look back toward the Sun, Earth's orbit subtends, or extends over, a smaller angle in the sky the farther you are from the Sun. As **Figure 1-14** shows, the distance at which 1 au subtends an angle of 1 arcsec is defined as 1 **parsec** (abbreviated **pc**):

$$1\ \text{pc} = 3.09 \times 10^{13}\ \text{km} = 3.26\ \text{ly}$$

The distance from Earth to Proxima Centauri can be stated as 1.3 pc or as 4.2 ly. Whether you choose to use parsecs or light-years is a matter of personal taste.

For even greater distances, astronomers commonly use **kiloparsecs** and **megaparsecs** (abbreviated **kpc** and **Mpc**). As we saw before, these prefixes simply mean "thousand" and "million," respectively:

$$1\ \text{kiloparsec} = 1\ \text{kpc} = 1000\ \text{pc} = 10^3\ \text{pc}$$
$$1\ \text{megaparsec} = 1\ \text{Mpc} = 1{,}000{,}000\ \text{pc} = 10^6\ \text{pc}$$

For example, the distance from Earth to the center of our Milky Way Galaxy is about 8 kpc, and the galaxy shown in Figure 1-9 is about 11 Mpc away.

Some astronomers prefer to talk about thousands or millions of light-years rather than kiloparsecs and megaparsecs. Once

BOX 1-3   TOOLS OF THE ASTRONOMER'S TRADE

Units of Length, Time, and Mass

To understand and appreciate the universe, we need to describe phenomena not only on the large scales of galaxies but also on the submicroscopic scale of the atom. Astronomers generally use units that are best suited to the topic at hand. For example, interstellar distances are conveniently expressed in either light-years or parsecs, whereas the diameters of the planets are more comfortably presented in kilometers.

Most scientists prefer to use a version of the metric system called the International System of Units, abbreviated SI (after the French name Système International). In **SI units**, length is measured in meters (m), time is measured in seconds (s), and mass (a measure of the amount of material in an object) is measured in kilograms (kg). How are these basic units related to other measures?

When discussing objects on a human scale, sizes and distances can also be expressed in millimeters (mm), centimeters (cm), and kilometers (km). These units of length are related to the meter as follows:

$$1 \text{ millimeter} = 0.001 \text{ m} = 10^{-3} \text{ m}$$
$$1 \text{ centimeter} = 0.01 \text{ m} = 10^{-2} \text{ m}$$
$$1 \text{ kilometer} = 1000 \text{ m} = 10^{3} \text{ m}$$

A useful set of conversions for the English system is

$$1 \text{ in.} = 2.54 \text{ cm}$$
$$1 \text{ ft} = 0.3048 \text{ m}$$
$$1 \text{ mi} = 1.609 \text{ km}$$

Each of these equalities can also be written as a fraction equal to 1. For example, you can write

$$\frac{0.3048 \text{ m}}{1 \text{ ft}} = 1 \quad \text{or} \quad \frac{1 \text{ ft}}{0.3048 \text{ m}} = 1$$

Fractions like these are useful for converting a quantity from one set of units to another. For instance, the Saturn V rocket used to send astronauts to the Moon stands about 363 ft tall. How can we convert this height to meters? The trick is to remember that a quantity does not change if you multiply it by 1. Expressing the number 1 by the fraction (0.3048 m)/(1 ft), we can write the height of the rocket as

$$363 \text{ ft} \times 1 = 363 \text{ ft} \times \frac{0.3048 \text{ m}}{1 \text{ ft}} = 111 \frac{\text{ft} \times \text{m}}{\text{ft}} = 111 \text{ m}$$

EXAMPLE: The diameter of Mars is 6794 km. Let's try expressing this in miles.

CAUTION!

You can get into trouble if you are careless in applying the trick of taking the number whose units are to be converted and multiplying it by 1. For example, if we multiply the diameter by 1 expressed as (1.609 km)/(1 mi), we get

$$6794 \text{ km} \times 1 = 6794 \text{ km} \times \frac{1.609 \text{ km}}{1 \text{ mi}} = 10{,}930 \frac{\text{km}^2}{\text{mi}}$$

The unwanted units of km did not cancel, so this answer cannot be right. Furthermore, a mile is larger than a kilometer, so the diameter expressed in miles should be a smaller number than when expressed in kilometers.

The correct approach is to write the number 1 so that the unwanted units *will* cancel. The number we are starting with is in kilometers, so we must write the number 1 with kilometers in the denominator ("downstairs" in the fraction). Thus, we express 1 as (1 mi)/(1.609 km):

$$6794 \text{ km} \times 1 = 6794 \text{ km} \times \frac{1 \text{ mi}}{1.609 \text{ km}}$$
$$= 4222 \text{ km} \times \frac{\text{mi}}{\text{km}} = 4222 \text{ mi}$$

Now the units of km cancel as they should, and the distance in miles is a smaller number than in kilometers (as it must be).

When discussing very small distances such as the size of an atom, astronomers often use the micrometer (μm) or the nanometer (nm). These are related to the meter as follows:

$$1 \text{ micrometer} = 1 \text{ μm} = 10^{-6} \text{ m}$$
$$1 \text{ nanometer} = 1 \text{ nm} = 10^{-9} \text{ m}$$

Thus, 1 μm = 10^3 nm. (Note that the micrometer is often called the micron.)

The basic unit of time is the second (s). It is related to other units of time as follows:

$$1 \text{ minute (min)} = 60 \text{ s}$$
$$1 \text{ hour (h)} = 3600 \text{ s}$$
$$1 \text{ day (d)} = 86{,}400 \text{ s}$$
$$1 \text{ year (y)} = 3.156 \times 10^{7} \text{ s}$$

(continued on the next page)

BOX 1-3 *(continued)*

In the SI system, speed is properly measured in meters per second (m/s). Quite commonly, however, speed is also expressed in km/s and mi/h:

$$1 \text{ km/s} = 10^3 \text{ m/s}$$
$$1 \text{ km/s} = 2237 \text{ mi/h}$$
$$1 \text{ mi/h} = 0.447 \text{ m/s}$$
$$1 \text{ mi/h} = 1.47 \text{ ft/s}$$

In addition to using kilograms, astronomers sometimes express mass in grams (g) and in solar masses ($M_\odot$), where the subscript $\odot$ is the symbol denoting the Sun. It is especially convenient to use solar masses when discussing the masses of stars and galaxies. These units are related to each other as follows:

$$1 \text{ kg} = 1000 \text{ g}$$
$$1 \text{ M}_\odot = 1.99 \times 10^{30} \text{ kg}$$

CAUTION!

You may be wondering why we have not given a conversion between kilograms and pounds (lb). The reason is that these units do not refer to the same physical quantity! A kilogram is a unit of *mass*, which is a measure of the amount of material in an object. By contrast, a pound is a unit of *weight*, which tells you how strongly gravity pulls on that object's material. Consider a person who weighs 110 pounds on Earth, corresponding to a mass of 50 kg. Gravity is only about one-sixth as strong on the Moon as it is on Earth, so on the Moon this person would weigh only one-sixth of 110 pounds, or about 18 pounds. But that person's mass of 50 kg is the same on the Moon; wherever you go in the universe, you take all of your material along with you. We will explore the relationship between mass and weight in Chapter 4.

again, the choice is a matter of personal taste. As a general rule, astronomers use whatever measuring sticks seem best suited for the issue at hand and do not restrict themselves to one system of measurement. For instance, an astronomer might say that the supergiant star Antares has a diameter of 860 million kilometers and is located at a distance of 185 parsecs from Earth. The *Cosmic Connections* figure on the following page shows where these different systems are useful.

1-8 Astronomy is an adventure of the human mind

An underlying theme of this book is that the behavior of the universe is governed by underlying laws of nature. It is not a hodgepodge of unrelated things behaving in unpredictable ways. Rather, we find strong evidence that fundamental laws of physics govern the nature of the universe and the behavior of everything in it. These unifying concepts enable us to explore realms far removed from our earthly experience. Thus, a scientist can do experiments in a laboratory to determine the properties of light or the behavior of atoms and then use this knowledge to investigate the structure of the universe. Through careful testing, the laws of physics discovered on Earth have been found to apply throughout the universe at the most distant locations and in the distant past. Indeed, Albert Einstein expressed the view that *"the eternal mystery of the world is its comprehensibility."*

> Studying the universe benefits our lives on Earth.

The discovery of fundamental laws of nature has had a profound influence on humanity. These laws have led to an immense number of practical applications that have fundamentally transformed commerce, medicine, entertainment, transportation, and other aspects of

our lives. In particular, space technology has given us instant contact with any point on the globe through communication satellites, precise navigation to any point on Earth using signals from the satellites of the Global Positioning System (GPS), and accurate weather forecasts from meteorological satellites (**Figure 1-15**).

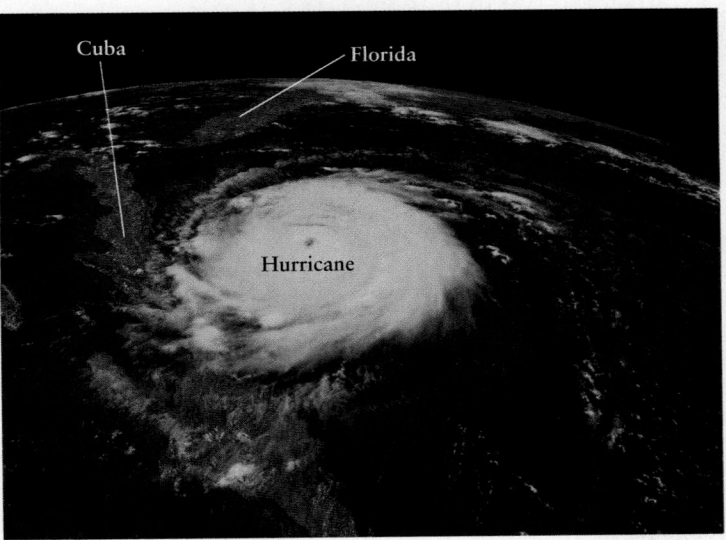

FIGURE 1-15 R I Ⓥ **U X G**

A Hurricane Seen from Space This image of Hurricane Frances was made on September 2, 2004, by GOES-12 (Geostationary Operational Environmental Satellite 12). A geostationary satellite like GOES-12 orbits around Earth's equator every 24 hours, the same length of time it takes the planet to make a complete rotation. Hence, this satellite remains over the same spot on Earth, from which it can monitor the weather continuously. By tracking hurricanes from orbit, GOES-12 makes it much easier for meteorologists to give early warning of these immense storms. The resulting savings in lives and property more than pay for the cost of the satellite. (NASA, NOAA)

COSMIC CONNECTIONS

Sizes in the Universe

ANIMATION
1.3

Powers-of-ten notation provides a convenient way to express the sizes of astronomical objects and distances in space. This illustration suggests the immense distances within our solar system, the far greater distances between the stars within our Milky Way Galaxy, and the truly cosmic distances between galaxies.

NSO/AURA/NSF

Sun: diameter = 1.39×10^6 km

The Sun, Earth, and other planets are members of our solar system.

Neptune

Uranus

Mercury   Saturn

Venus   Sun

Earth

Mars   Jupiter

Diameter of Neptune's orbit: 60 au
1 au (astronomical unit) =
1.50×10^8 km = average
Earth–Sun distance

The Sun is a typical star. Typical distances between our neighboring stars = 1 to 5 ly

1 ly = distance that light travels in one year = 6.32×10^4 au

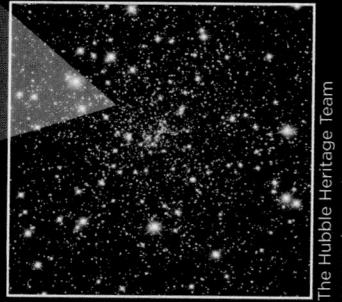

The Hubble Heritage Team (AURA/STScI/NASA)

NASA

Earth: diameter = 1.28×10^4 km

Galaxies are grouped into clusters, which can be up to 10^7 ly across.

NOAO/NASA

NASA/JPL-Caltech/R. Hurt (SSC)

Our Sun is one of more than 10^{11} stars in the Milky Way Galaxy.

Distance from the center of the Milky Way to the Sun = 2.8×10^4 ly

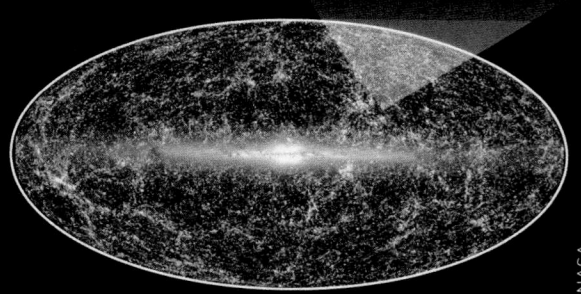

NASA

Each of the 1.6×10^6 dots in this map of the entire sky represents a relatively nearby galaxy. This is a tiny fraction of the number of galaxies in the observable universe.

As important as the applications of science are, the pursuit of scientific knowledge for its own sake is no less important. We are fortunate to live in an age in which this pursuit is in full flower. Just as explorers such as Columbus and Magellan discovered the true size of our planet in the fifteenth and sixteenth centuries, astronomers of the twenty-first century are exploring the universe to an extent that is unparalleled in human history. Indeed, even the voyages into space imagined by such great science fiction writers as Jules Verne and H. G. Wells pale in comparison to today's reality. Over a few short decades, humans have walked on the Moon, sent robotic spacecraft to dig into the Martian soil and explore the satellites of Saturn, and used the most powerful telescopes ever built to probe the limits of the observable universe. Never before has so much been revealed in so short a time.

As you proceed through this book, you will learn about the tools that scientists use to explore the natural world, as well as what they observe with these tools. But, most important, you will see how astronomers build from their observations an understanding of the universe in which we live. It is this search for understanding that makes science more than merely a collection of data and elevates it to one of the great adventures of the human mind. It is an adventure that will continue as long as there are mysteries in the universe—an adventure we hope you will come to appreciate and share.

KEY WORDS

Terms preceded by an asterisk (*) are discussed in the Boxes.

angle, p. 8
angular diameter (angular size), p. 8
angular distance, p. 8
arcminute (′, minute of arc), p. 9
arcsecond (″, second of arc), p. 9
astronomical unit (au), p. 13

Big Bang, p. 8
black hole, p. 6
degree (°), p. 8
exponent, p. 10
galaxy, p. 7
hypothesis, p. 2
kiloparsec (kpc), p. 13
laws of physics, p. 4
light-year (ly), p. 13

megaparsec (Mpc), p. 13
meteorite, p. 4
model, p. 2
nebula (*plural* nebulae), p. 6
Newtonian mechanics, p. 3
parsec (pc), p. 13
power of ten, p. 11
powers-of-ten notation, p. 10
pulsar, p. 6

quasar, p. 7
scientific method, p. 2
*SI units, p. 14
*small-angle formula, p. 9
solar system, p. 4
subtend (an angle), p. 8
supernova (*plural* supernovae), p. 6
theory, p. 2

KEY IDEAS

Astronomy, Science, and the Nature of the Universe: The universe is comprehensible. The scientific method is a procedure for formulating hypotheses about the universe. Hypotheses are tested by observation or experimentation in order to build consistent models or theories that accurately describe phenomena in nature.

Observations of the heavens have helped scientists discover some of the fundamental laws of physics. The laws of physics are in turn used by astronomers to interpret their observations (Section 1-1).

The Solar System: Exploration of the planets provides information about the origin and evolution of the solar system, as well as about the history and resources of Earth (Section 1-2).

adventtr/Getty Images

Stars and Nebulae: Studying the stars and nebulae helps us learn about the origin and history of the Sun and the solar system (Section 1-3).

2004–2013 R. Jay GaBany, Cosmotography.com

Galaxies: Observations of galaxies tell us about the origin and history of the universe (Section 1-4).

Angles: Astronomers use angles to denote the positions and sizes of objects in the sky. The size of an angle is measured in degrees, arcminutes, and arcseconds (Section 1-5).

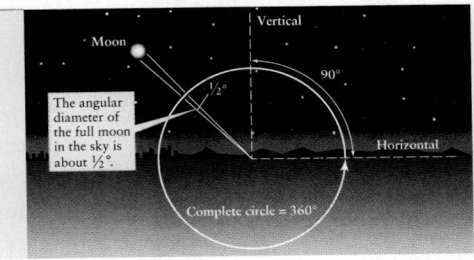

Moon

Vertical

90°

½°

Horizontal

The angular diameter of the full moon in the sky is about ½°.

Complete circle = 360°

The angular distance between the two pointer stars at the front of the Big Dipper is about 5°, roughly 10 times the angular diameter of the Moon.

5°

6°

The angular distance between the stars at the top and bottom of the Southern Cross is about 6°.

Powers-of-Ten Notation: This convenient shorthand system for writing numbers allows very large and very small numbers to be expressed in a compact form (Section 1-6).

Units of Distance: Astronomers use a variety of distance units. These include the astronomical unit (the average distance from Earth to the Sun), the light-year (the distance that light travels in one year), and the parsec (Section 1-7).

QUESTIONS

Problem-solving tips and tools

The small-angle formula, given in Box 1-1, relates the size of an astronomical object to the angle it subtends. Box 1-3 illustrates how to convert from one unit of measure to another. An object traveling at speed v for a time t covers a distance d given by $d = vt$; for example, a car traveling at 90 km/h (v) for 3 h (t) covers a distance $d = (90 \text{ km/h})(3 \text{ h}) = 270$ km. Similarly, the time t required to cover a given distance d at speed v is $t = d/v$; for example, if $d = 270$ km and $v = 90$ km/h, then $t = (270 \text{ km})/(90 \text{ km/h}) = 3$ h.

Section 1-1

1. **TUTORIAL 1.1** What is the difference between a hypothesis and a theory?
2. What is the difference between a theory and a law of physics?
3. How are scientific theories tested?
4. Describe the role that skepticism plays in science.

5. Describe one reason why it is useful to have telescopes in space.
6. What is the meaning of the letters **R I V U X G** that appear with some of the figures in this chapter? Why in each case is one of the letters highlighted? (*Hint:* See the Preface that precedes Chapter 1.)

Section 1-2

7. What caused the craters on the Moon?
8. What are meteorites? Why are they important for understanding the history of the solar system?

Section 1-3

9. What makes the Sun and stars shine?
10. What role do nebulae like the Orion Nebula play in the life stories of stars?

Section 1-4

11. What is the difference between a solar system and a galaxy?

Section 1-5

12. What are degrees, arcminutes, and arcseconds used for? What are the relationships among these units of measure?
13. How many arcseconds equal 1°?
14. With the aid of a diagram, explain what it means to say that the Moon subtends an angle of ½°.
15. **Box 1-1** Explain where the number 206,265 in the small-angle formula comes from.

16. **Box 1-1** At what distance would a person have to hold a European 2-euro coin (which has a diameter of about 2.6 cm) in order for the coin to subtend an angle of (**a**) 1°? (**b**) 1 arcmin? (**c**) 1 arcsec? Give your answers in meters.
17. **Box 1-1** A person with good vision can see details that subtend an angle of as small as 1 arcmin. If two dark lines on an eye chart are 2 mm apart, how far can such a person be from the chart and still be able to tell that there are two distinct lines? Give your answer in meters.
18. **Box 1-1** The average distance from Earth to the Moon is 384,000 km, and the Moon subtends an angle of ½°. Use this information to calculate the diameter of the Moon in kilometers.
19. **Box 1-1** Suppose your telescope can give you a clear view of objects and features that subtend angles of at least 2 arcsec. What is the diameter in kilometers of the smallest crater you can see on the Moon? (*Hint:* See the preceding question.)
20. **Box 1-1** On April 18, 2006, the planet Venus was a distance of 0.869 au from Earth. The diameter of Venus is 12,104 km. What was the angular size of Venus as seen from Earth on April 18, 2006? Give your answer in arcminutes.
21. **Box 1-1** (**a**) Use the information given in the caption to Figure 1-7 to determine the angular size of the Orion Nebula. Give your answer in degrees. (**b**) How does the angular diameter of the Orion Nebula compare to the angular diameter of the Moon?

Section 1-6

22. What is an exponent? How are exponents used in powers-of-ten notation?
23. What are the advantages of using powers-of-ten notation?
24. Write the following numbers using powers-of-ten notation: (**a**) ten million, (**b**) sixty thousand, (**c**) four one-thousandths, (**d**) thirty-eight billion, (**e**) your age in months.
25. The diameter of the Sun is 1.4×10^{11} cm, and the distance from Earth to the nearest star, Proxima Centauri, is 4.2 ly. Suppose you want to build an exact scale model of the Sun and Proxima Centauri, and you are using a ball 30 cm in diameter to represent the Sun. In your scale model, how far away would Proxima Centauri be from the Sun? Give your answer in kilometers, using powers-of-ten notation.
26. How many Suns would it take, laid side by side, to reach the nearest star? Use powers-of-ten notation. (*Hint:* See the preceding question.)
27. A hydrogen atom has a radius of about 5×10^{-9} cm. The radius of the observable universe is about 14 billion light-years. How many times larger than a hydrogen atom is the observable universe? Use powers-of-ten notation.

28. The Sun's mass is 1.99×10^{30} kg, three-quarters of which is hydrogen. The mass of a hydrogen atom is 1.67×10^{-27} kg. How many hydrogen atoms does the Sun contain? Use powers-of-ten notation.

29. The average distance from Earth to the Sun is 1.496×10^8 km. Express this distance (a) in light-years and (b) in parsecs. Use powers-of-ten notation. (c) Are light-years or parsecs useful units for describing distances of this size? Explain.

30. The speed of light is 3.00×10^8 m/s. How long does it take light to travel from the Sun to Earth? Give your answer in seconds, using powers-of-ten notation. (*Hint:* See the preceding question.)

31. When the Voyager 2 spacecraft sent back pictures of Neptune during its flyby of that planet in 1989, the spacecraft's radio signals traveled for 4 hours at the speed of light to reach Earth. How far away was the spacecraft? Give your answer in kilometers, using powers-of-ten notation. (*Hint:* See the preceding question.)

32. The star Altair is 5.15 pc from Earth. (a) What is the distance to Altair in kilometers? Use powers-of-ten notation. (b) How long does it take for light emanating from Altair to reach Earth? Give your answer in years. (*Hint:* You do not need to know the value of the speed of light.)

33. The age of the universe is about 13.7 billion years. What is this age in seconds? Use powers-of-ten notation.

Section 1-7

34. How is an astronomical unit (au) defined? Give an example of a situation in which this unit of measure would be convenient to use.

35. What is the advantage to the astronomer of using the light-year as a unit of distance?

36. What is a parsec? How is it related to a kiloparsec and to a megaparsec?

37. Give the word or phrase that corresponds to the following standard abbreviations: (a) km, (b) cm, (c) s, (d) km/s, (e) mi/h, (f) m, (g) m/s, (h) h, (i) y, (j) g, (k) kg. Which of these are units of speed? (*Hint:* You may have to refer to a dictionary. All of these abbreviations should be part of your working vocabulary.)

38. In the original (1977) *Star Wars* movie, Han Solo praises the speed of his spaceship by saying, "It's the ship that made the Kessel run in less than 12 parsecs!" Explain why this statement is obvious misinformation.

39. A reporter once described a light-year as "the time it takes light to reach us traveling at the speed of light." How would you correct this statement?

Section 1-8

40. How do astronomical observations differ from those of other sciences?

General Problems

41. Scientists assume that "reality is rational." Discuss what this means and the thinking behind it.

42. All scientific knowledge is inherently provisional. Discuss whether this is a weakness or a strength of the scientific method.

WEB/E-BOOK QUESTIONS

43. Use the links given in the *Universe* website or e-book, Chapter 1, to learn about the Orion Nebula (Figure 1-7). Can the nebula be seen with the naked eye? Does the nebula stand alone, or is it part of a larger cloud of interstellar material? What has been learned by examining the Orion Nebula with telescopes sensitive to infrared light?

44. Use the links given in the *Universe* website or e-book, Chapter 1, to learn more about the Crab Nebula (Figure 1-8). When did observers on Earth see the supernova that created this nebula? Does the nebula emit any radiation other than visible light? What kind of object is at the center of the nebula?

ACTIVITIES

Observing Projects

45. On a dark, clear, moonless night, can you see the Milky Way from where you live? If so, briefly describe its appearance. If not, what seems to be interfering with your ability to see the Milky Way?

46. Look up at the sky on a clear, cloud-free night. Is the Moon in the sky? If so, does it interfere with your ability to see the fainter stars? Why do you suppose astronomers prefer to schedule their observations on nights when the Moon is not in the sky?

47. Look up at the sky on a clear, cloud-free night and note the positions of a few prominent stars relative to such reference markers as rooftops, telephone poles, and treetops. Also note the location from where you make your observations. A few hours later, return to that location and again note the positions of the same bright stars that you observed earlier. How have their positions changed? From these changes, can you deduce the general direction in which the stars appear to be moving?

Starry Night™ Explorations

48. Install the *Starry Night*™ software and set up your home location. Familiarize yourself with the interface by completing the tutorial available by clicking the **Toggle SkyGuide** icon in the toolbar. Then set up a wide-angle view centered on the

Moon, as seen from your home location, one hour before sunrise on January 1, 2019.

(a) Use the HUD to identify the bright object in the sky near to the Moon. What type of object is this and what is its name? In which direction is the Moon in the sky relative to this object?

(b) Use the "D" keyboard shortcut to advance the time of the view by one calendar day. In which direction is the Moon now, relative to the bright object you identified earlier?

(c) What is the second bright object in the sky, about the same distance on the other side of the Moon as the bright object you have already identified?

(d) Click the **Quick Search Items** icon in the toolbar to open the **Planets…** panel. Use the features of this panel to apply labels to the bright objects on either side of the Moon and to see their magnified images. Observe the relative phases of these objects, as determined by the distribution of light and darkness on the object as seen from Earth. Do these objects exhibit the same or different phases? Formulate a hypothesis to explain this observation.

49. Use *Starry Night™* to measure the spacing between several stars in the sky to explore the concept of angular distance, also called angular separation. Open **Favorites > Explorations > Angular Distance** to view the northern sky from Key West, Florida, USA. The on-screen info shows the longitude and latitude of this location. Some of the brighter stars are indicated with their proper names, and the names and outlines of several asterisms (groupings of stars) are also shown. The stars in the Big Dipper asterism outline the shape of a "dipper," a tool that might be used for scooping water from a barrel. The two stars in the Big Dipper on the opposite side of the scoop from the handle, Merak and Dubhe, can be seen to point to the brightest star in the Little Dipper, Polaris.

(a) Use the **Angular Separation** cursor to measure the spacing between the two "pointer stars," Merak and Dubhe. What is the angular separation between these stars?

(b) What is the angular distance between Dubhe and Polaris?

(c) Approximately how many pointer-star spacings are there between Dubhe and Polaris; that is, what is the ratio of the angle separating Dubhe and Polaris compared to the angle between the pointer stars Dubhe and Merak?

(d) Click **Play** to see that, as time progresses, Polaris appears to remain fixed while the other stars circle it. This is because Polaris lies very close to the north celestial pole, the point in the sky directly above the north pole of Earth's spin axis. This coincidence makes Polaris, also called the Pole Star, a handy navigation aid for northern hemisphere observers because it indicates the approximate direction of true north on Earth. Click **Stop** and select **View > Celestial Guides > Celestial Poles** from the menu to show the position of the north celestial pole on the sky. Zoom in and measure the angular distance

between Polaris and this celestial pole. What angle separates Polaris, the Pole Star, from the north celestial pole?

(e) Use **File > Revert** to reset the view and then measure the angle between Polaris and the north point on the horizon. Compare this measurement with the latitude of Key West shown in the on-screen info. What strikes you about these two angles? If this relationship is general, how might it be used to estimate your location on Earth?

50. Imagine yourself in an open field, far from city lights beneath a clear, moonless night sky. Now use *Starry Night™* to make that image virtually real by opening **Favorites > Explorations > Orion**. Use the hand tool or keyboard shortcuts (N, S, W, E, and Z) to survey the sky. Click the **Quick Search Items** icon to open the **Options…** panel. In this panel, expand **Display Options > Constellations** and turn on **Labels** and **Illustrations**. Survey the sky once more to find the constellation Orion. In this illustration, the "belt of Orion" is a very recognizable line of three stars, as is the line of stars making up the sword hanging from his belt. As you will see, what appears as the central star in this sword is in fact something quite different!

(a) Use the HUD to identify the three stars of Orion's belt. What are the proper names of the three stars that comprise this recognizable line in the sky?

(b) Select **File > Revert** from the menu and use the angular separation cursor to draw a line from the upper star in Orion's belt, through the belt to the very bright star near the southeast horizon. What angle separates these two stars? What is the proper name of the bright star near the horizon?

(c) Zoom in slowly as you observe the middle star in the sword of Orion. As the view closes in, you will notice this star becomes fuzzy. Continue to zoom in to reveal a telescopic view of the detailed structure of this middle "star" in Orion's sword. From this vast cloud of glowing gas and dust in the sky, new stars are being born. Select **Labels > Messier Objects** from the menu. Designated M42, by what other name is this object known?

(d) Adjust the zoom and then use the angular separation cursor to obtain a rough measurement of the angular size of M42. Convert this measurement to arcseconds; use the result, along with the Distance from Observer for M42 from the HUD, to apply the small angle formula to estimate the physical size of this vast stellar nursery in our galaxy. Approximately how far across, in light years, is M42 in space?

Collaborative Exercises

51. A scientific theory is fundamentally different from the everyday use of the word *theory*. List and describe any three scientific theories of your choice and creatively imagine

three additional hypothetical theories that are not scientific. Briefly describe what is scientific and what is nonscientific about each of these theories.

52. Angles describe how far apart two objects appear to an observer. From where you are currently sitting, estimate the angular distance between the floor and the ceiling at the front of the room you are sitting in, the angular distance between the two people sitting closest to you, and the angular size of a clock or an exit sign on the wall. Draw sketches to illustrate each answer and describe how each of your answers would change if you were standing in the very center of the room.

53. Astronomers use powers of ten to describe the distances to objects. List an object or place that is located at very roughly each of the following distances from you: 10^{-2} m, 10^0 m, 10^1 m, 10^3 m, 10^7 m, 10^{10} m, and 10^{20} m.

ANSWERS

ConceptChecks

ConceptCheck 1-1: While a hypothesis is a testable idea that seems to explain an observation about nature, a scientific theory represents a set of well-tested and internally consistent hypotheses that are able to successfully and repeatedly predict the outcome of experiments and observations.

ConceptCheck 1-2: Shorter lifetime. The larger star has twice as much mass as the Sun. However, it consumes its nuclear fuel much more than twice as fast, so that its lifetime is shorter than the Sun's.

ConceptCheck 1-3: Yes. The apparent size of an object is given by how wide it appears (in other words, the angle that it subtends). By moving a basketball near and far, it can appear very large or very small.

Why Astronomy? by Sandra M. Faber

As you study astronomy, you may ask, "Why am I studying this subject? What good is it for people in general and for me in particular?" Admittedly, astronomy does not offer the same practical benefits as other sciences, so how can it be important to your life?

On the most basic level, I think of astronomy as providing the ultimate background for human history. Recorded history goes back about 3000 years. For knowledge of the time before that, we consult archeologists and anthropologists about early human history and paleontologists, biologists, and geologists about the evolution of life and of our planet—altogether going back some five billion years. Astronomy tells us about the time before that, the ten billion years or so when the Sun, solar system, and Milky Way Galaxy formed, and even about the origin of the universe in the Big Bang. Knowledge of astronomy is part of a well-educated person's view of history.

Astronomy challenges our belief system and impels us to put our "philosophical house" in order. For example, the Bible says the world and everything in it were created in six days by the hand of God. However, according to the ancient Egyptians, Earth arose spontaneously from the infinite waters of the eternal universe, called Nun. Alaskan legends teach that the world was created by the conscious imaginings of a deity named Father Raven.

Modern astronomy, supported by physics and observations, differs from these stories of the creation of Earth. Astronomers believe that the Sun formed about five billion years ago by gravitational collapse from a dense cloud of interstellar gas and dust. At the same time, and over a period of several hundred thousand years, the planets condensed within the swirling solar nebula. Astronomers have actually seen young stars form in this way.

At issue here, really, is the question of how we are to gain information about the nature of the physical world—whether by revelation and intuition or by logic and observation. Where science stops and faith begins is a thorny issue for everyone, but particularly for astronomers—and for astronomy students.

Astronomy cultivates our notions about cosmic time and cosmic evolution. Given the short span of human life, it is all too easy to overlook that the universe is a dynamic place. This idea implies fragility—if something can change, it might even some day disappear. For instance, in another five billion years or so, the Sun will swell up and brighten to 1000 times its present luminosity, incinerating Earth in the process. This is far enough in the future that neither you nor I need to feel any personal responsibility for preparing to meet this challenge. However, other cosmic catastrophes will inevitably occur before then. Earth will be hit by a sizable piece of space debris—craters show that this happens every few million years or so. Enormous volcanic eruptions have occurred in the past and will certainly occur again. Another Ice Age is virtually certain to begin within the next 20,000 years, unless we first cook Earth ourselves by burning too much fossil fuel.

Such common notions as the inevitability of human progress, the desirability of endless economic growth, and Earth's ability to support its human population are all based on limited experience—they will probably not prove viable in the long run. Consequently, we must rethink who we are as a species and what is our proper activity on Earth. These long-term problems involve the whole human race and are vital to our survival and well-being. Astronomy is essential to developing a perspective on human existence and its relation to the cosmos.

Many astronomers believe that the ultimate, proper concept of "home" for the human race is our universe. It seems increasingly likely that a large number of other universes exist, with the vast majority incapable of harboring intelligent life as we know it. The parallel with Earth is striking. Among the solar system planets, only Earth can support human life. Among the great number of planets in our galaxy, only a small fraction may be such that we can call them home. The fraction of hospitable universes is likely to be smaller still. It seems, then, that our universe is the ultimate "home," a sanctuary in a vast sea of inhospitable universes.

I began this essay by talking about history and ended with issues that border on the ethical and religious. Astronomy is like that: It offers a modern-day version of Genesis—and of the Apocalypse, too. I hope that during this course you will be able to take time out to contemplate the broader implications of what you are studying. This is one of the rare opportunities in life to think about who you are and where you and the human race are going. Don't miss it.

Sandra M. Faber is professor of astronomy at the University of California, Santa Cruz, and astronomer at Lick Observatory. She chaired the now legendary group of astronomers called the Seven Samurai, who surveyed the nearest 400 elliptical galaxies and discovered a new mass concentration, called the Great Attractor. Dr. Faber received the Bok Prize from Harvard University in 1978, was elected to the National Academy of Sciences in 1985, and in 1986 won the coveted Dannie Heineman Prize from the American Astronomical Society, in recognition of her sustained body of especially influential astronomical research.

LEARNING GOALS

It is a clear night on Mauna Kea, a dormant volcano on the island of Hawaii. As you gaze toward the north, as in this time-exposure photograph, you find that the stars are not motionless. Rather, they move in counterclockwise circles around a fixed point above the northern horizon. Stars close to this point never dip below the horizon, whereas stars farther from the fixed point rise in the east and set in the west. These motions fade from view when the Sun rises in the east and illuminates the sky. The Sun, too, arcs across the sky in the same manner as the stars. At day's end, when the Sun sets in the west, the panorama of stars is revealed for yet another night.

These observations are at the heart of *naked-eye astronomy*—the sort that requires no equipment but human vision. Naked-eye astronomy cannot tell us what the Sun is made of or how far away the stars are. But by studying naked-eye astronomy, you will learn the answers to equally profound questions, such as why there are seasons, why the night sky is different at different times of year, and why the night sky looks different in Australia than in North America. In discovering the answers to these questions, you will learn how Earth moves through space and will begin to understand our true place in the cosmos.

A time-exposure photograph captures stars tracing circles and cars driving on a mountain road. (Oliver Fluck)

R I V U X G

2-1 Naked-eye astronomy had an important place in civilizations of the past

Positional astronomy—the study of the positions of objects in the sky and how these positions change—has roots that extend far back in time. Four to five thousand years ago, the inhabitants of the British Isles erected stone structures, such as Stonehenge, that suggest a preoccupation with the motions of the sky. Alignments of these stones appear to show the changing locations where the Sun rose and set at key times during the year. Stoneworks of a different sort but with a similar astronomical purpose are found in the Americas, from the southwestern United States to the Andes of Peru. The ancestral Puebloans (also called Anasazi) of modern-day New Mexico, Arizona, Utah, and Colorado created stone carvings that were illuminated by the Sun on the first days of summer or winter (Figure 2-1). At the Incan city of Machu Picchu in Peru, a narrow window carved in a rock 2 meters (about 6.5 feet) thick looks out on the sunrise only on December 21 each year. It is thought that the need to keep track of seasons for farming—to know when to sow seeds—provided at least one motivation for early astronomical alignments.

> The astronomical knowledge of ancient peoples is the foundation of modern astronomy.

Other ancient peoples designed buildings with astronomical orientations. The great Egyptian pyramids, built around 3000 B.C.E., are oriented north–south and east–west with an accuracy much better than 1 degree. Similar alignments are found in the grand tomb of Shih Huang Ti (259 B.C.E.–210 B.C.E.), the first emperor of China.

Evidence of a highly sophisticated understanding of astronomy can be found in the written records of the Mayan civilization of Central America. Mayan astronomers deduced by observation that the apparent motions of the planet Venus follow a cycle that repeats every 584 days. They also developed a technique for calculating the position of Venus on different dates. The Maya believed that Venus was associated with war, so such calculations were important for choosing the most promising dates on which to attack an enemy.

These archaeological discoveries bear witness to an awareness of naked-eye astronomy by the peoples of many cultures. Many of the concepts of modern positional astronomy come to us from these ancients, including the idea of dividing the sky into constellations.

2-2 Eighty-eight constellations cover the entire sky

Looking at the sky on a clear, dark night, you might think that you can see millions of stars. Actually, the unaided human eye can detect only about 6000 stars. Because half of the sky is below the horizon at any one time, you can see at most about 3000 stars. When ancient peoples looked at these thousands of stars, they sometimes imagined that groupings of prominent stars

> The constellations provide a convenient framework for stating the position of an object in the heavens.

traced out pictures in the sky. Astronomers still refer to these groupings, called **constellations** (from the Latin for "group of stars").

You may already be familiar with some of these pictures or patterns in the sky, such as the Big Dipper, which is part of the large constellation Ursa Major (the Great Bear). Many constellations, such as Orion in Figure 2-2, have names derived from the myths and legends of antiquity. Although some star groupings vaguely resemble the figures they are supposed to represent (see Figure 2-2c), most require some imagination to recognize.

The term *constellation* has a broader definition in present-day astronomy. On modern star charts, the entire sky is divided into 88 regions, each of which is called a constellation. For example, the constellation Orion is now defined to be an irregular patch of sky whose borders are shown in Figure 2-2b. When astronomers refer to the "Great Nebula" M42 in Orion, they mean that, as seen from Earth, this nebula appears to be within Orion's patch of sky. Some constellations cover large areas of the sky (Ursa Major being one of the biggest) and others very small areas (Crux, the Southern Cross, being the smallest). But because the modern constellations cover the entire sky, every star lies in one constellation or another.

FIGURE 2-1 R I ⓥ U X G

The Sun Dagger at Chaco Canyon On the first day of winter, rays of sunlight passing between stone slabs bracket a spiral stone carving, or petroglyph, at Chaco Canyon in New Mexico. A single band of light strikes the center of the spiral on the first day of summer. This astronomically aligned petroglyph and others were carved by the ancestral Puebloan culture between 850 and 1250 C.E. (Karl Kernberger @ Solstice Project)

(a)

(b)

(c)

FIGURE 2-2 R I Ⓥ U X G

Three Views of Orion The constellation Orion is easily seen on nights from December through March. **(a)** This telescope image of Orion shows many more stars than can be seen with the naked eye. **(b)** A portion of a modern star atlas shows the distances in light-years (ly) to some of the stars in Orion. The yellow lines show the borders between Orion and its neighboring constellations (labeled in capital letters). **(c)** This fanciful drawing from a star atlas published in 1835 shows Orion the Hunter and his neighboring constellation, Taurus the Bull. (a: Eckhard Slawik/Science Source; c: © Historical Picture Archive/CORBIS/Corbis via Getty Images)

CAUTION!

When you look at a constellation's star pattern, it is tempting to conclude that you are seeing a group of stars that are all relatively close together. In fact, most of these stars are nowhere near one another. As an example, Figure 2-2b shows the distances in light-years from Earth to four stars in Orion. Although Bellatrix (Arabic for "the Amazon") and Mintaka ("the belt") appear to be close to each other, Mintaka is actually more than 600 light-years farther away from us. The two stars only *appear* to be close because they are in nearly the same direction as seen from Earth. The same illusion often appears when you see an airplane's lights at night. It is very difficult to tell how far away a single bright light is, which is why you can mistake an airplane a few kilometers away for a star trillions of times more distant.

LOOKING DEEPER 2.1 The star names shown in Figure 2-2b are from the Arabic language. For example, Betelgeuse means "armpit," which makes sense when you look at the star atlas drawing in Figure 2-2c. Other types of names are also used for stars. For instance, Betelgeuse is also known as α Orionis because it is the brightest star in Orion (α, or alpha, is the first letter in the Greek alphabet).

CAUTION!

A number of unscrupulous commercial firms offer to name a star for you for a fee. The money that they charge you for this "service" is real, but the star names are not; none of these names is recognized by astronomers. If you want to use astronomy to commemorate your name or the name of a friend or relative, consider making a donation to your local planetarium or science museum. The money will be put to much better use!

CONCEPT CHECK 2-1

If Jupiter is reported to be in the constellation of Taurus the Bull, does Jupiter need to be within the outline of the bull's body? Why or why not?

Answer appears at the end of the chapter.

2-3 The appearance of the sky changes during the course of the night and from one night to the next

Go outdoors soon after dark, find a spot away from bright lights, and note the patterns of stars in the sky. Do the same a few hours later. You will find that the entire pattern of stars (as well as the

Moon, if it is visible) has shifted its position. New constellations will have risen above the eastern horizon, and some will have disappeared below the western horizon. If you look again before dawn, you will see that the stars that were just rising in the east when the night began are now low in the western sky. This daily motion, or **diurnal motion**, of the stars is apparent in time-exposure photographs (see the photograph that opens this chapter).

By understanding the motions of Earth through space, we can understand why the Sun and stars appear to move in the sky.

If you repeat your observations on the following night, you will find that the motions of the sky are almost but not quite the same. The same constellations rise in the east and set in the west, but a few minutes earlier than on the previous night. If you look again after a month, the constellations visible at a given time of night (say, midnight) will be noticeably different, and after 6 months you will see an almost totally different set of constellations. Only after a year has passed will the night sky have the same appearance as when you began.

Why does the sky go through diurnal motion? Why do the constellations slowly shift from one night to the next? As we will see, the answer to the first question is that Earth *rotates* once a day around an axis running from the north pole to the south pole; the answer to the second question is that Earth also *revolves* once a year around the Sun.

Diurnal Motion and Earth's Rotation

To understand diurnal motion, note that at any given moment, it is daytime on the half of Earth illuminated by the Sun and nighttime on the other half (**Figure 2-3**). Earth rotates from west to east (counterclockwise if viewed looking down on the north pole), making one complete rotation every 24 hours, which is why there is a daily cycle

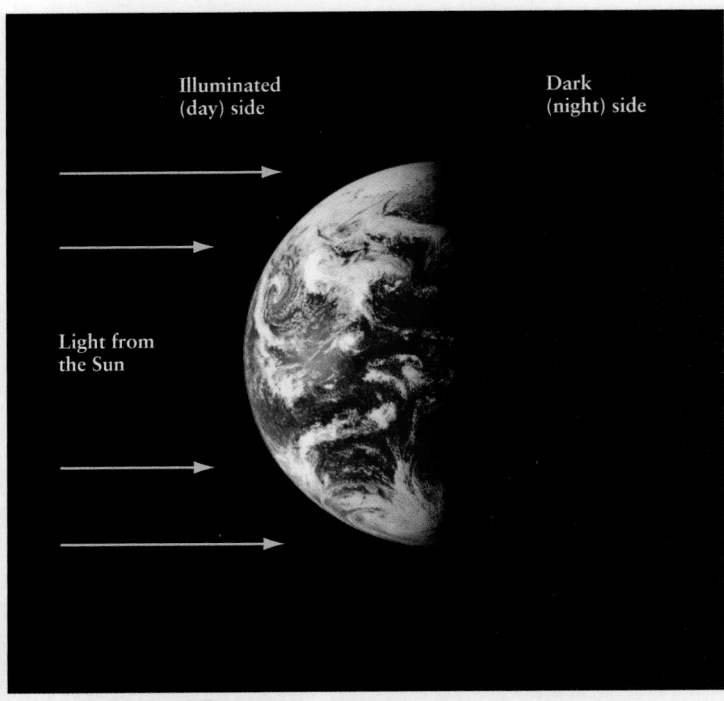

FIGURE 2-3 R I V U X G

Day and Night on Earth At any moment, half of Earth is illuminated by the Sun. As Earth rotates from west to east, your location moves from the dark (night) hemisphere into the illuminated (day) hemisphere and back again. This image was recorded in 1992 by the Galileo spacecraft as it was en route to Jupiter. (NASA/JPL)

of day and night. Because of this rotation, stars seem to rise in the east and set in the west, as do the Sun and Moon.

Figure 2-4 helps to further explain diurnal motion. It shows two views of Earth as seen from a point above the north pole.

View from above the north pole

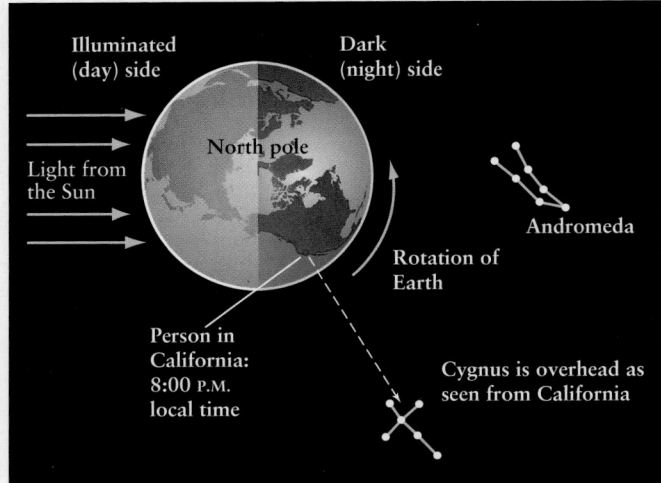

(a) Earth rotating counterclockwise, as seen from above the north pole

View from above the north pole

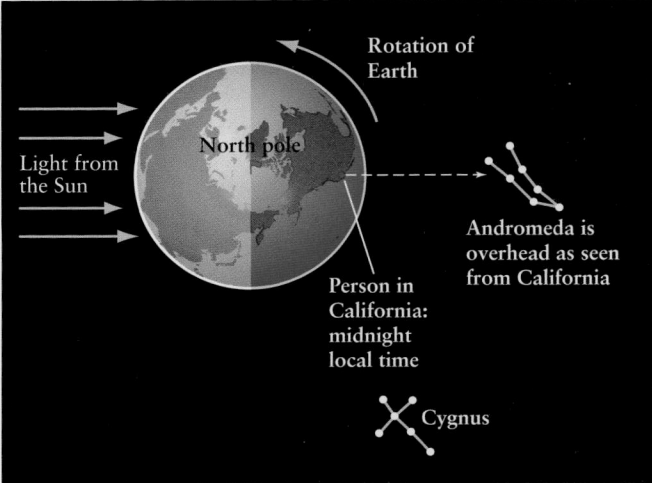

(b) 4 hours (one-sixth of a complete rotation) later

ANIMATION 2.1

FIGURE 2-4

Why Diurnal Motion Happens The diurnal (daily) motion of the stars, the Sun, and the Moon is a consequence of Earth's rotation. **(a)** This drawing shows Earth from a vantage point above the north pole. In this drawing, for a person in California, the local time is 8:00 P.M. and the constellation Cygnus is directly overhead. **(b)** Four hours later, Earth has

made one-sixth of a complete rotation to the east. As seen from Earth, the entire sky appears to have rotated to the west by one-sixth of a complete rotation. It is now midnight in California, and the constellation directly over California is Andromeda.

At the instant shown in Figure 2-4a, it is day in Asia but night in most of North America and Europe. Figure 2-4b shows Earth 4 hours later. Four hours is one-sixth of a complete 24-hour day, so Earth has made one-sixth of a rotation between Figures 2-4a and 2-4b. Europe is now in the illuminated half of Earth (the Sun has risen in Europe), while Alaska has moved from the illuminated to the dark half of Earth (the Sun has set in Alaska). For a person in California, in Figure 2-4a the time is 8:00 P.M. and the constellation Cygnus (the Swan) is directly overhead. Four hours later, the constellation over California is Andromeda (named for a mythological princess). Because Earth rotates from west to east, it appears to us on Earth that the entire sky rotates around us in the opposite direction, from east to west.

CONCEPT CHECK 2-2

People in which of the following cities in North America experience sunrise first: New York, San Francisco, Chicago, or Denver? Explain in terms of Earth's rotation.

CALCULATION CHECK 2-1

If the constellation Cygnus rises along the eastern horizon at sunset, at what time will it be highest above the southern horizon?

Answers appear at the end of the chapter.

Yearly Motion and Earth's Orbit

We described earlier that in addition to the diurnal motion of the sky, the constellations visible in the night sky also change slowly over the course of a year. This happens because Earth orbits, or revolves around, the Sun (Figure 2-5). Over the course of a year, Earth makes one complete *orbit*, and the darkened, nighttime side of Earth gradually turns toward different parts of the heavens. For example, as seen from the northern hemisphere, at midnight in late July the constellation Cygnus is close to overhead; at midnight in late September the constellation Andromeda is close to overhead; and at midnight in late November the constellation Perseus (commemorating a mythological hero) is close to overhead. If you follow a particular star on successive evenings, you will find that it rises approximately 4 minutes earlier each night, or 2 hours earlier each month.

CONCEPT CHECK 2-3

If Earth suddenly rotated on its axis 3 times faster than it does now, how many times would the Sun rise and set each year?

Answer appears at the end of the chapter.

Constellations and the Night Sky

Constellations can help you find your way around the sky. For example, if you live in the northern hemisphere, you can use the Big Dipper in Ursa Major to find the north direction by drawing a straight line through the two stars at the front of the Big Dipper's bowl (see the dashed red arrow in Figure 2-6). The first moderately bright star you come to is Polaris, also called the North Star because it is located almost directly over Earth's north pole. If you draw a line from Polaris straight down to the horizon, you will find the north direction.

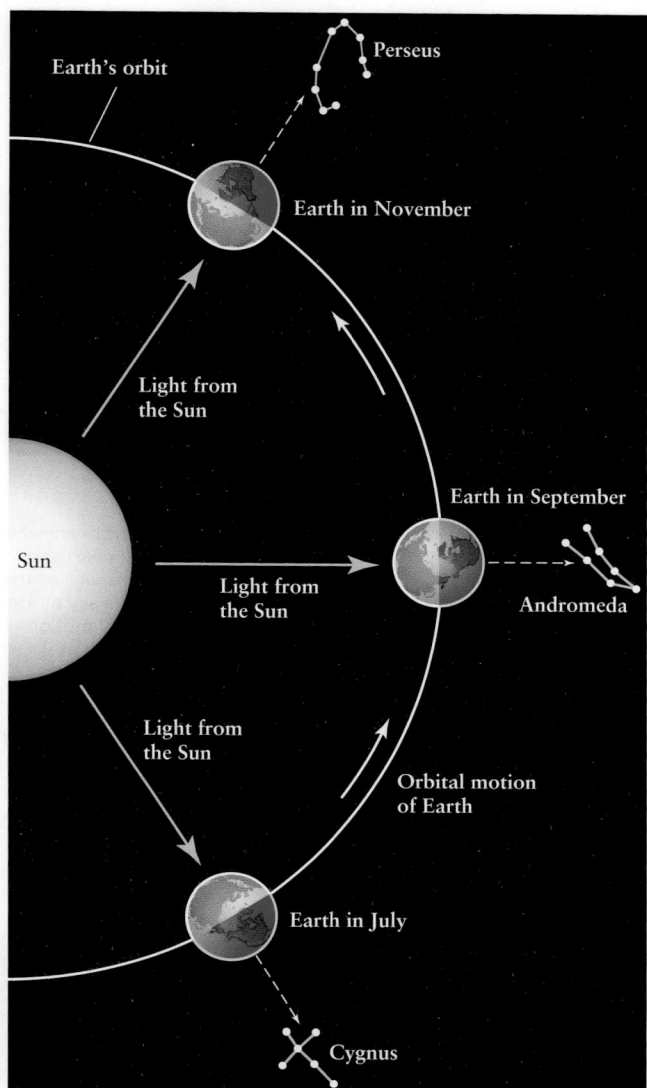

FIGURE 2-5

Why the Night Sky Changes During the Year As Earth orbits around the Sun, the nighttime side of Earth gradually turns toward different parts of the sky. Hence, the particular stars that you see in the night sky are different at different times of the year. This figure shows which constellation is overhead at midnight local time—when the Sun is on the opposite side of Earth from your location—during different months for observers at midnorthern latitudes (including the United States). If you want to view the constellation Andromeda, the best time of the year to do it is in late September, when Andromeda is nearly overhead at midnight.

As Figure 2-6 shows, by following the handle of the Big Dipper you can locate the bright reddish star Arcturus in Boötes (the Shepherd) and the prominent bluish star Spica in Virgo (the Virgin). The saying "Follow the arc to Arcturus and speed to Spica" may help you remember these stars, which are conspicuous in the evening sky during the spring and summer.

During winter in the northern hemisphere, you can see some of the brightest stars in the sky. Many of them are in the vicinity of the "winter triangle" (Figure 2-7), which connects bright stars in the constellations of Orion (the Hunter), Canis Major (the Large Dog), and Canis Minor (the Small Dog).

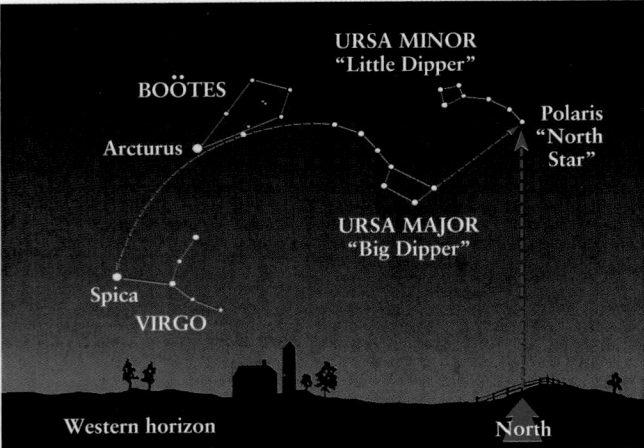

FIGURE 2-6

The Big Dipper as a Guide The North Star can be seen from anywhere in the northern hemisphere on any night of the year. This star chart shows how the Big Dipper can be used to point out the North Star as well as the brightest stars in two other constellations. The chart shows the sky at around 11 P.M. (daylight savings time) on August 1. Due to Earth's orbital motion around the Sun, you will see this same view at 1 A.M. on July 1 and at 9 P.M. on September 1. The angular distance from Polaris to Spica is 102°.

A similar feature, the "summer triangle," graces the summer sky in the northern hemisphere. This triangle connects the brightest stars in Lyra (the Harp), Cygnus (the Swan), and Aquila (the Eagle) (Figure 2-8). A conspicuous portion of the Milky Way forms a beautiful background for these constellations, which are nearly overhead during the middle of summer at midnight.

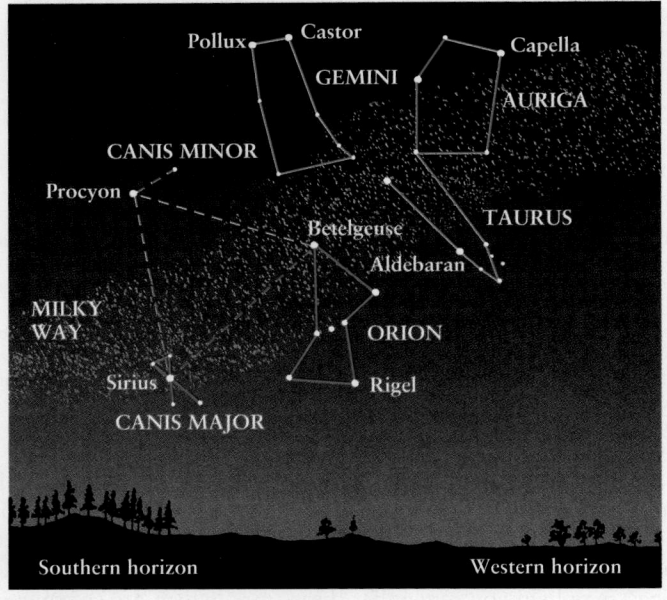

FIGURE 2-7

The "Winter Triangle" This star chart shows the view toward the southwest on a winter evening in the northern hemisphere (around midnight on January 1, 10 P.M. on February 1, or 8 P.M. on March 1). Three of the brightest stars in the sky make up the "winter triangle," which is about 26° on a side. In addition to the constellations involved in the triangle, the chart shows the prominent constellations Gemini (the Twins), Auriga (the Charioteer), and Taurus (the Bull).

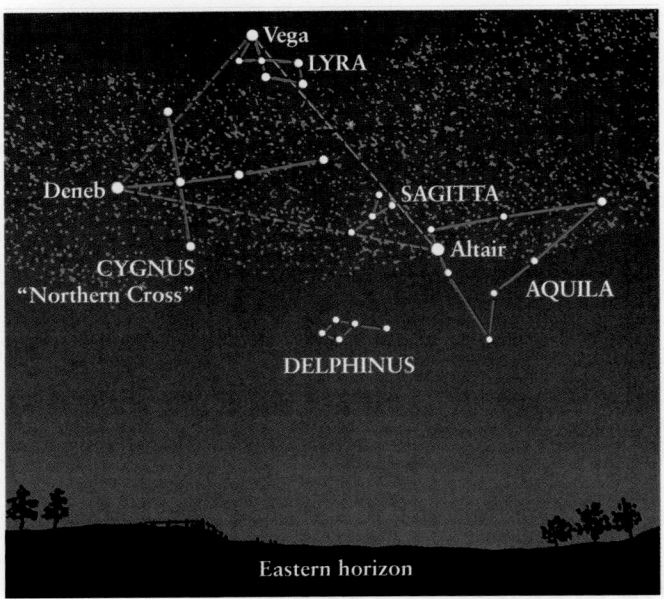

FIGURE 2-8

The "Summer Triangle" This star chart shows the eastern sky as it appears in the evening during spring and summer in the northern hemisphere (around 1 A.M. daylight savings time on June 1, around 11 P.M. on July 1, and around 9 P.M. on August 1). The angular distance from Deneb to Altair is about 38°. The summer triangle is made from the three brightest stars in that area of the sky: Vega, Altair, and Deneb.

A wonderful tool to help you find your way around the night sky is the planetarium program *Starry Night*™, which accompanies certain print copies of this book. (You can also obtain *Starry Night*™ separately.) In addition, at the end of this book you will find a set of selected star charts for the evening hours of all 12 months of the year. You may find stargazing an enjoyable experience, and *Starry Night*™ and the star charts will help you identify many well-known constellations.

Note that all the star charts in this section and at the end of this book are drawn for an observer in the northern hemisphere. If you live in the southern hemisphere, you can see constellations that are not visible from the northern hemisphere, and vice versa. In the next section, we will see why this is so.

CONCEPT CHECK 2-4

Suppose it is midnight in January and that you've sketched a diagram of the stars directly overhead. If you look directly overhead at midnight in May, do you see the same pattern of stars?

Answer appears at the end of the chapter.

2-4 It is convenient to imagine that the stars are located on a celestial sphere

TUTORIAL
2.1

Many ancient societies believed that all the stars are the same distance from Earth. They imagined the stars to be bits of fire embedded in the inner surface of an immense hollow sphere, called the **celestial sphere**, with Earth at its center. In this picture of the universe, Earth was fixed and did not rotate. Instead, the entire celestial sphere rotated once a day around Earth from east to west, thereby explaining the diurnal motion of the sky. The picture of a rotating celestial sphere fit well with naked-eye observations, and for its time was a useful model of how the universe works. (We discussed the role of models in science in Section 1-1).

> The concept of the imaginary celestial sphere helps us visualize the motions of stars in the sky.

Today's astronomers know that this simple model of the universe is not correct. Diurnal motion is due to the rotation of Earth, not the rest of the universe. Furthermore, as we learned when discussing the constellations in Section 2-2, the stars are not all at the same distance from Earth. Indeed, the stars that you can see with the naked eye range from 4.2 to more than 1000 light-years away, and telescopes allow us to see objects at distances of billions of light-years.

Thus, astronomers now recognize that the celestial sphere is an *imaginary* object that has no basis in physical reality. Nonetheless, the celestial sphere model remains a useful tool of positional astronomy. If we imagine, as did the ancients, that Earth is stationary and that the celestial sphere rotates around us, it is relatively easy to specify the directions to different objects in the sky and to visualize the motions of these objects. Figure 2-9 depicts the celestial sphere, with Earth at its center. (This is not to scale; just to contain the nearest star, a truly proportional drawing would leave

Earth as a miniscule dot.) We picture the stars as points of light that are fixed on the inner surface of the celestial sphere.

Other features on the celestial sphere result from **projections**. The projection of a point on Earth is made by extending an imaginary line perpendicular to the surface of Earth until it intersects the celestial sphere. If we project Earth's north and south poles into space, we obtain the **north celestial pole** and the **south celestial pole.** Thus, the two celestial poles are where Earth's axis of rotation intersects the celestial sphere (see Figure 2-9). The star Polaris is less than 1° away from the north celestial pole, which is why it is called the North Star or the Pole Star. If we project Earth's equator out into space, we obtain the **celestial equator.** The celestial equator divides the sky into northern and southern hemispheres, just as Earth's equator divides Earth into two hemispheres.

The point in the sky directly overhead an observer anywhere on Earth is called that observer's **zenith.** The zenith and celestial sphere are shown in Figure 2-10 for an observer located at 35° north latitude (that is, at a location on Earth's surface 35° north of the equator). The zenith is shown at the top of Figure 2-10, so Earth and the celestial sphere appear "tipped" compared to Figure 2-9. At any time, an observer can see only half of the celestial sphere; the other half is below the horizon, hidden by the body of Earth. The hidden half of the celestial sphere is darkly shaded in Figure 2-10.

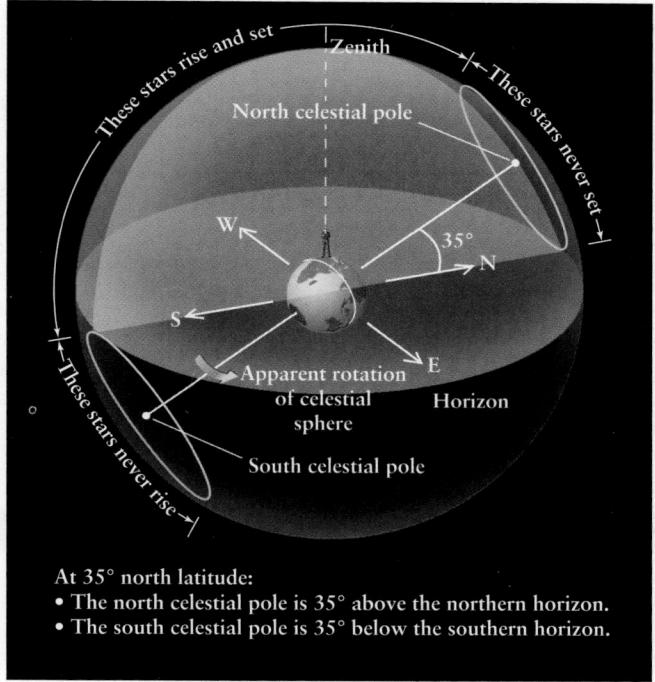

FIGURE 2-10

The View from 35° North Latitude To an observer at 35° north latitude (roughly the latitude of Los Angeles, Atlanta, Tel Aviv, and Tokyo), the north celestial pole is always 35° above the horizon. Stars within 35° of the north celestial pole are circumpolar; they trace out circles around the north celestial pole during the course of the night and are always above the horizon on any night of the year. Stars within 35° of the south celestial pole are always below the horizon and can never be seen from this latitude. Stars that lie between these two extremes rise in the east and set in the west. You can also see in the figure that heading toward the north celestial pole (into space!) is not the same as heading north in a car; for navigating Earth, north is only the portion of Earth's axis that points along the ground. Likewise for south.

ANIMATION 2.2 → **FIGURE 2-9**

The Celestial Sphere The celestial sphere is the apparent sphere of the sky encompassing Earth. The view in this figure is from the outside of this (wholly imaginary) sphere. Because Earth is at the center of the celestial sphere, our view is always from the inside of the sphere. The celestial equator and poles are the projections of Earth's equator and axis of rotation out into space. The celestial poles are therefore located directly over Earth's poles.

The stars move along paths parallel to the celestial equator . . .

To north celestial pole

(a) At middle northern latitudes

. . . at the north pole, the celestial equator lies all around you at the horizon . . .

To north celestial pole

(b) At the north pole

. . . and standing at the equator, the celestial equator is a semi-circle passing directly overhead.

To north celestial pole

(c) At the equator

 ANIMATION 2.3   **FIGURE 2-11** R I **V** U X G

The Apparent Motion of Stars at Different Latitudes As Earth rotates, stars appear to rotate around us along paths that are parallel to the celestial equator. **(a)** As shown in this long-time exposure, at most locations on Earth the rising and setting motions are at an angle to the horizon that depends on the latitude. This photo is looking west at setting stars. **(b)** At the north pole (latitude 90° north), the stars appear to move parallel to the horizon. **(c)** At the equator (latitude 0°), the stars rise and set along vertical paths. (a: © David Miller/David Malin Images)

Motions of the Celestial Sphere

For an observer anywhere in the northern hemisphere, including the observer in Figure 2-10, the north celestial pole is always above the horizon. As Earth turns from west to east, it appears to the observer that the celestial sphere turns from east to west. Stars sufficiently near the north celestial pole revolve around the pole, never rising or setting. Such stars are called **circumpolar**. For example, as seen from North America or Europe, Polaris is a circumpolar star and can be seen at any time of night on any night of the year. The long-exposure photograph that opens this chapter shows the circular trails of stars around the north celestial pole as seen from Hawaii (at 20° north latitude). Stars near the south celestial pole revolve around that pole but always remain below the horizon of an observer in the northern hemisphere. Hence, these stars can never be seen by the observer in Figure 2-10. Stars between those two limits rise in the east and set in the west.

CAUTION!

Keep in mind that which stars are circumpolar, which stars never rise, and which stars rise and set depends on the latitude on Earth from which you view the sky. As an example, for an observer at 35° south latitude (roughly the latitude of Sydney, Cape Town, and Buenos Aires), the roles of the north and south celestial poles are the opposite of those shown in Figure 2-10: Objects close to the *south* celestial pole are circumpolar, that is, they revolve around that pole and never rise or set. For an observer in the southern hemisphere, stars close to the *north* celestial pole are always below the horizon and can never be seen. Hence, astronomers in Australia, South Africa, and Argentina never see the North Star but are able to see other stars that are forever hidden from North American or European observers.

For observers at most locations on Earth, stars rise in the east and set in the west at an angle to the horizon. To see why this is so, notice that the rotation of the celestial sphere carries stars across the sky in paths that are parallel to the celestial equator (**Figure 2-11**).

CONCEPT CHECK 2-5

Where would you need to be standing on Earth for the celestial equator to pass through your zenith? (See Figure 2-10 for an illustration of the zenith.)

Answer appears at the end of the chapter.

How do we describe the location of a star? Using the celestial equator and poles, we can define a coordinate system based on angles to specify the position of any star on the celestial sphere. As **Box 2-1** describes, the most commonly used coordinate system uses two angles, *right ascension* and *declination*, that are analogous to longitude and latitude on Earth. These coordinates tell us in what direction we should look to see the star. To locate the star's true position in three-dimensional space, we must also know the distance to the star.

2-5 The seasons are caused by the tilt of Earth's axis of rotation

 TUTORIAL 2.2   In addition to rotating on its axis every 24 hours, Earth revolves around the Sun—that is, it orbits the Sun—in about 365¼ days (see Figure 2-5). As we travel with Earth around its orbit, we experience the annual cycle of seasons. But why *are* there seasons? Furthermore, the seasons are opposite in the northern and

Earth's tilt causes variations in the intensity, or concentration, of sunlight hitting the surface.

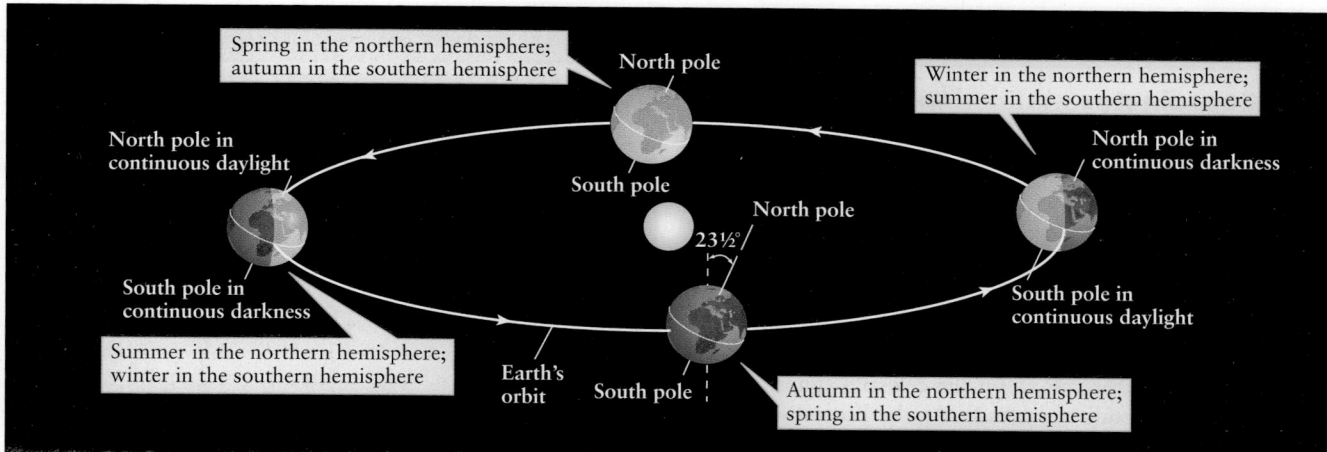

ANIMATION
2.4
FIGURE 2-12

The Seasons Earth's axis of rotation is inclined 23½° away from the perpendicular to the plane of Earth's orbit. The north pole points in the same direction as it orbits the Sun, approximately toward the star Polaris, the North Star. Consequently, the amount of solar illumination and the number of daylight hours at any location on Earth vary in a regular pattern throughout the year. This is the origin of the seasons.

southern hemispheres. For example, February is midwinter in North America but midsummer in Australia. Why should this be?

The Origin of the Seasons

As we know from common experience, there are warmer temperatures in summer. This warmth is due to an increased amount—that is, a higher concentration—of sunlight striking the ground beneath your feet and the surrounding areas. But what causes the concentration of sunlight to change through the seasons?

The reason why we have seasons, and why they are different in different hemispheres, is that Earth's axis of rotation is not perpendicular to the plane of Earth's orbit. Rather, as **Figure 2-12** shows, the axis is tilted about 23½° away from the perpendicular. Just as a top will not fall over as long as it is spinning, Earth maintains this tilt as it orbits the Sun, with Earth's north pole always pointing in the same direction throughout the year. As we saw earlier, this direction just happens to point to within 1° of a star named Polaris (the North Star).

During part of the year, when Earth is in the part of its orbit shown on the left side of Figure 2-12, the northern hemisphere is tilted toward the Sun. As Earth spins on its axis, a point in the northern hemisphere spends more than 12 hours in the sunlight. Thus, the days there are long and the nights are short, and it is summer in the northern hemisphere. The summer is hot not only because of the extended daylight hours but also because the Sun is high in the northern hemisphere's sky. As a result, sunlight strikes the ground at a nearly perpendicular angle that heats the ground efficiently (**Figure 2-13a**). During this same time of year in the southern hemisphere, the days are short and the nights are long, because a point in this hemisphere spends fewer than 12 hours a day in the sunlight. The Sun is low in the sky, so sunlight strikes the surface at a grazing angle that causes little heating (Figure 2-13b), and it is winter in the southern hemisphere.

(a) The Sun in summer

(b) The Sun in winter

FIGURE 2-13

Solar Energy in Summer and Winter At different times of the year, sunlight strikes the ground at different angles. **(a)** In summer, sunlight is concentrated and the days are also longer, which further increases the heating. **(b)** In winter, the sunlight is less concentrated, the days are short, and little heating of the ground takes place. This accounts for the low temperatures in winter.

BOX 2-1 TOOLS OF THE ASTRONOMER'S TRADE

Celestial Coordinates

Your *latitude* and *longitude* describe where on Earth's surface you are located. As shown in the top illustration, the latitude of your location denotes how far north or south of the equator you are, and the longitude of your location denotes how far west or east you are of an imaginary circle that runs from the north pole to the south pole through the Royal Observatory in Greenwich, England. In an analogous way, astronomers use coordinates called *declination* and *right ascension* to describe the position of a planet, star, or galaxy on the celestial sphere.

Declination is analogous to latitude. As the bottom illustration shows, the **declination** of an object is its angular distance north or south of the celestial equator, measured along a circle passing through both celestial poles. Like latitude, it is measured in degrees, arcminutes, and arcseconds (see Section 1-5).

Right ascension is analogous to longitude. It is measured from a line that runs between the north and south celestial poles and passes through a point on the celestial equator called the *vernal equinox* (shown as a red dot in the illustration). This point is one of two locations where the Sun crosses the celestial equator during its apparent annual motion, as we discuss in Section 2-5. In Earth's northern hemisphere, spring officially begins when the Sun reaches the vernal equinox in late March. The **right ascension** of an object is the angular distance from the vernal equinox eastward along the celestial equator to the circle used in measuring its declination (see illustration). Astronomers measure right ascension in *time* units (hours, minutes, and seconds), corresponding to the time required for the celestial sphere to rotate through this angle. For example, suppose there is a star at your zenith right now with right ascension $6^h 0^m 0^s$. Two hours and 30 minutes from now, there will be a different object at your zenith with right ascension $8^h 30^m 0^s$.

The coordinates of the bright star Rigel for the year 2000 are R.A. = $5^h 14^m 32.2^s$, Decl. = $-8° 12' 06''$. (R.A. and Decl. are abbreviations for right ascension and declination.) A minus sign on the declination indicates that the star is south of the celestial equator; a plus sign (or no sign at all) indicates that an object is north of the celestial equator. As we discuss in Box 2-2, right ascension helps determine the best time to observe a particular object.

It is important to state the year for which a star's right ascension and declination are valid. This is so because of precession, which we discuss in Section 2-6.

EXAMPLE: What are the coordinates of a star that lies exactly halfway between the vernal equinox and the south celestial pole?

Situation: Our goal is to find the right ascension and declination of the star in question.

Tools: We use the definitions depicted in the bottom figure.

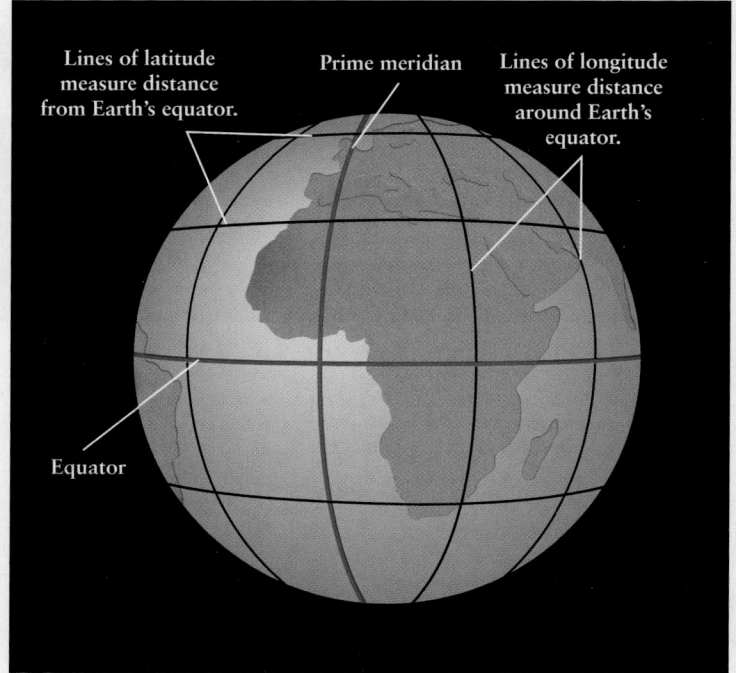

Lines of latitude measure distance from Earth's equator.

Prime meridian

Lines of longitude measure distance around Earth's equator.

Equator

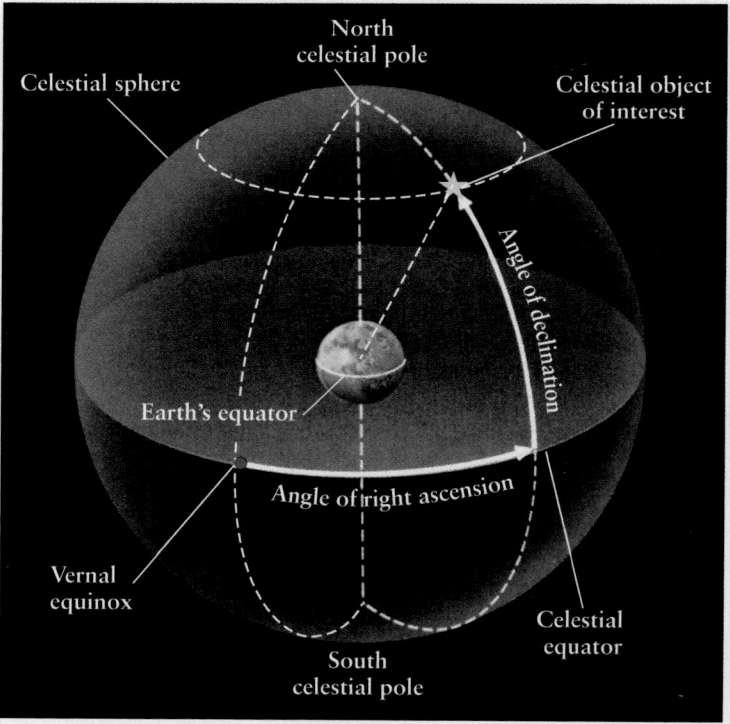

Celestial sphere

North celestial pole

Celestial object of interest

Angle of declination

Earth's equator

Angle of right ascension

Vernal equinox

Celestial equator

South celestial pole

BOX 2-1 (continued)

Answer: Since the circle used to measure this star's declination passes through the vernal equinox, this star's right ascension is R.A. = $0^h 0^m 0^s$. The angle between the celestial equator and south celestial pole is $90° 0' 0''$, so the declination of this star is Decl. = $-45° 0' 0''$.

Review: The declination in this example is negative because the star is in the southern half of the celestial sphere.

EXAMPLE: At midnight local time, you see a star with R.A. = $2^h 30^m 0^s$ at your zenith. When will you see a star at your zenith with R.A. = $21^h 0^m 0^s$?

Situation: If you held your finger stationary over a globe of Earth, the longitude of the point directly under your finger would change as you rotated the globe. In the same way, the right ascension of the point directly over your head (the zenith) changes as the celestial sphere rotates. We use this concept to determine the time in question.

Tools: We use the idea that a change in right ascension of 24^h corresponds to an elapsed time of 24 hours and a complete rotation of the celestial sphere.

Answer: The time required for the sky to rotate through the angle between the stars is the difference in their right ascensions: $21^h 0^m 0^s - 2^h 30^m 0^s = 18^h 30^m 0^s$. So the second star will be at your zenith 18½ hours after the first one, or at 6:30 P.M. the following evening.

Review: Our answer was based on the idea that the celestial sphere makes *exactly* one complete rotation in 24 hours. If this were so, each star would be in exactly the same position at a given time from one night to the next. But because of the way that we measure time, the celestial sphere makes slightly more than one complete rotation in 24 hours. (We explore the reasons for this in Box 2-2.) As a result, our answer is in error by about 3 minutes. For our purposes, this is a small enough error that we can ignore it.

Half a year later, Earth is in the part of its orbit shown on the right side of Figure 2-12. Now the situation is reversed, with winter in the northern hemisphere (which is now tilted away from the Sun) and summer in the southern hemisphere. During spring and autumn, the two hemispheres receive roughly equal amounts of illumination from the Sun, and daytime and nighttime are of roughly equal length everywhere on Earth.

CAUTION!

A common misconception is that the seasons are caused by variations in the distance from Earth to the Sun. According to this idea, Earth is closer to the Sun in summer and farther away in winter. But in fact, Earth's orbit around the Sun is very nearly circular, and the Earth–Sun distance varies only about 3 percent over the course of a year. (Earth's orbit only *looks* elongated in Figure 2-12 because this illustration shows the orbit from a side view; it would be circular if viewed overhead.) We are slightly closer to the Sun in January than in July, but this small variation has little influence on the cycle of the seasons. Also, if the seasons were really caused by variations in the Earth–Sun distance, the seasons would be the same in both hemispheres!

CONCEPT CHECK 2-6

If, instead of being tilted, Earth's axis were straight up and down compared to the plane created by the path of Earth's orbit around the Sun, would observers near Earth's north pole still observe periods where the Sun never rises and the Sun never sets?

Answer appears at the end of the chapter.

How the Sun Moves on the Celestial Sphere

The plane of Earth's orbit around the Sun is called the **ecliptic plane** (Figure 2-14a). The ecliptic plane is easy to picture if you imagine hovering above the solar system, as in Figure 2-14a, but it is described quite differently when viewed from Earth, as in Figure 2-14b. As a result of Earth's annual motion around the Sun, it appears to us (observing from Earth) that the Sun slowly changes its position on the celestial sphere over the course of a year. (Since we don't see the Sun and stars at the same time, this motion relative to the stars isn't readily familiar.) The circular path that the Sun appears to trace out against the background of stars is called the **ecliptic** (Figure 2-14b). The plane of this path is the same as the ecliptic plane. (The name *ecliptic* suggests that the path traced out by the Sun has something to do with eclipses. We will discuss the connection in Chapter 3.) Because there are 365¼ days in a year and 360° in a circle, the Sun appears to move along the ecliptic at a rate of about 1° per day. This motion is from west to east, that is, in the direction opposite to the apparent motion of the celestial sphere.

ANALOGY

Note that at the same time that the Sun is making its yearlong trip around the ecliptic, the entire celestial sphere is rotating around us once per day. As a very accurate analogy, you can envision the celestial sphere as a merry-go-round rotating clockwise and the Sun as a restless child who is walking slowly around the merry-go-round's rim in the counterclockwise direction. During the time it takes the child to make a round trip, the merry-go-round rotates 365¼ times.

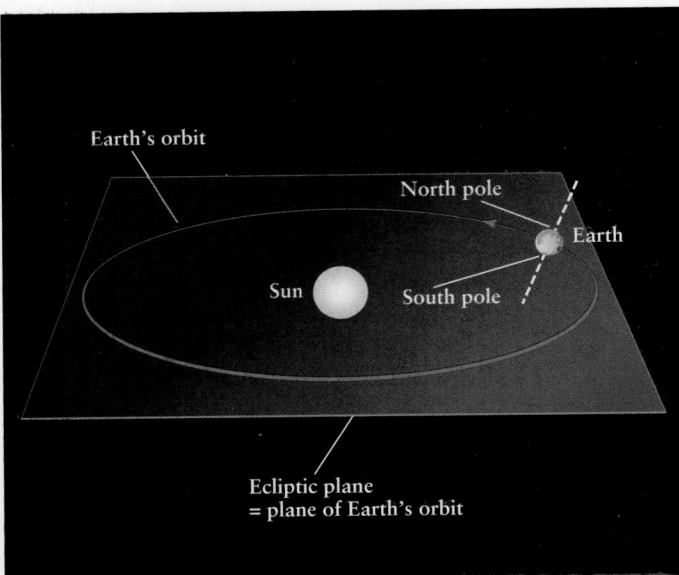

(a) In reality, Earth orbits the Sun once a year.

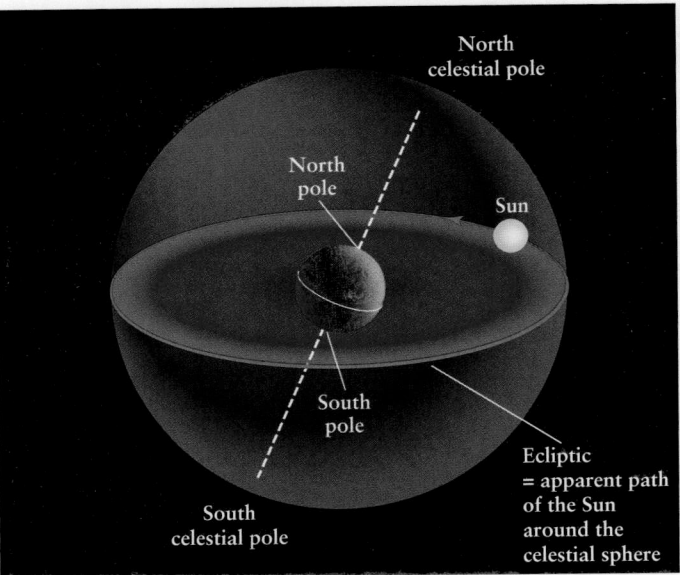

(b) It appears from Earth that the Sun travels around the celestial sphere once a year.

FIGURE 2-14

The Ecliptic Plane and the Ecliptic (a) The ecliptic plane is the plane in which Earth moves around the Sun. **(b)** As seen from Earth, the Sun appears to move around the celestial sphere along a circular path called the

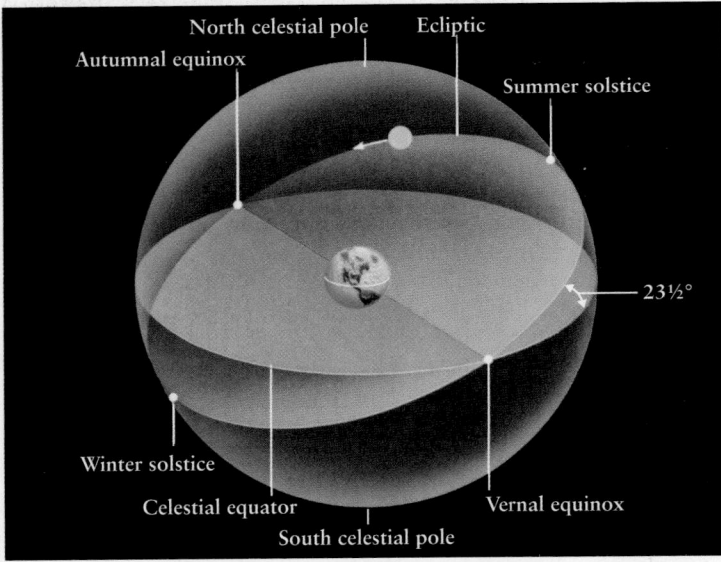

FIGURE 2-15

The Ecliptic, Equinoxes, and Solstices This illustration of the celestial sphere is similar to Figure 2-14b, but it is drawn with the north celestial pole at the top and the celestial equator oriented horizontally. The ecliptic is inclined to the celestial equator by 23½° because of the tilt of Earth's axis of rotation. It intersects the celestial equator at two points, called equinoxes. The northernmost point on the ecliptic is the summer solstice, and the southernmost point is the winter solstice. The Sun is shown in its approximate position for August 1.

The ecliptic plane is *not* the same as the plane of Earth's equator, thanks to the 23½° tilt of Earth's rotation axis shown in Figure 2-12.

ecliptic. Earth takes a year to complete one orbit around the Sun, so as seen by us the Sun takes a year to make a complete trip around the ecliptic.

As a result, the ecliptic and the celestial equator are inclined to each other by that same 23½° angle (**Figure 2-15**).

CONCEPT CHECK 2-7

How long does the Sun take to move from being next to a bright star all the way around the celestial sphere and back to that same bright star?

Answer appears at the end of the chapter.

Equinoxes and Solstices

As shown in Figure 2-15, the ecliptic and the celestial equator intersect at only two points, which are exactly opposite each other on the celestial sphere. Each point is called an **equinox** (from the Latin for "equal night"), because when the Sun appears at either of these points, day and night are each about 12 hours long at any location on Earth. The term *equinox* is also used to refer to the date on which the Sun passes through one of these special points on the ecliptic.

On about March 21 of each year, the Sun passes northward across the celestial equator at the **vernal equinox.** This marks the beginning of spring in the northern hemisphere (*vernal* is from the Latin for "spring"). On about September 22, the Sun moves southward across the celestial equator at the **autumnal equinox,** marking the moment when fall begins in the northern hemisphere (see Figure 2-15). Since the seasons are opposite in the northern and southern hemispheres, for Australians, South Africans, and South Americans the vernal equinox actually marks the beginning of autumn. The names of the equinoxes come from astronomers of the past who lived north of the equator.

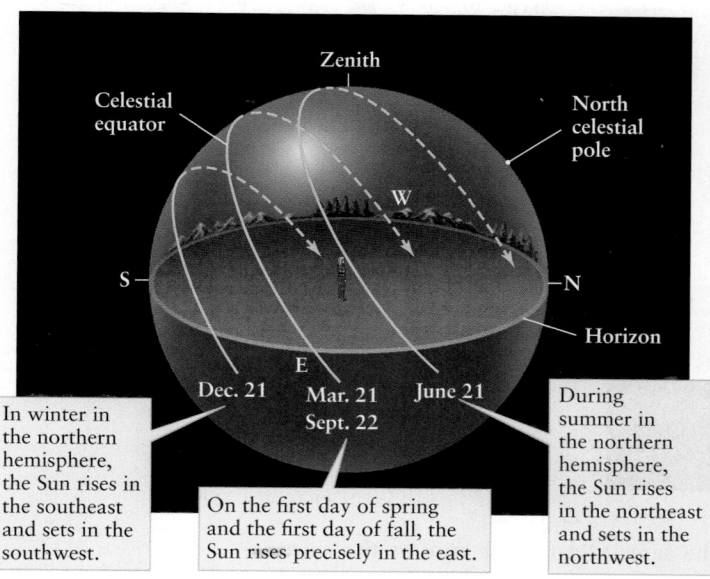

In winter in the northern hemisphere, the Sun rises in the southeast and sets in the southwest.

On the first day of spring and the first day of fall, the Sun rises precisely in the east.

During summer in the northern hemisphere, the Sun rises in the northeast and sets in the northwest.

ANIMATION 2.5   **FIGURE 2-16**

The Sun's Daily Path Across the Sky This drawing shows the apparent path of the Sun during the course of a day on four different dates. Like Figure 2-10, this drawing is for an observer at 35° north latitude. Note that the Sun is lower in the sky on December 21, which corresponds to the low winter position of the Sun in Figure 2-13b.

Between the vernal and autumnal equinoxes lie two other significant locations along the ecliptic. The point on the ecliptic farthest north of the celestial equator is called the **summer solstice.** *Solstice* is from the Latin for "solar standstill," and it is at the summer solstice that the Sun stops moving northward on

the celestial sphere. At this point, the Sun is as far north of the celestial equator as it can get. It marks the location of the Sun at the moment summer begins in the northern hemisphere (about June 21). At the beginning of the northern hemisphere's winter (about December 21), the Sun is farthest south of the celestial equator at a point called the **winter solstice** (see Figure 2-15).

Because the Sun's position on the celestial sphere varies slowly over the course of a year, its daily path across the sky (due to Earth's rotation) also varies with the seasons (**Figure 2-16**). On the first day of spring or the first day of fall, when the Sun is at one of the equinoxes, the Sun rises directly in the east and sets directly in the west.

When the northern hemisphere is tilted away from the Sun and it is winter in the northern hemisphere, the Sun rises in the southeast. Daylight lasts for fewer than 12 hours as the Sun skims low over the southern horizon and sets in the southwest. Northern hemisphere nights are longest when the Sun is at the winter solstice.

The closer you get to the north pole, the shorter the winter days and the longer the winter nights. In fact, anywhere within 23½° of the north pole (that is, north of latitude 90° − 23½° = 66½° N), the Sun is below the horizon for 24 continuous hours at least one day of the year. The circle around Earth at 66½° north latitude is called the **Arctic Circle** (**Figure 2-17**). The corresponding region around the south pole is bounded by the **Antarctic Circle** at 66½° south latitude. At the time of the winter solstice, explorers south of the Antarctic Circle enjoy "the midnight sun," or 24 hours of continuous daylight.

During summer in the northern hemisphere, when the northern hemisphere is tilted toward the Sun, the Sun rises in the northeast and sets in the northwest. The Sun is at its northernmost position

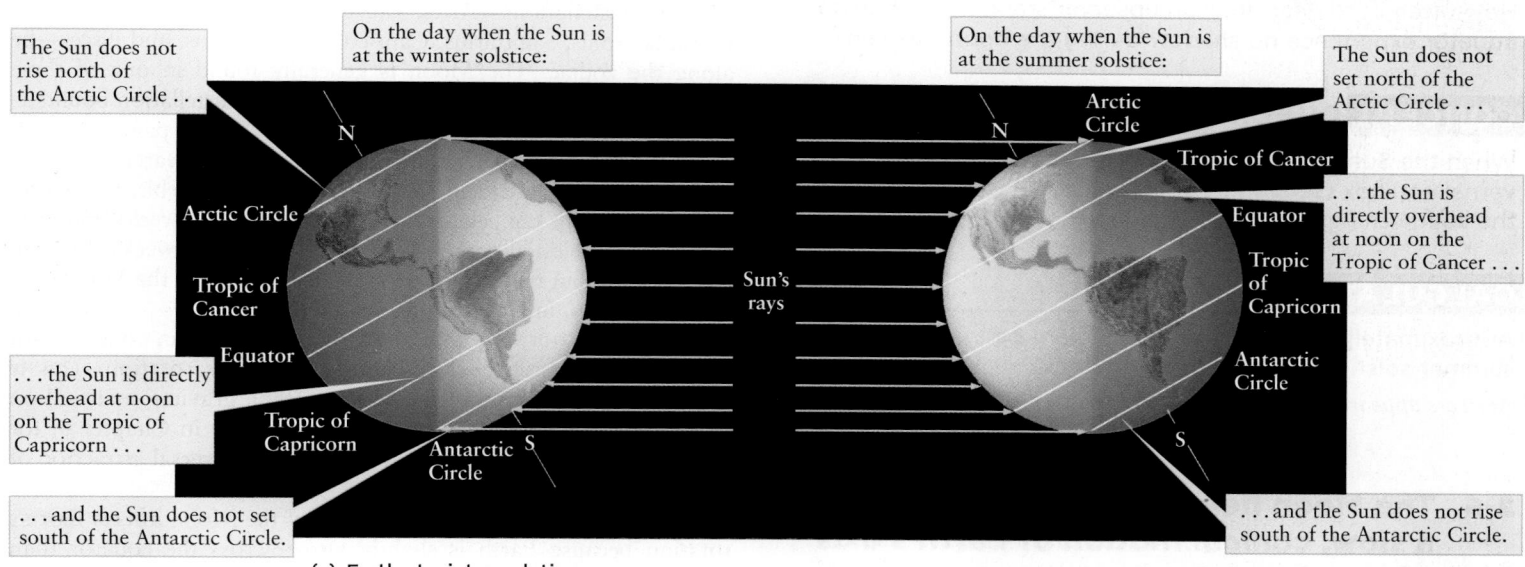

The Sun does not rise north of the Arctic Circle . . .

On the day when the Sun is at the winter solstice:

. . . the Sun is directly overhead at noon on the Tropic of Capricorn . . .

. . . and the Sun does not set south of the Antarctic Circle.

(a) Earth at winter solstice

On the day when the Sun is at the summer solstice:

The Sun does not set north of the Arctic Circle . . .

. . . the Sun is directly overhead at noon on the Tropic of Cancer . . .

. . . and the Sun does not rise south of the Antarctic Circle.

(b) Earth at summer solstice

FIGURE 2-17

Tropics and Circles Four important latitudes on Earth are the Arctic Circle (66½° north latitude), Tropic of Cancer (23½° north latitude), Tropic of Capricorn (23½° south latitude), and Antarctic Circle (66½° south latitude).

These drawings show the significance of these latitudes when the Sun is **(a)** at the winter solstice and **(b)** at the summer solstice.

FIGURE 2-18 R I V U X G

The Midnight Sun This time-lapse photograph was taken on July 19, 1985, at 69° north latitude in northeast Alaska. At this latitude, the Sun is above the horizon continuously, 24 hours a day (that is, it is circumpolar), from mid-May to the end of July. (Doug Plummer/Science Source)

11:40 P.M.

12:40 A.M.

1:40 A.M.

2:40 A.M.

3:40 A.M.

at the summer solstice, giving the northern hemisphere the greatest number of daylight hours. At the summer solstice, the Sun does not set at all north of the Arctic Circle (Figure 2-18) and does not rise at all south of the Antarctic Circle.

The variations of the seasons are much less pronounced close to the equator. Between the **Tropic of Capricorn** at 23½° south latitude and the **Tropic of Cancer** at 23½° north latitude, the Sun is directly overhead—that is, at the zenith—at high noon at least one day a year. Outside of the tropics, the Sun is never directly overhead, but is always either south of the zenith (as seen from locations north of the Tropic of Cancer) or north of the zenith (as seen from south of the Tropic of Capricorn).

CONCEPT CHECK 2-8

How often each year does an observer standing on Earth's equator experience no shadow during the noontime sun?

CONCEPT CHECK 2-9

When the Sun is between the winter solstice and the vernal equinox (see Figure 2-15), what is the season in the southern hemisphere?

CALCULATION CHECK 2-2

Approximately how many days are there between the summer solstice and the March equinox?

Answers appear at the end of the chapter.

2-6 The Moon helps to cause precession, a slow, conical motion of Earth's axis of rotation

The Moon is by far the brightest and most obvious naked-eye object in the nighttime sky. Like the Sun, the Moon slowly changes its position relative to the background stars; unlike the Sun, the Moon makes a complete trip around the celestial sphere in only about 4 weeks, or about a month. (The word *month* comes from

the same Old English root as the word *moon*.) Ancient astronomers realized that this motion occurs because the Moon orbits Earth in roughly 4 weeks. In 1 hour, the Moon moves on the celestial sphere by about ½° or roughly its own angular size.

The Moon's path on the celestial sphere is never far from the Sun's path (that is, the ecliptic). This is because the plane of the Moon's orbit around Earth is inclined only slightly from the plane of Earth's orbit around the Sun (the ecliptic plane shown in Figure 2-14a). The Moon's path varies somewhat from one month to the next, but always remains within a band called the **zodiac** that extends about 8° on either side of the ecliptic. Twelve famous constellations—Aries, Taurus, Gemini, Cancer, Leo, Virgo, Libra, Scorpius, Sagittarius, Capricornus, Aquarius, and Pisces—lie along the zodiac. The Moon is generally found in one of these 12 constellations. (Thanks to a redrawing of constellation boundaries in the mid-twentieth century, the zodiac actually passes through a thirteenth constellation—Ophiuchus, the Serpent Bearer—between Scorpius and Sagittarius.) As it moves along its orbit, the Moon appears north of the celestial equator for about 2 weeks and then south of the celestial equator for about the next 2 weeks. We will learn more about the Moon's motion, as well as why the Moon goes through phases, in Chapter 3.

The Moon not only moves around Earth but, in concert with the Sun, also causes a slow change in Earth's rotation. This is because both the Sun and the Moon exert a gravitational pull on Earth. We will learn much more about gravity in Chapter 4; for now, all we need is the idea that gravity is a universal attraction of matter for other matter.

The gravitational pull of the Sun and the Moon affects Earth's rotation because Earth is slightly fatter across the equator than it is from pole to pole: Its equatorial diameter is 43 kilometers (27 miles) larger than the diameter measured from pole to pole. Earth is therefore said to have an "equatorial bulge." Because of the gravitational pull of the Moon and the Sun on this bulge, the orientation of Earth's axis of rotation gradually changes, producing a motion called **precession**. Although the details are beyond the scope of this book, the combined actions of gravity and rotation cause

> Precession causes the apparent positions of the stars to slowly change over the centuries.

FIGURE 2-19

Precession Because Earth's rotation axis is tilted, the gravitational pull of the Moon and the Sun on Earth's equatorial bulge together cause Earth to precess. As Earth precesses, its axis of rotation slowly traces out a circle in the sky, like the shifting axis of a spinning top.

Earth's axis to trace out a circle in the sky (**Figure 2-19**). As the axis precesses, it remains tilted about 23½° to the perpendicular.

As Earth's axis of rotation slowly changes its orientation, the north and south celestial poles—which are the projections of that axis onto the celestial sphere—change their positions relative to the stars. At present, the north celestial pole lies within 1° of the star Polaris, which is why Polaris is the North Star. But 5000 years ago, the north celestial pole was closest to the star Thuban in the constellation of Draco (the Dragon). Thus, that star and not Polaris was the North Star. And 12,000 years from now, the North Star will be the bright star Vega in Lyra (the Harp). It takes 26,000 years for the north celestial pole to complete one full precessional circle around the sky (**Figure 2-20**). The south celestial pole executes a similar circle in the southern sky.

Precession also causes Earth's equatorial plane to change its orientation. Because this plane defines the location of the celestial equator in the sky, the celestial equator precesses as well. The intersections of the celestial equator and the ecliptic define the equinoxes (see Figure 2-15), so these key locations in the sky also shift slowly from year to year. For this reason, the precession of Earth is also called the **precession of the equinoxes**. The first person to detect the precession of the equinoxes, in the second century B.C.E., was the Greek astronomer Hipparchus, who compared his own observations with those of Babylonian astronomers three centuries earlier. Today, the vernal equinox is located in the constellation Pisces (the Fishes). Two thousand years ago, it was in Aries (the Ram). Around the year 2600 C.E., the vernal equinox will move into Aquarius (the Water Bearer).

FIGURE 2-20

Precession and the Path of the North Celestial Pole As Earth precesses, the north celestial pole slowly traces out a circle among the northern constellations. At present, the north celestial pole is near the moderately bright star Polaris, which serves as the North Star. Twelve thousand years from now, the bright star Vega will be the North Star.

CAUTION!

Astrological terms like the "Age of Aquarius" involve boundaries in the sky that are not recognized by astronomers and are generally not even related to the positions of the constellations. For example, most astrologers would call a person born on March 21, 1988, an "Aries" because the Sun was supposedly in the direction of that constellation on March 21. But due to precession, the Sun was actually in the constellation Pisces on that date! Indeed, astrology is not a science at all, but merely a collection of superstitions and unproven claims. Its practitioners use some of the terminology of astronomy but reject the logical thinking and rigorous testing that are at the heart of science. James Randi has more to say about astrology and other pseudosciences in his essay "Why Astrology Is Not Science" at the end of this chapter.

The astronomer's system of locating heavenly bodies by their right ascension and declination, discussed in Box 2-1, is tied to the positions of the celestial equator and the vernal equinox. Because of precession, these positions are changing, and thus the coordinates of stars in the sky are also constantly changing. These changes are very small and gradual, but they add up over the years. To cope with this difficulty, astronomers always make note of the date (called the **epoch**) for which a particular set of coordinates is precisely correct. Consequently, star catalogs and star charts are periodically updated. Most current catalogs and star charts are prepared for the epoch 2000. The coordinates in these reference books, which are precise for January 1, 2000, will require very little correction over the next few decades.

2-7 Positional astronomy plays an important role in keeping track of time

Astronomers have traditionally been responsible for telling time. This is because we want the system of timekeeping used in everyday life to reflect the position of the Sun in the sky. Thousands of years ago, the sundial was invented to keep track of **apparent solar time**. To obtain more accurate measurements, astronomers use the **meridian**. As Figure 2-21 shows, this is a north–south circle on the celestial sphere that passes through the zenith (the point directly overhead) and both celestial poles. *Local noon* is defined as when the Sun crosses the **upper meridian**, which is the half of the meridian above the horizon. At *local midnight*, the Sun crosses the **lower meridian**, the half of the meridian below the horizon; this crossing cannot be observed directly.

> Ancient scholars developed a system of timekeeping based on the Sun.

The crossing of the meridian by any object in the sky is called a **meridian transit** of that object. If the crossing occurs above the horizon, it is an *upper* meridian transit. When an object makes a meridian transit, it is not necessarily at the zenith directly overhead, but is at the *highest point above the horizon* that you will see it from your location.

An **apparent solar day** is the interval between two successive upper meridian transits of the Sun as observed from any fixed spot on Earth. Stated less formally, an apparent solar day is the time from one local noon to the next local noon, or from when the Sun is highest in the sky to when it is again highest in the sky.

The Sun as a Timekeeper

Unfortunately, the Sun is not a good timekeeper (Figure 2-22). The length of an apparent solar day (as measured by a device such as an hourglass) varies from one time of year to another. There are two main reasons why this is so, both having to do with the way in which Earth orbits the Sun.

The first reason is that Earth's orbit is not a perfect circle; rather, it is an ellipse, as Figure 2-22a shows in exaggerated form. As we will learn in Chapter 4, Earth moves more rapidly along its

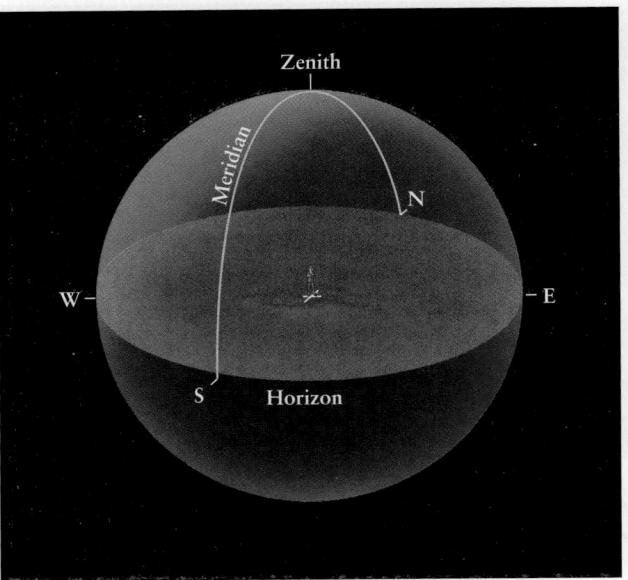

FIGURE 2-21

The Meridian The meridian is a circle on the celestial sphere that passes through the observer's zenith (the point directly overhead) and the north and south points on the observer's horizon. The passing of celestial objects across the meridian can be used to measure time. The upper meridian is the part above the horizon, and the lower meridian (not shown) is the part below the horizon.

orbit when it is near the Sun than when it is farther away. Hence, the Sun appears to us to move more than 1° per day along the ecliptic in January, when Earth is nearest the Sun, and less than 1° per day in July, when Earth is farthest from the Sun. By itself, this effect would cause the apparent solar day to be longer in January than in July.

The second reason why the Sun is not a good timekeeper is the 23½° angle between the ecliptic and the celestial equator (see Figure 2-15). As Figure 2-22b shows, this causes a significant part of the Sun's apparent motion when near the equinoxes to be in a north–south direction. The net daily eastward progress in the sky is then somewhat foreshortened. At the summer and winter solstices, by contrast, the Sun's motion is parallel to the celestial equator. Thus, there is no comparable foreshortening around the beginning of summer or winter. This effect by itself would make the apparent solar day shorter in March and September than in June or December. Combining these effects with those due to Earth's noncircular orbit, we find that the length of the apparent solar day varies in a complicated fashion over the course of a year.

To avoid these difficulties, astronomers invented an imaginary object called the **mean sun** that moves along the celestial equator at a uniform rate. (In science and mathematics, *mean* is a synonym for *average*.) The mean sun is sometimes slightly ahead of the real Sun in the sky, sometimes behind. As a result, mean solar time and apparent solar time can differ by as much as a quarter of an hour at certain times of the year.

Because the mean sun moves at a constant rate, it serves as a fine timekeeper. A **mean solar day** is the interval between successive upper meridian transits of the mean sun. It is exactly 24 hours

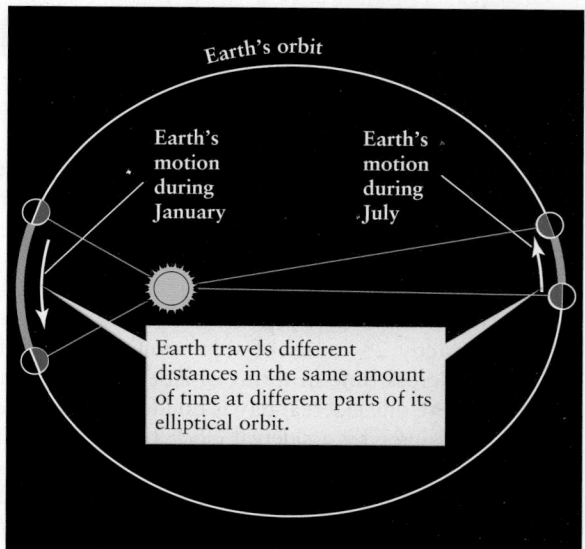

(a) A month's motion of Earth along its orbit

(b) A day's motion of the Sun along the ecliptic

FIGURE 2-22

Why the Sun Is a Poor Timekeeper There are two main reasons that the Sun is a poor timekeeper. **(a)** Earth's speed along its orbit varies during the year. It moves fastest when closest to the Sun in January and slowest when farthest from the Sun in July. Hence, the apparent speed of the Sun along the ecliptic is not constant. **(b)** Because of the tilt of Earth's rotation axis, the ecliptic is inclined with respect to the celestial equator. Therefore, the projection of the Sun's daily progress (shown in orange) along the ecliptic onto the celestial equator (shown in blue) varies during the year. This causes further variations in the length of the apparent solar day.

long, the average length of an apparent solar day. One 24-hour day as measured by your alarm clock or wristwatch is a mean solar day.

Time zones were invented for convenience in commerce, transportation, and communication. In a time zone, all clocks and watches are set to the mean solar time for a meridian of longitude that runs approximately through the center of the zone. Time zones around the world are generally centered on meridians of longitude at 15° intervals. In most cases, going from one time zone to the next requires you to change the time on your wristwatch by exactly 1 hour. The time zones for most of North America are shown in Figure 2-23.

In order to coordinate their observations with colleagues elsewhere around the globe, astronomers often keep track of time using Coordinated Universal Time, somewhat confusingly abbreviated UTC or UT. This is the time in a zone that includes Greenwich, England, a seaport just outside of London where the first internationally accepted time standard was kept. (UTC was formerly known as Greenwich Mean Time.) UTC is on a 24-hour system, with no A.M. or P.M. In North America, Eastern Standard Time (EST) is 5 hours different from UTC; 9:00 A.M. EST is 14:00 UTC. Coordinated Universal Time is also used by aviators and sailors, who regularly travel from one time zone to another.

Although it is natural to want our clocks and method of timekeeping to be related to the Sun, astronomers often use a system that is based on the apparent motion of the stars. This system, called **sidereal time,** is useful when aiming a telescope. Most observatories are therefore equipped with a clock that measures sidereal time, as discussed in Box 2-2.

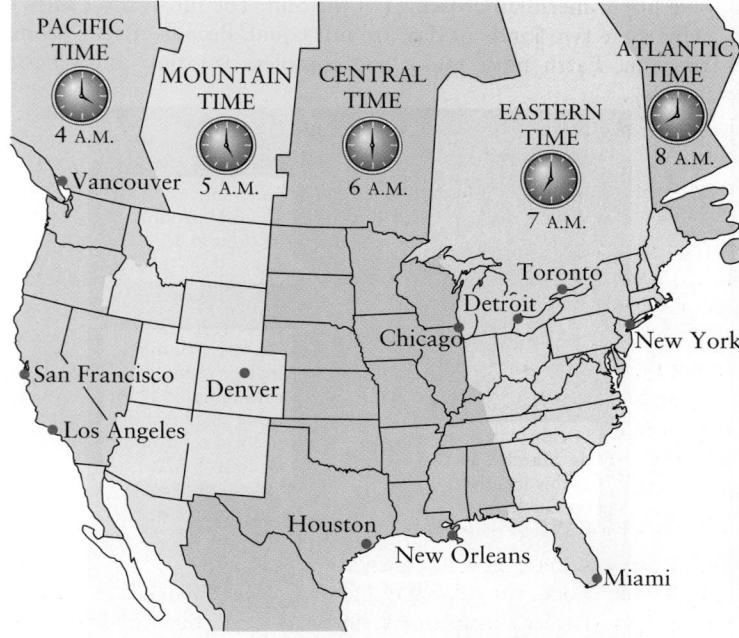

FIGURE 2-23

Time Zones in North America For convenience, Earth is divided into 24 time zones, generally centered on 15° intervals of longitude around the globe. There are four time zones across the continental United States, making for a 3-hour time difference between New York and California.

BOX 2-2 TOOLS OF THE ASTRONOMER'S TRADE

Sidereal Time

TUTORIAL 2.3

If you want to observe a particular object in the heavens, the ideal time to do so is when the object is high in the sky, on or close to the upper meridian.

Timing observations to occur when an object is high in the sky minimizes the distorting effects of Earth's atmosphere, which increase as you view objects closer to the horizon. For astronomers who study the Sun, this means making observations at local noon, which is not too different from noon as determined using mean solar time. For astronomers who observe planets, stars, or galaxies, however, the optimum time to observe depends on the particular object to be studied. Given the location of a given object on the celestial sphere, when will that object be on the upper meridian?

To answer this question, astronomers use *sidereal time* rather than solar time. It is different from the time on your wristwatch. In fact, a *sidereal clock* and an ordinary clock even tick at different rates, because they are based on different astronomical objects. Ordinary clocks are related to the position of the Sun, while sidereal clocks are based on the position of the vernal equinox, the location from which right ascension is measured. (See Box 2-1 for a discussion of right ascension.)

Regardless of where the Sun is, midnight sidereal time at your location is defined to be when the vernal equinox crosses your upper meridian. (Like solar time, sidereal time depends on where you are on Earth.) A **sidereal day** is the time between two successive upper meridian passages of the vernal equinox. By contrast, an apparent solar day is the time between two successive upper meridian crossings of the Sun. The illustration shows why these two kinds of day are not equal. Because Earth orbits the Sun, Earth must make one complete rotation plus about

1° to get from one local solar noon to the next. This extra 1° of rotation corresponds to 4 minutes of time, which is the amount by which a solar day exceeds a sidereal day. To be precise:

$$1 \text{ sidereal day} = 23^{\text{h}} 56^{\text{m}} 4.091^{\text{s}}$$

where the hours, minutes, and seconds are in mean solar time.

One day according to your wristwatch is one mean solar day, which is exactly 24 hours of solar time long. A **sidereal clock** measures sidereal time in terms of sidereal hours, minutes, and seconds, where one sidereal day is divided into 24 sidereal hours.

This explains why a sidereal clock ticks at a slightly different rate than your wristwatch. As a result, at some times of the year a sidereal clock will show a very different time than an ordinary clock. (At local noon on March 21, when the Sun is at the vernal equinox, a sidereal clock will say that it is midnight. Do you see why?)

We can now answer the question in the opening paragraph. The vernal equinox, whose celestial coordinates are R.A. = $0^{\text{h}} 0^{\text{m}} 0^{\text{s}}$, Decl. = $0° 0' 0''$, crosses the upper meridian at midnight sidereal time (0:00). The autumnal equinox, which is on the opposite side of the celestial sphere at R.A. = $12^{\text{h}} 0^{\text{m}} 0^{\text{s}}$, Decl. = $0° 0' 0''$, crosses the upper meridian 12 sidereal hours later at noon sidereal time (12:00). As these examples illustrate, *any* object crosses the upper meridian when the sidereal time is equal to the object's right ascension. That is why astronomers measure right ascension in units of time rather than degrees, and why right ascension is always given in sidereal hours, minutes, and seconds.

EXAMPLE: Suppose you want to observe the bright star Spica (Figure 2-6), which has epoch 2000 coordinates R.A. = $13^{\text{h}} 25^{\text{m}} 11.6^{\text{s}}$, Decl. = $-11° 9' 41''$. What is the best time to do this?

Situation: Our goal is to find the time when Spica passes through your upper meridian, where it can best be observed.

Tools: We use the idea that a celestial object is on your upper meridian when the sidereal time equals the object's right ascension.

Answer: Based on the given right ascension of Spica, it will be best placed for observation when the sidereal time at your location is about 13:25. (Note that sidereal time is measured using a 24-hour clock.)

Review: By itself, our answer doesn't tell you what time on your wristwatch (which measures mean solar time) is best for observing Spica. That's why most observatories are equipped with a sidereal clock.

While sidereal time is extremely useful in astronomy, mean solar time is still the best method of timekeeping for most earthbound purposes. All time measurements in this book are expressed in mean solar time unless otherwise stated.

To vernal equinox

Sun

Earth moves about 1° around its orbit in 1 day . . .

. . . so Earth must make a complete rotation plus 1° to bring this location to local solar noon on March 22.

Local solar noon on March 21 is at this location on Earth.

Earth on March 21

Earth on March 22

A month's motion of Earth along its orbit

ANIMATION 2.7

2-8 How should we define a "year"?

Deciding on the best way to define a year is not so easy, and the problems are very real without a good definition. For example, ancient calendars placed the start of winter (December 21) in the colder months, as we do today—but that date slowly crept along until it occurred in the hottest months. This created confusion for farmers, who could not rely on their calendar to know when to plant crops. In this section, we look at how this problem has been addressed.

> Our calendar is complex because a year does not contain a whole number of days.

Just as the day is a natural unit of time based on Earth's rotation, the year is a natural unit of time based on Earth's orbit around the Sun. However, nature has not arranged things for our convenience. The year does not divide into exactly 365 whole days. Ancient astronomers realized that the length of a year is approximately 365¼ days, motivating the Roman emperor Julius Caesar to establish the system of "leap years" to account for this extra quarter of a day. By adding an extra day to the calendar every 4 years, he hoped to ensure that seasonal astronomical events, such as the beginning of spring, would occur on the same date year after year.

Caesar's system would have been perfect if the year were exactly 365¼ days long and if there were no precession. Unfortunately, this is not the case. To be more accurate, astronomers now use several different types of years. For example, the **sidereal year** is defined to be the time required for the Sun to return to the same position with respect to the stars. It is equal to 365.2564 mean solar days, or $365^d\ 6^h\ 9^m\ 10^s$.

The sidereal year is the orbital period of Earth around the Sun, but it is *not* the year on which we base our calendar. Like Caesar, most people want annual events—in particular, the first days of the seasons—to fall on the same date each year. For example, we want the first day of spring to occur on March 21. But spring begins when the Sun is at the vernal equinox, and the vernal equinox moves slowly against the background stars because of precession. Therefore, to set up a calendar we use the **tropical year,** which is equal to the time needed for the Sun to return to the vernal equinox. This period is equal to 365.2422 mean solar days, or $365^d\ 5^h\ 48^m\ 46^s$. Because of precession, the tropical year is 20 minutes and 24 seconds shorter than the sidereal year.

Caesar's assumption that the tropical year equals 365¼ days was off by 11 minutes and 14 seconds. This tiny error adds up to about 3 days every 4 centuries. Although Caesar's astronomical advisers were aware of the discrepancy, they felt that it was too small to matter. However, by the sixteenth century the first day of spring was occurring on March 11.

The Roman Catholic Church became concerned because Easter kept shifting to progressively earlier dates. To straighten things out, Pope Gregory XIII instituted a calendar reform in 1582. He began by dropping 10 days (October 4, 1582, was followed by October 15, 1582), which brought the first day of spring back to March 21. Next, he modified Caesar's system of leap years.

Caesar had added February 29 to every calendar year that is evenly divisible by four. Thus, for example, 2020, 2024, and 2028 are all leap years with 366 days. But we have seen that this system produces an error of about 3 days every 4 centuries. To solve the problem, Pope Gregory decreed that only the century years evenly divisible by 400 should be leap years. For example, the years 1700, 1800, and 1900 (which would have been leap years according to Caesar) were not leap years in the improved Gregorian system, but the year 2000, which can be divided evenly by 400, *was* a leap year.

We use the Gregorian system today. It assumes that the year is 365.2425 mean solar days long, which is very close to the true length of the tropical year. In fact, the error is only 1 day in every 3300 years, which won't cause any problems for a long time.

KEY IDEAS

Ideas preceded by an asterisk (*) are discussed in the Boxes.

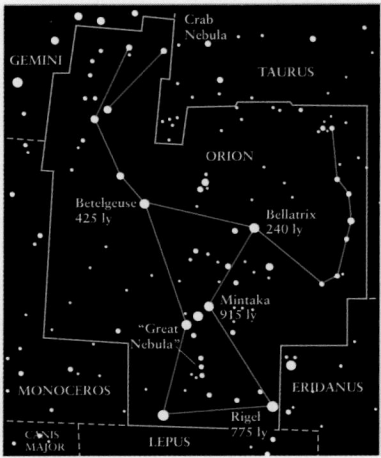

Constellations and the Celestial Sphere: The surface of the celestial sphere is divided into 88 regions called constellations (Section 2-2).

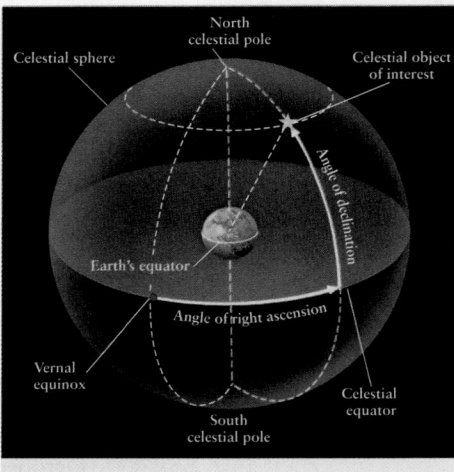

Diurnal (Daily) Motion of the Celestial Sphere: The celestial sphere appears to rotate around Earth once in each 24-hour period. In fact, it is actually Earth that is rotating (Sections 2-3 and 2-4).

- The poles and equator of the celestial sphere are determined by extending the axis of rotation and the equatorial plane of Earth out to the celestial sphere.
- *The positions of objects on the celestial sphere are described by specifying their right ascension (in time units) and declination (in angular measure).

Seasons and the Tilt of Earth's Axis: Earth's axis of rotation is tilted at an angle of about 23½° from the perpendicular to the plane of Earth's orbit (Section 2-5).

- The seasons are caused by the tilt of Earth's axis.
- Over the course of a year, the Sun appears to move around the celestial sphere along a path called the ecliptic. The ecliptic is inclined to the celestial equator by about 23½°.
- The ecliptic crosses the celestial equator at two points in the sky, the vernal and autumnal equinoxes. The northernmost point that the Sun reaches on the celestial sphere is the summer solstice, and the southernmost point is the winter solstice.

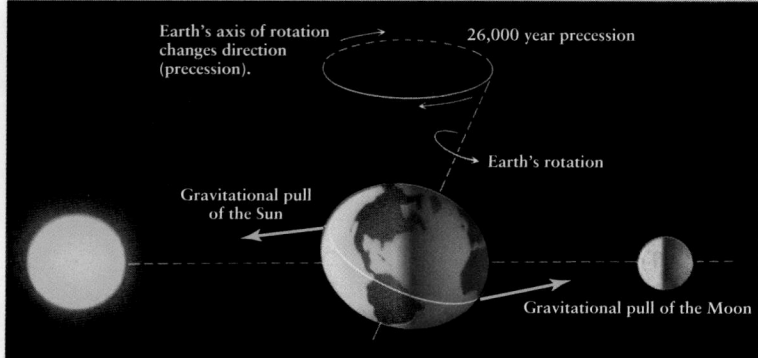

Precession: The orientation of Earth's axis of rotation changes slowly, a phenomenon called precession (Section 2-6).

- Precession is caused by the gravitational pull of the Sun and Moon on Earth's equatorial bulge.
- Precession of Earth's axis causes the positions of the equinoxes and celestial poles to shift slowly.
- *Because the system of right ascension and declination is tied to the position of the vernal equinox, the date (or epoch) of observation must be specified when giving the position of an object in the sky.

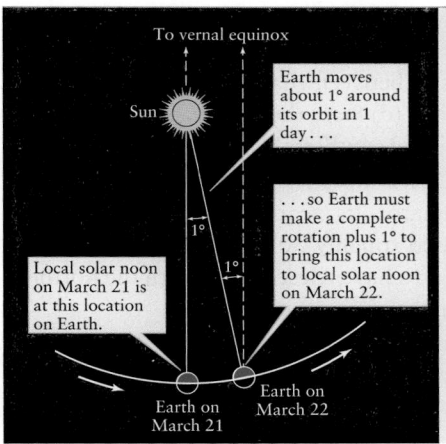

To vernal equinox

Earth moves about 1° around its orbit in 1 day...

...so Earth must make a complete rotation plus 1° to bring this location to local solar noon on March 22.

Local solar noon on March 21 is at this location on Earth.

Earth on March 21   Earth on March 22

Timekeeping: Astronomers use several different means of keeping time (Section 2-7).

- Apparent solar time is based on the apparent motion of the Sun across the celestial sphere, which varies over the course of the year.
- Mean solar time is based on the motion of an imaginary mean sun along the celestial equator, which produces a uniform mean solar day of 24 hours. Ordinary watches and clocks measure mean solar time.
- *Sidereal time is based on the apparent motion of the celestial sphere.

The Calendar: The tropical year is the period between two passages of the Sun across the vernal equinox. Leap year corrections are needed because the tropical year is not exactly 365 days. The sidereal year is the actual orbital period of Earth (Section 2-8).

QUESTIONS

Problem-solving tips and tools

To help you visualize the heavens, it is worth taking the time to become familiar with various types of star charts. These include the simple star charts at the end of this book, the monthly star charts published in such magazines as *Sky & Telescope* and *Astronomy*, and the more detailed maps of the heavens found in star atlases.

One of the best ways to understand the sky and its motions is to use the *Starry Night*™ computer program that accompanies certain printed copies of this book. This easy-to-use program allows you to view the sky on any date and at any time, as seen from any point on Earth, and to animate the sky to visualize its diurnal and annual motions.

You may also find it useful to examine a planisphere, a device consisting of two rotatable disks. The bottom disk shows all the stars in the sky (for a particular latitude), and the top one is opaque with a transparent oval window through which only some of the stars can be seen. By rotating the top disk, you can immediately see which constellations are above the horizon at any time of the year. A planisphere is a convenient tool to carry with you when you are out observing the night sky.

Section 2-1

1. Describe three structures or carvings made by past civilizations that show an understanding of astronomy.

Section 2-2

2. How are constellations useful to astronomers? How many stars are not part of any constellation?

3. A fellow student tells you that only those stars in Figure 2-2b that are connected by blue lines are part of the constellation Orion. How would you respond?

Section 2-3

4. Why is a particular star overhead at 10:00 P.M. on a given night rather than 2 hours later at midnight; how does this relate to Earth's orbit? Why are different stars overhead at midnight on June 1 rather than at midnight on December 1?

5. Suppose the Earth were to spin around its north–south axis twice as fast as it currently does. Would this change the effects of diurnal motion? Explain your answer.

6. Suppose the Earth were to take twice the time that it currently does to orbit the Sun. Would this change the effects of diurnal motion? Explain your answer.

7. Figure 2-4b shows a case where Andromeda is directly overhead at midnight. Is it ever possible for Cygnus to be directly overhead at midnight? (*Hint:* Also consult Figure 2-5.) Explain what kind of motion makes this possible, or prevents this from happening.

8. On November 1 at 8:30 P.M., you look toward the eastern horizon and see the bright star Bellatrix (shown in Figure 2-2b) rising. At approximately what time will Bellatrix rise 1 week later, on November 8?

9. Figure 2-4 shows the situation on September 21, when Cygnus is highest in the sky at 8:00 P.M. local time and Andromeda is highest in the sky at midnight. But as Figure 2-5 shows, on July 21 Cygnus is highest in the sky at midnight. On July 21, at approximately what local time is Andromeda highest in the sky? Explain your reasoning.

10. Figure 2-5 shows which constellations are high in the sky (for observers in the northern hemisphere) in the months of July, September, and November. From this figure, would you be able to see Perseus at midnight on May 15? Draw a picture to justify your answer.

11. Figure 2-6 shows the appearance of Polaris, the Little Dipper, and the Big Dipper at 11 P.M. (daylight saving time) on August 1. Sketch how these objects would appear on this same date at (a) 8 P.M. and (b) 2 A.M. Include the horizon in your sketches, and indicate the north direction.

12. Figure 2-6 shows the appearance of the sky near the North Star at 11 P.M. (daylight saving time) on August 1. Explain why the sky has this same appearance at 1 A.M. on July 1 and at 9 P.M. on September 1.

13. The time-exposure photograph that opens this chapter shows the trails made by individual stars as the celestial sphere appears to rotate around Earth. (a) For approximately what length of time was the camera shutter left open to take this photograph? (b) The stars in this photograph (taken in Hawaii, at roughly 20° north latitude) appear to rotate around one of the celestial poles. Which celestial pole is it? As seen from this location, do the stars move clockwise or counterclockwise around this celestial pole? (c) If you were at 20° south latitude, which celestial pole could you see? In which direction would you look to see it? As seen from this location, do the stars move clockwise or counterclockwise around this celestial pole?

Section 2-4

14. **TUTORIAL 2.1** What is the celestial sphere? Why is this ancient concept still useful today?

15. Imagine that someone suggests sending a spacecraft to land on the surface of the celestial sphere. How would you respond to such a suggestion?

16. What is the celestial equator? How is it related to Earth's equator? How are the north and south celestial poles related to Earth's axis of rotation?

17. Where would you have to look to see your zenith? Where on Earth would you have to be for the celestial equator to pass through your zenith? Where on Earth would you have to be for the south celestial pole to be at your zenith?

18. How many degrees is the angle from the horizon to the zenith? Does your answer depend on what point on the horizon you choose?

19. Why can't a person in Antarctica use the Big Dipper to find the north direction?

20. Is there any place on Earth where you could see the north celestial pole on the northern horizon? If so, where? Is there any place on Earth where you could see the north celestial pole on the western horizon? If so, where? Explain your answers.

21. How do the stars appear to move over the course of the night as seen from the north pole? As seen from the equator? Why are these two motions different?

22. (a) Redraw Figure 2-10 for an observer at the north pole. (*Hint:* The north celestial pole is directly above this observer.) (b) Redraw Figure 2-10 for an observer at the equator. (*Hint:* The celestial equator passes through this observer's zenith.) (c) Using Figure 2-10 and your drawings from (a) and (b), justify the following rule, long used by navigators: The latitude of an observer in the northern hemisphere is equal to the angle in the sky between that observer's horizon and the north celestial pole. (d) State the rule that corresponds to (c) for an observer in the southern hemisphere.

23. The Gemini North Observatory is located in Hawaii, roughly 20° north of the equator. Its near-twin, the Gemini South Observatory, is located roughly 30° south of the equator in Chile. Why is it useful to have telescopes in both the northern and southern hemispheres?

24. Is there any place on Earth where all the visible stars are circumpolar? If so, where? Is there any place on Earth where none of the visible stars is circumpolar? If so, where? Explain your answers.

Section 2-5

25. **TUTORIAL 2.2** Using a diagram, explain why the tilt of Earth's axis relative to Earth's orbit causes the seasons as we orbit the Sun.

26. Give two reasons why it is warmer in summer than in winter.

27. What is the ecliptic plane? What is the ecliptic?

28. Why is the ecliptic tilted with respect to the celestial equator? Does the Sun appear to move along the ecliptic, the celestial equator, or neither? By about how many degrees does the Sun appear to move on the celestial sphere each day?

29. Where on Earth do you have to be in order to see the north celestial pole directly overhead? What is the maximum possible elevation of the Sun above the horizon at that location? On what date can this maximum elevation be observed?

30. What are the vernal and the autumnal equinoxes? What are the summer and winter solstices? How are these four points related to the ecliptic and the celestial equator?

31. At what point on the horizon does the vernal equinox rise? Where on the horizon does it set? (*Hint:* See Figure 2-16.)

32. How does the daily path of the Sun across the sky change with the seasons? Why does it change?

33. Where on Earth do you have to be in order to see the Sun at the zenith? As seen from such a location, will the Sun be at the zenith every day? Explain your reasoning.

34. This image of Earth was made by the Galileo spacecraft while en route to Jupiter. South America is at the center of the image and Antarctica is at the bottom of the image. (a) In approximately which month of the year was this image made? Explain your reasoning. (b) When this image was made, was Earth relatively close to the Sun or relatively distant from the Sun? Explain your reasoning.

NASA/JPL

R I V U X G

35. Figure 2-16 shows the daily path of the Sun across the sky on March 21, June 21, September 22, and December 21 for an observer at 35° north latitude. Sketch drawings of this kind for (a) an observer at 35° south latitude; (b) an observer at the equator; and (c) an observer at the north pole.

36. Suppose that you live at a latitude of 40° N. What is the elevation (angle) of the Sun above the southern horizon at noon (a) at the time of the vernal equinox? (b) At the time of the winter solstice? Explain your reasoning. Include a drawing as part of your explanation.

37. In the northern hemisphere, houses are designed to have "southern exposure," that is, with the largest windows on the southern side of the house. But in the southern hemisphere houses are designed to have "northern exposure." Why are houses designed this way, and why is there a difference between the hemispheres?

38. The city of Mumbai (formerly Bombay) in India is 19° north of the equator. On how many days of the year, if any, is the Sun at the zenith at midday as seen from Mumbai? Explain your answer.

39. Describe how the seasons would be different if Earth's axis of rotation, rather than having its present 23½° tilt, were tilted (a) by 0° or (b) by 90°.

40. **Box 2-1** What is the right ascension of a star that is on the meridian at midnight at the time of the autumnal equinox? Explain your answer.

41. **Box 2-1** The coordinates on the celestial sphere of the summer solstice are R.A. = 6^h 0^m 0^s, Decl. = +23° 27′. What are the right ascension and declination of the winter solstice? Explain your answer.

42. **Box 2-1** Because 24 hours of right ascension takes you all the way around the celestial equator, $24^h = 360°$, what is the angle in the sky (measured in degrees) between a star with R.A. = 8^h 0^m 0^s, Decl. = 0° 0′ 0″ and a second star with R.A. = 11^h 20^m 0^s, Decl. = 0° 0′ 0″? Explain your answer.

43. **Box 2-1** On a certain night, the first star in Question 42 passes through the zenith at 12:30 A.M. local time. At what time will the second star pass through the zenith? Explain your answer.

Section 2-6

44. In Figure 2-12, the Earth's axis is seen to point in the same direction over the course of a year. Does this direction change over a timescale of thousands of years? Explain your answer.

45. Does precession of Earth's axis alter the seasons? Explain your answer.

46. When Polaris is no longer our North Star, about how many years must pass before it becomes our North Star once again?

47. Explain how the precession of Earth's axis leads to precession of the equinoxes.

48. Ancient records show that 2000 years ago, the stars of the constellation Crux (the Southern Cross) were visible in the southern sky from Greece. Today, however, these stars cannot be seen from Greece. What accounts for this change?

49. The Great Pyramid at Giza has a tunnel that points toward the north celestial pole. At the time the pyramid was built, around 2600 B.C.E., toward which star did it point? Toward which star does this same tunnel point today? (See Figure 2-20.)

50. The photo shows a statue of the Greek god Atlas. The globe that Atlas is holding represents the celestial sphere, with depictions of several important constellations and the celestial equator. Although the statue dates from around 150 C.E., it has been proposed that the arrangement of constellations depicts the sky as it was mapped in an early star atlas that dates from 129 B.C.E. Explain the reasoning that could lead to such a proposal.

Scala/Art Resource, NY

R I V U X G

51. In William Shakespeare's *Julius Caesar* (act 3, scene 1), Caesar says:

    > But I am constant as the northern star,
    > Of whose true-fix'd and resting quality
    > There is no fellow in the firmament.

    Translate Caesar's statement about the "northern star" into modern astronomical language. Is the northern star truly "constant"? Was the northern star the same in Shakespeare's time (1564–1616) as it is today?

Section 2-7

52. What is the (fictitious) mean sun? What path does it follow on the celestial sphere? Why is it a better timekeeper than the actual Sun in the sky?

53. Why is it convenient to divide Earth into time zones?

54. **Box 2-2** At local noon on March 21, when the Sun is at the vernal equinox, a sidereal clock will say that it is midnight. Explain why.

55. **Box 2-2** (a) What is the sidereal time when the vernal equinox rises? (b) On what date is the sidereal time nearly equal to the solar time? Explain your answer.

56. **Box 2-2** How would the sidereal and solar days change (a) if Earth's rate of rotation increased, (b) if Earth's rate of rotation decreased, and (c) if Earth's rotation were retrograde (that is, if Earth rotated about its axis opposite to the direction in which it revolves about the Sun)?

Section 2-8

57. Why is the time given by a sundial not necessarily the same as the time on your wristwatch?

58. What is the difference between the sidereal year and the tropical year? Why are these two kinds of year slightly different in length? Why are calendars based on the tropical year?

59. When is the next leap year? Was 2000 a leap year? Will 2100 be a leap year?

60. Unlike western Europe, Imperial Russia did not use the revised calendar instituted by Pope Gregory XIII. Explain why the Russian Revolution, which started on November 7, 1917, according to the modern calendar, is called the October Revolution in Russia. What was this date according to the Russian calendar at the time? Explain your answer.

General Problem

61. Examine a list of the 88 constellations. Are there any constellations whose names obviously date from modern times? Where are these constellations located? Why do you suppose they do not have archaic names?

WEB/E-BOOK QUESTIONS

62. Search the internet for information about the national flags of Australia, New Zealand, and Brazil and the state flag of Alaska. Which stars are depicted on these flags? Explain any similarities or differences among these flags.

63. Some people say that on the date that the Sun is at the vernal equinox, and only on this date, you can stand a raw egg on end. Others say that there is nothing special about the vernal equinox, and that with patience you can stand a raw egg on end on any day of the year. Search the internet for information about this story and for hints about how to stand an egg on end. Use these hints to try the experiment yourself on a day when the Sun is *not* at the vernal equinox. What do you conclude about the connection between eggs and equinoxes?

64. Use the U.S. Naval Observatory website to find the times of sunset and sunrise on (a) your next birthday and (b) the date this assignment is due. (c) Are the times the same for the two dates? Explain why or why not.

ACTIVITIES

Observing Projects

> **Observing tips and tools**
>
> Moonlight is so bright that it interferes with seeing the stars. For the best view of the constellations, do your observing when the Moon is below the horizon. You can find the times of moonrise and moonset in your local newspaper or on the internet. Each monthly issue of the magazines *Sky & Telescope* and *Astronomy* includes much additional observing information.

65. On a clear, cloud-free night, use the star charts at the end of this book to see how many constellations of the zodiac you can identify. Which ones were easy to find? Which were difficult? Are the zodiacal constellations the most prominent ones in the sky?

66. Examine the star charts that are published monthly in such popular astronomy magazines as *Sky & Telescope* and *Astronomy*. How do they differ from the star charts at the end of this book? On a clear, cloud-free night, use one of these star charts to locate the celestial equator and the ecliptic. Note the inclination of the Milky Way to the ecliptic and celestial equator. The Milky Way traces out the plane of our galaxy. What do your observations tell you about the orientation of Earth and its orbit relative to the galaxy's plane?

67. Suppose you wake up before dawn and want to see which constellations are in the sky. Explain how the star charts at the end of this book can be quite useful, even though chart times are given only for the evening hours. Which chart most closely depicts the sky at 4:00 A.M. on the morning that this assignment is due? Set your alarm clock for 4:00 A.M. to see if you are correct.

Starry Night™ Explorations

68. Use *Starry Night*™ to observe the diurnal motion of the sky. Select **Favorites > Explorations > Greenwich** from the menu to see the night sky looking south from Greenwich, England.

Play the animation to observe the motion of the stars and other objects in the sky as time passes. When daylight intervenes, select **View > Hide Daylight** from the menu.

(a) Toward which direction do the objects in the sky appear to move? (*Hint:* Notice the compass points along the horizon.)

(b) Use the "N" keyboard shortcut to change the view to look north. Do the stars appear to rotate clockwise or counterclockwise?

(c) Open **Favorites > Explorations > Sydney** to see the night sky looking north from Sydney, Australia, in the southern hemisphere and click **Play**. Toward which compass direction do the objects in the sky appear to move in this view?

(d) Use the "S" keyboard shortcut to change the view to look south from this location. Do the stars appear to rotate clockwise or counterclockwise?

(e) Open **Favorites > Explorations > Earth From Space** to see a view of Earth from a position in space and click **Play**. Use the location scroller to explore this view. How does it explain the motions of the objects in the sky observed in the views from the surface of Earth at Greenwich in the northern hemisphere and Sydney in the southern hemisphere?

69. If you go outside at the same time of night over the course of a year, you will notice that the entire sky appears to shift its position from night to night. Use *Starry Night™* to display this change and uncover its cause. Click the **Home** button in the toolbar and then the **Stop** button. Set the time in the toolbar to 1 A.M. Use the "S" key to direct the view to the south and the "K" key to display the constellations on the sky.

(a) Use the "D" key to advance time in steps of 1 calendar day per keypress. Toward which direction do the constellations appear to shift night by night?

(b) Note the date in the toolbar and the constellation in the central region of the view. Use the "M" keyboard shortcut to advance time in steps of 1 month until the selected constellation is back in the central region of the view again. How long does it take for the sky to complete a full cycle?

(c) Select **Favorites > Explorations > Year** from the menu to see a view of Earth from a position in space that remains on the line between Earth and the Sun, so that the Sun is always behind the observer and the view is always directed away from the Sun. Note the constellation in the sky behind Earth and use the "M" key to advance time 1 month per keypress until Earth returns to lie in front of the same constellation. To what do you attribute the yearly cycle of constellations in the night sky?

70. Use *Starry Night™* to explore the relationship between the altitude, the height above the horizon of the celestial poles in the sky, the latitude of the observer on Earth, and the concept of circumpolar stars. Begin by opening **Favorites > Explorations > Latitude** to display the northern sky from Calgary, Canada. On-screen info shows the name, latitude, and longitude of the observing location and the altitude and azimuth of the point in the sky that is at the center of the view. A red circle indicates the boundary of the circumpolar region. All the stars within this circle remain above the horizon at this latitude. Click **Play** to see the stars rotate around the north celestial pole. You will see that the star Polaris remains nearly stationary, indicating that it must be very close to this pole. Thus Polaris, the Pole Star, can be used as a beacon not only for the direction of true north but as an approximate measure of the angle above the horizon of the north celestial pole. As you will come to see, observers in the northern hemisphere are fortunate to have this relatively bright star so close to the celestial pole.

(a) Click the **Stop** button to halt time flow. Now use the position of the Pole Star to obtain an approximate measure for the altitude of the north celestial pole. To do this, click the name in the label indicating the Pole Star (Polaris) and select **Center**. Compare the altitude of the Pole Star to the latitude of the observing location in the on-screen info. Select **Options > Viewing Location...** to change the observing location to several other places in the northern hemisphere. Once you have changed to a new location, use the "N" keyboard shortcut to look north and then center Polaris in the view. What is the relationship between the altitude of the north celestial pole (or in this case, its close substitute, the Pole Star) and the latitude of an observer in the northern hemisphere? What is the relationship between the size of the circumpolar region in the sky with the observer's latitude?

(b) Now use **Options > Viewing Location...** to change your location to somewhere in the southern hemisphere, such as Perth, Western Australia. Use the "S" keyboard shortcut to look at the southern sky and click Play. Again, you will see that the stars rotate around a point in the sky. This point is the South Celestial Pole. You can display this exact position by selecting **View > Celestial Guides > Celestial Poles** from the menu. Use the angular separation tool to measure the angular distance from this point to the horizon. How does your measurement compare with the latitude of the location?

(c) From your observations, what general conclusion can you draw about the relationship between the elevation of the celestial pole and the latitude of the observing site?

(d) What will be the altitude of the north celestial pole as seen from the North Pole? What proportion of the stars in the sky will be circumpolar from this location?

(e) What will be the altitude of the south celestial pole in the sky for an observer at the equator of Earth? What will be the altitude of the north celestial pole for this observer? What will be the proportion of circumpolar stars in the sky from this location?

71. Use *Starry Night™* to demonstrate the reason for seasonal variations on Earth at mid-latitudes. A common misconception is

that summertime is warmer because Earth is closer to the Sun in the summer. The actual reason is that the tilt of the spin axis of Earth to its orbital plane places the Sun at a higher angle in the sky in the summer than in the winter. Thus, sunlight hits Earth's surface more directly in summertime than in winter, thereby depositing greater heat. Open **Favorites > Explorations > Seasons** to view the Sun in the sky from a location at 50°N on December 21, at a time near local noon when the Sun's angle above the horizon (i.e., its altitude) is at its highest. Use the HUD to obtain this altitude as well as the distance to the Sun on this date. Use the "M" key to advance time in steps of 1 month and compile a table noting the altitude of the Sun in the local sky and the Sun's distance from Earth (Distance from Observer) on the 21st day of December, March, June, and September.

(a) On which date does the Sun reach its highest altitude in the sky at midday?

(b) On which date does the Sun reach its lowest altitude in the sky at midday?

(c) How does the altitude of the Sun in March and September relate to the latitude of the observing location?

   (i) The Sun's altitude is equal to the latitude.

   (ii) The Sun's altitude is 90° minus the latitude.

   (iii) The Sun's altitude is not related to the latitude of the location of the observer.

(d) On which date is Earth nearest the Sun?

(e) On which date is Earth farthest from the Sun?

72. Use *Starry Night*™ to observe the Sun's motion on the celestial sphere. Select **Favorites > Explorations > Sun's Motion** from the menu. The view shows the Sun's position in the sky as seen from the North Pole of Earth. East is to the left and north is upwards, with the Pole Star out of the view at the top. Daylight has been removed, in order to reveal the background stars. The boundary and name of the constellation centered in the view at any time during the animation is also shown.

(a) Click **Play** to observe the motion of the Sun in the sky as you look toward a fixed point on the horizon, as seen from the surface of Earth. Toward which direction in the sky does the Sun appear to move relative to this point on the horizon?

(b) Select **File > Revert** from the menu to reset the view. Now select **Options > Orientation > Equatorial**. In this view, you will be looking toward a fixed point in the sky as Earth rotates (toward the left). Replay the animation. Hold down the "H" key on your keyboard for several moments to accelerate the time flow rate. What is the motion of the Sun relative to the background stars as seen from this nonrotating frame of reference?

(c) Open **Favorites > Explorations > Ecliptic** to observe the Sun again from the North Pole. In this case, the ecliptic

(green) and celestial equator (red) are shown and both the horizon and daylight have been removed. Click **Play** to observe the apparent motion of the Sun against the celestial sphere. Now use the time controls to find the most northerly and most southerly points on the Sun's path of apparent motion. At each point, use the angular separation cursor to measure the distance from the Sun to the celestial equator. What is the name of each of these points on the celestial sphere and on what dates does the Sun reach them? What is the angular distance of the Sun from the celestial equator at these points? What is the significance of this angle?

(d) Use the time controls to find the points on the celestial sphere when the Sun crosses the celestial equator. For each of these points, what is the name of this point on the celestial sphere, on which date does the Sun reach it, and in which direction is the Sun crossing the equator, from north to south or south to north? In which constellation would the Sun appear from Earth when it crosses the celestial equator from south to north, if our view were not blocked by daylight?

ANSWERS

ConceptChecks

ConceptCheck 2-1: No, Jupiter does not need to be one of the stars in the asterism that makes the outline of the bull's body. Jupiter would only need to be within the somewhat rectangular boundary of the constellation of Taurus.

ConceptCheck 2-2: Earth rotates from west to east. As locations on Earth rotate from the dark, nighttime side of the planet into the bright, daytime side of the planet, the easternmost cities experience sunrise first. New York is the farthest east of the cities listed, so the sun rises there first.

ConceptCheck 2-3: The Sun would still only rise and set once each day (since that defines what a day is), but each year would have 3 times as many days ($365 \times 3 = 1095$ days each year), and each day would be 3 times shorter.

ConceptCheck 2-4: No. Due to Earth's orbit around the Sun, the stars visible at night change slowly throughout the year. This is illustrated in Figure 2-5.

ConceptCheck 2-5: The celestial equator is a projection, or an extension, of Earth's equator out into the sky. In order for this imaginary line to pass directly overhead, one would need to be standing somewhere on Earth's equator.

ConceptCheck 2-6: No, they would not; in this imaginary scenario, even near the north pole the Sun would rise and set every 12 hours all year long. Earth's tilted axis means that observers near Earth's north pole will experience

6 months when the Sun never rises (when it is tilted away from the Sun) and then 6 months when the Sun never sets (when it is tilted toward the Sun).

ConceptCheck 2-7: It takes 1 year. The Sun slowly moves through the sky a little each day, taking 365¼ days to return to the same place it was 1 year earlier.

ConceptCheck 2-8: To cast no shadow, the Sun must be directly overhead. As the Sun's position on the celestial sphere slowly moves back and forth between the northern and southern solstice points over the course of a year, the noontime Sun will be directly overhead (and will cast no shadow) for an observer at Earth's equator only twice each year, on the March and September equinoxes.

ConceptCheck 2-9: Summer. When the Sun is between the winter solstice and the vernal equinox, it is winter in the northern hemisphere, and the tilt of the Earth's axis creates opposite seasons in the two hemispheres. You can also see that it would be summer in the southern hemisphere because sunlight would shine more directly (with a higher concentration on the ground) in the southern hemisphere.

CalculationChecks

CalculationCheck 2-1: Earth rotates once in 24 hours, so when the stars of Cygnus rise in the east at sunset, they take about 12 hours to go from one side of the sky to the other. As a result, it takes about half that time, or 6 hours, for stars of Cygnus to move halfway across the sky. Thus, Cygnus will be highest in the sky around midnight.

CalculationCheck 2-2: The summer solstice occurs on about June 21 and the March equinox occurs about March 21, so there are about 9 months or 270 days between these two events.

Why Astrology Is Not Science by James Randi

I'm involved in the strange business of telling folks what they should already know. I meet audiences who believe in all sorts of impossible things, often despite their education and intelligence. My job is to explain how science differs from the unproven, illogical assumptions of pseudoscience—and why it matters. Perhaps my best example is the difference between astronomy and astrology.

Both astrology and astronomy arose from the wonders of the night sky, from the stars to comets, planets, the Sun, and the Moon. Surely, humans have long reasoned, there must be some meaning in their motions. Surely the Moon's effect on tides hints at hidden "causes" for strange events. *Judiciary* (literally "judging") astrology therefore attempted to foretell the future—our earthly future. To serve it, *horary* (literally "hourly") astrology carefully tracked the heavens.

It is the latter that has become astronomy. Thanks to its process of careful measurement and testing, we now understand more about the true nature of the starry universe than astrologers could ever have imagined. With the birth of a new science, astronomers had a logical framework based on physical causes and systematic observations.

Astrology remains a popular delusion. Far too many believe today that patterns in the sky govern our lives. They accept the vague tendencies and portents of seers who cast horoscopes. They shouldn't. Just a glance at the tenets of astrology provides ample evidence of its absurdity.

An individual is said to be born under a sign. To the astrologer, the Sun was located "in" that sign at the moment of birth. (Stars are not seen in the daytime, but no matter—a calculation tells where the Sun is.) Each sign takes its name from a constellation, a totally imaginary figure invented for our convenience in referring to stars. Different cultures have different mythical figures up there, and so different schools of astrology assign different meanings to the signs they use.

In the spirit of equal-opportunity swindling, astrologers divide up the year fairly, ignoring variations in the size of constellations. Since Libra is tiny, while Virgo is huge, they chop some of the sky off Virgo and add it—along with bits of Scorpio—to bring Libra up to size. The Sun could well be declared "in" Libra when it is actually outside that constellation.

It gets worse. Science constantly challenges itself and changes. The rules of astrology could not, although they were made up thousands of years ago, and since then the "fixed" stars have moved. In particular, precession of the equinoxes has shifted objects in the sky relative to our calendar. The constellations have changed but astrology has not. If you were born August 7, you are said to be a Leo, but the Sun that day was really in the same part of the sky as the constellation Cancer.

With a theory like this to back it up, we should not be surprised at the bottom line: *A pseudoscience does not work.* Test after test has checked its predictions, and the result is always the same. One such investigator is Shawn Carlson of the University of California, San Diego. As he put it in *Nature* magazine, astrology is "a hopeless cause." Johannes Kepler, the pioneering astronomer, himself cast horoscopes, but they are little remembered today. Owen Gingerich, a historian of science at Harvard, puts it well: Kepler was the astrologer who destroyed astrology.

Astronomy works—it works very well indeed—which isn't easy. Because we humans tend to find what we want in any body of data, it takes science's careful process of observation, creative insight, and critical thinking to understand and predict changes in nature. As I write, a transit of Ganymede is due next Thursday at 21:47:20. At exactly that time, the satellite of Jupiter will cross in front of its planet as seen from Earth, and yet most of us will never know it. Still other moons of Jupiter may hold fresh clues to the formation of our entire solar system and the conditions for life elsewhere.

For most people, astronomy has too little fantasy or money in it, and they will never experience the beauty in its predictions. The dedicated labors of generations of scientists have enabled us to perform a genuine wonder.

James ("the Amazing") Randi works tirelessly to expose trickery so that others can relish the greater wonder of science. As a magician, he has had his own television show and an enormous public following. As a lecturer, he addresses teachers, students, and others worldwide. His newsletter and column for *The Skeptic* are key resources for educators. His many books include *Flim-Flam!, The Faith Healers,* and *The Mask of Nostradamus,* about a legendary con man with secrets of his own.

Mr. Randi is the founder of the James Randi Educational Foundation, and his $1 million prize for "the performance of any paranormal event . . . under proper observing conditions" has gone unclaimed for more than 25 years. An amateur archaeologist and astronomer as well, he lives in Florida with several untalented parrots and the occasional visiting magus.

3 Eclipses and the Motion of the Moon

LEARNING GOALS

Viewing a total solar eclipse is a very rare opportunity. On August 21, 2017, such an eclipse was visible within a narrow strip that ran across the United States. So moved by the event, it became known as The Great American Eclipse. In this figure, the very last bit of Sun is about to be blocked by the Moon, and the Sun's extended crownlike "corona" is easily visible. While this solar eclipse lasted only a couple of minutes, millions of viewers will remember it for a lifetime.

Such eclipses can be seen only on specific dates from special locations on Earth, so not everyone will ever see the Moon cover the Sun in this way. But anyone can find the Moon in the sky and observe how its appearance changes from night to night—going from new moon to full moon and back again—and how the times when the Moon rises and sets differ noticeably from one night to the next.

In this chapter, our subject is how the Moon moves as seen from Earth. We will explore why the Moon goes through a regular cycle of phases, and how the Moon's orbit around Earth leads to solar eclipses as well as lunar eclipses. We will also see how ancient astronomers used their observations of the Moon to determine the size and shape of Earth, as well as other features of the solar system. Thus, the Moon—which has always loomed large in the minds of poets, lovers, and dreamers—has also played a key role in the development of our modern picture of the universe.

A total solar eclipse behind a rock climber at Smith Rock, Oregon. (Andrew Studer Photography)

R I V U X G

3-1 The phases of the Moon are caused by its orbital motion

Ignoring the daily motion caused by Earth's rotation, both the Sun and the Moon as seen from Earth appear to move slowly from west to east relative to the background of stars, but they move at very different rates. The Sun takes 1 year to make a complete trip around the imaginary celestial sphere along the path we call the *ecliptic* (Section 2-5). By comparison, the Moon takes only about 4 weeks. In the past, these similar motions led people to believe that both the Sun and the Moon orbit around Earth. We now know that only the Moon orbits Earth, while the Earth–Moon system as a whole (Figure 3-1) orbits the Sun. (In Chapter 4 we will learn how this was discovered.)

> You can tell the Moon's position relative to Earth and the Sun by observing its phase.

One key difference between the Sun and the Moon is the nature of the light that we receive from them. The Sun emits its own light. So do the stars, which are objects like the Sun but much farther away, and so does an ordinary lightbulb. By contrast, the light that we see from the Moon is *reflected* light. This is sunlight that has struck the Moon's surface, bounced off, and ended up in our eyes here on Earth.

CAUTION!

You probably associate *reflection* with shiny objects like a mirror or the surface of a still lake. In science, however, the term refers to light bouncing off any object. You see most objects around you by reflected light. When you look at your hand, for example, you are seeing light from the Sun (or from a light fixture) that has been reflected from the skin of your hand into your eyes. In the same way, moonlight is really sunlight that has been reflected by the Moon's surface. This is what Shakespeare meant in the line "The Moon's an arrant thief, / And her pale fire she snatches from the Sun."

Understanding the Moon's Phases

 TUTORIAL 3.1

Figure 3-1 shows both the Moon and Earth as seen from a spacecraft. When this image was recorded, the Sun was far off to the right. Hence, only the right-hand hemispheres of both worlds were illuminated by the Sun; the left-hand hemispheres were in darkness and are not visible in the picture. In the same way, when we view the Moon from Earth, we see only the half of the Moon that faces the Sun and is illuminated. However, not all of the illuminated half of the Moon is necessarily facing us. As the Moon moves around Earth, from one night to the next, we see different amounts of the illuminated half of the Moon. These different appearances of the Moon are called **lunar phases.**

Figure 3-2 shows the relationship between the lunar phase visible from Earth and the position of the Moon in its orbit. For example, when the Moon is at position A, its location in the sky is in roughly the same direction as the Sun. Hence, the dark hemisphere of the Moon faces Earth. This phase, in which the Moon is barely visible, is called **new moon.** Since a new moon can only

FIGURE 3-1 R I V U X G

Earth and the Moon This picture of Earth and the Moon was taken in 1992 by the Galileo spacecraft on its way toward Jupiter. The Sun, which provides the illumination for both Earth and the Moon, was far to the right and out of the camera's field of view when this photograph was taken. (JPL/NASA)

be located near the Sun in the sky, it rises around sunrise and sets around sunset.

As the Moon continues its orbit from position A in Figure 3-2, more of its illuminated half becomes exposed to our view. The result, shown at position B, is a phase called **waxing crescent moon** (*waxing* is a synonym for "increasing"). About a week after new moon, the Moon is at position C; for a day, we then see half of the Moon's illuminated hemisphere and half of the dark hemisphere. This phase is called **first quarter moon.**

As seen from Earth, a first quarter moon is one-quarter of the way around the celestial sphere from the Sun. It rises and sets about one-quarter of an Earth rotation, or 6 hours, after the Sun does: Moonrise occurs around noon, moonset occurs around midnight, and the moon is highest in the sky around 6 P.M.

CAUTION!

Despite the name, a first quarter moon appears to be *half* illuminated, not one-quarter illuminated! The name indicates that this phase occurs one-quarter of the way through the complete cycle of lunar phases. Because the lunar disk is half illuminated as seen from Earth, it is commonly called a "half moon."

About 4 days later, the Moon reaches position D in Figure 3-2. Still more of the illuminated hemisphere can now be seen from Earth, giving us the phase called **waxing gibbous moon** (the name references the convex shape of the terminator, and *gibbous*

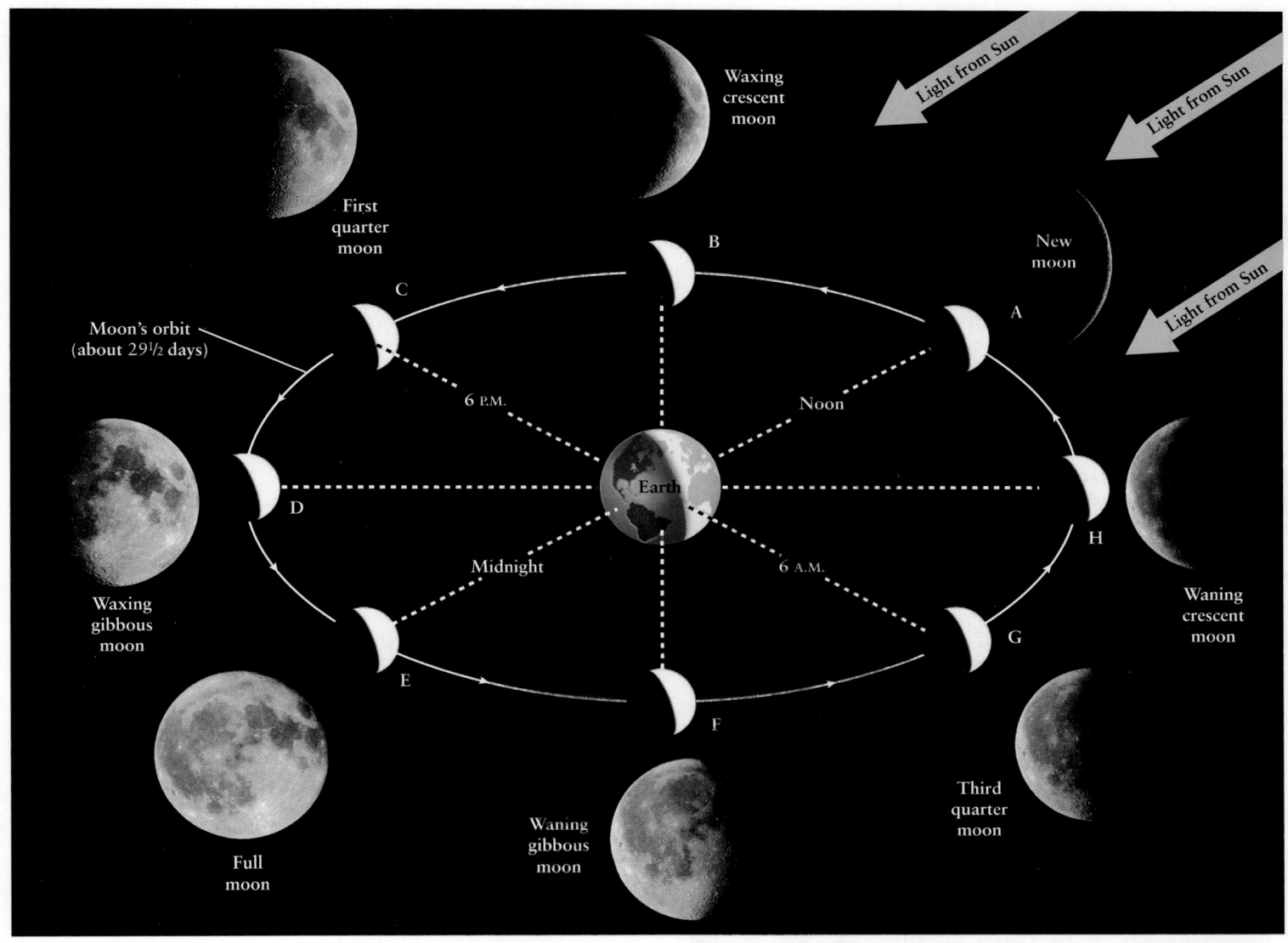

ANIMATION
3.1 ➤

FIGURE 3-2

Why the Moon Goes Through Phases This figure illustrates the Moon at eight positions in its orbit, along with photographs of what the Moon looks like at each position as seen from Earth. The changes in phase occur because light from the Sun illuminates one-half of the Moon, and as the Moon orbits Earth, we see varying amounts of the Moon's illuminated half. It takes about 29½ days for the Moon to go through a complete cycle of phases, and each of the phases shown is visible for 1 day during the complete cycle. The times given in the figure for each phase correspond to the time each phase is observed highest in the sky, halfway between its rise and set. (Larry Landolfi/Science Source)

is another word for "swollen"). When you look at the Moon in this phase, as in the waxing crescent and first quarter phases, the illuminated part of the Moon is on the western edge of the lunar disk. Two weeks after new moon, when the Moon stands opposite the Sun in the sky (position E), we see the fully illuminated hemisphere. This phase is called **full moon.** Because a full moon is opposite the Sun on the celestial sphere, it rises at sunset and sets at sunrise.

Over the following 2 weeks, we see less and less of the Moon's illuminated hemisphere as it continues along its orbit, and the Moon is said to be *waning* ("decreasing"). While the Moon is waning, its illuminated side is on the eastern edge of the lunar disk. The phases are called **waning gibbous moon** (position F),

third quarter moon (position G, also called *last quarter moon*), and **waning crescent moon** (position H). A third quarter moon appears one-quarter of the way around the celestial sphere from the Sun, but on the opposite side of the celestial sphere from a first quarter moon. Hence, a third quarter moon rises and sets about one-quarter Earth rotation, or 6 hours, *before* the Sun: Moonrise is around midnight and moonset is around noon.

The Moon takes about 4 weeks to complete one orbit around Earth, so it likewise takes about 4 weeks for a complete cycle of phases from new moon to full moon and back to new moon. Since the Moon's position relative to the Sun on the celestial sphere is constantly changing, and since our system of timekeeping is based on the Sun (see Section 2-7), the times of moonrise and moonset are different on different nights. *On average, the Moon rises and sets about an hour later each night.*

CONCEPT CHECK 3-1

What would the Moon look like a few days before the new moon to an observer looking up during a lunch break?

CONCEPT CHECK 3-2

At midnight, an observer sees a moon that is not quite a full moon because there is a little darkness on its right edge. Did this moon rise before or after sunset?

Answers appear at the end of the chapter.

Figure 3-2 also explains why the Moon is often visible in the daytime, as shown in Figure 3-3. From any location on Earth, about half of the Moon's orbit is visible at any time. For example, if it is midnight at your location, you are in the middle of the dark side of Earth that faces away from the Sun. At that time, you can easily see the Moon if it is at position C, D, E, F, or G (with the Moon near your horizon at midnight near position C and G). If it is midday at your location, you are in the middle of Earth's illuminated side, and the Moon will be easily visible if it is at position G, H, A, B, or C. (The Moon is so bright that it can be seen even against the bright

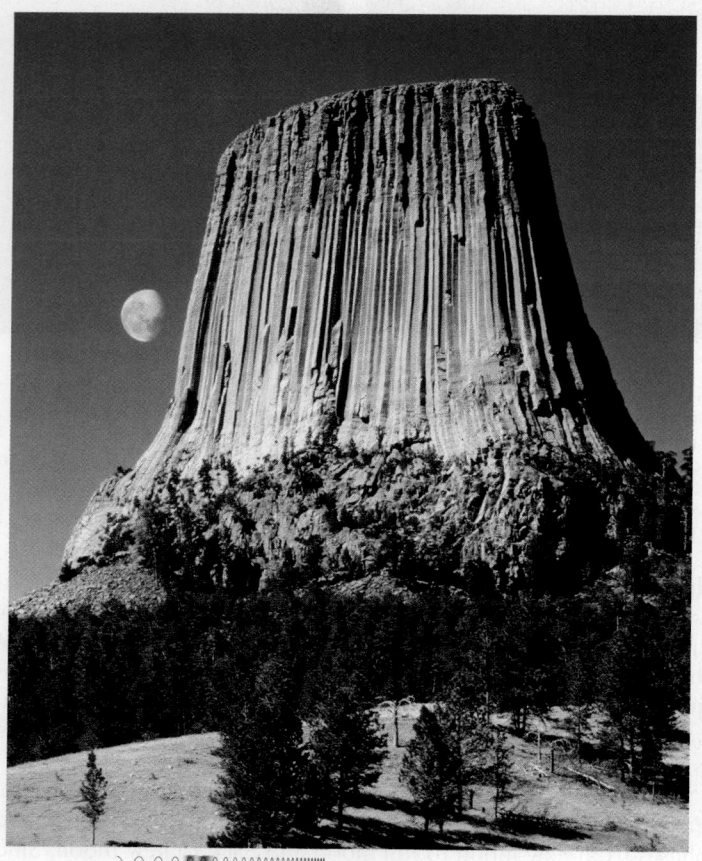

FIGURE 3-3 R I V U X G

The Moon During the Day The Moon can be seen during the daytime as well as at night. The time of day or night when it is visible depends on its phase. (ClassicStock/Alamy Stock Photo)

blue sky.) You can see that the Moon is prominent in the midnight sky for about half of its orbit, and prominent in the midday sky for the other half.

CAUTION!

A very common misconception about lunar phases is that they are caused by the shadow of *Earth* falling on the Moon. As Figure 3-2 shows, this is not the case at all. Instead, phases are simply the result of our seeing the illuminated half of the Moon at different angles as the Moon moves around its orbit. To help you better visualize how this works, **Box 3-1** describes how to simulate the cycle shown in Figure 3-2 using ordinary objects on Earth. (As we will learn in Section 3-3, Earth's shadow does indeed fall on the Moon on rare occasions. When this happens, we see a lunar eclipse.)

CONCEPT CHECK 3-3

If an observer on Earth sees just a tiny sliver of the crescent moon, how much of the Moon's total surface is being illuminated by the Sun?

CONCEPT CHECK 3-4

If the Moon appears in its waxing crescent phase, how will it appear in 2 weeks?

CONCEPT CHECK 3-5

The Moon in Figure 3-3 is not quite full, as can be seen by the slightly darker region on its right side. Consult Figure 3-2 to tell if it is rising or setting. (*Hint:* The Moon rises and sets about an hour later each night.)

Answers appear at the end of the chapter.

3-2 The Moon always keeps the same face toward Earth

Although the phase of the Moon is constantly changing, one aspect of its appearance remains the same: It always keeps essentially the same hemisphere, or face, toward Earth. Thus, you will always see the same craters and mountains on the Moon, no matter when you look at it; the only difference will be the angle at which these surface features are illuminated by the Sun. (You can verify this by carefully examining the photographs of the Moon in Figure 3-2.)

> The Moon rotates in a special way: It spins exactly once per orbit.

The Moon's Synchronous Rotation

Why is it that we only ever see one face of the Moon? You might think that it is because the Moon does not rotate (unlike Earth, which rotates around an axis that passes from its north pole

BOX 3-1   ASTRONOMY DOWN TO EARTH

Phases and Shadows

Figure 3-2 shows how the relative positions of Earth, the Moon, and the Sun explain the phases of the Moon. You can visualize lunar phases more clearly by doing a simple experiment here on Earth. All you need are a small round object, such as an orange or a baseball, and a bright source of light, such as a streetlamp.

In this experiment, you play the role of an observer on Earth looking at the Moon, and the round object plays the role of the Moon. The light source plays the role of the Sun. Hold the object in your right hand with your right arm stretched straight out in front of you, with the object directly between you and the light source (position A in the accompanying illustration). In this orientation, the illuminated half of the object faces away from you, like the Moon when it is in its new phase (position A in Figure 3-2).

Now, keeping your arm outstretched, slowly turn your body to the left so that the object in your hand "orbits" around you (toward positions C, E, and G in the illustration). As you turn, more and more of the illuminated side of the "moon" in your hand becomes visible, and it goes through the same cycle of

phases—waxing crescent, first quarter, and waxing gibbous—as does the real Moon. When you have rotated through half a turn so that the light source is directly behind you, you will be looking face on at the illuminated side of the object in your hand. This corresponds to a full moon (position E in Figure 3-2). Make sure your body does not cast a shadow on the "moon" in your hand—that would correspond to a lunar eclipse!

As you continue turning to the left, more of the unilluminated half of the object becomes visible as its phase moves through waning gibbous, third quarter, and waning crescent. When your body has rotated back to the same orientation that you were in originally, the unilluminated half of your handheld "moon" is again facing toward you, and its phase is again new. If you continue to rotate, the object in your hand repeats the cycle of "phases," just as the Moon does while orbiting around Earth.

The experiment works best when there is just one light source around. If there are several light sources, such as in a room with several lamps turned on, the different sources will create multiple shadows, and it will be difficult to see the phases of your handheld "moon."

to its south pole). To see that the Moon must rotate, consider Figure 3-4. This figure shows Earth and the orbiting Moon from a vantage point far above Earth's north pole. In this figure, two craters on the lunar surface have been colored, one in red and one in blue. If the Moon did not rotate on its axis, as in Figure 3-4a, sometimes the red crater would be visible from Earth, while at other times the blue crater would be visible. Thus, we would see different parts of the lunar surface over time, which does not happen in reality.

In fact, the Moon always keeps the same face toward us because it *is* rotating, but in a very special way: It takes exactly as long for the Moon to rotate on its axis as it does to make one orbit around Earth. This situation is called **synchronous rotation**. As Figure 3-4b shows, this keeps the crater shown in red always facing Earth, so that we always see the same face of the Moon. In Chapter 4, we will learn why the Moon's rotation and orbital motion are in step with each other.

Is there a permanently "dark side of the Moon?" Not at all. To understand this, consider the red crater in Figure 3-4b. Starting at the bottom and orbiting counterclockwise, the red crater would spend 2 weeks (half of a lunar orbit) in darkness, and the next 2 weeks in sunlight. Thus, no part of the Moon is perpetually in darkness. The side of the Moon that constantly faces away from Earth is properly called the *far* side.

CONCEPT CHECK 3-6

Viewed from space, the far side of the Moon is fully dark at only one point during the lunar orbit. Can you identify this point in Figure 3-4b? At what point in Figure 3-4b is the far side of the moon fully illuminated?

If the Moon did not rotate,
we could see all sides of the Moon.

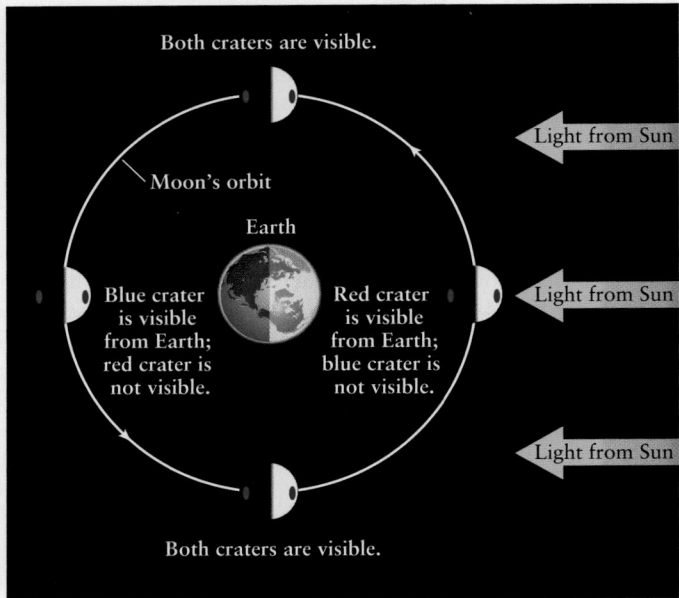

(a) Wrong model

In fact, the Moon does rotate,
and we see only one face of the Moon.

(b) Correct picture

 ANIMATION 3.2

FIGURE 3-4

The Moon's Rotation These diagrams show the Moon at four points in its orbit as viewed from high above Earth's north pole. **(a)** The wrong model: If the Moon did not rotate, then at various times the red crater would be visible from Earth while at other times the blue crater would be visible. Over a complete orbit, the entire surface of the Moon would be visible. **(b)** In reality, the Moon rotates on its north–south axis. Because the Moon makes one rotation in exactly the same time that it makes one orbit around Earth, we see only one face of the Moon.

CONCEPT CHECK 3-7

If astronauts landed on the Moon near the center of the visible surface at full moon, how many Earth days would pass before the astronauts experienced darkness on the Moon?

Answers appear at the end of the chapter.

Sidereal and Synodic Months

It takes about 4 weeks for the Moon to complete one cycle of its phases as seen from Earth. This regular cycle of phases inspired our ancestors to invent the concept of a month. For historical reasons, the calendar we use today has months of differing lengths. Astronomers find it useful to define two other types of months, depending on whether the Moon's motion is measured relative to the stars or to the Sun. Neither corresponds exactly to the familiar months of the calendar.

The **sidereal month** is the time it takes the Moon to complete one full orbit of Earth, as measured *with respect to the stars*. The sidereal period is what you would observe while hovering in space (like the stars), watching the Moon orbit Earth. This true orbital period is equal to about 27.32 days. The **synodic month**, or *lunar month*, is the time it takes the Moon to complete one cycle of phases (that is, from new moon to new moon or from full moon to full moon) and thus is measured *with respect to the Sun* rather than the stars.

The synodic month is longer than the sidereal month because Earth is also moving—that is, orbiting the Sun—while the Moon

goes through its phases. As Figure 3-5 shows, the Moon must travel *more* than 360° along its orbit to complete a cycle of phases (for example, from one new moon to the next). Because of this extra distance, the synodic month is equal to about 29.53 days, about 2 days longer than the sidereal month.

CONCEPT CHECK 3-8

If Earth were orbiting the Sun much faster than it is now, would the length of time between full moons increase, decrease, or stay the same?

Answer appears at the end of the chapter.

3-3 Eclipses occur only when the Sun and Moon are both on the line of nodes

TUTORIAL 3.2 From time to time the Sun, Earth, and Moon all happen to lie along a straight line. When this occurs, the shadow of Earth can fall on the Moon or the shadow of the Moon can fall on Earth. Such phenomena are called **eclipses**. They are perhaps the most dramatic astronomical events that can be seen with the naked eye.

> The tilt of the Moon's orbit makes lunar and solar eclipses rare events.

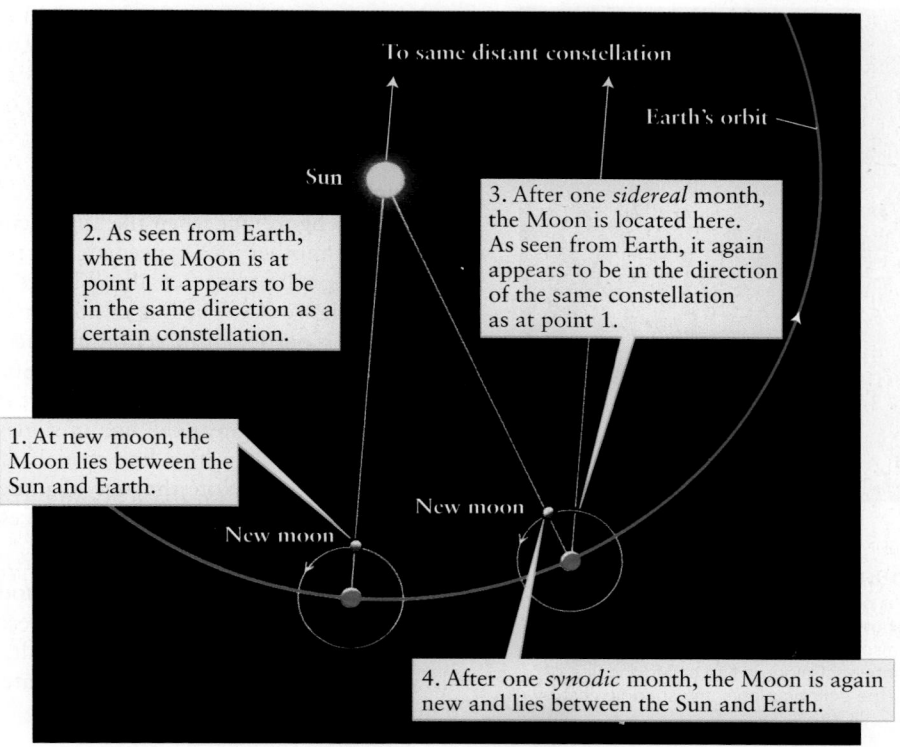

To same distant constellation

Earth's orbit

Sun

2. As seen from Earth, when the Moon is at point 1 it appears to be in the same direction as a certain constellation.

3. After one *sidereal* month, the Moon is located here. As seen from Earth, it again appears to be in the direction of the same constellation as at point 1.

1. At new moon, the Moon lies between the Sun and Earth.

New moon

New moon

4. After one *synodic* month, the Moon is again new and lies between the Sun and Earth.

ANIMATION 3.3 ➤ **FIGURE 3-5**

The Sidereal and Synodic Months The sidereal month is the time the Moon takes to complete one full revolution around Earth with respect to the background stars. However, because Earth is constantly moving along its orbit about the Sun, the Moon must travel through slightly more than 360° of its orbit to get from one new moon to the next. Thus, the synodic month—the time from one new moon to the next—is longer than the sidereal month.

A **lunar eclipse** occurs when the Moon passes through Earth's shadow. This occurs when the Sun, Earth, and Moon are in a straight line (**Figure 3-6a**), with Earth between the Sun and Moon so that the Moon is at full phase (position E in Figure 3-2). At this point in the Moon's orbit, the face of the Moon as seen from Earth would normally be fully illuminated by the Sun. Instead, it appears quite dim because Earth casts a shadow on the Moon.

A **solar eclipse** occurs when the Moon casts a shadow on Earth. As seen from Earth, the Moon moves in front of the Sun, blocking the Sun's bright disk. Once again, this can happen only when the Sun, Moon, and Earth fall into a straight line (Figure 3-6b). However, for a solar eclipse to occur, the Moon must be between Earth and the Sun. Therefore, a solar eclipse can occur only at new moon (position A in Figure 3-2).

CAUTION!

Both new moon and full moon occur at intervals of 29½ days. Hence, you might expect that there would be a solar eclipse every 29½ days, followed by a lunar eclipse about 2 weeks (half a lunar orbit) later. But in fact, there are only a few solar eclipses and lunar eclipses per year. Solar and lunar eclipses are so infrequent because the plane of the Moon's orbit and the plane of Earth's orbit are not exactly aligned, as **Figure 3-7** shows. The angle between the plane of Earth's orbit and the plane of the Moon's orbit is about 5°. Because of this tilt, new moon and full moon usually occur when the Moon is either above or below the plane of Earth's orbit. When the Moon is not in the plane of Earth's orbit, the Sun, Moon, and Earth cannot align perfectly, and an eclipse cannot occur.

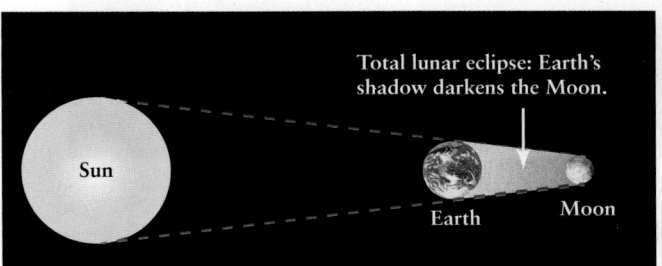

Total lunar eclipse: Earth's shadow darkens the Moon.

Sun

Earth

Moon

(a)

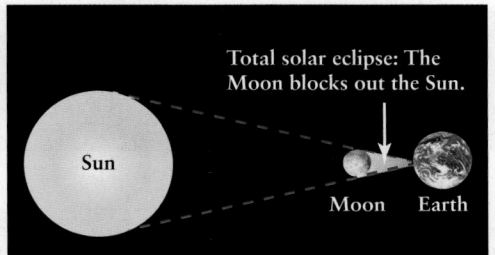

Total solar eclipse: The Moon blocks out the Sun.

Sun

Moon   Earth

(b)

FIGURE 3-6

Alignment for Lunar and Solar Eclipses For eclipses to occur, the Sun, Moon, and Earth must all be lined up. **(a)** For a lunar eclipse, the Moon is within Earth's shadow. **(b)** For a solar eclipse, as viewed from Earth, the Moon blocks out the Sun. For this to occur, an eclipse observer would have to be inside a small shadow created by the Moon.

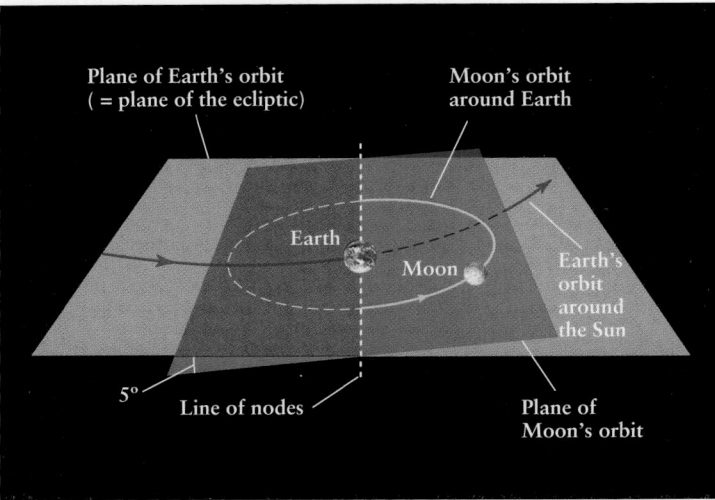

FIGURE 3-7

The Inclination of the Moon's Orbit This drawing shows the Moon's orbit around Earth (in yellow) and part of Earth's orbit around the Sun (in red). The plane of the Moon's orbit (shown in blue) is tilted by about 5° with respect to the plane of Earth's orbit, also called the plane of the ecliptic (shown in brown). These two planes intersect along a line called the line of nodes. Since the Sun and Earth are in the ecliptic plane, any shadow cast by Earth is also in this plane.

In order for the Sun, Earth, and Moon to be lined up for an eclipse, the Moon must lie in the same plane as Earth's orbit around the Sun. As we saw in Section 2-5, this plane is called the *ecliptic plane* because it is the same as the plane of the Sun's apparent path around the sky, or ecliptic (see Figure 2-14). *Thus, when an eclipse occurs, the Moon appears from Earth to be on the ecliptic—this is how the ecliptic gets its name.*

To help figure out when the alignment necessary for an eclipse is possible, another idea is needed. The planes of Earth's orbit and the Moon's orbit intersect along a line called the **line of nodes**, shown in Figure 3-7. The line of nodes passes through Earth and is oriented in a particular direction in space. Eclipses can occur only if the line of nodes points toward the Sun—that is, if the Sun lies on or near the line of nodes—and if, at the same time, the Moon *also* lies on or very near the line of nodes. Only then do the Sun, Earth, and Moon lie in a line straight enough for an eclipse to occur (Figure 3-8). Note that the alignment necessary for eclipses has nothing to do with either the solstices or the equinoxes.

LOOKING DEEPER 3.1 ➤ Anyone who wants to predict eclipses must know the orientation of the line of nodes. But the line of nodes is always gradually shifting because of the gravitational pull of the Sun on the Moon. As a result, the line of nodes rotates slowly westward. Astronomers calculate such details to fix the dates and times of upcoming eclipses.

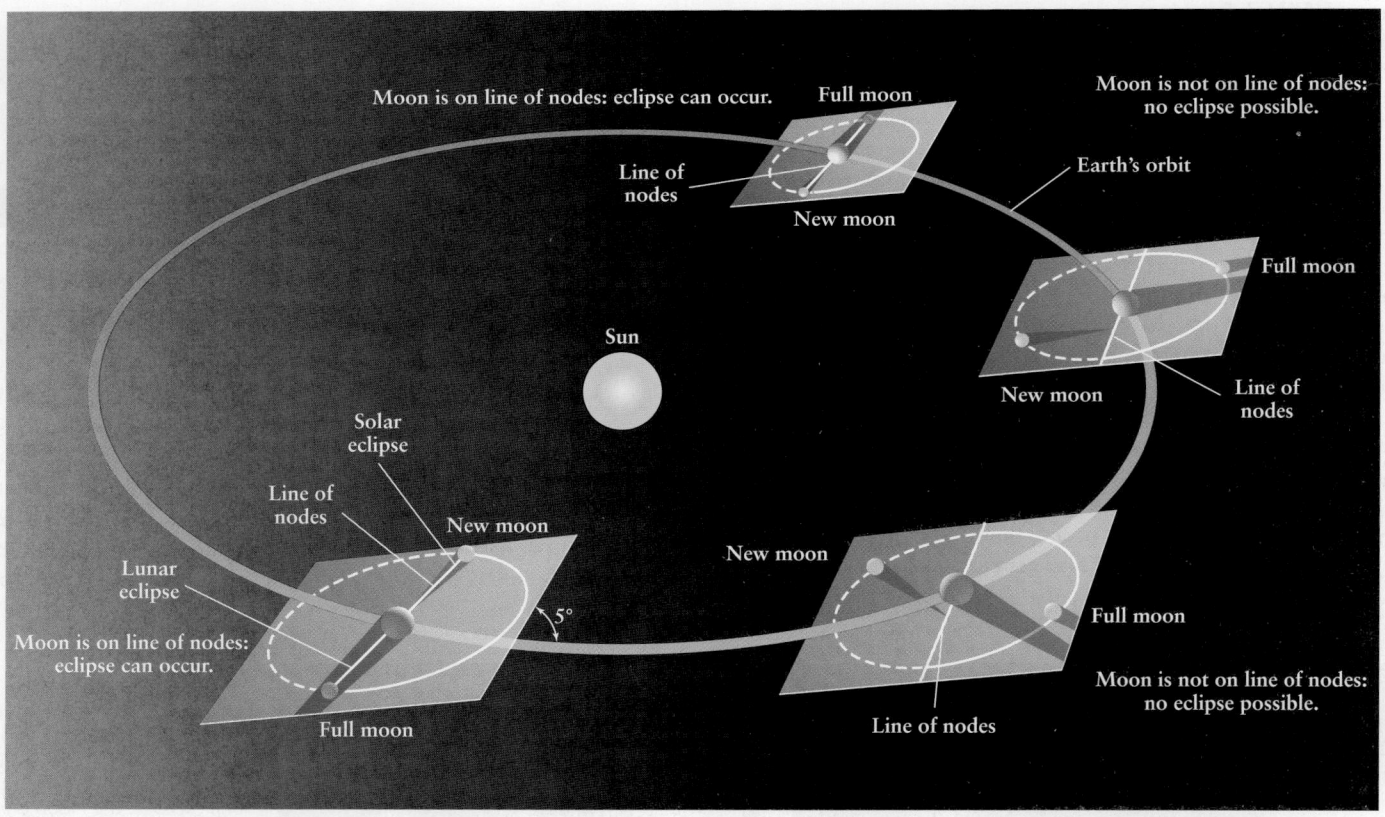

FIGURE 3-8

Conditions for Eclipses Eclipses can take place only if the Sun and Moon are both very near to or on the line of nodes. Only then can the Sun, Earth, and Moon all lie along a straight line. A solar eclipse occurs only if the Moon is very near the line of nodes at new moon; a lunar eclipse occurs only if the Moon is very near the line of nodes at full moon. If the Sun and Moon are not near the line of nodes, the Moon's shadow cannot fall on Earth and Earth's shadow cannot fall on the Moon.

There are at least two—but never more than five—solar eclipses each year. The last year in which five solar eclipses occurred was 1935. The fewest eclipses possible (two solar, zero lunar) happened in 1969. Lunar eclipses occur just about as frequently as solar eclipses, but the maximum possible number of eclipses (lunar and solar combined) in a single year is seven.

CONCEPT CHECK 3-9

Why don't lunar eclipses occur each time the Moon reaches its full moon phase?

Answer appears at the end of the chapter.

3-4 The character of a lunar eclipse depends on the alignment of the Sun, Earth, and Moon

The character of a lunar eclipse depends on exactly how the Moon travels through Earth's shadow. As Figure 3-9 shows, the shadow of Earth has two distinct parts. First, picture how the Moon's surface is illuminated as the Moon passes through the shadow. In the **umbra,** an observer on the Moon would not be able to see any portion of the Sun's disk: This is the darkest part of the shadow. On the other hand, a portion of the Sun's surface is visible to an observer in the **penumbra,** which therefore is not quite as dark. Most people notice a lunar eclipse only if the Moon passes into Earth's umbra, because the dimming of the Moon in the penumbra is not very strong. By contrast, when the umbral phase of the eclipse begins, a bite appears to have been taken out of the Moon.

The inset in Figure 3-9 shows the different ways in which the Moon can pass into Earth's shadow. When the Moon passes through only Earth's penumbra (Path 1), we see a **penumbral eclipse.** During a penumbral eclipse, Earth blocks only part of the Sun's light such

> The Sun's tenuous outer atmosphere is revealed during a total solar eclipse.

that none of the lunar surface is completely shaded. Because the Moon still looks full but only a little dimmer than usual, penumbral eclipses are easy to miss. If the Moon travels completely into the umbra (Path 2), a **total lunar eclipse** occurs. If only part of the Moon passes through the umbra (Path 3), we see a **partial lunar eclipse.**

If you were on the Moon during a total lunar eclipse, the Sun would be hidden behind Earth. But some sunlight would be visible through the thin ring of atmosphere around Earth, just as you can see sunlight through a person's hair if they stand with their head between your eyes and the Sun. As a result, a small amount of light reaches the Moon during a total lunar eclipse, so the Moon does not completely disappear from the sky as seen from Earth. Most of the sunlight that passes through Earth's atmosphere is red, and thus the eclipsed Moon glows faintly in reddish hues, as Figure 3-10 shows. (We'll see in Chapter 5 how our atmosphere causes this reddish color.)

Lunar eclipses occur at full moon, when the Moon is directly opposite the Sun in the sky. Hence, a lunar eclipse can be seen at any place where it is nighttime. A lunar eclipse has the maximum possible duration if the Moon travels directly through the center of the umbra. The Moon's speed through Earth's shadow is roughly 1 kilometer per second (3600 km/h, or 2280 mi/h), which means that **totality**—the period when the Moon is completely within Earth's umbra—can last for as long as 1 hour and 42 minutes.

On average, two or three lunar eclipses occur in a year. Table 3-1 lists all twelve lunar eclipses from 2018 to 2021. Of all lunar eclipses, roughly one-third are total, one-third are partial, and one-third are penumbral.

CONCEPT CHECK 3-10

Is it possible to have a total lunar eclipse without a penumbral eclipse?

CONCEPT CHECK 3-11

Suppose a lunar eclipse occurs in the early morning hours. When did the Moon rise on that day?

Answers appear at the end of the chapter.

FIGURE 3-9

Three Types of Lunar Eclipse People on the nighttime side of Earth see a lunar eclipse when the Moon moves through Earth's shadow. In the umbra, the darkest part of the shadow, the Sun is completely covered by Earth. The penumbra is less dark because only part of the Sun is covered by Earth. The three paths show the motion of the Moon if the lunar eclipse is penumbral (Path 1, in yellow), total (Path 2, in red), or partial (Path 3, in blue). The inset shows these same paths, along with the umbra and penumbra, as viewed from Earth.

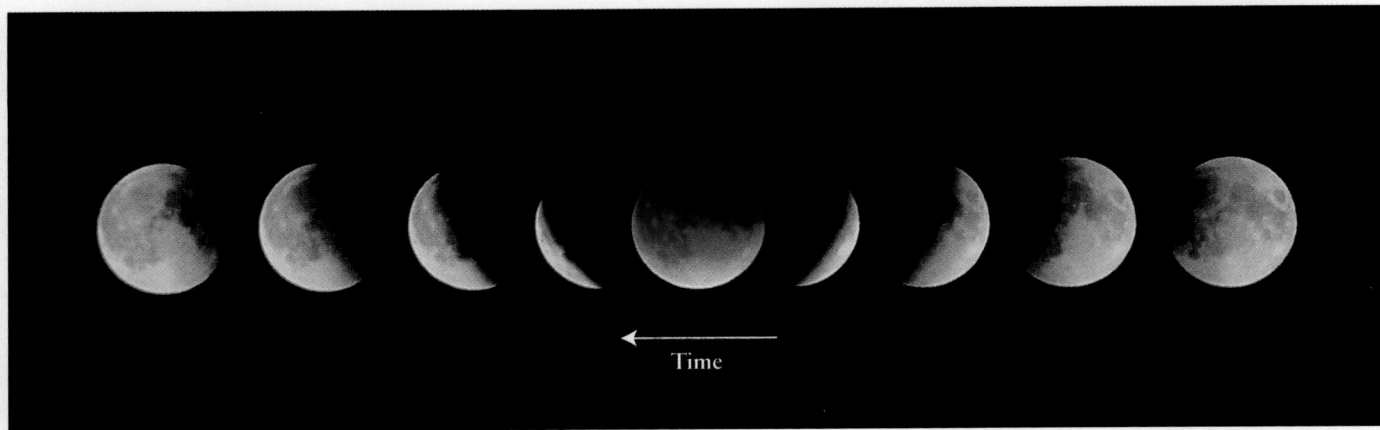

FIGURE 3-10 R I V U X G

A Total Lunar Eclipse This sequence of nine photographs was taken over a 3-hour period during the lunar eclipse of January 20, 2000. The sequence, which runs from right to left, shows the Moon moving through Earth's umbra. During the total phase of the eclipse (shown in the center), the Moon has a distinct reddish color. *Note:* While the moon can appear to have phases during an eclipse, the actual phase is a full moon. (Courtesy of Fred Espenak, www.mreclipse.com)

TABLE 3-1 Lunar Eclipses, 2018–2021

Date	Type	Where visible	Duration of totality (h = hours, m = minutes)
2018 Jan 31	Total	Asia, Australia, Pacific, western North America	01h 16m
2018 July 27	Total	South America, Europe, Africa, Asia, Australia	01h 43m
2019 Jan 21	Total	Central Pacific, Americas, Europe, Africa	01h 02m
2019 July 16	Partial	South America, Europe, Africa, Asia, Australia	—
2020 Jan 10	Penumbral	Europe, Africa, Asia, Australia	—
2020 June 5	Penumbral	Europe, Africa, Asia, Australia	—
2020 July 5	Penumbral	Americas, Southwest Europe, Africa	—
2020 Nov 30	Penumbral	Asia, Australia, Pacific, Americas	—
2021 May 26	Total	Asia, Australia, Pacific, Americas	00h 15m
2021 Nov 19	Penumbral	Americas, Europe, Asia, Australia, Pacific	—

Eclipse predictions by Fred Espenak, NASA/Goddard Space Flight Center.

3-5 Solar eclipses also depend on the alignment of the Sun, Earth, and Moon

TUTORIAL 3.3 As seen from Earth, the angular diameter of the Moon is almost exactly the same as the angular diameter of the far larger but more distant Sun—about 0.5°. Thanks to this coincidence of nature, the Moon just "fits" over the Sun during a **total solar eclipse.**

Total Solar Eclipses

A total solar eclipse is a dramatic event. The sky begins to darken, the air temperature falls, and winds increase as the Moon gradually covers more and more of the Sun's disk. All of nature responds: Birds go to roost, flowers close their petals, and crickets begin to chirp as if evening has already arrived. As the last

A lunar eclipse is most impressive when total, but can also be partial or penumbral.

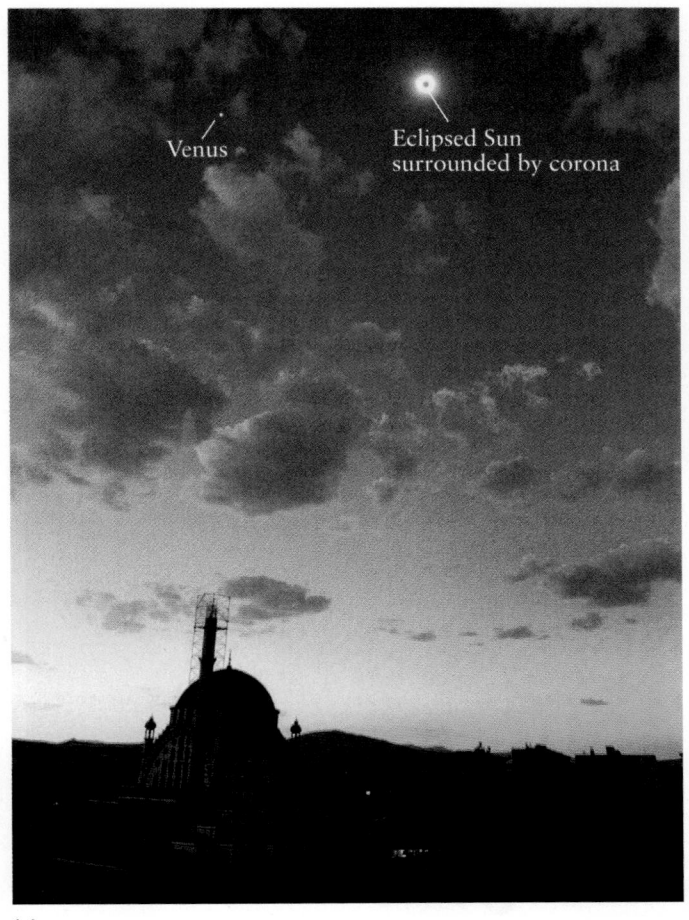

Venus

Eclipsed Sun
surrounded by corona

(a)

(b)

FIGURE 3-11 R I **V** U X G

A Total Solar Eclipse **(a)** This photograph shows the total solar eclipse of August 11, 1999, as seen from Elâziğ, Turkey. The sky is so dark that the planet Venus can be seen to the left of the eclipsed Sun. **(b)** When the Moon completely covers the Sun's disk during a total eclipse, the faint solar corona is revealed. (Courtesy of Fred Espenak, www.mreclipse.com)

few rays of sunlight peek out from behind the edge of the Moon and the eclipse becomes total, the landscape around you is bathed in an eerie gray or, less frequently, in shimmering bands of light and dark. Finally, for a few minutes the Moon completely blocks out the dazzling solar disk (Figure 3-11a). The **solar corona**—the Sun's thin, hot outer atmosphere, which is normally too dim to be seen—blazes forth in the darkened daytime sky (Figure 3-11b). It is an awe-inspiring sight.

CAUTION!

If you are fortunate enough to see a solar eclipse, keep in mind that the only time when it is safe to look at the Sun is during **totality,** when the solar disk is blocked by the Moon and only the solar corona is visible. Viewing this magnificent spectacle cannot harm you in any way. But you must *never* look directly at the Sun when even a portion of its intensely brilliant disk is exposed. *If you look directly at the Sun at any time without a special filter approved for solar viewing, you will suffer permanent eye damage or blindness.*

To see the remarkable spectacle of a total solar eclipse, you must be inside the darkest part of the Moon's shadow, also called the umbra, where the Moon completely blocks the Sun. Because the Sun and the Moon have nearly the same angular diameter as seen from Earth, only the tip of the Moon's umbra reaches Earth's surface (Figure 3-12). As Earth rotates, the tip of the umbra traces an **eclipse path** across Earth's surface. Only those locations within the eclipse path are treated to the spectacle of a total solar eclipse. The inset in Figure 3-12 shows the dark spot on Earth's surface produced by the Moon's umbra.

Partial Solar Eclipses

Immediately surrounding the Moon's umbra is the region of partial shadow called the penumbra. As seen from this area, the Sun's surface appears only partially covered by the Moon. During a solar eclipse, the Moon's penumbra covers a large portion of Earth's surface, and anyone standing inside the penumbra sees a **partial solar eclipse.** Such eclipses are much less interesting events than total solar eclipses, which is why astronomy enthusiasts strive to be inside the eclipse path. If you are within the eclipse path, you will still see a partial eclipse before and after the brief period of totality.

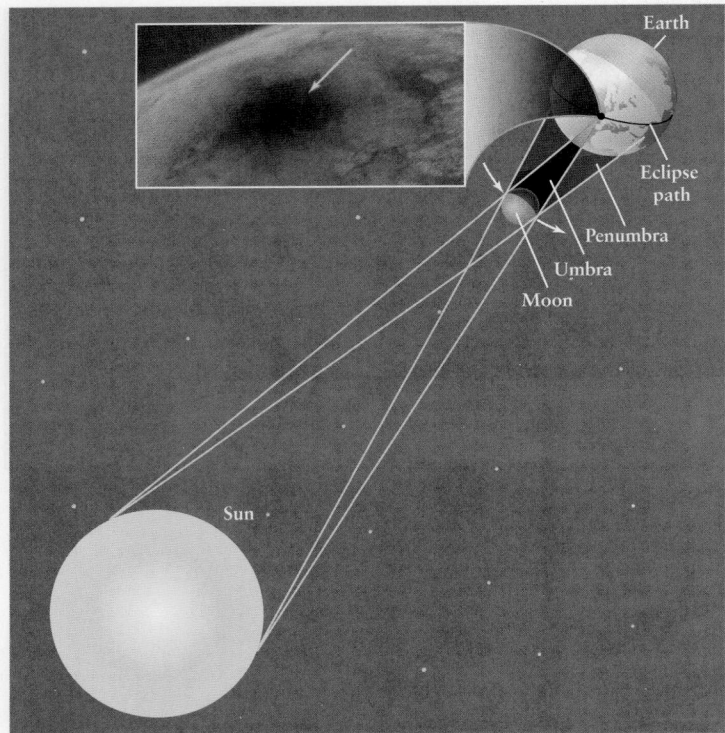

ANIMATION
3.4

FIGURE 3-12 R I V U X G

The Geometry of a Total Solar Eclipse During a total solar eclipse, the tip of the Moon's umbra reaches Earth's surface. As Earth and the Moon move along their orbits, this tip traces an eclipse path across Earth's surface. People within the eclipse path see a total solar eclipse as the tip moves over them. Anyone within the penumbra sees only a partial eclipse. The inset photograph was taken from the Mir space station during the August 11, 1999, total solar eclipse (the same eclipse shown in Figure 3-11). The tip of the umbra appears as a black spot on Earth's surface. At the time the photograph was taken, this spot was 105 km (65 mi) wide and was crossing the cloud-covered English Channel at 3000 km/h (1900 mi/h). (Jean-Pierre Haigneré, Centre National d'Etudes Spatiales, France/GSFS/NASA)

FIGURE 3-13 R I V U X G

An Annular Solar Eclipse This composite of six photographs taken at sunrise in Costa Rica shows the progress of an annular eclipse of the Sun on December 24, 1973. Note that at mideclipse the limb, or outer edge, of the Sun is visible around the Moon. (Photo Researchers/Getty Images)

The width of the eclipse path depends primarily on the Earth–Moon distance during totality. The eclipse path is widest if the Moon happens to be at **perigee,** the point in its orbit nearest Earth. In this case, the width of the eclipse path can be as great as 270 km (170 mi). In most eclipses, however, the path is much narrower.

CONCEPT CHECK 3-12

Why can a total lunar eclipse be seen by people all over the world whereas total solar eclipses can only be seen from a very limited geographic region?

Answer appears at the end of the chapter.

Annular Solar Eclipses

In some eclipses, the Moon's umbra does not reach all the way to Earth's surface. This can happen if the Moon is at or near **apogee,** its farthest position from Earth. In this case, the Moon appears too small to cover the Sun completely. The result is a third type of solar eclipse, called an **annular eclipse.** During an annular eclipse, a thin ring of the Sun is seen around the edge of the Moon (Figure 3-13). The length of the Moon's umbra is nearly 5000 km (3100 mi) less than the average distance between the Moon and Earth's surface. Thus, the Moon's shadow often fails to reach Earth even when the Sun, Moon, and Earth are properly aligned for an eclipse. Hence, annular eclipses are slightly more common—as well as far less dramatic—than total eclipses.

Even during a total eclipse, most people along the eclipse path observe totality for only a few moments. Earth's rotation, coupled with the orbital motion of the Moon, causes the umbra to race eastward along the eclipse path at speeds in excess of 1700 km/h (1060 mi/h). Because of the umbra's high speed, totality never lasts for more than 7½ minutes. In a typical total solar eclipse, the Sun–Moon–Earth alignment and the Earth–Moon distance are such that totality lasts much less than this maximum.

The details of solar eclipses can be calculated well in advance. Figure 3-14 shows the eclipse paths for all total solar eclipses from 1997 to 2021. Table 3-2 lists all the total, annular, and partial eclipses from 2018 to 2021, including the maximum duration of totality for total eclipses.

FIGURE 3-14

Eclipse Paths for Total Eclipses, 1997–2021 This map shows the eclipse paths for all 18 total solar eclipses occurring from 1997 through 2021. In each eclipse, the Moon's shadow travels along the eclipse path in a generally eastward direction across Earth's surface. (Courtesy of Fred Espenak, NASA/Goddard Space Flight Center)

TABLE 3-2 Solar Eclipses, 2018–2021

Date	Type	Where visible	Notes
2018 Feb 15	Partial	Antarctica, southern South America	60% eclipsed
2018 July 13	Partial	Southern Australia	34% eclipsed
2018 Aug 11	Partial	Northern Europe, northeast Asia	74% eclipsed
2019 Jan 6	Partial	Northeast Asia, north Pacific	72% eclipsed
2019 July 2	Total	South Pacific, South America [Total: south Pacific, Chile, Argentina]	Maximum duration of totality 04m 33s
2019 Dec 26	Annular	Asia, Australia [Annular: Saudi Arabia, India, Sumatra, Borneo]	97% eclipsed
2020 June 21	Annular	Africa, southeast Europe, Asia [Annular: central Africa, South Asia, China, Pacific]	99% eclipsed
2020 Dec 14	Total	Pacific, southern South America, Antarctica [Total: south Pacific, Chile, Argentina, South Atlantic]	Maximum duration of totality: 02m 10s
2021 Jun 10	Annular	North America, Europe, Asia [Annular: Canada, Greenland, Russia]	94% eclipsed
2021 Dec 4	Total	Antarctica, South Africa, south Atlantic [Total: Antarctica]	Maximum duration of totality: 01m 54s

Eclipse predictions by Fred Espenak, NASA/Goddard Space Flight Center.

BOX 3-2 TOOLS OF THE ASTRONOMER'S TRADE

Predicting Solar Eclipses

Suppose that you observe a solar eclipse in your hometown and want to figure out when you and your neighbors might see another eclipse. How would you begin?

First, remember that a solar eclipse can occur only if the line of nodes points toward the Sun at the same time that there is a new moon (see Figure 3-8). Second, you must know that it takes 29.53 days (one synodic month) to go from one new moon to the next. Because solar eclipses occur only during new moon, you must wait several whole lunar months for the proper alignment to occur again.

However, there is a complication: The line of nodes gradually shifts its position with respect to the background stars. It takes 346.6 days to move from one alignment of the line of nodes pointing toward the Sun to the next identical alignment. This period is called the **eclipse year.**

Therefore, to predict when you will see another solar eclipse, you need to know how many whole lunar months equal some whole number of eclipse years. This information will tell you how long you will have to wait for the next virtually identical alignment of the Sun, the Moon, and the line of nodes. By trial and error, you find that 223 lunar months is the same length of time as 19 eclipse years, because

$$223 \times 29.53 \text{ days} = 19 \times 346.6 \text{ days} = 6585 \text{ days}$$

This calculation is accurate to within a few hours. A more accurate calculation gives an interval, called the **saros,** that is about one-third of a day longer, or 6585.3 days (18 years, 11.3 days). Eclipses separated by the saros interval are said to form an *eclipse series.*

You might think that you and your neighbors would simply have to wait one full saros interval to go from one solar eclipse to the next. However, because of the extra one-third day, Earth will have rotated by an extra 120° (one-third of a complete rotation) when the next solar eclipse of a particular series occurs. The eclipse path will thus be one-third of the way around the world from you. Therefore, you must wait three full saros intervals (54 years, 34 days) before the eclipse path comes back around to your part of Earth. The illustration shows a series of six solar eclipse paths, each separated from the next by one saros interval.

There is evidence that ancient Babylonian astronomers knew about the saros interval. However, the discovery of the saros is more likely to have come from lunar eclipses than solar eclipses. If you are far from the eclipse path, there is a good chance that you could fail to notice a solar eclipse. Even if half the Sun is covered by the Moon, the remaining solar surface provides enough sunlight for the outdoor illumination not to be greatly diminished. By contrast, anyone on the nighttime side of Earth can see an eclipse of the Moon unless clouds block the view.

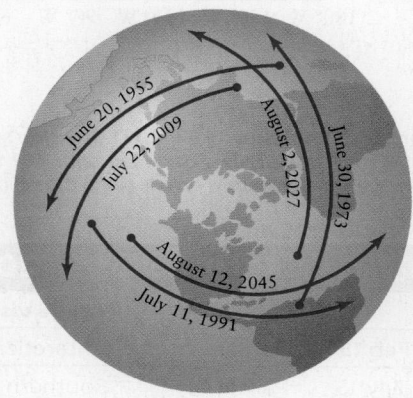

Ancient astronomers achieved a limited ability to predict eclipses. In those times, religious and political leaders who were able to predict such awe-inspiring events as eclipses must have made a tremendous impression on their followers. One of three priceless manuscripts to survive the devastating Spanish Conquest shows that the Mayan astronomers of Mexico and Guatemala had a fairly reliable method for predicting eclipses. The great Greek astronomer Thales of Miletus is said to have predicted the famous eclipse of 585 B.C.E., which occurred during the middle of a war. The sight was so unnerving that the soldiers put down their arms and declared peace.

In retrospect, it seems that what ancient astronomers actually produced were eclipse "warnings" of various degrees of reliability rather than true predictions. Working with historical records, these astronomers generally sought to discover cycles and regularities from which future eclipses could be anticipated. Box 3-2 describes how you might produce eclipse warnings yourself.

CONCEPT CHECK 3-13

If you had a chance to observe a total solar eclipse and a total lunar eclipse, in general, how much longer would you expect one type to last than the other?

Answer appears at the end of the chapter.

3-6 Ancient astronomers measured the size of Earth and attempted to determine distances to the Sun and Moon

The prediction of eclipses was not the only problem attacked by ancient astronomers. More than 2000 years ago, centuries before sailors of Columbus's era crossed the oceans, Greek astronomers were fully aware that Earth is not flat. They had come to this conclusion using a combination of observation and logical deduction, much like modern scientists. The Greeks noted that during lunar eclipses, when the Moon passes through Earth's shadow, the edge of the shadow is always circular. Because a sphere is the only shape that always casts a circular shadow from any angle, they concluded that Earth is spherical.

> The ideas of geometry made it possible for Greek scholars to estimate cosmic distances.

Eratosthenes and the Size of Earth

Around 200 B.C.E., the Greek astronomer Eratosthenes devised a way to measure the circumference of the spherical Earth. It was known that on the date of the summer solstice (the first day of summer; see Section 2-5) in the ancient town of Syene in Egypt, near present-day Aswan, the Sun shone directly down the vertical shafts of water wells. Hence, at local noon on that day, the Sun was at the zenith (see Section 2-4) as seen from Syene. Eratosthenes knew that the Sun never appeared at the zenith at his home in the Egyptian city of Alexandria, which is on the Mediterranean Sea almost due north of Syene. Rather, on the summer solstice in Alexandria, the position of the Sun at local noon was about 7° south of the zenith (**Figure 3-15**). This angle is about one-fiftieth of a complete circle, so he concluded that the distance from Alexandria to Syene must be about one-fiftieth of Earth's circumference.

In Eratosthenes's day, the distance from Alexandria to Syene was said to be 5000 stades. Therefore, Eratosthenes found Earth's circumference to be

$$50 \times 5000 \text{ stades} = 250{,}000 \text{ stades}$$

Unfortunately, no one today is sure of the exact length of the Greek unit called the stade. One guess is that the stade was about one-sixth of a kilometer, which would mean that Eratosthenes obtained a circumference for Earth of about 42,000 km. This is remarkably close to the modern-day value of 40,000 km.

Aristarchus and Distances in the Solar System

Eratosthenes was only one of several brilliant astronomers to emerge from the so-called Alexandrian school, which by his time had a distinguished tradition. One of the first Alexandrian astronomers, Aristarchus of Samos, had proposed a method of determining the relative distances to the Sun and Moon, perhaps as long ago as 280 B.C.E.

Aristarchus knew that the Sun, Moon, and Earth form a right triangle at the moment of first or third quarter moon, with the right angle at the location of the Moon (**Figure 3-16**). He estimated that, as seen from Earth, the angle between the Moon and the Sun at first and third quarters is 87°, or 3° less than a right angle. Using the rules of geometry, Aristarchus concluded that the Sun is about 20 times farther from us than is the Moon. We now know that Aristarchus erred in measuring angles and that the average distance to the Sun is about 390 times larger than the average distance to the Moon. It is nevertheless impressive that people were trying to measure distances across the solar system more than 2000 years ago.

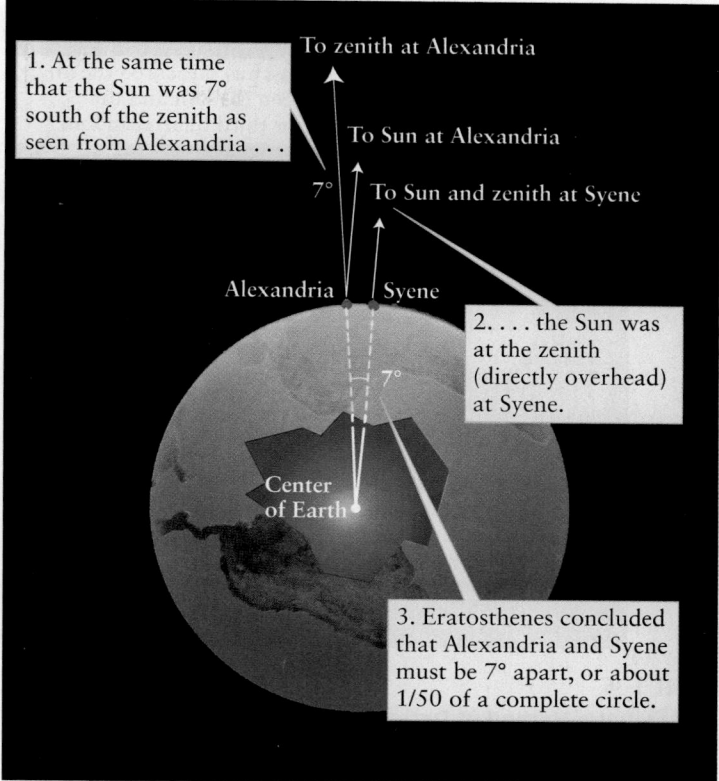

FIGURE 3-15

Eratosthenes's Method of Determining the Diameter of Earth Around 200 B.C.E., Eratosthenes used observations of the Sun's position at noon on the summer solstice to show that Alexandria and Syene were about 7° apart on the surface of Earth. This angle is about one-fiftieth of a circle, so the distance between Alexandria and Syene must be about one-fiftieth of Earth's circumference.

Aristarchus also made an equally bold attempt to determine the relative sizes of Earth, the Moon, and the Sun. From his observations of how long the Moon takes to move through Earth's shadow during a lunar eclipse, Aristarchus estimated the diameter of Earth to be about 3 times larger than the diameter of the Moon. To determine the diameter of the Sun, Aristarchus simply pointed out that the Sun and the Moon have the same angular size in the sky. Therefore, their diameters must be in the same proportion as their distances (see part a of the figure in Box 1-1). In other words, because Aristarchus thought the Sun to be 20 times farther from Earth than the Moon, he concluded that the Sun must be 20 times larger than the Moon. Once Eratosthenes had measured Earth's circumference, astronomers of the Alexandrian school could estimate the diameters of the Sun and Moon as well as their distances from Earth.

Table 3-3 summarizes some ancient and modern-day measurements of the sizes of Earth, the Moon, and the Sun and the distances between them. Some of these ancient measurements are far from the modern values. Yet the achievements of our ancestors still stand as impressive applications of observation and reasoning, important precursors to the development of the scientific method.

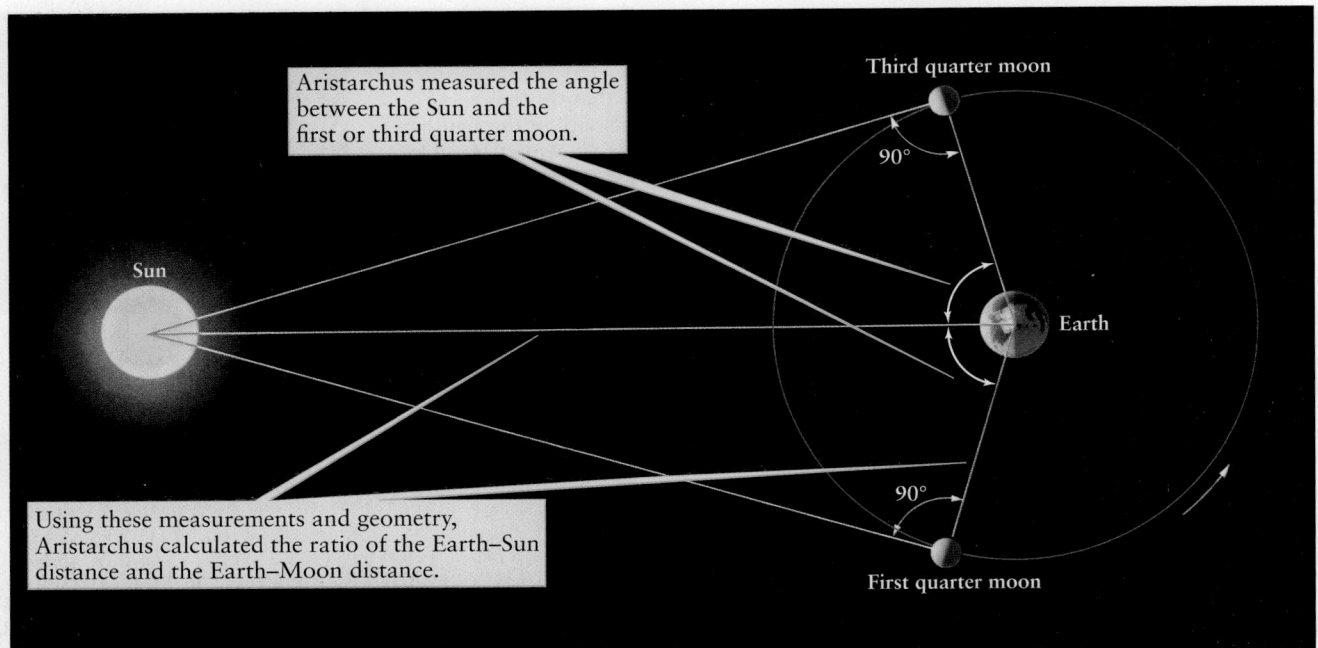

FIGURE 3-16
Aristarchus's Method of Determining Distances to the Sun and Moon Aristarchus knew that the Sun, Moon, and Earth form a right triangle at first and third quarter phases. Using geometrical arguments, he calculated the relative lengths of the sides of these triangles, thereby obtaining estimates of the distances to the Sun and Moon.

TABLE 3-3 Comparison of Ancient and Modern Astronomical Measurements

	Ancient (km)	Modern (km)
Earth's diameter	13,000	12,756
Moon's diameter	4300	3476
Sun's diameter	9×10^4	1.39×10^6
Earth–Moon distance	4×10^5	3.84×10^5
Earth–Sun distance	10^7	1.50×10^8

CONCEPT CHECK 3-14

If Eratosthenes had found that the Sun's noontime summer solstice altitude at Alexandria was much closer to directly overhead, would he then assume that Earth was larger, smaller, or about the same size?

CALCULATION CHECK 3-1

If Aristarchus had estimated the Sun to be 100 times farther from Earth than the Moon, how large would he have estimated the Sun to be?

Answers appear at the end of the chapter.

KEY WORDS

Terms preceded by an asterisk (*) are discussed in the Boxes.

annular eclipse, p. 62
apogee, p. 62
eclipse, p. 56
eclipse path, p. 61
*eclipse year, p. 64
first quarter moon, p. 52
full moon, p. 53
line of nodes, p. 58
lunar eclipse, p. 57

lunar phases, p. 52
new moon, p. 52
partial lunar eclipse, p. 59
partial solar eclipse, p. 61
penumbra (*plural* penumbrae), p. 59
penumbral eclipse, p. 59
perigee, p. 62
*saros, p. 64

sidereal month, p. 56
solar corona, p. 61
solar eclipse, p. 57
synchronous rotation, p. 55
synodic month, p. 56
third quarter moon, p. 53
totality (lunar eclipse), p. 59
totality (solar eclipse), p. 61
total lunar eclipse, p. 59

total solar eclipse, p. 60
umbra (*plural* umbrae), p. 59
waning crescent moon, p. 53
waning gibbous moon, p. 53
waxing crescent moon, p. 52
waxing gibbous moon, p. 52

KEY IDEAS

Ideas preceded by an asterisk (*) are discussed in the Boxes.

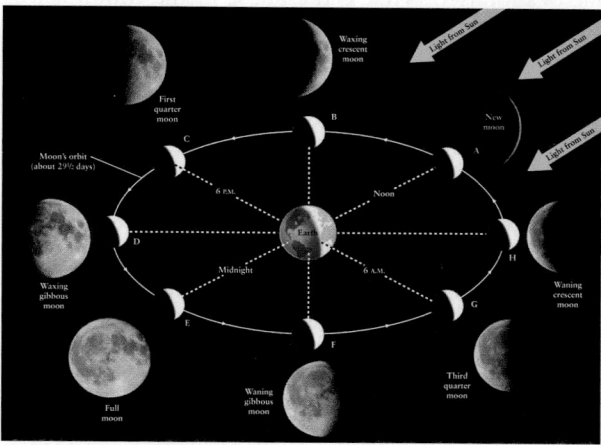

Lunar Phases: The phases of the Moon occur because light from the Moon is actually reflected sunlight. As the relative positions of Earth, the Moon, and the Sun change, we see more or less of the illuminated half of the Moon (Section 3-1).

Length of the Month: Two types of months are used in describing the motion of the Moon (Section 3-2).

- With respect to the stars, the Moon completes one orbit around Earth in a sidereal month, averaging 27.32 days.
- The Moon completes one cycle of phases (one orbit around Earth with respect to the Sun) in a synodic month, averaging 29.53 days.

The Moon's Orbit: The plane of the Moon's orbit is tilted by about 5° from the plane of Earth's orbit, or ecliptic (Section 3-3).

- The line of nodes is the line where the planes of the Moon's orbit and Earth's orbit intersect. The gravitational pull of the Sun gradually shifts the orientation of the line of nodes with respect to the stars.

Conditions for Eclipses: During a lunar eclipse, the Moon passes through Earth's shadow. During a solar eclipse, Earth passes through the Moon's shadow (Section 3-3).

- Lunar eclipses occur at full moon, while solar eclipses occur at new moon.
- Either type of eclipse can occur only when the Sun and Moon are both on or very near the line of nodes. If this condition is not met, Earth's shadow cannot fall on the Moon and the Moon's shadow cannot fall on Earth.

Umbra and Penumbra: The shadow of an object has two parts: the umbra, within which the light source is completely blocked, and the penumbra, where the light source is only partially blocked (Section 3-4).

Lunar Eclipses: Depending on the relative positions of the Sun, Moon, and Earth, lunar eclipses may be total (the Moon passes completely into Earth's umbra), partial (only part of the Moon passes into Earth's umbra), or penumbral (the Moon passes only into Earth's penumbra) (Section 3-4).

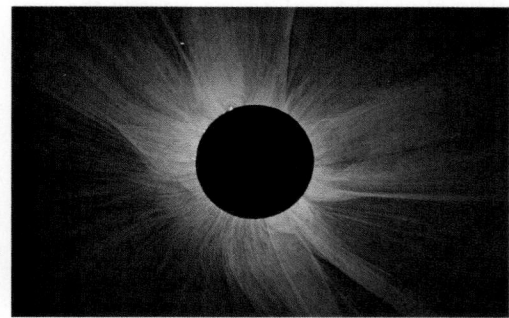

Solar Eclipses: Solar eclipses may be total, partial, or annular (Section 3-5).

- During a total solar eclipse, the Moon's umbra traces out an eclipse path over Earth's surface as Earth rotates. Observers outside the eclipse path but within the penumbra see only a partial solar eclipse.
- During an annular eclipse, the umbra falls short of Earth, and the outer edge of the Sun's disk is visible around the Moon at mideclipse.

The Moon and Ancient Astronomers: Ancient astronomers such as Aristarchus and Eratosthenes made great progress in determining the sizes and relative distances of Earth, the Moon, and the Sun (Section 3-6).

QUESTIONS

> ### Problem-solving tips and tools
>
> To estimate the average angular speed of the Moon along its orbit (that is, how many degrees around its orbit the Moon travels per day), divide 360° by the length of a sidereal month. It is helpful to know that the saros interval of 6585.3 days equals 18 years and 11⅓ days if the interval includes 4 leap years, but is 18 years and 10⅓ days if it includes 5 leap years.

Section 3-1

1. Explain the difference between sunlight and moonlight.

2. **TUTORIAL 3.1** (a) Explain why the Moon exhibits phases. (b) A common misconception about the Moon's phases is that they are caused by Earth's shadow. Use Figure 3-2 to explain why this is not correct.

3. How would the sequence and timing of lunar phases be affected if the Moon moved around its orbit (a) in the same direction, but at twice the speed; (b) at the same speed, but in the opposite direction? Explain your answers.

4. At approximately what time does the Moon rise when it is (a) a new moon, (b) a first quarter moon, (c) a full moon, and (d) a third quarter moon?

5. Astronomers sometimes refer to lunar phases in terms of the *age* of the Moon. This is the time that has elapsed since new moon phase. Thus, the age of a full moon is half of a 29½-day synodic period, or approximately 15 days. Find the approximate age of (a) a waxing crescent moon, (b) a third quarter moon, and (c) a waning gibbous moon.

6. If you lived on the Moon, would you see Earth go through phases? If so, would the sequence of phases be the same as those of the Moon as seen from Earth, or would the sequence be reversed? Explain using Figure 3-2.

7. The dividing line between the illuminated and unilluminated halves of the Moon is called the *terminator*. The terminator appears curved when there is a crescent or gibbous moon, but it appears straight when there is a first quarter or third quarter moon (see Figure 3-2). Describe how you could use these facts to explain to a friend why lunar phases cannot be caused by Earth's shadow falling on the Moon.

8. What is the phase of the Moon if it rises at (a) midnight, (b) sunrise, (c) halfway between sunset and midnight, and (d) halfway between noon and sunset? Explain your answers.

9. The Moon is highest in the sky when it crosses the meridian (see Figure 2-21), halfway between the time of moonrise and the time of moonset. At approximately what time does the Moon cross the meridian if it is (a) a new moon, (b) a first quarter moon, (c) a full moon, and (d) a third quarter moon? Explain your answers.

10. The Moon is highest in the sky when it crosses the meridian (see Figure 2-21), halfway between the time of moonrise and the time of moonset. What is the phase of the Moon if it is highest in the sky at (a) midnight, (b) sunrise, (c) noon, and (d) sunset? Explain your answers.

11. Suppose it is the first day of autumn in the northern hemisphere. What is the phase of the Moon if the Moon is located at (a) the vernal equinox, (b) the summer solstice, (c) the autumnal equinox, and (d) the winter solstice? Explain your answers. (*Hint:* Make a drawing showing the relative positions of the Sun, Earth, and Moon. Compare with Figure 3-2.)

12. This photograph of Earth was taken by the crew of the Apollo 8 spacecraft as they orbited the Moon. A portion of the lunar surface is visible at the right-hand side of the photo. In this photo, Earth is oriented with its north pole approximately at the top. When this photo was taken, was the Moon waxing or waning as seen from Earth? Explain your answer with a diagram.

NASA/JSC

R I **V** U X G

13. Describe the cycle of lunar phases that would be observed if the Moon moved around Earth in an orbit perpendicular to the plane of Earth's orbit. Would it be possible for both solar and lunar eclipses to occur under these circumstances? Explain your reasoning.

Section 3-2

14. Is the far side of the Moon (the side that can never be seen from Earth) the same as the dark side of the Moon? Explain your answer.

15. (a) If you lived on the Moon, would you see the Sun rise and set, or would it always be in the same place in the sky? Explain. (b) Would you see Earth rise and set, or would it always be in the same place in the sky? Explain using Figure 3-4.

16. What is the difference between a sidereal month and a synodic month? Which is longer? Why?

17. On a certain date, the Moon is in the direction of the constellation Gemini as seen from Earth. When will the Moon next be in the direction of Gemini: 1 sidereal month later, or 1 synodic month later? Explain your answer.

18. The Moon moves noticeably on the celestial sphere over the space of a single night. (**a**) To show this, calculate how long it takes the Moon to move through an angle equal to its own angular diameter (½°) against the background of stars. Give your answer in hours. (**b**) Through what angle (in degrees) does the Moon move during a 12-hour night? Can you notice an angle of this size? (*Hint:* See Figure 1-11.)

19. During an occultation, or "covering up," of Jupiter by the Moon, an astronomer notices that it takes the Moon's edge 90 seconds to cover Jupiter's disk completely. If the Moon's motion is assumed to be uniform and the occultation was "central" (that is, center over center), find the angular diameter of Jupiter. (*Hint:* Assume that Jupiter does not appear to move against the background of stars during this brief 90-second interval. You will need to convert the Moon's angular speed from degrees per day to arcseconds per second.)

20. The plane of the Moon's orbit is inclined at a 5° angle from the ecliptic, and the ecliptic is inclined at a 23½° angle from the celestial equator. Could the Moon ever appear at your zenith if you lived at (**a**) the equator, and (**b**) the south pole? Explain your answers.

21. How many more sidereal months than synodic months are there in a year? Explain your answer.

22. Suppose Earth moved a little faster around the Sun so that it took a bit less than 1 year to make a complete orbit. If the speed of the Moon's orbit around Earth were unchanged, would the length of the sidereal month be the same, longer, or shorter than it is now? What about the synodic month? Explain your answers.

23. If the Moon revolved about Earth in the same orbit but in the opposite direction, would the synodic month be longer or shorter than the sidereal month? Explain your reasoning.

24. One definition of a "blue moon" is the second full moon within the same calendar month. There is usually only one full moon within a calendar month, so the phrase "once in a blue moon" means "hardly ever." Why are blue moons so rare? Are there any months of the year in which it would be impossible to have two full moons? Explain your answer.

Section 3-3

25. **TUTORIAL 3.2** What is the difference between the umbra and the penumbra of a shadow?

26. Why doesn't a lunar eclipse occur at every full moon and a solar eclipse at every new moon?

27. What is the line of nodes? Why is it important to the subject of eclipses?

28. What is a penumbral eclipse of the Moon? Why do you suppose that it is easy to overlook such an eclipse?

Section 3-4

29. Why is the duration of totality different for different total lunar eclipses, as shown in Table 3-1?

30. Can one ever observe an annular eclipse of the Moon? Why or why not?

31. You are watching a lunar eclipse from somewhere on Earth's night side. Will you see the Moon enter Earth's shadow from the east or from the west? Explain your reasoning.

32. The total lunar eclipse of October 28, 2004, was visible from South America. The duration of totality was 1 hour, 21 minutes. Was this total eclipse also visible from Australia, on the opposite side of Earth? Explain your reasoning.

33. How would a lunar eclipse look if Earth had no atmosphere? Explain your reasoning.

Section 3-5

34. **TUTORIAL 3.3** If you were looking at Earth from the side of the Moon that faces Earth, what would you see during (**a**) a total lunar eclipse, and (**b**) a total solar eclipse? Explain your answers.

35. If there is a total eclipse of the Sun in April, can there be a lunar eclipse 3 months later in July? Why or why not?

36. Which type of eclipse—lunar or solar—do you think most people on Earth have seen? Why?

37. How is an annular eclipse of the Sun different from a total eclipse of the Sun? What causes this difference?

38. During a total solar eclipse, the Moon's umbra moves in a generally eastward direction across Earth's surface. Use a drawing like Figure 3-12 to explain why the motion is eastward, not westward.

39. A total solar eclipse was visible from Africa on March 29, 2006 (see Figure 3-11b). Draw what the eclipse would have looked like as seen from France, to the north of the path of totality. Explain the reasoning behind your drawing.

40. Figures 3-12 and 3-14 show that the path of a total eclipse is quite narrow. Use this to explain why a glow is visible all around the horizon when you are viewing a solar eclipse during totality (see Figure 3-11a).

41. (**a**) Suppose the diameter of the Moon were doubled, but the orbit of the Moon remained the same. Would total solar eclipses be more common, less common, or just as common as they are now? Explain. (**b**) Suppose the diameter of the Moon were halved, but the orbit of the Moon remained the same. Explain why there would be *no* total solar eclipses.

42. Just as the distance from Earth to the Moon varies somewhat as the Moon orbits Earth, the distance from the Sun to Earth changes as Earth orbits the Sun. Earth is closest to the Sun at its *perihelion;* it is farthest from the Sun at its *aphelion.* In order for a total solar eclipse to have the maximum duration of totality, should Earth be at perihelion or aphelion? Assume that the Earth–Moon distance is the same in both situations. As part of your explanation, draw two pictures like Figure 3-12, one with Earth relatively close to the Sun and one with Earth relatively far from the Sun.

43. In his 1885 novel *King Solomon's Mines*, H. Rider Haggard described a total solar eclipse that was seen in both South Africa and in the British Isles. Is such an eclipse possible? Why or why not?

44. Why do you suppose that total solar eclipse paths fall more frequently on oceans than on land? (You may find it useful to look at Figure 3-14.)

45. Examine Figure 3-14, which shows all of the total solar eclipses from 1997 to 2021. What are the chances that you might be able to travel to one of the eclipse paths? Do you think you might go through your entire life without ever seeing a total eclipse of the Sun?

46. Box 3-2 What is the saros? How did ancient astronomers use it to predict eclipses?

47. Box 3-2 On March 29, 2006, residents of northern Africa were treated to a total solar eclipse. (a) On what date and over what part of the world will the next total eclipse of that series occur? Explain your answer. (b) On what date might you next expect a total eclipse of that series to be visible from northern Africa? Explain your answer.

Section 3-6

48. How did Eratosthenes measure the size of Earth?

49. How did Aristarchus try to estimate the distance from Earth to the Sun and Moon?

50. How did Aristarchus try to estimate the diameters of the Sun and Moon?

WEB/E-BOOK QUESTIONS

51. ANIMATION 3.1 Access the animation "The Moon's Phases" in Chapter 3 of the *Universe* website or e-book. This shows the Earth–Moon system as seen from a vantage point looking down onto the north pole. (a) Describe where you would be on the diagram if you are on the equator and the time is 6:00 P.M. (b) If it is 6:00 P.M. and you are standing on Earth's equator, would a third quarter moon be visible? Why or why not? If it would be visible, describe its appearance.

52. Search the internet for information about the next total lunar eclipse. Will the total phase of the eclipse be visible from your location? If not, will the penumbral phase be visible? Draw a picture showing the Sun, Earth, and Moon when the totality is at its maximum duration, and indicate your location on the drawing of Earth.

53. Search the internet for information about the next total solar eclipse. Through which major cities, if any, does the path of totality pass? What is the maximum duration of totality? At what location is this maximum duration observed? Will this eclipse be visible (even as a partial eclipse) from your location? Draw a picture showing the Sun, Earth, and Moon when the totality is at its maximum duration, and indicate your location on the drawing of Earth.

54. ANIMATION 3.4 Access the animation "A Solar Eclipse Viewed from the Moon" in Chapter 3 of the *Universe* website or e-book. This shows the solar eclipse of August 11, 1999, as viewed from the Moon. Using a diagram, explain why the stars and the Moon's shadow move in the directions shown in this animation.

ACTIVITIES

Observing Projects

55. Observe the Moon on each clear night over the course of a month. On each night, note the Moon's location among the constellations and record that location on a star chart that also shows the ecliptic. After a few weeks, your observations will begin to trace the Moon's orbit. Identify the orientation of the line of nodes by marking the points where the Moon's orbit and the ecliptic intersect. On what dates is the Sun near the nodes marked on your star chart? Compare these dates with the dates of the next solar and lunar eclipses.

56. It is quite possible that a lunar eclipse will occur while you are taking this course. Look up the date of the next lunar eclipse in Table 3-1, on the internet, or in the current issue of a reference such as the *Astronomical Almanac* or *Astronomical Phenomena*. Then make arrangements to observe this lunar eclipse. You can observe the eclipse with the naked eye, but binoculars or a small telescope will enhance your viewing experience. If the eclipse is partial or total, note the times at which the Moon enters and exits Earth's umbra. If the eclipse is penumbral, can you see any changes in the Moon's brightness as the eclipse progresses?

Starry Night™ Explorations

57. Use *Starry Night*™ to explore the Moon's motion in the sky. Select **Favorites > Explorations > Moon's Motion** from the menu. The view shows the moon as seen from the north pole of Earth. In this view, eastward motion in the sky is toward the left and westward motion toward the right. Several of the brightest stars in the view are labeled with their proper names, including the bright star Aldebaran, just below the Moon. The time flow rate is set to 1 minute.

(a) Click **Play** and observe the minute-by-minute motion of the Moon across the sky as seen by an observer on the rotating Earth. In which direction does the Moon appear to move? In which direction do the stars appear to move? How do you explain this apparent motion of the stars and the Moon?

(b) Select **Edit > Undo Time Flow** from the menu to return to the initial view. Click **Play** and observe more carefully the relative position of the Moon with respect to the star Aldebaran and the other surrounding stars as time passes. How does the motion of the Moon compare with that of Aldebaran and the other stars in the sky?

(c) Select **File > Revert** from the menu. The apparent motion of the Moon across the sky for an observer on the surface of Earth as observed in part (a) is caused in part by the Moon's own orbital motion and in part by the eastward rotation of Earth. To remove the effect of Earth's rotation from the observation, select **Options > Orientation > Equatorial** from the menu and then click **Play**. Now, the stars remain stationary while Earth's rotation carries the horizon toward the left (east). Describe the Moon's motion in the sky as seen from this non-rotating point of view. What conclusions can you draw from these observations?

58. Use *Starry Night*™ to demonstrate the phases of the Moon and to explore the difference between a synodic and a sidereal month. Open **Favorites > Explorations > Moon Phases**. This view along the ecliptic plane from the center of Earth is centered upon the Moon, which is seen to be in the same direction in the sky as the Sun at this time and at the same location as the labeled double star Omicron Sagittarii.

(a) Point the cursor at the Moon to review the information in the HUD. What is the age of the Moon at the time of this observation? What is its phase? Where is the Sun in the sky relative to the Moon during this phase?

(b) Use the "D" keyboard shortcut or the toolbar time controls to step forward 1 day in time. In which direction is the Moon relative to the Sun and Omicron Sagittarii at the time of this observation? What is the phase of the Moon?

(c) Continue to step forward 1 calendar day at a time while monitoring the HUD for the age and phase of the Moon and stop when the phase reaches first quarter. What is the age of the Moon? Zoom out so that the Sun is visible in the view. Use the angular separation cursor to measure the angular distance separating the Sun and the Moon in the sky as seen from Earth. Which side of the Moon's disk, east or west, is illuminated during this phase? In which direction is the Moon relative to the Sun as seen from Earth at this phase of the Moon?

(d) Zoom in to about 10° and continue to advance time in steps of 1 day as you watch the phase and age of the Moon until the full phase. What is the Moon's age at this phase? Select **View > Center On . . .** from the menu. In the popup window, click the **Antipode** button followed by the **Center** button to reverse the direction of view 180° away from the Moon to the opposite point in the sky. Relative to the Sun, where is the Moon when it is full as seen from Earth? Is the Sun still in the same location in the sky as the star Omicron Sagittarii? If not, in which direction is the Sun now, relative to this star? Why?

(e) Use **View > Center on . . . > Antipode** again and zoom in to about 10°. Then click the name in the label for the Moon in the view and select **Center**. Resume time advance in steps of 1 day as you monitor the age and phase of the Moon. Stop when the Moon is at last quarter.

Where is the Moon in the sky, relative to the Sun as seen from Earth at this phase? What side of the Moon's disk is illuminated at this phase? What is the age of the Moon?

(f) Advance time in steps of 1 day while monitoring the age and phase of the Moon until the star Omicron Sagitarii is visible in the view. Now use the time controls to position the Moon in the same position relative to the reference star as in the initial view, so that the Moon has completed a full orbit around Earth relative to the stars. What is this period called? What is the approximate duration of this period?

(g) You will note that the Sun is now no longer in the same position relative to the stars, as in the initial view. So, although the Moon has completed one orbit relative to the stars, it has not yet caught up to the Sun in the sky. Bring the Moon to the same position in the sky as the Sun by using the time flow controls to find the age of the Moon in the HUD just before it turns over to 0.00 days and the Moon is again in its new phase. The Moon has completed a complete cycle of phases. What is this period called? What is its duration?

59. Use *Starry Night*™ to compare the view of the Moon from Earth with the view of Earth as seen from the Moon, and deduce from these observations the relationship between the rotation rate of the Moon and its orbital period around Earth.

(a) Open **Favorites > Explorations > Earth's Moon** to view the Moon as it appears from Earth and use the "D" keyboard shortcut to advance time in steps of 1 day per keypress while observing the Moon in the view. When the Moon sets below the horizon, use **File > Revert** to reset the view. Select **View > Hide Horizon** from the menu and then click the name in the label for the Moon and select **Magnify**. Once more, step time forward day by day and observe the features on the Moon. Do you see a different view of the Moon from night to night or does it always show the same face?

(b) Select **Favorites > Explorations > Taurus-Littrow** to view Earth from the Apollo 17 landing site on the Moon. Click **Play** and use the hand tool to adjust the view so that Earth stays in the view. Describe and explain your observations of the position of Earth in the lunar sky. Why does the lunar horizon not interfere with the visibility of Earth from this lunar viewing location? What do your observations indicate about the rotation rate of the Moon relative to its orbital period?

60. Use *Starry Night*™ to view the solar eclipse of August 21, 2017, from a unique perspective. Select **Favorites > Explorations > Aug 21** to see a view centered on Earth from a position near the Moon. The Moon's shadow is depicted as an orange cone comprised of the inner, more opaque umbra, and the fainter surrounding penumbra. From a position within the penumbral cone, part of the Sun will be obscured by the Moon while from a position within the umbral cone, the Sun is completely covered by the Moon.

(a) Where is the Sun in this view? Select **View > Center On . . .** from the menu, and in the window that opens, click the **Antipode** button followed by the **Center** button to reverse your view and look in the opposite direction.

(b) Select **File > Revert** from the menu to reset the view and then click **Play**. Observe the Moon's shadow as time progresses. When the Moon's penumbral shadow reaches Earth, a circle forms, showing the boundary of this shadow on Earth's surface. As time continues, another circle forms inside this penumbral circle that outlines the region of Earth that is within the umbra of the shadow. In which direction across Earth's surface does the Moon's shadow move? Does the shadow's motion across Earth's surface outpace or lag behind Earth's rate of rotation?

(c) Zoom in to a 5° field and use the time controls so that the complete umbral and penumbral circles are visible on Earth in the view and stop time flow. Right-click (Windows) or control-click (Mac) on a point on Earth's surface that is within the penumbral circle and select **Go to Surface**. Open the **Planets** panel, click the menu icon for the Sun in the list and select **Magnify** to see what this eclipse looks like from this location. Next, select **Edit > Undo** from the menu two or three times in succession until the view returns to that overlooking the Moon toward Earth. Then use the **Go to Surface** technique to move to a location on Earth within the umbra. Finally, use the **Planets** panel to **Center** the view on the Sun and zoom in to observe the eclipse from this location. Use the time controls to find the approximate duration of the total phase of the eclipse from this location.

Collaborative Exercise

61. Using a bright light source at the center of a darkened room, or a flashlight, use your fist held at arm's length to demonstrate the difference between a full moon and a lunar eclipse. (Use yourself or a classmate as Earth.) How must your fist "orbit" Earth so that lunar eclipses do not happen at every full moon? Create a simple sketch to illustrate your answers.

ANSWERS

ConceptChecks

ConceptCheck 3-1: In Figure 3-2, the last quarter of an orbit (which takes a little over 4 weeks) occurs between positions G and A, ending at position A with a new moon. A few days before that, the Moon would be near position H and would be visible only during the daytime. It would appear as a waning crescent with a slight illumination on the Moon's left side.

ConceptCheck 3-2: See Figure 3-2. First, recall that the full moon rises at sunset. This is because the Sun–Earth–Moon must be along a straight line for full illumination (at position E),

and while the full moon would be visible all night, it would first come into view rising in the east as the Sun sets in the west. Second, when an observer sees a little darkness on the right edge of the Moon, the Moon is near position F (or at least between positions E and F). Since the Moon rises and sets about an hour later each night, and this lunar phase occurs *after* the full moon, the Moon near position F must have risen after sunset. Note that the arrows labeling the white lunar orbit tell us the time ordering of the phases. Also note that while the observation was made at midnight in this question, the waning gibbous moon would look the same at any time it was visible.

ConceptCheck 3-3: Because the Sun is always shining on the Moon, the Moon is always half illuminated and half dark. Except in the rare events of eclipses, discussed later, the Moon's apparent changing phases are due to observers on Earth seeing differing amounts of the Moon's half illuminated surface.

ConceptCheck 3-4: The Moon goes through an entire cycle of phases in about 4 weeks, so if it is currently in the waxing gibbous phase, in 1 week it will be in the waxing gibbous phase, and after 2 weeks it will be in the waning gibbous phase.

ConceptCheck 3-5: In Figure 3-2, we can see that this waning gibbous moon occurs shortly after the full moon phase at position E. A full moon sets at sunrise, but the Moon sets about an hour later each day after that. Therefore, this Moon, like the Moon at position F, would set sometime shortly after sunrise. This Moon rises in the dark after sunset, so in the daylight here, it must be setting.

ConceptCheck 3-6: The Moon is shown at four points of its orbit in Figure 3-4b. In the left position, the far side is fully dark. At the position on the right, the far side is fully illuminated.

ConceptCheck 3-7: If the Moon is full, then after 1 week, it would reach the third quarter phase, and the point that used to be in the center of the Moon's visible surface at full moon would now fall into darkness that would last for 2 weeks. This scenario can be seen in two figures: In Figure 3-4b, this scenario is shown by the red dot on the leftmost depiction of the Moon, and the Moon's position a week later at the bottom of the figure. In Figure 3-2, this scenario occurs between positions E and G.

ConceptCheck 3-8: Increase. If Earth were moving around the Sun faster than it is now, Earth would move farther around the Sun during the Moon's orbit and it would take longer for the Moon to reach the position where it was in line with the Sun and Earth, increasing its synodic period. This is illustrated (for the new moon) in Figure 3-5.

ConceptCheck 3-9: Lunar eclipses occur only when the Sun, Earth, and Moon all lie exactly on a line; this line is called the line of nodes. As shown in Figure 3-8, during a full moon the Sun, Earth, and Moon are not necessarily lined up because the Moon orbits in a different plane than the ecliptic. Usually, the full moon is above or below the ecliptic plane of Earth's orbit around the Sun and misses being covered by Earth's shadow.

ConceptCheck 3-10: No. Before the Moon passes into the full shadow (the umbra), where the Sun is completely blocked by Earth, the Moon must pass through the penumbra where only part of the Sun is blocked. This is illustrated by Path 2 in Figure 3-9.

ConceptCheck 3-11: At sunset. Due to the required alignment, a lunar eclipse can only occur sometime during the night of a full moon. All full moons rise at sunset, no matter when the eclipse occurs later in the night.

ConceptCheck 3-12: Total solar eclipses are only observed in the small region where the Moon's tiny umbral shadow lands on Earth (see Figure 3-12), whereas when the Moon enters Earth's much larger umbral shadow, anyone on Earth who can see the Moon can observe it in Earth's shadow.

ConceptCheck 3-13: A total lunar eclipse can last for more than an hour, whereas a total solar eclipse lasts only a few minutes.

ConceptCheck 3-14: If the noontime position of the Sun on the summer solstice were much closer to being directly overhead, it would mean that on the curved surface of Earth, the overhead directions at both Syene and Alexandria are pointing in similar directions. This scenario does not occur on Earth and would only do so if Earth were much larger. (Keep in mind that the distance between these two cities is assumed to be constant for this question, no matter what the size of Earth.)

CalculationCheck

CalculationCheck 3-1: Because their diameters must be in the same proportion as their distances, if Aristarchus assumed the Sun to be 100 times farther away when the Sun and Moon appeared to be the same diameter in the sky, he would have proposed that the Sun is 100 times wider than the Moon. Today, we know that the Sun is about 400 times farther away from Earth and 400 times wider than the Moon.

Archaeoastronomy and Ethnoastronomy by Mark Hollabaugh

Many years ago I read astronomer John Eddy's *National Geographic* article about the Wyoming Medicine Wheel. A few years later I visited this archaeological site in the Bighorn Mountains and started on the major preoccupation of my career.

Archaeoastronomy combines astronomy and archaeology. You may be familiar with sites such as Stonehenge or the Mayan ruins of the Yucatán. You may not know that there are many archaeological sites in the United States that demonstrate the remarkable understanding a people, usually known as the Anasazi, had about celestial motions (see Figure 2-1). Chaco Canyon, Hovenweep, and Chimney Rock in the Four Corners area of the Southwest preserve ruins from this ancient Pueblo culture.

Moonrise at Chimney Rock in southern Colorado provides a good example of the Anasazi's knowledge of the lunar cycles. If you watched the rising of the Moon for many, many years, you would discover that the Moon's northernmost rising point undergoes an 18.6-year cycle. Dr. McKim Malville of the University of Colorado discovered that the Anasazi who lived there knew of the lunar standstill cycle and watched the northernmost rising of the Moon between the twin rock pillars of Chimney Rock.

My own specialty is the ethnoastronomy of the Lakota, or Teton Sioux, who flourished on the Great Plains of what are now Nebraska, the Dakotas, and Wyoming. Ethnoastronomy combines ethnography with astronomy. As an ethnoastronomer, I am less concerned with physical evidence in the form of ruins and more interested in myths, legends, religious belief, and current practices. In my quest to understand the astronomical thinking and customs of the nineteenth-century Lakota, I have traveled to museums, archives, and libraries in Nebraska, Colorado, Wyoming, South Dakota, and North Dakota. I frequently visit the Pine Ridge and Rosebud Reservations in South Dakota. The Lakota, and other Plains Indians, had a rich tradition of understanding celestial motions and developed an even richer explanation of why things appear the way they do in the sky.

My first professional contribution to ethnoastronomy was in 1996 at the Fifth Oxford International Conference on Astronomy in Culture held in Santa Fe, New Mexico. I had noticed that images of eclipses often appeared in Lakota winter counts—their method of making a historic record of events. The great Leonid meteor shower of 1833 appears in almost every Plains Indian winter count. As I looked at the hides or in ledger books recording these winter counts, I wondered why the Sun, the Moon, and the stars are so common among the Lakota of 150 years ago. My curiosity led me to look deeper: What did the Lakota think about eclipses? Why does their central ritual, the Sun Dance, focus on the Sun? Why do so many legends involve the stars?

The Lakota observed lunar and solar eclipses. Perhaps they felt they had the power to restore the eclipsed Sun or Moon. In August 1869, an Indian agency physician in South Dakota told the Lakota there would be an eclipse. When the Sun disappeared from view, the Lakota began firing their guns in the air. In their minds, they were more powerful than the white doctor because the result of their action was to restore the Sun.

The Lakota used a lunar calendar. Their names for the months came from the world around them. October was the moon of falling leaves. In some years, there actually are 13 new moons, and they often called this extra month the "Lost Moon." The lunar calendar often dictated the timing of their sacred rites. Although the Lakota were never dogmatic about it, they preferred to hold their most important ceremony, the Sun Dance, at the time of the full moon in June, which is when the summer solstice occurs.

Why did the Lakota pay attention to the night sky? Lakota elder Ringing Shield's statement about Polaris, recorded in the late nineteenth century, provides a clue: "One star never moves and it is *wakan*. Other stars move in a circle about it. They are dancing in the dance circle." For the Lakota, a driving force in their culture was a quest to understand the nature of the sacred, or *wakan*. Anything hard to understand or different from the ordinary was *wakan*.

Reaching for the stars, as far away as they are, was a means for the Lakota to bring the incomprehensible universe a bit closer to Earth. Their goal was the same as what Dr. Sandra M. Faber says in her essay "Why Astronomy?" at the end of Chapter 1, "a perspective on human existence and its relation to the cosmos."

Mark Hollabaugh taught physics and astronomy at Normandale Community College, St. Olaf College, Augsburg College, and the U.S. Air Force Academy. He majored in physics at St. Olaf College, earned an M.S. in astronomy at the University of Denver, and a Ph.D. in science education at the University of Minnesota. As a boy he watched the dance of the northern lights and the flash of meteors. Meeting Apollo 13 commander Jim Lovell and going for a ride in Roger Freedman's airplane are as close as he came to his dream of being an astronaut.

4 | Gravitation and the Waltz of the Planets

Not long ago, the idea of humans orbiting Earth or sending spacecraft to other worlds was regarded as science fiction. Today, science fiction has become commonplace reality. Literally thousands of artificial satellites orbit our planet to track weather, relay signals for communications and entertainment, and collect scientific data about Earth and the universe. Humans live and work in Earth orbit (as in the accompanying photograph), have ventured as far as the Moon, and have sent dozens of robotic spacecraft to explore the planets, asteroids, and comets.

While we think of spaceflight as an innovation of the twentieth century, we can trace its origins to a series of scientific revolutions that began in the 1500s. The first of these revolutions overthrew the ancient idea that Earth is an immovable object at the center of the universe, around which the Sun, the Moon, and the planets move. In this chapter, we will learn how Nicolaus Copernicus, Tycho Brahe, Johannes Kepler, and Galileo Galilei helped us understand that Earth is itself one of several planets orbiting the Sun.

An astronaut orbits Earth attached to the International Space Station. (NASA/JSC)

R I V U X G

We will learn, too, about Isaac Newton's revolutionary discovery of why the planets move in the way that they do. This was just one aspect of Newton's immense body of work, which included formulating the fundamental laws of physics and developing a precise mathematical description of the force of gravitation—the force that holds the planets in their orbits.

Gravitation proves to be a truly universal force: It guides spacecraft as they journey across the solar system and keeps satellites and astronauts in their orbits. In this chapter, you will learn more about this important force of nature, which plays a central role in all parts of astronomy.

4-1 Ancient astronomers invented geocentric models to explain planetary motions

Since the dawn of civilization, scholars have attempted to explain the nature of the universe. The ancient Greeks were the first to use the principle that still guides scientists today: *The universe can be described and understood logically.* For example, more than 2500 years ago, Pythagoras and his followers put forth the idea that nature can be described with mathematics. About 200 years later, Aristotle asserted that the universe is governed by physical laws. One of the most important tasks before the scholars of ancient Greece was to create a model (see Section 1-1) to explain the motions of objects in the heavens.

The Greek Geocentric Model

Most Greek scholars thought that the Sun, the Moon, the stars, and the planets revolve about a stationary Earth. A model of this kind, in which Earth is at the center of the universe, is called a **geocentric model**. Similar ideas were held by the scholars of ancient China.

> The scholars of ancient Greece imagined that the planets follow an ornate combination of circular paths.

Today we recognize that the stars are not merely points of light on an immense celestial sphere. But in fact, this is how the ancient Greeks regarded the stars in their geocentric model of the universe. To explain the diurnal motions of the stars, they assumed that the celestial sphere is *real*, and that it rotates around the stationary Earth once a day.

The Sun and Moon would both participate in this daily rotation of the sky, explaining their rising and setting motions. To explain why the Sun and Moon both move slowly with respect to the stars, the ancient Greeks imagined that both of these objects orbit around Earth.

ANALOGY

Imagine a merry-go-round that rotates clockwise as seen from above, as in **Figure 4-1a**. As it rotates, two children walk slowly counterclockwise at different speeds around the merry-go-round's platform. Thus, the children rotate along with the merry-go-round and also change their positions with respect to the merry-go-round's wooden horses. This scene is analogous to the way the ancient Greeks pictured the motions of the stars, the Sun, and the Moon. In their model, the celestial sphere rotated to the west around a stationary Earth (Figure 4-1b). The stars rotate along with the celestial sphere just as the wooden horses rotate along with the merry-go-round in Figure 4-1a. The Sun and Moon are analogous to the two children; they both turn westward with the celestial sphere, making one complete turn each day, and also move slowly eastward at different speeds with respect to the stars.

The geocentric model of the heavens also had to explain the motions of the planets. The ancient Greeks and other cultures of that time knew of five planets: Mercury, Venus, Mars, Jupiter, and Saturn, each of which is a bright object in the night sky. For example, when Venus is at its maximum brilliancy, it is 16 times brighter than the brightest star. (By contrast, Uranus and Neptune

Merry-go-round rotates clockwise.

Wooden horses fixed on merry-go-round.

(a) A rotating merry-go-round

Celestial sphere rotates to the west.

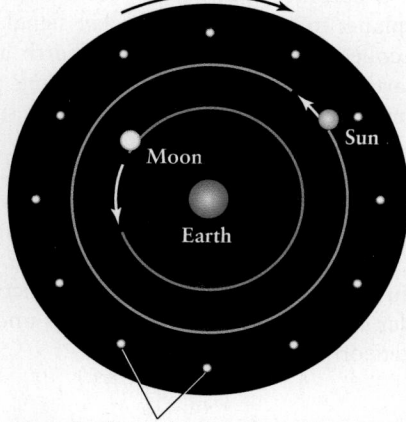

Stars fixed on celestial sphere.

(b) The Greek geocentric model

ANIMATION 4.1

FIGURE 4-1

A Merry-Go-Round Analogy (a) Two children walk at different speeds around a rotating merry-go-round with its wooden horses. **(b)** In an analogous way, the ancient Greeks imagined that the Sun and Moon move around the rotating celestial sphere with its fixed stars. Thus, the Sun and Moon move from east to west across the sky every day and also move slowly eastward from one night to the next, relative to the background of stars.

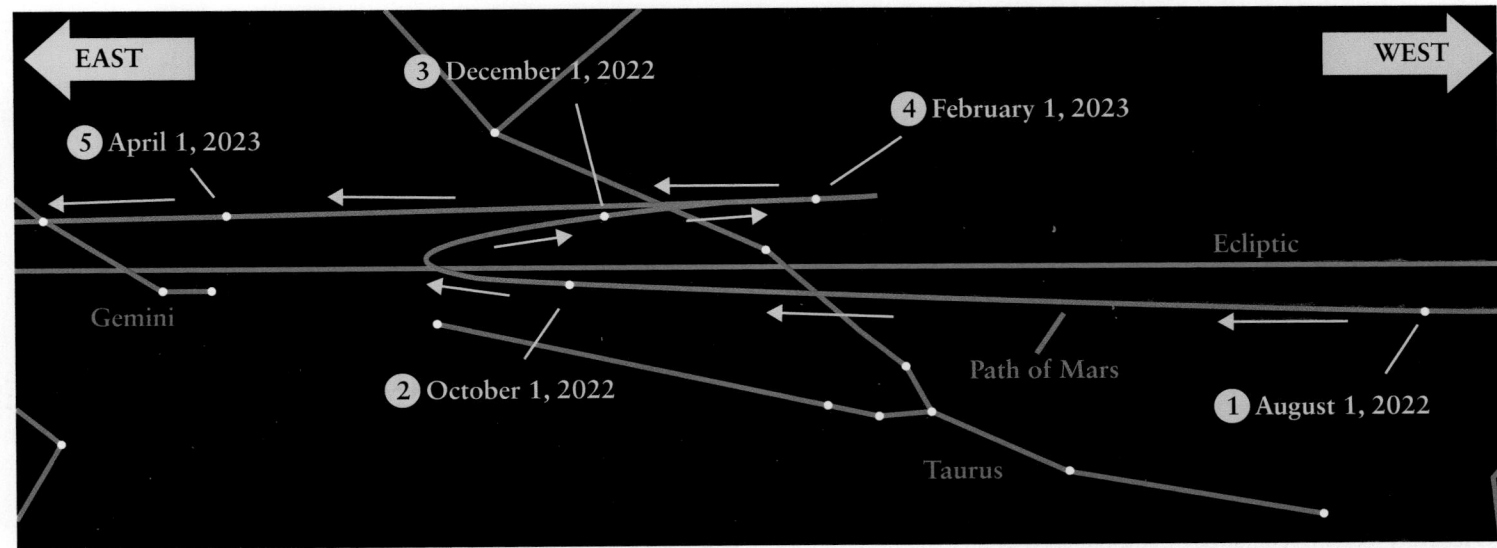

ANIMATION
4.2 **FIGURE 4-2**

The Path of Mars in 2022–2023 From August 2022 through April 2023, Mars traverses the constellations Taurus and Gemini. Mars's motion is direct (from west to east, or from right to left in this figure) most of the time, but it is retrograde (from east to west, or from left to right in this figure) during November and December 2022. Notice that the apparent speed of Mars is not constant: The planet travels farther across the sky during the 2 months from August 1, 2022, to October 1, 2022, than during the two months from December 1, 2022, to February 1, 2023.

are quite dim and were not discovered until after the invention of the telescope.)

Like the Sun and Moon, all the planets rise in the east and set in the west once a day. And like the Sun and Moon, from night to night the planets slowly move on the celestial sphere, that is, with respect to the background of stars. However, the character of this motion on the celestial sphere is quite different for the planets. Both the Sun and the Moon always move from west to east on the celestial sphere, that is, opposite the direction in which the celestial sphere appears to rotate. The Sun follows the path called the ecliptic (see Section 2-5), while the Moon follows a path that is slightly inclined to the ecliptic (see Section 3-3). Furthermore, the Sun and the Moon each move at relatively constant speeds around the celestial sphere. (The Moon's speed is faster than that of the Sun: It travels all the way around the celestial sphere in about a month whereas the Sun takes an entire year.) The planets, too, appear to move along paths that are close to the ecliptic. The difference is that each of the planets appears to wander back and forth on the celestial sphere with varying speed. As an example, Figure 4-2 shows the wandering motion of Mars with respect to the background of stars from August 2022 to April 2023. (This figure shows that the name *planet* is well deserved; it comes from a Greek word meaning "wanderer.")

CAUTION!

On a map of Earth with north at the top, west is to the left and east is to the right. Why, then, is *east* on the left and *west* on the right in Figure 4-2? The answer is that a map of Earth is a view looking downward at the ground from above, whereas a star map like Figure 4-2 is a view looking upward at the sky.

You can see how east and west are flipped if you first draw a set of north–south and east–west axes on paper, and then hold the paper up against the sky. For this reason, east and west are flipped in the star maps found at the end of this textbook.

Most of the time, planets move slowly eastward relative to the stars, just as the Sun and Moon do. This eastward progress is called **direct motion.** For example, Figure 4-2 shows that Mars will be in direct motion from August through October 2022 and from February through April 2023. Occasionally, however, the planet seems to stop and then back up for several weeks or months. This occasional westward movement is called **retrograde motion.** Mars will undergo retrograde motion from September 2022 to February 2023 (see Figure 4-2), and will do so again about every 22½ months. All the other planets go through retrograde motion, but at different intervals. In the language of the merry-go-round analogy in Figure 4-1, the Greeks imagined the planets as children walking around the rotating merry-go-round but who keep changing their minds about which direction to walk! As you can gather, this retrograde motion was very perplexing and required some explanation.

CAUTION!

Whether a planet is in direct or retrograde motion, over the course of a single night you will see it rise in the east and set in the west. The reason is that both direct and retrograde motions are much slower than the apparent daily rotation of the sky. Hence, they are best detected by mapping the position of a planet against the background stars from night to night over a long period. Figure 4-2 is a map of just this sort.

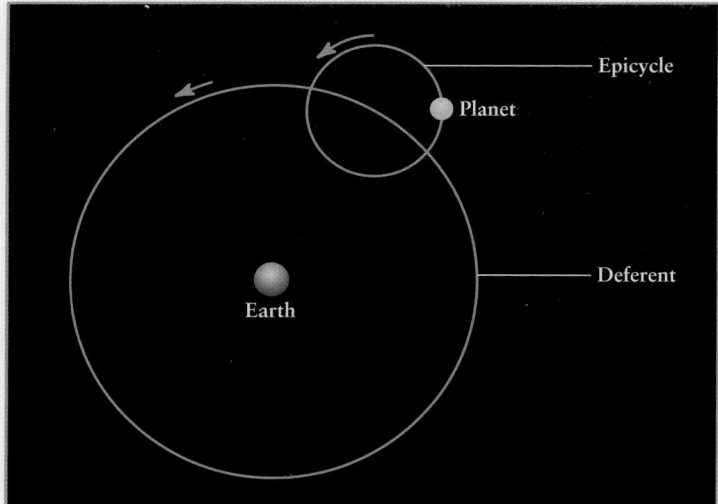

(a) Planetary motion modeled as a combination of circular motions

(b) Modeling direct motion

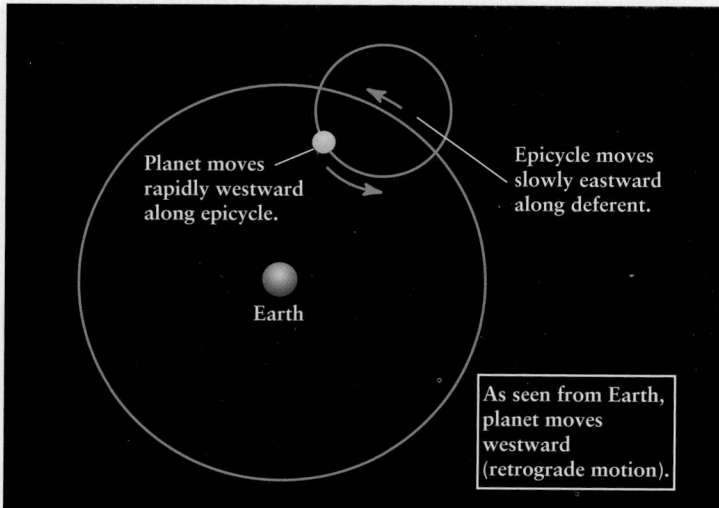

(c) Modeling retrograde motion

ANIMATION 4.3 **FIGURE 4-3**

A Geocentric Explanation of Retrograde Motion

(a) The ancient Greeks imagined that each planet moves along an epicycle, which in turn moves along a deferent centered approximately on Earth. The planet moves along the epicycle more rapidly than the epicycle moves along the deferent. **(b)** At most times, the eastward motion of the planet on the epicycle adds to the eastward motion of the epicycle on the deferent. Then the planet moves eastward in direct motion as seen from Earth. **(c)** When the planet is on the inside of the deferent, its motion along the epicycle is westward. Because this motion is faster than the eastward motion of the epicycle on the deferent, the planet appears from Earth to be moving westward in retrograde motion.

CONCEPT CHECK 4-1

If Mars is moving retrograde, will it rise above the eastern horizon or above the western horizon?

CONCEPT CHECK 4-2

How fast is Earth spinning on its axis in the Greek geocentric model?

Answers appear at the end of the chapter.

The Ptolemaic System

Explaining the nonuniform motions of the five planets was one of the main challenges facing the astronomers of antiquity. The Greeks developed many theories to account for retrograde motion and the loops that the planets trace out against the background stars. One of the most successful and enduring models was proposed by Apollonius of Perga and by Hipparchus in the second century B.C.E. and expanded upon by Ptolemy, the last of the great Greek astronomers, during the second century C.E. **Figure 4-3a** sketches the basic concept, usually called the **Ptolemaic system.** Each planet is assumed to move in a small circle called an **epicycle,** whose center in turn moves in a larger circle, called a **deferent,** which is centered approximately on Earth. Both the epicycle and deferent rotate in the same direction, shown as counterclockwise in Figure 4-3a.

As viewed from Earth, the epicycle moves eastward along the deferent. Most of the time, the eastward motion of the planet on its epicycle adds to the eastward motion of the epicycle on the deferent (Figure 4-3b). Then the planet is seen to be in direct (eastward) motion against the background stars. However, when the planet is on the inside of the deferent, the motion of the planet along the epicycle is opposite to the motion of the epicycle along the deferent. The planet therefore appears to slow down, halt its usual eastward movement among the constellations, and actually go backward in retrograde (westward) motion for a few weeks or months (Figure 4-3c). Thus, the concept of epicycles and deferents enabled Greek astronomers to explain the retrograde loops of the planets.

Using the wealth of astronomical data in the library at Alexandria, including records of planetary positions going back hundreds of years, Ptolemy deduced the sizes and rotation rates of the epicycles and deferents needed to reproduce the recorded paths of the planets. After years of tedious work, Ptolemy assembled his calculations into 13 volumes, collectively called the *Almagest.* His work was used to predict the positions and paths of the Sun,

Moon, and planets with unprecedented accuracy. In fact, the *Almagest* was so successful that for more than 1000 years, the Ptolemaic system endured as a useful description of the workings of the heavens.

A major problem posed by the Ptolemaic system was a philosophical one: It treated each planet as independent of the others. There was no rule in the *Almagest* that related the size and rotation speed of one planet's epicycle and deferent to the corresponding sizes and speeds for other planets. This problem made the Ptolemaic system very unsatisfying to many Islamic and European astronomers of the Middle Ages. They thought that a correct model of the universe should be based on a simple set of underlying principles applying to all the planets.

With only limited observational data, the astronomers of the Middle Ages needed a way to judge if a scientific model might be correct. To help in this assessment, a guiding principle called **Occam's razor** is often used (even today) to judge between competing scientific models. Named after the fourteenth-century English philosopher William of Occam, the principle states: *When two hypotheses can explain the available data, the hypothesis requiring the fewest new assumptions should be favored.* (The "razor" refers to shaving extraneous details or assumptions from an explanation.) In other words, the simplest explanation consistent with available data is considered most likely to be correct. The geocentric model required many parameters (or new assumptions) to characterize its epicycles and deferents, leaving open the possibility of a more simple theory. In the centuries that followed, Occam's razor would help motivate a simpler yet revolutionary view of the universe.

CONCEPT CHECK 4-3

Do the planets actually stop and change their direction of motion in the Ptolemaic model?

Answer appears at the end of the chapter.

4-2 Nicolaus Copernicus devised the first comprehensive heliocentric model

During the first half of the sixteenth century, a Polish lawyer, physician, canon of the Church, and gifted mathematician named Nicolaus Copernicus (Figure 4-4) began to construct a new model of the universe. His model, which placed the Sun at the center, explained the motions of the planets in a more simple way than the Ptolemaic system. As we will see, it also helped lay the foundations of modern physical science. But Copernicus was not the first to conceive of a Sun-centered model of planetary motion. In the third century B.C.E., the Greek astronomer Aristarchus suggested such a model as a way to explain retrograde motion.

A Heliocentric Model Explains Retrograde Motion

Imagine riding on a fast racehorse. As you pass a slowly walking pedestrian, he appears to move backward, even though he is traveling in the same direction as you and your horse. This sort of simple observation inspired Aristarchus to formulate a **heliocentric** (Sun-centered) **model** in which all the planets, including Earth, revolve about the Sun. Different planets take different lengths of

FIGURE 4-4

Nicolaus Copernicus (1473–1543) Copernicus was the first person to work out the details of a heliocentric system in which the planets, including Earth, orbit the Sun. (Erich Lessing/Art Resource, NY)

time to complete an orbit, so from time to time one planet will overtake another, just as a fast-moving horse overtakes a person on foot. When Earth overtakes Mars, for example, Mars appears to move backward in retrograde motion, as Figure 4-5 shows. Thus, in the heliocentric picture, the occasional retrograde motion of a planet is merely the result of Earth's fast motion.

> The retrograde motion of the planets in the sky results from our viewing these planets from a moving Earth.

As we saw in Section 3-6, Aristarchus demonstrated that the Sun is bigger than Earth (see Table 3-3), which made it sensible to imagine Earth orbiting the larger Sun. He also imagined that Earth rotates on its axis once a day, which explains the daily rising and setting of the Sun, Moon, and planets and the diurnal motions of the stars. To explain why the apparent motions of the planets never take them far from the ecliptic, Aristarchus proposed that the orbits of Earth and all the planets must lie in nearly the same plane. (Recall from Section 2-5 that the ecliptic is the projection onto the celestial sphere of the plane of Earth's orbit.)

This heliocentric model is conceptually much simpler than an Earth-centered system, such as that of Ptolemy, with all its "circles upon circles." In Aristarchus's day, however, the idea of an orbiting, rotating Earth seemed inconceivable, given Earth's apparent stillness and immobility. Nearly 2000 years would pass before a heliocentric model would gain broad acceptance.

CONCEPT CHECK 4-4

In the heliocentric model, could an imaginary observer on the Sun look out and see planets moving in retrograde motion?

Answer appears at the end of the chapter.

3. From point 6 to point 9, Mars again appears to move eastward against the background of stars as seen from Earth (direct motion).

2. As Earth passes Mars in its orbit from point 4 to point 6, Mars appears to move westward against the background of stars (retrograde motion).

1. From point 1 to point 4, Mars appears to move eastward against the background of stars as seen from Earth (direct motion).

ANIMATION 4.4   **FIGURE 4-5**

A Heliocentric Explanation of Retrograde Motion In the heliocentric model of Aristarchus, Earth and the other planets orbit the Sun. Earth travels around the Sun more rapidly than Mars. Consequently, as Earth overtakes and passes this slower-moving planet, Mars appears for a few months (from points 4 through 6) to fall behind and move backward with respect to the background of stars.

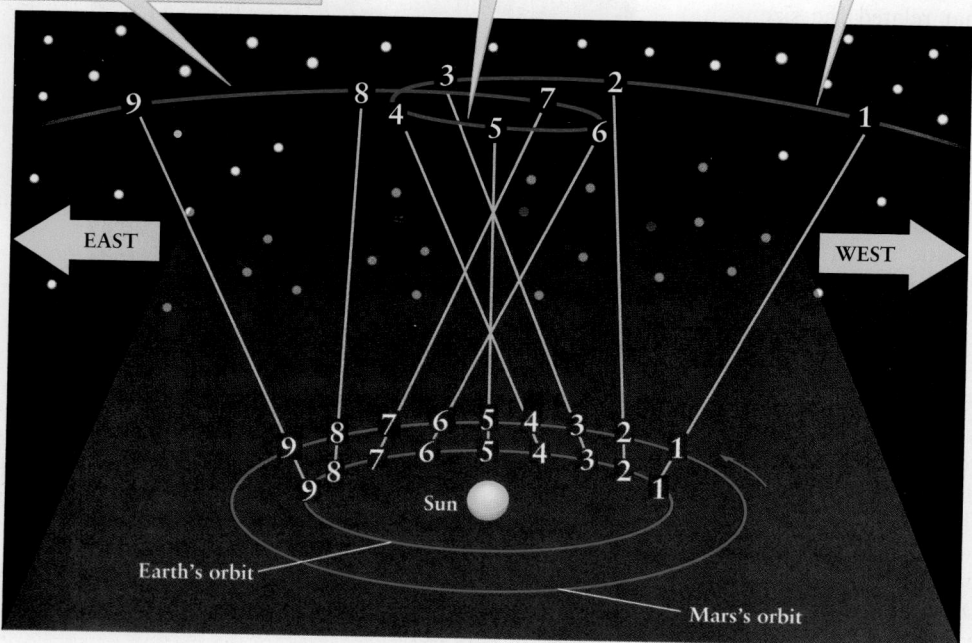

Copernicus and the Arrangement of the Planets

In the years after 1500, Copernicus came to realize that a heliocentric model has several advantages beyond providing a natural explanation of retrograde motion. In the Ptolemaic system, the arrangement of the planets—that is, which ones are close to Earth and which ones are far away—was chosen in large part by guesswork. By using a heliocentric model, Copernicus could determine the arrangement of the planets without ambiguity.

Copernicus realized that because Mercury and Venus are always observed fairly near the Sun in the sky, their orbits must be smaller than Earth's. Planets in such orbits are called **inferior planets** (Figure 4-6). The other visible planets—Mars, Jupiter, and Saturn—are sometimes seen on the side of the celestial sphere opposite the Sun, so these planets appear high above the horizon at midnight (when the Sun is far below the horizon). When this happens, Earth must lie between the Sun and these planets. Copernicus therefore concluded that the orbits of Mars, Jupiter, and Saturn must be larger than Earth's orbit. Hence, these planets are called **superior planets.**

Uranus and Neptune, as well as Pluto and a number of other small bodies called asteroids that also orbit the Sun, were discovered after the telescope was invented (and after the death of Copernicus). All of these can be seen at times in the midnight sky, far from the Sun, so these also have orbits larger than that of Earth.

The heliocentric model also explains why planets appear in different parts of the sky on different dates. Both inferior planets

(Mercury and Venus) go through cycles: The planet is seen in the west after sunset for several weeks or months, then for several weeks or months in the east before sunrise, and then in the west after sunset again.

Figure 4-6 shows the reason for this cycle. When Mercury or Venus is visible after sunset, it is near **greatest eastern elongation.** (The angle between the Sun and a planet as viewed from Earth is called the planet's **elongation.**) The planet's position in the sky is as far east of the Sun as possible, so it appears above the western horizon after sunset (that is, to the east of the Sun) and is often called an *evening star.* At **greatest western elongation,** Mercury or Venus is as far west of the Sun as it can possibly be. It then rises before the Sun, gracing the predawn sky as a *morning star* in the east. When Mercury or Venus is at **inferior conjunction,** it is between us and the Sun, and it moves from the evening sky into the morning sky over weeks or months. At **superior conjunction,** when the planet is on the opposite side of the Sun, it moves back into the evening sky.

A superior planet such as Mars, whose orbit is larger than Earth's, is best seen in the night sky when it is at **opposition.** At this point in its orbit, the planet is in the part of the sky opposite the Sun and is highest in the sky at midnight. This is also when the planet appears brightest, because it is closest to us. But when a superior planet like Mars is located behind the Sun at **conjunction,** it is above the horizon during the daytime and thus is not well placed for nighttime viewing.

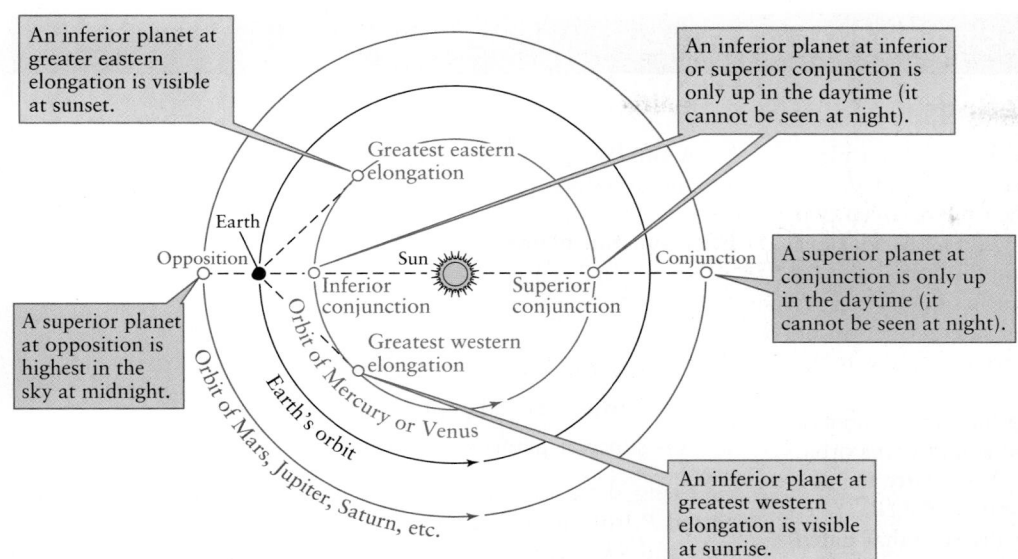

An inferior planet at greater eastern elongation is visible at sunset.

An inferior planet at inferior or superior conjunction is only up in the daytime (it cannot be seen at night).

A superior planet at conjunction is only up in the daytime (it cannot be seen at night).

A superior planet at opposition is highest in the sky at midnight.

An inferior planet at greatest western elongation is visible at sunrise.

FIGURE 4-6

Planetary Orbits and Configurations When and where in the sky a planet can be seen from Earth depends on the size of its orbit and its location on that orbit. Inferior planets have orbits smaller than Earth's, while superior planets have orbits larger than Earth's. (Note that in this figure you are looking down onto the solar system from a point far above Earth's northern hemisphere.)

CONCEPT CHECK 4-5

How many times is Mars at inferior conjunction during one orbit around the Sun?

Answer appears at the end of the chapter.

Planetary Periods and Orbit Sizes

The Ptolemaic system has no simple rules relating the motion of one planet to another. But Copernicus showed that there *are* such rules in a heliocentric model. In particular, he found a correspondence between the time a planet takes to complete one orbit—that is, its **period**—and the size of the orbit.

Determining the period of a planet takes some care, because Earth, from which we must make the observations, is also moving. Realizing this, Copernicus was careful to distinguish between two different periods of each planet. The **synodic period** is the time that elapses between two successive identical configurations as seen from Earth—from one opposition to the next, for example, or from one conjunction to the next. The **sidereal period** is the true orbital period of a planet, the time it takes the planet to complete one full orbit of the Sun relative to the stars.

The synodic period of a planet can be determined by observing the sky from Earth, but the sidereal period has to be found by calculation. Copernicus figured out how to do this (**Box 4-1**). Table 4-1 shows the results for all the planets.

To find a relationship between the sidereal period of a planet and the size of its orbit, Copernicus still had to determine the relative distances of the planets from the Sun. He devised a straightforward geometric method of determining the relative distances of the planets from the Sun using trigonometry. His answers turned out to be remarkably close to the modern values, as shown in Table 4-2.

The distances in Table 4-2 are given in terms of the astronomical unit (au), which is the average distance from Earth to the Sun (Section 1-7). The astronomical unit is 1 au = 1.496×10^8 km (92.96 million miles). However, Copernicus did not know the

TABLE 4-1 Synodic and Sidereal Periods of the Planets		
Planet	Synodic period	Sidereal period
Mercury	116 days	88 days
Venus	584 days	225 days
Earth	—	1.0 year
Mars	780 days	1.9 years
Jupiter	399 days	11.9 years
Saturn	378 days	29.5 years
Uranus	370 days	84.1 years
Neptune	368 days	164.9 years

TABLE 4-2 Average Distances of the Planets from the Sun		
Planet	Copernican value (au*)	Modern value (au)
Mercury	0.38	0.39
Venus	0.72	0.72
Earth	1.00	1.00
Mars	1.52	1.52
Jupiter	5.22	5.20
Saturn	9.07	9.55
Uranus	—	19.19
Neptune	—	30.07

1 au = 1 astronomical unit = average distance from Earth to the Sun.

precise value of this distance, so he could only determine the *relative* sizes of the orbits of the planets. (Using the astronomical unit, the average distance from the Sun to each of the planets in kilometers can be determined from Table 4-2. A table of these distances is given in Appendix 2.)

BOX 4-1   TOOLS OF THE ASTRONOMER'S TRADE

Relating Synodic and Sidereal Periods

We can derive a mathematical formula that relates a planet's sidereal period (the time required for the planet to complete one orbit) to its synodic period (the time between two successive identical configurations). To start with, let us consider an inferior planet (Mercury or Venus) orbiting the Sun as shown in the accompanying figure. Let P be the planet's sidereal period, S the planet's synodic period, and E Earth's sidereal period or sidereal year (see Section 2-8), which Copernicus knew to be nearly 365¼ days.

The rate at which Earth moves around its orbit is the number of degrees around the orbit divided by the time to complete the orbit, or $360°/E$ (equal to a little less than 1° per day). Similarly, the rate at which the inferior planet moves along its orbit is $360°/P$.

During a given time interval, the angular distance that Earth moves around its orbit is its rate ($360°/E$) multiplied by the length of the time interval. Thus, during a time S, or one synodic period of the inferior planet, Earth covers an angular distance of $(360°/E)S$ around its orbit. In that same time, the inferior planet covers an angular distance of $(360°/P)S$. Note, however, that the inferior planet has gained one full lap on Earth, and hence has covered 360° more than Earth has (see the figure). Thus, $(360°/P)S = (360°/E)S + 360°$. Dividing each term of this equation by $360°S$ yields the following.

For an inferior planet:

$$\frac{1}{P} = \frac{1}{E} + \frac{1}{S}$$

P = inferior planet's sidereal period

E = Earth's sidereal period = 1 year

S = inferior planet's synodic period

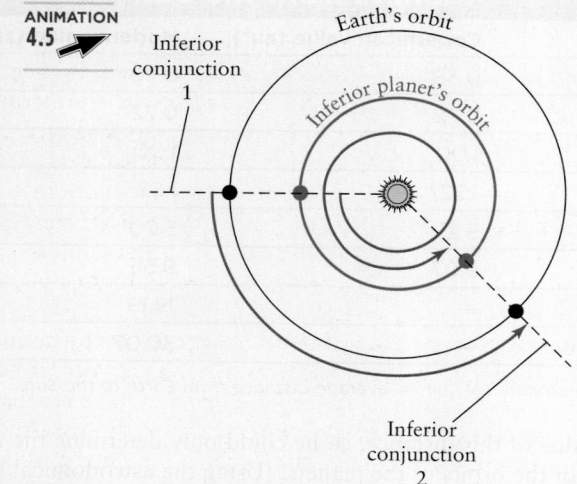

ANIMATION 4.5

Inferior conjunction 1

Earth's orbit

Inferior planet's orbit

Inferior conjunction 2

A similar analysis for a superior planet (for example, Mars, Jupiter, or Saturn) yields the following.

For a superior planet:

$$\frac{1}{P} = \frac{1}{E} - \frac{1}{S}$$

P = superior planet's sidereal period

E = Earth's sidereal period = 1 year

S = superior planet's synodic period

Using these formulas, we can calculate a planet's sidereal period P from its synodic period S. Often astronomers express P, E, and S in terms of years, by which they mean Earth years of approximately 365.26 days.

EXAMPLE: Jupiter has an observed synodic period of 398.9 days, or 1.092 years. What is its sidereal period?

Situation: Our goal is to find the sidereal period of Jupiter, a superior planet.

Tools: Since Jupiter is a superior planet, we use the second of the two equations given above to determine the sidereal period P.

Answer: We are given Earth's sidereal period $E = 1$ year and Jupiter's synodic period $S = 1.092$ years. Using the equation $1/P = 1/E - 1/S$

$$\frac{1}{P} = \frac{1}{1} - \frac{1}{1.092} = 0.08425$$

so

$$P = \frac{1}{0.08425} = 11.87 \text{ years}$$

Review: Our answer means that it takes 11.87 years for Jupiter to complete one full orbit of the Sun. This sidereal period is greater than Earth's 1-year sidereal period because Jupiter's orbit is larger than Earth's orbit. Jupiter's synodic period of 1.092 years is the time from one opposition to the next, or the time that elapses from when Earth overtakes Jupiter to when it next overtakes Jupiter. This synodic period is so much shorter than the sidereal period because Jupiter moves quite slowly around its orbit. Earth overtakes it a little less often than once per Earth orbit, that is, at intervals of a little bit more than a year.

By comparing Tables 4-1 and 4-2, you can see the unifying relationship between planetary orbits in the Copernican model: The farther a planet is from the Sun, the longer it takes to travel around its orbit (that is, the longer its sidereal period). That is so for *two* reasons: (1) the larger the orbit, the farther a planet must travel to complete an orbit; and (2) the larger the orbit, the slower a planet moves. For example, Mercury, with its small orbit, moves at an average speed of 47.9 km/s (107,000 mi/h). Saturn travels around its large orbit much more slowly, at an average speed of 9.64 km/s (21,600 mi/h). The older Ptolemaic model offers no such simple relations between the motions of different planets.

CONCEPT CHECK 4-6

What causes the planets to stop and change their direction of motion through the sky in the heliocentric model?

CONCEPT CHECK 4-7

Why is Jupiter's sidereal period longer than its synodic period?

Answers appear at the end of the chapter.

The Shapes of Orbits in the Copernican Model

At first, Copernicus assumed that Earth travels around the Sun along a circular path. He found, however, that perfectly circular orbits could not accurately describe the paths of the other planets, so he had to add an epicycle to each planet. These epicycles were *not* added to explain retrograde motion, which Copernicus realized was explained by the differences in orbital speeds of different planets, as shown in Figure 4-5. Rather, the small epicycles helped Copernicus account for slight variations in each planet's speed along its orbit.

Even though he clung to the old notion that orbits must be made up of circles, Copernicus had shown that a heliocentric model could explain the motions of the planets. He compiled his ideas and calculations into a book entitled *De revolutionibus orbium coelestium* (*On the Revolutions of the Celestial Spheres*), which was published in 1543, the year of his death.

For several decades after Copernicus, most astronomers saw little reason to change their allegiance from the older geocentric model of Ptolemy. The predictions that the Copernican model makes for the apparent positions of the planets are, on average, no better or worse than those of the Ptolemaic model. The test of Occam's razor does not really favor either model, because both use a combination of circles to describe each planet's motion.

More concrete evidence was needed to convince scholars to abandon the old, comfortable idea of a stationary Earth at the center of the universe. The story of how this evidence was accumulated begins nearly 30 years after the death of Copernicus, when a young Danish astronomer pondered the nature of a new star in the heavens.

CONCEPT CHECK 4-8

While Copernicus was alive, how did his model compare to the geocentric model of Ptolemy?

Answer appears at the end of the chapter.

4-3 Tycho Brahe's astronomical observations disproved ancient ideas about the heavens

On November 11, 1572, a bright star suddenly appeared in the constellation Cassiopeia. At first, it was even brighter than Venus, but then it began to grow dim. After 18 months, it had faded from view. Modern astronomers recognize this event as a supernova explosion, the violent death of a massive star.

> A supernova explosion and a comet revealed to Tycho that our universe is more dynamic than had been imagined.

In the sixteenth century, however, the vast majority of scholars held with the ancient teachings of Aristotle and Plato, who had argued that the heavens are permanent and unalterable. Consequently, the "new star" of 1572 could not really be a star at all, because the heavens do not change; it must instead be some sort of bright object quite near Earth, perhaps not much farther away than the clouds overhead.

The 25-year-old Danish astronomer Tycho Brahe (1546–1601) realized that straightforward observations might reveal the distance to the new star. It is common experience that when you walk from one place to another, nearby objects appear to change position against the background of more distant objects. This phenomenon, whereby the apparent position of an object changes because of the motion of the observer, is called **parallax.** If the new star was nearby, then its position should shift against the background stars over the course of a single night because Earth's rotation changes our viewpoint. Figure 4-7 (on the next page) shows this predicted shift. (Actually, Tycho believed that the heavens rotate about Earth, as in the Ptolemaic model, but the net effect is the same.)

Tycho's careful observations failed to disclose any parallax. The farther away an object is, the less it appears to shift against the background as we change our viewpoint, and the smaller the parallax. Hence, the new star had to be quite far away, farther from Earth than anyone had imagined. This was the first evidence that the "unchanging" stars were in fact changeable.

Tycho also attempted to measure the parallax of a bright comet that appeared in 1577 and, again, found it too small to measure. Thus, the comet also had to be far beyond Earth. Furthermore, because the comet's position relative to the stars changed from night to night, its motion was more like a planet than a star. But most scholars of Tycho's time taught that the motions of the planets had existed in unchanging form since the beginning of the universe. If new objects such as comets could appear and disappear within the realm of the planets, the conventional notions of planetary motions needed to be revised.

Tycho's observations showed that the heavens are by no means pristine and unchanging. This discovery flew in the face of nearly 2000 years of astronomical thought. In support of these revolutionary observations, the king of Denmark financed the construction of two magnificent observatories for Tycho's use on the island of Hven, just off the Danish coast. The two observatories—Uraniborg ("heavenly castle") and Stjerneborg ("star castle")—allowed Tycho to design and have built a set of naked-eye astronomical instruments vastly superior in quality to any earlier instruments (Figure 4-8).

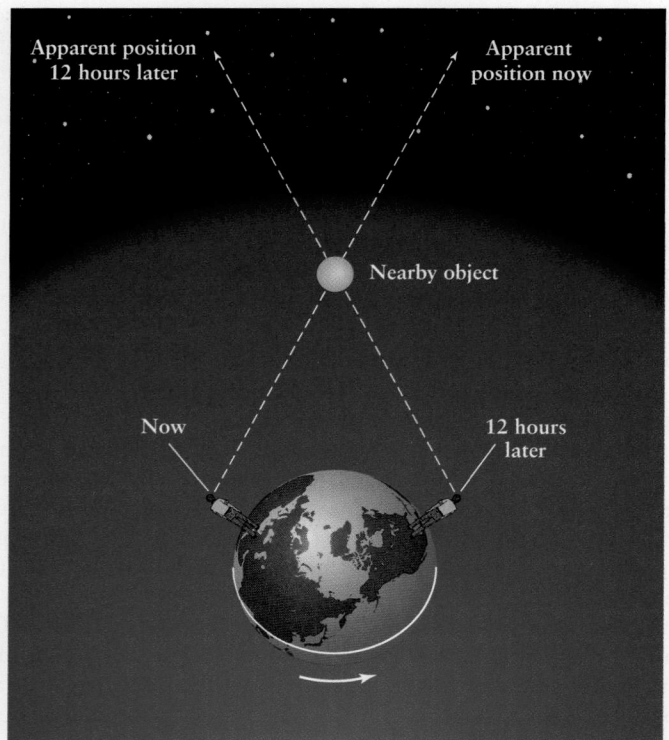

FIGURE 4-7

A Nearby Object Shows a Parallax Shift Tycho Brahe argued that if an object is near Earth, an observer would have to look in different directions to see that object over the course of a night and its position relative to the background stars would change. Tycho failed to measure such changes for a supernova in 1572 and a comet in 1577. He therefore concluded that these objects were far from Earth.

Tycho and the Positions of the Planets

With this state-of-the-art equipment, Tycho proceeded to measure the positions of stars and planets with unprecedented accuracy. In addition, he and his assistants were careful to make several observations of the same star or planet with different instruments, in order to identify any errors that might be caused by the instruments themselves. This painstaking approach to mapping the heavens revolutionized the practice of astronomy and is used by astronomers today.

A key goal of Tycho's observations during this period was to test the ideas Copernicus had proposed decades earlier about Earth going around the Sun. Tycho argued that if Earth was in motion, then nearby stars should appear to shift their positions with respect to background stars as we orbit the Sun. Tycho failed to detect any such parallax, and he concluded that Earth was at rest and the Copernican system was wrong.

On this point Tycho was in error, for nearby stars do in fact shift their positions as he had suggested. But even the nearest stars are so far away that the shifts in their positions are less than an arcsecond, too small to be seen with the naked eye. Tycho would have needed a telescope to detect the parallax that he was looking for, but the telescope was not invented until 1608, years after his death in 1601. Indeed, the first accurate determination of stellar parallax was not made until 1838.

Although he remained convinced that a stationary Earth was at the center of the universe, Tycho nonetheless made a tremendous

FIGURE 4-8

Tycho Brahe (1546–1601) Observing This contemporary illustration shows Tycho Brahe with some of the state-of-the-art measuring apparatus at Uraniborg, one of the two observatories that he built under the patronage of Frederik II of Denmark. (This magnificent observatory lacked a telescope, which had not yet been invented.) The data that Tycho collected were crucial to the development of astronomy in the years after his death. (Detlev van Ravenswaay/Science Source)

contribution toward putting the heliocentric model on a solid foundation. From 1576 to 1597, he used his instruments to make comprehensive measurements of the positions of the planets with an accuracy of 1 arcminute. These measurements are as well as can be done with the naked eye and were far superior to any earlier measurements. Within the reams of data that Tycho compiled lay the truth about the motions of the planets. The person who would extract this truth was a German mathematician who became Tycho's assistant in 1600, a year before the great astronomer's death. His name was Johannes Kepler (Figure 4-9).

4-4 Johannes Kepler proposed elliptical paths for the planets about the Sun

 TUTORIAL 4.1 The task that Johannes Kepler took on at the beginning of the seventeenth century was to find a model of planetary motion that agreed completely with Tycho's extensive and very accurate observations of planetary positions. To do this, Kepler found that he had to break with an ancient prejudice about planetary motions.

Kepler's ideas apply not just to the planets, but to all orbiting celestial objects.

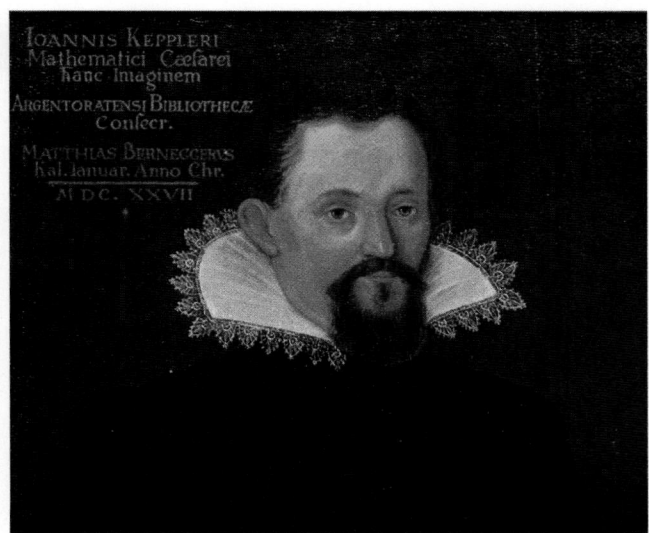

FIGURE 4-9

Johannes Kepler (1571–1630) By analyzing Tycho Brahe's detailed records of planetary positions, Kepler developed three general principles, called Kepler's laws, that describe how the planets move about the Sun. Kepler was the first to realize that the orbits of the planets are ellipses and not circles. (Erich Lessing/Art Resource, NY)

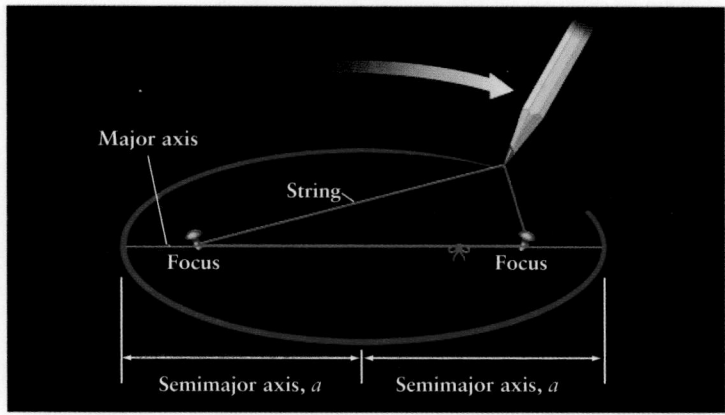

(a) The geometry of an ellipse

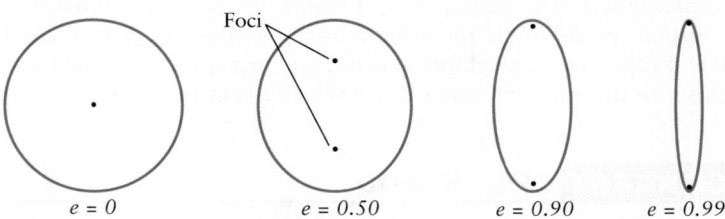

(b) Ellipses with different eccentricities

FIGURE 4-10

Ellipses (a) To draw an ellipse, use two thumbtacks to secure the ends of a piece of string, then use a pencil to pull the string taut. If you move the pencil while keeping the string taut, the pencil traces out an ellipse. The thumbtacks are located at the two foci of the ellipse. The major axis is the greatest distance across the ellipse; the semimajor axis is half of this distance. **(b)** A series of ellipses with the same major axis but different eccentricities. An ellipse can have any eccentricity from $e = 0$ (a circle) to just under $e = 1$ (virtually a straight line).

Elliptical Orbits and Kepler's First Law

Astronomers had long assumed that heavenly objects move in circles, which were considered the most perfect and harmonious of all geometric shapes. They believed that if a perfect God resided in heaven along with the stars and planets, then the motions of these objects must be perfect, too. Against this context, Kepler dared to try to explain planetary motions with noncircular curves. In particular, he found that he had the best success with a particular kind of curve called an **ellipse.**

You can draw an ellipse by using a loop of string, two thumbtacks, and a pencil, as shown in **Figure 4-10a.** Each thumbtack in the figure is at a **focus** (plural **foci**) of the ellipse; an ellipse has two foci. For any point on the ellipse, the sum of the distances to each foci is always the same—that's why the string and thumbtack exercise works. The longest diameter of an ellipse, called the **major axis,** passes through both foci. Half of that distance is called the **semimajor axis** and is usually designated by the letter *a*. A circle is a special case of an ellipse in which the two foci are at the same point (this corresponds to using only a single thumbtack in Figure 4-10b). The semimajor axis of a circle is equal to its radius.

By assuming that planetary orbits were ellipses, Kepler found, to his delight, that he could make his theoretical calculations match precisely to Tycho's observations. This important discovery, first published in 1609, is now called **Kepler's first law:**

The orbit of a planet about the Sun is an ellipse with the Sun at one focus.

The semimajor axis *a* of a planet's orbit is the average distance between the planet and the Sun.

CAUTION!

The Sun is at one focus of a planet's elliptical orbit, but there is *nothing* at the other focus. This "empty focus" has geometrical significance because it helps to define the shape of the ellipse, but it plays no other role.

Ellipses come in different shapes, depending on their elongation. The shape of an ellipse is described by its **eccentricity,** designated by the letter *e*. Figure 4-10b shows a few examples of ellipses with different eccentricities. The value of *e* can range from 0 (a circle) to just under 1 (nearly a straight line). The greater the eccentricity, the more elongated the ellipse. Because a circle is a special case of an ellipse, it is possible to have a perfectly circular orbit. But all of the objects that orbit the Sun have orbits that are at least slightly elongated. The most circular of any planetary orbit is that of Venus, with an eccentricity of just 0.007; Mercury's orbit has an eccentricity of 0.206, and a number of small bodies called comets move in very elongated orbits with eccentricities approaching 1.

CONCEPT CHECK 4-9

Which orbit is closer to being circular: Venus's orbit, with $e = 0.007$, or Mars's orbit, with $e = 0.093$?

Answer appears at the end of the chapter.

Orbital Speeds and Kepler's Second Law

Once he knew the shape of a planet's orbit, Kepler was ready to describe exactly *how* it moves on that orbit. As a planet travels in an elliptical orbit, its distance from the Sun varies. Kepler realized that the speed of a planet also varies along its orbit. A planet moves most rapidly when it is nearest the Sun, at a point on its orbit called **perihelion**. Conversely, a planet moves most slowly when it is farthest from the Sun, at a point called **aphelion** (Figure 4-11). After much trial and error, Kepler found a way to describe just how a planet's speed varies as it moves along its orbit. Figure 4-11 illustrates this discovery, referred to as **Kepler's second law:**

A line joining a planet and the Sun sweeps out equal areas in equal intervals of time.

This relationship is also called the **law of equal areas.** In the idealized case of a circular orbit, a planet would have to move at a constant speed around the orbit in order to satisfy Kepler's second law. In the case of an elliptical orbit, a planet speeds up when it is closer to the Sun, and slows down when it is farther away.

CONCEPT CHECK 4-10

According to Kepler's second law, at what point in a communications satellite's elliptical orbit around Earth will it move the slowest?

Answer appears at the end of the chapter.

Orbital Periods and Kepler's Third Law

Kepler's second law describes how the speed of a given planet changes as it orbits the Sun. Kepler also deduced from Tycho's data a relationship that can be used to compare the motions of *different* planets. Published in 1618 and now called **Kepler's third law**, it states a relationship between the size of a planet's orbit and the time the planet takes to go once around the Sun:

The square of the sidereal period of a planet is directly proportional to the cube of the semimajor axis of the orbit.

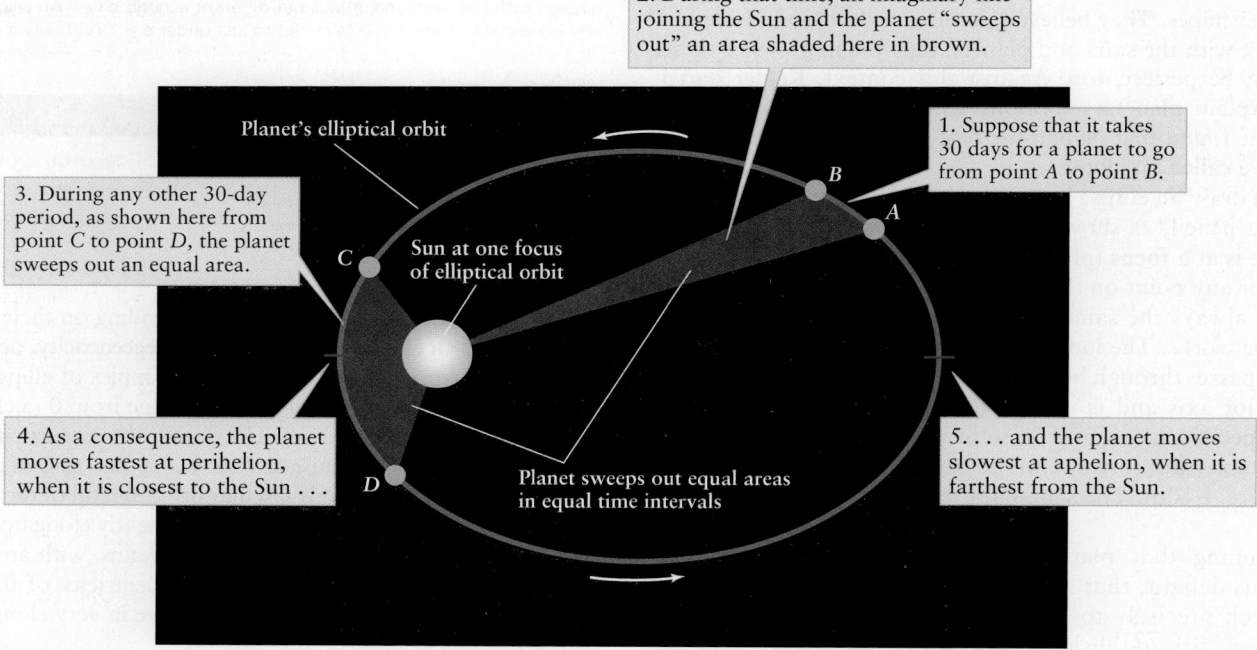

2. During that time, an imaginary line joining the Sun and the planet "sweeps out" an area shaded here in brown.

1. Suppose that it takes 30 days for a planet to go from point *A* to point *B*.

3. During any other 30-day period, as shown here from point *C* to point *D*, the planet sweeps out an equal area.

Planet's elliptical orbit

Sun at one focus of elliptical orbit

4. As a consequence, the planet moves fastest at perihelion, when it is closest to the Sun . . .

Planet sweeps out equal areas in equal time intervals

5. . . . and the planet moves slowest at aphelion, when it is farthest from the Sun.

ANIMATION
4.6

FIGURE 4-11

Kepler's First and Second Laws According to Kepler's first law, a planet travels around the Sun along an elliptical orbit with the Sun at one focus. According to his second law, a planet moves fastest when closest to the Sun (at perihelion) and slowest when farthest from the Sun (at aphelion). As the planet moves, an imaginary line joining the planet and the Sun sweeps out equal areas in equal intervals of time (from A to B or from C to D). By using these laws in his calculations, Kepler found an excellent fit to the apparent motions of the planets.

Kepler's third law says that the larger a planet's orbit—that is, the larger the semimajor axis, or average distance from the planet to the Sun—the longer the sidereal period, which is the time it takes the planet to complete an orbit. From Kepler's third law one can show that the larger the semimajor axis, the slower the average speed at which the planet moves around its orbit. (By contrast, Kepler's *second* law describes how the speed of a given planet is sometimes faster and sometimes slower than its average speed.) This qualitative relationship between orbital size and orbital speed is just what Aristarchus and Copernicus used to explain retrograde motion, as we saw in Section 4-2. Kepler's great contribution was to make this relationship a quantitative one.

It is useful to restate Kepler's third law as an equation. If a planet's sidereal period P is measured in years and its semimajor axis a is measured in astronomical units (au), then:

Kepler's third law

$$P^2 = a^3$$

P = planet's sidereal period, in years

a = planet's semimajor axis, in au

If you know either the sidereal period of a planet or the semimajor axis of its orbit, you can find the other quantity using this equation. Box 4-2 gives some examples of how this is done.

We can verify Kepler's third law for all the planets, including those that were discovered after Kepler's death, using data from Tables 4-1 and 4-2. If Kepler's third law is correct, for each planet the numerical values of P^2 and a^3 should be equal. This result is indeed true to a very high level of accuracy, as Table 4-3 shows.

BOX 4-2   TOOLS OF THE ASTRONOMER'S TRADE

Using Kepler's Third Law

Kepler's third law relates the sidereal period P of an object orbiting the Sun to the semimajor axis a of its orbit:

$$P^2 = a^3$$

You must keep two essential points in mind when working with this equation:

1. The period P *must* be measured in years, and the semimajor axis a *must* be measured in astronomical units (au). Otherwise you will get nonsensical results.
2. This equation applies *only* to the special case of an object, like a planet, that orbits the Sun. If you want to analyze the orbit of the Moon around Earth, of a spacecraft around Mars, or of a planet around a distant star, you must use a different, generalized form of Kepler's third law. We discuss this alternative equation in Section 4-7 and Box 4-4.

EXAMPLE: The average distance from Venus to the Sun is 0.72 au. Use this to determine the sidereal period of Venus.

Situation: The average distance from Venus to the Sun is the semimajor axis a of the planet's orbit. Our goal is to calculate the planet's sidereal period P.

Tools: To relate a and P, we use Kepler's third law, $P^2 = a^3$.

Answer: We first cube the semimajor axis (multiply it by itself twice):

$$a^3 = (0.72)^3 = 0.72 \times 0.72 \times 0.72 = 0.373$$

According to Kepler's third law this is also equal to P^2, the square of the sidereal period. So, to find P, we have to "undo" the square, that is, take the square root. Using a calculator, we find

$$P = \sqrt{P^2} = \sqrt{0.373} = 0.61$$

Review: The sidereal period of Venus is 0.61 year, or a bit more than 7 Earth months. This result makes sense: A planet with a smaller orbit than Earth's (an inferior planet) must have a shorter sidereal period than Earth.

EXAMPLE: A certain small asteroid (a rocky body a few tens of kilometers across) takes 8 years to complete one orbit around the Sun. Find the semimajor axis of the asteroid's orbit.

Situation: We are given the sidereal period $P = 8$ years, and are to determine the semimajor axis a.

Tools: As in the preceding example, we relate a and P using Kepler's third law, $P^2 = a^3$.

Answer: We first square the period:

$$P^2 = 8^2 = 8 \times 8 = 64$$

From Kepler's third law, 64 is also equal to a^3. To determine a, we must take the *cube root* of a^3, that is, find the number whose cube is 64. If your calculator has a cube root function, denoted by the symbol $\sqrt[3]{}$, you can use it to find that the cube root of 64 is 4: $\sqrt[3]{64}$. Otherwise, you can determine by trial and error that the cube of 4 is 64:

$$4^3 = 4 \times 4 \times 4 = 64$$

Because the cube of 4 is 64, it follows that the cube root of 64 is 4 (taking the cube root "undoes" the cube).

With either technique, you find that the orbit of this asteroid has semimajor axis $a = 4$ au.

Review: The period is greater than 1 year, so the semimajor axis is greater than 1 au. Note that $a = 4$ au is intermediate between the orbits of Mars and Jupiter (see Table 4-3). Many asteroids are known with semimajor axes in this range, forming a region in the solar system called the asteroid belt.

TABLE 4-3 A Demonstration of Kepler's Third Law ($P^2 = a^3$)

Planet	Sidereal period P (years)	Semimajor axis a (au)	P²	a³
Mercury	0.24	0.39	0.06	0.06
Venus	0.61	0.72	0.37	0.37
Earth	1.00	1.00	1.00	1.00
Mars	1.88	1.52	3.53	3.51
Jupiter	11.86	5.20	140.7	140.6
Saturn	29.46	9.55	867.9	871.0
Uranus	84.10	19.19	7072	7067
Neptune	164.86	30.07	27,180	27,190

Kepler's third law states that $P^2 = a^3$ for each of the planets. The last two columns of this table demonstrate that this relationship holds true to a very high level of accuracy.

CONCEPT CHECK 4-11

The space shuttle typically orbited Earth at an altitude of 300 km whereas the International Space Station orbits Earth at an altitude of 450 km. Although the space shuttle took less time to orbit Earth, which orbiter moved at a faster speed?

CALCULATION CHECK 4-1

If Pluto's orbit has a semimajor axis of 39.5 au, how long does it take Pluto to orbit the Sun once?

Answers appear at the end of the chapter.

The Significance of Kepler's Laws

Kepler's laws are a landmark in the history of astronomy. They made it possible to calculate the motions of the planets with better accuracy than any geocentric model ever had, and they helped to justify the idea of a heliocentric model. Kepler's laws also pass the test of Occam's razor, for they are simpler in every way than the schemes of Ptolemy or Copernicus, both of which used a complicated combination of circles.

But the significance of Kepler's laws goes beyond understanding planetary orbits. These same laws are also obeyed by spacecraft orbiting Earth, by two stars revolving about each other in a binary star system, and even by galaxies in their orbits about each other. Throughout this book, we shall use Kepler's laws in a wide range of situations.

As impressive as Kepler's accomplishments were, he did not prove that the planets orbit the Sun, nor was he able to explain *why* planets move in accordance with his three laws. These advances were made by two other figures who loom large in the history of astronomy: Galileo Galilei and Isaac Newton.

CONCEPT CHECK 4-12

Do Kepler's laws of planetary motion apply only to the planets?

Answer appears at the end of the chapter.

4-5 Galileo's discoveries with a telescope strongly supported a heliocentric model

When Dutch opticians invented the telescope during the first decade of the seventeenth century, astronomy was changed forever. The scholar who used this new tool to amass convincing evidence that the planets orbit the Sun, not Earth, was the Italian mathematician and physical scientist Galileo Galilei (**Figure 4-12**).

> Galileo used the cutting-edge technology of the 1600s to radically transform our picture of the universe.

Although Galileo did not invent the telescope, he was the first to point one of these new devices toward the sky and to publish his observations. Beginning in 1610, he saw sights of which no one had ever dreamed. He discovered mountains on the Moon, sunspots on the Sun, and the rings of Saturn, and he was the first to see that the Milky Way is not a featureless band of light but rather "a mass of innumerable stars."

The Phases of Venus

One of Galileo's most important discoveries with the telescope was that Venus exhibits phases like those of the Moon (**Figure 4-13**). Galileo also noticed that the apparent size of Venus as seen through his telescope was related to the planet's phase. Venus appears small at gibbous phase and largest at crescent phase. Galileo also noted a correlation between the phases of Venus and the planet's angular distance from the Sun.

FIGURE 4-12

Galileo Galilei (1564–1642) Galileo was one of the first people to use a telescope to observe the heavens. He discovered mountains on the Moon, sunspots on the Sun, the phases of Venus, and four moons orbiting Jupiter. His observations strongly suggested that Earth orbits the Sun, not vice versa. (Erich Lessing/Art Resource, NY) α = 58"   α = 42"

α = 24"   α = 15"   α = 10"

FIGURE 4-13 R I **V** U X G

The Phases of Venus This series of photographs shows how the appearance of Venus changes as it moves along its orbit. The number below each view is the angular diameter α of the planet in arcseconds. Venus has the largest angular diameter when it is a crescent, and the smallest angular diameter when it is gibbous (nearly full). (© New Mexico State University Observatory)

Figure 4-14 shows that these relationships are entirely compatible with a heliocentric model in which Earth and Venus both go around the Sun. They are also completely *incompatible* with the Ptolemaic system, in which the Sun and Venus both orbit Earth. To explain why Venus is never seen very far from the Sun, the

Ptolemaic model had to assume that the deferents of Venus and of the Sun move together in lockstep, with the epicycle of Venus centered on a straight line between Earth and the Sun (Figure 4-15). In this model, Venus was never on the opposite side of the Sun from Earth, and so it could never have shown the gibbous phases that Galileo observed.

CONCEPT CHECK 4-13

Which phase will Venus be in when it is at its maximum distance from Earth?

Answer appears at the end of the chapter.

The Moons of Jupiter

Galileo also found more unexpected evidence supporting the ideas of Copernicus. In 1610, Galileo discovered four moons, now called the Galilean satellites, orbiting Jupiter (Figure 4-16). He realized that they were orbiting Jupiter because they appeared to move back and forth from one side of the planet to the other. Figure 4-17 shows a page from Galileo's paper on Jupiter's moons. Astronomers soon realized that the larger the orbit of one of the moons around Jupiter, the slower that moon moves and the longer it takes that moon to travel around its orbit. These are the same relationships that Copernicus deduced for the motions of the planets around the Sun. Thus, the moons of Jupiter behave like a Copernican system in miniature.

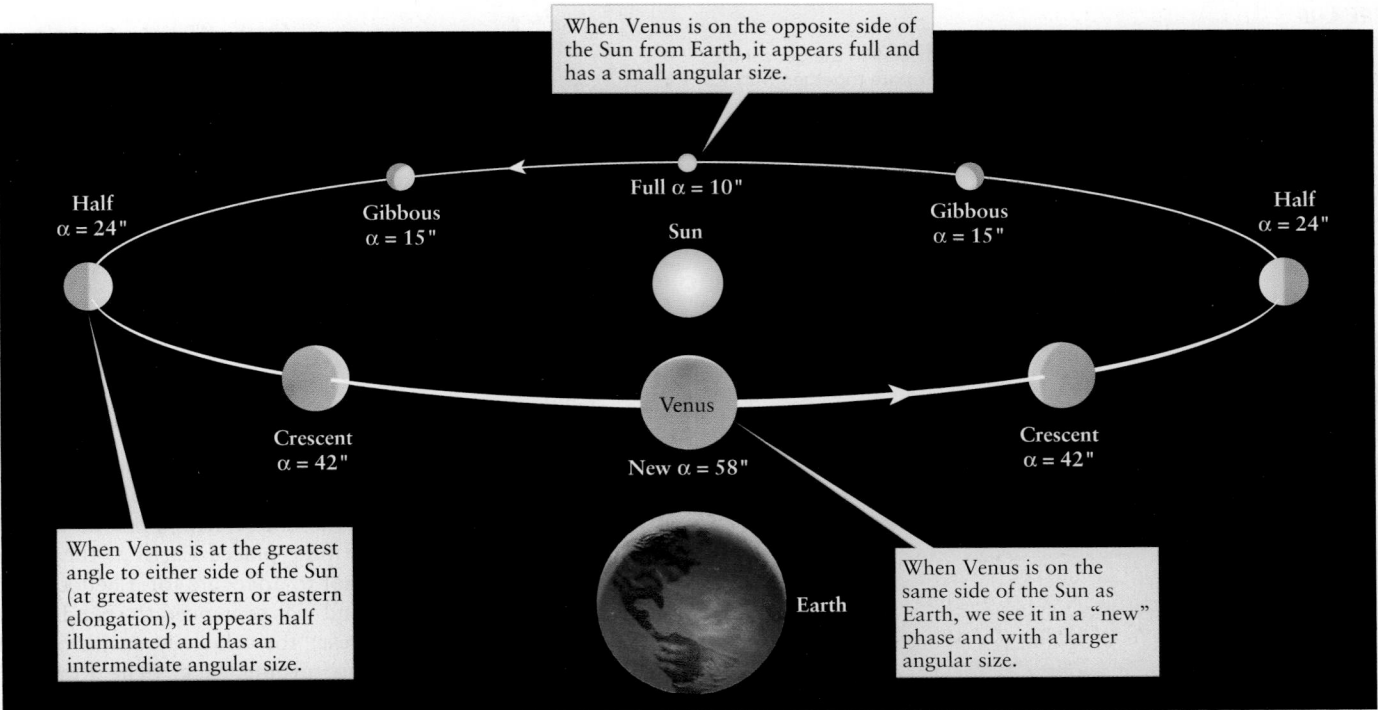

FIGURE 4-14

The Changing Appearance of Venus Explained in a Heliocentric Model A heliocentric model, in which Earth and Venus both orbit the Sun, provides a natural explanation for the changing appearance of Venus shown in Figure 4-13.

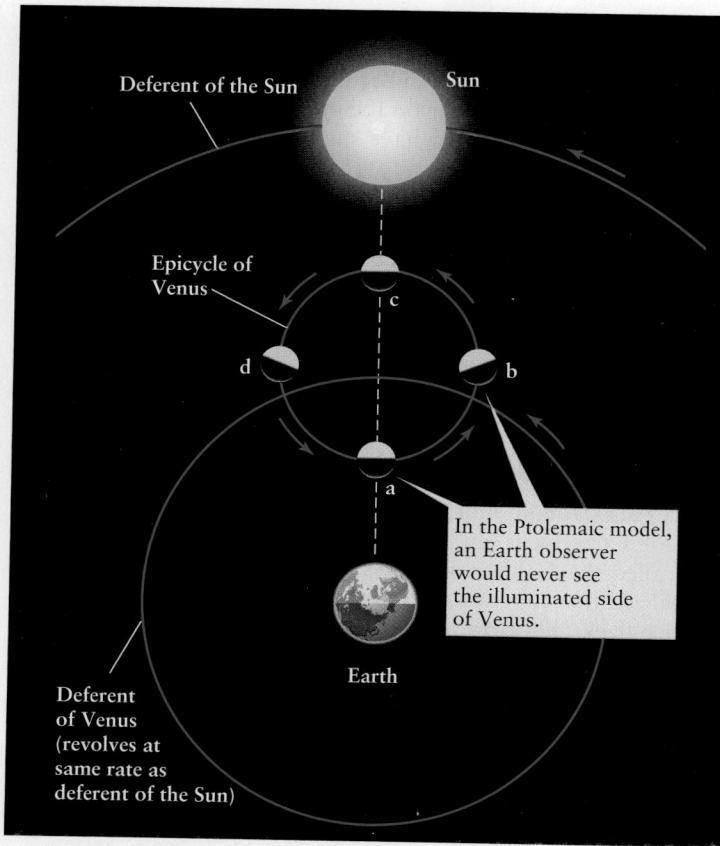

FIGURE 4-15

The Appearance of Venus in the Ptolemaic Model In the geocentric Ptolemaic model, the deferents of Venus and the Sun rotate together, with the epicycle of Venus (shown dashed) that connects the Sun and Earth. In this model, an Earth observer would never see Venus more than half illuminated. (At positions a and c, Venus appears in a "new" phase; at positions b and d, it appears as a crescent. Compare with Figure 3-2, which shows the phases of the Moon.) Because Galileo saw Venus in nearly fully illuminated phases, he concluded that the Ptolemaic model must be incorrect.

FIGURE 4-16   R I (V) U X G

Jupiter and Its Largest Moons This photograph, taken by an amateur astronomer with a small telescope, shows the four Galilean satellites alongside an overexposed image of Jupiter. Each satellite is bright enough to be seen with the unaided eye, were it not overwhelmed by the glare of Jupiter. (Rev. Ronald Royer/Science Source)

FIGURE 4-17

Early Observations of Jupiter's Moons In 1610, Galileo discovered four "stars" that move back and forth across Jupiter from one night to the next. He concluded that these are four moons that orbit Jupiter, much as our Moon orbits Earth. The circle represents Jupiter and the dots its moons. Compare the drawing labeled "Die 7" with the photograph in Figure 4-16. (Royal Astronomical Society/Science Source)

Galileo's telescopic observations constituted the first fundamentally new astronomical data in almost 2000 years. Contradicting prevailing opinion and religious belief, his discoveries strongly suggested a heliocentric structure of the universe. The Roman Catholic Church, which was a powerful political force in Italy and whose doctrine at the time placed Earth at the center of the universe, warned Galileo not to advocate for a heliocentric model. He nonetheless persisted and was sentenced to spend the last years of his life under house arrest "for vehement suspicion of heresy." Nevertheless, there was no turning back. (The Roman Catholic Church lifted its ban against Galileo's heliocentric ideas in the 1700s.)

While Galileo's observations showed convincingly that the Ptolemaic model was entirely wrong and that a heliocentric model is the more nearly correct one, he was unable to provide a complete explanation of why Earth should orbit the Sun and not vice versa. The first person who could provide that explanation was the Englishman Isaac Newton, born on Christmas Day of 1642, a dozen years after the death of Kepler and the same year that Galileo died. While Kepler and Galileo revolutionized our understanding of planetary motions, Newton's contribution went further: He deduced the basic laws that govern all motions on Earth as well as in the heavens.

CONCEPT CHECK 4-14

Why had Jupiter's moons not been observed prior to Galileo's time?

Answer appears at the end of the chapter.

4-6 Newton formulated laws of motion and gravity that describe fundamental properties of physical reality

TUTORIAL 4.2

Until the mid-seventeenth century, virtually all attempts to describe the motion of the heavens were based directly on data and observations of planetary positions. From Ptolemy to Kepler, astronomers would adjust their ideas and calculations by trial and error until they ended up with answers that agreed with observation.

> The same laws of motion that hold sway on Earth apply throughout the universe.

Isaac Newton (Figure 4-18) introduced a new approach. He began with three quite general statements, now called Newton's laws of motion. These laws, deduced from experimental observation, apply to all forces and all objects. Newton then showed that Kepler's three laws follow logically from these laws of motion and from a formula for the force of gravity that he derived from observation.

In other words, Kepler's laws are not just an empirical description of the motions of the planets, but a direct consequence of the fundamental laws of physical matter. Using this deeper insight into the nature of motions in the heavens, Newton and his successors were able to accurately describe not just the orbits of the planets but also the orbits of the Moon and comets.

Newton's First Law

Newton's laws of motion describe objects on Earth as well as in the heavens. Thus, we can understand each of these laws by considering the motions of objects around us. We begin with **Newton's first law of motion:**

An object remains at rest, or moves in a straight line at a constant speed, unless acted upon by a net outside force.

By **force** we mean any push or pull that acts on the object. An *outside* force is one that is exerted on the object by something other than the object itself. The net, or total, outside force is the combined effect of all the individual outside forces that act on the object.

Right now, you are demonstrating the first part of Newton's first law—remaining at rest. As you sit in your chair reading this passage, there are two outside forces acting on you: The force

FIGURE 4-18

Isaac Newton (1642–1727) Using mathematical techniques that he devised, Isaac Newton formulated the law of universal gravitation and demonstrated that the planets orbit the Sun according to simple mechanical rules. (DEA PICTURE LIBRARY/Getty Images)

of gravity pulls you downward, and the chair pushes up on you. These two forces are of equal strength but of opposite direction, so their effects cancel one another—there is no *net* outside force. Hence, your body remains at rest as stated in Newton's first law. If you try to lift yourself out of your chair by grabbing your knees and pulling up, you will remain at rest because this force is not an outside force: It comes from your own body.

CAUTION!

It is easy to confuse the *net* (or total) outside force on an object (central to Newton's first law) with *individual* outside forces on an object. If, for example, you want to make a book move across the floor in a straight line at a *constant speed*, you must continually push on it. You might therefore think that your push is a net outside force. But another force also acts on the book—the force of friction as the book rubs across the floor. The force of your push and the force of friction combine to make the *net* outside force. If you push the book at a constant speed, then the force of your push exactly balances the force of friction, so there is no net outside force. If you stop pushing, there will be nothing to balance the effects of friction. Then the friction will be a net outside force and the book will slow to a stop.

Newton's first law tells us that if no net outside force acts on a moving object, it can only move in a straight line and at a constant speed. This means that a net outside force *must* be acting on the planets since they move not in straight lines but rather move around elliptical paths. Another way to see that planetary orbits require a net outside force is that a planet would fly off into space at a constant speed along a straight line if there were no net outside force acting on it. Because planets don't fly off, Newton concluded that a force *must* act continuously on the planets to keep them in their elliptical orbits.

CONCEPT CHECK 4-15

Imagine a 10-kg rock speeding through empty space at 200 m/s, so far away from other objects that there is no gravitational force (or any other outside forces) exerted on the rock. Describe the rock's motion.

Answer appears at the end of the chapter.

Newton's Second Law

Newton's second law describes how the motion of an object *changes* if there is a net outside force acting on it. To appreciate Newton's second law, we must first understand three quantities that describe motion—speed, velocity, and acceleration.

Speed is a measure of how fast an object is moving. Speed and direction of motion together constitute an object's **velocity.** Compared with a car driving north at 100 km/h (62 mi/h), a car driving east at 100 km/h has the same speed but a different velocity. We can restate Newton's first law to say that an object has a constant velocity (its speed and direction of motion do not change) if no net outside force acts on the object.

Acceleration is the rate at which velocity changes. Because velocity involves both speed and direction, acceleration can result from changes in either. Contrary to popular use of the term, acceleration does not simply mean speeding up. A car is accelerating if it is speeding up, and it is also accelerating if it is slowing down or turning (that is, changing the direction in which it is moving).

You can verify these statements about acceleration if you think about the sensations of riding in a car. If the car is moving with a constant velocity (in a straight line at a constant speed), you feel the same, aside from vibrations, as if the car were not moving at all. But you can sense when the car accelerates in any way: You are pressed back in your seat when the car speeds up, pitched forward when the car slows down, and thrown sideways when the car changes direction in a tight turn. In Box 4-3, we discuss the reasons for these sensations, along with other applications of Newton's laws to everyday life.

An apple falling from a tree is a good example of acceleration that involves only an increase in speed. Initially, at the moment the stem breaks, the apple's speed is zero. After 1 second, its downward speed is 9.8 meters per second (m/s), or 32 feet per second (ft/s). After 2 seconds, the apple's speed is twice this, or 19.6 m/s. After 3 seconds, the speed is 29.4 m/s. Because the apple's speed increases by 9.8 m/s for each second of free fall, the rate of acceleration is 9.8 meters per second per second, or 9.8 m/s² (32 ft/s²).

Thus, Earth's gravity gives the apple a constant acceleration of 9.8 m/s² downward, toward the center of Earth.

A planet revolving about the Sun along a perfectly circular orbit is an example of acceleration that involves only a change of direction. As the planet moves along its orbit, its speed remains constant. Nevertheless, the planet is continuously accelerating because its direction of motion is continuously changing.

Newton's second law of motion says that in order to give an object an acceleration (that is, to change its velocity), a net outside force *must* act on the object. To be specific, this law says that the acceleration of an object is proportional to the net outside force acting on the object. That is, the harder you push on an object, the greater the resulting acceleration. This law can be succinctly stated as an equation. If a net outside force F acts on an object of mass m, the object will experience an acceleration a such that:

Newton's second law

$$F = ma$$

F = net outside force on an object

m = mass of object

a = acceleration of object

The **mass** of an object is a measure of the total amount of material in the object. It is usually expressed in kilograms (kg) or grams (g). For example, the mass of the Sun is 2×10^{30} kg, the mass of a hydrogen atom is 1.7×10^{-27} kg, and the mass of an average adult is 75 kg. The Sun, a hydrogen atom, and a person have these masses regardless of where they happen to be in the universe.

CAUTION!

It is important not to confuse the concepts of mass and weight. **Weight** is the force of gravity that acts on an object and, like any force, is usually expressed in pounds or newtons (1 newton = 0.225 pound). For example, astronauts feel lighter on the Moon because they weigh less in the Moon's weaker gravity, but an astronaut's mass on the Moon is the same as her mass on Earth.

We can use Newton's second law to relate mass and weight. We have seen that the acceleration caused by Earth's gravity is 9.8 m/s². When a 50-kg swimmer falls from a diving board, the only outside force acting on her as she falls is her weight. Thus, from Newton's second law ($F = ma$), her weight is equal to her mass multiplied by the acceleration due to gravity:

$$50 \text{ kg} \times 9.8 \text{ m/s}^2 = 490 \text{ newtons} = 110 \text{ pounds}$$

Note that this answer is correct only when the swimmer is on Earth. She would weigh less on the Moon, where the pull of gravity is weaker, and more on Jupiter, where the gravitational pull is stronger. Floating deep in space, she would have no weight at all; she would be "weightless." Nevertheless, in all these circumstances,

BOX 4-3 ASTRONOMY DOWN TO EARTH

Newton's Laws in Everyday Life

In our study of astronomy, we use Newton's three laws of motion to help us understand the motions of objects in the heavens. But you can see applications of Newton's laws every day in the world around you. By considering these everyday applications, we can gain insight into how Newton's laws apply to celestial events that are far removed from ordinary human experience.

Newton's *first* law, or principle of inertia, says that an object at rest naturally tends to remain at rest and that an object in motion naturally tends to remain in motion. This law explains the sensations that you feel when riding in an automobile. When you are waiting at a red light, your car and your body are both at rest. When the light turns green and you press on the gas pedal, the car accelerates forward but your body attempts to stay where it was. Hence, the seat of the accelerating car pushes forward into your body, and it feels as though you are being pushed back in your seat.

Once the car is up to cruising speed, your body wants to keep moving in a straight line at this cruising speed. If the car makes a sharp turn to the left, the right side of the car will move toward you. Thus, you will feel as though you are being thrown to the car's right side (the side on the outside of the turn). If you bring the car to a sudden stop by pressing on the brakes, your body will continue moving forward until the seat belt stops you. In this case, it feels as though you are being thrown toward the front of the car.

Newton's *second* law states that the net outside force on an object equals the product of the object's mass and its acceleration. You can accelerate a crumpled-up piece of paper to a pretty good speed by throwing it with a moderate force. But if you try to throw a heavy rock by using the same force, the acceleration

will be much less because the rock has much more mass than the crumpled paper. Because of the smaller acceleration, the rock will leave your hand moving at only a slow speed.

Automobile airbags are based on the relationship between force and acceleration. It takes a large force to bring a fast-moving object suddenly to rest because this requires a large acceleration. In a collision, the driver of a car not equipped with airbags is jerked to a sudden stop and the large forces that act can cause major injuries. But if the car has airbags that deploy in an accident, the driver's body will slow down more gradually as it contacts the airbag, and the driver's acceleration will be less. (Remember that *acceleration* can refer to slowing down as well as to speeding up.) Hence, the force on the driver and the chance of injury will both be greatly reduced.

Newton's *third* law, the principle of action and reaction, explains how a car can accelerate at all. It is not correct to say that the engine pushes the car forward, because Newton's second law tells us that it takes a force acting from outside the car to make the car accelerate. Rather, the engine makes the wheels and tires turn, and the tires push backward on the ground. From Newton's third law, the ground must exert an equally large forward force on the car, and this is the force that pushes the car forward. You can see this backward force in action when a car drives through wet ground and sprays mud backward from the tires.

You use the same principles when you walk: You push backward on the ground with your foot, and the ground pushes forward on you. Icy pavement or a freshly waxed floor have greatly reduced friction. In these situations, your feet and the surface under you can exert only weak forces on each other, and it is much harder to walk.

she would always have exactly the same mass, because mass is an inherent property of matter unaffected by details of the environment. Whenever we describe the properties of planets, stars, or galaxies, we speak of their masses, never of their weights.

We have seen that a planet is continually accelerating as it orbits the Sun. From Newton's second law, this means that there must be a net outside force that acts continually on each of the planets. As we will see in the next section, this force is the gravitational attraction of the Sun.

Newton's Third Law

The last of Newton's general laws of motion is called **Newton's third law of motion:**

Whenever one object exerts a force on a second object, the second object exerts an oppositely directed force of equal strength on the first object.

Newton's third law is sometimes described in terms of actions and reactions: When two objects interact by exerting forces on each other, these action and reaction forces are equal in magnitude but opposite in direction.

For example, consider your weight; Earth pulls down on you with a gravitational force equal to your weight. But you also pull back up on Earth with a gravitational force of equal strength. In another example, consider a bat hitting a ball. Clearly, the bat exerts a large force on the ball when hitting a home run (this can be called an action force). However, the ball exerts a force of *equal strength* on the bat as well (this can be called a reaction force). It might seem totally counterintuitive that a small, passive ball could exert an equally large force on a bat that has someone's full swing behind it. However, imagine a spring between the ball and bat during the hit; a spring is a very good approximation for the deformed ball. Regardless of which object moves in order to compress the spring, the spring pushes outward with an equal force on both ends (outward on the ball and bat).

CONCEPT CHECK 4-16

Two sumo wrestlers push against each other during a match. One wrestler is much larger than the other. The larger wrestler's feet remain on the floor while the smaller wrestler's feet slip as he is accelerated in a push right out of the ring. Compare the force that the larger wrestler exerts on the smaller wrestler to the force the smaller wrestler exerts on the larger wrestler.

CONCEPT CHECK 4-17

In midair after jumping off a diving board, a diver is pulled down to the water by her weight. However, the diver pulls up on Earth with a force of equal strength, so why doesn't Earth move an equal amount as the diver?

Answers appear at the end of the chapter.

In these examples, you can think of each force in these pairs as a reaction to the other force, which is the origin of the phrase "action and reaction."

Newton realized that because the Sun is exerting a force on each planet to keep it in orbit, each planet must also be exerting an equal and opposite force on the Sun. However, the planets are much less massive than the Sun (for example, Earth has only 1/300,000 of the Sun's mass). Therefore, although the Sun's force on a planet is the same as the planet's force on the Sun, the planet's much smaller mass allows for a much larger acceleration, according to Newton's second law. This is why the planets circle the Sun instead of vice versa. Thus, Newton's laws reveal the reason for our heliocentric solar system.

The Law of Universal Gravitation

Tie a ball to one end of a piece of string, hold the other end of the string in your hand, and whirl the ball around in a circle. As the ball "orbits" your hand, it is continuously accelerating because its velocity is changing—even though it moves at a constant speed, the ball is changing direction. In accordance with Newton's second law, this can happen only if the ball is continuously acted on by an outside force— the pull of the string. The pull is directed along the string toward your hand. In the same way, Newton saw, the force that keeps a planet in orbit around the Sun is a pull that always acts toward the Sun. That pull is **gravity, or gravitational force.**

> Newton's law of gravitation is truly universal: It applies to falling apples as well as to planets and galaxies.

Newton's discovery about the forces that act on planets led him to suspect that the force of gravity pulling a falling apple straight down to the ground is fundamentally the same as the force on a planet that is always directed straight at the Sun. In other words, gravity is the force that shapes the orbits of the planets. What is more, he was able to determine how the force of gravity depends on the distance between the Sun and the planet. His result was a law of gravitation that could apply to the motion of distant planets as well as to the flight of a football on Earth. Using this law, Newton achieved the remarkable goal of deducing Kepler's laws from fundamental principles of nature.

To see how Newton reasoned, think again about a ball attached to a string. If you use a short string, so that the ball orbits in a small circle, and whirl the ball around your hand at a high speed, you will find that you have to pull fairly hard on the string (Figure 4-19a). But if you use a longer string, so that the ball moves in a larger orbit, and if you make the ball orbit your hand at a slow speed, you only have to exert a light tug on the string (Figure 4-19b). The orbits of the planets behave in the same way: The larger the size of the orbit, the slower the planet's speed (Figure 4-19c, d). By analogy to the force of the string on the orbiting ball, Newton concluded that the force that attracts a planet toward the Sun must decrease with increasing distance between the Sun and the planet.

Using his own three laws and Kepler's three laws, Newton succeeded in formulating a general statement that describes the nature of the gravitational force. Newton's **law of universal gravitation** is as follows:

Two objects attract each other with a force that is directly proportional to the mass of each object and inversely proportional to the square of the distance between them.

This law states that *any* two objects exert gravitational pulls on each other. Normally, you notice only the gravitational force that Earth exerts on you, otherwise known as your weight. In fact, you are gravitationally attracted to *all* the objects around you. For example, a book exerts a gravitational force on you as you read it. But because the force exerted on you by this book is proportional to the book's mass, which is very small compared to Earth's mass, the force is too small to notice. (Although it can actually be measured with sensitive equipment.)

The farther apart two objects are, the weaker the gravitational force between them. Since the gravitational force weakens by the square of the distance between two objects, doubling their distance reduces their attraction by

$$\frac{1}{2^2} = \frac{1}{4}$$

and tripling their distance reduces their attraction by

$$\frac{1}{3^2} = \frac{1}{9}$$

 ANIMATION 4.9 Newton's law of universal gravitation can be stated as an equation. If two objects have masses m_1 and m_2 and are separated by a distance r, then the gravitational force F between these two objects is given by:

Newton's law of universal gravitation

$$F = G\left(\frac{m_1 m_2}{r^2}\right)$$

F = gravitational force between two objects

m_1 = mass of first object

m_2 = mass of second object

r = distance between objects

G = universal constant of gravitation

If the masses are measured in kilograms and the distance between them in meters, then the force is measured in newtons. In this formula, G is a number called the **universal constant of gravitation.** Laboratory experiments have yielded a value for G of

$$G = 6.67 \times 10^{-11} \text{ newton} \cdot \text{m}^2/\text{kg}^2$$

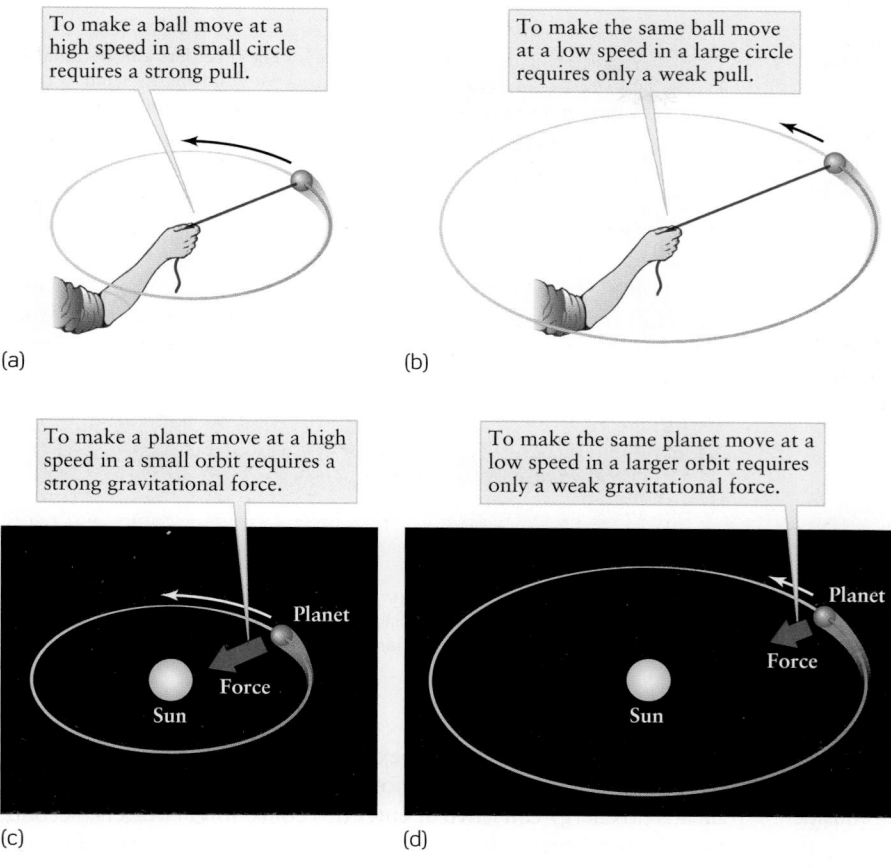

To make a ball move at a high speed in a small circle requires a strong pull.

To make the same ball move at a low speed in a large circle requires only a weak pull.

(a)   (b)

To make a planet move at a high speed in a small orbit requires a strong gravitational force.

To make the same planet move at a low speed in a larger orbit requires only a weak gravitational force.

Planet
Force
Sun

Planet
Force
Sun

(c)   (d)

FIGURE 4-19

An Orbit Analogy (a) To make a ball on a string move at high speed around a small circle, you have to exert a substantial pull on the string. **(b)** If you lengthen the string and make the same ball move at low speed around a large circle, much less pull is required. **(c)** Similarly, a planet that orbits close to the Sun moves at high speed and requires a substantial gravitational force from the Sun, while **(d)** a planet in a large orbit moves at low speed and requires less gravitational force to stay in orbit.

We can use Newton's law of universal gravitation to calculate the force with which any two objects attract each other. For example, to compute the gravitational force that the Sun exerts on Earth, we substitute values for Earth's mass ($m_1 = 5.98 \times 10^{24}$ kg), the Sun's mass ($m_2 = 1.99 \times 10^{30}$ kg), the distance between them ($r = 1$ au $= 1.5 \times 10^{11}$ m), and the value of G into Newton's equation. We get

$$F_{\text{Sun-Earth}} = 6.67 \times 10^{-11} \left[\frac{(5.98 \times 10^{24}) \times (1.99 \times 10^{30})}{(1.50 \times 10^{11})^2} \right]$$

$$= 3.53 \times 10^{22} \text{ newtons}$$

If we calculate the force that Earth exerts on the Sun, using the Sun's mass for m_1 and the Earth's mass for m_2, we find the same result (since the product of the two masses is the same). This is in accordance with Newton's third law: Any two objects exert *equal* gravitational forces on each other.

Your weight is just the gravitational force that Earth exerts on you, so we can calculate it using Newton's law of universal gravitation. Earth's mass is $m_1 = 5.98 \times 10^{24}$ kg, and the distance r to use is the distance between the *centers* of Earth and you. This distance is just the radius of Earth, which is $r = 6378$ km $= 6.378 \times 10^6$ m. If your mass is $m_2 = 50$ kg, your weight is

$$F_{\text{Earth-you}} = 6.67 \times 10^{-11} \left[\frac{(5.98 \times 10^{24}) \times (50)}{(6.378 \times 10^6)^2} \right]$$

$$= 490 \text{ newtons}$$

This value is the same as the weight of a 50-kg person that we calculated in Section 4-6. This example shows that your weight

would have a different value on a planet with a different mass m_1 and a different radius r.

CONCEPT CHECK 4-18

How much does the gravitational force of attraction change between two asteroids if the two asteroids drift 3 times closer together?

CALCULATION CHECK 4-2

How much would a 75-kg astronaut, weighing about 165 pounds on Earth, weigh in newtons and in pounds if he were standing on Mars, which has a mass of 6.4×10^{23} kg and a radius of 3.4×10^6 m?

Answers appear at the end of the chapter.

4-7 Describing orbits with energy and gravity

TUTORIAL
4.3
Imagine a cannonball shot into the air; it arcs through the sky before it crashes back to the ground. If you want to shoot the cannonball into orbit around Earth (like a satellite), you might guess that the ball must be shot faster to give it more energy. This guess is correct, and next we'll look at how to understand orbits in terms of gravity and energy.

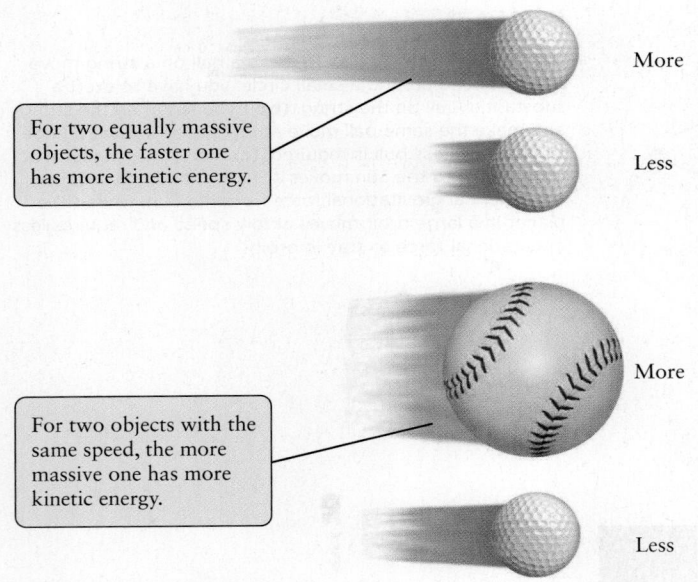

FIGURE 4-20

Kinetic Energy The kinetic energy of an object depends on both its speed and its mass.

FIGURE 4-21

Gravitational Potential Energy The higher the diver, the more gravitational potential energy she has. As the diver descends, some of her gravitational potential energy has transformed into kinetic energy.

Different Forms of Energy

Energy comes in a variety of familiar forms. For example, consider a car: The faster a car moves, the greater its energy. In general, the greater an object's speed, the more **kinetic energy** it has (Figure 4-20). The word *kinetic* refers to motion, so kinetic energy is the energy of motion. However, the mass of an object also contributes to its kinetic energy—a speeding car has more kinetic energy than a speeding bullet.

Now let's look at some other forms of energy. Both food and batteries store chemical energy, although they each contain very different chemicals. When energy is stored, we say it has the potential to be used, and this stored energy is often called *potential energy*. Consider a girl on the diving board of a swimming pool (Figure 4-21); gravity has the potential to pull her into the water (whenever she steps off the board). This is an example of stored gravitational energy, or **gravitational potential energy**. The higher the diving board, the more gravitational potential energy she has. After she steps off the board, she'll have both kinetic energy and gravitational potential energy. As she falls toward the water, her kinetic energy increases as she speeds up and her gravitational potential energy decreases.

While the concept of energy is easier to introduce with a diving board, the same principles apply to orbiting objects: Planets, moons, and satellites also have both kinetic and gravitational potential energy. For example, the faster a satellite orbits Earth, the greater its kinetic energy; if it falls back to Earth, it loses gravitational potential energy.

To understand another aspect of energy, let's think more about the diver in Figure 4-21. Standing high atop the board, the diver has more gravitational potential energy than she would have while standing on the ground. In midair during the dive, some of that gravitational potential energy has transformed into kinetic energy. This illustrates a law called the **conservation of energy**: While energy can change from one form to another, energy cannot be created or destroyed.

There's another way to express the law of conservation of energy. First consider a collection of objects that are *isolated*, which means no energy can leave or enter this collection of objects. Second, find the total energy by taking into account all the forms of energy the collection of objects has. This total energy is conserved, meaning that the total energy stays constant over time, even as objects interact with each other and the various forms of energy can be converted from one to another.

CONCEPT CHECK 4-19

Consider a cannonball shot vertically, straight up into the air. At the ball's highest point, it momentarily comes to a complete stop before beginning to fall back down. At its highest point, what form of energy does the ball have, and where did the energy come from?

Answer appears at the end of the chapter.

As a feather falls gently to the ground, it loses gravitational potential energy without gaining much kinetic energy. This is an example of **air drag**, where the kinetic energy of an object is transferred to the molecules in the surrounding air and the air gets hotter. Energy is still conserved—as it always is—and the nearby air molecules actually speed up. Since the kinetic energy of individual particles gives a gas its temperature and thermal energy, we can see that air drag converts kinetic energy into thermal energy (we will learn more about thermal energy in Section 5-3). Air drag can be good or bad, depending on the situation. Astronauts returning to Earth rely on this transfer of energy, as they use air drag during atmospheric reentry to reduce their speed for a safe landing. As we will see shortly, air drag is undesirable when it transfers energy away from an orbiting satellite, causing it to fall back to Earth.

Gravitational Force, Energy, and Orbits

Because there is a gravitational force between any two objects, Newton concluded that gravity is also the force that keeps the Moon in orbit around Earth. Gravity is also the force that keeps artificial satellites in orbit. But if the force of gravity attracts two objects to each other, why don't satellites immediately fall to Earth? Why doesn't the Moon fall into Earth? And, for that matter, why don't the planets fall into the Sun?

To see the answer, imagine (as Newton did) dropping a ball from a great height above Earth's surface, as in Figure 4-22. After you drop the ball, it, of course, falls straight down (path A in Figure 4-22). But if you *throw* the ball horizontally, it travels some distance across Earth's surface before hitting the ground (path B). If you throw the ball harder, it travels a greater distance (path C). If you could throw or shoot it at just the right speed, the curvature of the ball's path will exactly match the curvature of Earth's surface (path E). Although Earth's gravity is making the ball fall, Earth's surface is falling away under the ball at the same rate. Hence, the ball does not get any closer to the surface, and the ball is in circular orbit. So, the ball in path E is in fact falling, but it is falling *around* Earth rather than *toward* Earth.

As the hypothetical ball is launched faster and faster, you can also see that the paths (going from A to F) represent trajectories with more of both kinetic and gravitational potential energy; in other words, with increasing **orbital energy**. In general, for an object orbiting the Sun or a planet:

The greater the orbital energy, the greater the average orbital distance (that is, the greater the semimajor axis, a).

FIGURE 4-22

Orbits and the Escape Speed If a ball is dropped from a great height above Earth's surface, it falls straight down (A). If the ball is thrown with some horizontal speed, it follows a curved path before hitting the ground (B, C). If thrown with just the right speed (E), the ball goes into circular orbit; the ball's path curves but it never gets any closer to Earth's surface. If the ball is thrown with a speed that is slightly less (D) or slightly more (F) than the speed for a circular orbit, the ball's orbit is an ellipse. If thrown faster than the escape speed (G), the ball will leave Earth and never return.

A spacecraft is launched into orbit in just this way—by gaining enough speed and energy. Once the spacecraft is in a stable orbit, no more rocket fuel is needed and, if not for air drag, the spacecraft could orbit indefinitely. However, air drag from the thin outer wisps of Earth's atmosphere slowly removes orbital energy from the closer satellites, slowing them down and bringing them inward. Without occasional bursts of thrust, these satellites would fall back to Earth. The International Space Station is no exception, and air drag can decrease the station's altitude by a couple hundred feet each day!

CAUTION!

An astronaut performing a spacewalk outside of an orbiting spacecraft (like the one shown in the photograph that opens this chapter) feels "weightless." However, this is *not* because she is "beyond the pull of gravity." In fact, gravity on the Space Station is nearly 90% what we experience on Earth's surface. The astronaut is herself an independent satellite orbiting Earth, and Earth's gravitational pull is what holds her in orbit. She feels "weightless" because she and her spacecraft are falling *together* around Earth, so there is nothing pushing her against any of the spacecraft walls. You feel the same "weightless" sensation whenever you are falling, such as when you jump off a diving board or ride the free-fall ride at an amusement park.

CONCEPT CHECK 4-20

Suppose a spacecraft is initially shot into a circular orbit with path E in Figure 4-22. Which new path will the spacecraft likely take if air drag is present? Which new path will likely result if a rocket thruster is fired?

CONCEPT CHECK 4-21

As Earth orbits the Sun, does Earth's orbit decay due to air drag from its own atmosphere?

Answers appear at the end of the chapter.

The Escape Speed

If an object is hurled with enough speed, it can escape a planet altogether. For example, when a very large asteroid smashes into Earth, most of the debris ejected from the crater falls back to Earth's surface, but some rocks have enough speed to escape and roam the solar system. In Figure 4-22, the object on path G is shot with the **escape speed**, has more energy than any of the "bound" orbits shown in Figure 4-23, and never returns to Earth. The speed needed for escape depends on the size and mass of the planet, but *not* on the mass of the escaping object:

Escape speed

$$v_{\text{escape}} = \sqrt{\frac{2GM}{R}}$$

v_{escape} = escape speed

M = mass of the planet

R = radius of the planet

G = universal constant of gravitation

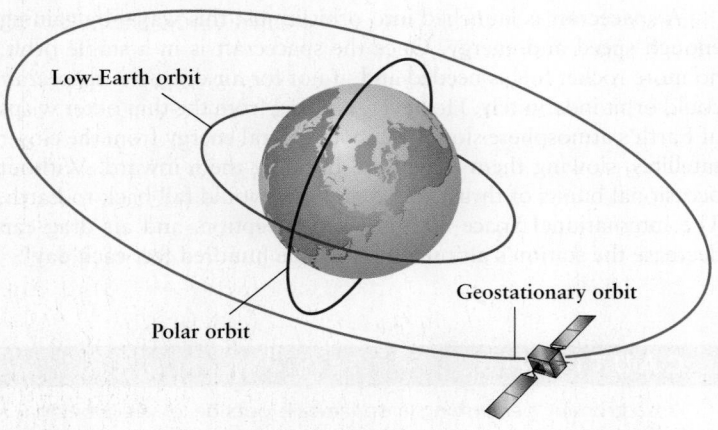

FIGURE 4-23

Many Possible Orbits Newton's form of Kepler's third law pertains to each of these orbits, and many more that are not shown. It also describes the orbits of any moon around a planet. Low-Earth orbits (called LEOs) are where most satellites are found. Polar orbits are a special LEO that can be used to map or monitor the entire Earth once each day. Geosynchronous, or geostationary, orbits have the unique property that they take 24 hours to orbit the Earth. This means they can orbit precisely in step with Earth's rotation and, as viewed from Earth, seemingly hover in the same place above Earth for constant monitoring of the same region. These orbits are not shown to scale, as the LEO orbits are quite close to Earth, and the geosynchronous orbit is about a quarter of the way to the Moon!

For Earth, the escape speed is 11.2 km/s. That's about 25,000 mi/h, or 33 times the speed of sound. This speed is so fast that, except in the largest asteroid impacts, very little ejected debris would have escaped Earth over time. However, the escape speed for Mars is only about 5 km/s, so compared to Earth, it is easier for impact debris from Mars to get ejected into the solar system. In fact, we know of about 100 rocks found on Earth that were originally ejected from Mars by asteroid impacts.

What about spacecraft; are they launched at the escape speed? The escape speed assumes a simple scenario where no air drag is present and where there is no rocket fuel to add energy during flight. However, even with fuel, spacecraft going to the Moon or beyond often leave Earth near the escape speed.

CONCEPT CHECK 4-22

Two objects, a large rock and a small rock, are shot with just enough speed to escape the Moon. Can both rocks be shot with equal speed? Can both rocks be shot with the same kinetic energy?

CONCEPT CHECK 4-23

New planets around other stars are constantly being discovered. Would it be possible to find a planet with the same mass as Earth but with a lower escape speed from its surface?

Answers appear at the end of the chapter.

Gravitation and Kepler's Laws

Using his three laws of motion and his law of gravity, Newton found that he could derive Kepler's three laws mathematically. (Newton also invented calculus, which helped in this endeavor!) While Kepler's laws describe the planets in the solar system specifically, Newton showed that similar behavior results from any orbital system, such as a planet and its moons. Kepler's first law, concerning the elliptical shape of planetary orbits, proved to be a direct consequence of the $1/r^2$ factor in the law of universal gravitation. The law of equal areas, or Kepler's second law, turns out to be a consequence of the Sun's gravitational force on a planet being directed straight toward the Sun.

Newton also demonstrated that Kepler's third law follows logically from his law of gravity. Specifically, he proved that if two objects with masses m_1 and m_2 orbit each other, the period P of their orbit and the semimajor axis a of their orbit (that is, the average distance between the two objects) are related by an equation that we call **Newton's form of Kepler's third law:**

$$P^2 = \left[\frac{4\pi^2}{G(m_1 + m_2)} \right] a^3$$

Newton's form of Kepler's third law is valid whenever two objects orbit each other because of their mutual gravitational attraction (**Figure 4-24**). It is invaluable in the study of binary star systems, in which two stars orbit each other. If the orbital period P and semimajor axis a of the two stars in a binary system are known, astronomers can use this formula to calculate the sum $m_1 + m_2$ of the masses of the two stars. Within our own solar system, Newton's form of Kepler's third law makes it possible to learn about the masses of planets. By measuring the period and semimajor axis for a satellite, astronomers can determine the sum of the masses of the planet and the satellite. (The satellite can be a moon of the planet or a spacecraft that we place in orbit around the planet. Newton's laws apply in either case.) Box 4-4 on page 101 gives an example of using Newton's form of Kepler's third law.

Newton also discovered new features of orbits around the Sun. For example, his equations soon led him to conclude that the trajectory of an object as it swings around the Sun need not follow an ellipse. It could be any one of a family of curves called conic sections.

A **conic section** is any curve that results from cutting a cone with a plane, as shown in **Figure 4-25**. You can get circles and ellipses by slicing all the way through the cone; these represent the familiar circular and elliptical orbits. You can also get two types of open curves called **parabolas** and **hyperbolas.** An object traveling at the escape speed follows a parabolic curve, and for objects traveling even faster than the escape speed, the path is hyperbolic. Some comets, such as Halley's comet, orbit the Sun on elliptical paths and keep returning, while other comets follow hyperbolic orbits and never return.

CONCEPT CHECK 4-24

A small celestial object on a hyperbolic orbit passes by a lonely moonless planet. Can the planet capture the object to acquire a moon?

Answer appears at the end of the chapter.

FIGURE 4-24 R I (V) U X G

In Orbit Around the Moon This photograph taken from the spacecraft *Columbia* shows the lunar lander *Eagle* after returning from the first human landing on the Moon in July 1969. Newton's form of Kepler's third law describes the orbit of a spacecraft around the Moon, as well as the orbit of the Moon around Earth (visible in the distance). (Michael Collins, Apollo 11, NASA)

The Triumph of Newtonian Mechanics

Newton's ideas turn out to be applicable to an incredibly wide range of situations. In fact, nearly all the details of the orbits of the planets and their satellites can be explained mathematically with what is called **Newtonian mechanics.**

Not only could Newtonian mechanics explain a variety of known phenomena in detail, but it could also predict new phenomena. For example, one of Newton's friends, Edmund Halley, was intrigued by three similar historical records of a comet that had been sighted at intervals of 76 years. Assuming these records to be accounts of the same comet, Halley used Newton's methods to work out the details of the comet's orbit and predicted its return in 1758. It was first sighted on Christmas night of 1757, and to this day the comet bears Halley's name (Figure 4-26).

Another dramatic success of Newton's ideas was their role in the discovery of the eighth planet from the Sun. The seventh planet, Uranus, was discovered accidentally by William Herschel in 1781 during a telescopic survey of the sky. Fifty years later, however, it was clear that Uranus was not following its predicted orbit. John Couch Adams in England and Urbain Le Verrier in France independently calculated that the gravitational pull of a yet unknown, more distant planet could explain the deviations of Uranus from its orbit. Le Verrier predicted that, at that time, the planet would be found at a certain location in the constellation of Aquarius. A brief telescopic search on September 23, 1846, revealed the planet Neptune within 1° of the calculated position. What a success! Before it was sighted with a telescope, the existence and location of Neptune was actually predicted with pencil and paper.

Because it has been so successful in explaining and predicting many important phenomena, Newtonian mechanics has become the cornerstone of modern physical science. Even today, as we send astronauts into Earth orbit and spacecraft to the outer planets, Newton's equations are used to calculate the orbits and trajectories of these spacecraft. The *Cosmic Connections* figure on the next page summarizes some of the ways that gravity plays an important role on scales from apples to galaxies.

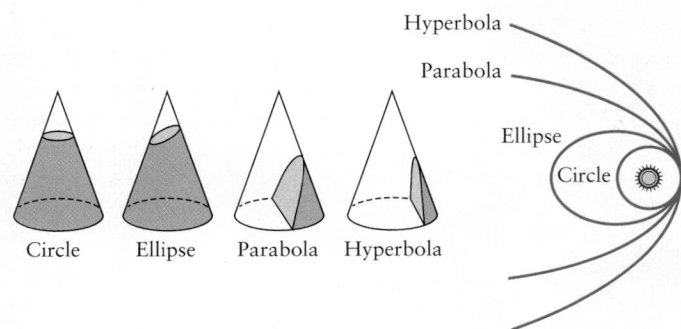

FIGURE 4-25

Conic Sections A conic section is any one of a family of curves obtained by slicing a cone with a plane. The orbit of one object about another can be any one of these curves: a circle, an ellipse, a parabola, or a hyperbola.

FIGURE 4-26 R I (V) U X G

Comet Halley This most famous of all comets orbits the Sun with an average period of about 76 years. During the twentieth century, the comet passed near the Sun in 1910 and again in 1986 (when this photograph was taken). It will next be prominent in the sky in 2061. (Harvard College Observatory/Science Source)

COSMIC CONNECTIONS
Universal Gravitation

Gravity is one of the fundamental forces of nature. We can see its effects here on Earth as well as in the farthest regions of the observable universe.

The weight of an ordinary object is just the gravitational force exerted on that object by Earth.

Earth is held together by the mutual gravitational attraction of its parts.

Gravitational forces exerted by the Sun keep the planets in their orbits. The farther a planet is from the Sun, the weaker the gravitational force that acts on the planet.

Milky Way Galaxy

Sun

26,000 light-years

The mutual gravitational attraction of all the matter in the Milky Way Galaxy holds it together. The gravitational force of the Galaxy on our Sun and solar system holds us in an immense orbit around the galactic center.

BOX 4-4   TOOLS OF THE ASTRONOMER'S TRADE

Newton's Form of Kepler's Third Law

Kepler's original statement of his third law, $P^2 = a^3$, is valid only for objects that orbit the Sun. (Box 4-2 shows how to use this equation.) But Newton's form of Kepler's third law is much more general: It can be used in *any* situation where two objects of masses m_1 and m_2 orbit each other. For example, Newton's form is the equation to use for a moon orbiting a planet or a satellite orbiting Earth. This equation is:

Newton's form of Kepler's third law

$$P^2 = \left[\frac{4\pi^2}{G(m_1 + m_2)} \right] a^3$$

P = sidereal period of orbit, in seconds

a = semimajor axis of orbit, in meters

m_1 = mass of first object, in kilograms

m_2 = mass of second object, in kilograms

G = universal constant of gravitation = 6.67×10^{-11} $\mathrm{m^3\,kg^{-1}\,s^{-2}}$

Notice that P, a, m_1, and m_2 must be expressed in these particular units. If you fail to use the correct units, your answer will be incorrect.

EXAMPLE: Io (pronounced "eye-oh") is one of the four large moons of Jupiter discovered by Galileo and shown in Figure 4-16. It orbits at a distance of 421,600 km from the center of Jupiter and has an orbital period of 1.77 days. Determine the combined mass of Jupiter and Io.

Situation: We are given Io's orbital period P and semimajor axis a (the distance from Io to the center of its circular orbit, which is at the center of Jupiter). Our goal is to find the sum of the masses of Jupiter (m_1) and Io (m_2).

Tools: Because this orbit is not around the Sun, we must use Newton's form of Kepler's third law to relate P and a. This relationship also involves m_1 and m_2, whose sum ($m_1 + m_2$) we are asked to find.

Answer: To solve for $m_1 + m_2$, we rewrite the equation in the form

$$m_1 + m_2 = \frac{4\pi^2 a^3}{GP^2}$$

To use this equation, we have to convert the distance a from kilometers to meters and convert the period P from days to seconds. There are 1000 meters in 1 kilometer and 86,400 seconds in 1 day, so

$$a = (421{,}600 \text{ km}) \times \frac{1000 \text{ m}}{1 \text{ km}} = 4.216 \times 10^8 \text{ m}$$

$$P = (1.77 \text{ days}) \times \frac{86{,}400 \text{ s}}{1 \text{ day}} = 1.529 \times 10^5 \text{ s}$$

We can now put these values and the value of G into the above equation:

$$m_1 + m_2 = \frac{4\pi^2 (4.216 \times 10^8)^3}{(6.67 \times 10^{-11})(1.529 \times 10^5)^2} = 1.90 \times 10^{27} \text{ kg}$$

Review: Io is very much smaller than Jupiter, so its mass is only a small fraction of the mass of Jupiter. Thus, $m_1 + m_2$ is very nearly the mass of Jupiter alone. We conclude that Jupiter has a mass of 1.90×10^{27} kg, or about 300 times the mass of Earth. This technique can be used to determine the mass of any object that has a second, much smaller object orbiting it. Astronomers use this technique to find the masses of stars, black holes, and entire galaxies of stars.

In the twentieth century, scientists found that Newton's laws do not apply in all situations. A new theory called *quantum mechanics* had to be developed to explain the behavior of matter on the very smallest of scales, such as within the atom and within the atomic nucleus. Albert Einstein developed the *theory of relativity* to explain what happens at very high speeds approaching the speed of light and in places where gravitational forces are very strong. For many purposes in astronomy, however, Newton's laws are as useful today as when Newton formulated them more than three centuries ago.

4-8 Gravitational forces between Earth and the Moon produce tides

We have seen how Newtonian mechanics explains why the Moon stays in orbit around Earth. It also explains why there are ocean tides, as well as why the Moon always keeps the same face toward Earth. Both

of these are the result of *tidal forces*— a consequence of gravity that deforms planets and reshapes galaxies.

Tidal forces arise when different parts of an object experience different gravitational pulls. As an illustration, imagine that three billiard balls are lined up in space at some distance from a planet, as in Figure 4-27a. According to Newton's law of universal gravitation, the force of attraction between two objects is greater the closer the two objects are to each other. Thus, the planet exerts more force on the 3-ball (in red) than on the 2-ball (in blue) and exerts more force on the 2-ball than on the 1-ball (in yellow). Now, imagine that the three balls are released and allowed to fall toward the planet. Figure 4-27b shows the situation a short time later. Because of the differences in gravitational pull, a short time

> Tidal forces reveal how gravitation can pull objects apart rather than drawing them together.

(a)

(b)

(c)

ANIMATION 4.8 **FIGURE 4-27**

The Origin of Tidal Forces **(a)** Imagine three identical billiard balls placed some distance from a planet and released. **(b)** The closer a ball is to the planet, the more gravitational force the planet exerts on it. Thus, a short time after the balls are released, the blue 2-ball has moved farther toward the planet than the yellow 1-ball, and the red 3-ball has moved farther still. **(c)** From the perspective of the 2-ball in the center, it appears that forces have pushed the 1-ball away from the planet and pulled the 3-ball toward the planet. These forces are called tidal forces.

later the 3-ball will have moved farther than the 2-ball, which will in turn have moved farther than the 1-ball. But now imagine that same motion from the perspective of the 2-ball. From this perspective, it appears as though the 3-ball is pulled toward the planet while the 1-ball is pushed away (Figure 4-27c). These apparent pushes and pulls are called tidal forces.

Tidal Forces on Earth

The Moon has a similar effect on Earth as the planet in Figure 4-27 has on the three billiard balls. The arrows in Figure 4-28a indicate the strength and direction of the gravitational force of the Moon at several locations on Earth. The side of Earth closest to the Moon feels a greater gravitational pull than does Earth's center, and the side of Earth that faces away from the Moon feels less gravitational pull than does Earth's center. This means that just as for the billiard balls in Figure 4-27, there are tidal forces acting on Earth (Figure 4-28b). These tidal forces exerted by the Moon try to elongate Earth along a line connecting the centers of Earth and the Moon and try to squeeze Earth inward in the direction perpendicular to that line.

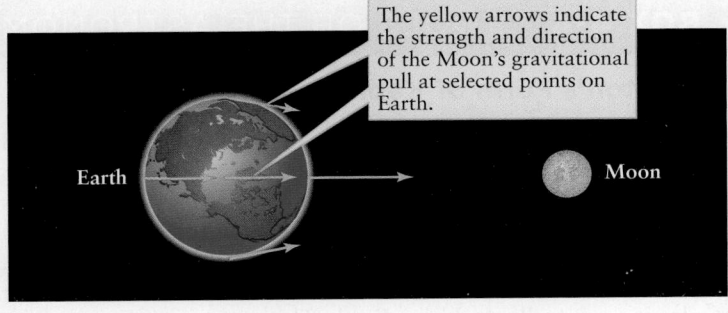

(a)

The yellow arrows indicate the strength and direction of the Moon's tidal forces acting on Earth.

From the perspective of the center of Earth

Moon

(b)

FIGURE 4-28

Tidal Forces on Earth **(a)** The Moon exerts different gravitational pulls at different locations on Earth. **(b)** At any location, the tidal force equals the Moon's gravitational pull at that location minus the gravitational pull of the Moon at the center of Earth. These tidal forces tend to deform Earth into a slightly elongated, or football-like, shape.

Even though the body of Earth is largely rigid, it deforms by about a meter in response to the tidal forces of the Moon. However, the water in the oceans deforms much more dramatically into a football shape, as Figure 4-29a shows. As Earth rotates, a point on its surface goes from where the water is shallow to where the water is deep and back again. This is the origin of low and high ocean tides. (In this simplified description, we have assumed that Earth is completely covered with water. The full story of the tides is much more complex, because the shapes of the continents and the effects of winds must also be taken into account.)

The Sun also exerts tidal forces on Earth's oceans, and the effects of these tidal forces are about half as great as those of the Moon. When the Sun, the Moon, and Earth are aligned, which happens at either new moon or full moon, the tidal effects of the Sun and Moon reinforce each other and the tidal distortion of the oceans is greatest. This produces large shifts in water level called **spring tides** (Figure 4-29b). At first quarter and last quarter, when the Sun and Moon form a right angle with Earth, the tidal effects of the Sun and Moon partially cancel each other. Hence, the tidal distortion of the oceans is the least pronounced, producing smaller tidal shifts called **neap tides** (Figure 4-29c).

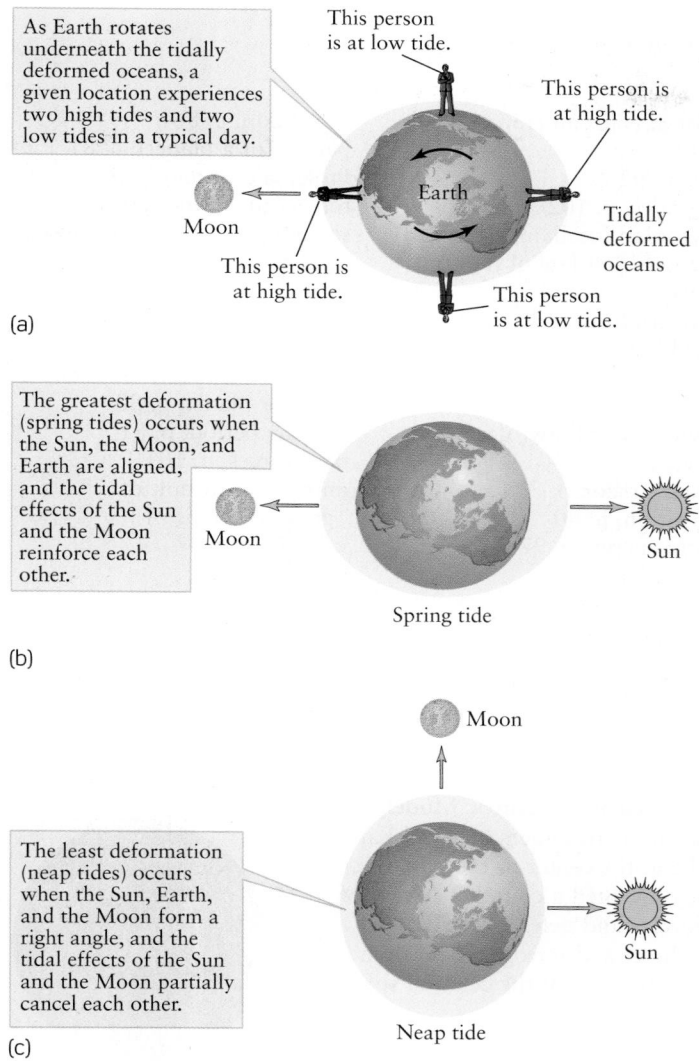

As Earth rotates underneath the tidally deformed oceans, a given location experiences two high tides and two low tides in a typical day.

This person is at low tide.

This person is at high tide.

Moon

Earth

Tidally deformed oceans

This person is at high tide.

This person is at low tide.

(a)

The greatest deformation (spring tides) occurs when the Sun, the Moon, and Earth are aligned, and the tidal effects of the Sun and the Moon reinforce each other.

Moon

Sun

Spring tide

(b)

Moon

The least deformation (neap tides) occurs when the Sun, Earth, and the Moon form a right angle, and the tidal effects of the Sun and the Moon partially cancel each other.

Sun

Neap tide

(c)

FIGURE 4-29

High and Low Tides **(a)** The gravitational forces of the Moon and the Sun deform Earth's oceans, giving rise to low and high tides. **(b), (c)** The strength of the tides depends on the relative positions of the Sun, the Moon, and Earth.

Tidal Forces on the Moon and Beyond

Just as the Moon exerts tidal forces on Earth, Earth exerts tidal forces on the Moon. Soon after the Moon formed some 4.56 billion years ago, it was semimolten throughout its volume. Earth's tidal forces deformed the molten Moon into a slightly elongated shape, with the long axis of the Moon pointed toward Earth. The Moon retained this shape when it cooled and solidified. With gravity's stronger force on the Moon's long axis (or bulge now facing Earth), the Moon's rotation has slowed and it has permanently settled into this orientation. This is why we always see the same face of the Moon (see Figure 3-4b). To keep its long axis pointed toward Earth, the Moon spins once on its axis as it makes one orbit around Earth—that is, it is in synchronous rotation (Section 3-2).

Compared to Earth's daily rotation, the month-long orbit of the Moon is slow enough that you can picture the Moon "standing still" while the Earth rotates. Therefore, people spending a long day at the beach pass through the higher and lower parts of the elongated oceans. These high and low tides are separated by approximately 6 hours. Due to tidal forces, most of the natural satellites in the solar system are in synchronous rotation, and thus always keep the same side facing their planet.

Tidal forces are also important on scales much larger than the solar system. Figure 4-30 shows two spiral galaxies, like the one in Figure 1-9, undergoing a near collision. During the millions of years that this close encounter has been taking place, the tidal forces of the larger galaxy have pulled an immense streamer of stars and interstellar gas out of the smaller galaxy.

On a very small scale, tidal forces can also play a role in the case of black holes. Rather than elongating a planet, a person falling into a black hole that is a few miles wide would be elongated to the point of being literally pulled apart.

IC 2163

NGC 2207

Stars and gas pulled out of IC 2163 by tidal forces from NGC 2207

FIGURE 4-30 R I V U X G

Tidal Forces on a Galaxy For millions of years, the galaxies NGC 2207 and IC 2163 have been moving ponderously past each other. The larger galaxy's tremendous tidal forces have drawn a streamer of material 100,000 light-years long out of IC 2163. Picture it this way: The larger galaxy on the left is pulling inward all of the matter within the smaller galaxy on the right. However, parts of the smaller galaxy that are closer to the larger one feel larger gravitational forces and are pulled in at a greater rate. As a result, the farther regions of the smaller galaxy are, in a sense, left behind and appear to be pulled out. (NASA and the Hubble Heritage Team, AURA/STScI)

KEY WORDS

Terms preceded by an asterisk (*) are discussed in the Boxes.

acceleration, p. 92
air drag, p. 96
aphelion, p. 86
conic section, p. 98
conjunction, p. 80
conservation of energy, p. 96
deferent, p. 78
direct motion, p. 77
eccentricity, p. 85
ellipse, p. 85
elongation, p. 80
epicycle, p. 78
escape speed, p. 97
focus (of an ellipse; *plural* foci), p. 85
force, p. 91
geocentric model, p. 76
gravitational force, p. 94

gravitational potential energy, p. 96
gravity, p. 94
greatest eastern elongation, p. 80
greatest western elongation, p. 80
heliocentric model, p. 79
hyperbola, p. 98
inferior conjunction, p. 80
inferior planet, p. 80
Kepler's first law, p. 85
Kepler's second law, p. 86
Kepler's third law, p. 86
kinetic energy, p. 96
law of equal areas, p. 86
law of universal gravitation, p. 94

major axis (of an ellipse), p. 85
mass, p. 92
neap tides, p. 102
Newtonian mechanics, p. 99
Newton's first law of motion, p. 91
Newton's form of Kepler's third law, p. 98
Newton's second law of motion, p. 92
Newton's third law of motion, p. 93
Occam's razor, p. 79
opposition, p. 80
orbital energy, p. 97
parabola, p. 98
parallax, p. 83

perihelion, p. 86
period (of a planet), p. 81
Ptolemaic system, p. 78
retrograde motion, p. 77
semimajor axis (of an ellipse), p. 85
sidereal period, p. 81
speed, p. 92
spring tides, p. 102
superior conjunction, p. 80
superior planet, p. 80
synodic period, p. 81
tidal forces, p. 101
universal constant of gravitation, p. 94
velocity, p. 92
weight, p. 92

KEY IDEAS

Apparent Motions of the Planets: Like the Sun and the Moon, the planets move on the celestial sphere with respect to the background of stars. Most of the time a planet moves eastward in direct motion, in the same direction as the Sun and the Moon, but from time to time it moves westward in retrograde motion (Section 4-1).

The Ancient Geocentric Model: Ancient astronomers believed Earth to be at the center of the universe. They invented a complex system of epicycles and deferents to explain the direct and retrograde motions of the planets on the celestial sphere (Section 4-1).

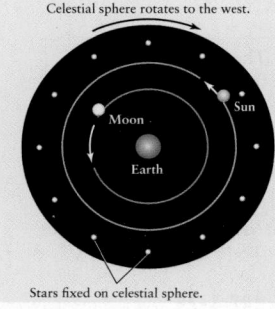

Celestial sphere rotates to the west.

Stars fixed on celestial sphere.

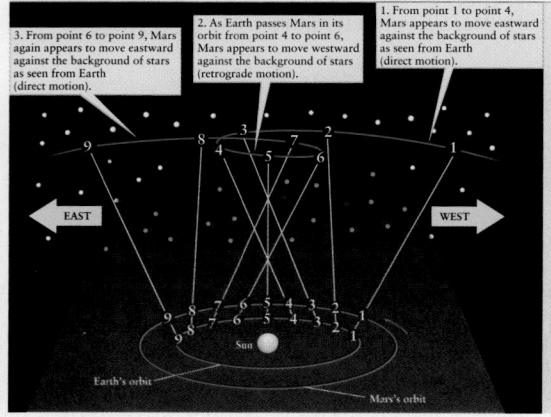

3. From point 6 to point 9, Mars again appears to move eastward against the background of stars as seen from Earth (direct motion).

2. As Earth passes Mars in its orbit from point 4 to point 6, Mars appears to move westward against the background of stars (retrograde motion).

1. From point 1 to point 4, Mars appears to move eastward against the background of stars as seen from Earth (direct motion).

EAST   WEST

Earth's orbit   Mars's orbit

Copernicus's Heliocentric Model: Copernicus's heliocentric (Sun-centered) theory simplified the general explanation of planetary motions (Section 4-2).

- In a heliocentric system, Earth is one of the planets orbiting the Sun.
- A planet undergoes retrograde motion as seen from Earth when Earth and the planet pass each other.
- The sidereal period of a planet, its true orbital period, is measured with respect to the stars. Its synodic period is measured with respect to Earth and the Sun (for example, from one opposition to the next).

Kepler's Improved Heliocentric Model and Elliptical Orbits:
Copernicus thought that the orbits of the planets were combinations
of circles. Using data collected by Tycho Brahe, Kepler deduced three
laws of planetary motion: (1) the orbits are in fact ellipses, (2) a
planet's speed varies as it moves around its elliptical orbit, and
(3) the orbital period of a planet is related to the size of its orbit
(Sections 4-3 and 4-4).

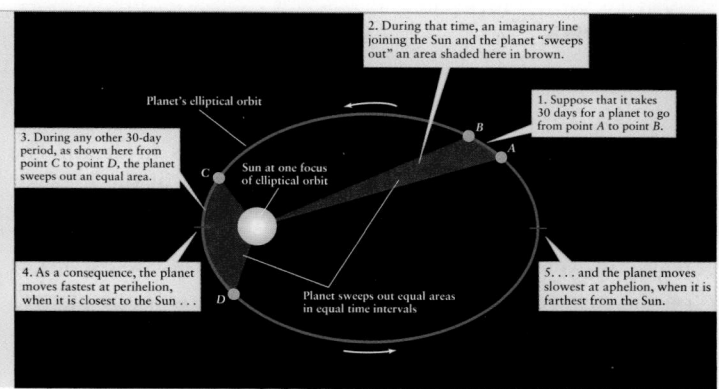

2. During that time, an imaginary line joining the Sun and the planet "sweeps out" an area shaded here in brown.

Planet's elliptical orbit

1. Suppose that it takes 30 days for a planet to go from point *A* to point *B*.

3. During any other 30-day period, as shown here from point *C* to point *D*, the planet sweeps out an equal area.

Sun at one focus of elliptical orbit

4. As a consequence, the planet moves fastest at perihelion, when it is closest to the Sun . . .

Planet sweeps out equal areas in equal time intervals

5. . . . and the planet moves slowest at aphelion, when it is farthest from the Sun.

Rev. Ronald Royer/ Science Source

Evidence for the Heliocentric Model:
The invention of the telescope led
Galileo to new discoveries that
supported a heliocentric model.
These included his observations of
the phases of Venus and of the
motions of four moons around
Jupiter (Section 4-5).

Newton's Laws of Motion: Isaac Newton developed three
principles, called the laws of motion, that apply to the motions
of objects on Earth as well as in space. These are (1) the tendency
of an object to maintain a constant velocity, (2) the relationship
between the net outside force on an object and the object's acceleration, and (3) the principle of action and reaction. These laws
and Newton's law of universal gravitation can be used to deduce
Kepler's laws. They lead to extremely accurate descriptions of planetary motions (Section 4-6).

- The mass of an object is a measure of the amount of matter in
  the object. Its weight is a measure of the force with which the
  gravity of some other object pulls on it.

- In general, the path of one object about another, such as that
  of a planet or comet about the Sun, is one of the curves called
  conic sections: circle, ellipse, parabola, or hyperbola.

Hyperbola

Parabola

Ellipse

Circle

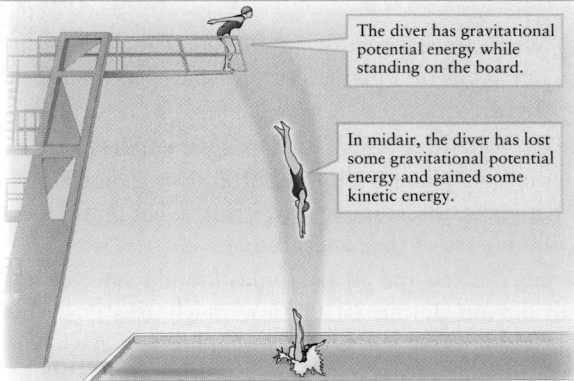

The diver has gravitational potential energy while standing on the board.

In midair, the diver has lost some gravitational potential energy and gained some kinetic energy.

Energy: There are different forms of energy an object can have. Kinetic
and gravitational potential energy are the most important for orbiting
objects (Section 4-7).

- **Kinetic energy:** The greater an object's speed, the larger its kinetic
  energy. For two objects with equal speed, the more massive object has
  a larger kinetic energy.

- **Gravitational potential energy:** The farther an object is located from
  the surface of a planet, the greater its gravitational energy.

- **Conservation of energy:** Energy can change forms, and be transferred
  from one object to another, but energy cannot be created or destroyed.

Escape Speed: The escape
speed of a planet is the ejection speed an object would
need to have near the surface
of the planet so that the
object can "break free" from
the planet and never return
(Section 4-7).

- The escape speed depends
  not on the mass of the
  object, but rather on both
  the size and mass of the
  planet.

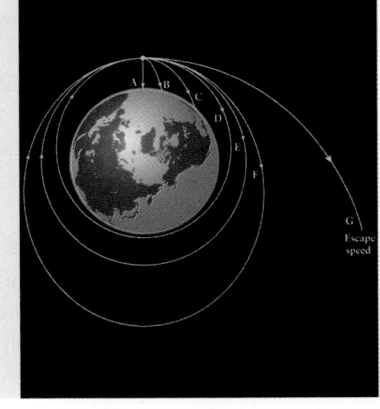

Tidal Forces: Tidal
forces are caused by
differences in the grav-
itational pull that one
object exerts on differ-
ent parts of a second
object (Section 4-8).

- The tidal forces of
  the Moon and the
  Sun produce tides
  in Earth's oceans.

- The tidal forces of
  Earth have locked
  the Moon into syn-
  chronous rotation.

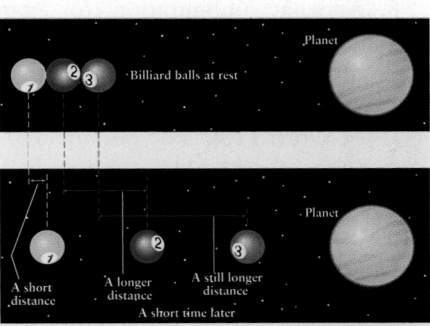

Billiard balls at rest

Planet

A short distance

A longer distance

A still longer distance

A short time later

Planet

From the perspective of the center ball

Planet

QUESTIONS

> ### Problem-solving tips and tools
>
> Box 4-1 explains sidereal and synodic periods in detail. The semimajor axis of an ellipse is half the length of the long, or major, axis of the ellipse. For data about the planets and their satellites, see Appendices 1, 2, and 3 at the back of this book. If you want to calculate the gravitational force that you feel on the surface of a planet, the distance r to use is the planet's radius (the distance between you and the center of the planet). Boxes 4-2 and 4-4 show how to use Kepler's third law in its original form and in Newton's form.

Section 4-1

1. How did the ancient Greeks explain why the Sun and the Moon slowly change their positions relative to the background stars?

2. In what direction does a planet move relative to the stars when it is in direct motion? When it is in retrograde motion? How do these compare with the direction in which we see the Sun move relative to the stars?

3. (a) In what direction does a planet move relative to the horizon over the course of one night? (b) The answer to (a) is the same whether the planet is in direct motion or retrograde motion. What does this tell you about the speed at which planets move on the celestial sphere?

4. What is an epicycle? How is it important in Ptolemy's explanation of the retrograde motions of the planets?

5. What is the significance of Occam's razor as a tool for analyzing theories?

Section 4-2

6. How did the models of Aristarchus and Copernicus explain the retrograde motion of the planets?

7. How did Copernicus determine that the orbits of Mercury and Venus must be smaller than Earth's orbit? How did he determine that the orbits of Mars, Jupiter, and Saturn must be larger than Earth's orbit?

8. At what configuration (for example, superior conjunction, greatest eastern elongation, and so on) would it be best to observe Mercury or Venus with an Earth-based telescope? At what configuration would it be best to observe Mars, Jupiter, or Saturn? Explain your answers.

9. Is it ever possible to see Mercury at midnight? Explain your answer.

10. Which planets can never be seen at opposition? Which planets can never be seen at inferior conjunction? Explain your answers.

11. What is the difference between the synodic period and the sidereal period of a planet?

12. Figure 4-2 shows the retrograde motion of Mars as seen from Earth. Sketch a similar figure that shows how Earth would appear to move against the background of stars during this same time period as seen by an observer on Mars.

13. In 2006, Mercury was at greatest western elongation on April 8, August 7, and November 25. It was at greatest eastern elongation on February 24, June 20, and October 17. Does Mercury take longer to go from eastern to western elongation, or vice versa? Explain why, using Figure 4-6.

14. **Box 4-1** The synodic period of Mercury (an inferior planet) is 115.88 days. Calculate its sidereal period in days.

15. **Box 4-1** Table 4-1 shows that the synodic period is *greater* than the sidereal period for Mercury, but the synodic period is *less* than the sidereal period for Jupiter. Draw diagrams like the one in Box 4-1 to explain why this is so.

16. **Box 4-1** A general rule for superior planets is that the greater the average distance from the planet to the Sun, the more frequently that planet will be at opposition. Explain how this rule comes about.

Section 4-3

17. What is parallax? What did Tycho Brahe conclude from his attempt to measure the parallax of a supernova and a comet?

18. What observations did Tycho Brahe make in an attempt to test the heliocentric model? What were his results? Explain why modern astronomers get different results.

Section 4-4

19. What are the foci of an ellipse? If the Sun is at one focus of a planet's orbit, what is at the other focus?

20. What are Kepler's three laws? Why are they important?

21. At what point in a planet's elliptical orbit does it move fastest? At what point does it move slowest? At what point does it sweep out an area at the fastest rate?

22. A line joining the Sun and an asteroid is found to sweep out an area of 6.3 au^2 during 2010. How much area is swept out during 2011? Over a period of 5 years?

23. The orbit of a spacecraft about the Sun has a perihelion distance of 0.1 au and an aphelion distance of 0.4 au. What is the semimajor axis of the spacecraft's orbit? What is its orbital period?

24. A comet with a period of 125 years moves in a highly elongated orbit about the Sun. At perihelion, the comet comes very close to the Sun's surface. What is the comet's average distance from the Sun? What is the farthest it can get from the Sun?

25. A certain comet is 2 au from the Sun at perihelion and 16 au from the Sun at aphelion. (a) Find the semimajor axis of the comet's orbit. (b) Find the sidereal period of the orbit.

26. A comet orbits the Sun with a sidereal period of 64.0 years. (a) Find the semimajor axis of the orbit. (b) At aphelion, the

comet is 31.5 au from the Sun. How far is it from the Sun at perihelion?

27. One trajectory that can be used to send spacecraft from Earth to Mars is an elliptical orbit that has the Sun at one focus, its perihelion at Earth, and its aphelion at Mars. The spacecraft is launched from Earth and coasts along this ellipse until it reaches Mars, when a rocket is fired to either put the spacecraft into orbit around Mars or cause it to land on Mars. (a) Find the semimajor axis of the ellipse. (*Hint:* Draw a picture showing the Sun and the orbits of Earth, Mars, and the spacecraft. Treat the orbits of Earth and Mars as circles.) (b) Calculate how long (in days) such a one-way trip to Mars would take.

28. Which planet would you expect to exhibit the greatest variation in apparent brightness as seen from Earth? Which planet would you expect to exhibit the greatest variation in angular diameter? Explain your answers. (*Hint:* Use Figure 4-6 and the information on the planets' eccentricities in Section 4-4.)

Section 4-5

29. What observations did Galileo make that reinforced the heliocentric model? Why did these observations contradict the older model of Ptolemy? Why could these observations not have been made before Galileo's time?

30. Why does Venus have its largest angular diameter when it is new and its smallest angular diameter when it is full?

Section 4-6

31. **TUTORIAL 4.2** ➤ What are Newton's three laws? Give an everyday example of each law.

32. How much force do you have to exert on a 3-kg brick to give it an acceleration of 2 m/s²? If you double this force, what is the brick's acceleration? Explain your answer.

33. What is the difference between weight and mass?

34. What is your weight in pounds and in newtons? What is your mass in kilograms?

35. The mass of the Moon is 7.35×10^{22} kg, while that of Earth is 5.98×10^{24} kg. The average distance from the center of the Moon to the center of Earth is 384,400 km. What is the size of the gravitational force that Earth exerts on the Moon? What is the size of the gravitational force that the Moon exerts on Earth? How do your answers compare with the force between the Sun and Earth calculated in the text?

36. The mass of Saturn is approximately 100 times that of Earth, and the semimajor axis of Saturn's orbit is approximately 10 au. To this approximation, how does the gravitational force that the Sun exerts on Saturn compare to the gravitational force that the Sun exerts on Earth? How do the accelerations of Saturn and Earth compare?

37. Suppose that you traveled to a planet with 4 times the mass and 4 times the diameter of Earth. Would you weigh more or less on that planet than on Earth? By what factor?

38. On Earth, a 50-kg astronaut weighs 490 newtons. What would she weigh if she landed on Jupiter's moon Callisto? What fraction is this of her weight on Earth? See Appendix 3 for relevant data about Callisto.

Section 4-7

39. **TUTORIAL 4.3** ➤ Suppose that Earth were moved to a distance of 3.0 au from the Sun. How much stronger or weaker would the Sun's gravitational pull be on Earth? Explain your answer.

40. How far would you have to go from Earth to be completely beyond the pull of its gravity? Explain your answer.

41. A cannonball is shot horizontally from a barrel off a building. Name two forms of energy the cannonball has as it exits the barrel. How do these two forms of energy increase or decrease during the cannonball's flight? Before the cannonball was fired, where was the energy stored?

42. A satellite is in circular orbit. What two forms of energy are part of the satellite's orbital energy? Would its orbital energy need to increase or decrease in order to orbit at a larger distance from Earth?

43. If an object loses orbital energy through air drag, is energy still conserved? If so, where does the energy go?

44. Calculate the escape speed for Earth and show that it is 11.2 km/s. (Consult Appendix 2 for planetary data.)

45. Including the effects of Earth's atmosphere, would you expect the real escape speed for a hypothetical cannonball to be greater or less than 11.2 km/s? Why?

46. What are conic sections? In what way are they related to the orbits of planets in the solar system?

47. Why was the discovery of Neptune an important confirmation of Newton's law of universal gravitation?

48. Except for some rare and exotic microbes, the Sun provides energy for life on Earth. Can you describe how energy in sunlight can end up in a kangaroo jumping through the air? In your description, can you name three forms of energy involved, other than sunlight?

49. You're talking with other students about how the energy of a satellite in circular orbit changes with altitude. One student says that lower orbits have more energy because Kepler's second law says that closer planets have higher speeds. Another student argues that orbits at larger distances must have more energy because rockets have to expend more energy to launch satellites farther into space. Which student reaches the right conclusion?

50. Some argue that life might have started on either Mars or Earth and spread to the other planet by microbe-carrying debris ejected during very large asteroid impacts. Consult Appendix 2 for planetary data and calculate the escape speed for both Earth and Mars. Assuming all else being equal, for which planet is it easier for rocks to escape?

51. Imagine a planet like Earth orbiting a star with 4 times the mass of the Sun. If the semimajor axis of the planet's orbit (*a*) is 1 au, what would be the planet's sidereal period? (*Hint:* Use Newton's form of Kepler's third law. Compared with the case of

Earth orbiting the Sun, by what factor has the quantity $m_1 + m_2$ changed? Has a changed? By what factor must P^2 change?)

52. A satellite is said to be in a "geosynchronous" orbit if it appears always to remain over the exact same spot on rotating Earth. (a) What is the period of this orbit? (b) At what distance from the center of Earth must such a satellite be placed into orbit? (*Hint:* Use Newton's form of Kepler's third law.) (c) Explain why the orbit must be in the plane of Earth's equator.

53. In July 1969, the lunar module *Eagle* orbited the Moon as humanity took its first steps on the lunar surface. The spacecraft orbited 111 km above the surface of the Moon. Calculate the period of the spacecraft's orbit. See Appendix 3 for relevant data about the Moon.

54. **Box 4-4** In Box 4-4, we analyze the orbit of Jupiter's moon Io. Look up information about the orbits of Jupiter's three other large moons (Europa, Ganymede, and Callisto) in Appendix 3. Demonstrate that these data are in agreement with Newton's form of Kepler's third law.

55. **Box 4-4** Suppose a newly discovered asteroid is in a circular orbit with synodic period 1.25 years. The asteroid lies between the orbits of Mars and Jupiter. (a) Find the sidereal period of the orbit. (b) Find the distance from the asteroid to the Sun.

Section 4-8

56. What is a tidal force? How do tidal forces produce tides in Earth's oceans?

57. What is the difference between spring tides and neap tides?

58. The average distance from the Moon to the center of Earth is 384,400 km, and the diameter of Earth is 12,756 km. Calculate the gravitational force that the Moon exerts on (a) a 1-kg rock at the point on Earth's surface closest to the Moon, and (b) a 1-kg rock at the point on Earth's surface farthest from the Moon. (c) Find the difference between the two forces you calculated in parts (a) and (b). This difference is the tidal force pulling these two rocks away from each other, like the 1-ball and 3-ball in Figure 4-27. Explain why tidal forces cause only a very small deformation of Earth.

WEB/E-BOOK QUESTIONS

59. Use two thumbtacks, a loop of string, and a pencil to draw several ellipses. Describe how the shape of an ellipse varies as the distance between the thumbtacks changes.

60. (a) Search the internet for information about Kepler. Before he realized that the planets move on elliptical paths, what other models of planetary motion did he consider? What was Kepler's idea of "the music of the spheres"? (b) Search the internet for information about Galileo. What were his contributions to physics? Which of Galileo's new ideas were later used by Newton to construct his laws of motion? (c) Search the internet for information about Newton. What were some of the contributions that he made to physics other than developing his laws of motion? What contributions did he make to mathematics?

61. ANIMATION 4.2 **Monitoring the Retrograde Motion of Mars.** Watching Mars night after night reveals that it changes its position with respect to the background stars. To track its motion, access and view the animation "The Path of Mars in 2016" in Chapter 4 of the *Universe* website or e-book. (a) Through which constellations does Mars move? (b) On approximately what date does Mars stop its direct (west-to-east) motion and begin its retrograde motion? (*Hint:* Use the "Stop" function on your animation controls.) (c) Over how many days does Mars move retrograde?

ACTIVITIES

Observing Projects

62. It is quite probable that within a few weeks of your reading this chapter, one of the planets will be near opposition or greatest eastern elongation, making it readily visible in the evening sky. Select a planet that is at or near such a configuration by searching the internet or by consulting a reference book, such as the current issue of the *Astronomical Almanac* or the pamphlet entitled *Astronomical Phenomena* (both published by the U.S. government). At that configuration, would you expect the planet to be moving rapidly or slowly from night to night against the background stars? Verify your expectations by observing the planet once a week for a month, recording your observations on a star chart.

63. If Jupiter happens to be visible in the evening sky, observe the planet with a small telescope on five consecutive clear nights. Record the positions of the four Galilean satellites by making nightly drawings, just as the Jesuit priests did in 1620 (see Figure 4-17). From your drawings, can you tell which moon orbits closest to Jupiter and which orbits farthest? Was there a night when you could see only three of the moons? What do you suppose happened to the fourth moon on that night?

64. If Venus happens to be visible in the evening sky, observe the planet with a small telescope once a week for a month. On each night, make a drawing of the crescent that you see. From your drawings, can you determine if the planet is nearer or farther from Earth than the Sun is? Do your drawings show any changes in the shape of the crescent from one week to the next? If so, can you deduce whether Venus is coming toward us or moving away from us?

Starry Night™ Explorations

65. Use the *Starry Night™* program to observe retrograde motion. Select **Favorites > Explorations > Retrograde** to see a view centered on Mars against the background of stars on June 23, 2010. Click **Play** and observe Mars as it moves against the background constellations for at least 2 years of simulated time. An orange line traces Mars's path in the sky from night to night. With the view centered on and tracking Mars, the sky appears to move, but the relative motion of Mars against this sky is obvious.

(a) For most of the time, does Mars move generally to the left (eastward) or to the right (westward) against the background stars?

(b) Select **File > Revert** from the menu to return to the original view. Use the zoom and time controls to determine when Mars's usual *direct* motion ends, as Mars comes to a momentary halt in the west–east direction, and *retrograde* motion begins. On what date does this occur and on what date does retrograde motion end and direct motion resume?

(c) To help understand the motion of Mars as seen from Earth, select **Favorites > Explorations > Heliocentric**. This view, from a position 5 au above the plane of the solar system, is centered on the Sun. The orbits and positions of Mars and Earth on June 23, 2010, are shown. Click **Play** and watch the motions of the planets for 2 years of simulated time. Note that Earth catches up with and overtakes Mars as time proceeds. This relative motion of the two planets leads to our observation of retrograde motion. On what date during this 2-year period is Earth directly between Mars and the Sun? How does this date compare to the dates you recorded in part b above? Explain the significance of this.

66. Use *Starry Night™* to observe the phases of Venus and of Mars as seen from Earth. Open **Favorites > Explorations > Venus Phases** to see an image of Venus.

(a) Draw the shape (phase) of Venus. With the time flow rate at 30 days, step time forward, sketching Venus to scale at each step. Make a total of 20 timesteps and drawings.

(b) From your drawings, determine when the planet is nearer or farther from Earth than is the Sun.

(c) Deduce from your drawings when Venus is coming toward Earth or is moving away from Earth.

(d) Explain why Venus goes through this cycle of phases.

(e) Select **Favorites > Explorations > Mars Phases** to see an image of Mars. Step time forward in 30-day intervals and observe the changing phase of Mars as seen from Earth. Compare this with the phases that you observed for Venus. Why are the cycles of phases as seen from Earth different for the two planets?

67. Use *Starry Night™* to observe the orbits of the planets of the inner solar system. Open **Favorites > Explorations > Kepler3**. The view is centered on the Sun from a position in space almost 2.5 au above the plane of the solar system and shows the Sun and the inner planets and their orbits, as well as many asteroids in the asteroid belt beyond the orbit of Mars. Three stars are also indicated that will be used as reference points in the last part of this exploration. Click **Play** and observe the motions of the planets from this unique location.

(a) List the planets visible in the view, first in the order of increasing distance from the Sun and then in the order of increasing orbital period. How do these lists compare?

(b) What might you conclude from this observation?

(c) Which of Kepler's laws accounts for this observation?

(d) To explore this law quantitatively, for each of the planets Mercury, Venus, and Mars in turn, use the time controls to place the planet next to the labeled star near the planet's orbit. Then use the angular separation cursor to measure the distance, in au, from the planet to the Sun to obtain an approximate value for the semimajor axis of the planet's orbit. Next, use the time controls to determine the number of days required for the planet to return to this position in its orbit. Convert your measured orbital period from days to years by dividing the value you obtain by 365.25. Finally, calculate the square of this orbital period in years and the cube of the approximate semimajor axis in au, then plot these values on a graph. Include Earth (period 1 year, semimajor axis 1 au) on the graph. Do your results confirm Kepler's law?

ANSWERS

ConceptChecks

ConceptCheck 4-1: When a planet is moving retrograde, the planet can be observed night after night to be slowly drifting from east to west compared to the very distant background stars (Figure 4-2). This is opposite to how a planet typically appears to move. However, regardless of this slow relative movement, all objects always appear to rise in the east and set in the west on a daily basis due to Earth's rotation.

ConceptCheck 4-2: The Greeks' ancient geocentric model pictured a nonspinning, stationary Earth in which the stars, planets, and the Sun all move around Earth.

ConceptCheck 4-3: In the Ptolemaic model, the planets are continuously orbiting in circles such that they appear to move backward for a brief time. However, the planets never actually stop and change their directions (Figure 4-3).

ConceptCheck 4-4: No. A planet only appears to move in retrograde motion when seen from another planet if the two planets move at different speeds and pass one another (Figure 4-5). An imaginary observer on the stationary Sun would only see planets moving in the same direction as they orbit the Sun.

ConceptCheck 4-5: Mars has an orbit around the Sun that is larger than Earth's orbit. As a result, Mars never moves to a position between Earth and the Sun, so Mars is never at inferior conjunction (Figure 4-6).

ConceptCheck 4-6: In Copernicus's heliocentric model, the more distant planets are moving slower than the planets closer to the Sun. As a faster-moving Earth moves past a slower-moving Mars, there is a brief time in which Mars appears to move backward through the sky. However, the planets never actually reverse their directions.

ConceptCheck 4-7: Slow-moving Jupiter does not move very far along its orbit in the amount of time it takes for Earth to pass by Jupiter, move around the Sun, and pass by Jupiter again, giving Jupiter a synodic period similar to Earth's orbital period around the Sun. However, slow-moving Jupiter takes more than a decade to move around the Sun back to its original starting place as measured by the background stars, giving it a large sidereal period.

ConceptCheck 4-8: During Copernicus's life, planetary data was limited to the positions of the planets. The models of Copernicus and Ptolemy made very similar predictions for positions, so there was no reason to favor one model over another at that time. Only much later, when other attributes of the planets became observable (such as the phases of Venus), did the models make radically different predictions that ruled out Ptolemy's model.

ConceptCheck 4-9: An ellipse with an eccentricity of zero is a perfect circle (see Figure 4-10b) and, compared to Mars's $e = 0.093$, the eccentricity of Venus's orbit is $e = 0.007$. Venus has a smaller eccentricity, so its orbit is closer to a perfect circle in shape.

ConceptCheck 4-10: Kepler's second law says that objects are moving slowest when they are farthest from the object they are orbiting, so an Earth-orbiting satellite will move slowest when it is farthest from Earth. Just as the Sun is at one focus for a planet's elliptical orbit, Earth is at one focus for a satellite's elliptical orbit (Figure 4-11).

ConceptCheck 4-11: According to Kepler's third law, planets closer to the Sun move faster than planets farther from the Sun. For objects orbiting Earth, the object closer to Earth is also moving the fastest, which, in this case, is the space shuttle.

ConceptCheck 4-12: No. Kepler's laws of planetary motion apply to any objects in space that orbit around another object, including comets orbiting the Sun, humanmade satellites and moons orbiting planets, and even stars orbiting other stars.

ConceptCheck 4-13: When Venus is on the opposite side of the Sun from Earth, it will be in a full or gibbous phase. The full phase can occasionally be observed because Earth and Venus do not orbit the Sun in the same exact plane.

ConceptCheck 4-14: Galileo was the first person to use and widely share what he learned from his telescope observations, and a telescope is necessary in order to observe Jupiter's tiny moons.

ConceptCheck 4-15: According to Newton's first law, the rock will continue to travel in the same direction and with the same speed, as long as there is no net outside force. Note that this principle applies no matter how massive the rock is, or how fast it's traveling.

ConceptCheck 4-16: As described by Newton's third law, the forces the sumo wrestlers exert on each other are of equal strength, but in opposite directions. This is true even though they are different in size and even though one wrestler is pushed out of the ring. Frictional forces from the floor and their body masses influence which wrestler moves the most, but the forces they exert onto each other are of equal strength.

ConceptCheck 4-17: Expressing Newton's second law in terms of acceleration, the acceleration of an object is given by $a = F/m$.

Since the gravitational forces on the diver and Earth are of equal strength, the smaller mass—the diver—will have the larger acceleration. Because Earth is so much more massive than the diver, Earth's acceleration in this case is nowhere near measurable.

ConceptCheck 4-18: According to Newton's universal law of gravitation, the gravitational attraction between two objects depends on the square of the distance between them. In this case, if the asteroids drift 3 times closer together, then the gravitational force of attraction between them increases 3^2 times, or, in other words, becomes 9 times greater.

ConceptCheck 4-19: At the cannonball's highest point, it is momentarily at rest and has gravitational potential energy but no kinetic energy. Starting from the chemical energy in gunpowder, that chemical energy is transferred to the cannonball, giving it kinetic energy. As the ball rises, kinetic energy is transformed into gravitational potential energy until the ball reaches its peak.

ConceptCheck 4-20: Air drag *removes* orbital energy and could put the spacecraft on the elliptical path D. Firing the rocket thruster *adds* orbital energy and could put the spacecraft on the elliptical path F. A circular orbit requires special circumstances, so it is not surprising that most moons and planets have at least slightly elliptical orbits.

ConceptCheck 4-21: In order for an object to feel air drag, it must pass through the air. Since Earth and our air orbit the Sun together, Earth feels no air drag and we are not slowed in our orbit.

ConceptCheck 4-22: An object's escape speed from a planet does not depend on the object's mass, so both rocks can be shot at the same speed, which is the escape speed from the Moon. As illustrated in Figure 4-20, for two objects with the same speed, the more massive object has more kinetic energy, so the kinetic energy of the two rocks will not be equal if they both leave at exactly the escape speed, and more energy is required to shoot the larger rock off of the Moon.

ConceptCheck 4-23: Yes. The escape speed $v_{escape} = \sqrt{\frac{2GM}{R}}$ depends on both the mass and radius of the planet. If another Earth-mass planet had a larger radius (which is certainly possible), its escape speed would be less than that of Earth.

ConceptCheck 4-24: The object—on its high-energy hyperbolic orbit—would have to lose orbital energy in order to enter a lower-energy elliptical or circular orbit as a moon. Special circumstances are required to lose this much energy, which is why we do not think Earth simply captured our Moon as it passed by.

CalculationChecks

CalculationCheck 4-1: According to Kepler's third law, $P^2 = a^3$. So, if $P^2 = (39.5)^3$, then $P = 39.5^{3/2} = 248$ years.

CalculationCheck 4-2: Using Newton's universal law of gravitation, $F_{\text{Mars-astronaut}} = G(m_{\text{Mars}}) \times (m_{\text{astronaut}}) \div (\text{radius})^2 = 6.67 \times 10^{-11} \times 6.4 \times 10^{23} \times 75 \div (3.4 \times 10^6)^2 = 277$ newtons, which we can convert to pounds because 277 newtons $\times$ 0.255 lb/N = 76 lb.

5 | The Nature of Light

LEARNING GOALS

In the early 1800s, the French philosopher Auguste Comte argued that because the stars are so far away, humanity would never know their nature and composition. Little did he know that this very information was already there for anyone to see, encoded in starlight. Just a few years after Comte's bold pronouncement, investigators began analyzing starlight to learn the very things that he had deemed unknowable.

We now know that atoms of each chemical element emit and absorb light at a unique set of wavelengths characteristic of that element alone. The red light in the chapter opener image of a gas cloud in space is of a wavelength emitted by nitrogen and no other element; the particular green light in this image is unique to oxygen, and the particular blue light is unique to helium. The light from nearby planets, distant stars, and remote galaxies also has characteristic "fingerprints" that reveal the chemical composition of these celestial objects.

In this chapter, we will learn about the basic properties of light. Light has a dual nature: It has properties of both waves and particles. The light emitted by

The Ring Nebula is a shell of glowing gases surrounding a dying star. The spectrum of the emitted light reveals which gases are present. (The Hubble Heritage Team, AURA/STScI/NASA)

R I V U X G

an object depends on the object's temperature; we can use this to determine the surface temperatures of stars. By studying the structure of atoms, we will learn why each element emits and absorbs light only at specific wavelengths and will see how astronomers determine what the atmospheres of planets and stars are made of. The motion of a light source also affects wavelengths, permitting us to deduce how fast stars and other objects are approaching or receding. These are but a few of the reasons why understanding light is a prerequisite to understanding the universe.

5-1   Light travels through empty space at a speed of 300,000 km/s

Galileo Galilei and Isaac Newton were among the first scientific thinkers to ask basic questions about light. Does light travel instantaneously from one place to another, or does it move with a measurable speed? Whatever the nature of light, it does seem to travel swiftly from a source to our eyes. We see a distant flash of lightning before we hear the accompanying thunderclap.

> The speed of light in a vacuum is a universal constant: It has the same value everywhere in the cosmos.

In the early 1600s, Galileo tried to measure the speed of light. He and an assistant stood at night on two hilltops a known distance apart, each holding a shuttered lantern. First, Galileo opened the shutter of his lantern; as soon as his assistant saw the flash of light, he opened his own. Galileo used his pulse as a timer to try to measure the time between opening his lantern and seeing the light from his assistant's lantern. From the distance and time, he hoped to compute the speed at which the light had traveled to the distant hilltop and back.

Galileo found that the measured time failed to increase noticeably, no matter how far away the assistant was stationed. Galileo therefore concluded that the speed of light is too high to be measured by slow human reactions. Thus, he was unable to tell whether or not light travels instantaneously.

The Speed of Light: Astronomical Measurements

The first evidence that light does *not* travel instantaneously was presented in 1676 by Olaus Rømer, a Danish astronomer. Rømer had been studying the orbits of the moons of Jupiter by carefully timing the moments when they passed into or out of Jupiter's shadow. To Rømer's surprise, the timing of these eclipses of Jupiter's moons seemed to depend on the relative positions of Jupiter and Earth. When Earth was far from Jupiter (that is, near conjunction; see Figure 4-6), the eclipses occurred several minutes later than when Earth was close to Jupiter (near opposition).

Rømer realized that this puzzling effect could be explained if light needs time to travel from Jupiter to Earth. When Earth is closest to Jupiter, the image of one of Jupiter's moons disappearing into Jupiter's shadow arrives at our telescopes a little sooner than it does when Jupiter and Earth are farther apart (Figure 5-1). The range of variation in the times at which such eclipses are observed is about 16.6 minutes, which Rømer interpreted as the length of time required for light to travel across the diameter of Earth's orbit

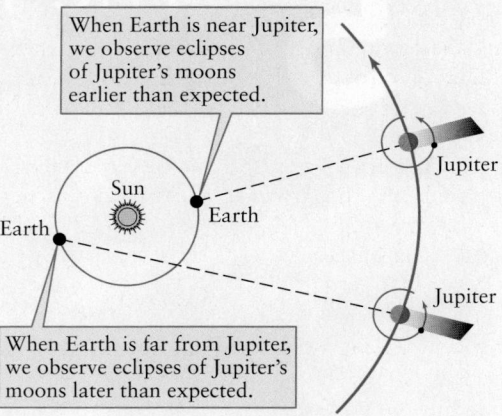

FIGURE 5-1

Rømer's Evidence That Light Does Not Travel Instantaneously The timing of eclipses of Jupiter's moons as seen from Earth depends on the Earth–Jupiter distance. Rømer correctly attributed this effect to variations in the time required for light to travel from Jupiter to Earth.

(a distance of 2 au). The size of Earth's orbit was not accurately known in Rømer's day, and he never actually calculated the speed of light. Today, using the modern value of 150 million kilometers for the astronomical unit, Rømer's method yields a value for the speed of light equal to roughly 300,000 km/s (186,000 mi/s).

The Speed of Light: A Universal Speed Limit

Almost two centuries after Rømer, the speed of light was measured very precisely using an ingenious experiment. In 1850, the French physicists Armand-Hippolyte Fizeau and Jean Foucault built the apparatus sketched in **Figure 5-2**. Light from a light source reflects from a rotating mirror toward a stationary mirror 20 meters away. The rotating mirror moves slightly while the light is making the round trip, so the returning light ray is deflected away from the source by a small angle. By measuring this angle and knowing the dimensions of their apparatus, Fizeau and Foucault could

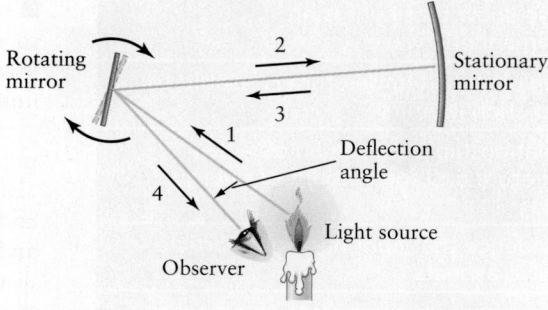

FIGURE 5-2

The Fizeau–Foucault Method of Measuring the Speed of Light Light from a light source (1) is reflected off a rotating mirror to a stationary mirror (2) and from there back to the rotating mirror (3). The ray that reaches the observer (4) is deflected away from the path of the initial beam because the rotating mirror has moved slightly while the light was making the round trip. The speed of light is calculated from the deflection angle and the dimensions of the apparatus.

deduce the speed of light. Once again, the answer was very nearly 300,000 km/s. Based on modern measurements, whether measuring the speed of light on Earth, in the solar system, or amongst galaxies, the speed of light appears to be the same throughout the known universe.

The speed of light in a vacuum (a space devoid of matter) is usually designated by the letter c (from the Latin *celeritas*, meaning "speed"). The modern value is $c = 299{,}792.458$ km/s ($186{,}282.397$ mi/s). In most calculations you can use

$$c = 3.00 \times 10^5 \text{ km/s} = 3.00 \times 10^8 \text{ m/s}$$

The most convenient set of units to use for c is different in different situations. The value in kilometers per second (km/s) is often most useful when comparing c to the speeds of objects in space, whereas the value in meters per second (m/s) is preferred when doing calculations involving the wave nature of light (which we will discuss in Section 5-2).

The speed of light is extremely fast. In 1 second, light travels a distance equal to about 7½ trips around Earth's equator! Since there are more than 31 million seconds in a year, 1 light-year—the distance light travels in 1 year—is about 6 trillion miles.

The speed of light in empty space is one of the most important numbers in modern physical science. This value appears in many equations that describe atoms, gravity, electricity, and magnetism. According to Einstein's special theory of relativity, there is a universal speed limit: *Nothing can travel faster than the speed of light.*

CAUTION!

Note that the quantity c is the speed of light *in a vacuum*. Light travels more slowly through air, water, glass, or any other transparent substance than it does in a vacuum. In our study of astronomy, however, we will almost always consider light traveling through the vacuum (or near-vacuum) of space.

CONCEPT CHECK 5-1

Why has the speed of light historically been so difficult to measure?

Answer appears at the end of the chapter.

5-2 Light is electromagnetic radiation and is characterized by its wavelength

TUTORIAL 5.1

Light is energy. This fact is apparent to anyone who has felt the warmth of the sunshine on a summer's day. But what exactly is light? How is it produced? What is it made of? How does it move through space? Scholars have struggled with these questions throughout history.

> Visible light, radio waves, and x rays are all the same type of wave: They differ only in their wavelength.

Newton and the Nature of Color

The first major breakthrough in understanding light came from a simple experiment performed by Isaac Newton around 1670. Newton was familiar with what he called the "celebrated Phenomenon of Colours," in which a beam of sunlight passing through a glass prism spreads out into the colors of the rainbow (Figure 5-3). This rainbow is called a **spectrum** (plural **spectra**).

Until Newton's time, it was thought that a prism somehow added colors to white light. To test this idea, Newton placed a second prism so that just one color of the spectrum passed through it. In Figure 5-4, only yellow light can pass through a slit to the second prism. According to the old theory, this should have caused a further change in the color of the light. But Newton found that each color of the spectrum was unchanged by the second prism;

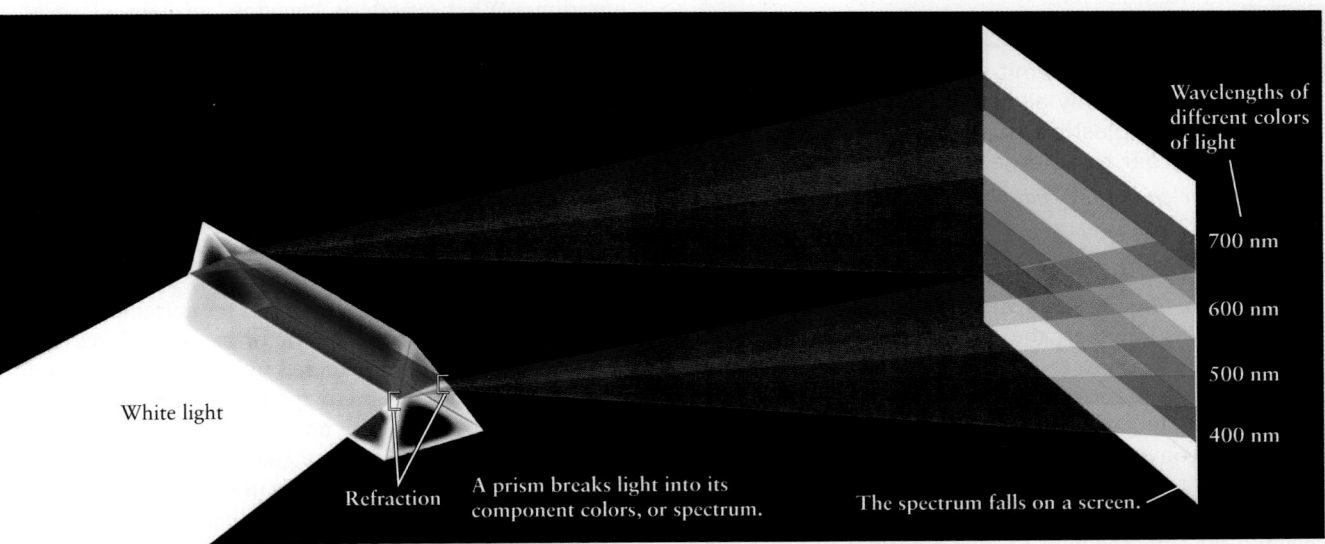

White light

Refraction

A prism breaks light into its component colors, or spectrum.

The spectrum falls on a screen.

Wavelengths of different colors of light

700 nm

600 nm

500 nm

400 nm

FIGURE 5-3

A Prism and a Spectrum When a beam of sunlight passes through a glass prism, the light is broken into a rainbow-colored band called a spectrum.

The wavelengths of different colors of light are shown on the right (1 nm = 1 nanometer = 10^{-9} m).

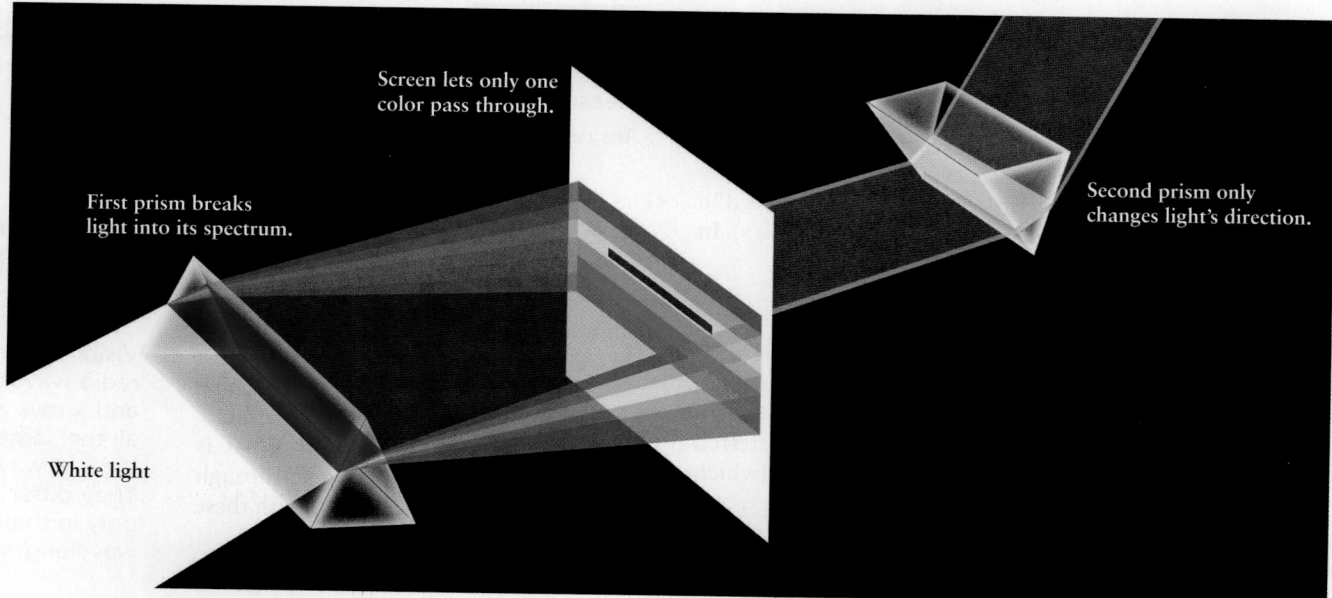

FIGURE 5-4

Newton's Experiment on the Nature of Light In a crucial experiment, Newton took sunlight that had passed through a prism and sent it through a second prism. Between the two prisms was a screen with a hole in it that allowed only one color of the spectrum to pass through. This same color emerged from the second prism. Newton's experiment proved that prisms do not add color to light but merely bend different colors through different angles. It also showed that white light, such as sunlight, is actually a combination of all the colors that appear in its spectrum.

yellow remained yellow, blue remained blue, and so on. He concluded that a prism merely separates colors and does not add color. Hence, the spectrum produced by the first prism shows that *sunlight is a mixture of all the colors of the rainbow*. We often call sunlight "white light," but keep in mind that white light is just our brain's perception of sunlight's mixed colors.

CONCEPT CHECK 5-2

The rear brake lights on a car emit white light, but are covered with plastic that only allows red light to pass through. What color of plastic would the red light have to pass through in order to emerge as green light?

Answer appears at the end of the chapter.

Young and the Wave Nature of Light

Newton suggested that light is made of particles too small to detect individually. In 1678, however, the Dutch physicist and astronomer Christiaan Huygens proposed a rival explanation. He suggested that light travels in the form of waves rather than particles.

Around 1801, the English scientist Thomas Young carried out an experiment that convincingly demonstrated the wavelike aspect of light. He passed a beam of light through two thin, parallel slits in a plate, as shown in Figure 5-5a. On a white screen some distance beyond the slits, the light formed a pattern of alternating bright and dark bands. Young reasoned that if a beam of light was a stream of particles (as Newton had suggested), the two beams of light from the slits would simply form bright images of the slits on the white surface. The pattern of bright and dark bands he observed, however, is just what would be expected if light had wavelike properties. An analogy with water waves demonstrates why (Figure 5-5b).

Maxwell and Light as an Electromagnetic Wave

The discovery of the wave nature of light posed some obvious questions. What exactly is "waving" in light? That is, what is it about light that goes up and down like water waves on the ocean? Because we can see light from the Sun, planets, and stars, light waves must be able to travel across empty space. Hence, whatever is "waving" cannot be any material substance. What, then, is it?

The answer came from a seemingly unlikely source—a comprehensive theory describing electricity and magnetism. Numerous experiments during the first half of the nineteenth century demonstrated an intimate connection between electric and magnetic forces. A central idea to emerge from these experiments is the concept of a *field*, an immaterial yet measurable disturbance of any region of space in which electric or magnetic forces are felt. Thus, an electric charge is surrounded by an electric field, and a magnet is surrounded by a magnetic field. Experiments in the early 1800s demonstrated that moving an electric charge produces a magnetic field; conversely, moving a magnet gives rise to an electric field.

In the 1860s, the Scottish mathematician and physicist James Clerk Maxwell succeeded in describing all the basic properties of electricity and magnetism in four equations. This mathematical achievement demonstrated that electric and magnetic forces are really two aspects of the same phenomenon, which we now call **electromagnetism.**

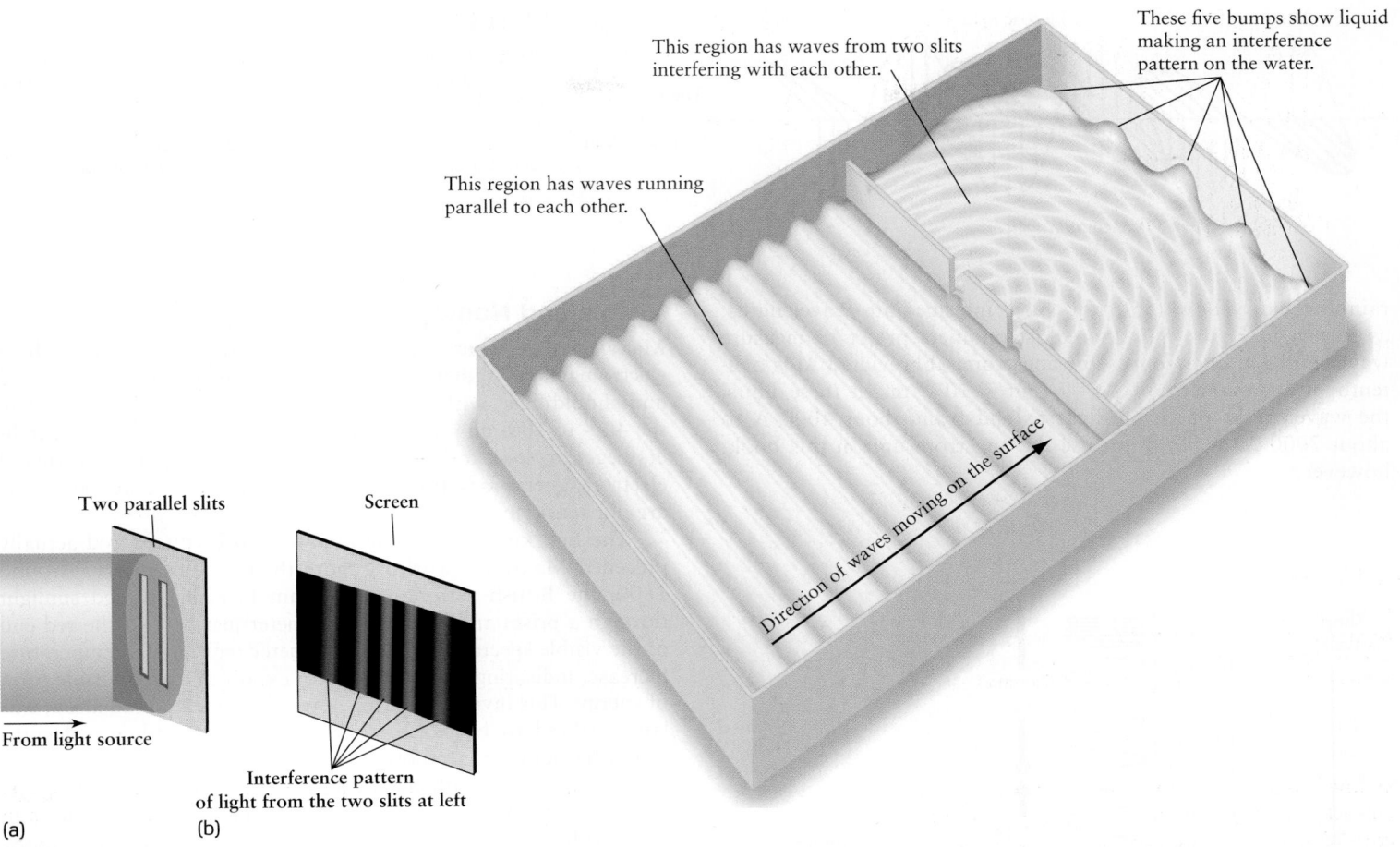

This region has waves from two slits interfering with each other.

These five bumps show liquid making an interference pattern on the water.

This region has waves running parallel to each other.

Direction of waves moving on the surface

Two parallel slits

Screen

From light source

Interference pattern of light from the two slits at left

(a)   (b)

FIGURE 5-5

Wave Interference **(a)** Electromagnetic radiation travels as waves. Thomas Young's interference experiment shows that light of a single color passing through a barrier with two slits behaves as waves that create alternating light and dark patterns on a screen. **(b)** Water waves passing through two slits creates two overlapping waves on the right side of a ripple tank; the result is an interference pattern. Constructive interference occurs where the crests (high points) from waves combine together. Destructive interference occurs where the crest from one wave combines with the trough (low point) of another, leaving the water's surface undisturbed. In the experiment with light, the bright bands result from constructive interference, and the dark bands from destructive interference, which shows that light is made of waves. (© University of Colorado, Center for Intergrated Plasma Studies, Boulder, CO)

By combining his four equations, Maxwell showed that electric and magnetic fields should travel through space in the form of waves at a speed of 3.0×10^5 km/s—a value equal to the best available value for the speed of light. Maxwell's suggestion that these waves do exist and are observed as light was soon confirmed by experiments. Because of its electric and magnetic properties, light is also called **electromagnetic radiation.**

CAUTION!

You may associate the term *radiation* with radioactive materials like uranium, but this term refers to anything that radiates, or spreads away, from its source. For example, scientists sometimes refer to sound waves as "acoustic radiation." Radiation does not have to be related to radioactivity!

Electromagnetic radiation consists of oscillating electric and magnetic fields, as shown in **Figure 5-6.** The distance between two successive wave crests is called the **wavelength** of the light, usually designated by the Greek letter λ (lambda). It is essential to keep in mind that no matter what the wavelength, all electromagnetic radiation travels at the same speed in a vacuum: $c = 3.0 \times 10^5$ km/s $= 3.0 \times 10^8$ m/s.

More than a century elapsed between Newton's experiments with a prism and the confirmation of the wave nature of light. One reason for this delay is that **visible light,** the light to which the human eye is sensitive, has extremely short wavelengths—less than a thousandth of a millimeter—that are not easily detectable. To express such tiny distances conveniently, scientists use a unit of length called the **nanometer** (abbreviated nm), where 1 nm $= 10^{-9}$ m. Experiments demonstrated that visible light has wavelengths covering the range from about 400 nm for violet light to about 700 nm for red light. Intermediate colors of the

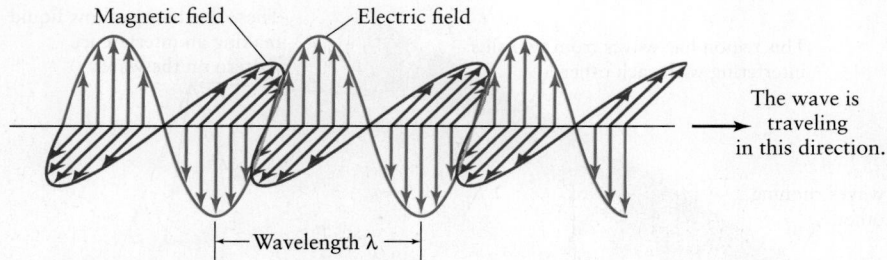

Magnetic field | Electric field

The wave is traveling in this direction.

←— Wavelength λ —→

FIGURE 5-6

Electromagnetic Radiation All forms of light consist of oscillating electric and magnetic fields that move through space at a speed of 3.00×10^5 km/s = 3.00×10^8 m/s. This figure shows a "snapshot" of these fields at one instant. The distance between two successive crests, called the wavelength of the light, is usually designated by the Greek letter λ (lambda).

rainbow like yellow (550 nm) have intermediate wavelengths, as shown in Figure 5-7. (Some astronomers prefer to measure wavelengths in *angstroms*. One angstrom, abbreviated Å, is one-tenth of a nanometer: 1 Å = 0.1 nm = 10^{-10} m. In these units, the wavelengths of visible light extend from about 4000 Å to about 7000 Å. We will not use the angstrom unit in this book, however.)

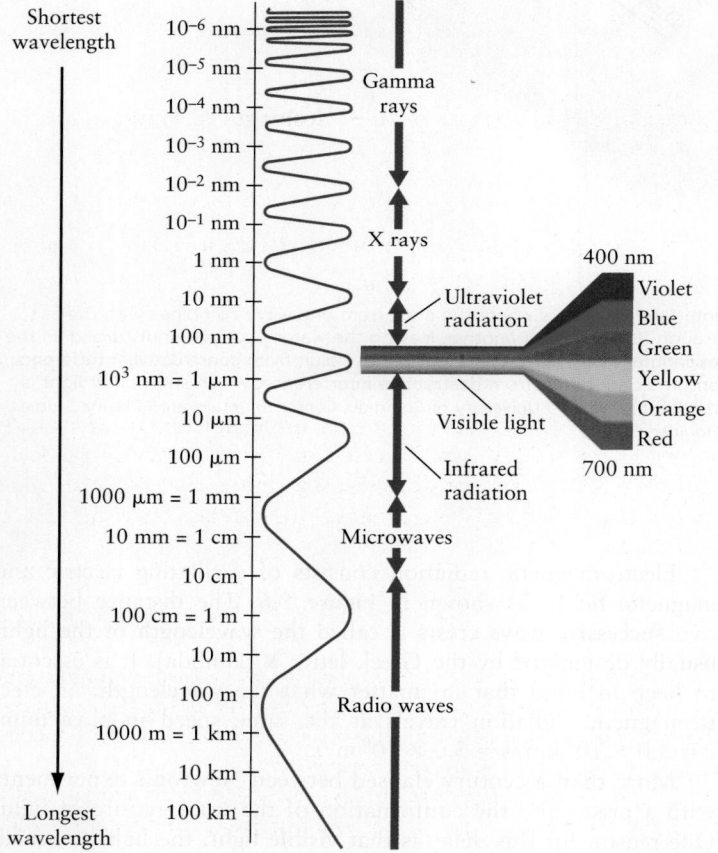

Shortest wavelength

10^{-6} nm
10^{-5} nm
10^{-4} nm Gamma rays
10^{-3} nm
10^{-2} nm
10^{-1} nm
1 nm X rays
10 nm
100 nm Ultraviolet radiation
10^3 nm = 1 μm
10 μm Visible light
100 μm Infrared radiation
1000 μm = 1 mm
10 mm = 1 cm Microwaves
10 cm
100 cm = 1 m
10 m
100 m Radio waves
1000 m = 1 km
10 km
100 km

Longest wavelength

400 nm
Violet
Blue
Green
Yellow
Orange
Red
700 nm

FIGURE 5-7

The Electromagnetic Spectrum The full array of all types of electromagnetic radiation is called the electromagnetic spectrum. It extends from the longest-wavelength radio waves to the shortest-wavelength gamma rays. Visible light occupies only a tiny portion of the full electromagnetic spectrum.

Visible and Nonvisible Light

Maxwell's equations place no restrictions on the wavelength of electromagnetic radiation. Hence, electromagnetic waves could and should exist with wavelengths both longer and shorter than the 400- to 700-nm range of visible light. Consequently, researchers began to look for *invisible* forms of light. These are forms of electromagnetic radiation to which the cells of the human retina are not sensitive.

The first kind of invisible radiation to be discovered actually preceded Maxwell's work by more than a half century. Around 1800, the British astronomer William Herschel passed sunlight through a prism and held a thermometer just beyond the red end of the visible spectrum. The thermometer registered a temperature increase, indicating that it was being exposed to an invisible form of energy. This invisible energy, now called **infrared radiation,** was later realized to be electromagnetic radiation with wavelengths somewhat longer than those of visible light.

In experiments with electric sparks in 1888, the German physicist Heinrich Hertz succeeded in producing electromagnetic radiation with even longer wavelengths of a few centimeters or more. These are now known as **radio waves.** In 1895, another German physicist, Wilhelm Röntgen, invented a machine that produces electromagnetic radiation with wavelengths shorter than 10 nm, now known as **x rays.** The x-ray machines in modern medical and dental offices are direct descendants of Röntgen's invention. Over the years, radiation has been discovered with many other wavelengths.

Thus, visible light occupies only a tiny fraction of the full range of possible wavelengths, collectively called the **electromagnetic spectrum.** As Figure 5-7 shows, the electromagnetic spectrum stretches from the longest-wavelength radio waves to the shortest-wavelength gamma rays. Figure 5-8 shows some applications of nonvisible light in modern technology.

On the long-wavelength side of the visible spectrum, infrared radiation covers the range from about 700 nm to 1 mm. Astronomers interested in infrared radiation often express wavelength in *micrometers* or *microns*, abbreviated μm, where 1 μm = 10^{-3} mm = 10^{-6} m. **Microwaves** have wavelengths from roughly 1 mm to 10 cm, while radio waves have even longer wavelengths.

At wavelengths shorter than those of visible light, **ultraviolet radiation** extends from about 400 nm down to 10 nm. Next are x rays, which have wavelengths between about 10 and 0.01 nm, and beyond them at even shorter wavelengths are **gamma rays.** Note that the rough boundaries between different types of radiation are simply arbitrary divisions in the electromagnetic spectrum.

Once again, in the vacuum of space, all light—from radio waves to gamma rays—travels at the speed of light, $c = 3 \times 10^8$ m/s.

(a) Mobile phone: radio waves

(b) Microwave oven: microwaves

(c) TV remote: infrared light

(d) Tanning booth: ultraviolet light

(e) Medical imaging: x rays

(f) Cancer radiotherapy: gamma rays

FIGURE 5-8

Uses of Nonvisible Electromagnetic Radiation (a) A mobile phone is actually a radio transmitter and receiver. The wavelengths used are in the range of 16 to 36 cm. **(b)** A microwave oven produces radiation with a wavelength near 10 cm. The water in food absorbs this radiation, thus heating the food. **(c)** A remote control sends commands to a television using a beam of infrared light. **(d)** Ultraviolet radiation in moderation gives you a suntan, but in excess can cause sunburn or skin cancer. **(e)** X rays can penetrate through soft tissue but not through bone, which makes them useful for medical imaging. **(f)** Gamma rays destroy cancer cells by breaking their DNA molecules, making them unable to multiply. (a: dpa picture alliance archive/Alamy Stock Photo; b: Michael Haegele/Corbis/Getty Images; c: Bill Losh/The Image Bank/Getty Images; d: Neil McAllister/Alamy Stock Photo; e: Ted Kinsman/Science Source; f: Will & Deni McIntyre/Science Source)

CONCEPT CHECK 5-3

Which form of electromagnetic radiation has a wavelength similar to the diameter of your finger?

Answer appears at the end of the chapter.

Frequency and Wavelength

Astronomers who work with radio telescopes often prefer to speak of *frequency* rather than wavelength. The **frequency** of a wave is the number of wave crests that pass a given point in 1 second. Equivalently, it is the number of complete *cycles* of the wave that pass per second (a complete cycle is from one crest to the next). Frequency is usually denoted by the letter f. The unit of frequency is the cycle per second, also called the *hertz* (abbreviated Hz) in honor of Heinrich Hertz, the physicist who first produced radio waves. For example, if 500 crests of a wave pass you in 1 second, the frequency of the wave is 500 cycles per second, or 500 Hz.

In working with frequencies, it is often convenient to use the prefix *mega-* (meaning "million," or 10^6, and abbreviated M) or *kilo-* (meaning "thousand," or 10^3, and abbreviated k). For example, AM radio stations broadcast at frequencies between 535 and 1605 kHz (kilohertz), while FM radio stations broadcast at frequencies in the range from 88 to 108 MHz (megahertz).

The relationship between the frequency and wavelength of an electromagnetic wave is a simple one. Because light moves at a constant speed $c = 3 \times 10^8$ m/s, if the wavelength (distance from one crest to the next) is made shorter, the frequency must increase (more of those closely spaced crests pass you each second). Thus, the shorter the wavelength of light, the higher the frequency; likewise, the longer the wavelength of light, the lower the frequency. The frequency f of light is related to its wavelength λ by the equation:

Frequency and wavelength of an electromagnetic wave

$$f = \frac{c}{\lambda}$$

f = frequency of an electromagnetic wave (in Hz)

c = speed of light = 3×10^8 m/s

λ = wavelength of the wave (in meters)

That is, the frequency of a wave equals the wave speed divided by the wavelength.

For example, hydrogen atoms in space emit radio waves with a wavelength of 21.12 cm. To calculate the frequency of this radiation, we must first express the wavelength in meters rather than centimeters: $\lambda = 0.2112$ m. Then we can use the above formula to find the frequency f:

$$f = \frac{c}{\lambda} = \frac{3 \times 10^8 \text{ m/s}}{0.2112 \text{ m}} = 1.42 \times 10^9 \text{ Hz} = 1420 \text{ MHz}$$

Visible light has a much shorter wavelength and higher frequency than radio waves. You can use the above formula to show that for yellow-orange light of wavelength 600 nm, the frequency is 5×10^{14} Hz, or 500 *million* megahertz!

While Young's experiment (Figure 5-5) showed convincingly that light has wavelike aspects, it was discovered in the early 1900s that light *also* has some of the characteristics of a stream of particles. We will explore light's dual nature in Section 5-5.

CONCEPT CHECK 5-4

How do the frequencies of the longest wavelengths of light compare to the frequencies of the shortest wavelengths of light?

CALCULATION CHECK 5-1

What is the wavelength of radio waves for the FM radio station 99.9 KTYD (which is 99.9 MHz)?

Answers appear at the end of the chapter.

5-3 An opaque object emits electromagnetic radiation according to its temperature

To learn about objects in the heavens, astronomers study the character of the electromagnetic radiation coming from those objects. Such studies can be very revealing because different kinds of electromagnetic radiation are typically produced in different ways. As an example, on Earth the most common way to generate radio waves is to make an electric current oscillate back and forth (as is done in the broadcast antenna of a radio station). By contrast, x rays for medical and dental purposes are usually produced by bombarding atoms in a piece of metal with fast-moving particles extracted from other atoms. Our own Sun emits radio waves from near its glowing surface and x rays from its corona (see the photo that opens Chapter 3). Hence, these observations indicate the presence of electric currents near the Sun's surface and of fast-moving particles in the Sun's outermost regions. (We will discuss the Sun at length in Chapter 18.)

> As an object is heated, it glows more brightly and its peak color shifts to shorter wavelengths.

Radiation from Heated Objects

TUTORIAL 5.2 The simplest and most common way to produce electromagnetic radiation, either on or off Earth, is to heat an object. The hot filament of wire inside an old-style incandescent lightbulb emits white light, and a neon sign has a characteristic red glow because neon gas within the tube is heated by an electric current. In like fashion, almost all the visible light that we receive from space comes from hot objects like the Sun and the stars. The kind and amount of light emitted by a hot object tell us not only how hot it is but also about other properties of the object.

We can tell whether the hot object is made of relatively dense or relatively thin material. Consider the difference between a lightbulb and a neon sign. The dense, solid filament of a lightbulb makes white light, which is a mixture of all different visible wavelengths, whereas the thin, transparent neon gas produces light of a rather definite red color and, hence, a rather definite wavelength. For now we will concentrate our attention on the light produced by dense, *opaque* objects. An *opaque* object does not allow light to travel through it without being scattered or absorbed—opaque is the opposite of transparent. Earth's atmosphere absorbs and scatters only some of the incoming sunlight, and we say that it is partially opaque, or partially transparent. Even though the Sun

and stars are gaseous, not solid, it turns out that because they are so large, light from the inside scatters many times before emerging from the surface. Therefore, stars are opaque and emit light with many of the same properties as light emitted by a hot, glowing, solid object.

Imagine a welder or blacksmith heating a bar of iron. As the bar becomes hot, it begins to glow deep red, as shown in Figure 5-9a. You can see this same glow from the coils of a toaster or from an electric range turned on "high." While the bar emits a range of wavelengths, the dominant wavelength is red in color. As the temperature rises further, the bar begins to give off a brighter orange light (Figure 5-9b). At still higher temperatures, it shines with a brilliant yellow color (Figure 5-9c); the dominant wavelength is yellow light. If the bar could be prevented from melting and vaporizing, at extremely high temperatures it would emit a dazzling blue-white light. As shown in Figure 5-9, the temperature of a star determines its color, and the hottest stars are blue.

As this example shows, the amount of energy emitted by the hot, dense object and the dominant wavelength of the emitted radiation both depend on the temperature of the object. The hotter the object, the more energy it emits and the shorter the wavelength at which most of the energy is emitted. Colder objects emit relatively little energy, and this emission is primarily at long wavelengths.

These observations explain why you cannot see in the dark. The temperatures of people, animals, and furniture are much less than even that of the iron bar in Figure 5-9a. So, while these objects emit radiation even in a darkened room, most of this emission is at wavelengths greater than those of red light, in the infrared part of the spectrum (see Figure 5-7). Your eye is not sensitive to infrared light, and you thus cannot see ordinary objects in a darkened room. But you can detect this radiation by using a camera that is sensitive to infrared light (Figure 5-10).

Temperature and Thermal Energy

To better understand the relationship between the temperature of a dense object and the radiation it emits, it is helpful to know just what "temperature" means. The temperature of a substance is directly related to the average speed of the tiny atoms or molecules that make up the substance. As shown in Figure 5-11, the hotter the substance, the faster its atoms or molecules are moving; the colder the substance, the slower the motion of its particles. The **thermal energy** of an object comes from the kinetic energy of its atoms and molecules. Therefore, the hotter a substance, the greater its thermal energy.

Scientists usually prefer to use the Kelvin temperature scale, on which temperature is measured in **kelvins** (K) upward from **absolute zero**. This is the coldest possible temperature, at which atoms move as slowly as possible (they can never quite stop completely). On the more familiar Celsius and Fahrenheit temperature scales, absolute zero (0 K) is −273°C and −460°F. Ordinary room temperature is 293 K, 20°C, or 68°F. Box 5-1 discusses the relationships among the Kelvin, Celsius, and Fahrenheit temperature scales.

(a) Hot: glows deep red   (b) Hotter: glows orange   (c) Even hotter: glows yellow

FIGURE 5-9 R I **V** U X G

Objects at Different Temperatures Have Different Colors and Brightnesses This sequence of photographs shows the changing appearance of a piece of iron as it is heated. As the temperature increases, the amount of energy radiated by the bar increases, and it appears brighter. The apparent color of the bar also changes because, as the temperature increases, the dominant or peak wavelength of light emitted by the bar decreases. The stars shown have roughly the same temperatures as the bars above them. (Top row: Richard Megna/Fundamental Photographs; bottom row: NASA)

FIGURE 5-10 R **I** V U X G

An Infrared Portrait In this image made with a camera sensitive to infrared radiation, the different colors represent regions of different temperature. That this image is made from infrared radiation is indicated by the highlighted I in the wavelength tab. Red areas (like the man's face) are the warmest and emit the most infrared light, while blue-green areas (including the man's hands and hair) are at the lowest temperatures and emit the least radiation. (Dr. Arthur Tucker/Science Source)

Gas   Gas

Solid or liquid   Solid or liquid

Colder: ## Hotter:

Slower vibration of   Faster vibration of
particles and lower   particles and higher
thermal energy   thermal energy

Ice water   Hot coffee

FIGURE 5-11

Thermal Energy The thermal energy of an object comes from the kinetic energy of its internal particles, and the hotter the material, the faster the particles move around. For hotter solids and liquids, particles vibrate faster and collide more intensely with their neighbors; for hotter gases, the particles stream around at higher average speeds.

Figure 5-12 depicts quantitatively how the radiation from a dense object depends on its Kelvin temperature. Each curve in this figure shows the intensity of light emitted at each wavelength by a dense object at a given temperature: 3000 K (the temperature of a "cool" star, and at which molten gold boils), 6000 K (about the temperature of our Sun, and of an iron-welding arc), and 12,000 K (the temperature of a "hot" star, and also found in special industrial furnaces). In other words, the curves show the spectrum of light emitted by such an object. At any temperature, a hot, dense object emits at all wavelengths, so its spectrum is a smooth, continuous curve with no gaps in it.

As we will see shortly, this is an example of *blackbody radiation*. Figure 5-12 shows that *the higher the temperature, the greater the intensity of light at all wavelengths*. One way to understand this response to temperature is that the emitted light results from the motion of the material's atoms and molecules, which, as we saw, increases with temperature (see Figure 5-11).

The temperature not only indicates the total intensity of the light, but also the shape of its spectrum. The most important feature in the spectrum—the dominant wavelength—is called the **wavelength of maximum emission,** at which the curve has its peak and the intensity of emitted energy is strongest (see Figure 5-12). The location of the peak also changes with temperature: *The higher the temperature, the shorter the wavelength of maximum emission.*

Figure 5-12 shows that for a dense object at a temperature of 3000 K, the wavelength of maximum emission is about 1000 nm (1 mm). Because this peak wavelength corresponds to the infrared range and is well outside the visible range, you might think that you cannot see the radiation from an object at this temperature.

BOX 5-1 TOOLS OF THE ASTRONOMER'S TRADE

Temperatures and Temperature Scales

Three temperature scales are in common use. Throughout most of the world, temperatures are expressed in **degrees Celsius** (°C). The Celsius temperature scale is based on the behavior of water, which freezes at 0°C and boils at 100°C at sea level on Earth. This scale is named after the Swedish astronomer Anders Celsius, who proposed it in 1742.

Astronomers usually prefer the Kelvin temperature scale. This is named after the nineteenth-century British physicist Lord Kelvin, who made many important contributions to our understanding of heat and temperature. Absolute zero, the temperature at which atomic motion is at the absolute minimum, is −273°C in the Celsius scale but 0 K in the Kelvin scale. Atomic motion cannot be any less than the minimum, so nothing can be colder than 0 K; hence, there are no negative temperatures on the Kelvin scale. Note that we do *not* use degrees (°) with the Kelvin temperature scale.

A temperature expressed in kelvins is always equal to the temperature in degrees Celsius plus 273. On the Kelvin scale, water freezes at 273 K and boils at 373 K. Water must be heated through a change of 100 K or 100°C to go from its freezing point to its boiling point. Thus, the "size" of a kelvin is the same as the "size" of a Celsius degree. When considering temperature changes, measurements in kelvins and Celsius degrees are the same. For extremely high temperatures, the 100°C difference between the Kelvin and Celsius scales becomes so tiny that they are essentially the same: For example, the Sun's core temperature is either 1.55×10^7 K or 1.55×10^{7}°C.

The now-archaic Fahrenheit scale, which expresses temperature in **degrees Fahrenheit** (°F), is used only in the United States. When the German physicist Gabriel Fahrenheit introduced this scale in the early 1700s, he intended 100°F to represent the temperature of a healthy human body. On the Fahrenheit scale, water freezes at 32°F and boils at 212°F. There are 180 Fahrenheit degrees between the freezing and boiling points of water, so a degree Fahrenheit is only $100/180 = 5/9$ as large as either a Celsius degree or a kelvin.

Two simple equations allow you to convert a temperature from the Celsius scale to the Fahrenheit scale and from Fahrenheit to Celsius:

$$T_F = \frac{9}{5}\, T_C + 32$$

$$T_C = \frac{5}{9}\,(T_F - 32)$$

T_F = temperature in degrees Fahrenheit
T_C = temperature in degrees Celsius

EXAMPLE: A typical room temperature is 68°F. We can convert this to the Celsius scale using the second equation:

$$T_C = \frac{5}{9}\,(68 - 32) = 20°C$$

To convert this to the Kelvin scale, we simply add 273 to the Celsius temperature. Thus,

$$68° = 20°C = 293 \text{ K}$$

The diagram displays the relationships among these three temperature scales.

However, the glow from such an object *is* visible; the curve shows that this object emits light within the visible range, although about 90% of the emitted energy comes out at longer, invisible wavelengths (to the right of the visible spectrum). This invisible light is also emitted by lightbulbs. Incandescent lightbulbs create visible light by electrically heating a thin wire filament to about 3000 K, but because most of the energy is wasted at longer wavelengths, more efficient lights have become popular.

The 3000-K curve in Figure 5-12 is quite a bit higher at the red end of the visible spectrum than at the violet end, and a dense object at this temperature will appear yellowish in color. At even lower temperatures below 1000 K, the object would appear red (see Figure 5-9a). For higher temperatures, the 12,000-K curve has its wavelength of maximum emission in the ultraviolet part of the spectrum, at a wavelength shorter than visible light. But such a hot, dense object also emits copious amounts of visible light

The higher the temperature of a blackbody, the shorter the wavelength of maximum emission (the wavelength at which the curve peaks).

The higher the temperature of a blackbody, the more light is emitted at *all* wavelengths.

ANIMATION 5.1

FIGURE 5-12

Blackbody Curves Each of these curves shows the intensity of light at every wavelength that is emitted by a blackbody (an idealized case of a dense object) at a particular temperature. The rainbow-colored band shows the range of visible wavelengths. The vertical scale has been compressed so that all three curves can be seen; the peak intensity for the 12,000-K curve is actually about 1000 times greater than the peak intensity for the 3000-K curve.

(much more than at 6000 K or 3000 K, for which the curves are lower) and thus will have a very visible glow. The curve for this temperature is higher for blue light than for red light, and so the color of a dense object at 12,000 K is a brilliant blue or blue-white. The same principles apply to stars of different temperatures, which accounts for the different colors of stars in Figure 5-9. A star that looks blue has a high surface temperature, while an orange star has a relatively cool surface.

To summarize these observations:

The higher an object's temperature, the more intensely the object emits electromagnetic radiation and the shorter the wavelength at which it emits most strongly.

We will make frequent use of this general rule to analyze the temperatures of celestial objects such as planets and stars.

The curves in Figure 5-12 are drawn for an idealized type of dense object called a **blackbody.** A perfect blackbody does not reflect any light at all; instead, it absorbs all radiation falling on it. Because it reflects no electromagnetic radiation, the radiation that it does emit is entirely the result of its temperature. Ordinary objects,

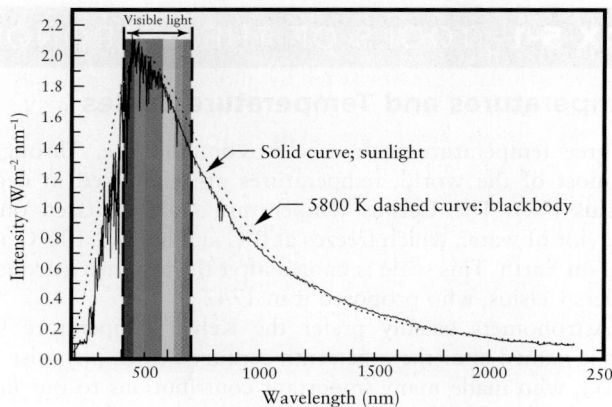

FIGURE 5-13

The Sun as a Blackbody This graph shows that the intensity of sunlight over a wide range of wavelengths (solid curve) is a remarkably close match to the intensity of radiation coming from a blackbody at a temperature of 5800 K (dashed curve). The measurements of the Sun's intensity were made above Earth's atmosphere (which absorbs and scatters certain wavelengths of sunlight). It is not surprising that the range of visible wavelengths includes the peak of the Sun's spectrum; the human eye evolved to take advantage of the most plentiful light available.

like tables, textbooks, and people, are not perfect blackbodies; they reflect light, which is why they are visible. A star such as the Sun, however, behaves very much like a perfect blackbody, because it absorbs almost completely any radiation falling on it from outside. The light emitted by a blackbody is called **blackbody radiation,** and the curves in Figure 5-12 are often called **blackbody curves.**

CAUTION!

Despite its name, a blackbody does not necessarily *look* black. The Sun, for instance, does not look black because its temperature is high (about 5800 K), and so it glows brightly. But a room-temperature (about 300 K) blackbody would appear very black indeed. Even if a 300-K blackbody were as large as the Sun, it would emit only about 1/100,000 as much energy: Its blackbody curve is far too low to graph in Figure 5-12. Furthermore, most of the radiation from a room-temperature object is emitted at infrared wavelengths that are too long for our eyes to perceive.

Figure 5-13 shows the blackbody curve for a temperature of 5800 K. It also shows the intensity curve for light from the Sun, as measured from above Earth's atmosphere. (This is necessary because Earth's atmosphere absorbs certain wavelengths.) The peak of both curves is at a wavelength of about 500 nm, near the middle of the visible spectrum. Note how closely the observed intensity curve for the Sun matches the blackbody curve. This is a strong indication that the temperature of the Sun's glowing surface is about 5800 K—a temperature that we can measure across a distance of 150 million kilometers! The close correlation between blackbody curves and the observed intensity curves for most stars is a key reason why astronomers are interested in the physics of blackbody radiation.

Blackbody radiation depends *only* on the temperature of the object emitting the radiation, not on the chemical composition of the object. The light emitted by molten gold at 2000 K is very nearly the same as that emitted by molten lead at 2000 K. Therefore, it might seem that analyzing the light from the Sun or from a star can tell astronomers the object's temperature but not what the star is made of. As Figure 5-13 shows, however, the intensity curve for the Sun (a typical star) is not precisely that of a blackbody. We will see later in this chapter that the *differences* between a star's spectrum and that of a blackbody allow us to determine the chemical composition of the star.

CONCEPT CHECK 5-5

An object is at 3000 K, emitting radiation that peaks at an infrared wavelength of about 1000 nm. Now the object is heated to 12,000 K and only emits a small fraction of its total energy at infrared wavelengths. At which of these temperatures does the object emit more infrared radiation?

CONCEPT CHECK 5-6

Does a 0°C blackbody object give off any radiation?

Answers appear at the end of the chapter.

5-4 Wien's law and the Stefan–Boltzmann law are useful tools for analyzing glowing objects like stars

The mathematical formula that describes the blackbody curves in Figure 5-12 is a rather complicated one. But there are two simpler formulas for blackbody radiation that prove to be very useful in many branches of astronomy. They are used by astronomers who investigate the stars as well as by those who study the planets (which are dense, relatively cool objects that emit infrared radiation). One of these formulas relates the temperature of a blackbody to its wavelength of maximum emission, and the other relates the temperature to the amount of energy that the blackbody emits. These formulas, which we will use throughout this book, restate in precise mathematical terms the qualitative relationships that we described in Section 5-3.

> Two simple mathematical formulas describing blackbodies are essential tools for studying the universe.

Wien's Law

Figure 5-12 shows that the higher the temperature (T) of a blackbody, the shorter its wavelength of maximum emission (λ_{max}). In 1893, the German physicist Wilhelm Wien used ideas about both heat and electromagnetism to make this relationship quantitative. The formula that he derived is now called **Wien's law**:

Wien's law for a blackbody

$$\lambda_{max} = \frac{0.0029 \text{ K m}}{T}$$

λ_{max} = wavelength of maximum emission of the object (in meters)

T = temperature of the object (in kelvins)

According to Wien's law, the wavelength of maximum emission of a blackbody is inversely proportional to its temperature in kelvins. In other words, if the temperature of the blackbody doubles, its wavelength of maximum emission is halved, and vice versa. For example, Figure 5-12 shows blackbody curves for temperatures of 3000 K, 6000 K, and 12,000 K. From Wien's law, a blackbody with a temperature of 6000 K has a wavelength of maximum emission $\lambda_{max} = (0.0029 \text{ K m})/(6000 \text{ K}) = 4.8 \times 10^{-7} \text{ m} = 480 \text{ nm}$, in the visible part of the electromagnetic spectrum. At 12,000 K, or twice the temperature, the blackbody has a wavelength of maximum emission half as great, or $\lambda_{max} = 240 \text{ nm}$, which is in the ultraviolet. At 3000 K, just half our original temperature, the value of λ_{max} is twice the original value—960 nm, which is an infrared wavelength. You can see that these wavelengths agree with the peaks of the curves in Figure 5-12.

CAUTION!

Remember that Wien's law involves the wavelength of maximum emission in *meters*. If you want to convert the wavelength to nanometers, you must multiply the wavelength in meters by $(10^9 \text{ nm})/(1 \text{ m})$.

Wien's law is very useful for determining the surface temperatures of stars. It is not necessary to know how far away the star is, how large it is, or how much energy it radiates into space. All we need to know is the dominant wavelength of the star's electromagnetic radiation.

CONCEPT CHECK 5-7

What single piece of information do astronomers need in order to determine if a star is hotter than our Sun?

CALCULATION CHECK 5-2

Which wavelength of light would our Sun emit most if its temperature were twice its current temperature of 5800 K?

Answers appear at the end of the chapter.

The Stefan–Boltzmann Law

The other useful formula for the radiation from a blackbody involves the total amount of energy the blackbody radiates at all wavelengths. (By contrast, the curves in Figure 5-12 show how much energy a blackbody radiates at each individual wavelength.)

Energy is usually measured in **joules** (J), named after the nineteenth-century English physicist James Joule. A joule is the amount of kinetic energy contained in the motion of a 2-kilogram mass moving at a speed of 1 meter per second. The joule is a convenient unit of energy because it is closely related to the familiar **watt** (W): 1 watt is 1 joule per second, or $1 \text{ W} = 1 \text{ J/s} = 1 \text{ J s}^{-1}$. (The superscript -1 means you are dividing by that quantity.) For example, a 100-watt lightbulb uses energy at a rate of 100 joules per second, or 100 J/s. The energy content of food is also often

BOX 5-2 TOOLS OF THE ASTRONOMER'S TRADE

Using the Laws of Blackbody Radiation

The Sun and stars behave like nearly perfect blackbodies. Wien's law and the Stefan–Boltzmann law can therefore be used to relate the surface temperature of the Sun or a distant star to the energy flux and wavelength of maximum emission of its radiation. The following examples show how to do this.

EXAMPLE: The maximum intensity of sunlight is at a wavelength of roughly 500 nm = 5.0×10^{-7} m. Use this information to determine the surface temperature of the Sun.

Situation: We are given the Sun's wavelength of maximum emission λ_{max}, and our goal is to find the Sun's surface temperature, denoted by $T_\odot$. (The symbol $\odot$ is the standard astronomical symbol for the Sun.)

Tools: We use Wien's law to relate the values of λ_{max} and $T_\odot$.

Answer: As written, Wien's law tells us how to find λ_{max} if we know the surface temperature. To find the surface temperature from λ_{max}, we first rearrange the formula, then substitute the value of λ_{max}:

$$T_\odot = \frac{0.0029 \text{ K m}}{\lambda_{max}} = \frac{0.0029 \text{ K m}}{5.0 \times 10^{-7} \text{ m}} = 5800 \text{ K}$$

Review: This temperature is very high by Earth standards, about the same as an iron-welding arc.

EXAMPLE: Using detectors above Earth's atmosphere, astronomers have measured the average flux of solar energy arriving at Earth. This value, called the **solar constant,** is equal to 1370 W m^{-2}. Use this information to calculate the Sun's surface temperature. (This calculation provides a check on our result from the preceding example.)

Situation: The solar constant is the flux of sunlight as measured at Earth. We want to use the value of the solar constant to calculate $T_\odot$.

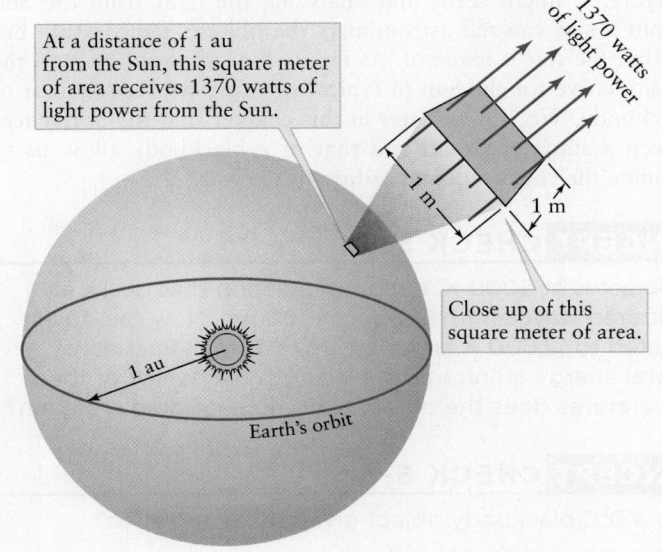

At a distance of 1 au from the Sun, this square meter of area receives 1370 watts of light power from the Sun.

1370 watts of light power

1 m

1 m

Close up of this square meter of area.

1 au

Earth's orbit

Tools: It may seem that all we need is the Stefan–Boltzmann law, which relates flux to surface temperature. However, the quantity F in this law refers to the flux measured at the Sun's surface, *not* at Earth. Hence, we will first need to calculate F from the given information.

Answer: To determine the value of F, we first imagine a huge sphere of radius 1 au with the Sun at its center, as shown in the figure. Each square meter of that sphere receives 1370 watts of power from the Sun, so the total energy radiated by the Sun per second is equal to the solar constant multiplied by the sphere's surface area. The result, called the **luminosity** of the Sun and denoted by the symbol $L_\odot$, is $L_\odot = 3.90 \times 10^{26}$ W. That is, in 1 second the Sun radiates 3.90×10^{26} joules of energy into space. Because we know the size of the Sun, we can compute

(continued on the next page)

measured in joules; in most of the world, diet soft drinks are labeled as "low joule" rather than "low calorie."

The amount of energy emitted by a blackbody depends both on its temperature and on its surface area. These characteristics make sense: A large burning log radiates much more heat than a burning match, even though the temperatures are the same. To consider the effects of temperature alone, it is convenient to look at the amount of energy emitted from each square meter of an object's surface in 1 second. This quantity is called the **energy flux** (F). Flux means "rate of flow," and thus F is a measure of how rapidly energy is flowing out of the object. It is measured in joules per square meter per second, usually written as J/m^2/s or J m^{-2} s^{-1}. Alternatively,

because 1 watt equals 1 joule per second, we can express flux in watts per square meter (W/m^2, or W m^{-2}).

The nineteenth-century Irish physicist John Tyndall performed the first careful measurements of the amount of radiation emitted by a blackbody. (He studied the light from a heated platinum wire, which behaves approximately like a blackbody.) By analyzing Tyndall's results, the Slovenian physicist Josef Stefan deduced in 1879 that the flux from a blackbody is proportional to the fourth power of the object's temperature (measured in kelvins). Five years after Stefan announced his law, Austrian physicist Ludwig Boltzmann showed how it could be derived mathematically from basic assumptions about atoms and molecules. For this reason,

BOX 5-2 *(continued)*

the energy flux (energy emitted per square meter per second) at its surface. The radius of the Sun is $R_\odot = 6.96 \times 10^8$ m, and the Sun's surface area is $4\pi R_\odot^2$. Therefore, its energy flux $F_\odot$ is the Sun's luminosity (total energy emitted by the Sun per second) divided by the Sun's surface area (the number of square meters of surface):

$$F_\odot = \frac{L_\odot}{4\pi R_\odot^2} = \frac{3.90 \times 10^{26} \text{ W}}{4\pi (6.96 \times 10^8 \text{ m})^2} = 6.41 \times 10^7 \text{ W m}^{-2}$$

Once we have the Sun's energy flux $F_\odot$, we can use the Stefan–Boltzmann law to find the Sun's surface temperature $T_\odot$:

$$T_\odot^4 = F_\odot/\sigma = 1.13 \times 10^{15} \text{ K}^4$$

Taking the fourth root (the square root of the square root) of this value, we find the surface temperature of the Sun to be $T_\odot = 5800$ K.

Review: Our result for $T_\odot$ agrees with the value we computed in the previous example using Wien's law. Notice that the solar constant of 1370 W m^{-2} is very much less than $F_\odot$, the flux at the Sun's surface. By the time the Sun's radiation reaches Earth, it is spread over a greatly increased area.

EXAMPLE: Sirius, the brightest star in the night sky, has a surface temperature of about 10,000 K. Find the wavelength at which Sirius emits most intensely.

Situation: Our goal is to calculate the wavelength of maximum emission of Sirius (λ_{max}) from its surface temperature T.

Tools: We use Wien's law to relate the values of λ_{max} and T.

Answer: Using Wien's law,

$$\lambda_{max} = \frac{0.0029 \text{ K m}}{T} = \frac{0.0029 \text{ K m}}{10,000 \text{ K}}$$
$$= 2.9 \times 10^{-7} \text{ m} = 290 \text{ nm}$$

Review: Our result shows that Sirius emits light most intensely in the ultraviolet. In the visible part of the spectrum, it emits more blue light than red light (like the curve for 12,000 K in Figure 5-11), so Sirius has a distinct blue color.

EXAMPLE: How does the energy flux from Sirius compare to the Sun's energy flux?

Situation: To compare the energy fluxes from the two stars, we want to find the *ratio* of the flux from Sirius to the flux from the Sun.

Tools: We use the Stefan–Boltzmann law to find the flux from Sirius and from the Sun, which from the preceding examples have surface temperatures 10,000 K and 5800 K, respectively.

Answer: For the Sun, the Stefan–Boltzmann law is $F_\odot = \sigma T_\odot^4$, and for Sirius we can likewise write $F_* = \sigma T_*^4$, where the subscripts $\odot$ and $*$ refer to the Sun and Sirius, respectively. If we divide one equation by the other to find the ratio of fluxes, the Stefan–Boltzmann constants cancel out and we get

$$\frac{F_*}{F_\odot} = \frac{T_*^4}{T_\odot^4} = \frac{(10,000 \text{ K})^4}{(5800 \text{ K})^4} = \left(\frac{10,000}{5800}\right)^4 = 8.8$$

Review: Because Sirius has such a high surface temperature, each square meter of its surface emits 8.8 times more energy per second than a square meter of the Sun's relatively cool surface. As it happens, Sirius is actually a larger star than the Sun, so it has more square meters of surface area and, hence, its *total* energy output is *more* than 8.8 times that of the Sun.

Stefan's law is commonly known as the **Stefan–Boltzmann law.** Written as an equation, the Stefan–Boltzmann law is:

Stefan–Boltzmann law for a blackbody

$$F = \sigma T^4$$

F = energy flux, in joules per square meter of surface per second

σ = a constant = 5.67×10^{-8} W m^{-2} K^{-4}

T = object's temperature, in kelvins

The value of the Stefan–Boltzmann constant σ (the Greek letter sigma) is known from laboratory experiments.

The Stefan–Boltzmann law says that if you double the temperature of an object (for example, from 300 K to 600 K), then the energy emitted from the object's surface each second increases by a factor of $2^4 = 16$. If you increase the temperature by a factor of 10 (for example, from 300 K to 3000 K), the rate of energy emission increases by a factor of $10^4 = 10,000$. Thus, a chunk of iron at room temperature (about 300 K) emits very little electromagnetic radiation (and essentially no visible light), but an iron bar heated to 3000 K glows quite intensely.

Box 5-2 gives several examples of applying Wien's law and the Stefan–Boltzmann law to typical astronomical problems.

Is it ever possible for a cooler object to emit more energy than a warmer object?

Answer appears at the end of the chapter.

5-5 Light has properties of both waves and particles

At the end of the nineteenth century, physicists mounted a valiant effort to explain *how* a material can produce blackbody radiation. To this end, they constructed theories based on Maxwell's description of light as electromagnetic waves. But while electromagnetic waves could explain many properties of light, it was not until the discovery of photons that all features of blackbody curves could be understood.

The Discovery of Photons

In 1905, the great physicist Albert Einstein developed a radically new model for the nature of light. Central to this new picture is Einstein's discovery that light is made out of particles! Each particle of light is called a **photon,** which is a distinct packet of electromagnetic energy.

> Light is made of smaller entities called photons, each one with a wavelength of its own.

Photons have a dual nature in that they are both particlelike and wavelike. They are particlelike in the sense that they are the small packets of energy that make up light. As expected with particles, you can count the number of photons in an electromagnetic wave. But photons are also wavelike because each one is itself a wave with a wavelength that matches that of the electromagnetic wave of which it is a small part. As expected with waves, photons exhibit interference patterns when passed through a double-slit experiment like the one in Figure 5-5.

The energy of each photon is related to the wavelength of light: the longer the wavelength, the lower the energy. Thus, a photon of red light (wavelength $\lambda = 700$ nm) has less energy than a photon of violet light ($\lambda = 400$ nm). Conversely, the shorter a photon's wavelength, the higher its energy.

It was quickly realized that photons explain more than just blackbody curves, and now the study of photons plays a key role in almost every branch of science. While scientific equipment is usually needed to detect individual photons, there is a somewhat familiar effect that photons can help to explain. When it comes to tanning, your skin must be exposed to ultraviolet light. For example, a low-intensity exposure to ultraviolet light causes tanning, but even a high-intensity exposure of visible light will not cause the skin to darken. (Recall that you do not get tanned with regular indoor lighting, which has negligible UV light). The reason is that tanning, at the molecular level, results from a chemical reaction in the skin that involves a single photon. A single high-energy, short-wavelength ultraviolet photon can trigger a tanning reaction. By contrast, even many lower-energy, longer-wavelength photons of visible light cannot trigger the reaction. In this situation, we see that it is not the total energy of light that matters, but how the energy is bundled into packets—and that is a photon.

The Energy of a Photon

The relationship between the energy E of a single photon and the wavelength of the electromagnetic radiation can be expressed in a simple equation:

Energy of a photon (in terms of wavelength)

$$E = \frac{hc}{\lambda}$$

$E =$ energy of a photon

$h =$ Planck's constant

$c =$ speed of light

$\lambda =$ wavelength of light

The value of the constant h in this equation, now called *Planck's constant*, has been shown in laboratory experiments to be

$$h = 6.625 \times 10^{-34} \text{ J s}$$

The units of h are joules multiplied by seconds, called "joule-seconds" and abbreviated J s.

Because the value of h is so tiny, a single photon carries a very small amount of energy. For example, a photon of red light with wavelength 633 nm has an energy of only 3.14×10^{-19} J (**Box 5-3**), which is why we ordinarily do not notice that light comes in the form of photons. Even a dim light source emits so many photons per second that it seems to be radiating a continuous stream of energy.

The energies of photons are sometimes expressed in terms of a small unit of energy called the **electron volt** (eV). One electron volt is equal to 1.602×10^{-19} J, so a 633-nm photon has an energy of 1.96 eV. If energy is expressed in electron volts, Planck's constant is best expressed in electron volts multiplied by seconds, abbreviated eV s:

$$h = 4.135 \times 10^{-15} \text{ eV s}$$

Because the frequency f of light is related to the wavelength λ by $f = c/\lambda$, we can rewrite the equation for the energy of a photon as:

Energy of a photon (in terms of frequency)

$$E = hf$$

$E =$ energy of a photon

$h =$ Planck's constant

$f =$ frequency of light

The equations $E = hc/\lambda$ and $E = hf$ are together called **Planck's law.** Both equations express a relationship between a particlelike property of light (the energy E of a photon) and a wavelike property (the wavelength λ or frequency f).

The photon picture of light is essential for understanding the detailed shapes of blackbody curves. As we will see, it also helps to explain how and why the spectra of the Sun and stars differ from those of perfect blackbodies.

If a photon's wavelength is measured to be longer than the wavelength of a green photon, will it have a greater or lower energy than a green photon?

Answer appears at the end of the chapter.

BOX 5-3 ASTRONOMY DOWN TO EARTH

Photons at the Supermarket

A beam of light can be regarded as a stream of tiny packets of energy called photons. The Planck relationships $E = hc/\lambda$ and $E = hf$ can be used to relate the energy E carried by a photon to its wavelength λ and frequency f.

As an example, the laser bar-code scanners used at stores and supermarkets emit orange-red light of wavelength 633 nm. To calculate the energy of a single photon of this light, we must first express the wavelength in meters. A nanometer (nm) is equal to 10^{-9} m, so the wavelength is

$$\lambda = (633 \text{ nm}) \left(\frac{10^{-9} \text{ m}}{1 \text{ nm}} \right) = 633 \times 10^{-9} \text{ m} = 6.33 \times 10^{-7} \text{ m}$$

Then, using the Planck formula $E = hc/\lambda$, we find that the energy of a single photon is

$$E = \frac{hc}{\lambda} = \frac{(6.625 \times 10^{-34} \text{ J s})(3 \times 10^{8} \text{ m/s})}{6.33 \times 10^{-7} \text{ m}} = 3.14 \times 10^{-19} \text{ J}$$

This amount of energy is very small. The laser in a typical bar-code scanner emits 10^{-3} joule of light energy per second, so the number of photons emitted per second is

$$\frac{10^{-3} \text{ joule per second}}{3.14 \times 10^{-19} \text{ joule per photon}} = 3.2 \times 10^{15} \text{ photons per second}$$

This number is so large that the laser beam seems like a continuous flow of energy rather than a stream of little energy packets.

5-6 Each chemical element produces its own unique set of spectral lines

In 1814, the German master optician Joseph von Fraunhofer repeated the classic experiment of shining a beam of sunlight through a prism (see Figure 5-3). But this time, Fraunhofer subjected the resulting rainbow-colored spectrum to intense magnification. To his surprise, he discovered that the solar spectrum contains hundreds of fine, dark lines, now called **spectral lines** (Figure 5-14). Yet light from a perfect blackbody would produce a smooth, continuous spectrum with no dark lines! A dark spectral line occurs when light at a specific wavelength is at least partially *absorbed* so that the spectrum appears darker; it is also called an *absorption line*. Fraunhofer counted more than 600 dark lines in the Sun's spectrum; today we know of more than 30,000. The photograph of the Sun's spectrum in Figure 5-14 shows hundreds of these spectral lines.

> Spectroscopy is the key to determining the chemical composition of planets and stars.

FIGURE 5-14 R I (V) U X G

The Sun's Spectrum Numerous dark spectral lines are seen in this image of the Sun's spectrum. The spectrum is spread out so much that it had to be cut into segments and stacked to fit on this page. (N.A. Sharp, NOAO/NSO/Kitt Peak FTS/AURA/NSF)

Spectral Analysis

Half a century later, chemists discovered that they could produce spectral lines in the laboratory and use these spectral lines to analyze what kinds of atoms different substances are made of. Chemists had long known that many substances emit distinctive colors when sprinkled into a flame. To facilitate study of these colors, around 1857 the German chemist Robert Bunsen invented a gas burner (today called a Bunsen burner) that produces a clean flame with no color of its own. Bunsen's colleague, the Prussian-born physicist Gustav Kirchhoff, suggested that the colored light produced when substances were added to the flame might best be studied by passing the resulting light through a prism (Figure 5-15). The two scientists promptly discovered that the spectrum from the flame consists of a pattern of thin, bright spectral lines against a dark background. A bright spectral line arises because at least some additional light is being emitted at a specific wavelength; it is also called an *emission line*.

1. Add a chemical substance to a flame.

2. Send light from the flame through a narrow slit, then through a prism.

3. Bright lines in the spectrum show that the substance emits light at only specific wavelengths.

FIGURE 5-15

The Kirchhoff–Bunsen Experiment In the mid-1850s, Gustav Kirchhoff and Robert Bunsen discovered that when a chemical substance is heated and vaporized, the spectrum of the emitted light exhibits a series of bright spectral lines. They also found that each chemical element produces its own characteristic pattern of spectral lines. (In an actual laboratory experiment, lenses would be needed to focus the image of the slit onto the screen.)

Kirchhoff and Bunsen then found that each chemical element produces its own unique pattern of spectral lines. Thus was born in 1859 the technique of **spectral analysis**: the identification of atoms and molecules by their unique patterns of spectral lines.

Atoms are the building blocks of substances on Earth and throughout the universe. Familiar examples are hydrogen, oxygen, carbon, iron, and gold. These atoms are also called chemical **elements** because a pure substance made from only one type of atom (such as pure gold) cannot be broken down into more basic chemicals. After Kirchhoff and Bunsen had recorded the prominent spectral lines of all the then-known elements, they soon began to discover other spectral lines in the spectra of vaporized mineral samples. In this way, they discovered elements on Earth that had never before been seen or inferred through any method.

Spectral analysis even allowed the discovery of new elements *outside* of Earth. During the solar eclipse of 1868, astronomers found a new spectral line in light coming from the hot gases at the upper surface of the Sun while the main body of the Sun was hidden by the Moon. This line was attributed to a new element that was named helium (from the Greek *helios*, meaning "sun"). It took almost another 30 years to discover helium on Earth!

Atoms can also combine to form **molecules**. For example, two hydrogen atoms (symbol H) can combine with an oxygen atom (symbol O) to form a water molecule (symbol H_2O). Substances like water, whose molecules include atoms of different elements, are called **compounds**. Just as each type of atom has its own unique spectrum, so does each type of molecule. Figure 5-16 shows the spectra of several types of atoms and molecules. You can easily see in Figure 5-16 that each substance produces a unique pattern of spectral lines; each pattern can be thought of as a spectral "fingerprint" for identification. This is enormously important in astronomy because it allows us to determine the detailed composition of distant planets and stars by analyzing the spectra of their light that reaches Earth.

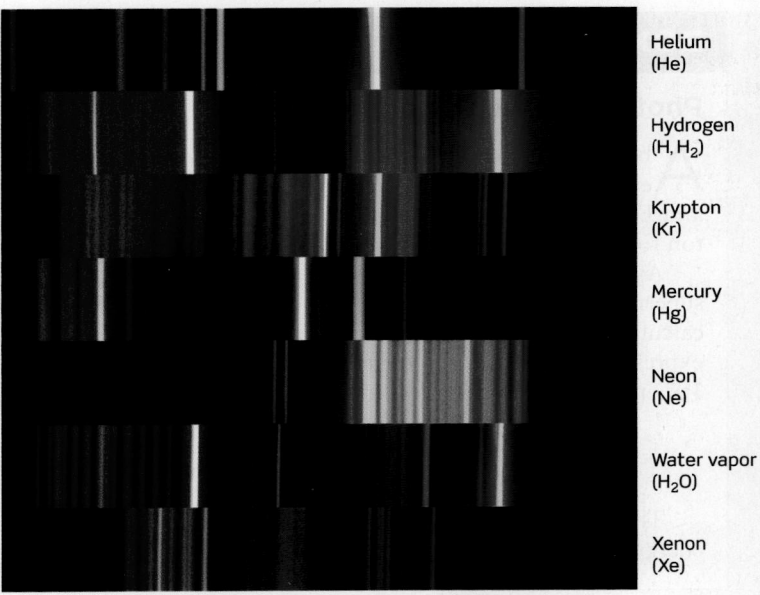

Helium (He)

Hydrogen (H, H_2)

Krypton (Kr)

Mercury (Hg)

Neon (Ne)

Water vapor (H_2O)

Xenon (Xe)

FIGURE 5-16 R I V U X G

Various Spectra These photographs show the spectra of different types of gases as measured in a laboratory on Earth. Each type of gas has a unique spectrum that is the same wherever in the universe the gas is found. Water vapor (H_2O) is a compound whose molecules are made up of hydrogen and oxygen atoms; the hydrogen molecule (H_2) is made up of two hydrogen atoms. For the hydrogen gas lamp, the bright lines are from single hydrogen atoms (H), and the fainter lines are emitted from hydrogen molecules (H_2). (Ted Kinsman/Science Source)

Kirchhoff's Laws

TUTORIAL 5.3 ➔ The spectrum of the Sun, with its dark spectral lines superimposed on a bright background (see Figure 5-14), may seem to be unrelated to the spectra of bright lines against a dark background produced by substances in a flame (see Figure 5-15). But by the early 1860s, Kirchhoff's experiments had revealed a direct connection between these two types of spectra. His conclusions are summarized in three important statements about spectra that are today called **Kirchhoff's laws**. These laws, which are illustrated in Figure 5-17, are as follows:

Law 1 A hot opaque body, such as a perfect blackbody or a hot, dense gas, produces a **continuous spectrum**—a complete rainbow of colors without any spectral lines.

Law 2 A hot, transparent gas produces an **emission line spectrum**—a series of bright spectral lines against a dark background.

Law 3 A cool, transparent gas in front of a source of a continuous spectrum (like a distant star) produces an **absorption line spectrum**—a series of dark spectral lines among the colors of the continuous spectrum. Furthermore, the dark lines in the absorption spectrum of a particular gas occur at exactly the *same* wavelengths as the bright lines in the emission spectrum of that same gas. As we will see in Section 5-8, absorption and emission lines correspond to distinct changes in an electron's orbit within an atom.

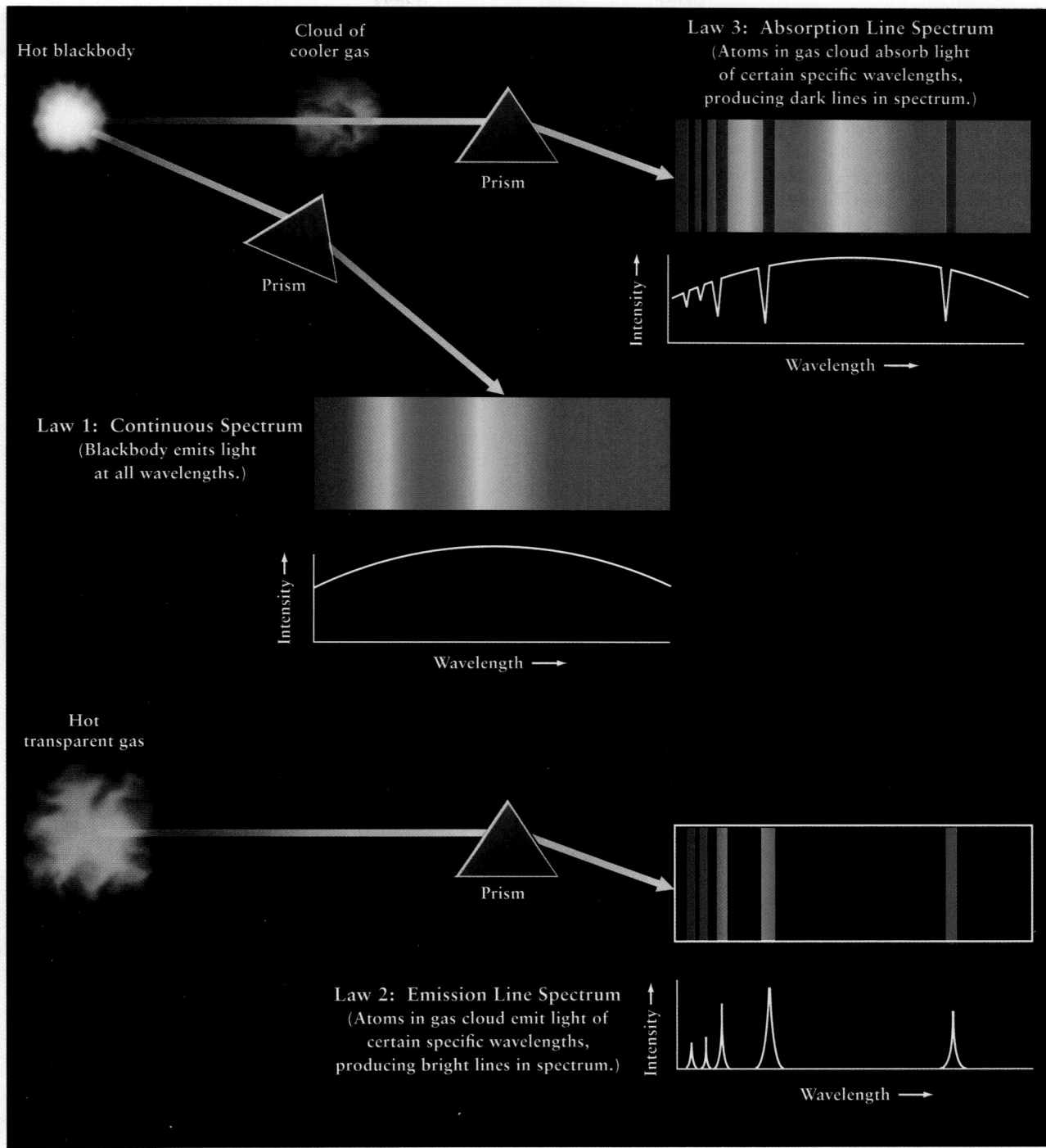

ANIMATION
5.2

FIGURE 5-17

Kirchhoff's Laws for Continuous, Absorption Line, and Emission Line Spectra **Law 1:** A hot, opaque body emits a continuous spectrum of light. **Law 3:** If this light is passed through a cloud of a cooler gas, the cloud absorbs light of certain specific wavelengths, leaving the light that passes directly through the cloud with an absorption line spectrum. **Law 2:** Atoms in a hot gas cloud collide energetically and emit light, creating an emission line spectrum that is unique to each type of atom. In each case, the spectrum is shown as it would appear with a prism, and just below it, by a graph of intensity versus wavelength.

CAUTION!

Figure 5-17 shows that light can either pass through a cloud of gas or be absorbed by the gas. But there is also a third possibility: The light can simply bounce off the atoms or molecules that make up the gas, a phenomenon called **light scattering.** In other words, photons passing through a gas cloud can miss the gas atoms altogether, be swallowed whole by the atoms (absorption), or bounce off the atoms like billiard balls colliding (scattering). **Box 5-4** describes how light scattering explains the blue color of the sky and the red color of sunsets.

BOX 5-4 ASTRONOMY DOWN TO EARTH

Why the Sky Is Blue

Light scattering is a process in which photons bounce off particles and change their direction of travel. The scattering particles can be atoms, molecules, or clumps of molecules. When you see regular objects like rocks or trees that do not produce their own intrinsic light, you see them because they scatter sunlight into your eyes.

An important fact about light scattering is that very small particles—ones that are smaller than a wavelength of visible light—are quite effective at scattering short-wavelength photons of blue light but are less effective at scattering long-wavelength photons of red light. This fact explains a number of phenomena that you can see here on Earth.

The light that comes from the daytime sky is sunlight that has been scattered by the molecules that make up our atmosphere (see part (a) of the accompanying figure). Air molecules are less than 1 nm across, far smaller than the wavelength of visible light, so they scatter blue light more than red light—which is why the sky looks blue. In other words, the sky is blue because while incoming white sunlight contains all the colors, air molecules scatter blue light around more than red light, literally spreading blue light all throughout the atmosphere.

During the day, distant mountains often appear blue thanks to sunlight being scattered from the atmosphere between the mountains and your eyes. The Blue Ridge Mountains, which extend from Pennsylvania to Georgia, and Australia's Blue Mountains derive their names from this effect.

Light scattering also explains why sunsets are red. Again, sunlight contains all the colors, but as this light passes through our atmosphere, the blue light is scattered away from the straight-line path from the Sun. On the other hand, red light undergoes relatively little scattering, so light coming directly from the Sun looks a bit redder once some blue has been removed. When you look toward the setting Sun (but don't ever look at the Sun!), the sunlight that reaches your eye has had to pass through a relatively thick layer of atmosphere (part (b) of the accompanying figure). Hence, a large fraction of the blue light from the Sun has been scattered, and the Sun appears quite red. If, at sunset, the sunlight shines onto clouds, the clouds also look red after much of the blue light has been scattered away.

The same effect also applies to sunrises, but sunrises seldom look as red as sunsets do. The reason is that dust is lifted into the atmosphere during the day by the wind (which, being driven by sunlight, typically blows stronger in the daytime), and dust particles in the atmosphere help to scatter even more blue light.

Finally, some scattered light appears white. If the small particles that scatter light are sufficiently concentrated, there will be almost as much scattering of red light as of blue light, and the scattered light will then appear white. This can occur when the scattering comes from dense ice crystals or water vapor and explains the white color of snow, clouds, and fog.

Light scattering has many applications to astronomy. For example, it explains why very distant stars in our Galaxy appear surprisingly red. The reason is that there are tiny dust particles throughout the space between the stars, and this dust scatters blue light. By studying how much scattering takes place, astronomers have learned about the tenuous material that fills interstellar space.

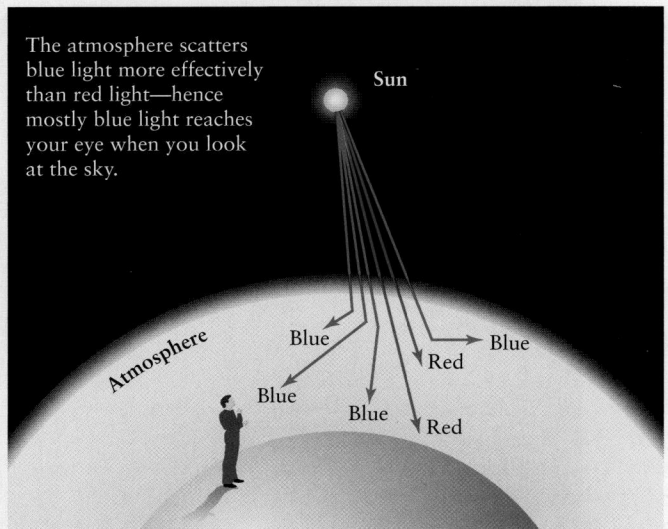

(a) Why the sky looks blue

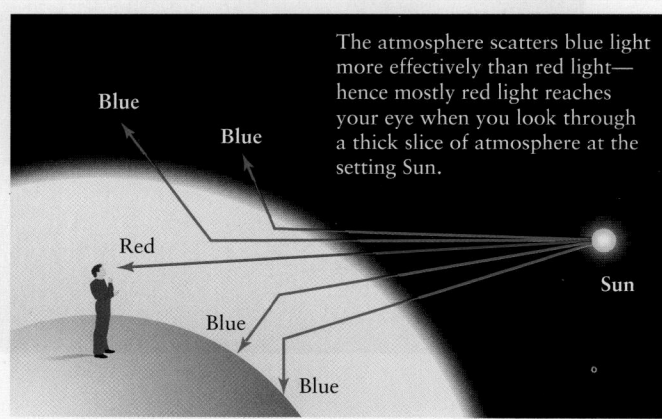

(b) Why the setting Sun looks red

In general, a gas cloud can both emit and absorb light. Whether an emission line spectrum or an absorption line spectrum is observed from an intervening gas cloud depends on the relative temperatures of the gas cloud and its background. Absorption lines are seen if the background is hotter than the gas, and emission lines are seen if the background is cooler. This is why the gas cloud illustrating absorption (Law 3) in Figure 5-17 is cooler than the star.

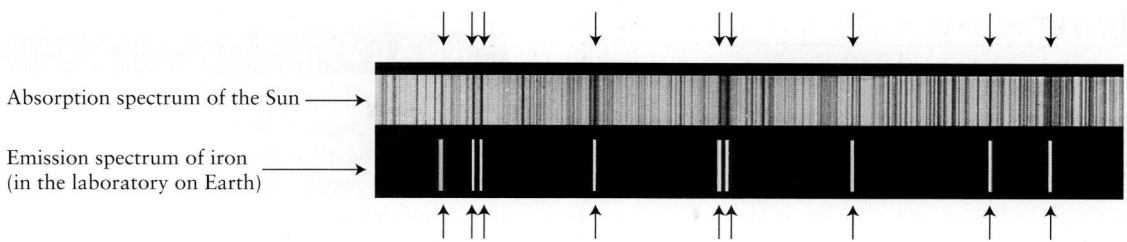

Absorption spectrum of the Sun ⟶

Emission spectrum of iron
(in the laboratory on Earth) ⟶

For each emission line of iron, there is a corresponding absorption
line in the solar spectrum; hence there must be iron in the Sun's atmosphere.

FIGURE 5-18 R I Ⓥ U X G

Iron in the Sun The upper part of this figure is a portion of the Sun's spectrum at violet wavelengths, showing numerous dark absorption lines. The lower part of the figure is a corresponding portion of the emission line spectrum of vaporized iron. The iron lines coincide with some of the solar lines, which proves that there is some iron (albeit a relatively small amount) in the Sun's atmosphere. (Carnegie Observatories)

Spectroscopy

Spectroscopy is the systematic study of spectra and spectral lines. As noted, spectral lines are tremendously important in astronomy, because they provide detailed evidence about the chemical composition of distant objects. As an example, the spectrum of the Sun shown in Figure 5-14 is an absorption line spectrum. The continuous spectrum comes from the hot surface of the Sun, which acts like a blackbody. The dark absorption lines are caused by this light passing through a cooler gas; this gas is the atmosphere that surrounds the Sun. (To explain the Sun's absorption lines in terms of Law 3 in Figure 5-17, the inner part of the Sun is the hot blackbody, and the Sun's outer atmosphere plays the role of the absorbing cloud of cooler gas.) Therefore, by identifying the spectral lines present in the solar spectrum, we can determine the chemical composition of the Sun's atmosphere.

Even when the spectral lines from more than one type of atom appear in the same spectrum, the different atoms can be identified. Figure 5-18 shows both a portion of the Sun's absorption line spectrum, which has numerous lines, and the emission line spectrum of iron vapor over the same wavelength range, measured on Earth. This pattern of bright spectral lines in the lower spectrum is iron's own distinctive "fingerprint," which no other substance can imitate. Because some absorption lines in the Sun's spectrum coincide with the iron lines, some vaporized iron must exist in the Sun's atmosphere.

Spectroscopy can also help us analyze gas clouds in space, such as the nebula surrounding the star cluster NGC 346 shown in Figure 5-19. The particular shade of red that dominates the color of this nebula is due to an emission line at a wavelength of 656 nm. This is one of the characteristic wavelengths emitted by hydrogen gas, so we can conclude that this nebula contains hydrogen. More detailed analyses of this kind show that hydrogen is the most common element in gaseous nebulae, and indeed in the universe as a whole. The spectra of other nebulae, such as the Ring Nebula shown in the image that opens this chapter, also reveal the presence of nitrogen, oxygen, helium, and other gases.

What is truly remarkable about spectroscopy is that it can determine chemical composition at any distance. The 656-nm red light produced by a sample of heated hydrogen gas on Earth (the bright red line in the hydrogen spectrum in Figure 5-16) is the same as that observed coming from the nebula shown in Figure 5-19, located about 210,000 light-years away. Throughout this book we will see many examples of how astronomers use spectra to determine the nature of celestial objects.

FIGURE 5-19 R I Ⓥ U X G

Analyzing the Composition of a Distant Nebula The glowing gas cloud in this Hubble Space Telescope image lies 210,000 light-years away in the constellation Tucana (the Toucan). The red light emitted by the nebula has a wavelength of 656 nm, characteristic of an emission line from hydrogen gas. Hot stars within the nebula emit high-energy, ultraviolet photons, which ionize hydrogen (removing the electron). When the electron returns to the atom, it jumps toward the center and gives off emission lines. (NASA, ESA, and A. Nota, STScI/ESA)

Why does an atom absorb light of only particular wavelengths? And why does it then emit light of only these same wavelengths? Maxwell's theory of electromagnetism (see Section 5-2) could not answer these questions. The answers did not come until scientists began to discover the structure and properties of atoms.

CONCEPT CHECK 5-10

Which one of Kirchhoff's laws describes the Sun's spectrum in Figure 5-14? If helium gas is heated up enough to produce the spectrum at the top of Figure 5-16, which of Kirchhoff's laws applies?

CONCEPT CHECK 5-11

What type of spectra would result from a glowing field of hot, dense lava as viewed by an orbiting satellite through Earth's atmosphere?

Answers appear at the end of the chapter.

5-7 An atom consists of a small, dense nucleus surrounded by electrons

By the early twentieth century, it was known that **atoms** contain negatively charged particles called **electrons** and some material with positive charge, but the structure of atoms was still a mystery. The "plum pudding" model popular at that time proposed that the negative electrons were like plums uniformly mixed into a positively charged pudding. However, in 1910 this model was overturned by Ernest Rutherford, an inquisitive physicist from New Zealand.

> To decode the information in the light from immense objects like stars and galaxies, we must understand the structure of atoms.

The Nucleus of an Atom

Rutherford discovered a more accurate model for the structure of an atom, shown in **Figure 5-20**. According to this model, a massive, positively charged **nucleus** at the center of the atom is orbited by tiny, negatively charged electrons. Rutherford concluded that at least 99.98% of the mass of an atom must be concentrated in its nucleus, whose diameter is only about 10^{-14} m. (The diameter of a typical atom is far larger, about 10^{-10} m.)

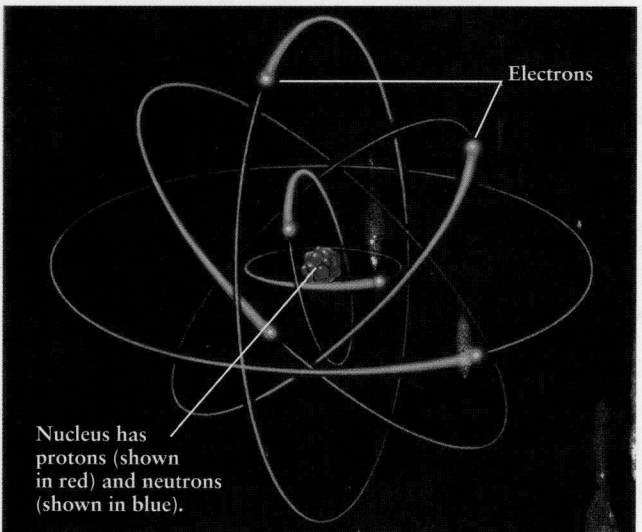

Electrons

Nucleus has protons (shown in red) and neutrons (shown in blue).

FIGURE 5-20

Rutherford's Model of the Atom In this model, electrons orbit the atom's nucleus, which contains most of the atom's mass. The nucleus contains two types of particles, protons and neutrons. The full behavior of electrons orbiting the nucleus is quite complicated and not represented in this simple illustration.

We know today that the nucleus of an atom contains two types of particles, **protons** and **neutrons**. A proton has a positive electric charge, equal in magnitude to that of the negatively charged electron. As its name suggests, a neutron has no electric charge—it is electrically neutral. As an example, the helium atom has two protons and two neutrons. Protons and neutrons are held together in a nucleus by the so-called *strong nuclear force,* whose great strength overcomes the electric repulsion between the positively charged protons. A proton and a neutron have almost the same mass, 1.7×10^{-27} kg, and each has about 2000 times as much mass as an electron (9.1×10^{-31} kg). In an ordinary atom, there are as many positive protons as there are negative electrons, so the atom has no net electric charge. Because the mass of the electron is so small, the mass of an atom is not much greater than the mass of its nucleus.

While the solar system is held together by gravitational forces, electrons are held in atoms by electrical force. Opposites attract: The negative charges on the orbiting electrons are attracted to the positive charges on the protons in the nucleus. Box 5-5 (on page 134) describes more about the connection between the structure of atoms and the chemical and physical properties of substances made of those atoms.

Rutherford's new model clarified the structure of the atom, but it did not explain how these tiny particles within the atom give rise to spectral lines. The task of reconciling Rutherford's atomic model with Kirchhoff's laws of spectral analysis was undertaken by the young Danish physicist Niels Bohr, who joined Rutherford's laboratory in 1912.

5-8 Spectral lines are produced when an electron jumps from one energy level to another within an atom

Niels Bohr began his study of the connection between atomic spectra and atomic structure by trying to understand the structure of hydrogen, the simplest and lightest of the elements. (As we discussed in Section 5-6, hydrogen is also the most common element in the universe.) When Bohr was done, he had found not only a way to explain this atom's spectrum, but also a justification for Kirchhoff's laws in terms of atomic physics.

Hydrogen and the Balmer Series

The most common type of hydrogen atom consists of a single electron and a single proton. Hydrogen atoms have a simple visible-light spectrum consisting of a pattern of lines that begins at a wavelength of 656.3 nm and ends at 364.6 nm. The first spectral line is called

Shorter wavelength

FIGURE 5-21   R I **V** U X G

Balmer Lines in the Spectrum of a Star This portion of the spectrum of the star Vega in the constellation Lyra (the Harp) shows eight Balmer lines, from H$_\alpha$ at 656.3 nm through H$_\theta$ (H-theta) at 388.9 nm. The series converges at 364.6 nm, slightly to the left of H$_\theta$. Note that the left end of this spectrum is transitioning into ultraviolet light, which is indicated by the highlighted U in the wavelength tab. (NOAO/AURA/NSF)

H$_\alpha$ (H-alpha), the second spectral line is called H$_\beta$ (H-beta), the third is H$_\gamma$ (H-gamma), and so forth (Figure 5-21). These are the bright lines in the spectrum of hydrogen shown in Figure 5-16. The fainter lines between these appear when hydrogen atoms form into molecules. The closer you get to the short-wavelength end of the spectrum at 364.6 nm, the more spectral lines you see.

The regularity in this spectral pattern was described mathematically in 1885 by Johann Jakob Balmer, a Swiss schoolteacher. In Figure 5-21, the spectral lines of hydrogen at visible wavelengths are today called **Balmer lines,** and the entire pattern from H$_\alpha$ onward is called the **Balmer series.** Stars in general, including the Sun, have Balmer absorption lines in their spectra, which shows they have atmospheres that contain hydrogen. Notice that the H$_\alpha$ line in Figure 5-21 is in the red portion of the visible spectrum, and when seen as an emission line, it accounts for much of the red seen in beautiful nebulae (including Figure 5-19).

Using trial and error, Balmer discovered a formula from which the wavelengths (λ) of hydrogen's spectral lines can be calculated. Balmer's formula is usually written

$$\frac{1}{\lambda} = R\left(\frac{1}{4} - \frac{1}{n^2}\right)$$

In this formula, R is the *Rydberg constant* ($R = 1.097 \times 10^7 \text{ m}^{-1}$), named in honor of the Swedish spectroscopist Johannes Rydberg, and n can be any integer (whole number) greater than 2. The meaning of n will become clear shortly when we discuss electron orbits in the hydrogen atom. To get the wavelength λ_α of the spectral line H$_\alpha$, first enter $n = 3$ into Balmer's formula:

$$\frac{1}{\lambda_\alpha} = (1.097 \times 10^7 \text{ m}^{-1})\left(\frac{1}{4} - \frac{1}{3^2}\right) = 1.524 \times 10^6 \text{ m}^{-1}$$

Then take the reciprocal:

$$\lambda_\alpha = \frac{1}{1.524 \times 10^6 \text{ m}^{-1}} = 6.563 \times 10^{-7} \text{ m} = 656.3 \text{ nm}$$

To get the wavelength of H$_\beta$, use $n = 4$, and to get the wavelength of H$_\gamma$ use $n = 5$. If you use $n = \infty$ (the symbol ∞ stands for infinity), you get the short-wavelength end of the hydrogen spectrum at 364.6 nm. (Note that 1 divided by infinity equals zero.)

Bohr's Model of Hydrogen

Bohr realized that to fully understand the structure of the hydrogen atom, he had to be able to derive Balmer's formula using the laws of physics. He first made the rather wild assumption that the electron in a hydrogen atom can orbit the nucleus only in certain

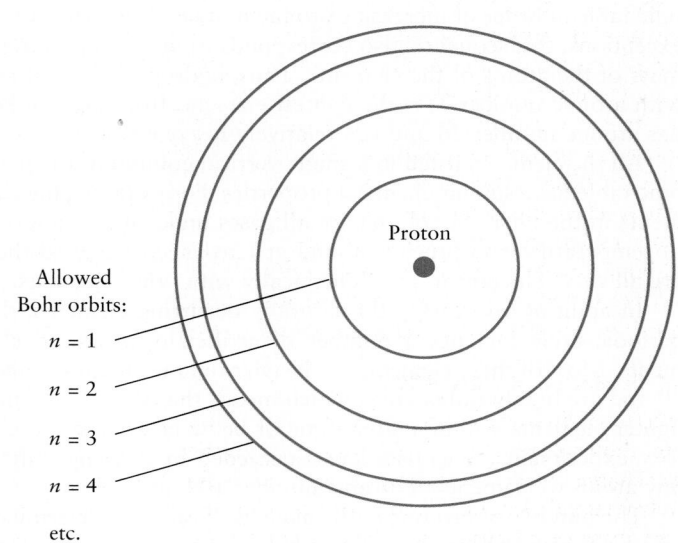

FIGURE 5-22

The Bohr Model of the Hydrogen Atom In this model, an electron circles the hydrogen nucleus (a proton) only in allowed orbits $n = 1, 2, 3$, and so forth. The first four Bohr orbits are shown here. This figure is not drawn to scale; in the Bohr model, the $n = 2, 3$, and 4 orbits are respectively 4, 9, and 16 times larger than the $n = 1$ orbit.

specific orbits. (This idea was a significant break with the ideas of Newton, in whose mechanics any orbit should be possible.) Figure 5-22 shows the four smallest of these **Bohr orbits,** labeled by the numbers $n = 1$, $n = 2$, $n = 3$, and so on.

Although confined to one of these allowed orbits while circling the nucleus, an electron can jump from one Bohr orbit to another. For an electron to jump between orbits, the hydrogen atom must gain or lose a specific amount of energy.

An atom must absorb energy for the electron to go from an inner to an outer orbit; an atom must release energy for the electron to go from an outer to an inner orbit.

As an example, Figure 5-23 shows an electron jumping between the $n = 2$ and $n = 3$ orbits of a hydrogen atom as the atom absorbs or emits an H$_\alpha$ photon. Figure 5-23 shows absorption and emission of a single photon, but in a gas with many atoms, these same jumps produce absorption lines and emission lines.

Bohr's orbital model also explains Kirchhoff's observation (see Figure 5-17) that an atom can show the same spectral lines

BOX 5-5   TOOLS OF THE ASTRONOMER'S TRADE

Atoms, the Periodic Table, and Isotopes

Each different chemical element is made of a specific type of atom. Each specific atom has a characteristic number of protons in its nucleus. For example, a hydrogen atom has 1 proton in its nucleus, an oxygen atom has 8 protons in its nucleus, and so on.

The number of protons in an atom's nucleus is the **atomic number** for that particular element. The chemical elements are most conveniently listed in the form of a **periodic table** (shown in the accompanying figure). Elements are arranged in the periodic table in order of increasing atomic number. With only a few exceptions, this sequence also corresponds to increasing average mass of the atoms of the elements. Thus, hydrogen (symbol H), with atomic number 1, is the lightest element. Iron (symbol Fe) has atomic number 26 and is a relatively heavy element.

All the elements listed in a single vertical column of the periodic table have similar chemical properties. For example, the elements in the far right column are all gases under the conditions of temperature and pressure found at Earth's surface, and they are all very reluctant to react chemically with other elements.

In addition to nearly 100 naturally occurring elements, the periodic table includes a number of artificially produced elements. Most of these elements are heavier than uranium (symbol U) and are highly radioactive, which means that they decay into lighter elements within a short time of being created in laboratory experiments. Scientists have succeeded in creating only a few atoms of elements of atomic number 104 and above.

The number of protons in the nucleus of an atom determines which element that atom is. Nevertheless, the same element may have different numbers of neutrons in its nucleus. For example, oxygen (O) has atomic number 8, so every oxygen nucleus has exactly 8 protons. But oxygen nuclei can have 8, 9, or 10

neutrons. These three slightly different kinds of oxygen are called **isotopes**. The isotope with 8 neutrons is by far the most abundant variety. It is written as ^{16}O (for the sum of 8 protons + 8 neutrons), or oxygen-16. The rarer isotopes with 9 and 10 neutrons are designated as ^{17}O and ^{18}O, respectively.

The superscript that precedes the chemical symbol for an element equals the total number of protons and neutrons in a nucleus of that particular isotope. For example, a nucleus of the most common isotope of iron, ^{56}Fe or iron-56, contains a total of 56 protons and neutrons. From the periodic table, the atomic number of iron is 26, so every iron atom has 26 protons in its nucleus. Therefore, the number of neutrons in an iron-56 nucleus is $56 - 26 = 30$. Most nuclei have more neutrons than protons, especially in the case of the heaviest elements.

It is extremely difficult to distinguish chemically between the various isotopes of a particular element. Ordinary chemical reactions involve only the electrons that orbit the atom, never the neutrons buried in its nucleus. But there are small differences in the wavelengths of the spectral lines for different isotopes of the same element. For example, the spectral line wavelengths of the hydrogen isotope ^{2}H are about 0.03% greater than the wavelengths for the most common hydrogen isotope, ^{1}H. Thus, different isotopes can be distinguished by careful spectroscopic analysis.

Isotopes are important in astronomy for a variety of reasons. By measuring the relative amounts of different isotopes of a given element in a Moon rock or meteorite, the age of that sample can be determined. The mixture of isotopes left behind when a star explodes into a supernova (see Section 1-3) tells

(continued on the next page)

Absorption

Incoming photon,
$\lambda = 656.3$ nm

Emission

Emitted photon,
$\lambda = 656.3$ nm

$n = 2$

$n = 3$

(a) Atom absorbs a 656.3-nm photon; absorbed energy causes electron to jump from the $n = 2$ orbit up to the $n = 3$ orbit.

(b) Electron falls from the $n = 3$ orbit to the $n = 2$ orbit; energy lost by atom goes into emitting a 656.3-nm photon.

ANIMATION 5.3

FIGURE 5-23

The Absorption and Emission of an H_α Photon This schematic diagram, drawn according to the Bohr model, shows what happens when a hydrogen atom **(a)** absorbs or **(b)** emits a photon whose wavelength is 656.3 nm.

BOX 5-5 (continued)

astronomers about the processes that led to the explosion. And knowing the properties of different isotopes of hydrogen and helium is crucial to understanding the nuclear reactions that make the Sun shine. Look for these and other applications of the idea of isotopes in later chapters.

Periodic Table of the Elements

1 **H** Hydrogen																	2 **He** Helium
3 **Li** Lithium	4 **Be** Beryllium											5 **B** Boron	6 **C** Carbon	7 **N** Nitrogen	8 **O** Oxygen	9 **F** Fluorine	10 **Ne** Neon
11 **Na** Sodium	12 **Mg** Magnesium											13 **Al** Aluminum	14 **Si** Silicon	15 **P** Phosphorus	16 **S** Sulfur	17 **Cl** Chlorine	18 **Ar** Argon
19 **K** Potassium	20 **Ca** Calcium	21 **Sc** Scandium	22 **Ti** Titanium	23 **V** Vanadium	24 **Cr** Chromium	25 **Mn** Manganese	26 **Fe** Iron	27 **Co** Cobalt	28 **Ni** Nickel	29 **Cu** Copper	30 **Zn** Zinc	31 **Ga** Gallium	32 **Ge** Germanium	33 **As** Arsenic	34 **Se** Selenium	35 **Br** Bromine	36 **Kr** Kryton
37 **Rb** Rubidium	38 **Sr** Strontium	39 **Y** Yttrium	40 **Zr** Zirconium	41 **Nb** Niobium	42 **Mo** Molybdenum	43 **Tc** Technetium	44 **Ru** Ruthenium	45 **Rh** Rhodium	46 **Pd** Palladium	47 **Ag** Silver	48 **Cd** Cadmium	49 **In** Indium	50 **Sn** Tin	51 **Sb** Antimony	52 **Te** Tellurium	53 **I** Iodine	54 **Xe** Xenon
55 **Cs** Cesium	56 **Ba** Barium	57 **La** Lanthanum	72 **Hf** Hafnium	73 **Ta** Tantaium	74 **W** Tungsten	75 **Re** Rhenium	76 **Os** Osmium	77 **Ir** Iridium	78 **Pt** Platinum	79 **Au** Gold	80 **Hg** Mercury	81 **Tl** Thallium	82 **Pb** Lead	83 **Bi** Bismuth	84 **Po** Polonium	85 **At** Astatine	86 **Rn** Radon
87 **Fr** Francium	88 **Ra** Radium	89 **Ac** Actinium	104 **Rf** Rutherfordium	105 **Db** Dubnium	106 **Sg** Seaborgium	107 **Bh** Bohrium	108 **Hs** Hassium	109 **Mt** Meitnerium	110 **Ds** Darmstadium	111 **Rg** Roentgenium							

58 **Ce** Cerium	59 **Pr** Praseodymium	60 **Nd** Neodymium	61 **Pm** Promethium	62 **Sm** Samarium	63 **Eu** Europium	64 **Gd** Gadolinium	65 **Tb** Terbium	66 **Dy** Dysprosium	67 **Ho** Holmium	68 **Er** Erbium	69 **Tm** Thulium	70 **Yb** Ytterbium	71 **Lu** Lutetium
90 **Th** Thorium	91 **Pa** Protactinium	92 **U** Uranium	93 **Np** Neptunium	94 **Pu** Plutonium	95 **Am** Americium	96 **Cm** Curium	97 **Bk** Berkelium	98 **Cf** Californium	99 **Es** Einsteinium	100 **Fm** Fermium	101 **Md** Mendelevium	102 **No** Nobelium	103 **Lr** Lawrencium

as either emission or absorption. Here's why: When the electron jumps from one orbit to another, the energy of the photon that is emitted or absorbed equals the difference in energy between these two orbits. This energy difference, and hence the photon energy, is the same whether the jump is from a low orbit to a high orbit (Figure 5-23a) or from the high orbit back to the low one (Figure 5-23b). If two photons have the same energy E, the relationship $E = hc/\lambda$ tells us that they must also have the same wavelength λ. Therefore, if an atom can emit photons of a given energy and wavelength, it can also absorb photons of precisely the same energy and wavelength.

The Bohr picture also helps us visualize how a hot, transparent gas produces an emission line spectrum (Kirchhoff's Law 2). When a gas is heated, its atoms move around rapidly and can collide forcefully with each other. These energetic collisions excite the atoms' electrons into various high orbits. The electrons then cascade back down to the innermost possible orbit, emitting photons whose energies are equal to the energy differences between different Bohr orbits. In this fashion, a hot gas produces an emission line spectrum with a variety of different wavelengths.

To produce an absorption line spectrum (Kirchhoff's Law 3), begin with a relatively cool gas, so that the electrons in most of the atoms are in inner, low-energy orbits. If a beam of light with a continuous spectrum is shone through the gas, most wavelengths will pass through undisturbed. *Only those photons will be absorbed whose energies are just right to excite an electron to an allowed outer orbit.* Hence, only certain wavelengths will be absorbed, and dark lines will appear in the spectrum at those wavelengths.

Using his picture of allowed orbits and the formula $E = hc/\lambda$, Bohr was able to prove mathematically that the wavelength λ of

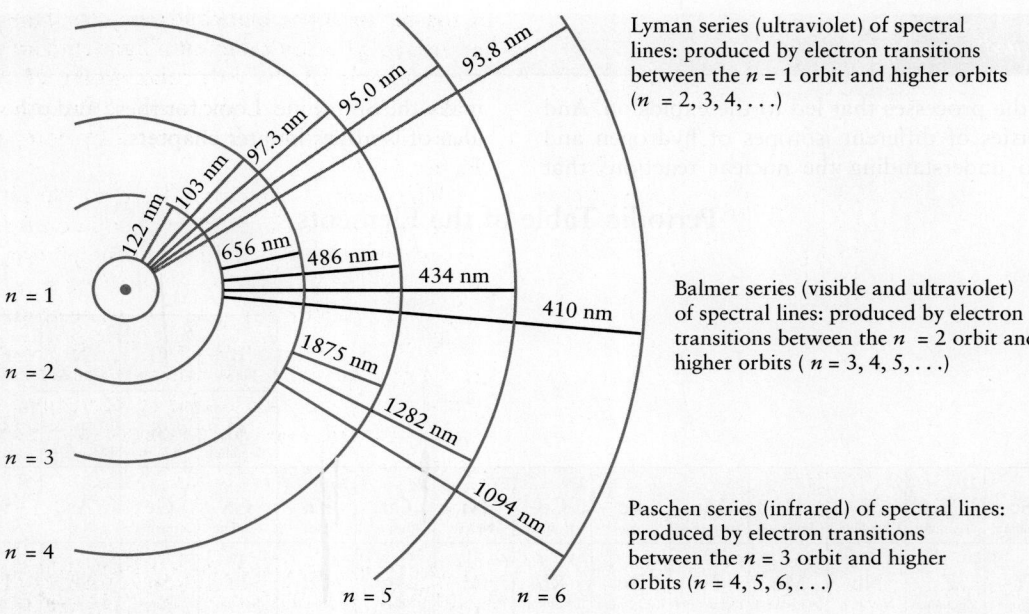

FIGURE 5-24

Electron Transitions in the Hydrogen Atom This diagram shows the photon wavelengths associated with different electron transitions in hydrogen. In each case, the same wavelength occurs whether a photon is emitted (when the electron drops from a high orbit to a low one) or absorbed (when the electron jumps from a low orbit to a high one). The orbits are not shown to scale.

the photon emitted or absorbed as an electron jumps between an inner orbit N and an outer orbit n is:

Bohr formula for hydrogen wavelengths

$$\frac{1}{\lambda} = R\left(\frac{1}{N^2} - \frac{1}{n^2}\right)$$

N = number of inner orbit

n = number of outer orbit

R = Rydberg constant = $1.097 \times 10^7 \text{ m}^{-1}$

λ = wavelength (in meters) of emitted or absorbed photon

Letting $N = 2$ for the inner orbit in this formula yielded the formula that Balmer discovered by trial and error. Hence, Bohr deduced the meaning of the Balmer series: All the Balmer lines are produced by electrons jumping between the second Bohr orbit ($N = 2$) and higher orbits ($n = 3, 4, 5$, and so on).

Bohr's formula also correctly predicts the wavelengths of other series of spectral lines that occur at nonvisible wavelengths (**Figure 5-24**). Using $N = 1$ for the inner orbit results in the **Lyman series,** which is entirely in the ultraviolet. All the spectral lines in this series involve electron transitions between the lowest Bohr orbit and all higher orbits ($n = 2, 3, 4$, and so on). This pattern of spectral lines begins with L_α (Lyman alpha) at 122 nm and converges on L_∞ at 91.2 nm. Using $N = 3$ gives a series of infrared wavelengths called the **Paschen series.** This series, which involves transitions between the third Bohr orbit and all higher orbits, begins with P_α (Paschen alpha) at 1875 nm and converges on P_∞ at 822 nm. Additional series exist at still longer wavelengths.

Atomic Energy Levels

Today's view of the atom owes much to the Bohr model, but it is different in certain ways. The modern picture is based on **quantum mechanics,** a branch of physics developed during the 1920s that deals with photons and subatomic particles. As a result of this work, physicists no longer picture electrons as moving in specific orbits about the nucleus. Instead, electrons are now known to have both particle and wave properties and are said to occupy only certain **energy levels** in the atom.

An extremely useful way to display the structure of an atom is with an **energy-level diagram. Figure 5-25** shows such a diagram for hydrogen. The lowest energy level, called the **ground state,** corresponds to the $n = 1$ Bohr orbit. Higher energy levels, called **excited states,** correspond to successively larger Bohr orbits.

An electron can jump from the ground state ($n = 1$) up to the $n = 2$ level if the atom absorbs a Lyman-alpha photon with a wavelength of 122 nm. Such a photon has energy $E = hc/\lambda = 10.2$ eV (electron volts; see Section 5-5). That is why the energy level of $n = 2$ is shown in Figure 5-25 as having an energy 10.2 eV above that of the ground state (which is usually assigned a value of 0 eV). Similarly, the $n = 3$ level is 12.1 eV above the ground state, and so forth. Electrons can make transitions to higher energy levels by absorbing a photon or in a collision between atoms; they can make transitions to lower energy levels by emitting a photon.

In the energy-level diagram for hydrogen, the $n = \infty$ level has an energy of 13.6 eV. (This corresponds to an infinitely large "orbit" in the Bohr model.) If the electron is initially in the ground state and the atom absorbs a photon of any energy greater than 13.6 eV, the electron will be removed completely from the atom. This process is called **ionization.** A 13.6-eV photon has a wavelength of 91.2 nm, equal to the shortest wavelength in the ultraviolet Lyman series (L_∞). So any photon with a wavelength of 91.2 nm or less can ionize hydrogen. (Recall that the Planck formula $E = hc/\lambda$ tells us that the shorter the wavelength, the higher the photon energy.)

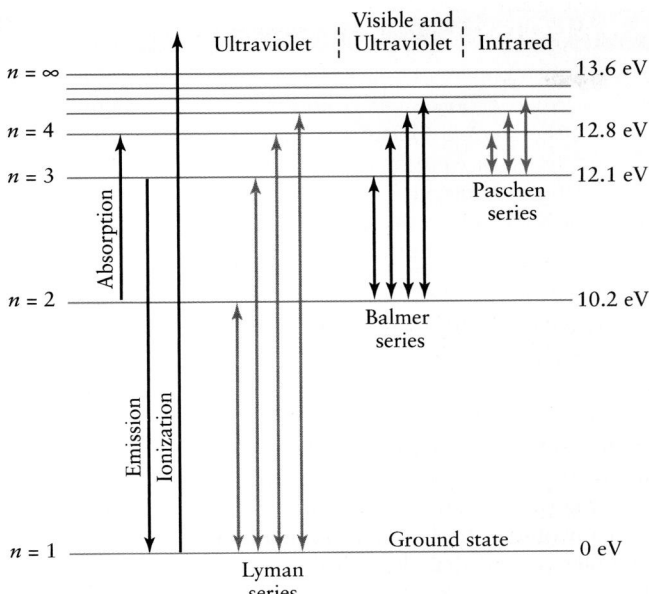

FIGURE 5-25

Energy-Level Diagram of Hydrogen A convenient way to display the structure of the hydrogen atom is in a diagram like this, which shows the allowed energy levels. The diagram shows a number of possible electron jumps, or transitions, between energy levels. An upward transition occurs when the atom absorbs a photon; a downward transition occurs when the atom emits a photon. (Compare with Figure 5-24.)

As an example, the gaseous nebula shown in Figure 5-19 surrounds a cluster of hot stars that produce copious amounts of ultraviolet photons with wavelengths less than 91.2 nm. Hydrogen atoms in the nebula that absorb these photons become ionized and lose their electrons. When the electrons recombine with the nuclei, they cascade down the energy levels to the ground state and emit visible light in the process. This process is what makes the nebula glow. Note that this emission is due to the ionization of the atoms, not the temperature of the gas, and illustrates that Kirchhoff's Law 2 is not the only way to produce an emission line spectrum.

The Spectra of Other Elements

LOOKING DEEPER 5.1 The same basic principles that explain the hydrogen spectrum also apply to the atoms of other elements. Electrons in each kind of atom can only be in certain energy levels, so only photons of certain wavelengths can be emitted or absorbed. Because each kind of atom has its own unique arrangement of electron energy levels, the pattern of spectral lines is likewise unique to that particular type of atom. This is what is meant by the statement that each atom has its own unique spectral fingerprint. For example, the unique spectra in Figure 5-16 entail emission lines of different colors, and different colors result from the unique energy levels of each atom. Because some telescopes are designed to measure the precise energy of emission lines, we can identify specific atoms throughout the universe!

Continuous Spectra

The properties of energy levels explain the emission line spectra and absorption line spectra of gases. But what about the continuous spectra produced by dense objects like the glowing coils of

a toaster or other blackbody object? These objects are made of atoms, so why don't they emit light with an emission line spectrum characteristic of the particular atoms of which they are made? (*Note:* You can see the stark differences between continuous spectra and emission line spectra by comparing Laws 1 and 2 in Figure 5-17.)

The reason is directly related to the difference between a gas on the one hand and a liquid or solid on the other. In a gas, atoms are widely separated and can emit photons without interference from other atoms. But in a liquid or a solid, atoms are so close that they almost touch, and thus these atoms interact strongly with each other. In addition, each of the many possible interactions in a liquid or solid have their own energy levels. As a result, the pattern of distinctive bright spectral lines that the atoms would emit in isolation becomes crowded with additional lines forming a continuous spectrum.

ANALOGY

Think of atoms as being like tuning forks. If you strike a single tuning fork, it produces a sound wave with a single clear frequency and wavelength, just as an isolated atom emits light of definite wavelengths. But if you shake a box packed full of tuning forks, you will hear a clanging noise that is a mixture of sounds of all different frequencies and wavelengths. This is directly analogous to the continuous spectrum of light emitted by a dense object with closely packed atoms.

With the work of such people as Planck, Einstein, Rutherford, and Bohr, the interchange between astronomy and physics came full circle. Modern physics was born when Newton set out to understand the motions of the planets. Two and a half centuries later, physicists in their laboratories probed the properties of light and the structure of atoms. Their labors had immediate applications in astronomy. Armed with this new understanding of light and matter, astronomers were able to investigate in detail the chemical and physical properties of planets, stars, and galaxies.

CONCEPT CHECK 5-12

Two photons are emitted from a hydrogen atom. One comes from the electron jumping down from the n = 3 level to the n = 1 level. The other photon comes from an electron transition from n = 2 to n = 1. Which photon has the highest frequency?

CONCEPT CHECK 5-13

Hydrogen is, by far, the most abundant element in the universe. Are any hydrogen emission lines visible to humans? If so, what color is most common?

Answers appear at the end of the chapter.

5-9 The wavelength of a spectral line is affected by the relative motion between the source and the observer

In addition to telling us about temperature and chemical composition, the spectrum of a planet, star, or galaxy can also reveal something about that object's motion through space. This idea dates from 1842, when Christian Doppler, a professor of mathematics in Prague, pointed out that the observed wavelength of light must be affected by motion.

The Doppler Effect

In Figure 5-26 a light source is moving from right to left; the circles represent the crests of waves emitted from the moving source at various positions. Each successive wave crest is emitted from a position slightly closer to the observer on the left, so she sees a shorter wavelength—the distance from one crest to the next—than she would if the source were stationary. What is the effect on a full spectrum of light? All the lines in the spectrum of an approaching source are shifted toward the short-wavelength (blue) end of the spectrum. This phenomenon is called a **blueshift**.

> The Doppler effect makes it possible to tell whether astronomical objects are moving toward us or away from us.

The source is receding from the observer on the right in Figure 5-26. The wave crests that reach him are stretched apart, so that he sees a longer wavelength than he would if the source were stationary. All the lines in the spectrum of a receding source are shifted toward the longer-wavelength (red) end of the spectrum, producing a **redshift**. In general, the effect of relative motion on wavelength is called the **Doppler effect**. Police radar guns use the Doppler effect to check for cars exceeding the speed limit: The radar gun sends a radio wave toward the car, and measures the wavelength shift of the reflected wave. During reflection of the wave, the car acts like a moving source so the Doppler shift measures the speed of the car.

Suppose that λ_0 is the wavelength of a particular spectral line from a light source that is not moving. It is the wavelength that you might look up in a textbook or determine in a laboratory experiment for this spectral line. If the source is moving, this particular spectral line is shifted to a different wavelength λ. The size of the wavelength shift is usually written as $\Delta\lambda$, where $\Delta\lambda = \lambda - \lambda_0$. Thus, $\Delta\lambda$ is the difference between the wavelength listed in textbooks and the wavelength that you actually observe in the spectrum of a moving star or galaxy.

Doppler proved that the wavelength shift ($\Delta\lambda$) is governed by the following simple equation:

Doppler shift equation

$$\frac{\Delta\lambda}{\lambda_0} = \frac{v}{c}$$

$\Delta\lambda$ = wavelength shift

λ_0 = wavelength if source is not moving

v = velocity of the source measured along the line of sight

c = speed of light = 3.0×10^5 km/s

Wave crest 1: emitted when light source was at S_1.

Wave crest 2: emitted when light source was at S_2.

Wave crests 3 and 4: emitted when light source was at S_3 and S_4, respectively.

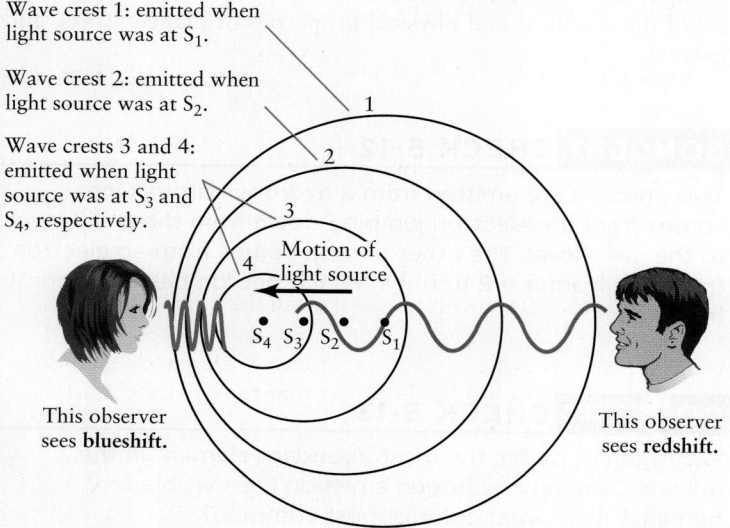

Motion of light source

This observer sees **blueshift**.

This observer sees **redshift**.

FIGURE 5-26

The Doppler Effect The wavelength of light is affected by motion between the light source and an observer. The light source shown here is moving, so wave crests 1, 2, etc., emitted when the source was at points S_1, S_2, etc., are crowded together in front of the source but are spread out behind it. Consequently, wavelengths are shortened (blueshifted) if the source is moving toward the observer and lengthened (redshifted) if the source is moving away from the observer. Motion perpendicular to an observer's line of sight does not affect wavelength.

Interpreting the Doppler Effect

The velocity determined from the Doppler effect is called **radial velocity**, because v is the component of the star's motion parallel to our line of sight, or along a "radius" drawn from Earth to the star. Of course, a sizable fraction of a star's motion might be perpendicular to our line of sight, but this sideways or transverse movement across the sky does not cause Doppler shifts. Box 5-6 includes two examples of calculations with radial velocity using the Doppler formula.

CAUTION!

The redshifts and blueshifts of stars visible to the naked eye, or even through a small telescope, are only a small fraction of a nanometer. These tiny wavelength changes are far too small to detect visually. Astronomers were able to detect the tiny Doppler shifts of starlight only after they had developed highly sensitive equipment for measuring wavelengths. This was done around 1890, a half-century after Doppler's original proposal. So, if you see a star with a red color, it means that the star really is red; it does *not* mean that it is moving rapidly away from us.

BOX 5-6   TOOLS OF THE ASTRONOMER'S TRADE

Applications of the Doppler Effect

Doppler's formula relates the radial velocity of an astronomical object to the wavelength shift of its spectral lines. Here are two examples that show how to use this remarkably powerful formula.

EXAMPLE: As measured in the laboratory, the prominent H_α spectral line of hydrogen has a wavelength $\lambda_0 = 656.285$ nm. But in the spectrum of the star Vega (Figure 5-21), this line has a wavelength $\lambda = 656.255$ nm. What can we conclude about the motion of Vega?

Situation: Our goal is to use the ideas of the Doppler effect to find the velocity of Vega toward or away from Earth.

Tools: We use the Doppler shift formula $\Delta\lambda/\lambda_0 = v/c$ to determine Vega's velocity v.

Answer: The wavelength shift is

$$\Delta\lambda = \lambda - \lambda_0 = 656.255 \text{ nm} - 656.285 \text{ nm} = -0.030 \text{ nm}$$

The negative value means that we see the light from Vega shifted to shorter wavelengths—that is, there is a blueshift. (Note that the shift is very tiny and can be measured only using specialized equipment.) From the Doppler shift formula, the star's radial velocity is

$$v = c\frac{\Delta\lambda}{\lambda_0} = (3.00 \times 10^5 \text{ km/s})\left(\frac{-0.030 \text{ nm}}{656.285 \text{ nm}}\right) = -14 \text{ km/s}$$

Review: The minus sign indicates that Vega is coming toward us at 14 km/s. *The star might also have some motion perpendicular to the line from Earth to Vega, but such motion produces no Doppler shift.*

By plotting the motions of stars such as Vega toward and away from us, astronomers have been able to learn how the Milky Way Galaxy (of which our Sun is a part) is rotating. From this knowledge, and aided by Newton's universal law of gravitation (see Section 4-6), they have made the surprising discovery that the Milky Way contains roughly 10 times more matter than had once been thought! The nature of this unseen *dark matter* is still one of the great unsolved mysteries in astronomy.

EXAMPLE: In the radio region of the electromagnetic spectrum, hydrogen atoms emit and absorb photons with a wavelength of 21.12 cm, giving rise to a spectral feature commonly called the *21-centimeter line*. The galaxy NGC 3840 in the constellation Leo (the Lion) is receding from us at a speed of 7370 km/s, or about 2.5% of the speed of light. At what wavelength do we expect to detect the 21-cm line from this galaxy?

Situation: Given the velocity of NGC 3840 away from us, our goal is to find the wavelength as measured on Earth of the 21-centimeter line from this galaxy.

Tools: We use the Doppler shift formula to calculate the wavelength shift $\Delta\lambda$, then use this to find the wavelength λ measured on Earth.

Answer: The wavelength shift is

$$\Delta\lambda = \lambda_0\left(\frac{v}{c}\right) = (21.12 \text{ cm})\left(\frac{7370 \text{ km/s}}{3.00 \times 10^5 \text{ km/s}}\right) = 0.52 \text{ cm}$$

Therefore, we will detect the 21-cm line of hydrogen from this galaxy at a wavelength of

$$\lambda = \lambda_0 + \Delta\lambda = 21.12 \text{ cm} + 0.52 \text{ cm} = 21.64 \text{ cm}$$

Review: The 21-cm line has been redshifted to a longer wavelength because the galaxy is receding from us. In fact, most galaxies are receding from us. This observation is one of the key pieces of evidence that the universe is expanding, and has been doing so since the Big Bang that took place almost 14 billion years ago.

The Doppler effect is an important tool in astronomy because it uncovers basic information about the motion of planets, stars, and galaxies. For example, the rotation of the planet Venus was deduced from the Doppler shift of radar waves reflected from its surface. Small Doppler shifts in the spectrum of sunlight have shown that the entire Sun is vibrating like an immense gong. Astronomers also use the Doppler effect along with Kepler's third law to measure a galaxy's mass. The Doppler effect will also play a major role in Chapter 8 for the discovery of planets around other stars; planets cause a star to wobble back and forth, which produces a Doppler shift in the star's spectral lines. These are but a few examples of how Doppler's discovery has empowered astronomers in their quest to understand the universe.

In this chapter, we have glimpsed how much can be learned by analyzing light from the heavens. To analyze this light, however, it is first necessary to collect as much of it as possible, because most light sources in space are very dim. Collecting the faint light from distant objects is the key purpose of telescopes. In the next chapter, we will describe how telescopes work, as well as how they are used.

CONCEPT CHECK 5-14

Compare the absorption spectrum from an approaching star to the spectrum it would have if it were stationary.

CALCULATION CHECK 5-3

How fast and in what direction is a star moving if it has a line that shifts from 486.2 nm to 486.3 nm?

Answers appear at the end of the chapter.

KEY WORDS

Terms preceded by an asterisk (*) are discussed in the Boxes.

absolute zero, p. 118
absorption line spectrum, p. 128
atom, p. 132
*atomic number, p. 134
Balmer line, p. 133
Balmer series, p. 133
blackbody, p. 122
blackbody curve, p. 122
blackbody radiation, p. 122
blueshift, p. 138
Bohr orbits, p. 133
compound, p. 128
continuous spectrum, p. 128
*degrees Celsius, p. 121
*degrees Fahrenheit, p. 121
Doppler effect, p. 138
electromagnetic radiation, p. 115

electromagnetic spectrum, p. 116
electromagnetism, p. 114
electron, p. 132
electron volt, p. 126
element, p. 128
emission line spectrum, p. 128
energy flux, p. 124
energy level, p. 136
energy-level diagram, p. 136
excited state, p. 136
frequency, p. 117
gamma rays, p. 116
ground state, p. 136
infrared radiation, p. 116
ionization, p. 136
*isotope, p. 134
joule, p. 123

kelvin, p. 118
Kirchhoff's laws, p. 128
light scattering, p. 129
*luminosity, p. 124
Lyman series, p. 136
microwaves, p. 116
molecule, p. 128
nanometer, p. 115
neutron, p. 132
nucleus, p. 132
opaque, p. 118
Paschen series, p. 136
*periodic table, p. 134
photon, p. 126
Planck's law, p. 126
proton, p. 132
quantum mechanics, p. 136
radial velocity, p. 139
radio waves, p. 116

redshift, p. 138
*solar constant, p. 124
spectral analysis, p. 128
spectral line, p. 127
spectroscopy, p. 131
spectrum (*plural* spectra), p. 113
Stefan–Boltzmann law, p. 125
thermal energy, p. 118
ultraviolet radiation, p. 116
visible light, p. 115
watt, p. 123
wavelength, p. 115
wavelength of maximum emission, p. 120
Wien's law, p. 123
x rays, p. 116

KEY IDEAS

The Nature of Light:
Light is electromagnetic radiation. It has wavelike properties described by its wavelength λ and frequency f, and travels through empty space at the constant speed $c = 3.0 \times 10^8$ m/s $= 3.0 \times 10^5$ km/s (Sections 5-1 and 5-2).

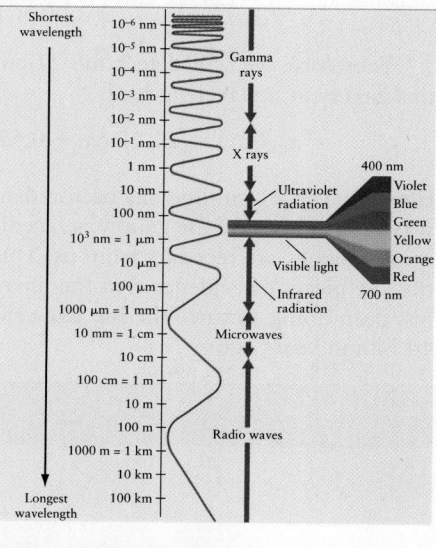

Thermal Energy:
The thermal energy of a material comes from the kinetic energy of its microscopic particles (atoms and molecules). The hotter a material, the faster its particles move, and the greater its thermal energy (Section 5-3).

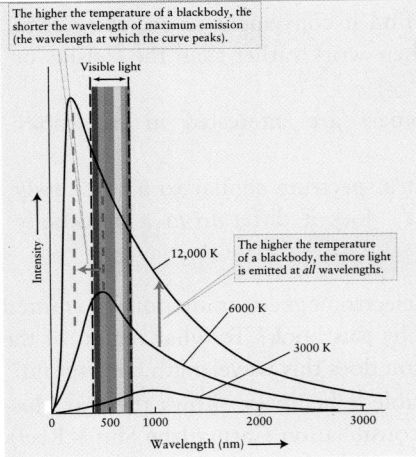

The higher the temperature of a blackbody, the shorter the wavelength of maximum emission (the wavelength at which the curve peaks).

Visible light

12,000 K

The higher the temperature of a blackbody, the more light is emitted at *all* wavelengths.

6000 K

3000 K

Intensity

Wavelength (nm)

Blackbody Radiation: A blackbody is a hypothetical object that emits a continuous spectrum; the hotter the object, the greater the emission. Stars closely approximate the behavior of blackbodies, as do other hot, dense objects (Section 5-4).

- The intensities of radiation emitted at various wavelengths by a blackbody at a given temperature are shown by a blackbody curve.
- Wien's law states that the dominant wavelength at which a blackbody emits electromagnetic radiation is inversely proportional to the Kelvin temperature of the object: λ_{max} (in meters) = $(0.0029 \text{ K m})/T$.
- The Stefan–Boltzmann law states that a blackbody radiates electromagnetic waves with a total energy flux F directly proportional to the fourth power of the Kelvin temperature T of the object: $F = \sigma T^4$.

Photons: Light is made of particles called photons. Each photon has a wavelength equal to the wavelength of the light that the photons make up (Section 5-5).

- Planck's law relates the energy E of a photon to its frequency f or wavelength λ: $E = hf = hc/\lambda$, where h is Planck's constant. The shorter the wavelength a photon has, the higher its frequency and the larger its energy.

Kirchhoff's Laws: Kirchhoff's three laws of spectral analysis describe conditions under which different kinds of spectra are produced (Section 5-6).

- A hot, dense object such as a blackbody emits a continuous spectrum covering all wavelengths.
- A hot, transparent gas produces a spectrum that contains bright (emission) lines.
- A cool, transparent gas in front of a light source that itself has a continuous spectrum produces dark (absorption) lines in the continuous spectrum.

Absorption

Incoming photon, $\lambda = 656.3$ nm

Emission

Emitted photon, $\lambda = 656.3$ nm

$n = 2$

$n = 3$

(a) Atom absorbs a 656.3-nm photon; absorbed energy causes electron to jump from the $n = 2$ orbit up to the $n = 3$ orbit.

(b) Electron falls from the $n = 3$ orbit to the $n = 2$ orbit; energy lost by atom goes into emitting a 656.3-nm photon.

Atomic Structure: An atom has a small dense nucleus composed of protons and neutrons. The nucleus is surrounded by electrons that only occupy certain orbits or energy levels (Sections 5-7 and 5-8).

- When an electron jumps from one energy level to another, it emits or absorbs a photon of corresponding energy (and hence of a specific wavelength).
- The spectral lines of a particular atom correspond to various electron orbital transitions between energy levels in the atom. Each atom has a unique "spectral fingerprint."

The Doppler Shift: The Doppler shift enables us to determine the line-of-sight (radial) velocity of a light source from the displacement of its spectral lines (Section 5-9).

- The spectral lines of an approaching light source are shifted toward short wavelengths (a blueshift); the spectral lines of a receding light source are shifted toward long wavelengths (a redshift).
- The size of a wavelength shift is proportional to the radial velocity of the light source relative to the observer.

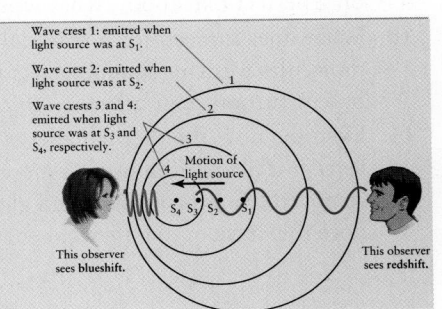

Wave crest 1: emitted when light source was at S_1.

Wave crest 2: emitted when light source was at S_2.

Wave crests 3 and 4: emitted when light source was at S_3 and S_4, respectively.

Motion of light source

This observer sees blueshift.

This observer sees redshift.

QUESTIONS

> ### Problem-solving tips and tools
>
> You can find formulas in Box 5-1 for converting between temperature scales. Box 5-2 discusses how a star's radius, luminosity, and surface temperature are related. Box 5-3 shows how to use Planck's law to calculate the energy of a photon. To learn how to do calculations using the Doppler effect, see Box 5-6.

Section 5-1

1. When Jupiter is undergoing retrograde motion as seen from Earth, would you expect the eclipses of Jupiter's moons to occur several minutes early, several minutes late, or neither? Explain your answer.

2. Approximately how many times around Earth could a beam of light travel in 1 second?

3. How long does it take light to travel from the Sun to Earth, a distance of 1.50×10^8 km?

Section 5-2

4. **TUTORIAL 5.1** How did Newton show that a prism breaks white light into its component colors but does not add any color to the light?

5. For each of the following wavelengths, state whether it is in the radio, microwave, infrared, visible, ultraviolet, x-ray, or gamma-ray portion of the electromagnetic spectrum. Explain your reasoning. (a) 2.6 μm, (b) 34 m, (c) 0.54 nm, (d) 0.0032 nm, (e) 0.620 μm, (f) 310 nm, (g) 0.012 m

6. What is meant by the frequency of light? How is frequency related to wavelength?

7. A cellular phone is actually a radio transmitter and receiver. You receive an incoming call in the form of a radio wave of frequency 880.65 MHz. What is the wavelength (in meters) of this wave?

8. A light source emits infrared radiation at a wavelength of 1150 nm. What is the frequency of this radiation?

Section 5-3

9. **TUTORIAL 5.2** (a) What is a blackbody? (b) In what way is a blackbody black? (c) If a blackbody is black, how can it emit light? (d) If you were to shine a flashlight beam on a perfect blackbody, what would happen to the light?

10. What does it mean for a material to be opaque? Can water molecules form an opaque object in some forms and a transparent substance in others? (*Hint:* Think of clouds.)

11. Describe how the thermal energy changes at a microscopic level as a blacksmith heats a piece of iron. If the iron is then removed from the furnace and glows bright red, what effect does this have on the thermal energy?

12. Why do astronomers find it convenient to use the Kelvin temperature scale in their work rather than the Celsius or Fahrenheit scales?

13. Explain why astronomers are interested in blackbody radiation.

14. In what way is the Sun's spectrum similar to a blackbody spectrum? In what way does it differ from a blackbody spectrum?

15. What wavelength of electromagnetic radiation is emitted with greatest intensity by this book? To what region of the electromagnetic spectrum does this wavelength correspond?

16. The accompanying visible-light image shows the star cluster NGC 3293 in the constellation Carina (the Ship's Keel). What can you say about the surface temperatures of most of the bright stars in this cluster? In what part of the electromagnetic spectrum do these stars emit most intensely? Are your eyes sensitive to this type of radiation? If not, how is it possible to see these stars at all? There is at least one bright star in this cluster with a distinctly different color from the others; what can you conclude about its surface temperature?

R I **V** U X G   (David Malin/Australian Astronomical Observatory)

Section 5-4

17. Using Wien's law and the Stefan–Boltzmann law, explain the color and intensity changes that are observed as the temperature of a hot, glowing object increases.

18. If you double the Kelvin temperature of a hot piece of steel, how much more energy will it radiate per second?

19. The bright star Bellatrix in the constellation Orion has a surface temperature of 21,500 K. What is its wavelength of maximum emission in nanometers? What color is this star?

20. The bright star Antares in the constellation Scorpius (the Scorpion) emits the greatest intensity of radiation at a

wavelength of 853 nm. What is the surface temperature of Antares? What color is this star?

21. Your normal body temperature is 98.6°F. What kind of radiation do you predominantly emit? At what wavelength (in nm) do you emit the most radiation?

22. Black holes are objects whose gravity is so strong that not even an object moving at the speed of light can escape from their surfaces. Hence, black holes do not themselves emit light. But it is possible to detect radiation from material falling *toward* a black hole. Calculations suggest that as this matter falls, it is compressed and heated to temperatures around 10^6 K. Calculate the wavelength of maximum emission for this temperature. In what part of the electromagnetic spectrum does this wavelength lie?

23. **Box 5-2** Use the value of the solar constant given in Box 5-2 and the distance from Earth to the Sun to calculate the luminosity of the Sun.

24. **Box 5-2** The star Alpha Lupi (the brightest in the constellation Lupus the Wolf) has a surface temperature of 21,600 K. How much more energy is emitted each second from each square meter of the surface of Alpha Lupi than from each square meter of the Sun's surface?

25. **Box 5-2** Jupiter's moon Io has an active volcano named Pele whose temperature can be as high as 320°C. (a) What is the wavelength of maximum emission for the volcano at this temperature? In what part of the electromagnetic spectrum is this? (b) The average temperature of Io's surface is −150°C. Compared with a square meter of surface at this temperature, how much more energy is emitted per second from each square meter of Pele's surface?

26. **Box 5-2** The bright star Sirius in the constellation of Canis Major (the Large Dog) has a radius of 1.67 R$_\odot$ and a luminosity of 25 L$_\odot$. (a) Use this information to calculate the energy flux at the surface of Sirius. (b) Use your answer in part (a) to calculate the surface temperature of Sirius. How does your answer compare to the value given in Box 5-2?

Section 5-5

27. (a) Describe an experiment in which light behaves like a wave. (b) Describe an experiment in which light behaves like a particle.

28. How is the energy of a photon related to its wavelength? What kind of photons carry the most energy? What kind of photons carry the least energy?

29. To emit the same amount of light energy per second, which must emit more photons per second: a source of red light or a source of blue light? Explain your answer.

30. Instruments on board balloons and spacecraft detect 511-keV photons coming from the direction of the center of our galaxy. (The prefix k means "kilo," or "thousand," so 1 keV = 10^3 eV.) What is the wavelength of these photons? To what part of the electromagnetic spectrum do these photons belong?

31. The equation that relates the frequency, wavelength, and speed of a light wave, $f = c/\lambda$, can be rewritten as $c = f\lambda$. A friend who has studied mathematics but not much astronomy or physics might look at this equation and say: "This equation tells me that the higher the frequency f, the greater the wave speed c. Since visible light has a higher frequency than radio waves, this means that visible light goes faster than radio waves." How would you respond to your friend?

Section 5-6

32. Certain interstellar clouds contain a very cold, very thin gas of hydrogen atoms. Ultraviolet radiation with any wavelength shorter than 91.2 nm cannot pass through this gas; instead, it is absorbed. Explain why.

33. Which one of Kirchhoff's laws best describes the spectrum from the red-hot filament in a toaster? After the toaster has been off for a while and has cooled down to room temperature, do any of Kirchhoff's laws apply to the filament?

34. The Sun's spectrum is altered as it passes through Earth's atmosphere. Which of Kirchhoff's laws would describe the alteration of the Sun's spectrum?

35. Some streetlamps operate by emitting light from sodium vapor. For many of these lamps, about 90% of the light is emitted by emission lines at 589.0 nm and 589.6 nm. Which of Kirchhoff's laws best describes this emission?

Section 5-8

36. (a) Describe the spectrum of hydrogen at visible wavelengths. (b) Explain how Bohr's model of the atom accounts for the Balmer lines.

37. Why do different elements display different patterns of lines in their spectra?

38. (a) Calculate the wavelength of P$_\Delta$ (P-delta), the fourth wavelength in the Paschen series. (b) Draw a schematic diagram of the hydrogen atom and indicate the electron transition that gives rise to this spectral line. (c) In what part of the electromagnetic spectrum does this wavelength lie?

39. (a) Calculate the wavelength of H$_\eta$ (H-eta), the spectral line for an electron transition between the $n = 7$ and $n = 2$ orbits of hydrogen. (b) In what part of the electromagnetic spectrum does this wavelength lie? Use this to explain why Figure 5-21 is labeled R I V U X G.

40. (a) Can a hydrogen atom in the ground state absorb an H-alpha (H$_\alpha$) photon? Explain why or why not. (b) Can a hydrogen atom in the $n = 2$ state absorb a Lyman-alpha (L$_\alpha$) photon? Explain why or why not.

41. An imaginary atom has just three energy levels: 0 eV, 1 eV, and 3 eV. Draw an energy-level diagram for this atom. Show all possible transitions between these energy levels. For each transition, determine the photon energy and the photon wavelength. Which transitions involve the emission or absorption of visible light?

Section 5-9

42. What is the Doppler effect? Why is it important to astronomers?

43. If you see a blue star, what does its color tell you about how the star is moving through space? Explain your answer.

44. The star cluster NGC 346 and nebula shown in Figure 5-19 are located within the Small Magellanic Cloud (SMC), a small galaxy that orbits our Milky Way Galaxy. The SMC and the stars and gas within it are moving away from us at 158 km/s. At what wavelength does the red H_α line of hydrogen (which causes the color of the nebula) appear in the nebula's spectrum?

45. The wavelength of H_β in the spectrum of the star Megrez in the Big Dipper (part of the constellation Ursa Major the Great Bear) is 486.112 nm. Laboratory measurements demonstrate that the normal wavelength of this spectral line is 486.133 nm. Is the star coming toward us or moving away from us? At what speed?

46. You are given a traffic ticket for going through a red light (wavelength 700 nm). You tell the police officer that because you were approaching the light, the Doppler effect caused a blueshift that made the light appear green (wavelength 500 nm). How fast would you have had to be going for this to be true? Would the speeding ticket be justified? Explain your answer.

General Problems

47. Which part of the electromagnetic spectrum contains light with a higher frequency: microwaves or radio waves?

48. Can a satellite convert some of its orbital energy into thermal energy? If yes, describe a context in which this might occur.

49. (a) If you could see ultraviolet radiation, how might the night sky appear different? Would ordinary objects appear different in the daytime? (b) What differences might there be in the appearance of the night sky and in the appearance of ordinary objects in the daytime if you could see infrared radiation?

50. The human eye is most sensitive over the same wavelength range at which the Sun emits the greatest intensity of radiation. Suppose creatures were to evolve on a planet orbiting a star somewhat hotter than the Sun. To what wavelengths would their vision most likely be sensitive?

51. Why do you suppose that ultraviolet light can cause skin cancer but ordinary visible light does not?

WEB/E-BOOK QUESTION

52. Search the internet for information about rainbows. Why do rainbows form? Why do they appear as circular arcs? Why can you see different colors?

ACTIVITIES

Observing Projects

53. Turn on an electric stove or toaster oven and carefully observe the heating elements as they warm up. Relate your observations to Wien's law and the Stefan–Boltzmann law.

Starry Night™ Explorations

54. *Starry Night™* can be used to measure the speed of light by simulating the observation of an event from two sites at different distances away from this event and measuring the difference of arrival times at the two sites. The event is the emergence of Jupiter's moon Io from Jupiter's shadow. The two observing locations from which you will time the arrival of that light signal are at the north poles of the Earth and the planet Mercury, respectively. The difference in the distances between Io and these two locations at the time the signal arrives divided by the difference between the arrival times of the signal at the two locations will give you a measure of the speed of light. Open **Favorites** > **Explorations** > **Io from Earth**. The view is centered on Io, which at this time is hidden in Jupiter's shadow. Click **Play** to watch the moon emerge from this shadow. Then zoom in to fill the view with Io and use the time controls to freeze time at the point where half of Io's disk has emerged from eclipse. Record the universal time from the on-screen info for this event and then use the HUD to find the Distance from Observer of Io at this time. Now open **Favorites** > **Exploration** > **Io from Mercury** to observe this event from the north pole of Mercury. Again, use the time controls to find the time at which half of Io's disk has emerged from the shadow and record the universal time of the event and the distance to Io from this location. Use these observations to calculate the speed of light. First, calculate the difference of the distance, in au, of the two locations from Io. Then calculate the difference, in seconds, between the arrival times of the light signal at the two locations. Divide the difference in the distance by the difference in time to calculate the speed of light in au per second. Finally, convert this value to kilometers per second by multiplying the result by 1.496×10^8 (the number of kilometers in one au).

   (a) What result do you obtain for the speed of light and how does it compare to the accepted value of the speed of light of 2.9979×10^5 km/s?

   (b) Your calculated value might not agree with the accepted value for the speed of light. Estimate what an error of 1 second in the timing difference would make to your value for this speed by recalculating your values with 1 fewer or 1 extra second for the time difference. Will such an observational error explain the inaccuracy in your measured value for the speed of light?

55. Use *Starry Night*™ to examine the colors and temperatures of several relatively nearby stars. Select **Favorites** > **Explorations** > **Atlas** and then open the **Observing Lists** panel. In this panel, expand the list named **Stellar Temperatures**. For each star in the list, click the contextual menu icon for the star, select **Center** and then click the checkbox to apply a label to the star in the view. Zoom in on the star to estimate the approximate color of the star and then use the HUD to obtain the star's temperature. Record your color estimate and the temperature for each star.
    (a) Which of the stars have a longer wavelength of maximum emission λ_{max} than that of the Sun?
    (b) Which of the stars have a shorter λ_{max} than the Sun?
    (c) Which of the stars has a definite reddish color?

56. Use *Starry Night*™ to compare the brightness of two similarly sized stars in the constellation Auriga. Select **Favorites** > **Explorations** > **Auriga**. The two stars Capella and Delta Aurigae are indicated in this view. Use the HUD to find the temperature and radius for both of these stars. Note that these two stars have the same radius but differ in temperature. From these data, which of these stars is intrinsically brighter and by what proportion?

57. Use *Starry Night*™ to examine several different celestial objects and consider the type of spectrum you would expect each object to radiate. Select **Favorites** > **Explorations** > **Spectra** from the menu to show a view of the sky as seen from Bozeman, Montana. Three celestial objects, M31 (the Andromeda galaxy), M8 (the Lagoon nebula) and the Moon are indicated. For each object, click the name in the label and select **Magnify** to examine the object's appearance and then use **File** > **Revert** to reset to the wide-angle view. For each object decide whether you think it will have a continuous spectrum, an absorption line spectrum, or an emission line spectrum, and explain your reasoning. Keep in mind the facts that the light coming from a galaxy is the combined light of hundreds of billions of individual stars and moonlight is simply reflected sunlight.

Collaborative Exercise

58. The Doppler effect describes how relative motion impacts wavelength. With a classmate, stand up and demonstrate each of the following: (a) a blueshifted source for a stationary observer; (b) a stationary source and an observer detecting a redshift; and (c) a source and an observer both moving in the same direction, but the observer is detecting a redshift. Create simple sketches to illustrate what you and your classmate did.

ANSWERS

ConceptChecks

ConceptCheck 5-1: The speed of light is incredibly fast, which made it very difficult to measure light's speed except over enormous distances.

ConceptCheck 5-2: As Newton found when passing sunlight through a series of prisms, when one color is isolated from white light, there are no longer any other colors present in the remaining light. As a result, you cannot turn pure red into any other color. White light contains all of the colors, but once any of those colors are absorbed, they cannot be recovered. The red plastic absorbs all the visible wavelengths other than red, so no green can be obtained.

ConceptCheck 5-3: The width of your finger is about 1 cm, which falls in the range of the wavelength of microwaves (1 mm–10 cm).

ConceptCheck 5-4: Because the relationship between wavelength and frequency is $c = \lambda \times f$, as one increases the other decreases (wavelength and frequency are inversely related). Thus, the longest wavelengths have the lowest frequencies and the shortest wavelengths have the highest frequencies.

ConceptCheck 5-5: As shown in Figure 5-12, at every wavelength, including infrared, the higher the temperature of a blackbody, the more energy it emits.

ConceptCheck 5-6: Yes. An object at 0°C still has a temperature of 237 K, and kelvins are the temperature unit for working with blackbodies. At 273 K, the blackbody would peak in the infrared, although humans cannot see this light.

ConceptCheck 5-7: A star is hotter than our Sun if the star's peak wavelength of blackbody emission is a shorter wavelength than the peak from our Sun.

ConceptCheck 5-8: Yes. If the cooler object is much larger than the warmer object, the cooler object can emit a greater total energy. The energy flux F refers to the energy emitted per second for each square meter of surface area, so even for a cool object, the larger the object, the more energy it radiates. This is why some stars (called red giants) that are much larger than our Sun are actually brighter than our Sun, even though they are cooler.

ConceptCheck 5-9: Photons with longer wavelengths will have lower energy than those with shorter wavelengths because the greater the wavelength, the lower the energy of a photon associated with that wavelength.

ConceptCheck 5-10: While still hot, the Sun's outer atmosphere is cooler than deeper down where most of the Sun's continuous blackbody emission originates. Thus, Figure 5-14 is an absorption line spectrum described by Kirchhoff's third law. Kirchhoff's second law would describe the emission line spectrum from hot helium gas, as in Figure 5-16.

ConceptCheck 5-11: An absorption spectrum results when the light from a hot, dense object passes through the cooler, transparent gas of our atmosphere. At visible wavelengths, the atmosphere does not produce a significant absorption spectrum, but at infrared wavelengths (which are also emitted by hot lava), the atmosphere has many absorption lines.

ConceptCheck 5-12: The greater a photon's energy, the higher its frequency. Since the change in energy for the $n = 3$ to $n = 1$ jump is the larger of the two transitions (see Figure 5-25), its emitted photon has the highest frequency. You can also compare the wavelengths of these two transitions in Figure 5-24, and the shorter the wavelength of light, the higher its frequency.

ConceptCheck 5-13: Yes. The most common visible-light hydrogen transition emits red photons of wavelength 656 nm. It is an emission line (see Figure 5-24) with a transition from $n = 3$ to $n = 2$ and is referred to as H_α (H-alpha). This single transition accounts for most of the red seen in pretty nebulae such as Figure 5-19.

ConceptCheck 5-14: When the distance between an observer and a source is decreasing, the source's entire spectrum will be shifted toward shorter wavelengths; alternatively, when the distance between an object and a source is increasing, the emissions lines will be shifted toward longer wavelengths. Thus, the star's absorption lines will be observed at shorter wavelengths.

CalculationCheck

CalculationCheck 5-1: This classic rock station in Santa Barbara, California, emits waves with a frequency of 99.9 MHz, but this can be rounded to 100 MHz. To calculate the wavelength of these radio waves, we rearrange the equation $c = f \div \lambda$ to get $\lambda = 3 \times 10^8$ m/s $\div 100 \times 10^6$ Hz = 3.00 m.

CalculationCheck 5-2: Wien's law can be rearranged to calculate the temperature of a star as $T = 0.0029$ K m $\div$ (5800 K $\times$ 2) = 250 nm, which is ultraviolet.

CalculationCheck 5-3: $f = c \times \Delta\lambda \div \lambda_0 = 3 \times 10^5$ km/s $\times$ (486.3 nm $-$ 486.2 nm) $\div$ 486.2 nm = 61.7 km/s, and because it is moving toward longer wavelengths, the distance between the observer and the star must be increasing.

Visible-light image of galaxy M82

Infrared image of galaxy M82

LEARNING GOALS

There is more to the universe than meets the eye—literally! As seen through a small telescope, the galaxy M82 (shown in the above image) appears as a bright patch that glows with the light of its billions of stars. But when observed with a telescope sensitive to infrared light—with wavelengths longer than your eye can detect—M82 displays an immense halo that extends for tens of thousands of light-years.

The spectrum of this halo reveals it to be composed of tiny dust particles, which are ejected by newly formed stars. The halo's immense size shows that new stars are forming in M82 at a far greater rate than within our own galaxy.

These observations are just one example of the tremendous importance of telescopes to astronomy. Whether a telescope detects visible or nonvisible light, its fundamental purpose is the same: to gather more light than the unaided human eye. Telescopes are used to gather the feeble light from distant objects to make bright, sharp images. Telescopes also produce finely detailed spectra of objects in space. These spectra reveal the chemical compositions of nearby planets as well as of distant galaxies and help astronomers understand the nature and evolution of astronomical objects of all kinds.

Modern telescopes can view the universe in every range of electromagnetic radiation, although some must be placed above Earth's atmosphere. (NASA/JPL-Caltech/C. Engelbracht [University of Arizona]; NASA)

R I V U X G

As the opening images of the galaxy M82 show, telescopes that collect nonvisible light reveal otherwise hidden aspects of the universe. In addition to infrared telescopes, radio telescopes have mapped out the structure of our Milky Way Galaxy, ultraviolet telescopes have revealed the workings of the Sun's outer atmosphere, and gamma-ray telescopes have detected the most powerful explosions in the universe. The telescope, in all its variations, is by far astronomers' most useful tool for collecting data about the universe.

6-1 A refracting telescope uses a lens to concentrate incoming light at a focus

The **optical telescope**—that is, a telescope designed for use with visible light—was invented in the Netherlands in the early seventeenth century. Soon after, Galileo used one of these new inventions for his groundbreaking astronomical observations (see Section 4-5). These first telescopes used carefully shaped pieces of glass, or **lenses**, to make distant objects

> Refracting telescopes gave humans the first close-up views of the Moon and planets.

appear larger and brighter. Telescopes of this same basic design are used today by many amateur astronomers. To understand telescopes of this kind, we need to understand how lenses work.

Refraction of Light

Lenses used in telescopes gather and focus light. Larger lenses can gather more light, which is why professional telescopes must be big. Focusing light requires that the light's path be bent. The ability of a lens to bend light is based on a universal fact: *Light travels at a slower speed in a dense substance.*

Thus, although the speed of light *in a vacuum* is 3.00×10^8 m/s, its speed in glass is less than 2×10^8 m/s. Imagine you are walking briskly down a boardwalk before stepping onto a sandy beach. Just as your pace slows suddenly as you hit the sand, so light slows abruptly as it enters a piece of glass. When you step back onto the boardwalk, you can easily resume your original pace; in the same way, light resumes its original speed upon exiting the glass.

A material through which light travels is called a **medium** (*plural* **media**). As a beam of light passes from one transparent medium into another—say, from air into glass, or from glass back into air—the direction of the light can change. This path-bending phenomenon, called **refraction**, is caused by the change in the speed of light.

Figure 6-2a shows the refraction of a beam of light passing through a piece of flat glass. As the beam enters the upper surface of the glass, refraction takes place, and the beam is bent to a direction

(a) How cars behave

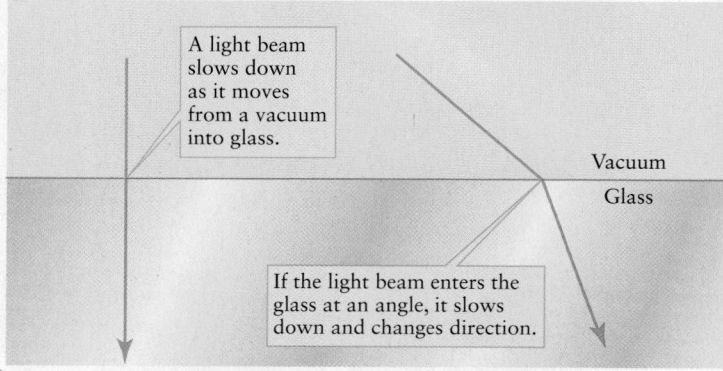

(b) How light beams behave

FIGURE 6-1

Refraction (a) When a car drives from smooth pavement into soft sand, it slows down. If it enters the sand at an angle, the front wheel on one side feels the drag of the sand before the other, causing the car to veer to the side and change direction. **(b)** Similarly, light slows down when it passes from a vacuum into glass and changes direction if it enters the glass at an angle.

more nearly perpendicular to the surface of the glass. As the beam exits from the glass back into the surrounding air, a second refraction takes place, and the beam bends in the opposite sense. The amount of bending depends on the speed of light in the glass, so different kinds of glass produce slightly different amounts of refraction. Because the two surfaces of the glass are parallel, the beam emerges from the glass traveling in the same direction at which it entered.

Lenses and Refracting Telescopes

Something more useful happens if the glass is curved into a convex shape (one that is fatter in the middle than at the edges), like the lens in Figure 6-2b. When a beam of incoming light rays passes through the lens, refraction causes all the outgoing rays to converge at a point called the **focus**. If the light rays entering the lens are all parallel, the focus occurs at a special point called the **focal point**. The distance from the lens to the focal point is called the **focal length** of the lens.

The case of incoming parallel light rays, shown in Figure 6-2b, is not merely a theoretical ideal. The stars are so far away that light rays

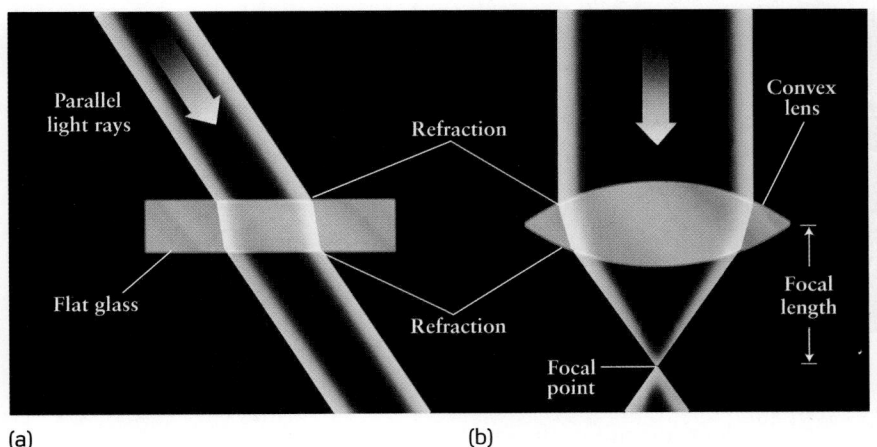

FIGURE 6-2

Refraction and Lenses (a) Refraction is the change in direction of a light ray when it passes into or out of a transparent medium such as glass. When light rays pass through a flat piece of glass, the two refractions bend the rays in opposite directions. There is no overall change in the direction in which the light travels. **(b)** If the glass is in the shape of a convex lens, parallel light rays converge to a focus at a special point called the focal point. The distance from the lens to the focal point is called the focal length of the lens.

from them are essentially parallel, as Figure 6-3 shows. Consequently, a lens always focuses light from an astronomical object to the focal point. If the object has a very small angular size, like a distant star, all the light entering the lens from that object converges onto the focal point. The resulting image is just a single bright dot.

However, if the object is *extended*—that is, if it has a relatively large angular size, like the Moon or a nearby planet—then light coming from each point on the object is brought to a focus at its own individual point. The result is a clearly focused and extended image that lies in the **focal plane** of the lens (Figure 6-4), which is a plane that

FIGURE 6-3

Light Rays from Distant Objects Are Parallel Light rays travel away in all directions from an ordinary light source. If a lens is located very far from the light source, only a few of the light rays will enter the lens, and these rays will be essentially parallel. This observation is why we drew parallel rays entering the lens in Figure 6-2b.

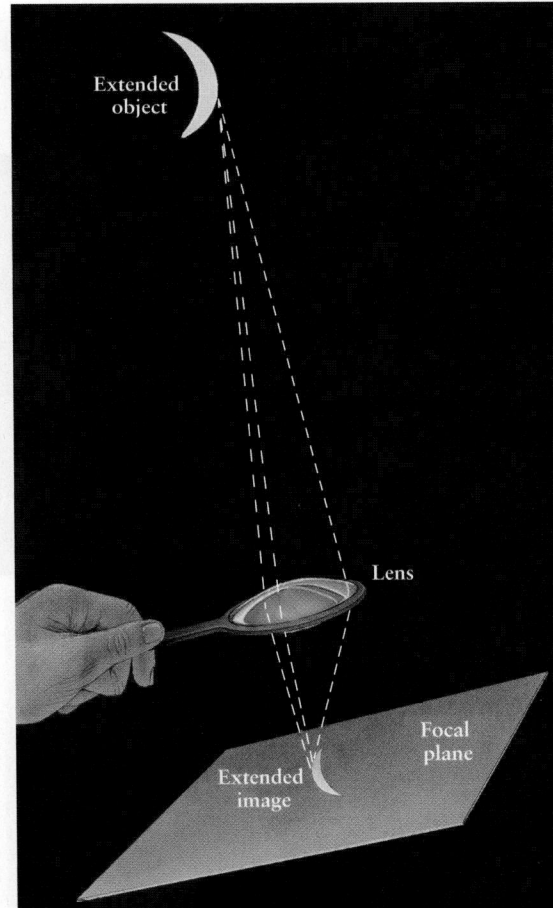

FIGURE 6-4

A Lens Creates an Extended Image of an Extended Object Light coming from each point on an extended object passes through a lens and produces an image of that point. All of these tiny images put together make an extended image of the entire object. The image of a very distant object is formed in a plane called the focal plane. The distance from the lens to the focal plane is called the focal length.

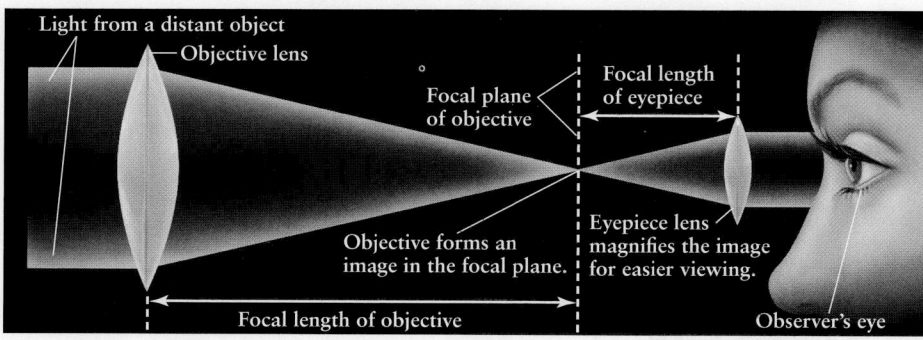

ANIMATION
6.1 ➜

FIGURE 6-5

A Refracting Telescope A refracting telescope consists of a large-diameter objective lens with a long focal length and a small eyepiece lens of short focal length. The eyepiece lens magnifies the image formed by the objective lens in its focal plane (shown as a dashed line). To take a photograph, the eyepiece is removed and an electronic detector is placed in the focal plane.

includes the focal point. You can use an ordinary magnifying glass in this way to make an image of the Sun on the ground.

To use a lens to make a permanent picture of an astronomical object, you would place an electronic detector in the focal plane—the same type of detector used in digital cameras. In fact, an ordinary digital camera works in a very similar way for photographing objects here on Earth. However, many amateur astronomers want to view the image with their eye, not a camera, so they add a second lens to magnify the image formed in the focal plane. Such an arrangement of two lenses is called a **refracting telescope,** or **refractor** (Figure 6-5). The large-diameter, long-focal-length lens at the front of the telescope, called the **objective lens,** forms the image; the smaller, shorter-focal-length lens at the rear of the telescope, called the **eyepiece lens,** magnifies the image for the observer.

Light-Gathering Power

In addition to the focal length, the other important dimension of the objective lens of a refractor is its diameter. Compared with a small-diameter lens, a large-diameter lens captures more light, produces brighter images, and allows astronomers to detect fainter objects. (For the same reason, the iris of your eye opens when you go into a darkened room to allow you to see dimly lit objects.)

The **light-gathering power** of a telescope is directly proportional to the area of the objective lens, which in turn is proportional to the square of the lens diameter (Figure 6-6). Thus, if you

Small-diameter objective lens:
dimmer image, less detail

Large-diameter objective lens:
brighter image, more detail

FIGURE 6-6 R I V U X G

Light-Gathering Power These two photographs of the galaxy M31 in Andromeda were taken using the same exposure time and at the same magnification, but with two different telescopes with objective lenses of different diameters. The right-hand photograph is brighter and shows more detail because it was made using the large-diameter lens, which intercepts more starlight than a small-diameter lens. This same principle applies to telescopes that use curved mirrors rather than lenses to collect light (see Section 6-2). (Association of Universities for Research in Astronomy) double the diameter of the lens, the light-gathering power increases by a factor of $2^2 = 2 \times 2 = 4$. Box 6-1 describes how to compare the light-gathering power of different telescopes.

Because light-gathering power is so important for seeing faint objects, the lens diameter is almost always given when describing a telescope. For example, the Lick telescope on Mount Hamilton in California is a 90-cm refractor, which means that it is a refracting telescope whose objective lens is 90 cm in diameter. By comparison, Galileo's telescope of 1610 was a 3-cm refractor. The Lick telescope has an objective lens 30 times larger in diameter, and so has 30^2 or $30 \times 30 = 900$ times the light-gathering power of Galileo's instrument.

CONCEPT CHECK 6-1

If someone says he or she is using an 8-in. telescope, which dimension of the telescope—length or tube diameter—is he or she most likely referring to?

CONCEPT CHECK 6-2

If a thick lens is able to bend light more than a thin lens, which lens has a greater focal length?

Answers appear at the end of the chapter.

BOX 6-1   TOOLS OF THE ASTRONOMER'S TRADE

Magnification and Light-Gathering Power

The magnification of a telescope is equal to the focal length of the objective divided by the focal length of the eyepiece. Telescopic eyepieces are usually interchangeable, so the magnification of a telescope can be changed by using eyepieces of different focal lengths.

EXAMPLE: A small refracting telescope has an objective focal length of 120 cm. If the eyepiece has a focal length of 4 cm, what is the magnification of the telescope?

Situation: We are given the focal lengths of the telescope's objective and eyepiece lenses. Our goal is to calculate the magnification provided by this combination of lenses.

Tools: We use the relationship that the magnification equals the focal length of the objective (120 cm) divided by the focal length of the eyepiece (4 cm).

Answer: Using this relationship,

$$\text{Magnification} = \frac{120 \text{ cm}}{4 \text{ cm}} = 30 \text{ (usually written as } 30\times)$$

Review: A magnification of $30\times$ means that, as viewed through this telescope, a large lunar crater that subtends an angle of 1 arcminute to the naked eye will appear to subtend an angle 30 times greater, or 30 arcminutes (one-half of a degree). This magnification makes the details of the crater much easier to see.

If a 2-cm-focal-length eyepiece is used instead, the magnification will be (120 cm)/(2 cm) = $60\times$. The shorter the focal length of the eyepiece, the greater the magnification.

The light-gathering power of a telescope depends on the diameter of the objective lens; it does not depend on the focal length. The light-gathering power is proportional to the square of the diameter. As an example, a fully dark adapted human eye has a pupil diameter of about 5 mm. By comparison, a small telescope whose objective lens is 5 cm in diameter has 10 times the diameter and $10^2 = 100$ times the light-gathering power

of the eye. (Recall that there are 10 mm in 1 cm.) Hence, this telescope allows you to see objects 100 times fainter than you can see without a telescope.

EXAMPLE: The same relationships apply to reflecting telescopes, discussed in Section 6-2. Each of the two Keck telescopes on Mauna Kea in Hawaii (discussed in Section 6-3; see Figure 6-16) uses a concave mirror 10 m in diameter to bring starlight to a focus. How many times greater is the light-gathering power of either Keck telescope compared to that of the human eye?

Situation: We are given the diameters of the pupil of the human eye (5 mm) and of the mirror of either Keck telescope (10 m). Our goal is to compare the light-gathering powers of these two optical instruments.

Tools: We use the relationship that light-gathering power is proportional to the square of the diameter of the mirror (or pupil) that gathers light. Hence, the *ratio* of the light-gathering powers is equal to the square of the ratio of the diameters.

Answer: We first calculate the ratio of the diameter of the Keck mirror to the diameter of the pupil. To determine this ratio, we must first express both diameters in the same units. Because there are 1000 mm in 1 meter, the diameter of the Keck mirror can be expressed as

$$10 \text{ m} \times \frac{1000 \text{ mm}}{1 \text{ m}} = 10,000 \text{ mm}$$

Thus, the light-gathering power of either of the Keck telescopes is greater than that of the human eye by a factor of

$$\frac{(10,000 \text{ mm})^2}{(5 \text{ mm})^2} = (2000)^2 = 4 \times 10^6 = 4,000,000$$

Review: Each Keck telescope can gather *4 million* times as much light as a human eye. When it comes to light-gathering power, the bigger the telescope, the better!

Magnification

In addition to their light-gathering power, telescopes are useful because they magnify distant objects. As an example, the angular diameter of the Moon as viewed with the naked eye is about 0.5°. But when Galileo viewed the Moon through his telescope, its apparent angular diameter was 10°, large enough that he could identify craters and mountain ranges. The **magnification,** or **magnifying power,** of a telescope is the ratio of an object's angular diameter seen through the telescope to its naked-eye angular diameter. Thus, the magnification of Galileo's telescope was 10°/0.5° = 20 times, usually written as 20×.

The magnification of a refracting telescope depends on the focal lengths of both of its lenses:

$$\text{Magnification} = \frac{\text{focal length of objective lens}}{\text{focal length of eyepiece lens}}$$

This formula shows that using a long-focal-length objective lens with a short-focal-length eyepiece gives a large magnification. Box 6-1 illustrates how this formula is used.

Disadvantages of Refracting Telescopes

If you were to build a telescope like the one shown in Figure 6-5 using only the instructions given so far, you would probably be disappointed with the results. The problem is that a lens bends different colors of light through different angles, just as a prism does (recall Figure 5-3). As a result, different colors do not focus at the same point, and stars viewed through a telescope that uses a simple lens are surrounded by fuzzy, rainbow-colored halos. Figure 6-7a shows this optical defect, which is called **chromatic aberration.**

One way to correct for chromatic aberration is to use an objective lens that is not just a single piece of glass. Different types of glass can be manufactured by adding small amounts of chemicals to the glass when it is molten. Because of the different properties of these added chemicals, the speed of light varies slightly from one kind of glass to another, and the refractive properties vary as well. If a thin lens is mounted just behind the main objective lens of a telescope, as shown in Figure 6-7b, and if the telescope designer carefully chooses two different kinds of glass for these two lenses, different colors of light can be brought to a focus at the same point.

Chromatic aberration is only the most severe of a host of optical problems that must be solved in designing a high-quality

(a) The problem: chromatic aberration.

(b) The solution: Use two lenses.

FIGURE 6-7

Chromatic Aberration (a) A single lens suffers from a defect called chromatic aberration, in which different colors of light are brought to a focus at different distances from the lens. While the aberration occurs for all colors, only red and blue are shown here. **(b)** This problem can be corrected by adding a second lens made from a different kind of glass.

Objective lens made of two different types of glass (see Figure 6-7b)

(a)

(b)

FIGURE 6-8 R I 🅥 U X G

A Large Refracting Telescope (a) This giant refractor, built in 1897, is housed at Yerkes Observatory near Chicago. The telescope tube is 19.5 m (64 ft) long; it has to be this long because the focal length of the objective is just under 19.5 m (see Figure 6-5). As Earth rotates, the motor drive rotates the telescope in the opposite direction in order to keep the object of interest within the telescope's field of view. **(b)** This historical photograph shows the astronomer George van Biesbrock with the objective lens of the Yerkes refractor. This lens, the largest refracting lens still in use, is 102 cm (40 in.) in diameter. (a: ALAN SOLOMON/Tribune News Service/LAKE GENEVA/WI/USA/Newscom; b: Yerkes Observatory)

refracting telescope. Master opticians of the nineteenth century devoted their careers to solving these problems, and several magnificent refractors were constructed in the late 1800s (Figure 6-8).

Unfortunately, there are several negative aspects of refractors that even the finest optician cannot overcome:

1. Because faint light must readily pass through the objective lens, the glass from which the lens is made must be totally free of defects, such as the bubbles that frequently form when molten glass is poured into a mold. Such defect-free glass is extremely expensive.

2. Glass is opaque to certain kinds of light. Ultraviolet light is absorbed almost completely, and even visible light is dimmed substantially as it passes through the thick slab of glass that makes up the objective lens.

3. It is impossible to produce a large lens that is entirely free of chromatic aberration.

4. Because a lens can be supported only around its edges, a large lens tends to sag and distort under its own weight as it tracks objects through the night. This distortion has negative effects on the image clarity.

For these reasons and more, few major refractors have been built since the beginning of the twentieth century. Instead, astronomers have avoided all the limitations of refractors by building telescopes that use mirrors instead of lenses to form images.

CONCEPT CHECK 6-3

How does the eyepiece focal length affect a telescope's overall magnification?

Answer appears at the end of the chapter.

6-2 A reflecting telescope uses a mirror to concentrate incoming light at a focus

Almost all modern telescopes form an image using the principle of **reflection**. To understand reflection, imagine drawing a dashed line perpendicular to the surface of a flat mirror at the point where a light ray strikes the mirror (**Figure 6-9a**). The angle $\theta_{incidence}$ between the *incident* (arriving) light ray and the perpendicular is always equal to the angle $\theta_{reflection}$ between the *reflected* ray and the perpendicular.

All of the largest professional telescopes—and most amateur telescopes—are reflecting telescopes.

In 1663, the Scottish mathematician James Gregory first proposed a telescope using reflection from a concave mirror—one that is fatter at the edges than at the middle. Such a mirror makes parallel light rays converge to a focus (Figure 6-9b). The distance

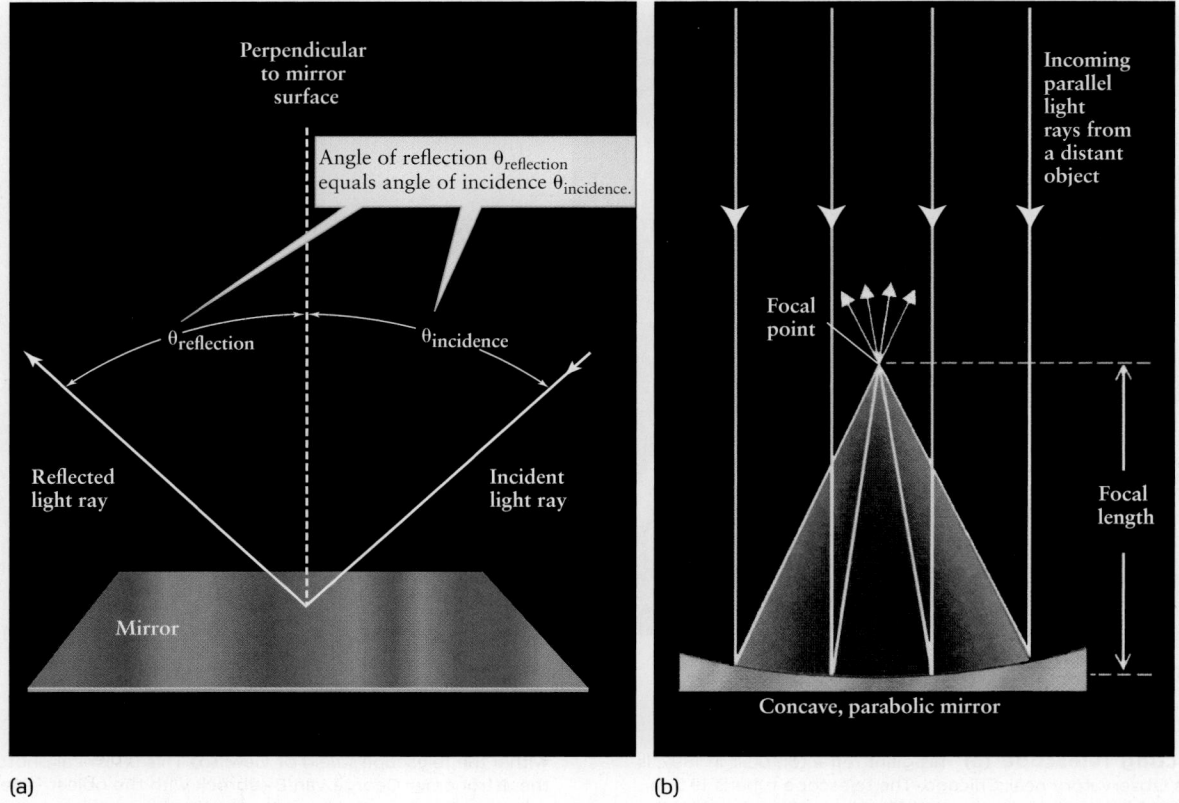

FIGURE 6-9

Reflection (a) The angle at which a beam of light approaches a mirror, called the angle of incidence ($\theta_{incidence}$), is always equal to the angle at which the beam is reflected from the mirror, called the angle of reflection ($\theta_{reflection}$).

(b) A concave mirror causes parallel light rays to converge to a focus at the focal point. The distance between the mirror and the focal point is the focal length of the mirror.

between the reflecting surface and the focus is the focal length of the mirror. A telescope that uses a curved mirror to make an image of a distant object is called a **reflecting telescope,** or **reflector.** Using terminology similar to that used for refractors, the mirror that forms the image is called the **objective mirror** or **primary mirror.**

To make a reflector mirror, an optician grinds and polishes a large slab of glass into the appropriate concave shape. The glass is then coated with silver, aluminum, or a similar highly reflective substance. Because light reflects off the surface of the coated glass rather than passing through it, defects within the glass—which would have very negative consequences for the objective lens of a refracting telescope—have no effect on the optical quality of a reflecting telescope.

Another advantage of reflector mirrors is that they do not suffer from the chromatic aberration that plagues refractors. This is because reflection is not affected by the wavelength of the incoming light (only the angle at which it hits the mirror), so all wavelengths are reflected to the same focus. (A small amount of chromatic aberration may arise if the image is viewed using an eyepiece lens.) Furthermore, the mirror can be fully supported by braces on its back, so that a large, heavy mirror can be mounted and well-supported without its shape warping much from gravity.

Designs for Reflecting Telescopes

Although a reflecting telescope has many advantages over a refractor, the arrangement shown in Figure 6-9a is not ideal. One problem is that the focal point is in front of the objective mirror. If you try to view the image formed at the focal point, your head will block part or all of the light from reaching the mirror.

To get around this problem, in 1668 Isaac Newton simply placed a small, flat mirror at a 45° angle in front of the focal point, as sketched in **Figure 6-10a.** This secondary mirror deflects the light rays to one side, where Newton placed an eyepiece lens to magnify the image. A reflecting telescope with this optical design is appropriately called a **Newtonian reflector** (Figure 6-10b). The magnifying power of a Newtonian reflector is calculated in the same way as for a refractor: The focal length of the objective mirror is divided by the focal length of the eyepiece (see Box 6-1).

Later astronomers modified Newton's original design. The objective mirrors of some modern reflectors are so large that an astronomer could actually sit in an "observing cage" at the undeflected focal point directly in front of the objective mirror. (In practice, riding in this cage on a winter's night is a remarkably cold and uncomfortable experience.) This arrangement is called a **prime focus** (Figure 6-11a). It usually provides the highest-quality image,

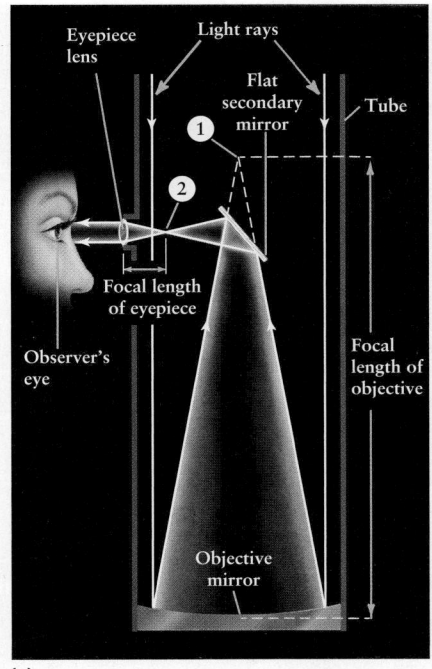

(a)

FIGURE 6-10 R I **V** U X G

A Newtonian Telescope (a) In a Newtonian telescope, the image made by the objective is moved from point 1 to point 2 by means of a flat mirror called the secondary mirror. An eyepiece magnifies this image, just as for a refracting telescope (Figure 6-5). **(b)** This is a replica of a Newtonian

(b)

telescope built by Isaac Newton in 1672. The objective mirror is 3 cm (1.3 in.) in diameter and the magnification is 40×. (Right: Royal Greenwich Observatory/ Science Source)

Incoming parallel rays from a distant object

(a) Prime focus

(b) Cassegrain focus

(c) Coudé focus

because there is no need for a secondary mirror (which might have imperfections).

Another popular optical design, called a **Cassegrain focus** after the French contemporary of Newton who first proposed it, also has a convenient, accessible focal point. A hole is drilled directly through the center of the primary mirror, and a convex secondary mirror placed in front of the original focal point reflects the light rays back through the hole (Figure 6-11b).

A fourth design is useful when there is optical equipment too heavy or bulky to mount directly on the telescope. Instead, a series of mirrors channels the light rays away from the telescope to a remote focal point where the equipment is located. This design is called a **coudé focus,** from a French word meaning "bent like an elbow" (Figure 6-11c).

ANIMATION 6.2

FIGURE 6-11

Designs for Reflecting Telescopes Three common optical designs for reflecting telescopes are shown here. **(a)** Prime focus is used only on some large telescopes; an observer or instrument is placed directly at the focal point, within the barrel of the telescope. **(b)** The Cassegrain focus is used on reflecting telescopes of all sizes, from 90-mm (3.5-in.) reflectors used by amateur astronomers to giant research telescopes on Mauna Kea (see Figure 6-14). **(c)** The coudé focus is useful when large and heavy optical apparatus is to be used at the focal point. Light reflects off the objective mirror to a secondary mirror, then back down to a third, angled mirror. With this arrangement, the heavy apparatus at the focal point does not have to move when the telescope is repositioned.

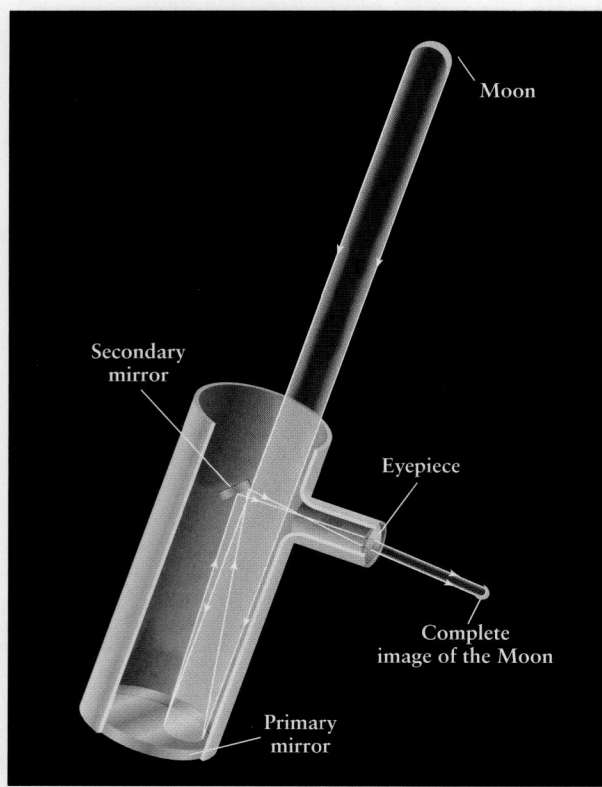

FIGURE 6-12

The Secondary Mirror Does Not Cause a Hole in the Image This illustration shows how even a small portion of the primary (objective) mirror of a reflecting telescope can make a complete image of the Moon. Thus, the secondary mirror does not cause a black spot or hole in the image. (It does, however, make the image a bit dimmer by reducing the total amount of light that reaches the primary mirror.)

> **CAUTION!**
>
> You might think that the secondary mirror in the Newtonian design and the Cassegrain and coudé designs shown in Figure 6-11 would cause a black spot or hole in the center of the telescope image. But this spot does not form. The reason is that light from every part of the object lands on every part of the primary, objective mirror. Hence, any portion of the mirror can itself produce an image of the distant object, as **Figure 6-12** shows. The only effect of the secondary mirror is that it prevents part of the light from reaching the objective mirror, which reduces somewhat the total light-gathering power of the telescope.

Spherical Aberration

A reflecting telescope must be designed to minimize a defect called **spherical aberration** (Figure 6-13). At issue is the precise shape of a mirror's concave surface. A spherical surface is easy to grind and polish, but different parts of a spherical mirror focus light onto slightly different spots (Figure 6-13a). This results in a fuzzy image.

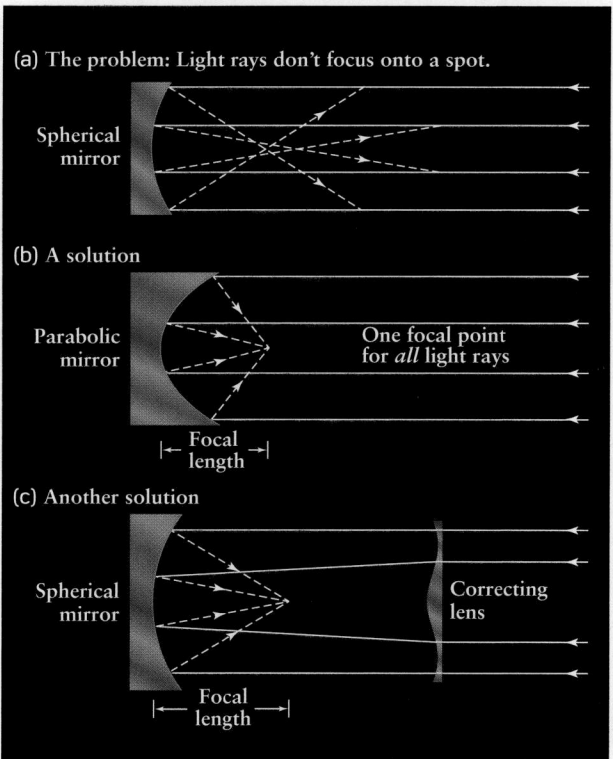

FIGURE 6-13

Spherical Aberration **(a)** Different parts of a spherically concave mirror reflect light to slightly different points. This effect, called spherical aberration, causes image blurring, but the difficulty can be corrected by either **(b)** using a parabolic mirror or **(c)** using a correcting lens in front of the mirror.

One common way to eliminate spherical aberration is to polish the mirror's surface to a parabolic shape, because a parabola reflects parallel light rays to a common focus (Figure 6-13b). Unfortunately, the astronomer then no longer has a wide-angle view. Furthermore, unlike spherical mirrors, parabolic mirrors suffer from a defect called **coma,** wherein star images far from the center of the field of view are elongated to look like tiny teardrops. A different approach is to use a spherical mirror, thus minimizing coma, and to place a thin correcting lens at the front of the telescope to eliminate spherical aberration (Figure 6-13c). This approach is only used on relatively small reflecting telescopes for amateur astronomers.

The Largest Reflectors

There are over a dozen optical reflectors in operation with primary mirrors between 8 m (26.2 ft) and 11 m (36.1 ft) in diameter. Figure 6-14a shows the objective mirror of one of the four Very Large Telescope (VLT) units in Chile, and Figure 6-14b shows the objective and secondary mirrors of the Gemini North telescope in Hawaii. A near-twin of Gemini North, called Gemini South, is in Cerro Pachón, Chile. These twins allow astronomers to observe both the northern and southern parts of the celestial sphere with essentially the same state-of-the-art instrument. Two other "twins" are the side-by-side 8.4-m objective mirrors of the Large Binocular Telescope

(a) A large objective mirror

FIGURE 6-14   R   I   ⓥ   U   X   G

Reflecting Telescopes **(a)** This photograph shows technicians preparing an objective mirror 8.2 m in diameter for the European Southern Observatory in Chile. The mirror was ground to a curved shape with a remarkable precision of 8.5 nm. **(b)** This view of the Gemini North telescope shows its 8.1-m objective mirror (1). Light incident on this mirror is reflected toward the 1-m secondary mirror (2), then through the hole in the objective mirror (3) to the Cassegrain focus (See FIGURE 6-11b). (a: SAGEM and the European Southern Observatory; b: NOAO/AURA/NSF)

(b) A large Cassegrain telescope

in Arizona. Combining the light from these two mirrors gives double the light-gathering power, equivalent to a single 11.8-m mirror.

Several other reflectors around the world have objective mirrors between 3 and 6 m in diameter, and dozens of smaller but still powerful telescopes have mirrors in the range of 1 to 3 m. There are thousands of professional astronomers, each of whom has several ongoing research projects; thus the demand for all these telescopes is high. On any night of the year, virtually every research telescope in the world is being used to explore the universe.

CONCEPT CHECK 6-4

Does the angle of light reflected off a mirrored surface depend on wavelength? How does the answer to this question help or hurt a reflecting telescope?

Answer appears at the end of the chapter.

6-3 Telescope images are degraded by the blurring effects of the atmosphere and by light pollution

TUTORIAL 6.1 In addition to providing a brighter image, a large telescope also helps achieve a second major goal: It produces star images that are sharp and crisp. A quantity called **angular**

resolution gauges how well fine details can be seen. Poor angular resolution causes star images to be fuzzy and blurred together.

To determine the angular resolution of a telescope, pick out two adjacent stars whose separate images are just barely discernible (Figure 6-15). The angle θ (the Greek letter theta) between these stars is the telescope's angular resolution; the *smaller* that angle, the finer the details that can be seen and the sharper the image.

When you are asked to read the letters on an eye chart, what's being measured is the angular resolution of your eye. If you have 20/20 vision, the angular resolution θ of your eye is about 1 arcmin, or 60 arcsec. (You may want to review the definitions of angles and their units in Section 1-5.) Hence, with the naked eye it is impossible to distinguish two stars less than 1 arcminute apart or to see details on the Moon with an angular size smaller than this. All the planets have angular sizes (as seen from Earth) of 1 arcmin or less, which is why they appear as featureless points of light to the naked eye.

Limits to Angular Resolution

One factor limiting angular resolution is **diffraction,** which is the tendency of light waves to spread out when they are confined to a small area like the lens or mirror of a telescope. (A rough analogy is the way water exiting a garden hose sprays out in a wider angle when you cover part of the end of the hose with your thumb.) As a result of diffraction, a narrow beam of light tends to spread out

(a)

Two light sources with angular separation greater than angular resolution of telescope: two sources easily distinguished

(b)

Light sources moved closer so that angular separation equals angular resolution of telescope: just barely possible to tell that there are two sources

FIGURE 6-15 R I **V** U X G

Angular Resolution The angular resolution of a telescope indicates the sharpness of the telescope's images. **(a)** This telescope view shows two sources of light whose angular separation is greater than the angular resolution. **(b)** The light sources have been moved together so that their angular separation is equal to the angular resolution. If the sources were moved any closer together, the telescope image would show them as a single source. (Courtesy of John D. Monnier)

within a telescope's optics, thus blurring the image. If diffraction were the only limit, the angular resolution of a telescope would be given by the formula:

Diffraction-limited angular resolution

$$\theta = 2.5 \times 10^5 \frac{\lambda}{D}$$

θ = diffraction-limited angular resolution of a telescope, in arcseconds

λ = wavelength of light, in meters

D = diameter of telescope objective, in meters

For a given wavelength of light, using a telescope with an objective of *larger* diameter D *reduces* the amount of diffraction and makes the angular resolution θ *smaller* (and hence better). For example, with red light of wavelength 640 nm, or 6.4×10^{-7} m, the diffraction-limited resolution of an 8-m telescope (see Figure 6-15) would be

$$\theta = (2.5 \times 10^5) \frac{6.4 \times 10^{-7} \text{ m}}{8 \text{ m}} = 0.02 \text{ arcsec}$$

In practice, however, ordinary optical telescopes cannot achieve such fine angular resolution. The problem is that turbulence in the air causes star images to jiggle around and twinkle. Even through the largest telescopes, a star still looks like a tiny blob rather than a pinpoint of light. A measure of the limit that atmospheric turbulence places on a telescope's resolution is called the **seeing disk.** This disk is the angular diameter of a star's image broadened by turbulence. The size of the seeing disk varies from one observatory site to another and from one night to another. For example, at the observatories on Kitt Peak in Arizona and Cerro Tololo in Chile, the seeing disk is typically around 1 arcsec. But some of the very best conditions in the world can be found at the observatories atop Mauna Kea in Hawaii, where the seeing disk is often as small as 0.5 arcsec. These great conditions are one reason why so many telescopes have been built there (**Figure 6-16**).

University of Hawaii 0.6-m Telescope

Caltech Submillimeter Observatory

James Clerk Maxwell Telescope

Submillimeter Array

Subaru

Keck I

Keck II

NASA Infrared Telescope Facility

United Kingdom Infrared Telescope

University of Hawaii 2.2-meter Telescope

Gemini North

Canada-France-Hawaii Telescope

FIGURE 6-16 R I **V** U X G

The Telescopes of Mauna Kea The summit of Mauna Kea—a dormant Hawaiian volcano that reaches more than 4100 m (13,400 ft) above the waters of the Pacific—has nighttime skies that are unusually clear, still, and dark. To take advantage of these superb viewing conditions, Mauna Kea has become the home of many powerful telescopes. (Richard J. Wainscoat, University of Hawaii)

CONCEPT CHECK 6-5

A small-diameter telescope collects less light than a larger-diameter telescope. If a telescope with a larger-diameter lens or mirror is used for observations of dim, barely illuminated features on Jupiter, does the larger diameter worsen or improve the image resolution compared to the smaller telescope?

Answer appears at the end of the chapter.

Active Optics and Adaptive Optics

In many cases, the angular resolution of a telescope is even worse than the limit imposed by the seeing disk. This occurs if the objective mirror deforms even slightly due to variations in air temperature or slight gravitational sagging of the telescope mount. To combat this, many large telescopes are equipped with an **active optics** system. Such a system slowly adjusts the mirror shape every few seconds to help keep the telescope in optimum focus and properly aimed at its target. These adjustments can be determined once during a testing phase for each orientation of the telescope and applied for all future observations. However, the adjustments discussed next are more challenging because they change rapidly throughout each observation.

Changing the mirror shape is also at the heart of a more refined technique called **adaptive optics**. The goal of this technique is to compensate for atmospheric turbulence, so that the angular resolution can be smaller than the size of the seeing disk and can even approach the theoretical limit set by diffraction. Turbulence causes the image of a star to "dance" around erratically. While

> Adaptive optics produces sharper images by "undoing" atmospheric turbulence.

this turbulence produces twinkling stars seen with the naked eye, it also produces blurry images in large telescopes. In an adaptive optics system, sensors monitor this dancing motion 10 to 100 times per second, and a powerful computer rapidly calculates the mirror shape needed to compensate. Fast-acting mechanical devices called *actuators* then deform the mirror accordingly, to produce a sharp, focused image. In some adaptive optics systems, the actuators deform a small secondary mirror rather than the large objective mirror.

One difficulty with adaptive optics is that a fairly bright star must be in or near the field of the telescope's view to serve as a "calibration target" for the sensors that track atmospheric turbulence. This is seldom the case, since the field of view of most telescopes is rather narrow, lowering the chances of viewing a sufficiently bright star. With no lack of creative problem-solving, astronomers get around this limitation by shining a laser beam toward a spot in the sky near the object to be observed (Figure 6-17). The laser beam causes atoms in Earth's upper atmosphere to glow, making an artificial "star." The light that comes down to Earth from this "star" travels through the same part of our atmosphere as the light from the object being observed, so its image in the telescope will dance around in the same erratic way as the image of a real star. By observing how the changing atmosphere distorts the artificial star, computers apply opposing adjustments to cancel the atmosphere's effects on the real star.

Figure 6-18 shows the dramatic improvement in angular resolution possible with adaptive optics. Images made with adaptive

FIGURE 6-17 R I (V) U X G

Creating an Artificial "Star" A laser beam shines upward from Yepun, an 8.2-m telescope at the European Southern Observatory in the Atacama Desert of Chile. (Figure 6-14a shows the objective mirror for this telescope.) The beam strikes sodium atoms that lie about 90 km (56 mi) above Earth's surface, causing them to glow and make an artificial "star." Tracking the twinkling of this "star" makes it possible to undo the effects of atmospheric turbulence on telescope images. (ESO/G.Hüdepohl)

optics are nearly as sharp as if the telescope were in the vacuum of space, where there is no atmospheric distortion whatsoever and the only limit on angular resolution is diffraction. A number of large telescopes are now being used with adaptive optics systems.

CAUTION!

The images in Figure 6-18 are **false-color** images: They do not represent the true color of the stars shown. False color is often used when the image is made using wavelengths that the eye cannot detect, as with the infrared images in Figure 6-18. A different use of false color is to indicate the relative brightness of different parts of the image, as in the infrared image of a person in Figure 5-10. Throughout this book, we'll always point out when false color is used in an image.

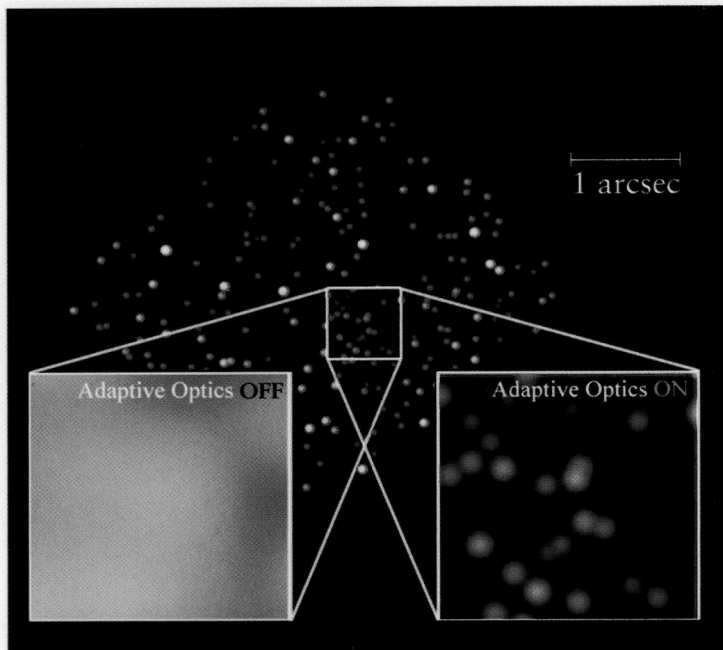

FIGURE 6-18 R **I** V U X G

Using Adaptive Optics to "Unblur" Telescope Images The two false-color, inset images show the same 1-arcsecond-wide region of the sky at infrared wavelengths as observed with the 10-m Keck II telescope on Mauna Kea (see Figure 6-16). Without adaptive optics, it is impossible to distinguish individual stars in this region. With adaptive optics turned on, more than two dozen stars can be distinguished. (These images/animations were created by Prof. Andrea Ghez and her research team at UCLA and are from data sets obtained with the W. M. Keck Telescopes.)

Interferometry

Several large observatories are developing a technique called **interferometry** that promises to further improve the angular resolution of telescopes. The idea is to have two widely separated telescopes observe the same object simultaneously, then use fiber optic cables to "pipe" the light signals from each telescope to a central location where they "interfere" or blend together. This method makes the combined signal sharp and clear. The effective resolution of such a combination of telescopes is equivalent to that of one giant telescope with a diameter equal to the **baseline,** or distance between the two telescopes. For example, the Keck I and Keck II telescopes atop Mauna Kea (Figure 6-16) are 85 m apart, so when used as an interferometer, the angular resolution is the same as a single 85-m telescope.

Interferometry has been used for many years with radio telescopes (which we will discuss in Section 6-6), but it is still under development with telescopes for visible light or infrared wavelengths. Astronomers are devoting a great deal of effort to this development because the potential rewards are great. For example, the Keck I and II telescopes used together should give an angular resolution as small as 0.005 arcsec, which corresponds to being able to read the bottom row on an eye chart 36 km (22 mi) away!

Light Pollution

Light from city street lamps and from buildings also degrades telescope images. This **light pollution** illuminates the sky, making

it more difficult to see the stars. You can appreciate the problem if you have ever looked at the night sky from a major city. Only a few of the very brightest stars can be seen, as opposed to the thousands that can be seen with the naked eye from the desert or the mountains. To avoid light pollution, observatories are built in remote locations far from any city lights.

Unfortunately for astronomers, the expansion of cities has brought light pollution to observatories that in former times had none. As an example, the growth of Tucson, Arizona, has had deleterious effects on observations at the nearby Kitt Peak National Observatory. Efforts have been made to have cities adopt light fixtures that provide safe illumination for their citizens but produce little light pollution. These efforts have met with only mixed success.

One factor over which astronomers have absolutely no control is the weather. Optical telescopes cannot see through clouds, so it is important to build observatories where the weather is usually clear. One advantage of mountaintop observatories such as Mauna Kea is that most clouds form at altitudes below the observatory, giving astronomers a better chance of having clear skies.

In many ways, the best location for a telescope is in orbit around Earth, where it is unaffected by weather, light pollution, or atmospheric turbulence. We will discuss orbiting telescopes in Section 6-7.

CONCEPT CHECK 6-6

If astronomers are using an adaptive optics system on a night when the atmosphere is unusually turbulent, will the adaptive optics actuators deform the telescope's mirror more rapidly or less rapidly than on a typical night?

Answer appears at the end of the chapter.

6-4 A charge-coupled device is commonly used to record the image at a telescope's focus

 TUTORIAL 6.2

Telescopes provide astronomers with detailed pictures of distant objects. The process of recording these pictures is called **imaging.**

Astronomical imaging really began in the nineteenth century with the invention of photography. It was soon realized that this new invention was a boon to astronomy. By mounting a camera at the focus of a telescope, long exposures can gather light from an object for an entire evening. Such long exposures can reveal details in galaxies, star clusters, and nebulae that would not be visible to an astronomer by simply looking through a telescope. Indeed, most large, modern telescopes do not have eyepieces at all.

Originally, cameras and telescopes used photographic film, but film has now been replaced by electronic detectors. Photographic film is not a very efficient light detector, as only 2% of the photons that strike photographic film actually trigger the chemical reaction needed to produce an image. And, of course, to analyze images on a computer we need them in some electronic form. For much greater efficiency, and to capture images digitally, we use charge-coupled devices (CCDs).

The same technology that makes digital cameras possible has revolutionized astronomy.

(a) Pan-STARRS detector CCDs

(b) An image made with photographic film

(c) An image of the same region of the sky made with a CCD

FIGURE 6-19   R I Ⓥ U X G

Charge-Coupled Devices (CCDs) and Imaging (a) These CCDs are part of the Pan-STARRS detector developed at MIT's Lincoln Laboratory. The full detector is made from many of these wafers, has about 1.4 billion pixels, and is nearly 40 cm wide. **(b)** This negative print (black stars and white sky) shows a portion of the sky as imaged with a 4-m telescope and photographic film. **(c)** This negative image of the same region of the sky was made with the same telescope, but using the photographic film replaced by a CCD. Many more stars and galaxies are visible. (a: LBNL/Roy Kaltschmidt/Science Source; b, c: Patrick Seitzer, National Optical Astronomy Observatories)

Charge-Coupled Devices

The most sensitive light detector currently available to astronomers is the **charge-coupled device (CCD).** At the heart of a CCD is a semiconductor wafer divided into an array of small, light-sensitive squares called picture elements or, more commonly, **pixels.** A personal digital camera or cell phone uses a CCD chip (or similar device) and might have about 10 million pixels. A large astronomical camera combines many CCDs to capture a wide field of view at high resolution (Figure 6-19a). For example, many CCDs like those shown in Figure 6-19a are used together to give the Pan-STARRS detector about 1.4 billion pixels.

When an image from a telescope is focused on the CCD, an electric charge builds up in each pixel in proportion to the number of photons falling on that pixel. When the exposure is finished, the amount of charge on each pixel is read by a computer, where the resulting image can be analyzed and stored in digital form. Compared with photographic film, CCDs are some 35 times more sensitive to light, can record much finer details, and respond more uniformly to light of different colors. Figures 6-19b and 6-19c show the dramatic difference between photographic and CCD images. The great sensitivity of CCDs also makes them useful for **photometry,** which measures the brightness of a star or other astronomical object.

In the modern world of CCD astronomy, astronomers no longer need to spend the night in the unheated dome of a telescope. Instead, they operate the telescope electronically from a separate control room, where the electronic CCD images can be viewed on a computer monitor. By viewing the CCD's electronic image during the evening's observations, an astronomer gets immediate feedback and often makes adjustments to get the best results. The control room need not even be adjacent to the telescope. Although the Keck I and II telescopes (see Figure 6-16) are at an altitude of 4100 m (13,500 ft), astronomers can now make observations from a facility elsewhere on the island of Hawaii that is much closer to sea level.

This saves the laborious drive to the summit of Mauna Kea and eliminates the need for astronomers to acclimate to the high altitude.

Most of the images you will see in this book were made with CCDs. Because of their extraordinary sensitivity and their ability to be used in conjunction with computers, CCDs have attained a role of central importance in astronomy.

CONCEPT CHECK 6-7

Why can CCDs more efficiently observe faint stars than photographic film or photographic plates?

Answer appears at the end of the chapter.

6-5 Spectrographs record the spectra of astronomical objects

We saw in Section 5-6 how the spectrum of an astronomical object provides a tremendous amount of information about that object, including its chemical composition and temperature. Thus, measuring spectra, or **spectroscopy,** is one of the most important uses of telescopes. Indeed, some telescopes are designed solely for measuring the spectra of distant, faint objects; they are never used for imaging.

> The spectrum of a planet, star, or galaxy can reveal more about its nature than an image.

Spectrographs and Diffraction Gratings

An essential tool of spectroscopy is the **spectrograph,** a device that records spectra. This optical device is mounted at the focus of a telescope. Figure 6-20a shows one design for a spectrograph, in which a **diffraction grating** is used to form the spectrum of a planet, star, or galaxy. A diffraction grating is a piece of glass on which thousands

(a)

(b)

FIGURE 6-20 R I V U X G

A Grating Spectrograph (a) This optical device uses a diffraction grating to break up the light from a source into a spectrum. The spectrum is then recorded on a CCD. **(b)** A diffraction grating has a large number of parallel lines in its surface that reflect light of different colors in different directions. A compact disc, which stores information in a series of closely spaced pits, reflects light in a similar way. (Right: Dale E. Boyer/Science Source)

of very regularly spaced parallel lines have been cut. Some of the finest diffraction gratings have more than 10,000 lines per centimeter, which are usually cut by drawing a diamond back and forth across the glass. When light is shone on a diffraction grating, light waves leaving different parts of the grating interfere with each other, producing a spectrum. (This same effect produces the rainbow of colors you see reflected from a compact disc or DVD, as shown in Figure 6-20b. Information is stored on the disc in a series of closely spaced pits, which acts as a diffraction grating.)

Older types of spectrographs used a prism rather than a diffraction grating to form a spectrum, but using a prism has several drawbacks. A prism does not disperse the colors of the rainbow evenly: Blue and violet portions of the spectrum are spread out more than the red portion. In addition, because the blue and violet wavelengths must pass through more of the prism's glass than do the red wavelengths (examine Figure 5-3), light is absorbed unevenly across the spectrum. Indeed, a glass prism is opaque to near-ultraviolet light. For these reasons, diffraction gratings are preferred in modern spectrographs.

In Figure 6-20a the spectrum of a planet, star, or galaxy formed by the diffraction grating is recorded on a CCD. Light from a hot gas (such as helium, neon, argon, iron, or a combination of these) is then focused on the spectrograph slit. The result is a *comparison spectrum* alongside the spectrum of the celestial object under study. The wavelengths of the bright spectral lines of the comparison spectrum are known from laboratory experiments and can therefore serve as reference markers. (See Figure 5-18, which shows the spectrum of the Sun and a comparison spectrum of iron.)

When the exposure is finished, electronic equipment measures the charge that has accumulated in each pixel. These data are used to graph light intensity versus wavelength. Dark absorption lines in the spectrum appear as depressions or valleys on the graph, while bright emission lines appear as peaks. **Figure 6-21** compares two ways of

exhibiting spectra with absorption lines and emission lines. Later in this book, we shall see spectra presented in both of these ways.

CONCEPT CHECK 6-8

A planet is viewed through a spectrograph like the one illustrated in Figure 6-20a. Does the CCD also record an image of the planet?

Answer appears at the end of the chapter.

6-6 A radio telescope uses a large concave dish to reflect radio waves to a focus

For thousands of years, all the information that astronomers gathered about the universe was based on ordinary visible light. In the twentieth century, however, astronomers first began to explore the nonvisible electromagnetic radiation coming from astronomical objects. In this way, they have discovered aspects of the cosmos that are forever hidden to optical telescopes.

Astronomers have used ultraviolet light to map the outer regions of the Sun and the clouds of Venus, and used infrared radiation to see new stars and perhaps new planetary systems in the process of formation. By detecting radio waves from Jupiter and Saturn, they have mapped the intense magnetic fields that surround those giant planets; by detecting curious bursts of x rays from space, they have learned about the utterly alien

Observing at radio wavelengths reveals aspects of the universe hidden from optical telescopes.

(a) Two representations of an absorption line spectrum

(b) Two representations of an emission line spectrum

FIGURE 6-21

Two Ways to Represent Spectra When a CCD is placed at the focus of a spectrograph, it records the rainbow-colored spectrum. A computer can convert the recorded data into a graph of intensity versus wavelength.

(a) Absorption lines appear as dips on such a graph, while **(b)** emission lines appear as peaks. The dark absorption lines and bright emission lines in this example are the Balmer lines of hydrogen (see Section 5-8).

conditions in the vicinity of a black hole. It is no exaggeration to say that today's astronomers learn as much about the universe using telescopes for nonvisible wavelengths as they do using visible light.

Radio Astronomy

Radio waves were the first part of the electromagnetic spectrum beyond the visible to be exploited for astronomy. Their usefulness was discovered as a result of a research project seemingly unrelated to astronomy. In the early 1930s, Karl Jansky, a young electrical engineer at Bell Telephone Laboratories, was trying to locate the source of interference with the then-new transatlantic radio link. By 1932, he realized that one kind of radio noise is strongest when the constellation Sagittarius is high in the sky. The center of our galaxy is located in the direction of Sagittarius; Jansky concluded that he was detecting radio waves from an astronomical source.

At first, only Grote Reber, a radio engineer living in Illinois, took up Jansky's research. In 1936, Reber built in his backyard the first **radio telescope,** a radio-wave detector dedicated to astronomy. He modeled his design after an ordinary reflecting telescope, with a parabolic metal "dish" (reflecting antenna) measuring 10 m (31 ft) in diameter and a radio receiver at the focal point of the dish.

Reber spent the years from 1938 to 1944 mapping radio emissions from the sky at wavelengths of 1.9 m and 0.63 m. He found radio waves coming from the entire Milky Way, with the greatest emission from the center of the galaxy. These results, together with the development of improved radio technology during World War II, encouraged the growth of radio astronomy and the construction of new radio telescopes around the world. Today, radio observatories are as common as major optical observatories.

Modern Radio Telescopes

Like Reber's prototype, a typical modern radio telescope has a large parabolic dish (**Figure 6-22**). An antenna tuned to the desired frequency is located at the focus (like the prime focus design for

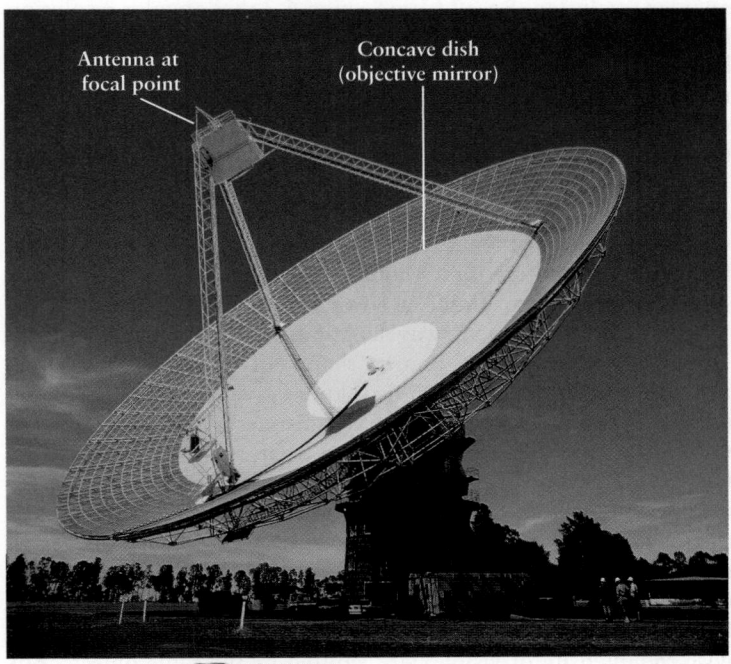

FIGURE 6-22 R I ⓥ U X G

A Radio Telescope The dish of the Parkes radio telescope in New South Wales, Australia, is 64 m (210 ft) in diameter. Radio waves reflected from the dish are brought to a focus and collected by an antenna at the focal point. (David Nunuk/Science Source)

optical reflecting telescopes shown in Figure 6-11a). The incoming signal is relayed from the antenna to amplifiers and recording instruments, typically located in a room at the base of the telescope's pier.

Many radio telescope dishes, like the one in Figure 6-22, have visible gaps in them like a wire mesh. This does not affect their reflecting power because the holes are much smaller than the wavelengths of the radio waves; the radio waves reflect from the wire mesh as if from a solid surface. The same idea is used in the design of microwave ovens. The glass window in the oven door would allow the microwaves to leak out, so the window is covered by a metal screen with small holes. These holes are much smaller than the 12.2-cm (4.8-in.) wavelength of the microwaves, so the screen reflects the microwaves back into the oven while still allowing you to watch your food heat up.

> **CAUTION!**
>
> The radio telescope in Figure 6-22 looks like a radar dish but is used in a different way. In active radar, the dish is used to send out a narrow beam of radio waves. If this beam encounters an object like an airplane, some of the radio waves will be reflected back to the radar dish and detected by a receiver at the focus of the dish. Thus, an active radar dish looks for radio waves *reflected* by distant objects. By contrast, a radio telescope is designed to receive radio waves *emitted* by objects in space.

Radio Telescopes: Limits to Angular Resolution

One great drawback of early radio telescopes was their very poor angular resolution. Recall from Section 6-3 that angular resolution is the smallest angular separation between two stars that can just barely be distinguished as separate objects. Unlike visible light, radio waves are only slightly affected by turbulence in the atmosphere, so the factor limiting the angular resolution of a radio telescope is diffraction. The problem is that diffraction-limited angular resolution is directly proportional to the wavelength being observed: The longer the wavelength, the larger (and hence worse) the angular resolution and the fuzzier the image. (See the formula for angular resolution θ in Section 6-3.)

VIDEO 6.1 As an example, a 1-m radio telescope detecting radio waves of 5-cm wavelength has an angular resolution 100,000 times poorer than a 1-m optical telescope. Because radio radiation has very long wavelengths, *small* radio telescopes can produce only blurry, indistinct images. A very large radio telescope can produce a somewhat sharper radio image, because as the diameter of the telescope increases, the angular resolution decreases. In other words, the bigger the dish, the better the resolution. For this reason, most modern radio telescopes have dishes more than 30 m (100 ft) in diameter. A large dish is also useful for increasing light-gathering power, because radio signals from astronomical objects are typically very weak in comparison with the intensity of visible light emitted by the same objects. But even the largest single radio dish in existence, the 305-m (1000-ft) Arecibo radio telescope in Puerto Rico, cannot come close to the resolution of the best optical instruments.

To improve angular resolution at radio wavelengths, astronomers combine observations from two or more widely separated radio telescopes using the interferometry technique that we described in Section 6-3. Interferometry using radio waves is much easier than for visible or infrared light because radio signals can be carried over electrical wires. Consequently, two radio telescopes observing the same astronomical object can be hooked together, even if they are separated by a large distance, or baseline, of many

FIGURE 6-23 **R** **I** (**V**) **U** **X** **G**

The Very Large Array (VLA) The 27 radio telescopes of the VLA in central New Mexico are arranged along the arms of a Y. The north arm of the array is 19 km long; the southwest and southeast arms are each 21 km long. By spreading out the telescopes along the legs and combining the signals received, the VLA can give the same angular resolution as a single dish many kilometers in radius. (Courtesy of NRAO/AUI)

kilometers. The connected radio dishes can act as a single large-diameter dish with vastly improved angular resolution.

One of the largest arrangements of radio telescopes for interferometry is the Very Large Array (VLA), located in the desert near Socorro, New Mexico (Figure 6-23). The VLA consists of 27 parabolic dishes, each 25 m (82 ft) in diameter. These 27 telescopes are arranged along the arms of a gigantic Y that covers an area 27 km (17 mi) in diameter. By pointing all 27 telescopes at the same object and combining the 27 radio signals, this system can produce radio views of the sky with an angular resolution as small as 0.05 arcsec, comparable to that of the very best optical telescopes.

Dramatically better angular resolution can be obtained by combining the signals from radio telescopes at different observatories thousands of kilometers apart. This technique is called **very-long-baseline interferometry (VLBI)**. VLBI is used by a system called the Very Long Baseline Array (VLBA), which consists of ten 25-m dishes at different locations between Hawaii and the Caribbean. Although the 10 dishes are not physically connected, they are all used to observe the same object at the same time. The data from each telescope are recorded electronically and processed later. By carefully synchronizing the 10 recorded signals, they can be combined just as if the telescopes had been linked together during the observation. With the VLBA, features smaller than 0.001 arcsec can be distinguished at radio wavelengths. This angular resolution is 100 times better than a large optical telescope with adaptive optics.

Even better angular resolution can be obtained by adding radio telescopes in space; the baseline is then the distance from the VLBA to the orbiting telescope. The first such space radio telescope, the Japanese HALCA spacecraft, operated from 1997 to 2003. Orbiting at a maximum distance of 21,400 km (13,300 mi) above Earth's surface, HALCA gave a baseline 3 times longer—and thus an angular resolution 3 times better—than can be obtained with

Earthbound telescopes alone. A Russian-led mission launched in 2011, Spektr-R, continues the effort for space-based interferometry.

Figure 6-24 shows how optical and radio images of the same object can give different and complementary kinds of information. The visible-light image of Saturn (Figure 6-24a) shows clouds in the planet's atmosphere and the structure of the rings. Like the visible light from the Moon, the light used to make this image is just reflected sunlight. By contrast, the false-color radio image of Saturn (Figure 6-24b) is a record of waves *emitted* by the planet and its rings. Analyzing such images provides information about the structure of Saturn's atmosphere and rings that could never be obtained from a visible-light image such as Figure 6-24a.

CONCEPT CHECK 6-9

What radio telescope would make an image with the better resolution: a single radio antenna dish with a diameter of 100 m, or two 25-m dishes operating together at a separation of 200 m?

Answer appears at the end of the chapter.

(a) R I **V** U X G

(b) **R** I V U X G

FIGURE 6-24

Optical and Radio Views of Saturn (a) This picture was taken by a spacecraft 18 million km from Saturn. The view was produced by sunlight reflecting from the planet's cloudtops and rings. **(b)** This image was made using radio waves, which is indicated by the highlighted R in the wavelength tab. This VLA image shows radio emission from Saturn at a wavelength of 2 cm. In this false-color image, the most intense radio emission is shown in red, the least intense in blue. Yellow and green represent intermediate levels of radio intensity; black indicates no detectable radio emission. Note the radio "shadow" caused by Saturn's rings where they lie in front of the planet. (a: NASA/JPL/Space Science Institute; b: I. de Pater, J.R. Dickel; NRAO/AUI/NSF)

6-7 Telescopes in orbit around Earth detect radiation that does not penetrate the atmosphere

TUTORIAL 6.3 The many successes of radio astronomy show the value of observations at nonvisible wavelengths. But Earth's atmosphere is opaque to many wavelengths. Other than visible light and radio waves, very little radiation from space manages to reach Earth's surface. To overcome this, astronomers have placed a variety of telescopes in orbit around the planet.

> Space telescopes make it possible to study the universe across the entire electromagnetic spectrum.

Figure 6-25 shows the transparency of Earth's atmosphere to different wavelengths of electromagnetic radiation. The atmosphere is most transparent in two wavelength regions, the **optical window** (which includes the entire visible spectrum) and the **radio window** (which includes part, but not all, of the radio spectrum). There are also several relatively transparent regions at infrared wavelengths between 1 and 40 μm. Infrared radiation within these wavelength intervals can penetrate Earth's atmosphere somewhat and can be detected with ground-based telescopes. This wavelength range is called the *near-infrared*, because it lies just beyond the red end of the visible spectrum.

CONCEPT CHECK 6-10

Look at Figure 6-25, which shows the transparency of Earth's atmosphere. Would astronomers most prefer to have a new ground-based telescope constructed that is most sensitive in the x-ray region, the ultraviolet wavelength region, or the microwave region?

Answer appears at the end of the chapter.

Infrared Astronomy

Water vapor is the main absorber of infrared radiation from space, which is why infrared observatories are located at sites with exceptionally low humidity. The site must also be at high altitude to get above as much of the atmosphere's water vapor as possible. One site that meets both criteria is the summit of Mauna Kea in Hawaii, shown in Figure 6-16. (The complete lack of vegetation on the summit attests to its extreme dryness.) Some of the telescopes on Mauna Kea are designed exclusively for detecting infrared radiation. Others, such as the Keck I and Keck II telescopes, are used for both visible and near-infrared observations (see Figure 6-18).

Even at the elevation of Mauna Kea, water vapor in the atmosphere restricts the kinds of infrared observations that astronomers can make. This situation can be improved by carrying telescopes on board high-altitude balloons or aircraft. But the ultimate solution is to place a telescope in Earth's orbit and radio its data back to astronomers on the ground. The first such orbiting infrared observatory, the Infrared Astronomical Satellite (IRAS), was launched in 1983. During its 9-month mission, IRAS used its 57-cm (22-in.) telescope to map almost the entire sky at wavelengths from 12 to 100 μm.

The IRAS data revealed the presence of dust disks around nearby stars. Planets are thought to coalesce from disks of this kind, so this was the first observational indication that there might

FIGURE 6-25

The Transparency of Earth's Atmosphere This graph shows the percentage of radiation that can penetrate Earth's atmosphere at different wavelengths. Regions in which the curve is high are called "windows," because the atmosphere is relatively transparent at those wavelengths.

There are also three wavelength ranges in which the atmosphere is opaque and the curve is near zero. At wavelengths where the atmosphere is opaque, ground-based telescopes cannot receive light from space because it is either absorbed or reflected by the atmosphere.

be planets orbiting other stars. The dust that IRAS detected is warm enough to emit infrared radiation but too cold to emit much visible light, so it remained undetected by ordinary optical telescopes. IRAS also discovered distant, ultraluminous galaxies that emit almost all their radiation at infrared wavelengths.

In 1995, the Infrared Space Observatory (ISO), a more advanced 60-cm reflector with better light detectors, was launched into orbit by the European Space Agency (ESA). During its 2.5-year mission, ISO made a number of groundbreaking observations of very distant galaxies and of the thin, cold material between the stars of our own galaxy. Like IRAS, ISO had to be cooled by liquid helium to temperatures just a few degrees above absolute zero. Had this not been done, the infrared blackbody radiation from the telescope itself would have outshone the infrared radiation from astronomical objects. The ISO mission came to an end when the last of the helium evaporated into space.

One of the largest orbiting infrared observatories is the Herschel Space Observatory, a 3.5-m infrared telescope designed to survey the infrared sky with unprecedented resolution (**Figure 6-26**). Placed in orbit in 2009, the Herschel Space Observatory is being used to study galaxy formation in the early universe, star formation, and the chemical composition of atmospheres around planets, moons, and comets in our solar system.

Ultraviolet Astronomy

Astronomers are also very interested in observing at ultraviolet wavelengths. These observations can reveal a great deal about hot stars, ionized clouds of gas between the stars, and the Sun's high-temperature corona (see Section 3-5), all of which emit copious amounts of ultraviolet light. The spectrum of ultraviolet sunlight reflected from a planet can also reveal the composition of the planet's atmosphere. However, Earth's atmosphere is opaque to ultraviolet light except for the narrow *near-ultraviolet* range, which extends from about 400 nm (the violet end of the visible spectrum) down to 300 nm.

To see shorter-wavelength, *far-ultraviolet* light, astronomers must again make their observations from space. The first ultraviolet telescope was placed in orbit in 1962, and several others have since followed it into space. Small rockets have also lifted ultraviolet cameras briefly above Earth's atmosphere. **Figure 6-27** shows an ultraviolet view of the constellation Orion, along with infrared and visible views.

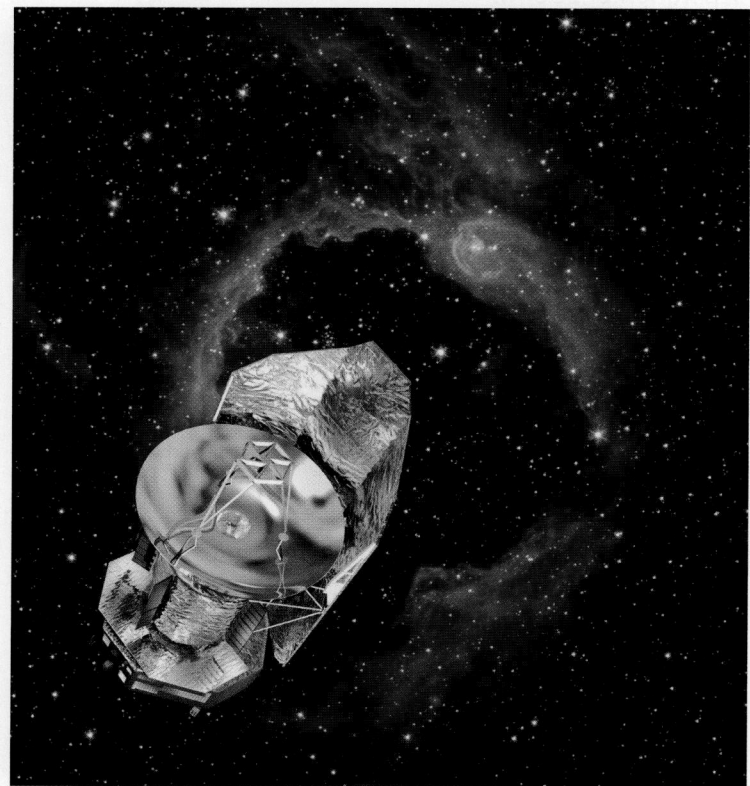

FIGURE 6-26

Herschel Space Observatory This is an artist's concept of the Herschel Space Observatory. Herschel and the Planck observatory lifted off together on May 14, 2009, aboard an Ariane ECA rocket. (ESA - D. Ducros, 2009/NASA)

The Far Ultraviolet Spectroscopic Explorer (FUSE), which went into orbit in 1999 and remained until 2007, specialized in measuring spectra at wavelengths from 90 to 120 nm. Highly ionized oxygen atoms, which can exist only in an extremely high-temperature gas, have a characteristic spectral line in this range. By looking for this spectral line in various parts of the sky, FUSE confirmed that our Milky Way Galaxy (Section 1-4) is surrounded by an immense "halo" of gas at temperatures in excess of 200,000 K. Only an ultraviolet

Immense clouds of dust are heated by hot, luminous, newly formed stars; the clouds glow at (a) ultraviolet and (b) infrared wavelengths.

R I **U** X G
(a) Ultraviolet Orion

R **I** V U X G
(b) Infrared Orion

At visible wavelengths the dust is opaque, obscuring the newly formed stars.

R I **V** U X G
(c) Visible Orion

(d) A star chart of Orion

FIGURE 6-27

Orion Seen at Ultraviolet, Infrared, and Visible Wavelengths
(a) An ultraviolet view of the constellation Orion was obtained during a brief rocket flight in 1975. This 100-s exposure covers the wavelength range 125–200 nm. **(b)** The false-color view from the Infrared Astronomical Satellite displays emission at different wavelengths in different colors: red for 100–μm

radiation, green for 60–μm radiation, and blue for 12–μm radiation. Compare these images with **(c)** an ordinary visible-light photograph and **(d)** a star chart of Orion. (a: G. R. Carruthers, Naval Research Laboratory; b: NASA/JPL-Caltech/IRAS/H. McCallon; c: mike black photography/Getty Images)

telescope could have detected this "halo," which is thought to have been produced by exploding stars called supernovae (Section 1-3).

The Hubble Space Telescope

Infrared and ultraviolet satellites give excellent views of the heavens at selected wavelengths. But since the 1940s, astronomers had dreamed of having one large telescope that could be operated at any wavelength from the near-infrared through the visible range and out into the ultraviolet. This is the mission of the Hubble Space Telescope (HST), which was placed in a 600-km-high orbit by the space shuttle *Discovery* in 1990 (**Figure 6-28**). HST has a 2.4-m (7.9-ft) objective mirror and was designed to observe at wavelengths from 115 nm to 1 μm. Like most ground-based telescopes, HST uses a CCD to record images. (In fact, the development of HST helped drive advances in CCD technology.) The images are then radioed back to Earth in digital form.

The great promise of HST was that from its vantage point high above the atmosphere, its angular resolution would be limited only by diffraction. But soon after HST was placed in orbit, astronomers discovered that a manufacturing error had caused the telescope's objective mirror to suffer from spherical aberration. The mirror should have been able to concentrate 70% of a star's light into an image with an angular diameter of 0.1 arcsec. Instead, only 20% of the light was focused into this small area. The remainder was smeared out over an area about 1 arcsec wide, giving images little better than those achieved at major ground-based observatories.

On an interim basis, astronomers used only the 20% of incoming starlight that was properly focused and, with computer processing, discarded the remaining poorly focused 80%. This was practical only for brighter objects, but many of the observing projects scheduled for HST involved extremely dim galaxies and nebulae.

These problems were resolved by a second space shuttle mission in 1993. Astronauts installed a set of small secondary mirrors whose curvature exactly compensated for the error in curvature of the primary mirror. Once these were in place, HST was able to make truly sharp images of extremely faint objects. Astronomers have used the repaired HST to make discoveries about the nature of planets, the evolution of stars, the inner workings of galaxies, and the expansion of the universe. You will see many HST images in later chapters.

FIGURE 6-28 R **I** V U X G

The Hubble Space Telescope The largest optical telescope yet placed in orbit, HST is a joint project of NASA and the ESA. HST has helped discover new moons of Pluto, probed the formation of stars, and found evidence that the universe is now expanding at a faster rate than several billion years ago. This photograph was taken from the space shuttle *Discovery* during a 1997 mission to service HST. (NASA)

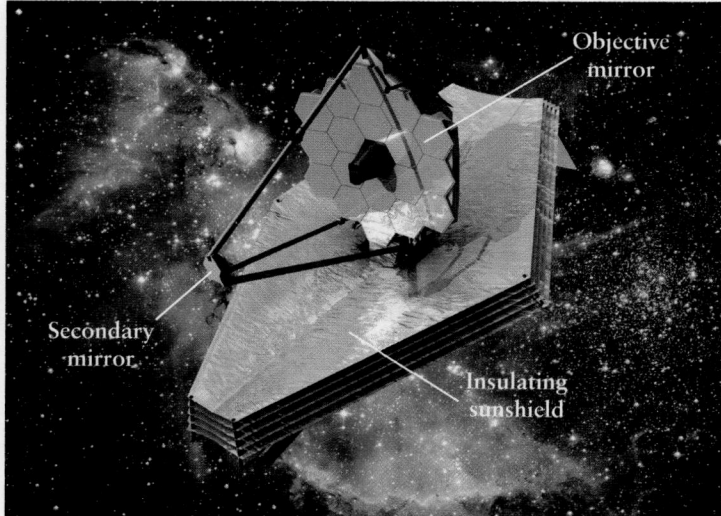

ANIMATION
6.3

FIGURE 6-29

The James Webb Space Telescope (JWST) The
successor to the Hubble Space Telescope, JWST has an objective
mirror 6.5 m (21 ft) in diameter. Like many Earth-bound telescopes, JWST
is of Cassegrain design (see Figures 6-11b and 6-14b). Rather than using
liquid helium to keep the telescope and instruments at the low temperatures
needed to observe at infrared wavelengths, JWST stays cool using a
multilayer sunshield the size of two tennis courts. (ESA, C. Carreau)

The success of HST has inspired plans for its larger successor,
the James Webb Space Telescope, or JWST (Figure 6-29). Planned
for a 2020 launch, JWST will observe at visible and infrared
wavelengths from 600 nm to 28 μm. With its 6.5-m objective
mirror—2.5 times the diameter of the HST objective mirror, with
6 times the light-gathering power—JWST will study faint objects
such as planetary systems forming around other stars and galaxies
near the limit of the observable universe. Unlike HST, which is in a
relatively low-altitude orbit around Earth, JWST will orbit the Sun
some 1.5 million km beyond Earth. In this orbit, the telescope's
view will not be blocked by Earth. Furthermore, by remaining far
from the radiant heat of Earth it will be easier to keep JWST at the
very cold temperatures required by its infrared detectors.

X-Ray Astronomy

Space telescopes have also made it possible to explore objects
whose temperatures reach the almost inconceivable values of 10^6
to 10^8 K. Atoms in such a high-temperature gas move so fast that
when they collide, they emit x-ray photons of very high energy and
very short wavelengths less than 10 nm. x-ray telescopes designed
to detect these photons must be placed in orbit, because Earth's
atmosphere is totally opaque at these wavelengths.

Astronomers got their first quick look at the x-ray sky from
brief rocket flights during the late 1940s. These observations confirmed that the Sun's corona (see Section 3-5) is a source of x rays,
and must therefore be at a temperature of millions of kelvins. In
1962, a rocket experiment revealed that objects beyond the solar
system also emit x rays.

Since 1970, a series of increasingly sensitive and sophisticated
x-ray observatories have been placed in orbit, including NASA's
Einstein Observatory, the ESA's Exosat, and the German-British-
American ROSAT. These telescopes have shown that other stars also
have high-temperature coronae and have found hot, x-ray–emitting

gas clouds so immense that hundreds of galaxies fit inside them. They
also discovered unusual stars that emit x rays in erratic bursts. These
bursts are now thought to be coming from heated gas swirling around
a small but massive object—possibly a black hole.

> **CAUTION!**
>
> X-ray telescopes work on a very different principle from the x-ray
> devices used in medicine and dentistry. If you have your foot
> "x-rayed" to check for a broken bone, a piece of photographic film
> (or an electronic detector) sensitive to x rays is placed under your
> foot and an x-ray beam is directed at your foot from above. The
> radiation penetrates through soft tissue but not as much through
> bone, so the bones cast an "x-ray shadow" on the film. A fracture
> will show as a break in the shadow. X-ray telescopes, by contrast,
> do *not* send beams of x rays toward astronomical objects in an
> attempt to see inside them. Rather, these telescopes detect x rays
> that the objects emit on their own.

X-ray astronomy took a giant leap forward in 1999 with the
launch of NASA's Chandra X-ray Observatory and the ESA's
XMM-Newton (for *X-ray Multi-mirror Mission*). Named for
the Indian-American Nobel laureate astrophysicist Subrahmanyan
Chandrasekhar, Chandra can view the x-ray sky with an angular
resolution of 0.5 arcsec (Figure 6-30a). This is comparable to the
best ground-based optical telescopes and more than 1000 times better than the resolution of the first orbiting x-ray telescope. Chandra
can also measure x-ray spectra 100 times more precisely than any
previous spacecraft and can detect variations in x-ray emissions on
time scales as short as 16 ms. This latter capability is essential for
understanding how x-ray bursts are produced around black holes.

XMM-Newton is actually three x-ray telescopes that all
point in the same direction (Figure 6-30b). Their combined light-
gathering power is 5 times greater than that of Chandra, which enables
XMM-Newton to observe fainter objects. (For reasons of economy,
the mirrors were not ground as precisely as those on Chandra, so the
angular resolution of XMM-Newton is only about 6 arcseconds.) It
also carries a small but highly capable telescope for ultraviolet and visible observations. Hot x-ray sources are usually accompanied by cooler
material that radiates at these longer wavelengths, so XMM-Newton
can observe these hot and cool regions simultaneously.

The latest x-ray telescope in space is NuSTAR, with an incredible
10-m long mast (Figure 6-30c). NuSTAR searches for enormous black
holes, studies particles moving at almost the speed of light, and analyzes how atoms are produced during supernova explosions.

Gamma-Ray Astronomy

Gamma rays, the shortest-wavelength photons of all, help us to
understand phenomena even more energetic than those that produce
x rays. As an example, when a massive star explodes in a supernova,
it produces radioactive atomic nuclei that are strewn across interstellar
space. Observing the gamma rays emitted by these nuclei helps astronomers understand the nature of supernova explosions.

Like x rays, gamma rays do not penetrate Earth's atmosphere,
so space telescopes are required. One of the first gamma-ray telescopes placed in orbit was the Compton Gamma Ray Observatory
(CGRO). One particularly important task for CGRO was the study
of gamma-ray bursts, which are brief, unpredictable, and very
intense flashes of gamma rays that are found in all parts of the sky.
By analyzing data from CGRO and other orbiting observatories,

Incoming x rays

Solar panels provide electric power.

X rays graze the polished inner surface of a gently curved cylinder.

Reflected x rays come to a focus in the focal plane.

(a) Chandra X-ray Observatory

Apertures for the three x-ray telescopes

(b) XMM-Newton

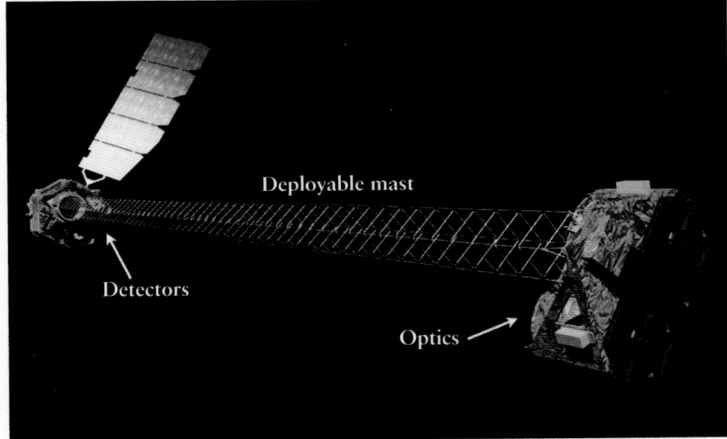

Deployable mast

Detectors

Optics

(c) NuSTAR

FIGURE 6-30

Three Orbiting X-Ray Observatories **(a)** X rays are absorbed by ordinary mirrors like those used in optical reflectors, but they can be reflected if they graze the mirror surface at a very shallow angle. In the Chandra X-ray Observatory, x rays are focused in this way onto a focal plane 10 m (33 ft) behind the mirror. **(b)** XMM-Newton is about the same size as Chandra, and its three x-ray telescopes form images in the same way. **(c)** NuSTAR has a 33-ft (10-m) mast that deployed after launch to separate the optics modules (right) from the detectors in the focal plane (left) (a: NASA; b: ESA/D. Ducros; c: NASA)

astronomers have shown that the sources of these gamma-ray bursts are billions of light-years away. For these bursts to be visible across such great distances, their sources must be among the most energetic objects in the universe. By combining these gamma-ray observations with images made by optical telescopes, astronomers have found that at least some of the gamma-ray bursts emanate from stars that explode catastrophically. Launched in 2008, the Fermi Gamma-ray Space Telescope (Figure 6-31) continues the study of gamma-ray bursts and other highly energetic phenomena.

The *Cosmic Connections* figure on page 171 shows the wavelengths at which Earth-orbiting telescopes are particularly useful, as well as summarizing the design of refracting and reflecting Earth-based telescopes. The advantages and benefits of Earth-orbiting observatories cannot be overemphasized. We are no longer limited to the narrow ranges of whatever wavelengths manage to leak through our shimmering, hazy atmosphere (Figure 6-32). For the first time, we are really *seeing* the universe.

CONCEPT CHECK 6-11

What is the primary advantage of an orbiting space telescope, compared to a ground-based telescope?
Answer appears at the end of the chapter.

FIGURE 6-31 R I V U X G

The Fermi Gamma-Ray Space Telescope This photograph shows the telescope placed inside half of the launching rocket's nose cone. For scale: Several engineers in white cleanroom gowns are at the base of the nose cone. (NASA/Jim Grossmann)

FIGURE 6-32

The Entire Sky at Five Wavelength Ranges These five views show the entire sky at visible, radio, infrared, x-ray, and gamma-ray wavelengths. The entire celestial sphere is mapped onto an oval, with the Milky Way stretching horizontally across the center. **(a)** In the visible portion, the constellation Orion is at the right, Sagittarius is in the middle, and Cygnus is toward the left. Many of the dark areas along the Milky Way are locations where interstellar dust is sufficiently thick to block visible light. **(b)** The radio view shows the sky at a wavelength of 21 cm. This wavelength is emitted by hydrogen atoms in interstellar space. The brightest regions (shown in red) are in the plane of the Milky Way, where the hydrogen is most concentrated. **(c)** This is a mosaic of the images covering the entire sky as observed by the Wide-field Infrared Survey Explorer (WISE). Most of the emission is from dust particles in the plane of the Milky Way that have been warmed by starlight. **(d)** This image is made by using x rays, as indicated by the highlighted X in the wavelength tab. The x-ray map from

ROSAT shows about 50,000 x-ray sources whose intensity runs from red for the less intense sources through yellow and blue for the brightest ones. Extremely high temperature gas emits these x rays. The white regions, which emit strongly at all x-ray wavelengths, are remnants of supernovae. **(e)** This image is made by using gamma rays, as indicated by the highlighted G in the wavelength tab. The gamma-ray view from the Compton Gamma Ray Observatory includes all wavelengths less than about 1.2×10^{-5} nm (photon energies greater than 10^8 eV). The diffuse radiation from the Milky Way is emitted when fast-moving subatomic particles collide with the nuclei of atoms in interstellar gas clouds. The bright spots above and below the Milky Way are distant, extremely energetic galaxies. (a: Axel Mellinger; b: Max Planck Institute for Radio Astronomy/Science Photo Library/ Science Source; c: NASA/JPL-Caltech/UCLA; d: Max Planck Institute for Extraterrestrial Physics/Science Photo Library/Science Source; e: NASA/Compton Gamma Ray Observatory/Science Source)

scopes Across the EM Spectrum

Today, telescopes can view the universe in every range of electromagnetic (EM) radiation, although some must be above Earth's atmosphere to receive radiation without interference.

Light from distant object

Objective lens

A refracting telescope uses a lens to form an image. An eyepiece is used to magnify the image.

Image of distant object

Eyepiece for magnifying image

Light from distant object

Secondary mirror

A reflecting telescope uses a curved mirror to form an image. An eyepiece can be used to magnify the image, or the image can be recorded on a CCD.

Objective mirror

Image

CCD detector

Earth's atmosphere is transparent to visible wavelengths, so visible-light telescopes (refracting or reflecting) can be used from Earth's surface.

Infrared telescopes are best placed in orbit since most infrared wavelengths do not penetrate the atmosphere.

Infrared telescope

Ultraviolet telescope

ay, x-ray, and
telescopes
laced in
use these
hs do
ate the
re.

X-ray telescope

Radio telescope

Gamma-ray

The atmosphere is transparent to radio waves, so radio telescopes can be used from Earth's surface.

KEY WORDS

active optics, p. 159
adaptive optics, p. 159
angular resolution, p. 157
baseline, p. 160
Cassegrain focus, p. 155
charge-coupled device (CCD), p. 161
chromatic aberration, p. 152
coma, p. 156
coudé focus, p. 155
diffraction, p. 157
diffraction grating, p. 161
eyepiece lens, p. 150
false color, p. 159

focal length, p. 148
focal plane, p. 149
focal point, p. 148
focus (of a lens or mirror), p. 148
imaging, p. 160
interferometry, p. 160
lens, p. 148
light-gathering power, p. 150
light pollution, p. 160
magnification (magnifying power), p. 152
medium (*plural* media), p. 148
Newtonian reflector, p. 154

objective lens, p. 150
objective mirror (primary mirror), p. 154
optical telescope, p. 148
optical window (in Earth's atmosphere), p. 165
photometry, p. 161
pixel, p. 161
prime focus, p. 154
radio telescope, p. 163
radio window (in Earth's atmosphere), p. 165
reflecting telescope (reflector), p. 154

reflection, p. 153
refracting telescope (refractor), p. 150
refraction, p. 148
seeing disk, p. 158
spectrograph, p. 161
spectroscopy, p. 161
spherical aberration, p. 156
very-long-baseline interferometry (VLBI), p. 164

KEY IDEAS

Refracting Telescopes: Refracting telescopes, or refractors, produce images by bending light rays as they pass through glass lenses (Section 6-1).

- Chromatic aberration is an optical defect whereby light of different wavelengths is bent in different amounts by a lens.

- Glass impurities, chromatic aberration, opacity to certain wavelengths, and structural difficulties make it inadvisable to build extremely large refractors.

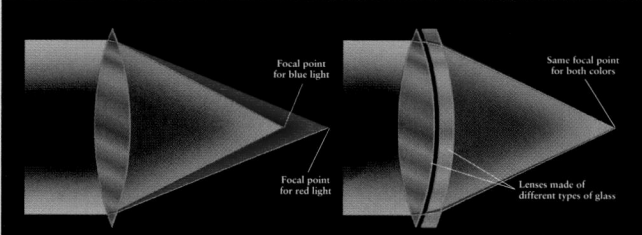

Reflecting Telescopes: Reflecting telescopes, or reflectors, produce images by reflecting light rays to a focus point from curved mirrors (Section 6-2).

- Reflectors are not subject to most of the problems that limit the useful size of refractors.

Courtesy of John D. Monnier

Two light sources with angular separation greater than angular resolution of telescope: two sources easily distinguished

Courtesy of John D. Monnier

Light sources moved closer so that angular separation equals angular resolution of telescope: just barely possible to tell that there are two sources

Angular Resolution: A telescope's angular resolution, which indicates ability to see fine details, is limited by two key factors (Section 6-3).

- Diffraction is an intrinsic property of light waves. Its effects can be minimized by using a larger objective lens or mirror.

- The blurring effects of atmospheric turbulence can be minimized by placing the telescope atop a tall mountain with very smooth air. They can be dramatically reduced by the use of adaptive optics and can be eliminated entirely by placing the telescope in orbit.

Charge-Coupled Devices: Sensitive electronic light detectors called charge-coupled devices (CCDs) are often used at a telescope's focus to digitally record faint images (Section 6-4).

Spectrographs: A spectrograph uses a diffraction grating to form the spectrum of an astronomical object (Section 6-5).

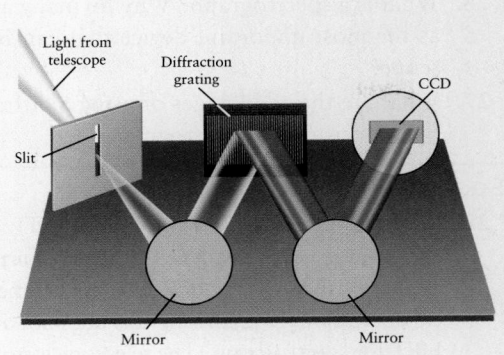

Radio Telescopes: Radio telescopes use large reflecting dishes to focus radio waves onto a detector (Section 6-6).

- Very large dishes provide reasonably sharp radio images. Higher resolution is achieved with interferometry techniques that link smaller dishes together.

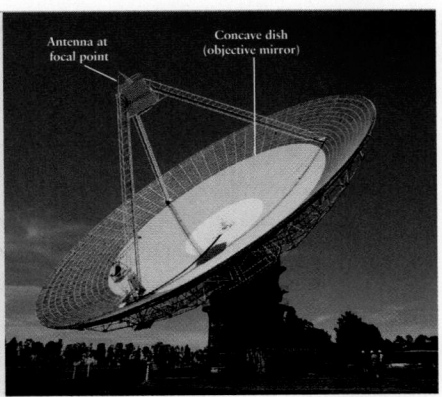

David Nunuk/Science Source

Transparency of Earth's Atmosphere: Earth's atmosphere absorbs much of the radiation that arrives from space (Section 6-7).

- The atmosphere is transparent chiefly in two wavelength ranges known as the optical window and the radio window. A few wavelengths in the near-infrared also reach the ground.

Telescopes in Space: For observations at wavelengths to which Earth's atmosphere is opaque, astronomers depend on telescopes carried above the atmosphere by rockets or spacecraft (Section 6-7).

- Satellite-based observatories provide new information about the universe and permit coordinated observation of the sky at all wavelengths.

QUESTIONS

Problem-solving tips and tools

You may find it useful to review the small-angle formula discussed in Box 1-1. The area of a circle is proportional to the square of its diameter. Data on the planets can be found in the appendices at the end of this book. Section 5-2 discusses the relationship between frequency and wavelength. Box 6-1 gives examples of how to calculate magnifying power and light-gathering power.

Section 6-1

1. Describe refraction and reflection. Explain how these processes enable astronomers to build telescopes.

2. Explain why a flat piece of glass does not bring light to a focus while a curved piece of glass can.

3. Explain why the light rays that enter a telescope from an astronomical object are essentially parallel.

4. With the aid of a diagram, describe a refracting telescope. Which dimensions of the telescope determine its light-gathering power? Which dimensions determine the magnification?

5. What is the purpose of a telescope eyepiece? What aspect of the eyepiece determines the magnification of the image? In what circumstances would the eyepiece not be used?

6. Do most professional astronomers actually look through their telescopes? Why or why not?

7. Quite often advertisements appear for telescopes that extol their magnifying power. Is this a good criterion for evaluating telescopes? Explain your answer.

8. What is chromatic aberration? For what kinds of telescopes does it occur? How can it be corrected?

9. Show by means of a diagram why the image formed by a simple refracting telescope is upside down.

10. Consider replacing the standard lens of a camera with a telephoto lens. The images made with the telephoto lens will make distant objects appear closer. How does the focal length of the telephoto lens compare with the standard lens?

Section 6-2

11. With the aid of a diagram, describe a reflecting telescope. Describe four different ways in which an astronomer can access the focal plane.

12. Explain some of the disadvantages of refracting telescopes compared to reflecting telescopes.

13. What kind of telescope would you use if you wanted to take a color photograph entirely free of chromatic aberration? Explain your answer.

14. Explain why a Cassegrain reflector can be substantially shorter than a refractor of the same focal length.

15. No major observatory has a Newtonian reflector as its primary instrument, whereas Newtonian reflectors are extremely popular among amateur astronomers. Explain why this is so.

16. What is spherical aberration? How can it be corrected?

17. The observing cage in which an astronomer can sit at the prime focus of the 5-m telescope on Palomar Mountain is about 1 m in diameter. Calculate what fraction of the incoming starlight is blocked by the cage.

18. (a) Compare the light-gathering power of the Keck I 10-m telescope with that of the Hubble Space Telescope (HST), which has a 2.4-m objective mirror. (b) What advantages does Keck I have over HST? What advantages does HST have over Keck I?

Section 6-3

19. **TUTORIAL 6.1** What is diffraction? Why does it limit the angular resolution of a telescope? What other physical phenomenon is often a more important restriction on angular resolution?

20. What is active optics? What is adaptive optics? Why are they useful? Would either of these be a good feature to include on a telescope to be placed in orbit?

21. Explain why combining the light from two or more optical telescopes can give dramatically improved angular resolution.

22. What is light pollution? What effects does it have on the operation of telescopes? What can be done to minimize these effects?

23. Suppose your Newtonian reflector has an objective mirror 20 cm (8 in.) in diameter with a focal length of 2 m. What magnification do you get with eyepieces whose focal lengths are (a) 9 mm, (b) 20 mm, and (c) 55 mm? (d) What is the telescope's diffraction-limited angular resolution when used with orange light of wavelength 600 nm? (e) Would it be possible to achieve this angular resolution if you took the telescope to the summit of Mauna Kea? Why or why not?

24. Several groups of astronomers are making plans for large ground-based telescopes. (a) What would be the diffraction-limited angular resolution of a telescope with a 40-m objective mirror? Assume that yellow light with wavelength 550 nm is used. (b) Suppose this telescope is placed atop Mauna Kea. How will the actual angular resolution of the telescope compare to that of the 10-m Keck I telescope? Assume that adaptive optics is not used.

Section 6-4

25. **TUTORIAL 6.2** What is a charge-coupled device (CCD)? Why have CCDs replaced photographic film for recording astronomical images?

26. What is a spectrograph? Why do many astronomers regard it as the most important device that can be attached to a telescope?

27. What are the advantages of using a diffraction grating rather than a prism in a spectrograph?

Section 6-5

28. The Hobby-Eberly Telescope (HET) at the McDonald Observatory in Texas has a spherical mirror, which is the least expensive shape to grind. Consequently, the telescope has spherical aberration. Explain why this does not affect the usefulness of HET for spectroscopy. (The telescope is not used for imaging.)

29. Consider the spectrum in Figure 6-21a. Does the distance between the blue absorption lines on the left and the red absorption line on the right depend on the width of the astronomical object being observed? Explain your reasoning.

Section 6-6

30. Compare an optical reflecting telescope and a radio telescope. What do they have in common? How are they different?

31. Why can radio astronomers make observations at any time during the day, whereas optical astronomers are mostly limited to observing at night? (*Hint:* Does your radio work any better or worse in the daytime than at night?)

32. Why are radio telescopes so large? Why does a single radio telescope have poorer angular resolution than a large optical telescope? How can the resolution be improved by making simultaneous observations with several radio telescopes?

33. The Institute of Space and Astronautical Science in Japan proposes to place a radio telescope into an even higher orbit than the HALCA telescope. Using this telescope in concert with a ground-based radio telescope, baselines as long as 25,000 km may be obtainable. Astronomers want to use this combination to study radio emission at a frequency of 43 GHz from the molecule silicon monoxide, which is found in the interstellar clouds from which stars form. (1 GHz = 1 gigahertz = 10^9 Hz.) (a) What is the wavelength of this emission? (b) Taking the baseline to be the effective diameter of this radio-telescope array, what angular resolution can be achieved?

Section 6-7

34. **TUTORIAL 6.3** What are the optical window and the radio window? Why isn't there an x-ray window or an ultraviolet window?

35. Why is it necessary to keep an infrared telescope at a very low temperature?

36. How are the images made by an x-ray telescope different from those made by a medical x-ray machine?

37. Why must astronomers use satellites and Earth-orbiting observatories to study the heavens at x-ray and gamma-ray wavelengths?

38. The four largest moons of Jupiter are roughly the same size as our Moon and are about 628 million (6.28×10^8) km from Earth at opposition. What is the size (in kilometers) of the smallest surface features that the Hubble Space Telescope (resolution of 0.1 arcsec) can detect? How does this compare with the smallest features that can be seen on the Moon with the unaided human eye (resolution of 1 arcmin)?

39. The Hubble Space Telescope (HST) has been used to observe the galaxy M100, some 70 million light-years from Earth. (a) If the angular resolution of the HST image is 0.1 arcsec, what is the diameter in light-years of the smallest detail that can be discerned in the image? (b) At what distance would a U.S. dime (diameter 1.8 cm) have an angular size of 0.1 arcsec? Give your answer in kilometers.

40. At its closest to Earth, Pluto is 28.6 au from Earth. Can the Hubble Space Telescope distinguish any features on Pluto? Justify your answer using calculations.

41. The mission of the Submillimeter Wave Astronomy Satellite (SWAS), launched in 1998, was to investigate interstellar clouds within which stars form. One of the frequencies at which it observed these clouds is 557 GHz (1 GHz = 1 gigahertz = 10^9 Hz), characteristic of the emission from interstellar water molecules. (a) What is the wavelength (in meters) of this emission? In what part of the electromagnetic spectrum is this? (b) Why was it necessary to use a satellite for these observations? (c) SWAS had an angular resolution of 4 arcmin. What was the diameter of its primary mirror?

42. To search for ionized oxygen gas surrounding our Milky Way Galaxy, astronomers aimed the ultraviolet telescope of the FUSE spacecraft at a distant galaxy far beyond the Milky Way. They then looked for an ultraviolet spectral line of ionized oxygen in that galaxy's spectrum. Were they looking for an emission line or an absorption line? Explain your answer.

43. A sufficiently thick interstellar cloud of cool gas can absorb low-energy x rays but is transparent to high-energy x rays and gamma rays. Explain why both Figure 6-32b and Figure 6-32d reveal the presence of cool gas in the Milky Way. Could you infer the presence of this gas from the visible-light image in Figure 6-32a? Explain your answer.

General Problems

44. If you were in charge of selecting a site for a new observatory, what factors would you consider important?

45. Discuss the advantages and disadvantages of using a small telescope in Earth's orbit versus a large telescope on a mountaintop.

WEB/E-BOOK QUESTIONS

46. Several telescope manufacturers build telescopes with a design called a Schmidt–Cassegrain. These use a correcting lens in an arrangement like that shown in Figure 6-13c. Consult advertisements on the internet to see the appearance of these telescopes and find out their cost. Why do you suppose they are very popular among amateur astronomers?

47. The Large Zenith Telescope (LZT) in British Columbia, Canada, uses a 6-m *liquid* mirror made of mercury. Use the internet to investigate this technology. How can a liquid metal be formed into the necessary shape for a telescope mirror? What are the advantages of a liquid mirror? What are the disadvantages?

48. Three of the telescopes shown in Figure 6-16—the James Clerk Maxwell Telescope (JCMT), the Caltech Submillimeter Observatory (CSO), and the Submillimeter Array (SMA)—are designed to detect radiation with wavelengths close to 1 mm. Search for current information about JCMT, CSO, and SMA on the internet. What kinds of celestial objects emit radiation at these wavelengths? What can astronomers see using JCMT, CSO, and SMA that cannot be observed at other wavelengths? Why is it important that they be at high altitude? How large are the primary mirrors used in JCMT, CSO, and SMA? What are the differences among the three telescopes? Which can be used in the daytime? What recent discoveries have been made using JCMT, CSO, or SMA?

49. In 2003, an ultraviolet telescope called GALEX (Galaxy Evolution Explorer) was placed into orbit. Use the internet to learn about GALEX and its mission. What aspects of galaxies was GALEX designed to investigate? Why is it important to make these observations using ultraviolet wavelengths?

50. At the time of this writing, NASA's plans for the end of the Hubble Space Telescope's mission were uncertain. Consult the Space Telescope Science Institute website to learn about plans for HST's final years of operation. Are future space shuttle missions planned to service HST? If so, what changes will be made to HST on such missions? What will become of HST at the end of its mission lifetime?

ACTIVITIES

Observing Projects

51. Obtain a telescope during the daytime, along with several eyepieces of various focal lengths. If you can determine the telescope's focal length, calculate the magnifying powers of the eyepieces. Focus the telescope on some familiar object, such as a distant lamppost or tree. **DO NOT FOCUS ON THE SUN! Looking directly at the Sun can cause blindness.** Describe the image you see through the telescope. Is it upside down? How does the image move as you slowly and gently shift the telescope left and right or up and down? Examine the eyepieces, noting their focal lengths. By changing the eyepieces, examine the distant object under different magnifications. How do the field of view and the quality of the image change as you go from low power to high power?

52. Many towns and cities have amateur astronomy clubs. If you are so inclined, attend a "star party" hosted by your local club. People who bring their telescopes to such gatherings are delighted to show you their instruments and take you on a telescopic tour of the heavens. Such an experience can lead to a very enjoyable, lifelong hobby.

Starry Night™ Explorations

53. Use *Starry Night*™ to explore the effect of light pollution on the night sky. Click the **Home** button to view the sky from your home location. Click on the **Quick Search Items** button and open the **Options** panel, then expand the **Local View** layer and turn off the **Daylight** option to view the sky under ideal conditions of darkness. Survey the sky and use the HUD to find the Apparent Magnitude of the dimmest stars you can discern.

    (a) Return to the Options panel and click on the **Local Light Pollution** label in the Local View layer to open the Light Pollution options window. In this window, turn on **Local** light pollution, set the level to its maximum limit, and click **OK**. Look around the sky again. What is the apparent magnitude of the dimmest discernible stars in this sky? How does this compare to the apparent magnitude limit under ideal conditions?

    (b) Estimate the fraction of the visible stars you can see under ideal conditions that are obscured by severe nearby light pollution. Open the **FOV Indicators** panel, expand the **All Charts** layer and turn on the 20° × 20° indicator. Drag the view so that the square indicator encompasses a reasonable number of stars. Without changing the zoom, count the number of stars visible within this 20° × 20° area of the sky. Now reopen the **Options** panel, turn off **Local Light Pollution**, and count the stars visible within the indicator under ideal conditions of darkness. Try this for one or two more areas in the sky and then turn off the 20° × 20° box in the **FOV Indicators** panel. Approximately what fraction of the stars that are visible under ideal conditions are obscured by severe local light pollution?

54. Use *Starry Night*™ to explore the difference between magnification and resolution. Select **Favorites > Explorations > Orion**. The view is centered on the Great Orion Nebula.

    (a) Zoom in to a field of view about 7° wide. Would you say that the view has been magnified? What has happened to the size of the field of view? Has the resolution improved?

    (b) Note the two bright spots near the center of the nebula and observe them closely as you slowly zoom in to a field of view about 1° wide. Is more detail discernible in the nebula as you zoom in? What has become of the lower of the two bright spots? During the zoom, has magnification increased? Has the resolution improved?

    (c) Zoom in to a field of view about 25° wide. How did the zoom affect the image of the other bright spot in the nebula? Use the HUD to identify the components of this bright spot. With this zoom, has the magnification increased? Has the resolution improved?

    (d) Continue to zoom in on the image. As you zoom closer, does the clarity of the image improve, deteriorate, or remain unchanged? Explain your observations in terms of magnification and resolution.

Collaborative Exercises

55. Stand up and have everyone in your group join hands, making as large a circle as possible. If a telescope mirror were built as big as your circle, what would be its diameter? What would be your telescope's diffraction-limited angular resolution for blue light? Would atmospheric turbulence have a noticeable effect on the angular resolution?

56. Are there enough students in your class to stand and join hands and make two large circles that recreate the sizes of the two Keck telescopes? Explain how you determined your answer.

ANSWERS

ConceptChecks

ConceptCheck 6-1: The diameter, because it is the width of the opening that determines how much light can be captured by the telescope.

ConceptCheck 6-2: The thin lens bends light less, so, according to Figure 6-5, it must focus light at a more distant location. Thus, the thinner lens has the greater focal length.

ConceptCheck 6-3: The larger the eyepiece focal length, the smaller the magnification.

ConceptCheck 6-4: Reflection off a mirror does not depend on wavelength. This is a key advantage for a reflecting telescope, because it prevents the chromatic aberration that blurs images in telescopes made from lenses.

ConceptCheck 6-5: The angular resolution is decreased for larger telescope diameters. This improves the sharpness of the image, by reducing the angle over which light is spread out and blurred by the telescope. Therefore, a larger telescope is better in two ways: sharper images of even dimmer features.

ConceptCheck 6-6: Adaptive optics actuators slightly deform the telescope's mirror to match the apparent movement of a star due to distortions in Earth's atmosphere. The actuators must deform the mirror further on nights when the atmosphere is fluctuating more rapidly.

ConceptCheck 6-7: Photographic film captures only 2% of the light it receives. For a faint star, there might be very little light captured by the film. CCDs, on the other hand, are 35 times more sensitive, capturing $35 \times 2\% = 70\%$ of the light they receive.

ConceptCheck 6-8: No. Instead of the CCD capturing a focused image of the planet, the spectrograph device spreads out light from the planet in wavelength to form a spectrum. A choice must be made to either look at an image using imaging equipment, or to look at a spectrum using a spectrograph.

ConceptCheck 6-9: If more than one radio dish is connected together, the distance between the two dishes—not the size of the dishes—determines the telescope's overall diameter when considering the angular resolution. Thus, two dishes at 200 m produce sharper images than a single 100-m dish.

ConceptCheck 6-10: Given these three choices, astronomers would much prefer to have a new telescope in the microwave region because x rays and ultraviolet wavelengths rarely pass through Earth's atmosphere to the ground. While the atmosphere absorbs some portion of microwaves, not all microwaves are absorbed.

ConceptCheck 6-11: Orbiting space telescopes are placed far above most of Earth's atmosphere. Without Earth's atmosphere to absorb infrared, ultraviolet, x-ray, and gamma-ray light, astronomers can study the universe at these wavelengths.

7 | Comparative Planetology I: Our Solar System

LEARNING GOALS

As recently as the mid-1900s, astronomers knew precious little about the other worlds that orbit the Sun. Even the best telescopes provided images of the planets that were frustratingly hazy and indistinct. Of asteroids, comets, and the satellites (or moons) of the planets, we knew even less.

Today, our knowledge of the solar system has grown exponentially, due almost entirely to robotic spacecraft. Spacecraft have been sent to all the planets at close range, revealing details unimagined by astronomers of an earlier generation. Most recently, we visited Pluto in 2015 to get our first look at this much-loved dwarf planet. As you can see in the image, Pluto seems to have a heart of its own. We have landed spacecraft on the Moon, Venus, and Mars and dropped a probe into the immense atmosphere of Jupiter. The Cassini spacecraft spent 13 years orbiting Saturn before passing through its rings to make its final plunge into the thick atmosphere on September 15, 2017. This is truly the golden age of solar system exploration.

In this chapter, we paint in broad outline our present understanding of the solar system. We will see that the planets come in a variety of sizes and chemical compositions. A rich variety also exists among the moons (or satellites) of the planets and among smaller bodies we call asteroids, comets, and trans-Neptunian objects. We will investigate

July 14, 2015: New Horizons, the first spacecraft to visit Pluto, reaches its closest approach to the dwarf planet. (NASA/Johns Hopkins University Applied Physics Laboratory/Southwest Research Institute)

R I V U X G

the nature of craters on the Moon and other worlds of the solar system. And by exploring the magnetic fields of planets, we will be able to peer inside Earth and other worlds and learn about their interior compositions.

An important reason to study the solar system is to search for our own origins. In Chapter 8, we will see how astronomers have used evidence from the present-day solar system to understand how the Sun, Earth, and the other planets formed some four and a half billion years ago, and how they have evolved since then. But for now, we invite you to join us on a guided tour of the worlds that orbit our Sun.

CONCEPT CHECK 7-1

How many stars are in our solar system?

Answer appears at the end of the chapter.

7-1 The solar system has two broad categories of planets: Earthlike and Jupiterlike

Each of the planets that orbit the Sun is unique. Only Earth has liquid water and an atmosphere that humans can breathe; only Venus has a perpetual cloud layer made of sulfuric acid droplets; and only Jupiter has immense storm systems that persist for centuries. But there are also striking similarities among planets. Volcanoes are found not only on Earth, but also on Venus and Mars; rings encircle Jupiter, Saturn, Uranus, and Neptune; and impact craters dot the surfaces of Mercury, Venus, Earth, and Mars, showing that all of these planets have been bombarded by interplanetary debris.

How can we make sense of the many similarities and differences among the planets? An important first step is to organize our knowledge of the planets in a systematic way. There are two different approaches we can take. First, we can contrast the orbits of different planets around the Sun; second, we can compare the planets' physical properties such as size, mass, average density, and chemical composition.

We can understand the most important similarities and differences among the planets by comparing their orbits, masses, and diameters.

Comparing the Planets: Orbits

TUTORIAL
7.1

A planet falls naturally into one of two categories according to the size of its orbit. As **Figure 7-1** shows, the orbits of the four inner planets (Mercury, Venus, Earth, and Mars) are crowded in close to the Sun. In contrast, the orbits of the next four planets (Jupiter, Saturn, Uranus, and Neptune) are widely spaced at great distances from the Sun. **Table 7-1** lists the orbital characteristics of these eight planets.

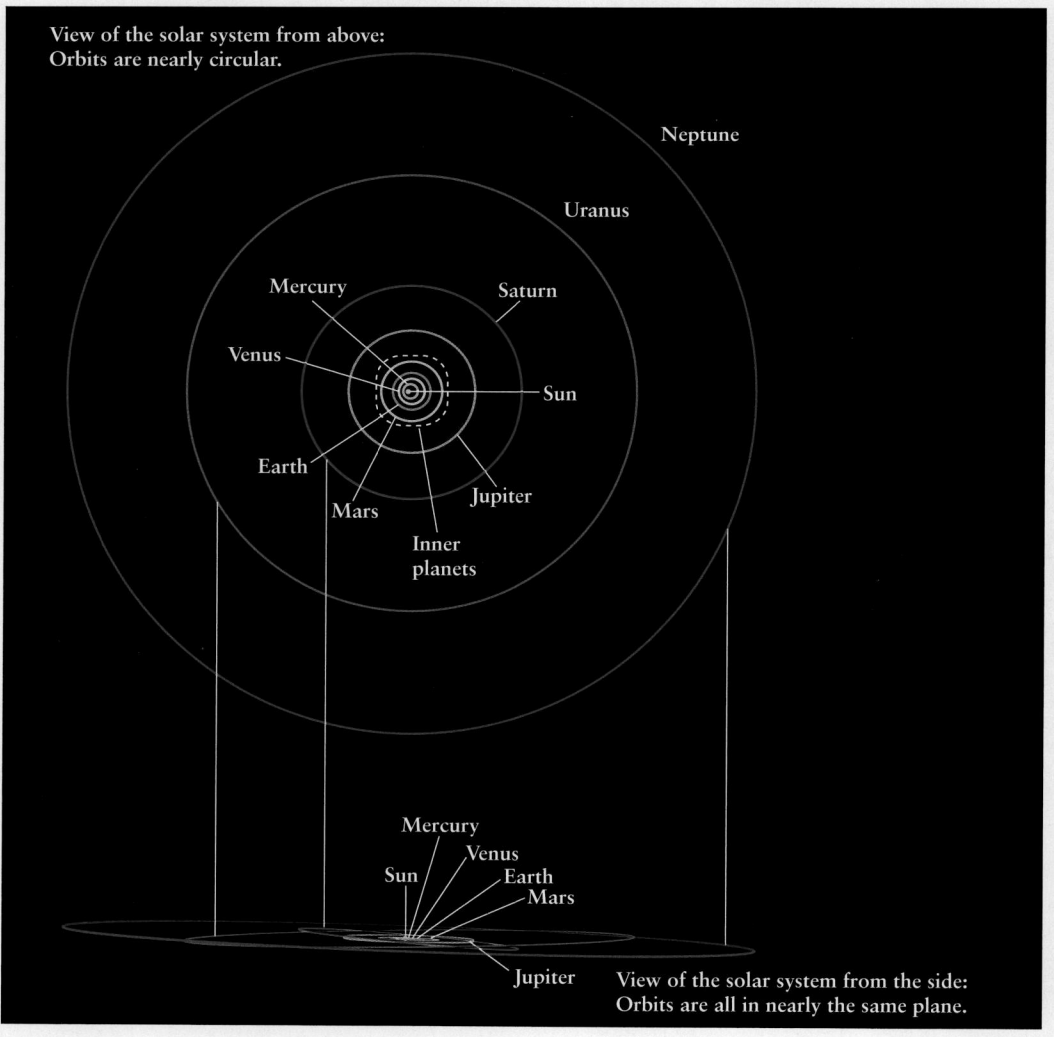

View of the solar system from above:
Orbits are nearly circular.

Neptune

Uranus

Mercury   Saturn

Venus

Sun

Earth

Mars

Jupiter

Inner
planets

Mercury
Venus
Sun   Earth
Mars

Jupiter   View of the solar system from the side:
Orbits are all in nearly the same plane.

ANIMATION
7.1

FIGURE 7-1

The Solar System to Scale This scale drawing shows the orbits of the planets around the Sun. The four inner planets are crowded in close to the Sun, while the four outer planets orbit the Sun at much greater distances. On the scale of this drawing, the planets themselves would be much smaller than the diameter of a human hair and too small to see.

TABLE 7-1 **Characteristics of the Planets**

	The Inner (Terrestrial) Planets			
	Mercury	**Venus**	**Earth**	**Mars**
Average distance from the Sun (10^6 km)	57.9	108.2	149.6	227.9
Average distance from the Sun (au)	0.387	0.723	1.000	1.524
Orbital period (years)	0.241	0.615	1.000	1.88
Orbital eccentricity	0.206	0.007	0.017	0.093
Inclination of orbit to the ecliptic	7.00°	3.39°	0.00°	1.85°
Tilt of rotational axis	0.01°	2.64°	23.44°	25.19°
Equatorial diameter (km)	4880	12,104	12,756	6794
Equatorial diameter (Earth = 1)	0.383	0.949	1.000	0.533
Mass (kg)	3.302×10^{23}	4.868×10^{24}	5.974×10^{24}	6.418×10^{23}
Mass (Earth = 1)	0.0553	0.8150	1.0000	0.1074
Average density (kg/m^3)	5430	5243	5515	3934

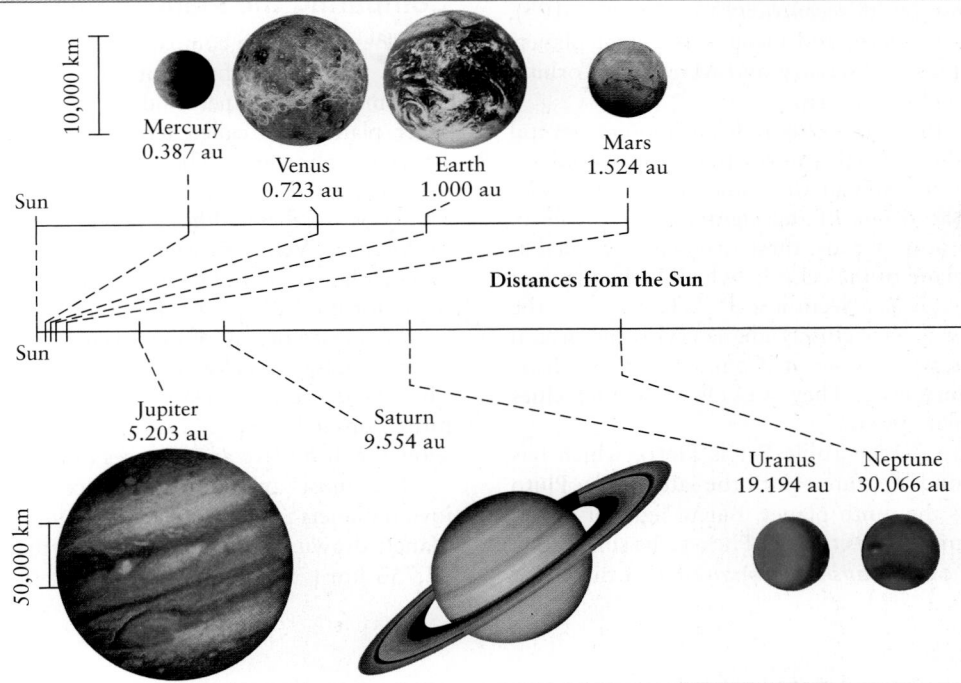

10,000 km

Mercury
0.387 au

Venus
0.723 au

Earth
1.000 au

Mars
1.524 au

Sun

Distances from the Sun

Sun

Jupiter
5.203 au

Saturn
9.554 au

Uranus
19.194 au

Neptune
30.066 au

50,000 km

NASA/Johns Hopkins
University Applied Physics
Laboratory/Carnegie
Institution of Washington;
NASA/JPL; NASA/JSC;
NASA/USGS; NASA;
NASA/JPL/USGS; NASA/
JPL-Caltech; NASA/JPL

R I V U X G

	The Outer (Jovian) Planets			
	Jupiter	**Saturn**	**Uranus**	**Neptune**
Average distance from the Sun (10^6 km)	778.3	1429	2871	4498
Average distance from the Sun (au)	5.203	9.554	19.194	30.066
Orbital period (years)	11.86	29.46	84.10	164.86
Orbital eccentricity	0.048	0.053	0.043	0.010
Inclination of orbit to the ecliptic	1.30°	2.48°	0.77°	1.77°
Tilt of rotational axis	3.12°	26.73°	82.23°	28.33°
Equatorial diameter (km)	142,984	120,536	51,118	49,528
Equatorial diameter (Earth = 1)	11.209	9.449	4.007	3.883
Mass (kg)	1.899×10^{27}	5.685×10^{26}	8.682×10^{25}	1.024×10^{26}
Mass (Earth = 1)	317.8	95.16	14.53	17.15
Average density (kg/m^3)	1326	687	1318	1638

CAUTION!

While Figure 7-1 shows the orbits of the planets, it does not show the planets themselves. The reason is simple: If Jupiter, the largest of the planets, were to be drawn to the same scale as the rest of this figure, it would be a dot just 0.0002 cm across—about 1/300 of the width of a human hair and far too small to be seen without a microscope. The planets themselves are *very* small compared to the distances between them. While an airplane traveling at 1000 km/h (620 mi/h) can fly around Earth in less than 2 days, at this speed it would take 17 *years* to fly from Earth to the Sun. The solar system is a very large and very empty place!

Most of the planets have orbits that are nearly circular. As we learned in Section 4-4, Kepler discovered in the seventeenth century that these orbits are actually ellipses. Astronomers denote the elongation of an ellipse by its *eccentricity* (see Figure 4-10b). The eccentricity of a circle is zero, and most of the eight planets (with the notable exceptions of Mercury and Mars) have orbital eccentricities that are very close to zero.

If you could observe the solar system from a point several astronomical units (au) above Earth's north pole, you would see that all the planets orbit the Sun in the same counterclockwise direction. Furthermore, the orbits of the eight planets all lie in nearly the same plane. In other words, these orbits are *inclined* at only slight angles to the plane of the ecliptic, which is the plane of Earth's orbit around the Sun (see Section 2-5). What's more, the plane of the Sun's equator is very closely aligned with the orbital planes of the planets. As we will see in Chapter 8, these near-alignments are not a coincidence. They provide important clues about the origin of the solar system.

Not included in Figure 7-1 or Table 7-1 is Pluto, which has an orbit that reaches beyond Neptune. Until the late 1990s, Pluto was generally regarded as the ninth planet. But in light of recent discoveries, most astronomers now consider Pluto to be simply one member of a large collection of *trans-Neptunian objects* that orbit

far from the Sun. Pluto is not even the largest of this new class of objects! Trans-Neptunian objects orbit the Sun in the same counterclockwise direction as the eight planets, though many of them have orbits that are steeply inclined to the plane of the ecliptic and have high eccentricities (that is, the orbits are quite elongated and noncircular). We will discuss trans-Neptunian objects, along with other small bodies that orbit the Sun, in Section 7-5.

CONCEPT CHECK 7-2

Is Mars classified as an inner planet or an outer planet?

CALCULATION CHECK 7-1

Using Table 7-1, which of the planets has an orbital path that is most nearly a perfect circle in shape?

Answers appear at the end of the chapter.

Comparing the Planets: Physical Properties

TUTORIAL 7.2 When we compare the physical properties of the planets, we again find that they fall naturally into two classes—four small inner planets and four large outer ones. The four small inner planets are called **terrestrial planets** because they resemble Earth (in Latin, *terra*). They all have hard, rocky surfaces with mountains, craters, valleys, and volcanoes. You could stand on the surface of any one of them, although you would need a protective spacesuit on Mercury, Venus, or Mars. The four large outer planets are called **jovian planets** because they resemble Jupiter. (Jove was another name for the Roman god Jupiter.) An attempt to land a spacecraft on the surface of any of the jovian planets would be futile, because the materials that make up these planets are mostly gaseous or liquid. The visible "surface" features of a jovian planet are actually cloud formations in the planet's atmosphere. The photographs in Table 7-1 show the distinctive appearances of the two classes of planets.

The most apparent difference between the terrestrial and jovian planets is their *diameters*. Figure 7-2 shows the Sun and the planets drawn to the same scale. Earth, with its diameter of about 12,756 km (7926 mi), is the largest of the four inner, terrestrial

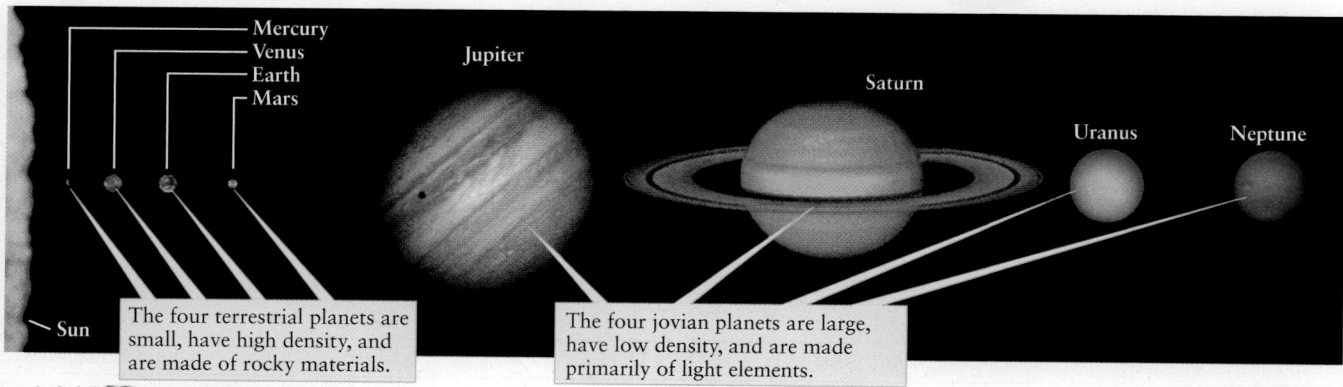

FIGURE 7-2   R I V U X G

The Planets to Scale This figure shows the planets from Mercury to Neptune to the same scale, but their distances apart are not drawn to scale.

The four terrestrial planets have orbits nearest the Sun, and the jovian planets are the next four planets from the Sun. (Calvin J. Hamilton and NASA/JPL)

BOX 7-1 ASTRONOMY DOWN TO EARTH

Average Density

Average density—the mass of an object divided by that object's volume—is a useful quantity for describing the differences between planets in our solar system. This same quantity has many applications here on Earth.

A rock tossed into a lake sinks to the bottom, while an air bubble produced at the bottom of a lake (for example, by the air tanks of a scuba diver) rises to the top. These are examples of a general principle: An object sinks in a fluid if its average density is greater than that of the fluid; it rises if its average density is less than that of the fluid. The average density of water is 1000 kg/m^3, which is why a typical rock (with an average density of about 3000 kg/m^3) sinks, while an air bubble (average density of about 1.2 kg/m^3) rises.

At many summer barbecues, cans of soft drinks are kept cold by putting them in a container full of ice. When the ice melts, the cans of diet soda always rise to the top, while the cans of regular soda sink to the bottom. Why is this? The average density of a can of diet soda—which includes water, flavoring, artificial sweetener, and the trapped gas that makes the drink fizzy—is slightly less than the density of water, and so the can floats. A can of regular soda contains sugar instead of artificial sweetener, and the sugar is a bit heavier than the sweetener. The extra weight is just enough to make the average density of a can of regular soda slightly more than that of water, making the can sink. (You can test these statements for yourself by putting unopened cans of diet soda and regular soda in a sink or bathtub full of water.)

The concept of average density provides geologists with important clues about the early history of Earth. The average density of surface rocks on Earth, about 3000 kg/m^3, is less than Earth's average density of 5515 kg/m^3. The simplest explanation is that in the ancient past, Earth was partially—maybe even completely—molten throughout its volume, so that low-density materials rose to the surface and high-density materials sank deep into Earth's interior in a process called *chemical differentiation*. This series of events also suggests that Earth's core must be made of relatively dense materials, such as iron and nickel. A tremendous amount of other geological evidence has convinced scientists that this picture is correct.

---

planets. In sharp contrast, the four outer, jovian planets are much larger than the terrestrial planets. First place goes to Jupiter, whose equatorial diameter is more than 11 times that of Earth. On the other end of the scale, Mercury's diameter is less than two-fifths that of Earth. The diameters of the planets are given in Table 7-1.

The *masses* of the terrestrial and jovian planets are also dramatically different. If a planet has a moon, you can calculate the planet's mass from the moon's period and semimajor axis by using Newton's form of Kepler's third law (see Section 4-7 and Box 4-4). Astronomers have also measured the mass of each planet by sending a spacecraft to pass near the planet. The planet's gravitational pull (which is proportional to its mass) deflects the spacecraft's path, and the amount of deflection tells us the planet's mass. Even the single flyby of a comet past a planet can be used to measure that planet's mass. Using these techniques, astronomers have found that the four jovian planets have masses that range from tens to hundreds of times greater than the mass of any of the terrestrial planets. Again, first place goes to Jupiter, whose mass is 318 times greater than Earth's.

Once we know the diameter and mass of a planet, we can learn something about what that planet is made of. The trick is to calculate the planet's **average density,** or mass divided by volume, measured in kilograms per cubic meter (kg/m^3). The average density of any substance depends in part on that substance's composition. For example, air near sea level on Earth has an average density of 1.2 kg/m^3, water's average density is 1000 kg/m^3, and a piece of concrete has an average density of 2000 kg/m^3. Box 7-1 describes some applications of the idea of average density to everyday phenomena on Earth.

The four inner, terrestrial planets have very high average densities (see Table 7-1); the average density of Earth, for example, is 5515 kg/m^3. By contrast, a typical rock found on Earth's surface has a lower average density, about 3000 kg/m^3. Thus, Earth must contain a large amount of material that is denser than rock. This information provides our first clue that terrestrial planets have dense iron cores.

In sharp contrast, the outer, jovian planets have quite low densities. Saturn has an average density less than that of water. This information strongly suggests that the giant outer planets are composed primarily of light elements such as hydrogen and helium. All four jovian planets probably have large cores of mixed rock and highly compressed hydrogen buried beneath low-density outer layers tens of thousands of kilometers thick.

We can conclude that the following general rule applies to the planets:

The terrestrial planets are made of rocky materials and have dense iron cores. These planets have solid surfaces and high average densities. The jovian planets are composed primarily of light elements such as hydrogen and helium, resulting in low average densities. These planets have interiors made mostly of gas and liquid, and they have no solid surface.

CONCEPT CHECK 7-3

A planet's average density can be estimated by measuring its size and how much the planet's gravity deflects a nearby spacecraft's path. If the density of rocks recovered from a planet's surface is lower than the planet's average density, what can one infer about the density of the planet's core?

CALCULATION CHECK 7-2

If Earth's diameter is 12,756 km and Saturn's diameter is 120,536 km, how many Earths could fit across the diameter of Saturn?

Answers appear at the end of the chapter.

7-2 Seven large satellites are almost as big as the terrestrial planets

All the planets except Mercury and Venus have moons (also called satellites). At least 170 satellites are known: Earth has 1 (the Moon), Mars has 2, Jupiter has at least 69, Saturn at least 62, Uranus at least 27, and Neptune at least 14. Dozens of other small satellites probably remain to be discovered as our telescope technology continues to improve. Like the terrestrial planets, all of the satellites of the planets have solid surfaces.

You can see that there is a striking difference between the terrestrial planets, with few or no satellites, and the jovian planets, each of which has so many moons that it resembles a miniature solar system. In Chapter 8, we will explore this evidence (as well as other evidence) that the jovian planets formed in a manner similar to the solar system as a whole, but on a smaller scale.

> The various moons of the planets are not simply copies of Earth's Moon.

Seven of the jovian satellites are roughly as big as the planet Mercury! Table 7-2 lists these satellites and shows them to the same scale. Note that Earth's Moon and Jupiter's satellites Io and Europa have relatively high average densities, indicating that these moons are made primarily of rocky materials. By contrast, the average densities of Ganymede, Callisto, Titan, and Triton are all relatively low. Planetary scientists conclude that the interiors of these four moons also contain substantial amounts of water ice,

which is less dense than rock. (In Section 7-4, we will learn about types of frozen "ice" made of substances other than water.)

CAUTION!

Water ice may seem like a poor material for building a satellite, since the ice you find in your freezer can easily be cracked or crushed. But under high pressure, as is found in the interior of a large satellite, water ice becomes as rigid as rock. (It also becomes denser than the ice found in ice cubes, although not as dense as rock.) Note that water ice is an important constituent only for satellites in the outer solar system, where the Sun is far away and temperatures are very low. For example, the surface temperature of Titan is a frigid 95 K (−178°C = −288°F). In Section 7-4, we will learn more about the importance of temperature in determining the composition of a planet or satellite.

The satellites listed in Table 7-2 are unusually large. Most of the known satellites have diameters less than 2000 km, and many are irregularly shaped and just a few kilometers across.

Interplanetary spacecraft have made many surprising and fascinating discoveries about the satellites of the solar system. We now know that Jupiter's satellite Io is the most geologically active world in the solar system, with numerous volcanoes that continually belch forth sulfur-rich compounds. The fractured surface of Europa, another of Jupiter's large satellites, was an early clue that a worldwide ocean of liquid water lies beneath its icy surface. On Saturn's moon Titan, lakes of hydrocarbons, similar to the natural gas used as a source of energy on Earth, have been found near the poles—some the size of Lake Superior! In 2015, the space probe Cassini made its last flight through the plumes of geysers on Saturn's moon Enceladus as they ejected water ice, gases, and even the chemical food sources needed for some microbial life forms.

TABLE 7-2 The Seven Giant Satellites

	Moon	Io	Europa	Ganymede	Callisto	Titan	Triton
Parent planet	Earth	Jupiter	Jupiter	Jupiter	Jupiter	Saturn	Neptune
Diameter (km)	3476	3642	3130	5268	4806	5150	2706
Mass (kg)	7.35×10^{22}	8.93×10^{22}	4.80×10^{22}	1.48×10^{23}	1.08×10^{23}	1.34×10^{23}	2.15×10^{22}
Average density (kg/m³)	3340	3530	2970	1940	1850	1880	2050
Substantial atmosphere?	No	No	No	No	No	Yes	No

Moon   Io   Europa   Ganymede   Callisto   Titan   Triton

R I V U X G (NASA/JPL/USGS; NASA/JPL/DLR; NASA/JPL/Space Science Institute)

While Jupiter and Saturn lack a solid surface, their many moons contain a rich variety of features.

How many moons in the solar system are larger than Earth's Moon (see Table 7-2)?

Answer appears at the end of the chapter.

7-3 Spectroscopy reveals the chemical composition of the planets

As we have seen, the average densities of the planets and satellites give us a crude measure of their **chemical compositions—** that is, what substances they are made of. For example, the low average density of the Moon (3340 kg/m^3) compared with Earth (5515 kg/m^3) tells us that the Moon contains relatively little iron or other dense metals. But to truly understand the nature of the planets and satellites, we need to know their chemical compositions in much greater detail than we can learn from average density alone.

> The light we receive from a planet or satellite is reflected sunlight—but with revealing differences in its spectrum.

The most accurate way to determine chemical composition is by directly analyzing samples taken from a planet's atmosphere and surface. Unfortunately, of all the planets and satellites, we have such direct information only for Earth and the five worlds on which spacecraft have landed or descended into—Venus, the Moon, Mars, Jupiter, and Titan. In all other cases, astronomers must analyze sunlight reflected from the distant planets and their satellites. To do that, astronomers bring to bear one of their most powerful tools, **spectroscopy,** the systematic study of spectra and spectral lines. (We discussed spectroscopy in Sections 5-6 and 6-5.) On Earth, spectroscopy is already a very useful tool for assessing vegetation on a global scale, as well as greenhouse gases in our atmosphere.

Determining Atmospheric Composition

As on Earth, spectroscopy gives us a sensitive probe of the composition of a planet's *atmosphere*. If a planet has an atmosphere, then the reflected light that we see must have passed through some of its atmosphere before making its way to us. During this passage, some of the wavelengths of sunlight will have been absorbed. Hence, the spectrum of this reflected sunlight will have dark absorption lines. Astronomers look at the particular wavelengths absorbed and the amount of light absorbed at those wavelengths. Both of these depend on the kinds of chemicals present in the planet's atmosphere and the abundance of those chemicals.

For example, astronomers have used spectroscopy to analyze the atmosphere of Saturn's largest satellite, Titan (see **Figure 7-3a** and Table 7-2). The graph in Figure 7-3b shows the spectrum of

(a) Saturn's satellite Titan

FIGURE 7-3 R I V U X G

Analyzing a Satellite's Atmosphere through Its Spectrum (a) Titan is the only satellite in the solar system with a substantial atmosphere. **(b)** The dips in the spectrum of sunlight reflected from Titan are due to absorption by hydrogen atoms (H), oxygen molecules (O₂), and methane molecules (CH₄). Of these, only methane is actually present in Titan's atmosphere. **(c)** This illustration shows the path of the light that reaches us from Titan. To interpret the spectrum of this light as shown in (b), astronomers must account for the absorption that takes place in the atmospheres of the Sun and Earth. (a: NASA/JPL/Space Science Institute)

(b) The spectrum of sunlight reflected from Titan

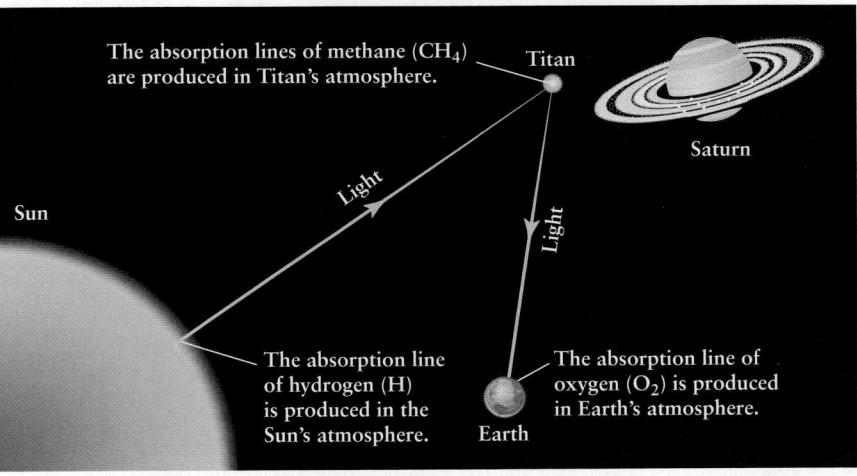

(c) Interpreting Titan's spectrum

visible sunlight reflected from Titan. (We first saw this method of displaying spectra in Figure 6-21.) The dips in this curve of intensity versus wavelength represent absorption lines. However, not all of these absorption lines are produced in the atmosphere of Titan (Figure 7-3c). Before reaching Titan, light from the Sun's glowing surface must pass through the Sun's own hydrogen-rich atmosphere. This produces the hydrogen absorption line in Figure 7-3b at a wavelength of 656 nm. After being reflected from Titan, the light must pass through Earth's atmosphere before reaching the observing telescope; this is where the oxygen absorption line in Figure 7-3b is produced. Only the two dips near 620 nm and 730 nm are caused by gases in Titan's atmosphere.

These two absorption lines are caused not by individual atoms in the atmosphere of Titan but by atoms combined to form molecules. (We introduced the idea of molecules in Section 5-6.) Molecules, like atoms, also produce unique patterns of lines in the spectra of astronomical objects. The absorption lines in Figure 7-3b indicate the presence in Titan's atmosphere of methane molecules (CH_4, a molecule made of one carbon atom and four hydrogen atoms). Titan must be a curious place indeed, because on Earth, methane—the primary ingredient in natural gas—is a rather rare substance! When we examine other planets and satellites with atmospheres, we find that all of their spectra have absorption lines of molecules of various types.

In addition to visible-light measurements such as those shown in Figure 7-3b, it is very useful to study the *infrared* and *ultraviolet* spectra of planetary atmospheres. Many molecules have much stronger spectral lines in these nonvisible wavelength bands than in the visible. As an example, the ultraviolet spectrum of Titan shows that nitrogen molecules (N_2) are the dominant constituent of Titan's atmosphere. Furthermore, Titan's infrared spectrum includes spectral lines of a variety of molecules that contain carbon and hydrogen, indicating that Titan's atmosphere has a very complex chemistry. None of these molecules could have been detected by visible light alone. Because Earth's atmosphere is largely opaque to infrared and ultraviolet wavelengths (see Section 6-7), telescopes in space are important tools for these spectroscopic studies of the solar system.

It is a testament to the power of spectroscopy that when the robotic spacecraft Huygens landed on Titan in 2005, its onboard instruments confirmed the presence of methane and nitrogen in Titan's atmosphere—just as had been predicted years before by spectroscopic observations.

CONCEPT CHECK 7-5

If planets reflect some of the Sun's light rather than emitting visible light of their own, how can spectroscopy reveal information about a planet's atmosphere?

Answer appears at the end of the chapter.

Determining Surface Composition

Spectroscopy can also provide useful information about the *solid surfaces* of planets and satellites without atmospheres. When light shines on a solid surface, some wavelengths are absorbed while others are reflected. (For example, a plant leaf absorbs red and violet light but reflects green light—which is why leaves look green.) Unlike a gas, a solid illuminated by sunlight does not produce sharp, definite spectral lines. Instead, only broad absorption features appear in the spectrum. By comparing such a spectrum with the spectra of samples of different substances on Earth, astronomers can infer the chemical composition of the surface of a planet or satellite.

As an example, **Figure 7-4a** shows Jupiter's satellite Europa (see Table 7-2), and Figure 7-4b shows the infrared spectrum of light reflected from the surface of Europa. Because this spectrum is so close to that of water ice—that is, frozen water—astronomers conclude that water ice is the dominant constituent of Europa's surface.

(a) Jupiter's moon Europa

R I **V** U X G

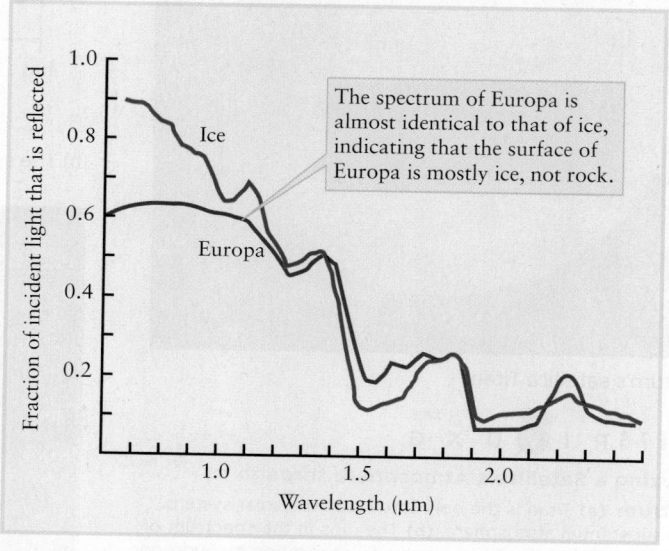

The spectrum of Europa is almost identical to that of ice, indicating that the surface of Europa is mostly ice, not rock.

(b) The spectum of light reflected from Europa

FIGURE 7-4

Analyzing a Satellite's Surface from Its Spectrum (a) Unlike Titan (Figure 7-3a), Jupiter's satellite Europa has no atmosphere. (b) Infrared light from the Sun that is reflected from the surface of Europa has nearly the same spectrum as sunlight reflected from ordinary water ice. (NASA)

(We saw in Section 7-2 that water ice cannot be the dominant constituent of Europa's *interior*, because this satellite's density is too high.)

Unfortunately, spectroscopy tells us little about what the material is like just below the surface of a satellite or planet. For this purpose, there is simply no substitute for sending a spacecraft to a planet and examining its surface directly.

CONCEPT CHECK 7-6

How is the spectrum of reflected sunlight from a solid planetary surface different from the spectrum produced when sunlight passes through a planet's gaseous atmosphere?

Answer appears at the end of the chapter.

7-4 The jovian planets are made of lighter elements than the terrestrial planets

Spectroscopic observations from Earth and spacecraft show that the outer layers of the jovian planets are composed primarily of the lightest gases, hydrogen and helium (see Box 5-5). In contrast, chemical analysis of soil samples from Venus, Earth, and Mars demonstrate that the terrestrial planets are made mostly of heavier elements, such as iron, oxygen, silicon, magnesium, nickel, and sulfur. Spacecraft images such as Figure 7-5 and Figure 7-6 only hint at

> Differences in distance from the Sun, and hence in temperature, explain many distinctions between terrestrial and jovian planets.

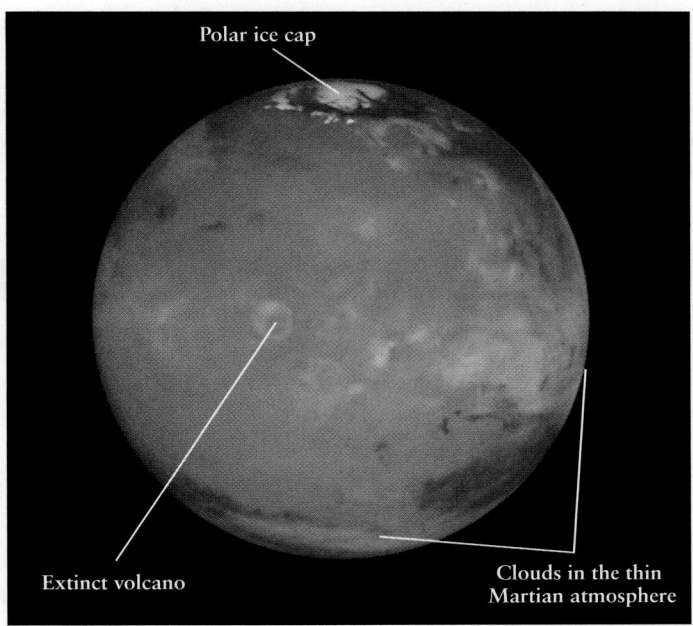

FIGURE 7-6 R I V U X G

A Terrestrial Planet Mars is composed mostly of heavy elements such as iron, oxygen, silicon, magnesium, nickel, and sulfur. The planet's red surface can be seen clearly in this Hubble Space Telescope image because the Martian atmosphere is thin and nearly cloudless. Olympus Mons, the extinct volcano to the left of center, is nearly 3 times the height of Mount Everest. (Steve Lee, University of Colorado; Jim Bell, Cornell University; Mike Wolff, Space Science Institute; and NASA)

these striking differences in chemical composition, which are summarized in Table 7-3.

Temperature plays a major role in determining whether the materials of which planets are made exist as solids, liquids, or gases. Hydrogen (H_2) and helium (He) are gaseous except at extremely low temperatures and extraordinarily high pressures. By contrast, rock-forming substances such as iron and silicon are solids except at temperatures well above 1000 K. (You may want to review the discussion of temperature scales in Box 5-1.) Between these two extremes are substances such as water (H_2O), carbon dioxide (CO_2), methane (CH_4), and ammonia (NH_3),

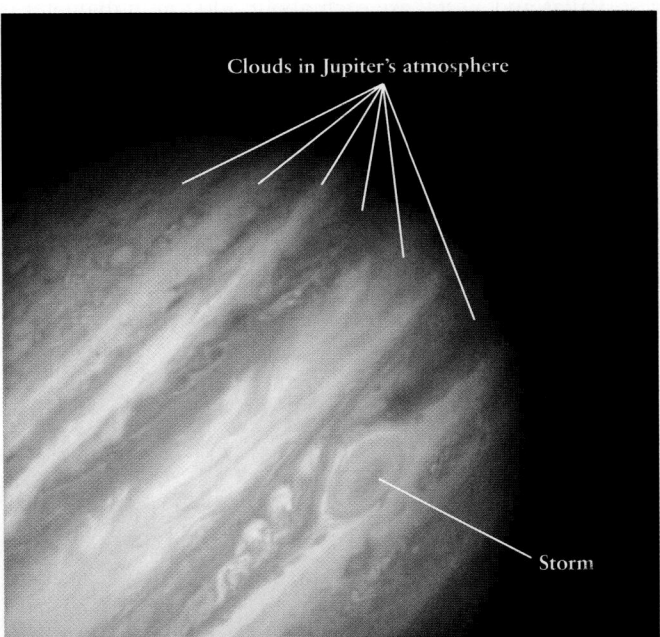

FIGURE 7-5 R I V U X G

A Jovian Planet This Hubble Space Telescope image gives a detailed view of Jupiter's cloudtops. Jupiter is composed mostly of the lightest elements, hydrogen and helium, which are colorless; the colors in the atmosphere are caused by trace amounts of other substances. The giant storm at lower right, called the Great Red Spot, has been raging for as many as 300 years. (Amy Simon, Cornell U.; Reta Beebe, NMSU; Heidi Hammel, Space Science Institute, MIT)

TABLE 7-3 Comparing Terrestrial and Jovian Planets

	Terrestrial planets	Jovian planets
Distance from the Sun	Less than 2 au	More than 5 au
Size	Small	Large
Composition	Mostly rocky materials containing iron, oxygen, silicon, magnesium, nickel, and sulfur	Mostly light elements such as hydrogen and helium
Density	High	Low

which solidify at low temperatures (from below 100 to 300 K) into solids called **ices**. (In astronomy, frozen water is just one kind of "ice.") At somewhat higher temperatures, they can exist as liquids or gases. For example, clouds of ammonia ice crystals are found in the cold upper atmosphere of Jupiter, but within Jupiter's warmer interior, ammonia exists primarily as a liquid.

CAUTION!

The jovian planets are sometimes called "gas giants." It is true that their primary constituents, including hydrogen, helium, ammonia, and methane, are gases under normal conditions on Earth. But in the interiors of these planets, pressures are so high that these substances are *liquids*, not gases. The jovian planets might be better described as "liquid giants"!

As you might expect, a planet's surface temperature is related to its distance from the Sun. The four inner planets are quite warm. For example, midday temperatures on Mercury may climb to 700 K (= 427°C = 801°F), and during midsummer on Mars, it is sometimes as warm as 290 K (= 17°C = 63°F). The outer planets, which receive much less solar radiation, are cooler. Typical temperatures range from about 125 K (= −148°C = −234°F) in Jupiter's upper atmosphere to about 55 K (= −218°C = −360°F) at the tops of Neptune's clouds.

How Temperature Affects Atmospheres

The higher surface temperatures of the terrestrial planets help to explain the following observation: The atmospheres of the terrestrial planets contain virtually *no* hydrogen molecules or helium atoms. Instead, the atmospheres of Venus, Earth, and Mars are composed of heavier molecules such as nitrogen (N_2, 14 times more massive than a hydrogen molecule, H_2), oxygen (O_2, 16 times more massive), and carbon dioxide (22 times more massive). To understand the connection between surface temperature and the absence of hydrogen and helium, we need to know a few basic facts about gases.

The temperature of a gas is directly related to the speeds at which the atoms or molecules of the gas move: The higher the gas temperature, the greater the speed of its atoms or molecules. Furthermore, for a given temperature, lightweight atoms and molecules move more rapidly than heavy ones. On the four inner, terrestrial planets, where atmospheric temperatures are high, low-mass hydrogen molecules and helium atoms move so swiftly that they can escape from the relatively weak gravity of these planets. Hence, the atmospheres that surround the terrestrial planets are composed primarily of more massive, slower-moving molecules such as CO_2, N_2, O_2, and water vapor (H_2O). On the four jovian planets, low temperatures and relatively strong gravity prevent even lightweight hydrogen and helium gases from escaping into space, and so their atmospheres are much more extensive. The combined mass of Jupiter's atmosphere, for example, is about a million (10^6) times greater than that of Earth's atmosphere. This is comparable to the entire mass of Earth! **Box 7-2** (page 190) describes more about the ability of a planet's gravity to retain gases.

CONCEPT CHECK 7-7

If the average speed of hydrogen molecules in Earth's atmosphere is below Earth's escape speed, why do Earth's atmospheric hydrogen molecules slowly leak into space?

Answer appears at the end of the chapter.

7-5 Small chunks of rock and ice also orbit the Sun

In addition to the eight planets, many smaller objects orbit the Sun. These objects fall into three broad categories: asteroids, which are rocky objects found in the inner solar system; trans-Neptunian objects, which are found beyond Neptune in the outer solar system and contain both rock and ice; and comets, which are mixtures of rock and ice that originate in the outer solar system but can venture close to the Sun.

> The smaller bodies of the solar system contain important clues about its origin and evolution.

Asteroids

Within the orbit of Jupiter are hundreds of thousands of rocky objects called **asteroids**. There is no sharp dividing line between planets and asteroids, which is why asteroids are also called **minor planets**. The largest asteroid, Ceres, has a diameter of about 900 km (about one-quarter of our Moon's diameter). The two next largest, Pallas and Vesta, are each about 500 km in diameter. Still smaller ones, like the asteroid shown in close-up in **Figure 7-7**, are more numerous. Hundreds of thousands of kilometer-sized

FIGURE 7-7 R I V U X G

An Asteroid The asteroid shown in this image, 433 Eros, is only 33 km (21 mi) long and 13 km (8 mi) wide—about the same size as the island of Manhattan. Because Eros is so small, its gravity is too weak to have pulled it into a spherical shape. This image was taken in 2000 by NEAR Shoemaker, the first spacecraft to orbit around and land on an asteroid. (NASA/JHUAPL) asteroids are known, and there are probably hundreds of thousands more that are boulder sized or smaller. All of these objects orbit the Sun in the same direction as the planets.

Most (although not all) asteroids orbit the Sun at distances of 2 to 3.5 au. This region of the solar system between the orbits of Mars and Jupiter is called the **asteroid belt.**

CAUTION!

One common misconception about asteroids is that they are the remnants of an ancient planet that somehow broke apart or exploded, like the fictional planet Krypton in the comic book adventures of Superman. In fact, the combined mass of all the asteroids (including Ceres, Pallas, and Vesta) is much less than the mass of the Moon, and they were probably never part of any planet-sized body. The early solar system is thought to have been filled with asteroidlike objects, most of which were incorporated into the planets. The "leftover" objects that missed out on this process make up our present-day population of asteroids.

CONCEPT CHECK 7-8

Is the largest asteroid, Ceres, about the same size as a large city, a large U.S. state, a large country, or Earth itself?

Answer appears at the end of the chapter.

Trans-Neptunian Objects

While asteroids are the most important small bodies in the inner solar system, the outer solar system is the realm of the **trans-Neptunian objects.** As the name suggests, these are small bodies whose orbits lie beyond the orbit of Neptune. The first of these to be discovered (in 1930) was Pluto, with a diameter of only 2274 km. Pluto is larger than any asteroid, but smaller than any planet or any of the moons listed in Table 7-2. The orbit of Pluto has a greater semimajor axis (39.54 au), is more steeply inclined to the ecliptic (17.15°), and has a greater eccentricity (0.250) than that of any of the planets (Figure 7-8). In fact, Pluto's noncircular orbit sometimes takes it nearer the Sun than Neptune. (Happily, the orbits of Neptune and Pluto are such that they will never collide.) Pluto's density is just under 2000 kg/m³, about the same as Neptune's moon Triton, shown in Table 7-2. Hence, its composition is thought to be a mixture of about 70% rock and 30% ice.

Since 1992 astronomers have discovered more than 1800 other trans-Neptunian objects, and are discovering more each year. Like asteroids, the vast majority of trans-Neptunian objects orbit the Sun in the same direction as the planets. A handful of trans-Neptunian objects are comparable in size to Pluto; at least one, Eris, is 27% more massive than Pluto, as well as being in an orbit that is much larger, more steeply inclined, and more eccentric (Figure 7-8). Table 7-4 lists the seven largest trans-Neptunian objects known as of this writing. (Note that Charon is actually a satellite of Pluto.) Haumea and its two moons are shown. While Haumea's shape has not been observed directly, modeling of its reflected light suggests that it is unusually elongated, possibly from a collision.

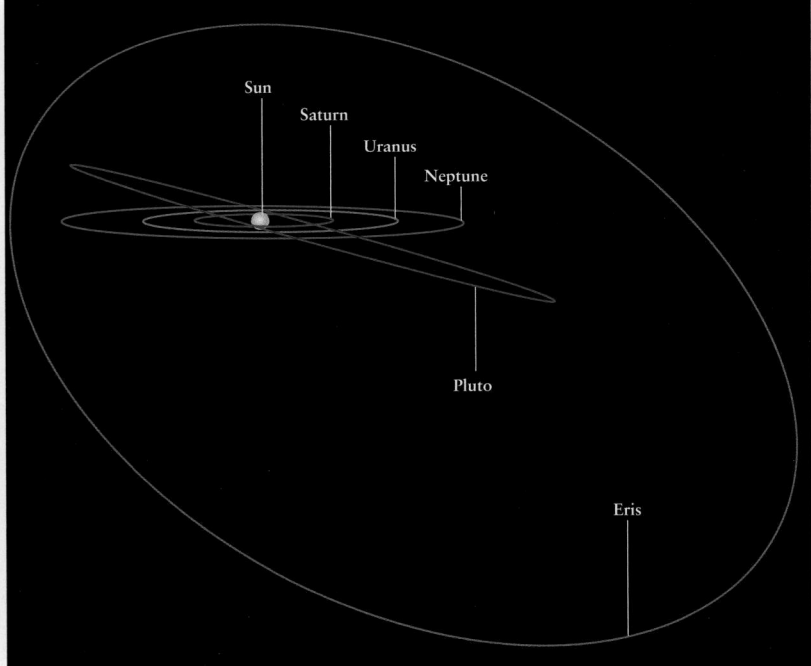

FIGURE 7-8

Trans-Neptunian Objects **(a)** Pluto and Eris are the two largest trans-Neptunian objects, small worlds of rock and ice that orbit beyond Neptune. **(b)** Unlike the orbits of the planets, the orbits of these two objects are steeply inclined to the ecliptic: Pluto's orbit is tilted by about 17°, and that of Eris is tilted by 44°.

BOX 7-2 TOOLS OF THE ASTRONOMER'S TRADE

Kinetic Energy, Temperature, and Whether Planets Have Atmospheres

TUTORIAL 7.3

A moving object possesses energy, and the faster it moves, the more energy it has (see Figure 4-20). Energy of this type is called **kinetic energy**. If an object of mass m is moving with a speed v, its kinetic energy E_k is given by:

Kinetic energy of an object

$$E_k = \frac{1}{2}mv^2$$

E_k = kinetic energy of an object

m = mass of object

v = speed of object

This expression for kinetic energy is valid for all objects, both big and small, from atoms and molecules to planets and stars, as long as their speeds are slow in relation to the speed of light. If the mass is expressed in kilograms and the speed in meters per second, the kinetic energy is expressed in joules (J).

EXAMPLE: A self-driving car of mass 1000 kg traveling at a typical freeway speed of 30 m/s (= 108 km/h = 67 mi/h) has a kinetic energy of

$$E_k = \frac{1}{2}(1000 \times 30^2) = 450{,}000 \text{ J} = 4.5 \times 10^5 \text{ J}$$

Upper atmosphere

Smaller, faster molecules are more likely to reach the escape speed and leave the atmosphere.

On average, more massive molecules move slower than lighter molecules.

Interior

Consider a gas, such as the atmosphere of a star or planet. Some of the gas atoms or molecules will be moving slowly, with little kinetic energy, while others will be moving faster and have more kinetic energy. The temperature of the gas is a direct measure of the *average* amount of kinetic energy per atom

or molecule. The hotter the gas, the faster atoms or molecules move, on average, and the greater the average kinetic energy of an atom or molecule (see Figure 5-11).

If the gas temperature is sufficiently high (typically several thousand kelvins), molecules move so fast that when they collide with one another, the energy of the collision can break the molecules apart into their constituent atoms. Thus, the Sun's atmosphere, where the temperature is 5800 K, consists primarily of individual hydrogen atoms rather than hydrogen molecules. By contrast, the hydrogen atoms in Earth's atmosphere (temperature 290 K) are combined with oxygen atoms into molecules of water vapor (H_2O).

The physics of gases tells us that in a gas of temperature T (in kelvins), the average kinetic energy of an atom or molecule is:

Kinetic energy of a gas atom or molecule

$$E_k = \frac{3}{2}kT$$

E_k = average kinetic energy of a gas atom or molecule, in joules

$k = 1.38 \times 10^{-23}$ J/K

T = temperature of gas, in kelvins

The quantity k is called the Boltzmann constant. Note that the higher the gas temperature, the greater the average kinetic energy of an atom or molecule of the gas. This average kinetic energy becomes zero at absolute zero, or $T = 0$, the temperature at which molecular motion is at a minimum.

At a given temperature, all kinds of atoms and molecules will have the same average kinetic energy. But the average *speed* of a given kind of atom or molecule depends on the particle's mass. To see this dependence on mass, note that the average kinetic energy of a gas atom or molecule can be written in two equivalent ways:

$$E_k = \frac{1}{2}mv^2 = \frac{3}{2}kT$$

where v represents the average speed of an atom or molecule in a gas with temperature T. Rearranging this equation, we obtain:

Average speed of a gas atom or molecule

$$v = \sqrt{\frac{3kT}{m}}$$

v = average speed of a gas atom or molecule, in m/s

$k = 1.38 \times 10^{-23}$ J/K

T = temperature of gas, in kelvins

m = mass of the atom or molecule, in kilograms

(continued on the next page)

BOX 7-2 (*continued*)

For a given gas temperature, the greater the mass of a given type of gas atom or molecule, the slower its average speed. (*Note:* The value v in this equation is actually the root-mean-square speed, which is slightly higher than the average speed of the atoms or molecules in the gas, but it is close enough for our purposes here.)

EXAMPLE: What is the average speed of the oxygen molecules that you breathe at a room temperature of 20°C (= 68°F)?

Situation: We are given the temperature of a gas and are asked to find the average speed of the gas molecules.

Tools: We use the relationship $v = \sqrt{3kT/m}$ where T is the gas temperature in kelvins and m is the mass of a single oxygen molecule in kilograms.

Answer: To use this equation to calculate the average speed v, we must express the temperature T in kelvins (K) rather than degrees Celsius (°C). As we learned in Box 5-1, we do this by adding 273 to the Celsius temperature, so 20°C becomes $(20 + 273) = 293$ K. The mass m of an oxygen molecule is not given, but you can easily find that the mass of an oxygen *atom* is 2.66×10^{-26} kg. The mass of an oxygen molecule (O_2) is twice the mass of an oxygen atom, or $2(2.66 \times 10^{-26}$ kg$) = 5.32 \times 10^{-26}$ kg. Thus, the average speed of an oxygen molecule in 20°C air is

$$v = \sqrt{\frac{3(1.38 \times 10^{-23})(293)}{5.32 \times 10^{-26}}} = 478 \text{ m/s} = 0.478 \text{ km/s}$$

Review: This speed is about 1700 kilometers per hour, or about 1100 miles per hour. Hence, atoms and molecules move rapidly in even a moderate-temperature gas.

In some situations, atoms and molecules in a gas may be moving so fast that they can overcome the attractive force of a planet's gravity and escape into interplanetary space. The minimum speed that an object at a planet's surface must have in order to permanently leave the planet is called the planet's **escape speed** (see Figure 4-22). The escape speed for a planet of mass M and radius R is given by

$$v_{\text{escape}} = \sqrt{\frac{2GM}{R}}$$

where $G = 6.67 \times 10^{-11}$ N m²/kg² is the universal constant of gravitation.

The accompanying table gives the escape speed for the Sun, the planets, and the Moon. For example, to escape Earth, a cannonball would have to be shot with an implausible speed greater than 11.2 km/s (25,100 mi/h).

A good rule of thumb is that a planet can retain a gas if the escape speed is at least 6 times greater than the average speed of the molecules in the gas. Some molecules are moving slower than average and others are moving faster, but very few are moving more than 6 times faster than average.

EXAMPLE: Consider Earth's atmosphere. We saw that the average speed of oxygen molecules is 0.478 km/s at room temperature. The escape speed from Earth (11.2 km/s) is much more than 6 times the average speed of the oxygen molecules, so Earth has no trouble keeping oxygen in its atmosphere.

A similar calculation for hydrogen molecules (H_2) gives a different result, however. At 293 K, the average speed of a hydrogen molecule is 1.9 km/s. Six times this speed is 11.4 km/s, which is about the escape speed from Earth. Thus, Earth does not retain hydrogen in its atmosphere. Any hydrogen released into the air slowly leaks away into space. On Jupiter, by contrast, the escape speed is so high that even the lightest gases such as hydrogen are retained in its atmosphere. But on Mercury the escape speed is low and the temperature high (so that gas molecules move faster), and Mercury cannot retain any significant atmosphere at all. Impacts and the solar wind constantly replenish what atmosphere there is, but these molecules never stick around for very long.

Object	Escape speed (km/s)
Sun	618.0
Mercury	4.3
Venus	10.4
Earth	11.2
Moon	2.4
Mars	5.0
Jupiter	59.5
Saturn	35.5
Uranus	21.3
Neptune	23.5

TABLE 7-4 Seven Large Trans-Neptunian Objects

	Sedna	Quaoar	Haumea	Makemake	Eris	Pluto	Charon (satellite of Pluto)
Average distance from the Sun (au)	489	43.54	43.34	45.71	67.67	39.54	39.54
Orbital period (years)	10,800	287	285	309	557	248.6	248.6
Orbital eccentricity	0.844	0.035	0.189	0.155	0.442	0.250	0.250
Inclination of orbit to the ecliptic	11.9°	8.0°	28.2°	29.0°	44.2°	17.15°	17.15°
Approximate diameter (km)	995	1110	1240	1430	2326	2372	1212

3476 km = diameter of Earth's Moon

Sedna   Quaoar   Haumea   Makemake   Eris   Pluto   Charon

(Note: To date, astronomers have been able to capture images of the surfaces of Pluto and Charon only.)

R I V U X G

Alan Stern [Southwest Research Institute]/Marc Buie [Lowell Observatory]/NASA/ESA; A. Field [STScI]/NASA; Alan Stern [Southwest Research Institute]/Marc Buie [Lowell Observatory]/NASA/ESA; NASA/JHUAPL/SwRI; NASA/Johns Hopkins University Applied Physics Laboratory/Southwest Research Institute

Just as most asteroids lie in the asteroid belt, most trans-Neptunian objects orbit within a band called the **Kuiper belt** (pronounced "ki-per") that extends from 30 au to 50 au from the Sun and is centered on the plane of the ecliptic. Like asteroids, many more trans-Neptunian objects are yet to be discovered: Astronomers estimate that there are 35,000 or more such objects with diameters greater than 100 km. If so, the combined mass of all trans-Neptunian objects is comparable to the mass of Jupiter and is several hundred times greater than the combined mass of all the asteroids found in the inner solar system.

Like asteroids, trans-Neptunian objects are thought to be debris left over from the formation of the solar system. In the inner regions of the solar system, rocky fragments have endured continuous exposure to the Sun's heat, but any ice originally present at the surface would have evaporated. Far from the Sun, ice has survived for billions of years. Thus, *debris* in the solar system naturally divides into two families—rocky asteroids and partially icy trans-Neptunian objects—which can be arranged according to distance from the Sun, just like the two categories of planets (terrestrial and jovian).

To study Pluto and the Kuiper belt in more detail, NASA launched the New Horizons mission in 2006, which arrived at Pluto in 2015. As we'll see when discussing craters in the next section, Pluto has some unexpected and unexplained features.

CONCEPT CHECK 7-9

Is Pluto an asteroid, planet, Kuiper belt object, or trans-Neptunian object?

CALCULATION CHECK 7-3

Which has a larger average orbital radius, the asteroid belt or the Kuiper belt?

Answers appear at the end of the chapter.

Comets

Two objects in the Kuiper belt can collide if their orbits cross each other. When this happens, a fragment can be knocked off one of the colliding objects and diverted into an elongated orbit that brings it close to the Sun. Such small objects, each a combination of rock and ice, are called **comets.** When a comet comes close enough to the Sun, the Sun's radiation vaporizes some of the comet's ices, producing long flowing tails of gas and dust particles (Figure 7-9). Astronomers deduce the composition of comets by studying the spectra of these tails.

CAUTION!

Science-fiction movies and television programs sometimes show comets tearing across the night sky like a rocket, which would be a pretty impressive sight. However, comets do not zoom across the sky. Like the planets, comets orbit the Sun. And like the planets, comets move hardly at all against the background of stars over the course of a single night (see Section 4-1). If you are lucky enough to see a bright comet, it will not zoom dramatically from horizon to horizon. Instead, it will seem to hang majestically among the stars, so you can admire it at your leisure.

To attain this stunning view, the Cassini spacecraft was carefully lined up to watch Saturn eclipse the Sun. With the Sun directly behind Saturn, otherwise faint and diffuse ring particles appear bright in this color-enhanced image. The outermost ring contains Saturn's moon Enceladus, and the diffuse ring itself is produced by ejecta from that moon's ice volcanoes. At about the "10 o'clock" position in the upper left, Earth can be seen just above the bright inner rings. (CICLOPS, JPL, ESA, NASA)

R I V U X G

II Planets and Moons

For 85 years after its discovery, Pluto's appearance remained a mystery. Too far away for clear telescopic views of its surface, it took NASA's New Horizons spacecraft to finally reveal the dwarf planet's many features in 2015. The darker regions on the surface arise when methane and nitrogen in Pluto's atmosphere are broken apart by ultraviolet light. These molecules then combine into larger particles that fall back to the surface and give it a red hue. Something very different is going on with Pluto's smooth white region. It's a glacier made of nitrogen ice, and it's surprisingly young at no more than 10 million years. Scientists still haven't figured out what energy source could have powered this recent glacial activity. While Pluto is quite small (only half the diameter of Mercury), the mountains surrounding the glacial plains can reach an impressive 11,000 feet. (NASA/Johns Hopkins University Applied Physics Laboratory/Southwest Research Institute)

R I **V** U X G

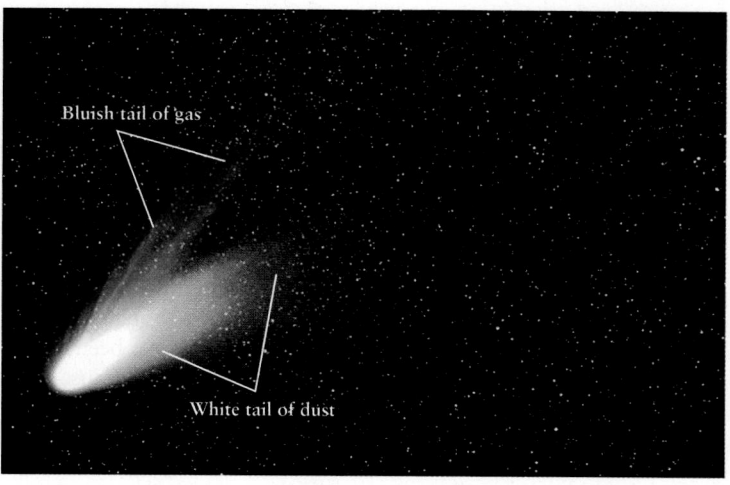

FIGURE 7-9 R I **V** U X G

A Comet This photograph shows Comet Hale-Bopp as it appeared in April 1997. The solid part of a comet like this is a chunk of dirty ice a few tens of kilometers in diameter. When a comet passes near the Sun, solar radiation vaporizes some of the icy material, forming a bluish tail of gas and a white tail of dust. Both tails can extend for tens of millions of kilometers. (Agencia el Universal/AP Images)

Some comets appear to originate from locations far beyond the Kuiper belt. The source of these is thought to be a swarm of comets that forms a spherical "halo" around the solar system called the **Oort comet cloud** (also known as **Oort cloud**). This hypothesized "halo" extends to 50,000 au from the Sun (about one-fifth of the way to the nearest other star). The spherical shape of the Oort cloud means that comets arriving from this region can follow orbits with very high inclinations away from the ecliptic plane. Because the Oort cloud is so distant, it has not yet been possible to detect objects in the Oort cloud directly.

CONCEPT CHECK 7-10

Consider a comet that has an orbital plane that is perpendicular to the plane of Earth's orbit around the Sun. Did this comet most likely originate from the Kuiper belt or the Oort cloud?

Answer appears at the end of the chapter.

7-6 Craters on planets and their moons are the result of impacts from asteroids and comets

One of the great challenges in studying planets and their moons, or satellites, is determining their internal structures. Are they solid or liquid inside? Because planets are opaque, we cannot see directly into their interiors to answer these questions. But we can gather important clues about the interiors of terrestrial planets and satellites by studying the extent to which their surfaces are covered with craters (Figure 7-10). To see how information about the interiors is gathered, we first need to understand where craters come from.

> Scientists study craters on a planet or satellite to learn the age and geologic history of the surface.

The Origin of Craters

The planets orbit the Sun in roughly circular orbits. But many asteroids and comets are in more elongated orbits. Slight disturbances to such an elongated orbit can send these small objects on a collision course with a planet or satellite. If the object collides with a jovian planet, it is swallowed up by the planet's thick atmosphere.

(a) A crater on the Moon

(b) A crater on Earth

(c) A crater on Mars

FIGURE 7-10 R I **V** U X G

Impact Craters These images, all taken from spacecraft, show impact craters on three different worlds. **(a)** The Moon's surface has craters of all sizes. The large crater near the middle of this image is about 80 km (50 mi) in diameter, equal to the length of San Francisco Bay. **(b)** Manicouagan Reservoir in Quebec is the relic of a crater formed by an impact more than 200 million years ago. The crater was eroded over the ages by the advance and retreat of glaciers, leaving a ring lake 100 km (60 mi) across. A planet or moon's internal heat can power geologic activity and erosion processes that slowly erase craters. **(c)** Lowell Crater in the southern highlands of Mars is 201 km (125 mi) across. Like the image of the Moon in part (a), there are craters on top of craters. Note the light-colored frost formed by condensation of carbon dioxide from the Martian atmosphere. (a: NASA; b: NASA; c: NASA/JPL/MSSS)

(Astronomers actually saw an event of this kind in 1994, when a comet crashed into Jupiter.) But if the object collides with the solid surface of a terrestrial planet or a satellite, the result is an **impact crater** (see Figure 7-10). Such impact craters, found throughout the solar system, offer stark evidence of these violent collisions.

The easiest way to view impact craters is to examine the Moon through a telescope or binoculars. Some 30,000 lunar craters are visible, with diameters ranging from 1 km to several hundred km. Close-up photographs from lunar orbit have revealed millions of craters too small to be seen from Earth (Figure 7-10a). These smaller craters are thought to have been caused by impacts of relatively small objects called **meteoroids,** which range in size from a few hundred meters across to the size of a pebble or smaller. Meteoroids are the result of collisions between asteroids, whose orbits sometimes cross. The chunks of rock that result from these collisions go into independent orbits around the Sun, which can lead them to collide with the Moon or another world.

When British astronomer Richard Proctor revived the idea in 1873 that lunar craters were the result of impacts, a major sticking point was the observation that nearly all craters are circular. If craters were merely gouged out by high-speed rocks, a rock striking the Moon in any direction except straight downward would have created a noncircular crater. In the early twentieth century, it was realized that a meteoroid colliding with the Moon generates a shock wave in the lunar surface that spreads out from the point of impact. Such a shock wave produces a circular crater no matter what direction the meteoroid was moving. (In a similar way, the craters made by artillery shells are almost always circular.) Many of the larger lunar craters also have a central peak, which is characteristic of a high-speed impact (see Figure 7-10a). Craters made by other processes, such as volcanic action, would not have central peaks of this sort.

Comparing Cratering on Different Worlds

Not all planets and satellites show the same amount of cratering. The Moon is heavily cratered over its entire surface, with craters on top of craters, as shown in Figure 7-10a. On Earth, by contrast,

craters are very rare. Geologists have identified fewer than 200 impact craters on our planet (Figure 7-10b). Our understanding is that both Earth and the Moon formed at nearly the same time and have been bombarded at comparable rates over their histories. Why, then, are craters so much rarer on Earth than on the Moon?

The answer is that Earth is a *geologically active* planet: Through **plate tectonics**—the motion of rocky plates over Earth's surface—new material wells up from the interior to form new crust, and old surface material is pushed back into the interior. These processes, which literally recycle Earth's surface, coupled with erosion from wind and water, cause craters on Earth to be erased over time. The few craters found on Earth today must be relatively young, or they would have been erased already. (However, there are a few older craters around, because recycling of the surface isn't a perfectly even process.)

The Moon, by contrast, is geologically *inactive*. There are no volcanoes and no movements of continents (and, indeed, no continents). Furthermore, the Moon has neither liquid water nor an atmosphere, so there is no erosion as we know it on Earth. With none of the processes that tend to erase craters on Earth, the Moon's surface remains pockmarked with the scars of billions of years of impacts.

In order for a planet to be geologically active, its interior must be hot. Heat is the energy source that drives plate tectonics. And even if a planet has volcanoes but no plate tectonics, heat rising to the surface powers these volcanoes. We'll discuss this more in Chapter 9, but the process has similarities to a pot of boiling water. Therefore, we would expect geologically inactive (and hence heavily cratered) worlds like the Moon to have less heat in their interiors than does Earth. Investigation of these geologically inactive worlds bears this out. But, *why is there less heat in the Moon's interior than in Earth's?*

ANALOGY

To understand the answer to this question, let's compare two hot objects of different sizes. Consider a single turkey leg and an entire turkey, both cooked to the same temperature and taken out of the oven together. It won't take long for the turkey leg to cool down, but it will take the entire turkey several hours to cool down.

This difference in cooling times happens because heat is contained in the volume of the object, but escapes through the surface. In fact, smaller objects cool down faster than larger objects because they have more surface area relative to their volume. Both Earth and the Moon could have been similarly heated when they first formed (by gravity and radioactivity, processes discussed in Chapter 8), but because the Moon is so much smaller than Earth, the Moon has lost much of its internal heat (**Figure 7-11**), and with it the energy to power geologic processes necessary to erase craters.

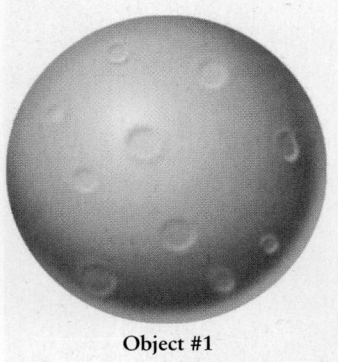

Object #1

Compared to object #1, object #2:
— Has 1/2 the radius.
— Has 1/4 the surface area (so it can lose heat only 1/4 as fast).
— But has only 1/8 the volume (so it has only 1/8 as much heat to lose).

Hence compared to object #1, object #2:
— Will cool off more rapidly.
— Will sustain less geologic activity.
— Will have more craters.

Object #2

FIGURE 7-11

Object Size and Cratering Of these two hypothetical objects, which could be planets or moons, the smaller one (#2) has less volume and less internal heat, as well as less surface area from which to radiate heat into space. But the ratio of surface area to volume is greater for the smaller object. Hence, in the absence of other heat sources, the smaller object will lose heat faster, have a colder interior, and be less geologically active. It will also have a more heavily cratered surface, since it takes geologic activity to erase craters.

A planet's internal heat can also be important in less obvious ways. As we'll see in Chapter 11, volcanism and a global magnetic field—both dependent on internal heat—help to preserve a planet's atmosphere. While Mars lost much of its atmosphere as its heat dissipated, Earth's atmosphere is still going strong. In Figure 7-10b, glaciers have ground down a massive crater, and wind and water continue to erode the terrain. Without our atmosphere, Earth would not experience erosion processes like these.

Cratering Measures Geologic Activity

By considering these differences between Earth and the Moon, we have uncovered a general rule for worlds with solid surfaces:

The smaller the terrestrial world, the less internal heat it is likely to have retained, and, thus, the less geologic activity it will display on its surface. The less geologically active the world, the older and hence more heavily cratered its surface.

This rule means that we can use the amount of cratering visible on a planet or satellite to estimate the age of its surface and how geologically active it is. As an example, Mercury has a heavily cratered surface, which means that the surface is very old. This geologically inactive surface is consistent with Mercury being the smallest of the terrestrial planets (see Table 7-1). Due to its small size, it has lost the internal heat required to sustain geologic activity. On Venus, by comparison, there are only about a thousand craters larger than a few kilometers in diameter. This is many more craters than have been found on Earth, but only a small fraction of the number on the Moon or Mercury. Venus is only slightly smaller than Earth, and it has enough internal heat to power the geologic activity required to erase most of its impact craters.

Mars is an unusual case, in that extensive cratering (Figure 7-10c) is found only in the higher terrain; the lowlands of Mars are remarkably smooth and free of craters. Thus, it follows that the Martian highlands are quite old, while the lowlands have a younger surface from which most craters have been erased. Considering the planet as a whole, the amount of cratering on Mars is intermediate between that on Mercury and Earth. This agrees with our general rule, because Mars is intermediate in size between Mercury and Earth. The interior of Mars was once hotter than it is now, such that geologic processes were able to erase some of the impact craters. A key piece of evidence that supports this picture is that Mars has a number of immense volcanoes (Figure 7-12). These volcanoes were active when Mars was young, but as this relatively small planet cooled down, its geological activity tapered off. All of the volcanoes of Mars are now inactive.

As with all rules, however, there are limitations and exceptions to the rule relating a world's size to its geologic activity. One limitation is that the four terrestrial planets all have slightly different compositions, which affects the types and extent of geologic activity that can take place on their surfaces. The different compositions also complicate the relationship between the number of craters and the age of the surface. An important exception to our rule is Jupiter's satellite Io, which, despite its small size, is the most volcanic world in the solar system (see Section 7-2). Something must be supplying Io with energy to keep its interior hot; this energy comes from Jupiter, which exerts powerful tidal forces on Io as it moves in a relatively small orbit around its planet. These tidal forces cause Io to flex like a ball of clay being kneaded between your fingers, and this flexing heats up the satellite's

FIGURE 7-12 R I V U X G

A Martian Volcano Olympus Mons is the largest of the inactive volcanoes of Mars and the largest volcano in the solar system. The base of Olympus Mons measures 600 km (370 mi) in diameter, and the scarps (cliffs) that surround the base are 6 km (4 mi) high. The caldera, or volcanic crater, at the summit is approximately 70 km across, large enough to contain the state of Rhode Island. (Kees Veenenbos/Science Source)

interior. Another example of heating due to tidal forces occurs within Saturn's moon Enceladus. But despite these limitations and exceptions, the relationships between a world's size, internal heat, geologic activity, and amount of cratering are powerful tools for understanding the terrestrial planets and satellites.

Pluto is another exception to the rule that a small object should be geologically inactive. Pluto is certainly small: Its radius is less than 20% that of Earth, and it only has about 0.2% the mass of Earth. However, part of Pluto's surface shows very recent geological activity in the form of smooth nitrogen ice. Figure 7-13 shows

FIGURE 7-13

Pluto Has a Heart This heart-shaped feature recently revealed on Pluto is named Tombaugh Regio. (NASA/Johns Hopkins University Applied Physics Laboratory/Southwest Research Institute.)

a heart-shaped feature named Tombaugh Regio; its western lobe is devoid of craters. At this time, there is no consensus as to where Pluto's internal heat comes from that powers this recent flow or melting of nitrogen ice over the surface.

CONCEPT CHECK 7-11

Io, which is a *moon,* is an exception to the rule that smaller worlds should have less geologic activity. Do you expect that some small *planets* might break this rule as well?

Answer appears at the end of the chapter.

7-7 A planet with a magnetic field indicates a fluid interior in motion

The amount of impact cratering on a terrestrial planet or moon provides indirect evidence about its interior. But another, more direct tool for probing the interior of *any* planet or moon is an ordinary compass, which senses the magnetic field outside the planet or moon. Magnetic field measurements prove to be an

extremely powerful way to investigate the internal structure of a world without having to actually dig into its interior. To illustrate how this works, consider the behavior of a compass on Earth.

> Global magnetic fields around a planet indicate that it must have a liquid interior.

The needle of a compass on Earth points north because it aligns with Earth's *magnetic field.* Such magnetic fields arise whenever electrically charged particles are in motion. For example, a loop of wire carrying an electric current generates a magnetic field in the space around it. The source of the magnetic field that surrounds an ordinary bar magnet (Figure 7-14a) is complicated, but part of it is created by the motion of negatively charged electrons within the iron atoms of which the magnet is made. Earth's magnetic field is similar to that of a bar magnet, as Figure 7-14b shows. The consensus among geologists is that this magnetic field is caused by the motion of liquid iron in the outer portion of Earth's core. Because this molten iron conducts electricity, these motions give rise to electric currents, which in turn produce Earth's magnetic field. Our planet's rotation helps to

(a)

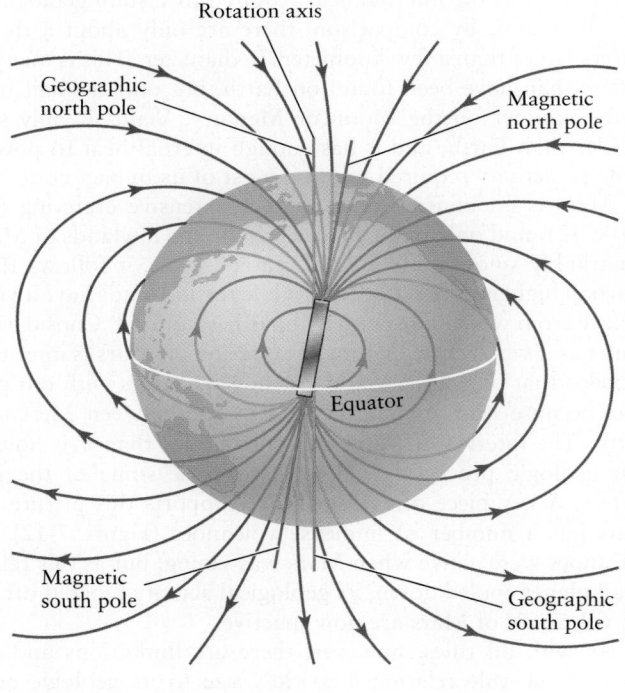

(b)

FIGURE 7-14

The Magnetic Fields of a Bar Magnet and of Earth **(a)** This picture was made by placing a piece of paper on top of a bar magnet, then spreading thin iron filings on the paper. Each elongated iron filing acts like a compass needle and aligns with the magnetic field at its location. The pattern of the filings show the magnetic field lines, which appear to stream from one of the magnet's poles to the other. **(b)** Earth's magnetic field lines have a similar pattern. Although Earth's field is

produced in a different way—by electric currents in the liquid portion of our planet's core—the field is much the same as if there were a giant bar magnet inside Earth. This "bar magnet" is not exactly aligned with Earth's rotation axis, which is why the magnetic north and south poles are not at the same locations as the true, or geographic, poles. A compass needle points toward the north magnetic pole, not the true north pole. (a: Jules Bucher/Science Source)

CAUTION!

While Earth's magnetic field is similar to that of a giant bar magnet, you should not take this picture too literally. Earth is *not* simply a magnetized ball of iron. In an iron bar magnet, the electrons of different atoms orbit their nuclei in the same general direction, so that the magnetic fields generated by individual atoms add together to form a single, strong field. But at temperatures above 770°C (1418°F = 1043 K), the orientations of the electron orbits become randomized. The fields of individual atoms tend to cancel each other out, and the iron loses its magnetism. Geological evidence shows that almost all of Earth's interior is hotter than 770°C, so the iron there cannot be extensively magnetized. The correct picture is that Earth acts as a dynamo: The liquid iron carries electric currents, and these currents create Earth's magnetic field.

sustain these motions and hence the magnetic field. This process for producing a magnetic field is called a **dynamo.**

If a planet or moon has a solid interior, then the dynamo mechanism cannot work: Electrically conductive material in the interior cannot flow, there are no electric currents, and the planet or moon does not generate a magnetic field. One example of this is the Moon. Measurements made during the Apollo missions, in which 12 humans visited the lunar surface between 1969 and 1972, showed that the present-day Moon indeed has no global magnetic field, which means no active dynamo with electric currents. However, careful magnetic measurements of lunar rocks returned by the Apollo astronauts indicate that the Moon *did* have a magnetic field when the rocks solidified. These rocks, like the rest of the lunar surface, are very old. Hence, in the distant past the Moon may have had a dynamo of its own, generated by motions within a liquid iron core. As the Moon cooled, the energy powering this dynamo declined, and the lunar magnetic field disappeared.

In addition to magnetism, extensive cratering on the Moon's surface is consistent with a hot interior that cooled down long before volcanism could erase the craters. However, it is important to understand that while magnetism and volcanism both require heat, each of these phenomena is complex and one can exist without the other.

We have now identified another general rule about planets and moons:

> *A planet or moon with a global magnetic field has liquid material in its interior that conducts electricity and is in motion, generating the magnetic field.*

Thus, by studying the magnetic field of a planet or moon, we can learn about that world's interior. So, many spacecraft carry devices called **magnetometers** to measure magnetic fields. Measurements made with magnetometers on spacecraft have led to a number of striking discoveries throughout the solar system, which we now discuss.

Mercury

It has been found that Mercury has a planetwide magnetic field with a shape similar to Earth's, although it's only about 1% as strong as Earth's field. If Mercury's magnetic field were simply due to a magnetized crust, variations in the field due to crust-altering craters would have been detected, yet no such variations were observed. Instead, Mercury's smooth magnetic field—created beneath the surface—suggests that currents within a liquid outer core are sustaining a dynamo. A magnetic dynamo requires enough heat for a liquid core, and this was unexpected. First of all, Mercury is the smallest planet and should have cooled off too much to have a molten liquid iron core. This cooling is certainly consistent with Mercury's heavily cratered surface, where a lack of volcanic activity has left its craters intact.

Nonetheless, as mentioned in Section 7-6, tidal forces can produce heat in a planet or moon. As a result, the Sun's tidal forces on Mercury produce just enough heat for a liquid iron core, but not enough to produce volcanism at present.

Venus

By contrast, Venus has no measurable planetwide magnetic field, even though the paucity of craters on its surface indicates the presence of geologic activity and hence a hot interior of the planet. One possible reason for the lack of a magnetic field on Venus is that the planet turns on its axis very slowly, taking 243 days for a complete rotation. Because of this slow rotation, the fluid material within the planet is hardly agitated at all, and so may not move in the fashion that generates a magnetic field.

Mars

Mars has a very interesting magnetic field that sheds light on its early history. The magnetometer aboard the Mars Global Surveyor spacecraft, which went into orbit around Mars in 1997, found that most of the planet's crust is magnetized (Figure 7-15): Mars has no planetwide magnetic field like Earth, but portions of its rocky surface are magnetized. This information actually tells us a lot about the Martian past. During the planet's first billion years or so, the interior was still hot and molten. Electric currents in the flowing molten material could then produce a planetwide magnetic field. As surface material cooled and solidified, the crust became magnetized by the planetwide field, and remained magnetized even after the internally generated planetwide field shut down.

The alternating stripes of magnetism on the Martian crust—seen most clearly in red and blue in the middle of Figure 7-15—tell us there is even more to learn about the Martian past. Far from being "alien" features, these stripes resemble magnetic stripes found in places on Earth's ocean seafloor. However, these Martian magnetic stripes remain a mystery because on Earth they were produced in part by plate tectonics, which Mars seems to lack.

FIGURE 7-15

Relic Magnetism on Mars For almost a billion years after Mars formed, molten material with electric currents generated a planetwide magnetic field. Most of the surface was magnetized during this time, which is the crustal field we measure today. Regions without magnetism probably experienced higher temperatures that can remove magnetism at a later time, or were formed by lava flows after the planetwide field shut down. Some magnetic stripes are visible in the lower middle of the image, where the field alternately points up and down. These stripes, which resemble magnetic stripes on Earth's ocean floors, could indicate that Mars went through a period with plate tectonics, but their origin remains a mystery. (NASA)

Jupiter and Saturn

The most intense planetary magnetic field in the solar system is that of Jupiter: At the tops of Jupiter's clouds, the magnetic field is about 14 times stronger than the field at Earth's surface. It is thought that Jupiter's field, like Earth's, is produced by a dynamo acting deep within the planet's interior. Unlike Earth, however, Jupiter is composed primarily of hydrogen and helium, not substances like iron that conduct electricity. How, then, can Jupiter have a dynamo that generates such strong magnetic fields?

To answer this question, recall from Section 5-8 that a hydrogen atom consists of a single proton orbited by a single electron. Deep inside Jupiter, the pressure is so great and hydrogen atoms are squeezed so close together that electrons can hop from one atom to another. This hopping motion creates an electric current, just as the ordered movement of electrons in the copper wires of a flashlight constitutes an electric current. In other words, the highly compressed hydrogen deep inside Jupiter behaves like an electrically conducting metal; thus, it is called **liquid metallic hydrogen.**

Laboratory experiments show that hydrogen becomes a liquid metal when the pressure is more than about 1.4 million times ordinary atmospheric pressure on Earth. Recent calculations suggest that this transition occurs about 7000 km below Jupiter's cloudtops. Most of the planet's enormous bulk lies below this level, so there is a tremendous amount of liquid metallic hydrogen within Jupiter. Since Jupiter rotates rapidly—a "day" on Jupiter is just less than 10 hours long—this liquid metal moves rapidly, generating the planet's powerful magnetic field. Saturn also has a magnetic field produced by dynamo action in liquid metallic hydrogen. (The field is weaker than Jupiter's because Saturn is a smaller planet with less internal pressure, so there is less of the liquid metal available.)

Uranus and Neptune

Uranus and Neptune also have magnetic fields, but they cannot be produced in the same way: Because these planets are relatively small, the internal pressure is not great enough to turn liquid hydrogen into a metal. Instead, it is thought that both Uranus and Neptune have large amounts of liquid water in their interiors and that this water has molecules of ammonia and other substances dissolved in it. (The fluid used for washing windows has a similar chemical composition.) Under the pressures found in this interior water, the dissolved molecules lose one or more electrons and become electrically charged (that is, they become ionized; see Section 5-8). Water is a good conductor of electricity when it has such electrically charged molecules dissolved in it, and electric currents in this fluid would be the source of the magnetic fields of Uranus and Neptune. However, there is no direct evidence of currents in liquid water for these planets, and an alternative model proposes an electrically conductive mixture of hydrogen and rocky material.

As you can see in these cases throughout our solar system, the mere presence of a global magnetic field measured around a planet or moon compels us to consider major questions about its internal composition. Where does the object get the heat necessary to produce a liquid interior? What is the electrically conductive material in its interior? The answers to these questions have broadened our understanding of the diversity within our solar system.

CONCEPT CHECK 7-12

By comparing parts (a) and (b) of Figure 7-14, can you conclude that there is a large bar magnet inside Earth?

CONCEPT CHECK 7-13

Does every planet with a molten interior produce a magnetic field? Do any of the jovian planets create their magnetic fields through electrically conductive molten iron?

Answers appear at the end of the chapter.

7-8 The diversity of the solar system is a result of its origin and evolution

Our brief tour of the solar system has revealed its almost dizzying variety. No two planets are alike, satellites come in all sizes, the extent of cratering varies from one terrestrial planet to another, and the magnetic fields of different planets vary dramatically in their strength and in how they are produced. (The *Cosmic Connections* that closes this chapter summarizes these properties of the planets.) All this variety leads us to a simple yet profound question: *Why* are the planets and satellites of the solar system so different from each other?

> The similarities and differences among the planets can be logically explained by a model of the solar system's origin and evolution.

Among humans, the differences from one individual to another result from heredity (the genetic traits passed on from an individual's parents) and environment (the circumstances under which the individual matures to an adult). As we will find in the following chapter, much the same is true for the worlds of the solar system.

In Chapter 8, we will see evidence that the entire solar system shares a common "heredity," in that the planets, satellites, comets, asteroids, and the Sun itself formed from the same cloud of interstellar gas and dust. The composition of this cloud was shaped by cosmic processes, including nuclear reactions that took place within stars that died long before our solar system was formed. We will see how different planets formed in different environments, depending on their distance from the Sun, and will discover how these environmental variations gave rise to the planets and satellites of our present-day solar system. And we will see how we can test these ideas of solar system origin and evolution by studying planetary systems orbiting other stars.

Our journey through the solar system is just beginning. In this chapter, we have explored space to examine the variety of the present-day solar system; in Chapter 8, we will examine other planetary systems and journey through time to see how our own solar system came to be.

CALCULATION CHECK 7-4

Using the *Cosmic Connections* figure, *Characteristics of the Planets*, how many Earth masses does it take to equal Jupiter's mass? How many Saturn masses does it take to equal Jupiter's mass?

Answer appears at the end of the chapter.

KEY WORDS

Terms preceded by an asterisk (*) are discussed in the Boxes.

asteroid, p. 188	dynamo, p. 197	Kuiper belt, p. 192	Oort comet cloud (Oort
asteroid belt, p. 189	*escape speed, p. 191	liquid metallic hydrogen,	cloud), p. 193
average density, p. 183	ices, p. 188	p. 198	plate tectonics, p. 194
chemical composition,	impact crater, p. 194	magnetometer, p. 197	spectroscopy, p. 185
p. 185	jovian planet, p. 182	meteoroid, p. 194	terrestrial planet, p. 182
comet, p. 192	*kinetic energy, p. 190	minor planet, p. 188	trans-Neptunian object, p. 189

The Inner (Terrestrial) Planets Close to the Sun—small diameter, small mass—high density

	Mercury	Venus	Earth	Mars
Average distance from the Sun (au)	0.387	0.723	1.000	1.524
Equatorial diameter (Earth = 1)	0.383	0.949	1.000	0.533
Mass (Earth = 1)	0.0553	0.8150	1.0000	0.1074
Average density (kg/m³)	5430	5243	5515	3934

Mercury **Venus** **Earth** **Mars**

Atmosphere
None

Atmosphere
Carbon dioxide

Atmosphere
Nitrogen, oxygen

Atmosphere
Carbon dioxide

Magnetic field
Weak

Magnetic field
None

Magnetic field
Moderate, due to
liquid iron core

Magnetic field
Weak, due to
magnetized crust

Interior
Iron–nickel core,
rocky shell

Interior
Iron–nickel core,
rocky shell

Interior
Iron–nickel core,
rocky shell

Interior
Iron–nickel core,
rocky shell

NASA/Johns Hopkins University Applied Physics Laboratory/Carnegie Institution of Washington; NASA/JPL · NASA/JSC; NASA/USGS

The Outer (Jovian) Planets
Far from the Sun—large diameter, large mass—low density

	Jupiter	Saturn	Uranus	Neptune
Average distance from the Sun (au)	5.203	9.554	19.194	30.066
Equatorial diameter (Earth = 1)	11.209	9.449	4.007	3.883
Mass (Earth = 1)	317.8	95.16	14.53	17.15
Average density (kg/m³)	1326	687	1318	1638

Jupiter

Saturn

Uranus

Neptune

Atmosphere
Hydrogen, helium

Atmosphere
Hydrogen, helium

Atmosphere
Hydrogen, helium

Atmosphere
Hydrogen, helium

Magnetic field
Strong, due to
liquid metallic hydrogen

Magnetic field
Strong, due to
liquid metallic hydrogen

Magnetic field
Moderate, due to
dissolved ions

Magnetic field
Moderate, due to
dissolved ions

Interior
Rocky core, liquid
hydrogen and helium

Interior
Rocky core, liquid
hydrogen and helium

Interior
Rocky core, liquid
water and ammonia

Interior
Rocky core, liquid
water and ammonia

NASA; NASA/JPL/USGS; NASA/JPL-Caltech; NASA/JPL

KEY IDEAS

Properties of the Planets: All the planets orbit the Sun in the same direction and in almost the same plane. Most of the planets have nearly circular orbits (Section 7-1).

- The four inner planets are called terrestrial planets. They are relatively small (with diameters of 5000 to 13,000 km), have high average densities (4000 to 5500 kg/m^3), and are composed primarily of rocky materials.

- The four giant outer planets are called jovian planets. They have large diameters (50,000 to 143,000 km) and low average densities (700 to 1700 kg/m^3) and are composed primarily of light elements such as hydrogen and helium.

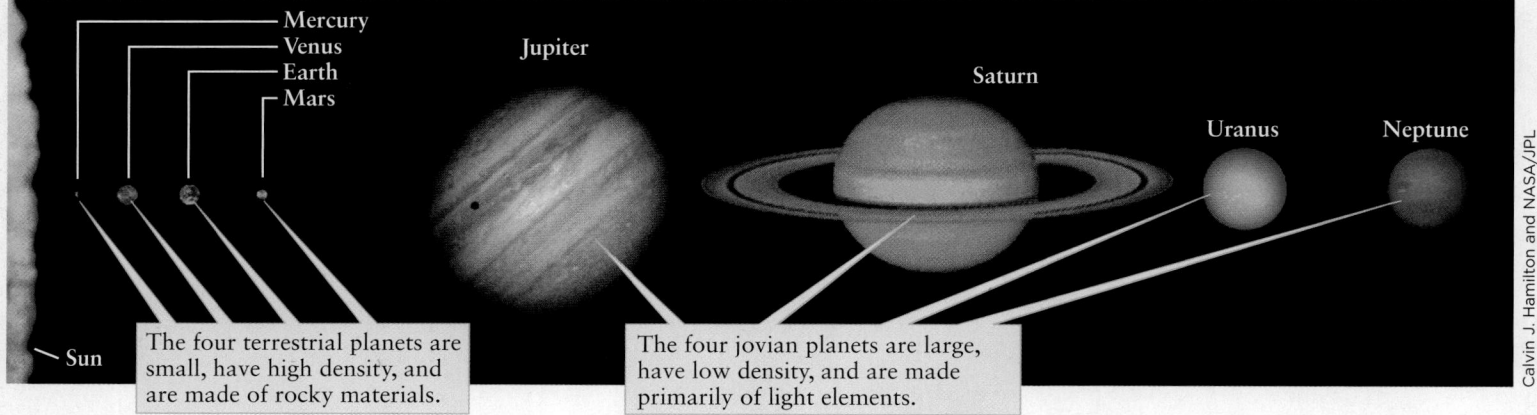

The four terrestrial planets are small, have high density, and are made of rocky materials.

The four jovian planets are large, have low density, and are made primarily of light elements.

Calvin J. Hamilton and NASA/JPL

Satellites and Small Bodies in the Solar System: Besides the planets, the solar system includes satellites of the planets, asteroids, comets, and trans-Neptunian objects (Section 7-2).

- Seven large planetary satellites (one of which is the Moon) are comparable in size to the planet Mercury. The remaining satellites of the solar system are much smaller.

- Asteroids are small, rocky objects, whereas comets and trans-Neptunian objects are made of ice and rock. All are remnants left over from the formation of the planets.

- Most asteroids are found in the asteroid belt between the orbits of Mars and Jupiter, and most trans-Neptunian objects lie in the Kuiper belt outside the orbit of Neptune. Pluto is one of the largest members of the Kuiper belt.

Moon   Io   Europa   Ganymede   Callisto   Titan   Triton

NASA/JPL/USGS, NASA/JPL/Space Science Institute, NASA/JPL/Space Science Institute, NASA/JPL

Spectroscopy and the Composition of the Planets: Spectroscopy, the study of spectra, provides information about the chemical composition of objects in the solar system (Section 7-3).

- The spectrum of a planet or satellite with an atmosphere reveals the atmosphere's composition. If there is no atmosphere, the spectrum indicates the composition of the surface.

- The substances that make up the planets can be classified as gases, ices, or rock, depending on the temperatures at which they solidify.

Impact Craters: When an asteroid, comet, or meteoroid collides with the surface of a terrestrial planet or satellite, the result is an impact crater (Section 7-6).

- Geologic activity renews the surface and erases craters, so a terrestrial world with extensive cratering has an old surface and little or no geologic activity.

- Because geologic activity is powered by internal heat, and smaller worlds lose heat more rapidly, as a general rule smaller terrestrial worlds are more extensively cratered.

Central peak

50 km

NASA

50 km

NASA

50 km

NASA/JPL/MSSS

Magnetic Fields and Planetary Interiors: Planetary magnetic fields are produced by the motion of electrically conducting liquids inside the planet. This mechanism is called a dynamo. A planet having no magnetic field is evidence that there is little such liquid material in the planet's interior or that the liquid is not in a state of motion (Section 7-7).

- The magnetic fields of terrestrial planets are produced by metals such as iron in the liquid state. The stronger fields of the jovian planets are generated by liquid metallic hydrogen or by water with ionized molecules dissolved in it.

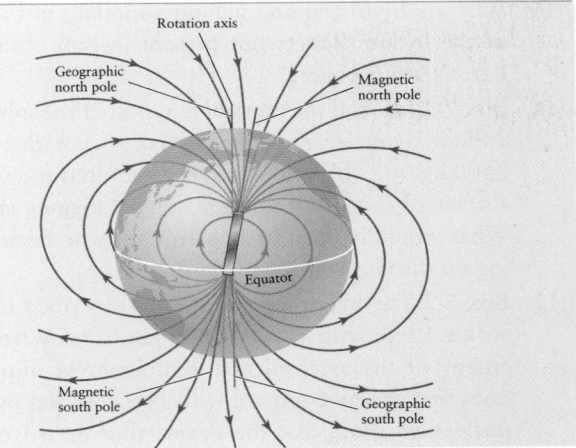

Rotation axis

Geographic north pole

Magnetic north pole

Equator

Magnetic south pole

Geographic south pole

QUESTIONS

Problem-solving tips and tools

The volume of a sphere of radius r is $4\pi r^3/3$, and the surface area of a sphere of radius r is $4\pi r^2$. The surface area of a circle of radius r is πr^2. The average density of an object is its mass divided by its volume. To calculate escape speeds, you will need to review Box 7-2. Be sure to use the same system of units (meters, seconds, kilograms) in all your calculations involving escape speeds, orbital speeds, and masses. Appendix 6 gives conversion factors between different sets of units, and Box 5-1 has formulas relating various temperature scales.

Section 7-1

1. Do all the planets orbit the Sun in the same direction? Are all of the orbits circular?

2. What are the characteristics of a terrestrial planet?

3. What are the characteristics of a jovian planet?

4. **TUTORIAL 7.2** What is meant by the average density of a planet? What does the average density of a planet tell us?

5. **TUTORIAL 7.3** What are the differences in chemical composition between the terrestrial and jovian planets?

6. Mars has two small satellites, Phobos and Deimos. Phobos circles Mars once every 0.31891 day at an average altitude of 5980 km above the planet's surface. The diameter of Mars is 6794 km. Using this information, calculate the mass and average density of Mars. (*Hint:* Consider using Newton's form of Kepler's third law.)

Section 7-2

7. In what ways are the largest satellites similar to the terrestrial planets? In what ways are they different? Which satellites are largest?

Section 7-3

8. **TUTORIAL 7.3** The absorption lines in the spectrum of a planet or satellite do not necessarily indicate the composition of the planet or satellite's atmosphere. Why not?

9. Figure 7-3 shows the spectrum of Saturn's largest satellite, Titan. Can you think of a way that astronomers can tell which absorption lines are due to Titan's atmosphere and which are due to the atmospheres of the Sun and Earth? Explain your answer.

Section 7-4

10. Why are hydrogen and helium abundant in the atmospheres of the jovian planets but present in only small amounts in Earth's atmosphere?

11. **Box 7-2** (a) Find the mass of a hypothetical spherical asteroid 2 km in diameter and composed of rock with average density 2500 kg/m^3. (b) Find the speed required to escape from the surface of this asteroid. (c) A typical jogging speed is 3 m/s. What would happen to an astronaut who decided to go for a jog on this asteroid?

12. **Box 7-2** The hypothetical asteroid described in Question 11 strikes Earth with a speed of 25 km/s. (a) What is the kinetic energy of the asteroid at the moment of impact? (b) How does this energy compare with that released by a 20-kiloton nuclear weapon, like the device that destroyed Hiroshima, Japan, on August 6, 1945? (*Hint:* 1 kiloton of TNT releases 4.2×10^{12} joules of energy.)

13. **Box 7-2** Suppose a spacecraft landed on Jupiter's moon Europa (see Table 7-2), which moves around Jupiter in an orbit of radius 670,900 km. After collecting samples from the satellite's surface, the spacecraft prepares to return to Earth. (a) Calculate the escape speed from Europa. (b) Calculate the escape speed from Jupiter at the distance of Europa's orbit. (c) In order to begin its homeward journey, the spacecraft must leave Europa with a speed greater than either your answer to (a) or your answer to (b). Explain why.

14. **Box 7-2** A hydrogen atom has a mass of 1.673×10^{-27} kg, and the temperature of the Sun's surface is 5800 K. What is the average speed of hydrogen atoms at the Sun's surface?

15. **Box 7-2** The Sun's mass is 1.989×10^{30} kg, and its radius is 6.96×10^8 m. (a) Calculate the escape speed from the Sun's surface. (b) Using your answer to Question 14, explain why the Sun has lost very little hydrogen over its entire 4.56-billion-year history.

16. **Box 7-2** Saturn's satellite Titan has an appreciable atmosphere, yet Jupiter's satellite Ganymede—which is about the same size and mass as Titan—has no atmosphere. Explain why there is a difference.

17. **Box 7-2** There are no asteroids with an atmosphere. Discuss why not.

Section 7-5

18. What is an asteroid? What is a trans-Neptunian object? In what ways are these minor members of the solar system like or unlike the planets?

19. What are the asteroid belt, the Kuiper belt, and the Oort cloud? Where are they located? How do the objects found in these three regions compare?

20. In what ways is Pluto similar to a terrestrial planet? In what ways is it different?

21. What is the connection between comets and the Kuiper belt? Between comets and the Oort cloud?

22. The distance from the asteroid 433 Eros (Figure 7-7) to the Sun varies between 1.13 and 1.78 au. (a) Find the period of Eros's orbit. (b) Does Eros lie in the asteroid belt? How can you tell?

23. Imagine a trans-Neptunian object with roughly the same mass as Earth but located 50 au from the Sun. (a) What do you think this object would be made of? Explain your reasoning. (b) On the basis of this speculation, assume a reasonable density for this object and calculate its diameter. How many times bigger or smaller than Earth would it be?

24. Consider a hypothetical trans-Neptunian object located 100 au from the Sun. (a) What would be the orbital period (in years) of this object? (b) There are 360 degrees in a circle, and 60 arcminutes in a degree. How long would it take this object to move 1 arcminute across the sky? (c) Trans-Neptunian objects are discovered by looking for "stars" that move on the celestial sphere. Use your answer from part (b) to explain why these discoveries require patience. (d) Discovering trans-Neptunian objects also requires large telescopes equipped with sensitive detectors. Explain why.

Section 7-6

25. What is one piece of evidence that impact craters are actually caused by impacts?

26. What is the relationship between the extent to which a planet or satellite is cratered and the amount of geologic activity on that planet or satellite?

27. How do we know that the surface of Venus is older than Earth's surface but younger than the Moon's surface?

28. Why do smaller worlds retain less of their internal heat?

29. How does the size of a terrestrial planet influence the number of craters visible on the planet's surface?

30. The surfaces of Mercury, the Moon, and Mars are riddled with craters formed by the impact of space debris. Many of these craters are billions of years old. By contrast, there are only a few conspicuous craters on Earth's surface, and these are generally less than 500 million years old. What do you suppose explains the difference?

31. During the period of most intense bombardment by space debris, a new 1-km-radius crater formed somewhere on the Moon about once per century. During this same period, what was the probability that such a crater would be created within 1 km of a certain location on the Moon during a 100-year period? During a 10^6-year period? (*Hint:* If you drop a coin onto a checkerboard, the probability that the coin will land on any particular one of the board's 64 squares is $\frac{1}{64}$.)

32. When an impact crater is formed, material (called *ejecta*) is sprayed outward from the impact. (The accompanying photograph of the Moon shows light-colored ejecta extending outward from the crater Copernicus.) While ejecta are found surrounding the craters on Mercury, they do not extend as far from the craters as do ejecta on the Moon. Explain why, using the difference in surface gravity between the Moon (surface gravity = 0.17 that on Earth) and Mercury (surface gravity = 0.38 that on Earth).

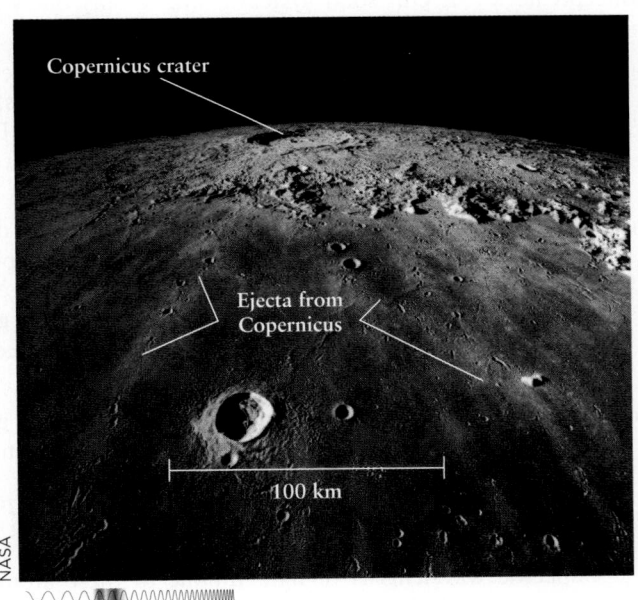

Copernicus crater

Ejecta from Copernicus

100 km

NASA

R I V U X G

Section 7-7

33. How is the magnetic field of a planet different from that of a bar magnet? Why is a large planet more likely to have a magnetic field than a small planet?

34. Could you use a compass to find your way around Venus? Why or why not?

35. If Mars has no planetwide magnetic field, why does it have magnetized regions on its surface?

36. What is liquid metallic hydrogen? Why is it found only in the interiors of certain planets?

37. Mercury rotates once on its axis every 58.646 days, compared to 1 day for Earth. Use this information to argue why Mercury's magnetic field should be much smaller than Earth's.

38. As you can see in Figure 7-15, Mars Global Surveyor did not find a significant magnetic field in the northern region of Utopia Planitia, which is a large lava field. Based on this observation, would you expect the lava field to have formed before or after Mars ceased to have a molten core?

39. Liquid metallic hydrogen is the source of the magnetic fields of Jupiter and Saturn. Explain why liquid metallic hydrogen cannot be the source of Earth's magnetic field.

40. The Galileo spacecraft that orbited Jupiter from 1995 to 2003 discovered that Ganymede (Table 7-2) has a magnetic field twice as strong as that of Mercury. Does this discovery surprise you? Why or why not?

General Problems

41. **TUTORIAL 7.2** On March 16, 2007, Venus was 1.97×10^8 km from Earth and had an angular diameter of 12.7 arcsec. Using the small-angle formula from Box 1-1, calculate the diameter of Venus.

WEB/E-BOOK QUESTION

42. **ANIMATION 7.1** Search the internet for information about impact craters on Earth. Where is the largest crater located? How old is it estimated to be? Which crater is closest to where you live?

43. **Determining Terrestrial Planet Orbital Periods.** Access the animation "Planetary Orbits" in Chapter 7 of the *Universe* website or e-book. Focus on the motions of the inner planets at the last half of the animation. Using the stop and start buttons, determine how many days it takes Mars, Venus, and Mercury to orbit the Sun once if Earth takes approximately 365 days.

ACTIVITIES

Observing Projects

44. Use a telescope or binoculars to observe craters on the Moon. Make a drawing of the Moon, indicating the smallest and largest craters that you can see. Can you estimate their sizes? For comparison, the Moon as a whole has a diameter of 3476 km. (*Hint:* You can see craters most distinctly when the Moon is near first quarter or third quarter [see Figure 3-2]. At these phases, the Sun casts long shadows across the portion of the Moon in the center of your field of view, making the variations in elevation between the rims and centers

of craters easy to identify. You can determine the phase of the Moon by looking at a calendar, the weather page of the newspaper, on the internet, or by using the *Starry Night*™ program.)

Starry Night™ Explorations

45. Open **Favorites > Explorations > Inner Planet Orbits** to investigate the orbits of the major terrestrial planets around the Sun.

    (a) Which of these four inner planets has the least circular orbit? How do the other three planets compare? (You may want to use the angular separation cursor to measure distances from the Sun to various parts of each orbit. If the distance is the same at all points, then the orbit is circular; otherwise, it is noncircular.)

    (b) Use the location scroller cursor to adjust the view so that Earth's orbit (green) is edge on. Do the orbits of the other three planets lie in the same plane as the plane of Earth's orbit (the ecliptic plane)?

    (c) Which of these planets has the orbit with the greatest tilt from Earth's orbit?

    (d) If you were on any of these other planets in the solar system, would the Sun appear to follow exactly the same path across the celestial sphere as is seen from Earth?

46. Use *Starry Night*™ to examine magnified images of the terrestrial major planets: Mercury, Venus, Earth, and Mars, and also the dwarf planet Ceres. Select **Favorites > Explorations > Terrestrial Planets** and click the **Quick Search Items** icon in the toolbar to open the **Observing Lists** panel. In this panel, locate and expand the **Terrestrial Planets** list. For each planet in the list, click the menu icon and select **Magnify**. Adjust the zoom and use keyboard shortcuts and time flow controls to allow you to inspect each planet's appearance and rotation. In the case of Venus, take the additional step of clicking its menu icon in the observing list and select **Surface Image/ Model > Magellan Color** to see the surface of the planet as seen by the Magellan radar-imaging satellite.

    (a) Describe each planet's appearance. From what you observe in each case, is there any way of knowing whether you are looking at a planet's surface or at complete cloud cover over the planet?

    (b) Which planet or planets have clouds?

    (c) Which of these selected planets shows the heaviest cratering?

    (d) Do any of these planets show evidence of liquid water on the planet's surface today? Do any show evidence of liquid water on the surface in ancient times?

    (e) What do you notice about Venus's rotation compared to the rotation of the other planets?

47. Use *Starry Night*™ to examine the gas giant planets Jupiter, Saturn, Uranus, and Neptune. Select **Favorites > Explorations > Atlas** and click the **Quick Search Items** icon in the toolbar to open the **Observing Lists** panel and expand the **Gas Giants**

list. For each planet in the list, click its contextual menu icon and select **Magnify**. Adjust the zoom and time flow as needed to explore the appearance and rotation of the planet as seen from Earth. For Saturn, take the additional step of using the "Y" keyboard shortcut to advance time by 1 year per keypress to observe the changing appearance of its rings over time. For the planet Uranus, take the additional step of selecting **Orbit** from its contextual menu and then zoom out to observe the orientation of Uranus's rotation relative to its orbital plane.

    (a) Which planet shows the greatest color contrast in its cloudtops?

    (b) Which planet has the least color contrast in its cloudtops?

    (c) What can you say about the thickness of Saturn's rings compared to their diameter?

    (d) Which planet's axis is most inclined to its orbital plane?

ANSWERS

ConceptChecks

ConceptCheck 7-1: Our solar system contains only one star, the Sun. The other stars are very far away.

ConceptCheck 7-2: Mars is classified as an inner planet. The inner planets (Mercury, Venus, Earth, and Mars) orbit within 2 au of the Sun, while the outer planets (Jupiter, Saturn, Uranus, and Neptune) all orbit farther out, more than 5 au from the Sun.

ConceptCheck 7-3: If the rocks on the surface have a density lower than the planet's average density, then the planet's core has a density greater than the planet's average. This is not surprising, since gravity causes material of greater density to sink toward the center of a planet. (The concept of average density is discussed in Box 7-1.)

ConceptCheck 7-4: Table 7-2 lists four having diameters greater than our Moon's 3476 km: Io, Ganymede, Callisto, and Titan.

ConceptCheck 7-5: As sunlight passes through a planet's atmosphere, the atoms and molecules in the atmosphere absorb specific wavelengths of light unique to these atoms and molecules. By looking at this absorption spectrum in the reflected sunlight, astronomers can infer the composition of the atmosphere as illustrated in Figure 7-3.

ConceptCheck 7-6: The reflected spectrum from a solid surface shows broad absorption features, whereas the spectrum observed from light passing through a gaseous atmosphere shows sharper spectral lines.

ConceptCheck 7-7: Hydrogen molecules in a gaseous atmosphere have a range of speeds based on the temperature of the gas, with some moving about six times the average speed. These faster molecules exceed Earth's escape speed and leave Earth entirely as described in Box 7-2. Continuously, some of the remaining slower hydrogen molecules speed up through molecular collisions and steadily escape.

ConceptCheck 7-8: Ceres has a diameter of about 900 km. This is about the same size as a large U.S. state, such as the length of California.

ConceptCheck 7-9: Pluto orbits beyond Neptune. All such objects are called trans-Neptunian objects. One group of trans-Neptunian objects—including Pluto—orbits in the Kuiper belt, so Pluto is both a trans-Neptunian object and a Kuiper belt object. Pluto is neither a planet nor an asteroid.

ConceptCheck 7-10: Whereas the Kuiper belt lies in the same plane as Earth's orbit around the Sun, the Oort cloud is a spherical distribution of comets that completely surrounds the solar system. If a comet has an orbit that is considerably different from that of the flat plane of the solar system, it most likely came from the spherical Oort cloud that exists in all directions around our solar system.

ConceptCheck 7-11: No, but planets can experience some heating. Tidal forces on planets decrease with their distance from our Sun. Only Mercury is close enough for any significant heating, and this does seem to occur, as evidenced by its magnetic field (discussed in Section 7-7). Nonetheless, Mercury has only minimal geologic activity. In other solar systems with planets very close to their stars, tidal-force heating could produce strong geological activity.

ConceptCheck 7-12: No. Figure 7-14a shows the field created by a bar magnet. Figure 7-14b shows that these fields are similar in structure, but the source of the magnetic field is not a bar magnet. Instead, Earth's magnetic field is due to electric currents flowing in a molten iron interior.

ConceptCheck 7-13: The answer to both questions is no. Based on its similar size to Earth, Venus probably has a molten interior. However, it exhibits no magnetic field, probably due to its very slow rotation: Earth rotates about 243 times for each single rotation of Venus. In the cases of Jupiter, Saturn, Uranus, and Neptune, some electrically conducting material other than iron is needed to explain the observed magnetic fields.

CalculationChecks

CalculationCheck 7-1: The shape of a planet's orbit is given by the value of its eccentricity. The closer this value is to zero, the closer the orbit's shape is to that of a perfect circle. According to the table, the orbit of Venus has the eccentricity closest to zero (0.007), making it the most circlelike of all planetary orbits.

CalculationCheck 7-2: If we divide Saturn's 120,536-km diameter by Earth's 12,756-km diameter—we find that 120,536 km ÷ 12,756 km = 9.449, so about 9½ Earth's would fit across Saturn's diameter.

CalculationCheck 7-3: The asteroid belt is located between Mars and Jupiter at about 3 au from the Sun, whereas the much larger Kuiper belt is beyond the orbit of Neptune and is located between about 30 and 50 au from the Sun.

CalculationCheck 7-4: The masses in this figure are all given as multiples of Earth's mass, so it would take just over 317 Earth masses to equal Jupiter's mass. The number of Saturn masses that would equal Jupiter's mass is 317.8/95.16 = 3.33.

8 Comparative Planetology II: The Origin of Our Solar System

Size of
Neptune's orbit

Size of our
solar system

LEARNING GOALS

What did our solar system look like before the planets were fully formed? The answer may lie in these remarkable images from the Hubble Space Telescope (left) and the Atacama Large Millimeter/submillimeter Array (ALMA, right). Each image shows an immense disk of gas and dust surrounding a young star. (In the first image, the light from the star itself was blocked out within the telescope to make the rather faint disk more visible.) Astronomers strongly suspect that in its infancy our own Sun was surrounded by a similar disk from which the planets of our solar system eventually coalesced. In the second image, the gaps in the disk are due to forming planets that have begun to accumulate material. This protostar, HL Tauri, is only 1 million years old, demonstrating that planet formation begins very quickly.

Planets are thought to form within the disks surrounding young stars such as au Microscopii (left, seen edge on) and HL Tauri (right). Neptune's orbit, shown for scale, is about 60 au across. (Science History Images/Alamy Stock Photo; ALMA [ESO/NAOJ/NRAO])

R I V U X G

In this chapter, we will examine the evidence that led astronomers to this model of the origin of the planets. We will see how the abundances of different chemical elements in the solar system indicate that the Sun and planets formed from a thin cloud of interstellar matter some 4.54 billion years ago, an age determined by measuring the radioactivity of meteorites. We will learn how the nature of planetary orbits provides important clues to what happened as this cloud contracted and how evidence from meteorites reveals the chaotic conditions that existed within the cloud. And we will see how this cloud eventually evolved into the solar system that we see today.

In the past few decades, astronomers have been able to test this picture of planetary formation by examining disks around young stars (like the ones in the chapter-opening images). Most remarkably, they have discovered planets in orbit around hundreds of other stars. These recent observations provide valuable information about how our own system of planets came to be.

8-1 Any model of solar system origins must explain the present-day Sun and planets

How did the Sun and planets form? In other words, where did the solar system come from? This question has tantalized astronomers for centuries. Our goal in this chapter is to examine our current understanding of how the solar system came to be—that is, our current best *theory* of the origin of the solar system.

Recall from Section 1-1 that a theory is not merely a set of wild speculations, but a self-consistent collection of ideas that must pass the test of providing an accurate description of the real world. Since no humans were present to witness the formation of the planets, scientists must base their theories of solar system origins on their observations of the present-day solar system. (In an analogous way, paleontologists base their understanding of the lives of dinosaurs on the evidence provided by fossils that have survived to the present day.) In so doing, they are following the steps of the scientific method that we described in Section 1-1.

What key attributes of the solar system should guide us in building a theory of solar system origins? Among the many properties of the planets that we discussed in Chapter 7, three of the most important are listed in Table 8-1. Any theory that attempts to describe the origin of the solar system must be able to explain how these attributes came to be. We begin by considering what

Property 1 tells us; we will return to Properties 2 and 3 and the orbits of the planets later in this chapter.

CONCEPT CHECK 8-1

Is Earth an exception to any of the three key properties of our solar system in Table 8-1?
Answer appears at the end of the chapter.

8-2 The cosmic abundances of the chemical elements are the result of how stars evolve

The small sizes of the terrestrial planets compared to the jovian planets (Property 1 in Table 8-1) suggest that some chemical elements are quite common in our solar system, while others are quite rare. The tremendous masses of the jovian planets—Jupiter alone has more mass than all of the other planets combined—means that the elements of which they are made, primarily hydrogen and helium, are very abundant. The Sun, too, is made almost entirely of hydrogen and helium. Its average density of 1410 kg/m^3 is in the same range as the densities of the jovian planets (see Table 7-1). Hydrogen, the most abundant element, makes up nearly three-quarters of the combined mass of the Sun and planets. Helium is the second most abundant element. Together, hydrogen and helium account for about 98% of the mass of all the material in the solar system. All of the other chemical elements are relatively rare; combined, they make up the remaining 2% (Figure 8-1).

> The terrestrial planets are small because they are made of less abundant elements.

The dominance of hydrogen and helium is not merely a characteristic of our local part of the universe. By analyzing the spectra of stars and galaxies, astronomers have observed essentially the same pattern of chemical abundances even at the farthest distance attainable by the most powerful telescopes. Hence, the vast majority of the atoms in the universe are hydrogen and helium atoms. The elements that make up the bulk of Earth—mostly iron, oxygen, and silicon—are relatively rare in the universe as a whole, as are the elements of which living organisms are made: carbon, oxygen, nitrogen, and phosphorus, among others. (You may find it useful to review the periodic table of the elements, described in Box 5-5.)

TABLE 8-1 Three Key Properties of Our Solar System	
Any theory of the origin of the solar system must be able to account for these properties of the planets.	
Property 1: Sizes and compositions of terrestrial planets versus jovian planets	The terrestrial planets, which are composed primarily of rocky materials, are relatively small, while the jovian planets, which are composed primarily of hydrogen and helium, are relatively large (see Sections 7-1 and 7-4).
Property 2: Directions and orientations of planetary orbits	All of the planets orbit the Sun in the same direction, and all of their orbits are in nearly the same plane (see Section 7-1).
Property 3: Distances of terrestrial planets versus jovian planets from the Sun	The terrestrial planets orbit close to the Sun, while the jovian planets orbit far from the Sun (see Section 7-1).

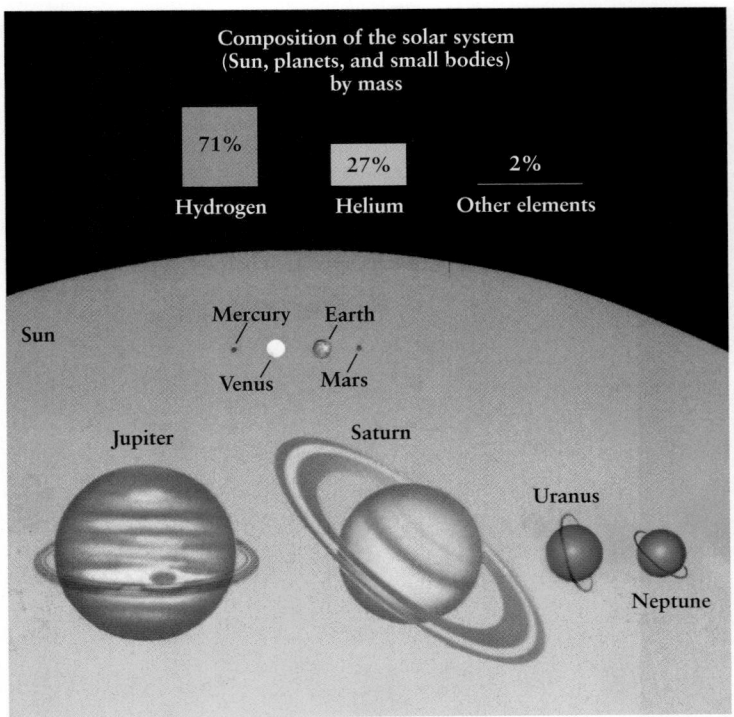

FIGURE 8-1

Composition of the Solar System Hydrogen and helium make up almost all of the mass of our solar system. Other elements such as carbon, oxygen, nitrogen, iron, gold, and uranium constitute only 2% of the total mass.

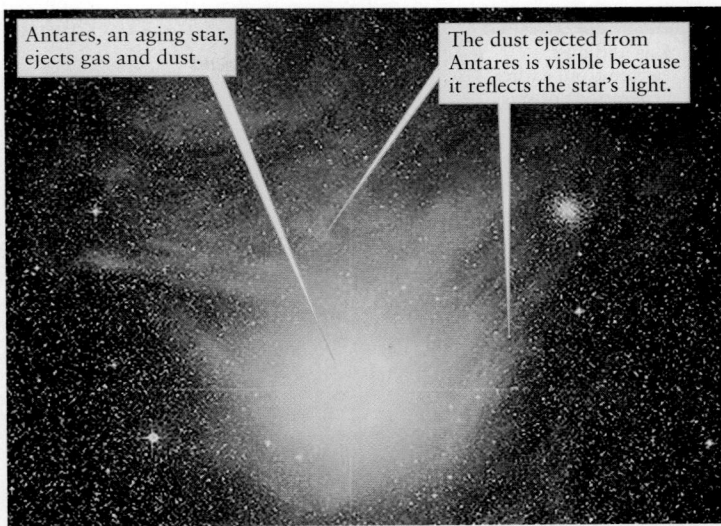

FIGURE 8-2 R I **V** U X G

A Mature Star Ejecting Gas and Dust The star Antares is shedding material from its outer layers, forming a thin cloud around the star. We can see the cloud because some of the ejected material has condensed into tiny grains of dust that reflect the star's light. (Dust particles in the air around you reflect light in the same way, which is why you can see them within a shaft of sunlight in a darkened room). Antares lies some 600 ly from Earth in the constellation Scorpius. (© Australian Astronomical Observatory/David Malin Images)

The Origin of the Elements and Cosmic "Recycling"

There is a good reason for this overwhelming abundance of hydrogen and helium. A wealth of evidence has led astronomers to conclude that the universe began some 13.7 billion years ago with a violent event called the Big Bang (Chapter 26). Only the lightest elements—hydrogen and helium, as well as tiny amounts of lithium and perhaps beryllium—emerged from the enormously high temperatures following this cosmic event. All the heavier elements were later manufactured by stars, either by thermonuclear fusion reactions deep in their interiors, by the violent explosions that mark the end of massive stars, or by collisions between ultradense stellar remnants called *neutron stars*. Were it not for these processes that take place only in stars, there would be no heavy elements in the universe, no planet like our Earth, and no humans to contemplate the nature of the cosmos.

Because our solar system contains heavy elements, at least some of its material must have once resided inside other stars. But how did this material become available to help build our solar system? The answer is that near the ends of their lives, stars cast much of their matter back out into space. For most stars, this process is a comparatively gentle one, in which the star's outer layers are gradually expelled. Figure 8-2 shows a star losing material in this fashion. This ejected material appears as the cloudy region, or **nebulosity** (from *nubes*, Latin for "cloud"), that surrounds the star and is illuminated by it. A small fraction of all stars eject matter much more dramatically at the very end of their lives, in a spectacular detonation called a *supernova*, which blows the star apart (see Figure 1-8).

No matter how it escapes, the ejected material contains heavy elements dredged up from the star's interior, where they were formed. This material becomes part of the **interstellar medium,** a tenuous collection of gas and dust that pervades the spaces between the stars. As stars die, they increasingly enrich the interstellar medium with heavy elements. Observations show that new stars form as condensations in the interstellar medium (Figure 8-3). Thus, these new stars have an adequate supply of heavy elements from which to develop a system of planets, moons, comets, and asteroids. Our own solar system must have formed from enriched material in just this way. Thus, our solar system contains "recycled" material that was produced long ago inside now-dead stars. This "recycled" material includes all of the carbon in your body, all of the oxygen that you breathe, and all of the iron and silicon in the soil beneath your feet. Scientifically accurate, a rock song from 1970 proclaimed: "We are stardust."

The Abundances of the Elements

Stars create different heavy elements in different amounts. For example, oxygen (as well as carbon, silicon, and iron) is readily produced in the interiors of massive stars, whereas gold (as well as silver, platinum, and uranium) is created only under special circumstances. Consequently, gold is rare in our solar system and in the universe as a whole, whereas oxygen is relatively abundant (although still much less abundant than hydrogen or helium).

A convenient way to express the relative abundances of the various elements is to say how many atoms of a particular element are found for every trillion (10^{12}) hydrogen atoms. For example, for every 10^{12} hydrogen atoms in space, there are about 100 billion (10^{11}) helium atoms. From spectral analysis of stars and chemical analysis of Earth rocks, Moon rocks, and bits of interplanetary

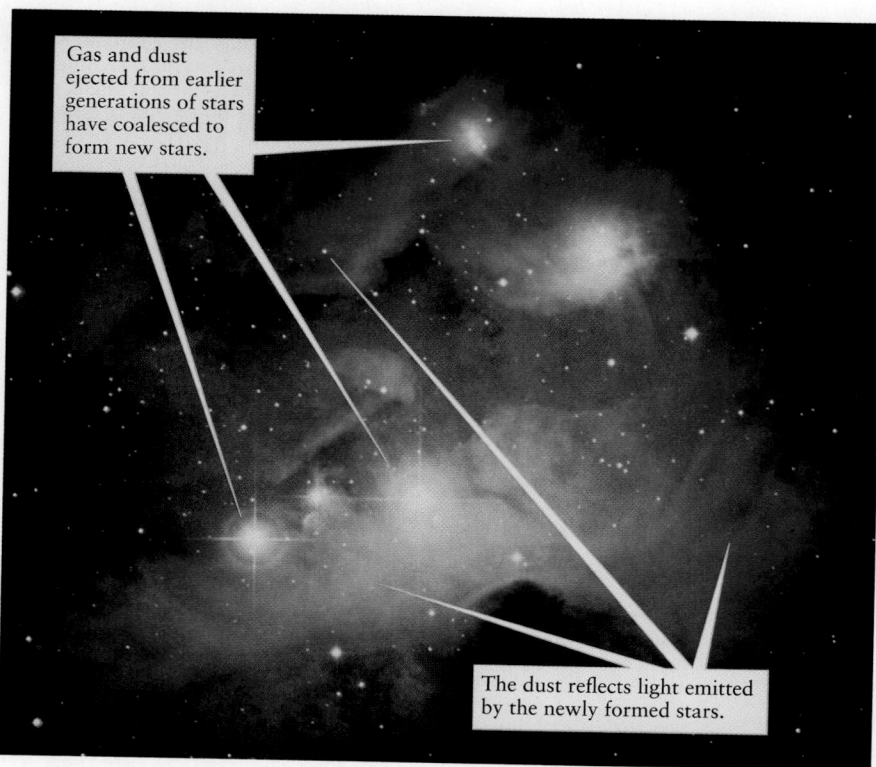

Gas and dust ejected from earlier generations of stars have coalesced to form new stars.

The dust reflects light emitted by the newly formed stars.

FIGURE 8-3 R I **V** U X G

New Stars Forming from Gas and Dust Unlike Figure 8-2, which depicts an old star that is ejecting material into space, this image shows young stars in the constellation Orion (the Hunter) that have only recently formed from a cloud of gas and dust. The bluish, wispy appearance of the cloud (called NGC 1973-1975-1977) is caused by starlight reflecting off interstellar dust grains within the cloud (see Box 5-4). The grains are made of heavy elements produced by earlier generations of stars. (© Australian Astronomical Observatory/David Malin Images)

debris called meteorites, scientists have determined the relative abundances of the elements in our part of the Milky Way Galaxy today. Figure 8-4 shows the relative abundances of the 30 lightest elements, arranged in order of their **atomic numbers.** An element's

atomic number is the number of protons in the nucleus of an atom of that element. It is also equal to the number of electrons orbiting the nucleus (see Box 5-5). In general, the greater the atomic number of an atom, the greater its mass.

The 10 most abundant elements:

1. Hydrogen (H)   5. Neon (Ne)   8. Magnesium (Mg)
2. Helium (He)   6. Nitrogen (N)   9. Silicon (Si)
3. Oxygen (O)   7. Iron (Fe)   10. Sulfur (S)
4. Carbon (C)

FIGURE 8-4

Abundances of the Lighter Elements This graph shows the abundances in our part of the Milky Way Galaxy of the 30 lightest elements (listed in order of increasing atomic number) compared to a value of 10^{12} for hydrogen. The inset lists the 10 most abundant of these elements, which are also indicated in the graph. Notice that the vertical scale is not linear; each division on the scale corresponds to a tenfold increase in abundance. After hydrogen and helium, the next most abundant element is oxygen, which is why water (H_2O) is a common molecule found in our galaxy. All elements heavier than zinc (Zn) have abundances of fewer than 1000 atoms per 10^{12} atoms of hydrogen.

CAUTION!

Figure 8-4 shows that there is about 10 times more hydrogen than helium in our galaxy when comparing the *number of atoms*. Figure 8-1 shows that our solar system is made of 71% hydrogen versus 27% helium when comparing *mass*. The explanation of this seeming inconsistency lies in the difference between comparing the mass of atoms versus their numbers: Each helium atom has about 4 times the mass of a hydrogen atom, which makes helium's contribution to total mass larger than its contribution to the total number of atoms.

The box inset in Figure 8-4 lists the 10 most abundant elements. Note that even oxygen (chemical symbol O), the third most abundant element, is quite rare relative to hydrogen (H) and helium (He): There are only 8.5×10^8 oxygen atoms for each 10^{12} hydrogen atoms and each 10^{11} helium atoms. Expressed another way, for each oxygen atom in our region of the Milky Way Galaxy, there are about 1200 hydrogen atoms and 120 helium atoms.

In addition to the 10 most abundant elements listed in Figure 8-4, five elements are moderately abundant: sodium (Na), aluminum (Al), argon (Ar), calcium (Ca), and nickel (Ni). These elements have abundances in the range of 10^6 to 10^7 relative to the standard 10^{12} hydrogen atoms. Most other elements are much rarer. For example, for every 10^{12} hydrogen atoms in the solar system, there are only 6 atoms of gold. The extreme rarity of many of these heavier elements results from the different processes by which they are created. As we'll see in Chapter 21, gold is produced in the violent merger of objects called *neutron stars*.

The small cosmic abundances of elements other than hydrogen and helium help to explain why the terrestrial planets are so small (Property 1 in Table 8-1). Because the heavier elements required to make a terrestrial planet are rare, only relatively small planets can form out of them. By contrast, hydrogen and helium are so abundant that it was possible for these elements to form large jovian planets.

CONCEPT CHECK 8-2

What is meant by the phrase "We are stardust"?

CONCEPT CHECK 8-3

From the abundances in Figure 8-4, do you expect that water (H_2O) might be a common substance in the galaxy?

Answers appear at the end of the chapter.

8-3 The abundances of radioactive elements reveal the solar system's age

The heavy elements can tell us even more about the solar system: They also help us determine its age. The particular heavy elements that provide us with this information are *radioactive*. Their atomic nuclei are unstable because they contain too many protons or too many neutrons. A radioactive nucleus therefore ejects particles until it becomes stable. In doing so, a nucleus may change from one element to another. Physicists refer to this transmutation as **radioactive decay**. For example, a radioactive form of the element rubidium (atomic number 37) spontaneously decays into the element strontium (atomic number 38) when one of the neutrons in the rubidium nucleus decays into a proton and an electron (which is ejected from the nucleus).

Experiments show that each type of radioactive nucleus decays at its own characteristic rate, which can be measured in the laboratory. Furthermore, the older a solid rock is, the more of its bulk has decayed and the less of its original radioactive material remains. This behavior is the key to a technique called **radioactive dating**, which is used to determine how many years ago a rock cooled and solidified, or simply, to determine the "ages" of rocks. For example, if a rock contained a certain amount of radioactive rubidium when it first solidified, over time more and more of the atoms of rubidium within the rock will decay into strontium atoms. The ratio of the number of strontium atoms the rock contains to the number of rubidium atoms it contains then gives a measure of the age of the rock. Box 8-1 describes radioactive dating in more detail.

> Our solar system, which formed 9 billion years after the Big Bang, is a relative newcomer to the universe.

Dating the Solar System

Scientists have applied techniques of radioactive dating to rocks taken from all over Earth. The results show that most rocks are tens or hundreds of millions of years old, but that some rocks are as much as 4 billion (4×10^9) years old. These results confirm that geologic processes—for example, lava flows—have produced new surface material over Earth's history, as we concluded from the small number of impact craters found on Earth (see Section 7-6). They also show that Earth must be at least 4×10^9 years old.

Radioactive dating has also been applied to rock samples brought back from the Moon by the Apollo astronauts. The oldest Apollo specimen, collected from one of the most heavily cratered—and hence most ancient—regions of the Moon, is 4.5×10^9 years old. But the oldest rocks found anywhere in the solar system are **meteorites**, bits of interplanetary debris that survive passing through Earth's atmosphere and land on our planet's surface (Figure 8-5). Radioactive dating of meteorites reveals that they are all nearly the *same* age, about 4.54 billion years old. The absence of any younger or older meteorites indicates that these are all remnants of objects that formed around the same time when rocky material in the early solar system—which was initially hot—first cooled and solidified. We conclude that the age of the oldest meteorites, about 4.54×10^9 years, is the age of the solar system itself. Note that this almost inconceivably long span of time is only about one-third of the current age of the universe, 13.7×10^9 years.

Thus, by studying the abundances of radioactive elements, we are led to a remarkable insight: Some 4.54 billion years ago, a collection of hydrogen, helium, and a much smaller amount of heavy elements came together to form the Sun and all of the objects that orbit around it. All those heavy elements, including the carbon atoms in your body and the oxygen atoms that you

FIGURE 8-5 R I V U X G

A Meteorite Although it resembles an ordinary Earth rock, this is actually a meteorite that fell from space. The proof of its extraterrestrial origin is the meteorite's composition and its surface. Searing heat melted the surface as the rock slammed into our atmosphere. Some meteorites are the oldest objects in the solar system (see Section 15-5). (Ted Kinsman/Science Source)

breathe, were created and cast off by stars that lived and died long before our solar system formed, during the first 9 billion years of the universe's existence. We are literally made of old stardust, and our solar system is relatively young.

CONCEPT CHECK 8-4

What is meant by the "age" of a rock? Is it the age of the rock's atoms?

Answer appears at the end of the chapter.

BOX 8-1 TOOLS OF THE ASTRONOMER'S TRADE

Radioactive Dating

How old are the rocks found on Earth and other planets? Are rocks that are found at different locations the same age or different ages? How old are meteorites? Questions like these are important to scientists who wish to reconstruct the history of our solar system. The age of a rock indicates how long ago it formed, but we cannot tell by simply looking at a rock whether it was formed a million or a billion years ago. Fortunately, most rocks contain trace amounts of radioactive elements such as uranium. By measuring the relative abundances of various radioactive isotopes and their decay products within a rock, scientists can determine the rock's age.

As we saw in Box 5-5, every atom of a particular element has the same number of protons in its nucleus. However, different isotopes of the same element have different numbers of neutrons in their nuclei. For example, the common isotopes of uranium are ^{235}U and ^{238}U. Each isotope of uranium has 92 protons in its nucleus (correspondingly, uranium is element 92 in the periodic table; see Box 5-5). However, a ^{235}U nucleus contains 143 neutrons, whereas a ^{238}U nucleus has 146 neutrons.

A radioactive nucleus with too many protons or too many neutrons is unstable; it *decays* by ejecting particles until it becomes stable. If the number of protons (the atomic number) changes in this process, the nucleus changes from one element to another.

Some radioactive isotopes decay rapidly, while others decay slowly. Physicists find it convenient to talk about the decay rate in terms of an isotope's **half-life.** The half-life of an isotope is the time interval in which one-half of the nuclei decay. For example, the half-life of ^{238}U is 4.5 billion (4.5×10^9) years. Uranium's half-life means that if you start out with 1 kg of ^{238}U, after 4.5 billion years, you will have only ½ kg of ^{238}U remaining; the other ½ kg will have turned into other elements. If you wait another half-life, so that a total of 9 billion years has elapsed, only ¼ kg of ^{238}U—one-half of one-half of the original amount—will remain. Several isotopes useful for determining the ages of rocks are listed in the accompanying table.

To see how geologists date rocks, consider the slow conversion of radioactive rubidium (^{87}Rb) into strontium (^{87}Sr). (The periodic table in Box 5-5 shows that the atomic numbers for these elements are 37 for rubidium and 38 for strontium, so in the decay a neutron is transformed into a proton. In this process, an electron is ejected from the nucleus.) Over the years, the amount of ^{87}Rb in a rock decreases, whereas the amount of ^{87}Sr increases. Because the ^{87}Sr accumulates in the rock due to radioactive decay, this isotope is called *radiogenic.* Dating the rock is not simply a matter of measuring its ratio of rubidium to strontium, however, because the rock already contained some strontium when it was formed. Geologists must therefore determine how much fresh strontium came from the decay of rubidium after the rock's formation.

To make this determination, geologists use as a reference another isotope of strontium whose concentration has remained constant. In this case, they use ^{86}Sr, which is stable and is not created by radioactive decay; it is said to be *nonradiogenic.* Dating a rock thus entails comparing the ratio of radiogenic and nonradiogenic strontium ($^{87}Sr/^{86}Sr$) in the rock

(continued on the next page)

BOX 8-1 (continued)

to the ratio of radioactive rubidium to nonradiogenic strontium ($^{87}Rb/^{86}Sr$). Because the half-life for converting ^{87}Rb into ^{87}Sr is known, the rock's age can then be calculated from these ratios (see the table).

Radioactive isotopes decay with the same half-life no matter where in the universe they are found. Hence, scientists have used the same techniques to determine the ages of rocks from the Moon and of meteorites.

Original radioactive isotope	Final stable isotope	Half-life (years)	Range of ages that can be determined (years)
Rubidium (^{87}Rb)	Strontium (^{87}Sr)	47 billion	10 million–4.54 billion
Uranium (^{238}U)	Lead (^{206}Pb)	4.5 billion	10 million–4.54 billion
Potassium (^{40}K)	Argon (^{40}Ar)	1.3 billion	50,000–4.54 billion
Carbon (^{14}C)	Nitrogen (^{14}N)	5730	100–70,000

8-4 The Sun and planets formed from a solar nebula

We have seen how processes in the Big Bang and within ancient stars produced the raw ingredients for our solar system. But how were these ingredients combined to make the Sun and planets? Astronomers have developed a variety of models for the origin of the solar system. The test of these models is whether they explain the properties of the present-day system of Sun and planets.

Astronomers see young stars that may be forming planets today in the same way that our solar system did billions of years ago.

The Failed Tidal Hypothesis

Any model of the origin of the solar system must explain why all the planets orbit the Sun in the same direction and in nearly the same plane (Property 2 in Table 8-1). One model that was devised explicitly to address this issue was the *tidal hypothesis,* proposed in the early 1900s. As we saw in Section 4-8, two nearby planets, stars, or galaxies exert tidal forces on each other that cause the objects to elongate. In the tidal hypothesis, another star happened to pass close by the Sun, and tidal forces induced by the star drew a long filament out of the Sun. In this scenario, the filament material would then have entered into orbit around the Sun, and the planets would have condensed from it, potentially explaining why they all orbit in the same direction and in the same plane. However, it was shown in the 1930s that tidal forces strong enough to pull a filament out of the Sun would also cause the filament to disperse before it could condense to form any planets. Hence, the tidal hypothesis cannot be correct.

The Successful Nebular Hypothesis

An entirely different model is now thought to describe the most likely series of events that produced our present solar system (Figure 8-6). The central idea of this model dates to the late 1700s,

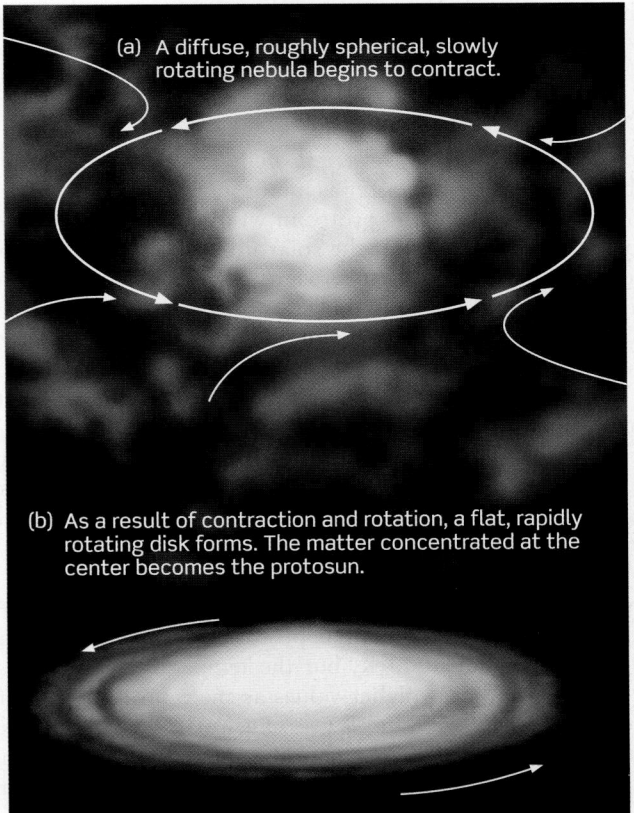

(a) A diffuse, roughly spherical, slowly rotating nebula begins to contract.

(b) As a result of contraction and rotation, a flat, rapidly rotating disk forms. The matter concentrated at the center becomes the protosun.

FIGURE 8-6

The Birth of the Solar System (a) A cloud of interstellar gas and dust begins to contract because of its own gravity. (b) As the cloud flattens and spins more rapidly around its rotation axis, a central concentration develops and evolves into a glowing protosun. The planets will form out of the surrounding disk of gas and dust.

when the German philosopher Immanuel Kant and the French scientist Pierre-Simon de Laplace each turned their attention to the manner in which the planets orbit the Sun. Both concluded that the arrangement of the orbits—all in the same direction and in nearly the same plane—could not be mere coincidence. To explain the orbits, Kant and Laplace independently proposed that our entire solar system, including the Sun as well as all of its planets and satellites, formed from a vast, rotating cloud of gas and dust called the **solar nebula** (Figure 8-6a). This model is called the **nebular hypothesis**.

The consensus among today's astronomers is that Kant and Laplace were on exactly the right track. In the modern version of the nebular hypothesis, at the outset the solar nebula was similar in character to the cloud of gas and dust shown in Figure 8-3 and had a mass somewhat greater than that of our present-day Sun. Also as seen in Figure 8-3, it is very likely that other stars formed from such a large cloud.

Each part of the nebula exerted a gravitational attraction on the other parts, and these mutual gravitational pulls tended to make the nebula contract. As it contracted, the greatest concentration of matter built up at the center of the nebula, forming a relatively dense region called the **protosun**. As its name suggests, this part of the solar nebula eventually developed into the Sun. The planets formed from the much sparser material in the outer regions of the solar nebula. Indeed, the mass of all the planets together is only 0.1% of the Sun's mass.

Evolution of the Protosun

When you drop a ball, the gravitational attraction of Earth makes the ball move faster and faster as it falls; in the same way, material falling inward toward the protosun would have gained speed as it approached the center of the solar nebula. As this fast-moving material collided with the protosun, the kinetic energy of each collision was converted into thermal energy, causing the temperature deep inside the solar nebula to climb. This process, in which the gravitational energy of a contracting gas cloud is converted into thermal energy, is called **Kelvin–Helmholtz contraction,** after the nineteenth-century physicists who first described it. This heating process will be discussed several times in future chapters so let's summarize it here: *Kelvin–Helmholtz contraction is the conversion of gravitational energy into heat as a gaseous cloud contracts.*

As the newly created protosun continued to contract and become denser, its temperature continued to climb as well. After about 10^5 (100,000) years, the protosun's *surface* temperature leveled off at about 6000 K, but the temperature in its *interior* kept increasing to ever higher values as the central regions of the protosun became denser and denser. Eventually, after perhaps 10^7 (10 million) years had passed since the solar nebula first began to contract, the gas at the center of the protosun reached a density of about 10^5 kg/m^3 (about 13 times denser than typical iron) and a temperature of a few million kelvins (that is, a few times 10^6 K). Under these extreme conditions, nuclear reactions that convert hydrogen into helium began in the protosun's interior. These nuclear reactions released energy that significantly increased the pressure in the protosun's core. When the pressure built up enough, it stopped further contraction of the protosun and a true

star was born. *In fact, the onset of nuclear reactions defines the end of a protostar and beginning of a star.* Nuclear reactions continue to the present day in the interior of the Sun and are the source of all the energy that the Sun radiates into space.

The Protoplanetary Disk

If the solar nebula had not been rotating at all, everything would have fallen directly into the protosun, leaving nothing behind to form the planets. Instead, the solar nebula must have had an overall slight rotation, which caused its evolution to follow a different path. As the slowly rotating nebula collapsed inward, it would naturally have tended to rotate faster. This relationship between the size of an object and its rotation speed is an example of a general principle called the **conservation of angular momentum.**

ANALOGY

Figure skaters make use of the conservation of angular momentum. When a spinning skater pulls her arms and legs in close to her body, the rate at which she spins automatically increases (**Figure 8-7**). Even if you are not a figure skater, you can demonstrate this by sitting on a rotating office chair. Sit with your arms outstretched and hold a weight, like a brick or a full water bottle, in either hand. Now use your feet to start your body and the chair rotating, lift your feet off the ground, and then pull your arms inward. Your rotation will speed up quite noticeably.

Astronomers see young stars that may be forming planets today in the same way that our solar system did billions of years ago. As the solar nebula began to rotate more rapidly, it also tended to flatten out (Figure 8-6b)—but why? From the

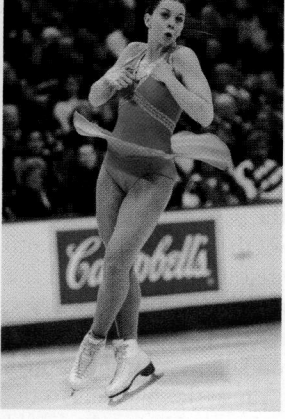

(a)   (b)

FIGURE 8-7 R I V U X G

Conservation of Angular Momentum A figure skater who **(a)** spins slowly with her limbs extended will naturally speed up when **(b)** she pulls her limbs in. In the same way, the solar nebula spun more rapidly as its material contracted toward the center of the nebula. (Amy Sancetta/AP Images)

(a)

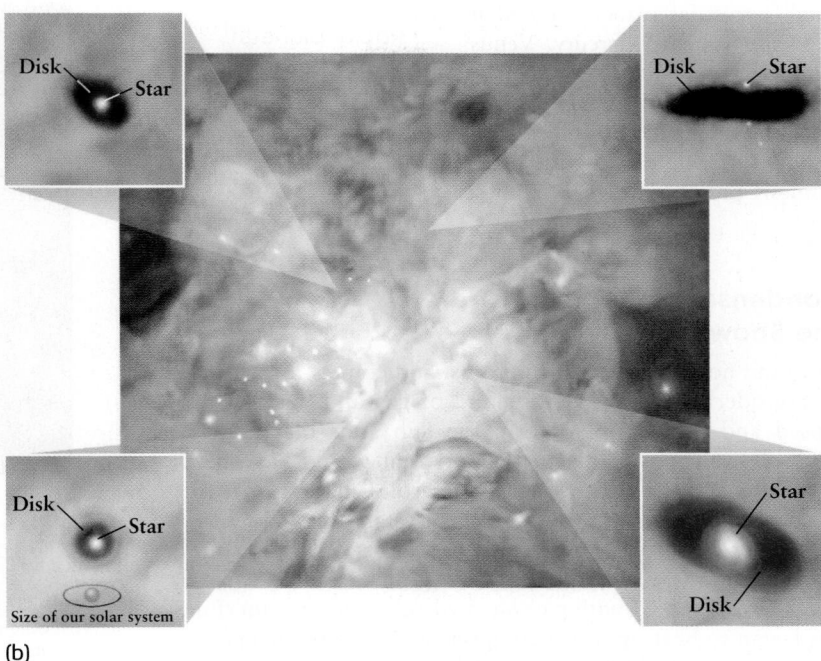

(b)

VIDEO
8.1

FIGURE 8-8 R I ◉ **U X G**

Protoplanetary Disks (a) The Orion Nebula is a star-forming region located some 1500 ly from Earth. It is the middle "star" in Orion's "sword" (see Figure 2-2a). The smaller, bluish nebula is the object shown in Figure 8-3. **(b)** This view of the center of the Orion Nebula is a mosaic of Hubble Space Telescope images. The four insets are false-color close-ups of four protoplanetary disks that lie within the nebula. A young, recently formed star is at the center of each disk. (The disk at upper right is seen nearly edge on.) The inset at the lower left shows the size of our own solar system for comparison. (a: © Australian Astronomical Observatory/ David Malin Images; b: C. R. O'Dell and S. K. Wong, Rice University; NASA)

perspective of a particle on its orbit around the center of mass of the nebula, it felt as though there were a force pushing the particle away from the nebula's axis of rotation. (Likewise, passengers on a merry-go-round or spinning carnival ride seem to feel a force pushing them outward and away from the ride's axis of rotation.) This apparent force was directed opposite to the inward pull of gravity, so it tended to slow the contraction of material toward the nebula's rotation axis. But there was no such effect opposing contraction in a direction parallel to the rotation axis. Some 10^5 (100,000) years after the solar nebula first began to contract, it had developed the structure shown in Figure 8-6b, with a rotating, flattened disk surrounding what would become the protosun. This disk is called the **protoplanetary disk,** since planets formed from its material. This model explains why their orbits all lie in essentially the same plane and why they all orbit the Sun in the same direction.

There were no humans to observe these processes taking place during the formation of the solar system. But Earth astronomers have seen disks of material surrounding other stars that formed only recently. These, too, are called protoplanetary disks, because it is thought that planets can eventually form from these disks around other stars. Hence, these disks are planetary systems that are still "under construction." By studying these disks around other stars, astronomers are able to examine what our solar nebula may have been like some 4.5×10^9 years ago.

Figure 8-8 shows a number of protoplanetary disks in the Orion Nebula, a region of active star formation. A star is visible at the center of each disk, which reinforces the idea that our Sun began to shine before the planets were fully formed. (The images that open this chapter show even more detailed views of disks surrounding young stars.) A study of 110 young stars in the Orion Nebula detected protoplanetary disks around 56 of them, which suggests that systems of planets may form around a substantial fraction of stars. Later in this chapter, we will see direct evidence for planets that have formed around stars other than the Sun.

CONCEPT CHECK 8-5

If the nebular hypothesis is correct, what must it explain about the orbits of the planets?

CONCEPT CHECK 8-6

Was the solar nebula cold until nuclear reactions began powering the Sun?

Answers appear at the end of the chapter.

8-5 The terrestrial planets formed by the accretion of planetesimals

We have seen how a contracting molecular cloud would have evolved into a spinning protoplanetary disk with a young Sun at its center. But how did the material in this disk form into planets? Why are there small terrestrial planets located in the inner solar system (Mercury, Venus, Earth, and Mars), and giant jovian planets in the outer solar system (Jupiter, Saturn, Uranus, and Neptune)? In this section and the next, we will see how the nebular hypothesis provides answers to these questions.

> Rocky planets formed in the inner solar nebula as a consequence of the high temperatures close to the protosun.

The Condensation Temperature and the Snow Line

To understand how the planets, asteroids, and comets formed, we start by considering the vast cold cloud of gas and dust, before it collapsed and heated up to form the Sun and solar system. Figure 8-9 shows a dust grain of the sort that would have been present throughout the cold cloud. Other solid materials in the cold cloud would have formed small ice crystals (like snow). In the low pressures of space, no liquids can exist, only gases and solids. Together, these solids—ices, dust grains, and ice-coated dust grains—were mixed with gaseous hydrogen and helium. But, as things began to heat up, these components began to change.

Due to Kelvin–Helmholtz contraction—the conversion of gravitational energy into thermal energy—a hot protosun began to form in the center of the nebula, and it was actually quite a bit more luminous than the present-day Sun. The heat from Kelvin–Helmholtz contraction was so great that nearly the entire cloud was vaporized into a gaseous state. The product of this transition from a giant cold cloud of gas and dust to a more dense and disk-shaped swirl of heated gas is now called the *solar nebula*.

With most of the contraction over, the solar nebula began to cool, especially in the outer regions. The temperature in the nebula varied significantly, rising above 2000 K closer to the hot protosun, and dropping below 50 K in the outermost regions (Figure 8-10a).

$10 \ \mu m = 0.01 \ mm$

FIGURE 8-9

A Grain of Cosmic Dust This highly magnified image shows a microscopic dust grain that came from interplanetary space. It entered Earth's upper atmosphere and was collected by a high-flying aircraft. Dust grains of this sort are abundant in star-forming regions like that shown in Figure 8-3. These tiny grains were also abundant in the solar nebula and served as the building blocks of the planets. (NASA)

Our first goal in determining a model for the origin of the solar system is to understand why the inner four planets are small and rocky, whereas the outer jovian planets are large and made of lighter elements (recall Properties 1 and 3 in Table 8-1). To answer this, we first need to understand the **condensation temperature**, which determines whether a substance forms a solid, liquid, or gas under given conditions.

ANIMATION
8.2

FIGURE 8-10

Temperature Distribution and Snow Line in the Solar Nebula (a) This graph shows how temperatures probably varied across the solar nebula as the planets were forming, and the present-day position of the planets (red arrows). Note the general decline in temperature with increasing distance from the center of the nebula. Between the present-day distances of Mars and Jupiter, the snow line marked where

Beyond the snow line, where water freezes, ices add to planet-building material, leading to the large jovian planets.

temperatures were low enough for water to condense and form ice; beyond about 16 au, methane (CH_4) could also condense into ice. **(b)** Terrestrial planets formed inside the snow line, where the low abundance of solid dust grains kept these planets small. The jovian planets formed beyond the snow line where solid ices of water, methane, and ammonia added their mass to build larger cores, and attract surrounding gas.

If the temperature of a substance in the solar nebula is above its condensation temperature, the substance is a gas. On the other hand, if the temperature is below the condensation temperature, the substance solidifies (or condenses) into tiny specks of dust or icy frost. You can often see similar behavior on a cold morning. The morning air temperature can be above the condensation temperature of water, while the cold windows of parked cars may have temperatures below the condensation temperature. Thus, water molecules in the warmer air remain as a gas (water vapor) but condense to form solid ice particles (frost) on the colder car windows.

With temperatures decreasing over time, and decreasing with distance from the protosun, the solar nebula evolved into two distinct regions. These regions—inner and outer—had very different properties and eventually formed very different planets:

- **Inner Region: rock and metal.** In the hot inner region of the nebula, materials with the highest condensation temperatures would have been the first to condense into solids as the nebula began to cool down. These rocky and metallic materials, in the form of solid dust grains (Figure 8-9), eventually formed much of the rock and metal of the terrestrial planets. On the other hand, hydrogen compounds—such as water (H_2O), methane (CH_4), and ammonia (NH_3)—remained as gas; they could not condense into solids in the warm inner solar system. (Hydrogen and helium could not condense into solids anywhere in the solar system.) The terrestrial planets that formed in this inner region are Mercury, Venus, Earth, and Mars, and now we understand why they are composed primarily of rock and metal.

- **Outer Region: ices beyond the snow line.** In the region between the present-day orbits of Mars and Jupiter, the temperature would have fallen below about 170 K. In the low pressures of the solar nebula, this is the condensation temperature at which water vapor (H_2O) forms ice (see Figure 8-10a). One consequence of this **snow line**—the distance from the Sun where it is cool enough for water vapor to condense, or to solidify into ice or frost—is that beyond this line, *solid ice particles can join with rock and metal grains to provide more mass for building planets* (Figure 8-10b). Farther out, at even cooler temperatures, methane (CH_4), and ammonia (NH_3) also condense to form ices. (Recall from Section 7-4 that "ice" can refer to frozen carbon dioxide, methane, or ammonia as well as to frozen water.) To summarize: Beyond the snow line, rocky and metallic dust grains were mixed with ices, resulting in more material available to build even larger planets. The large jovian planets that formed in this outer region are Jupiter, Saturn, Uranus, and Neptune.

The abundances of different materials in the solar nebula provide the final insight about planet sizes. With a mixture similar to the Sun, the solar nebula was composed of about 98% hydrogen and helium and about 1.4% hydrogen compounds (H_2O, CH_4, and NH_3), with the remaining 0.6% made up of rocks and metals. Inside the snow line, the only solid material available to build planets consisted of rock and metal. Thus, with rock and metal having the lowest abundances, the inner terrestrial planets that formed from these dust grains are the smallest. On the other hand, beyond the snow line there was more solid mass available in the form of ice-coated dust grains to help build larger planets.

The phenomena described so far help us understand why smaller terrestrial planets formed close to the Sun, and larger jovian planets formed farther out. However, there is more to this story, and some big mysteries remain. Next, we consider the process of building planets in more detail, beginning with the terrestrial planets.

Planetesimals, Protoplanets, and Terrestrial Planets

ANIMATION
8.1

In the inner part of the solar nebula, the grains of rocks and metals would have collided and merged into small clumps. Initially, electric forces—that is, chemical bonds—held these clumps together, in the same way that chemical bonds hold an ordinary rock together. Over a few million years, these clumps continued to collide and coalesced into roughly a billion asteroidlike objects called **planetesimals,** with diameters of 1 km or so. These larger planetesimals were massive enough to be held together by their own gravity, resulting from the mutual gravitational attraction of all the material within them.

During the next stage, gravitational attraction between different planetesimals caused them to collide and accumulate into about a hundred still-larger objects called **protoplanets** (also called planetary embryos), each of which was roughly the size and mass of our Moon. This accumulation of material to form larger and larger objects is called **accretion**. During a period lasting a few tens of millions of years, these Moon-sized protoplanets collided to form the inner planets. This final episode must have involved some truly spectacular, world-shattering collisions. In fact, after Earth formed, detailed modeling suggests that a leftover protoplanet collided with Earth and the debris created our Moon! Snapshots from a computer simulation of accreting planetesimals are shown in Figure 8-11.

In the higher temperatures of the inner solar nebula, only materials that can remain solid at high temperatures—which excludes water ice—could form solid dust grains to build protoplanets. Therefore, we would expect to form a set of inner planets made predominantly of materials such as iron, silicon, magnesium, and nickel. In strong support for this model of solar system formation, these rocky and metallic materials match the composition of the present-day terrestrial planets. There is also supporting evidence from meteorites (dated from their radioactivity) that are thought to have come from smashed-up planetesimals during this early period. As expected for planetesimals from the inner solar system, these meteorites contain metallic grains mixed with rocky material (Figure 8-12).

At first, the material that coalesced to form protoplanets in the inner solar nebula remained largely in solid form, despite the high temperatures close to the protosun. But as the protoplanets grew, they were heated by violent impacts as they collided with other planetesimals, as well as by the energy released from the decay of radioactive elements, and all this heat caused melting. Thus, the terrestrial planets began their existence as spheres of at least partially molten rocky materials. Material was free to move within these molten spheres, so the denser, iron-rich minerals sank to the centers of the planets while the less dense silicon-rich rocky minerals floated to their surfaces. This process is called **chemical differentiation** (see Box 7-1). In this way, the terrestrial planets developed their dense iron cores.

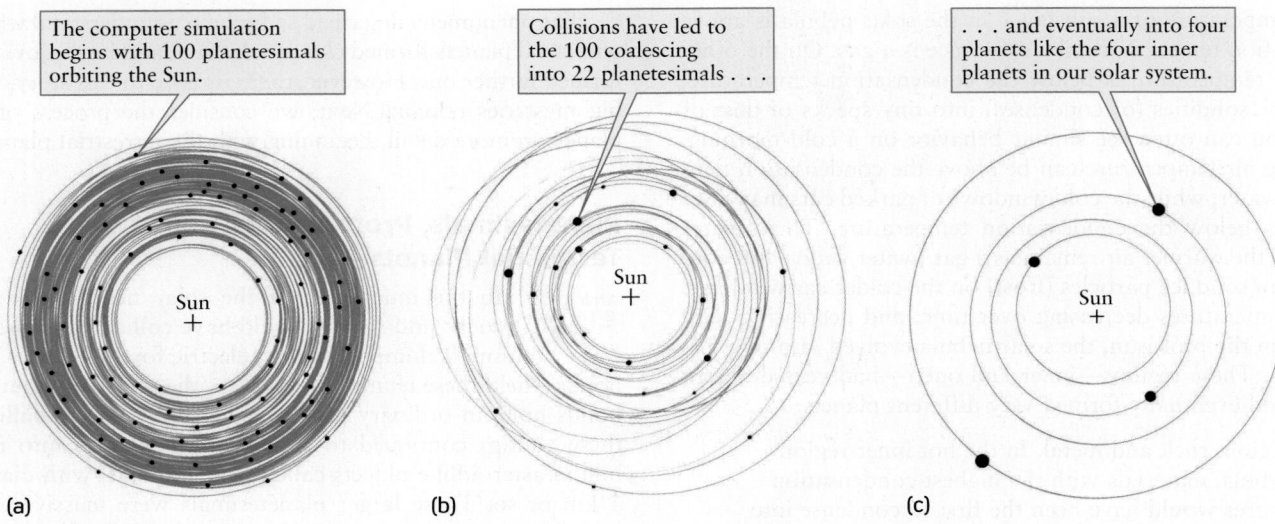

The computer simulation begins with 100 planetesimals orbiting the Sun.

Collisions have led to the 100 coalescing into 22 planetesimals . . .

. . . and eventually into four planets like the four inner planets in our solar system.

(a)   (b)   (c)

FIGURE 8-11

Accretion of the Inner Planets This computer simulation shows the formation of the inner planets over time. (Data from George W. Wetherill)

FIGURE 8-12 R I V U X G

Primitive Meteorite This is a cross-section from a fragment of the Allende meteorite that landed in Chihuahua, Mexico in 1969. A fireball was observed at 1:05 A.M. local time, and over the next 25 years, about 3 tons of meteorite fragments were collected over an 8 km by 50 km (5 mi by 31 mi) area. Radioactive dating yields an age of 4.57 billion years, corresponding to the very early solar system. The meteorite contains rocks and metals that are consistent with the inner solar system and is thought to come from a smashed-up planetesimal. In 2009, astronomers where shocked to discover that the meteorite is also magnetized in a manner that suggests the original planetesimal had a chemically differentiated iron core with a magnetic dynamo (similar to the source of Earth's magnetic field). (Matteo Chinellato/Shutterstock)

CONCEPT CHECK 8-7

How might the solar system be different if the location at which water could freeze in the solar nebula was much more distant from the Sun than it was in the solar nebula that formed our solar system?

CONCEPT CHECK 8-8

How does a rocky planet develop a core that has a higher density than rocks on its surface?

Answers appear at the end of the chapter.

8-6 Gases in the outer solar nebula formed the jovian planets, and planetary migration reshaped the solar system

We have seen how the low abundance of solid material in the inner solar nebula led to the formation of the small, rocky terrestrial planets. To explain the very different properties of the jovian planets, we need to consider the conditions that prevailed in the relatively cool outer regions of the solar nebula.

> Low temperatures in the outer solar nebula made it possible for planets to grow to titanic size.

The Core Accretion Model

The large jovian planets *initially* formed through a similar process as the terrestrial planets—through the accretion of planetesimals. As discussed in Section 8-5, a key difference is that ices, which are solid—in addition to solid rock and metal grains—were able to form in the cooler outer regions of the solar nebula (see Figure 8-10). *Crucially, the elements of which ices are made are much more abundant than those that form rocky grains.* Thus, more solid material would have been available to form planetesimals in the outer solar nebula than in the inner part. As a result, solid objects larger than any of the terrestrial planets could have formed in the outer solar nebula. Each such object could have become the core of a jovian planet, a "seed" around which the rest

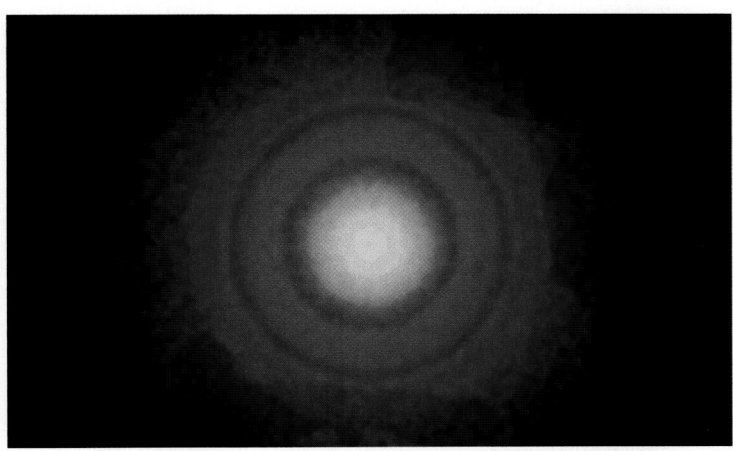

Formation of Planets in a Protoplanetary Disk About 175 light-years away, the protostar TW Hydrae is only about 10 million years old and obtains its energy through gravitational contraction. The protoplanetary disk has several dark gaps due to the formation of jovianlike planets that have accreted material from their orbital region. The two prominent dark rings are at distances similar to the orbits of Uranus and Pluto around our Sun. In additional analysis, astronomers even see evidence for a planet forming around the star at a similar distance as Earth is from the Sun. (S. Andrews, Harvard-Smithsonian CfA; B. Saxton, NRAO/AUI/NSF; ALMA, ESO/NAOJ/NRAO)

of the planet eventually grew. For example, the mass of Jupiter's rock and metal core is estimated to equal about 10 Earth masses. While the addition of icy material beyond the snow line played a role in forming the larger cores of jovian planets, astronomers do not fully understand how their cores get as large as they do, and this is an active area of research.

For Jupiter, a large seed mass of rock, metal, and ice is only the beginning: Most of Jupiter's mass is hydrogen and helium. Recall that for a planet, retaining a gaseous atmosphere depends on both the planet's mass and on the gas temperature (see Box 7-2). Accordingly, due to the lower temperatures in the outer solar system, Jupiter's large seed mass could capture and retain hydrogen and helium gas. This picture—where a jovian protoplanet core captures gas and grows by accretion—is called the **core accretion model**. In Figure 8-13, evidence of the formation of large jovianlike planets can be seen in the gaps of the protoplanetary disk as matter is accreted.

As Jupiter grew, its gravitational pull increased, allowing it to capture more gases and grow even larger (further enhancing its gravitational pull in a feedback cycle) until most of the available gas in its region had been captured. You can see this actually taking place in a disk around another star in the chapter-opening figure (right figure, page 209); the two dark gaps in the disk are due to jovianlike planets capturing gas as they grow. In our solar system, because hydrogen and helium were so abundant (98% of the solar nebula), Jupiter quickly grew to more than 300 Earth masses. You can see Jupiter forming before the other planets in Figure 8-14, which summarizes the formation of the solar system. Farther out in the solar nebula, Saturn would have gone through a similar process. About one-third the mass of Jupiter, Saturn's 95 Earth masses would have taken longer to accumulate. Uranus and Neptune formed well beyond the snow line, where temperatures were cold enough for additional ices of carbon dioxide, methane, and ammonia to form the bulk of these planets.

Like the protosun, each jovian planet would have been surrounded by a disk—a solar nebula in miniature (see Figure 8-6). The large moons of the jovian planets, including those shown in Table 7-2, are thought to have formed from ice particles and dust grains within these disks. Furthermore, since the moons formed from a rotating disk, these large satellites all orbit in the same direction. However, the jovian planets also have smaller bodies—called irregular satellites—that orbit in the *opposite* direction, and these were probably captured after the planets formed.

Observations of protoplanetary disks around other protostars (such as those in Figure 8-8b) suggest that high-energy photons and a solar wind (mostly made of electrons and protons) would eventually disperse any remaining hydrogen and helium gas, and this gas ends up in the space between the stars. With hydrogen and helium gone from the disk, formation of the jovian planets came to an end after a few million years. However, the accumulation of planetesimals that built the terrestrial planets did not stop when gas was expelled from the protoplanetary disk, and the terrestrial planets took longer to form than the jovian planets. It is thought that Earth formed within a few tens of millions of years, whereas Mars, being smaller, was formed in less than 10 million years.

Early Migration of Jovian Planets Shapes the Inner Solar System

We now look at **migration,** which refers to changes in orbital distances of the planets. Migration of the planets results from interactions—gravitational tugs and pulls—between planets and different parts of the forming solar system. The details of planetary migration are far from certain, but the broad picture that follows is emerging from computer simulations.

Within the first few hundred thousand years, long before the terrestrial planets formed, Jupiter's interaction with the gaseous disk of hydrogen and helium led to an inward migration of the growing giant, followed by an outward migration. At its closest distance to the protosun, Jupiter migrated inward to about 1.5 au, which is near the current orbit of Mars. With its large mass, Jupiter gravitationally deflected many of the planetesimals near the current Martian orbit. As a result, when Mars eventually formed in this region, it ended up with a low mass of about one-tenth of Earth's mass. In fact, it was the mysteriously low mass of Mars that prompted this analysis of Jupiter's inward migration. Moreover, inward-then-outward migration of all the jovian planets—called the **Grand Tack model**—also solves a long-standing mystery involving the asteroid belt.

The asteroid belt (Section 7-5) contains rocky objects typical of the inner solar system, but, surprisingly, also contains icy objects expected to have formed well beyond the asteroid belt where ices can form. In the Grand Tack model, as both Jupiter and the other jovian planets migrate outward, they deflect planetesimals inward to form the asteroid belt. Some of these planetesimals come from the inner solar system, but some also come from much farther out beyond the snow line, providing a very natural explanation for the icy objects in the asteroid belt.

This early migration, along with jovian planet formation, ended within the solar system's first few million years or so. Another type of migration may have occurred over the next few hundred million years and reshaped the outer solar system.

Vast, rotating cloud of cool gas and dust begins to collapse.

Protosun forms within a solar nebula and begins to grow.

Increased density, temperature, rotation rate

Decreased overall size

About a million years

Dusty debris

Snow line

Dust and ice debris

Protosun

Orbiting gas and dust forms a disk. Heat from protosun prevents ices from forming inside the snow line.

About a million years

Collisions between dust particles create larger-size debris and planetesimals. Jupiter quickly accretes gas and grows very large. The other jovian planets form soon after.

Protosun

Snow line

Jupiter forms

Planetesimals and protoplanets

Inner planets

Snow line

About a million years

Saturn, Neptune, and Uranus are nearly finished and formed closer than they are today. Icy material beyond the snow line helps to make these planets large. Terrestrial planets are still building, taking tens of millions of years to reach their present size.

Oort cloud

Jupiter

Saturn

Uranus   Neptune

Dust, ice, and small particles

Hundreds of millions of years

Kuiper belt

Asteroid belt

The jovian planets migrate outward through gravitational interactions with planetesimals. Jupiter continues to fling objects out to form the Oort cloud. Neptune flings objects out to form the Kuiper belt.

Jupiter

Saturn

Uranus

Neptune

ANIMATION 8.9

FIGURE 8-14

The Formation of the Solar System This sequence of drawings shows stages in the formation of the solar system.

Late Migration of Jovian Planets Reshapes the Outer Solar System

Astronomers have long suspected that some planetary migration must have taken place in the outer solar system. To see why, consider Neptune. At Neptune's current large orbital distance (about 30 au), the timescale necessary to build a planet of Neptune's large size by core accretion is much longer than the timeframe during which the protoplanetary disk was around. However, planets can grow much faster if closer to the protosun, where the protoplanetary disk is denser. Therefore, astronomers suspect that Neptune formed closer in, and then migrated outward.

The leading theory described here for late migration of the jovian planets is based on the **Nice** (pronounced "niece") **model**. This model was developed in Nice, France, around 2005 using detailed computer simulations. It is important to emphasize that the Nice model is a work in progress and has already evolved with new ideas and better simulations.

In the Nice model, the jovian planets would have all originally formed within about 20 au of the protosun, even though the outermost planet today, Neptune, is at 30 au. Furthermore, the outermost planet during this early time could have been Uranus, at 20 au, since gravitational interactions could have switched the orbital ordering of Neptune and Uranus. In the Nice model, both initial scenarios can lead to the arrangement of our present-day solar system.

Initially, there was also a disk of planetesimals beyond the outermost jovian planet at 20 au, and through gravitational encounters, these planetesimals, on average, were scattered inward. When a big planet knocks a small planetesimal inward, the planet itself is kicked slightly outward. Over a few hundred million years, as numerous planetesimals were knocked inward by Saturn, Neptune, and Uranus, these jovian planets slowly migrated outward to their current locations.

Most of the inward-moving planetesimals even made it close to Jupiter. With its much greater mass, Jupiter did not migrate inward very much, but did gravitationally fling most of these planetesimals clear out of the solar system. However, a small fraction of these icy objects would not have made it all the way out of the solar system and are thought to currently orbit at about 50,000 au from the Sun—about 0.8 ly away and one-fifth the distance to the nearest star. These planetesimals would be very loosely bound to our Sun, and gravitational deflections from passing stars would spread many of their orbits into a spherical "halo." We call this hypothesized distribution of icy planetesimals the Oort cloud (Figure 8-15a), which lies beyond the objects of our next topic—the Kuiper belt.

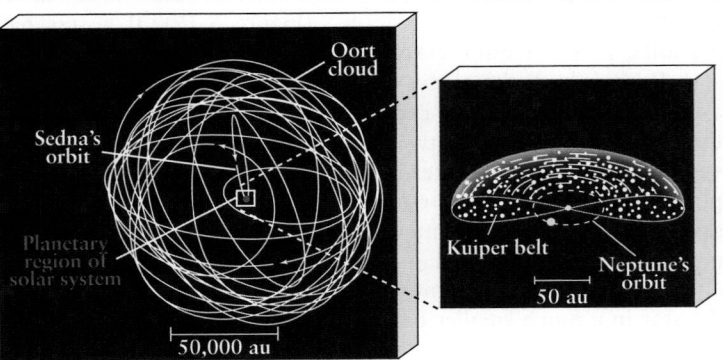

(a) Orbits of some Oort and Kuiper belt objects

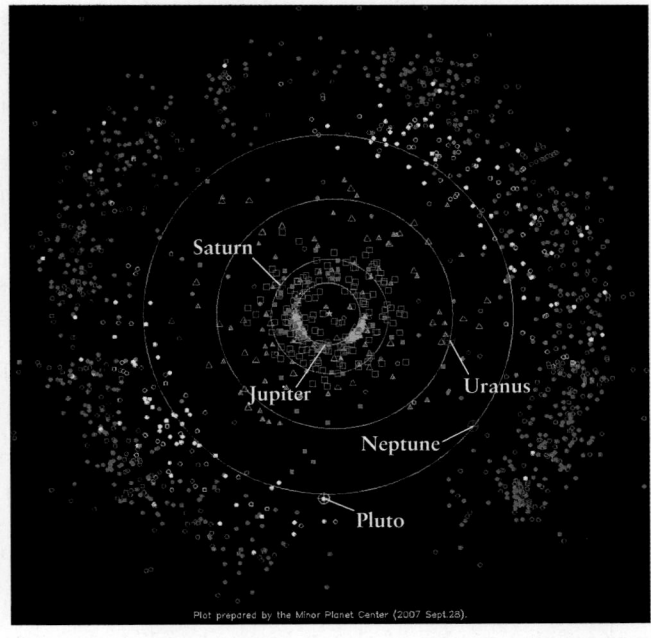

(b)

ANIMATION 8.4

FIGURE 8-15

The Kuiper Belt and Oort Cloud **(a)** The classical Kuiper belt of comets spreads from Neptune out to 50 au from the Sun. Most of the estimated 200 million belt comets are believed to orbit in or near the plane of the ecliptic. The spherical Oort cloud extends from beyond the Kuiper belt. Note from the scale bars that the Oort cloud extends about 1000 times farther out than the Kuiper belt. **(b)** Orbits of identified bodies in the Kuiper belt. (b: Minor Planet Center)

Unstable Orbits, the Kuiper Belt, and the Late Heavy Bombardment

During the slow outward migration of the jovian planets, their orbits remained mostly circular; in other words, they had very low eccentricity. The slowly expanding jovian orbits were also quite stable and allowed the migration to continue for about 600 million years or so. However, gravitational interactions with the planetesimals eventually destabilized planetary orbits, leading to elongated, or eccentric, orbits.

With elongated orbits, the planets gravitationally interacted with each other more strongly. In many computer simulations, this nearly *doubles* the orbital distance of Neptune, sending Neptune out to its current distance of about 30 au. As Neptune moved outward, its gravity flung nearby planetesimals to greater distances as well, creating an orbiting collection of icy objects called the Kuiper belt (Figure 8-15b). (The images that open this chapter show two young stars surrounded by dusty disks that resemble the Kuiper belt, one seen edge on and the other seen face on.) Taken together, the Oort cloud and Kuiper belt contain icy planetesimals that could not have formed at these great distances where matter was sparse, but instead were gravitationally deflected outward by Jupiter and then by Neptune.

The Nice model seems to solve another mystery of the solar system. Many astronomers think there was a cataclysmic period when the planets and moons of the inner solar system were subjected to a short but intense period of large impacts; this hypothesis is referred to as the **Late Heavy Bombardment.** The evidence comes from radioactive dating of craters on the Moon, but more lunar samples are needed to firmly establish a spike in the timing of impact events. This bombardment is considered "late" because the frequent planetesimal collisions of planet-building had long since passed.

So, what could cause a big jump in impacts? There are several ideas, but here is the most favored scenario. In the Nice model, the destabilized and elongated planetary orbits also led to deflections of planetesimals throughout the inner solar system. Very significantly, this widespread scattering of the planetesimals takes about 600 million years, which is consistent with the surprisingly late timing of the Late Heavy Bombardment.

ANIMATION 8.3 ➤ To a *much* lesser extent, gravitational deflections continue today. Planetesimals from either the Kuiper belt or the Oort cloud are occasionally knocked into new orbits. If one of these icy objects is deflected on a path through the inner solar system, it begins to evaporate, producing a visible tail, and is seen as a comet (see Figure 7-9).

Prior to 1995, the only fully formed planetary system to which we could apply this model was our own. As we will see in the next section, astronomers can now further test this model on an ever-growing number of planets known to orbit other stars.

CONCEPT CHECK 8-9

Why is it unlikely that Neptune formed by core accretion at its current location?

CONCEPT CHECK 8-10

In the Nice model, how did Neptune get to its present location?

CONCEPT CHECK 8-11

Does accretion refer to the accumulation of matter by gravitational attraction or the formation of chemical bonds?

Answers appear at the end of the chapter.

8-7 Exoplanets: The search for planets around other stars

If planets formed around our Sun, have they formed around other stars? The answer is yes, and they are called **extrasolar planets,** or **exoplanets.** The discovery of exoplanets brings up several questions: What are the properties of these exoplanets—are they similar to the planets of our solar system? What do they tell us about the formation of other solar systems? And perhaps the biggest question of all: Are there other Earthlike planets suitable for life, or that perhaps already contain alien life?

> Many planets have been discovered around other stars, and some are Earth size.

Since exoplanets were first detected in the 1990s, this field of astronomy has exploded with new discoveries. At the time of this writing, over 3700 exoplanets have been discovered—each one a new world! Some of these exoplanets are not alone: Over 600 stars have been discovered with multiple planets. There are even exoplanets orbiting pairs of stars (or binary star systems). Some exoplanets are larger than Jupiter, but orbit their stars even closer than Mercury orbits our Sun. Other planets are Earth sized and might have conditions suitable for life. To appreciate how remarkable these discoveries are, we must look at the methods used to find exoplanets.

Methods to Search for Exoplanets

It is very difficult to make direct observations of planets orbiting other stars. The problem is that planets are small and dim compared with stars; for example, the Sun is a billion times brighter than Jupiter at visible wavelengths. Exoplanets can sometimes be imaged directly, as in Figure 8-16a, but a star's glare makes this method the least productive means of detection. By the time the exoplanets in Figure 8-16b were imaged in 2010, hundreds of others had been detected through other means.

One very powerful method of exoplanet detection is to search for stars that appear to "wobble" (Figure 8-17). If a star has a planet, it is not quite correct to say that the planet orbits the star. Rather, both the planet and the star move in elliptical orbits around a point called the **center of mass** (Figure 8-17a). Imagine the planet and the star as two objects sitting at opposite ends of a very long seesaw; the center of mass is where you would have to place the support point (or fulcrum) in order to balance the seesaw. Because of the star's much greater mass, the center of mass is much closer to the star than to the planet. Thus, both the planet and the star orbit this point—the planet in a very large orbit and the star in a much smaller orbit. The small orbit of the star is the star's wobble.

FIGURE 8-16 R I V U X G

Imaging Extrasolar Planets (a) Four planets can be seen orbiting the star HR 8799. The star is about 129 ly from Earth and can be seen with the naked eye on a clear night. In capturing this image, special techniques make the star less visible to reduce its glare and reveal its planets. The white arrows indicate possible trajectories that each planet might take over the next 10 years. **(b)** About 170 ly away, the star 2M1207 and a planet with about 1.5 times the diameter of Jupiter are captured in this infrared image. First observed in 2004, this extrasolar planet was the first to be visible in a telescopic image. At infrared wavelengths, a star outshines a Jupiter-sized planet by only about 100 to 1, compared to 10^9 to 1 at visible wavelengths, making imaging possible in the infrared. (a: NASA/NRC-HIA, Christian Marois, Keck Observatory/NASA; b: ESO/VLT/NACO)

As an example, the Sun and Jupiter both orbit their common center of mass with an orbital period of 11.86 years. Because Jupiter has more mass than the other seven planets put together, it is a reasonable approximation to assume that the Sun's wobble is due to Jupiter alone. The Sun's orbit around the common center of mass has a semimajor axis only slightly greater than the Sun's radius, so the Sun slowly orbits, or wobbles, around a point not far outside its surface. If astronomers elsewhere in the galaxy could detect the Sun's wobbling motion, they would be able to tell that there is a large planet (Jupiter) orbiting the Sun. As we'll see, they

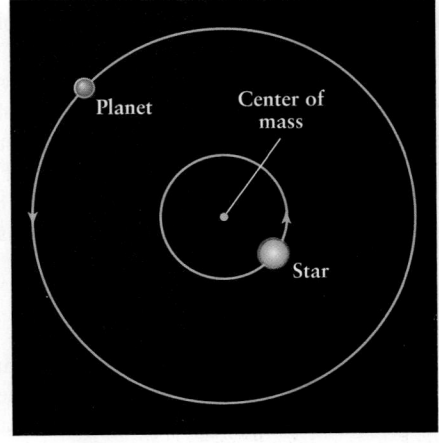

(a) A planet and its star

(b) The astrometric method

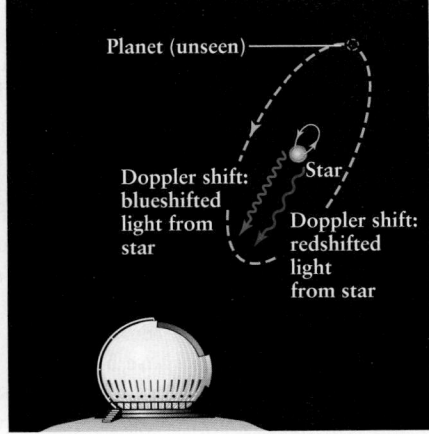

(c) The radial velocity method

FIGURE 8-17

Detecting a Planet by Measuring Its Parent Star's Motion (a) A planet and its star both orbit around their common center of mass, always staying on *opposite sides* of this point. Even if the planet cannot be seen, its presence can be inferred if the star's motion can be detected. **(b)** The astrometric method of detecting the unseen planet involves making direct measurements of the star's orbital motion during its orbit. **(c)** In the radial velocity method, astronomers measure the Doppler shift of the star's spectrum as it moves alternately toward and away from Earth. Analyzing the Doppler shift of the star's light leads to the orbital period, orbital size (semimajor axis), and mass of the unseen planet. The radial velocity method works for circular and elliptical orbits and can even determine the orbital eccentricity.

could even determine the planet's mass and the size of its orbit, without seeing the planet itself!

Detection Based on Changes in Position

Detecting the wobble of other stars is not an easy task. One approach to the problem, called the **astrometric method**, involves making very precise measurements of a star's position in the sky relative to other stars. The goal is to find stars whose positions change in a cyclic way (Figure 8-17b). The measurements must be made with very high accuracy and over a long enough time to span at least one entire orbital period of the star's motion.

While very few exoplanets have been discovered by the astrometric method so far, the most extensive search for extrasolar planets will be carried out with this method by the European Space Agency (ESA)'s Gaia telescope, which was launched in 2013. Gaia will survey a billion stars—1% of all the stars in the Milky Way Galaxy!—and will be able to detect tiny stellar wobbles. To detect these wobbles, Gaia measures changes in a star's position as small as 2.4×10^{-5} arcsec, which is fine enough to see something move 1 cm on the Moon all the way from Earth!

Astronomers estimate that Gaia will find about 10 exoplanets a day, leading to about 15,000 discoveries over the life of its mission. At the time of this writing, the first batch of data has been released for astronomers to analyze, but the results are not yet available.

Initially, Gaia finds exoplanet candidates based on their changing positions. This is the side-to-side motion of the star shown in Figure 8-17b. Then, Gaia performs a second type of measurement to see whether the star also moves toward and away from us during its orbit. This brings us to our second method for detecting exoplanets.

Detection Based on the Doppler Effect

ANIMATION 8.5 A star's wobble can also be detected by the **radial velocity method** (Figure 8-17c). This method is based on the Doppler effect, which we discussed in Section 5-9. Unless its orbit is perfectly oriented to face Earth, a wobbling star will alternately move toward and away from us over repeated orbits. Only this motion along a radius or line between Earth and the star—in other words, the star's radial velocity—contributes to the Doppler shift. The star's wobble causes the wavelengths of its entire spectrum to shift in a periodic fashion, and this can be measured in the star's dark absorption lines (for example, the absorption lines in Figure 5-14). When the star is moving away from us, its spectrum will undergo a shift to longer wavelengths. When the star is approaching, there will be a shift of the spectrum to shorter wavelengths (see Figure 5-26).

These shifts in color are represented by the red and blue colors used for the starlight in Figure 8-17c. These wavelength shifts are very small because the star's motion around its orbit is quite slow. As an example, the Sun moves around its small orbit at only 12.5 m/s (45 km/h, or 28 mi/h)—slow enough that someone could beat the Sun on a fast bike ride! If the Sun were moving directly toward an observer at this speed, the hydrogen absorption line at a wavelength of 656 nm in the Sun's spectrum would shift by only about 1 part in 25 million. Detecting these tiny shifts requires extraordinarily careful measurements and painstaking data analysis.

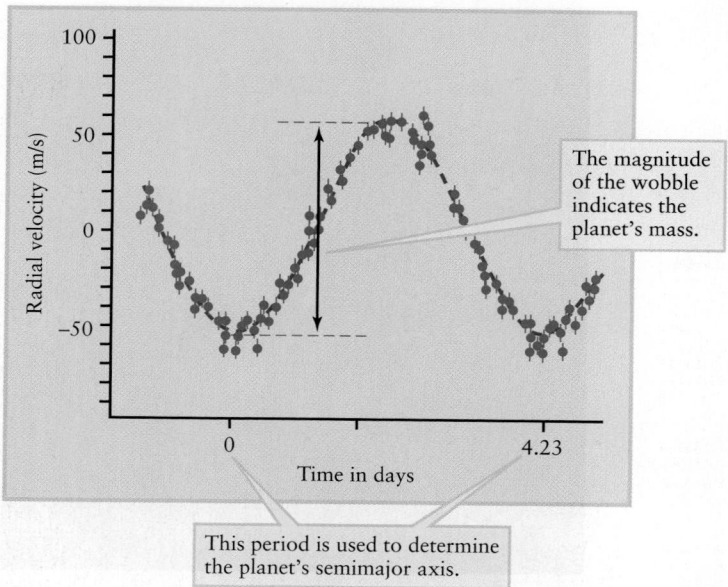

The magnitude of the wobble indicates the planet's mass.

This period is used to determine the planet's semimajor axis.

FIGURE 8-18

Discovering Planets from a Star's Wobble: 51 Pegasi As a star wobbles, it moves toward us and away from us, with a varying radial velocity that is measured through the Doppler shift. The shape of the curve indicates the planet's semimajor axis and also the planet's mass. However, the planetary mass is actually a minimum mass because an orbital plane tilted out of our line of sight would require an even larger mass to produce the observed Doppler shift.

In 1995, the radial velocity method was used to discover a planet orbiting the star 51 Pegasi, which is 47.9 ly away (see Figure 8-18). For the very first time, solid evidence had been found for a planet orbiting another star like our Sun. Since that groundbreaking discovery, techniques have advanced to the point that we can now measure stars that move *slower than we walk*. These improvements have led to the discovery of nearly 600 exoplanets using the radial velocity method.

The radial velocity method also has a significant drawback. Naturally, we are interested in discovering Earth-sized exoplanets at orbital distances similar to our own. However, to generate a large enough radial velocity to be detected, exoplanets need to be quite massive and must be **orbiting** close to their star (Figure 8-19). To detect exoplanets more similar to Earth, we need to use the method discussed next.

Detection Based on the Transit Method

ANIMATION 8.7 While the radial velocity method was the first technique used to find exoplanets, the **transit method** has been the most successful, both in finding the most planets and in determining many of their physical properties. Even better, the transit method can discover terrestrial Earth-sized planets far enough away from their star that temperatures might be just right for water to be stable in the liquid state.

ANIMATION 8.8 The transit method takes advantage of the rare situation in which a planet passes between us and its parent star, an event called a **transit**. Similar to a partial solar eclipse (Section 3-5), a planet passing in front of a star during

FIGURE 8-19

A Selection of Extrasolar Planets This figure summarizes what we know about planets orbiting a number of other stars. The star name is given at the left of each line. Each planet is shown at its average distance from its star (equal to the semimajor axis of its orbit), and some stars are found with multiple planets. The mass of each planet—actually a lower limit—is given as a multiple of Jupiter's mass (M_J), equal to 318 Earth masses. Comparison with our own solar system (at the top of the figure) shows how closely many of these extrasolar planets orbit their stars—many planets about the size of Jupiter orbit closer to their star than Mercury does to our Sun! Due to higher temperatures closer to the Sun, many of these exoplanets are called "hot Jupiters." However, the radial velocity method detects large close-in planets most easily, and if all planets could be detected, the majority would be smaller and farther away. (Data from the California and Carnegie Planet Search)

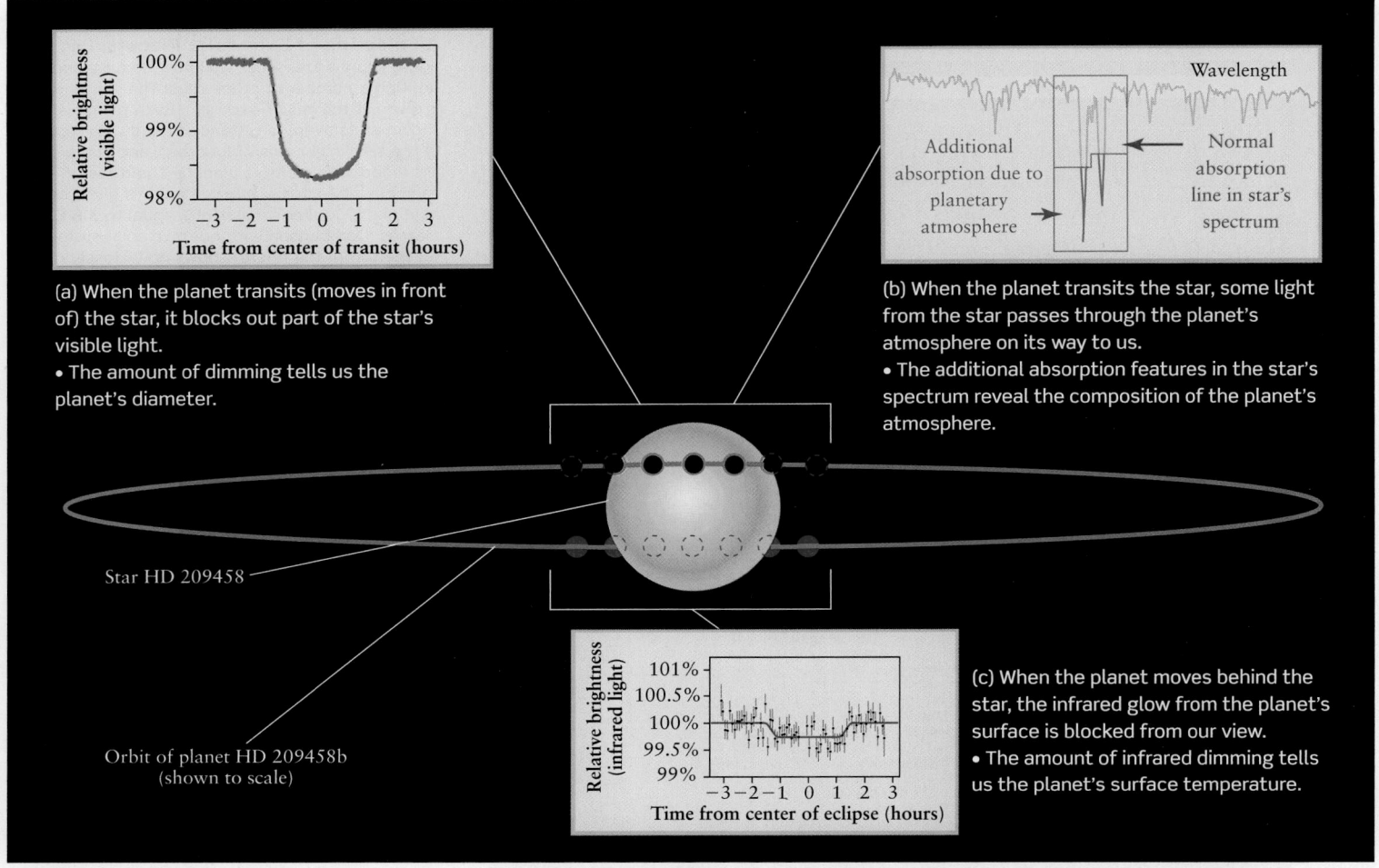

(a) When the planet transits (moves in front of) the star, it blocks out part of the star's visible light.
• The amount of dimming tells us the planet's diameter.

(b) When the planet transits the star, some light from the star passes through the planet's atmosphere on its way to us.
• The additional absorption features in the star's spectrum reveal the composition of the planet's atmosphere.

(c) When the planet moves behind the star, the infrared glow from the planet's surface is blocked from our view.
• The amount of infrared dimming tells us the planet's surface temperature.

Star HD 209458

Orbit of planet HD 209458b (shown to scale)

FIGURE 8-20

A Transiting Extrasolar Planet If the orbit of an extrasolar planet is nearly edge on to our line of sight, like the planet that orbits the star HD 209458, we can learn about the planet's **(a)** diameter, **(b)** atmospheric composition, and **(c)** surface temperature.

its orbit causes a small and repeating dimming of the star's light (Figure 8-20). For a transit to be seen, the orbit must be nearly edge on to our line of sight. While only 1% of exoplanets are expected to have this ideal orbital orientation, with over 200 billion stars in our galaxy, there are plenty of transits to observe.

To find Earth-sized exoplanets, and many other types as well, NASA launched the Kepler spacecraft in 2009 to search for transits among 145,000 stars. Figure 8-21 illustrates two Earth-sized exoplanets detected by Kepler; many more have been discovered. Kepler is so sensitive that it can detect transit-caused dips in brightness of 0.01%. Despite mechanical failures that altered the original mission plan, Kepler has found over 2330 confirmed exoplanets (as of this writing).

Furthering searches based on the transit method, NASA's TESS and ESA's CHEOPS are expected to launch in 2018. TESS is designed to monitor about 500,000 stars for transits, while CHEOPS will analyze known transits in detail to accurately determine each exoplanet's radius.

Determining Physical Properties of Exoplanets

In our own solar system, Jupiter is more than 28 times larger than Mercury, yet it is only one-quarter as dense. Venus has a scorching hot and toxic atmosphere, whereas Earth's temperature supports liquid water and it has an oxygen-rich atmosphere. To search for Earthlike planets—and to understand if our solar system represents a typical planetary system—we need to be able to determine the physical properties of exoplanets. This process begins by determining orbital properties such as period and distance.

• **Orbital period.** Most methods of detecting planets provide direct information about the orbital period. This is no coincidence! Repeating patterns in the brightness or motions of a star are good indications that a planet (or more than one) is there. For example, detections based on changes in position, or the wobble of the star, directly observe *how long it takes* for the star to go through one complete wobble. Since the star's wobble is produced by the gravitational tug of

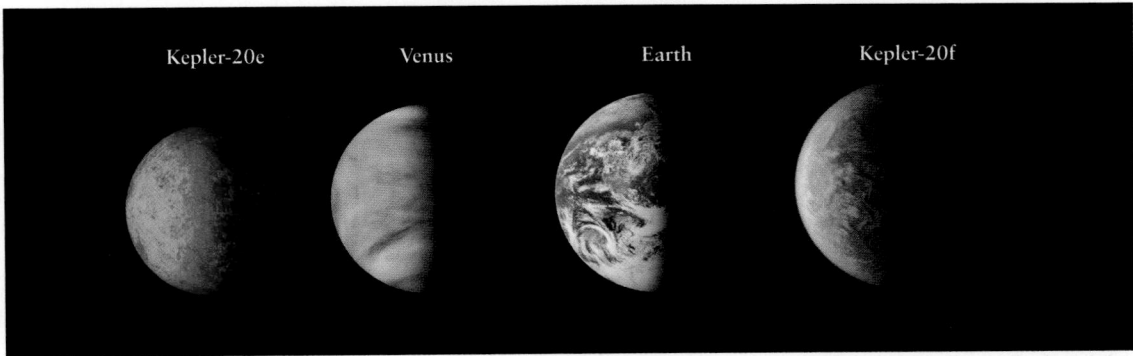

FIGURE 8-21

Earth-Sized Exoplanets Kepler-20 is a star with five known exoplanets. The exoplanets Kepler-20e and Kepler-20f are illustrated in comparison to Earth and Venus. Kepler-20e (at 0.05 au) has a radius just 0.87 times smaller than Earth, while Kepler-20f (at 0.1 au) is 1.03 times larger than Earth. However, these objects orbit extremely close to their star and are much too hot for liquid water to be present on their surfaces. (NASA/Ames/JPL-Caltech)

a planet, this observation gives a direct measurement of the planet's orbital period. Likewise, the radial velocity method produces graphs like Figure 8-18, which reveal the planet's orbital period—the time it takes for the star's wobble to repeat. In Figure 8-20a, you can see that the transit method would also reveal the planet's orbital period as the time between successive dips in the brightness of the host star.

- **Orbital distance.** Once a planet's orbital period p is known (from any of the methods above), the average orbital distance a can be found using Newton's form of Kepler's third law (from Section 4-7):

$$p^2 = \left[\frac{4\pi^2}{G(m_{star} + m_{planet})} \right] a^3$$

However, we need to consider the masses in the denominator. First, we can safely assume that the planet's mass is much smaller than the star's mass and simply ignore m_{planet}. Then, the star's mass can be determined from other methods covered later in Chapter 17 (the mass–luminosity relation). With this information, the equation can be solved for the average orbital distance a.

- **Planetary mass.** For a given orbital distance (as determined above), a larger planetary mass will produce a larger wobble of the star, because the wobble of the star is produced by the gravitational tugs from the planet. The radial velocity method measures these gravitational tugs in Doppler shifts of the star's spectrum. From these shifts, the star's velocity as it moves toward and away from us can be inferred, after which the equations that describe orbiting bodies can be solved to determine the unseen planet's mass. However, there's a subtlety here: *The Doppler shift only results from motion along your line of sight.* Depending on the planetary system's orientation with respect to Earth, we might be observing an orbital plane exactly edge on, nearly perpendicular, or somewhere in between—with no way of knowing the orientation direction from the Doppler shift method.

To understand the effects of this uncertainty, first consider a case where the unseen planet's orbital plane is perpendicular

to the line of sight to the star. Imagine the star and its planet as a distant clock, with the star at the center and the planet at the tip of the second-hand arm as it circles around. With the clock oriented face on in this way, the tug-induced wobble of the star at the center would remain in the plane of the clock, never directed toward or away from you, and thus would not produce any Doppler shift.

Conversely, if the clock were oriented edge on to your line of sight so that you couldn't even see the clock's face, the observed Doppler shift of the star would be the maximum that this star-planet system could produce. When we don't know the orientation of the unseen planet's orbit, we can do the calculations as if the orbital plane is edge on, but *that only gives us a lower limit to the planet's mass,* since even more mass would be required to produce a given Doppler shift if the plane of the wobble is tilted away from us. Fortunately, the thousands of planets discovered by the transit method all have to have edge on orbits for a transit to occur! For these planets, after doing follow-up measurements of their Doppler shifts, we have precise determinations of planetary masses.

- **Planetary size.** During a transit, the larger the planet, the more of the star's disk it covers and the greater the dip in starlight (see Figure 8-20a). To turn this effect into a numerical value, you first determine the size of the star using methods described later in Chapter 17. Then, the planet's radius can be determined by the size of the starlight dip during the transit.

- **Planetary density.** The average density of any object is its mass divided by its volume. A planet's volume is easily computed once you know its radius (from $V = \frac{4}{3}\pi r^3$), and methods for determining a planet's radius and mass have already been described. Knowing the density can quickly tell you if a planet is a low-density gas giant like our jovian planets or a dense rocky planet like Earth.

- **Planetary temperature.** Consider the orbit of the transiting planet in Figure 8-20c. When the planet is *behind* the star, any light emitted by the planet is blocked out. But what light from the planet would this be? Recall that every object with a temperature emits blackbody radiation, and for planets this is infrared radiation. By watching the infrared emission go up and down very slightly during the planet's orbit, we can isolate the portion that is coming from the planet and determine the planet's temperature. In the absence of this difficult measurement, we can still estimate a temperature for any planet simply based on the orbital distance to its star. This is quite accurate if the planet has no atmosphere, but atmospheric gases such as water or methane would lead to higher temperatures (as we will see in Chapter 9, these are called *greenhouse gases*.)

- **Planetary atmosphere.** Learning about a planet's atmosphere is very important in assessing its suitability for life, but also very difficult. Once again, we turn to the transit method, which offers the opportunity to collect a spectrum as illustrated in Figure 8-20b. While the body of the planet blocks the starlight behind it, a thin atmosphere, if present, adds absorption lines to the star's spectrum. The physical scenario is accurately described by Kirchhoff's third law at the top of Figure 5-17, where in this case the "Hot blackbody" is the star, and the "Cloud of cooler gas" is the thin atmosphere around the planet. The resulting absorption spectrum is a unique spectral fingerprint of the atoms and molecules (and their concentrations) that make up the atmosphere.

The planet that orbits the star HD 209458 in Figure 8-20 was the first transiting exoplanet to be discovered; it is a good example of the physical properties that can be determined for exoplanets. The mass of this planet is 0.69 times that of Jupiter, but its diameter is 1.32 times larger. Hence, this lower-mass, larger-volume planet has only one-quarter the density of Jupiter. Because this planet, called HD 209458b, orbits just 0.047 au from its star, its surface temperature is a torrid 1130 K (860°C, or 1570°F)—about twice the temperature needed to melt lead! Thanks to its large mass, extended radius, and close orbit, this planet is known as a **hot Jupiter.** Observations of the spectrum during a transit of this hot Jupiter reveal the presence of hydrogen, carbon, oxygen, sodium, and possibly water—all in a planetary atmosphere that is literally evaporating away in the intense starlight. Discoveries such as these provide insights into the exotic circumstances that can be found in other planetary systems.

What Have We Discovered?

What do we find in these other solar systems? The planet orbiting 51 Pegasi has a mass of at least 0.47 times that of Jupiter and orbits only 0.052 au away from its star, with a very short orbital period of just 4.23 days. Transferred to our own solar system, this planet would orbit about *7 times closer to the Sun than Mercury!* In fact, as shown in Figure 8-18, most of the extrasolar planets discovered by the radial velocity method have large masses (comparable with Jupiter) and also orbit very close to their stars. Why are there so many? This strange result is an example of a **selection effect,** where an observational technique is more likely to detect some systems and not others. For example, very massive planets that *also* have very close orbits tug on their stars to a greater degree, creating a larger Doppler shift. With the radial velocity method, it is hence

much easier to detect this type of extrasolar system—and that shows in the numbers. Overall, hot Jupiters form a minority of exoplanets, but, due to the selection effect of the radial velocity method, they were the most common exoplanet discovered until Kepler's discoveries using the transit method.

Hot Jupiters like HD 209458b and 51 Pegasi b orbit so close to their stars that the star's intense light must produce temperatures high enough to melt lead. These high temperatures also present a mystery. According to the picture of our own solar system's formation (Section 8-5), jovian planets would be expected to have formed relatively far from their stars, where low temperatures permit ices to form. The abundance of extra solid material available for planet building would then allow the protoplanet to grow quickly and attract a massive envelope of hydrogen and helium gas. As Figure 8-18 shows, many extrasolar planets are hot Jupiters orbiting very *close* to their stars, which contradicts this picture.

Another surprising result is that many of the extrasolar planets found so far have noncircular orbits with very large eccentricities (see Figure 4-10b). As an example, the planet around the star 16 Cygni B has an orbital eccentricity of 0.67; its distance from the star varies between 0.55 au and 2.79 au. This eccentricity is quite unlike planetary orbits in our own solar system, where no planet has an orbital eccentricity greater than 0.2. Recall that solar system formation according to the nebular hypothesis begins with matter mostly in circular motion. Therefore, to explain the highly eccentric exoplanet orbits requires some phenomena different than our understanding of our own solar system's formation.

Do these observations mean that our picture of how planets form is incorrect? If Jupiterlike extrasolar planets such as 51 Pegasi b formed close to their stars, their formation processes might have been very different from those that produced our solar system.

But another possibility is that extrasolar planets actually did form at large distances from their stars, just like our jovian planets, and later *migrated* inward after their formation. Recall that in the Grand Tack model of our solar system formation (Section 8-6), Jupiter migrates inward (although to a much lesser extent than the extreme migration implied by these hot Jupiters).

This mystery has fueled extensive computer modeling, which indicates that Jupiter-sized planets can indeed migrate to orbits very close to their stars. In simulations, this migration is due to interactions with the disk of gas and dust (formed from the solar nebula) in cases where the planet forms and migrates *before* the gas is dispersed (either blown away by intense radiation from the star or gathered up by forming planets).

ANIMATION 8.6 Simulations even suggest that planets can migrate *all the way* inward without stopping, ultimately getting swallowed by the central star. One piece of evidence that young planets may sometimes spiral into their parent stars is the spectrum of the star HD 82943, which has at least two planets orbiting it currently. The spectrum shows that this star's atmosphere contains a rare form of lithium that is found in planets but is very rare in stars, because it is destroyed by nuclear reactions within 30 million years. The presence of this exotic form of lithium, known as ^{6}Li, indicates that in the past HD 82943 was orbited by at least one other planet. It is thought that this planet was consumed by the star when it spiraled too close, leaving behind a ^{6}Li smoking gun in the star's atmosphere.

Are exoplanets intriguingly rare in a galaxy that contains over 200 billion stars? Not at all, as estimates indicate that over 70% of stars have at least one planet, with the most typical planet being about twice the size of Earth. When the selection effects of exoplanet detection are taken into account, we can calculate what the true average exoplanet system looks like in our galaxy. *Compared to our solar system, exoplanet systems tend to orbit closer to their stars and have higher eccentricities.*

One of the most exciting multiplanet systems discovered is TRAPPIST-1 (Figure 8-22a). This system contains *seven* Earth-sized exoplanets, more than any other discovered so far. The star is a cool red dwarf, is only 8% the mass of our Sun, and is just under half the temperature of the Sun. These planets would be very cold if they orbited at 1 au like our Earth. However, the largest orbit (that of TRAPPIST-h) is only 6% of Earth's orbital distance, and all seven planets orbit their star closer than Mercury orbits our Sun. Because the star is so faint, the farthest planet is likely cold enough to freeze any water on its surface, while any water on the closest planet would quickly boil away. However, models of possible conditions on three of the planets (e, f, and g) produce temperatures that allow for liquid water. Earth-sized planets that could have liquid water are of tremendous interest in our search for planets suitable for life, and we discuss them next.

The Search for Earth-Sized Planets in the Habitable Zone

The ultimate goal of planetary searches is to find planets with conditions suitable for life and, if possible, to search for alien life. Since we do not know what conditions would be required for some unknown form of alien life, we focus on two properties: liquid water and planetary size. We assume that liquid water significantly improves the chances that a planet could support life. Even the most exotic forms of life on Earth require liquid water, and many biochemists argue that an alien life-form based on some alternative biochemistry would probably still require liquid water. Fortunately, as you can see in Figure 8-4, hydrogen and oxygen are the first and third most abundant atoms in the universe. Since these atoms spontaneously bond together, water is fairly common in the universe. Most of the universe is too cold or too hot for these molecules to exist in the liquid state, so conditions need to be "just right."

A rocky planet in the so-called **habitable zone** of a stellar system might have liquid water on its surface (Figure 8-22b). In the habitable zone, a planet is close enough to its star—a direct source of heat—to melt water ice, yet not close enough to boil the water away. But the habitable zone is only a rough guide; planetary greenhouse gases and other conditions can warm up a cold planet

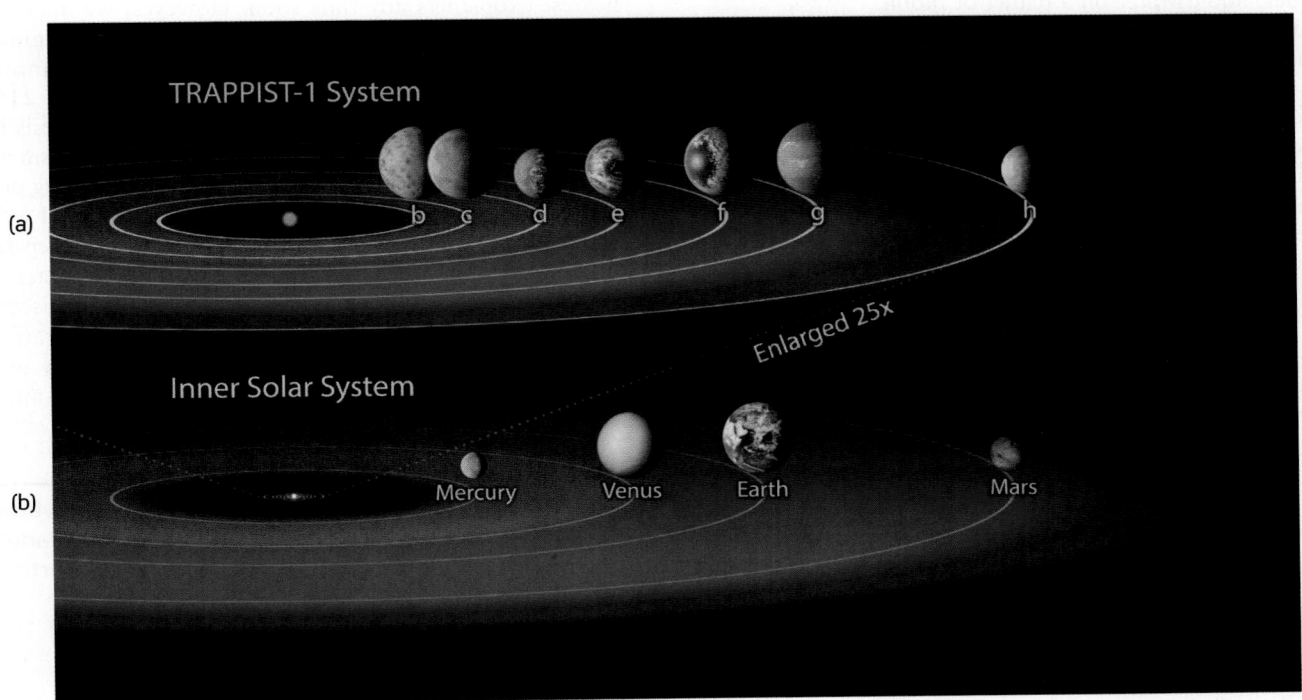

FIGURE 8-22

TRAPPIST-1 Exoplanets and the Habitable Zone This illustration compares the presently known properties of the star's seven exoplanets to our four terrestrial planets orbiting the Sun. **(a)** Illustrations of the seven planets are labeled b through h in order of the distance from their star, named TRAPPIST-1. In galactic terms, this star system is very close at only 40 ly away, making additional observations much easier. **(b)** Three of the exoplanets, TRAPPIST-1 e, f, and g, could be in the so-called habitable zone—shaded in green for both systems—where the planets' distance from their star and models of possible atmospheric greenhouse gases could result in temperatures compatible with liquid water, and thus, conditions possibly suitable for life. For TRAPPIST-1 f and g, even greenhouse gases are not likely to create enough warming for liquid water to be present on their surfaces. However, the same tidal forces that keep our Moon facing Earth could keep one side of these exoplanets facing their star to create enough warmth to allow liquid water. (NASA)

FIGURE 8-23

Potentially Habitable Exoplanets Each of
these exoplanets is in the habitable zone around its
star. These exoplanets are ranked by how similar they
might be to Earth, given by the number below them
in the square brackets, which is based on both the
size of the planet and the distance to its star. Thus,
Earth has a value of [1.00], and Proxima Cen b might
be similar to Earth with a value of [0.87]. These
values do not include the effects of a possible planet-
warming atmosphere that might be discovered
with additional observations, nor the effects of
atmosphere-depleting stellar winds for exoplanets
close to their stars. (PHL@UPR Arecibo [phl.upr.edu],
Nov 15, 2017)

or even (as in the case of Venus) warm it so much that any water
boils away. Furthermore, as in the case of Jupiter's Europa and
other moons (which are far beyond our habitable zone), tidal
forces can warm up an icy moon to support liquid water in a subsurface ocean. Therefore, the habitable zone neither guarantees,
nor excludes, liquid water on a planet or moon.

In addition to an ideal range of orbital distances, we also seek
planets that are rocky and roughly "Earth sized." As in the case of
Mars (discussed later in Chapter 11), small planets can lose their
heat too quickly, ultimately leading to a loss of their atmosphere
and the conditions required to maintain liquid water. Planets that
are too large also seem less likely to support life. For radii larger
than about 1.6 R_{Earth}, planets begin to retain thick atmospheres,
and, rather than having a distinct rocky surface, transition to have
a more gradual change in density with depth. Life is not ruled out
on these larger planets lacking a solid surface, but we exclude them
from lists of Earthlike planets. Taking a planet's size and distance
from its star into account, Figure 8-23 ranks the best candidates
for habitable planets at the time of this writing.

Within our galaxy, how many Earth-sized planets could we
expect to orbit in a habitable zone? This question was answered
only recently with the Kepler telescope. If we consider exoplanets
with a radius between 1 and 1.6 times Earth's, and only those
orbiting stars like our Sun, there should be a whopping 1 billion
exoplanets in habitable zones! If you include stars as cool as half
our Sun's temperature, there are about 14 billion Earth-sized
exoplanets in habitable zones. *Roughly speaking, about one in*

*five stars in our galaxy should have an Earth-sized exoplanet in a
habitable zone!*

With so many stars in habitable zones, the search for life is in
full swing. Unfortunately, even with recent technological advances,
we do not expect to be able to *directly* observe life on even the
nearest exoplanets any time soon. However, we might observe a
signature byproduct of some forms of life. For example, oxygen
is considered a "biosignature" if found in large quantities within
an atmosphere. Here on Earth, oxygen makes up 21% of our
atmosphere and mostly originated from photosynthesis in oceanic
microbes. Other than life, there is no known mechanism to produce
that much atmospheric oxygen, so it would be a strong piece of evidence (though not *proof*) for life on another planet. If we get lucky
and there is a nearby exoplanet with a thick oxygen-rich atmosphere, we could detect it with the James Webb Space Telescope
(JWST), expected to launch in late 2018. While the discovery of an
oxygen-rich atmosphere may require the next generation of telescopes following JWST, humanity is entering a truly unique time
when we could find strong indications of life around another star!

CONCEPT CHECK 8-12

Why will the radial velocity method fail to discover an
extrasolar planet if the plane of the extrasolar planet's orbit
is oriented perpendicular to the line between Earth and the
wobbling star?

Answer appears at the end of the chapter.

KEY WORDS

Term preceded by an asterisk (*) is discussed in Box 8-1.

accretion, p. 219
astrometric method (for
   detecting extrasolar planets),
   p. 226
atomic number, p. 212

center of mass, p. 224
chemical differentiation,
   p. 219
condensation temperature,
   p. 218

conservation of angular
   momentum, p. 216
core accretion model, p. 221
exoplanet, p. 224
extrasolar planet, p. 224

Grand Tack model, p. 221
habitable zone, p. 231
*half-life, p. 214
hot Jupiter, p. 230
interstellar medium, p. 211

KEY IDEAS

The Nebular Hypothesis: The most successful model of the origin of the solar system is called the nebular hypothesis. According to this hypothesis, the solar system formed from a cloud of interstellar material called the solar nebula. This occurred 4.54 billion years ago (as determined by radioactive dating) (Section 8-4).

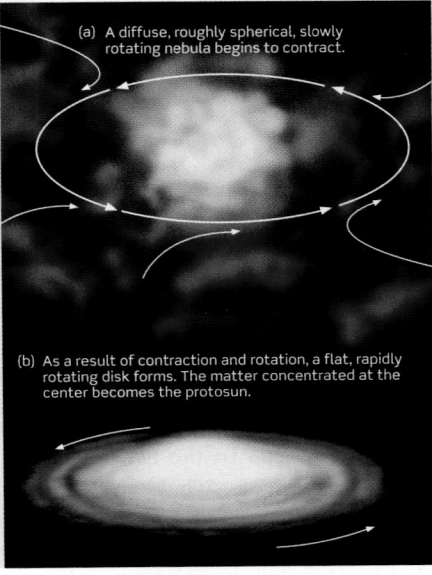

(a) A diffuse, roughly spherical, slowly rotating nebula begins to contract.

(b) As a result of contraction and rotation, a flat, rapidly rotating disk forms. The matter concentrated at the center becomes the protosun.

The Solar Nebula and Its Evolution: The chemical composition of the solar nebula, by mass, was 98% hydrogen and helium (elements that formed shortly after the beginning of the universe) and 2% heavier elements (produced much later in the centers of stars and cast into space when the stars died). The heavier elements were in the form of ice and dust particles (Section 8-4).

- The nebula flattened into a disk in which all the material orbited the center in the same direction, just as do the present-day planets.

Formation of the Planets and Sun: The terrestrial planets, the jovian planets, and the Sun followed different pathways to formation (Sections 8-5 and 8-6).

- The four terrestrial planets formed through the accretion of rock and metal dust grains into planetesimals, then into larger protoplanets.

- Closer to the Sun—inside the snow line—water could not freeze into ice to help build larger planets. With rocks and metals making up only a small fraction of mass throughout the nebula, the terrestrial planets made from these materials in the inner solar system are small.

- Jovian planets form beyond the snow line, where ices of water and methane contribute to make larger cores as these planets form. These larger cores can then attract copious amounts of hydrogen and helium gas, becoming large jovian planets.

- Gravitational interactions between the jovian planets and a gaseous disk, and much later with a disk of planetesimals, lead to planetary migration. These migrations can explain the small size of Mars, the composition of the asteroid belt, the locations of the jovian planets, aspects of the Kuiper belt and Oort cloud, and even account for a Late Heavy Bombardment.

- The Sun formed by gravitational contraction of the center of the nebula. After about 10^8 years, temperatures at the protosun's center became high enough to ignite nuclear reactions that convert hydrogen into helium, thus forming a true star.

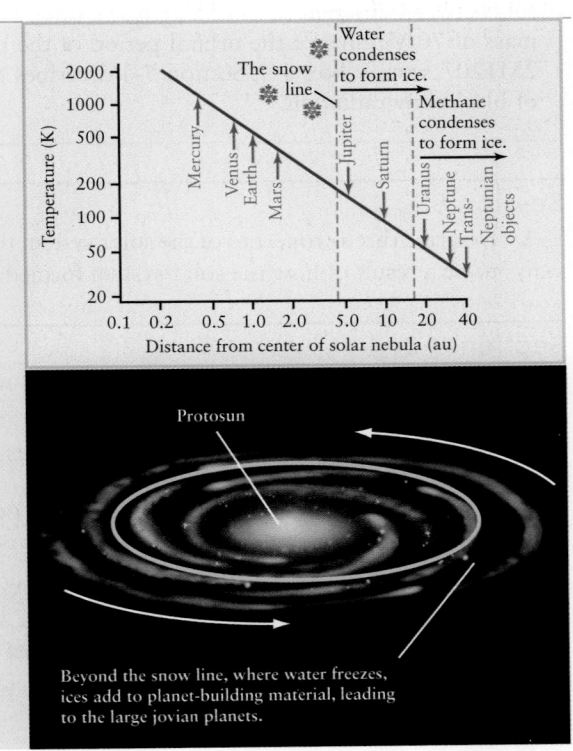

Beyond the snow line, where water freezes, ices add to planet-building material, leading to the large jovian planets.

Extrasolar Planets: Astronomers have discovered many planets orbiting other stars (Section 8-7).

- Many of these planets are detected by the "wobble" of the stars around which they orbit. The radial velocity method detects this wobble through Doppler shifts. The astrometric method directly observes a star's change in position as it wobbles.

- Many extrasolar planets have been discovered by the transit method, and a lesser number by direct imaging.

- Most of the extrasolar planets discovered to date are quite massive and have orbits that are very different from planets in our solar system. This is expected because big planets close to stars are easier to detect. We are just beginning to detect Earth-sized planets at distances that might allow liquid water on their surfaces.

NASA/NRC-HIA, Christian Marois, Keck Observatory and ESO/VLT/NACO

QUESTIONS

Problem-solving tips and tools

The volume of a disk of radius r and thickness t is $\pi r^2 t$. Box 1-1 explains the relationship between the angular size of an object and its actual size. An object moving at speed v for a time t travels a distance $d = vt$; Appendix 6 includes conversion factors between different units of length and time. To calculate the mass of 70 Virginis or the orbital period of the planet around 2M1207, review Box 4-4. Section 5-4 describes the properties of blackbody radiation.

Section 8-1

1. Describe three properties of the solar system that are thought to be a result of how the solar system formed.

Section 8-2

2. The graphite in your pencil is a form of carbon. Where were these carbon atoms formed?

3. What is the interstellar medium? How does it become enriched over time with heavy elements?

4. What is the evidence that other stars existed before our Sun was formed?

5. Figure 8-4 shows that carbon, nitrogen, and oxygen are among the most abundant elements (after hydrogen and helium). In our solar system, the atoms of these elements are found primarily in the molecules CH_4 (methane), NH_3 (ammonia), and H_2O (water). Explain why you suppose this is.

6. (a) If Earth had retained hydrogen and helium in the same proportion to the heavier elements that exist elsewhere in the universe, what would its mass be? Give your answer as a multiple of Earth's actual mass. Explain your reasoning. (b) How does your answer to (a) compare with the mass of Jupiter, which is 318 Earth masses? (c) Based on your answer to (b), would you expect Jupiter's rocky core to be larger, smaller, or the same size as Earth? Explain your reasoning.

Section 8-3

7. How do radioactive elements make it possible to determine the age of the solar system?

8. What are the oldest objects that have been found in the solar system?

9. Consider a 4.54×10^9 year old meteorite. Are the atoms in that meteorite—which were created in stars—4.54×10^9 years old? Explain your reasoning.

10. **Box 8-1** The half-life for uranium-238, discussed in Box 8-1, is 4.5 billion years. There is plenty of uranium-238 in Earth, and much of Earth's heat comes from the radioactive decay of this isotope. Compared to today, about how much more uranium-238 was there when Earth formed? (*Hint:* See Box 8-1.)

11. **Box 8-1** If you start with 0.80 kg of radioactive potassium (^{40}K), how much will remain after 1.3 billion years? After 2.6 billion years? After 3.9 billion years? How long would you have to wait until there was *no* ^{40}K remaining?

12. **Box 8-1** Three-quarters of the radioactive potassium (^{40}K) originally contained in a certain volcanic rock has decayed into argon (^{40}Ar). How long ago did this rock form?

Section 8-4

13. What is the tidal hypothesis? What aspect of the solar system was it designed to explain? Why was this hypothesis rejected?

14. What is the nebular hypothesis? Why is this hypothesis accepted?

15. What was the protosun? What caused it to shine? Into what did it evolve?

16. Why is it thought that the solar nebula developed into a rotating disk?

17. What are protoplanetary disks? What do they tell us about the plausibility of our model of the solar system's origin?

18. Suppose you were to use the Hubble Space Telescope to monitor one of the protoplanetary disks shown in Figure 8-8b. Over the course of 10 years, would you expect to see planets forming within the disk? Why or why not?

19. The protoplanetary disk at the upper right of Figure 8-8b is seen edge on. The diameter of the disk is about 700 au. (a) Make measurements on this image to determine the thickness of the disk in au. (b) Explain how this disk must have changed over time to get to this state.

20. The image accompanying Question 52 shows a dark, opaque disk of material surrounding the young star IRAS 04302+2247. The disk is edge on to our line of sight, so it appears as a dark band running vertically across this image. The material to the left and right of this band is still falling onto the disk. (a) Make measurements on this image to determine the diameter of the disk in au. Use the scale bar at the lower left of this image. (b) If the thickness of the disk is 50 au, find its volume in cubic meters. (c) The total mass of the disk is perhaps 2×10^{28} kg (0.01 of the mass of the Sun). How many atoms are in the disk? Assume that the disk is all hydrogen. A single hydrogen atom has a mass of 1.673×10^{-27} kg. (d) Find the number of atoms per cubic meter in the disk. Is the disk material thick or thin compared to the air that you breathe, which contains about 5.4×10^{25} atoms per cubic meter?

Section 8-5

21. What is meant by a substance's condensation temperature? What role did condensation temperatures play in the formation of the planets?

22. At distances within the snow line, what is the state of water (solid, liquid, or gas)? How does this affect the formation of terrestrial planets?

23. What is a planetesimal? How did planetesimals give rise to the terrestrial planets?

24. (a) What is meant by accretion? (b) Why are the terrestrial planets denser at their centers than at their surfaces?

Section 8-6

25. If hydrogen and helium account for 98% of the mass of all the atoms in the universe, why aren't Earth and the Moon composed primarily of these two gases?

26. Why did the terrestrial planets form close to the Sun while the jovian planets formed far from the Sun?

27. How did the jovian planets form?

28. In the Grand Tack model, why is Mars much smaller than Earth or Venus?

29. How does the Grand Tack model help us understand the composition of the asteroid belt?

30. Why do we think that Neptune must have formed closer to the Sun, and later migrated outward to its present position?

31. In the Nice model of the jovian planets, which planet migrates more, Jupiter or Neptune?

32. In the Nice model, how does the Oort cloud and the Kuiper belt form?

33. What is the Late Heavy Bombardment? How does the Nice model explain this event?

34. Explain why most of the satellites of Jupiter orbit that planet in the same direction that Jupiter rotates.

35. Propose an explanation for why the jovian planets are orbited by terrestrial-like satellites.

Section 8-7

36. What is the radial velocity method used to detect planets orbiting other stars? Why is it difficult to use this method to detect planets like Earth?

37. What type of planets and orbits are easiest to detect with the radial velocity method? What are "hot Jupiters"? Explain your answer.

38. Summarize the differences between the planets of our solar system and those found orbiting other stars.

39. Is there evidence that planets have fallen into their parent stars? Explain your answer.

40. What does it mean for a planet to transit a star? What can we learn from such events?

41. What combination of methods are required to determine the average density of an exoplanet? Explain your reasoning. (*Hint:* The average density is the planet's mass divided by its volume.)

42. What is so special about the habitable zone? Is every planet in this zone like Earth?

43. A 1999 news story about the discovery of three planets orbiting the star Upsilon Andromedae ("Ups And" is near the top of stars listed in Figure 8-19) stated that "the newly discovered galaxy, with three large planets orbiting a star known as Upsilon Andromedae, is 44 ly away from Earth." What is wrong with this statement?

44. The planet discovered orbiting the star 70 Virginis (70Vir is near the middle of the stars listed in Figure 8-19), 59 ly from Earth, moves in an orbit with semimajor axis 0.48 au and eccentricity

0.40. The period of the orbit is 116.7 days. Find the mass of 70 Virginis. Compare your answer with the mass of the Sun. (*Hint:* The planet has far less mass than the star.)

45. Because of the presence of Jupiter, the Sun moves in a small orbit of radius 742,000 km with a period of 11.86 years. (a) Calculate the Sun's orbital speed in meters per second. (b) An astronomer on a hypothetical planet orbiting the star Vega, 25 ly from the Sun, wants to use the astrometric method to search for planets orbiting the Sun. What would be the angular diameter of the Sun's orbit as seen by this alien astronomer? Would the Sun's motion be discernible if the alien astronomer could measure positions to an accuracy of 0.001 arcsec? (c) Repeat part (b), but now let the astronomer be located on a hypothetical planet in the Pleiades star cluster, 360 ly from the Sun. Would the Sun's motion be discernible to this astronomer?

46. (a) Figure 8-20c shows how astronomers determine that the planet of HD 209458 has a surface temperature of 1130 K. Treating the planet as a blackbody, calculate the wavelength at which it emits most strongly. (b) The star HD 209458 itself has a surface temperature of 6030 K. Calculate its wavelength of maximum emission, assuming it to be a blackbody. (c) If a high-resolution telescope were to be used in an attempt to record an image of the planet orbiting HD 209458, would it be better for the telescope to use visible or infrared light? Explain your reasoning.

47. (a) The star 2M1207 shown in Figure 8-16b is 170 light-years from Earth. Find the angular distance between this star and its planet as seen from Earth. Express your answer in arcseconds. (b) The mass of 2M1207 is 0.025 that of the Sun; the mass of the planet is very much smaller. Calculate the orbital period of the planet, assuming that the distance between the star and planet shown in Figure 8-16b is the semimajor axis of the orbit. Is it possible that an astronomer could observe a complete orbit in one lifetime?

48. Suppose that a planetary system is now forming around some protostar in the sky. In what ways might this planetary system turn out to be similar to or different from our own solar system? Explain your reasoning.

49. Suppose astronomers discovered a planetary system in which the planets orbit a star along randomly inclined orbits. How might a theory for the formation of that planetary system differ from that for our own?

General Problems

50. Why are terrestrial planets smaller than jovian planets?

51. Explain how our current understanding of the formation of the solar system can account for the following characteristics of the solar system: (a) All planetary orbits lie in nearly the same plane. (b) All planetary orbits are nearly circular. (c) The planets orbit the Sun in the same direction in which the Sun itself rotates.

52. The accompanying infrared image shows IRAS 04302+2247, a young star that is still surrounded by a disk of gas and dust. The scale bar at the lower left of the image shows that at the distance of IRAS 04302+2247, an angular size of 2 arcsec corresponds to a linear size of 280 au. Use this information to find the distance to IRAS 04302+2247.

(Courtesy of D. Padgett and W. Brandner, IPAC/Caltech; K. Stapelfeldt, JPL; and NASA)

2"
280 au

R I V U X G

WEB/E-BOOK QUESTIONS

53. Search the internet for information about recent observations of protoplanetary disks. What insights have astronomers gained from these observations? Is there any evidence that planets have formed within these disks?

54. In 2000, extrasolar planets with masses comparable to that of Saturn were first detected around the stars HD 16141 (also called 79 Ceti) and HD 46375. Search the internet for information about these "lightweight" planets. Do these planets move around their stars in the same kind of orbit as Saturn follows around the Sun? Why do you suppose this is? How does the discovery of these planets reinforce the model of planet formation described in this chapter?

55. In 2006, a planet called XO-1b was discovered using the transit method. Search the internet for information about this planet and how it was discovered. What unusual kind of telescope was used to make this discovery? Have other extrasolar planets been discovered using the same kind of telescope?

ACTIVITIES

Starry Night™ Explorations

56. Use *Starry Night*™ to make observations of the different components of the solar system. Select **Favorites > Explorations > Solar System**. The view shows the orbits of

the major planets of the solar system against the backdrop of the stars of the Milky Way Galaxy, from a location 64 au from the Sun.

(a) Use the location scroller to look at the solar system from different angles and observe the general distribution and motion of the major planets. Make a list of your observations.

(b) To examine the inner solar system in more detail, zoom in to fill the screen with Jupiter's orbit and reduce the time step to 1 day. The many smaller objects moving in the region between the orbits of Mars and Jupiter are the asteroids in the asteroid belt. Open the **Planets** panel and expand **Our Solar System > Main Belt Asteroids**. Click the radio button beside the names of a sample of asteroids in the list to show their orbits in the view. Use the location scroller to examine the shape and inclination of these orbits. How do these characteristics differ from those of the planets?

(c) Select **Favorites > Explorations > Solar System-dwarf planets** to show the locations and orbits of several of the outer dwarf planets such as Pluto. Use the location scroller to examine these orbits. Describe why you think these objects are not assigned to the planets category but rather to this separate dwarf planets grouping within our solar system.

(d) How does the nebular hypothesis of solar system formation account for your observations?

57. Use *Starry Night*™ to examine stars that have planets. Very recent observations from space and ground-based telescopes suggest that most stars in our galaxy probably have at least one planet. Select **Favorites > Explorations > Exoplanets**. The view shows the Kepler space telescope, a spacecraft dedicated to the search for these extrasolar planets by detecting the very small reduction in light that occurs when the planet passes between its star and the telescope, an event known as a transit. All the stars in the view marked with a light blue halo are known to have at least one orbiting planet. One such star, Tau Ceti, is labeled. Use the location scroller to look around and judge the extent of the growth in our knowledge over the past few years. You will note an intense concentration of stars with exoplanets in one area, not because there are more exoplanets in this region but because this is the selected observing area for the Kepler space telescope.

(a) Apparent magnitude uses an "inverse" scale: The greater the apparent magnitude, the dimmer the star. Most of the brighter stars you can see from Earth with the naked eye have apparent magnitudes between 0 and +1, whereas the faintest star you can see from a dark location has an apparent magnitude of about +6. Use the HUD to sample the apparent magnitudes of a number of the marked stars in the view. Are most of the circled stars in your sample visible from Earth with the naked eye? List

at least two stars that are visible to the unaided eye from Earth and that are known to have extrasolar planets.

(b) Select **File > Revert** to reset the view. Tau Ceti, a close neighbor to our Sun, is a star with known exoplanets. Use the HUD to find this star's apparent magnitude. Is it visible from Earth with the unaided eye?

(c) Careful measurements have allowed astronomers to determine the orbits of the known planets in the Tau Ceti system. Select **Favorites > Explorations > Tau Ceti** to see the orbits of the exoplanets in the Tau Ceti system. Move the cursor to the position indicated by the label for the central star, Tau Ceti, and use the HUD to obtain the distance of this star from the Sun, as well as its temperature for comparison with that of the Sun (surface temperature 5800 K). To compare the planets in this system to the planets orbiting the Sun, click the info icon in each exoplanet's label to open an info window from which you can compile a table of the following important planetary properties for each exoplanet orbiting Tau Ceti: radius, mass, semi-major axis and period of its orbit, and whether the planet is considered to be a mesoplanet—that is, potentially habitable. (For reference, Earth's radius is 6400 km, masses are in units of Earth mass, Earth's semi-major axis is 1 au, its period is 365.26 days, and Earth is habitable!) From your data, answer the following questions:

(i) How many of these planets are more massive than Earth?

(ii) How many Tau Ceti planets are larger than Earth?

(iii) How many of these planets have larger orbits than does Earth?

(iv) How many planets have a period longer than that of Earth?

(v) Although Tau Ceti has a similar temperature to that of the Sun, it has a lower luminosity, or total energy output. On this basis, would you expect the habitable zone around this star to be larger or smaller than that of the Sun?

(vi) On the basis of your answer in (v), do you consider the habitability designations for the Tau Ceti planets to be correct?

ANSWERS

ConceptChecks

ConceptCheck 8-1: No. Earth is not an exception to any of the three properties in Table 8-1, and in fact, all the planets are consistent with these properties.

ConceptCheck 8-2: The carbon atoms that make up much of our human bodies and the very oxygen atoms we breathe were made inside of stars. Stars have made most of the atoms other than hydrogen.

ConceptCheck 8-3: Yes. Hydrogen (H) is the first most abundant element, and oxygen (O) is the third most abundant element. Water (H_2O) is, in fact, quite common even though it is usually frozen.

ConceptCheck 8-4: No. The age of a rock refers to how much time has passed since the rock cooled and solidified. It is not the age of the rock's atoms, which were made in stars, and possibly different stars at different times.

ConceptCheck 8-5: To be correct, the nebular hypothesis must explain why all the planets orbit the Sun in the same direction and in nearly the same plane. This arrangement is unlikely to have arisen purely by chance.

ConceptCheck 8-6: No. Through Kelvin–Helmholtz contraction, gravitational energy is converted into heat—a lot of heat. The protosun, which formed before nuclear reactions began, had a surface temperature of 6000 K, slightly higher than our present-day Sun. Thus, the solar nebula was warmed by the protosun.

ConceptCheck 8-7: Frozen water helped to make Jupiter large enough to hold onto hydrogen and helium. The closest

that a large, Jupiterlike planet could form would be much farther away in a much hotter solar nebula.

ConceptCheck 8-8: Called chemical differentiation, denser, iron-rich materials migrate to the center of a planet while the less dense, silicon-rich minerals float to the outside surface before the initially molten planet solidifies in the early solar system.

ConceptCheck 8-9: At Neptune's present distance, the time to build a planet the size of Neptune is much longer than the time that the protoplanetary disk is around.

ConceptCheck 8-10: In the Nice model, once planetary orbits became unstable, Neptune was deflected to about twice its original orbital distance.

ConceptCheck 8-11: Accretion refers to the gravitational accumulation of matter. Chemical bonds might, or might not, be formed once matter has accumulated.

ConceptCheck 8-12: An extrasolar planet with an orbit that causes a star to wobble side to side will not exhibit any Doppler shifted spectra as seen from Earth because the star will not be moving alternately toward and away from Earth, and therefore the radial velocity method will fail to detect such an orbiting extrasolar planet.

Exoplanets

by Prof. Sara Seager

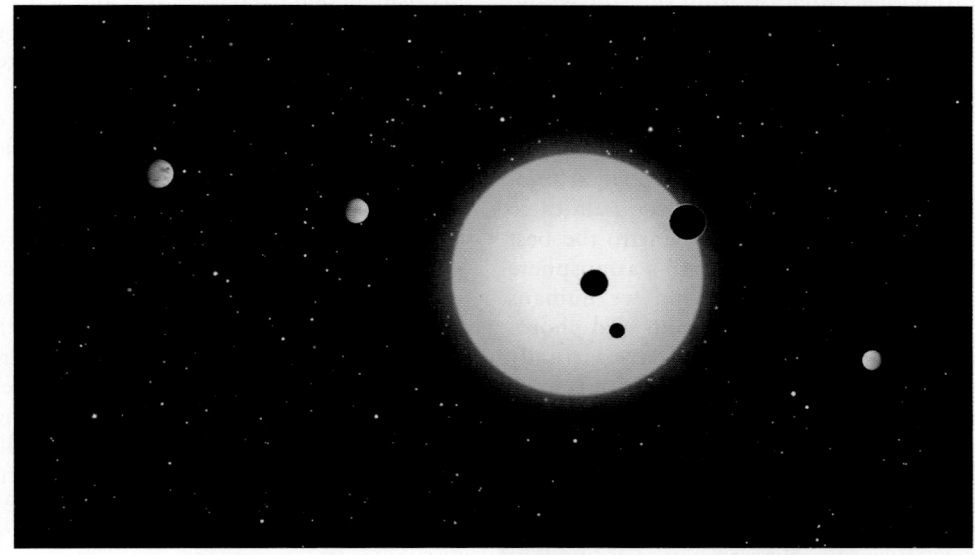

NASA/Ames/JPL-Caltech

For thousands of years, since the time of the Greek philosophers, humans have wondered, "Do other Earths exist?" "Are other Earths common?" "Do any have signs of life?" There are records of Italian philosophers in the 1600s, America's founding fathers in the 1700s, American school children in the 1800s who believed in other worlds and life forms on them. Today, Hollywood movies show the continuing human desire to wonder, "What is out there?" "Who is out there?"

To our huge surprise, evidence suggests that every star in the sky has planets. Moreover, planetary systems are incredibly diverse. Some stars have planets the size of Earth orbiting more than twenty times closer to their star than Mercury is to our Sun. Blasted by radiation, such hot Earths may have liquid lava lakes. A "circumbinary planet" orbits two suns—a planet on which there are two shadows and two sunsets. Some planetary systems consist of several planets tightly packed together—all orbiting interior to where Venus should be. The most common type of planet appears to be two to three times the size of Earth, not only with no solar system counterpart but one whose formation astronomers are struggling to explain. The range of exoplanet sizes, masses, and orbits highlights that the planet formation process is both more random and more robust than previously thought.

Despite the discovery of thousands of exoplanets, we know relatively little about most of them, beyond their sizes or masses and orbits. To learn more, we want to know what the planet is made of, including both the planet interior and atmosphere.

Driving the field of exoplanets is the key question "Do any exoplanets have the right conditions to support life?" The answer is, "We do not know—yet." The first step is the search for planets in the host star's habitable zone, the region around a star where a planet with a thin atmosphere heated by the star could have surface temperatures not too hot, not too cold, but just right for life. Because all life on Earth requires liquid water, a suitable surface temperature for life is synonymous with a suitable surface temperature for supporting liquid water.

While dozens of planets in their host star's habitable zone are known, there is no guarantee that any of them actually have a suitable surface temperature for liquid water. This is because of the unknown properties of the atmosphere—we do not know what sets the mass and composition of a planet atmosphere. Like a suffocating blanket, a massive atmosphere would trap heat, making the surface far warmer than a comparable planet with a thin atmosphere. For example, with current exoplanet discovery techniques, both Venus and Earth would appear to be the same size and mass, and both estimated to be in the habitable zone of the Sun. Yet, Venus is completely hostile to life, owing to a massive atmosphere with a high carbon dioxide content, leading to a strong greenhouse effect causing high surface temperatures (>700 K). In contrast, Earth has the right surface temperature for liquid water oceans and is teeming with life.

We would like to detect water vapor in a rocky exoplanet's atmosphere—directly indicating the presence of liquid water oceans. Without an ocean reservoir and a water cycle, water vapor molecules would diffuse to the upper

(continued on the next page)

Exoplanets *(continued)*

atmosphere. There, ultraviolet radiation from the star would split the water vapor into hydrogen and oxygen. Hydrogen, a very light gas, would escape to space with oxygen disappearing by reacting with the planet surface or other atmospheric molecules.

If we find a predominately rocky world with atmospheric water vapor, we will move on to our most lofty goal, a "biosignature gas"—a gas produced by life that accumulates to detectable levels in a planet atmosphere. On Earth, the best example of a biosignature gas is oxygen. Earth's atmosphere is filled to 20% by volume with oxygen, and we humans breathe in oxygen to survive. Yet, without plants and photosynthetic bacteria, Earth's atmosphere would have virtually no oxygen. In fact, oxygen is so reactive it should not exist in our atmosphere at all. There are so many more gases to

consider than oxygen; life on Earth produces thousands of different gases. But most of these gases are problematic because they are either also produced geologically (i.e., volcanoes) or are produced in too small of a quantity to register in an atmosphere as viewed from a great distance. Nonetheless, the research field of biosignature gases on exoplanets is a burgeoning field.

In the near term, to make the job of finding and identifying a habitable world easier, astronomers are focusing on small red dwarf stars where both the planet discovery and atmosphere follow-up have larger signals than planets orbiting Sunlike stars. The next generation telescopes, both the James Webb Space Telescope and the large 20 to 40 m mirror ground-based telescopes of the future, will be able to observe the atmospheres of rocky, habitable-zone exoplanets orbiting red dwarf stars.

Simultaneous to the fixation on small stars, we are investing in the longer-term search for a true Earth twin: an Earth-sized planet in the habitable zone of a Sunlike star. We are designing very sophisticated telescopes to operate in space, above the blurring effect of Earth's atmosphere. We will develop a coronagraph and/or a starshade (a giant specially shaped screen) to block out the starlight so we can directly image a planet billions of times fainter than its host star.

It is breathtaking to consider that, after thousands of years of wondering, we are finally poised to begin answering the ancient questions on the existence and properties of Earthlike exoplanets.

Sara Seager is an astrophysicist and planetary scientist at MIT. Her research focuses on theoretical models of atmospheres and interiors of exoplanets, as well as novel space science missions. She has introduced many new ideas to the field of exoplanet characterization, including work that led to the first detection of an exoplanet atmosphere. She was part of a team that co-discovered the first detection of light emitted from an exoplanet and the first spectrum of an exoplanet. The exceedingly surprising diversity of exoplanets has led to Seager's maxim, "For exoplanets, anything is possible under the laws of physics and chemistry."

9 The Living Earth

LEARNING GOALS

When astronauts made the first journeys from Earth to the Moon between 1968 and 1972, they often reported that our planet is the most beautiful sight visible from space. Of all the worlds of the solar system, only Earth has the distinctive green color of vegetation, and only Earth has liquid water on its surface. Compared with the Moon, whose lifeless, dry surface bears evidence of billions of years of impacts by interplanetary debris, Earth seems an inviting and tranquil place.

But the appearance of tranquility is deceiving. The seemingly solid surface of our planet is in a state of slow but constant motion, driven by the flow of hot, deformable rock within Earth's interior. Sunlight provides Earth's atmosphere with the energy that drives our planet's sometimes violent weather, including all the cloud patterns visible in the images shown here.

Unseen in these images are immense clouds of subatomic particles that drift around the planet in two giant belts, held in thrall by Earth's magnetic field. And equally unseen are trace gases in Earth's atmosphere, whose abundance may determine the future of our climate, and on which the survival of entire species may depend.

The two hemispheres of Earth. (NASA Goddard Space Flight Center. Image by Reto Stöckli)

R I V U X G

In this chapter, our goal is to learn about Earth's components—its dynamic oceans and atmosphere, its ever-changing surface, and its hot, active interior—and how they interact to make up our planetary home. By learning about Earth, we also learn about some of the same processes shaping the planets and moons that we have seen in Chapter 7 and will look at in more detail in Chapters 10 through 14. Farther from home, by understanding Earth's processes, we can better interpret data gathered on exoplanets and assess a distant world's prospects for habitability from light-years away.

9-1 Earth's atmosphere, oceans, and surface are extraordinarily active

The crew of an alien spacecraft exploring our solar system might overlook Earth altogether. Although it is the largest of the terrestrial planets, with a mass greater than that of Mercury, Venus, and Mars put together, Earth is far smaller than any of the giant jovian planets (see Table 7-1 and Table 9-1). But a more careful inspection would reveal that Earth is unique among all the planets that orbit the Sun.

Dynamic Earth: Water, Air, and Land

Unlike the arid surfaces of Venus and Mars, Earth's surface is very wet. Indeed, nearly 71% of Earth's surface is covered with water (Figure 9-1). Water is also locked up inside many of the minerals that make Earth's rocks, including those found in the driest deserts. Furthermore, Earth's liquid water is in constant motion. The oceans ebb and flow with the tide, streams and rivers flow down to the sea, and storms whip lake waters into a frenzy.

Earth's atmosphere is also in a state of perpetual activity. Winds blow at all altitudes, with speeds and directions that change from one hour to the next. The atmosphere is at its most dynamic when water evaporates to form clouds, then returns to the surface as rain or snow (Figure 9-2). While other worlds of the solar system have atmospheres, only Earth's atmosphere contains the oxygen that animals (including humans) need to live.

The combined effects of water and wind cause erosion of mountains and beaches. But these are by no means the only forces reshaping the face of our planet. All across Earth, we find evidence that the surface has been twisted, deformed, and folded (Figure 9-3). What is more, new material is continually being added to Earth's surface as lava pours forth from volcanoes and from immense cracks in the ocean floors. At the same time, other geologic activity slowly buries old surface material, or pulls material back into our planet's interior. These processes and others work together to renew and refresh our planet's exterior. Thus, although Earth is some 4.54 billion (4.54×10^9) years old, most of

To understand our dynamic planet, we must understand the energy sources that power its activity.

TABLE 9-1 Earth Data

Average distance from Sun:	1.000 au = 1.496×10^8 km
Maximum distance from Sun:	1.017 au = 1.521×10^8 km
Minimum distance from Sun:	0.983 au = 1.471×10^8 km
Eccentricity of orbit:	0.017
Average orbital speed:	29.79 km/s
Orbital period:	365.256 days
Rotation period:	23.9345 hours
Inclination of equator to orbit:	23.45°
Diameter (equatorial):	12,756 km
Mass:	5.974×10^{24} kg
Average density:	5515 kg/m³
Escape speed:	11.2 km/s
Albedo:	0.31
Surface temperature range:	Maximum: 60°C = 140°F = 333 K
	Mean: 14°C = 57°F = 287 K
	Minimum: −90°C = −130°F = 183 K
Atmospheric composition (by number of molecules):	78.08% nitrogen (N_2)
	20.95% oxygen (O_2)
	0.035% carbon dioxide (CO_2)
	about 1% water vapor

(NASA Goddard Space Flight Center. Image by Reto Stöckli)

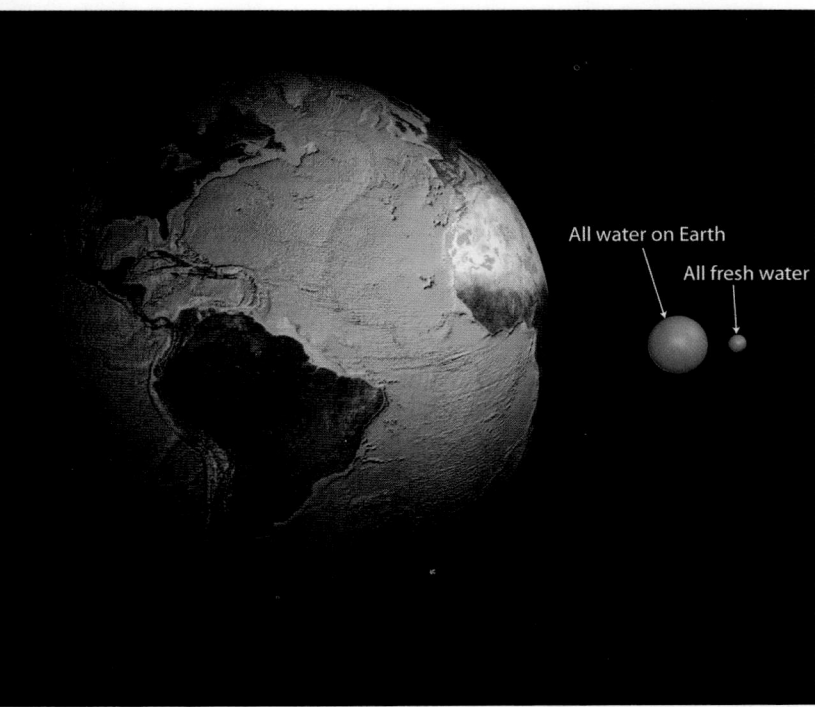

FIGURE 9-1 R I Ⓥ U X G

Earth's Dynamic Oceans This image shows the relative size of Earth compared to a sphere 860 miles in diameter containing all of Earth's water: fresh water, oceans, ice, and even water in the atmosphere. For comparison, the smaller sphere, with a diameter of 167.5 miles, contains just Earth's fresh water. Nearly three-quarters of Earth's surface is covered with water, a substance that is essential to the existence of life. In contrast, there is no liquid water at all on Mercury, Venus, Mars, or the Moon. (Illustration by Jack Cook. © Woods Hole Oceanographic Institution.)

its surface is less than 200 million years old (Figure 9-4). (We saw in Section 8-3 how scientists use radioactive dating to determine the ages of rocks and of Earth as a whole.)

Solar Energy and Convection

TUTORIAL 9.1 ➔ What powers all of this activity in Earth's oceans, atmosphere, and surface? There are three energy sources:

FIGURE 9-2 R I Ⓥ U X G

Earth's Dynamic Atmosphere This space shuttle image shows thunderstorm clouds over the African nation of Zaire. The tops of thunderstorm clouds frequently reach altitudes of 10,000 m (33,000 ft) or higher. At any given time, nearly 2000 thunderstorms are in progress over Earth's surface. (JSC/NASA)

radiation from the Sun, the tidal effects of the Moon and Sun, and Earth's own internal heat (Table 9-2).

The Sun is the principal source of energy for the atmosphere. Earth's surface is warmed by sunlight, which in turn warms the air next to the surface. Hot air is less dense than cool air, so it tends to rise (see the discussion of density in Box 7-1). As the air rises, it transfers heat to its surroundings. As a result, the rising air cools and becomes denser. It then sinks downward to be heated again, and the process starts over. This up-and-down motion is called **convection**, and the overall pattern of circulation is called a **convection current**. (You can see convection currents in action by heating water on a stove, as Figure 9-5 shows.) Convection currents also play a key role in reshaping Earth's surface, as warmer low-density material rises up from Earth's interior, cools and increases its density, and then sinks back down.

Solar energy also powers atmospheric activity by evaporating water from the surface. The energy it takes to change liquid water into water vapor is released later when the water vapor condenses back into liquid to form water droplets, like those droplets that make up clouds. In a typical thunderstorm (see Figure 9-2), the amount of energy released when this water condenses is as much as a city of 100,000 people consume in a month!

FIGURE 9-3 R I Ⓥ U X G

Earth's Dynamic Surface These tan-colored ridges, or hogbacks, in Colorado's Rocky Mountains were once horizontal layers of sediment at the bottom of an ancient body of water. Forces within Earth folded this terrain and rotated the layers into a vertical orientation. The layers were revealed when wind and rain eroded away the surrounding material. (Mr. Klein/Shutterstock)

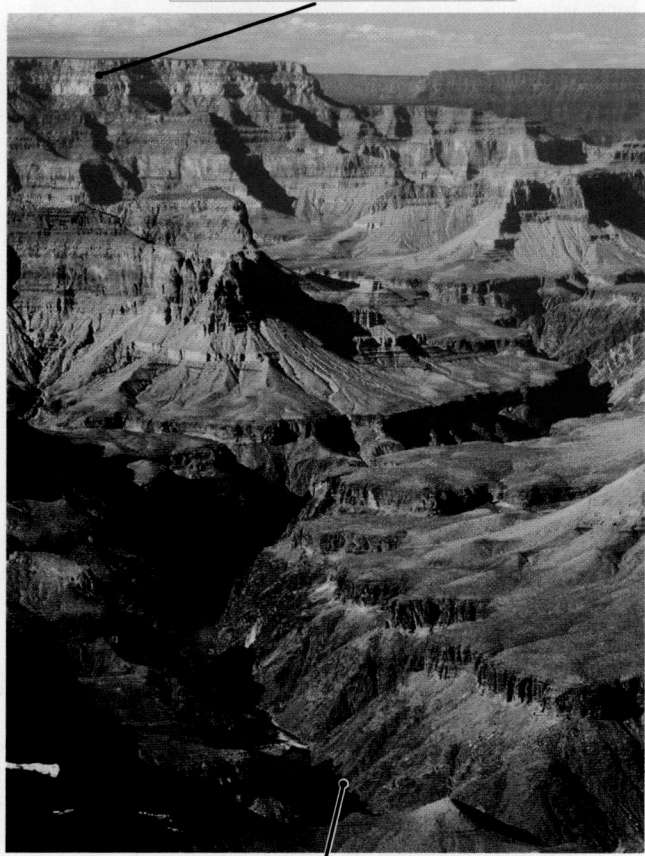

2. Over millions of years, layers of sediments built up over that rock. The most recent layer—the top—is about 250 million years old.

1. The rocks at the bottom of the Grand Canyon are 1.7–2.0 billion years old.

FIGURE 9-4 R I V U X G

Old and Young Rocks in the Grand Canyon The ages of rocks in Arizona's Grand Canyon demonstrate that geologic processes take place over very long timescales. (John Wang/Getty Images)

Solar energy also helps to power the oceans. Warm water from near the equator moves toward the poles, while cold polar water returns toward the equator. As we saw in Section 4-8, however, there is back-and-forth motion of the oceans due to the tidal forces of the Moon and Sun. Sometimes these two influences can reinforce each other, as when a storm (caused by solar energy) reaches a coastline at high tide (caused by tidal forces), producing waves strong enough to seriously erode beaches and sea cliffs.

Neither solar energy nor tidal forces can explain the reshaping of Earth's surface suggested by Figure 9-3. Rather, most geologic activity is powered by heat flowing from the interior of Earth itself. Roughly half of the heat is caused by the nuclear decay of radioactive elements such as uranium and thorium deep inside Earth. The rest is heat left over from the energetic collisions among planetesimals that created Earth (see Section 8-5).

TABLE 9-2 Earth's Energy Sources

Activity	Energy Sources
Motion of water in oceans, lakes, rivers	Solar energy, tidal forces
Motion of the atmosphere	Solar energy
Reshaping of surface	Earth's internal heat
Life	Solar energy (a few species that live on the ocean floor make use of Earth's internal heat)

The heat flow from Earth's interior to its surface is minuscule—just 1/6000 as great as the flow of energy we receive from the Sun—but it has a profound effect on the face of our planet. In Sections 9-2 and 9-3, we will explore the interior structure of Earth and discover that the heat-driven convection seen in a pot of boiling water (Figure 9-5) also takes place in our Earth.

CONCEPT CHECK 9-1

Consider the convection illustrated in Figure 9-5. Is gravity necessary for convection to occur? Can convection occur if the temperature is the same above and below the fluid?

Answer appears at the end of the chapter.

Earth's Surface Temperature and the Greenhouse Effect

Since so little heat comes from inside our planet, the average surface temperature of Earth depends almost entirely on the amount of energy that reaches us from the Sun in the form of visible light and other wavelengths of electromagnetic radiation. (In an analogous way, whether you feel warm or cool outdoors on a summer

1. Convection moves hot water from the bottom to the top . . .

2. . . . where it cools, becomes more dense, moves laterally, sinks, . . .

3. . . . warms, becomes less dense, and rises again.

FIGURE 9-5

Convection in the Kitchen Heat supplied to the bottom of a pot warms the water there, making the water expand and lowering its average density. This low-density water rises and transfers heat to its cooler surroundings. The water that began at the bottom thus cools down, becomes denser, and sinks back to the bottom to repeat the process. (Data source: F. Press, R. Siever, T. Grotzinger, and T. H. Jordan, *Understanding Earth,* 4th ed., New York, NY: W. H. Freeman, 2004) day depends on whether you are in the sunlight, receiving lots of solar energy, or in the shade, receiving little of this energy.)

If Earth did nothing but *absorb* radiation from the Sun, it would get hotter and hotter until the surface temperature became high enough to melt rock. Happily, this does not happen. And, even before incoming sunlight gets absorbed by the Earth, several processes reduce the amount of energy absorbed. Clouds, snow, ice, and sand *reflect* about 31% of the incoming sunlight back into space. The fraction of incoming sunlight that a planet reflects is called its **albedo** (from the Latin for "whiteness"); thus, Earth's albedo is about 0.31. As a result, only 69% of the incoming solar energy is absorbed by Earth. Earth also *emits* radiation into space because of its temperature, in accordance with the laws that describe heated dense objects (see Section 5-3). Earth's average surface temperature is nearly constant, which means that on the whole it is neither gaining nor losing energy. It is in equilibrium. Thus, the rate at which Earth emits energy into space must equal the rate at which it absorbs energy from the Sun.

To better understand this balance between absorbed and emitted radiation, remember that Wien's law tells us that the wavelength at which such an object emits most strongly (λ_{max}) is inversely proportional to its temperature (T) on the Kelvin scale (see Section 5-4 and Box 5-2). For example, the Sun's surface temperature is about 5800 K, and sunlight has its greatest intensity at a wavelength λ_{max} of 500 nm, which is green light in the middle of the visible spectrum. Earth's average surface temperature is far lower than the Sun's, so Earth radiates most strongly at longer wavelengths in the infrared portion of the electromagnetic spectrum. The Stefan–Boltzmann law tells us that temperature also determines the *amount* of radiation that Earth emits: The higher the temperature, the more energy it radiates.

Given the amount of energy reaching us from the Sun each second, together with Earth's albedo, we can calculate the amount of solar energy that Earth should *absorb* each second. Since this amount must equal the amount of electromagnetic energy that Earth *emits* each second, which in turn depends on Earth's average surface temperature, we can calculate what Earth's average surface temperature should be. The result is a very chilly 255 K ($-18°C = -0.7°F$), so cold that oceans and lakes around the world should be frozen over. In fact, Earth's actual average surface temperature is 287 K ($14°C = 57°F$). What is wrong with our model? Why is Earth so much warmer than we would expect?

The explanation for this discrepancy is called the **greenhouse effect**: Our atmosphere prevents some of the outgoing radiation emitted by Earth's surface from escaping into space. Certain gases in our atmosphere called **greenhouse gases,** among them water vapor (H_2O) and carbon dioxide (CO_2), are transparent to visible light but not to infrared radiation. Consequently, visible sunlight has no trouble entering our atmosphere and warming the surface. But some of the infrared radiation given off by Earth's heated surface is trapped by the atmosphere, raising the temperatures of both the atmosphere and the surface. As the surface and atmosphere become hotter, they both emit more infrared radiation, part of which is able to escape into space. The temperature levels off when the amount of infrared energy that escapes just balances the amount of solar energy reaching the surface (**Figure 9-6**). The result is that our planet's surface is some 33°C (59°F) warmer

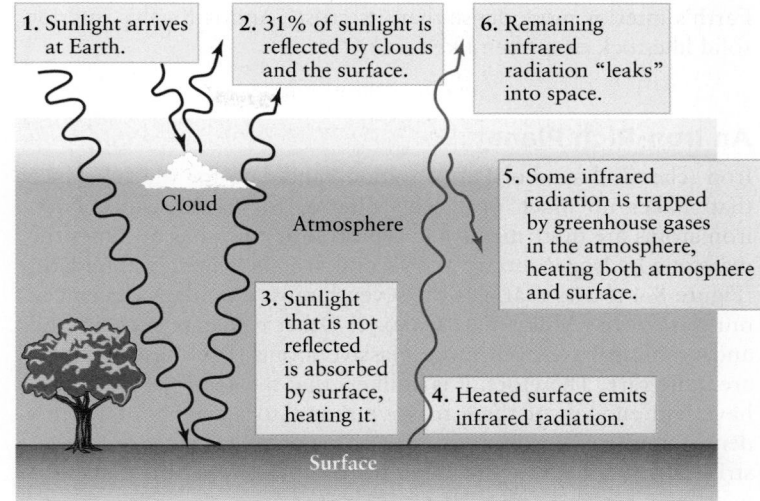

ANIMATION 9.1 ➤   **FIGURE 9-6**

The Greenhouse Effect Sunlight warms Earth's surface, which due to its temperature emits infrared radiation. Much of this radiation is absorbed by atmospheric water vapor and carbon dioxide, helping to raise the average temperature of the surface. Some infrared radiation does penetrate the atmosphere and leaks into space. In a state of equilibrium, the rate at which Earth loses energy to space in this way is equal to the rate at which it absorbs energy from the Sun.

than it would be without the greenhouse effect, and water remains unfrozen over most of Earth.

The warming caused by the greenhouse effect gives our planet the moderate temperatures needed for the existence of life. For more than a century, however, our technological civilizations have been adding greenhouse gases to the atmosphere at an unprecedented rate. As we will see in Section 9-7, the likely consequences for our climate are extremely grave.

CONCEPT CHECK 9-2

What is the positive impact that the greenhouse effect has on our planet?

Answer appears at the end of the chapter.

9-2 Studies of earthquakes reveal Earth's layered interior structure

TUTORIAL 9.2 ➤   The kinds of rocks found on and near Earth's surface provide an important clue about our planet's interior. The densities of typical surface rocks, including the rocky seafloor, are around 3000 kg/m³, but the average density of Earth as a whole (that is, its mass divided by its volume) is 5515 kg/m³. The interior of Earth must therefore be composed of substances much denser than those found near the surface. But what is this substance? Why is

> Like an exotic dessert, our planet's interior has layers of both solid and liquid material.

Figure labels (Figure 9-6):
1. Sunlight arrives at Earth.
2. 31% of sunlight is reflected by clouds and the surface.
3. Sunlight that is not reflected is absorbed by surface, heating it.
4. Heated surface emits infrared radiation.
5. Some infrared radiation is trapped by greenhouse gases in the atmosphere, heating both atmosphere and surface.
6. Remaining infrared radiation "leaks" into space.
Cloud
Atmosphere
Surface

Earth's interior more dense than its crust? And is Earth's interior solid like rock or molten like lava?

An Iron-Rich Planet

Iron (chemical symbol Fe) is a good candidate for the substance that makes up much of Earth's interior for two reasons. First, iron atoms are quite massive (a typical iron atom has 56 times the mass of a hydrogen atom), and second, iron is relatively abundant. (Figure 8-4 shows that it is the seventh most abundant element in our part of the Milky Way Galaxy.) Other elements such as lead and uranium have even more massive atoms, but these elements are quite rare. Therefore, it is unlikely that the solar nebula would have had enough of these massive atoms to account for Earth's dense interior. Furthermore, iron is common in meteoroids that strike Earth, which suggests that it was abundant in the planetesimals from which Earth formed.

Earth was almost certainly molten throughout its volume soon after its formation, about 4.54×10^9 years ago. Energy released by the violent impacts of numerous meteoroids and asteroids and by the decay of radioactive isotopes likely melted the solid material collected from the earlier planetesimals. Gravity caused abundant, dense iron to sink toward Earth's center, forcing less dense material to the surface. **Figure 9-7a** shows this process of *chemical differentiation*. (We discussed chemical differentiation in Section 8-5.) The result was a planet with the layered structure shown in Figure 9-7b—a central **core** composed mostly of iron, surrounded by a **mantle** of dense, iron-rich minerals. The mantle, in turn, is surrounded by a thin **crust** of relatively light silicon-rich minerals. We live on the surface of this crust.

Seismic Waves as Earth Probes

How do we know that this layered structure is correct? The challenge in testing this model is that Earth's interior is as inaccessible as the most distant galaxies in space. The deepest wells go down only a few kilometers, barely penetrating the surface of our planet. Despite these difficulties, geologists have learned basic properties of Earth's interior by studying earthquakes and the seismic waves that they produce.

Over decades, forces build up in Earth's crust. Occasionally, these forces are relieved with a sudden motion called an **earthquake**. Most earthquakes occur within 20 km of Earth's surface. The point on Earth's surface directly over an earthquake's location is called the **epicenter**.

Earthquakes produce three different kinds of **seismic waves**, which travel around or through Earth in different ways and at different speeds. Geologists use a sensitive instrument called a **seismometer** to detect and record these vibratory motions. The first type of wave, which is analogous to ocean waves, causes the rolling motion that people feel around an epicenter. These waves are called **surface waves** because they travel only over Earth's surface. The two other kinds of waves, called **P waves** (for "primary") and **S waves** (for "secondary"), travel through the interior of Earth. P waves are called *longitudinal* waves because their oscillations are parallel to the direction of wave motion, like a spring that is alternately pushed and pulled. In contrast, S waves are called *transverse* waves because their vibrations are perpendicular to the direction in which the waves move. S waves are analogous to waves produced by a person shaking a rope up and down (**Figure 9-8**).

What makes seismic waves useful for learning about Earth's interior is that they do not travel in straight lines when they pass from one type of material to another. Instead, the paths that they

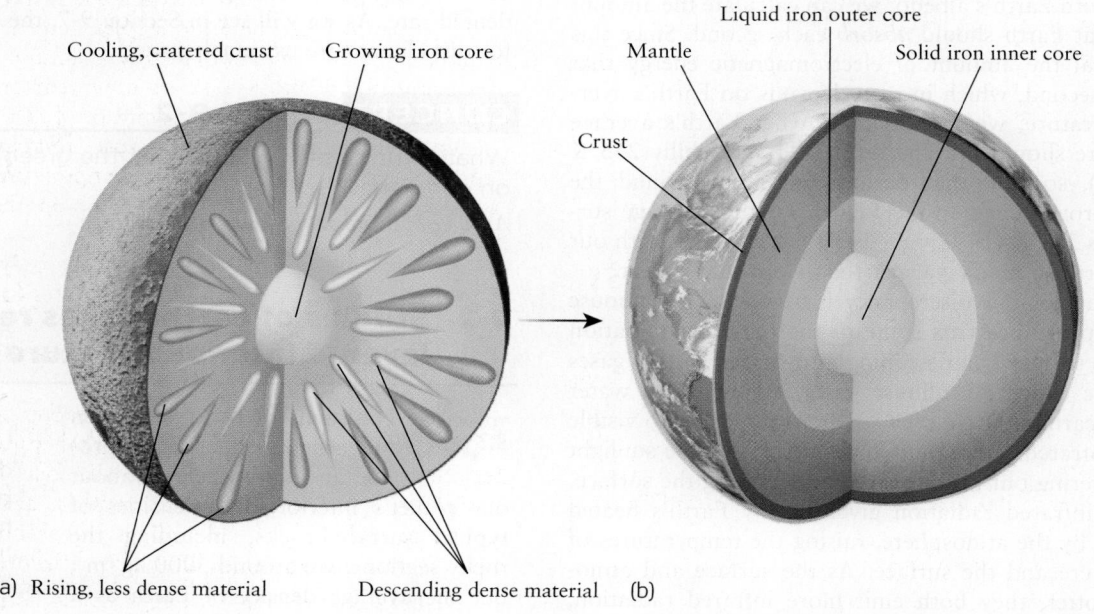

(a) Rising, less dense material   Descending dense material   (b)

FIGURE 9-7

Chemical Differentiation and Earth's Internal Structure **(a)** The newly formed Earth was likely molten throughout its volume. Dense materials such as iron (shown in yellow) sank toward the center, while low-density materials (shown in orange) rose toward the surface. **(b)** The present-day

Earth is no longer completely molten inside. A dense, solid iron core is surrounded by a less dense liquid core and an even less dense mantle. The crust, which includes the continents and ocean floors, is the least dense of all.

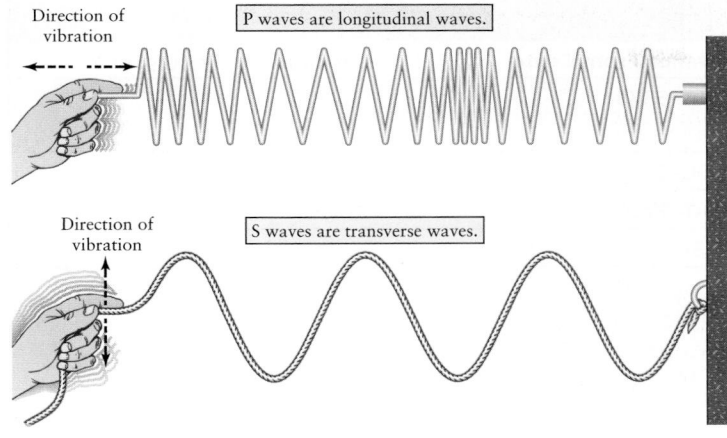

ANIMATION 9.2 **FIGURE 9-8**

Seismic Waves Earthquakes produce two kinds of waves that travel through the body of our planet. One kind, called P waves, are longitudinal waves. They are analogous to those produced by pushing a spring in and out. The other kind, S waves, are transverse waves analogous to the waves produced by shaking a rope up and down.

follow through the body of Earth are bent because of the varying density and composition of Earth's interior. We saw in Section 6-1 that light waves behave in a very similar way. Just as light waves bend, or refract, when they pass from air into glass or vice versa

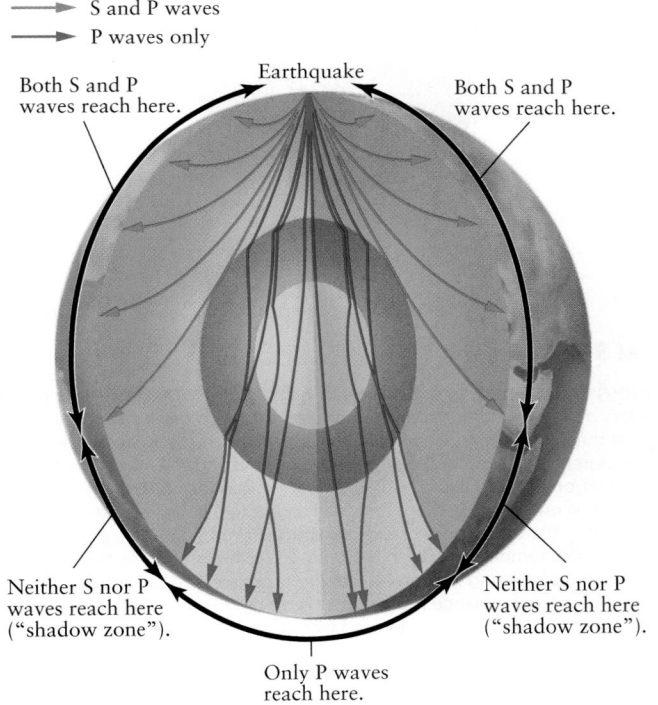

FIGURE 9-9

Earth's Internal Structure and the Paths of Seismic Waves Seismic waves follow curved paths because of differences in the density and composition of the material in Earth's interior. The paths curve gradually where there are gradual changes in density and composition. Sharp bends occur only where there is an abrupt change from one kind of material to another, such as at the boundary between the outer core and the mantle. Only P waves can pass through Earth's liquid outer core.

(see Figure 6-2), seismic waves refract as they pass through different parts of Earth's interior. By studying how the paths of these waves bend, geologists can map out the general interior structure of Earth.

One key observation about seismic waves and how they bend has to do with the differences between S and P waves. When an earthquake occurs, seismometers relatively close to the epicenter record both S and P waves, but seismometers on the opposite side of Earth record only P waves. The absence of S waves was first explained in 1906 by British geologist Richard Dixon Oldham, who noted that transverse vibrations such as S waves cannot travel far through liquids. Oldham therefore concluded that our planet has a molten core. Furthermore, there is a region in which neither S waves nor P waves from a distant earthquake can be detected (**Figure 9-9**). This "shadow zone" results from the specific way in which P waves are deflected at the boundary between the solid mantle and the liquid iron outer core (the molten core). By measuring the size of the shadow zone, geologists have concluded that the radius of the molten core is about 3500 km (2200 mi), about 55% of our planet's overall radius and about double the radius of the Moon (1738 km = 1080 mi).

As the quality and sensitivity of seismometers improved, geologists discovered faint traces of P waves in an earthquake's shadow zone. In 1936, the Danish seismologist Inge Lehmann explained that some of the P waves passing through Earth are deflected into the shadow zone by a small **inner core**, later determined to be solid, at the center of our planet. The radius of this solid iron inner core is about 1300 km (800 mi).

Earth's Major Layers

LOOKING DEEPER 9.1 The seismic evidence reveals that our planet has a curious internal structure—a liquid **outer core** of iron sandwiched between a solid inner iron core and a solid mantle. **Table 9-3** summarizes this structure. To understand this arrangement, we must look at how temperature and pressure inside Earth affect the melting point of rock.

Both temperature and pressure increase with increasing depth below Earth's surface. The temperature of Earth's interior rises steadily from about 14°C (57°F) on the surface to nearly 5000°C (9000°F) at our planet's center (**Figure 9-10**).

Earth's outermost layer, the crust, is only about 5 to 35 km (3 to 22 mi) thick. It is composed of rocks for which the **melting point,** or temperature at which the rock changes from solid to liquid, is far higher than the temperatures actually found in the crust. Thus, the crust is solid.

Earth's mantle, which extends to a depth of about 2900 km (1800 mi), is largely composed of substances rich in iron and magnesium. Most of Earth's mantle is solid, but there is a layer closer to Earth's surface where the temperature and pressure cause the mantle to melt (**Figure 9-11**). This transition between solid and liquid is very important, so let's summarize it:

Without any change in temperature, solid rock can become liquid when the pressure on it decreases.

As Figure 9-10 shows, the temperature throughout most of the mantle falls below the melting point of mantle materials. Hence, the mantle is primarily solid. However, the upper level of the mantle is liquid. Not all liquids can flow as easily as water, and a very thick fluid like this is called **plastic.**

TABLE 9-3 Earth's Internal Structure

Region	Depth Below Surface (km)	Distance From Center (km)	Average Density (kg/m³)
Crust (solid)	0–5 (under oceans)	6343–6378	3500
	0–35 (under continents)		
Mantle (plastic, solid)	from bottom of crust to 2900	3500–6343	3500–5500
Outer core (liquid)	2900–5100	1300–3500	10,000–12,000
Inner core (solid)	5100–6400	0–1300	13,000

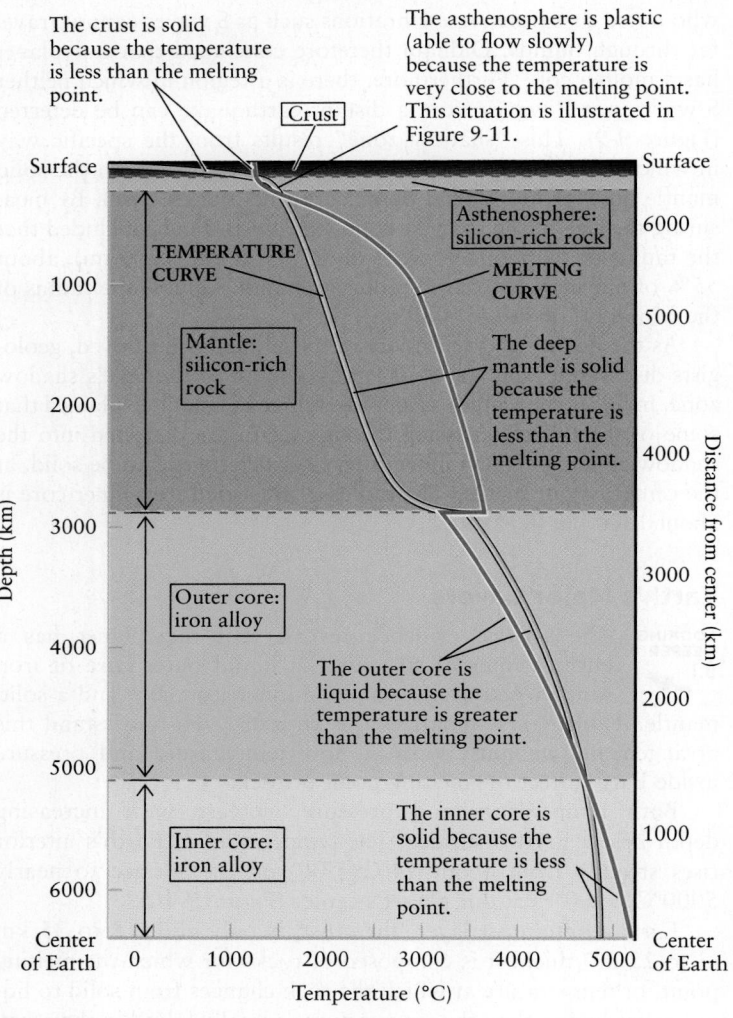

FIGURE 9-10

Temperature and Melting Point of Rock Inside Earth The temperature (yellow curve) rises steadily from Earth's surface to its center. The temperature at the center of Earth is about the same as the surface of the Sun! The melting point of Earth's material (red curve) is also shown on this graph. Where the temperature is below the melting point, as in the mantle and inner core, the material is solid; where the temperature is above the melting point, as in the asthenosphere and outer core, the material is liquid. (Data from T. Grotzinger, T. H. Jordan, F. Press, and R. Siever, *Understanding Earth*, 5th ed., New York, NY: W. H. Freeman, 2007)

CAUTION!

Why use the term *plastic*? It may seem strange to think of a plastic material as one that is able to flow. Normally we think of plastic objects as being hard and solid, like the rigid plastic

FIGURE 9-11

Melting: From Solid to Liquid with a Drop in Pressure Rocky material in Earth's mantle is under less and less pressure toward Earth's surface. This is because there is less mass above pressing down under the force of gravity. With less pressure, molecules can begin to move around and the mantle material transitions from solid to liquid (although this liquid does not flow easily). However, low pressure can't turn any solid rock into a liquid, or else asteroids would melt in the miniscule pressures of space. For rock to melt, it must also have a sizeable temperature, as in the case of Earth's interior. As the pressure drops, when considering material closer to Earth's surface, the melting point temperature of rocky mantle material can drop below the actual temperature, and the mantle melts. This occurs in a layer called the asthenosphere.

exterior of a computer or cell phone. But to make these objects, the plastic material is heated so that it can flow into a mold, then cooled so that it appears to solidify. Strictly speaking, the material only has a "plastic" property when it is able to flow. Maple syrup and tree sap are two other substances that flow at warm temperatures but become more solid when cooled. Furthermore, as with a plastic spoon or the asthenosphere, a material might seem solid to us when we touch it, but it actually flows when enormous forces act on it over millions of years.

The thin plastic layer of the upper mantle is called the **asthenosphere** (from the Greek *asthenia*, meaning "weakness"). As we'll soon see, the ability for material in the asthenosphere to flow, or move around, is essential for plate tectonics.

At the boundary between the mantle and the outer core, there is an abrupt change in chemical composition, from silicon-rich materials to a mixture of iron (about 80%) with some nickel. Because this iron-nickel mixture has a lower melting point than the silicon-rich material above it, the melting curve in Figure 9-10 changes abruptly as it crosses from the mantle down into the outer core. As a result, the outer core is liquid.

At depths greater than about 5100 km, the pressure is more than 10^{11} newtons per square meter. This is about the same pressure that 3000 midsize cars would produce if stacked and balanced on a quarter! Because the melting point of the iron-nickel mixture under this pressure is higher than the actual temperature (see Figure 9-10), Earth's inner core is solid.

LOOKING DEEPER 9.2 ➤ Until the 1980s, geologists knew little more about the inner core than that it is solid and dense. Since then, evidence from seismic waves has accumulated that suggests the inner core is a collection of iron **crystals**—that is, where each crystal is made of iron atoms arranged in orderly rows, like carbon atoms are arranged within a diamond. If this is indeed the case, the inner core contains the largest known crystalline objects anywhere in the solar system. Furthermore, there are strong indications that the inner core is rotating at a slightly faster rate than the rest of Earth! These remarkable discoveries suggest that even more surprises may lurk deep within Earth's interior.

Heat naturally flows from where the temperature is high to where it is low. Figure 9-10 thus explains why heat flows from the center of Earth outward. We will see in the next section how this heat flow acts as the "engine" that powers our planet's geologic activity.

CONCEPT CHECK 9-3

Hypothetically, what would have changed about Earth's interior if S waves from earthquakes suddenly started being observed on the side of Earth opposite an earthquake?

Answer appears at the end of the chapter.

9-3 Plate movement produces earthquakes, mountain ranges, and volcanoes that shape Earth's surface

One of the most important geological discoveries of the twentieth century was the realization that Earth's crust is constantly changing. We have learned that the crust (which includes ocean seafloors) is divided into huge **plates** whose motions produce earthquakes, volcanoes, mountain ranges, and oceanic trenches. Furthermore, material in most of these plates is slowly but constantly replenished about every few hundred million years, so that most of Earth's surface is young compared to the planet's 4.54-billion-year chronologic age. The motion of Earth's

> Evidence from the bottom of the ocean confirmed the theory of moving continents.

plates has come to be the central unifying theory of geology, much as the theory of evolution has become the centerpiece of modern biology. By understanding the effects of plate motion and what causes it, we also learn how to read the surface of other planets and moons to learn about their interiors.

Continental Drift

Anyone who carefully examines a map of Earth might come up with the idea of moving continents. South America, for example, would fit snugly against Africa were it not for the Atlantic Ocean. As Figure 9-12 shows, the fit between land masses on either side of the Atlantic Ocean is quite remarkable. This observation inspired the German meteorologist Alfred Wegener to find evidence for "continental drift"—the idea that the continents on either side of the Atlantic Ocean have simply drifted apart. After much research, in 1915 Wegener published the theory that there had originally been a single gigantic supercontinent, which he called Pangaea (meaning "all lands"), that began to break up and drift apart roughly 200 million years ago. Other geologists refined this theory over time, arguing that Pangaea must have split up in stages to make the continents we see today (Figure 9-13). (*Note:* No

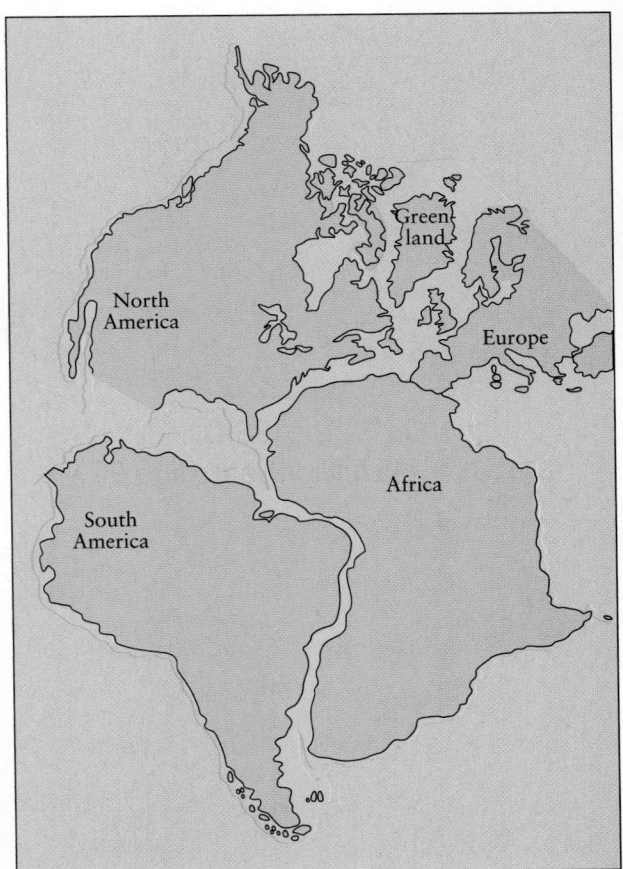

FIGURE 9-12

Fitting the Continents Together Africa, Europe, Greenland, North America, and South America fit together remarkably well. The fit is especially convincing if the edges of the continental shelves (shown in yellow) are used, rather than today's shorelines. This strongly suggests that these continents were in fact joined together at some point in the past. (Data source: Hurley, P. M. (1968). The Confirmation of Continental Drift. *Scientific American, 218*(4), 52–68.)

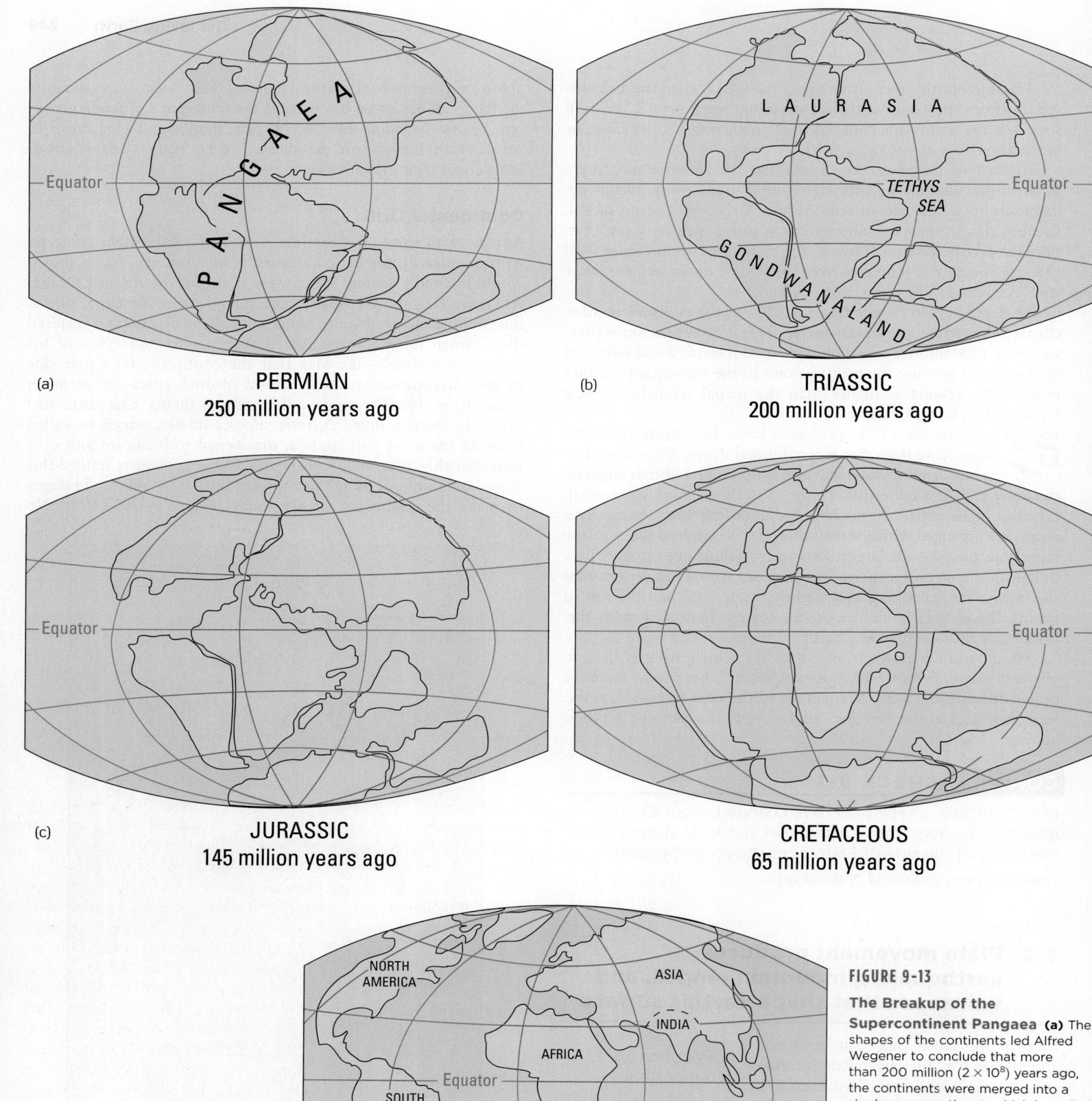

(a) **PERMIAN**
250 million years ago

(b) **TRIASSIC**
200 million years ago

(c) **JURASSIC**
145 million years ago

(d) **CRETACEOUS**
65 million years ago

(d) **PRESENT DAY**

FIGURE 9-13

The Breakup of the Supercontinent Pangaea **(a)** The shapes of the continents led Alfred Wegener to conclude that more than 200 million (2×10^8) years ago, the continents were merged into a single supercontinent, which he called Pangaea. **(b)** Pangaea first split into two smaller land masses, Laurasia and Gondwanaland. **(c)** Over millions of years, the continents continued to move apart, **(d)** ultimately reaching their present-day locations. Among the evidence confirming this picture are nearly identical rock formations 200 million years in age that today are thousands of kilometers apart but would have been side by side on Pangaea. (U.S. Geological Survey) humans were present to witness the break up; the earliest humans did not appear on Earth until about 2 million years ago. At the time that Pangaea broke up, *small* dinosaurs were the dominant land animals.)

Sadly, most geologists ridiculed Wegener's ideas and he died before they were accepted. Although it was generally acknowledged that the continents do "float" on top of the denser mantle beneath them, few geologists could accept the idea that entire continents could move around Earth at speeds of centimeters per year. We now know that Wegener was right, and GPS satellites currently track all the continents, which range in travel speed from 2.5 cm/yr to as many as 15 cm/yr (about 1 to 6 in./yr). The difficulty in accepting Wegener's fragmentary evidence of continental drift stemmed from a lack of understanding what forces could possibly move entire continents. The clue they needed was hidden beneath the ocean.

Plate Tectonics

Beginning in the mid-1950s, geologists found evidence that new material is being forced upward to the crust from deep within Earth. Completely submerged mountain ranges on the ocean floors were discovered, including the Mid-Atlantic Ridge, which is the longest mountain chain in the world and stretches all the way from Iceland to Antarctica (Figure 9-14). Through further analysis, it was found that mantle rock melts as it reaches lower pressures near the surface, and rises to emerge along the Mid-Atlantic Ridge, which is in essence a long chain of underwater volcanoes that forms new seafloor.

As newly risen material cools and hardens at the underwater ridge, this new seafloor material also *moves away* from the ridge, leading to **seafloor spreading** as even more material rises and hardens. For example, because of seafloor spreading from the Mid-Atlantic Ridge, South America and Africa are moving apart at a speed of roughly 3 cm per year, with new seafloor forming in between. Working backward, these two continents would have been next to each other some 200 million years ago—just as Wegener suggested (see Figure 9-12).

In the early 1960s, geologists began to find additional evidence supporting the existence of large, moving plates. Thus was born the modern theory of crustal motion, which came to be known as **plate tectonics** (from the Greek *tekton*, meaning "builder").

One such piece of evidence is that earthquakes and volcanoes tend to occur at the boundaries of Earth's tectonic plates, where the plates are colliding, separating, or sliding against each other. The boundaries of the plates therefore stand out clearly when the epicenters of earthquakes are plotted on a map, as in Figure 9-15. The boundary in Figure 9-15 that surrounds the Pacific plate is called the Ring of Fire: More than 75% of the world's volcanoes are found here, along with 90% of the world's earthquakes.

Convection and Plate Motion

What makes the plates move? The answer is twofold. First, heat flows outward from Earth's hot interior to its cool crust, where it

FIGURE 9-14

The Mid-Atlantic Ridge This artist's rendition shows the Mid-Atlantic Ridge, an immense 40,000 km–long mountain ridge that rises up from the floor of the North Atlantic Ocean. It is caused by lava seeping up from Earth's interior along a rift that extends from Iceland to Antarctica. Despite solid data, not everyone believed in the existence of such a massive underwater mountain chain because it was totally unexpected and made little sense. (Courtesy of M. Tharp and B. C. Heezen)

is radiated to space. Second, Earth accomplishes this heat flow by convection, which is far more efficient at carrying heat than other processes like conduction or radiation. In Earth's mantle, hotter interior temperatures cause rocky material to expand, *lowering* the material's density compared to its surroundings, which then causes it to rise. This material then cools off at or near the surface, *increasing* the material's density compared to its surroundings, which then causes it to sink. Figure 9-5 shows that convection can cause material to circulate in convection currents. While convection involves many layers within the mantle, it is the motion of Earth's outer layer that directly reshapes our planet's surface. This outer layer is called the **lithosphere** (from the Greek word *litho* for "rock"), which is physically divided into the plates described by plate tectonics. The lithosphere is a cooled rocky outer layer that contains the seafloor in the oceans, and the crust that forms continents.

Figure 9-16 shows the basic features of an oceanic seafloor plate. Hot, molten subsurface rock seeps upward along **oceanic rifts**, hardening to form the seafloor plates. (The Mid-Atlantic Ridge, shown in Figure 9-14, is an oceanic rift.) Where a seafloor plate meets a continental plate (on the right in Figure 9-16), the higher-density

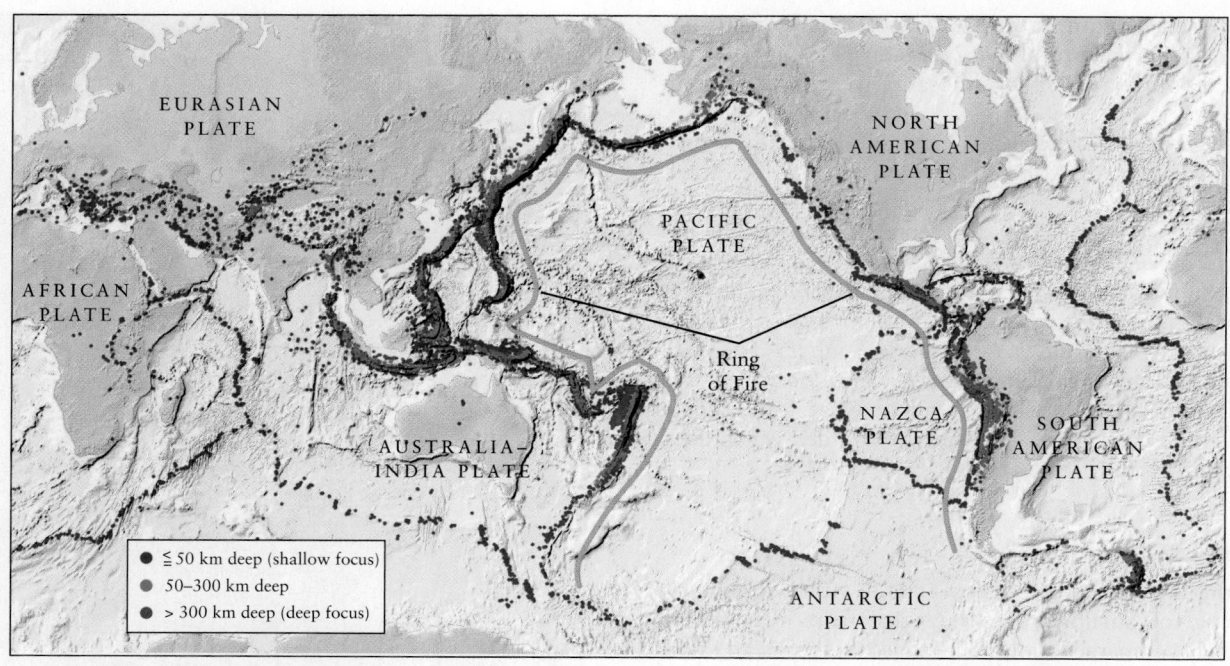

ANIMATION
9.3

FIGURE 9-15

Earth's Major Plates The boundaries of Earth's major plates are the scenes of violent seismic and geologic activity. Most earthquakes occur where plates separate, collide, or rub together. Plate boundaries are therefore easily identified simply by plotting earthquake epicenters (shown here as dots) on a map. The colors of the dots indicate the depths at which the earthquakes originate. Typical travel speeds of plates are between 2 cm and 10 cm (1 to 5 in.) per year, where some separate from each other and some move closer together. (Data from Harvard CMT catalog; plot by M. Boettcher and T. Jordan)

ANIMATION
9.4

FIGURE 9-16

The Mechanism of Plate Tectonics Convection currents transport rock to a rift where it oozes toward the surface as magma, hardens in the cold ocean, and spreads away from the rift to create new seafloor. Each side of the rift exhibits seafloor spreading and is a separate tectonic plate. When higher-density seafloor meets lower-density continental crust at a subduction zone, the seafloor plunges downward. Slab pull, due to gravity acting on this slab, is the dominant force moving the plate across Earth. Collision of plates at a subduction zone also leads to mountain-building, earthquakes, and volcanoes. Continental material is actually built up over time as seafloor material is scraped off during subduction, and altered over time to a lower density.

seafloor plate sinks back down into the mantle, creating a **sub-duction zone**. In most cases, the greatest force that moves seafloor plates comes from **slab pull** at the sinking edge of the plate: Due to gravity, the cooler high-density seafloor slab falls into the warmer lower-density mantle, pulling the rest of the plate behind it.

Another force on the oceanic plate is **ridge push**, where molten rock emerging from the oceanic rift is pulled down the ridge's slopes by gravity, just as a ball rolls down a hill. Thus, gravity gives a push to the plate, contributing to its horizontal slide. With mountainous ridges typically a couple of miles high, this can be a large force.

CAUTION!

The plates are not merely riding on top of deeper convection currents. Sometimes convection in lower layers drags the plates along, and other times the plates drag lower layers along, but both of these cases tend to create weaker forces than slab pull and ridge push. While convection currents also circulate deeper in the mantle, plate tectonics results from the participation of Earth's surface layer in convection. A further caution is that rocky material only becomes liquid (or magma) very near the surface at the rifts; there isn't molten lavalike material deep inside the mantle. Most of the convection current is very slow plastic flow within mantle rock that experiences great forces, and this mantle rock would feel very solid to the touch.

A rift between two plates is not always found at the ocean seafloor, although continental rifts occur much less often than oceanic rifts. As shown in Figure 9-17, the valley created by a continental rift can be filled with adjacent seawater. If the continental spreading persists, it can slowly transition into a mid-ocean rift, as happened between the continental plates spreading from Pangaea. However, continental plates are thicker and are not subducted nearly as often as oceanic plates, and most continental plate motion comes from pushing by adjacent seafloor plates.

ANIMATION 9.5 ➤   ANIMATION 9.6 ➤ The boundaries between plates are the sites of some of the most impressive geological activity

on our planet. Great mountain ranges, such as the Sierra Nevada and the Cascade Range along the western coast of North America and the Andes along South America's west coast, are thrust up by ongoing collisions between continental plates and the plates of the ocean floor. The Himalayas, due to the collision of plates containing India and China, contain the highest mountains on Earth (Figure 9-18).

(a)

(b)

ANIMATION 9.8 ➤ **FIGURE 9-18**
The Collision of Two Plates Collisions between the plates that carry India and China have created Mount Everest and K2 in the Himalayas. **(a)** While plates typically move a few cm/yr, the Indian plate has sped along at up to 20 cm/yr. **(b)** Millions of years ago, the ocean seafloor subducted neatly beneath the Eurasian plate, but now there is a buildup of continental crust on both plates that forms the Himalayas.

The Red Sea is formed by the spreading apart of the African plate (at lower left) and the Arabian plate (at upper right).

ANIMATION 9.7 ➤ **FIGURE 9-17 R I V U X G**
The Separation of Two Plates The plates that carry Africa and Arabia are moving apart, leaving a great rift that has been flooded to form the Red Sea. This view from orbit shows the northern Red Sea, which splits into the Gulf of Suez and the Gulf of Aqaba. (Gemini 12, NASA)

CONCEPT CHECK 9-4

What are two forces that move tectonic seafloor plates?

CALCULATION CHECK 9-1

If seafloor spreading opens South America and Africa apart at a speed of 2 cm per year, how long would it have taken for the Atlantic to open to a current width of 600 million (6×10^8) cm?

Answers appear at the end of the chapter.

Plate Tectonics and the Varieties of Rocks

ANIMATION 9.9

Rocks are solid material made from minerals. A **mineral** can be made out of a single type of atom, such as diamond made from carbon, but most minerals contain molecules made from several different types of atoms. For example, quartz—which can form sandstone and sand at the beach and desert—is made from molecules of silicon and oxygen (SiO_2). Plate tectonics plays a major role in forming rocks and moving them around, as seen in the three major categories of rocks (Figure 9-19). **Igneous rocks** result when minerals cool from a molten state. (Molten rock is called **magma** when it is buried below the surface and **lava** when it flows out upon the surface, as in a volcanic eruption.) The ocean floor is made predominantly of a type of igneous rock called basalt (Figure 9-19a). Igneous rock is just what we expect for the spreading ocean seafloor, which is produced by material welling up from the mantle. Taken together, igneous rock in the seafloor and from volcanic flows on land comprise about 95% of Earth's crust.

Sedimentary rocks are produced by the accumulation of smaller particles. For example, winds and water can pile up layer upon layer of sand grains. Other minerals present amid the sand can gradually cement the grains together to produce sandstone (Figure 9-19b). In another process, the skeletal remains of tiny marine organisms can cover the ocean floor to form a sedimentary rock called limestone. The motion of tectonic plates can move such rocks to places far from where they were formed. This movement explains why the pyramids of Egypt are made of limestone that formed on the ocean floor. Plate tectonics can also upturn horizontally formed sedimentary rock, which explains the vertical layers of the hogback ridges in Colorado (see Figure 9-3).

Sometimes igneous or sedimentary rocks become buried far beneath the surface, where they are subjected to enormous pressure and high temperatures without melting. These severe conditions can change the minerals and structure of the rocks, producing **metamorphic rock** (Figure 9-19c). The presence of metamorphic rocks at Earth's surface tells us that tectonic activity sometimes lifts up material from deep within the crust. This activity can happen when two plates collide, as shown in Figure 9-18.

The Cycle of Supercontinents

Plate tectonic theory offers insight into geology on the largest of scales, that of an entire supercontinent. In recent years, geologists have uncovered evidence that points to a whole succession of supercontinents that once broke apart and then reassembled. Pangaea is only the most recent supercontinent in this cycle, which repeats about every 500 million years. As a result, intense episodes of mountain building have occurred at roughly 500-million-year intervals.

Once assembled, a supercontinent sows the seeds of its own destruction because it disrupts the even flow of heat from Earth's interior to the surface. In one scenario, as soon as a supercontinent forms, temperatures beneath it rise, much as they do when a lid is placed on a heated cooking pot. As heat accumulates in one place, the lithosphere domes upward and cracks. The effect of these cracks is like taking the lid off of a pressure cooker, and the pressure drops. Rising rocky material, suddenly at lower pressures, melts and wells up to fill the resulting fractures, which continue to widen as pieces of the fragmenting supercontinent are pushed apart.

The changes wrought by plate tectonics are very slow on the scale of a human lifetime, but have occurred on Earth's surface for more than 4 billion years. Ocean seafloor recycles back into the mantle every few hundred million years as continents are built up and broken apart. The lesson of plate tectonics is that the seemingly permanent face of Earth is in fact dynamic and ever-changing.

(a) Igneous rock (basalt)

(b) Sedimentary rock (sandstone)

(c) Metamorphic rock (marble)

FIGURE 9-19 R I **V** U X G

Igneous, Sedimentary, and Metamorphic Rocks **(a)** Igneous rocks such as basalt are created when molten materials solidify. This example contains iron-rich minerals that give it its dark color. **(b)** Sedimentary rocks such as sandstone are typically formed when loose particles of soil or sand are compacted and fused into rock as groundwater leaves behind other minerals, which act as a cement. **(c)** Metamorphic rocks are produced when igneous or sedimentary rocks are subjected to high temperatures and pressures deep within Earth's crust. Marble is formed from sedimentary limestone; gneiss is formed from igneous granite. (a: Doug Martin/Science Source; b: Dorling Kindersley/Getty Images; c: Juniors/Superstock)

What might cause a supercontinent to break apart?

If supercontinents are formed and break up in a 500-million-year cycle, about how many supercontinents could have come and gone over Earth's roughly 4.5-billion-year existence?

Answers appear at the end of the chapter.

9-4 Earth's magnetic field produces a magnetosphere and reverses direction

We saw in Section 7-7 that Earth's magnetic field provides evidence that our planet's interior is partially molten. While the magnetic field is generated deep inside Earth, it has effects on our surface, our atmosphere, and even in our solar system neighborhood.

The Magnetosphere

Earth's magnetic field extends far above the atmosphere, where it interacts dramatically with charged particles from the Sun. This **solar wind** is a flow of mostly protons and electrons that streams constantly outward from the Sun's upper atmosphere. Near Earth, the particles in the solar wind move at speeds of roughly 450 km/s, or about a million miles per hour. Because this is considerably faster than sound waves can travel in the very thin gas between the planets, the solar wind is said to be *supersonic*. (Because the gas between the planets is so thin, interplanetary sound waves carry too little energy to be heard by astronauts.)

If Earth had no magnetic field, we would be continually bombarded by the solar wind. But our planet *does* have a magnetic field, and the forces that this field can exert on charged particles are strong enough to deflect them away from us. The region of space around a planet in which the motion of charged particles is dominated by the planet's magnetic field is called the planet's

magnetosphere. Figure 9-20 is a scale drawing of Earth's magnetosphere, which was discovered in the late 1950s by the first satellites placed in orbit.

When the supersonic particles in the solar wind first encounter Earth's magnetic field, they abruptly slow to subsonic speeds. The boundary where this sudden decrease in velocity occurs is called the **bow shock**. Still closer to Earth lies another boundary, called the **magnetopause**, where the outward magnetic pressure of Earth's field is exactly counterbalanced by the impinging pressure of the solar wind. Most of the particles of the solar wind (mostly electrons and protons) are deflected around the magnetopause, just as water is deflected to either side of the bow of a ship.

Some charged particles of the solar wind manage to leak through the magnetopause. When they do, they are trapped by Earth's magnetic field in two huge, doughnut-shaped rings around Earth called the **Van Allen belts**. These belts were discovered in 1958 during the flight of the first successful U.S. Earth-orbiting satellite. They are named after the physicist James Van Allen, who insisted that the satellite carry a Geiger counter to detect charged particles. The inner Van Allen belt, which extends over altitudes of about 2000 to 5000 km, contains mostly protons. The outer Van Allen belt, about 6000 km thick, is centered at an altitude of about 16,000 km above Earth's surface and contains mostly electrons. Electric charges of the same type, such as positively charged protons, repel each other. Negatively charged electrons also repel each other. This repulsion limits how many particles each of the Van Allen belts can hold.

Aurorae

Sometimes the Van Allen belts become overloaded with charged particles. The particles then leak through the magnetic fields at their weakest points and cascade down into Earth's upper atmosphere, usually in a ring-shaped pattern (**Figure 9-21a**). As these high-speed charged particles collide with atoms in the upper atmosphere, they excite the atoms to high energy levels. The atoms then emit visible light as they drop down to their ground states, like the excited gas atoms in a neon light (see Section 5-8). The result is a beautiful, shimmering display called an **aurora** (*plural* **aurorae**). These are also called the **northern lights (aurora borealis)** or **southern lights (aurora australis)**, depending on the hemisphere in which the phenomenon

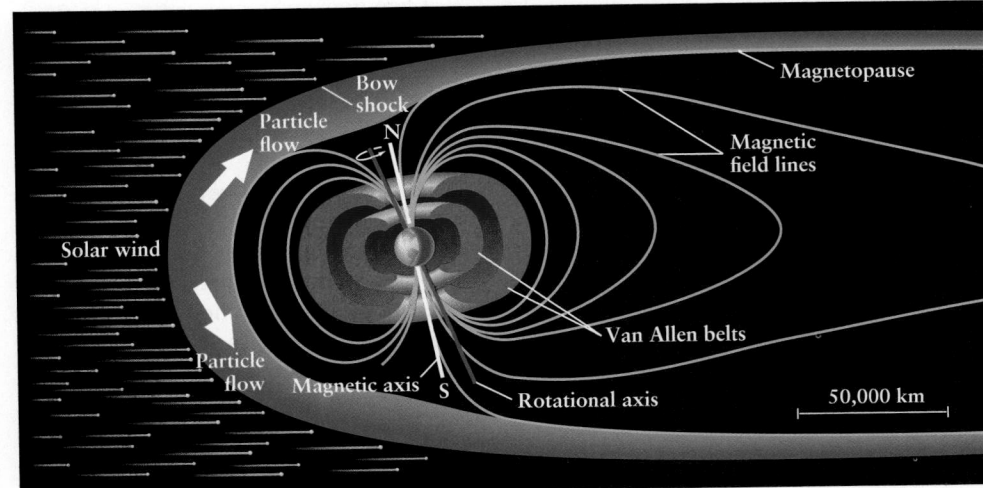

FIGURE 9-20

Earth's Magnetosphere Earth's magnetic field carves out a cavity in the solar wind, shown here in cross section. The bow shock marks the boundary where the supersonic solar wind is abruptly slowed to subsonic speeds. Most of the particles of the solar wind are deflected around Earth in a turbulent region (colored blue in this drawing). Earth's magnetic field also traps some charged particles in two huge, doughnut-shaped rings called the Van Allen belts (shown in red). This figure shows only a slice through the Van Allen belts.

(a) R I V **U** X G   (b) R I **V** U X G   (c) R I **V** U X G

FIGURE 9-21

VIDEO
9.1
→

The Aurorae An increased flow of charged particles from the Sun can overload the Van Allen belts and cascade toward Earth, producing aurorae. **(a)** This false-color ultraviolet image from the Dynamics Explorer 1 spacecraft shows a glowing oval of auroral emission about 4500 km in diameter centered on the north magnetic pole. **(b)** This photograph shows the aurora australis over Antarctica as seen from NASA's IMAGE satellite. **(c)** This view from Alaska shows aurorae at their typical altitudes of 100 to 400 km. The green color shows that the light is emitted by excited oxygen atoms in the upper atmosphere. (a: NASA/GSFC/Univ. of Iowa; b: NASA; c: J. Finch/Science Source)

is observed. Figures 9-21b and 9-21c show the aurorae as seen from orbit and from Earth's surface. Aurorae have also been seen on Jupiter and Saturn, and detailed studies of aurorae on Earth have led to a clear interpretation of these distant light-rings on other worlds. For example, by spotting an aurora, we can quickly say that a planet probably has an electrically conductive fluid interior that is in motion and produces a global magnetic field!

Occasionally, a violent event on the Sun's surface sends a particularly intense burst of protons and electrons toward Earth. The resulting auroral display can be exceptionally bright and can often be seen over a wide range of latitudes. Such events also disturb radio transmissions and can damage communication satellites and transmission lines.

It is remarkable that Earth's magnetosphere, including its vast belts of charged particles, was entirely unknown to us until a few decades ago. Such discoveries remind us of how little we truly understand and how much remains to be learned even about our own planet.

Earth's Magnetic Field Reverses Direction

Earth's magnetic field is produced by a **dynamo**. While a full description is beyond the scope of this book, there are several key ingredients to a **magnetic dynamo**. A dynamo requires an electrically conductive rotating fluid. All electricity produces magnetic fields, and the rotation is necessary to keep the electricity flowing. Far from being solely relevant to Earth's geology, understanding the magnetic dynamo will help us understand the Sun, Mars, Mercury, and moons around Jupiter.

In the dynamo, heat is required for convection in the liquid iron core. Recall from Section 9-1 that heat is released when water condenses from a vapor to

> Like the motions of tectonic plates, Earth's magnetic field results from our planet's internal heat.

a liquid, and this adds to a storm's energy. (Imagining the reverse is very easy: Heat is required to turn liquid water into steam.) Similarly, energy is released when material from our planet's liquid core cools and solidifies onto the solid core. This released energy powers convection in the liquid core, which combines with Earth's rotation to generate Earth's magnetic field. (While there is also convective motion in the mantle, it does not produce an appreciable magnetic field: The silicon-rich rocks that comprise the mantle are poor conductors of electricity, and their motions are a million times slower than those in the outer core.)

If Earth were a permanent magnet, like the small magnets used to attach notes to refrigerators, it would be hard to imagine how our magnetic field could spontaneously reverse direction. But in a magnetic dynamo, computer simulations show that fields produced by electrical currents in Earth's liquid iron core should reverse direction from time to time (**Figure 9-22**). The flip of Earth's field is irregular, but occurs on average about every 300,000 years. The most recent reversal occurred 780,000 years ago—before that flip, a compass needle would have pointed south, not north! While Earth's present-day magnetic field varies, the evidence does not suggest that we are heading toward another field flip anytime soon.

A historical record of Earth's magnetic field is found in the magnetized crust of the ocean seafloor. As illustrated in **Figure 9-23**, the seafloor indicates repeated reversals of Earth's magnetic field. As lava is cooled by seawater at a mid-ocean rift, the newly formed rocky crust is magnetized by Earth's magnetic field. The magma is too hot to hold a magnetic field before it solidifies, and upon cooling, the direction of the magnetic field it acquires is preserved in the oceanic crust. As seafloor spreading continues over time (Section 9-3), this magnetized rock is carried away from the ridge, in a somewhat symmetric pattern on both sides of the rift. Now imagine what happens if Earth's magnetic field reverses direction. While the older magnetized rock is carried away, new lava emerges and cools at the rift, recording Earth's

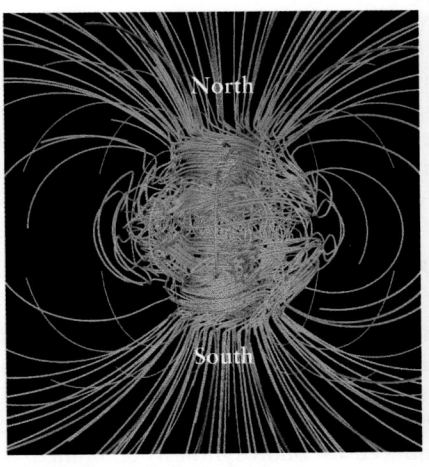

(a) Similar to today's field

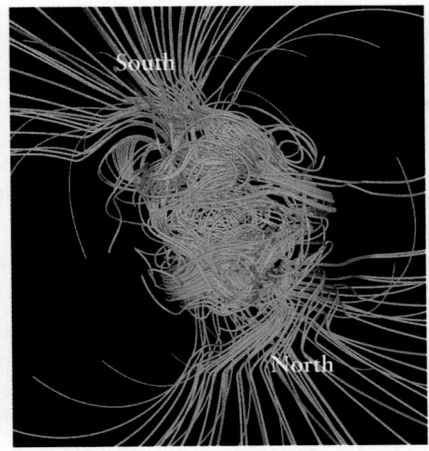

(b) After a field reversal

(c) During a reversal

FIGURE 9-22

Simulations of Earth's Magnetic Dynamo Running for almost 3 months, a supercomputer was used to simulate the complex behavior of Earth's liquid iron core and the magnetic field it produces. During the simulation, the computed magnetic field spontaneously flipped, consistent with reversals by Earth's magnetic field. **(a)** Blue indicates magnetic field lines pointed toward Earth, as they do today at our north magnetic pole. **(b)** The field has reversed. **(c)** During field reversals, which can take a few thousand years to complete, the magnetic field structure can be very complicated. (a, b, and c: The computer model was developed and run by Gary A. Glatzmaier, University of California, Santa Cruz; and Paul H. Roberts, University of California, Los Angeles)

newly reversed field. The result is magnetic stripes, where the magnetic field alternately points mostly up and down (see the magnetic field penetrating Earth in Figure 7-14b). These striations, discovered by magnetic surveys of the seafloor, provide a key piece of evidence in support of the theory of plate tectonics.

Magnetic stripes—and all they imply about a molten interior and plate tectonics—are a compelling example of how phenomena discovered on Earth shed light on other worlds in the solar system. Consider the magnetic stripes on Mars represented in Figure 7-15. Generations of scientists have pieced together the story that these stripes tell on Earth, and with one orbiting magnetometer, a similar story *might* be revealed on Mars! However, Mars shows evidence of only minor plate motion between two plates, and the magnetic stripes on Mars are still a mystery.

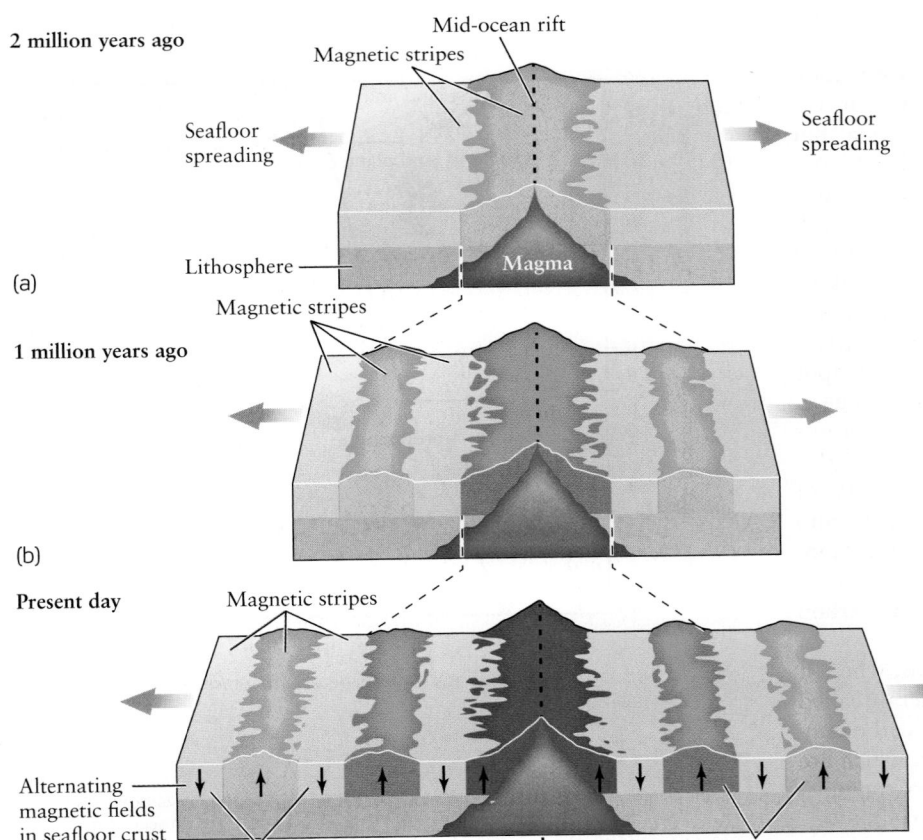

FIGURE 9-23

Magnetic Stripes on the Ocean Seafloor Magnetic measurements, illustrated here, clearly show that the seafloor preserves a record of past magnetic field reversals. New material is magnetized as it comes up at the mid-ocean rift and solidifies into the crust. Then, the newly magnetized crust is carried away by seafloor spreading as freshly solidified material repeats this process at the rift. Over millions of years, Earth's magnetic field flips several times, leaving stripes of alternating field directions in the seafloor. For simplicity, the magnetic field is drawn pointing up or down, but the actual direction depends on the latitude of the rift.

CONCEPT CHECK 9-6

If Earth's iron-rich core solidified, would a magnetic compass still point north?

CALCULATION CHECK 9-3

If the seafloor spreads away from a rift at 2 cm/yr, and Earth's magnetic field points in the same direction for 300,000 years before flipping, how many km wide is the resulting magnetic stripe?

Answers appear at the end of the chapter.

9-5 Earth's atmosphere has changed substantially over our planet's history

Both plate tectonics and Earth's magnetic field paint a picture of a planet with a dynamic, evolving surface and interior. Our atmosphere also has changed and evolved substantially over Earth's 4.54-billion-year history. As we will see, this evolution explains why we have an atmosphere totally unlike that of any other world in the solar system.

> The composition of our atmosphere has been dramatically altered by the evolution of life.

Earth's Early Atmosphere

When Earth first formed by accretion of planetesimals (see Section 8-5), gases were probably trapped within Earth's interior in the same proportions that they were present in the solar nebula. But since the early Earth was hot enough to be molten throughout its volume, most of these trapped gases were released. As we discussed in Section 7-4 and Box 7-2, Earth's gravity was too weak to prevent hydrogen and helium—the two most common kinds of atoms in the universe, but also the least massive—from leaking away into space. The atmosphere that remained still contained substantial amounts of hydrogen, but in the form of relatively massive molecules of water vapor (H_2O). In fact, water vapor was probably the dominant constituent of the early atmosphere, which is thought to have been about 100 times denser than our present-day atmosphere.

In addition to releasing water, intense volcanic activity of this early period would have released carbon dioxide (CO_2) and ammonia (NH_3). This process, in which volcanoes inject the atmosphere with gas, is called **outgassing**. As Earth cooled, much of the water condensed and fell as rain. But the rain lowered carbon dioxide levels as well.

Carbon dioxide dissolves in rainwater and the oceans, where it combines with other substances to form a class of minerals called *carbonates*. (Limestone and marble are examples of carbonate-bearing rock.) These form sediments on the ocean floor, which are eventually recycled into the crust by subduction. As an example, marble (Figure 9-19c) is a metamorphic rock formed deep within the crust from limestone, a carbonate-rich sedimentary rock.

At the same time that levels of water and carbon dioxide were decreasing, ultraviolet sunlight broke ammonia (NH_3) apart to create two gases in the atmosphere: hydrogen (H_2) and nitrogen (N_2). The hydrogen, being too light, escaped into space. Through the loss or relocation of water vapor, carbon dioxide, and hydrogen, the remaining nitrogen became the dominant component in the atmosphere. Today, nitrogen still forms the bulk of our atmosphere—about 78%—but it is the other main component that matters the most—oxygen.

Life's Impact on Earth's Atmosphere

The appearance of life on Earth set into motion a radical transformation of the atmosphere. Early single-cell organisms converted energy from sunlight into chemical energy using **photosynthesis,** a chemical process that consumes CO_2 and water and releases oxygen (O_2). Oxygen molecules are very reactive, so originally most of the O_2 produced by photosynthesis combined with other substances to form minerals called oxides. (Evidence for this can be found in rock formations of various ages. The oldest rocks have very low oxide content, while oxides are prevalent in rocks that formed after the appearance of organisms that used photosynthesis.) But as life proliferated, the amount of photosynthesis increased dramatically. Eventually, so much oxygen was being produced that it could not all be absorbed to form oxides, and O_2 began to accumulate in the atmosphere. Figure 9-24 shows how the amount of O_2 in the atmosphere has increased over the history of Earth.

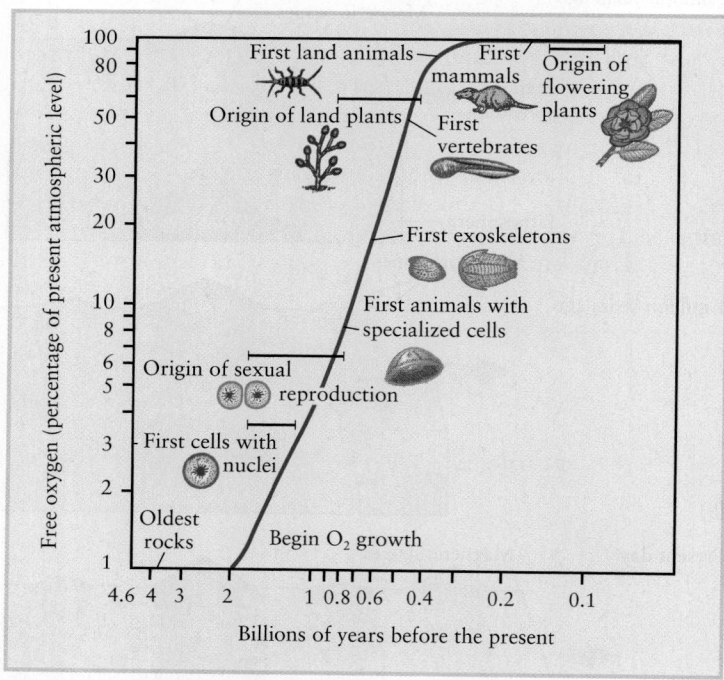

FIGURE 9-24

The Increase in Atmospheric Oxygen This graph shows how the amount of oxygen in the atmosphere (expressed as a percentage of its present-day value) has evolved with time. Note that the atmosphere contained essentially no oxygen until about 2 billion years ago. (Data from Preston Cloud, "The Biosphere," *Scientific American*, September 1983, p. 176)

About 2 billion (2×10^9) years ago, a new type of life evolved to take advantage of the newly abundant oxygen. These new organisms produced energy by consuming oxygen and releasing carbon dioxide—a process called **respiration** that is used by modern animals, including humans. Such organisms thrived because photosynthetic plants continued to add even more oxygen to the atmosphere. The abundance of atmospheric oxygen is due almost exclusively to the presence of life—a situation that has no parallel anywhere else in the solar system. For this reason, finding oxygen in an atmosphere around a planet in another solar system would signal the possible presence of life beyond Earth. And, as we saw in Section 8-7, determining the composition of atmospheres around exoplanets is now possible, so astronomers are looking for this oxygen "biosignature."

Comparing Atmospheres: Earth, Venus, and Mars

Table 9-4 shows the dramatic differences between Earth's atmosphere and those of Venus and Mars. (Mercury, the other terrestrial planet, is too small and has too little gravity to hold an appreciable atmosphere.) The greater intensity of sunlight on Venus caused higher temperatures, which boiled any liquid water and made it impossible for CO_2 to be taken out of the atmosphere and put back into rocks. Venus's atmosphere thus became far denser than our own and rich in greenhouse gases. The result was a very strong greenhouse effect that raised temperatures on Venus to their present value of about 460°C (733 K = 855°F).

Just the opposite happened on Mars, where sunlight is less than half as intense as it is on Earth. The lower temperatures drove CO_2 from the atmosphere into Martian rocks and froze water into large ice caps at its poles, as well as a permafrost beneath its surface. The atmosphere that remains on Mars has a similar composition to that of Venus but is less than 1/10,000 as dense. On neither Venus nor Mars was life able to blossom and transform the atmosphere as it did here on Earth.

Clearly, all else being equal, closer proximity to the Sun would lead to higher temperatures. However, the amount of heat-trapping greenhouse gases in a planet's atmosphere also raises a planet's temperature. Therefore, the liquid water—upon which all life on Earth depends—is a result of our location in the solar system, as well as the composition and the density of our atmosphere.

CONCEPT CHECK 9-7

Which, if any, of the two most abundant gases in Earth's present atmosphere came directly from volcanoes?

Answer appears at the end of the chapter.

TABLE 9-4 Chemical Compositions of Three Planetary Atmospheres

	Venus	Earth	Mars
Nitrogen (N_2)	3.5%	78.08%	2.7%
Oxygen (O_2)	almost zero	20.95%	almost zero
Carbon dioxide (CO_2)	96.5%	0.035%	95.3%
Water vapor (H_2O)	0.003%	about 1%	0.03%
Other gases	almost zero	almost zero	2%

9-6 Like Earth's interior, our atmosphere has a layered structure

While life has shaped our atmosphere's chemical composition, the structure of the atmosphere is controlled by the influence of sunlight. Scientists describe the structure of any atmosphere in terms of two properties: temperature and atmospheric pressure, both of which vary with altitude.

> Circulation in our atmosphere results from convection and Earth's rotation.

Pressure, Temperature, and Convection in the Atmosphere

LOOKING DEEPER 9.3 ➤ **Atmospheric pressure** at any height in the atmosphere is caused by the weight of all the air above that height. For a useful analogy, if 10 cars are stacked one on top of another, it is the bottom car that must bear all the weight from those above. It is similar for air: The pressure is greatest at Earth's surface. The average atmospheric pressure at sea level is defined to be 1 **atmosphere** (1 atm), equal to 1.01×10^5 N/m², or 14.7 pounds per square inch. As you go higher in the atmosphere, there is less air above to weigh down on you. Hence, atmospheric pressure decreases smoothly with increasing altitude.

ANIMATION 9.10 ➤ Unlike atmospheric pressure, temperature varies with altitude in a complex way. Figure 9-25 shows that temperature decreases with increasing altitude in some layers of the atmosphere, but in other layers, it actually *increases* with increasing altitude. These differences result from the individual ways in which each layer is heated. The lowest layer, called the **troposphere,** extends from the surface to an average altitude of 12 km (roughly 7.5 mi, or 39,000 ft). It is heated only indirectly by the Sun. Sunlight warms Earth's surface, which heats the lower part of the troposphere. By contrast, the upper part of the troposphere

FIGURE 9-25

Temperature Profile of Earth's Atmosphere This graph shows how the temperature in Earth's atmosphere varies with altitude. In the troposphere and mesosphere, temperature decreases with increasing altitude; in the stratosphere and thermosphere, temperature actually increases with increasing altitude.

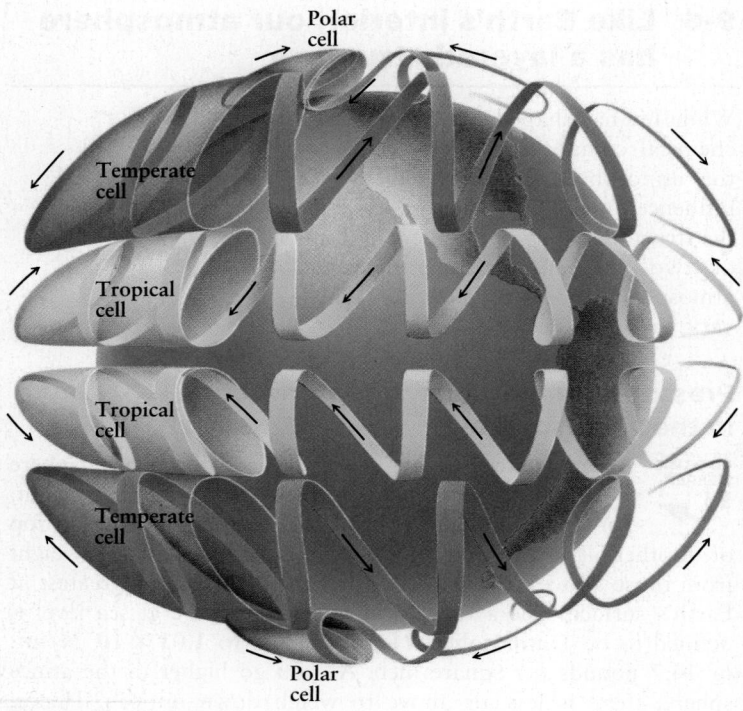

Polar
cell

Temperate
cell

Tropical
cell

Tropical
cell

Temperate
cell

Polar
cell

FIGURE 9-26

Circulation Patterns in Earth's Atmosphere The dominant circulation in our atmosphere consists of six convection cells, three in the northern hemisphere and three in the southern hemisphere. In the northern temperate region (including the continental United States), the prevailing winds at the surface are from the southwest toward the northeast. Farther south, within the northern tropical region (for example, Hawaii), the prevailing surface winds are from northeast to southwest.

remains at cooler temperatures. This vertical temperature variation causes convection currents that move up and down through the troposphere (see Section 9-1). Much of Earth's weather is a consequence of this convection.

Convection on a grand scale is caused by the temperature difference between Earth's equator and its poles. If Earth did not rotate, heated air near the equator would rise upward and flow at high altitude toward the poles. There it would cool and sink to lower altitudes, at which it would flow back to the equator. On Venus, which rotates extremely slowly, this pattern is very nearly what we see. Earth's rotation, on the other hand, breaks up this simple convection pattern into a series of **convection cells**. The structure of these cells explains why the prevailing winds blow in different directions at different latitudes (**Figure 9-26**).

CONCEPT CHECK 9-8

Is the atmospheric layer called the troposphere heated from above or below?

Answer appears at the end of the chapter.

Upper Layers of the Atmosphere

Almost all the oxygen in the troposphere is in the form of O_2, a molecule made of two oxygen atoms. But in the **stratosphere**,

which extends from about 12 to 50 km (about 7.5 to 31 mi) above the surface, an appreciable amount of oxygen is in the form of **ozone**, a molecule made of *three* oxygen atoms (O_3). Unlike oxygen (O_2), ozone is very efficient at absorbing ultraviolet radiation from the Sun, which means that the stratosphere can directly absorb solar energy. The result is that the temperature actually increases as you move upward in the stratosphere. Convection requires that the temperature must decrease, not increase, with increasing altitude, so there are essentially no convection currents in the stratosphere. This lack of convection leads to low levels of air turbulence, which is why commercial jets fly in the stratosphere.

Above the stratosphere lies the **mesosphere.** Very little ozone is found there, so solar ultraviolet radiation is not absorbed within the mesosphere, and atmospheric temperature again declines with increasing altitude.

LOOKING DEEPER 9.4 The temperature of the mesosphere reaches a minimum of about $-75°C$ ($= -103°F = 198$ K) at an altitude of about 80 km (50 mi). This minimum marks the bottom of the atmosphere's thinnest and uppermost layer, the **thermosphere,** in which temperature once again rises with increasing altitude. This temperature increase is not due to the presence of ozone because, in this very low-density region, oxygen and nitrogen are found as individual atoms rather than in molecules. Instead, the thermosphere is heated because these isolated atoms absorb very-short-wavelength solar ultraviolet radiation (which oxygen and nitrogen molecules cannot absorb).

CAUTION!

At altitudes near 300 km (about 186 mi), the temperature of the thermosphere is about 1000°C (1800°F). This altitude is near the altitude at which the space shuttle and satellites orbit Earth. Nonetheless, a satellite in orbit does *not* risk being burned up as if in a hot oven. The reason is that the thermosphere is far less dense than the atmosphere at sea level. The high temperature simply means that an average atom in the thermosphere is moving very fast (see Box 7-2). But because the thermosphere is so thin (only about 10^{-11} as dense as the air at sea level), these fast-moving atoms are few and far between. Hence, the thermosphere contains very little thermal energy. When meteors burn up in the thermosphere, it is because they move much faster than satellites, which generates heat regardless of the thermosphere's average temperature.

The *Cosmic Connections* figure compares the layered structure of our atmosphere with that of Earth's interior.

CONCEPT CHECK 9-9

What would happen to the temperature in the stratosphere if there were an absence of ozone?

Answer appears at the end of the chapter.

COSMIC CONNECTIONS

Comparing Earth's Atmosphere and Interior

 Earth's atmosphere (which is a gas) and Earth's interior (which is partly solid, partly liquid) both decrease in density and pressure as you go farther away from Earth's center.

Thermosphere:
above 80 km
(50 mi)

Aurorae

Meteors
burn up in
thermosphere.

Mesosphere:
50–80 km
(31–50 mi)

Stratosphere:
12–50 km
(7.5–31 mi)

Ozone layer

Airplanes fly
in stratosphere.

Troposphere:
ground to 12 km
(7.5 mi)

Weather occurs
in troposphere.

Structure of Earth's Atmosphere

Crust:
solid rock,
rich in silicon

Lithosphere:
rigid plates

6378 km

6343 km

Asthenosphere:
plastic flow

Mantle:
primarily
solid rock

3500 km

Outer core:
liquid iron

1300 km

Inner core:
solid iron

0 km

Earth's center

Structure of Earth's Interior

9-7   A burgeoning human population is profoundly altering Earth's biosphere

One of the extraordinary characteristics of Earth is that it is covered with life, from the floors of the oceans to the tops of mountains and from frigid polar caps to blistering deserts. So far in this chapter, we have hinted at how our planet and living organisms interact: The greenhouse effect has given Earth a suitable temperature for the evolution

> Global warming and its consequences pose a major challenge to our civilization.

of life (Section 9-1), and over billions of years that evolution has transformed the chemical composition of the atmosphere (Section 9-5). Let's explore in greater depth how Earth and the organisms that live on it, especially humans, affect each other.

The Biosphere and Natural Climate Variation

All life on Earth subsists in a relatively thin layer called the **biosphere,** which includes the land, the oceans, the crust a kilometer beneath our feet, and the stratosphere 30 km overhead where microbes have been found. Figure 9-27 is a portrait of Earth's biosphere based on NASA satellite data. The biosphere, which has taken billions of years to evolve to its present state, is a delicate, highly complex system in which plants and animals depend on each other for their mutual survival.

The state of the biosphere depends crucially on the temperatures of the oceans and atmosphere. Even small temperature changes can have dramatic consequences. An example that recurs every 3 to 7 years is the El Niño phenomenon, in which temperatures at the surface of the equatorial Pacific Ocean rise by 2 to 3°C (3.6 to 5.4°F). Ordinarily, water from the cold depths of the ocean is able to well upward, bringing with it nutrients that are used by microscopic marine organisms called phytoplankton that live near the surface (see Figure 9-27). But during an El Niño, the warm surface water suppresses this upwelling, and the phytoplankton starve. This wreaks havoc on organisms such as mollusks that feed on phytoplankton, on the fish that feed on the mollusks, and on the birds and mammals that eat the fish. During the 1982–1983 El Niño, one-quarter of the adult sea lions off the Peruvian coast starved, along with all of their pups.

Many different factors can change the surface temperature of our planet. One is that the amount of energy radiated by the Sun can vary up or down by a few tenths of a percent. Reduced solar brightness may explain the period from 1450 to 1850, when European winters were substantially colder than they are today.

Even the gravitational influences of the Moon and the planets can affect the extent of ice that covers Earth. Thanks to these astronomical influences, the eccentricity of Earth's orbit varies with a period of 90,000 to 100,000 years, the tilt of its rotation axis varies between 22.1° and 24.5° with a 40,000-year period, and the orientation of its rotation axis changes due to precession (see Section 2-6) with a 26,000-year period. These effects

Chlorophyll concentration (mg/m³)

| 0.01 | 0.1 | 1 | 10 | 50 |

Vegetation index

| 0 | 0.2 | 0.4 | 0.6 | 0.8 |

FIGURE 9-27

Earth's Biosphere This image, based on data from the SeaWiFS space-borne instrument, shows the distribution of plant life over Earth's surface. Note that a different color scale is being used for the ocean and for land masses. Phytoplankton, a microorganism in fresh and ocean water, is a plant and contains the chlorophyll mapped in this image. Phytoplankton is more abundant where cold water brings nutrients to the surface. About equal parts of Earth's atmospheric oxygen come from land-based plants and phytoplankton. (NASA) can reinforce or cancel each other and lead to variations called *Milankovitch cycles.*

The Milankovitch cycles affect the prevalence of glaciers by altering the seasonal contrast between winter (when glaciers grow) and summer (when glaciers retreat). While commonly called "ice ages," these times of periodically advancing ice sheets are actually called *glacial periods* or *glaciations.* During the most recent glaciation, ice sheets covering what is now the state of Wisconsin were over 3 km thick, and their retreat from North America began only about 10,000 years ago.

As we saw in Sections 9-1 and 9-5, one of the most important factors affecting global temperatures is the abundance of greenhouse gases such as CO_2. Geologic processes can alter this abundance, either by supplying new CO_2 (mostly through volcanism), or by removing CO_2 from the atmosphere. Removing CO_2 starts with rain, because CO_2 easily dissolves in water and is absorbed by the ground. As this groundwater comes into contact with rocky minerals, chemical bonds can bind CO_2 into the rock. From time to time in our planet's history, natural events have caused dramatic increases in the amount of greenhouse gases in the atmosphere. One such event may have taken place 250 million years ago, when Siberia went through a period of intense volcanic activity. (The evidence for this is a layer of 250-million-year-old solidified lava in Siberia that covers an area about the size of Europe!) The tremendous amounts of CO_2 released by this activity could have elevated the global temperature by several degrees. Remarkably, the fossil record reveals that 95% of all species on Earth became extinct at this same time (this is called the Permian extinction), although this event was likely caused by a combination of factors.

CONCEPT CHECK 9-10

How can a tiny 3°C increase in ocean surface temperature disrupt the food chain?

Answer appears at the end of the chapter.

Human Effects on the Biosphere: Deforestation

Our species is having an increasing effect on the biosphere because our population is skyrocketing. Figure 9-28 shows the sharp rise in the human population that began in the late 1700s with the Industrial Revolution and the spread of modern ideas about hygiene. This rise accelerated in the twentieth century thanks to medical and technological advances ranging from antibiotics to high-yield grain crops. In 1960, there were 3 billion people on Earth; in 1975, 4 billion; and in 1999, 6 billion. Projections by the United Nations Population Division show that there will be more than 8 billion people on Earth by the year 2030 and more than 9 billion by 2050.

Every human being has basic requirements: food, clothing, and housing. We all need fuel for cooking and heating. To meet these demands, we burn fossil fuel, cut down forests, and build sprawling cities. A striking example of this activity is occurring in the Amazon rain forest of Brazil. Tropical rain forests are vital to our planet's ecology because they absorb significant amounts of CO_2 and release O_2 through photosynthesis. When plant material burns or decays, the reverse is true; oxygen is consumed and CO_2 is released. Although rain forests occupy only 7% of the world's land areas, they are home to at least 50% of all plant and animal

FIGURE 9-28

The Human Population Data and estimates from the U.S. Bureau of the Census, the Population Reference Bureau, and the United Nations Population Fund were combined to produce this graph showing the human population from 500 B.C.E. to 2000 C.E. The population began to rise in the eighteenth century and has been increasing at an astonishing rate since 1900. The world's population surpassed 7.5 billion in 2017, and there is great uncertainty as to when, and at what level, it will reach its peak.

species on Earth. Nevertheless, to make way for farms and grazing land, people systematically cut down the trees and set them on fire—a process called slash-and-burn agriculture. Such deforestation, along with extensive lumbering operations, is occurring in Malaysia, Indonesia, and Papua New Guinea. The rain forests that once thrived in Central America, India, and the western coast of Africa are almost gone (Figure 9-29).

FIGURE 9-29 R I V U X G

The Deforestation of Amazonia The Amazon, the world's largest rain forest, is being destroyed at a rate of 20,000 square kilometers (nearly 8000 square miles) per year in order to provide land for grazing and farming and as a source for lumber. About 80% of the logging is being carried out illegally. (Martin Wendler/Science Source)

CONCEPT CHECK 9-11

In what two ways does deforestation through slash-and-burn agriculture increase Earth's atmospheric carbon dioxide?

Answer appears at the end of the chapter.

Human Effects on the Biosphere: The Ozone Layer

Human activity is also having potentially disastrous effects on the upper atmosphere. Certain chemicals released into the air—in particular chlorofluorocarbons (CFCs), which have been used in refrigeration and electronics, and methyl bromide, which is used in fumigation—are destroying the ozone in the stratosphere. We saw in Section 9-6 that ozone molecules absorb ultraviolet radiation from the Sun. Without this high-altitude **ozone layer,** solar ultraviolet radiation would beat down on Earth's surface with greatly increased intensity. Such radiation breaks apart most of the delicate molecules that form living tissue. Hence, a complete loss of the ozone layer would lead to a catastrophic ecological disaster.

In 1985, scientists discovered an **ozone hole,** a region with an abnormally low concentration of ozone, over Antarctica. Since then this hole has generally expanded from one year to the next (**Figure 9-30**), and only recently has shown signs of recovering, thanks in large part to regulatory efforts. Smaller but still serious effects have been observed in the stratosphere above other parts of Earth. As a result, there has been a worldwide increase in the number of deaths due to skin cancer caused by solar ultraviolet radiation. By international agreement, CFCs are being replaced by compounds that do not deplete stratospheric ozone, and sunlight naturally, but slowly, produces more ozone in the stratosphere. Agreements to limit CFCs are hailed as some of the greatest successes of the international community in addressing global environmental threats.

CONCEPT CHECK 9-12

If ozone is not actually leaking out of the ozone hole, what is moving through this hole?

Answer appears at the end of the chapter.

Human Effects on the Biosphere: Global Warming and Climate Change

The most troubling influence of human activity on the biosphere is a consequence of burning fossil fuels (natural gas, petroleum, and coal) in automobiles, airplanes, and power plants, as well as burning forests and brushland for agriculture and cooking (Figure 9-29). This burning releases carbon dioxide into Earth's atmosphere—and we are now adding CO_2 to the atmosphere faster than plants and geological processes can extract it. **Figure 9-31** shows how the carbon dioxide content of the atmosphere has increased since the Industrial Revolution.

To put the values shown in Figure 9-31 into perspective, we need to know the atmospheric CO_2 concentration in earlier eras. Scientists have learned this by analyzing air bubbles trapped at various depths in the ice that blankets the Antarctic and Greenland. Each winter a new ice layer is deposited, so the depth of a bubble indicates the year in which it was trapped. **Figure 9-32** uses data obtained in this way to show how the atmospheric CO_2 concentration has varied since 1000 C.E. While there has been some natural variation in the concentration, its value has skyrocketed since the beginning of the Industrial Revolution around 1800. Data from older, deeper bubbles of trapped air show that in the 650,000 years before the Industrial Revolution, the CO_2 concentration

October 1979   0 100 200 300 400 500 600 700   September 2016
                        Total ozone (Dobson units)

FIGURE 9-30

The Antarctic Ozone Hole These two false-color images show that there was a net decrease of 50% in stratospheric ozone over Antarctica between October 1979 and September 2016. The amount of ozone at midlatitudes, where most of the human population lives, decreased by 10% to 20% over the same period. (NASA's Goddard Space Flight Center)

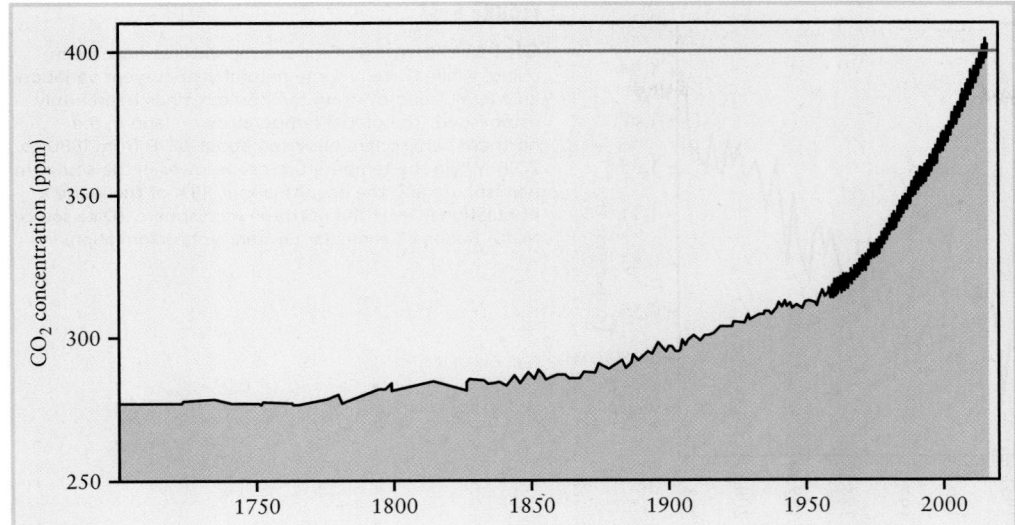

FIGURE 9-31

Atmospheric Carbon Dioxide Is Rising

Fast This graph shows a dramatic increase in atmospheric carbon dioxide that begins around the time of the Industrial Revolution (1880). Carbon dioxide levels are measured in parts per million, or ppm. In 2016, carbon levels exceeded 400 ppm (green line), which is a 40% increase from the preindustrial era. This increase is very significant, as carbon dioxide levels haven't exceeded 400 ppm in 4 million years! This graph is made from two distinct methods of measurement. Since 1958, direct measurements of carbon dioxide have been made at Mauna Loa, Hawaii. These ongoing measurements are frequent enough that they capture natural seasonal variations, leading to a thicker line in the graph after 1958. Prior to 1958, atmospheric carbon dioxide measurements were made from ice cores. (Data from the Scripps CO₂ Program)

was never greater than 300 parts per million. The 400 parts per million level was exceeded in 2016, which is a 40% increase from the year 1800.

Increasing atmospheric CO_2 is of concern because this strengthens the greenhouse effect and raises Earth's average surface temperature. Figure 9-32 also shows how this average temperature has varied since 1000 C.E. (Like the CO_2 data, the temperature data from past centuries comes from analyzing trapped air bubbles. Scientists measure the relative amounts of two different types of atmospheric oxygen called ^{16}O and ^{18}O, described in Box 5-5, in a bubble. The ratio of these is a sensitive measure of the temperature at that time the air was trapped inside the ice.) The recent dramatic increase in atmospheric CO_2 concentration has produced an equally dramatic increase in the average surface temperature (**Figure 9-33**). This temperature increase is called **global warming.** Other explanations for

global warming have been proposed, such as changes in the Sun's brightness, but only greenhouse gases produced by human activity can explain the steep temperature increase shown in Figure 9-32.

CONCEPT CHECK 9-13

Is the greenhouse effect the same as global warming?

CONCEPT CHECK 9-14

As shown in Figure 9-33, the average temperature from about 2002 to 2010 did not rise much. Are these recent measurements consistent with a long-term rise in temperatures?

Answers appear at the end of the chapter.

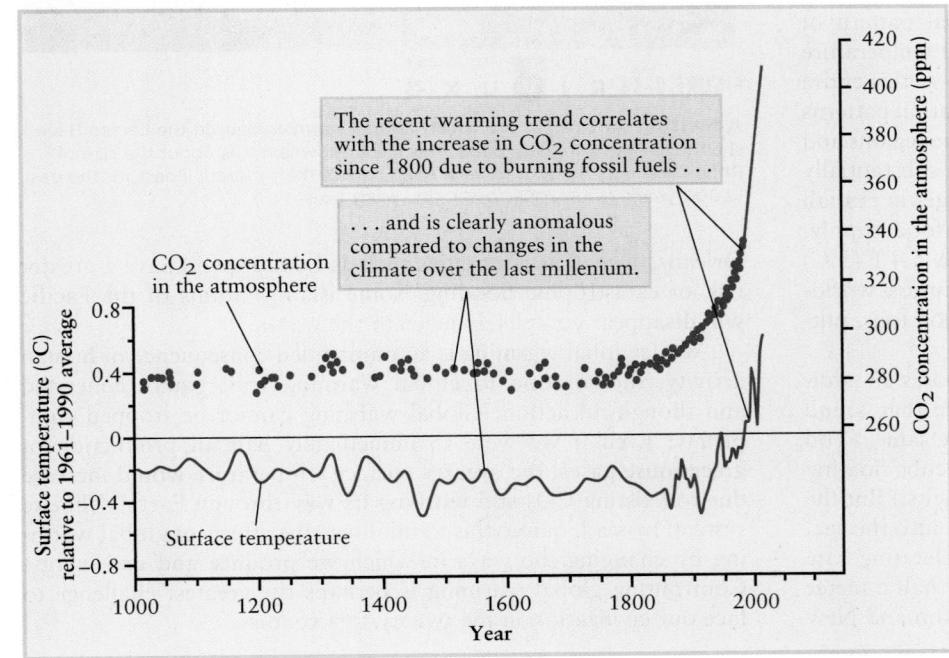

The recent warming trend correlates with the increase in CO_2 concentration since 1800 due to burning fossil fuels . . .

. . . and is clearly anomalous compared to changes in the climate over the last millenium.

CO_2 concentration in the atmosphere

Surface temperature

FIGURE 9-32

Atmospheric CO₂ and Changes in Global

Temperature This figure shows how the carbon dioxide concentration in our atmosphere (upper curve) and Earth's average surface temperature (lower curve) have changed since 1000 C.E. The increase in CO₂ since 1800 due to burning fossil fuels has strengthened the greenhouse effect and caused a dramatic temperature increase. (Data from Intergovernmental Panel on Climate Change and Hadley Centre for Climate Prediction and Research, UK Meteorological Office)

FIGURE 9-33

Global Warming Global temperatures have been rising. While there is some natural year-to-year variation, an overall trend of rising temperatures has been firmly established. The global temperature on land in the northern hemisphere has risen about 3.7°F from 1880 to 2016. While the temperature rise is lower in the southern hemisphere and the oceans, about 88% of the world's population lives in the northern hemisphere. (Data source: NOAA National Centers for Environmental information)

Impact of Global Warming

The effects of global warming can be seen around the world. Glaciers worldwide are receding, the area of sea ice cover around the north pole has decreased by over a third since 1979, and a portion of the Antarctic ice shelf has broken off (Figure 9-34).

Unfortunately, global warming is predicted to intensify in the decades to come. The United Nations Intergovernmental Panel on Climate Change predicts that if nothing is done to decrease the rate at which we add greenhouse gases to our atmosphere, the average surface temperature will continue to rise by an additional 3.6 to 7.2°F (2 to 4°C) during the twenty-first century. As of 2017, the United States has pulled out of the international Paris Agreement to reduce CO_2 emissions, making the higher estimate for the temperature increase more likely.

What is worse, the temperature increase is predicted to be greater at the poles than at the equator. The global pattern of atmospheric circulation (Figure 9-26) depends on the temperature difference between the warm equator and cold poles, so this entire pattern will be affected. The same is true for the circulation patterns in the oceans. As a result, temperatures will rise in some regions and decline in others, and the patterns of rainfall will be substantially altered. Agriculture depends on rainfall, so these changes in rainfall patterns can cause major disruptions in the world food supply. Studies suggest that the climate changes caused by a 5.4°F (3°C) increase in the average surface temperature could cause a world-wide drop in wheat and corn crops of 20 to 400 million tons, putting hundreds of millions of people at risk of hunger.

The melting of polar ice due to global warming poses an additional risk to our civilization. When floating ice like that found near the north pole melts, the ocean level remains the same. (You can see this by examining a glass of water with an ice cube floating in it. The water level does not change when the ice melts.) But the ocean level rises when ice on land melts and runs off into the sea. The Greenland ice cap has been melting at an accelerating rate since 2000 and has the potential to raise sea levels by half a meter or more. Low-lying coastal communities such as Boston and New

FIGURE 9-34 R I V U X G

A Melting Antarctic Ice Shelf Global warming caused the Larsen B ice shelf to break up in early 2002. This ice shelf, which was about the size of Rhode Island, is thought to have been part of the Antarctic coast for the past 12,000 years. (NASA/GSFC/LaRC/JPL, MISR Team)

Orleans, as well as river cities such as London, will have a greater risk of catastrophic flooding. Some island nations of the Pacific will disappear completely beneath the waves.

While global warming is an unintended consequence of human activity, the solution to global warming will require concerted and thoughtful action. Global warming cannot be stopped completely: Even if we were to immediately halt all production of greenhouse gases, the average surface temperature would increase due to existing CO_2 still working its way through Earth's climate system. Instead, our goal is to minimize the effects of global warming by changing the ways in which we produce and use energy. Confronting global warming is perhaps the greatest challenge to face our civilization in the twenty-first century.

KEY WORDS

albedo, p. 245
asthenosphere, p. 249
atmosphere (atm), p. 259
atmospheric pressure, p. 259
aurora (*plural* aurorae),
  p. 255
aurora australis, p. 255
aurora borealis, p. 255
biosphere, p. 262
bow shock, p. 255
convection, p. 243
convection cell, p. 260
convection current, p. 243
core, p. 246
crust (of Earth), p. 246
crystal, p. 249

dynamo, p. 256
earthquake, p. 246
epicenter, p. 246
global warming, p. 265
greenhouse effect, p. 245
greenhouse gas, p. 245
igneous rock, p. 254
inner core (of Earth), p. 247
lava, p. 254
lithosphere, p. 251
magma, p. 254
magnetic dynamo, p. 256
magnetopause, p. 255
magnetosphere, p. 255
mantle, p. 246
melting point, p. 247

mesosphere, p. 260
metamorphic rock, p. 254
mineral, p. 254
northern lights, p. 255
oceanic rift, p. 251
outer core (of Earth), p. 247
outgassing, p. 258
ozone, p. 260
ozone hole, p. 264
ozone layer, p. 264
P wave, p. 246
photosynthesis, p. 258
plastic, p. 247
plate (lithospheric), p. 249
plate tectonics, p. 251
respiration, p. 259

ridge push, p. 252
rock, p. 254
S wave, p. 246
seafloor spreading, p. 251
sedimentary rock, p. 254
seismic wave, p. 246
seismometer, p. 246
slab pull, p. 252
solar wind, p. 255
southern lights, p. 255
stratosphere, p. 260
subduction zone, p. 252
surface wave, p. 246
thermosphere, p. 260
troposphere, p. 259
Van Allen belts, p. 255

KEY IDEAS

Earth's Energy Sources: All activity in Earth's atmosphere, oceans, and surface is powered by three sources of energy (Section 9-1).

- Solar energy is the energy source for the atmosphere. In the greenhouse effect, some of this energy is trapped by infrared-absorbing gases in the atmosphere, raising Earth's surface temperature.

- Tidal forces from the Moon and Sun help to power the motion of the oceans.

- The internal heat of Earth is the energy source for geologic activity.

Earth's Interior: Studies of seismic waves (vibrations produced by earthquakes) show that Earth has a small, solid inner core surrounded by a liquid outer core. The outer core is surrounded by the dense mantle, which in turn is surrounded by the thin low-density crust (Section 9-2).

- Seismologists deduce Earth's interior structure by studying how longitudinal P waves and transverse S waves travel through Earth's interior.

- Earth's inner and outer cores are composed of almost pure iron with some nickel mixed in. The mantle is composed of iron-rich minerals.

- Both temperature and pressure increase with depth inside Earth.

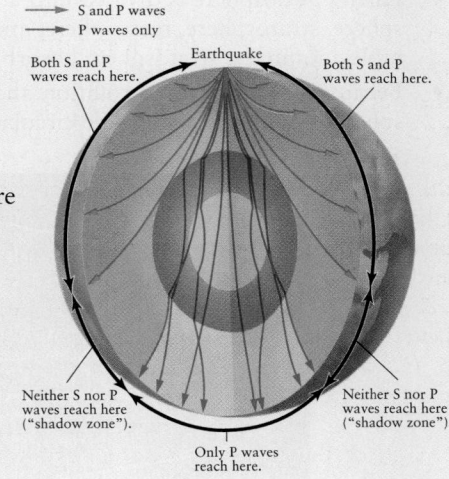

Plate Tectonics: Earth's crust and a small part of its upper mantle form a rigid layer called the lithosphere. The lithosphere is divided into huge plates that move about over the plastic layer called the asthenosphere in the upper mantle (Section 9-3).

- Plate tectonics, or movement of the plates, is driven by convection within the asthenosphere. Molten material wells up at oceanic rifts, producing seafloor spreading, and is returned to the asthenosphere in subduction zones. As one end of a plate is subducted back into the asthenosphere, it helps to pull the rest of the plate along.

- Plate tectonics is responsible for most of the major features of Earth's surface, including mountain ranges, volcanoes, and the shapes of the continents and oceans.

- Plate tectonics is involved in the formation of the three major categories of rocks: igneous rocks (cooled from molten material), sedimentary rocks (formed by sediments in water), and metamorphic rocks (altered by extreme heat and pressure).

Doug Martin/Science Source

Dorling Kindersley/Getty Images

Juniors/Superstock

Earth's Magnetic Field and Magnetosphere: Electric currents in the liquid outer core generate a magnetic field. This magnetic field produces a magnetosphere that surrounds Earth (Section 9-4).

- Most of the particles of the solar wind are deflected around Earth by the magnetosphere.

- Some charged particles from the solar wind are trapped in two huge, doughnut-shaped rings called the Van Allen belts. An excess of these particles can initiate an auroral display.

- Earth's magnetic field flips irregularly, but with an average of about 300,000 years between reversals. These field flips are preserved in magnetized rock on the ocean seafloor.

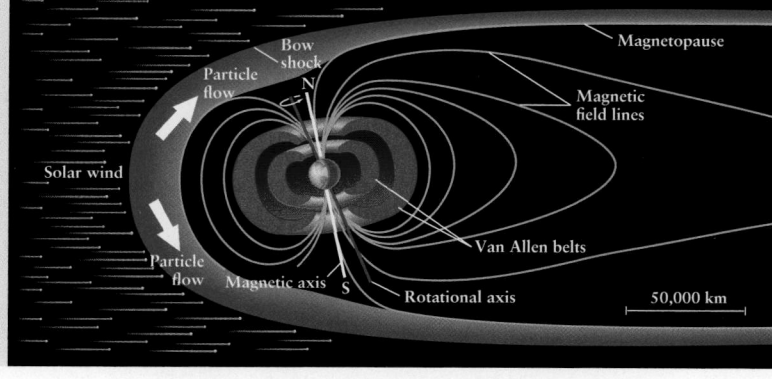

Earth's Atmosphere: Earth's atmosphere differs from those of the other terrestrial planets in its chemical composition, circulation pattern, and temperature profile (Sections 9-5 and 9-6).

- Earth's atmosphere evolved from being mostly water vapor to its present composition of 78% nitrogen and 21% oxygen.

- Earth's atmosphere is divided into layers called the troposphere, stratosphere, mesosphere, and thermosphere. Ozone molecules in the stratosphere absorb ultraviolet light.

- Because of Earth's rapid rotation, the circulation in its atmosphere is complex, with three circulation cells in each hemisphere.

The Biosphere: Human activity is changing Earth's biosphere, on which all living organisms depend (Sections 9-7).

- Industrial chemicals released into the atmosphere have damaged the ozone layer in the stratosphere.

- Deforestation and the burning of fossil fuels are increasing the greenhouse effect in our atmosphere and causing a global warming of the planet.

QUESTIONS

> **Problem-solving tips and tools**
>
> You can find most of the Earth data that you need in Tables 9-1 and 9-3. You will need to know that a sphere has surface area $4\pi r^2$ and volume $4\pi r^3/3$, where r is the sphere's radius. You may have to consult an atlas to examine the geography of the South Atlantic. Also, remember that the average density of an object is its mass divided by its volume. Sections 5-3 and 5-4 and Box 5-2 describe how to solve problems involving blackbody radiation, Section 2-6 discusses precession, and Section 4-4 describes the properties of elliptical orbits.

Section 9-1

1. Why are typical rocks found on Earth's surface much younger than Earth itself?

2. **TUTORIAL 9.1** What is convection? What causes convection in Earth's atmosphere?

3. Consider a variation of the pot of water in Figure 9-5. If the source of heat were placed at the top of the pot (instead of the bottom), would convection currents flow in the opposite direction? Explain your answer.

4. Describe how energy is transferred from Earth's surface to the atmosphere by both convection and radiation.

5. If heat flows to Earth's surface from both the Sun and Earth's interior, why do we say that the motions of the atmosphere are powered by the Sun?

6. How does solar energy help power the motion of water in Earth's oceans?

7. How does the greenhouse effect influence the temperature of the atmosphere? Which properties of greenhouse gases in the atmosphere cause this effect?

8. The total power in sunlight that reaches the top of our atmosphere is 1.75×10^{17} W. (a) How many watts of power are reflected back into space due to Earth's albedo? (b) If Earth had no atmosphere, all of the solar power that was not reflected would be absorbed by Earth's surface. In equilibrium, the heated surface would act as a blackbody that radiates as much power as it absorbs from the Sun. How much power would the entire Earth radiate? (c) How much power would one square meter of the surface radiate? (d) What would be the average temperature of the surface? Give your answer in both the Kelvin and Celsius scales. (e) Why is Earth's actual average temperature higher than the value you calculated in (d)?

Section 9-2

9. **TUTORIAL 9.2** How do we know that part of the Earth is molten?

10. What are the different types of seismic waves? Why are seismic waves useful for probing Earth's interior structure?

11. Describe the interior structure of Earth.

12. What is a plastic material? Which parts of Earth's interior are described as being plastic?

13. The deepest wells and mines go down only a few kilometers. What then is the evidence that iron is abundant in Earth's core? That Earth's outer core is molten but the inner core is solid?

14. The inner core of Earth is at a higher temperature than the outer core. Why, then, is the inner core solid and the outer core molten instead of the other way around?

15. On average, the temperature beneath Earth's crust increases at a rate of 20°C per kilometer. At what depth would water boil? (Assume the surface temperature is 20°C and ignore the effect of the pressure of overlying rock on the boiling point of water.)

16. What fractions of Earth's total volume are occupied by the core, the mantle, and the crust?

17. What fraction of the total mass of Earth lies within the inner core?

18. (a) Using the mass and diameter of Earth listed in Table 9-1, verify that the average density of Earth is 5500 kg/m^3. (b) Assuming that the average density of material in Earth's mantle is about 3500 kg/m^3, what must the average density of the core be? Is your answer consistent with the values given in Table 9-3?

Section 9-3

19. Describe the process of plate tectonics. Give specific examples of geographic features created by plate tectonics.

20. Explain how convection in Earth's interior drives the process of plate tectonics.

21. Why do you suppose that active volcanoes, such as Mount St. Helens in Washington State, are usually located in mountain ranges that border on subduction zones?

22. What is the difference between a rock and a mineral?

23. What are the differences among igneous, sedimentary, and metamorphic rocks? What do these rocks tell us about the sites at which they are found?

24. Why do geologists suspect that Pangaea was the most recent in a succession of supercontinents?

25. Identical fossils of the reptile *Mesosaurus*, which lived 300 million years ago, are found in eastern South America and western Africa and nowhere else in the world. Explain how these fossils provide evidence for the theory of plate tectonics.

26. Measurements of the seafloor show that the Eurasian and North American plates have moved 60 km apart in the past 3.3 million years. How far apart (in millimeters) do they move in 1 year? (By comparison, your fingernails grow at a rate of about 50 mm/year.)

27. The oldest rocks found on the continents are about 4 billion years old. By contrast, the oldest rocks found on the ocean floor are only about 200 million years old. Explain why there is such a large difference in ages.

Section 9-4

28. Why do geologists think that Earth's magnetic field is produced in the liquid outer core rather than in the mantle?

29. Describe Earth's magnetosphere. If Earth did not have a magnetic field, do you think aurorae would be more common or less common than they are today?

30. What are the Van Allen belts?

31. Explain how magnetic stripes are created around a rift on the ocean seafloor.

32. If the seafloor spreads from a rift at a speed of 2.5 cm/yr and a magnetic stripe is produced that is 20 km wide, for how many years is Earth's magnetic field in the same orientation without flipping?

33. It was stated in Section 7-7 that iron loses its magnetism at temperatures above 770°C. Use this fact and Figure 9-10 to explain why Earth's magnetic field cannot be due to a magnetized solid core, but must instead be caused by the motion of electrically conducting material in the liquid core.

34. Most auroral displays (like those shown in Figure 9-21c) have a green color dominated by emission from oxygen atoms at a wavelength of 557.7 nm. (a) What minimum energy (in electron volts) must be imparted to an oxygen atom to make it emit this wavelength? (b) Why is your answer in (a) a *minimum* value?

Section 9-5

35. Summarize the history of Earth's atmosphere. What role has biological activity played in this evolution?

36. Describe how the present-day atmosphere and surface temperature of Earth might be different (a) if carbon dioxide had never been released into the atmosphere; (b) if carbon dioxide had been released, but life had never evolved on Earth.

37. The following photograph shows the soil at Bryce Canyon National Park, Utah. The color of the whitish layer is due to a lack of iron oxide. More recent soils typically contain iron oxide and have a darker color. Explain what this tells us about the history of Earth's atmosphere.

(Frank Vetere/Alamy Stock Photo) R I **V** U X G

Section 9-6

38. Earth's atmospheric pressure decreases by a factor of one-half for every 5.5-km increase in altitude above sea level. Construct a plot of pressure versus altitude, assuming the pressure at sea level is 1 atmosphere (1 atm). Discuss the characteristics of your graph. At what altitude is the atmospheric pressure equal to 0.001 atm?

Section 9-7

39. Ozone and carbon dioxide each make up only a fraction of a percent of our atmosphere. Why, then, should we be concerned about small increases or decreases in the atmospheric abundance of these gases?

40. How are scientists able to measure what the atmospheric CO_2 concentration and average surface temperature were in the distant past?

41. Does global warming increase the surface temperature of all parts of Earth by equal amounts or by different amounts? What consequences does this have?

42. Earth is at perihelion on January 3 and at aphelion on July 4. Because of precession, in 15,000 C.E. the amount of sunlight in summer will be more than at present in the northern hemisphere but less than at present in the southern hemisphere. Explain why.

43. Antarctica has an area of 13 million square km and is covered by an ice cap that varies in thickness from 300 m near the coast to 1800 m in the interior. Estimate the volume of this ice cap. Assuming that water and ice have roughly the same density, estimate the amount by which the water level of the world's oceans would rise if Antarctica's ice were to melt completely (see Figure 9-34). What portions of Earth's surface would be inundated by such a deluge?

44. The human population on Earth is currently doubling about every 30 years. Describe the various pressures placed on Earth by uncontrolled human population growth. Can such growth continue indefinitely? If not, what natural and human controls might arise to curb this growth? It has been suggested that overpopulation problems could be solved by colonizing the Moon or Mars. Do you think this is a reasonable solution? Explain your answer.

45. In order to alleviate global warming, it will be necessary to dramatically reduce the amount of carbon dioxide that we release into the atmosphere by burning petroleum. What changes in technology and society do you think will be needed to bring this about?

General Problems

46. Describe the various ways in which Earth is unique among the planets of our solar system.

47. Describe the various ways in which Earth's surface is reshaped over time.

WEB/E-BOOK QUESTIONS

48. Search the internet for information about Rodinia, the supercontinent that predated Pangaea. When did this supercontinent form? When did it break apart? Do we have any evidence of supercontinents that predated Rodinia? Why or why not?

49. Because of plate tectonics, the arrangement of continents in the future will be different from today. Search the internet for information about "Pangaea Ultima," a supercontinent that may form in the distant future. When is it predicted to form? How will it compare to the supercontinent that existed 200 million years ago (see Figure 9-13a)?

50. Use the internet to investigate the current status of the Antarctic ozone hole. Is the situation getting better or worse? Is there a comparable hole over the north pole? Why do most scientists blame the chemicals called CFCs for the existence of the ozone hole?

51. Use the internet to research predictions of Earth's future surface temperature. What are some predictions for the surface temperature in 2050? In 2100? What effects may the increased temperature have on human health?

52. The Kyoto Protocol is an agreement made by a number of nations around the world to reduce their production of greenhouse gases. Use the internet to investigate the status of the Kyoto Protocol. How many nations have signed the protocol? What fraction of the world's greenhouse gas production comes from these nations? Have any developed nations failed to sign? How effective has the protocol been?

53. **Observing Mountain Range Formation.** Access the two animations "The Collision of Two Plates: South America" and "The Collision of Two Plates: The Himalayas" in Chapter 9 of the *Universe* website or e-book. (a) In which case are the mountains formed by tectonic uplift? In which case are the mountains formed by volcanoes from rising lava? (b) For each animation, describe which plate is moving in which direction.

ACTIVITIES

Starry Night™ Explorations

54. Use *Starry Night*™ to view Earth from space. Select **Favorites > Explorations > Earth** from the menu to see a view of Earth from a position 12,000 km above the surface. Use the location scroller, zoom, and time controls to explore the view. Is there any evidence for the presence of an atmosphere on Earth? Explain your answer.

55. Use *Starry Night*™ to view Earth from space. Open the view named **Favorites > Explorations > Earth**. Click the name in the label for Earth and select **Surface Image/Model > No clouds** to examine the surface of Earth without the interference of its atmosphere. Use the location scroller, zoom, and time flow controls to inspect the entire surface of our home planet.
    (a) What evidence, if any, can you see of plate tectonics on Earth?
    (b) Can you see any evidence of life or artificial structures or objects?

ANSWERS

ConceptChecks

ConceptCheck 9-1: For convection to occur, gravity is necessary, as well as a cooler temperature on top of the fluid. Convection depends on lower-density material rising when surrounded by higher-density material, and this requires gravity. But the density depends on temperature, and if the entire fluid were at the same temperature, there would be no changes in fluid density; thus, a cooler temperature on top is needed.

ConceptCheck 9-2: By trapping heat, the greenhouse effect keeps Earth's temperature warmer than it would otherwise be, allowing liquid water and hence life to exist on Earth.

ConceptCheck 9-3: Because longitudinal S waves are unable to travel through liquid, the sudden appearance of S waves would suggest that the planet's interior has solidified.

ConceptCheck 9-4: The largest force is slab pull, as the dense sinking portion of the plate pulls on the entire seafloor plate. Another force is due to the portion of the plate on the side of the underwater mountains forming the mid-ocean ridge. The weight of this portion of the plate pulls

it down the ridge and pushes the entire plate; this force is called ridge push.

ConceptCheck 9-5: In one scenario, a supercontinent traps heat beneath it, causing that region to expand and crack. This process allows fissures to be filled by rapidly upwelling material from underneath, breaking the continent apart.

ConceptCheck 9-6: No. Earth's magnetic field is predominantly created by a dynamo effect caused by the movement of iron-rich material in the liquid core.

ConceptCheck 9-7: Nitrogen and oxygen are the most abundant gases in our atmosphere, but neither came directly from volcanoes. Ammonia (NH_3) came directly from volcanoes, but after sunlight broke this gas apart, only nitrogen gas remained. Later, photosynthesizing microorganisms produced oxygen.

ConceptCheck 9-8: As sunlight warms Earth's surface, energy in the form of heat increases the temperature of the troposphere near Earth's surface such that the troposphere is heated from below.

ConceptCheck 9-9: Ozone in the stratosphere absorbs ultraviolet radiation from the Sun and helps warm the stratosphere. So a deficit of ozone would make the stratosphere cooler in temperature.

ConceptCheck 9-10: A 3°C change in ocean surface temperature is sufficient to starve plankton, the base of the food chain, by preventing cooler, nutrient-rich water from moving to the surface.

ConceptCheck 9-11: Living plants remove CO_2 from the atmosphere through photosynthesis, and burning plants release the CO_2 they have captured back into the atmosphere.

ConceptCheck 9-12: Atmospheric ozone prevents much of the Sun's harmful ultraviolet light from entering Earth's atmosphere. The ozone hole is a region allowing more ultraviolet light to enter Earth's atmosphere.

ConceptCheck 9-13: No. Our planet's natural greenhouse effect has kept the planet warm for billions of years. Global warming refers to the sudden and significant rise in temperatures since the 1800s.

ConceptCheck 9-14: Yes. As can be seen in Figure 9-33, global temperatures do not rise every single year, or every decade, and global warming refers to the overall rising trend since the 1800s.

CalculationChecks

CalculationCheck 9-1: Using the relationship that distance equals rate times $(6 \times 10^8 \text{ cm}) \div 2 \text{ cm/year} = 3 \times 10^8$ years or 300 million years.

CalculationCheck 9-2: $(5 \times 10^9 \text{ years}) \div (5 \times 10^8 \text{ years/cycle}) = 10$ cycles.

CalculationCheck 9-3: The width of the stripe is $(\text{speed}) \times (\text{time}) = (2 \text{ cm/yr}) \times (300{,}000 \text{ yr}) = 600{,}000 \text{ cm}$. $1 \text{ km} = 10^5 \text{ cm}$, so the width is $(6 \times 10^5 \text{ cm}) \times (1 \text{ km}/10^5 \text{ cm}) = 6 \text{ km}$ wide (or about 3.7 mi).

Climate Change: Clearing Up Misconceptions That Cloud The Issue

by Jay Phelan

Misconceptions about climate change arise from its scale and complexity. And addressing it requires cooperation among many countries, corporations, and individuals, with varied interests and priorities. Nonetheless, the science is straightforward, as are the necessary courses of action. The misconceptions can (and must) be clarified and corrected.

Misconception: The climate changes naturally. It has changed before and will change again. Analyses of ice cores, tree rings, and lake sediments have helped us understand long-term historical patterns of climate change. These reveal changes that occurred over thousands or even millions of years, rather than decades, as we're observing now. Also, past climate changes were not accompanied by the large increase in atmospheric carbon dioxide seen over the past 100 years. Recent increases in CO_2 levels are due to human activities—80% due to fossil fuel burning and 20% to deforestation—with no changes in natural sources of CO_2 emission.

Misconception: We had more snow the other year than ever before, so there can't be any global warming. Anecdotal observations are appealing, but also misleading. Weather events are short-term phenomena, influenced by regional factors. Climate, on the other hand, is the long-term average of the weather. Cold spells or unusually heavy snowfalls or storms do occur, but the benchmark measure is *average* temperature. This has increased consistently since the 1950s.

Jay Phelan is a professor of biology at UCLA, where he has taught introductory biology in large lectures for majors and nonmajors for 15 years. He received his PhD in evolutionary biology from Harvard in 1995, and his master's and bachelor's degrees from Yale and UCLA, respectively. His primary area of research is evolutionary genetics, and his original research has been published in *Evolution, Experimental Gerontology,* and the *Journal of Integrative and Comparative Biology*, among others. His research has been featured on Nightline, CNN, the BBC, and NPR; in *Science Times* and *Elle* magazine; and in more than 100 newspapers. He is the recipient of more than a dozen teaching awards. With Terry Burnham, Jay is the coauthor of the international best-seller *Mean Genes: From Sex to Money to Food—Taming Our Primal Instincts*. Written for the general reader, *Mean Genes* explains in simple terms how knowledge of the genetic basis of human nature can empower individuals to lead more satisfying lives. Writing for a nonscientific audience has honed Phelan's writing style to one that is casual and inviting to students but also scientifically precise.

Misconception: It's the ozone hole that causes global warming, and we've fixed that. Depletion of ozone in the atmosphere—much of which was human-caused—increased the UV radiation reaching Earth. This radiation causes harmful mutations in organisms, but has little impact on temperatures. Several other human-caused environmental issues—including acid rain, nuclear waste, and heat from car exhausts—also don't have a significant impact on Earth's climate.

Misconception: Plants will grow faster, winters will be milder. Global warming isn't such a bad thing. Yes, the impact of global warming will vary by location, and in some places may seem positive. But globally, the negative consequences will vastly outweigh any positive changes. Coral reef die-offs, for example, will damage the entire food webs of the oceans. Melting polar ice will flood coastal cities and small islands around the world, displacing many millions of people. Though not obvious, rising sea levels will also lead to longer droughts and altered wind patterns that may produce intense storms, tornadoes, and other extreme weather events.

Misconception: So, maybe climate change is occurring, but there's nothing we can do. We can reduce greenhouse gas emissions by burning fewer fossil fuels. We can develop new, renewable energy sources. However, because CO_2 emissions diffuse evenly throughout Earth's atmosphere, all nations must work together. The Kyoto Protocol is an international agreement created to address climate change. Lack of full commitment from the countries responsible for most greenhouse gas emissions (including the United States) highlights the difficulty in securing international cooperation.

Misconception: The Sun is to blame. It has gotten hotter, or gotten closer to Earth. Solar output—which has been closely monitored by NASA and other independent organizations for several decades—has not increased. A peer-reviewed summary of published research on solar energy concluded that fluctuations in solar output have had a negligible impact on climate and cannot explain the magnitude or direction (consistently upward) of observed climate change over the past hundred years.

Misconception: Not everyone agrees; there's no consensus among the experts. There is debate among scientists about how best to model the climate and predict changes. And in the 1970s, there were even some predictions of "global cooling." As data accumulate, hypotheses are refined and improved. Today, there is near-universal agreement among climate scientists that climate change is occurring and is the result of human activities. In a recent analysis of the publications of 1372 climate researchers, more than 97% of the researchers were in agreement about the existence and cause of global climate change.

10 | Our Barren Moon

LEARNING GOALS

By reading the sections of this chapter, you will learn

10-1 The nature of the Moon's surface

10-2 The story of human exploration of the Moon

10-3 How we have learned about the Moon's interior

10-4 How Moon rocks differ from rocks on Earth

10-5 Why scientists think the Moon formed as the result of a violent collision between worlds

Other than Earth itself, the most familiar world in the solar system is the Moon. It is so near to us that many of its surface features are visible to the naked eye. But the Moon is also a strange and alien place, with dramatic differences from our own Earth. It has no substantial atmosphere, no global magnetic field, and no liquid water of any kind. On the airless Moon, the sky is black even at midday. And unlike the geologically active Earth, the Moon has a surface that changes very little with time. When Apollo 16 astronauts Charlie Duke (shown in the photograph) and John Young explored the region of the lunar highlands known as the Descartes Formation, the youngest rocks that they found were virtually the same age as the oldest rocks found on Earth.

Because the Moon has changed so little over the eons, it has preserved the early history of the inner solar system and the terrestrial planets. In this chapter, we will see evidence that the Moon's formation and early evolution was violent and chaotic. The Apollo missions revealed that the Moon may have formed some 4.5 billion years ago as the result of a titanic collision between the young Earth and a rogue protoplanet the size of Mars. The young Moon was then pelted incessantly for hundreds of millions of years by large chunks of interplanetary debris,

Walking on the lunar surface, April 1972. (NASA/John W. Young/Science Source)

R I V U X G

producing a cratered landscape. We will discover that the largest of these impacts formed immense circular basins that then flooded with molten lava. You can see the solidified lava as dark patches that cover much of the face of the Moon. In later chapters, we will find other evidence that catastrophic collisions have played an important role in shaping the solar system.

10-1 Lunar motions and surface

Compared with the distances between planets, which are measured in tens or hundreds of millions of kilometers, the average distance from Earth to the Moon is just 384,400 km (238,900 mi). Despite being so close, the Moon has an angular diameter of just ½° on the sky, which tells us that it is a rather small world. The diameter of the Moon is 3476 km (2160 mi), just 27% of Earth's diameter and about the same as the distance from Las Vegas to Philadelphia or from London to Cairo (**Figure 10-1**). **Table 10-1** lists some basic data about the Moon.

Motions of the Earth–Moon System

TUTORIAL 10.1 Although the Moon has a small mass (just 1.23% of Earth's mass), Newton's third law tells us that the Moon exerts just as much gravitational force on Earth as Earth does on the Moon (see Section 4-6). Hence, it is not strictly correct to say that the Moon orbits Earth; rather, they both orbit around a point between their centers called the **center of mass** of the Earth–Moon system. Because Earth is so much more massive than the Moon, the center of mass is close to Earth's center: Indeed, it actually lies within Earth's interior. The center of mass of the system follows an elliptical orbit around the Sun (**Figure 10-2a**). In a similar way, a spinning wrench sliding along a table actually rotates around its center of mass (Figure 10-2b).

The Moon's Rotation and Libration

Even without binoculars or a telescope, you can easily see that dark gray and light gray areas cover vast expanses of the Moon's rocky surface. If you observe the Moon over many nights, you will see the Moon go through its phases as the dividing line between the illuminated and darkened portion of the Moon (the **terminator**) shifts. But you will always see the same pattern of dark and light areas, because the Moon keeps the same side turned toward Earth; you never see the **far side** (or back side) of the Moon. We saw in Section 3-2 that this is the result of the Moon's **synchronous rotation**: It takes precisely the same amount of time for the Moon to rotate on its axis (one lunar "day") as it does to complete one orbit around Earth.

> ### CAUTION!
> There is no permanent "dark side" of the moon. The same side of the moon always faces Earth, and this is called the near side. The side we never see is the far side, but it is not always in the dark. Just like the near side, the far side goes through cycles of light and dark during its monthly orbit. (The enormous popularity of *Dark Side of the Moon* by the band Pink Floyd has perpetuated this misconception.)

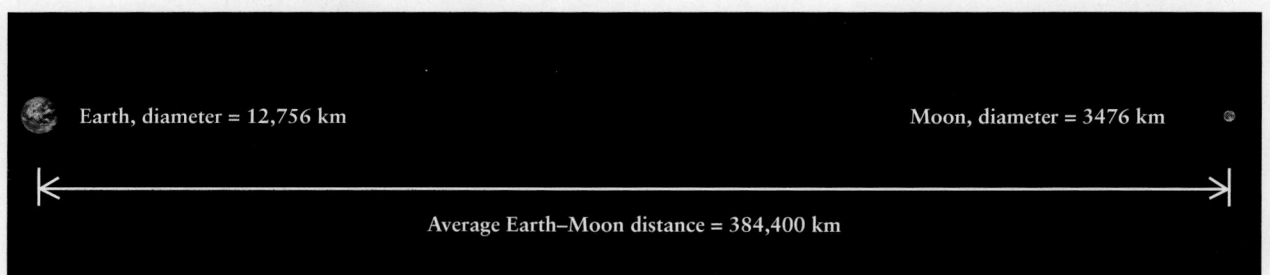

Earth, diameter = 12,756 km   Moon, diameter = 3476 km

Average Earth–Moon distance = 384,400 km

(a) The Earth–Moon system. Both objects and the distance between them are to scale.

Earth

Moon

(b) Earth and the Moon to scale, shown 10 times larger than in part (a) (distance not to scale).

FIGURE 10-1 R I V U X G

Comparing Earth and the Moon (a) The Moon is a bit more than one-quarter the diameter of Earth. The average Earth–Moon distance is about 30 times Earth's diameter. **(b)** Earth has blue oceans, an atmosphere streaked with white clouds, and continents continually being reshaped by plate tectonics. By contrast, the Moon has no oceans, no atmosphere, and no evidence of plate tectonics. (a: NASA/JPL/Arizona State University; b: NASA Goddard Space Flight Center. Image by Reto Stöckli; NASA/GSFC/Arizona State Univ./Lunar Reconnaissance Orbiter)

TABLE 10-1 Moon Data

Distance from Earth (center to center):	Average: 384,400 km = 238,900 mi
	Maximum (apogee): 405,500 km
	Minimum (perigee): 363,300 km
Eccentricity of orbit:	0.0549
Average orbital speed:	3680 km/h
Sidereal period (relative to fixed stars):	27.322 days
Synodic period (new moon to new moon):	29.531 days
Inclination of lunar equator to orbit:	6.68°
Inclination of orbit to ecliptic:	5.15°
Diameter (equatorial):	3476 km = 2160 mi = 0.272 Earth diameter
Mass:	7.349×10^{22} kg = 0.0123 Earth mass
Average density:	3344 kg/m³
Escape speed:	2.4 km/s
Surface gravity (Earth = 1):	0.17
Albedo:	0.11
Average surface temperatures:	Day: 130°C = 266°F = 403 K
	Night: −180°C = −292°F = 93 K (Larry Landolfi, Science Source)
Atmosphere:	Essentially none

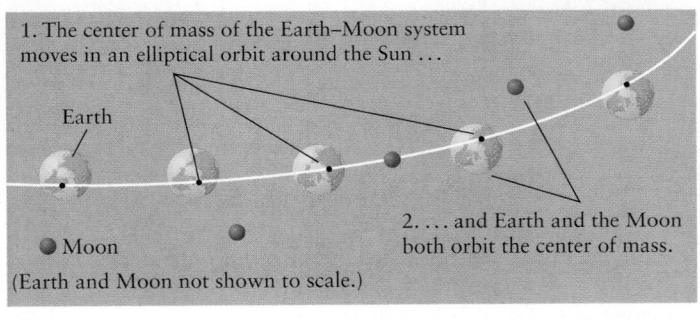

1. The center of mass of the Earth–Moon system moves in an elliptical orbit around the Sun …

Earth

Moon
(Earth and Moon not shown to scale.)

2. … and Earth and the Moon both orbit the center of mass.

(a)

FIGURE 10-2 R I V U X G

Motion of the Earth–Moon System **(a)** Earth and the Moon both orbit around their center of mass, which in turn follows an elliptical orbit around the Sun. **(b)** The illustration in part (a) is analogous to this time-lapse image of a spinning wrench sliding across a table. (b: Berenice Abbott/Science Source)

1. The center of mass of this wrench (shown as a red cross) moves in a straight line …

2. … and the wrench rotates around the center of mass.

(b)

ANIMATION
10.1

With patience, however, you can see slightly more than half the lunar surface, because the Moon appears to wobble slightly over the course of an orbit. It seems to rock back and forth around its north–south axis and to nod up and down in a north–south direction. This apparent periodic wobbling, called **libration,** permits us to view 59% of our satellite's surface from Earth. We stress that the Moon only *appears* to be wobbling. In fact, its rotation axis keeps nearly the same orientation in space as the Moon goes around its orbit. The reason why the Moon seems to rock back and forth is that

LOOKING
DEEPER
10.1

its orbit around Earth is slightly elliptical. As a result, the Moon's orbital motion does not keep pace with its rotation at all points around the orbit. Furthermore, because the Moon's rotation axis is not exactly perpendicular to the plane of its orbit, it also appears to us to nod up and down.

An Airless World

Clouds always cover some portion of Earth's surface. However, clouds are never seen on the Moon (see Figure 10-1), because the Moon is too small a world to have an atmosphere. Its surface gravity is low, only about one-sixth as great as that of Earth, and thus any gas molecules can easily escape into space (see Box 7-2). Because there is no atmosphere to scatter sunlight, the daytime sky on the Moon appears as black as the nighttime sky on Earth (see the photograph that opens this chapter). For the same reason, shadows on the Moon are much darker than they appear on Earth, where surrounding air bounces light into crevices.

The absence of an atmosphere means that the Moon can have no liquid water on its surface. Here on Earth, water molecules are kept in the liquid state by the pressure of the atmosphere pushing down on them. To see what happens to liquid water on the airless Moon, consider what happens when the pressure of Earth's atmosphere is reduced. The air pressure in Denver, Colorado (elevation 1650 m, or 5400 ft), is only 83% as great as at sea level. Therefore, in Denver, less energy has to be added to the molecules in a pot of water to make them evaporate, and water boils at only 95°C (= 368 K = 203°F), as compared with 100°C (= 373 K = 212°F) at sea level. (This temperature dependence on pressure is why some foods require longer cooking times at high elevations, where the boiling water used for cooking is not as hot as at sea level.) On the Moon, the atmospheric pressure is essentially zero, and the boiling temperature is lower than the temperature of the lunar surface. Under these conditions, water can exist as a solid (ice) or a vapor but not as a liquid.

CAUTION!

If astronauts were to venture out onto the lunar surface without wearing spacesuits, the lack of atmosphere would *not* cause their blood to boil. That's because our skin is strong enough to contain bodily fluids. In the same way, if we place an orange in a sealed container and remove all the air, the orange retains all of its fluid and does not explode. It even tastes the same afterward! Astronauts wore pressurized spacesuits on the Moon (see the photograph that opens this chapter) to provide oxygen for breathing, to regulate body temperature, and to protect against ultraviolet radiation from the Sun. Since the Moon has no atmosphere, and thus no ozone layer like that of Earth to screen out ultraviolet wavelengths (Section 9-7), an unprotected astronaut would suffer a sunburn after just 10 seconds of exposure!

Lunar Craters

ANIMATION 10.2 Since the Moon has no atmosphere to obscure its surface, you can get a clear view of several different types of lunar features with a small telescope or binoculars. These include *craters,* the dark-colored *maria,* and the light-colored *lunar highlands* or *terrae* (Figure 10-3). Taken together, these various features show the striking differences between the surfaces of Earth and the Moon.

With an Earth-based telescope, some 30,000 **craters** are visible, with diameters ranging from 1 km to several hundred km. Close-up photographs from lunar orbit have revealed millions of craters too small to be seen from Earth. Following a tradition established in the seventeenth century, the most prominent craters are named after famous philosophers, mathematicians, and scientists, such as Plato, Aristotle, Pythagoras, Copernicus, and Kepler. Virtually all lunar craters, both large and small, are the result of

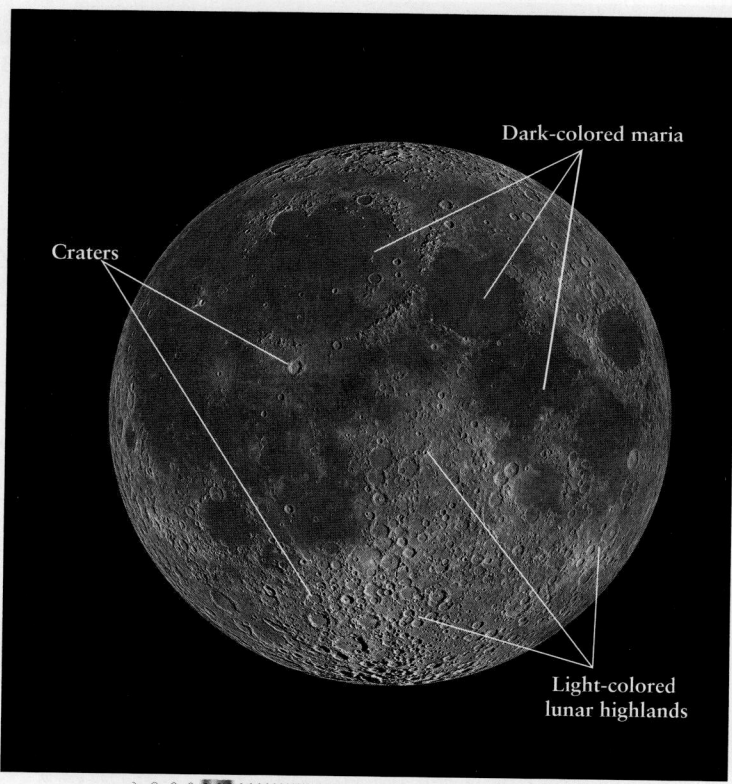

FIGURE 10-3 R I V U X G

The Near Side of the Moon The key features of the lunar surface can be seen with binoculars, a small telescope, or even the naked eye. This composite image was made with more than 1300 images from the Lunar Reconnaissance Orbiter. (NASA/GSFC/Arizona State Univ./Lunar Reconnaissance Orbiter)

the Moon's having been bombarded by meteoroids and asteroids. (Meteoroids are just the smaller version of these space rocks.) For this reason, they are also called **impact craters** (see Section 7-6). High-speed impacts have been recreated in laboratories to study what happens for different impacting materials and different angles of impact. Based on experiments, two key pieces of evidence for the impact-origin of lunar craters are the central peak found in some craters (see Figure 7-10a), and the fact that different angles of impact still produce circular craters.

An asteroid striking the lunar surface generates a shock wave over the surface outward from the point of impact. Such shock waves can spread over an area much larger than the size of the asteroid that generated the wave. Very large craters more than 100 km across, such as the crater Clavius shown in Figure 10-4, were probably created by the impact of a fast-moving piece of rock only a few kilometers in diameter.

Maria: Dark and Relatively Young

The large dark areas on the lunar surface shown in Figure 10-3 are called **maria** (pronounced "MAR-ee-uh"). They form a pattern that looks crudely like a human face, sometimes called "the man in the Moon," or perhaps a rabbit, as Chinese tradition holds. The singular form of maria is **mare** (pronounced "MAR-ay"), which means "sea" in Latin. These terms date from the seventeenth century, when observers using early telescopes thought these

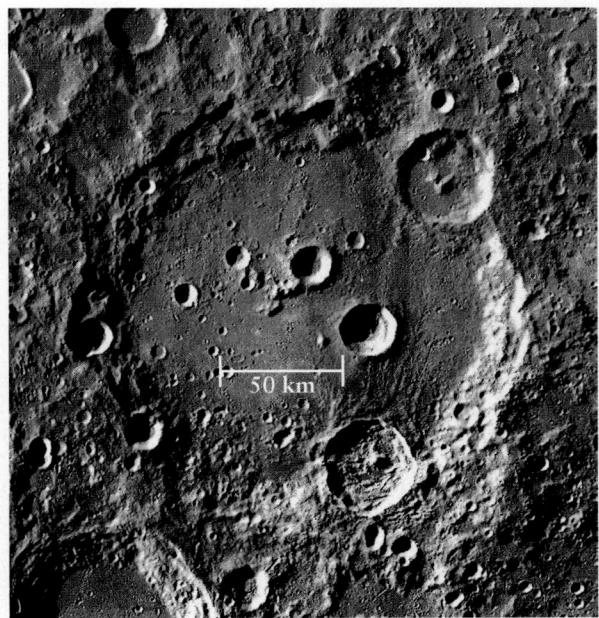

FIGURE 10-4 R I **V** U X G

The Crater Clavius This crater is one of the largest on the Moon. Clavius has a diameter of 232 km (144 mi) and a depth of 4.9 km (16,000 ft), measured from the crater's floor to the top of the surrounding rim. Later impacts formed the smaller craters inside Clavius. (NASA/GSFC/Arizona State University)

FIGURE 10-5 R I **V** U X G

Mare Imbrium from Earth Mare Imbrium is the largest of 14 dark plains that dominate the Earth-facing side of the Moon. Its diameter of 1100 km (700 mi) is about the same as the distance from Chicago to Philadelphia or from London to Rome. The maria formed after the surrounding light-colored terrain, so they have not been exposed to meteoritic bombardment for as long and hence have fewer craters. (Carnegie Observatories)

dark features were large bodies of water. They gave the maria fanciful and romantic names such as Mare Tranquillitatis (Sea of Tranquility), Mare Nubium (Sea of Clouds), Mare Nectaris (Sea of Nectar), Mare Serenitatis (Sea of Serenity), and Mare Imbrium (Sea of Showers). While modern astronomers still use these names, they now understand that the maria are not bodies of water (which, as we have seen, could not exist on the airless Moon) but rather the remains of huge lava flows. The maria get their distinctive appearance from the dark color of the solidified lava.

On the whole, the maria have fewer craters than the surrounding terrain. Craters are caused by meteoritic bombardment, so the maria must not have been exposed to that bombardment for as long as the surrounding terrain. Hence, the maria must be young in comparison, and the lava flows that formed them must have occurred at a later stage in the Moon's geologic history (**Figure 10-5**). The craters that are found in the maria were caused by the relatively fewer and smaller impacts that have occurred since the maria solidified (**Figure 10-6**).

Although they are much larger than craters, maria also tend to be circular in shape (see Figure 10-5). This suggests that, like craters, these depressions in the lunar surface were also created by large impacts. It is thought that very large asteroids some tens of kilometers across struck the Moon, forming basins. These depressions subsequently flooded with lava that flowed out from the Moon's interior through cracks in the lunar crust. The solidified lava formed the maria that we see today. However, the large basin-creating impacts probably did not directly cause the lava flows because the lava is typically dated to about 500 million years after the impacts. The *Cosmic Connections* figure depicts our understanding of how lunar craters and maria formed.

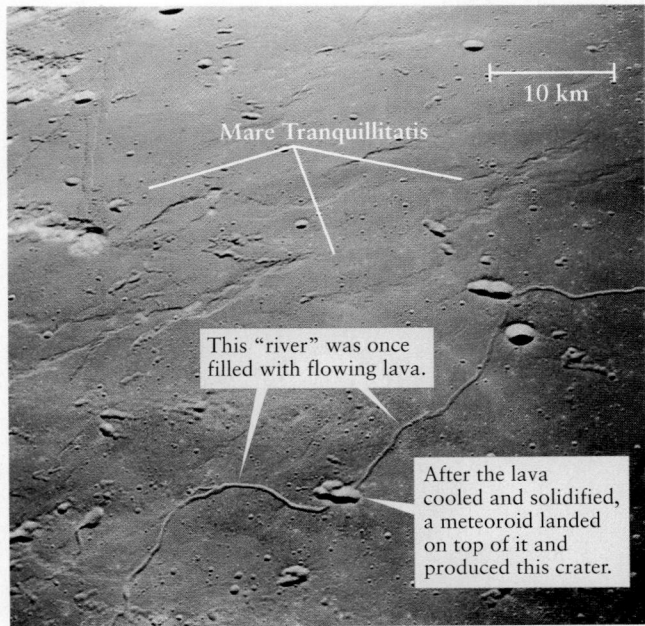

FIGURE 10-6 R I **V** U X G

Details of Mare Tranquillitatis This color photograph from lunar orbit reveals numerous tiny craters in the surface of a typical mare. The curving "river," or rille, was carved out by flowing lava that later solidified; similar features are found in volcanic areas on Earth. The Apollo 11 landing site is near the center of the top edge of this image. (NASA/JPL)

COSMIC CONNECTIONS

The Formation of Craters and Maria on the Moon

The physical features of lunar craters and maria show that they were formed by asteroids impacting the Moon's surface at high speed. The crater labeled in the photograph, called Aristillus after a Greek astronomer who lived around 300 B.C.E., is 55 km (34 mi) across and 3600 m (12,000 ft) deep. This crater is only shown for illustrative purposes, as the actual impact basins that make up the maria are much larger in scale.

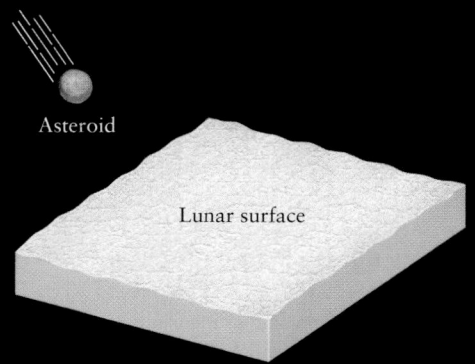

1. An asteroid approaches the lunar surface.

2. When the asteroid strikes, it ejects rocks and boulders and produces a shock wave in the lunar surface.

3. What remains is a crater with a central peak. The ejected material forms a blanket around the crater and can cause secondary craters.

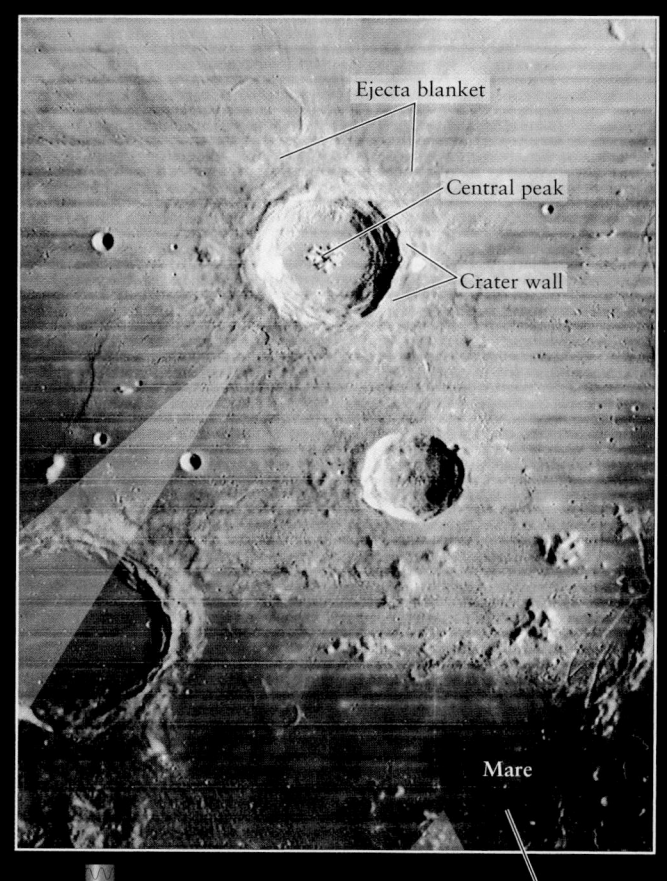

R I V U X G

Lava fills crater and solidifies to form the dark mare floor.

NASA LOIV-110-H1 courtesy of Lunar and Planetary Institute/USRA

4. About 500 million years later, lava flows into the crater to form a mare.

An impact large enough to create a mare basin must have thrust upward tremendous amounts of material around the impact site. This explains the mountain ranges that curve in circular arcs around mare basins. Analysis of lunar rocks as well as observations made from lunar orbit have validated this picture of the origin of lunar mountains.

Lunar Highlands: Light-Colored and Ancient

The light-colored terrain that surrounds the maria is called **terrae**, from the Latin for "lands." (Seventeenth-century astronomers speculated that this terrain was the dry land that surrounded lunar oceans.) Detailed measurements by astronauts in lunar orbit demonstrated that the maria on the Moon's Earth-facing side are 2 to 5 km below the average lunar elevation, while the terrae extend to several kilometers above the average elevation. For this reason terrae are more commonly called **lunar highlands** (see Figure 10-3). The flat, low-lying, dark gray maria cover only 15% of the lunar surface; the remaining 85% is light gray lunar highlands. Since the lunar highlands are more heavily cratered than the maria, we can conclude that the highlands are older.

Remarkably, when the Moon's far side was photographed for the first time by the Soviet Luna 3 spacecraft in 1959, scientists were surprised to find almost no maria there. Lunar highlands predominate on the side of the Moon that faces away from Earth

(Figure 10-7). One factor contributing to this asymmetry is a big difference in crustal thickness; thicker crust on the far side may have impeded volcanic activity. The Moon is not unique in having major portions with different crustal thicknesses: Earth's northern hemisphere has much more continental crust than its southern hemisphere, and the southern hemisphere of Mars has a thicker crust than the northern hemisphere.

Although the Moon's far side has fewer maria than the near side, the highlands on the far side are as heavily pockmarked by impacts as those on the near side. The most prominent feature on the far side is the South Pole–Aitken Basin, the largest known impact crater in the solar system. This basin is about 2500 km (1600 mi) across—about the same size as the continent of Europe. Surprisingly, the South Pole–Aitken Basin isn't as dark as the maria because very little lava has filled in the basin, even though it has an extraordinary depth of 12 km and very thin crust beneath its floor.

A World Without Plate Tectonics

Something is entirely missing from our description of the Moon's surface—any mention of plate tectonics. Indeed, no evidence for plate tectonics can be found on the Moon. Recall from Section 9-3 that Earth's major mountain ranges were created by collisions between lithospheric plates. In contrast, it appears that the Moon is a *one-plate world:* Its entire lithosphere is a single, solid plate.

(a) Near side: highlands and maria

(b) Far side: highlands but almost no maria

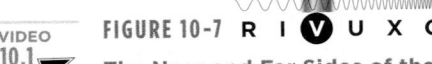

VIDEO
10.1

FIGURE 10-7 R I **V** U X G

The Near and Far Sides of the Moon (a) The iron-rich darker maria are a younger surface made from ancient lava flows and are 2 to 5 km below the average surface elevation. (b) The cratered highlands are an older surface several kilometers above the average elevation.

The South Pole–Aitken Basin reaches depths of 12 km, but was never filled in by darker lava flows. These images were made by the Lunar Reconnaissance Orbiter. (a, b: NASA/GSFC/Arizona State Univ./Lunar Reconnaissance Orbiter)

There are no rifts where plates are moving apart, no mountain ranges of the sort formed by collisions between plates, and no subduction zones. Thus, while lunar mountain ranges bear the names of terrestrial ranges, such as the Alps, Carpathians, and Apennines, they are entirely different in character. They were formed by material ejected from impact sites, not by plate tectonics. Geologically, the Moon is a nearly dead world. In Section 7-6, we discussed the general rule that a world with abundant craters has an old, geologically inactive surface. The Moon is a prime example of this rule. Without an atmosphere to cause erosion and without plate tectonics, the surface of the Moon changes only very slowly. That is why the Moon's craters, many of which are now known to be billions of years old, remain visible today. On Earth, by contrast, the continents and seafloor are continually being reshaped by seafloor spreading, mountain building, and subduction—not to mention the wind and the rain. Thus, any craters that were formed by impacts on the ancient Earth's surface have long since been erased. Fewer than 200 impact craters are known to exist on Earth today; these were formed within the last few hundred million years and have not yet been erased.

In 2010, evidence for some surprisingly recent geological activity on the Moon was found. The Moon has long cliffs called *scarps* that develop as the entire Moon cools and shrinks (**Figure 10-8**). A good analogy for the formation of these scarps is the wrinkling

> In contrast to Earth and its active geology, the Moon's surface is geologically dead.

seen on a fruit's skin as the fruit dries out and contracts. It had been thought that the Moon cooled quickly after formation, and that all scarps were billions of years old. However, the relatively young craters disturbed by a scarp must have formed first, providing evidence of the scarp's age and suggesting that some of these geologic features might be as young as a few hundred million years old. The Moon has probably shrunk about 600 ft (about 180 m) in diameter since it first formed, and this new evidence suggests that it might still be shrinking today.

CONCEPT CHECK 10-1

Why does darker rock appear at lower elevations on the Moon?

Answer appears at the end of the chapter.

10-2 Human exploration of the Moon in the 1960s and 1970s has been continued by robotic spacecraft

 TUTORIAL 10.2 For thousands of years, storytellers have invented fabulous tales of voyages to the Moon. The seventeenth-century French satirist Cyrano de Bergerac wrote of a spacecraft whose motive power was provided by spheres of morning dew. (After all, reasoned Cyrano, dew rises with the morning sun, so a spacecraft would rise along with it.) In his imaginative tale *Somnium*, Cyrano's contemporary Johannes Kepler (the same Kepler who deduced the laws of planetary motion) used the astronomical knowledge of his time to envision what it would be like to walk on the Moon's surface. In the eighteenth century, the Moon became a popular setting for satire and adventure stories. In 1865, the French author Jules Verne published *From the Earth to the Moon*, a story of a spacecraft sent to the Moon from a launching site in Florida.

Almost exactly 100 years later, Florida was indeed the starting point of the first voyages made by humans to other worlds. Between 1969 and 1972, 12 astronauts walked on the Moon, and fantasy became magnificent reality.

> The Moon is the only world other than Earth to be visited by humans.

The First Spacecraft to the Moon

As in Verne's story, politics had much to do with the motivation for going to the Moon. But in fact, there were excellent scientific reasons to do so as well. Because Earth is a geologically active planet, typical terrestrial surface rocks are only a few hundred million years old, which is only a small fraction of Earth's age. Thus, all traces of Earth's origins have been erased. By contrast, rocks on the geologically inactive surface of the Moon have been essentially undisturbed for billions of years. Samples of lunar rocks have helped answer many questions about the Moon and have shed light on the origin of Earth and the birth of the entire solar system. (It has also been suggested that some Earth rocks would have been ejected all the way to the Moon from large impacts in the solar system's early years, holding clues to Earth's oldest geology.)

Lunar missions began in 1959, when the Soviet Union sent three small spacecraft—Lunas 1, 2, and 3—toward the Moon.

FIGURE 10-8 R I **V** U X G

Recent Geologic Activity of Scarps This image shows a long cliff called a scarp—formed as the Moon cools, shrinks, and wrinkles—that is cutting across several small craters (see the inset, and the two arrows). Until this discovery in 2010, it was thought these scarps were billions of years old. But small craters allow for a unique way to estimate ages. Small craters (less than about 400 m) typically do not last very long on the lunar surface because they are disturbed by ongoing micrometeorite impacts. Furthermore, when an image shows that a scarp cuts across a crater, it means the scarp is even younger than the crater. By analyzing scarps and their underlying craters, some scarps are estimated to be less than a billion years old, and could be as young as 100 million years old. (NASA/Goddard/Arizona State University/Smithsonian)

American attempts to reach the Moon began in the early 1960s with Project Ranger. Equipped with six television cameras, each of the Ranger 7, 8, and 9 spacecraft transmitted close-up views of the lunar surface taken in the few minutes before it crashed onto the Moon. These images showed far more detail than even the best Earth-based telescope.

In 1966 and 1967, five Lunar Orbiter spacecraft were placed in orbit around the Moon and made a high-resolution photographic survey of 99.5% of the lunar surface. Some of these photographs show features as small as 1 m across and were essential in helping NASA scientists choose suitable landing sites for the astronauts.

Even after the Ranger missions, it was unclear whether the lunar surface was a safe place to land. One leading theory was that billions of years of impacts by meteoroids and the solar wind had ground the Moon's surface to a fine powder. Perhaps a spacecraft attempting to land would simply sink into a quicksand-like sea of dust! Thus, before a crewed landing on the Moon could be attempted, it was necessary to perform a trial landing of a robotic spacecraft. This was the purpose of the Surveyor program. Between June 1966 and January 1968, five Surveyors successfully touched down on the Moon, sending back pictures and data directly from the lunar surface. These missions demonstrated convincingly that the Moon's surface was as solid as that of Earth. Figure 10-9 shows an astronaut visiting Surveyor 3, 2 and a half years after it had landed on the Moon.

"That's One Small Step for Man ..."

VIDEO
10.2

The first of six crewed lunar landings took place on July 20, 1969, when the Apollo 11 lunar module *Eagle* set down in Mare Tranquillitatis (Sea of Tranquility), seen in Figure 10-6. Astronauts Neil Armstrong and Edwin "Buzz" Aldrin were the first humans to set foot on the Moon. Apollo 12 also landed in a mare, Oceanus Procellarum (Ocean of Storms). The Apollo 13 mission experienced a nearly catastrophic inflight explosion that made it impossible for it to land on the Moon. Fortunately, titanic efforts by the astronauts and ground personnel brought it safely home. The remaining four missions, Apollo 14 through Apollo 17, were made in progressively more challenging terrain, chosen to permit the exploration of a wide variety of geologically interesting features. The first era of human exploration of the Moon culminated in December 1972, when Apollo 17 landed in a rugged region of the lunar highlands. Figure 10-10 shows one of the Apollo bases.

In a curious postscript to the Apollo missions, in the 1990s an "urban legend" arose that the Moon landings had never taken place and were merely an elaborate hoax by NASA. While amusing, this legend does not hold up under even the slightest scientific scrutiny. For example, believers in the "Moon hoax" cannot explain why the rocks brought back from the Moon are substantially different in their chemistry from Earth rocks, as we will see later in this chapter.

Between 1966 and 1976, a series of robotic Soviet spacecraft also orbited the Moon and landed on its surface. The first landing, by Luna 9, came 4 months before the U.S. Surveyor 1 arrived. The first lunar satellite, Luna 10, orbited the Moon 4 months before Lunar Orbiter 1. In the 1970s, robotic spacecraft named Luna 17 and Luna 21 landed vehicles that roamed the lunar surface, and Lunas 16, 20, and 24 brought samples of rock back to Earth. The Luna samples complement those returned by the Apollo astronauts because they were taken from different parts of the Moon's surface.

VIDEO
10.3

FIGURE 10-9 R I ⊙V U X G

Visiting a Robotic Pioneer Surveyor 3 landed on the Moon in April 1967. It was visited in November 1969 by Apollo 12 astronaut Alan Bean, who took pieces of the spacecraft back to Earth for analysis. (The Apollo 12 lunar module is visible in the distance.) The retrieved pieces showed evidence of damage by micrometeoroid impacts, demonstrating that small bits of space debris are still colliding with the Moon. (NASA)

FIGURE 10-10 R I ⊙V U X G

The Apollo 15 Base This photograph shows astronaut James Irwin and the Apollo 15 landing site at the foot of the Apennine Mountains near the eastern edge of Mare Imbrium. The hill in the background is about 5 km away and rises about 3 km high. At the right is a Lunar Rover, an electric-powered vehicle used by astronauts to explore a greater radius around the landing site. The last three Apollo missions carried Lunar Rovers. (Dave Scott, Apollo 15, NASA)

Recent Lunar Exploration and the Search for Water

While robots and probes can pursue scientific research in space more efficiently than humans, there might be other reasons for people to venture into space some day. Indeed, some countries have already outlined initial steps toward a lunar base in the coming decades. However, it is an expensive endeavor—it costs over $100,000 to transport the mass of a gallon of water to the Moon. Contemplation of a lunar base and the resources required to maintain it has added extra importance to the question, "Is there water on the moon?" To help answer this question, in 1994 a small NASA spacecraft named Clementine spent more than 2 months observing the Moon from lunar orbit. It carried an array of very high-resolution cameras sensitive to ultraviolet and infrared light as well as to the visible spectrum. Different types of materials on the lunar surface reflect different wavelengths, so analysis of Clementine images made in different parts of the electromagnetic spectrum has revealed the composition of large areas of the Moon's surface. For example, the unique reflection signature of iron led to the map of the Moon's iron distribution, shown in **Figure 10-11**.

One remarkable result of the Clementine mission was evidence for a patch of water ice tens or hundreds of kilometers across at the Moon's southern pole, in deep craters where sunlight never reaches and where temperatures are low enough to prevent ice from evaporating. Using a remarkable technique, radio waves beamed from Clementine were reflected off these areas and were then detected by radio telescopes on Earth! The reflected radio waves showed the unique signature of reflection from water ice. Then, data from the Lunar Prospector spacecraft, which went into orbit around the Moon in 1998, reinforced the Clementine findings, indicating that even more water ice might lie in deep craters at the Moon's northern pole.

Encouragingly, the Lunar Prospector used a different measurement technique than Clementine, which strengthens the overall case for water. However, the Lunar Prospector observations show only that there is an excess of hydrogen atoms at both lunar poles. The assumption is that this hydrogen is most likely contained within water molecules (H_2O), but the possibility also exists that the hydrogen could be within other, more exotic minerals. Providing further support for the case for water, in 2017 NASA's Lunar Reconnaissance Orbiter (LRO) reported that the lowest-temperature areas near the south pole also reflect light in a manner consistent with a thin layer of frost. Overall, this evidence for water on the moon is highly suggestive but not overwhelming. Next, we discuss a different approach to truly settle this question.

At the end of the Lunar Prospector mission in July 1999, the spacecraft was intentionally crashed into a crater at the lunar south pole in the hope that a plume of water vapor might rise from the impact site. Although more than a dozen large Earth-based telescopes watched the Moon during the impact, no such plume or debris of any kind was observed. The advantage of seeing water kicked up in a plume is that you can detect the unique infrared spectrum of H_2O directly, not just hydrogen atoms alone, which would *confirm* the presence of water. However, this attempt was unsuccessful.

Designed specifically for impact and analysis, the Lunar Crater Observation and Sensing Satellite (LCROSS) tried again to detect water directly. The impact target was suspected to contain water in a permanently shadowed region of a crater only 100 meters from the lunar south pole, and should have been cold enough to retain water ice for billions of years. In 2009, after ferrying the LCROSS probe from Earth, a spent fuel canister weighing more than 2 tons was used as an artificial meteoroid to crash into the suspected ice. Traveling at about 8 times the speed of sound, the impact created a large plume of debris (**Figure 10-12**). Then, a few minutes later, the probe itself traveled through the plume, analyzing the debris before smashing into the surface. Playing a supporting role before the mission, the LRO was also able to estimate the mass of material ejected.

The results were spectacular and lunar water was finally confirmed! As shown in **Figure 10-13**, the characteristic absorption features due to H_2O were found in the plume's infrared spectrum. Ammonia and methane were also found, matching the abundances expected for comets, indicating that comets are the most likely source of the water.

About 155 kg (41 gallons) of water were ejected from LCROSS's 30-meter impact crater. However, the concentration of water in the crater is similar to that found in Earth's Sahara Desert. As a resource for supporting human activity on the Moon, the water discovered so far on the Moon is quite scarce. On the other hand, the ice has scientific value analogous to ice cores that have been obtained on Earth: By preserving a record in ice over time, craters may hold clues to the origins and evolution of our solar system.

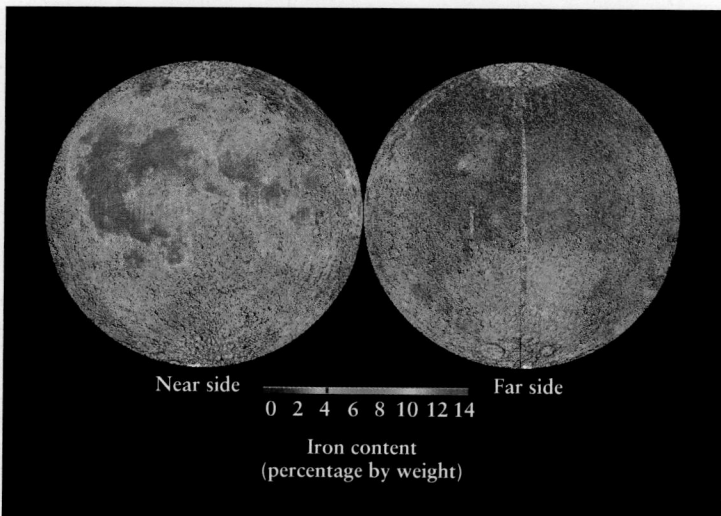

Near side ————————— Far side

0 2 4 6 8 10 12 14

Iron content
(percentage by weight)

FIGURE 10-11

Iron on the Moon Images at different wavelengths from the Clementine spacecraft were used to make this map of the concentration of iron on the lunar surface. The areas of highest concentration (red) coincide closely with the maria on the near side (compare with Figure 10-7), confirming that the maria formed from iron-rich lavas. The lowest iron concentration (blue) is found in the lunar highlands. (No data were collected in the gray areas.) The large green region of intermediate iron concentration on the far side is the South Pole–Aitken Basin (see Figure 10-7). The impact that created this basin may have excavated all the way to the Moon's mantle, in which iron is more abundant than in the crust. (The Clementine Project/JSC/NASA)

FIGURE 10-12 R I V U X G

Water in the Plume Created by LCROSS The permanent shadow in Cabeus crater is visible, in addition to the plume that rose 20 seconds after the impact of a spent fuel canister. About 5% of the ejected material is water. Other evidence suggests that the water may have originally come from comets. (NASA)

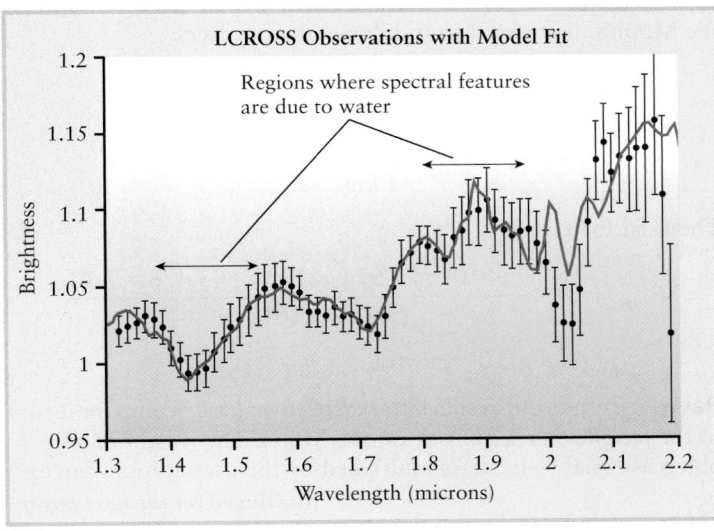

FIGURE 10-13

Evidence for Lunar Water The infrared spectrum of the plume shown in Figure 10-12 indicates both water vapor and water ice. The collected data are shown in black with their vertical error bars. The red curve is a model based on the known spectrum of water vapor and ice. There is a very good fit between the model and data in the regions where the spectral features are due to water.

CONCEPT CHECK 10-2

What provides better evidence of water on the Moon: a clear excess of hydrogen atoms or the spectrum shown in Figure 10-13?

CALCULATION CHECK 10-1

The material ejected from the LCROSS impact contained 5% water. How many pounds of this material would be needed for an 8-oz (237-g) glass of water? (*Note:* 1 kg = 2.2 lb.)
Answers appear at the end of the chapter.

10-3 The Moon has no global magnetic field but has a small molten core

We saw in Section 7-6 and Section 7-7 that a small planet or satellite will have little remaining internal heat, and hence is likely to have a mostly solid interior. As we learned in Section 7-7, Earth's magnetic field is generated by fluid motions in its interior. Because it lacks substantial amounts of moving electrically conducting material in its interior, a small world is unlikely to generate a global magnetic field. However, a small world can still have a small molten core, which is just what has been found for the Moon.

> Compared to earthquakes, moonquakes are few and feeble—but they reveal key aspects of the Moon's interior.

No Global Magnetic Field

The Moon landings gave scientists an unprecedented opportunity to explore an alien world. The Apollo spacecraft were therefore packed with an assortment of apparatus that the astronauts deployed around the landing sites (see Figure 10-9). For example, all the missions carried magnetometers, which confirmed that the present-day Moon has no global magnetic field. However, careful magnetic measurements of lunar rocks returned by the Apollo astronauts indicated that the Moon did have a weak magnetic field when the rocks solidified billions of years ago. Hence, the Moon must originally have had a small, molten, iron-rich core.

The Moon's Interior

The Apollo astronauts also set up four seismometers on the Moon's surface, which made it possible for scientists to investigate the Moon's interior using seismic waves, just as is done for Earth (Section 9-2). It was found that the Moon exhibits far less seismic activity than Earth. Roughly 3000 **moonquakes** were detected per year, whereas a similar seismometer on Earth would record hundreds of thousands of earthquakes per year. Furthermore, typical moonquakes are very weak. A major terrestrial earthquake is in the range from 6 to 8 on the Richter scale (where a magnitude of 8 is 100 times stronger than a magnitude 6), while a major moonquake measures only 0.5 to 1.5 on that scale and would go unnoticed by a person standing near the epicenter.

Analysis of moonquakes reveals that the Moon has a thick mantle surrounding an iron-rich core. The Moon also has a crust with an average thickness of about 60 km on the Earth-facing side and up to 100 km on the far side. For comparison, the thickness of Earth's crust ranges from 5 km under the oceans to about 70 km under major mountain ranges on the continents (see

BOX 10-1   TOOLS OF THE ASTRONOMER'S TRADE

Calculating Tidal Forces

The gravitational force between any two objects decreases with increasing distance between the objects. This principle leads to a simple formula for estimating the tidal force that Earth exerts on parts of the Moon.

Consider two small objects, each of mass m, on opposite sides of the Moon. Because one object is farther from Earth's center than the other, Earth exerts different forces F_{near} and F_{far} on them (see the accompanying illustration). The closer object is pulled harder toward Earth than the more distant one; as a result, they are pulled away from each other and away from the center of the Moon (see Section 4-8, especially Figure 4-27). The *tidal force* on these two objects, which is the force that pulls them apart, is the difference between the forces acting on the individual objects:

$$F_{tidal} = F_{near} - F_{far}$$

Both F_{near} and F_{far} are inversely proportional to the square of the distance from Earth (see Section 4-7). If this distance were doubled, both forces would decrease to $\frac{1}{2}^2 = \frac{1}{4}$ of their original values. But the tidal force F_{tidal}, which is the *difference* between these two forces, decreases even more rapidly with distance: It is inversely proportional to the *cube* of the distance from Earth to the Moon. That is, doubling the Earth–Moon distance decreases the tidal force to $\frac{1}{2}^3 = \frac{1}{8}$ of its original value. If we use the symbols M_{Earth} for Earth's mass (5.974×10^{24} kg), d for the Moon's diameter, r for the center-to-center distance from Earth to the Moon, and G for the universal constant of gravitation (6.67×10^{-11} newton m^2/kg^2), the tidal force is given approximately by the formula

$$F_{tidal} = \frac{2GM_{Earth}md}{r^3}$$

EXAMPLE: Calculate the tidal force for two 1-kg lunar rocks at the locations shown in the figure.

Situation: We are asked to calculate a tidal force. We know the masses of Earth (M_{Earth}) and of each of the two

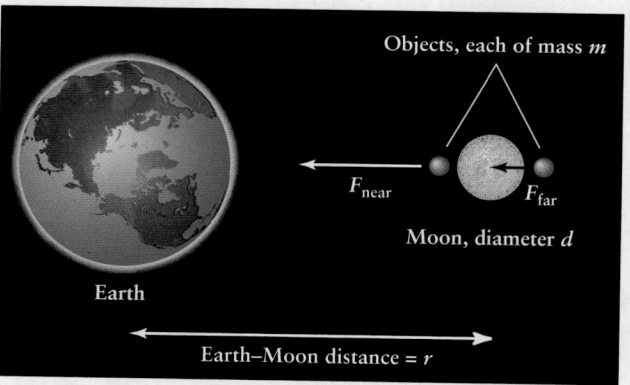

rocks ($m = 1$ kg). The Earth–Moon distance (r) and the diameter of the Moon (d, equal to the separation between the two rocks) are given in Table 10-1.

Tools: We use the equation given for the tidal force F_{tidal}.

Answer: To use the given values in the equation, we must first convert the values for the average Earth–Moon distance r and the Moon's diameter d from kilometers to meters:

$$d = 3476 \text{ km} \times \frac{10^3 \text{ m}}{1 \text{ km}} = 3.476 \times 10^6 \text{ m}$$

$$r = 384{,}400 \text{ km} \times \frac{10^3 \text{ m}}{1 \text{ km}} = 3.844 \times 10^8 \text{ m}$$

The tidal force is then

$$F_{tidal} = \frac{2(6.67 \times 10^{-11})(5.974 \times 10^{24})(1)(3.476 \times 10^6)}{(3.844 \times 10^8)^3}$$

$$= 4.88 \times 10^{-5} \text{ newton}$$

Review: To put our result into perspective, let us compare it to the *weight* of a 1-kg rock on the Moon. The weight of an object is equal to its mass multiplied by the acceleration due to

(continued on the next page)

Table 9-3). This slightly lopsided distribution of mass, together with tides raised in the solid Moon by Earth, helps to keep the Moon in synchronous rotation, so that we always see the same side of the Moon (see Section 4-8 and Box 10-1).

More than 30 years after the original Apollo data shed light on the Moon's internal structure, the seismic data were reanalyzed in 2011 and revealed a partially molten core. Like Earth, the Moon's iron-rich core has a solid inner core surrounded by a liquid outer core (Figure 10-14, on page 286). However, the liquid core is too small for the convection that is needed to generate a global magnetic field.

The Origin of Moonquakes

Earthquakes tend to occur at the boundaries between tectonic plates, where plates move past each other, collide, or undergo subduction (see Figure 9-15). But there are no plate motions on the Moon. What, then, causes moonquakes? The answer turns out to be Earth's tidal forces. Just as the tidal forces of the Sun and Moon deform Earth's oceans and give rise to ocean tides (see Figure 4-29), Earth's tidal forces deform the solid body of the Moon. The amount of the deformation changes as the Moon moves around its elliptical orbit and the Earth–Moon distance varies. As a result, the body of

BOX 10-1 (continued)

gravity (see Section 4-6). This acceleration is equal to 9.8 m/s² on Earth, but is only 0.17 as great on the Moon. So, the weight of a 1-kg rock on the Moon is

$$0.17 \times 1 \text{ kg} \times 9.8 \text{ m/s}^2 = 1.7 \text{ newtons}$$

Thus, the tidal force on lunar rocks is *much* less than their weight, which is a good thing. If the tidal force (which tries to pull a rock away from the Moon's center) were greater than the weight (which pulls it toward the Moon's center), loose objects would levitate off the Moon's surface and fly into space! Other objects aren't so lucky. As we'll see in Chapter 12, Saturn's rings are thought to be created by objects experiencing tidal forces larger than the gravity holding them together. As a result, they are pulled apart into fragments that form the rings.

The tidal force of Earth on the Moon deforms the solid body of the Moon, so that it acquires *tidal bulges* on its near and far sides. The mass of each bulge must be proportional to the tidal force that lifts lunar material away from the Moon's center. So the above formula tells us that the mass of each bulge is inversely proportional to the cube of the Earth–Moon distance r:

$$m_{\text{bulge}} = \frac{A}{r^3}$$

(Determining the value of the constant A requires knowing how easily the Moon deforms, which is a complicated problem beyond our scope.) We can then use our formula for the tidal force F_{tidal} to tell us the *net* tidal force on the Moon. To make this calculation, we think of the objects of mass m on opposite sides of the Moon as being the tidal bulges. So we substitute m_{bulge} for m in the tidal force formula

$$F_{\text{tidal-net}} = \frac{2GM_{\text{Earth}}m_{\text{bulge}}d}{r^3} = \frac{2GM_{\text{Earth}}}{r^3}\frac{A}{r^3}d$$
$$= \frac{2GM_{\text{Earth}}Ad}{r^6}$$

Thus, the net tidal force on the Moon is inversely proportional to the *sixth* power of the Earth–Moon distance. Even a

small decrease in this distance can cause a substantial increase in the net tidal force.

As we saw in Chapter 7, Jupiter's moon Io is the most volcanically active body in the solar system (including Earth). The energy powering Io's volcanism is called *tidal heating*. Io is the closest of four large moons around Jupiter, and frequently gets pulled away from Jupiter by the other moons. As a result, the distance between Jupiter and Io changes, leading to large changes in tidal forces distorting Io. These frequent changes in Io's shape generate tremendous internal heat due to the friction from shifting rocky material. This occurs to a lesser degree on Jupiter's other moons and helps generate warmth for subsurface oceans.

EXAMPLE: Compare the net tidal force on the Moon at perigee (when it is closest to Earth) and at apogee (when it is farthest away).

Situation: We are asked to *compare* the tidal forces on the Moon at two different distances from Earth. To make this comparison, we will find the *ratio* of the tidal force at perigee to the tidal force at apogee.

Tools: We use the equation given above for the net tidal force on the Moon.

Answer: The Earth–Moon distances at perigee and apogee are given in Table 10-1. If we take the ratio of the tidal forces at these two distances, the constants (including the unknown quantity A) cancel out:

$$\frac{F_{\text{tidal-net-perigee}}}{F_{\text{tidal-net-apogee}}} = \frac{2GM_{\text{Earth}}Ad/r^6_{\text{perigee}}}{2GM_{\text{Earth}}Ad/r^6_{\text{apogee}}} = \left(\frac{r_{\text{apogee}}}{r_{\text{perigee}}}\right)^6$$
$$= \left(\frac{405,500 \text{ km}}{363,000 \text{ km}}\right)^6 = (1.116)^6 = 1.93$$

Review: Our result tells us that the net tidal force at perigee is 1.93 times as great (that is, almost double) as at apogee. Tidal forces on the Moon help create the stresses that generate moonquakes, so it is not surprising that these quakes are more frequent at perigee than at apogee.

---

the Moon flexes slightly, triggering moonquakes. The tidal stresses are greatest when the Moon is nearest Earth—that is, at perigee (see Section 3-5)—which is just when the Apollo seismometers reported the most moonquakes. Box 10-1 has more information about how the tidal stresses on the Moon depend on its distance from Earth.

Without plate tectonics, and without the erosion caused by an atmosphere or oceans, the only changes occurring today on the lunar surface are those due to meteoroid impacts. These impacts were also monitored by the seismometers left on the Moon by the Apollo astronauts. These devices, which remained in operation for several years,

were sensitive enough to detect a hit by a grapefruit-sized meteoroid anywhere on the lunar surface. The data show that every year, the Moon is struck by 80 to 150 meteoroids having masses between 0.1 and 1000 kg (roughly ¼ lb to 1 ton).

CONCEPT CHECK 10-3

Does every world with a liquid iron-rich core contain a magnetic dynamo?

Answer appears at the end of the chapter.

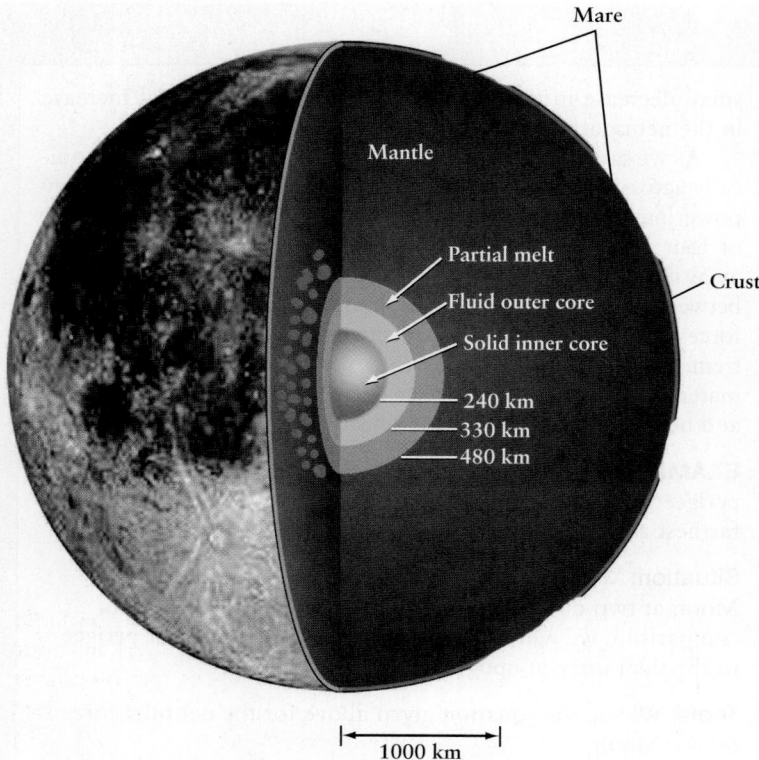

FIGURE 10-14

The Internal Structure of the Moon Like Earth, the Moon has a crust, a mantle, and a core. The red dots indicate deep earthquakes detected by seismometers left by the Apollo missions in the 1970s. Reanalysis of this data in 2011 uncovered a liquid outer core surrounded by a solid inner core. (NASA)

FIGURE 10-15 R I V U X G

The Regolith Billions of years of bombardment by space debris have pulverized the uppermost layer of the Moon's surface into powdered rock. This layer, called the regolith, is utterly bone dry. It nonetheless sticks together like wet sand when compressed, as shown by the sharp outline of an astronaut's bootprint. (NASA)

10-4 Lunar rocks reveal a geologic history quite unlike that of Earth

One of the most important scientific goals of the Apollo missions was to learn as much as possible about the lunar surface. Although only 1 of the 12 astronauts who visited the Moon was a professional geologist, all of the astronauts received intensive training in field geology, and they were in constant communication with planetary scientists on Earth as they explored the lunar surface.

Weathering by Micrometeorites and the Solar Wind

The Moon has no atmosphere or oceans to cause "weathering" of the surface. But the Apollo astronauts found that the lunar surface has undergone a kind of erosion due to billions of years of relentless micrometeorite bombardment. When lunar astronauts retrieved equipment left on the moon two years earlier, micrometeorite impacts showed that this weathering is still taking place. Another significant form of weathering comes from bombardment by charged particles in the solar wind.

On longer timescales, craters fewer than 400 m wide could be erased by weathering over a period of hundreds of millions of years to perhaps a billion years (this is used to date the scarps

shown in Figure 10-8). This bombardment has pulverized the surface rocks into a layer of fine powder and rock fragments called the **regolith,** from the Greek for "blanket of stone" (**Figure 10-15**). This layer varies in depth from about 2 to 20 m. Just as porous foam material helps to absorb sound waves in a recording studio, the regolith absorbs most of the light waves, or sunlight, that falls on it. This absorption of light helps to explain why the Moon only reflects about 11% of incident sunlight (see albedo, Table 10-1), compared to a typical Earth desert that reflects about 25% of incident light.

> Although no wind or rain disturbs the lunar surface, it is "weathered" over the ages by the impact of tiny meteoroids.

There is also concern that the regolith might be dangerous for lunar explorers. As the Apollo astronauts found, even after wearing sealed space suits when venturing about, fine regolith dust eventually makes its way into living quarters. The dust is sharp like glass, and some of the particles are very small. Small particles are particularly dangerous for the lungs, where they can lodge deep into lung tissue and cause a range of serious health problems.

While the effects of weathering can be dangerous to the touch, the impacts that cause weathering might provide an intriguing visual display. Meteorites smaller than 1 mm in size are referred to as micrometeorites, but less frequent impacts by larger objects still occur. These larger impacts might solve a mystery that is more than 500 years old—a sparkling moon. Observers have long noted occasional flashes of light from the moon, although their origin and even their very existence, was in dispute. In 2012, researchers found that when meteorites as small as 10 cm hit the lunar surface, enough heat is generated to produce light that is visible from Earth. Such events—called **transient lunar phenomena**—can occur hundreds of times a year, and are more frequent during meteor showers.

Moon Rocks

In addition to their observations of six different landing sites, the Apollo astronauts brought back 2415 individual samples of lunar material totaling 382 kg (842 lb). In addition, the robotic Soviet spacecraft Luna 16, 20, and 24 returned 300 g (²/₃ lb) from three other sites on the Moon. This collection of lunar material has provided important information about the early history of the Moon that could not have been obtained any other way.

All of the lunar samples are igneous rocks; there are no metamorphic or sedimentary rocks (see Section 9-3, especially Figure 9-19). The presence of only igneous rocks suggests that most or all of the Moon's surface was once molten (though not necessarily at the same time). Indeed, Moon rocks are composed mostly of the same minerals that are found in volcanic rocks on Earth.

Astronauts who visited the maria discovered that these dark regions of the Moon are covered with basaltic rock quite similar to the dark-colored rocks formed by lava from volcanoes on Hawaii and Iceland. The rock of these low-lying lunar plains is called **mare basalt** (Figure 10-16).

In contrast to the dark maria, the lunar highlands are composed of a light-colored rock called **anorthosite** (Figure 10-17). On Earth, anorthositic rock is found only in such very old mountain ranges as the Adirondacks in the eastern United States. Anorthosite is less dense than basalt. As a result, during the period when the Moon's surface was molten, the less dense anorthosite rose to the top, forming the majority of the present-day lunar surface. Mare basalt later covered some portions, filling in the large basins.

FIGURE 10-17 R I **V** U X G

Anorthosite The light-colored lunar terrae (highlands) are composed of this ancient type of rock, which is thought to be the material of the original lunar crust. Lunar anorthosites vary in color from dark gray to white; this sample from the Apollo 16 mission is a medium gray. It measures 18 × 16 × 7 cm. (NASA)

Although anorthosite is the dominant rock type in the lunar highlands, most of the rock samples collected there were not pure anorthosite but rather **impact breccias.** These rocks are composites of different rock types that were broken apart, mixed, and fused together by a series of meteoritic impacts (Figure 10-18). The prevalence of breccias is evidence that the lunar highlands have undergone eons of bombardment from space.

By carefully measuring the abundances of trace amounts of radioactive elements in lunar samples and applying the principles of radioactive dating (see Section 8-3), geologists have determined that lunar rocks formed more than 3 billion years ago. Of these ancient rocks, however, anorthosite is much older than the mare basalts. Typical anorthositic specimens from the highlands are

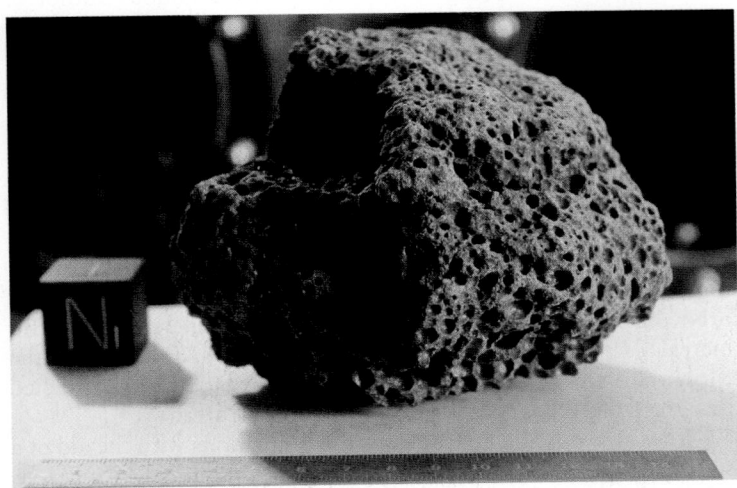

FIGURE 10-16 R I **V** U X G

Mare Basalt This 1.53-kg (3.38-lb) specimen of mare basalt was brought back by the crew of Apollo 15. Gas must have dissolved under pressure in the lava from which this rock solidified. When the lava reached the airless lunar surface, the pressure on the lava dropped and the gas formed bubbles. Some of the bubbles were frozen in place as the rock cooled, forming the holes on its surface. (Figure 9-19a shows a sample of basalt from Earth.) (NASA)

FIGURE 10-18 R I **V** U X G

Impact Breccias These rocks are evidence of the Moon's long history of bombardment from space. Parts of the Moon's original crust were shattered and strewn across the surface by meteoritic impacts. Later impacts compressed and heated this debris, welding it into the type of rock shown here. Such impact breccias are rare on Earth but abundant on the Moon. (NASA) between 4.0 and 4.3 billion years old. All of these extremely ancient specimens are probably samples of the Moon's original crust. In sharp contrast, all the mare basalts are between 3.1 and 3.8 billion years old. With no plate tectonics to bring fresh material to the surface, no new anorthosites or basalts have formed on the Moon for more than 3 billion years.

The History of Lunar Cratering

By correlating the ages of Moon rocks with the density of craters at the sites where the rocks were collected, geologists have found that the rate of impacts on the Moon has changed over the ages. The ancient, heavily cratered lunar highlands are evidence of an intense bombardment that dominated the Moon's early history. For around half a billion years, rocky debris left over from the formation of the planets rained down on the Moon's young surface. As Figure 10-19 shows, this barrage extended from 4 billion years ago, when the Moon's surface solidified, until about 3.8 billion years ago. At its peak, this bombardment from space would have produced a new crater of about 1 km radius somewhere on the Moon about once per century. If 100 years seems like a long time interval, remember that we are talking about a bombardment that lasted hundreds of millions of years and produced millions of craters over that time. Smaller impacts would have occurred much more frequently and larger ones more occasionally.

As meteoroids, asteroids, and the even larger planetesimals were swept up by the newly formed planets, you might expect that the rate of impacts would have tapered off. In fact, radioactive dating indicates that the impact rate may have spiked upward between 4.1 and 3.8 billion years ago (see Figure 10-19), and this hypothetical period is referred to as the Late Heavy Bombardment.

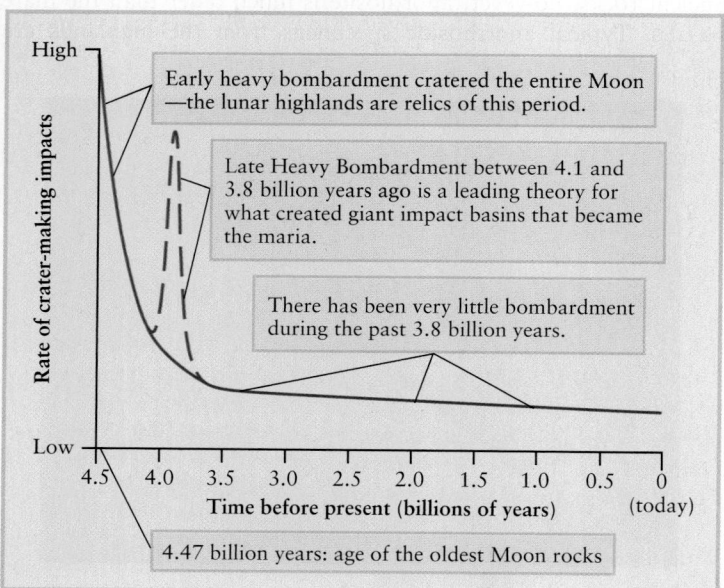

FIGURE 10-19

The Rate of Crater Formation on the Moon This graph shows the rate at which impact craters formed over the Moon's history. The impact rate today is only about 1/10,000 as great as during the most intense bombardment. (Data from T. Grotzinger, T. H. Jordan, F. Press, and R. Siever, *Understanding Earth*, 5th ed., W. H. Freeman, 2007)

This spike in impacts is theorized to have occurred throughout the inner solar system as the jovian planets scattered planetesimals—before the solar system settled into its present configuration (see Section 8-6).

Whatever the explanation for the epoch of heavy bombardment, major impacts near the end of solar system formation gouged out the mare basins, which later filled with lava to form the maria (see the *Cosmic Connections* figure). The relative absence of craters in the maria (see Figure 10-5) tells us that the impact rate has been quite low since the lava solidified. During these past 3.8 billion years, the Moon's surface has changed very little.

CONCEPT CHECK 10-4

Astronauts have left footprints on the Moon. With no wind or rain, do you expect that these features will last on the surface for a billion years?

Answer appears at the end of the chapter.

10-5 The Moon probably formed from debris cast into space when a huge protoplanet struck the young Earth

 TUTORIAL 10.3   The lunar rocks collected by the Apollo and Luna missions have provided essential information about the history of the Moon. In particular, they have helped astronomers come to a consensus about one of the most important questions in lunar science: Where did the Moon come from?

The Receding Moon

Seventy years before the first spacecraft landed on the Moon, the British astronomer George Darwin (son of the famous author of the theory of evolution, Charles Darwin) deduced that the Moon must be slowly moving away from Earth. He began with the notion that the Moon's tidal forces elongate the oceans into a football shape (see Section 4-8, especially Figure 4-28). However, the long axis of this "football" does not point precisely at the Moon. Earth spins on its axis more rapidly than the Moon revolves around Earth, and this rapid rotation carries the tidal bulge about 10° ahead of a line connecting Earth and the Moon (Figure 10-20). This misaligned bulge produces a small but steady gravitational force that tugs the Moon forward, adding energy to the Moon's orbit. As we saw in Section 4-7, increasing the energy of an orbit increases its semimajor axis. Therefore, Earth's tidal bulge tugs the Moon into an ever larger orbit. In other words, Darwin predicted that the Moon must be spiraling away from Earth.

It became possible to test Darwin's predictions with the aid of a simple device that the Apollo 11, 14, and 15 and Luna 21 missions left on the Moon—a set of reflectors, similar to the orange and red ones found in automobile taillights. Since 1969, scientists on Earth have been firing pulses of laser light at these reflectors and measuring very accurately the length of time it takes each pulse to return to Earth. Knowing the speed of light, they can use these data to calculate the distance to the Moon to an accuracy of just 3 cm. They have found that the Moon is moving away from Earth at a

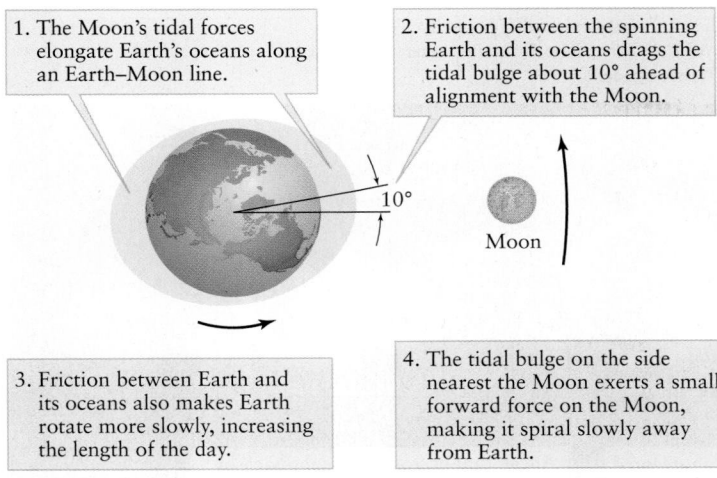

1. The Moon's tidal forces elongate Earth's oceans along an Earth–Moon line.

2. Friction between the spinning Earth and its oceans drags the tidal bulge about 10° ahead of alignment with the Moon.

3. Friction between Earth and its oceans also makes Earth rotate more slowly, increasing the length of the day.

4. The tidal bulge on the side nearest the Moon exerts a small forward force on the Moon, making it spiral slowly away from Earth.

10°

Moon

FIGURE 10-20

The Moon's Tidal Recession Earth's rapid rotation drags the tidal bulge of the oceans about 10° ahead of a direct alignment with the Moon. The bulge on the side nearest the Moon exerts more gravitational force than the other, more distant bulge. The net effect is a small forward force on the Moon that makes it spiral slowly away from Earth.

very gradual rate of 3.8 cm per year—which is just what Darwin predicted!

As the Moon moves away from Earth, its sidereal period is getting longer in accordance with Kepler's third law (see Section 4-7). At the same time, friction between the oceans and the body of Earth is gradually slowing Earth's rotation. This dance between Earth and the Moon demonstrates the conservation of angular momentum: As the Moon moves outward, it gains orbital angular momentum, and to balance this increase, Earth gradually loses angular momentum as its rotation slows. The length of Earth's day is therefore increasing by approximately 0.002 second per century. In fact, 400 million years ago, the length of Earth's day was only 21 hours!

Theories of the Moon's Origins

These observations mean that in the distant past, Earth was spinning faster than it is now and the Moon was much closer. Darwin theorized that early Earth may have been spinning so fast that a large fraction of its mass tore away, and this material coalesced to form the Moon. This is called the **fission theory** of the Moon's origin. Prior to the Apollo program, two other theories were in competition with the fission theory. The **capture theory** posits that the Moon was formed elsewhere in the solar system and then drawn into orbit about Earth by gravitational forces. By contrast, the **co-creation theory** proposes that Earth and the Moon were formed at the same time but coalesced from the solar nebula separately.

One fact used to support the fission theory was that the Moon's average density (3344 kg/m^3) is comparable to that of Earth's outer layers, as would be expected if the Moon had been flung out of a rapidly rotating proto-Earth. However, the fission theory predicts that lunar and terrestrial rocks should be very similar. In fact, samples returned from the Moon by the Apollo and Luna missions show that the material forming lunar rocks

has experienced higher temperatures than rocks on Earth. These higher temperatures are inferred by the presence of fewer **volatile elements** in lunar rock, which are elements that boil (and escape as gas) at temperatures under 900°C (about 1650°F).

LOOKING DEEPER 10.2 ➤ Arguing that the fission theory could not account for a temperature difference, some scientists maintained that the Moon formed elsewhere and was later captured by Earth. They also noted that the plane of the Moon's orbit is close to the plane of the ecliptic, as would be expected if the Moon had originally been in orbit about the Sun but was later captured by Earth.

There are, however, difficulties with the capture theory. If Earth did capture the Moon intact, then some necessary conditions must have been satisfied entirely by chance. It is easiest to understand the required coincidence in terms of orbital energy and speed. Again, from Section 4-7, each semimajor axis of an orbiting object has a certain energy, with larger energies corresponding to lower speeds and larger orbital distances. If an object were passing by too quickly, there would be nothing to slow it down into a captured orbit. If an object came in at a lower speed, nothing could speed it up into a steady orbit before it merged with Earth. Therefore, the only way to simply capture a passing Moon would be if it happened to pass by at just the right speed for its orbital distance from Earth; this seems very unlikely.

Another difficulty with the capture theory has to do with the specific chemistry of Earth and Moon rocks. The proportion of different oxygen isotopes (oxygen atoms with the same number of protons but a different number of neutrons) is specific to each solar system body whose rocks we can sample: Earth's ratio is not the same as that found in Mars meteorites or other meteorite classes. But the Moon's oxygen isotope ratio matches Earth's very closely, ruling out a capture origin.

Proponents of the co-creation theory argued that it is easier for a planet to capture swarms of tiny rocks. In this theory, the Moon formed from just such rocky debris. Great numbers of rock fragments orbited the newborn Sun in the plane of the ecliptic along with the protoplanets. Heat generated in collisions could have baked the volatile elements out of these smaller rock fragments. Then, just as planetesimals accreted to form the proto-Earth in orbit about the Sun, the fragments in orbit about Earth accreted to form the Moon.

The Collisional Ejection Theory

VIDEO 10.4 ➤ The present scientific consensus is that *none* of these three theories—fission, capture, or co-creation—is correct. Instead, the evidence points toward an idea proposed in 1975 by William Hartmann and Donald Davis and independently by Alastair Cameron and William Ward. In this **collisional ejection theory**, the proto-Earth was struck off-center by a Mars-sized object, and this collision ejected debris from which the Moon formed.

The collisional ejection theory agrees with what we know about the early history of the solar system. We saw in Section 8-5 that smaller objects collided and fused together to form the inner

The difference in chemical composition between lunar rocks and Earth rocks helps to constrain theories of the Moon's origin.

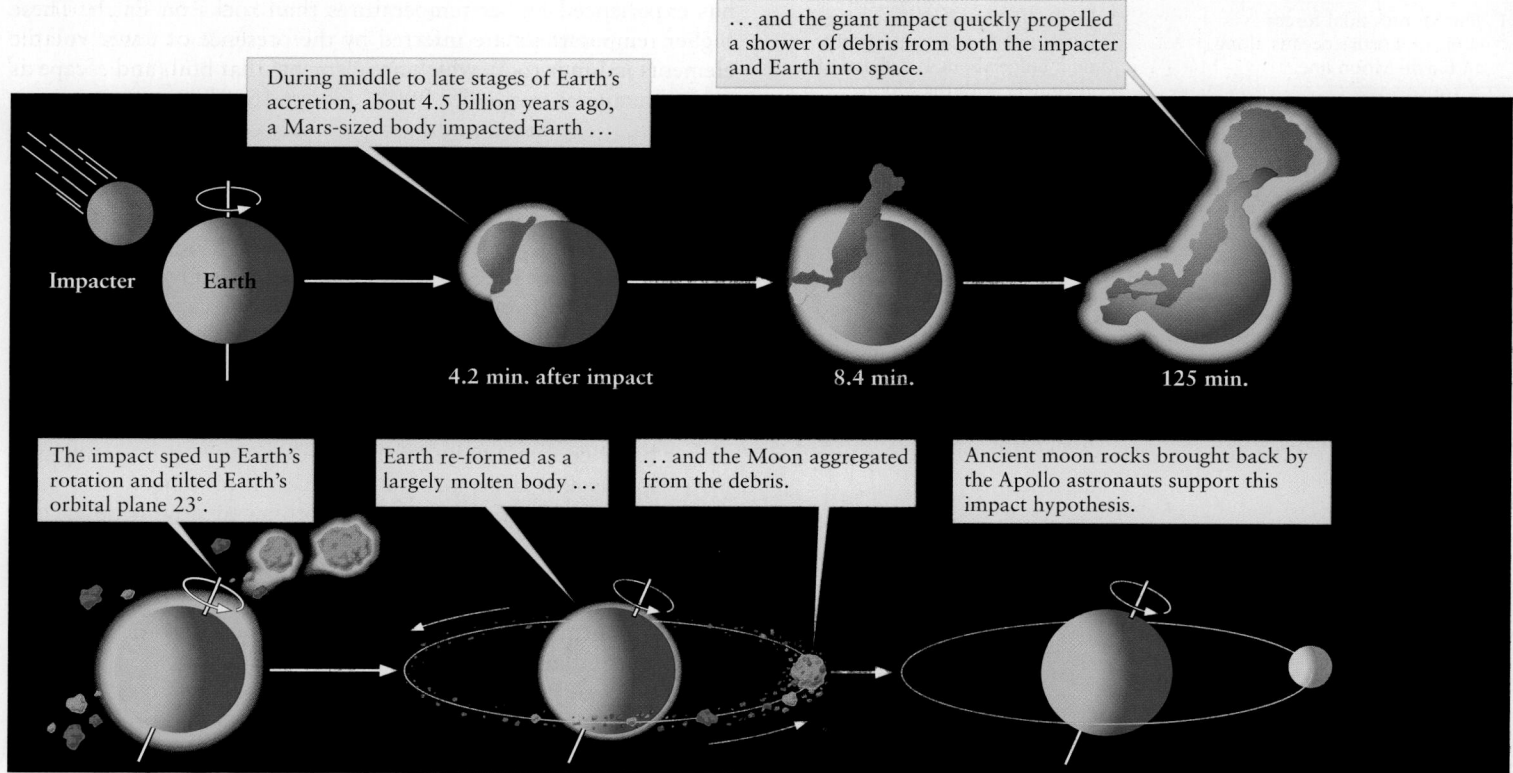

During middle to late stages of Earth's accretion, about 4.5 billion years ago, a Mars-sized body impacted Earth ...

... and the giant impact quickly propelled a shower of debris from both the impacter and Earth into space.

Impacter   Earth

4.2 min. after impact

8.4 min.

125 min.

The impact sped up Earth's rotation and tilted Earth's orbital plane 23°.

Earth re-formed as a largely molten body ...

... and the Moon aggregated from the debris.

Ancient moon rocks brought back by the Apollo astronauts support this impact hypothesis.

FIGURE 10-21

The Formation of the Moon This figure shows how the Moon could have formed in the aftermath of a collision between a Mars-sized protoplanet and the proto-Earth. (Data from T. Grotzinger, T. H. Jordan, F. Press, and R. Siever, *Understanding Earth*, 5th ed., New York, NY: W. H. Freeman, 2007)

planets. Some of these collisions must have been quite spectacular, especially near the end of planet formation, when most of the mass of the inner solar system was contained in a few dozen large protoplanets. The collisional ejection theory proposes that one such object, a Mars-sized protoplanet, struck the proto-Earth about 4.5 billion years ago, about halfway between the age of the solar system (4.56 billion years) and the age of the oldest known Moon rocks (4.47 billion years). Figure 10-21 shows the results of a supercomputer simulation of this cataclysm. In the simulation, energy released during the collision produces a huge plume of vaporized rock that shoots out from the point of impact. As this ejected material cools, it coalesces to form the Moon. The tidal effects depicted in Figure 10-20 then cause the size of the Moon's orbit to gradually increase.

The collisional ejection theory also agrees with many properties of lunar rocks and of the Moon as a whole. For example, rock vaporized by the impact would have been depleted of volatile elements and water, leaving behind the Moon rocks we know today. If the off-center collision took place after chemical differentiation had occurred on Earth, so that our planet's iron had sunk to its center, then relatively little iron would have been ejected to become part of the Moon. The lack of iron on Earth's surface explains the Moon's low density and the small size of its iron core. It also explains the curiously low iron content found in the South Pole–Aitken Basin (see Figure 10-7), where an ancient impact excavated the surface down to a depth of 12 km. Data from the Clementine spacecraft show that iron concentration in this basin is only 10% (see Figure 10-11), which is far less than the 20%–30% concentration at a corresponding depth below Earth's surface.

Detailed computer simulations support the collisional ejection theory. However, in order to create the Moon with the properties we observe, the collision would have had to occur at just the right angle and speed. Some researchers are unsatisfied with this special requirement, and have explored models that involve more than one impact and smaller bodies. Recent isotope measurements of oxygen and other elements also suggest that a more violent collision may have vaporized more of Earth than previously thought. Thus, the origin of the Moon is still an active area of research.

An Overview of the Moon's History

We can now summarize the geologic history of the Moon. The surface of the newborn Moon was probably molten for many years, both from heat produced during the formation of the Moon and from the decay of short-lived radioactive isotopes. As the Moon gradually cooled, low-density lava floating on the Moon's surface began to solidify into the anorthositic crust that exists today. A barrage of impacts then produced the craters that cover the lunar highlands.

As the stray debris was swept up by the newly formed planets, the impact rate declined. A Late Heavy Bombardment between 4.1 and 3.8 billion years ago (see Figure 10-19) may have been caused by perturbations from the giant outer planets, and more than a dozen objects measuring at least 100 km across may have struck the Moon and blasted out the vast mare basins. Meanwhile, heat from the decay of long-lived radioactive elements like uranium and thorium began to accumulate inside of the Moon. Molten rock seeped up from the lunar interior, filling the impact basins and creating the dark basaltic maria we see today.

Very little has happened on the Moon since those ancient times. The most recent geologic activity appears to be wrinklelike scarps forming as the cooling Moon shrinks in size, perhaps within the past few hundred million years. A few relatively fresh craters have been formed, but the astronauts visited a world that has remained largely unchanged for more than 3 billion years. During that same period on Earth, by contrast, the continents have been totally transformed time and time again (see Section 9-3). Earth, too, would have experienced a violent past full of impact events, but plate tectonics and our oceans and atmosphere have erased this early record.

Many questions and mysteries still remain. The six American and three Soviet lunar landings have brought back samples from only nine locations, barely scratching the lunar surface. We still know very little about the Moon's far side and poles. How many craters contain ice deposits at the poles? How much of the Moon's interior is molten? Just how large is its iron core? How old are the youngest lunar rocks? Did lava flows occur over western Oceanus Procellarum only 2 billion years ago, as crater densities there suggest? Could there still be active volcanoes on the Moon? There is still much to be learned by exploring the Moon.

CONCEPT CHECK 10-5

In the collisional ejection theory, does it matter if Earth has undergone chemical differentiation before the impact?

Answer appears at the end of the chapter.

KEY WORDS

anorthosite, p. 287
capture theory (of Moon's formation), p. 289
center of mass, p. 274
co-creation theory (of Moon's formation), p. 289

collisional ejection theory (of Moon's formation), p. 289
crater, p. 276
far side (of the Moon), p. 274
fission theory (of Moon's formation), p. 289
impact breccia, p. 287

impact crater, p. 276
libration, p. 275
lunar highlands, p. 279
mare (*plural* maria), p. 276
mare basalt, p. 287
moonquake, p. 283
regolith, p. 286

synchronous rotation, p. 274
terminator, p. 274
terrae, p. 279
transient lunar phenomena, p. 287
volatile element, p. 289

KEY IDEAS

Appearance of the Moon: The Earth-facing side of the Moon displays light-colored, heavily cratered highlands and dark-colored, smooth-surfaced maria. The Moon's far side has almost no maria (Section 10-1).

- Virtually all lunar craters were caused by space debris striking the surface. There is no evidence of plate tectonic activity on the Moon.

NASA/GSFC/Arizona State Univ./Lunar Reconnaissance Orbiter

Internal Structure of the Moon: Much of our knowledge about the Moon has come from human exploration in the 1960s and early 1970s and from more recent observations by robotic spacecraft (Sections 10-2 and 10-3).

- Analysis of seismic waves and other data indicates that the Moon has a crust thicker than that of Earth (and thickest on the far side of the Moon), a thick mantle, and a small iron core.
- The Moon has a solid inner core surrounded by a liquid outer core.
- The Moon has no global magnetic field today, although it had a weak magnetic field billions of years ago.

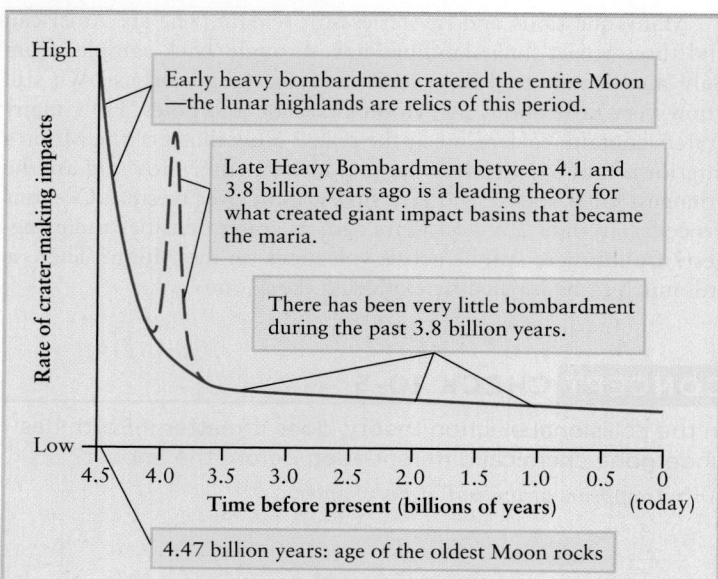

Early heavy bombardment cratered the entire Moon—the lunar highlands are relics of this period.

Late Heavy Bombardment between 4.1 and 3.8 billion years ago is a leading theory for what created giant impact basins that became the maria.

There has been very little bombardment during the past 3.8 billion years.

4.47 billion years: age of the oldest Moon rocks

Geologic History of the Moon: The anorthositic crust exposed in the highlands was formed between 4.3 and 4.0 billion years ago. The Late Heavy Bombardment is thought to have formed the maria basins between 4.1 and 3.8 billion years ago, and the mare basalts solidified between 3.8 and 3.1 billion years ago (Section 10-4).

- The Moon's surface has undergone very little change over the past 3 billion years.

- Meteoroid impacts and the solar wind have been the only significant "weathering" agents on the Moon. These weathering processes formed the Moon's regolith, or surface layer of powdered and fractured rock.

- All of the lunar rock samples are igneous rocks formed largely of the same minerals found in terrestrial rocks on Earth. However, material in lunar rocks appears to have been exposed to higher temperatures than terrestrial rocks.

Origin of the Moon: The collisional ejection theory of the Moon's origin holds that the proto-Earth was struck by a Mars-sized proto-planet and that debris from this collision coalesced to form the Moon. This theory successfully explains most properties of the Moon (Section 10-5).

- The Moon was molten in its early stages, and the anorthositic crust solidified from low-density magma that floated to the lunar surface. The mare basins were created later by the impact of planetesimals and later filled with lava from the lunar interior.

- Tidal interactions between Earth and the Moon are slowing Earth's rotation and pushing the Moon away from Earth. This also causes the length of Earth's day to slowly increase.

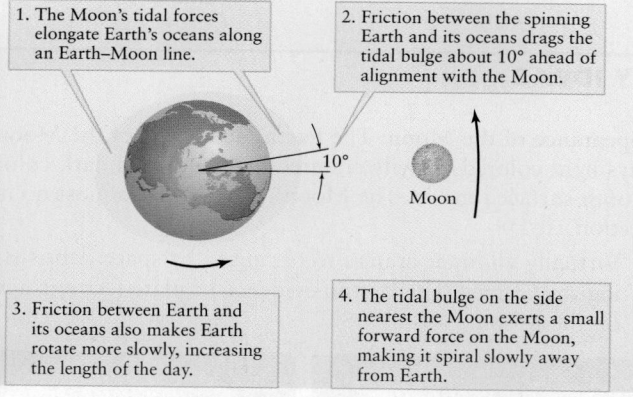

QUESTIONS

Problem-solving tips and tools

Recall that the average density of an object is its mass divided by its volume. The volume of a sphere is $4\pi r^3/3$, where r is the sphere's radius. The surface area of a sphere of radius r is $4\pi r^2$, while the surface area of a circle of radius r is πr^2. Recall also that the acceleration of gravity on Earth's surface is 9.8 m/s^2. You may find it useful to know that a 1-lb weight presses down on Earth's surface with a force of 4.448 newtons. You might want to review Newton's law of universal gravitation in Section 4-6. The time to travel a certain distance is equal to the distance traveled divided by the speed of motion. Consult Table 10-1 and the Appendices for any additional data.

Section 10-1

1. **TUTORIAL 10.1** Is it correct to say that the Moon orbits Earth? If not, what is a more correct description?

2. If the Moon always keeps the same face toward Earth, how is it possible for Earth observers to see more than half of the Moon's surface?

3. Why does the sky look black on the Moon even during daytime?

4. Why is it impossible for liquid water to exist on the surface of the Moon?

5. Describe two reasons why astronauts needed to wear space-suits on the lunar surface.

6. Describe the kinds of features that can be seen on the Moon with a small telescope.

7. Are impact craters on the Moon the same size as the meteoroids that made the impact? Explain your answer.

8. Describe the differences between the maria and the lunar highlands. Which kind of terrain covers more of the Moon's surface? Which kind of terrain is more heavily cratered? Which kind of terrain was formed later in the Moon's history? How do we know?

9. Describe the differences between the near and far sides of the Moon. What is thought to be the likely explanation for these differences?

10. What does it mean to say the Moon is a "one-plate world"? What is the evidence for this statement?

11. Calculate the round-trip travel time for a pulse of laser light that is fired from a point on Earth nearest the Moon, hits a reflector at the point on the Moon nearest Earth, and returns to its point of origin. Assume that Earth and the Moon are at their average separation from each other.

Section 10-2

12. TUTORIAL **10.2** Why was it necessary to send robotic spacecraft to land on the Moon before sending humans there?

13. What is the evidence that water ice exists at the lunar poles? Is this evidence definitive?

14. Why was it useful for the Apollo astronauts to bring magnetometers and seismometers to the Moon?

15. No Apollo mission landed on the far side of the Moon. Why do you suppose this was? What would have been the scientific benefits of a mission to the far side?

16. NASA is planning a new series of crewed missions to the Moon. Compare the advantages and disadvantages of exploring the Moon with astronauts as opposed to using mobile, robotic instrument packages.

17. Imagine that you are planning a lunar landing mission. What type of landing site would you select? Where might you land to search for evidence of recent volcanic activity?

Section 10-3

18. Could you use a magnetic compass to navigate on the Moon? Why or why not?

19. Does the absence of a global magnetic field around the Moon prove there is no liquid iron core?

20. **Box 10-1** Explain why moonquakes occur more frequently when the Moon is at perigee than at other locations along its orbit.

21. **Box 10-1** Suppose two worlds (say, a planet and its satellite) have masses m_1 and m_2, and the center-to-center distance between the worlds is r. The distance d_{cm} from the center of world 1 to the center of mass of the system of two worlds is given by the formula

$$d_{cm} = \frac{m_2 r}{m_1 + m_2}$$

(a) Suppose world 1 is Earth and world 2 is the Moon. If Earth and the Moon are at their average center-to-center distance, find the distance from the center of Earth to the center of mass of the Earth–Moon system. (b) Is the Earth–Moon system's center of mass within Earth? How far below Earth's surface is it located? (c) Find the distance from the center of the Sun (mass 1.989×10^{30} kg) to the center of mass of the Sun–Earth system. How does this compare to the diameter of the Sun? Is it a good approximation to say that Earth orbits around the center of the Sun?

22. **Box 10-1** In Box 10-1, we calculated the tidal force that Earth exerts on two 1-kg rocks located on the near and far sides of the Moon. We assumed that the Earth–Moon distance was equal to its average value. Repeat this calculation (a) for the Moon at perigee and (b) for the Moon at apogee. (c) What is the ratio of the tidal force on the rocks at perigee to the tidal force at apogee?

23. **Box 10-1** When the Moon originally coalesced, it may have been only one-tenth as far from Earth as it is now. (a) When the Moon first coalesced, was Earth's tidal force strong enough to lift rocks off the lunar surface? Explain your answer. (b) Compared with the net tidal force that Earth exerts on the Moon today, how many times larger was the net tidal force on the newly coalesced Moon? (This strong tidal force kept the one axis of the Moon oriented toward Earth, and the Moon kept that orientation after it solidified.)

Section 10-4

24. What is the regolith? What causes its powdery character?

25. Why are there no sedimentary rocks on the Moon?

26. On the basis of Moon rocks brought back by the astronauts, explain why the maria are dark-colored but the lunar highlands are light-colored.

27. Briefly describe the main differences and similarities between Moon rocks and Earth rocks.

28. Rocks found on the Moon are between 3.1 and 4.47 billion years old. By contrast, the majority of Earth's surface is made of oceanic crust that is less than 200 million years old, and the very oldest Earth rocks are about 4 billion years old. If Earth and the Moon are essentially the same age, why is there such a disparity in the ages of rocks on the two worlds?

29. The youngest lunar anorthosites are 4.0 billion years old, and the youngest mare basalts are 3.1 billion years old. Would you expect to find any impact breccias on the Moon that formed less than 3.1 billion years ago? Explain your answer.

30. In the maria, the lunar regolith is about 2 to 8 meters deep. In the lunar highlands, by contrast, it may be more than 15 meters deep. Explain how the different ages of the maria and highlands can account for these differences.

31. The mare basalts are volcanic rock. Is it likely that active volcanoes exist anywhere on the Moon today? Explain your answer.

32. Before the Apollo missions to the Moon, there were two diametrically opposite schools of thought about the history of lunar geology. The "cold moon" theory held that all lunar surface features were the result of impacts. The most violent impacts melted the surface rock, which then solidified to form the maria. The opposite "hot moon" theory held that all lunar features, including maria, mountains, and craters, were the result of volcanic activity. Explain how lunar rock samples show that neither of these theories is entirely correct.

Section 10-5

33. **TUTORIAL 10.3** If Earth's tidal bulge pointed directly toward the Moon, would the Moon still be receding from Earth? Explain your answer.

34. Why do most scientists favor the collisional ejection theory of the Moon's formation?

35. Some people who supported the fission theory proposed that the Pacific Ocean basin is the scar left when the Moon pulled away from Earth. Explain why this idea is probably wrong.

36. How would our theories of the Moon's history have been affected if astronauts had discovered sedimentary rock on the Moon?

General Problems

37. Why is Earth geologically active while the Moon is not?

38. If you view the Moon through a telescope, you will find that details of its craters and mountains are more visible when the Moon is near first quarter phase or third quarter phase than when it is at full phase. Explain why.

39. In a whimsical moment during the Apollo 14 mission, astronaut Alan Shepard hit two golf balls over the lunar surface. Give two reasons why they traveled much farther than golf balls do on Earth.

40. Temperature variations between day and night are much more severe on the Moon than on Earth. Explain why.

41. How much would an 80-kg person weigh on the Moon? How much does that person weigh on Earth?

42. Using the diameter and mass of the Moon given in Table 10-1, verify that the Moon's average density is about $3344 \ kg/m^3$. Explain why this average density implies that the Moon's interior contains much less iron than the interior of Earth.

43. Comment on the idea that without the presence of the Moon in our sky, astronomy would have developed far more slowly.

44. Describe how you would empirically test the idea that human behavior is related to the phases of the Moon. What problems are inherent in such testing?

WEB/E-BOOK QUESTION

45. In 2005, the SMART-1 spacecraft detected calcium on the lunar surface. Search the internet for information about the SMART-1 mission and this discovery. How was the presence of calcium detected? What does this tell astronomers about the origin of the Moon?

ACTIVITIES

Observing Projects

> **Observing tips and tools**
>
> If you do not have access to a telescope, you can learn a lot by observing the Moon through binoculars. Note that the Moon will appear right side up through binoculars but inverted through a telescope; if you are using a map of the Moon to aid your observations, you will need to take this into account. Inexpensive maps of the Moon can be purchased from most good bookstores or educational supply stores. You can determine the phase of the Moon either by looking at a calendar (most of which tell you the dates of new moon, first quarter, full moon, and third quarter), by checking the weather page of your newspaper, by consulting the current issue of *Sky & Telescope* or *Astronomy* magazine, by using the *Starry Night*™ program, or by using the internet.

46. Observe the Moon through a telescope every few nights over a period of 2 weeks between new moon and full moon. Make sketches of various surface features, such as craters, mountain ranges, and maria. How does the appearance of these features change with the Moon's phase? Which features are most easily seen at a low angle of illumination? Which features show up best with the Sun nearly overhead?

Starry Night™ Explorations

47. Use *Starry Night*™ to observe the apparent size of the Moon as seen from Earth. Select **Favorites > Explorations > Moon Size**. The view shows the full Moon as it would appear from the North Pole of Earth. Surrounding the Moon is a yellow circle that is 30 arcmin in diameter. As you can see, at the time and date of this view, the Moon fills this circle. In order for you to follow the apparent size of the Moon over time, daylight and the horizon have been removed from the view and the dark hemisphere of the Moon is artificially brightened to prevent the phases of the Moon from interfering with your observations.

(a) With the time flow rate at 1 day, step time forward and observe the size of the Moon relative to the 30′ reference circle. Does the apparent size of the Moon remain constant? If not, why not? Explain what this tells us about the shape of the Moon's orbit around Earth.

(b) Approximately how many days does it take for the Moon to go through a complete cycle of variation of its apparent size? (*Hint:* Count the number of 1-day time steps between two successive occasions when the Moon appears to fit precisely inside the 30′ reference circle.)

(c) As you make your observations, note that the lunar surface features do not remain stationary in the view. The Moon appears to rock back and forth, shifting its profile slightly from day to day. What term is used to describe this phenomenon and how do you explain it? (*Hint:* An internet search on the word *libration* [not libation!] may help.)

48. Open *Starry Night*™ and select **Favorites > Explorations > Luna** to display the familiar face of the Moon that we always see from Earth. The dark hemisphere of the Moon is artificially brightened in this view so that the features on the complete surface of the Moon can be observed clearly.

(a) First, consider the overall appearance of the lunar surface by using the location scroller to allow you to compare the Earth-facing hemisphere with the hemisphere facing away from Earth. You can ensure that you are looking directly at this other hemisphere by positioning the label for Earth behind the center of the image of the Moon. What is the primary difference between these two hemispheres? Use the HUD to identify any maria on the normally hidden hemisphere. Suggest a hypothesis to account for the appearance of the two hemispheres of the Moon.

(b) Based on your observations, does the presence or absence of specific types of features indicate that the Moon is geologically inactive now? Explain your answer.

(c) From your observations, do you see evidence to support the claim that the darker lunar maria are geologically younger than the brighter highland regions of the Moon?

(d) Spreading outward from some of the largest craters on the Moon are straight lines of slightly lighter-colored material called rays. These result from material ejected outward by the impacts that caused these craters. Name at least three craters on the lunar surface that have rays.

(e) Use the zoom controls and the location scroller to examine these ray patterns and follow them across the lunar surface. Is there any evidence that this ejected material disturbed crater walls during the material outbursts following the violent impacts that caused these craters?

49. The image of Crater Clavius in Figure 10-4 reveals numerous craters. Using the idea that the Moon's landscape can only be changed by impacts, make a rough sketch showing 10 of the largest craters and label them from oldest (those that showed up first) to youngest (the most recent ones). Explain your reasoning and any uncertainties.

ANSWERS

ConceptChecks

ConceptCheck 10-1: The darker color of the maria comes from solidified iron-rich lava that has partially filled large impact basins. These impact basins are deep (with the South Pole–Aitken Basin reaching depths of 12 km) and partial filling by lava still leaves the maria surfaces at lower elevations.

ConceptCheck 10-2: Figure 10-13 shows conclusive evidence of water on the Moon because the observed spectrum is such a good match to the known spectrum of water. An excess of hydrogen atoms is likely due to water, but is less certain because it could also arise from unusual minerals.

ConceptCheck 10-3: No. The Moon's liquid iron-rich core is too small for the convection that creates a magnetic dynamo. This is why the Moon today has no global magnetic field.

ConceptCheck 10-4: No. Impacting micrometeorites constantly "weather" the surface of the Moon. After a billion years, this weathering might even erase a 400-meter-wide crater.

ConceptCheck 10-5: Yes. The Moon has a lower average density than Earth, indicating a lower iron fraction. If Earth's iron sank before the impact, then outer material forming the Moon would be lower in iron.

CalculationCheck

CalculationCheck 10-1: 10.4 lb. Since 5% of ejected material x is water: $(0.05)x = 0.237$ kg. Then $x = 4.74$ kg $\times$ (2.2 lb/kg) = 10.4 lb.

11 | Mercury, Venus, and Mars: Terrestrial, Yet Unique

Mercury

R I **V** U X G

Venus

R I V U X G

Mars

R I **V** U X G

LEARNING GOALS

These images of Mercury, Venus, and Mars show the three planets that share the inner solar system with our Earth. All three have solid surfaces, and, in principle, a properly protected astronaut could stand on any of them. The differences between these three worlds, however, are pronounced.

Mercury is small, airless, and extensively cratered. It is also mysterious: Mercury has a perplexing and unexpected magnetic field. It also rotates in a manner unique in the solar system, spinning 3 times on its axis for every two orbits around the Sun.

Venus is nearly the same size as Earth, but it is shrouded by a perpetual cloud cover that hides its surface from view. Robotic spacecraft have used radar to map Venus at close range and even land on its surface. The false-color topographic map shown in the image reveals that Venus has no true continents but merely highlands (shown in red) that rise gently above the planet's lower-lying areas (shown in blue). We have also learned that Venus's atmosphere is thick and searing hot, and that its clouds contain droplets of corrosive sulfuric acid.

Mars has captivated the popular imagination like no other planet. But rather than being the home of warlike aliens, Mars proves to be an enigmatic world. Some

Three terrestrial planets: Mercury, Venus, and Mars. (NASA; NASA/JPL/USGS; NASA/J. Bell [Cornell U.] and M. Wolff [SSI])

parts of its surface are drier than any desert on Earth, while other locations show evidence of having been underwater for extended periods. Our challenge is to understand how these three terrestrial planets evolved to be so unique and so different from Earth.

11-1 Mercury, Venus, and Mars can all be seen with the naked eye

At various times Mercury, Venus, and Mars are among the brightest objects in the sky. At their greatest brilliance, each of these planets appears brighter than any star, which is why they have been known since ancient times and why their motions played a role in the religious beliefs of many ancient cultures.

Mercury and Venus are *inferior* planets, orbiting closer to the Sun than Earth does. As a result, Mercury and Venus never appear very far from the Sun in our sky (Figure 11-1). Table 11-1 and Table 11-2 summarize some basic information about these planets and their orbits.

We get the best view of either Mercury or Venus when it appears as far from the Sun in the sky as it can be, at its greatest eastern or western elongation (Section 4-2). Of course, these planets cannot be viewed with the naked eye unless the Sun is below the horizon. On dates near its **greatest eastern elongation**, Mercury or Venus appears after sunset over

> Viewing Mercury is difficult because it is so close to the Sun; Venus is easier to see but covered with clouds; Mars is best seen at opposition.

the western horizon as an "evening star." On dates near its **greatest western elongation**, Mercury or Venus appears as a "morning star" that rises before the Sun in the eastern sky.

Unlike Mercury and Venus, Mars is a *superior* planet, which means that Mars has a semimajor axis larger than Earth's orbit. As Table 11-3 shows, the Martian orbital semimajor axis is $a = 1.524$ au. Hence, Mars and the Sun are sometimes on opposite sides of the sky as seen from Earth—that is when Mars is at opposition (see Figure 4-6)—and Mars appears high in our night sky.

Because Mercury's orbit is so close to the Sun, its maximum elongation is only 28° and it is quite difficult to observe. However, the orbit of Venus is about twice the size of Mercury's orbit. As a result, Venus can be seen about 47° away from the Sun and is easy to observe.

11-2 While Mars rotates much like Earth, Mercury's rotation is coupled to its orbital motion and Venus's rotation is slow and retrograde

Mars is relatively easy to observe with visible-light telescopes, and by tracking patches of color on its surface, its rotation has been measured. By contrast, only a few faint, hazy features can be seen on Mercury's surface, and the surface of Venus is perpetually hidden by

> Radar technology revealed the curious rotation of Mercury and Venus.

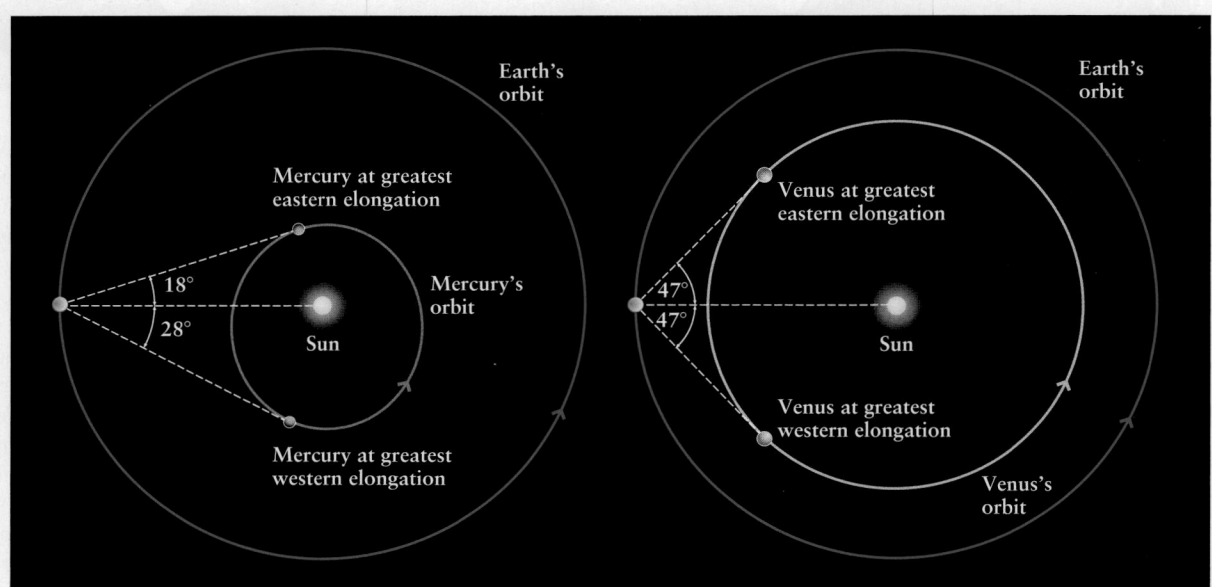

FIGURE 11-1

The Orbits of Mercury and Venus Mercury moves around the Sun every 88 days in a rather eccentric orbit. As seen from Earth, the angle between Mercury and the Sun at greatest eastern or western elongation can be as large as 28° (when Mercury is near aphelion) or as small as 18° (near perihelion). By contrast, Venus follows a larger, nearly circular orbit with a 224.7-day period. The angle between Venus and the Sun at eastern or western elongation is 47°.

TABLE 11-1 **Mercury Data**	
Average distance from the Sun:	0.387 au = 5.79×10^7 km
Maximum distance from the Sun:	0.467 au = 6.98×10^7 km
Minimum distance from the Sun:	0.307 au = 4.60×10^7 km
Eccentricity of orbit:	0.206
Average orbital speed:	47.9 km/s
Orbital period:	87.969 days
Rotation period:	58.646 days
Inclination of equator to orbit:	0.5°
Inclination of orbit to ecliptic:	7° 00′ 16″
Diameter (equatorial):	4880 km = 0.383 Earth diameter
Mass:	3.302×10^{23} kg = 0.0553 Earth mass
Average density:	5430 kg/m³
Escape speed:	4.3 km/s
Surface gravity (Earth = 1):	0.38
Albedo:	0.12
Average surface temperatures:	Day: 350°C = 662°F = 623 K Night: −170°C = −274°F = 103 K
Atmosphere:	Essentially none

R I **V** U X G
NASA

clouds. These differences explain why the rotation of Mars has been well understood since the mid-seventeenth century, while it was not until the 1960s that astronomers accurately measured the rotation of Mercury and Venus. As we will see, Mars rotates in a very Earthlike way, while Mercury and Venus rotate like no other objects in the solar system.

The Rotation of Mars

ANIMATION 11.2
In 1659, the first reliable record of surface features on Mars was made by the Dutch scientist Christiaan Huygens, who observed a dark feature appear and disappear for several weeks. Huygens concluded that the rotation period of Mars—1 day on Mars—is approximately 24 hours, surprisingly similar to a day on Earth.

Also like Earth, Mars's axis of rotation is not perpendicular to the plane of the planet's orbit, but is tilted by about 25° away from the perpendicular. This tilt is very close to Earth's 23½° tilt (see Figure 2-12). This striking coincidence means that Mars experiences Earthlike seasons, with opposite seasons in the northern and southern Martian hemispheres (see Section 2-5). Because Mars takes nearly 2 (Earth) years to orbit the Sun, the Martian seasons last nearly twice as long as on Earth.

The Challenge of Observing Mercury's Rotation

During the 1880s, the Italian astronomer Giovanni Schiaparelli attempted to make the first map of Mercury. Unfortunately,

Schiaparelli's telescopic views of Mercury were so indistinct that he made a major error, which went uncorrected for more than half a century. He erroneously concluded that Mercury always keeps the same side facing the Sun. As we will see in Section 11-4, it would not be the last time that Schiaparelli's judgment would be overturned.

Many objects in our solar system are in synchronous rotation, so that their rotation period equals their period of revolution—a situation also called **1-to-1 spin-orbit coupling**. We saw in Section 4-8 how Earth's tidal forces keep the Moon in synchronous rotation, so that it always keeps the same side toward Earth. Tidal forces also keep the two moons of Mars with the same side facing their parent planet, and likewise many of the satellites of Jupiter and Saturn.

It had been suggested as early as 1865 that the Sun's tidal forces would keep Mercury in synchronous rotation (resulting in one side being perpetually illuminated by the Sun). But this is not how Mercury rotates. Because Mercury's surface features are very muted in visible light, measuring the planet's rotation had to wait nearly a century for the development of advanced radar technology.

In 1965, Rolf B. Dyce and Gordon H. Pettengill used the giant 1000-ft radio telescope at the Arecibo Observatory in Puerto Rico to bounce powerful radar pulses off Mercury. The outgoing radiation consisted of microwaves of a very specific wavelength. In the reflected signal that echoed back from the planet, the wavelength had shifted as a result of the Doppler effect (see Section 5-9, especially Figure 5-26). As Mercury rotates, one side of the

TABLE 11-2   Venus Data

Average distance from the Sun:	$0.723 \text{ au} = 1.082 \times 10^8 \text{ km}$
Maximum distance from the Sun:	$0.728 \text{ au} = 1.089 \times 10^8 \text{ km}$
Minimum distance from the Sun:	$0.718 \text{ au} = 1.075 \times 10^8 \text{ km}$
Eccentricity of orbit:	0.0068
Average orbital speed:	35.0 km/s
Orbital period:	224.70 days
Rotation period:	243.01 days (retrograde)
Inclination of equator to orbit:	177.4°
Inclination of orbit to ecliptic:	3.39°
Diameter (equatorial):	12,104 km = 0.949 Earth diameter
Mass:	$4.868 \times 10^{24} \text{ kg} = 0.815$ Earth mass
Average density:	5243 kg/m^3
Escape speed:	10.4 km/s
Surface gravity (Earth = 1):	0.91
Albedo:	0.59
Average surface temperature:	460°C = 860°F = 733 K
Atmosphere composition (by number of molecules):	96.5% carbon dioxide (CO_2) 3.5% nitrogen (N_2) 0.003% water vapor (H_2O)

R I V U X G
NASA/Ames/JPL-Caltech

planet approaches Earth, while the other side recedes from Earth. Microwaves reflected from the planet's approaching side were shortened in wavelength, whereas those from its receding side were lengthened (Figure 11-2). By analyzing these variations in the Doppler-shifted microwaves, a rotation period of approximately 58.6 days was found for Mercury.

Spin-Orbit Coupling and the Curious Rotation of Mercury

Giuseppe Colombo, an Italian physicist with a long-standing interest in Mercury, was intrigued by Dyce and Pettengill's results. Colombo noticed that their result of 58.6 days for the rotation

FIGURE 11-2

Measuring Mercury's Rotation Period As Mercury rotates, one side of the planet moves away from Earth and the other side toward Earth. If radio waves of a single wavelength are beamed toward Mercury, waves reflected from these two sides will be Doppler shifted to longer and shorter wavelengths, respectively. By measuring the wavelength shift of the reflected radiation, astronomers deduced how rapidly Mercury rotates and thus determined its rotation period.

period is very close to two-thirds of Mercury's accurately measured sidereal (orbital) period of 87.969 days:

$$\tfrac{2}{3} (87.969 \text{ days}) = 58.646 \text{ days}$$

Colombo therefore boldly speculated that Mercury's true rotation period is exactly 58.646 days, or 58 days and 15½ hours. He realized that this value would mean that Mercury is locked into a **3-to-2 spin-orbit coupling**: The planet makes *three* complete rotations on its axis for every *two* complete orbits around the Sun. No other planet or satellite in the solar system currently has this curious relationship between its rotation and its orbital motion.

Figure 11-3 shows how gravitational forces from the Sun maintain Mercury's 3-to-2 spin-orbit coupling. The force of gravity decreases with increasing distance, which explains how the Moon is able to raise a tidal bulge in Earth's oceans (see Section 4-8). Mercury has no oceans, but it has a natural bulge of its own; thanks to the Sun's tidal forces, the planet is slightly elongated along one axis. The stronger gravitational force that the Sun exerts on the near side of the planet tends to twist the long axis to point toward the Sun, as Figure 11-3a shows. Indeed, Mercury's long axis would always point toward the Sun if its orbit were circular or nearly so. In this case, the same side of Mercury would always face the Sun, and there would be synchronous rotation (Figure 11-3b). However, Mercury's orbit has a rather high eccentricity, which leads to greater twisting forces at closer approaches. As the planet continues in its orbit, moving farther from the Sun, the twisting effect on Mercury decreases rapidly. As a result, Mercury's long axis points toward the Sun only at perihelion (Figure 11-3c). From one perihelion to the next, alternate sides of Mercury face the Sun.

TABLE 11-3 Mars Data

Average distance from the Sun:	1.524 au $= 2.279 \times 10^8$ km
Maximum distance from the Sun:	1.666 au $= 2.492 \times 10^8$ km
Minimum distance from the Sun:	1.381 au $= 2.067 \times 10^8$ km
Eccentricity of orbit:	0.093
Average orbital speed:	24.1 km/s
Orbital period:	686.98 days = 1.88 years
Rotation period:	24h 37m 22s
Inclination of equator to orbit:	25.19°
Inclination of orbit to ecliptic:	1.85°
Diameter (equatorial):	6794 km = 0.533 Earth diameter
Mass:	6.418×10^{23} kg = 0.107 Earth mass
Average density:	3934 kg/m³
Escape speed:	5.0 km/s
Surface gravity (Earth = 1):	0.38
Albedo:	0.15
Surface temperatures:	Maximum: 20°C = 70°F = 293 K Mean: −23°C = −10°F = 250 K Minimum: −140°C = −220°F = 133 K
Atmosphere composition (by number of molecules):	95.3% carbon dioxide (CO_2) 2.7% nitrogen (N_2) 0.03% water vapor (H_2O) 2% other gases

R I V U X G

NASA/JPL-Caltech

(a) The shape of Mercury's equator is slightly elongated: The different gravitational forces on Mercury's two ends tend to twist the long axis to point toward the Sun.

(b) If Mercury were in a circular orbit, its long axis would always point toward the Sun: Mercury would be in synchronous rotation (1-to-1 spin-orbit coupling).

(c) In fact, Mercury is in an elliptical orbit, and its long axis points toward the Sun only at perihelion: Mercury spins on its axis 1½ times during each complete orbit (3-to-2 spin-orbit coupling).

ANIMATION
11.3

FIGURE 11-3

Mercury's Spin-Orbit Coupling **(a)** The Sun's gravitational force on the near side of the long axis of Mercury's equator is greater than the force on the far side. This tends to twist the long axis to point toward the Sun. **(b)** The twisting effect shown in (a) would always keep the same side of Mercury (shown by a red dot) facing the Sun. **(c)** Because Mercury's orbit is rather elongated, its rotation is more complex: The end of Mercury's long axis (marked by a red dot) faces the Sun at one perihelion (point 1), but at the next perihelion (point 15) the opposite end—marked by a blue dot—faces the Sun. Since the poles of the long axis face the Sun when Mercury is closest to the Sun (points 1 and 15), these are the hottest places on Mercury—about 430°C (about 800°F) and *hot enough to melt lead!* Surprisingly, nighttime temperatures can sink below −170°C (about −270°F), which is cold enough to freeze carbon dioxide and methane.

(a) Prograde rotation

(b) Retrograde rotation

ANIMATION
11.4

FIGURE 11-4

Prograde and Retrograde Rotation **(a)** If you could view the solar system from a point several astronomical units above Earth's north pole, you would see all the planets orbiting the Sun in a counterclockwise direction. Most of the planets also rotate counterclockwise on their axes in prograde (forward) rotation. **(b)** Venus is an exception; from your perspective high above the plane of the solar system, you would see Venus rotating clockwise in retrograde (backward) rotation.

Observing Venus's Rotation: Penetrating the Clouds with Radio

Venus's perpetual cloud cover makes it impossible to measure the planet's rotation using visible-light telescopes. To penetrate the planet's perpetual cloud cover, astronomers applied to Venus the same technique of using radio waves and the Doppler effect that revealed the rotation of Mercury.

Clouds may contain gas, dust, haze, water droplets, or other small particles. In general, electromagnetic radiation can pass easily through such a cloud only if the wavelength is large compared to the size of the particles. We then say that the cloud is *transparent* to the radiation. For example, clouds in Earth's atmosphere are made of water droplets with an average diameter of about 20 μm (2×10^{-5} m, or 20,000 nm). Visible light has wavelengths between 400 and 700 nm, which is less than the droplet size. Hence, visible light cannot easily pass through such clouds, which is why cloudy days are darker than sunny days. But radio waves, with wavelengths of 0.1 m or more, and microwaves, with wavelengths from 10^{-3} m to 0.1 m, can pass through such a cloud with ease. So, your cell phone or radio works just as well on a cloudy or foggy day as on a clear one.

Like clouds in Earth's atmosphere, the clouds of Venus are transparent to radio waves and microwaves. Just as for Mercury (see Figure 11-2), the Doppler shift of waves reflected from the two sides of Venus reveals the speed and direction of the planet's rotation.

Venus: Where the Sun Rises in the West

These measurements show that Venus rotates very slowly; the rotation period of the planet is 243.01 days, even longer than the planet's year, which is 224.7 days. To an imaginary inhabitant of Venus who could somehow see through the cloud cover, stars would move across the sky at a rate of only 1½° per Earth day. As seen from Earth, by contrast, stars move 1½° across the sky in just 6 minutes.

Not only does Venus rotate slowly, it also rotates in an unusual direction. Most of the planets in the solar system spin on their axes in **prograde rotation.** *Prograde* means "forward," and planets with prograde rotation spin on their axes in the same direction in which they orbit the Sun (**Figure 11-4a**). Venus is an exception to the rule: It spins in **retrograde rotation.** That is, the direction in which it spins on its axis is backward compared to the direction in which it orbits the Sun, as Figure 11-4b shows. (Recall that *retrograde* means "backward.")

> **CAUTION!**
>
> Be careful not to confuse retrograde *rotation* with retrograde *motion*. When Venus is near inferior conjunction, so that it is passing Earth on its smaller, faster orbit around the Sun, it appears to us to move from east to west on the celestial sphere from one night to the next. This apparent motion is called retrograde motion, because it is backward to the apparent west-to-east motion that Venus displays most of the time. (You may want to review the discussion of retrograde motion in Sections 4-1 and 4-2.) By contrast, the retrograde rotation of Venus means that it spins on its axis from east to west, rather than west to east like Earth.

Venus's retrograde rotation adds to the puzzle of our solar system's origins. If you could view our solar system from a great distance above Earth's north pole, you would see all the planets orbiting the Sun counterclockwise. Closer examination would reveal that the Sun and most of the planets also *rotate* counterclockwise on their axes; the exceptions are Venus and Uranus. Most of the satellites of the planets also move counterclockwise along their orbits and rotate in the same direction. Thus, the rotation of Venus is opposite to most of the orbital and rotational motions in the solar system!

As we saw in Section 8-4, the Sun and planets formed from a rotating solar nebula. Because the planets formed from the material of this rotating cloud, they tend to orbit and rotate in that same direction. It is difficult to imagine how Venus and Uranus could have bucked this trend. One theory suggests that a huge impact billions of years ago could have reversed Venus's direction of rotation. Another idea holds that the Sun's gravitational pull on the dense atmosphere could have generated strong tides and affected the planet's rotation.

CONCEPT CHECK 11-1

If we build bigger visible-light telescopes, will astronomers finally be able to see the surface of Venus?

Answer appears at the end of the chapter.

11-3 Mercury is cratered like the Moon but has a surprisingly strong magnetic field

Imagine an alien astronomer observing Earth from Mercury, with the same telescope technology available to our own astronomers. Such an astronomer would be unable to resolve any features on Earth less than a few hundred kilometers across. It would be impossible for her to learn about Earth's mountain ranges or volcanoes. Earth-based astronomers face the same limitations in studying Mercury. To truly understand Mercury, it is essential to study this world at close range using spacecraft.

> Spacecraft observations at close range were necessary to reveal Mercury's unusual properties.

Mercury Spacecraft: Mariner 10 and MESSENGER

LOOKING DEEPER 11.1 The first spacecraft to provide some knowledge about Mercury was Mariner 10, which flew close to Mercury on three occasions in 1974 and 1975. Decades later, two brief flybys in 2008 produced the first images of Mercury from the spacecraft MESSENGER (MErcury Surface, Space ENvironment, GEochemistry, and Ranging) before it settled into orbit in 2011. MESSENGER could distinguish features as small as 18 m across—a vast improvement over Mariner 10's 2-km resolution. Figure 11-5 shows a real-color image of Mercury from MESSENGER.

FIGURE 11-5 R I **V** U X G

MESSENGER Image of Mercury This real-color image shows the natural gray color of Mercury's surface. Much like the Moon, Mercury's surface contains many craters, but none of the Moon's dark maria. The lack of darker maria is consistent with a low iron content measured on Mercury's surface. (NASA/Johns Hopkins University Applied Physics Laboratory/Carnegie Institution of Washington)

Mercury's Cratered Surface

As Mariner 10 closed in on Mercury, scientists were struck by the Moonlike pictures appearing on their television screens. It was obvious that Mercury, like the Moon, is a barren, heavily cratered world, with no evidence of plate tectonics. These surface features are what we would expect based on Mercury's small size (see Table 7-1 and the figure that opens this chapter). As we saw in Section 7-6, a small world is expected to have little internal heat and, hence, little or no geologic activity to erase an ancient, cratered surface (see Figure 7-11). But impact craters are not the only features of Mercury's surface; there are also gently rolling plains, long, meandering cliffs, and an unusual sort of jumbled terrain. As we will see, Mercury's distinctive surface shows that this planet is *not* merely a somewhat larger version of the Moon.

Figure 11-6 shows a typical close-up view of Mercury's surface. The consensus among astronomers is that most of the craters on both Mercury and the Moon were produced during the first 700 million years after the planets formed. Debris remaining after planet formation rained down on these young worlds, gouging out most of the craters we see today. Recall from Section 10-4 that the strongest evidence for this chronology comes from analysis and dating of Moon rocks. No spacecraft has yet landed on Mercury, so we are not able to make the same kind of direct analysis of rocks from the planet's surface. Likewise, no meteorites delivered

VIDEO
11.1
FIGURE 11-6 R I **V** U X G

Mercurian Craters and Plains MESSENGER took this true-color image of a region near Mercury's equator at a range of 2800 km (1700 mi). Lava flows formed the plains inside and to the west of the crater on the right. (NASA/Johns Hopkins University Applied Physics Laboratory/Carnegie Institution of Washington)

to Earth have yet been identified to have come from Mercury, although if one is discovered, it will provide a wealth of valuable information.

Scarps: Evidence of a Shrunken Planet

Images also reveal numerous long cliffs, called **scarps**, meandering across Mercury's surface (Figure 11-7). Some scarps rise as high as 3 km (2 mi) above the surrounding plains and are 20 to 500 km (10 to 300 mi) long. These cliffs probably formed as Mercury

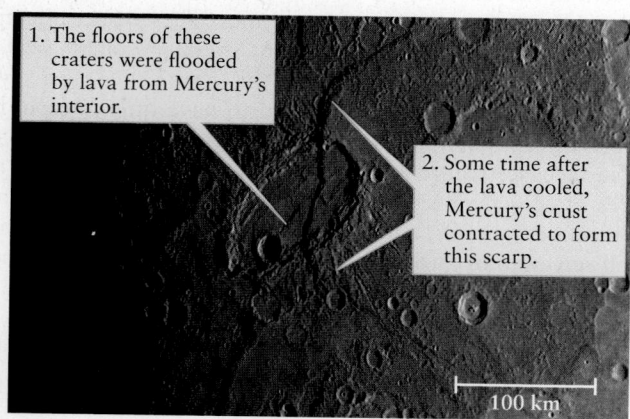

1. The floors of these craters were flooded by lava from Mercury's interior.

2. Some time after the lava cooled, Mercury's crust contracted to form this scarp.

ANIMATION
11.6
FIGURE 11-7 R I **V** U X G

A Scarp A long cliff, or scarp, called Beagle Rupes runs across this MESSENGER image. This scarp is almost a kilometer high and extends for more than 600 km (370 mi) across a region near Mercury's equator. (NASA/Johns Hopkins University Applied Physics Laboratory/Carnegie Institution of Washington)

cooled and contracted by a few kilometers in diameter. Scarps are analogous to wrinkles on fruit—as the fruit dries out and shrinks in size, its skin is deformed. Just as we saw on the Moon (Figure 10-8), when a scarp cuts through a crater, it means that the scarp formed after the crater. Using age estimates related to craters, astronomers have started to date the scarps of Mercury, which began to form early in the planet's history. Even considering the scarps, there are no features on Mercury that resemble the boundaries of tectonic plates. Thus, we can regard Mercury's crust, like that of the Moon, as a single plate.

Fire and Ice: The Caloris Basin and Mercury's Poles

The most impressive feature discovered by Mariner 10 was a huge impact scar called the Caloris Basin (*calor* is the Latin word for "heat"). The Sun is directly over the Caloris Basin during alternating perihelion passages, making it the hottest place on the planet once every 176 days. In Figure 11-8, enhanced colors help to identify different chemical compositions present on the surface of the Caloris Basin. The human eye would see only shades of gray, as seen in the other Mercury images. With enhanced coloring, slight variations can be emphasized, like the bright spots shown in orange near the rim of the basin. These spots are thought to be made up of volcanic material, while the blue areas indicate an above-average titanium oxide content. Material that appears blue is thought to have originated a few kilometers below the surface before being brought to the surface by impact events.

The Caloris Basin, which measures 1500 km (960 mi) in diameter, is both filled with and surrounded by smooth lava plains. The basin was initially gouged out by the impact of a large meteoroid, and later volcanic eruptions flooded the depression with lava. Because fewer craters pockmark these lava flows than elsewhere on the surface, the Caloris impact must be relatively young, forming toward the end of the intense bombardment period that dominated the first 700 million years of our solar system's history.

The impact that created the huge Caloris Basin must have been so violent that it shook the entire planet. On the side of Mercury *opposite* the Caloris Basin, Mariner 10 discovered a possible consequence of this shake-up—a jumbled, hilly region covering about 500,000 square km, about twice the size of the state of Wyoming. Geologists theorize that seismic waves from the Caloris impact became focused as they passed through Mercury. Perhaps as this concentrated seismic energy reached the opposite side of the planet, called the **antipode**, it deformed the surface and created jumbled terrain. This same process may also have taken place on the Moon: There is a region of chaotic hills on the side of the Moon directly opposite from the large impact basin called Mare Orientale. What about antipodal effects on Earth? Detailed calculations of Earth's dinosaur-killing asteroid impact (Chapter 15) indicate that the antipodal region, on the opposite side of Earth, would have been raised 3 to 5 m!

Even though temperatures at Mercury's equator are high enough to melt lead, there is the possibility of finding water ice at the poles. In 1991, radar waves that bounced back to Earth from the surface showed an unexpectedly high reflectivity near the poles, and ice is highly reflective. Further analysis showed that the radar reflections came precisely from regions of permanent

FIGURE 11-8   R I ⓥ U X G

The Caloris Basin About 4 billion years ago, an impact on Mercury created an immense basin, shown in orange in this enhanced-color MESSENGER image, along with a close-up in the second panel. The crater is about 1500 km (960 mi) in diameter. The impact fractured the surface extensively, forming several concentric chains of mountains. The mountains

in the outermost ring are up to 2 km (6500 ft) high. Craters inside the basin (shown in blue) were formed later by smaller impacts. (NASA/Johns Hopkins University Applied Physics Laboratory/Carnegie Institution of Washington; NASA/APL)

shadow within craters near the pole, which are well below the freezing point of water. This idea gained even more credibility in 2010 with the discovery of water in a permanently shadowed crater on the Moon (Section 10-2).

Finally, in 2012, the MESSENGER spacecraft confirmed water ice by detecting excess hydrogen in these shaded craters, where the hydrogen is part of water molecules. Analysis of hydrogen concentrations indicates a layer of almost pure water ice more than 30 cm thick beneath a layer about 15 cm thick containing little water. Some astronomers speculate that the water was delivered by comets, but the source is unknown.

Mercury's Interior and Iron Core

By measuring how much Mariner 10 was deflected in its trajectory by Mercury's gravity, scientists were able to make a precise determination of Mercury's mass. Given the mass and the diameter of Mercury, they could then calculate that the planet has an average density of 5430 kg/m³—a value quite similar to Earth's average density of 5515 kg/m³. However, as shown in Figure 11-9, a much larger fraction of Mercury is composed of iron.

CAUTION!

You might wonder how Mercury and Earth can have similar densities when Mercury's iron fraction is twice as large as Earth's. In Earth, which is 18 times more massive than Mercury, lighter material is compressed by gravity to create an overall high density. Mercury gets its own high density from its abundance of intrinsically heavy iron.

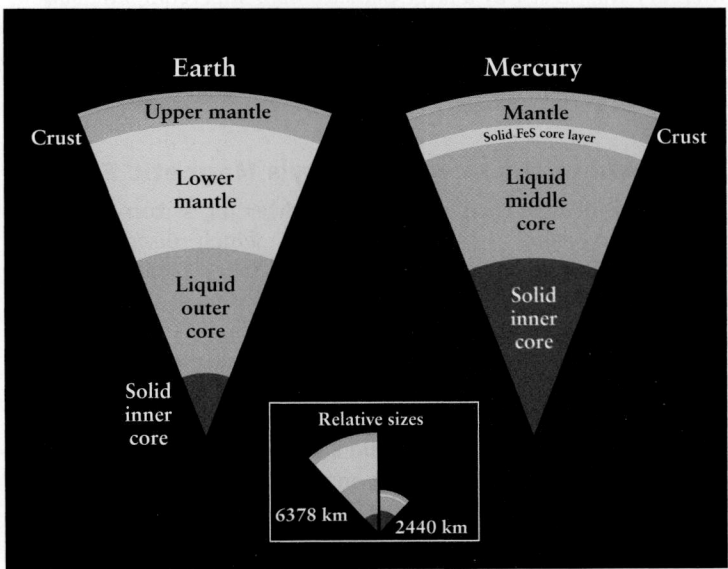

FIGURE 11-9

The Internal Structures of Mercury and Earth Compared to Earth's core, Mercury's iron-rich core takes up a much larger fraction of the planet's volume. Indeed, Mercury is the most iron-rich planet in the solar system. Mercury's iron core covers 74% of its radius and 40% of its volume, whereas Earth's core is only 53% of its radius and 15% of its volume. (NASA/Johns Hopkins University Applied Physics Laboratory/Carnegie Institution of Washington)

Several theories have been proposed to account for Mercury's high iron content. According to one theory, the inner regions of the primordial solar nebula were so warm that only those substances with high condensation temperatures—like iron-rich minerals—could have condensed into solids. Another theory suggests that

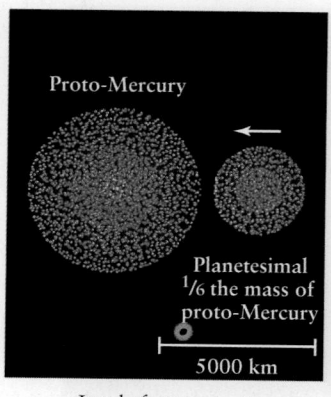

Proto-Mercury

Planetesimal ⅙ the mass of proto-Mercury

5000 km

Just before contact

Planetesimal plows into proto-Mercury's interior.

5000 km

3 minutes after contact

Impact vaporizes the rocky mantles of both worlds.

5000 km

6 minutes after contact

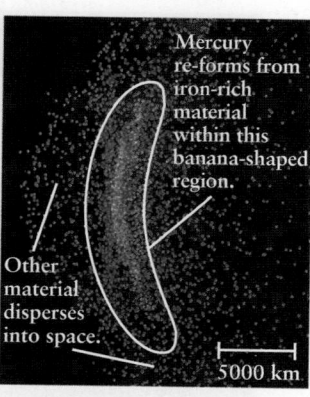

Mercury re-forms from iron-rich material within this banana-shaped region.

Other material disperses into space.

5000 km

30 minutes after contact

FIGURE 11-10

Stripping Mercury's Mantle by a Collision To account for Mercury's high iron content, one theory proposes that a collision with a planet-sized object stripped Mercury of most of its rocky mantle. These four images show a supercomputer simulation of a head-on collision between proto-Mercury and a planetesimal one-sixth its mass. (Courtesy of W. Benz, A. G. W. Cameron, and W. Slattery)

a brief episode of very powerful solar winds could have stripped Mercury of its low-density mantle shortly after the Sun formed. A third possibility is that during the final stages of planet formation, Mercury was struck by a large planetesimal. Supercomputer simulations show that this cataclysmic collision would have ejected much of the lighter mantle, leaving a disproportionate amount of iron to reaccumulate to form the planet we see today (Figure 11-10).

Clues About the Core: Mercury's Magnetic Field

An important clue to the structure of Mercury's iron core came from the Mariner 10 magnetometers, which discovered that Mercury has a magnetic field similar to that of Earth but only about 1% as strong. MESSENGER has confirmed the presence of a global magnetic field, which indicates that Mercury contains moving liquid material in its interior that conducts electricity. As in Earth, this moving material would act as a dynamo to generate the magnetic field. Hence, we conclude that at least part of Mercury's core must be in a liquid state.

The discovery of Mercury's magnetic field comes as something of a surprise. The ancient, cratered surface of Mercury is evidence for a lack of geologic activity capable of erasing craters, which in turn suggests that Mercury must have lost the internal heat that powers such activity. Again, we would expect that such a small planet would quickly lose much of its internal heat (see Section 7-6). But the existence of a global magnetic field shows that Mercury is still partially molten inside, which is evidence that the planet has some internal heat. Thus, Mercury's small size and dense cratering are in contradiction to its global magnetic field.

Another mystery is that a planet must rotate fairly rapidly in order to stir its molten interior into the kind of motion that generates a magnetic field. But as we saw in Section 11-2, Mercury rotates very *slowly*. It takes 58.646 days to spin on its axis. We still have much to learn about the interior of Mercury.

Mercury's Unsolved Mysteries

A very unusual and unexplained property of Mercury's magnetic field is that it does not originate at the planet's center. Instead, MESSENGER found that Mercury's magnetic field originates about 488 km north of the planet's center—an offset equal to a whopping 20% of the planet's radius! No other planet or moon has shown such a large offset and this is currently one of Mercury's greatest mysteries. Yet, another related mystery is perhaps the most perplexing of all.

Mercury presumably formed close to the Sun under high temperatures. (The early migration expected in our solar system was mostly carried out by the jovian planets.) It is generally found that under higher temperatures, certain atomic elements—referred to as **volatile elements**—"boil off" out of planetary material. We saw in Section 10-5 that the Moon has fewer volatile elements than Earth, just as would be expected if the Moon were created in a fiery impact with Earth. However, *Mercury has an unexpectedly high abundance of volatile elements*, as if it formed at lower temperatures.

Not only is Mercury's high volatile content a problem in trying to explain its formation so close to the hot Sun, but it also presents a challenge in explaining subsequent large impacts. For example, we saw that one explanation for Mercury's large iron core is that a large impact may have blasted away some of the lighter outer material (Figure 11-10), but this hot impact event is inconsistent with a high volatile content; volatile elements should have been evaporated into space when heated. It is also tempting to invoke a large impact to explain Mercury's off-center magnetic field, but this too runs against the observed high abundance of volatile elements on Mercury's surface.

Simply put, Mercury holds unsolved mysteries that are certain to keep astronomers busy. While additional data from MESSENGER (which ended its mission in 2015) and the European Space Agency's BepiColumbo mission will help to answer some questions, new questions are bound to arise as well.

CONCEPT CHECK 11-2

Why is it thought that Caloris Basin would have formed near the end of the planet-forming period?

CONCEPT CHECK 11-3

In Figure 11-10, rocky mantle is vaporized about 6 minutes after Mercury is hit by a large planetesimal. Would this be more consistent with a low or high abundance of volatile elements on Mercury?

Answers appear at the end of the chapter.

11-4 The first missions to Venus and Mars demolished decades of speculations about those planets

Prior to the 1960s, scientists developed a number of intriguing ideas about Venus and Mars. While these ideas were consistent with observations made using Earth-based telescopes, they needed to be radically revised with the advent of spacecraft data.

Speculations About Venus: A Lush Tropical Paradise?

Venus is bathed in more intense sunlight than Earth because it is closer to the Sun. If Venus had no atmosphere, and if its surface had an albedo like that of Mercury or the Moon, heating by the Sun would bring its surface temperature to about 45°C (113°F)—comparable to that found in the hottest regions on Earth. What was unclear until the 1960s was how much effect Venus's atmosphere has on the planet's surface temperature.

> Venus was once thought to be steamy and tropical, and Mars was thought to have vegetation that varied with the seasons.

We saw in Section 9-1 that gases such as water vapor (H_2O) and carbon dioxide (CO_2) in Earth's atmosphere trap some of the infrared radiation that is emitted from our planet's surface. This trapped radiation elevates the temperatures of both the atmosphere and the surface, a phenomenon called the *greenhouse effect*. In 1932, Walter S. Adams and Theodore Dunham Jr. at the Mount Wilson Observatory found absorption lines of CO_2 in the spectrum of sunlight reflected from Venus, indicating that carbon dioxide is also present in the Venusian atmosphere. (Section 7-3 describes how astronomers use spectroscopy to determine the chemical compositions of atmospheres.) Thus, Venus's surface, like Earth's, should be warmed by the greenhouse effect.

However, Venus's perpetual cloud cover reflects back into space a substantial amount of the solar energy reaching the planet. This effect by itself acts to cool Venus's surface, just as clouds in our atmosphere can make overcast days cooler than sunny days. Some scientists suspected that this cooling kept the greenhouse effect in

check, leaving Venus with surface temperatures below the boiling point of water. Venus might then have warm oceans and possibly even life in the form of tropical vegetation. But if the greenhouse effect were strong enough, surface temperatures could be so high that any liquid water would boil away. Then Venus would have a dry, desertlike surface with no oceans, lakes, or rivers.

Close-Up Observations of Venus: Revealing a Broiled Planet

This scientific question was resolved in 1962 when the U.S. spacecraft Mariner 2 (Figure 11-11) made the first close flyby of Venus. As we learned in Sections 5-3 and 5-4, a dense object (like a planet's surface) emits blackbody radiation: light whose intensity and spectrum depend on the temperature of the object. Mariner 2 carried instruments that measured radiation coming from the planet at two microwave wavelengths, 1.35 cm and 1.9 cm. Venus's atmosphere is transparent to both of these wavelengths, so Mariner 2 saw radiation that had been emitted by the planet's surface and then passed through the atmosphere like visible light through glass.

From the amount of radiation that Mariner 2 detected at these two wavelengths, astronomers concluded that the surface temperature of Venus was more than 400°C—well above the boiling point of water and even higher than the daytime temperatures on Mercury. Thus, there cannot be any liquid water on the planet's surface. Moreover, water vapor absorbs microwaves at 1.35 cm,

FIGURE 11-11

The Mariner 2 Spacecraft Mariner 2 was the first of all spacecraft from Earth to make a successful flyby of another planet. After coming within 34,773 km of Venus on December 14, 1962, it went into a permanent orbit around the Sun. (GSFC/NASA)

so if there were substantial amounts of water vapor in Venus's atmosphere, it would have blocked this wavelength from reaching the detectors on board Mariner 2. In fact, the spacecraft did detect strong 1.35-cm emissions from Venus. Thus, the atmosphere, too, must be all but devoid of water.

This picture of Venus as a dry, hellishly hot world was confirmed in 1970 by the Soviet spacecraft Venera 7, which survived a descent through the Venusian atmosphere and managed to transmit data for a few seconds directly from the planet's surface. Venera 7 and other successful Soviet landers during the early 1970s finally determined the surface temperature to be a nearly constant 460°C (860°F), far higher than the temperature it would have with no atmosphere at all. Thus, Venus has an extraordinarily strong greenhouse effect, and the planet's surface is no one's idea of a tropical paradise.

Speculations About Mars: A World with Plant Life and Canals?

As we saw in Section 11-2, Mars's axis of rotation is tilted by almost the same angle as Earth's axis, so Mars goes through seasons much like those on Earth. Perhaps biased by expectations of Earthlike conditions, in 1877 the Italian astronomer Giovanni Schiaparelli reported seeing 40 straight-line features crisscrossing the Martian surface (Figure 11-12). He called these dark linear

FIGURE 11-12

The Mirage of the Martian Canals Giovanni Schiaparelli studied the Martian surface using a 20-cm (8-in.) telescope, the same size used by many amateur astronomers today. He recorded his observations in drawings like this one, which shows a network of linear features that Percival Lowell and others interpreted as irrigation canals. Later observations with larger telescopes revealed that the "canals" were mere illusions. (G.V. Schiaparelli)

features *canali,* an Italian word for "channels," which was soon mistranslated into English as "canals." The alleged discovery of canals implied that there were intelligent creatures on Mars capable of substantial engineering feats. This speculation helped motivate the American millionaire Percival Lowell to finance a major new observatory near Flagstaff, Arizona, primarily to study Mars. By the end of the nineteenth century, Lowell had reported observations of 160 Martian canals.

Although many astronomers observed Mars, not all saw the canals. In 1894, the American astronomer Edward Barnard, working at Lick Observatory in California, complained that "to save my soul I can't believe in the canals as Schiaparelli draws them." But objections by Barnard and others did little to sway the proponents of the canals.

As the nineteenth century drew to a close, speculation about Mars grew more and more fanciful. Perhaps the reddish-brown color of the planet meant that Mars was a desert world, and perhaps the Martian canals were an enormous planetwide irrigation network. From these ideas, it was a small leap to envision Mars as a dying world with canals carrying scarce water from melting polar caps to farmlands near the equator. And of special concern, some suggested that the Martians might be a warlike race who schemed to invade Earth for its abundant resources. (In Roman mythology, Mars was the god of war.) Stories of Martian invasions, from H. G. Wells's 1898 novel *The War of the Worlds* down to the present day, all owe some inspiration to the *canali* of Schiaparelli.

Observations of Mars: Revealing a Windblown, Cratered World

Another feature of Mars that gained attention is the changes in color that its surface goes through on a seasonal cycle. This observation led to claims that Mars must be covered in vegetation that varied seasonally, as on Earth. In the 1960s, three American spacecraft flew past Mars and sent back the first close-up pictures of the planet's surface. (Figure 11-13 shows even better images from a subsequent mission to Mars.) These images showed that the dark surface markings are just different-colored rocky terrain.

The seasonal variations in color can be explained by winds in the Martian atmosphere that blow in different directions in different seasons. These seasonal winds blow fine dust across the Martian surface, frequently producing planetwide dust storms that can be seen from Earth (Figure 11-14). The motion of dust covers some parts of the Martian terrain and exposes others. Thus, Earth observers see large areas of the planet vary from dark to light with the passage of the seasons, mimicking the color changes that would be expected from vegetation.

Instead of showing irrigation canals and vegetation, the surface of Mars is dry and pockmarked with craters (Figure 11-13b). During the 1960s and 1970s, scientists learned that most of the craters on the Moon were formed by interplanetary debris that struck the lunar surface more than 3 billion years ago. The Martian craters, too, are the result of long-ago impacts from space. Since so many craters survive to the present day, at least part of the Martian surface must be extremely ancient (see Section 7-6). Yet the surface of Mars is more dynamic than the Moon's, and these ancient craters contend with geological forces. The spectacular Victoria Crater is shown in Figure 11-15, with its heavily eroded rim and sand dunes on its floor.

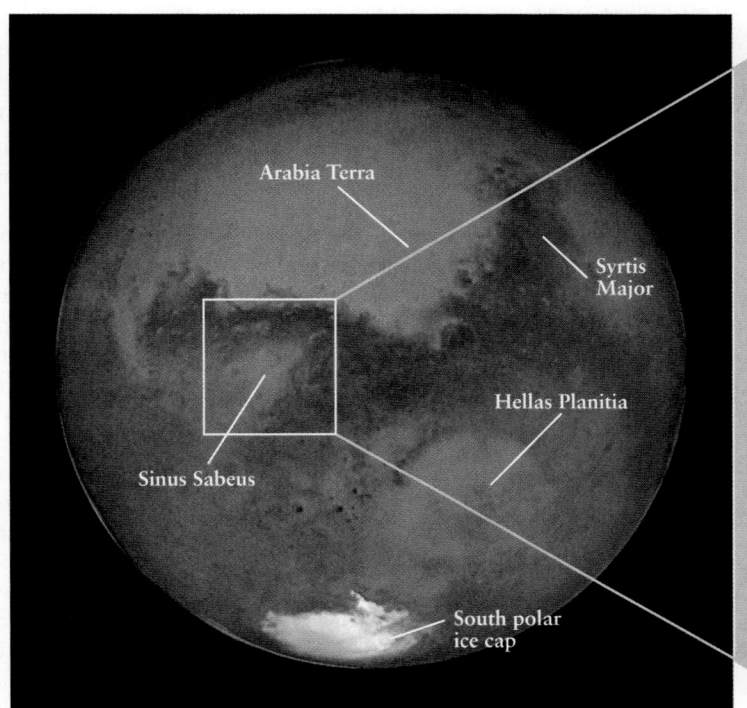

(a) Mars from the Hubble Space Telescope

(b) Closeup of Sinus Sabeus region

FIGURE 11-13 R I V U X G

Martian Craters (a) This image, made during the favorable opposition of 2003, shows cratered regions on the side of Mars opposite to that in the image that opens this chapter. Arabia Terra is dotted with numerous flat-bottom craters. Syrtis Major was first identified by Christiaan Huygens in 1659. A single titanic impact carved out Hellas Planitia, which is 5 times the size of Texas. **(b)** This mosaic of images from the Viking Orbiter 1 and 2 spacecraft shows an extensively cratered region located south of the Martian equator. (a: NASA, ESA, and The Hubble Heritage Team [STScI/AURA]; and b: NASA/JPL/USGS)

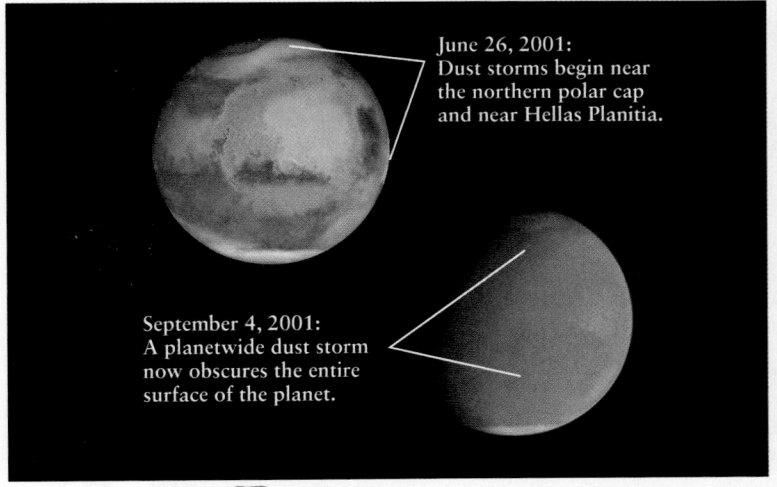

FIGURE 11-14 R I V U X G

A Martian Dust Storm Dust storms on Mars can expand to cover the entire planet. These Hubble Space Telescope images show a particularly intense storm that took place in 2001, when it was spring in the Martian northern hemisphere and autumn in the southern hemisphere. (NASA, James Bell [Cornell Univ.], Michael Wolff [Space Science Inst.], and the Hubble Heritage Team [STScI/AURA])

FIGURE 11-15 R I V U X G

Victoria Crater This crater is about 800 m wide (half a mile) and about 50 m deep. The false-coloring of this image brings out sand dunes at the bottom. The robotic Opportunity rover just arrived when this image was taken, barely visible as a speck indicated by the arrow. This image was made by the HiRISE camera aboard the Mars Reconnaissance Orbiter. (NASA)

CONCEPT CHECK 11-4

Microwaves are emitted from Venus's hot surface. What do 1.35-cm microwaves tell us about the presence of water on the surface of Venus and in its thick atmosphere?

Answer appears at the end of the chapter.

11-5 Both Venus and Mars have volcanoes—and Mars has signs of ancient plate tectonics

TUTORIAL 11.1 From a distance, Venus and Mars appear radically different: Venus is nearly the size of Earth and has a thick atmosphere, whereas Mars is much smaller and has only a thin atmosphere. But spacecraft observations reveal that these two worlds have many surface features in common, including volcanoes, faults, and impact craters. Mars even shows some evidence for ancient plate tectonic activity. By comparing the surfaces of these two worlds to that of Earth, we can gain insight into the similarities and differences between the three largest terrestrial planets.

> Venus has no plate tectonics because its crust is too thin, whereas on Mars the crust is too thick.

Observing Venus and Mars from Orbit

Because additional fuel is required to slow down a spacecraft, it is much easier to send a spacecraft whizzing past a planet (a "flyby") than it is to park one in orbit around the planet. But the trouble is worth it! A much more detailed study can be made from orbit. Since the 1970s, a number of spacecraft have been placed in orbit around Venus and Mars.

ANIMATION 11.5 In order to map the surface of Venus through the perpetual cloud layer, several of the Venus orbiters carried radar devices. A beam of microwave radiation from the orbiter easily penetrates Venus's clouds and reflects off the planet's surface; a receiver on the orbiter then detects the reflected beam. By measuring the time it takes for reflected waves to return to the orbiter, astronomers can determine the height and depth of Venus's terrain. The most recent spacecraft to orbit Venus, Magellan, produced the topographic map in Figure 11-16. The same radar method can be used on Mars using visible light, and the topographic map of Mars in Figure 11-17 was produced by the Mars Global Surveyor spacecraft (which entered Mars's orbit in 1997).

The topographies of Mars and Venus differ in important ways from that of Earth. Our planet has two broad classes of terrain: About 71% of Earth's surface is oceanic crust and about 27% is continental crust that rises above the ocean floor by about 4 to 6 km on average. On Venus, by contrast, about 60% of the terrain lies within 500 m of the average elevation, with only a few localized highlands (shown in yellow and red in Figure 11-16). Mars is

FIGURE 11-16

A Topographic Map of Venus Radar altimeter measurements by Magellan were used to produce this topographic map of Venus. Color indicates elevations above (positive numbers) or below (negative numbers) the planet's average radius. (The blue areas are not oceans!) Gray areas were not mapped by Magellan. Flat plains of volcanic origin cover most of the planet's surface, with only a few continentlike highlands. (NASA/JPL)

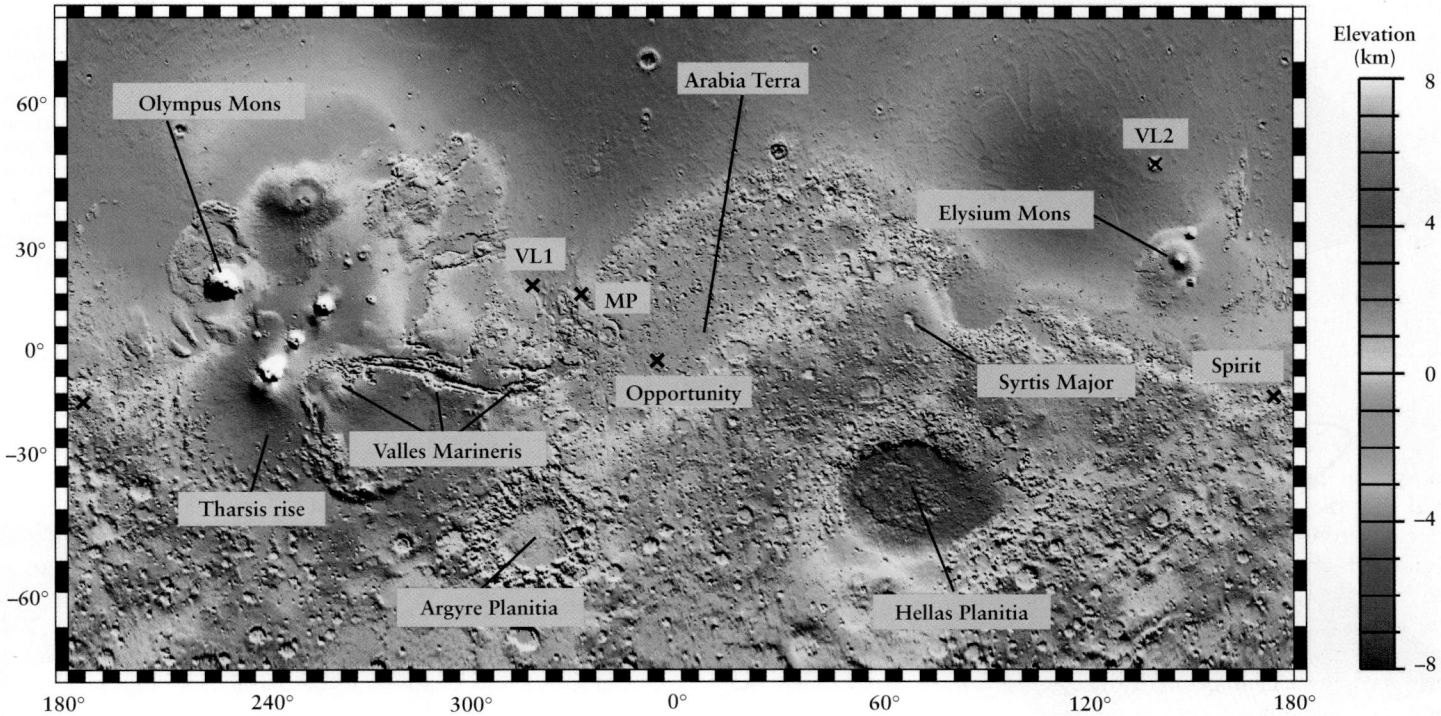

FIGURE 11-17

A Topographic Map of Mars This map was generated from measurements made by the laser altimeter on board the Mars Global Surveyor spacecraft. As in Figure 11-15, color indicates elevations above or below the planet's average radius. Most of the southern hemisphere lies several kilometers above the northern hemisphere, with the exception of the immense impact feature called Hellas Planitia. The landing sites for Viking Landers 1 and 2 (VL1 and VL2), Mars Pathfinder (MP), and the Mars Exploration Rovers (Spirit and Opportunity) are each marked with an X. (NASA/JPL-Caltech)

different from both Earth and Venus: Rather than having elevated continents scattered among low-lying ocean floors, all of the high terrain on Mars (shown in red and orange in Figure 11-17) is in the southern hemisphere. Hence, planetary scientists refer to Mars as having **northern lowlands** and **southern highlands.** This difference indicates that Mars has a **crustal dichotomy**—that is, the Martian crust is about 5 km thicker in the southern hemisphere than in the northern hemisphere. In this sense, Mars has something in common with the Moon, which has a thicker crust on the far side than on the side that faces Earth (see Section 10-1).

CONCEPT CHECK 11-5

In Figure 11-17, the northern hemisphere is in the top half of the map, with the southern hemisphere in the bottom half. Which hemisphere has the younger surface, and why?

Answer appears at the end of the chapter.

Tectonics on Venus: A Light, Flaky Crust

Before radar maps like Figure 11-16 were available, scientists wondered whether Venus had plate tectonics like the process that has remolded the face of Earth. Venus is only slightly smaller than Earth and should have retained enough heat to sustain the convection currents that drive tectonic activity on Earth (see Section 9-3,

especially Figure 9-16). If this were the case, then the same tectonic effects might also have shaped the surface of Venus. As we saw in Section 9-3, Earth's hard outer shell, or lithosphere, is broken into about a dozen large plates that move slowly across the globe.

Radar images from Magellan show *no* evidence of Earthlike plate tectonics on Venus. On Earth, long chains of volcanic mountains (like the Cascades in North America or the Andes in South America) form along plate boundaries where subduction is taking place. Mountainous features on Venus, by contrast, do not appear in chains. There are also no structures like Earth's Mid-Atlantic Ridge (Figure 9-14), which suggests that there is no seafloor spreading on Venus. With no subduction or seafloor spreading, there has been only limited horizontal displacement of Venus's lithosphere. Thus, like the Moon (see Section 10-1) and Mercury (see Section 11-3), Venus has a one-plate crust.

Unlike the Moon, however, Venus has had local, small-scale deformations and reshaping of the surface. One piece of evidence for this is that roughly one-fifth of Venus's surface is covered by folded and faulted ridges. Venus also has very unique formations called *coronae* which might have helped to reshape its surface (**Figure 11-18**). Further evidence that Venus's surface has evolved over time comes from close-up Magellan images showing that Venus has about a thousand craters that are larger than a few kilometers in diameter, many more than have been found on Earth but only a small fraction of the number on the Moon or Mercury.

Coronae formation on Venus

Increasing time

(a)

Concentric
fracture rings

Radial fractures

(b)

FIGURE 11-18 Coronae: The Crowns of Venus Originally mistaken as impact craters, these round and oval-shaped crown, or corona-like, features originate from a very different mechanism. **(a)** The leading model is that hot magma rises up from Venus's interior, pushing the crust upward to create a dome above the surface. As the magma cools, and as some magma oozes out the sides, the center of the dome can solidify above the surface or drop below the surface, leaving a ring-shaped or crown-like ridge. **(b)** Some coronae have prominent radial and concentric fractures. (a: Courtesy of Michael Manga; b: NASA)

R I V U X G

We saw in Section 7-6 that the number of impact craters is a clue to the age of a planet's surface. Such impact craters formed at a rapid rate during the early history of the solar system, when considerable interplanetary debris still orbited the Sun, and have formed at a much slower rate since then. Consequently, the more craters a planet has, the older its surface. The number of craters on Venus indicates that the Venusian surface is roughly 500 million years old. This is about twice the age of Earth's surface but much younger than the surfaces of the Moon or Mercury, each of which is billions of years old. No doubt Venus was more heavily cratered in its youth, but localized activity in its crust has erased the older craters (Figure 11-19).

Surprisingly, Venus's craters are uniformly scattered across the planet's surface. We would expect that older regions on the surface—which have been exposed to bombardment for a longer time—would be more heavily cratered, while younger regions would be relatively free of craters. For example, the ancient highlands on the Moon are much more heavily cratered than the younger maria (see Section 10-1). Because such variations are not found on Venus,

scientists conclude that the entire surface of the planet has essentially the *same* age. This is very different from Earth, where geological formations of widely different ages can be found.

One model that can explain these features suggests that the convection currents in Venus's interior are actually more vigorous than inside Earth, but that the Venusian crust is much thinner than the continental crust on Earth. Rather than sliding around like the plates of Earth's crust, the thin Venusian crust stays in roughly the same place but undergoes wrinkling and flaking (Figure 11-20). However, the current and past tectonic activity on Venus is still highly debated, and we have much to learn about our neighboring planet.

LOOKING DEEPER 11.2 Although Venus almost certainly has a partially molten core in its interior, it has no planetwide magnetic field. As we discussed in Section 7-7, Venus may have no magnetic field because it rotates too slowly to generate the kind of internal motions that would produce a magnetic field. With no such field, Venus has no magnetosphere. The planet is nonetheless shielded from the solar wind by ions in its upper atmosphere.

1. This circular region is the remnant of a crater formed by a meteoroid impact.

2. Later, a fault in the crust formed this deep valley, destroying half of the crater.

3. These "streaks" are other fractures or faults that have occurred since the crater was formed.

50 km

FIGURE 11-19 Ⓡ I V U X G

A Partially Obliterated Crater on Venus   This Magellan image shows how the right half of an old impact crater 37 km (20 mi) in diameter was erased by a fault in the crust. This crater lies in Beta Regio (see the left side of Figure 11-16). Most features on Venus are named for women in history and legend; this crater commemorates Emily Greene Balch, an American economist and sociologist who won the 1946 Nobel Peace Prize. (NASA/JPL)

CONCEPT CHECK 11-6

How was it determined that Venus's surface is older than parts of Earth's surface, yet younger than the surface of Mercury?

Answer appears at the end of the chapter.

Plate Tectonics on Mars

Like Venus, Mars lacks the global network of ridges and subduction zones like those produced by the seven major plates on Earth. However, evidence has been discovered for the existence of two plates on Mars separated along the 4000-km-long canyon named Valles Marineris (**Figure 11-21**). This enormous chasm was first imaged in 1971, and in 2012 astronomers realized that the two sides of this canyon have slid past each other horizontally by 150 km (93 mi). So far, this is the only known evidence for a plate boundary on Mars, giving Mars a total of two plates.

Another piece of evidence about ancient Mars comes from the pattern of crustal magnetism resembling stripes that display alternating north–south polarities of the magnetic field. These magnetic stripes are shown as red and blue bands in the lower middle of Figure 7-15. If these stripes on Mars formed in a similar way to magnetic stripes on Earth, as illustrated in Figure 9-23, that would tell us quite a bit about Mars's early magnetic field and tectonics. On Earth, the stripes arise from a global magnetic field that flips in polarity over time and magnetizes lava cooling at a rift. As the newly formed crust spreads away from the rift, stripes of magnetism with alternating directions spread across the surface. Two mechanisms are required to form Earth's magnetic stripes that would also need to take place on Mars—plate tectonics and a magnetic dynamo in a molten core with an alternating magnetic field. It is a remarkable achievement of geology that stripes of magnetism observed on Earth today might tell us so much about another planet's early history. However, if Martian plate tectonics helped to create these magnetic stripes, another mystery arises:

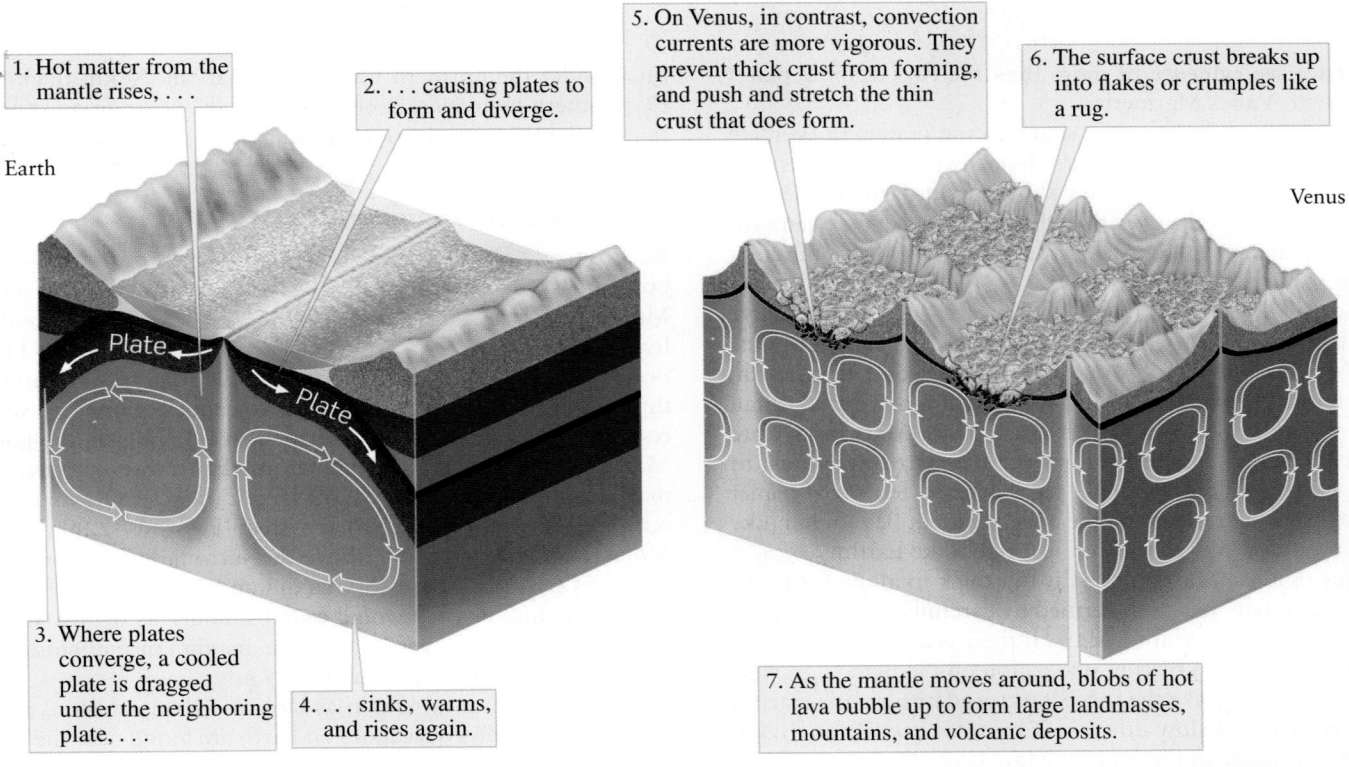

1. Hot matter from the mantle rises, . . .

2. . . . causing plates to form and diverge.

5. On Venus, in contrast, convection currents are more vigorous. They prevent thick crust from forming, and push and stretch the thin crust that does form.

6. The surface crust breaks up into flakes or crumples like a rug.

Earth

Venus

Plate

Plate

3. Where plates converge, a cooled plate is dragged under the neighboring plate, . . .

4. . . . sinks, warms, and rises again.

7. As the mantle moves around, blobs of hot lava bubble up to form large landmasses, mountains, and volcanic deposits.

FIGURE 11-20

Tectonics on Earth Versus Tectonics on Venus   This illustration shows the difference between plate tectonics on Earth and a model of tectonics on Venus. (© Courtesy of John Grotzinger)

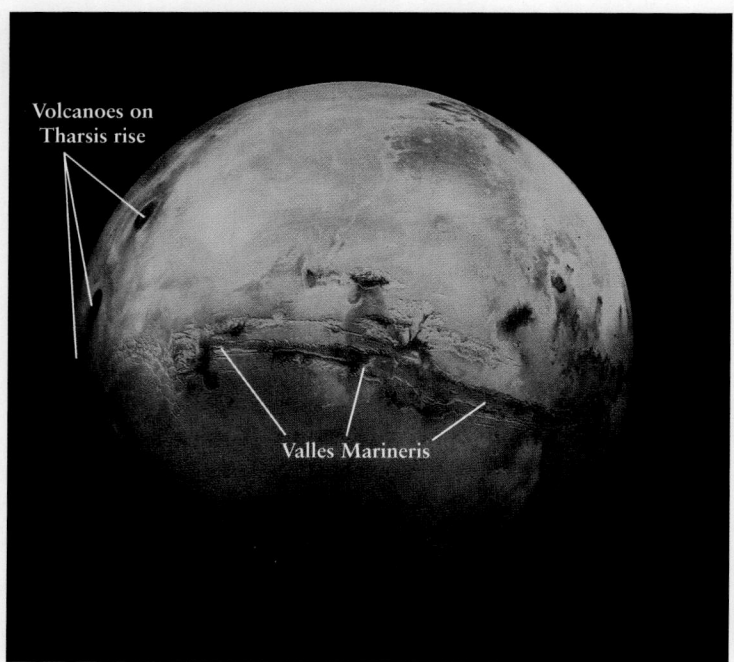

(a) Mars and Valles Marineris

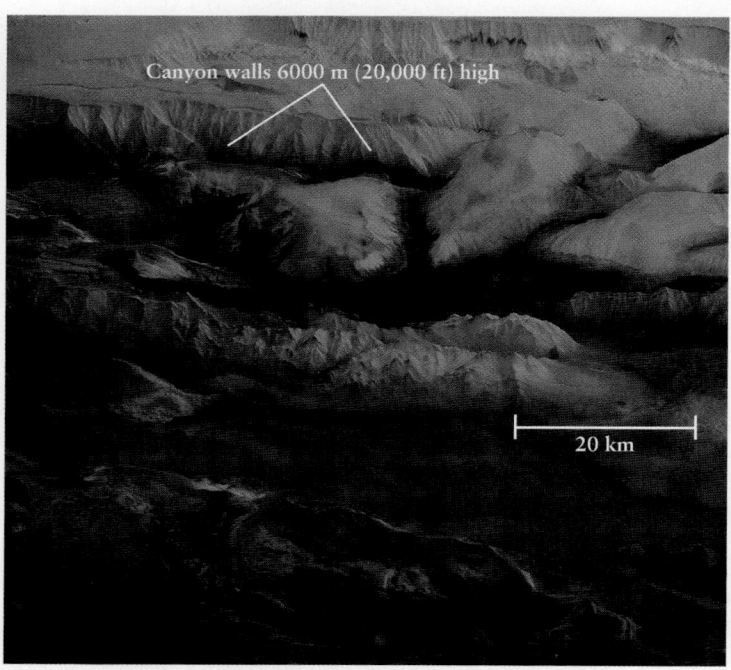

(b) The central region of Valles Marineris

FIGURE 11-21 R I Ⓥ U X G

Valles Marineris **(a)** This mosaic of Viking Orbiter images shows the huge rift valley of Valles Marineris, which extends from west to east for more than 4000 km (2500 mi), which is approximately the width of the United States, and is 600 km (400 mi) wide at its center. It reaches depths of 8 km (5 mi) beneath the surrounding plateau, which is over 4 times as deep as the Grand

Canyon, Arizona. At its western end is the Tharsis rise. **(b)** This perspective image from the Mars Express spacecraft shows what you would see from a point high above the central part of Valles Marineris. (a: USGS/NASA; b: ESA/DLR/FU Berlin (G. Neukum))

Why can't we see evidence of more than the two tectonic plates associated with Valles Marineris?

We can also look at magnetic signatures in craters to estimate when the magnetic dynamo of Mars shut down, since craters that formed after this point would not show any crustal magnetism. Using age estimates of the craters without magnetism, we can estimate that the dynamo might have shut down about 500 million years after Mars formed.

Why didn't Mars experience more plate tectonic activity? Recall that smaller objects cool off more quickly than larger objects (Section 7-6). Because Mars is a much smaller world than Earth, the outer layers of the red planet have cooled more quickly than on Earth, which has led to a thicker crust. Thus, Mars lacks plate tectonics because its crust is too thick for one part of the crust to be subducted beneath another. We see that for a terrestrial planet to have extensive plate tectonics, the crust must not be too thin (like Venus) or too thick (like Mars), but just right (like Earth).

The idea that the Martian crust is too thick to allow for much plate tectonic activity has been verified by carefully monitoring the motion of a spacecraft orbiting Mars. If there is a concentration of mass (such as a thicker crust) in one region on the planet, gravitational attraction will make the spacecraft speed up as it approaches the concentration and slow down as it moves away. By analyzing these small deviations in the orbit of Mars Global Surveyor in just this way, scientists found that unlike Earth's crust, which varies in thickness from 5 to 35 km, the Martian crust is about 40 km

thick under the northern lowlands but about 70 km thick under the southern highlands. Both regions of the Martian crust are too thick to undergo subduction, making plate tectonics very difficult.

Volcanoes on Venus and Mars

Radar images of Venus and visible-light images of Mars show that both planets have a number of large volcanoes (**Figure 11-22**). Magellan observed more than 1600 major volcanoes and volcanic features on Venus, two of which are shown in Figure 11-22a. Both of these volcanoes have gently sloping sides. A volcano with this characteristic is called a **shield volcano,** because in profile it resembles an ancient Greek warrior's shield lying on the ground. Martian volcanoes are less numerous than those on Venus, but they are also shield volcanoes; the largest of these, Olympus Mons, is the largest volcano in the solar system (Figure 11-22b). Olympus Mons rises 24 km (15 mi) above the surrounding plains. In fact, Olympus Mons is about the maximum size a mountain can get on Mars before its rock would crumble under its own weight. On Earth, with our stronger gravity, the maximum mountain size is about Mount Everest (8.8 km, or 5.5 mi).

But how does a mountain as large as Olympus Mons form in the first place? Most volcanoes on Earth are found near the boundaries of tectonic plates, where subducted material becomes molten magma and rises upward to erupt from the surface. This cannot explain the volcanoes of Venus and Mars, since there are no

(a) Volcanoes and lava flows on Venus   R I V U X G

(b) Cloud-topped volcanoes on Mars   R I V U X G

FIGURE 11-22

Volcanoes on Venus and Mars (a) The false color in this radar image approximates the real color of sunlight that penetrates Venus's thick clouds. The brighter color of the extensive lava flows indicates that they reflect radio waves more strongly. To emphasize the gently sloping volcanoes, the vertical scale has been exaggerated 10 times. **(b)** The volcanoes of Mars also have gently sloping sides. In this view, looking down from Mars orbit, you can see bluish clouds topping the summits of the volcanoes. These clouds, made of water ice crystals, form on most Martian afternoons. (a: NASA/JPL; b: NASA/JPL/MSSS)

subduction zones on those planets. Instead, Venusian and Martian volcanoes were probably formed by **hot-spot volcanism.** In this process, magma wells upward from a hot spot in a planet's mantle, elevating the overlying surface and producing a shield volcano.

On Earth, hot-spot volcanism is the origin of the Hawaiian Islands. These islands are part of a long chain of shield volcanoes that formed in the middle of the Pacific tectonic plate as that plate moved over a long-lived hot spot (Figure 11-23). On Venus and Mars, by contrast, the absence of significant plate tectonics means that the crust remains stationary over a hot spot. On Mars, a single hot spot under Olympus Mons probably pumped magma upward through the same vent for millions of years, producing one giant volcano rather than a long chain of smaller ones. The Tharsis rise and its volcanoes (see Figure 11-17, Figure 11-21a, and Figure 11-22b) may have formed from the same hot spot as gave rise to

Olympus Mons; a different hot spot on the opposite side of Mars produced a smaller bulge centered on the volcano Elysium Mons, shown near the right-hand side in Figure 11-17. The same process of hot-spot volcanism likely gave rise to large shield volcanoes on Venus like those shown in Figure 11-22a.

Volcanic Activity on Venus

About 80% of the surface of Venus is composed of flat plains of volcanic origin. In other words, essentially, the entire planet is covered with lava! This observation shows the tremendous importance of volcanic activity in Venusian geology.

To verify the volcanic nature of the Venusian surface, it was necessary to visit the surface and examine rock samples. Figure 11-24 is a panoramic view taken in 1981 by the Soviet spacecraft Venera 13, one of 10 robotic spacecraft that the Soviet Union

FIGURE 11-23

Hot-Spot Volcanoes on Earth A hot spot under the Pacific plate has remained essentially stationary for 70 million years while the plate has moved some 6000 km to the northwest. The upwelling magma has thus produced a long chain of volcanoes. The Hawaiian Islands are the newest of these; the oldest, the Emperor Seamount Chain, have eroded so much that they no longer protrude above the ocean surface. This obscured chain also indicates that the plate was moving in a different direction when it was formed. (National Geophysical Data Center/NOAA)

(a) Image from Venera 13

(b) Color-corrected image

FIGURE 11-24   R I ⓥ U X G

A Venusian Landscape (a) This wide-angle color photograph from Venera 13 shows the rocky surface of Venus. The thick atmosphere absorbs the blue component of sunlight, giving the image an orange tint: The stripes on the spacecraft's color calibration bar that appear yellow are actually white in color. **(b)** This color-corrected view shows that the rocks are actually gray in color. The rocky plates covering the ground may be fractured segments of a thin layer of lava. (a: NSSDC/NASA; b: GSFC/NASA)

landed successfully on the surface of Venus. Russian scientists hypothesize that this region was covered with a thin layer of lava that fractured upon cooling to create the rounded, interlocking shapes seen in the photograph. This hypothesis agrees with information obtained from chemical analyses of surface material made by the spacecraft's instruments. These analyses indicate that the surface composition is similar to lava rocks called basalt, which are common on Earth (see Figure 9-19a) and in the maria of the Moon (see Figure 10-16). The results from Venera 13 and other landers are consistent with the picture of Venus as a world whose surface and atmosphere have been shaped by volcanic activity.

Most of the volcanoes on Venus are probably inactive at present, just as is the case with most volcanoes on Earth. But in 2010, evidence was found for very recent volcanic activity on Venus. By analyzing infrared emission, the European Space Agency's Venus Express orbiter found material on volcanoes showing signs that it had not experienced significant weathering. However, due to weathering by Venus's hot and thick atmosphere, the surface of fresh volcanic rock should show weathering fairly quickly. This analysis points to very young lava flows, perhaps a few hundred years old, or as many as 2.5 million years old (which is still geologically young). The presence of such young lava flows suggests that Venus, like Earth, has some present-day volcanic activity.

Another piece of evidence for ongoing volcanic activity on Venus comes from the planet's atmosphere. An erupting volcano on Earth ejects substantial amounts of sulfur dioxide, sulfuric acid, and other sulfur compounds into the air. Many of these substances are highly reactive and short-lived, forming sulfate compounds that become part of the planet's surface rocks. For these substances to be relatively abundant in a planet's atmosphere, they must be constantly replenished by new eruptions. Sulfur compounds make up about 0.015% of the Venusian atmosphere, compared to less than 0.0001% of Earth's atmosphere. This evidence suggests that ongoing volcanic eruptions on Venus are ejecting sulfur compounds into the atmosphere to sustain the high sulfur content.

Volcanic Activity on Mars

Spectroscopic observations from Mars's orbit confirm that the planet's rocks and sands are made almost entirely of the three minerals feldspar, pyroxene, and olivine. These are the components of basalt, or solidified lava. Thus, Mars, like Venus, had a volcanic past.

Unlike lava flows on Venus, however, most of the lava flows on Mars have impact craters on them. These craters suggest that most Martian lava flows are very old and that most of the volcanoes on the red planet are no longer active. This is what we would expect from a small planet whose crust has cooled significantly to a greater depth than on Earth, making it difficult for magma to travel from the Martian mantle to the surface.

However, a few Martian lava flows are crater free, which suggests that they are only a few million years old. If volcanoes erupted on Mars within the past few million years, are they erupting now? Scientists have used infrared telescopes to search for telltale hot spots on the Martian surface, but have yet to discover any. Perhaps volcanism on Mars is rare but not yet wholly extinct.

Spacecraft that have landed on Mars have taught us a great deal about the geology and history of the red planet. Before we explore the Martian surface in detail, however, it is useful to examine the unique atmospheres of both Mars and Venus.

11-6 The dense atmosphere of Venus and the thin Martian atmosphere are dramatically different but have similar chemical compositions

We have seen that the three large terrestrial planets—Earth, Venus, and Mars—display very different styles of geology. As we will discover, these differences help to explain why each of these three worlds has a distinctively different atmosphere.

Recall from Chapter 9 (Section 9–7) that Earth's atmosphere is so sensitive to carbon dioxide, that even the CO_2 released from human activity is enough to raise the temperature of our planet. Likewise, as we will see in this section and the next, carbon dioxide will also play a major role in the resulting temperatures on Venus and Mars.

On Venus, sulfuric acid rain evaporates before reaching the ground; on Mars, it snows frozen carbon dioxide.

To explore a planet's atmosphere in detail, it is best to send a spacecraft to make measurements as it descends through that atmosphere. Both Soviet and U.S. spacecraft have done this for Venus (see Figure 11-24), and seven U.S. spacecraft have successfully landed on Mars. The results of these missions are summarized in **Figure 11-25**, which shows how pressure and temperature vary with altitude in the atmospheres of Earth, Venus, and Mars.

The Venusian Atmosphere: Dense, Hot, and Corrosive

Figure 11-25b depicts just how dense the Venusian atmosphere is. At the surface, the pressure is 90 atmospheres. This pressure is 90 times greater than the average air pressure at sea level on Earth and is about the same as the water pressure at a depth of about 1 km below the surface of Earth's oceans. The density of the atmosphere at the surface of Venus is likewise high, more than 50 times greater than the sea-level density of our atmosphere. The atmosphere is so massive that once heated by the Sun, it retains its heat throughout the long Venusian night. As a result, temperatures on the day and night sides of Venus are almost identical.

The composition of atmosphere explains why Venus's surface temperatures are so high: 96.5% of the molecules in Venus's atmosphere are carbon dioxide, with nitrogen (N_2) making up most of the remaining 3.5% (see Table 9-4 for a comparison of the atmospheres of Venus, Earth, and Mars). Because Venus's atmosphere is so thick, and because most of the atmosphere is the greenhouse gas CO_2, the greenhouse effect (Section 9-1) has run wild. On Earth, the greenhouse gases H_2O and CO_2 together make up only about 1% of our relatively sparse atmosphere, and the greenhouse effect has elevated the surface temperature by an additional 33°C

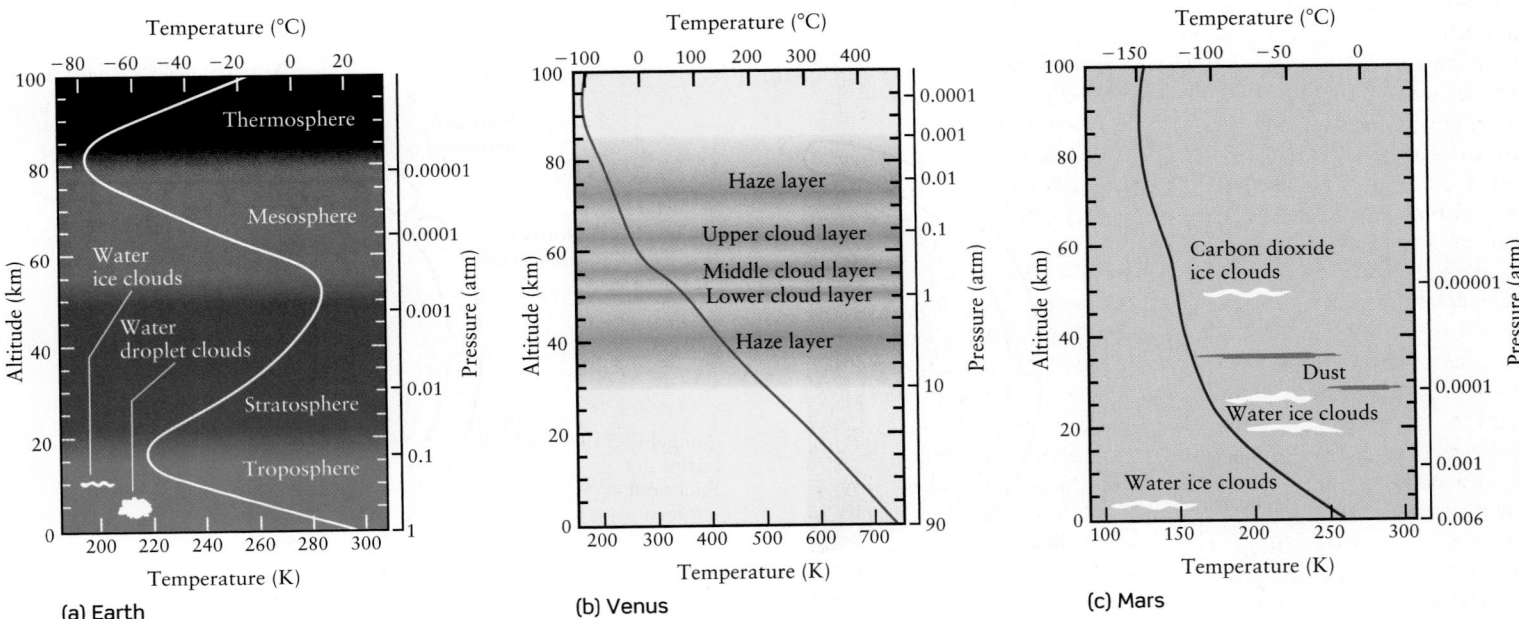

(a) Earth   (b) Venus   (c) Mars

FIGURE 11-25

Atmospheres of the Terrestrial Planets In each graph, the curve shows how temperature varies with altitude from 0 to 100 km above the planet's surface. The scale on the right-hand side of each graph shows how pressure varies with altitude. **(a)** Clouds in Earth's atmosphere are seldom found above 12 km (40,000 ft). **(b)** Venus's perpetual cloud layers lie at much higher altitudes. The atmosphere is so dense that the pressure 50 km above the Venusian surface is 1 atm, the same as at sea level on Earth. **(c)** By contrast, the Martian atmosphere is so thin that the surface pressure is the same as the pressure at an altitude of 35 km on Earth. Wispy clouds can be found at extreme altitudes.

(59°F). But on Venus, the dense shroud of CO_2 traps infrared radiation from the surface so effectively that the surface temperature is more than 400°C (720°F) greater than it would have been without the CO_2.

Soviet and U.S. spacecraft also discovered that Venus's clouds are primarily confined to three high-altitude layers. An upper cloud layer lies at altitudes between 68 and 58 km, a denser and more opaque cloud layer from 58 to 52 km, and an even more dense and opaque layer between 52 and 48 km. Above and below the clouds are 20-km-thick layers of haze. Below the lowest haze layer, the atmosphere is remarkably clear all the way down to the surface of Venus.

We saw in Section 11-5 that sulfur, which is not found in any appreciable amount in our atmosphere, plays an important role in the Venusian atmosphere. Sulfur combines with other elements to form gases such as sulfur dioxide (SO_2) and hydrogen sulfide (H_2S), along with sulfuric acid (H_2SO_4), the same acid used in automobile batteries. While Earth's clouds are composed of water droplets, Venusian clouds contain almost no water. Instead, they are composed of droplets of concentrated, corrosive sulfuric acid. Due to the high temperatures on Venus, these droplets never rain down on the planet's surface; they simply evaporate at high altitude. (You can see a similar effect on Earth. On a hot day in the desert of the U.S. Southwest, streamers of rain called *virga* appear out of the bottoms of clouds but evaporate before reaching the ground.)

Convection in the Venusian Atmosphere

We saw in Section 9-6 that Earth's atmosphere is in continuous motion: Hot gases rise and cooler gases sink, forming convection cells in the atmosphere. The same is true on Venus, but with some important differences. On Venus, gases in the equatorial regions are warmed by the Sun, then rise upward and travel in the upper cloud layer toward the cooler polar regions. At the polar latitudes, the cooled gases sink to the lower cloud layer, in which they are transported back toward the equator. The result is two huge convection cells, one in the northern hemisphere and another in the southern hemisphere (**Figure 11-26a**). These cells are almost entirely contained within the main cloud layers shown in Figure 11-25b. The circulation is so effective at transporting heat around Venus's atmosphere that there is almost no temperature difference between the planet's equator and its poles. By contrast, Earth's atmosphere has a more complicated convection pattern (see Figure 11-26b and Figure 9-26) because Earth's rotation—which is far more rapid than Venus's—distorts the convection cells.

In addition to the north–south motions due to convection, high-altitude winds with speeds of 350 km/h (220 mi/h) blow from east to west, the same direction as Venus's retrograde rotation. Thus, the upper atmosphere rotates around the planet once every 4 days. The fast-moving winds stretch out the convection cells and produce V-shaped, chevronlike patterns in the clouds (**Figure 11-27**). In a similar way, Earth's rotation stretches out the convection cells in our atmosphere to produce the complicated circulation patterns shown in Figure 9-26.

On Earth, friction between the atmosphere and the ground causes wind speeds at the surface to be much lower than at high altitude. The same effect occurs on Venus: The greatest wind speed measured by spacecraft on Venus's surface is only about 5 km/h (3 mi/h). Thus, only slight breezes disturb the crushing pressures and infernal temperatures found on Venus's dry, lifeless surface.

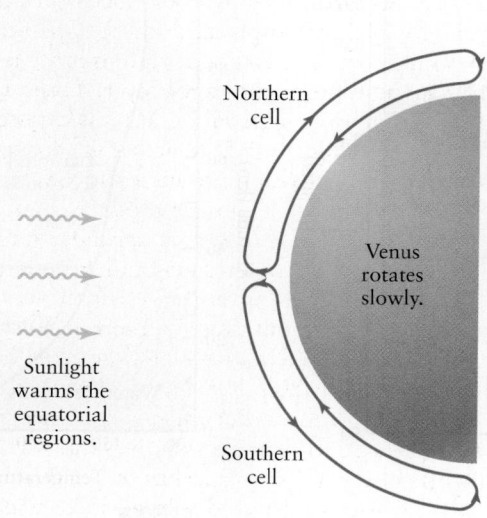

(a) Atmospheric circulation on Venus:
two convection cells

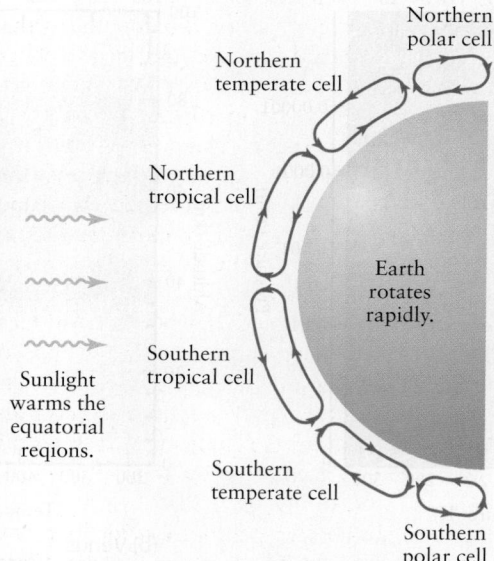

(b) Atmospheric circulation on Earth:
six convection cells

FIGURE 11-26

Atmospheric Circulation on Venus and Earth Solar heating causes convection in the atmospheres of both Venus and Earth, with warm air rising at the equator and cold air descending at the poles. **(a)** Because Venus rotates very slowly, it has little effect on the circulation. **(b)** Earth's rapid rotation distorts the atmospheric circulation into a more complex pattern (compare Figure 9-26).

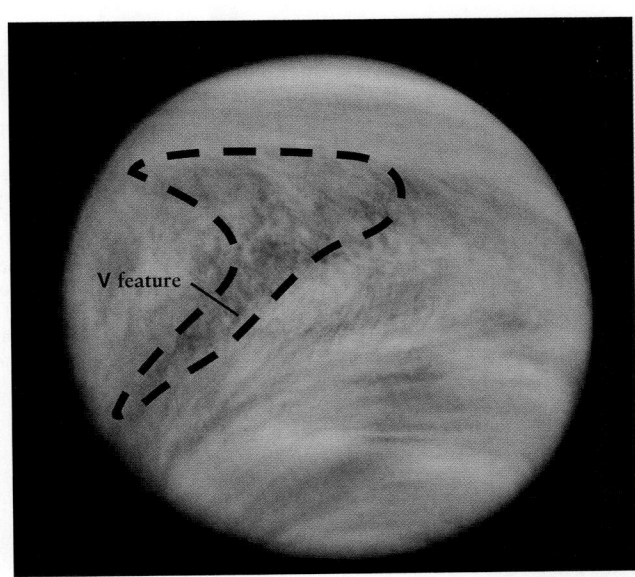

FIGURE 11-27 R I **V** U X G

Venus's Cloud Patterns This false-color image of Venus was made at ultraviolet wavelengths, at which the planet's atmospheric markings stand out best. The dark V feature is produced by the rapid motion of clouds around the planet's equator. The clouds move in the same retrograde direction as the rotation of the planet itself but at a much greater speed. (GSFC/NASA)

The Martian Atmosphere: Not a Drop to Drink

As Figure 11-25c shows, the atmosphere of Mars is very cold and thin compared to the atmosphere of either Earth or Venus: The surface pressure is a mere 0.006 atmosphere. But its chemical composition is very close to that of Venus: The Martian atmosphere is 95.3% carbon dioxide (versus 96.5% for Venus) and 2.7% nitrogen (versus 3.5% for Venus). The predominance of carbon dioxide means that Mars, like Earth and Venus, is warmed by the greenhouse effect. However, the Martian atmosphere is so thin that the greenhouse effect is very weak and warms the Martian surface by only 5°C (versus 33°C on Earth). While the perpetual cloud cover and large convection cells on Venus maintain a steady surface temperature, the thin atmosphere on Mars provides very little thermal insulation. While daytime highs on Mars can be a pleasant 20°C (68°F), at night the temperature can plummet to −140°C (−220°F).

Water vapor makes up about 0.03% of the Martian atmosphere (versus about 1% for Earth), and this vapor can form clouds. Figure 11-22b shows clouds that form near the tops of Martian volcanoes when air moves up the volcanic slopes. The rising air cools until the water vapor condenses into ice crystals. The same process occurs on Earth, but the clouds that form in this way are made of droplets of liquid water rather than ice crystals. Why is there a difference?

The explanation is that liquid water cannot exist for long anywhere on the Martian surface or in the Martian atmosphere. Water is liquid over only a limited temperature range: If the temperature is too low, water becomes ice, and if the temperature is too high, it becomes water vapor. What determines this temperature range is the atmospheric pressure above a body of water or around a water drop. If the pressure is very low, molecules easily escape from the liquid's surface, causing the water to vaporize. Thus, at low pressures, water more easily becomes water vapor. The average surface temperature on Mars is only about 250 K (−23°C or −10°F), and the average pressure is only 0.006 atmosphere. With this combination of temperature and pressure, water can exist as a solid (ice) and as a gas (water vapor) but not as a liquid. (You can see this same situation inside a freezer, where water vapor swirls around over ice cubes.) Hence, it never rains on Mars and there are no bodies of standing water anywhere on the planet.

In order to keep water on Mars in the liquid state, it would be necessary to increase both the temperature (to keep water from freezing) and the pressure (to keep the liquid water from evaporating). Later in this chapter, we will see evidence that higher temperatures and pressures did exist on Mars billions of years ago, and that there was once liquid water on the surface.

Methane in the Martian Atmosphere

The discovery of large quantities of methane (CH_4) in the Martian atmosphere came as a big surprise in 2003. More specifically, while some methane is expected, the observed *variation* (changes over time, and for different regions) in the methane abundance is not understood. Because methane is destroyed in the Martian atmosphere after about 300 years, the discovery of large variations in methane abundance indicate an ongoing variable process for producing and removing methane. The source for more than 95% of Earth's methane is well known: *life!* Naturally, these observations raise some hopes for a microbial source of the Martian methane, possibly varying with the seasons. Another proposed source is geologic activity (such as volcanism), which would imply significantly more ongoing geologic activity than observed on Mars so far. Either way, solving the methane mystery seemed to promise breakthroughs in either Martian biology or geology.

However, in 2012, scientists found evidence for a third scenario to explain Martian methane—it involves meteorites. Small micrometeorites have showered Mars for billions of years, and, unlike on Earth, most of these meteorites do not burn up in the thin Martian atmosphere. Also, compared to Earth, ultraviolet (UV) light on the Martian surface is more intense without absorption in an atmospheric ozone layer. Taken together, a large quantity of meteoritic material on Mars is exposed to intense UV light.

By exposing UV light to actual meteorites on Earth with a similar carbon composition to meteoritic material on Mars, scientists found that the UV exposure produces methane. Furthermore, the amount of methane produced depends on temperature and would vary with the Martian seasons and latitude. And, by exposing new surfaces, even the movement of small meteorites by dust storms would produce significant variations in methane production. These results do not rule out contributions from biology or volcanism to methane production, but they weaken the case for these causes.

The Mars rover Curiosity is equipped to detect methane and was expected to shed light on the source of Martian methane. But, as often occurs, more data has only led to more questions! In late 2013, Curiosity detected a large spike with a tenfold increase in methane. The source of this spike, along with the methane previously detected on Mars, is still unknown.

The Martian Poles: Freezing the Atmosphere

Figure 11-25c shows that at very high altitudes above the Martian surface, atmospheric carbon dioxide can freeze into crystals and form clouds. (Frozen carbon dioxide can be made on Earth and

FIGURE 11-28 R I **V** U X G

Changing Seasons Reveal Water Ice at the Martian North Pole **(a)** During the Martian winter, the temperature drops so low that carbon dioxide freezes out of the Martian atmosphere to make a large polar ice cap of carbon dioxide. During summer, the carbon dioxide returns to the atmosphere, revealing a residual ice cap made of water. **(b)** The dark canyon, Chasma Boreale, is more than 300 km long—similar in length and depth to the Grand Canyon, Arizona. Spectra reveal the residual ice is made of water and not CO_2. (a: JPL/NASA/STScI; b: G. Neukum [FU Berlin] et al., Mars Express, DLR, ESA/NASA; NASA/JPL-Caltech/MSSS)

is called *dry ice.*) Such frozen carbon dioxide is also found in large quantities in ice caps at the north and south poles of Mars (**Figure 11-28**). Although no spacecraft has successfully landed on the ice caps, their composition can be measured from orbit by comparing the spectrum of sunlight reflected from the ice caps with the known reflection spectrum of frozen carbon dioxide. (Figure 7-4b shows another application of this idea.)

As shown in Figure 11-28a, the size of the ice caps varies substantially with the seasons. When it is winter at one of the Martian poles, temperatures there are so low that atmospheric carbon dioxide freezes and the ice cap grows. (Think of it: It gets so cold at the Martian poles that the atmosphere freezes!) With the coming of warmer spring temperatures, much of the carbon dioxide returns to the gaseous state and the ice cap shrinks. With the arrival of summer, however, the rate of shrinkage slows abruptly and a **residual ice cap** remains solid throughout the summer. Scientists had long suspected that the north and south residual polar caps contain a large quantity of frozen water that is less easily evaporated (Figure 11-28b). The ESA's Mars Express spacecraft confirmed this idea by detecting the characteristic spectrum of water ice.

The wintertime freezing of atmospheric carbon dioxide was observed by the Viking Lander 1 and Viking Lander 2 spacecraft, which made the first successful landings on Mars in 1976. Both of these spacecraft touched down in the northern hemisphere of Mars (you can see the landing sites in Figure 11-17). **Figure 11-29** compares the springtime and wintertime views seen from the Viking

Landers. Closer to the north pole, seasonal warming of carbon dioxide has lead to a very strange landscape (**Figure 11-30**). The dark areas in this image are dust patches that only become exposed after carbon dioxide ice evaporates off the surface. On the steeper slopes, the dust cascades down, creating the dark treelike streaks.

Snowfall on Mars

Similar to the way that frost can form on a car's windshield on a cold morning, water and carbon dioxide in the Martian atmosphere can form frost, as in Figure 11-29b. However, actual *snowfall* from clouds indicates a more complex climate system that also occurs on Mars. In 2008, NASA's Phoenix lander detected water snow from clouds about 4 km above the lander. As the Martian winter approached and temperatures dropped, reflections from a green laser shot into the clouds above the lander revealed large ice crystals, large enough to assume they had fallen to the ground. The snow was assumed to be water snow because it was not yet cold enough for carbon dioxide to freeze.

Then, in 2012, astronomers reported that Mars Reconnaissance Orbiter found evidence of falling snow made from carbon dioxide (CO_2) during the deep cold of winter—the only known case of CO_2 snowfall anywhere in our solar system. The snow was detected above the Martian south pole through its visible and infrared spectrum, and it was observed extending all the way to the ground.

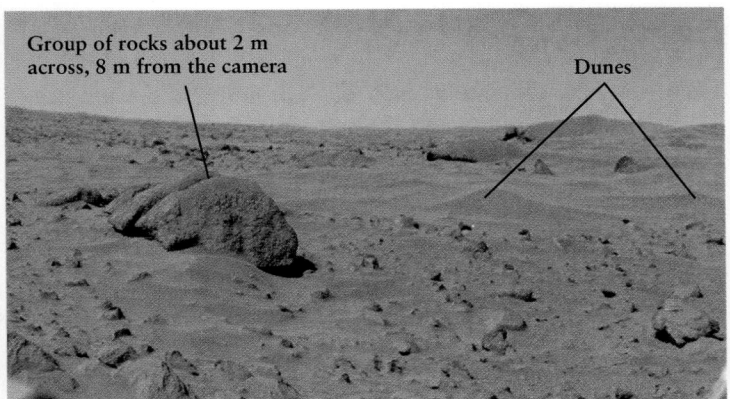

Group of rocks about 2 m
across, 8 m from the camera

Dunes

(a) A view from Viking Lander 1

FIGURE 11-29 R I ⓥ U X G

Seasons on the Martian Surface (a) This springtime view from Viking
Lander 1 shows rocks that resemble volcanic rocks on Earth (see Figure
9-19a). They are thought to be part of an ancient lava flow that was broken
apart by crater-forming asteroid impacts. Fine-grained debris has formed
sand dunes. **(b)** This picture was taken during midwinter at the Viking

(b) A wintertime view from Viking Lander 2

Lander 2 site. Freezing carbon dioxide condenses onto water-ice crystals or
dust grains in the atmosphere, causing them to fall to the ground and coat
the surface with frost. This frost lasted for about 100 days. (a: NASA/Viking 1/
USGS; b: NASA/JPL)

FIGURE 11-30 R I ⓥ U X G

Spring on Mars The dark areas in this enhanced-color image are dust patches
uncovered by evaporating carbon dioxide ice near the north pole in springtime.
On the steeper slopes, the dust cascades down, creating the dark treelike streaks.
The white arrow points to a plume being kicked up by one of these dust-slides.
(NASA/JPL/University of Arizona)

Martian Dust, Dunes, and Devils

Figure 11-29a shows that the landscape around the Viking Lander 1
is covered with a fine-grained dust. On much larger scales, wind
can pile up the dust particles to form sand dunes (**Figure 11-31**).
The dust particles are much smaller than ordinary household dust
on Earth: They are only about 1 μm (10^{-6} m) across, about the
size of the particles in cigarette smoke.

Because Martian dust is so fine, it can be carried aloft and
spread over the surface by Martian winds even though the

atmosphere is so tenuous. (A wind of 100 km/h, or 60 mi/h, on
Mars is only about as strong as an Earth breeze of 10 km/h or
6 mi/h.) Windblown dust can sometimes be seen from Earth as a
yellow haze that obscures Martian surface features. Figure 11-14
shows a dust storm that covered the entire planet, triggered by the
flow of carbon dioxide evaporating from the north polar ice cap
with the coming of northern spring. On Earth, stormy weather
means cloudy skies, wind, and falling rain; on Mars, where no
raindrop falls, stormy weather means wind and dust.

Dust also plays a role in the ordinary daily weather on Mars.
Each afternoon, columns of warm air rise from the heated surface
and form whirlwinds called **dust devils** (**Figure 11-32**). A similar
phenomenon with the same name occurs in dry or desert terrain on
Earth. Martian dust devils can reach an altitude of 6 km (3.79 mi)
and help spread dust particles across the planet's surface. Martian
landers have measured changes in air pressure as dust devils swept
past; sometimes several of these would pass in a single afternoon.
As the sun sets and afternoon turns to evening, the dust devils
subside until the following day.

CONCEPT CHECK 11-7

How might Mars's environment become more hospitable to
life if more CO_2 could be added to the atmosphere?

CONCEPT CHECK 11-8

What three processes could lead to variable amounts of
methane on Mars?

CONCEPT CHECK 11-9

Daytime temperatures on Mars near the equator can reach a
warm 68°F (20°C). Why wouldn't an ice cube on Mars at this
temperature melt?

Answers appear at the end of the chapter.

FIGURE 11-31 R I V̲ U X G

Martian Sand Dunes **(a)** Enhanced coloring brings out the composition of these dunes. Blue indicates sand that came from volcanic rock. The lighter colors on top are probably very fine-grained dust. **(b)** Sand from clay minerals form dunes that partially cover sedimentary rock. (NASA/JPL/University of Arizona)

(b)

FIGURE 11-32 R I V̲ U X G

A Martian Dust Devil This Mars Global Surveyor image shows a dust devil as seen from almost directly above. This tower of swirling air and dust casts a long shadow in the afternoon sun. Spotted in 2012, this dust devil is about 100 ft wide and half a mile in height. About a month later, another dust devil was seen reaching a height of 12 mi! (NASA/JPL-Caltech/University of Arizona)

11-7 The atmospheres of Venus and Mars were very different billions of years ago

TUTORIAL 11.2

Why do Earth, Venus, and Mars have such dramatically different atmospheres? As we will see, the original atmospheres of all three worlds were essentially the same.

However, each atmosphere evolved in a unique way determined by the planet's size and distance from the Sun. The *Cosmic Connections* figure on page 324 summarizes the processes that led to the present-day atmospheres of the three large terrestrial planets.

> The early atmospheres of Earth, Venus, and Mars were predominantly water vapor and carbon dioxide.

The Origin of Atmospheres on Earth, Venus, and Mars

Hydrogen in the solar nebula formed the bulk of the earliest atmospheres on Earth, Venus, and Mars, but none of these planets could hold onto this light gas—so it escaped (see Box 7-2). The next atmospheres of Earth, Venus, and Mars all derived from gases that were emitted, or *outgassed*, from volcanoes (Figure 11-33). The gases released by present-day Earth volcanoes are predominantly water vapor (H_2O), followed by carbon dioxide (CO_2), sulfur dioxide (SO_2), and nitrogen (N_2). These gases should therefore have been important parts of the original atmospheres of Earth, Venus, and Mars. It is also possible—but far from certain—that icy comets from the outer solar system delivered significant quantities of water to these planets.

Due in part to intense volcanic activity that took place when the young planets had warmer interiors, water probably once dominated the atmospheres of Earth, Venus, and Mars. Studies of how stars evolve suggest that the early Sun was only about 70% as luminous as it is now, so the temperature in Venus's early atmosphere must have been quite a bit lower. Water vapor may actually have been able to liquefy and form oceans on Venus. Mars

FIGURE 11-33 R I V U X G

A Volcanic Eruption on Earth The eruption of Mount St. Helens in Washington State on May 18, 1980, released a plume of ash and gas 40 km (25 mi) high. Eruptions of this kind on Earth, Venus, and Mars probably gave rise to those planets' original atmospheres. (USGS)

is smaller than either Earth or Venus, but its large volcanoes could also have emitted substantial amounts of water vapor. With a thicker atmosphere and hence greater pressure, liquid water could also have existed on Mars.

The Evolution of Earth's Atmosphere

On present-day Earth, most of the water is in the oceans. Nitrogen, which is not very reactive, is still in the atmosphere. By contrast, carbon dioxide dissolves in water, which can fall as rain; as a result, rain removed most of the H_2O and CO_2 from our planet's atmosphere long ago. The dissolved CO_2 reacted with rocks in rivers and streams, and the residue was ultimately deposited on the ocean floors, where it turned into carbonate rocks, such as limestone. Consequently, most of Earth's carbon dioxide is tied up in Earth's crust, and only about 1 in every 3000 molecules in our atmosphere is a CO_2 molecule. If Earth were to become as hot as Venus, much of its CO_2 would be boiled out of the oceans and baked out of the crust, and our planet would soon develop a thick, oppressive carbon dioxide atmosphere much like that of Venus.

The amount of carbon dioxide in our atmosphere is maintained in part by plate tectonics. Tectonic activity causes carbonate rocks to cycle through volcanoes, where they are heated and forced to liberate their trapped CO_2. This liberated gas then rejoins the atmosphere. Without volcanoes, rainfall would remove all of the

CO_2 from our atmosphere in only a few thousand years. This lack of CO_2 would dramatically reduce the greenhouse effect, our planet's surface temperature would drop precipitously, and the oceans would freeze. Plate tectonics is thus essential for maintaining our planet's livable temperatures.

The oxygen (O_2) in our atmosphere is the result of photosynthesis by plant life, which absorbs CO_2 and releases O_2. The net result is an atmosphere that is predominantly N_2 and O_2, with enough pressure and a high enough temperature (thanks to the greenhouse effect) for water to remain a liquid—and hence for life, which (as far as we can tell) requires liquid water, to exist.

The Evolution of the Venusian Atmosphere: A Runaway Greenhouse Effect

When Venus was young and had liquid water, the amount of atmospheric CO_2 was probably kept in check much as on Earth today: CO_2 was released by volcanoes, then dissolved in the oceans and bound up in carbonate rocks. Enough of the liquid water would have vaporized to create a thick cover of water vapor clouds. Since water vapor is a greenhouse gas, this humid atmosphere—perhaps denser than Earth's present-day atmosphere, but far less dense than the atmosphere that surrounds Venus today—would have efficiently trapped heat from the Sun. At first, this would have had little effect on the oceans of Venus. Although the temperature would have climbed above 100°C, the boiling point of water at sea level on Earth, the added atmospheric pressure from water vapor would have kept the water in Venus's oceans in the liquid state.

This hot and humid state of affairs may have persisted for several hundred million years. But as the Sun's energy output slowly increased over time, the temperature at the surface would eventually have risen above 374°C (647 K or 705°F). Above this temperature, no matter what the atmospheric pressure, Venus's oceans would have begun to evaporate. The added water vapor in the atmosphere would have increased the greenhouse effect, raising the temperature even higher and causing the oceans to evaporate faster. This temperature increase would have added even more water vapor to the atmosphere, further intensifying the greenhouse effect and making the temperature climb higher still. This is an example of a **runaway greenhouse effect**, in which an increase in temperature causes a further increase in temperature, and so on.

ANALOGY

A runaway greenhouse effect is like a house in which the thermostat has accidentally been connected backward. Normally, when the temperature rises too high in a house, the heater shuts off. However, if a thermostat is connected backward, the hotter the house, the greater the heater output, raising the temperature even further, and so on.

Once Venus's oceans disappeared, so did the mechanism for removing carbon dioxide from the atmosphere. With no oceans to dissolve it, outgassed CO_2 began to accumulate in the atmosphere, making the greenhouse effect "run away" even faster. Temperatures eventually became high enough to "bake out" any

COSMIC CONNECTIONS

Evolution of Terrestrial Atmospheres

ANIMATION
11.8

Earth, Venus, and Mars formed with similar original atmospheres. However, these atmospheres changed dramatically over time due to factors such as planetary size and distance from the Sun.

Earth: Active plate tectonics "recycles" gases and maintains a moderate atmosphere.

1. H_2O, CO_2, SO_2, and N_2 are outgassed from Earth volcanoes on an ongoing basis.

2. H_2O falls as rain; CO_2 and SO_2 dissolve in the oceans; N_2 goes into nitrates.

6. Plants convert CO_2 to O_2; bacteria convert nitrates to N_2 and O_2.

Clouds

Oceans

5. The liberated H_2O, CO_2, and SO_2 rise upward by volcanic action.

3. CO_2, SO_2, and some H_2O are incorporated into sedimentary rocks.

Sediment

Subduction

4. Sedimentary rocks are eventually subducted to great depths, where heat liberates the gases.

NASA

Venus: No plate tectonics and more intense sunlight produce a thick atmosphere.

Space

Ultraviolet light

1. H_2O, CO_2, SO_2, and N_2 are outgassed from Venusian volcanoes on an ongoing basis.

2. Sunlight warms Venus more than Earth, so H_2O does not fall as rain. There are no oceans, so CO_2 remains in the atmosphere and intensifies the greenhouse effect.

3. Ultraviolet light from the sun breaks apart some H_2O molecules; hydrogen escapes into space.

Atmosphere

Clouds

4. H_2O that does not escape combines with SO_2 to make H_2SO_4 (sulfuric acid) in Venus's clouds.

Surface

5. Remaining SO_2 is locked up in sulfate minerals.

NASA/JPL

Mars: No plate tectonics, weak gravity, and feeble sunlight produce a thin atmosphere.

Space

Ultraviolet light

1. H_2O, CO_2, SO_2, and N_2 were outgassed from Martian volcanoes; this stopped billions of years ago as the planet's interior cooled.

2. Sunlight warms Mars less than Earth, so H_2O fell as rain; CO_2 and SO_2 dissolved in the water.

3. Ultraviolet light from the sun broke apart some H_2O and N_2 molecules; the weak gravity allowed H, N, and O atoms to escape into space.

Regolith

Permafrost

4. Removing CO_2 and H_2O from the atmosphere reduced the greenhouse effect, lowering temperatures. H_2O froze beneath the surface, while CO_2 and SO_2 went into rocks and the regolith.

5. There is no plate tectonics on Mars due to its small size. Gases that go into the surface are not recycled into the atmosphere.

NASA

CO_2 that was trapped in carbonate rocks. This liberated carbon dioxide formed the thick atmosphere of present-day Venus.

Sulfur dioxide (SO_2) is also a greenhouse gas released by volcanoes. Although present in far smaller amounts than carbon dioxide, it would have contributed to Venus's rising temperatures. Like CO_2, it dissolves in water, a process that helps moderate the amount of sulfur dioxide in our atmosphere. But when the oceans evaporated on Venus, the amount of atmospheric SO_2 would have increased.

The temperature within Venus's atmosphere would not have increased indefinitely. In time, solar ultraviolet radiation striking molecules of water vapor—the dominant cause of the runaway greenhouse effect—would have broken them into hydrogen and oxygen atoms. The lightweight hydrogen atoms would have then escaped into space (see Box 7-2 for a discussion of why lightweight atoms can escape a planet more easily than heavy atoms). The remaining atoms of oxygen, which is one of the most chemically active elements, would have readily combined with other substances in Venus's atmosphere. Eventually, almost all of the water vapor would have been irretrievably lost from Venus's atmosphere. With all of the water vapor gone, and essentially all of the carbon dioxide removed from surface rocks, the greenhouse effect would no longer have "run away," and the rising temperature would have leveled off.

Today, the infrared-absorbing properties of CO_2 have stabilized Venus's surface temperature at its present value of 460°C (860°F). Only minuscule amounts of water vapor—about 30 parts per million, or 0.003%—remain in the atmosphere. As on Earth, volcanic outgassing might still add small amounts of water vapor to the atmosphere, along with carbon dioxide and sulfur dioxide. But on Venus, these water molecules either combine with sulfur dioxide to form sulfuric acid clouds or break apart due to solar ultraviolet radiation. The great irony is that this state of affairs is the direct result of an earlier Venusian atmosphere that was predominantly water vapor!

The Evolution of the Martian Atmosphere: A Runaway Icehouse Effect

Mars probably had a thicker atmosphere 4 billion years ago. Thanks to Mars's greater distance from the Sun, and hence less intense sunlight, temperatures would have been lower than on young Earth, and any water in the atmosphere would more easily have fallen as rain or snow. This precipitation would have washed much of the planet's carbon dioxide from its atmosphere, perhaps creating carbonate minerals in which the CO_2 is today chemically bound. Measurements from Mars orbit show only small amounts of carbonate materials on the surface, suggesting that the amount of atmospheric CO_2 that rained out was small. Hence, even the original Martian atmosphere was relatively thin, though thicker than the present-day atmosphere.

Because Mars is so small, it cooled early in its history and volcanic activity came to an end. Thus, any carbonate rocks were not recycled through volcanoes as they are on Earth. The depletion of carbon dioxide from the Martian atmosphere into the surface would therefore have been permanent.

As the amount of atmospheric CO_2 declined, the greenhouse effect on Mars would have weakened and temperatures begun to fall. This temperature decrease would have caused more water vapor to condense into rain or snow and fall to the surface, taking even more CO_2 with them and further weakening the greenhouse effect. Thus, a decrease in temperature would have caused a further decrease in temperature—a phenomenon sometimes called a **runaway icehouse effect**. (This is the reverse of the runaway greenhouse effect that has taken place on Venus.) Ultimately, both water vapor and most of the carbon dioxide would have been removed from the Martian atmosphere. With only a very thin CO_2 atmosphere remaining, surface temperatures on Mars eventually stabilized at their present frigid values.

As both water vapor and carbon dioxide became depleted, ultraviolet light from the Sun could penetrate the thinning Martian atmosphere to strip it of nitrogen. Nitrogen molecules (N_2) normally do not have enough thermal energy to escape from Mars, but they can acquire that energy from ultraviolet photons, which break the molecules in two. Ultraviolet photons can also split carbon dioxide and water molecules, giving their atoms enough energy to escape. Indeed, in 1971 the Soviet Mars 2 spacecraft found a stream of oxygen and hydrogen atoms (from the breakup of water molecules) escaping into space. Oxygen atoms that did not escape into space could have combined with iron-bearing minerals in the surface, forming rustlike compounds. Such compounds have a characteristic reddish-brown color and may be responsible for the overall color of the planet; if this is correct, Mars gained its color by losing its atmosphere!

A Magnetic Field Protects the Martian Atmosphere

Some scientists think a large asteroid impact triggered the loss of the Martian atmosphere. To understand the possible effect of an impact, we need to understand how the Martian atmosphere might have been protected. Ancient magnetic stripe-like features on Mars's surface indicate that early on, Mars may have had a magnetic field arising from a magnetic dynamo in a molten core, similar to Earth's magnetic field (see Section 11-5). As with Earth's magnetic field, the Martian field would have protected the planet's atmosphere by deflecting the solar wind away from Mars (see Figure 9-20). We say "protected" because if they are not deflected, high-speed particles in the solar wind can break apart atmospheric CO_2 and H_2O, allowing their lighter constituents to escape into space. Furthermore, the timeline of magnetic signatures in Martian craters suggests that this protective magnetic field persisted for about Mars's first 500 million years, which coincides with the onset of the Late Heavy Bombardment (see Section 8-6).

Putting all of these processes together, a new idea proposes that a very large asteroid impact on Mars may have disrupted the Martian magnetic dynamo, shut down its magnetic field, and led to the loss of the planet's atmosphere before its interior cooled down. Advocates of this scenario estimate that there are several large craters of the right age that indicate enough impact energy to significantly disrupt the core convection that produced a magnetic field. However, others are not convinced that such an impact could significantly alter the magnetic field.

Due to Mars's small size, it only contained enough heat for a short period of volcanism. Therefore, Mars may have been destined to lose its atmosphere anyway. Since planetary cooling shuts down the core convection that creates a global magnetic field (which protects CO_2 and H_2O), an atmosphere cannot outlast its planet's internal heat for very long. Whether or not a large impact brought an early end to the Martian atmosphere remains unknown.

Before it was lost, was the Martian atmosphere thick enough and warm enough for water to remain as a liquid on the planet's surface? As we will see in the next section, scientists have found convincing evidence that the answer is yes.

CONCEPT CHECK 11-10

What is "runaway" about Venus's runaway greenhouse?

CONCEPT CHECK 11-11

How would a global magnetic field on a young Mars help to keep the planet warm?

Answers appear at the end of the chapter.

11-8 Extensive evidence exists for water on Mars

Just as ancient fossils on Earth provide evidence of how life has evolved on our planet, ancient rocks and geological formations help us understand how our planet's surface and atmosphere have evolved. (For example, the oldest rocks on Earth have a chemical structure that shows that they formed during a time when there was little oxygen in the atmosphere, and hence before the appearance of photosynthetic planet

> Parts of Mars are far drier than a bone, while others had standing water for extended periods.

life.) Planetary scientists who want to learn about the history of Venus and Mars are therefore very interested in studying ancient rocks on those worlds.

On Venus we cannot look very far into the past, since the planet's volcanic activity and tectonic activity have erased features more than about 500 million years old (see Section 11-5). On Mars, by contrast, we expect to find rocks that are billions of years old: The thin atmosphere causes very little erosion, there has been little recent volcanic activity to cover ancient rocks, and there is no subduction to drag old surface features back into the planet's interior. (The heavy cratering of the southern highlands bears testament to the great age of much of the Martian surface.) In a quest to find these ancient rocks and learn about the early history of Mars, scientists have sent spacecraft to land on the red planet and explore it at close range.

As we will see, by understanding rocks found on Mars, we also learn about water and gain insight into some of the biggest Mars questions of all: Did Mars have liquid water that could support life? Did Mars ever have life? These questions shape NASA's missions to search for evidence of past microbial Martian life.

Landing on Mars

As of 2012, seven spacecraft have successfully landed on the Martian surface. Viking Lander 1 and Viking Lander 2, which we first discussed in Section 11-6, touched down in 1976. They had no capability to move around on the surface, and so each Viking Lander could study only objects within reach of the scoop at the end of its mechanical arm (**Figure 11-34a**). Figure 11-29 shows the view from the two Viking Lander sites.

Three more recent robotic explorers, Mars Pathfinder (landed 1997) and two Mars Exploration Rovers named Spirit and Opportunity (landed 2004), were equipped with electrically driven wheels that allow travel over the surface. Unlike the Viking

Cameras   Mars Exploration Rover   Mars Pathfinder rover (Sojourner)

(a) Two generations of rovers

(b) Curiosity rover

FIGURE 11-34 R I V U X G

Martian Rovers (a) The technicians in this photograph pose with one of the Mars Exploration Rovers and with a replica of its predecessor, the Mars Pathfinder rover. **(b)** The rover Curiosity is much bigger than previous rovers;

you can compare to the height of a person in parts (a) and (b). Curiosity is searching for areas that may have been suitable for life. (a: JPL/NASA; b: NASA/JPL-Caltech)

Landers, these rovers did not have a heavy and expensive rocket engine to lower them gently to the Martian surface. Instead, the spacecraft was surrounded by airbags like those used in automobiles. Had there been anyone to watch Mars Pathfinder or one of the Mars Exploration Rovers land, they would have seen an oversized beach ball hit the surface at 50 km/h (30 mi/h), then bounce for a kilometer before finally rolling to a stop. Unharmed by its wild ride, the spacecraft then deflated its airbags and began to study the geology of Mars. In 2008, the Phoenix Mars Lander sampled a challenging north polar region over a period of a few months; it did not rove around.

Figure 11-17 shows the landing sites of the Mars Pathfinder rover (named Sojourner) and the two Mars Exploration Rovers (Spirit and Opportunity). Following directions from Earth, Sojourner spent 3 months exploring the rocks around its landing site. The Mars Exploration Rovers were also designed for a 3-month lifetime, and both remained operational years after landing. In fact, as of this writing, Opportunity has traveled more than 44 km over the Martian surface and continues to collect data. However, Spirit became stuck in soft sand in 2009 and communication was lost a year later. In addition to carrying cameras, most landers carried a device called an *x-ray spectrometer* to measure the chemical composition of rock samples. Spirit and Opportunity also carried a tool designed to grind holes into rocks, thus revealing parts of the rock's interior that have not been weathered by exposure to the Martian atmosphere.

The most recent Mars landing was by the rover Curiosity (Figure 11-34b). The stunning landing, engineered by NASA's Mars Science Laboratory mission, caught the public imagination in August 2012. Curiosity is too massive (900 kg) for the airbag landing scenario used to land earlier rovers. Instead, a parachute followed by rocket-thrusters provided the necessary speed reduction, but not all the way to the surface. To avoid stirring up too much dust onto the rover, a "sky crane" began lowering Curiosity down the last 20 m.

Curiosity was not designed to look for life directly, but to analyze the habitability of Mars—the past or present conditions favorable to Martian microbial life. Curiosity has found the most important chemical ingredients for life: carbon, hydrogen, nitrogen, oxygen, phosphorous, and sulfur (often referred to as CHNOPS). In addition to these essential elements, organic molecules—the building blocks of life—were found as well. Curiosity has also found evidence that all these ingredients occurred in a region with liquid water and low levels of salt.

While the past habitability of Mars looks promising, Curiosity found a challenge for the future human habitability of Mars. One of the greatest risks for astronauts in getting to Mars and staying for an extended visit is radiation. Some of these high-energy particles come from supernovas in our galaxy, and some come from the Sun. Curiosity measured radiation levels jumping high enough, on occasion, that safety shelters would be needed for astronauts during these times. The radiation is strong enough that signs of previous microbial life within the top meter of the Martian surface might have been destroyed, and this guides the strategy of NASA's next rover: the Mars 2020.

The Mars 2020 uses the same landing system and rover-robotics as Curiosity but carries different instruments. It will have some limited capability to discover biosignatures—the past or present signs of life. Its most intriguing feature is the ability to collect promising samples and prepare them for a return to Earth by a future mission. With samples of Mars in our best laboratories on Earth, we will have tremendous power to search for signs of life. You can read more about the Mars 2020 rover in the essay at the end of this chapter.

The Martian Regolith

The Viking Landers were the first to reveal the nature of the Martian regolith, or surface material. The scoop on the extendable arm (Figure 11-34a) was used to dig into the regolith and retrieve samples for analysis. Bits of the regolith clung to a magnet mounted on the scoop, indicating that the regolith contains iron. The x-ray spectrometers showed that samples collected at both Viking Lander sites were rich in iron, silicon, and sulfur.

The Viking Landers each carried a miniature chemistry laboratory for analyzing the regolith. Curiously, when a small amount of water was added to a sample of regolith, the sample began to fizz and release oxygen. This experiment indicated that the regolith contains unstable chemicals called *peroxides* and *superoxides*, which break down in the presence of water to release oxygen gas.

The unstable chemistry of the Martian regolith probably comes from solar ultraviolet radiation that beats down on the planet's surface. (Unlike Earth, Mars has no ozone layer to screen out ultraviolet light.) Ultraviolet photons easily break apart molecules of carbon dioxide (CO_2) and water vapor (H_2O) by knocking off oxygen atoms, which then become loosely attached to chemicals in the regolith.

Among the most important scientific goals of the Viking Landers was to search for evidence of life on Mars. Although it was clear that Mars has neither civilizations nor vast fields of plants, it still seemed possible that simple microorganisms could have evolved on Mars during its early brief epoch of Earthlike climate. If so, perhaps some species had survived to the present day. To test this idea, the Viking Landers looked for biologically significant chemical reactions in samples of the Martian regolith, but they failed to detect any reactions of this type. Perhaps life never existed on Mars, or perhaps it existed once but has since vanished. (In Chapter 27, we discuss the Viking Lander experiments in detail.)

Exploring Martian Rocks

Images made from Mars orbit show a number of features that suggest water flowed there in the past (**Figure 11-35**). To investigate these, Mars Pathfinder landed in an area of the northern lowlands called Ares Valles, which appears to be an ancient flood plain. Figure 11-35b shows a portion of this plain. Flood waters can carry rocks of all kinds great distances from their original locations, so scientists expected that this landing site (**Figure 11-36**) would show great geologic diversity. They were not disappointed. Many of the rocks appeared to be igneous rock produced by volcanic action. Other rocks, which resemble the impact breccias found in the ancient highlands of the Moon (see Figure 10-18), appear to have come from areas that were struck by crater-forming meteoroids. Still others have a layered structure like sedimentary rocks on Earth (see Figure 9-19b). Such rocks form gradually at the bottom of bodies of water, suggesting that liquid water was stable in at least some regions of Mars for a substantial period of time.

(a)

(b)

FIGURE 11-35 R I **V** U X G

Signs of Ancient Martian Water **(a)** This Mars Reconnaissance Orbiter image shows patterns from flowing water almost identical to those caused by rainfall in deserts on Earth. No rain falls in today's sparse Martian atmosphere, so these must date from an earlier era when the atmosphere was thicker.

(b) These features rise above the floor of Ares Valles. They were carved out by a torrent of water that flowed from the bottom of the image toward the top. Similar flood-carved islands are found on Earth in eastern Washington State. (a: NASA/JPL/Univ. of Arizona; b: NASA/USGS)

FIGURE 11-36 R I **V** U X G

Roving the Martian Surface This view from the stationary part of Mars Pathfinder shows the rover, called Sojourner, deploying its x-ray spectrometer against a rock named Moe. Scientists gave other rocks whimsical names such as Jiminy Cricket and Pooh Bear. (NASA/JPL)

Evidence for Ancient Martian Water

  In the search for life beyond Earth, the search for signs of ancient water on Mars has been one of NASA's primary missions. If Mars ever supported life, biologists expect Martian life to have been dependent on water, just as it is for life on Earth. By finding where liquid water existed on Mars, rovers such as Curiosity can narrow the search further by looking for organic compounds, which are the building blocks of life (Chapter 27).

The two Mars Exploration Rovers landed in sites that were suspected to have been under water in the distant past. Spirit touched down in Gusev Crater, a large impact basin. A valley opens into this crater, resembling a river that runs into a lake. The destination for Opportunity was the flat plain of Sinus Meridianii, where orbiting spectrometers had spotted evidence of hematite—a type of iron oxide that, on Earth, tends to form in wet locations.

Contrary to expectations, most of the rocks encountered by Spirit show *no* evidence that they have been exposed to water. The volcanic rocks found in and around Gusev Crater (Figure 11-37a) are composed of the minerals olivine and pyroxene, which break down if they are exposed to liquid water. In a few locations, Spirit found deposits of sulfur salts, which can be formed by the action of water (Figure 11-37b). However, olivine and pyroxene were still found among the sulfur salts. While it remains a possibility that Gusev Crater was once a river-fed lake, the results from Spirit show that for the past billion years this region has been drier than any desert on Earth.

The results from Opportunity were entirely different. This rover discovered outcrops of sedimentary rocks on the plains of Sinus Meridianii (Figure 11-38). On these plains, Opportunity found millimeter-sized spheres of hematite that, like analogous spheres found on Earth, probably formed from iron-rich water circulating through sediment. These observations suggest Sinus Meridianii, which is among the flattest places known in the solar system, was once a lake bed. It is not known how long ago this area was under water, but the density of craters found there suggests that the surface formed 3 to 4 billion years ago.

When combined with observations from Mars orbit, the Spirit and Opportunity results show that ancient Mars was a very diverse planet. Most of the planet was probably very dry, like Gusev Crater, but there were isolated areas like Sinus Meridianii that had liquid water for some period of time. It remains a challenge to planetary scientists to explain why liquid water appeared on Mars only in certain places and at certain times.

(a) Volcanic rocks

(b) Sulfur salts

FIGURE 11-37 R I ⊙ U X G

Dry Mars At the Spirit landing site, both **(a)** the dusty volcanic rocks that dominate the site and **(b)** these light-colored sulfur salts contain minerals that break apart upon contact with liquid water. Hence, this area must have been dry for billions of years. (a: NASA/JPL-Caltech/Cornell; b: NASA/JPL/Cornell)

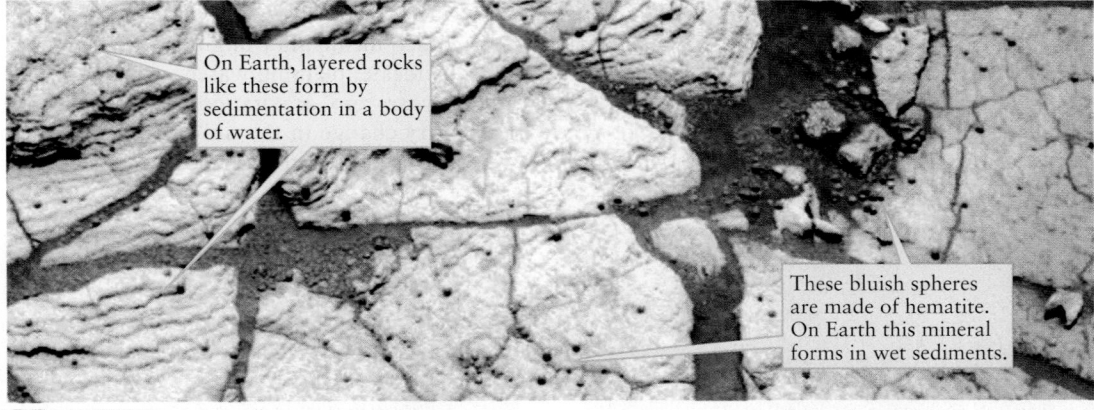

On Earth, layered rocks like these form by sedimentation in a body of water.

These bluish spheres are made of hematite. On Earth this mineral forms in wet sediments.

FIGURE 11-38 R I ⊙ U X G

Wet Mars This false-color image shows two signatures of water found by Opportunity. Both the sedimentary rock (exposed by erosion) and the hematite spheres (appearing blue) require standing water to form. So this area must have been under water for an extended period. The small spheres are typically a few millimeters in size and in these false-color images are often referred to as "blueberries." (NASA/JPL/Cornell)

Where Is the Water Now?

Whatever liquid water once coursed over parts of the Martian surface is presumably now frozen, either at the polar ice caps or beneath the planet's surface. To check this idea, scientists used a device on board the Mars Odyssey orbiter to measure the result of cosmic rays hitting Mars.

Cosmic rays are fast-moving subatomic particles that enter our solar system from interstellar space. When these particles strike and penetrate the Martian soil, they collide with the nuclei of atoms in the soil and knock out neutrons. (Recall from Section 5-7 that atomic nuclei contain both positively charged protons and electrically neutral neutrons.) If the ejected neutrons originate within a

meter or so of the surface, they can penetrate through the regolith and escape into space. However, if there are water molecules (H_2O) in the surface, the nuclei of the hydrogen atoms in these molecules tend to absorb the neutrons and prevent them from escaping. An instrument on board Mars Odyssey detects neutrons coming from Mars; where the instrument finds a *deficiency* of neutrons is where the soil contains hydrogen and, presumably, water.

The results from the Mars Odyssey neutron measurements show that there is abundant water beneath the surface at both Martian poles (Figure 11-39). But they also show a surprising amount of subsurface water in two regions near the equator, Arabia Terra (see Figure 11-13a and Figure 11-17) and another

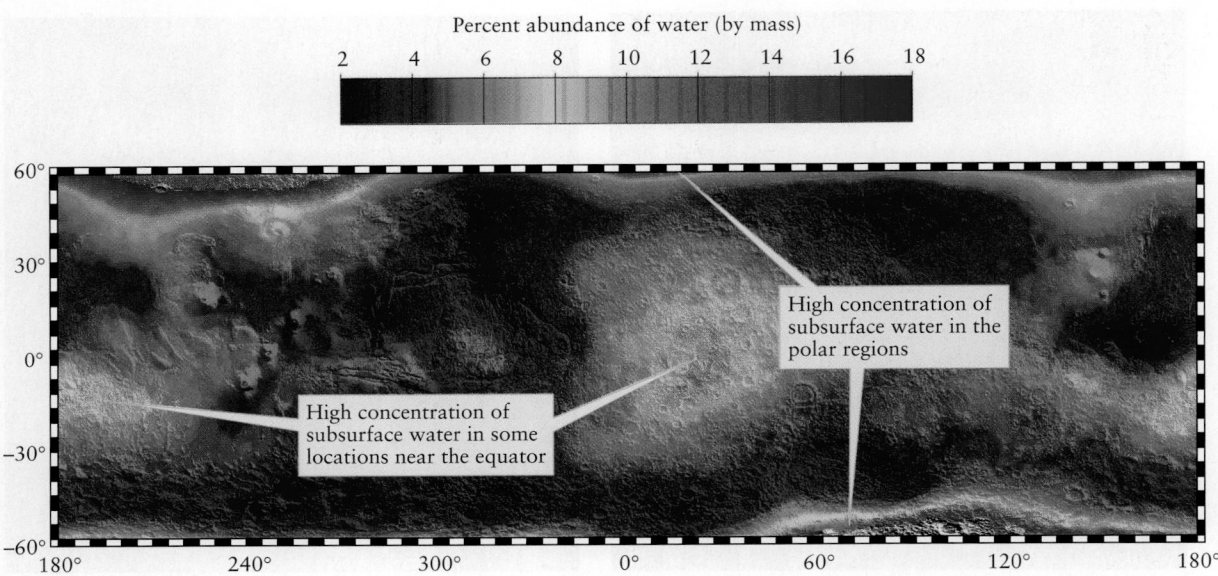

Percent abundance of water (by mass)

High concentration of subsurface water in the polar regions

High concentration of subsurface water in some locations near the equator

FIGURE 11-39

Water Beneath the Martian Surface This map of the abundance of water (probably in the form of ice) just below the surface of Mars is based on data collected from orbit by the Mars Odyssey spacecraft. These

measurements can reveal the presence of water or water ice only to a depth of about 1 m; there may be much more water at greater depths. (Los Alamos National Laboratory)

region on the opposite side of the planet. Recently, even small impacts have directly revealed ice just beneath the surface, but far away from the poles (Figure 11-40). A much larger impact appears to have revealed subsurface water as well (Figure 11-41). And in 2018, HiRISE images revealed 100-m-deep glacier-like ice formations in areas where erosion cut into the Martian surface. In

some cases, the glaciers are only buried by a few feet of Martian dust, which means they might be accessible for missions on Mars.

One idea explaining water ice near the equator is that about a million years ago, gravitational forces from other planets caused a temporary change in the tilt of Mars's rotation axis. The ice caps were then exposed to more direct sunlight, causing some of the

October 2008

January 2009

FIGURE 11-40 R I V U X G

Water Revealed by Recent Impacts A fresh impact observed in 2008 revealed water ice. The small crater is only 6 m wide and shows that ice is just beneath the Martian surface about halfway between the north pole and the equator. A spectrum taken of the white material shows that it is

water, not carbon dioxide. By January 2009, several months later, the water had evaporated. Quite surprisingly, the white material is 99% pure water; a mixture with about 50% dirt was expected, and this remains a mystery. (NASA/JPL-Caltech/University of Arizona)

FIGURE 11-41 R I **V** U X G

Impact Melts Subsurface Ice The flood channels of Hephaestus Fossae run 600 km from a 20-km-wide impact crater. Heat from the impact is thought to have caused frozen subsurface water ice to melt. (ESA/DLR/FU Berlin [G. Neukum])

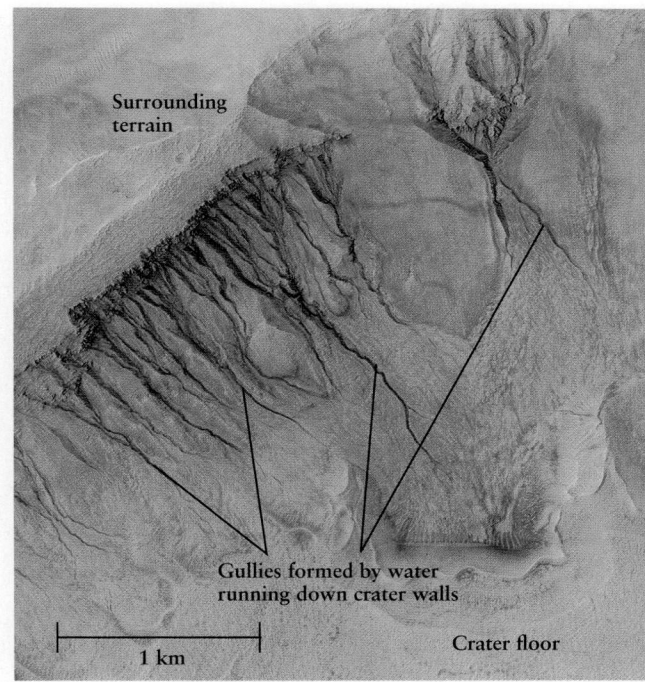

FIGURE 11-42 R I **V** U X G

Martian Gullies When the Mars Global Surveyor spacecraft looked straight down into this crater in the Martian southern hemisphere, it saw a series of gullies along the crater wall. These gullies may have formed by subsurface water seeping out to the surface, or by the melting of snow that fell on crater walls. Gullies like these have been found in many places near the equator and show little to no disruption by impact craters. Therefore, these gullies are considered as likely evidence for the *recent* flow of water. (NASA/JPL/MSSS)

water to evaporate and eventually refreeze elsewhere on the planet. A big change in the tilt of Mars's axis would not be surprising for two reasons. First, Earth's large Moon stabilizes our rotation axis, but the Martian moons are too small to provide the planet with stabilizing forces. And second, Mars is close enough to Jupiter to be significantly influenced by Jupiter's strong gravity.

The significance of water on Mars in the search for life is so strong that researchers have put in great effort to find signs of water as ice or liquid, in the past or present. Water near the warmer equator could also be valuable for any longer-term human missions on Mars. Do we have additional evidence for water beneath the dusty surface of Mars? We do!

The steep walls of some craters in the warmer equatorial regions of Mars have "gullies" that appear to have been formed by liquid water (**Figure 11-42**). And, in 2010, an undergraduate student named Lujendra Ojha, who was studying HiRISE images, noticed dark streaks running down slopes that only appeared during warmer times. Shaped like lines, or "lineae," these features came to be called *recurring slope lineae*, or RSL. While their formation isn't well understood, there is strong evidence that these features arise from liquid water (**Figure 11-43**). Thus, an undergraduate student discovered the only known present-day liquid water on Mars!

Clues from Curiosity

In 2012, Curiosity found strong evidence for the *steady flow* of water in an ancient stream or river, as opposed to intermittent flow. **Figure 11-44** shows rounded pebbles, just as we find in dry

streambeds here on Earth. The pebbles are too big to have been moved by wind and indicate water depths between ankle and knee high. Other nearby sites show similar rounding of pebbles, and based on images from orbit, the entire region has long been suspected as the outlet of an ancient river.

The landing site for Curiosity, inside Gale Crater, was chosen by NASA because it appeared to be an ancient lake bed. In 2015, Curiosity confirmed that sedimentary layers at the bottom of Gale Crater formed in a lake that lasted hundreds of millions of years (**Figure 11-45**).

We should expect to discover more ancient rivers and lakes because the ice on Mars today is enough to cover the planet 35 km (21.7 miles) deep in water. During Mars's first few hundred million years—with a thicker, higher pressure atmosphere and warmer temperatures—at least some of that water would have been in liquid form.

CONCEPT CHECK 11-12

What are two different lines of evidence that point to present-day subsurface ice on Mars?

Answer appears at the end of the chapter.

(a)

(b)

FIGURE 11-43 R I **V** U X G

RSL and the Seasonal Flow of Water **(a)** The arrows point to some of the dark streaks that recur seasonally during warmer times on the Sun-facing wall of Newton crater. These features are called recurring slope lineae (RSL), and researchers have concluded they form from liquid saltwater. The source of the water is uncertain, and may emerge from below the surface, or come from the atmosphere. Even with warmer temperatures, this water can only form a liquid due to its high salt concentration (just as salt on icy roads is applied to melt the ice). Note that 3D modeling of satellite images was used to create a view as if looking at the crater's wall. **(b)** A 100-m-long RSL running along the wall of Hale crater. (NASA/JPL-Caltech/Univ. of Arizona)

1 cm

FIGURE 11-44 R I **V** U X G

Dry Streambed The rounding of the pebbles at this site indicate they were exposed to flowing water for a period lasting at least thousands of years. The pebbles, and certainly the largest rock (in the white circle), could not have been rounded or transported by winds in the thin Martian atmosphere. The jagged overhanging rock ledge (running down the middle of the image) appears to be the bank of the stream, undercut by erosion from the stream itself. Even the rocky ledge appears to be made of a sedimentary rock from prior exposure to water. (NASA/JPL-Caltech/MSSS)

FIGURE 11-45 R I **V** U X G

Liquid Water Formed Sedimentary Layers Curiosity has provided data to confirm that these sedimentary layers were produced within a lake and stream system within Gale crater. While it might look obvious that these broken-up slabs formed in water, wind-blown sediments can (and do) create similar features in this crater. A stream-fed lake probably existed here between about 3.8 and 3.3 billion years ago. (NASA/JPL-Caltech/MSSS)

11-9 The two Martian moons resemble asteroids

Two moons move around Mars in orbits close to the planet's surface. These satellites are so tiny that they were not discovered until the favorable opposition of 1877. While Schiaparelli was seeing canals, the American astronomer Asaph Hall spotted the two moons, which orbit almost directly above its equator. He named them Phobos ("fear") and Deimos ("panic"), after the mythical horses that drew the chariot of the Greek god of war.

The Orbits of Phobos and Deimos

Phobos is the inner and larger of the Martian moons. It circles Mars in just 7 hours and 39 minutes at an average distance of only 9378 km (5828 mi) from the center of Mars. (Recall that our Moon orbits at an average distance of 384,400 km, or 238,855 mi from Earth's center.) This orbital period is much shorter than the Martian day; hence, Phobos rises in the *west* and gallops across the sky in only 5½ hours, as viewed by an observer near the Martian equator. During this time, Phobos appears several times brighter in the Martian sky than Venus does from Earth.

> As seen from Mars, Deimos rises in the east but Phobos rises in the west.

Deimos, which is farther from Mars and somewhat smaller than Phobos, appears about as bright from Mars as Venus does from Earth. Deimos orbits at an average distance of 23,460 km (14,580 mi) from the center of Mars, only slightly beyond the right distance for a synchronous orbit—an orbit in which a satellite seems to hover above a single location on the planet's equator. As seen from the Martian surface, Deimos rises in the east and takes about 3 full days to creep slowly from one horizon to the other (Figure 11-46a).

Images from spacecraft in Mars's orbit reveal Phobos and Deimos to be jagged and heavily cratered (Figure 11-46b). Both moons are somewhat elongated from a spherical shape: Phobos is approximately 28 by 23 by 20 km in size, while Deimos is slightly smaller, roughly 16 by 12 by 10 km. Phobos and Deimos are both in synchronous rotation: The tidal forces that Mars exerts on these elongated moons cause them to always keep the same sides toward Mars, just as Earth's tidal forces ensure that the same side of the Moon always faces us (see Section 4-8).

We saw in Section 10-5 that the Moon raises a tidal bulge on Earth. Because Earth rotates on its axis more rapidly than the Moon orbits Earth, this bulge is dragged ahead of a line connecting Earth and the Moon (see Figure 10-20). The gravitational force of this bulge pulls the Moon into an ever-larger orbit. In the same way, and for the same reason, Deimos is slowly moving away from Mars. But for fast-moving Phobos, the situation is reversed: Mars rotates on its axis more slowly than Phobos moves around Mars. This slow rotation holds back the tidal bulge that Phobos raises in the solid body of Mars; this bulge is thus *behind* a line connecting the centers of Mars and Phobos. (Imagine that the bulge in Figure 10-20 is tilted clockwise rather than counterclockwise.) This bulge exerts a gravitational force on Phobos that pulls it back, slowing the moon's motion and making it slowly spiral inward toward Mars. This inward spiral will cease in about 40 million years when Phobos smashes into the Martian surface!

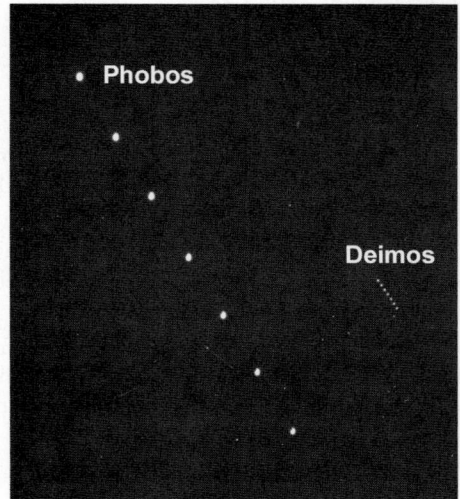

(a) As seen from Mars

(b) As seen from Mars orbit

FIGURE 11-46 R I V U X G

The Moons of Mars (a) The Spirit rover recorded seven images of the Martian night sky at 15-minute intervals to produce this composite view of the motions of Deimos and Phobos. **(b)** Two Mars Reconnaissance Orbiter images have been combined to show Deimos and Phobos and their heavily cratered surfaces to the same scale. The gravity on these moons is only about 1/1000 of Earth's, and not strong enough to shape them into spheres. (a: NASA/JPL/Cornell/Texas A&M; b: NASA/JPL Caltech/University of Arizona)

The Nature of the Martian Moons

The origin of the Martian moons is unknown. One idea is that they are asteroids that Mars captured from the nearby asteroid belt (see Section 7-5). Like Phobos and Deimos, many asteroids reflect very little sunlight (typically less than 10%) because of their high carbon content. Phobos and Deimos also have very low densities (1900 kg/m³ and 1760 kg/m³, respectively), which are also characteristic of carbon-rich asteroids. Perhaps two of these asteroids wandered close enough to Mars to become permanently trapped by the planet's gravitational field. Alternatively, Phobos and Deimos may have formed in orbit around Mars out of debris left over from the formation

of the planets. Some of this debris might have come from the asteroid belt, giving these two moons an asteroidlike character. Yet another theory proposes that the Martian moons may be debris left over from a gigantic impact, similar to the formation of our own Moon. Future measurements of the composition of the Martian moons—perhaps by landing spacecraft on their surfaces—may shed light on their origin.

CONCEPT CHECK 11-13

What causes Phobos to slowly spiral in toward Mars?

Answer appears at the end of the chapter.

KEY WORDS

1-to-1 spin-orbit coupling, p. 299
3-to-2 spin-orbit coupling, p. 300
antipode, p. 304
cosmic rays, p. 329
crustal dichotomy, p. 311

dust devil, p. 321
greatest eastern elongation, p. 298
greatest western elongation, p. 298
hot-spot volcanism, p. 315
northern lowlands, p. 311

prograde rotation, p. 302
residual ice cap, p. 320
retrograde rotation, p. 302
runaway greenhouse effect, p. 323
runaway icehouse effect, p. 325

scarp, p. 304
shield volcano, p. 314
southern highlands, p. 311
volatile elements, p. 306

KEY IDEAS

Motions of Mercury, Venus, and Mars in Earth's Sky: Mercury and Venus can be seen only in the morning or evening sky, whereas it is possible to see Mars at any time of night, depending on its position in its orbit (Section 11-1).

- At their greatest eastern and western elongations, Mercury is only 28° from the Sun and Venus is only 47° from the Sun.

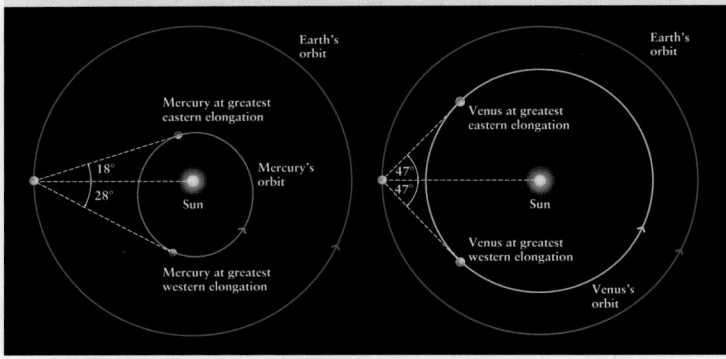

Mercury's Surface, Interior, and Magnetic Field: Mercury's surface is pocked with craters, but there are extensive smooth plains between these craters (Section 11-3).

- Long cliffs called scarps meander across the surface of Mercury. These probably formed as the planet's crust cooled, solidified, and shrank.

- Mercury has an iron core with a diameter equal to about 83% of the planet's diameter. By contrast, the diameter of Earth's core is only slightly more than one-half of Earth's diameter.

- Mercury has a weak magnetic field, which indicates that at least part of the iron core is liquid. However, the field is off-center by 20%, which remains unexplained.

- The high abundance of volatile elements that should have "boiled off" from Mercury's material during formation presents a mystery considering the planet's close proximity to the Sun.

Rotation of Mercury, Venus, and Mars: Poor telescopic views of Mercury's surface led to the mistaken impression that the planet always keeps the same face toward the Sun (1-to-1 spin-orbit coupling) (Section 11-2).

- Radio and radar observations revealed that Mercury in fact has 3-to-2 spin-orbit coupling: The planet rotates on its axis three times every two orbits.

- Venus rotates very slowly in a retrograde direction. Its rotation period is longer than its orbital period.

- Mars rotates at almost the same rate as Earth, and its rotation axis is tilted by almost the same angle as Earth's axis.

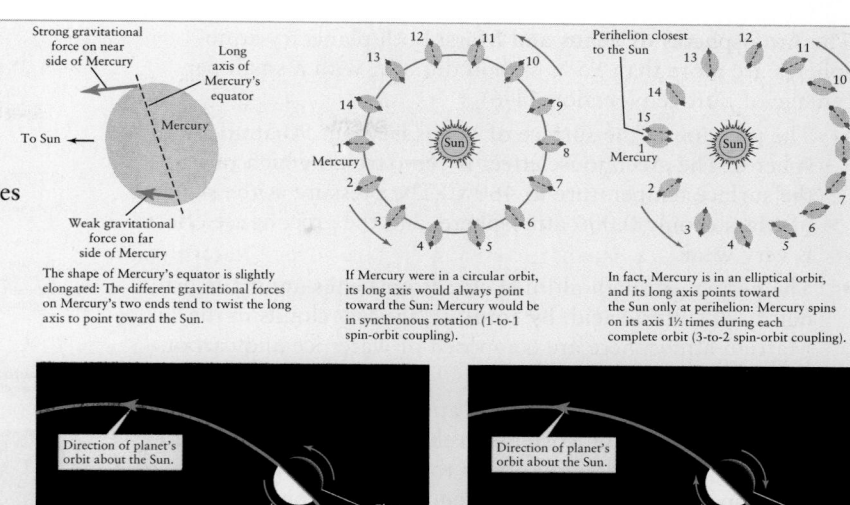

The shape of Mercury's equator is slightly elongated: The different gravitational forces on Mercury's two ends tend to twist the long axis to point toward the Sun.

If Mercury were in a circular orbit, its long axis would always point toward the Sun: Mercury would be in synchronous rotation (1-to-1 spin-orbit coupling).

In fact, Mercury is in an elliptical orbit, and its long axis points toward the Sun only at perihelion: Mercury spins on its axis 1½ times during each complete orbit (3-to-2 spin-orbit coupling).

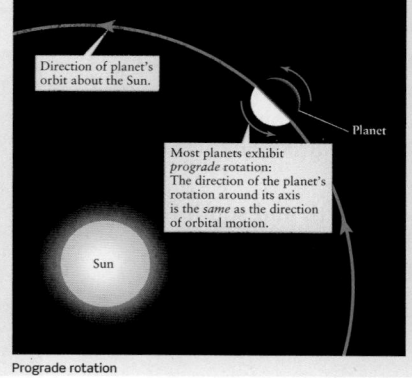

Direction of planet's orbit about the Sun.

Most planets exhibit *prograde* rotation: The direction of the planet's rotation around its axis is the *same* as the direction of orbital motion.

Prograde rotation

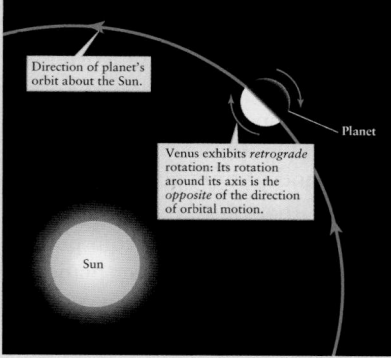

Direction of planet's orbit about the Sun.

Venus exhibits *retrograde* rotation: Its rotation around its axis is the *opposite* of the direction of orbital motion.

Retrograde rotation

Comparing Venus and Mars: Most of the surface of Venus is at about the same elevation, with just a few elevated regions. On Mars, the southern highlands rise several kilometers above the northern lowlands (Sections 11-4 and 11-5).

- Venus has a thick atmosphere and a volcanically active surface. Mars has a very thin atmosphere and little or no current volcanism.

- There is no evidence of plate tectonics on Venus, but Valles Marineris shows there may have been at least two plates on Mars. On Venus, vigorous convection is predicted in the planet's interior, with tectonic activity leading to a somewhat wrinkled surface. On Mars, the planet's smaller size means the crust cooled long ago and became too thick to allow widespread plate tectonic activity.

- Volcanoes on both Venus and Mars were produced by hot spots in the planet's interior.

- The entire Venusian surface is about 500 million years old and has relatively few craters. By contrast, most of the Martian surface is cratered and is probably billions of years old. The southern highlands on Mars are the most heavily cratered and hence the oldest part of the planet's surface.

Venus

Canyon walls 6000 m (20,000 ft) high

20 km

NASA; NASA/JPL/USGS; NASA/J. Bell (Cornell U.) and M. Wolff (SSI)

ESA/DLR/FU Berlin (G. Neukum)

The Atmospheres of Venus and Mars: Both planetary atmospheres are more than 95% carbon dioxide, with a small percentage of nitrogen (Section 11-6).

- The pressure at the surface of Venus is about 90 atmospheres. The greenhouse effect is very strong, which raises the surface temperature to 460°C. The pressure at the surface of Mars is only 0.006 atmosphere, and the greenhouse effect is very weak.

- The permanent high-altitude clouds on Venus are made primarily of sulfuric acid. By contrast, the few clouds in the Martian atmosphere are composed of water ice and carbon dioxide ice.

- The circulation of the Venusian atmosphere is dominated by two huge convection currents in the cloud layers, one in the northern hemisphere and one in the southern hemisphere. The upper cloud layers of the Venusian atmosphere move rapidly around the planet in a retrograde direction, with a period of only about 4 Earth days.

- Weather on Mars is dominated by the north and south flow of carbon dioxide from pole to pole with the changing seasons. This can trigger planetwide dust storms.

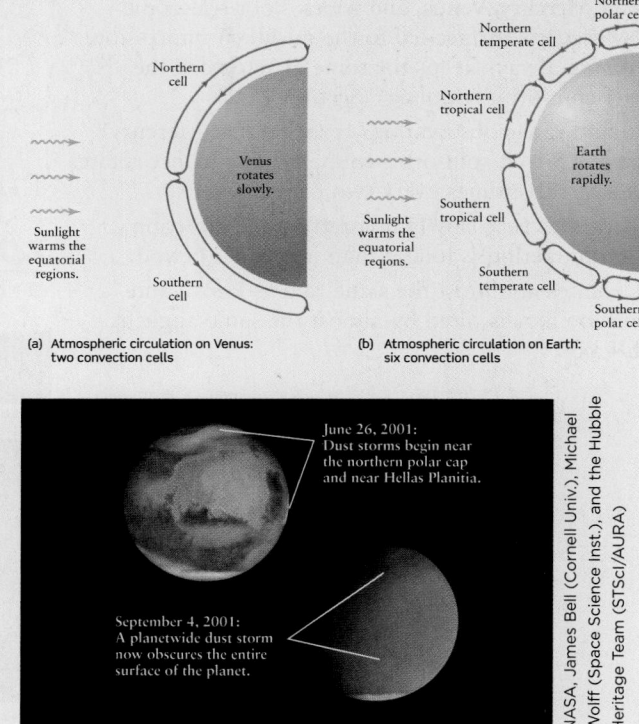

(a) Atmospheric circulation on Venus: two convection cells

(b) Atmospheric circulation on Earth: six convection cells

June 26, 2001: Dust storms begin near the northern polar cap and near Hellas Planitia.

September 4, 2001: A planetwide dust storm now obscures the entire surface of the planet.

NASA, James Bell (Cornell Univ.), Michael Wolff (Space Science Inst.), and the Hubble Heritage Team (STScI/AURA)

Evolution of Atmospheres: Earth, Venus, and Mars all began with relatively thick atmospheres of carbon dioxide, water vapor, and sulfur dioxide (Section 11-7).

- On Earth, most of the carbon dioxide went into carbonate rocks and most of the water into the oceans. Ongoing plate tectonics recycles atmospheric gases through the crust.

- On Venus, more intense sunlight and the absence of plate tectonics led to a thick carbon dioxide atmosphere and a runaway greenhouse effect.

- On Mars, a runaway icehouse effect resulted from weaker sunlight and a lack of strong plate tectonic activity.

Venus: No plate tectonics and more intense sunlight produce a thick atmosphere.

1. H_2O, CO_2, SO_2, and N_2 are outgassed from Venusian volcanoes on an ongoing basis.

2. Sunlight warms Venus more than Earth, so H_2O does not fall as rain. There are no oceans, so CO_2 remains in the atmosphere and intensifies the greenhouse effect.

3. Ultraviolet light from the sun breaks apart some H_2O molecules; hydrogen escapes into space.

4. H_2O that does not escape combines with SO_2 to make H_2SO_4 (sulfuric acid) in Venus's clouds.

5. Remaining SO_2 is locked up in sulfate minerals.

Atmosphere

Clouds

Surface

Ultraviolet light

Space

NASA/JPL

Water on Mars: Liquid water cannot exist on present-day Mars because the atmosphere is too thin and cold. But there is evidence for frozen water at the polar ice caps and beneath the surface of the regolith (Section 11-8).

- Geological evidence from robotic rovers shows that much of the Martian surface has been dry for billions of years, but some regions had substantial amounts of liquid water in the past.

NASA/JPL-Caltech/Univ. of Arizona

The Moons of Mars: Mars has two small, football-shaped satellites that move in orbits close to the surface of the planet. They may be captured asteroids or may have formed in orbit around Mars out of solar system debris (Section 11-9).

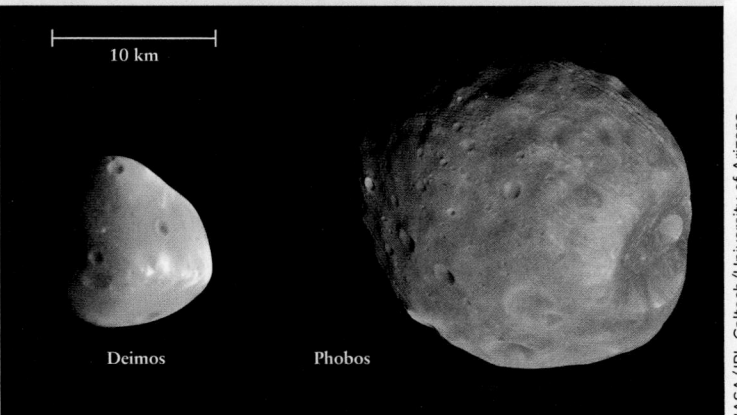

Deimos   Phobos

NASA/JPL Caltech/University of Arizona

QUESTIONS

Problem-solving tips and tools

We discussed the relationship between angular distance and linear distance in Box 1-1 and the idea of angular resolution in Section 6-3. You may need to refresh your memory about Kepler's third law, described in Section 4-4. You may also need to review the form of Kepler's third law that explicitly includes mass (see Section 4-7 and Box 4-4). Box 4-1 describes the relationship between a planet's sidereal orbital period and its synodic period (the time from one inferior conjunction to the next). You should recall that Wien's law (Section 5-4) relates the temperature of a blackbody to λ_{max}, its wavelength of maximum emission. Section 5-9 and Box 5-6 explain the Doppler effect and how to do calculations using it. The linear speed of a point on a planet's equator is the planet's circumference divided by its rotation period; recall that the circumference of a circle of radius r is $2\pi r$. The volume of a sphere of radius r is $4\pi r^3/3$. The speed of light is given in Appendix 7.

Section 11-1

1. Why is it impossible to see Mercury or Venus in the sky at midnight?

2. Figure 11-1 shows Mercury with a greatest eastern elongation of 18° and a greatest western elongation of 28°. On November 25, 2006, Mercury was at a greatest western elongation of 20°. Was Mercury at perihelion, aphelion, or some other point on its orbit? Explain your answer.

3. Venus takes 440 days to move from greatest western elongation to greatest eastern elongation, but it needs only 144 days to go from greatest eastern elongation to greatest western elongation. With the aid of a diagram like Figure 11-1, explain why.

4. This time-lapse photograph was taken on May 7, 2003, during a *solar transit* of Mercury. Over a period of 5 hours and 19 minutes, Mercury appeared to move across the face of the Sun. Such solar transits of Mercury occur 13 or 14 times each century; they do *not* happen each time that Mercury is at inferior conjunction. Explain why not. (*Hint:* For a solar transit to occur, the Sun, Mercury, and Earth must be in a nearly perfect alignment. Does the orbit of Mercury lie in the plane of the ecliptic?)

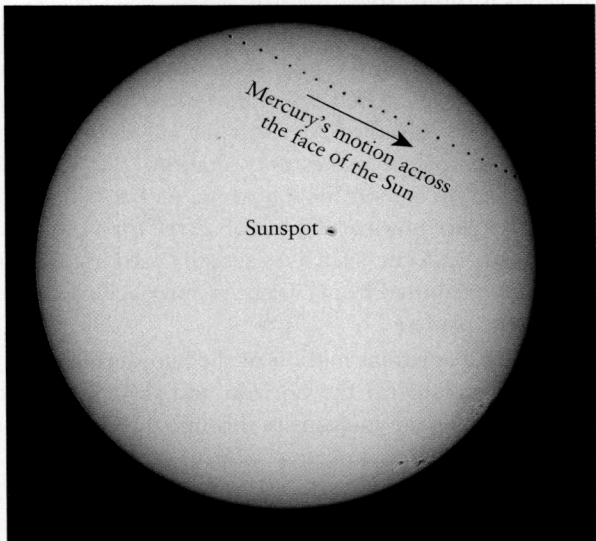

Mercury's motion across the face of the Sun

Sunspot

R I **V** U X G   (Dominique Dierick)

5. Find the largest angular size that Mercury can have as seen from Earth. In order for Mercury to have this apparent size, at what point in its orbit must it be? Consult Table 11-1.

6. Before about 350 B.C.E., the ancient Greeks did not realize that Mercury seen in the morning sky (which they called Apollo)

and seen in the evening sky (which they called Hermes) were actually the same planet. Discuss why you think it took some time to realize this.

Section 11-2

7. In his 1964 science fiction story "The Coldest Place," author Larry Niven described the "dark side" of Mercury as the coldest place in the solar system. What assumption did he make about the rotation of Mercury? Did this assumption turn out to be correct?

8. What is 3-to-2 spin-orbit coupling? How is the rotation period of an object exhibiting 3-to-2 spin-orbit coupling related to its orbital period? What aspects of Mercury's orbit cause it to exhibit 3-to-2 spin-orbit coupling? What telescopic observations proved this?

9. Why was it so difficult to determine the rate and direction of Venus's rotation? How were these finally determined? What is one proposed explanation for the slow, retrograde rotation of Venus?

10. Venus's sidereal rotation period is 243.01 days and its orbital period is 224.70 days. Use these data to prove that a solar day on Venus lasts 116.8 days. (*Hint:* Develop a formula relating Venus's solar day to its sidereal rotation period and orbital period similar to the first formula in Box 4-1.)

11. In Section 11-2, we described the relationship between the length of Venus's synodic period and the length of an apparent solar day on Venus. Using this relationship and a diagram, explain why at each inferior conjunction the same side of Venus is turned toward Earth.

12. (a) Mercury has a 58.646-day rotation period. What is the speed at which a point on the planet's equator moves due to this rotation? (*Hint:* Remember that speed is distance divided by time. What distance does a point on Mercury's equator travel as the planet makes one rotation?) (b) Use your response to (a) to answer the following: As a result of rotation, what difference in wavelength is observed for a radio wave of wavelength 12.5 cm (such as is actually used in radar studies of Mercury) emitted from either the approaching or receding edge of the planet?

13. Describe the apparent motion of the Sun during a "day" on Venus relative to (a) the horizon and (b) the background stars. (Assume that you can see through the cloud cover.)

Section 11-3

14. What kind of surface features are found on Mercury? How do they compare to surface features on the Moon? Why are they probably much older than most surface features on Earth?

15. How do we know that the scarps on Mercury are younger than the lava flows? How can you tell that the scarp in Figure 11-7 is younger than the vertically distorted crater at the center of the figure?

16. If Mercury is the closest planet to the Sun and has such a high average surface temperature, how is it possible that ice might exist on its surface?

17. Why do astronomers think that Mercury has a very large iron core?

18. Why is it surprising that Mercury has a global magnetic field? Why does the 58.646-day rotation period of Mercury imply that the planet can have only a weak magnetic field?

19. Is there a consistent explanation for why Mercury's magnetic field is off-center? If Earth's magnetic field was off-center by the same percentage, how many kilometers would that be?

20. Why are high concentrations of volatile elements a mystery on Mercury?

21. Consider the idea that Mercury has a solid iron-bearing mantle that is permanently magnetized like a giant bar magnet. Using the fact that iron demagnetizes at temperatures above 770°C, present an argument against this explanation of Mercury's magnetic field.

22. What evidence do we have that the surface features on Mercury were not formed during recent geological history?

23. If you could examine rock samples from the surface of Venus, would you expect them to be the same as rock samples from Earth? Would you expect to find igneous, sedimentary, and metamorphic rocks like those found on Earth (see Section 9-3)? Explain your answers.

Section 11-4

24. Why was it difficult to determine Venus's surface temperature from Earth? How was this finally determined?

25. The Mariner 2 spacecraft did not enter Venus's atmosphere, but it was nonetheless able to determine that the atmosphere contains little water vapor. How was this done?

26. How did winds trick early observers of Mars into thinking that Mars had vegetation?

27. (a) At what wavelength does Venus's surface emit the most radiation? (b) Do astronomers have telescopes that can detect this radiation? (c) Why can't we use such telescopes to view the planet's surface?

28. The Mariner 2 spacecraft detected more microwave radiation when its instruments looked at the center of Venus's disk than when it looked at the edge, or limb, of the planet. (This effect is called *limb darkening*.) Explain how these observations show that the microwaves are emitted by the planet's surface rather than its atmosphere.

Section 11-5

29. Do Venus and Mars have continents like those on Earth?

30. What is the Martian crustal dichotomy? What is the evidence that the southern highlands are older than the northern lowlands?

31. What is the evidence that the surface of Venus is only about 500 million years old?

32. **TUTORIAL 11.1** ➤ What geologic features (or lack thereof) on Mars have convinced scientists that extensive plate tectonics did not significantly shape the present Martian surface? Does Mars have any plates now?

33. What magnetic features indicate that Mars might have experienced plate tectonics in the past?

34. Compare the volcanoes of Venus, Earth, and Mars. Cite evidence that hot-spot volcanism is or was active on all three worlds.

35. Describe the evidence that there has been recent volcanic activity (a) on Venus and (b) on Mars.

36. Suppose all of Venus's volcanic activity suddenly stopped. (a) How would this affect Venus's clouds? (b) How would this affect the overall Venusian environment?

37. Earth's northern hemisphere is 39% land and 61% water, while its southern hemisphere is only 19% land and 81% water. Thus, the southern hemisphere could also be called the "water hemisphere." The Moon also has two distinct hemispheres: the near side (which has a number of maria) and the far side (which has almost none). How are these hemispheric differences on Earth and on the Moon similar to the Martian crustal dichotomy? How are they different?

38. For a group of properly attired astronauts equipped with oxygen tanks, a climb to the summit of Olympus Mons would actually be a relatively easy (albeit long) hike rather than a true mountain climb. Give two reasons why.

39. On Mars, the difference in elevation between the highest point (the summit of Olympus Mons) and the lowest point (the bottom of the Hellas Planitia basin) is 30 km. On Earth, the corresponding elevation difference (from the peak of Mount Everest to the bottom of the deepest ocean) is only 20 km. Discuss why the maximum elevation difference is so much greater on Mars.

40. The Mars Global Surveyor (MGS) spacecraft was in a nearly circular orbit around Mars with an orbital period of 117 minutes. (a) Using the data in Table 11-3, find the radius of the orbit. (b) What is the average altitude of MGS above the Martian surface? (c) The orbit of MGS passed over the north and south poles of Mars. Explain how this makes it possible for the spacecraft to observe the entire surface of the planet.

41. The elevations in Figure 11-17 were measured using an instrument called MOLA (Mars Orbiter Laser Altimeter) on board Mars Global Surveyor. MOLA fired a laser beam downward, then measured how long it took for the beam to return to the spacecraft after reflecting off the surface. Suppose MOLA measured this round-trip time for the laser beam to reflect off the summit of Olympus Mons, as well as the round-trip time to reflect off the bottom of Hellas Planitia (see Question 39). Which round-trip time is longer? How much longer is it? Do you need to know the distance from Mars Global Surveyor to the surface to answer this question?

42. (a) The Grand Canyon in Arizona was formed over 15 to 20 million years by the flowing waters of the Colorado River, as well as by rain and wind. Contrast this formation scenario to that of Valles Marineris on Mars. (b) Valles Marineris is sometimes called the "Grand Canyon of Mars." Is this an appropriate description? Why or why not?

43. In 1978, the Pioneer Venus Orbiter spacecraft arrived at Venus. It carried an ultraviolet spectrometer to measure the chemical composition of the Venusian atmosphere. This instrument recorded unexpectedly high levels of sulfur dioxide and sulfuric acid, which steadily declined over the next several years. Discuss how this observation suggests that volcanic eruptions occurred on Venus not long before Pioneer Venus Orbiter arrived there.

Section 11-6

44. Why are the patterns of convection in the Venusian atmosphere so different from those in our atmosphere?

45. Why is it impossible for liquid water to exist on Mars today? If liquid water existed on Mars in the past, what must have been different then?

46. Why does the atmospheric pressure on Mars vary with the seasons? What is the relationship between this pressure variation and Martian dust storms?

47. What is a dust devil? Why would you feel much less breeze from a Martian dust devil than from a dust devil on Earth?

48. A hypothetical planet has an atmosphere that is opaque to visible light but transparent to infrared radiation. How would this affect the planet's surface temperature? Contrast and compare this hypothetical planet's atmosphere with the greenhouse effect in Venus's atmosphere.

49. Water has a density of 1000 kg/m^3, so a column of water n meters tall and 1 meter square at its base has a mass of $n \times 1000$ kg. On either Earth or Venus, which have nearly the same surface gravity, a mass of 1 kg weighs about 9.8 newtons (2.2 lb). Calculate how deep you would have to descend into Earth's oceans for the pressure to equal the atmospheric pressure on Venus's surface, 90 atm or 9×10^6 newtons per square meter. Give your answer in meters.

50. Marine organisms produce sulfur-bearing compounds, some of which escape from the oceans into Earth's atmosphere. (These compounds are largely responsible for the characteristic smell of the sea.) Even more sulfurous gases are injected into our atmosphere by the burning of sulfur-rich fossil fuels, such as coal, in electric power plants. Both of these processes add more sulfur compounds to the atmosphere than do volcanic eruptions. On lifeless Venus, by contrast, volcanoes are the only source for sulfurous atmospheric gases. Why, then, are sulfur compounds so much rarer in our atmosphere than in the Venusian atmosphere?

51. The classic 1950 science fiction movie *Rocketship X-M* shows astronauts on the Martian surface with oxygen masks

for breathing but wearing ordinary clothing. Would this be a sensible choice of apparel for a walk on Mars? Why or why not?

Section 11-7

52. **TUTORIAL** Why is it reasonable to assume that the primordial atmospheres of Earth, Venus, and Mars were roughly the same?

53. Carbon dioxide accounts for about 95% of the present-day atmospheres of both Mars and Venus. Why, then, is there a strong greenhouse effect on Venus but only a weak greenhouse effect on Mars?

54. (a) What is a runaway greenhouse effect? (b) What is a runaway icehouse effect?

55. In the classic Ray Bradbury science fiction story "All Summer in a Day," human colonists on Venus are subjected to continuous rainfall except for 1 day every few years when the clouds part and the Sun comes out for an hour or so. Discuss how our understanding of Venus's atmosphere has evolved since this story was first published in 1954.

Section 11-8

56. (a) Why is Mars red? (b) Why is the Martian sky the color of butterscotch?

57. Were the Viking Landers able to determine whether life currently exists on Mars or whether it once existed there? Why or why not?

58. (a) The Spirit rover found the minerals olivine and pyroxene at its landing site on Mars. Explain how this shows that there has been no liquid water at that site for billions of years. (b) What evidence did the Opportunity rover find at its landing site to suggest that liquid water had once been present there?

59. How was the Mars Odyssey spacecraft able to detect water beneath the Martian surface without landing on the planet?

60. How do craters reveal evidence for water ice beneath the Martian surface?

61. Biologists think that an extended period of time with liquid present water is more promising for life on Mars than a brief period of liquid water. Explain one of the pieces of evidence for water that implies an extended period of time with liquid water.

Section 11-9

62. Why do you suppose that Phobos and Deimos are not round like our Moon?

63. The orbit of Phobos has a semimajor axis of 9378 km. Use this information and the orbital period given in the text to calculate the mass of Mars. How does your answer compare with the mass of Mars given in Table 11-3?

64. Calculate the angular sizes of Phobos and Deimos as they pass overhead, as seen by an observer standing on the Martian equator. How do these sizes compare with that of the Moon

seen from Earth's surface? Would Phobos and Deimos appear as impressive in the Martian sky as the Moon does in our sky?

65. You are to put a spacecraft into a synchronous circular orbit around the Martian equator, so that its orbital period is equal to the planet's rotation period. Such a spacecraft would always be over the same part of the Martian surface. (a) Find the radius of the orbit and the altitude of the spacecraft above the Martian surface. (b) Suppose Mars had a third moon that was in a synchronous orbit. Would tidal forces make this moon tend to move toward Mars, away from Mars, or neither? Explain your answer.

General Problems

66. Explain why Mercury does not have a substantial atmosphere.

67. Explain two ways that heat within Mars helped to maintain a greenhouse effect.

68. (a) Suppose you have a telescope with an angular resolution of 1 arcsec. What is the size (in kilometers) of the smallest feature you could have seen on the Martian surface during the opposition of 2005, when Mars was 0.464 au from Earth? (b) Suppose you had access to the Hubble Space Telescope (HST), which has an angular resolution of 0.1 arcsec. What is the size (in kilometers) of the smallest feature you could have seen on Mars with the HST during the 2005 opposition?

69. For a planet to appear to the naked eye as a disk rather than as a point of light, its angular size would have to be 1 arcmin, or 60 arcsec. (This is the same as the angular separation between lines in the bottom row of an optometrist's eye chart.) (a) How close would you have to be to Mars in order to see it as a disk with the naked eye? Does Mars ever get this close to Earth? (b) Would Earth ever be visible as a disk to an astronaut on Mars? Would she be able to see Earth and the Moon separately, or would they always appear as a single object? Explain your answers.

70. Imagine that you are part of the scientific team monitoring a spacecraft that has landed on Mars. At 5:00 P.M. in your control room on Earth, the spacecraft reports that the Sun is highest in the sky as seen from its location on Mars. When the Sun is next at its highest point as seen from the spacecraft, what time will it be in the control room on Earth?

71. The orbital period of Mariner 10 is twice that of Mercury. Use this fact to calculate the length of the semimajor axis of the spacecraft's orbit.

72. Suppose that Venus had no atmosphere at all. How would the albedo of Venus then compare with that of Mercury or the Moon? Explain your answer.

73. Suppose that the only information you had about Mars was the images of the surface in Figure 11-29. Describe at least two ways that you could tell from these images that Mars has an atmosphere.

74. Although the Viking Lander 1 and Viking Lander 2 landing sites are 6500 km apart and have different geologic histories, the chemical compositions of the dust at both sites are nearly identical. (a) What does this suggest about the ability of the Martian winds to transport dust particles? (b) Would you expect that larger particles such as pebbles would also have identical chemical compositions at the two Viking Lander sites? Why or why not?

75. If you were planning a new mission to Mercury, what features and observations would be of particular interest to you?

WEB/E-BOOK QUESTIONS

77. ANIMATION **Elongations of Mercury.** Access the animation **11.1** "Elongations of Mercury" in Chapter 11 of the *Universe* website or e-book. (a) View the animation and notice the dates of the greatest eastern and greatest western elongations. Which time interval is greater: from a greatest eastern elongation to a greatest western elongation, or vice versa? (b) Based on what you observe in the animation, draw a diagram to explain your answer to the question in (a).

78. Just as Mercury can pass in front of the Sun as seen from Earth (see Question 44), so can Venus. Transits of Venus are quite rare. The dates of the only transits in the twenty-first century are June 8, 2004, and June 6, 2012; the next ones will occur in 2117 and 2125. A number of European astronomers traveled to Asia and the Pacific islands to observe the transits of Venus in 1761 and 1769. Search the internet for information about these expeditions. Why were these events of such interest to astronomers? How definitive were the results of these observations?

79. Search the internet for information about possible crewed missions to Mars. How long might such a mission take? How expensive would such a project be? What would be the advantages of a crewed mission compared to a robotic one?

80. ANIMATION **Conjunctions of Mars.** Access and view the animation **11.2** "The Orbits of Earth and Mars" in Chapter 11 of the *Universe* website or e-book. (a) The animation highlights three dates when Mars is in opposition, so that Earth lies directly between Mars and the Sun. By using the "Stop" and "Play" buttons in the animation, find two times during the animation when Mars is in *conjunction,* so that the Sun lies directly between Mars and Earth (see Figure 4-6). For each conjunction, make a drawing showing the positions of the Sun, Earth, and Mars, and record the month and year when the conjunction occurs. (b) When Mars is in conjunction, at approximately what time of day does it rise as seen from Earth? At what time of day does it set? Is Mars suitably placed for telescopic observation when in conjunction?

ACTIVITIES

Observing Projects

> **Observing tips and tools**
>
> Mercury and Venus are visible in the morning sky when at or near greatest western elongation and in the evening sky when at or near greatest eastern elongation. These are also the times when Venus can be seen at the highest altitude above the horizon during the hours of darkness. Mars is most easily seen around an opposition. At other times, Mars may be visible only in the early morning hours before sunrise or in the early evening just after sunset. Consult such magazines as *Sky & Telescope* and *Astronomy* or their websites for more detailed information about when and where to look for Mercury, Venus, and Mars during a given month. You can also use the *Starry Night*™ program that accompanies certain printed copies of this textbook.

81. Refer to the *Universe* website or e-book for a link to a website that calculates the dates of upcoming greatest elongations of Mercury. Consult such magazines as *Sky & Telescope* and *Astronomy,* or the websites for these magazines, to determine if any of these greatest elongations is going to be a favorable one. If so, make plans to be one of those rare individuals who has actually seen the innermost planet of the solar system. Set aside several evenings (or mornings) around the date of the favorable elongation to reduce the chances of being "clouded out." Select an observing site that has a clear, unobstructed view of the horizon where the Sun sets (or rises). If possible, make arrangements to have a telescope at your disposal. Search for the planet on the dates you have selected, and make a drawing of its appearance through your telescope.

82. *This observing project should be performed only under the direct supervision of an astronomer who knows how to point a telescope safely at Mercury.* Make arrangements to view Mercury during broad daylight. This is best done by visiting an observatory where the coordinates (right ascension and declination) of Mercury's position can be used to point the telescope. **DO NOT LOOK AT THE SUN! Looking directly at the Sun can cause blindness.**

83. Refer to the *Universe* website or e-book for a link to a website that calculates the dates of greatest elongations of Venus. If Venus is near a greatest elongation, view the planet through a telescope. Make a sketch of the planet's appearance. From your sketch, can you determine if Venus is closer to us or farther from us than the Sun?

84. Using a small telescope, observe Venus once a week for a month and make a sketch of the planet's appearance on each occasion. From your sketches, can you determine whether Venus is approaching us or moving away from us?

85. *This observing project should be performed only under the direct supervision of an astronomer who knows how to point a telescope safely at Venus.* Make arrangements to view Venus during broad daylight. This is best done by visiting an observatory where the coordinates (right ascension and declination) of Venus's position can be used to point the telescope. **DO NOT LOOK AT THE SUN! Looking directly at the Sun can cause blindness.**

86. If Mars is suitably placed for observation, arrange to view the planet through a telescope. Draw a picture of what you see. What magnifying power seems to give you the best image? Can you distinguish any surface features? Can you see a polar cap or dark markings? If not, can you offer an explanation for Mars's bland appearance?

Starry Night™ Explorations

87. Use the *Starry Night™* program to observe the apparent motion of the inferior planets Mercury and Venus. Open **Favorites > Explorations > Elongations** to see a view along the ecliptic plane from the center of a transparent Earth showing the positions in the sky of the Sun, Mercury, and Venus. The date and universal time are shown in the upper right of the main view. The view is locked on the Sun with north at the top and east to the left. **Play** the animation and observe the motions of these two inferior planets against the background stars as seen from Earth.

    (a) Select **File > Revert** from the menu to reset the animation and then use the time controls to find the first date in 2019 when Mercury is at its farthest point to the left of the Sun, and the first date in 2019 when Mercury is at its farthest point to the right of the Sun in the view. What is your interpretation of these two dates and how would you label them?

    (b) If you wanted to make observations of Mercury on these two dates, what would be the best time of day to do so for each date?

    (c) Select **File > Revert** from the menu to reset the date to the beginning of 2019. Click the name in Venus's label and select **Orbit** from the menu to show the extent of this planet's orbit and then use the "D" keyboard shortcut to advance time by 1 day. Notice that Venus is very close to the Moon in the sky. If you wished to see this conjunction of the Moon and Venus, at what time of day would you plan to make your observation and in which direction would you need to look? Click the **Home** button in the toolbar to go to your home location and then adjust the date, time, and direction of the view to confirm your answer.

88. Use *Starry Night™* to explore the characteristics and surface features of the planet Mercury. Open **Favorites > Explorations > Mercury** to see this planet from a distance above the planet surface about equal to the diameter of the planet.

    (a) Begin your exploration of Mercury by obtaining some basic data about this planet from the HUD. What are the planet's radius and mass? How does the average density of this planet compare to that of Earth? What does this suggest about the planet's composition?

    (b) Mercury's surface shows clear evidence of impact craters. Use the HUD to identify the bright crater visible near the center of the view. What name is given to this crater? What are the bright striations emanating from this bright object called?

    (c) Use the location scroller to survey the surface of Mercury and describe your observations. To aid in this examination, the dark side of the planet has been brightened.

    (d) Reset the view with **File > Revert** and then zoom in to a field about 7° wide to see the crater Lu Hsun. Use the angular separation cursor to measure the angular diameter of this crater. Using this measurement and the height of the observer from the planet surface as given on the toolbar, what is the approximate diameter, in km, of this crater?

89. Use *Starry Night™* to observe a transit of the Sun by Venus. Select **Favorites > Explorations > Transit** to open a view of the sky from Rome, Italy, at 5:30:00 A.M. on June 8, 2004. The view is centered on the planet Venus.

    (a) Click **Play** to observe Venus transit the disk of the Sun. What is the name given to the planetary configuration of Venus relative to Earth during this transit?

    (b) Select **File > Revert** to reset the view and then use the "F" keyboard shortcut to display the ecliptic in the view. During the transit, is Venus precisely on the ecliptic?

    (c) Use the zoom and time controls to find the approximate duration of this transit. (Use the times when the center of Venus crosses the limb of the Sun.) How long does it take for Venus to complete the transit as seen from Rome?

    (d) Select **Favorites > Explorations > Fairbanks** to see this transit from Fairbanks, Alaska. Use the zoom and time controls to observe the details of this transit. What is the duration of the transit at this location? Based on the transit duration, from which location, Fairbanks or Rome, does Venus appear closer to the ecliptic?

    (e) Change the date in the toolbar to June 5, 2012, and the time to 12:30 P.M., the date of the most recent transit of the Sun by Venus. Use the time controls to find the approximate duration of this transit from this location. Is Venus closer to or farther from the ecliptic on this date compared to that seen in 2004 from this location?

90. Use *Starry Night™* to examine Mars. Open **Favorites > Explorations > Mars Surface** and use the zoom controls and the location scroller to explore this planet's surface features. You will notice that four volcanoes (Mons) and the Valles Marineris have been labeled.

    (a) Which of the volcanoes appears to be the largest?

    (b) Click the name in the label for Mars and select **Markers and Outlines...** In the window that opens, type "Crater" into the Filter box and then click the **Check all Shown** button. Describe the distribution of craters on the Martian surface.

(c) What does the distribution of craters in the region around the large volcanoes suggest about the time at which these volcanoes formed on Mars?

Collaborative Exercises

91. Figures 11-3b and 11-3c show a planet in synchronous rotation and Mercury with a 3-to-2 spin-orbit coupling, respectively. Stand up and demonstrate how planets move in each of these rotations by "orbiting" around a stationary classmate who represents our Sun. How would Mercury's motion be different if it had a 4-to-2 spin-orbit coupling instead?

92. In the nineteenth century, French mathematician and astronomer U. J. J. Le Verrier led a failed search for a hypothetical planet named Vulcan (after the mythical blacksmith of the gods) orbiting closer to our Sun than Mercury. If Vulcan had an orbit with the same eccentricity as Mercury's orbit but only one-half the size, what would have been its maximum eastern and western elongations?

93. Figure 11-4 shows prograde and retrograde rotation. Stand up and demonstrate how Venus rotates as it orbits our Sun, using a classmate as our stationary Sun. How is this different from how Earth rotates in its orbit?

94. The image of Mars that opens this chapter is from the Hubble Space Telescope. Draw a circle on your paper roughly 5 cm in diameter and, taking turns, have each person in your group sketch a different region of Mars. How is your collaborative sketch different from Schiaparelli's drawing shown in Figure 11-12?

ANSWERS

ConceptChecks

ConceptCheck 11-1: No. Venus is perpetually shrouded in a thick layer of clouds and, because visible light does not penetrate the clouds, no light escapes for us to observe from Earth.

ConceptCheck 11-2: When Caloris Basin formed, all existing craters in that area would have been wiped out, and the craters we see today occurred afterward. The density of craters inside Caloris Basin is low compared to typical regions that still show high densities from the Late Heavy Bombardment. Therefore, it appears that Caloris Basin would have formed later.

ConceptCheck 11-3: The high temperatures of such an impact would "boil off" the volatile elements, and would be consistent with a low abundance on Mercury. However, we observe a high abundance of volatile elements, and this is a great mystery.

ConceptCheck 11-4: 1.35-cm microwaves are absorbed by water molecules. If the surface of Venus contained a significant amount of water, it would absorb these wavelengths. The same is true for the Venusian atmosphere. Since microwaves at 1.35 cm were readily detected, there can be little water on or around Venus.

ConceptCheck 11-5: The northern hemisphere has a younger surface. Compared to the southern hemisphere, the northern hemisphere is almost free of impact craters. This indicates that craters must have been erased on the northern hemisphere by a more recent process.

ConceptCheck 11-6: The age of a planetary surface is indicated by how many craters it has. Venus has more craters than Earth, yet fewer than Mercury. Detailed analysis estimates the age of Venus's surface at 500 million years old.

ConceptCheck 11-7: If CO_2 could be added to the atmosphere, the temperature on Mars might be increased through the greenhouse effect.

ConceptCheck 11-8: Methane-producing microbes, volcanic activity, and the interaction of UV sunlight with small meteorites could all produce variable amounts of methane on Mars.

ConceptCheck 11-9: There needs to be sufficient atmospheric pressure for ice to melt. Mars has a very thin atmosphere that produces only 0.6% of the atmospheric pressure on Earth found at sea level. Raising the temperature of ice on Mars will cause it to skip the liquid phase and go directly to the gas phase: The ice will evaporate.

ConceptCheck 11-10: The "runaway" term refers to the fact that an increase in temperature creates changes in the atmosphere, which causes the temperature to rise even higher and higher.

ConceptCheck 11-11: A global magnetic field can deflect the solar wind, protecting the Martian atmosphere from high-energy collisions. These collisions can break apart the greenhouse gases CO_2 and H_2O, allowing the constituent atoms to escape into space.

ConceptCheck 11-12: It appears that hydrogen, presumed to reside in frozen water, is absorbing neutrons on the surface of Mars (Figure 11-39). Also, impact craters have directly revealed water ice (Figure 11-40).

ConceptCheck 11-13: Phobos raises a tidal bulge on Mars that lags behind Phobos due to Phobos orbiting faster than Mars rotates. Through gravity, this bulge pulls back on Phobos, slowing it down. As Phobos slowly loses orbital energy, it spirals in toward Mars.

NASA's Next Great Mars Rover

By Jay Bennett (from *Popular Mechanics,* January 17, 2018)

Of all the places in the solar system aside from Earth and our moon, humans have a more complete understanding of Mars than any other world. NASA's robotic rovers have been combing the calm sands of our red neighbor for more than two decades to uncover the secret history of the planet, and two-and-a-half short years from now, NASA's Jet Propulsion Laboratory will launch its fifth rover to Mars.

The first, Sojourner, was barely more than two feet long. It roved the Red Planet for about three months in 1997, equipped with a couple cameras and a small spectrometer. Two more rovers that landed in 2004, Spirit and Opportunity, found mineralogical evidence of past water on Mars and even stumbled upon evidence of ancient hydrothermal vents that could have supported life. Curiosity, officially called the Mars Science Laboratory, truly is a rolling laboratory of science instruments about the size of a small car. It has been probing geological and chemical processes on Mars for more than five years, finding evidence of water around just about every corner.

Now, Curiosity's twin is almost ready to join the Mars rover fleet. Mars 2020 will be the most advanced rolling science experiment yet. The rover, which will receive a common name like "Curiosity" before its trip, is a major step forward in Mars exploration, advancing environmental studies of the planet to experiments that will test how the martian environment could be put to use by future human explorers. The mission is all about life on Mars: the possibility of finding ancient or modern microbial martians, and preparing for earthlings to visit in the future.

Mars 2020, scheduled to launch from Cape Canaveral in July of that year, has four primary science objectives: determine whether life ever arose on Mars, characterize the planet's climate, study its geology, and prepare for human exploration.

The Mars 2020 rover leverages many successful features of NASA's Curiosity rover, which landed on Mars in 2012, but also adds new science instruments and a sampling system to carry out new goals.

NASA/JPL–Caltech

The rover will carry seven primary science instruments, some of which will search for signs of life known as biosignatures.

"Biosignatures is a big umbrella term for basically any kind of structure, pattern, texture, or composition that really only life can form," says Katie Stack Morgan, deputy scientist for Mars 2020, whom we met during a visit to NASA's Jet Propulsion Laboratory in Pasadena, California. "Determining whether something is truly a biosignature can be challenging. Obviously in a best case scenario, you run into a fossil."

While it's possible that Mars 2020 finds a microfossil in a rock sample, Morgan says it's far more likely the rover will stumble upon subtler clues of past life—if, in fact, life ever existed on Mars. Chemical and mineral remnants could be left behind in the martian rock record, and there could be other clues left behind in the shape of the rock itself.

"It may be that it's not just one of those things that will tell us. It could be the assemblage of the composition plus the texture that may give us a clue that something we're looking at is indeed a biosignature," Morgan says. Organic compounds, amino acids, hydrogen, carbon—all these clues could help scientists put together a picture of past life on Mars.

Mars 2020 will carry two tools to study martian samples on a microscopic scale. The first is the Planetary Instrument for X-Ray Lithochemistry, or PIXL, which is an x-ray spectrometer designed to determine the chemical composition of lithified rock. In addition to collecting powdered rock, Mars 2020 will be able to drill out solid cores of rock to analyze with PIXL, something Curiosity cannot do.

The other instrument to search for signs of life is SHERLOC, or Scanning Habitable Environments with Raman & Luminescence for Organics & Chemicals. This tool uses spectrometers, a UV laser, and a camera to look for chemicals that are indicative of past watery environments or aquatic organisms.

Then there are the cameras. In addition to Mars 2020's primary camera, Mastcam-Z, the rover also will have an instrument called SuperCam that will scan the landscape with a laser and generate a detailed composition map from a distance. Whereas Curiosity can look at a rock outcrop and detect aluminum, Morgan says, Mars 2020 would be able to locate aluminum in a specific vein.

Perhaps the most likely way Mars 2020 could help scientists find evidence of life, however, is by collecting rock samples and leaving them on the surface for a future mission to collect. If NASA could bring samples of martian rock back to labs on Earth, then scientists could go far beyond the analysis capabilities of a rover. For example, they could search for isotope ratios and isotopic remnants of past life. An imbalance in naturally occurring isotope ratios is one of the most convincing biosignatures, though the technology to search for such an imbalance cannot yet be packaged onto a Mars rover.

From little Sojourner to Mars 2020, the tradition of NASA's plucky Mars rovers rolls on. Mars 2020 will be long into its chilly retirement years before the first boots imprint the red sand, but its mission will go a long way toward getting them there.

12 | Jupiter and Saturn: Lords of the Planets

LEARNING GOALS

By reading the sections of this chapter, you will learn

12-1 What observations from Earth reveal about Jupiter and Saturn

12-2 How Jupiter and Saturn rotate differently from terrestrial planets like Earth

12-3 The nature of the immense storms seen in the clouds of Jupiter and Saturn

12-4 How the internal heat of Jupiter and Saturn drives activity in their atmospheres

12-5 What the Galileo space probe revealed about Jupiter's atmosphere

12-6 How the shapes of Jupiter and Saturn indicate the sizes of their rocky cores

12-7 How Jupiter and Saturn's intense magnetic fields are produced by an exotic form of hydrogen

12-8 The overall structure and appearance of Saturn's ring system

12-9 What kinds of particles form the rings of Jupiter and Saturn

12-10 How spacecraft observations revealed the intricate structure of Saturn's rings

12-11 How Saturn's rings are affected by the presence of several small satellites

Among the most remarkable sights in our solar system are the colorful, turbulent atmosphere of Jupiter and the ethereal beauty of Saturn's rings. Both of these giant worlds dwarf our own planet: More than 1200 Earths would fit inside Jupiter's immense bulk, and more than 700 Earths inside Saturn. Unlike Earth, both Jupiter and Saturn are composed primarily of the lightweight elements hydrogen and helium, in abundances very similar to those in the Sun.

As these images show, the rapid rotations of both Jupiter and Saturn stretch their weather systems into colorful bands that extend completely around each planet. Both planets also have intense magnetic fields that are generated in their interiors, where hydrogen is so highly compressed that it becomes a metal.

Saturn, however, is not merely a miniature version of Jupiter. Its muted colors and more flattened shape are clues that Saturn's atmosphere and interior have important differences from those of Jupiter. The most striking difference is Saturn's elaborate system of rings, composed of countless numbers of icy fragments orbiting in the plane of the planet's equator. The rings display a complex and elegant structure, which is shaped by subtle gravitational influences from Saturn's retinue of

Jupiter (left) and Saturn (right). (NASA/JPL/Space Science Institute)

R I V U X G

TABLE 12-1 Jupiter Data

Average distance from the Sun:	5.203 au = 7.783×10^8 km
Maximum distance from the Sun:	5.455 au = 8.160×10^8 km
Minimum distance from the Sun:	4.950 au = 7.406×10^8 km
Eccentricity of orbit:	0.048
Average orbital speed:	13.1 km/s
Orbital period:	11.86 years
Rotation period:	$9^h\ 50^m\ 28^s$ (equatorial) $9^h\ 55^m\ 29^s$ (internal)
Inclination of equator to orbit:	3.12°
Inclination of orbit to ecliptic:	1.30°
Diameter:	142,984 km = 11.209 Earth diameters (equatorial) 133,708 km = 10.482 Earth diameters (polar)
Mass:	1.899×10^{27} kg = 317.8 Earth masses
Average density:	1326 kg/m^3
Escape speed:	60.2 km/s
Surface gravity (Earth = 1):	2.36
Albedo:	0.44
Average temperature at cloudtops:	−108°C = −162°F = 165 K
Atmosphere composition: (by number of molecules)	86.2% hydrogen (H_2), 13.6% helium (He), 0.2% methane (CH_4), ammonia (NH_3), water vapor (H_2O), and other gases

R I **V** U X G

(Ocean/Corbis)

moons. Jupiter, too, has rings, but they are made of dark, dustlike particles that reflect little light. These systems of rings, along with the distinctive properties of the planets themselves, make Jupiter and Saturn highlights of our tour of the solar system.

12-1 Jupiter and Saturn are the most massive planets in the solar system

Jupiter and Saturn are respectively the largest and second largest of the planets, and the largest objects in the solar system other than the Sun itself. Astronomers have known for centuries about the huge diameters and immense masses of these two planets. Given the distance to each planet and their angular sizes, scientists used the small-angle formula (Box 1-1) to calculate that Jupiter is about 11 times larger in diameter than Earth, while Saturn's diameter is about 9 times larger than Earth's (see Figure 7-2). By observing the orbits of Jupiter's four large moons and Saturn's large moon Titan (see Table 7-2) and applying Newton's form of Kepler's third law (see Section 4-7 and Box 4-4), astronomers also determined that Jupiter and Saturn are, respectively, 318 times and 95 times more massive than Earth. In fact, Jupiter has 2½ times the combined mass of all the other planets, satellites, asteroids, meteoroids, and

comets in the solar system. A visitor from interstellar space might well describe our solar system as the Sun, Jupiter, and some debris! Table 12-1 and Table 12-2 list some basic data about Jupiter and Saturn.

Observing Jupiter and Saturn

As for any planet whose orbit lies outside Earth's orbit, the best time to observe Jupiter or Saturn is when they are at opposition. At opposition, Jupiter can appear nearly 3 times brighter than Sirius, the brightest star in the sky. Only the Moon and Venus can outshine Jupiter at opposition. Saturn's brightness at opposition is only about one-seventh that of Jupiter, but it still outshines all of the stars except Sirius and Canopus.

Through a telescope, Jupiter at opposition presents a disk nearly 50 arcsec in diameter, approximately twice the angular diameter of Mars under the most favorable conditions (Figure 12-1a). Because Jupiter takes almost a dozen Earth years to orbit the Sun, it appears to meander slowly across the 12 constellations of the zodiac at the rate of approximately one constellation per year. Successive oppositions occur at intervals of about 13 months.

Saturn's orbital period is even longer, more than 29 Earth years, so it moves even more slowly across the zodiac. If Saturn did not move at all, successive oppositions would occur exactly

TABLE 12-2 **Saturn Data**

Average distance from the Sun:	9.572 au = 1.432 × 10⁹ km
Maximum distance from the Sun:	10.081 au = 1.508 × 10⁹ km
Minimum distance from the Sun:	9.063 au = 1.356 × 10⁹ km
Eccentricity of orbit:	0.053
Average orbital speed:	9.64 km/s
Orbital period:	29.37 years
Rotation period:	10^h 13^m 59^s (equatorial)
	10^h 39^m 25^s (internal)
Inclination of equator to orbit:	26.73°
Inclination of orbit to ecliptic:	2.48°
Diameter:	120,536 km = 9.449 Earth diameters (equatorial)
	108,728 km = 8.523 Earth diameters (polar)
Mass:	5.685 × 10²⁶ kg = 95.16 Earth masses
Average density:	687 kg/m³
Escape speed:	35.5 km/s
Surface gravity (Earth = 1):	0.92
Albedo:	0.46
Average temperature at cloudtops:	−180°C = −292°F = 93 K
Atmosphere composition (by number of molecules):	96.3% hydrogen (H₂), 3.3% helium (He), 0.4% methane (CH₄), ammonia (NH₃), water vapor (H₂O), and other gases

R I V U X G
(NASA, ESA, and the Hubble Heritage Team [STScI/AURA])

one Earth year apart; thanks to its slow motion, Saturn's oppositions occur at intervals of about 1 year and 2 weeks. At opposition, Saturn appears as a disk about 20 arcsec in diameter, with a dramatic difference from Jupiter: Saturn is surrounded by a magnificent system of rings (Figure 12-1b). Later in this chapter, we will examine these rings in detail.

(a)

(b)

FIGURE 12-1 R I V U X G

Jupiter and Saturn as Viewed from Earth **(a)** The disk of Jupiter at opposition appears about 2½ times larger than **(b)** the disk of Saturn at opposition. Both planets display dark and light bands, though these are fainter on Saturn. (a: © Damian Peach; b: NASA)

Zones (light-colored)

Belts (dark-colored)

Shadow of Europa (a moon of Jupiter)

Great Red Spot

(a) Jupiter

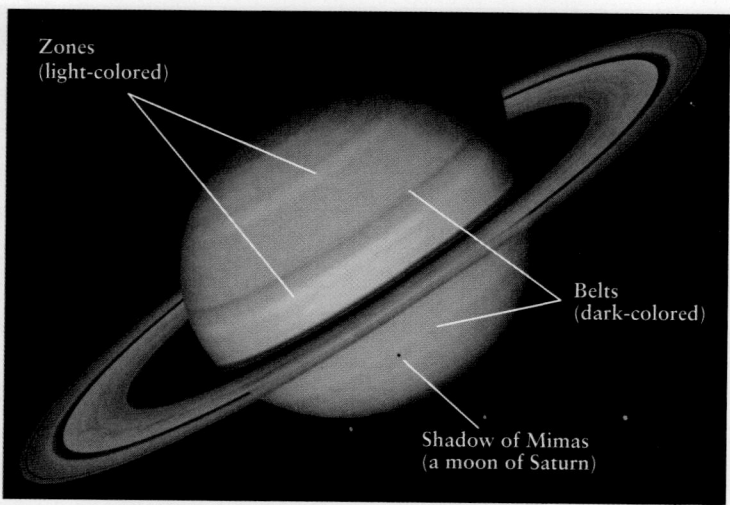

Zones (light-colored)

Belts (dark-colored)

Shadow of Mimas (a moon of Saturn)

(b) Saturn

VIDEO 12.1   VIDEO 12.2

FIGURE 12-2   R I **V** U X G

Jupiter and Saturn as Viewed from Space
(a) This view of Jupiter is a composite of four images made by the Cassini spacecraft as it flew past Jupiter in 2000. **(b)** Images from the Voyager 2 spacecraft as it flew past Saturn in 1981 were combined to show the planet in approximately natural color. (a: NASA/JPL/University of Arizona; b: JPL/NASA)

Dark Belts and Light Zones

As seen through an Earth-based telescope, both Jupiter and Saturn display colorful bands that extend around each planet parallel to its equator (Figure 12-1). More detailed close-up views of Jupiter from a spacecraft (**Figure 12-2a**) show alternating dark and light bands parallel to Jupiter's equator in subtle tones of red, orange, brown, and yellow. The dark, reddish bands are called **belts,** and the light-colored bands are called **zones.** The image of Saturn in Figure 12-2b shows similar features, but with colors that are much less pronounced. (We will see in Section 12-4 how differences between the atmospheres of Jupiter and Saturn can explain Saturn's washed-out appearance.)

In addition to these conspicuous stripes, a huge, red-orange oval called the **Great Red Spot** is often visible in Jupiter's southern hemisphere. This remarkable feature was first seen by the English scientist Robert Hooke in 1664 but may be much older. It appears to be an extraordinarily long-lived storm in the planet's dynamic atmosphere. Saturn has no long-lived storm systems of this kind that we have ever observed. However, Saturn has a remarkable hexagonal pattern around its north pole that we'll discuss in Section 12-3.

> Even a small telescope reveals colored bands on Jupiter and rings around Saturn.

12-2 Unlike the terrestrial planets, Jupiter and Saturn exhibit differential rotation

Observations of features like the Great Red Spot and smaller storms allow astronomers to determine how rapidly Jupiter and Saturn rotate. At its equator, Jupiter completes a full rotation in only 9 hours, 50 minutes, and 28 seconds, making it not only the largest and most massive planet in the solar system but also the one with the fastest rotation. However, Jupiter rotates in a strikingly different way from Earth, the Moon, Mercury, Venus, or Mars.

Differential Rotation

If Jupiter were a solid body like a terrestrial planet (or, for that matter, a billiard ball), all parts of Jupiter's surface would rotate through one complete circle in this same amount of time (**Figure 12-3a**). But by watching features in Jupiter's cloud cover, Giovanni Domenico Cassini discovered in 1690 that the polar regions of the planet rotate a little more slowly than do the equatorial regions. Near the poles, the rotation period of Jupiter's atmosphere is about 5 minutes longer than at the equator (about 9 hours, 55 minutes, and 41 seconds). Saturn, too, has a longer rotation period near its poles, by about 25 minutes (10 hours, 39 minutes, and 24 seconds near the poles compared to 10 hours, 13 minutes, and 59 seconds at the equator).

ANALOGY

You can see this kind of rotation, called **differential rotation,** in the kitchen. If you drop some bread crumbs into a pot of water and give it a stir, different parts of the liquid take different amounts of time to make one "rotation" around the center of the pot (Figure 12-3b). Differential rotation shows that neither Jupiter nor Saturn can be completely solid throughout their volumes: They must be at least partially fluid, like water in a pot.

Solid rotation typifies the terrestrial planets: Every part of the object takes exactly the same time to complete one rotation.

Differential rotation typifies Jupiter and Saturn: Particles at different locations in the fluid take different lengths of time to complete one rotation.

(a)   (b)

FIGURE 12-3

Solid Rotation versus Differential Rotation **(a)** All parts of a solid object rotate together, but **(b)** a rotating fluid displays differential rotation. To see differential rotation, put some grains of sand, bread crumbs, or other small particles in a pot of water. Stir the water with a spoon to start it rotating, then take out the spoon. The particles near the center of the pot take less time to make a complete rotation than those farther away from the center.

CONCEPT CHECK 12-1

Will Jupiter's moons of rock and ice, such as Ganymede, exhibit differential rotation?

Answer appears at the end of the chapter.

The Compositions of Jupiter and Saturn

If Jupiter and Saturn have partially fluid interiors, they cannot be made only of the rocky materials that constitute the terrestrial planets. An important clue to the compositions of Jupiter and Saturn is their average densities: only 1326 kg/m^3 for Jupiter and 687 kg/m^3 for Saturn. (By comparison, Earth's average density is 5515 kg/m^3.) To explain these low average densities, Rupert Wildt of the University of Göttingen in Germany suggested in the 1930s that Jupiter and Saturn are composed mostly of hydrogen and helium atoms—the two lightest elements in the universe—held together by their mutual gravitational attraction to form a planet. Wildt was motivated in part by his observations of prominent absorption lines of methane and ammonia in Jupiter's spectrum. (We saw in Section 7-3 how spectroscopy plays an important role in understanding the planets.) A molecule of methane (CH_4) contains four hydrogen atoms, and a molecule of ammonia (NH_3) contains three. The presence of these hydrogen-rich molecules was strong, but indirect, evidence of abundant hydrogen in Jupiter's atmosphere.

Direct evidence for hydrogen and helium in Jupiter's atmosphere, however, was slow in coming. The problem was that neither gas produces prominent spectral

> Spacecraft observations were needed to determine the compositions of Jupiter and Saturn.

lines in the visible sunlight reflected from the planet. To show the presence of these elements conclusively, astronomers had to look for spectral lines in the ultraviolet part of the spectrum. These lines are very difficult to measure from Earth, because almost no ultraviolet light can penetrate our atmosphere (see Figure 6-25). Astronomers first detected the weak spectral lines of hydrogen molecules in Jupiter's spectrum in 1960. The presence of helium on Jupiter and Saturn was finally confirmed in the 1970s and 1980s, when spacecraft first flew past these planets and measured their hydrogen spectra in detail.

Today, we know that the chemical composition of Jupiter's atmosphere is 86.2% hydrogen molecules (H_2), 13.6% helium atoms, and 0.2% methane, ammonia, water vapor, and other gases. The percentages in terms of *mass* are somewhat different because a helium atom is twice as massive as a hydrogen molecule. Hence, by mass, Jupiter's atmosphere is approximately 75% hydrogen, 24% helium, and 1% other substances, quite similar to that of the Sun. We will see evidence in Section 12-6 that Jupiter has a rocky core—large compared to Earth, but only a small fraction of Jupiter's volume—made of heavier elements. It is estimated that the breakdown by mass of the planet as a whole (atmosphere plus interior) is approximately 71% hydrogen, 24% helium, and 5% all heavier elements.

Saturn's Missing Helium

Like Jupiter, Saturn is thought to have a large rocky core. But unlike Jupiter, data from Earth-based telescopes and spacecraft show that the atmosphere of Saturn has a serious helium deficiency: Its chemical composition is 96.3% hydrogen molecules, 3.3% helium, and 0.4% other substances (by mass, 92% hydrogen, 6% helium, and 2% other substances). Saturn's atmosphere is a puzzle because Jupiter and Saturn are thought to have formed in similar ways from the gases of the solar nebula (see Section 8-4), thus both planets (and the Sun) should have essentially the same abundances of hydrogen and helium. So where did Saturn's helium go?

The explanation may be simply that Saturn is smaller than Jupiter, and as a result, Saturn probably cooled more rapidly. (We saw in Section 7-6 why a small world cools down faster than a large one.) This cooling would have triggered a process analogous to the way rain develops here on Earth. When the air is cool enough, humidity in Earth's atmosphere condenses into raindrops that fall to the ground. On Saturn, however, it is droplets of liquid helium that condense within the planet's cold, hydrogen-rich outer layers. In this scenario, helium is deficient in Saturn's upper atmosphere simply because it has fallen deeper into the planet. By contrast, Jupiter's helium has not rained out because its upper atmosphere is warmer and the helium does not form droplets.

ANALOGY

An analogy to helium "rainfall" within Saturn is what happens when you try to sweeten tea by adding sugar. If the tea is cold, the sugar does not dissolve well and tends to sink to the bottom of the glass even if you stir the tea with a spoon. But if the tea is hot, the sugar dissolves with only a little stirring. In the same way, it is thought that the descending helium droplets once again dissolve in hydrogen once they reach the warmer depths of Saturn's interior.

In this scenario, Jupiter and Saturn both have about the same overall chemical composition. But Saturn's smaller mass, less than one-third that of Jupiter, results in less gravitational force tending to compress its hydrogen and helium. This lack of compression explains why Saturn's density is only about half that of Jupiter, is even less than the density of water, and is in fact the lowest of any planet in the solar system.

CAUTION!

Because Jupiter and Saturn are almost entirely hydrogen and helium, it would be impossible to land a spacecraft on either planet. An astronaut foolish enough to try would notice the hydrogen and helium around the spacecraft becoming denser, the temperature rising, and the pressure increasing as the spacecraft descended. But the hydrogen and helium would never solidify into a surface on which the spacecraft could touch down. Long before reaching the planet's rocky core, the pressure of the hydrogen and helium would reach such unimaginably high levels that any spacecraft, even one made of the strongest known materials, would be crushed.

CONCEPT CHECK 12-2

If Jupiter and Saturn formed with nearly the same chemical composition, why might Jupiter be observed to have more helium than Saturn?

Answer appears at the end of the chapter.

12-3 Spacecraft images show remarkable activity in the clouds of Jupiter and Saturn

Most of our detailed understanding of Jupiter and Saturn comes from a series of robotic spacecraft that have examined these remarkable planets at close range. They found striking evidence of stable, large-scale weather patterns in both planets' atmospheres, as well as evidence of dynamic changes on smaller scales.

Spacecraft to Jupiter and Saturn

The first several spacecraft to visit Jupiter and Saturn each made a single flyby of the planet. Pioneer 10 flew past Jupiter in December 1973; it was followed a year later by the nearly identical Pioneer 11, which went on to make the first-ever flyby of Saturn in 1979. Also in 1979, another pair of spacecraft, Voyager 1 and Voyager 2, sailed past Jupiter. These spacecraft sent back spectacular close-up color pictures of Jupiter's dynamic atmosphere. Both Voyagers subsequently flew past Saturn.

The first spacecraft to go into Jupiter orbit was Galileo, which carried out an extensive program of observations from 1995 to 2003. The Cassini spacecraft went into orbit around Saturn in 2004. A cooperative project of NASA, the European Space Agency, and the Italian Space Agency, this spacecraft is named for astronomer Giovanni Domenico Cassini. On the way to its destination, it viewed Jupiter at close range, recording detailed images such as Figure 12-2a. The latest spacecraft, Juno, was launched in 2011

FIGURE 12-4   R I V U X G

Juno Spacecraft and a Whole New View of Jupiter Juno produces the highest resolution images of Jupiter ever taken, and the spacecraft orbits the poles to give views of Jupiter never seen before. This is an enhanced-color view, looking directly down at Jupiter's south pole (appearing blue). Cyclones (or hurricanes) were not expected at Jupiter's poles, but are seen here as circular or oval in shape and appear white, or a whiter shade within blue. The cyclones range in size, but can reach widths of 1000 km or so. Only longer-term observations will tell us if these cyclones are a persistent feature, or if they come and go. (NASA/JPL-Caltech/SwRI/MSSS/Betsy Asher Hall/Gervasio Robles)

and reached Jupiter in 2016 (Figure 12-4). By measuring Jupiter's composition, structure, and magnetic field, Juno is shedding light on planet formation, the early solar system, and magnetic dynamos. But as is often the case, data from Juno are also revealing more mysteries and raising more questions than expected.

Observing Jupiter's Dynamic Atmosphere

While the general pattern of Jupiter's atmosphere stayed the same during the 4 years between the Pioneer and Voyager flybys, there were some remarkable changes in the area surrounding the centuries-old Great Red Spot. During the Pioneer flybys, the Great Red Spot was embedded in a broad white zone that dominated the planet's southern hemisphere (Figure 12-5a). By the time of the Voyager missions, a dark belt had broadened and encroached on the Great Red Spot from the north (Figure 12-5b). In the same way that colliding weather systems in our atmosphere can produce strong winds and turbulent air, the interaction in Jupiter's atmosphere between the belt and the Great Red Spot embroiled the entire region in turbulence (Figure 12-6). By 1995, the Great Red Spot was once again centered within a white

> Immense storms on Jupiter and Saturn can last for months or years.

The Great Red Spot lies within a white zone.

(a) Pioneer 11, December 1974

A dark belt has encroached on the Great Red Spot.

(b) Voyager 2, July 1979

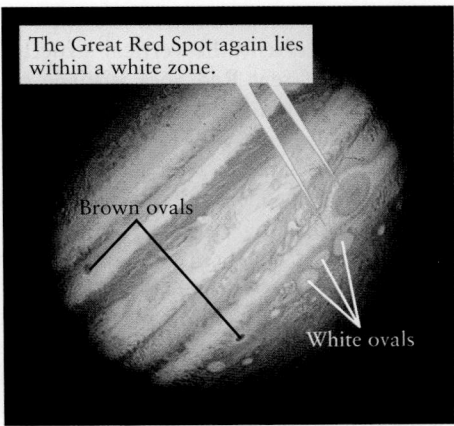

The Great Red Spot again lies within a white zone.

Brown ovals

White ovals

(c) HST, February 1995

FIGURE 12-5 R I **V** U X G

Jupiter's Changing Appearance These images from **(a)**, **(b)** spacecraft and **(c)** the Hubble Space Telescope (HST) show major changes in the planet's upper atmosphere over a 20-year period. (a and b: JPL/NASA; c: Reta Beebe, Amy Simon, New Mexico State University, and NASA)

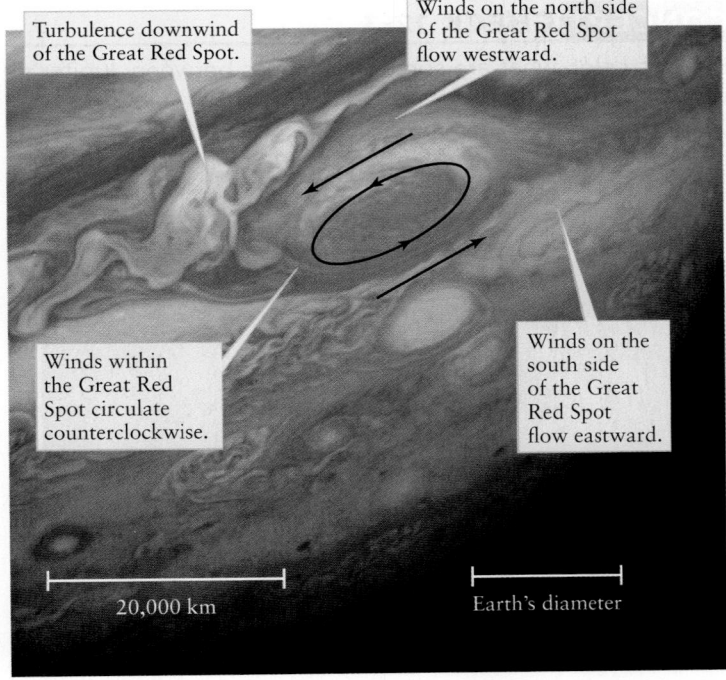

Turbulence downwind of the Great Red Spot.

Winds on the north side of the Great Red Spot flow westward.

Winds within the Great Red Spot circulate counterclockwise.

Winds on the south side of the Great Red Spot flow eastward.

20,000 km

Earth's diameter

(a)

(b)

VIDEO 12.3

FIGURE 12-6 R I **V** U X G

The Great Red Spot (a) This image of the Great Red Spot shows the counterclockwise circulation of gas in the Great Red Spot that takes about 6 days to make one rotation. The clouds that encounter the spot are forced to pass around it, and when other oval features are near it, the entire system becomes particularly turbulent, like batter in a two-bladed blender. **(b)** A close-up of the Great Red Spot made by the Juno spacecraft in 2017. It is about 1.3 times as wide as Earth. (a: NASA/JPL; b: NASA/JPL-Caltech/SwRI/MSSS/Jason Major)

zone (Figure 12-5c); when Cassini flew past Jupiter in 2000, the dark belt to the north of the Great Red Spot embroiled the entire region in turbulence.

Over the past three centuries, Earth-based observers have reported many long-term variations in the Great Red Spot's size and color. At its largest, it measured 40,000 by 14,000 km—so large that three Earths could fit side by side across it! At other times (as in 1976 and 1977), the spot almost faded from view. During the Voyager flybys of 1979, the Great Red Spot was comparable in size to Earth (see Figure 12-6).

Clouds at different heights within Jupiter's atmosphere reflect different wavelengths of infrared light. Using this effect, astronomers used an infrared telescope on board the Galileo space-craft to help clarify the vertical structure of the Great Red Spot. Most of the Great Red Spot is made of clouds at relatively high altitudes, surrounded by a collar of very low-level clouds about 50 km (30 mi, or 160,000 ft) below the high clouds at the center of the spot. This same kind of structure is seen in high-pressure areas in Earth's atmosphere, although on a much smaller scale.

Cloud motions in and around the Great Red Spot reveal that the spot rotates counterclockwise and takes about 6 days to complete a full rotation. Furthermore, winds to the north of the spot blow to the west, and winds south of the spot move toward the east. The circulation around the Great Red Spot is thus like a wheel spinning between two oppositely moving surfaces (see Figure 12-6). Weather patterns on Earth tend to change charac-ter and eventually dissipate when they move between plains and mountains or between land and sea. But because no solid surface or ocean underlies Jupiter's clouds, no such changes can occur for the Great Red Spot—which may explain how this wind pattern has survived for at least three centuries.

Other persistent features in Jupiter's atmosphere are the **white ovals.** Several white ovals are visible in Figure 12-5c. As in the Great Red Spot, the wind flow in white ovals is counterclockwise. White ovals are also apparently long lived; Earth-based observers have reported seeing them in the same location since 1938.

Most of the white ovals are observed in Jupiter's southern hemisphere, whereas **brown ovals** are more common in Jupiter's northern hemisphere. Brown ovals appear dark in a visible-light image like Figure 12-5c, but they appear bright in an infrared image. For this reason, brown ovals are understood to be holes in Jupiter's cloud cover. They permit us to see into the depths of the jovian atmosphere, where the temperature is higher and the atmosphere emits infrared light more strongly. White ovals, by contrast, have relatively low temperatures. They are areas with cold, high-altitude clouds that block our view of the lower levels of the atmosphere.

Between 1998 and 2000, three white ovals merged in Jupiter's southern hemisphere. The merger of these ovals, each of which had been observed for 60 years, led to the formation of a massive storm about half the size of the Great Red Spot (**Figure 12-7**). Time will tell whether this new feature—called Red Spot Jr.—is as long lived as its larger cousin, and this storm was still brewing when Juno viewed it in 2017 (Figure 12-7e). Even Jupiter's large belts can disappear. In 2009, Jupiter's dark southern belt began to fade and was completely gone by June 2010 (**Figure 12-8**). This disappearing act has occurred about 14 times in the last century.

CONCEPT CHECK 12-3

Besides color, what are some differences between Jupiter's white ovals and brown ovals?

Answer appears at the end of the chapter.

FIGURE 12-7 R I **V** U X G

Creating Red Spot Jr. **(a)** Three white ovals (FA, DE, BC); **(b)** DE and BC merge to form BE; **(c)** FA moves closer to BE; **(d)** FA and BE merge to form BA. For 60 years prior to 1998, the three white ovals labeled FA, DE, and BC traveled together at the same latitude on Jupiter. Between 1998 and 2000, they combined into one white oval, labeled BA, which **(e)** became a red spot, named Red Spot Jr., in 2006. **(f)** In 2017, Juno confirmed that Red Spot Jr. is still going strong. (a–d: NASA/JPL/WFPC2; e: NASA/ESA/Amy Simon-Miller, NASA/ESA/Amy Simon-Miller; f: NASA/JPL-Caltech/SwRI/MSSS/Roman Tkachenko)

July 23, 2009   June 7, 2010

FIGURE 12-8 R I V U X G

Dark Belt Fades on Jupiter The south equatorial belt completely faded from view between mid-2009 and mid-2010. White-colored ammonia ice crystals appeared at higher elevations, blocking the darker belt from view, but the full mechanism is not understood. By late 2010, the belt began to reappear. (NASA, ESA, M.H. Wong [University of California, Berkeley], H.B. Hammel [Space Science Institute, Boulder, Colo.], A.A. Simon-Miller [Goddard Space Flight Center], and the Jupiter Impact Science Team)

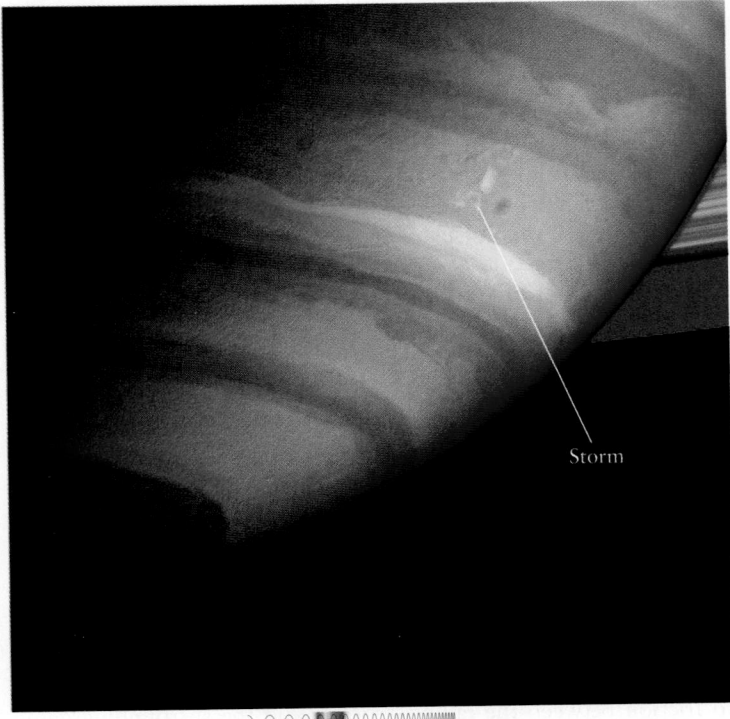

Storm

VIDEO 12.5

FIGURE 12-9 R I V U X G

A New Storm on Saturn This infrared image from the Cassini spacecraft shows a storm the size of Earth that appeared in Saturn's southern hemisphere in 2004. Cassini has observed several storms in this same region of Saturn. (NASA/JPL/Space Science Institute)

Storms and Patterns on Saturn

While Saturn has belts and zones like Jupiter, it has no storm systems as long lived as Jupiter's Great Red Spot. But about every 30 years—roughly the orbital period of Saturn—Earth-based observers have reported seeing storms in Saturn's clouds that last for several days or even months. Figure 12-9 shows one such storm that appeared in Saturn's southern hemisphere in 2004.

VIDEO 12.4

Saturn's storms are thought to form when warm atmospheric gases rise upward and cool, causing gaseous ammonia to solidify into crystals and form white clouds. On Earth, similar rapid lifting of air occurs within thunderstorms and can cause water droplets to solidify into hailstones. Earth thunderstorms produce electrical discharges (lightning) that generate radio waves, which you hear as "static" on an AM radio. Radio receivers on board Cassini have recorded similar static emitted by storms on Saturn, which strongly reinforces the idea that these storms are like thunderstorms on Earth.

Around Saturn's north pole, an enigmatic hexagonal pattern was glimpsed by Voyager in 1980. Cassini captured this pattern in greater detail, establishing that it has persisted for more than 26 years and survives the change of seasons on Saturn (Figure 12-10a). Amazingly, experiments on Earth that simulate Saturn's atmosphere have recreated this pattern (Figure 12-10b).

(a)

(b)

FIGURE 12-10 R V U X G

Hexagonal Pattern Around Saturn's North Pole **(a)** An infrared image made by Cassini. Each side of the hexagonal pattern is about the diameter of Earth. Inside the hexagon is a rapidly circulating jet stream. **(b)** Experiments with fluids have recreated the green hexagonal pattern in a cylinder. The cylinder, filled with fluid, spins slowly to simulate Saturn's overall rotation, while an inner ring is spun faster than the fluid to simulate the jet stream observed around Saturn's pole. Simulating Saturn's gaseous atmosphere can be performed with fluids because under the right conditions, fluids and gas behave similarly. (a: NASA/JPL/University of Arizona; b: Ana Claudia Barbosa Aguiar and Peter Read, Atmospheric, Oceanic and Planetary Physics, Clarendon Laboratory, University of Oxford)

12-4 The internal heat of Jupiter and Saturn has a major effect on the planets' atmospheres

TUTORIAL 12.1 Weather patterns on Earth are the result of the motions of air masses. Winter storms in the midwestern United States occur when cold, moist air moves southward from Canada; tropical storms in the southwest United States are caused by the northeastward motion of hot, moist air from Hawaii. All such motions, as well as the overall global circulation of our atmosphere, are powered by the Sun. Sunlight is absorbed by Earth's surface, and the heated surface in turn warms the atmosphere and stirs it into motion (see Sections 9-1 and 9-5). The energy of sunlight also powers the motions of the atmospheres of Venus and Mars (see Section 11-6). On both Jupiter and Saturn, however, atmospheric motions are powered both by solar energy and by the internal energy of the planet.

Jupiter and Saturn: Radiating Energy into Space

In the late 1960s, astronomers using Earth-based telescopes made the remarkable discovery that Jupiter emits more energy in the form of infrared radiation than it absorbs from sunlight—in fact, about twice as much. (By comparison, the internal heat that Earth radiates into space is only about 0.005% of what it absorbs from the Sun.) About 4.56 billion years ago when Jupiter formed, it would have been about twice as large. Since then, it has contracted under its own gravity, converting gravitational energy into heat (see Kelvin–Helmholtz contractions, Section 8-4). More specifically, Jupiter shrinks as its outer layers are pulled in closer, but as this gas picks up speed, collisions between gas molecules randomize their motion, thereby producing heat (see Figure 5-11). Because Jupiter is so large, it has retained substantial heat, or thermal energy, even after billions of years of cooling, so it still has plenty of energy to emit as infrared radiation. Even today, the contraction continues so that Jupiter shrinks by about 2 cm per year.

Saturn, too, radiates into space more energy than it receives as sunlight. Because Saturn is smaller than Jupiter, it should have begun with less internal heat trapped inside, and it should have radiated that heat away more rapidly. Hence, we would expect Saturn to be radiating very little energy today. Yet, when we take Saturn's smaller mass into account, we find that Saturn in fact releases about 25% *more* energy from its interior on a per-kilogram basis than does Jupiter. The explanation for this seeming paradox may be that helium is raining out of Saturn's upper atmosphere, as we described in Section 12-2. The helium condenses into relatively heavy droplets at cold upper altitudes, and generates heat at lower altitudes due to friction between the falling droplets and the surrounding gases. Thus, friction converts gravitational potential energy of falling droplets into heat. This heat eventually escapes from the planet's surface as infrared radiation. This "raining out" of helium from Saturn's upper layers is calculated to have begun 2 billion years ago. The amount of thermal energy released by this process adequately accounts for the extra heat radiated by Saturn since that time.

Heat naturally flows spontaneously from a hot place to a colder place, never the other way around. Hence, in order for heat to flow upward through the atmospheres of Jupiter and Saturn and radiate out into space, the temperature must be warmer deep inside the atmosphere than it is at the cloudtops. Infrared measurements have confirmed that the temperature within the atmospheres of both planets does indeed increase with increasing depth (**Figure 12-11**).

ANALOGY

You can understand the statement that "heat naturally flows from a hot place to a colder place" by visualizing an ice cube placed on a hot frying pan. Experience tells you that the hot frying pan cools down a little while the ice warms up and melts. If heat were to flow the other way, the ice would get colder and the frying pan would get hotter—which never happens in the real world.

Zones and Convection

When a fluid is warm at the bottom and cool at the top, like water being warmed in a pot, one effective way for energy to be distributed through the fluid is by the up-and-down motion called *convection*. (We introduced the idea of convection in Section 9-1; see, in particular, Figure 9-5.) For many years, it was thought that Jupiter's light-colored zones are regions where warm gas from low levels is rising and cooling, forming high-altitude clouds, while the dark-colored belts are regions where cool high-altitude gas is descending and being heated (**Figure 12-12**). However, this picture has been called into question with images made by Cassini during its 2000–2001 flyby of Jupiter, which show evidence for rising material in the dark belts.

At the time of this writing (2018), the Juno spacecraft has already collected data that further muddles our picture of Jupiter's belts and zones. It was expected that structure associated with the bands would disappear by a depth of about 100 km. However, Juno has found variations as far down as the spacecraft can measure, which is about 350 km. Clearly, there is more to learn about Jupiter's behavior.

The Fastest Winds in the Solar System

 In addition to the vertical motion of gases within the belts and zones, the very rapid rotation of both Jupiter and Saturn creates a global pattern of eastward and westward **zonal winds**. Wind speeds on Jupiter can exceed 500 km/h (300 mi/h). Earth's atmospheric circulation has a similar pattern of eastward and westward flow (see Figure 9-26), but with slower wind speeds. The faster winds on Jupiter presumably result from the planet's more rapid rotation, as well as from the substantial flow of heat from the planet's interior.

The zonal winds on Jupiter are generally strongest at the boundaries between belts and zones. Within a given belt or zone, the wind reverses direction between the northern and southern boundaries. As Figure 12-6 shows, such reversals in wind direction are associated with circulating storms such as the Great Red Spot.

Saturn rotates at about the same rate as Jupiter, but its interior releases less heat and it receives only one-quarter as much energy in sunlight as does Jupiter. With less energy available to power the motions of Saturn's atmosphere, we would expect its zonal winds to be slower than Jupiter's. Surprisingly, Saturn's winds are *faster*! In 1980, the Voyager spacecraft measured wind speeds near Saturn's equator that approach 1800 km/h (1100 mi/h), approximately two-thirds the speed of sound in Saturn's atmosphere and

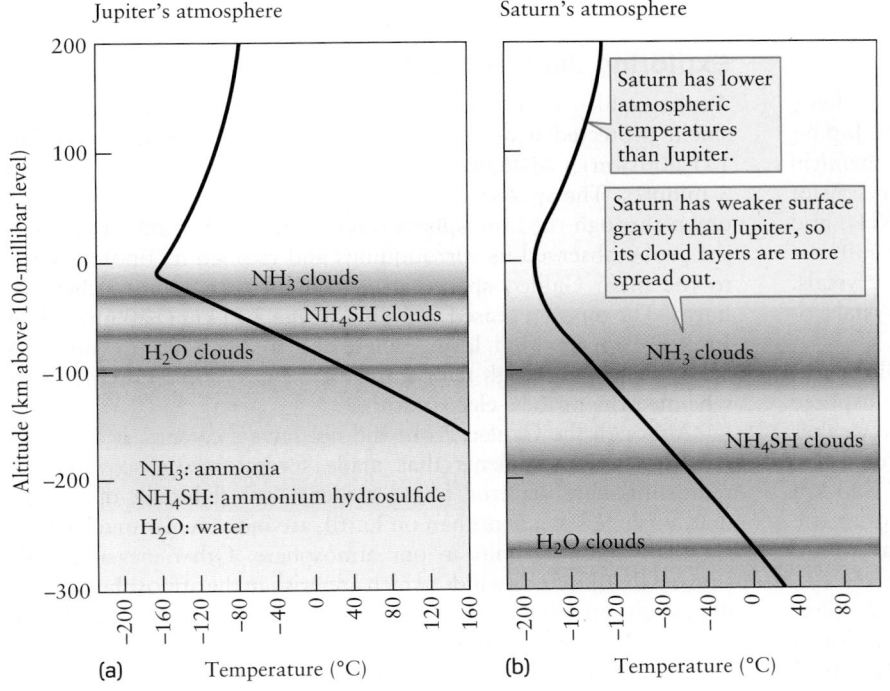

FIGURE 12-11

The Upper Atmospheres of Jupiter and Saturn The black curves in these graphs show temperature versus altitude in each atmosphere, as well as the probable arrangements of the cloud layers. Compared to Jupiter, the clouds that give these planets their colors are spread out over a larger range of altitude on Saturn. This reduces the appearance of any one color and gives Saturn a more washed-out appearance. Zero altitude in each atmosphere is chosen to be the point where the pressure is 100 millibars, or one-tenth of Earth's atmospheric pressure. Beneath both planets' cloud layers, the atmosphere is composed almost entirely of hydrogen and helium. (Data from A. P. Ingersoll)

East/west flow of belts and zones

North

South

Belt

Zone

?
?
?
?

Cool material sinking?
Warm material rising?

Upper atmosphere

To Jupiter's center

FIGURE 12-12

Original Model of Jupiter's Belts and Zones The light-colored zones and dark-colored belts in Jupiter's atmosphere were believed until recently to be regions of rising and descending gases, respectively. In the zones, gases warmed by heat from Jupiter's interior were thought to rise upward and cool, forming high-altitude clouds. In the belts, cooled gases were thought to descend and undergo an increase in temperature; the cloud layers seen there are at lower altitudes than in the zones. Observations by the Cassini spacecraft on its way to Saturn suggest that just the opposite might be correct, with the darker belts rising (these new doubts are indicated by the question marks in the figure above). Cassini was able to spot numerous small clouds—too small to be identified from Earth—rising within the dark belts, suggesting that the belts themselves might be rising.

faster than winds on any other planet. Since then, the equatorial winds seem to have abated somewhat. Between 1996 and 2004, astronomers using the Hubble Space Telescope showed that the wind speed had slowed to about 1000 km/h, and in 2005 Cassini measured wind speeds of about 1350 km/h at the equator. (These changes may be due to seasonal variations or changes in Saturn's clouds.) Why Saturn's winds should be even faster than Jupiter's is not yet completely understood.

Jupiter and Saturn's Cloud Layers

From spectroscopic observations and calculations of atmospheric temperature and pressure, scientists conclude that both Jupiter and Saturn have three main cloud layers of differing chemical composition. The uppermost cloud layer is composed of crystals of frozen ammonia. Deeper in the atmosphere, ammonia (NH_3) and hydrogen sulfide (H_2S)—a compound of hydrogen and sulfur—combine to produce ammonium hydrosulfide (NH_4SH) crystals. At even greater depths, the clouds are composed of crystals of frozen water.

Jupiter's strong surface gravity compresses these cloud layers into a region just 75 km deep in the planet's upper atmosphere (Figure 12-11a). But Saturn has a smaller mass and hence weaker surface gravity, so the atmosphere is less compressed and the same three cloud layers are spread out over a range of nearly 300 km (Figure 12-11b). The colors of Saturn's clouds are less dramatic than Jupiter's (see Figure 12-2) because deeper cloud layers are partially obscured by the hazy atmosphere above them.

The colors of clouds on Jupiter and Saturn depend on the temperatures of the clouds and, therefore, on the depth of the clouds within the atmosphere. Brown clouds are the warmest and are thus

the deepest layers that we can see. Whitish clouds form the next layer up, followed by red clouds in the highest layer. The whitish zones on each planet are therefore somewhat higher than the brownish belts, while the red clouds in Jupiter's Great Red Spot are among the highest found anywhere on that planet.

Despite all the data from various spacecraft, we still do not know what gives the clouds of Jupiter and Saturn their colors. The crystals of NH_3, NH_4SH, and water ice in the three main cloud layers are all white. Thus, other chemicals must cause the browns, reds, and oranges. The search continues, and progress in 2016 suggests that the Sun's ultraviolet radiation might break apart NH_3 and hydrocarbons, which in turn produce a new compound that creates Jupiter's colors.

CONCEPT CHECK 12-4

Why are Jupiter's cloud layers more compressed than Saturn's?

Answer appears at the end of the chapter.

12-5 The Galileo space probe explored Jupiter's deep atmosphere

By observing Jupiter's and Saturn's cloud layers from a distance, we have been able to learn a great deal about the atmospheres beneath the clouds. But there is no substitute for making measurements on site, and for many years scientists planned to send a spacecraft to explore deep into Jupiter's atmosphere. While still 81.52 million kilometers (50.66 million miles) from Jupiter, the Galileo spacecraft released the Galileo Probe, a cone-shaped body about the size of an office desk. While Galileo itself later fired its rockets to place itself into an orbit around Jupiter, in 1995 the Galileo Probe continued on a course that led it to a point in Jupiter's clouds just north of the planet's equator.

Exploring Jupiter's Atmosphere

As the Galileo Probe descended into Jupiter's atmosphere, a heat shield protected it as air friction slowed its descent speed from 171,000 km/h (106,000 mi/h) to 40 km/h (25 mi/h) in just 3 minutes. The spacecraft then deployed a parachute and floated down through the atmosphere (**Figure 12-13**). For the next hour, the probe observed its surroundings and radioed its findings back to the main Galileo spacecraft, which in turn radioed them to Earth. The mission ceased at a point some 200 km (120 mi) below Jupiter's upper cloud layer, where the tremendous pressure (24 atmospheres) and high temperature (152°C = 305°F) finally overwhelmed the probe's electronics.

Although the Galileo Probe did not have a camera, it did carry a variety of instruments that made several new discoveries. A radio-emissions detector found evidence for lightning discharges that, while less frequent than on Earth, are individually much stronger than lightning bolts in our atmosphere. Other measurements showed that Jupiter's winds, which are brisk in the atmosphere above the clouds, are even stronger beneath the clouds. The Galileo Probe measured a nearly constant wind speed of 650 km/h (400 mi/h) throughout its descent. This shows convincingly that the energy

FIGURE 12-13

The Galileo Probe Enters Jupiter's Atmosphere This artist's impression shows the Galileo Probe descending under a parachute through Jupiter's clouds. The jettisoned heat shield (shown to the right of the probe) protected the probe during its initial high-speed entry into the atmosphere. The Galileo Probe returned data for 58 minutes before it was crushed and melted by the pressure and temperature of the atmosphere. (NASA)

source for the winds is Jupiter's interior heat. If the winds were driven primarily by solar heating, as is the case on Earth, the wind speed would have decreased with increasing depth.

A central purpose of the Galileo Probe mission was to test the three-layer cloud model depicted in Figure 12-11a by making direct measurements of Jupiter's clouds. But the probe saw only traces of the NH_3 and NH_4SH cloud layers and found no sign at all of the low-lying water clouds. (One explanation of this surprising result is that the probe by sheer chance entered an unusually warm and cloud-free part of Jupiter's atmosphere.)

The Galileo Probe also made measurements of the chemical composition of Jupiter's atmosphere. These data revealed that the relative proportions of hydrogen and helium are almost exactly the same as in the Sun, a finding in line with the accepted picture of how Jupiter formed from the solar nebula (see Section 8-4). A number of heavy elements, including carbon, nitrogen, and sulfur, were found to be significantly more abundant on Jupiter than in the Sun, which was also an expected result. Over the past several billion years, Jupiter's strong gravity should have pulled in many pieces of interplanetary debris, thus building up a small but appreciable abundance of heavy elements. What happens to these heavy elements—whether they settle into the interior or become evenly mixed into the atmosphere—remains to be determined.

Noble Gases and the Origin of Jupiter

Another surprising and potentially important result from the Galileo Probe concerns three elements: argon (Ar), krypton (Kr), and xenon (Xe). These elements are three of the six so-called **noble gases,** which do not combine with other atoms to form molecules. (The noble gases, which also include helium, appear in the rightmost column of the periodic table in Box 5-5.) All three of these elements appear in tiny (or trace) amounts in Earth's atmosphere and in the Sun, and so must also have been present in the solar nebula. But the Galileo Probe found that argon, krypton, and xenon are about 3 times as abundant in Jupiter's atmosphere as in the Sun's atmosphere.

If these elements were incorporated into Jupiter directly from the gases of the solar nebula, they should be equally as abundant in Jupiter as in the Sun (which formed from the gas at the center of the solar nebula; see Section 8-4). So the excess amounts of argon, krypton, and xenon must have entered Jupiter in the form of solid planetesimals, just as did carbon, nitrogen, and sulfur. But

> Measuring trace amounts of rare gases on Jupiter provides clues about the origins of the planets.

at Jupiter's distance from the Sun, the presumed temperature of the solar nebula was too high to permit argon, krypton, or xenon to solidify. Left in their warmer gaseous form, it is unclear how Jupiter could have accumulated so much of these gases. How, then, did the excess amounts of these noble gases become part of Jupiter?

One hypothesis proposes that Jupiter originally formed much farther from the Sun. Out in the cold, remote reaches of the solar nebula, argon, krypton, and xenon could have solidified into icy particles and fallen into the young Jupiter, thus enriching it in those elements. Interactions between Jupiter and the material of the solar nebula could then have forced the planet to spiral inward, eventually reaching its present distance from the Sun. Although we saw in Section 8-6 that some migration of planets is expected, current models do not support Jupiter forming so far away. It's also possible that the noble gases were delivered within icy planetesimals. With Juno's arrival at Jupiter in 2016, we may soon learn more about the origin of planet's noble gases.

CONCEPT CHECK 12-5

Why is it difficult to understand that Jupiter has greater abundances of noble gases than the Sun?

Answer appears at the end of the chapter.

12-6 The oblateness of Jupiter and Saturn might reveal rocky cores

The Galileo Probe penetrated only a few hundred kilometers into Jupiter's atmosphere before it was crushed. To learn about Jupiter's structure at greater depths, as well as that of Saturn, astronomers must use indirect clues. The shapes of Jupiter and Saturn are important clues, because they indicate the size of the rocky cores at the centers of the planets. Now that the Juno spacecraft has reached Jupiter, we are learning even more about the planet's internal structure.

Oblateness and Rotation

Even a casual glance through a small telescope shows that Jupiter and Saturn are not spherical but slightly flattened or **oblate**. (You can see this in Figures 12-1 and 12-2.) The diameter across Jupiter's equator (142,980 km) is 6.5% larger than its diameter from pole to pole (133,700 km). Thus, Jupiter is said to have an **oblateness** of 6.5%, or 0.065. Saturn has an even larger oblateness of 9.8%, or 0.098, making it the most oblate of all the planets. By comparison, the oblateness of Earth is just 0.34%, or 0.0034.

If Jupiter and Saturn did not rotate, both would be perfect spheres. A massive, nonrotating object naturally settles into a spherical shape so that every atom on its surface experiences the same gravitational pull aimed directly at the object's center. Because Jupiter and Saturn do rotate, however, the body of each planet tends to bulge outward and away from the axis of rotation.

ANALOGY

You can demonstrate this effect for yourself. Stand with your arms hanging limp at your sides, then spin yourself around. Your arms will naturally tend to fly outward, away from the vertical axis of rotation of your body. Likewise, if you drive your car around a circular road, you feel thrown toward the outside of the circle; the car as a whole is rotating around the center of the circle, and you tend to move away from the rotation axis. (We discussed the physical principles behind this in Box 4-3.) For Jupiter and Saturn, this same effect deforms the planets into their nonspherical, oblate shapes.

Modeling Jupiter and Saturn's Interiors

The oblateness of a planet depends both on its rotation rate and on how the planet's mass is distributed throughout its volume. In general, the more a planet's mass is concentrated in a dense, central core, the less oblate (or more spherical) the planet's shape. The challenge to planetary scientists is to create a model of a planet's mass distribution that is consistent with the observed oblateness and rotation rate. One such model for Jupiter suggests that 4% of its mass is concentrated in a dense (but not necessarily solid) core made of rock and ice (Figure 12-14a). Although a core of this size range would have about 10 times the mass of Earth, the crushing weight of the remaining bulk of Jupiter would compress it to a diameter similar to the size of Earth! The pressure at the center of such a core is estimated to be about 70 million atmospheres, and the central temperature is probably about 22,000 K. By contrast, the temperature at Jupiter's cloudtops is only 165 K. Results from the Juno spacecraft are expected to allow for more accurate modeling of Jupiter's interior.

If Jupiter formed by accretion of gases onto a rocky "seed," any present-day core is presumably that very seed. Additional rocky material was presumably added later by meteoritic material that fell into the planet and sank to the center. If, however, Jupiter formed directly from the gases of the solar nebula, then the present-day core is more likely to be made of meteoritic material that later fell in. Material from icy planetesimals, too, would have

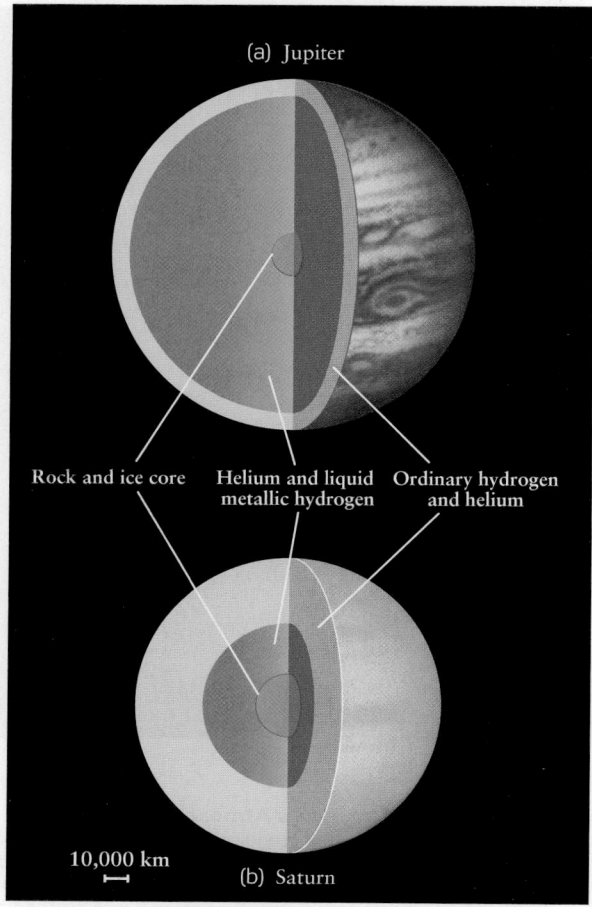

FIGURE 12-14

The Internal Structures of Jupiter and Saturn These diagrams of the interiors of Jupiter and Saturn are drawn to the same scale. Each planet's rock and ice core is surrounded by a layer of helium and liquid metallic hydrogen and a layer of helium and ordinary molecular hydrogen (H_2). Saturn's rock and ice core contains a larger fraction of the planet's mass than does Jupiter's, while Saturn has a smaller volume of liquid metallic hydrogen than does Jupiter.

sunk deep within Jupiter and added to the core. Despite the high pressures in the core, the temperature is so high that the "ices" are probably in a liquid state.

Analyzing Jupiter's core is one of the goals for the Juno spacecraft, and early results hint at a larger core than described here. With additional data collected by Juno in 2017, better models of Jupiter's core are expected.

Saturn rotates at about the same rate as Jupiter but has less mass, and therefore less gravity to pull its material inward. Hence, Saturn's rotation should cause material at its equator to bulge outward more than on Jupiter, giving Saturn a greater oblateness. In fact, if Saturn and Jupiter were to have the same internal structure, we would actually expect Saturn to be even more oblate than it really is. We can account for this discrepancy if Saturn has a different mass distribution than Jupiter. Detailed calculations suggest that about 10% of Saturn's mass is contained in its central core, compared to 4% for Jupiter's central core (see Figure 12-14b). Like Jupiter, the central core of Saturn is probably not solid and contains rock and ices.

If Jupiter's rotation rate was made large enough, could it be more oblate than Saturn?

Answer appears at the end of the chapter.

12-7 Metallic hydrogen inside Jupiter and Saturn endows the planets with strong magnetic fields

The oblateness of Jupiter and Saturn gives information about the size of the planets' dense cores. However, oblateness does not provide much insight into the dynamics of the liquid layer surrounding the planets' cores. Just as with Earth, a window into a planet's inner workings can come from a global magnetic field.

Jupiter's Magnetic Field

Measurements by the Pioneer and Voyager spacecraft detected a magnetic field and indicate that the field at Jupiter's equator is 14 times stronger than at Earth's equator. Like Earth's magnetic field, Jupiter's field is thought to be generated by motions of an electrically conducting fluid in the planet's interior. But wholly unlike Earth, the moving fluid within Jupiter is an exotic form of hydrogen called **liquid metallic hydrogen.**

As we discussed in Section 7-7, when hydrogen is squeezed under tremendous pressures, the atoms are pushed so close together that they behave like a liquid, conductive metal. This transition occurs when the pressure exceeds about 1.4 million atmospheres. This pressure occurs at depths more than about 7000 km below Jupiter's cloudtops. Thus, the internal structure of Jupiter likely consists of three distinct regions, as shown in Figure 12-14a: a central core of rock and ice, a layer of helium and liquid metallic hydrogen about 56,000 km thick, and a layer of helium and ordinary hydrogen about 7000 km thick. The colorful cloud patterns that we can see through telescopes are located in the outermost 100 km of the outer layer.

Figure 12-14a shows that much of the enormous bulk of Jupiter is composed of liquid metallic hydrogen. Jupiter's rapid rotation sets the liquid metallic hydrogen into motion, giving rise to a powerful magnetic dynamo 20,000 times stronger than Earth's and generating an intense magnetic field.

Once again, preliminary data from the Juno spacecraft has complicated our understanding of Jupiter. Not only is Jupiter's magnetic field stronger than we thought, but it appears to arise closer to the surface than expected. In particular, it might be originating from the outer layer that is currently modeled to contain ordinary hydrogen rather than from the electrically conductive metallic hydrogen. As of this writing (2017), Juno still has a lot of data to collect that will shed light on this topic.

Saturn's Magnetic Field

Saturn also has a substantial magnetic field, but its strength at the equator is only 5% as strong as Jupiter's equatorial field. This difference between the planets suggests that even though Saturn rotates nearly as rapidly as Jupiter, there must be some difference between their resulting magnetic dynamos. Cassini presented us with another mystery, which may or may not be related to Saturn's weak magnetic field. Theoretical modeling of magnetic dynamos indicates that the resulting magnetic field axis should not be aligned with the planet's rotational axis. This is consistent with the 11° difference between Earth's magnetic and rotational axes. At this time, the meager 0.06° difference between the axes measured by Cassini is not understood.

Jupiter's Magnetosphere

Jupiter's strong magnetic field surrounds the planet with a magnetosphere so large that it envelops the orbits of many of its moons (Figure 12-15). The Pioneer and Voyager spacecraft that carried instruments to detect charged particles

Both Jupiter and Saturn have magnetospheres that are millions of kilometers across.

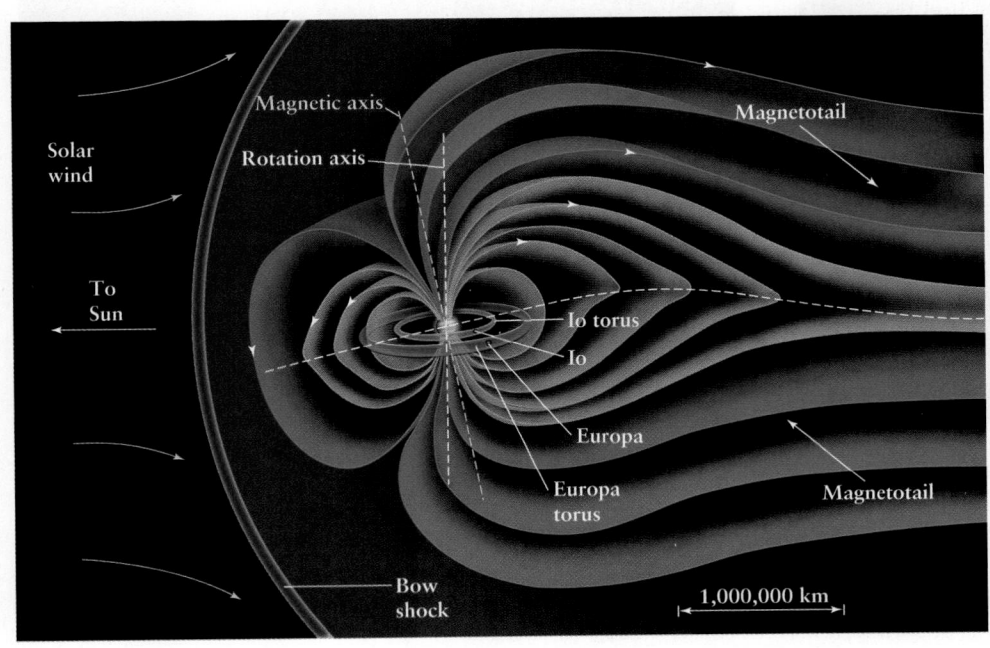

FIGURE 12-15

Jupiter's Magnetosphere Created by the planet's rotation, the ion-trapping regions of Jupiter's magnetosphere (in orange, analogous to the Van Allen belts) extend into the realm of the Galilean moons. Gases originating from Jupiter's moons Io and Europa form tori (doughnut-shaped regions) in the magnetosphere. Some of Io's particles are pulled by the field onto Jupiter. Swept back by the solar wind, the magnetosphere has a "magnetotail" pointing away from the Sun. The magnetotail is often over 500 million km long and sometimes it reaches all the way to Saturn. (J. Clarke, University of Michigan, and NASA) and magnetic fields revealed the awesome dimensions of the jovian magnetosphere. If you could see Jupiter's magnetosphere from Earth, it would cover an area in the sky 16 times larger than the Moon!

As charged particles in the solar wind collide with the jovian magnetosphere, they slow down, creating a region called a *bow shock* (see Section 9-4). The bow shock marks the transition between the magnetosphere and the surrounding space. Deeper within Jupiter's magnetosphere, charged particles—also referred to as ions—are trapped in belts that are analogous to Earth's Van Allen belts (recall Figure 9-20). An astronaut venturing into these belts would quickly receive a lethal dose of this charged-particle radiation.

A major source for charged particles in Jupiter's magnetosphere is the moon Io (see Table 7-2). Io's active volcanoes eject about a ton of material into Jupiter's magnetosphere each second. As shown in Figure 12-15, these particles form a torus.

Saturn's Magnetosphere

Saturn's magnetosphere is similar to Jupiter's, but is only about 10% to 20% as large and contains far fewer charged particles than Jupiter's. There are two reasons for this deficiency. First, Saturn lacks a highly volcanic moon like Io to inject particles into the magnetosphere. Second, and more important, many charged particles are absorbed by the icy particles that make up Saturn's rings (see Section 12-10). As with Jupiter, the charged particles that do exist in Saturn's magnetosphere are concentrated in radiation belts similar to the Van Allen belts in Earth's magnetosphere.

Aurorae on Jupiter and Saturn

On Earth, charged particles trapped in our planet's magnetic field stream onto the north and south magnetic poles, creating the aurora (see Section 9-4). The same effect takes place on both Jupiter and Saturn. **Figure 12-16** shows images from the Hubble Space Telescope that reveal glowing auroral rings centered on the magnetic poles of Jupiter and of Saturn. As on Earth, the emission comes from a region high in each planet's atmosphere, hundreds of kilometers above the cloudtops. However, preliminary data from Juno indicates that Jupiter's aurorae are created with different types of charged particles than those on Earth.

CONCEPT CHECK 12-7

Consider Figures 12-15 and 12-16a. Could charged particles from volcanoes on the moon Io lead to brighter aurorae on Jupiter?

Answer appears at the end of the chapter.

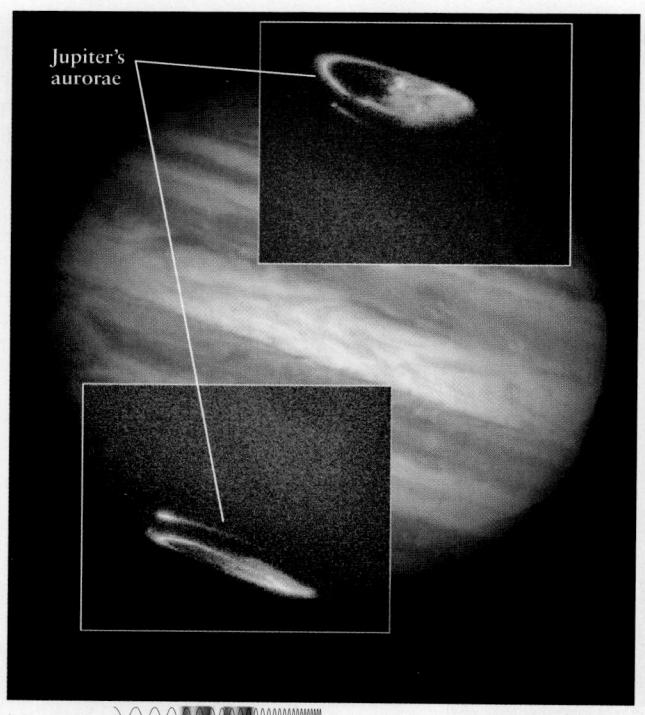

(a) Jupiter R I V U X G

Jupiter's aurorae

(b) Saturn R I V U X G

FIGURE 12-16

Aurorae on Jupiter and Saturn These images combine visible wavelengths of the planets and ultraviolet light from the aurorae. On both Jupiter and Saturn, charged particles from the magnetosphere are funneled onto the planet's magnetic poles. When the particles collide with and excite molecules in the upper atmosphere, the molecules emit ultraviolet light. In **(b)**, a large disturbance in the solar wind approaching Saturn coincided with the brightening of the aurora in the middle image of the three. (a: JPL/NASA/STScI, JPL/NASA/STScI, JPL/NASA/STScI; b: NASA/Hubble/Z. Levay and J. Clarke)

12-8 Earth-based observations reveal three broad rings encircling Saturn

TUTORIAL 12.2 Compared to Saturn, Jupiter is unquestionably larger, more massive, more colorful, and more dynamic in both its atmosphere and magnetosphere. But as seen through even a small telescope, Saturn stands out as perhaps the most beautiful of all the worlds of the solar system, thanks to its majestic rings (Figure 12-17). As we will see in Section 12-9, Jupiter also has rings, but they are so faint that they were not discovered until spacecraft could observe Jupiter at close range.

Discovering Saturn's Rings

In 1610, Galileo became the first person to see Saturn through a telescope. He saw few details, but he did notice two puzzling lumps protruding from opposite edges of the planet's disk. Curiously, these lumps disappeared in 1612, only to reappear in 1613. Other observers saw similar appearances and disappearances over the next several decades.

In 1655, the Dutch astronomer Christiaan Huygens began to observe Saturn with a better telescope than was available to any of his predecessors. (We discussed in Section 11-2 how Huygens used this same telescope to observe Mars.) On the basis of his observations, Huygens suggested that Saturn was surrounded by a thin, flattened ring. At times, this ring was edge on as viewed from Earth, making it almost impossible to see. At other times, Earth observers viewed Saturn from an angle either above or below the plane of the ring, and the ring was visible, as in Figure 12-1. (The lumps that Galileo saw were the parts of the ring to either side of Saturn, blurred by the poor resolution of his rather small telescope.) Astronomers confirmed this brilliant deduction over the next several years as they watched the ring's appearance change just as Huygens had predicted.

As the quality of telescopes improved, astronomers realized that Saturn's "ring" is actually a *system* of rings, as Figure 12-17 shows. In 1675, the Italian astronomer Giovanni Domenico Cassini discovered a dark division in the ring. The **Cassini division** is a gap about 4800 km wide that separates the outer **A ring** from the brighter **B ring** closer to the planet. In the mid-1800s, astronomers managed to identify the faint **C ring**, or *crepe* ring, that lies just inside the B ring. To appreciate the enormous scale of the ring system, consider that Mercury could fit snugly within the Cassini division!

Saturn and its rings are best seen when the planet is at or near opposition. A modest telescope gives a good view of the A and B rings, but a large telescope and excellent observing conditions are needed to see the C ring.

How the Rings Appear from Earth

Earth-based views of the Saturnian ring system change as Saturn slowly orbits the Sun. Huygens was the first to understand that this change occurs because the rings lie in the plane of Saturn's equator, and this plane is tilted 27° from the plane of Saturn's orbit. As Saturn orbits the Sun, its rotation axis and the plane of its equator maintain the same orientation in space. (The same is true for Earth; see Figure 2-12.) Hence, over the course of a Saturnian year, an Earth-based observer views the rings from various angles (Figure 12-18). At certain times, Saturn's north pole is tilted toward Earth and the observer looks "down" on the "top side" of the rings. Half a Saturnian year later, Saturn's south pole is tilted toward us and the "underside" of the rings is exposed to our Earth-based view.

> As seen from Earth, Saturn's rings change their orientation as the planet orbits the Sun.

When our line of sight to Saturn is in the plane of the rings, the rings are viewed edge on and seem to disappear entirely (see the images at the upper left and lower right corners of Figure 12-18). This disappearance indicates that the rings are very thin. In fact, they are thought to be only about 10 m thick (about 30 ft). In proportion to their diameter, Saturn's rings are thousands of times thinner than the sheets of paper used to print this book.

The last edge-on presentation of Saturn's rings of the twentieth century was in 1995–1996; the first two of the twenty-first century are in 2008–2009 and in 2025. From 1996 to 2008, we Earth observers saw the "underside" of the rings (as in the image

FIGURE 12-17 R I **V** U X G

Saturn's System of Rings This Cassini image shows many details of Saturn's rings. The C ring is so faint that it is almost invisible in this view. At 274,000 km, the outer diameter of the A ring is almost twice the diameter of Jupiter. Saturn is visible through the rings (look near the very bottom of the image), which shows that the rings are not solid. The ring system is so large that Mercury could fit inside the Cassini division. (NASA/cassini3d.com/images)

A ring

B ring

C ring

Cassini division

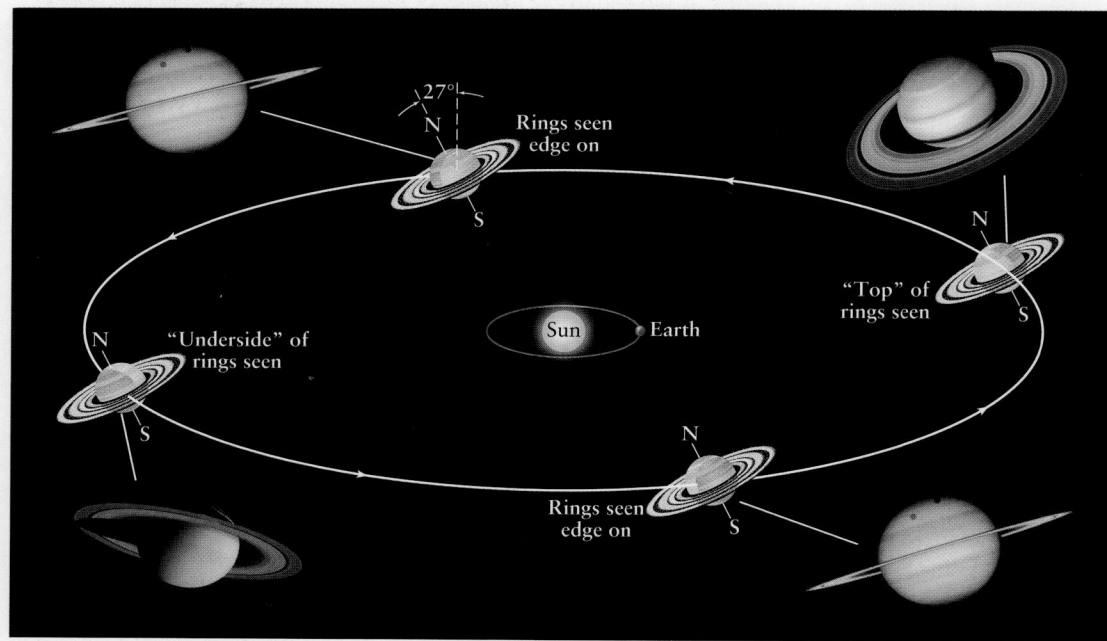

ANIMATION **FIGURE 12-18**
12.1
R I V U X G
**The Changing Appearance of
Saturn's Rings** Saturn's rings are
aligned with its equator, which is
tilted 27° from Saturn's orbital plane.
As Saturn moves around its orbit, the
rings maintain the same orientation
in space, so observers on Earth
see the rings at various angles. The
accompanying images of Saturn are
made from the "underside" of the
rings, the "top side," and nearly edge
on. (Top left & bottom right: NASA, ESA,
and the Hubble Heritage Team [STScI/
AURA]; top right & bottom left: NASA/
JPL/Space Science Institute)

at the lower left corner of Figure 12-18; since 2009, we have been seeing the "top" of the rings, and will continue to get this view until 2025.

CONCEPT CHECK 12-8

What causes Saturn's brilliant rings to be sometimes nearly invisible and at other times easily observable with the smallest of telescopes?

Answer appears at the end of the chapter.

12-9 Saturn's rings are composed of numerous icy fragments, while Jupiter's rings are made of small rocky particles

Astronomers have long known that Saturn's rings could not possibly be solid sheets of matter. In 1857, the Scottish physicist James Clerk Maxwell (whom we last encountered in Section 5-2) proved theoretically that if the rings were solid, differences in Saturn's gravitational pull on different parts of the rings would cause the rings to tear apart. He concluded that Saturn's rings are made of "an indefinite number of unconnected particles."

Ring Particles

In 1895, James Keeler at the Allegheny Observatory in Pittsburgh became the first to confirm by observation that the rings are not rigid. He arrived at this conclusion by photographing the spectrum of sunlight reflected from Saturn's rings. As the rings orbit Saturn, the spectral lines from the side approaching us are blueshifted

by the Doppler effect (recall Section 5-9, especially Figure 5-26). At the same time, spectral lines from the receding side of the rings are redshifted. Keeler noted that the size of the wavelength shift increases inward across the rings—the closer to the planet, the greater the shift. This variation in Doppler shift proves that the inner portions of Saturn's rings are moving around the planet more rapidly than the outer portions. Indeed, the orbital speeds across the rings are in complete agreement with Kepler's third law: The square of the orbital period about Saturn at any place in the rings is proportional to the cube of the distance from Saturn's center (see Section 4-7 and Box 4-4). This result is exactly what would be expected if the rings consisted of numerous tiny "moonlets," or **ring particles**, each individually circling Saturn.

Saturn's rings are quite bright; they reflect 80% of the sunlight that falls on them. (By comparison, Saturn itself reflects 46% of incoming sunlight.) Astronomers therefore have long suspected that the ring particles are made of ice and ice-coated rock. This hunch was confirmed in the 1970s, when the American astronomers Gerard P. Kuiper and Carl Pilcher identified absorption features of frozen water in the rings' near-infrared spectrum. The Voyager and Cassini spacecraft have made even more detailed infrared measurements that indicate the temperature of the rings ranges from −180°C (−290°F) in the sunshine to less than −200°C (−330°F) in Saturn's shadow. Water ice is in no danger of melting or evaporating at these temperatures.

To determine the sizes of the particles that make up Saturn's rings, astronomers have analyzed the radio signals received from a spacecraft as it passed behind the rings. How easily radio waves can travel through the rings depends on the relationship between the wavelength and the particle size. The results show that **most of the particles range in size from pebble-sized fragments about 1 cm in diameter to chunks about 5 m across,** the size of large boulders. Most abundant are snowball-sized particles about 10 cm in diameter. Recall that Mercury could fit inside the Cassini division, a gap that appears quite narrow compared to the rings themselves, so seeing even the largest boulder-sized ring particles is not possible in images such as Figure 12-17.

The Roche Limit

All of Saturn's ring material may be ancient debris that failed to accrete (fall together) into satellites. The total amount of material in the rings is quite small. If Saturn's entire ring system were compressed together to make a satellite, it would be no more than 100 km (60 mi) in diameter. But, in fact, the ring particles are so close to Saturn that they will never be able to form moons.

To see why, imagine a collection of small particles orbiting a planet. Gravitational attraction between neighboring particles tends to pull the particles together. However, because the various particles are at differing distances from the parent planet, they also experience different amounts of gravitational pull from the planet. This difference in gravitational pull is a **tidal force** that tends to keep the particles separated. (We discussed tidal forces in detail in Section 4-8. You may want to review that section, and in particular Figure 4-27.)

> Tidal forces prevent the material of Saturn's rings from coalescing into a moon.

The closer a pair of particles is to the planet, the greater the tidal force that tries to pull the pair apart. At a certain distance from the planet's center, called the **Roche limit,** the disruptive tidal force is just as strong as the gravitational force between the particles. (The concept of this limit was developed in the mid-1800s by the French mathematician Edouard Roche.) Inside the Roche limit, the tidal force overwhelms the gravitational pull between neighboring particles, and these particles cannot accrete to form a larger body. Instead, they tend to spread out into a ring around the planet. (The *Cosmic Connections* figure depicts this process.) Indeed, most of Saturn's system of rings visible in Figure 12-17 lies within the planet's Roche limit.

All large planetary moons are found outside their planet's Roche limit. If any large moon were to come inside its planet's Roche limit, the planet's tidal forces would cause the moon to break up into fragments. Such a catastrophic tidal disruption may be the eventual fate of the Martian moon Phobos and of Neptune's moon Triton.

CAUTION!

It may seem that it would be impossible for any object to hold together inside a planet's Roche limit. But the ring particles inside Saturn's Roche limit survive and do not break apart. The reason is that the Roche limit applies only to objects held together by the *gravitational* attraction of each part of the object for the other parts. By contrast, the forces that hold a rock or a chunk of ice together are *chemical* bonds between the object's atoms and molecules. These chemical forces are much stronger than the disruptive tidal force of a nearby planet, so a rock or chunk of ice does not break apart. In the same way, people walking around on Earth's surface (which is inside Earth's Roche limit) are in no danger of coming apart, because we are held together by comparatively strong chemical forces rather than gravity.

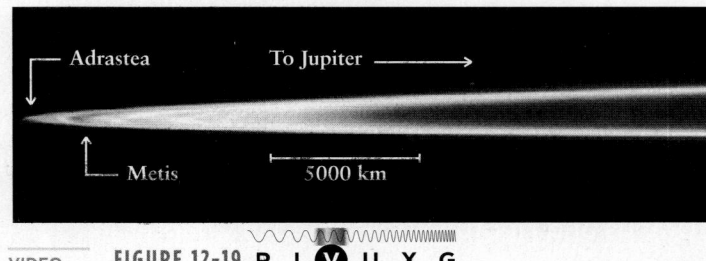

VIDEO 12.7

FIGURE 12-19 R I **V** U X G

Jupiter's Main Ring This Galileo image shows Jupiter's main ring almost edge on. The ring's outer radius, about 129,000 km from the center of Jupiter, is close to the orbit of Adrastea, the second closest (after Metis) of Jupiter's moons. Not shown is an even larger and even more tenuous pair of "gossamer rings," one of which extends out to 181,000 km and the other out to 222,000 km. (NASA/JPL/Cornell University)

Jupiter's Ring and the Roche Limit

When the Voyager 1 spacecraft flew past Jupiter in 1979, it trained its cameras not just on the planet but also on the space around the planet's equator. It discovered that Jupiter, too, has a system of rings that lies within its Roche limit (Figure 12-19). These rings differ from Saturn's in two important ways. First, Jupiter's rings are composed of tiny particles of rock with an average size of only about 1 μm (= 0.001 mm = 10^{-6} m) that reflect less than 5% of the sunlight that falls on them. Second, there is very little material in the rings of Jupiter, less than 1/100,000 (10^{-5}) the amount of material in Saturn's rings. As a result, Jupiter's rings are extremely faint, which explains why their presence was first revealed by a spacecraft rather than an Earth-based telescope. The ring particles are thought to originate from meteorite impacts on Jupiter's four small, inner moons, two of which are visible in Figure 12-19.

We will see in Chapter 14 that the rings of Uranus and Neptune are also made of many individual particles orbiting inside each planet's Roche limit. Like the rings of Jupiter, these rings are quite dim and difficult to see from Earth.

CONCEPT CHECK 12-9

If an asteroid were to enter a low-Earth orbit inside Earth's Roche limit, what would become of it?

Answer appears at the end of the chapter.

12-10 Saturn's rings consist of thousands of narrow, closely spaced ringlets

The photograph in Figure 12-1b, which was taken in 1974, indicates what astronomers understood about the structure of Saturn's rings in the mid-1970s. Each of the A, B, and C rings appeared to be rather uniform, with little or no evidence of any internal structure. Within a few years, however, close-up observations from spacecraft revealed the true complexity of the rings, as well as their chemical composition.

COSMIC CONNECTIONS

Planetary Rings and the Roche Limit

ANIMATION 12.2 If a small moon wanders too close to a planet, tidal forces will eventually tear the moon into smaller particles. These particles will form a ring around the planet. The critical distance from the planet at which this happens is called the Roche limit. This helps us understand why the rings of Jupiter and Saturn are made of small particles, as are the rings of Uranus and Neptune (discussed in Chapter 14).

If a satellite is outside the planet's Roche limit, the tidal forces exerted by the planet may deform the satellite but will not tear it apart.

If the satellite crosses the planet's Roche limit, the tidal forces overwhelm the gravitational forces that hold the satellite together and the satellite eventually fragments.

Fragments that are closer to the planet orbit faster (in accordance with Kepler's third law). This spreads the fragments around the planet . . .

. . . and the result is a ring.

Jupiter, Saturn, Uranus, and Neptune all have systems of rings that lie mostly within the Roche limit. This diagram shows each of the four ring systems scaled to the radius of the planet.

A planet's Roche limit is about 2.4 times the radius of the planet, provided the material orbiting the planet has the same density as the planet itself. For denser material, the Roche limit is closer to the planet.

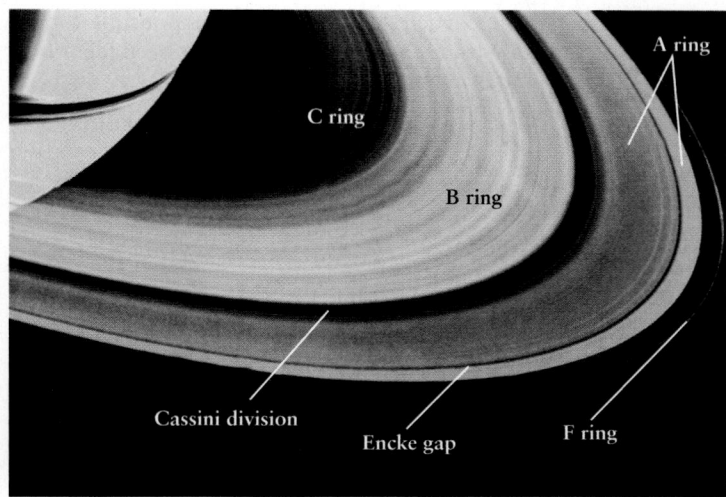

FIGURE 12-20 R I (V) U X G

Details of Saturn's Rings The colors in this Voyager 1 image have been enhanced to emphasize small differences between different portions of the rings. The broad Cassini division is clearly visible, as is the narrow Encke gap in the outer A ring. The very thin F ring lies just beyond the outer edge of the A ring. (NASA/JPL)

Spacecraft Views of Saturn's Rings

Pioneer 11, Voyager 1, and Voyager 2 recorded images of Saturn's rings during their flybys in 1979, 1980, and 1981, respectively. Pioneer 11 had only a relatively limited capability to make images, but cameras on board the two Voyager spacecraft sent back a number of pictures showing the detailed structure of Saturn's rings. Since entering Saturn orbit in 2004, the Cassini spacecraft has provided scientists with even higher-resolution images of the rings.

Figure 12-20 shows a Voyager 1 image. Some of the features seen in this image were expected, including the **Encke gap,** a 270-km-wide division in the outer A ring observed by the German astronomer Johann Franz Encke in 1838. But to the amazement of scientists, images like the one in Figure 12-20 revealed that the broad A, B, and C rings are not uniform at all; instead, they consist of hundreds upon hundreds of closely spaced bands or **ringlets.** These ringlets are arrangements of ring particles that have evolved from the combined gravitational forces of neighboring particles, of Saturn's moons, and of the planet itself.

Pioneer 11 first detected the narrow **F ring,** which you can see in Figure 12-20. It is only about 100 km wide and lies 4000 km beyond the outer edge of the A ring. Close-up views from the Voyagers and Cassini show that the F ring is made up of several narrow ringlets. Small moons that orbit close to the F ring exert gravitational forces on the ring particles, thus warping and deforming these ringlets (**Figure 12-21**).

Exploring Saturn's Ring Particles

Spacecraft have done more than show Saturn's rings in greater detail; they have also viewed the rings from perspectives not possible from Earth. Through Earth-based telescopes, we can see only the sunlit side of the rings. From this perspective, the B ring

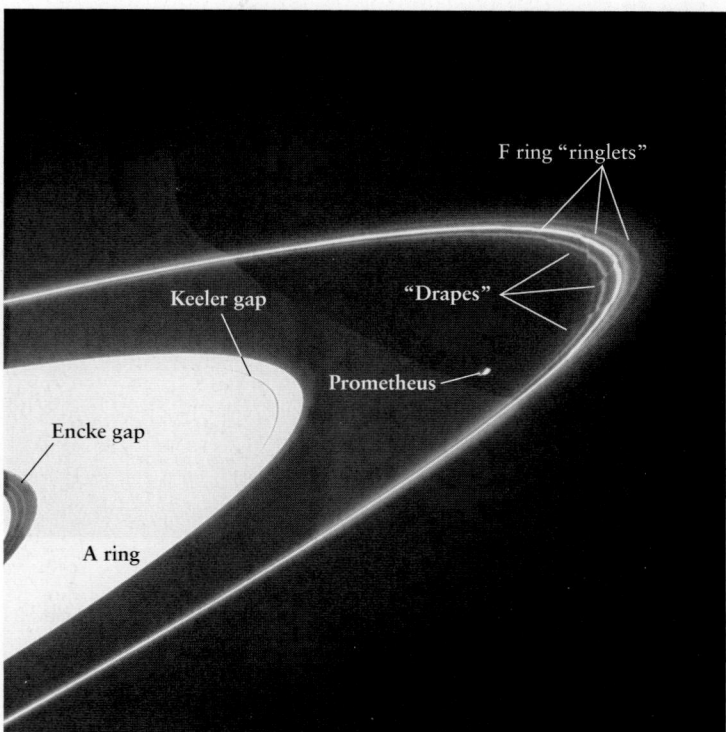

VIDEO 12.8

FIGURE 12-21 R I (V) U X G

Details of Saturn's F Ring This Cassini image shows Saturn's F ring (compare Figure 12-20). The bright central ringlet is about 50 km wide. As the small moon Prometheus orbits just inside the F ring, its gravitational pull causes disturbances called *drapes* in the ringlets. The narrowness of the F ring is the result of gravitational forces exerted by Prometheus and another small moon called Pandora (see Section 12-11). This image also shows two divisions in the A ring, the Encke gap and the Keeler gap. (NASA/JPL/Space Science Institute)

appears very bright, the A ring moderately bright, the C ring dim, and the Cassini division dark (see Figure 12-20). The fraction of sunlight reflected back toward the Sun (the *albedo*) is directly related to the concentration and size of the particles in the ring. The B ring is bright because it has a high concentration of relatively large, icy, reflective particles, whereas the darker Cassini division has a lower concentration of such particles.

The Voyagers and Cassini have expanded our knowledge of the ring particles by imaging the shaded side of the rings, which cannot be seen from Earth. Figure 12-22 shows such an image. Because the spacecraft was looking back toward the Sun, this picture shows sunlight that has passed through the rings. The B ring looks darkest in this image because little sunlight gets through its dense collection of particles. If the Cassini division were completely empty, it would look black, because we would then see through it to the blackness of empty space. But the Cassini division looks *bright* in Figure 12-22. This shows that the Cassini division is not empty but contains a relatively small number of particles. (In the same way, dust particles in the air in your room become visible when a shaft of sunlight passes through them.)

The process by which light bounces off particles in its path is called **light scattering.** The proportion of light scattered in various

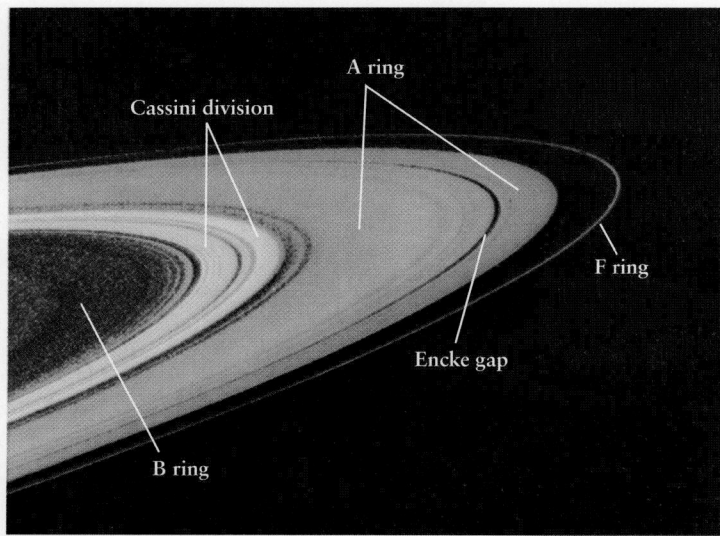

FIGURE 12-22 R I V U X G

The View from the Far Side of Saturn's Rings This false-color view of the side of the rings away from the Sun was taken by Voyager 1. The Cassini division appears white, not black like the empty gap between the A and F rings. Hence, the Cassini division cannot be empty, but must contain a number of relatively small particles that scatter sunlight like dust motes. (NASA/JPL/Space Science Institute)

directions depends on both the size of the particles and the wavelength of the light. (See Box 5-4 for examples of light scattering on Earth.) Because visible light has a much smaller wavelength than radio waves, light scattering allows scientists to measure the sizes of particles too small to detect using the radio technique described in Section 12-10. As an example, by measuring the amount of light scattered from the rings at various wavelengths and from different angles as the two Voyager spacecraft sped past Saturn, scientists showed that the F ring contains a substantial number of tiny particles about 1 μm (= 10^{-6} m) in diameter. This size is about the same as the size of the particles found in smoke.

There are also subtle differences in color from one ring to the next, as the computer-enhanced image in **Figure 12-23** shows. Although the main chemical constituent of the ring particles is frozen water, trace amounts of other chemicals—perhaps coating

the surfaces of the ice particles—are probably responsible for the different colors. These trace chemicals have not yet been identified.

The color variations in Figure 12-23 suggest that the icy particles do not migrate substantially from one ringlet to another. Had such migration taken place, the color differences would have been smeared out over time. The color differences may also indicate that different sorts of material were added to the rings at different times. In this scenario, the rings did not all form at the same time as Saturn but were added to over an extended period. New ring material could have come from small satellites that shattered after being hit by a stray asteroid or comet.

> Different colors in the rings reveal variations in chemical composition.

Discovering New Ring Systems

In addition to revealing new details about the A, B, C, and F rings, the Voyager cameras also discovered three new ring systems: the D, E, and G rings. The drawing in **Figure 12-24** shows the layout of all of Saturn's known rings, along with the orbits of some of Saturn's satellites. The **D ring** is Saturn's innermost ring system. It consists of a series of extremely faint ringlets located between the inner edge of the C ring and the Saturnian cloudtops. The **E ring** and the **G ring** both lie far from the planet, well beyond the narrow F ring. Both of these outer ring systems are extremely faint and fuzzy, but can be seen in **Figure 12-25**. Each lacks the ringlet structure so prominent in the main ring systems. The E ring encloses the orbit of Enceladus, one of Saturn's icy moons. Some scientists suspected that water ejected from Enceladus is the source of ice particles in the E ring, much as Io's volcanoes supply material to Jupiter's magnetosphere (see Section 12-8). Then, in 2010, the Cassini spacecraft observed directly that water geysers on Enceladus feed the E ring (Chapter 13).

CONCEPT CHECK 12-10

How would Figure 12-23 look different if ring particles migrated quickly through Saturn's rings?

Answer appears at the end of the chapter.

FIGURE 12-23 R I V U X G

Color Variations in Saturn's Rings This mosaic of Cassini images shows the rings in natural color. The color variations are indicative of slight differences in chemical composition among particles in different parts of the rings. (NASA/JPL/Space Science Institute)

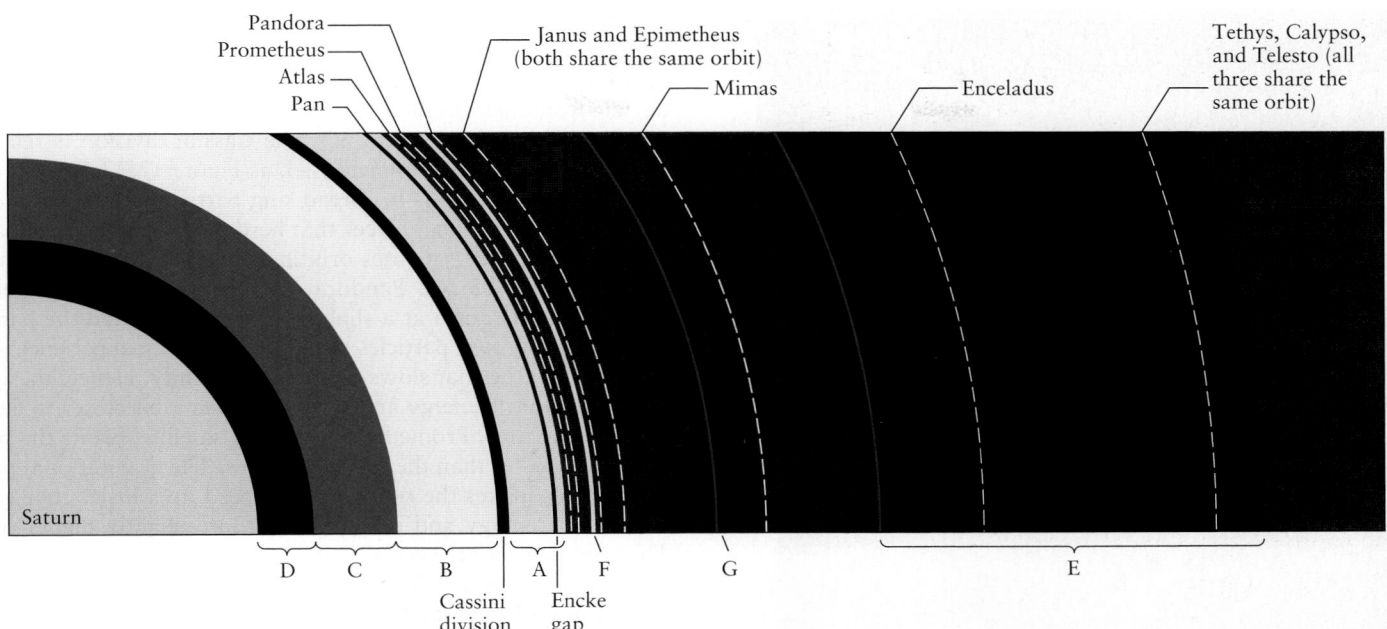

Pandora
Prometheus
Atlas
Pan
Janus and Epimetheus
(both share the same orbit)
Mimas
Enceladus
Tethys, Calypso,
and Telesto (all
three share the
same orbit)

Saturn

D   C   B   A   F   G   E

Cassini
division
Encke
gap

FIGURE 12-24

The Arrangement of Saturn's Rings This scale drawing shows the location of Saturn's rings, along with the orbits of the 11 inner satellites. Only the A, B, and C rings can readily be seen from Earth. The remaining rings are very faint, and their existence was only confirmed by spacecraft flybys.

Cassini
division   G ring   E ring

FIGURE 12-25 R I **V** U X G

Backlit Saturn With Saturn eclipsing the Sun, Cassini took 165 images to form this breathtaking color-enhanced portrait. The contrast was also increased to reveal fainter rings, and the outermost E ring is visible. The E ring contains water-ice particles *as well as* the moon Enceladus that ejected this water through ice volcanoes. The ring labels point to the prominent feature of each ring, although the rings are noticeably wide. (NASA/JPL/Space Science Institute)

12-11 Saturn's inner satellites affect the appearance and structure of its rings

Why do Saturn's rings have such a complex structure? The answer is that, like any object in the universe, the particles that make up the rings are affected by the force of gravity. Saturn's gravitational pull keeps the ring particles in orbit around the planet. But the satellites of Saturn also exert gravitational forces on the rings. These forces can shape the orbits of the ring particles and help give rise to the rings' structure.

Resonance and Shepherd Satellites

An arrangement in which one orbiting object has twice the orbital period of another is an example of an **orbital resonance.** (More generally, an orbital resonance occurs whenever the orbital periods of two objects are related by a ratio of small integers.) In an orbital resonance, gravitational interactions occur with just the right rhythm to transfer the greatest amount of energy. As an analogy, you naturally push someone on a swing in resonance with the swing's motion; in this way, a little energy is added with each push, and the swing gets very high with

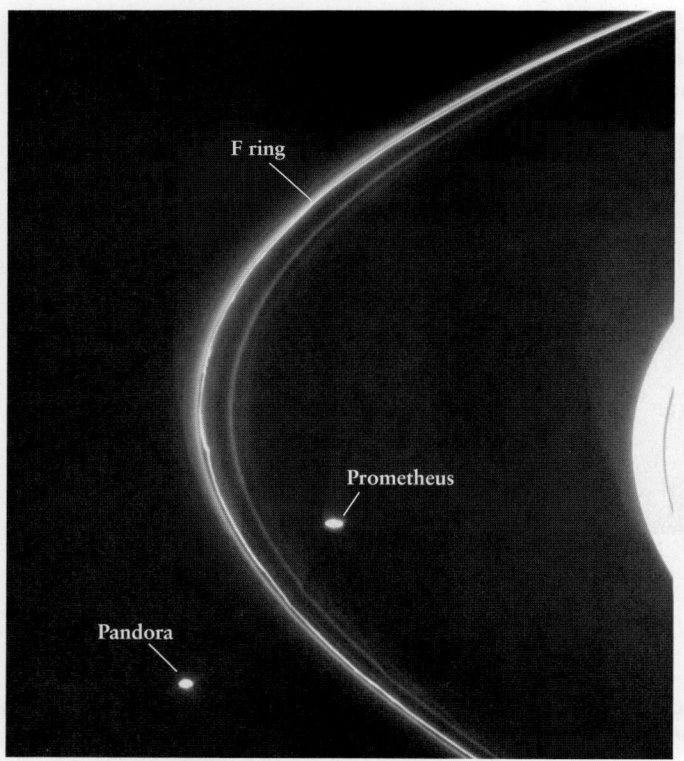

FIGURE 12-26 R I **V** U X G

Saturn's F Ring and Its Two Shepherds This Cassini image shows the tiny satellites Prometheus (102 km across) and Pandora (84 km across). (Compare with Figure 12-21.) These orbit Saturn on either side of its F ring, shepherding the ring particles into a narrow band. The satellites lap each other every 25 days. (NASA/JPL/Space Science Institute)

repeated pushes. Now let's see how the Cassini division results from an orbital resonance.

Astronomers have long known that one of Saturn's moons, Mimas, has a gravitational effect on Saturn's ring system. Mimas is a moderate-sized satellite that orbits Saturn every 22.6 hours. According to Kepler's third law, ring particles in the Cassini division—if they existed—would orbit Saturn approximately every 11.3 hours. Consequently, a given set of ring particles in the Cassini division would find itself between Saturn and Mimas on every second orbit (that is, every 22.6 hours). In other words,

particles in the Cassini division are in a 2-to-1 orbital resonance with Mimas. Because of these repeated alignments, the combined gravitational forces of Saturn and Mimas cause small particles to deviate from their original orbits, sweeping them out of the Cassini division. This helps explain why the Cassini division is relatively empty (although not completely so, as Figure 12-22 shows).

While Mimas tends to spread ring particles apart, other satellites exert gravitational forces that herd particles together. Cassini observed two such satellites orbiting on either side of the narrow F ring (Figure 12-26). Pandora, the outer of the two satellites, moves around Saturn at a slightly slower speed than the particles in the ring. As ring particles pass Pandora, they experience a tiny gravitational tug that slows them down slightly. Hence, these particles lose a little energy and fall into orbits a bit closer to Saturn. At the same time, Prometheus, the inner satellite, orbits the planet somewhat faster than the F ring particles. The gravitational pull of Prometheus makes the ring particles speed up a little, giving them some extra energy and nudging them into slightly higher orbits. The competition between the pulls exerted by these two satellites focuses the F ring particles into a well-defined, narrow band.

Because of their confining influence, Prometheus and Pandora are called **shepherd satellites.** We will see in Chapter 14 that Uranus is encircled by narrow rings that are likewise kept confined by shepherd satellites.

An effect similar to shepherding explains two divisions in the A ring, the Encke gap and the even narrower Keeler gap (Figure 12-21). In 2005, Cassini images revealed a miniature moon, Daphnis, which actually orbits inside the Keeler gap (Figure 12-27). In a fashion similar to the F ring's shepherds, Daphnis's gravitational forces cause ring particles outside its orbit to move farther out and cause those inside its orbit to move farther inward toward Saturn. In this way, Daphnis helps to maintain a gap in the rings that is 42 km (26 mi) wide, many times wider than Daphnis's 7-km (4-mi) diameter. In a similar way, a 20-km-wide moon called Pan orbits within and maintains the 270-km-wide Encke gap. Both of these moons cause wavelike disturbances in the ring material on either side of the gap through which they travel (see Figure 12-26).

In addition to their dynamic ring systems, Jupiter and Saturn together have an exotic collection of moons. These include

> Gravitational effects carve gaps in the rings and focus ring particles into narrow ringlets.

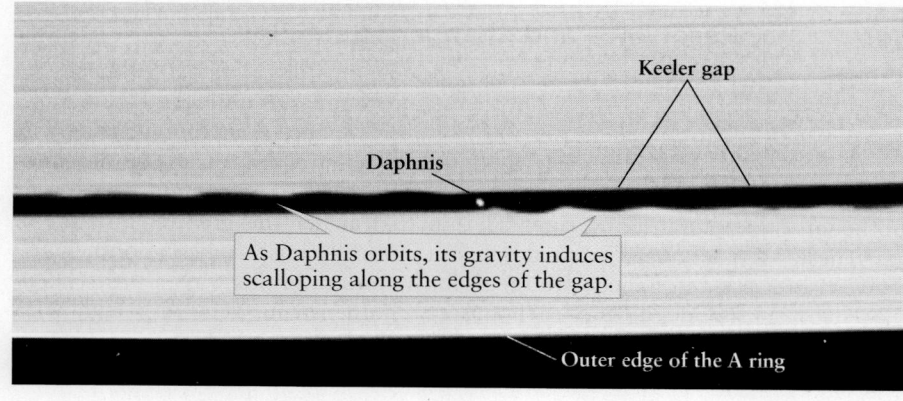

As Daphnis orbits, its gravity induces scalloping along the edges of the gap.

Keeler gap

Daphnis

Outer edge of the A ring

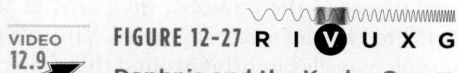

VIDEO 12.9 **FIGURE 12-27** R I **V** U X G

Daphnis and the Keeler Gap This Cassini spacecraft image shows Saturn's tiny moon Daphnis orbiting within the Keeler gap (see also Figure 12-21). Although it is just 7 km across, Daphnis exerts enough gravitational force to keep the 42-km-wide gap open and induce wavelike disturbances in the two edges of the gap. (NASA/JPL/Space Science Institute)

Saturn's moon Titan, with a thick atmosphere rich in hydrocarbons; Jupiter's satellite Io, the most volcanic world in the solar system; and Saturn's satellite Enceladus, which has exotic geysers that spew a mixture of water vapor and ice. We will examine these alien worlds in Chapter 13.

CONCEPT CHECK 12-11

If Saturn's moon Mimas orbited somewhat farther than it currently does, how would this affect the Cassini division?
Answer appears at the end of the chapter.

KEY WORDS

A ring, p. 361
B ring, p. 361
belts, p. 348
brown oval, p. 352
C ring, p. 361
Cassini division, p. 361
D ring, p. 366

differential rotation, p. 348
E ring, p. 366
Encke gap, p. 365
F ring, p. 365
G ring, p. 366
Great Red Spot, p. 348
light scattering, p. 365

liquid metallic hydrogen, p. 359
noble gases, p. 357
oblate, oblateness, p. 358
orbital resonance, p. 367
ring particles, p. 362
ringlets, p. 365

Roche limit, p. 363
shepherd satellite, p. 368
tidal force, p. 363
white oval, p. 352
zonal winds, p. 355
zones, p. 348

KEY IDEAS

Composition and Structure: Jupiter and Saturn are both much larger than Earth. Each is composed of 71% hydrogen, 24% helium, and 5% all other elements by mass. Both planets have a higher percentage of heavy elements than does the Sun (Sections 12-1, 12-2, and 12-6).

- Jupiter probably has a central core several times more massive than Earth, made of rock and "ices" (water, ammonia, methane, and associated compounds). On top of this rock and ice layer is a layer of helium and liquid metallic hydrogen, as well as an outermost layer composed primarily of ordinary hydrogen and helium. All of Jupiter's visible features are near the top of this outermost layer.

- Saturn's internal structure is similar to that of Jupiter, but its core makes up a larger fraction of its volume and its liquid metallic hydrogen mantle is shallower than that of Jupiter.

- Jupiter and Saturn both rotate so rapidly that these planets are noticeably flattened, or oblate.

- Both Jupiter and Saturn emit more energy than they receive from the Sun. Presumably both planets are still cooling.

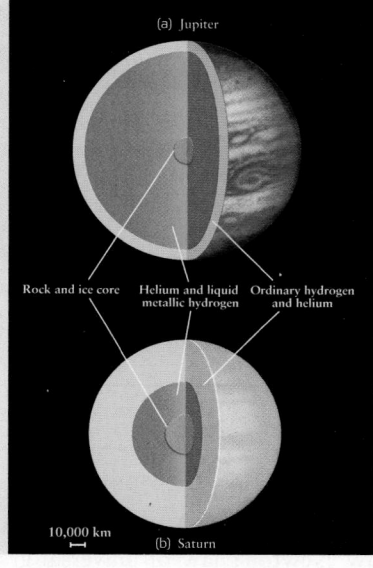

Atmospheres: The visible "surfaces" of Jupiter and Saturn are actually the tops of their clouds. The rapid rotation of the planets twists the clouds into dark belts and light zones that run parallel to the equator. Strong zonal winds run along the belts and zones (Sections 12-3, 12-4, and 12-5).

- The outer layers of both planets' atmospheres show differential rotation: The equatorial regions rotate slightly faster than the polar regions.

- The colored ovals visible in the jovian atmosphere represent gigantic storms. Some, such as the Great Red Spot, are quite stable and persist for many years. Storms in Saturn's atmosphere seem to be shorter lived.

- There are presumed to be three cloud layers in the atmospheres of Jupiter and Saturn. The reasons for the distinctive colors of these different layers are not yet known. The cloud layers in Saturn's atmosphere are spread out over a greater range of altitude than those of Jupiter, giving Saturn a more washed-out appearance.

- Saturn's atmosphere contains less helium than Jupiter's atmosphere. This lower abundance may be the result of helium raining downward into the planet.

NASA/JPL-Caltech/SwRI/MSSS/Jason Major

Magnetic Fields and Magnetospheres: Jupiter has a strong magnetic field created by currents in a metallic hydrogen layer. With less liquid metallic hydrogen, Saturn's magnetic field is much weaker than Jupiter's (Section 12-7).

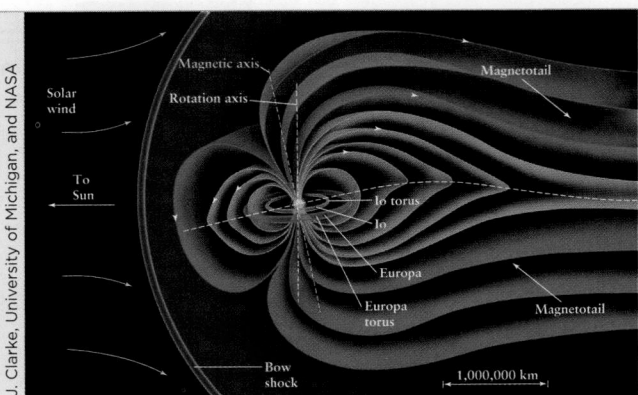

J. Clarke, University of Michigan, and NASA

Rings: Saturn is circled by a system of thin, broad rings lying in the plane of the planet's equator. This system is tilted away from the plane of Saturn's orbit, which causes the rings to be seen at various angles by an Earth-based observer over the course of a Saturnian year (Section 12-8).

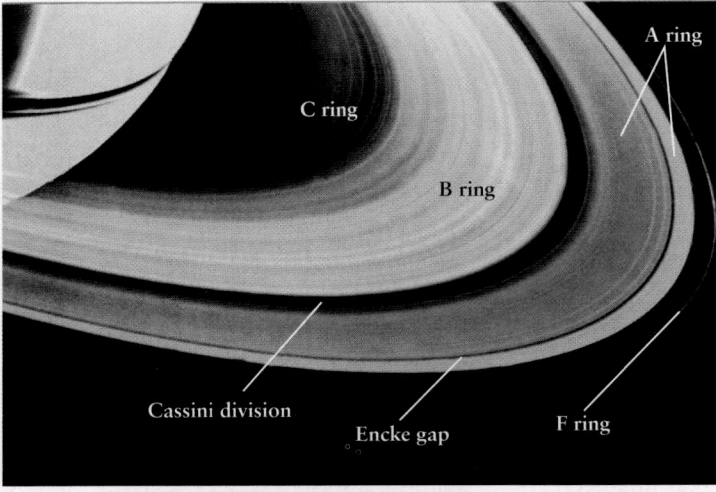

NASA/JPL

Structure of the Rings: Saturn has three major, broad rings (A, B, and C) that can be seen from Earth. Other, fainter rings were found by the Voyager spacecraft (Sections 12-9, 12-10, and 12-11).

- The principal rings of Saturn are composed of numerous particles of ice and ice-coated rock ranging in size from a few micrometers to about 10 m. Most of the rings exist inside the Roche limit of Saturn, where disruptive tidal forces are stronger than the gravitational forces attracting the ring particles to each other.

- Each of Saturn's major rings is composed of a great many narrow ringlets. The faint F ring, which is just outside the A ring, is kept narrow by the gravitational pull of shepherd satellites.

- Jupiter's faint rings are composed of a relatively small amount of small, dark, rocky particles that reflect very little light.

QUESTIONS

Problem-solving tips and tools

Box 4-4 describes how to use Newton's form of Kepler's third law. Newton's law of universal gravitation, discussed in Section 4-6, is the basic equation from which you can calculate a planet's surface gravity. Box 7-2 discusses escape speed, and Sections 5-3 and 5-4 discuss the properties of thermal radiation (including the Stefan–Boltzmann law, which relates the temperature of a body to the amount of thermal radiation that it emits).

Section 12-1

1. Mars passes closer to Earth than Jupiter does, but with an Earth-based telescope it is easier to see details on Jupiter than on Mars. Why is this?

2. Saturn is the most distant of the planets visible without a telescope. Is there any way we could infer this from naked-eye observations? Explain your answer. (*Hint:* Think about how

Saturn's position on the celestial sphere must change over the course of weeks or months.)

3. As seen from Earth, does Jupiter or Saturn undergo retrograde motion more frequently? Explain your answer.

4. The angular diameter of Jupiter at opposition varies little from one opposition to the next. By contrast, the angular diameter of Mars at opposition is quite variable. Explain why there is a difference between these two planets.

5. Jupiter was at opposition on May 9, 2018. On that date, Jupiter appeared to be in the constellation Ophiuchus. Approximately when will Jupiter next be at opposition in this same region of the celestial sphere? Explain your answer.

Section 12-2

6. Jupiter's equatorial diameter and the rotation period at Jupiter's equator are both given in Table 12-1. Use these data to calculate the speed at which an object at the cloudtops along Jupiter's equator moves around the center of the planet.

7. Suppose a cloud in Saturn's atmosphere could be observed rotating around Saturn's south pole (that is, near the pole,

far from the equator). About how long would it take for one rotation (rounded to the nearest minute)?

8. How could Saturn have approximately half the density of Jupiter if the two planets have similar compositions?

Section 12-3

9. What are the belts and zones in the atmospheres of Jupiter and Saturn? Is the Great Red Spot more like a belt or a zone? Explain your answer.

10. Give one possible explanation why weather systems on Jupiter are longer lived than weather systems on Earth.

11. What are white ovals and brown ovals? What can we infer about them from infrared observations?

12. Use the information given in Section 12-3 to estimate the wind velocities in the Great Red Spot, which rotates with a period of about 6 days.

13. If Jupiter emitted just as much energy per second (as infrared radiation) as it receives from the Sun, the average temperature at the outer reaches of the planet's atmosphere, where Jupiter's own blackbody radiation is emitted, would be about 107 K. Given that Jupiter actually emits twice this much energy per second from the same altitude, calculate what the average temperature at this height must actually be.

14. Describe some of the semipermanent features in Jupiter's atmosphere. Compare and contrast these long-lived features with some of the transient phenomena seen in Jupiter's clouds.

Section 12-4

15. **TUTORIAL 12.1** In what ways are the motions of Jupiter's atmosphere like the motion of water stirred in a pot (see Figure 12-3b)? In what ways are they different?

16. On a warm, humid day, water vapor remains in the atmosphere. But if the temperature drops suddenly, the water vapor forms droplets, clouds appear, and it begins to rain. Relate this observation to why there is relatively little helium in Saturn's atmosphere compared to the atmosphere of Jupiter.

17. Compare and contrast the source of energy for motions in Earth's atmosphere with the energy source for motions in the atmospheres of Jupiter and Saturn.

18. Both Jupiter and Saturn emit more energy than they receive from the Sun in the form of sunlight. Compare the internal energy sources of the two planets that produce this infrared emission.

19. What observations from the Cassini spacecraft contradict the accepted picture of zone and belt convection in Jupiter's atmosphere?

20. Compare the atmospheres of Jupiter and Saturn. Why does Saturn's atmosphere look "washed out" in comparison to that of Jupiter?

21. (a) From Figure 12-11a, by how much does the temperature increase as you descend from the 100-millibar level in Jupiter's atmosphere to an altitude 100 km below that level? (b) Use Figure 12-11b to answer the same question for Saturn's atmosphere. (c) In Earth's troposphere (see Section 9-5), the air temperature increases by 6.4°C for each kilometer that you descend. In which planet's atmosphere—Earth, Jupiter, or Saturn—does the temperature increase most rapidly with decreasing altitude?

Section 12-5

22. Which data from the Galileo Probe were in agreement with astronomers' predictions? Which data were surprising?

23. Fewer than one in every 10^5 atoms in Jupiter's atmosphere is an argon atom, and fewer than one in 10^8 is an atom of krypton or xenon. If these atoms are such a tiny portion of Jupiter's composition, why are scientists confused about their abundances? How do the abundances of these elements in Jupiter's atmosphere compare to the abundances in the Sun? What hypothesis has been offered to explain these observations?

Section 12-6

24. **TUTORIAL 12.1** Is the chemical composition of Jupiter as a whole the same as that of its atmosphere? Explain any differences.

25. Why is Jupiter oblate? What do astronomers learn from the value of Jupiter's oblateness?

26. Describe the internal structures of Jupiter and Saturn, and compare them with the internal structure of Earth.

27. Explain why Saturn is more oblate than Jupiter, even though Saturn rotates more slowly.

28. From the model described in Section 12-6, calculate the radius of Jupiter's central core if its density is 5 times Earth's average density. How does this compare with the radius of Earth? (See Table 9-1 for data about Earth.)

Section 12-7

29. What is liquid metallic hydrogen? What is its significance for Jupiter?

30. Compare and contrast Jupiter's magnetosphere with the magnetosphere of a terrestrial planet like Earth.

31. Why is Saturn's magnetosphere less extensive than Jupiter's?

32. In the outermost part of Jupiter's outer layer (shown in yellow in the upper part of Figure 12-14), hydrogen is principally in the form of molecules (H_2). Deep within the liquid metallic hydrogen layer (shown in orange in Figure 12-14), hydrogen is in the form of single atoms. Recent laboratory experiments suggest that there is a gradual transition between these two states, and that the transition layer overlaps the boundary between the ordinary hydrogen and liquid metallic hydrogen layers. Use this information to redraw the upper part of

Figure 12-13 and to label the following regions in Jupiter's interior: (i) ordinary (nonmetallic) hydrogen molecules; (ii) nonmetallic hydrogen with a mixture of atoms and molecules; (iii) liquid metallic hydrogen with a mixture of atoms and molecules; (iv) liquid metallic hydrogen atoms.

Section 12-8

33. **TUTORIAL 12.2** What observations of Saturn's rings proved that they are not solid?

34. If Saturn's rings are not solid, why do they look solid when viewed through a telescope?

35. When Saturn is at different points in its orbit, we see different aspects of its rings because the planet has a 27° tilt. If the tilt angle were different, would it be possible to see the upper and lower sides of the rings at all points in Saturn's orbit? If so, what would the tilt angle have to be? Explain your answers.

36. As seen from Earth, the intervals between successive edge-on presentations of Saturn's rings alternate between about 13 years, 9 months, and about 15 years, 9 months. Why do you think these two intervals are not equal?

Section 12-9

37. **TUTORIAL 12.2** Although the Voyager and Cassini spacecraft did not collect any samples of Saturn's ring particles, measurements from these spacecraft allowed scientists to determine the sizes of the particles. Explain how this was done.

38. The Space Shuttle and other spacecraft orbit Earth well within Earth's Roche limit. Explain why these spacecraft are not torn apart by tidal forces.

39. How do Jupiter's rings differ from those of Saturn?

40. (a) Use Newton's form of Kepler's third law to calculate the orbital periods of particles at the outer edge of Saturn's A ring and at the inner edge of the B ring. (b) Saturn's rings orbit in the same direction as Saturn's rotation. If you were floating along with the cloudtops at Saturn's equator, would the outer edge of the A ring and the inner edge of the B ring appear to move in the same or opposite directions? Explain your answer.

41. **VIDEO 12.10** This Cassini close-up image of Saturn's rings shows a number of bright, straight features elongated outward called *spokes*. As these features orbit around Saturn, they tend to retain their shape like the rigid spokes on a rotating bicycle wheel. The spokes rotate at the same rate as Saturn's magnetic field and are thought to be clouds of tiny, electrically charged particles kept in orbit by magnetic forces. Explain why the spokes could *not* maintain their shape if they were kept in orbit by gravitational forces alone.

R I V U X G
(NASA/JPL/Space Science Institute)

42. Explain how the Roche limit helps us understand Saturn's rings.

Section 12-10

43. Describe the structure of Saturn's rings. What evidence is there that ring particles do not migrate significantly between ringlets?

44. During the planning stages for the Pioneer 11 mission, when relatively little was known about Saturn's rings, it was proposed to have the spacecraft fly through the Cassini division. Why might this have been a bad idea?

Section 12-11

45. What is the relationship between Saturn's satellite Mimas and the Cassini division?

46. Why is the term *shepherd satellite* appropriate for the objects so named? Explain how a shepherd satellite operates.

47. The Cassini division involves a 2-to-1 resonance with Mimas. Does the location of the Encke gap—133,500 km from Saturn's center—correspond to a resonance with one of the other satellites? If so, which one? (See Appendix 3.)

General Problems

48. **TUTORIAL 12.1** Astronomers can detect the presence of hydrogen in stars by looking for the characteristic absorption lines of hydrogen in the star's visible spectrum (Figure 5-21). They can also detect hydrogen in glowing gas clouds by looking for hydrogen's characteristic emission lines (Figure 5-18).

Explain why neither of these techniques helped Earth-based astronomers to detect hydrogen in Jupiter's atmosphere. (*Hint:* Consider the conditions required for emission and absorption.)

49. What would happen if you tried to land a spacecraft on the surface of Jupiter?

50. Using orbital data for a jovian satellite of your choice (see Appendix 3), calculate the mass of Jupiter. How does your answer compare with the mass quoted in Table 12-1?

51. It has been claimed that Saturn would float if one had a large enough bathtub. Using the mass and size of Saturn given in Table 12-2, confirm that the planet's average density is about 690 kg/m^3, and comment on this somewhat fanciful claim.

52. Roughly speaking, Jupiter's composition (by mass) is three-quarters hydrogen and one-quarter helium. The mass of a single hydrogen atom is given in Appendix 7; the mass of a single helium atom is about 4 times greater. Use these numbers to calculate how many hydrogen atoms and how many helium atoms there are in Jupiter.

53. An astronaut floating above Saturn's cloudtops would see a blue sky, even though Saturn's atmosphere has a very different chemical composition than Earth's. Explain why. (*Hint:* See Box 5-4.)

54. Consider a hypothetical future spacecraft that would float, suspended from a balloon, for extended periods in Jupiter's upper atmosphere. If we want this spacecraft to return to Earth after completing its mission, calculate the speed at which the spacecraft's rocket motor would have to accelerate it in order to escape Jupiter's gravitational pull. Compare with the escape speed from Earth, equal to 11.2 km/s.

55. The Galileo Probe had a mass of 339 kg. On Earth, its weight (the gravitational force exerted on it by Earth) was 3320 newtons, or 747 lb. What was the gravitational force that Jupiter exerted on the Galileo Probe when it entered Jupiter's clouds?

56. Suppose you were asked to design a mission to Jupiter involving an unmanned airplanelike vehicle that would spend many days (months?) flying through the jovian clouds. What observations, measurements, and analyses should this aircraft be prepared to make? What dangers might the aircraft encounter, and what design problems would you have to overcome?

57. What sort of experiment or space mission would you design in order to establish definitively whether Jupiter has a rocky core?

58. The classic science fiction films *2001: A Space Odyssey* and *2010: The Year We Make Contact* both involve crewed spacecraft in orbit around Jupiter. What kinds of observations could humans make on such a mission that cannot be made by robotic spacecraft? What would be the risks associated with such a mission? Do you think that a crewed Jupiter mission would be as worthwhile as a crewed mission to Mars? Explain your answers.

59. Suppose that Saturn were somehow moved to an orbit around the Sun with semimajor axis 1 au, the same as Earth's. Discuss what long-term effects this would have on the planet and its rings.

WEB/E-BOOK QUESTIONS

60. On Jupiter, the noble gases argon, krypton, and xenon provide important clues about Jupiter's past. On Earth, xenon is used in electronic strobe lamps because it emits a very white light when excited by an electric current. Argon, by contrast, is one of the gases used to fill ordinary incandescent lightbulbs. Search the internet for information about why argon is used in this way. Why do premium, long-life lightbulbs use krypton rather than argon?

61. Search the internet, especially the websites for NASA's Jet Propulsion Laboratory and the European Space Agency, for information about the current status of the Juno mission. What recent discoveries has Juno made?

62. The two Voyager spacecraft were launched from Earth along a trajectory that took them directly to Jupiter. The force of Jupiter's gravity then gave the two spacecraft a "kick" that helped push them onward to Saturn. The much larger Cassini spacecraft, by contrast, was first launched on a trajectory that took it past Venus. Search the internet, especially the websites for NASA's Jet Propulsion Laboratory and the European Space Agency, for information about the trajectory that Cassini took through the solar system to reach Saturn. Explain why this trajectory was so different from that of the Voyagers.

63. **VIDEO 12.5** **The Rotation Rate of Saturn.** Access and view the video "Saturn from the Hubble Space Telescope" in Chapter 12 of the *Universe* website or e-book. The total amount of time that actually elapses in this video is 42.6 hours. Using this information, identify and follow an atmospheric feature and determine the rotation period of Saturn. How does your answer compare with the value given in Table 12-2?

ACTIVITIES

Observing Projects

> #### Observing tips and tools
>
> Like Mars, Jupiter and Saturn are most easily seen around opposition. (The dates of Jupiter's and Saturn's oppositions are available on the internet.) Each planet is visible in the night sky for several months before or after its opposition. At other times, Jupiter and Saturn may be visible in either the predawn morning sky or the early evening sky. Consult such magazines as *Sky & Telescope* and *Astronomy* or their websites for more detailed information about when and where to look for Jupiter during a given month. You can also use the *Starry Night™* program that accompanies selected copies of this textbook. A relatively small telescope with an objective diameter of 15 to 20 cm (6 to 8 inches), used with a medium-power eyepiece to give a magnification of 25× or so, should enable you to see some of the dark belts on Jupiter and the rings of Saturn.

64. If Jupiter is visible in the night sky, make arrangements to view the planet through a telescope. What magnifying power seems to give you the best view? Draw a picture of what you see. Can you see any belts and zones? How many? Can you see the Great Red Spot?

65. Make arrangements to view Jupiter's Great Red Spot through a telescope. Consult the *Sky & Telescope* website, which lists the times when the center of the Great Red Spot passes across Jupiter's central meridian. The Great Red Spot is well placed for viewing for 50 minutes before and after this happens. You will need a refractor with an objective of at least 15-cm (6-in.) diameter or a reflector with an objective of at least 20-cm (8-in.) diameter. Using a pale blue or green filter can increase the color contrast and make the spot more visible. For other useful hints, see the article "Tracking Jupiter's Great Red Spot" by Alan MacRobert (*Sky & Telescope*, September 1997).

66. View Saturn through a small telescope. Make a sketch of what you see. Estimate the angle at which the rings are tilted to your line of sight. Can you see the Cassini division? Can you see any belts or zones in Saturn's clouds? Is there a faint, starlike object near Saturn that might be Saturn's satellite Titan? What observations could you perform to test whether the starlike object is a Saturnian satellite?

Starry Night™ Explorations

67. Use *Starry Night*™ to examine Jupiter. Open **Favorites > Explorations > Jupiter** and use the location scroller to study the planet's atmosphere.
    (a) Describe the appearance of the atmosphere of the planet.
    (b) Compare the number of white ovals and brown ovals in Jupiter's southern hemisphere (the hemisphere in which the Great Red Spot is located) with the number of white ovals and brown ovals visible in the northern hemisphere. What general conclusion can you draw about the abundance of these storms in the two hemispheres?

68. Use *Starry Night*™ to observe the rotation and shape of Jupiter. Open **Favorites > Explorations > Jupiter**.
    (a) Determine the rotation period of the planet. (*Hint:* You may want to track the motion of an easily recognizable feature, such as the Great Red Spot.) What is the approximate rotation period of Jupiter?
    (b) Use the angular separation cursor to measure Jupiter's equatorial radius and its polar radius. Measure from the center of the planet to its east or west limb (edge) along the equator to determine its equatorial radius and then measure from the center of the planet to its north or south pole to measure the polar radius. Are these two radii the same? If not, which is larger? Offer an explanation for this observation.

69. Use *Starry Night*™ to observe the changing appearance of Saturn as seen from Earth. Select **Favorites > Explorations > Atlas** and then open the **Planets** panel. Locate **Saturn** in the list and double-click its name to center and label this planet

in the view. Zoom in until Saturn and its rings are clearly visible. Now use the "Y" keyboard shortcut to move forward in time in steps of 1 year per keypress and observe the orientation of Saturn's rings as time advances.
    (a) Describe qualitatively how the rings change orientation, as seen from Earth. What causes this changing orientation?
    (b) Approximately how long does it take for Saturn's rings, as seen from Earth, to go through a complete cycle from edge on to fully open, to edge on, to fully open and then finally to edge on again?
    (c) During which of the next 30 years will Saturn's rings appear edge on as seen from Earth?

70. Use *Starry Night*™ to examine the rings of Saturn. Open **Favorites > Explorations > Rings** to see a view of Saturn from above the planet's north pole, so that you are looking straight down on the plane of the rings.
    (a) Sketch what you observe and label the different rings and divisions.
    (b) Use the location scroller to adjust the view so that Saturn's rings appear edge on, and then rotate the image until the Sun comes into view. Which of the following is correct?
        (i) The Sun is in the same plane as the rings of Saturn.
        (ii) The rings are in Saturn's equatorial plane.
        (iii) Neither (i) nor (ii) is correct.
        (iv) Both (i) and (ii) are correct.
    (c) Zoom in to a field of view of about 85° and use the location scroller to examine the structure of the rings in detail, watching particularly the obscuration of the light from bright stars as the rings pass in front of them. Which ring appears to block the most light from distant stars?
    (d) Which of the following models for Saturn's rings is the most likely, from your observations?
        (i) The ring system is one solid disk whose thickness varies with radius.
        (ii) The rings are composed of millions of small rocks and pieces of ice, each orbiting independently around the planet.
        (iii) The rings are composed of several separate, solid, semi-transparent rings orbiting around the planet.

Collaborative Exercises

71. Using a ruler with millimeter markings on five various images of Jupiter in the text (for example, Figures 12-1a, 12-2a, 12-5a–c, and the image in Table 12-1), determine the ratio of the longest width of the Great Red Spot to the full diameter of Jupiter. Each group member should measure a different image and all values should be averaged.

72. The text provides different years that spacecraft have flown by Jupiter and Saturn. List these dates and create a time line by listing one important event that was occurring on Earth during each of those years.

73. If the largest circle you can draw on a piece of paper represents the largest diameter of Saturn's rings, about how large would Saturn be if scaled appropriately? Which item in a group member's backpack is closest to this size?

ANSWERS

ConceptChecks

ConceptCheck 12-1: No. Solid objects cannot undergo differential rotation because all parts of a solid object rotate together.

ConceptCheck 12-2: If the planets formed with the same chemical composition, then helium in Saturn's colder atmosphere could have condensed and moved deeper inside, resulting in Saturn appearing to be mostly hydrogen.

ConceptCheck 12-3: White ovals are long-lasting, low-temperature, high-altitude features that obscure views of what lies underneath. Brown ovals are openings in Jupiter's cloud cover that emit more infrared energy from higher-temperature material at lower altitudes.

ConceptCheck 12-4: Jupiter has significantly more mass, and as a result the cloud layers are gravitationally compressed much more on Jupiter than on less massive Saturn.

ConceptCheck 12-5: We expect the Sun and Jupiter to have formed from the same material—the solar nebula—each with the same abundances of noble gases. Because the noble gases do not undergo chemical reactions with other atoms and molecules, their observed abundance today is thought to be representative of the abundance when the Sun and Jupiter formed.

ConceptCheck 12-6: Yes. All else being equal, increasing the rotation rate increases the oblateness.

ConceptCheck 12-7: Yes. Some charged particles ejected by Io's volcanism can join the ion-trapping belts around Jupiter. Just as on Earth, when these ions stream along the magnetic field, they can cause aurorae at Jupiter's magnetic poles.

ConceptCheck 12-8: As Saturn travels around the Sun, the tilted and quite thin rings are sometimes seen somewhat "face on" and are easily observed, whereas at other times the view from Earth makes the rings "edge on" and nearly invisible.

ConceptCheck 12-9: An asteroid orbiting Earth within Earth's Roche limit for enough time would be ripped apart by Earth's gravitational attraction differing on the near side and the far side of the asteroid (which is a tidal force), resulting in a tiny ring of debris orbiting Earth.

ConceptCheck 12-10: Slight color variations indicate differences in material making the rings. If this material migrated quickly, color variations would be smeared out; instead, there are some very narrow ring features with different colors in Figure 12-23.

ConceptCheck 12-11: The Cassini division is created by a destabilizing 2-to-1 orbital resonance with Saturn's moon Mimas. Due to this resonance, repeated gravitational tugs clear out any particles at this location. If Mimas orbited farther away, its period would increase, and the Cassini division would appear farther out as well (where its dislodged particles also had a larger period).

13 Jupiter and Saturn's Satellites of Fire and Ice

LEARNING GOALS

To many astronomers, the moons of Jupiter and Saturn are so unique and complex that they are best thought of as *worlds*. Io, shown in the above image, is the most geologically active world in the solar system. Its surface is covered with colorful sulfur compounds deposited by dozens of active volcanoes. Europa, another of Jupiter's large satellites, has no volcanoes but does have a weirdly cracked surface of ice. Beneath Europa's ice there is probably a worldwide ocean, within which small, single-celled organisms could possibly live.

No less exotic is Saturn's large satellite Titan, which is enveloped by a thick, nitrogen-rich atmosphere from which liquid hydrocarbons fall as rain and form massive lakes. Another moon of Saturn, Enceladus, shocks us with its ice volcanoes, which Cassini has caught in the act of erupting. In this chapter, we will explore these and other intriguing aspects of Jupiter and Saturn's extensive collection of satellites.

This is Jupiter's moon Io, the most volcanically active body in the solar system. The faint plume on the top is an active volcano. This image was taken by the Galileo spacecraft. (Galileo Project/JPL/NASA)

R I V U X G

13-1 Jupiter's Galilean satellites are easily seen with Earth-based telescopes

TUTORIAL 13.1 When Galileo trained his telescope on Jupiter in January 1610, he became the first person to observe satellites, or moons, orbiting another planet (see Section 4-5). He called them the Medicean stars in order to curry the favor of a wealthy Florentine patron of the arts and sciences. We now call them the **Galilean satellites**. Individually, these moons are named Io, Europa, Ganymede, and Callisto, after four mythical lovers and companions of the god whom the Greeks called Zeus and the Romans called Jupiter.

When viewed through an Earth-based telescope, the Galilean satellites look like pinpoints of light (see Figure 4-16). All four satellites orbit Jupiter in nearly the same plane as the planet's equator (Figure 13-1). From Earth, we always see that plane nearly edge on, so the satellites appear to move back and forth relative to Jupiter (see Figure 4-17). The orbital periods are fairly short, ranging from 1.8 (Earth) days for Io to 16.7 days for Callisto, so we can easily see the satellites move from one night to the next and even over the course of a single night. These motions led Galileo to realize that he was seeing objects orbiting around Jupiter, in much the same way that Copernicus said that planets move around the Sun (see Section 4-2).

Each of the four Galilean satellites is bright enough to be visible to the naked eye. Why, then, did Galileo need a telescope to discover them? The reason is that as seen from Earth, the angular separation between Jupiter and the satellites is quite small—never more than 10 arcminutes for Callisto and even less for the other three Galilean satellites. To the naked eye, these satellites are lost in the overwhelming glare of Jupiter. But a small telescope or even binoculars increases the apparent angular separation by enough to make the Galilean satellites visible.

Synchronous Rotation and Orbital Resonance

The brightness of each satellite varies slightly as it orbits Jupiter because, as it moves along its orbit, the satellite also spins on its axis, and dark and light areas on the surface are alternately exposed and hidden from our view. Remarkably, each Galilean satellite goes through one complete cycle of brightness during one orbital period. This observation tells us that each satellite is in *synchronous*

rotation, so that it rotates exactly once on its axis during each trip around its orbit (see Figure 3-4b). As we saw in Section 4-8, our Moon's synchronous rotation is the result of gravitational forces exerted on the Moon by Earth. Likewise, Jupiter's gravitational forces keep the Galilean satellites in synchronous rotation.

Much more important, however, are the effects of an **orbital resonance**. Recall that an orbital resonance occurs when bodies exert a regular, repeating gravitational influence on each other. As we will see in this chapter, heat generated by this effect has a profound influence on Io and Europa.

ANIMATION 13.1 Synchronous rotation means that the *rotation* period and orbital period are in a 1:1 ratio for each Galilean satellite. Notably, there is also a simple ratio of the different *orbital* periods of the three inner Galilean satellites, Io, Europa, and Ganymede, which leads to an orbital resonance. Io orbits Jupiter in 1.77 days, while Europa has twice the orbital period of Io, and Ganymede has 4 times the orbital period of Io. Thus, the orbital periods of Io, Europa, and Ganymede are in the ratio 1:2:4, which you can verify from the data in Table 13-1. This is an example of a stabilizing orbital resonance. (In Section 12-11, we saw that a destabilizing resonance cleared out the Cassini division.)

This orbital resonance—a special relationship among the satellites' orbits—is maintained by the gravitational forces that they exert on one another. Those forces can be quite strong because the three inner Galilean satellites pass relatively close to one another; at their closest approach, Io and Europa are separated by only two-thirds the distance from Earth to the Moon. Such a close approach occurs once for every two of Io's orbits, so Europa's gravitational pull acts on Io in a rhythmic way. Indeed, Io, Europa, and Ganymede all exert rhythmic gravitational tugs on one another. Just as a drummer's rhythm keeps musicians on the same beat, this gravitational rhythm maintains the simple ratio of orbital periods among the satellites. As we will see, this "rhythmic drumming" provides a significant source of energy that heats the interiors of Io and Europa.

> Gravitational interactions between Io, Europa, and Ganymede induce a rhythmic relationship among their orbits.

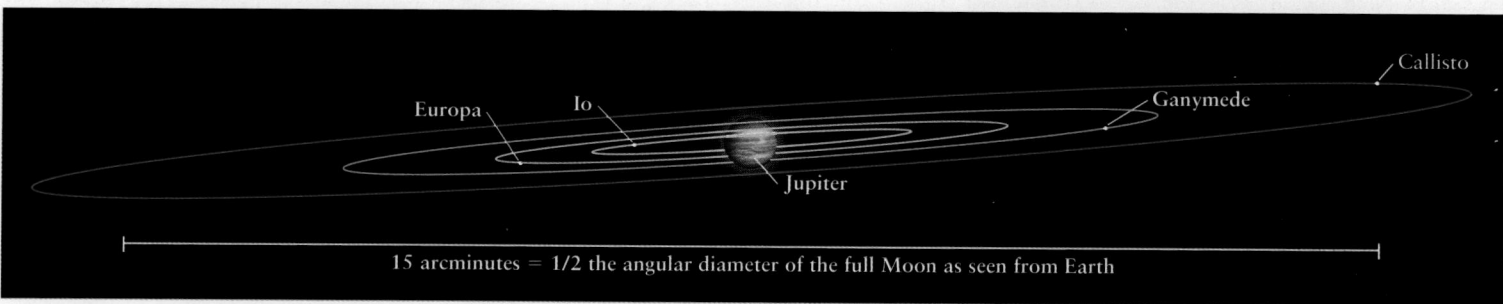

15 arcminutes = 1/2 the angular diameter of the full Moon as seen from Earth

FIGURE 13-1

The Orbits of the Galilean Satellites This illustration shows the orbits of the four Galilean satellites as seen from Earth. All four orbits lie in nearly the same plane as Jupiter's equator. The apparent angular size and orientation of the orbits depends on the relative positions of Earth and Jupiter.

TABLE 13-1 Jupiter's Galilean Satellites Compared with the Moon, Mercury, and Mars

	Average distance from Jupiter (km)	Orbital period (days)	Diameter (km)	Mass (kg)	Mass (Moon = 1)	Average density (kg/m³)	Albedo
Io	421,600	1.769	3642	8.932×10^{22}	1.22	3529	0.63
Europa	670,900	3.551	3120	4.791×10^{22}	0.65	3018	0.64
Ganymede	1,070,000	7.155	5268	1.482×10^{23}	2.02	1936	0.43
Callisto	1,883,000	16.689	4800	1.077×10^{23}	1.47	1851	0.17
Moon	—	—	3476	7.349×10^{22}	1.00	3344	0.11
Mercury	—	—	4880	3.302×10^{23}	4.49	5430	0.12
Mars	—	—	6794	6.419×10^{23}	8.73	3934	0.15

5000 km

Io   Europa   Ganymede   Callisto

Jupiter   421,600 km   670,900 km   1,070,000 km   1,883,800 km

Average distance from Jupiter

Note: Jupiter is shown to the same scale as the distances of the satellites from Jupiter. Compared to this scale, the images of the satellites themselves have been enlarged 74×.

R I V U X G

(NASA/JPL/Ames Research Center; NASA/JPL/University of Arizona; NASA/JPL/DLR; NASA/JPL)

By contrast, Callisto orbits at a relatively large distance from the other three large satellites. Hence, the gravitational forces on Callisto from the other satellites are relatively weak, and there is no simple relationship between Callisto's orbital period and the period of Io, Europa, or Ganymede.

Transits, Eclipses, and Occultations

Because the orbital planes of the Galilean satellites are nearly edge on to our line of sight, we see these satellites undergoing transits, eclipses, and occultations. In a *transit*, a satellite passes between us and Jupiter, and we see the satellite's shadow as a black dot against the planet's colorful cloudtops. (Figure 12-2a shows Europa's shadow on Jupiter.) In an *eclipse*, one of the satellites disappears and then reappears as it passes into and out of Jupiter's enormous shadow. In an **occultation**, a satellite passes completely behind Jupiter, and the satellite is temporarily blocked from our Earth-based view. (The word *occultation* comes from a Latin verb meaning "to cover.")

Before spacecraft first ventured to Jupiter, astronomers used eclipses in a very clever way to estimate the diameters of the Galilean satellites. When a satellite emerges from Jupiter's shadow, it does not blink on instantly. Instead, there is a brief interval during which the satellite gets progressively brighter as more of its surface is exposed to sunlight. The satellite's diameter is calculated from the duration of this interval and the satellite's orbital speed (which is known from Kepler's third law).

As Table 13-1 shows, the smallest Galilean satellite (Europa) is slightly smaller than our Moon; the largest satellite (Ganymede) is larger than Mercury and more than three-quarters the size of Mars. Thus, the Galilean satellites truly are worlds in their own right.

CONCEPT CHECK 13-1

If you know the speed of a moon or satellite (from Kepler's third law), and you measure how long it takes to disappear during its orbit behind its planet (which is an occultation), what can you learn about the satellite?

Answer appears at the end of the chapter.

13-2 Data from spacecraft reveal the unique properties of the Galilean satellites

Even the finest images made with Earth-based telescopes have revealed relatively few details about the Galilean satellites. Almost everything we know about these satellites has come from observations made at close range by spacecraft.

The first close-range observations of the Galilean satellites were made by the Pioneer 10 and Pioneer 11 spacecraft as they flew past Jupiter in 1973 and 1974. The images from these missions were of relatively low resolution, however, and much more information has come from three subsequent missions. Voyager 1 and Voyager 2 flew past Jupiter in 1979 and recorded tens of thousands of images, one of which is shown in Figure 13-2. The Galileo spacecraft, which made 35 orbits around Jupiter between 1995 and 2003, carried out an even more extensive investigation of Jupiter's satellites. The Cassini and New Horizons spacecraft also imaged Jupiter on their way to Saturn and Pluto, respectively, and Juno entered into orbit in July 2016. Juno will make only limited observations of Jupiter's satellites because the spacecraft will remain relatively distant from these moons.

Measuring Satellite Densities

One key goal of these missions was to determine the densities of the Galilean satellites to very high accuracy. Given the density of a planet or satellite, astronomers can draw conclusions about its chemical composition and internal structure. To find the densities, scientists first determined the satellite masses. They did this by measuring how the gravity of Io, Europa, Ganymede, and Callisto deflected the trajectories of the Voyager and Galileo spacecraft. Given the masses and diameters of the satellites, they then calculated each satellite's average density (mass divided by volume). Table 13-1 lists our current knowledge about the sizes and masses of the Galilean satellites.

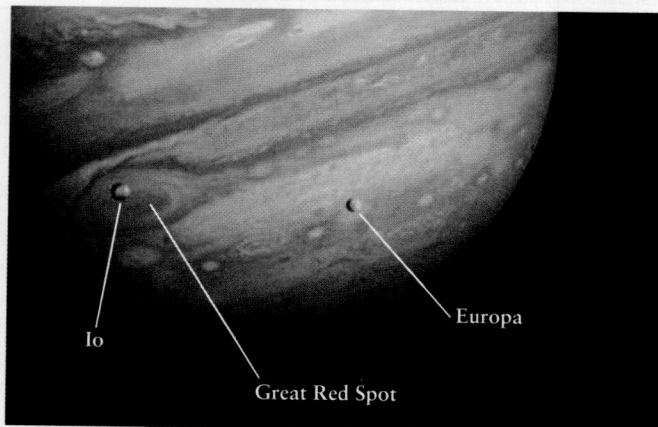

FIGURE 13-2 R I V U X G

Io and Europa from Voyager 1 Voyager 1 recorded this image of Jupiter and the Galilean satellites Io and Europa in February 1979, when it was 20 million km (12.4 million mi) from the giant planet. The satellites are actually smaller than some of Jupiter's cloud features, such as the Great Red Spot (see Section 12-3). Voyager 1 and Voyager 2 also recorded many images of the Galilean satellites at closer range. (NASA/JPL)

Of the four Galilean satellites, Europa proves to be the least massive: Its mass is only two-thirds that of our Moon. Ganymede is by far the most massive of the four, with more than double the mass of our Moon. In fact, Ganymede is the most massive satellite anywhere in the solar system. In second place is Callisto, with about 1½ times the Moon's mass.

However, Ganymede and Callisto are *not* merely larger versions of our Moon. The most obvious difference is that both Ganymede and Callisto have very low average densities of less than 2000 kg/m^3. By comparison, typical rocks in Earth's crust have densities around 3000 kg/m^3, and the average density of our Moon is 3344 kg/m^3. The low densities of Ganymede and Callisto indicate that these satellites cannot be composed primarily of rock. Instead, they are probably roughly equal parts of rock and water ice. Water ice is substantially less dense than rock, but it becomes as rigid as rock under high pressure and at low temperatures, such as is found in the interiors of the Galilean satellites. Furthermore, water molecules are relatively common in the solar system. Enough water would have been available in the early solar system to make up a substantial portion of a large satellite such as Ganymede or Callisto.

Water ice, however, cannot be a major constituent of the two inner Galilean satellites. The innermost satellite, Io, has the highest average density, 3529 kg/m^3, slightly greater than the density of our Moon. The next satellite out, Europa, has an average density of 3018 kg/m^3. Both of these values are close to the densities of typical rocks in Earth's crust. Hence, it is reasonable to suppose that both Io and Europa are made primarily of rocky material.

> Three of the Galilean satellites are made of a mixture of ice and rock; only Io is ice free.

The most definitive evidence for water ice on three of Jupiter's Galilean satellites has come from spectroscopic observations. The spectra of sunlight reflected from Europa, Ganymede, and Callisto show absorption at the infrared wavelengths characteristic of water ice molecules. You can see this in Figure 7-4, which compares the spectrum of Europa to the spectrum of water ice. Europa's density shows that it is composed mostly of rock, so its ice must be limited to the satellite's outer regions. Among the Galilean satellites, only Io shows no trace of water ice on its surface or in its interior.

CONCEPT CHECK 13-2

By using Kepler's laws on any of the moons orbiting Jupiter, the mass of Jupiter is easily determined from Earth-based observations. Can the masses of Jupiter's moons be determined from Earth as well?

Answer appears at the end of the chapter.

13-3 The Galilean satellites formed like a solar system in miniature

As Table 13-1 shows, the more distant a Galilean satellite is from Jupiter, the greater the proportion of ice within the satellite and the lower its average density. But why should this be? An important clue is that the average densities of the planets show a similar trend: Moving outward from the Sun, the average density of the inner six planets steadily decreases, from more than 5000 kg/m³ for Mercury to less than 1000 kg/m³ for Saturn (recall Table 7-1). The best explanation for this similarity is that the Galilean satellites formed around Jupiter in much the same way that the

> The differences in composition among the Galilean satellites mimic those among planets orbiting the Sun.

planets formed around the Sun, although on a much smaller scale.

We saw in Section 8-5 that dust grains in the solar nebula played an important role in planet formation. In the inner parts of the nebula, close to the hot protosun, only dense rock and metal grains were able to remain solid. These grains accumulated over time into the dense, rocky inner planets. But in the cold outer reaches of the nebula (beyond the "snow line"), dust grains were able to retain icy coatings of low-density material like water and ammonia. Hence, when the jovian planets formed in the outer solar nebula and incorporated these ice-coated grains, the planets ended up with both rock and ices in their cores, as well as substantial amounts of ammonia and water throughout their interiors and atmospheres.

As Jupiter coalesced, the gas accumulating around it formed a rotating "jovian nebula." The central part of this nebula became the huge envelope of hydrogen and helium that makes up most of Jupiter's bulk. But in the outer parts of this nebula, dust grains could have accreted to form small solid bodies. These grew to become the Galilean satellites.

Although the "jovian nebula" was far from the protosun, not all of it was cold. Jupiter, like the protosun, must have emitted substantial amounts of radiation due to Kelvin–Helmholtz contraction, which we discussed in Section 8-4. (We saw in Section 12-4 that even today, Jupiter emits more energy due to its internal heat than it absorbs from sunlight.) Hence, temperatures very close to Jupiter must have been substantially higher than at locations farther from the planet. Calculations show that only rocky material would condense at the orbital distances of Io and Europa, but frozen water could be retained and incorporated into satellites at the distances of Ganymede and Callisto. In this way, Jupiter ended up with two classes of Galilean satellites: primarily rocky satellites versus satellites made of a mixture of rock and water (**Figure 13-3**).

The analogy between the solar nebula and the "jovian nebula" is not exact. Unlike the Sun, Jupiter can be thought of as a "failed star." Its internal temperatures and pressures never became high enough to ignite nuclear reactions that convert hydrogen into helium. (Jupiter's mass would have had to be about 80 times larger for these reactions to have begun.) Furthermore, the icy worlds that formed in the outer region of the "jovian nebula"—namely, Ganymede and Callisto—were of relatively small mass. Hence, they were unable to attract gas from the nebula and become jovian planets in their own right. Nonetheless, it is remarkable how the main processes of planet formation seem to have occurred twice in our solar system—once in the solar nebula around the Sun, and once in microcosm in the gas cloud around Jupiter.

The diverse densities of the Galilean satellites suggest that these four worlds may be equally diverse in their geology. This

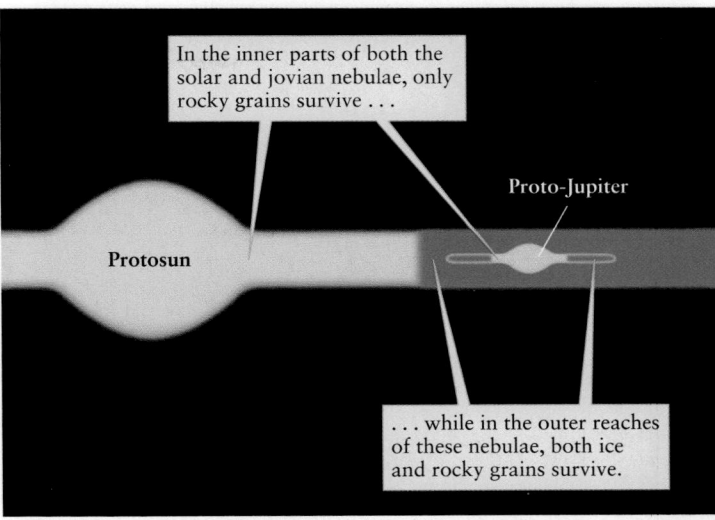

FIGURE 13-3

Formation of the Galilean Satellites Heat from the protosun made it impossible for icy grains to survive within the innermost 2.5 to 4 au of the solar nebula. In the same way, Jupiter's heat evaporated any icy grains that were too close to the center of the "jovian nebula." Hence, the two inner Galilean satellites were formed primarily from rock, while the outer two incorporated both rock and ice.

conclusion turns out to be entirely correct. Because each of the Galilean satellites is unique, we devote the next several sections to each satellite in turn.

CONCEPT CHECK 13-3

Why do the average densities of the Galilean moons suggest a "jovian nebula" developing like the formation of a miniature solar system?

Answer appears at the end of the chapter.

13-4 Io is covered with colorful sulfur compounds ejected from active volcanoes

Before the two Voyager spacecraft flew past Jupiter, the nature of Io was largely unknown. On the basis of Io's size and density, however, it was expected that Io would be a world much like our own Moon—geologically dead, with little internal heat available to power tectonic or volcanic activity (see Section 7-6). Therefore, it was thought that Io's surface would be extensively cratered, because there would have been little geologic activity to erase those craters over the satellite's history.

Io proved to be utterly different from these naive predictions, as you can see in **Figure 13-4** and the opening figure of this chapter. For our first views, Voyager 1 came within 21,000 km (13,000 mi) of Io and began sending back a series of bizarre and unexpected pictures of the satellite. These images showed that Io has *no* impact craters at all! Instead, the surface is pockmarked by irregularly shaped pits and blotched with vibrant colors.

(a)

(b)

VIDEO
13.1

FIGURE 13-4 R I **V** U X G

The Colors of Io **(a)** This true-color image of Io shows a striking yellow surface. Numerous volcanoes dot Io's surface, spewing out sulfur compounds. Sulfur (S) appears yellow, sulfur dioxide (SO_2) appears white, and S_3 and S_4 appear red. **(b)** Here on Earth, we find yellow and white sulfur formations around volcanic vents. (a: NASA/JPL/University of Arizona; b: Cubo Images/Superstock)

The Voyager mission scientists were seeing confirmation of a prediction published in the journal *Science* just 3 days before Voyager 1 flew past Io. In this article, two astrophysicists reported their conclusion that Io's interior must be kept hot by Jupiter's tidal forces and that this should lead to intense volcanic activity.

Tidal Heating on Io

As we saw in Section 4-8, tidal forces arise from differences in the gravitational pull on different parts of a planet or satellite. These forces tend to deform the shape of the planet or satellite (see Figure 4-29). Io, for example, is somewhat distorted from a spherical shape by tidal forces from Jupiter. If Io were in a perfectly circular orbit so that it traveled around Jupiter at a steady speed, Io's rotation would always be in lockstep with its orbital motion. In this case, Io would be in perfect synchronous rotation, with its long axis always pointed directly toward Jupiter. But as Io moves around its orbit, Europa and Ganymede exert gravitational tugs on it in a regular, rhythmic fashion, thanks to the 1:2:4 ratio among the orbital periods of these three satellites. These tugs distort Io's orbit into an ellipse, so its speed varies as it moves around its orbit. Consequently, Io's long axis "nods" back and forth by about ½° as seen from Jupiter. Due to this nodding, the tidal stresses that Jupiter exerts on Io vary rhythmically as Io moves around its orbit. These varying tidal stresses alternately squeeze and flex Io. Just as a ball of clay or bread dough gets warm as you knead it between your fingers, this squeezing and flexing causes **tidal heating** of Io's interior. (Earth's Moon also "nods" as it orbits our planet, but

> The theory of tidal forces helped predict Io's volcanic nature.

experiences almost no tidal heating because Earth's tidal forces are very weak compared to those of massive Jupiter.)

Tidal heating adds energy to Io at a rate of about 10^{14} watts, equivalent to 24 tons of TNT exploding every second. As this energy makes its way to the satellite's surface, it provides about 2.5 watts of power to each square meter of Io's surface. By comparison, the average global heat flow through Earth's crust is 0.06 watt per square meter. Only in volcanically active areas on Earth do we find heat flows that even come close to Io's average. As a result of Io's heat flow, and just as predicted in the journal *Science*, this moon has "widespread and recurrent surface volcanism."

Io's Active Volcanoes

No one expected that Voyager 1 would obtain images of erupting volcanoes on Io. After all, a spacecraft making a single trip past Earth would be highly unlikely to catch a large volcano actually erupting. But, in fact, Voyager 1 images of Io revealed *eight* different giant plumes of gas from volcanic eruptions. (**Figure 13-5a** shows two of these.) More recently, the Galileo and New Horizons spacecraft returned detailed images of several such plumes (Figure 13-5b, c). These observations resoundingly confirm the prediction that Io is volcanically active: In fact, this small world is by far the most volcanic body in the solar system.

Io's volcanic plumes rise to astonishing heights of 70 to 290 km (45 to 180 mi/h) above Io's surface. To reach such altitudes, the material must emerge from volcanic vents with speeds between 300 and 1000 m/s (1100 to 3600 km/h, or 700 to 2200 mi/h). Even the most violent volcanoes on Earth have eruption speeds of only around 100 m/s (360 km/h, or 220 mi/h). Scientists thus began to suspect that Io's volcanic activity must be fundamentally different from volcanism here on Earth.

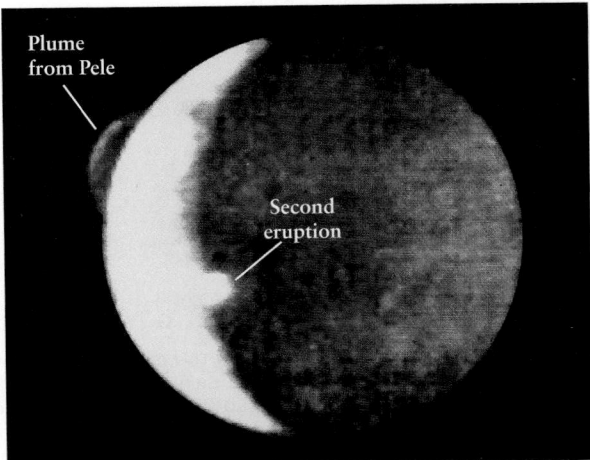

(a) Voyager 1, March 1979

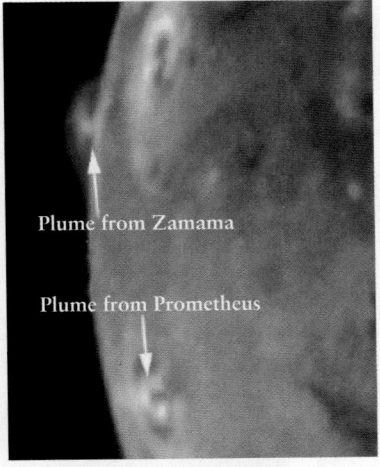

(b) Galileo, November 1997

FIGURE 13-5 R I V U X G

Volcanic Plumes on Io **(a)** Io's volcanic eruptions were first discovered on this Voyager 1 image. The plume at upper left, which extends 260 km (150 mi) above Io's surface, came from the volcano Pele. **(b)** This Galileo image shows plumes from the volcanoes Prometheus and Zamama. The plumes rise to about 100 km (60 mi) above Io's surface and are about 250 km (150 mi) wide. **(c)** At the top of this New Horizons image, the plume from the volcano Tvashtar reaches 290 km (180 mi) high. Another plume is visible at the far left edge. The mountains near the center, just to the left of the terminator (the line between day and night) are about as tall as Mount Everest, nearly 9 km (5.5 mi). (a: NASA/ JPL; b: NASA/JPL/University of Arizona; c: NASA/Johns Hopkins University Applied Physics Laboratory/Southwest Research Institute)

(c) New Horizons, 2007

The Nature of Io's Volcanic Eruptions

An important clue about volcanism on Io came from the infrared spectrometers aboard Voyager 1, which detected abundant sulfur and sulfur dioxide in Io's volcanic plumes. This observation led to the idea that the plumes are actually more like geysers than volcanic eruptions. In a geyser on Earth, water seeps down to volcanically heated rocks, where it changes to steam and erupts explosively through a vent. Some have suggested that sulfur dioxide rather than water could be the principal propulsive agent driving volcanic plumes on Io. Sulfur dioxide is a solid at the frigid temperatures found on most of Io's surface, but it should be molten at depths of only a few kilometers. Just as the explosive conversion of water into steam produces a geyser on Earth, the conversion of liquid sulfur dioxide into a high-pressure gas could result in eruption velocities of up to 1000 m/s.

VIDEO 13.2 ▸ Io's dramatic coloration (see Figure 13-4) is probably due to sulfur and sulfur dioxide, which are ejected in volcanic plumes and later fall back to the surface. Sulfur is normally bright yellow, which explains the dominant color of Io's surface. But if sulfur is heated and suddenly cooled, as would

happen if it were ejected from a volcanic vent and allowed to fall to the surface, it can assume a range of colors from orange and red to black. Indeed, these colors are commonly found around active volcanic vents (**Figure 13-6**). Whitish surface deposits (examine Figure 13-4 and Figure 13-6a), by contrast, are probably due to sulfur dioxide (SO_2). Volcanic vents on Earth commonly discharge SO_2 in the form of an acrid gas. But on Io, when hot SO_2 gas is released by an eruption into the cold vacuum of space, it crystallizes into white snowflakes. This sulfur dioxide "snow" then falls back down onto Io's surface.

The plumes are only part of the story of volcanism on Io, however. The sources of the plumes appear in Voyager images as black spots 10 to 50 km (6–30 mi) in diameter, which are actually volcanic vents (see Figure 13-4a). Many of these black spots, which cover 5% of Io's surface, are surrounded by dark lava flows. **Figure 13-7** is a Galileo image of one of Io's most active volcanic regions, where in 1999 lava was found spouting to altitudes of thousands of meters along a fissure 40 km (25 mi) in length. The glow from this lava fountain was so intense that it was visible with Earth-based telescopes. Thus, Io has two distinct styles of volcanic

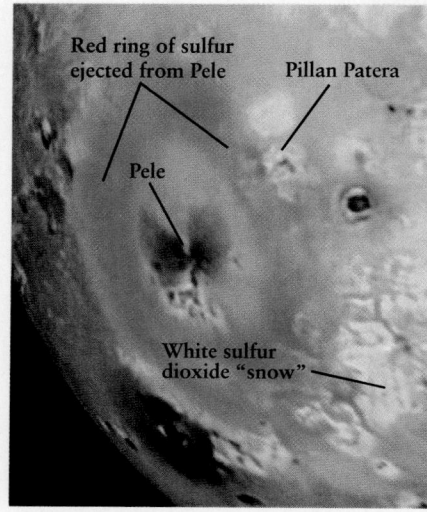

Red ring of sulfur
ejected from Pele

Pillan Patera

Pele

White sulfur
dioxide "snow"

(a)

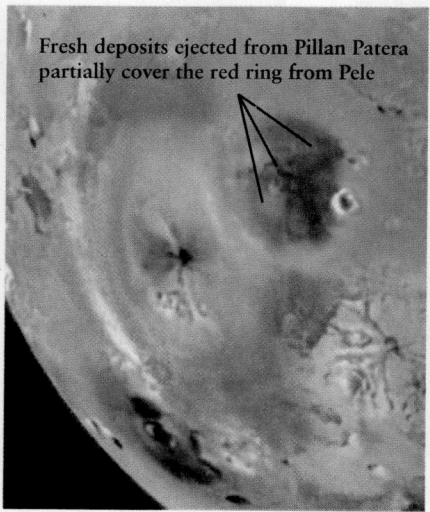

Fresh deposits ejected from Pillan Patera
partially cover the red ring from Pele

(b)

FIGURE 13-6 R I V U X G

Rapid Changes on Io **(a)** This Galileo image shows a portion of Io's southern hemisphere. Sulfur ejected from the active volcanic vent Pele produced a red ring that resembles paint sprayed from a can. **(b)** A few months later, a volcanic plume erupted from a vent called Pillan Patera, spraying dark material over an area some 400 km (250 mi) in diameter (about the size of Arizona). Other than Earth, Io is the only world in the solar system that shows such noticeable changes over short time spans. (NASA/JPL/University of Arizona)

Lava flows

Volcanic
calderas

Curtain of lava

50 km

FIGURE 13-7 R I V U X G

Io's Lava Flows and a Curtain of Fire This mosaic of Galileo spacecraft images shows a nested chain of volcanic calderas (pits) with black lava flows. The red streak at upper left is a fissure in the surface about 40 km (25 mi) long, through which lava is erupting upward in a vertical sheet or curtain. The erupting lava reaches the amazing height of 1500 m (5000 ft). (NASA/JPL/University of Arizona)

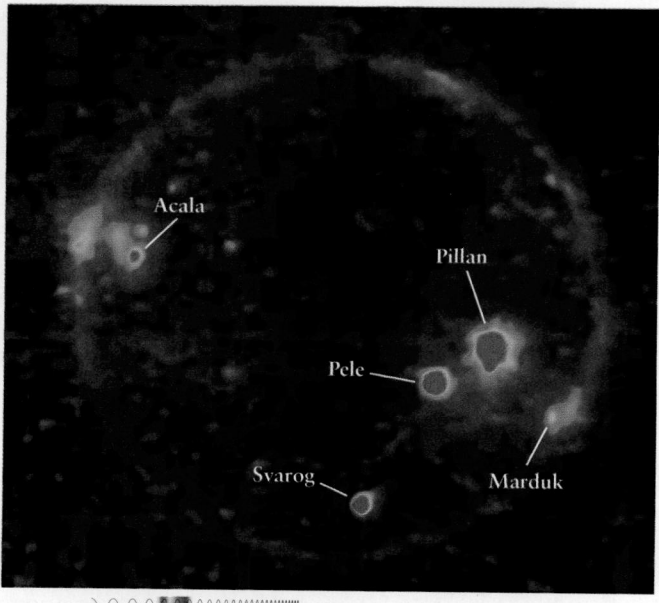

Acala

Pillan

Pele

Svarog

Marduk

FIGURE 13-8 R I V U X G

Io's Glowing Volcanoes This false-color Galileo image (made while Io was in the darkness of Jupiter's shadow) shows the infrared and visible glow from several volcanic regions, including Pele and Pillan (see Figure 13-6). These regions are substantially hotter than typical lava flows on Earth. (NASA/JPL/University of Arizona)

activity: unique geyserlike plumes, and lava flows that are more dramatic versions of volcanic flows on Earth.

While sulfur and sulfur compounds are the key constituents of Io's volcanic plumes, the lava flows must be made of something else. Infrared measurements from the Galileo spacecraft (Figure 13-8) show that lava flows on Io are at temperatures of 1700 K to 2000 K (about 1450°C to 1750°C, or 2600°F to 3150°F). At these temperatures, sulfur could not remain molten but would evaporate almost instantly. Io's lavas are also unlikely to have the same chemical composition as typical lavas on Earth, which have temperatures of only 1300 K to 1450 K. Instead, lavas on Io are probably enriched in magnesium and iron, which give the lava a higher melting temperature.

Similar lavas enriched in magnesium and iron are found on Earth, but primarily in solidified lava beds that formed billions of

years ago when Earth's interior was much hotter than today. The presence of this enriched lava on Io suggests that its interior, too, is substantially hotter than that of the present-day Earth. One model proposes that Io has a 100-km (60-mi) thick crust floating atop a worldwide ocean of liquid magma 800 km (500 mi) deep. (As we saw in Section 9-2, the mantle that lies underneath Earth's crust is a plastic solid rather than a true liquid.) The global extent of this "magma ocean" would explain why volcanic activity is found at all points on Io's surface, rather than being concentrated in pockets as on Earth (see Section 9-3). It could also explain the origin of Io's mountains, which reach heights up to 10 km (30,000 ft). In this model, these mountains are blocks of Io's crust that have tilted before sinking into the depths of the magma ocean.

Io's worldwide volcanic activity is remarkably persistent. When Voyager 2 flew through the jovian system in July 1979, 4 months after Voyager 1, almost all of the volcanic plumes seen by Voyager 1 were still active. More recently, 48 volcanoes were tracked for 29 months ending in 2015. This data (from the Keck II telescope) confirmed that Io is still the most active site of volcanism in the solar system.

Io may have as many as 425 active volcanoes, each of which ejects an estimated 10,000 tons of material per second. Altogether, volcanism on Io may eject as much as 10^{13} tons of matter each year. This amount of material is sufficient to cover the satellite's entire surface to a depth of 1 meter in a century, or to cover an area of 1000 square km in a few weeks (see Figure 13-6). Thanks to this continual "repaving" of the surface, there are probably no long-lived features on Io, and any impact craters are quickly covered up.

CONCEPT CHECK 13-4

The strongest tidal forces on Io are exerted by Jupiter. Then why are the orbital properties of Europa and Ganymede important for generating the heat that powers Io's volcanism?

Answer appears at the end of the chapter.

13-5 Jupiter's magnetic field makes electric currents flow through Io

Most of the material ejected from Io's volcanoes falls back onto the satellite's surface. But some material goes on a remarkable journey into space, by virtue of Io's location deep within Jupiter's magnetosphere.

The Io Torus

The jovian magnetosphere contains charged particles, or ions, some of which collide with Io and its volcanic plumes. The impact of these collisions knocks more ions out of the plumes and off the surface; these additional charged particles become part of Jupiter's magnetosphere. Indeed, material from Io is the main source of material in Jupiter's magnetosphere as a whole. Some of the ejected material forms the **Io torus**, a huge doughnut-shaped ring of ionized gas that circles Jupiter at the distance of Io's orbit. A gas that is hot enough for electrons to dissociate from atoms forms a gas of charged particles called a **plasma**. The plasma's glow can be detected from Earth (Figure 13-9a).

Jupiter's magnetic field has other remarkable effects on Io. As Jupiter rotates, its magnetic field rotates with it. This field sweeps over Io at high speed and generates a voltage of 400,000 volts across the satellite, causing 5 million amperes of electric current to flow through Io.

> Jupiter and Io together act like an immense electric generator.

ANALOGY

Whenever you use a credit card with a magnetic strip, you use the same physical principle that generates an electric current within Io. The credit card's number is imprinted as a magnetic code in the dark stripe on the back of the card. The salesperson who swipes your card through the card reader is actually moving the card's magnetic field past a coil of wire within the reader. This action generates within the coil an electric current that carries the same coded information about your credit card number. That information is transmitted to the credit card company, which (it is hoped) approves your purchase. An electric generator at a power plant works in the same way. A coil of wire is moved through a strong magnetic field, which makes current flow in the coil. This current is delivered to transmission lines and eventually to your home.

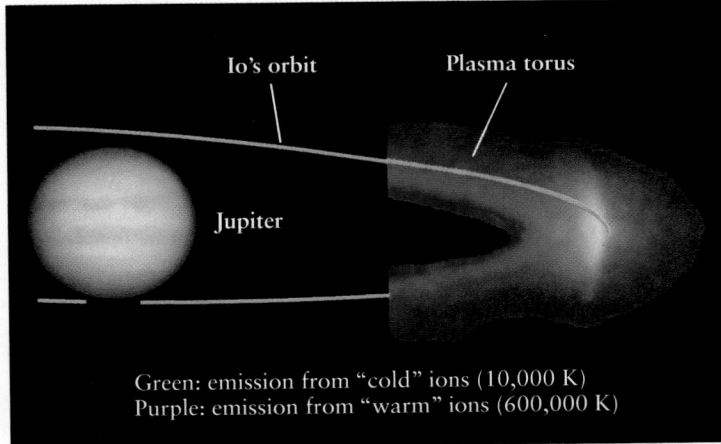

(a)

Green: emission from "cold" ions (10,000 K)
Purple: emission from "warm" ions (600,000 K)

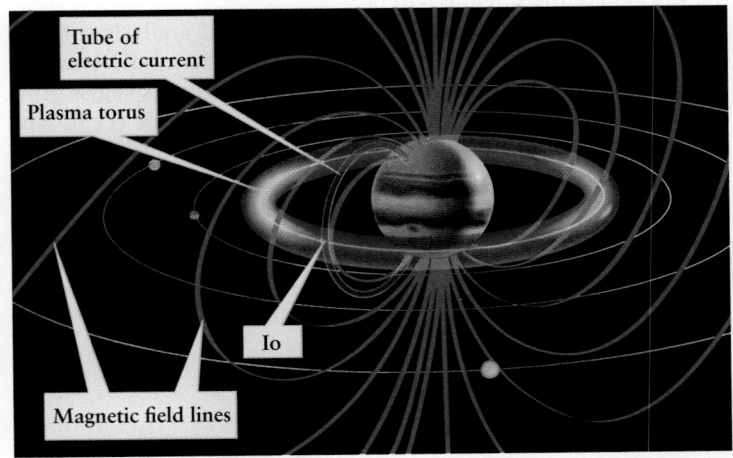

(b)

VIDEO 13.3   **FIGURE 13-9** R **I V** U X G

The Io Torus (a) An Earth-based telescope recorded this false-color infrared image of Io's plasma torus of charged particles. Because of Jupiter's glare, only the outer edge of the torus could be photographed; an artist added the visible-light picture of Jupiter and the yellow line indicating the rest of the torus. **(b)** A tube of electric current carries a charged-particle plasma between Jupiter and Io, forming an immense electric circuit. (a: Courtesy of John Trauger; b: Alfred T. Kamajian)

In fact, a current flows not only through Io but also through the sea of charged particles in Jupiter's magnetosphere and through Jupiter's atmosphere. This forms a gigantic, oval-shaped electric circuit that connects Io and Jupiter in the same way that wires connect the battery and the lightbulb in a flashlight (Figure 13-9b). Jupiter's aurora, which we discussed in Section 12-7, is particularly strong at the locations where this current strikes the upper atmosphere of Jupiter.

This important phenomenon occurs on all the Galilean moons, leading to profound insights and discoveries, so let us summarize what we have learned for further use in the chapter. As Jupiter's magnetic field sweeps over an electrically conductive medium (in this case, a plasma concentrated around Io), there is an induced flow of electricity in the conductive medium called an **induced electric current**. In addition to inducing the flow of electricity, all electricity produces its own magnetic field, which, in this scenario, is called an **induced magnetic field**.

Probing Io's Interior

The electric current within and near Io also creates a weak magnetic field, which was first detected by Voyager 1. But what about an intrinsic magnetic field generated by Io's interior? Galileo data determined that Io does not generate a magnetic field. This means that Io does not contain a convecting conducting core.

If Io has undergone enough tidal heating to melt its interior, chemical differentiation must have taken place and the satellite should have a dense core. (We described chemical differentiation in Box 7-1 and Section 8-5.) To test for a dense core, scientists measured how Io's gravity deflected the trajectory of Galileo as the spacecraft flew past. From these measurements, they could determine not only Io's mass but also the satellite's oblateness (which is how much it deviates from a spherical shape, bulging out along the equator). The amount of oblateness indicates the size of the core. A rotating body will tend to assume an oblate shape, but the more of its mass that is concentrated in the center (in a more rigid core), the less oblate the satellite will be.

The Galileo observations allowed a detailed analysis combining the moon's rotation with its oblate shape to model how Io's density changes toward the moon's center. The results suggest that Io has a dense core composed of iron and iron sulfide (FeS), with a radius of about 900 km (about half the satellite's overall radius). Surrounding the core is a mantle of partially molten rock, on top of which is Io's visible crust.

CONCEPT CHECK 13-5

Is Jupiter's magnetic field generated by the induced electric currents flowing between Jupiter and Io?

Answer appears at the end of the chapter.

13-6 Europa is covered with a smooth layer of ice that may cover a worldwide ocean

Europa, the second closest of the Galilean satellites, is the smoothest body in the solar system. There are no mountains and no surface features greater than a few hundred meters high (Figure 13-10). There

FIGURE 13-10 R **I** V U X G

Europa Dark lines crisscross Europa's smooth, icy surface in this false-color composite of visible and infrared images from Galileo. These lines are fractures in Europa's crust that can be as much as 20 to 40 km (12 to 25 mi) wide. Only a few impact craters are visible on Europa, which indicates that this satellite has a very young surface on which all older craters have been erased. (NASA/JPL/University of Arizona)

are almost no craters, indicating a young surface, between 40 and 90 million years old, that has been reprocessed by geologic activity. The dominant surface feature is a worldwide network of stripes and cracks (Figure 13-11). Like Io, Europa is an exception to the general rule that a small world should be cratered and geologically dead (see Section 7-6).

Europa's Surface: Icy but Active

Neither Voyager 1 nor Voyager 2 flew very near Europa, so most of our knowledge of this satellite comes from the close passes made by Galileo. (Both Figure 13-10 and Figure 13-11 are Galileo images.) But even before these spacecraft visited Jupiter, spectroscopic observations from Earth indicated that Europa's surface is almost pure frozen water (see Figure 7-4). This observation was confirmed by instruments on board Galileo, which showed that Europa's infrared spectrum is a close match to that of a thin layer of fine-grained water-ice frost on top of a surface of pure water ice. (The brown areas in Figure 13-10 show where the icy surface contains deposits of rocky material from meteoritic impacts, from Europa's interior, or from a combination of these sources.)

> Geology on Europa is based on solid ice and liquid water, not rock and molten magma.

The purity of Europa's ice suggests that water is somehow brought upward from the moon's interior to the surface, where it

Smooth ice plains with a network of fractures.

Linear ridges (dark colors are caused by minerals in the ice).

500 km

FIGURE 13-11   R I **V** U X G

Europa's Fractured Crust False colors in this Galileo composite image emphasize the difference between the linear ridges and the surrounding plains. The smooth ice plains (shown in blue) are the basic terrain found on Io. (NASA/JPL)

Ridges produced by folding and faulting of the surface.

5 km

Smooth area formed by fluid erupting onto the surface.

Rugged patch created by a subsurface disturbance.

FIGURE 13-12   R I **V** U X G

Europa in Close-up This high-resolution Galileo image shows a network of overlapping ridges, part of which has been erased to leave a smooth area and part of which has been jumbled into a rugged patch of terrain. Europa's interior must be warm enough to power this complex geologic activity. (NASA/JPL/ASU)

solidifies to make a fresh, smooth layer of ice. Indeed, some Galileo images show what appear to be lava flows on Europa's surface, although the "lava" in this case is mostly ice. This idea helps to explain why Europa has very few craters (any old ones have simply been covered up) and why its surface is so smooth. Europa's surface may thus represent a water-and-ice version of plate tectonics.

Although Europa's surface is almost pure water ice, keep in mind that Europa is *not* merely a giant ice ball. The satellite's density shows that rocky material makes up about 85% to 90% of Europa's mass. Hence, only a small fraction of the mass, about 10% to 15%, is water ice. Because the surface is icy, we can conclude that the rocky material is found within Europa's interior.

Europa is too small to have retained much of the internal heat that it had when it first formed. But there must be internal heat nonetheless to power the geologic processes that erase craters and bring fresh water to Europa's surface. What keeps Europa's interior warm? The most likely answer, just as for Io, is heating by Jupiter's tidal forces. The rhythmic gravitational tugs exerted by the orbital resonance of Io, Ganymede, and Europa deform Europa's orbit, causing variations in the tidal stresses from Jupiter that make Europa flex. But because Europa is farther from Jupiter, tidal effects on Europa are only about one-fourth as strong as those on Io, which may explain why ongoing volcanic activity has yet to be confirmed on Europa.

Some features on Europa's surface, such as the fracture patterns shown in Figure 13-11, may be the direct result of the crust being stretched and compressed by tidal flexing. **Figure 13-12** shows other features, such as networks of ridges and a young, very smooth circular area, that were probably caused by the internal heat that tidal flexing generates. The rich variety of terrain depicted in Figure 13-12, with stress ridges going in every direction, shows that Europa has a complex geologic history.

Among the unique structures found on Europa's surface are **ice rafts**. The area shown in **Figure 13-13a** was apparently subjected to folding, producing the same kind of linear features as those in Figure 13-12. But a later tectonic disturbance broke the surface into small chunks of crust a few kilometers across, which then "rafted" into new positions. A similar sort of rafting happens in Earth's Arctic Ocean every spring, when the winter's accumulation of surface ice breaks up into drifting ice floes (Figure 13-13b). The existence of such structures on Europa suggests that there is a subsurface layer of liquid water or soft ice over which the ice rafts can slide with little resistance.

But questions remain: Is Europa's heat mostly gone, leaving behind a frozen interior today? Is there a large subsurface ocean, or just thin pockets of liquid water? To answer these questions, we return to measurements of magnetic fields, which act like faithful messengers from a planet's interior.

An Underground Ocean?

Magnetic field measurements made by the Galileo spacecraft have provided key evidence for a large saltwater ocean within Europa. Unlike Earth or Jupiter, Europa does not seem to create a steady magnetic field of its own. But as Europa moves through Jupiter's intense magnetic field, electric currents are induced within the satellite's interior, just as they are around Io (see Section 13-5), and these currents generate a weak but measurable induced magnetic field. The strength and direction of this induced magnetic field varies, and can even reverse within hours, as Europa moves through different parts of the jovian magnetosphere. These quick reversals could not occur if Europa generated its field by itself.

To explain these observations, there must be an electrically conducting fluid beneath Europa's crust—a perfect description of an underground ocean of water with dissolved minerals.

(a) Ice rafts on Europa

Ice rafts (fractured segments of Europa's icy crust)

Linear features show how rafts once fit together.

10 km

(b) Ice floes on Earth

Ice floes

Akpatok Island

10 km

FIGURE 13-13 R I V U X G

Moving Ice on Europa and Earth (a) Some time after a series of ridges formed in this region of Europa's surface, the icy crust broke into "rafts" that were moved around by an underlying liquid or plastic layer. The colors in this Galileo image may be due to minerals that were released from beneath the surface after the crust broke apart. **(b)** Europa's ice rafts are analogous to ice floes created when pack ice breaks up, as in this spacecraft view of part of the Canadian arctic. (a: NASA/JPL/University of Arizona; b: NASA/USGS)

(Pure water is a very poor conductor of electricity, but electrically charged particles, such as dissolved minerals, would make the ocean water conducting.) The dissolved minerals might be the familiar salt found in Earth's saltwater oceans, but the actual minerals are not yet known.

If some of this water should penetrate upward to Europa's surface through cracks in the crust, it would vaporize and leave the dissolved minerals on the surface. In fact, these minerals might explain the dark linear features in Figure 13-11.

By combining measurements of Europa's induced magnetic field, gravitational pull, and oblateness, scientists estimate that Europa has a frozen ice crust on the order of 10 km thick on top of a liquid water ocean about 100 km deep (Figure 13-14). For comparison, the total liquid water in Europa's ocean is more than all of Earth's oceans combined! Beneath the ocean, models suggest a rocky mantle surrounding a metallic core some 600 km (400 mi) in radius. Since water is known to partially dissolve some of the minerals that make rock, contact with this mantle would be a likely source of the ocean's electrical conductivity.

Interestingly, Europa also has an extremely thin atmosphere of oxygen (producing an atmospheric pressure only a trillionth of Earth's). We saw in Section 9-5 that oxygen in Earth's atmosphere is produced by plants through photosynthesis. But Europa's oxygen atmosphere is probably the result of ions from the solar wind and Jupiter's magnetosphere striking the satellite's icy surface. These collisions break apart water molecules, liberating atoms of hydrogen (which escape into space) and oxygen.

The existence of a warm, subsurface ocean on Europa, if proved, would make Europa one of the few worlds in the solar system other than Earth with liquid water. The evidence for a subsurface ocean is very strong, and confirming the presence of liquid

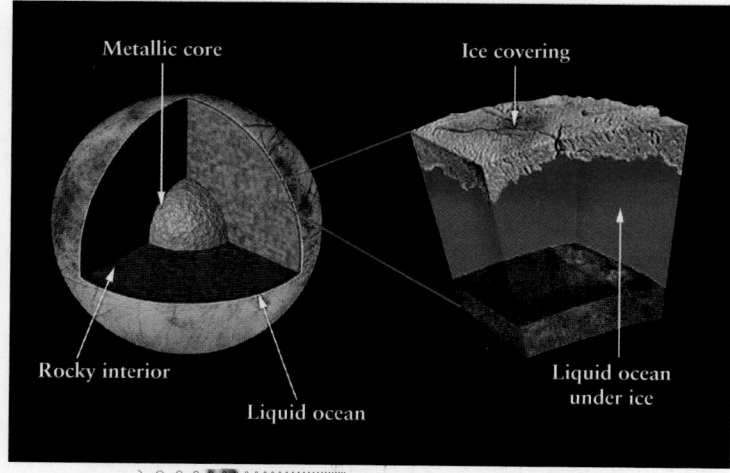

Metallic core

Ice covering

Rocky interior

Liquid ocean

Liquid ocean under ice

FIGURE 13-14 R I V U X G

Europa's Ocean While small compared to Europa's overall size, its ocean is still enormous. Estimated at about 100 km deep, it would be over 10 times deeper than Earth's oceans and contain more than twice the volume of water. (NASA/JPL)

water on Europa would have dramatic implications. On Earth, water and warmth are essentials for the existence of life. Perhaps single-celled organisms have evolved in the water beneath Europa's crust, where they would use dissolved minerals and organic compounds as food sources. In light of this possibility, NASA has taken steps to prevent biological contamination of Europa. At the end of the Galileo mission in 2003, the spacecraft (which may have carried traces of organisms from Earth) was sent to burn up in

Jupiter's atmosphere, rather than remaining in orbit where it might someday crash into Europa.

Does Europa Emit Plumes of Water Vapor?

In 2012, the Hubble telescope found evidence consistent with water plumes erupting from Europa's surface. Emission lines from water's components—hydrogen and oxygen—were observed above the south polar region (at levels above those found in Europa's very thin atmosphere). Then, in 2016, Hubble observed Europa 10 times and saw evidence for plumes on three occasions. Both observations led to estimates of similar masses of the plumes, coming from similar areas, and shooting water to a height of about 200 km (125 mi).

These plumes might result from **cryovolcanoes**. Instead of erupting molten rock, a cryovolcano (or ice volcano) ejects water. (Another term for this could be *cryogeyser*.)

If confirmed, such water plumes would be enormously significant in the search for life beyond Earth. Penetrating through Europa's approximately 10-km-thick layer of ice, in order to sample the ocean directly, would be very difficult. We can't even penetrate ice that deeply on Earth yet using people and heavy machinery. But if Europa is ejecting water from its ocean into space, sampling this ocean through the plumes becomes much easier.

As we'll see in Section 13-10, cryovolcanoes on Europa share the spotlight with Saturn's moon Enceladus, as two places with plumes of water from a subsurface ocean that might contain life.

A new NASA mission called the Europa Clipper is under development. This spacecraft might launch in 2022 (there is no firm date presently) and will carry instruments to more accurately determine the thickness of Europa's icy surface at various locations, as well as the depth of its ocean. The mission will also assess where conditions might be suitable for microbial life forms. And, if Europa indeed emits plumes from its liquid interior, this spacecraft will be able to fly through and sample these plumes for signs of microbial life.

CONCEPT CHECK 13-6

If the orbital resonance of Io, Europa, and Ganymede contributes to the heating of their interiors, why isn't Europa hot enough to have volcanism like Io?

CONCEPT CHECK 13-7

Why does an induced magnetic field coming from Europa indicate a large subsurface ocean?

Answers appear at the end of the chapter.

13-7 Liquid water may also lie beneath the cratered surfaces of Ganymede and Callisto

Unlike Io and Europa, the two outer Galilean satellites have cratered surfaces. In this respect, Ganymede and Callisto bear a superficial resemblance to our Moon (see the illustration with Table 13-1). But unlike the Moon's craters, the craters on both Ganymede and Callisto are made primarily of ice rather than rock, and Ganymede has a number of surface features that indicate a geologically active past.

> Both Ganymede and Callisto show evidence of past geologic activity.

The Two Faces of Ganymede's Surface

Ganymede is the largest satellite in our solar system and is even larger than the planet Mercury. Like our Moon, Ganymede has two distinct kinds of terrain, called **dark terrain** and **bright terrain** (**Figure 13-15**). On the Moon, the dark maria are younger than the light-colored lunar highlands (see Section 10-1). On Ganymede, by contrast, the dark terrain is older, as indicated by its higher density of craters. The bright terrain is much less cratered and therefore younger. Because ice is more reflective than rock, even Ganymede's dark terrain is substantially brighter than the lunar surface.

Voyager and Galileo images show noticeable differences between young and old impact craters on Ganymede. The youngest craters are surrounded by bright rays of freshly exposed ice that has been excavated by the impact (see the lower left of Figure 13-15). Older craters are darker, perhaps because of chemical processes triggered by exposure to sunlight.

The close-up Galileo image in **Figure 13-16** shows the contrast between dark and bright terrain. The area on the left-hand side of this image is strewn with large and small craters, reinforcing the idea that the dark terrain is billions of years old. The dark terrain is also marked by long, deep furrows, which are deformations of

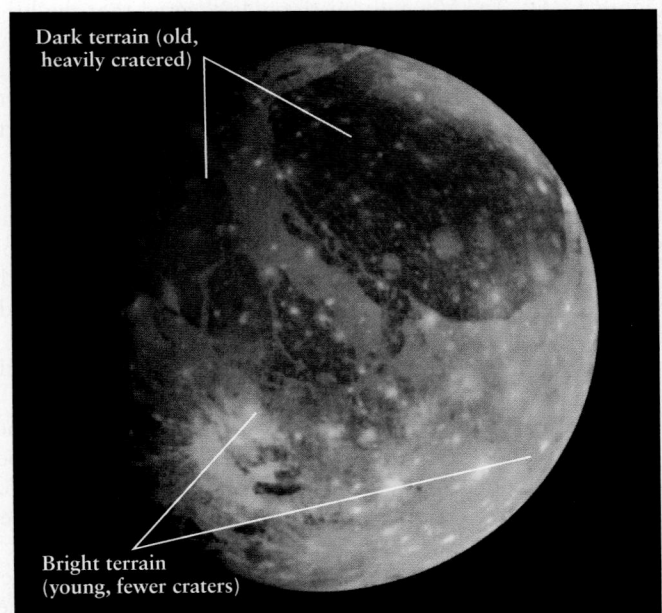

Dark terrain (old, heavily cratered)

Bright terrain (young, fewer craters)

FIGURE 13-15 R I Ⓥ U X G

Ganymede Two distinct types of terrain—one dark, heavily cratered, and hence old; the other bright, less cratered, and hence younger—are visible in this Galileo image of the hemisphere of Ganymede that always faces away from Jupiter. Craters in general appear bright, suggesting that the impacts that formed the craters excavated the surface to reveal ice underneath. (NASA/JPL)

Dark terrain: more craters

Bright terrain: fewer craters

Long grooves

Long, deep furrows

50 km

FIGURE 13-16 R I **V** U X G

Evidence for Geologic Activity on Ganymede Stresses in Ganymede's crust have produced linear features that extend for hundreds of kilometers across both the dark and bright terrain. These features must be quite ancient, since numerous impact craters have formed on top of them. (NASA/JPL/DLR)

Ganymede's crust that have partially erased some of the oldest craters. Since other craters lie on top of the furrows, the stresses that created the furrows must have occurred long ago. Because geologic activity is a result of a planet or satellite's internal heat, this shows that Ganymede must have had a warm interior in the distant past.

Linear features are also seen in the bright terrain on the right in Figure 13-16. The Voyager missions discovered that the bright terrain is marked by long grooves, some of which extend for hundreds of kilometers and are as much as a kilometer deep. Images from Galileo show that some adjacent grooves run in different directions, indicating that the bright terrain has been subjected to a variety of tectonic stresses. Galileo also found a number of small craters overlying the grooves, some of which are visible in Figure 13-16. The density of craters shows that the bright terrain, while younger than the dark terrain, is at least a billion years old.

Images from the Voyager missions led scientists to hypothesize that the bright terrain represented fresh ice that had flooded through cracks in Ganymede's crust, analogous to the maria on our Moon (see Section 10-1). However, high-resolution Galileo images do not show the kinds of surface features that would be expected if this terrain were produced by ice volcanoes. Yet there are several low-lying areas within the bright terrain that appear to have been flooded by a watery fluid that subsequently froze. These low-lying areas may be *rift valleys* where the crust has been pulled apart, like Valles Marineris on Mars (see Section 11-5). In order to flood the floors of these valleys, there must have been liquid water beneath Ganymede's surface when the valleys formed. This is additional evidence that Ganymede must have had a warm interior in the past, and that this internal warmth persisted at least until the era when the bright terrain formed a billion years ago. The persistence of internal warmth is rather surprising, because Ganymede is too small to have retained much of the heat left over from its formation or from radioactive elements, and it orbits too far from Jupiter to have any substantial tidal heating.

Ganymede's Interior and Bizarre Magnetic Field

An equally great surprise from Galileo observations was the discovery that Ganymede has its own magnetic field, and that this field is twice as strong as that of Mercury. This field is sufficiently strong to trap charged particles, giving Ganymede its own "mini-magnetosphere" within Jupiter's much larger magnetosphere. The presence of a magnetic field shows that electrically conducting material must be in motion within Ganymede to act as a dynamo, which means that the satellite must have substantial internal heat even today. A warm interior causes differentiation, and Galileo measurements of how its mass is distributed indeed showed that Ganymede is highly differentiated. It has a metallic core some 500 km in radius, surrounded by a rocky mantle and by an outer shell of ice 800 km thick.

An even more surprising result is that Ganymede's magnetic field varies somewhat as it orbits Jupiter, and these variations are correlated with Ganymede's location within Jupiter's magnetic field. One explanation for this variation is that Ganymede has a layer of salty, electrically conducting liquid water some 170 km beneath its surface. As the satellite moves through Jupiter's magnetic field, electric currents are induced in this layer, just as happens within Europa (see Section 13-6), and these currents generate a weak field that adds to the stronger field produced by Ganymede's dynamo.

In 2015, The Hubble telescope strengthened the case for a subsurface ocean on Ganymede. By observing changes in the aurorae on Ganymede, which would be affected by a salty, electrically conductive ocean, astronomers found evidence for a 100-km-deep ocean, containing 10 times the volume of Earth's oceans.

A possible explanation for Ganymede's high internal heat is that gravitational forces from the other Galilean satellites may have affected its orbit in the past. While Ganymede's present-day orbit around Jupiter is quite circular, calculations show that the orbit could once have been more elliptical. Just as for Io and

Europa, such an elliptical orbit would have given rise to substantial tidal heating of Ganymede's interior. In that case, it may have retained a substantial amount of that heat to the present day.

Callisto's Perplexing Surface

Standing in marked contrast to Io, Europa, and Ganymede is Callisto, Jupiter's outermost Galilean satellite. Images from Voyager showed that Callisto has numerous impact craters scattered over an icy crust (Figure 13-17). Callisto's ice is not as reflective as that of Europa or Ganymede, however; the surface appears to be covered with some sort of dark mineral deposit.

CAUTION!

Figure 13-17, as well as the images that accompany Table 13-1, shows that Callisto is the darkest of the Galilean satellites. Nonetheless, its surface is actually more than twice as reflective as that of our own Moon. Callisto's surface may be made of "dirty" ice, but even dirty ice reflects more light than the gray rock found on the Moon. If you could somehow replace our Moon with Callisto, a full moon as seen from Earth would be 4 times brighter than it is now (thanks also in part to Callisto's larger size).

The Voyager images led scientists to conclude that Callisto is a dead world, with an ancient surface that has never been reshaped by geologic processes. There is no simple relationship between Callisto's orbital period and those of the other Galilean satellites, so it would seem that Callisto never experienced any tidal heating and hence was unable to power any geologic processes at its surface. But once the Galileo spacecraft went into orbit around Jupiter and began returning high-resolution images, scientists realized that Callisto's nature is not so simple. In fact, Callisto proves to be the most perplexing of the Galilean satellites.

One curious discovery from Galileo is that while Callisto has numerous large craters, there are very few with diameters less than 1 km. Such small craters are abundant on Ganymede, which has probably had the same impact history as Callisto. The lack of small craters on Callisto implies that most of these craters have been eroded away. How this erosion could have happened is not known. Another surprising result is that Callisto's surface is covered by a blanket of dark, dusty material (see the inset in Figure 13-17). Where this material came from, and how it came to be distributed across the satellite's surface, is not understood.

Callisto's Interior: Hot or Cold?

Callisto's most unexpected feature is that, like Europa and Ganymede, it appears to have an induced magnetic field that varies as the satellite moves through different parts of Jupiter's magnetosphere. Callisto must have a layer of electrically conducting material in order to generate such a field. One model suggests that this material is in the form of a subsurface ocean like Europa's but only about 10 km (6 mi) deep. However, without tidal heating, Callisto's interior is probably so cold that a layer of liquid water would quickly freeze. One proposed explanation is that Callisto's ocean is a mixture of water and a form of "antifreeze." In cold climates on Earth, antifreeze is added to the coolant in automobile radiators to lower the liquid's freezing temperature and prevent it from solidifying. Ammonia (NH_3) in Callisto's proposed subsurface ocean could play the same role: There was plenty of ammonia available at these far reaches of the solar system, and the ability for ammonia to act like an antifreeze in water is well-established. Another example of a substance making it harder for water to freeze is salt; rock salt is spread onto icy roads to lower the temperature at which water melts, allowing the ice to melt for safer driving conditions.

The presence of a liquid layer beneath Callisto's surface is difficult to understand. Even with ammonia acting as an antifreeze, this layer cannot be too cold or it would solidify. This argues for a relatively *warm* interior. But Galileo measurements suggest that Callisto's interior is not highly differentiated. The lack of differentiation implies that the interior is *cold* and may never have been heated enough to undergo complete chemical differentiation. The full story of Callisto's interior, it would seem, has yet to be told.

Although Ganymede and Callisto have many differences, one feature they share is an extremely thin and tenuous atmosphere. Like Europa, Ganymede has a thin oxygen (O_2) atmosphere thought to be created when radiation splits water ice (H_2O) on the surface into oxygen and hydrogen, with the lighter H_2 escaping Ganymede's weak gravity. On Callisto, the atmosphere is known to contain carbon dioxide (CO_2). This gas would escape in several days and is thought to be replenished by evaporation of CO_2 ice from Callisto's cold polar regions.

The diagrams in Figure 13-18 summarize the probable interior structures of the Galilean satellites. They demonstrate the remarkable variety of these terrestrial worlds. We'll learn much more about Jupiter's icy moons from the European Space Agency (ESA)'s spacecraft JUICE (Jupiter's Icy Moon Explorer). Expected to be launched in 2022, JUICE would arrive at Jupiter in 2030.

Close-up from Galileo

Callisto from Voyager 1

5 km

FIGURE 13-17 R I **V** U X G

Callisto Numerous craters pockmark Callisto's icy surface. The inset shows a portion of a huge impact basin called Valhalla. Most of the very smallest craters in this close-up view have been completely obliterated, and the terrain between the craters is blanketed by dark, dusty material. (NASA/JPL; NASA/JPL/ASU)

CONCEPT CHECK 13-8

Why is it thought that Callisto's interior might contain ammonia?

Answer appears at the end of the chapter.

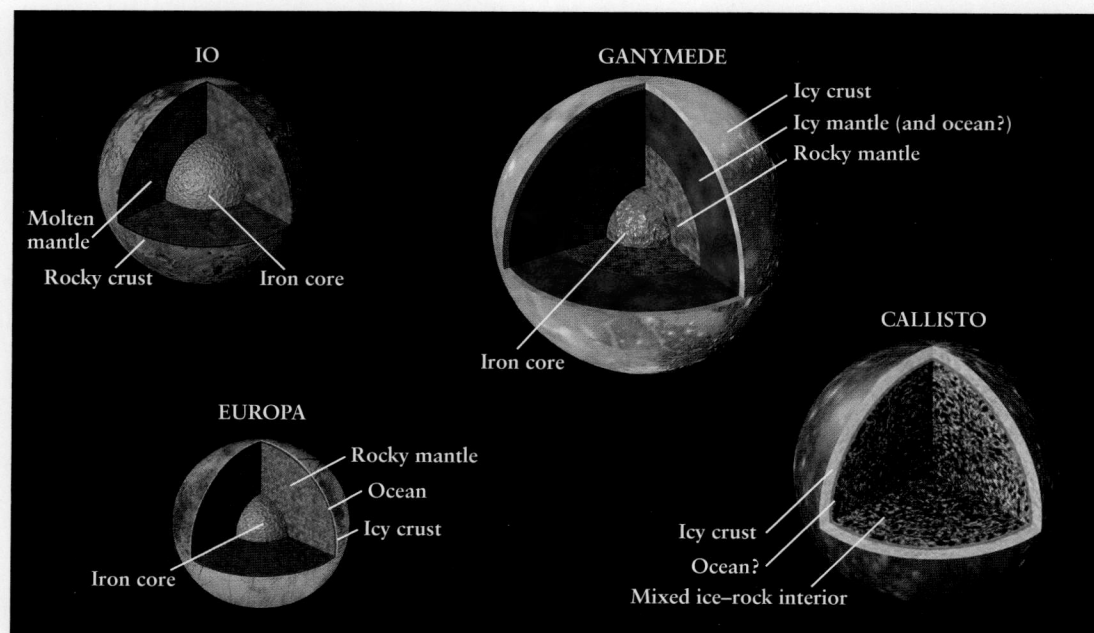

FIGURE 13-18

**Interiors of the Galilean
Satellites** These cross-sectional
diagrams show the probable internal
structures of the four Galilean satellites
of Jupiter, based on information from the
Galileo mission. (NASA/JPL)

13-8 Titan has a thick atmosphere and hydrocarbon lakes

TUTORIAL
13.2 ➤ Unlike Jupiter, Saturn has only one large satellite that is comparable in size to our Moon. This satellite, Titan, was discovered by Christiaan Huygens in 1665. Titan's diameter of 5150 km makes it second in size among satellites to Ganymede (Section 13-7). Like Ganymede, Titan has a low density that suggests it is made of a mixture of ice and rock. (See Table 7-2 for a comparison between Titan, the Moon, and the Galilean satellites.) But unlike Ganymede or any other satellite in the solar system, Titan has a thick atmosphere with a unique chemical composition.

Titan's Atmosphere

By the early 1900s, scientists had begun to suspect that Titan (Figure 13-19) might have an atmosphere, because it is cool enough and massive enough to retain heavy gases. (Box 7-2 explains the criteria for a planet or satellite to retain an atmosphere.) These suspicions were confirmed in 1944, when Gerard Kuiper discovered the spectral lines of methane in sunlight reflected from Titan. (We discussed Titan's spectrum and its interpretation in Section 7-3.)

Because of its atmosphere, Titan was a primary target for the Voyager missions. To the chagrin of mission scientists and engineers, however, Voyager 1 spent hour after precious hour of its limited, preprogrammed observation time sending back featureless

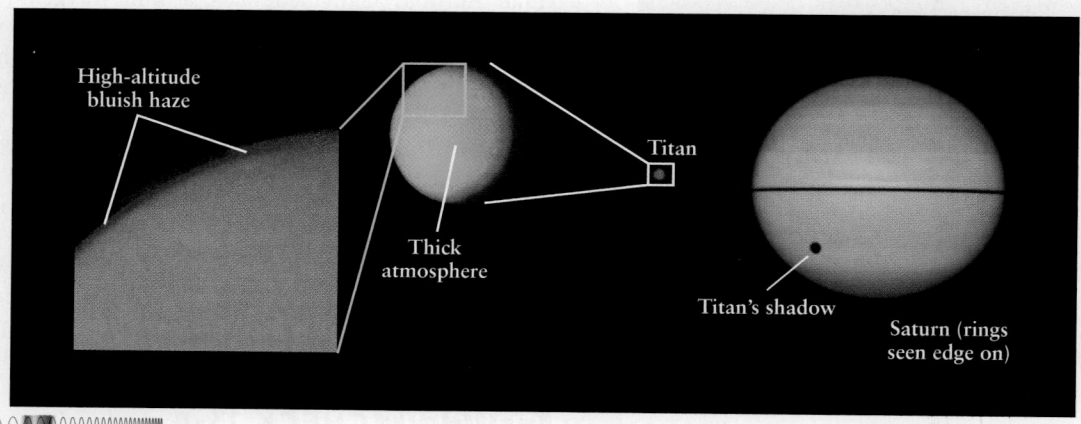

FIGURE 13-19 R I (V) U X G

Titan The Hubble Space Telescope image at far right shows Titan and the shadow it casts on Saturn's clouds. The Voyager image in the middle shows how Titan's atmosphere gives it a nearly featureless appearance. At far left is a color-enhanced Voyager image that shows a haze extending well above the satellite's visible edge, or limb. (Left: NASA/JPL; center: NASA/JPL; right: Eich Karkoschka, LPL/STSSc1/NASA) images like the one at the center of Figure 13-19. Titan's atmosphere is so thick—about 200 km (120 mi) deep, or about 10 times the depth of Earth's troposphere—that it blocked Voyager's view of the surface. The haze surrounding Titan is so dense that little sunlight penetrates to the ground; noon on Titan is only about 1/1000 as bright as noon on Earth, though still 350 times brighter than night on Earth under a full moon.

By examining how the radio signal from Voyager 1 was affected by passing through Titan's atmosphere, scientists inferred that the atmospheric pressure at Titan's surface is 50% greater than the atmospheric pressure at sea level on Earth. Titan's surface gravity is weaker than that of Earth, so to produce this pressure, Titan must have considerably more gas in its atmosphere than Earth. Indeed, calculations show that about 7 times more gas (by mass) lies above a square meter of Titan's surface than above a square meter of Earth. With Titan's combination of low gravity and thick atmosphere, estimates even suggest that humans could literally fly through the atmosphere by flapping special "wings" attached to their arms!

Voyager data show that more than 95% of Titan's atmosphere is nitrogen gas, although it is not clear where this nitrogen came from. One idea suggests that most of this nitrogen came from ammonia (NH_3), a compound of nitrogen and hydrogen that is quite common in the outer solar system. In this scenario, the ammonia is released from Titan's interior, and then is easily broken into hydrogen and nitrogen by the Sun's ultraviolet (UV) radiation. Titan's gravity is too weak to retain hydrogen atoms (see Box 7-2), so these atoms escape into space and leave the more massive nitrogen gas molecules behind. However, not all of the details for this scenario fit well with recent Cassini data, prompting another model.

A 2011 experiment at the University of Tokyo indicates that *comets smashing into Titan's surface* might have created its nitrogen atmosphere. The experiment involved two steps. First, a laser was shot into water ice that was rich in ammonia. The energy of the laser blast simulates a cometary impact, and the ammonia-rich ice approximates Titan's crust. Second, measurements were made of the amount of nitrogen gas liberated by the simulated impact. Finally, by assuming the frequency of impacts expected during the Late Heavy Bombardment about 3.9 billion years ago (see Section 8-6), researchers could account for the nitrogen seen in Titan's present-day atmosphere. Nonetheless, both of these models (based on UV light or impacts) for Titan's atmospheric nitrogen have shortcomings and the origin of this atmosphere remains a mystery.

The remaining 5% of Titan's atmosphere is predominantly methane (CH_4), which is the principal component of the "natural gas" used on Earth as fuel. However, in the absence of oxygen, Titan's atmosphere is in no danger of catching on fire. Methane is an example of a **hydrocarbon**, which is any compound made only from hydrogen and carbon atoms. The interaction of methane in Titan's atmosphere with ultraviolet light from the Sun produces other hydrocarbons, including ethane (C_2H_6). As we will see, methane and ethane play a very significant role on Titan beyond its atmosphere.

What gives Titan its reddish-brown color (seen in Figure 13-19)? Nitrogen in Titan's atmosphere can combine with airborne hydrocarbons to produce other compounds, such as hydrogen cyanide (HCN). Hydrogen cyanide, along with other molecules, can join together in long, repeating molecular chains—some of which remain suspended in Titan's atmosphere and might be responsible for its unique color.

Titan's Methane Cycle

On Earth, water can be a gas (like water vapor in the atmosphere), a liquid (as in the oceans), or a solid (as in snow and ice). On Titan, the atmospheric pressure is so high and the surface temperature is so low—about 95 K ($-178°C$, or $-288°F$)—that any water is frozen solid. But temperatures and pressures on Titan are just right for methane to exist as a gas, liquid, or solid. This single revelation by the Voyager missions—that Titan could support three phases of matter for methane—led to the bold prediction that vapor clouds of methane might rain down onto Titan's surface to form the only lakes known outside of Earth (**Figure 13-20**). Are these predictions correct? Does Titan have something analogous to Earth's hydrologic cycle, with methane playing the role of water? As we will see, the answer is definitely yes!

There is one difference, though. While Titan's surface consists mostly of water ice, methane ice is only expected at high altitudes in Titan's atmosphere, where this frozen methane snow melts to rain before hitting the surface. Even though methane ice appears to play little role in shaping Titan, it can shape plans for future spacecraft orbits that might want to avoid this methane snow.

> A faint mist of methane rain falls sporadically on Titan.

FIGURE 13-20

Methane Cycle on Titan On Titan, small variations in temperature and pressure can change methane between its gas and liquid states, just as water can change states to form the hydrological cycle on Earth. Even though Titan receives only about 1% as much sunlight as Earth does, it is enough to drive a methane cycle.

Bright region covered with icy volcanic deposits

Dark terrain near Titan's equator

Clouds near Titan's south pole

(a) R **I** **V** U X G

1. Liquid flowed in a number of small streams . . .

2. . . . which merged into a river . . .

3. . . . which emptied into a large, dark-colored outflow channel.

4. Dark hydrocarbon polymers from Titan's atmosphere fell onto the surface, then were washed by methane rains into the streams, river, and outflow channel. Hence these surface features have a dark color.

(b) R **I** **V** U X G

15 cm

(c) R **I** **V** U X G

VIDEO 13.4   VIDEO 13.5

FIGURE 13-21

Beneath Titan's Clouds (a) This mosaic of Titan was constructed from Cassini images made at visible and infrared wavelengths during a close flyby in December 2005. **(b)** From 6.5 km (4 mi) high, the descending Huygens lander recorded this view of features on Titan. Carved by flowing liquid, rivers and streams seem to have flowed between ridges about 200 m (650 ft) high. It is too cold on Titan for water to be a liquid, so these streams and rivers must have carried liquid methane or ethane. **(c)** This nearly true-color Huygens image shows the view from Titan's surface. The 15-cm (6-in.) wide "rock" is about 85 cm (33 in.) from the camera. Unlike Earth rocks, Titan "rocks" are chunks of water ice. (a: NASA/JPL/University of Arizona; b: NASA/JPL/ESA/University of Arizona; c: NASA/JPL/ESA/University of Arizona)

Exploring Titan with Cassini and Huygens

The designers of the Cassini mission to Saturn—which ended in September 2017—made the exploration of Titan a central goal. Carried with Cassini was the Huygens probe, which landed successfully on Titan; Huygens was the first landing on any object in the outer solar system. By penetrating Titan's atmosphere, Cassini and Huygens lifted a veil on the largest unexplored surface in the solar system.

Titan's atmosphere is more transparent to infrared wavelengths than to visible light, so Cassini has used its infrared telescope to obtain the first detailed images of Titan's surface (Figure 13-21a). Most of the surface is light colored, but a swath of dark terrain extends around most of Titan's equator.

Cassini used its atmosphere-penetrating radar to map the dark terrain. (Like infrared light, radio waves can pass through Titan's atmosphere.) These images reveal a series of long, parallel lines of sand dunes about 1 to 2 km apart (Figure 13-22). These are aligned west to east, in the same direction that winds blow at Titan's equator, and so presumably formed by wind action. The dunes can be 100 m high and stretch for hundreds of kilometers.

The most remarkable images of Titan's surface came from Huygens, the small landing craft that Cassini carried on its journey from Earth. Named for the discoverer of Titan, Huygens entered Titan's atmosphere on January 14, 2005. The lander took 2½ hours to descend to the surface under a parachute, during which time it recorded detailed images of the terrain (see Figure 13-21b).

FIGURE 13-22 R I V U X G

Sand Dunes on Titan Unlike sand on Earth, which is made of small particles of silicate rock, sand on Titan is probably made from a mixture of water ice and organic compounds. The organic compounds give the dark sand their color, which might look something like coffee grounds. A mountainous region in the center of this image prevents formation of dunes. Concentrated around the equator, at least 13% of Titan's total surface area is covered in dunes. (NASA/JPL-Caltech/ASI)

Chunks of water ice appear on the surface like rocks (see Figure 13-21c). These are thought to come from the breakup of Titan's mostly water-ice crust. After touchdown, Huygens continued to return data for another 70 minutes before its batteries succumbed to Titan's low temperatures.

Huygens images such as Figure 13-21b show that liquids have indeed flowed on Titan like water on Earth. Yet none of the images made by the lander during its descent or after touchdown showed any evidence of standing liquid. However, dampness in the ground detected by Huygens indicated recent methane rainfall or flooding at the landing site.

What would a rainstorm be like on Titan? First, consider that the strength of a downpour on Earth depends on how much water vapor builds up in the atmosphere before larger droplets form and fall as rain. Compared to Earth, measurements indicate that Titan's atmosphere holds much more moisture before droplets of methane would precipitate out as rain. However, the evaporation rate of methane from Titan's surface is slower than the evaporation of water into Earth's atmosphere. As a result, it can take 100 to 1000 years before enough methane builds up to rain down from Titan's skies. But, when it does rain, so much methane has built up that flash floods are the likely result. The methane raindrops are also expected to be several times larger than the water drops on Earth, and to fall more slowly in Titan's lower gravity and thicker atmosphere. Grape-sized, slow-falling raindrops are nice to imagine, but keep in mind that they are similar in composition to liquefied natural gas.

Cassini Discovers Vast Lakes

Huygens's view of Titan was limited to the equatorial region where it landed. While evidence near the equator pointed to the presence of liquid in the past, it took Cassini's radar maps of the poles to find present-day liquid on Titan. The radar maps (Figure 13-23) have revealed the *first standing bodies of liquid* in the solar system beyond Earth. The terrain includes vast lakes that are most likely methane and ethane. If correct, evaporation from these large reservoirs could help explain how Titan's atmosphere contains so much methane even though ultraviolet sunlight breaks it apart.

How Titan's lakes were discovered is more subtle than the images imply. To analyze Titan's terrain through its atmosphere, radar waves are reflected like echoes off its surface. This works for imaging typical surfaces with mountains and canyons, but very flat surfaces reflect radar at only one angle—like a ball bouncing off a wall—and no radar echo is returned (except when Cassini is directly overhead). Therefore, Titan's lakes are inferred by their *flatness:* They are surface regions where very little radar echo is detected. In fact, the measurements are so precise that the lakes present another mystery: *They are too flat.*

(a)

(b)

FIGURE 13-23 🅡 I V U X G

Hydrocarbon Seas on Titan Cassini used its onboard radar to record these false-color views of the region near Titan's north pole. **(a)** The dark regions show where the surface does not reflect radio waves at all, which leads to the interpretation that these are flat lakes (presumably filled with liquid methane and ethane). **(b)** Lake Ligeia Mare is one of the larger lakes and is about the size of Lake Superior (about 350 mi, or 560 km wide). In this high-resolution image, many tributaries can be seen around the lake's perimeter. (a: NASA/JPL/USGS; b: NASA/JPL-Caltech/ASI/Cornell)

Most large lakes have wind-driven waves, and with Titan's thick atmosphere and low gravity, wind is expected even at the poles. Indeed, we mentioned that near the equator, windblown sand dunes can reach 100 m in height, and stretch for hundreds of kilometers (see Figure 13-22). Thus, the flatness of the polar lakes remains a mystery.

Overall, there are many more lakes near Titan's north pole than its south pole. Why? Titan's northern hemisphere is currently in its winter season. Since lower temperatures lead to more condensation, more lakes are expected. Over the course of a Titan year (which lasts 29.37 Earth years) the supply of liquid will migrate with the seasons from pole to pole, much as frozen carbon dioxide migrates seasonally on Mars (see Section 11-6).

More than the seasons however, it is expected that an even larger effect accounts for a long-term predominance of lakes in the north. Due to the shape of Saturn's orbit around the Sun, during the Titan summer the entire Saturn–Titan system is 12% farther from the Sun than during the winter, making Titan's north cooler, on average, than the south. Over long timescales, when you consider the combined orbital properties of Saturn and Titan, lakes are expected to predominate at one pole or another for about 32,000 years before reversing, and we just happen to be in a northern part of that cycle.

Cryovolcanoes on Titan?

Combining infrared and radar images, Cassini also detected surface features that could indicate a cryovolcano (**Figure 13-24**). Instead of erupting molten rock, a cryovolcano (or ice volcano) on Titan might erupt or ooze icy mixtures of water, ammonia, and methane.

Deep crater

FIGURE 13-24 Ⓡ Ⓘ V U X G

Cryovolcanism The Sorta Facula region contains Titan's best candidate for a cryovolcano made of water ice. Some of the taller peaks are more than 1000 m high, next to a deep 1500-m volcanic crater. There are also examples on Earth where a crater is deeper than the height of the volcano. At the moment, the potential cryovolcano does not appear to be active. The blue areas in this false-colored radar map are thought to be exposed water ice, with other colors possibly due to a thin layer of organic compounds covering water ice. (NASA/JPL-Caltech/ASI/USGS/University of Arizona)

Acting as an "antifreeze," the ammonia (NH_3) is key, allowing the mixture to remain a liquid within Titan's interior. If correct, this scenario could also explain the source of Titan's atmospheric nitrogen: ammonia from the interior broken down by ultraviolet light.

In addition to evaporation from lakes, the presence of ice volcanoes on Titan could explain why there is so much methane in the satellite's atmosphere. Methane molecules are broken apart by ultraviolet photons from the Sun, and the hydrogen atoms escape into space. This process would destroy all of Titan's methane within 10 million years unless a fresh supply is added to the atmosphere, perhaps aided by cryovolcanoes. While entirely fascinating, Titan's cryovolcanism isn't entirely unique: We saw evidence for plumes from possible cryovolcanos on Europa in Section 13-6, and we'll see in Section 13-10 that another moon of Saturn (Enceladus) shows ongoing cryovolcanism.

CONCEPT CHECK 13-9

Explain two roles that ammonia (NH_3) seems to play on Titan: one role on or above the surface, and one below.

Answer appears at the end of the chapter.

Titan's Geology

There are some indications that Titan might have a subsurface ocean of liquid water. Electrical measurements of the surface made by Huygens are consistent with the top of such an ocean about 45 km beneath the surface. Furthermore, models with a subsurface ocean might provide a better fit, compared to a solid body, for Titan's carefully measured shape and gravitational field. However, due to Titan's atmosphere, Cassini cannot orbit closely enough to make the additional magnetic field measurements that would more clearly reveal a subsurface ocean on Titan. Why is water so interesting? As we will see in Chapter 27 on the search for extraterrestrial life, liquid water either in a subsurface ocean or associated with cryovolcanoes means that Titan could be habitable to simple forms of life.

The predicted interior of present-day Titan is shown in **Figure 13-25**, but what was Titan like in the past? We have seen that Titan was probably volcanically active to provide its atmosphere with methane, but what has powered the volcanic activity? Titan is too small to have retained the internal heat from its formation. Furthermore, although Titan's rotation is synchronous like that of Io, its orbit around Saturn is so circular that it cannot be heated by tidal flexing in the way that Io is (see Section 13-4).

One model to power cryovolcanism proposes that about 500 million years ago, the last of Titan's internal heat from radioactivity caused convection to take place in the satellite's icy crust. The resulting ice volcanoes would have repaved the entire surface, explaining Titan's unexpectedly low crater density. The amount of methane released during this proposed episode is enough to sustain the gas in the atmosphere, but—very importantly—not necessarily enough to create the extensive hydrocarbon lakes. Thus, we have much to learn about Titan's present condition, as well as its past.

What already seems clear from the data is that the early predictions were correct: Titan's methane does indeed take liquid and gas forms, producing a cycle similar to water's hydrological cycle on Earth. With its storms, lakes, and volcanoes, Titan is a truly fascinating world.

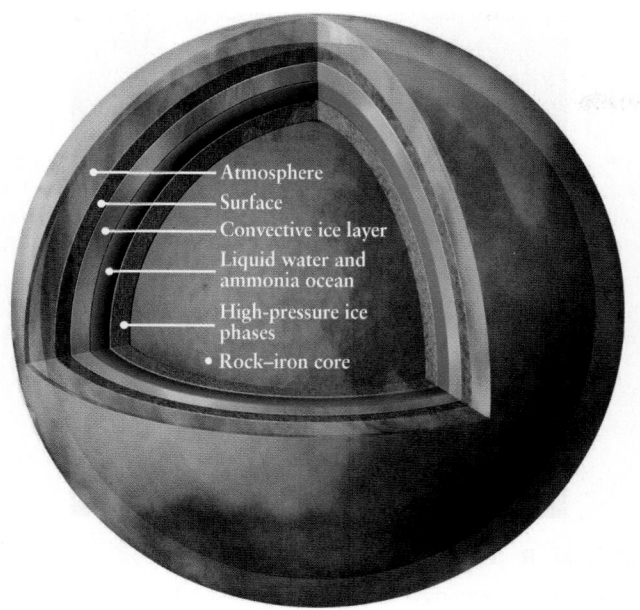

FIGURE 13-25

Titan's Interior This cross-sectional diagram is based on models of Titan's interior. As shown, there could be an outer layer of convective ice, and a liquid water subsurface ocean could extend hundreds of kilometers deep. Based on Titan's density, about half the moon is made of iron, with the other half water in various states. (Kevin Hand/NASA)

CONCEPT CHECK 13-10

If lakes were not found in images made by the Huygens probe, and Cassini cannot see through Titan's thick haze, how were lakes discovered?

CONCEPT CHECK 13-11

What insight from Voyager data led scientists to predict that Titan's methane might be involved in a cycle similar to Earth's hydrological cycle?

Answers appear at the end of the chapter.

13-9 Jupiter has dozens of small satellites that have different origins

Besides the four Galilean satellites, Jupiter has at least 75 other confirmed satellites (as of 2018). Each is named for a mythological character associated with Jupiter. Appendix 3 gives details of some of the satellites' orbits. It is helpful to sort Jupiter's satellites into three groups according to their sizes and orbits.

Four planet-sized satellites, Io, Europa, Ganymede, and Callisto, can be thought of as four of the terrestrial worlds of the solar system. All are comparable in size to the planet Mercury (see Table 13-1). The semimajor axes of their orbits range from 5.9 to 26 times the radius of Jupiter. (By comparison, the distance from Earth to the Moon is about 60 times Earth's radius.)

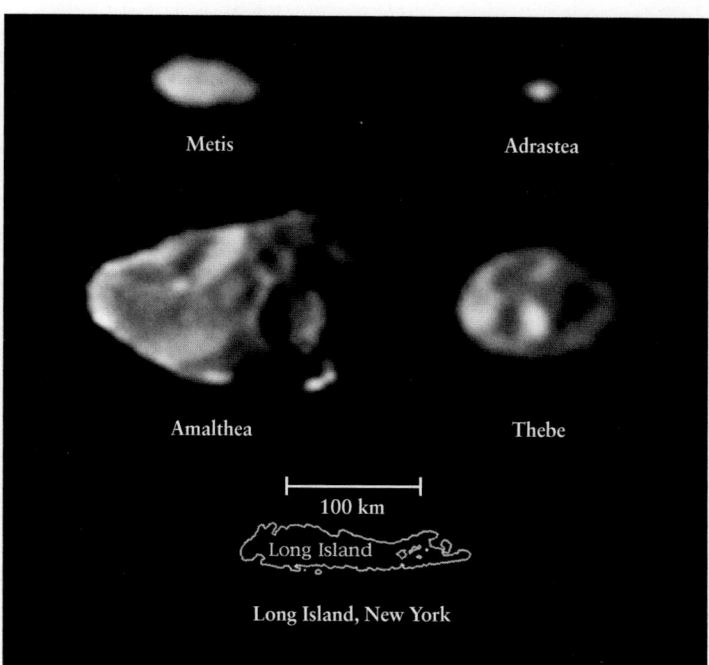

FIGURE 13-26 R I V U X G

Jupiter's Four Inner Satellites First observed in 1892, Amalthea was the last satellite of any planet to be discovered without the aid of photography. Metis, Adrastea, and Thebe, by contrast, are so small that they were not discovered until 1979. These images from the Galileo spacecraft are the first to show these three satellites as anything more than points of light. (NASA/JPL/Cornell University)

Four small inner satellites, Metis, Adrastea, Amalthea, and Thebe, have orbital semimajor axes between 1.8 and 3.1 times the radius of Jupiter. As Figure 13-26 shows, these satellites have irregular shapes, much like a potato (or an asteroid). The largest of the inner satellites, Amalthea, measures about 270 by 150 km (170 by 95 mi)—about 10 times larger than the moons of Mars, but only about one-tenth the size of the Galilean satellites. At these small sizes, gravity is too weak to pull the moons into a spherical shape. Discovered in 1892, Amalthea has a distinct reddish color due to sulfur that Jupiter's magnetosphere removed from Io's volcanic plumes, then deposited onto Amalthea's surface. The three other inner satellites are even smaller than Amalthea. They were first discovered in images from Voyager 1 and Voyager 2.

Seventy-one small outer satellites have diameters from 184 km for the largest to about 1 km for the smallest. All orbit far beyond Callisto, with semimajor axes that range from about 100 to 360 times the radius of Jupiter. (Twenty-three of these satellites were discovered in 2003 alone, and even more may be discovered as telescope technology continues to evolve.)

The Galilean satellites and the four inner satellites all orbit Jupiter in the plane of the planet's equator. (The same is true of Jupiter's rings, discussed in Section 12-9.) Furthermore, these orbits are all **prograde**, which means that these

Only a few of Jupiter's satellites orbit the planet in the same direction as Jupiter rotates.

objects orbit Jupiter in the same direction as Jupiter's rotation. We would expect orbits in the same direction if these satellites formed from the same primordial, rotating cloud as Jupiter itself. (In a similar way, the planets all orbit the Sun in nearly the same plane and in the same direction as the Sun's rotation. As we saw in Section 8-4, this reinforces the idea that the Sun and planets formed from the same solar nebula.)

In contrast, the 71 outer satellites all circle Jupiter along orbits that are inclined at steep angles to the planet's equatorial plane. These satellites are thought not to have formed along with Jupiter; rather, they are probably wayward asteroids captured by Jupiter's powerful gravitational field. One piece of evidence for this is that most of the outer satellites are in **retrograde orbits**: They orbit Jupiter in a direction opposite the planet's rotation. Calculations show that it is easier for Jupiter to capture a satellite into a retrograde orbit than a prograde one.

CONCEPT CHECK 13-12

Why does a retrograde orbit suggest that a moon is a captured asteroid?

Answer appears at the end of the chapter.

13-10 Saturn has moderate-sized moons, including Enceladus and its watery plumes

Saturn, too, has an extensive collection of satellites: 62 are known as of 2018. As we did for Jupiter's satellites in Section 13-9, we can subdivide the satellites of Saturn into three categories based on their sizes and their orbits. (Appendix 3 has information about the orbits of all of Saturn's satellites.) We will see that while Saturn does not have a set of large moons like the Galilean satellites of Jupiter, it does have a remarkably diverse collection of moderate-sized satellites.

Saturn's Retinue of Moons

One planet-sized satellite, Titan, is actually one of the terrestrial worlds of our solar system. With its diameter of 5150 km, Titan is intermediate in size between Mercury and Mars. Its prograde orbit has a semimajor axis of just over 20 times the radius of Saturn and lies in nearly the same plane as Saturn's equator and rings.

Six moderate-sized satellites were all discovered before 1800. These six satellites have properties and diameters that lead us to group them in pairs: Mimas and Enceladus (400 to 500 km in diameter), Tethys and Dione (roughly 1000 km in diameter), and Rhea and Iapetus (about 1500 km in diameter). The semimajor axes of their orbits range from 3.1 to 59 times the radius of Saturn. Like Titan, all of their orbits are prograde and close to the plane of Saturn's equator.

Fifty-five small satellites have sizes from 3 to 266 km. Some are in prograde orbits as small as 2.2 times the radius of Saturn, and lie within or close to the rings (see Figure 12-26). These satellites may be jagged fragments of ice and rock produced by impacts and collisions. The shepherd satellites that shape Saturn's F ring (see Figure 12-25) may be such fragments. Others are in retrograde

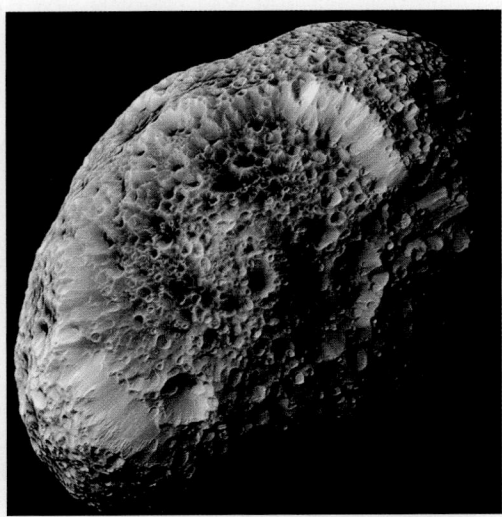

FIGURE 13-27   R I V U X G

Hyperion This heavily pockmarked moon is about 410 km across its elongated axis. Its density is about *half* that of water, so it is probably very porous. Rather than rotating around a single axis, the gravitational tugs on its elongated body by Titan and Saturn lead to a non-repeating chaotic tumble. (NASA, ESA, JPL, SSI and Cassini Imaging Team)

orbits with semimajor axes as large as 380 times Saturn's radius and may be captured asteroids. A little farther out than Titan, Cassini made a close flyby of one strange moon named Hyperion (Figure 13-27).

CONCEPT CHECK 13-13

In Figure 13-27, Hyperion appears almost like a sponge on the outside. Why would we, in fact, expect a lot of empty space throughout Hyperion?

Answer appears at the end of the chapter.

Six Strange Satellites

The most curious and mysterious of these worlds may be Saturn's six moderate-sized satellites; four of these are shown in Figure 13-28. The six satellites have very low average densities of less than 1400 kg/m³. This implies that unlike the satellites of the inner solar system or the Galilean satellites of Jupiter, they are composed primarily of ices (mostly frozen water and ammonia) with very little rock. The motions of these midsized satellites are not surprising: All six orbit the planet in the plane of Saturn's equator and in the direction of the planet's rotation, and all six exhibit synchronous rotation. What is surprising is the curious variety of surface features that the Voyager and Cassini spacecraft discovered on these satellites.

Mimas, the innermost of the six middle-sized moons, has the heavily cratered surface we would expect on such a small world. Its most remarkable feature is a crater so large (Figure 13-28a) that the impact that formed it must have come close to shattering Mimas into fragments.

Most of Tethys, like Mimas, is heavily cratered. But Tethys also has an immense rift valley called Ithaca Chasma that extends for 2000 km, or about ¾ of the way around Tethys (Figure 13-28b). This rift may have formed when the water inside Tethys cooled

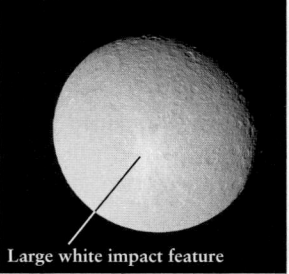

(a) Mimas
(diameter 392 km)

(b) Tethys
(diameter 1060 km)

(c) Dione
(diameter 1120 km)

(d) Rhea
(diameter 1530 km)

FIGURE 13-28 R I (V) U X G

Saturn's Mid-Sized Satellites (a–d) These images from Cassini show the variety of surface features found on four of Saturn's six moderate-sized satellites. The satellites depicted are not shown to scale. (NASA/JPL/Space Science Institute)

and solidified, or it may be the result of a shock wave that traveled through Tethys after an immense asteroid impact. (A large impact crater lies at the antipode, which is on the side of Tethys opposite Ithaca Chasma.)

Dione, slightly larger than Tethys, has a strange surface dichotomy: The surfaces of its leading hemisphere (that is, the hemisphere facing toward its direction of orbital motion) and trailing hemisphere are quite different. Dione's leading hemisphere is heavily cratered, but its trailing hemisphere is covered by a network of bright stress cracks caused by some sort of tectonic activity (Figure 13-28c). Like Dione, the moon Rhea also has more craters on its leading hemisphere (Figure 13-28d).

Another moon of Saturn, Iapetus, is also associated with a newly discovered ring of Saturn (**Figure 13-29**). Iapetus has been known to be unusual ever since its discovery in 1671: Unlike the other satellites, its brightness varies greatly as it orbits Saturn. Images from Cassini show extreme differences between the leading and trailing hemispheres of Iapetus (Figure 13-29a); the leading side faces in the direction of Iapetus's orbit. Most of the leading hemisphere is as black as asphalt, but the trailing hemisphere is highly reflective, like the surfaces of the other moderate-sized moons.

What could account for the two-sided nature of Iapetus? Models for Iapetus suggest that a small initial difference in coloration

(a) R (I V) U X G

(b) R (I) V U X G

FIGURE 13-29

The Two Faces of Iapetus (a) This color-enhanced Cassini image shows nonreflective material on the leading face of Iapetus. The contrast is not very strong in true color, but the contrast is very strong when comparing reflectivity: The lighter side reflects like somewhat dirty ice, whereas the darker side reflects like black charcoal. **(b)** The leading model to explain the two faces requires debris to create some initial color differences, which is thought to come from a very large ring observed in the infrared by the Spitzer Space Telescope. Discovered in 2009, this ring has a diameter about 300 times larger than Saturn; if it were visible to the naked eye, it would appear wider than two of Earth's full moons. Saturn's moon Phoebe orbits within the ring, making this moon the likely source of the ring's material. This is an artist's illustration of the ring, with an inset of Saturn in the infrared. (a: NASA/JPL/Space Science Institute; b: NASA/JPL-Caltech/Keck) between the two sides would be amplified. A slightly darker surface (less reflective) would absorb more sunlight, warm up more, and evaporate more ice. The residue left behind would leave the surface even darker and therefore warmer, amplifying the effect even further. In 2009, this idea gained support when a new ring of Saturn was discovered near Iapetus that could provide the initial debris to start the runaway process of darkening (Figure 13-29b).

Enceladus Erupts with Plumes of Water

 Enceladus has become one of the most interesting moons in the entire solar system (Figure 13-30). Cassini discovered that Enceladus is volcanically active, though with ice taking the place of lava. While the moon Titan may show evidence of ice volcano formations (Figure 13-24), Cassini has caught icy eruptions on

> Although just 500 km across, Enceladus has powerful ice volcanoes at its south pole.

Enceladus in the act! Figure 13-30b shows several distinct plumes ejecting water vapor and ice from the features referred to as "tiger stripes" near the satellite's south pole. By measuring the distribution of speeds for water-ice particles ejected from Enceladus, high-speed ejecta were found that could escape Enceladus's gravity to leave behind a ring in its orbit (Figure 13-30c). This is Saturn's E ring (Figure 12-25) and solves a mystery about how this ring formed.

The discovery of these icy jets raised questions about the source of the ejected water. Could it have come from a global subsurface ocean, or a localized sea beneath the tiger stripes? Cassini has shed light on this question when passing directly through the plumes and analyzing their content. By detecting dissolved minerals (or "salts") within the water, Cassini data indicate that the water steadily dissolves a rocky mineral interior or mantle (just as Earth's oceans do), which strongly supports the model of a large subsurface ocean.

With ocean water conveniently shot into space by Enceladus for our sampling and analysis, some scientists argue that Enceladus is the solar system's best place to search for microbial life beyond Earth (in a future mission). Recall that Europa's oceans are buried under 10 km of ice, making the liquid water very difficult to sample. Fortunately, evidence is growing that Europa, too, ejects plumes into space, making the search for life much easier on this moon as well.

What could be powering the geologic activity on Enceladus? One guess is tidal heating, which provides the internal heat for the inner Galilean satellites of Jupiter (see Section 13-4). The satellite Dione orbits Saturn in 65.7 hours, almost exactly twice Enceladus's orbital period of 32.9 hours, and the 1:2 ratio of orbital periods should set Enceladus up to be caught in a tidal tug-of-war between Saturn and Dione. Calculations suggest that the amount of tidal heating produced in this way is modest, but should be enough to melt the ices of Enceladus at least part of the time.

(a)

(b)

(c)

FIGURE 13-30 R I **V** U X G

Cryovolcanoes on Enceladus (a) This Cassini image reveals an icy Enceladus. The lack of significant cratering suggests that some geologic process resurfaced the moon within the past 100 million years. **(b)** Plumes of water vapor and ice are erupting from cryovolcanoes associated with the "tiger stripes" near the south pole of Enceladus. The faint plumes are best seen when they are backlit by the Sun, so Enceladus appears as a crescent. **(c)** A backlit view from Cassini shows how particles ejected from Enceladus become part of Saturn's E ring, within which Enceladus orbits (see Figure 12-24). (NASA/JPL/Space Science Institute)

CONCEPT CHECK 13-14

Some early models to explain water ejected from Enceladus avoided reliance on a global subsurface ocean (although a large ocean is certainly the more exciting case), and instead considered smaller pockets of melted water. What subsequent data shifted models to a large ocean, and why?

Answer appears at the end of the chapter.

KEY WORDS

bright terrain (Ganymede), p. 389
cryovolcano, p. 389
dark terrain (Ganymede), p. 389

Galilean satellites, p. 378
hydrocarbon, p. 393
ice rafts (Europa), p. 387
induced electric current, p. 386

induced magnetic field, p. 386
Io torus, p. 385
occultation, p. 379
orbital resonance, p. 378
plasma, p. 385

prograde orbit, p. 397
retrograde orbit, p. 398
tidal heating, p. 382

KEY IDEAS

Nature of the Galilean Satellites: The four Galilean satellites orbit Jupiter in the plane of its equator. All are in synchronous rotation (Sections 13-1, 13-2, and 13-3).

- The orbital periods of the three innermost Galilean satellites, Io, Europa, and Ganymede, are in the ratio 1:2:4. This forms an orbital resonance.

- The two innermost Galilean satellites, Io and Europa, have roughly the same size and density as our Moon. They are composed principally of rocky material. The two outermost Galilean satellites, Ganymede and Callisto, are roughly the size of Mercury. Lower in density than either the Moon or Mercury, they are made of roughly equal parts ice and rock.

- The Galilean satellites probably formed in a similar fashion to planets in our solar system but on a smaller scale.

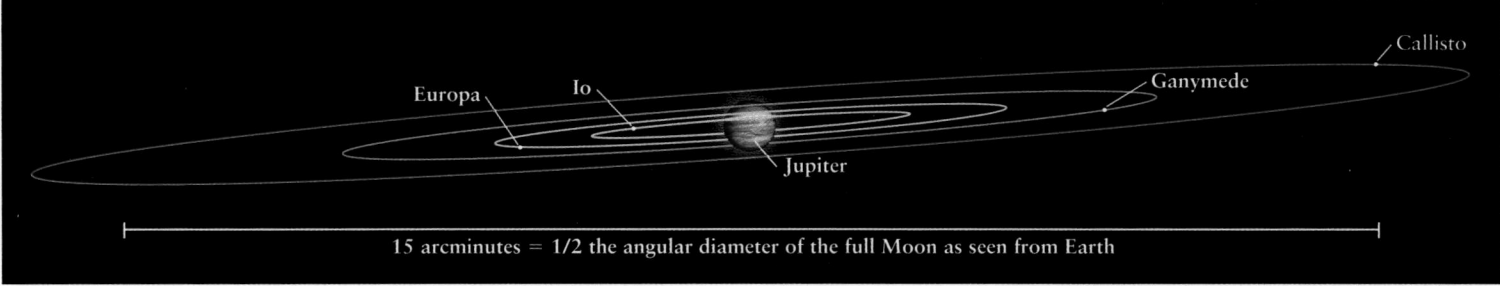

15 arcminutes = 1/2 the angular diameter of the full Moon as seen from Earth

Io: Io is covered with a colorful layer of sulfur compounds deposited by frequent explosive eruptions from volcanic vents. These eruptions resemble terrestrial geysers (Sections 13-4 and 13-5).

- The energy to heat Io's interior and produce the satellite's volcanic activity comes from tidal forces that flex the satellite. This tidal flexing is aided by an orbital resonance: the 1:2:4 ratio of orbital periods among the inner three Galilean satellites.

- The Io torus is a ring of electrically charged particles circling Jupiter at the distance of Io's orbit.

- When Jupiter's magnetic field sweeps over the Galilean satellites, it induces electric currents, which in turn create an induced magnetic field.

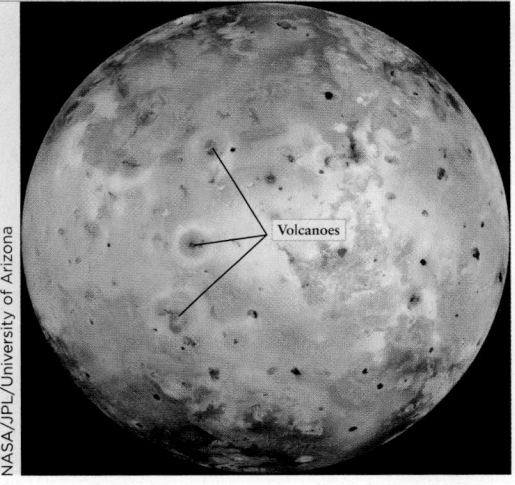

Europa: While composed primarily of rock, Europa is covered with a smooth layer of water ice (Section 13-6).

- The surface has hardly any craters, indicating a geologically active history. Other indications are a worldwide network of long cracks and ice rafts that suggest a subsurface layer of liquid water or soft ice. As with Io, tidal heating and an orbital resonance is responsible for Europa's internal heat.

- There is probably an ocean beneath Europa's frozen surface. Minerals dissolved in this ocean may explain Europa's induced magnetic field.

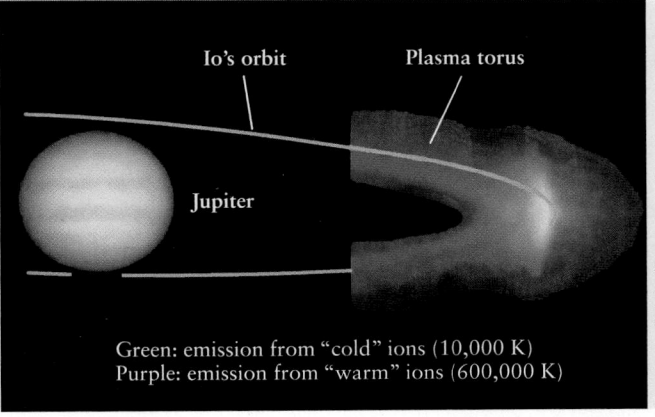

Green: emission from "cold" ions (10,000 K)
Purple: emission from "warm" ions (600,000 K)

Ganymede: Two types of terrain are found on the icy surface of Ganymede: areas of dark, ancient, heavily cratered surface and regions of heavily grooved, lighter-colored, younger terrain (Section 13-7).

- Ganymede is highly differentiated and probably has a metallic core. It has a surprisingly strong magnetic field and a magnetosphere of its own.

- While there is at present little tidal heating of Ganymede, it may have been heated in this fashion in the past. An induced magnetic field suggests that it, too, has a layer of liquid water beneath the surface.

Ice rafts (fractured segments of Europa's icy crust)

Linear features show how rafts once fit together.

10 km

NASA/JPL/University of Arizona

Callisto: Callisto has a heavily cratered crust of water ice. The surface shows little sign of geologic activity because there was never any significant tidal heating of Callisto. However, some unknown processes have erased the smallest craters and blanketed the surface with a dark, dusty substance (Section 13-7).

- Magnetic field data seem to suggest that Callisto has a shallow subsurface ocean.

CALLISTO

Icy crust
Ocean?
Mixed ice–rock interior

NASA/JPL

Titan: The largest Saturnian satellite, Titan, is a terrestrial world with a dense atmosphere consisting mostly of nitrogen and an important 5% in the form of methane. Temperatures and pressures on Titan are near the point where methane can be found in a solid, liquid, or gas phase (Section 13-8).

- Titan shows evidence for flooding expected from methane rain, and many lakes near its north pole. Methane appears to work in a cycle similar to water in Earth's hydrological cycle.

Methane clouds
Precipitation
Evaporation
Methane lakes

Other Satellites: As of 2018, Jupiter has a total of 79 confirmed satellites and Saturn has a total of 62 (Sections 13-9 and 13-10).

- Beyond the Galilean satellites, Jupiter has many small satellites that move in much larger orbits that are noticeably inclined to the plane of Jupiter's equator. Many of these orbit in the direction opposite to Jupiter's rotation. These do not appear to have formed with Jupiter, and they are thought to be captured asteroids.

- In addition to Titan, six moderate-sized moons circle Saturn in regular orbits: Mimas, Enceladus, Tethys, Dione, Rhea, and Iapetus. They are probably composed largely of ice, but their surface features and histories vary significantly. The other, smaller moons include shepherd satellites that control the shapes of Saturn's rings and captured asteroids in large retrograde orbits.

- Enceladus ejects plumes of water from a region near its south pole. Minerals detected in these plumes indicate a subsurface ocean in contact with a rocky mantle.

NASA/JPL-Caltech/Keck

QUESTIONS

Problem-solving tips and tools

Newton's form of Kepler's third law (Box 4-4) relates the masses of two objects in orbit around each other to the period and size of the orbit. The small-angle formula is discussed in Box 1-1. The best seeing conditions on Earth give a limiting angular resolution of ¼ arcsec. Because the orbits of the Galilean satellites are almost perfect circles, you can easily calculate the orbital speeds of these satellites from the data listed in Table 13-1. Data about Jupiter itself are given in Table 12-1 and data about all of the satellites of Jupiter and Saturn can be found in Appendix 3. For a discussion of escape speed and how planets retain their atmospheres, see Box 7-2.

Section 13-1

1. If you observed the Galilean satellites through a telescope for a single night, could you notice their motions around Jupiter?

2. Why can't the Galilean satellites be seen with the naked eye?

3. During the time it takes Ganymede to complete one orbit, how many orbits do Io and Europa complete?

4. In what ways does the system of Galilean satellites resemble our solar system? In what ways is it different?

5. Using the orbital data in Table 13-1, demonstrate that the Galilean satellites obey Kepler's third law.

6. What is the size of the smallest feature you should be able to see on a Galilean satellite through a large telescope under conditions of excellent seeing when Jupiter is near opposition? How does this compare with the best Galileo images, which have resolutions around 25 m?

7. How long does it take for Ganymede to entirely enter or entirely leave Jupiter's shadow? Assume that the shadow has a sharp edge.

Section 13-2

8. No spacecraft from Earth has ever landed on any of the Galilean satellites. How, then, can we know if they are made of rock, water, or a combination of the two?

9. What would the average density of a moon be if it contained solid rock with a density of 3500 kg/m³ for the inner 50% of its radius, and water with 1000 kg/m³ for the rest of its volume?

Section 13-3

10. In what ways did the formation of the Galilean satellites mimic the formation of the planets? In what ways were the two formation processes different?

11. In the classic science-fiction film *2010: The Year We Make Contact,* an alien intelligence causes Jupiter to contract so much that nuclear reactions begin at its center. As a result, Jupiter becomes a star like the Sun. Is this possible in principle? Explain your answer.

12. All of the Galilean satellites orbit Jupiter in the same direction. Furthermore, the planes of their orbits all lie within 0.5° of Jupiter's equatorial plane. Explain why these observations are consistent with the idea that the Galilean satellites formed from a "jovian nebula."

13. Most of Jupiter's smaller moons orbit in a direction opposite to Jupiter's rotation. What does this tell us about those moons?

Section 13-4

14. What is the source of energy that powers Io's volcanoes? How is it related to the orbits of Io and the other Galilean satellites?

15. Long before the Voyager flybys, Earth-based astronomers reported that Io appeared brighter than usual for the few hours after it emerged from Jupiter's shadow. From what we know about the material ejected from Io's volcanoes, suggest an explanation for this brief brightening of Io.

16. How do lavas on Io differ from typical lavas found on Earth? What does this difference tell us about Io's interior?

17. What accounts for the different colors found on Io's surface?

18. Using the diameter of Io (3642 km) as a scale, estimate the height to which the plume of Pele rises above the surface of Io in Figure 13-5a. (You will need to make measurements on this figure using a ruler.) Compare your answer to the value given in the figure caption.

19. Explain why the volcanic plumes in Figure 13-5b have a bluish color. (*Hint:* See Box 5-4.)

20. If you could replace our Moon with Io, and if Io could maintain its present amount of volcanic activity, what changes would this cause in our nighttime sky? Do you think that Io could in fact remain volcanically active in this case? Why or why not?

Section 13-5

21. What is the Io torus? What is its source?

22. What is the origin of the electric current that flows through Io?

23. Jupiter, its magnetic field, and the charged particles that are trapped in the magnetosphere all rotate together once every 10 hours. Io takes 1.77 days to complete one orbit. Using a diagram, explain why particles from Jupiter's magnetosphere hit Io primarily from behind (that is, on the side of Io that trails as it orbits the planet).

24. Assuming material is ejected from Io into Jupiter's magnetosphere at the rate of 1 ton per second (1000 kg/s), how long will it be before Io loses 10% of its mass? How does your answer compare with the age of the solar system?

25. Volcanic eruptions from Io add about 1 ton of material to Jupiter's magnetosphere every second. Based on the information

given in Section 13-4, what fraction of the material ejected from Io's volcanoes goes into Jupiter's magnetosphere?

Section 13-6

26. How was the Galileo spacecraft used to determine that the surface of Europa consists of frozen water?

27. What visible surface features on Europa provide evidence for geologic activity?

28. What is the evidence for an ocean of liquid water beneath Europa's icy surface? What is the evidence that substances other than water are dissolved in this ocean?

29. What aspects of Europa lead scientists to speculate that life may exist there? What precautions have been taken to prevent contaminating Europa with Earth organisms?

Section 13-7

30. Why is ice an important constituent of Ganymede and Callisto, but not of Earth's Moon?

31. How do scientists know that the dark terrain on Ganymede is younger than the bright terrain?

32. In what ways is Ganymede like our own Moon? In what ways is it different? What are the reasons for the differences?

33. Why were scientists surprised to learn that Ganymede has a magnetic field? What does this field tell us about Ganymede's history?

34. What leads scientists to suspect that there is liquid water beneath Ganymede's surface?

35. Why are numerous impact craters found on Ganymede and Callisto but not on Io or Europa?

36. Describe the surprising aspects of Callisto's surface and interior that were revealed by the Galileo spacecraft. Why did these come as a surprise?

37. Figure 13-16 shows that Ganymede's dark terrain has both more craters and larger craters than the bright terrain. Explain what this tells you about the sizes of meteoroids present in the solar system in the distant past and in the more recent past.

38. Put a few cubes of ice in a glass, then fill the glass with warm water to make the ice cubes crack. The fracture lines in each cube are clearly visible because they reflect light. Use this observation to suggest why Ganymede's icy bright terrain, which may have been badly fractured by the stresses that produced the grooves (Figure 13-16), is 25% more reflective than the dark terrain.

39. Speculate on the possibility that Europa, Ganymede, or Callisto might harbor some sort of life. Explain your reasoning.

Section 13-8

40. **TUTORIAL 13.2** Describe Titan's atmosphere. Explain how the Sun's ultraviolet radiation might have altered Titan's original atmosphere.

41. In what ways do hydrocarbons like methane on Titan behave like water on Earth?

42. Why does the presence of methane in Titan's atmosphere imply that Titan has had recent volcanic activity, or has some other source?

43. What is a cryovolcano? How can water be anything but frozen solid in Titan's low temperatures?

44. What is the evidence that there were liquid hydrocarbons on Titan in the past? That there are liquid hydrocarbons now?

45. Is the bulk of Titan's surface made of water ice or methane ice?

46. Many of the gases in the atmosphere of Titan, such as methane, ethane, and acetylene, are highly flammable. Why, then, doesn't Titan's atmosphere catch fire? (*Hint:* What gas in our atmosphere is needed to make wood, coal, or gasoline burn?)

47. Suppose Earth's Moon were removed and replaced in its orbit by Titan. What changes would you expect to occur in Titan's atmosphere? Would solar eclipses be more or less common as seen from Earth? Explain your answers.

48. At infrared wavelengths, the Hubble Space Telescope can see details on Titan's surface as small as 580 km (360 mi) across. Determine the angular resolution of the Hubble Space Telescope using infrared light. If visible light is used, is the angular resolution better, worse, or the same? Explain your answer.

49. Comment on the suggestion that Titan may harbor life-forms.

Section 13-9

50. How would you account for the existence of the satellites of Jupiter other than the Galilean ones?

51. What is a retrograde orbit? What does a moon's retrograde orbit suggest about that moon's origins?

Section 13-10

52. Which of Saturn's moderate-sized satellites show evidence of geologic activity? What might be the energy source for this activity?

53. Explain the present-day difference in the reflectivity on parts of Iapetus. Explain what might cause this if Saturn's largest ring—the newly discovered ring around Saturn in Figure 13-29—produces an initial difference on Iapetus's surface.

54. Saturn's equator is tilted by 27° from the ecliptic, while Jupiter's equator is tilted by only 3°. Use these data to

explain why we see fewer transits, eclipses, and occultations of Saturn's satellites than of the Galilean satellites.

55. (a) To an observer on Enceladus, what is the time interval between successive oppositions of Dione? Explain your answer. (b) As seen from Enceladus, what is the angular diameter of Dione at opposition? How does this compare to the angular diameter of the Moon as seen from Earth (about ½°)?

56. Jupiter's satellite Io and Saturn's satellite Enceladus are both geologically active, and both are in 2:1 resonances with other satellites. However, the amount of geologic activity of Enceladus is far less than on Io. Discuss some possible reasons for this difference.

57. Imagine that you are in charge of planning a successor to the Cassini spacecraft to further explore the Saturnian system. In your opinion, which satellites in the system should be examined more closely? What data should be collected? What kinds of questions should the new mission attempt to answer?

58. What does the presence of salts in the plumes from Enceladus tell us about the liquid water within this moon?

59. What is the connection between Enceladus and Saturn's E ring?

60. What role do the plumes on Enceladus play in the search for life?

General Problems

61. Io has no impact craters on its surface, while our Moon is covered with craters. What is the explanation for this difference?

62. Despite all the gases released from its interior by volcanic activity, Io does not possess a thick atmosphere. Explain why not.

63. Compare and contrast the surface features of the four Galilean satellites. Discuss their relative geological activity and the evolution of these four satellites.

64. The larger the orbit of a Galilean satellite, the less geologic activity that satellite has. Explain why this might be.

65. Explain how orbital resonance from the 1:2:4 ratio of the orbital periods of Io, Europa, and Ganymede is related to the geologic activity on these satellites.

66. Find the escape speed on Titan. What is the limiting molecular weight of gases that could be retained by Titan's gravity? (*Hint:* Use the ideas presented in Box 7-2 and assume an average atmospheric temperature of 95 K.)

WEB/E-BOOK QUESTIONS

67. Various spacecraft missions have been proposed to explore Europa in greater detail. Search the internet for information about these. How would these missions test for the presence of an ocean beneath Europa's surface?

68. Two new satellites of Jupiter were discovered in 2011. Search the internet for information about these satellites. What do the orbital characteristics of these moons tell you about their origins?

ACTIVITIES

Observing Projects

Observing tips and tools

You can easily find the apparent positions of the Galilean satellites for any date and time using the *Starry Night*™ software if you have access. For even more detailed information about satellite positions, consult the "Satellites of Jupiter" section in the *Astronomical Almanac* for the current year. If your goal is to view Saturn's satellites, consult the section entitled "Satellites of Saturn" in the *Astronomical Almanac* for the current year. This includes a diagram showing the orbits of Mimas, Enceladus, Tethys, Dione, Rhea, Titan, and Hyperion. Plan your observing session by looking up the dates and times of the most recent greatest eastern elongations of the various satellites. You will have to convert from universal time (UT), also known as Greenwich Mean Time, to your local time zone. Then, using the tick marks along the orbits in the diagram, estimate the positions of the satellites relative to Saturn at the time you will be at the telescope. Another useful resource is the "Celestial Calendar" section of *Sky & Telescope*. During months when Saturn is visible in the night sky, this section of the magazine includes a chart of Saturn's satellites.

69. Observe Jupiter through a pair of binoculars. Can you see all four Galilean satellites? Make a drawing of what you observe. If you look again after an hour or two, can you see any changes?

70. Observe Jupiter through a small telescope on three or four consecutive nights. Make a drawing each night showing the positions of the Galilean satellites relative to Jupiter. Record the time and date of each observation. Consult the sources listed above in the "Observing tips and tools" to see if you can identify the satellites by name.

71. If you have access to a moderately large telescope, make arrangements to observe several of Saturn's satellites. At the telescope, you should have no trouble identifying Titan. Tethys, Dione, and Rhea are about one-sixth as bright as Titan and should be the next easiest satellites to find. Can you confidently identify any of the other satellites?

Starry Night™ Explorations

72. Use *Starry Night*™ to observe the moons of Jupiter. Select **Favorites > Explorations > Galilean Moons** from the menu to see a view of Jupiter and the four Galilean moons Callisto, Io, Europa, and Ganymede.
    (a) Click **Edit > Select None** in the menu to remove the labels and then **Play** the animation to observe the motion of the Galilean moons. Which moon orbits closest to Jupiter and which orbits farthest away?

(b) Use the time controls to determine the periods of the orbits of each of the Galilean moons. Express these periods in days (e.g., 1.23 days).

73. Use *Starry Night*™ to explore the orbits of some of Saturn's moons. Open **Favorites > Explorations > Saturn**. The view is centered on Saturn and shows the positions of its six largest moons. Use the location scroller and elevation controls to see the view from various perspectives, especially in the plane of Saturn's rings.

    (a) Do these moons orbit Saturn in the same or in different directions?

    (b) Which, if any, of these moons orbit in nearly the same plane as that of the planet's rings?

    (c) Stop time flow, open the **Planets** panel, and locate and expand **Our Solar System > Saturn**. Click the menu icon in the list for each of the moons and select **Magnify** to see an image of the moon or use the slide bar next to each moon's name to magnify its image. Which of these six largest moons of Saturn have appearances most similar to Earth's moon?

    (d) Use **File > Revert** to reset the view and in the **Planets** panel, expand **Saturn > Smaller Moons**. Click the radio buttons for a number of these moons to display their orbits in the view. Use the location scroller and elevation controls to examine the appearance of these orbits. Do these orbits fall mostly in the plane of Saturn's rings?

74. In *Starry Night*™, open **Favorites > Explorations > Enceladus** to view Saturn as it might appear if you were observing it from the crater named Ahmad on the surface of Enceladus, one of Saturn's moons. North is at the top and east is to the left. Click **Play** to observe Saturn in Enceladus's sky.

    (a) How do the stars appear to move?

    (b) How does Saturn appear to move in the sky?

    (c) What do these observations suggest about the relationship between the orbital and rotational periods of Enceladus?

    (d) Notice the appearance of Saturn's rings in this view from Enceladus. What does this suggest about the plane of Enceladus's orbit compared to the plane of Saturn's rings?

Collaborative Exercises

75. Imagine that scientists are proposing to send a robotic lander to Jupiter's satellite Callisto. Create a 100-word written proposal describing a robotic lander mission to another of the Galilean satellites, explaining why your group found it to be the most interesting and why the government should allocate the money for your alternative project. In your proposal, be sure to demonstrate your knowledge of Callisto and the other satellite.

76. From the data and the accompanying images for Table 13-1: Jupiter's Galilean Satellites Compared with the Moon, Mercury, and Mars, use someone's shoe to represent the 150,000-km diameter of Jupiter and determine about how many "shoes" away each of the Galilean satellites would be from Jupiter.

ANSWERS

ConceptChecks

ConceptCheck 13-1: You can determine the diameter of the satellite, even though you cannot see the physical extent of the satellite. As the satellite is increasingly blocked or occulted by the planet (from our view), we receive less and less of the satellite's reflected light. The time it takes for the satellite's light to completely fade away tells you the transit time, and, combined with the satellite's speed, the diameter can be calculated.

ConceptCheck 13-2: No. In order to measure the masses of Jupiter's moons, the gravitational deflection of a spacecraft by these moons had to be measured.

ConceptCheck 13-3: Because the average density of each moon decreases with distance from Jupiter (Table 13-1). This implies a warmer interior in an early jovian nebula, where rock and metal grains would form the bulk of solid moon-forming material. Farther away from Jupiter, water would freeze with solid grains and help build moons with a lower average density. This is analogous to the formation of the solar system discussed in Section 8-5.

ConceptCheck 13-4: It is not enough for Jupiter to exert a constant tidal force on Io; that would be like squeezing a ball into one deformed shape and keeping it there. A varying tidal force is necessary, and this arises due to Europa and Ganymede rhythmically tugging the long axis of Io's orbit back and forth by about ½°.

ConceptCheck 13-5: No. Jupiter has its own magnetic field that comes from electric currents within its interior. As Jupiter's field sweeps over Io, electric currents and a magnetic field are induced. This induced magnetic field is a separate and much weaker field than Jupiter's.

ConceptCheck 13-6: The orbital resonance of Io, Europa, and Ganymede causes variations in the tidal forces exerted onto these moons by Jupiter. These tidal forces generate heat, but they are stronger on Io because it is closer to Jupiter than to Europa.

ConceptCheck 13-7: The induced magnetic field comes from the induced electric current in Europa created by Jupiter. For this current to flow through Europa's water interior, the water must be liquid and therefore not solid ice. When water comes in contact with rock (probably Europa's rocky mantle), the water dissolves some of the rock's minerals, making the water electrically conductive. Modeling of Europa's induced magnetic field points to a large subsurface ocean.

ConceptCheck 13-8: Callisto has an unexpected induced magnetic field: unexpected because this indicates a liquid ocean, yet at Callisto's temperature any interior water should be frozen. However, ammonia allows water to remain a liquid at temperatures where pure water would freeze.

ConceptCheck 13-9: Ninety-five percent of Titan's atmosphere is nitrogen, which is thought to come from either ultraviolet sunlight breaking down ammonia or from comet impacts generating nitrogen from ammonia. Ammonia in water also lowers the temperature at which that water will freeze, which allows water below the surface to act like lava in cryovolcanoes.

ConceptCheck 13-10: Cassini discovered lakes using radar. By reflecting radio waves, radar can be used to map the height of surface features. The lakes show up as very flat regions, and tributary channels support this interpretation.

ConceptCheck 13-11: Voyager found that temperatures and pressures on Titan are just right for methane to exist as a gas, liquid, or solid. Titan's methane cycle depends on these changes between liquid and gas, just as Earth's hydrologic cycle depends on changes between liquid and gas phases of water.

ConceptCheck 13-12: If Jupiter formed like a mini solar system as expected, any moon that formed with Jupiter would have a prograde orbit (in the same directions as Jupiter's rotation), and would also orbit approximately in Jupiter's equatorial plane. The retrograde moons (orbiting in the direction opposite to Jupiter's rotation at high inclinations to Jupiter's equator) must have a different origin, and the capture of asteroids is most likely.

ConceptCheck 13-13: Hyperion's average density is less than half that of water. The only obvious way to account for Hyperion's size, then, is to assume it is very "porous" and has a lot of empty space inside.

ConceptCheck 13-14: Cassini passed through the plumes and detected dissolved minerals (or salts). The simplest interpretation is that a large subsurface ocean partially dissolves minerals from a rocky mantle.

14

Uranus, Neptune, Pluto, and the Kuiper Belt: Remote Worlds

Earth Uranus Neptune Pluto Eris

LEARNING GOALS

Beyond Saturn, in the cold, dark recesses of the solar system, orbit three worlds that have long been shrouded in mystery. These worlds—Uranus, Neptune, and Pluto (all shown here to the same scale as Earth)—are so distant, so dimly lit by the Sun, and so slow in their motion against the stars that they were unknown to ancient astronomers and were discovered only after the invention of the telescope. Even then, little was known about Uranus and Neptune until Voyager 2 flew past these planets during the 1980s. Future missions may reveal much more about these icy giants.

Surrounded by a system of small moons and thin, dark rings, Uranus is tipped on its side so that its axis of rotation lies nearly in its orbital plane. This remarkable orientation suggests that Uranus may have been the victim of a staggering impact by a massive planetesimal. Neptune is, at first glance, a denser, more massive version of Uranus, but it is a far more active world. It has an internal energy source that Uranus seemingly lacks, as well as atmospheric bands and storm activity resembling those on Jupiter.

Earth, Uranus, Neptune, Pluto, and Eris to scale. (NASA/NOAA/GSFC/Suomi NPP/VIIRS/Norman Kuring; NASA; Sromovsky and P. Fry [Univ. Wisconsin-Madison] et al., NASA; NASA/Johns Hopkins University Applied Physics Laboratory/Southwest Research Institute; NASA, ESA, and A. Schaller [for STScI])

R I **V** U X G

Neptune also has dark rings; small, icy moons; and an intriguing large satellite, Triton, which is nearly devoid of impact craters and has geysers that squirt nitrogen-rich vapors. Triton's retrograde orbit suggests that this strange world may have been gravitationally captured by Neptune.

Close cousins of Triton are Pluto and thousands of other trans-Neptunian objects, among the most remote members of the solar system. These objects harbor many mysteries that are only beginning to be explored by spacecraft.

14-1 Uranus was discovered by chance, but Neptune's existence was predicted by applying Newtonian mechanics

 TUTORIAL 14.1   Before the eighteenth century, only six planets were known to orbit the Sun: Mercury, Venus, Earth, Mars, Jupiter, and Saturn. Another planet was discovered by William Herschel, a German-born musician who moved to England in 1757 and became fascinated by astronomy.

Discovering the "Georgian Star"

Using a telescope that he built himself, Herschel was systematically surveying the sky when, on March 13, 1781, he noticed a faint, fuzzy object that he first thought to be a distant comet. By the end of 1781, however, his observations had revealed that the object's orbit was relatively circular and was larger than Saturn's orbit. Comets, by contrast, can normally be seen only when they follow elliptical orbits that bring them much closer to the Sun.

Herschel had discovered the seventh planet from the Sun. In doing so, he had doubled the radius of the known solar system from 9.5 au (the semimajor axis of Saturn's orbit) to 19.2 au (the average distance from the Sun to Uranus). Herschel originally named his discovery Georgium Sidus (Latin for "Georgian star") in honor of the reigning British monarch, George III. The name Uranus—in Greek mythology, the personification of Heaven—came into currency only some decades later, to be consistent with other myth-inspired planet names.

Although Herschel received the credit for discovering Uranus, he was by no means the first person to have seen it. At opposition, Uranus is just barely bright enough to be seen with the naked eye under good observing conditions, so it was probably seen by the ancients. Many other astronomers with telescopes had sighted this planet before Herschel; it is plotted on at least 20 star charts drawn between 1690 and 1781. But all these other observers mistook Uranus for a dim star. Herschel was the first to track its motion relative to the stars and recognize it as a planet. Tracking Uranus was no small task, because it moves very slowly on the celestial sphere, just over 4° in the space of a year (compared to about 12° for Saturn and about 30° for Jupiter).

The Discovery of Neptune

It was by carefully tracking Uranus's slow motions that astronomers were led to discover Neptune. By the beginning of the nineteenth century, it had become painfully clear to astronomers that they could not accurately predict the orbit of Uranus using Newtonian mechanics. By 1830, the discrepancy between the planet's predicted and observed positions had become large enough (2 arcmin) that some scientists suspected that Newton's law of gravitation might not be accurate at great distances from the Sun.

By the mid-1840s, two scientists working independently—the French astronomer Urbain Jean Joseph Le Verrier and the English mathematician John Couch Adams—were exploring an earlier suggestion that would leave Newton's law intact. Perhaps the gravitational pull of an as-yet-undiscovered planet was causing Uranus to deviate slightly from its predicted orbit. Calculations by both scientists indicated that Uranus had indeed caught up with and then passed a more distant planet. On approaching the unknown planet, they reasoned, Uranus had accelerated slightly; then, as it moved away from the planet, Uranus decelerated slightly, explaining the observations.

Inspired by Le Verrier's results, astronomers at Cambridge University Observatory undertook a 6-week search for the proposed new planet in the summer of 1846. They were unsuccessful, in part because they lacked accurate star maps of the part of the sky being searched.

Meanwhile, Le Verrier wrote to Johann Gottfried Galle at the Berlin Observatory with detailed predictions of where to search for the new planet. Galle received the letter on September 23, 1846. That very night, aided by more complete star maps and after just a half hour of searching, Galle and Heinrich d'Arrest located an uncharted star with the expected brightness in the predicted location. Subsequent observations confirmed that this new "star" showed a planetlike motion with respect to other stars.

Le Verrier proposed that the planet be called Neptune. After years of debate between English and French astronomers, the credit for its discovery came to be divided between Adams and Le Verrier. As of this writing, Neptune is the only planet in our solar system whose existence was predicted by calculation rather than revealed by chance discovery.

Is There Another Planet to Discover?

Beginning in 2014, astronomers have been gathering evidence for a ninth planet. Recall that the trail leading to the discovery of Neptune was inspired by the unexplained deviations from Uranus's expected orbit. In a similar manner, a ninth planet is being proposed to account for deviations in the orbital properties of comet-like objects in the outer reaches of the solar system. These objects orbit at distances beyond Neptune and will be discussed in Section 14-10. By observing thousands of these objects, it becomes clear what their typical orbital properties are, but it also appears that 10 or so of them have different orbital properties (such as the orientations of their orbital planes) that cannot be easily explained.

One likely reason is that a ninth planet exists at about 10 times the distance of Neptune, between about 300 and 400 au. This ninth planet would also need to have about 10 times the mass of Earth for its gravity to cause the observed deviations.

Again, similar to Neptune, deviations in expected orbits provide a clue, but discovering a new planet requires a direct telescopic observation. As you can imagine, the race is on, and a number of astronomers are searching hard to discover a ninth planet. With predictions putting the planet so far away, it could be a very difficult search.

TABLE 14-1 Uranus Data

Average distance from the Sun:	19.201 au $= 2.871 \times 10^9$ km
Maximum distance from the Sun:	20.024 au $= 2.995 \times 10^9$ km
Minimum distance from the Sun:	18.275 au $= 2.748 \times 10^9$ km
Eccentricity of orbit:	0.0457
Average orbital speed:	6.83 km/s
Orbital period:	84.011 years
Rotation period (internal):	17.24 hours
Inclination of equator to orbit:	82.23°
Inclination of orbit to ecliptic:	0.77°
Diameter:	51,118 km $= 4.007$ Earth diameters (equatorial)
Mass:	8.681×10^{25} kg $= 14.54$ Earth masses
Average density:	1271 kg/m^3
Escape speed:	21.3 km/s
Surface gravity (Earth $= 1$):	0.90
Albedo:	0.49
Average temperature at cloudtops:	$-218°C = -360°F = 55$ K
Atmosphere composition (by number of molecules):	82.5% hydrogen (H$_2$), 15.2% helium (He), 2.3% methane (CH$_4$)

R I V U X G
NASA/JPL

TABLE 14-2 Neptune Data

Average distance from the Sun:	30.047 au $= 4.498 \times 10^9$ km
Maximum distance from the Sun:	30.304 au $= 4.543 \times 10^9$ km
Minimum distance from the Sun:	29.630 au $= 4.453 \times 10^9$ km
Eccentricity of orbit:	0.011
Average orbital speed:	5.5 km/s
Orbital period:	164.79 years
Rotation period (internal):	16.11 hours
Inclination of equator to orbit:	28.32°
Inclination of orbit to ecliptic:	1.77°
Diameter:	49,528 km $= 3.883$ Earth diameters (equatorial)
Mass:	1.024×10^{26} kg $= 17.15$ Earth masses
Average density:	1638 kg/m^3
Escape speed:	23.5 km/s
Surface gravity (Earth $= 1$):	1.1
Albedo:	0.44
Average temperature at cloudtops:	$-218°C = -360°F = 55$ K
Atmosphere composition (by number of molecules):	79.5% hydrogen (H$_2$), 19% helium (He), 1.5% methane (CH$_4$)

R I V U X G
NASA/JPL

Observing Uranus and Neptune

At opposition, Neptune can be bright enough to be visible through small telescopes. The first person thus to have seen it was probably Galileo. His drawings from January 1613, when he was using his telescope to observe Jupiter and its four large satellites, show a "star" less than 1 arcmin from Neptune's location. Galileo even noted in his observation log that on one night this star seemed to have moved in relation to the other stars. But Galileo would have been hard pressed to identify Neptune as a planet, because its motion against the background stars is so slow (just over 2° per year).

Uranus and Neptune are in such large orbits that they move very slowly on the celestial sphere.

Through a large, modern telescope, both Uranus and Neptune are dim, uninspiring sights. Each planet appears as a hazy, featureless disk with a faint greenish-blue tinge. Although Uranus and Neptune are both about 4 times larger in diameter than Earth, they are so distant that their angular diameters as seen from Earth are tiny—no more than 4 arcsec for Uranus and just over 2 arcsec for Neptune. To an Earth-based observer, Uranus is roughly the size of a golf ball seen at a distance of 1 km. **Table 14-1** and **Table 14-2** give basic data about Uranus and Neptune, respectively.

CONCEPT CHECK 14-1

How was Neptune predicted to exist at the right location before it was ever observed?

Answer appears at the end of the chapter.

14-2 Uranus is nearly featureless and has an unusually tilted axis of rotation

TUTORIAL 14.2

Scientists had hoped that Voyager 2 would reveal cloud patterns in Uranus's atmosphere when it flew past the planet in January 1986. But even images recorded at close range showed Uranus to be remarkably featureless (**Figure 14-1**). Faint cloud markings became visible in images of Uranus only after computer enhancement (**Figure 14-2**).

Uranus's Atmosphere

Voyager 2 data confirmed that the Uranian atmosphere is dominated by hydrogen (82.5%) and helium (15.2%), similar to the atmospheres of Jupiter and Saturn. Uranus differs, however, in that 2.3% of its atmosphere is methane (CH_4), which is 5 to 10 times the percentage found on Jupiter and Saturn. In fact, Uranus has a higher percentage of all heavy elements—including carbon atoms, which are found in molecules of methane—than Jupiter and Saturn. (In Section 14-4, we will investigate how this could have come about.)

Methane preferentially absorbs the longer wavelengths of visible light, so sunlight reflected from Uranus's upper atmosphere is depleted of its reds and yellows. Fewer reds and yellows in the reflected light give the planet its distinct greenish-blue

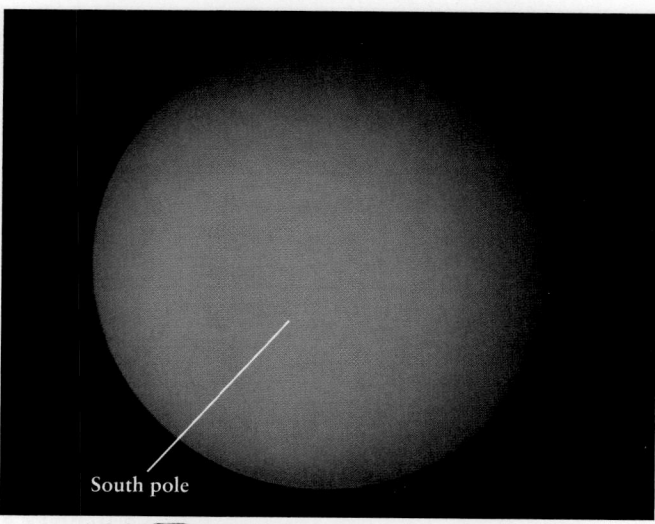
FIGURE 14-1 R I **V** U X G

Uranus from Voyager 2 This image looks nearly straight down onto Uranus's south pole, which was pointed almost directly at the Sun when Voyager 2 flew past in 1986. None of the Voyager 2 images of Uranus shows any pronounced cloud patterns. The color is due to methane in the planet's atmosphere, which absorbs red light but reflects green and blue. (NASA/JPL)

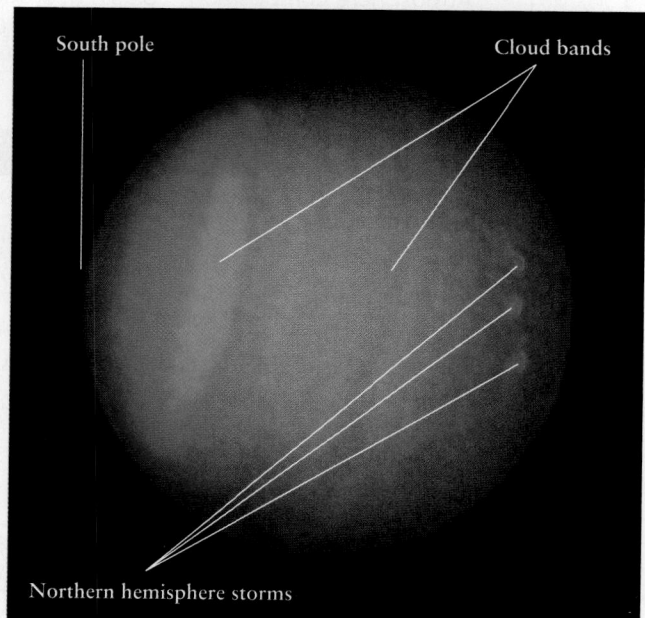
FIGURE 14-2 R **I** **V** **U** X G

Uranus from the Hubble Space Telescope Images made at infrared, visible, and ultraviolet wavelengths were combined and enhanced to give this false-color view of cloud features on Uranus. These images were captured in August 2004 (18 years after the image in Figure 14-1), during springtime in Uranus's northern hemisphere. (NASA/Erich Karkoschka, University of Arizona)

appearance. As on Saturn's moon Titan (see Section 13-8), ultraviolet light from the Sun turns some of the methane gas into a hydrocarbon haze, making it difficult to see the lower levels of the atmosphere.

Ammonia (NH_3), which constitutes 0.01% to 0.03% of the atmospheres of Jupiter and Saturn, is almost completely absent from the Uranian atmosphere. The reason is that Uranus is colder than Jupiter or Saturn: The temperature in its upper atmosphere is only 55 K ($-218°C = -360°F$). Ammonia freezes at these very low temperatures, so any ammonia has long since precipitated out of the atmosphere and into the planet's interior. For the same reason, Uranus's atmosphere is also lacking in water. Hence, the substances that make up the clouds on Jupiter and Saturn—ammonia, ammonium hydrosulfide (NH_4SH), and water—are not available in Uranus's atmosphere. This helps to explain the bland, uniform appearance of the planet shown in Figure 14-1.

The few clouds found on Uranus are made primarily of methane, which condenses into droplets only if the pressure is sufficiently high. Hence, methane clouds form only at lower levels within the atmosphere, where they are difficult to see.

An Oddly Tilted Planet

The rotation period of Uranus's atmosphere is about 16 hours. Like Jupiter and Saturn, Uranus rotates differentially, so this period depends on the latitude and can be measured by tracking the motions of clouds. To determine the rotation period for the underlying body of the planet, scientists looked to Uranus's magnetic field, which is presumably anchored in the planet's interior, or at least in the deeper and denser layers of its atmosphere. Data from Voyager 2 indicate that Uranus's internal period of rotation is 17.24 hours.

Voyager 2 also confirmed that Uranus's rotation axis is tilted in a unique and bizarre way. Herschel found the first evidence of this in 1787, when he discovered two moons orbiting Uranus in a plane that is almost perpendicular to the plane of the planet's orbit around the Sun. Thus, compared to the other planets, Uranus is nearly tipped on its side (**Figure 14-3**).

Careful measurement shows that Uranus's axis of rotation is tilted by 98°, as compared to 23½° for Earth (compare Figure 14-3 with Figure 2-12). A tilt angle greater than 90° means that Uranus exhibits retrograde (backward) rotation like that of Venus, shown in Figure 11-4b. Astronomers suspect that Uranus might have acquired its large tilt angle billions of years ago, when one or more massive bodies collided with Uranus while the planet was still forming. It turns out that the moons of Uranus place constraints on such axis-tilting impacts, a scenario we will discuss further in Section 14-7.

The radical tilt of its axis means that as Uranus moves along its 84-year orbit, its north and south poles alternately point toward or away from the Sun. This tilt produces highly exaggerated seasonal changes. For example, when Voyager 2 flew by in 1986, Uranus's south pole was pointed toward the Sun. Most of the planet's southern hemisphere was bathed in continuous sunlight, while most of the northern hemisphere was subjected to a continuous, frigid winter night. But over the following quarter of a Uranian year later (21 of our years), sunlight has gradually been returning to the northern hemisphere, triggering immense storms there. Figure 14-2 shows some of these storms, which are much more visible at infrared wavelengths than with visible light.

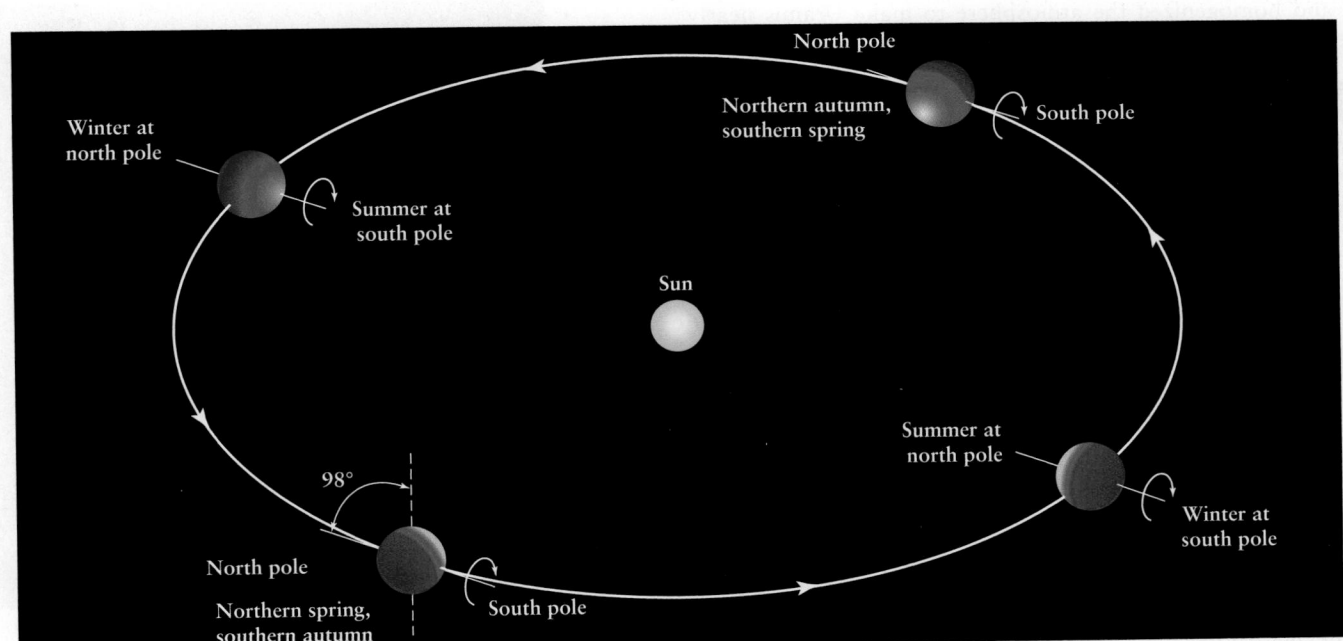

FIGURE 14-3

Exaggerated Seasons on Uranus For most planets, the rotation axis is roughly perpendicular to the plane of the planet's orbit around the Sun. But for Uranus the rotation axis is tilted by 98° from the perpendicular. This tilt causes severely exaggerated seasons. For example, during midsummer at Uranus's south pole, the Sun appears nearly overhead for many Earth years, while the planet's northern regions are in continuous darkness. Half an orbit later, the seasons are reversed.

Atmospheric Motions on Uranus

By following the motions of clouds and storm systems on Uranus, scientists find that the planet's winds flow to the east—that is, in the same direction as the planet's rotation—at northern and southern latitudes, but to the west near the equator. This wind pattern is quite unlike the situation on Jupiter and Saturn, where the zonal winds alternate direction many times between the north and south poles (see Section 12-4). The fastest Uranian winds (about 700 km/h, or 440 mi/h) are found at the equator.

We saw in Section 12-4 that the internal heat of Jupiter and Saturn plays a major role in driving atmospheric activity on those worlds. Uranus is different: It appears to have little or no internal source of thermal energy. Measurements to date show that unlike Jupiter or Saturn, Uranus radiates into space only as much energy as it receives from the Sun. With only feeble sunlight to provide energy to its atmosphere, and little internal thermal energy, Uranus lacks the dramatic wind and cloud dynamics found on Jupiter and Saturn. It is not known why Uranus lacks a significant source of internal thermal energy.

> Unlike the other jovian planets, Uranus's interior heat has no effect on its atmosphere's motions.

Although Uranus's equatorial region was receiving little sunlight at the time of the Voyager 2 flyby, the atmospheric temperature there (about 55 K = −218°C = −360°F) was not too different from that at the sunlit pole. Heat must therefore be efficiently transported from the poles to the equator. This north–south heat transport, which is perpendicular to the wind flow, may have mixed and homogenized the atmosphere to make Uranus nearly featureless.

CONCEPT CHECK 14-2

If methane were absent from Uranus's atmosphere, what color would it appear?

Answer appears at the end of the chapter.

14-3 Neptune is a cold, bluish world with Jupiter-like atmospheric features

At first glance, Neptune appears to be the twin of Uranus. (See the image that opens this chapter, and compare Tables 14-1 and 14-2.) But these two planets are by no means identical. While Neptune and Uranus have almost the same diameter, Neptune is 18% more massive. Neptune's axis of rotation also has a more modest 29½° tilt. When Voyager 2 flew past Neptune in August 1989, it revealed that the planet has a more active and dynamic atmosphere than Uranus. This activity suggests that Neptune, unlike Uranus, has a powerful source of energy in its interior.

Neptune's Atmosphere

VIDEO 14.1 The Voyager 2 data showed that Neptune has essentially the same atmospheric composition as Uranus: 79.5% hydrogen (by number of molecules), 19% helium, 1.5% methane, and almost no ammonia or water vapor. As for Uranus, the presence

of methane gives Neptune a characteristic bluish-green color. The temperature in the upper atmosphere is also the same as on Uranus, about 55 K. That these temperatures should be so similar, even though Neptune is much farther from the Sun, is further evidence that Neptune has a strong internal heat source.

Unlike Uranus, however, Neptune has clearly visible cloud patterns in its atmosphere. At the time that Voyager 2 flew past Neptune, the most prominent feature in the planet's atmosphere was a giant storm called the Great Dark Spot (Figure 14-4). The Great Dark Spot had a number of similarities to Jupiter's Great Red Spot (see Sections 12-1 and 12-3). The storms on both planets were comparable in size to Earth's diameter, both appeared at about the same latitude in the southern hemisphere, and the winds in both storms circulated in a counterclockwise direction (see Figure 12-6). But Neptune's Great Dark Spot appears not to have been as long lived as the Great Red Spot on Jupiter. When the Hubble Space Telescope (HST) first viewed Neptune in 1994, the Great Dark Spot had disappeared. Other dark vortices—high-pressure systems with bright "companion clouds"—have since been observed by HST.

> Unlike its close relative Uranus, Neptune has truly immense storms.

Voyager 2 also saw a few conspicuous whitish clouds on Neptune. These clouds are thought to be produced when winds carry methane gas into the cool, upper atmosphere, where it condenses into crystals of methane ice. Voyager 2 images show these high-altitude clouds casting shadows onto lower levels of

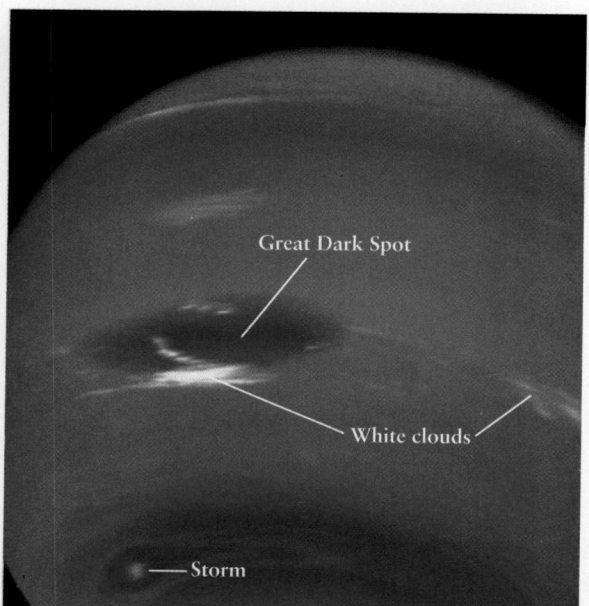

VIDEO 14.2 **FIGURE 14-4** R I **V** U X G

Neptune from Voyager 2 When this picture of Neptune's southern hemisphere was taken in 1989, the Great Dark Spot measured about 12,000 by 8000 km, comparable in size to Earth. Wind speeds in the Great Dark Spot (up to 1500 mi/h) can reach 3 times the speeds in Jupiter's Great Red Spot. (A smaller storm appears at the lower left.) The white clouds are thought to be composed of crystals of methane ice. The color contrast in this image has been exaggerated to emphasize the differences between dark and light areas. (NASA/JPL)

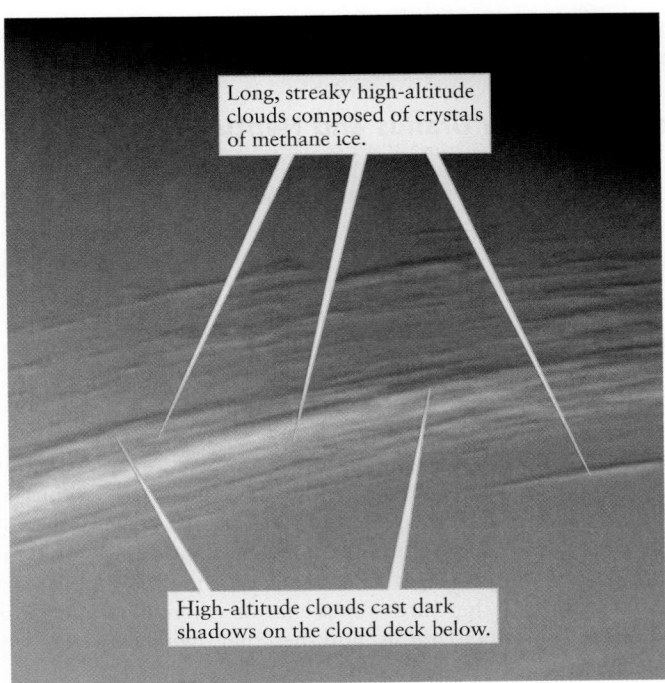

FIGURE 14-5 R I **V** U X G

Cirrus Clouds over Neptune Voyager 2 recorded this image of clouds near Neptune's terminator (the border between day and night on the planet). Like wispy, high-altitude cirrus clouds in Earth's atmosphere, these clouds are thought to be made of ice crystals. The difference is that Neptune's cirrus clouds are probably methane ice, not water ice as on Earth. (NASA/JPL)

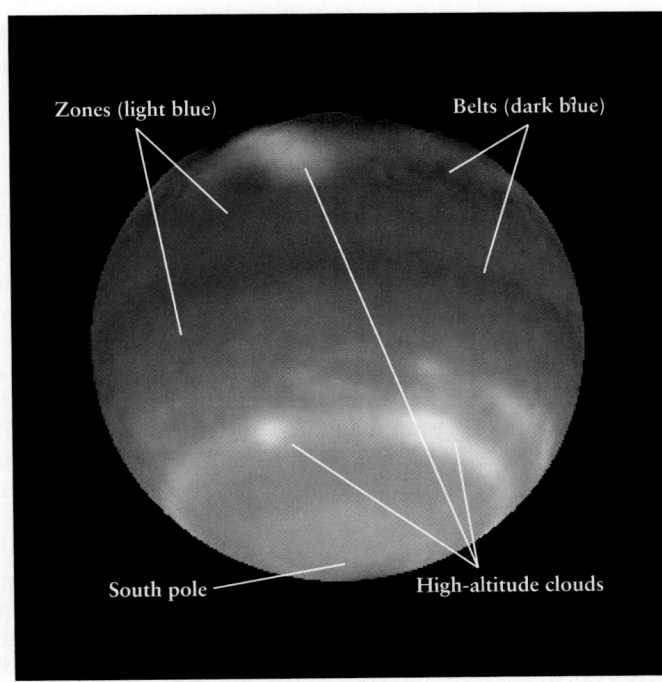

FIGURE 14-6 R I **V** U X G

Neptune's Banded Structure This enhanced-color Hubble Space Telescope image shows Neptune's belts and zones. White areas denote high-altitude clouds; the very highest clouds (near the top of the image) are shown in yellow-red. The green belt near the south pole is a region where the atmosphere absorbs blue light, perhaps indicating a different chemical composition there. (NASA/JPL/STScI)

Neptune's atmosphere (Figure 14-5). Images from HST also show the presence of high-altitude clouds (Figure 14-6).

Neptune's Internal Heat

Due to its greater distance from the Sun, Neptune receives less than half as much energy from the Sun as Uranus does. But with less solar energy available to power atmospheric motions, why are there high-altitude clouds and huge, dark storms on Neptune but not on Uranus? At least part of the answer is probably that Neptune is still slowly contracting, thus converting gravitational energy into thermal energy that heats the planet's core. (The same is true for Jupiter and Saturn, as we saw in Section 12-4.) The evidence for this contraction is that Neptune, like Jupiter and Saturn but unlike Uranus, emits more energy than it receives from the Sun. The combination of a warm interior and a cold outer atmosphere can cause convection in Neptune's atmosphere, producing the up-and-down motion of gases that generates clouds and storms. Neptune also resembles Jupiter in having belts and zones parallel to the planet's equator, although Neptune's are much fainter (see Figure 14-6).

Like those on Uranus, most of Neptune's clouds are probably made of droplets of liquid methane. Because these droplets form fairly deep within the atmosphere, the clouds are more difficult to see than the ones on Jupiter. Hence, Neptune's belts and zones are less pronounced than Jupiter's, although more so than those on Uranus (thanks to the extra cloud-building energy from Neptune's interior). As described above, Neptune's high-altitude clouds (see Figure 14-5) are probably made of *frozen* methane.

Although Neptune displays more evidence of up-and-down motion in its atmosphere than Uranus, the global pattern of east and west winds is almost identical on the two planets. This similarity is rather strange. The two planets are heated very differently by the Sun, thanks to their different distances from the Sun and the different tilts of their axes of rotation, so we might have expected the wind patterns on Uranus and Neptune also to be different. Perhaps the explanation of these wind patterns will involve understanding how heat is transported not only within the atmospheres of Uranus and Neptune but within their interiors as well.

CONCEPT CHECK 14-3

With Neptune even farther from the Sun than Uranus, where does Neptune get the energy to power more atmospheric activity than Uranus?

Answer appears at the end of the chapter.

14-4 Uranus and Neptune contain a higher proportion of heavy elements than Jupiter and Saturn

At first glance, Uranus and Neptune might seem to be simply smaller and less massive versions of Jupiter or Saturn (see Table 7-1). But like many first impressions, this one is misleading because it fails to take into account what lies beneath the surface. We saw in Section 12-2 that the interiors of both Jupiter and Saturn are composed primarily of hydrogen and helium, in nearly the same abundance as the Sun. But the average densities of Uranus and Neptune indicate that they must have a different composition, which we look at next.

Uranus and Neptune: Curiously Dense

If Uranus and Neptune formed from gas with similar abundances of elements as Jupiter and Saturn, the smaller masses of Uranus and Neptune would produce less gravitational compression, and Uranus and Neptune would both have lower average densities than Jupiter or Saturn. In fact, however, Uranus and Neptune have average densities (1271 kg/m^3 and 1638 kg/m^3, respectively) that are comparable to or greater than those of Jupiter (1330 kg/m^3) or Saturn (690 kg/m^3). Therefore, we can conclude that Uranus and Neptune contain greater proportions of the heavier elements than do Jupiter or Saturn.

This picture is not what we might expect. According to our discussion in Section 8-5 of how the solar system formed, hydrogen and helium should be relatively more abundant as distance from the vaporizing heat of the Sun increases. But Uranus and Neptune contain a greater percentage of heavy elements, and thus a *smaller* percentage of hydrogen and helium, than Jupiter and Saturn.

A related problem is how to explain the masses of Uranus and Neptune. At the locations of Uranus (19 au from the Sun) and Neptune (30 au from the Sun), the solar nebula was probably so sparse that these planets would have taken billions of years to grow to their present sizes around a core of icy planetesimals. But protoplanetary disks observed around other stars (see Section 8-4) do not appear to survive for that long before they dissipate. In other words, the problem is not how to explain why Uranus and Neptune are smaller than Jupiter and Saturn; it is how to explain why Uranus and Neptune should exist at all!

Models of How Uranus and Neptune Formed

In the Nice model (see Section 8-6) for the origin of Uranus and Neptune, the planets formed in denser regions of the solar nebula closer to the Sun. In this region, they could have grown rapidly to their present sizes. Had the planets remained at these locations, they could eventually have accumulated enough hydrogen and helium to become as large as Jupiter or Saturn. But before that could happen, gravitational interactions with Jupiter, Saturn, and the disk of planetesimals would have deflected Uranus and Neptune outward into their present large orbits. Because the solar nebula was very sparse at those greater distances, Uranus and Neptune stopped growing and remained at the sizes we see today. In the Nice model, there are even scenarios where Neptune formed closer to the Sun than Uranus, and then they switched places when deflected outward.

> Explaining the origins of Uranus and Neptune poses a challenge to planetary scientists.

Figure 14-7 shows one model for the present-day internal structures of Uranus and Neptune. (Compare this to the internal structures of Jupiter and Saturn, shown in Figure 12-14.) In this model, each planet has a rocky core roughly the size of Earth, although more massive. Each planet's core is surrounded by a mantle of liquid water and ammonia. (This mixture means that the mantle is chemically similar to household window cleaning fluid.) Around the mantle is a layer of liquid molecular hydrogen and liquid helium, with a smaller portion (about 10%) of liquid methane.

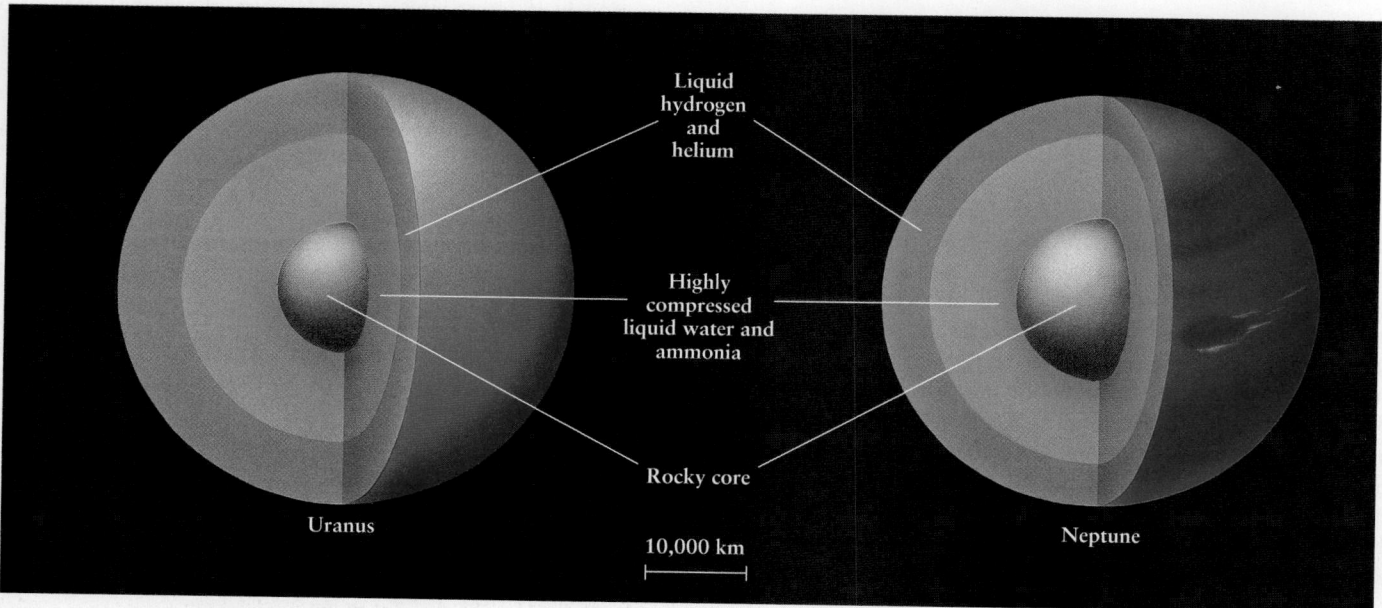

Liquid hydrogen and helium

Highly compressed liquid water and ammonia

Rocky core

Uranus

10,000 km

Neptune

FIGURE 14-7

The Internal Structures of Uranus and Neptune In the model shown here, both Uranus and Neptune have a rocky core (resembling a terrestrial planet), a mantle of liquid water with ammonia dissolved in it, and an outer layer of liquid molecular hydrogen and liquid helium. The atmospheres are very thin shells on top of the hydrogen-helium layer. Uranus and Neptune have about the same diameter, but since Neptune is more massive, it may have a somewhat larger core.

Unlike Jupiter and Saturn, the pressure is not high enough to convert the liquid hydrogen into liquid metallic hydrogen.

Weather Forecast: Raining Diamonds!

Both Uranus and Neptune contain methane (CH_4), which might actually lead to solid diamonds raining down onto a liquid carbon ocean. Recall that diamonds are made of pure carbon, and this carbon would come from the methane that has broken apart in the planets' high-temperature interiors.

The formation of diamonds has been suspected at depths within Neptune and Uranus, where temperatures reach 50,000°F and pressures reach more than 10 million times Earth's atmospheric pressure. However, these conditions couldn't be recreated in the lab—that is, until 2017, when scientists used lasers to compress and heat carbon to the desired conditions. To their own astonishment, scientists saw solid diamonds form, where they floated like icebergs on liquid carbon!

Verifying this process within these giant planets won't be easy, but the ocean of carbon could have a measurable effect on their magnetic fields.

CONCEPT CHECK 14-4

Why is it thought that Neptune had to form closer to the Sun than at its present distance?

Answer appears at the end of the chapter.

14-5 The magnetic fields of both Uranus and Neptune are oriented at unusual angles

Astronomers were quite surprised by the data sent back from Voyager 2's magnetometer as the spacecraft sped past Uranus and Neptune. These data, as well as radio emissions from charged particles in their magnetospheres, showed that the magnetic fields of both Uranus and Neptune are tilted at steep angles to their axes of rotation (Figure 14-8). Uranus's **magnetic axis**, the line connecting its north and south magnetic poles, is inclined by 59° from its axis of rotation; Neptune's magnetic axis is tilted by 47°. By contrast, the magnetic axes of Earth, Jupiter, and Saturn are all nearly aligned with their rotation axes; the angle between their magnetic and rotation axes is 12° or less. Scientists were also surprised to find that the magnetic fields of Uranus and Neptune are offset from the centers of the planets.

CAUTION!

The drawing of Earth's magnetic field at the far left of Figure 14-8 may seem to be mislabeled, because it shows the *south* pole of a magnet at Earth's *north* pole. But, in fact, this is correct, as you can understand by thinking about how magnets work. On a magnet that is free to swivel, like the magnetized needle in a compass, the "north pole" is called that because it tends to point north on Earth. Likewise, a compass needle's south pole points toward the south on Earth. Furthermore, opposite magnetic poles attract. If you take two magnets and put them next to each other, they try to align themselves so that one magnet's north pole is next to the other magnet's south pole. Now, if you think of a compass needle as one magnet and the entire Earth as the other magnet, it makes sense that the compass needle's north pole is being drawn toward a magnetic *south* pole—which happens to be located near Earth's geographic north pole. Earth's magnetic pole nearest its geographic North Pole is called the "magnetic north pole." Note that the magnets drawn inside Jupiter, Saturn, and Neptune are oriented opposite to Earth's. On any of these planets, the north pole of a compass needle would point south, not north!

FIGURE 14-8

The Magnetic Fields of Five Planets This illustration shows how the magnetic fields of Earth, Jupiter, Saturn, Uranus, and Neptune are tilted relative to their rotation axes (shown in red). For both Uranus and Neptune, the magnetic fields are offset from the planet's center and steeply inclined to the rotation axis.

Explaining Misaligned Magnetism

Why are the magnetic axes and axes of rotation of Uranus and Neptune so badly misaligned? And why are the magnetic fields offset from the centers of the planets? One possibility is that their magnetic fields might be undergoing a reversal; geological data show that Earth's magnetic field has switched north to south and back again many times in the past. Another possibility is that the misalignments resulted from catastrophic collisions with planet-sized bodies. As we discussed in Section 14-2, the tilt of Uranus's rotation axis and its system of moons suggest that such collisions may have occurred long ago. As we will see in Section 14-7, Neptune may have gravitationally captured its largest moon, Triton, but no one knows if that incident was responsible for offsetting Neptune's magnetic axis.

Another possibility has garnered interest since 2017, when scientists showed in the lab that conditions within Uranus and Neptune could lead to a liquid ocean of carbon. Liquid carbon is expected to have electrical properties that could alter these planets' magnetic fields, possibly misaligning their magnetic axes.

Because neither Uranus nor Neptune is massive enough to compress hydrogen to a metallic state, their magnetic fields cannot be generated in the same way as those of Jupiter and Saturn. Instead, under the high pressures found in the watery mantles of Uranus and Neptune, dissolved molecules such as ammonia lose one or more electrons and become electrically charged (that is, they become *ionized*; see Section 5-8). Water is a good conductor of electricity when it has such electrically charged molecules dissolved in it, and electric currents in this fluid are probably the source of the magnetic fields of Uranus and Neptune.

> Unlike any other planets, Uranus and Neptune have magnetic fields likely caused by ionized ammonia.

The *Cosmic Connections* figure on the next page summarizes the key properties of Uranus and Neptune and how they compare with those of Jupiter and Saturn.

CONCEPT CHECK 14-5

In order for a magnetic dynamo to create a global magnetic field around Uranus and Neptune, these planets must have an electrically conducting fluid. What is that conducting fluid thought to be?

Answer appears at the end of the chapter.

14-6 Uranus and Neptune each have a system of thin, dark rings

On March 10, 1977, Uranus was scheduled to move in front of a faint star, as seen from the Indian Ocean. A team of astronomers headed by James L. Elliot of Cornell University observed this event, called an **occultation**, from a NASA airplane equipped with a telescope. They hoped that by measuring how long the star was hidden, they could accurately measure Uranus's size. In addition, by carefully measuring how the starlight faded when Uranus passed in front of the star, they planned to deduce important properties of Uranus's upper atmosphere.

The Surprising Rings of Uranus

To everyone's surprise, the background star briefly blinked on and off several times just before the star passed behind Uranus and again immediately after (Figure 14-9). The astronomers concluded that Uranus must be surrounded by a series of nine narrow rings. In addition to these nine rings, Voyager 2 discovered two others that lie even closer to the cloudtops of Uranus. Another two extremely faint rings, much larger in diameter than any of the others, were found in 2005 using the Hubble Space Telescope (Figure 14-10).

> Like the planet itself, the rings of Uranus were discovered by accident.

Unlike Saturn's rings, the rings of Uranus are dark and narrow: Most are less than 10 km wide. Typical particles in Saturn's rings are chunks of reflective water ice the size of snowballs (a few centimeters across), but typical particles in Uranus's rings—thought to be mostly water ice—are 0.1 to 10 m in size and are no more reflective than lumps of coal. It is not surprising that these narrow, dark rings escaped detection for so long. Figure 14-11 is a Voyager 2 image of the rings from the side of the planet away from the Sun, where light scattering from exceptionally small ring particles makes them more visible. (We discussed light scattering in Section 5-6 and Box 5-4.) Figure 14-10 shows the Sun-facing side of the rings.

The Uranian rings are so dark because sunlight at Uranus is only one-quarter as intense as at Saturn. As a result, Uranus's ring particles are so cold that they can retain some methane ice. Scientists speculate that eons of impacts by electrons trapped in the magnetosphere of Uranus have converted this methane ice into dark carbon compounds. This **radiation darkening** can account for the low reflectivity of the rings.

Uranus's major rings are located less than 2 Uranian radii from the planet's center, well within the planet's Roche limit (see Section 12-9). Some sort of mechanism, possibly one involving shepherd satellites, efficiently confines particles to their narrow orbits. (In Section 12-11, we discussed how shepherd satellites help keep planetary rings narrow.) Voyager 2 searched for shepherd satellites but found only two. The others may be so small and dark that they have simply escaped detection.

The two faint rings discovered in 2005 (see Figure 14-10) are different: They lie well *outside* Uranus's Roche limit. The outer of these two rings owes its existence to a miniature satellite called Mab (named for a diminutive fairy in English folklore) that orbits within this ring. Meteorites colliding with Mab knock dust off the satellite's surface. Since Mab is so small (just 36 km in diameter) and hence has little gravity, the ejected dust escapes from the satellite and goes into orbit around Uranus, forming a ring. The inner of the two faint rings may be formed in the same way; however, no small satellite has yet been found within this ring. Recall from Section 13-10 that Saturn's E ring is produced from the moon Enceladus—orbiting within the ring itself—spewing out material from cryovolcanoes.

The Decaying Rings of Neptune

Like Uranus, Neptune is surrounded by a system of thin, dark rings that were first detected in stellar occultations. Figure 14-12 is a Voyager 2 image of Neptune's rings. The ring particles vary

COSMIC CONNECTIONS

The Outer Planets: A Comparison

ANIMATION 14.4 Uranus and Neptune are not simply smaller versions of Jupiter and Saturn. This table summarizes the key differences among the four jovian planets.

	Interior	Surface	Rings	Atmosphere	Magnetic Field*
Jupiter	Terrestrial core, thick liquid metallic hydrogen layer, molecular hydrogen	No solid surface, atmosphere gradually thickens to liquid state, belt and zone structure, hurricane-like features	Yes	Primarily H, He	19,000× Earth's total field; at its cloud layer, 14× stronger than Earth's surface field
Saturn	Similar to Jupiter, with bigger terrestrial core and less metallic hydrogen	No solid surface, less distinct belt and zone structure than Jupiter	Yes	Primarily H, He	570× Earth's total field; at its cloud layer, ⅔× Earth's surface field
Uranus	Terrestrial core, liquid water shell, liquid hydrogen and helium mantle	No solid surface, weak belt and zone system, hurricane-like features, color from methane absorption of red, orange, yellow	Yes	Primarily H, He, some CH_4	50× Earth's total field, at its cloud layer, 0.7× Earth's surface field
Neptune	Similar to Uranus	Similar to Uranus	Yes	Primarily H, He, some CH_4	35× Earth's total field, at its cloud layer, 0.4× Earth's surface field

For detailed comparisons between planets, see Appendices 1 and 2

*To see the orientations of these magnetic fields relative to the rotation axes of the planets, see Figure 14-8.

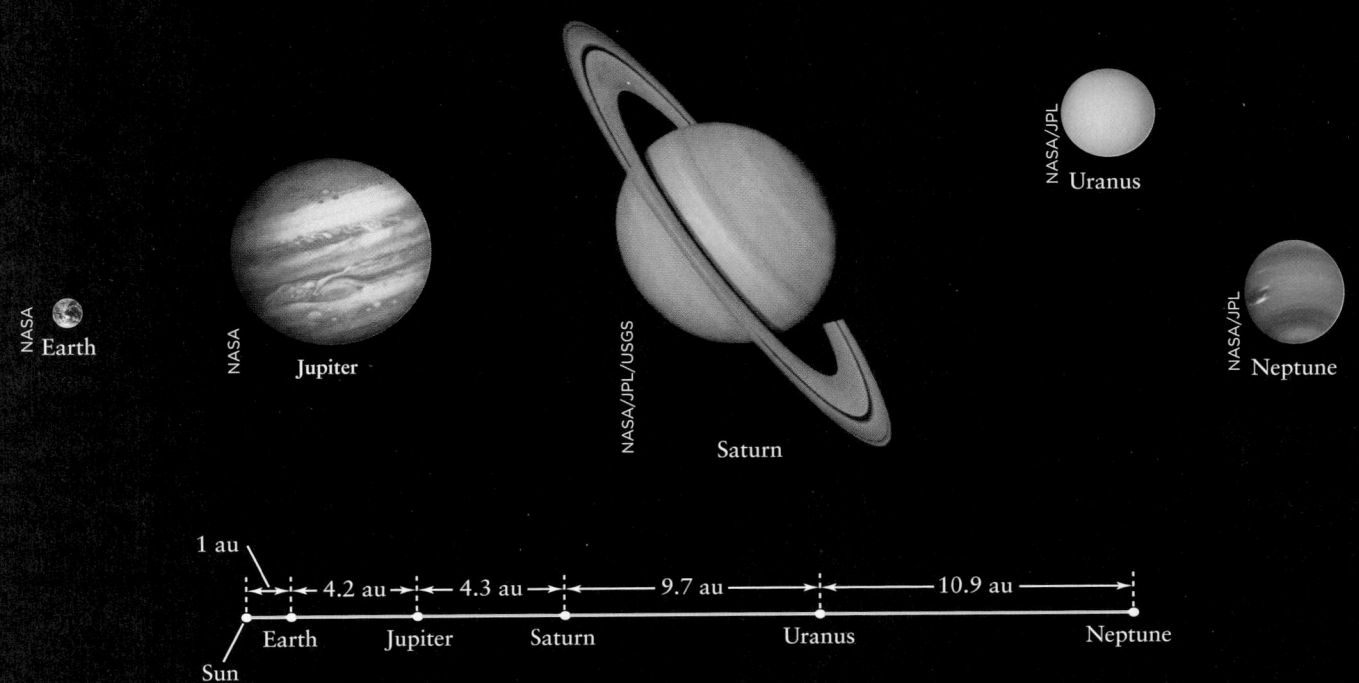

NASA — Earth
NASA — Jupiter
NASA/JPL/USGS — Saturn
NASA/JPL — Uranus
NASA/JPL — Neptune

1 au
Sun — Earth ← 4.2 au → Jupiter ← 4.3 au → Saturn ← 9.7 au → Uranus ← 10.9 au → Neptune

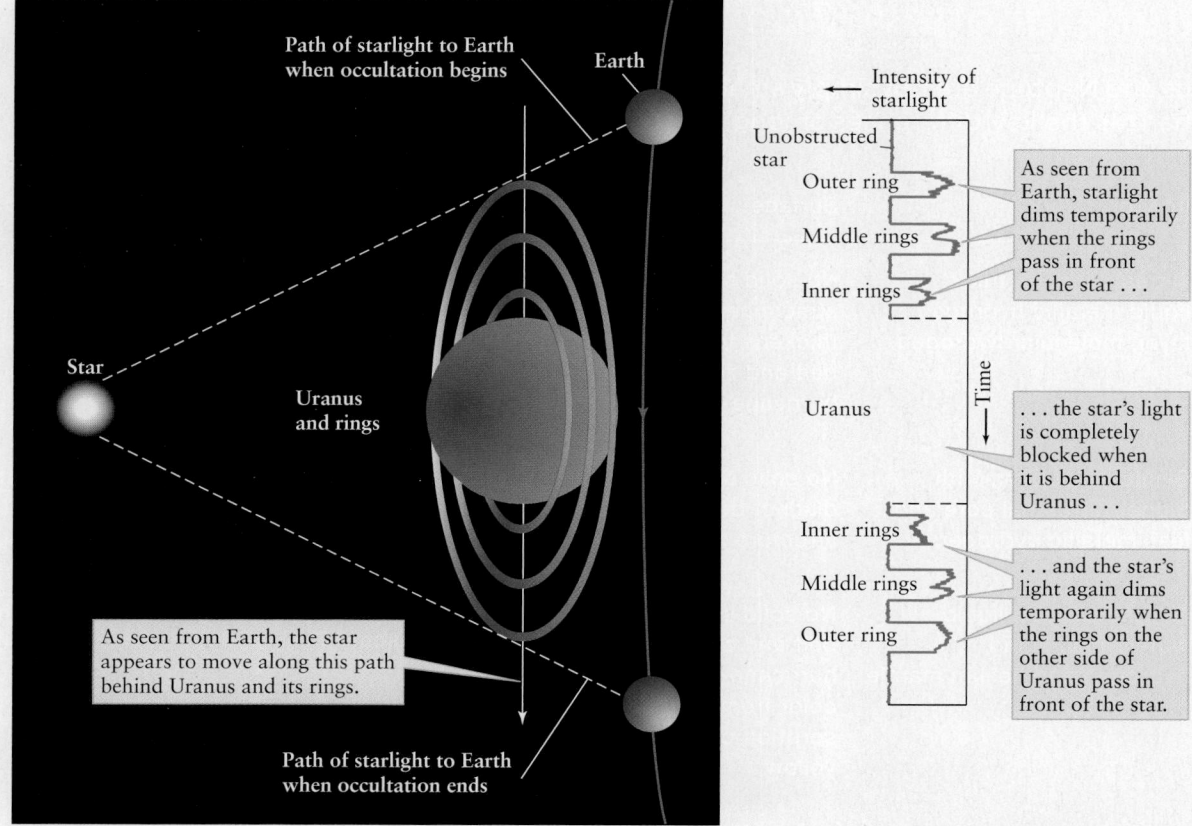

ANIMATION 14.2 **FIGURE 14-9**

How the Rings of Uranus Were Discovered As seen from Earth, Uranus occasionally appears to move in front of a distant star. Such an event is called an occultation. The star's light is completely blocked when it is behind the planet. But before and after the occultation by Uranus, the starlight dims temporarily as the rings pass in front of the star. Computer monitors weren't in common use at the time, and these

occultations—the intensity versus time shown in the figure—were printed out on a long piece of paper. When the paper was folded back on itself, the dips in intensity matched up, suggesting that they corresponded to rings. An editor for this book actually took her first astronomy class from James Elliot in 2004, and she remembers being shown the original 1977 readout showing the occultations!

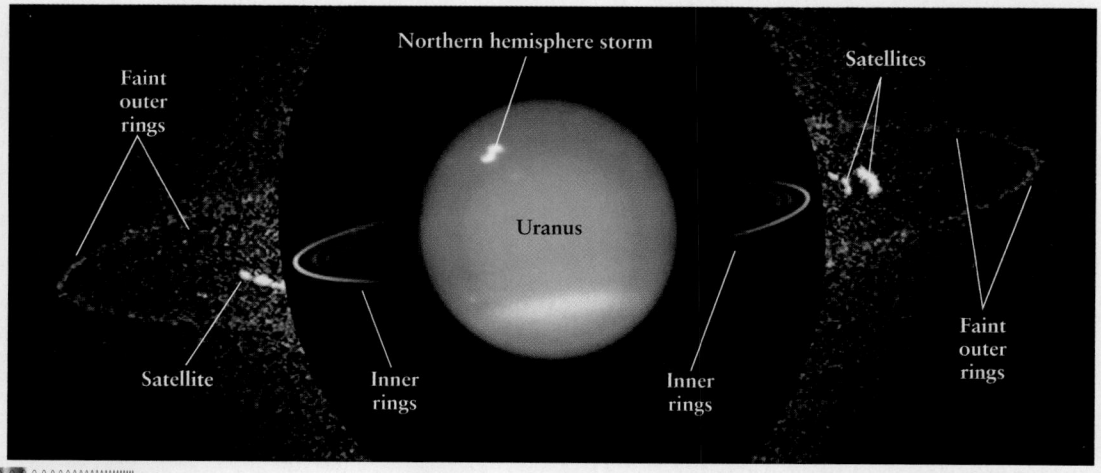

FIGURE 14-10 **R I V U X G**

Uranus's Rings from the Hubble Space Telescope Several HST images were assembled into this view of Uranus's rings. Relatively short exposure times reveal the planet and the inner rings, while much longer

exposure times were needed to reveal the outer rings. During these long exposures, several of Uranus's satellites moved noticeably, leaving bright trails on the image. (NASA, ESA, and M. Showalter, SETI Institute)

FIGURE 14-11 R I (V) U X G

The Shaded Side of Uranus's Rings This view from Voyager 2, taken when the spacecraft was in Uranus's shadow, looks back toward the Sun. Numerous fine-grained dust particles gleam in the spaces between the main rings. This dust is probably debris from collisions between larger particles in the main rings. The short horizontal streaks are star images blurred by the spacecraft's motion during the exposure. (NASA/JPL)

VIDEO
14.3

FIGURE 14-12 R I (V) U X G

Neptune's Rings The two main rings of Neptune and a faint, inner ring can easily be seen in this composite of two Voyager 2 images. There is also a faint sheet of particles whose outer edge is located between the two main rings and extends inward toward the planet. The overexposed image of Neptune itself has been blocked out. (NASA)

in size from a few micrometers ($1 \ \mu m = 10^{-6}$ m) to about 10 m. As for Uranus's rings, the particles that make up Neptune's rings reflect very little light because they have undergone radiation darkening.

In 2002 and 2003, a team led by Imke de Pater of the University of California, Berkeley, used the 10-m Keck telescope in Hawaii

to observe in detail the rings of Neptune. Remarkably, they found that all of the rings had become fainter since the Voyager 2 flyby in 1989. Apparently, the rings are losing particles faster then new ones are being added. If the rings continue to decay at the same rate, one of them may vanish completely within a century. More research into the nature of Neptune's rings is needed to explain this curious and rapid decay.

CONCEPT CHECK 14-6

Why are Uranus's rings so much more difficult to see than Saturn's rings?

Answer appears at the end of the chapter.

14-7 Uranus's larger and smaller satellites

Before the Voyager 2 flyby of Uranus, five moderate-sized satellites—Titania, Oberon, Ariel, Umbriel, and Miranda—were known to orbit the planet (Figure 14-13). These are Uranus's largest moons. Most are named after sprites and spirits in Shakespeare's plays. They range in diameter from about 1600 km (1000 mi) for Titania and Oberon to less than 500 km (300 mi) for Miranda. These moons have average densities around 1500 kg/m³, which is consistent with a mixture of half water ice and half rock (although Miranda is mostly ice).

Voyager 2 discovered 11 other small Uranian satellites, most of which have diameters of less than 100 km (60 mi); 11 more were found using ground-based telescopes and the HST between 1997 and 2005. Only Jupiter and Saturn have more known satellites than Uranus. In the infrared, Uranus, its rings, and some of its moons are easily visible (Figure 14-14).

The moons of Uranus are generally divided into three groups: the larger moons, the smaller inner moons orbiting near the rings, and the smaller outer moons that are probably captured asteroids or comets.

FIGURE 14-13 R I (V) U X G

Uranus's Principal Satellites This "family portrait" (a montage of five Voyager 2 images) shows Uranus's five largest moons to the same scale and correctly displays their respective reflectivities. (The darkest satellite, Umbriel, is actually more reflective than Earth's Moon.) All five satellites have grayish surfaces, with only slight variations in color. (NASA/JPL; NASA)

(a)

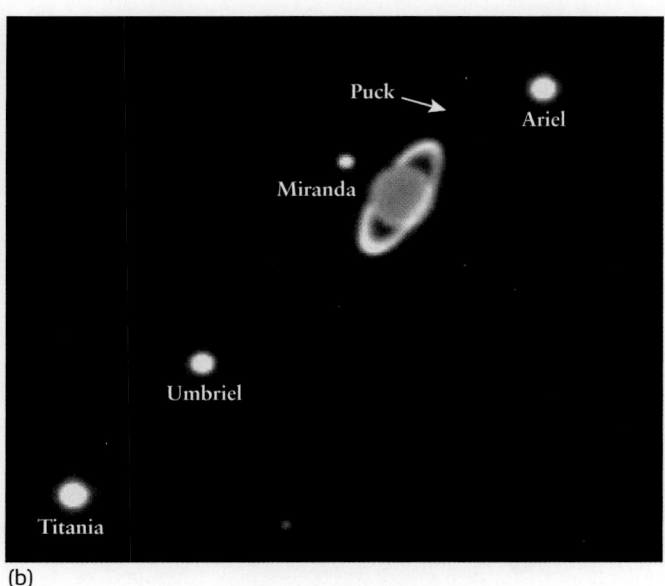

(b)

FIGURE 14-14 R I V U X G

Uranus's Rings and Moons (a) This false-color infrared image from the HST shows eight of Uranus's inner moons, all of which were discovered by Voyager 2 when it flew past Uranus in 1986. They all lie within 86,000 km of the planet's center (only about one-fifth of the distance from Earth to our Moon). The arcs show how far each moon moves around its orbit in 90 minutes. Note the pole and equator; unlike the other planets, Uranus nearly lies on its side with its rotational axis close to its orbital plane around the Sun. **(b)** This infrared image from the Very Large Telescope shows five of Uranus's larger moons. All the moons visible in this image are in prograde orbit, orbiting in the same direction as Uranus rotates. The rings in this image appear even brighter than Uranus: Gaseous methane in Uranus's atmosphere absorbs infrared light, while ice in the rings reflects this wavelength. (a: NASA/ JPL/STScI; b: ESO)

Uranus's Larger Satellites

Umbriel and Oberon both appear to be geologically dead worlds, with surfaces dominated by impact craters. By contrast, Ariel's surface appears to have been cracked at some time in the past, allowing some sort of ice lava to flood low-lying areas. A similar process appears to have taken place on Titania. This geologic activity may be due to a combination of the satellites' internal heat and tidal heating like that which powers volcanism on Jupiter's satellite Io (see Section 13-4).

Unique among Uranus's satellites is Miranda, which has a landscape unlike that of any other world in the solar system. Much of the surface is heavily cratered, as we would expect for a satellite only 470 km in diameter, but several regions have unusual and dramatic topography (Figure 14-15). Detailed analysis of Miranda's geology suggests that this satellite's orbital period was once in a whole-number ratio with that of more massive Umbriel or Ariel. The resulting tidal heating melted Miranda's interior, causing dense rocks in some locations on the surface to settle toward the satellite's center as blocks of less dense ice were forced upward toward the surface, thus creating Miranda's resurfaced terrain. Tidal heating must have ceased before this process could run its full course, which explains why some ancient, heavily cratered regions remain on Miranda's surface.

> Uranus's satellite Miranda has cliffs twice as high as Mount Everest.

VIDEO
14.4

FIGURE 14-15 R I V U X G

Miranda This composite of Voyager 2 images shows that part of Miranda's surface is ancient and heavily cratered, while other parts are dominated by parallel networks of valleys and ridges. At the very bottom of the image—where a "bite" seems to have been taken out of Miranda—is a range of enormous cliffs that jut upward to an elevation of 20 km (12 mi), twice as high as Mount Everest. (NASA/JPL)

Uranus's Larger Satellites Constrain Models of an Axis-Tilting Impact

The five large moons of Uranus help us understand the collision that is proposed to have tilted the rotation axis of Uranus. A collision large enough to turn a planet on its side would be quite a violent event. Such an impactor is estimated to be several times the mass of Earth—quite large!—though there might instead have been several smaller impacts. While several impacts might seem less likely to occur, modeling indicates that a single large impact would have resulted in retrograde orbits for Uranus's large moons. On the other hand, two or more smaller impactors, each the size of Earth, would better explain the observed prograde orbits of the large Uranian moons.

Another constraint on this impact model comes from the orientation of the moons' orbital plane. If these impacts occurred after the large moons formed, the moons would have remained in their original orbital plane, which is assumed to be in Uranus's original equatorial plane. Instead, what we see today is both Uranus *and* the orbital plane of its moons tilted together (see Figure 14-14b). In order to tilt Uranus and its moons together, models suggest that impacts might have rotated a disk of protoplanetary material from which both Uranus and its moons formed. Another hypothesis suggests that gravitational interactions with a large moon might have tipped Uranus on its side. This satellite later could have been lost after a close encounter with another gas giant. The origin of Uranus's tilt is still an open question.

Uranus's Perplexing Small Satellites

While Uranus's small satellites are unlikely to exhibit the kind of geology found in Miranda, some of them move in curious orbits. Eight of the nine smaller outer satellites beyond Oberon are in retrograde orbits, moving around Uranus opposite to the direction in which Uranus rotates. Many of the small outer satellites of Jupiter and Saturn have retrograde orbits and are probably captured asteroids (see Section 13-9 and Section 13-10); the same is probably true of the outer satellites of Uranus.

The 13 inner satellites that orbit closest to Uranus (inside the orbit of Miranda) are all in prograde orbits, so they travel around Uranus in the same direction as the planet rotates (Figure 14-14a). Just like the rings of Uranus, these inner moons are very dark; this is probably due to the same process of radiation darkening. In fact, it is thought that the Uranian rings were probably created by the fragmentation of one or more inner moons.

Will the inner moons orbit Uranus indefinitely? Probably not. When Mark Showalter and Jack Lissauer compared 11 of these satellites' orbits in 2005 with their orbits in 1994, they found surprisingly large differences. (The other two inner satellites were only discovered in 2003, so it is not known whether their orbits underwent similar changes.) These satellites can pass within a few thousand to a few hundred kilometers from each other, so they can exert strong gravitational forces on each other. Over time, these forces can modify their orbits. Computer simulations of these gravitational interactions indicate that the inner satellites might begin colliding with each other within 100 million years. If these inner satellites are in such unstable orbits, have they really been in orbit around Uranus since it formed more than 4.56 billion years ago? Or did Uranus somehow acquire these satellites in the relatively recent past?

Why is there doubt about most of Uranus's inner moons being roughly as old as Uranus?

Answer appears at the end of the chapter.

14-8 Neptune's satellite Triton is an icy world with a young surface and a tenuous atmosphere

Neptune has 13 known satellites, listed in Appendix 3. They are named for mythological beings related to bodies of water. (Neptune itself is named for the Roman god of the sea.) Most of these worlds are small, icy bodies, probably similar to the smaller satellites of Uranus. The one striking exception is Triton, Neptune's largest satellite (see Table 7-2). In many ways, Triton is quite unlike any other world in the solar system.

Triton's Backward Orbit and Young Surface

Like many of the small satellites of Jupiter, Saturn, Uranus, and Neptune, Triton is in a retrograde orbit, such that it moves around Neptune opposite to the direction of Neptune's rotation. Furthermore, the plane of Triton's orbit is inclined by 23° from the plane of Neptune's equator. It is difficult to imagine how a satellite might form out of the same cloud of material as a planet and end up orbiting in a direction opposite the planet's rotation and in such a tilted plane. One hypothesis is that Triton formed elsewhere in the solar system, collided long ago with a now-vanished satellite of Neptune, and was captured by Neptune's gravity. With a diameter of 2706 km, a bit smaller than our Moon but much larger than any other satellite in a retrograde orbit, Triton would certainly be the largest captured satellite. Another hypothesis for the origin of Triton is that it was part of a binary system with another object. In this scenario, as Neptune's gravity attracted the binary system, Triton could have been slowed down enough to capture it, while the other object was ejected away.

Figure 14-16 shows the icy, reflective surface of Triton as imaged by Voyager 2. Most of Triton's ice is water, with some additional nitrogen and methane ice. There is a conspicuous absence of large craters, which immediately tells us that Triton has a young surface on which the scars of ancient impacts have largely been erased by tectonic activity. There are areas that resemble frozen lakes and may be the calderas of extinct ice volcanoes. Some of Triton's surface features resemble the long cracks seen on Europa (Section 13-6) and Ganymede (Section 13-7). Still other features are unique to Triton. For example, in the upper portion of Figure 14-16, you can see dimpled, wrinkled terrain that resembles the skin of a cantaloupe.

Triton's tectonically active history is probably related to its having been captured into orbit around Neptune. After its capture, Triton most likely started off in a highly elliptical orbit, but today the satellite's orbit is quite circular. The satellite's original elliptical orbit would have been made circular by tidal forces exerted on Triton by Neptune's gravity. These tidal forces would also have stretched and flexed Triton, causing enough tidal heating to melt much of the interior. The resulting volcanic activity (with

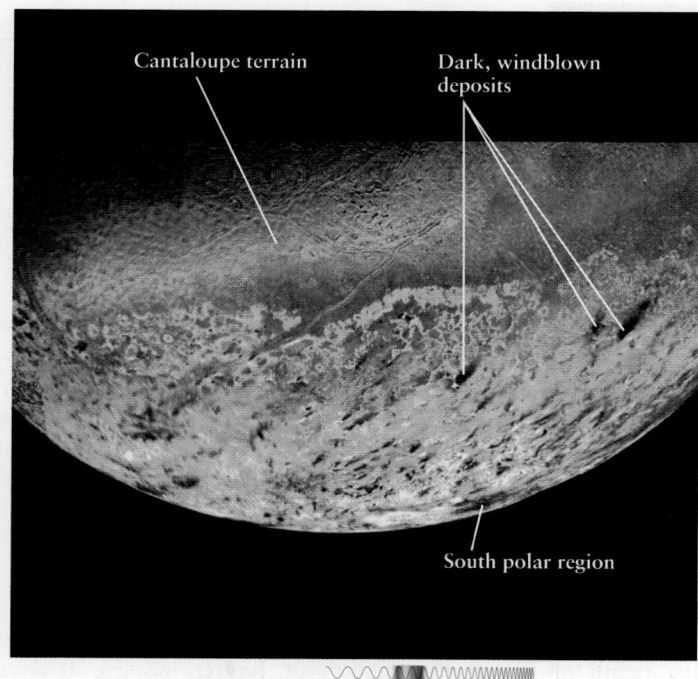

Cantaloupe terrain

Dark, windblown deposits

South polar region

FIGURE 14-16   R I V U X G

Triton Several high-resolution Voyager 2 images were combined to create this mosaic. The pinkish material surrounding Triton's south polar region is probably nitrogen frost. Some of this presumably evaporates when summer comes to the south pole; the northward flow of the evaporated gas may cause the dark surface markings. Farther north is a brown area of "cantaloupe terrain" that resembles the skin of a melon. (NASA/JPL)

lavas made of ice rather than molten rock) would have obliterated Triton's original surface features, including craters.

There is still enough warmth in Triton's interior to keep it active with cryovolcanoes (where the lava consists of water that is prevented from freezing due to dissolved ammonia). Voyager 2 observed plumes of dark material being ejected from the surface to a height of 8 km (5 mi). The plumes contain invisible nitrogen gas, along with small and dark dust particles. These plumes may have been generated from a hot spot far below Triton's surface, similar to geysers on Earth. Alternatively, the energy source for the plumes may be sunlight that warms the surface, producing subsurface pockets of gas and creating fissures in the icy surface through which the gas can escape.

Triton's surface temperature is only 38 K ($-235°$C = $-391°$F), the lowest of any world yet visited by spacecraft, aside from Pluto. This temperature is low enough to solidify nitrogen, and indeed the spectrum of sunlight reflected from Triton's surface shows absorption lines due to nitrogen ice as well as methane ice. But Triton is also warm enough to allow some nitrogen to evaporate from the surface, like the steam that rises from ice cubes when you first take them out of the freezer. Voyager 2 confirmed that Triton has a very thin nitrogen atmosphere with a surface pressure of only 1.6×10^{-5} atmosphere, about the same as at an altitude of 100 km above Earth's surface. Despite its thinness, Triton's atmosphere was revealed by its winds. Voyager 2 saw areas on Triton's surface where dark dusty material has been blown downwind by a steady breeze (see Figure 14-16). Dark dusty material ejected from the geyserlike plumes was carried as far as 150 km by high-altitude winds.

Tidal Forces and Triton's Doom

Just as tidal forces presumably played a large role in Triton's history, they also determine its future. Triton raises a tidal bulge on Neptune, just as our Moon distorts Earth (recall Figure 10-20). In the case of the Earth–Moon system, the gravitational pull of Earth's tidal bulge causes the Moon to spiral away from Earth. But because Triton's orbit is retrograde, the tidal bulge on Neptune exerts a force on Triton that makes the satellite slow down rather than speed up. (In Figure 10-20, imagine that the moon is orbiting toward the bottom of the figure rather than toward the top.) This is causing Triton to spiral gradually in toward Neptune. In approximately 100 million years, Triton will be inside Neptune's Roche limit, and the satellite will eventually be torn to pieces by tidal forces. When this happens, the planet will develop a spectacular ring system—overshadowing by far its present-day set of narrow rings—as rock fragments gradually spread out along Triton's former orbit.

> Within 100 million years, Triton will be torn apart by Neptune's tidal forces.

Prior to Voyager 2, only one other satellite was known to orbit Neptune. Nereid, which was first sighted in 1949, is in a prograde orbit. Hence, it orbits Neptune in the direction opposite to Triton. Nereid also has the most eccentric orbit of any satellite in the solar system; its distance from Neptune varies from 1.4 million to 9.7 million km. One possible explanation is that when Triton was captured by Neptune's gravity, the interplay of gravitational forces exerted on Nereid by both Neptune and Triton moved Nereid from a relatively circular orbit (like those of Neptune's other, smaller moons) into its present elliptical one.

CONCEPT CHECK 14-8

How was an atmosphere detected on Triton?

Answer appears at the end of the chapter.

14-9 Pluto is a small, icy, mysterious world

Speculations about worlds beyond Neptune date back to the late 1800s, when a few astronomers suggested that Neptune's orbit was being perturbed by an unknown object. Encouraged by the fame of Adams and Le Verrier, several people set out to discover "Planet X." Two prosperous Boston gentlemen, William Pickering and Percival Lowell, were prominent in this effort. (Lowell also enthusiastically promoted the idea of canals on Mars; see Section 11-4.) Modern calculations show that there are, in fact, no unaccounted perturbations of Neptune's orbit. It is thus not surprising that no orbiting body was found at the positions predicted by Pickering, Lowell, and others. Yet the search continued.

The Discovery of Pluto

Before he died in 1916, Lowell urged that a special camera with a wide field of view be constructed to help search for Planet X. After many delays, the camera was finished in 1929 and installed at the

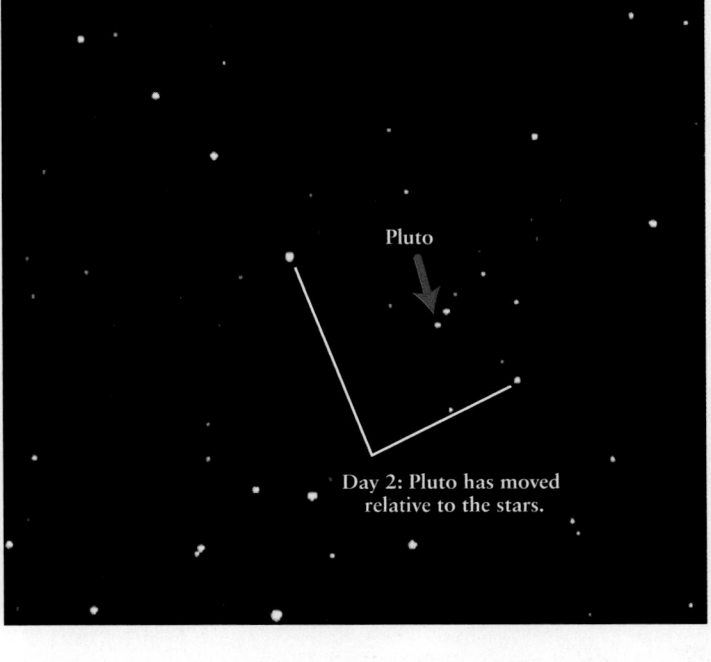

FIGURE 14-17 R I V U X G

Pluto's Motion Across the Sky Pluto was discovered in 1930 by searching for a dim, starlike object that moves slowly in relation to the background stars. These two photographs were taken 3 days apart. Even when its apparent motion on the celestial sphere is fastest, Pluto moves only about 1 arcmin per day relative to the stars. (John Chumack/Science Source)

Lowell Observatory in Flagstaff, Arizona, where a young astronomer, Clyde W. Tombaugh, had recently joined the project. On February 18, 1930, Tombaugh finally discovered the long-sought object. It was disappointingly faint—1000 times dimmer than the dimmest object visible with the naked eye and 250 times dimmer than Neptune at opposition—and presented no discernible disk (**Figure 14-17**). The new world, originally labeled a planet, was named by the 11-year-old student Venetia Burney for Pluto, the mythological god of the underworld. In 2006, Pluto was reclassified as a **dwarf planet**, which is an object orbiting the Sun (not a satellite of another body) with enough mass to gravitationally pull itself into a spherical shape, yet not enough gravity to clear out planetesimals from its surroundings. As of this writing, there are 10 dwarf planets that are nearly confirmed. (We will further discuss Pluto's reclassification in the next section.)

Pluto's orbit (see Figure 7-8) around the Sun is more elliptical (eccentricity 0.2501) and more steeply inclined to the plane of the ecliptic (17.15°) than the orbit of any of the planets. In fact, Pluto's orbit is so eccentric that it is sometimes closer to the Sun than Neptune is. This was the case from 1979 until 1999; indeed, when Pluto was at perihelion in 1989, it was more than 10^8 km closer to the Sun than was Neptune. Happily, the orbits of Neptune and Pluto are such that the two worlds will never collide.

New Horizons Visits Pluto

Eighty-five years after the discovery of Pluto, the New Horizons spacecraft visited this distant icy world. It took 9 years for New Horizons to reach Pluto in 2015, but not for lack of speed. New Horizons was launched fast enough to pass our Moon 9 hours after launch—a trip that took the Apollo 11 mission more than 3 days to complete in 1969. New Horizons had the greatest launch speed of any manmade object and is currently traveling so fast that it has enough speed to escape our solar system, which is expected to happen sometime around 2040. However, before leaving our solar system, New Horizons has another mission to accomplish. At the time of this writing, New Horizons is headed for a close flyby of a Kuiper belt object on January 1, 2019.

Due to the tremendous speed needed for New Horizons to reach Pluto in only 9 years, it wasn't feasible to weigh the spacecraft down with the additional fuel that would have been needed to slow it down into an orbit around Pluto. As a result, the New Horizons mission had to be accomplished as it zipped past Pluto.

Figure 14-18 shows the first clear view of Pluto ever captured. Two features of Pluto jump out at first sight—the heart-shaped region and the darker colors surrounding it. The darker colors are thought to originate when ultraviolet light from the Sun breaks apart methane (CH_4) in Pluto's atmosphere. The broken-up methane then recombines into a variety of molecules called *tholins*, which then fall down to Pluto's surface, giving it a dark reddish appearance. Pluto's heart, on the other hand, is a recent glacial flow of nitrogen ice, as shown in **Figure 14-19**.

Pluto's Mysteriously Young Surface

Upon seeing Pluto, astronomers were immediately struck by its lack of significant cratering. Recall that smaller objects (such as our Moon and Mercury) are expected to be heavily cratered. Since

FIGURE 14-18 R I V U X G

Pluto Has a Heart! This image, captured on July 13, 2015, stunned the world and revealed for all to see that Pluto has a heart. The heart-shaped feature is named Tombaugh Regio, after Pluto's discoverer, Clyde Tombaugh. He died in 1997, and a small container with some of his ashes was carried away on New Horizons. The left lobe of the heart, called Sputnik Planum, is about the size of Texas. This photo was taken when the distance between New Horizons and Pluto was a little farther than the Moon is from Earth, yet its resolution could just discern features the size of San Francisco's Golden Gate Bridge. New Horizons sped past Pluto so quickly that this is the only full view we'll get until some future mission; we don't know what Pluto's other side looks like. (NASA/Johns Hopkins University Applied Physics Laboratory/ Southwest Research Institute)

smaller bodies contain less heat-releasing radioactive material, and lose heat faster than larger objects, they have a shorter period of heat-powered volcanic or cryovolcanic activity to repave their surfaces. Thus, while volcanism on smaller objects fills in some of the impact craters formed during the *early* years of the solar system, the surface then experiences billions of years of continual bombardment, leaving numerous unfilled craters.

And Pluto is small! Pluto's volume, where its heat is generated, is only about one-third the volume of our Moon (and less than one-eighth of Mercury's volume). We have seen examples of small objects like Io and Europa with surfaces free of impact craters, but those moons are heated by tidal forces from Jupiter—Pluto has no source for significant tidal heating. Therefore, astronomers expected an old, cratered surface, yet found a relatively smooth young surface.

By counting the number of impacts for different crater sizes, the age of a surface can be estimated. This analysis indicates that Pluto has been geologically active for the last 4 billion years and that the smooth plains of Sputnik Planum (the left side of the

(a)

(b)

FIGURE 14-19 R I V U X G

Glacial Flows of Nitrogen Ice Pluto's heart is made of nitrogen ice, which can flow on Pluto because it forms a much softer ice than water does at Pluto's cold average temperature of around −228°C (−378°F, or only 45 K). **(a)** In the upper left of Pluto's heart, very smooth nitrogen ice flows up against hard mountains made of water ice. Analysis of the rounded features on the smooth surface indicates that they are convection cells. While the overall lack of cratering suggests a surface age of no more than 10 million years, modeling of the convection points to a surface age of about 200,000 years. **(b)** This is an enhanced-color image of a lower-left region of the heart. The smooth terrain is nitrogen ice. The numerous small pits are due to nitrogen sublimation, where the ice transforms directly into gas without melting first. The reddish rugged terrain (named Krun Macula) is mountains of water ice covered in tholins. These mountains reach about 2.5 km (1.5 mi) in height. (NASA/JHUAPL/SwRI)

FIGURE 14-20 R I **V** U X G

Mountains and Volcanoes of Ice These mountains of water ice on Pluto in the upper right of the image reach heights of about 3.5 km (about 2 mi). (The tallest mountain in the contiguous United States is Mount Whitney, at 3.1 km.) The circular feature in the middle of the image with a dark shadow in its center might be a water-ice volcano (or cryovolcano) about 100 km wide. While 98% of Pluto is probably nitrogen ice, water ice is less dense and floats on top of nitrogen ice. Nitrogen ice is too soft to form tall mountains, and the water-ice mountains probably float atop the nitrogen ice like icebergs. The lack of cratering in this region indicates that it's no more than 100 million years old, and could be much younger. (NASA, Johns Hopkins Univ./APL, Southwest Research Institute)

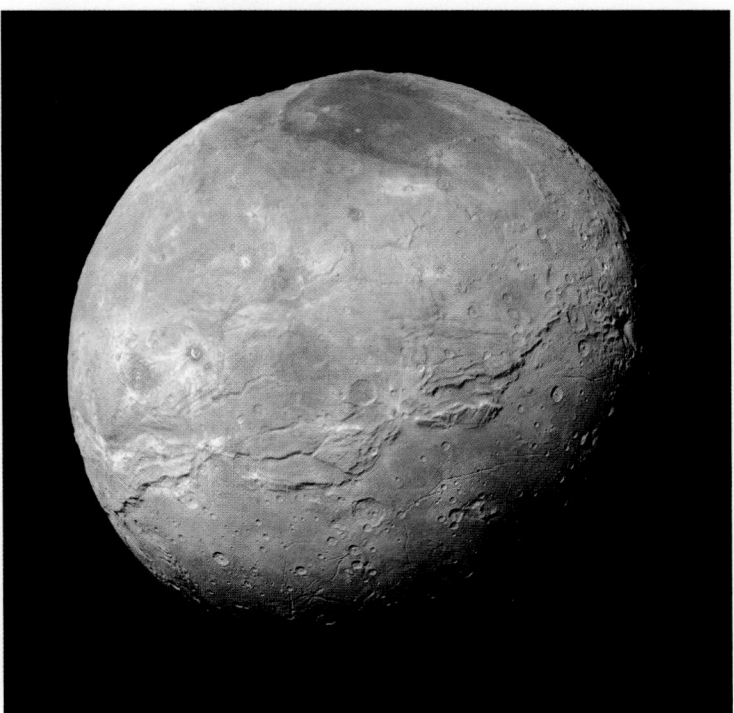

FIGURE 14-21 R I **V** U X G

Pluto's Moon Charon This is an enhanced-color image of Charon taken by New Horizons in July 2015. Charon's width is about 1200 km, which is similar in size to the state of Texas. Charon is quite close to Pluto, at only 5% the Earth–Moon distance. Gravitational forces keep the same sides of Charon and Pluto facing each other at all times. The reddish color around Charon's north pole comes from methane gas escaping Pluto's thin atmosphere, landing on Charon, then settling at the colder poles and getting processed by ultraviolet (UV) sunlight to make reddish-colored tholins. As expected, evidence from reflected light (not from direct imaging) indicates a similar reddish patch around Charon's southern pole. (NASA/JHUAPL/SwRI)

heart) are no more than 10 million years old! Even Pluto's mountain ranges appear to be young (**Figure 14-20**).

One idea to explain the unexpectedly young surface involves Pluto's largest moon, Charon. Similar to Earth's Moon, Charon is expected to have formed from an impact with Pluto. If that impact occurred much later than expected, then that more recent heating event for Pluto could explain its ongoing heat-powered geologic activity. A major problem with this late-impact idea, however, is that it seems inconsistent with our understanding that the era of large impact events would have long since ended. Furthermore, as we'll see next, impact craters on Charon imply an old age for the moon. For now, Pluto's minimally cratered surface—and the heat that it implies is still under the surface—remains a major unsolved mystery.

Pluto's Moons

In 1978, while examining some photographs of Pluto, James W. Christy of the U.S. Naval Observatory noticed that the image of Pluto on a photographic plate appeared slightly elongated, as though Pluto had a lump on one side. He promptly examined a number of other photographs of Pluto and found a series that

showed the lump moving clockwise around Pluto, with a period of about 6 days.

VIDEO 14.7 ➤ Christy concluded that the lump was actually a satellite of Pluto. He proposed that the newly discovered moon be christened Charon (pronounced "SHA-ron"), after the mythical boatman who ferried souls across the River Styx to Hades, the domain ruled by Pluto.

New Horizons captured an image of Charon showing a surprising complexity of surface features (**Figure 14-21**). Unlike Pluto, Charon shows a mixture of young and old surface features. There are fewer craters below Charon's equator than above, indicating a large-scale resurfacing event. As with Pluto, the source of the heat needed to power geology on Charon is an unsolved mystery.

> Pluto and its satellite Charon both rotate in lockstep, always keeping the same faces toward each other.

While Pluto's surface is mostly nitrogen ice, Charon's surface is mostly water ice. Water might even help to explain the crack-like feature around Charon's equator in Figure 14-21. Charon may have contained a liquid water ocean when it first formed and was still relatively warm. As Charon cooled down and its water froze, the water ice would have expanded. This familiar expansion is why we don't put

Styx   Nix   Kerberos   Hydra

Charon

10 miles
10 km

FIGURE 14-22 R I V U X G

VIDEO
14.8   **Pluto's Smaller Moons** Images
        from New Horizons are shown here
        approximately to scale. You can see that
Charon is much larger than all other moons. Except
for Charon, all the moons are irregularly
shaped—not surprising, because smaller objects don't
have enough gravity to pull themselves into spheres.
These are the best images ever captured of these
moons. (NASA/JHUAPL/SwRI)

glass bottles of water in the freezer—they'll break. Likewise, Charon's surface may have broken around the equator when its ocean froze.

Beyond Charon, Pluto has four smaller moons with increasing orbital distance: Styx, Nix, Kerberos, and Hydra (**Figure 14-22**). One goal of the New Horizons mission was to help understand the origins of Pluto's moons. The favored idea was that all of the moons were created in a single violent impact with Pluto, similar to the origin of Earth's Moon. New Horizons found support for this impact origin. For the moons seen clearly enough to count their craters, these moons have the same distribution of craters, and are therefore roughly the same age. This is consistent with all the moons forming in a single impact.

Another piece of evidence for the impact origin leads us naturally to our next topic. Pluto is not the only object beyond Neptune; one idea was that Pluto managed to capture some of these objects as its moons. However, New Horizons established that the reflectivity of Pluto's moons was very similar to each other, while being quite different from the other objects beyond Neptune. This, too, points to an impact origin.

For this scenario to be feasible, there must have been many Pluto-like objects in the outer regions of the solar system. Astronomers estimate that in order for a collision or close encounter between two of them to have occurred at least once since the solar system formed some 4.56 billion years ago, there must have been at least a thousand Plutos. As we will see, astronomers have begun to discover this population of Pluto-like objects (called *plutoids*) in the dark recesses of the solar system beyond Neptune.

CONCEPT CHECK 14-9

What is the difference between a planet and a dwarf planet like Pluto?

CONCEPT CHECK 14-10

Explain why astronomers can't estimate the age of the surfaces of Styx and Kerberos. (*Hint*: See Figure 14-22.)

Answers appear at the end of the chapter.

14-10 Trans-Neptunian Objects and Pluto's Reclassification

For many years, astronomers attempted to find other worlds beyond Neptune and Pluto. In 1992, astronomers found an object with a semimajor axis of 42 au and a diameter estimated to be only 240 km. This object, named 1992 QB1, has a reddish color like Pluto, possibly because frozen methane has been degraded by eons of radiation exposure. The first object was discovered in 1992, and as of 2017, more than 2300 **trans-Neptunian objects** (TNO)—icy worlds whose orbits have semimajor axes larger than that of Neptune—have been discovered. Most of these are relatively small, but a number are larger than Charon and one is nearly as large as Pluto itself (**Figure 14-23**). In light of these recent discoveries, Pluto is best understood as a particularly large trans-Neptunian object, rather than as a planet.

Reclassifying Pluto as both a dwarf planet and a trans-Neptunian object (both new terms) instead of as a planet was controversial. However, Pluto was already known to be somewhat of an oddball among planets, as it is much smaller than the true planets and has an orbit inclined from the ecliptic plane significantly more than the planets. Furthermore, compared to the terrestrial planets, Pluto also has a significantly lower density, suggesting it has a much larger proportion of water ice. However, once large trans-Neptunian objects were discovered, a new class of objects came into view, of which Pluto was only a member. The main decision then facing astronomers was whether to *add* several more oddball objects like Pluto to the list of planets, or to consider Pluto and these new objects members of a different category. The solution adopted by the International Astronomical Union in 2006 was to create a category called trans-Neptunian objects and, in addition, to determine that a larger object like Pluto anywhere in the solar system could also be called a dwarf planet if it met certain requirements (see Section 14-9).

With improvements in the sensitivity of telescopes, new objects beyond Neptune are being discovered at a rapid pace. Based on these observations, it is thought that there could be 35,000 or more objects beyond Neptune with diameters greater than

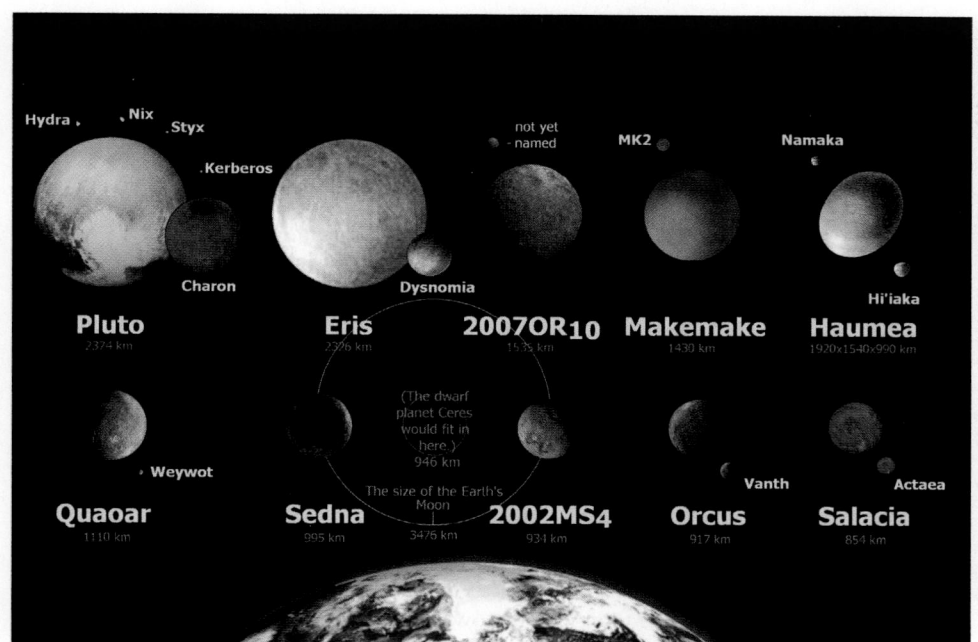

FIGURE 14-23

The Largest Trans-Neptunian Objects This artist's impression depicts Earth and the largest objects known beyond Neptune (as of 2016) to the same scale. Eris has a diameter of about 2300 km, while Pluto's diameter is now known to be 2377 km, thanks to New Horizons. Note the differences in color among these objects and that five of them have satellites of their own. Table 7-4 in Section 7-5 lists the properties of seven of these objects. (Rescaled and Updated Oct 21, 2016, R.L. McNish, RASC Calgary Centre)

100 km. One of the larger objects could have collided with Pluto in the past, giving rise to Pluto's retinue of satellites (see Section 14-9). At least 40 trans-Neptunian objects have satellites, which suggests that such collisions have taken place many times.

The Kuiper Belt and Beyond

Most of the trans-Neptunian objects lie within the **Kuiper belt**, which extends from about 30 to 50 au from the Sun and is relatively close to the ecliptic (see Figure 14-24). When the solar system was young, a large number of icy planetesimals formed in the region beyond Jupiter. Over time, the gravitational forces of the massive jovian planets deflected most of these planetesimals beyond Neptune's orbit, concentrating them into a belt centered on the plane of the ecliptic. Most of the trans-Neptunian objects within the belt are in orbits that are only slightly inclined to the ecliptic; these objects are thought to have formed beyond Neptune and to be in roughly their original orbits. Other objects such as Makemake and Haumea (see Figure 14-23 and Table 7-4) are in orbits that are inclined by about 30° from the ecliptic. These objects are thought to have been pushed into their steeply inclined orbits

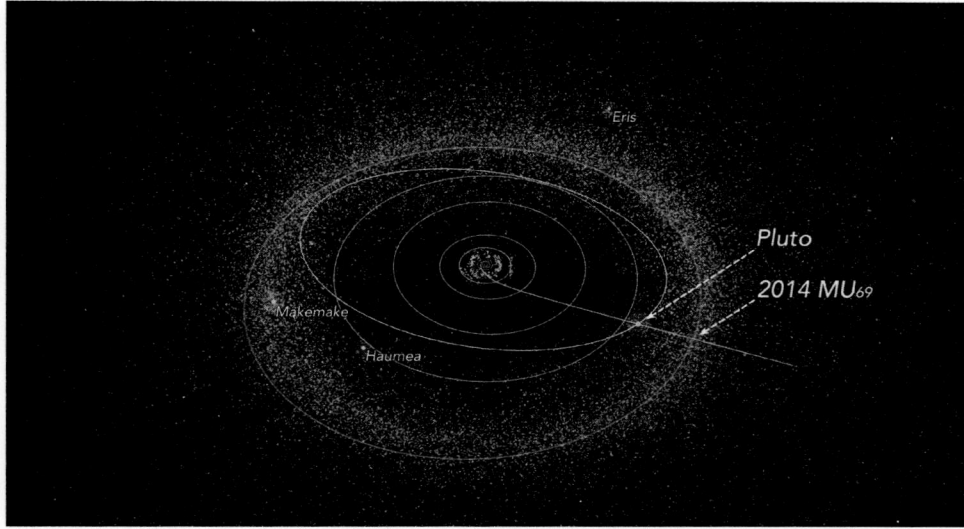

FIGURE 14-24

The Kuiper Belt and New Horizons This illustration shows the thousands of objects (in yellow) that make up the Kuiper belt. Orbits of the gas giants are shown in green and surround the inner solar system. The New Horizons spacecraft has already flown past Pluto and is en route to inspect the Kuiper belt object 2014-MU69 on January 1, 2019. All of the objects past Neptune's orbit are trans-Neptunian objects, just as the name implies. However, Eris is not a member of the Kuiper belt. (Alex Parker)

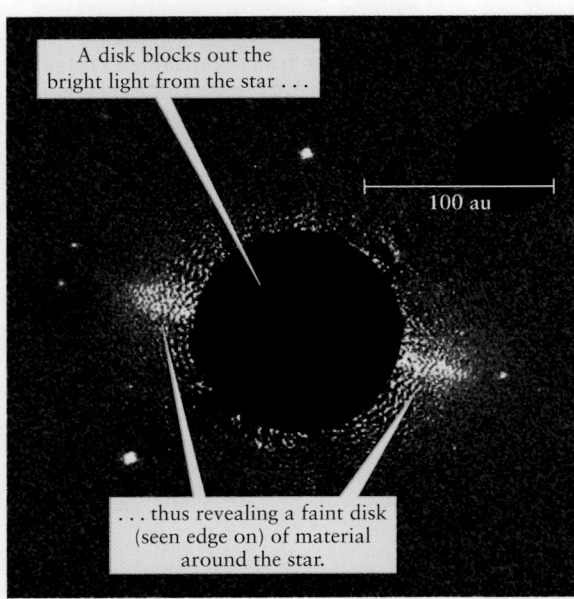

A disk blocks out the bright light from the star . . .

100 au

. . . thus revealing a faint disk (seen edge on) of material around the star.

FIGURE 14-25   R **I** V U X G

A "Kuiper Belt" Around Another Star The star HD 139664 is only slightly more massive than the Sun but is thought to be just 300 million years old. This HST image shows a ring of material around HD 139664 that is similar in size to the Kuiper belt in our solar system. HD 139664 is 57 ly (17 pc) from Earth in the constellation Lupus (the Wolf). (NASA, ESA, and P. Kalas [University of California, Berkeley])

by gravitational interactions with Neptune. If all goes according to plan, we will learn a lot more about Kuiper belt objects when New Horizons gets a close look at 2014-MU69 in 2019 (Figure 14-24).

The processes that gave rise to the Kuiper belt in our solar system also appear to have taken place around other stars. Figure 14-25 shows a disk of material surrounding the young star HD 139664. This disk has dimensions comparable to our Kuiper belt. A number of other young stars have been found with disks of this same type.

There are relatively few members of the Kuiper belt between the orbits of Neptune and Pluto. Remarkably, there are about 100 objects that have nearly the same semimajor axis as Pluto. These so-called **plutinos**, which include Pluto itself, have the property that they complete nearly two orbits around the Sun in the same time that Neptune completes three orbits. The plutinos thus

experience rhythmic gravitational pulls from Neptune, and these pulls keep them in their orbits. (In Section 13-1, we saw a similar relationship among the orbital periods of Jupiter's satellites Io, Europa, and Ganymede, though the ratio of their orbital periods is 1:2:4 rather than the 2:3 ratio for Neptune and the plutinos.) Most Kuiper belt objects orbit at distances beyond the plutinos but within about 50 au from the Sun, at which distance an object completes one orbit for every two orbits of Neptune. At this distance, the rhythmic gravitational forces of Neptune appear to pull small objects inward, giving the Kuiper belt a relatively sharp outer edge.

Neptune's gravity shapes the orbits of many icy worlds, including Pluto.

An example of a trans-Neptunian object that is not a member of the Kuiper belt is shown in Figure 14-23. Eris, the largest known trans-Neptunian object aside from Pluto, has a semimajor axis of more than 68 au and an orbital eccentricity of 0.442. This orbit takes Eris from inside the orbit of Pluto to far beyond the Kuiper belt. Eris is also the most reflective body in the solar system—more reflective than ice or snow, and even more reflective than a mirror! Eris reflects about 96% of the light striking its surface, whereas a mirror reflects about 90%–95%. It is thought that this unusual reflectivity comes from a very thin layer (less than 1 mm thick) of nitrogen-rich ice mixed with methane frost. Due to Eris's highly elongated orbit, as it moves farther from the Sun, its atmosphere can freeze onto the surface to produce this reflective layer.

The discoverer of Eris, astronomer Michael Brown, has been labeled as the one who "killed Pluto." It was his discovery of Eris that led to our understanding that Pluto is just one of many objects orbiting beyond Neptune (what we now call trans-Neptunian objects, or TNOs).

ANIMATION 14.3 ▶   Figure 14-23 also shows an even more extreme case with Sedna: Its orbital semimajor axis is 518 au, giving it an orbital period of more than 10,000 years, and the orbital eccentricity has the remarkably high value of 0.85. It is not well understood how Sedna could have been moved into such an immense and elongated orbit.

CONCEPT CHECK 14-11

How do Pluto's properties differ from the planets?

Answer appears at the end of the chapter.

KEY WORDS

dwarf planet, p. 425
Kuiper belt, p. 429

magnetic axis, p. 417
occultation, p. 418

plutino, p. 430
radiation darkening, p. 418

trans-Neptunian object
   (TNO), p. 428

KEY IDEAS

Discovery of the Outer Planets: Uranus was discovered by chance, whereas Neptune was discovered at a location predicted by applying Newtonian mechanics (Section 14-1).

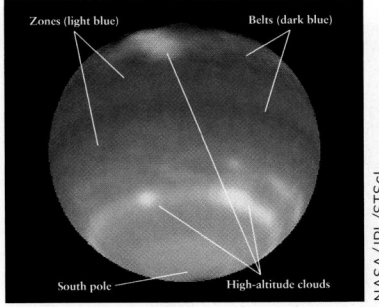

Atmospheres of Uranus and Neptune: Both Uranus and Neptune have atmospheres composed primarily of hydrogen, helium, and a small percentage of methane (Sections 14-2 and 14-3).

- Methane absorbs red light, giving Uranus and Neptune their greenish-blue color.
- No white ammonia clouds are seen on Uranus or Neptune. Presumably the low temperatures have caused almost all the ammonia to precipitate into the interiors of the planets. All of these planets' clouds are composed of methane.
- Much more cloud activity is seen on Neptune than on Uranus. This is because Uranus lacks a substantial internal heat source.

Interiors and Magnetic Fields of Uranus and Neptune: Both Uranus and Neptune may have a rocky core surrounded by a mantle of water and ammonia. Each of these planets might even contain a liquid carbon ocean (Sections 14-4 and 14-5).

- The magnetic axes of both Uranus and Neptune are steeply inclined from their axes of rotation. The magnetic and rotational axes of all the other planets are more nearly parallel. The magnetic fields of Uranus and Neptune are also offset from the centers of the planets.

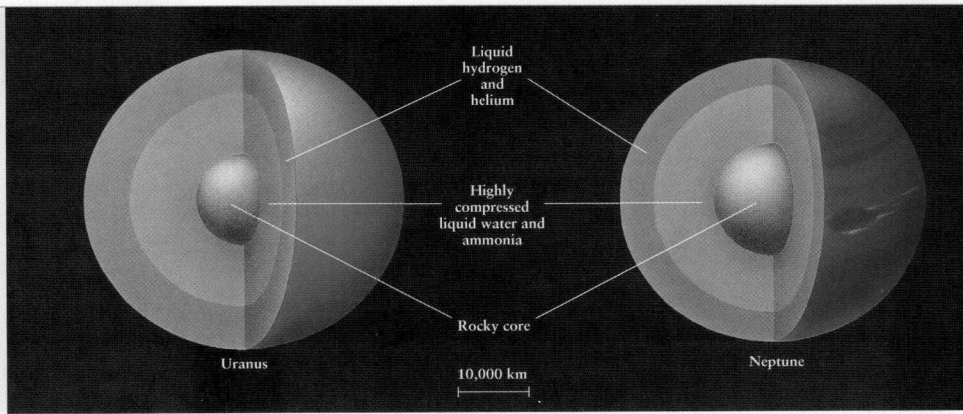

Uranus's Unusual Rotation: Uranus's axis of rotation lies nearly in the plane of its orbit, producing greatly exaggerated seasonal changes on the planet (Section 14-7).

- This unusual orientation may be the result of one or more collisions with planetlike objects early in the history of our solar system. Such collisions could have knocked Uranus on its side.

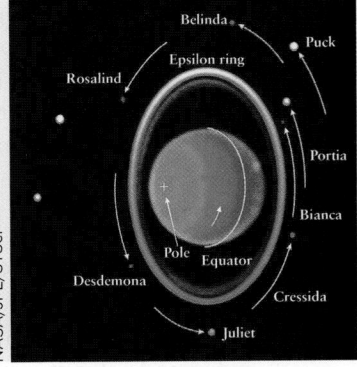

Ring Systems of Uranus and Neptune: Uranus and Neptune are both surrounded by systems of thin, dark rings. The low reflectivity of the ring particles may be due to radiation-darkened methane ice (Section 14-6).

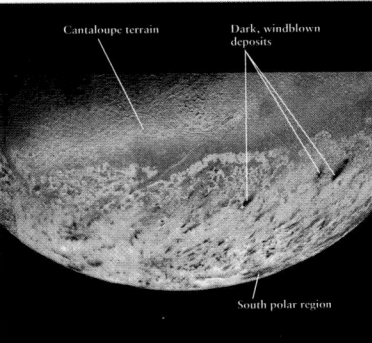

Satellites of Uranus and Neptune: Uranus has five satellites similar to the moderate-sized moons of Saturn, plus at least 22 more small satellites. Neptune has 13 satellites, one of which (Triton) is comparable in size to our Moon or the Galilean satellites of Jupiter (Section 14-8).

- Triton has a young, icy surface indicative of tectonic activity. The energy for this activity may have been provided by tidal heating that occurred when Triton was captured by Neptune's gravity into a retrograde orbit.
- Triton has a tenuous nitrogen atmosphere.

Dwarf Planet: An object orbiting the Sun (but is not a moon) with enough mass to gravitationally pull itself into a spherical shape, yet not enough gravity to clear out planetesimals from its surroundings (Section 14-9).

- This term was introduced in 2006 to help reclassify Pluto as a dwarf planet.

Trans-Neptunian Object: An object that orbits the Sun (other than planets and comets) and on average orbits at a distance greater than Neptune's (Section 14-10).

- More than 1000 icy worlds have been discovered beyond Neptune. Pluto and Charon are part of this population.

- Most trans-Neptunian objects lie in a band called the Kuiper belt that extends from 30 to 50 au from the Sun. Neptune's gravity shapes the orbits of objects within the Kuiper belt.

QUESTIONS

Problem-solving tips and tools

See Box 1-1 for the small-angle formula. Recall that the volume of a sphere of radius r is $4\pi r^3/3$. Section 4-7 discusses the gravitational force between two objects and Newton's form of Kepler's third law. You will find discussions of the original form of Kepler's third law in Section 4-4, tidal forces in Section 4-8, Wien's law for blackbody radiation in Section 5-4, and the transparency of Earth's atmosphere to various wavelengths of light in Section 6-7.

Section 14-1

1. **TUTORIAL 14.1** Could astronomers in antiquity have seen Uranus? If so, why was it not recognized as a planet?

2. Why do you suppose that the discovery of Neptune is rated as one of the great triumphs of science, whereas the discoveries of Uranus and Pluto are not?

3. For which configuration of the Sun, Uranus, and Neptune is the gravitational force of Neptune on Uranus at a maximum? For this configuration, calculate the gravitational force exerted by the Sun on Uranus and by Neptune on Uranus. Then calculate the fraction by which the sunward gravitational pull on Uranus is reduced by Neptune at that configuration. Based on your calculations, do you expect that Neptune has a relatively large or relatively small effect on Uranus's orbit?

Section 14-2

4. Why is it so difficult to see features in the atmosphere of Uranus?

5. (a) Draw a figure like Figure 14-3, indicating where Uranus was in 1986 and 2004. Explain your reasoning. (*Hint:* See Figure 14-1 and Figure 14-2.) (b) In approximately what year will the Sun next be highest in the sky as seen from Uranus's south pole? Explain your reasoning.

6. Why do you suppose the tilt of Uranus's rotation axis was deduced from the orbits of its satellites and not by observing the rotation of the planet itself?

7. Describe the seasons on Uranus. In what ways are the Uranian seasons different from those on Earth?

8. **TUTORIAL 14.2** Explain the statement "Methane is to Uranus's atmosphere as water is to Earth's atmosphere."

9. A number of storms in the Uranian atmosphere can be seen in Figure 14-2, but none are visible in Figure 14-1. How can you account for the difference?

10. Why are Uranus and Neptune distinctly blue-green in color, while Jupiter and Saturn are not?

11. How does the energy source for Uranus's atmospheric motions differ from those from Jupiter, Saturn, and Neptune?

Section 14-3

12. Why are fewer white clouds seen on Uranus and Neptune than on Jupiter and Saturn?

13. What source of energy is thought to still be heating Neptune's core today?

Section 14-4

14. Why do Uranus and Neptune have higher densities than Jupiter and Saturn?

15. Discuss the main reason why Uranus and Neptune are substantially smaller than Jupiter and Saturn.

16. According to one model for the internal structure of Uranus, the rocky core and the surrounding shell of water ices and methane ices together make up 80% of the planet's mass. This interior region extends from the center of Uranus to about 70% of the planet's radius. (a) Find the average density of this interior region. (b) How does your answer to (a) compare with the average density of Uranus as a whole? Is this what you would expect? Why?

Section 14-5

17. How do the orientations of Uranus's and Neptune's magnetic axes differ from those of other planets?

18. Briefly describe why it is thought that Uranus was struck by at least one large planetlike object several billion years ago.

Section 14-6

19. Compare the rings that surround Jupiter, Saturn, Uranus, and Neptune. Briefly discuss their similarities and differences.

20. The 1977 occultation that led to the discovery of Uranus's rings was visible from the Indian Ocean. Explain why it could not be seen from other parts of Earth during nighttime.

Section 14-7

21. Why do astronomers think that the energy needed to resurface parts of Miranda came from tidal heating rather than the satellite's own internal heat?

22. Using the data in Appendix 3, explain why Uranus's satellites Caliban and Stephano were probably captured from space rather than having formed at the same time as the planet itself.

Section 14-8

23. Briefly describe the evidence supporting the idea that Triton was captured by Neptune.

24. If you were floating in a balloon in Neptune's upper atmosphere, in what part of the sky would you see Triton rise? Explain your reasoning.

25. Why is it reasonable to suppose that Neptune will someday be surrounded by a broad system of rings, perhaps similar to those that surround Saturn?

26. If Triton had been formed along with Neptune rather than having been captured, would you expect it to be in a prograde or retrograde orbit? Would you expect the satellite to show signs of tectonic activity? Explain your answers.

Section 14-9

27. How can astronomers distinguish a faint solar system object like Pluto from background stars within the same field of view?

28. What is the evidence that the five moons of Pluto might have a common origin? What is the leading hypothesis for the origin of Pluto's moons?

29. How did the presence of Pluto's moons suggest that there should be other worlds beyond Neptune?

30. What is the difference between a planet and a dwarf planet?

31. What process leads to Pluto's reddish colors?

32. Why did we expect Pluto to have a heavily cratered surface? What did New Horizons actually see?

33. Pluto appears to have a young surface. Why does this imply more heat within Pluto than expected? Have we identified the source of this heat?

34. Is Pluto's "heart" younger or older than the rest of the surface? Explain your answer. What makes up the white material of Pluto's heart?

35. Why is it reasonable to describe Pluto's mountains as "icebergs"?

36. Does the entire surface of Charon appear to have the same age?

37. How can the properties of water help to explain the cracklike feature around Charon's equator?

38. What is the leading hypothesis for the origin of Pluto's moons, and what evidence did New Horizons find to support this idea?

Section 14-10

39. Why are there a large number of objects with the same semimajor axis as Pluto?

40. Are there any trans-Neptunian objects that are not members of the Kuiper belt? Are there any members of the Kuiper belt that are not trans-Neptunian objects? Explain your answers.

41. If Earth-based telescopes can resolve angles down to 0.25 arcsec, how large could a trans-Neptunian object be at Pluto's average distance from the Sun and still not present a resolvable disk?

42. (a) Find the semimajor axis of the orbit of an object whose period is $3/2$ of the orbital period of Neptune. How does your result compare to the semimajor axis of Pluto's orbit? (b) A number of Kuiper belt objects called plutinos have been discovered with the same orbital period and hence the same semimajor axis as Pluto. Explain how these objects can avoid colliding with Pluto.

43. Find the semimajor axis of the orbit of an object whose period is twice the orbital period of Neptune. How does your result compare to the outer limit of the Kuiper belt?

44. Suppose you wanted to search for trans-Neptunian objects. Why might it be advantageous to do your observations at infrared rather than visible wavelengths? (*Hint:* At visible wavelengths, the light we see from planets is reflected sunlight. At

what wavelengths would you expect distant planets to *emit* their own light most strongly? Use Wien's law to calculate the wavelength range best suited for your search.) Could such observations be done at an observatory on Earth's surface? Explain your answer.

45. Why do astronomers argue that it makes more sense to classify Pluto as a trans-Neptunian object than as a planet?

46. If Pluto were still classified as a planet, would there likely be 9 planets, 10 planets, or even more? Explain your reasoning.

47. If the largest trans-Neptunian objects in Figure 14-23 were all put together into a new object, would it have more mass than Earth?

General Problems

48. As Voyager 2 flew past Uranus, it produced images only of the southern hemispheres of the planet's satellites. Why do you suppose this was?

49. At certain points in its orbit, a stellar occultation by Uranus would *not* reveal the existence of the rings. What points are those? How often does this circumstance arise? Explain using a diagram.

50. Uranus's epsilon (ε) ring has a radius of 51,150 km. (a) How long does it take a particle in the ε ring to make one complete orbit of Uranus? (b) If you were riding on one of the particles in the ε ring and watching a cloud near Uranus's equator, would the cloud appear to move eastward or westward as Uranus rotates? Explain your answer.

51. Suppose you were standing on Pluto. Describe the motions of Charon relative to the Sun, the stars, and your own horizon. Would you ever be able to see a total eclipse of the Sun? (*Hint:* You will need to calculate the angles subtended by Charon and by the Sun as seen by an observer on Pluto.) In what circumstances would you *never* see Charon?

52. It is thought that Pluto's tenuous atmosphere may become even thinner as the planet moves toward aphelion (which it will reach in 2113), then regain its present density as it again moves toward perihelion. Why should this be?

53. The brightness of sunlight is inversely proportional to the square of the distance from the Sun. For example, at a distance of 4 au from the Sun, sunlight is only $(1/4)^2 = 1/16 = 0.0625$ as bright as at 1 au. Compared with the brightness of sunlight on Earth, what is its brightness (a) on Pluto at perihelion (29.649 au from the Sun) and (b) on Pluto at aphelion (49.425 au from the Sun)? (c) How much brighter is it on Pluto at perihelion compared with aphelion? (Even this brightness is quite low. Noon on Pluto is about as dim as it is on Earth a half hour after sunset on a moonless night.)

54. Pluto and Charon always face each other, with the same faces, throughout their orbits and rotations. Presumably Pluto and Charon raise tidal bulges on each other. Explain why the average distance between Pluto and Charon is probably constant, rather than increasing like the Earth–Moon distance or decreasing like the Neptune–Triton distance. Include a diagram like Figure 10-20 as part of your answer.

55. The New Horizons spacecraft swung by Jupiter to get a boost from that planet's gravity, enabling it to reach Pluto relatively quickly. To see what would happen if this technique were not used, consider a spacecraft trajectory that is an elliptical orbit around the Sun. The perihelion of this orbit is at 1 au from the Sun (at Earth) and the aphelion is at 30 au (at Pluto's position). Calculate how long it would take a spacecraft in this orbit to make the one-way trip from Earth to Pluto. Based on the information in Section 14-10, how much time is saved by making a swing by Jupiter instead?

56. Discuss the evidence presented by the outer planets suggesting that catastrophic impacts of planetlike objects occurred during the early history of our solar system.

57. Some scientists are discussing the possibility of placing spacecraft in orbit about Uranus and Neptune. What kinds of data should be collected, and what questions would you like to see answered by these missions?

58. Would you expect the surfaces of Pluto and Charon to be heavily cratered? Explain why or why not.

59. In 2006, the International Astronomical Union changed Pluto's designation from *planet* to *dwarf planet*. One criterion that Pluto failed to meet was that a planet must have "cleared the neighborhood" around its orbit. In what sense has Pluto not done so? In what sense have the eight planets (Mercury through Uranus) cleared their neighborhoods? Do you agree with this criterion?

WEB/E-BOOK QUESTIONS

60. **VIDEO 14.4** **Miranda.** Access and view the video "Uranus's Moon Miranda" in Chapter 14 of the *Universe* website or e-book. Discuss some of the challenges that would be involved in launching a spacecraft from Earth to land on the surface of Miranda.

61. Charon was discovered by an astronomer at the U.S. Naval Observatory. Why do you suppose the U.S. Navy carries out work in astronomy? Search the internet for the answer.

62. Search the internet for a list of trans-Neptunian objects. What are the largest and smallest objects of this sort that have so far been found, and how large are they? Have any objects larger than Eris been found?

63. Calculate the maximum possible angular separation between Pluto and Charon as seen from Earth. Assume that Pluto is at perihelion (29.649 au from the Sun) and that Pluto is at opposition as seen from Earth. You might need to search the internet for other details.

ACTIVITIES

Observing Projects

> **Observing tips and tools**
>
> You can find Uranus and Neptune with binoculars if you know where to look (a good star chart is essential), but Pluto is so dim that it can be a challenge to spot even with a 25-cm (10-in.) telescope. Each year, star charts that enable you to find these planets are printed in the issue of *Sky & Telescope* for the month in which each planet is first visible in the nighttime sky. You can also locate the outer planets using the *Starry Night™* program if you have access.

64. Make arrangements to view Uranus through a telescope. The planet is best seen at or near opposition. Use a star chart at the telescope to find the planet. Are you certain that you have found Uranus? Can you see a disk? What is its color?

65. If you have access to a large telescope, make arrangements to view Neptune. Like Uranus, Neptune is best seen at or near opposition and can most easily be found using a star chart. Can you see a disk? What is its color?

66. If you have access to a large telescope (at least 25 cm in diameter), make arrangements to view Pluto. Using the star chart from *Sky & Telescope*, view the part of the sky where Pluto is expected to be seen and make a careful sketch of all of the stars that you see. Repeat this process on a later night. Can you identify the "star" that has moved?

Starry Night™ Explorations

67. Use *Starry Night™* to observe the planet Uranus and the orbits of its five largest moons: Umbriel, Ariel, Miranda, Titania, and Oberon. Open **Favorites > Explorations > Uranus** to see a view of this planet as it might have looked (apart from the labels) through a telescope to an astronomer at the South Pole in June 1986. Choose **Edit > Select None** from the menu to remove the labels and then click **Play** to watch the planet and its moons move against the background stars.

    (a) Select **File > Revert** to reset the view. Click the name in the label of each of the moons in turn, select **Orbit** from the menu that appears, and then click **Play**. Describe how the satellites move and relate your observations to Kepler's third law.

    (b) As the animation progresses, use the "Y" keyboard shortcut to advance time by 1 year per keypress to change the year to **2007** and observe the motion of the moons at this time. Explain how and why the orbits look different in this view.

68. Use the *Starry Night™* program to observe Neptune and two of its moons, Triton and Nereid. Select **Favorites > Explorations > Neptune**. The view shows Neptune and its moon Triton, from a location in space over Neptune's equator. This equator, as well as a meridian line and pole sticks, are shown on Neptune.

    (a) Click **Play** and observe the motion of Triton around Neptune. Does Triton orbit Neptune in the same direction as Neptune's direction of rotation or in the opposite direction? How will this motion affect this moon's fate?

    (b) Nereid is in a rather elongated orbit that takes it far from Neptune. Zoom out until the complete orbit of Nereid is visible in the view. Use the location scroller to adjust the view to obtain different perspectives and use the "D" key to move time forward in 1-day steps. Does Nereid orbit Neptune in the same direction as Neptune's direction of rotation or in the opposite direction?

    (c) Zoom in on Neptune and use the time flow controls and the position of the meridian on the planet to determine the approximate rotational period of Neptune. What is this period?

69. Use *Starry Night™* to explore the relationship between Pluto's rotation rate and the orbital period of its moon, Charon. Select **Favorites > Explorations > Pluto** to see a view centered on Pluto from above its north pole, with a meridian line marked upon the planet. Pluto's moon Charon is also shown, along with its orbit.

    (a) Use the angular separation cursor to draw a line from the center of Pluto to the center of Charon and note that the labeled star Sadal Suud is on this line from the observer's viewpoint. Now zoom in to a field of view of about 15° and note the alignment of the red meridian line drawn on Pluto with the star Kitalpha. Determine Pluto's rotational period by using the keyboard shortcuts or time controls to adjust the number of days, hours, and minutes required for Pluto to make one complete rotation so that the meridian line is again aligned with Kitalpha in the view. What is the rotational period of Pluto? Now, select **File > Revert** to return to the original view, then measure the orbital period of Charon, using the time controls to advance this moon through a complete orbit, using the position of the star Sadal Suud as the position marker. What is the relationship of this orbital period of Charon to the rotational period of Pluto?

    (b) To explore the consequences of this relationship between planet rotation and moon revolution, which is relatively common in the solar system, open **Favorites > Explorations > Charon**. The view, from a location on Pluto's surface slightly north of the equator of this dwarf planet, shows Charon lying just below the star Brachium in Pluto's eastern sky at this time. The view is not locked on to Charon but set to look directly east. Make a note of the date and the time in the toolbar and then click **Play**. As time progresses, Pluto's rotation carries the observing location toward the east, resulting in the apparent westward motion of the stars in the sky. Charon, however, remains stationary relative to the horizon. Click **Stop** when the star Brachium reappears in the view and then use the time controls to bring this star back to a position just above Charon in the sky, indicating that Charon has completed a full orbit around Pluto relative to the stars. Thus, the difference between these two times is a measure of Charon's sidereal month as seen from Pluto. How does this compare with the period you observed for Pluto's rotation?

(c) If you were to observe Pluto in Charon's sky, would you expect it to remain stationary, move progressively westward, progressively eastward, or some other direction in the sky? To answer this question, click the name in the label for Charon and select **Go There** to move to Charon. Click and hold the decrease elevation button in the toolbar to descend to Charon's surface. Then adjust the view to find Pluto in the sky (or open **Quick Search > Planets...** and click on **Pluto**) and click **Play**. Does Pluto appear to move in Charon's sky? What does your observation indicate about the relationship between Charon's orbital period around Pluto and Charon's rotational period around its own axis?

(d) Zoom in on Pluto and watch the planet's surface as you advance time. Apart from the phase changes, does the face of Pluto that is visible from Charon change as time progresses?

(e) If the Earth and the Moon were to have the same orbital and rotational relationship between them that Pluto and Charon have, which of the following statements would *not* be true? (i) The Moon would always appear above one particular location on Earth. (ii) Everyone on Earth would see the Moon pass through their sky every month. (iii) The Earth would rotate with a period of 27.3 days.

Collaborative Exercise

70. Sir William Herschel, a British astronomer, discovered Uranus in 1781 and named it Georgium Sidus, after the reigning monarch, George III. What name might Uranus have been given in 1781 if an astronomer in your country had discovered it? Why? What if it had been discovered in your country in 1881? In 1991?

ANSWERS

ConceptChecks

ConceptCheck 14-1: The trajectory of Uranus did not show slow changes in speed predicted by its orbit around the Sun. Instead, Uranus deviated by slightly speeding up and slowing down, and this could be attributed to a new planet at just the right location.

ConceptCheck 14-2: Methane in Uranus's atmosphere absorbs the Sun's longer wavelengths of light, so if it were absent, Uranus would have more reds and, as a result, appear more yellow-white, like Saturn.

ConceptCheck 14-3: It is thought that Neptune is still contracting, which converts gravitational energy into thermal energy to power its atmosphere.

ConceptCheck 14-4: At Neptune's present distance, there would not have been enough matter to form such a large planet. Furthermore, even if there were more matter, Neptune has a higher proportion of heavy elements than expected in the outer reaches of the solar system.

ConceptCheck 14-5: Ammonia dissolved in water under high pressure loses electrons, which become free to move, and this makes the fluid electrically conductive.

ConceptCheck 14-6: Due to radiation darkening, Uranian rings are black as coal, unlike Saturn's icy light-reflecting rings. Uranus's rings are also very narrow, at about 10 km wide, compared to rings on Saturn that can be more than 1000 times wider.

ConceptCheck 14-7: The orbits of some moons were seen to change between observations made in 1994 and 2005. Based on the degree of observed change, it might only have taken a few hundred million years before these moons would have collided, so they might have been acquired somewhat recently.

ConceptCheck 14-8: Geyserlike plumes observed by Voyager 2 showed streaking, indicating that they were carried downwind by a thin atmosphere.

ConceptCheck 14-9: Pluto's gravity is too weak to clear its neighborhood of nearby planetesimals in the Kuiper belt. According to new classifications adopted in 2006—that a planet must clear its neighborhood—this makes Pluto a dwarf planet instead of a planet.

ConceptCheck 14-10: Estimating the age of an object's surface requires counting the distribution of its craters, and images of these moons are too poor to see craters.

ConceptCheck 14-11: Pluto is much smaller than any planet, and has a much higher fraction of water ice.

LEARNING GOALS

On the Tuesday night after Easter in the year 1066, a strange new star appeared in the European sky. Seemingly trailing fire, this "star" hung in the night sky for weeks. Some, including the English king Harold, may have taken it as an ill omen; others, such as Harold's rival William, Duke of Normandy, may have regarded it as a sign of good fortune. Perhaps they were both right, because by the end of the year Harold was dead and William was installed on the English throne.

Neither man ever knew that he was actually seeing the trail of a city-sized ball of ice and dust, part of which had evaporated as it rounded the Sun to produce a long tail. They were seeing a comet—remarkably, the same Comet Halley that last appeared in 1986 and will next be seen in 2061. Although we now understand that comets are natural phenomena rather than supernatural omens, they still have the power to awe and inspire us, as did comet 67P/Churyumov–Gerasimenko when the Rosetta spacecraft got this close-up view in 2014. As a comet gets closer to the Sun, dusty ice evaporates to form jets, as in this image, as well as the long tails associated with comets.

Jets from Comet 67P/Churyumov-Gerasimenko. (ESA/Rosetta/MPS for OSIRIS Team MPS/UPD/LAM/
IAA/SSO/INTA/UPM/DASP/IDA)

R I V U X G

This chapter is about comets, asteroids, and the impacts they create when colliding with Earth. Comets are made of ice and dust, and they sometimes shoot out jets of this material, as can be seen in the opening photo. When the dust from a comet enters Earth's atmosphere, it creates "shooting stars"—which we all love! Asteroids, by contrast, are mostly rock and are made from bits of the solar nebula that never formed into planets. We don't always like it when asteroids hit Earth, as the larger impacts can release the tremendous amount of energy. In fact, such an asteroid impact probably led to the extinction of the dinosaurs about 65 million years ago.

15-1 A search for a planet between Mars and Jupiter led to the discovery of asteroids

TUTORIAL 15.1 After William Herschel's discovery of Uranus in 1781 (which we described in Section 14-1), many astronomers began to wonder if there were other, as yet undiscovered, planets. If these planets were too dim to be seen by the naked eye, they might still be visible through telescopes. These planets would presumably be found close to the plane of the ecliptic, because all the other planets orbit the Sun in or near that plane (see Section 7-1). But how far from the Sun might these additional planets be found?

The Hunt for the "Missing Planet"

LOOKING DEEPER 15.1 Astronomers in the late eighteenth century had a simple rule of thumb, called the *Titius–Bode law,* which relates the sizes of planetary orbits. The law states that from one planet to the next, the semimajor axis of the orbit increases by a factor between approximately 1.4 and 2. For example, the semimajor axis of Mercury's orbit is 0.39 au. Venus's orbit has a semimajor axis of 0.72 au, which is larger by a factor of $(0.72)/(0.39) = 1.85$. (Unlike Newton's laws, the Titius–Bode "law" is not a fundamental law of nature, but merely a numerical coincidence.)

The glaring exception is Jupiter (semimajor axis 5.20 au), which is more than *3 times* farther from the Sun than Mars (semimajor axis 1.52 au). Table 7-1 depicts this large space between the orbits of Mars and Jupiter. If the rule of thumb is correct, astronomers reasoned, there should be a "missing planet" with an orbit about 1.4 to 2 times larger than that of Mars. This planet should therefore have a semimajor axis between 2 and 3 au.

Six German astronomers, who jokingly called themselves the "Celestial Police," organized an international group to begin a careful search for this missing planet. Before their search even began, however, surprising news reached them from Giuseppe Piazzi, a Sicilian astronomer. Piazzi had been carefully mapping faint stars in the constellation of Taurus when on January 1, 1801, the first night of the nineteenth century, he noticed a dim, previously uncharted star. This star's position shifted slightly over the next several nights. Suspecting that he might have found the "missing planet," Piazzi excitedly wrote to Johann Bode, the director of the Berlin Observatory and a member of the Celestial Police.

Unfortunately, Piazzi's letter did not reach Bode until late March. By that time, Earth had moved around its orbit so that Piazzi's object appeared too near the Sun to be visible in the night sky. With no way of knowing where to look after it emerged from the Sun's glare, astronomers feared that Piazzi's object might have been lost.

Upon hearing of this dilemma, the brilliant young German mathematician Karl Friedrich Gauss took up the challenge. He developed a general method of computing an object's orbit from only three separate measurements of its position on the celestial sphere. (With slight modifications, this same method is used by astronomers today.) In November 1801, Gauss predicted that Piazzi's object would be found in the predawn sky in the constellation of Virgo. And indeed Piazzi's object was sighted again on December 31, 1801, only a short distance from the position Gauss had calculated. Piazzi named the object Ceres (pronounced "SEE-reez"), after the patron goddess of Sicily in Roman mythology.

Ceres orbits the Sun once every 4.6 years at an average distance of 2.77 au, just where astronomers had expected to find the missing planet. But Ceres is very small; its equatorial diameter is a mere 974.6 km (Figure 15-1). (The pole-to-pole diameter is even smaller, just 909.4 km.) Hence, Ceres reflects only a little sunlight, which is why it cannot be seen with the naked eye even at opposition; it can be seen with binoculars, but it looks like just another faint star.

On March 28, 1802, Heinrich Olbers discovered another faint, starlike object that moved against the background stars. He called it Pallas, after the Greek goddess of wisdom. Like Ceres, Pallas orbits the Sun every 4.6 years at an average distance of 2.77 au, but its orbit is more steeply inclined from the plane of the ecliptic and is somewhat more eccentric. Pallas is even smaller and dimmer than Ceres, with an estimated diameter of only 522 km. Obviously, Pallas was also not the missing planet. Ceres and Pallas are **asteroids**—small objects with various mixtures of rock,

FIGURE 15-1 R I V U X G

Ceres Compared with Earth and Moon Ceres, the largest of the asteroids and the first one discovered, is shown here to the same scale as Earth and the Moon. Ceres is too small to be considered a planet, and it cannot be regarded as a moon because it does not orbit a planet. To denote their status, asteroids are also called minor planets. (NASA Image Collection/Alamy Stock Photo) ice, and metal that orbit the Sun between Mars and Jupiter. (Like all asteroids, Ceres and Pallas are classified as *small solar system bodies*, or **minor planets**; these are all objects other than planets and comets that are in direct orbit around the Sun.)

A new question then faced the astronomers of the early nineteenth century: Did a missing planet even exist? Some speculated that perhaps there had once been such a full-sized planet, but it had somehow broken apart or exploded to produce the asteroids. The search was on to discover this population. Two more such asteroids were discovered in the next few years, Juno in 1804 and Vesta in 1807. As telescope technology improved, hundreds more asteroids were discovered. Going beyond telescopes, in 2007—200 years after Vesta's discovery—NASA's spacecraft Dawn was launched on a mission to visit Vesta and Ceres (more on this in Section 15-3).

An Abundance of Asteroids

The next major breakthrough came in 1891, when the German astronomer Max Wolf began using photographic techniques to search for asteroids. Before this, asteroids had to be painstakingly discovered by scrutinizing the skies for faint, uncharted, starlike objects whose positions moved slightly from one night to the next. With photography, an astronomer simply aimed a camera-equipped telescope at the stars and took a long exposure. If an asteroid happened to be in the field of view, its orbital motion during the long exposure would leave a distinctive trail on the photographic plate (Figure 15-2). Using this technique, Wolf alone discovered 228 asteroids.

Today, more than a million asteroids have been discovered, of which more than 400,000 have been studied well enough so that their orbits are known. Thousands more are discovered every month, some by amateur astronomers. After the discoverer reports a new find to the Minor Planet Center of the Smithsonian Astrophysical Observatory, the new asteroid is given a provisional

designation. For example, asteroid 1980 JE was the fifth ("E") to be discovered during the second half of May (the tenth half-month of the year, "J") 1980. The same scheme for provisional designations is also used for small objects found in the outer solar system beyond Neptune, or trans-Neptunian objects.

If the asteroid is located again on at least four succeeding oppositions—a process that can take decades—the asteroid is assigned an official sequential number (Ceres is 1, Pallas is 2, and so forth), and the discoverer is given the privilege of suggesting a name for the asteroid. The suggested name must then be approved by the International Astronomical Union. For example, the asteroid 1980 JE was officially named 3834 Zappafrank (after the American musician Frank Zappa) in 1994.

Ceres (formally called "1 Ceres") is by far the largest asteroid at 975 km; it has one-half the mass of all other asteroids combined. In recognition of its size, Ceres is also called a dwarf planet—large enough to pull itself into a sphere, but unable to gravitationally clear out its surroundings (see Section 14-10). Note that, because this term describes only the size and shape of an object, an asteroid like Ceres and a trans-Neptunian object like Pluto can both be designated as dwarf planets, even though they reside in different parts of the solar system.

Besides Ceres, only two asteroids—Pallas and Vesta—have diameters greater than 500 km. A general feature of asteroids is that smaller ones are more numerous. Thus, we find other asteroids have diameters between 200 and 300 km, and 200 more are bigger than 100 km across. Infrared observations indicate that there are between 700,000 and 1.7 million asteroids greater than 1 km in size. And while a million asteroids might sound like a lot, the vast majority of asteroids are smaller than 1 km across.

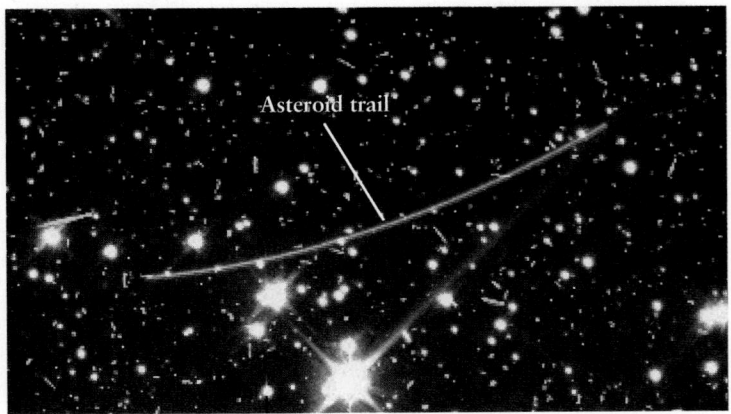

FIGURE 15-2 R I V U X G

The Trail of an Asteroid Telescopes used to photograph the stars are motorized to follow the apparent motion of the celestial sphere. Because asteroids orbit the Sun, their positions change with respect to the stars, and they leave blurred trails on time exposure images of the stars. This bluish asteroid trail was recorded by accident by the Hubble Space Telescope. The shorter streaks (mostly yellow) are imaging artifacts. (R. Evans and K. Stapelfeldt [JPL] and NASA)

> ### CAUTION!
>
> Be careful not to confuse asteroids with the trans-Neptunian objects that we discussed in Section 14-10. Asteroids are found in the inner solar system, are made primarily of rocky materials, and are generally quite small. By contrast, trans-Neptunian objects are found primarily in the Kuiper belt beyond Neptune, are a mixture of ices and rock, and range in size from a few kilometers across to 2900 km in diameter—more than 3 times the diameter of Ceres. The combined mass of all trans-Neptunian objects is not known, but it is probably thousands of times greater than the combined mass of all asteroids. Asteroids make up only a tiny fraction of the total mass of the solar system!

Interestingly, the dwarf planets Ceres and Pluto have a common story line. Like Pluto, Ceres was considered a planet at first and retained that designation for about 50 years. Just as Pluto recently went from being the smallest planet to being one of the largest trans-Neptunian objects, Ceres had been reclassified and went from being the smallest planet to the largest asteroid. Ceres and Pluto were simply the first objects discovered in these new categories.

The Asteroid Belt

Like Ceres, Pallas, Vesta, and Juno, most asteroids orbit the Sun at distances between 2 and 3.5 au. This region of our solar system between the orbits of Mars and Jupiter is called the **asteroid belt** (Figure 15-3).

> ### CAUTION!
>
> You may wonder how spacecraft such as Voyager 1, Voyager 2, Galileo, and Cassini were able to cross the asteroid belt to reach Jupiter or Saturn. Aren't asteroids hazards to space navigation, as they are sometimes depicted in science-fiction movies? Wouldn't these spacecraft have been likely to run into an asteroid? Happily, the answer to both questions is no! While it is true that there are more than 10^6 asteroids, they are spread over a belt with a total area (as seen from above the plane of the ecliptic) of about 10^{17} square km—about 10^8 times greater than Earth's entire surface area. Hence, the average distance between asteroids in the ecliptic plane is about 10^6 km, about twice the distance between Earth and the Moon. Furthermore, many asteroids have orbits that are tilted out of the ecliptic. The Galileo spacecraft did pass within a few thousand kilometers of two asteroids, but these passes were intentional and required that the spacecraft be carefully aimed.
>
> *Science-fiction movies to the contrary, asteroids are not hazards to space navigation.*

What about the nineteenth-century notion of a missing planet? This idea has long since been discarded, because the combined matter of all the asteroids (including an estimate for those not yet observed) would produce an object barely 1500 km in diameter—considerably smaller than anything that could be considered a

FIGURE 15-3

The Asteroid Belt Most asteroids, including the large asteroids Ceres, Pallas, and Juno, have roughly circular orbits that lie within the asteroid belt. By contrast, some asteroids, such as Apollo and Icarus, have eccentric orbits that carry them inside the orbit of Earth.

missing planet. We now understand that asteroids are actually debris left over from the formation of the solar system and, as we'll see next, why that debris couldn't easily build up into larger objects.

CONCEPT CHECK 15-1

Why was a "missing planet" suspected to exist around the orbital distance of the asteroid belt?

CONCEPT CHECK 15-2

Is Ceres an asteroid or a dwarf planet?

Answers appear at the end of the chapter.

15-2 Jupiter's gravity helped shape the asteroid belt

Where did the asteroid belt come from? Why did asteroids, rather than a planet, form in the solar nebula (see Section 8-5) in the region between Mars and Jupiter? Important insights into these questions come from supercomputer simulations of the formation of the terrestrial planets, like the one depicted in Figure 8-11 but more elaborate.

Jupiter and the Formation of the Asteroid Belt

A typical simulation of the inner solar system starts off with a billion (10^9) or so planetesimals, each with a mass of 10^{15} kg or more, so that their combined mass equals that of the terrestrial planets. The computer then follows these planetesimals as they collide and accrete to form the planets. By selecting different speeds and positions for the planetesimals at the start of the simulation, a scientist can study alternative scenarios for the development of the inner solar system.

If the effects of Jupiter's gravity are not included in a simulation, an Earth-sized planet usually forms in the asteroid belt, giving us five terrestrial planets instead of four. But when Jupiter's gravity is taken into account, this fifth terrestrial planet is less likely to form. Jupiter's strong pull "clears out" the asteroid belt by disrupting the orbits of planetesimals in the belt, ejecting most of them from the solar system altogether. As a result, the asteroid belt becomes depleted of planetesimals within about 1 million years, before a planet has a chance to form. The few planetesimals that remain in the simulation—only about 0.1% of those that originally orbited between Mars and Jupiter—become the asteroids that we see today.

The asteroids were never part of a larger planet.

Kirkwood Gaps

Jupiter's gravity continues to influence the asteroid belt down to the present day. The American astronomer Daniel Kirkwood found the first evidence for this in 1867, when he discovered gaps in the asteroid belt. These features, today called **Kirkwood gaps**, can best be seen in a histogram like Figure 15-4, which shows the

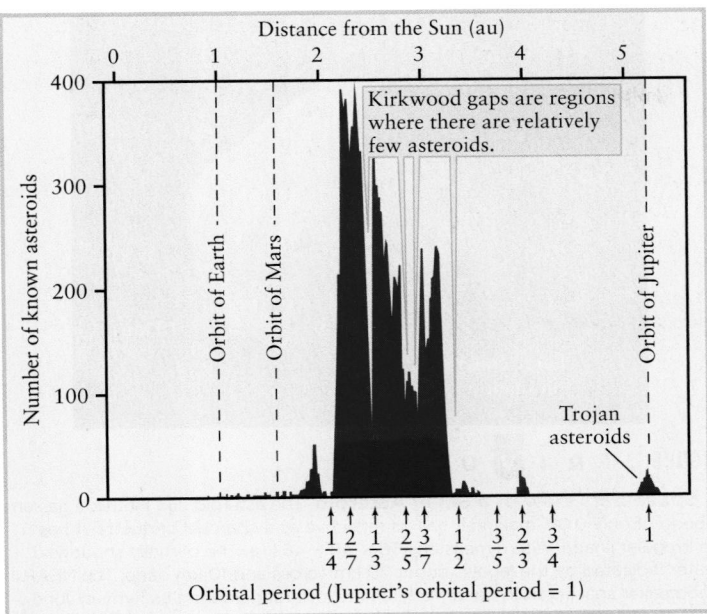

FIGURE 15-4

The Kirkwood Gaps This histogram displays the numbers of asteroids at various distances from the Sun. Very few asteroids are found in orbits whose orbital periods are equal to the simple fractions 1/3, 2/5, 3/7, and 1/2 of Jupiter's orbital period. The Trojan asteroids (at far right) follow the same orbit as Jupiter and thus have the same orbital period. We describe these in Box 15-1 and Section 15-4.

number of asteroids at each increment of distance from the Sun. The gaps in this histogram show regions where there are relatively few asteroids. Curiously, these gaps occur for asteroid orbits whose periods are simple fractions (such as 1/3, 2/5, 3/7, and 1/2) of Jupiter's orbital period.

To understand why the Kirkwood gaps exist, imagine an asteroid within the belt that circles the Sun once every 5.93 years, exactly half of Jupiter's orbital period. On every *second* trip around the Sun, the asteroid finds itself lined up between Jupiter and the Sun again, always at the same location and with the same orientation. Because of these repeated alignments, called a *2-to-1 orbital resonance*, Jupiter's gravity deflects the asteroid from its original 5.93-year orbit, ultimately ejecting it from the asteroid belt (this is the gap labeled "1/2" in Figure 15-4). By contrast, for an object not in resonance—whose orbital period does not result in repeated Sun-asteroid-Jupiter alignments—Jupiter's repeated tugs on the object occur at different locations around the Sun and do not add up to eventually pull it out of orbit.

Another Kirkwood gap corresponds to an orbital period of 1/3 Jupiter's period, or 3.95 years (a *3-to-1 orbital resonance*). Although not necessarily as strong, additional gaps exist for other simple ratios between the periods of asteroids and Jupiter. Elsewhere in the solar system, similar orbital resonances help clear out the Cassini division in Saturn's rings (see Section 12-11) and give an outer edge to the Kuiper belt beyond Neptune (see Section 14-10). These are examples of *unstable* resonances, but under special conditions a resonance can actually be a *stabilizing* influence (see the clump of asteroids with a period of 2/3 in Figure 15-4).

CONCEPT CHECK 15-3

What prevented a planet from forming where the asteroid belt currently exists? Did this effect eject mass from the region or just keep the existing mass from forming a planet?

Answer appears at the end of the chapter.

15-3 Astronomers use a variety of tools, including the Dawn spacecraft, to study asteroids

Although the asteroid belt is mostly empty space, collisions between asteroids should occur from time to time. Asteroids move in orbits with a variety of different eccentricities and inclinations (see Figure 15-3), and some of these orbits intersect each other. A collision between asteroids must be an awe-inspiring event. Typical collision speeds are estimated to be from 1 to 5 km/s (3600 to 18,000 km/h, or 2000 to 11,000 mi/h), which is more than sufficient to shatter rock. Recent observations show that these titanic impacts play an important role in determining the nature of asteroids.

> Many asteroids have been battered by collisions with other asteroids.

Observing Asteroids with Telescopes, Radar, and Spacecraft

VIDEO 15.1 State-of-the-art optical telescopes barely have the resolution to reveal surface details on large asteroids such as Ceres, Pallas, and Vesta (**Figure 15-5**). These asteroids resemble miniature terrestrial planets, and one of them (Ceres) has enough mass, and hence enough gravity, to pull itself into a roughly spherical shape. These large asteroids are likely to have extensively cratered surfaces due to collisions with other asteroids in the asteroid belt.

Optical telescopes reveal much less detail about the smaller asteroids. But by studying how the brightness of an asteroid changes as it rotates, astronomers can infer the asteroid's shape. An elongated asteroid appears dimmest when its long axis points toward the Sun so that it presents less of its surface to reflect sunlight. By contrast, a rotating spherical asteroid will have a more constant brightness. Such observations show that most smaller asteroids are not spherical at all. Instead, they retain the odd shapes produced by previous collisions with other asteroids.

An important recent innovation in studying asteroids has been the use of radar. An intense radio beam is sent toward an asteroid, and the reflected signal is detected and analyzed to determine the asteroid's shape (**Figure 15-6**). The amount of the radio signal that is reflected also gives clues about the texture of the surface.

VIDEO 15.2 The most powerful technique for studying asteroids is to send a spacecraft to make observations at close range. The Jupiter-bound Galileo spacecraft gave us our first close-up view of an asteroid in 1991, when it flew past 951 Gaspra. Two years later, Galileo made a close pass by 243 Ida. The NEAR Shoemaker spacecraft made a close approach to 253 Mathilde in 1997 (**Figure 15-7**). It later went into orbit around 433 Eros (see Figure 7-7) and touched down on Eros in 2001. It was the first spacecraft to orbit and land on an asteroid. A number

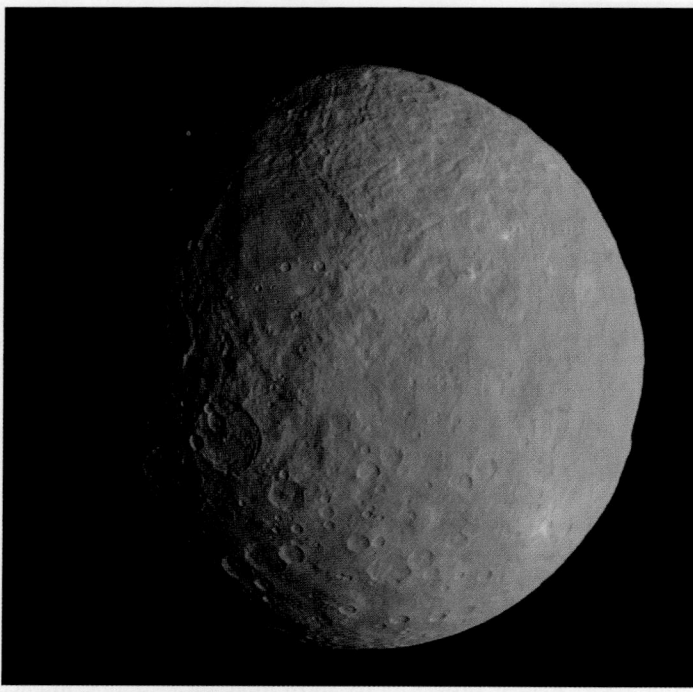

FIGURE 15-5 R I V U X G

A Dawn Spacecraft View of the Large Asteroid Ceres With an equatorial diameter of 974.6 km (605.6 mi), Ceres is the largest of all the asteroids. Its very nearly spherical shape suggests that it was once molten throughout its volume, in which case it probably underwent chemical differentiation like the terrestrial planets (see Box 7-1 and Section 8-5). The bright crater on the right is called Haulani. (NASA/JPL-Caltech/UCLA/MPS/DLR/IDA)

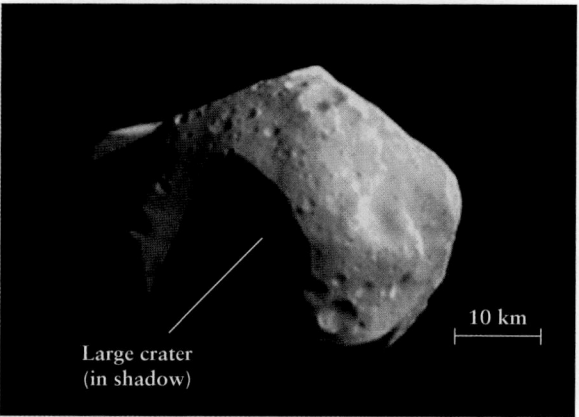

FIGURE 15-7 R I V U X G

A Spacecraft View of a Small Asteroid The asteroid 253 Mathilde has an albedo of only 0.04, making it half as reflective as a charcoal briquette. It has an irregular shape, with dimensions 66 × 48 × 46 km. The partially shadowed crater indicated by the label is about 20 km across and 10 km deep. The NEAR Shoemaker spacecraft imaged Mathilde at close range during its flyby in June 1997. (NASA/Johns Hopkins University Applied Physics Laboratory/Science Source)

25143 Itokawa, where it gathered small samples of dust from the asteroid's surface (Figure 15-8). In 2010, the dust samples were safely returned to Earth, completing the first-ever asteroid sample return. Analysis indicates that Itokawa may have once been part of a larger asteroid, and that the dust has been exposed to space for

of other spacecraft have since made close flybys of asteroids, and Dawn reached Vesta in 2011.

One of the most ambitious asteroid missions to date was Hayabusa ("Falcon"), a project of the Japan Aerospace Exploration Agency (JAXA). In 2005, it touched down briefly on the surface of

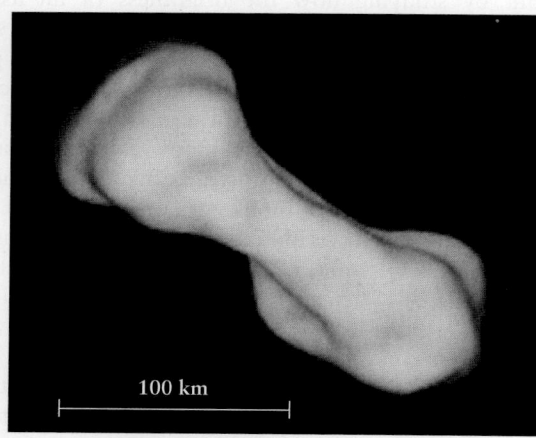

FIGURE 15-6 R I V U X G

A Radar View of a Medium-Sized Asteroid This radar image reveals the curious dog-bone shape of the asteroid 216 Kleopatra. The 300-m Arecibo radio telescope in Puerto Rico was used both to send radio waves toward Kleopatra and to detect the waves that were reflected back. Scientists then created this image by computer analysis of the reflected waves. (NASA/JPL)

FIGURE 15-8 R I V U X G

Sample Returned from Itokawa This is Itokawa photographed from a distance of only 4.7 km. The surface of Itokawa is rough and studded with boulders. It has been described as a loosely held-together "rubble pile." Compared to Mathilde and other asteroids, Itokawa has a surprising lack of craters, which indicates a relatively young surface. The asteroid may even be the result of two asteroids merging together. The sample returned to Earth contained about 1500 small grains, with properties indicating that Itokawa may have been part of a larger asteroid at one time. (Japan Aerospace Exploration Agency)

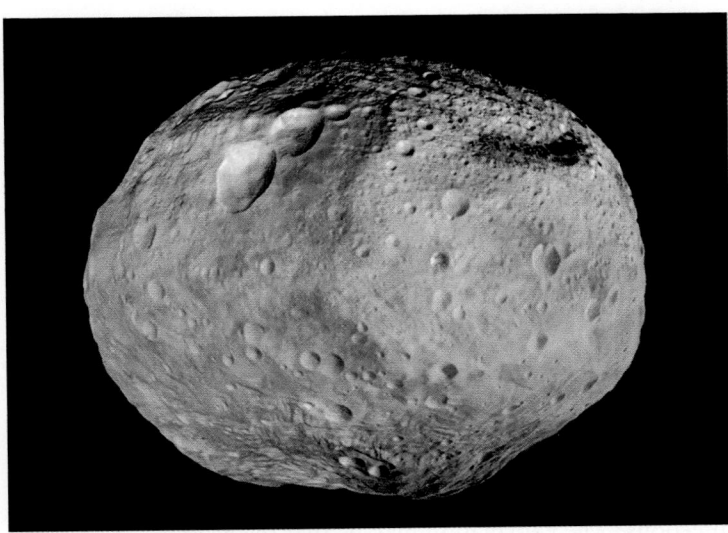

FIGURE 15-9   R I **V** U X G

Vesta This view of Vesta is a mosaic composed of the best views from the Dawn spacecraft. Many craters pockmark the surface. Vesta's average diameter is 525 km. (NASA/JPL-Caltech/UCAL/MPS/DLR/IDA)

about 8 million years. As expected, the composition of the asteroid dust also matches some of the meteoritic material that makes its way to Earth (more on meteorites in Section 15-5).

Hayabusa also contained a mini-lander (about half a kilogram in weight and 12 cm wide) designed to hop around in Itokawa's low gravity. Unfortunately, the mini-lander did not deploy correctly and floated off into space. A follow-up mission, Hayabusa 2,

reached the asteroid 162173 Ryugu in 2018 and will return samples to Earth in 2020.

Dawn Visits Vesta and Ceres

In 2011, NASA's Dawn spacecraft went into orbit around Vesta, the second largest asteroid after Ceres (**Figure 15-9**). With an average diameter of 525 km, Vesta is not quite massive enough to gravitationally pull itself into a sphere, so it is not a dwarf planet like Ceres. However, Vesta is large enough that its own radioactivity is thought to have melted the asteroid early on, allowing iron to sink to the core (we will learn more about *differentiated asteroids* in Section 15-5).

Estimates suggest that Vesta lost about 1% of its mass in a huge collision about one or two billion years ago. Amazingly, debris from this impact with Vesta has actually made it to Earth as meteorites. In fact, about 6% of meteorites landing on Earth originated from Vesta. The only other known bodies in the solar system whose rocky debris has been identified on Earth are the Moon and Mars.

Remarkably, the Dawn spacecraft has now imaged the crater produced by the impact that ejected these rocks from Vesta (**Figure 15-10**). By analyzing infrared spectra, Dawn has confirmed that the chemical composition around Vesta's south pole crater matches the meteorites long-suspected to have come from this asteroid. The collision at Vesta's south pole was so energetic that grooves formed around the equator as the entire asteroid vibrated like a ringing bell.

The Dawn mission did not end there. After leaving Vesta, the spacecraft traveled to Ceres and entered into orbit around this asteroid in March 2015 (see Figure 15-5). Dawn found a surface covered in water-infused clay minerals, along with evidence of a layer of water ice

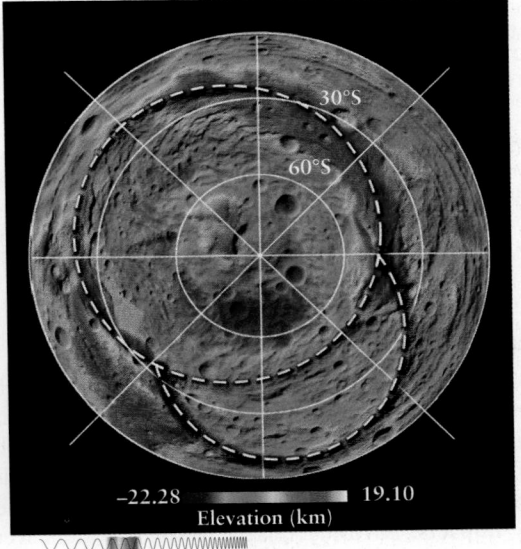

FIGURE 15-10

Impact on Vesta (a) The impact at Vesta's south pole appears to have created its unique equatorial grooves. **(b)** This topographic map looking directly over the south pole shows the outlines of a massive crater named Rheasilvia. The crater is so wide that it covers most of the southern hemisphere and is about 90% the width of the entire asteroid. The central

peak (shown in red, about 14 mi (22.5 km) high) is similar in size to the largest mountain in the solar system—Olympus Mons on Mars. There is speculation about another large crater underneath Rheasilvia (jutting out in the lower right), but this is uncertain. (a, b: NASA/JPL-Caltech/UCLA/MPS/DLR/IDA)

(a)

(b)

FIGURE 15-11

Occator Crater on Ceres This 90-km-wide crater is one of many craters on Ceres that contain bright reflective material. **(a)** This false-color image indicates different materials by their coloring. Blue indicates the presence of salts containing sulfur. Silicates are also expected, which is a component of sandstone rock here on Earth. **(b)** An enhanced-color close-up of the bright spot in Occator Crater. At a surprisingly young 4 million years old, this bright material is about 30 million years younger than the crater in which it sits. The brightest features are due to salts containing sodium and carbon. This salt is thought to seep up from below in briny liquid water that is melted (or vaporized) by the heat of the impact. Once exposed to the negligible pressure on the surface, the water sublimates away, leaving behind a coating of reflective salty minerals. (a: NASA/JPL-Caltech/UCLA/MPS/DLR/IDA; b: NASA/JPL-Caltech/UCLA/MPS/DLR/IDA/PSI/LPI)

just meters below. Enigmatic bright spots in a crater named Occator are thought to be salty, reflective deposits (Figure 15-11).

Collisions and the Nature of Asteroids

What have spacecraft and Earth-based radar told us about the asteroids? To appreciate the answer, we should first consider what scientists expected to find with these tools. Until rather recently, the consensus among planetary scientists was that the smaller asteroids were solid, rocky objects. The surfaces were presumed to be cratered by impacts, but only relatively small craters were expected. The reason is that an impact forceful enough to make a very large crater would probably fracture a rocky asteroid into two or more smaller asteroids.

The truth proves to be rather different. It is now suspected that most asteroids larger than about 1 km are monolithic, compact bodies that are actually many smaller solid pieces held together by gravity, with a lot of air pockets in between. An extreme example

is Mathilde (see Figure 15-7). This asteroid has such a low density (1300 kg/m³) that it cannot be made of solid rock, which typically has densities of 2500 kg/m³ or higher. Itokawa is also well below the density of solid rock. Therefore, a good description of an asteroid like this is a "rubble pile" of small fragments that fit together loosely. Most likely, billions of years of impacts have totally shattered this asteroid. But the impacts were sufficiently gentle that the resulting fragments only drifted apart relatively slowly, only to be pulled back onto the asteroid by gravitational attraction.

Another surprising discovery is that some asteroids have extremely large craters, comparable in size to the asteroid itself. In addition to Vesta's massive south pole crater (Figure 15-10b), an example is the immense crater on Mathilde, shown in Figure 15-7. The object that formed this crater probably collided with Mathilde at a speed of 1 to 5 km/s (2000 to 11,000 mi/h), which could have easily shattered solid rock. Why didn't this collision break the asteroid completely apart, sending the pieces flying in all directions? The explanation may be that the asteroid had already been broken into a loose collection of fragments before this major impact took place. Such a fragmented asteroid can more easily absorb the energy of a collision than can a solid, rocky one.

ANALOGY

If you fire a rifle bullet at a wine glass, the glass will shatter into tiny pieces. But if you fire the same bullet into a bag full of sand (which has much the same chemical composition as glass), the main damage will be a hole in the bag. In a similar way, an asteroid that is a collection of small pieces can survive a major impact more easily than a solid asteroid can.

In some cases, a collision that produces a rubble-pile asteroid can eject a piece that goes into orbit around the asteroid. This may explain why a handful of asteroids are known to have satellites. There are even six known cases of an asteroid with two satellites, like a miniature planetary system (Figure 15-12). If someone were lucky enough to observe an asteroid soon after an impact, the short-lived debris from head-on collisions could be observed, as in Figure 15-13.

If a collision is sufficiently energetic, it may shatter an asteroid permanently. Evidence for this was first pointed out in 1918 by the Japanese astronomer Kiyotsugu Hirayama. He drew attention to families of asteroids that share nearly identical orbits. Each of these groupings, now called **Hirayama families**, presumably resulted from a parent asteroid that was broken into fragments by a high-speed, high-energy collision with another asteroid.

CONCEPT CHECK 15-4

How do we know that some meteoritic dust comes from asteroids?

CONCEPT CHECK 15-5

Without a returned sample, how did the Dawn spacecraft verify that some meteors on Earth came from the asteroid Vesta?

Answers appear at the end of the chapter.

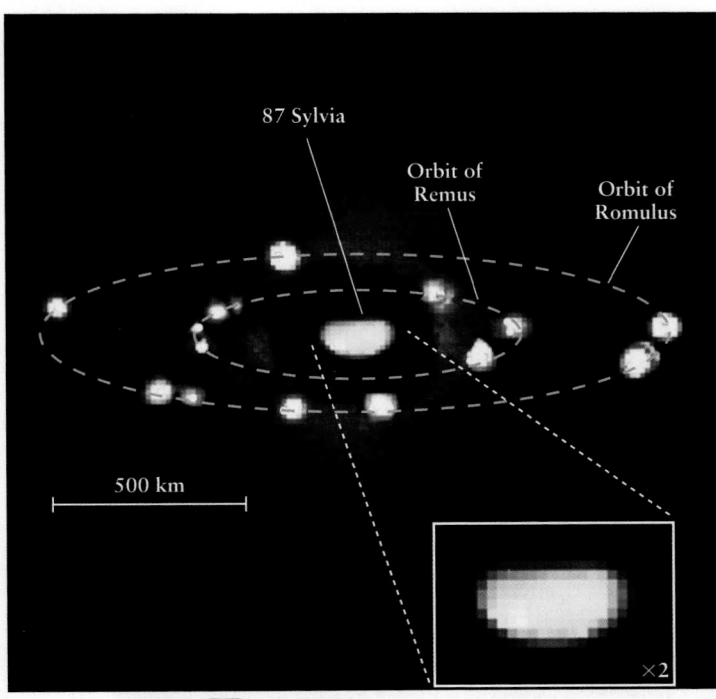

FIGURE 15-12 R **I** V U X G

An Asteroid with Two Satellites Although it is only about 280 km across, the potato-shaped asteroid 87 Sylvia has two satellites of its own. The outer satellite, Romulus, is 18 km across; the inner satellite, Remus, is just 7 km across. This image is actually a composite of nine observations made with the adaptive optics system at the Yepun telescope of the European Southern Observatory (see Figure 6-17). (ESO)

15-4 The dinosaur-killing asteroid, the Tunguska object, and other impacting asteroids are found outside the asteroid belt

While the majority of asteroids lie within the asteroid belt, there are some notable exceptions. Of course, for an asteroid to hit Earth, its orbit must not be confined to the asteroid belt (see **Box 15-1**), and we look at some of the resulting impacts here. The most notable impact of all is the dinosaur-killing asteroid event, and there have been several other significant impacts as well.

Near-Earth Objects and Impacts from Space

Some asteroids have highly elliptical orbits that bring them into the inner regions of the solar system. Asteroids that *cross* Mars's orbit, or whose orbits lie completely within that of Mars, are called **near-Earth objects** or **NEOs**. These are shown in red in Figure 15-14b.

Occasionally, a near-Earth object may pass relatively close to Earth. For example, in 2009 the 35-m-wide asteroid 2009 DD45 came within 72,000 km of Earth, which is about one-third of the distance to the Moon. Estimates suggest that several dozen asteroids ranging in size from 6 to 12 m pass by closer than our Moon each year, with most going undetected. On February 15, 2013, the 30-m-wide asteroid DA14 missed hitting us by only 27,000 km. This is closer to us than some communications satellites (more on that historic day later in this section). In 2017, an asteroid flew by even closer, at just over 8000 km, with a size estimated between 10 and 31 m. More than 16,000 near-Earth objects are known, and there may be thousands more that we have not yet detected. Fortunately, space is a big place and Earth is a rather small target, so large asteroids rarely strike Earth.

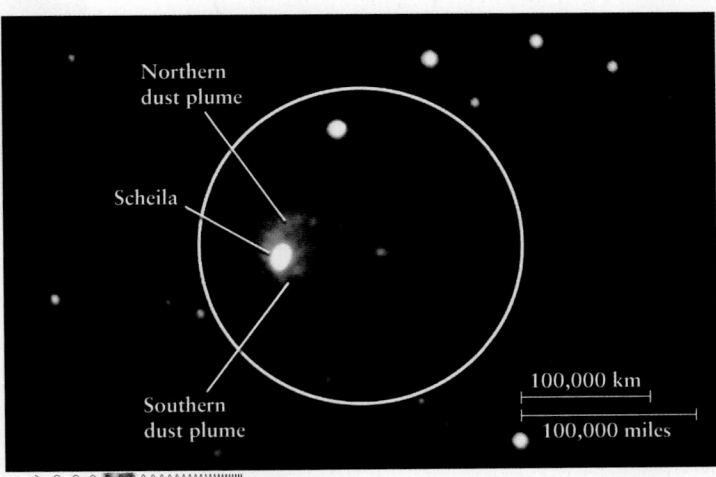

(a) R I **V** U X G

FIGURE 15-13

(b) R I **V** **U** X G

Asteroid Collisions **(a)** This X-shaped profile of debris is suspected to result from the collision of two asteroids. The debris tails are made of dust and gravel, and this object (named P/2010 A2) is seen in false color orbiting within the asteroid belt. **(b)** The asteroid Scheila unexpectedly doubled in brightness, which prompted a closer look at UV wavelengths. While UV images reveal dust plumes kicked up by a collision, they also indicate

Scheila is not a comet (due to a lack of any water vapor emission lines in the plumes). Weeks after the collision's plume was imaged, it faded from view. Within the white circle, images are combined at UV and visible wavelengths, and other than Scheila, the bright spots are stars. (a: NASA, ESA, and D. Jewitt [UCLA]; b: NASA/Swift/DSS/D. Bodewits [UMD])

BOX 15-1   TOOLS OF THE ASTRONOMER'S TRADE

The Lagrange Points and Trojan Asteroids

We saw in Section 15-2 that Jupiter's gravitational pull by itself depletes certain orbits in the asteroid belt, forming the Kirkwood gaps. But the gravitational forces of the Sun and Jupiter work together to *capture* asteroids at two locations outside the asteroid belt called the **stable Lagrange points** of Jupiter. These points of orbital stability exist for any pair of orbiting bodies, like Earth and the Moon or Jupiter and the Sun. One of these points is one-sixth of the way around Jupiter's orbit ahead of the planet, while the other point is the same distance behind the planet (Figure 15-14a). The French mathematician Joseph Louis Lagrange predicted the existence of these points in 1772; asteroids were first discovered there in 1906.

The asteroids trapped at Jupiter's Lagrange points are called **Trojan asteroids**, with some named individually after heroes of the Trojan War. More than 5000 Trojan asteroids have been catalogued so far (see Figure 15-14b). Nine objects have also been found at the Lagrange points of Neptune. These are icy trans-Neptunian objects rather than rocky asteroids. In 2017, NASA's asteroid sample return mission OSIRIS-REx began a search for Earth-Trojan asteroids on its way to its target, asteroid 101955 Bennu. One such asteroid, 2010 TK7 (discovered by the Wide-field Infrared Survey Explorer [WISE] satellite) has been confirmed as an Earth-Trojan.

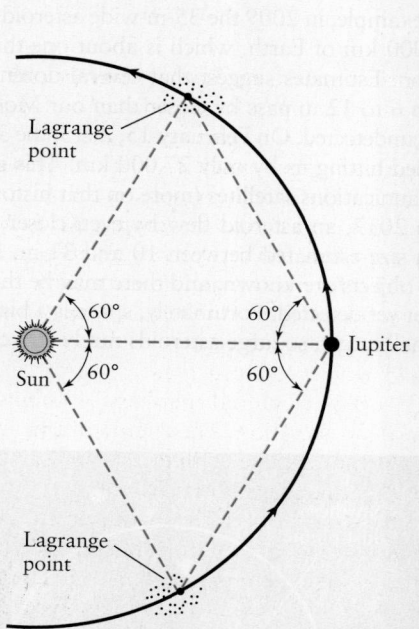

(a) A clump of asteroids is found at each of Jupiter's Lagrange points.

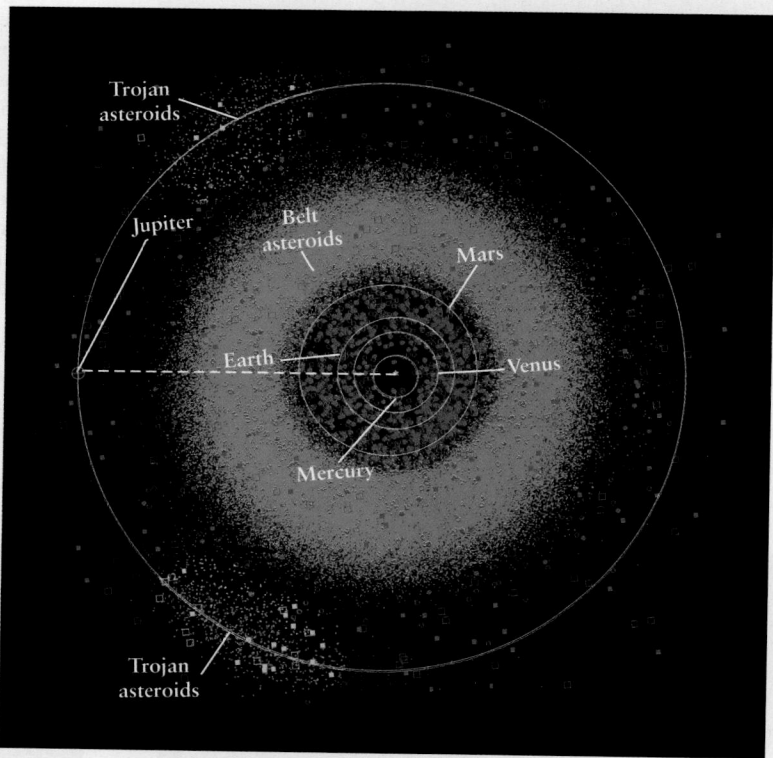

(b) A map of all asteroids within Jupiter's orbit.

FIGURE 15-14

Asteroids Outside the Belt (a) The combined gravitational forces of Jupiter and the Sun tend to trap asteroids near the two stable Lagrange points along Jupiter's orbit. Note that the Sun, Jupiter, and either one of these points lie at the vertices of an equilateral triangle. **(b)** This plot shows the actual positions of all known asteroids at Jupiter's orbit or closer. Green dots denote belt asteroids, deep blue squares denote Trojan asteroids, and red circles denote asteroids that come within 1.3 au of the Sun. Most of the red dots are called near-Earth objects (NEOs). Comets are filled and unfilled light blue squares. (b: Gareth Williams/Minor Planet Center)

However, collisions between asteroids produce numerous smaller chunks of rock, and many of these do eventually rain down on the terrestrial planets. Fortunately for us, the majority of these fragments are quite small. But on rare occasions, a large fragment collides with our planet. When such a collision takes place, the result is an impact crater whose diameter depends on the mass and speed of the impinging object.

A relatively young, pristine terrestrial impact crater is the Barringer Crater near Winslow, Arizona (Figure 15-15). Also known as Meteor Crater, this feature formed 50,000 years ago

FIGURE 15-15 R I **V** U X G

The Barringer Crater This 1.2-km (3/4-mi) wide crater was formed by an asteroid impact about 50,000 years ago and is one of Earth's best preserved impact features. Almost 200 other impact craters—typically tens of kilometers across—have been identified on Earth. Few of these are more than 500 million years old, because the reshaping of Earth's surface by erosion and by tectonic processes tends to erase craters over time. (© Meteor Crater Enterprises)

FIGURE 15-16 R I **V** U X G

Iridium-Rich Clay This photograph of strata in Italy's Apennine Mountains shows a dark-colored layer of iridium-rich clay. This clay is sandwiched between older layers of white limestone (lower right) and younger layers of grayish limestone (upper left). The iridium-rich layer is thought to be the result of an asteroid impact 65 million years ago. The coin is the size of a U.S. quarter. (Prof. Walter Alvarez/Science Source)

when an iron-rich asteroid approximately 50 m across struck the ground at more than 11 km/s (40,000 km/h, or 25,000 mi/h). The resulting blast was equivalent to the detonation of a 20-megaton hydrogen bomb and left a crater 1.2 km (3/4 mi) wide and 200 m (650 ft) deep at its center. This energy was also large enough to vaporize most of the asteroid, and only small fragments of the impacting object have been found. (Figure 7-10b shows a much larger but much older crater in Canada.)

Earth's atmosphere filters out small asteroids, which burn up before reaching the surface. As a rough rule of thumb, asteroids initially larger than about 10 m in size can make it through the atmosphere, with about half of the asteroid burning up or breaking up in the process. The resulting crater can be 10 or even 20 times larger than the main fragment impacting Earth's surface, depending on the asteroid's composition and impact speed, as well as the kind of material being impacted.

Iridium and the Demise of the Dinosaurs

The element iridium plays a special role in deciphering the history of asteroid impacts on Earth. Iridium is common in asteroids, but rare in Earth's crustal rocks, so iridium's presence is a reliable indication of a material's extraterrestrial origins. Measurements of iridium in Earth's crust can therefore tell us the rate at which meteoritic material has been deposited on Earth over the ages. The geologist Walter Alvarez and his physicist father, Luis Alvarez, from the University of California, Berkeley, made such measurements in the late 1970s.

Working at a site of exposed marine limestone in the Apennine Mountains in Italy, the Alvarez team discovered an *exceptionally high abundance of iridium* in a dark-colored layer of clay between limestone strata (**Figure 15-16**). Since this discovery in 1979, a comparable layer of iridium-rich material has been uncovered

An asteroid impact probably ended the age of dinosaurs 65 million years ago.

at numerous sites around the world. In every case, geological dating reveals that this apparently worldwide layer of iridium-rich clay was deposited about 65 million years ago.

Paleontologists were quick to realize the significance of this date, because it was 65 million years ago that the dinosaurs rather suddenly became extinct. In fact, more than 85% of all species on Earth disappeared within a relatively brief span of time.

The Alvarez discovery suggests a startling explanation for this dramatic mass extinction: Perhaps an asteroid or comet hit Earth at that time. An asteroid 10 km in diameter slamming into our planet would have produced a fiery plume that ignited tremendous wildfires, killing untold numbers of animals and consuming much of Earth's vegetation. The energy of this impact would have been a *billion* times stronger than the nuclear bomb dropped on Hiroshima, Japan, during World War II.

In this scenario, the dust thrown up from the impact and soot from wildfires hung in the atmosphere, blackening the sky for about a decade after the impact. When the dust eventually drifted to the ground, it formed the worldwide layer of iridium-rich clay that the Alvarez team identified in 1979. A layer of soot 65 million years old is also found at various sites around the world, supporting this theory. From the thickness of this layer, scientists calculate that the wildfires produced nearly 70 *billion* tons of soot. Temperatures would have plummeted as sunlight was blocked out by dust and soot, and the loss of many plants would have killed larger animals higher up in the food chain. The *Cosmic Connections* figure depicts the sequence of events that may have followed the asteroid impact.

Tiny rodentlike creatures capable of ferreting out seeds and nuts were among the animals that managed to survive this catastrophe, setting the stage for the rise of mammals in the eons that followed. As the dominant mammals on the planet, we may owe our very existence to an ancient asteroid.

In 1992, a team of geologists suggested that this asteroid impact occurred at a site in Mexico. They based this conclusion

...ler Asteroid

...compelling evidence that 65 million years ago, Earth was struck by an asteroid about ...bout 6 mi) in diameter. This impact triggered a biological cataclysm that wiped out 85% ...nt and animal species on our planet. [Chris Butler]

The Day Before

...e Cretaceous swamps and rivers in North America ...d a mix of coniferous, broad-leaved evergreen, and ...ciduous trees. They formed canopied forests and ...en woodlands with understories of ferns, aquatic ...nts, and flowering shrubs.

Impact

The Chicxulub Impact occurred in a shall... sea and immediately lofted rocky, molten... vaporous debris into the atmosphere. The... of the debris rained down on nearby cont... regions, but much of it rose all the way in...

40 Minutes

...e vapor-rich plume of material expanded to ...velop Earth. As material in that plume fell back ...he ground, it streaked through the atmosphere ...trillions of meteors, heating it in some places by ...ndreds of degrees.

5 Years

After fires had ravaged the landscape, only... stark trunks and skeletons remained. Soot f... fires and dust from the impact slowly settle... ground. Sunlight was dramatically, if not tot... attenuated for years.

10 Years

...e postimpact environment was less diverse. Ferns ...d algae were the first to recover. Plant species in ...amps and swamp margins generally survived better ...n species in other types of ecosystems. Conifers

50 Years

Shrubs took advantage of the vacant lands... began to cover it. Species pollinated by the... did better than those that relied on insects. ... began to grow, but it took years for forest ... to rebuild. The recovery time is uncertain...

FIGURE 15-17

Confirming an Extinction-Level Impact Site   Measuring
the gravitational field from slight variations in density can reveal
a buried crater. Concentric rings of the underground Chicxulub
Crater (right inset) lie under a portion of the Yucatán Peninsula.
This crater has been dated to 65 million years ago and is believed
to be the site of the impact that led to the extinction of the
dinosaurs. A piece of 65-million-year-old meteorite discovered
in the middle of the Pacific Ocean in 1998 is believed to be a
fragment of that meteorite. The fragment, about 0.3 cm (0.1 in.)
long, was cut into two pieces for study (left inset). (Top left: Frank
T. Kyte, UCLA; top right: Mark Pilkington/Geological Survey of Canada/
Science Source; bottom: Peter W. Sloss, NOAA-NESDIS-NGDCD)

on glassy debris and violently shocked grains of rock ejected from
the 180-km-diameter Chicxulub Crater on the Yucatán Peninsula
(Figure 15-17). Using radioactive dating techniques (see Section
8-3 and Box 8-1), the scientists found that the asteroid struck
64.98 million years ago, in remarkable agreement with the age of
the iridium-rich layer of clay.

In 2010, a special international panel of 41 scientists, spanning fields ranging from astronomy to geology to paleontology,
reviewed 20 years of data and several different hypotheses for
extinction of the dinosaurs. It was their strong conclusion that
the current evidence points to an asteroid or comet impact that
triggered the mass extinction 65 million years ago.

The Tunguska Event

On June 30, 1908, a spectacular explosion occurred over the
Tunguska region of Siberia that released about 10^{15} joules of
energy, equivalent to a nuclear detonation of several hundred kilo-
tons. The blast knocked a man off his porch some 60 km away
and could be heard more than 1000 km away. Millions of tons of
dust were injected into the atmosphere, darkening the sky as far
away as California.

Preoccupied with political upheaval and World War I, a successful expedition from Russia did not reach the site until 1927.
Researchers found that trees had been seared and felled radially
outward in an area about 50 km in diameter (Figure 15-18). There
was no clear evidence of a crater. In fact, the trees at "ground
zero" were left standing upright, although they were completely
stripped of branches and leaves.

Several teams of modern astronomers have argued that the
Tunguska explosion was caused by an asteroid or comet traveling
at supersonic speed. They arrived at this conclusion after assessing
the effects of various impactor sizes, speeds, and compositions.
Even data from above-ground nuclear detonations of the 1940s
and 1950s were worked into the calculations. The Tunguska event
is well-matched by an asteroid or comet about 80 m (260 ft) in
diameter entering Earth's atmosphere at 22 km/s (50,000 mi/h)
and exploding about 7 km high in the atmosphere.

The Historic Asteroid Impact of 2013

On February 15, 2013, astronomers were eagerly awaiting the
flyby of asteroid DA14. Discovered a year earlier, its predictable
and historic near miss—passing closer than some communications

FIGURE 15-18 R I V U X G

Aftermath of the Tunguska Event In 1908, an asteroid or comet with a mass of about 10⁸ kilograms entered Earth's atmosphere over the Tunguska region of Siberia. Its passage through the atmosphere made a blazing trail in the sky some 800 km (500 mi) long. The object apparently exploded before reaching the surface, blowing down trees for hundreds of kilometers around "ground zero." (Sovfoto)

The Chelyabinsk impact is the largest since the Tunguska event in 1908, and impacts of this size are expected about once every 100 years. Unlike any other impact event, it caused more than 1000 human injuries and damaged thousands of buildings. Drawn to

(a)

satellites—was sure to make headlines on astronomy news sites. Instead, by sheer coincidence, an unexpected asteroid with nearly the opposite trajectory struck Earth first, and it made headlines around the world (Figure 15-19).

Named after the province in Russia where the meteor exploded, the Chelyabinsk asteroid was about 15 m wide and weighed about 10,000 tons. The asteroid slammed into Earth's atmosphere at about 60 times the speed of sound; at that tremendous speed, an object could orbit Earth in less than 40 minutes. Most of the asteroid vaporized in a bright fireball about 20 km above Earth's surface. Even though it didn't reach the surface, the fireball produced a damaging blast wave when it released about as much energy as 20 nuclear bombs of the size detonated in World War II.

(b)

FIGURE 15-19 R I V U X G

Russian Asteroid Impact of 2013 (a) The fireball lasted about 30 seconds, leaving a smoke trail about 400 km (250 mi) across the sky over the Russian town of Chelyabinsk. (b) Impact with the atmosphere produced a powerful shockwave that shattered more than 1 million square feet of glass, and damaged many buildings, such as this zinc plant. With winter temperatures around −15°C (5°F), broken windows posed an ongoing threat. (c) This 6-m-wide hole in Lake Chebarkul is a likely impact site for a larger piece of meteoritic debris, although divers could not find the suspected piece in the first few weeks after the impact. Many smaller meteorites were found around this hole and, to a lesser degree, throughout the city. The collected meteorites are mostly rock, consisting of about 10% iron. (a: Xinhua/RIA Xinhua News Agency/Newscom; b: Jiang Kehong/Xinhua News Agency/Newscom; c: Itar-Tass/ABACA/Newscom)

(c)

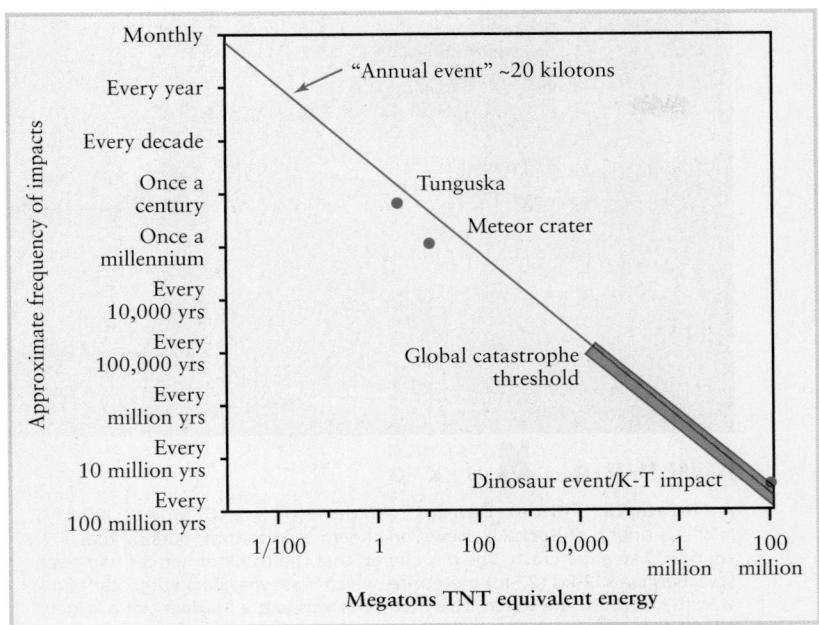

Megatons TNT equivalent energy

FIGURE 15-20

Frequency of Impacts on Earth This is a plot of how frequently, on average, an impact with a given energy occurs on Earth. These are only estimates because this kind of information is not known in detail. The energy is given in units of millions of tons of TNT, which is a high explosive. The lower region corresponds to rare events, but includes events with energies leading to global catastrophe. The upper region indicates that it is as if a World War II–sized nuclear blast (about 20,000 tons of TNT) is detonated each year somewhere in our atmosphere. (Data from C. R. Chapman and D. Morrison, *Cosmic Catastrophes*, 1989)

their windows to view a fireball that outshone the Sun, more than 1000 people were injured by flying shards of glass when the slower-traveling shock wave hit their windows nearly 3 minutes later.

The Dangers of Asteroid Impacts

Large objects, such as that which formed the Chicxulub Crater and may have led to the demise of the dinosaurs, occasionally strike Earth with the destructive force of as much as 1 million megatons of TNT. An asteroid 1 or 2 km in diameter, striking Earth at 30 km/s (more than 67,000 mi/h), would devastate an area the size of California. Such an impact would throw more than 10^{13} kg of microscopic dust particles into the atmosphere, which would decrease the amount of sunlight reaching Earth's surface by enough to threaten the health of the world's agriculture. The loss of a year's crops could lead to the demise of one-quarter of the world's population and would place our civilization—although perhaps not the survival of our species—in grave jeopardy.

The good news is that dangerously large objects strike Earth *very* infrequently (**Figure 15-20**). On average, an asteroid about 5 to 10 m wide harmlessly strikes Earth once a year, releasing as much energy in the upper atmosphere as the atomic bomb on Hiroshima. A more serious impact from an object greater than 50 m, like the one that caused the Tunguska event, is expected every 300 years or so. Getting hit by a 1-km asteroid would certainly be dangerous, but is expected only about once every 500,000 years. For asteroids 10 km or larger, impacts are so rare that the last known case is the dinosaur-killer asteroid 65 million years ago.

Smaller objects, on the other hand, strike Earth more frequently. Basketball-sized objects enter Earth's atmosphere about once a day. Even refrigerator-sized objects hit us about once a month. These smaller objects mostly burn up in the atmosphere, but next we consider the debris that occasionally reaches Earth's surface from objects large and small.

CONCEPT CHECK 15-6

What are two pieces of evidence for a massive asteroid impact 65 million years ago?

CALCULATION CHECK 15-1

If an impact crater on Earth is 2000 m wide, how might you estimate the size of the asteroid fragment that made it through Earth's atmosphere?

Answers appear at the end of the chapter.

15-5 Meteorites are classified as stones, stony irons, or irons, depending on their composition

TUTORIAL 15.2 Asteroids can only be studied at close range using spacecraft. But from time to time, pieces of asteroids fall to Earth, where they can be examined by scientists. These asteroid fragments are called *meteorites*.

Meteoroids, Meteors, and Meteorites

A **meteoroid**, like an asteroid, is a chunk of rock or metal in space. There is no official dividing line between meteoroids and asteroids, but many astronomers think of meteoroids as objects measuring less than 1 m across.

A **meteor** is the brief flash of light (sometimes called a *shooting star*) that is visible at night when a meteoroid enters Earth's atmosphere (**Figure 15-21**). The shooting star begins with an incoming rock about 50 km high in Earth's atmosphere. As the fast-moving

FIGURE 15-21 R I **V** U X G

A Meteor A meteor is produced when a meteoroid—a piece of interplanetary rock or dust—strikes Earth's atmosphere at high speed (typically 5 to 30 km/s). The light comes from heat as the air in front of the meteor is highly compressed. This heat also vaporizes rock, and most meteoroids burn up completely at altitudes of 100 km (60 mi) or so. This long exposure (notice the star trails) shows an exceptionally bright meteor called a fireball. (SPL/Science Source)

FIGURE 15-22 R I **V** U X G

A Meteorite "Fender-Bender" On the evening of October 9, 1992, Michelle Knapp of Peekskill, New York, heard a noise from outside that sounded like a car crash. She discovered that the trunk of her car had been smashed by a 12-kg (27-lb) meteorite, which was lying beside the car and was still warm to the touch. This meteorite was just a fragment of a larger object that produced a fireball that was seen streaking across several states. Meteorite damage is not covered by automobile insurance, but Ms. Knapp was offered several tens of thousands of dollars for the meteorite (and the car). One of the editors for this textbook grew up in Peekskill, where this event is well-known. Occurring on the night of the Homecoming football game, it was captured on several home videos that can now be found online. (Wolfgang Kehmeier/dapd/AP Photo)

meteoroid rams into the air in front of it, the compressed air heats up to about 3000 K (= 4940°F = 2730°C) and produces the bright trail of light we see. The width of the bright streak might only be 1 m or so wide, and cools in less than a second. However, due to the rock's high speed, the bright trail can be several kilometers long.

While an impressive sight, the size of a typical meteoroid entering Earth's atmosphere is quite small, between a sand grain and a pebble. For incoming rocks less than 10 m wide, the same heat that produces its bright trail vaporizes most of the object. Therefore, the typical meteoroid is completely burned up in the atmosphere. On the other hand, small particles of cosmic dust (see Figure 8-9) slow down quickly, produce little heat, and continuously drift down to Earth's surface without being noticed.

If a piece of rock or metal survives its fiery descent through the atmosphere and reaches the ground, it is called a **meteorite**. As incredible as it may seem, an estimated total of 80 tons of extraterrestrial matter falls on Earth *each day*. However, most of this matter is cosmic dust rather than meteorites. It is fortunate that large meteorite falls are rather rare events—direct hits can cause damage (Figure 15-22). Because they are so rare, meteorites are prized by scientists and collectors alike. Almost all meteorites are asteroid fragments and thus provide information about the chemical composition of asteroids. A small number of meteorites have been identified as pieces of the Moon, Mars, or the asteroid Vesta, that were blasted off the surfaces of those worlds by violent impacts. Perhaps someday a meteorite from Mercury will turn up.

People have been finding specimens of meteorites for thousands of years, and descriptions of them appear in ancient Chinese, Greek, and Roman literature. Many civilizations have regarded meteorites as objects of veneration. The sacred Black

> Studying meteorites reveals the properties of the asteroids from which they came.

Stone, enshrined at the Ka'aba in the Great Mosque at Mecca, may be a relic of a meteorite impact. More recently, selling meteorites has become profitable enough that when scientists show up to investigate an impact, oftentimes the fragments have already been taken.

The extraterrestrial origin of meteorites was hotly debated until as late as the eighteenth century. Upon hearing a lecture by two Yale professors, President Thomas Jefferson is said to have remarked, "I could more easily believe that two Yankee professors could lie than that stones could fall from Heaven." Although several falling meteorites had been widely witnessed and specimens from them had been collected, many scientists were reluctant to accept the idea that rocks could fall to Earth from outer space. Conclusive evidence for the extraterrestrial origin of meteorites came on April 26, 1803, when thousands of fragments pelted the French town of L'Aigle. This event was analyzed by the noted physicist Jean-Baptiste Biot, and his conclusions helped to finally convince scientists that meteorites were indeed extraterrestrial.

Types of Meteorites

Meteorites are classified into three broad categories: stones, stony irons, and irons. As their name suggests, **stony meteorites**, or **stones**, look like ordinary rocks at first glance, but they are sometimes covered with a **fusion crust** (Figure 15-23a). This crust is produced by the melting of the meteorite's outer layers during its fiery descent through the atmosphere. When a stony meteorite

Many stony meteorites are coated with dark fusion crusts . . .

. . . but when cut and polished they reveal tiny specks of iron in the rock.

(a)   (b)

FIGURE 15-23 R I (V) U X G

Stony Meteorites Of all meteorites that fall on Earth, 95% are stones. **(a)** Many freshly discovered specimens, like the one shown here, are coated with dark fusion crusts. This particular stone fell in Texas. **(b)** Most stony meteorites contain small grains of nickel-rich iron distributed throughout the rock. These distinctive grains are also evident in this specimen from Australia. (a: Tom McHugh/Science Source; b: © Photo by Daniel Ball/School of Earth and Space Exploration/ASU. Courtesy of the Center for Meteorite Studies.)

is cut in two and polished, tiny flecks of iron can sometimes be found in the rock (Figure 15-23b). Meteorites collected from the 2013 Chelyabinsk impact event are of this type and contain about 10% iron.

Although stony meteorites account for about 95% of all the meteoritic material that falls to Earth, they are the most difficult meteorite specimens to find. If they lie undiscovered and become exposed to the weather for a few years, they look almost indistinguishable from common terrestrial rocks. (The asteroid that caused the Tunguska explosion, described in Section 15-4, was probably of stony composition, which would account for the lack of obvious meteoritic debris on the ground.) Meteorites with high iron content can be found much more easily because they can be located with a metal detector. Consequently, iron and stony iron meteorites dominate most museum collections.

Stony iron meteorites consist of roughly equal amounts of rock and iron. Figure 15-24, for example, shows the mineral olivine suspended in a matrix of iron. Only about 1% of the meteorites that fall to Earth are stony irons.

Iron meteorites (Figure 15-25), or **irons**, account for about 4% of the material that falls on Earth. Iron meteorites usually contain no stone, but many contain from 10% to 20% nickel. Before humans began extracting iron from ores around 2000 B.C.E., the only source of the metal was from iron meteorites. Because these are so rare, iron was regarded as a precious metal like gold or silver.

In 1808, Count Alois von Widmanstätten (pronounced "VIT-mahn-shtetten"), director of the Imperial Porcelain Works in Vienna, discovered a conclusive test for the most common type of iron meteorite. About 75% of all iron meteorites have a unique crystalline structure. These **Widmanstätten patterns** become visible when the meteorite is cut, polished, and briefly dipped into a dilute solution of acid (Figure 15-25b). Nickel–iron crystals of this type can form only if the molten metal cools slowly over many millions of years. Therefore, Widmanstätten patterns are never found in counterfeit meteorites, or "meteorwrongs."

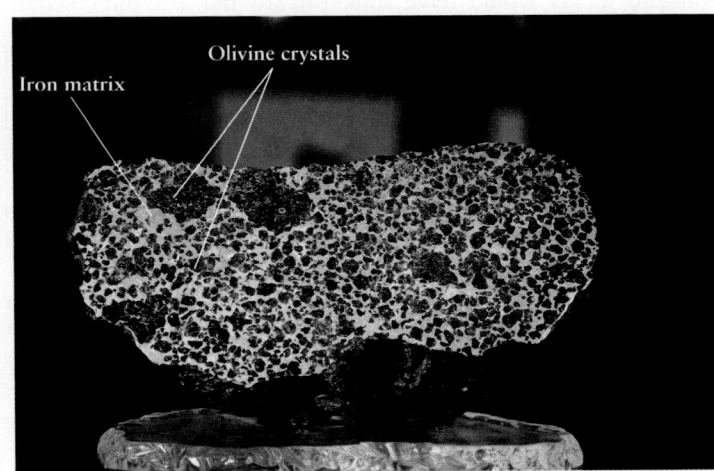

Olivine crystals

Iron matrix

FIGURE 15-24 R I (V) U X G

A Stony Iron Meteorite Stony irons account for about 1% of all the meteorites that fall to Earth. This particular specimen, a variety of stony iron called a pallasite, fell in China. The dark greenish-amber pockets are crystals of the mineral olivine (also known as peridot, August's birthstone) embedded in an iron matrix that fills in the gaps. Meteorites like this one are called stony irons because these two components—rocky crystals and iron matrix—have separated from each other, forming a very distinctive and beautiful pattern. The leading idea for the origins of this very rare type of meteorite is that it forms from material in the boundary region between the core and mantle of an asteroid that was large and hot enough for chemical differentiation to take place. In that boundary region, iron from the core would be present, as well as silicates in the mantle to form olivine. (© 2014 Symeon Platts/Arizona Board of Regents)

Iron meteorites are composed of nickel–iron minerals and are characterized by a surface covered with depressions . . .

(a)

. . . and when cut and polished, by interlocking crystals in a Widmanstätten pattern.

(b)

FIGURE 15-25   R I **V** U X G

Iron Meteorites (a) The surface of a typical iron is covered with small depressions caused by ablation (removal by melting) during its high-speed descent through the atmosphere. **(b)** When cut, polished, and etched with a weak acid solution, most iron meteorites exhibit interlocking crystals in designs called Widmanstätten patterns. (a: Detlev van Ravenswaay/ Science Source; b: The Natural History Museum/Alamy Stock Photo)

Meteorites and the Early History of Asteroids

Widmanstätten patterns suggest that some asteroids were originally at least partly molten inside and remained partly molten long after they were formed. As soon as an asteroid accreted from planetesimals some 4.56 billion years ago, rapid decay of short-lived radioactive isotopes could have heated the asteroid's interior to temperatures above the melting point of rock. If the asteroid was large enough—at least 200 to 400 km in diameter—to insulate the interior and prevent it from cooling, the interior would have remained molten over the next few million years and chemical differentiation would have occurred (see Box 7-1 and Section 8-5). Iron and nickel would have sunk toward the asteroid's center, forcing less dense rock upward toward the asteroid's surface.

Like a terrestrial planet, such a **differentiated asteroid** would have a distinct core and crust. After a differentiated asteroid cooled and its core solidified, collisions with other asteroids would have fragmented the parent body into meteoroids. Iron meteorites are therefore specimens from a differentiated asteroid's core, while stony irons come from regions between a differentiated asteroid's core and its crust. Vesta is an example of a differentiated asteroid that has not been completely blasted apart by collision.

Unlike irons and stony irons, stony meteorites may or may not come from differentiated asteroids. Smaller asteroids would not have retained their internal heat long enough for chemical differentiation to occur. Some stony meteorites are fragments of such **undifferentiated asteroids** (also called *primitive* asteroids), while others are relics of the crust of differentiated asteroids.

CALCULATION CHECK 15-2

If a meteor passes through Earth's atmosphere at about 5 km/s, how long is the tail if the atmosphere heated by the meteor only remains hot enough to be visible for about 1 second?

CONCEPT CHECK 15-7

Why does a Widmanstätten pattern ensure that a sample is truly a meteorite?

Answers appear at the end of the chapter.

15-6 Some meteorites retain traces of the early solar system

A class of rare stony meteorites called **carbonaceous chondrites** shows no evidence of ever having been melted while in space. As suggested by their name, carbonaceous chondrites contain substantial amounts of carbon and carbon compounds, including complex organic molecules and as much as 20% water bound into the minerals. These compounds would have been broken down and the water driven out if these meteorites had been subjected to heating and melting. Carbonaceous chondrites may therefore be samples of *the original material from which our solar system was created*. The asteroid 253 Mathilde, shown in Figure 15-7, has a very dark gray color and the same sort of spectrum as a carbonaceous chondrite. It, too, is likely composed of material that dates back to the formation of the solar system.

Amino acids, the building blocks of proteins on which terrestrial life is based, are among the organic compounds occasionally found inside carbonaceous chondrites. Interstellar organic material has thus probably been falling on our planet since its formation. Some scientists suspect that carbonaceous chondrites may have provided some of the original building blocks in the development of life on Earth.

Shortly after midnight on February 8, 1969, the night sky around Chihuahua, Mexico, was illuminated by a brilliant blue-white light moving across the heavens. As the light crossed the sky, it disintegrated in a spectacular, noisy explosion that dropped thousands of rocks and pebbles over terrified onlookers. Within hours, teams of scientists were on their way to collect specimens of a carbonaceous chondrite, collectively named the Allende meteorite after the locality (Figure 15-26).

Specimens were scattered in an elongated ellipse approximately 50 km long by 10 km wide. Most fragments were coated with a fusion crust, but minerals immediately beneath the crust showed no signs of damage. As it heats up on its flight through the atmosphere, surface material is peeled away; this process forms the fusion crust. Because the heat has little time to penetrate the meteorite's interior, compounds there are left intact.

One of the most striking discoveries to come from the Allende meteorite was evidence suggesting that a nearby star exploded

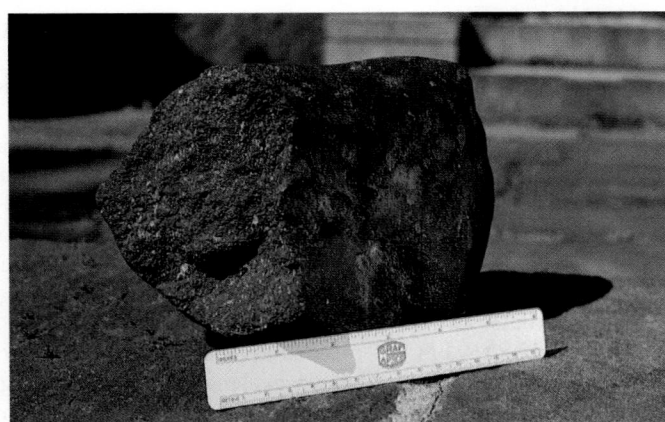

FIGURE 15-26 R I V U X G

A Piece of the Allende Meteorite This carbonaceous chondrite fell near Chihuahua, Mexico, in February 1969. The meteorite's dark color is due to its high abundance of carbon. Radioactive age dating indicates that this meteorite is 4.56 billion years old, suggesting that this meteorite is a specimen of primitive planetary material that predates the formation of the planets. The ruler is 15 cm (6 in.) long. (Courtesy of J. A. Wood)

into a supernova about 4.6 billion years ago. In a **supernova**, a massive star reaches the end of its life cycle and blows itself apart in a cataclysm that hurls matter outward at tremendous speeds (see Figure 1-8). During this detonation, violent collisions between nuclei produce a host of radioactive elements, including ^{26}Al, a radioactive isotope of aluminum.

Researchers found clear evidence for the former presence of ^{26}Al in the Allende meteorite: Chemical analyses revealed a high abundance of a stable isotope of magnesium (^{26}Mg), which is produced by the radioactive decay of ^{26}Al. Some astronomers interpret this as evidence for a supernova in our vicinity at about the time the Sun was born. Indeed, by compressing interstellar gas and dust, the supernova's shock wave may have triggered the birth of our solar system.

CONCEPT CHECK 15-8

Why are carbonaceous chondrites thought to preserve some of the primal material that formed our solar system?

Answer appears at the end of the chapter.

15-7 A comet is a chunk of ice and dust that partially vaporizes as it passes near the Sun

TUTORIAL 15.3 Just as heat from the protosun produced two classes of planets, the terrestrial and the jovian, two main types of small bodies formed in the solar system. Near the Sun, inter-planetary debris consists of the rocky objects called asteroids or meteoroids. Far from the Sun, where temperatures in the early solar system were low enough to permit ices of water, methane, ammonia, and carbon dioxide to form, interplanetary debris took the form of loose collections of ices and small rocky particles.

These objects are like icy versions of "rubble pile" asteroids such as 253 Mathilde, which we described in Section 15-3. The Harvard astronomer Fred Whipple, who devised this model in 1950, called them "icy conglomerates" or "dirty snowballs." Some of these snowballs are trans-Neptunian objects that orbit forever in the outer regions of the solar system. Others, however, are in noncircular orbits that take them inward toward the Sun, where the ice begins to evaporate. These objects are known as **comets**.

The Structure of a Comet

Kuiper belt objects and asteroids travel around the Sun along roughly circular orbits that lie close to the plane of the ecliptic (see Figure 2-14). In sharp contrast, comets travel around the Sun along highly elliptical orbits inclined at random angles to the ecliptic. Comets are just a few kilometers across, so when they are far from the Sun they are difficult to spot even with a large telescope. As a comet approaches the Sun, however, solar heat begins to vaporize the comet's ices, liberating gases as well as dust particles. The liberated gases begin to glow, producing a fuzzy, luminous ball called a *coma* that is typically 1 million km in diameter. Some of these luminous gases stream outward into a long, flowing **tail**. This tail, which can be more than 100 million km in length—comparable to the distance from Earth to the Sun—is one of the most awesome sights that can be seen in the night sky (see **Figure 15-27**).

LOOKING DEEPER 15.2 Many comets are discovered when they are still far from the Sun, long before their tails grow to full size. A number of these discoveries are made by amateur astronomers, many of whom use only binoculars or small telescopes to aid in their search.

FIGURE 15-27 R I V U X G

Comet Hyakutake Using binoculars, Japanese amateur astronomer Yuji Hyakutake first noticed this comet on the morning of January 30, 1996. Two months later, Comet Hyakutake came within 0.1 au (15 million km, or 9 million mi) of Earth and became one of the brightest comets of the twentieth century. This image was captured in April 1996, when the comet's tail extended more than 30° across the sky (more than 60 times the diameter of the full moon). The stars appear as short streaks because the camera followed the comet's motion during the exposure. (Gordon Garradd/SPL/ Science Source)

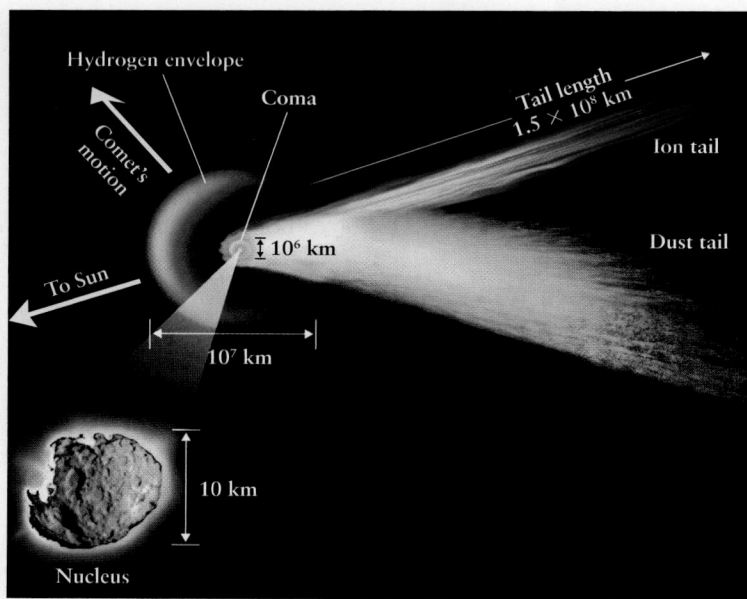

FIGURE 15-28

The Structure of a Comet The solid part of a comet is the nucleus. It is typically about 10 km in diameter. The coma typically measures 1 million km in diameter, and the hydrogen envelope usually spans 10 million km. A comet's tail can be as long as 1 au. Comet Wild 2 (inset) is examined further in Figure 15-30. (This drawing is not to scale.) (NASA/JPL)

Figure 15-28 depicts the structure of a comet. The solid part of the comet, from which the coma and tail emanate, is called the **nucleus**. It is a mixture of ice and dust that typically measures a few kilometers across. Before 1986, no one had seen the nucleus of a comet. It is small, dim, and buried in the glare of the bright coma. In that year, the first close-up pictures of a comet's nucleus were obtained by Soviet spacecraft Vega 1 and 2 and the European Space Agency's Giotto mission, all of which flew past and photographed Comet Halley (Figure 15-29a). Comet Halley's potato-shaped nucleus is darker than coal, reflecting only about 4% of the light that strikes it. This dark color is probably caused by a

layer of carbon-rich compounds and dust that is left behind as the comet's ice evaporates.

Several 15-km-long jets of dust particles were seen emanating from bright areas on Comet Halley's nucleus. These jets seem to be active only when exposed to the Sun. The bright areas, which probably cover about 10% of the surface of the nucleus, are presumably places where the dark layer covering the nucleus is particularly thin. When exposed to the Sun, these areas evaporate rapidly, producing the jets. The jets build the coma, which feeds the comet's tails. When the nucleus's rotation brings the bright areas into darkness, away from the Sun, the jets shut off.

In 2010, NASA's EPOXI mission flew by Comet Hartley 2 (Figure 15-29b) and measured the wavelength spectrum of light coming from the jets that stream off its nucleus. Earlier data indicated that within the jets, water evaporates at a rate of about 200 kg/s (440 lb/s), and water's evaporation was thought to power the jets. However, the spectrum measured by EPOXI reveals that there is a lot of carbon dioxide (CO_2) in the jets as well. Carbon dioxide undergoes the same transformation as water, changing from solid form (familiar as "dry ice") directly to vapor, when exposed to sunlight. Surprisingly, it appears that the evaporation of carbon dioxide powers the jets even more than water.

After imaging comets and analyzing their composition with spectra, the next steps in exploration are to sweep up material from their jets, smash projectiles into their surfaces, and even land on their surfaces for detailed analysis.

Sweep, Smash, and Land

In 2006, NASA's Stardust probe crashed to Earth. Far from being an accident, Stardust's landing was the final phase of a huge success after having swept up dust surrounding the nucleus of Comet Wild 2 (Figure 15-30). Some of the most interesting materials collected from the comet are amino acids. We already saw in Section 15-6 that amino acids are present in some meteorites as well; both comets *and* asteroids contain these basic building blocks of life.

The Stardust mission also contradicted the long-held notion that comets never get warm enough to contain liquid water. In 2011, analysis of samples revealed minerals that could only form

(a) Comet Halley

(b) Comet Hartley 2

FIGURE 15-29 R I **V** U X G

Comet Nuclei and Jets When sunlight reaches exposed ice on the surface of the nuclei, evaporation produces jets. **(a)** This image from the European Space Agency spacecraft Giotto shows the dark, potato-shaped nucleus of Comet Halley in March 1986. Sunlight from the left illuminates half of the nucleus, which is about 15 km long and 8 km wide. Two bright jets of dust extend from the nucleus toward the Sun. **(b)** Comet Hartley 2 imaged by NASA's EPOXI spacecraft. This small comet is only about 1.2 mi (1.9 km) wide from end to end. A spectrum of the jets indicates that they emit about 200 kg/s of water, and even more carbon dioxide. (a: Max Planck Institute for Solar System Research; b: NASA/JPL-Caltech/UMD)

(a)

Ejecta in aerogel

Tiny crater in aluminum
foil holding aerogel

(b)

(c)

FIGURE 15-30 R I **V** U X G

Comet Wild 2 (a) This picture shows two images combined. The inner portion is a high-resolution photograph showing the surprisingly heavily cratered comet. The bright outer portion shows gas and dust jetting away from the comet. Its tails are millions of kilometers long. **(b)** A substance called aerogel was used to capture particles from Comet Wild 2's dust tail.

A piece of space debris pierced the aluminum foil holding the aerogel, embedding itself in it, along with pieces of the foil. **(c)** A 2-μm piece of comet dust, composed of a mineral called forsterite. On Earth, this mineral is used to make gems called peridot. (a, b: NASA/JPL; c: NASA/JPL-Caltech/University of Washington)

in liquid water, indicating that the full story on comets is still far from clear.

VIDEO
15.3

To assess the structural properties of a comet, a mission was designed to smash a projectile into a comet's surface and analyze the debris. To carry out this bold mission, the Deep Impact spacecraft flew to within 500 km of Comet Tempel 1

in 2005 (**Figure 15-31**). Then, Deep Impact launched a 372-kg copper projectile that slammed into the surface of the nucleus at 37,000 km/h (23,000 mi/h). The collision excavated about 11,000 tons of material from the nucleus for analysis. The character of the ejected material confirmed that the nucleus of Tempel 1 is not held together by chemical forces (the way a solid block of ice is held together).

Smooth regions

Impact site

(a)

Impact site

(b)

(c)

(d)

(e)

FIGURE 15-31 R I **V** U X G

Comet Tempel 1 (a) This composite image of Comet Tempel 1 has higher resolution at the bottom, as the projectile from Deep Impact headed in that direction. The smooth regions on the comet have yet to be explained. **(b)** Thirty seconds before the projectile struck the comet. **(c)** Seconds after impact, hot debris explodes away from the comet nucleus. The horizontal white streak through the center is where the camera's charge-coupled device became overloaded with light from the event. **(d)** Moments later, the gases and dust were expanding outward. **(e)** An image taken 67 s after impact. Within minutes, the cloud of debris became much larger than the entire nucleus. (a: NASA/JPL/UMD; b–e: NASA/JPL-Caltech/UMD)

Instead, the nucleus is like a "rubble pile" and is held together by gravitational forces between numerous parts.

The Deep Impact mission also measured the mass and hence the density of the comet nucleus. The very low density of 600 kg/m^3 (60% of the density of water) shows that *the nucleus cannot be a solid block*. It is best regarded as a porous jumble of rock, fine dust, and ice.

An even more ambitious mission to explore a comet is the European Space Agency's Rosetta. This spacecraft was launched in 2004, just in time to observe Deep Impact's collision with Comet Tempel 1 the following year on its way to the outer solar system. Completing a 10-year journey, Rosetta reached Comet 67P/Churyumov-Gerasimenko in 2014 (**Figure 15-32**). After

entering into orbit to map the comet from an altitude of 29 km (18 mi), Rosetta released a special probe—named Philae—to land on the nucleus. Due to the low gravity around a comet, Philae was equipped with tethered harpoons that were to be fired into the surface to prevent the lander from bouncing off into space. Unfortunately, the probe bounced twice upon landing and the harpoons never fired. Stranded in a shadow, Philae could not collect enough solar power to carry out its primary science mission, which included detecting organic molecules (such as amino acids).

Even with the landing probe trouble, Rosetta was able to pursue scientific objectives with its own instruments. One long-standing question is where Earth got its water (because any water present when Earth formed would have been boiled away); a

(a)

(b)

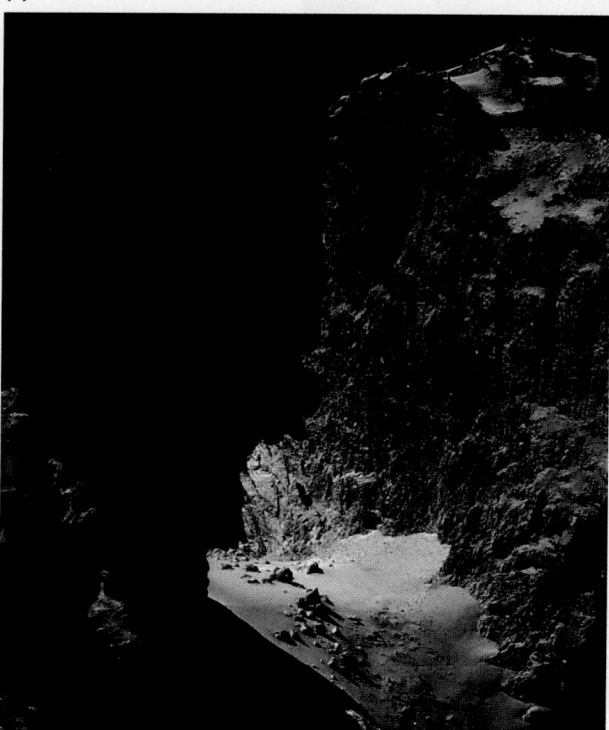

(c)

FIGURE 15-32

Rosetta Visits Comet 67P **(a)** This is a small comet whose length is only 4.3 km (2.6 mi). Rather than the usual description of comets as dusty snowballs, Comet 67P's higher dust content makes it more of a "snowy dustball." When comets are closer to the Sun, they emit gas and dust as ice evaporates, but Rosetta was able to observe even more. Farther from the Sun, Rosetta released rocks of up to about 1 cm in size ejected from the surface. But as 67P got closer to the Sun, it started to release rocks up to 1 m in size. Estimates indicate that 67P could lose about 1 m of its surface material with each pass around the Sun. **(b)** A single jet, which emits both dust and water ice. It is many jets together, like this one, that emit the material that make a comet's tail. **(c)** This cliff on the comet is about 1 km tall, with the larger boulders at the bottom around 20 m across. With so little gravity on this small object, you could jump this cliff to a safe landing. (a: ESA/Rosetta/MPS for OSIRIS Team MPS/UPD/LAM/IAA/SSO/INTA/UPM/DASP/IDA; b: ESA, Rosetta, MPS, OSIRIS; UPD/LAM/IAA/SSO/INTA/UPM/DASP/IDA; c: ESA, Rosetta spacecraft, NAVCAM/NASA) natural hypothesis is that comet impacts delivered water to Earth. Recall that an *isotope* refers to the differing numbers of neutrons that an atom can have, and Rosetta found hydrogen isotopes in Comet 67P's water that don't match Earth's water. Therefore, either comets weren't the main source of Earth's water or, if they were, they weren't like Comet 67P.

Comet Envelopes and Comet Tails

A part of a comet that is not visible to the human eye is the **hydrogen envelope**, a huge sphere of tenuous gas surrounding the nucleus (Figure 15-28). This hydrogen comes from water molecules (H_2O) that escape from the comet's evaporating ice and then break apart when they absorb ultraviolet photons from the Sun. The hydrogen atoms also absorb solar ultraviolet photons, which excite the atoms from the ground state into an excited state (see Section 5-8). When the atoms return to the ground state, they emit more ultraviolet photons.

Unfortunately for astronomers, Earth-based telescopes cannot detect the emission from a comet's hydrogen envelope because our atmosphere absorbs ultraviolet light (see Section 6-7). Instead, cameras above Earth's atmosphere must be used. Figure 15-33 shows two views of the same comet, one as it appeared in visible light to Earth-based observers and one as photographed by an ultraviolet camera aboard a rocket. The ultraviolet image shows the enormous extent of the hydrogen envelope, which can span 10 million km.

As the diagram in Figure 15-28 suggests, comet tails always point away from the Sun. This observation is true regardless of the direction of the comet's motion (Figure 15-34). The implication that something from the Sun was "blowing" the comet's gases radially outward led Ludwig Biermann to predict the existence of a *solar wind*, a stream of particles rushing away from the Sun (see

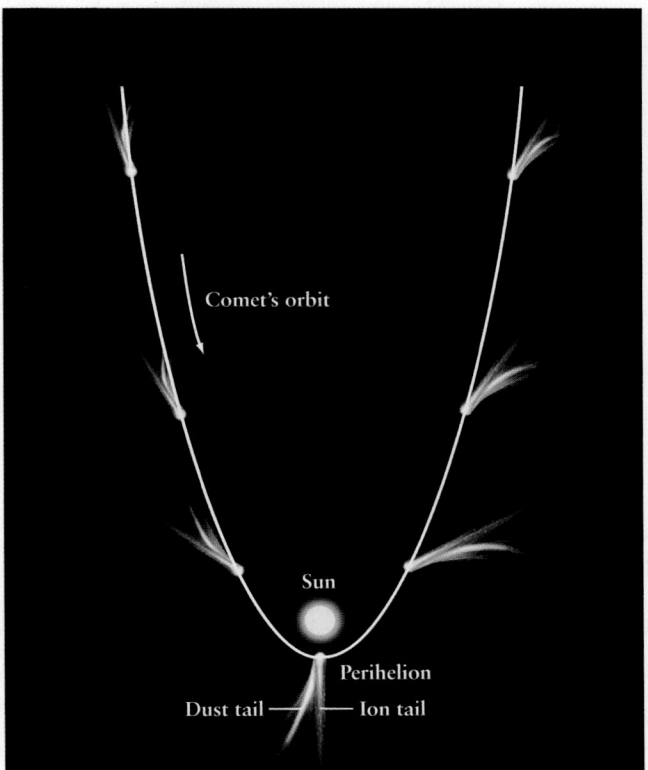

ANIMATION 15.1 **FIGURE 15-34**

The Orbit and Tail of a Comet The solar wind and the pressure of sunlight blow a comet's dust particles and ionized atoms away from the Sun. Consequently, a comet's tail points generally away from the Sun. In particular, the tail does not always stream behind the nucleus. At the upper right of this figure, the comet is moving upward along its orbit and is literally chasing its own tail.

Section 8-5). A decade later, in 1962, Biermann's prediction was confirmed when the solar wind was detected for the first time by instruments on the Mariner 2 spacecraft (see Figure 11-11).

In fact, the Sun usually produces *two* comet tails—an **ion tail** and a **dust tail** (Figure 15-35). Ionized atoms and molecules—that is, atoms and molecules missing one or more electrons—are swept directly away from the Sun by the solar wind to form the relatively straight ion tail. The distinct blue color of the ion tail is caused by emissions from ionized carbon-bearing molecules such as CN and C_2. The dust tail is formed when photons of light strike dust particles freed from the evaporating nucleus. (Figure 8-9 shows a dust particle of this type.) Light exerts a pressure on any object that absorbs or reflects it. This pressure, called **radiation pressure**, is quite weak but is strong enough to make fine-grained dust particles in a comet's coma drift away from the comet, thus producing a dust tail. The solar wind has less of an effect on dust particles than on ions, so the dust tail ends up being curved rather than straight. On rare occasions, a comet is oriented in such a

> A comet's tail does not stream behind it; instead, it points generally away from the Sun.

A visible-light image shows the comet's tail.

R I **V** U X G

An ultraviolet image reveals the comet's immense hydrogen envelope.

R I V **U** X G

FIGURE 15-33

A Comet and Its Hydrogen Envelope These visible-light and ultraviolet images of Comet Kohoutek, which made an appearance in 1974, are reproduced to the same scale. The hydrogen envelope is only visible in the ultraviolet image. (John Hopkins University, Naval Research Laboratory)

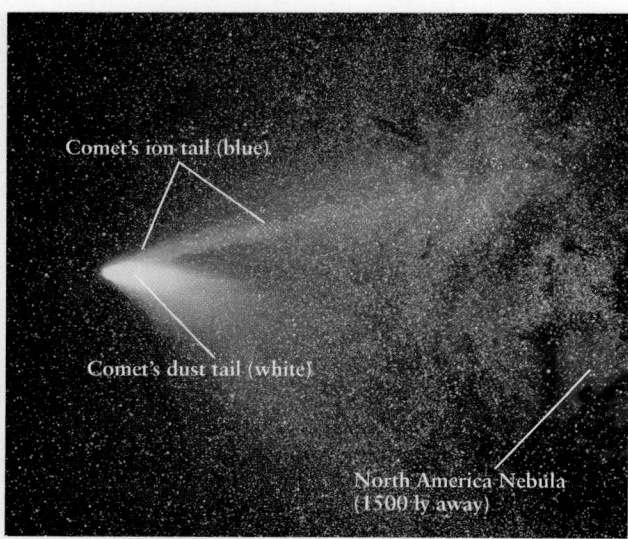

FIGURE 15-35 R I **V** U X G

The Two Tails of Comet Hale-Bopp A comet's dust tail is white because it reflects sunlight, while the molecules in the ion tail emit their own light with a characteristic blue color. When this picture was taken on March 8, 1997, the ion tail extended more than 10° across the sky. The red object to the right is the North America Nebula, a star-forming region some 1500 light-years beyond the solar system. (Tony & Daphne Hallas/Science Source)

way that its dust tail appears to stick out in front of the comet (**Figure 15-36**).

After the comet passes perihelion, it recedes from the Sun back into the cold regions of the outer solar system. The ices stop vaporizing, the coma and tail disappear, and the comet goes back into an inert state—until the next time its orbit takes it toward the inner

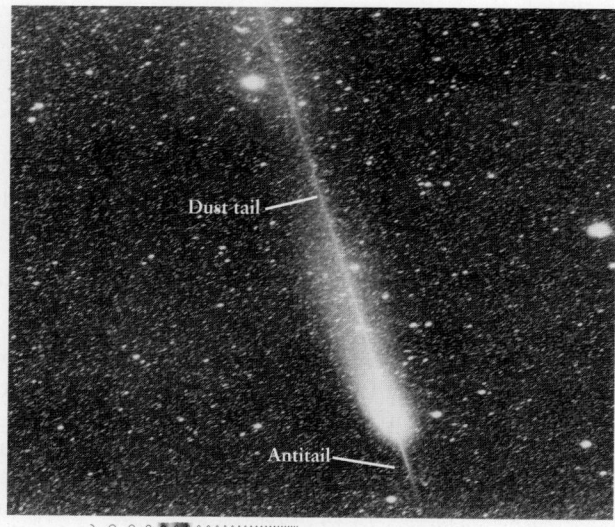

FIGURE 15-36 R I **V** U X G

The Antitail of Comet Hale-Bopp In January 1998, 9 months after passing perihelion, Comet Hale-Bopp was oriented in such a way that the end of its arched dust tail looked like a spike sticking out of the comet's head. This spike is called an "antitail" because it appears to protrude in front of the comet. (ESO)

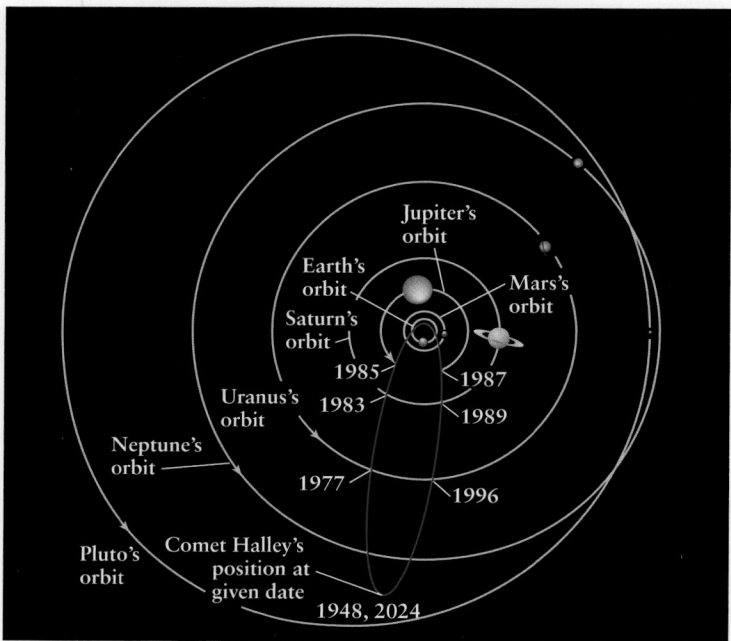

ANIMATION 15.2 ➤ **FIGURE 15-37**

Comet Halley's Eccentric Orbit Like most comets, Comet Halley has an elongated orbit and spends most of its time far from the Sun. Figure 4-26 shows the comet's tail, which appears only when the comet is close to perihelion. Comet Halley has been observed at intervals of about 76 years—the period of its orbit—since 88 B.C.E. (see Section 4-7).

solar system. An object in an elliptical orbit spends most of its time far from the Sun, so it is only during a relatively brief period before and after perihelion that a comet can have a prominent tail. An example is Comet Halley, which orbits the Sun along a highly elliptical path that stretches from just inside Earth's orbit to slightly beyond the orbit of Neptune (**Figure 15-37**). The comet's tail is visible to the naked eye or with binoculars only during a few months around perihelion, which last occurred in 1986 and will next occur in 2061.

Other comets have orbits that are larger and more elongated, with even longer periods. The orbit of Comet Hyakutake (see Figure 15-27) takes it to a distance of about 2000 au from the Sun, 50 times the size of Pluto's orbit, with an orbital period of about 70,000 years.

CONCEPT CHECK 15-9

What is the evidence that comets are not mostly solid chunks of ice?

Answer appears at the end of the chapter.

15-8 Comets originate either from the Kuiper belt or from the Oort cloud

Both comets and Kuiper belt objects are mixtures of ice and rock that must have formed in the outer solar system. (Had they formed in the warm inner solar system, they would not have included ices.) But most Kuiper belt objects have relatively circular orbits that lie close to the plane of the ecliptic, whereas comets have very elongated orbits that are often steeply inclined to the ecliptic. What gives comets their distinctive orbits? As we will see, the answer to this question depends on whether the comet's orbital period is relatively short or relatively long.

Jupiter-Family Comets and the Kuiper Belt

As we saw in Section 14-10, the gravitational effects of Neptune shape the orbits of objects in the **Kuiper belt**. Most of these objects are found between the orbit of Pluto (where an object makes two orbits for every three orbits of Neptune) and about 50 au from the Sun (where an object makes one orbit for every two orbits of Neptune).

However, collisions between Kuiper belt objects can break off relatively small chunks. Gravitational perturbations from Neptune can occasionally launch one of these chunks into a highly elliptical orbit that takes it close to the Sun. When one of these chunks approaches the Sun, it develops a visible tail and appears as a comet. These are called **Jupiter-family comets** because their orbits tend to be influenced strongly by Jupiter's gravitational pull. Jupiter-family comets orbit the Sun in fewer than 20 years. One example is Comet Tempel 1, which has an orbital period of 5.5 years.

There may be tens of thousands of objects in the Kuiper belt that could eventually be perturbed into an elongated orbit and become comets. It is also thought that a few of these refugees from the Kuiper belt have been captured by the gravitational pull of Saturn, becoming that planet's small outer satellites (see Section 13-10).

Short-Period Comets, Long-Period Comets, and the Oort Cloud

The majority of comets are **short-period comets**, with orbital periods between 20 and 200 years, or **long-period comets**, which take more than 200 years to complete one orbit around the Sun. Comet Halley, shown in Figure 15-29a, has an orbital period of 76 years and is a short-period comet; Comet Hyakutake (see Figure 15-27) is a long-period comet that takes about 70,000 years to complete one orbit.

The short-period comets are thought to come from the Kuiper belt (Figure 15-38). Due to their longer periods, the long-period comets are thought to come from much farther out—a reservoir of icy objects that extends from the Kuiper belt to some 50,000 au from the Sun, or about one-fifth of the way to the nearest star. This reservoir, first hypothesized by Dutch astronomer Jan Oort in 1950 and now called the **Oort cloud**, is thought to contain two vast arrangements of comets around the Sun—an inner doughnut-shaped distribution and an outer spherical distribution. The outer spherical distribution was proposed in order to account for the observation that long-period comets can have large orbital inclinations relative to the ecliptic (the orbital plane of the planets).

Because astronomers discover long-period comets at the rate of about one per month, it is reasonable to suppose that there is an enormous population of comets in the Oort cloud. The number of Oort cloud objects greater than 1 km in size range is as high as 5 trillion (5×10^{12}). Only such a large reservoir of comet nuclei would explain why we see so many long-period comets, even though each one takes tens of thousands of years to travel once around its

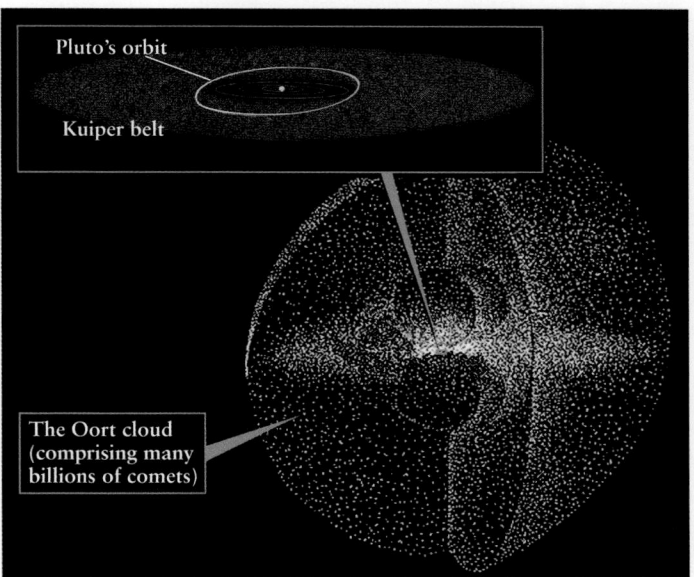

FIGURE 15-38

The Kuiper Belt and Oort Cloud Short-period comets are thought to originate in the Kuiper belt, which begins just beyond the orbit of Neptune at 30 au, and extends to about 50 au from the Sun. Most of the Kuiper belt objects lie in a disk close to the plane of the solar system. The Oort cloud, by contrast, contains a doughnut-shaped distribution in the plane of the ecliptic, as well as a spherical distribution of icy objects whose orbits are randomly oriented relative to the ecliptic. The inner edge of the doughnut-shaped distribution is about 2000 au; the Kuiper belt is so small compared to the Oort cloud that it would appear as a dot on the scale used in this figure. Long-period comets are thought to originate in the Oort cloud. (NASA, JPL)

orbit. Even with trillions of "dirty snowballs" in the Oort cloud, their average separation from each other is about 10 million km. Taken together, the total mass in the Oort cloud is equal to about 5 Earth masses.

Because the Oort cloud is so distant, direct detection of objects in the Oort cloud is a daunting challenge. The best candidate Oort cloud object is the trans-Neptunian object Sedna (Figure 14-23), with an aphelion of 960 au.

The Oort cloud was probably created 4.56 billion years ago from numerous icy planetesimals that orbited the Sun in the vicinity of the newly formed jovian planets. When the outward migration of Saturn, Uranus, and Neptune deflected these icy objects inward toward Jupiter, Jupiter's large gravity catapulted most of these objects clear out of the solar system. However, a small portion of them failed to leave the solar system entirely and formed the Oort cloud (Section 8-6). Gravitational perturbations from nearby stars could have tilted the planes of the orbits in all directions, giving the Oort cloud its spherical shape. Computer simulations even suggest the possibility that the Sun could have captured a substantial fraction of the existing Oort cloud objects from the protoplanetary disks of *other* stars.

Changing a Comet's Orbit

The distinction between long-period, short-period, and Jupiter-family comets can be blurred by the effects of gravitational perturbations. During a return trip toward the Sun, an encounter

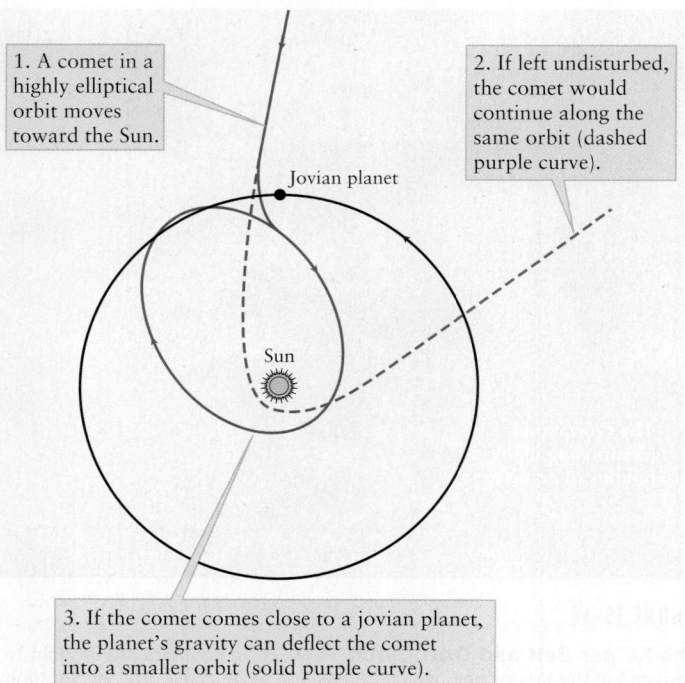

1. A comet in a highly elliptical orbit moves toward the Sun.

2. If left undisturbed, the comet would continue along the same orbit (dashed purple curve).

Jovian planet

Sun

3. If the comet comes close to a jovian planet, the planet's gravity can deflect the comet into a smaller orbit (solid purple curve).

FIGURE 15-39

Transforming a Comet's Orbit The gravitational force of a planet can deflect a comet from a highly elliptical orbit into a less elliptical one. In some cases, a comet ends up with an orbit that keeps it within the inner solar system.

with a jovian planet may force a comet into a much larger orbit. Alternatively, such an encounter can move a long-period comet into a smaller orbit (**Figure 15-39**). Several comets have been perturbed in this way into orbits that always remain within the inner solar system. One example is Comet Tempel, which now orbits between Mars and Jupiter.

Comet Breakup and Meteor Showers

Because they evaporate away part of their mass each time they pass near the Sun, comets cannot last forever. A typical comet may lose about 0.5% to 1% of its ice each time it passes near the Sun. Hence, the ice completely vaporizes after about 100 or 200 perihelion passages, leaving only a swarm of dust and pebbles. Astronomers have observed some comet nuclei in the process of fragmenting (**Figure 15-40**).

> Comets eventually break apart, and their fragments give rise to meteor showers.

One remarkable comet, called Shoemaker-Levy 9, broke apart for a different reason. As the comet swung by Jupiter in July 1992, the giant planet's tidal forces tore the nucleus into more than 20 fragments. Jupiter's gravity then deflected the trajectories of these fragments so that they plummeted into the planet's atmosphere (**Figure 15-41**).

As a comet's nucleus evaporates, residual dust and rock fragments form a **meteoritic swarm**, a loose collection of debris that continues to circle the Sun along the comet's orbit (**Figure 15-42**).

(a)

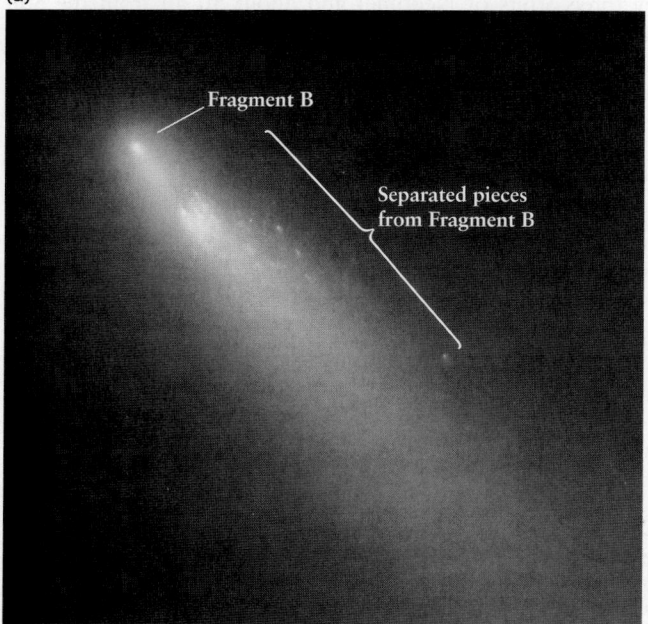

Fragment B

Separated pieces from Fragment B

(b)

FIGURE 15-40 R I ⬤ U X G

The Fragmentation of Comets (a) Comet LINEAR was discovered by the Lincoln Near Earth Asteroid Research (LINEAR) project in 1999. As it passed the perihelion of its orbit in July 2000, Comet LINEAR broke into more than a dozen small fragments. These continue to orbit the Sun along the same trajectory that the comet followed prior to its breakup. This image was made using the Very Large Telescope (see Figure 6-14b). **(b)** This comet, with a 5.4-year orbit, has been coming apart for decades. In 2006, it further fragmented after passing perihelion. One piece, Fragment B, shed at least 30 smaller pieces, shown here. (a: ESO; b: NASA; ESA; H. Weaver [APL/JHU]; M. Mutchler and Z. Levay [STScI])

If Earth's orbit happens to pass through this swarm, a **meteor shower** is seen as the dust particles strike Earth's upper atmosphere. Nearly a dozen meteor showers can be seen each year (**Table 15-1**). Two of the bigger meteor showers—Perseids in summer and Geminids in winter—have almost a meteor per minute, which makes for great viewing.

CONCEPT CHECK 15-10

What motivated the Oort cloud hypothesis of a distant collection of icy objects?

Answer appears at the end of the chapter.

(a) R I **V** U X G

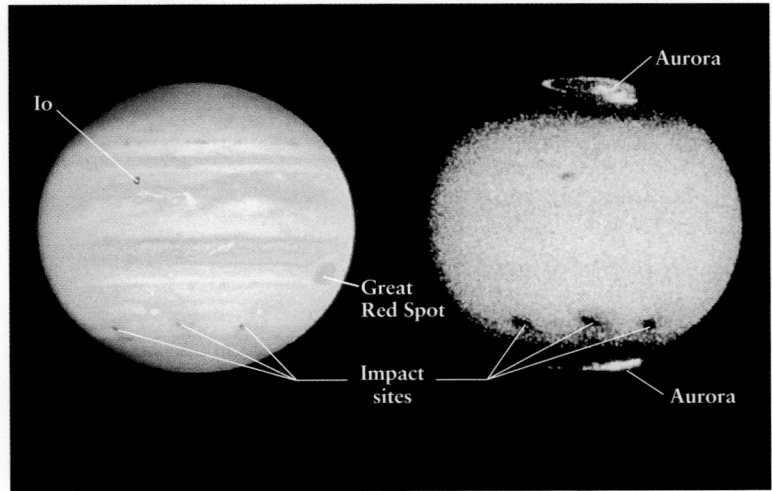

(b) R I **V** U X G   R I V **U** X G

FIGURE 15-41

Comet Shoemaker-Levy 9 and Its Encounter with Jupiter
(a) This comet, originally orbiting Jupiter, was torn apart by the planet's gravitational force on July 7, 1992, fracturing into at least 21 pieces. Its returning debris, shown here in May 1994, struck the planet between July 16 and July 22, 1994. **(b)** Shown here are visible (left) and ultraviolet (right) images of Jupiter taken by the Hubble Space Telescope after three pieces of Comet Shoemaker-Levy 9 struck the planet. Astronomers had expected white remnants (the color of condensing ammonia or water vapor); the darkness of the impact sites may have come from carbon compounds in the comet debris. Note the aurorae in the ultraviolet image. (a: NASA, ESA, and H. Weaver and E. Smith [STScI]; b: JPL/NASA/STScI)

(a)

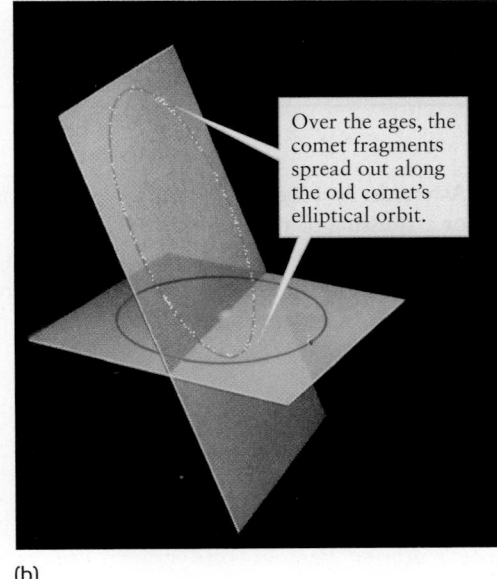

(b)

FIGURE 15-42

Meteoritic Swarms Rock fragments and dust from "burned out" comets continue to circle the Sun. **(a)** The most spectacular meteor showers occur when Earth passes through compact swarms of debris from recently extinct comets. **(b)** Meteor showers associated with older comets are more predictable, because Earth passes through the evenly distributed swarm on each trip around the Sun.

TABLE 15-1 Prominent Yearly Meteor Showers

Shower name	Date of maximum intensity*	Typical hourly rate	Average speed (km/s)	Radiant constellation
Quadrantids	January 3	40	40	Boötes
Lyrids	April 22	15	50	Lyra
Eta Aquarids	May 4	20	64	Aquarius
Delta Aquarids	July 30	20	40	Aquarius
Perseids	August 12	50	60	Perseus
Orionids	October 21	20	66	Orion
Taurids	November 4	15	30	Taurus
Leonids	November 16	15	70	Leo
Geminids	December 13	50	35	Gemini
Ursids	December 22	15	35	Ursa Minor

*The date of maximum intensity is the best time to observe a particular shower, although good displays can often be seen a day or two before or after the maximum. The typical hourly rate is given for an observer under optimum viewing conditions. The average speed refers to how fast the meteoroids are moving when they strike the atmosphere. The radiant indicates which constellation, and thus the direction, the meteors appear to come from.

KEY WORDS

amino acids, p. 454
asteroid, p. 438
asteroid belt, p. 440
carbonaceous chondrite, p. 454
comet, p. 455
differentiated asteroid, p. 454
dust tail, p. 459
fusion crust, p. 452
Hirayama family, p. 444

hydrogen envelope, p. 459
iron meteorite (iron), p. 453
ion tail, p. 459
Jupiter-family comet, p. 461
Kirkwood gaps, p. 440
Kuiper belt, p. 461
long-period comet, p. 461
meteor, p. 452
meteor shower, p. 462
meteorite, p. 452

meteoritic swarm, p. 462
meteoroid, p. 451
minor planet, p. 439
near-Earth object (NEO), p. 445
nucleus (of a comet), p. 456
Oort cloud, p. 461
radiation pressure, p. 459
short-period comet, p. 461
stable Lagrange points, p. 446

stony iron meteorite, p. 453
stony meteorite (stone), p. 452
supernova, p. 455
tail (of a comet), p. 455
Trojan asteroid, p. 446
undifferentiated asteroid, p. 454
Widmanstätten patterns, p. 453

KEY IDEAS

Discovery of the Asteroids: Astronomers first discovered the asteroids while searching for a "missing planet" (Section 15-1).

• Thousands of asteroids with diameters ranging from a few kilometers up to about 1000 km orbit within the asteroid belt between the orbits of Mars and Jupiter.

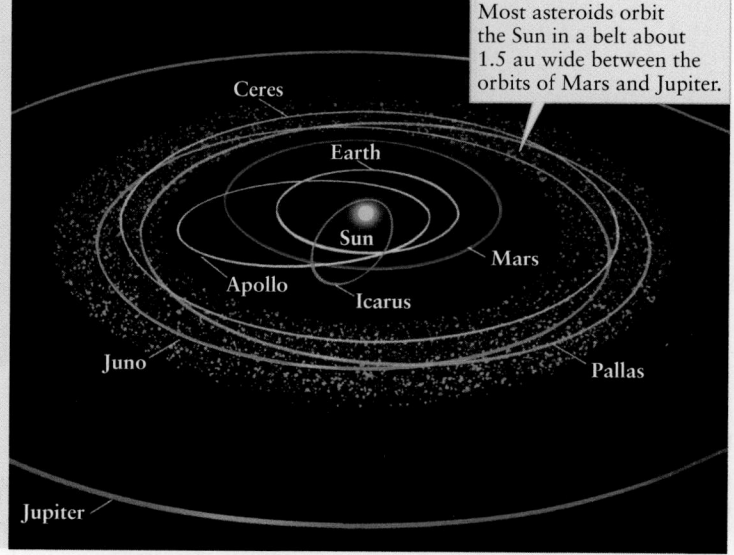

Most asteroids orbit the Sun in a belt about 1.5 au wide between the orbits of Mars and Jupiter.

Origin of the Asteroids: The asteroids are the relics of planetesimals that failed to accrete into a full-sized planet, thanks to the effects of Jupiter and other Mars-sized objects (Section 15-2).

- Even today, gravitational perturbations by Jupiter deplete certain orbits within the asteroid belt. The resulting gaps, called Kirkwood gaps, occur at simple fractions of Jupiter's orbital period.

- Jupiter's gravity also captures asteroids in two locations, called Lagrangian points, along Jupiter's orbit.

Asteroid Collisions: Asteroids collide with one another, causing them to break into smaller fragments (Section 15-4).

- Some asteroids, called near-Earth objects, move in elliptical orbits that cross the orbits of Mars and Earth. If such an asteroid strikes Earth, it forms an impact crater whose diameter depends on both the mass and the speed of the asteroid.

- An asteroid struck Earth 65 million years ago, probably causing the extinction of the dinosaurs and many other species.

Meteoroids, Meteors, and Meteorites: Small rocks in space less than 1 m in size are called meteoroids. If a meteoroid enters Earth's atmosphere, it produces a bright trail called a meteor (or shooting star). A rock that does not burn up in the atmosphere and reaches Earth's surface is called a meteorite; most of the meteoroids are too small to make it to Earth before being completely vaporized (Sections 15-5 and 15-6).

- Meteorites are grouped into three major classes, according to composition: iron, stony iron, and stony meteorites. Irons and stony irons are fragments of the core of an asteroid that was large enough and hot enough to have undergone chemical differentiation, just like a terrestrial planet. Some stony meteorites come from the crust of such differentiated meteorites, while others are fragments of small asteroids that never underwent differentiation.

© Photo by Daniel Ball/School of Earth and Space Exploration/ASU. Courtesy of the Center for Meteorite Studies

Copyright 2014 Symeon Platts/Arizona Board of Regents

The Natural History Museum/Alamy Stock Photo

- Rare stony meteorites called carbonaceous chondrites may be relatively unmodified material from the solar nebula. These meteorites often contain organic material and may have provided building blocks for the origin of life on Earth.

- Analysis of isotopes in carbonaceous chondrites suggests that a nearby supernova may have triggered the formation of the solar system 4.56 billion years ago.

Courtesy of J. A. Wood

Comets: A comet is a chunk of ice with embedded rock fragments that generally moves in a highly elliptical orbit about the Sun (Section 15-7).

- As a comet approaches the Sun, its icy nucleus develops a luminous coma, surrounded by a vast hydrogen envelope. An ion tail and a dust tail extend from the comet, pushed away from the Sun by the solar wind and radiation pressure, respectively.

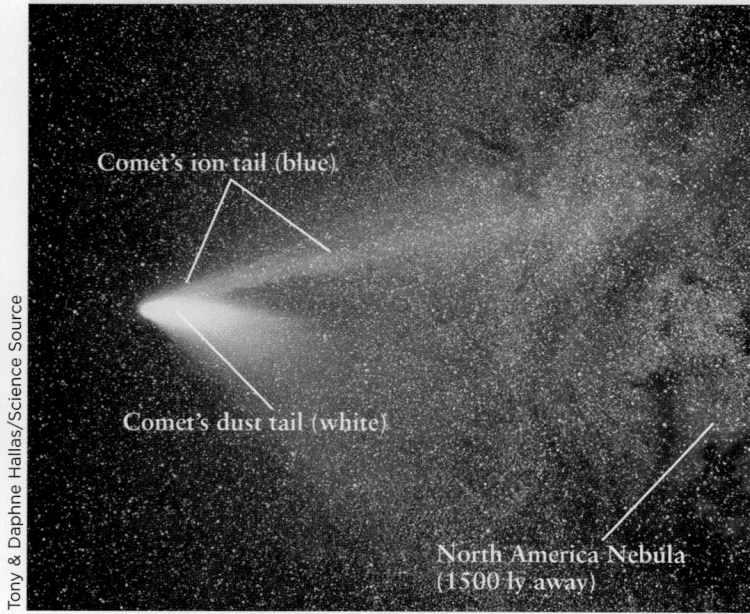

Comet's ion tail (blue)

Comet's dust tail (white)

North America Nebula (1500 ly away)

Tony & Daphne Hallas/Science Source

Origin and Fate of Comets: Comets are thought to originate from two regions, the Kuiper belt and the Oort cloud (Section 15-8).

- The Kuiper belt lies in the plane of the ecliptic at distances between 30 and 50 au from the Sun. Its icy objects are thought to be the source of short-period comets, whose periods are less than 200 years.

- The Oort cloud is thought to contain billions of icy objects in a spherical distribution that extends out to 50,000 au from the Sun. The long-period comets, with periods greater than 200 years, are thought to originate in the Oort cloud.

- Dust from comet tails due to evaporation of ice by the Sun produces meteoritic swarms. A meteor shower is seen when Earth passes through the trail of dust left behind by the comet.

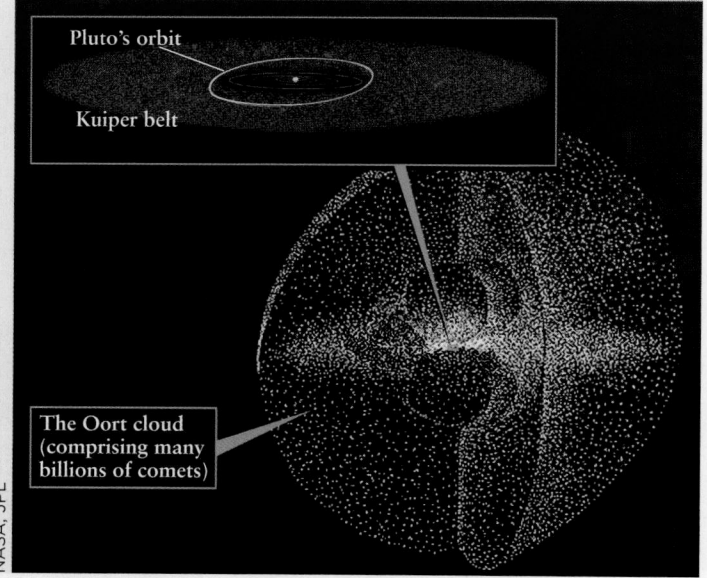

Pluto's orbit

Kuiper belt

The Oort cloud (comprising many billions of comets)

NASA, JPL

QUESTIONS

> ### Problem-solving tips and tools
>
> The small-angle formula is described in Box 1-1. We discussed retrograde motion in Section 4-1 and described its causes in Section 4-2. You will need to use Kepler's third law, described in Section 4-4 and Box 4-2, in some of the problems below. Box 7-2 discusses the concept of escape speed. A spherical object of radius r intercepts an amount of sunlight proportional to its cross-sectional area, equal to πr^2. The volume of a sphere of radius r is $4\pi r^3/3$.

Section 15-1

1. What led astronomers to suspect that there were members of the solar system that orbit between Mars and Jupiter?

2. How did the first asteroid come to be discovered? What role did theoretical calculations play in confirming the discovery? How did this discovery differ from what astronomers had expected to find?

3. How do modern astronomers discover new asteroids?

4. Describe the asteroid belt. Does it lie completely within the plane of the ecliptic? What are its inner and outer radii?

5. When Olbers discovered Pallas in March 1802, the asteroid was moving from east to west relative to the stars. At what time of night was Pallas highest in the sky over Olbers's observatory? Explain your reasoning.

6. Discuss the idea that Ceres should be regarded as the smallest dwarf planet rather than the largest asteroid. What are the advantages of this scheme? What are the disadvantages?

7. In the 1998 movie *Armageddon,* an asteroid "the size of Texas" is on a collision course with Earth. The asteroid is first discovered by astronomers just 18 days prior to impact. To avert disaster, a team of astronauts blasts the asteroid into two pieces just 4 hours before impact. Discuss the plausibility of this scenario. (*Hint:* On average, the state of Texas extends for about 750 km from north to south and from east to west. How does this compare with the size of the largest known asteroids?)

Section 15-2

8. If Jupiter was not present in our solar system, would the asteroid belt exist? Why or why not?

9. What are Kirkwood gaps? What causes them?

10. Consider the early solar system when Jupiter's semimajor axis changed as the planet migrated. Would the effect of the unstable orbital resonances (called Kirkwood gaps in the present-day asteroid belt) lead to more disruption of the original belt objects, or less?

11. Compare the explanation of the Kirkwood gaps in the asteroid belt to the way in which Saturn's moons help produce divisions in that planet's rings (see Section 12-11).

12. Consider the Kirkwood gap whose orbital period is two-fifths of Jupiter's 11.86-year period. Calculate the distance from the Sun to this gap. Does your answer agree with Figure 15-4?

Section 15-3

13. How is it possible to tell that some asteroids are nonspherical even though we do not have images of those asteroids?

14. **TUTORIAL 15.1** What is the evidence that some asteroids are made of a loose conglomeration of smaller pieces?

15. What evidence links some meteorites on Earth to the asteroid Vesta?

16. The asteroid 243 Ida, which was viewed by the Galileo spacecraft, is a member of a Hirayama family. Discuss what this tells us about the history of this asteroid. Where might you look to find other members of the same family?

17. What explains the bright spots on Ceres in Figure 15-11?

18. What do Ceres and Pluto have in common?

19. When this image was made, the asteroid Ceres was 1.63 au, or 2.44×10^8 km, from the Hubble Space Telescope. (a) What was the angular size of the asteroid as a whole? (b) You can see individual pixels in the image shown here. Using a ruler and the scale bar in the figure, determine how many kilometers on the surface of Ceres are contained in the width of one pixel. (c) What is the angular width of each pixel in arcseconds?

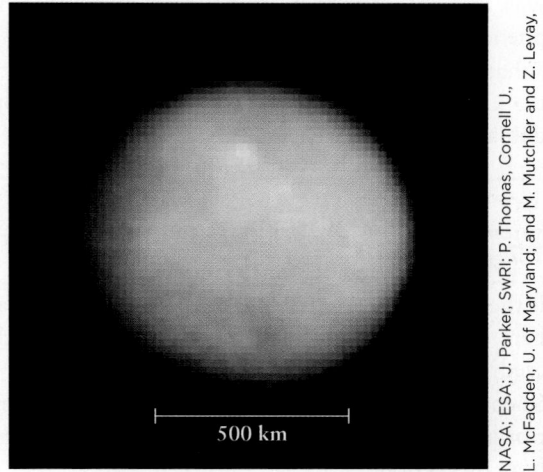

500 km

NASA: ESA: J. Parker, SwRI; P. Thomas, Cornell U., L. McFadden, U. of Maryland; and M. Mutchler and Z. Levay, STScI

20. Suppose that a binary asteroid (two asteroids orbiting each other) is observed in which one member is 16 times brighter than the other. Suppose that both members have the same albedo and that the larger of the two is 120 km in diameter. What is the diameter of the other member?

21. The accompanying image from the Galileo spacecraft shows the asteroid 243 Ida, which has dimensions $56 \times 24 \times 21$ km. Galileo discovered a tiny moon called Dactyl, just 1.6×1.2 km in size, which orbits Ida at a distance of about 100 km. (In Greek mythology, the Dactyli were beings who

lived on the slopes of Mount Ida.) Describe a scenario that could explain how Ida came to have a moon.

Ida

Dactyl

(NASA/JPL)

R I **V** U X G

Section 15-4

22. What are near-Earth objects? What is the evidence that Earth has been struck by these objects?

23. What is the evidence that an asteroid impact could have contributed to the demise of the dinosaurs?

24. For many years, it was thought that the Tunguska event was caused by a comet striking Earth. This idea was criticized because a small comet would have broken up too high in the atmosphere to cause significant damage on the ground. Explain why, using your knowledge of a comet's structure.

25. Summarize the evidence suggesting that the Tunguska event was due to an impact, as opposed to some other natural phenomena.

26. What was the greatest source of damage to structures from the Chelyabinsk asteroid? Why did so many people get eye damage from this event?

Section 15-5

27. **TUTORIAL 15.2** What is the difference between a meteoroid, a meteor, and a meteorite?

28. Is there anywhere on Earth where you might find large numbers of stony meteorites that are not significantly weathered? If so, where? If not, why not?

29. Scientists can tell that certain meteorites came from the interior of an asteroid rather than from its outer layers. Explain how this is done.

30. Why are some asteroids differentiated while others are not?

31. Suppose you found a rock you suspect might be a meteorite. Describe some of the things you could do to determine whether it was a meteorite or a "meteorwrong."

32. Use the percentages of stones, irons, and stony iron meteorites that fall to Earth to estimate what fraction of their parent asteroids' interior volume consisted of an iron core. Assume that the percentages of stones and irons that fall to Earth indicate the fractions of a parent asteroid's interior volume occupied by rock and iron, respectively. How valid do you think this assumption is?

Section 15-6

33. What is the evidence that carbonaceous chondrites are essentially unaltered relics of the early solar system? What do they suggest about how the solar system may have formed?

34. What is a possible role that asteroids might have played in the origins of life on Earth?

35. How do some astronomers explain the high abundance of ^{26}Mg in the Allende meteorite?

Section 15-7

36. **TUTORIAL 15.3** With the aid of a drawing, describe the structure of a comet.

37. Why is the phrase "dirty snowball" an appropriate characterization of a comet's nucleus?

38. What process produces jets that stream off of a comet's nucleus?

39. What did we learn from the Stardust mission about the possible connection between comets and the origins of life on Earth?

40. What did scientists learn about the structure of comet nuclei from the Deep Impact mission?

41. Why do the ion tail and dust tail of a comet point in different directions?

42. Why do comets have prominent tails for only a short time during each orbit?

43. On March 8, 1997, Comet Hale-Bopp was 1.39 au from Earth and 1.00 au from the Sun. Use this information and that given in the caption to Figure 15-35 to estimate the length of the comet's ion tail on that date. Give your answer in kilometers and astronomical units.

44. *Sun-grazing comets* come so close to the Sun that their perihelion distances are essentially zero. Find the orbital periods of Sun-grazing comets whose aphelion distances are (**a**) 100 au, (**b**) 1000 au, (**c**) 10,000 au, and (**d**) 100,000 au. Assuming that these comets can survive only 100 perihelion passages, calculate their lifetimes. (*Hint:* Remember that the semimajor axis of an orbit is one-half the length of the orbit's long axis.)

45. Comets are generally brighter a few weeks after passing perihelion than a few weeks before passing perihelion. Explain why might this be. (*Hint:* Water, including water ice, does an excellent job of retaining heat.)

46. The hydrogen envelopes of comets are especially bright at an ultraviolet wavelength of 122 nm. Use Figure 5-24 to explain why.

47. A *very* crude model of a typical comet nucleus is a cube of ice (density 1000 kg/m^3) 10 km on a side. (**a**) What is the mass of this nucleus? (**b**) Suppose 1% of the mass of the nucleus evaporates away to form the comet's tail. Suppose further that the tail is 100 million (10^8) km long and 1 million (10^6) km wide. Estimate the average density of the tail (in kg/m^3). For comparison, the density of the air you breathe is about 1.2 kg/m^3. (**c**) In 1910, Earth actually passed through the tail

of Comet Halley. At the time, there was some concern among the general public that this could have deleterious effects on human health. Was this concern justified? Why or why not?

48. What is the relationship between the Kuiper belt and comets?

49. What is the Oort cloud? How might it be related to planetesimals left over from the formation of the solar system?

50. Why are comets more likely to break apart at perihelion than at aphelion?

51. Why do astronomers think that meteorites come from asteroids, while meteor showers are related to comets?

52. Use Kepler's third law to calculate the orbital period of comets just beyond Neptune at 35 au. Then use Kepler's third law to calculate Comet Halley's average distance from the Sun. Explain why Comet Halley's period is much shorter than short-period comets in the Kuiper belt. (*Hint:* See Figure 15-37.)

53. TUTORIAL 15.1 **Box 15-1** What are the Trojan asteroids, and where are they located? What holds them in this location?

54. **Box 15-1** Imagine that you are an astronaut standing on the surface of a Trojan asteroid. How will you see the phase of Jupiter change with the passage of time? How will you see Jupiter move relative to the distant stars? Explain your answers.

General Problems

55. What are the differences between asteroids and trans-Neptunian objects?

56. Why is it that Jupiter and Saturn can be seen in the night sky every year, while seeing specific comets such as Halley and Hyakutake is a once-in-a-lifetime event?

57. Why are asteroids, meteorites, and comets all of special interest to astronomers who want to understand the early history and subsequent evolution of the solar system?

58. Assume that Ida's tiny moon Dactyl (see Question 21) has a density of 2500 kg/m^3. (**a**) Calculate the mass of Dactyl in kilograms. For simplicity, assume that Dactyl is a sphere 1.4 km in diameter. (**b**) Calculate the escape speed from the surface of Dactyl. If you were an astronaut standing on Dactyl's surface, could you throw a baseball straight up so that it would never come down? Professional baseball pitchers can throw at speeds around 40 m/s (140 km/h, or 90 mi/h); your throwing speed is probably a bit less.

59. From the abundance of craters on the Moon and Mercury, we know that numerous asteroids and meteoroids struck the inner planets early in the history of our solar system. Is it reasonable to suppose that numerous comets also pelted the planets 3.5 to 4.5 billion years ago? Speculate about the effects of such a cometary bombardment, especially with regard to the evolution of the primordial atmospheres on the terrestrial planets.

60. Suppose astronomers discover that a near-Earth object the size of 1994 XM1 is on a collision course with Earth. Describe what humanity could do within the framework of present technology to counter such a catastrophe.

WEB/E-BOOK QUESTIONS

61. Search the internet for information about the present status of NASA's Dawn mission. Describe one result from Vesta that is not described in this textbook.

62. Search the internet to find out why some scientists disagree with the idea that a tremendous impact led to the demise of the dinosaurs. (They do not dispute that the impact took place, only what its consequences were.) What are their arguments? From what you learn, what is your opinion?

63. Several scientific research programs are dedicated to the search for near-Earth objects (NEOs), especially those that might someday strike our planet. Search the internet for information about at least one of these programs. How does this program search for NEOs? How many NEOs has this program discovered? Will any of these pose a threat in the near future?

64. VIDEO 15.4 **Estimating the Speed of a Comet.** Access and view the video "Two Comets and an Active Sun" in Chapter 15 of the *Universe* website or e-book. (**a**) Why don't the comets reappear after passing the Sun? (**b**) The white circle shows the size of the Sun, which has the diameter 1.39×10^6 km. Using this to set the scale, step through the video and measure how the position of one of the comets changes. Use your measurements and the times displayed in the video to estimate the comet's speed in km/h and km/s. (Assume that the comet moved in the same plane as that shown in the video.) As part of your answer, explain the technique and calculations you used. (**c**) How does your answer in (b) compare with the orbital speed of Mercury, the innermost and fastest-moving planet (see Table 11-1)? Why is there a difference? (**d**) If the comet's motion was not in the same plane as that shown in the video, was its actual speed more or less than your estimate in (b)? Explain your answer.

ACTIVITIES

Observing Project

Observing tips and tools

Meteors: Informative details concerning upcoming meteor showers appear on the websites for *Sky & Telescope* and *Astronomy* magazines.

Comets: Because astronomers discover dozens of comets each year, there is usually a comet visible somewhere in the sky. Unfortunately, they are often quite dim, and you will need to have access to a moderately large telescope (at least 35 cm, or 14 in.). You can get up-to-date information from the Minor Planet Center website. Also, if there is an especially bright comet in the sky, useful information about it might be found at the websites for *Sky & Telescope* and *Astronomy* magazines.

Asteroids: At opposition, some of the largest asteroids are bright enough to be seen through a modest telescope. Check the Minor Planet Center website to see if any bright asteroids are near opposition. If so, check the current issue as well as the most recent January issue of *Sky & Telescope* magazine for a star chart showing the asteroid's path among the constellations. You will need such a chart to distinguish the asteroid from background stars. Also, you can locate Ceres, Pallas, Juno, and Vesta using the *Starry Night*™ program if you have access.

65. Make arrangements to view a meteor shower. Table 15-1 lists the dates of major meteor showers. Choose a shower that will occur near the time of a new moon. Set your alarm clock for the early morning hours (1 to 3 A.M.). Get comfortable in a reclining chair or lie on your back so that you can view a large portion of the sky. Record how long you observe, how many meteors you see, and what location in the sky they seem to come from. How well does your observed hourly rate agree with published estimates, such as those in Table 15-1? Is the radiant of the meteor shower apparent from your observations?

66. If a comet is visible with a telescope at your disposal, make arrangements to view it. Can you distinguish the comet from background stars? Can you see its coma? Can you see a tail?

67. Make arrangements to view an asteroid. Observe the asteroid on at least two occasions, separated by a few days. On each night, draw a star chart of the objects in your telescope's field of view. Has the position of one starlike object shifted between observing sessions? Does the position of the moving object agree with the path plotted on published star charts? Do you feel confident that you have in fact seen the asteroid?

Starry Night™ Explorations

68. Use *Starry Night*™ to observe the motions of two smaller objects in the solar system, the dwarf planet Ceres, and the asteroid Pallas as seen from Earth. Open **Favorites > Explorations > Ceres and Pallas**. The view shows the sky from the center of a transparent Earth, looking along the ecliptic plane, shown as a green line. The positions of the dwarf planet Ceres and the asteroid Pallas in the sky are shown. Click the name in the label for Ceres and select **Center**. Click **Play** and observe the motion of Ceres over the course of at least two calendar years.

(a) Describe the motion of Ceres across the sky, including its direction of motion relative to the background stars and its proximity to the ecliptic. Keep in mind that the left side of the view is eastward and that the background stars will appear to move in a direction opposite to the motion of Ceres because the view is centered and locked onto Ceres.

(b) Select **File > Revert** from the menu and now center the view on Pallas. Click **Play** to observe the motion of this asteroid over a period of at least two calendar years. How does the motion of Pallas compare with that of Ceres?

(c) Of the two objects Ceres and Pallas, which has the more steeply inclined orbit with respect to the orbit of Earth?

(d) Based on your observations of their motions in the sky as seen from Earth, do you expect the orbits of Ceres and Pallas around the Sun to be of similar or very different sizes? Why?

(e) One distinguishing feature that identifies a dwarf planet from an asteroid is shape. A dwarf planet is approximately spherical in shape, whereas an asteroid has a more irregular shape. Select **File > Revert** to return to the initial view and then click the name on the label for Ceres and select **Magnify** from the menu to see the appearance of this object. Click **Play** to see changes in appearance as the dwarf planet rotates. Now return to the initial view (**File > Revert**) and **Magnify** the object Pallas. Adjust the time controls to see the whole of the asteroid's surface. Describe the appearance of each of these objects. Is their appearance consistent with their classification?

69. Use *Starry Night*™ to locate the largest asteroid Ceres, now designated a dwarf planet, and to examine the asteroid belt. Select **Favorites > Explorations > Asteroids**. The view, centered on the Sun, shows the orbits of the major planets and that of the dwarf planet Ceres, moving among the rich field of objects that make up the asteroid belt. Use the location scroller to examine the extent of the asteroid belt and observe the orbits of some of the asteroids from different perspectives, particularly along the plane of Earth's orbit, the ecliptic plane, and perpendicular to this plane.

(a) Between the orbits of which two major planets is the major concentration of asteroids?

(b) How would you describe the plane of Ceres's orbit relative to the ecliptic plane?

(c) Are the orbits of the asteroids generally concentrated on the ecliptic plane and close to the orbital planes of the major planets, or do their orbits carry them well above and below the ecliptic plane?

70. Use *Starry Night*™ to explore some of the dwarf planets of the solar system. Select **Favorites > Explorations > Dwarf Planets** from the menu. This view, from a position in space 100 au from the Sun, shows the orbit of Neptune as an example of the orbits of the major planets of the solar system, as well as the orbits of several dwarf planets. Click **Play** and use the location scroller to explore this view from different angles.

(a) Do the dwarf planets revolve about the Sun in the same direction as the major planets or in the opposite direction?

(b) Adjust the view so that the plane of Neptune's orbit appears edge on. How do the orbital planes of the dwarf planets compare to those of the major planets?

71. Use *Starry Night*™ to study the motion and orbits of comets. Open **Favorites > Explorations > Comet SOHO**. The view shows the Comet SOHO (P/2008 Y12) along with the position of Earth in its orbit, from a stationary position in space directly above the Sun. Click **Play** and watch the changing position and appearance of the comet as it approaches and recedes from the Sun over a few days in October 2019. Select **File > Revert** to return to the original view and use the location scroller, zoom controls, elevation buttons, and time-flow controls to view the animation from different perspectives and for longer time spans.

    (a) In which direction does the tail of this comet point with respect to the Sun?

    (b) Adjust the date and time controls to place the comet at its closest point of approach to the Sun, known as perihelion. What is the time of perihelion for this comet in October 2019?

    (c) Select **File > Revert** and click on the label for Comet SOHO (P/2008 Y12). In the pop-up menu that appears, select **Orbit** to show the complete orbit of this comet. Manipulate the view, using the location scroller, zoom controls, and elevation buttons to study this orbit. Is the comet's orbit in about the same plane as Earth's orbit or is it steeply inclined to that plane?

    (d) How does the shape of this comet's orbit compare to the shape of the orbit of Earth?

    (e) How does the comet's speed vary as it moves along its orbit?

    (f) Select **File > Revert** and click the **Quick Search Items** button in the toolbar to open the **Planets...** panel to expand **Our Solar System > Comets**. Click the circular **Orbit** icon next to several comets in the list to display the orbits of these comets in the main view. Use the view controls to study these orbits. In general, how do the orbits of comets differ from the orbits of the planets?

72. Use *Starry Night*™ to explore the outer regions of the solar system, from which most comets appear to originate. Open **Favorites > Explorations > Oort Cloud** to see the orbits of the major planets from a position 60 au away from the Sun. You will also see two spheres centered on the Sun, which represent the inner and outer boundaries of the region in the solar system where the Kuiper Belt is found. This belt is in fact pancake-shaped, with most objects moving in orbits close to the ecliptic plane. These rocky and icy objects were probably formed closer to the Sun where the density of the solar nebula was relatively high but migrated outward to occupy these orbits early in the evolution of the solar system.

    (a) Use the elevation control to move inward until you are passing through the outer boundary. What is the approximate radius of this outer boundary, as given by your distance from the Sun?

    (b) What type of comet originates from this belt?

    (c) Use the location scroller to view the inner boundary of the Kuiper belt region. Which major planet has its orbit at about the distance of this inner boundary?

    (d) Lower the elevation again until you appear to pass though this inner boundary. What is the distance of this boundary from the Sun?

    (e) Now use the elevation control to move much farther from the Sun, where you will encounter the outer boundary of the much larger Oort cloud. At what distance is this outer boundary?

    (f) Compare this distance with that to the nearest star. At what fraction of this distance to this star do we find the outer regions of the Oort cloud?

    (g) What type of comet arises from this region of the solar system?

ANSWERS

ConceptChecks

ConceptCheck 15-1: The solar system is filled with planets that have semimajor axes increasing from one to the next by about a factor between 1.4 and 2. The exception to this pattern is Jupiter, but the pattern would hold up if there was a planet where the asteroid belt currently exists.

ConceptCheck 15-2: Ceres is an asteroid and also a dwarf planet—large enough to pull itself into a sphere but unable to gravitationally clear out its surroundings.

ConceptCheck 15-3: Jupiter's gravity prevented a planet from forming where the asteroid belt currently exists. The planet would have probably been about the size of Earth, but mass was ejected and all the mass in the asteroid belt today could only form a very small object 1500 km in diameter (less than the distance from Los Angeles to Chicago).

ConceptCheck 15-4: The Hayabusa mission returned samples of the asteroid Itokawa, where analysis on Earth showed a match to meteoritic dust.

ConceptCheck 15-5: The Dawn spacecraft was able to assess Vesta's composition by measuring the infrared spectra of the asteroid's surface. Analysis finds a match with about 6% of the meteorites found on Earth.

ConceptCheck 15-6: Iridium is common in asteroids, but rare in Earth's rocks. A thick layer of iridium is found around the world dating back 65 million years, implying a massive impact at that time. More evidence comes from a massive impact event dated at 65 million years old near the Yucatán Peninsula.

ConceptCheck 15-7: If a nickel–iron sample cools too quickly, it will not contain a Widmanstätten pattern, which takes millions of years to form. This is not a feature that can be artificially produced on Earth.

ConceptCheck 15-8: The buildup of larger objects in the formation of the solar system involved heating (due to radioactivity and collisions) leading to alteration of their chemical composition. However, the carbonaceous chondrites do not show chemical alteration that would have occurred if they were heated. Therefore, this class of meteorites appears to contain some original unaltered material present in the early solar nebula.

ConceptCheck 15-9: The Deep Impact mission measured the density of Comet Tempel 1 and found it be significantly less dense than water—only 60%. Because the comet is mostly water (with smaller amounts of other material), the comet must be porous with empty spaces to have a density lower than water.

ConceptCheck 15-10: From Kepler's third law, the longer the orbital period, the larger the semimajor axis of the orbit.

The short-period comets already orbit beyond Neptune in the Kuiper belt, and the long-period comets require a collection of objects much farther out; this is called the Oort cloud.

CalculationChecks

CalculationCheck 15-1: The size of the crater formed in an impact depends on the size of the fragment hitting Earth and how fast it is moving, as well as the kind of material being struck. If the crater can be as much as 20 times the size of the fragment, then the smallest the fragment might be is (2000 m)/20 = 100 m across.

CalculationCheck 15-2: 5 km. (5 km/s) × (1 s) = 5 km.

Pluto and the Kuiper Belt

The "what is a planet" controversy began with the discovery of Ceres and several other objects between Mars and Jupiter in the early 1800s. At first, Ceres and then Pallas, Juno, and Vesta were considered planets, even though they were orders of magnitude less massive than the other known planets and all had similar orbits. The definition of a planet at this time was simple. Any object in orbit about the Sun that did not show cometary effects was a planet.

By the mid-1800s, the discovery of more and more objects between Mars and Jupiter saw the words "asteroid" and "minor planet" start to be used in order to signify these objects as an ensemble of bodies. That is, astronomers felt a major planet should be a rare and unique thing. The major planet club thus was left with eight members: Mercury, Venus, Earth, Mars, Jupiter, Saturn, Uranus, and Neptune.

The planet controversy arose once again in 1930 with the discovery of Pluto, which was quickly regarded as the ninth planet. During the ensuing years, new observations downgraded Pluto's size from being near that of Earth to being smaller than our Moon. Its orbit also proved to be peculiar because it was highly inclined and eccentric, having it even crossing Neptune's orbit. Its uniqueness kept Pluto in the planet club with relatively few detractors until 1992, when the first Kuiper belt object was discovered. As it became apparent that hundreds of thousands of Kuiper belt objects existed on stable orbits just beyond Neptune, Pluto's planet status was becoming less and less certain. With the discovery of Eris (2003 UB313) in 2003, an object larger than Pluto, Pluto's last unique claim of being the largest Kuiper belt object was lost. Thus, the planet controversy had finally reached a point where a major planet's minimum size must be precisely defined.

No matter how you count them, there are no longer nine major planets in our solar system. The same situation of the main belt asteroids in the mid-1800s is now apparent with Pluto, Eris, and the rest of the Kuiper belt objects. Although the largest Kuiper belt objects are significantly bigger than the largest main belt asteroids, they are still an ensemble of bodies with a continuous size distribution and similar formation and evolution histories. The reason Pluto's status as a planet is a controversy now is simply because it took 62 years between the discovery of Pluto and the next Kuiper belt object in 1992. Within that time span, Pluto became known as the ninth planet through several generations and was ingrained in society as such.

Classification is not that important to scientists, but it is the understanding of how the object formed and its history. Even so, any classification should have a scientific base and not be based on tradition, as seems to be the case for Pluto in the recent past. Eris as well as Pluto and the rest of the Kuiper belt objects had very similar histories and formation scenarios and thus they should be considered as an ensemble of objects just like the main belt asteroids between Mars and Jupiter. As technology advances, our knowledge and thus understanding advances and so should our classification.

Several minimum-size planet definitions were debated recently. The first is that anything larger than a certain minimum size is a planet. The chosen size would likely be arbitrary and may be a nice round number like 1000 km or the size of Pluto. This stems from the traditional belief that Pluto has been considered a planet in the past and should continue as such. This would give us 10 current planets including Eris. A second possible definition is that anything that is spherical should be a planet. This definition is based on physics in which the gravitational force of an object is large enough to overcome any other forces in determining the overall shape of the object. Determining if an object meets this criteria or not is complicated. This definition would also significantly change the status quo by increasing the number of known planets by several factors overnight. A third definition is that any object with a unique orbit and that gravitationally dominates its local orbital environment should be a planet. This definition is the most scientifically based, in that objects with similar formation and evolutionary histories would be grouped together. Using this definition would give us eight known major planets, dropping Pluto from the list.

A final proposal as to what is a planet takes into account that the current known major planets themselves are remarkably different and can be split into individual categories themselves. Mercury, Venus, Earth, and Mars are terrestrial planets whose compositions are dominated by rock. Jupiter and Saturn are gas giant planets dominated by their hydrogen and helium envelopes. Uranus and Neptune are ice giant planets dominated by gases other than hydrogen and helium. The trans-Neptunian objects or "ice dwarf planets," as some are calling them, are probably composed of large amounts of volatiles such as methane ice and water ice.

The International Astronomical Union (IAU) recently chose to use a combination of the second and third definitions above to define a planet. That is, a planet is an object that is both spherical and has a unique orbit in which it is gravitationally dominant. An object that only satisfies definition two above and not three is now being called a dwarf planet, of which Ceres, Pluto, and Eris qualify, as well as several more objects in the Kuiper belt. In reality, it is the public that must accept

Scott S. Sheppard is an astronomer at the Carnegie Institution of Washington in the Department of Terrestrial Magnetism located in Washington, DC. His main research currently involves discovering small bodies in our solar system to understand their dynamical and physical properties that help constrain solar system and planet formation. He has discovered several small moons around all the giant planets, been a pioneer in the discovery of Neptune Trojans, and found tens of Kuiper Belt objects.

(continued on the next page)

this definition, and we will only know if that is the case a few generations from now.

Defining the word "planet" is like defining what an ocean is. At first it appears to be a simple term, but is very hard to precisely define. Further planet discoveries will be made and there are sure to be borderline cases. This is just the way nature

is; it does not have little bins to nicely classify objects, but there is usually a continuous array of objects. The important thing to take away from all of this is to understand this array of objects by using our rapidly expanding scientific knowledge and understanding our place in the solar system.

16 | Our Star, the Sun

LEARNING GOALS

T he Sun is by far the brightest object in the sky. By earthly standards, the temperature of its glowing surface is remarkably high, about 5800 K (nearly 10,000°F), but there are regions of the Sun that reach far higher temperatures of tens of thousands or even millions of kelvins. Gases at such temperatures emit ultraviolet light or x rays, which makes the Sun's outer regions prominent in the accompanying image from an ultraviolet telescope in space. Far from being smooth and placid, the Sun is a dynamic object with many features. Some of the hottest and most energetic regions on the Sun spawn immense disturbances and can propel solar material across space to reach Earth and other planets.

In recent decades, we have learned that the Sun shines because at its core hundreds of millions of tons of hydrogen are converted to helium every second. We have confirmed this picture by detecting the by-products of this transmutation—strange, ethereal particles called *neutrinos*—streaming outward from the Sun into space. We have discovered that the Sun has a surprisingly violent atmosphere, with a host of features such as sunspots whose numbers rise and fall on a predictable 11-year cycle. By studying the Sun's vibrations, we have begun to probe beneath its

This ultraviolet image of the Sun, taken in 2010 by the Solar Dynamics Observatory, shows a prominence (upper left) in which magnetic fields carry gases above the Sun's surface in a loop that can extend upward hundreds of thousands of kilometers. (NASA/GSFC/AIA)

R I V U X G

surface into hitherto unexplored realms. And we have just begun to investigate how changes in the Sun's activity can affect Earth's environment as well as our technological society.

16-1 The Sun's energy is generated by thermonuclear reactions in its core

TUTORIAL
16.1

The Sun is the largest member of the solar system. It has almost 1000 times more mass than all of the planets, moons, asteroids, comets, and meteoroids put together. But the Sun is also a star. In fact, it is a remarkably typical star, with a mass, size, surface temperature, and chemical composition that are roughly midway between the extremes exhibited by the myriad other stars in the heavens. Table 16-1 lists essential data about the Sun.

Solar Energy

For most people, what matters most about the Sun is the energy that it radiates into space. Without the Sun's warming rays, our oceans would freeze. Furthermore, since most life is part of a food chain dependent on sunlight for energy, an Earth without sunlight would be nearly dead. To understand the nature of life on Earth, we must understand the Sun.

The Sun's surface is hot enough to emit visible light.

Why is the Sun such a strong source of energy? Recall that the hotter an object is, the more light it emits (see Figure 5-9). The Sun's spectrum is close to that of an idealized blackbody with a temperature of 5800 K (see Sections 5-3 and 5-4, especially Figure 5-13). Thanks to this high temperature, each square meter of the Sun's surface emits a tremendous amount of radiation, principally at visible wavelengths. Indeed, the Sun is the only object in the solar system that *emits* substantial amounts of visible light. The surface temperatures of the Moon and planets are very low compared to the Sun; the visible light that we see from bodies in the solar system is actually sunlight that struck those worlds and was *reflected* toward Earth. It is extremely helpful in astronomy to see sunlight, but it is no coincidence; our eyes evolved to see the predominant form of light around us.

The bright Sun emits a tremendous amount of energy. The total amount of energy emitted by the Sun each second, called its **luminosity**, is about 3.9×10^{26} watts, or joules of energy emitted per second. (We discussed the relationship between the Sun's surface temperature, radius, and luminosity in Box 5-2.) Astronomers denote the Sun's luminosity by the symbol $L_\odot$. A circle with a dot in the center, once used by ancient astronomers, is the astronomical symbol for the Sun.

Does the Sun emit so much energy because it is hot or because it is big? To put the Sun's temperature and size into perspective, we can make a comparison with ovens and lightbulbs. The solar surface is about 10 times hotter than the maximum temperature of a household oven. A factor of 10 might not sound like a great amount between the Sun and an oven, but we saw in Section 5-4, for the Stefan–Boltzmann law, F (energy flux) $= \sigma T^4$, that raising an object's temperature 10 times will increase the energy it emits per second by a factor of 10,000! But the Sun is also much bigger than an oven, and to appreciate the effect of size, let us consider lightbulbs.

TABLE 16-1 Sun Data

Distance from Earth:	Mean: 1 au = 149,598,000 km
	Maximum: 152,000,000 km
	Minimum: 147,000,000 km
Light travel time to Earth:	8.32 min
Mean angular diameter:	32 arcmin
Radius:	696,000 km = 109 Earth radii
Mass:	1.9891×10^{30} kg = 3.33×10^5 Earth masses
Composition (by mass):	74% hydrogen, 25% helium, 1% other elements
Composition (by number of atoms):	92.1% hydrogen, 7.8% helium, 0.1% other elements
Mean density:	1410 kg/m^3
Mean temperature:	Surface: 5800 K; Center: 1.55×10^7 K
Luminosity:	3.9×10^{26} W
Distance from center of galaxy:	8000 pc = 26,000 ly
Orbital period around center of galaxy:	220 million years
Orbital speed around center of galaxy:	220 km/s

R I V U X G
NOAO

Suppose you stood in a room that contained an object at the same temperature as the Sun but with only a surface area of 1 square millimeter. Would it burn you up? No. Such an object would emit around 100 watts, similar to a household lightbulb. Just like a lightbulb, you would have to put your hand quite close to feel any warmth. What helps make the Sun so strong is that each square meter patch of the solar surface emits about as much as a million 100-watt lightbulbs, and the Sun's surface area has more than a million trillion of these 1-square-meter patches. Taken together, it is the high temperature *and* large size of the Sun that make it so bright.

CONCEPT CHECK 16-1

The Sun emits light at nearly all possible wavelengths. At which range of wavelengths is most of the Sun's energy emitted?

Answer appears at the end of the chapter.

The Source of the Sun's Energy: Early Ideas

The bright Sun leads us to a more fundamental question: What keeps the Sun's visible surface so hot? Or, put another way, what is the fundamental source of the tremendous energy that the Sun radiates into space? For centuries, this question was one of the greatest mysteries in science. The mystery deepened in the nineteenth century, when geologists and biologists found convincing evidence that life has existed on Earth for at least several hundred million years. Since life as we know it depends crucially on sunlight, the conclusion was that the Sun was at least several hundred million years old. (We now know that both the Sun and Earth are 4.54 billion years old and that life has existed on our planet for most of its history.) The source of the Sun's energy posed a severe problem for physicists. What source of energy could have kept the Sun shining for so long (**Figure 16-1**)?

One attempt to explain solar energy was made in the mid-1800s by the English physicist Lord Kelvin (for whom the temperature scale is named) and the German scientist Hermann von Helmholtz. They argued that the tremendous weight of the Sun's outer layers should cause the Sun to contract gradually, compressing its interior gases. Whenever a gas is compressed, its temperature rises. (You can demonstrate this with a bicycle pump: As you pump air into a tire, the temperature of the air increases and the pump becomes warm to the touch.) Kelvin and Helmholtz thus suggested that gravitational contraction could cause the Sun's gases to become compressed and hot enough to radiate energy out into space.

This process, called *Kelvin–Helmholtz contraction*, actually does occur during the earliest stages of the birth of a star like the Sun (see Section 8-4). But Kelvin–Helmholtz contraction cannot be the major source of the Sun's energy today. If it were, the Sun would have had to be much larger in the relatively recent past. Helmholtz's own calculations showed that the Sun could have started its initial collapse from the solar nebula no more than about 25 million years ago. But the geological and fossil record shows that Earth is far older than that, and so the Sun must be as well. Hence, this model of a Sun that shines because it shrinks cannot be correct.

On Earth, a common way to produce heat and light is by burning fuel, such as a log in a fireplace or charcoal in a barbecue grill.

FIGURE 16-1 R I V U X G

The Sun The Sun's visible surface has a temperature of about 5800 K. At this temperature, all solids and liquids vaporize to form gases. It was only in the twentieth century that scientists discovered what has kept the Sun so hot for billions of years—the thermonuclear fusion of hydrogen nuclei in the Sun's core. (Jeremy Woodhouse/Getty Images)

Is it possible that a similar process explains the energy released by the Sun? The answer is no, because this process could not continue for a long enough time to explain the age of Earth. It might seem difficult to estimate how long the Sun could last if its energy came from burning, but the calculation only requires a few steps. To begin with, the chemical reactions involved in burning release roughly 10^{-19} joules of energy per atom. Therefore, the number of atoms that would have to undergo chemical reactions each second to generate the Sun's luminosity of 3.9×10^{26} joules per second is approximately

$$\frac{3.9 \times 10^{26} \text{ joules per second}}{10^{-19} \text{ joules per atom}} = 3.9 \times 10^{45} \text{ atoms per second}$$

From its mass and chemical composition, we know that the Sun contains about 10^{57} atoms. Thus, the length of time that would be required to consume the entire Sun by burning is

$$\frac{10^{57} \text{ atoms}}{3.9 \times 10^{45} \text{ atoms per second}} = 3 \times 10^{11} \text{ seconds}$$

There are about 3×10^7 seconds in a year. Hence, in this model, the Sun would burn itself out in a mere 10,000 (10^4) years! This period of time is far shorter than the known age of Earth, so chemical reactions cannot explain how the Sun shines. This is a very powerful result because even though there are numerous types of chemical reactions that release energy (anything from fire to dynamite), the maximum possible chemical energy in the Sun's total mass could not keep it burning for very long.

The Source of the Sun's Energy: Discovering Thermonuclear Fusion

The source of the Sun's luminosity could be explained if there were a process that was like burning but released much more energy per atom. Then the rate at which atoms would have to be consumed would be far less, and the lifetime of the Sun could be long enough to be consistent with the known age of Earth. Albert Einstein discovered the key to such a process in 1905. According to his *special theory of relativity,* a quantity m of mass can in principle be converted into an amount of energy E according to a now-famous equation:

Einstein's mass-energy equation

$$E = mc^2$$

m = quantity of mass, in kg

c = speed of light = 3×10^8 m/s

E = amount of energy into which the mass can be converted, in joules

The speed of light c is a large number, so c^2 is huge. Therefore, a small amount of matter can release an awesome amount of energy. For example, if the mass in just one penny (about 2.5 g) were completely converted into energy, it would release about 5 times the energy of the nuclear bomb detonated over Hiroshima, Japan, during World War II. (While the energy in a penny's worth of mass is impressive, the largest nuclear bombs release the energy contained in the mass of 10 rolls of quarters; see Figure 1-6.)

Inspired by Einstein's ideas, astronomers began to wonder if the Sun's energy output might come from the conversion of matter into energy. Thanks to a key discovery by Cecilia Payne-Gaposchkin in 1925, we understand today that the Sun is made up primarily of hydrogen and helium. The Sun's low density of 1410 kg/m^3 supports this idea, since it must be made of the very lightest atoms. In the 1920s, the British astronomer Arthur Eddington showed that temperatures near the center of the Sun must be so high that atoms become completely ionized. Hence, at the Sun's center we expect to find hydrogen nuclei and electrons flying around independent of each other.

Another British astronomer, Robert Atkinson, suggested that under these conditions hydrogen nuclei could fuse together to produce helium nuclei in a *nuclear reaction* that transforms a tiny amount of mass into a large amount of energy. Experiments in the laboratory using individual nuclei show that such reactions can indeed take place. The process of converting hydrogen into helium is called **hydrogen fusion.** (It is also sometimes called *hydrogen burning,* even though nothing is actually burned in the conventional sense. Ordinary burning involves chemical reactions that rearrange the outer electrons of atoms but have no effect on the atoms' nuclei.)

> Ideas from relativity and nuclear physics led to an understanding of how the Sun shines.

The fusing together of nuclei is also called **thermonuclear fusion,** because it can take place only at extremely high temperatures. This is because all nuclei have a positive electric charge and so tend to repel one another. But in the extreme heat and pressure at the Sun's center, hydrogen nuclei (protons) are moving so fast that they can overcome their electric repulsion and actually touch one another. When that happens, thermonuclear fusion can take place.

ANALOGY

You can think of protons as tiny electrically charged spheres that are coated with a very powerful glue. If the spheres are not touching, the repulsion between their charges pushes them apart. But if the spheres are forced into contact, the strength of the glue "fuses" them together. In this fusion, a little bit of mass is lost, which gets converted into energy.

CAUTION!

Be careful not to confuse thermonuclear fusion with the similar-sounding process of *nuclear fission.* In nuclear fusion, energy is released by joining or fusing together nuclei of lightweight atoms such as hydrogen. In nuclear fission, by contrast, the nuclei of very massive atoms such as uranium or plutonium release energy by fragmenting into smaller nuclei. Nuclear power plants produce energy using fission, not fusion. (Generating power using fusion has been a goal of researchers for decades, but no one has yet devised a commercially viable way to do this.)

CONCEPT CHECK 16-2

If hydrogen nuclei are positively charged, under what conditions can two hydrogen nuclei overcome the repulsive force of their electric charges and fuse together, thus releasing energy according to Einstein's equation, $E = mc^2$?

CALCULATION CHECK 16-1

How much energy is released when just 5 kg of mass is converted into energy?

Answers appear at the end of the chapter.

Converting Hydrogen to Helium

We learned in Section 5-8 that the nucleus of a hydrogen atom (H) consists of a single proton. The nucleus of a helium atom (He) consists of two protons and two neutrons. In the nuclear process that Atkinson described, four hydrogen nuclei combine to form one helium nucleus, with a concurrent release of energy:

$$4\,H \rightarrow He + energy$$

In several separate reactions, two of the four protons are changed into neutrons and eventually combine with the remaining protons to produce a helium nucleus. This sequence of reactions is called the **proton–proton chain.** The *Cosmic Connections* figure on the next page depicts the proton–proton chain in detail.

Each time this process takes place, a small fraction (0.7%) of the initial combined mass of the hydrogen nuclei does not show up in the final mass of the helium nucleus. This "lost" mass is converted into energy. **Box 16-1** describes how to use Einstein's mass–energy equation to calculate the amount of energy released.

COSMIC CONNECTIONS

The Proton–Proton Chain

The most common form of hydrogen fusion in the Sun involves three steps, each of which releases energy.

Hydrogen fusion in the Sun usually takes place in a sequence of steps called the proton–proton chain. Each of these steps releases energy that heats the Sun and gives it its luminosity.

STEP 1

(a) Two protons (hydrogen nuclei, ^{1}H) collide.

(b) One of the protons changes into a neutron (shown in blue). The proton and neutron form a hydrogen isotope (^{2}H).

(c) One by-product of converting a proton to a neutron is a neutral, nearly massless neutrino (ν), which escapes from the Sun.

(d) The other by-product of converting a proton to a neutron is a positively charged electron, or positron (e^+). This encounters an ordinary electron (e^-), annihilating both particles and converting them into gamma-ray photons (γ). The energy of these photons goes into sustaining the Sun's internal heat.

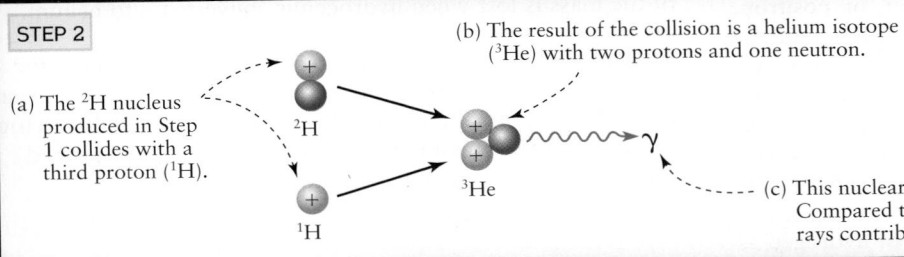

STEP 2

(a) The ^{2}H nucleus produced in Step 1 collides with a third proton (^{1}H).

(b) The result of the collision is a helium isotope (^{3}He) with two protons and one neutron.

(c) This nuclear reaction releases another gamma-ray photon (γ). Compared to other sources in these three steps, these gamma rays contribute the most energy to the Sun's internal heat.

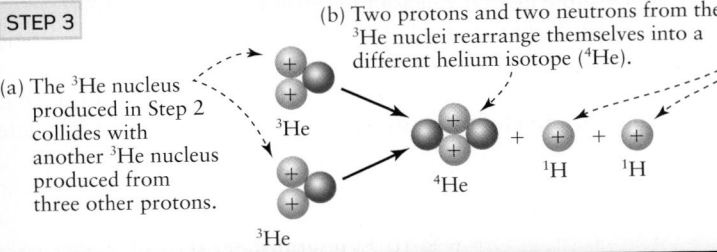

STEP 3

(a) The ^{3}He nucleus produced in Step 2 collides with another ^{3}He nucleus produced from three other protons.

(b) Two protons and two neutrons from the two ^{3}He nuclei rearrange themselves into a different helium isotope (^{4}He).

(c) The two remaining protons are released. The energy of their motion contributes to the Sun's internal heat.

(d) Six ^{1}H nuclei went into producing the two ^{3}He nuclei, which combine to make one ^{4}He nucleus. Since two of the original ^{1}H nuclei are returned to their original state, we can summarize the three steps as:

$$4\ ^1\text{H} \longrightarrow {}^4\text{He} + \text{energy}$$

 ANIMATION
16.1

 ANIMATION
16.2

Hydrogen fusion also takes place in all of the stars visible to the naked eye. (Fusion follows a different sequence of steps in the most massive stars, but the net result is the same.)

NSO/AURA/NSF; Walter Pacholka, Astropics/Science Source

BOX 16-1 TOOLS OF THE ASTRONOMER'S TRADE

Converting Mass into Energy

The *Cosmic Connection: The Proton–Proton Chain* figure shows the steps involved in the thermonuclear fusion of hydrogen at the Sun's center. In these steps, four protons are converted into a single nucleus of ^{4}He, the most common isotope of helium with two protons and two neutrons. (As we saw in Box 5-5, different isotopes of the same element have the same number of protons but different numbers of neutrons.) The reaction depicted in the *Cosmic Connections* Step 1 also produces a neutral, nearly massless particle called the *neutrino*. Neutrinos respond hardly at all to ordinary matter, so they travel almost unimpeded through the Sun's massive bulk. Hence, the energy that neutrinos carry is quickly lost into space. This loss is not great, however, because each neutrino carries relatively little energy. (See Section 16-4 for more about these curious particles.)

Most of the energy released by thermonuclear fusion appears in the form of gamma-ray photons. The energy of these photons remains trapped within the Sun for a long time, thus maintaining the Sun's intense internal heat. Some gamma-ray photons are produced by the reaction shown as Step 2 in the *Cosmic Connections* figure. Others appear when an electron in the Sun's interior annihilates a positively charged electron, or **positron**, which is a by-product of the reaction shown in Step 1 in the *Cosmic Connections* figure. An electron and a positron are respectively matter and antimatter, and they convert their mass entirely into energy when they meet. (You may have thought that "antimatter" was pure science fiction. In fact, tremendous amounts of antimatter are being created and annihilated in the Sun as you read these words.)

We can summarize the thermonuclear fusion of hydrogen as follows:

$$4 \, ^1\text{H} \rightarrow \, ^4\text{He} + \text{neutrinos} + \text{gamma-ray photons}$$

To calculate how much energy is released in this process, we use Einstein's mass–energy formula: The energy released is equal to the amount of mass consumed multiplied by c^2, where c is the speed of light. To see how much mass is consumed, we compare the combined mass of four hydrogen atoms (the ingredients) to the mass of one helium atom (the product):

$$
\begin{aligned}
4 \text{ hydrogen atoms} &= 6.693 \times 10^{-27} \text{ kg} \\
- \; 1 \text{ helium atom} &= 6.645 \times 10^{-27} \text{ kg} \\
\hline
\text{Mass lost} &= 0.048 \times 10^{-27} \text{ kg}
\end{aligned}
$$

Thus, a small fraction (0.7%) of the mass of the hydrogen going into the nuclear reaction does not show up in the mass of the helium. This lost mass is converted into an amount of energy $E = mc^2$:

$$
\begin{aligned}
E = mc^2 &= (0.048 \times 10^{-27} \text{ kg})(3 \times 10^8 \text{ m/s})^2 \\
&= 4.3 \times 10^{-12} \text{ joules}
\end{aligned}
$$

This amount of energy is released by the formation of a single helium atom. It would light a 10-watt lightbulb for almost one-half of a trillionth of a second.

EXAMPLE: How much energy is released when 1 kg of hydrogen is converted to helium?

Situation: We are given the initial mass of hydrogen. We know that a fraction of the mass is lost when the hydrogen undergoes fusion to make helium; our goal is to find the quantity of energy into which this lost mass is transformed.

Tools: We use the equation $E = mc^2$ and the result that 0.7% of the mass is lost when hydrogen is converted into helium.

Answer: When 1 kg of hydrogen is converted to helium, the amount of mass lost is 0.7% of 1 kg, or 0.007 kg. (This means that 0.993 kg of helium is produced.) Using Einstein's equation, we find that this missing 0.007 kg of matter is transformed into an amount of energy equal to

$$E = mc^2 = (0.007 \text{ kg})(3 \times 10^8 \text{ m/s})^2 = 6.3 \times 10^{14} \text{ joules}$$

Review: The energy released by the fusion of 1 kg of hydrogen is the same as that released by burning *20,000 metric tons* (2×10^7 kg) of coal! Hydrogen fusion is a *much* more efficient energy source than ordinary burning.

The Sun's luminosity is 3.9×10^{26} joules per second. To generate this much power, hydrogen must be consumed at a rate of

$$
\frac{3.9 \times 10^{26} \text{ joules per second}}{6.3 \times 10^{14} \text{ joules per kilogram}}
$$
$$= 6 \times 10^{11} \text{ kilograms per second}$$

That is, the Sun converts 600 million metric tons of hydrogen into helium every second.

CAUTION!

You may have heard the idea that mass is always conserved (that is, it is neither created nor destroyed), or that energy is always conserved in a reaction. Einstein's ideas show that neither of these statements is quite correct, because mass can be converted into energy and vice versa. A more accurate statement is that the total amount of mass *plus* energy is conserved. Hence, the destruction of mass in the Sun does not violate any laws of nature.

For every four hydrogen nuclei converted into a helium nucleus, 4.3×10^{-12} joules of energy is released—only enough energy to power a small Christmas light for a trillionth of a second. This amount of energy may seem tiny, but it is about 10^7 times larger than the amount of energy released in a typical chemical reaction, such as occurs in ordinary burning. Thus, thermonuclear fusion explains how the Sun could have been shining for billions of years.

To produce the Sun's luminosity of 3.9×10^{26} joules per second, 6×10^{11} kg (600 million metric tons) of hydrogen must be converted into helium each second—this is about 100 times the mass of Egypt's largest pyramid. This rate is prodigious, but there is, literally, an astronomical amount of hydrogen in the Sun. In particular, the Sun's core contains enough hydrogen to have been giving off energy at the present rate for as long as the solar system has existed, about 4.56 billion years, and to continue doing so for more than 5 billion years into the future.

 LOOKING DEEPER 16.1 The proton–proton chain is also the energy source for many of the stars in the sky. In stars with central temperatures that are much hotter than that of the Sun, however, hydrogen fusion proceeds according to a different set of nuclear reactions, called the **CNO cycle**, in which carbon, nitrogen, and oxygen nuclei absorb protons to produce helium nuclei. Still other thermonuclear reactions, such as helium fusion, carbon fusion, and oxygen fusion, occur late in the lives of many stars.

CONCEPT CHECK 16-3

If 1 kg of hydrogen combines to form helium in the proton-proton chain, why is only 0.007 kg (0.7%) available to be converted into energy?

CONCEPT CHECK 16-4

How do astronomers estimate that our Sun can continue to shine for more than 5 billion years?

Answers appear at the end of the chapter.

16-2 A theoretical model of the Sun shows how energy gets from its center to its surface

TUTORIAL 16.2 While thermonuclear fusion is the source of the Sun's energy, this process cannot take place everywhere within the Sun. As we have seen, extremely high temperatures—in excess of 10^7 K—are required for atomic nuclei to fuse together to form larger nuclei. The temperature of the Sun's visible surface, about 5800 K, is too low for these reactions to occur there. Hence, thermonuclear fusion can be taking place only within the Sun's interior. But precisely where does it take place? And how does the energy produced by fusion make its way to the surface, where it is emitted into space in the form of photons?

To answer these questions, we must understand conditions in the Sun's interior. Ideally, we would send an exploratory spacecraft to probe deep into the Sun; in practice, the Sun's intense heat would vaporize even the sturdiest spacecraft. Instead, astronomers use the laws of physics to construct a theoretical model of the Sun, and then compare predictions of the model to a variety of observations. (We discussed the use of models in science in Section 1-1.) Let's see what ingredients go into building a model of this kind.

Hydrostatic Equilibrium

Note first that the Sun is not undergoing any dramatic changes. The Sun is not exploding or collapsing, nor is it significantly heating or cooling. The Sun is thus said to be in both *hydrostatic equilibrium* and *thermal equilibrium*.

To understand what is meant by **hydrostatic equilibrium**, imagine a slab of material in the solar interior (Figure 16-2a). In equilibrium, the slab on average will move neither up nor down. (In fact, there are upward and downward motions of material inside the Sun, but these motions average out.) Equilibrium is maintained by a balance among three forces that act on this slab:

1. The downward pressure of the layers of solar material above the slab
2. The upward pressure of the hot gases beneath the slab
3. The slab's weight—that is, the downward gravitational pull it feels from the rest of the Sun

The pressure from below must balance both the slab's weight and the pressure from above. Hence, the pressure below the slab must be greater than that above the slab. In other words, pressure has to increase with increasing depth. For the same reason, pressure increases as you dive deeper into the ocean (Figure 16-2b) or as you move toward lower altitudes in our atmosphere.

ANALOGY

Imagine you are one of five acrobats standing on each other's shoulders. Each acrobat lower in the stack must bear the weight of all the acrobats above them, so the pressure on each of them increases with depth.

Hydrostatic equilibrium also tells us about the density of the slab of solar material. If the slab is too dense, its weight will be too large and it will sink; if the density is too low, the slab will rise. To prevent this, the density of solar material must have a certain value at each depth within the solar interior. (The same principle applies to objects that float beneath the surface of the ocean. Scuba divers wear weight belts and inflatable vests to adjust their average density so that they will neither rise nor sink but will stay submerged at the same level.)

Pressure from gases above the slab

Slab of solar material

Weight of the slab

Pressure from gases below the slab

(a) Material inside the Sun is in hydrostatic equilibrium, so forces balance.

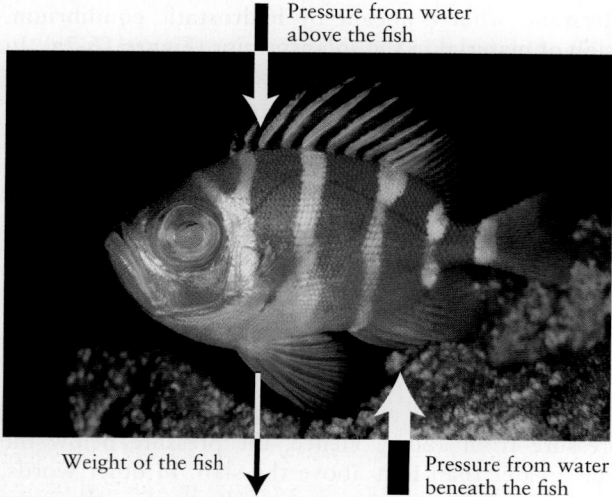

Pressure from water above the fish

Weight of the fish

Pressure from water beneath the fish

(b) A fish floating in water is in hydrostatic equilibrium, so forces balance.

FIGURE 16-2

Hydrostatic Equilibrium (a) Material in the Sun's interior tends to move neither up nor down. The upward forces on a slab of solar material (due to pressure of gases below the slab) must balance the downward forces (due to the slab's weight and the pressure of gases above the slab). Hence, the pressure must increase with increasing depth. (b) The same principle applies to a fish floating in water. In equilibrium, the forces balance and the fish neither rises nor sinks. (Ken Usami/Getty Images)

Thermal Equilibrium

Another consideration is that the Sun's interior is so hot that it is completely gaseous. Gases compress and become more dense when you apply greater pressure to them, so density must increase along with pressure as you go to greater depths within the Sun. Furthermore, when you compress a gas, its temperature tends to increase, so the temperature must also increase as you move toward the Sun's center.

While the temperature in the solar interior is different at different depths, the temperature at each depth remains constant in time. This principle is called **thermal equilibrium**. For the Sun to be in thermal equilibrium, all of the energy generated by thermonuclear reactions in the Sun's core must be transported to

the Sun's glowing surface, where it can be radiated into space. If too much energy were flowing from the core to the surface to be radiated away, the Sun's interior would cool down; alternatively, the Sun's interior would heat up if too little energy were flowing to the surface.

CONCEPT CHECK 16-5

If our Sun were much less massive, how would the pressure at the Sun's center be different?

Answer appears at the end of the chapter.

Transporting Energy Outward from the Sun's Core

But exactly how is energy transported from the Sun's center to its surface? There are three methods of energy transport: *conduction*, *convection*, and *radiative diffusion*. Only the last two are important inside the Sun.

If you heat one end of a metal bar with a blowtorch, energy flows to the other end of the bar so that it too becomes warm. The efficiency of this method of energy transport, called **conduction**, varies significantly from one substance to another. For example, metal is a good conductor of heat, but plastic is not (which is why metal cooking pots often have plastic handles, to keep you from burning your hand). Conduction is *not* an efficient means of energy transport in substances with low average densities, including the gases inside stars like the Sun.

Inside stars like our Sun, energy moves from center to surface by two other means: convection and radiative diffusion. **Convection** is the circulation of fluids—gases or liquids—between hot and cool regions. Hot gases (with lower density) rise toward a star's surface, while cool gases (with higher density) sink back down toward the star's center. This physical movement of gases transports heat energy outward in a star, just as the physical movement of water boiling in a pot transports energy from the bottom of the pot (where the heat is applied) to the cooler water at the surface.

In **radiative diffusion**, photons created in the thermonuclear inferno at a star's center diffuse outward toward the star's surface. Individual photons are absorbed and reemitted by atoms and electrons inside the star, transferring their energy as they go. The overall result is an outward migration from the hot core, where photons are constantly created, toward the cooler surface, where they escape into space.

CONCEPT CHECK 16-6

Why is the energy transport process of conduction relatively unimportant when studying how energy moves toward the Sun's surface?

Answer appears at the end of the chapter.

Modeling the Sun

To construct a model of a star like the Sun, astrophysicists express the ideas of hydrostatic equilibrium, thermal equilibrium, and energy transport as a set of equations. To ensure that the model applies to the particular star under study, they also make use of astronomical observations of the star's surface. (For example, to

TABLE 16-2 A Theoretical Model of the Sun

Distance from the Sun's center (solar radii)	Fraction of luminosity	Fraction of mass	Temperature (× 10⁶ K)	Density (kg/m³)	Pressure relative to pressure at center
0.0	0.00	0.00	15.5	160,000	1.00
0.1	0.42	0.07	13.0	90,000	0.46
0.2	0.94	0.35	9.5	40,000	0.15
0.3	1.00	0.64	6.7	13,000	0.04
0.4	1.00	0.85	4.8	4,000	0.007
0.5	1.00	0.94	3.4	1,000	0.001
0.6	1.00	0.98	2.2	400	0.0003
0.7	1.00	0.99	1.2	80	4×10^{-5}
0.8	1.00	1.00	0.7	20	5×10^{-6}
0.9	1.00	1.00	0.3	2	3×10^{-7}
1.0	1.00	1.00	0.006	0.00030	4×10^{-13}

Note: The distance from the Sun's center is expressed as a fraction of the Sun's radius (R$_\odot$). Thus, 0.0 is at the center of the Sun and 1.0 is at the surface. The fraction of luminosity is that portion of the Sun's total luminosity produced within each distance from the center; this is equal to 1.00 for distances of 0.25 R$_\odot$ or more, which means that all of the Sun's nuclear reactions occur within 0.25 solar radius from the Sun's center. The fraction of mass is that portion of the Sun's total mass lying within each distance from the Sun's center. The pressure is expressed as a fraction of the pressure at the center of the Sun.

construct a model of the Sun, they use the Sun's surface temperature of 5800 K, its luminosity of 3.9×10^{26} watts, and the gas pressure and density at the surface, which are almost zero.) The astrophysicists then use a computer to solve their set of equations and calculate conditions layer by layer in toward the star's center. The result is a model of how temperature, pressure, and density increase with increasing depth below the star's surface.

Table 16-2 and Figure 16-3 show a theoretical model of the Sun that was calculated in just this way. Different models of the Sun use slightly different assumptions, but all models give essentially the same results as those shown here. From such computer models we have learned that at the Sun's center the density is 160,000 kg/m³ (14 times the density of lead!), the temperature is 1.55×10^7 K, and the pressure is 3.4×10^{11} atm. (One atmosphere, or 1 atm, is the average atmospheric pressure at sea level on Earth.)

Table 16-2 and Figure 16-3 show that the solar luminosity rises to 100% at about one-quarter of the way from the Sun's center to its surface. In other words, the Sun's energy production occurs within a volume that extends out only to 0.25 R$_\odot$. (The symbol R$_\odot$

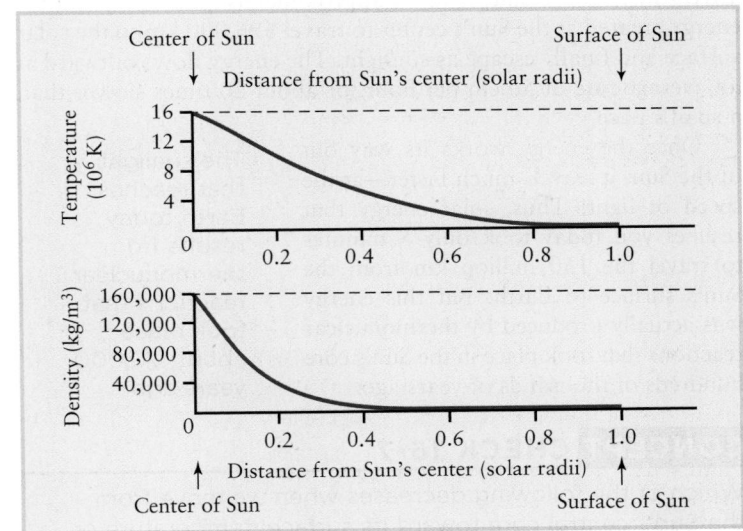

FIGURE 16-3

A Theoretical Model of the Sun's Interior These graphs depict what percentage of the Sun's total luminosity is produced within each distance from the center (upper left), what percentage of the total mass lies within each distance from the center (lower left), the temperature at each distance (upper right), and the density at each distance (lower right). (See Table 16-2 for a numerical version of this model.)

denotes the solar radius, or radius of the Sun as a whole, equal to 696,000 km.) Outside 0.25 $R_\odot$, the density and temperature are too low for thermonuclear reactions to take place. Also note that 94% of the total mass of the Sun is found within the inner 0.5 $R_\odot$. Hence, the outer 0.5 $R_\odot$ contains only a relatively small amount of material.

How energy flows from the Sun's center toward its surface depends on how easily photons move through the gas. If the solar gases are relatively transparent, photons can travel moderate distances before being scattered or absorbed, and energy is thus transported by radiative diffusion. You can think of radiative diffusion as the "free flight" of photons, and energy is transported in these photons. If the gases are relatively *opaque*, photons are frequently scattered or absorbed and cannot easily get through the gas. In an opaque gas, heat builds up and convection then becomes the most efficient means of energy transport. In convection, the energy is in the form of heat; the movement of hot gas transports the energy. The gases start to churn, with hot lower-density gas moving outward and cooler gas sinking inward.

From the center of the Sun out to about 0.71 $R_\odot$, energy is transported by radiative diffusion. Hence, this region is called the **radiative zone**. Beyond about 0.71 $R_\odot$, the temperature is low enough (a mere 2×10^6 K or so) for electrons and hydrogen nuclei to join into hydrogen atoms. These atoms are very effective at absorbing photons, much more so than free electrons or nuclei, and this absorption chokes off the outward flow of photons. Therefore, beyond about 0.71 $R_\odot$, radiative diffusion is not an effective way to transport energy. Instead, convection dominates the energy flow in this outer region, which is why it is called the **convective zone**. Figure 16-4 shows these aspects of the Sun's internal structure.

Although energy travels through the radiative zone in the form of photons, the photons have a difficult time of it. Table 16-2 shows that the material in this zone is extremely dense, so photons from the Sun's core take a long time to diffuse through the radiative zone. As a result, it takes approximately 170,000 years for energy created at the Sun's center to travel 696,000 km to the solar surface and finally escape as sunlight. The energy flows outward at an average rate of 50 cm per hour, or about 20 times slower than a snail's pace.

Once the energy works its way out of the Sun, it travels much faster—at the speed of light! Thus, solar energy that reaches you today took only 8 minutes to travel the 150 million km from the Sun's surface to Earth. But this energy was actually produced by thermonuclear reactions that took place in the Sun's core hundreds of thousands of years ago.

> The sunlight that reaches Earth today results from thermonuclear reactions that took place about 170,000 years ago.

CONCEPT CHECK 16-7

Which of the following decreases when we move from the Sun's central core toward its surface: temperature or luminosity?

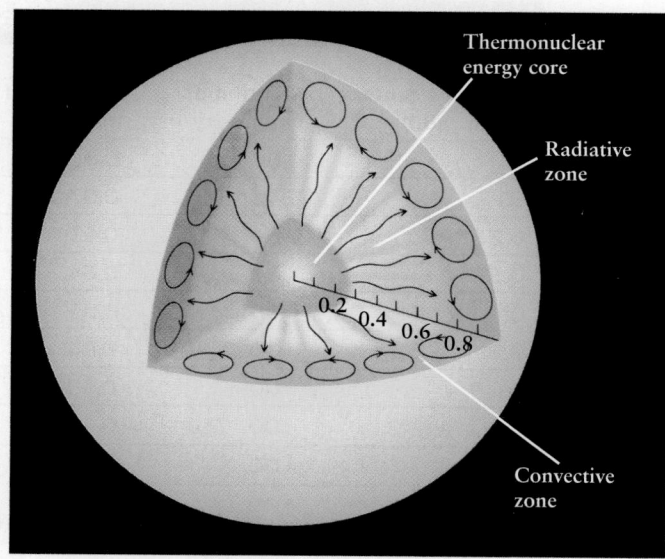

FIGURE 16-4

The Sun's Internal Structure Thermonuclear reactions occur in the Sun's core, which extends out to a distance of 0.25 $R_\odot$ from the center. Energy is transported outward, via radiative diffusion, to a distance of about 0.71 $R_\odot$. In the outer layers between 0.71 $R_\odot$ and 1.00 $R_\odot$, energy flows outward by convection.

CALCULATION CHECK 16-2

By how much (in percent) does the Sun's temperature drop from its temperature at the central core to its temperature at 0.5 $R_\odot$?

Answers appear at the end of the chapter.

16-3 Astronomers probe the solar interior using the Sun's own vibration

We have described how astrophysicists construct models of the Sun. But since we cannot see into the Sun's opaque interior, how can we check these models to see if they are accurate? What is needed is a technique for probing the Sun's interior. A very powerful technique of just this kind involves measuring vibrations of the Sun as a whole. This field of solar research is called **helioseismology**.

Vibrations are a useful tool for examining the hidden interiors of all kinds of objects. Food shoppers test whether melons are ripe by tapping on them and listening to the sound made by the resulting vibrations. Since the vibrations of a melon depend on its density profile, the sounds from tapping provide clues about the melon's interior. Similarly, geologists can determine the structure of Earth's interior by using seismographs to record vibrations from earthquakes that traveled through Earth's interior (and along its surface; see Section 9-2).

Although there are no true "sunquakes," the Sun does vibrate at a variety of frequencies, somewhat like a ringing bell. These vibrations were first noticed in 1960 by Robert Leighton of the California Institute of Technology, who made high-precision Doppler shift observations of the solar surface. These measurements revealed that parts of the Sun's surface move up and down about 10 m every 5 minutes. Since the mid-1970s, several astronomers have reported slower vibrations, having periods ranging from 20 to 160 minutes. The detection of extremely slow vibrations has inspired astronomers to organize networks of telescopes around and in orbit above Earth to monitor the Sun's vibrations on a continuous basis.

The vibrations of the Sun's surface can be compared with sound waves. If you could somehow survive within the Sun's outermost layers, you would first notice a deafening roar, somewhat like a jet engine, produced by turbulence in the Sun's gases. But superimposed on this noise would be a variety of nearly pure tones. You would need greatly enhanced hearing to detect these tones, however; the strongest has a frequency of just 0.003 hertz, 13 octaves below the lowest frequency audible to humans. (Recall from Section 5-2 that one hertz is one oscillation per second.)

While it might not seem obvious, sound waves moving upward from the solar interior would be reflected back inward after reaching the solar surface. However, as a reflected sound wave descends back into the Sun, the increasing density and pressure bend the wave severely, turning it around and aiming it back toward the solar surface. In other words, sound waves bounce back and forth between the solar surface and layers deep within the Sun. These sound waves can reinforce each other if their wavelength is the right size, just as sound waves of a particular wavelength resonate inside an organ pipe.

The Sun oscillates in millions of ways as a result of waves resonating in its interior. Figure 16-5 is a computer-generated illustration of one such mode of vibration. Helioseismologists can deduce information about the solar interior from measurements of these oscillations. In addition to providing information about interior temperatures and densities, helioseismology sets limits on the amount of helium in the Sun's core and convective zone and determines the thickness of the transition region between the radiative zone and the convective zone.

> The Sun's oscillations resemble those inside an organ pipe but are much more complex.

CONCEPT CHECK 16-8

Do solar sound waves travel exclusively on the surface of the Sun—like ripples across a pond—or do they travel through the Sun's interior as well?

CONCEPT CHECK 16-9

What can be determined from carefully monitoring the Sun's vibrations?

Answers appear at the end of the chapter.

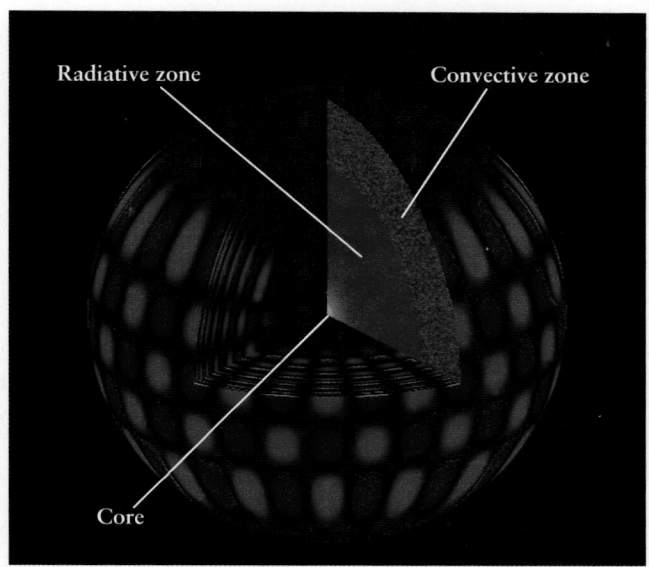

FIGURE 16-5

A Sound Wave Resonating in the Sun This computer-generated image shows one of the millions of ways in which the Sun's interior vibrates. The regions that are moving outward are colored blue, those moving inward, red. As the cutaway shows, these oscillations are thought to extend into the Sun's radiative zone (compare Figure 16-4). (GONG/NSO/AURA/NSF)

16-4 Neutrinos reveal information about the Sun's core—and have surprises of their own

We have seen circumstantial evidence that thermonuclear fusion is the source of the Sun's power. To be certain, however, we need more definitive evidence. How can we show that thermonuclear fusion really is taking place in the Sun's core?

What Sunlight Cannot Tell Us

Although the light energy that we receive from the Sun originates in the core, it provides few clues about conditions there. The problem is that this energy has changed form repeatedly during its passage from the core: It appeared first as photons diffusing through the radiative zone, then as heat transported through the outer layers by convection, and then again as photons emitted from the Sun's glowing surface. As a result of these transformations, much of the information that the Sun's radiated energy once carried about conditions in the core has been lost.

ANALOGY

If you make a photocopy of a photocopy of a photocopy of an original document, the final result may be so blurred as to be unreadable. In an analogous way, because solar energy is transformed many times while en route through the Sun, the story it could tell us about the Sun's core is hopelessly blurred.

Solar Neutrinos

Happily, there is a way for scientists to learn about conditions in the Sun's core and to get direct evidence that thermonuclear fusion really does happen there. The trick is to detect the subatomic by-products of thermonuclear fusion reactions.

> Scientists use the most ethereal of subatomic particles to learn about the Sun, and vice versa.

As part of the process of hydrogen fusion, protons change into neutrons and release **neutrinos** (see the *Cosmic Connections* figure in Section 16-1, as well as Box 16-1). Like photons, neutrinos are particles that have no electric charge. Unlike photons, however, neutrinos interact only very weakly with matter. Even the vast bulk of the Sun offers little impediment to their passage, so neutrinos must be streaming out of the core and into space. Indeed, the conversion of hydrogen into helium at the Sun's center produces 10^{38} neutrinos each second. Some of these neutrinos hit Earth, and about 10^{14} neutrinos from the Sun—that is, **solar neutrinos**—must pass through each square meter of Earth every second. Perhaps the exceedingly weak interaction between neutrinos and matter is a good thing, since about 100 trillion neutrinos pass through our bodies each second!

If it were possible to detect these solar neutrinos, we would have direct evidence that thermonuclear reactions really do take place in the Sun's core. Beginning in the 1960s, scientists began to build neutrino detectors for precisely this purpose. The challenge is that neutrinos are exceedingly difficult to detect. Just as neutrinos pass unimpeded through the Sun, they also pass through Earth almost as if it were not there. We stress the word "almost," because neutrinos can and do interact with matter, albeit infrequently.

On rare occasions, a neutrino will strike a neutron and convert it into a proton. This effect was the basis of the original solar neutrino detector, designed and built by Raymond Davis of the Brookhaven National Laboratory in the 1960s. This device used 100,000 gallons of perchloroethylene (C_2Cl_4), a fluid used in dry cleaning, in a huge tank buried deep underground. Most of the solar neutrinos that entered Davis's tank passed right through it with no effect whatsoever. But occasionally, a neutrino would strike the nucleus of one of the chlorine atoms (^{37}Cl) in the cleaning fluid and convert one of its neutrons into a proton, creating a radioactive atom of argon (^{37}Ar). Fortunately for experimenters, what individual neutrinos lack in barely interacting with matter, the Sun partially compensates for by producing neutrinos in extraordinarily large numbers.

The rate at which argon is produced is related to the neutrino flux—that is, the number of solar neutrinos arriving at Earth per square meter per second. By counting the number of newly created argon atoms, Davis was able to determine the neutrino flux from the Sun. (Other subatomic particles besides neutrinos can also induce reactions that create radioactive atoms. By placing the experiment deep underground, however, the overlying Earth absorbs essentially all such particles—with the exception of neutrinos.)

The Solar Neutrino Problem

Davis and his collaborators found that solar neutrinos created one radioactive argon atom in the tank every 3 days. But this rate corresponded to only one-third of the neutrino flux predicted from standard models of the Sun. This troubling discrepancy between theory and observation, called the **solar neutrino problem**, motivated scientists around the world to conduct further experiments to measure solar neutrinos.

One key question was whether the neutrinos that Davis had detected had really come from the Sun. (The Davis experiment had no way to determine the direction from which neutrinos had entered the tank of cleaning fluid.) The direction was resolved by an experiment in Japan called Kamiokande, which was designed by the physicist Masatoshi Koshiba. A large underground tank containing 3000 tons of water was surrounded by 1100 light detectors. From time to time, a high-energy solar neutrino would strike an electron in one of the water molecules, dislodging it and sending it flying like a pin hit by a bowling ball. Each time, the recoiling electron produced a streaking flash of light, which was sensed by the detectors. By analyzing the flashes, scientists could tell the direction from which the neutrinos were coming and confirmed that they emanated from the Sun. These results in the late 1980s gave direct evidence that thermonuclear fusion is indeed occurring in the Sun's core. (Davis and Koshiba both received the 2002 Nobel Prize in Physics for their pioneering research on solar neutrinos.)

Like Davis's experiment, however, Koshiba and his colleagues at Kamiokande detected only a fraction of the expected flux of neutrinos. Where, then, were the missing solar neutrinos?

LOOKING DEEPER 16.1 ➤ One proposed solution had to do with the energy of the detected neutrinos. The vast majority of neutrinos from the Sun are created during the first step in the proton–proton chain, in which two protons combine to form a heavy isotope of hydrogen (see the *Cosmic Connections* figure in Section 16-1). But these neutrinos have too little energy to convert chlorine into argon. Both Davis's and Koshiba's experiments responded only to high-energy neutrinos produced by reactions that only occasionally occur near the end of the proton–proton chain. (The *Cosmic Connections* figure does not show these reactions.) Could it be that the discrepancy between theory and observation could be explained if the flux of low-energy neutrinos could be measured?

To test this idea, two teams of physicists constructed neutrino detectors using several tons of gallium (a liquid metal) rather than cleaning fluid. Low-energy neutrinos convert gallium (^{71}Ga) into a radioactive isotope of germanium (^{71}Ge). By chemically separating the germanium from the gallium and counting the radioactive atoms, the physicists were able to measure the flux of low-energy solar neutrinos. These experiments—GALLEX (Gallium Experiment) in Italy and SAGE (Soviet-American Gallium Experiment) in Russia—detected only 50% to 60% of the expected neutrino flux. Hence, the solar neutrino problem extended to neutrinos of all energies.

Another proposed solution to the neutrino problem suggested that the Sun's core might be cooler than predicted by solar models. If the Sun's central temperature were only 10% lower than the current estimate, fewer neutrinos would be produced and the neutrino flux would agree with experiments. However, a lower central temperature would cause other obvious features, such as the Sun's size and surface temperature, to be different from what we observe.

Finding the Missing Neutrinos

Only very recently has the solution to the neutrino problem been found. The answer lies not in how neutrinos are produced, but rather in what happens to them between the Sun's core and detectors on Earth. Motivated by the lack of observed solar neutrinos, physicists have discovered that there are actually three types of neutrinos. Only one of these types is produced in the Sun, and it is only this type that can be detected by the experiments we have described. But if some of the solar neutrinos change *in flight* into a different type of neutrino, the detectors in these experiments would record only a fraction of the total neutrino flux. This effect, where one type of neutrino can spontaneously transform into another type, is called **neutrino oscillation**. In June 1998, scientists at the Super-Kamiokande neutrino observatory (a larger and more sensitive device than Kamiokande) revealed evidence that neutrino oscillation does indeed take place.

The best confirmation of this idea has come from the Sudbury Neutrino Observatory (SNO) in Canada. Like Kamiokande and Super-Kamiokande, SNO uses a large tank of water placed deep underground (**Figure 16-6**). But unlike those earlier experiments, SNO can detect all three types of neutrinos, and it does this using *heavy water*. In ordinary, or "light," water, each hydrogen atom in the H₂O molecule has a solitary proton as its nucleus. In heavy water, by contrast, each of the hydrogen nuclei has a nucleus made up of a proton and a neutron. (This isotope ²H is shown in the *Cosmic Connections* figure in Section 16-1.) If a high-energy solar

neutrino of any type passes through SNO's tank of heavy water, it can knock the neutron out of one of the ²H nuclei. The ejected neutron can then be captured by another nucleus, and this capture releases energy that manifests itself as a tiny burst of light. As in Kamiokande, detectors around SNO's water tank record these light flashes.

Scientists using SNO have found that the combined flux of all three types of neutrinos coming from the Sun is *equal* to the theoretical prediction. Together with the results from earlier neutrino experiments, this result strongly suggests that the Sun is indeed producing neutrinos at the predicted rate as a by-product of thermonuclear reactions. But before these neutrinos can reach Earth, about two-thirds of them undergo spontaneous oscillation and change their type. Thus, there is really no lack of neutrinos, and hence no solar neutrino problem—scientists merely needed the right kind of detectors to observe all of the neutrinos, including the ones that transformed in flight.

The story of the solar neutrino problem illustrates how two different branches of science—in this case, studies of the solar interior and investigations of subatomic particles—can sometimes interact, to the mutual benefit of both. While there is still much we do not understand about the Sun and about neutrinos, a new generation of neutrino detectors in Japan, Canada, and elsewhere promises to further our knowledge of these exotic realms of astronomy and physics.

CONCEPT CHECK 16-10

Why did the SNO experiment find 3 times as many neutrinos as the Kamiokande experiment?

Answer appears at the end of the chapter.

16-5 The photosphere is the lowest of three main layers in the Sun's atmosphere

Although the Sun's core is hidden from our direct view, we can easily see sunlight coming from the high-temperature gases that make up the Sun's atmosphere. These outermost layers of the Sun prove to be the sites of truly dramatic activity, much of which has a direct impact on our planet. By studying these layers, we gain further insight into the character of the Sun as a whole.

Observing the Photosphere

A visible-light photograph like **Figure 16-7** makes it appear that the Sun has a definite surface. This surface is actually an illusion; because of its high temperature, the Sun is gaseous throughout its volume, and the gases simply become less and less dense as you move farther away from the Sun's center.

Why, then, does the Sun appear to have a sharp, well-defined surface in Figure 16-7? The reason is that essentially all of the Sun's visible light emanates from a single, thin layer of gas called the **photosphere** ("sphere of light"). Just as you can see only a certain distance through Earth's atmosphere before objects vanish in the haze, we can see only about 400 km into the photosphere. This

FIGURE 16-6   R I ⓥ U X G

A Solar Neutrino Experiment Located 2073 m (6800 ft) underground in the Creighton nickel mine in Sudbury, Canada, the Sudbury Neutrino Observatory is centered around a tank that contains 1000 tons of water. Occasionally, a neutrino entering the tank interacts with one or another of the particles already there. Such interactions create flashes of light, called Cherenkov radiation. Some 9600 light detectors sense this light. The numerous silver protrusions are the back sides of the light detectors prior to their being wired and connected to electronics in the lab (seen at the bottom of the photograph). (LBNL/Roy Kaltschmidt/Science Source)

FIGURE 16-7 R I V U X G

The Photosphere The photosphere is the layer in the solar atmosphere from which the Sun's visible light is emitted. Note that the Sun appears darker around its limb, or edge; here we are seeing the upper photosphere, which is relatively cool and thus glows less brightly. (The dark sunspots, which we discuss in Section 16-8, are also relatively cool regions.) (Celestron International)

FIGURE 16-8

The Origin of Limb Darkening Light from the Sun's limb and light from the center of its disk both travel about the same straight-line distance through the photosphere to reach us. Because of the Sun's curved shape, light from the limb comes from a greater height within the photosphere, where the temperature is lower and the gases glow less brightly. Hence, the limb appears darker and more orange.

distance is so small compared with the Sun's radius of 696,000 km that the photosphere appears to be a definite surface.

The photosphere is actually the lowest of the three layers that together constitute the solar atmosphere. Above it are the *chromosphere* and the *corona*, both of which are transparent to visible light. We can see them only using special techniques, which we discuss later in this chapter. Everything below the photosphere is called the *solar interior*.

The Lesson of Limb Darkening

The photosphere is heated from below by energy streaming outward from the solar interior. Hence, temperature should decrease as you go upward in the photosphere, just as in the solar interior (see Table 16-2 and Figure 16-3). We know this is the case because the photosphere appears darker around the edge, or *limb*, of the Sun than it does toward the center of the solar disk, an effect called **limb darkening** (examine Figure 16-7). This happens because when we look near the Sun's limb, we do not see as deeply into the photosphere as we do when we look near the center of the disk (Figure 16-8). The high-altitude gas we observe at the limb is not as hot and thus does not glow as brightly as the deeper, hotter gas seen near the disk center.

The spectrum of the Sun's photosphere confirms how its temperature varies with altitude. As we saw in Section 16-1, the photosphere shines like a nearly perfect blackbody with an average temperature of about 5800 K. However, superimposed on this spectrum are many dark absorption lines (see Figure 5-13). As discussed in Section 5-6, we see an *absorption line spectrum* of this sort whenever we view a hot, glowing object through a relatively cool gas. In this case, the hot object is the lower part of the photosphere; the cooler gas is in the upper part of the photosphere, where the temperature declines to about 4400 K. All the absorption lines in the Sun's spectrum are produced in this relatively cool layer, as atoms selectively absorb photons of various wavelengths streaming outward from the hotter layers below.

> **CAUTION!**
>
> You may find it hard to think of 4400 K as "cool." But keep in mind that the ratio of 4400 K to 5800 K, the temperature in the lower photosphere, is the same as the ratio of the temperature on a Siberian winter night to that of a typical day in Hawaii.

Granules and Supergranules in the Photosphere

We can learn still more about the photosphere by examining it with a telescope—but only when using special dark filters to prevent eye damage. *Looking directly at the Sun without the correct filter, whether with the naked eye or with a telescope, can cause permanent blindness!* Under good observing conditions, astronomers using such filter-equipped telescopes can often see a blotchy pattern in the photosphere, called **granulation** (Figure 16-9). Each light-colored **granule** measures about 1000 km (600 mi) across—equal in size to the areas of Texas and Oklahoma combined—and is surrounded by a darkish boundary. The difference in brightness between the center and the edge of a granule corresponds to a temperature drop of about 300 K (80°F).

Granulation is caused by convection of the gas in the photosphere. The inset in Figure 16-9 shows how hot gas from lower levels rises upward in granules, cools off, spills over the edges of the granules, and then plunges back down into the Sun. Convection can occur only if the gas is heated from below, like a

Like water boiling on a stove, the photosphere bubbles with convection cells.

VIDEO 16.1   VIDEO 16.2

FIGURE 16-9 R I **V** U X G

Solar Granulation High-resolution photographs of the Sun's surface reveal a blotchy pattern called granulation. Granules are convection cells about 1000 km (600 mi) wide in the Sun's photosphere. **Inset:** Rising hot gas produces bright granules. Cooler gas sinks downward along the boundaries between granules; this gas glows less brightly, giving the boundaries their dark appearance. This convective motion transports heat from the Sun's interior outward to the solar atmosphere. (Hinode JAXA/NASA/PPARC)

pot of water being heated on a stove (see Section 16-2). Along with limb darkening and the Sun's absorption line spectrum, granulation shows that the upper part of the photosphere must be cooler than layers below.

Time-lapse photography reveals more of the photosphere's dynamic activity. Granules form, disappear, and reform in cycles lasting only a few minutes. At any one time, about 4 million granules cover the solar surface.

Superimposed on the pattern of granulation are even larger convection cells called **supergranules** (Figure 16-10). As in granules, gases rise upward in the middle of a supergranule, move horizontally outward toward its edge, and descend back into the Sun. The difference is that a typical supergranule is about 35,000 km in diameter, large enough to enclose several hundred granules. This large-scale convection moves at only about 0.4 km/s (1400 km/h, or 900 mi/h), about one-tenth the speed of gases churning in a regular-sized granule. A given supergranule lasts about a day.

ANALOGY

Similar patterns of large-scale and small-scale convection can be found in Earth's atmosphere. On the large scale, air rises gradually at a low-pressure area, then sinks gradually at a high-pressure area, which might be hundreds of kilometers away. This pattern is analogous to the flow in a supergranule. Thunderstorms in our atmosphere are small but intense convection cells within which air moves rapidly up and down. Like granules, they last only a relatively short time before they dissipate.

The Photosphere: Hot, Thin, and Opaque

Although the photosphere is a very active place, it actually contains relatively little material. Careful examination of the spectrum shows that it has a density of only about 10^{-4} kg/m³, roughly 0.01% the density of Earth's atmosphere at sea level. The

■ Blue: areas of rising gas
■ Red: areas of sinking gas

FIGURE 16-10 R I **V** U X G

Supergranules and Large-Scale Convection Supergranules display relatively little contrast between their center and edges, so they are hard to observe in ordinary images. But they can be seen in a false-color Doppler image like this one. Light from gas that is approaching us (that is, rising) is shifted toward shorter "bluer" wavelengths, while light from receding gas (that is, descending) is shifted toward longer "redder" wavelengths (see Section 5-9). (David Hathaway, MSFC/NASA)

photosphere is made primarily of hydrogen and helium, the most abundant elements in the solar system (see Figure 8-4).

Despite being such a thin gas, the photosphere is surprisingly opaque to visible light. (This means that visible light is scattered numerous times as it passes through the photosphere.) If it were not so opaque, we could see into the Sun's interior to a depth of hundreds of thousands of kilometers, instead of the mere 400 km that we can see down into the photosphere. What makes the photosphere so opaque is that its hydrogen atoms sometimes acquire an extra electron, becoming **negative hydrogen ions**. The scattering of photons can be pictured in two steps. In the first step, the extra electron of a negative hydrogen ion is only loosely attached and can be dislodged if it absorbs a photon of any visible wavelength. In the second step, a similar photon can be emitted when the dislodged electron once again combines with hydrogen to form another negative hydrogen ion, but the emitted photon *heads in a random direction each time.* Hence, negative hydrogen ions absorb light very efficiently, and there are enough of these light-absorbing ions in the photosphere to make it quite opaque. Because it is so opaque, the photosphere's spectrum is close to that of an ideal blackbody (blackbodies are discussed in Section 5-3).

CONCEPT CHECK 16-11

What causes the photosphere to bubble like water boiling on the stove?

Answer appears at the end of the chapter.

16-6 Spikes of rising gas extend through the Sun's chromosphere

An ordinary visible-light image such as Figure 16-7 gives the impression that the Sun ends at the top of the photosphere. But during a total solar eclipse, the Moon blocks the photosphere from our view, revealing a glowing, pinkish layer of gas above the photosphere (**Figure 16-11**). This layer is the tenuous **chromosphere** ("sphere of color"), the second of the three major levels in the Sun's atmosphere. The chromosphere is only about one ten-thousandth (10^{-4}) as dense as the photosphere, or about 10^{-8} as dense as our own atmosphere. No wonder it is normally invisible!

Comparing the Chromosphere and Photosphere

Unlike the photosphere, which has an absorption line spectrum, the chromosphere has a spectrum dominated by emission lines. An emission line spectrum is produced by the atoms of a hot, thin gas (see Section 5-6 and Section 5-8). As their electrons fall from higher to lower energy levels, the atoms emit photons.

One of the strongest emission lines in the chromosphere's spectrum is the H_α line at 656.3 nm, which is emitted by a hydrogen atom when its single electron falls from the $n = 3$ level to the $n = 2$ level (recall Figure 5-23b). This wavelength is in the red part of the spectrum, which gives the chromosphere its characteristic pinkish color. The spectrum also contains emission lines of singly ionized calcium, as well as lines due to ionized helium and ionized metals. In fact, helium—named after Helios, a Greek reference to the Sun—was originally discovered through detection in the chromospheric spectrum in 1868, almost 30 years before helium was discovered on Earth.

Analysis of the chromospheric spectrum shows that temperature *increases* with increasing height in the chromosphere. This trend is just the opposite of the situation in the photosphere, where

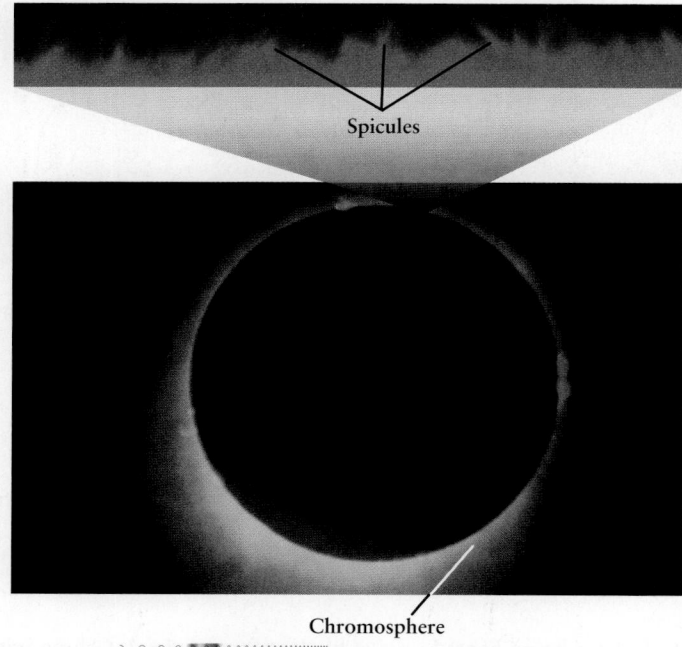

FIGURE 16-11   R I V U X G

The Chromosphere During a total solar eclipse, the Sun's glowing chromosphere can be seen around the edge of the Moon. It appears pinkish because its hot gases emit light at only certain discrete wavelengths, principally the H_α emission of hydrogen at a red wavelength of 656.3 nm. The expanded area above shows spicules, jets of chromospheric gas that surge upward into the Sun's outer atmosphere. (Bill Livingston/NSO/AURA/NSF)

temperature decreases with increasing height. The temperature is about 4400 K at the top of the photosphere; 2000 km higher, at the top of the chromosphere, the temperature is nearly 25,000 K. This temperature difference is very surprising, since it may seem like temperatures should decrease as you move away from the Sun's interior. In Section 16-10, we will see how solar scientists explain this seeming paradox.

The photosphere's spectrum is dominated by absorption lines at certain wavelengths, while the spectrum of the chromosphere has emission lines at these same wavelengths. In other words, the photosphere appears dark at the specific wavelengths at which the chromosphere emits most strongly, such as the H_α wavelength of 656.3 nm. By viewing the Sun through a special filter that is transparent to light only at the wavelength of H_α, astronomers can screen out light from the photosphere and make the chromosphere visible. Filters make it possible to see the chromosphere at any time, not just during a solar eclipse.

Spicules

The top photograph in Figure 16-11 is a high-resolution image of the Sun's chromosphere taken through an H_α filter. This image shows numerous vertical spikes, which are actually jets of rising gas called **spicules**. A typical spicule lasts just 15 minutes or so, rising about

Jets of gas thousands of kilometers in height rise through the chromosphere.

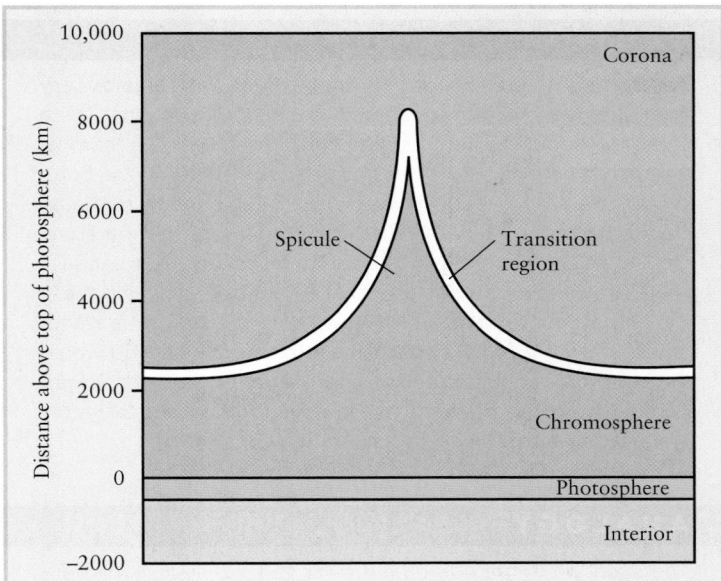

FIGURE 16-12

The Solar Atmosphere This schematic diagram shows the three layers of the solar atmosphere. The lowest, the photosphere, is about 400 km thick. The chromosphere extends about 2000 km higher, with spicules jutting up to nearly 10,000 km above the photosphere. Above a transition region is the Sun's outermost layer, the corona, which we discuss in Section 16-7. It extends many millions of kilometers out into space. (Data from J. A. Eddy)

18,000 km and then collapsing and fading away (**Figure 16-12**). Approximately 300,000 spicules exist at any one time, covering about 1% of the Sun's surface. When viewed from above, spicules can surround a sunspot, as in **Figure 16-13** (we will learn more about sunspots in Section 16-8). While the Sun's magnetic field appears to be involved in the production of spicules, the details are not understood.

CONCEPT CHECK 16-12

Why do the spicules in Figure 16-11 appear red? Do spicules emit light at other wavelengths?

Answer appears at the end of the chapter.

16-7 The corona ejects mass into space to form the solar wind

The **corona**, or outermost region of the Sun's atmosphere, begins at the top of the chromosphere. It extends out to a distance of several million kilometers. Despite its tremendous extent, the corona is only about one-millionth (10^{-6}) as bright as the photosphere—no brighter than the full moon. Hence, the corona can be viewed only when the light from the photosphere is blocked out, either by use of a specially designed telescope or during a total solar eclipse.

Figure 16-14 is an exceptionally detailed photograph of the Sun's corona taken during a solar eclipse. It shows that the corona

FIGURE 16-13 R I **V** U X G

Spicules from Above Surrounding this sunspot are numerous spicules, also called *fibrils* because of their long and thin fiberlike shape. These spicules are about as long as Earth's diameter and contain very hot gas confined by magnetic fields. This is one of the highest-resolution images of the Sun ever made. (K. Reardon [Arcetri Astrophysical Observatory/INAF; NSO]—IBIS, DST)

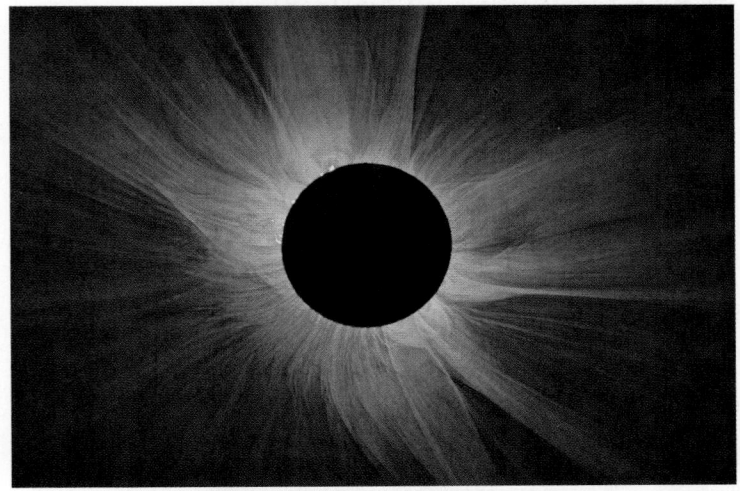

FIGURE 16-14 R I **V** U X G

The Solar Corona This striking photograph of the corona was taken during the total solar eclipse of August 21, 2017. Numerous streamers extend for millions of kilometers above the solar surface. The unearthly light of the corona is one of the most extraordinary aspects of experiencing a solar eclipse. The pink features jutting above the disk in the upper left are called solar prominences (Section 16-10). (S. Habbal, M. Druckmuller, P. Aniol/ NASA) is not merely a spherical shell of gas surrounding the Sun. Rather, numerous streamers extend in different directions far above the solar surface. The shapes of these streamers vary on timescales of days or weeks. (For other views of the corona during a solar eclipse, see Figure 3-11b and the image that opens Chapter 3.)

Comparing the Corona, Chromosphere, and Photosphere

Like the chromosphere that lies below it, the corona has an emission line spectrum characteristic of a hot, thin gas. When the spectrum of the corona was first measured in the nineteenth century, astronomers found a number of emission lines at wavelengths that had never been seen in the laboratory. Their explanation was that the corona contained elements that had not yet been detected on Earth. However, laboratory experiments in the 1930s revealed that these unusual emission lines were in fact caused by the same atoms found elsewhere in the universe—but in highly ionized states. For example, a prominent green line at 530.3 nm is caused by highly ionized iron atoms, each of which has been stripped of 13 of its 26 electrons. In order to strip that many electrons from atoms, temperatures in the corona must reach 2 million kelvins (2×10^6 K) or even higher—far greater than the temperatures in the chromosphere. Figure 16-15 shows how temperature varies in the chromosphere and corona.

The low density of the corona explains why it is so dim compared with the photosphere. In general, the higher the temperature of a gas, the brighter it glows. But because there are so few atoms in the corona, the net amount of light that it emits is very feeble compared with the light from the much cooler, but also much denser, photosphere.

CAUTION!

The corona is actually not very "hot"—that is, it contains very little thermal energy—because it is nearly a vacuum. There are only about 10^{11} atoms per cubic meter in the corona, compared with about 10^{23} atoms per cubic meter in the Sun's photosphere and about 10^{25} atoms per cubic meter in the air that we breathe. Because of the corona's high temperature, the atoms there are moving at very high speeds. But because there are so few of them, the total amount of energy in these moving atoms (a measure of how "hot" the gas is) is rather low. If you were to fly a spaceship into the corona, you would have to worry about becoming overheated by the intense light coming from the photosphere, but you would notice hardly any heating from the corona's ultrathin gas.

ANALOGY

The situation in the corona is similar to that inside a conventional oven that is being used for baking. Both the walls of the oven and the air inside the oven are at the same high temperature, but the air contains very few atoms and thus carries little energy. If you put your hand in the oven momentarily, the lion's share of the heat you feel is radiation from the oven walls.

CONCEPT CHECK 16-13

Why is the corona so difficult to see if it is so much hotter than the photosphere?

Answer appears at the end of the chapter.

In this narrow transition region between the chromosphere and corona, the temperature rises abruptly by about a factor of 100.

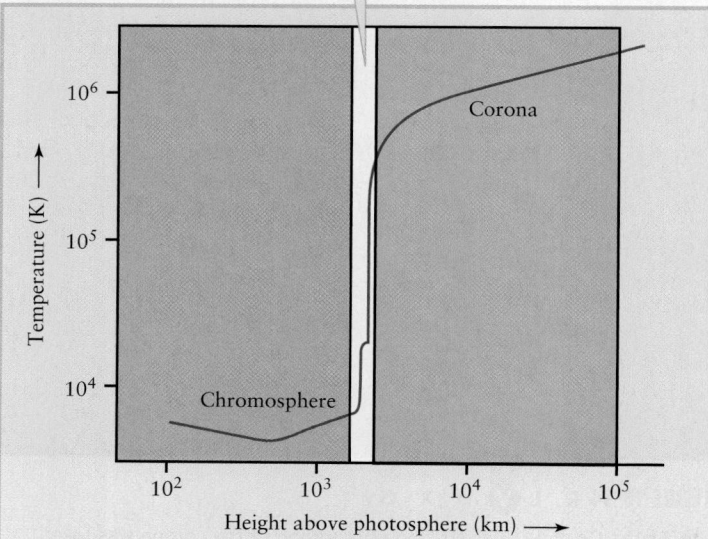

(a)

FIGURE 16-15 R I V U X G

Temperatures in the Sun's Upper Atmosphere **(a)** This graph shows how temperature varies with altitude in the Sun's chromosphere and corona and in the narrow transition region between them. In order to show a large range of values, both the vertical and horizontal scales are nonlinear. **(b)** This composite portrait shows the different features that appear as wavelengths

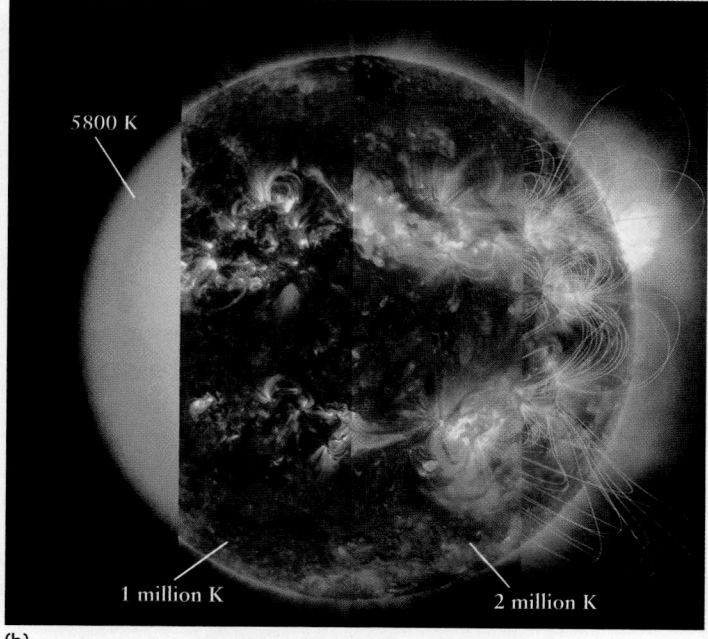

(b)

corresponding to higher temperatures are imaged. At visible wavelengths, the 5800 K photosphere is visible, while ultraviolet light captures features at millions of degrees. On the far right, modeling has been used to draw lines of the Sun's magnetic field in the corona. (a: Data from A. Gabriel; b: NASA)

The Solar Wind and Coronal Holes

Earth's gravity keeps our atmosphere from escaping into space. In the same way, the Sun's powerful gravitational attraction keeps most of the gases of the photosphere, chromosphere, and corona from escaping. But the corona's high temperature means that its atoms and ions are moving at very high speeds of about 1 million kilometers per hour. As a result, some of the coronal gas can and does escape. This outflow of gas, which we first encountered in Section 8-6, is called the **solar wind.**

Each second, the Sun ejects about 1 million tons (10^9 kg) of material into the solar wind. But the Sun is so massive that, even over its entire lifetime, it will eject only a few tenths of a percent of its total mass. The solar wind is composed almost entirely of electrons and nuclei of hydrogen and helium. About 0.1% of the solar wind is made up of ions of more massive atoms, such as silicon, sulfur, calcium, chromium, nickel, iron, and argon. The aurorae seen at far northern or southern latitudes on Earth are produced when electrons and ions from the solar wind enter our upper atmosphere.

Special telescopes enable astronomers to see the point of origin of the solar wind. To appreciate what sort of telescopes are needed, note that because the temperature of the coronal gas is so high, ions in the corona have a lot of kinetic energy, so they are moving very fast (see Box 7-2). When ions collide, the energy of the impact is so great that the ion's electrons are boosted to very high energy levels. As the electrons fall back to lower levels, they emit high-energy photons in the ultraviolet and x-ray portions of the spectrum—wavelengths at which the photosphere and chromosphere are relatively dim. Hence, telescopes sensitive to these short wavelengths are ideal for studying the corona and the flow of the solar wind.

Earth's atmosphere absorbs most ultraviolet light and x rays, so telescopes for these wavelengths must be placed above the atmosphere on spacecraft (see Section 6-7, especially Figure 6-25). Figure 16-16 shows an ultraviolet view of the corona from the SOHO spacecraft (Solar and Heliospheric Observatory), a joint project of the European Space Agency (ESA) and the National Aeronautics and Space Administration (NASA).

Figure 16-16 reveals that the corona is not uniform in temperature or density. The densest, highest-temperature regions appear bright, while the thinner, lower-temperature regions are dark. Note the large dark area, called a **coronal hole** because it is almost devoid of luminous gas. Particles streaming away from the Sun can most easily flow outward through these particularly thin regions. Therefore, it is thought that coronal holes are the main corridors through which particles of the solar wind escape from the Sun.

> Unlike the lower levels of the Sun's atmosphere, the corona has immense holes that shift and reshape.

Evidence in favor of this picture has come from the Ulysses spacecraft, another joint ESA/NASA mission. In 1994 and 1995, Ulysses became the first spacecraft to fly over the Sun's north and south poles, where there are apparently permanent coronal holes. The spacecraft indeed measured a stronger solar wind emanating from these holes.

Coronal hole

VIDEO 16.3 **FIGURE 16-16** R I V Ⓤ X G
The Ultraviolet Corona The SOHO spacecraft recorded this false-color ultraviolet view of the solar corona. The dark feature running across the Sun's disk from the top is a coronal hole, a region where the coronal gases are thinner than elsewhere. Such holes are often the source of strong gusts in the solar wind. (SOHO/EIT/ESA/NASA)

The temperatures in the corona and the chromosphere are not at all what we would expect. Just as you feel warm if you stand close to a campfire but cold if you move away, we would expect that the temperature in the corona and chromosphere would *decrease* with increasing altitude—that is, increasing distance from the warmth of the Sun's photosphere. Why, then, does the temperature in these regions *increase* with increasing altitude? This question, also referred to as the **coronal heating problem**, has remained one of the major unsolved mysteries in astronomy for the last half-century.

In 2011, the Solar Dynamics Observatory found evidence that spicules (Section 16-6) play a role in heating the corona. Some spicules can get quite large—the width of a typical state in the United States and as tall as Earth—and shoot gas at about 150,000 mi/h. Observations show that some of the gas ejected by spicules is heated to millions of degrees as it is shot into the corona. With so many spicules present at any given time (about 300,000), estimates indicate that this ejected gas makes a significant contribution to coronal heating.

However, spicules are only one contribution to coronal heating, and as astronomers have tried to resolve this dilemma, they have found important clues in one of the Sun's most familiar features—sunspots.

CONCEPT CHECK 16-14

What is the connection between the solar wind and coronal holes?

Answer appears at the end of the chapter.

16-8 Sunspots are low-temperature regions in the photosphere

Granules, supergranules, spicules, and the solar wind occur continuously. These features are said to be aspects of the *quiet* Sun. But other, more dramatic features appear periodically, including massive eruptions and regions of concentrated magnetic fields. When these features are present, astronomers refer to the *active* Sun. The features of the active Sun that can most easily be seen with even a small telescope are sunspots (although only with a safety filter attached).

Observing Sunspots

Sunspots are irregularly shaped dark regions in the photosphere. Sometimes sunspots appear in isolation (Figure 16-17a), but frequently they are found in sunspot groups (Figure 16-17b; see also Figure 16-7). Although sunspots vary greatly in size, typical ones measure a few tens of thousands of kilometers across—comparable to the diameter of Earth. Sunspots are not permanent features of the photosphere but last between a few hours and a few months.

Each sunspot has a dark central core, called the *umbra,* and a brighter border called the *penumbra.* We used these same terms in Section 3-4 to refer to different parts of Earth's shadow or the Moon's shadow. But a sunspot is not a shadow: It is a region in the photosphere where the temperature is relatively low, which makes it appear darker than its surroundings. As we saw in Section 5-4, Wien's law relates the color of a blackbody, such as the photosphere, to the blackbody's temperature; the cooler the blackbody, the longer the wavelength at which it emits the most light. If the surrounding hot photosphere is blocked from view, a sunspot's cooler umbra appears red and the warmer penumbra appears orange. The colors of a sunspot indicate that the temperature of the umbra is typically 4300 K and that of the penumbra is typically 5000 K. While high by earthly standards, these temperatures are quite a bit lower than the average photospheric temperature of 5800 K.

The Stefan–Boltzmann law (see Section 5-4) tells us that the energy flux from a blackbody is proportional to the fourth power of its temperature. This law lets us compare the amounts of light energy emitted by a square meter of a sunspot's umbra and by a square meter of undisturbed photosphere. The ratio is

$$\frac{\text{flux from umbra}}{\text{flux from photosphere}} = \left(\frac{4300 \text{ K}}{5800 \text{ K}}\right)^4 = 0.30$$

That is, the umbra emits only 30% as much light as an equally large patch of undisturbed photosphere, which is why sunspots appear so dark.

Sunspots and the Sun's Rotation

Occasionally, a sunspot group is large enough to be seen without a telescope. Chinese astronomers recorded such sightings 2000 years ago, and huge sunspot groups visible to the naked eye (with an appropriate filter) were seen in 1989 and 2003. But it was not until Galileo introduced the telescope into astronomy (see Section 4-5) that anyone was able to examine sunspots in detail.

(a)

(b)

VIDEO 16.4

FIGURE 16-17 R I Ⓥ **U X G**

Sunspots (a) This high-resolution photograph of the photosphere shows a mature sunspot. The dark center of the spot is called the umbra and has the lowest temperature. It is bordered by the penumbra, which is less dark, is hotter than the umbra, and has a featherlike appearance. (Even the umbra emits visible light, but not enough to capture in an image against the glare of the Sun.) **(b)** In this view of a typical sunspot group, several sunspots are close enough to overlap. In both images, you can see granulation in the surrounding, undisturbed photosphere. (a: Scharmer et al, Royal Swedish Academy of Sciences/Science Source; b: NSO/NASA)

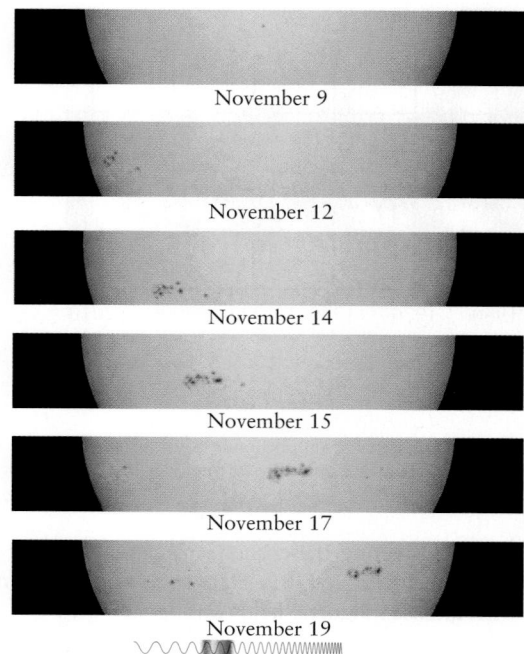

November 9

November 12

November 14

November 15

November 17

November 19

VIDEO 16.5 **FIGURE 16-18** R I **V** U X G

Tracking the Sun's Rotation with Sunspots This series of photographs taken in 1999 shows the rotation of the Sun. By observing the same group of sunspots from one day to the next, Galileo found that the Sun rotates once in about 4 weeks. (The equatorial regions of the Sun actually rotate somewhat faster than the polar regions.) Notice how the sunspot group shown here changed its shape. (Carnegie Observatories)

Galileo discovered that he could determine the Sun's rotation rate by tracking sunspots as they moved across the solar disk (Figure 16-18). He found that the Sun rotates once in about 4 weeks. A typical sunspot group lasts about 2 months, so a specific one can be followed for two solar rotations.

Further observations by the British astronomer Richard Carrington in 1859 demonstrated that the Sun does not rotate as a rigid body. Instead, the equatorial regions rotate more rapidly than the polar regions. This phenomenon is known as **differential rotation**. Thus, while a sunspot near the solar equator takes only 25 days to go once around the Sun, a sunspot at 30° north or south of the equator takes 27½ days. The rotation period at 75° north or south is about 33 days, while near the poles it may be as long as 35 days.

CONCEPT CHECK 16-15

If the center of a sunspot has a temperature of about 4300 K, why does it appear dark?

Answer appears at the end of the chapter.

The Sunspot Cycle

The average number of sunspots on the Sun is not constant, but varies in a predictable **sunspot cycle** (Figure 16-19a). This phenomenon was first reported by the German astronomer Heinrich Schwabe in 1843 after many years of observing. As Figure 16-19a shows, the average number of sunspots varies with a period of about 11 years. A period of exceptionally many sunspots is a **sunspot maximum** (Figure 16-19b), as occurred in 1979, 1989, and 2000. Conversely, the Sun is almost devoid of sunspots at a **sunspot minimum** (Figure 16-19c), as occurred in 1976, 1986, 1996, and 2008. During the 2008 minimum in sunspot activity, there were fewer sunspots observed than in any other year since 1913.

The locations of sunspots also vary with the same 11-year sunspot cycle. At the beginning of a cycle, just after a sunspot

> The number of sunspots increases and decreases on an 11-year cycle.

(a)

(b) Near sunspot maximum

(c) Near sunspot minimum

FIGURE 16-19 R I **V** U X G

The Sunspot Cycle (a) The number of sunspots on the Sun varies with a period of about 11 years. The most recent sunspot maximum occurred in 2000. **(b)** This photograph, taken near sunspot maximum in 1989, shows a number of sunspots and large sunspot groups. The sunspot group visible near the bottom of the Sun's disk has about the same diameter as the planet Jupiter. **(c)** Near sunspot minimum, as in this 1986 photograph, essentially no sunspots are visible. (NSO/AURA/NSF)

FIGURE 16-20

Variations in the Average Latitude of Sunspots The dots in this graph (sometimes called a "butterfly diagram") record how far north or south of the Sun's equator sunspots were observed. At the beginning of each sunspot cycle, most sunspots are found near latitudes 30° north or south. As the cycle goes on, sunspots typically form closer to the equator. (NASA, Marshall Space Flight Center)

minimum, sunspots first appear at latitudes around 30° north and south of the solar equator (Figure 16-20). Over the succeeding years, the sunspots occur closer and closer to the equator.

16-9 Sunspots are produced by a 22-year cycle in the Sun's magnetic field

TUTORIAL 16.3

Why should the number of sunspots vary with an 11-year cycle? Why should their average latitude vary over the course of a cycle? And why should sunspots exist at all? The

first step toward answering these questions came in 1908, when the American astronomer George Ellery Hale discovered that sunspots are associated with intense magnetic fields on the Sun.

Probing Solar Magnetism

LOOKING DEEPER 16.2

When Hale focused a spectroscope on sunlight coming from a sunspot, he found that many spectral lines, which normally appear as a single band, appear to be split into several closely spaced lines (Figure 16-21). This "splitting" of spectral lines is called the **Zeeman effect**, after the Dutch physicist Pieter

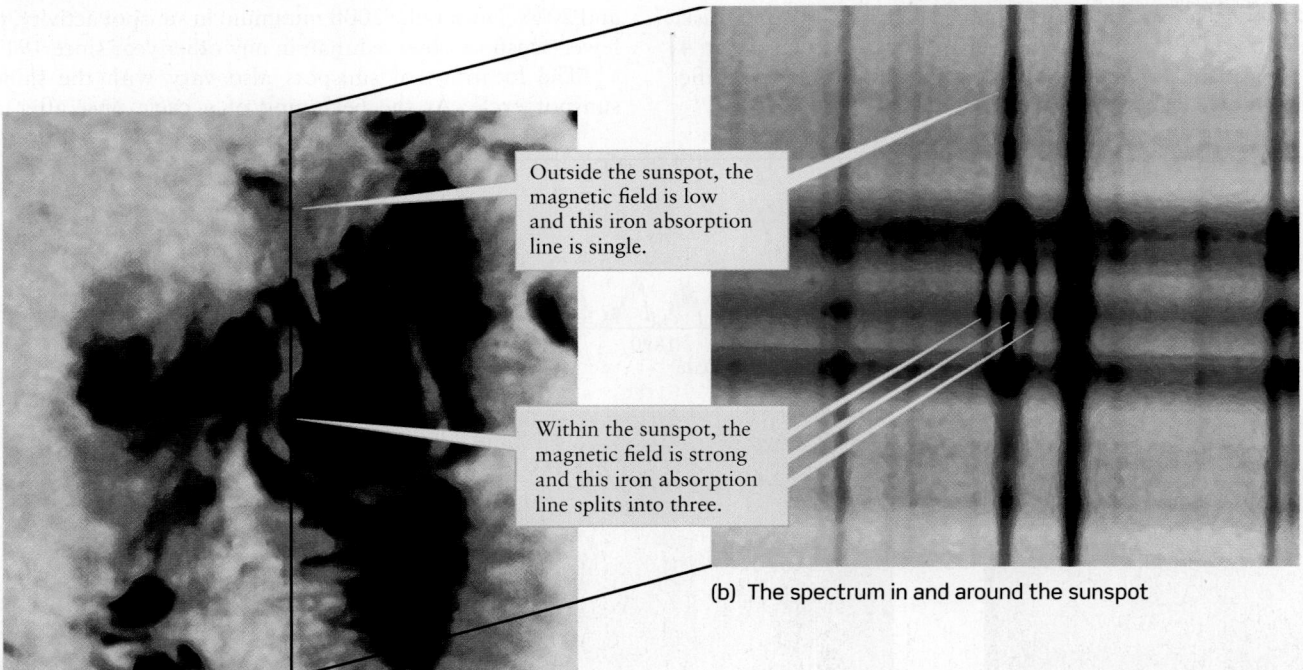

Outside the sunspot, the magnetic field is low and this iron absorption line is single.

Within the sunspot, the magnetic field is strong and this iron absorption line splits into three.

(a) A sunspot

(b) The spectrum in and around the sunspot

FIGURE 16-21 R I V U X G

Sunspots Have Strong Magnetic Fields (a) A black line in this image of a sunspot shows where the slit of a spectrograph was aimed. **(b)** This is a portion of the resulting spectrum, including a dark absorption line caused by iron atoms in the photosphere. The splitting of this line by the sunspot's magnetic field can be used to calculate the field strength. Typical sunspot magnetic fields are over 5000 times stronger than Earth's field at its north and south poles. (NSO/AURA/NSF) gas called a **plasma**, in which electrically charged ions and electrons can move freely. Like any moving, electrically charged objects, they can be deflected by magnetic fields. **Figure 16-22** shows how a magnetic field in the laboratory bends a beam of fast-moving electrons into a curved trajectory. Similarly, the paths of moving ions and electrons in the photosphere are deflected by the Sun's magnetic field. In particular, magnetic forces act on the hot plasma that rises from the Sun's interior due to convection. Where the magnetic field is particularly strong, these forces push the hot plasma away. The result is a localized region where the gas is relatively cool and thus glows less brightly—in other words, a sunspot.

To get a fuller picture of the Sun's magnetic fields, astronomers take images of the Sun at two wavelengths, one just less than and one just greater than the wavelength of a magnetically split spectral line. From the difference between these two images, they can construct a picture called a **magnetogram**, which reveals key details about the magnetic fields in the solar atmosphere. **Figure 16-23a** is an ordinary white-light photograph of the Sun taken at the same time as the magnetogram in Figure 16-23b. In the magnetogram, dark blue indicates areas of the photosphere with one magnetic polarity (north), and yellow indicates areas with the opposite (south) magnetic polarity. This image shows that many sunspot groups have roughly comparable areas covered by north and south magnetic polarities (see also Figure 16-23c). Thus, a sunspot group resembles a giant bar magnet, with a north magnetic pole at one end and a south magnetic pole at the other.

If different sunspot groups were unrelated to one another, their magnetic poles would be randomly oriented, like a bunch of compass needles all pointing in random directions. As Hale discovered, however, there is a striking regularity in the magnetization of sunspot groups. As a given sunspot group moves with the Sun's rotation, the sunspots in front are called the "preceding members" of the group. The spots that follow behind are referred to as the "following members." Hale compared the sunspot groups in the two solar hemispheres, north or south of the Sun's equator. He found that the preceding members in one solar hemisphere all have the same magnetic polarity, while the preceding members in the other hemisphere have the opposite polarity. Furthermore, in the hemisphere where the Sun has its north magnetic pole, the preceding members of all sunspot groups have north magnetic polarity.

FIGURE 16-22 R I **V** U X G

Magnetic Fields Deflect Moving, Electrically Charged Objects
In this laboratory experiment, a beam of negatively charged electrons (shown by a blue arc) is aimed straight upward from the center of the apparatus. The entire apparatus is inside a large magnet, and the magnetic field deflects the beam into a curved path. (Andrew Lambert Photography/Science Source)

Zeeman, who first observed it in his laboratory in 1896. Zeeman showed that a spectral line splits when the atoms are subjected to an intense magnetic field. The more intense the magnetic field, the wider the separation of the split lines.

Hale's discovery showed that sunspots are places where the hot gases of the photosphere are bathed in a concentrated magnetic field. Many of the atoms of the Sun's atmosphere are ionized due to the high temperature. The solar atmosphere is thus a special type of

(a) Visible-light image

(b) Magnetogram

(c) Magnetogram of a sunspot group

FIGURE 16-23 R I **V** U X G

Mapping the Sun's Magnetic Field **(a)** This visible-light image and **(b)** this false-color magnetogram were recorded at the same time. Dark blue and yellow areas in the magnetogram have north and south magnetic polarity, respectively; blue-green regions have weak magnetic fields. The highly magnetized regions in (b) correlate with the sunspots in (a). **(c)** The two ends of this large sunspot group have opposite magnetic polarities (colored blue and yellow), like the ends of a giant bar magnet. (a: NSO/AURA/NSF; b: NOAO/AURA/NSF; c: M. Sigwarth, J. Elrod, K. S. Balasubramaniam, S. Fletcher NSO/AURA/NSF)

In the opposite hemisphere, where the Sun has its south magnetic pole, the preceding members all have south magnetic polarity.

Along with his colleague Seth B. Nicholson, Hale also discovered that the Sun's polarity pattern completely reverses itself every 11 years—the same interval as the time from one solar maximum to the next. The hemisphere that has preceding north magnetic poles during one 11-year sunspot cycle will have preceding south magnetic poles during the next 11-year cycle, and vice versa. The north and south magnetic poles of the Sun itself also reverse every 11 years. Thus, the Sun's magnetic pattern repeats itself only after two sunspot cycles, which is why astronomers speak of a **22-year solar cycle**.

The Magnetic-Dynamo Model

In 1960, the American astronomer Horace Babcock proposed a description that seems to account for many features of this 22-year solar cycle. Babcock's scenario, called a **magnetic-dynamo model**, makes use of two basic properties of the Sun's photosphere—differential rotation and convection. Differential rotation causes the magnetic field in the photosphere to become wrapped around the Sun (**Figure 16-24**). As a result, the magnetic field becomes concentrated at certain latitudes on either side of the solar equator. Convection in the photosphere creates tangles in the concentrated magnetic field, and "kinks" erupt through the solar surface. Sunspots appear where the magnetic field protrudes through the photosphere. This theory suggests that sunspots should appear first at northern and southern latitudes and later form nearer to the equator, which is just what is observed (see Figure 16-20). Note also that as shown on the far right in Figure 16-24, the preceding member of a sunspot group has the same polarity (N or S) as the Sun's magnetic pole in that hemisphere, which is just as Hale observed.

Differential rotation eventually undoes the twisted magnetic field. The preceding members of sunspot groups move toward the Sun's equator, while the following members migrate toward the poles. Because the preceding members from the two hemispheres have opposite magnetic polarities, their magnetic fields cancel each other out when they meet at the equator. The following members in each hemisphere have the opposite polarity to the Sun's pole in that hemisphere; hence, when they converge on the pole, the following members first cancel out and then reverse the Sun's overall magnetic field. The fields are now completely relaxed. Once again, differential rotation begins to twist the Sun's magnetic field, but now with all magnetic polarities reversed. In this way, Babcock's model helps to explain the change in field direction every 11 years.

> The Sun's differential rotation makes the magnetic field twist like a rubber band.

Recent discoveries in helioseismology (Section 16-3) offer new insights into the Sun's magnetic field. By comparing the speeds of sound waves that travel with and against the Sun's rotation, helioseismologists have been able to determine the Sun's rotation rate at different depths and latitudes. As shown in **Figure 16-25**, the Sun's surface pattern of differential rotation persists through the convective zone. Farther in, within the radiative zone, the Sun seems to rotate like a rigid object with a period of 27 days at all latitudes. Astronomers suspect that the Sun's magnetic field originates in a relatively thin layer where the radiative and convective zones meet and slide past each other due to their different rotation rates.

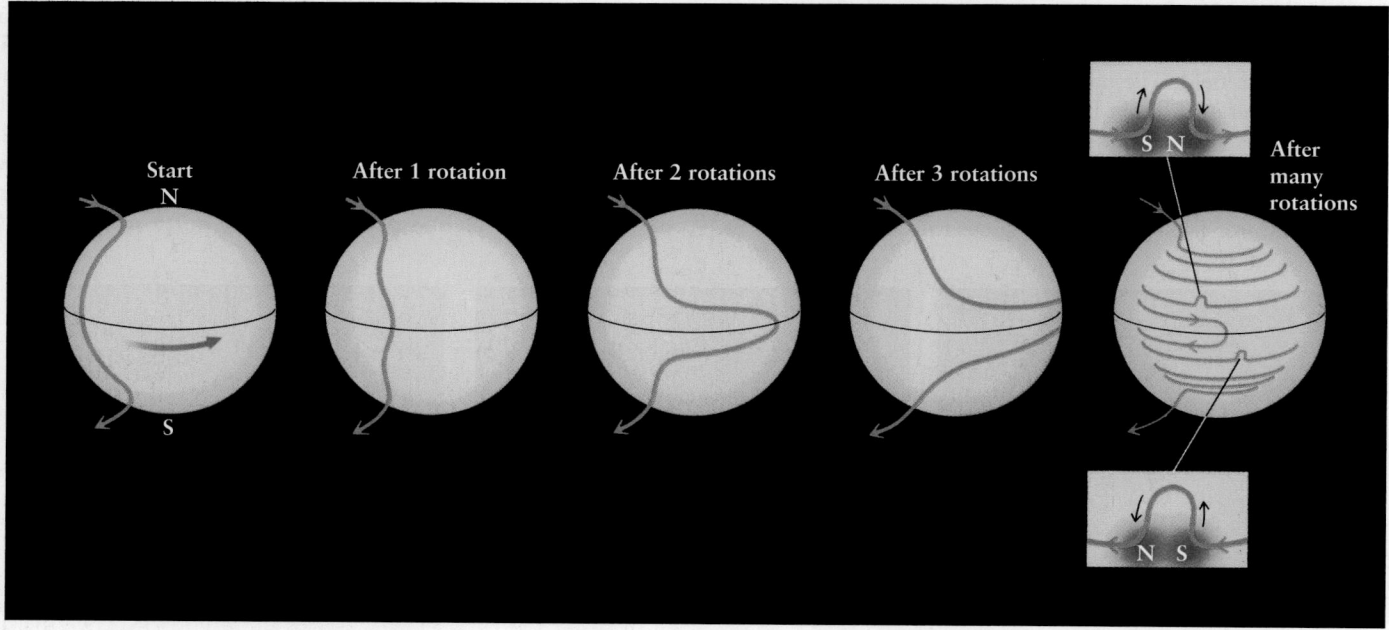

FIGURE 16-24

Babcock's Magnetic Dynamo Model Magnetic field lines tend to move along with the plasma in the Sun's outer layers. Because the Sun rotates faster at the equator than near the poles, a field line that starts off running from the Sun's north magnetic pole (N) to its south magnetic pole (S) ends up wrapped around the Sun like twine wrapped around a ball. The insets on the far right show how sunspot groups appear where the concentrated magnetic field rises through the photosphere.

FIGURE 16-25

Rotation of the Solar Interior This cutaway picture of the Sun shows how the solar rotation period (shown by different colors) varies with depth and latitude. The surface and the convective zone have differential rotation (a short period at the equator and longer periods near the poles). Deeper within the Sun, the radiative zone seems to rotate like a rigid sphere. (Courtesy of K. Libbrecht, Big Bear Solar Observatory)

One dilemma about sunspots is that compressed magnetic fields tend to push themselves apart, which means that we would expect sunspots to dissipate rather quickly. Yet observations show that sunspots can persist for many weeks. The resolution of this paradox may have been found using helioseismology (see Section 16-3). Analysis of the vibrations of the Sun around sunspots shows that beneath the surface of the photosphere, the gases surrounding each sunspot are circulating at high speed—rather like a hurricane as large as Earth. The circulation of charged gases around the magnetic field holds the fields in place, thus stabilizing the sunspot.

Helioseismology—the analysis of solar sound waves—can also take advantage of the relationship between sunspots and the solar magnetic field. Compared to typical solar material, sunspots absorb solar sound waves more strongly. When sunspots occur on the *opposite* side of the Sun, their increased absorption means that interior sound waves do not reflect back as strongly through the body of the Sun. Amazingly, this effect is used to estimate sunspot activity on the *opposite side* of the Sun. Knowing the number of sunspots facing away from Earth is actually useful: Added to the sunspots facing Earth, the *total* sunspot activity is continuously monitored to help forecast **space weather**—these are variations in the solar wind and magnetic field that can affect satellites and astronauts.

Much about sunspots and solar activity remains mysterious. There are perplexing irregularities in the solar cycle. For example, the overall reversal of the Sun's magnetic field is often piecemeal and haphazard. One pole may reverse polarity long before the other. For several weeks, the Sun's surface may have two north magnetic poles and no south magnetic pole at all.

Furthermore, there seem to be times when all traces of sunspots and the sunspot cycle vanish for many years. For example, virtually no sunspots were seen from 1645 through 1715. Curiously, during these same years Europe experienced record low temperatures, often referred to as the Little Ice Age, whereas the western United States was subjected to severe drought. By contrast, there was apparently a period of increased sunspot activity during the eleventh and twelfth centuries, during which Earth was warmer than it is today. Thus, variations in solar activity appear to affect climates on Earth. The origin of this Sun–Earth connection is a topic of ongoing research.

CONCEPT CHECK 16-16

Is the sunspot cycle an 11-year cycle or a 22-year cycle?

CONCEPT CHECK 16-17

How might the Sun's sunspot cycle change if the Sun were rotating much faster than it is now?

Answers appear at the end of the chapter.

16-10 The Sun's magnetic field heats the corona, produces flares, and causes massive eruptions

Astronomers now understand that the Sun's magnetic field does more than just explain the presence of sunspots. It is also responsible for the existence of spicules, as well as a host of other dramatic phenomena in the chromosphere and corona.

Magnetic Reconnection

In a plasma, magnetic field lines and the material of the plasma tend to move together. This means that as convection pushes material toward the edge of a supergranule, it pushes magnetic field lines as well. The result is that vertical magnetic field lines pile up around a supergranule. Plasma that "sticks" to these magnetic field lines thus ends up lifted upward, forming a spicule (see Figures 16-12 and 16-13).

The tendency of plasma to follow the Sun's magnetic field can also explain why the temperature of the chromosphere and corona is so high. Spacecraft observations show magnetic field arches extending tens of thousands of kilometers into the corona, with streamers of electrically charged particles moving along each arch (**Figure 16-26a**). If the magnetic fields of two arches come into proximity, their magnetic fields can rearrange in a process called **magnetic reconnection** (Figure 16-26b). The tremendous amount of energy stored in the magnetic field is then released into the solar atmosphere. (A single arch contains as much energy as a hydroelectric power plant would generate in a million years.) The amount of energy released in this way is significant and could be a major contributor to the corona's unexpectedly high temperature. In other words, additional research could establish that magnetic reconnections help to solve the coronal heating problem.

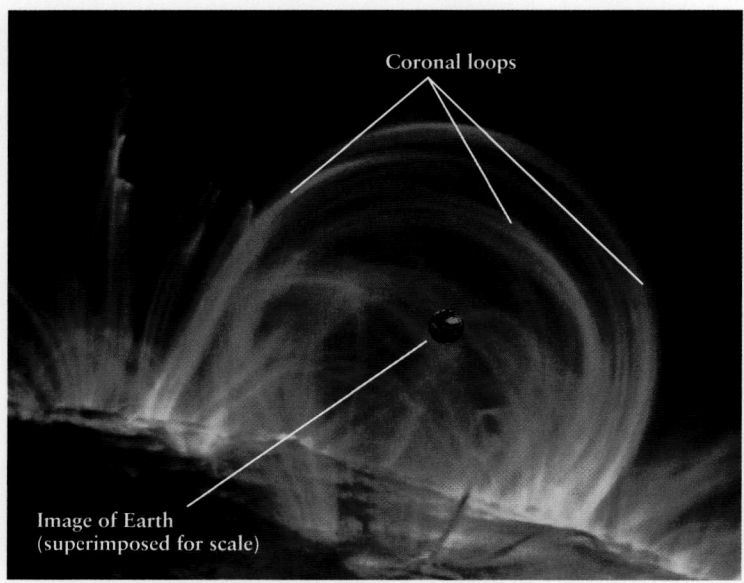

(a)

FIGURE 16-26 R I V **U** X G

Magnetic Arches and Magnetic Reconnection **(a)** This false-color ultraviolet image from the TRACE spacecraft (Transition Region and Coronal Explorer) shows magnetic field loops suspended high above the solar surface. The

1. If magnetic field loops begin to pinch together . . .

2. . . . the field lines of adjacent loops can reconnect, causing a release of energy.

3. The upper helix or "coil" of magnetic field can break loose, carrying material with it into space.

Sun's surface

Before magnetic reconnection   After magnetic reconnection

(b)

loops are made visible by the glowing gases trapped within them. **(b)** When the magnetic fields in these loops change their arrangement, a tremendous amount of energy is released and solar material can be ejected upward. (NASA/LMSAL)

ANALOGY

The idea that a magnetic field can heat gases has applications on Earth as well as on the Sun. In an automobile engine's ignition system, an electric current is set up in a coil of wire, which produces a magnetic field. When the current is shut off, the magnetic field collapses and its energy is directed to a spark plug in one of the engine's cylinders. The released energy heats the mixture of air and gasoline around the plug, causing the mixture to ignite. This drives the piston in that cylinder and makes the automobile go.

VIDEO
16.6

Magnetic heating can also explain why the parts of the corona that lie on top of sunspots are often the most prominent in ultraviolet images. (Some examples are the bright regions in Figure 16-16.) The intense magnetic field of the sunspots helps trap and compress hot coronal gas, giving it such a high temperature that it emits copious amounts of high-energy ultraviolet photons and even more energetic x-ray photons.

CONCEPT CHECK 16-18

Why does glowing plasma on the Sun appear to arch up above the Sun's photosphere?

Answer appears at the end of the chapter.

Prominences, Flares, and Coronal Mass Ejections

Spicules and coronal heating occur even when the Sun is quiet. But magnetic fields can also explain many aspects of the active Sun in addition to sunspots. **Figure 16-27** is an image of the chromosphere made with an H_α filter during a sunspot maximum. The bright areas are called **plages** (from the French word for "beach").

These plages are bright, hot regions in the chromosphere that tend to form just before the appearance of new sunspots. They are probably created by magnetic fields that push upward from the Sun's interior, compressing and heating a portion of the chromosphere. The dark streaks, called **filaments**, are relatively cool and dense parts of the chromosphere that have been pulled along with magnetic field lines as they arch to high altitudes. Images with an

Prominences

Plages

Filaments

FIGURE 16-27 R I V **U** X G

The Active Sun Seen through an H_α Filter This image was made using a red filter that only passes light at a wavelength of 656 nm. The spectrum of the photosphere has an absorption line at this wavelength and so appears dark. Hence, this filter reveals the photosphere and corona. Prominences, plages, and filaments are associated with strong magnetic fields. (NASA)

FIGURE 16-28 R I V **U** X G

The Sun in Ultraviolet Light This false-color image was made in ultraviolet light by the Solar Dynamics Observatory. A very long filament is seen cutting across the Sun's southern hemisphere; it is longer than the distance from Earth to the Moon! The filament is confined by a strong magnetic field, and above the filament is a bright sunspot group. (NASA/SDO)

H_α filter are made from visible light, and images using ultraviolet light can highlight filaments and sunspots (**Figure 16-28**).

When seen from the side, so that they are viewed against the dark background of space, filaments appear as bright, arching columns of gas called **prominences** (**Figure 16-29**). They can extend for tens of thousands of kilometers above the photosphere. Some prominences last for only a few hours, while others persist for many months. The most energetic prominences break free of the magnetic fields that confined them and burst into space.

VIDEO 16.7 ➤ Magnetic reconnection also causes violent, eruptive events on the Sun, called **solar flares**. Rooted to the same magnetic phenomena, the solar flares occur in complex sunspot groups. Within only a few minutes, temperatures in a compact region may soar to 5×10^6 K and vast quantities of particles and radiation—including as much material as is in the prominence shown in Figure 16-29—are blasted out into space. These eruptions can also cause disturbances that spread outward in the solar atmosphere, like the ripples that appear when you drop a rock into a pond.

While the external effects of the Sun's magnetism are on full display with solar flares, the advancing magnetic fields can also be discerned below the surface. Observations from NASA's Solar Dynamics Observatory employed helioseismology to detect magnetic fields about 65,000 km below the surface (that distance is about 2½ times Earth's circumference). As predicted, the effects of these magnetic fields emerged several days later as sunspots.

The most energetic flares carry as much as 10^{30} joules of energy, equivalent to 10^{14}

> A solar flare can have as much energy as 100 trillion nuclear bombs.

FIGURE 16-29 R I V **U** X G

A Solar Prominence A huge prominence arches above the solar surface in this ultraviolet image from the SOHO spacecraft. The image was recorded using ultraviolet light at a wavelength of 30.4 nm, emitted by singly ionized helium atoms at a temperature of about 60,000 K. By comparison, the material within the arches in Figure 16-25 reaches temperatures in excess of 2×10^6 K. (SOHO/EIT/ESA/NASA)

one-megaton nuclear weapons being exploded at once! However, the energy of a solar flare does not come from thermonuclear fusion in the solar atmosphere; instead, it appears to be released from the intense magnetic field around a sunspot group.

FIGURE 16-30 R I V **U** X G

Coronal Mass Ejection The Sun produced a huge coronal mass ejection over several hours on April 16, 2012. Not all of the gas was shot into the solar system; some material from this event was observed crashing back onto the solar surface. (This ejection was not directed toward Earth.) (NASA)

As energetic as solar flares are, they are dwarfed by **coronal mass ejections**. One such event is shown in **Figure 16-30**. If directed toward Earth, these events are large enough to affect us directly (**Figure 16-31**). In a coronal mass ejection, more than 10^{12} kilograms of high-temperature coronal gas is blasted into space at speeds of hundreds of kilometers per second. That is about 200 times the mass of the largest Egyptian pyramid, and a typical coronal mass ejection lasts a few hours. These explosive events seem to be related to large-scale alterations in the Sun's magnetic field, like the magnetic reconnection shown in Figure 16-26b. Coronal mass ejections occur every few months; smaller eruptions may occur almost daily.

If a solar flare or coronal mass ejection happens to be aimed toward Earth, a stream of high-energy electrons and nuclei reaches us a few days later (Figure 16-31b); this is part of space weather (Section 16-9). When this plasma arrives, it can interfere with satellites, pose a health hazard to astronauts in orbit, and disrupt electrical and communications equipment on Earth's surface. Telescopes on Earth and on board spacecraft now monitor the Sun continuously to provide warnings of dangerous levels of solar particles. By design, the walls of the International Space Station are thick enough to protect astronauts from material ejected during regular solar activity, but extra precautions are needed during a stronger solar event.

The numbers of plages, filaments, solar flares, and coronal mass ejections all vary with the same 11-year cycle as sunspots. But unlike sunspots, coronal mass ejections never completely cease, even when the Sun is at its quietest. Astronomers are devoting substantial effort to understanding these and other aspects of our dynamic Sun.

CONCEPT CHECK 16-19

Which of the following phenomena is the most energetic: prominences, solar flares, or coronal mass ejections?

Answer appears at the end of the chapter.

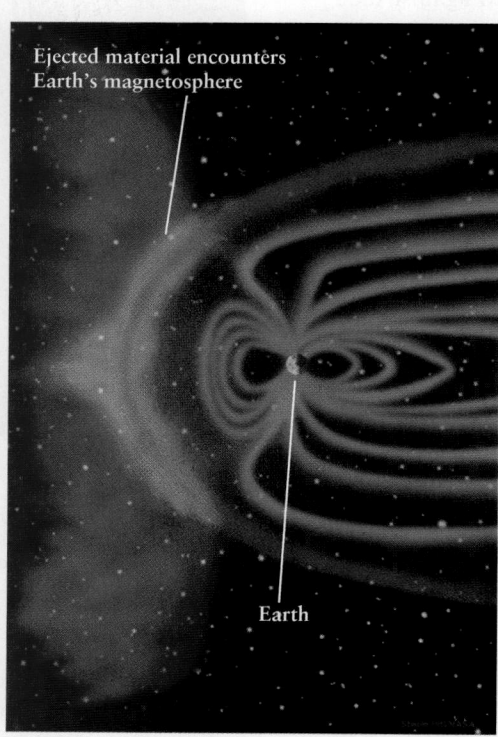

(a) A coronal mass ejection

(b) Two to four days later

FIGURE 16-31 R I V **U** X G

A Sun–Earth Connection (a) SOHO recorded this coronal mass ejection in an x-ray image. (The image of the Sun itself was made at ultraviolet wavelengths.) **(b)** In this artist's illustration, we see that within 2 to 4 days the fastest-moving ejected material reaches a distance of 1 au from the Sun. Most particles are deflected by Earth's magnetosphere, but some are able to reach Earth, where they can damage satellites and our electrical power grid. [The ejection shown in (a) was not aimed toward Earth and did not affect us.] (Steele Hill/NASA)

KEY WORDS

The term preceded by an asterisk (*) is discussed in Box 16-1.

KEY IDEAS

Hydrogen Fusion in the Sun's Core: The Sun's energy is produced by hydrogen fusion, a sequence of thermonuclear reactions in which four hydrogen nuclei combine to produce a single helium nucleus (Section 16-1).

- The energy released in a nuclear reaction corresponds to a slight reduction of mass according to Einstein's equation $E = mc^2$.

- Thermonuclear fusion occurs only at very high temperatures; for example, hydrogen fusion occurs only at temperatures in excess of about 10^7 K. In the Sun, fusion occurs only in the dense, hot core.

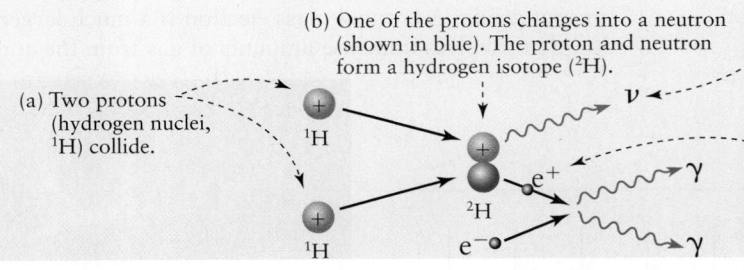

(a) Two protons (hydrogen nuclei, ^{1}H) collide.

(b) One of the protons changes into a neutron (shown in blue). The proton and neutron form a hydrogen isotope (^{2}H).

(c) One by-product of converting a proton to a neutron is a neutral, nearly massless neutrino (ν), which escapes from the Sun.

(d) The other by-product of converting a proton to a neutron is a positively charged electron, or positron (e^+). This encounters an ordinary electron (e^-), annihilating both particles and converting them into gamma-ray photons (γ). The energy of these photons goes into sustaining the Sun's internal heat.

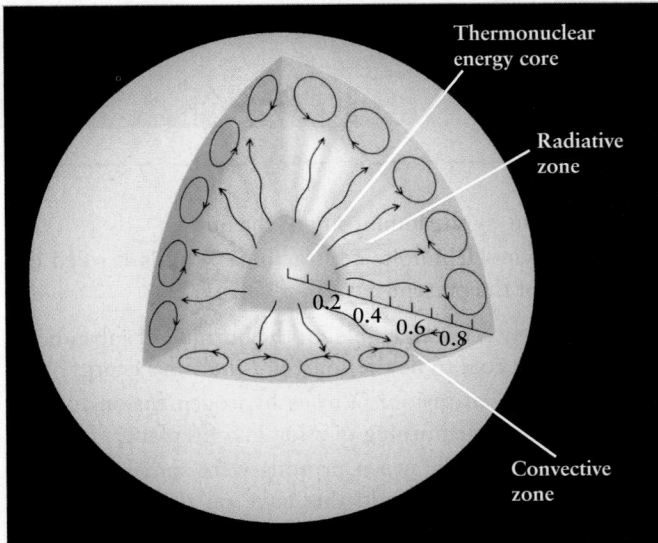

Thermonuclear energy core

Radiative zone

Convective zone

Models of the Sun's Interior: A theoretical description of a star's interior can be calculated using the laws of physics (Section 16-2).

- The standard model of the Sun suggests that hydrogen fusion takes place in a core extending from the Sun's center to about 0.25 solar radius.

- The core is surrounded by a radiative zone extending to about 0.71 solar radius. In this zone, energy travels outward through radiative diffusion.

- The radiative zone is surrounded by a rather opaque convective zone of gas at relatively low temperature and pressure. In this zone, energy travels outward primarily through convection.

Solar Neutrinos and Helioseismology: Conditions in the solar interior can be inferred from measurements of solar neutrinos and of solar vibrations (Sections 16-3 and 16-4).

- Neutrinos emitted in thermonuclear reactions in the Sun's core have been detected, but in smaller numbers than expected. Recent neutrino experiments explain why this is so.

- Helioseismology is the study of how the Sun vibrates. These vibrations have been used to infer pressures, densities, chemical compositions, and rotation rates within the Sun.

The Sun's Atmosphere: The Sun's atmosphere has three main layers: the photosphere, the chromosphere, and the corona. Everything below the solar atmosphere is called the solar interior (Sections 16-5, 16-6, and 16-7).

- The visible surface of the Sun, the photosphere, is the lowest layer in the solar atmosphere. Its spectrum is similar to that of a blackbody at a temperature of 5800 K. Convection in the photosphere produces granules.

- Above the photosphere is a layer of less dense but higher-temperature gases called the chromosphere. Spicules extend upward from the chromosphere into the corona.

- The outermost layer of the solar atmosphere, the corona, is made of very high-temperature gases at extremely low density. Activity in the corona includes coronal mass ejections and coronal holes. The solar corona blends into the solar wind at great distances from the Sun.

S. Habbal, M. Druckmuller, P. Aniol/NASA; Celestron International

The Active Sun: The Sun's surface features vary in an 11-year cycle. This is related to a 22-year cycle in which the surface magnetic field increases, decreases, and then increases again with the opposite polarity (Sections 16-8, 16-9, and 16-10).

- Sunspots are relatively cool regions produced by local concentrations of the Sun's magnetic field. The average number of sunspots increases and decreases in a regular cycle of approximately 11 years, with reversed magnetic polarities from one 11-year cycle to the next. Two such cycles make up the 22-year solar cycle.

- The magnetic-dynamo model suggests that many features of the solar cycle are due to changes in the Sun's magnetic field. These changes are caused by convection and the Sun's differential rotation.

- A solar flare is a brief eruption of hot, ionized gases from a sunspot group. A coronal mass ejection is a much larger eruption that involves immense amounts of gas from the corona.

NASA

NASA, Marshall Space Flight Center

QUESTIONS

Problem-solving tips and tools

You may have to review Wien's law and the Stefan–Boltzmann law, which are the subjects of Section 5-4. Section 5-5 discusses the properties of photons. As we described in Box 5-2, you can simplify calculations by taking ratios, such as the ratio of the flux from a sunspot to the flux from the undisturbed photosphere. When you do this, all the cumbersome constants cancel out. Figure 5-7 shows the various parts of the electromagnetic spectrum. We introduced the Doppler effect in Section 5-9 and Box 5-6. For information about the planets, see Table 7-1.

Section 16-1

1. What is meant by the luminosity of the Sun?

2. What is Kelvin–Helmholtz contraction? Why is it ruled out as a source of the present-day Sun's energy?

3. **TUTORIAL 16.1** Why is it impossible for the burning of substances like coal to be the source of the Sun's energy?

4. What is hydrogen fusion? Why is hydrogen fusion fundamentally unlike the burning of a log in a fireplace?

5. If the electric force between protons were somehow made stronger, what effect would this have on the temperature required for thermonuclear fusion to take place?

6. Why do thermonuclear reactions occur only in the Sun's core, not in its outer regions?

7. Describe how the net result of the reactions shown in the *Cosmic Connections* figure (Section 16-1) is the conversion of four protons into a single helium nucleus. What other particles are produced in this process? How many of each particle are produced?

8. Calculate how much energy would be released if each of the following masses were converted *entirely* into their equivalent energy: (a) a carbon atom with a mass of 2×10^{-26} kg, (b) 1 kg, and (c) a planet as massive as Earth (6×10^{24} kg).

9. Use the luminosity of the Sun (given in Table 16-1) and the answers to the previous question to calculate how long the Sun must shine in order to release an amount of energy equal to that produced by the complete mass-to-energy conversion of (a) a carbon atom, (b) 1 km, and (c) Earth.

10. Assuming that the current rate of hydrogen fusion in the Sun remains constant, what fraction of the Sun's mass will be converted into helium over the next 5 billion years? How will this affect the overall chemical composition of the Sun?

11. (a) Estimate how many kilograms of hydrogen the Sun has consumed over the past 4.56 billion years, and estimate the amount of mass that the Sun has lost as a result. Assume that the Sun's luminosity has remained constant during that time. (b) In fact, the Sun's luminosity when it first formed was only about 70% of its present value. With this in mind, explain whether your answers to part (a) are an overestimate or an underestimate.

12. To convert 1 kg of hydrogen (^{1}H) into helium (^{4}He) as described in Box 16-1, you must start with 1.5 kg of hydrogen. Explain why, and explain what happens to the other 0.5 kg. (*Hint*: How many ^{1}H nuclei are used to make the two ^{3}He nuclei shown in the *Cosmic Connections* figure in Section 16-1? How many of these ^{1}H nuclei end up being incorporated into the ^{4}He nucleus shown in the figure?)

13. (a) A positron has the same mass as an electron (see Appendix 7). Calculate the amount of energy released by the annihilation of an electron and positron. (b) The products of this annihilation are two photons, each of equal energy. Calculate the wavelength of each photon, and confirm from Figure 5-7 that this wavelength is the gamma-ray range.

14. Sirius is the brightest star in the night sky. It has a luminosity of 23.5 $L_\odot$, that is, it is 23.5 times as luminous as the Sun and burns hydrogen at a rate 23.5 times greater than the Sun. How many kilograms of hydrogen does Sirius convert into helium each second?

15. (Refer to the preceding question.) Sirius has 2.3 times the mass of the Sun. Do you expect that the lifetime of Sirius will be longer, shorter, or the same length as that of the Sun? Explain your reasoning.

Section 16-2

16. **TUTORIAL 16.2** Give an everyday example of hydrostatic equilibrium. Give an example of thermal equilibrium. Explain how these equilibrium conditions apply to each example.

17. If thermonuclear fusion in the Sun were suddenly to stop, what would eventually happen to the overall radius of the Sun? Justify your answer using the ideas of hydrostatic equilibrium and thermal equilibrium.

18. Give some everyday examples of conduction, convection, and radiative diffusion.

19. Describe the Sun's interior. Include references to the main physical processes that occur at various depths within the Sun.

20. (a) If the Sun were not in a state of hydrostatic equilibrium, would its diameter remain the same? Explain your reasoning. (b) If the Sun were not in a state of thermal equilibrium, would its luminosity remain the same? What about its surface temperature? Explain your reasoning.

21. Using the mass and size of the Sun given in Table 16-1, verify that the average density of the Sun is 1410 kg/m^3. Compare your answer with the average densities of the jovian planets.

22. Use the data in Table 16-2 to calculate the average density of material within 0.1 $R_\odot$ of the center of the Sun. (You will need to use the mass and radius of the Sun as a whole, given in Table 16-1.) Explain why your answer is not the same as the density at 0.1 $R_\odot$ given in Table 16-2.

Section 16-3

23. Explain how studying the oscillations of the Sun's surface can give important, detailed information about physical conditions deep within the Sun.

24. In a typical solar oscillation, the Sun's surface moves up or down at a maximum speed of 0.1 m/s. An astronomer sets out to measure this speed by detecting the Doppler shift of an absorption line of iron with wavelength 557.6099 nm. What is the maximum wavelength shift that she will observe?

Section 16-4

25. What is a neutrino? Why is it useful to study neutrinos coming from the Sun? What do they tell us that cannot be learned from other avenues of research?

26. Unlike all other types of telescopes, neutrino detectors are placed deep underground. Why?

27. What was the solar neutrino problem? What solution to this problem was suggested by the results from the Sudbury Neutrino Observatory?

28. Explain why the results from the Sudbury Neutrino Observatory (SNO) only provide an answer to the solar neutrino problem for relatively high-energy neutrinos. (*Hint:* Can SNO detect solar neutrinos of all energies?)

Section 16-5

29. Briefly describe the three layers that make up the Sun's atmosphere. In what ways do they differ from one another?

30. What is solar granulation? Describe how convection gives rise to granules.

31. High-resolution spectroscopy of the photosphere reveals that absorption lines are blueshifted in the spectrum of the central, bright regions of granules but are redshifted in the spectrum of the dark boundaries between granules. Explain how these observations show that granulation is due to convection.

32. What is the difference between granules and supergranules?

33. The amount of energy required to dislodge the extra electron from a negative hydrogen ion is 1.2×10^{-19} J. (a) The extra electron can be dislodged if the ion absorbs a photon of sufficiently short wavelength. (Recall from Section 5-5 that the higher the energy of a photon, the shorter its wavelength.) Find the longest wavelength (in nm) that can accomplish this. (b) In what part of the electromagnetic spectrum does this wavelength lie? (c) Would a photon of visible light be able to dislodge the extra electron? Explain your answer. (d) Explain why the photosphere, which contains negative hydrogen ions, is quite opaque to visible light but is less opaque to light with wavelengths longer than the value you calculated in (a).

Section 16-6

34. What are spicules? Where are they found? How can you observe them? What causes them?

35. Astronomers often use an H$_\alpha$ filter to view the chromosphere. Explain why this can also be accomplished with filters that are transparent only to the wavelengths of the H and K lines of ionized calcium. (*Hint:* The H and K lines are dark lines in the spectrum of the photosphere.)

36. Spicules can attain high speeds as they rise above the chromosphere. Use values from Section 16-6 to estimate the speed of spicules in km/h and mi/h.

Section 16-7

37. How do astronomers know that the temperature of the corona is so high?

38. If the conditions within the corona were recreated in a lab for 1 hour, including its high temperature of 2 million kelvins, why wouldn't you be able to bake a potato in this simulated corona?

39. What particles make up the majority of the solar wind? Do these particles ever reach Earth?

40. What is the connection between the solar wind and coronal holes?

41. What is the coronal heating problem?

Section 16-8

42. **TUTORIAL 16.3** How do astronomers know when the next sunspot maximum and minimum will occur?

43. Find the wavelength of maximum emission of the umbra of a sunspot and the wavelength of maximum emission of a sunspot's penumbra. In what part of the electromagnetic spectrum do these wavelengths lie?

44. (a) Find the ratio of the energy flux from a patch of a sunspot's penumbra to the energy flux from an equally large patch of undisturbed photosphere. Which patch is brighter? (b) Find the ratio of the energy flux from a patch of a sunspot's penumbra to the energy flux from an equally large patch of umbra. Again, which patch is brighter?

45. Suppose that you want to determine the Sun's rotation rate by observing its sunspots. Is it necessary to take Earth's orbital motion into account? Why or why not?

Section 16-9

46. **TUTORIAL 16.3** Why do astronomers say that the solar cycle is really 22 years long, even though the number of sunspots varies over an 11-year period?

47. Explain how the magnetic-dynamo model accounts for the solar cycle.

48. The moving images on a television set that uses a tube (as opposed to an LCD or plasma flat-screen TV) are made by a fast-moving electron beam that sweeps over the back of the screen. Explain why placing a strong magnet next to the screen distorts the picture. (*Caution:* Do not try this with a tube television; it can cause permanent damage.)

49. (a) Using a ruler to make measurements of Figure 16-26, determine how far the arches in that figure extend above the Sun's surface. The diameter of Earth is 12,756 km. (b) In Figure 16-26 (an ultraviolet image) the photosphere appears dark compared to the arches. Explain why.

50. The amount of visible light emitted by the Sun varies only a little over the 11-year sunspot cycle. But the number of x rays emitted by the Sun can be 10 times greater at solar maximum than at solar minimum. Explain why these two types of radiation should be so different in their variability.

Section 16-10

51. Describe one explanation involving magnetism for why the corona has a higher temperature than the chromosphere. Could this mechanism help to solve one of the long-standing problems in astronomy?

52. Why should solar flares and coronal mass ejections be a concern for businesses that use telecommunication satellites?

53. Discuss some of the difficulties in correlating solar activity with changes in Earth's climate.

General Problems

54. Suppose thermonuclear fusion in the Sun stopped abruptly. Would the intensity of sunlight decrease just as abruptly? Why or why not?

55. Describe the dangers in attempting to observe the Sun. How have astronomers learned to circumvent these observational problems?

56. Calculate the wavelengths at which the photosphere, chromosphere, and corona emit the most radiation. Explain how the results of your calculations suggest the best way to observe these regions of the solar atmosphere. (*Hint:* Treat each part of the atmosphere as a perfect blackbody. Assume average temperatures of 50,000 K and 1.5×10^6 K for the chromosphere and corona, respectively.)

57. On November 15, 1999, the planet Mercury passed in front of the Sun as seen from Earth. The TRACE spacecraft made these time-lapse images of this event using ultraviolet light (top) and visible light (bottom). (Mercury moved from left to right in these images. The time between successive views of Mercury is 6 to 9 minutes.) Explain why the Sun appears somewhat larger in the ultraviolet image than in the visible-light image.

R I **V** U X G

R I V U X G

(K. Schrijver, Stanford-Lockheed Institute for Space Research; TRACE; and NASA)

58. Discuss the extent to which cultures around the world have worshipped the Sun as a deity throughout history. Why do you suppose there has been such widespread veneration?

59. In the movie *Star Trek IV: The Voyage Home*, the starship *Enterprise* flies on a trajectory that passes close to the Sun's surface. What features should a real spaceship have to survive such a flight? Why?

60. Describe some of the advantages and disadvantages of observing the Sun (**a**) from space and (**b**) from Earth's south pole. What kinds of phenomena and issues might solar astronomers want to explore from these locations?

WEB/E-BOOK QUESTIONS

61. Search the internet for information about features in the solar atmosphere called *sigmoids*. What are they? What causes them? How might they provide a way to predict coronal mass ejections?

62. **VIDEO 16.1** **Determining the Lifetime of a Solar Granule.** Access and view the video "Granules on the Sun's Surface" in Chapter 16 of the *Universe* website or e-book. Your task is to determine the approximate lifetime of a solar granule on the photosphere. Select an area, then slowly and rhythmically repeat "start, stop, start, stop" until you can consistently predict the appearance and disappearance of granules. While keeping your rhythm, move to a different area of the video and continue monitoring the appearance and disappearance of granules. When you are confident you have the timing right, move your eyes (or use a partner) to look at the clock shown in the video. Determine the length of time between the appearance and disappearance of the granules and record your answer.

ACTIVITIES

Observing Projects

Observing tips and tools

At the risk of repeating ourselves, we remind you to ***never look directly at the Sun, because it can easily cause permanent blindness.*** You can view the Sun safely without a telescope just by using two pieces of white cardboard. First, use a pin to poke a small hole in one piece of cardboard; this will be your "lens," and the other piece of cardboard will be your "viewing screen." Hold the "lens" piece of cardboard so that it is face on to the Sun and sunlight can pass through the hole. With your other hand, hold the "viewing screen" so that the sunlight from the "lens" falls on it. Adjust the distance between the two pieces of cardboard so that you see a sharp image of the Sun on the "viewing screen." This image is perfectly safe to view. It is actually possible to see sunspots with this low-tech apparatus.

For a better view, use a telescope with a solar filter that fits on the front of the telescope. A standard solar filter is a piece of glass coated with a thin layer of metal to give it a mirrorlike appearance. This coating reflects almost all the sunlight that falls on it, so that only a tiny, safe amount of sunlight enters the telescope. An H$_\alpha$ filter, which looks like a red piece of glass, keeps the light at a safe level by admitting only a very narrow range of wavelengths. (Filters that fit on the back of the telescope are *not* recommended. The telescope focuses concentrated sunlight on such a filter, heating it and making it susceptible to cracking—and if the filter cracks when you are looking through it, your eye will be ruined instantly and permanently.)

To use a telescope with a solar filter, first aim the telescope away from the Sun, then put on the filter. Keep the lens cap on the telescope's secondary wide-angle "finder scope" (if it has one), because the heat of sunlight can fry the finder scope's optics. Next, aim the telescope toward the Sun, using the telescope's shadow to judge when you are pointed in the right direction. You can then safely look through the telescope's eyepiece. When you are done, make sure you point the telescope away from the Sun before removing the filter and storing the telescope.

Note that the amount of solar activity that you can see (sunspots, filaments, flares, prominences, and so on) will depend on where the Sun is in its 11-year sunspot cycle.

63. Use a telescope with a solar filter to observe the surface of the Sun. Do you see any sunspots? Sketch their appearance. Can you distinguish between the umbrae and penumbrae of the sunspots? Can you see limb darkening? Can you see any granulation?

64. If you have access to an H$_\alpha$ filter attached to a telescope especially designed for viewing the Sun safely, use this instrument to examine the solar surface. How does the appearance of the Sun differ from that in white light? What do sunspots look like in H$_\alpha$? Can you see any prominences? Can you see any filaments? Are the filaments in the H$_\alpha$ image near any sunspots seen in white light? (Note that the amount of activity that you see will be much greater at some times during the solar cycle than at others.)

Starry Night™ Explorations

65. Use *Starry Night™* to examine simulations of various features that appear on the surface of the Sun. Select **Favorites > Explorations > Solar Surface** to show a simulated view of the visible surface of the Sun. Use the location scroller to examine this surface.
    (a) Which layer of the Sun's atmosphere is shown in this part of the simulation?
    (b) List the different features that are visible in this view of the Sun's surface, zooming in if necessary to examine the details of this surface.
    (c) Select **Favorites > Explorations > Chromosphere** to see a view from the "surface" of the Sun. Use the view controls to look around this view. This simulated view of the chromosphere is at the color of the wavelength of hydrogen light. The opacity of the gas at this wavelength means that you can see the structure of the hot chromosphere that lies above the visible surface. Provide a detailed

description of the various features visible in this simulation of the Sun's surface.

66. Use *Starry Night™* to measure the Sun's rotation. Select **Favorites > Explorations > Solar Rotation** to display the Sun as seen from about 0.008 au above its surface, well inside the orbit of Mercury. Use the time controls to stop the Sun's rotation at a time when a line of longitude on the Sun makes a straight line between the solar poles, preferably a line crossing a recognizable solar feature. Note the date and time and then use the time-flow controls to adjust the date and time to place this selected meridian in this position again.
    (a) What is the rotation rate of the Sun as shown in *Starry Night™*? This demonstration does not show one important feature of the Sun, namely its differential rotation, where the equator of this fluid body rotates faster than the polar regions.
    (b) To which region of the real Sun does your measured rotation rate refer? In the real Sun, this differential rotation is thought to generate the magnetic fields and active regions that make the Sun an active star. Occasional emission of high-energy particles from these active regions can disturb the Earth's environment and disrupt electrical transmission systems.
    (c) Compare this simulated image of the Sun with real images in your textbook or on the internet. In particular, how does the distribution of sunspots and active regions on this image compare to the distribution of these regions on the real Sun?

Collaborative Exercises

67. Figure 16-20 shows variations in the average latitude of sunspots. Estimate the average latitude of sunspots in the year you were born and estimate the average latitude on your twenty-first birthday. Make rough sketches of the Sun during those years to illustrate your answers.

68. Create a diagram showing a sketch of how limb darkening on the Sun would look different if the Sun had either a thicker or thinner photosphere. Be sure to include a caption explaining your diagram.

69. Solar granules, shown in Figure 16-9, are about 1000 km across. What city is about that distance away from where you are right now? What city is that distance from the birthplace of each group member?

70. Magnetic arches in the corona are shown in Figure 16-26a. How many Earths high are these arches, and how many Earths could fit inside one arch?

ANSWERS

ConceptChecks

ConceptCheck 16-1: The Sun emits most of its energy in the form of visible light.

ConceptCheck 16-2: At the extremely high temperatures and pressures existing in the Sun's core, hydrogen nuclei can move fast enough to overcome the repulsive force from their positive electric charges and fuse together.

ConceptCheck 16-3: When 1 kg of hydrogen combines to form helium, the vast majority of the mass ends up in helium atoms, with only 0.7% of the original mass converted into energy.

ConceptCheck 16-4: Astronomers use the current energy output of the Sun to estimate how fast the Sun is consuming its usable fuel (hydrogen). Then the amount of hydrogen fuel remaining in the Sun's core indicates how much longer the Sun will shine.

ConceptCheck 16-5: Because pressure in the Sun's core is due to the downward pushing weight of the overlying mass of material, having less mass pressing down would result in a lower pressure at the core.

ConceptCheck 16-6: The energy transport process of conduction occurs when energy moves through a relatively dense material by hot material transferring its kinetic energy to cooler material through direct contact. The Sun's density is simply too low for conduction to be an important process in transferring energy from one part of the Sun to another part.

ConceptCheck 16-7: Only the temperature decreases with increasing distance from the Sun's central core.

ConceptCheck 16-8: As shown in Figure 16-5, sound waves do penetrate into the Sun, which allows helioseismologists to study the solar interior.

ConceptCheck 16-9: By carefully monitoring how sound waves move through the Sun, astronomers are able to deduce the amount of helium in the Sun's core and convective zone, as well as the thickness of various zones.

ConceptCheck 16-10: Kamiokande could detect only one of three possible types of neutrinos, and this was the type emitted by the Sun. However, along the way, two-thirds of these neutrinos transformed into the other types and were no longer detectable by Kamiokande. SNO, on the other hand, was able to detect all three types of neutrinos.

ConceptCheck 16-11: The energy transport process of convection causes warmer material to rise to the surface and cooler material to sink back into the photosphere, giving the photosphere a granulelike appearance.

ConceptCheck 16-12: Figure 16-11 is taken with an H_α filter that only allows red light to pass through and form the image. This method is used to block out the Sun's other light that would overwhelm the camera. Since the H_α line is red (at 656.3 nm), the spicules appear red, but they emit other wavelengths as well.

ConceptCheck 16-13: The corona has an extremely low density and can be observed only when the more dominant photosphere is blocked, such as during a solar eclipse.

ConceptCheck 16-14: The solar wind must escape through the corona, and this occurs in greater abundance through holes in the corona that contain less gas.

ConceptCheck 16-15: The photosphere surrounding a sunspot has a temperature of about 5800 K, which far outshines the relatively cooler sunspot region, resulting in the sunspots appearing to be quite dark in comparison.

ConceptCheck 16-16: The length of time between large numbers of sunspots to few numbers of sunspots and back to large numbers of sunspots averages about 11 years. However, the magnetic character of sunspots flips every 11-year cycle, suggesting an overarching 22-year sunspot cycle.

ConceptCheck 16-17: According to the magnetic-dynamo model, the sunspot cycle is a result of twisting magnetic fields. In the event the Sun was turning faster, the twisting would occur more quickly and the length of the sunspot cycle would decrease.

ConceptCheck 16-18: The glowing plasma tends to follow the pathways created by the Sun's invisible magnetic field lines, which form curved arches above the Sun's surface.

ConceptCheck 16-19: Coronal mass ejections are many times more energetic than any other event on the Sun and, when directed at Earth, can cause severe problems with telecommunications, electrical power distribution, and radiation health hazards for astronauts working in space.

CalculationChecks

CalculationCheck 16-1: According to Einstein's equation that $E = mc^2$, a mass of 5 kg is equivalent to $5 \text{ kg} \times (3 \times 10^8 \text{ m/s})^2 = 45 \times 10^{16}$ joules, which is equivalent to burning about 100,000 metric tons of coal!

CalculationCheck 16-2: According to Figure 16-3, the Sun's core temperature is about 16 million kelvins. At a distance of 50% of the Sun's radius, the Sun's temperature has dropped to 4 million kelvins, which is a drop of about 75%.

17 | The Nature of the Stars

LEARNING GOALS

To the unaided eye, the night sky is spangled with several thousand stars, each appearing as a bright pinpoint of light. With a pair of binoculars, you can see some 10,000 other, fainter stars; with a 15-cm (6-in.) telescope, the total rises to more than 2 million. Astronomers now know that there are more than 100 billion (10^{11}) stars in our Milky Way Galaxy alone.

But what are these distant pinpoints? To the great thinkers of ancient Greece, the stars were bits of light embedded in a vast sphere, with Earth at the center. They thought the stars were composed of a mysterious "fifth element," quite unlike anything found on Earth.

Today, we know that the stars are made of the same chemical elements found on Earth. We know their sizes, their temperatures, their masses, and something of their internal structures. We understand, too, why the stars in the accompanying image come in a range of beautiful colors: Blue stars have high surface temperatures, whereas the surface temperatures of red and yellow stars are relatively low.

How have we learned these things? How can we know the nature of the stars, objects so distant that their light takes years or even centuries to reach us? In this

Stars come in various sizes, brightnesses, and colors. Different temperatures lead to different colors from red to blue. This is the central region of the star cluster Omega Centauri, which is about 15,800 ly from Earth. (NASA, ESA, and the Hubble SM4 ERO Team)

R I V U X G

chapter, we will learn about the measurements and calculations that astronomers make to determine the properties of stars. We will also take a first look at the Hertzsprung–Russell diagram, an important tool that helps astronomers systematize the wealth of available information about the stars. In later chapters, we will use this diagram to help us understand how stars are born, evolve, and eventually die.

17-1 Careful measurements of the parallaxes of stars reveal their distances

The vast majority of stars are objects very much like the Sun. This understanding followed from the discovery that the stars are tremendously far from us, at distances so great that their light takes years to reach us. Because the stars at night are clearly visible to the naked eye despite these huge distances, it must be that the **luminosity** of the stars—that is, how much light energy they emit into space per second—is comparable to or greater than that of the Sun. Just as for the Sun, the only explanation for such tremendous luminosities is that thermonuclear reactions are occurring within the stars (see Section 16-1).

Clearly, then, it is important to know how distant the stars are from Earth. But how do we measure these distances? You might think these distances are determined by comparing how bright different stars appear. Perhaps the star Betelgeuse in the constellation Orion appears bright because it is relatively close, while the dimmer and less conspicuous star Polaris (the North Star, in the constellation Ursa Minor) is farther away.

But this line of reasoning is misleading—Polaris is actually closer to us than Betelgeuse! How bright a star appears is *not* a good indicator of its distance because the intrinsic brightness of the star also plays a role. If you see a light on a darkened road, it could be a motorcycle headlight a kilometer away or a person holding a flashlight just a few meters away. In the same way, a bright star might be extremely far away but have an unusually high luminosity, and a dim star might be relatively close but have a rather low luminosity. Astronomers must use other techniques to determine the distances to the stars.

CONCEPT CHECK 17-1

Which appears brighter, a small handheld flashlight 10 cm away or a large spotlight atop a lighthouse 10 km away?

Answer appears at the end of the chapter.

Parallax and the Distances to the Stars

 TUTORIAL 17.1 The most straightforward way of measuring stellar distances uses an effect called **parallax**, which is the apparent displacement of an object because of a change in the observer's point of view (Figure 17-1). To see how parallax works, hold your arm out straight in front of you. Now look at the hand on your outstretched arm, first with your left eye closed, then with your right eye closed. When you close one eye and open the other, your hand appears to shift back and forth against the background of more distant objects.

FIGURE 17-1

Parallax Imagine looking at some nearby object (a tree) against a distant background (mountains). When you move from one location to another, (from position A to position B), the nearby object appears to shift with respect to the distant background scenery. This familiar phenomenon is called parallax.

The closer the object you are viewing, the greater the parallax shift. To see this increased shift, repeat the experiment with your hand held closer to your face. Your brain analyzes such parallax shifts constantly as it compares the images from your left and right eyes, and in this way determines the distances to objects around you. This analysis is the basis for depth perception.

To measure the distance to a star, astronomers measure the parallax shift of the star using two points of view that are as far apart as possible—at opposite sides of Earth's orbit. The direction from Earth to a nearby star changes as our planet orbits the Sun, and the nearby star appears to move back and forth against the background of more distant stars (Figure 17-2). This motion is called **stellar parallax**. The parallax (p) of a star is equal to half the angle through which the star's apparent position shifts as Earth moves from one side of its orbit to the other. The larger the parallax p, the smaller the distance d to the star (compare Figure 17-2a with Figure 17-2b).

> You measure distances around you by comparing the images from your left and right eyes, and we find the distances to stars using the same principle.

It is convenient to measure the distance d in **parsecs**. A star with a parallax angle of 1 second of arc ($p = 1$ arcsec) is at a

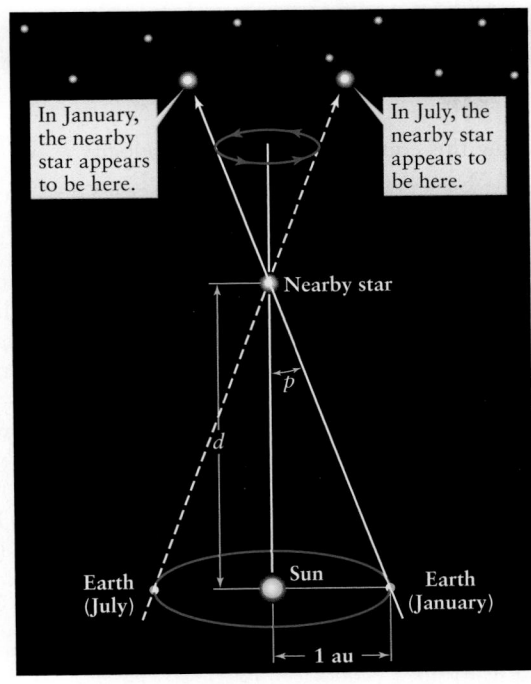

(a) Parallax of a nearby star

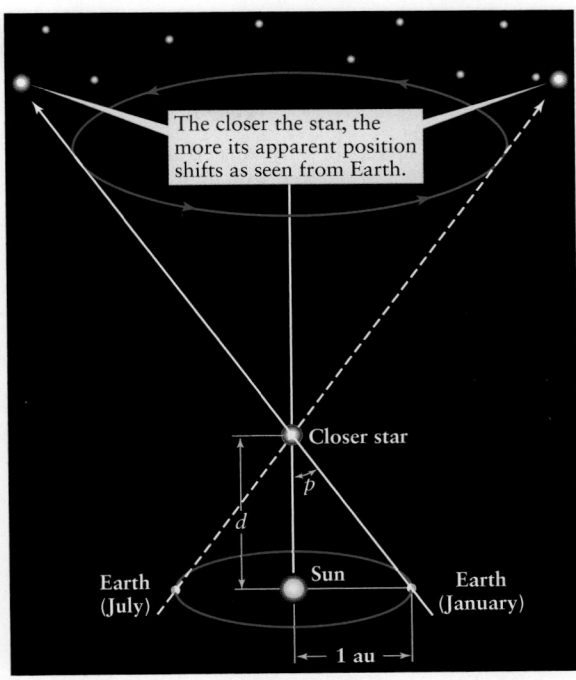

(b) Parallax of an even closer star

ANIMATION **FIGURE 17-2**

17.1

Stellar Parallax **(a)** As Earth orbits the Sun, a nearby star appears to shift its position against the background of distant stars. The parallax (*p*) of the star is equal to half the angle that the star appears to shift as Earth moves from one side of its orbit to the other. **(b)** The closer the star is to us, the greater the parallax angle *p*. The distance *d* to the star (in parsecs) is equal to the reciprocal of the parallax angle *p* (in arcseconds): *d* = 1/*p*.

distance of 1 parsec ($d = 1$ pc). (The word *parsec* is a contraction of the phrase "the distance at which a star has a *par*allax of one arc*sec*ond." Recall from Section 1-7 that 1 parsec equals 3.26 ly, 3.09×10^{13} km, or 206,265 au; see Figure 1-14.)

If the angle p is measured in arcseconds, then the distance d to the star in parsecs is given by the following equation:

Relation between a star's distance and its parallax

$$d = \frac{1}{p}$$

d = distance to a star, in parsecs

p = parallax angle of that star, in arcseconds

This simple relationship between parallax and distance in parsecs is one of the main reasons that astronomers usually measure cosmic distances in parsecs rather than light-years. For example, a star whose parallax is $p = 0.1$ arcsec is at a distance

$d = 1/(0.1) = 10$ pc from Earth. Barnard's star, named for the American astronomer Edward E. Barnard, has a parallax of 0.547 arcsec. Hence, the distance to this star is

$$d = \frac{1}{p} = \frac{1}{0.547} = 1.83 \text{ pc}$$

Because 1 pc is 3.26 ly, this distance can also be expressed as

$$d = 1.83 \text{ pc} \times \frac{3.26 \text{ ly}}{1 \text{ pc}} = 5.97 \text{ ly}$$

All known stars have parallax angles of less than 1 arcsec. In other words, the closest star is more than 1 pc away. Such small parallax angles are difficult to detect, so it was not until 1838 that the first successful parallax measurements were made by the German astronomer and mathematician Friedrich Wilhelm Bessel. He found the parallax angle of the star 61 Cygni to be just $\frac{1}{3}$ arcsec—equal to the angular diameter of a dime at a distance of 11 km, or 7 mi. He thus determined that this star is about 3 pc from Earth. (Modern measurements give a slightly smaller parallax angle, which means that 61 Cygni is actually more than 3 pc away.) The star Proxima Centauri has the largest known parallax angle, 0.772 arcsec, and hence is the closest known star (other than the Sun); its distance is 1/(0.772) = 1.30 pc.

Appendix 4 at the back of this book lists all the stars within 4 pc of the Sun, as determined by parallax measurements. Most of these stars are far too dim to be seen with the naked eye, which is why their names are probably unfamiliar to you. By contrast, the

BOX 17-1 TOOLS OF THE ASTRONOMER'S TRADE

Stellar Motions

Stars can move through space in any direction. The **space velocity** of a star describes how fast and in what direction it is moving. As the accompanying figure shows, a star's space velocity v can be broken into components parallel and perpendicular to our line of sight.

The component of velocity perpendicular to our line of sight—that is, across the plane of the sky—is called the star's **tangential velocity** (v_t). To determine it, astronomers must know the distance to a star (d) and its **proper motion** (μ, the Greek letter mu), which is the number of arcseconds that the star appears to move per year on the celestial sphere. Proper motion does not repeat itself yearly, so it can be distinguished from the apparent back-and-forth motion due to parallax. In terms of a star's distance and proper motion, its tangential velocity (in km/s) is

$$v_t = 4.74 \, \mu \, d$$

where μ is in arcseconds per year and d is in parsecs. For example, Barnard's star (Figure 17-3) has a proper motion of 10.358 arcsec per year and a distance of 1.83 pc. Hence, its tangential velocity is

$$v_t = 4.74(10.358)(1.83) = 89.8 \text{ km/s}$$

The component of a star's motion parallel to our line of sight—that is, either directly toward us or directly away from us—is its **radial velocity** (v_r). It can be determined from measurements of the Doppler shifts of the star's spectral lines (see Section 5-9 and Box 5-6). If a star is approaching us, the wavelengths of all of its spectral lines are decreased (blueshifted); if the star is receding from us, the wavelengths are increased (redshifted). The radial velocity v_r is related to the wavelength shift by the equation

$$\frac{\lambda - \lambda_0}{\lambda_0} = \frac{v_r}{c}$$

In this equation, λ is the wavelength of light coming from the star, λ_0 is what the wavelength would be if the star were not moving, and c is the speed of light. As an illustration, a particular spectral line of iron in the spectrum of Barnard's star has a wavelength (λ) of 516.445 nm. As measured in a laboratory on Earth, the same spectral line has a wavelength (λ_0) of 516.629 nm. Thus, for Barnard's star, our equation becomes

$$\frac{516.445 \text{ nm} - 516.629 \text{ nm}}{516.629 \text{ nm}} = -0.000356 = \frac{v_r}{c}$$

Solving this equation for the radial velocity v_r, we find

$$v_r = (-0.000356) \, c = (-0.000356)(3.00 \times 10^5 \text{ km/s})$$
$$= -107 \text{ km/s}$$

The minus sign means that Barnard's star is moving toward us. You can check this interpretation by noting that the wavelength $\lambda = 516.445$ nm received from Barnard's star is less than the laboratory wavelength $\lambda_0 = 516.629$ nm; hence, the light from the star is blueshifted, which indeed means that the star is approaching. If the star were receding, its light would be redshifted, and its radial velocity would be positive.

The illustration shows that the tangential velocity and radial velocity form two sides of a right triangle. The long side (hypotenuse) of this triangle is the space velocity (v). From the Pythagorean theorem, the space velocity is

$$v = \sqrt{v_t^2 + v_r^2}$$

For Barnard's star, the space velocity is

$$v = \sqrt{(89.4 \text{ km/s})^2 + (-107 \text{ km/s})^2} = 140 \text{ km/s}$$

Therefore, Barnard's star is moving through space at a speed of 140 km/s (503,000 km/h, or 312,000 mi/h) relative to the Sun.

Determining the space velocities of stars is essential for understanding the structure of the galaxy. Studies show that the stars in our local neighborhood are moving in wide orbits around the center of the galaxy, which lies some 8000 pc (26,000 ly) away in the direction of the constellation Sagittarius (the Archer). While many of the orbits are roughly circular and lie in nearly the same plane, others are highly elliptical or steeply inclined to the galactic plane. We will see in Chapter 22 how the orbits of stars and gas clouds reveal the galaxy's spiral structure.

v = the star's space velocity
v_r = the star's radial velocity
v_t = the star's tangential velocity

This side is also equal to v_r.
This side is also equal to v_t.
d = distance from Earth to the star
Star
Earth

majority of the familiar, bright stars in the nighttime sky (listed in Appendix 5) are so far away that their parallaxes cannot be measured from Earth's surface. They appear bright not because they are close, but because they are far more luminous than the Sun. The brightest stars in the sky are *not* necessarily the nearest stars!

CONCEPT CHECK 17-2

If an astronomer erroneously measures a parallax angle for a star to be smaller than it actually is, does the measurement suggest that the star is farther away or closer to Earth than it actually is?

CALCULATION CHECK 17-1

How many light-years away is Alpha Centauri if it has a parallax angle of 0.772 arcsec?

Answers appear at the end of the chapter.

Measuring Parallax from Space

Parallax angles smaller than about 0.01 arcsec are extremely difficult to measure from Earth, in part because of the blurring effects of the atmosphere. Therefore, the parallax method used with ground-based telescopes can give fairly reliable distances only for stars nearer than about $1/0.01 = 100$ pc. But an observatory in space is unhampered by the atmosphere. Observations made from spacecraft therefore permit astronomers to measure even smaller parallax angles and thus determine the distances to more remote stars.

In 1989, the European Space Agency (ESA) launched the satellite Hipparcos, an acronym for *hi*gh *p*recision *par*allax *co*llecting *s*atellite (and a commemoration of the ancient Greek astronomer Hipparchus, who created one of the first star charts). Over more than 3 years of observations, the telescope aboard Hipparcos was used to measure the parallaxes of 118,000 stars, some with an accuracy of 0.001 arcsec. This telescope has enabled astronomers to determine stellar distances out to several hundred parsecs, and with much greater precision than has been possible with ground-based observations. In the years to come, astronomers will increasingly turn to space-based observations to determine stellar distances.

Unfortunately, most of the stars in the galaxy are so far away that their parallax angles are too small to measure even with an orbiting telescope. Later in this chapter, we will discuss a technique that can be used to find the distances to these more remote stars. In Chapter 23, we will learn about other techniques that astronomers use to determine the much larger distances to galaxies beyond the Milky Way. These techniques also help us understand the overall size, age, and structure of the universe.

The Importance of Parallax Measurements

Because stellar parallax can be used only on relatively close stars, it might seem to be of limited usefulness. But parallax measurements are the cornerstone of all other methods used to find the distances to remote objects. These other methods require a precise and accurate knowledge of the distances to nearby stars, as determined by stellar parallax. Hence, any inaccuracies in the parallax angles for

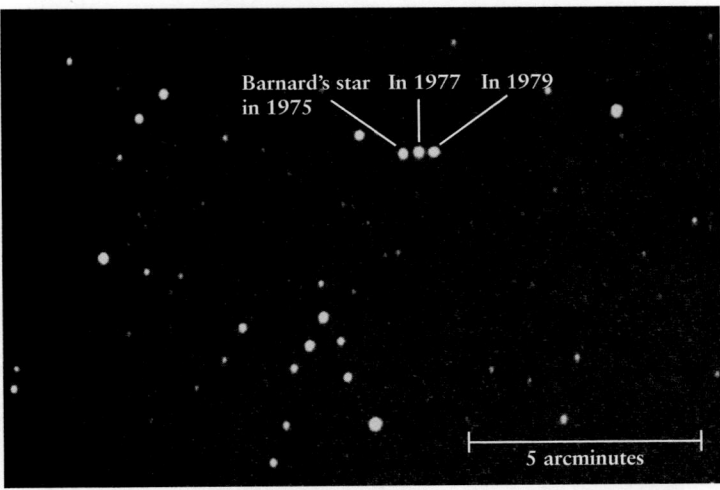

FIGURE 17-3 R I V U X G

The Motion of Barnard's Star Three photographs taken over a four-year period were combined to show the motion of Barnard's star, which lies 1.82 pc away in the constellation Ophiuchus. Over this time interval, Barnard's star moved more than 41 arcsec on the celestial sphere (about 0.69 arcmin, or 0.012°), more than any other star. (John Sanford/Science Source)

nearby stars can translate into substantial errors in measurement for the whole universe. To minimize these errors, astronomers are continually trying to perfect their parallax-measuring techniques.

Stellar parallax is an *apparent* motion of stars caused by Earth's orbital motion around the Sun. But stars are not fixed objects, and they actually do move through space. As a result, stars change their positions on the celestial sphere (Figure 17-3), and they move either toward or away from the Sun. These motions are sufficiently slow, however, that changes in the positions of the stars are hardly noticeable over a human lifetime. Box 17-1 describes how astronomers study these motions and what insights they gain from these studies.

CONCEPT CHECK 17-3

Why can't the distances to most stars in our galaxy be measured using parallax?

Answer appears at the end of the chapter.

17-2 If a star's distance is known, its luminosity can be determined from its apparent brightness

TUTORIAL 17.2 All the stars you can see in the nighttime sky shine by thermonuclear fusion, just as the Sun does (see Section 16-1). But they are by no means merely identical copies of the Sun. Stars differ in their luminosity (L), the amount of light energy they emit each second. Luminosity is usually measured either in watts (1 watt, or 1 W, is 1 joule per second) or as a multiple of the Sun's luminosity ($L_\odot$, equal to 3.83×10^{26} W). Most stars are less luminous than the Sun, but some blaze forth with a million times the Sun's luminosity. Knowing a star's luminosity is essential for determining the star's history, its present-day internal structure, and its future evolution.

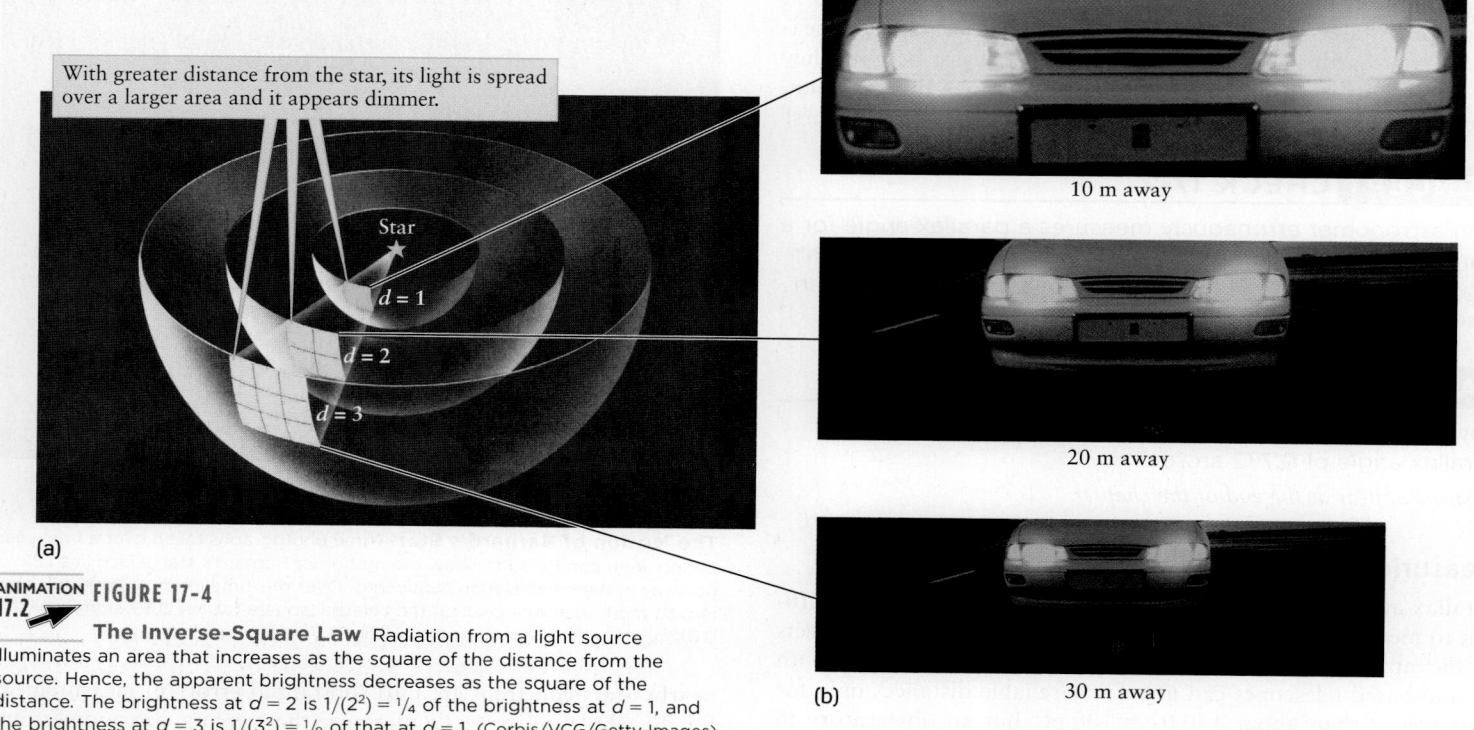

With greater distance from the star, its light is spread over a larger area and it appears dimmer.

Star
d = 1
d = 2
d = 3

(a)

10 m away

20 m away

(b)   30 m away

ANIMATION
17.2 FIGURE 17-4
     The Inverse-Square Law Radiation from a light source illuminates an area that increases as the square of the distance from the source. Hence, the apparent brightness decreases as the square of the distance. The brightness at $d = 2$ is $1/(2^2) = 1/4$ of the brightness at $d = 1$, and the brightness at $d = 3$ is $1/(3^2) = 1/9$ of that at $d = 1$. (Corbis/VCG/Getty Images)

Luminosity, Apparent Brightness, and the Inverse-Square Law

Imagine a star radiating light into space. As the light energy moves away from the star, it spreads out over increasingly larger regions of space—the energy thins out. Now imagine a sphere of radius d centered on the star, as in **Figure 17-4**. The amount of energy that passes each second through a square meter of the sphere's surface area is called the **brightness** (b). The brightness is the total luminosity of the source (L) divided by the total surface area of the sphere (equal to $4\pi d^2$). This quantity is also called the **apparent brightness** (b), because how bright a light source appears depends on how far away it is being observed. Apparent brightness is measured in watts per square meter (W/m^2).

Written in the form of an equation, the relationship between apparent brightness and luminosity is:

> Apparent brightness is a measure of how faint a star looks to us; luminosity is a measure of the star's total light output.

Inverse-square law relating apparent brightness and luminosity

$$b = \frac{L}{4\pi d^2}$$

b = apparent brightness of a star's light, in W/m^2

L = star's luminosity, in watts

d = distance to star, in meters

This relationship is called the **inverse-square law** because the apparent brightness of light that an observer can see or measure is inversely proportional to the square of the observer's distance (d) from the source. If you double your distance from a light source, its radiation is spread out over an area 4 times larger, so the apparent brightness you see is decreased by a factor of 4. Similarly, at triple the distance, the apparent brightness is $1/9$ as great (see Figure 17-4).

We can apply the inverse-square law to the Sun, which is 1.50×10^{11} m from Earth. Its apparent brightness ($b_\odot$) is

$$b_\odot = \frac{3.83 \times 10^{26}\ \text{W}}{4\pi(1.50 \times 10^{11}\ \text{m})^2} = 1362\ \text{W/m}^2$$

That is, a solar panel with an area of 1 square meter receives 1362 watts of power from the Sun. This quantity is called the **solar constant** and corresponds to solar maximum, when the number of sunspots peaks. At solar minimum, solar constant is 1361 W.

When astronomers measure the apparent brightness of a star, this process is called **photometry**.

CAUTION!

No matter what your distance is from a star, the star's luminosity (L) remains unchanged. Luminosity is the intrinsic energy output of the star. Only the apparent brightness (b) changes with distance as the star's energy thins out. Furthermore, a big telescope and a small telescope on Earth will both measure the same apparent brightness; while a larger telescope collects more light, its larger lens or mirror is taken into account when computing b.

CALCULATION CHECK 17-2

How many times less light falls on a book illuminated by a lightbulb if the book is moved 3 times farther away?

Answer appears at the end of the chapter.

Calculating a Star's Luminosity

The inverse-square law says that we can find a star's luminosity if we know its distance and its apparent brightness. For convenience, this law can be expressed in a somewhat different form. We first rearrange the inverse-square law equation as follows:

$$L = 4\pi d^2 b$$

We then apply this equation to the Sun. That is, we write a similar equation relating the Sun's luminosity ($L_\odot$), the distance from Earth to the Sun ($d_\odot$, equal to 1 au), and the Sun's apparent brightness ($b_\odot$):

$$L_\odot = 4\pi d_\odot^2 b_\odot$$

If we take the ratio of these two equations, the unpleasant factor of 4π drops out and we are left with the following:

Determining a star's luminosity from its apparent brightness

$$\frac{L}{L_\odot} = \left(\frac{d}{d_\odot}\right)^2 \frac{b}{b_\odot}$$

$L/L_\odot$ = ratio of the star's luminosity to the Sun's luminosity

$d/d_\odot$ = ratio of the star's distance to the Earth–Sun distance

$b/b_\odot$ = ratio of the star's apparent brightness to the Sun's apparent brightness

We need to know just two things to find a star's luminosity: the distance to a star as compared to the Earth–Sun distance (the ratio $d/d_\odot$), and how that star's apparent brightness compares to the brightness of the Sun (the ratio $b/b_\odot$). Then we can use the earlier equation to determine how luminous that star is compared to the Sun (the ratio $L/L_\odot$).

In other words, this equation gives us a general rule relating the luminosity, distance, and apparent brightness of a star:

We can determine the luminosity of a star from its distance and apparent brightness. For a given distance, the brighter the star appears, the more luminous that star must be. For a given apparent brightness, the more distant the star, the more luminous it must be to be seen at that distance.

Box 17-2 shows how to use the above equation to determine the luminosity of the nearby star ε (epsilon) Eridani, the fifth brightest star in the constellation Eridanus (named for a river in Greek mythology). Parallax measurements indicate that ε Eridani is 3.23 pc away, and photometry shows that the star appears only 6.73×10^{-13} as bright as the Sun. Using the above equation, we find that ε Eridani has only 0.30 times the luminosity of the Sun.

BOX 17-2   TOOLS OF THE ASTRONOMER'S TRADE

Luminosity, Distance, and Apparent Brightness

The inverse-square law (Section 17-2) relates a star's luminosity, distance, and apparent brightness to the corresponding quantities for the Sun:

$$\frac{L}{L_\odot} = \left(\frac{d}{d_\odot}\right)^2 \frac{b}{b_\odot}$$

We can use a similar equation to relate the luminosities, distances, and apparent brightnesses of any two stars, which we call star 1 and star 2:

$$\frac{L_1}{L_2} = \left(\frac{d_1}{d_2}\right)^2 \frac{b_1}{b_2}$$

EXAMPLE: The star ε (epsilon) Eridani is 3.23 pc from Earth. As seen from Earth, this star appears only 6.73×10^{-13} as bright as the Sun. What is the luminosity of ε Eridani compared with that of the Sun?

Situation: We are given the distance to ε Eridani ($d = 3.23$ pc) and this star's brightness compared to that of the Sun ($b/b_\odot = 6.73 \times 10^{-13}$). Our goal is to find the ratio of the luminosity of ε Eridani to that of the Sun, that is, the quantity $L/L_\odot$.

Tools: Since we are asked to compare this star to the Sun, we use the first of the two equations given, $L/L_\odot = (d/d_\odot)^2(b/b_\odot)$, to solve for $L/L_\odot$.

Answer: Our equation requires the ratio of the star's distance to the Sun's distance, $d/d_\odot$. The distance from Earth to the Sun is $d_\odot = 1$ au. To calculate the ratio $d/d_\odot$, we must express both distances in the same units. There are 206,265 au in 1 pc, so we can write the distance to ε Eridani as $d = (3.23 \text{ pc})(206{,}265 \text{ au/pc}) = 6.66 \times 10^5$ au. Hence, the ratio of distances is $d/d_\odot = (6.66 \times 10^5 \text{ au})/(1 \text{ au}) = 6.66 \times 10^5$. Then we find that the ratio of the luminosity of ε Eridani (L) to the Sun's luminosity ($L_\odot$) is

$$\frac{L}{L_\odot} = \left(\frac{d}{d_\odot}\right)^2 \frac{b}{b_\odot} = (6.66 \times 10^5)^2 \times (6.73 \times 10^{-13}) = 0.30$$

Review: This result means that ε Eridani is only 0.30 as luminous as the Sun; that is, its power output is only 30% as great.

EXAMPLE: Suppose star 1 is at half the distance of star 2 (that is, $d_1/d_2 = \frac{1}{2}$) and that star 1 appears twice as bright as star 2 (that is, $b_1/b_2 = 2$). How do the luminosities of these two stars compare?

(continued on the next page)

BOX 17-2 (continued)

Situation: For these two stars, we are given the ratio of distances (d_1/d_2) and the ratio of apparent brightnesses (b_1/b_2). Our goal is to find the ratio of their luminosities (L_1/L_2).

Tools: Since we are comparing two stars, neither of which is the Sun, we use the second of the two equations above: $L_1/L_2 = (d_1/d_2)^2(b_1/b_2)$.

Answer: Plugging values into our equation, we find

$$\frac{L_1}{L_2} = \left(\frac{d_1}{d_2}\right)^2 \frac{b_1}{b_2} = \left(\frac{1}{2}\right)^2 \times 2 = \frac{1}{2}$$

Review: This result says that star 1 has only one-half the luminosity of star 2. Despite this, star 1 appears brighter than star 2 because it is closer to us.

The two equations above are also useful in the method of *spectroscopic parallax,* which we discuss in Section 17-8. It turns out that a star's luminosity can be determined simply by analyzing the star's spectrum. If the star's apparent brightness is also known, the star's distance can be calculated. The inverse-square law can be rewritten as an expression for the ratio of the star's distance from Earth (d) to the Earth–Sun distance ($d_\odot$):

$$\frac{d}{d_\odot} = \sqrt{\frac{(L/L_\odot)}{(b/b_\odot)}}$$

We can also use this formula as a relation between the properties of any two stars, 1 and 2:

$$\frac{d_1}{d_2} = \sqrt{\frac{(L_1/L_2)}{(b_1/b_2)}}$$

EXAMPLE: The star Pleione in the constellation Taurus is 190 times as luminous as the Sun but appears only 3.19×10^{-13} as bright as the Sun. How far is Pleione from Earth?

Situation: We are told the ratio of Pleione's luminosity to that of the Sun ($L/L_\odot = 190$) and the ratio of their apparent brightnesses ($b/b_\odot = 3.19 \times 10^{-13}$). Our goal is to find the distance d from Earth to Pleione.

Tools: Since we are comparing Pleione to the Sun, we use the first of the two equations above.

Answer: Our equation tells us the ratio of the Earth–Pleione distance to the Earth–Sun distance:

$$\frac{d}{d_\odot} = \sqrt{\frac{(L/L_\odot)}{(b/b_\odot)}} = \sqrt{\frac{190}{3.19 \times 10^{-13}}}$$
$$= \sqrt{5.95 \times 10^{14}} = 2.44 \times 10^7$$

Hence, the distance from Earth to Pleione is 2.44×10^7 times greater than the distance from Earth to the Sun. The Earth–Sun distance is $d_\odot = 1$ au and 206,265 au = 1 pc, so we can express the star's distance as $d = (2.44 \times 10^7 \text{ au}) \times (1 \text{ pc}/206,265 \text{ au}) = 118$ pc.

Review: We can check our result by comparing it with the above example about the star ε Eridani. Pleione has a much greater luminosity than ε Eridani (190 times the Sun's luminosity versus 0.30 times), but Pleione appears dimmer than ε Eridani (3.19×10^{-13} times as bright as the Sun compared to 6.73×10^{-13} times). For this to be true, Pleione must be much farther away from Earth than is ε Eridani. This is just what our results show: $d = 118$ pc for Pleione compared to $d = 3.23$ pc for ε Eridani.

EXAMPLE: The star δ (delta) Cephei, which lies 300 pc from Earth, is thousands of times more luminous than the Sun. Thanks to this great luminosity, stars like δ Cephei can be seen in galaxies millions of parsecs away. As an example, the Hubble Space Telescope has detected stars like δ Cephei within the galaxy NGC 3351, which lies in the direction of the constellation Leo. These stars appear only 9×10^{-10} as bright as δ Cephei. What is the distance to NGC 3351?

Situation: To determine the distance we want, we need to find the distance to a star within NGC 3351. We are told that certain stars within this galaxy have the same luminosity as δ Cephei but appear only 9×10^{-10} as bright.

Tools: We use the equation $d_1/d_2 = \sqrt{(L_1/L_2)/(b_1/b_2)}$ to relate two stars, one within NGC 3351 (call this star 1) and the identical star δ Cephei (star 2). Our goal is to find d_1.

Answer: Since the two stars are identical, they have the same luminosity ($L_1 = L_2$, or $L_1/L_2 = 1$). The brightness ratio is $b_1/b_2 = 9 \times 10^{-10}$, so our equation tells us that

$$\frac{d_1}{d_2} = \sqrt{\frac{(L_1/L_2)}{(b_1/b_2)}} = \sqrt{\frac{1}{9 \times 10^{-10}}} = \sqrt{1.1 \times 10^9} = 33,000$$

Hence, NGC 3351 is 33,000 times farther away than δ Cephei, which is 300 pc from Earth. The distance from Earth to NGC 3351 is therefore $(33,000)(300 \text{ pc}) = 10^7$ pc, or 10 megaparsecs (10 Mpc).

Review: This example illustrates one technique that astronomers use to measure extremely large distances. We will learn more about stars like δ Cephei in Chapter 19, and in Chapter 23 we will further explore how they are used to determine the distances to remote galaxies.

CONCEPT CHECK 17-4

A woman sees her rideshare coming to pick her up. In terms of the car's headlights, how does the luminosity change? How does the apparent brightness change?

CONCEPT CHECK 17-5

Two stars appear equally bright from Earth. However, one of them is 100 times more luminous than the other. Which star is farther away?

CALCULATION CHECK 17-3

Star A is 8 times more luminous than Star B. Star B appears twice as bright. How much farther away from Earth is Star A than Star B?

Answers appear at the end of the chapter.

The Stellar Population

Calculations of this kind show that stars come in a wide variety of luminosities, with values that range from about $10^6 L_\odot$ (a million times the Sun's luminosity) to only about $10^{-4} L_\odot$ (a mere ten-thousandth of the Sun's light output). The most luminous star emits roughly 10^{10} times more energy each second than the least luminous! (To put this number in perspective, about 10^{10} human beings have lived on Earth since our species first evolved.)

As stars go, our Sun is neither extremely luminous nor extremely dim; it is a rather ordinary, garden-variety star. It is somewhat more luminous than most stars, however. Of more than 30 stars within 4 pc of the Sun (see Appendix 4), only 3 (α Centauri, Sirius, and Procyon) have a greater luminosity than the Sun.

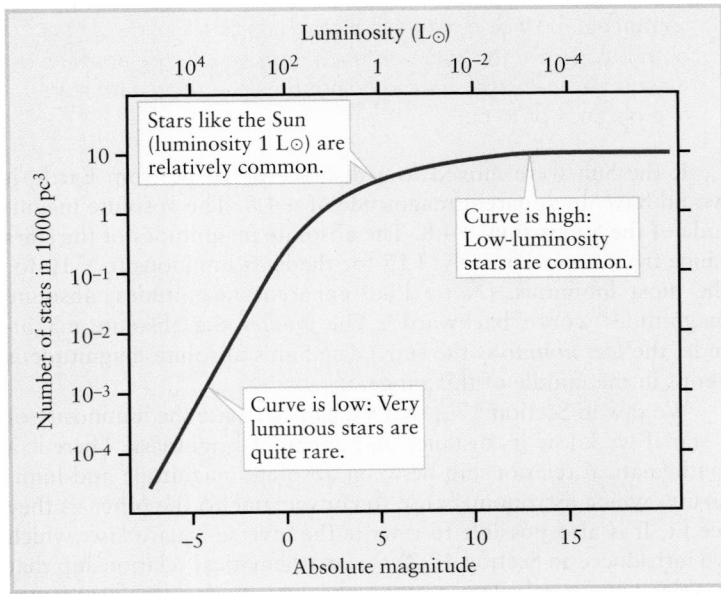

FIGURE 17-5

The Luminosity Function This graph shows how many stars of a given luminosity lie within a representative 1000–cubic-parsec volume. The scale at the bottom of the graph shows absolute magnitude, an alternative measure of a star's luminosity (described in Section 17-3). (Data from J. Bahcall and R. Soneira)

To better characterize a typical population of stars, astronomers count the stars out to a certain distance from the Sun and plot the number of stars that have different luminosities. The resulting graph is called the **luminosity function**. Figure 17-5 shows the luminosity function for stars in our part of the Milky Way Galaxy. The curve declines very steeply for the most luminous stars toward the left side of the graph (absolute magnitudes are discussed in the next section), indicating that they are quite rare. For example, this graph shows that stars like the Sun are about 10,000 times more common than stars like Spica (which has a luminosity of 2100 $L_\odot$).

The exact shape of the curve in Figure 17-5 applies only to the vicinity of the Sun and similar regions in our Milky Way Galaxy. Other locations have somewhat different luminosity functions. In stellar populations in general, however, low-luminosity stars are much more common than high-luminosity ones.

17-3 Astronomers often use the magnitude scale to denote brightness

Because astronomy is among the most ancient of sciences, some of the tools used by modern astronomers are actually many centuries old. One such tool is the **magnitude scale**, which astronomers frequently use to denote the brightness of stars. This scale was introduced in the second century B.C.E. by the Greek astronomer Hipparchus, who called the brightest stars first-magnitude stars. Stars about half as bright as first-magnitude stars were called second-magnitude stars, and so forth, down to sixth-magnitude stars, the dimmest ones he could see. After telescopes came into use, astronomers extended Hipparchus's magnitude scale to include even dimmer stars.

Apparent Magnitudes

The magnitudes in Hipparchus's scale are properly called **apparent magnitudes**, because they describe how bright an object *appears* to an Earth-based observer. Apparent magnitude is directly related to apparent brightness.

> **CAUTION!**
>
> The magnitude scale can be confusing because it works "backward." Keep in mind that the *greater* the apparent magnitude, the *dimmer* the star. A star of apparent magnitude +3 (a third-magnitude star) is dimmer than a star of apparent magnitude +2 (a second-magnitude star).

In the nineteenth century, astronomers developed better techniques for measuring the light energy arriving from a star. These measurements showed that a first-magnitude star is about 100 times brighter than a sixth-magnitude star. In other words, it would take 100 stars of magnitude +6 to provide as much light energy as we receive from a single star of magnitude +1. To make computations easier, the magnitude scale was redefined so that a magnitude difference of 5 corresponds exactly to a factor of 100 in brightness. A magnitude difference of 1 corresponds to a factor of 2.512 in brightness, because

$$2.512 \times 2.512 \times 2.512 \times 2.512 \times 2.512 = (2.512)^5 = 100$$

(a) Some apparent magnitudes

FIGURE 17-6 R I **V** U X G

The Apparent Magnitude Scale (a) Astronomers denote the apparent brightness of objects in the sky by their apparent magnitudes. The greater the apparent magnitude, the dimmer the object. **(b)** This photograph of the

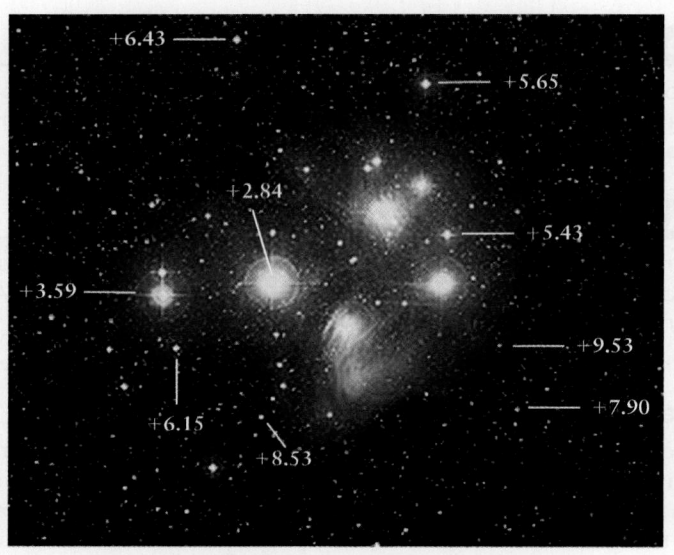

(b) Apparent magnitudes of stars in the Pleiades

Pleiades cluster, located about 120 pc away in the constellation Taurus, shows the apparent magnitudes of some of its stars. Most are too faint to be seen by the naked eye. (b: © Australian Astronomical Observatory/David Malin Images)

Thus, it takes 2.512 third-magnitude stars to provide as much light as we receive from a single second-magnitude star. Stated another way, a second-magnitude star is 2.512 times brighter than a third-magnitude star.

If one star is brighter than another star by 1 magnitude, it is 2.512 times brighter.

Figure 17-6 illustrates the modern apparent magnitude scale. The dimmest stars visible through a pair of binoculars have an apparent magnitude of +10, and the dimmest stars that can be photographed in a 1-hour exposure with the Keck telescopes (see Section 6-2) or the Hubble Space Telescope have apparent magnitude +30. Modern astronomers also use negative numbers to extend Hipparchus's scale to include very bright objects. For example, Sirius, the brightest star in the sky, has an apparent magnitude of −1.43. The Sun, the brightest object in the sky, has an apparent magnitude of −26.7.

Absolute Magnitudes

Apparent magnitude is a measure of a star's apparent brightness as seen from Earth. A related quantity that measures a star's true energy output—that is, its luminosity—is called **absolute magnitude**. This is the apparent magnitude a star would have if it were located exactly 10 parsecs from Earth.

> Apparent magnitude measures a star's brightness; absolute magnitude measures its luminosity.

ANALOGY

If you wanted to compare the light output of two different lightbulbs, you would naturally place them side by side so that both bulbs were the same distance from you. In the absolute magnitude scale, we imagine doing the same thing with stars at a distance of 10 pc.

If the Sun were moved to a distance of 10 pc from Earth, it would have an apparent magnitude of +4.8. The absolute magnitude of the Sun is thus +4.8. The absolute magnitudes of the stars range from approximately +15 for the least luminous to −10 for the most luminous. (*Note:* Like apparent magnitudes, absolute magnitudes work "backward": The *greater* the absolute magnitude, the *less luminous* the star.) The Sun's absolute magnitude is about in the middle of this range.

We saw in Section 17-2 that we can calculate the luminosity of a star if we know its distance and apparent brightness. There is a mathematical relationship between absolute magnitude and luminosity, which astronomers use to convert one to the other as they see fit. It is also possible to rewrite the inverse-square law, which we introduced in Section 17-2, as a mathematical relationship that enables you to calculate a star's absolute magnitude (a measure of its luminosity) from its distance and apparent magnitude (a measure of its apparent brightness). **Box 17-3** describes these relationships and how to use them.

Because the backward magnitude scales can be confusing, we will use them only occasionally in this book. We will usually speak of a star's luminosity rather than its absolute magnitude and will

BOX 17-3   TOOLS OF THE ASTRONOMER'S TRADE

Apparent Magnitude and Absolute Magnitude

Astronomers commonly express a star's apparent brightness in terms of apparent magnitude (denoted by a lowercase *m*), and the star's luminosity in terms of absolute magnitude (denoted by a capital *M*). While we do not use these quantities extensively in this book, it is useful to know a few simple relationships involving them.

Consider two stars, labeled 1 and 2, with apparent magnitudes m_1 and m_2 and brightnesses b_1 and b_2, respectively. The *ratio* of their apparent brightnesses (b_1/b_2) corresponds to a *difference* in their apparent magnitudes $(m_2 - m_1)$. As we learned in Section 17-3, each step in magnitude corresponds to a factor of 2.512 in brightness; we receive 2.512 times more energy per square meter per second from a third-magnitude star than from a fourth-magnitude star. This idea was used to construct the following table:

Apparent magnitude difference ($m_2 = m_1$)	Ratio of apparent brightness (b_1/b_2)
1	2.512
2	$(2.512)^2 = 6.31$
3	$(2.512)^3 = 15.85$
4	$(2.512)^4 = 39.82$
5	$(2.512)^5 = 100$
10	$(2.512)^{10} = 10^4$
15	$(2.512)^{15} = 10^6$
20	$(2.512)^{20} = 10^8$

A simple equation relates the difference between two stars' apparent magnitudes to the ratio of their brightnesses:

Magnitude difference related to brightness ratio

$$m_2 - m_1 = 2.5 \log\left(\frac{b_1}{b_2}\right)$$

m_1, m_2 = apparent magnitudes of stars 1 and 2

b_1, b_2 = apparent brightnesses of stars 1 and 2

In this equation, $\log (b_1/b_2)$ is the logarithm of the brightness ratio. The logarithm of $1000 = 10^3$ is 3, the logarithm of $10 = 10^1$ is 1, and the logarithm of $1 = 10^0$ is 0.

EXAMPLE: At their most brilliant, Venus has a magnitude of about −4 and Mercury has a magnitude of about −2. How many times brighter are these planets than the dimmest stars visible to the naked eye, with a magnitude of +6?

Situation: In each case, we want to find a ratio of two apparent brightnesses (the brightness of Venus or Mercury compared to that of the dimmest naked-eye stars).

Tools: In each case, we will convert a *difference* in apparent magnitude between the planet and the naked-eye star into a *ratio* of their brightnesses.

Answer: The magnitude difference between Venus and the dimmest stars visible to the naked eye is $+6 - (-4) = 10$. From the table, this difference corresponds to a brightness ratio of $(2.512)^{10} = 10^4 = 10,000$, so Venus at its most brilliant is 10,000 times brighter than the dimmest naked-eye stars.

The magnitude difference between Mercury and the dimmest naked-eye stars is $+4 - (-4) = 8$. While this value is not in the table, you can see that the corresponding ratio of brightnesses is $(2.512)^8 = (2.512)^{5+3} = (2.512)^5 \times (2.512)^3$. From the table, $(2.512)^5 = 100$ and $(2.512)^3 = 15.85$, so the ratio of brightnesses is $100 \times 15.85 = 1585$. Hence, Mercury at its most brilliant is 1585 times brighter than the dimmest stars visible to the naked eye.

Review: Can you show that when at their most brilliant, Venus is 6.31 times brighter than Mercury? (*Hint:* No multiplication or division is required—just notice the difference in apparent magnitude between Venus and Mercury, and consult the table.)

EXAMPLE: The variable star RR Lyrae in the constellation Lyra (the Harp) periodically doubles its light output. By how much does its apparent magnitude change?

Situation: We are given a ratio of two brightnesses (the star at its maximum is twice as bright as at its minimum). Our goal is to find the corresponding difference in apparent magnitude.

Tools: We let 1 denote the star at its maximum brightness and 2 denote the same star at its dimmest, so the ratio of brightnesses is $b_1/b_2 = 2$. We then use the equation $m_2 - m_1 = 2.5 \log (b_1/b_2)$ to solve for the apparent magnitude difference $m_2 - m_1$.

Answer: Using a calculator, we find $m_2 - m_1 = 2.5 \log (2) = 2.5 \times 0.30 = 0.75$. RR Lyrae therefore varies periodically in brightness by 0.75 magnitude.

Review: Our answer means that at its dimmest, RR Lyrae has an apparent magnitude m_2 that is 0.75 *greater* than its apparent magnitude m_1 when it is brightest. (Remember that a greater value of apparent magnitude means the star is dimmer, not brighter!)

The inverse-square law relating a star's apparent brightness and luminosity can be rewritten in terms of the star's apparent magnitude (*m*), absolute magnitude (*M*), and distance from Earth (*d*). This can be expressed as an equation:

(continued on the next page)

BOX 17-3 *(continued)*

Relation between a star's apparent magnitude and absolute magnitude

$$m - M = 5 \log d - 5$$

m = star's apparent magnitude

M = star's absolute magnitude

d = distance from Earth to the star in parsecs

In this expression, $m - M$ is called the **distance modulus**, and $\log d$ means the logarithm of the distance d in parsecs. For convenience, the following table gives the values of the distance d corresponding to different values of $m - M$.

Distance modulus $m - M$	Distance d (pc)
−4	1.6
−3	2.5
−2	4.0
−1	6.3
0	10
1	16
2	25
3	40
4	63
5	100
10	10^3
15	10^4
20	10^5

This table shows that if a star is less than 10 pc away, its distance modulus $m - M$ is negative. That is, its apparent magnitude (m) is less than its absolute magnitude (M). If the star is more than 10 pc away, $m - M$ is positive and m is greater than M. As an example, the star ε (epsilon) Indi, which is in the direction of the southern constellation Indus, has apparent magnitude $m = +4.7$. It is 3.6 pc away, which is less than 10 pc, so its apparent magnitude is less than its absolute magnitude.

EXAMPLE: Find the absolute magnitude of ε Indi.

Situation: We are given the distance to ε Indi ($d = 3.6$ pc) and its apparent magnitude ($m = 14.7$). Our goal is to find the star's absolute magnitude M.

Tools: We use the formula $m - M = 5 \log d - 5$ to solve for M.

Answer: Since $d = 3.6$ pc, we use a calculator to find $\log d = \log 3.6 = 0.56$. Therefore, the star's distance modulus is $m - M = 5(0.56) - 5 = -2.2$, and the star's absolute magnitude is $M = m - (-2.2) = +4.7 + 2.2 = +6.9$.

Review: As a check on our calculations, note that this star's distance modulus $m - M = -2.2$ is less than zero, as it should be for a star less than 10 pc away. Note that our Sun has absolute magnitude $+4.8$; ε Indi has a greater absolute magnitude, so it is less luminous than the Sun.

EXAMPLE: Suppose you were viewing the Sun from a planet orbiting another star 100 pc away. Could you see it without using a telescope?

Situation: We learned in the preceding examples that the Sun has absolute magnitude $M = +4.8$ and that the dimmest stars visible to the naked eye have apparent magnitude $m = +6$. Our goal is to determine whether the Sun would be visible to the naked eye at a distance of 100 pc.

Tools: We use the relationship $m - M = 5 \log d - 5$ to find the Sun's apparent magnitude at $d = 100$ pc. If this is greater than $+6$, the Sun would not be visible at that distance. (Remember that the greater the apparent magnitude, the dimmer the star.)

Answer: The table indicates that at $d = 100$ pc, the distance modulus is $m - M = 5$. So, as seen from this distant planet, the Sun's apparent magnitude would be $m = M + 5 = +4.8 + 5 = +9.8$. This is greater than the naked-eye limit $m = +6$, so the Sun could not be seen.

Review: The Sun is by far the brightest object in Earth's sky. But our result tells us that to an inhabitant of a planetary system 100 pc away—a rather small distance in a galaxy that is thousands of parsecs across—our own Sun would be just another insignificant star, visible only through binoculars or a telescope.

The magnitude system is also used by astronomers to express the colors of stars as seen through different filters, as we describe in Section 17-4. For example, rather than quantifying a star's color by the color ratio b_V/b_B (a star's apparent brightness as seen through a V filter divided by the brightness through a B filter), astronomers commonly use the *color index* B–V, which is the difference in the star's apparent magnitude as measured with these two filters. We will not use this system in this book, however (but see Questions 32 and 33).

describe a star's appearance in terms of apparent brightness rather than apparent magnitude. But if you go on to study more about astronomy, you will undoubtedly make frequent use of apparent magnitude and absolute magnitude.

CONCEPT CHECK 17-6

If we were observing our Sun from a distance of 10 pc, what would be its apparent and absolute magnitudes?

CONCEPT CHECK 17-7

The star Tau Ceti has an apparent magnitude of about +3 and an absolute magnitude of about +6. Is it much closer or much farther from Earth than 10 pc?

Answers appear at the end of the chapter.

17-4 A star's color depends on its surface temperature

The image that opens this chapter shows that stars come in different colors. You can see these colors even with the naked eye. For example, you can easily see the reddish color of Betelgeuse (the star in the "armpit" of the constellation Orion) and the blue tint of Bellatrix at Orion's other "shoulder" (see Figure 2-2). Colors are most evident for the brightest stars, because human color vision works poorly at low light levels.

CAUTION!

It is true that the light from a star will appear redshifted if the star is moving away from you and blueshifted if it is moving toward you. But for even the fastest stars, these color shifts are so tiny that it takes sensitive instruments to measure them. The red color of Betelgeuse and the blue color of Bellatrix are not due to their motions; they are the actual colors of the stars, governed by temperature.

Color and Temperature

A star's surface temperature is a very important characteristic that is integral to understanding other features of stars discussed later in this chapter. If there were no difficulties collecting data, then measuring the blackbody curves of a star, as illustrated in Figure 17-7, would be one way to determine a star's temperature. However, it is not so easy to directly measure the blackbody curves for large numbers of stars, and we next look at other methods to determine a star's surface temperature.

We saw in Section 5-3 that a star's color is directly related to its surface temperature. The intensity of light from a relatively cool star peaks at long wavelengths, making the star look red (Figure 17-7a). A hot star's intensity curve peaks at shorter wavelengths, so the star looks blue (Figure 17-7c). For a star with an intermediate temperature, such as the Sun, the intensity peak is near the middle of the visible spectrum. The human visual system interprets an object with this spectrum of wavelengths as yellowish in color (Figure 17-7b). This leads to an important general rule about star colors and surface temperatures:

Red stars are relatively cold, with low surface temperatures; blue stars are relatively hot, with high surface temperatures.

Figure 17-7 shows that astronomers can accurately determine the surface temperature of a star by carefully measuring its color. To measure color, the star's light is collected by a telescope and passed through various color filters. For example, a red filter passes red light while blocking other wavelengths. The filtered light is then collected by a light-sensitive

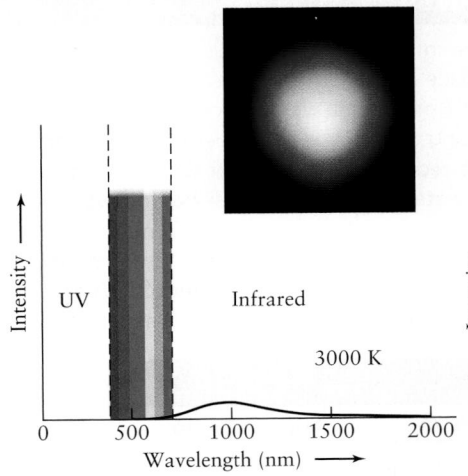

(a) A **cool star** with surface temperature 3000 K emits much more red light than blue light, and so appears red.

(b) A **warmer star** with surface temperature 5800 K (like the Sun) emits similar amounts of all visible wavelengths, and so appears yellow-white.

(c) A **hot star** with surface temperature 10,000 K emits much more blue light than red light, and so appears blue.

FIGURE 17-7

Temperature and Color The black curves in these graphs show the intensity of light emitted by three hypothetical stars plotted against wavelength. These are the same blackbody curves shown in Figure 5-12 for stars of different temperatures. The rainbow band indicates the range of visible wavelengths. The star's apparent color depends on whether the intensity curve has larger values at the short-wavelength (blue) or long-wavelength (red) end of the visible spectrum. However, the human brain combines all of the colors in a spectrum, so where the intensity peaks is just one factor shaping the color perceived by the eye. The image insets show stars of about these surface temperatures. UV stands for the ultraviolet wavelength range, which extends from 10 to 400 nm. See Figure 5-12 for more on wavelengths of the spectrum. (Insets a: Andrea Dupree [Harvard-Smithsonian CFA], Ronald Gilliland [STScI], NASA and ESA; b: NSO/AURA/NSF; c: NASA, ESA, H. Bond [STScI] and M. Barstow [University of Leicester])

FIGURE 17-8

U, B, and V Filters This graph shows the wavelengths to which the standard filters are transparent. The U filter is transparent to near-ultraviolet light. The B filter is transparent to violet, blue, and green light, while the V filter is transparent to green and yellow light. By measuring the apparent brightness of a star with each of these filters and comparing the results, an astronomer can determine the star's surface temperature.

device such as a charge-coupled device (CCD) (see Section 6-4). The process is then repeated with each of the filters in the set. The star's image will have a different brightness through each colored filter, and by comparing these brightnesses, astronomers can find the wavelength at which the star's intensity curve has its peak—and hence the star's temperature.

> Astronomers use a set of filters in their telescopes to measure the surface temperatures of stars.

UBV Photometry

Let's look at this procedure in more detail. The most commonly used filters are called U, B, and V, and the technique that uses them is called **UBV photometry**. Each filter is transparent to a different band of wavelengths: the ultraviolet (U), the blue (B), and the yellow-green (V, for visual) region in and around the visible spectrum (**Figure 17-8**). The transparency of the V filter mimics the sensitivity of the human eye.

To determine a star's temperature using UBV photometry, the astronomer first measures the star's brightness through each of the filters individually. This gives three apparent brightnesses for the star, designated b_U, b_B, and b_V. The astronomer then compares the intensity

of starlight in neighboring wavelength bands by taking the ratios of these brightnesses: b_V/b_B and b_B/b_U. **Table 17-1** gives values for these **color ratios** for several stars with different surface temperatures.

If a star is very hot, its radiation is skewed toward short, ultraviolet wavelengths as in Figure 17-7c. This makes the star dim through the V filter, brighter through the B filter, and brightest through the U filter. Hence, for a hot star, b_V is less than b_B, which in turn is less than b_U, and the ratios b_V/b_B and b_B/b_U are both less than 1. One such star is Bellatrix (see Table 17-1), which has a surface temperature of 21,500 K.

In contrast, if a star is cool, its radiation peaks at long wavelengths, as in Figure 17-7a. Such a star appears brightest through the V filter, dimmer through the B filter, and dimmest through the U filter (see Figure 17-8). In other words, for a cool star, b_V is greater than b_B, which in turn is greater than b_U. Hence, the ratios b_V/b_B and b_B/b_U will both be greater than 1. The star Betelgeuse (surface temperature 3500 K) is an example.

You can see these differences between hot and cool stars in parts a and c of Figure 6-27, which show the constellation Orion at ultraviolet wavelengths (a bit shorter than those transmitted by the U filter) and at visible wavelengths that approximate the transmission of a V filter. The hot star Bellatrix is brighter in the ultraviolet image (Figure 6-27a) than at visible wavelengths (Figure 6-27c). (Figure 6-27d shows the names of the stars.) The situation is reversed for the cool star Betelgeuse: It is bright at visible wavelengths, but at ultraviolet wavelengths it is too dim to show up in the image.

Figure 17-9 graphs the relationship between a star's b_V/b_B color ratio and its temperature. If you know the value of the b_V/b_B color ratio for a given star, you can use this graph to find the star's surface temperature. As an example, b_V/b_B for the Sun equals 1.87, which corresponds to a surface temperature of 5800 K.

CAUTION!

As we will see in Chapter 18, tiny dust particles that pervade interstellar space cause distant stars to appear redder than they really are. (In the same way, particles in Earth's atmosphere make the setting Sun look redder; see Box 5-4.) Astronomers must take this reddening into account whenever they attempt to determine a star's surface temperature from its color ratios. A star's spectrum provides a more precise measure of a star's surface temperature, as we will see next. But it is quicker and easier to observe a star's colors with a set of U, B, and V filters than it is to take the star's spectrum with a spectrograph.

TABLE 17-1 Colors of Selected Stars

Star	Surface temperature (K)	b_V/b_B	b_B/b_U	Apparent color
Bellatrix (γ Orionis)	21,500	0.81	0.45	Blue
Regulus (α Leonis)	12,000	0.90	0.72	Blue-white
Sirius (α Canis Majoris)	9400	1.00	0.96	Blue-white
Megrez (δ Ursae Majoris)	8630	1.07	1.07	White
Altair (α Aquilae)	7800	1.23	1.08	Yellow-white
Sun	5800	1.87	1.17	Yellow-white
Aldebaran (α Tauri)	4000	4.12	5.76	Orange
Betelgeuse (α Orionis)	3500	5.55	6.66	Red

Source: J.-C. Mermilliod, B. Hauck, and M. Mermilliod, University of Lausanne

FIGURE 17-9

Temperature, Color, and Color Ratio The b_V/b_B color ratio is the ratio of a star's apparent brightnesses through a V filter and through a B filter. This ratio is small for hot, blue stars but large for cool, red stars. After measuring a star's brightness with the B and V filters, an astronomer can estimate the star's surface temperature from a graph like this one.

CONCEPT CHECK 17-8

Which is hotter: a star with an orange color or one with a blue color?

Answer appears at the end of the chapter.

17-5 The spectra of stars reveal their chemical compositions as well as their surface temperatures

We have seen how the color of a star's light helps astronomers determine its surface temperature. To determine the other properties of a star, astronomers must analyze the spectrum of its light in more detail. This technique of *stellar spectroscopy* began in 1817 when Joseph Fraunhofer, a German instrument maker, attached a spectroscope to a telescope and pointed it toward the stars. Fraunhofer had earlier observed that the Sun has an absorption line spectrum—that is, a continuous spectrum with dark absorption lines (see Section 5-6). He found that stars have the same kind of spectra, which reinforces the idea that our Sun is a rather typical star. But Fraunhofer also found that the pattern of absorption lines is different for different stars.

Classifying Stars: Absorption Line Spectra and Spectral Classes

We see an absorption line spectrum when a cool gas lies between us and a hot, glowing object (recall Figure 5-17). The light from the hot, glowing object itself has a continuous spectrum. In the case of a star, light with a continuous spectrum is produced at low-lying levels of the star's atmosphere where the gases are hot and dense. The absorption lines are created when this light flows outward

through the upper layers of the star's atmosphere. Atoms in these cooler, less dense layers absorb radiation at specific wavelengths, which depend on the specific kinds of atoms present—hydrogen, helium, or other elements—and on whether or not the atoms have lost an electron (leaving them ionized). Absorption lines in the Sun's spectrum are produced in this same way (see Section 16-5).

Some stars have spectra in which the Balmer absorption lines of hydrogen are prominent. But in the spectra of other stars, including the Sun, the Balmer lines are nearly absent and the dominant absorption lines are those of heavier elements such as calcium, iron, and sodium. Still other stellar spectra are dominated by broad absorption lines caused by molecules, such as titanium oxide, rather than single atoms. To cope with this diversity, astronomers group similar-appearing stellar spectra into **spectral classes**. In a popular classification scheme that emerged in the late 1890s, a star was assigned a letter from A through O according to the strength or weakness of the hydrogen Balmer lines in the star's spectrum.

Nineteenth-century science could not explain why or how the spectral lines of a particular chemical are affected by the temperature and density of the gas. Nevertheless, a team of astronomers at the Harvard College Observatory forged ahead with a monumental project of examining the spectra of hundreds of thousands of stars. Their goal was to develop a system of spectral classification in which all spectral features, not just Balmer lines, change smoothly from one spectral class to the next.

> Deciphering the information in starlight took the painstaking work of generations of astronomers.

The Harvard project was financed by Mary Ann Palmer Draper, the widow of Henry Draper, a wealthy New York physician and amateur astronomer who in 1872 became the first person to photograph stellar absorption lines. Researchers on the project included Edward C. Pickering, Williamina Fleming, Antonia Maury, and Annie Jump Cannon (Figure 17-10). As a result of their efforts, many of the original A-through-O classes were dropped and others were consolidated. The remaining spectral classes were reordered in the sequence **OBAFGKM**. They remembered this sequence (and you can, too) with the mnemonic "Oh, *Be A Fine Girl* (or *Guy*), *Kiss Me!*"

CONCEPT CHECK 17-9

What principal characteristic of a star's spectrum most dominates which spectral class letter of the alphabet it is assigned?

Answer appears at the end of the chapter.

Refining the Classification: Spectral Types

Cannon refined the original OBAFGKM sequence into smaller steps called **spectral types**. These steps are indicated by attaching an integer from 0 through 9 to the original letter. For example, the spectral class F includes spectral types F0, F1, F2, . . . , F8, F9, which are followed by the spectral types G0, G1, G2, . . . , G8, G9, and so on.

(a)

(b)

FIGURE 17-10   R I **V** U X G

Classifying the Spectra of the Stars The modern classification scheme for stars, based on their spectra, was developed at the Harvard College Observatory in the late nineteenth century. Female astronomers, initially led by Edward C. Pickering (not shown) and **(a)** Williamina Fleming, standing, and then by **(b)** Annie Jump Cannon, analyzed hundreds of thousands of spectra. Social conventions of the time prevented most female astronomers from using research telescopes or receiving salaries comparable to those of men. (a: Harvard College Observatory; b: Bettmann/Getty Images)

Figure 17-11 shows representative spectra of several spectral types. The strengths of spectral lines change gradually from one spectral type to the next. For example, the Balmer absorption lines of hydrogen become increasingly prominent as you go from spectral type B0 to A0. From A0 onward through the F and G classes, the hydrogen lines weaken and almost fade from view. The Sun, whose spectrum is dominated by calcium and iron, is a G2 star.

The Harvard project culminated in the *Henry Draper Catalogue*, published between 1918 and 1924. It listed 225,300 stars, each of which Cannon had personally classified. Meanwhile, physicists had been making important discoveries about the structure of atoms. Ernest Rutherford had shown that atoms have nuclei (recall Figure 5-20), and Niels Bohr made the remarkable hypothesis that electrons circle atomic nuclei along discrete orbits (see Figure 5-22). These advances gave scientists the conceptual and mathematical tools needed to understand stellar spectra.

In the 1920s, the Harvard astronomer Cecilia Payne and the Indian physicist Meghnad Saha demonstrated that the OBAFGKM spectral sequence is actually a sequence in temperature. The hottest stars are O stars. Their absorption lines can occur only if these stars have surface temperatures above 25,000 K. M stars are the coolest stars. The spectral features of M stars are consistent with stellar surface temperatures of about 3000 K.

CONCEPT CHECK 17-10

Is a spectral class F2 star more similar to an A-spectral class star or a G-spectral class star?

CONCEPT CHECK 17-11

As spectral type numbers increase within the G-spectral class of stars, do the larger numbers represent higher- or lower-temperature stars?

Answers appear at the end of the chapter.

Why Surface Temperature Affects Stellar Spectra

To see why the appearance of a star's spectrum is profoundly affected by the star's surface temperature, consider the Balmer lines of hydrogen. Hydrogen is by far the most abundant element in the universe, accounting for about three-quarters of the mass of a typical star. Yet the Balmer lines do not necessarily show up in a star's spectrum. As we saw in Section 5-8, Balmer absorption lines are produced when an electron in the $n = 2$ orbit of hydrogen is lifted into a higher orbit by absorbing a photon with the right amount of energy (see Figure 5-24). If the star is much hotter than 10,000 K, the photons pouring out of the star's interior have such high energy that they easily knock electrons out of hydrogen atoms in the star's atmosphere. This process ionizes the gas. With its only electron torn away, a hydrogen atom cannot produce absorption lines. Hence, the Balmer lines will be relatively weak in the spectra of such hot stars, such as the hot O and B2 stars in Figure 17-11.

Conversely, if the star's atmosphere is much cooler than 10,000 K, almost all of the hydrogen atoms are in the lowest ($n = 1$) energy state. Most of the photons passing through the star's atmosphere possess too little energy to boost electrons up from the $n = 1$ to the $n = 2$ orbit of the hydrogen atoms. Hence, very few of these atoms will have electrons in the $n = 2$ orbit, and only these few can absorb the photons characteristic of the Balmer lines. As a result, these lines are nearly absent from the spectrum of a cool star. (You can see this in the spectra of the cool M0 and M2 stars in Figure 17-11.)

For the Balmer lines to be prominent in a star's spectrum, the star must be hot enough to excite the electrons out of the ground state but not so hot that all of the hydrogen atoms become ionized. A stellar surface temperature of about 9000 K produces the strongest hydrogen lines; this is the case for the stars of spectral types A0 and A5 in Figure 17-11.

 LOOKING DEEPER 17.1 Every other type of atom or molecule also has a characteristic temperature range in which it produces prominent absorption lines in the observable part of the spectrum. Figure 17-12 shows the relative strengths of absorption lines produced

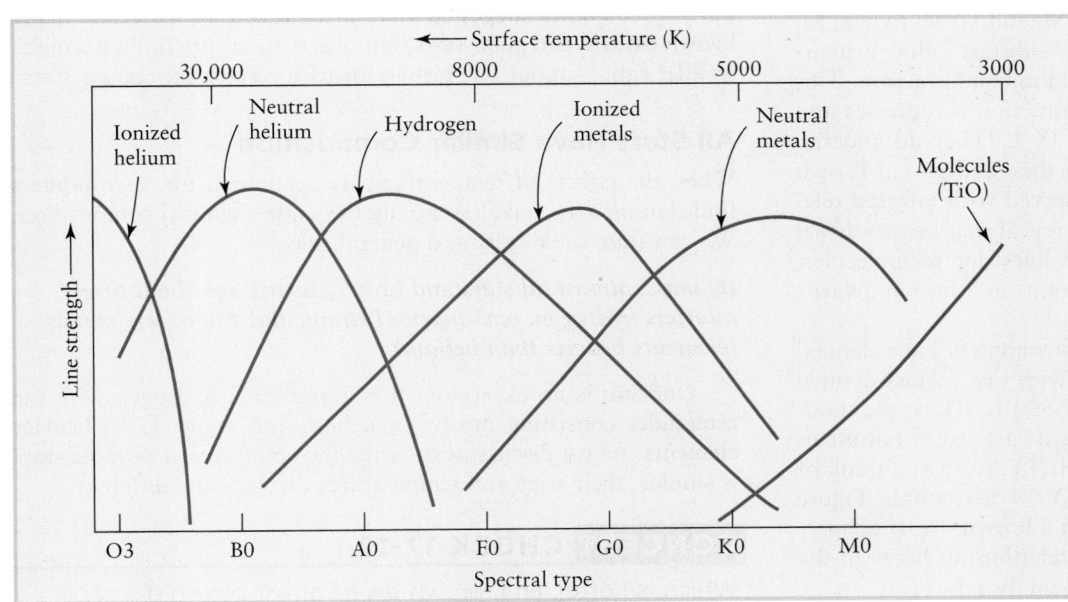

FIGURE 17-11 R I Ⓥ U X G

Principal Types of Stellar Spectra Stars of different spectral classes and different surface temperatures have spectra dominated by different absorption lines. Notice how the Balmer lines of hydrogen (H_α, H_β, H_γ, and H_δ) are strongest for hot stars of spectral class A, while absorption lines due to calcium (Ca) are strongest in cooler K and M stars. The spectra of M stars also have broad, dark bands caused by molecules of titanium oxide (TiO), which can exist only at relatively low temperatures. A Roman numeral after a chemical symbol shows whether the absorption line is caused by un-ionized atoms (Roman numeral I) or by atoms that have lost one electron (Roman numeral II). The Sun, a G2 star, has a spectrum between G0 and G5. (R. Bell, University of Maryland, and M. Briley, University of Wisconsin at Oshkosh)

FIGURE 17-12

The Strengths of Absorption Lines
Each curve in this graph peaks at the stellar surface temperature for which that chemical's absorption line is strongest. For example, hydrogen (H) absorption lines are strongest in A stars with surface temperatures near 10,000 K. Roman numeral I denotes neutral, un-ionized atoms; II, III, and IV denote atoms that are singly, doubly, or triply ionized (that is, have lost one, two, or three electrons).

by different chemicals. By measuring the details of these lines in a given star's spectrum, astronomers can accurately determine that star's surface temperature.

For example, the spectral lines of neutral (that is, un-ionized) helium are strong around 25,000 K. At this temperature, photons have enough energy to excite helium atoms without tearing away the electrons altogether. In stars hotter than about 30,000 K, helium atoms become singly ionized; that is, they lose one of their two electrons. The remaining electron produces a set of spectral lines that is recognizably different from those of neutral helium. Hence, when the spectral lines of singly ionized helium appear in a star's spectrum, we know that the star's surface temperature is greater than 30,000 K.

Astronomers use the term **metals** to refer to all elements other than hydrogen and helium. This idiosyncratic use of the term "metal" is quite different from the definition used by chemists and other scientists. To a chemist, sodium and iron are metals but carbon and oxygen are not; to an astronomer, all of these substances are metals. In this terminology, metals dominate the spectra of stars cooler than 10,000 K. Ionized metals are prominent for surface temperatures between 6000 and 8000 K, whereas neutral metals are strongest between approximately 5500 and 4000 K.

Below 4000 K, certain atoms in a star's atmosphere combine to form molecules. (At higher temperatures, atoms move so fast that when they collide, they bounce off each other rather than "sticking together" to form molecules.) As these molecules vibrate and rotate, they produce bands of spectral lines that dominate the star's spectrum. Most noticeable are the lines of titanium oxide (TiO), which are strongest for surface temperatures of about 3000 K.

Spectral Classes for Brown Dwarfs

Since 1995, astronomers have found a number of stars with surface temperatures even lower than those of spectral class M. Strictly speaking, these are not stars at all but **brown dwarfs**. Brown dwarfs are too small to sustain thermonuclear fusion in their cores; they have masses between 13 M_J and 80 M_J (where M_J stands for mass of Jupiter). Instead, these "substars" glow primarily from the heat released by Kelvin–Helmholtz contraction. This is the process by which gravitational contraction compresses and heats up gas, as we described in Section 16-1. (They do undergo fusion reactions for a brief period during their evolution.) Brown dwarfs are so cold that they are best observed with infrared telescopes (Figure 17-13). Such observations reveal that brown dwarf spectra have a rich variety of absorption lines due to molecules, some of which actually form into solid grains in a brown dwarf's atmosphere.

To describe brown dwarf spectra, astronomers have defined three new spectral classes, L, T, and Y, with the Y class defined for brown dwarfs below temperatures of 600 K. Thus, the modern spectral sequence of stars and brown dwarfs from hottest to coldest surface temperature is OBAFGKMLTY. (Can you think of a new mnemonic that includes L, T, and Y?) For example, Figure 17-13 shows a star of spectral class K and a brown dwarf of spectral class T. Table 17-2 summarizes the relationship between the temperatures and spectra of stars and brown dwarfs.

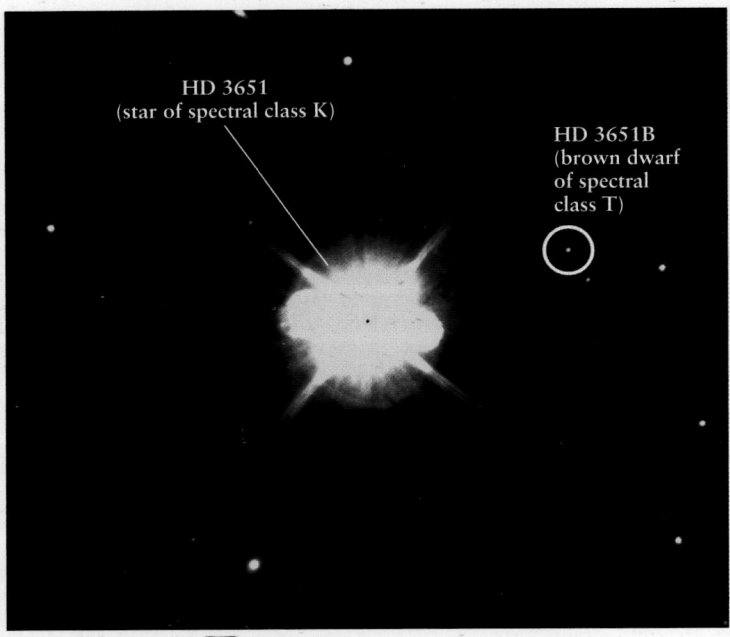

FIGURE 17-13 R I V U X G

An Infrared Image of Brown Dwarf HD 3651B The star HD 3561 is of spectral class K, with a surface temperature of about 5200 K. ("HD" refers to the *Henry Draper Catalogue*.) In 2006, it was discovered that HD 3651 is orbited by a brown dwarf named HD 3651B with a surface temperature between 800 and 900 K and a luminosity just 1/300,000 that of the Sun. The brown dwarf emits most of its light at infrared wavelengths, so an infrared telescope was used to record this image. The hotter and more luminous star HD 3651 is greatly overexposed in this image and appears much larger than its true size. HD 3651 and HD 3651B are both 11 pc (36 ly) from Earth in the constellation Pisces (the Fish); the other stars in this image are much farther away. (M. Mugrauer and R. Neuhauser, U. of Jena; A. Seifahrt, ESO; T. Mazeh, Tel Aviv, U.)

NASA's Wide-field Infrared Satellite (WISE) telescope discovered more than 100 new brown dwarfs. The coolest known brown dwarf is a mere 300 K (80°F), which is about room temperature. Early predictions suggested that there might be twice as many brown dwarfs as regular stars, but observations are finding a much smaller ratio—about one brown dwarf for every six regular stars.

All Stars Have Similar Composition

When the effects of temperature are accounted for, astronomers find that *all* stars have essentially the same chemical composition. We can state the results as a general rule:

By mass, almost all stars and brown dwarfs are about three-quarters hydrogen, one-quarter helium, and 1% or less metals (elements heavier than helium).

Our Sun is no exception: It is about 75% hydrogen with the remainder consisting mostly of helium and about 1.7% heavier elements. As we discuss next, while the composition of most stars is similar, their sizes and temperatures can be quite different.

CONCEPT CHECK 17-12

Which is hotter, an L brown dwarf or a T brown dwarf?

Zooming in on a small portion of the Carina Nebula, this feature, which is 3 light-years long, is called Mystic Mountain. The dense hydrogen body of this feature is slowly being eroded by ultraviolet light from surrounding stars. In the coolest and densest regions of this feature, new stars are forming. New stars can form a pair of oppositely directed jets, which are visible at the peak of this mystic mountain. On the flat-topped structure to the left of the peak, another jet can be seen, while its other half is presumably obscured by the nebula. (NASA, ESA, and M. Livio and the Hubble 20th Anniversary Team [STScI])

R I V U X G

III Stars and Stellar Evolution

Our own Milky Way Galaxy has a galactic neighbor called the Small Magellanic Cloud. Within this nearby galaxy, stars can be seen that are surrounded by a dense cloud of gas. The dense gas is a stellar nursery that gives birth to these stars. The bright blue stars are the youngest and most massive, with temperatures several times hotter than our Sun. The light emitted by these blue stars is so strong that it causes the surrounding gas to glow, while simultaneously eroding and clearing the gas to form a somewhat spherical cavity. The densest gas and dust resists the erosion, leaving small fingerlike columns of gas around the inner edge that point toward the center of the spherical cavity. Within these dense structures, a new generation of stars is born. (NASA/CXC/JPL-Caltech/STScI)

R I V U X G

TABLE 17-2   The Spectral Sequence

Spectral class	Color	Temperature (K)	Spectral lines	Examples
O	Blue-violet	30,000–50,000	Ionized atoms, especially helium	Naos (ζ Puppis), Mintaka (δ Orionis)
B	Blue-white	11,000–30,000	Neutral helium, some hydrogen	Spica (α Virginis), Rigel (β Orionis)
A	White	7500–11,000	Strong hydrogen, some ionized metals	Sirius (α Canis Majoris), Vega (α Lyrae)
F	Yellow-white	5900–7500	Hydrogen and ionized metals such as calcium and iron	Canopus (α Carinae), Procyon (α Canis Minoris)
G	Yellow	5200–5900	Both neutral and ionized metals, especially ionized calcium	Sun, Capella (α Aurigae)
K	Orange	3900–5200	Neutral metals	Arcturus (α Boötis), Aldebaran (α Tauri)
M	Red-orange	2500–3900	Strong titanium oxide and some neutral calcium	Antares (α Scorpii), Betelgeuse (α Orionis)
L	Red	1300–2500	Neutral potassium, rubidium, and cesium, and metal hydrides	Brown dwarf Teide 1
T	Red	700–1300	Methane, strong neutral potassium and some water (H_2O)	Brown dwarfs Gliese 229B, HD 3651B
Y	Red	below 600	Possibly ammonia	Brown dwarfs CFBDS J005910.90–011401.3, WISE 1828+2650

CONCEPT CHECK 17-13

Aside from size and surface temperature, what is the main difference between a regular star like our Sun and a brown dwarf?

Answers appear at the end of the chapter.

17-6 Stars come in a wide variety of sizes and temperatures

With even the best telescopes, stars appear as nothing more than bright points of light. On a photograph or CCD image, brighter stars appear larger than dim ones (see Figures 17-3, 17-6b, and 17-13), but these apparent sizes give no indication of the star's actual size. To determine the size of a star, astronomers combine information about its luminosity (determined from its distance and apparent brightness) and its surface temperature (determined from its spectral type). In this way, they find that some stars are quite a bit smaller than the Sun, whereas others are 1000 times larger.

> A star's radius can be calculated if we know its luminosity and surface temperature.

Calculating the Radii of Stars

The key to finding a star's radius from its luminosity and surface temperature is the Stefan–Boltzmann law (see Section 5-4). This law says that the amount of energy radiated per second from a square meter of a blackbody's surface—that is, the energy flux (F)—is proportional to the fourth power of the temperature of that surface (T), as given by the equation $F = \sigma T^4$. This equation applies very well to stars, whose spectra are quite similar to that of a perfect blackbody. (Absorption lines, while important for determining the star's chemical composition and surface temperature, make only relatively small modifications to a star's blackbody spectrum.)

A star's luminosity is the amount of energy emitted per second from its entire surface. This quantity equals the energy flux F multiplied by the total number of square meters on the star's surface (that is, the star's surface area). We expect that most stars are nearly spherical, like the Sun, so we can use the formula for the surface area of a sphere. The formula is $4\pi R^2$, where R is the star's radius (the distance from its center to its surface). Multiplying together the formulas for energy flux and surface area, we can write the star's luminosity as follows:

Relationship between a star's luminosity, radius, and surface temperature

$$L = 4\pi R^2 \sigma T^4$$

L = star's luminosity, in watts

R = star's radius, in meters

σ = Stefan–Boltzmann constant = 5.67×10^{-8} W m^{-2} K^{-4}

T = star's surface temperature, in kelvins

This equation says that a relatively cool star (low surface temperature T), for which the energy flux is quite low, can nonetheless be very luminous if it has a large enough radius R. Alternatively, a relatively hot star (large T) can have a very low luminosity if the star has only a little surface area (small R).

Box 17-4 describes how to use the preceding equation to calculate a star's radius if its luminosity and surface temperature are known.

We can express the idea behind these calculations in terms of the following general rule:

We can determine the radius of a star from its luminosity and surface temperature.

For a given luminosity: The greater the surface temperature, the smaller the radius must be.

For a given surface temperature: The greater the luminosity, the larger the radius must be.

ANALOGY

In a similar way, a roaring campfire can emit more light than a welder's torch. The campfire is at a lower temperature than the torch, but has a much larger surface area from which it emits light.

The Range of Stellar Radii

Using this general rule, as shown in Box 17-4, astronomers find that stars come in a wide range of sizes. The smallest stars visible through ordinary telescopes, called *white dwarfs,* are about the same size as Earth. Although their surface temperatures can be very high (25,000 K or more), white dwarfs have so little surface area that their luminosities are very low (less than $0.01\ L_\odot$). The largest stars, called *supergiants,* are 1000 times larger in radius than the Sun and 10^5 times larger than white dwarfs. If our own Sun were replaced by one of these supergiants, Earth's orbit would lie completely inside the star!

Figure 17-14 summarizes how astronomers determine the distance from Earth, luminosity, surface temperature, chemical composition, and radius of a star close enough to us so that its parallax can be measured. Remarkably, all of these properties can be deduced from just a few measured quantities: the star's parallax angle, apparent brightness, and spectrum.

CONCEPT CHECK 17-14

Why is it that sitting by a campfire warms you up, but sitting next to a candle of the same temperature does not?

CONCEPT CHECK 17-15

Based on the flowchart in Figure 17-14, do you need to know the distance of a star in order to determine its surface temperature? Do you need to know the distance of a star from Earth in order to determine its radius? Assume you know the star's apparent brightness.

CALCULATION CHECK 17-4

If two stars are at the same temperature, but one is 3 times larger, how many times more luminous is it?

Answers appear at the end of the chapter.

17-7 Hertzsprung–Russell (H-R) diagrams reveal different kinds of stars

Astronomers have collected a wealth of data about the stars, but merely having tables of numerical data is not very useful by itself. Like all scientists, astronomers want to analyze their data to look for trends and underlying principles. One of the best ways to look for trends in any set of data, whether it comes from astronomy,

FIGURE 17-14

Finding Key Properties of a Nearby Star This flowchart shows how astronomers determine the properties of a relatively nearby star (one close enough that its parallax can be measured). The rounded purple boxes show the measurements that must be made of the star, the blue ovals show the key equations that are used (from Sections 17-2, 17-5, and 17-6), and the green rectangles show the inferred properties of the stars. A different procedure is followed for more distant stars (see Section 17-8, especially Figure 17-17).

BOX 17-4   TOOLS OF THE ASTRONOMER'S TRADE

Stellar Radii, Luminosities, and Surface Temperatures

Because stars emit light in almost exactly the same fashion as blackbodies, we can use the Stefan–Boltzmann law to relate a star's luminosity (L), surface temperature (T), and radius (R). The relevant equation is

$$L = 4\pi R^2 \sigma T^4$$

As written, this equation involves the Stefan–Boltzmann constant σ, which is equal to 5.67×10^{-8} W m^{-2} K^{-4}. In many calculations, it is more convenient to relate everything to the Sun, which is a typical star. Specifically, for the Sun we have $L_\odot = 4\pi R_\odot^2 \sigma T_\odot^4$, where $L_\odot$ is the Sun's luminosity, $R_\odot$ is the Sun's radius, and $T_\odot$ is the Sun's surface temperature (equal to 5800 K). Dividing the general equation for L by this specific equation for the Sun, we obtain

$$\frac{L}{L_\odot} = \left(\frac{R}{R_\odot}\right)^2 \left(\frac{T}{T_\odot}\right)^4$$

This formula is easier to use because the constant σ has canceled out. We can also rearrange terms to arrive at a useful equation for the radius (R) of a star:

Radius of a star related to its luminosity and surface temperature

$$\frac{R}{R_\odot} = \left(\frac{T_\odot}{T}\right)^2 \sqrt{\frac{L}{L_\odot}}$$

$R/R_\odot$ = ratio of the star's radius to the Sun's radius

$T_\odot/T$ = ratio of the Sun's surface temperature to the star's surface temperature

$L/L_\odot$ = ratio of the star's luminosity to the Sun's luminosity

EXAMPLE: The bright reddish star Betelgeuse in the constellation Orion (see Figure 2-2) is 60,000 times more luminous than the Sun and has a surface temperature of 3500 K. What is its radius?

Situation: We are given the star's luminosity $L = 60,000\ L_\odot$ and its surface temperature $T = 3500$ K. Our goal is to find the star's radius R.

Tools: We use the above equation to find the ratio of the star's radius to the radius of the Sun, $R/R_\odot$. Note that we also know the Sun's surface temperature, $T_\odot = 5800$ K.

Answer: Substituting these data into the above equation, we get

$$\frac{R}{R_\odot} = \left(\frac{5800\ \text{K}}{3500\ \text{K}}\right)^2 \sqrt{6 \times 10^4} = 670$$

Review: Our result tells us that Betelgeuse's radius is 670 times larger than that of the Sun. The Sun's radius is 6.96×10^5 km, so we can also express the radius of Betelgeuse as $(670)(6.96 \times 10^5\ \text{km}) = 4.7 \times 10^8$ km, which is more than 3 au. If Betelgeuse were located at the center of our solar system, it would extend beyond the orbit of Mars!

EXAMPLE: Sirius, the brightest star in the sky, is actually two stars orbiting each other (a binary star). The less luminous star, Sirius B, is a white dwarf that is too dim to see with the naked eye. Its luminosity is $0.0025\ L_\odot$ and its surface temperature is 10,000 K. How large is Sirius B compared to Earth?

Situation: Again, we are asked to find a star's radius from its luminosity and surface temperature.

Tools: We use the same equation as in the preceding example.

Answer: The ratio of the radius of Sirius B to the Sun's radius is

$$\frac{R}{R_\odot} = \left(\frac{5800\ \text{K}}{10,000\ \text{K}}\right)^2 \sqrt{0.002.5} = 0.017$$

Since the Sun's radius is $R_\odot = 6.96 \times 10^5$ km, the radius of Sirius B is $(0.017)(6.96 \times 10^5\ \text{km}) = 12,000$ km. As indicated in Table 7-1, Earth's radius (half its diameter) is 6378 km. Hence, this star is only about twice the radius of Earth.

Review: Sirius B's radius would be large for a terrestrial planet, but it is minuscule for a star. The name *dwarf* is well deserved!

The radii of some stars have been measured with other techniques (see Section 17-11). These other methods yield values consistent with those calculated by the methods just described.

---

finance, medicine, or meteorology, is to create a graph showing how one quantity depends on another. For example, investors consult graphs of stock market values versus dates, and weather forecasters make graphs of temperature versus altitude to determine whether thunderstorms are likely to form. Astronomers have found that a particular graph of stellar properties shows that stars fall naturally into just a few categories. This graph, one of the most important in all of astronomy, will in later chapters help us understand how stars form, evolve, and eventually die.

H-R Diagrams

Which properties of stars should we include in a graph? Most stars have about the same chemical composition, but two properties of stars—their luminosities and surface temperatures—differ substantially from one star to another. Stars also come in a wide range of radii, but a star's radius is a secondary property that can be found from the luminosity and surface temperature (as we saw in Section 17-6 and Box 17-4). We also relegate the positions and space velocities of stars to secondary importance. (In a similar way, a physician is more interested in your weight and blood pressure than in where you live or how fast you drive.) We can then ask the following question: What do we learn when we graph the luminosities of stars versus their surface temperatures?

> A graph of luminosity versus surface temperature reveals that stars fall into a few basic categories.

The first answer to this question was given in 1911 by the Danish astronomer Ejnar Hertzsprung. He pointed out that a regular pattern appears when the absolute magnitudes of stars (which measure their luminosities) are plotted against their colors (which measure their surface temperatures). Two years later, the American astronomer Henry Norris Russell independently discovered a similar regularity in a graph using spectral types (another measure of surface temperature) instead of colors. In recognition of their originators, graphs of this kind are today known as **Hertzsprung–Russell diagrams**, or **H-R diagrams** (Figure 17-15).

Figure 17-15a is a typical Hertzsprung–Russell diagram. Each dot represents a star whose spectral type and luminosity have been determined. The most luminous stars are near the top of the diagram, the least luminous stars near the bottom. Hot stars of spectral classes O and B are toward the left side of the graph and cool stars of spectral class M are toward the right.

CAUTION!

You are probably accustomed to graphs in which the numbers on the horizontal axis increase as you move to the right. (For example, the business section of a newspaper includes a graph of stock market values versus dates, with later dates to the right of earlier ones.) But in an H-R diagram, the temperature scale on the horizontal axis increases toward the *left*. This practice stems from the original diagrams of Hertzsprung and Russell, who placed hot O stars on the left and cool M stars on the right. This arrangement is a tradition that no one has seriously tried to change.

CONCEPT CHECK 17-16

Where on the H-R diagram are the stars with the greatest luminosity and highest temperatures plotted?

Answer appears at the end of the chapter.

Star Varieties: Main-Sequence Stars, Giants, Supergiants, White Dwarfs, and Brown Dwarfs

The most striking feature of the H-R diagram is that the data points are not scattered randomly over the graph but are grouped in a few distinct regions. The luminosities and surface temperatures of stars do *not* have random values; instead, these two quantities are related!

The band stretching diagonally across the H-R diagram includes about 90% of the stars in the night sky. This band, called the **main sequence**, extends from the hot, luminous, blue stars in the upper left corner of the diagram to the cool, dim, red stars in the lower right corner. A star whose properties place it in this region of an H-R diagram is called a **main-sequence star**. The Sun (spectral type G2, luminosity 1 $L_\odot$, absolute magnitude +4.8) is such a star. We will find that all main-sequence stars are like the Sun in that *hydrogen fusion*—thermonuclear reactions that convert hydrogen into helium (see Section 16-1)—is taking place in their cores. Thus, we have a very important rule:

Main-sequence stars gain their energy through conversion of hydrogen to helium in their cores.

The upper right side of the H-R diagram shows a second major grouping of data points. Stars represented by these points are both luminous and cool. From the Stefan–Boltzmann law, we know that a cool star radiates much less light per unit of surface area than a hot star. In order for these stars to be as luminous as they are, they must be huge (see Section 17-6)—therefore, they are called **giants**. These stars are about 10 to 100 times larger than the Sun. You can see this size difference in Figure 17-15b, which is an H-R diagram to which dashed lines have been added to represent stellar radii. Most giant stars are about 100 to 1000 times more luminous than the Sun and have surface temperatures of about 3000 to 6000 K. Cooler members of this class of stars (those with surface temperatures from about 3000 to 4000 K) are often called **red giants** because they appear reddish. In the collection of stars in Figure 17-15c, the brightest orange and red stars are red giants. However, while these giant stars also appear larger in this image, their true size is much too small to see; it is really their brightness that overexposes the camera and makes them appear large. (A number of red giants can easily be seen with the naked eye, including Aldebaran in the constellation Taurus and Arcturus in Boötes.)

A few rare stars in the night sky are considerably bigger and brighter than typical red giants, with radii up to 1000 $R_\odot$. Appropriately enough, these superluminous stars are called **supergiants**. Betelgeuse in Orion (Figure 17-16) and Antares in Scorpius are two supergiants you can find in the nighttime sky. Together, giants and supergiants make up about 1% of the stars in the sky.

Both giants and supergiants have thermonuclear reactions occurring in their interiors, but the character of those reactions and where in the star they occur can be quite different than for a main-sequence star like the Sun. We will study these stars in more detail in Chapter 21.

The remaining 9% of stars in the H-R diagram of Figure 17-15a form a distinct grouping of data points toward the lower left corner. Although these stars are hot (making them somewhat blue-white in appearance), their luminosities are quite low; hence, they must be small. They are appropriately called **white dwarfs**. These stars, which are so dim that they can be seen only with a telescope, are approximately the same size as Earth. As we will learn in Chapter 20, no thermonuclear reactions take place within white dwarf stars. Rather, like hot embers left from a fire, they are the still-glowing remnants of what were once giant stars. Technically, since true stars are powered by thermonuclear reactions, white dwarfs are not true stars.

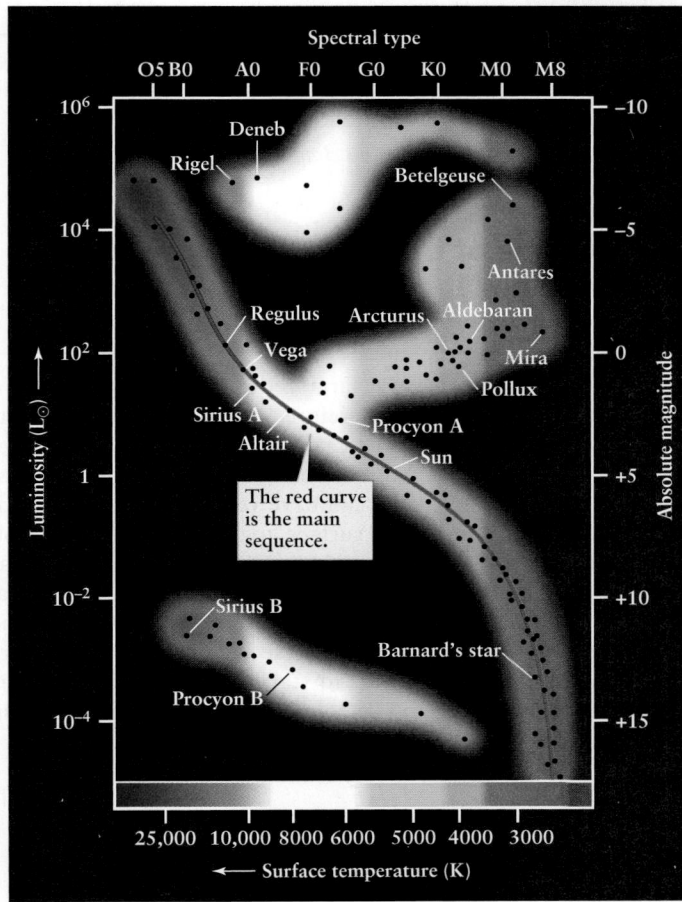

(a) A Hertzsprung–Russell (H-R) diagram

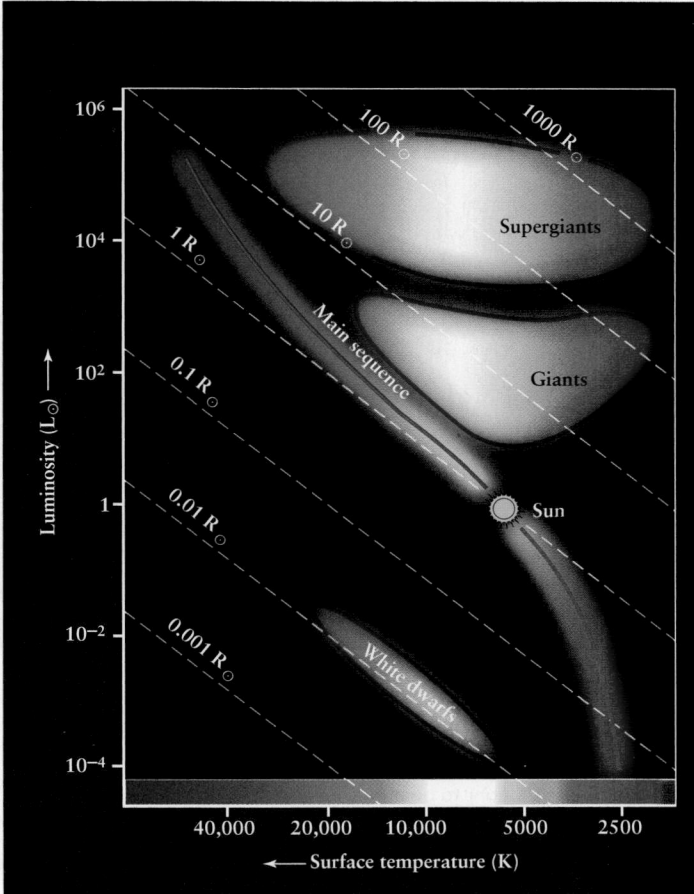

(b) The sizes of stars in an H-R diagram

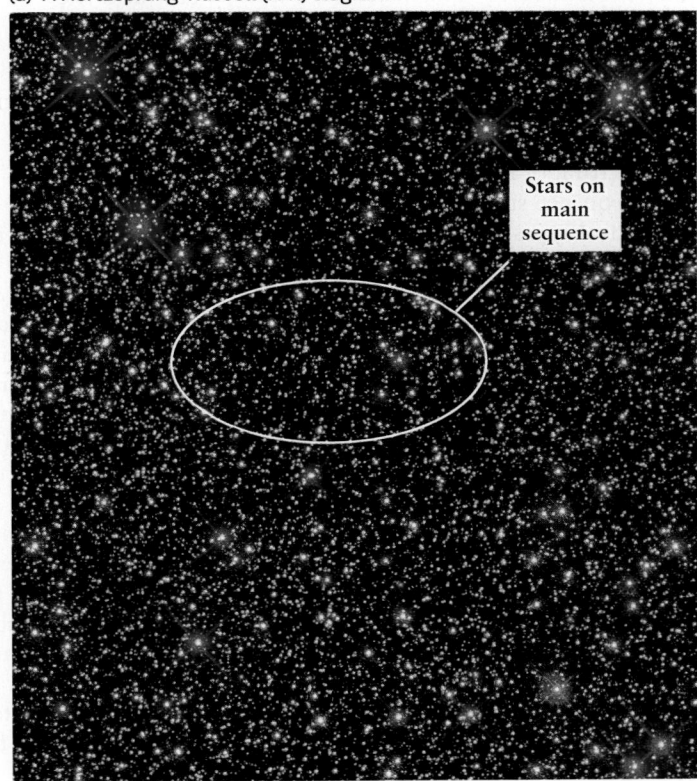

(c) A portion of the star cluster Omega Centauri

~~~~~~~~~~ᴡᴡᴡᴡᴡᴡᴡᴡᴡᴡᴡ

FIGURE 17-15 R I V U X G

Hertzsprung–Russell (H-R) Diagrams In an H-R diagram, the luminosities (or absolute magnitudes) of stars are plotted against their spectral types (or surface temperatures). **(a)** The data points are grouped in just a few regions on the graph, showing that luminosity and spectral type are correlated. Most stars lie along the red curve called the main sequence. Giants like Arcturus, as well as supergiants like Rigel and Betelgeuse, are above the main sequence, and white dwarfs like Sirius B are below it. **(b)** The blue curves in this H-R diagram enclose the regions of the diagram in which different types of stars are found. The dashed diagonal lines indicate different stellar radii. For a given stellar radius, as the surface temperature increases (that is, moving from right to left in the diagram), the star glows more intensely and the luminosity increases (that is, moving upward in the diagram). Note that the Sun is intermediate in luminosity, surface temperature, and radius. **(c)** Some stars in Omega Centauri are burning hydrogen in their cores and would appear on the main sequence in an H-R diagram. The orange- and red-colored stars are red giants, and the blue stars have begun burning helium in their cores. (c: NASA, ESA, and the Hubble SM4 ERO Team)

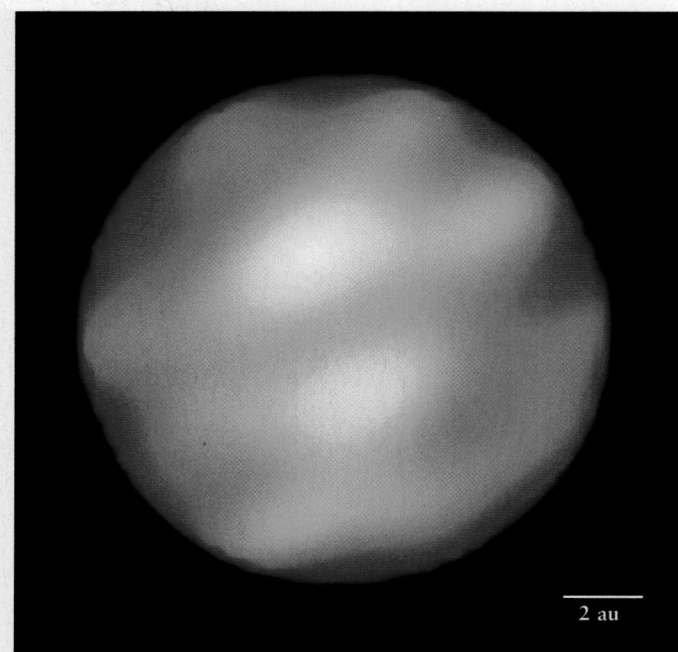

FIGURE 17-16 R I̶ V U X G

Supergiant Betelgeuse Betelgeuse is just close enough and big enough that its surface features can be reconstructed using special techniques (interferometry at infrared wavelengths). The two brighter regions near the center are thought to be hotter gas rising by convection. The size of Betelgeuse is uncertain because its precise distance from Earth is not known, and its radius changes as the star expands and contracts. In this image, the radius is about 5.5 au—slightly larger than Jupiter's orbital distance. (Xavier Haubois [Observatoire de Paris] et al.)

By contrast, *brown* dwarfs (which lie at the extreme lower right of the main sequence, off the bottom and right-hand edge of Figure 17-15a or Figure 17-15b) are objects that will never become stars. They are not massive enough for significant nuclear fusion of hydrogen in their cores.

ANALOGY

You can think of white dwarfs as "has-been" stars whose days of glory have passed. In this analogy, a brown dwarf is a "never-will-be" and is sometimes called a "failed star."

The existence of fundamentally different types of stars is the first important lesson to come from the H-R diagram. In later chapters, we will find that these different types represent various stages in the lives of stars. For example, as their temperatures and luminosities change over time, stars can move through the H-R diagram as they leave the main sequence and become red giants. We will use the H-R diagram as an essential tool for understanding how stars evolve.

CONCEPT CHECK 17-17

Betelgeuse and Barnard's star are both in the red portion of Figure 17-15a. What can you deduce about their relative sizes from the H-R diagram?

CONCEPT CHECK 17-18

In Figure 17-15a, Regulus and Mira are both about 100 times more luminous than our Sun. Which one is powered through conversion of hydrogen to helium in its core?

Answers appear at the end of the chapter.

17-8 Details of a star's spectrum reveal whether it is a giant, a white dwarf, or a main-sequence star

A star's surface temperature largely determines which lines are prominent in its spectrum. Therefore, classifying stars by spectral type is essentially the same as categorizing them by surface temperature. But as the H-R diagram in Figure 17-15b shows, stars of the same surface temperature can have very different luminosities. As an example, a star with surface temperature 5800 K could be a white dwarf, a main-sequence star, a giant, or a supergiant, depending on its luminosity. By examining the details of a star's spectrum, however, astronomers can determine to which of these categories a star belongs. As we will see, this gives astronomers a tool to determine the distances to stars millions of parsecs away, far beyond the maximum distance that can be measured using stellar parallax.

Determining a Star's Size from Its Spectrum

Figure 17-17 compares the spectra of two stars of the same spectral type but different luminosity (and hence different size): a B8 supergiant and a B8 main-sequence star. Note that the Balmer lines of hydrogen are narrow in the spectrum of the very large, very luminous supergiant but are quite broad in the spectrum of the small, less luminous main-sequence star. In general, for stars of spectral types B through F, the larger and more luminous the star, the narrower its hydrogen lines.

> The smaller a star and the denser its atmosphere, the broader the absorption lines in its spectrum.

Fundamentally, these differences between stars of different luminosity are due to differences between the stars' atmospheres, where absorption lines are produced. Hydrogen lines in particular are affected by the density and pressure of the gas in a star's atmosphere. The higher the density and pressure, the more frequently hydrogen atoms collide and interact with other atoms and ions in the atmosphere. These collisions shift the energy levels in the hydrogen atoms and thus broaden the hydrogen spectral lines.

In the atmosphere of a luminous giant star, the density and pressure are quite low because the star's mass is spread over a huge volume. Atoms and ions in the atmosphere are relatively far apart; hence, collisions between them are sufficiently infrequent that hydrogen atoms can produce narrow Balmer lines. A main-sequence star, however, is much more compact than a giant or supergiant. In the denser atmosphere of a main-sequence star, frequent interatomic collisions perturb the energy levels in the hydrogen atoms, thereby producing broader Balmer lines.

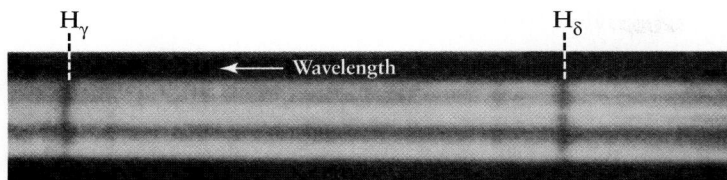

(a) A supergiant star has a low-density, low-pressure atmosphere: Its spectrum has narrow absorption lines.

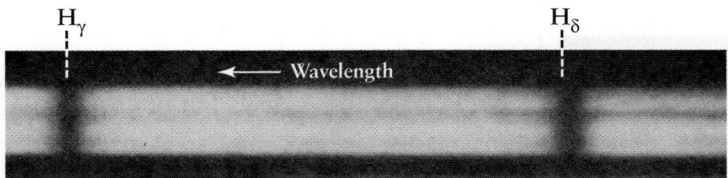

(b) A main-sequence star has a denser, higher-pressure atmosphere: Its spectrum has broad absorption lines.

FIGURE 17-17 R I **V** U X G

How a Star's Size Affects Its Spectrum These are the spectra of two stars of the same spectral type (B8) and surface temperature (13,400 K) but different radii and luminosities: **(a)** the B8 supergiant Rigel (luminosity 58,000 $L_\odot$) in Orion, and **(b)** the B8 main-sequence star Algol (luminosity 100 $L_\odot$) in Perseus. (© From *An Atlas of Stellar Spectra*, W. W. Morgan, P. C. Keenan, and E. Kellman. University of Chicago Press, © 1943)

Luminosity Classes

In the 1930s, W. W. Morgan and P. C. Keenan of the Yerkes Observatory of the University of Chicago developed a system of **luminosity classes** based on the subtle differences in spectral lines. When these luminosity classes are plotted on an H-R diagram (Figure 17-18), they provide a useful subdivision of the star types in the upper right of the diagram. Luminosity classes Ia and Ib are composed of supergiants; luminosity class V includes all the main-sequence stars. The intermediate classes distinguish giant stars of various luminosities. Note that for stars of a given surface temperature (that is, a given spectral type), the *higher* the number of the luminosity class, the *lower* the star's luminosity.

As we will see in Chapters 19 and 20, different luminosity classes represent different stages in the evolution of a star. White dwarfs are not always given a luminosity class of their own; as we mentioned in Section 17-7, they represent a final stage in stellar evolution in which no thermonuclear reactions take place.

Astronomers commonly use a shorthand description that combines a star's spectral type and its luminosity class. For example, the Sun is said to be a G2 V star. The spectral type indicates the star's surface temperature, and the luminosity class indicates its luminosity. Thus, an astronomer knows immediately that any G2 V star is a main-sequence star with a luminosity of about 1 $L_\odot$ and a surface temperature of about 5800 K. Similarly, a description of Aldebaran as a K5 III star tells an astronomer that it is a red giant with a luminosity of around 370 $L_\odot$ and a surface temperature of about 4000 K.

Spectroscopic Parallax

A star's spectral type and luminosity class, combined with the information in the H-R diagram, enable astronomers to estimate the star's distance from Earth. As an example, consider the star Pleione in the constellation Taurus. Its spectrum reveals Pleione

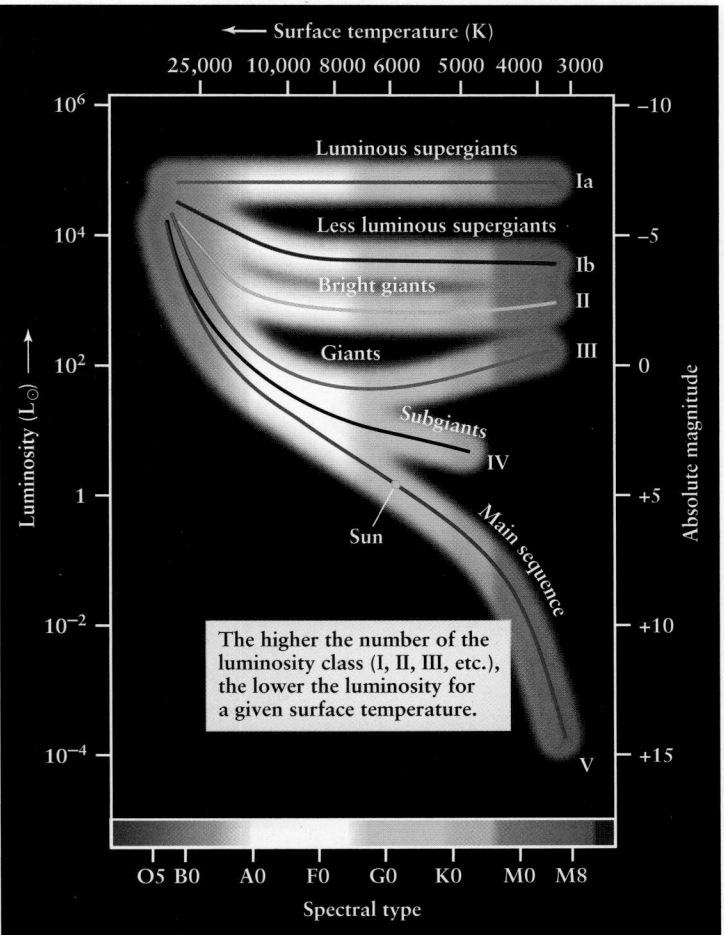

FIGURE 17-18

Luminosity Classes The H-R diagram is divided into regions corresponding to stars of different luminosity classes. (White dwarfs do not have their own luminosity class.) A star's spectrum reveals both its spectral type and its luminosity class; from these, the star's luminosity can be determined.

to be a B8 V star (a hot, blue, main-sequence star, like the one in Figure 17-15b). Using Figure 17-18, we can read off that such a star's luminosity is 190 $L_\odot$. Given the star's luminosity and its apparent brightness—in the case of Pleione, 3.9×10^{-13} of the apparent brightness of the Sun—we can use the inverse-square law to determine its distance from Earth. The mathematical details are worked out in Box 17-2.

This method for determining distance, in which the luminosity of a star is found using spectroscopy, is called **spectroscopic parallax.** Figure 17-19 summarizes the method of spectroscopic parallax.

> **CAUTION!**
>
> The term *spectroscopic parallax* is a bit misleading, because no parallax angle is involved! The idea is that measuring the star's spectrum takes the place of measuring its parallax as a way to find the star's distance. A better name for this method, although not the one used by astronomers, would be *spectroscopic distance determination.*

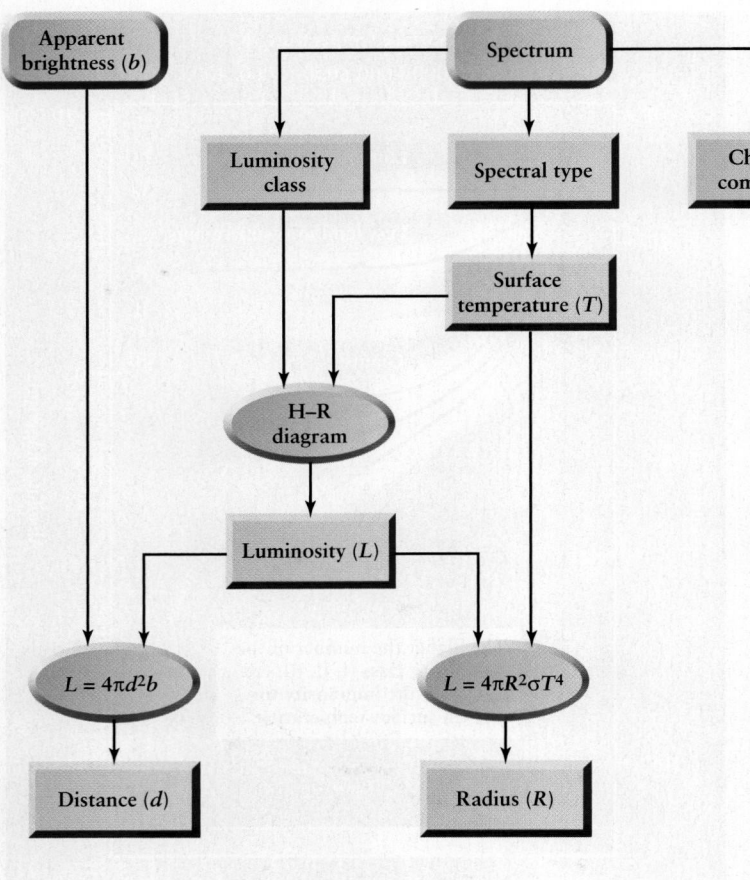

FIGURE 17-19

The Method of Spectroscopic Parallax If a star is too far away, its parallax angle is too small to allow a direct determination of its distance. This flowchart shows how astronomers deduce the properties of such a distant star. Note that the H-R diagram plays a central role in determining the star's luminosity from its spectral type and luminosity class. Just as for nearby stars (see Figure 17-14), the star's chemical composition is determined from its spectrum, and the star's radius is calculated from the luminosity and surface temperature.

CONCEPT CHECK 17-20

How does a K5 V star compare to a K5 II star in terms of temperature, luminosity, and radius?

Answers appear at the end of the chapter.

17-9 Observing binary star systems reveals the masses of stars

TUTORIAL 17.3 We now know something about the sizes, temperatures, and luminosities of stars. To complete our picture of the physical properties of stars, we need to know their masses. In this section, we will see that stars come in a wide range of masses. We will also discover an important relationship between the mass and luminosity of main-sequence stars. This relationship is crucial to understanding why some main-sequence stars are hot and luminous, whereas others are cool and dim. It will also help us understand what happens to a star as it ages and evolves.

> For main-sequence stars, there is a direct correlation between mass and luminosity.

Determining the masses of stars is not trivial, however. The problem is that there is no practical, direct way to measure the mass of an isolated star. Fortunately for astronomers, about half of the visible stars in the night sky are not isolated individuals. Instead, they are *multiple-star systems,* in which two or more stars orbit each other. By carefully observing the motions of these stars, astronomers can glean important information about their masses.

Binary Stars

A pair of stars located at nearly the same position in the night sky is called a **double star.** The Anglo-German astronomer William Herschel made the first organized search for such pairs. Between 1782 and 1821, he published three catalogs listing more than 800 double stars. Late in the nineteenth century, his son, John Herschel, discovered 10,000 more doubles. Some of these double stars are **optical double stars,** which are two stars that lie along nearly the same line of sight but are actually at very different distances from us. Yet many double stars are true **binary stars,** or **binaries**—pairs of stars that actually orbit each other. Figure 17-20 shows an example of this orbital motion.

When astronomers can actually see the two stars orbiting each other, a binary is called a **visual binary.** By observing the binary over an extended period, astronomers can plot the orbit that one star appears to trace around the other, as shown in the center diagram in Figure 17-20.

Spectroscopic parallax is an incredibly powerful technique. No matter how remote a star is, this technique allows astronomers to determine its distance, provided only that its spectrum and apparent brightness can be measured. Box 17-2 gives an example of how spectroscopic parallax has been used to find the distance to stars in other galaxies tens of millions of parsecs away. By contrast, we saw in Section 17-1 that "real" stellar parallaxes can be measured only for stars within a few hundred parsecs.

Unfortunately, spectroscopic parallax has its limitations; distances to individual stars determined using this method often have error margins greater than 10%. The reason is that the luminosity classes shown in Figure 17-18 are not thin lines on the H-R diagram but are moderately broad bands. Hence, even if a star's spectral type and luminosity class are known, there is still some uncertainty in the luminosity that we read off an H-R diagram. Nonetheless, spectroscopic parallax is often the only means that an astronomer has to estimate the distance to remote stars.

What has been left out of this discussion so far is *why* different stars have different spectral types and luminosities. One key factor, as we shall see, turns out to be the mass of the star.

CONCEPT CHECK 17-19

What is the luminosity class of the main sequence (see Figure 17-18)? Are all stars in this luminosity class less luminous than the stars in luminosity class III?

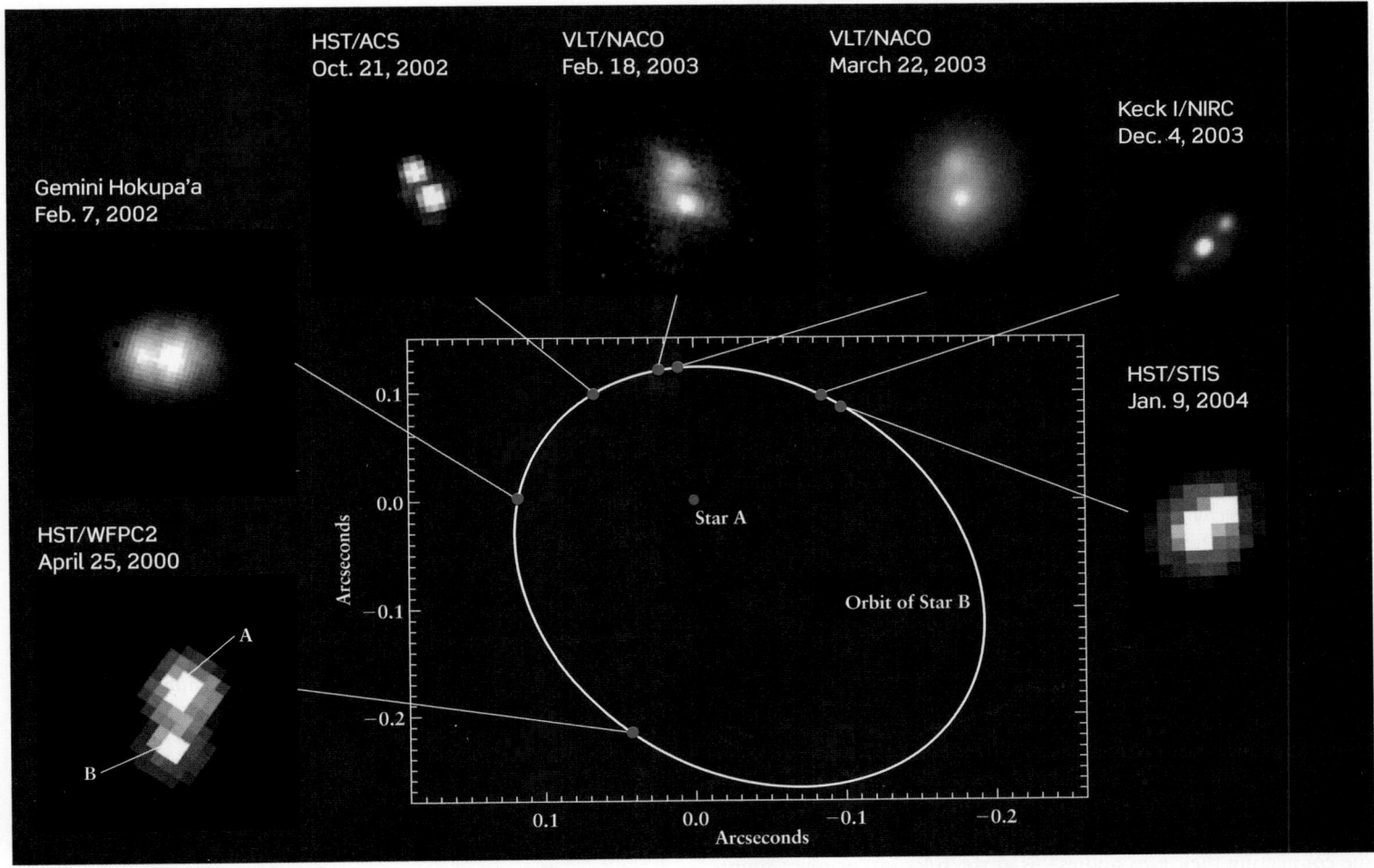

FIGURE 17-20 R **I** V U X G

A Binary Star System As seen from Earth, the two stars that make up the binary system called 2MASSW J0746425+2000321 are separated by less than ¹/₃ arcsec. The images surrounding the center diagram show the relative positions of the two stars over a 4-year period. These images were made by the Hubble Space Telescope (HST), the European Southern Observatory's Very Large Telescope (VLT), and Keck I and Gemini North in Hawaii (see Figure 6-16). For simplicity, the diagram shows one star as remaining stationary; in reality, both stars move around their common center of mass. (H. Bouy et al., MPE and ESO)

In fact, *both* stars in a binary system are in motion. They orbit each other because of their mutual gravitational attraction, and their orbital motions obey Kepler's third law as formulated by Isaac Newton (see Section 4-7 and Box 4-4). This law can be written as follows:

Kepler's third law for binary star systems

$$M_1 + M_2 = \frac{a^3}{P^2}$$

M_1, M_2 = masses of two stars in binary system, in solar masses

  a = semimajor axis of one star's orbit around the other, in au

  P = orbital period, in years

Here, a is the semimajor axis of the elliptical orbit that one star appears to follow around the other, plotted as in the center diagram in Figure 17-20. As the equation indicates, if we can measure this semimajor axis (a) and the orbital period (P), we can learn something about the masses of the two stars.

In principle, the orbital period of a visual binary is easy to determine. All you have to do is see how long it takes for the two stars to revolve once about each other. The two stars shown in Figure 17-20 are relatively close, about 2.5 au on average, and their orbital period is only 10 years. Many binary systems have much larger separations, however, and the period may be so long that more than one astronomer's lifetime is needed to complete the observations.

Determining the semimajor axis of an orbit can also be a challenge. The *angular* separation between the stars can be determined by observation. To convert this angle into a physical distance between the stars, we need to know the distance between the binary and Earth. This distance can be found from parallax measurements or by using spectroscopic parallax. The astronomer must also take into account how the orbit is tilted to our line of sight.

The center of mass of the system of two children is nearer to the more massive child.

Fulcrum

(a) A "binary system" of two children

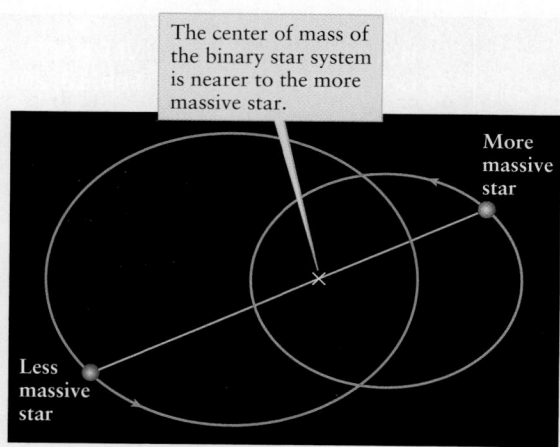

The center of mass of the binary star system is nearer to the more massive star.

More massive star

Less massive star

(b) A binary star system

ANIMATION 17.3 ➤ **FIGURE 17-21**

Center of Mass in a Binary Star System **(a)** A seesaw balances if the fulcrum is at the center of mass of the two children. **(b)** The members of a binary star system orbit around the center of mass of the two stars. Although their elliptical orbits cross each other, the two stars are always on opposite sides of the center of mass and thus never collide.

Once both P and a have been determined, Kepler's third law can be used to calculate $M_1 + M_2$, the sum of the masses of the two stars in the binary system. But this analysis tells us nothing about the *individual* masses of the two stars. To obtain these masses, more information about the motions of the two stars is needed.

Each of the two stars in a binary system actually moves in an elliptical orbit about the **center of mass** of the system. Imagine two children sitting on opposite ends of a seesaw (Figure 17-21a). For the seesaw to balance properly, they must position themselves so that their center of mass—an imaginary point that lies along a line connecting their two bodies—is at the fulcrum, or pivot point of the seesaw. If the two children have the same mass, the center of mass lies midway between them, and they should sit equal distances from the fulcrum. If their masses are different, the center of mass is closer to the heavier child.

Just as the seesaw naturally balances at its center of mass, the two stars that make up a binary system naturally orbit around their center of mass (Figure 17-21b). The center of mass always lies along the line connecting the two stars and is closer to the more massive star.

The center of mass of a visual binary is located by plotting the separate orbits of the two stars, as in Figure 17-21b, using the background stars as reference points. The center of mass lies at the common focus of the two elliptical orbits. Comparing the relative sizes of the two orbits around the center of mass yields the ratio of the two stars' masses, M_1/M_2. With the sum $M_1 + M_2$ already known from Kepler's third law, the individual masses of the two stars can then be determined using algebra.

CONCEPT CHECK 17-21

If two stars in a binary system were moved farther apart, how would their masses and orbital periods change?

Answer appears at the end of the chapter.

Main-Sequence Masses and the Mass–Luminosity Relation

Years of careful, patient observations of binaries have slowly yielded the masses of many stars. As the data accumulated, an important trend began to emerge: For main-sequence stars, there is a direct correlation between mass and luminosity. The more massive a main-sequence star, the more luminous it is. Figure 17-22 depicts this **mass–luminosity relation** as a graph. The range of

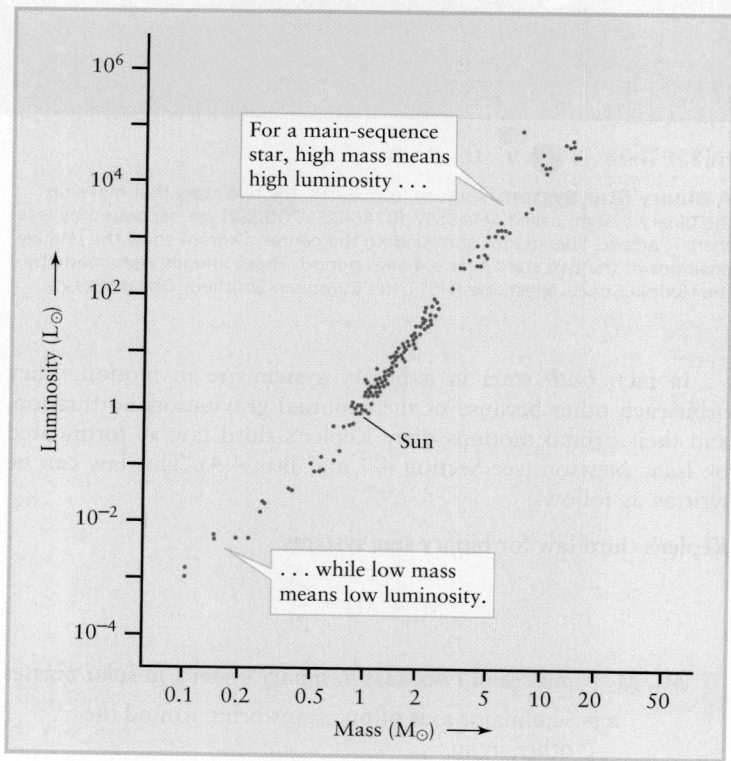

For a main-sequence star, high mass means high luminosity . . .

Sun

. . . while low mass means low luminosity.

FIGURE 17-22

The Mass–Luminosity Relation For main-sequence stars, there is a direct correlation between mass and luminosity—the more massive a star, the more luminous it is. A main-sequence star of mass 10 $M_\odot$ (that is, 10 times the Sun's mass) has roughly 3000 times the Sun's luminosity (3000 $L_\odot$); one with 0.1 $M_\odot$ has a luminosity of only about 0.001 $L_\odot$.

stellar masses extends from less than 0.1 of a solar mass to more than 50 solar masses. The Sun's mass lies between these extremes.

The *Cosmic Connections* figure on the next page depicts the mass–luminosity relation for main-sequence stars in an H-R diagram. This figure shows that the main sequence in an H-R diagram is a progression in mass as well as in luminosity and surface temperature. The hot, bright, bluish stars in the upper left corner of an H-R diagram are the most massive main-sequence stars. Likewise, the dim, cool, reddish stars in the lower right corner of an H-R diagram are the least massive. Main-sequence stars of intermediate temperature and luminosity also have intermediate masses.

The mass of a main-sequence star also helps determine its radius. Referring back to Figure 17-15b, we see that if we move along the main sequence from low luminosity to high luminosity, the radius of the star increases. Thus, we have the following general rule for main-sequence stars:

The greater the mass of a main-sequence star, the greater its luminosity, its surface temperature, and its radius.

Mass and Main-Sequence Stars

Why is mass the controlling factor in determining the properties of a main-sequence star? The answer is that all main-sequence stars are objects like the Sun, with essentially the same chemical composition as the Sun but with different masses. Like the Sun, all main-sequence stars shine because thermonuclear reactions at their cores convert hydrogen to helium and release energy. The greater the total mass of the star, the greater the pressure and temperature at the core, the more rapidly thermonuclear reactions take place in the core, and the greater the energy output—that is, the luminosity—of the star. In other words, the greater the mass of a main-sequence star, the greater its luminosity. This statement is just the mass–luminosity relation, which we can now recognize as a natural consequence of the nature of main-sequence stars.

Like the Sun, main-sequence stars are in a state of both hydrostatic equilibrium and thermal equilibrium. Calculations using models of a main-sequence star's interior (like the solar models we discussed in Section 16-2) show that as a star settles into equilibrium, a more massive star must have a larger radius and a higher surface temperature. This result is just what we see when we plot the curve of the main sequence on an H-R diagram (see Figure 17-15b). As you move up the main sequence from less massive stars (at the lower right in the H-R diagram) to more massive stars (at the upper left), the radius and surface temperature both increase.

Calculations using hydrostatic and thermal equilibrium also show that if a star's mass is less than about $0.08 M_\odot$, the core pressure and temperature are too low for thermonuclear reactions to take place. The "star" (or so-called failed star) is then a brown dwarf. Brown dwarfs are still luminous; as they slowly contract, gravitational energy is converted into heat.

CAUTION!

The mass–luminosity relation we have discussed applies to main-sequence stars only. There are *no* simple mass–luminosity relations for giant, supergiant, or white dwarf stars. Why these stars lie where they do on an H-R diagram will become apparent when we study the evolution of stars in Chapters 19 and 20. We will find that main-sequence stars evolve into giant and supergiant stars, and that some of these eventually end their lives as white dwarfs.

CONCEPT CHECK 17-22

From the mass–luminosity relation in the Cosmic Connections figure, what can you say about the masses, temperatures, and radii for the most luminous stars?

Answer appears at the end of the chapter.

17-10 Spectroscopy makes it possible to study binary systems in which the two stars are close together

We have described how the masses of stars can be determined from observations of visual binaries, in which the two stars can be distinguished from each other. But if the two stars in a binary system are too close together, the images of the two stars can blend to produce the semblance of a single star. Happily, in many cases we can use spectroscopy to decide whether a seemingly single star is in fact a binary system. Spectroscopic observations of binaries provide additional useful information about star masses.

Some binaries are discovered when the spectrum of a star shows incongruous spectral lines. For example, the spectrum of what appears to be a single star may include both strong hydrogen lines (characteristic of a type A star) and strong absorption bands of titanium oxide (typical of a type M star). Because a single star cannot have the differing physical properties of these two spectral types, such a star must actually be a binary system that is too far away for us to resolve its individual stars. A binary system detected in this way is called a **spectrum binary**.

Other binary systems can be detected using the Doppler effect. If a star is moving toward Earth, its spectral lines are displaced toward the short-wavelength (blue) end of the spectrum. Conversely, the spectral lines of a star moving away from us are shifted toward the long-wavelength (red) end of the spectrum. The upper portion of Figure 17-23 applies these ideas to a hypothetical binary star system with an orbital plane that is edge on to our line of sight.

As the two stars move around their orbits, they periodically approach and recede from us. Hence, the spectral lines of the two stars are alternately blueshifted and redshifted. The two stars in this hypothetical system are so close together that they appear through a telescope as a single star with a single spectrum. Because one star shows a blueshift while the other is showing a redshift, the spectral lines of the binary system appear to split apart and rejoin periodically. Stars whose binary character is revealed by such shifting spectral lines are called **spectroscopic binaries**.

Exploring Spectroscopic Binary Stars

To analyze a spectroscopic binary, astronomers measure the wavelength shift of each star's spectral lines and use the Doppler shift formula (introduced in Section 5-9 and Box 5-6) to determine the *radial velocity* of each star—that is, how fast and in what direction it is moving (toward us or away) along our line of sight. The lower portion of Figure 17-23 shows a graph of the radial velocity versus time, called a **radial velocity curve**,

Stars in close binary systems move so rapidly that we can detect their motion using the Doppler effect.

COSMIC CONNECTIONS

The Main Sequence and Masses

The main sequence is an arrangement of stars according to their mass. The most massive main-sequence stars have the greatest luminosity, greatest radius, and greatest surface temperature. These characteristics are consequences of the behavior of thermonuclear reactions at the core of a main-sequence star.

Each dot represents a main-sequence star. The number next to each dot is the mass of that star in solar masses.

60-solar-mass star shown to scale

Increasing mass
Increasing luminosity
Increasing radius
Increasing surface temperature

0.21-M$_\odot$ star shown to scale

Sun (a 1-M$_\odot$ star) shown to scale

Luminosity (L$_\odot$)

Surface temperature (K)

- A star with 60 M$_\odot$ has much higher pressure and temperature at its core than does the Sun.
- This heat and pressure cause thermonuclear reactions in the core to occur much more rapidly and release energy at a much faster rate—790,000 times faster than in the Sun.
- Energy is emitted from the star's surface at the same rate that it is released in the core, so the star has 790,000 times the Sun's luminosity.
- The tremendous rate of energy release also heats the star's interior tremendously, increasing the star's internal pressure. This pressure inflates the star to 15 times the Sun's radius.
- The star's surface must be at a high temperature (about 44,500 K) in order for it to emit energy into space at such a rapid rate.

- A star with 0.21 M$_\odot$ has much lower pressure and temperature at its core than does the Sun.
- This causes thermonuclear reactions in the core to occur much more slowly and release energy at a much slower rate—0.011 times as fast as in the Sun.
- Energy is emitted from the star's surface at the same rate that it is released in the core, so the star has 0.011 of the Sun's luminosity.
- The low rate of energy release supplies relatively little heat to the star's interior, so the star's internal pressure is low. Hence the star's radius is only 0.33 times the Sun's radius.
- The star's surface need be at only a low temperature (about 3200 K) to emit energy into space at such a relatively slow rate.

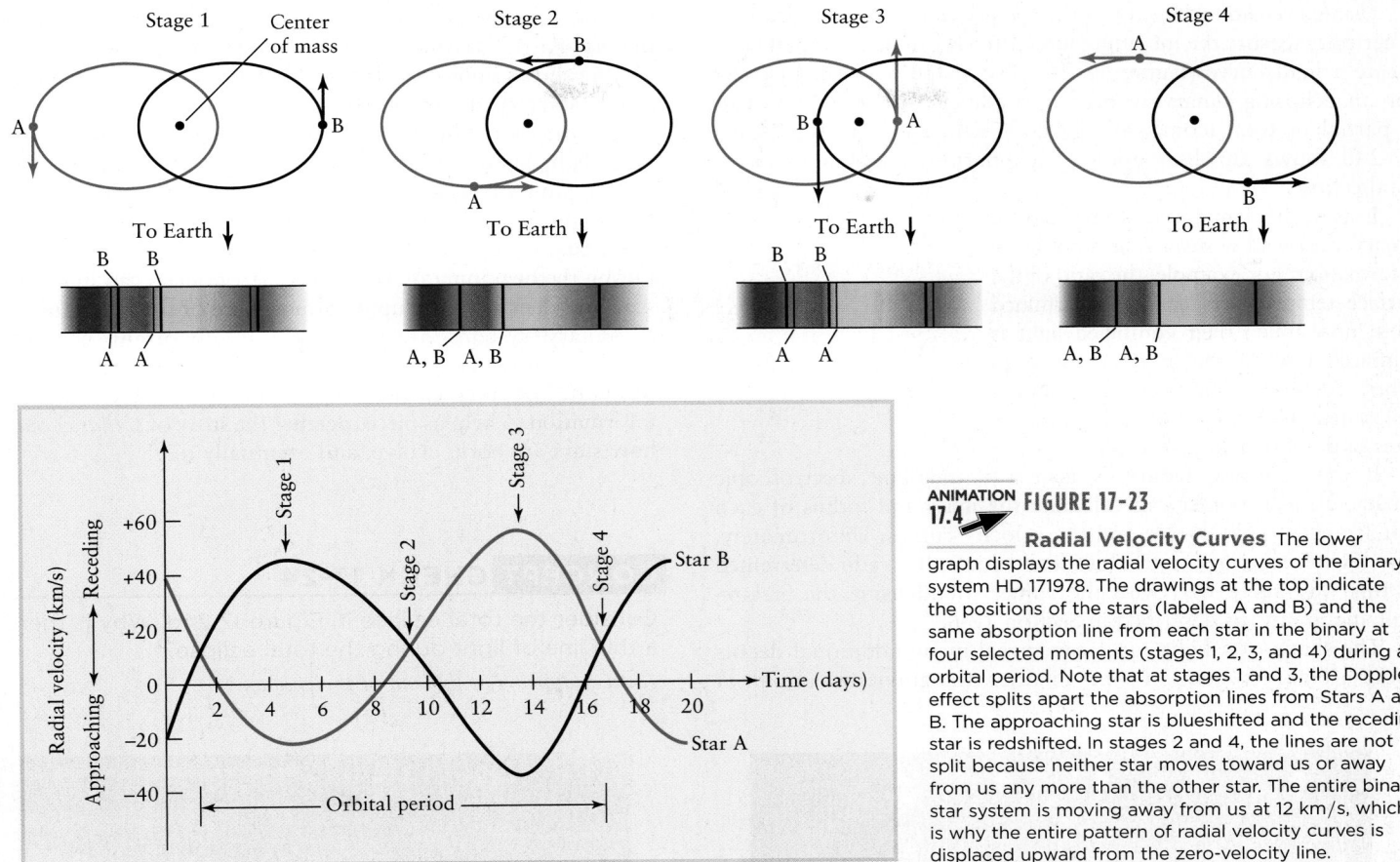

ANIMATION
17.4 **FIGURE 17-23**

Radial Velocity Curves The lower graph displays the radial velocity curves of the binary system HD 171978. The drawings at the top indicate the positions of the stars (labeled A and B) and the same absorption line from each star in the binary at four selected moments (stages 1, 2, 3, and 4) during an orbital period. Note that at stages 1 and 3, the Doppler effect splits apart the absorption lines from Stars A and B. The approaching star is blueshifted and the receding star is redshifted. In stages 2 and 4, the lines are not split because neither star moves toward us or away from us any more than the other star. The entire binary star system is moving away from us at 12 km/s, which is why the entire pattern of radial velocity curves is displaced upward from the zero-velocity line.

for the binary system HD 171978. Each of the two stars alternately approaches and recedes as it orbits around the center of mass. The pattern of the curves repeats every 15 days, which is the orbital period of the binary.

It is important to emphasize that the Doppler effect applies only to motion along the line of sight. Motion perpendicular to the line of sight does not affect the observed wavelengths of spectral lines. Hence, the ideal orientation for a spectroscopic binary is to have the stars orbit in a plane that is edge on to our line of sight. (By contrast, a *visual* binary is best observed if the orbital plane is face on to our line of sight.) For the Doppler shifts to be noticeable, the orbital speeds of the two stars should be at least a few kilometers per second.

As for visual binaries, spectroscopic binaries allow astronomers to learn about stellar masses. From a radial velocity curve, one can find the *ratio* of the masses of the two stars in a binary. The *sum* of the masses is related to the orbital speeds of the two stars by Kepler's laws and Newtonian mechanics. If both the ratio of the masses and their sum are known, the individual masses can be determined using algebra. However, determining the sum of the masses requires that we know how the binary orbits are tilted from our line of sight. This is because the Doppler shifts reveal only the radial velocities of the stars rather than their true orbital speeds. This tilt is often impossible to determine, because we cannot see the individual stars in the binary. Thus, the masses of stars in spectroscopic binaries tend to be uncertain.

There is one important case in which we can determine the orbital tilt of a spectroscopic binary. If the two stars are observed to eclipse each other periodically, then we must be viewing the orbit nearly edge on. As we will see next, individual stellar masses—as well as other useful data—can be determined if a spectroscopic binary also happens to be such an *eclipsing* binary.

CONCEPT CHECK 17-23

In Figure 17-23, describe the Doppler shift observed at Stage 2 of the binary orbit.

Answer appears at the end of the chapter.

17-11 Light curves of eclipsing binaries provide detailed information about the two stars

Some binary systems are oriented so that the two stars periodically eclipse each other as seen from Earth. These **eclipsing binaries** can be detected even when the two stars cannot be resolved visually as two distinct images in the telescope. The apparent brightness of the image of the binary dims briefly each time one star blocks the light from the other.

Using a sensitive detector at the focus of a telescope, an astronomer can measure the incoming light intensity quite accurately and create a **light curve** (Figure 17-24). The shape of the light curve for an eclipsing binary reveals at a glance whether the eclipse is partial or total (compare Figures 17-24a and 17-24b). Figure 17-24d shows an observation of a binary system undergoing a total eclipse.

In fact, the light curve of an eclipsing binary can yield a surprising amount of information. For example, the ratio of the surface temperatures can be determined from how much their combined light is diminished when the stars eclipse each other. Also, the duration of a mutual eclipse tells astronomers about the relative sizes of the stars and their orbits.

> Eclipsing binaries can reveal the sizes and shapes of stars.

If the eclipsing binary is also a double-line spectroscopic binary, an astronomer can calculate the mass and radius of each star from the light curves and the velocity curves. Unfortunately, very few binary stars are of this ideal type. Stellar radii determined in this way agree well with the values found using the Stefan–Boltzmann law, as described in Section 17-6.

The shape of a light curve can reveal many additional details about a binary system. In some binaries, the gravitational pull of one star distorts the other through tidal forces, much as the Moon distorts Earth's oceans in producing tides (see Figure 4-29). Figure 17-24c shows how such tidal distortion gives the light curve a different shape than in Figure 17-24b.

Information about stellar atmospheres can also be derived from light curves. Suppose that one star of a binary is a luminous main-sequence star and the other is a bloated red giant. By observing exactly how the light from the bright main-sequence star is gradually cut off as it moves behind the edge of the red giant during the beginning of an eclipse, astronomers can infer the pressure and density in the upper atmosphere of the red giant.

Binary systems are tremendously important because they enable astronomers to measure stellar masses as well as other key properties of stars. In the next several chapters, we will use this information to help us piece together the story of *stellar evolution*—how stars are born, evolve, and eventually die.

CONCEPT CHECK 17-24

Consider the total eclipse in Figure 17-24d. Why is there still a thin line of light during the total eclipse?

Answer appears at the end of the chapter.

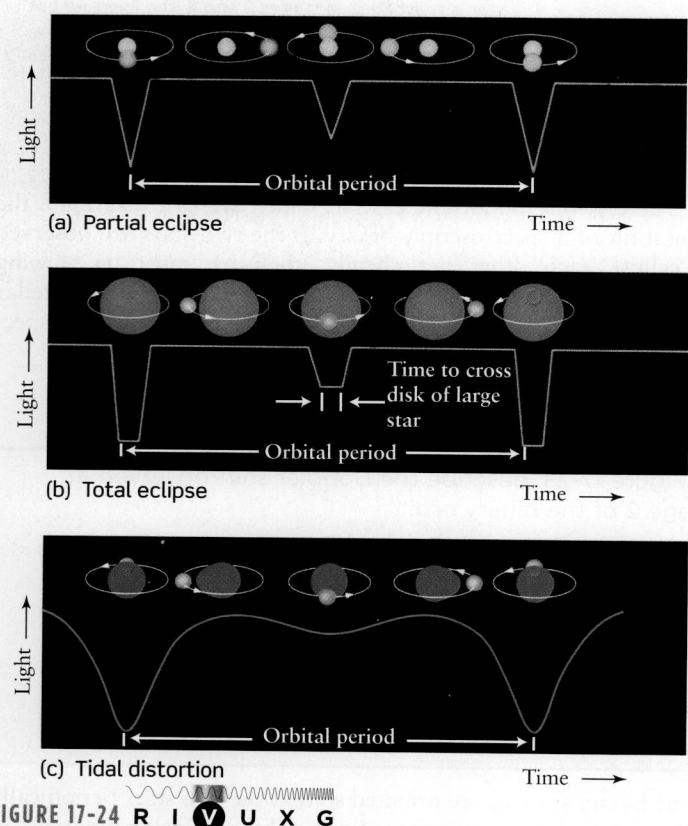

(a) Partial eclipse

(b) Total eclipse

(c) Tidal distortion

FIGURE 17-24   R I **V** U X G

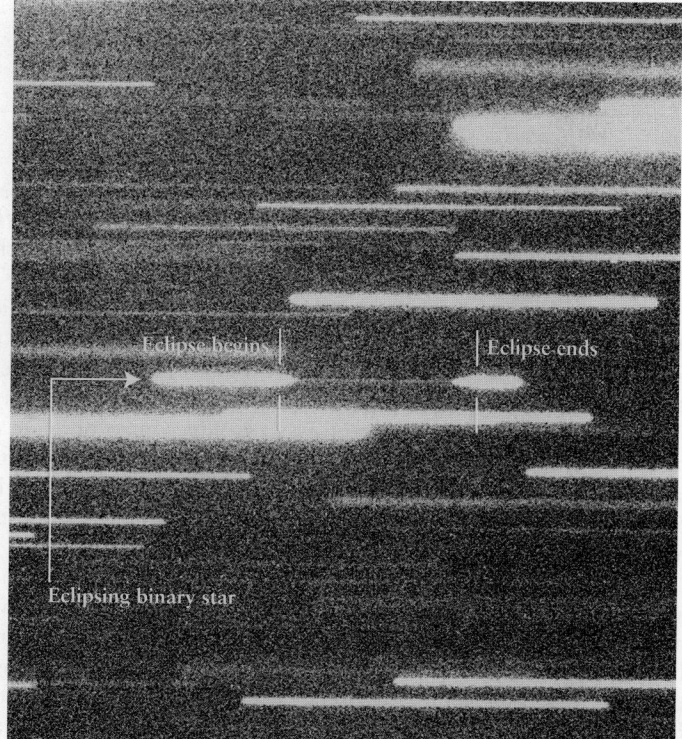

(d) Eclipse of a binary star

Representative Light Curves of Eclipsing Binaries **(a), (b), (c)**
The shape of the light curve of an eclipsing binary can reveal many details about the two stars that make up the binary. **(d)** This image shows the binary star NN Serpens (indicated by the arrow) undergoing a total eclipse. The telescope was moved during the exposure so that the sky drifted slowly from left to right across the field of view. During the 10.5-minute duration of the eclipse, the dimmer star of the binary system (an M6 main-sequence star) passed in front of the other, more luminous star (a white dwarf). The binary became so dim that it almost disappeared. (d: European Southern Observatory)

KEY WORDS

Terms preceded by an asterisk (*) are discussed in the Boxes.

absolute magnitude,
  p. 520
apparent brightness
  (brightness), p. 516
apparent magnitude, p. 519
binary star (binary), p. 536
brown dwarf, p. 528
center of mass, p. 538
color ratio, p. 524
*distance modulus, p. 522
double star, p. 536
eclipsing binary, p. 541
giant, p. 532

Hertzsprung–Russell diagram
  (H-R diagram), p. 532
inverse-square law, p. 516
light curve, p. 542
luminosity, p. 512
luminosity class, p. 535
luminosity function, p. 519
magnitude scale, p. 519
main sequence, p. 532
main-sequence star, p. 532
mass–luminosity relation,
  p. 538
metals, p. 528

OBAFGKM, p. 525
optical double star, p. 536
parallax, p. 512
parsec, p. 512
photometry, p. 516
*proper motion, p. 514
*radial velocity, p. 514
radial velocity curve, p. 539
red giant, p. 532
solar constant, p. 516
*space velocity, p. 514
spectral classes, p. 525
spectral types, p. 525

spectroscopic binary, p. 539
spectroscopic parallax,
  p. 535
spectrum binary, p. 539
stellar parallax, p. 512
supergiant, p. 532
*tangential velocity, p. 514
UBV photometry, p. 524
visual binary, p. 536
white dwarf, p. 532

KEY IDEAS

Measuring Distances to Nearby Stars: Distances to the nearer stars can be determined by parallax, the apparent shift of a star against the background stars observed as Earth moves along its orbit (Section 17-1).

- Parallax measurements made from orbit, above the blurring effects of the atmosphere, are much more accurate than those made with Earth-based telescopes.

- Stellar parallaxes can be measured only for stars within a few hundred parsecs.

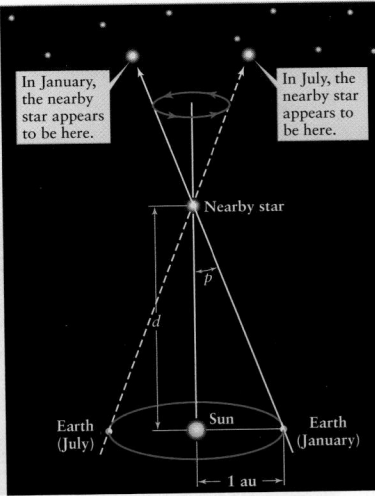

The Inverse-Square Law: A star's luminosity (total light output), apparent brightness, and distance from Earth are related by the inverse-square law. If any two of these quantities are known, the third can be calculated (Section 17-2).

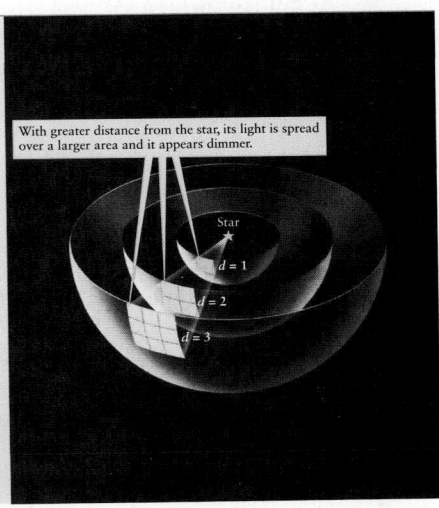

The Population of Stars: Stars of relatively low luminosity are more common than more luminous stars. Our own Sun is a rather average star of intermediate luminosity (Section 17-2).

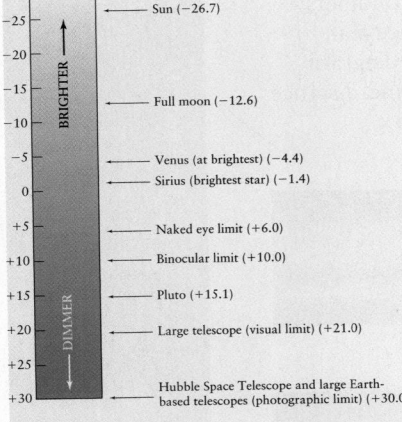

The Magnitude Scale: The apparent magnitude scale is an alternative way to measure a star's apparent brightness (Section 17-3).

- The absolute magnitude of a star is the apparent magnitude it would have if viewed from a distance of 10 pc. A version of the inverse-square law relates a star's absolute magnitude, apparent magnitude, and distance.

Photometry and Color Ratios: Photometry measures the apparent brightness of a star. The color ratios of a star are the ratios of brightness values obtained through different standard filters, such as the U, B, and V filters. These ratios are a measure of the star's surface temperature (Section 17-4).

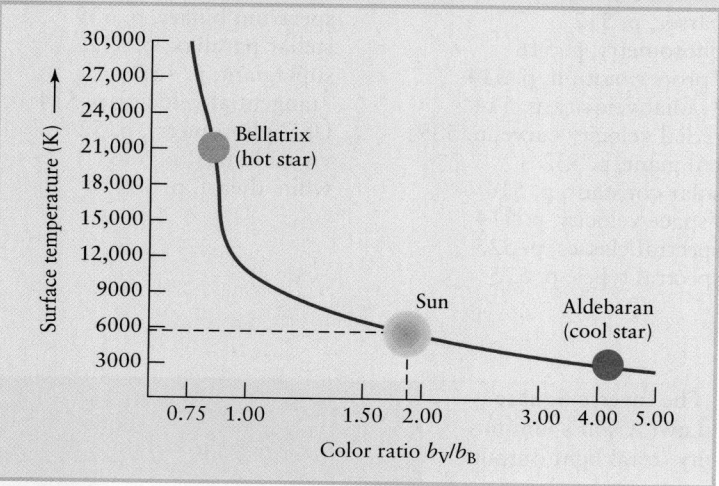

Spectral Types: Stars are classified into spectral types (subdivisions of the spectral classes O, B, A, F, G, K, and M), based on the major patterns of spectral lines in their spectra. The spectral class and type of a star is directly related to its surface temperature: O stars are the hottest and M stars are the coolest (Section 17-5).

- Most brown dwarfs are in even cooler spectral classes called L, T, and Y. Unlike true stars, brown dwarfs are too small to sustain thermonuclear fusion.

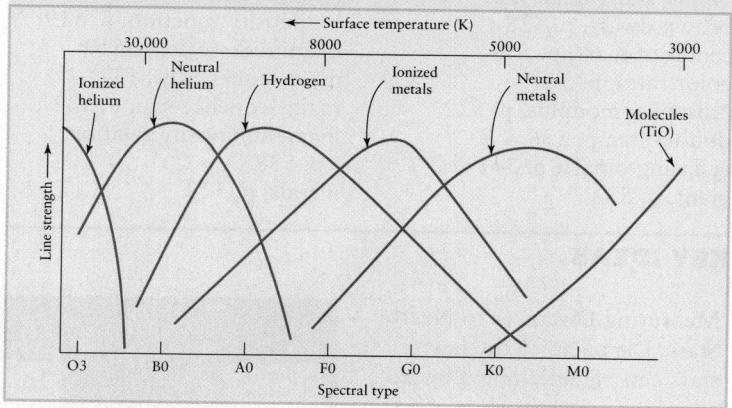

Hertzsprung–Russell Diagram: The Hertzsprung–Russell (H-R) diagram is a graph plotting the absolute magnitudes of stars against their spectral types—or, equivalently, their luminosities against surface temperatures (Sections 17-7 and 17-8).

- Most stars plotted on the H-R diagram are positioned along the main sequence, a band that extends from high luminosity and high surface temperature to low luminosity and low surface temperature.

- In the H-R diagram, giant and supergiant stars lie above the main sequence, while white dwarfs are below the main sequence.

- By carefully examining a star's spectral lines, astronomers can determine whether that star is a main-sequence star, giant, supergiant, or white dwarf. Using the H-R diagram and the inverse-square law, the star's luminosity and distance can be found without measuring its stellar parallax.

© From *An Atlas of Stellar Spectra*, W. W. Morgan, P. C. Keenan, and E. Kellman. University of Chicago Press, © 1943.

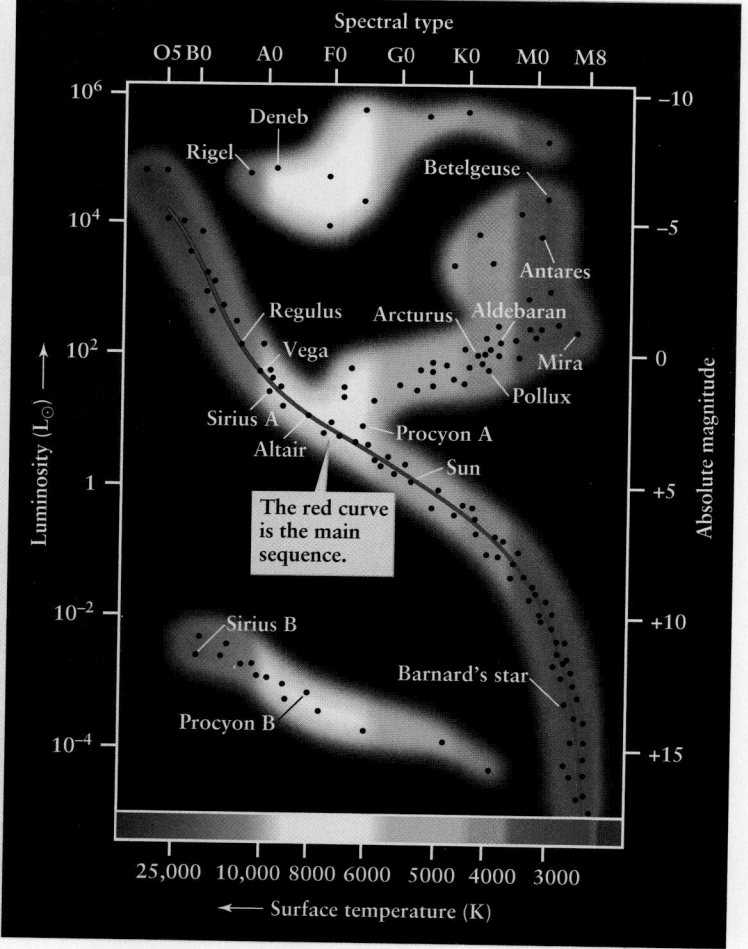

Binary Stars: Binary stars, in which two stars are held in orbit around each other by their mutual gravitational attraction, are surprisingly common. Those that can be resolved into two distinct star images by a telescope are called visual binaries (Section 17-9).

- Each of the two stars in a binary system moves in an elliptical orbit about the center of mass of the system.

- Binary stars are important because they allow astronomers to determine the masses of the two stars in a binary system. The masses can be computed from measurements of the orbital period and orbital dimensions of the system.

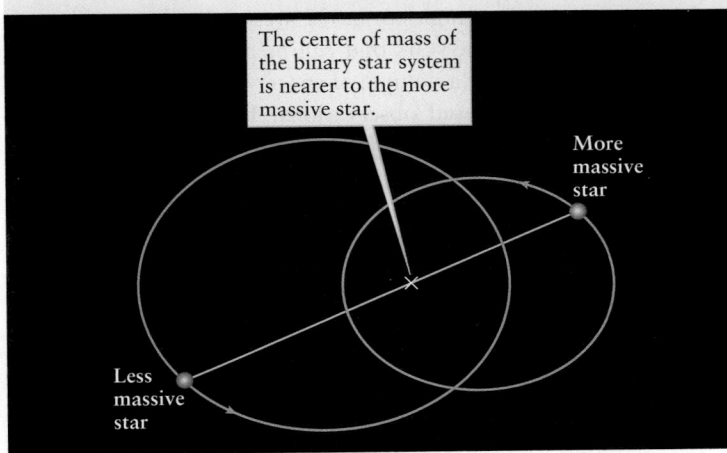

The center of mass of the binary star system is nearer to the more massive star.

More massive star

Less massive star

Mass–Luminosity Relation for Main-Sequence Stars: Main-sequence stars are stars like the Sun but with different masses (Section 17-9).

- The mass–luminosity relation expresses a direct correlation between mass and luminosity for main-sequence stars. The greater the mass of a main-sequence star, the greater its luminosity (and also the greater its radius and surface temperature).

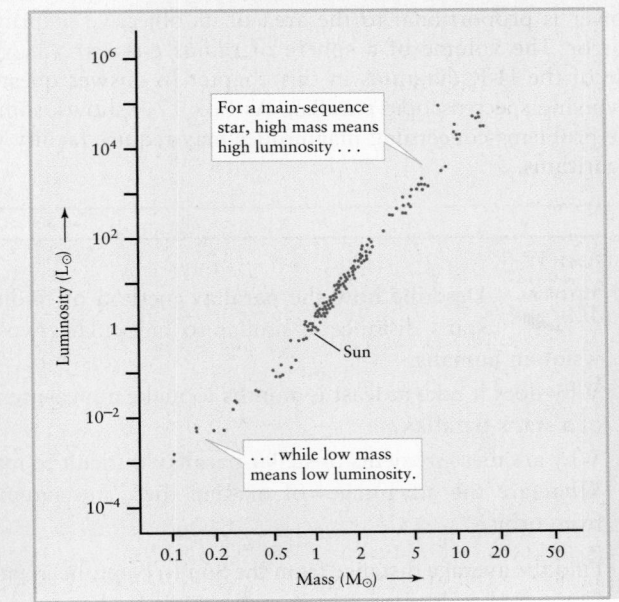

For a main-sequence star, high mass means high luminosity . . .

Sun

. . . while low mass means low luminosity.

Spectroscopic Observations of Binary Stars: Some binaries can be detected and analyzed, even though the system may be so distant or the two stars so close together that the two star images cannot be resolved (Sections 17-10 and 17-11).

- A spectrum binary appears to be a single star but has a spectrum with the absorption lines for two distinctly different spectral types.

- A spectroscopic binary has spectral lines that shift back and forth in wavelength. This is caused by the Doppler effect, as the orbits of the stars carry them first toward and then away from Earth.

- An eclipsing binary is a system whose orbits are viewed nearly edge on from Earth, so that one star periodically eclipses the other. Detailed information about the stars in an eclipsing binary can be obtained from a study of the binary's radial velocity curve and its light curve.

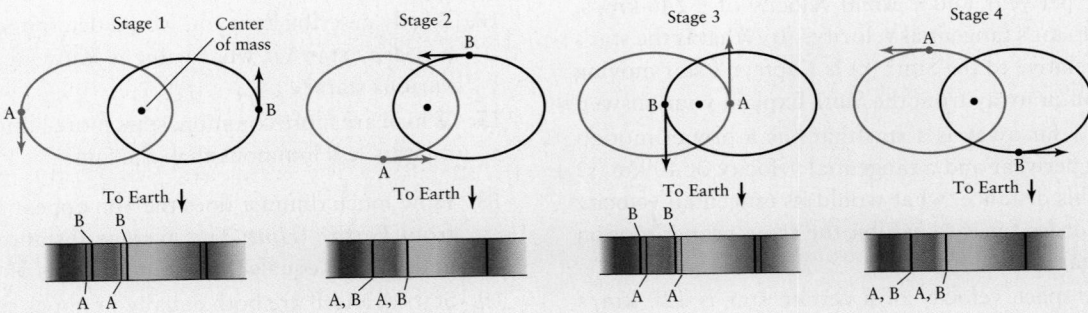

QUESTIONS

> **Problem-solving tips and tools**
>
> Look carefully at the worked examples in Boxes 17-1, 17-2, 17-3, and 17-4 before attempting these exercises. For data on the planets, see Table 7-1 or Appendices 1 and 2 at the back of this book. Remember that a telescope's light-gathering power is proportional to the area of its objective or primary mirror. The volume of a sphere of radius r is $4\pi r^3/3$. Make use of the H-R diagrams in this chapter to answer questions involving spectroscopic parallax. As Box 17-3 shows, some of the problems concerning magnitudes may require facility with logarithms.

Section 17-1

1. **TUTORIAL 17.1** Describe how the parallax method of finding a star's distance is similar to binocular (two-eye) vision in humans.

2. Why does it take at least 6 months to make a measurement of a star's parallax?

3. Why are measurements of stellar parallax difficult to make? What are the advantages of making these measurements from orbit?

4. Find the average distance from the Sun to Neptune in parsecs. Compared to Neptune, how many times farther away from the Sun is Proxima Centauri?

5. Suppose that a dim star were located 2 million au from the Sun. Find (**a**) the distance to the star in parsecs and (**b**) the parallax angle of the star. Would this angle be measurable with present-day techniques?

6. The star GJ 1156 has a parallax angle of 0.153 arcsec. How far away is the star?

7. **Box 17-1** Kapteyn's star (named after the Dutch astronomer who found it) has a parallax of 0.255 arcsec, a proper motion of 8.67 arcsec per year, and a radial velocity of +246 km/s. (**a**) What is the star's tangential velocity? (**b**) What is the star's actual speed relative to the Sun? (**c**) Is Kapteyn's star moving toward the Sun or away from the Sun? Explain your answer.

8. **Box 17-1** How far away is a star that has a proper motion of 0.08 arcsec per year and a tangential velocity of 40 km/s? For a star at this distance, what would its tangential velocity have to be in order for it to exhibit the same proper motion as Barnard's star?

9. **Box 17-1** The space velocity of a certain star is 120 km/s and its radial velocity is 72 km/s. Find the star's tangential velocity.

10. **Box 17-1** In the spectrum of a particular star, the Balmer line H_α has a wavelength of 656.15 nm. The laboratory value for the wavelength of H_α is 656.28 nm. (**a**) Find the star's radial velocity. (**b**) Is this star approaching us or moving away? Explain your answer. (**c**) Find the wavelength at which you would expect to find H_α in the spectrum of this star, given that the laboratory wavelength of H_α is 486.13 nm. (**d**) Do your answers depend on the star's distance from the Sun? Why or why not?

11. **Box 17-1** Derive the equation given in Box 17-1 relating proper motion and tangential velocity. (*Hint:* See Box 1-1.)

12. From its orbit around Earth, the Hipparcos satellite could measure stellar parallax angles with acceptable accuracy only if the angles were larger than about 0.002 arcsec. Discuss the advantages or disadvantages of making parallax measurements from a satellite in a large solar orbit, say, at the distance of Jupiter from the Sun. If this satellite can also measure parallax angles of 0.002 arcsec, what is the distance of the most remote stars that can be accurately determined? How much bigger a volume of space would be covered compared to Earth-based observations? How many more stars would you expect to be contained in that volume?

13. **Box 17-1** As seen from the starship *Enterprise* in the *Star Trek* television series and movies, stars appear to move across the sky due to the starship's motion. How fast would the *Enterprise* have to move in order for a star 1 pc away to appear to move 1° per second? (*Hint:* The speed of the star as seen from the *Enterprise* is the same as the speed of the *Enterprise* relative to the star.) How does this compare with the speed of light? Do you think the stars appear to move as seen from an orbiting space shuttle, which moves at about 8 km/s?

Section 17-2

14. Explain the difference between a star's apparent brightness and its luminosity.

15. **TUTORIAL 17.2** What is the inverse-square law? Use it to explain why an ordinary lightbulb can appear brighter than a star, even though the lightbulb emits far less light energy per second.

16. Briefly describe how you would determine the luminosity of a nearby star. Of what value is knowing the luminosity of various stars?

17. Which are more common, stars more luminous than the Sun or stars less luminous than the Sun?

18. How much dimmer does the Sun appear from Neptune than from Earth? (*Hint:* The average distance between a planet and the Sun equals the semimajor axis of the planet's orbit.)

19. Stars A and B are both equally bright as seen from Earth, but A is 120 pc away while B is 24 pc away. Which star has the greater luminosity? How many times greater is it?

20. Stars C and D both have the same luminosity, but C is 32 pc from Earth while D is 128 pc from Earth. Which star appears brighter as seen from Earth? How many times brighter is it?

21. Suppose two stars have the same apparent brightness, but one star is 8 times farther away than the other. What is the

ratio of their luminosities? Which one is more luminous, the closer star or the farther star?

22. The *solar constant*, equal to 1370 W/m^2, is the amount of light energy from the Sun that falls on 1 square meter of Earth's surface in 1 second (see Section 17-2). What would the distance between Earth and the Sun have to be in order for the solar constant to be 1 W/m^2?

23. The star Procyon in Canis Minor (the Small Dog) is a prominent star in the winter sky, with an apparent brightness 1.3×10^{-11} that of the Sun. It is also one of the nearest stars, being only 3.50 pc from Earth. What is the luminosity of Procyon? Express your answer as a multiple of the Sun's luminosity.

24. The star HIP 92403 (also called Ross 154) is only 2.97 pc from Earth but can be seen only with a telescope, because it is 60 times dimmer than the dimmest star visible to the unaided eye. How close to us would this star have to be in order for it to be visible without a telescope? Give your answer in parsecs and in au. Compare with the semimajor axis of Pluto's orbit around the Sun.

25. **Box 17-2** The star HIP 72509 has an apparent magnitude of +12.1 and a parallax angle of 0.222 arcsec. (a) Determine its absolute magnitude. (b) Find the approximate ratio of the luminosity of HIP 72509 to the Sun's luminosity.

26. **Box 17-2** Suppose you can just barely see a twelfth-magnitude star through an amateur's 6-in. telescope. What is the magnitude of the dimmest star you could see through a 60-in. telescope?

27. **Box 17-2** A certain type of variable star is known to have an average absolute magnitude of 0.0. Such stars are observed in a particular star cluster to have an average apparent magnitude of +14.0. What is the distance to that star cluster?

28. **Box 17-2** (a) Find the absolute magnitudes of the brightest and dimmest of the labeled stars in Figure 17-6b. Assume that all of these stars are 110 pc from Earth. (b) If a star in the Pleiades cluster is just bright enough to be seen from Earth with the naked eye, what is its absolute magnitude? Is such a star more or less luminous than the Sun? Explain your answer.

Section 17-3

29. **TUTORIAL 17.2** Why is the magnitude scale called a "backward" scale? What is the difference between apparent magnitude and absolute magnitude?

30. Consider two stars, A and B. If Star A has absolute magnitude 3.0 and Star B has absolute magnitude 5.0, which star is brighter? How many times more luminous is the brighter star compared to the dimmer star?

31. The star Zubenelgenubi (from the Arabic for "scorpion's southern claw") has apparent magnitude +2.75, while the star Sulafat (Arabic for "tortoise") has apparent magnitude +3.25. Which star appears brighter? From this information

alone, what can you conclude about the luminosities of these stars? Explain your answer.

32. **Box 17-3** Astronomers usually express a star's color using apparent magnitudes. The star's apparent magnitude as viewed through a B filter is called m_B, and its apparent magnitude as viewed through a V filter is m_V. The difference $m_B - m_V$ is called the *B–V color index* ("B minus V"). Is the B–V color index positive or negative for very hot stars? What about very cool stars? Explain your answers.

33. **Box 17-3** (See Question 32.) The B–V color index is related to the color ratio b_V/b_B by the equation

$$m_B - m_V = 2.5 \log\left(\frac{b_V}{b_B}\right)$$

(a) Explain why this equation is correct. (b) Use the data in Table 17-1 to calculate the B–V color indices for Bellatrix, the Sun, and Betelgeuse. From your results, describe a simple rule that relates the value of the B–V color index to a star's color.

Section 17-4

34. Explain why the color ratios of a star are related to the star's surface temperature.

35. Would it be possible for a star to appear bright when viewed through a U filter or a V filter, but dim when viewed through a B filter? Explain your answer.

36. (a) On a copy of Figure 17-8, sketch the intensity curve for a blackbody at a temperature of 3000 K. Note that this figure shows a smaller wavelength range than Figure 17-7a. (b) Repeat part (a) for a blackbody at 12,000 K (see Figure 17-7c). (c) Use your sketches from parts (a) and (b) to explain why the color ratios b_V/b_B and b_B/b_U are less than 1 for very hot stars but greater than 1 for very cool stars.

37. Why doesn't a star with its intensity peaking in the green portion of the visible spectrum appear green?

38. Star A has a bluish color; Star B is much more luminous with a reddish color. Which star has a higher surface temperature?

Section 17-5

39. Which gives a more accurate measure of a star's surface temperature, its color ratios or its spectral lines? Explain your answer.

40. Menkalinan (Arabic for "shoulder of the rein-holder") is an A2 star in the constellation Auriga (the Charioteer). What is its spectral class? What is its spectral type? Which gives a more precise description of the spectrum of Menkalinan?

41. What are the most prominent absorption lines you would expect to find in the spectrum of a star with a surface temperature of (a) 35,000 K, (b) 2800 K, and (c) 5800 K (like the Sun)? Briefly describe why these stars have such different spectra even though they have essentially the same chemical composition.

42. (See Figure 17-12.) What temperature and spectral classification would you give to a star with equal line strengths of hydrogen (H) and neutral helium (He I)? Explain your answer.

43. The Sun's surface temperature is 5800 K. Using Figure 17-12, arrange the following absorption lines in the Sun's spectrum from the strongest to the weakest, and explain your reasoning: (i) neutral calcium, (ii) singly ionized calcium, (iii) neutral iron, and (iv) singly ionized iron.

Section 17-6

44. If a red star and a blue star both have the same radius and both are the same distance from Earth, which one looks brighter in the night sky? Explain why.

45. If a red star and a blue star both appear equally bright and both are the same distance from Earth, which one has the larger radius? Explain why.

46. If a red star and a blue star both have the same radius and both appear equally bright, which one is farther from Earth? Explain why.

47. The bright star Rigel in the constellation Orion has a surface temperature about 1.6 times that of the Sun. Its luminosity is about 64,000 $L_\odot$. What is Rigel's radius compared to the radius of the Sun?

48. **Box 17-4** Star P has one-half the radius of Star Q. Stars P and Q have surface temperatures 4000 K and 8000 K, respectively. Which star has the greater luminosity? How many times greater is it?

49. **Box 17-4** Star X has 12 times the luminosity of Star Y. Stars X and Y have surface temperatures 3500 K and 7800 K, respectively. Which star has the larger radius? How many times larger is it?

50. **Box 17-4** Suppose a star experiences an outburst in which its surface temperature doubles but its average density (its mass divided by its volume) decreases by a factor of 8. The mass of the star stays the same. By what factors do the star's radius and luminosity change?

51. **Box 17-4** The Sun experiences solar flares (see Section 16-10). The amount of energy radiated by even the strongest solar flare is not enough to have an appreciable effect on the Sun's luminosity. But when a flare of the same size occurs on a main-sequence star of spectral class M, the star's brightness can increase by as much as a factor of 2. Why should there be an appreciable increase in brightness for a main-sequence M star but not for the Sun?

52. **Box 17-4** The bright star Zubeneschmali (β Librae) is of spectral type B8 and has a luminosity of 130 $L_\odot$. What is the star's approximate surface temperature? How does its radius compare to that of the Sun?

53. **Box 17-4** Castor (α Geminorum) is an A1 V star with an apparent brightness of 4.4×10^{-12} that of the Sun. Determine the approximate distance from Earth to Castor (in parsecs).

54. **Box 17-4** A brown dwarf called CoD-33°7795 B has a luminosity of $0.0025 L_\odot$. It has a relatively high surface temperature of 2550 K, which suggests that it is very young and has not yet had time to cool down by emitting radiation. (a) What is this brown dwarf's spectral class? (b) Find the radius of CoD-33°7795 B. Express your answer in terms of the Sun's radius and in kilometers. How does this compare to the radius of Jupiter? Is the name "dwarf" justified?

Section 17-7

55. Sketch a Hertzsprung–Russell diagram. Indicate the regions on your diagram occupied by (a) main-sequence stars, (b) red giants, (c) supergiants, (d) white dwarfs, and (e) the Sun.

56. Explain why the dashed lines in Figure 17-15b slope down and to the right.

57. Some giant and supergiant stars are of the same spectral type (G2) as the Sun. What aspects of the spectrum of a G2 star would you concentrate on to determine the star's luminosity class? Explain what you would look for.

Section 17-8

58. Briefly describe how you would determine the distance to a star whose parallax is too small to measure.

59. Consider a gas at a fixed temperature that produces hydrogen absorption lines. Other than temperature, what conditions of the gas would influence how wide (in a wavelength spectrum) the absorption lines would be?

60. For a given surface temperature, explain why the radius of a star influences the width of its absorption lines.

61. Does every star within the same luminosity class have approximately the same luminosity?

Section 17-9

62. What information about stars do astronomers learn from binary systems that cannot be learned in any other way? What measurements do they make of binary systems to garner this information?

63. What is the mass–luminosity relation? Does it apply to stars of all kinds?

64. Use Figure 17-22 to (a) estimate the mass of a main-sequence star that is 1000 times as luminous as the Sun, and (b) estimate the luminosity of a main-sequence star whose mass is one-fifth that of the Sun. Explain your answers.

65. Which is more massive, a red main-sequence star or a blue main-sequence star? Which has the greater radius? Explain your answers.

66. How do white dwarfs differ from brown dwarfs? Which are more massive? Which are larger in radius? Which are denser?

67. The star HD 3651 shown in Figure 17-13 has a mass of 0.79 $M_\odot$. Its brown dwarf companion, HD 3651B, has about 40 times the mass of Jupiter. The average distance between

the two stars is about 480 au. How long does it take the two stars to complete one orbit around each other?

68. The visual binary 70 Ophiuchi (see the accompanying figure) has a period of 87.7 years. The parallax of 70 Ophiuchi is 0.2 arcsec, and the apparent length of the semimajor axis as seen through a telescope is 4.5 arcsec. (a) What is the distance to 70 Ophiuchi in parsecs? (b) What is the actual length of the semimajor axis in au? (c) What is the sum of the masses of the two stars? Give your answer in solar masses.

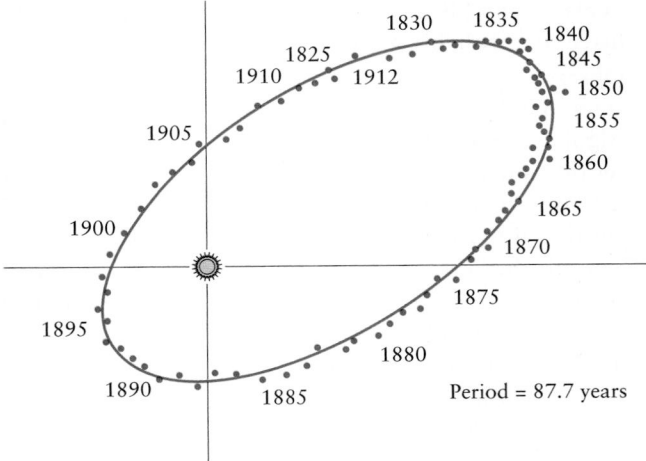

69. An astronomer observing a binary star finds that one of the stars orbits the other once every 5 years at a distance of 10 au. (a) Find the sum of the masses of the two stars. (b) If the mass ratio of the system is $M_1/M_2 = 0.25$, find the individual masses of the stars. Give your answers in terms of the mass of the Sun.

70. Why are more massive stars on the main sequence also more luminous?

Section 17-10

71. **TUTORIAL 17.3** Sketch the radial velocity curves of a binary consisting of two identical stars moving in circular orbits that are (a) perpendicular to our line of sight and (b) parallel to our line of sight.

72. Give two reasons why a visual binary star is unlikely to also be a spectroscopic binary star.

73. Explain the role of the Doppler shift in analyzing spectroscopic binaries.

Section 17-11

74. Sketch the light curve of an eclipsing binary consisting of two identical stars in highly elongated orbits oriented so that (a) their major axes are pointed toward Earth and (b) their major axes are perpendicular to our line of sight.

75. To determine the relative sizes of two stars using the eclipsing binary method, do astronomers need to see both stars separated on the sky as they orbit each other?

76. Stars are typically round. Explain why a star might not be round, and how the light curve from an eclipsing binary might reveal a star's shape.

General Problems

77. A fellow student expresses the opinion that since the Sun's spectrum has only weak absorption lines of hydrogen, this element cannot be a major constituent of the Sun. How would you enlighten this person?

78. Most of the bright stars in the night sky (see Appendix 5) are giants and supergiants. How can this be, if giants and supergiants make up only 1% of the population of stars?

79. Suppose that you want to determine the temperature, diameter, and luminosity of an isolated star (not a member of a binary system). Which of these physical quantities require you to know the distance to the star? Explain your answer.

80. It is desirable to be able to measure the radial velocity of stars (using the Doppler effect) to an accuracy of 1 km/s or better. One complication is that radial velocities refer to the motion of the star relative to the Sun, while the observations are made using a telescope on Earth. Is it important to take into account the motion of Earth around the Sun? Is it important to take into account Earth's rotational motion? To answer this question, you will have to calculate Earth's orbital speed and the speed of a point on Earth's equator (the part of Earth's surface that moves at the greatest speed because of the planet's rotation). If one or both of these effects are of importance, how do you suppose astronomers compensate for them?

WEB/E-BOOK QUESTIONS

81. Search the internet for information about Gaia, a European Space Agency (ESA) spacecraft planned to extend the work carried out by Hipparcos. When is the spacecraft planned to be launched? How will Gaia compare to Hipparcos? For how many more stars will it be able to measure parallaxes? What other types of research will it carry out?

82. Search the internet for recent discoveries about brown dwarfs. Are all brown dwarfs found orbiting normal stars, or are they also found orbiting other brown dwarfs? Are any found in isolation (that is, not part of a binary system)? The Sun experiences flares (see Section 16-10), as do other normal stars; is there any evidence that brown dwarfs also experience flares? If so, is there anything unusual about these flares? What is the lowest temperature for a brown dwarf found so far?

ACTIVITIES

Observing Projects

> **Observing tips and tools**
>
> Even through a telescope, the colors of stars are sometimes subtle and difficult to see. To give your eye the best chance of seeing color, use the "averted vision" trick: When looking through the telescope eyepiece, direct your vision a little to one side of the star you are most interested in. This places the light from that star on a more sensitive part of your eye's retina.

83. The following table lists five well-known red stars. It includes their right ascension and declination (celestial coordinates described in Box 2-1), apparent magnitudes, and color ratios. As their apparent magnitudes indicate, all these stars are somewhat variable. Observe at least two of these stars both by eye and through a small telescope. Is the reddish color of the stars readily apparent, especially in contrast to neighboring stars? (The Jesuit priest and astronomer Angelo Secchi named Y Canum Venaticorum "La Superba," and μ Cephei is often called William Herschel's "Garnet Star.")

Star	Right ascension	Declination	Apparent magnitude	b_V/b_B
Betelgeuse	$5^h 55.2^m$	$+7° 24'$	0.4–1.3	5.5
Y Canum Venaticorum	$12^h 45.1^m$	$+45° 26'$	5.5–6.0	10.4
Antares	$16^h 29.4^m$	$−26° 26'$	0.9–1.8	5.4
μ Cephei	$21^h 43.5^m$	$+58° 47'$	3.6–5.1	8.7
TX Piscium	$23^h 46.4^m$	$+3° 29'$	5.3–5.8	11.0

Note: The right ascensions and declinations are given for epoch 2000.

84. The table of double stars shown at the bottom of the page includes vivid examples of contrasting star colors. The table lists the angular separation between the stars of each double. Observe at least four of these double stars through a telescope. Use the spectral types listed to estimate the difference in surface temperature of the stars in each pair you observe. Does the double with the greatest difference in temperature seem to present the greatest color contrast? From what you see through the telescope and based on what you know about the H-R diagram, explain why all the cool stars (spectral types K and M) listed are probably giants or supergiants.

85. Observe the eclipsing binary Algol (β Persei), using nearby stars to judge its brightness during the course of an eclipse. Algol has an orbital period of 2.87 days, and, with the onset of primary eclipse, its apparent magnitude drops from 2.1 to 3.4. It remains this faint for about 2 hours. The entire eclipse, from start to finish, takes about 10 hours. Consult the "Celestial Calendar" section of the current issue of *Sky & Telescope* for the predicted dates and times of the minima of Algol. Note that the schedule is given in Universal Time (the same as Greenwich Mean Time), so you will have to convert the time to that of your own time zone. Algol is normally the second brightest star in the constellation of Perseus. Because of its position on the celestial sphere (R.A. = $3^h 08.2^m$, Decl. = $40° 57'$), Algol is readily visible from northern latitudes during the fall and winter months.

Star	Right ascension	Declination	Apparent magnitude	Angular separation (arcseconds)	Spectral types
55 Piscium	$0^h 39.9^m$	$+21° 26'$	5.4 and 8.7	6.5	K0 and F3
γ Andromedae	$2^h 03.9^m$	$+42° 20'$	2.3 and 4.8	9.8	K3 and A0
32 Eridani	$3^h 54.3^m$	$−2° 57'$	4.8 and 6.1	6.8	G5 and A
ι Cancri	$8^h 46.7^m$	$+28° 46'$	4.2 and 6.6	30.5	G5 and A5
γ Leonis	$10^h 20.0^m$	$+19° 51'$	2.2 and 3.5	4.4	K0 and G7
24 Coma Berenicis	$12^h 35.1^m$	$+18° 23'$	5.2 and 6.7	20.3	K0 and A3
ν Boötis	$14^h 45.0^m$	$+27° 04'$	2.5 and 4.9	2.8	K0 and A0
α Herculis	$17^h 14.6^m$	$+14° 23'$	3.5 and 5.4	4.7	M5 and G5
59 Serpentis	$18^h 27.2^m$	$+0° 12'$	5.3 and 7.6	3.8	G0 and A6
β Cygni	$19^h 30.7^m$	$+27° 58'$	3.1 and 5.1	34.3	K3 and B8
δ Cephei	$22^h 29.2^m$	$+58° 25'$	4* and 7.5	20.4	F5 and A0

Note: The right ascensions and declinations are given for epoch 2000.

*The brighter star in the δ Cephei binar system is a variable star of approximately the fourth magnitude.

Starry Night™ Explorations

86. Use *Starry Night*™ to explore the proper motion of a star, defined as its angular motion in our sky in arcseconds per year. Open **Favorites > Explorations > Stellar Motion** to see a view of the constellations against the celestial sphere.

    (a) Click the **CE** value in the date shown in the toolbar to toggle between the common era and the age before common era. In other words, each click of this value changes the time back and forth over a time span of over 4000 years. Do the constellation patterns remain static over this time?

    (b) With the time flow rate set to 100 years, click **Play**. How do you account for the changing appearance of the constellations?

    (c) Select **Favorites > Explorations > Barnard's Star** to see a view of the sky from a location hovering 1000 km above the surface of Earth. The view is centered on the star Tyc421-2053-1. In the same position on the celestial sphere as this star is Barnard's star, which is also labeled in the view. Click **Play** to observe the motion of Barnard's star relative to the other stars in the view. Is Barnard's star's motion unique in this view or do other stars show a similar change in position?

    (d) Select **File > Revert** from the menu and zoom in to a field of view 15′ wide. Use the "U" keyboard shortcut to advance time by 10 years and then use the angular separation cursor to measure the angular distance between Barnard's star and the star Tyc421-2053-1. What is the velocity, in arcseconds per year, of Barnard's star across the sky, its proper motion?

87. Use *Starry Night*™ to examine the effect of distance on the apparent brightness of a star, using the Sun as an example and observing it from different locations in the solar system. The amount of energy emitted per second by a star is its luminosity, L. This energy spreads into space and, at a distance d, will be distributed over the surface of a sphere of area $4\pi d^2$. Thus, the brightness per unit area is $b = L/4\pi d^2$ and the ratio of two brightnesses at two different distances is $b_1/b_2 = (d_2/d_1)^2$, the inverse-square law of light intensity. The historically based astronomical scale of brightness uses apparent magnitude, m, in which a ratio of brightnesses (b_1/b_2) is represented in terms of a difference in magnitudes by the number 2.512 raised to the power of $(m_2 - m_1)$. Thus, $\left(\dfrac{b_1}{b_2}\right) = \left(\dfrac{d_2}{d_1}\right)^2 = 2.512^{(m_2 - m_1)}$. Taking the logarithm of both sides of this equation results in the relation $2 \cdot \log d_2 - 2 \cdot \log d_1 = (m_2 - m_1) \cdot \log 2.512$. If we hold d_1 constant while choosing different values of d_2, then m_1 will also be constant, and, with $\log 2.512 = 0.4$, we have the final equation representing m in terms of d: $m = 5 \cdot \log d + constant$. To demonstrate that the inverse-square law holds over a range of distance, one simply has to show that this relationship is correct for a set of measured data. Prepare a table with four

columns: Location, Apparent Magnitude, Distance from Observer (d), and Log d. Open **Favorites > Explorations > Sun from 1-Mercury.** The view is centered on the Sun from the planet Mercury. Move the cursor over the Sun and use the HUD to find the apparent magnitude of the Sun and its distance from Mercury. Record these data in the appropriate columns of the first row of your data table. Next, open **Favorites > Explorations** and in the list of files, find the series labeled **Sun from *n-location*.** Open each of these views in turn to obtain the apparent magnitude of the Sun and its distance from each of the observing locations, then record these data in your table. Calculate the logarithm of the distance for each location and plot a graph or use a spreadsheet to graph the apparent magnitude of the Sun as a function of log d.

    (a) What can you say about this plot and what is your interpretation?

    (b) Calculate the slope of the graph. How does this relate to the above equation?

88. Use *Starry Night*™ to examine the 10 brightest stars in Earth's night sky. Select **Favorites > Explorations > Atlas** and open the **Observing Lists** panel. In this panel, expand the **10 Brightest Stars** list. Double-click on each of the stars in this list in turn to center and label the star in the view. Click on the label for the star, select **Show Info...** and open Position in Space to obtain the star's distance from the Sun; then open the **Other Data** page in the info window to obtain the star's apparent magnitude, temperature, and luminosity. Compile a table of these values.

    (a) Which is the brightest star in Earth's night sky? Why?

    (b) Which of these brightest stars has the highest temperature? What would you expect to be the color of this star compared to others in the list?

    (c) Which of these stars is intrinsically the most luminous?

89. Use *Starry Night*™ to study the Hertzsprung–Russell (H-R) diagram for stars. Select **Favorites > Explorations > H-R Diagram** to display a view of the Sun against the background stars, along with an H-R diagram that plots the absolute magnitudes of the stars in the view as a function of their surface temperatures. Absolute magnitude is an inverse, logarithmic parameter related to the star's luminosity; the smaller the value of absolute magnitude, the larger the luminosity. The surface temperature is plotted in an inverse direction, with the hottest stars appearing to the left of the diagram. You can point the cursor at any star and display its position on the H-R diagram.

    (a) Place the cursor on the Sun. What are the approximate surface temperature and absolute magnitude values for the Sun, as read from the H-R diagram? (Note that temperature is plotted in units of 1000 K; that is, "10" represents 10,000 K.)

    (b) Scroll around the sky and watch the H-R diagram change as different stars enter and leave the view. Does the distribution of stars in the H-R diagram change drastically from one part of the sky to another, or are all types of stars approximately equally represented in all directions from Earth?

(c) In which region of the H-R diagram do you see most of the stars?

(d) Select **File > Revert** from the menu to reset the view and click the **Quick Search Items** icon in the toolbar to open the **Observing Lists . . .** panel. In this panel, find and expand the **H-R Diagram** list. For each star in this list, double-click its name to center and label it in the view. Then use the cursor to determine its position in the H-R diagram. Tabulate the values of absolute magnitude and temperature for each of these stars as determined from the H-R diagram. (Note the nonlinear nature of the temperature axis when estimating these values.)

(e) Based on its absolute magnitude-temperature classification, which of these stars is most similar to the Sun? In which region of the H-R diagram do we find both this star and the Sun?

(f) If these two physical properties are similar for the Sun and this star and absolute magnitude is a measure of the luminosity of a star, which other two parameters will necessarily be similar? Explain your answer.

ANSWERS

ConceptChecks

ConceptCheck 17-1: A very nearby flashlight will *appear* much brighter than a very distant spotlight. Although a large spotlight emits more light than a small flashlight, which light appears brighter also depends on their distances away.

ConceptCheck 17-2: The parallax angle is smallest for the most distant stars, so if a parallax angle were measured to be too small, then the astronomer would assume the star is farther away than it actually is.

ConceptCheck 17-3: Parallax only works for stars that are relatively nearby, and the majority of stars in our galaxy are much too far away to exhibit any apparent shift in position as Earth orbits the Sun.

ConceptCheck 17-4: The luminosity refers to the intrinsic energy output of the headlights, which does not change no matter how far away the car is observed. On the other hand, apparent brightness refers to how much light energy is observed by the woman, and the apparent brightness increases as the headlights get closer (see Figure 17-4).

ConceptCheck 17-5: For both stars to have the same apparent brightness, the more luminous star must be farther away. (In fact, due to the inverse-square law, the more luminous star must be 10 times farther away, which you can determine using the techniques in Box 17-2.)

ConceptCheck 17-6: Because a star's apparent magnitude when viewed from a distance of 10 pc is the same as its absolute magnitude, both magnitudes would be the same, +4.8.

ConceptCheck 17-7: From Earth, Tau Ceti appears to be relatively bright at +3. However, if it were viewed from a distance of 10 pc, it would appear much dimmer, at +6. The only way this could be the case is if Tau Ceti is much closer to Earth than 10 pc.

ConceptCheck 17-8: The shorter the peak wavelength, the higher the temperature. The blue star emits light that peaks at blue wavelengths, and these are shorter wavelengths than the other star's peak at orange wavelengths. Thus, the blue star is hotter. (Figure 17-7 illustrates these effects.)

ConceptCheck 17-9: The prominence of certain hydrogen absorption lines visible in the star, with A class being the greatest and O class being the least.

ConceptCheck 17-10: In the sequence OBAFGKM, F-spectral class stars with designations F1, F2, F3, and F4 are more similar to A-spectral class stars, and F6, F7, F8, and F9 are more similar to G-spectral class stars.

ConceptCheck 17-11: Within a particular spectral class, the larger numbers correspond to cooler stars. For example, a G2 star is hotter than a G8 star because a G2 is closer to the hotter F-spectral class stars in the sequence OBAFGKM.

ConceptCheck 17-12: Given that the modern spectral sequence of stars and brown dwarfs from hottest to coolest is OBAFGKMLTY, L-spectral class brown dwarfs are hotter than T-spectral class brown dwarfs.

ConceptCheck 17-13: Stars like our Sun derive their energy from nuclear reactions in the core, whereas brown dwarfs derive their energy from gravitational contraction. Brown dwarfs are not true stars.

ConceptCheck 17-14: Even though the campfire and the candle have the same surface temperature, the larger surface area of a campfire results in a greater energy output.

ConceptCheck 17-15: As indicated in Figure 17-14, the surface temperature of a star can be determined from its spectral type, and does not require knowing the star's distance. On the other hand, the star's distance is needed to determine the star's radius, because solving for the radius requires knowing the star's luminosity, which, in turn, requires knowing the star's distance and apparent brightness.

ConceptCheck 17-16: The H-R diagram is plotted as increasing luminosity on the vertical axis and decreasing temperature along the horizontal axis. The stars with the greatest luminosity and the highest temperatures are found in the upper left-hand corner of the H-R diagram.

ConceptCheck 17-17: You can read from the vertical axis of the H-R diagram that Betelgeuse is much more luminous than Barnard's star. Since they both have similar surface temperatures (around 3500 K), Betelgeuse must be much larger to emit its larger luminosity.

ConceptCheck 17-18: Regulus. All main-sequence stars, and only these stars (such as Regulus, but not Mira), convert hydrogen to helium in their cores.

ConceptCheck 17-19: The main sequence and all its stars have luminosity class V. At higher temperatures, the main-sequence stars can have a greater luminosity than the lower-temperature stars in luminosity class III.

ConceptCheck 17-20: Main-sequence stars are luminosity class "V" whereas luminosity class "II" are giant stars. In this instance, the two stars have the same temperature because they are in the same spectral class, but the luminosity class II star is larger and has a greater luminosity than the much smaller main-sequence star.

ConceptCheck 17-21: If two stars orbiting a common center of mass were moved farther apart, their masses would not change, but the period would increase following Kepler's third law.

ConceptCheck 17-22: According to the mass–luminosity relation, the most luminous main-sequence stars are also the most massive, have the largest radii, and have the greatest temperatures.

ConceptCheck 17-23: At Stage 2, neither of the stars moves toward or away from Earth. The only radial velocity Doppler shift arises from the pair of stars as they drift together through space, which in this case, can be read near the bottom of Figure 17-23 for Stage 2 at about +12 km/s away from Earth. Since both stars have this same radial velocity at Stage 2, their absorption lines are not split.

ConceptCheck 17-24: When the bright white dwarf passes behind the larger and dimmer star, the light coming from the dimmer star is still visible.

CalculationChecks

CalculationCheck 17-1: Given that distance in parsecs is the inverse of the parallax angle, $d = 1/0.7772$, then $d = 1.35$ pc. Because 1 pc is 3.26 ly, we find that 1.35 pc $\times$ (3.26 ly/1pc) = 4.4 ly.

CalculationCheck 17-2: According to the inverse-square law nature of light, the intensity of the light received decreases with the square of the distance. In this instance, increasing the distance 3 times means the newspaper receives only $1/(3)^2$ or one-ninth the amount of light it originally received.

CalculationCheck 17-3: Using the equation at the beginning of Box 17-2, Star A must be 4 times farther away than Star B.

CalculationCheck 17-4: With luminosity given by the Stefan–Boltzmann law, where $L = 4\pi R^2 \sigma T^4$, if radius R is tripled, then L must increase by the R^2, which is, in this instance, 3^2 or 9 times, for a star 3 times larger but at the same temperature.

LEARNING GOALS

By reading the sections of this chapter, you will learn

The stars that illuminate our nights seem eternal and unchanging. But this permanence is an illusion. Each of the stars visible to the naked eye shines due to thermonuclear reactions and has only a finite amount of fuel available for these reactions. Hence, stars cannot last forever: They form from material in interstellar space, evolve over millions or billions of years, and eventually die. In this chapter, our concern is with how stars are born and become part of the main sequence.

Stars form within cold, dark clouds of gas and dust that are scattered abundantly throughout our galaxy. One such cloud appears as a dark area on the far right-hand side of the photograph on this page. Perhaps a dark cloud like this encounters one of the galaxy's spiral arms, or perhaps a supernova detonates nearby. From the shock of events like these, the cloud begins to contract under the pull of gravity, forming protostars—the fragments that will one day become stars. As we saw in Chapter 8, our own Sun was likely formed in exactly this way. As a protostar develops, its internal pressure builds and its temperature rises. In time, hydrogen fusion begins, and a star is born. The hottest, bluest, and brightest young stars, like those

A region of star formation about 1400 pc (4000 ly) from Earth in the southern constellation Ara (the Altar). (European Southern Observatory)

R I V U X G

in the accompanying image, emit ultraviolet radiation that excites atoms in the surrounding interstellar gas. The result is a beautiful glowing nebula, which typically has the red color characteristic of excited hydrogen (as shown in the photograph).

In Chapters 19 and 20, we will see how stars mature and grow old. Some even blow themselves apart in death throes that enrich interstellar space with the material for future generations of stars. Thus, like the mythical phoenix, new stars arise from the ashes of the old.

18-1 Understanding how stars evolve requires observation as well as ideas from physics

Over the past several decades, astronomers have labored to develop an understanding of **stellar evolution**, that is, how stars are born, live their lives, and finally die. Our own Sun provides evidence that stars are not permanent. The energy radiated by the Sun comes from thermonuclear reactions in its core, which consume 6×10^{11} kg of hydrogen each second and convert it into helium (see Section 16-1). While the amount of hydrogen in the Sun's core is vast, it is not infinite; therefore, the Sun cannot always have been shining, nor can it continue to shine forever. The same is true for all other main-sequence stars, which are fundamentally the same kinds of objects as the Sun but with different masses (see Section 17-8). Thus, stars must have a beginning as well as an end.

> Stars consume the material of which they are made, and so cannot last forever.

Stars last very much longer than the lifetime of any astronomer—indeed, far longer than the entire history of human civilization. Thus, it is impossible to watch a single star go through its formation, evolution, and eventual demise. Rather, astronomers have to piece together the evolutionary history of stars by studying different stars at different stages in their life cycles.

ANALOGY

To see the magnitude of this task, imagine that you are a biologist from another planet who sets out to understand the life cycles of human beings. You send a spacecraft to fly above Earth and photograph humans in action. Unfortunately, the spacecraft fails after collecting only 20 seconds of data, but during that time its sophisticated equipment sends back observations of thousands of different humans. From this brief snapshot of life on Earth—only 10^{-8} (a hundred-millionth) of a typical human lifetime—how would you decide which were the young humans and which were the older ones? Without a look inside our bodies to see the biological processes that shape our lives, could you tell how humans are born and how they die? And how could you deduce the various biological changes that humans undergo as they age?

Astronomers, too, have data spanning only a tiny fraction of any star's lifetime. A star like the Sun can last for about 10^{10} years, whereas astronomers have been observing stars in detail for only about a century—as in our analogy, roughly 10^{-8} of the life span of a typical star. Astronomers are also frustrated by being unable to see the interiors of stars. For example, we cannot see the thermonuclear reactions that convert hydrogen into helium. But astronomers have an advantage over the biologist in our story. Unlike humans, stars are made of relatively simple substances, primarily hydrogen and helium, that are found almost exclusively in the form of gases. Of the three phases of matter—gas, liquid, and solid—gases are by far the simplest to understand.

Astronomers use our understanding of gases to build theoretical models of the interiors of stars, like the model of the Sun we saw in Section 16-2. Models help to complete the story of stellar evolution. In fact, like all great dramas, the story of stellar evolution can be regarded as a struggle between two opposing and unyielding forces: Gravity continually tries to make a star shrink, while the star's internal pressure tends to make the star expand. When these two opposing forces are in balance, the star is in a state of hydrostatic equilibrium (see Figure 16-2).

But what happens when changes within the star cause either pressure or gravity to predominate? The star must then either expand or contract until it reaches a new equilibrium. In the process, it will change not only in size but also in luminosity and color.

In the following chapters, we will find that giant and supergiant stars are the result of pressure gaining the upper hand over gravity. Both giants and supergiants turn out to be aging stars that have become tremendously luminous and ballooned to hundreds or thousands of times their previous size. White dwarfs, by contrast, are the result of the balance tipping in gravity's favor. These dwarfs are even older stars that have collapsed to a fraction of their former size while on the main sequence. In this chapter, however, we will see how the opposing influences of gravity and pressure explain the birth of stars. We start our journey within the diffuse clouds of gas and dust that permeate our galaxy.

18-2 Interstellar gas and dust pervade the galaxy

Where do stars come from? As we saw in Section 8-4, our Sun condensed from a solar nebula, a collection of gas and dust in interstellar space—the space between the stars. Observations suggest that other stars originate in a similar way (see Figure 8-8). To understand the formation of stars, we must first understand the nature of the interstellar matter from which the stars form.

> Different types of nebulae emit, absorb, or reflect light.

Nebulae and the Interstellar Medium

At first glance, the space between the stars seems to be empty. On closer inspection, we find that it is filled with a thin gas laced with microscopic dust particles. This combination of gas and dust is called the **interstellar medium**.

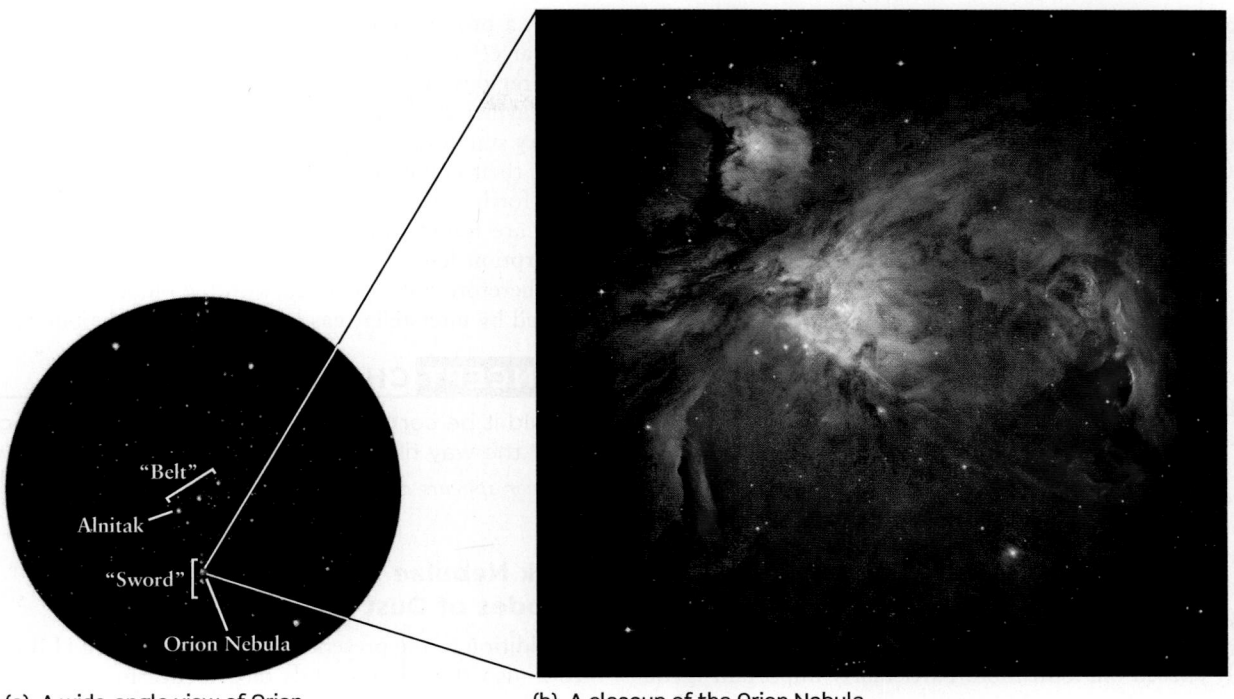

(a) A wide-angle view of Orion   (b) A closeup of the Orion Nebula

FIGURE 18-1 R I Ⓥ U X G

The Orion Nebula **(a)** The middle "star" of the three that make up Orion's sword is actually an interstellar cloud called the Orion Nebula. **(b)** The nebula is about 450 pc (1500 ly) from Earth and contains about 300 solar masses of material. Most of the ultraviolet light that makes the nebula glow comes from just five hot massive stars. (a: © Akira Fujii/David Malin Images; b: NASA, ESA, M. Robberto [Space Telescope Science Institute/ESA] and the Hubble Space Telescope Orion Treasury Project Team)

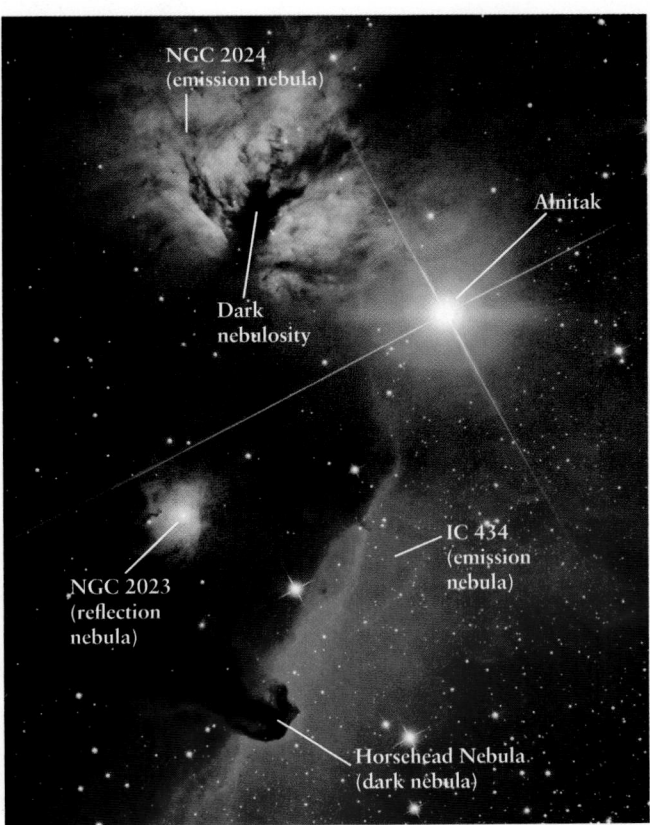

FIGURE 18-2 R I Ⓥ U X G

Emission, Reflection, and Dark Nebulae in Orion A variety of different nebulae appear in the sky around Alnitak, the easternmost star in Orion's belt (see Figure 18-1a). All the nebulae lie approximately 500 pc (1600 ly) from Earth. They are actually nowhere near Alnitak, which is only 250 pc (820 ly) distant. This photograph shows an area of the sky about 1.5° across. (Stocktrek Images/Roth Ritter/Stocktrek Images/Corbis)

You can see a dense concentration of the interstellar medium with the naked eye. Look carefully at the constellation Orion (Figure 18-1a), visible on winter nights in the northern hemisphere and summer nights in the southern hemisphere. While most of the stars in the constellation appear as sharply defined points of light, the middle "star" in Orion's sword has a fuzzy appearance. As Figure 18-1b shows, this "star" is actually not a star at all, but the Orion Nebula—a cloud in interstellar space. Any interstellar cloud is called a **nebula** (plural **nebulae**) or **nebulosity**.

Emission Nebulae: Clouds of Excited Gas

The Orion Nebula emits its own light, with the characteristic emission line spectrum of a hot, thin gas. For this reason, it is called an **emission nebula**. Many emission nebulae can be seen with a small telescope. Figure 18-2 shows some of these nebulae in a different part of the constellation Orion. Emission nebulae are direct evidence of gas atoms in the interstellar medium.

Typical emission nebulae have masses that range from about 100 to about 10,000 solar masses. Because this mass is spread over a huge volume that is light-years across, the density is quite low by Earth standards, only a few thousand hydrogen atoms per cubic centimeter. (By comparison, the air you are breathing contains more than 10^{19} atoms per cm^3.)

Emission nebulae are found near hot, luminous stars of spectral types O and B. Such stars emit copious amounts of ultraviolet radiation. When atoms in the nearby interstellar gas absorb these energetic ultraviolet photons, the atoms become ionized. Indeed, emission nebulae are composed primarily of ionized hydrogen atoms—that is, free protons (hydrogen nuclei) and electrons. Astronomers use the notation H I for neutral, un-ionized hydrogen atoms and H II for ionized hydrogen atoms, which is why emission nebulae are also called **H II regions**.

H II regions emit red visible light when some of the free protons and electrons get back together to form hydrogen atoms, a process called **recombination** (Figure 18-3). When an atom forms by recombination, the electron is typically captured into a high-energy orbit. As the electron cascades downward through the atom's energy levels toward the ground state, the atom emits photons with lower energies and longer wavelengths than the photons that originally caused the ionization. Particularly important is the transition from $n = 3$ to $n = 2$. It produces H$_\alpha$ photons with a wavelength of 656 nm, in the red portion of the visible spectrum (see Section 5-8, especially Figure 5-23b). These photons give H II regions their distinctive reddish color.

For each high-energy, ultraviolet photon absorbed by a hydrogen atom to ionize it, several photons of lower energy are emitted when a proton and electron recombine. As Box 18-1 describes, a similar effect takes place in a fluorescent lightbulb. In this sense, H II regions are immense fluorescent light fixtures!

Further evidence of interstellar gas comes from the spectra of binary star systems. As the two stars that make up a binary system orbit their common center of mass, their spectral lines shift back and forth (see Figure 17-23). But certain calcium and sodium lines are found to remain at fixed wavelengths. These **stationary absorption lines**, which have captured the curiosity of observers, are therefore not associated with the binary star. Instead, they are caused by interstellar gas between us and the binary system.

CONCEPT CHECK 18-1

Would it be correct to describe a reddish H II region as "red hot," the way filaments in an electric toaster glow red?
Answer appears at the end of the chapter.

Dark Nebulae and Reflection Nebulae: Abodes of Dust

In addition to the presence of gas atoms in red H II regions, Figure 18-2 also shows two kinds of evidence for larger bits of matter, called **dust grains**, in the interstellar medium. These dust grains make their appearance in *dark nebulae* and *reflection nebulae*.

A **dark nebula** is so opaque that it blocks any visible light coming from stars that lie behind it. One such dark nebula, called the Horsehead Nebula for its shape, protrudes in front of one of the H II regions in Figure 18-2. Figure 18-4 shows another dark

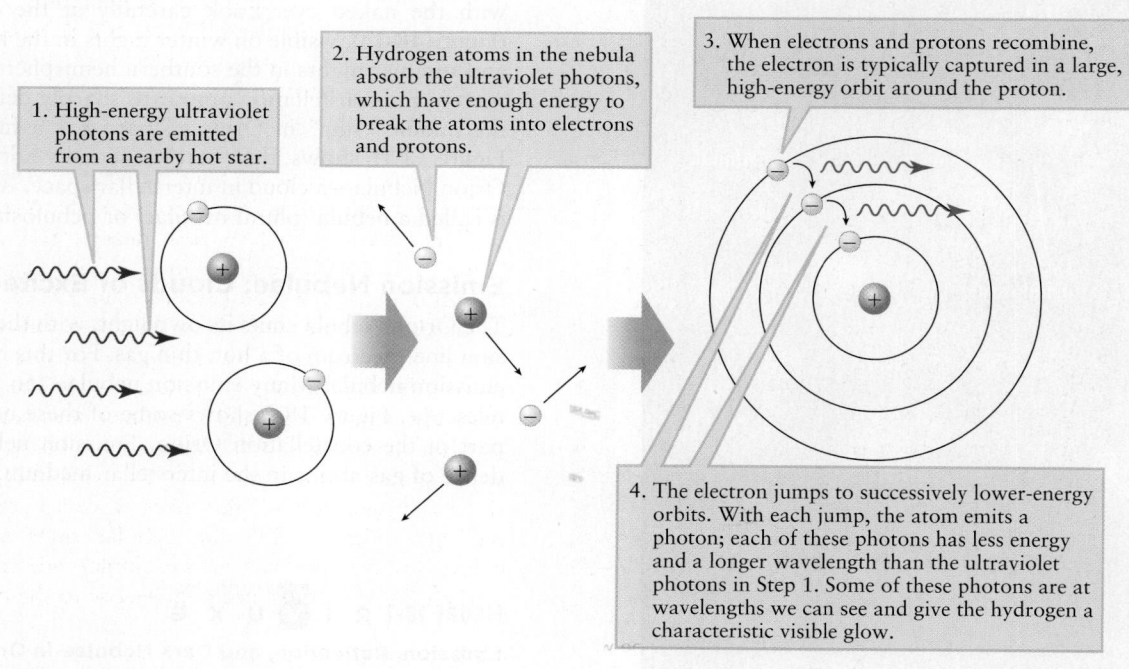

1. High-energy ultraviolet photons are emitted from a nearby hot star.

2. Hydrogen atoms in the nebula absorb the ultraviolet photons, which have enough energy to break the atoms into electrons and protons.

3. When electrons and protons recombine, the electron is typically captured in a large, high-energy orbit around the proton.

4. The electron jumps to successively lower-energy orbits. With each jump, the atom emits a photon; each of these photons has less energy and a longer wavelength than the ultraviolet photons in Step 1. Some of these photons are at wavelengths we can see and give the hydrogen a characteristic visible glow.

FIGURE 18-3

Ionization and Recombination In an H II region, the characteristic red glow of emission nebulae (like those shown in Figure 18-1 and Figure 18-2) comes from gas atoms that are excited by ultraviolet radiation from nearby hot stars. While many photons are emitted during recombination, only some are visible, and red is the most common.

Fluorescent Lights

The light that comes from glowing interstellar clouds is, quite literally, otherworldly. But the same principles that explain how such clouds emit light are also at the heart of light phenomena that we see here on Earth.

A fluorescent lamp produces light in a manner not too different from an emission nebula (H II region). In both cases, the physical effect is called **fluorescence:** High-energy ultraviolet photons are absorbed, and the absorbed energy is reradiated as lower-energy photons of visible light.

Within the glass tube of a fluorescent lamp is a small amount of the element mercury. When you turn on the lamp, an electric current passes through the tube, vaporizing the mercury and exciting its atoms. This excited mercury vapor radiates light with an emission-line spectrum, including lines in the ultraviolet. The white fluorescent coating on the inside of the glass tube absorbs these ultraviolet photons, exciting electrons in the coating's molecules to high energy levels.

The electrons in the molecules then cascade down through a number of lower levels before reaching the ground state. During this cascade, visible-light photons of many different wavelengths are emitted, giving an essentially continuous spectrum and a very white light. (By comparison, the hydrogen atoms in an H II region emit at only certain discrete wavelengths, because the spectrum of hydrogen is much simpler than that of the fluorescent coating's molecules. Another difference is that the molecules in the fluorescent tube never become ionized.)

Many common materials display fluorescence. Among them are teeth, fingernails, and certain minerals. When illuminated with ultraviolet light, these materials glow with a blue or green color. (Most natural history museums and science museums have an exhibit showing fluorescent minerals.) Laundry detergent also contains fluorescent material. After washing, your laundry absorbs ultraviolet light from the Sun and glows faintly, making it appear "whiter than white."

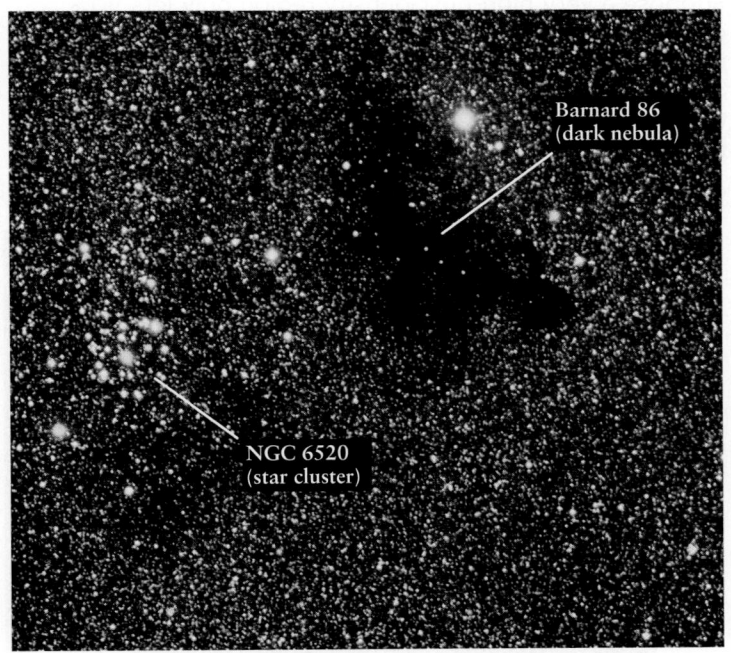

FIGURE 18-4 R I ⓥ U X G

A Dark Nebula When first discovered in the late 1700s, dark nebulae were thought to be "holes in the heavens" where very few stars are present. In fact, they are opaque regions that block out light from the stars beyond them. The few stars that appear to be within Barnard 86 lie between us and the nebula. Barnard 86 is in the constellation Sagittarius and has an angular diameter of 4 arcmin, about 1/7 the angular diameter of the full moon. (© Australian Astronomical Observatory/David Malin Images)

nebula, called Barnard 86. These nebulae have a relatively dense concentration of microscopic dust grains, which scatter and absorb light much more efficiently than single atoms. Typical dark nebulae have temperatures between 10 and 100 K, which is low enough for hydrogen atoms to form molecules. Such nebulae contain from 10^4 to 10^9 particles (atoms, molecules, and dust grains) per cubic centimeter. Although tenuous by Earth standards, dark nebulae are large enough—typically many light-years deep—that they block the passage of light. In the same way, a sufficient depth of haze or smoke in our atmosphere can make it impossible to see distant mountains.

The other evidence for dust in Figure 18-2 is the bluish haze, labeled NGC 2023, surrounding the star immediately above and to the left of the Horsehead Nebula. **Figure 18-5** shows a similar haze around a different set of stars. A haze of this kind, called a **reflection nebula**, is caused by fine grains of dust in a lower concentration than that found in dark nebulae. The light we see coming from the nebula is starlight that has been scattered and reflected by these dust grains. The grains are only about 500 nm across, no larger than a typical wavelength of visible light, and they scatter short-wavelength blue light more efficiently than long-wavelength red light. Hence, reflection nebulae have a characteristic blue color. Box 5-4 explains how a similar process—the scattering of sunlight in our atmosphere—gives rise to the blue color of the sky.

CONCEPT CHECK 18-2

Does the blue color in Figure 18-5 come from hot gas similar to the surfaces of massive hot blue stars?

Answer appears at the end of the chapter.

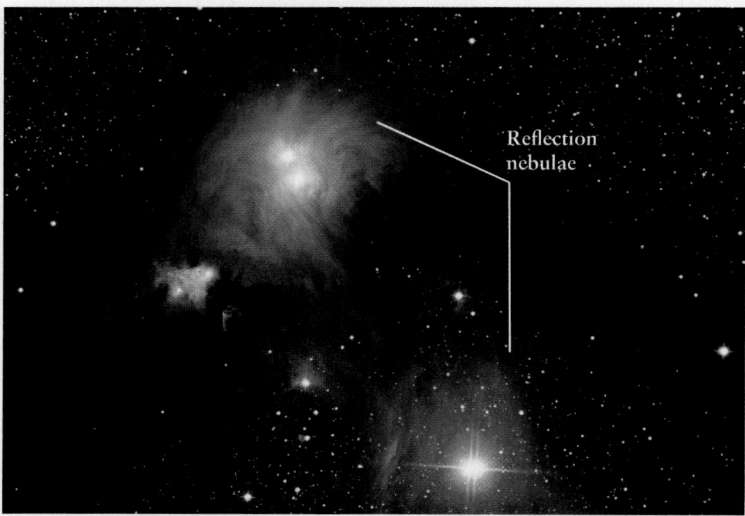

FIGURE 18-5 R I V U X G

Reflection Nebulae Wispy reflection nebulae called NGC 6726-27-29 surround several stars in the constellation Corona Australis (the Southern Crown). Unlike emission nebulae, reflection nebulae do not emit their own light, but scatter and reflect light from the stars that they surround. This scattered starlight is quite blue in color. The region shown here is about 23 arcmin across. (Dr. Stefan Binneweis and Josef Popsel/www.capellaobservatory.com)

Interstellar Extinction and Reddening

In the 1930s, the American astronomer Robert Trumpler discovered two other convincing pieces of evidence for the existence of interstellar matter—*interstellar extinction* and *interstellar reddening*. While studying the brightness and distances of certain star clusters, Trumpler noticed that remote clusters seem to be dimmer than would be expected from their distance alone. His observations demonstrated that the intensity of light from remote stars is reduced as the light passes through material in interstellar space. This process is called **interstellar extinction**. (In the same way, the

headlights of oncoming cars appear dimmer when there is smoke or fog in the air.) The light from remote stars is also reddened as it passes through the interstellar medium, because the blue component of their starlight is scattered and absorbed by interstellar dust. This effect, shown in Figure 18-6, is called **interstellar reddening**. The same effect makes the setting Sun look red (see Box 5-4).

CAUTION!

It is important to understand the distinction between interstellar reddening and the Doppler effect. If an object is moving away from us, the Doppler effect *shifts* all of that object's light toward longer wavelengths (a redshift). Interstellar reddening, by contrast, makes objects appear red not by shifting wavelengths but by filtering out shorter blue wavelengths. The effect is the same as if we looked at an object through red-colored glasses, which let red light pass but block out blue light. The one similarity between Doppler shifts and interstellar reddening is that neither has any discernible effect on what you see with the naked eye. Both of these effects cause only subtle color changes that your eye (which does a poor job of seeing colors in faint objects like stars) cannot detect. If you see a star with a red color, the reason is *not* interstellar reddening or the Doppler effect; it is because the star really is red due to its low surface temperature (see Figure 17-7a).

CALCULATION CHECK 18-1

Suppose a very high-dust-density region only lets 90% of starlight through for each light-year that the starlight passes through this interstellar medium. After starlight is diminished due to interstellar extinction by traveling 2 ly through this medium, what percentage of the light makes it through?

Answer appears at the end of the chapter.

(a) How dust causes interstellar reddening

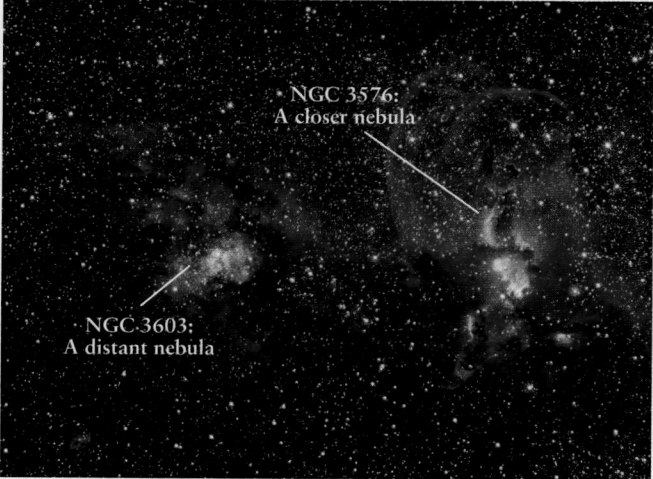

(b) Reddening depends on distance

FIGURE 18-6

Interstellar Reddening (a) Dust grains in interstellar space scatter or absorb blue light more than red light. Thus, light from a distant object appears redder than it really is. **(b)** The emission nebulae NGC 3603 and NGC 3576 are different distances from Earth. Light from the more distant

nebula must pass through more interstellar dust to reach us, so more interstellar reddening occurs and NGC 3603 is a deeper shade of red. The two nebulae are about 1° apart in the sky. (b: Stocktrek Images/Corbis)

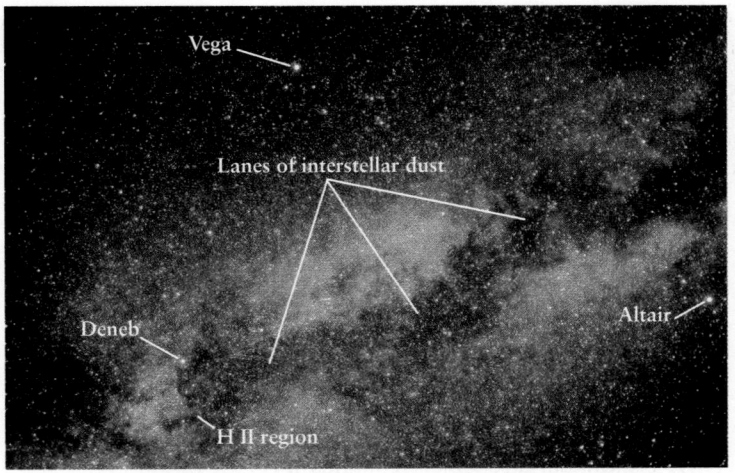

FIGURE 18-7 R I (V) U X G

Gas and Dust in the Milky Way There are so many stars that they aren't easy to see individually, resulting in the diffuse whitish band that gives the Milky Way its name. Red glowing gas clouds (emission nebulae or H II regions) can be seen in the foreground, and dark, dusty regions that block starlight are concentrated close to the midplane of the Milky Way Galaxy. This wide-angle photograph also shows the three bright stars that make up the "summer triangle" (see Figure 2-8). (Jerry Lodriguss/Science Source)

Interstellar Gas and Dust in Spiral Galaxies

Observations indicate that interstellar gas and dust are largely confined to the plane of the Milky Way—that faint, hazy band of stars you can see stretching across the sky on a dark, moonless night. Figure 18-7 shows glowing emission nebulae and dark lanes of dust along the Milky Way.

As we will see in Chapter 23, the band of the Milky Way is actually our inside view of our galaxy, which is a flat, pancake-shaped collection of several hundred billion stars about 50,000 pc (160,000 ly) in diameter. The Sun is located inside this pancake shape and is about 8000 pc (26,000 ly) from our galaxy's center. Astronomers know these dimensions because they can map our galaxy from the locations of bright stars and nebulae, as well as by using radio telescopes. Observations indicate that bright stars, gas, and glowing nebulae are mostly located within a few hundred parsecs of the midplane of our galaxy, form somewhat of a pancake shape, and are concentrated along arching spiral arms that wind outward from the galaxy's center. If we could view our galaxy from a great distance, it would look somewhat like one of the galaxies shown in Figure 18-8.

Interstellar gas and dust are the raw material from which stars are made. The disk of our galaxy, where most of this matter is concentrated, is therefore the site of ongoing star formation. Our next goal is to examine the steps by which stars are formed.

18-3 Protostars form in cold, dark nebulae

TUTORIAL 18.1 How do stars form in the interstellar medium? In this section, we discuss how a contracting cloud gives birth to a clump called a *protostar*—a future main-sequence star.

Protostars: Initial Gravitational Collapse

In order for interstellar material to condense and form a star, the force of gravity—which tends to pull interstellar material

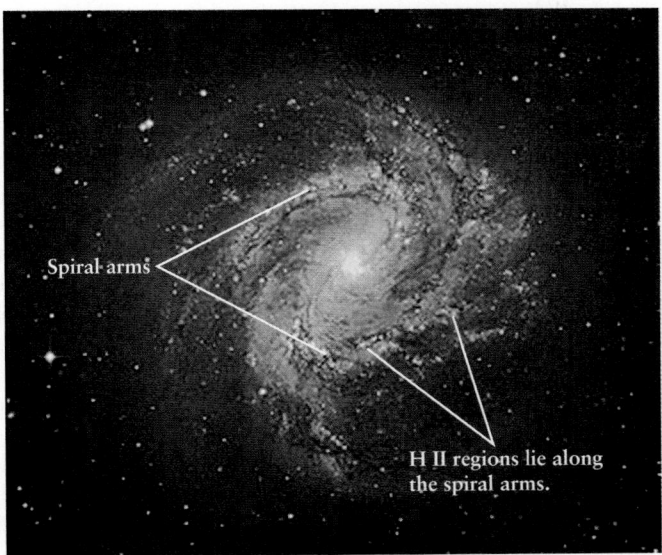

(a) We see spiral galaxy M83 nearly face on.

FIGURE 18-8 R I (V) U X G

Spiral Galaxies Spiral galaxies, like our own Milky Way Galaxy, consist of stars, gas, and dust that are largely confined to a flattened, rotating disk. **(a)** This face-on view of M83 shows luminous stars and H II regions along the spiral arms. **(b)** This edge-on view of NGC 891 shows a dark band caused

(b) We see spiral galaxy NGC 891 nearly edge on.

by dust in this galaxy's interstellar medium. Although in different parts of the sky, both galaxies are about 7 million pc (23 million ly) from Earth and have angular diameters of about 13 arcmin. (a: © Australian Astronomical Observatory/David Malin Images; b: © Malin/IAC/RGO)

FIGURE 18-9   R **I** V U X G

Bok Globules The dark blobs in this photograph of a glowing H II region are clouds of dust and gas called Bok globules. A typical Bok globule is 1 parsec or less in size and contains from 1 to 1000 solar masses of material. The Bok globules and H II region in this image are part of a much larger star-forming region called NGC 281, which lies about 2900 pc (9500 ly) from Earth in the constellation Cassiopeia. The image shows an area about 2.7 pc (8.8 ly) across. (NASA, ESA, and The Hubble Heritage Team [STScI/AURA])

together—must overwhelm the internal pressure pushing the material apart. This means that stars will most easily form in regions where the interstellar material is relatively dense, so that atoms and dust grains are close together and gravitational attraction is enhanced.

To assist star formation, the pressure of the interstellar medium should be relatively low, to allow for the gas and dust to contract. This means that the star-forming region of the interstellar medium should be cold, because the pressure of a gas goes down as the gas temperature decreases. (Conversely, increasing the temperature of a gas makes the pressure increase. For example, the air pressure increases inside a covered pot of boiling water, which often causes steam to shoot out.) While a higher gas density can lead to higher pressure, the actual densities found in star-forming regions are not high enough to cause a significant pressure increase.

The only parts of the interstellar medium with high enough density and low enough temperature for stars to form are the dark

nebulae. Many of these nebulae, such as the Horsehead Nebula (see Figure 18-2) and Barnard 86 (see Figure 18-4), were discovered and catalogued around 1900 by Edward Barnard and are known as **Barnard objects.** (The Horsehead Nebula is also known as Barnard 33.) Other relatively small, nearly spherical dark nebulae are known as **Bok globules,** after the Dutch-American astronomer Bart Bok, who first called attention to them during the 1940s (**Figure 18-9**). A Bok globule resembles the inner core of a Barnard object with the outer, less dense portions stripped away. The density of the gas and dust within a Barnard object or a Bok globule is indeed quite high by cosmic standards, in the range from 100 to 10,000 particles per cm^3; by comparison, most of the interstellar medium contains only 0.1 to 20 particles per cm^3. Radio emissions from molecules within these clouds indicate that their internal temperatures are very low, only about 10 K.

A typical Barnard object contains a few thousand solar masses of gas and dust spread over a volume roughly 10 pc (30 ly) across; a typical Bok globule is about one-tenth as large. The chemical composition of this material is the standard "cosmic abundance" of about 74% (by mass) hydrogen, 25% helium, and 1% heavier elements (review Figure 8-4). Within these clouds, the densest portions can contract under their own mutual gravitational attraction and form clumps called **protostars.** Each protostar will eventually evolve into a main-sequence star. Because dark nebulae contain many solar masses of material, it is possible for a large number of protostars to form out of a single such nebula. Thus, we can think of dark nebulae as "stellar nurseries."

> In the first stage of star formation, a protostar coalesces and contracts due to the mutual gravitational attraction of its parts.

The Evolution of a Protostar

In the 1950s, astrophysicists such as Louis Henyey in the United States and Chushiro Hayashi in Japan performed calculations that enabled them to describe the earliest stages of a protostar. At first, a protostar is merely a cool blob of gas several times larger than our solar system. The pressure inside the protostar is too low to support all this cool gas against the mutual gravitational attraction of its parts, so the protostar collapses. As the protostar collapses, gravitational energy is converted into thermal energy, making the gases heat up and start glowing. (We discussed this process, called Kelvin–Helmholtz contraction, in Section 8-4 and Section 16-1.) Energy from the interior of the protostar is transported outward by convection, warming its surface.

After only a few thousand years of gravitational contraction, the protostar's surface temperature reaches 2000 to 3000 K and shines brightly by emitting blackbody radiation (see Section 5-3). At this point, the protostar is still quite large, so its glowing gases produce substantial luminosity. (The greater a star's radius and surface temperature, the greater its luminosity. You can review the details in Section 17-6 and Box 17-4.) For example, after only 1000 years of contraction, a protostar of 1 solar mass (1 $M_\odot$) is 20 times larger in radius than the Sun and has 100 times the Sun's luminosity. Unlike the Sun, however, the luminosity of a young protostar is not the result of thermonuclear fusion, because the protostar's core is not yet hot enough for fusion reactions to begin. Instead, the radiated energy comes exclusively from the heating of the protostar as it contracts. The development of a protostar is illustrated in steps 1 through 4 of the *Cosmic Connections* figure on page 576.

In order to determine the conditions inside a contracting protostar, astrophysicists use computers to solve equations similar to those used for calculating the structure of the Sun (see Section 16-2). The results tell us how the protostar's luminosity and surface temperature change at various stages in its contraction. This information, when plotted on a Hertzsprung–Russell (H-R) diagram, provides a protostar's **evolutionary track**. The track shows us how the protostar's appearance changes because of changes in its interior. These theoretical tracks agree quite well with actual observations of protostars.

CAUTION!

When astronomers refer to a star "following an evolutionary track" or "moving on the H-R diagram," they are not talking about how the star physically moves through space. Over a star's lifetime, its luminosity, surface temperature, or both change. Hence, the point that represents the star in an H-R diagram changes its position.

Figure 18-10 shows evolutionary tracks for protostars of seven different masses, ranging from 0.5 $M_\odot$ to 15 $M_\odot$. Because protostars are relatively cool when they begin to shine at visible wavelengths, these tracks all begin near the right (low-temperature)

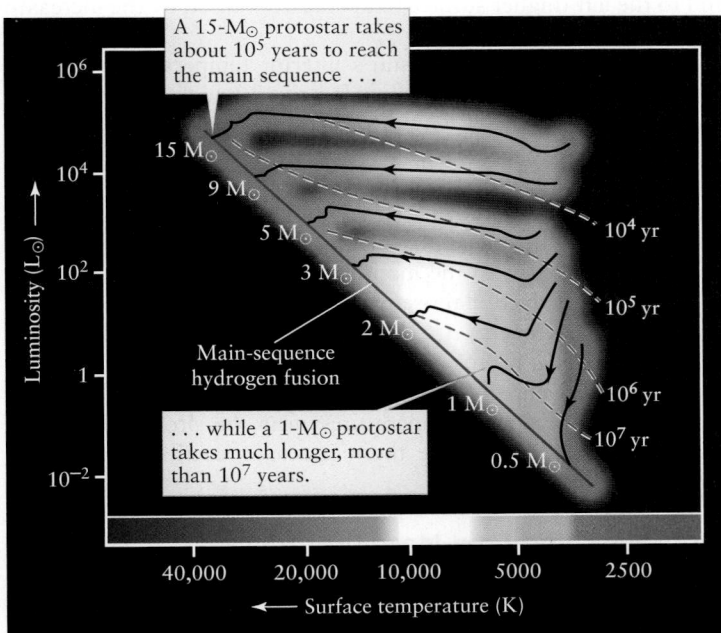

FIGURE 18-10

Pre-Main-Sequence Evolutionary Tracks in an H-R Diagram
As a protostar evolves, its luminosity and surface temperature both change. The tracks shown here depict these changes for protostars of seven different masses. Each dashed red line shows the age of a protostar when its evolutionary track crosses that line. (We will see in Section 18-5 that protostars lose quite a bit of mass as they evolve: The mass shown for each track is the value when the protostar finally settles down as a main-sequence star.)

side of the H-R diagram. The subsequent evolution is somewhat different, however, depending on the protostar's mass. Note that the greater a star's mass, the more rapidly it moves along its evolutionary track: A 15-$M_\odot$ protostar takes only about 10^5 years to reach the main sequence, while a 1-$M_\odot$ protostar takes more than 100 times longer (more than 10^7 years).

CONCEPT CHECK 18-3

Is it possible for a protostar to shine brighter than the main-sequence star it becomes?

CONCEPT CHECK 18-4

About how long does it take for a 2-$M_\odot$ protostar to reach the main sequence?
Answers appear at the end of the chapter.

Observing Protostars

Observing the evolution of a protostar can be quite a challenge. Indeed, while Figure 18-10 shows that young protostars are much more luminous than the Sun, it is quite unlikely that you have ever seen a protostar shining in the night sky. The reason is that protostars form within clouds that contain substantial amounts of interstellar dust. The dust in a protostar's immediate surroundings, called its **cocoon nebula**, absorbs the vast amounts of visible light emitted by the protostar and makes it very hard to detect using visible wavelengths.

Protostars can be seen, however, using infrared wavelengths. Because a cocoon nebula absorbs so much energy from the protostar that it surrounds, the nebula becomes heated to a few hundred kelvins. The warmed dust then reradiates its thermal energy at infrared wavelengths, to which the cocoon nebula is relatively transparent. So, by using infrared telescopes, astronomers can see protostars within the "stellar nursery" of a dark nebula.

Figure 18-11 shows visible-light and infrared views of one such stellar nursery. The visible-light view (Figure 18-11a) shows a dark, dusty nebula (or Bok globule) that appears completely opaque. The infrared image (Figure 18-11b) allows us to see through the dust, revealing a newly formed protostar within the dark nebula. The properties of this protostar agree well with the theoretical ideas outlined in this section.

18-4 Protostars evolve into main-sequence stars

Evolutionary tracks in Figure 18-10 show how a protostar matures into a star as its gases contract. The details of this evolution depend on the star's mass. To follow these details, keep in mind the basic principle that a star's luminosity, radius, and surface temperature are intimately related. The luminosity, which is the total energy output per second, depends on how hot and how

The course of a protostar's evolution depends on its mass.

(a) A dark nebula R I **V** U X G   (b) A hidden protostar within the dark nebula R **I** V U X G

FIGURE 18-11

Revealing a Hidden Protostar (a) This visible-light view shows a dark nebula (or Bok globule) called L1014 in the constellation Cygnus (the Swan). No stars are visible within the nebula. **(b)** The Spitzer Space Telescope was used to make this false-color infrared image of the outlined area in (a). The bright red-yellow spot is a protostar within the dark nebula. (NASA/JPL-Caltech/N. Evans [University of Texas at Austin])

large the star is. Specifically, the luminosity is proportional to the square of a star's radius and to the fourth power of its surface temperature (see Section 17-6).

A One-Solar-Mass Protostar

For a protostar with the same mass as our Sun (1 $M_\odot$), the outer layers are relatively cool and quite opaque (for the same reasons that the Sun's photosphere is opaque; see the discussion at the end of Section 16-5). Due to frequent scattering or absorption, light does not pass easily through an opaque material. This means that energy released from the shrinking inner layers in the form of radiation cannot easily reach the protostar's surface. Instead, energy flows outward by the slower and less effective method of convection. The result is that for a contracting 1-$M_\odot$ protostar, the surface temperature stays roughly constant, the luminosity decreases as the radius decreases, and the protostar's evolutionary track initially moves downward on the H-R diagram in Figure 18-10.

Although its surface temperature changes relatively little, the internal temperature of the shrinking protostar increases. After a time, the interior becomes more ionized, which makes it less opaque. Energy is then conveyed outward by radiation in the interior and by convection in the opaque outer layers, just as in the present-day Sun (see Section 16-2, especially Figure 16-4). This makes it easier for energy to escape from the protostar, so the luminosity—the rate at which energy is emitted from the protostar's surface—increases. As a result, the evolutionary track for a 1-$M_\odot$ protostar in Figure 18-10 bends upward (higher luminosity)

and to the left (higher surface temperature, caused by the increased energy flow).

In time, the 1-$M_\odot$ protostar's interior temperature reaches a few million kelvins, hot enough for thermonuclear reactions to begin converting hydrogen into helium. As we saw in Section 16-1 and Box 16-1, these reactions release enormous amounts of energy. Eventually, these reactions provide enough heat and internal pressure to stop the star's gravitational contraction, and hydrostatic equilibrium is reached. The protostar's evolutionary track has now led it to the main sequence, and the protostar has become a full-fledged main-sequence star.

In summary, energy transport is initially dominated by convection as a 1-$M_\odot$ protostar shrinks and dims. Next, the protostar transports more energy through radiation and its surface temperature increases. Finally, temperatures near the protostar's center climb high enough to cause thermonuclear reactions, and a star is born.

High-Mass and Low-Mass Protostars

More massive protostars evolve a bit differently. If its mass is more than about 4 $M_\odot$, a protostar contracts and heats more rapidly, and hydrogen fusion begins quite early. As a result, the luminosity quickly stabilizes at nearly its final value, while the surface temperature continues to increase as the star shrinks. Thus, the evolutionary tracks of massive protostars traverse the H-R diagram roughly horizontally (signifying approximately constant luminosity) in the direction from right to left (from low to high surface temperature). You can see this most easily for the 9-$M_\odot$ and 1-$M_\odot$ evolutionary tracks in Figure 18-10.

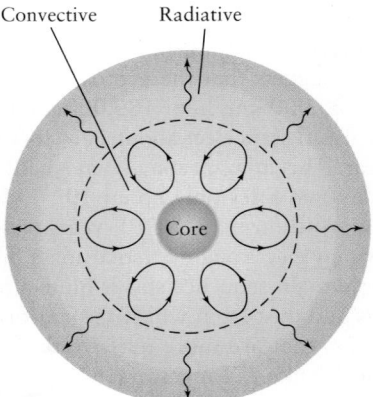

(a) **Mass more than about 4 M⊙:** Energy flows by convection in the inner regions and by radiation in the outer regions.

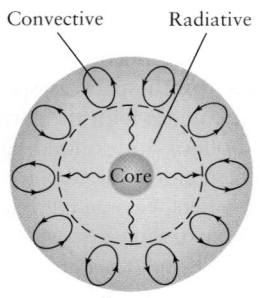

(b) **Mass between about 4 M⊙ and 0.4 M⊙:** Energy flows by radiation in the inner regions and by convection in the outer regions.

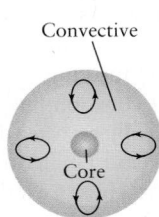

(c) **Mass less than 0.4 M⊙:** Energy flows by convection throughout the star's interior.

FIGURE 18-12

Main-Sequence Stars of Different Masses Stellar models show that when a protostar evolves into a main-sequence star, its internal structure depends on its mass. *Note:* The three stars shown here are not drawn to scale. Compared with a 1-M⊙ main-sequence star like that shown in (b), a 6-M⊙ main-sequence star like that in (a) has more than 4 times the radius, and a 0.2-M⊙ main-sequence star like that in (c) has only one-third the radius.

Greater mass means greater pressure and temperature in the interior, which means that a massive star has an even larger temperature difference between its core and its outer layers than the Sun. This difference causes convection deep in the interior of a massive star (**Figure 18-12**). By contrast, a massive star's outer layers are of such low density that energy flows through them more easily by radiation than by convection. Therefore, main-sequence stars with masses more than about 4 M⊙ have convective interiors but radiative outer layers (Figure 18-12a). By contrast, less massive main-sequence stars such as the Sun have radiative interiors and convective outer layers (Figure 18-12b).

The internal structure is also different for main-sequence stars of very low mass. When such a star forms from a protostar, the interior temperature is never high enough to fully ionize the interior. The interior remains too opaque for radiation to flow efficiently, so energy is transported by convection throughout the volume of the star (Figure 18-12c).

Arriving on the Main Sequence

The main sequence represents stars in which thermonuclear reactions are converting hydrogen into helium, and all of the protostar evolutionary tracks in Figure 18-10 end on the main sequence. For most stars, these thermonuclear reactions are part of a stable situation. For example, our Sun will remain on or very near the main sequence with a roughly constant temperature and luminosity, quietly fusing hydrogen into helium at its core, for a total of some 10^{10} years. The point along the main sequence where each evolutionary track ends depends on the star's mass. The most massive stars are the most luminous and their evolutionary tracks end at the upper left of the main sequence, while the least massive stars are the least luminous and their evolutionary tracks end at the lower right of the main sequence. The formation of a main sequence star is illustrated in step 5 of the *Cosmic Connections* figure on page 576. Because the most massive protostars evolve the fastest, many massive blue stars are seen in young clusters, as in **Figure 18-13**.

The theory of how protostars evolve helps explain why the main sequence has both an upper mass limit and a lower mass limit. As we saw in Section 17-5, protostars less massive than about 0.08 M⊙ can never develop the necessary pressure and temperature to start hydrogen fusion in their cores. Instead, such "failed stars" end up as *brown dwarfs*, which shine faintly by Kelvin–Helmholtz contraction (see Figure 17-13).

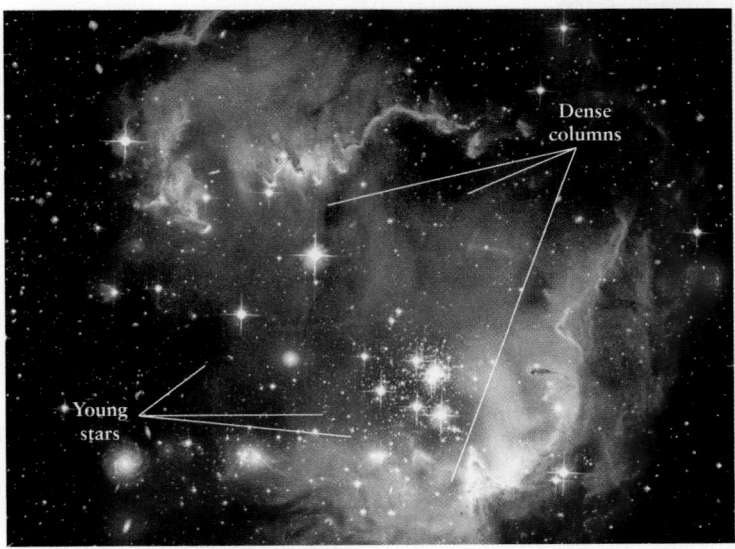

FIGURE 18-13 R I V U X G

Stellar Nursery with Young Main-Sequence Stars Before its first stars formed, this cloud of gas and dust would have originally looked like the Bok globule in Figure 18-9. With star formation, the nebula undergoes significant change. The most massive protostars reach the main sequence first. These are the bright blue stars near the center. Ultraviolet light from these bright stars also causes the gas to glow as an emission nebula (see Figure 18-3). Their intense emission of light erodes the surrounding gas, creating the cavity that hollowed out this nebula. However, the densest regions of the nebula block some of this emission, leaving small dense columns of gas around the inner edge that point approximately toward the center of the spherical cavity. Within these dense structures, compression of gas will lead to a new generation of stars. This nebula is in a nearby galactic neighbor called the Small Magellanic Cloud. (NASA/CXC/JPL-Caltech/STScI)

On the other hand, protostars with masses greater than about 200 solar masses also do not become main-sequence stars. Such a protostar rapidly becomes very luminous, resulting in tremendous internal pressures. This pressure is so great that it overwhelms the effects of gravity, expelling the outer layers into space and disrupting the star. Main-sequence stars therefore have masses between about 0.08 and 200 $M_\odot$, although the high-mass stars are extremely rare.

> **CAUTION!**
>
> A few words of caution are in order here. First, while the evolutionary tracks of protostars begin in the red giant region of the H-R diagram (the upper right), protostars are *not* red giants. As we will see in Chapter 19, red giant stars represent a stage in the evolution of stars that comes *after* being a main-sequence star. Second, it is worth remembering that stars live out most of their lives on the main sequence, after only a relatively brief period as protostars. A 15-$M_\odot$ protostar takes only 20,000 years to become a main-sequence star, and a 1-$M_\odot$ protostar takes about 2×10^7 years. By contrast, the Sun has been a main-sequence star for about 4.56×10^9 years. By astronomical standards, pre-main-sequence stars are quite transitory. Furthermore, during the last 4.56×10^9 years, the Sun has moved very little from its spot on the main sequence.

CONCEPT CHECK 18-5

A 9-$M_\odot$ protostar can have a somewhat constant luminosity, and this holds up even when it reaches the main sequence (see Figure 18-10). Does that mean it has the same energy source while it is a protostar as when it is a main-sequence star?

Answer appears at the end of the chapter.

18-5 During the birth process, stars both gain and lose mass

After reading the previous section, you may think that a main-sequence star forms simply by collapsing inward. In fact, much of the material of a cold, dark nebula is ejected into space and never incorporated into stars. As it is ejected, this material may help sweep away the dust surrounding a young star, making the star observable at visible wavelengths.

> Before settling down, protostars eject more than half of their initial mass into space.

T Tauri Stars

Mass ejection into space is a hallmark of **T Tauri stars**. These objects are protostars with emission lines as well as absorption lines in their spectra and whose luminosity can change irregularly on timescales of a few days. The namesake of this class of stars, T Tauri, is a protostar in the constellation Taurus (the Bull).

T Tauri stars have final masses less than about 3 $M_\odot$ and ages around 10^6 years, so in an H-R diagram such as Figure 18-10, they appear above the right-hand end of the main sequence. (The stellar masses shown in Figure 18-10 are those of the final,

main-sequence stars.) The emission lines show that these protostars are surrounded by a thin, hot gas. The Doppler shifts of these emission lines suggest that the protostars eject gas at speeds around 80 km/s (300,000 km/h, or 180,000 mi/h).

On average, T Tauri stars eject about 10^{-8} to 10^{-7} solar masses of material per year. This amount may seem small, but by comparison the present-day Sun loses only about 10^{-14} $M_\odot$ per year. The T Tauri phase of a protostar may last 10^7 years or so, during which time the protostar may eject roughly a solar mass of material. Thus, the mass of the final main-sequence star is quite a bit less than that of the cloud of gas and dust from which the star originated.

Young protostars that are more massive than about 3 $M_\odot$ do not vary in luminosity like T Tauri stars. They do lose mass, however, because the pressure of radiation at their surfaces is so strong that it blows gas into space (Figure 18-14).

Bipolar Outflows and Herbig–Haro Objects

ANIMATION 18.1 In the early 1980s, it was discovered that many young stars, including T Tauri stars, lose mass by ejecting gas along two narrow, oppositely directed jets—a phenomenon called **bipolar outflow**. As this material is ejected into space at speeds of several hundred kilometers per second, it collides with the surrounding interstellar medium and produces knots of hot, ionized gas that glow with an emission-line spectrum. These glowing knots are called **Herbig–Haro objects** after the two astronomers, George Herbig in the United States and Guillermo Haro in Mexico, who discovered them independently. Thus, a Herbig–Haro object is the observational manifestation of bipolar outflow around a star. Figure 18-15 is a Hubble Space Telescope (HST) image of the Herbig–Haro objects HH 1 and HH 2, which are produced by the two jets from a single young star in the constellation Orion. Herbig–Haro objects like these change noticeably in position, size, shape, and brightness from year to year, indicating the dynamic character of bipolar outflows.

Observations suggest that most protostars eject material in the form of jets at some point during their evolution. These bipolar outflows are very short-lived by astronomical standards, a mere 10^4 to 10^5 years, but they are so energetic that they typically eject into space more mass than ends up in the final protostar.

FIGURE 18-14

Mass Loss from a Supermassive Star The Quintuplet Cluster is 25,000 ly from Earth. *Inset:* Within the cluster is one of the brightest known stars, called the Pistol Star. Astronomers calculate that the Pistol Star formed nearly 3 million years ago and originally had 100–200 $M_\odot$. The structure of the gas cloud suggests the star ejected the gas we see in two episodes, 6000 and 4000 years ago. The gas from any previous ejections is so thinly spread now that we cannot see it. The nebula shown in the inset is more than 1.25 pc (4 ly) across—it would stretch from the Sun nearly to the closest star, Proxima Centauri. (NASA and Don F. Figer [UCLA]; NASA)

HH 2

HH 1

Protostar
(hidden by dark, dusty nebula)

FIGURE 18-15 R I V U X G

Bipolar Outflow and Herbig–Haro Objects The two bright knots of glowing, ionized gas called HH 1 and HH 2 are Herbig–Haro objects. They are created when fast-moving gas ejected from a protostar slams into the surrounding interstellar medium, heating the gas to high temperature. HH 1 and HH 2 are 0.34 pc (1.1 ly) apart and lie 470 pc (1500 ly) from Earth in the constellation Orion. (J. Hester [ASU], WFPC2 Team, NASA)

In one object, water has been observed flowing out in the jets along with other material. The amount of water ejected each second is about 100 million times the amount of water flowing out of the Amazon River each second. The water in the jets moves at about 80 times the speed of a machine gun's bullet and eventually slams into surrounding material about 5000 au away.

Accretion Disks

Protostars slowly add mass to themselves at the same time that they rapidly eject it into space. In fact, the two processes are related. As a protostar's nebula contracts, it spins faster and flattens into a disk with the protostar itself at the center. The same flattening took place in the solar nebula from which the Sun and planets formed (see Section 8-4). Particles orbiting the protostar within this disk collide with each other, causing them to lose energy, spiral inward onto the protostar, and add to the protostar's mass. This process is called **accretion**, and the disk of material being added to the protostar in this way is called a **circumstellar accretion disk**. Figure 18-16 is an edge-on view of a circumstellar accretion disk, showing two oppositely directed jets emanating from a point at or near the center of the disk (where the protostar is located).

What causes some of the material in the disk to be blasted outward in a pair of jets? One model involves the magnetic field of the dark nebula in which the star forms (Figure 18-17). As material in the circumstellar accretion disk falls inward, it drags the magnetic field lines along with it. (We saw in Section 16-9 how a similar mechanism may

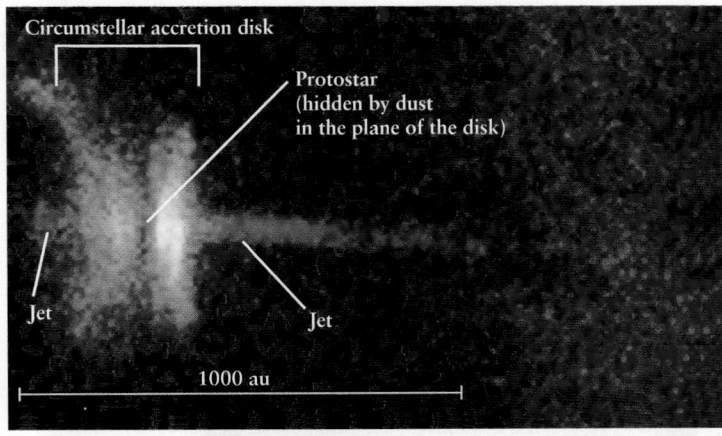

Circumstellar accretion disk

Protostar
(hidden by dust
in the plane of the disk)

Jet

Jet

1000 au

FIGURE 18-16 R I V U X G

A Circumstellar Accretion Disk and Jets This false-color image shows a T Tauri star surrounded by an accretion disk, which we see nearly edge on. Red denotes emission from ionized gas, while green denotes starlight scattered from dust particles in the disk. The midplane of the accretion disk is so dusty and opaque that it appears dark. Two oppositely directed jets flow away from the star, perpendicular to the disk and along the disk's rotation axis. This star, Herbig–Haro object HH 30, lies 140 pc (460 ly) from Earth. (NASA/Hubble/STScI)

explain the Sun's 22-year cycle.) Parts of the disk at different distances from the central protostar orbit at different speeds, and this can twist

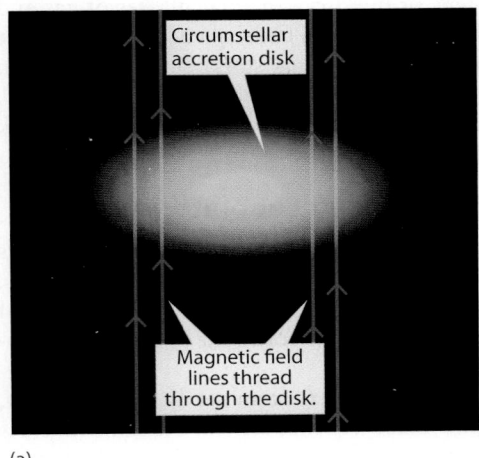

Circumstellar accretion disk

Magnetic field lines thread through the disk.

(a)

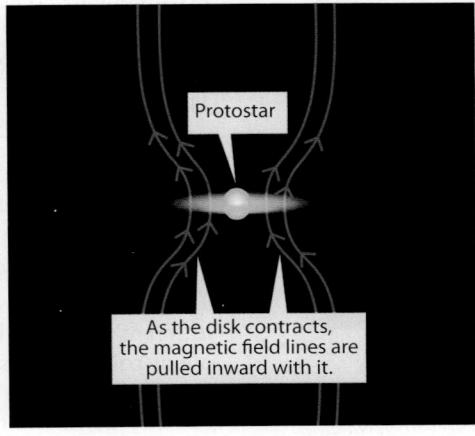

Protostar

As the disk contracts, the magnetic field lines are pulled inward with it.

(b)

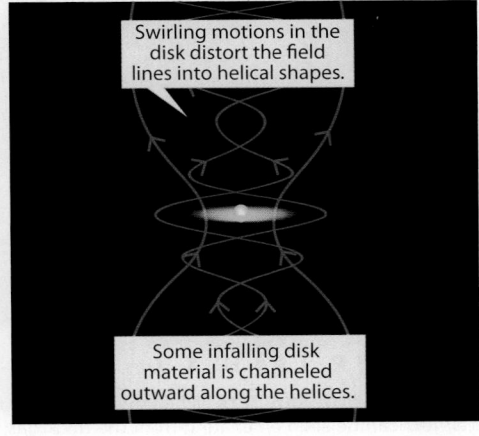

Swirling motions in the disk distort the field lines into helical shapes.

Some infalling disk material is channeled outward along the helices.

(c)

FIGURE 18-17

A Magnetic Model for Bipolar Outflow (a) Observations suggest that circumstellar accretion disks are threaded by magnetic field lines, as shown here. **(b)** Magnetic field lines move with the material they thread. **(c)** The contraction and rotation of the disk make the magnetic field lines distort

and twist into helices. These helices steer some of the disk material into jets that stream perpendicular to the plane of the disk, as in Figure 18-16. (Source: Alfred T. Kamajian/Thomas P. Ray, "Fountain of Youth: Early Days in the Life of a Star," *Scientific American*, August 2000, p. 17.)

the magnetic field lines into two helix shapes, one on each side of the disk. The helices then act as channels that guide initially infalling material *away* from the protostar, forming two opposing jets. While it's clear that jets do form, the details of their formation are still uncertain. The star-forming region in **Figure 18-18** shows two spectacular outflows from jets.

Protoplanetary Disks

ANIMATION 18.2 In the 1990s, astronomers using the HST discovered many examples of disks around newly formed stars in the Orion Nebula (see Figure 8-8), one of the most prominent star-forming regions in the northern sky. These disks are called **protoplanetary disks**, or **proplyds**, and they surround young stars. As the name suggests, protoplanetary disks are thought to contain the material from which planets form around stars, which is visible in **Figure 18-19**. They are what remains of a circumstellar accretion disk after much of the material has either fallen onto the star or been ejected by bipolar outflows. The formation of a protoplanetary disk is illustrated in step 8 of the *Cosmic Connections* figure on page 576.

FIGURE 18-19 R I **V** U X G

Protoplanetary Disk with a Forming Planet (a) The ALMA telescope captured this protoplanetary disk around the young star HL Tauri using millimeter wavelengths. The darker gaps are thought to result from material getting pulled into a forming planet in the gap. The disk is about 235 au across. **(b)** Using radio wavelengths, the Very Large Array (VLA) was able to image the inner portion of the disk. Here, a clump is observed to be between 3 and 8 Earth masses in size. This clump is thought to be a forming planet, but it is so early in the process that it hasn't yet cleared out its orbital region to create a dark gap. (Carrasco-Gonzalez, et al.; Bill Saxton, NRAO/AUI/NSF)

CONCEPT CHECK 18-6

Protostars gain and lose mass on their way to the main sequence, but which effect is dominant in the formation of main-sequence stars with less than 3 $M_\odot$?

Answer appears at the end of the chapter.

18-6 Young star clusters give insight into star formation and evolution

Dark nebulae contain tens or hundreds of solar masses of gas and dust, enough to form many stars. As a consequence, these nebulae tend to form groups or **clusters** of young stars. One cluster still in **TUTORIAL 18.2** the act of star formation is M16, shown in **Figure 18-20**. As a cluster ages, intense radiation transforms the dark nebula, producing a reddish H II region as in **Figure 18-21**.

Star Clusters as Evolutionary Laboratories

In addition to being objects of great natural beauty, star clusters give us a unique way to compare the evolution of different stars. The reason is that clusters typically include stars with a range of different masses, all of which began to form out of the parent nebula at roughly the same time.

All the stars in a cluster may begin to form nearly simultaneously, but they do not all become main-sequence stars at the same time. As you can see from their evolutionary tracks (Figure 18-10), high-mass stars evolve more rapidly than low-mass stars. The more massive the protostar, the sooner it develops the central pressures and temperatures needed for steady hydrogen fusion to begin, thus joining the main sequence.

FIGURE 18-18 R I **V** U X G

Star Formation and Jets While the protostars aren't visible, some of their jets can be seen emanating from the thick gas and dust from which the stars are forming. As new stars form, their strong winds and intense ultraviolet radiation blow away lower-density material, leaving behind distinct regions with greater density. The denser material makes an ideal star-forming region to produce stars; the first few bright stars are already visible. This is only part of the 30 ly-wide Carina Nebula, which is about 7500 ly away. (NASA, ESA, and M. Livio and the Hubble 20th Anniversary Team [STScI])

1. This emission nebula (about 2200 pc away and about 20 pc across) surrounds the star cluster M16.

2. Star formation is still taking place within this dark, dusty nebula.

3. Hot, luminous stars (beyond the upper edge of the closeup image) emit ultraviolet radiation: This makes the dark nebula evaporate, leaving these pillars.

4. At the tip of each of these "fingers" is a cocoon nebula containing a young star.

5. Eventually the cocoon nebulae evaporate, revealing the stars.

FIGURE 18-20 R I **V** U X G

ANIMATION 18.3 **Formation of a Star Cluster** The star cluster M16 is thought to be no more than 800,000 years old, and star formation is still taking place within adjacent dark, dusty globules. Massive young stars in the cluster power red emission lines of the H II region (called the Eagle Nebula for its shape). The inset zooms in on three dense, cold pillars of gas with ongoing star formation. The pillar at the upper left extends about 0.3 pc (1 ly) from base to tip, and each of its "fingers" is somewhat broader than our entire solar system. (Johannes Schedler [Panther Observatory]; NASA/ESA/STScI/Arizona State University)

Upon reaching the main sequence, *high-mass* protostars become hot, ultraluminous stars of spectral types O and B. As we saw in Section 18-2, these types of stars have ultraviolet radiation that ionizes the surrounding interstellar medium to produce an H II region. Figure 18-20 shows such an H II region, called the Eagle

FIGURE 18-21 R I **V** U X G

A Mature Star Cluster NGC 3603 is one the closest star clusters at only 20,000 ly away. Star formation has ended and ultraviolet light from its stars power the largest visible reddish H II region in our galaxy. (NASA, ESA, and the Hubble Heritage)

(a) The star cluster NGC 2264

FIGURE 18-22 R I $\mathbf{V}$ U X G

A Young Star Cluster and Its H-R Diagram **(a)** This image shows the young star cluster NGC 2264 in a reddish H II region known as the Fox Fur Nebula. Stars within the cluster provide the ultraviolet light that powers this emission nebula. The cluster lies about 800 pc (2600 ly) from Earth. **(b)** Each dot plotted on this H-R diagram represents a star in NGC 2264

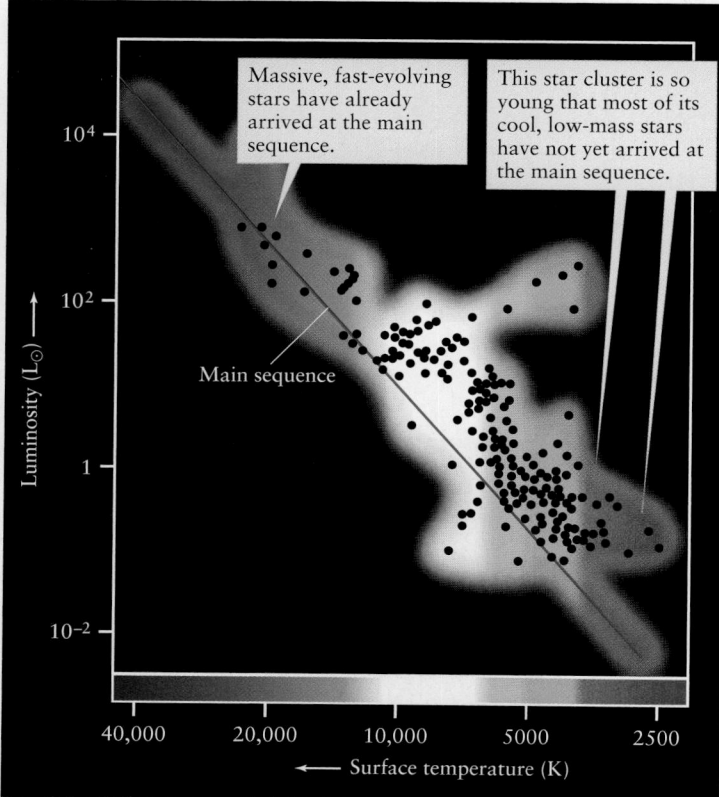

(b) An H-R diagram of the stars in NGC 2264

whose luminosity and surface temperature have been determined. This star cluster probably started forming only 2 million years ago, and its lower-mass stars have not yet reached the main sequence. (a: © 2004–2013 R. Jay GaBany, Cosmotography.com)

Nebula, surrounding the young star cluster M16. A few hundred thousand years ago, this region of space would have had a far less dramatic appearance. It was then a dark nebula, with protostars just beginning to form. Over the intervening millennia, mass ejection from these evolving protostars swept away the obscuring dust. The exposed young, hot stars heated the relatively thin remnants of the original dark nebula, creating the H II region that we see today.

When the most massive protostars to form out of a dark nebula have reached the main sequence, other *low-mass* protostars are still evolving nearby within their dusty cocoons. The evolution of these low-mass stars can be disturbed by their more massive neighbors. As an example, the inset in Figure 18-20 is a close-up of part of the Eagle Nebula. Within these opaque pillars of cold gas and dust, protostars are still forming. At the same time, however, the pillars are being eroded by intense ultraviolet light from hot, massive stars that have already shed their cocoons. As each pillar evaporates, the embryonic stars within have their surrounding material stripped away prematurely, limiting the total mass that these stars can accrete.

Analyzing Young Clusters Using H-R Diagrams

Star clusters tell us still more about how high-mass and low-mass stars evolve. **Figure 18-22a** shows the young star cluster NGC 2264

and its associated emission nebula. Astronomers have measured each star's apparent brightness and color ratio. Knowing the distance to the cluster, they have deduced the luminosities and surface temperatures of the stars (see Section 17-2 and Section 17-4). Figure 18-22b shows all these stars in an H-R diagram. Note that the hottest and most massive stars, with surface temperatures around 20,000 K, are on the main sequence. Stars cooler than about 10,000 K, however, have not yet quite arrived at the main sequence. These are less massive stars in the final stages of pre–main-sequence contraction and are just now beginning to ignite thermonuclear reactions at their centers. To find the ages of these stars, we can compare Figure 18-22b with the theoretical calculations of protostar evolution in Figure 18-10. It turns out that this particular cluster is probably about 2 million years old.

Figure 18-23a shows another young star cluster called the Pleiades. The photograph shows gas that must once have formed an H II region around this cluster and has dissipated into interstellar space, leaving only traces of dusty material that forms reflection nebulae around the

> The H-R diagram of a young cluster reveals how much time has elapsed since its stars began to form.

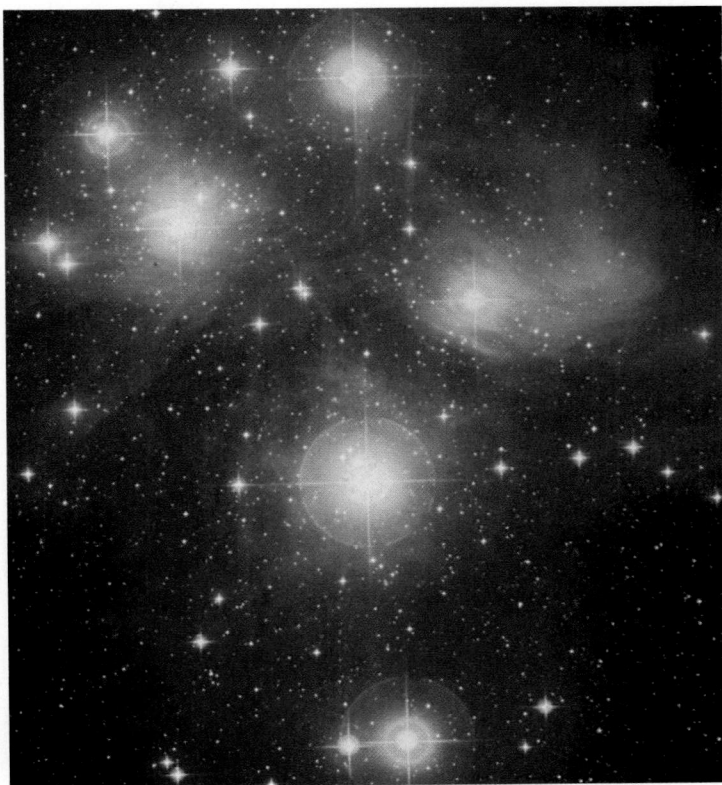

(a) The Pleiades star cluster

FIGURE 18-23 R I **V** U X G

The Pleiades and Its H-R Diagram (a) The Pleiades star cluster is 117 pc (380 ly) from Earth in the constellation Taurus and can be seen with the naked eye. (b) Each dot plotted on this H-R diagram represents a star in the Pleiades whose luminosity and surface temperature have been measured.

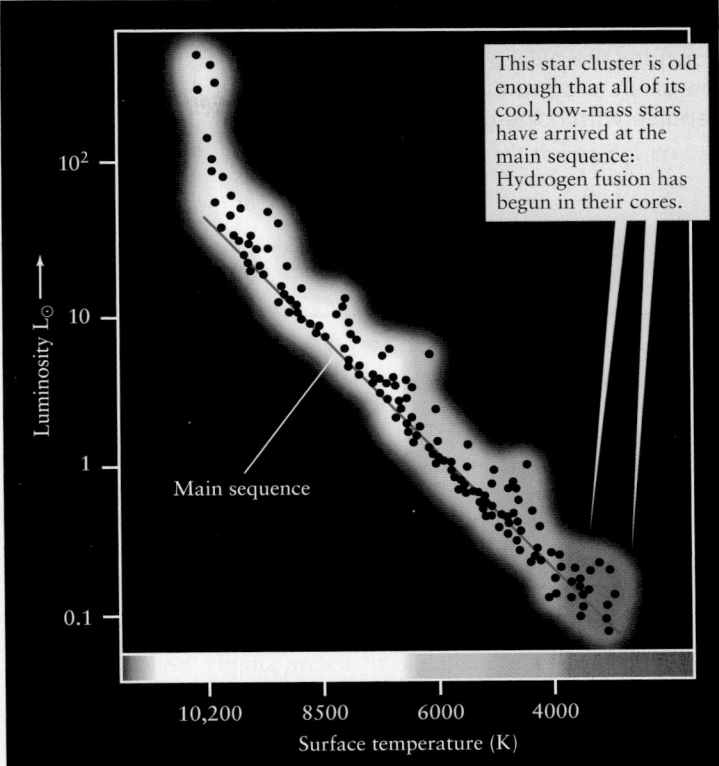

(b) An H-R diagram of the stars in the Pleiades

> This star cluster is old enough that all of its cool, low-mass stars have arrived at the main sequence: Hydrogen fusion has begun in their cores.

Main sequence

(*Note:* The scales on this H-R diagram are different from those in Figure 18-20b.) The Pleiades is about 50 million (5×10^7) years old. (© Australian Astronomical Observatory/David Malin Images)

cluster's stars. This implies that the Pleiades must be older than NGC 2264, the cluster in Figure 18-22a, which is still surrounded by an H II region. The H-R diagram for the Pleiades in Figure 18-23b bears out this idea. In contrast to the H-R diagram for NGC 2264, nearly all the stars in the Pleiades are on the main sequence. The cluster's age is about 50 million years, which is how long it takes for the least massive stars to finally begin hydrogen fusion in their cores.

CAUTION!

Note that the data points for the most massive stars in the Pleiades (at the upper left of the H-R diagram in Figure 18-23b) lie above the main sequence. This is *not* because these stars have yet to arrive at the main sequence. Rather, these stars were the first members of the cluster to arrive at the main sequence some time ago and are now the first members to leave it. They have used up the hydrogen in their cores, so the steady process of core hydrogen fusion that characterizes main-sequence stars cannot continue. In Chapter 19, we will see why massive stars spend a rather short time as main-sequence stars and will study what happens to stars after the main-sequence phase of their lives.

There are two ways to get the ages of clusters with H-R diagrams. What we have seen in Figure 18-22 and Figure 18-23 is for *young* clusters. The age is estimated by figuring out how long it takes the less massive, slower-evolving protostars to *arrive* at the main sequence. In the next chapter (Section 19-4), we will estimate the ages of *old* clusters by determining how long it has taken more massive stars to *leave* the main sequence.

A loose collection of stars such as NGC 2264 or the Pleiades is referred to as an **open cluster** (or *galactic cluster*, since such clusters are usually found in the plane of the Milky Way Galaxy). Open clusters possess barely enough mass to hold themselves together by gravitation. Occasionally, a star moving faster than average will escape, or "evaporate," from an open cluster. Indeed, by the time the stars are a few billion years old, they may be so widely separated that a cluster no longer exists.

If a group of stars is gravitationally unbound from the very beginning—that is, if the stars are moving away from one another so rapidly that gravitational forces cannot keep them together—then the group is called a **stellar association**. Because young stellar associations are typically dominated by luminous O and B main-sequence stars, they are also called **OB associations.** The image that opens this chapter shows part of an OB association in the southern constellation Ara (the Altar).

CONCEPT CHECK 18-7

Consider the pillars shown in the inset of Figure 18-20. Are the pillars eroded away by the stars forming in their cocoon nebulae within the pillars?

Answer appears at the end of the chapter.

18-7 Star birth can begin in giant molecular clouds

We have seen that star formation takes place within dark nebulae. But where within our galaxy are these dark nebulae found? Does star formation take place everywhere within the Milky Way, or only in certain special locations? The answers to such questions can enhance our understanding of star formation and of the nature of our home galaxy.

Exploring the Interstellar Medium at Millimeter Wavelengths

Dark nebulae are a challenge to locate simply because they *are* dark—they do not emit visible light. Nearby dark nebulae can be seen silhouetted against background stars or reddish H II regions (see Figure 18-2), but sufficiently distant dark nebulae are impossible to see in contrast with background visible light because of interstellar extinction from dust grains. They can, however, be detected using longer-wavelength radiation that can pass unaffected through interstellar dust. In fact, dark nebulae actually emit radiation at millimeter wavelengths.

Such emission takes place because in the cold depths of interstellar space, atoms combine to form molecules. The laws of quantum mechanics predict that just as electrons within atoms can occupy only certain specific energy levels (see Section 5-8), molecules can vibrate and rotate only at certain specific rates. When a molecule goes from one vibrational state or rotational state to another, it either emits or absorbs a photon. (In the same way, an atom emits or absorbs a photon as an electron jumps from one energy level to another.) Most molecules are strong emitters of radiation with wavelengths of around 1 to 10 millimeters (mm). Consequently, observations with radio telescopes tuned to millimeter wavelengths make it possible to detect interstellar molecules of different types. More than 200 different kinds of molecules have so far been discovered in interstellar space, and the list is constantly growing.

Hydrogen is by far the most abundant element in the universe. Unfortunately, in cold nebulae, much of it is in a molecular form (H_2) that is difficult to detect. The reason is that the hydrogen molecule is symmetric, with two atoms of equal mass joined together, and such molecules do not emit many photons at radio frequencies. In contrast, asymmetric molecules that consist of two atoms of unequal mass joined together, such as carbon monoxide (CO), are easily detectable at radio frequencies. When a carbon

> Observing the galaxy at millimeter wavelengths reveals the cold gas that spawns new stars.

monoxide molecule makes a transition from one rate of rotation to another, it emits a photon at a wavelength of 2.6 mm or shorter.

The ratio of carbon monoxide to hydrogen in interstellar space is reasonably constant: For every CO molecule, there are about 10,000 H_2 molecules. As a result, carbon monoxide is an excellent "tracer" for molecular hydrogen gas. Wherever astronomers detect strong emission from CO, they know molecular hydrogen gas must be abundant.

Giant Molecular Clouds

The first systematic surveys of our galaxy looking for 2.6-mm CO radiation were undertaken in 1974 by the American astronomers Philip Solomon and Nicholas Scoville. In mapping the locations of CO emission, they discovered huge clouds, now called **giant molecular clouds**, that must contain enormous amounts of hydrogen. These clouds have masses in the range of 10^5 to 2×10^6 solar masses and diameters that range from about 15 to 100 pc (50 to 300 ly). Inside one of these clouds, there are about 200 hydrogen molecules per cubic centimeter. This density is several thousand times greater than the average density of matter in the disk of our galaxy, yet only 10^{-17} as dense as the air we breathe. Astronomers now estimate that our galaxy contains about 5000 of these enormous clouds.

Figure 18-24 is a map of radio emissions from carbon monoxide in the constellations Orion and Monoceros. Note the extensive

FIGURE 18-24 Ⓡ I V U X G

Mapping Molecular Clouds A radio telescope was tuned to a wavelength of 2.6 mm to detect emissions from carbon monoxide (CO) molecules in the constellations Orion and Monoceros. The result was this false-color map, which shows a 35° × 40° section of the sky. The Orion and Horsehead star-forming nebulae are located at sites of intense CO emission (shown in red and yellow), indicating the presence of a particularly dense molecular cloud at these sites of star formation. The molecular cloud is much thinner at the positions of the Cone and Rosette nebulae, where star formation is less intense. (Courtesy of R. Maddalena, M. Morris, J. Moscowitz, and P. Thaddeus) areas of the sky covered by giant molecular clouds. This part of the sky is of particular interest because it includes several star-forming regions. By comparing the radio map with the star chart overlay, you can see that the areas where CO emission is strongest, and thus where giant molecular clouds are densest, are sites of star formation. Therefore, giant molecular clouds are associated with the formation of stars. Particularly dense regions within these clouds form dark nebulae, and within these stars are born.

By using CO emissions to map out giant molecular clouds, astronomers can find the locations in our galaxy where star formation occurs. These investigations reveal that molecular clouds clearly outline our galaxy's spiral arms, as **Figure 18-25** shows. These clouds lie roughly 1000 pc (3000 ly) apart and are strung along the spiral arms like beads on a string. This arrangement resembles the spacing of H II regions along the arms of other spiral galaxies, such as the galaxy shown in Figure 18-8a. The presence of both molecular clouds and H II regions shows that spiral arms are sites of ongoing star formation.

Star Formation in Spiral Arms

In Chapter 23, we will learn that spiral arms are locations where matter "piles up" temporarily as it orbits the center of the galaxy. You can think of matter in a spiral arm as analogous to a freeway traffic jam. Just as cars are squeezed close together when they enter a traffic jam, a giant molecular cloud is compressed when it passes through a spiral arm. When this happens, vigorous star formation begins in the cloud's densest regions.

As soon as massive O and B stars form, they emit ultraviolet light that ionizes the surrounding hydrogen, and an H II region is born. An H II region is thus a small, bright "hot spot" in a giant molecular cloud (but keep in mind that light from an H II region comes from emission lines, and not blackbody radiation from hot

matter). An example is the Orion Nebula, shown in Figure 18-1b. Four hot, luminous O and B stars at the heart of the nebula produce the ionizing radiation that makes the surrounding gases glow. The Orion Nebula is embedded on the edge of a giant molecular cloud whose mass is estimated at 500,000 $M_\odot$. The H II regions in Figure 18-2 are located at a different point on the edge of the same molecular cloud, some 25 pc (80 ly) from the Orion Nebula.

Once star formation has begun and an H II region has formed, the massive O and B stars at the core of the H II region induce star formation in the rest of the giant molecular cloud. Ultraviolet radiation and vigorous stellar winds from the O and B stars carve out a cavity in the cloud, and the reddish H II region, powered by the stars, expands into it. These winds travel faster than the speed of sound in the gas—that is, they are **supersonic.** Just as an airplane creates a shock wave (a sonic boom) if it flies faster than sound waves in our atmosphere, a shock wave forms where the expanding H II region pushes at supersonic speed into the rest of the giant molecular cloud. This shock wave compresses the gas through which it passes, stimulating more star birth (**Figure 18-26**).

Newborn O and B stars further expand the H II region into the giant molecular cloud. Meanwhile, the older O and B stars, which were left behind, begin to disperse (**Figure 18-27**). In this way, an OB association "eats into" a giant molecular cloud, leaving stars in its wake.

FIGURE 18-26 R Ⓘ V U X G

A Star-Forming Bubble Radiation and winds from the hot, young O and B stars at the center of this Spitzer Space Telescope image have carved out a bubble about 20 pc (70 ly) in diameter in the surrounding gas and dust. The material around the surface of the bubble has been compressed and heated, making the dust glow at the infrared wavelengths used to record this image. The compressed material is so dense that new stars have formed within that material. This glowing cloud, called RCW 79, lies about 5300 pc (17,200 ly) from Earth in the constellation Centaurus. (NASA/JPL-Caltech/E. Churchwell [University of Wisconsin])

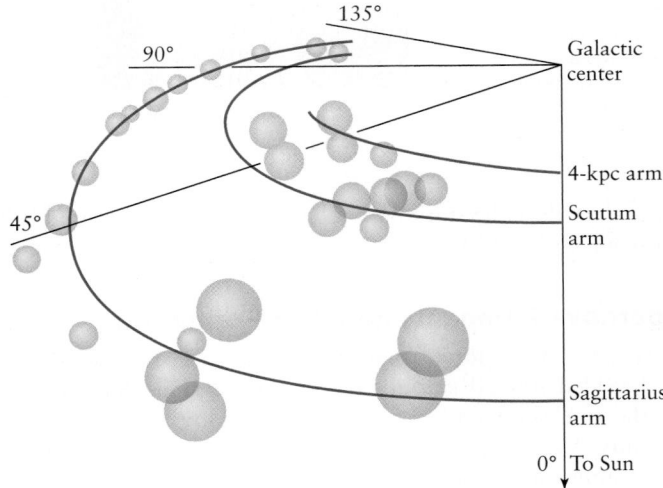

FIGURE 18-25

Giant Molecular Clouds in the Milky Way This perspective drawing shows the locations of giant molecular clouds in an inner part of our galaxy as seen from a vantage point above the Sun. These clouds lie primarily along the galaxy's spiral arms, shown by red arcs. The distance from the Sun to the galactic center is about 8000 pc (26,000 ly). (Data from T. M. Dame and colleagues)

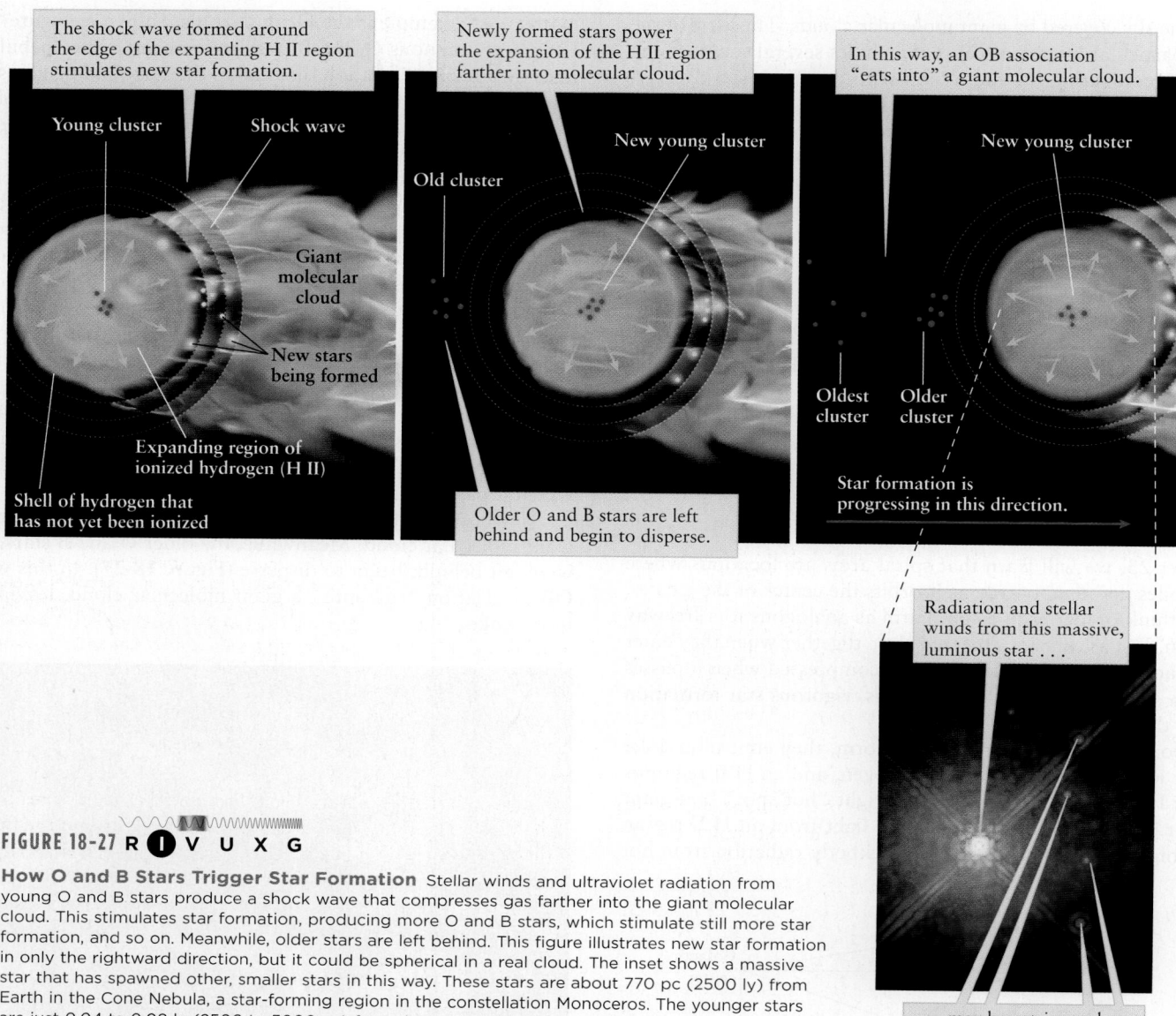

The shock wave formed around the edge of the expanding H II region stimulates new star formation.

Young cluster   Shock wave

Giant molecular cloud

New stars being formed

Expanding region of ionized hydrogen (H II)

Shell of hydrogen that has not yet been ionized

Newly formed stars power the expansion of the H II region farther into molecular cloud.

Old cluster   New young cluster

Older O and B stars are left behind and begin to disperse.

In this way, an OB association "eats into" a giant molecular cloud.

New young cluster

Oldest cluster   Older cluster

Star formation is progressing in this direction.

Radiation and stellar winds from this massive, luminous star . . .

. . . may have triggered the formation of these stars.

FIGURE 18-27 R **I** V U X G

How O and B Stars Trigger Star Formation Stellar winds and ultraviolet radiation from young O and B stars produce a shock wave that compresses gas farther into the giant molecular cloud. This stimulates star formation, producing more O and B stars, which stimulate still more star formation, and so on. Meanwhile, older stars are left behind. This figure illustrates new star formation in only the rightward direction, but it could be spherical in a real cloud. The inset shows a massive star that has spawned other, smaller stars in this way. These stars are about 770 pc (2500 ly) from Earth in the Cone Nebula, a star-forming region in the constellation Monoceros. The younger stars are just 0.04 to 0.08 ly (2500 to 5000 au) from the central star. (R. Thompson, M. Rieke and G. Schneider [Univ. Arizona], NASA)

CONCEPT CHECK 18-8

How can O and B stars initiate star formation in a giant molecular cloud?

Answer appears at the end of the chapter.

18-8 Supernovae compress the interstellar medium and can trigger star birth

Spiral arms are not the only mechanism for triggering the birth of stars. Presumably, anything that compresses interstellar clouds will do the job. Perhaps the most dramatic is a *supernova*, caused by the violent death of a massive star after it has left the main sequence. As we will see in Chapter 20, the core of the doomed star collapses suddenly, releasing vast quantities of particles and

energy that blow the star apart. The star's outer layers are blasted into space at speeds of several thousand kilometers per second.

Supernova Remnants and Star Formation

Astronomers have found many nebulae across the sky that are the shredded funeral shrouds of these dead stars. Such nebulae, like the one shown in Figure 18-28, are known as **supernova remnants**. Many supernova remnants have a distinctly circular or arched appearance, as would be expected for an expanding shell of gas. This wall of gas is typically moving away from the dead star faster than sound waves can travel through the interstellar medium. As we saw in Section 18-7, such supersonic motion produces a shock wave that abruptly compresses the medium through which it passes. When a gas is compressed rapidly, its temperature rises, and this temperature rise causes the gas to glow as shown in Figure 18-28.

A Supernova Remnant This composite image shows
Cassiopeia A, the remnant of a supernova that occurred
about 3000 pc (10,000 ly) from Earth. In the roughly 300 years since
the supernova explosion, a shock wave has expanded about 3 pc (10 ly)
outward in all directions from the explosion site. The shock wave has warmed
interstellar dust to a temperature of about 300 K (Spitzer Space Telescope
infrared image in red), and has heated interstellar gases to temperatures that
range from 10^4 K (Hubble Space Telescope visible-light image in yellow) to
10^7 K (Chandra X-ray Observatory x-ray image in green and blue). (NASA/JPL-
Caltech/Univ. of Ariz./STScI/CXC/SAO)

The Canis Major R1 Association These luminous arcs of gas
are studded with numerous young stars. Both the luminous arcs
and the young stars within can be traced to the same source—a supernova
explosion. The shock wave from the supernova explosion is exciting the gas
and making it glow; the same shock wave also compresses the interstellar
medium through which it passes, triggering star formation. (© Davide De
Martin)

When the expanding shell of a supernova remnant slams into an interstellar cloud, it compresses the cloud, stimulating star birth. This kind of star birth is taking place in the stellar association seen in Figure 18-29. This stellar nursery is located along reddish luminous arcs of gas extending about 30 pc (100 ly) in length that is presumably the remnant of a supernova explosion. (In fact, these arcs are part of an almost complete ring of glowing gas.) Since a supernova remnant does not last very long, this interpretation requires that the stars triggered by the supernova are young. But are the stars young? Spectroscopic observations of stars along this arc reveal substantial T Tauri activity. This activity results from young protostars undergoing mass loss in their final stages of contraction before they become main-sequence stars.

A supernova a few light-years away may have triggered the formation of our own Sun. Supernovae produce a variety of radioactive atomic nuclei, including some that are not produced in any other way. These nuclei are dispersed into space by the explosion and act as telltale signs of their supernova origin. Traces

Our Sun may have formed in association with a number of other stars.

of these telltale nuclei have been discovered in meteorites that have fallen to Earth. Since meteorites formed very early in the history of our solar system (see Section 8-3), this suggests that a supernova occurred nearby when our solar system was very young. Detailed modeling suggests this interpretation is plausible, with a supernova about 15 ly away triggering the collapse of a cloud that became our solar system.

Many other processes can also trigger star formation. For example, a collision between two interstellar clouds can create new stars. Compression occurs at the interface between the two colliding clouds and vigorous star formation follows. Similarly, stellar winds from a group of O and B stars may exert strong enough pressure on interstellar clouds to cause compression, followed by star formation. (This process is similar to the one depicted in Figure 18-27.)

Star Birth in Perspective

Our understanding of star birth has improved dramatically in recent years, primarily through infrared- and millimeter-wavelength observations. The *Cosmic Connections* figure summarizes our present state of knowledge about the formation of stars.

ATION If a clump of interstellar matter is cold and dense enough, it will begin to collapse thanks to the mutual gravitational attraction of its parts. If the clump is massive enough, it will evolve into a main-sequence star through the sequence of events shown here.

1

In this cold, dark nebula, gas atoms and dust particles move so slowly that gravity can draw them together.

2

Gas and dust begin to condense into clumps, forming the cores of protostars.

3 Protostar cores within the dark nebula

As the cores condense, their density and temperature both increase.

4

As the protostars continue to heat up and accrete matter from the nebula, they begin to glow due to their increasing temperature.

6

In the T Tauri stage, the young star ejects mass into space in a bipolar outflow. A stellar wind blows away the remaining parts of the nebula that surround the star, exposing the star to space.

5

Once the temperature at the center of a protostar becomes sufficiently high, thermonuclear fusion of hydrogen into helium begins. The protostar has now become a main-sequence star. The mass that is continuing to fall onto the star forms an accretion disk.

7

The ejected mass can induce a shock wave in the surrounding interstellar material, triggering the formation of additional stars.

8

Processes that cause the star to lose or gain mass come to an end, and the star stabilizes as a main-sequence star in hydrostatic equilibrium. The remnants of the accretion disk may remain as a protoplanetary disk, from which a system of

Nevertheless, many puzzles and mysteries remain. One problem is that different modes of star birth tend to produce different percentages of different kinds of stars. For example, the passage of a spiral arm through a giant molecular cloud tends to produce an abundance of massive O and B stars. In contrast, the shock wave from a supernova seems to produce fewer O and B stars but many more of the less massive A, F, G, and K stars. We do not yet know why this is so.

Despite these unanswered questions, it is now clear that star birth involves mechanisms on a colossal scale, from the deaths of massive stars to the rotation of an entire galaxy. In many respects, we have just begun to appreciate these cosmic processes. The study of cold, dark stellar nurseries will be an active and exciting area of astronomical research for many years to come.

In the chapters that follow, we will learn that the interstellar medium is both the birthplace of new stars and a dumping ground for dying stars. At the end of its life, a star can shed most of its mass in an outburst that enriches interstellar space with new chemical elements. The interstellar medium is therefore both nursery and graveyard. Because of this intimate relationship with stars, the interstellar medium evolves as successive generations of stars live out their lives. Understanding the details of this cosmic symbiosis is one of the challenges of modern astronomy.

CONCEPT CHECK 18-9

Several processes have been considered in this chapter that initiate star formation in clouds. What do they have in common?

Answer appears at the end of the chapter.

KEY WORDS

The term preceded by an asterisk (*) is discussed in Box 18-1.

accretion, p. 567
Barnard object, p. 562
bipolar outflow, p. 566
Bok globule, p. 562
circumstellar accretion disk, p. 567
cluster (of stars), p. 568
cocoon nebula, p. 563
dark nebula, p. 558

dust grains, p. 558
emission nebula, p. 557
evolutionary track, p. 563
*fluorescence, p. 559
giant molecular cloud, p. 572
H II region, p. 558
Herbig–Haro object, p. 566
interstellar extinction, p. 560
interstellar medium, p. 556

interstellar reddening, p. 560
nebula (*plural* nebulae), p. 557
nebulosity, p. 557
OB association, p. 571
open cluster, p. 571
protoplanetary disk (proplyd), p. 568
protostar, p. 562

recombination, p. 558
reflection nebula, p. 559
stationary absorption lines, p. 558
stellar association, p. 571
stellar evolution, p. 556
supernova remnant, p. 574
supersonic, p. 573
T Tauri star, p. 566

KEY IDEAS

Stellar Evolution: Because stars shine by thermonuclear reactions, they have a finite life span. The theory of stellar evolution describes how stars form and change during that life span (Section 18-1).

The Interstellar Medium: Interstellar gas and dust, which make up the interstellar medium, are concentrated in the disk of the galaxy. Clouds within the interstellar medium are called nebulae (Section 18-2).

- Dark nebulae are so dense that they are opaque. They appear as dark blots against a background of distant stars.
- Emission nebulae, or H II regions, are reddish glowing, ionized clouds of gas. Emission nebulae are powered by ultraviolet light that they absorb from nearby hot stars.
- Reflection nebulae are produced when starlight is reflected from dust grains in the interstellar medium, producing a characteristic bluish glow.

Jerry Lodriguss/Science Source

NASA, ESA, and The Hubble Heritage Team (STScI/AURA)

Protostars: Star formation begins in dense, cold nebulae, where gravitational attraction causes a clump of material to condense into a protostar (Section 18-3).

- As a protostar grows by the gravitational accretion of gases, Kelvin–Helmholtz contraction causes it to heat and begin glowing. Its relatively low temperature and high luminosity place it in the upper right region of an H-R diagram.
- Further evolution of a protostar causes it to move toward the main sequence in the H-R diagram. When its core temperatures become high enough to ignite steady hydrogen burning, it becomes a main-sequence star.
- The more massive the protostar, the more rapidly it evolves.

Protostar

NASA/JPL–Caltech/N. Evans (University of Texas at Austin)

Mass Loss by Protostars: In the final stages of pre–main-sequence contraction, when thermonuclear reactions are about to begin in its core, a protostar may eject large amounts of gas into space (Sections 18-4 and 18-5).

- Low-mass stars that vigorously eject gas are called T Tauri stars.
- A circumstellar accretion disk provides material that a young star ejects as jets. Clumps of glowing gas called Herbig–Haro objects are sometimes found along these jets and at their ends.

HH 2

HH 1

Protostar (hidden by dark, dusty nebula)

J. Hester (ASU), WFPC2 Team, NASA

Star Clusters: Newborn stars may form an open or galactic cluster. Stars are held together in such a cluster by gravity. Occasionally, a star moving more rapidly than average will escape, or "evaporate," from such a cluster (Section 18-6).

- A stellar association is a group of newborn stars that are moving apart so rapidly that their gravitational attraction for one another cannot pull them into orbit about one another.

© 2004–2013 R. Jay GaBany, Cosmotography.com

O and B Stars and Their Relation to H II Regions: The most massive protostars to form out of a dark nebula rapidly become main-sequence O and B stars. They emit strong ultraviolet radiation that ionizes hydrogen in the surrounding cloud, thus creating the reddish emission nebulae called H II regions (Section 18-7).

- Ultraviolet radiation and stellar winds from the O and B stars at the core of an H II region create shock waves that move outward through the gas cloud, compressing the gas and triggering the formation of more protostars.

Giant Molecular Clouds: The spiral arms of our galaxy are laced with giant molecular clouds, immense nebulae so cold that their constituent atoms can form into molecules (Sections 18-7 and 18-8).

- Star-forming regions appear when a giant molecular cloud is compressed. This can be caused by the cloud's passage through one of the spiral arms of our galaxy, by a supernova explosion, or by other mechanisms.

NASA/JPL–Caltech/E. Churchwell (University of Wisconsin)

QUESTIONS

Problem-solving tips and tools

You may find it helpful to review Box 17-4, which describes the relationship among a star's luminosity, radius, and surface temperature. The small-angle formula is described in Box 1-1. Orbital periods are described by Kepler's third law, which we discussed in Box 4-2 and Box 4-4. Remember that the Stefan–Boltzmann law (Box 5-2) relates the temperature of a blackbody to its energy flux. Remember, too, that the volume of a sphere of radius r is $4\pi r^3/3$.

Section 18-1

1. If no one has ever seen a star go through the complete formation process, how are we able to understand how stars form?

2. Why is it more difficult to observe the life cycles of stars than the life cycles of plants or animals?

Section 18-2

3. If an interstellar medium fills the space between the stars, how is that we are able to see the stars at all?

4. Summarize the evidence showing that interstellar space contains (a) gas and (b) dust.

5. What are H II regions? Near what kinds of stars are they found? Why do only these stars give rise to H II regions?

6. What are stationary absorption lines? In what sort of spectra are they seen? How do they give evidence for the existence of the interstellar medium?

7. In Figure 18-2, what makes the Horsehead Nebula dark? What makes IC 434 glow?

8. Why is the daytime sky blue? Why are distant mountains purple? Why is the Sun red when seen near the horizon at sunrise or sunset? In what ways are your answers analogous to the explanations for the bluish color of reflection nebulae and the process of interstellar reddening?

9. To see the constellation Coma Berenices (Berenice's Hair) you must look perpendicular to the plane of the Milky Way. By contrast, the Milky Way passes through the constellation Cassiopeia (named for a mythical queen). Would you expect H II regions to be more abundant in Coma Berenices or in Cassiopeia? Explain your reasoning.

10. The interior of a dark nebula is billions of times less dense than the air that you breathe. How, then, are dark nebulae able to block out starlight?

11. If you looked at the spectrum of a reflection nebula, would you see absorption lines, emission lines, or no lines? Explain your answer. As part of your explanation, describe how the spectrum demonstrates that the light was reflected from nearby stars.

12. In the direction of a particular star cluster, interstellar extinction allows only 15% of a star's light to pass through each kiloparsec (1000 pc) of the interstellar medium. If the star cluster is 3.0 kpc away, what percentage of its photons survive the trip to Earth?

Section 18-3

13. **TUTORIAL 18.1** Why are low temperatures necessary in order for protostars to form inside dark nebulae?

14. Compare and contrast Barnard objects and Bok globules. How many Sun-sized stars could you make out of a Barnard object? Out of a Bok globule?

15. Describe the energy source that causes a protostar to shine. How does this source differ from the energy source inside a main-sequence star?

16. What is an evolutionary track? How can evolutionary tracks help us interpret the H-R diagram?

17. Why are protostars more easily seen with an infrared telescope than with a visible-light telescope?

18. Find the density (in atoms per cubic centimeter) of a Bok globule having a radius of 1 ly and a mass of 100 $M_\odot$. How does your result compare with the density of a typical H II region, between 80 and 600 atoms per cm^3? (Assume that the globule is made purely of hydrogen atoms.)

19. The *Becklin–Neugebauer object* is a newly formed star within the Orion Nebula. It is substantially more luminous than the other newly formed stars in that nebula. Assuming that all these stars began the process of formation at the same time, what can you conclude about the mass of the Becklin–Neugebauer object compared with those of the other newly formed stars? Does your conclusion depend on whether or not the stars have reached the main sequence? Explain your reasoning.

Section 18-4

20. What happens inside a protostar to slow and eventually halt its gravitational contraction?

21. Why are the evolutionary tracks of high-mass stars different from those of low-mass stars? For which kind of star is the evolution more rapid? Why?

22. In what ways is the internal structure of a 1-$M_\odot$ main-sequence star different from that of a 5-$M_\odot$ main-sequence star? From that of a 0.5-$M_\odot$ main-sequence star? What features are common to all these stars?

23. What sets the limits on the maximum and minimum masses of a main-sequence star?

Section 18-5

24. What are T Tauri stars? How do we know that they eject matter at high speed? How does their rate of mass loss compare to that of the Sun?

25. What are Herbig–Haro objects? Why are they often found in pairs?

26. Why do disks form around contracting protostars? What is the connection between disks and bipolar outflows?

27. A newly formed protostar and a red giant are both located in the same region of the H-R diagram. Explain how you could distinguish between these two, assuming that you could observe their surroundings in detail.

28. (**a**) Determine the radius of the circumstellar accretion disk in Figure 18-16. (You will need to measure this image with a ruler. Note the scale bar in this figure.) Give your answer in astronomical units and in kilometers. (**b**) Assume that the young star at the center of this disk has a mass of 1 M$_\odot$. What is the orbital period (in years) of a particle at the outer edge of the disk? (**c**) Using your ruler again, determine the length of the jet that extends to the right of the circumstellar disk in Figure 18-16. At a speed of 200 km/s, how long does it take gas to traverse the entire visible length of the jet?

29. The star cluster NGC 2264 (Figure 18-22) contains numerous T Tauri stars, while the Pleiades (Figure 18-23) contains none. Explain why there is a difference.

Section 18-6

30. **TUTORIAL 18.2** Young open clusters like those shown in Figures 18-22a and 18-23 are found only in the plane of the galaxy. Explain why this should be.

31. How does a star cluster allow us to learn which kinds of stars evolve the fastest?

32. Consider the H-R diagrams in Figures 18-22b and 18-23b. How can you tell from these H-R diagrams which cluster is older?

33. The most luminous stars in the Pleiades cluster are not on the main sequence in Figure 18-23b. Are they protostars that are going to arrive on the main sequence, or have they already left? Explain your reasoning.

Section 18-7

34. Why are observations at millimeter wavelengths so much more useful in exploring interstellar clouds than observations at visible wavelengths?

35. What are giant molecular clouds? What role do these clouds play in the birth of stars?

36. Giant molecular clouds are among the largest objects in our galaxy. Why, then, were they discovered only relatively recently?

37. Consider the following stages in the evolution of a young star cluster: (i) H II region; (ii) dark nebula; (iii) formation of O and B stars; (iv) giant molecular cloud. Put these stages in the correct chronological order and discuss how they are related.

38. The concentration or abundance of ethyl alcohol in a typical molecular cloud is about 1 molecule per 10^8 cubic meters. What volume of such a cloud would contain enough alcohol to make a martini (about 10 grams of alcohol)? A molecule of ethyl alcohol has 46 times the mass of a hydrogen atom (that is, ethyl alcohol has a molecular weight of 46).

39. What do you think would happen if our solar system were to pass through a giant molecular cloud? Do you think Earth has ever passed through such clouds?

Section 18-8

40. Briefly describe four mechanisms that compress the interstellar medium and trigger star formation.

41. From the information given in the caption to Figure 18-28, calculate the angular diameter in arcminutes of Cassiopeia A as seen from Earth.

42. From the information given in the caption for Figure 18-28, calculate the average speed at which the shock wave has spread away from the site of the supernova explosion. Give your answer in kilometers per second and as a fraction of the speed of light. (*Hint:* There are 3.16×10^7 seconds in a year and the speed of light is 3.00×10^5 km/s.)

43. Many of the molecules found in giant molecular clouds are *organic* molecules (that is, they contain carbon). Speculate about the possibility of life-forms and biological processes occurring in giant molecular clouds. In what ways might the conditions existing in giant molecular clouds favor or hinder biological evolution?

General Problems

44. The visible-light photograph below shows the Trifid Nebula in the constellation Sagittarius. Label the following features on this photograph: (**a**) reflection nebulae (and the star or stars whose light is being reflected); (**b**) dark nebulae; (**c**) H II regions; (**d**) regions where star formation may be occurring. Explain how you identified each feature.

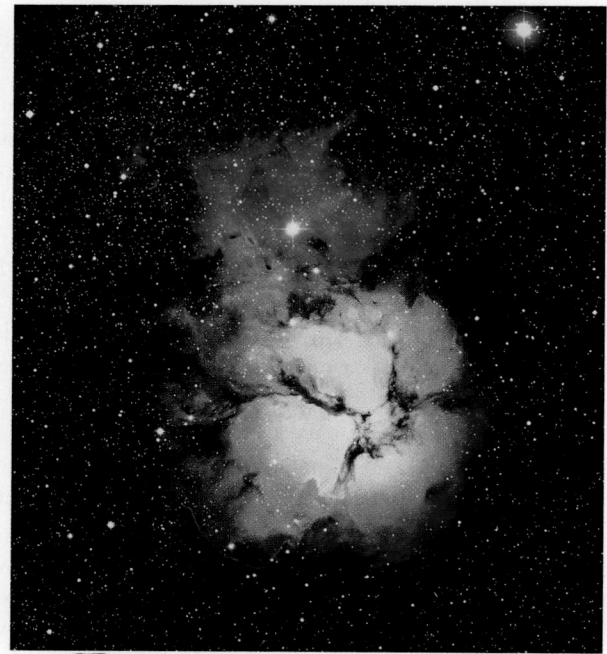

R I V U X G

(Australian Astronomical Observatory/David Malin Images)

45. The two false-color images that follow show a portion of the Trifid Nebula (see Question 44). The reddish-orange view is a false-color infrared image, while the bluish picture (shown to the same scale) was made with visible light. Explain why the dark streaks in the visible-light image appear bright in the infrared image.

R I **V** U X G

R **I** V U X G

(ESA/ISO, ISOCAM, and J. Cernicharo et. al.; IAC, Observatorio del Teide, Tenerife)

46. At one stage during its birth, the protosun had a luminosity of 1000 $L_\odot$ and a surface temperature of about 1000 K. At this time, what was its radius? Express your answer in three ways: as a multiple of the Sun's present-day radius, in kilometers, and in astronomical units. (*Hint:* See Box 17-4.)

47. Some science-fiction movies show stars suddenly becoming dramatically brighter when they are "born" (that is, when thermonuclear fusion reactions begin in their cores). Discuss whether this is a reasonable depiction.

48. Suppose that the electrons in hydrogen atoms were not as strongly attracted to the nuclei of those atoms, so that these atoms were easier to ionize. What consequences might this have for the internal structure of main-sequence stars? Explain your reasoning.

49. Speculate on why a shock wave from a supernova seems to produce relatively few high-mass O and B stars, compared to the lower-mass A, F, G, and K stars.

WEB/E-BOOK QUESTIONS

50. In recent years, astronomers have been able to learn about the character of the interstellar medium in the vicinity of the Sun. Search the internet for information about aspects of the nearby interstellar medium, including features called the Local Interstellar Cloud and the Local Bubble. How do astronomers study the nearby interstellar medium? What makes these studies difficult? Is the interstellar medium relatively uniform in our neighborhood, or is it clumpy? If the latter, is our solar system in a relatively thin or thick part of the interstellar medium? How is our solar system moving through the interstellar medium?

51. Search the internet for recent discoveries about how brown dwarfs form. Do they tend to form in the same locations as "real" stars? Do they form in relatively small or relatively large numbers compared to "real" stars? What techniques are used to make these discoveries?

52. ANIMATION 18.1 **Measuring a Stellar Jet.** Access the animation "A Stellar Jet in the Trifid Nebula" in Chapter 18 of the *Universe* website or e-book. (**a**) The Trifid Nebula as a whole has an angular diameter of 28 arcmin. By stepping through the animation, estimate the angular size of the stellar jet shown at the end of the animation. (**b**) The Trifid Nebula is about 2800 pc (9000 ly) from Earth. Estimate the length of the jet in light-years. (**c**) If gas in the jet travels at 200 km/s, how long does it take to traverse the length of the jet? Give your answer in years. (*Hint:* There are 3.16×10^7 seconds in a year.)

ACTIVITIES

Observing Projects

Observing tips and tools

After looking at the beautiful color photographs of nebulae in this chapter, you may find the view through a telescope a bit disappointing at first, but fear not. You can see a great deal with even a small telescope. To get the best view of a dim nebula using a telescope, use the same "averted vision" trick we described in "Observing tips and tools" in Chapter 17: If you direct your vision a little to one side of the object that you are looking at, the light from that object will go onto a more sensitive part of the retina.

53. Use a telescope to observe at least two of the H II regions listed in the following table. In each case, can you guess which stars are probably responsible for the ionizing radiation that causes the nebula to glow? Can you see any obscuration or silhouetted features that suggest the presence of interstellar dust? Draw a picture of what you see through the telescope and compare it with a photograph of the object. Take note of which portions of the nebula were not visible through your telescope.

Nebula	Right ascension	Declination
M42 (Orion)	5^h 35.4^m	−5° 27′
M43	5^h 35.6^m	−5° 16′
M20 (Trifid)	18^h 02.6^m	−23° 02′
M8 (Lagoon)	18^h 03.8^m	−24° 23′
M17 (Omega)	18^h 20.8^m	−16° 11′

Note: The right ascensions and declinations are given for epoch 2000.

54. On an exceptionally clear, moonless night, use a telescope to observe at least one of the dark nebulae listed in the following table. These nebulae are very difficult to find, because they are recognizable only by the absence of stars in an otherwise starry part of the sky. Are you confident that you actually saw the dark nebula? Does the pattern of background stars suggest a particular shape to the nebula?

Nebula	Right ascension	Declination
Barnard 72 (the Snake)	17^h 23.5^m	−23° 38′
Barnard 86	18^h 02.7^m	−27° 50′
Barnard 133	19^h 06.1^m	−6° 50′
Barnard 142 and 143	19^h 40.7^m	−10° 57′

Note: The right ascensions and declinations are given for epoch 2000.

55. A few fine objects cover such large regions of the sky that they are best seen with binoculars. If you have access to a high-quality pair of binoculars, observe the North American Nebula in Cygnus and the Pipe Nebula in Ophiuchus. Both nebulae are quite faint, so you should attempt to observe them only on an exceptionally dark, clear, moonless night. The North American Nebula is a cloud of glowing hydrogen gas located about 3° east of Deneb, the brightest star in Cygnus. While searching for the North American Nebula, you may glimpse another diffuse H II region, the Pelican Nebula, located about 2° southeast of Deneb. The Pipe Nebula is a 7°-long, meandering, dark nebula to the south and to the east of the star θ (theta) Ophiuchi, which is in a section of Ophiuchus that extends southward between the constellations of Scorpius and Sagittarius. Located about 12° east of the bright red star Antares, you can locate θ Ophiuchi using the *Starry Night™* software.

Starry Night™ Explorations

56. Use *Starry Night™* to explore the distribution and appearances of a sample of the several different types of nebulae found in the Milky Way Galaxy. Open **Favorites > Explorations > Nebulae** to see a view of the celestial sphere from Earth with the Milky Way distinctly visible and a blue line indicating the plane of the disk of this galaxy.

(a) Look around this view and observe the mottled appearance of the Milky Way caused by regions of opaque dust and gas. Click the **Quick Search Items** icon to open the **Options** panel. In this panel, locate and expand **Display Options > Options for Other Objects > Nebulae**. Turn on the option to display **Dark Nebula** and use the hand tool and zoom controls to look around the view again. Describe the distribution of these objects in the galaxy.

(b) Return to the **Options** panel, turn off the **Dark Nebula** option, turn on the display of **Emission Nebula** and survey the view again. What is the distribution of these nebulae in the galaxy?

(c) Use the "K" keyboard shortcut to display the constellations on the celestial sphere and then locate the tight concentration of emission nebulae between Draco and Ursa Major. Center and zoom in on this region to a field of view about 1° wide. You will notice that the outlined emission nebulae belong to the galaxy M101 (the Pinwheel Galaxy), which is about 27 million ly from Earth. Describe the distribution of these emission nebulae within this galaxy and compare it to your observations of the distribution of emission nebulae in the Milky Way.

(d) Given the distance to M101, what does the fact that these emission nebulae in this galaxy are visible from Earth suggest about their properties?

(e) Open the **Observing Lists** panel and locate and expand the **Nebulae** list. Click the menu icon next to the name of each nebula in the list and select **Magnify** to see telescopic images of some of these nebulae and describe your observations of the details of some of them, such as color, shape, and structure.

57. Use *Starry Night™* to observe the characteristics of several nebulae in the Milky Way Galaxy. Open **Favorites > Explorations > Atlas** and then select **View > Galactic Guides > Equator** from the menu to indicate the plane of the disk of the Milky Way Galaxy. Then open the **Observing Lists** panel and expand the **Nebulae** list. For each of the nebulae in the list, use its contextual menu to **Magnify** it in the view; then click the check box to label it and examine the appearance of each nebula. Place the cursor on the point of the indicating

arrow for each nebula to activate the HUD and note the nebula's distance from Earth. Zoom out if necessary and measure the approximate angular distance from the nebula to the galactic equator as seen from Earth. Is the nebula an example of a dark nebula, a reflection nebula, an emission nebula, or perhaps some combination of these? Are all these nebulae found relatively near to the galactic plane?

58. Use *Starry Night™* to examine the H-R diagram of the Pleiades star cluster. This group of stars was formed relatively recently in astronomical time. Select **Favorites > Explorations > Pleiades** to display this cluster of young stars in the sky, along with an H-R diagram of the stars that comprise it. Moving the cursor over individual stars will indicate the positions of these stars on the diagram.
    (a) Where on the H-R diagram do you find the majority of the stars of this cluster?
    (b) In view of the existence within this cluster of very hot stars with high output of energy, what does this tell you about the age of this cluster compared to the general population of stars?

Collaborative Exercises

59. Imagine that your group walks into a store that specializes in selling antique clothing. Prepare a list of observable characteristics that you would look for to distinguish which items were from the early, middle, and late twentieth century. Also, write a paragraph that specifically describes how this task is similar to the way astronomers understand the evolution of stars.

60. Consider advertisement signs visible at night in your community and provide specific examples of ones that represent the three different types of nebulae that astronomers observe and study. If an example doesn't exist in your community, creatively design an advertisement sign that could serve as an example.

61. The pre–main-sequence evolutionary tracks shown in Figure 18-10 describe the tracks of seven protostars of different masses. Imagine a new sort of H-R diagram that plots a human male's increasing age versus decreasing hair density on the head instead of increasing luminosity versus decreasing temperature. Create and carefully label a sketch of this imaginary H-R diagram showing both the majority of the U.S. male population and a few oddities. Finally, draw a line that clearly labels your sketch to show how a typical male undergoing male-pattern baldness might slowly change position in the H-R diagram over the course of a human life span.

ANSWERS

ConceptChecks

ConceptCheck 18-1: No. Toaster filaments glow red because they are hot blackbodies. The red light from an H II region comes from a single red emission line. The gas in an H II region is too thin to emit as a blackbody.

ConceptCheck 18-2: No. The blue gas is simply reflecting starlight, but with more reflection at blue wavelengths.

ConceptCheck 18-3: Yes. The evolutionary track for a 1-$M_\odot$ solar star like our Sun in Figure 18-10 indicates that such a protostar is almost 100 times more luminous when it forms than when it reaches the main sequence.

ConceptCheck 18-4: The red dashed lines indicate how long it takes for protostars to cross those lines on their way to the main sequence. A 2-$M_\odot$ star reaches the main sequence after about 10^7 (10 million) years.

ConceptCheck 18-5: No. The energy source of a protostar is gravitational contraction compressing and heating the protostar's interior. The energy source for all main-sequence stars are nuclear reactions converting hydrogen to helium in the star's core.

ConceptCheck 18-6: For a main-sequence star that ends up with less than 3 $M_\odot$, it was more massive as a protostar; these protostars lose more mass than they gain. The masses shown in Figure 18-10 are for the final main-sequence stars.

ConceptCheck 18-7: No. The pillars are eroded away by hot luminous stars outside of the pillars that emit ultraviolet radiation.

ConceptCheck 18-8: Ultraviolet light from hot O and B stars creates an H II region that expands into the surrounding molecular cloud. This compresses the cloud, causing star formation.

ConceptCheck 18-9: Protostars begin to form when gravity overwhelms a gas cloud's internal pressure. This is initiated when an external process compresses the cloud.

CalculationCheck

CalculationCheck 18-1: After the first light-year, only 90% passes through. In the second light-year, 90% of what emerged from the first light-year passes through. Therefore, $0.9 \times 0.9 = 0.81$ passes through 2 ly, or 81%.

19 | Stellar Evolution: On and After the Main Sequence

LEARNING GOALS

I magine a world like Earth, but orbiting a star more than 100 times larger and 2000 times more luminous than our Sun. Bathed in the star's intense light, the surface of this world is utterly dry, airless, and hot enough to melt iron. If you could somehow survive on the daytime side of this world, you would see the star filling almost the entire sky.

This bizarre planet is not a creation of science fiction—it is our own Earth some 5 billion years from now. The bloated star is our own Sun, which in that remote era will have become a red giant.

In this chapter, we will learn how a main-sequence star evolves into a red giant when all the hydrogen in its core is consumed. The star's core contracts and heats up, but its outer layers expand and cool. In the hot, compressed core, helium fusion becomes a new energy source. The more massive a star, the more rapidly it consumes its core's hydrogen and the sooner it evolves into a giant.

The interiors of stars are hidden from our direct view, so much of the story in this chapter is based on theory. We back up those calculations with observations of star clusters, which contain stars of different masses but roughly the same age.

The red stars in this image of open cluster NGC 290 are red giants, a late stage in stellar evolution. (ESA & NASA/Edward W. Olszewski, U. of Arizona)

R I V U X G

(An example is the cluster shown in the chapter-opening image, many of whose stars have evolved into luminous red giants.) Other observations show that some red giant stars actually pulsate and that stars can evolve along very different paths if they are part of a binary star system.

19-1 During a star's main-sequence lifetime, it expands and becomes more luminous

In their cores, main-sequence stars are all fundamentally alike. As we saw in Section 16-1 and Section 18-4, it is in their cores that all such stars convert hydrogen into helium by thermonuclear reactions (**Figure 19-1**). This process is called **core hydrogen fusion**. The total time that a star will spend fusing hydrogen into helium in its core, and thus the total time that it will spend as a main-sequence star, is called its **main-sequence**

> Over the past 4.56 billion years, thermonuclear reactions have caused an accumulation of helium in our Sun's core.

lifetime. For a star of a given mass, its main-sequence lifetime can be estimated by modeling the nuclear reactions within its core. For our Sun, the main-sequence lifetime is about 10 billion (10^{10}) years. Hydrogen fusion has been going on in the Sun's core for

the past 4.56 billion (4.56×10^9) years, so our Sun is less than halfway through its main-sequence lifetime.

What happens to a star like the Sun after the core hydrogen has been used up, so that it is no longer a main-sequence star? As we will see, it expands dramatically to become a red giant. To understand why this happens, it is useful to first look at how a star evolves *during* its main-sequence lifetime. The nature of that evolution depends on whether its mass is less than or greater than about 0.4 $M_\odot$.

Main-Sequence Stars of 0.4 $M_\odot$ or Greater: Consuming Core Hydrogen

A protostar becomes a main-sequence star when steady hydrogen fusion begins in its core and it achieves *hydrostatic equilibrium*—a balance between the inward force of gravity and the outward pressure produced by hydrogen fusion (see Section 16-2 and Section 18-4). Such a freshly formed main-sequence star is called a **zero-age main-sequence star**.

We make the distinction between "main sequence" and "zero-age main sequence" because a star undergoes noticeable changes in luminosity, surface temperature, and radius during its main-sequence lifetime. These changes are a result of core hydrogen fusion, which alters the chemical composition of the core. As an example, when our Sun first formed, its composition was the same at all points throughout its volume: by mass, about 74% hydrogen, 25% helium, and 1% heavy elements. But as **Figure 19-2**

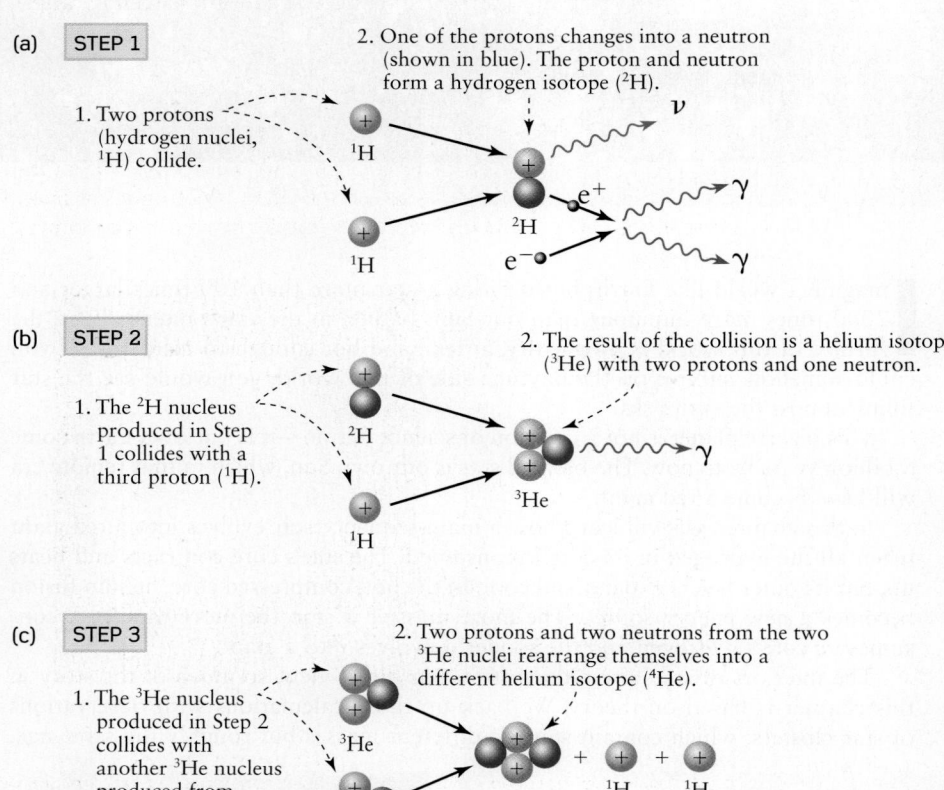

(a) **STEP 1**

1. Two protons (hydrogen nuclei, ^{1}H) collide.

2. One of the protons changes into a neutron (shown in blue). The proton and neutron form a hydrogen isotope (^{2}H).

^{1}H

^{1}H

^{2}H

ν

e^+

γ

e^-

γ

(b) **STEP 2**

1. The ^{2}H nucleus produced in Step 1 collides with a third proton (^{1}H).

^{2}H

^{1}H

2. The result of the collision is a helium isotope (^{3}He) with two protons and one neutron.

^{3}He

γ

(c) **STEP 3**

1. The ^{3}He nucleus produced in Step 2 collides with another ^{3}He nucleus produced from three other protons.

^{3}He

^{3}He

2. Two protons and two neutrons from the two ^{3}He nuclei rearrange themselves into a different helium isotope (^{4}He).

^{4}He

^{1}H

^{1}H

FIGURE 19-1

Hydrogen Fusion via the Proton–Proton Chain Most hydrogen fusion in the Sun occurs through the following three steps, all of which release energy. **(a)** Step 1: Two protons combine to form a proton and neutron bound together, which is just an isotope of hydrogen. **(b)** Step 2: The hydrogen isotope combines with another proton, and the result is a helium isotope with 2 protons and 1 neutron. Many of these helium isotopes are produced in this manner. **(c)** Step 3: Two helium isotopes combine to make the most common form of helium with 2 protons and 2 neutrons. There are also 2 hydrogen atoms produced, which can participate anywhere hydrogen is used in further proton–proton chain reactions. For more information on these reactions, see the *Cosmic Connections* figure in Chapter 16 (page 479).

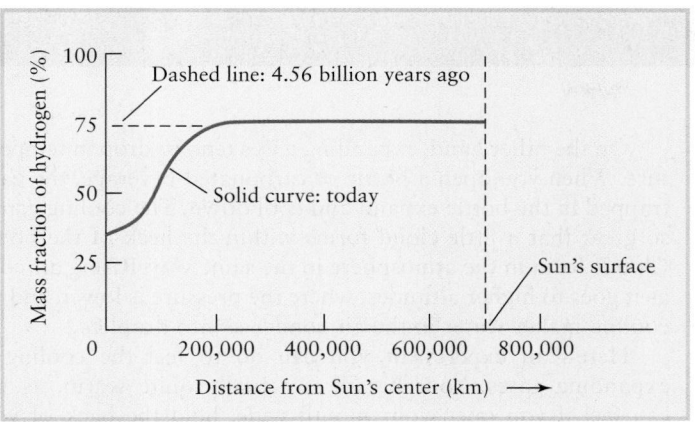

(a) Hydrogen in the Sun's interior

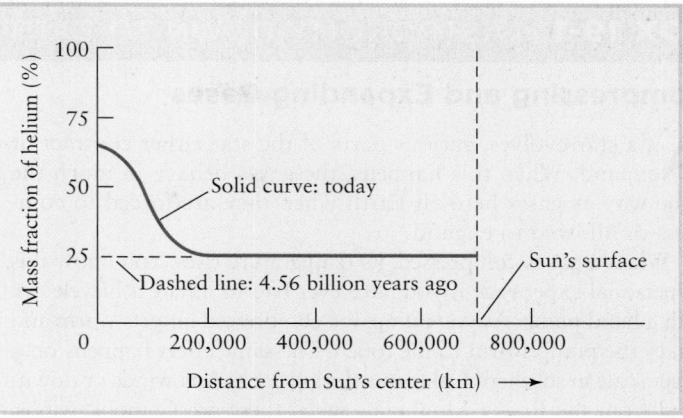

(b) Helium in the Sun's interior

FIGURE 19-2

Changes in the Sun's Chemical Composition These graphs show the percentage by mass of **(a)** hydrogen and **(b)** helium at different points within the Sun's interior. The dashed horizontal lines show that these percentages were the same throughout the Sun's volume when it first formed. As the solid curves show, over the past 4.56×10^9 years, thermonuclear reactions at the core have depleted hydrogen in the core and increased the amount of helium in the core.

shows, the Sun's core now contains a greater mass of helium than of hydrogen. (There is still enough hydrogen in the Sun's core for another 5 billion years or so of core hydrogen fusion.)

CAUTION!

Although the outer layers of the Sun are also predominantly hydrogen, there are two reasons why this hydrogen cannot undergo fusion. The first reason is that while the temperature and pressure in the core are high enough for thermonuclear reactions to take place, the temperatures and pressure in the outer layers are not. The second reason is that there is no flow of material between the Sun's core and outer layers, so the hydrogen in the outer layers cannot move into the hot, high-pressure core to undergo fusion. The same is true for main-sequence stars with masses of about 0.4 M$_\odot$ or greater. (We will see below that the outer layers *can* undergo fusion in main-sequence stars with a mass less than about 0.4 M$_\odot$.)

Thanks to core hydrogen fusion, the total number of atomic nuclei in a star's core decreases with time: In each reaction, four hydrogen nuclei are converted to a single helium nucleus (see the *Cosmic Connections* figure in Section 16-1, as well as Box 16-1). With fewer particles bouncing around to provide the core's internal pressure, the core contracts slightly under the weight of the star's outer layers. Compression makes the core denser and increases its temperature. (Box 19-1 gives some everyday examples of how the temperature of a gas changes when it compresses or expands.) As a result of these changes in density and temperature, the pressure in the compressed core is actually higher than before.

As the star's core shrinks, its outer layers expand and shine more brightly. Here's why: As the core's density and temperature increase, hydrogen nuclei in the core collide with one another more frequently, and the rate of core hydrogen fusion increases. Hence, the star's release of energy increases, which increases the star's luminosity. The radius of the star as a whole also increases slightly, because increased core pressure pushes outward on the star's outer layers. The star's surface temperature changes as well, because it is related to the luminosity and radius (see Section 17-6 and Box 17-4). As an example, theoretical calculations indicate that over the past 4.56×10^9 years, our Sun has become 40% more luminous, grown in radius by 6%, and increased in surface temperature by 300 K (Figure 19-3).

Main-Sequence Stars of Less than 0.4 M$_\odot$: Consuming All Their Hydrogen

The story is somewhat different for the least massive main-sequence stars, with masses between 0.08 M$_\odot$ (the minimum mass for sustained thermonuclear reactions to take place in a star's core)

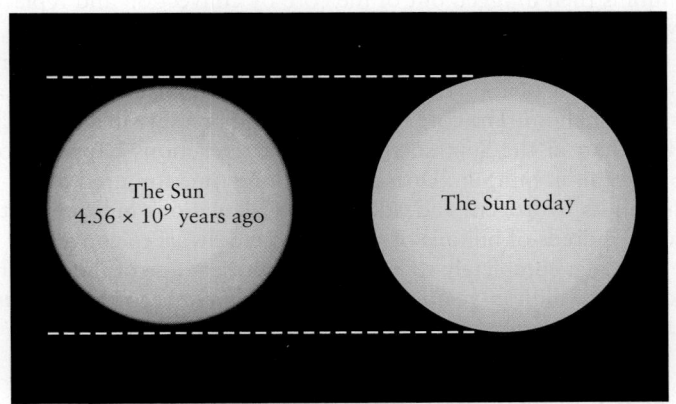

FIGURE 19-3

The Zero-Age Sun and Today's Sun Over the past 4.56×10^9 years, about half of the hydrogen in the Sun's core has been converted into helium, the core has contracted a bit, and the Sun's luminosity has gone up by about 40%. These changes in the core have made the Sun's outer layers expand in radius by 6% and increased the surface temperature from 5500 K to 5800 K.

BOX 19-1   ASTRONOMY DOWN TO EARTH

Compressing and Expanding Gases

As a star evolves, various parts of the star either contract or expand. When this happens, the gases behave in much the same way as gases here on Earth when they are forced to compress or allowed to expand.

When a gas is compressed, its temperature rises. You know this by personal experience if you have ever had to inflate a bicycle tire with a hand pump. As you pump, the compressed air gets warm and makes the pump warm to the touch. The same effect happens on a larger scale in southern California during Santa Ana winds or downwind from the Rocky Mountains when there are Chinook winds. Both of these strong winds blow from the mountains down to the lowlands. Even though the mountain air is cold, the winds that reach low elevations can be very hot. (Chinook winds have been known to raise the temperature by as much as 27°C, or 49°F, in only 2 minutes!) The explanation is compression. Air blown downhill by the winds is compressed by the greater air pressure at lower altitudes, and this compression raises the temperature of the air.

On the other hand, expanding gases tend to drop in temperature. When you open a bottle of carbonated beverage, the gases trapped in the bottle expand and cool down. The cooling can be so great that a little cloud forms within the neck of the bottle. Clouds form in the atmosphere in the same way. Rising air cools as it goes to higher altitudes, where the pressure is lower, and the cooling makes water in the air condense into droplets.

Here's an experiment you can do to feel the cooling of expanding gases. Your breath is actually quite warm, as you can feel if you open your mouth wide, hold the back of your hand next to your mouth, and exhale. But if you bring your lips together to form a small "o" and again blow on your hand, your breath feels cool. In this second case, your exhaled breath has to expand as it passes between your lips to the outside, which makes its temperature drop.

---

and about 0.4 M_☉. These stars, of spectral class M, are called **red dwarfs** because they are small in size and have a red color due to their low surface temperature. They are also very numerous; about 85% of all stars in the Milky Way Galaxy are red dwarfs.

In a red dwarf, helium does *not* accumulate in the core to the same extent as in the Sun's core. The reason is that in a red dwarf, there are convection cells of rising and falling gas that extend throughout the star's volume and penetrate into the core (see Figure 18-12c). These convection cells drag helium outward from the core and replace it with hydrogen from the outer layers (Figure 19-4). The fresh hydrogen can undergo thermonuclear fusion that releases energy and makes additional helium. This helium is then dragged out of the core by convection and replaced by even more hydrogen from the red dwarf's outer layers.

As a consequence, over a red dwarf's main-sequence lifetime, essentially all of the star's hydrogen can be consumed and converted to helium. The core temperature and pressure in a red dwarf is less than in the Sun, so thermonuclear reactions happen more slowly than in our Sun. Does it take a long time for a red dwarf to burn all its hydrogen? Indeed it does! Calculations indicate that it takes hundreds of billions of years for a red dwarf to convert all of its hydrogen completely to helium. The present age of the universe is only 13.7 billion years, so there has not yet been time for *any* red dwarfs—not one—to become pure helium.

A Star's Mass Determines Its Main-Sequence Lifetime

The main-sequence lifetime of a star depends critically on its mass. As Table 19-1 shows, *massive stars have short main-sequence lifetimes.* The greater core pressure created by a larger total mass causes nuclear reactions to occur more rapidly, and the star burns through the available hydrogen fuel more quickly. Hence, even though a

massive main-sequence star contains much more hydrogen fuel in its core than is in the entire volume of a red dwarf, a massive star exhausts its hydrogen much sooner. More rapid burning also produces a greater luminosity, so that *massive stars are more luminous* (see Section 17-9, and particularly the *Cosmic Connections* figure

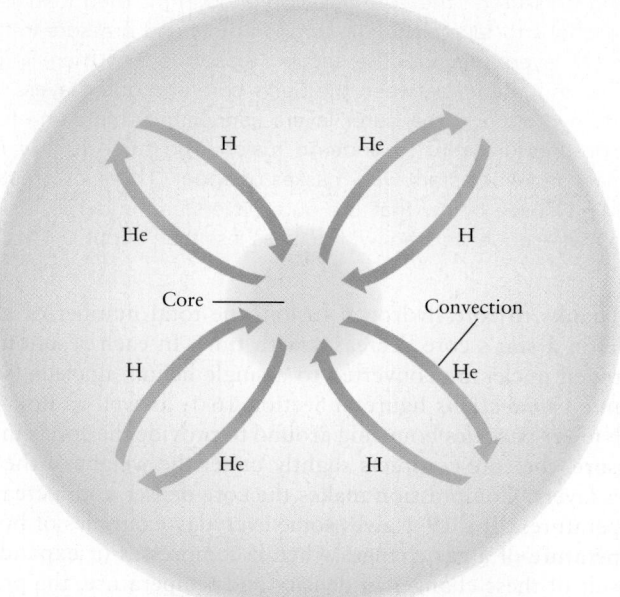

FIGURE 19-4

A Fully Convective Red Dwarf In a red dwarf—a main-sequence star with less than about 0.4 solar masses—helium (He) created in the core by thermonuclear reactions is carried to the star's outer layers by convection. Convection also brings fresh hydrogen (H) from the outer layers into the core. This process continues until the entire star is helium.

TABLE 19-1 Approximate Main-Sequence Lifetimes

Mass (M☉)	Surface temperature (K)	Spectral class	Luminosity (L☉)	Main-sequence lifetime (10^6 years)
25	35,000	O	80,000	4
15	30,000	B	10,000	15
3	11,000	A	60	800
1.5	7000	F	5	4500
1.0	6000	G	1	12,000
0.75	5000	K	0.5	25,000
0.50	4000	M	0.03	700,000

Main-sequence lifetimes were estimated using the relationship $t \propto 1/M^{2.5}$ (see Box 19-2).

in Chapter 17). Thus, a main-sequence star's mass determines not only its luminosity, but also how long it can remain a main-sequence star (see Box 19-2 for details).

We saw in Section 18-4 how more massive stars evolve more quickly through the protostar phase to become main-sequence stars (see Figure 18-10). In general, the more massive the star, the more rapidly it goes through *all* the phases of its life. Still, most of the stars we are able to detect are in their main-sequence phase, because this phase lasts so much longer than other luminous phases.

ANALOGY

A large powerful car can reach high speeds, but it requires a lot more gasoline per mile to do so than a smaller, less powerful car does. Similarly, a massive star shines brighter but burns through its nuclear fuel faster than a smaller star.

In the remainder of this chapter, we will look at the luminous phases that can take place after the end of a star's main-sequence lifetime. (In Chapters 20 and 21, we will explore the final phases of a star's existence, when it ceases to have an appreciable luminosity.)

CONCEPT CHECK 19-1

Why are there no red dwarfs that have already left the main sequence?

Answer appears at the end of the chapter.

19-2 When core hydrogen fusion ceases, a main-sequence star like the Sun becomes a red giant

TUTORIAL 19.1 Like so many properties of stars, what happens at the end of a star's main-sequence lifetime depends on its mass. If the star is a red dwarf of less than about 0.4 M☉, after hundreds of billions of years, the star will have converted all of its hydrogen to

helium. It is possible for helium to undergo thermonuclear fusion, but this requires temperatures and pressures far higher than those found within a red dwarf. Thus, this red dwarf will end its life as a ball of helium, which has no further nuclear reactions but still glows due to its internal heat. As it radiates energy into space, it slowly cools and shrinks. This slow, quiet demise is the ultimate fate of the red dwarfs, which are 85% of the stars in the Milky Way. (As we have seen, there has not yet been time in the history of the universe for any red dwarf to reach this final stage in its evolution.)

What is the fate of stars more massive than about 0.4 M☉, including the Sun? As we will see, the late stages of their evolution are far more dramatic. Studying these stages will give us insight into the fate of our solar system and of life on Earth.

Stars of 0.4 M☉ or Greater: From Main-Sequence Star to Red Giant

When a star of at least 0.4 M☉ reaches the end of its main-sequence lifetime, all of the hydrogen in its core has been used up and hydrogen fusion ceases there. In this new stage, hydrogen fusion continues only in the hydrogen-rich material just outside the core, a situation called **shell hydrogen fusion**. At first, this process occurs only in the hottest region just outside the core, where the hydrogen fuel has not yet been exhausted. Outside this region, no fusion reactions take place.

It may seem counterintuitive, but the end of the core hydrogen fusion process leads to an *increase* in the core's temperature. Here's why: When thermonuclear reactions first cease in the core, nothing remains to generate heat there. Hence, the core starts to cool and the pressure in the core starts to decrease. This pressure decrease allows the star's core to again compress under the weight of the outer layers. As the core contracts, its temperature again increases, and heat begins to flow outward from the core even though no nuclear reactions are taking place there. (Technically, gravitational energy is converted into

In the transition from main-sequence star to red giant, the star's core contracts while its outer layers expand.

BOX 19-2   TOOLS OF THE ASTRONOMER'S TRADE

Main-Sequence Lifetimes

Hydrogen fusion converts a portion of a star's mass into energy. We can use Einstein's famous equation relating mass and energy to calculate how long a star will remain on the main sequence.

Suppose that M is the mass of a star and f is the fraction of the star's mass that is converted into energy by hydrogen fusion. During its main-sequence lifetime, the total energy E supplied by the hydrogen fusion can be expressed as

$$E = fMc^2$$

In this equation, c is the speed of light.

This energy from hydrogen fusion is released gradually over millions or billions of years. If L is the star's luminosity (energy released per unit time) and t is the star's main-sequence lifetime (the total time over which the hydrogen fusion occurs), then we can write

$$L = \frac{E}{t}$$

(Actually, this equation is only an approximation. A star's luminosity is not quite constant over its entire main-sequence lifetime. But the variations are not important for our purposes.) We can rewrite this equation as

$$E = Lt$$

From this equation and $E = fMc^2$, we see that

$$Lt = fMc^2$$

We can rearrange this equation as

$$t = \frac{fMc^2}{L}$$

Thus, a star's lifetime on the main sequence is proportional to its mass (M) divided by its luminosity (L). Using the symbol $\propto$ to denote "is proportional to," we write

$$t \propto \frac{M}{L}$$

We can carry this analysis further by recalling that main-sequence stars obey the mass–luminosity relation. The distribution of data on the graph in the *Cosmic Connections* figure in Section 17-9 tells us that a star's luminosity is roughly proportional to the 3.5 power of its mass:

$$L \propto M^{3.5}$$

Substituting this relationship into the previous proportionality, we find that

$$t \propto \frac{M}{M^{3.5}} = \frac{1}{M^{2.5}} = \frac{1}{M^2 \sqrt{M}}$$

This approximate relationship can be used to obtain rough estimates of how long a star will remain on the main sequence. It is often convenient to relate these estimates to the Sun (a typical 1-$M_\odot$ star), which will spend 10^{10} years on the main sequence.

EXAMPLE: How long will a star whose mass is 4 $M_\odot$ remain on the main sequence?

Situation: Given the mass of a star, we are asked to determine its main-sequence lifetime.

Tools: We use the relationship $t \propto 1/M^{2.5}$.

Answer: The star has 4 times the mass of the Sun, so it will be on the main sequence for approximately

$$\frac{1}{4^{2.5}} = \frac{1}{4^2 \sqrt{4}} = \frac{1}{32}$$

times the Sun's main-sequence lifetime.

Thus, a 4-$M_\odot$ main-sequence star will fuse hydrogen in its core for about $(1/32) \times 10^{10}$ years, or about 3×10^8 (300 million) years.

Review: Our result makes sense: A star more massive than the Sun must have a shorter main-sequence lifetime.

---

thermal energy, as in Kelvin–Helmholtz contraction; see Section 8-4 and Section 16-1).

This new flow of heat warms the gases around the core, increasing the rate of shell hydrogen fusion and causing the shell of fusion to "eat" further outward into the surrounding matter. Helium produced by reactions in the shell falls down onto the core, which continues to contract and heat up as it gains mass. Over the course of hundreds of millions of years, the core of a 1-$M_\odot$ star compresses to about one-third of its original radius, while its

central temperature increases from about 15 million (1.5×10^7) K to about 100 million (10^8) K.

During this post–main-sequence phase, the star's outer layers expand just as dramatically as the core contracts. As the hydrogen-fusing shell works its way outward, egged on by heat from the contracting core, the star's luminosity increases substantially. This increases the star's internal pressure and makes the star's outer layers expand to many times its original radius. This tremendous expansion causes those outer layers to cool down, and the star's surface

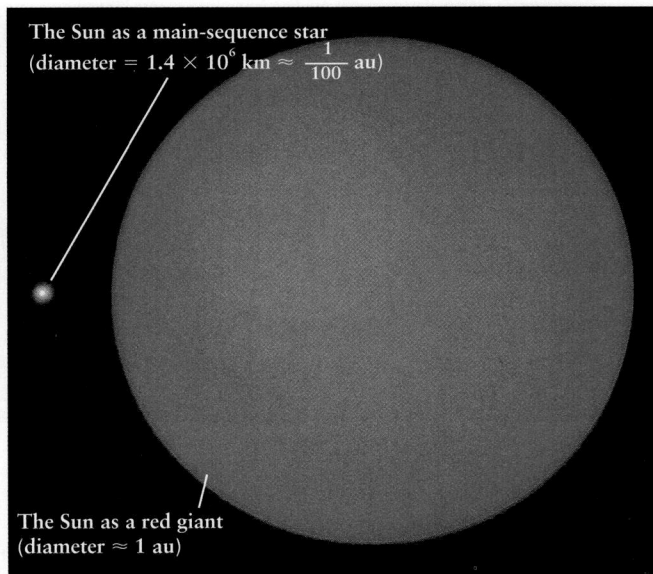

The Sun as a main-sequence star
(diameter = 1.4×10^6 km $\approx \frac{1}{100}$ au)

The Sun as a red giant
(diameter $\approx$ 1 au)

(a) The Sun today and as a red giant

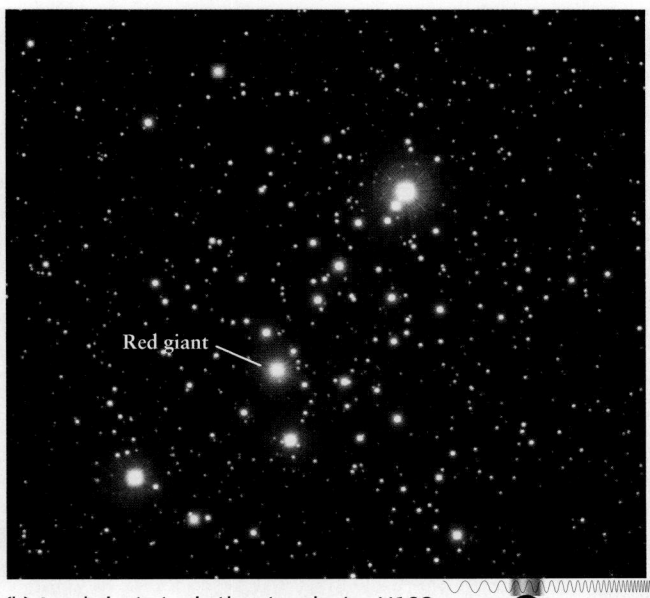

Red giant

(b) A red giant star in the star cluster M103   **R I V U X G**

FIGURE 19-5

Red Giants **(a)** The present-day Sun produces energy in a hydrogen-fusing core about 100,000 km in diameter. Some 5 billion years from now, when the Sun becomes a red giant, its energy source will be a shell only about 30,000 km in diameter within which hydrogen fusion will take place at a furious rate. The Sun's luminosity will be about 2000 times greater than today, and the increased luminosity will make the Sun's outer layers expand to approximately 100 times their present size. **(b)** The cluster M103 contains a red giant star. M103 is 14 ly across and lies about 8000 ly from Earth in the constellation Cassiopeia. (Hillary Mathis/N.A. Sharp/NOAO/AURA/NSF/Science Source)

temperature drops (see Box 19-1). Once the temperature of the star's bloated surface falls to about 3500 K, the gases glow with a reddish hue, in accordance with Wien's law (see Figure 17-7a). The star is then appropriately called a **red giant** (Figure 19-5). Thus, we see that red giant stars are former main-sequence stars that have evolved into a new stage of existence. We can summarize these observations as a general rule:

Stars join the main sequence when they begin hydrogen fusion in their cores. They leave the main sequence and become giant stars when the core hydrogen is depleted.

Red giant stars undergo substantial **mass loss** because of their large diameters and correspondingly weak surface gravity. This makes it relatively easy for gases to escape from the red giant into space. Mass loss can be detected in a star's spectrum, because gas escaping from a red giant toward a telescope on Earth produces narrow absorption lines that are slightly blueshifted by the Doppler effect (review Figure 5-26). Typical observed blueshifts correspond to a speed of about 10 km/s. A typical red giant loses roughly $10^{-7} \, M_\odot$ of matter per year. For comparison, the Sun's present-day mass loss rate is only $10^{-14} \, M_\odot$ per year. Hence, an evolving star loses a substantial amount of mass as it becomes a red giant. Figure 19-6 shows a star losing mass in this way.

The Distant Future of Our Solar System

We can use these ideas to peer into the future of our planet and our solar system. The Sun's luminosity will continue to increase as it goes through its main-sequence lifetime, and the temperature of Earth will increase with it. One and a half billion years from now, Earth's average surface temperature will be 50°C (122°F). By 3½ billion years from now, the surface temperature of Earth will exceed the boiling temperature of water. All the oceans will boil away, and Earth will become utterly incapable of supporting life (at least, near the surface). These increasingly hostile conditions will pose the ultimate challenge to whatever intelligent beings might inhabit Earth in the distant future.

About 5 billion years from now, our Sun will finish converting hydrogen into helium at its core. As the Sun's core contracts, its atmosphere will expand to envelop Mercury and perhaps reach to the orbit of Venus. Roughly 700 million years after leaving the main sequence, our red giant Sun will have swollen to a diameter of about 1 au—roughly 100 times larger than its present size—and its surface temperature will have dropped to about 3500 K. Shell hydrogen fusion will proceed at such a furious rate that our star will shine with the brightness of 2000 present-day Suns. Finding themselves *inside* the bloated Sun, some of the inner planets, possibly including Earth, will be vaporized. Even the thick atmospheres of the outer planets will evaporate away to reveal tiny, rocky cores. Thus, in its later years, the aging Sun may destroy the planets that have accompanied it since its birth.

CONCEPT CHECK 19-2

When core hydrogen burning ceases in a red giant, why does the core's temperature increase?

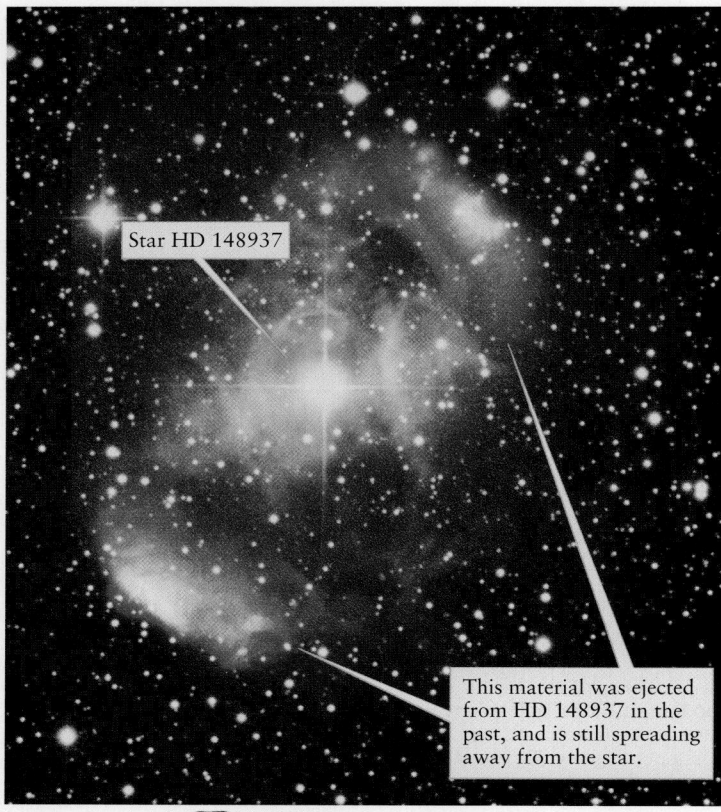

Star HD 148937

This material was ejected from HD 148937 in the past, and is still spreading away from the star.

FIGURE 19-6 R I V U X G

A Mass-Loss Star As stars age and become giant stars, they expand tremendously and shed matter into space. This star, HD 148937, is losing matter at a high rate. Other strong outbursts in the past ejected the clouds that surround HD 148937. These clouds absorb ultraviolet radiation from the star, which excites the atoms in the clouds and causes them to glow. The characteristic red color of the clouds reveals the presence of hydrogen (see Section 5-6) that was ejected from the star's outer layers. (© Australian Astronomical Observatory/David Malin Images)

CONCEPT CHECK 19-3

Why are red giants so large?

Answers appear at the end of the chapter.

19-3 Fusion of helium into carbon and oxygen begins at the center of a red giant

When a star with a mass greater than 0.4 M☉ first changes from a main-sequence star (Figure 19-7a) to a red giant (Figure 19-7b), its hydrogen-fusing shell surrounds a small, compact core of almost pure helium. In a red giant of moderately low mass, like the Sun will be 5 billion years from now, the dense helium core is about twice the diameter of Earth. Most of this core helium was *produced* by thermonuclear reactions during the star's main-sequence lifetime; during the red giant era, this helium will itself *undergo* thermonuclear reactions.

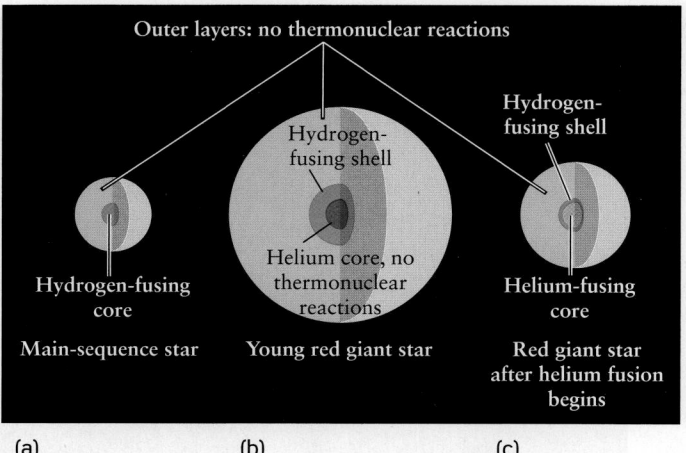

Outer layers: no thermonuclear reactions

Hydrogen-fusing shell

Hydrogen-fusing shell

Hydrogen-fusing core

Helium core, no thermonuclear reactions

Helium-fusing core

Main-sequence star   Young red giant star   Red giant star after helium fusion begins

(a)   (b)   (c)

FIGURE 19-7

Stages in the Evolution of a Star with More than 0.4 M☉
(a) During the star's main-sequence lifetime, hydrogen is converted into helium in the star's core. **(b)** When the core hydrogen is exhausted, hydrogen fusion continues in a shell, and the star expands to become a red giant. **(c)** When the temperature in the red giant's core becomes high enough because of contraction, core helium fusion begins. (These three pictures are not drawn to scale. The star is about 100 times larger in its red giant phase than in its main-sequence phase, then shrinks somewhat when core helium fusion begins.)

Core Helium Fusion

Helium, which is the left over remains from hydrogen fusion, is a potential nuclear fuel: **Helium fusion**, the thermonuclear fusion of helium nuclei to make heavier nuclei, releases energy. But this reaction cannot take place within the core of our present-day Sun because the temperature there is too low. Each helium nucleus contains two protons, so it has twice the positive electric charge of a hydrogen nucleus, and there is a much stronger electric repulsion between two helium nuclei than between two hydrogen nuclei. For helium nuclei to overcome this repulsion and get close enough to fuse together, they must be moving at very high speeds, which means that the temperature of the helium gas must be very high. (For more on the relationship between the temperature of a gas and the speed of atoms in the gas, see Box 7-2.)

When a star first becomes a red giant, the temperature of its contracted helium core is still too low for helium nuclei to fuse. But as the hydrogen-fusing shell adds mass to the helium core, the core contracts even more, further increasing the star's central temperature. When the central temperature finally reaches 100 million (10^8) K, **core helium fusion**—that is, thermonuclear fusion of helium in the core—begins. As a result, the aging star again has a central energy source for the first time since leaving the main sequence (see Figure 19-7c).

Helium fusion occurs in two steps. First, two helium nuclei combine to form a beryllium nucleus:

$$^4He + {^4He} \rightarrow {^8Be}$$

This particular beryllium isotope, which has four protons and four neutrons, is very unstable and would normally break up again into two helium nuclei soon after it forms. However, in the star's dense core, a third helium nucleus may strike the ^{8}Be nucleus before it has a chance to fall apart. Such a collision creates a stable isotope of carbon and releases energy in the form of a gamma-ray photon (γ):

$$^8\text{Be} + {}^4\text{He} \rightarrow {}^{12}\text{C} + \gamma$$

This process of fusing three helium nuclei to form a carbon nucleus is called the **triple alpha process**, because helium nuclei (^{4}He) are also called **alpha particles** by nuclear physicists. Some of the carbon nuclei created in this process can fuse with an additional helium nucleus to produce a stable isotope of oxygen and release more energy:

$$^{12}\text{C} + {}^4\text{He} \rightarrow {}^{16}\text{O} + \gamma$$

Thus, both carbon and oxygen make up the left over remains of helium fusion. The second step in the triple alpha process, and the process of oxygen formation, both release energy. The pressure resulting from this energy prevents any further gravitational contraction of the star's core. The *Cosmic Connections* figure summarizes the reactions involved in helium fusion.

It is interesting to note that ^{12}C and ^{16}O are the most common isotopes of carbon and oxygen, respectively; the vast majority of the carbon atoms in your body are ^{12}C, and almost all the oxygen you breathe is ^{16}O. We will discuss the significance of this in Section 19-5.

A mature red giant fuses helium in its core for a much shorter time than it spent fusing hydrogen in its core as a main-sequence star. For example, in the distant future the Sun will sustain helium core fusion for only about 100 million years—a period only about 1% as long as the time that hydrogen fusion occurs. While this is going on, hydrogen fusion is still continuing in a shell around the core.

CONCEPT CHECK 19-4

What is different about the core of a red giant that allows helium to fuse even though this does not occur in the core of our present-day Sun?

CONCEPT CHECK 19-5

What part of a red giant is hotter than it was during its main-sequence phase? What part is cooler?

Answers appear at the end of the chapter.

The Helium Flash and Electron Degeneracy

How helium fusion begins at a red giant's center depends on the mass of the star. In high-mass red giants (greater than about 2 to 3 M$_\odot$), helium fusion begins gradually as temperatures in the star's core approach 10^8 K. In red giants with a mass less than about 2 to 3 M$_\odot$, helium fusion begins explosively and suddenly, in what is called the **helium flash**. **Table 19-2** summarizes these differences.

The helium flash occurs because of unusual conditions that develop in the core of a moderately low-mass star as it becomes a red giant. To appreciate these conditions, we must first understand how an ordinary gas behaves. Then we can explore how the densely packed electrons at the star's center alter this behavior.

TABLE 19-2	How Helium Core Fusion Begins in Stars of Different Masses	
Mass of star	**Onset of helium fusion in core**
Less than about 0.4 M$_\odot$	None (These low-mass stars never begin helium fusion.)
More than about 0.4 but less than 2–3 M$_\odot$	Explosive (helium flash)
More than 2–3 M$_\odot$	Gradual

When a gas is compressed, it usually becomes denser and warmer. To describe this process, scientists use the convenient concept of an **ideal gas**, which has a simple relationship between pressure, temperature, and density. Specifically, the pressure exerted by an ideal gas is directly proportional to both the density and the temperature of the gas. Many real gases actually behave like an ideal gas over a wide range of temperatures and densities.

Under most circumstances, the gases inside a star act like an ideal gas. If the gas expands, it cools down, and if it is compressed, it heats up (see Box 19-1). This behavior—like a safety valve—has a stabilizing influence, ensuring that the star neither collapses nor explodes (see Section 16-2). For example, if the rate of thermonuclear reactions in the star's core should increase, the additional energy releases heat and expands the core. This expansion cools the core's gases and slows the rate of thermonuclear reactions back to the original value. Conversely, if the rate of thermonuclear reactions should decrease, the core will cool down and compress under the pressure of the overlying layers. The compression of the core will make its temperature increase, thus speeding up the thermonuclear reactions and returning them to their original rate.

In a red giant with a mass between about 0.4 M$_\odot$ and 2–3 M$_\odot$, however, the core behaves very differently from an ideal gas. The core must be compressed tremendously in order to become hot enough for helium fusion to begin. At these extreme pressures and temperatures, the atoms are completely ionized, and most of the core consists of nuclei and detached electrons. Eventually, the free electrons become so closely crowded that a limit to further compression is reached, as predicted by a remarkable law of quantum mechanics called the **Pauli exclusion principle**. Formulated in 1925 by the Austrian physicist Wolfgang Pauli, this principle states that two electrons cannot simultaneously occupy the same quantum state. A quantum state is a particular set of circumstances concerning locations and speeds that are available to a particle. In the quantum world of electrons, the Pauli exclusion principle is analogous to saying that you can't have two things in the same place at the same time.

Just before the onset of helium fusion, the electrons in the core of a low-mass star are so closely crowded together that any further compression would violate the Pauli exclusion principle. Because the electrons cannot be squeezed any closer together, they produce a powerful pressure that resists further core contraction.

This phenomenon, in which closely packed particles resist compression as a consequence of the Pauli exclusion principle,

> Before helium fusion begins in a red giant's core, the helium gas there behaves like a metal.

Helium Fusion in a Red Giant

ION 7 A star becomes a red giant after the fusion of hydrogen into helium in its core has come to an end. As the red giant's core shrinks and heats up, a new cycle reactions can occur that create the even heavier elements on and oxygen.

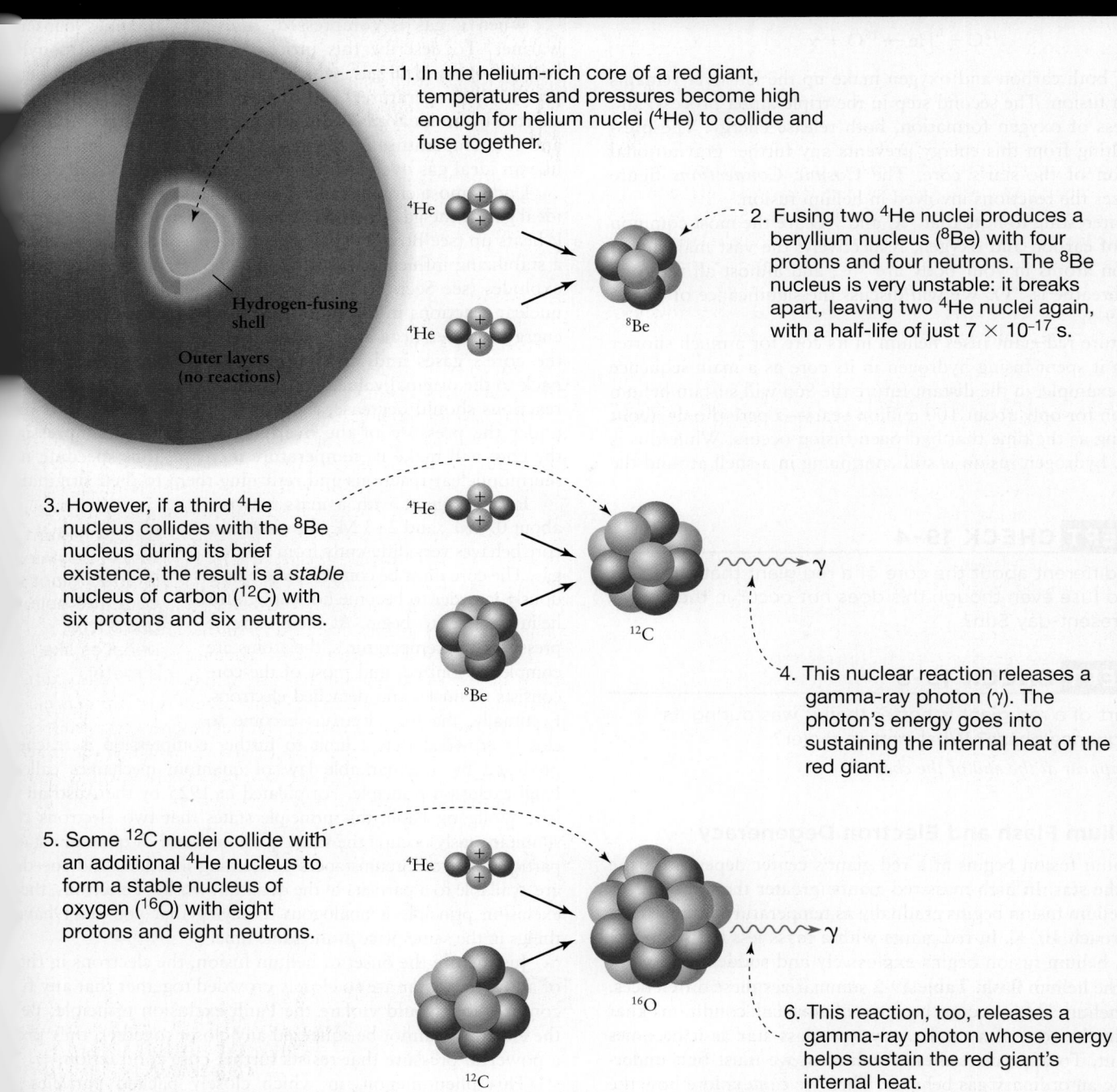

1. In the helium-rich core of a red giant, temperatures and pressures become high enough for helium nuclei (^{4}He) to collide and fuse together.

Hydrogen-fusing shell

Outer layers (no reactions)

^{4}He

^{4}He

^{8}Be

2. Fusing two ^{4}He nuclei produces a beryllium nucleus (^{8}Be) with four protons and four neutrons. The ^{8}Be nucleus is very unstable: it breaks apart, leaving two ^{4}He nuclei again, with a half-life of just 7×10^{-17} s.

3. However, if a third ^{4}He nucleus collides with the ^{8}Be nucleus during its brief existence, the result is a *stable* nucleus of carbon (^{12}C) with six protons and six neutrons.

^{4}He

^{8}Be

^{12}C

γ

4. This nuclear reaction releases a gamma-ray photon (γ). The photon's energy goes into sustaining the internal heat of the red giant.

5. Some ^{12}C nuclei collide with an additional ^{4}He nucleus to form a stable nucleus of oxygen (^{16}O) with eight protons and eight neutrons.

^{4}He

^{12}C

^{16}O

γ

6. This reaction, too, releases a gamma-ray photon whose energy helps sustain the red giant's internal heat.

is called **degeneracy**. Astronomers say that the electrons in the helium-rich core of a low-mass red giant are "degenerate" and that the core is supported by **degenerate-electron pressure**. This degenerate pressure, unlike the pressure of an ideal gas, does not depend on temperature. Remarkably, you can find degenerate electrons on Earth in an ordinary piece of metal (Figure 19-8).

When the temperature in the core of a low-mass red giant reaches the high level required for the triple alpha process, energy begins to be released. The helium heats up, which makes the triple alpha process happen even faster. However, the pressure provided by the degenerate electrons is independent of the temperature, so the pressure does not change. Recall that an ideal gas's increase in pressure with temperature acted as a stabilizing influence in the core. Now, without the "safety valve" of increasing pressure, the star's core cannot expand and cool. The rising temperature causes the helium to fuse at an ever-increasing rate, producing the helium flash.

At this stage, the pressure (arising from degeneracy) is roughly constant, but the temperature is rising due to helium fusion. However, electrons also *lose* their degeneracy at even higher temperatures. Eventually, the temperature becomes so high that the electrons in the core are no longer degenerate. The electrons then behave like an ideal gas and the star's core expands, terminating the helium flash. These events occur so rapidly that the helium flash is over in seconds, after which the star's core settles down to a steady rate of helium fusion.

CAUTION!

The term "helium flash" might give you the impression that a star emits a sudden flash of light when the helium flash occurs. If this were true, it would be an incredible sight. During the brief time interval when the helium flash occurs, the helium-fusing core is 10^{11} times more luminous than the present-day Sun, which is similar to the total luminosity of all the stars in the Milky Way Galaxy! But, in fact, the helium flash has no immediately visible consequences—for two reasons. First, much of the energy released during the helium flash goes into heating the core and terminating the degenerate state of the electrons. Second, the energy that does escape the core is largely absorbed by the star's outer layers and must work its way slowly to the surface (just like the Sun's present-day interior; see Section 16-2). Therefore, the explosive drama of the helium flash takes place where it cannot be seen directly.

The Continuing Evolution of a Red Giant

Whether a helium flash occurs or not, the onset of core helium fusion actually causes a *decrease* in the luminosity of the star. This decrease is the opposite of what you might expect—after all, turning on a new energy source should make the luminosity greater, not smaller, right? What happens is that after the onset of core helium fusion, a star's superheated core expands like an ideal gas. (If the star is of sufficiently low mass to have had a degenerate core, the increased temperature after the helium flash makes the core too hot to remain degenerate. Hence, these stars also end up with cores that behave like ideal gases.) Temperatures drop

FIGURE 19-8 R I ⓥ U X G

Degenerate Electrons In an ordinary piece of metal, like the chrome grille on this classic car, the electrons are affected by the Pauli exclusion principle. The resulting degenerate electron pressure helps make metals strong and difficult to compress. A more powerful version of this same effect happens inside the cores of low-mass red giant stars. (Santokh Kochar/Getty Images)

around the expanding core, so the hydrogen-fusing shell reduces its energy output and the star's luminosity decreases. This temperature decrease allows the star's outer layers to contract and heat up. Consequently, a post–helium-flash star is less luminous, hotter at the surface, and smaller than a red giant.

Core helium fusion lasts for only a relatively short time. Calculations suggest that a 1-$M_\odot$ star like the Sun sustains core hydrogen fusion for about 10 billion (10^{10}) years, followed by about 250 million (2.5×10^8) years of shell hydrogen fusion leading up to the helium flash. After the helium flash, such a star can fuse helium in its core (while simultaneously fusing hydrogen in a shell around the core) for only 100 million (10^8) years, a mere 1% of its main-sequence lifetime. Figure 19-9 summarizes these evolutionary stages in the life of a 1-$M_\odot$ star. In Chapter 20, we will take up the story of what happens after a star has consumed all the helium in its core.

Here is the story of post–main-sequence evolution in its briefest form: Before the beginning of core helium fusion, the star's core compresses and the outer layers expand, and just after core helium fusion begins, the core expands and the outer layers compress. We will see in Chapter 20 that this behavior, in which the inner and outer regions of the star change in opposite ways, occurs again and again in the final stages of a star's evolution.

CONCEPT CHECK 19-6

What is the "safety valve" in a main-sequence star's nuclear reactions, and how is it not in place for a red giant with a core supported by degenerate-electron pressure?

CONCEPT CHECK 19-7

What happens to the luminosity when the helium-fusing core is no longer supported by degenerate-electron pressure?

Answers appear at the end of the chapter.

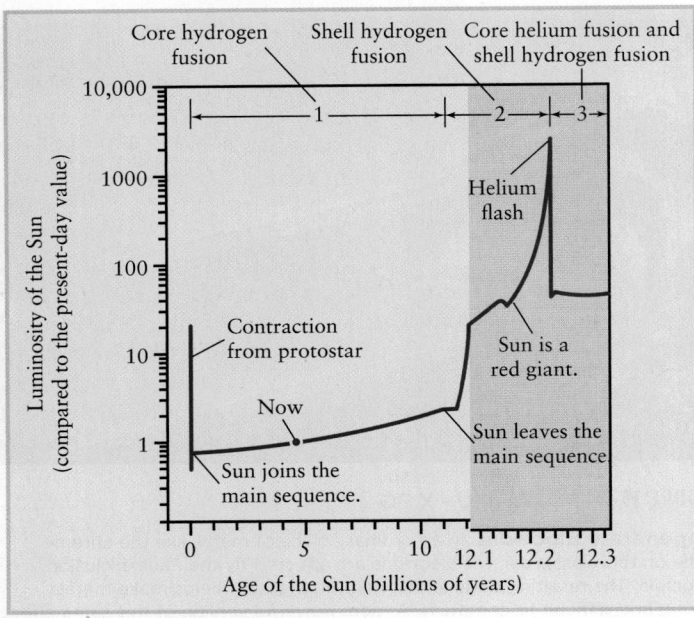

FIGURE 19-9

Stages in the Evolution of the Sun This diagram shows how the luminosity of the Sun (a 1-M☉ star) changes over time. The Sun began as a protostar whose luminosity decreased rapidly as the protostar contracted. Once established as a main-sequence star with core hydrogen fusion, the Sun's luminosity increased slowly over billions of years. The post–main-sequence evolution was much more rapid, so a different timescale is used in the right-hand portion of the graph. (Data from Mark A. Garlick, based on calculations by I.-Juliana Sackmann and Kathleen E. Kramer)

19-4 H-R diagrams and observations of star clusters reveal how red giants evolve

TUTORIAL 19.2 To see how stars evolve during and after their main-sequence lifetimes, it is helpful to follow them on a Hertzsprung–Russell (H-R) diagram. In such a diagram, zero-age main-sequence stars lie along a line called the **zero-age main sequence**, or **ZAMS** (Figure 19-10a). These stars have just emerged from their protostar stage, are steadily fusing hydrogen into helium in their cores, and have attained hydrostatic equilibrium. With the passage of time, hydrogen in a main-sequence star's core is converted to helium, the luminosity slowly increases, the star slowly expands, and the star's position on the H-R diagram inches away from the ZAMS. As a result, the main sequence on an H-R diagram is a fairly broad band rather than a narrow line (Figure 19-10b).

Post–Main-Sequence Evolution on an H-R Diagram

The dashed line in Figure 19-10a denotes stars whose cores have been exhausted of hydrogen and in which core hydrogen fusion has ceased. These stars have reached the ends of their main-sequence lifetimes. From there, the points representing high-mass stars (3 M☉, 5 M☉, and 9 M☉) move rapidly from left to right across the H-R diagram. This means that, although the star's surface temperature is decreasing, its surface area is increasing at a rate that keeps its overall luminosity roughly constant. During this transition, the star's core contracts and its outer layers expand as energy flows outward from the hydrogen-fusing shell.

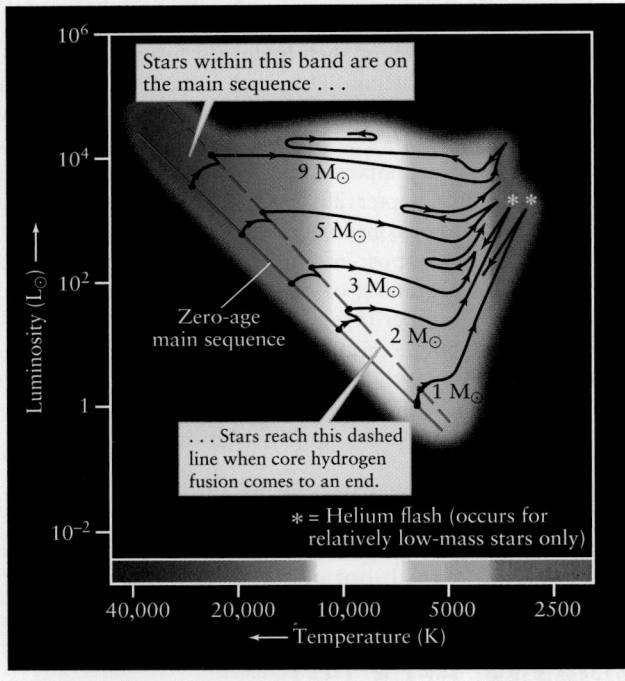

(a) Post–main-sequence evolutionary tracks of five stars with different mass

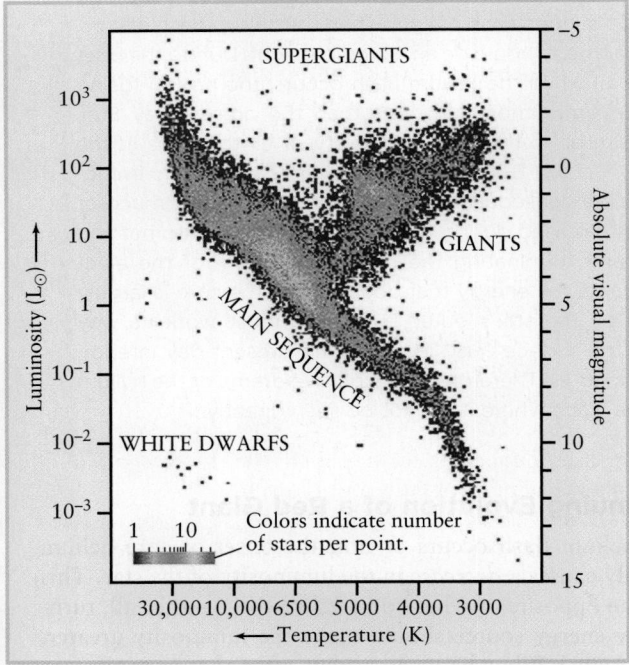

(b) H-R diagram of 20,853 stars—note the width of the main sequence

FIGURE 19-10

H-R Diagrams of Stellar Evolution On and Off the Main Sequence
(a) The two lowest-mass stars shown here (1 M☉ and 2 M☉) undergo a helium flash at their centers, as shown by the asterisks. In the high-mass stars, core helium fusion ignites more gradually where the evolutionary tracks make a sharp downward turn in the red giant region on the right-hand side of the H-R diagram. **(b)** Data from the Hipparcos satellite (see Section 17-1) was used to create this H-R diagram. The thickness of the main sequence is due in large part to stars evolving during their main-sequence lifetimes. (a: Data from I. Iben; b: Data from M. A. C. Perryman)

Just before core helium fusion begins, the evolutionary tracks of high-mass stars turn upward in the red giant region of the H-R diagram (to the upper right of the main sequence). After core helium fusion begins, however, the cores of these stars expand, the outer layers contract, and the evolutionary tracks back away from these temporary peak luminosities. The tracks then wander back and forth in the red giant region while the stars readjust to their new energy sources.

Figure 19-10a also shows the evolutionary tracks of two stars of moderately low mass (1 $M_\odot$ and 2 $M_\odot$). The onset of core helium fusion in these stars occurs with a helium flash, indicated by the red asterisks in the figure. As we saw in the previous section, after the helium flash, these stars shrink and become less luminous. The decrease in size is proportionally greater than the decrease in luminosity, so the surface temperatures increase. Hence, after the helium flash, the evolutionary tracks for the 1-$M_\odot$ and 2-$M_\odot$ stars move down and to the left.

A Simulated Star Cluster: Tracking 4½ Billion Years of Stellar Evolution

We can summarize our understanding of stellar evolution from birth through the onset of helium fusion by following the evolution of a hypothetical cluster of stars. We saw in Section 18-6 that the stars that make up a cluster all begin to form at essentially the same time but have different initial masses. Hence, studying star clusters allows us to compare how stars of different masses evolve.

The eight H-R diagrams in **Figure 19-11** are from a computer simulation of the evolution of 100 stars that all form at the same moment and differ only in initial mass. All 100 stars begin as cool protostars on the right side of the H-R diagram (see Figure 19-11a). The protostars are spread out on the diagram according to their masses, and the greater the mass, the greater the protostar's initial luminosity. As we saw in Section 18-3, the source of a protostar's luminosity is its gravitational energy. As the protostar contracts, this gravitational energy is converted to thermal energy and radiated into space.

The most massive protostars contract and heat up very rapidly. After only 5000 years, they have already moved across the H-R diagram toward the main sequence (see Figure 19-11b). After 100,000 years, these massive stars have ignited hydrogen fusion in their cores and have settled down on the main sequence as O stars (see Figure 19-11c). After 3 million years, stars of moderate mass have also ignited core hydrogen fusion and become main-sequence stars of spectral classes B and A (see Figure 19-11d). Meanwhile, low-mass protostars continue to inch their way toward the main sequence as they leisurely contract and heat up.

After 30 million years (see Figure 19-11e), the most massive stars have depleted the hydrogen in their cores and become red giants. These stars have moved from the upper left end of the main sequence to the upper right corner of the H-R diagram. (This simulation follows stars only to the red giant stage, after which they are simply deleted from the diagram.) Intermediate-mass stars lie on the main sequence, while the lowest-mass stars are still in the protostar stage and lie above the main sequence.

After 66 million years (see Figure 19-11f), even the lowest-mass protostars have finally ignited core hydrogen fusion and have settled down on the main sequence as cool, dim M stars. These lowest-mass stars can continue to fuse hydrogen in their cores for hundreds of billions of years.

In the final two H-R diagrams (Figures 19-11g and 19-11h), the main sequence stars get "peeled" or "eaten away" from the upper left to the lower right as stars exhaust their core supplies of hydrogen and evolve into red giants. The stars that leave the main sequence between Figure 19-11g and Figure 19-11h have masses between about 1 $M_\odot$ and 3 $M_\odot$ and undergo the helium flash in their cores.

For all stars in this simulation, the giant stage lasts only a brief time compared to the star's main-sequence lifetime. Compared to a 1-$M_\odot$ star (see Figure 19-9), a more massive star has a shorter main-sequence lifetime *and* spends a shorter time as a giant star. Thus, at any given time, only a small fraction of the stellar population is passing through the giant stage. Hence, most of the stars we can see through telescopes are main-sequence stars. As an example, of the stars within 4.00 pc (13.05 ly) of the Sun listed in Appendix 4, only one—Procyon A—is presently evolving from a main-sequence star into a giant. Two other nearby stars are white dwarfs, an even later stage in stellar evolution that we will discuss in Chapter 20. (By contrast, most of the *brightest* stars listed in Appendix 5 are giants and supergiants. Although they make up only a small fraction of the stellar population, these stars stand out due to their extreme luminosity.)

CONCEPT CHECK 19-8

In Figure 19-11, is there an age when the most massive stars in the cluster have become red giants, yet the least massive stars are finally approaching the main sequence?

CONCEPT CHECK 19-9

Suppose you happened to observe the simulated cluster in Figure 19-11 just as its most massive stars were joining the main sequence. How old would the cluster be?

Answers appear at the end of the chapter.

Real Star Clusters: Cluster Ages and Turnoff Points

The evolution of the hypothetical cluster displayed in Figure 19-11 helps us interpret what we see in actual star clusters. We can observe the early stages of stellar evolution in **open clusters**, which typically contain a few hundred to a few thousand stars. Many open clusters are just a few million years old, so their H-R diagrams resemble Figure 19-11d, 19-11e, or 19-11f (see Section 18-6, especially Figure 18-22 and Figure 18-23).

Star clusters tend to form in a very short period of time compared to how long their stars shine, which means each cluster has an age—the time since its formation. **Figure 19-12** shows two open clusters of different ages. The nearer cluster, called M35, must be relatively young because it contains several dozen luminous, blue, high-mass main-sequence stars. These stars lie in the upper part of the main sequence on an H-R diagram. They have main-sequence lifetimes of only a few

As a cluster ages and its stars end their main-sequence lifetimes, the cluster's color changes from blue to red.

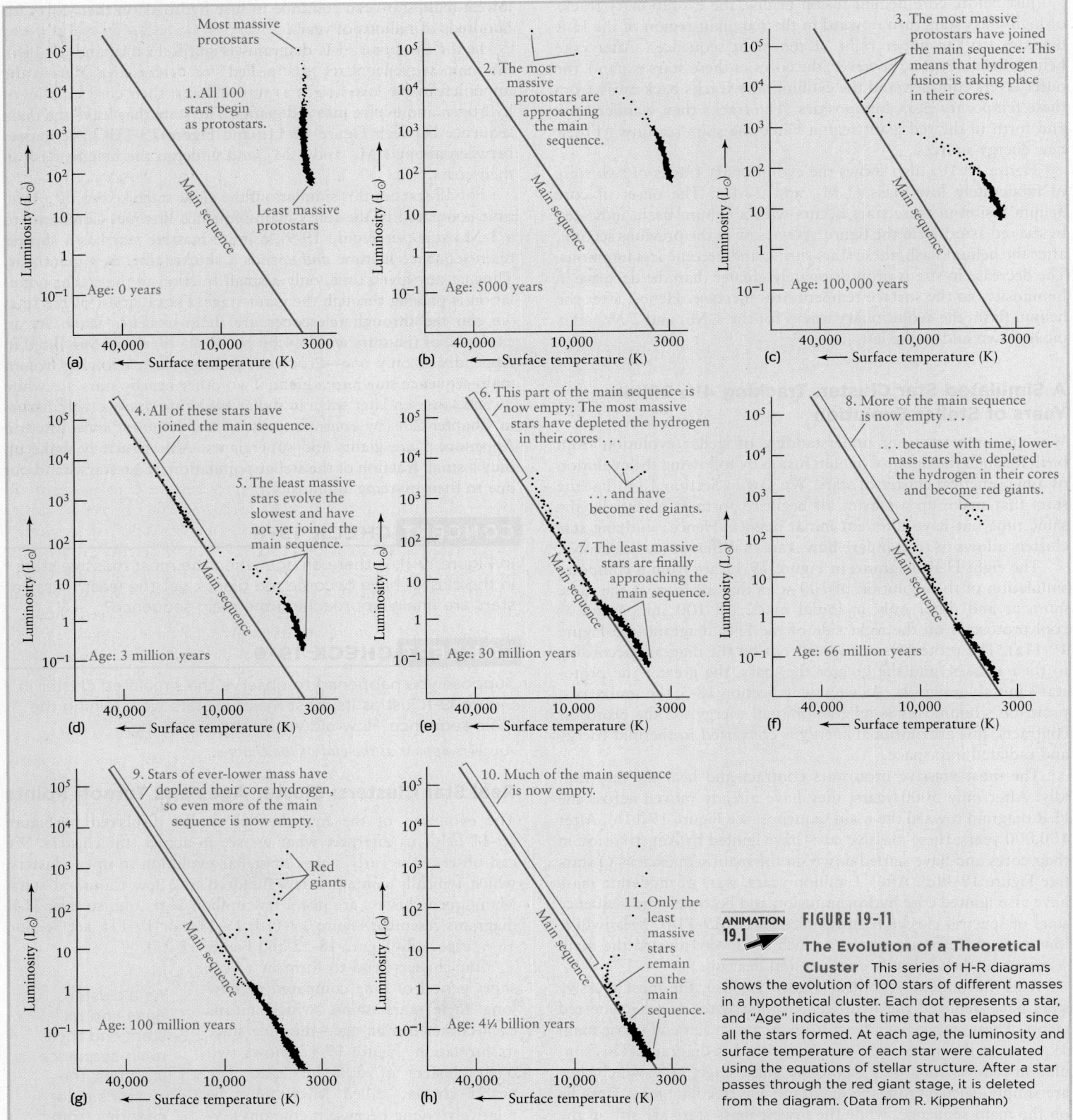

ANIMATION 19.1

FIGURE 19-11

The Evolution of a Theoretical Cluster This series of H-R diagrams shows the evolution of 100 stars of different masses in a hypothetical cluster. Each dot represents a star, and "Age" indicates the time that has elapsed since all the stars formed. At each age, the luminosity and surface temperature of each star were calculated using the equations of stellar structure. After a star passes through the red giant stage, it is deleted from the diagram. (Data from R. Kippenhahn)

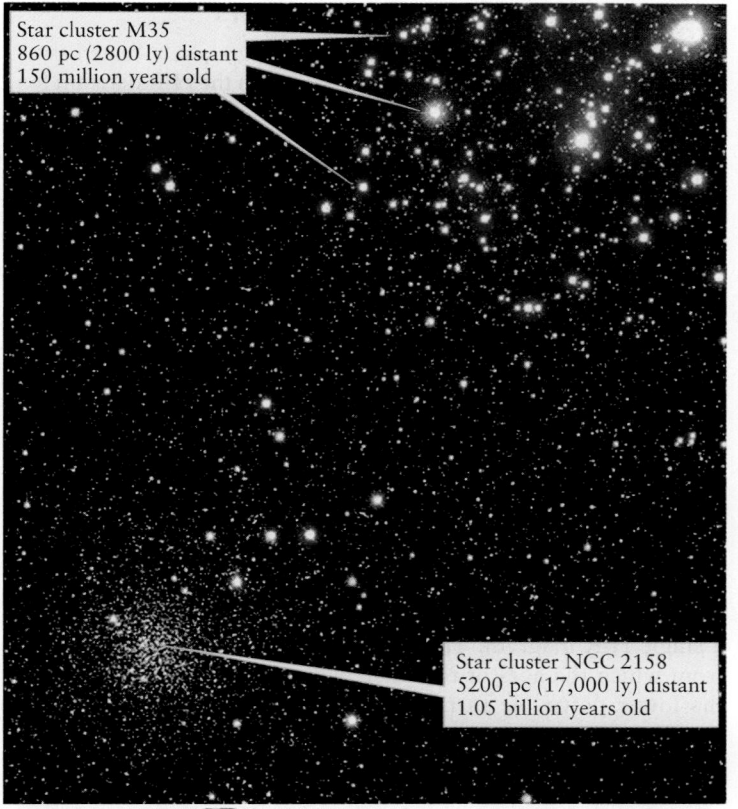

Star cluster M35
860 pc (2800 ly) distant
150 million years old

Star cluster NGC 2158
5200 pc (17,000 ly) distant
1.05 billion years old

FIGURE 19-12 R I **V** U X G

Two Open Clusters The two clusters in this image, M35 and NGC 2158, lie in almost the same direction in the constellation Gemini. The nearer cluster, M35, has a number of luminous blue main-sequence stars with surface temperatures around 10,000 K as well as a few red giants. Hence, its H-R diagram resembles that shown in Figure 19-11g, and its age is around 100 million years (more accurately, around 150 million). The more distant cluster, NGC 2158, has no blue main-sequence stars; long ago, all of these massive stars came to the end of their main-sequence lifetimes and became giants. The H-R diagram for NGC 2158 is intermediate between Figure 19-11g and Figure 19-11h, and its age (1.05 billion years) is as well. (© Jean-Charles Cuillandre, Hawaiian Starlight, CFHT)

hundred million years, so M35 can be no older than that. Some of the most luminous stars in M35 are red or yellow in color; these are stars that ended their main-sequence lifetimes some time ago and have evolved into red giants. The H-R diagram for this cluster resembles Figure 19-11g.

There are no high-mass blue main-sequence stars at all in NGC 2158, the more distant cluster shown in Figure 19-12. Any such stars that were once in NGC 2158 have long since come to the end of their main-sequence lifetimes. As a result, the main sequence in this cluster has been "eaten away" more than that of M35, leaving only stars that are yellow or red in color. This tells us that NGC 2158 must be older than M35 (compare Figures 19-11g and 19-11h). This example shows that as a cluster ages, it generally becomes redder in its average color.

We can see even later stages in stellar evolution by studying **globular clusters**, so called because of their spherical shape. A typical globular cluster contains up to 1 million stars in a volume less than 100 pc across (Figure 19-13). Among these are many highly evolved post–main-sequence stars.

Globular clusters must be old, because they contain no high-mass stars on the main-sequence. To determine that these clusters are old, you would measure the apparent magnitude (a measure of apparent brightness, which we introduced in Section 17-3) and color ratio of many stars in a globular cluster, then plot the data as shown in Figure 19-14. Such a **color-magnitude diagram** for a cluster is equivalent to an H-R diagram. The color ratio of a star tells you its surface temperature (as described in Section 17-4), and because all the stars in the cluster are at essentially the same distance from us, their relative brightnesses indicate their relative luminosities. What you would discover is that a globular cluster's main sequence has been "peeled" or "eaten away" even more extensively than the open cluster NGC 2158 in Figure 19-12. Hence, globular clusters must be even older than NGC 2158. In a typical globular cluster, all the main-sequence stars with masses more than about 1 M☉ or 2 M☉ evolved long ago into red giants. Only low-mass, slowly evolving stars still have core hydrogen fusion. (Compare Figure 19-14 with Figure 19-11h.) As we will see next, not only can we tell that globular clusters are old, we can also figure out just how old they are.

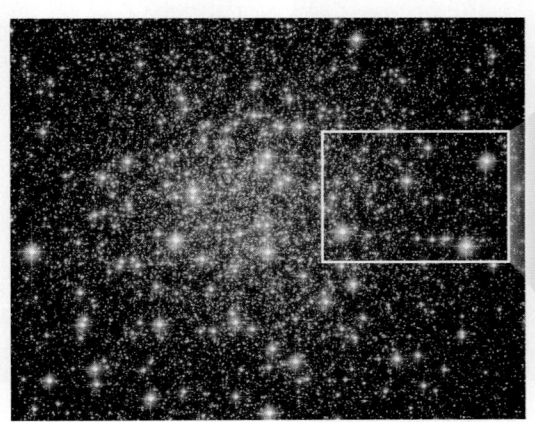

FIGURE 19-13 R **I** V U X G

A Globular Cluster This cluster, called M10, contains a few hundred thousand stars within a diameter of only 20 pc (70 ly). It lies approximately 5000 pc (16,000 ly) from Earth in the constellation Ophiuchus (the Serpent

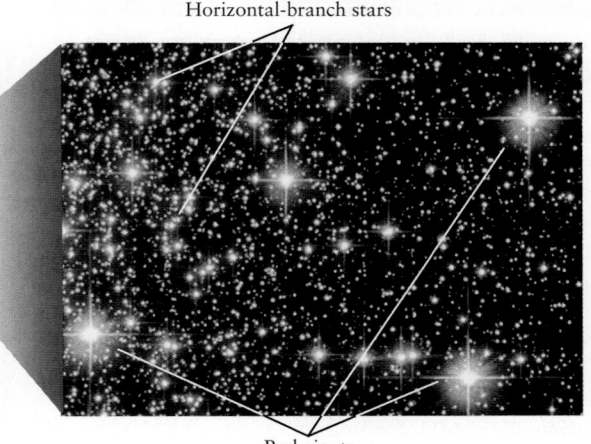

Horizontal-branch stars

Red giants

Holder). Most of the stars in this image are either red giants or blue, horizontal-branch stars with both core helium fusion and shell hydrogen fusion. (ESA/Hubble & NASA)

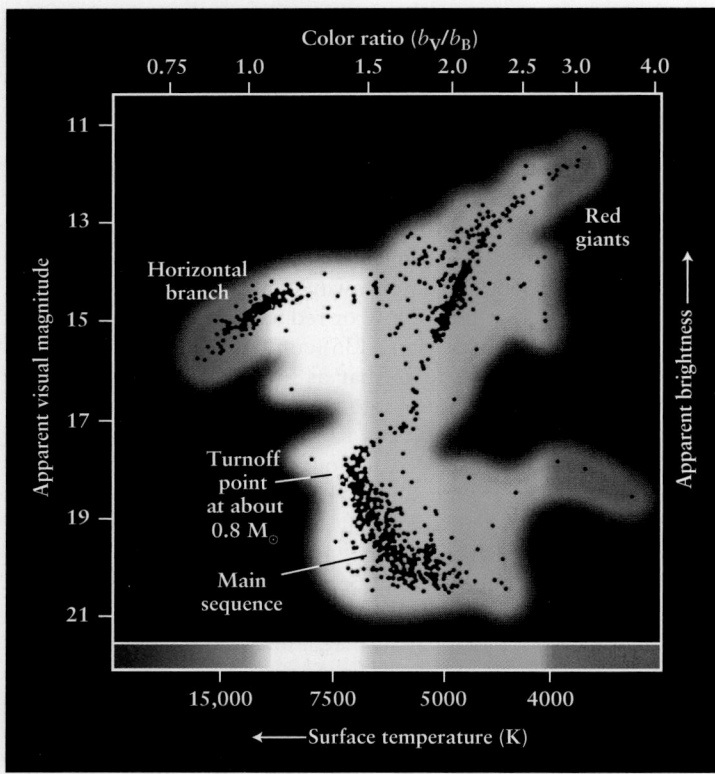

FIGURE 19-14

Age of a Globular Cluster This color-magnitude diagram is equivalent to an H-R diagram and reveals the age of the cluster. Below the turnoff point, stars are still on the main sequence. Above the turnoff point, higher-mass stars have already moved off the main sequence. The turnoff point occurs for stars around 0.8 $M_\odot$; since stars of this mass are only on the main sequence for about 10 billion years, this tells us the age of the cluster. Each dot in this diagram represents the apparent visual magnitude (a measure of the brightness as seen through a V filter) and surface temperature (as measured by the color ratio b_V/b_B) of a star in the globular cluster M55 in Sagittarius. (Data from D. Schade, D. VandenBerg, and F. Hartwick)

CAUTION!

The inset in Figure 19-13 shows something surprising: There are luminous *blue* stars in the ancient globular cluster M10. This seems to contradict our earlier statements that blue main-sequence stars evolve into red giants after just a few hundred million years. The explanation is that these are *not* main-sequence stars, but rather **horizontal-branch stars**. These stars get their name because in the color-magnitude diagram of a globular cluster, they form a horizontal grouping in the left-of-center portion of the diagram (see Figure 19-14). Horizontal-branch stars are relatively low-mass stars that have already become red giants and undergone a helium flash, so there is both core helium fusion and shell hydrogen fusion taking place in their interiors. After the helium flash, their luminosity decreased to about 50 $L_\odot$ (compared to about 1000 $L_\odot$ before the flash) and their outer layers contracted and heated, giving these stars their blue color. In years to come, these stars will move back toward the red giant region as their fuel is devoured. Our own Sun will go through a horizontal-branch phase in the distant future; this is the phase labeled by the number 3 at the far right of Figure 19-9.

Measuring the Ages of Star Clusters

The idea that a cluster's main sequence is progressively "eaten away" is the key to determining the age of a cluster. In the H-R diagram for a very young cluster, all the stars are on or near the main sequence. (An example is the open cluster NGC 2264, shown in Figure 18-22.) As a cluster gets older, however, stars begin to leave the main sequence. The high-mass, high-luminosity stars are the first to consume their core hydrogen and become red giants. As time passes, fewer and fewer stars remain on the main sequence (see parts d through h of Figure 19-11).

The age of a cluster can be found from the **turnoff point**, which is at the peak of the surviving portion of the main-sequence stars on the cluster's H-R diagram (see Figure 19-14). The stars at the turnoff point are just at the stage of exhausting the hydrogen in their cores, so their main-sequence lifetime is equal to the age of the cluster. Since we know the main-sequence lifetimes of stars with differing masses through modeling, we are able to determine how long ago the cluster formed. For example, in the case of the globular cluster M55 plotted in Figure 19-14, 0.8-$M_\odot$ stars have just left the main sequence, indicating that the cluster's age is more than 10 billion (10^{10}) years (see Table 19-1).

Figure 19-15 shows data for several star clusters plotted on a single H-R diagram. This graph also shows turnoff-point times from which the ages of the clusters can be estimated. Notice that the lower down the turnoff point occurs (which corresponds to lower masses moving off the main sequence), the older the cluster.

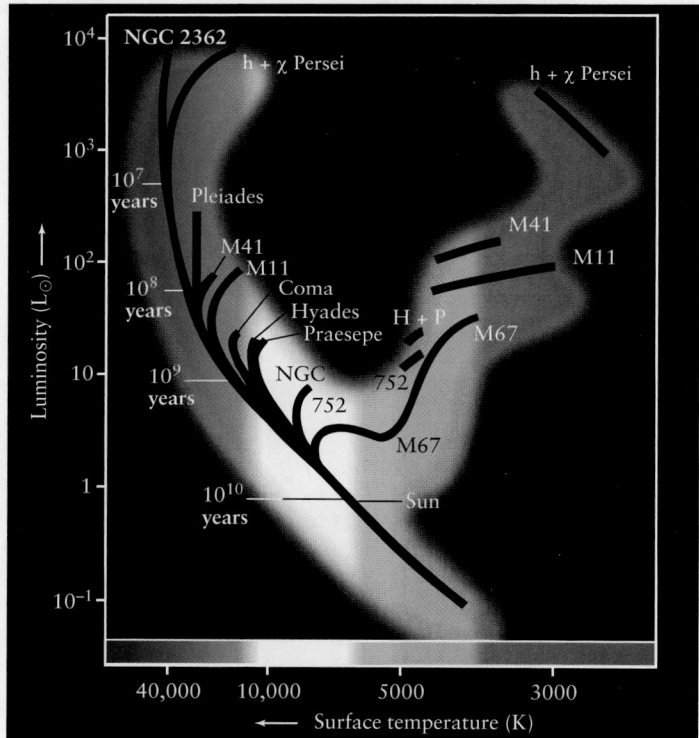

FIGURE 19-15

An H-R Diagram for Open Star Clusters The black bands indicate where stars from various open clusters fall on the H-R diagram. The age of a cluster can be estimated from the location of the cluster's turnoff point, where the cluster's most massive stars are just now leaving the main sequence. The times for these turnoff points are listed alongside the main sequence. For example, the Pleiades cluster turnoff point is near the 10^8-year point, so this cluster is about 10^8 years old. (Data from A. Sandage)

Suppose you knew that certain birthday candles took 10 minutes to burn down. If you walked into a room and saw that these candles were burned halfway down on a birthday cake, you would know they were lit 5 minutes ago. In this analogy, knowing the rate that the candle burns is analogous to knowing how long a star of a given mass will remain on the main sequence.

More than just finding the age of a cluster, we have seen how each main-sequence star has a lifetime for how long it will remain on the main sequence. Thus, lifetime is additional information that can be indicated in an H-R diagram. The *Cosmic Connections* figure on the next page summarizes the stellar properties we have already covered since Chapter 17 that can be noted in an H-R diagram.

CONCEPT CHECK 19-10

The turnoff point in Figure 19-14 corresponds to main-sequence stars that are about 7500 K. Suppose that the turnoff point was instead for stars of around 10,000 K. For a 10,000-K turnoff point, would the cluster be younger or older than 10 billion years?

CONCEPT CHECK 19-11

Using Figure 19-15, explain why turnoff points at lower luminosities correspond to older clusters.

Answers appear at the end of the chapter.

19-5 Different populations of stars have developed over time

Studies of star clusters reveal a curious difference between the youngest and oldest stars in our galaxy. Stars in the youngest clusters (those with most of their main sequences still intact) are said to be **metal rich**, because their spectra contain many prominent spectral lines of heavy elements. (Recall from Section 17-5 that astronomers use the term *metal* to denote any element other

than hydrogen and helium, which are the two lightest elements.) Such stars are also called **Population I stars**. The Sun is a relatively young, metal-rich, Population I star.

By contrast, the spectra of stars in the oldest clusters show only weak lines of heavy elements. These ancient stars are thus said to be **metal poor**, because heavy elements are only about 3% as abundant in these stars as in the Sun. They are also called Population II **stars**. The stars in globular clusters are metal-poor, Population II stars. **Figure 19-16** shows the difference in spectra between a metal-poor, Population II star and the Sun (a metal-rich, Population I star).

Even metal-poor Population II stars have more heavier metals than would have been around for the very first population of stars, since only the stars create these metals. This very first generation of metal-free stars is called **Population III**, yet none have been convincingly detected. It's expected that these stars were as large as hundreds of solar masses—much larger than any stars around today—and that they had very brief lifetimes.

Note that "metal rich" and "metal poor" are relative terms. In even the most metal-rich star known, metals make up just a few percent of the total mass of the star.

Stellar Populations and the Origin of Heavy Elements

To explain why there are two distinct populations of stars, we must go back to the Big Bang, the explosive origin of the universe that took place some 13.7 billion years ago. As we will discuss in Chapter 26, the early universe consisted almost exclusively of hydrogen and helium, with almost no heavy elements (metals). The first stars to form were likewise metal poor. The least massive of these stars have survived to the present day and are now the ancient stars of Population II.

The more massive of the original stars evolved more rapidly and no longer shine. But as these stars evolved, helium fusion in their cores produced metals—carbon and oxygen. In the most massive stars, as we will learn in Chapter 20, further thermonuclear reactions produced even heavier elements. As these massive original

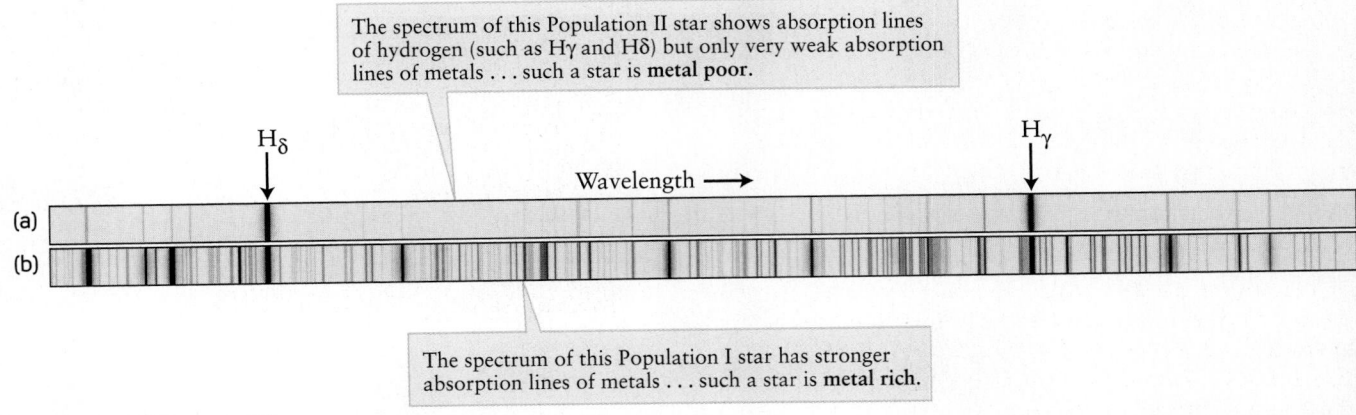

The spectrum of this Population II star shows absorption lines of hydrogen (such as Hγ and Hδ) but only very weak absorption lines of metals . . . such a star is **metal poor**.

H$_\delta$ Wavelength ⟶   H$_\gamma$

(a)

(b)

The spectrum of this Population I star has stronger absorption lines of metals . . . such a star is **metal rich**.

FIGURE 19-16 R I **V** U X G

Spectra of a Metal-Poor Star and a Metal-Rich Star The abundance of metals (elements heavier than hydrogen and helium) in a star can be inferred from its spectrum. These spectra compare **(a)** a metal-poor, Population II star and **(b)** a metal-rich, Population I star (the Sun) of the same surface temperature. We described the hydrogen absorption lines H$_\gamma$ (wavelength 434 nm) and H$_\delta$ (wavelength 410 nm) in Section 5-8.

COSMIC CONNECTIONS

Stellar Lifetimes on the Main Sequence

A star's location on the H-R diagram reveals more than its temperature and luminosity. The farther up the main sequence, the more massive the star and the shorter its lifetime before leaving the main sequence to become a red giant. Also, the radius of the star increases as you move toward the upper right. Although clearly not part of any single cluster, several familiar stars are labeled.

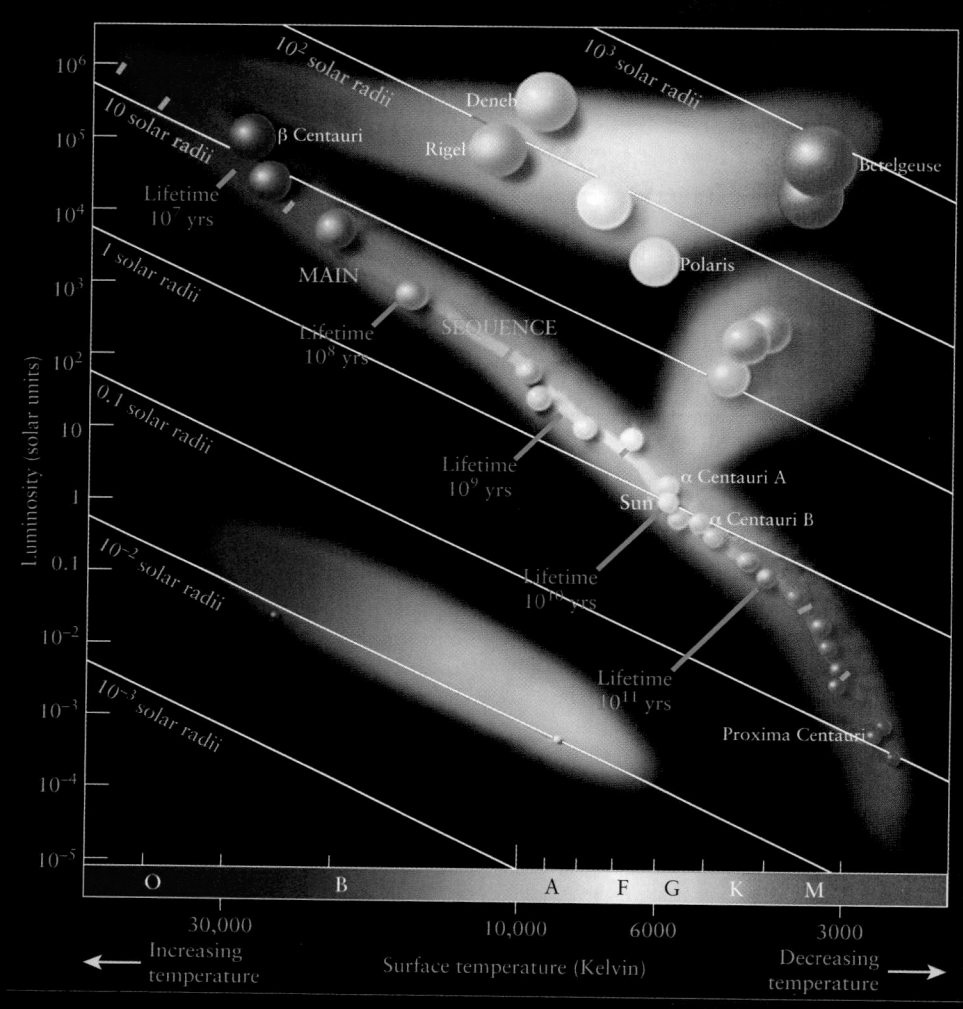

stars aged and died, they expelled their metal-enriched gases into space. (The star shown in Figure 19-6 is going through such a mass-loss phase late in its life.) This expelled material joined the interstellar medium and was eventually incorporated into a second generation of stars that have a higher concentration of heavy elements. These metal-rich members of the second stellar generation are the Population I stars, of which our Sun is an example.

> Stars like the Sun contain material that was processed through an earlier generation of stars.

CAUTION!

Be careful not to let the designations of the three stellar populations confuse you. Population *I* stars are members of a *second* stellar generation, while Population *II* stars belong to an older generation. And the hypothesized Population *III* stars are the *first* generation. Unfortunately, the names were assigned in the order in which these populations of stars were discovered or proposed, and not the order in which they arise in a galaxy.

The relatively high concentration of heavy elements in the Sun means that the solar nebula, from which both the Sun and planets formed (see Section 8-4), must likewise have been metal rich. Earth is composed almost entirely of heavy elements, as are our bodies. Thus, our very existence is intimately linked to the Sun's being a Population I star. A planet like Earth probably could not have formed from the metal-poor gases that went into making Population II stars.

The concept of different stellar populations provides insight into our own origins. Recall from Section 19-3 that helium fusion in red giant stars produces the same isotopes of carbon (^{12}C) and oxygen (^{16}O) that are found most commonly on Earth. The reason is that Earth's carbon and oxygen atoms, including all of those in your body, actually *were* produced by helium fusion. These reactions occurred billions of years ago within an earlier generation of stars that died and gave up their atoms to the interstellar medium—the same atoms that later became part of our solar system, our planet, and our bodies. We are literally children of the stars.

CONCEPT CHECK 19-12

How do we know that Population II stars had to form before Population I stars?

Answer appears at the end of the chapter.

19-6 Many mature stars pulsate

ANIMATION 19.2

We saw in Section 16-3 that the surface of our Sun vibrates in and out, although by only a small amount. But other stars undergo substantial changes in size, alternately swelling and shrinking. As these stars pulsate, their surface temperatures also vary. As a result, pulsating stars can vary dramatically in brightness. We now understand that these **pulsating variable stars** are actually evolved, post–main-sequence stars.

Long-Period Variables

Pulsating variable stars were first discovered in 1595 by David Fabricius, a Dutch minister and amateur astronomer. He noticed that the star o (omicron) Ceti is sometimes bright enough to be easily seen with the naked eye but at other times fades to invisibility (Figure 19-17). By 1660, astronomers realized that these brightness variations repeated with a period of 332 days. Seventeenth-century astronomers were so enthralled by this variable star that they renamed it Mira ("wonderful").

Mira is an example of a class of pulsating stars called **long-period variables**. These stars are cool red giants that vary in brightness by a factor of 100 or more over a period of months or years. With surface temperatures of about 3500 K and average luminosities that range from about 10 to 10,000 $L_\odot$, they occupy the upper right side of the H-R diagram (Figure 19-18). Some, like Mira, are periodic, but others are irregular. Many eject large amounts of gas and dust into space.

Astronomers do not fully understand why some cool red giants become long-period variables. It is difficult to calculate accurate stellar models to describe such huge stars with extended, tenuous atmospheres.

Cepheid Variables

Astronomers have a much better understanding of other pulsating stars, called **Cepheid variables**, or simply Cepheids. As we will see, Cepheid stars play a crucial role in measuring the large-scale behavior of the universe and were some of the first objects targeted by the Hubble Space Telescope.

A Cepheid variable is recognized by the characteristic way in which its light output varies—rapid brightening followed by gradual dimming. They are named for δ (delta) Cephei, an example of this type of star discovered in 1784 by John Goodricke, a 19-year-old English amateur astronomer who happened to be deaf. He found that at its most brilliant, δ Cephei is 2.3 times as bright as at its dimmest. The cycle of brightness variations repeats every 5.4 days. (Sadly, Goodricke

Mira at minimum

Mira at maximum

(a)   (b)

FIGURE 19-17 R I **V** U X G

Mira—A Long-Period Variable Star Mira, or o (omicron) Ceti, is a variable star whose luminosity varies with a 332-day period. Luminosity variations are due to changes in both surface temperature and radius. At its dimmest, as in **(a)** (photographed in December 1961), Mira is less than 1% as bright as when it is at maximum, as in **(b)** (January 1965). The physical size of the star is too small to discern with the telescope used, but variations in brightness change how wide it appears in the image. (Lowell Observatory)

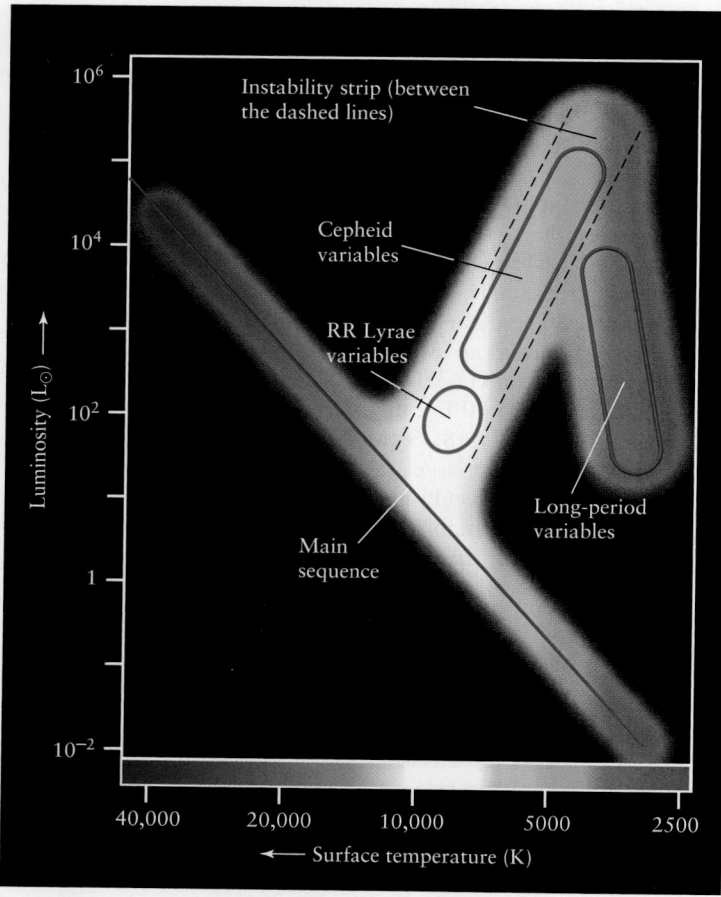

FIGURE 19-18

Variable Stars on the H-R Diagram Pulsating variable stars are found in the upper right of the H-R diagram. Long-period variables like Mira are cool red giant stars that pulsate slowly, changing their brightness in a semiregular fashion over months or years. Cepheid variables and RR Lyrae variables are located in the instability strip, which lies between the main sequence and the red giant region. A star passing through this strip along its evolutionary track becomes unstable and pulsates.

caught pneumonia and died before his twenty-second birthday.) The surface temperatures and luminosities of the Cepheid variables place them in the upper middle of the H-R diagram (see Figure 19-18).

> By studying variable stars, astronomers gain insight into late stages of stellar evolution.

After core helium fusion begins, mature stars move across the middle of the H-R diagram. Figure 19-10a shows the evolutionary tracks of high-mass stars crisscrossing the H-R diagram. Post–helium-flash stars of moderate mass also cross the middle of the H-R diagram between the red giant region and the horizontal branch.

During these transitions across the H-R diagram, a star can become unstable and pulsate. In fact, there is a region on the H-R diagram between the upper main sequence and the red giant branch called the **instability strip** (see Figure 19-18). When an evolving star's temperature and luminosity fall within this region, the star pulsates and its brightness varies periodically. **Figure 19-19a** shows the brightness variations of δ Cephei, which lies within the instability strip.

A Cepheid variable brightens and fades because the star's outer envelope cyclically expands and contracts. The first to observe this

was the Russian astronomer Aristarkh Belopolsky, who noticed in 1894 that spectral lines in the spectrum of δ Cephei shift back and forth with the same 5.4-day period as that of the magnitude variations. From the Doppler effect, we can translate these wavelength shifts into radial velocities and draw a velocity curve (Figure 19-19b). Negative velocities mean that the star's surface is expanding toward us; positive velocities mean that the star's surface is receding. Note that the light curve and velocity curve are mirror images of each other. The star is brighter than average while it is expanding and dimmer than average while contracting.

When a Cepheid variable pulsates, the star's surface oscillates up and down like a spring. During these cyclical expansions and contractions, the star's gases alternately heat up and cool down. Figure 19-19c shows the resulting changes in the star's surface temperature. Figure 19-19d graphs the periodic changes in the star's diameter.

Just as a bouncing ball eventually comes to rest, a pulsating star would soon stop pulsating without something to keep its oscillations going. In 1914, the British astronomer Arthur Eddington suggested that a Cepheid pulsates because the star is more opaque when compressed than when expanded. When the star is compressed, trapped heat increases the internal pressure, which pushes the star's surface outward. When the star expands, the heat escapes, the internal pressure drops, and the star's surface falls inward. This process is analogous to a boiling pot with a lid that lifts on and off the pot as the pressure varies (**Figure 19-20**).

In the 1960s, the American astronomer John Cox followed up on Eddington's idea and proved that helium is what keeps Cepheids pulsating. Normally, when a star's helium is compressed, the gas increases in temperature and becomes more transparent. But in certain layers near the star's surface, compression may ionize helium (remove one of its electrons) instead of raising its temperature. Ionized helium gas is quite opaque, so these layers effectively trap heat and make the star expand, as Eddington suggested. This expansion cools the outer layers and makes the helium ions recombine with electrons, which makes the gas more transparent and releases the trapped energy. The star's surface then falls inward, recompressing the helium, and the cycle begins all over again.

CAUTION!

In our discussion of the behavior of gases (see Box 19-1, Section 19-1, and Section 19-3) we saw that a gas cools when it expands and heats up when it is compressed. Hence, you would expect that the gases in a pulsating star like δ Cephei would reach their maximum temperature when the star is at its smallest diameter, so that the gases are most compressed. By this reasoning, the hotter the gas, the more brightly it glows, so δ Cephei should also have its maximum brightness when its diameter is smallest. But Figure 19-19 shows that the star's brightness and the temperature of the gases at the surface reach their maximum values when the star is expanding, some time *after* the star has contracted to its smallest diameter. How can this be? The explanation is again related to how opaque the gases are inside the star. The rate at which energy is emitted from the central regions of the star is indeed greatest when the star is at its minimum diameter, but the opaque gases in the star's outer layers impede the flow of energy to the surface. Hence, δ Cephei reaches its maximum brightness and maximum surface temperature about half a day after the star is at its smallest size.

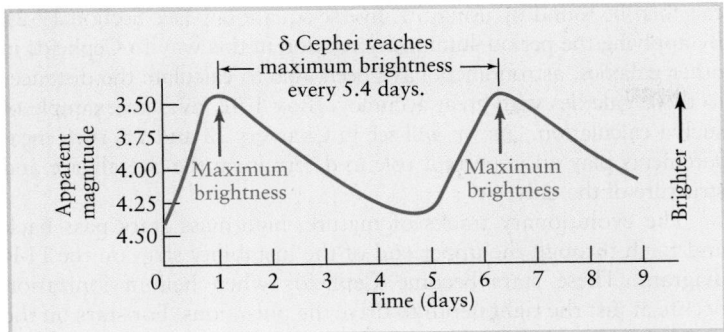

(a) The light curve of δ Cephei (a graph of brightness versus time)

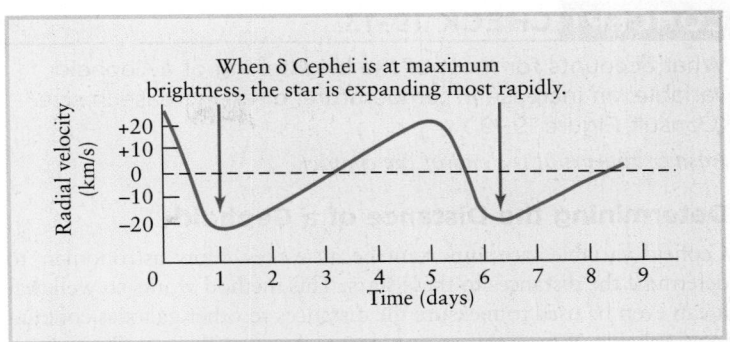

(b) Radial velocity versus time for δ Cephei
(positive: star is contracting; negative: star is expanding)

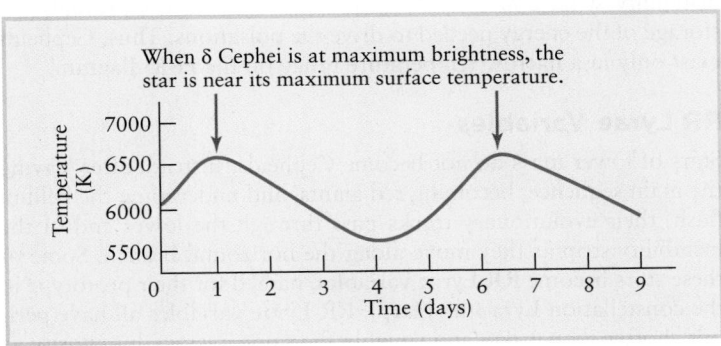

(c) Surface temperature versus time for δ Cephei

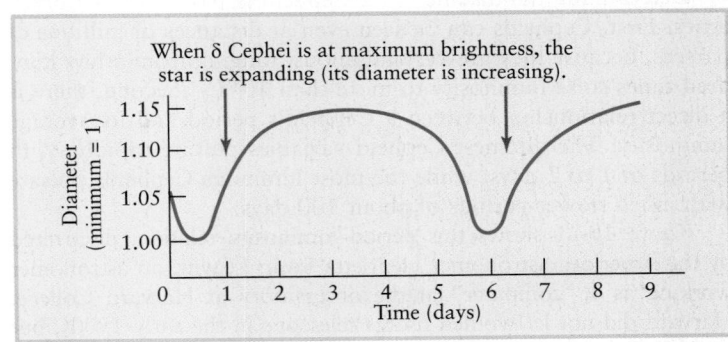

(d) Diameter versus time for δ Cephei

FIGURE 19-19

δ Cephei—A Pulsating Star **(a)** As δ Cephei pulsates, it brightens quickly (the light curve moves upward sharply) but fades more slowly (the curve declines more gently). The increases and decreases in brightness are nearly in step with variations in **(b)**, the star's radial velocity (positive when the star contracts and the surface moves away from us, negative when the star expands and the surface approaches us), as well as in **(c)**, the star's surface temperature. **(d)** The star is still expanding when it is at its brightest and hottest (compare with parts a and b).

(a)

(b)

(c)

(d)

FIGURE 19-20 R I ⊙ U X G

Analogy for Cepheid Variability **(a)** As pressure builds up in this pot, the force on the lid (analogous to a Cepheid's outer layers) increases. **(b)** When the pressure inside the pot is sufficient, it lifts the lid off (expands the star's outer layers) and thereby allows some of the energy inside to escape. This process goes through cycles (two are shown), as do the luminosity and temperature of Cepheid stars. (© Janet Horton)

CONCEPT CHECK 19-13

What accounts for most of the brightening of a Cepheid variable: an increase in temperature, or an increase in size? (Consult Figure 19-19.)

Answer appears at the end of the chapter.

Determining the Distance of a Cepheid

Cepheid variables are important because they allow astronomers to determine the distances to these stars. This method works so well that it can even be used to measure the distances to other galaxies containing Cepheids. Why is this so important? As we will see in Chapter 23, mapping out distances to other galaxies has lead to profound discoveries about the very structure and nature of the universe.

Determining the distance to a Cepheid is based on two properties. First, Cepheids can be seen even at distances of millions of parsecs, because they are very luminous, ranging from a few hundred times solar luminosity to more than 10^4 $L_\odot$. Second, there is a direct relationship between a Cepheid's period and its average luminosity: The dimmest Cepheid variables pulsate rapidly, with periods of 1 to 2 days, while the most luminous Cepheids pulsate with much slower periods of about 100 days.

Figure 19-21 shows this **period–luminosity relation**, discovered by the American astronomer Henrietta Swan Leavitt, an astronomer working as a "computer" at the observatory at Harvard College. Harvard did not let women use its telescope in the early 1900s, but in 1912 Leavitt analyzed photographic plates containing thousands of nearby Cepheids to establish a relationship between their period and luminosity. By first measuring the period of a distant Cepheid's brightness variations, and then using a graph like Figure 19-21, an astronomer can determine the star's intrinsic luminosity. By also measuring the star's apparent brightness, the distance to the Cepheid

can then be found by using the inverse-square law (see Section 17-2). By applying the period–luminosity relation in this way to Cepheids in other galaxies, astronomers have been able to calculate the distances to those galaxies with great accuracy. (Box 17-2 gives an example of such a calculation.) As we will see in Chapters 23 and 25, such measurements play an important role in determining the overall size and structure of the universe.

The evolutionary tracks of mature, high-mass stars pass back and forth through the upper end of the instability strip on the H-R diagram. These stars become Cepheids when helium ionization occurs at just the right depth to drive the pulsations. For stars on the high-temperature (left) side of the instability strip, helium ionization occurs too close to the surface and involves only an insignificant fraction of the star's mass. For stars on the cool (right) side of the instability strip, convection in the star's outer layers prevents the storage of the energy needed to drive the pulsations. Thus, Cepheids exist only in a narrow temperature range on the H-R diagram.

RR Lyrae Variables

Stars of lower mass do not become Cepheids. Instead, after leaving the main sequence, becoming red giants, and undergoing the helium flash, their evolutionary tracks pass through the lower end of the instability strip as they move along the horizontal branch. Some of these stars become **RR Lyrae variables**, named for their prototype in the constellation Lyra (the Harp). RR Lyrae variables all have periods shorter than 1 day and roughly the same average luminosity as horizontal-branch stars, about 100 $L_\odot$. In fact, the RR Lyrae region of the instability strip (see Figure 19-18) is actually a segment of the horizontal branch. RR Lyrae stars are all metal-poor, Population II stars. Many have been found in globular clusters, and they have been used to determine the distances to those clusters in the same way that Cepheids are used to find the distances to other galaxies. In Chapter 22, we will see how RR Lyrae stars helped astronomers determine the size of the Milky Way Galaxy.

In some cases, the expansion speed of a pulsating star exceeds the star's escape speed. When this happens, the star's outer layers are ejected completely. We will see in Chapter 20 that dying stars eject significant amounts of mass in this way, renewing and enriching the interstellar medium for future generations of stars.

CONCEPT CHECK 19-14

If a Cepheid variable star was observed to vary in brightness with a period of 30 days, how would you use Figure 19-21 to determine the distance to this star?

Answer appears at the end of the chapter.

19-7 Mass transfer can affect the evolution of stars in a close binary system

We have outlined what happens when a main-sequence star evolves into a red giant. What we have ignored so far is that more than half of all stars are members of multiple-star systems, including binaries. If the stars in such a system are widely separated, the individual stars follow the same course of evolution as if they were isolated. In a **close binary**, however, when one star expands to become a red giant, its

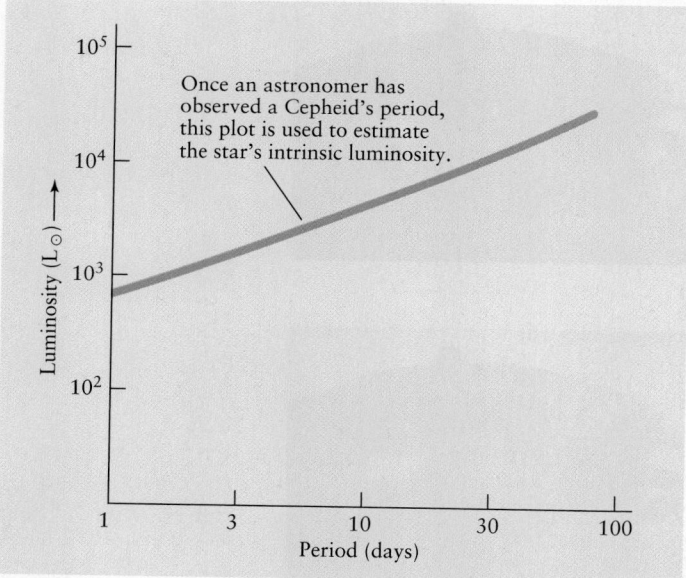

FIGURE 19-21

Period–Luminosity Relation for Cepheids The greater the average luminosity of a Cepheid variable, the longer its period and the slower its pulsations. (Data from H. C. Arp)

outer layers can be gravitationally captured by the nearby companion star. In other words, a bloated red giant in a close binary system can dump gas onto its companion, a process called **mass transfer**.

Roche Lobes and Lagrangian Points

Our modern understanding of mass transfer in close binaries is based on the work of the French mathematician Edouard Roche. In the mid-1800s, Roche studied how rotation and mutual tidal interaction affect the stars in a binary system. Tidal forces cause the two stars in a close binary to keep the same sides facing each other, just as our Moon keeps its same side facing Earth (see Section 4-8). But because stars are gaseous, not solid, rotation and tidal forces can have significant effects on their shapes.

> If the stars in a binary system are sufficiently close, tidal forces can pull gases off one star and onto the other.

In widely separated binaries, the stars are so far apart that tidal effects are small; therefore, the stars are nearly perfect spheres. In close binaries, where the separation between the stars is not much greater than their sizes, tidal effects are strong, causing the stars to be somewhat egg-shaped.

Roche discovered a mathematical surface that marks the gravitational domain of each star in a close binary. (This surface is not a real physical one, like the surface of a balloon, but a mathematical construct.) **Figure 19-22a** shows the outline of this surface as a

dashed line. The two halves of this surface, each of which encloses one of the stars, are known as **Roche lobes**. The more massive star is always located inside the larger Roche lobe. If gas from a star leaks over its Roche lobe, it is no longer bound by gravity to that star. This escaped gas is free either to fall onto the companion star or to escape from the binary system.

The point where the two Roche lobes touch, called the **inner Lagrangian point**, is a kind of balance point between the two stars in a binary. It is here that the effects of gravity and rotation cancel each other. When mass transfer occurs in a close binary, gases flow through the inner Lagrangian point from one star to the other.

In many binaries, the stars are so far apart that even during their red giant stages the surfaces of the stars remain well inside their Roche lobes. As a result, little mass transfer can occur and each star lives out its life as if it were single and isolated. A binary system of this kind is called a **detached binary** (Figure 19-22a).

However, if the two stars are close enough, when one star expands to become a red giant, it may fill or overflow its Roche lobe. Such a system is called a **semidetached binary** (Figure 19-22b). If both stars fill their Roche lobes, the two stars actually touch and the system is called a **contact binary** (Figure 19-22c). It is quite unlikely, however, that both stars exactly fill their Roche lobes at the same time. (This would only be the case if the two stars had identical masses, so that they both evolved at exactly the same rate.) It is more likely that they overflow their Roche lobes, giving rise to a common envelope of gas. Such a system is called an **overcontact binary** (Figure 19-22d).

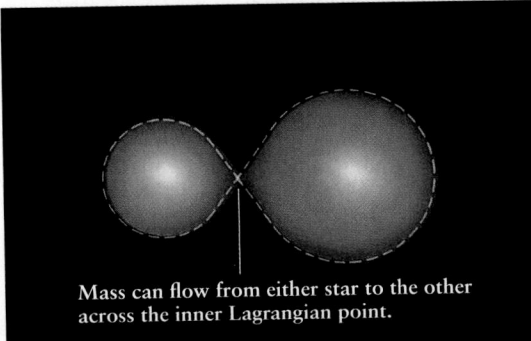

(a) Detached binary: Neither star fills its Roche lobe.

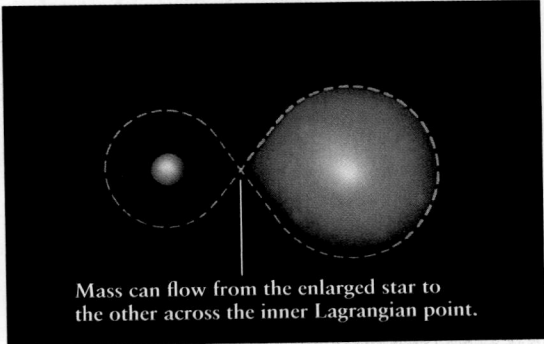

(b) Semidetached binary: One star fills its Roche lobe.

(c) Contact binary: Both stars fill their Roche lobes.

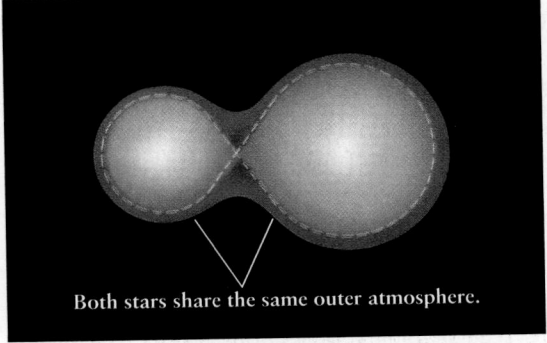

(d) Overcontact binary: Both stars overfill their Roche lobes.

FIGURE 19-22

Close Binary Star Systems The gravitational domain of a star in a close binary system is called its Roche lobe. The two Roche lobes meet at the inner Lagrangian point. The sizes of the stars relative to their Roche lobes determine whether the system is **(a)** a detached binary, **(b)** a semidetached binary, **(c)** a contact binary, or **(d)** an overcontact binary.

(a) A semidetached binary

Small star eclipses the large one.

2.87 days

Large star eclipses the small one.

Bright — Apparent brightness — Dim

(b) A semidetached binary with mass transfer

Mass flows from the large star onto the small one, forming an accretion disk.

12.9 days

Small star eclipses the large one.

Large star eclipses the small one.

Bright — Apparent brightness — Dim

(c) An overcontact binary

Both stars are about the same size, so all eclipses are roughly the same.

8.0 hours

Bright — Apparent brightness — Dim

FIGURE 19-23

Three Eclipsing Binaries Compare the light curves of these three eclipsing binaries to those in Figure 17-24. **(a)** Algol is a semidetached binary. The deep eclipse occurs when the large red giant star blocks the light from the smaller but more luminous main-sequence star. (Compare with Figure 17-24b.) **(b)** β Lyrae's light curve is also at its lowest when the larger star completely eclipses the smaller one. Half an orbital period later, the smaller star partially eclipses the larger one, making a shallower dip in the light curve. **(c)** W Ursae Majoris is an overcontact binary in which both stars overfill their Roche lobes. The extremely short period of this binary indicates that the two stars are very close to each other.

CONCEPT CHECK 19-15

What happens if the mass from one star enters the Roche lobe of another star?

Answer appears at the end of the chapter.

Observations of Mass Transfer

The binary star system Algol (from an Arabic term for "demon") provided the first clear evidence of mass transfer in close binaries. Also called β (beta) Persei, Algol can easily be seen with the naked eye. Ancient astronomers knew that Algol varies periodically in brightness by a factor of more than 2. In 1782, John Goodricke (the discoverer of δ Cephei's variability) first suggested that these brightness variations take place because Algol is an *eclipsing* binary. (We discussed this type of binary in Section 17-11.) The orbital plane of the two stars that make up the binary system happens to be nearly edge on to our line of sight, so one star periodically eclipses the other. Algol's light curve (**Figure 19-23a**) and spectrum show that Goodricke's brilliant hypothesis is correct, and that Algol is a semidetached binary. The detached star (on the right in Figure 19-23a) is a luminous blue main-sequence star, while its less massive companion is a dimmer red giant that fills its Roche lobe.

CAUTION!

According to stellar evolution theory, the more massive a star, the more rapidly it should evolve. Since the two stars in a binary system form simultaneously and thus are the same age, the more massive star should become a red giant before the less massive one. But in Algol and similar binaries, the *more* massive star (on the right in Figure 19-23a) is still on the main sequence, whereas the *less* massive star (on the left in Figure 19-23a) has evolved to become a red giant. How can we explain this apparent contradiction? The answer is that the red giant in Algol-type binaries was *originally* the more massive star. As it left the main sequence to become a red giant, this star expanded until it overflowed its Roche lobe and dumped gas onto its originally less massive companion. Because of the resulting mass transfer, that companion (which is still on the main sequence) became the more massive star.

Mass transfer is also important in another class of semidetached binaries, called β (beta) Lyrae variables, after their prototype in the constellation Lyra. As with Algol, the less massive star in β Lyrae (on the left in Figure 19-23b) fills its Roche lobe. Unlike Algol, however, the more massive detached star (on the right in Figure 19-23b) is the dimmer of the two stars. Apparently, this detached star is enveloped in a rotating *accretion disk* of gas captured from its bloated companion. This disk partially blocks the light coming from the detached star, making it appear dimmer.

What is the fate of an Algol or β Lyrae system? If the detached star is massive enough, it will evolve rapidly, expanding to also fill its Roche lobe. The result will be an overcontact binary in which

the two stars share the gases of their outer layers. Such binaries are sometimes called W Ursae Majoris stars, after the prototype of this class (Figure 19-23c).

Mass transfer can also continue even after nuclear reactions cease in one of the stars in a close binary. In Chapters 20 and 21, we will see how mass transfer onto different types of dead stars called *white dwarfs*, *neutron stars*, and *black holes* can produce some of the most unusual and dramatic objects in the sky.

KEY WORDS

alpha particle, p. 593
Cepheid variable, p. 603
close binary, p. 606
color-magnitude diagram, p. 599
contact binary, p. 607
core helium fusion, p. 592
core hydrogen fusion, p. 586
degeneracy, p. 595
degenerate-electron pressure, p. 595
detached binary, p. 607

globular cluster, p. 599
helium flash, p. 593
helium fusion, p. 592
horizontal-branch star, p. 600
ideal gas, p. 593
inner Lagrangian point, p. 607
instability strip, p. 604
long-period variable, p. 603
main-sequence lifetime, p. 586
mass loss, p. 591
mass transfer, p. 607
metal-poor star, p. 601

metal-rich star, p. 601
open cluster, p. 597
overcontact binary, p. 607
Pauli exclusion principle, p. 593
period–luminosity relation, p. 606
Population I, Population II, and Population III stars, p. 601
pulsating variable star, p. 603
red dwarf, p. 588

red giant, p. 591
Roche lobe, p. 607
RR Lyrae variable, p. 606
semidetached binary, p. 607
shell hydrogen fusion, p. 589
triple alpha process, p. 593
turnoff point, p. 600
zero-age main sequence (ZAMS), p. 596
zero-age main-sequence star, p. 586

KEY IDEAS

The Main-Sequence Lifetime: The duration of a star's main-sequence lifetime depends on the amount of hydrogen available to be consumed in the star's core and the rate at which this hydrogen is consumed (Section 19-1).

- The more massive a star, the shorter is its main-sequence lifetime. The Sun has been a main-sequence star for about 4.56 billion years and should remain one for about another 5 billion years.

- During a star's main-sequence lifetime, the star expands somewhat and undergoes a modest increase in luminosity.

- If a star's mass is greater than about 0.4 M☉, only the hydrogen present in the core can undergo thermonuclear fusion during the star's main-sequence lifetime. If the star is a red dwarf with a mass less than about 0.4 M☉, over time convection brings all of the star's hydrogen to the core where it can undergo fusion.

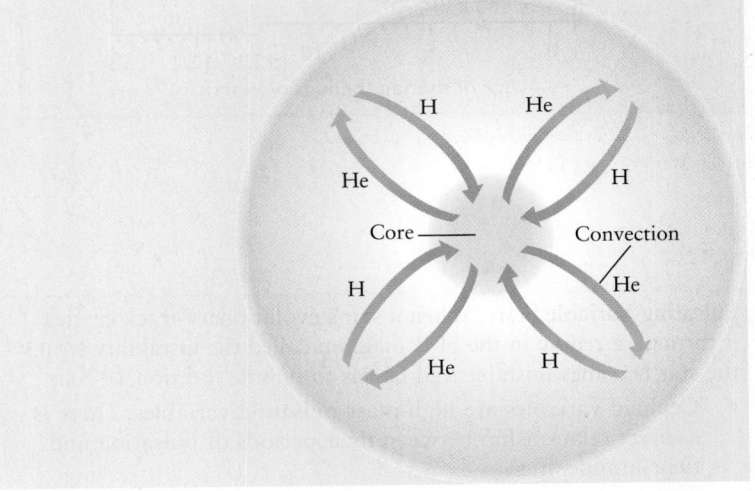

Becoming a Red Giant: Core hydrogen fusion ceases when the hydrogen has been exhausted in the core of a main-sequence star with mass greater than about 0.4 M☉. This leaves a core of nearly pure helium surrounded by a shell through which hydrogen fusion works its way outward in the star. The core shrinks and becomes hotter, while the star's outer layers expand and cool. The result is a red giant star (Section 19-2).

- As a star becomes a red giant, its evolutionary track moves rapidly from the main sequence to the red giant region of the H-R diagram. The more massive the star, the more rapidly this evolution takes place.

Helium Fusion: When the central temperature of a red giant reaches about 100 million K, helium fusion begins in the core. This process, also called the triple alpha process, converts helium to carbon and oxygen (Section 19-3).

- In a more massive red giant, helium fusion begins gradually; in a less massive red giant, it begins suddenly, in a process called the helium flash.

- After the helium flash, a low-mass star moves quickly from the red giant region of the H-R diagram to the horizontal branch.

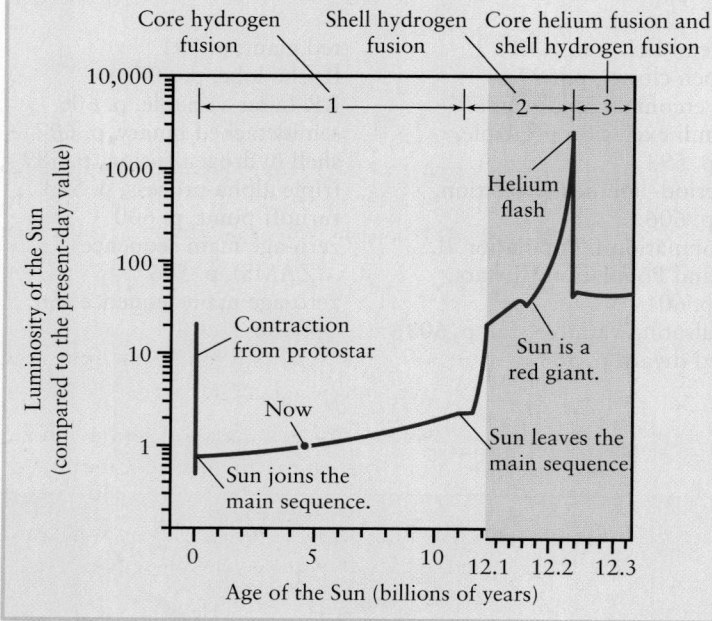

Star Clusters and Stellar Populations: The age of a star cluster can be estimated by plotting its stars on an H-R diagram (Sections 19-4 and 19-5).

- The cluster's age is equal to the age of the main-sequence stars at the turnoff point (the upper end of the remaining main sequence).

- As a cluster ages, the main sequence is "eaten away" from the upper left as stars of progressively smaller mass evolve into red giants.

- Relatively young Population I stars are metal rich; ancient Population II stars are metal poor. The very first generation of metal-free stars is called Population III. The metals (heavy elements) in Population I stars were manufactured by thermonuclear reactions in an earlier generation of Population II stars, then ejected into space and incorporated into a later stellar generation.

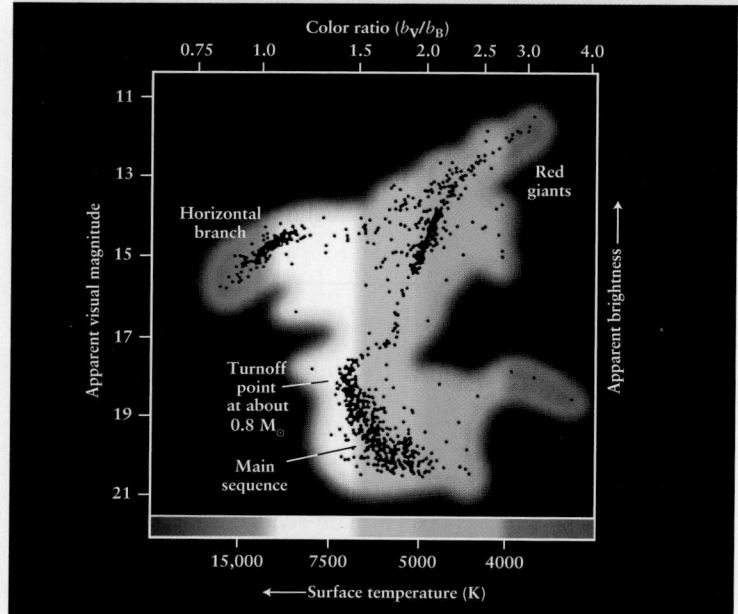

Pulsating Variable Stars: When a star's evolutionary track carries it through a region in the H-R diagram called the instability strip, the star becomes unstable and begins to pulsate (Section 19-6).

- Cepheid variables are high-mass pulsating variables. There is a direct relationship between their periods of pulsation and their luminosities.

- RR Lyrae variables are low-mass, metal-poor pulsating variables with short periods.

- Long-period variable stars also pulsate but in a fashion that is less well understood.

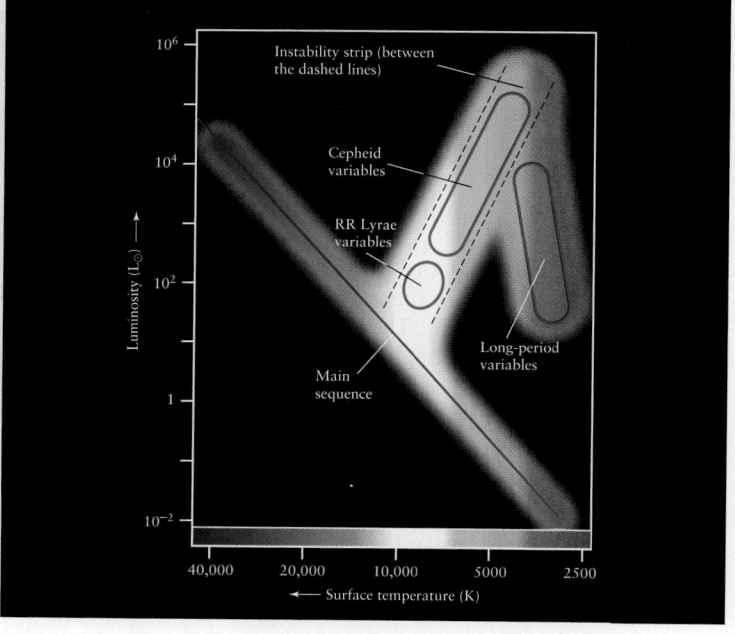

Close Binary Systems: Mass transfer in a close binary system occurs when one star in a close binary overflows its Roche lobe. Gas flowing from one star to the other passes across the inner Lagrangian point. This mass transfer can affect the evolutionary history of the stars that make up the binary system (Section 19-7).

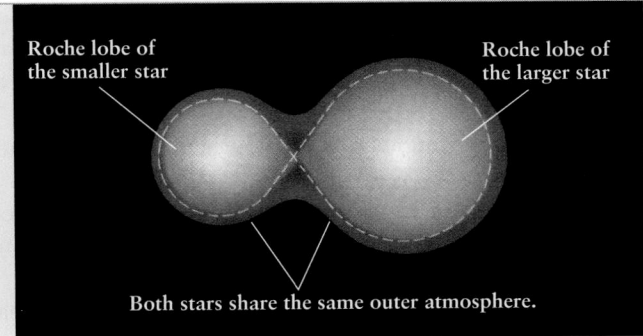

Roche lobe of the smaller star

Roche lobe of the larger star

Both stars share the same outer atmosphere.

QUESTIONS

Section 19-1

1. How does the chemical composition of the present-day Sun's core compare to the core's composition when the Sun formed? What caused the change?

2. Which regions of the Sun are denser today than they were a billion years ago? Which regions are less dense? What has caused these changes?

3. What is a red dwarf? How are thermonuclear reactions in the core of a red dwarf able to consume hydrogen from the star's outer layers?

4. Why do high-mass main-sequence stars have shorter lifetimes than those of lower mass?

5. There is a good deal of evidence that our universe is about 13.7 billion years old. Explain why no main-sequence stars with masses less than 0.4 $M_\odot$ have yet converted all of their hydrogen to helium.

6. **Box 19-2** Explain why the quantity f in Box 19-2 has a different value for stars with masses less than 0.4 $M_\odot$ than for stars with masses greater than 0.4 $M_\odot$. In which case does f have a greater value?

7. **Box 19-2** Calculate the main-sequence lifetimes of (**a**) a 9-$M_\odot$ star and (**b**) a 0.25-$M_\odot$ star. Compare these lifetimes with that of the Sun.

8. **Box 19-2** The earliest fossil records indicate that life appeared on Earth about a billion years after the formation of the solar system. What is the most mass that a star could have in order that its lifetime on the main sequence is long enough to permit life to form on one or more of its planets? Assume that the evolutionary processes would be similar to those that occurred on Earth.

Section 19-2

9. What will happen inside the Sun 5 billion years from now, when it begins to mature into a red giant?

10. **TUTORIAL 19.1** Explain why Earth is expected to become inhospitable to life long before the Sun becomes a red giant.

11. Explain how it is possible for the core of a red giant to contract at the same time that its outer layers expand.

Section 19-3

12. Why does helium fusion require much higher temperatures than hydrogen fusion?

13. How is a degenerate gas different from ordinary gases?

14. What is the helium flash? Why does it happen in some stars but not in others?

15. Why does a star's luminosity decrease after helium fusion begins in its core?

16. The half-life of the ^{8}Be nucleus, 2.6×10^{-16} second, is the average time that elapses before this unstable nucleus decays into two alpha particles. How would the universe be different if instead the ^{8}Be half-life were zero? How would the universe be different if the ^{8}Be nucleus were stable and did not decay?

Section 19-4

17. **TUTORIAL 19.2** Explain why the majority of the stars visible through telescopes are main-sequence stars.

18. On an H-R diagram, main-sequence stars do not lie along a single narrow line but are spread out over a band (see Figure 19-10b). On the basis of how stars evolve during their main-sequence lifetimes, explain why this should be so.

19. Explain how and why the turnoff point on the H-R diagram of a cluster is related to the cluster's age. Why can't this method be used with any collection of stars plotted on an H-R diagram?

20. How do astronomers know that globular clusters are made of old stars?

21. Red giant stars appear more pronounced in composites of infrared images and visible-light images, like those in Figure 19-5b and Figure 19-13. Explain why.

22. The horizontal-branch stars in Figure 19-13 appear blue. (a) Explain why this is consistent with the color-magnitude diagram shown in Figure 19-14. (b) All horizontal-branch stars were once red giants. Explain what happened to these stars to change their color.

23. (a) The main-sequence stars Sirius (spectral type A1), Vega (A0), Spica (B1), Fomalhaut (A3), and Regulus (B7) are among the 20 brightest stars in the sky. Explain how you can tell that all these stars are younger than the Sun. (b) The third-brightest star in the sky, although it can be seen only south of 29° north latitude, is α (alpha) Centauri A. It is a main-sequence star of spectral type G2, the same as the Sun. Can you tell from this whether α Centauri A is younger than the Sun, the same age, or older? Explain your reasoning.

24. Using the same horizontal and vertical scales as in Figure 19-10a, draw points on an H-R diagram for each of the stars listed in Table 19-1. Label each point with the star's mass and its main-sequence lifetime. Which of these stars will remain on the main sequence after 10^9 years? After 10^{11} years?

25. Discuss how H-R diagrams of star clusters could be used to set limits on the age of the universe. Could they be used to set lower limits on the age? Could they be used to set upper limits? Explain your reasoning.

Section 19-5

26. What is the difference between Population I and Population II stars? In what sense can the stars of one population be regarded as the "children" of the other population?

27. Both diamonds and graphite (the material used in pencils to make marks on paper) are crystalline forms of carbon. Most of the carbon atoms in these substances have nuclei with 6 protons and 6 neutrons (^{12}C). Where did these nuclei come from?

28. The star whose spectrum is shown in Figure 19-16a has a lower percentage of heavy elements than the Sun, whose spectrum is shown in Figure 19-16b. Hence, the star in Figure 19-16a has a higher percentage of hydrogen. Why, then, isn't the H_δ absorption line of hydrogen noticeably darker for the star in Figure 19-16a?

29. If Population III stars have never been observed, why do astronomers suspect they existed in the past? Why wouldn't this population of stars be observable today?

Section 19-6

30. Why do astronomers attribute the observed Doppler shifts of a Cepheid variable to pulsation rather than to some other cause, such as orbital motion?

31. Why do Cepheid stars pulsate? Why are these stars important to astronomers who study galaxies beyond the Milky Way?

32. Would you expect the color of a Cepheid variable star (consider Figure 19-19) to change during the star's oscillation period? If not, why not? If so, describe why the color should change, and describe the color changes you would expect to see during an oscillation period.

33. The brightness of a certain Cepheid variable star changes from maximum brightness to minimum brightness in 26 days. In 4 more days, the star returns to maximum brightness. (a) What is this star's period? (b) What is this star's approximate luminosity?

34. The star X Arietis is an RR Lyrae variable. Its apparent brightness varies between 2.0×10^{-15} and 4.9×10^{-15} that of the Sun with a period of 0.65 day. Interstellar extinction dims the star by 37%. Approximately how far away is the star?

35. The apparent brightness of δ Cephei (a Cepheid variable star) varies with a period of 5.4 days. Its average apparent brightness is 5.1×10^{-13} that of the Sun. Approximately how far away is δ Cephei? (Ignore interstellar extinction.)

Section 19-7

36. What is a Roche lobe? What is the inner Lagrangian point? Why are Roche lobes important in close binary star systems?

37. What is the difference between a detached binary, a semi-detached binary, a contact binary, and an overcontact binary?

38. Massive main-sequence stars turn into red giants before less massive stars do. Why, then, is the more massive star in Algol a main-sequence star and the less massive star a red giant?

39. Suppose you find a binary star system in which the more massive star is a red giant and the less massive star is a main-sequence star. Would you expect that mass transfer between the stars has played an important role in the evolution of these stars? Explain your reasoning.

40. The larger star in the Algol binary system (see Figure 19-23a) is of spectral class K, while the smaller star is of spectral class B. Discuss how the color of Algol changes as seen through a small telescope (through which Algol appears as a single star). What is the color during a deep eclipse, when the large star eclipses the small one? What is the color when the small star eclipses the large one?

41. Suppose the detached star in β Lyrae (Figure 19-23b) did not have an accretion disk. Would the deeper dips in the light curve be deeper, shallower, or about the same? What about the shallower dip? Explain your answers.

42. The two stars that make up the overcontact binary W Ursae Majoris (Figure 19-23c) have estimated masses of 0.99 $M_\odot$

and 0.62 M$_\odot$. (**a**) Find the average separation between the two stars. Give your answer in kilometers. (**b**) The radii of the two stars are estimated to be 1.14 R$_\odot$ and 0.83 R$_\odot$. Show that these values and your result in part (**a**) are consistent with the statement that this is an overcontact binary.

General Problems

43. **TUTORIAL 19.2** What does it mean when an astronomer says that a star "moves" from one place to another on an H-R diagram?

44. The radius of the Sun has increased over the past several billion years. Over the same time period, the size of the Moon's orbit around Earth has also increased. A few billion years ago, were annular eclipses of the Sun (see Figure 3-13) more or less common than they are today? Explain your answer.

45. The Sun has increased in radius by 6% over the past 4.56 billion years. Its present-day radius is 696,000 km. What was its radius 4.56 billion years ago? (*Hint:* The answer is *not* 654,000 km.)

46. Calculate the escape speed from (**a**) the surface of the present-day Sun and (**b**) the surface of the Sun when it becomes a red giant, with essentially the same mass as today but with a radius that is 100 times larger. (**c**) Explain how your results show that a red giant star can lose mass more easily than a main-sequence star.

47. Calculate the average speed of a hydrogen atom (mass 1.67×10^{-27} kg) (**a**) in the atmosphere of the present-day Sun, with temperature 5800 K, and (**b**) in the atmosphere of a 1-M$_\odot$ red giant, with temperature 3500 K. (**c**) Compare your results with the escape speeds that you calculated in Question 46. Use this comparison to discuss how well the present-day Sun and a 1-M$_\odot$ red giant can retain hydrogen in their atmospheres.

48. Use the value of the Sun's luminosity (3.90×10^{26} watts, or 3.90×10^{26} joules per second) to calculate what mass of hydrogen the Sun will convert into helium during its entire main-sequence lifetime of 1.2×10^{10} years. (Assume that the Sun's luminosity remains nearly constant during the entire 1.2×10^{10} years.) What fraction does this represent of the total mass of hydrogen that was originally in the Sun?

49. As a red giant, the Sun's luminosity will be about 2000 times greater than it is now, so the amount of solar energy falling on Earth will increase to 2000 times its present-day value. Hence, to maintain thermal equilibrium, each square meter of Earth's surface will have to radiate 2000 times as much energy into space as it does now. Use the Stefan–Boltzmann law to determine what Earth's surface temperature will be under these conditions. (*Hint:* The present-day Earth has an average surface temperature of 14°C.)

50. When the Sun becomes a red giant, its luminosity will be about 2000 times greater than it is today. Assuming that this luminosity is caused *only* by fusion of the Sun's remaining hydrogen, calculate how long our star will be a red giant. (In fact, only a fraction of the remaining hydrogen will be consumed, and the luminosity will vary over time, as shown in Figure 19-9.)

51. What observations would you make of a star to determine whether its primary source of energy is hydrogen fusion or helium fusion?

52. The stars that make up the binary system W Ursae Majoris (see Figure 19-23c) have particularly strong magnetic fields. Explain how astronomers could have discovered this. (*Hint:* See Section 16-9.)

53. Eventually the Sun's luminosity will increase to the point where Earth can no longer sustain life. Discuss what measures a future civilization might take to preserve itself from such a calamity.

WEB/E-BOOK QUESTIONS

54. Suppose that an oxygen nucleus (^{16}O) were fused with a helium nucleus (^{4}He). What element would be formed? Look up the relative abundance of this element in, for example, the *Handbook of Chemistry and Physics* or on the internet. Based on the abundance, comment on whether such a process is likely. (*Hint:* See Figure 8-4.)

55. Although Polaris, the North Star, is a Cepheid variable, it pulsates in a somewhat different way than other Cepheids. Search the internet for information about this star's pulsations and how they have been measured by astronomers at the U.S. Naval Observatory. How does Polaris pulsate? How does this differ from other Cepheids?

56. **ANIMATION 19.1** **Observing Stellar Evolution.** Step through the animation "The Hertzsprung–Russell Diagram and Stellar Evolution" in Chapter 19 of the *Universe* website or e-book. Use this animation to answer the following questions. (**a**) How does a 1-M$_\odot$ star move on the H-R diagram during its first 4.56 billion (4560 million) years of existence? Compare this with the discussion in Section 19-1 of how the Sun has evolved over the past 4.56 billion years. (**b**) What is the zero-age spectral class of a 2-M$_\odot$ star? At what age does such a star evolve into a red giant of spectral class K? (**c**) What is the approximate zero-age luminosity of a 1.3-M$_\odot$ star? What is its approximate luminosity when it becomes a red giant? (**d**) Suppose a star cluster has no main-sequence stars of spectral classes O or B. What is the approximate age of the cluster? (**e**) Approximately how long do the most massive stars of spectral class B live before leaving the main sequence? What about the most massive stars of spectral class F?

ACTIVITIES

Observing Projects

> #### Observing tips and tools
>
> An excellent resource for learning how to observe variable stars is the website of the American Association of Variable Star Observers. A wealth of data about specific variable stars can be found on the *Sky & Telescope* website and in the three volumes of *Burnham's Celestial Handbook: An Observer's Guide to the Universe Beyond the Solar System* (Dover, 1978). This book also provides useful information about observing star clusters.

Starry Night™ Explorations

57. Search the internet to find out when Mira will next reach maximum brightness. Look up the star's location in the sky using the *Starry Night™* program if you have access. (Use the **Find...** command in the **Edit** menu to search for Omicron Ceti.) Why is it unlikely that you will be able to observe Mira at maximum brightness?

58. Use *Starry Night™* to demonstrate your knowledge of the evolution of a main sequence star. Select **Favorites > Explorations > Stellar Evolution** to see a view looking toward the constellation Orion. Point the cursor at each of the five labeled stars in the view to see the star's position in the H-R diagram. List these five stars in order of increasing evolutionary age.

59. Use *Starry Night™* to look for signs of stellar evolution in M101, the Pinwheel Galaxy. Select **Favorites > Explorations > M101.**
    (a) What is the color of the central part of this galaxy?
    (b) What is the color of the outer regions of this galaxy?
    (c) Based on your observations, what type of stars would you expect to predominate in each of these two regions of this galaxy?

Collaborative Exercises

60. The inverse relationship between a star's mass and its main-sequence lifetime is sometimes likened to automobiles in that the more massive vehicles, such as commercial semi–tractor-trailer trucks, need to consume significantly more fuel to travel at highway speeds than more lightweight and economical vehicles. As a group, create a table called "Maximum Vehicle Driving Distances" (much like Table 19-1, "Main-Sequence Lifetimes") by making estimates for any five vehicles of your group's choosing. The table's column headings should be (1) vehicle make and model; (2) estimated gas tank size; (3) cost to fill tank; (4) estimated mileage (in miles per gallon); and (5) number of miles driven on a single fill-up.

61. Consider Figure 19-21, showing the period–luminosity relation for Cepheids. If a certain Cepheid star has an average luminosity of 4×10^{30} W, what is the variable star's approximate period?

62. Figure 19-19a shows a light curve of apparent magnitude versus time in days for δ Cephei—a pulsating star that reaches maximum brightness every 5.4 days. Create a new sketch of apparent magnitude versus time in days showing three different stars: (1) δ Cephei; (2) a slightly smaller pulsating star; and (3) a slightly larger pulsating star, all of which have about the same total change in apparent magnitude.

ANSWERS

ConceptChecks

ConceptCheck 19-1: To leave the main sequence, a red dwarf has to finish converting all of its hydrogen to helium. This takes hundreds of billions of years for a red dwarf—longer than the age of the universe.

ConceptCheck 19-2: When main-sequence hydrogen burning stops, core pressure decreases, and the core contracts. As the core contracts, gravitational energy is converted into heat, and the core's temperature increases well beyond the temperature it had during its main-sequence burning.

ConceptCheck 19-3: The higher temperatures in a red giant's core lead to an increased rate of shell hydrogen fusion. The additional energy released by this fusion produces a greater pressure outside the core than before, which greatly expands the star.

ConceptCheck 19-4: In a red giant, the core is at a significantly higher temperature and density that allows helium atoms to combine and fuse together, releasing energy.

ConceptCheck 19-5: Compared to when it was on the main sequence, a red giant is hotter in the core and cooler on the surface.

ConceptCheck 19-6: If fusion in a main-sequence star increases its energy output, the higher temperatures increase the internal pressure. This in turn expands the core's gas, lowers its temperature, and reduces the rate of nuclear reactions. The "safety valve" aspect is that increased nuclear reactions end up reducing further reactions so that the rate of reactions remains stable. When a red giant's core is supported by degenerate-electron pressure, an increase in temperature from an increase in nuclear reactions does not increase the pressure to expand the core and does not reduce further reaction rates.

ConceptCheck 19-7: At high enough temperatures, the core is no longer supported by degenerate-electron pressure and the core begins to expand. By expanding, the core cools somewhat, nuclear reaction rates slow down, and luminosity of the star decreases. This is shown on the right of Figure 19-9.

ConceptCheck 19-8: Yes. After 30 million years, you can see in Figure 19-11e a few red giants just as low-mass stars are finally approaching the main sequence.

ConceptCheck 19-9: In Figure 19-11c, showing the cluster after 100,000 years, the most massive stars are just joining the main sequence.

ConceptCheck 19-10: The turnoff point in Figure 19-14 corresponds to 0.8-$M_\odot$ stars that stay on the main sequence for about 10 billion years. If the turnoff point were instead at around 10,000 K, these stars would be more massive and spend *less* time on the main sequence, and the cluster would be younger.

ConceptCheck 19-11: Recall that lower-mass stars are less luminous. Therefore, the lower the luminosity of the turnoff point, the lower the mass of the main-sequence star at the turnoff point. Lower-mass stars remain on the main sequence longer, indicating an older cluster when these stars finally leave the main sequence.

ConceptCheck 19-12: Population I stars are enriched with heavier atomic elements (they are metal rich). The heavier elements can only be made by an earlier generation of stars through supernova explosions. The earlier generation lacks the heavier elements (they are metal poor) and are called Population II stars.

ConceptCheck 19-13: You can see that the Cepheid in Figure 19-19a peaks in luminosity on day 1. This is about when the temperature peaks as well in Figure 19-19c. However, when the star reaches its maximum size in Figure 19-19d, the luminosity is already declining. Therefore, surface temperature contributes more to the peak luminosity.

ConceptCheck 19-14: Figure 19-21 indicates that a variable star with a 30-day period has an intrinsic luminosity (L) about 10,000 times brighter than our Sun. Direct observations of the star also tell you its apparent brightness (b). Section 17-2 describes how these two quantities are directly related to the distance (d) of the star, so the distance is easily determined.

ConceptCheck 19-15: If a star expands enough that some of its mass enters the Roche lobe of its binary companion, the matter becomes gravitationally bound to the binary companion. Over time, this can transfer a significant amount of mass to the companion.

20

Stellar Evolution: The Deaths of Stars

1 pc

10 pc

When a star of 0.4 $M_\odot$ or more reaches the end of its main-sequence lifetime and becomes a red giant, it has a compressed core and a bloated atmosphere. Finally, the star devours its remaining nuclear fuel and begins to die. As we will learn in this chapter, the character of the star's death depends crucially on the value of its mass.

A star of relatively low mass—such as our own Sun—ends its evolution by gently expelling its outer layers into space. These ejected gases form a glowing cloud called a *planetary nebula*, such as the one shown above in the left-hand image. The burned-out core that remains is called a *white dwarf*.

In contrast, a high-mass star ends its life in almost inconceivable violence. At the end of its short life, the core of such a star collapses suddenly, which triggers a powerful *supernova* explosion that can be as luminous as an entire galaxy of stars. A white dwarf, too, can become a supernova if it accretes gas from a companion star in a close binary system.

In supernovae, nuclear reactions produce a wide variety of heavy elements, which are ejected into the interstellar medium. (The supernova remnant shown

Left: The planetary nebula NGC 7293 (the Helix Nebula). Right: The supernova remnant LMC N49. (Left: C. R. O'Dell [Vanderbilt] et al., ESA, NOAO, NASA; Right: NASA/JPL/Hubble Heritage Team [STScI/AURA])

R I V U X G

in the right-hand image is rich in these elements.) Such heavy elements are essential building blocks for terrestrial worlds like our Earth. Supernovae are also associated with the formation of neutron stars. And, as we'll see in Chapter 21, the violent collision of neutron stars also produces heavy elements. Thus, supernovae play a major role in the origin of most of the elements in the periodic table.

20-1 Stars of between 0.4 and 8 M_⊙ go through two distinct red giant stages

TUTORIAL 20.1 All main-sequence stars convert hydrogen to helium in their cores via a series of energy-releasing nuclear reactions. As we saw in Section 19-1, convection within a low-mass main-sequence star—a so-called *red dwarf* with a mass between 0.08 and 0.4 M_⊙—will eventually bring all of the star's hydrogen into the core. Thus, over hundreds of billions of years, a red dwarf will end up as a ball of helium. Convection is less important in main-sequence stars with masses greater than 0.4 M_⊙, so these stars are able to consume only the hydrogen that is present within the core. These stars of greater mass then leave the main sequence. Let's examine what happens next for a star of moderately low mass, between 0.4 and 4 M_⊙. One example of such a star is our own Sun, with a mass of 1 M_⊙. We will begin by reviewing what we learned in Chapter 19 about the first stages of post–main-sequence evolution for such a star. (Later in this chapter, we will study the evolution of more massive stars.)

Note: Technically, the nuclear reactions referred to above are **thermonuclear**. The *thermo-* in thermonuclear indicates that high-speed collisions, due to very high temperatures, help fuse atoms together despite electric repulsion between protons. In this chapter, we will often refer to these simply as nuclear reactions.

The Red Giant and Horizontal-Branch Stages: A Review

We can describe a star's post–main-sequence evolution using an evolutionary track in an H-R diagram. Figure 20-1 shows the track for a 1-M_⊙ star like the Sun. Once core hydrogen fusion ceases, the core shrinks, heating the nearby surrounding hydrogen and triggering shell hydrogen fusion (see Section 19-2). The new outpouring of energy causes the star's outer layers to expand and cool, and the star becomes a red giant. As the luminosity increases and the surface temperature drops, the post–main-sequence star moves up and to the right along the **red giant branch** in an H-R diagram (Figure 20-1a).

Next, the helium-rich core of the star shrinks and heats until eventually **core helium fusion** begins. This second post–main-sequence stage begins gradually in stars more massive than about 2–3 M_⊙, but for less massive stars it comes suddenly—in a *helium flash*. During core helium fusion, the surrounding hydrogen-fusing shell still provides most of the red giant's luminosity and the helium flash is not observed.

As we learned in Section 19-3, the core expands when core helium fusion begins, which makes the core cool down a bit. (We

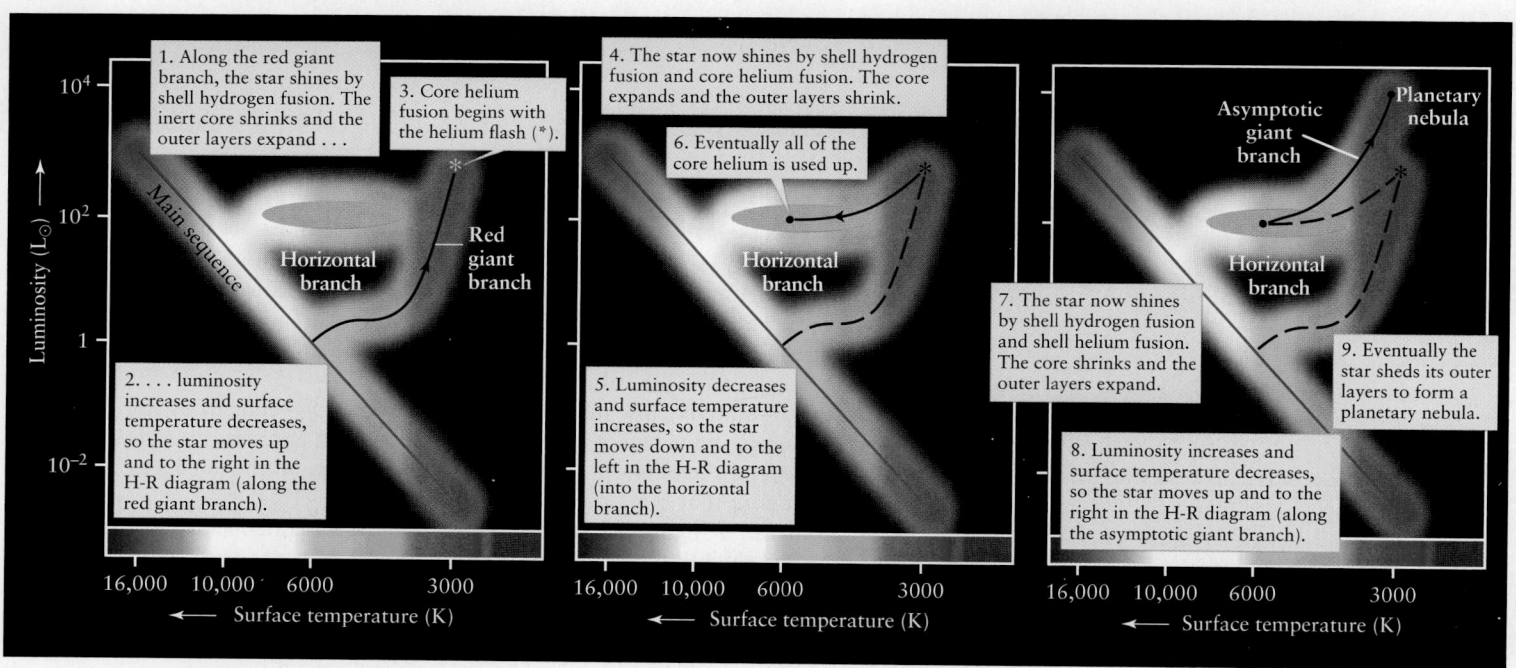

(a) Before the helium flash: A red giant star

(b) After the helium flash: A horizontal-branch star

(c) After core helium fusion ends: An AGB star

ANIMATION 20.1

FIGURE 20-1

The Post–Main-Sequence Evolution of a 1-M_⊙ Star These H-R diagrams show the evolutionary track of a star like the Sun as it goes through the stages of being **(a)** a red giant star, **(b)** a horizontal-branch star, and **(c)** an asymptotic giant branch (AGB) star. The star eventually evolves into a planetary nebula (described in Section 20-3).

saw in Box 19-1 that letting a gas expand tends to lower its temperature, whereas compressing a gas tends to increase its temperature.) The cooling of the core also cools the surrounding hydrogen-fusing shell, so that the shell releases energy more slowly. Hence, the luminosity goes down a bit after core helium fusion begins.

The slower rate of energy release also lets the star's outer layers contract. As they contract, they heat up, so the star's surface temperature increases and its evolutionary track moves to the left in the H-R diagram in Figure 20-1b. The luminosity changes relatively little during this stage, so the evolutionary track moves almost horizontally, along a path called the **horizontal branch.** Horizontal-branch stars have helium-fusing cores surrounded by hydrogen-fusing shells. Figure 19-13 shows horizontal-branch stars in a globular cluster, and Figure 19-9 shows the evolution of the luminosity of a 1-$M_\odot$ star up to this point in its history. (Technically, a star ceases to be a red giant as soon as it begins core helium fusion, even though it is still red and giant for a while; we did not make this distinction in Chapter 19.)

> ### CAUTION!
> It is easy to get confused about the relationship between fusion inside and outside the core of a star versus the star's overall size, temperature, and luminosity. Compared to fusion in the core, shell fusion releases energy at a greater rate. This increased energy puffs up the star's outer layers. Shell fusion results in a larger star with a larger luminosity, even though its surface temperature decreases.

AGB Stars: The Second Red Giant Stage

Helium fusion produces nuclei of carbon and oxygen. After about 100 million (10^8) years of core helium fusion, essentially all the helium in the core of a 1-$M_\odot$ star has been converted into carbon and oxygen, and the fusion of helium in the core ceases. (This corresponds to the

right-hand end of the graph in Figure 19-9.) Without nuclear reactions to maintain the core's internal pressure, the core again contracts, until it is stopped by degenerate electron pressure (described in Section 19-3). This contraction releases heat into the surrounding helium-rich gases, and a new stage of helium fusion begins in a thin shell around the core. This process is called **shell helium fusion.**

History now repeats itself—the star enters a *second* red giant phase. A star first becomes a red giant at the end of its main-sequence lifetime, when the outpouring of energy from shell hydrogen fusion makes the star's outer layers expand and cool. In the same way, the outpouring of energy from shell *helium* fusion causes the outer layers to expand again. The low-mass star ascends into the red giant region of the H-R diagram for a second time (Figure 20-1c), but now with even greater luminosity than during its first red giant phase.

Stars in this second red giant phase are commonly called **asymptotic giant branch stars,** or **AGB stars,** and their evolutionary tracks follow what is called the **asymptotic giant branch.** (*Asymptotic* means "approaching"; the name means that a star on the asymptotic giant branch approaches the red giant branch from the left in an H-R diagram.)

When a low-mass star first becomes an AGB star, it consists of a carbon–oxygen core (supported by degenerate electron pressure) with no nuclear reactions, and a helium-fusing shell, both inside a hydrogen-fusing shell, all within a volume not much larger than Earth. This small, dense central region is surrounded by an enormous hydrogen-rich envelope about as big as Earth's orbit around the Sun. After a while, the expansion of the star's outer layers causes the hydrogen-fusing shell to also expand and cool, and nuclear reactions in this shell temporarily cease. This leaves the aging star's structure as shown in Figure 20-2.

> Stars like the Sun go through a second red giant phase, during which helium fusion takes place in a shell around an inert core.

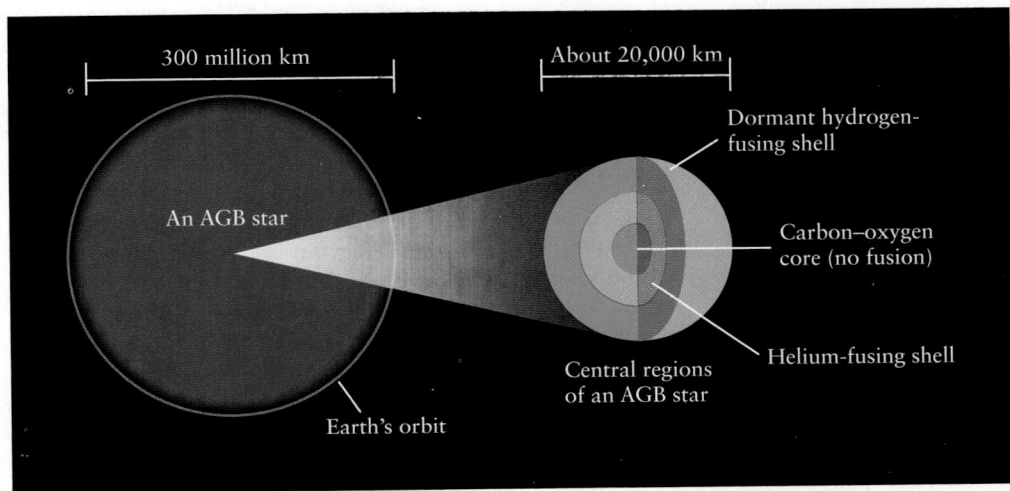

FIGURE 20-2

The Structure of an Old, Moderately Low-Mass AGB Star Near the end of its life, a star like the Sun becomes an immense, red, asymptotic giant branch (AGB) star. The star's inert core, active helium-fusing shell, and dormant hydrogen-fusing shell are all contained within a volume roughly the size of Earth. Nuclear reactions in the helium-fusing shell are so rapid that the star's luminosity is thousands of times that of the present-day Sun. (The relative sizes of the shells in the star's interior are not shown to scale.)

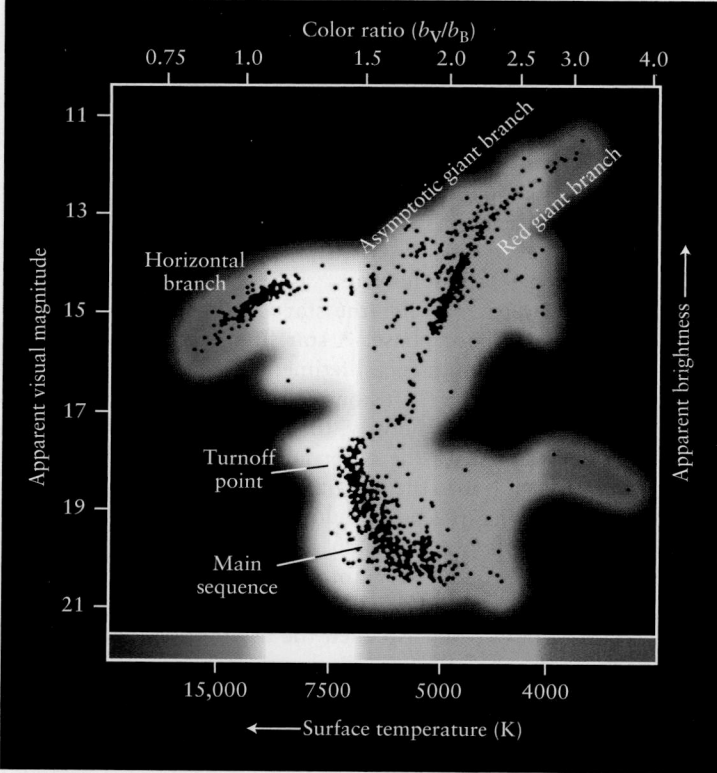

Color ratio (b_V/b_B)

0.75   1.0   1.5   2.0   2.5 3.0   4.0

Asymptotic giant branch

Red giant branch

Horizontal branch

Turnoff point

Main sequence

15,000   7500   5000   4000

Surface temperature (K)

Apparent visual magnitude

Apparent brightness

FIGURE 20-3

Stellar Evolution in a Globular Cluster In the old globular cluster M55, stars with masses less than about 0.8 M$_\odot$ are still on the main sequence, converting hydrogen into helium in their cores. Slightly more massive stars have consumed their core hydrogen and are ascending the red giant branch; even more massive stars have begun core helium fusion and are found on the horizontal branch. The most massive stars (which still have less than 4 M$_\odot$) have consumed all the helium in their cores and are ascending the asymptotic giant branch. (Data from D. Schade, D. VandenBerg, and F. Hartwick)

We saw in Section 19-1 that the more massive a star, the shorter the amount of time it remains on the main sequence. Similarly, the greater the star's mass, the more rapidly it goes through the stages of post–main-sequence evolution. Hence, we can see all of these stages by studying star clusters, which contain stars that are all the same age but that have a range of masses (see Section 19-4). Figure 20-3 shows a color–magnitude diagram for the globular cluster M55, which is about 12.3 billion years old. The least massive stars in this cluster are still on the main sequence. Progressively more massive stars have evolved to the red giant branch, the horizontal branch, and the asymptotic giant branch.

A 1-M$_\odot$ AGB star can reach a maximum luminosity of nearly 10^4 L$_\odot$, as compared with approximately 10^3 L$_\odot$ when it reached the helium flash and a relatively paltry 1 L$_\odot$ during its main-sequence lifetime. When the Sun becomes an AGB star some 5 billion years from now, this tremendous increase in luminosity will cause Mars and the jovian planets to largely evaporate away. The Sun's bloated outer layers will reach Earth's orbit. Mercury and perhaps Venus will simply be swallowed whole.

CONCEPT CHECK 20-1

Why does helium fusion in a star's core, which releases energy, actually lead to a decrease in its luminosity? What are these stars called in the H-R diagram?

Answer appears at the end of the chapter.

20-2 Dredge-ups bring the products of nuclear fusion to a giant star's surface

As we saw in Section 16-2, energy is transported outward from a star's core by one of two processes—radiative diffusion or convection. The first is the passage of energy in the form of electromagnetic radiation, and it dominates only when a star's gases are relatively transparent. The second involves up-and-down movement of the star's gases. Convection plays a very important role in giant stars, and it helps supply the cosmos with the elements essential to life.

Convection, Dredge-ups, and Carbon Stars

In the Sun, convection dominates only the outer layers, from around 0.71 solar radius (measured from the center of the Sun) up to the photosphere (recall Figure 16-4). During the final stages of a star's life, however, the convective zone can become so broad that it extends down to the star's core. At these times, convection can "dredge up" the heavy elements produced in and around the core by nuclear fusion, transporting them all the way to the star's surface.

The *first* dredge-up takes place after core hydrogen fusion stops, when the star becomes a red giant for the first time. Convection dips so deeply into the star that material processed by the CNO cycle of hydrogen fusion (see Section 16-1) is carried up to the star's surface, changing the relative abundances of carbon, nitrogen, and oxygen. A *second* dredge-up occurs after core helium fusion ceases, further altering the abundances of carbon, nitrogen, and oxygen. Still later, during the AGB stage, a *third* dredge-up can occur if the star has a mass greater than about 2 M$_\odot$. This third dredge-up transports large amounts of freshly synthesized carbon to the star's surface, and the star's spectrum thus exhibits prominent absorption bands of carbon-rich molecules like C_2, CH, and CN. For this reason, an AGB star that has undergone a third dredge-up is called a **carbon star**.

All AGB stars have very strong stellar winds that cause them to lose mass at very high rates, up to 10^{-4} M$_\odot$ per year (1000 times greater than that of a red giant, and 10^{10} times greater than the rate at which our present-day Sun loses mass). The surface temperature of AGB stars is relatively low, around 3000 K, so any ejected carbon-rich molecules can condense to form tiny grains of soot. Indeed, carbon stars are commonly found to be obscured in sooty cocoons of ejected matter (Figure 20-4).

The carbon that forms the basis of all life on Earth was ejected billions of years ago from giant stars.

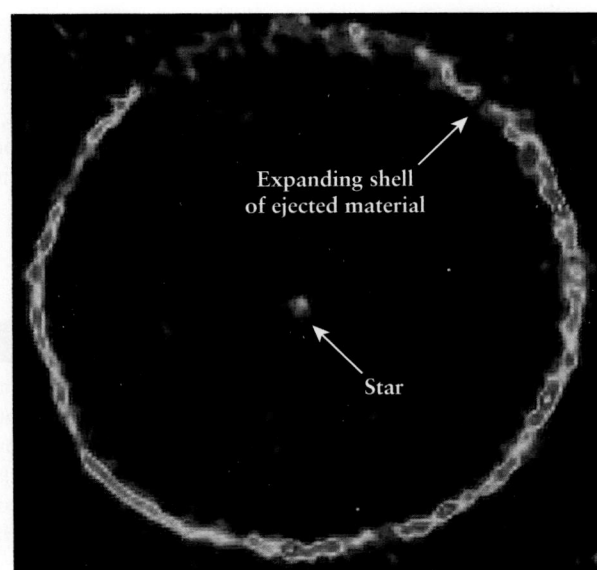

FIGURE 20-4 ⓡ I V U X G

FIGURE 20-4 Ⓡ I V U X G

A Carbon Star TT Cygni is an AGB star in the constellation Cygnus that ejects some of its carbon-rich outer layers into space. Some of the ejected carbon combines with oxygen to form molecules of carbon monoxide (CO), whose emissions can be detected with a radio telescope. This radio image shows the CO emissions from a shell of material that TT Cygni ejected some 7000 years ago. Over that time, the shell has expanded to a diameter of about ½ ly. (H. Olofsson [Stockholm Observatory] et al./NASA)

Carbon stars are important because they enrich the interstellar medium with carbon and some nitrogen and oxygen. The triple alpha process that occurs in helium fusion is the *only* way that carbon can be made, and carbon stars are the primary avenue by which this element is dispersed into interstellar space. Indeed, most of the carbon in your body was produced long ago inside a star by the triple alpha process (see Section 19-3). This carbon was later dredged up to the star's surface and ejected into space. Some 4.56 billion years ago, a clump of the interstellar medium that contained this carbon coalesced into the solar nebula from which our Earth—and all of the life on it—eventually formed. In this sense, you can think of your body as containing "recycled" material—substances that were once in the heart of a star that formed and evolved long before our solar system existed.

CONCEPT CHECK 20-2

How did most of the carbon in your body leave the stars where these atoms formed—through a supernova explosion, or a stellar wind?

Answer appears at the end of the chapter.

20-3 Stars of moderately low mass die by gently ejecting their outer layers, creating planetary nebulae

For a star that, like our Sun, began with a moderately low mass (between about 0.4 and 4 M⊙), the AGB stage in its evolution is a dramatic turning point. Before this stage, a star loses mass only

gradually through steady stellar winds. But as it evolves during its AGB stage, a star divests itself completely of its outer layers. The aging star undergoes a series of bursts in luminosity, and in each burst it ejects a shell of material into space. (The shell around the AGB star TT Cygni, shown in Figure 20-4, was probably created in this way.) Eventually, all that remains of a low-mass star is a fiercely hot, exposed core, surrounded by glowing shells of ejected gas. This late stage in the life of a star is called a **planetary nebula**; see **Figure 20-5**. (The left-hand image on the opening page of this chapter also shows a planetary nebula called the Helix Nebula.)

> An aging AGB star casts off much of the mass that it possesses and makes a beautiful glowing nebula.

Making a Planetary Nebula

To understand how an AGB star can eject its outer layers in shells, consider the internal structure of such a star as shown in Figure 20-2. As the helium in the helium-fusing shell is used up, the pressure that holds up the dormant hydrogen-fusing shell decreases. Hence, the dormant hydrogen shell contracts and heats up, and

FIGURE 20-5 R I V U X G

The Cat's Eye Nebula Named after its shape, this nebula was first discovered in 1786. The exposed core of the central star—clearly visible in the image—is about 10,000 times more luminous than the Sun, and its intense ultraviolet radiation powers the nebula by ionizing its gas. While the nebula is hotter than the Sun, the gas is too diffuse to shine by thermal radiation as the Sun does. Instead, like all planetary nebula, the visible light comes from emission lines: After electrons are excited to higher energies, photons are emitted as the electrons cascade down to lower energies in the gas's atoms. As with many planetary nebulae, the mechanisms shaping the intricate visible structures are not well understood. (NASA, MAST, STScI, AURA and Vicent Peris [OAUV/PTeam]) hydrogen fusion begins anew. This revitalized hydrogen fusion creates helium, which rains downward onto the temporarily dormant helium-fusing shell. As the helium shell gains mass, it shrinks and heats up. When the temperature of the helium shell reaches a certain critical value, it reignites in a **helium shell flash** that is similar to (but less intense than) the helium flash that occurred earlier in the evolution of a low-mass star (see Section 19-3). The released energy pushes the hydrogen-fusing shell outward, making it cool off, so that hydrogen fusion ceases and this shell again becomes dormant. The process then starts over again.

When a helium shell flash occurs, the luminosity of an AGB star increases substantially in a relatively short-lived burst called a **thermal pulse**. Figure 20-6, which is based on a theoretical calculation of the evolution of a 1-$M_\odot$ star, shows that thermal pulses begin when the star is about 12.365 billion years old. The calculations predict that thermal pulses occur at ever-shorter intervals of about 100,000 years.

During these thermal pulses, the dying star's outer layers can separate completely from its carbon–oxygen core. As the ejected material expands into space, dust grains condense out of the cooling gases. Radiation pressure from the star's hot, burned-out core acts on the specks of dust, propelling them farther outward, and the star sheds its outer layers altogether. In this way, an aging 1-$M_\odot$ star loses as much as 40% of its mass. More massive stars eject even greater fractions of their original mass.

As a dying star ejects its outer layers, the star's hot core becomes exposed. With a surface temperature of about 100,000 K, this exposed core emits ultraviolet radiation intense enough to ionize

and excite atoms in the expanding shell of ejected gases. These gases therefore glow and emit visible light through the process of fluorescence (see Figure 18-3 and Box 18-1). In fluorescence, visible photons are emitted after electrons knocked up to higher energy levels fall back down to lower energy levels. This mechanism produces the light seen in Figure 20-5, as well as more colorful emission in planetary nebula like those shown in **Figure 20-7**.

CAUTION!

Despite their name, planetary nebulae have nothing to do with planets. This misleading term was introduced in the nineteenth century because these glowing objects looked like distant jovian planets—a small colored blur—when viewed through the small telescopes available at the time. The difference between planets and planetary nebulae first became obvious with the advent of spectroscopy: Planets have *absorption* line spectra (see Section 7-3), but the excited gases of planetary nebulae have *emission* line spectra.

The Properties of Planetary Nebulae

Planetary nebulae are quite common. Astronomers estimate that there are 20,000 to 50,000 planetary nebulae in our galaxy alone. However, in another sense, planetary nebulae are quite rare. With about 200 billion stars in our galaxy, even 50,000 planetary nebulae would constitute only 1 in every 4 million stars. This rarity of planetary nebulae is due to *time*. Since a planetary nebula typically lasts less than 50,000 years, yet low-mass stars last many billions of years, the chance that any given star is observed during its planetary nebula stage is very low.

Many planetary nebulae, such as those in Figure 20-7, are more or less spherical in shape. This shape is a result of the symmetrical way in which the gases were ejected. But if the rate of expansion is not the same in all directions, the resulting nebula can take on an hourglass or dumbbell appearance (**Figure 20-8**).

Spectroscopic observations of planetary nebulae show emission lines of ionized hydrogen, oxygen, and nitrogen. From the Doppler shifts of these lines, astronomers have concluded that the expanding shell of gas moves outward from a dying star at speeds from 10 to 30 km/s. For a shell expanding at such speeds to have attained the typical diameter of a planetary nebula (about 1 ly), it must have begun expanding about 10,000 years ago. Thus, by astronomical standards, the planetary nebulae we see today were created only very recently.

We do not observe planetary nebulae that are more than about 50,000 years old. After this length of time, the shell has spread out so far from the cooling central star that its gases cease to glow and simply fade from view. The nebula's gases then mix with the surrounding interstellar medium.

Astronomers estimate that all the planetary nebulae in the galaxy return a total of about 5 $M_\odot$ of matter to the interstellar medium each year. This amount is about 15% of all the matter expelled by the various sorts of stars in the galaxy each year.

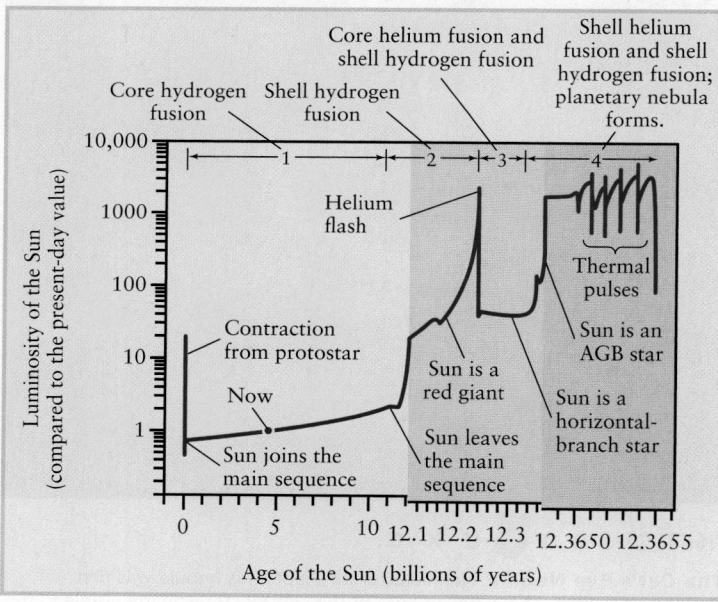

FIGURE 20-6

Further Stages in the Evolution of the Sun This diagram, which shows how the luminosity of the Sun (a 1- $M_\odot$ star) changes over time, is an extension of Figure 19-9. We use different scales for the final stages because the evolution is so rapid. During the AGB stage, there are brief periods of runaway helium fusion, causing spikes in luminosity called thermal pulses. Planetary nebulae form during these thermal pulses. (Data from Mark A. Garlick, based on calculations by I.-Juliana Sackmann and Kathleen E. Kramer)

(a)

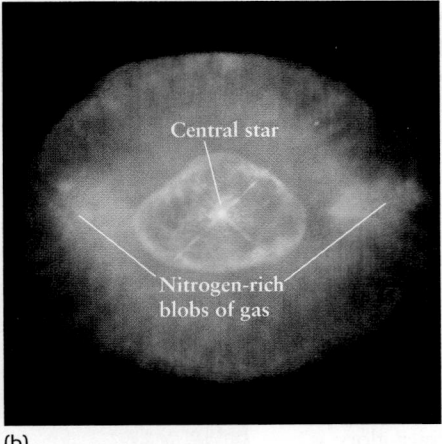

(b)

FIGURE 20-7 R I **V** U X G

Some Shapes of Planetary Nebulae The outer shells of dying low-mass stars are ejected in a wonderful variety of patterns. **(a)** This shell of expanding gas, in the globular cluster M15, is about 2150 pc (7000 ly) away from Earth. The almost perfectly spherical shell that comprises the nebula is about 1.5 pc (5 ly) in diameter; the thickness of the shell is only about 0.1 pc (0.3 ly). **(b)** NGC 6826 shows, among other features, lobes of nitrogen-rich gas (red). The process by which they were ejected is as yet unknown. **(c)** Mz 3 (Menzel 3) is 900 pc (3000 ly) from Earth. The dying star, creating these bubbles of gas, may be part of a binary system. **(d)** The spectacular Butterfly Nebula is thought to have been channeled by a doughnut-shaped cloud at the center of the disk that separates the two "wings." (a: WIYN/NOAO/NSF; b: B. Balick, J. Alexander [University of Washington] et al., NASA; c: NASA/Space Telescope Science Institute; d: NASA/ESA/Hubble)

(c)

(d)

(a)

(b)

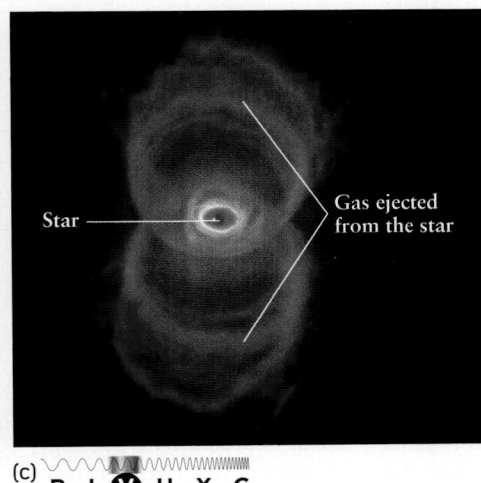

(c) R I **V** U X G

FIGURE 20-8

Making an Elongated Planetary Nebula **(a)**, **(b)** These illustrations show one proposed explanation for why many planetary nebulae have an elongated shape. **(c)** The planetary nebula MyCn18, shown here in false color, may have acquired its elongated shape in this way. It lies some 2500 pc (8000 ly) from Earth in the constellation Musca. (NASA/JPL-Caltech/ESA, the Hubble Heritage Team [STScI/AURA])

However, planetary nebulae make up less than one-fifth of the total mass that low-mass stars expel into the space between the stars. While a high-mass star ejects more matter than a low-mass star, there are many more lower-mass stars, and they make up 80% of the ejected mass.

CONCEPT CHECK 20-3

What provides the energy for a planetary nebula's emission? Does a planetary nebula fade from view when its gases cool?

Answer appears at the end of the chapter.

20-4 The burned-out core of a moderately low-mass star cools and contracts until it becomes a white dwarf

We have seen that after a moderately low-mass star (from about 0.4 to about 4 M$_\odot$) consumes all the hydrogen in its core, it is able to ignite nuclear reactions that convert helium to carbon and oxygen. Given sufficiently high temperature and pressure, carbon and oxygen can also undergo fusion reactions that release energy. But for such a moderately low-mass star, the core temperature and pressure never reach the extremely high values needed for these reactions to take place. Instead, as we have seen, the process of mass ejection just strips away the star's outer layers and leaves behind the hot carbon–oxygen core. With no nuclear reactions taking place, the core simply cools down like a dying ember. This burned-out relic of a star's former glory is called a **white dwarf.** Such white dwarfs prove to have exotic physical properties that are wholly unlike any object found on Earth.

CAUTION!

Unfortunately, the word *dwarf* is used in astronomy for several very different kinds of small objects. Here's a review of the three kinds that we have encountered so far in this book. A *white* dwarf is the relic that remains at the very end of the evolution of a star of initial mass between about 0.4 M$_\odot$ and 4 M$_\odot$. Nuclear reactions are no longer taking place in its interior; it emits light simply because it is still hot. A *red* dwarf, discussed in Section 19-1, is a cool main-sequence star with a mass between about 0.08 M$_\odot$ and 0.4 M$_\odot$. The energy emitted by a red dwarf in the form of light comes from its core, where fusion reactions convert hydrogen into helium. Finally, a *brown* dwarf (see Section 17-5) is an object like a main-sequence star but with a mass less than about 0.08 M$_\odot$. Because its mass is so small, its internal pressure and temperature are too low to sustain nuclear reactions. Instead, a brown dwarf emits light because it is slowly contracting, a process that releases energy (see Section 16-1). White dwarfs are comparable in size to Earth (see Section 17-7); by contrast, brown dwarfs are larger than the planet Jupiter, and red dwarfs are even larger.

Properties of White Dwarfs

You might think that without nuclear reactions to provide internal heat and pressure, a white dwarf should keep on shrinking under the influence of its own gravity as it cools. In actuality, a cooling white dwarf maintains its size because the burned-out stellar core is so dense that most of its electrons are degenerate (see Section 19-3). Thus, degenerate electron pressure supports the star against further collapse. This pressure does not depend on temperature, so it continues to hold up the star even as the white dwarf cools and its temperature drops.

> A white dwarf is kept from collapsing by the pressure of its degenerate electrons.

Many white dwarfs are found in the solar neighborhood, but all are too faint to be seen with the naked eye. One of the first white dwarfs to be discovered is a companion to Sirius, the brightest star in the night sky. In 1844, the German astronomer Friedrich Bessel noticed that Sirius was moving back and forth slightly,

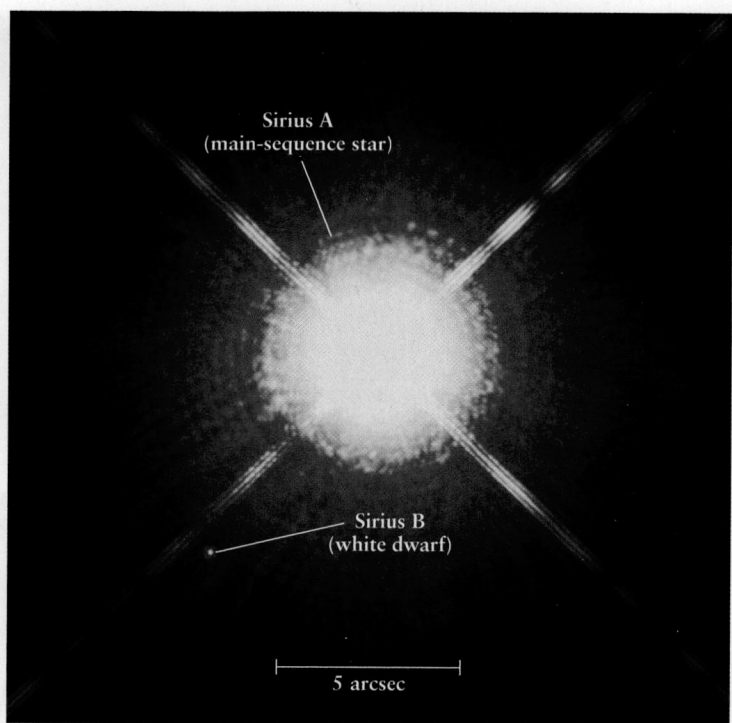

FIGURE 20-9 R I V U X G

Sirius A and Its White Dwarf Companion Sirius, the brightest-appearing star in the sky, is actually a binary star: The secondary star, called Sirius B, is a white dwarf. In this Hubble Space Telescope image, Sirius B is almost obscured by the glare of the overexposed primary star, Sirius A, which is about 10^4 times more luminous than Sirius B. The halo and rays around Sirius A are the result of optical effects within the telescope. (NASA, ESA, H. Bond [STScI] and M. Barstow [University of Leicester])

as if it was being orbited by an unseen object. This companion, designated Sirius B (Figure 20-9), was first glimpsed in 1862 by the American astronomer Alvan Clark. Recent Hubble Space Telescope observations at ultraviolet wavelengths, where hot white dwarfs emit most of their light, show that the surface temperature of Sirius B is 25,200 K. (By contrast, the main-sequence star Sirius A has a surface temperature of 10,500 K, while the Sun's surface temperature is a relatively frosty 5800 K.)

The Density of White Dwarfs

Observations of white dwarfs in binary systems like Sirius allow astronomers to determine the mass, radius, and density of these stars (see Sections 17-9, 17-10, and 17-11). Such observations show that the density of the degenerate matter in a white dwarf is typically 10^9 kg/m^3 (a million times denser than water). A teaspoonful of white dwarf matter brought to Earth would weigh nearly 5.5 tons—as much as an elephant!

Compressing matter so that an elephant could fit into a teaspoon might seem impossible until you remember that most of an atom is empty space. We mentioned in Section 5-7 that if you imagine a typical atom expanded to the size of a football field, the nucleus at the 50-yard line would only be 1 cm wide. That leaves a lot of space between the nucleus and its electrons, which would be out near the goal posts; even farther out would be the nucleus of the next nearest atom. The incredible density of a white dwarf leaves much less empty space between its atoms, but as we will see in Chapter 21, objects called *neutron stars* are even denser than white dwarfs.

CONCEPT CHECK 20-4

Do all atoms consist mostly of empty space, or is empty space a property of only the carbon and oxygen atoms that lead to a white dwarf?

Answer appears at the end of the chapter.

The Mass–Radius Relation for White Dwarfs

As we learned in Section 19-3, degenerate matter has a very different relationship among its pressure, density, and temperature than ordinary gases do. Consequently, white dwarf stars have an unusual **mass–radius relation:** The more massive a white dwarf star, the *smaller* it is. Figure 20-10 displays the mass–radius relation for white dwarfs. Note that the more degenerate matter you pile onto a white dwarf, the smaller it becomes. However, there is a limit to how much pressure degenerate electrons can produce. As a result, there is an upper limit to the mass that a white dwarf can have. This maximum mass is called the **Chandrasekhar limit,** after the Indian-American scientist Subrahmanyan Chandrasekhar, who pioneered theoretical studies of white dwarfs in the 1930s. (The orbiting Chandra X-ray Observatory, described in Section 6-7, is named in his honor.) The Chandrasekhar limit is equal to 1.4 $M_\odot$, meaning that all white dwarfs must have masses less than 1.4 $M_\odot$. As we will see in Section 20-6, when the Chandrasekhar limit is exceeded, the star is no longer a white dwarf, and the results are explosive.

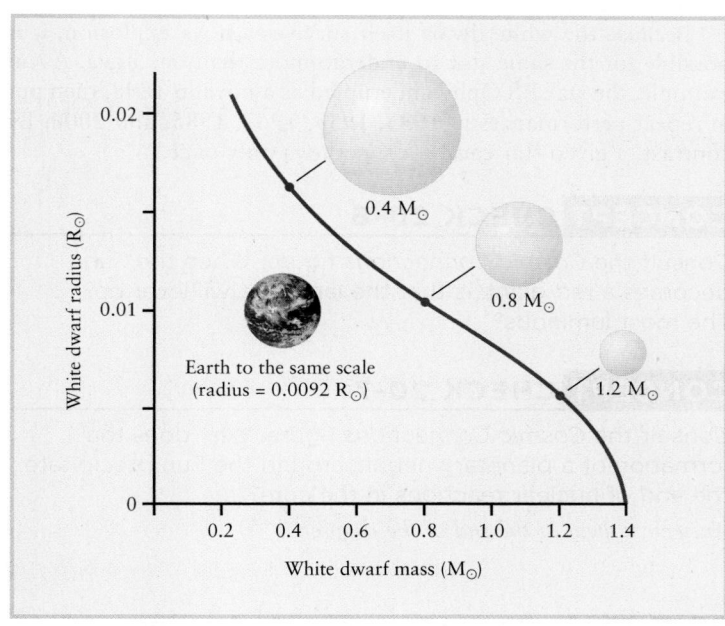

FIGURE 20-10

The Mass–Radius Relationship for White Dwarfs The more massive a white dwarf is, the smaller its radius. (The drawings of white dwarfs of different mass are drawn to the same scale as the image of Earth.) This unusual relationship is a result of the degenerate-electron pressure that supports the star. The maximum mass of a white dwarf, called the Chandrasekhar limit, is 1.4 $M_\odot$.

It is worth pointing out that the Chandrasekhar limit represents a huge scientific success. Electron degeneracy pressure is based on the quantum physics of electrons on the smallest of imaginable scales, yet the Chandrasekhar limit applies this physics to a solar mass of material about the size of Earth. Observations verify that white dwarfs do not exceed the Chandrasekhar limit: Numerous white dwarfs are in binary systems (which allows their masses to be measured; see Section 4-7), and amazingly, none are above 1.4 $M_\odot$!

The material inside a white dwarf consists mostly of ionized carbon and oxygen atoms floating in a sea of degenerate electrons. As the dead star cools, the carbon and oxygen ions slow down, and electric forces between the ions begin to prevail over the random thermal motions. About 5×10^9 years after the star first becomes a white dwarf, when its luminosity has dropped to about 10^{-4} $L_\odot$ and its surface temperature is a mere 4000 K, the ions no longer move freely. Instead, they arrange themselves in orderly rows, like an immense crystal lattice. From this time on, you could say that the star is "solid." The degenerate electrons move around freely in this crystalline material, just as electrons move freely through an electrically conducting metal like copper or silver. A diamond is also crystalline (with carbon ordered in a crystal lattice), so a cool carbon–oxygen white dwarf has some similarities to a giant diamond in the sky!

CONCEPT CHECK 20-5

Could electron degeneracy pressure support a 1.2-$M_\odot$ white dwarf? If so, how would the white dwarf compare in size to Earth?

Answer appears at the end of the chapter.

From Red Giant to Planetary Nebula to White Dwarf

Figure 20-11 shows the evolutionary tracks followed by three burned-out stellar cores as they pass through the planetary nebula stage and become white dwarfs. When these three stars were red giants, they had masses of 0.8, 1.5, and 3.0 $M_\odot$. Mass ejection strips these dying stars of up to 60% of their matter. During their final spasms, the luminosity and surface temperature of these stars change quite rapidly. The points representing these stars in an H-R diagram race along their evolutionary tracks, sometimes executing loops corresponding to thermal pulses (see Track B and Track C in Figure 20-11). Finally, as the ejected nebulae fade and the stellar cores cool, the evolutionary tracks of these dying stars take a sharp turn toward the white dwarf region of the H-R diagram. As the table accompanying Figure 20-11 shows, the final white dwarf has only a fraction of the mass of the giant star from which it evolved.

Although a white dwarf maintains the same size as it cools, its luminosity and surface temperature both decrease with time. Consequently, the evolutionary tracks of aging white dwarfs point toward the lower right corner of the H-R diagram. You can see this in the lower right portion of Figure 20-11. After ejecting much of its mass into space, our own Sun will eventually evolve into a white dwarf star about the size of Earth and with perhaps one-tenth of its present luminosity. It will become even dimmer as it cools. After 5 billion years as a white dwarf, the Sun will radiate with no more than one ten-thousandth of its present brilliance. With the passage of eons, our Sun will simply fade into obscurity.

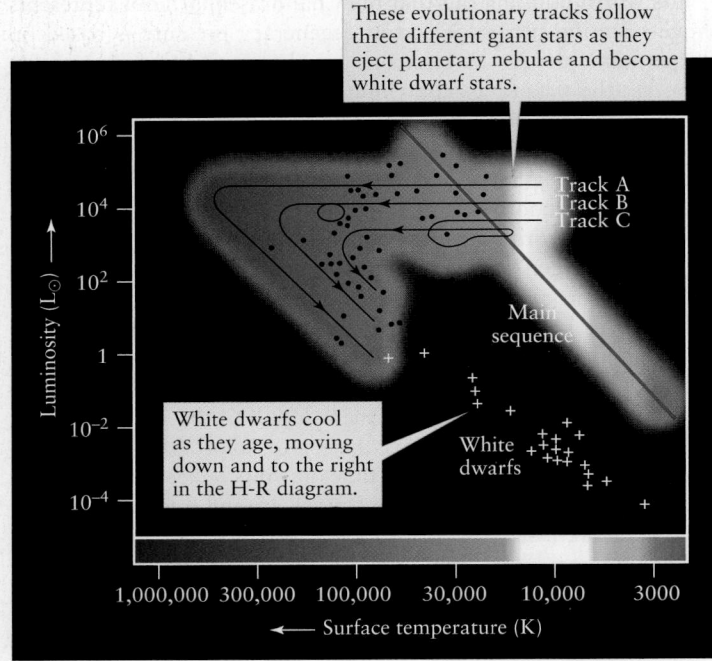

These evolutionary tracks follow three different giant stars as they eject planetary nebulae and become white dwarf stars.

White dwarfs cool as they age, moving down and to the right in the H-R diagram.

	Mass ($M_\odot$)		
Evolutionary track	**Giant star**	**Ejected nebula**	**White dwarf**
A	3.0	1.8	1.2
B	1.5	0.7	0.8
C	0.8	0.2	0.6

FIGURE 20-11

Evolution from Giants to White Dwarfs This H-R diagram shows the evolutionary tracks of three low-mass giant stars as they eject planetary nebulae. The table gives the extent of mass loss in each case. The dots represent the central stars of planetary nebulae whose surface temperatures and luminosities have been determined; the crosses represent white dwarfs of known temperature and luminosity. (Data from B. Paczynski)

The *Cosmic Connections* figure on the next page summarizes the full evolutionary cycle of a 1-$M_\odot$ star like the Sun, from its birth as a main-sequence star to its demise as a white dwarf.

Still other exotic phenomena occur when a stellar corpse is part of a close binary system. One example is a **nova** (plural **novae**), in which a faint star suddenly brightens by a factor of 10^4 to 10^8 over a few days or hours, reaching a peak luminosity of about $10^5\ L_\odot$. By contrast, the *super*nova events we discuss in Section 20-6 have a peak luminosity of about $10^9\ L_\odot$.

Novae and White Dwarfs

In the 1950s, painstaking observations led to the conclusion that all novae are members of close binary systems containing a white dwarf. Gradual mass transfer from the ordinary companion star deposits fresh hydrogen onto the white dwarf (see Figure 19-22b for a schematic diagram of this sort of mass transfer). Because of the white dwarf's strong gravity, this hydrogen is compressed into

1. Material from a star accretes onto a companion white dwarf.

2. When enough accreted material builds up, thermonuclear reactions occur on the white dwarf's surface, creating a burst of visible light.

3. The nova fades over several weeks.

FIGURE 20-12

The Light Curve of a Nova This illustration and graph show the history of Nova Cygni 1975, a typical nova. Its rapid rise and gradual decline in apparent brightness are characteristic of all novae. This nova, also designated V1500 Cyg, was easily visible to the naked eye (that is, was brighter than an apparent magnitude of +6) for nearly a week. (Illustration: NASA/CXC/M. Weiss)

a dense layer covering the hot surface of the white dwarf. As more gas is deposited and compressed, the temperature in the hydrogen layer increases. When the temperature reaches about 10^7 K, hydrogen fusion ignites throughout the gas layer, embroiling the white dwarf's surface in a nuclear detonation that we see as a nova (**Figure 20-12**).

Because the white dwarf itself survives a nova explosion, it is possible for the same star to undergo more than one nova. As an example, the star RS Ophiuchi erupted as a nova in 1898, then put in repeat performances in 1933, 1958, 1967, 1985, and 2006. By contrast, a given star can be a supernova only once.

CONCEPT CHECK 20-6

Consult the *Cosmic Connections* figure. When the Sun becomes a red giant, is that the largest it will ever be? The most luminous?

CONCEPT CHECK 20-7

Consult the *Cosmic Connections* figure. Why does the formation of a planetary nebula around the Sun precipitate the end of nuclear reactions in the core?

Answers appear at the end of the chapter.

20-5 High-mass stars create heavy elements in their cores

During the entire lifetime of a low-mass red dwarf star (with an initial mass less than about 0.4 $M_\odot$), the only nuclear reaction that takes place is the fusion of hydrogen nuclei to form helium nuclei. In stars with initial masses from about 0.4 $M_\odot$ to about 4 $M_\odot$, a

COSMIC CONNECTIONS

The Future of Our Sun

ANIMATION 20.9 The Sun is presently less than halfway through its lifetime as a main-sequence star. The H-R diagram and cross sections on this page summarize the dramatic changes that will take place when the Sun's main-sequence lifetime comes to an end.

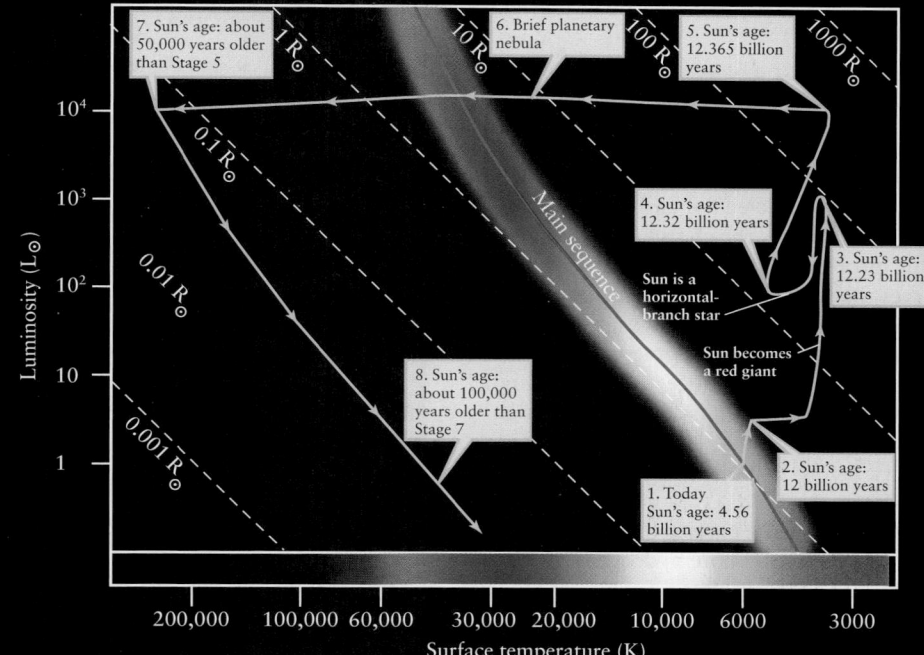

7. Sun's age: about 50,000 years older than Stage 5

6. Brief planetary nebula

5. Sun's age: 12.365 billion years

4. Sun's age: 12.32 billion years

Sun is a horizontal-branch star

3. Sun's age: 12.23 billion years

Sun becomes a red giant

8. Sun's age: about 100,000 years older than Stage 7

2. Sun's age: 12 billion years

1. Today Sun's age: 4.56 billion years

Main sequence

Luminosity ($L_\odot$)

Surface temperature (K)

NOTE: The illustrations below do *not* show the dramatic changes in the Sun's radius as it evolves. The sizes of the various layers are not shown to scale.

---

1. On the main sequence

The present-day Sun is a main-sequence star—in its core, hydrogen fuses to produce helium.

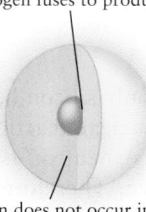

Fusion does not occur in the outer layers (which contain predominantly hydrogen and helium).

2. Becoming a red giant

At the end of the Sun's main-sequence lifetime, fusion stops in the core (which has been converted to helium).

Fusion of hydrogen into helium continues in a shell around the core. The core shrinks, accelerating the fusion reactions in the shell and making the outer layers expand and cool.

3. The helium flash

As the core contracts and heats, the core helium begins to fuse to make carbon and oxygen. The core expands and the rate of energy release slows.

Hydrogen fusion continues in a shell around the core.

The outer layers (where there are still no fusion reactions) contract and get hotter due to the slower rate of energy release.

4. Beginning the second red giant phase

Once the core helium is consumed, what remains is an inert core of carbon and oxygen. The core again shrinks and gets hotter.

Helium-fusing shell

Hydrogen-fusing shell

The shrinkage of the core accelerates fusion reactions, and a second shell burns—now with the fusion of helium. Outer layers expand and cool. The Sun becomes an AGB star.

---

5. The Sun reaches its maximum size

Inert carbon–oxygen core

Hydrogen-fusing shell

Helium-fusing shell

Outer layers (still no fusion reactions)

The Sun is more than 100 times larger in radius than when it was a main-sequence star. Part of the outer layers escapes into space in a stellar wind.

6. A planetary nebula

Thermal pulses cause spikes in luminosity that eject the star's outer layers.

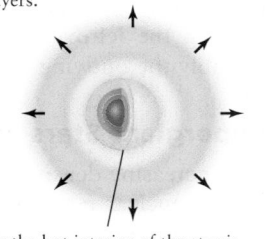

As the hot interior of the star is exposed, we observe an increase in the star's surface temperature.

7. The end of nuclear reactions

With the outer layers gone, the pressure on the shells around the core is too little to sustain nuclear reactions.

The star still glows intensely because of its high temperature. As energy is lost in the form of electromagnetic radiation, the star slowly cools.

8. A white dwarf

The core is now a white dwarf star, and the former shells around the core become its thin atmosphere.

The carbon–oxygen interior of the white dwarf is degenerate, so it does not contract as it cools. Hence the white dwarf's radius no longer changes.

Hydrogen and helium, no fusion

Hydrogen fusion producing helium

Helium, no fusion

Helium fusion producing carbon and oxygen

Carbon and oxygen, no fusion

second kind of nuclear reaction takes place—helium fusion. The heaviest elements manufactured by helium fusion are carbon and oxygen.

The life story of a *high-mass* star (with a main-sequence mass greater than about 4 $M_\odot$) begins with these same reactions. But calculations show that high-mass stars can also go through several additional stages of nuclear reactions involving the fusion of carbon, oxygen, and other heavier nuclei. As a result, high-mass stars end their lives quite differently from low-mass stars.

Heavy-Element Fusion in Massive Stars

Why is stellar fusion of these heavier nuclei possible only in a high-mass star? The reason is that the electric charge of a nucleus increases with its size: For example, a nucleus of carbon has 6 positively charged protons and hence 6 times the charge of a hydrogen nucleus (which has a single proton). This means that there are strong repulsive electric forces that tend to keep these nuclei apart. Only at the great speeds associated with extremely high temperatures can the nuclei travel fast enough to overcome their mutual electric repulsion and fuse together. To produce these very high temperatures at a star's center, the pressure must also be very high. Hence, the star must have a very large mass, because only such a star has strong enough gravity trying to pull it together and thus strong enough pressure at its center.

As we discussed in Section 19-2, when a main-sequence star with a mass greater than about 0.4 $M_\odot$ uses up its core hydrogen, it begins shell hydrogen fusion and enters a red giant phase. Such a star then begins core helium fusion when the core temperature becomes high enough. The differences between moderately low-mass (from about 0.4 $M_\odot$ to about 4 $M_\odot$) and high-mass stars (more than about 4 $M_\odot$) become pronounced after helium core fusion ends, when the core is composed primarily of carbon and oxygen.

Let us consider how the late stages in the evolution of a high-mass star differ from those of a low-mass star. In low-mass stars, as we saw in Section 20-4, fusion stops in the carbon–oxygen core; it eventually becomes a white dwarf. But in stars whose overall mass is more than about 4 $M_\odot$, the temperature and pressure of the carbon–oxygen core is enough to trigger further fusion reactions. The first of the new nuclear reactions is **carbon fusion,** which consumes carbon nuclei and produces oxygen, neon, sodium, and magnesium.

If a star has an even larger main-sequence mass of about 8 $M_\odot$ or so (before mass ejection), even more nuclear reactions can take place. After the cessation of carbon fusion, the core will again contract, and the star's central temperature exceeds 1 billion kelvins (10^9 K). At about this temperature, **neon fusion** begins. This uses up the neon accumulated from carbon fusion and further increases the amount of oxygen and magnesium in the star's core.

After neon fusion ends, the core will again contract, and **oxygen fusion** begins. The principal product of oxygen fusion is silicon. Once oxygen fusion is over, the core will contract yet again. With this contraction, the central temperature rises and **silicon fusion** begins, producing a variety of nuclei from sulfur to iron and nickel. While all of this fusion is going on in the star's interior, at the surface the star is losing mass at a rapid rate (**Figure 20-13**).

As a high-mass star consumes increasingly heavier nuclei, the nuclear reactions produce a wider variety of products. For

FIGURE 20-13 R I **V** U X G

Mass Loss from a Supergiant Star At the heart of this nebulosity, called SMC N76, lies a supergiant star with a mass of at least 18 $M_\odot$. This star is losing mass at a rapid rate in a strong stellar wind. As this wind collides with the surrounding interstellar gas and dust, it creates the "bubble" shown here. SMC N76, which has an angular diameter of 130 arcsec, lies within the Small Magellanic Cloud, a small galaxy that orbits our Milky Way. It is about 60,600 pc (198,000 ly) distant. (European Southern Observatory)

example, oxygen fusion produces not only silicon but also magnesium, phosphorus, and sulfur. Some nuclear reactions that create heavy elements also release neutrons. A neutron is like a proton except that it carries no electric charge. Therefore, neutrons are not repelled by positively charged nuclei, and so can easily collide and combine with them. This absorption of neutrons by nuclei, called **neutron capture,** creates many elements and isotopes that are not produced directly in fusion reactions.

Each stage of nuclear reactions in a high-mass star helps to trigger the succeeding stage. In each stage, when the star exhausts a given type of nuclear fuel in its core, gravitational contraction takes the core to ever-higher densities and temperatures, thereby igniting the leftover remains of the previous fusion stage—and possibly the outlying shell of unburned fuel as well.

Supergiant Stars and Their Evolution

The increasing density and temperature of the core make each successive nuclear reaction more rapid than the one that preceded it. As an example, **Table 20-1** shows a theoretical calculation of the evolutionary stages for a star with an initial main-sequence mass of 25 $M_\odot$. This calculation indicates that carbon fusion in such a star

TABLE 20-1 Evolutionary Stages of a 25-M$_\odot$ Star

Stage	Core temperature (K)	Core density (kg/m^3)	Duration of stage
Hydrogen fusion	7×10^7	10×10^3	10^7 years
Helium fusion	2×10^8	2×10^6	10^6 years
Carbon fusion	8×10^8	10^9	1000 years
Neon fusion	1.6×10^9	10^{10}	3 years
Oxygen fusion	1.8×10^9	10^{10}	3.6 months
Silicon fusion	2.5×10^9	10^{11}	5 days
Core collapse	10^{10}	10^{13}	¼ second
Explosive (supernova)	about 10^9	varies	10 seconds

Based on calculations by Stanford Woosley (University of California, Santa Cruz) and H.-T. Janka.

lasts for 1000 years, neon fusion for 3 years, and oxygen fusion for only 3.6 months. The last, and briefest, stage of nuclear reactions is silicon fusion. The entire core supply of silicon in a 25-M$_\odot$ star is used up in only 5 days!

Each stage of core fusion in a high-mass star generates a new shell of material around the core. After several such stages, the internal structure of a truly massive star—say, 25 to 30 M$_\odot$ or greater—resembles that of an onion (Figure 20-14). Notice the order of the layers in Figure 20-14—the earlier the fusion reaction begins, the farther it is from the center. Because nuclear reactions can take place simultaneously in several shells, energy is released at such a rapid rate that the star's outer layers expand tremendously. The result is a

> A star of 8 M$_\odot$ evolves into a supergiant 100 times (or more) larger than the Sun.

supergiant star, whose luminosity and radius are much larger than those of a giant (see Section 17-7).

Several of the brightest stars in the sky are supergiants, including Betelgeuse and Rigel in the constellation Orion and Antares in the constellation Scorpius. (Figure 17-15 shows the locations of these stars in an H-R diagram.) They appear bright not because they are particularly close, but because they are extraordinarily luminous.

The End of Nuclear Fusion

A supergiant star cannot keep adding shells to its "onion" structure forever, because the sequence of nuclear fusion reactions cannot go on indefinitely. In order for atoms to serve as a nuclear fuel, energy must be given off when they collide and fuse.

Let's consider a common case in which a helium nucleus fuses with a larger atomic nucleus, although the results apply beyond helium. When they are close enough (due to high temperatures and pressures), the helium atom and the larger atom are pulled together by strong nuclear forces and energy is released. However, to get close enough, the atoms must overcome a repulsive force because all protons repel each other. As a result, fusion with larger atomic nuclei becomes increasingly difficult as the larger atom's greater number of protons leads to a greater *repulsive* electric force against the protons in the smaller helium nucleus. Very large atoms can be fused with helium, but they require an *input* of energy, rather than releasing energy. Everything in a star changes when a fusion reaction requires an input of energy in order to occur.

To act as a fuel, nuclear fusion reactions must release energy; this is the energy that powers a star. As larger and larger atoms are built up inside a star's core, they will eventually approach a size where they can no longer release energy through fusion and would instead require an input of energy to fuse. At this point, the star has run out of nuclear fuel in its core: **Iron is the first atom produced for which further fusion does not release energy.** As silicon fusion produces iron, the iron atoms in the core steadily grow in number without fusing into even larger atoms. Hence, the sequence of energy-producing fusion stages ends with silicon fusion, and no new layers in the fusing "onion" shells of Figure 20-14 will be produced. Shell fusion in the layers surrounding the iron-rich core consumes the star's remaining reserves of fuel. At this stage, the entire energy-producing region

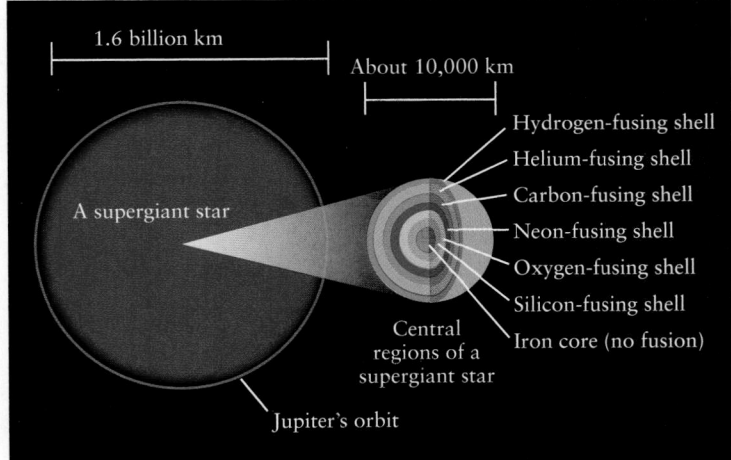

1.6 billion km

About 10,000 km

A supergiant star

Hydrogen-fusing shell
Helium-fusing shell
Carbon-fusing shell
Neon-fusing shell
Oxygen-fusing shell
Silicon-fusing shell
Iron core (no fusion)

Central regions of a supergiant star

Jupiter's orbit

FIGURE 20-14

The Structure of an Old High-Mass Star Near the end of its life, a star with an initial mass greater than about 8 M$_\odot$ becomes a red supergiant. The star's overall size can be as large as Jupiter's orbit around the Sun. The stars energy comes from a series of concentric fusing shells, all combined within a volume roughly the same size as Earth. Nuclear reactions do not occur within the iron core because fusion reactions that involve iron require an input of energy rather than release it.

of the star is contained in a volume no bigger than Earth, some 10^6 times smaller in radius than the overall size of the star. This state of affairs will soon come to an end because the buildup of the iron-rich core signals the impending violent death of a massive supergiant star.

CONCEPT CHECK 20-8

Is it possible for helium nuclei to fuse together with atoms larger than iron to make even larger atoms?

Answer appears at the end of the chapter.

20-6 High-mass stars violently blow apart in core-collapse supernova explosions

Our present understanding is that stars of about 8 $M_\odot$ or less shed most of their mass in the form of planetary nebulae. The burned-out core that remains settles down to become a white dwarf star. But the truly massive stars—stellar heavyweights that begin their lives with more than 8 $M_\odot$ of material—do not pass through a planetary nebula phase. Instead, they die in spectacular *core-collapse supernova explosions*.

The Violent End of a High-Mass Star

To understand what happens in a core-collapse supernova explosion, we rely on theoretical models based on the known behavior of gases and atomic nuclei. The following model accounts for the observed properties of supernovae fairly well, which lends support to the overall picture. And, as we will see in Section 20-8, a special kind of "telescope" has allowed us to glimpse the interior of at least one relatively nearby supernova.

The core of an aging, massive star gets progressively hotter as it contracts and ignites successive stages of nuclear fusion (see Stage 1 in Figure 20-15). But even as more of the star's core is converted into iron, this iron is unable to undergo further fusion. Without this heat-producing nuclear energy, the usual pressure that supports a star's core from collapsing is absent. Nonetheless, the core is held up briefly by degenerate electron

> By forming a dense core of iron, a massive star sows the seeds of its own destruction.

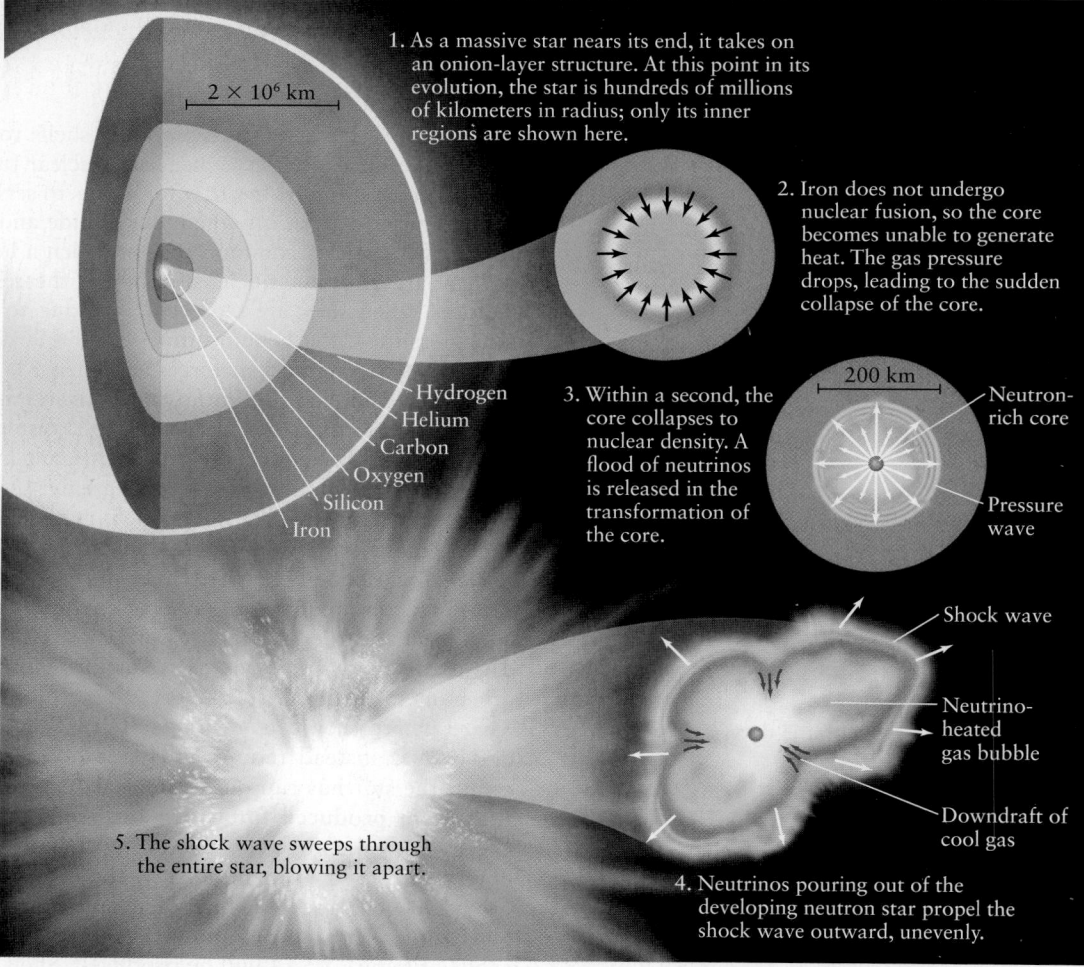

FIGURE 20-15

A Core-Collapse Supernova This series of illustrations depicts our understanding of the last day in the life of a star of more than about 8 $M_\odot$. (Art © Don Dixon) pressure, which is the same pressure that keeps a white dwarf from collapsing. However, degenerate electron pressure can only hold up a 1.4-$M_\odot$ core against its own tremendous inward gravitational pull (this is the Chandrasekhar limit, discussed in Section 20-4). When enough iron builds up to cross this threshold of mass, the result is a cataclysmic core collapse (Stage 2 in Figure 20-15). The collapse is so sudden and gravity is so strong that the outer core comes rushing in at over 20% the speed of light!

Even though steady fusion reactions in the core end with iron, the collapsing core soars in temperature due to the gravitational implosion. Temperatures skyrocket up to about 10 billion K within a fraction of a second. As a result, gamma-ray photons emitted by the intensely hot core have so much energy that when they collide with iron nuclei, they begin to break down the iron nuclei into much smaller helium nuclei. This process is called **photodisintegration**. As Table 20-1 shows, it takes a high-mass star millions of years and several stages of nuclear reactions to build up an iron core—but within a fraction of a second, photo-disintegration blasts apart much of that core.

Within another fraction of a second, the core becomes so dense that the negatively charged electrons within it are forced to combine with the positively charged protons to produce neutrons. In fact, most of the core is converted into neutrons. This process also releases a flood of *neutrinos* (see Section 16-4), denoted by the Greek letter ν (nu):

$$e^- + p \rightarrow n + \nu$$

There are other processes that produce neutrinos as well, and because these neutrinos carry away a substantial amount of energy from the core, the core cools down and condenses even further.

At about 0.25 s after its rapid contraction begins, the core is less than 20 km in diameter and its density is in excess of 4×10^{17} kg/m^3. This is **nuclear density**, the density with which neutrons and protons are packed together inside nuclei. (If Earth were compressed to this density, it would be only 300 m, or 1000 ft, in diameter!)

Matter at nuclear density or higher is extraordinarily difficult to compress. Just as compressed electrons experience a degenerate electron pressure, these neutrons in the core attain a **degenerate neutron pressure**. Thus, when the density of the neutron-rich core begins to exceed nuclear density, the core suddenly becomes very stiff and rigid. The core's contraction comes to a sudden halt, and the inner part of the core actually bounces back and expands somewhat. This *core bounce* sends a powerful wave of pressure, like an unimaginably intense sound wave, outward into the outer core (see Stage 3 in Figure 20-15).

Initially, it was thought that the core bounce was sufficient to generate a shock wave that propelled a star's matter outward during a supernova. While there is evidence for a shock wave, more energy is needed than can be provided by the core bounce alone. One of the mechanisms that can provide additional energy to the shock wave appears to involve neutrinos.

At first glance, you might not expect neutrinos to have an explosive effect because they interact so weakly with matter. Recall that after a trillion or so neutrinos pass through your head each second—which is our Sun's normal output—most of them pass right through Earth without getting absorbed. In a collapsing core, however, the density of matter is so high during the collapse that

enough neutrinos are absorbed to significantly heat matter around the core.

These neutrinos are essential to what follows. The absorbed neutrinos are thought to carry enough energy to produce hot inflating bubbles (Stage 4 in Figure 20-15). As the outer boundary of these bubbles pushes on matter around the core, the resulting shock wave ejects material from the star altogether. **The result is an explosive supernova.**

When the shock wave breaks out of the star, a portion of the explosive energy escapes in a torrent of light (see Stage 5 in Figure 20-15). At this point, the star has become a **supernova** (plural **supernovae**). Specifically, what we have described is the formation of a **core-collapse supernova**. We use this term because, as we will see in Section 20-9, it is possible for a different type of supernova explosion to occur that does not involve the collapse of the core in a massive star.

The details of the shock wave are not fully understood, and neither is the mechanism that drives it, nor all the forms of energy that power it. Supernova explosions are still an active area of investigation, and as one researcher put it: "Astrophysicists still don't know how to blow up a star."

CAUTION!

The energy released in a core-collapse supernova is an incomprehensibly large 10^{46} joules—100 times more energy than the Sun has emitted due to nuclear reactions over its *entire* 4.56-billion-year history. However, it is important to recognize that the source of the supernova's energy release is *not* nuclear reactions. Rather, it is the *gravitational* energy released by the collapse of the core. (You release gravitational energy when you step off a diving board, and this released energy goes into making a big splash in the swimming pool.) The nuclear reactions that occur in a supernova do not occur unless the atoms are pushed very close together, and this is powered by gravity. In a similar way, gravitational energy can power a bouncing ball: While the elastic (or springlike) energy pushes the ball back up when it is compressed against the ground, it was gravitational energy that powered the compression.

CONCEPT CHECK 20-9

Our Sun cannot create the pressures necessary to fuse silicon into iron. However, would a 1-$M_\odot$ core of iron collapse and produce a supernova?

CONCEPT CHECK 20-10

What can provide pressure to halt the collapse of an iron core that exceeds the Chandrasekhar limit?

CONCEPT CHECK 20-11

Hot, expanding bubbles might produce a shock wave during a supernova explosion. What delivers energy to heat up these bubbles?

Answers appear at the end of the chapter.

The Big Picture of Core Collapse

One way to think about the evolution of a massive star's core is that it undergoes an unrelenting gravitational contraction. Before the star arrives on the main sequence, hydrogen contracts until hydrogen fusion produces enough heat and pressure to temporarily halt further contraction. But as the resulting helium builds up, the core contracts until helium fusion turns on and produces a temporary halt to further gravitational contraction. This pattern continues all the way through to the production of iron in the core, with each stage of fusion—each "onion" layer in Figure 20-14—providing only a "pause" in gravity's unrelenting contraction.

But these pauses in gravitational contraction required pressure that was produced by the release of nuclear energy. While that was possible when fusing elements lighter than iron, the iron core does not release nuclear energy and offers nothing to support the surrounding layers against gravity.

With nothing to hold off gravity, the surrounding layers rush inward, gaining tremendous speed and kinetic energy. The core is also compressed further as a variety of processes unfold. However, more than anything else, the kinetic energy of the infalling matter is ultimately converted into neutrinos. In this sense, *a supernova can be seen as a gravity-powered neutrino explosion.*

For a few seconds during the collapse, the neutrino energy output is unimaginably intense: While a supernova can outshine an entire galaxy with its visible light, the energy in neutrinos is 100 times greater. By depositing only a tiny fraction of this energy in surrounding layers of the star, the neutrinos produce hot inflating bubbles that propel most of the star's matter into space.

The Wreckage of a Core-Collapse Supernova

Supercomputer simulations provide many insights into the violent, complex, and rapidly changing conditions deep inside a star as it is torn apart by a supernova explosion (Figure 20-16). For example, Figure 20-16a shows a computer simulation of a high-mass star several hours after the core bounce that initiates the supernova. The simulation produces turbulent swirls and eddies. Evidence in favor of such turbulence comes from images of the remnants of long-ago supernovae (Figure 20-16b). Such images show that material is ejected from the supernova not in uniform shells but in irregular clumps. These clumps are just what would be expected from a turbulent explosion. (We discuss supernova remnants further in Section 20-10.)

Detailed computer calculations suggest that a 25-$M_\odot$ star ejects about 96% of its material to the interstellar medium for use in producing future generations of stars. Less massive stars eject a smaller percentage of their mass into space when they become core-collapse supernovae.

Before this material is ejected into space, it is compressed so much that **a new wave of nuclear reactions sets in.** These reactions produce many of the atoms on the periodic table of the elements (see Box 5-5)

(a) A simulated supernova 2.8 hours after the core bounce.

ANIMATION 20.3

FIGURE 20-16

Turbulence in a Core-Collapse Supernova **(a)** This image from a supercomputer simulation shows a cross section of a massive star 2.8 hours into the supernova explosion. Density increases from black through red and orange to white. The darker inner region of the core is lower density because its heavier elements have been ejected into the outer layers. These dense metals appear almost as white flames as they undergo turbulence on the outermost shells of helium and hydrogen. The simulated supernova is SN 1987A. **(b)** Turbulence causes material to be ejected from the supernova in irregular "blobs," as shown by these images of the supernova remnant Cassiopeia A. Each image was made using an x-ray wavelength emitted by a particular element. (Figure 18-28 shows a false-color image of Cassiopeia A made using visible, infrared, and x-ray wavelengths.) (a: Konstantinos Kifonidis/Science Photo Library/Science Source; b: NASA/GSFC U. Hwang et al./CXC)

Silicon

Calcium

Iron

(b) Material was ejected in "blobs" from the supernova that produced the Cassiopeia A supernova remnant.   R I V U Ⓧ G that are heavier than iron. Reactions producing the heavier elements require a tremendous *input* of energy (recall Section 20-5) and thus cannot take place during the star's pre-supernova lifetime. Hence, much of the heavy elemental material that makes up our solar system, Earth, and even our bodies was produced in violent, explosive supernovae. In the next chapter, we will discover another source of heavy elements: collisions between neutron stars.

CONCEPT CHECK 20-12

Uranium atoms are much heavier than iron atoms. During what stage of a star's life are uranium atoms produced? Do they release energy when produced?

Answer appears at the end of the chapter.

20-7 In 1987, a nearby supernova gave us a close-up look at the death of a massive star

Typical supernovae have peak luminosities of visible light as great as $10^9\,L_\odot$, rivaling the light output of an entire galaxy. These large luminosities make it possible to see supernovae in galaxies far beyond our own Milky Way Galaxy, and indeed hundreds of these distant supernovae are observed each year. In fact, the vast majority of supernovae occur so far beyond our galaxy that astronomers cannot tell from earlier images the type of star that exploded. However, in a handful of nearby cases, images made before the explosion have allowed astronomers to identify the star that subsequently "went supernova," called the **progenitor star**. One significant case occurred in 1987.

A Supernova in the Galaxy Next Door

On February 23, 1987, a supernova was discovered in the Large Magellanic Cloud (LMC), a companion galaxy to our Milky Way some 51,500 pc (168,000 ly) from Earth. The supernova,

designated SN 1987A (because it was the first discovered that year), was so bright that observers in the southern hemisphere could see it without a telescope (**Figure 20-17**).

Such a bright supernova is a rare event. In the past 1000 years, only five other supernovae—in 1006, 1054, 1181, 1572, and 1604—have been bright enough to be seen with the naked eye, and all of these occurred in the Milky Way Galaxy. As the brightest supernova with modern telescopes in place, SN 1987A has given astronomers an unprecedented view of the violent death of a massive star.

> Supernova 1987A was the first nearby supernova to be seen since the invention of the telescope.

The light from a supernova such as SN 1987A does not all come in a single brief flash; the outer layers continue to glow as they expand into space. For the first 20 days after the detonation of SN 1987A, its glow was powered primarily by the explosion's tremendous heat. As the expanding gases cooled, the light energy began to come from a different source—the decay of radioactive atoms that were produced during the supernova explosion.

Why SN 1987A Was Unusual

Ideally, SN 1987A would have confirmed the theories of astronomers about typical supernovae. But SN 1987A was *not* typical. Its luminosity peaked at roughly $10^8\,L_\odot$, only one-tenth of the maximum luminosity observed for typical supernovae. Fortunately, the doomed star was close enough to be observed prior to becoming a supernova, and these observations helped explain why SN 1987A was an exceptional case.

Figure 20-14 indicates that a massive star is usually in a red supergiant stage when the iron core collapses and the star becomes a supernova. The progenitor star of SN 1987A was indeed a high-mass star: Its estimated main-sequence mass was about 20 $M_\odot$, although by the time it exploded—some 10^7 years after it first formed—it probably had shed a few solar masses. However, this

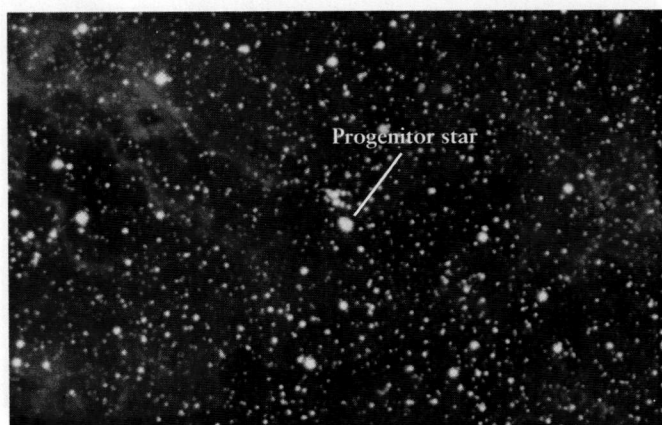

(a) Before the star exploded

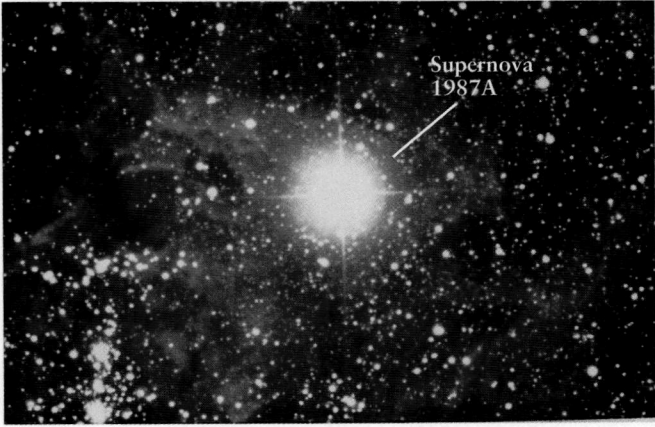

(b) After the star exploded

FIGURE 20-17 R I V U X G

SN 1987A—Before and After (a) This photograph shows a small section of the Large Magellanic Cloud as it appeared before the explosion of SN 1987A. The supernova's progenitor star was a B3 blue supergiant.

(b) This image shows a somewhat larger region of the sky a few days after the supernova exploded into brightness. (© Australian Astronomical Observatory/David Malin Images) progenitor star was identified not as a red supergiant, but as a *blue* B3 I supergiant.

The explanation of this seeming contradiction is that stars in the Large Magellanic Cloud, including the progenitor of SN 1987A, are Population II stars with a very low percentage of metals—that is, elements heavier than hydrogen and helium (see Section 19-5). A small difference in the amount of metals present can cause a star to follow a somewhat different evolutionary track on the H-R diagram. Apparently, when the progenitor star of SN 1987A developed an iron core and became a supernova, it was in the blue supergiant stage.

When in a blue supergiant phase, a star's radius may be less than one-tenth as large as when it is in a red supergiant phase. Hence, for its large mass, the progenitor star was relatively small when its core collapsed (though its radius was still more than 10 times that of our Sun). This means that the star's outer layers were held more strongly by the core's gravitational attraction. When the detonation occurred, a relatively large fraction of the supernova's energy had to be used against this gravitational attraction to push the outer layers into space. Hence, the amount of energy available to be converted into light was smaller than for most supernovae. This explains why SN 1987A was only one-tenth as bright as an exploding red supergiant would have been.

The Aftermath of SN 1987A

Three and a half years after SN 1987A exploded, astronomers used the newly launched Hubble Space Telescope to obtain a picture of the supernova. To their surprise, the image showed a ring of glowing gas around the exploded star. After the optics of the Hubble Space Telescope were repaired in 1994 (see Section 6-7),

SN 1987A was observed again and a set of *three* glowing rings was revealed (**Figure 20-18a**).

These rings are relics of a hydrogen-rich outer atmosphere that was ejected by gentle stellar winds from the progenitor star when it was a red supergiant, about 20,000 years ago. This diffuse gas expanded in an hourglass shape (Figure 20-18b) because it was blocked from expanding around the star's equator either by a pre-existing ring of gas or by the orbit of an as yet unseen companion star. (Figure 20-8 shows a similar model used to explain the shapes of certain planetary nebulae.) The outer rings in Figure 20-18a are parts of the hourglass that were ionized by the initial flash of ultraviolet radiation from the supernova; as electrons recombine with the ions, the rings emit visible-light photons. (We described this process of *recombination* in Section 18-2.)

ANIMATION
20.5 → By the early years of the twenty-first century, the explosion debris from the supernova was beginning to collide with the "waist" of the hourglass shown in Figure 20-18b. This collision is making the hourglass glow more brightly in visible wavelengths—though not enough, unfortunately, to make the supernova again visible to the naked eye—and emit copious radiation at x-ray and ultraviolet wavelengths. Because SN 1987A provides a unique laboratory for studying the evolution of a supernova, astronomers will monitor it carefully for decades to come.

CONCEPT CHECK 20-13

Why was SN 1987A about one-tenth as bright as initially expected?

Answer appears at the end of the chapter.

(a) Supernova 1987A seen in 1996

(b) An explanation of the rings

Outer ring at edge of swept-up gas from earlier mass loss

Inner ring of swept-up red supergiant gas

Supernova remnant; a dark, invisible outer portion surrounds the brighter inner region lit by radioactive decay.

SN 1987A

Outer rings

Inner ring, about 1.3 ly (0.4 pc) in diameter

ANIMATION
20.4 → FIGURE 20-18 R I V U X G

SN 1987A and Its "Three-Ring Circus" **(a)** This true-color view from the Hubble Space Telescope shows three bright rings around SN 1987A. **(b)** This drawing shows the probable origin of the rings. A wind from the progenitor star, shown in Figure 20-17a, formed an hourglass-shaped shell surrounding the star. (Compare with Figure 20-8.) Ultraviolet light from

the supernova explosion ionized ring-shaped regions in the shell, causing them to glow. (a: Rober Kirshner and Peter Challis, Harvard-Smithsonian Center for Astrophysics; STScI; b: © Rober Kirshner and Peter Challis, Harvard-Smithsonian Center for Astrophysics; STScI/NASA)

20-8 Neutrinos emanate from supernovae like SN 1987A

In addition to electromagnetic radiation, supernovae also emit a brief but intense burst of neutrinos from their collapsing cores. In fact, theory suggests that *most* of the energy released by the exploding star is in the form of neutrinos. If it were possible to detect the flood of neutrinos from an exploding star, astronomers would have direct evidence of the nuclear processes that occur within the star during its final seconds before becoming a supernova.

> Observations of SN 1987A confirm that a core-collapse supernova emits most of its energy in the form of neutrinos.

Unfortunately, detecting neutrinos is difficult because under most conditions, matter is transparent to neutrinos (see Section 16-4). Consequently, when supernova neutrinos encounter Earth, almost all of them pass completely through the planet as if it were not there. The challenge to scientists is to detect the tiny fraction of neutrinos that *do* interact with the matter through which they pass.

Neutrino Telescopes

During the 1980s, two "neutrino telescopes" designed uniquely for detecting supernova neutrinos went into operation—the Kamiokande detector in Japan (a joint project of the University of Tokyo and the University of Pennsylvania), and the IMB detector (a collaboration of the University of California at Irvine, the University of Michigan, and Brookhaven National Laboratory). Both detectors consisted of large tanks containing thousands of tons of water. On the rare occasion when a neutrino collided with one of the water molecules, it produced a brief flash of light. Any light flash was recorded by photomultiplier tubes lining the walls of the tank; photomultiplier tubes simply convert any detected light into an electronic signal that can be recorded. Because other types of subatomic particles besides neutrinos could also produce similar light flashes, the detectors were placed deep underground so that hundreds of meters of Earth would screen out almost all particles except neutrinos. (The more recent Sudbury Neutrino Observatory, shown in Figure 16-6, has a similar design.)

What causes the light flashes? And how do we know whether a neutrino comes from a supernova? The key is that a single supernova neutrino carries a relatively large amount of energy, typically 20 MeV or more. (We introduced the unit of energy called the electron volt, or eV, in Section 5-5. One MeV is equal to 10^6 electron volts. Only the most energetic nuclear reactions produce particles with energies of more than 1 MeV.) If such a high-energy neutrino hits a proton in the water-filled tank of a neutrino telescope, the collision produces a positron (see Box 16-1). The positron is emitted at a speed greater than the speed of light in water, which is 2.3×10^5 km/s. (Such motion does not violate the ultimate speed limit in the universe, 3×10^5 km/s, which is the speed of light in a *vacuum*.)

Just as an airplane that flies faster than sound produces a shock wave (a sonic boom), a positron that moves through a substance such as water faster than the speed of light in that substance produces a shock wave of light. This shock wave is called **Cherenkov radiation**, after the Russian physicist Pavel A. Cherenkov who first observed it in 1934. It is this radiation that is detected by the photomultiplier tubes that line the detector walls.

By measuring the properties of the Cherenkov radiation from an emitted positron, scientists can determine the positron's energy and, therefore, the energy of the neutrino that created the positron. Determining the energy allows them to tell the difference between the high-energy neutrinos from supernovae and neutrinos from the Sun, which typically have energies of 1 MeV or less.

Neutrinos from SN 1987A

Both the Kamiokande and IMB detectors were operational on February 23, 1987, when SN 1987A was first observed in the Large Magellanic Cloud. Soon afterward, the physicists working with these detectors excitedly reported that they had detected Cherenkov flashes from a 12-second burst of neutrinos that reached Earth 3 hours before astronomers saw the light from the exploding star. Only a few neutrinos were seen: The Kamiokande detector saw flashes from 11 neutrinos at about the same time that 8 were recorded by the IMB detector. But when the physicists factored in the sensitivity of their detectors, they calculated that Kamiokande and IMB had actually been exposed to a torrent of more than 10^{16} neutrinos.

Given the flux of neutrinos measured by the detectors and the distance of 168,000 ly from SN 1987A to Earth, physicists used the inverse-square law to determine the total number of neutrinos that had been emitted from the supernova. (This law applies to neutrinos just as it does to electromagnetic radiation; see Section 17-2.) They found that over a 10-second period, SN 1987A emitted 10^{58} neutrinos with a total energy of 10^{46} joules. This energy is more than 100 times as much energy as the Sun has emitted in its entire history and more than 100 times the amount of energy that the supernova emitted in the form of electromagnetic radiation. Indeed, for a few seconds, the supernova's neutrino luminosity—that is, the *rate* at which it emitted energy in the form of neutrinos—was 10 times greater than the total luminosity in electromagnetic radiation of all of the stars in the observable universe! Such comparisons give a hint of the incomprehensible violence with which a supernova explodes.

Why did the neutrinos from SN 1987A arrive 3 hours *before* the first light was seen? As we saw in Section 20-6, neutrinos are produced when nuclear reactions cease in the core of a massive star and the core collapses. The neutrinos that are not absorbed around the core pass easily through the volume of the star. However, the tremendous increase in the star's light output occurs only when the shock wave reaches the star's outermost layers (see Figure 20-16a). For SN 1987A, it took 3 hours for this shock wave to travel outward from the star's core to its surface, by which time the neutrino burst was already billions of kilometers beyond the dying star. Traveling toward Earth for the next 168,000 years, the neutrinos that would eventually produce light flashes in Kamiokande and IMB remained in front of the photons emitted from the star's surface, so the neutrinos were detected before the supernova's light. Thus, the neutrino data from SN 1987A gave astronomers direct confirmation of theoretical ideas about how supernova explosions take place.

Kamiokande and IMB have both been replaced by a new generation of neutrino telescopes (see Section 16-4). While these new detectors are intended primarily to observe neutrinos from the Sun, they are also fully capable of measuring neutrino bursts from nearby supernovae. Astronomers have identified a number of supergiant stars in our galaxy that are likely to explode into supernovae, among them the bright red supergiant Betelgeuse in the constellation Orion (see Figure 2-2 and Figure 6-27).

Unfortunately, astronomers do not yet know how to predict precisely when such stars will explode into supernovae. It may be many thousands of years before Betelgeuse explodes. Then again, it could happen tomorrow. If it does, neutrino telescopes will be ready to record the collapse of its massive core. However, you will not need a telescope to observe this spectacular event, as it would be about 10 times as bright as the full Moon!

How do high-energy neutrinos produce detectable light?
Answer appears at the end of the chapter.

20-9 White dwarfs in close binary systems can also become supernovae

Astronomers discover dozens of supernovae in distant galaxies every year, but not all of these are the result of massive stars dying violently. A totally different type of supernova occurs when a white dwarf star in a binary system (possibly with a second white dwarf) blows itself completely apart. We now look at the different types of supernovae (Figure 20-19).

In just a few seconds, a thermonuclear supernova completely destroys an entire white dwarf star.

(a) Type Ia supernova
- The spectrum has no hydrogen or helium lines, but does have a strong absorption line of ionized silicon (Si II).
- Produced by runaway carbon fusion in a white dwarf in a close binary system (the ionized silicon is a by-product of carbon fusion).

SN 1987N (Type Ia)
Si II

Red giant   White dwarf explodes
White dwarf

(b) Type Ib supernova
- The spectrum has no hydrogen lines, but it does have a strong absorption line of un-ionized helium (He I).
- Produced by core collapse in a massive star that lost the hydrogen from its outer layers.

SN 1984L (Type Ib)
He I

Supergiant star with outer hydrogen removed → Core collapse, explosion

(c) Type Ic supernova
- The spectrum has no hydrogen lines or helium lines.
- Produced by core collapse in a massive star that lost the hydrogen and the helium from its outer layers.

SN 1987M (Type Ic)

Supergiant star with outer hydrogen, helium removed → Core collapse, explosion

(d) Type II supernova
- The spectrum has prominent hydrogen lines such as H_α.
- Produced by core collapse in a massive star whose outer layers were largely intact.

SN 1992H (Type II)
H_α

Supergiant star with outer layers largely intact → Core collapse, explosion

Wavelength (nm)

FIGURE 20-19

Supernova Types These illustrations show the characteristic spectra and the probable origins of supernovae of **(a)** Type Ia, **(b)** Type Ib, **(c)** Type Ic, and **(d)** Type II. (Data from Alexei V. Filippenko, University of California, Berkeley)

Types of Supernovae

There are two main mechanisms for a supernova explosion. We have already discussed one mechanism, which is a core-collapse supernova. A second mechanism involves at least one white dwarf in a binary system. Had astronomers understood these processes all along, they might have named the different supernovae according to their underlying mechanisms. However, the first indication that there were different types of supernova came from their *differing spectral lines*. As a result, supernovae are labeled not by their underlying mechanism, but by their spectra.

Supernovae with hydrogen emission lines are called **Type II supernovae;** these are core-collapse supernovae of the sort we described in Section 20-6. They are caused by the deaths of highly evolved massive stars that still have ample hydrogen in their atmospheres when they explode. When the star explodes, the hydrogen atoms are excited and glow prominently, producing hydrogen emission lines. SN 1987A (the topic of Sections 20-7 and 20-8) was a Type II supernova.

Hydrogen lines are missing in the spectrum of a **Type I supernova** (Figure 20-19), which tells us that little or no hydrogen is left in the debris from the explosion. Type I supernovae are further divided into three important subclasses. **Type Ia supernovae** have spectra that include a strong absorption line of ionized silicon. **Type Ib** and **Type Ic supernovae** both lack the ionized silicon line. The difference between them is that the spectra of Type Ib supernovae have a strong helium absorption line, while those of Type Ic supernovae do not.

Astronomers suspect that Type Ib and Ic supernovae are caused by core collapse in dying massive stars, just like Type II supernovae. The difference is that the progenitor stars of Type Ib and Ic supernovae have been stripped of their outer layers before they explode. A star can lose its outer layers to a strong stellar wind (see Figure 20-13) or, if it is part of a close binary system, by transferring mass to its companion star (see Figure 19-22b). If enough mass remains for the star's core to collapse, the star dies as a Type Ib supernova. Because the outer layers of hydrogen are absent, the supernova's spectrum exhibits no hydrogen lines but many helium lines (Figure 20-19b). Type Ic supernovae have apparently undergone even more mass loss prior to their explosion; their spectra show that they have lost much of their helium as well as their hydrogen (Figure 20-19c).

As Figure 20-19 shows, dividing supernovae by their underlying explosive mechanism leaves us with supernovae involving white dwarfs (Type Ia), and those involving core collapse (Type II, Type Ib, and Type Ic). We now look at how to detonate a white dwarf.

Type Ia Supernovae: Detonating a White Dwarf

Type Ia supernovae are *not* the death throes of massive supergiant stars. Instead, Type Ia supernovae are thought to result from the thermonuclear explosion of a white dwarf star. Recall that the *thermo-* in thermonuclear indicates that high-speed collisions at very high temperatures fuse the atoms together. Furthermore, a supernova resulting from this rapid release of energy is called a **thermonuclear supernova.** As we will see, a thermonuclear supernova begins with a white dwarf.

This may seem contradictory, because we saw in Section 20-4 that white dwarf stars have no nuclear reactions going on in their interiors. But these reactions *can* occur if a carbon-oxygen–rich white dwarf gains mass in a close, semidetached binary system with a red giant star (see Figure 19-22b and Figure 20-19a).

Figure 20-20 shows the likely series of events that lead to a Type Ia supernova involving only one white dwarf. Stage 1 in this figure shows a close binary system in which both stars have less than 4 $M_\odot$. The more massive star on the left evolves more rapidly than its less massive companion and eventually becomes a white dwarf. As the companion evolves and its outer layers expand, it overflows its Roche lobe and dumps gas from its outer layers onto the white dwarf (see Stage 2 in Figure 20-20). When the total mass of the white dwarf approaches the Chandrasekhar limit (1.4 $M_\odot$), the increased pressure applied to the white dwarf's interior causes carbon fusion to begin there (Stage 3 in Figure 20-20). Hence, the interior temperature of the white dwarf increases.

If the white dwarf were made of ordinary matter, the temperature increase would cause a further increase in pressure, the white dwarf would expand and cool, and the carbon-fusing reactions would abate. But because the white dwarf is composed of degenerate matter, this "safety valve" between temperature and pressure does not operate. Instead, the increased temperature just makes the reactions proceed at an ever-increasing rate, in a catastrophic runaway process reminiscent of the helium flash in low-mass stars (Stage 4 in Figure 20-20). The reaction spreads rapidly outward from the white dwarf's *center*, with its leading edge (called the *flame front*) being propelled by convection and turbulence, in a manner analogous to what happens to the shock wave in a core-collapse supernova. Within seconds, the white dwarf blows apart, dispersing 100% of its mass into space (Stage 5 in Figure 20-20).

Before exploding, the white dwarf contained primarily carbon and oxygen and almost no hydrogen or helium, which explains the absence of hydrogen and helium lines in the spectrum of the resulting supernova. Silicon is a by-product of the carbon-fusing reaction and gives rise to the silicon absorption line characteristic of Type Ia supernovae.

CAUTION!

Different types of supernovae have fundamentally different energy sources. We saw in Section 20-6 that core-collapse supernovae (which we have now sorted into Types II, Ib, and Ic) are powered by *gravitational* energy released as the star's iron-rich core and outer layers fall inward. The core undergoes nuclear reactions during the supernova, but these reactions *consume* energy provided by gravity. Therefore, the nuclear reactions that occur are gravity powered; gravity "fuels" the explosion. Type Ia supernovae, by contrast, are powered by *nuclear* energy released in the explosive thermonuclear fusion of a white dwarf star. In these supernovae, gravity only acts as a *trigger*—just as a spark can light a candle—but the source of energy is thermonuclear carbon-fusion reactions; nuclear energy contained in the carbon atoms is the "fuel."

To highlight the source of the energy, we also use the term *thermonuclear supernova* to refer to a Type Ia supernova. The different types of supernova also differ in their energy output. While Type Ia supernovae typically emit more energy in the form of visible light than supernovae of other types, they do not emit copious numbers of neutrinos because there is no core collapse. If we include the energy emitted in the form of neutrinos, the most energetic supernovae by far are those of Type II. A Type Ia supernova consumes its nuclear fuel in much the same way as a fusing star, but instead of lasting a stellar lifetime, the thermonuclear reactions are over within seconds.

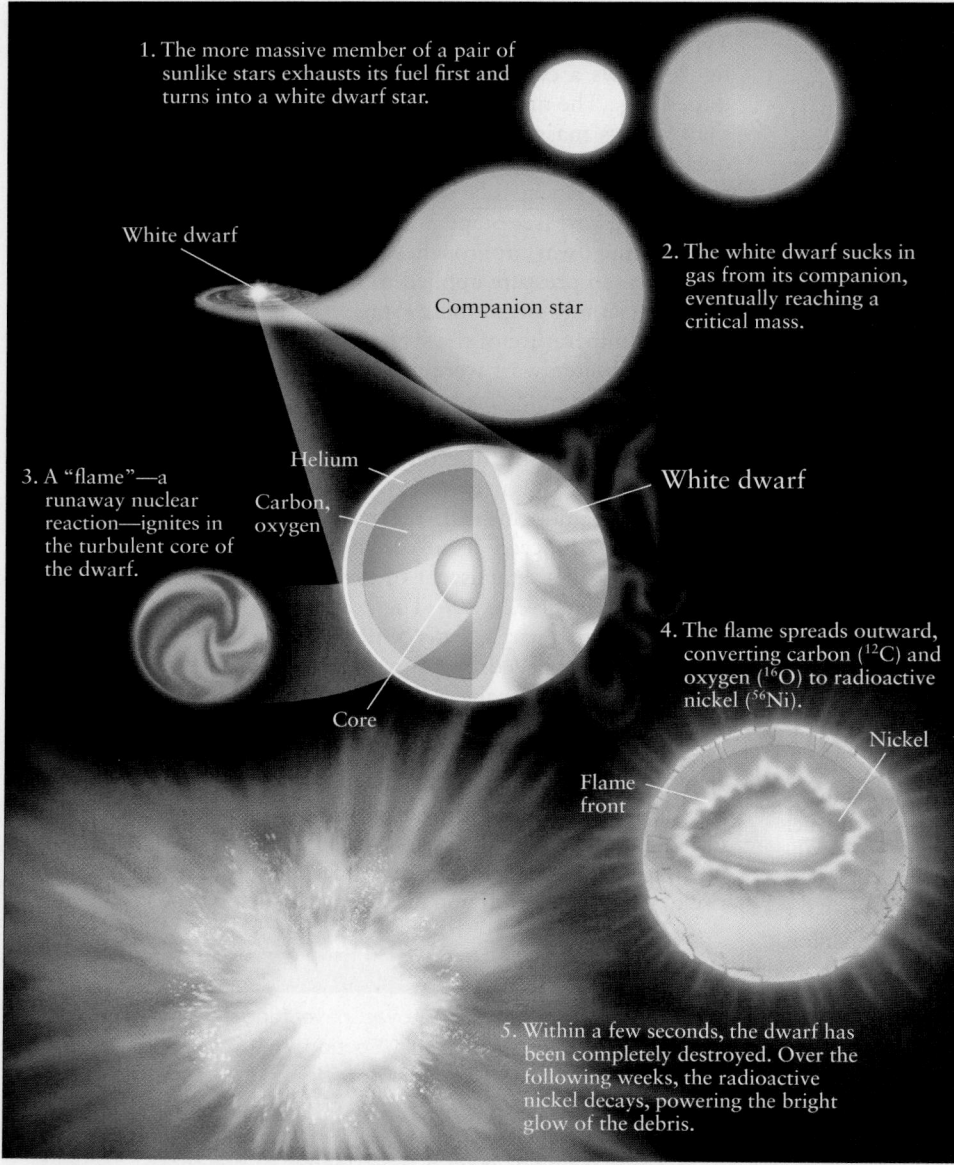

1. The more massive member of a pair of sunlike stars exhausts its fuel first and turns into a white dwarf star.

White dwarf

Companion star

2. The white dwarf sucks in gas from its companion, eventually reaching a critical mass.

3. A "flame"—a runaway nuclear reaction—ignites in the turbulent core of the dwarf.

Helium

Carbon, oxygen

White dwarf

Core

4. The flame spreads outward, converting carbon (^{12}C) and oxygen (^{16}O) to radioactive nickel (^{56}Ni).

Nickel

Flame front

5. Within a few seconds, the dwarf has been completely destroyed. Over the following weeks, the radioactive nickel decays, powering the bright glow of the debris.

ANIMATION 20.6 **ANIMATION 20.7** **FIGURE 20-20**
A Type Ia Supernova This series of illustrations depicts our understanding of how a white dwarf in a close binary system can undergo a sudden nuclear detonation that destroys it completely. Such a cataclysmic event is called a Type Ia supernova or thermonuclear supernova. (Art © Don Dixon)

The Decay of a Supernova: Light Curves

In addition to the differences in their spectra, different types of supernovae can be distinguished by their light curves (Figure 20-21). All supernovae begin with a sudden rise in brightness that occurs in less than a day. The energy powering this initial burst of light comes from heat released during the explosion. After reaching peak luminosity, Type Ia, Ib, and Ic supernovae settle into a steady, gradual decline in luminosity. By contrast, most Type II light curves have a long plateau, where they remain at a similar brightness for about 100 days, before rapidly fading. For all supernova types, during the period of declining brightness—after the peak—energy comes from the decay of radioactive isotopes. These decays heat the surrounding material enough to emit visible light.

Merging White Dwarfs

There is another possible mechanism for a Type Ia supernova that involves the merger of two white dwarfs, where enough mass is concentrated in the merger to ignite a nuclear explosion. While the initial scenario is quite different from a binary system consisting of a white dwarf and a red giant, under certain circumstances, the predicted outcome of merging white dwarfs can be quite similar. One motivation for this mechanism is that some Type Ia supernovae fail to show evidence, before or after the event, for any red giant companion.

An example of a Type Ia supernova with strong evidence that it lacked a companion is SN 2011fe. This supernova event occurred quite close to us, and inspection of Hubble Space Telescope (HST) images captured prior to the supernova

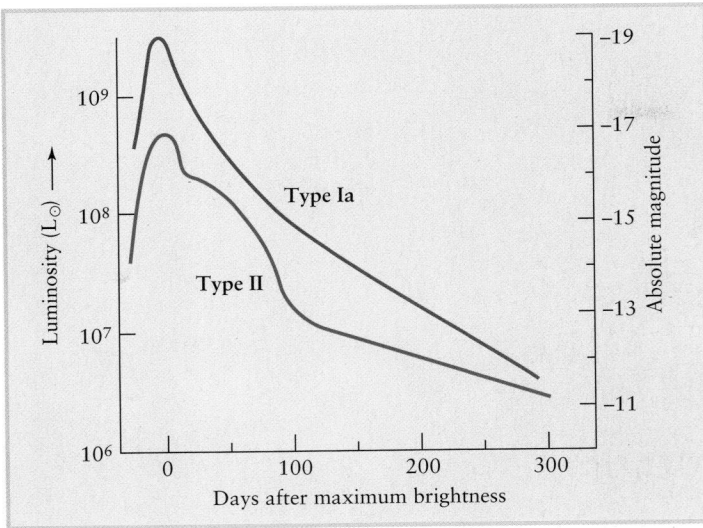

FIGURE 20-21

Supernova Light Curves A Type Ia supernova reaches maximum brightness in about 1 day, followed by a gradual decline in brightness. A Type II supernova reaches a maximum brightness only about one-fourth that of a Type Ia supernova and usually has alternating intervals of steep and gradual declines. The initial peak in brightness results from heat produced in the supernova explosion; after the peak, the energy comes from radioactive decay of elements that were produced by the supernova.

rule out that it had a red giant or main-sequence companion. Furthermore, the presence of a companion would have also produced observable effects that were ruled out by observations made during and after the supernova. Whether or not SN 2011fe resulted from merging white dwarfs is unknown, but it certainly appears to have resulted from a mechanism different from what is described in Figure 20-19.

Type Ia Supernova Are Used to Measure Distances to Other Galaxies

A number of astronomers are now measuring the distances to remote galaxies by looking for Type Ia supernovae in those galaxies. This is possible because there is a simple relationship between the rate at which a Type Ia supernova fades away and its peak luminosity: The slower it fades, the greater its luminosity. Hence, by observing how rapidly a distant Type Ia supernova fades, astronomers can determine its peak luminosity—not just its apparent brightness, but its actual intrinsic luminosity at its peak.

Then, a measurement of the supernova's peak apparent brightness tells us (through the inverse-square law) the distance to the supernova and, therefore, the distance to the supernova's host galaxy. In this way, the Type Ia supernova is being used as a "standard candle" of known luminosity, and by comparison to its apparent brightness, its distance is determined.

The tremendous luminosity of Type Ia supernovae allows this method to be used for galaxies more than 10^9 ly distant. In Chapter 25, we will learn what such studies tell us about the size

and evolution of the universe as a whole. In fact, we will see that distances determined from Type Ia supernovae underlie such discoveries as "dark energy."

CONCEPT CHECK 20-15

If a supernova is observed with no lines in emission or absorption of hydrogen, helium, or silicon, did it result from core collapse, or involve a white dwarf (consult Figure 20-19)? What "type" is it?

CONCEPT CHECK 20-16

Why is a Type Ia supernova brighter than a nova, even though both involve white dwarfs? (See the discussion at the end of Section 20-4.)

CONCEPT CHECK 20-17

Why does a Type Ia supernova occur only once for a given object, whereas novae can occur repeatedly from the same object?

Answers appear at the end of the chapter.

20-10 A supernova remnant can be detected at many wavelengths for centuries after the explosion

Astronomers find the debris of supernova explosions, called **supernova remnants**, scattered across the sky. A beautiful example of a supernova remnant is the Veil Nebula, shown in Figure 20-22. The doomed star's outer layers were blasted into space with such violence that they are still traveling through the interstellar medium at supersonic speeds 8000 years later. As this expanding shell of gas plows through space, it collides with atoms in the interstellar medium, exciting the gas and making it glow. We saw in Section 18-8 that the passage of a supernova remnant through the interstellar medium can trigger the formation of new stars, so the death of a single massive star (in a core-collapse supernova) or white dwarf (in a thermonuclear supernova) can cause a host of new stars to be born.

> Many supernovae in our galaxy are hidden from us by the obscuring interstellar medium.

A few nearby supernova remnants cover sizable areas of the sky. The largest is the Gum Nebula, named after the astronomer Colin Gum, who first noticed its faint glowing wisps on photographs of the southern sky. Its astonishingly wide 40° angular diameter is centered on the constellation Vela (the Ship's Sail).

The Gum Nebula looks big because it is quite close to us and has had a long time to expand—its center is only about 1000 ly

FIGURE 20-23 R I **V** U X G

The Vela Nebula—A Supernova Remnant The Vela Nebula is embedded in an even larger remnant called the Gum Nebula. The Vela supernova explosion occurred about 11,000 years ago, and the remnant now has a diameter of about 700 pc (2300 ly). (Axel Mellinger)

FIGURE 20-22 R I **V** U X G

The Veil Nebula—A Supernova Remnant The Cygnus Loop, or Veil Nebula, is a roughly spherical remnant of a supernova that exploded about 8000 years ago. The distance to the nebula is about 1500 ly and the overall diameter of the loop is about 100,000 ly. This supernova remnant is so large that it covers an area of about 45 full moons! The image contains more than 1.6 Gb of data and is one of the largest astronomical images ever made. (National Optical Astronomy Observatory [NOAO] and WIYN partners)

ANIMATION 20.8 **FIGURE 20-24** R I **V** U X G

Cassiopeia A—A Supernova Remnant This false-color radio image of Cassiopeia A was produced by the Very Large Array (see Figure 6-23). The impact of supernova material on the interstellar medium causes ionization, and the liberated electrons generate radio waves as they move. Cassiopeia A is roughly 3300 pc (11,000 ly) from Earth. (NRAO/AUI)

from Earth, and it originated from a supernova explosion about 1 million years ago. Embedded in the Gum Nebula is another large supernova remnant called the Vela Nebula (Figure 20-23), which is about 16° wide. Studies of the nebula's expansion rate suggest that this supernova exploded around 9000 B.C.E. At maximum brilliance, the exploding star probably was as bright as the Moon at first quarter. Like the first quarter moon, it would have been visible in the daytime!

Many supernova remnants are nearly invisible at optical wavelengths. However, when the expanding gases collide with the interstellar medium, they radiate energy at a wide range of wavelengths, from x rays through radio waves. For example, Figure 20-24 shows a radio image of the supernova remnant Cassiopeia A. (Compare to Figure 18-28, which is a composite of observations of Cassiopeia A at x-ray, visible, and infrared wavelengths.) As a rule, radio searches for supernova remnants are more fruitful than optical searches. Only two dozen supernova remnants have been found in visible-light images, but more than 100 remnants have been discovered by radio astronomers.

From the expansion rate of Cassiopeia A, astronomers conclude that this supernova explosion occurred about 300 years ago. Although telescopes were in wide use by the late 1600s, no one saw the outburst (and no one today knows why). The last supernova seen in our galaxy, which occurred in 1604, was observed by Johannes Kepler. In 1572, Tycho Brahe also recorded the sudden appearance of an exceptionally bright star in the sky. To find any other accounts of nearby bright supernovae, we must delve into astronomical records that are almost 1000 years old.

(a)

FIGURE 20-25

A Summary of Stellar Evolution **(a)** The evolution of an isolated star (one that is not part of a close multiple-star system) depends on the star's mass. The more massive the star, the more rapid its evolution. If the star's initial mass is less than about 0.4 $M_\odot$, it evolves slowly over the eons into an inert ball of helium. If the initial mass is in the range from about 0.4 $M_\odot$ to about 8 $M_\odot$, it ejects enough mass over its lifetime so that what remains is a white dwarf with a mass less than the Chandrasekhar limit of 1.4 $M_\odot$. If the star's initial mass is more than about 8 $M_\odot$, it ends as a core-collapse supernova, leaving behind a neutron star or black hole. (Timelines are not to scale.) **(b)** These images summarize the key stages in the cycle of stellar evolution. (b: *top:* ESA ISO/ISOCAM/Alain Abergel (Abergel et al. (1996) A&A 315, L329); *right:* NASA/ESA/STScI/Arizona State University; *bottom:* NASA; *left:* NASA; *center:* © Australian Astronomical Observatory/David Malin Images)

R I **V** U X G

At first glance, this apparent lack of nearby supernovae may seem puzzling. From the frequency with which supernovae occur in distant galaxies, it is expected that we should have about two supernovae per century. Where have they been?

As we will learn when we study galaxies in Chapters 22 and 23, the plane of our galaxy is where massive stars are born and supernovae explode. This disklike region is so rich in interstellar dust, however, that we simply cannot see very far into space when looking in the plane of the disk, which is in the directions occupied by the Milky Way (see Section 20-2). In other words, supernovae probably do in fact erupt every few decades in remote parts of our galaxy, but their detonations are hidden from our view by intervening interstellar matter.

Relics of the Fall: White Dwarfs, Neutron Stars, and Black Holes

A supernova remnant may be all that is left after some supernovae explode. But for core-collapse supernovae of Types II, Ib, and Ic, the core itself may also remain. If there is a relic of the core, it may be either a *neutron star* or a *black hole*, depending on the mass of the core and the conditions within it during the collapse. Neutron stars, as the name suggests, are made primarily of neutrons. Wholly unlike anything we have studied so far, these exotic objects are one of the subjects of the next chapter. We will also study black holes, which are far stranger than even neutron stars.

CAUTION!

Although neutron stars and black holes can be part of the debris from a supernova explosion, they are *not* called "supernova remnants." That term is applied exclusively to the gas and dust that spreads away from the site of the supernova explosion.

We have seen in this chapter that only the most massive stars end their lives as core-collapse supernovae. We learned earlier that mass plays a central role in determining the speed with which a star forms and joins the main sequence (see Figure 18-10), the star's luminosity and surface temperature while on the main sequence (see Figure 17-22 and the *Cosmic Connections* figure in Chapter 17), and how long a star can remain on the main sequence (see Table 19-1). Now we see that a star's initial mass also determines its eventual fate (**Figure 20-25a**). We also see that a star's fate is connected to the formation of future stars and the matter from which even we are made (Figure 20-25b).

CONCEPT CHECK 20-18

How can astronomers estimate how long ago a supernova exploded by its remnant?

Answer appears at the end of the chapter.

KEY WORDS

asymptotic giant branch, p. 619
asymptotic giant branch star (AGB star), p. 619
carbon fusion, p. 628
carbon star, p. 620
Cherenkov radiation, p. 635
Chandrasekhar limit, p. 625
core-collapse supernova, p. 631
core helium fusion, p. 618
degenerate neutron pressure, p. 631

dredge-up, p. 620
helium shell flash, p. 622
horizontal branch, p. 619
mass–radius relation, p. 625
neon fusion, p. 628
neutron capture, p. 628
nova (*plural* novae), p. 626
nuclear density, p. 631
oxygen fusion, p. 628
photodisintegration, p. 631
planetary nebula, p. 621

progenitor star, p. 633
red giant branch, p. 618
shell helium fusion, p. 619
silicon fusion, p. 628
supergiant, p. 629
supernova (*plural* supernovae), p. 631
supernova remnant, p. 639
thermal pulse, p. 622
thermonuclear, p. 618

thermonuclear supernova, p. 637
Type I supernova, p. 637
Type Ia supernova, p. 637
Type Ib supernova, p. 637
Type Ic supernova, p. 637
Type II supernova, p. 637
white dwarf, p. 624

KEY IDEAS

Late Evolution of Low-Mass Stars: A star of moderately low mass (about 0.4 $M_\odot$ to about 4 $M_\odot$) becomes a red giant when shell hydrogen fusion begins, becomes a horizontal-branch star when core helium fusion begins, and becomes an asymptotic giant branch (AGB) star when the helium in the core is exhausted and shell helium fusion begins (Section 20-1).

• As a moderately low-mass star ages, convection occurs over a larger portion of its volume. This takes heavy elements formed in the star's interior and distributes them throughout the star.

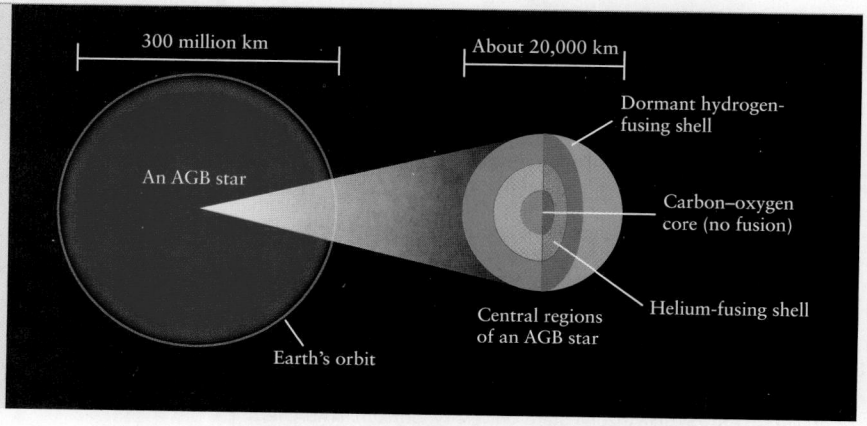

Planetary Nebulae and White Dwarfs: Helium shell flashes in an old, moderately low-mass star produce thermal pulses during which more than half the star's mass may be ejected into space. This exposes the hot carbon–oxygen core of the star (Sections 20-3 and 20-4).

- Ultraviolet radiation from the exposed core ionizes and excites the ejected gases, producing a planetary nebula.
- No further nuclear reactions take place within the exposed core. Instead, it becomes a degenerate, dense sphere about the size of Earth and is called a white dwarf. It glows from thermal radiation; as a white dwarf cools, it becomes dimmer.

Star / Gas ejected from the star

NASA/JPL–Caltech/ESA, the Hubble Heritage Team (STScI/AURA)

Novae: Material from an ordinary star in a close binary can fall onto the surface of the companion white dwarf to produce a thin surface layer in which explosive hydrogen fusion may occur, producing the sudden increase in luminosity that we call a nova (Section 20-4).

Late Evolution of High-Mass Stars: Unlike a moderately low-mass star, a high-mass star (initial mass more than about 4 $M_\odot$) undergoes an extended sequence of nuclear reactions in its core and shells. These include carbon fusion, neon fusion, oxygen fusion, and silicon fusion (Section 20-5).

- In the last stages of its life, a high-mass star has an iron-rich core surrounded by concentric shells hosting the various nuclear reactions. The sequence of nuclear reactions stops here, because the formation of elements heavier than iron requires an input of energy rather than causing energy to be released.

1.6 billion km / About 10,000 km / A supergiant star / Hydrogen-fusing shell / Helium-fusing shell / Carbon-fusing shell / Neon-fusing shell / Oxygen-fusing shell / Silicon-fusing shell / Iron core (no fusion) / Central regions of a supergiant star / Jupiter's orbit

Core-Collapse Supernovae: A star with an initial mass greater than 8 $M_\odot$ dies in a violent cataclysm in which its core collapses and most of its matter is ejected into space at high speeds. The luminosity of the star increases suddenly by a factor of about 10^8 during this explosion, producing a supernova (Sections 20-6 and 20-8).

- More than 99% of the energy from a core-collapse supernova is emitted in the form of neutrinos from the collapsing core.
- The matter ejected from the supernova, moving at supersonic speeds through interstellar gases and dust, glows as a nebula called a supernova remnant.
- A Type II supernova is the result of the collapse of the core of a massive star, as are supernovae of Type Ib and Type Ic.

Konstantinos Kifonidis/Science Photo Library/Science Source

White Dwarf Supernovae: An accreting white dwarf in a close binary system can also become a supernova when carbon fusion ignites explosively throughout such a degenerate star. This is called a thermonuclear supernova (Section 20-9).

- One scenario for Type Ia supernovae is an accreting white dwarf in a close binary with a star; another scenario involves the merger of two white dwarfs.

SN 1987N (Type Ia) / Relative intensity / Si II

Red giant / White dwarf explodes / White dwarf

QUESTIONS

Problem-solving tips and tools

The small-angle formula is given in Box 1-1. You may find it useful to review Box 17-4, which discusses stellar radii and their relationship to temperature and luminosity. Section 4-7 explains the formula for gravitational force, and Box 7-2 explains the concept of escape speed. Sections 5-2 and 5-4 describe some key properties of light, especially blackbody radiation. We discussed the relationship among luminosity, apparent brightness, and distance in Box 17-2. The relationship among absolute magnitude, apparent magnitude, and distance was the topic of Box 17-3. Section 17-6 gives the formula relating a star's luminosity, surface temperature, and radius (see Box 17-4 for worked examples). In our discussion of binary stars in Section 17-9, we saw how the masses of the stars, the orbital period, and the average distance between the two stars are all related. The volume of a sphere of radius r is $4\pi r^3/3$. Appendix 6 gives the conversion between seconds and years, as well as the radius of the Sun.

Section 20-1

1. **TUTORIAL 20.1** What is the horizontal branch? Where is it located in an H-R diagram? How do stars on the horizontal branch differ from red giants or main-sequence stars?

2. Horizontal-branch stars are sometimes referred to as "helium main-sequence stars." In what sense is this true?

3. What is the asymptotic giant branch? Where is it located in an H-R diagram? How do asymptotic giant branch stars differ from red giants or main-sequence stars?

4. Some blue main-sequence stars in our region of the galaxy have the same luminosity and surface temperature as the horizontal-branch stars in the globular cluster M55 (see Figure 20-3). How do we know that the horizontal-branch stars in M55 are not main-sequence stars?

Section 20-2

5. Is a carbon star a star that is made of carbon? Explain your answer.

6. What is the connection between dredge-ups in old stars and life on Earth?

7. Stellar winds from an AGB star can cause it to lose mass at a rate of up to 10^{-4} M$_\odot$ per year. (**a**) Express this rate in metric tons per second. (One metric ton equals 1000 kg.) (**b**) At this rate, how long would it take an AGB star to eject an amount of mass equal to the mass of Earth? Express your answer in days.

8. Suppose the convective zone in AGB stars did *not* reach all the way down into their carbon-rich cores. How might this have affected the origin and evolution of life on Earth?

Section 20-3

9. What are thermal pulses in AGB stars? What causes them? What effect do they have on the luminosity of the star?

10. How is a planetary nebula formed?

11. How can an astronomer tell the difference between a planetary nebula and a planet?

12. What is the evidence that typical planetary nebulae are only a few thousand years old?

13. Why do we not observe planetary nebulae that are more than about 50,000 years old?

14. The globular cluster M15 depicted in Figure 20-7a contains 30,000 old stars, but only one of these stars is presently in the planetary nebula stage of its evolution. Explain why planetary nebulae are not more prevalent in M15.

15. You want to determine the age of a planetary nebula. What observations should you make, and how would you use the resulting data?

16. Figure 20-7a shows the planetary nebula Abell 39. Use the information given in the figure caption to calculate the angular diameter of the nebula as seen from Earth.

17. The Ring Nebula is a planetary nebula in the constellation Lyra. It has an angular size of 1.4 arcmin × 1.0 arcmin and is expanding at the rate of about 20 km/s. Approximately how long ago did the central star shed its outer layers? Assume that the nebula is 2700 ly from Earth.

18. The accompanying image shows the planetary nebula IC 418 in the constellation Lepus (the Hare). (**a**) The image shows a small shell of glowing gas (shown in blue) within a larger glowing gas shell (shown in orange). Discuss how IC 418 could have acquired this pair of gas shells. (**b**) Explain why the outer shell looks thicker around the edges than near the middle.

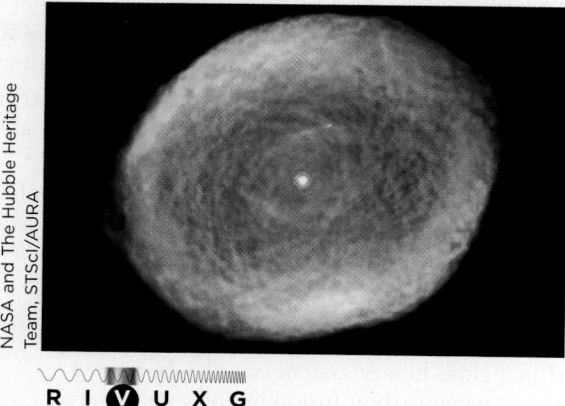

NASA and The Hubble Heritage Team, STScI/AURA

R I **V** U X G

19. Suppose that you discover a small, glowing disk of light while searching the sky with a telescope. How would you decide if this object is a planetary nebula? Could your object be something else? Explain your answer.

Section 20-4

20. What is a white dwarf? Does it produce light in the same way as a star like the Sun?

21. What is nuclear density? Why is it significant when a star's core reaches this density?

22. How does the radius of a white dwarf depend on its mass? How is this different from other types of stars?

23. What is the significance of the Chandrasekhar limit?

24. In an H-R diagram, sketch the evolutionary track that the Sun will follow from when it leaves the main sequence to when it becomes a white dwarf. Approximately how much mass will the Sun have when it becomes a white dwarf? Where will the rest of the mass go?

25. What prevents nuclear reactions from occurring at the center of a white dwarf? If no nuclear reactions are occurring in its core, why doesn't the star collapse?

26. A white dwarf has a greater mass than either a red dwarf or a brown dwarf. Yet a white dwarf has a smaller radius than either a red dwarf or a brown dwarf. Explain why, in terms of the types of pressure that keep the different kinds of dwarfs from collapsing under their own gravity.

27. The central star in a newly formed planetary nebula has a luminosity of $1000 \ L_\odot$ and a surface temperature of 100,000 K. What is the star's radius? Give your answer as a multiple of the Sun's radius.

28. (a) Calculate the wavelength of maximum emission of the white dwarf Sirius B. In what part of the electromagnetic spectrum does this wavelength lie? (b) In a visible-light photograph such as Figure 20-9, Sirius B appears much fainter than its primary star. But in an image made with an x-ray telescope, Sirius B is the brighter star. Explain the difference.

29. Sirius is 2.63 pc from Earth. By making measurements on Figure 20-9, calculate the distance between the centers of Sirius A and Sirius B at the time that this image was made in October 2003. Give your answer in astronomical units (au). (Note that your result is the true distance only if Sirius A and Sirius B were exactly the same distance from Earth at the time the image was made. If one of the stars was closer to us than the other, the actual distance between them is greater than what you calculate.)

30. (a) Find the average density of a $1\text{-}M_\odot$ white dwarf having the same diameter as Earth. (b) Do your own calculation to check the statement that a white dwarf has a density of about one elephant per teaspoon. (c) What is the escape speed (Section 4-7) required to eject gas from the white dwarf's surface? (This is also the speed with which interstellar gas falling from a great distance would strike the star's surface.)

31. In the classic 1960s science-fiction comic book *The Atom*, a physicist discovers a basketball-sized meteorite (about 10 cm in radius) that is actually a fragment of a white dwarf star. With some difficulty, he manages to hand-carry the meteorite back to his laboratory. Estimate the mass of such a fragment, and discuss the plausibility of this scenario.

32. Imagine that our Sun was somehow replaced by a recently formed $1\text{-}M_\odot$ white dwarf star, and that our Earth continued in an orbit of semimajor axis 1 au around this star. Discuss what effects this would have on our planet. What would the white dwarf look like as seen from Earth? Could you look at it safely with the unaided eye? Would Earth's surface temperature remain the same as it is now?

33. The similar names *white dwarf, red dwarf,* and *brown dwarf* describe three very different kinds of objects. Suggest better names for these three kinds of objects, and describe how your names more accurately describe the objects' properties.

34. What are the similarities between a nova and a Type Ia supernova? What are the differences?

Section 20-5

35. What is the difference between a red giant and a red supergiant?

36. Why does the evolutionary track of a high-mass star move from left to right and back again in the H-R diagram?

37. In what way does the structure of an aging supergiant resemble that of an onion?

38. Can stars fuse iron as nuclear fuel? How does this affect the core of a supergiant?

39. What happens if the iron core of a star exceeds the Chandrasekhar limit? About how large must the star be for this to occur?

40. (a) Use the information in the caption to Figure 20-13 to calculate the diameter of the nebula SMC N76. Express your answer in parsecs. (b) How does your answer to part (a) compare to the diameters of the planetary nebulae depicted in Figure 20-7? Explain how this is consistent with the observation that gases ejected from a supergiant travel faster than gases ejected from an AGB star.

41. The supergiant star depicted in Figure 20-13 is actually one member of a binary star system. The masses of the two stars are $18 \ M_\odot$ and $34 \ M_\odot$, and the orbital period is 19.56 days. (a) What is the average separation between the two stars? Give your answer in au. (b) Compare your answer in part (a) to the sizes of the orbits of Mercury, Venus, and Earth around the Sun.

42. The major final product of silicon fusion is ^{56}Fe, an isotope of iron with 26 protons and 30 neutrons. This is also the most common isotope of iron found on Earth. Discuss what this tells you about the origin of the solar system.

Section 20-6

43. Supernovae can produce uranium that, billions of years later, releases nuclear energy through radioactive decay. In what form was this energy during the supernova event?

44. (a) What kinds of stars would you monitor if you wished to observe a core-collapse supernova explosion from its very beginning? (b) Examine Appendices 4 and 5, which list the nearest and brightest stars, respectively. Which, if any, of these stars are possible supernova candidates? Explain your answer.

45. Consider a high-mass star just prior to a supernova explosion, with a core of diameter 20 km and density 4×10^{17} kg/m^3. (a) Calculate the mass of the core. Give your answer in kilograms and in solar masses. (b) Calculate the force of gravity on a 1-kg object at the surface of the core. How many times larger is this than the gravitational force on such an object at the surface of Earth, which is about 10 newtons? (c) Calculate the escape speed from the surface of the star's core. Give your answer in meters per second and as a fraction of the speed of light. What does this tell you about how powerful a supernova explosion must be in order to blow material away from the star's core?

46. Iron is produced by nuclear reactions and can undergo nuclear reactions. Why, then, is it unsuitable as a star's nuclear fuel? (*Hint:* Think about the role that *fuel* plays in a process.)

47. In a supernova, why doesn't degenerate electron pressure hold up a star's core against collapse?

48. Describe what part of a supernova produces neutrinos.

49. Describe what is meant by a *core bounce*. How do neutrinos contribute to the effects of a core bounce?

50. Describe how one can view a supernova as a gravity-powered neutrino explosion.

Section 20-7

51. Why is SN 1987A so interesting to astronomers? In what ways was it not a typical supernova?

52. The shock wave that traveled through the progenitor star of SN 1987A took 2.8 hours to reach the star's surface (see Figure 20-16a). (a) Given the size of a blue supergiant star (see Section 20-7), estimate the speed with which the front end of the shock wave traveled through the star's outer layers. (The core of the progenitor star was very small, so you may consider the shock wave to have started at the very center of the star.) Give your answer in meters per second.

53. Compared to SN 1987A (see Figure 20-17), a supernova observed in 1993 called SN 1993J had a maximum apparent brightness only 9.1×10^{-4} as great. SN 1993J was 3.6 Mpc away in the galaxy M81. Using the distances from Earth to each of these supernovae, determine the ratio of the maximum luminosity of SN 1993J to that of SN 1987A. Which of the two supernovae had the greater maximum luminosity?

54. Suppose that the brightness of a star becoming a supernova increases by 20 magnitudes. Show that this corresponds to an increase of 10^8 in luminosity.

55. SN 1987A did not agree with the theoretical picture outlined in Section 20-6. Does this mean that the theory was wrong? Discuss.

56. What two processes account for the bright light given off by SN 1987A?

Section 20-8

57. Why are neutrinos emitted by core-collapse supernovae? How can these neutrinos be detected? How can they be distinguished from solar neutrinos?

58. The neutrinos from SN 1987A arrived 3 hours before the visible light. While they were en route to Earth, what was the distance between the neutrinos and the first photons from SN 1987A? Assume that neutrinos are massless and thus travel at the speed of light. Give your answers in kilometers and in au.

59. Why did neutrinos from SN 1987A arrive hours before this supernova's visible light?

Section 20-9

60. What causes a thermonuclear supernova? How does a thermonuclear supernova compare with a core-collapse supernova?

61. What are the differences among Type Ia, Type Ib, Type Ic, and Type II supernovae? Which type is most unlike the other three, and why?

62. Suppose that the red supergiant star Betelgeuse, which lies some 425 ly from Earth, becomes a Type II supernova. (a) At the height of the outburst, how bright would it appear in the sky? Give your answer as a fraction of the brightness of the Sun ($b_\odot$). (b) How would it compare with the brightness of Venus (about 10^{-9} $b_\odot$)?

63. In July 1997, a supernova named SN 1997cw exploded in the galaxy NGC 105 in the constellation Cetus (the Whale). It reached an apparent magnitude of $+16.5$ at maximum brilliance, and its spectrum showed an absorption line of ionized silicon. Use this information to find the distance to NGC 105. (*Hint:* Inspect the light curves in Figure 20-21 to find the *absolute* magnitudes of typical supernovae at peak brightness.)

64. After 200 days, what is the source of light for all types of supernovae? What is the initial source of light in the first day?

65. What is it about Type Ia supernovae that allows them to be used as "standard candles" for measuring distances?

Section 20-10

66. How can a supernova continue to shine for many years after it explodes?

67. How do supernova remnants produce radiation at nonvisible wavelengths?

68. Figure 20-22 shows the Veil Nebula in Cygnus. Use the information given in the caption to find the average speed at which material has been moving away from the site of the supernova explosion over the past 15,000 years. Express your answer in km/s and as a fraction of the speed of light.

General Problems

69. Why does the mass of a star play such an important role in determining the star's evolution?

70. Why is the temperature in a star's core so important in determining which nuclear reactions can occur there?

71. There may have been recent supernovae in our galaxy that have not been observable even though they are incredibly luminous. How is this possible?

72. Is our own Sun likely to become a supernova? Why or why not?

73. The chapter-opening images show two kinds of glowing gas clouds: a planetary nebula and a supernova remnant.

(a) Explain what makes the planetary nebula glow and what makes the supernova remnant glow. (*Hint:* The explanations are different for the two kinds of gas clouds.) (b) Which of these two kinds of gas clouds continues to glow for a longer time? Why?

74. The planetary nebula and supernova remnant shown in the chapter-opening images are about the same age. Both objects consist of glowing gases that have expanded away from a central star. Based on these images, in which of these objects have the gases expanded more rapidly? Explain your reasoning.

75. There are many more main-sequence stars of low mass (less than 8 $M_\odot$) than of high mass (8 $M_\odot$ or more). Use this fact to explain why white dwarf stars are far more common than neutron stars.

WEB/E-BOOK QUESTIONS

76. It has been claimed that the Dogon tribe in western Africa has known for thousands of years that Sirius is a binary star. Search the internet for information about these claims. What is the basis of these claims? Why are scientists skeptical, and how do they refute these claims?

77. Search the internet for information about SN 1994I, a supernova that occurred in the galaxy M51 (NGC 5194). Why was this supernova unusual? Was it bright enough to have been seen by amateur astronomers?

78. **ANIMATION 20.2** **Convection Inside a Giant Star.** Access and view the animation "Convection Inside a Giant Star" in Chapter 20 of the *Universe* website or e-book. Describe the motion of material in the interior of the star. In what ways is this motion similar to convection within the present-day Sun (see Section 16-2)? In what ways is it different? Is a dredge-up taking place in this animation? How can you tell?

79. **ANIMATION 20.3** **ANIMATION 20.6** **Types of Supernovae.** Access and view the animations "In the Heart of a Core-Collapse (Type II, Ib, or Ic) Supernova" and "A ThermoNuclear (Type Ia) Supernova" in Chapter 20 of the *Universe* website or e-book. Describe how these two types of supernova are fundamentally different in their origin.

80. Search the internet for information about the latest observations of the stellar remnant at the center of SN 1987A. Has a neutron star been detected? Has a pulsar been detected?

Has the supernova's debris thinned out enough to give a clear view of the neutron star?

ACTIVITIES

Observing Projects

> **Observing tips and tools**
>
> While planetary nebulae are rather bright objects, their brightness is spread over a relatively large angular size, which can make it a challenge for the beginning observer to see them. For example, the Helix Nebula (shown in the left-hand image on the chapter-opening page) has the largest angular size of any planetary nebula but is also one of the most difficult to see. To improve your view, make your observations on a dark, moonless night from a location well shielded from city lights. Another useful trick, mentioned in Chapters 17 and 18, is to use "averted vision." Once you have the nebula centered in the telescope, you will get a brighter and clearer image if you look at the nebula out of the corner of your eye. The so-called Blinking Planetary in Cygnus affords an excellent demonstration of this effect; the nebula seems to disappear when you look straight at it, but it reappears as soon as you look toward the side of your field of view.
>
> Another useful tip is to view the nebula through a green filter (a #58, or O III, filter available from telescope supply houses). Green light is emitted by excited, doubly ionized oxygen atoms, which are common in planetary nebulae but not in most other celestial objects. Using such a filter can make a planetary nebula stand out more distinctly against the sky. As a side benefit, it also helps to block out stray light from street lamps. The same tips also apply to observing supernova remnants.

81. Although they represent a fleeting stage at the end of a star's life, planetary nebulae are found all across the sky. Some of the brightest are listed in the accompanying table. Note that the distances to most of these nebulae are quite uncertain. Observe as many of these planetary nebulae as you can on a clear, moonless night using the largest telescope at your disposal. Note and compare the various shapes of the different nebulae. In how many cases can you see the central star? The central star in the Eskimo Nebula is supposed to be the "nose" of an Eskimo wearing a parka. Can you see this pattern?

Planetary nebula	Distance (ly)	Angular size	Constellation	Right ascension	Declination
Dumbbell (M27, NGC 6853)	490–3500	8.0′ × 5.7′	Vulpecula	19h 59.6m	+22° 43′
Ring (M57, NGC 6720)	1300–4100	1.4′ × 1.0′	Lyra	18h 53.6m	+33° 02′
Little Dumbbell (M76, NGC 650)	1700–15,000	2.7′ × 1.8′	Perseus	01h 42.4m	+51° 34′
Owl (M97, NGC 3587)	1300–12,000	3.4′ × 3.3′	Ursa Major	11h 14.8m	+55° 01′
Saturn (NGC 7009)	1600–3900	0.4′ × 1.6′	Aquarius	21h 04.2m	−11° 22′
Helix (NGC 7293)	450	41′ × 41′	Aquarius	22h 29.6m	−20° 48′
Eskimo (NGC 2392)	1400–10,000	0.5′ × 0.5′	Gemini	7h 29.2m	+20° 55′
Blinking Planetary (NGC 6826)	3300(?)	2.2′ × 0.5′	Cygnus	19h 44.8m	+50° 31′

82. Northern hemisphere observers with modest telescopes can see two supernova remnants, one in the winter sky and the other in the summer sky. Both are quite faint, however, so you should schedule your observations for a moonless night. The winter sky contains the Crab Nebula, which is discussed in detail in Chapter 23. The coordinates are R.A. = $5^h 34.5^m$ and Decl. = $+22° 00'$, which places the object near the star marking the eastern horn of Taurus (the Bull). Whereas the entire Crab Nebula easily fits in the field of view of an eyepiece, the Veil or Cirrus Nebula in the summer sky is so vast that you can see only a small fraction of it at a time. The easiest way to find the Veil Nebula is to aim the telescope at the star 52 Cygni (R.A. = $20^h 45.7^m$ and Decl. = $+30° 43'$), which lies on one of the brightest portions of the nebula. If you then move the telescope slightly north or south until 52 Cygni is just out of the field of view, you should see faint wisps of glowing gas.

83. Use a telescope to observe the remarkable triple star 40 Eridani, whose coordinates are R.A. = $4^h 15.3^m$ and Decl. = $-7° 39'$. The primary, a 4.4-magnitude yellowish star like the Sun, has a 9.6-magnitude white dwarf companion, the most easily seen white dwarf in the sky. On a clear, dark night with a moderately large telescope, you should also see that the white dwarf has an 11th-magnitude companion, which completes this most interesting trio.

Starry Night™ Explorations

84. Use *Starry Night*™ to examine two planetary nebulae in detail—M57, the Ring Nebula; and M27, the Dumbbell Nebula—and compare their shapes and sizes. Select **Favorites > Explorations > Planetary Nebulae** to see the positions of these two planetary nebulae against the background sky, with the constellations displayed and labeled. For each of these nebulae in turn, click the name in its label and select **Magnify** to see a telescopic image of the nebula. Then use the HUD to obtain each object's distance from Earth and use the angular separation cursor to measure its angular radius in Earth's sky. Next, use **File > Revert** to reset the view and repeat these steps with the other labeled nebula.

    (a) From the different shapes of these two planetary nebulae, which appears to have the more symmetrical shape?

    (b) What is the nature of the central star in each of these nebulae?

    (c) Using the data for the distance and the angular radius in the sky for each of these nebulae, use the small angle formula to calculate their size, in light-years. By assuming that both of these nebulae have been expanding at the same rate, determine which of the stars at the cores of these nebulae reached the end of its life cycle first (i.e., which of these nebulae is the oldest)?

    (d) Assuming an average rate of expansion of 20 km/s for the shell of gas that forms M57, approximately how long ago, in years, did the star that formed this nebula expel the outer shell of its atmosphere?

85. The red supergiant star Betelgeuse in the constellation Orion will explode as a supernova at some time in the future. Use *Starry Night*™ to investigate how the supernova might appear from Earth. Select **Favorites > Explorations > Betelgeuse** to see the sky looking south from Edmonton, Canada, at 4:00 A.M. on September 7, 2020, with the Moon and the planet Venus in the sky and Betelgeuse indicated.

    (a) Use the HUD to find the apparent magnitude of a sample of the brightest stars in the view, including Betelgeuse. What is the apparent magnitude of Betelgeuse?

    (b) If Betelgeuse became a supernova, then at peak brightness it would be 11 magnitudes brighter than it is now. What would be the apparent magnitude of this supernova at its peak brightness? (Remember that the magnitude scale is inverted, so that the magnitude value goes down as brightness rises.)

    (c) For comparison, the apparent magnitude of Venus on this night is -4.17 and that of the Moon is -12.34. How would the supernova Betelguese appear at night compared to bright planets and to the Moon? Do you think the supernova would be bright enough to cast shadows at night?

    (d) Use the "H" keyboard shortcut to advance time to 9 A.M. If Betelgeuse went supernova, do you think it would be visible from Earth in the daytime?

86. Use *Starry Night*™ to examine the Veil nebula, a large supernova remnant. Open **Favorites > Explorations > Veil** to see a view of this object in the night sky.

    (a) What significant feature do you notice about this supernova remnant in this 5° field of view?

    (b) Use the angular separation cursor to measure the approximate angular distance between the components of this nebula. What angular distance separates these components? What form of optical aid is best suited to observing this object?

    (c) It is believed that the supernova that produced this nebula would have been visible to people on Earth about 6500 years ago and that its distance from Earth is about 1500 light-years. From these data, what is the approximate speed of expansion of this nebula in kilometers per second? Zoom in on each component of the nebula to examine the fine details of this expanding remnant of a violent stellar explosion.

Collaborative Exercises

87. Imagine that a supernova originating from a close binary star system, both of whose stars have less than 4 $M_\odot$, began (as seen from Earth) on the most recent birthday of the youngest person in your group. Using the light curves in Figure 20-21, what would its new luminosity be today and how bright would it appear in the sky (apparent magnitude)

if it were located 10 pc (32.6 ly) away? How would your answers change if you were to discover that the supernova actually originated from an isolated star with a mass 15 times greater than our Sun?

ANSWERS

ConceptChecks

ConceptCheck 20-1: The energy released from core helium fusion expands the star's core, thereby slowing down reactions for both the core helium fusion and the hydrogen-fusing shell. With a decreased rate of energy released in and around the star's core, the luminosity of the star decreases. These stars are called horizontal-branch stars.

ConceptCheck 20-2: Most of the carbon in your body came from stellar winds around carbon stars.

ConceptCheck 20-3: The visible colors of a planetary nebula do not come from the temperature of the nebula's gas. Rather, a planetary nebula's emission is powered by the ultraviolet light emitted by the central star.

ConceptCheck 20-4: For any atom with electrons, the electrons orbit very far from the nucleus, compared to the size of the nucleus itself. Thus, any atom on the periodic table of the elements consists mostly of empty space.

ConceptCheck 20-5: Yes. Electron degeneracy pressure can support white dwarf masses up to the Chandrasekhar limit of 1.4 $M_\odot$. There can be no white dwarfs beyond this limit. As shown in Figure 20-10, a 1.2-$M_\odot$ white dwarf is smaller than Earth.

ConceptCheck 20-6: No. The Sun's size can be read in the *Cosmic Connections* figure from the diagonal dashed lines. The Sun expands to about 100 $R_\odot$ as a red giant, but this phase ends when the Sun is about 12.23 billion years old (Stage 3 in the figure). At 12.365 billion years, the Sun is even larger (Stage 5). The Sun is also more luminous by this time, and remains so until nuclear reactions come to an end.

ConceptCheck 20-7: It is the loss of the star's mass going into the nebula that relieves pressure on the core. The pressure becomes too low for further nuclear reactions to occur (Stages 6 and 7 in the *Cosmic Connections* figure).

ConceptCheck 20-8: Yes. However, these reactions do not release energy; energy must be added for these reactions to take place.

ConceptCheck 20-9: No. In order for the iron core to collapse, it must overcome degenerate electron pressure, which requires over 1.4 $M_\odot$ (this is the Chandrasekhar limit).

ConceptCheck 20-10: For stars that had main-sequence masses over 8 $M_\odot$, iron cores larger than the Chandrasekhar limit can develop and collapse. During collapse, iron is converted into neutrons, which then exert a degenerate-neutron pressure that can halt further collapse.

ConceptCheck 20-11: Neutrinos deliver energy to these bubbles. These neutrinos were produced in the collapse when electrons combined with protons to produce neutrons and neutrinos.

ConceptCheck 20-12: Atoms heavier than iron (such as uranium) are produced during a supernova. Nuclear reactions that produce uranium require an *input* of energy, and there is plenty of gravitational energy in the collapse that ultimately ends up producing atomic elements heavier than iron. Uranium does not release energy when produced.

ConceptCheck 20-13: The typical progenitor star for a core-collapse supernova is a red supergiant. The blue supergiant progenitor star that led to SN 1987A was still large in mass but smaller in size than a red supergiant. That means some of the supernova energy that would have normally gone into light was used just to push the smaller star's matter outward against its own gravity.

ConceptCheck 20-14: When a high-energy neutrino hits a proton in a big water tank, it can cause a positron to be emitted. The positron can travel faster than light travels within that same water, and this produces a shock wave of light that can be detected.

ConceptCheck 20-15: By comparison with the spectra in Figure 20-19, the lack of hydrogen, helium, or silicon lines matches the core collapse of a Type Ic supernova.

ConceptCheck 20-16: In a supernova, the entire white dwarf undergoes nuclear reactions; whereas in a nova, only a surface layer undergoes nuclear fusion.

ConceptCheck 20-17: A Type Ia supernova blows the object apart. For a nova, only an outer layer of transferred mass undergoes nuclear reactions. When enough mass is again transferred from the binary partner, the detonation repeats itself.

ConceptCheck 20-18: By observing the expansion rate of the remnant (its speed), and how far it has expanded (its approximate radius), the time since the explosion can be estimated.

21 Stellar Remnants: Neutron Stars and Black Holes

LEARNING GOALS

N ow our tour of the universe brings us to the strange physics of Dr. Albert Einstein. So far, the effects of gravity we have described have been accurately characterized by Newton's law of gravity. However, Einstein discovered that when the pull of gravity is strong enough, additional and unexpected behavior takes place. The most extreme example of Einstein's description of gravity is a black hole. And the most recent confirmation of black holes comes from another prediction of Einstein's: gravitational waves.

Before we jump into black holes, however, we need to cover several key ideas. First on that list is to describe the conditions in which a black hole can be created. These conditions occur in the collapsed core of a star undergoing a supernova explosion. Since the collapsed core can instead lead to a neutron star, we begin with these strange objects. In fact, the illustration that opens this chapter depicts an event observed in 2017 in which two neutron stars merged together to form a black hole! We will then develop Einsten's ideas on relativity, black holes, and gravitational waves. With these concepts in hand, we will be able to see where black holes are found and how to discover them.

An artist's impression of two neutron stars merging into a black hole. (NSF/LIGO/Sonoma State University/A. Simonnet)

21-1 Neutron stars are the remnants of massive stars

On the morning of July 4, 1054, Yang Wei-T'e, the imperial astronomer to the Chinese court, made a startling discovery. Just a few minutes before sunrise, a new and dazzling object ascended above the eastern horizon. This "guest star," as Yang called it, was far brighter than Venus and more resplendent than any star he had ever seen. Yang's records show that the "guest star" was so brilliant that it could easily be seen during broad daylight for the rest of July. Records from Constantinople (now Istanbul, Turkey) also describe this object, and works of art made by the Chacoans in the American Southwest suggest that they may have seen it as well (Figure 21-1). Over the next 21 months, however, the "guest star" faded to invisibility.

We now know that the "guest star" of 1054 was actually a remarkable stellar transformation: A massive star some 6500 ly away perished in a supernova explosion, leaving behind both a supernova remnant and a bizarre object called a **neutron star**—an incredibly dense sphere composed primarily of neutrons. (We learned in Section 5-7 that a neutron has about the same mass as a proton, but has no electric charge.) Today, the remnant of this supernova that was seen nearly 1000 years ago is called the Crab Nebula. The neutron star at its center is known as the Crab pulsar for reasons we will see shortly, and it was discovered at radio wavelengths (Figure 21-2a). As we'll see, neutron stars, when merging together (as illustrated in the chapter-opening figure), can also be detected by their gravitational waves.

We saw in Section 20-6 that under the very high pressures within a core-collapse supernova, a proton and an electron can combine to form a neutron (as well as a neutrino). Far from being a rare transformation, *a core collapse can convert over a solar mass of material into almost pure neutrons*. Recall that white dwarfs are supported by degenerate electron pressure; by a similar effect, neutron stars are supported against further gravitational collapse by degenerate *neutron* pressure. (Like electrons, neutrons obey the Pauli exclusion principle that we described in Section 19-3.)

Although the possibility for how neutron stars form was proposed in 1934, most scientists politely ignored the idea for years. After all, a neutron star must be a rather weird object. If brought to Earth's surface, a single teaspoonful of neutron star matter would weigh the same as about 20 million elephants (about 100 million tons)!

Neutron stars can exist for only a narrow range of masses. Supported by degenerate electron pressure, stellar cores below the Chandrasekhar limit of $1.4 \, M_\odot$ form white dwarfs. Above the Chandrasekhar limit, a neutron star forms, but the maximum possible neutron star mass is only about 2 to $3 \, M_\odot$. Beyond a stellar core mass of 2 to $3 \, M_\odot$, gravity overwhelms internal pressure, and a black hole is formed. Keep in mind that these are only the masses of the stellar cores. Stars with total masses between 8 and $25 \, M_\odot$ are expected to develop into neutron stars, and those above $25 \, M_\odot$ will become black holes.

A couple of solar masses is a lot of material, but at extreme densities, these objects are surprisingly small. A $1.4\text{-}M_\odot$ neutron star would have a diameter of only 20 km (12 mi). Thus, a neutron star is like a giant atomic nucleus made only of neutrons—the size of a city! Its surface gravity would be so strong that its escape speed—the speed required for an object to escape into space (Section 4-7)—would be one-half the speed of light. Even climbing a 1-mm "mountain" on a neutron star would require about the same energy as a person on Earth jumping straight to the Moon! These conditions seemed outrageous until the late 1960s, when astronomers discovered pulsating radio sources.

The Discovery of Pulsars

As a young graduate student at Cambridge University, Jocelyn Bell (later Jocelyn Bell Burnell) spent many months helping construct an array of radio antennae covering 4½ acres of the English countryside. The instrument was completed by the summer of 1967, and Bell and her colleagues began using it to scrutinize radio emissions from the sky. While searching for something quite different, Bell noticed that the antennae had detected regular pulses of radio noise from one particular location in the sky. These radio pulses were arriving at regular intervals of 1.3373011 s—much more rapid than those of any other astronomical object known at that time. Indeed, they were so rapid and regular that the Cambridge team first wondered if they might be signals from an advanced alien civilization.

That possibility had to be discarded within a few months after several more of these pulsating radio sources, which came to be called **pulsars**, were discovered across the sky. In all cases, the periods were extremely regular, ranging from about 0.25 s for the fastest to about 1.5 s for the slowest (Figure 21-3). After ruling out other mechanisms to generate such rapid periodic pulses (e.g., rapidly orbiting binary systems), it appeared that a "hot spot" on a rotating white dwarf might account for pulsars. However, the discovery of a pulsar at the center of the Crab Nebula—called the Crab pulsar (Figure 21-2b)—was about to rule out white dwarfs and fundamentally change astronomy.

> The rapid flashing of radiation from a pulsar gave evidence that white dwarfs are not the only endpoint of stellar evolution.

FIGURE 21-1

A Supernova Pictograph This drawing in an eleventh-century structure in New Mexico shows a 10-pointed star next to a crescent. It may depict the scene on the morning of July 5, 1054, when a "guest star" appeared next to the waning crescent moon. (Efrain Padro/Alamy Stock Photo)

(a) The Crab Nebula

R I V U X G

(b) The Crab pulsar in visible light

R I V U X G

(c) The Crab pulsar in x rays

(d) Composite image of the Crab Nebula

R I V U X G

FIGURE 21-2

The Crab Pulsar (a) A pulsar is located at the center of the Crab Nebula, which is about 2000 pc (6500 ly) from Earth and about 3 pc (10 ly) across. The boxed area is shown in close-up in part (b). **(b)** We see a flash of visible light when the rotating pulsar's beam is directed toward us (the "on" state). The pulsar fades (the "off" state) when the beam is aimed elsewhere. **(c)** The Crab pulsar also pulses "on" and "off" at x-ray wavelengths. The radio pulses, visual flashes, and x-ray flashes all have a period of 0.033 s. **(d)** False-color image of the Crab Nebula with visible light in purple, infrared in pink, and x rays in blue-white. Jets of matter and antimatter moving near the speed of light emit x rays and can be seen shooting out from the pulsar toward the lower left. From the scale bar, you can see that this image is similar in size to part (a). (a: ESO; b: The FORS Team, VLT, European Southern Observatory; c: Harvard-Smithsonian Center for Astrophysics; d: NASA—x ray: CXC, Optical: STScI, Infrared: JPL-Caltech)

FIGURE 21-3

A Recording of a Pulsar This chart recording shows the intensity of radio emission from one of the first pulsars to be discovered. (The designation PSR B0329+54 means pulsar at a right ascension of 03h 29m and a declination of +54°.) Note that the interval between pulses is very regular, even though some pulses are weak and others are strong. (Data from R. N. Manchester and J. H. Taylor)

At the time of its discovery, the Crab pulsar was the fastest pulsar known to astronomers. Its period is 0.0333 s per rotation, which means that it flashes or rotates about $1/0.0333 = 30$ times each second. It was immediately apparent to astrophysicists that white dwarfs are too big and bulky to generate 30 signals per second; calculations demonstrated that a white dwarf could not rotate that fast without tearing itself apart! Hence, the Crab pulsar indicated that pulsars had to be much smaller and more compact than a white dwarf. We now understand that pulsars arise from rapidly rotating neutron stars, which were created in supernova explosions. This was a truly remarkable revelation, because most astronomers in the mid-1960s thought that *all* stellar corpses are white dwarfs.

Because neutron stars are very small, they should also rotate rapidly. A typical star, such as our Sun, takes nearly a full month to rotate once about its axis. But just as a spinning ice skater speeds up when she pulls in her arms, a collapsing star also speeds

FIGURE 21-4

Analogy for How Magnetic Field Strengths Increase When growing, these wheat stalks cover a much larger area than when they are harvested and bound together. A star's magnetic field behaves similarly. The collapsing star carries the field inward, thereby increasing its strength. (left: Mitch Diamond/Getty Images; right: Oscar Burriel/Science Source)

up as its size shrinks. (We introduced this principle, called the *conservation of angular momentum*, in Section 8-4; see Figure 8-7.) If our Sun were compressed to the size of a neutron star, it would spin about 1000 times per second! Because neutron stars are so small and dense, they can spin this rapidly without flying apart.

The neutron star model of pulsars had to pass several stringent tests to be accepted by astronomers.

The small size of neutron stars also leads to intense magnetic fields. The magnetic field of a main-sequence star is spread out over billions of square kilometers of the star's surface. However, if such a star collapses down to a neutron star, its surface area shrinks by a factor of about 10^{10}. The magnetic field, which weaves through and is connected to the star's ionized gases (see Section 16-9), becomes concentrated over a smaller area than before the collapse and increases by a factor of 10^{10} (Figure 21-4).

How can astronomers determine the strength of a neutron star's magnetic field? In the presence of a magnetic field, certain spectral lines can be split apart—after initially appearing as a single line—and the greater the magnetic field, the farther apart the lines move. This method reveals that magnetic fields on neutron stars are over a million times stronger than the strongest fields produced in labs on Earth. The magnetic field of a neutron star makes it possible for the star to radiate pulses of energy toward our telescopes. As a neutron star rotates, the magnetic axis of a neutron star—the imaginary line passing through the north and south magnetic poles—is likely to be inclined at an angle to the rotation axis (Figure 21-5). After all, there is no fundamental reason for these two axes to coincide. (Indeed, these two axes do not coincide for many of the planets of our solar system.)

Furthermore, nearby electrically charged particles are accelerated by the magnetic fields, and this motion produces electromagnetic radiation. The result is that two narrow beams of radiation pour out of the neutron star's north and south magnetic polar regions. While pulsars were originally discovered at radio wavelengths, some also pulse in visible light and x rays (Figure 21-2b,c).

Note that a neutron *star* is not a star in the usual sense: It isn't powered by nuclear reactions.

ANALOGY

A rotating, magnetized neutron star is somewhat like a lighthouse beacon. As the star rotates, the beams of radiation sweep around the sky. If, at some point during the rotation, one of those beams happens to point toward Earth, as shown in Figure 21-5a, we will detect a brief flash as the beam sweeps over us. At other points during the rotation, the beam will be pointed away from Earth, and the radiation from the neutron star will appear to have turned off (Figure 21-5b). Hence, a radio telescope will detect regular pulses of radiation, with one pulse being received for each rotation of the neutron star.

CAUTION!

The name *pulsar* may lead you to think that the source of radio waves is actually pulsing. But in the model just described, this is not the case at all. Instead, beams of radiation are emitted continuously from the magnetic poles of the neutron star. The pulsing that astronomers detect here on Earth is simply a result of the rapid rotation of the neutron star, which brings one of the beams periodically into our line of sight, as Figure 21-5 shows. In this sense, the analogy between a pulsar and a lighthouse beacon is a very close one.

CONCEPT CHECK 21-1

What prevents a neutron star from collapsing under its own tremendous gravitational attraction?

CONCEPT CHECK 21-2

How can a rotating neutron star lead to radio pulses?
Answers appear at the end of the chapter.

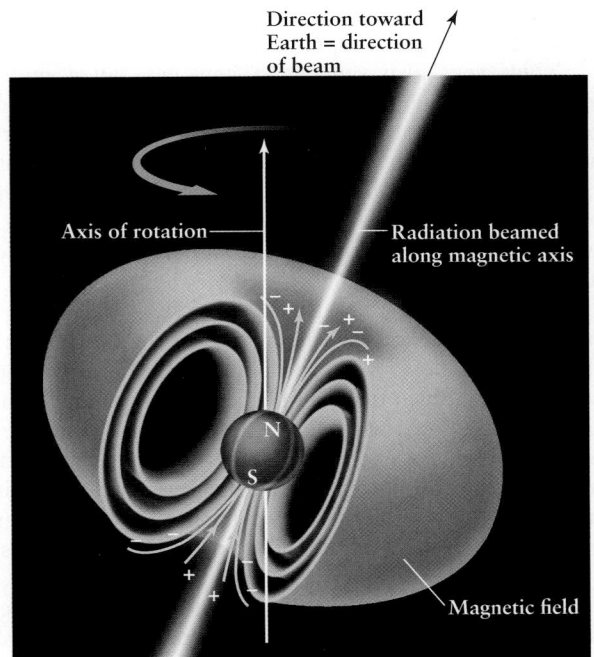

Direction toward
Earth = direction
of beam

Axis of rotation

Radiation beamed
along magnetic axis

N
S

Magnetic field

(a) One of the beams from the rotating neutron star is
aimed toward Earth: We detect a pulse of radiation.

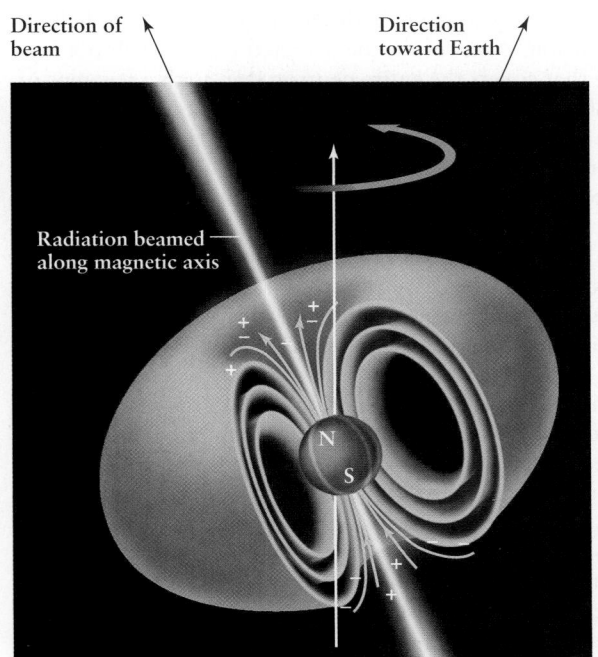

Direction of
beam

Direction
toward Earth

Radiation beamed
along magnetic axis

N
S

(b) Half a rotation later, neither beam is aimed toward
Earth: We detect that the radiation is "off."

FIGURE 21-5

A Rotating, Magnetized Neutron Star Charged particles are
accelerated near a magnetized neutron star's magnetic poles (labeled N and
S), producing two oppositely directed beams of radiation. This radiation is
emitted from charged particles in the jets, which are made of matter and
antimatter. If the star's magnetic axis (a line that connects the north and

south magnetic poles) is tilted at an angle from the axis of rotation, as shown
here, the beams sweep around the sky as the star rotates. If Earth happens to
lie in the path of one of the beams, we detect radiation that appears to pulse
(a) on and **(b)** off.

X-Ray Bursters and Neutron Stars

Recall from Section 20-4 that a nova is an explosion on the surface
of a white dwarf. A surface explosion similar to a nova also occurs
with neutron stars. In 1975, it was discovered that some objects in
the sky emit sudden, powerful bursts of x rays. Figure 21-6 shows
the record of a typical burst. The source emits x rays at a constant
low level until suddenly, without warning, there is an abrupt
increase in x rays, followed by a gradual decline. An entire burst
typically lasts for only 20 seconds, and the same object can repeat
these bursts, although at irregular intervals. Sources that behave
in this fashion are known as **x-ray bursters**. Several dozen x-ray
bursters have been discovered in our galaxy.

X-ray bursters, like novae, are thought to involve close binaries
whose stars are engaged in mass transfer. With a burster, however,
the stellar corpse is a neutron star rather than a white dwarf. Gases
escaping from the ordinary companion star fall onto the neutron
star. The energy released as these gases crash down onto the neutron star's surface produces the low-level x rays that are continuously emitted by the burster.

Most of the gas falling onto the neutron star is hydrogen,
which the star's powerful gravity compresses against its hot surface. In fact, temperatures and pressures in this accreting layer

1. Material from a star accretes
onto a companion neutron star.

2. When enough accreted material builds up,
thermonuclear reactions occur on the neutron
star's surface, creating a burst of x rays.

3. The x-ray burster
fades within seconds.

FIGURE 21-6

The Light Curve of an X-Ray Burster This illustration and graph show
the history of a typical x-ray burster. A burster emits a constant low intensity
of x rays interspersed with occasional powerful x-ray bursts. This burst
was recorded on September 28, 1975 by an Earth-orbiting x-ray telescope.
Contrast this figure with Figure 20-12, which shows a typical nova. (Data from
W. H. G. Lewin; illustration courtesy CXC/M. Weiss)

become so high that the arriving hydrogen is converted into helium by hydrogen fusion. As a result, the accreted gases develop a layered structure that covers the entire neutron star, with a few tens of centimeters of hydrogen lying atop a similar thickness of helium. The structure is reminiscent of the layers within an evolved giant star (see Figure 20-2), although the layers atop a neutron star are much more compressed, thanks to the star's tremendous surface gravity.

When the helium layer grows to about 1 m thick, helium fusion ignites explosively and heats the neutron star's surface to about 3×10^7 K. At this temperature, the surface predominantly emits x rays, but the emission ceases within a few seconds as the surface cools. Hence, we observe a sudden burst of x rays only a few seconds in duration. New hydrogen then flows onto the neutron star, and the whole process starts over. Indeed, x-ray bursters typically emit a burst every few hours or days.

Whereas explosive hydrogen fusion on a white dwarf produces a nova, explosive helium fusion on a neutron star produces an x-ray burster. In both cases, the process is explosive, because the fuel is compressed so tightly against the star's surface that it becomes degenerate, like the star itself. As with the helium flash inside red giants (described in Section 19-3), the ignition of a degenerate nuclear fuel involves a sudden thermal runaway. This is because an increase in temperature does not produce a corresponding increase in pressure that would otherwise relieve compression of the gases and slow the nuclear reactions.

Pulsars and x-ray bursters are just some of the intriguing phenomena associated with neutron stars. However, some of the most interesting objects of all—black holes—are created when there is so much matter that a neutron star cannot form. The gravity associated with a neutron star is so strong that the escape speed from it is roughly one-half the speed of light. But if a stellar corpse has a mass greater than 3 $M_\odot$, so much matter is crushed into such a small volume that the escape speed actually exceeds the speed of light. Because nothing can travel faster than light, nothing—not even light—can leave this dead star. Its gravity is so powerful that it leaves a hole in the fabric of space and time. We begin to take up the story of black holes in Section 21-3.

CONCEPT CHECK 21-3

Why does a Type Ia supernova occur only once for a given object, whereas x-ray bursters can occur repeatedly from the same object?

Answer appears at the end of the chapter.

The Incredible Merger of Two Neutron Stars

Something incredible was observed on August 17, 2017: the merger of two neutron stars. This merger, which is depicted in the figure that opens this chapter, involved two neutron stars that then quickly collapsed into a black hole, but the science and excitement didn't stop there. The rapidly spinning black hole shot out jets that led to a burst of gamma rays and solved a long-standing mystery about the origin of this type of burst. For about 100 s,

these neutron stars produced intense gravitational waves that were detected on Earth—a topic we'll discuss further in Section 21-4.

This event also signaled a new era called *multi-messenger astronomy*, in which many telescopes around the world are used to probe an event (see **Box 21-1** for more on multi-messenger astronomy). While this event was discovered through its gravitational waves, more than 70 telescopes were used, encompassing all wavelengths of light from gamma rays to radio waves, to decipher what took place. It was the most studied event in the history of astronomy!

At the end of their stellar lives, 130 million ly away, two neutron stars locked in a binary orbit began their death spiral. As they spiraled closer and closer, they radiated gravitational waves, somewhat like two children playing in a pool and creating ripples that spread out across the water. However, in the case of orbiting neutron stars, the energy radiated away in gravitational waves is lost from the orbital energy of the neutron stars, bringing them into an even closer orbit that radiates even stronger gravitational waves. This inward spiral continues until the merger is complete.

Meanwhile, other effects of gravity are still in play. Recall that tidal forces (see Section 4-8) tend to elongate an object, and the tidal forces from each neutron star upon the other are so intense that they literally rip neutron star material free and create a swirl of debris. After stellar lives lasting about 11 billion years, we detected this death spiral in its last 100 s. At only 200 mi or so (about 320 km) apart, the neutron stars only had about 1500 more orbits before merging together into a black hole. We'll introduce black holes in Section 21-3; for now you just need to know that spinning black holes can cause jets to shoot out from material surrounding them at nearly the speed of light. These jets slammed into the debris that was ripped off the neutron stars by their tidal forces. The impact of the jets slamming into the debris was so energetic that it produced a powerful burst of gamma rays lasting about 2 s. These mysterious bursts had been seen before and are called **short gamma ray bursts** (short GRBs). However, it was this event that provided the proof that short GRBs indeed come from the merger of two neutron stars.

Here on Earth, we depend on these neutron star mergers! Each of the two neutron stars had masses somewhat greater than 1 $M_\odot$, and about 1% of their mass was ejected, escaping a fate trapped within the resulting black hole. Furthermore, recall that neutron stars already have densities similar to that of the nucleus of atoms, so perhaps it's not surprising that the violently ejected neutron star debris could fuse together to produce the nuclei of heavier elements. While stars make up the lighter half of the periodic table (roughly speaking), the majority of the remaining heavy elements are produced in neutron star mergers. This is a very new insight, as it was previously thought that most of the heavier elements were produced in supernovae. In this merger, an entire Earth mass of gold was produced! A similar amount of silver was formed, and about one-third of an Earth mass of uranium was produced, along with many other less familiar heavier elements (see **Figure 21-7**).

As we'll see in Section 21-4, the merger of two neutron stars is not the only event that can be detected with gravitational waves: The merging of black holes has also been detected! However, before we can get into black holes, we need to discuss Albert Einstein's special and general theories of relativity.

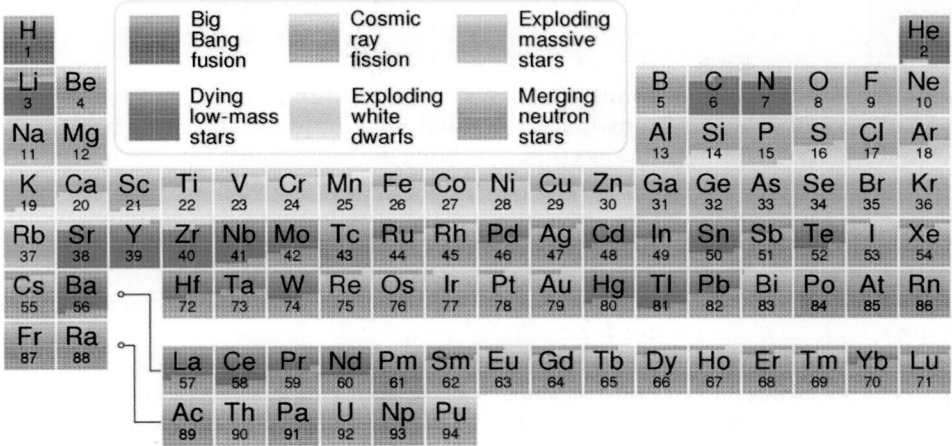

FIGURE 21-7

Origin of the Elements This chart indicates our latest understanding of what process produced the different elements. Starting at the light end, hydrogen, helium, and some lithium emerged from the Big Bang (in blue). Beryllium (Be) and boron (B) are created when fast-moving particles break apart larger atoms; these are created in our own atmosphere by cosmic rays (in pink). Supernovae from massive stars produce a wide range of elements (in yellow). Novae from white dwarfs in a binary system produce a narrower range of elements (in light gray). Dying low-mass stars go through an asymptotic giant branch (AGB) stage. During this stage, iron produced from previous generations of supernovae acts as a seed for a slow process (over thousands of years) of building larger and larger nuclei (in green). Some atoms are produced by stars but are so unstable that they radioactively decay before leaving significant amounts to observe (in dark gray). The majority of the heavier half of the periodic table (approximately) was produced by the merger of neutron stars (in purple). For example, elements such as silver (Ag) and gold (Au) come from neutron star mergers. These mergers also produce our uranium (U) and thorium (Th), which account for about 75% of the heat from radioactivity in Earth's mantle; along with heat from Earth's core, these sources of heat power plate tectonics. (NASA/Wikipedia: Cmglee; Data: Jennifer Johnson [OSU])

BOX 21-1   ASTRONOMY DOWN TO EARTH

A New Era of Multi-Messenger Astronomy *by Andy Howell*

On August 17, 2017, physicists detected the ripples in spacetime from two neutron stars spiraling together at one-third the speed of light (for more on the Laser Interferometer Gravitational-Wave Observatory [LIGO] and Virgo collaborations and the method of detecting these ripples, see Section 21-4). Meanwhile, NASA's Fermi satellite sensed a coincident burst of gamma rays. The teams could localize the event to a region on the sky near the constellation Hydra about 150 times the size of the full moon, and roughly 130 million ly distant. While the merger of two black holes had been detected before, the smashing together of two neutron stars had not. Such a death spiral was theorized to culminate in a never-before-seen explosion called a *kilonova*.

Astronomers and physicists from all over the world sprang into action, looking for signals in their neutrino detectors, radio dishes, satellites, and telescopes. About 11 hours after the merger, six teams, including my own group at Las Cumbres Observatory and the University of California, Santa Barbara, used telescopes in Chile to detect a new source of light in the sky. Scientists shared their findings in a mailing list, while simultaneously arranging to observe the phenomenon with telescopes all over the globe, and in space.

At Las Cumbres Observatory, we were especially well prepared, as we had built a new kind of observatory specifically designed to observe phenomena like this. It is a global network of 20 robotic telescopes wired together to act as a single instrument. This combines the telescope and the internet into a new kind of tool, allowing nearly instantaneous and continuous access to the sky. It is made possible only by repeatedly solving a complex, changing schedule of observations and continuously shuffling them around to the ever-changing night side of Earth. As dawn rose over Chile, night fell over Australia, and we kept observing the kilonova there. Then, as night shifted over to South Africa, we picked up the kilonova a third time. The explosion rose and faded in brightness so quickly that our globally distributed telescopes were essential to catching the action.

This discovery ushered in a new era, called *multi-messenger astronomy*. As long as humans have been around, we've used light to survey the cosmos. Now we have an entirely new way of sensing—gravitational waves—and by combining the two, we can learn more than we could from either technique alone.

This discovery brought together not just physicists and astronomers, but scientists from all over the globe. Each held only one piece of a much larger puzzle. It is a remarkable triumph of science that we all came to essentially the same conclusion regardless of cultural background.

(continued on the next page)

BOX 21-1 *(continued)*

This discovery also required instantaneous communication and cooperation. In that sense, it is a direct consequence of the third revolution in information sharing: the internet. The first revolution, writing, allowed the storage of knowledge beyond death. The second, the printing press, allowed knowledge to disperse widely and rapidly. Finally, the internet allowed easy access to nearly all accumulated human knowledge. Even more importantly, it made possible immediate global cooperation.

Humans are composed of the hydrogen created in the Big Bang, combined with the lighter elements, such as carbon and oxygen, synthesized in stars—thus, we are the algorithm for the elements of the universe to contemplate their own creation. Our computers, critical to this discovery, require the heavier rare Earth elements. So in a sense, the heavy elements are now joining us in helping to work out the details of their own synthesis.

Adding a new chapter to the story of creation is the rarest and most thrilling of scientific findings. Yet the idea that we can bring the entire planet together to make breakthroughs that were elusive at the national level is even more profound. All parts of Earth had to come together to find the last piece of the puzzle of our origin. Humans now have a new and exciting tool in their arsenal—global scientific cooperation—and I can't wait to see what secrets of the universe we unlock with it next.

Dr. Andrew Howell is a staff scientist at Las Cumbres Observatory and Adjunct Faculty in Physics at the University of California, Santa Barbara. He leads a global collaboration of more than 150 scientists called the Global Supernova Project. He has helped to discover and characterize new kinds of supernovae and has helped produce some of the best measurements of the properties of Dark Energy by using supernovae as standard candles.

21-2 The special theory of relativity changes our conceptions of space and time

Black holes, whose existence was predicted by Einstein's general theory of relativity (our best description of what gravity is and how it behaves), are both strange and simple. If something enters a black hole, the speed it would need to escape would be greater than the speed of light. Because nothing can travel faster than light, nothing can escape from a black hole—and even light is trapped. Perhaps strangest of all is that black holes actually exist at all. In recent years, astronomers have found that certain binary star systems contain a black hole. They have also found evidence that extremely massive stars form black holes when they die, producing immense bursts of gamma rays in the process. Even more extraordinary is the discovery that black holes of more than a million solar masses lie at the centers of most galaxies, including our own Milky Way. Perhaps most remarkable of all, pairs of black holes have been observed orbiting each other and merging into an even larger black hole!

The idea of a black hole is a truly strange one. To appreciate it, we must discard some "commonsense" notions about the nature of space and time.

The Special Theory of Relativity

According to the classical physics of Newton, space is perfectly uniform and fills the universe like a rigid framework. Similarly, time passes at a monotonous, unchanging rate. It is always possible to know exactly how fast you are moving through this rigid fabric of space and time.

Those ideas were upset forever in 1905 when Albert Einstein proposed his **special theory of relativity**. This theory describes how motion affects our measurements of distance and time. It is "special" in the sense of being specialized. In particular, it does not include the effects of gravity. The word *relativity* is used because one of the key ideas

of the theory is that all measurements are made relative to an observer. Contrary to Newton, there is *no* absolute framework of space and time.

In particular, the distance between two points is not an absolute, nor is the time interval between two events. Instead, the values that you measure for these quantities depend on how you are moving, and these values are relative to you. Thus, one of Einstein's key discoveries is that someone moving in a different way would measure different lengths for objects and different durations of events. In our everyday experience, we do not notice that time and distance measurements depend on the speed between an observer and the events being observed because, as we will see, these effects are most noticeable when that speed approaches the speed of light.

Remarkably, Einstein's theory is based on just two basic principles. The first is quite simple:

Your description of physical reality is the same, regardless of the constant velocity at which you move.

In other words, if you are moving in a straight line at a constant speed, you experience the same laws of physics as you would if you were moving at any other constant speed and in any other direction. As an example, suppose you were inside a closed railroad car moving due north in a straight line at 100 km/h. Any measurements you make inside the car—for example, how long your thumb is or how long it takes for a ball to drop to the floor—will give exactly the same results as if the car were moving in any other direction or at any other speed, or were not moving at all. This first principle is what we experience in everyday life when riding on a bus or flying on an airplane: Phenomena inside a flying airplane (such as dropping a book) appear the same as when the plane is on the ground even though we are moving, and this would still be true if the plane could fly near the speed of light.

> To understand black holes, we must first grasp the nature of space and time as described by Einstein's special theory of relativity.

You might think there is a contradiction between the two statements that (1) measurements of the same event (such as dropping a ball) inside the railroad car give the same results regardless of the train's speed, and that (2) motion between an observer and events changes the times and distances being observed, and that someone moving in a different way could measure different values. To understand these statements, think of the train car as a laboratory, and *one observer* is inside moving right along with the experiments being conducted. The first statement says that you can move the laboratory at any speed and it will not change what the observer inside measures. The second statement—a primary result of special relativity—refers to differences between *two observers*. If a second observer standing outside the train on the tracks takes measurements of the experiments taking place inside the train (perhaps looking through a window), the two observers are moving at different speeds with respect to the experiment, and they can measure different values for some of the phenomena. Again, this discrepancy between observers is most noticeable when the speed between them is near the speed of light.

(a)

(b)

FIGURE 21-8

The Speed of Light Is the Same to All Observers (a) The speed you measure for ordinary objects depends on how you are moving. Thus, the batter sees the ball moving at 30 m/s, but the outfielder sees it moving at 40 m/s relative to her. **(b)** Einstein showed that this commonsense principle does not apply to light. No matter how fast or in what direction the astronaut in the spaceship is moving, she and the astronaut holding the flashlight will always measure light to have the same speed.

Einstein's second principle is much more bizarre than the first:

Regardless of your speed or direction of motion, you always measure the speed of light to be the same.

To see what this implies, imagine you are in a spaceship that is moving toward a flashlight. Even if you are moving at 99% of the speed of light, you will measure the photons from the flashlight to be moving at the same speed ($c = 3 \times 10^8$ m/s $= 3 \times 10^5$ km/s) as if your spaceship were motionless. This statement is in direct conflict with the Newtonian view that a stationary person and a moving person should measure different speeds (**Figure 21-8**).

Speed involves both distance and time. Since speed has a very different behavior in the special theory of relativity than in Newton's physics, it follows that both space and time behave differently as well. Indeed, in relativity, time proves to be so intimately intertwined with the three dimensions of space that we regard them as a single *four*-dimensional entity called **spacetime**.

Length Contraction

Einstein expressed his ideas about spacetime in a mathematical form and used this description to make a number of predictions about nature. All these predictions have been verified in innumerable experiments. One prediction is that the length you measure an object to have depends on how that object is moving; the faster it moves, the shorter its length along its direction of motion (**Figure 21-9**). This is called **length contraction**. In other words, if a railroad car moves past you at high speed, from your perspective on the ground you will actually measure it to be shorter than if it were at rest (Figure 21-9a). However, if you are on board the railroad car and moving with it, you will measure its length to be the same as measured on the ground when it was at rest. The word *relativity* emphasizes the importance of the relative speed between the observer and the object being measured.

If the idea of length contraction seems outrageous, it is because this effect is noticeable only at very high speeds, near the speed of light. But even the fastest spacecraft ever built by humans travels at a mere 1/25,000 of the speed of light. At this speed, a spacecraft 10 m long would be contracted in length by only 8 nm—a distance equal to the width of a single protein molecule! For moving cars, trains, and airplanes, length contraction is far too small to measure. As **Box 21-2** describes, however, we can easily see the effects of length contraction for subatomic particles that do indeed travel at nearly the speed of light.

Time Dilation

A second result of relativity is no less strange: A clock moving past you runs more slowly than a clock that is at rest. Like length contraction, this **time dilation** is a very small effect unless the clock is moving at extremely high speeds (Figure 21-9b). Nonetheless, physicists have confirmed the existence of time dilation by using an extremely accurate atomic clock carried on a jet airliner. When the airliner landed, they found that the on-board clock had actually ticked off slightly less time than an identical, stationary clock on the ground. For the passengers on board the airliner during this experiment, however, time flowed at a normal rate: The on-board clock ticked off the seconds as usual, their hearts beat at a normal rate, and so on. You only measure a clock (or a beating heart) to

BOX 21-2 TOOLS OF THE ASTRONOMER'S TRADE

Time Dilation and Length Contraction

The special theory of relativity describes how motion affects measurements of time and distance. By using the two basic principles of his theory—that no matter how fast you move, the laws of physics and the speed of light are the same—Einstein concluded that measurements of time and distance must depend on how the person making the measurements is moving.

To describe how measurements depend on motion, Einstein derived a series of equations named the **Lorentz transformations,** after the famous Dutch physicist Hendrik Antoon Lorentz. (Lorentz was a contemporary of Einstein who developed these equations independently but did not grasp all of their implications.) These equations tell us exactly how moving clocks slow down and how moving objects shrink.

To appreciate the Lorentz transformations, imagine two observers named Sergio and Majeeda. Sergio is at rest on Earth, while Majeeda is flying past in her spaceship at a speed v. Sergio and Majeeda both observe the same phenomenon on Earth—say, the beating of Sergio's heart or the ticking of Sergio's watch—which appears to occur over an interval of time. According to Sergio's clock (which is not moving relative to the phenomenon), the phenomenon lasts for T_0 seconds. This time period is called the **proper time** of the phenomenon. But according to Majeeda's clock (which *is* moving relative to the phenomenon), the same phenomenon lasts for a different length of time, T seconds. These two time intervals are related as follows:

Lorentz transformation for time

$$T = \frac{T_0}{\sqrt{1 - (v/c)^2}}$$

T = time interval measured by an observer moving relative to the phenomenon

T_0 = time interval measured by an observer *not* moving relative to the phenomenon (proper time)

v = speed of the moving observer relative to the phenomenon

c = speed of light

EXAMPLE: Sergio heats a cup of water in a microwave oven for 1 minute. According to Majeeda, who is flying past Sergio at 98% of the speed of light, how long does it take to heat the water?

Situation: The phenomenon in question is heating the water in the microwave oven. Sergio is not moving relative to this phenomenon (so he measures the *proper* time interval T_0). Majeeda is moving at speed $v = 0.98c$ relative to this phenomenon (so she measures a different time interval T). Our goal is to determine the value of T.

Tools: We use the Lorentz transformation for time to calculate T.

Answer: We have $v/c = 0.98$, so

$$T = \frac{T_0}{\sqrt{1 - (v/c)^2}} = \frac{1 \text{ minute}}{\sqrt{1 - (0.98)^2}} = 5 \text{ minutes}$$

Review: A phenomenon that lasts for $T_0 = 1$ minute on Sergio's clock is stretched out to $T = 5T_0 = 5$ minutes as measured on Majeeda's clock moving at 98% of the speed of light. Other phenomena are affected in the same way: As measured by *Majeeda,* it takes 5 seconds for Sergio's wristwatch to tick off one second, and the minute hand on Sergio's wristwatch takes 5 hours to make a complete sweep. The converse is also true. As measured by *Sergio,* the minute hand on Majeeda's wristwatch will take 5 hours to make a complete sweep. This phenomenon, in which events moving relative to an observer happen at a slower pace, is called *time dilation.*

The Lorentz transformation for time is plotted in Figure 21-9b, which shows how 1 s measured on a stationary clock (say, Sergio's) is stretched out when measured using a clock carried by a moving observer (such as Majeeda). For speeds less than about half the speed of light, the mathematical factor $\sqrt{1 - (v/c)^2}$ is not too different from 1, and there is little difference between the recordings of the stationary and moving clocks. At the fastest speeds that humans have ever traveled (en route from Earth to the Moon), the difference between 1 and the factor $\sqrt{1 - (v/c)^2}$ is less than 10^{-9}. So, for any speed associated with human activities, stationary and slowly moving clocks tick at essentially the same rate. As the next example shows, however, time dilation is important for subatomic particles that travel at speeds comparable to c.

EXAMPLE: When unstable particles called *muons* (pronounced "mew-ons") are produced in experiments on Earth, they decay into other particles in an average time of 2.2×10^{-6} s. Muons are also produced by fast-moving protons from interstellar space when they collide with atoms in Earth's upper atmosphere. These muons typically move at 99.9% of the speed of light and are formed at an altitude of 10 km. How long do such muons last before they decay?

Situation: The phenomenon in question is the life of a muon, which lasts a time $T_0 = 2.2 \times 10^{-6}$ s as measured by an observer not moving with respect to the muon. Our goal is to find the time interval T measured by an observer on Earth, which is moving at $v = 0.999c$ relative to the muon (the same speed at which the muon is moving relative to Earth).

Tools: As in the preceding example, we use the Lorentz transformation for time.

(continued on the next page)

BOX 21-2 *(continued)*

Answer: Using $v/c = 0.999$,

$$T = \frac{2.2 \times 10^{-6}\,\text{s}}{\sqrt{1 - (0.999)^2}} = 4.9 \times 10^{-5}\,\text{s}$$

Review: At this speed, the muon's lifetime is slowed down by time dilation by a factor of more than 22. Note that as measured by an observer on Earth, the time that it takes a muon produced at an altitude of 10 km = 10,000 m to reach Earth's surface is

$$\frac{\text{distance}}{\text{speed}} = \frac{10,000\,\text{m}}{0.999 \times 3.00 \times 10^8\,\text{m/s}} = 3.3 \times 10^{-5}\,\text{s}$$

Were there no time dilation, such a muon would decay in just $2.2 \times 10^{-6}\,\text{s}$ and would never reach Earth's surface. But in fact, these muons *are* detected by experiments on Earth's surface, because a muon moving at $0.999c$ lasts more than $3.3 \times 10^{-5}\,\text{s}$. The detection at Earth's surface of muons from the upper atmosphere is compelling evidence for the reality of time dilation.

In the same terminology as "proper time," we say that a ruler at rest measures **proper length** or **proper distance** (L_0). According to the Lorentz transformations, distances perpendicular to the direction of motion are unaffected. However, a ruler of proper length L_0 held parallel to the direction of motion shrinks to a length L, given by:

Lorentz transformation for length

$$L = L_0\sqrt{1 - (v/c)^2}$$

L = length of a moving object along the direction of motion
L_0 = length of the same object at rest (proper length)
v = speed of the moving object
c = speed of light

EXAMPLE: Again, imagine that Majeeda is traveling at 98% of the speed of light relative to Sergio. If Majeeda holds a 1-m ruler parallel to the direction of motion, how long is this ruler as measured by Sergio?

Situation: The ruler is at rest relative to Majeeda, so she measures its *proper* length $L_0 = 1$ m. Our goal is to determine its length L as measured by Sergio, who is moving at $v = 0.98c$ relative to Majeeda and her ruler.

Tools: We use the Lorentz transformation for length.

Answer: With $v/c = 0.98$, we find

$$L = L_0\sqrt{1 - (v/c)^2} = (1\,\text{m})\sqrt{1 - (0.98)^2}$$
$$= (1\,\text{m}) \times (0.20) = 0.20\,\text{m} = 20\,\text{cm}$$

Review: We saw in the first example that according to *Sergio*, Majeeda's clocks are ticking only one-fifth as fast as his. This example shows that he also measures Majeeda's 1-m ruler to be only one-fifth as long (20 cm) when held parallel to the direction of motion. Note that the converse is also true: If Sergio holds a 1-m ruler parallel to the direction of relative motion, *Majeeda* measures it to be only 20 cm long. This shrinkage of length is called *length contraction*.

EXAMPLE: Consider again the above example about muons created 10 km above Earth's surface. If a muon is traveling straight down, what is the distance to the surface as measured by an observer riding along with the muon?

Situation: Imagine a ruler that extends straight up from Earth's surface to where the muon is formed. This ruler is at rest relative to Earth, so its length of 10 km is the proper length L_0. Our goal is to find the length L of this ruler as measured by the observer riding with the muon.

Tools: As in the previous example, we use the Lorentz transformation for length.

Answer: With $v/c = 0.999$, we calculate

$$L = (1\,\text{km})\sqrt{1 - (0.999)^2} = 0.45\,\text{km} = 450\,\text{m}$$

Review: The distance is contracted tremendously as measured by an observer riding with the muon. This result gives us another way to explain how such muons are able to reach Earth's surface. As measured by the muon, the time required to travel the contracted distance is

$$\frac{\text{distance}}{\text{speed}} = \frac{450\,\text{m}}{0.999 \times 3.00 \times 10^8\,\text{m/s}} = 1.5 \times 10^{-6}\,\text{s}$$

This time is less than the $2.2 \times 10^{-6}\,\text{s}$ that an average muon takes to decay. Hence, muons can successfully reach Earth's surface.

be running slow if it is *moving* relative to you. Box 21-2 discusses time dilation in more detail.

As an example of the time dilation in Figure 21-9b, suppose a rider on the train has a heart that beats once per second. Any method used to measure the rider's heartbeat on the train gives this value—1 s between beats. Perhaps watching through a window, consider a nurse on the tracks observing the rider's heartbeat as the hypothetical train speeds by at 98% of the speed of light (labeled $0.98c$ in the figure). In Figure 21-9b, we see that from the tracks,

the nurse would measure that the rider's heart only beat once every 5 s on the nurse's own clock. In other words, the stationary observer would see a slower heartbeat.

The special theory of relativity predicts that the astronaut with the flashlight in Figure 21-8b sees the astronaut flying in the spaceship as shortened along the direction of motion and as having slowly ticking clocks. Remarkably, it also says that the *flying* astronaut sees the astronaut with the flashlight (who is moving relative to her) as being shortened and as having slowly

This train is at rest relative to you.

The same train is now moving relative to you.

(a) Length contraction

(b) Time dilation

FIGURE 21-9

Length Contraction and Time Dilation (a) The faster an object moves, the shorter it becomes along its direction of motion. Speed has no effect on the object's dimensions perpendicular to the direction of motion. **(b)** The faster a clock moves, the slower it runs. This graph shows how many seconds

(as measured on your clock) it takes a clock that moves relative to you to tick off 1 second. The effect is pronounced only for speeds near the speed of light c.

ticking clocks! These observations may seem contradictory, but they are not: Each spaceship is moving relative to the other, and in the special theory of relativity, only relative motion matters. Each astronaut's measurements are correct relative to his or her frame of reference.

Now we see the significance of relative motion between observers: Relative motion changes what observers measure for certain quantities, such as length and time. **Objects observed in motion are shorter in their direction of motion, and clocks observed in motion tick more slowly.** However, relative motion does not change the speed of light (through empty space) and it does not change the laws used to describe nature.

CAUTION!

You may have the idea that the special theory of relativity implies that there are *no* absolutes in nature and that everything is relative. But, in fact, the special theory is based on the principles that the laws of physics and the speed of light in a vacuum *are* absolutes. Only certain quantities, such as distance and time, depend on your state of motion. You may also have the idea that length contraction and time dilation are just optical illusions caused by high speeds. But these effects are *real*. A moving spaceship does not just look shorter or seem to be shorter—it really *is* shorter. Likewise, a moving clock does not merely look like it is ticking slowly, or seem to be ticking slowly—it really *does* tick more slowly. Einstein's theory is supported by every experiment designed to test it. Relativity is very strange, but it is also very real!

CONCEPT CHECK 21-4

Suppose you are on the hypothetical train in Figure 21-9a. If the train sped along at half the speed of light, would it appear to you that the train contracts in length? Would your heart beat more slowly?

CONCEPT CHECK 21-5

Suppose two spaceships travel directly toward each other at 90% the speed of light relative to an observer watching from a planet right in the middle of the spaceships. Each spaceship turns on its lights for everyone to see. At what speed does a traveler from one spaceship see light arriving from the other spaceship? At what speed does the observer on the planet see light arriving from the spaceship?

Answers appear at the end of the chapter.

Mass and Energy

Einstein's special theory of relativity predicts another famous relationship: Any object with mass (m) has energy (E) embodied in its mass. This idea is expressed by the well-known equation $E = mc^2$, where c is the speed of light. In thermonuclear reactions, part of the energy embodied in the mass of atomic nuclei is converted into other forms of energy. This released energy is what makes the Sun and the stars shine (see Section 16-1). The conversion of mass into energy results in a real loss of mass; as the Sun shines, it loses about 100 times the mass of the largest Egyptian pyramid each second. The sunlight we see by day and the starlight that graces our nights are a brilliant verification of the special theory of relativity.

The special theory of relativity also explains why it is impossible for a spaceship to move at the speed of light. If it could, then a light beam traveling in the same direction as the spaceship would appear to the ship's crew to be stationary. But this would contradict the second of Einstein's principles, which says that all observers, including those on a spaceship, must see light traveling at the speed of light. Therefore, it cannot be possible for a spaceship to travel at the speed of light. **In fact, no object with mass can travel at the speed of light.** This conclusion has also been verified by experiments. Subatomic particles can be made to travel at 99.99999999% of the speed of light *c*, but they can never make it all the way to *c*.

21-3 The general theory of relativity predicts black holes

TUTORIAL 21.1 Einstein's special theory of relativity is a *special* case in the sense that it does not include the effects of gravity. Einstein's next goal was to develop an even more comprehensive theory that includes gravity and its effects on space and time. This is Einstein's **general theory of relativity**, which he published in 1915.

In the special theory of relativity, a relative velocity between observers leads to different observations of distance and time. In addition to all of the effects of the special theory, the general theory of relativity describes how observers in gravitational fields of different strengths also measure differences in distance and time.

The Equivalence Principle

According to Newton's theory of gravity, an apple falls to the floor because the force of gravity pulls the apple down. But Einstein pointed out that the apple would appear to behave in exactly the same way in space, far from the gravitational influence of any planet or star, if the floor were to accelerate upward (in other words, if the floor came up to meet the apple).

In **Figure 21-10**, two famous gentlemen are watching an apple fall toward the floor of their closed compartments. They have no way of telling who is at rest on Earth and who is in the hypothetical elevator moving upward through empty space at a constantly increasing speed. This is an example of Einstein's **equivalence principle**, which states that in a small volume of space, the downward pull of gravity can be accurately and completely duplicated by an upward acceleration of the observer.

The equivalence principle is the key to the general theory of relativity. It allowed Einstein to focus entirely on motion, rather than force, in discussing gravity. A hallmark of gravity is that it causes the same acceleration no matter the mass of the object. For example, a baseball and a cannon ball have very different masses, but if you drop them side by side in a vacuum, they accelerate downward at exactly the same rate. To explain this observation, Einstein envisioned gravity as being caused by a *curvature of space*. In fact, his general theory of relativity describes gravity entirely in terms of the geometry of both space *and* time, that is, of spacetime. Far from a source of gravity, like a planet or a star, spacetime is "flat" and clocks tick at their normal rate. Closer to a source of gravity, however, space is curved and clocks slow down.

ANALOGY

A useful analogy is to picture the spacetime near a massive object such as the Sun as being curved like the surface in **Figure 21-11**. Imagine a ball rolling along this surface. Far from the "well" that represents the Sun, the surface is fairly flat and the ball moves in a straight line. If the ball passes near the well, however, it curves in toward it. If the ball is moving at an appropriate speed, it might move in an orbit around the sides of the well. The curvature of the well has the same effect on a ball of any size, which explains why gravity produces the same acceleration on objects of different mass.

(a) The apple hits the floor of the compartment because Earth's gravity accelerates the apple downward.

This compartment is at rest in Earth's gravitational field.

(b) The apple hits the floor of the compartment because the compartment accelerates upward.

This compartment is accelerating in a gravity-free environment.

ANIMATION 21.1 **FIGURE 21-10**

The Equivalence Principle The equivalence principle asserts that you cannot distinguish between **(a)** being at rest in a gravitational field and **(b)** being accelerated upward in a gravity-free environment. This idea was an important step in Einstein's quest to develop the general theory of relativity.

(a)

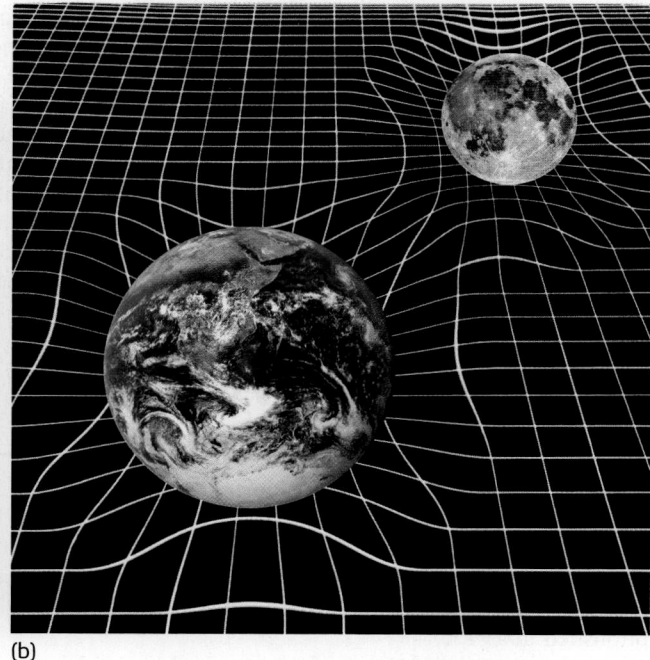

(b)

FIGURE 21-11

The Gravitational Curvature of Spacetime According to Einstein's general theory of relativity, spacetime becomes curved near a massive object. **(a)** To help you visualize the curvature of four-dimensional spacetime, this figure shows the curvature of a two-dimensional space around a massive object. The curvature of spacetime has been measured directly around Earth, where Earth was the massive object in the center (location 1), and a satellite called Gravity Probe B was orbiting at location 3. **(b)** If two objects are massive enough, they can each warp or distort the spacetime around them, with the larger mass curving spacetime the most. The greater the curvature of spacetime, the slower clocks will tick and the more contracted lengths become, compared to distant regions with little spacetime curvature. (b: Victor de Schwanberg/Science Source)

Testing the General Theory

Einstein's general theory of relativity and its picture of curved spacetime have been tested in a variety of different ways. In what follows, we discuss some key experimental tests of the theory.

Experimental Test 1: The gravitational bending of light. Light rays naturally travel in straight lines. But if the space through which the rays travel is curved, as happens when light passes near the surface of a massive object like the Sun, the paths of the rays will likewise be curved (**Figure 21-12**). In other words, gravity should bend light rays. This prediction was first tested in 1919 during a total solar eclipse. During totality, when the Moon blocked out the Sun's disk, astronomers photographed the stars around the Sun. Careful measurements afterward revealed that the stars around the Sun appeared to have shifted from their usual positions by an amount consistent with Einstein's theory.

> Careful experiments have verified the key ideas of Einstein's theory of gravity.

Experimental Test 2: The precession of Mercury's orbit. As the planet Mercury moves along its elliptical orbit, the orbit itself slowly changes orientation, or *precesses* (**Figure 21-13**). Most of Mercury's precession is caused by the gravitational pull of the other planets, as explained by Newtonian mechanics. But once the effects of all the other planets had been accounted for, there remained an unexplained excess rotation of Mercury's major axis of 43 arcsec per century. Although this discrepancy may seem very small, it frustrated astronomers for half a century. Some

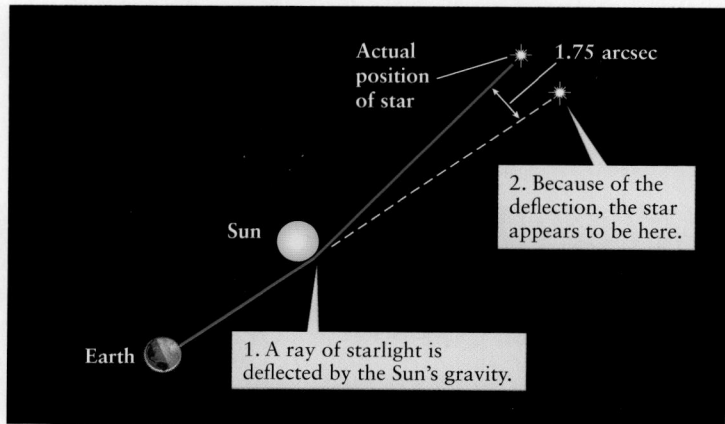

FIGURE 21-12

The Gravitational Deflection of Light Light rays are deflected by the curved spacetime around a massive object like the Sun. General relativity predicts a small deflection of 1.75 arcsec for a light ray grazing the Sun's surface, and this deflection could not be explained by Newton's theory of gravity. The deflection of starlight by the Sun was confirmed during a solar eclipse in 1919.

astronomers searched for a missing planet even closer to the Sun that might be tugging on Mercury; none has been found. Einstein showed that at Mercury's position close to the Sun, the general theory of relativity predicts a small correction to Newton's description of gravity. This correction is exactly enough to account for the excess precession.

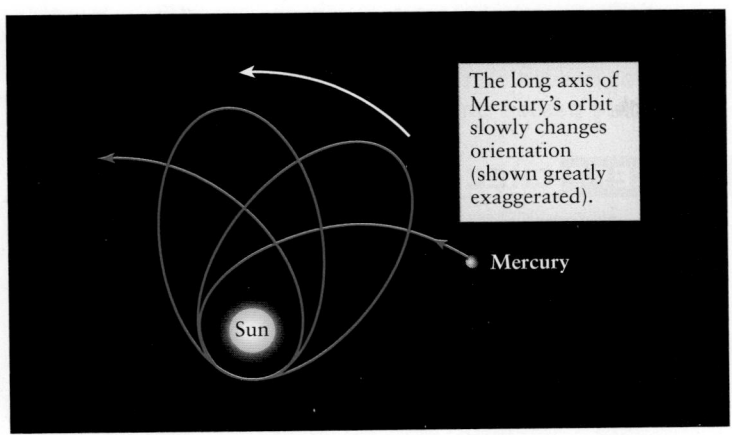

FIGURE 21-13

The Precession of Mercury's Orbit Mercury's orbit changes orientation at a rate of 574 arcsec (about one-sixth of a degree) per century. Newtonian mechanics predicts that the gravitational influences of the other planets should make the orientation change by only 531 arcsec per century. Einstein was able to explain the excess 43 arcsec per century using his general theory of relativity.

Experimental Test 3: The gravitational slowing of time and the gravitational red shift (Figure 21-14). In the general theory of relativity, a massive object such as Earth warps time as well as space. Einstein predicted that clocks on the ground floor of a building should tick slightly more slowly than clocks on the top floor, which are farther from Earth (Figure 21-14a).

A light wave can be thought of as a clock; just as a clock makes a steady number of ticks per minute, an observer sees a steady number of complete cycles of a light wave passing by each second. If a light beam is aimed straight up from the ground floor of a building, an observer at the top floor will measure a "slow-ticking" light wave with a lower frequency, and thus a longer wavelength, than will an observer on the ground floor (Figure 21-14b). Because photons with longer wavelengths have lower energy, the increase in wavelength means that a photon reaching the top floor has less energy than when it left the ground floor. (You may want to review the discussion of frequency and wavelength in Section 5-2 as well as the description of photons in Section 5-5.) These effects, which have no counterpart in Newton's theory of gravity, are called the **gravitational redshift**.

> **CAUTION!**
>
> Be careful not to confuse the gravitational redshift with a Doppler shift. In the Doppler effect, redshifts are caused by a light source moving away from an observer. Gravitational redshifts, by contrast, are caused by time flowing at different rates at different locations. No motion is involved.

The American physicists Robert Pound and Glen Rebka first measured the gravitational redshift in 1960 using gamma rays fired between the top and bottom of a shaft 20 m tall. Because Earth's gravity is relatively weak, the redshift that they measured was very small ($\Delta\lambda/\lambda = 2.5 \times 10^{-15}$) but was in complete agreement with Einstein's prediction.

Much larger shifts are seen in the spectra of white dwarfs, whose spectral lines are redshifted as light climbs out of the white dwarf's intense surface gravity. As an example, the gravitational redshift of the spectral lines of the white dwarf Sirius B (see Figure 20-9) is $\Delta\lambda/\lambda = 3.0 \times 10^{-4}$, which also agrees with the general theory of relativity.

Compared to a clock on the top floor, a clock on the ground floor is deeper in Earth's gravitational field and so ticks more slowly.

(a) The gravitational slowing of time

As a light wave climbs in a gravitational field, its frequency decreases and its wavelength increases.

Light source

(b) The gravitational redshift

FIGURE 21-14

The Gravitational Slowing of Time and the Gravitational Redshift **(a)** Clocks at different heights in a gravitational field tick at different rates. Assume that identical clocks were first brought together and set to the same time. After being placed on different floors and left to tick for a while, the stronger gravity on the lower floor (closer to Earth) results in less time passing for that clock. In stronger gravity, time passes more slowly. **(b)** The oscillations of a light wave constitute a type of clock. As a light wave climbs from the ground floor toward the top floor, its oscillation frequency becomes lower and its wavelength becomes longer.

Experimental Test 4: Gravitational waves. Electric charges oscillating up and down in a radio transmitter's antenna produce electromagnetic radiation. In a similar way, the general theory of relativity predicts that oscillating massive objects should produce **gravitational radiation,** or **gravitational waves.** (Newton's theory of gravity makes no such prediction.)

Gravitational radiation is exceedingly difficult to detect, because it is by nature much weaker than electromagnetic radiation. As we will see in the next section, exciting detections of gravitational waves by the Laser Interferometer Gravitational-Wave Observatory (LIGO) and partner instruments have recently confirmed the prediction made by Albert Einstein almost exactly 100 years earlier. In addition to these recent discoveries, compelling *indirect* evidence for the existence of gravitational radiation has come from a binary system of two neutron stars. Russell Hulse and Joseph Taylor at the University of Massachusetts at Amherst discovered that these stars are slowly spiraling toward each other and losing energy in the process. The rate at which they lose energy is just what would be expected if the two neutron stars are emitting gravitational radiation as predicted by Einstein. Hulse and Taylor shared the 1993 Nobel Prize in Physics for their discovery.

Recent Evidence for Einstein's Gravity Theory

The curvature of spacetime around Earth was measured in 2011. Analogous to Figure 21-11, a scientific satellite called Gravity Probe B carrying sensitive gyroscopes was sent to measure the curvature produced by Earth's gravity. Spacetime curvature would produce a subtle precession of a gyroscope's axis as it orbits Earth, which is an effect that is unique to Einstein's theory. Amazingly, scientists reported precise agreement between general relativity and Gravity Probe B's detailed measurements.

Using the world's most accurate clocks in 2010, the gravitational slowing of time has been measured at two vertical positions *separated by only 1 foot* in a laboratory. Just as predicted, time slows down enough that the lower clock would read about 90 billionths of a second less over a 79-year period.

Another context in which the general theory of relativity shows up is when certain technology, by its very nature, is highly sensitive. For example, global positioning system (GPS) satellites must use extremely accurate clocks in order to triangulate an accurate position on Earth. The effects of gravity slow down time on Earth's surface compared to the satellite's clocks, so general relativity *must* be taken into account for an accurate GPS result. Without general relativity included, GPS would accumulate errors of more than 10 km (6.2 mi) per day!

The general theory of relativity has never made an incorrect prediction. It now stands as our most accurate and complete description of gravity. Einstein demonstrated that Newtonian mechanics is accurate only when applied to low speeds and weak gravity. If extremely high speeds or powerful gravity are involved, only a calculation using relativity will give correct answers. As we will see next, the bizarre phenomenon of a black hole is an example of strong gravity that Newton's theory does not describe.

CONCEPT CHECK 21-6

Suppose two identical clocks are synchronized far out in space and away from any strong sources of gravity. You learn what time they measure by radio waves sent from each clock. Then one clock is placed near a massive planet, while you stay floating in space with the other clock. After 1 hour has passed on the clock next to you, has more time or less time passed on the clock near the planet?

CONCEPT CHECK 21-7

Consider the same set of clocks as in ConceptCheck 21-6. When radio waves from the clock near the planet reach you out in space, are they the same wavelength as when they were emitted near the planet?

Answers appear at the end of the chapter.

The Formation of Black Holes

Perhaps the most dramatic prediction of the general theory of relativity concerns what happens when a large amount of matter is concentrated in a small volume. We have seen that if a dying star is not too massive, it ends up as a white dwarf star. If the dying star is more massive than the Chandrasekhar limit of about 1.4 $M_\odot$, it cannot exist as a stable white dwarf star and, instead, shrinks down to form a neutron star. But if the dying star has more mass than the maximum permissible for a neutron star, about 2 to 3 $M_\odot$, not even the internal pressure of neutrons (the degenerate neutron pressure) can hold the star up against its own gravity, and the core collapses.

As the star's core becomes compressed to enormous densities, the strength of gravity at the surface of this rapidly shrinking sphere also increases dramatically. According to the general theory of relativity, the space immediately surrounding the core becomes so highly curved that it closes on itself (**Figure 21-15**). Photons flying outward at an angle from such an object's surface would arc back inward, while photons that fly straight outward undergo such a strong gravitational redshift that they lose all their energy and cease to exist.

Ordinary matter can never travel as fast as light. Hence, if light cannot escape from the collapsing core, neither can anything else. An object from which neither light nor matter can escape is called a **black hole. Figure 21-16** illustrates the extreme curvature of spacetime due to a black hole.

The mass of the star's collapsing core forms the black hole; this mass gives the spacetime around the black hole its strong curvature. Thanks to this curvature, the black hole's gravitational influence can still be felt by other objects. However, most of the original mass of the dying star does not make it into the black hole. For example, a 5-$M_\odot$ black hole formed by core collapse would probably come from a star of about 20 $M_\odot$ or more.

CAUTION!

Some low-quality science-fiction movies and books suggest that black holes are evil things that go around gobbling up everything in the universe. Not so! The bizarre effects created by highly warped spacetime are limited to a region quite near the hole. For example, the effects of the general theory of relativity predominate only within 1000 km of a 10-$M_\odot$ black hole. Beyond 1000 km, gravity is weak enough that Newtonian physics can adequately describe everything. If our own Sun somehow turned into a black hole (an event that, happily, seems to be quite impossible), the orbits of the planets would hardly be affected at all.

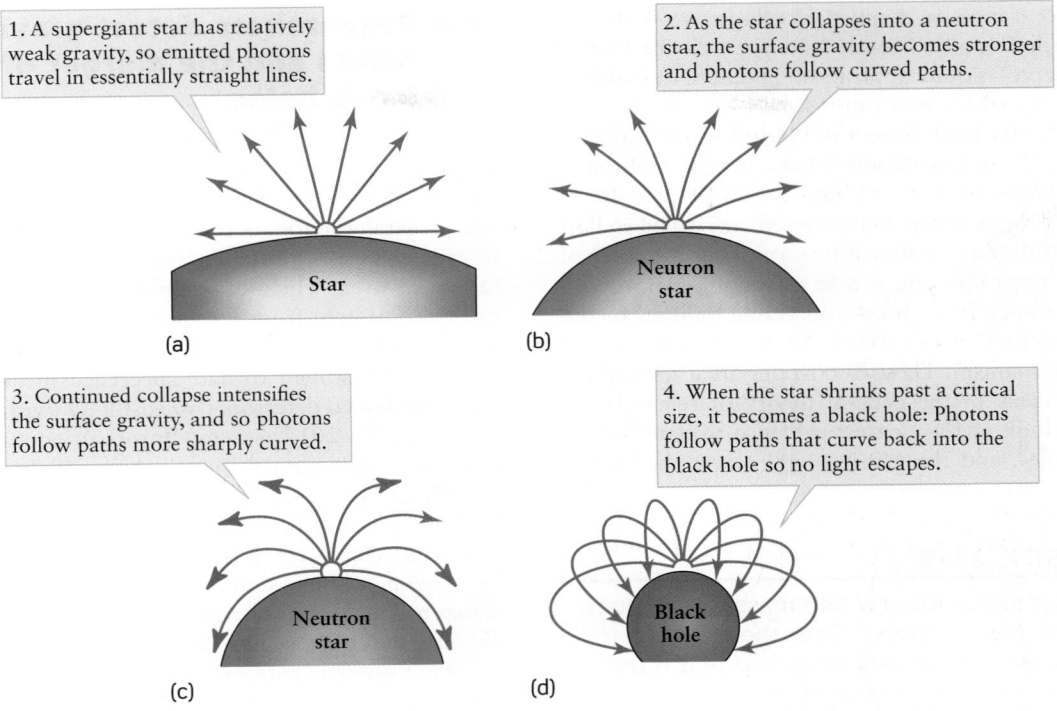

FIGURE 21-15

The Formation of a Black Hole (a)–(c) These illustrations show four steps leading up to the formation of a black hole from a dying star. **(d)** When the star becomes a black hole, not even photons emitted directly upward from the surface can escape; they undergo an infinite gravitational redshift and disappear.

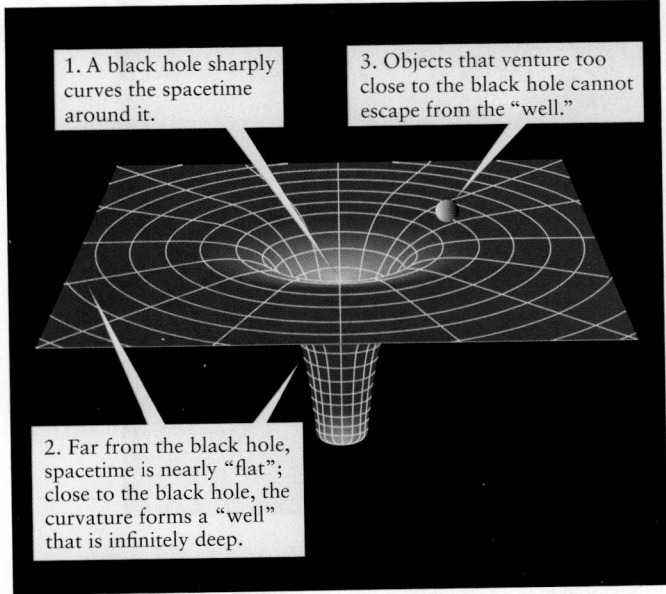

FIGURE 21-16

Curved Spacetime around a Black Hole This diagram suggests how spacetime is distorted by a black hole's mass. As in Figure 21-11, spacetime is represented as a two-dimensional surface. Unlike the situation in Figure 21-11, a black hole's gravitational "well" is infinitely deep.

CONCEPT CHECK 21-8

The exact number is not known, but approximately what is the lowest-mass black hole you might expect from a stellar core collapse?

Answer appears at the end of the chapter.

The Escape Speed from Black Holes Is the Speed of Light

One way to understand black holes is to first consider the escape speed from a planet (see Section 4-7). The escape speed of a planet is the minimum ejection speed an object must have at the planet's surface in order to escape the planet and never return.

Very importantly, if you imagine compressing a planet's mass into a smaller radius, the planet's escape speed increases. It is even possible, if an object were compressed into a small enough radius, that its calculated escape speed would exceed the speed of light. Since nothing—not even light—can travel faster than the speed of light, nothing can escape such an unusual object. *An object with an escape speed exceeding the speed of light is a black hole.* Surrounding the black hole, where the escape speed from the hole just equals the speed of light, is the **event horizon**.

You can think of the event horizon as the "surface" of the black hole, although it is just a spherical boundary and not a physical surface. Once a massive dying star collapses to within its event horizon, it disappears permanently from the universe's view. The term "event horizon" is quite appropriate, because this surface is like a horizon beyond which we cannot see any events.

Perhaps the most remarkable aspect of black holes is that they really exist! As we will see, astronomers have located a number of black holes with masses between about 5 and 30 M_⊙. Black holes in this range (or even greater, if discovered) are called **stellar-mass black holes**, and they are assumed to form from the cores of supernovae. What is truly amazing is that astronomers have also discovered many enormous black holes containing millions or billions of solar masses; these were formed through a different and poorly understood mechanism. These discoveries are a resounding confirmation of the ideas of the general theory of relativity. In the next section, we look at the clearest evidence to date for the existence of black holes, and the gravitational waves that led to these discoveries.

CONCEPT CHECK 21-9

Suppose that aliens could somehow add mass to our Moon without changing the Moon's radius. Does the Moon's escape speed increase or decrease? What happens if the Moon's escape speed exceeds the speed of light?

Answer appears at the end of the chapter.

21-4 Researchers detect gravitational waves and the merger of two black holes

One hundred years after Einstein published his theory of gravity (the general theory of relativity), the world of astronomy was forever changed. Einstein's theory led to the prediction of gravitational waves, which are ripples in spacetime radiating away at the speed of light from accelerating masses. Finally, in 2015, after half a century of trying, the National Science Foundation's largest financial investment to date succeeded in detecting gravitational waves with two detectors named LIGO (with the "I" pronounced as "eye"). By 2017, several European countries finished a third detector in Italy named Virgo, which collaborates with LIGO.

What is a gravitational wave or ripple in spacetime? Recall from Section 21-2 that the stronger the gravity, the more that time slows down and lengths contract. Similarly, as a gravitational wave travels by a region, it briefly slows down time and causes lengths to contract in that region. Since it's harder to measure tiny changes in time than in distance, the challenge is to measure these length contractions.

What event is extreme enough to create gravitational waves? The most likely event to generate observable gravitational waves is the merging of two large black holes (see **Figure 21-17**). In this

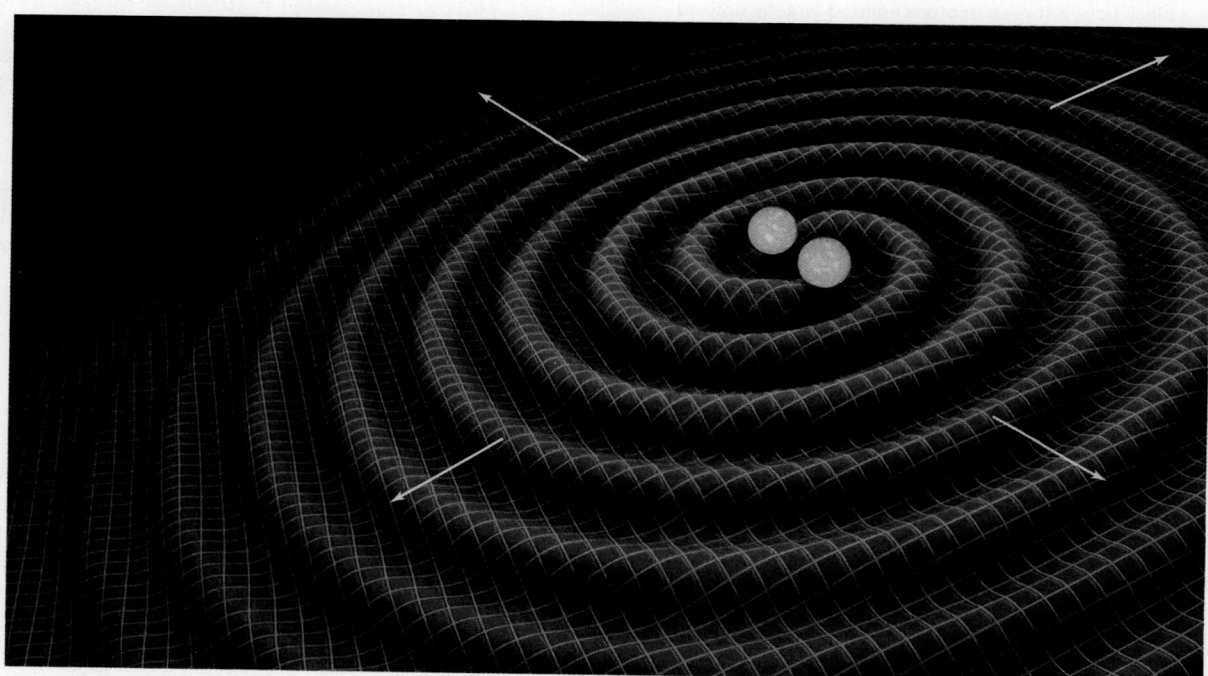

FIGURE 21-17

Gravitational Waves Are Ripples in Spacetime As two black holes orbit each other, they radiate gravitational waves. This also occurs for any accelerating mass, but black holes orbiting each other make some of the strongest waves. The "grid" represents spacetime, and the gravitational waves, or ripples, are distortions in spacetime. The ripples travel outward at the speed of light (indicated by the yellow arrows); within each ripple, time slows down and lengths contract (or get shorter). In Figure 21-11, it is assumed that the masses don't have any significant acceleration, so the curvatures of spacetime in those figures are not examples of radiating gravitational waves. (R. Hurt/Caltech-JPL/NASA) process (which we'll discuss in more detail), the black holes accelerate as they orbit each other and finally merge into an even larger black hole. Even the merger of two neutron stars generates observable gravitational waves, as discussed in Section 21-1.

In this section, we'll look at how gravitational waves are detected and what events have been observed.

LIGO, Virgo, and the Detection of Gravitational Waves

Tiny changes in length need to be measured in order to detect a gravitational wave passing by. The details of how this measurement is carried out are beyond the scope of this textbook, but the basic layout is shown in **Figure 21-18**. Each detector is like a "telescope" that is pointed everywhere; they can detect gravitational waves coming from anywhere, including those passing right through Earth! This is extremely fortunate and greatly increases the chances of capturing an event.

Even with 4-km arms and the strongest gravitational waves, the change in length of an arm is miniscule. An arm changes its length by only about one-ten-thousandth the size of a proton! Measuring such a tiny change over 4 km is equivalent to accurately measuring changes in the Earth–Moon distance by the size of one hydrogen atom! Expressed in another way, it is equivalent to measuring changes in the distance to the nearest star, Proxima Centauri (4.25 ly away), by the width of a human hair! These are the most precise measurements humankind has ever made.

Merging Black Holes

It is hard to imagine a more extreme scenario than the collision and merger of two black holes. LIGO's very first detection of gravitational waves came from the merger of two unexpectedly large black holes estimated at 29 $M_\odot$ and 36 $M_\odot$. The mass of the resulting black hole is 62 $M_\odot$, and the "missing" 3 $M_\odot$ was radiated away as gravitational waves! In a fraction of a second, three Suns' worth of mass simply disappeared, converting mass into gravitational wave energy according to $E = mc^2$. In fact, during the merger, the peak energy output from the gravitational waves was 50 times greater than all the stars in the observable universe!

The original mass of each star is estimated to be between 40 $M_\odot$ and 100 $M_\odot$. In a galaxy 1.4 billion ly away, these two stars in a binary system evolved very rapidly, due to their large masses, and left behind a binary black hole system. Just as with the merger of two neutron stars discussed in Section 21-1, as gravitational waves carry energy away from the binary system, the orbital distance of the black holes shrinks, leading to even stronger radiation of gravitational waves. Therefore, gravitational waves beget even stronger waves in a system of positive feedback. Very slowly at first, this draws the two black holes closer and closer until the gravitational radiation begins to skyrocket, and the black holes collide and merge in a fraction of a second (see **Figure 21-19**).

The LIGO data match theoretical predictions of black hole mergers so well that GW150914 is the strongest proof yet for the existence of black holes! Immediately after the black holes merged, the resulting larger black hole wasn't spherical. This asymmetric black hole was predicted to generate a brief signal called *ringdown* as it quickly became spherical, and even this detail was confirmed by LIGO. The data and the models are so precise for merger events that the masses can be determined before and after the merger, and also if one or both of the objects is a black hole or neutron star.

As of this writing (August 2018), LIGO has detected five black hole mergers and one neutron star merger. Many more merger events will be discovered soon with upgrades to the detectors. Gravitational waves open up a whole new window through which to view the universe and have ushered in a new era in astronomy. Gravitational

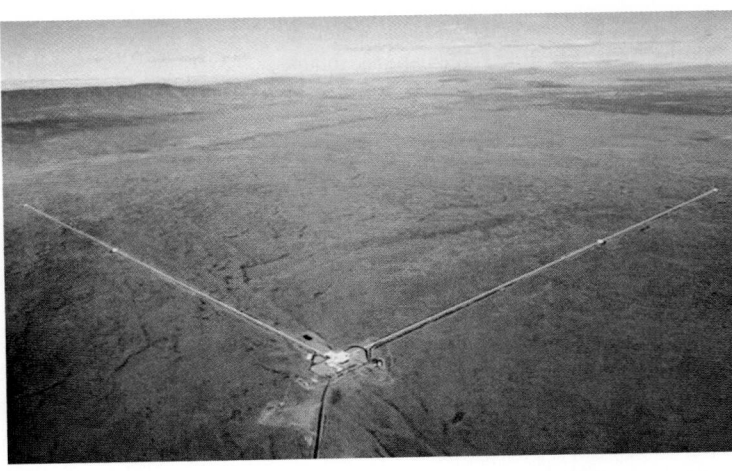

(a)

R I **V** U X G

FIGURE 21-18

LIGO and Virgo: Two Arms with Lasers to Measure Small Changes in Distance (a) The LIGO detector in Hanford, Washington. The second LIGO detector is in Livingston, Louisiana, and there's a similar European detector named Virgo near Pisa, Italy. The three teams work together and can triangulate the approximate location of an event, allowing visible-light telescopes to know where to point for additional observations. **(b)** The detectors use lasers to make extremely precise measurements of the length of each 4-km (2.5-mi) arm. In the "normal situation" when no gravitational wave is

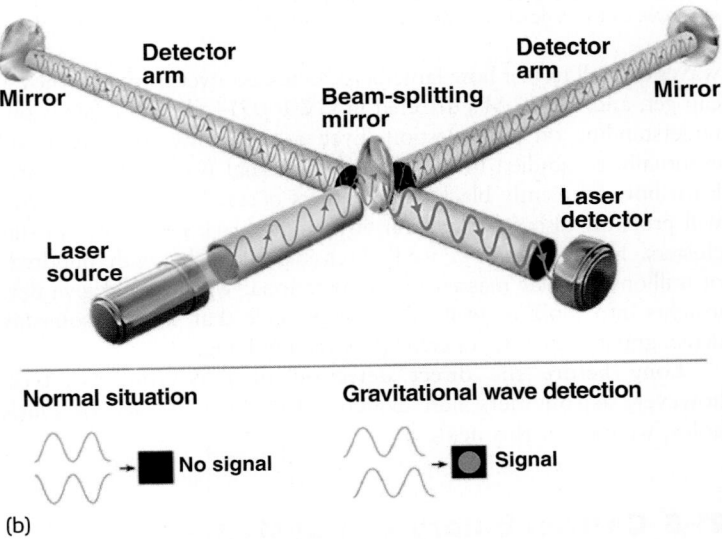

(b)

present, each arm is the same length and the laser light will cancel out, leaving no signal. However, it is the unique property of a passing gravitational wave that it increases the length of one arm while shrinking the other. These arm-length changes disrupt the perfect cancellation of the laser light; this produces a signal and indicates the detection of a gravitational wave. By having several detectors spread around Earth, it is easy to tell the difference between the false signal created by a truck driving by and a signal that all three detectors receive. (a: LIGO, Caltech, MIT, NSF/NASA; b: Nicolle Rager Fuller/Science News)

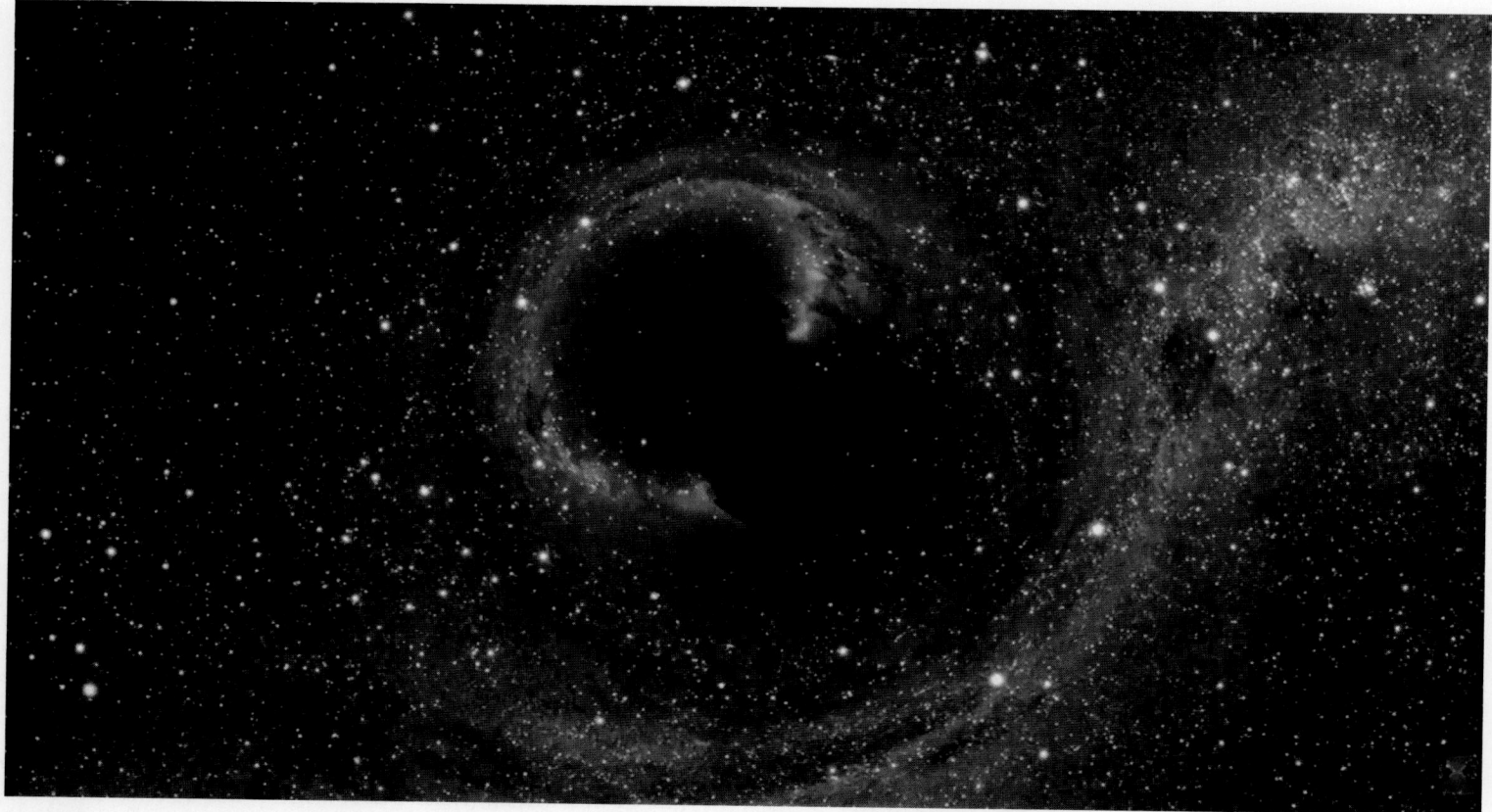

FIGURE 21-19

Merging Black Holes A snapshot from a simulation of the first two merging black holes observed by LIGO. Even the gravity from a single black hole distorts background light; the merging pair does that here as well, creating a nearly circular outer ring. The event horizons of each black hole are indicated by the outer border of the black regions. Normally spherical in shape, as the black holes merge, their event horizons get distorted. Only the last 0.2 s of this merger produced detectable gravitational waves. By the time of the collision and merger, the orbital distance of the black holes was about 350 km (217 mi), and their orbital speed was about half the speed of light. Like "objects" in the sky, this event is named GW150914, where the "GW" refers to gravitational waves, and the number refers to the date it was detected: 2015, September 14. (SXS/NASA)

waves can tell us just how large black holes left over from supernovae can get, and the 36 $M_\odot$ detected in GW150914 already extends our understanding on this question. Even a supernova event itself will eventually be studied through its gravitational waves. We will also learn how frequently black hole mergers occur. Gravitational waves will probably also shed light on how many black holes form in star clusters. In Section 21-6, we'll discuss black holes with hundreds of millions of solar masses, and gravitational waves could give new insights into those as well. It's even possible that we will someday detect gravitational waves created by the Big Bang.

Long before the direct detection of gravitational waves, however, astronomers had evidence for the existence of black holes; we look at this next.

21-5 Certain binary star systems probably contain black holes

TUTORIAL 21.2 Finding black holes is a difficult business. Because light cannot escape from inside the black hole, you cannot observe one directly in the same way that you can observe a star or a planet. The best you can hope for is to detect the effects of a black hole's powerful gravity.

Close binary star systems offer a unique way to find stellar-mass black holes. Imagine that one of the stars in a binary system evolves into a black hole. The binary system is not destroyed when one star undergoes a supernova to produce a black hole, leaving the binary partner to aid in the detection of the black hole itself. If the black hole orbits close enough to the remaining ordinary star, tidal forces can draw matter from the star onto the black hole. Fortunately, we can detect this "stolen" matter as it accretes around the black hole.

One key to detecting black holes is to search for the radiation emitted by material as it falls into a hole.

Given the tremendous strength of gravity around stellar-mass black holes, their accreted matter would be compressed and heated up to very high temperatures before being sucked into the black hole. In fact, the temperatures are so high that the accreted matter would give off x rays, so the search for these black holes is conducted with x-ray telescopes.

The Case of Cygnus X-1

The first sign of such emissions from a binary system with a black hole came shortly after the launch of the Uhuru x-ray–detecting satellite in 1971. Astronomers became intrigued with an x-ray

2. As gases spiral toward a black hole, they are heated by friction and magnetic fields: Just outside the black hole, they are hot enough to emit x rays.

HDE 226868
(blue-white supergiant)

X rays

Black hole

1. Gases from the supergiant are captured into an accretion disk around the black hole.

FIGURE 21-20

X rays Generated by Accretion of Matter Near a Black Hole
Stellar-remnant black holes, such as Cygnus X-1, V404 Cygni, and probably A0620-00, are detected in close binary star systems. This drawing (of the Cygnus X-1 system) shows how gas from the 30-$M_\odot$ companion star, HDE 226868, transfers to the black hole, which has about 14.8 $M_\odot$. This process creates an accretion disk. As the gas spirals inward, friction and compression heat it so much that the gas emits x rays, which astronomers can detect. (NASA, ESA, Martin Kornmesser [ESA/Hubble])

source designated Cygnus X-1 (**Figure 21-20**). This source is quite unlike pulsating x-ray sources, which emit regular bursts of x rays every few seconds (see Section 21-10). Instead, the x-ray emissions from Cygnus X-1 are highly variable and irregular; they flicker on timescales as short as one-hundredth of a second.

Is Cygnus X-1 big, like a regular star, or very compact, like a stellar remnant core? Amazingly, this can be answered very well without directly seeing the physical extent of Cygnus X-1. One of the fundamental concepts in physics is that nothing can travel faster than the speed of light (recall Section 21-2). Because of this limitation, different parts of an object cannot flicker in unison faster than the time required for light to travel across the object. Because light travels 3000 km in one-hundredth of a second, Cygnus X-1 can be no more than 3000 km across, or about one-quarter the size of Earth. This places a fundamental upper limit on the size of Cygnus X-1 and clearly rules it out as a star.

Cygnus X-1 occasionally emits bursts of radio emission, and in 1971 radio astronomers used these bursts to show that the source was nearly at the same location in the sky as the star HDE 226868. Spectroscopic observations revealed that HDE 226868 is a B0 supergiant with a surface temperature of about 31,000 K. Such stars do not emit significant amounts of x rays, so HDE 226868 alone cannot be the x-ray source for Cygnus X-1. Because binary stars are very common, astronomers began to suspect that the visible star and the x-ray source are in orbit about each other.

Further spectroscopic observations soon showed that the lines in the spectrum of HDE 226868 shift back and forth with a period of 5.6 days; this is the Doppler shift, as illustrated in Figure 17-23. The clear implication is that HDE 226868 and Cygnus X-1 are the two members of a binary star system. From what we know about the masses of other supergiant stars, HDE 226868 is estimated to have a mass of roughly 30 $M_\odot$.

The rest of the puzzle involves estimating the mass of Cygnus X-1 based on the properties of the binary system. For example, the *magnitude* of the measured Doppler shifts—how much the spectral lines of the ordinary star shift back and forth—is influenced by the mass of Cygnus X-1. After decades of careful observation, the mass of Cygnus X-1 is estimated to be 14.8 $M_\odot$. Keep in mind that Cygnus X-1 is very compact—smaller than Earth. With a maximum possible white dwarf mass of 1.4 $M_\odot$ and a maximum neutron star mass of 2–3 $M_\odot$, it is very likely that Cygnus X-1 is a 14.8 $M_\odot$ black hole.

If Cygnus X-1 does contain a black hole, the x rays do not come from the black hole itself. Gas captured from HDE 226868 goes into orbit about the hole, forming an accretion disk observed to be about 4 million km in diameter (see Figure 21-20). As material in the disk gradually spirals in toward the hole, friction heats the gas to temperatures approaching 2×10^6 K. In the final 200 km above the hole, these extremely hot gases emit the x rays that our satellites detect. Presumably, the x-ray flickering is caused by small hot spots on the rapidly rotating inner edge of the accretion disk. In this way, the black hole's existence is revealed by hot gases just before they plunge to oblivion.

But is there really a black hole at the center of Cygnus X-1? With Cygnus X-1 so far away (about 6100 ly), astronomers do not have the capability to detect the emission signatures unique to black holes expected near the event horizon. Furthermore, the size constraints mentioned above for Cygnus X-1 do not place its mass of 14.8 $M_\odot$ within its event horizon. Nonetheless, theoretical calculations provide two compelling results. First, there is no known type of object or state of matter that could contain 14.8 $M_\odot$ within the observed upper size limit of Cygnus X-1. Second, even if the observed 14.8 $M_\odot$ were due to some unknown object (or collection of objects), it would very quickly collapse into a black hole anyway. More than 40 years after its discovery, astronomers think the case for a black hole at the center of Cygnus X-1 is very strong.

CONCEPT CHECK 21-10

There are stars that are 14.8 $M_\odot$, so why is Cygnus X-1 strongly considered to be a black hole?

Answer appears at the end of the chapter.

Other Black Hole Candidates

Astronomers have found more than 20 other black hole candidates like Cygnus X-1. All are compact x-ray sources orbiting ordinary stars in a spectroscopic binary system. One of the best candidates is V404 Cygni in the constellation Cygnus. Doppler shift measurements reveal that as the visible star moves around its 6.47-day orbit, its radial velocity (see Section 17-10) varies by more than 400 km/s. These data indicate a black hole mass of 12 $M_\odot$ for the unseen companion of the visible star. It is probably impossible for a neutron star to be more massive than 3 $M_\odot$, so V404 Cygni must almost certainly be a black hole.

Another particularly convincing case is the flickering x-ray source A0620-00; at only 3000 ly away, it is the black hole candidate closest to Earth. (The A refers to the British satellite Ariel 5, which discovered the source; the numbers denote its position in the sky.) The visible companion to A0620-00 is an orange K5 main-sequence star called V616 Monocerotis, which orbits the x-ray source every 7.75 hours. Because this star is relatively faint, it is possible to observe the shifting spectral lines from *both* the visible companion and the x-ray source as they orbit each other. With this more complete information about the orbits, astronomers have determined the black hole mass of A0620-00 to be about 11 $M_\odot$.

Astronomers have seen jets of hot, glowing material extending several light-years from some black hole candidates. The ejected material emerges from the vicinity of the black hole at speeds approaching the speed of light. It is thought that these jets are formed by strong electric and magnetic fields in the material around a *rotating* black hole, much as occurs for rotating protostars (see Figure 18-17) and for rotating neutron stars (see Figure 21-5). **Figure 21-21** is an artist's impression of the immediate vicinity of such a rotating black hole. (We will discuss the curious features of rotating black holes in Section 21-9.)

If there really are black holes in close binary systems, how did they get there? One possibility already mentioned is that one member of the binary explodes as a Type II supernova, leaving a burned-out core whose mass exceeds 3 $M_\odot$. This core undergoes gravitational collapse to form a black hole.

A white dwarf or a neutron star in a binary system could also become a black hole if it accretes enough matter from its companion star. This transformation can occur when the companion star becomes a red giant and dumps a significant part of its mass over its Roche lobe (see Figure 19-22b).

FIGURE 21-21

The Environment of an Accreting Black Hole If a black hole is rotating, it can generate strong electric and magnetic fields in its immediate vicinity. These fields draw material from the accretion disk around the black hole and accelerate it into oppositely directed jets along the black hole's rotation axis. This illustration also shows other features of the material surrounding such a black hole. (NASA/Chandra X-ray Observatory/M. Weiss)

(labels in figure:)
This black hole is surrounded by a disk of hot gas.

Torus (doughnut) of cooler gas and dust (shown cut away).

X rays from the hot disk excite iron atoms in the torus, making them glow.

Fast-moving jets of subatomic particles are formed and ejected by electric and magnetic fields.

Finally, another possibility is two dead stars coalescing to form a black hole. For example, imagine two neutron stars orbiting each other, like the binary system discussed in Section 21-1. Because the two stars emit gravitational radiation, they gradually spiral in toward each other and can eventually merge. If their total mass exceeds 2 to 3 $M_\odot$, the entire system may become a black hole.

21-6 The most intense radiation bursts in the universe may be caused by the formation of black holes

As we have seen, bursts of x rays from binary systems such as Cygnus X-1 result from black holes and show that certain stars can evolve into black holes. Even more remarkable is that astronomers may actually be able to observe the moment that a black hole forms from a dying star. The telltale signals are called **gamma-ray bursts**. These are from mysterious objects that emit the most powerful bursts of high-energy photons ever measured. To understand the connection between these objects and black holes, we begin by looking at how gamma-ray bursts were discovered and how astronomers struggled to determine their true nature.

> Gamma-ray bursts are incredibly bright flashes of radiation from remarkably distant galaxies.

A Gamma-Ray Mystery

Gamma-ray bursts were discovered in the late 1960s by the orbiting Vela satellites, whose detectors noticed flashes of gamma rays coming from random parts of the sky at random intervals. This discovery was an unexpected consequence of the Cold War. The Vela satellites were originally placed in orbit by the United States to look for high-energy photons coming from aboveground tests of nuclear weapons by the Soviet Union, tests that had been banned by treaty since 1963. (No such tests were ever detected.) With a new generation of gamma-ray telescopes currently in orbit (see Section 6-7), new gamma-ray bursts are being found at a rate of about one per day.

Gamma-ray bursts fall into two types. *Long* gamma-ray bursts, which are more common, last from about 2 to about 1000 seconds before fading to invisibility. The less common *short* gamma-ray bursts last from a few hundredths of a second to about 2 seconds, and tend to emit photons of shorter wavelength and hence higher energy. Unlike x-ray bursts (see Section 21-1), gamma-ray bursts of both types appear to emit only one burst in their entire history.

What are gamma-ray bursts, and how far away are they? These questions plagued astronomers for almost 30 years. One important clue is that gamma-ray bursts are seen with roughly equal probability in all parts of the sky, as **Figure 21-22a** shows. This suggests that they are not in the disk of our galaxy, because then most gamma-ray bursts would be found in the plane of the Milky Way (Figure 21-22b). One idea to explain their uniformity on the sky was that gamma-ray bursts are relatively close to us and lie in a spherical halo surrounding the Milky Way. Alternatively, gamma-ray bursts could be strewn throughout space like galaxies, with some of them billions of light-years away.

The source of the short gamma-ray bursts was finally solved in 2017 by LIGO's detection of gravitational waves from merging

Gamma-ray bursts are found in all parts of the sky . . .

. . . and are *not* concentrated in the plane of the Milky Way Galaxy.

+90°

+180°   −180°

−90°

(a) Map of the entire sky showing the positions of gamma-ray bursts

R I V U X **G**

+90°

+180°   −180°

−90°

(b) Map of the entire sky at visible wavelengths

R I **V** U X G

FIGURE 21-22

Gamma-Ray Bursts **(a)** This map shows the locations of 2704 gamma-ray bursts detected by the Compton Gamma Ray Observatory (see Section 6-7). The entire celestial sphere is mapped onto an oval. The colors in the order of the rainbow indicate the total amount of energy detected from each burst; bright bursts appear in red and weak bursts appear in violet. **(b)** This map shows the same sky as in part (a) but at visible wavelengths. Comparing part (b) with part (a) shows that unlike x-ray bursts, which originate in the disk of the Milky Way Galaxy, gamma-ray bursts are seen in all parts of the sky. (a, b: NASA)

neutron stars (see Section 21-1). Following the LIGO detection, many telescopes looked for light coming from the event, including gamma rays. For the first time, astronomers observed a short gamma-ray burst from an astrophysical system they could piece together. Thus, it appears that these bursts arise from high-energy jets slamming into debris, giving off a short burst of gamma rays. The jets are produced by the black hole that is formed after the two neutron stars merge together.

Long Gamma-Ray Bursts and Supernovae

For many years, there was no convincing way to decide between the two competing models, which placed the long gamma-ray bursts either quite close to or very far away from Earth. This state of affairs changed dramatically in 1997, thanks to the Italian-Dutch BeppoSAX satellite. Unlike previous gamma-ray satellites, which could determine the position of a burst only to within a few degrees, BeppoSAX could pin down its position to within 1 arcmin. This made it far easier for astronomers using optical telescopes to search for a counterpart to the burst. Just 21 hours after BeppoSAX observed a long gamma-ray burst on February 28, 1997, astronomers located a visible-light afterglow of the burst. Since then, astronomers using large visible-light and infrared telescopes have detected the afterglow of many gamma-ray bursts and found that they occur within galaxies (Figure 21-23). As of this writing, the most distant gamma-ray burst yet seen lies within a galaxy some 13 billion (1.3×10^{10}) ly away. (In Chapter 23, we will see how astronomers determine the distances to remote galaxies.) To be seen at such immense distances, a gamma-ray burst must release a truly stupendous amount of energy in the form of radiation.

The relationship between gamma-ray bursts and galaxies strongly suggests that a burst is a cataclysmic event involving one of a galaxy's stars. Observations of a relatively bright (and, hence, relatively close) long gamma-ray burst have confirmed this idea. On March 29, 2003, the orbiting gamma-ray telescope HETE-2 (short for High Energy Transient Explorer) detected this burst—denoted

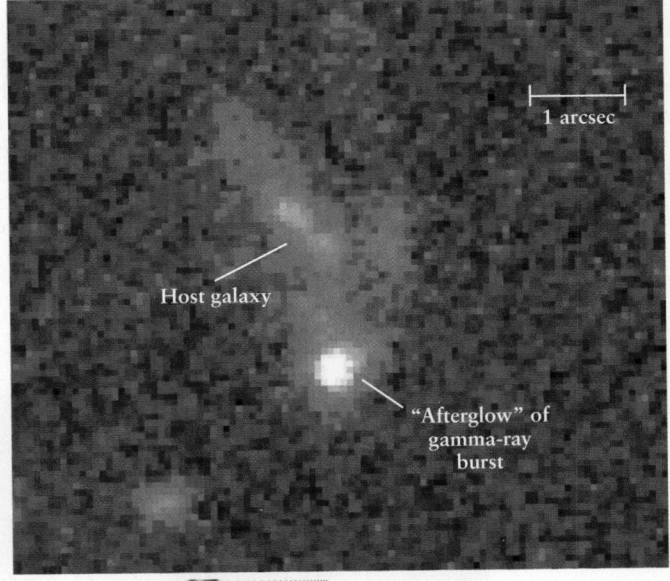

1 arcsec

Host galaxy

"Afterglow" of gamma-ray burst

FIGURE 21-23 R I **V** U X G

The Host Galaxy of a Gamma-Ray Burst This false-color image was captured on February 8, 1999, 16 days after a gamma-ray burst was observed at this location. The host galaxy of the gamma-ray burst has a very blue color (not shown in this image), indicating the presence of many recently formed stars. The gamma-ray burst may have been produced when one of the most massive of these stars became a supernova. (HST GRB Collaboration, STIS, HST, NASA)

GRB 030329 for the date it was observed—in the constellation Leo. (GRB 030329 was so intense that its high-energy photons temporarily ionized part of Earth's atmosphere—the only event of its kind ever caused by an object beyond our own galaxy.) Within 90 minutes, a visible-light afterglow of GRB 030329 was found by astronomers in Australia and Japan. It was bright enough that astronomers in the United States and Chile were able to measure its detailed spectrum, which turned out to be that of a Type Ic supernova at a distance of some 2.65 billion (2.65×10^9) ly. We saw in Section 20-9 that a Type Ic supernova results when a massive star undergoes core collapse after having first shed the hydrogen and helium from its outer layers (see Figure 20-19c). Subsequent observations with the Hubble Space Telescope revealed an expanding shell of gas at the position of the gamma-ray burst. This shell is just what we would expect from a supernova explosion. Another long gamma-ray burst with an associated Type Ic supernova was seen in 2006.

Beamed Radiation

A long gamma-ray burst cannot simply be an ordinary Type Ic core-collapse supernova. Such supernovae release about 10^{46} joules of energy, of which only about 0.03% is released as electromagnetic radiation. (Most of the remaining energy goes into neutrinos, while some goes into accelerating the debris that expands away from the supernova.) This radiation streams outward from the supernova in all directions equally, like light from the Sun. If we assume that a gamma-ray burst also emits light equally in all directions, the inverse-square law that relates brightness, luminosity, and distance (see Section 17-2) tells us that the most energetic bursts would have to emit about 3×10^{47} joules of radiation in less than a minute. It would take 100,000 ordinary supernovae going off simultaneously to release this much radiation!

This dilemma can be resolved if we imagine a type of supernova that emits most of its radiation in narrow beams. If such a beam happened to be aimed toward Earth, we would detect a far more intense burst of radiation than we would from an ordinary supernova.

Collapsars and the Birth of a Black Hole

One theoretical model that would produce beamed radiation invokes a special type of supernova called a **collapsar** (also called a *hypernova*). These objects are thought to result from progenitor stars that are very massive (more than about 30 $M_\odot$), have lost their outer layers of hydrogen and helium, and are rotating rapidly. The core of such a star is too massive to be a white dwarf or a neutron star, so when thermonuclear reactions cease, the core will become a black hole. **Figure 21-24** depicts what happens when thermonuclear reactions cease in the core of such a star. The black hole forms before the material outside the core has a chance to contract very much (Figure 21-24a). As a result, the black hole quickly forms an accretion disk from the surrounding stellar material, as shown in Figure 21-24b. This process is aided because the progenitor star was rotating rapidly, which makes it easier for the infalling material to form a rotating disk around the black hole. Some of the infalling material does not fall into the black hole, but is ejected in powerful, back-to-back jets along the rotation axis of

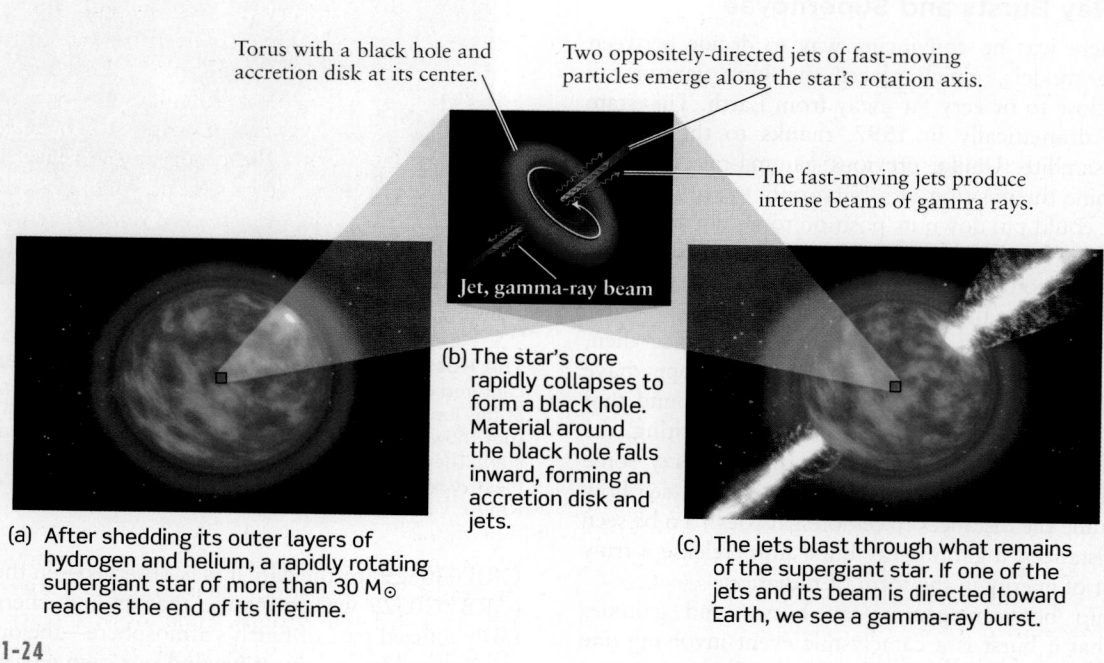

Torus with a black hole and accretion disk at its center.

Two oppositely-directed jets of fast-moving particles emerge along the star's rotation axis.

The fast-moving jets produce intense beams of gamma rays.

Jet, gamma-ray beam

(a) After shedding its outer layers of hydrogen and helium, a rapidly rotating supergiant star of more than 30 $M_\odot$ reaches the end of its lifetime.

(b) The star's core rapidly collapses to form a black hole. Material around the black hole falls inward, forming an accretion disk and jets.

(c) The jets blast through what remains of the supergiant star. If one of the jets and its beam is directed toward Earth, we see a gamma-ray burst.

ANIMATION 21.2

FIGURE 21-24

The Collapsar Model of a Long Gamma-Ray Burst
These illustrations show the final few seconds in the life of a massive, rapidly rotating supergiant star that has lost its outer layers of hydrogen and helium. After the jets have subsided, what remains is a Type Ic supernova remnant with a black hole at its center. (a, c: NASA/SkyWorks Digital; b: NASA and A. Feild [STScI])

the accretion disk. (Much the same mechanism acts in close binary star systems in which one member is a black hole, as we described in Section 21-5.)

The jets are so energetic that they reach and break through the surface of the star within 5 to 10 s after being formed, as Figure 21-24c shows. (It helps that the star has already lost much of its outer layers, which makes it easier for the jets to break through what remains of the star.) As they travel outward, the jets produce powerful shock waves that blow the star to pieces.

If one of the jets from a collapsar happens to be aimed toward Earth, we see an intense burst of gamma rays. These are produced as the relativistic particles in the jet slow down and convert their kinetic energy (energy of motion) into radiation. The burst is short-lived because the jets have only a brief existence: It takes the black hole only about 20 s to accrete the entire inner core of the star, after which there is no longer an accretion disk to produce the jets. We do not see a burst if the jets are aimed away from Earth. Hence, for each gamma-ray burst that we observe, there may be hundreds or thousands of collapsars that go undetected.

The newest tool in the search for answers about gamma-ray bursts is a NASA satellite called Swift, which was launched in 2004 and has observed over 800 gamma-ray bursts. In addition to sensitive gamma-ray detectors to identify bursts, Swift carries telescopes that are sensitive to x rays, ultraviolet light, and visible light. Thus, instead of having to wait for ground-based telescopes to make follow-up observations of gamma-ray bursts, Swift makes these observations itself as swiftly as possible (hence the space-craft's name). Observations from Swift may help us better understand the nature of long bursts, as well as provide important clues about the still enigmatic short bursts.

CONCEPT CHECK 21-11

Why do supernova models associated with gamma-ray bursts require jets?

Answer appears at the end of the chapter.

21-7 Supermassive black holes exist at the centers of most galaxies

Stellar evolution makes it possible to form black holes with masses several times that of the Sun. But calculations suggest other ways to form black holes that are either extremely large or extraordinarily small.

Supermassive Black Holes

As we will discuss in Chapter 25, galaxies formed in the early universe by the coales-cence of gas clouds. During this formation process, some of a galaxy's gas might plunge straight toward the center of the galaxy, where it collects and compresses together under its own gravity. If this central clump has enough mass, it can form a black hole. Typical galaxies have masses of 10^{11} $M_\odot$, so even a tiny percentage of the galaxy's gas collected at its center would give rise to a **supermassive black hole** with a truly stupendous mass.

> The largest black holes contain billions of times the mass of the Sun.

CAUTION!

While stellar-mass-sized black holes require incomprehensibly high-density material to form, supermassive black holes require much lower-density material to form. For example, if you *imagine filling our solar system with water* out to the orbital distance of Earth, the total mass of water in this sphere would be about 500 million $M_\odot$, and a supermassive black hole would form! In other words, a supermassive black hole can be formed with an average density equal to that of water. Because larger black holes require lower average densities to form compared to smaller black holes, supermassive black holes might form through *less extreme* processes than stellar-mass-sized black holes.

Since the 1970s, astronomers have found evidence suggesting the existence of supermassive black holes in other galaxies. With the advent of a new generation of optical telescopes in the 1990s, it became possible to see the environments of these suspected black holes with unprecedented detail (**Figure 21-25**). Figure 21-25a shows radio emissions from two jets that extend some 15,000 pc (50,000 ly) from the center of the galaxy NGC 4261. These resemble an immensely magnified version of the jets seen emerging from some stellar-mass black holes (see Section 21-5). The highly magnified Hubble Space Telescope image in Figure 21-25b shows a disk some 250 pc (800 ly) in diameter at the very center of this galaxy. The jets are perpendicular to the plane of the disk, just as in Figure 21-21.

By measuring the Doppler shifts of light coming from the two sides of the disk, astronomers found that the disk material was orbiting around the bright object at the center of the disk at speeds of hundreds of kilometers per second. (By comparison, Earth orbits the Sun at 30 km/s.) Given the size of the material's orbit, and using Newton's form of Kepler's third law (see Section 4-7 and Box 4-4), they were able to calculate the mass of the central bright object. The answer is an amazing 1.2 *billion* (1.2×10^9) solar masses! What's more, the observations show that this object can be no larger than our solar system. The only known explanation is that the object at the center of NGC 4261 is a black hole.

Dozens of other black holes in the centers of galaxies have been identified by their gravitational effect on surrounding gas and dust. Surveys of galaxies have shown that supermassive black holes are not at all unusual; most large galaxies appear to have them at their centers. As we will see in the next chapter, a black hole with several million solar masses lies at the center of our own Milky Way Galaxy, some 8000 pc (26,000 ly) from Earth.

Other Black Holes

Theoretically, a black hole can have any mass, but the masses of black holes that exist reflect what can be produced through natural processes. The black hole masses we do observe seem to fall into two very different groups. Supermassive black holes have masses from 10^6 to 10^9 $M_\odot$, while stellar-mass black holes like those discussed in Section 21-5 have masses of about 10 $M_\odot$. In 2000, a team of astronomers from Japan, the United States, and Great Britain reported finding a "missing link" between these two groups.

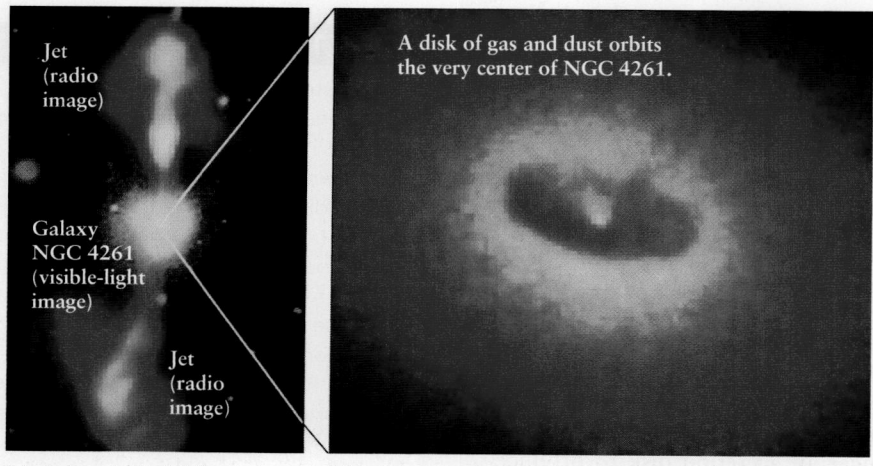

Jet
(radio
image)

Galaxy
NGC 4261
(visible-light
image)

Jet
(radio
image)

A disk of gas and dust orbits
the very center of NGC 4261.

 (a) Galaxy NGC 4261   (b) Evidence for a supermassive black hole in NGC 4261

ANIMATION
21.3

FIGURE 21-25

A Supermassive Black Hole The galaxy NGC 4261 lies some
30 million pc (100 million ly) from Earth in the constellation Virgo.
(a) This composite view superimposes a visible-light image of the galaxy
(white) with a radio image of the galaxy's immense jets (orange).

(b) This Hubble Space Telescope image shows a disk of gas and dust about
250 pc (800 ly) across at the center of NGC 4261. Observations indicate that a
supermassive black hole is at the center of the disk. (a, b: NASA/HST/WFPC2)

The new discovery was a fluctuating x-ray source located some
200 pc (650 ly) from the center of the galaxy M82 (**Figure 21-26**).
Images do not reveal material orbiting around this source, so
we cannot determine the source's mass using Newton's form of

Kepler's third law (see Box 4-4). But by relating the source's x-ray
luminosity to the rate at which matter would have to fall onto the
source to produce that luminosity, it can be shown that the source
must have a mass of at least 500 $M_\odot$ or so. The rapid fluctuations

Center of galaxy

X-ray
source

VIDEO
21.1

FIGURE 21-26

An Intermediate-Mass Black Hole? The largest black holes
have billions of times the mass of the Sun. M82 is an unusual
galaxy in the constellation Ursa Major. The inset is an image of the central
region of this galaxy from the Chandra X-ray Observatory. The bright,
compact x-ray source varies in its light output over a period of months.
The properties of this source suggest that it may be a black hole of roughly

500 $M_\odot$ or more. If correct, the intermediate-mass black hole is about 600 ly
away from a 30-million-$M_\odot$ supermassive black hole at the center of the
galaxy. (inset: NASA/CXC/Tsinghua Univ./H. Feng et al.; main: X-ray: NASA/CXC/
JHU/D. Strickland; optical: NASA/ESA/STScI/AURA/The Hubble Heritage Team;
infrared: NASA/JPL-Caltech/Univ. of AZ/C. Engelbracht)

of the source show that it has a very small diameter, which suggests that it is a black hole (see Section 21-5). Hence, the source in M82 may be an **intermediate-mass black hole** (also called a **mid-mass black hole**). There is also some evidence for a 2300 $M_\odot$ black hole in the globular cluster 47 Tuc, based on its gravitational influence on nearby pulsars. Additional intermediate-mass black hole candidates have been found as well. Such black holes might have formed by the coalescence of many normal stars or by the direct merger of stellar-mass black holes.

During the early 1970s, Stephen Hawking of Cambridge University proposed the existence of yet another type of black hole. He suggested that during the Big Bang some 13.7 billion years ago, local pockets of the universe could have been dense enough and under sufficient pressure to form small black holes. According to his theory, these **primordial black holes** could have had masses as large as Earth or as small as 5×10^{-8} kg, about 1/40 the mass of a single raindrop. To date, however, no evidence has been found for the existence of primordial black holes in nature.

CONCEPT CHECK 21-12

Do all black holes require extreme densities of matter to form?

CONCEPT CHECK 21-13

Can black holes only exist within certain mass ranges?
Answers appear at the end of the chapter.

21-8 A nonrotating black hole has only a "center" and a "surface"

Understanding in detail how black holes form is a challenging task. But once a black hole forms, it has a remarkably simple structure.

Recall from Section 21-3 that a black hole is an object with an escape speed greater than the speed of light. Since nothing can travel faster than the speed of light, nothing can escape a black hole. Even light can't escape, which is how black holes got their name. A black hole has no surface, but rather a spherical boundary called the *event horizon*, where the escape speed just equals the speed of light. Once the core of a massive dying star collapses to within its event horizon, it disappears permanently from view.

CAUTION!

A black hole does *not* suck in everything from its surroundings. While anything that crosses the event horizon is lost forever, material outside the event horizon traveling at high speeds can pass by without getting trapped. Far away from a black hole, the black hole is like any other object. As mentioned earlier, if the Sun became a black hole, the planets are all too far away for any noticeable orbital effects.

If the dying star is not rotating before it collapses, the black hole will likewise not be rotating. We will consider these nonrotating black holes first, and then consider rotating black holes

in the next section. The distance from the center of a nonrotating black hole to its event horizon is called the **Schwarzschild radius** (denoted R_{Sch}), after the German physicist Karl Schwarzschild who first determined its properties. Box 21-3 describes how to calculate the Schwarzschild radius, which depends only on the black hole's mass. The more massive the black hole, the larger its event horizon.

Inside a Black Hole

Once an object (such as a stellar core) has contracted enough to form an event horizon, nothing can prevent the matter's further collapse. Equations describing the matter within the event horizon indicate that the object's entire mass is crushed nearly to a single point, known as the **singularity**, at the center of the black hole. We now can see that the structure of a nonrotating black hole is

> All of the mass of a nonrotating black hole is concentrated nearly to a point at the center called the singularity.

quite simple. As drawn in **Figure 21-27**, it has only two parts: a singularity at the center surrounded by a spherical event horizon. Note that while a black hole's effects on space and time away from the singularity are fairly well understood, the singularity itself—with its nearly infinite density—poses both mathematical and conceptual problems in physics.

As seen in Box 21-3, the size—called the Schwarzschild radius—of a black hole is determined by its mass. For larger mass black holes, this size corresponds to lower average densities than for smaller mass black holes. As noted in Section 21-7, a supermassive black hole could be formed by a solar system filled with water. However, once a black hole forms, its mass rapidly collapses toward the center. So, while less extreme conditions are

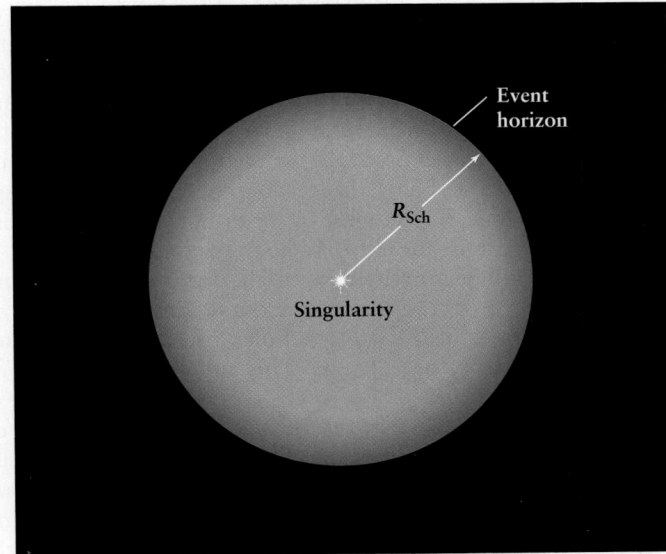

FIGURE 21-27

The Structure of a Nonrotating (Schwarzschild) Black Hole A nonrotating black hole has only two parts: a singularity, where all of the mass is located, and a surrounding event horizon. The distance from the singularity to the event horizon is called the Schwarzschild radius (R_{Sch}). The event horizon is a one-way surface: Things can fall in, but nothing can get out.

BOX 21-3   TOOLS OF THE ASTRONOMER'S TRADE

The Schwarzschild Radius

The Schwarzschild radius (R_{Sch}) is the distance from the center of a nonrotating black hole to its event horizon. You can think of this radius as the "size" of the black hole. For an object of a given mass, the Schwarzschild radius is the size that mass would need to be concentrated within, in order to make the escape speed from that object equal to the speed of light. The Schwarzschild radius is related to the mass M of the black hole by:

Schwarzschild radius

$$R_{Sch} = \frac{2GM}{c^2}$$

R_{Sch} = Schwarzschild radius of a black hole
  G = universal constant of gravitation
  M = mass of black hole
  c = speed of light

When using this equation, be careful to express the mass in kilograms, not solar masses, because of the units in which G and c are commonly expressed ($G = 6.67 \times 10^{-11}$ Nm²/kg² and $c = 3.00 \times 10^8$ m/s). If you use M in kilograms, the answer that you get for R_{Sch} will be in meters.

EXAMPLE: Find the Schwarzschild radius (in kilometers) of a black hole with 10 times the mass of the Sun.

Situation: We are given the mass of the black hole (10 M$_\odot$) and wish to determine its Schwarzschild radius R_{Sch}.

Tools: We use the preceding equation for R_{Sch}, being careful to first convert the mass from solar masses to kilograms.

Answer: One solar mass is 1.99×10^{30} kg, so in this case $M = 10 \times 1.99 \times 10^{30}$ kg $= 1.99 \times 10^{31}$ kg. The Schwarzschild radius of this black hole is then

$$R_{Sch} = \frac{2GM}{c^2} = \frac{2(6.67 \times 10^{-11})(1.99 \times 10^{31})}{(3.00 \times 10^8)^2}$$

$$= 3.00 \times 10^4 \text{ m}$$

One kilometer is 10^3 m, so the Schwarzschild radius of this 10-M$_\odot$ black hole is

$$R_{Sch} = (3.0 \times 10^4 \text{ m}) \times \frac{1 \text{ km}}{10^3 \text{ m}} = 30 \text{ km, or 19 mi}$$

Review: The Schwarzschild radius is directly proportional to the mass of the black hole. Thus, a black hole with 10 times the mass (100 M$_\odot$) would have a Schwarzschild radius 10 times larger ($R_{Sch} = 10 \times 30$ km = 300 km); a black hole with half the mass (5 M$_\odot$) would have half as large a Schwarzschild radius ($R_{Sch} = 1/2 \times 30$ km = 15 km), and so on.

---

required to *form* supermassive black holes, the conditions inside all black holes are extreme.

To understand why the complete collapse of such a doomed star is inevitable, first think about your own life on Earth, far from any black holes. You have the freedom to move as you wish through the three dimensions of space: up and down, left and right, or forward and back. But you do not have the freedom to move at will through the dimension of time. Whether we like it or not, we are all carried inexorably through time from the cradle to the grave.

Inside a black hole, gravity distorts the structure of spacetime so severely that the directions of space and time become interchanged. In a sense, inside a black hole, you acquire a limited ability to affect the passage of time. This apparent gain does you no good, however, because you lose a corresponding amount of freedom to move through space. Whether you like it or not, you will be dragged inexorably from the event horizon toward the singularity. Just as no force in the universe can prevent the forward march of time from past to future outside a black hole, no force in the universe can prevent the inward march of matter from the event horizon to the singularity inside a black hole. Once an object dropped into a black hole crosses the event horizon, it is gone forever, like an object dropped into a bottomless pit.

The same is true for a light beam aimed at the black hole. Because all this light will be absorbed and none reflected back, a black hole is indeed black. By contrast, even the blackest object on Earth reflects *some* light when you shine a flashlight on it—a black hole would

reflect none. (In Section 21-11, we will see that black holes are expected to emit some particles and light through a quantum physics process, but this radiation arises just *outside* the event horizon and would be negligible for black holes of stellar-mass size or larger.)

Although black holes are really very intriguing objects, there are some common misconceptions about their nature. The *Cosmic Connections* figure exposes two of these misconceptions.

CONCEPT CHECK 21-14

What is the connection between a black hole's Schwarzschild radius and its event horizon?

CONCEPT CHECK 21-15

Consult the *Cosmic Connections: Urban Legend #1* figure. Consider being 10 million km from the Sun, and then the same distance from a black hole of 1 M$_\odot$. From which object do you feel a greater gravitational force?

CONCEPT CHECK 21-16

Consult the *Cosmic Connections: Urban Legend #2* figure. If you had a personal rocket spacecraft that was low on gas, would it be safer to fly by the event horizon of a small black hole or a large black hole?

Answers appear at the end of the chapter.

COSMIC CONNECTIONS

Black Hole "Urban Legends"

ANIMATION
21.4 Black holes are such intriguing objects that a folklore has developed about them. As often is the case with folklore, myth and reality can become confused. These illustrations depict two misconceptions or "urban legends" about black holes.

Urban Legend #1:
When a star becomes a black hole, its gravitational pull becomes stronger.

Reality: At great distances from a black hole, the gravitational force that it exerts is exactly the *same* as that exerted by an ordinary star of the same mass. The truly stupendous gravitational effects of a black hole appear only if you venture close to the black hole's event horizon.

— Consider gravitational forces on a 60-kg person (not shown to scale). The gravitational force of Earth on this person when standing on our planet's surface— that is, this person's weight on Earth—is 590 newtons (132 pounds).

— 10,000,000 km from the center of the Sun: Force of the Sun on the person is 80 newtons.

Sun

Radius = 696,000 km

Same distance, same force

— 10,000,000 km from the center of 1-$M_\odot$ black hole: Force of the black hole on the person is 80 newtons.

Black hole with the same mass as the Sun

Schwarzschild radius = 3 km

1000 km from the center of a 1-$M_\odot$ black hole: Force of the black hole on the person is 8,000,000,000 (8×10^9) newtons!

Urban Legend #2:
The larger a black hole, the more powerful the gravity near its Schwarzschild radius.

Reality: If you increase the mass of a black hole, its Schwarzschild radius increases by the same factor. Newton's law of universal gravitation, $F = Gm_1m_2/r^2$, tells us that the force that an object of mass m_1 exerts on a second object of mass m_2 is directly proportional to mass m_1 but *inversely* proportional to the *square* of the distance r between the two objects. Hence the force that a black hole exerts on an object a given distance outside its Schwarzschild radius actually *decreases* as the black hole's mass and Schwarzschild radius increase.

Black hole with the same mass as the Sun

Schwarzschild radius = 3 km

Black hole with 1,000,000,000 (10^9) solar masses: Schwarzschild radius = 3×10^9 = 20 au

1000 km outside the Schwarzschild radius of a 1-$M_\odot$ black hole: Force of the black hole on the person is 8,000,000,000 (8×10^9) newtons.

1000 km outside the Schwarzschild radius of a 10^9-$M_\odot$ black hole: Force of the black hole on the person is only 900,000 (9×10^5) newtons.

21-9 Just three numbers completely describe the structure of a black hole

Because light and matter cannot escape the event horizon, we are prevented from ever knowing much about what has already fallen into a black hole. In this sense, a black hole would appear to be an "information sink." Many properties of an object that falls into a black hole—such as the object's chemical composition, texture, color, shape, and size—would appear to vanish as soon as it crosses the event horizon.

Because this information has seemingly vanished, it cannot affect the basic structure or properties of the hole. For example, consider two hypothetical black holes, one made from the gravitational collapse of 10 $M_\odot$ of iron and the other made from the gravitational collapse of 10 $M_\odot$ of peanut butter. Obviously, quite different substances went into the creation of the two holes. However, once the event horizons of these two black holes have formed, both the iron and the peanut butter will have permanently disappeared from the universe. As seen from the outside, the two holes are identical, making it impossible for us to tell which ate the peanut butter and which ate the iron. A black hole is thus unaffected by the type of matter it consumes.

> When an object falls into a black hole, all information about that object disappears except its mass, electric charge, and angular momentum.

The Three Properties of a Black Hole

Because a black hole appears to be an information sink, it is reasonable to wonder whether we can determine anything at all about a black hole. What properties *do* characterize a black hole?

First, we can measure the *mass* of a black hole. One way to do this would be by placing a satellite into orbit around the hole. After measuring the size and period of the satellite's orbit, we could use Newton's form of Kepler's third law (recall Section 4-7 and Box 4-4) to determine the mass of the black hole. This mass is equal to the total mass of all the material that has gone into the hole.

Second, we can also measure the total *electric charge* possessed by a black hole. Like gravity, the electric force acts over long distances—it is a long-range interaction that is felt in the space around the hole. Appropriate equipment on a space probe passing near the hole could measure the strength of the electric force, and the electric charge could thus be determined.

In actuality, we would not expect a black hole to possess much, if any, electric charge. For example, if a hole did happen to start off with a sizable positive charge, it would vigorously attract vast numbers of negatively charged electrons from the interstellar medium, which would soon neutralize the hole's charge. For this reason, astronomers usually neglect electric charge when discussing real black holes.

Although a black hole might theoretically have a tiny electric charge, it can have no magnetic field of its own whatsoever. We discussed in Section 21-5 how magnetic fields are involved in producing jets from black holes. However, these fields are associated with the accretion disk around the black hole, not the black hole itself. When a black hole is created, however, the collapsing star from which it forms may possess an appreciable magnetic field. The star must therefore radiate this magnetic field away before it can settle down inside its event horizon. Theory predicts that the star does this by emitting electromagnetic and gravitational waves. As we saw in Section 21-3, gravitational waves are ripples in the overall geometry of space. While gravitational waves from merging black holes and neutron stars have been detected, gravitational waves from a collapsing star have not been seen yet.

Third, we can detect the effects of a black hole's rotation, that is, measure its *angular momentum*. An object's angular momentum is related to how fast it rotates and how the object's mass is distributed over its volume. As a dead star collapses into a black hole, its rotation naturally speeds up as its mass moves toward the center, just as a figure skater rotates faster when she pulls her arms and legs in. This same effect explains the rotation of the solar nebula (see Section 8-4), as well as why neutron stars spin so fast (see Section 21-1). We expect a black hole that forms from a rotating star to be spinning very rapidly.

Rotating Black Holes

When the matter that collapses to form a black hole is rotating, that matter does not compress to a point. Instead, it collapses into a ring-shaped singularity located between the center of the hole and the event horizon (Figure 21-28). The structure of such rotating black holes was first worked out in 1963 by the New Zealand mathematician Roy Kerr. Most rotating black holes should be spinning thousands of times per second, even faster than the most rapid pulsars.

If a rotating black hole is surrounded by an accretion disk with a magnetic field, it may be possible for the magnetic field to steal some of the rotational energy and angular momentum from the black hole and transfer it to the disk. The magnetic field acts as a "brake" that retards the black hole's rotation and makes the disk's rotation speed up. (On Earth, magnetic braking is used to slow locomotives, amusement park rides, and hybrid cars to a smooth stop.) Figure 21-29 shows an arching magnetic field that connects an accretion disk with a rotating black hole in just this way. This process may be taking place with the supermassive black hole at the center of the galaxy MCG–6-15-30. Observations with the XMM-Newton telescope (see Section 6-7) indicate that unusually intense x rays are coming from the accretion disk around this black hole. The suspicion is that the energy to power this radiation may be extracted from the black hole's rotation.

Even a rotating black hole without an accretion disk can transfer energy to other objects. This energy transfer is possible according to Einstein's general theory of relativity, which makes the startling prediction that a rotating body drags spacetime around it. (Very precise spacecraft measurements indicate that the rotating Earth drags spacetime in just this way.) Surrounding the event horizon of every rotating black hole is a region where this dragging of space and time is so severe that it is impossible to stay in the same place. No matter what you do, you get pulled around the hole, along with the rotating geometry of space and time. This

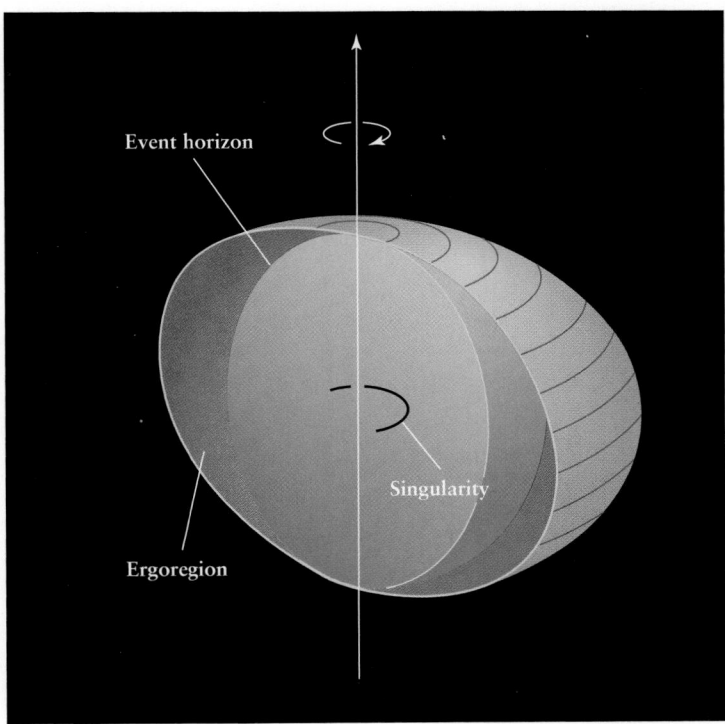

FIGURE 21-28

The Structure of a Rotating (Kerr) Black Hole The singularity of a rotating black hole is an infinitely thin ring centered on the geometrical center of the hole. (It appears as an arc in this cutaway diagram.) Outside the spherical event horizon is the doughnut-shaped ergoregion, where the dragging of spacetime around the hole is so severe that nothing can remain at rest. The tan-colored surface marks the outer boundary of the ergoregion.

region, where it is impossible to be at rest, is called the **ergoregion** (see Figure 21-28).

To measure a black hole's angular momentum, we could hypothetically place two satellites in orbit about the hole. Suppose that one satellite circles the hole in the same direction the hole rotates while the other circles in the opposite direction. One satellite is thus carried along by the geometry of space and time, but the other is constantly fighting its way "upstream." The two satellites will thus have different orbital periods. By comparing these two periods, astronomers can deduce the total angular momentum of the hole.

Because the ergoregion is outside the event horizon, this bizarre region is accessible to us, and spacecraft could travel through it without disappearing into the black hole. According to detailed calculations, objects grazing the ergoregion could be catapulted back into space at tremendous speeds. In other words, the ejected object could leave the ergoregion with more energy than it initially had, having extracted added energy from the hole's rotation. This process is called the *Penrose process,* after Roger Penrose, the British mathematician who proposed it.

Mass, charge, and angular momentum are the only three properties that a black hole possesses. This simplicity is the essence of the **no-hair theorem**, first formulated in the early 1970s: "Black holes have no hair." In other words, "hair" would be some property of a black hole other than mass, charge, or angular momentum, yet

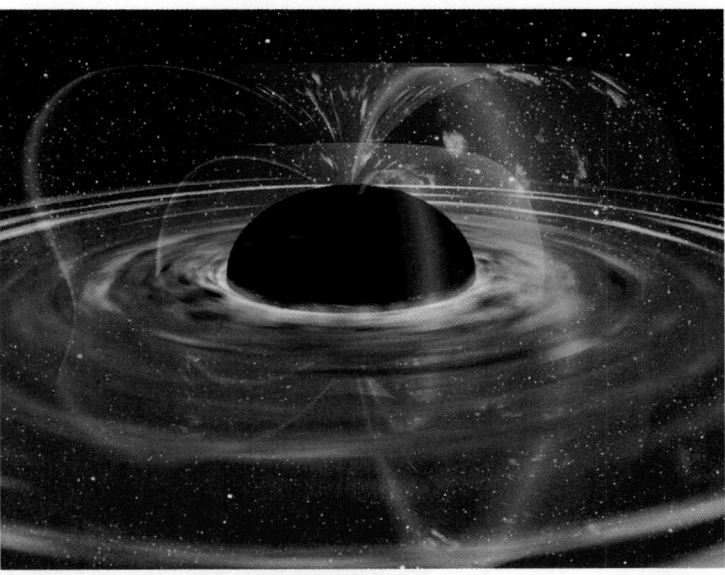

FIGURE 21-29

A Rotating Supermassive Black Hole This artist's impression shows the accretion disk around the supermassive black hole at the heart of the galaxy MCG−6-15-30. The arching magnetic field allows the accretion disk to extract energy and angular momentum from the black hole. While it might appear that the magnetic fields come from within the black hole, they only get close and are still *outside* of the black hole. (NASA/D. Berry)

another property is not possible. Therefore, once an object has fallen into a black hole, except for the object's mass, charge, and angular momentum, any additional properties carried by the object disappear from the universe. However, the idea that black holes can completely remove information from the universe is contentious, as some scientists do not think information is ever truly lost. We will revisit black hole information in Section 21-11, where subtle quantum effects might help resolve the issue.

CONCEPT CHECK 21-17

Consider two black holes. They have the same mass and angular momentum, and the total electric charge for each of them is zero. According to the no-hair theorem, could we tell if one of these black holes was made from iron and the other from peanut butter?

Answer appears at the end of the chapter.

21-10 Falling into a black hole is an infinite voyage

TUTORIAL 21.3 Imagine that you are on board a spaceship at a safe distance from a 5-$M_\odot$ black hole. A distance of 1000 Schwarzschild radii, or 15,000 km, would suffice. You now release a space probe with a video camera and let it fall into the black hole. What will you see on the video as the probe falls?

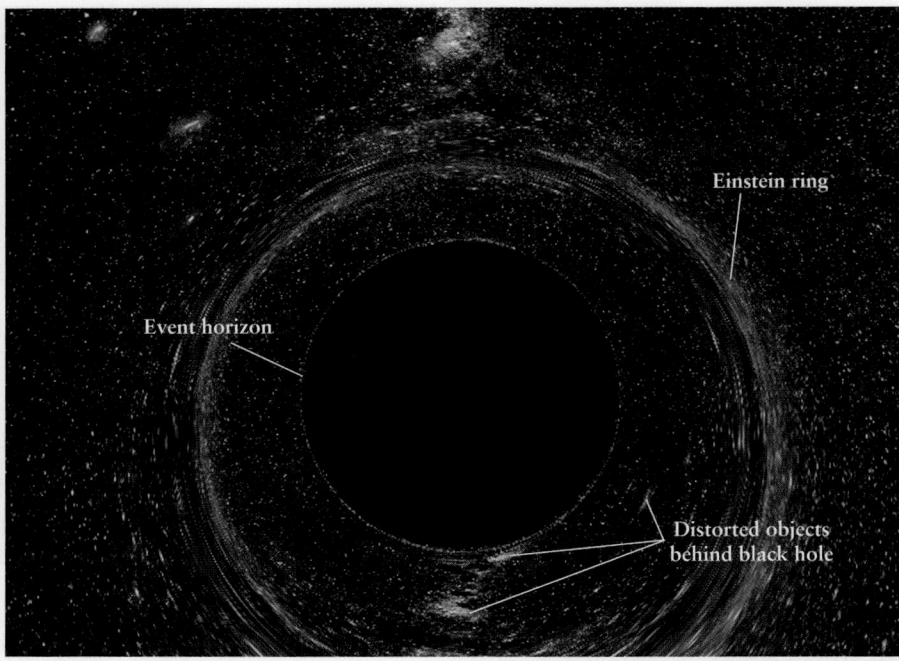

FIGURE 21-30

The View Approaching a Black Hole An image from a computer simulation of the appearance of the sky behind a black hole. For typical objects behind a black hole, lines of sight that are closer to the edge of the black hole result in greater light bending and distortion of the object's image.

In the rare case of a bright object right behind a black hole, a bright "Einstein ring" is formed. For illustration, this simulation places the extensive Milky Way behind the black hole, which leads to an unusually strong Einstein ring. (Andrew J. S. Hamilton/University of Colorado, Boulder)

Initially, at 1000 Schwarzschild radii from the black hole, the video camera would send back a rather normal view of space. But as the probe approaches the black hole, the bending of light by the black hole becomes more pronounced (Figure 21-30). Light rays passing close to the black hole are deflected so much that background stars are severely distorted. For the black hole in Figure 21-30, light from a bright object directly behind the black hole is warped into a feature called an "Einstein ring."

The gravitational field of a black hole distorts the image of other stars.

How will the probe itself look from your vantage point on the spaceship at a safe distance? To discuss changes in the probe, let's assume it emits blue light, and it has a clock that was initially synchronized with a clock on your spaceship (Figure 21-31). You might expect that, as the probe falls, its speed should continue to increase. This expectation is true up to a point. But as the probe approaches the event horizon, where the black hole's gravity is extremely strong, the *gravitational slowing of time* becomes so pronounced that the probe will appear to slow down! From your point of view on the spaceship, the probe takes an infinite amount of time to reach the event horizon, where it will appear to remain suspended for all eternity. You can see the slower passage of time by comparing the clocks on the spaceship to the clocks on the probe in Figure 21-31.

To watch these effects, however, you will need special equipment. The reason for this is the gravitational redshift: As the probe falls, the light reaching you from the probe is shifted to longer and longer wavelengths. The probe's blue light will turn green, then yellow, then red, and eventually fade into infrared wavelengths that your unaided eye cannot detect. Therefore, while the object appears to never cross the event horizon to an outside observer, any light emitted by the object still disappears from view.

If you could view the falling probe with goggles that tracked light of ever-increasing wavelengths, you would see it come to an unpleasant end. Near the event horizon, the strength of the black hole's gravity increases dramatically as the probe moves just a short distance closer to the hole. In fact, the side of the probe nearest the black hole feels a much stronger gravitational pull than the side farther from the hole. These *tidal forces* are like those that the Moon exerts on Earth (see Section 4-8), but they are tremendously strong. Furthermore, the sides of the probe are pulled together, since the hole's gravity makes them fall in straight lines aimed at the center of the hole. The net effect is that the probe will be stretched out along the line pointing toward the hole, and squeezed together along the perpendicular directions. The stretching—sometimes called *spaghettification*—is so great that it can rip even the strongest materials apart.

Someone foolhardy enough to ride along with the probe would have a *very different view* of the fall than your view from the spaceship. From the point of view while riding the probe, there is no slowing of time. The probe and rider experience spaghettification and continue to pass through the event horizon into the singularity.

Clock on a spaceship
far from the black hole

Black
hole

Event
horizon

Probe close to black hole

Probe far from
black hole

FIGURE 21-31

Effect of a Black Hole's Tidal Force on Infalling Matter A cube-shaped probe 1500 km from a 5-$M_\odot$ black hole. Near the Schwarzschild radius, the probe is pulled long and thin by the difference in the gravitational forces felt by its different sides. This tidal effect is a greatly magnified version of the Moon's gravitational force on Earth. The probe changes color as its photons undergo extreme gravitational redshift and time slows down on the probe, as seen from far away.

The size of the black hole matters: Tidal forces causing spaghettification are greater for smaller black holes. For stellar-mass black holes (like the 5-$M_\odot$ black hole considered here), tidal forces would squeeze and stretch a person to death even before they reach the event horizon. However, a person could fall right through the event horizon of a supermassive black hole, with tidal forces only pulling them apart closer to the center. Therefore, a person really could see what is inside a supermassive black hole—but at great cost, for he could never share what he learned.

CONCEPT CHECK 21-18

If a person fell into a black hole feet first, would she feel a strong tidal force pulling on her feet before crossing the event horizon?

Answer appears at the end of the chapter.

21-11 Hawking radiation and black hole evaporation

With all the mass of a black hole hidden behind its event horizon and collapsed into a singularity, it may seem that there is no way of getting mass from the black hole back into the universe. But in fact there is, as Stephen Hawking figured out in the 1970s. To understand how this is possible, we must look at the quantum mechanical behavior of matter at the microscopic scale of nuclei and electrons.

The **Heisenberg uncertainty principle** is a basic tenet of quantum physics. This principle states that you cannot determine precisely both the position and the speed of a subatomic particle. Over extremely short distances or times, a certain amount of "fuzziness" is built into the nature of reality.

The Heisenberg uncertainty principle leads to the concept of **virtual pairs**: *At every point in space, pairs of particles and antiparticles are constantly and spontaneously being created and*

destroyed. An **antiparticle** is quite like an ordinary particle except that it has an opposite electric charge and can annihilate an ordinary particle so that both particles disappear. In the case of virtual pairs, the process of creation and annihilation occurs over such incredibly brief time intervals that these virtual particles and antiparticles cannot be observed directly. With some added energy, a virtual particle can become a real and observable particle, and a black hole helps this to happen.

With virtual particle pairs everywhere, now let's think about a black hole. Furthermore, think about the momentary creation of a virtual pair of an electron and a positron (the antiparticle of an electron) just *outside* the black hole's event horizon (**Figure 21-32**). It may happen that one of the virtual particles falls into the black hole. Its partner is then deprived of a counterpart with which to annihilate and ends up becoming a real particle. However, some energy is required for a virtual particle to become a real particle. The energy to accomplish this conversion comes from the black hole's mass according to $E = mc^2$. This decreases the mass of the black hole by a tiny amount, and the newly created real particle is free to radiate away from the black hole. In this way, real particles can quantum mechanically "leak" out from a microscopic region just outside the event horizon of a black hole, carrying some of the hole's mass with them. These particles, called **Hawking radiation**, would appear as a form of steady emission by a black hole.

> A fundamental uncertainty in physical knowledge makes it possible for black holes to emit particles and radiation into space.

Hawking radiation might also have implications for the information loss implied by the no-hair theorem. Some scientists suggest that the detailed information about objects entering a black hole is not lost, but is actually preserved in slight quantum variations of the Hawking radiation. While that might answer an important question within theoretical physics, for all practical purposes, the information is unobservable through any foreseeable techniques.

1. Pairs of virtual particles spontaneously appear and annihilate everywhere in the universe.

2. If a pair appears just outside a black hole's event horizon, tidal forces can pull the pair apart, preventing them from annihilating each other.

e⁻ • • e⁺   γ   γ

Event horizon

3. If one member of the pair crosses the event horizon, the other can escape into space, carrying energy away from the black hole.

FIGURE 21-32

Hawking Radiation and Evaporation of a Black Hole This illustration shows two pairs of virtual particles—an electron (e^-) and a positron (e^+), and a pair of virtual photons (γ)—appearing just outside the event horizon of a black hole. If one member of the pair becomes real and escapes, it carries a little energy away from the black hole, the black hole decreases in mass, and the event horizon shrinks. The real particles radiating away are referred to as Hawking radiation, and the process is described as black hole evaporation.

The Temperature of a Black Hole

Hawking radiation led to another remarkable discovery: Black holes have a temperature. Recall that the warmer an object, the more radiation it emits (Section 5-3). Detailed calculations show that Hawking radiation has the same properties as the radiation that is emitted from objects due to their temperature: Black holes emit blackbody radiation. The result is that black holes actually have a temperature, and *the smaller a black hole's mass, the higher its temperature.*

As an example, a 1-trillion-ton (10^{15} kg) black hole would emit energy as if it were a blackbody with a temperature of nearly 10^9 K. That is hotter than the center of our Sun. However, the black holes that nature seems to produce are much larger and cooler. A 1-M$_\odot$ (2×10^{30} kg) black hole would have a miniscule temperature around 10^{-7} K. This temperature is barely above absolute zero. In other words, stellar mass black holes (and larger) have such low temperatures that they would emit very little Hawking radiation.

The story is different for very low-mass black holes, such as the proposed primordial black holes we discussed in Section 21-7. As particles escape from a very small black hole, the mass of the black hole decreases significantly, making its temperature go up. As its temperature rises, still more particles escape, further decreasing the hole's mass and forcing the temperature still higher. This runaway process of **black hole evaporation** finally causes the hole to vanish altogether! During its final seconds of evaporation, the hole gives up the last of its mass in a violent burst of energy equal to the detonation of a billion megaton hydrogen bombs!

A 10^{10}-kg primordial black hole (comparable to the mass of Mount Everest) would take about 15 billion years to evaporate.

This period is close to the present age of the universe. If primordial black holes were formed in the Big Bang, we would expect to see some of them today going through the explosive final stages of evaporation. In 2008, the Fermi Gamma-ray Space Telescope was launched on a 10-year mission with the capability to detect the nearest of these hypothesized bursts; so far, none have been found.

By contrast, a 5-M$_\odot$ black hole would take more than 10^{62} years to evaporate, and a supermassive black hole of 5 million M$_\odot$ would take more than 10^{80} years. These time spans are far longer than the age of the universe. We can safely predict that the black holes we have observed to date will remain as black holes for the foreseeable future.

So far, we have considered black holes in isolation. As we will see in Chapter 25, the universe contains numerous photons left over from the Big Bang (giving the universe a background temperature of 2.7 K). As a result, larger black holes that have cooler temperatures *absorb* more energy from their surroundings than they emit through Hawking radiation; these black holes would slowly grow! At the present universal temperature of 2.7 K, only black holes with masses less than that of the Moon (which would have a diameter about the width of a human hair) are hot enough to shrink by emitting more radiation than they absorb.

While Hawking radiation and black hole evaporation have not been observationally verified (yet), these predictions—based on established laws of physics—are being pursued. The Fermi Gamma-ray Space Telescope is already looking for evidence of evaporating black holes as noted above, and even particle accelerators have joined the search. The Large Hadron Collider at CERN (which discovered the Higgs particle in 2012) has the energy to look for certain miniature black holes. They have not been detected, but if created by the collider, they would immediately evaporate by emitting observable Hawking radiation.

From supermassive black holes to stellar mass and primordial black holes, the physical parameters of black holes stretch across a wide range—from large, massive, and cold to small, hot, and evaporating away with a burst. As new instruments are designed to detect these black holes, their exotic properties can be explored. As we will see in Chapter 24, supermassive black holes are already known to play a central role in powering objects called *quasars*, which are some of the brightest objects in the universe.

CONCEPT CHECK 21-19

Which object radiates more intensely, a supermassive black hole or a stellar-mass black hole?

CONCEPT CHECK 21-20

Do black holes remain the same temperature forever?

CONCEPT CHECK 21-21

Suppose for a moment that a 1-kg black hole has been detected. Is it likely that this miniature black hole was produced billions of years ago in the Big Bang or created more recently?

Answers appear at the end of the chapter.

KEY WORDS

Terms preceded by an asterisk (*) are discussed in the Boxes.

antiparticle, p. 683
black hole, p. 666
black hole evaporation,
    p. 684
collapsar, p. 674
equivalence principle, p. 663
ergoregion, p. 681
event horizon, p. 667
gamma-ray burst, p. 672
general theory of relativity,
    p. 663

gravitational radiation
    (gravitational waves), p. 666
gravitational redshift, p. 665
Hawking radiation, p. 683
Heisenberg uncertainty
    principle, p. 683
intermediate-mass (mid-mass)
    black hole, p. 677
length contraction, p. 659
*Lorentz transformations,
    p. 660

neutron star, p. 652
no-hair theorem, p. 681
primordial black hole, p. 677
*proper length (proper
    distance), p. 661
*proper time, p. 660
pulsar, p. 652
Schwarzschild radius (R_{Sch}),
    p. 677
short gamma-ray bursts,
    p. 656

singularity, p. 677
spacetime, p. 659
special theory of relativity,
    p. 658
stellar-mass black hole,
    p. 668
supermassive black hole,
    p. 675
time dilation, p. 659
virtual pairs, p. 683
x-ray bursters, p. 655

KEY IDEAS

Neutron Stars: A neutron star is a dense stellar corpse consisting primarily of closely packed degenerate neutrons (Section 21-1).

- Neutron stars form in a Type II core-collapse supernova. They are held up from further collapse by degenerate neutron pressure.

- A neutron star typically has a diameter of about 20 km, a mass less than 3 $M_\odot$, a magnetic field 10^{12} times stronger than that of the Sun, and a rotation period of roughly 1 s.

Pulsars: A pulsar is a source of periodic pulses of radio emission. These pulses are produced as beams of radio waves from a neutron star's magnetic poles sweep past Earth (Section 21-1).

The Special Theory of Relativity: This theory, published by Einstein in 1905, describes how measurements of lengths and durations of time depend on the motion of an observer and can appear different to different observers (Section 21-2).

- The speed of light is the same to all observers, no matter how fast they are moving.

- An observer will note a slowing of clocks and a shortening of rulers that are moving with respect to the observer. This effect becomes significant only if the clock or ruler is moving at a substantial fraction of the speed of light.

- Space and time are not wholly independent of each other, but are aspects of a single entity called spacetime.

The General Theory of Relativity: Published by Einstein in 1915, this is a theory of gravity. Any massive object causes space to curve and time to slow down, and these effects manifest themselves as a gravitational force. These distortions of space and time are most noticeable in the vicinity of large masses or compact objects (Section 21-3).

- The general theory of relativity is our most accurate description of gravitation. It predicts a number of phenomena, including the bending of light by gravity and the gravitational redshift, whose existence has been confirmed by observation and experiment.

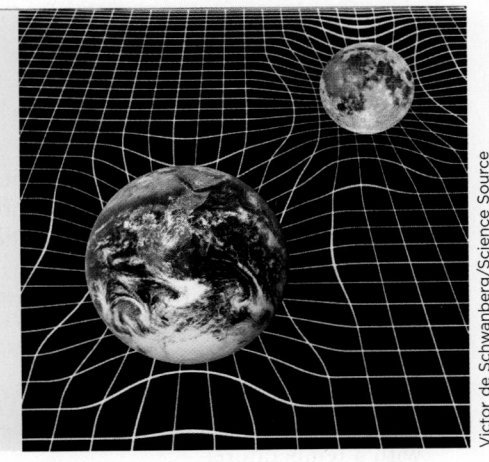

Victor de Schwanberg/Science Source

Black Holes: If a stellar corpse has a mass greater than about 2 to 3 $M_\odot$, gravitational compression will overwhelm any and all forms of internal pressure (Section 21-3).

- A black hole has an escape speed greater than the speed of light.

- Nothing, not even light, can escape a black hole once it crosses into a black hole's event horizon.

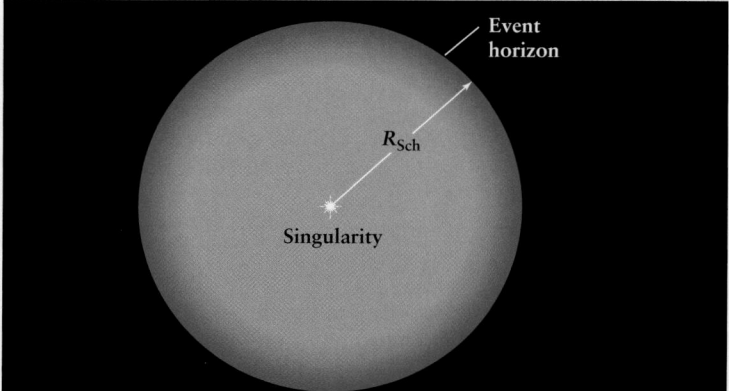

Gravitational Waves: The general theory of relativity also predicts the existence of gravitational waves, which are ripples in the overall geometry of space and time produced by moving masses (Sections 21-3 and 21-4).

- Gravitational waves have been detected by the Laser Interferometer Gravitational-Wave Observatory (LIGO) and its collaborators.

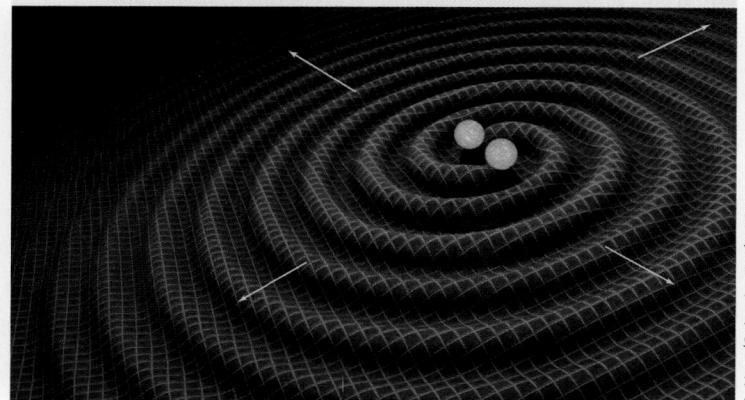

R. Hurt/Caltech-JPL/NASA

Observing Black Holes: Black holes have been detected using indirect methods (Sections 21-5 and 21-7).

- Some binary star systems contain a black hole. In such a system, gases captured from the companion star by the black hole emit detectable x rays. The emission comes from outside the event horizon.

- Many galaxies have supermassive black holes at their centers. These are detected by observing the motions of material around the black hole.

Gamma-Ray Bursts: Short, intense bursts of gamma rays are observed at random times coming from random parts of the sky (Section 21-6).

- By observing the afterglow of long gamma-ray bursts, astronomers find that these objects have very large cosmological redshifts and appear to be located within distant galaxies. The bursts are correlated with supernovae and may be due to an exotic type of supernova called a collapsar.

- The origin of short gamma-ray bursts is now understood to arise from the merger of neutron stars.

NSF/LIGO/Sonoma State University/A. Simonnet

Properties of Black Holes: The entire mass of a black hole gets quickly concentrated in a nearly infinitely dense singularity (Sections 21-9, 21-10, and 21-11).

- A black hole has only three physical properties: mass, electric charge, and angular momentum.

- A rotating black hole (one with angular momentum) has an ergoregion around the outside of the event horizon. In the ergoregion, space and time themselves are dragged along with the rotation of the black hole.

- Black holes emit Hawking radiation from a quantum physics process that occurs just outside of their event horizons.

- Black holes have temperatures and their Hawking radiation is emitted like the blackbody radiation that is emitted by any object with a temperature.

- Black holes can evaporate, but they do so at extremely slow rates unless the black holes are very small.

NASA/D. Berry

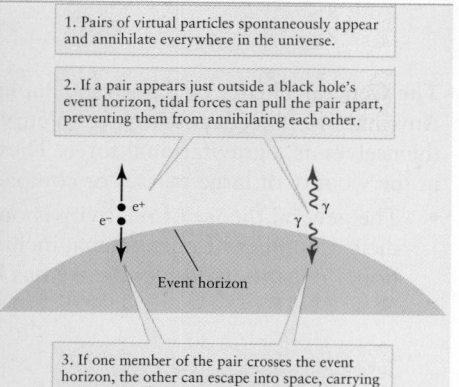

1. Pairs of virtual particles spontaneously appear and annihilate everywhere in the universe.

2. If a pair appears just outside a black hole's event horizon, tidal forces can pull the pair apart, preventing them from annihilating each other.

3. If one member of the pair crosses the event horizon, the other can escape into space, carrying energy away from the black hole.

QUESTIONS

> **Problem-solving tips and tools**
>
> Remember that the time to travel a certain distance is equal to the distance divided by the speed, and the density of an object is its mass divided by its volume. The volume of a sphere of radius r is $4\pi r^3/3$. Section 4-7 describes Newton's law of universal gravitation. Box 4-4 shows how to use Newton's formulation of Kepler's third law, which explicitly includes masses; when using this equation, note that the period P must be expressed in seconds, the semimajor axis a in meters, and the masses in kilograms. For another version of this formula, in which period is in years, semimajor axis in au, and masses in solar masses, see Section 17-9.

Section 21-1

1. What are neutron stars? Where do the neutrons come from?

2. In what type of supernova can a neutron star form? Why?

3. What is degenerate neutron pressure? How is it related to degenerate electron pressure?

4. What does the radio emission from a pulsar look like?

5. During the weeks immediately following the discovery of the first pulsar, one suggested explanation was that the pulses might be signals from an extraterrestrial civilization. Why did astronomers soon discard this idea?

6. How are rotating neutron stars able to produce pulses of radiation as seen by an observer on Earth?

7. Why do neutron stars rotate so much more rapidly than ordinary stars? Why do they have such strong magnetic fields?

8. Is our Sun likely to end up as a neutron star? Why or why not?

9. Why does a neutron star need to have a mass above 1.4 $M_\odot$? What effects contribute to the pressure as a neutron star approaches its maximum mass?

10. The distance to the Crab Nebula is about 2000 pc. In what year did the star actually explode? Explain your answer.

11. The Crab Nebula has an apparent size of about 5 arcmin, and this size is increasing at a rate of 0.23 arcsec per year. (a) Assume that the expansion rate has been constant over the entire history of the Crab Nebula. Based on this assumption, in what year would Earth observers have seen the supernova explosion that formed the nebula? (b) Does your answer to part (a) agree with the known year of the supernova, 1054 C.E.? If not, can you point to assumptions you made in your computations that led to the discrepancies? Or do you think your calculations suggest that additional physical effects are at work in the Crab Nebula, over and above a constant rate of expansion?

12. Emission lines in the spectrum of the Crab Nebula exhibit a Doppler shift, which indicates that gas in the part of the nebula closest to us is moving toward us at 1450 km/s. (a) Assume that the expanding gas has been moving at the same speed since the original supernova explosion observed in 1054 C.E., and calculate what radius and what diameter (in light-years) we should observe the nebula to have today. (b) Compare your result in part (a) to the actual size of the nebula, given in the caption to Figure 21-2.

13. A neutron has a mass of about 1.7×10^{-27} kg and a radius of about 10^{-15} m. (a) Compare the density of matter in a neutron with the average density of a neutron star. (b) If the neutron star's density is more than that of a neutron, the neutrons within the star are overlapping; if it is less, the neutrons are not overlapping. Which of these seems to be the case for average neutrons within the star? Which do you think is the case at the center of the neutron star, where densities are higher than average?

14. Imagine that we are somehow able to stand (and survive) at one of the magnetic poles of the Crab pulsar. Describe what you would see. How would the stars appear to move in the sky? What would you see if you looked straight up? What factors make this location a very unhealthy place to visit?

15. When neutrons are very close together, they repel one another through the strong nuclear force. If this repulsion were made even stronger, what effect might this have on the maximum mass of a neutron star? Explain your answer.

16. What causes neutron stars in a binary system to orbit closer and closer together over time?

17. What process leads to short gamma-ray bursts?

18. Are there materials around today that are the result of merging neutron stars? Explain your answer.

19. What are the similarities between novae (Section 20-4) and x-ray bursters? What are the differences?

Section 21-2

20. You drop a ball inside a car traveling at a steady 50 km/h in a straight line on a smooth road. Does it fall in the same way as it does inside a stationary car? How does this question relate to Einstein's special theory of relativity?

21. In Einstein's special theory of relativity, two different observers moving at different speeds will measure the same value of the speed of light. Will these same observers measure the same value of, say, the speed of an airplane? Explain your answer.

22. A friend summarizes the special theory of relativity by saying "Everything is relative." Explain why this statement is inaccurate.

23. Serena flies past Michael in her spaceship at nearly the speed of light. According to Michael, Serena's clock runs slow. According to Serena, does Michael's clock run slow, fast, or at the normal rate? Explain your answer.

24. Ole flies past Lena in a spherical spaceship at nearly the speed of light. According to Lena, how does the distance from the front to the back of Ole's spaceship (that is, measured along the direction of motion) compare to the distance from the top to the bottom (that is, measured perpendicular to the direction of motion)? Explain your answer.

25. Why does the speed of light represent an ultimate speed limit?

26. The speed of light is the same for all observers, regardless of their motion. Discuss why this requires us to abandon the Newtonian view of space and time.

27. **Box 21-1** A spaceship flies from Earth to a distant star at a constant speed. Upon arrival, a clock on board the spaceship shows a total elapsed time of 8 years for the trip. An identical clock on Earth shows that the total elapsed time for the trip was 10 years. What was the speed of the spaceship relative to Earth?

28. **Box 21-1** An unstable particle called a positive pion (pronounced "pie-on") decays in an average time of 2.6×10^{-8} s. On average, how long will a positive pion last if it is traveling at 95% of the speed of light?

29. **Box 21-1** How fast should a meter stick be moving in order to appear to be only 60 cm long?

30. **Box 21-1** An astronaut flies from Earth to a distant star at 80% of the speed of light. As measured by the astronaut, the one-way trip takes 15 years. (a) How long does the trip take as measured by an observer on Earth? (b) What is the distance from Earth to the star (in light-years) as measured by an Earth observer? As measured by the astronaut?

Section 21-3

31. **TUTORIAL 21.1** Why is Einstein's general theory of relativity a better description of gravity than Newton's universal law of gravitation? Under what circumstances is Newton's description of gravity adequate?

32. Describe two different predictions of the general theory of relativity and how these predictions were tested experimentally. Do the results of the experiments agree with the theory?

33. How does a gravitational redshift differ from a Doppler shift?

34. According to the general theory of relativity, why can't some sort of yet-undiscovered degenerate pressure prevent the matter inside a black hole from collapsing all the way down to a singularity?

35. In the binary system of two neutron stars discovered by Hulse and Taylor (Section 21-3), one of the neutron stars is a pulsar. The distance between the two stars varies between 1.1 and 4.8 times the radius of the Sun. The time interval between pulses from the pulsar is not constant: It is greatest when the two stars are closest to each other and least when the two stars are farthest apart. Explain why this is consistent with the gravitational slowing of time (Figure 21-14a).

36. Find the total mass of the neutron star binary system discovered by Hulse and Taylor (Section 21-3), for which the orbital period is 7.75 hours and the average distance between the neutron stars is 2.8 solar radii. Is your result reasonable for a pair of neutron stars? Explain your answer.

37. Estimate how long it will be until the two neutron stars that make up the binary system discovered by Hulse and Taylor collide with each other. Assume that the distance between the two stars will continue to decrease at its present rate of 3 mm every 7.75 hours, and use the data given in Question 35. (You can assume that the two stars are very small, so they will collide when the distance between them is equal to zero.)

38. The orbital period of the binary system containing A0620-00 is 0.32 day, and Doppler shift measurements reveal that the radial velocity of the x-ray source peaks at 457 km/s (about 1 million mi/h). (a) Assuming that the orbit of the x-ray source is a circle, find the radius of its orbit in kilometers. (This is actually an estimate of the semimajor axis of the orbit.) (b) By using Newton's form of Kepler's third law, prove that the mass of the x-ray source must be at least 3.1 times the mass of the Sun. (*Hint:* Assume that the mass of the K5V visible star—about 0.5 $M_\odot$ from the mass–luminosity relationship—is negligible compared to that of the invisible companion.)

Section 21-4

39. Gravity affects both space and time. Explain the idea behind the method that LIGO uses to detect gravitational waves.

40. Why do orbiting black holes get closer and closer together over time?

41. When black holes merge, the mass of the resulting (larger) black hole is less than the sum of the two masses that merged to create it. Where does that mass go?

42. Why can't LIGO detect gravitational waves radiated by the binary star system that eventually lead to the binary black hole system?

Section 21-5

43. Should we worry about Earth being pulled into a black hole? Why or why not?

44. How does rapid flickering of an x-ray source provide evidence that the source is small (see discussion of Cygnus X-1)?

45. **TUTORIAL 21.2** All the stellar-mass black hole candidates mentioned in the text are members of very short-period binary systems. Explain how this makes it possible to detect the presence of the black hole.

46. Astronomers cannot actually see the black hole candidates in close binary systems. How, then, do they know that these candidates are not white dwarfs or neutron stars?

Section 21-6

47. Describe two ways in which a member of a binary star system could become a black hole.

48. What is a gamma-ray burst? What is the evidence that gamma-ray bursts are not located in the disk of our galaxy or in a halo surrounding our galaxy?

49. Summarize the evidence that gamma-ray bursts result from a process involving a star in a distant galaxy.

50. What is a collapsar? How does the collapsar model account for the existence of long gamma-ray bursts?

51. Contrast gamma-ray bursts with x-ray bursts (discussed in Section 21-1). From our models of what causes these energetic phenomena, explain why x-ray bursts emit repeated pulses but gamma-ray bursts apparently emit just once.

52. Long gamma-ray bursts are only observed in galaxies where there is ongoing star formation. Explain how this is consistent with the collapsar model of how these bursts occur.

53. The spectrum of a Type Ic supernova lacks absorption lines of hydrogen and helium. This means that when a black hole formed at the center of the progenitor star, the resulting jets were more easily able to escape into space. Explain why this is so.

Section 21-7

54. How do astronomers locate supermassive black holes in galaxies?

55. What is an intermediate-mass black hole? How are such objects thought to form?

56. Find the orbital period of a star moving in a circular orbit of radius 500 au around the supermassive black hole in the galaxy NGC 4261 (Section 21-7).

Section 21-8

57. When we say that the Moon has a radius of 1738 km, we mean that this is the smallest radius that encloses all of the Moon's material. In this sense, is it correct to think of the Schwarzschild radius as the radius of a black hole? Why or why not?

58. Why is it unlikely that a black hole has a large electric charge?

59. **Box 21-2** Find the Schwarzschild radius for an object having a mass equal to that of the planet Saturn.

60. **Box 21-2** What is the Schwarzschild radius of a black hole whose mass is that of (**a**) Earth, (**b**) the Sun, (**c**) the supermassive black hole in NGC 4261 (Section 21-7)? In each case, also calculate what the density would be if the matter were spread uniformly throughout the volume of the event horizon.

61. **Box 21-2** What is the mass in kilograms of a black hole whose Schwarzschild radius is 11 km?

62. **Box 21-2** To what density must the matter of a dead 8-$M_\odot$ star be compressed in order for the star to disappear inside its event horizon? How does this compare with the density at the center of a neutron star, about 3×10^{18} kg/m^3?

63. **Box 21-2** Prove that the density of matter needed to produce a black hole is inversely proportional to the square of the mass of the hole. If you wanted to make a black hole from matter compressed to the density of water (1000 kg/m^3), how much mass would you need?

Section 21-9

64. What kind of black hole is surrounded by an ergoregion? What happens inside the ergoregion?

65. What is the no-hair theorem?

Section 21-10

66. As seen by a distant observer, how long does it take an object dropped from a great distance to fall through the event horizon of a black hole? Explain your answer.

67. Speculate on the effects you might encounter on a trip to the center of a black hole (assuming that you could survive the journey).

Section 21-11

68. **TUTORIAL 21.3** A twenty-third-century instructor at Starfleet Academy tells her students, "If someday your starship falls into a black hole, it'll be your own fault." Explain why it would require careful piloting to direct a spacecraft into a black hole.

69. Summarize what leads to Hawking radiation.

70. If even light cannot escape from a black hole, how is it possible for black holes to evaporate?

71. Why do smaller black holes evaporate more quickly than larger black holes?

General Problems

72. In what way is a black hole blacker than black ink or a black piece of paper?

73. If the Sun suddenly became a black hole, how would Earth's orbit be affected? Explain your answer.

74. Is it possible to tell which chemical elements went into a black hole? Why or why not?

75. Describe the kinds of observations you might make in order to locate and identify black holes.

WEB/E-BOOK QUESTIONS

76. Search the internet for information about a stellar-mass black hole candidate named V4641 Sgr. In what ways does it resemble other black hole candidates such as Cygnus X-1 and V404 Cygni? In what ways is it different and more dramatic? How do astronomers explain why V4641 Sgr is different?

77. Search the internet for information about supernova SN 2006aj, which was associated with gamma-ray burst GRB 060218. In what ways were this supernova and gamma-ray burst unusual? Are the observations of these objects consistent with the collapsar model?

78. Search the internet for information about the intermediate-mass black hole candidate in M82. Is this still thought to be an intermediate-mass black hole? What new evidence has been used to either support or oppose the idea that this object is an intermediate-mass black hole?

79. **ANIMATION 21.1** **The Equivalence Principle.** Access the animation "The Equivalence Principle" in Chapter 21 of the *Universe* website or e-book. View the animation and answer the following questions. (a) Describe what is happening as viewed from the frame of reference of the elevator. What causes the apple to fall to the floor of each elevator? (b) Describe what is happening as viewed from the frame of reference of the stars. What causes the apple to fall to the floor of each elevator? (c) Think of another experiment you could perform with the apple. Describe what would happen during this experiment as seen by Newton (in the left-hand box) and by Einstein (in the right-hand box).

ACTIVITIES

Observing Project

80. You cannot see a black hole with a telescope. Nevertheless, you might want to observe the visible companion of Cygnus X-1. The epoch 2000 coordinates of this ninth-magnitude star are R.A. = $19^h 58.4^m$ and Decl. = $+35° 12'$, which is quite near the bright star η (eta) Cygni. Compare what you see with the photograph in Figure 21-20.

Starry Night™ Explorations

81. Use *Starry Night™* to observe the sky in July 1054, when the supernova that spawned the Crab Nebula (M1) was recorded as a "guest" star, visible in daylight by Chinese astronomers. The daytime appearance and position of this "star" near to a crescent Moon may even have been recorded as a pictograph on an overhanging cliff face at this time by inhabitants of Chaco Canyon, New Mexico. Open **Favorites > Explorations > Pictograph** to position yourself in Chaco Canyon, at latitude 36° N and longitude 108° W at 5 A.M. on July 5, 1054, looking east just before sunrise.

(a) Zoom in to display a field of view of about 10° centered on the Crab Nebula, where you can see the position of the Moon near to the nebula. You may find it helpful to turn daylight on or off (select **Show Daylight** or **Hide Daylight** in the **View** menu). What is the phase of the Moon?

(b) Investigate how the relative positions of the Moon and the Crab Nebula change when you set the date to July 4, 1054, or July 6, 1054. On which date do the relative positions of the Moon and the Crab Nebula give the best match to the pictograph shown in the textbook or on internet sites? (N.B. The image of the pictograph in the textbook is upside down.)

(c) From the HUD, obtain the Distance from Observer for the Crab Nebula. Taking light travel time into account, in which year did the star actually explode?

(d) Zoom in closer to a field of view of about 30 arcmin and remove daylight to see the detailed structure of this exploding cloud of gas from the supernova that produced the Crab Nebula. Briefly describe the appearance of this gas cloud.

82. Einstein formulated his general theory of relativity in 1915 and predicted several observations that would confirm the correctness of this theory. One of these predictions was that light would be deflected by the curvature of space around a massive object. He suggested that a total solar eclipse would provide an ideal opportunity to test this prediction using the Sun as the massive body. This would be done by comparing the positions of background stars close to the Sun's position when the Sun's light is blocked by the Moon with star positions on equivalent nighttime photographs outside of the eclipse. This prediction was verified only 3 years later. You can use *Starry Night™* to demonstrate the geometry and observing conditions of this experiment. Of course, the relativistic effects cannot be simulated, but you can view the total eclipse and assess the limitations of this measurement. Open **Favorites > Explorations > Eclipse** to show the Moon covering most of the Sun during the August 21, 2017, solar eclipse over Idaho, USA. Click **Play** to watch the period of totality at 30 times normal speed as the Moon covers the visible Sun to reveal the outer corona, the Sun's hot but tenuous surrounding atmosphere. Background stars appear as the sky darkens, but you can see that fainter stars close to the Sun's limb are hidden by the corona. Adjust the time to central eclipse at about 11:34:50 A.M. Use **View > Hide Daylight** to remove the remaining effects of daylight, then select **Options > Solar System > Planets-Moons...** and turn off the "Show Corona During Eclipse" option to remove the effect of the corona, privileges that observers would not have during real measurements.

(a) Would the bending of the starlight as it passes through the curved space generated by the Sun cause the star images to appear farther inward toward the Sun or outward away from the Sun?

(b) The maximum predicted shift of a star's position by gravitational bending caused by the Sun for stars close to the solar limb is 1.75 arcsec. In these observations, relative star positions would be compared in photos in and out of eclipse. Stars on opposite sides of the Sun near the solar limb should shift position in opposite directions, so the stars should shift relative to each other by 3.5 arcsec. Use the angular separation tool to measure the relative spacing between two stars near the solar limb on opposite sides of the Sun, translate this measurement into arcseconds, and calculate roughly the precision of star position measurement that would be needed to see the above gravitational effect by dividing the expected relative shift by the star spacing.

83. Use *Starry Night*™ to examine in some detail the central regions of two galaxies that contain supermassive black holes at their centers. The Milky Way is one such galaxy. Select **Favorites > Explorations > Chandra** to view the central region of our galaxy.

    (a) Note that at visible wavelengths, this central region of the Milky Way appears to be dark. What explanation can you give for this dark band across the galactic plane? Zoom in on the center of the galaxy to examine the region surrounding the black hole using the highly penetrating x rays detected by the Chandra X-ray space telescope.

    (b) M87 is an active galaxy for which evidence is strong for the presence of a supermassive black hole at its center. Select **Favorites > Explorations > Black Hole** to see two separate images of this galaxy offset from each other for clarity. The large image shows this galaxy as it might appear in a ground-based telescope. The smaller image centered in the view is the high-resolution Hubble Space Telescope image of the gas jet emanating from the black hole at the center of this galaxy. Zoom in to a field of view of about 1 arcmin to see this spectacular jet as it interacts violently with the interstellar medium above the black hole. Zoom out again, click on **Quick Search Items**, and open the **Options** panel. Expand the **Deep Space** layer and turn off **Messier Objects** and toggle **Chandra Images** on to display the Chandra X-ray image of this galaxy. Comment on the suggestion that supermassive black holes were discovered only after relatively recent advances were made in telescope and detector technology.

Collaborative Exercises

84. Consider the graph showing a recording of a pulsar in Figure 21-3. Sketch and label similar graphs that your group estimates for (a) a rapidly spinning, professional ice skater holding a flashlight, and (b) a siren on an emergency ambulance.

85. As stars go, pulsars are tiny, only about 20 km across. Name three specific things or places that have a size or a separation of about 20 km.

86. Using Einstein's special theory of relativity, estimate (1) the length of your pencil or pen at constant speed at the speeds of a bicycle rider, a car traveling on the highway, and a commercial jet liner at cruising altitude; and (2) the speed of a light beam emitted by a spaceship traveling at 200,000 km/s toward another spaceship traveling at the same speed.

ANSWERS

ConceptChecks

ConceptCheck 21-1: Just as a white dwarf is held up by degenerate electron pressure, a neutron star is held up by degenerate neutron pressure.

ConceptCheck 21-2: By shrinking during formation, neutron stars spin rapidly and have very large magnetic fields. Charged particles in these magnetic fields emit the light (often radio waves) observed at Earth. The light arrives in pulses the way a lighthouse beacon repeatedly sweeps by an observer.

ConceptCheck 21-3: A type Ia supernova blows the object apart. For an x-ray burster, as for a nova (see Section 20-4), only an outer layer of transferred mass undergoes nuclear reactions. When enough mass is again transferred from the binary partner, the detonation repeats itself.

ConceptCheck 21-4: For anyone riding on the speeding train, the train is not moving relative to the riders. They observe no length contraction or time dilation, so their hearts do not beat more slowly.

ConceptCheck 21-5: All observers, whether moving or not, see light travel at the same speed through the vacuum of space: They all see light travel at $c = 3 \times 10^8$ m/s.

ConceptCheck 21-6: Less time has passed on the clock near the planet—the planet's gravity slows down time; the stronger the gravity, the more time slows down.

ConceptCheck 21-7: No. As the radio waves "climb" out of the gravitational field of the planet, they experience a gravitational redshift—the frequency decreases and the wavelength increases.

ConceptCheck 21-8: Neutron stars can support up to 2 to 3 $M_\odot$ without collapsing further. Masses larger than this can collapse into black holes, so these are the lowest black hole masses expected from stellar core collapse.

ConceptCheck 21-9: If an object is compressed, its escape speed increases, and the escape speed also increases if more mass is added to the object while maintaining its size. When an object's escape speed exceeds the speed of light, no light can escape, and the object becomes a black hole.

ConceptCheck 21-10: Stars do not emit strongly in x rays, as Cygnus X-1 does. More importantly, fast variations in the x-ray brightness tell us that the mass of Cygnus X-1 is smaller than Earth—far too small for any type of star or other object.

ConceptCheck 21-11: The collapsar supernova model for gamma-ray bursts involves the special case in which jets are formed during the collapse, and one jet of gamma rays points toward Earth. Without the formation and enhanced intensity of jets, a regular supernova (of Type I or II) would be too weak to account for the strength of the observed gamma rays.

ConceptCheck 21-12: No. The larger the mass of a black hole, the lower the average density required to form the black hole. If it could be carried out, filling our solar system with water would produce a very large black hole.

ConceptCheck 21-13: No. Theoretically speaking, the laws of gravity allow a black hole to exist with any mass. However, *natural processes* in the cosmos might only *produce* black holes in certain mass ranges.

ConceptCheck 21-14: A black hole's event horizon is a spherical boundary with a radius given by the Schwarzschild radius. Objects getting closer to the Schwarzschild radius have fallen irreversibly within the event horizon.

ConceptCheck 21-15: At such large distances, the gravitational force is the same from each object. Stronger forces and other effects only show up much closer to a black hole.

ConceptCheck 21-16: It would be safer to fly by the event horizon of the larger black hole, where the gravitational force is actually weaker.

ConceptCheck 21-17: No. Once matter enters a black hole, the only properties of the black hole are its mass, charge, and angular momentum.

ConceptCheck 21-18: Maybe, as it depends on the mass of the black hole. For a supermassive black hole with weak tidal forces, they would not notice their feet being tugged with a greater force than any other part of their body. However, for a stellar-mass black hole, their feet could be pulled relative to their legs by deadly forces.

ConceptCheck 21-19: The smaller the black hole, the higher its temperature. Therefore, the stellar-mass black hole is hotter and radiates more intensely.

ConceptCheck 21-20: No. As black holes lose mass through Hawking radiation, they heat up to even greater temperatures. This causes them to lose even more mass in a process of evaporation, which is more noticeable for smaller black holes.

ConceptCheck 21-21: Small black holes evaporate more quickly and a black hole of this extraordinarily small size would have evaporated long ago if it were produced during the Big Bang. Therefore, the black hole would have been created more recently.

Gravitational Waves: A Bright Future

By Shane L. Larson

Astronomy is a spectator sport. We are largely confined to remain near Earth, staring out into the cosmos, and learning what we can from the meager signals that wash across our shores. For the better part of the history of the human race, the messages we received from space have been carried on beams of light. Originally, we only knew of the few kinds of light our eyes could see, the colors of the visible rainbow. But beginning in the late 1800s, physicists and astronomers became aware that there were other kinds of light that our eyes could not see. Using technology, we were able to build instruments to see what our eyes and telescopes could not: an invisible universe detectable only in radio waves, infrared radiation, ultraviolet light, x rays, and gamma rays. Astonishingly, every kind of light we know of is useful in some way to study astronomy.

At the start of the twentieth century, when we were first figuring out how to measure these different kinds of light, Albert Einstein was working on general relativity (GR), the description of gravity used today in modern astrophysics. As part of his early work on GR, Einstein made a remarkable prediction: Dynamic motion of masses instigates changes in the gravitational field around them, and those changes must move through space at the speed of light. Today, we call these propagating disturbances *gravitational waves*. Like all good physicists of the day, Einstein thought carefully about how we might go about detecting the new waves he had predicted. He wanted to know how strong would they be? What influence would they have on an apparatus that one could build in the laboratory to detect them? What he discovered was the waves were vanishingly

weak. For any practical mass you could shake in a laboratory, attempting to make the waves, the strength simply would not be large enough to measure. Even considering the most massive astrophysical systems known in Einstein's day, the stars and planets, the overall effect of gravitational waves on Earth would be vanishingly small. It seemed unlikely that they would ever be detected by experiment.

Fast forward 100 years to today. We have amazing technology at our disposal, and have discovered that the universe is home to extreme astrophysical objects, like neutron stars and black holes, with tremendous gravitational influence. After more than two decades of design and construction, we have built the world's first successful gravitational wave interferometers. They are called the Laser Interferometer Gravitational-Wave Observatory (LIGO) and Virgo and they have definitively detected gravitational waves. The technology is exquisite, and has added to the astronomer's toolbox a new and provocative method for probing the distant cosmos.

Gravitational waves are created by accelerating huge amounts of mass, which surprisingly happens quite often in the cosmos. There is an entire spectrum of gravitational waves covering a vast range of frequencies. Roughly speaking, the timescale that defines the motion of the masses is approximately the inverse frequency of the gravitational waves generated. The figure here illustrates the entire gravitational wave spectrum, noting the parts of the spectrum where we have good observing techniques, as well as different types of astrophysical systems that might be observable in gravitational waves.

(continued on the next page)

Gravitational Waves: A Bright Future (continued)

The first gravitational wave detection, made by LIGO in September 2015, was of a merging black hole binary—two black holes, each roughly 30 times the mass of the Sun, merged to create a new, bigger black hole. There have been several other binary black hole mergers detected, but in August of 2017, the LIGO-Virgo network made the first detection of a merging neutron star binary. This event was significant because the gravitational wave detection was followed 1.7 seconds later by a short gamma-ray burst observed by the Fermi satellite. This coincidence of signals, with the confident identification of neutron stars using gravitational wave data, confirmed a long-held theoretical belief that short gamma-ray bursts (GRBs) are created by merging neutron star binaries. For several weeks after the event, more than 70 observatories around the world conducted follow-up observations. The combination of gravitational wave observations with electromagnetic data provided new insights into the "kilonova" mechanism, the gamma-ray burst afterglow, and the nucleosynthesis of heavy elements. This *multi-messenger* detection, with both gravitational waves and light, illustrated the powerful ways that astronomers will be able to use gravitational waves to probe the distant universe.

Shane L. Larson is a Research Associate Professor of Physics and Astronomy at Northwestern University and the Associate Director of the Center for Interdisciplinary Exploration and Research in Astrophysics (CIERA). A member of both the Laser Interferometer Gravitational-Wave Observatory (LIGO) Collaboration and the Laser Interferometer Space Antenna (LISA) Study Team, Larson studies how the evolutionary history of the galaxy is imprinted in the stellar graveyard and how gravitational waves can be used to understand aspects of the universe that light cannot reveal.

This is only the beginning. Around 2030, the European Space Agency, in collaboration with NASA, will fly a gravitational wave observatory in space, called LISA. Roughly a million times larger than LIGO or Virgo, LISA will be sensitive to much lower-frequency gravitational waves. In an astrophysical context, lower frequencies correspond to systems with much longer orbital periods of tens to tens of thousands of seconds.

There are still *many* discoveries to be made, but the most common question asked is "Is this really that important?" In the history of science, there are many turning points—pivotal discoveries that transformed the knowledge we use and the way we think about the universe around us. The discovery of gravitational waves, and the subsequent development of our ability to *observe the universe* in gravitational waves, will, in hindsight, be one of the pivotal moments in the history of astronomy. Gravitational waves are one of the few observational methods we have that do not map onto some other sensory experience connected to our long history as a biological species; in this way, they are revealing to us a cosmos that has never been perceived before. New ideas require new ways of thinking, and new ways of looking at old ideas. History has shown us, time and again, that turning traditional thinking on its head can lead to new discoveries. The discovery of gravitational waves is as important and as transformative as other big discoveries in astronomy, like the fact that the Milky Way is not the only galaxy, or that the universe is expanding, or the discovery of the cosmic microwave background. Students beginning their studies in astronomy today will never know a world without gravitational wave data, and will look back on the times before gravitational wave observations in the same way we look back on the early days of telescopes or the times before the Space Age.

22 | Our Galaxy

20°

0.5° = 30 arcmin
= angular size of
the full Moon

LEARNING GOALS

By reading the sections of this chapter, you will learn

22-1 How astronomers discovered the solar system's location within the Milky Way Galaxy

22-2 The shape and size of our galaxy

22-3 How the Milky Way's spiral structure was discovered

22-4 The evidence for the existence of dark matter in our galaxy

22-5 What causes the Milky Way's spiral arms to form and persist

22-6 How astronomers discovered a supermassive black hole at the galactic center

O n a clear, moonless night, away from the glare of city lights, you can often see a hazy, luminous band stretching across the sky. This band, called the Milky Way, extends all the way around the celestial sphere. The upper of the two opening photographs is centered on the brightest part of the Milky Way, in the constellation Sagittarius.

Galileo, the first person to view the Milky Way with a telescope, discovered that it is composed of countless dim stars. Today, we realize that the Milky Way is actually a disk tens of thousands of parsecs across containing hundreds of billions of stars. (It is hard to picture hundreds of billions of stars, but it is about the number of grains in a few trashcans full of beach sand.) Between the stars, there are also large quantities of gas and dust. One of these stars is our Sun, and this vast assemblage of matter—our home galaxy—is collectively called the Milky Way Galaxy.

Just as Galileo's telescope revealed aspects of the Milky Way that the naked eye could not, modern astronomers use telescopes at nonvisible wavelengths to peer through our galaxy's obscuring dust and observe what visible-light telescopes never could. For example, the lower of the two opening photographs is an infrared

Two views of the Milky Way: a wide-angle infrared image (upper) and a close-up infrared image (lower). (top: ESO/VVV Survey/D. Minniti; bottom: NASA/JPL-Caltech/S. Stolovy [SSC/Caltech])

R I V U X G

image that shows hundreds of thousands of stars near the center of the galaxy. As we will see, radio, infrared, and x-ray observations reveal that at the very center of the galaxy lies a black hole with a mass of 4.1 million Suns.

Modern astronomers have also discovered that most of the Milky Way's mass is not in its stars, gas, or dust, but in a halo of *dark matter* that emits no measurable radiation. What the character of this dark matter could be remains one of the greatest unanswered questions in astronomy and physics.

The Milky Way is just one of myriad *galaxies,* or systems of stars and interstellar matter, that are spread across the observable universe. By studying our home galaxy, the Milky Way Galaxy, we begin to explore the universe on a grand scale. Instead of focusing on individual stars, we look at the overall arrangement and history of a huge stellar community of which the Sun is a member. In this way, we gain insights into galaxies in general and prepare ourselves to ask fundamental questions about the cosmos.

22-1 The Sun is located in the disk of our galaxy, about 8000 pc from the galactic center

TUTORIAL 22.1 Eighteenth-century astronomers were the first to suspect that because the Milky Way completely encircles us, all the stars in the sky are part of an enormous disk of stars—the **Milky Way Galaxy**. As we learned in Section 1-4, a *galaxy* is an immense collection of stars—as well as gas and dust referred to as *interstellar matter.*

Our galaxy is both large and mostly empty. To put our galaxy into perspective, consider this scale model suggested by astronomer Mark Whittle: Suppose that our galaxy was shrunk down to

the size of the continental United States. Then, the average distance between the shrunken stars would be about the size of a football field. But what about the stars themselves? The average size of the shrunken stars would be about the size of human cells! Clearly, the stars of the Milky Way are widely separated, and there are a lot of them.

We are within our own Milky Way Galaxy, which makes it hard to view. Our solar system is located inside the pancake-like disk of our galaxy, which is why the Milky Way appears as a band around the sky (**Figure 22-1**).

Locating the Sun Within the Galaxy: Early Attempts

But where within this disk is our own Sun? Until the twentieth century, the prevailing opinion was that the Sun and planets lie at the galaxy's center. One of the first to come to this conclusion was the eighteenth-century English astronomer William Herschel, who discovered the planet Uranus and was a pioneering cataloger of binary star systems (see Section 17-9). Herschel's approach to determining the Sun's position within the galaxy was to count the number of stars in each of 683 regions of the sky. He reasoned that he should see the greatest number of stars toward the galaxy's center and a lesser number toward the galaxy's edge.

Herschel found approximately the same density of stars all along the Milky Way. Therefore, he concluded that we are at the center of our galaxy (**Figure 22-2**). In the early 1900s, the Dutch astronomer Jacobus Kapteyn came to essentially the same conclusion by analyzing the brightness and proper motions of a large number of stars. According to Kapteyn, the Milky Way is about 17 kpc (17 kiloparsecs = 17,000 pc or 55,000 ly) in diameter, with the Sun near its center.

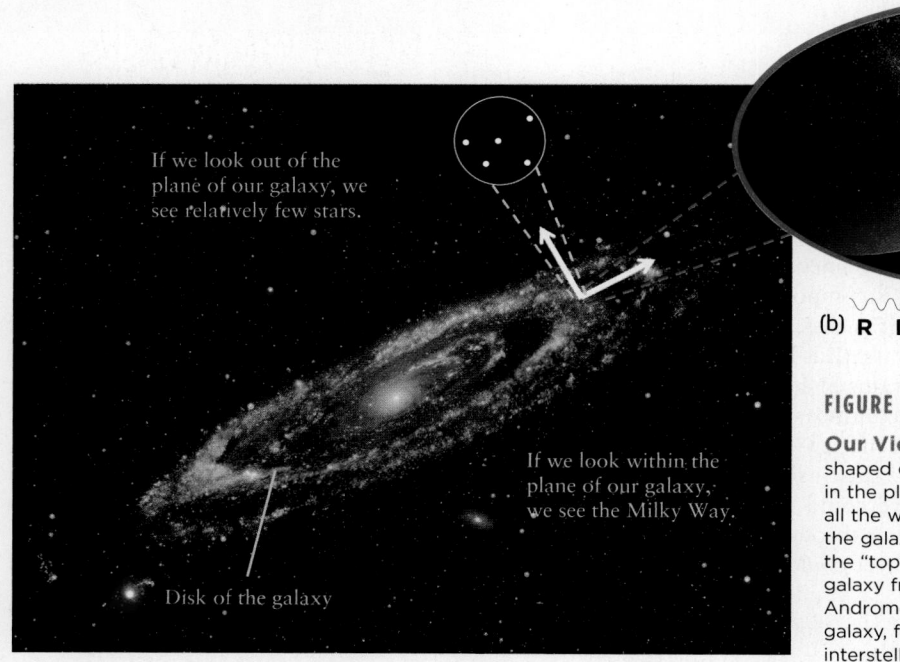

(a)

(b) R I **V** U X G

FIGURE 22-1

Our View of the Milky Way (a) The Milky Way Galaxy is a disk-shaped collection of stars like this one. When we look out at the night sky in the plane of the disk, the stars appear as a band of light that stretches all the way around the sky. When we look perpendicular to the plane of the galaxy, we see only those relatively few stars that lie between us and the "top" or "bottom" of the disk. (Since we can't take a picture of our galaxy from the outside, this is not the Milky Way but our neighbor, the Andromeda Galaxy.) **(b)** The Milky Way, which is our view of our own galaxy, from the inside. The dark streaks across the Milky Way are due to interstellar dust in the plane of our galaxy. (a: NASA/JPL-Caltech; b: Denis Belitsky/Shutterstock.com)

If we look out of the plane of our galaxy, we see relatively few stars.

If we look within the plane of our galaxy, we see the Milky Way.

Disk of the galaxy

FIGURE 22-2

Herschel's Map of Our Galaxy In a paper published in 1785, the English astronomer William Herschel presented this map of the Milky Way Galaxy. He determined the galaxy's shape by counting the numbers of stars in various parts of the sky. Herschel's conclusions were flawed because interstellar dust blocked his view of distant stars, leading him to the erroneous idea that the Sun is at the center of the galaxy. (Dr. Jeremy Burgess/Science Source)

The Problem: Interstellar Extinction

Both Herschel and Kapteyn were wrong about the Sun being at the center of our galaxy. The reason for their mistake was discerned in 1930 by Robert J. Trumpler of Lick Observatory. While studying star clusters, Trumpler discovered that the more remote clusters appear unusually dim—more so than would be expected from their distances alone. As a result, Trumpler concluded that interstellar space must not be a perfect vacuum: It must contain dust that absorbs or scatters light from distant stars.

Like the stars themselves, interstellar dust is concentrated in the plane of the galaxy (see Section 18-2). As a result, it obscures our view within the plane and makes distant objects appear dim, an effect called **interstellar extinction**. Great patches of interstellar dust are clearly visible in wide-angle photographs such as the ones that open this chapter. Thanks to interstellar extinction, Herschel and Kapteyn were actually seeing only the nearest stars in the galaxy. Hence, they had no idea of either the enormous size of the galaxy or of the vast number of stars concentrated around the galactic center.

The same principle applies to our galaxy. While interstellar dust in the plane of our galaxy hides the sky covered by the Milky Way, we have an almost unobscured view out of the plane (that is, to either side of the Milky Way). To find our location in the galaxy, we need to locate bright objects that are part of the galaxy but lie outside its plane in unobscured regions of the sky.

The Breakthrough: Globular Clusters and Variable Stars

Fortunately, bright objects of the sort we need do in fact exist. They are the **globular clusters**, a class of star clusters associated with the galaxy but which lie outside its plane (Figure 22-3b). As we saw in Section 19-4, a typical globular cluster is a spherical distribution of roughly 10^6 stars packed in a volume only a few hundred light-years across (see Figure 19-13).

However, to use globular clusters to determine our location in the galaxy, we must first determine the distances from Earth to

(a) Determining your position in the fog

(b) Determining your position in the galaxy

FIGURE 22-3

Finding the Center of the Galaxy **(a)** A motorist lost on a foggy night can determine his location by looking for tall buildings that extend above the fog. **(b)** In the same way, astronomers determine our location in the galaxy by observing globular clusters that are part of the galaxy but lie outside the obscuring material in the galactic disk. The globular clusters form a spherical halo centered on the center of the galaxy.

these clusters. (Think again of our lost motorist—glimpsing the lights of a sky-scraper through the fog may be useful to the motorist, but only if he can tell how far away the skyscraper is.) Pulsating variable stars in globular clusters provide the distances, giving astronomers the key to the dimensions of our galaxy.

In 1912, the American astronomer Henrietta Leavitt reported her important discovery of the period–luminosity relation for Cepheid variables. As we saw in Section 19-6, Cepheid variables are pulsating stars that vary periodically in brightness (see Figure 19-19). Leavitt studied numerous Cepheids in the Small Magellanic Cloud (a small galaxy near the Milky Way) and found their periods to be directly related to their average luminosities. Figure 22-4 shows that the longer a Cepheid's period, the greater its average luminosity.

The period–luminosity law is an important tool in astronomy because it can be used to determine distances. For example, suppose you find a Cepheid variable in the sky. By measuring its period and using a graph like Figure 22-4, you can determine the star's average luminosity. Knowing the star's average luminosity, you can find out how far away the star must be in order to give the observed brightness. (Box 17-2 explains how this is done.)

Shortly after Leavitt's discovery of the period–luminosity law, Harlow Shapley, a young astronomer at the Mount Wilson Observatory in California, began studying a family of pulsating stars closely related to Cepheid variables called **RR Lyrae variables.** The light curve of an RR Lyrae variable is similar to

> Observations of pulsating variable stars revealed the immense size of the Milky Way.

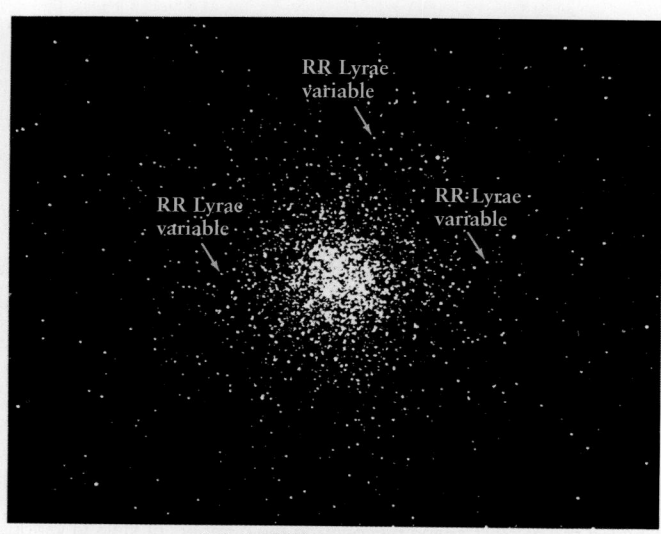

FIGURE 22-5 R I V U X G

VIDEO 22.1

RR Lyrae Variables in a Globular Cluster The arrows point to three RR Lyrae variables in the globular cluster M55, located in the constellation Sagittarius. From the average apparent brightness (as seen in this photograph) and average luminosity (known to be roughly 100 L$_\odot$) of these variable stars, astronomers have deduced that the distance to M55 is 6500 pc (20,000 ly). (Harvard-Smithsonian Center for Astrophysics)

that of a Cepheid, but RR Lyrae variables have shorter pulsation periods and lower peak luminosities (see Figure 22-4).

The tremendous importance of RR Lyrae variables lies in the fact that they are commonly found in globular clusters (Figure 22-5). By using the period–luminosity relationship for these stars, Shapley was able to determine the distances to the 93 globular clusters then known. He found that some of them were more than 100,000 ly from Earth. The large values of these distances immediately suggested that the galaxy was much larger than Herschel or Kapteyn had thought.

Another striking property of globular clusters is how they are distributed across the sky. Ordinary stars are rather uniformly spread along the Milky Way. However, the majority of the 93 globular clusters that Shapley studied are located in one-half of the sky, widely scattered around the portion of the Milky Way that is in the constellation Sagittarius.

From the directions to the globular clusters and their distances from us, Shapley mapped out the three-dimensional distribution of these clusters in space. In 1920, he concluded that the globular clusters form a huge spherical distribution centered not on Earth but rather about a point in the Milky Way several kiloparsecs away in the direction of Sagittarius (see Figure 22-3b). This point, reasoned Shapley, must coincide with the center of our galaxy, because of gravitational forces between the disk of the galaxy and the "halo" of globular clusters. Therefore, by locating the center of the distribution of globular clusters, Shapley was in effect measuring the location of the galactic center.

Modern-Day Measurements

Since Shapley's pioneering observations, many astronomers have measured the distance from the Sun to the **galactic nucleus,** the center of our galaxy. Shapley's estimate of this distance was too

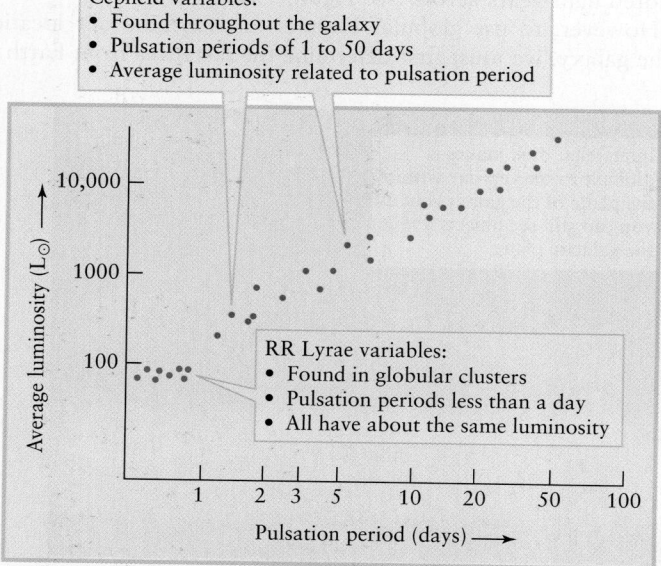

FIGURE 22-4

Period and Luminosity for Cepheid and RR Lyrae Variables This graph shows the relationship between period and average luminosity for Cepheid variables and RR Lyrae variables. Cepheids come in a broad range of luminosities: The more luminous the Cepheid, the longer its pulsation period. By contrast, RR Lyrae variables are horizontal-branch stars that all have roughly the same average luminosity of about 100 L$_\odot$.

large by about a factor of 2, because he did not take into account the effects of interstellar extinction (which were not well understood at the time). Today, the generally accepted distance to the galactic nucleus is about 8 kpc (26,000 ly); the actual distance could be greater or less than that value by about 1 kpc (3300 ly).

Just as Copernicus and Galileo showed that Earth was not at the center of the solar system, Shapley and his successors showed that the solar system lies nowhere near the center of the galaxy. We see that Earth indeed occupies no special position in the universe.

CONCEPT CHECK 22-1

If astronomers observed that globular clusters were evenly spread across all parts of the sky, what would astronomers assume about the location of the Sun and Earth in our galaxy?

Answer appears at the end of the chapter.

22-2 Observations at nonvisible wavelengths reveal the shape of the galaxy

At visible wavelengths, light suffers so much interstellar extinction that the center of the galaxy is totally obscured from view. But the amount of interstellar extinction is roughly inversely proportional to wavelength of light. In other words, the longer the wavelength of light, the farther that light can travel through interstellar dust without being scattered or absorbed. As a result, we can see farther into the plane of the Milky Way at infrared wavelengths than at visible wavelengths, and radio waves can travel all the way through the galaxy! For this reason, telescopes sensitive to these nonvisible wavelengths are important tools for studying the structure of our galaxy.

Exploring the Milky Way in the Infrared

Infrared light is particularly useful for tracing the location of interstellar dust in the galaxy. Starlight warms the dust grains to temperatures in the range of 10 to 90 K; thus, in accordance with Wien's law (see Section 5-4), the dust emits radiation predominately at wavelengths from about 30 to 300 μm. These are called **far-infrared wavelengths**, because they lie in the part of the infrared spectrum most different in wavelength from visible light (see Figure 5-7). At these wavelengths, interstellar dust radiates more strongly than stars, so a far-infrared view of the sky is principally a view of where the dust is. In 1983, the Infrared Astronomical Satellite (IRAS) scanned the sky with a 60-cm reflecting telescope at far-infrared wavelengths, giving the panoramic view of the Milky Way's dust shown in Figure 22-6a.

> Our galaxy's dust and stars—including the Sun—lie mostly in a relatively thin disk.

In 1990, an instrument on the Cosmic Background Explorer (COBE) satellite scanned the sky at **near-infrared wavelengths,** that is, relatively short wavelengths closer to the visible spectrum. Figure 22-6b shows the resulting near-infrared view of the plane of the Milky Way. At near-infrared wavelengths, interstellar dust

Dust lies mostly in the plane of the galaxy (seen edge on).

(a) Infrared emission from dust at wavelengths of 25, 60, and 100 μm

Central bulge

Stars lie mostly in the plane of the galaxy and in the central bulge.

(b) Infrared emission from dust at wavelengths of 1.2, 2.2, and 3.4 μm

FIGURE 22-6 R I V U X G

The Infrared Milky Way **(a)** This view was constructed from observations made at far-infrared wavelengths by the IRAS spacecraft. Interstellar dust, which is mostly confined to the plane of the galaxy, is the principal source of radiation in this wavelength range. **(b)** Observing at near-infrared wavelengths, as in this composite of COBE data, allows us to see much farther through interstellar dust than we can at visible wavelengths. Light in this wavelength range comes mostly from stars in the plane of the galaxy and in the bulge at the galaxy's center. (a: NASA/JPL-Caltech; b: DIRBE, COBE, NASA)

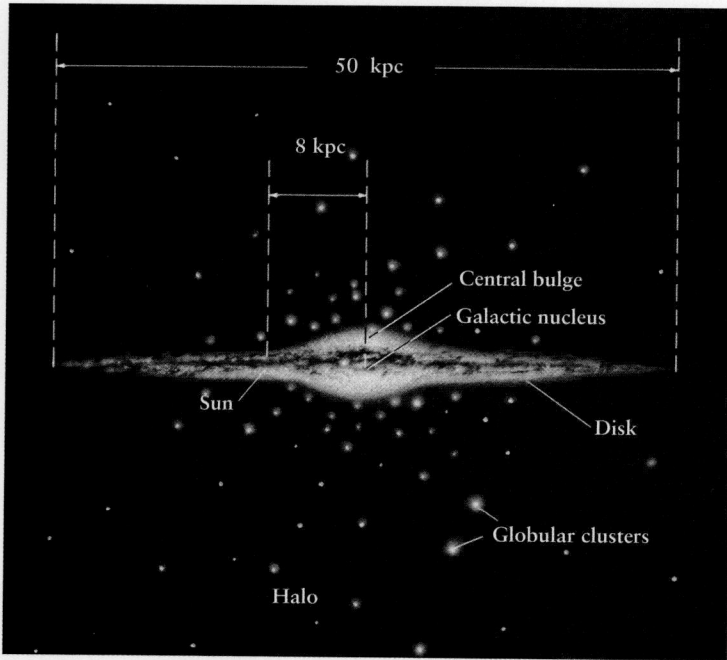

FIGURE 22-7

Our Galaxy (Schematic Edge-on View) There are three major components of our galaxy: a disk, a central bulge, and a halo. The disk contains gas and dust along with metal-rich (Population I) stars. The halo is composed almost exclusively of old, metal-poor (Population II) stars. The central bulge is a mixture of Population I and Population II stars.

does not emit very much light. Hence, the light sources in Figure 22-6b are stars, which do emit strongly in the near-infrared (especially the cool stars, such as red giants and supergiants). Because interstellar dust causes little interstellar extinction in the near-infrared, many of the stars whose light is recorded in Figure 22-6b lie deep within the Milky Way.

Observations such as those shown in Figure 22-6, along with the known distance to the center of the galaxy, have helped astronomers establish the dimensions of the galaxy. The **disk** of our galaxy is about 50 kpc (160,000 ly) in diameter and about 0.6 kpc (2000 ly) thick, as shown in Figure 22-7. The center of the galaxy is surrounded by a distribution of stars, called the **central bulge**, which is about 2 kpc (6500 ly) in diameter. This central bulge is clearly visible in Figure 22-6b. The spherical distribution of globular clusters traces the **halo** of the galaxy.

This structure is not unique to our Milky Way Galaxy. Figure 22-8 shows another galaxy whose dust and stars lie in a disk and that has a central bulge of stars, just like the Milky Way. In the same way that our Sun is a rather ordinary member of the stellar community that makes up the Milky Way, the Milky Way turns out to be a fairly common variety of galaxy.

CONCEPT CHECK 22-2

If an astronomer wanted to study the nature of cool stars hiding inside dust clouds, which wavelength would be the best choice, near-infrared (with shorter wavelengths that are nearer to the visible spectrum) or far-infrared (with longer wavelengths farther from the visible spectrum)?

CONCEPT CHECK 22-3

Would an imaginary space traveler have a better view of our Sun and its planets by standing at the galactic center or in the halo?

Answers appear at the end of the chapter.

The Milky Way's Distinct Stellar Populations

It is estimated that our galaxy contains about 200 billion (2×10^{11}) stars. Remarkably, different kinds of stars are found in the various components of the galaxy. The globular clusters in the halo are

(a) Infrared emission from dust in NGC 7331 at 5.8 and 8.0 μm

FIGURE 22-8 R I V U X G

NGC 7331: A Near-Twin of the Milky Way If we could view our galaxy from a great distance, it would probably look like this galaxy in the constellation Pegasus. As in Figure 22-6, the far-infrared image **(a)** reveals the presence of dust in the galaxy's plane, while the near-infrared image

(b) Infrared emission from stars in NGC 7331 at 3.6 and 4.5 μm

(b) shows the distribution of stars. These images of NGC 7331, which is about 15 million pc (50 million ly) from Earth, were made with the Spitzer Space Telescope. (NASA/JPL-Caltech/M. Regan [STScI], and the SINGS Team)

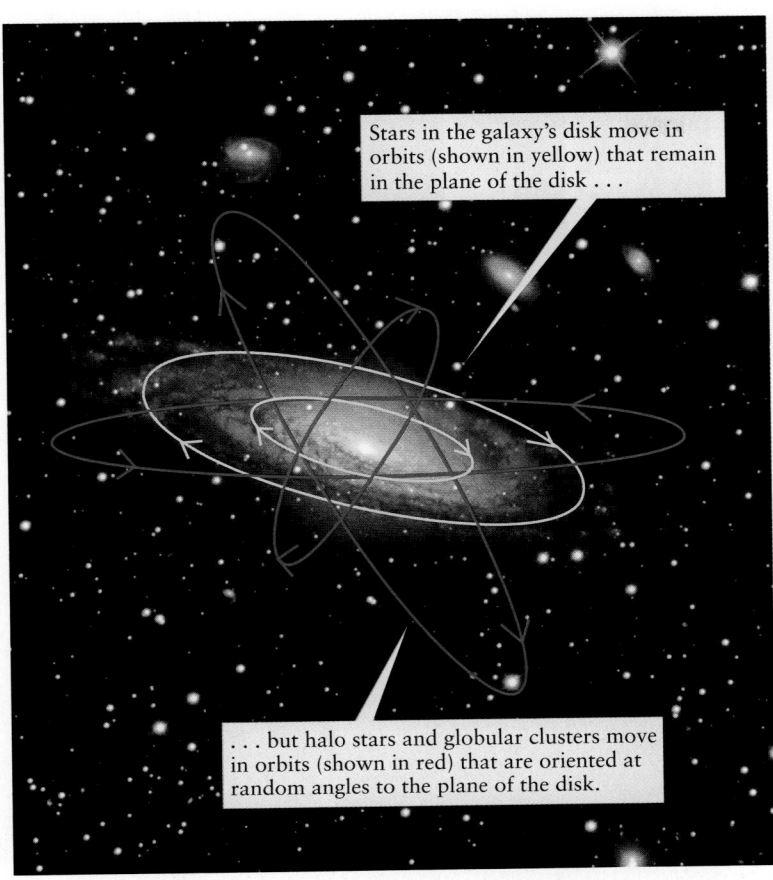

Stars in the galaxy's disk move in orbits (shown in yellow) that remain in the plane of the disk . . .

. . . but halo stars and globular clusters move in orbits (shown in red) that are oriented at random angles to the plane of the disk.

FIGURE 22-9 R I **V** U X G

Star Orbits in the Milky Way The different populations of stars in our galaxy travel along different sorts of orbits. The galaxy in this visible-light image is the Milky Way's near-twin NGC 7331, the same galaxy shown at infrared wavelengths in Figure 22-8. (Russell Croman/Science Source)

must be active star formation in the galactic disk. By contrast, no O or B stars are present in the halo, which implies that star formation ceased there long ago.

The central bulge contains *both* Population I stars and metal-poor Population II stars. Since Population II stars are thought to have formed early in the history of the universe, this suggests that some of the stars in the bulge are quite ancient while others were created more recently. The central bulge looks yellowish or reddish because it contains many red giants and red supergiants (see Figure 1-7), but does *not* contain luminous, short-lived, blue O or B stars. Hence, there cannot be ongoing star formation in the central bulge. The same is true for other galaxies whose structure is similar to that of the Milky Way (Figure 22-10).

Why are there such different populations of stars in the halo, disk, and central bulge? Why has star formation stopped in some regions of the galaxy but continues in other regions? The answers to these questions are related to the way that stars, as well as the gas and dust from which stars form, move within the galaxy.

CONCEPT CHECK 22-4

If you were looking to find the most recently formed stars in the galaxy, where might you look?

Answer appears at the end of the chapter.

composed of old, metal-poor, Population II stars (see Section 19-5). Although globular clusters are conspicuous, they contain only about 1% of the total number of stars in the halo; most halo stars are single Population II stars in isolation. These ancient stars orbit the galaxy along paths tilted at random angles to the disk of the Milky Way, as do the globular clusters. By contrast, stars in the disk travel along orbits that remain in the disk (Figure 22-9).

Unlike the halo, the stars in the disk are mostly young, metal-rich, Population I stars like the Sun. The disk of a galaxy like the Milky Way appears bluish because its light is dominated by radiation from hot O and B main-sequence stars. Such stars have very short main-sequence lifetimes (see Section 19-1, especially Table 19-1), so they must be quite young by astronomical standards. Hence, their presence shows that there

FIGURE 22-10 R I **V** U X G

Stellar Populations: Disk Versus Central Bulge The disk and central bulge of the Milky Way contain rather different populations of stars. The same is true for the galaxy NGC 1309, which has a similar structure to the Milky Way Galaxy and happens to be oriented face on to us. NGC 1309 is about 30 million pc (100 million ly) from us in the constellation Eridanus. (NASA, ESA, The Hubble Heritage Team [STScI/AURA], and A. Riess [STScI])

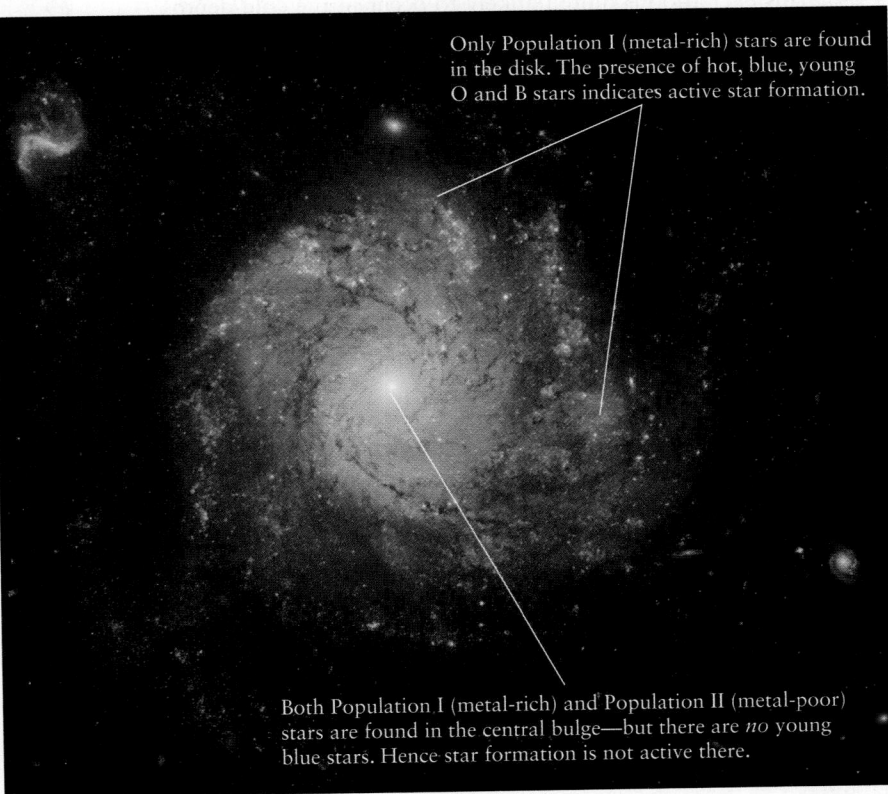

Only Population I (metal-rich) stars are found in the disk. The presence of hot, blue, young O and B stars indicates active star formation.

Both Population I (metal-rich) and Population II (metal-poor) stars are found in the central bulge—but there are *no* young blue stars. Hence star formation is not active there.

22-3 Observations of cold hydrogen clouds and star-forming regions reveal that our galaxy has spiral arms

The galaxies shown in Figure 22-8 and Figure 22-10 both have **spiral arms**, spiral-shaped concentrations of gas and dust that extend outward from the center in a shape reminiscent of a pinwheel. Assuming our galaxy is similar to other galaxies, observations of spiral arms would lead us to suspect that our own Milky Way Galaxy has this feature. However, because interstellar dust obscures our visible-light view in the pancake-like plane of our galaxy, a detailed understanding of the structure of our galactic disk had to wait until the development of radio astronomy. Thanks to their long wavelengths, radio waves can penetrate the interstellar medium even more easily than infrared light and can travel without being scattered or absorbed. As we shall see in this section, both radio and optical observations reveal that our galaxy does indeed have spiral arms.

Mapping Hydrogen in the Milky Way

LOOKING DEEPER 22.1 → Hydrogen is by far the most abundant element in the universe (see Figure 8-4 in Section 8-2). Hence, by looking for concentrations of hydrogen gas, we should be able to detect important clues about the distribution of matter in our galaxy. Unfortunately, ordinary visible-light telescopes are of little use in this quest,

> Our galaxy's dust and stars—including the Sun—lie mostly in a relatively thin disk.

because hydrogen atoms can only emit visible light if they are first excited to high energy levels (see Section 5-8, and especially Figure 5-24). This excitation is quite unlikely to occur in the cold depths of interstellar space. Furthermore, even if there are some hydrogen atoms that glow strongly at visible wavelengths, interstellar extinction due to dust (see Section 22-1) would make it impossible to see this glow from distant parts of the galaxy.

What makes it possible to map out the distribution of hydrogen in our galaxy is that even cold hydrogen clouds emit *radio* waves. As we saw in Section 22-2, radio waves can easily penetrate the interstellar medium, so we can detect the radio emission from such cold clouds no matter where in the galaxy they lie. The hydrogen in these clouds is neutral—that is, not ionized—and is called **H I**. (This distinguishes it from ionized hydrogen, which is designated H II.) To understand how H I clouds can emit radio waves, we must probe a bit more deeply into the structure of protons and electrons, the particles of which hydrogen atoms are made.

In addition to having mass and charge, particles such as protons and electrons possess a tiny amount of angular momentum (that is, rotational motion) commonly called **spin**. Very roughly, you can visualize a proton or electron as a tiny, electrically charged sphere that spins on its axis. Because electric charges in motion generate magnetic fields, a proton or electron behaves like a tiny magnet with a north pole and a south pole (**Figure 22-11**).

If you have ever played with magnets, you know that two magnets attract when the north pole of one magnet is next to the south

(a) The magnetic energy of two bar magnets depends on their relative orientation.

(b) The magnetic energy of a proton and electron depends on their relative spin orientation.

FIGURE 22-11

Magnetic Interactions in the Hydrogen Atom (a) The energy of a pair of magnets is high when their north poles or their south poles are near each other, and low when they have opposite poles near each other. **(b)** Thanks to their spin, electrons and protons are both tiny magnets. When the electron flips from the higher-energy configuration (with its spin in the same direction as the proton's spin) to the lower-energy configuration (with its spin opposite to the proton's spin), the atom loses a tiny amount of energy and emits a radio photon with a wavelength of 21 cm.

pole of the other and repel when two like poles (both north or both south) are next to each other (Figure 22-11a). This behavior can also be described in terms of magnetic energy: The energy of the two magnets is least when opposite poles are together and highest when like poles are together. Hence, as shown in Figure 22-11b, the energy of a hydrogen atom is slightly different, depending on whether the spins of the proton and electron are in the same direction or opposite directions. (According to the laws of quantum mechanics, these are the only two possibilities; the spins cannot be at random angles.)

If the spin of the electron changes its orientation from the higher-energy configuration to the lower-energy one—called a **spin-flip transition**—a photon is emitted. The energy difference between the two spin configurations is very small, only about 10^{-6} as great as those between different electron orbits (see Figure 5-24). Therefore, the photon emitted in a spin-flip transition between these configurations has only a small energy, and thus its wavelength is a relatively long 21 cm—a radio wavelength. These spin-flip transitions occur spontaneously in our Milky Way's diffuse hydrogen gas, which allows the possibility of mapping hydrogen through this method.

The spin-flip transition in neutral hydrogen was first predicted in 1944 by the Dutch astronomer Hendrik van de Hulst. His calculations suggested that it should be possible to detect the **21-cm radio emission** from interstellar hydrogen, although a very sensitive radio telescope would be required. In 1951, Harold Ewen and Edward Purcell at Harvard University first succeeded in detecting this faint emission from hydrogen between the stars.

Figure 22-12 shows the results of a more recent 21-cm survey of the entire sky. Neutral hydrogen gas (H I) in the plane of the Milky Way stands out prominently as a bright band across the middle of this image.

LOOKING DEEPER 22.2 The distribution of gas in the Milky Way is not uniform but is actually quite frothy. In fact, our Sun lies near the edge of an irregularly shaped region within which the interstellar medium is very thin but at very high temperatures (about 10^6 K, but so thin it would feel cold). This region, called the **Local Bubble**, is several hundred parsecs across. The Local Bubble may have been carved out by a supernova that exploded nearby some 300,000 years ago.

Remarkably, spin-flip transitions are used not only to map our galaxy but also to map the internal structure of the human body. Box 22-1 discusses this application, called *magnetic resonance imaging*.

CONCEPT CHECK 22-5

If a spin-flip transition were to release substantially more energy, such that a photon is emitted in the far-infrared range, would the Milky Way stretching across the night sky appear brighter or dimmer?

CALCULATION CHECK 22-1

If a 21-cm photon is observed when a single hydrogen atom undergoes a spin-flip transition, what wavelength is observed when 10 hydrogen atoms undergo spin-flip transitions?

Answers appear at the end of the chapter.

Detecting Our Galaxy's Spiral Arms

The detection of 21-cm radio emission was a major breakthrough that permitted astronomers to reveal the presence of spiral arms in the galactic disk. Figure 22-13 shows how spiral arms were detected. Suppose that you aim a radio telescope along a particular line of sight across the galaxy. Your radio receiver, located at point S (the position of the solar system), picks up 21-cm emission from H I clouds at points 1, 2, 3, and 4. However, the radio waves from these various clouds are Doppler shifted by slightly different amounts because the clouds are moving at different speeds as they travel with the rotating galaxy.

It is important to remember that the Doppler shift reveals only motion along the line of sight (review Figure 5-26). In Figure 22-13, cloud 2 has the highest speed along our line of sight because it is moving directly toward us. Consequently, the radio waves from cloud 2 exhibit a larger Doppler shift than those from the other three clouds along our line of sight. Because clouds 1 and 3 are at the same distance from the galactic center, they have the same orbital speed. The fraction of their velocity parallel to our line of sight is also the same, so their radio waves exhibit the same

21-cm emission shows that hydrogen gas is concentrated along the plane of the galaxy.

FIGURE 22-12 R I V U X G

The Sky at 21 Centimeters This image was made by mapping the sky with radio telescopes tuned to the 21-cm wavelength emitted by neutral interstellar hydrogen (H I). The entire sky has been mapped onto an oval, and the plane of the galaxy extends horizontally across the image as in Figure 22-6. Black and blue represent the weakest emission, and red and white the strongest. (Courtesy of C. Jones and W. Forman, Harvard-Smithsonian Center for Astrophysics)

BOX 22-1 ASTRONOMY DOWN TO EARTH

Spin-Flip Transitions in Medicine

Thanks to their spin, protons and electrons act like microscopic bar magnets. In a hydrogen atom, the interaction between the magnetism of the electron and that of the proton gives rise to the 21-cm radio emission. But these particles can also interact with outside magnetic fields, such as that produced by a large electromagnet. This physical principle is behind **magnetic resonance imaging**, an important diagnostic tool of modern medicine.

Much of the human body is made of water. Every water molecule has two hydrogen atoms, each of which has a nucleus made of a single proton. If a person's body is placed in a strong magnetic field, the spins of the protons in the hydrogen atoms making up his or her body can either be in the same direction as the field ("aligned") or in the direction opposite to the field ("opposed"), and these directions correspond to different energies. The majority of protons end up with their spins in the lower-energy orientation. But if a radio wave of just the right wavelength is now sent through the person's body, a proton can absorb a radio photon and flip its spin into the higher-energy orientation. How much of the radio wave is absorbed depends on the number of protons in the body, which in turn depends on how much water (and, thus, how much water-containing tissue) is in the body.

In magnetic resonance imaging, a magnetic field is used whose strength varies from place to place. The difference in energy between the lower- and higher-energy orientations of a proton depends on the strength of the magnetic field, so radio waves will be absorbed only at places where this energy difference is equal to the energy of a radio photon. (This equality is called *resonance*, which is how magnetic resonance imaging gets its name.) By varying the magnetic field strength over the body

and the wavelength of the radio waves, and by measuring how much of the radio wave is absorbed by different parts of the body, it is possible to map out the body's tissues. The accompanying image shows such a map of a patient's head.

Unlike x-ray images, which show only the densest parts of the body, such as bones and teeth, magnetic resonance imaging can be used to view less dense (but water-containing) soft tissue. Just as the 21-cm radio emission has given astronomers a clear view of what were hidden regions of our galaxy, magnetic resonance imaging allows modern medicine to see otherwise invisible parts of the human body.

R I V U X G

Doppler shift, which is less than the Doppler shift of cloud 2. Finally, cloud 4 is the same distance from the galactic center as the Sun. This cloud is thus orbiting the galaxy at the same speed as the Sun, resulting in no net motion along the line of sight. Radio waves from cloud 4, as well as from hydrogen gas near the Sun, are not Doppler shifted at all.

These various Doppler shifts cause radio waves from gases in different parts of the galaxy to arrive at our radio telescopes with wavelengths slightly different from 21 cm. It is therefore possible to sort out the various gas clouds and thus produce a map of the galaxy like that shown in Figure 22-14.

Figure 22-14 shows that neutral hydrogen gas is not spread uniformly around the disk of the galaxy but is concentrated into

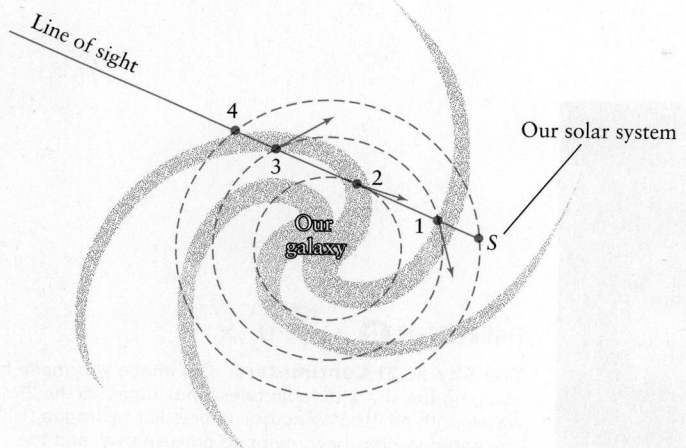

- Hydrogen clouds 1 and 3 are approaching us: They have a moderate blueshift.
- Hydrogen cloud 2 is approaching us at a faster speed: It has a larger blueshift.
- Hydrogen cloud 4 is the same distance from the galactic center as the Sun: No Doppler shift observed.

FIGURE 22-13

A Technique for Mapping Our Galaxy If we look within the plane of our galaxy from our position at point S, hydrogen clouds at different locations (shown as 1, 2, 3, and 4) along our line of sight are moving at slightly different speeds relative to us. As a result, radio waves from these various gas clouds are subjected to slightly different Doppler shifts. This permits radio astronomers to sort out the gas clouds and thus map the galaxy.

You are here

FIGURE 22-14 Ⓡ I V U X Ⓖ

A Map of Neutral Hydrogen in Our Galaxy This map, constructed from radio-telescope surveys of 21-cm radiation, shows the distribution of hydrogen gas in a reconstructed (or hypothetical) face-on view of our galaxy. The map suggests a spiral structure. Details in the blank, wedge-shaped region at the bottom of the map are unknown. Gas in this part of the galaxy is moving perpendicular to our line of sight and thus does not exhibit a detectable Doppler shift. (Image courtesy Leo Blitz, PhD)

numerous arched lanes. Similar features are seen in other galaxies beyond the Milky Way. Unlike the view from within our own galaxy, we can view other galaxies face on to easily see their

distribution of stars, gas, and dust. As an example, the galaxy in **Figure 22-15a** has prominent spiral arms outlined by hot, luminous, blue main-sequence stars and the red emission nebulae (H II regions) found near many such stars. Stars of this sort are very short-lived, so these features indicate that spiral arms are sites of active, ongoing star formation. The 21-cm radio image of this same galaxy, shown in Figure 22-15b, shows that spiral arms are also regions where neutral hydrogen gas is concentrated, similar to the structures in our own galaxy visible in Figure 22-14. This similarity is a strong indication that our galaxy also has spiral arms.

CAUTION!

Photographs such as Figure 22-15a can lead to the impression that there are very few stars between the spiral arms of a galaxy. Nothing could be further from the truth! In fact, stars are distributed rather uniformly throughout the disk of a galaxy like the one in Figure 22-15a; the density of stars in the spiral arms is only about 5% higher than in the rest of the disk. The spiral arms stand out nonetheless because they are where hot, blue O and B stars are found. One such star is about 10^4 times more luminous than an average star in the disk, so the light from O and B stars completely dominates the visible appearance of a spiral galaxy.

CONCEPT CHECK 22-6

When looking at neutral hydrogen gas that is moving away from you, how is its wavelength changed compared to neutral hydrogen gas that is not moving relative to you?

Answer appears at the end of the chapter.

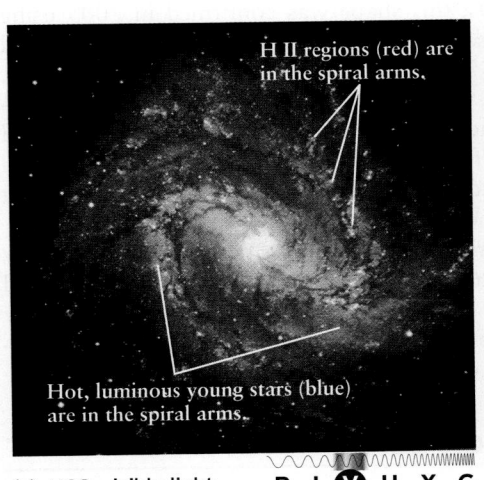

H II regions (red) are in the spiral arms.

Hot, luminous young stars (blue) are in the spiral arms.

(a) M83, visible light   Ⓡ I Ⓥ U X Ⓖ

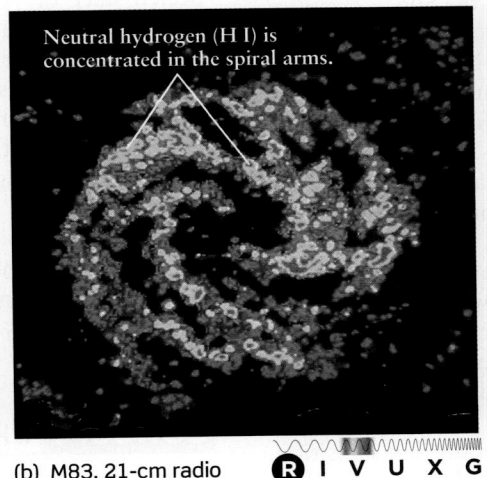

Neutral hydrogen (H I) is concentrated in the spiral arms.

(b) M83, 21-cm radio   Ⓡ I V U X Ⓖ

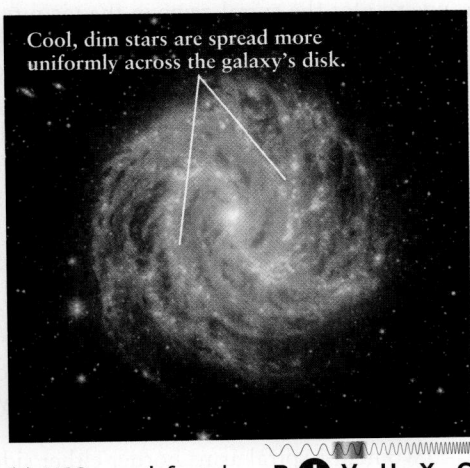

Cool, dim stars are spread more uniformly across the galaxy's disk.

(c) M83, near-infrared   Ⓡ Ⓘ V U X Ⓖ

FIGURE 22-15

A Spiral Galaxy The galaxy M83 lies in the southern constellation Hydra about 5 million pc (15 million ly) from Earth. **(a)** This visible-light image clearly shows the spiral arms. The presence of young stars and H II regions indicates that star formation takes place in spiral arms. **(b)** This radio view at a wavelength of 21 cm shows the emission from neutral interstellar hydrogen gas (H I). Note that essentially the same pattern of spiral arms is traced

out in this image as in the visible-light photograph. **(c)** M83 has a different appearance in this near-infrared view. The starlight has been removed in this image to reveal the infrared emission of dust throughout the galaxy. (a: © Australian Astronomical Observatory/David Malin Images; b: NRAO/AUI/NSF/ Science Photo Library/Science Source; c: NASA/JPL-Caltech)

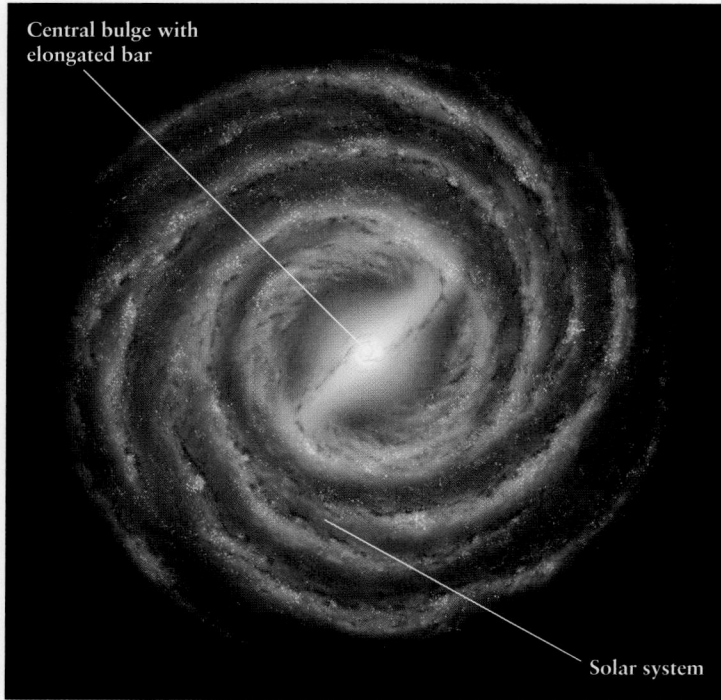

Central bulge with elongated bar

Solar system

(a) The structure of the Milky Way's disk

To the center of the galaxy

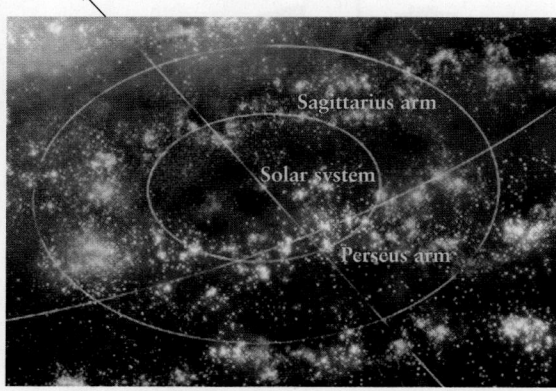

Sagittarius arm

Solar system

Perseus arm

(b) Closeup of the Sun's galactic neighborhood

FIGURE 22-16

Our Galaxy Seen Face On: Artist's Impressions **(a)** The galaxy's diameter is about 50,000 pc (160,000 ly), and our solar system is about 8000 pc (26,000 ly) from the galactic center. The elongated central bulge is about 8300 pc (27,000 ly) long and is oriented at approximately 45° to a line running from the solar system to the galactic center. **(b)** Our solar system is located between the Sagittarius and Perseus arms, two of the major spiral arms in the Milky Way. (a: NASA/JPL-Caltech; b: Ng Maps/National Geographic Creative)

Mapping the Spiral Arms and the Central Bulge

Figure 22-15a suggests that we can confirm the presence of spiral structure in our own galaxy by mapping the locations of star-forming regions. Such regions are marked by OB associations, H II regions, and molecular clouds (see Section 18-7). Unfortunately, the first two of these are best observed using visible light, and interstellar extinction limits the range of visual observations in the plane of the galaxy to less than 3 kpc (10,000 ly) from Earth. But there are enough OB associations and H II regions within this range to plot the spiral arms in the vicinity of the Sun.

Molecular clouds are easier to observe at great distances, because molecules of carbon monoxide (CO) in these clouds emit radio waves that are relatively unaffected by interstellar extinction. Hence, the positions of molecular clouds have been plotted even in remote regions of the galaxy, as Figure 18-25 shows. (We saw in Section 18-7 that CO molecules in molecular clouds emit more strongly than the hydrogen atoms do, even though hydrogen is the principal constituent of these clouds.)

Taken together, all these observations demonstrate that our galaxy has at least four major spiral arms as well as several short arm segments (Figure 22-16). The Sun is located just outside a relatively short arm segment called the Orion arm, which includes the Orion Nebula and neighboring sites of vigorous star formation in that constellation.

Two major spiral arms border either side of the Sun's position. The Sagittarius arm is on the side toward the galactic center. You see this arm on June and July nights when you look at the portion of the Milky Way stretching across Scorpius and Sagittarius, near the center of the upper photograph that opens this chapter.

In December and January, when our nighttime view is directed away from the galactic center, we see the Perseus arm. The other major spiral arms cannot be seen at visible wavelengths due to the obscuring effects of dust.

Figure 22-16a also shows that the central bulge of the Milky Way is not spherical, but is elongated like a bar. The Milky Way's bulge is unlike the bulge of the galaxy NGC 7331 shown in Figure 22-8, but similar to the bulge of the galaxy M83 shown in Figure 22-15. The elongated shape of the central bulge had been suspected since the 1980s; this shape was confirmed in 2005 using the Spitzer Space Telescope, which was used to survey the infrared emissions from some 3 million stars in the central bulge. Thus, the artist's impression shown in Figure 22-16a is based on observations using both radio wavelengths (for the spiral arms) and infrared wavelengths (for the central bulge). We will see in Section 22-5 that this elongated shape may play a crucial role in sustaining the galaxy's spiral structure.

To get a full picture of our galaxy, we need to know not only where our stars and gas are located, but we must also understand their motion. We now look at the rotational motion of our galaxy and how it reveals one of the greatest mysteries in all of science—dark matter.

CONCEPT CHECK 22-7

If you traveled from Earth in a direction *away* from the center of the galaxy, what is the first major arm you would reach?

Answer appears at the end of the chapter.

22-4 The rotation of our galaxy reveals the presence of dark matter

The Doppler shifts observed in the disk of our galaxy tell us that the disk rotates. This rotation shows that the stars, gas, and dust in our galaxy are all orbiting the galactic center. Indeed, if this were not the case, mutual gravitational attraction would cause the entire galaxy to collapse into the galactic center. In the same way, the Moon is kept from crashing into Earth and the planets from crashing into the Sun because of their orbital motion (see Section 4-7).

Measuring the rotation of our galaxy accurately is a difficult business. But such challenging measurements have been made, as we shall see, and the results led to a remarkable conclusion: Most of the mass of the galaxy is in the form of *dark matter*, a mysterious sort of material that emits no light at all.

Measuring How the Milky Way Rotates

Radio observations of 21-cm radiation from hydrogen gas provide important clues about our galaxy's rotation. Doppler shift measurements of this radiation indicate that stars and gas all orbit in the same direction around the galactic center, just as the planets all orbit in the same direction around the Sun.

Measurements also show that the orbital speed of stars and gas around the galactic center is fairly uniform throughout much of the galaxy's disk (Figure 22-17). As a result, stars orbiting between the Sun and the galactic center complete a trip around the galactic center more quickly than the Sun because the stars have a shorter distance to travel. Conversely, stars outside the Sun's orbit take longer to go once around the galactic center because they have farther to travel. As seen by Earth-based astronomers moving along with the Sun, stars inside the Sun's orbit overtake and pass us, while we overtake and pass stars outside the Sun's orbit (Figure 22-17a).

CAUTION!

Note that when we say that objects in different parts of the galaxy orbit at the same speed, we do *not* mean that the galaxy rotates like a solid disk. All parts of a rotating solid disk—a CD or DVD, for example—take the same time to complete one rotation. Because the outer part of the disk has to travel around a larger circle than the inner part, the speed (distance per time) is greater in the outer part (Figure 22-17b). By contrast, the orbital speed of material in our galaxy is roughly the *same* at all distances from the galactic center.

FIGURE 22-17

The Rotation of Our Galaxy Each schematic diagram follows three stars (the Sun and two others) orbiting the center of the galaxy at different distances from the galactic center. **(a)** This case of stars with similar, nearly uniform, orbital speeds is what we observe in our galaxy. Although they start off lined up in this illustration, the stars become increasingly separated as they move along their orbits. With a uniform speed for all stars, stars inside the Sun's orbit overtake and move ahead of the Sun, while stars far from the galactic center lag behind the Sun. **(b)** The stars would remain lined up if the galaxy rotated like a solid disk. This orientation is not what is observed. **(c)** If stars orbited the galactic center in the same way that planets orbit the Sun, stars inside the Sun's orbit would overtake us faster than they are observed to do.

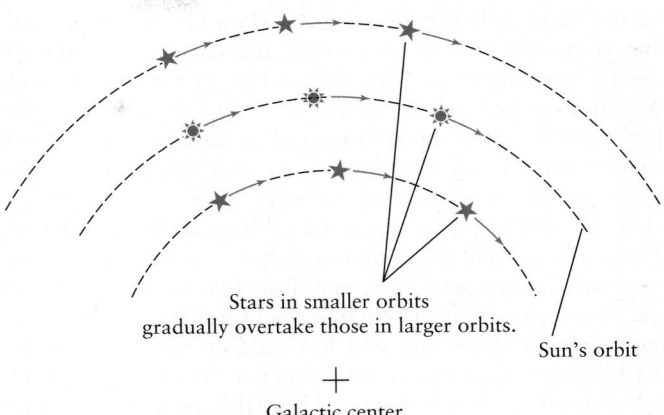

(a) The orbital speed of stars and gas around the galactic center is nearly uniform throughout most of our galaxy.

Stars in smaller orbits gradually overtake those in larger orbits.

Sun's orbit

Galactic center

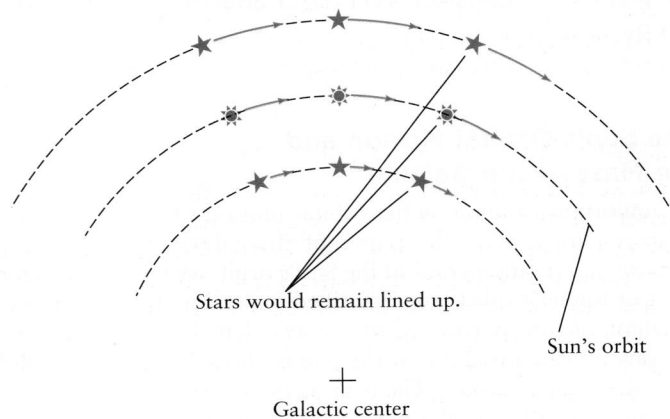

(b) If our galaxy rotated like a solid disk, the orbital speed would be greater for stars and gas in larger orbits.

Stars would remain lined up.

Sun's orbit

Galactic center

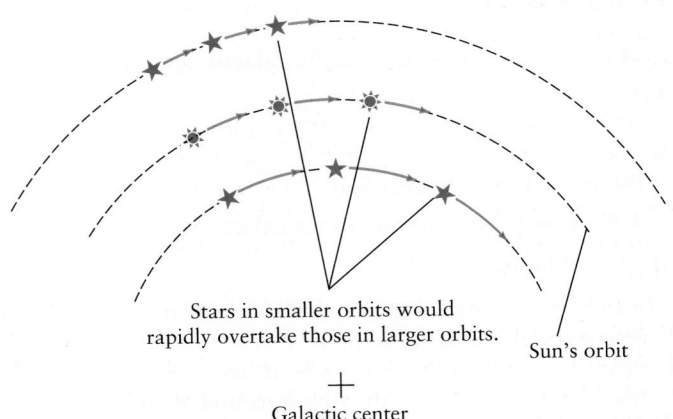

(c) If the Sun and stars obeyed Kepler's third law, the orbital speed would be less for stars and gas in larger orbits.

Stars in smaller orbits would rapidly overtake those in larger orbits.

Sun's orbit

Galactic center

The most familiar examples of orbital motion are the motions of the planets around the Sun. As we saw in Section 4-7, the farther a planet is from the Sun, the less gravitational force it experiences and the slower the speed it needs to have to remain in orbit. The same would be true for the orbits of stars and gas in the galaxy *if* they were held in orbit by a single, massive object at the galactic center (see Figure 22-17c). Hence, the 21-cm H I wavelength observations of our galaxy, which show that the speed of orbiting objects does *not* decrease with increasing distance from the galactic center, demonstrate that there is no such single, massive object holding objects in their galactic orbits.

Instead, what keeps a star in its orbit around the center of galaxy is the combined gravitational force exerted on it by all of the mass (including stars, gas, and dust) that lies *within* the star's orbit. (It turns out that the gravitational force from matter *outside* a star's orbit balances out and has little or no net effect on the star's motion around the galactic center.) This gives us a tool for determining the galaxy's mass and how that mass is distributed.

CONCEPT CHECK 22-8

Which would have a greater impact on changing the length of time it takes the Sun to orbit the galactic center: adding more stars to the galactic center or adding more stars to the galactic halo?

Answer appears at the end of the chapter.

The Sun's Orbital Motion and the Mass of the Galaxy

An important example is the orbital motion of the Sun (and the solar system) around the center of the galaxy. If we know the semimajor axis and period of the Sun's orbit, we can use Newton's form of Kepler's third law (described in Section 4-7) to determine the mass of that portion of the galaxy that lies within the orbit. We saw in Section 22-2 that the Sun is about 8000 pc (26,000 ly) from the galactic center. The orbit is in fact nearly circular, so we can regard 8000 pc as the radius r of the orbit. (This radius is also the semimajor axis of the Sun's orbit.) In one complete trip around the galaxy, the Sun travels a distance equal to the circumference of its orbit, which is $2\pi r$. The time required for one orbit, or orbital period P, is equal to the distance traveled divided by the Sun's orbital speed v:

Period of the Sun's orbit around the galactic center

$$P = \frac{2\pi r}{v}$$

P = orbital period of the Sun

r = distance from the Sun to the galactic center

v = orbital speed of the Sun

Unfortunately, we cannot tell the Sun's orbital speed from 21-cm observations because these reveal only how fast things are moving relative to the Sun. Instead, we need to measure how the Sun is moving relative to a background that is not rotating along with the rest of the galaxy. Such a background is provided by distant galaxies beyond the Milky Way and by the globular clusters. (Since globular clusters lie outside the plane of the Milky

Way, they do not take part in the rotation of our galaxy's disk.) By measuring the Doppler shifts of these objects and averaging their velocities, astronomers deduce that the Sun is moving along its orbit around the galactic center at about 220 km/s—about 790,000 km/h or 490,000 mi/h!

Using this information, we find that the Sun's orbital period is

$$P = \frac{2\pi \times 8000 \text{ pc}}{220 \text{ km/s}} \times \frac{3.09 \times 10^{13} \text{ km}}{1 \text{ pc}} = 7.1 \times 10^{15} \text{ s}$$

$$= 2.2 \times 10^8 \text{ years}$$

Traveling at 790,000 km/h, it takes the Sun about 220 million years to complete one trip around the galaxy. (In the 65 million years since the demise of the dinosaurs, our solar system has traveled less than a third of the way around its orbit.) The galaxy is a very large place!

Box 22-2 shows how to combine the radius and period of the Sun's orbit to calculate the total mass of all the matter that lies inside the Sun's orbit. Such calculations give an answer of $9.0 \times 10^{10} \text{ M}_\odot$ (90 billion solar masses). As Figure 22-7 shows, the galaxy extends well beyond the Sun's orbit, so the mass of the entire galaxy must be larger than this mass.

CALCULATION CHECK 22-2

Since it formed 4.5 billion years ago, how many orbits around the galaxy has the Sun made?

Answer appears at the end of the chapter.

Rotation Curves and the Mystery of Dark Matter

In recent years, astronomers have been astonished to discover how much matter may lie outside the Sun's orbit. The clues come from 21-cm radiation emitted by hydrogen in spiral arms that extend to the outer reaches of the galaxy. Because we know the true speed of the Sun, we can convert the Doppler shifts of this radiation into actual speeds for the spiral arms. This calculation gives us a **rotation curve**, a graph of the speeds of galactic rotation measured outward from the galactic center (**Figure 22-18**). We would expect that for gas clouds beyond the confines of most of the galaxy's mass, the orbital speed should decrease with increasing distance from the galaxy's center, just as the orbital speeds of the planets decrease with increasing distance from the Sun (see Figure 22-17c). But as Figure 22-18 shows, the galaxy's rotation curve is quite flat, indicating roughly uniform orbital speeds well beyond the visible edge of the galactic disk.

To explain these nearly uniform orbital speeds in the outer parts of the galaxy, astronomers conclude that a large amount of matter must lie outside the Sun's orbit. When this matter is included, the total mass of our galaxy could exceed $10^{12} \text{ M}_\odot$ or more, of which only about 10% is regular matter (mostly in the form of stars).

Unlike the galaxy's stars and dust, its dark matter forms a roughly spherical halo.

These observations lead to a profound mystery. Stars, gas, and dust account for only about 10% of the galaxy's total mass. What, then, makes up the remaining 90% of the matter in our galaxy? Whatever it's made of, it's dark. It does not show up on photographs nor indeed in images made in any part of the

BOX 22-2   TOOLS OF THE ASTRONOMER'S TRADE

Estimating the Mass Inside the Sun's Orbit

The force that keeps the Sun in orbit around the center of the galaxy is the gravitational pull of all the matter *interior* to the Sun's orbit—this force is described by Newton's law of gravity. We can estimate the total mass of all of this matter using Newton's form of Kepler's third law (see Section 4-7 and Box 4-4):

$$P^2 = \frac{4\pi^2 a^3}{G(M + M_\odot)}$$

In this equation, P is the orbital period of the Sun, a is the semimajor axis of the Sun's orbit around the galactic center, G is the universal constant of gravitation, M is the amount of mass inside the Sun's orbit, and $M_\odot$ is the mass of the Sun.

Because the Sun is only one of more than 10^{11} stars in the galaxy, the Sun's mass is minuscule compared to M. Hence, we can safely replace the sum $M + M_\odot$ in the above equation with just M. If we now assume that the Sun's orbit is a circle, the semimajor axis a of the orbit is just the radius of this circle, which we call r. The period P of the orbit is equal to $2\pi r/v$, where v is the Sun's orbital speed. You can then show that

$$M = \frac{rv^2}{G}$$

(We leave the derivation of this equation as an exercise at the end of this chapter.)

Now we can insert known values to obtain the mass inside the Sun's orbit. Being careful to express distance in meters and speed in meters per second, we have $v = 220$ km/s $= 2.2 \times 10^5$ m/s, $G = 6.67 \times 10^{-11}$ newton·m²/kg², and

$$r = 26{,}000 \text{ ly} \times \frac{9.46 \times 10^{12} \text{ km}}{1 \text{ ly}} \times \frac{10^3 \text{ m}}{1 \text{ km}}$$

$$= 2.5 \times 10^{20} \text{ m}$$

Hence, we find that

$$M = \frac{2.5 \times 10^{20} \times (2.2 \times 10^5)^2}{6.67 \times 10^{-11}} = 1.8 \times 10^{41} \text{ kg}$$

or, in terms of the mass of the Sun,

$$M = 1.8 \times 10^{41} \text{ kg} \times \frac{1 \text{ M}_\odot}{1.99 \times 10^{30} \text{ kg}}$$

$$= 9.0 \times 10^{10} \text{ M}_\odot$$

This estimate involves only mass that is interior to the Sun's orbit. Matter outside the Sun's orbit has no net gravitational effect on the Sun's motion and thus does not enter into Kepler's third law. (This is strictly true only if the matter outside our orbit is distributed over a sphere rather than a disk. In fact, the dark matter that dominates our galaxy seems to have a spherical distribution.)

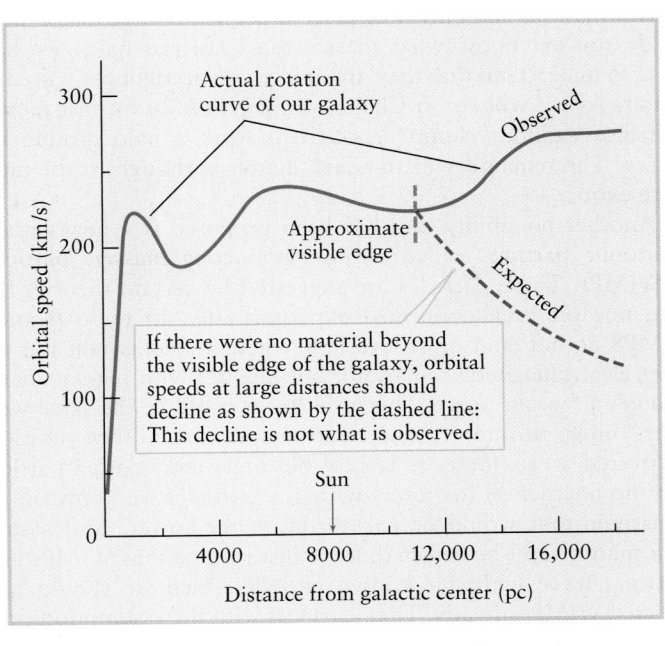

FIGURE 22-18

The Galaxy's Rotation Curve   The blue curve shows the orbital speeds of stars and gas in the disk of the galaxy out to a distance of 18,000 pc from the galactic center. (Very few stars are found beyond this distance.) The dashed red curve indicates how this orbital speed should decline beyond the confines of most of the galaxy's visible mass. Because there is no such decline, there must be an abundance of invisible dark matter that extends to great distances from the galactic center.

electromagnetic spectrum. This unseen material, which is by far the predominant constituent of our galaxy, is called **dark matter**. We sense its presence in our galaxy only through its gravitational influence on the orbits of stars and gas clouds.

CAUTION!

Be careful not to confuse dark *matter* with dark *nebulae*. A dark nebula like the one in Figure 18-4 emits no visible light, but it does radiate at longer wavelengths. By contrast, no electromagnetic radiation of any kind has yet been discovered coming from dark matter.

Our galaxy's dark matter forms a spherical halo, with the galactic nucleus at its center, like the halo stars and globular clusters shown in Figure 22-7. However, the dark matter halo is much larger; it may extend to a distance of 100–200 kpc from the center of our galaxy, some 2 to 4 times the extent of the visible halo (**Figure 22-19**). Analysis of the rotation curve in Figure 22-18 shows that the density of the dark matter halo decreases with increasing distance from the center of the galaxy.

CONCEPT CHECK 22-9

How would the motion of stars at the edge of our galaxy be different if there were no dark matter present?

Answer appears at the end of the chapter.

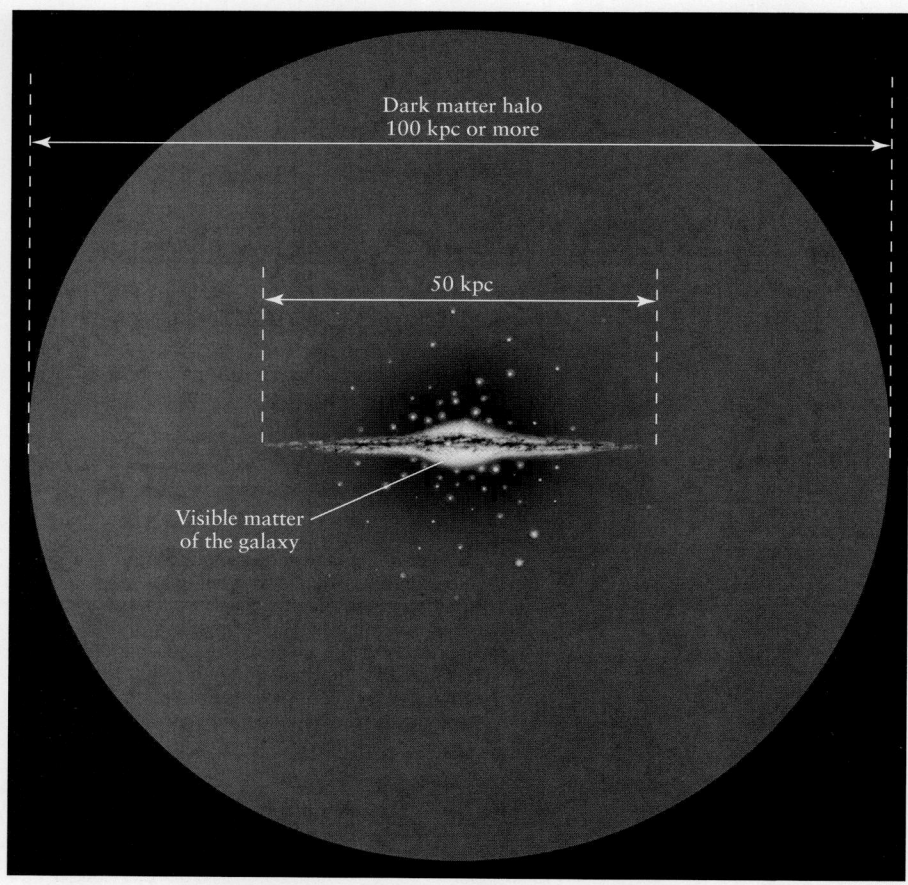

Dark matter halo
100 kpc or more

50 kpc

Visible matter
of the galaxy

FIGURE 22-19

The Galaxy and Its Dark Matter Halo The dark matter in our galaxy forms a spherical halo whose center is at the center of the visible galaxy. The extent of the dark matter halo is unknown, but its diameter is at least 100 kpc. The total mass of the dark matter halo is at least 10 times the combined mass of all of the stars, dust, gas, and planets in the Milky Way.

Dark Matter Speculations

What is the nature of this mysterious dark matter? One proposal is that the dark matter halo is composed, at least in part, of dim objects with masses less than 1 $M_\odot$. These objects, which could include brown dwarfs, white dwarfs, or black holes, are called **massive compact halo objects**, or **MACHOs**. While some brown dwarfs and white dwarfs emit measurable light, it is possible that many more are simply too dim to detect, making them candidates for dark matter. Astronomers have searched for MACHOs by monitoring the light from distant stars. If a MACHO passes between us and the star, its gravity will bend the light coming from the star. (In Section 23-8, we describe how gravity can bend starlight.) As **Figure 22-20** shows, the MACHO's gravity acts like a lens that focuses the light from the star. This effect, called **microlensing**, makes the star appear to brighten substantially for a few days.

Astronomers have indeed detected MACHOs in this way, but not nearly enough to solve the dark matter mystery. MACHOs with very low mass (10^{-6} to 0.1 $M_\odot$ each) do not appear to be a significant part of the dark matter halo. MACHOs of roughly 0.5 $M_\odot$ are more prevalent, but account for less than 10% of the dark matter halo.

The remainder of the dark matter is thought to be more exotic than the regular atoms (consisting of electrons, protons, and neutrons) that make up known astrophysical objects. One idea proposed for dark matter is the *neutrino*. As we saw in Section 16-4, one type of neutrino can transform into another, and these transformations can only take place if neutrinos have mass. While we do not yet know what masses the neutrinos have, we have come to understand that they are unlikely to account for halo dark matter. As we will see in Chapter 26 (Section 26-6), fast-moving neutrinos cannot "clump" enough to form a halo around our galaxy. The remainder of the dark matter is thought to be much more exotic.

Another possibility that has been proposed is a new class of subatomic particles called **weakly interacting massive particles**, or **WIMPs**. These particles are suggested by certain theories that have not yet been confirmed experimentally. In these theories, WIMPS do not emit or absorb light (neither visible light nor any other electromagnetic radiation), but interact with regular matter through a "weak" force. This weak force could be the same weak force known to interact with neutrinos, or some other yet-to-be-discovered weak force. It is possible that dark matter particles have no interaction (or force) with any particles we know of, but to assume that would be to "throw in the towel" and assume dark matter can't be detected. If we just assume that WIMPs have a strong force with our known particles, then we should have detected WIMPs already. That leaves us with the assumption—and hope!—that dark matter particles might have some weak interaction (or weak force) with the particles we know of.

Based on models that suggest likely masses, a billion or so WIMPs might be passing through you each second (compared to a trillion or so neutrinos). While dark matter makes up 90% of the matter in our galaxy, dark matter particles, such as WIMPs, would

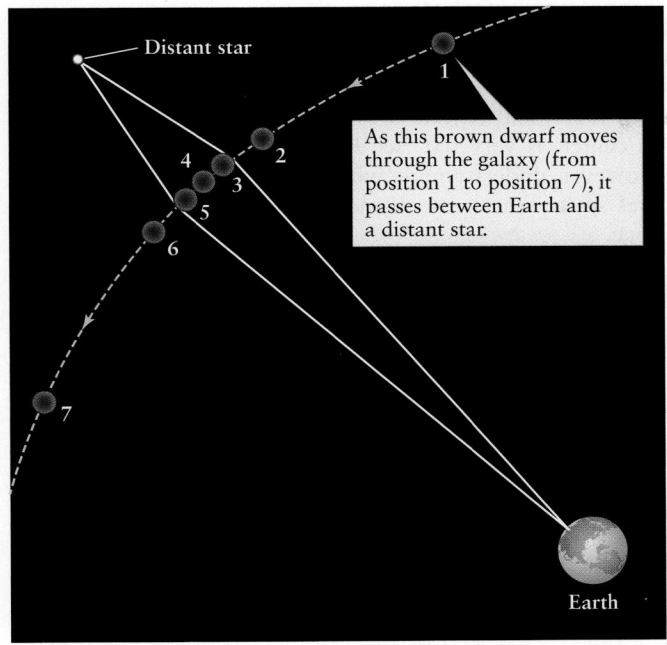

(a)

When the brown dwarf is directly between us and the distant star [near position 4 in (a)], it acts as a gravitational lens and makes the distant star appear brighter.

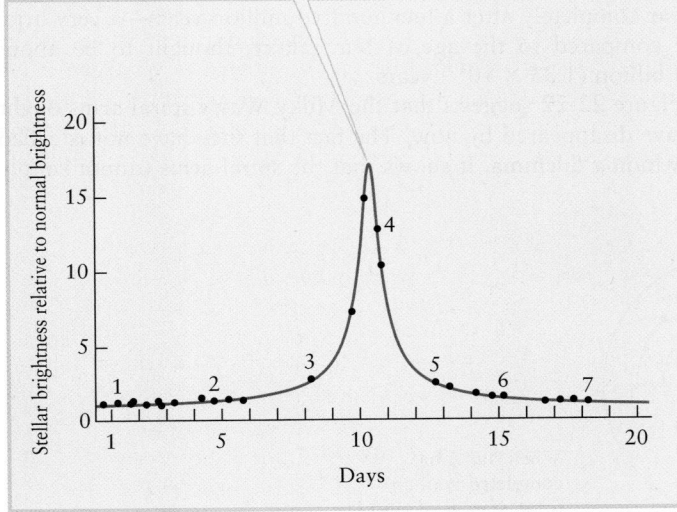

(b)

FIGURE 22-20

Microlensing by Dark Matter in the Galactic Halo **(a)** If a dense object such as a brown dwarf or black hole passes between Earth and a distant star, the gravitational curvature of spacetime around the dense object deflects the starlight and focuses it in our direction. This effect is called microlensing. **(b)** This light curve shows the gravitational microlensing of light from a star in the galaxy's central bulge. Astronomers do not know the nature of the object that passed between Earth and this star to cause the microlensing. (Courtesy of the MACHO and GMAN Collaborations)

be spread out much more than regular matter. In fact, the amount of dark matter passing through Earth at any given time would only be about 1 kg!

Physicists are attempting to detect these curious particles, which might have masses 10 to 10,000 times greater than a proton or neutron, by using a large crystal cooled almost to absolute zero. If a WIMP should enter this crystal and collide with one of its atoms, the collision will deposit a tiny but measurable amount of heat in the crystal.

Another approach for detecting WIMPs searches for the light (and electrons) that should be produced after the WIMPs (if they exist at all) collide with xenon atoms. One such experiment that is expected to begin collecting data in 2020 is called LZ and is at the bottom of the Homestake Mine in South Dakota (Figure 22-21). The earth above the experiment shields out unwanted particles coming in from space, while the WIMPs should easily pass through

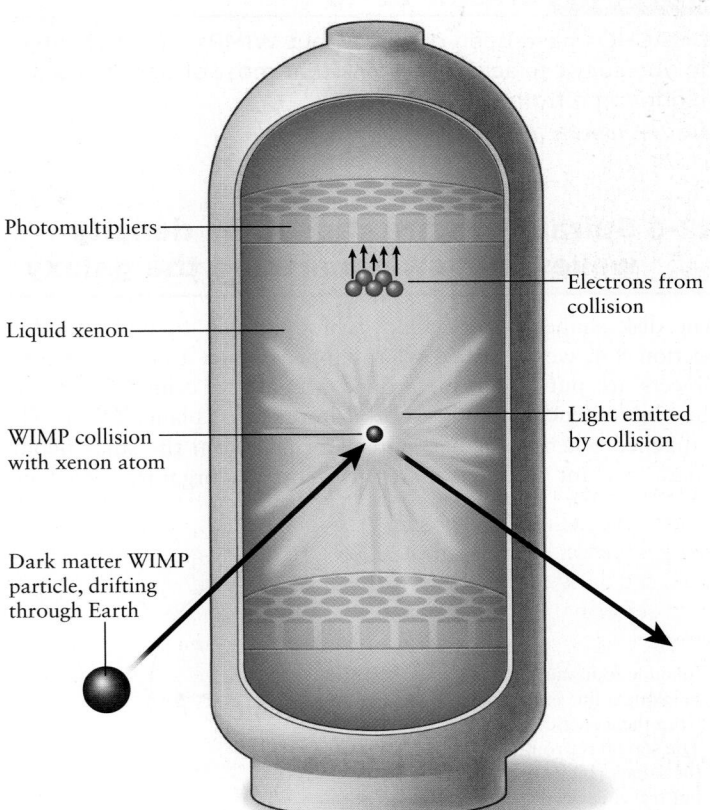

FIGURE 22-21

Detecting Dark Matter WIMPs This detector sits near the bottom of an old gold mine 4850 ft underground. If a WIMP collides with a xenon particle, the so-called weak interaction ends up producing light that can be detected by photomultipliers. Electrons are also produced in collisions to varying degrees, which helps to determine if the collisions were actually due to WIMPs, rather than some other particle. While Earth shields out cosmic rays that would otherwise overwhelm the many fewer WIMP detections, the tank shown here is also surrounded by 72,000 gallons of water that helps to shield neutrons (also a source of noise) that are emitted by natural radioactivity from the surrounding rock.

toward the xenon-filled detector. With WIMPs as the main hypothesis for dark matter, even larger experiments are being built for their discovery.

There are many searches for dark matter underway. Like the LZ experiment, some are conducted underground. But particle accelerators such as those under development by the European Organization for Nuclear Research (CERN) are also in on the hunt, and there's also a dark matter experiment on the International Space Station.

As yet, the true nature of dark matter remains a mystery. Furthermore, this mystery is not confined to our own galaxy. In Chapter 23, we will find that other galaxies have the same sort of rotation curve as in Figure 22-18, indicating that they also contain vast amounts of dark matter. We will also discuss additional ways to measure dark matter, such as a method called gravitational lensing. Overall, dark matter appears to make up most of the mass in our universe. Hence, dark matter is one of the most important unsolved problems in physics and astronomy.

CONCEPT CHECK 22-10

If MACHOs have been observed but WIMPs have not, why do you suppose scientists think that most of dark matter might come from WIMPs?

Answer appears at the end of the chapter.

22-5 Spiral arms are caused by density waves that sweep around the galaxy

The disk shape of our galaxy is not difficult to understand. In Section 8-4, we described what happens when a large number of objects are put into orbit around a common center: Over time, the objects tend naturally to orbit in the same plane. This is what happened when our solar system formed from the solar nebula. There, a giant cloud of material eventually organized itself into

planets, all of which orbit in nearly the same plane. Similarly, the disk of our galaxy, which is also made up of a large number of individual objects orbiting a common center, is very flat (see Figure 22-7). However, understanding why our galaxy has spiral arms presents more of a challenge.

> Our Sun and its solar system were spawned when a cloud of gas and dust passed through a spiral arm.

The Winding Dilemma

One early explanation for the galaxy's spiral structure was that the material in the galaxy had somehow condensed into a spiral pattern from the very start. In this view, once stars, gas, and dust had become concentrated within the spiral arms, the pattern would have remained fixed. This fixed pattern would be possible only if the galaxy rotated like a solid disk (see Figure 22-17b); the fixed pattern would be like the spokes on a rotating bicycle wheel. But the reality is that the galaxy is not a solid disk. As we have seen, stars, gas, and dust all orbit the galactic center with approximately the same speed, as shown in Figure 22-17a. Let us see why this makes it impossible for a rigid spiral pattern to persist.

Imagine four stars, A, B, C, and D, that originally lie on a line extending outward from the galactic center (**Figure 22-22a**). In a given amount of time, each of the stars travels the same distance around its orbit. But because the innermost star has a smaller orbit than the others, it takes less time to complete one orbit. As a result, a line connecting the four stars is soon bent into a spiral (Figure 22-22b). Moreover, the spiral becomes tighter and tighter with the passage of time (Figures 22-22c and 22-22d). This "winding up" of the spiral arms causes the spiral structure to disappear completely after a few hundred million years—a very brief time compared to the age of our galaxy, thought to be about 13.5 billion (1.35×10^{10}) years.

Figure 22-22 suggests that the Milky Way's spiral arms ought to have disappeared by now. The fact that they have not is called the **winding dilemma**. It shows that the spiral arms cannot simply

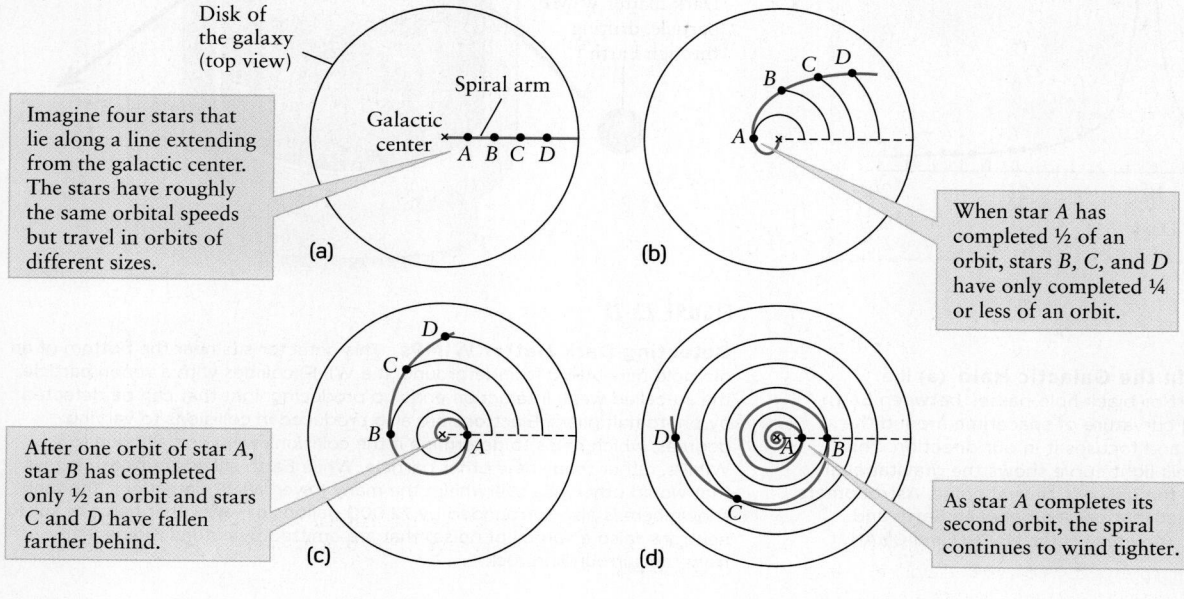

Imagine four stars that lie along a line extending from the galactic center. The stars have roughly the same orbital speeds but travel in orbits of different sizes.

(a)

Disk of the galaxy (top view)

Spiral arm

Galactic center ×   A B C D

(b)

When star A has completed ½ of an orbit, stars B, C, and D have only completed ¼ or less of an orbit.

After one orbit of star A, star B has completed only ½ an orbit and stars C and D have fallen farther behind.

(c)

As star A completes its second orbit, the spiral continues to wind tighter.

(d)

FIGURE 22-22

The Winding Dilemma
This series of drawings shows that spiral arms in galaxies like the Milky Way cannot simply be assemblages of stars. If they were, the spiral arms would "wind up" and disappear in just a few hundred million years.

be assemblages of stars and interstellar matter that travel around the galaxy together, like a troop of soldiers marching in formation around a flagpole. What, then, can the spiral arms be?

The Density-Wave Model

ANIMATION 22.2 In the 1940s, the Swedish astronomer Bertil Lindblad proposed that the spiral arms of a galaxy are actually a pattern that moves through the galaxy like ripples on water. This idea was elaborated on greatly in the 1960s by the American astronomers Chia Chiao Lin and Frank Shu. In this picture, spiral arms are a kind of wave, like the waves that move across the surface of a pond when you toss a stone into the water. Water molecules pile up at a crest of the wave but spread out again when the crest passes. By analogy, Lindblad, Lin, and Shu pictured a pattern of **density waves** sweeping around the galaxy. These waves make matter pile up in the spiral arms, which are the crests of the waves. Individual parts of the galaxy's material are compressed only temporarily when they pass through a spiral arm. The pattern of spiral arms persists, however, just as the waves made by a stone dropped in the water can persist for quite awhile after the stone has sunk.

To understand better how a density wave operates in a galaxy, think again about a water wave in a pond. If one part of the pond is disturbed by dropping a stone into it, the molecules in that part will be displaced a bit. They will nudge the molecules next to them, causing those molecules to be displaced and to nudge the molecules beyond them. In this way, the wave disturbance spreads throughout the pond.

In a galaxy, stars play the role of water molecules. Although stars and interstellar clouds of gas and dust are separated by vast distances, they can nonetheless exert forces on each other because they are affected by each other's gravity. If a region of above-average density should form, its gravitational attraction will draw nearby material into it. The displacement of this material will change the gravitational force that it exerts on other parts of the galaxy, causing additional displacements. In this way a spiral-shaped density wave can travel around the disk of a galaxy.

ANALOGY

A key feature of density waves is that they move more slowly around a galaxy than do stars or interstellar matter. To visualize this movement, imagine workers painting a line down a busy freeway (**Figure 22-23**). The cars normally cruise along the freeway at high speed, but the crew of painters is moving much more slowly. When the cars come up on the painters, they must slow down temporarily to avoid hitting anyone. As seen from the air, cars are jammed together around the painters. An individual car spends only a few moments in the traffic jam before resuming its usual speed, but the traffic jam itself lasts all day, inching its way along the road as the painters advance.

A similar crowding takes place when the matter between the stars, called the interstellar medium, enters a spiral arm. This crowding plays a key role in the formation of stars and the recycling of the interstellar medium. As interstellar gas and dust moves through a spiral arm, it is compressed into new nebulae (**Figure 22-24**). This

1. A crew of painters moves slowly along the highway, creating a moving traffic jam.

2. Cars enter the traffic jam from the right and exit to the left: While in the traffic jam, they are compressed together.

FIGURE 22-23

A Density Wave on the Highway   A density wave in a spiral galaxy is analogous to a crew of painters moving slowly along the highway, creating a moving traffic jam. Like such a traffic jam, a density wave in a spiral galaxy is a slow-moving region where stars, gas, and dust are more densely packed than in the rest of the galaxy. As the material of the galaxy passes through the density wave, it is compressed. This triggers star formation, as Figure 22-24 shows.

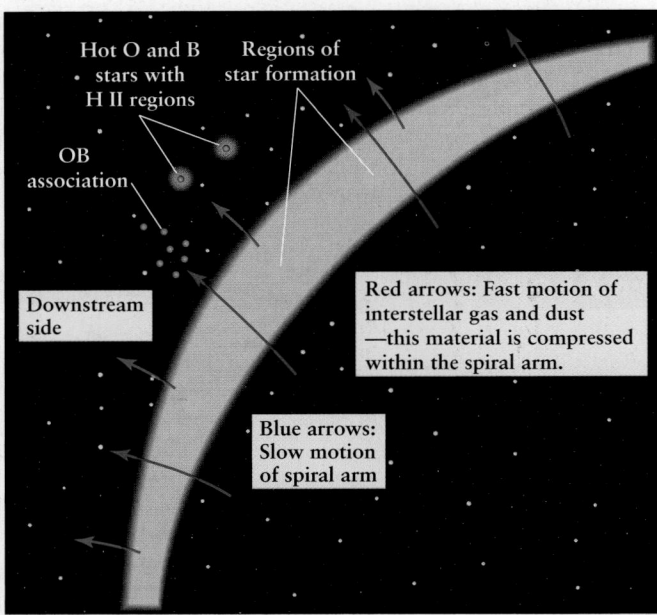

Hot O and B
stars with
H II regions

Regions of
star formation

OB
association

Downstream
side

Red arrows: Fast motion of
interstellar gas and dust
—this material is compressed
within the spiral arm.

Blue arrows:
Slow motion
of spiral arm

FIGURE 22-24

Star Formation in the Density-Wave Model A spiral arm is a region where the density of material is higher than in the surrounding parts of a galaxy. Interstellar matter moves around the galactic center rapidly (shown by the red arrows) and is compressed as it passes through the slow-moving spiral arms (whose motion is shown by the blue arrows). This compression triggers star formation in the interstellar matter, so that new stars appear on the "downstream" side of the densest part of the spiral arms.

compression begins the process by which new stars form, which we described in Section 18-3.

These freshly formed stars continue to orbit around the center of their galaxy, just like the matter from which they formed. The most luminous among these are the hot, massive, blue O and B stars, which may have emission nebulae (H II regions) associated with them. These stars have main-sequence lifetimes of only 3 to 15 million years (see Table 19-1), which is very short compared to the 220 million years required for the Sun to make a complete orbit around the galaxy.

As a result, these luminous O and B stars can travel only a relatively short distance before dying off. Therefore, these stars, and their associated H II regions, are seen only in or slightly "downstream" of the spiral arm in which they formed. **Figure 22-25** illustrates this for the spiral galaxy M51. Less massive stars have much longer main-sequence lifetimes, thus their orbits are able to carry them all around the galactic disk. These less luminous stars are found throughout the disk, including between the spiral arms.

The density-wave model of spiral arms explains why the disk of our galaxy is dominated by metal-rich Population I stars. Because the material left over from the death of ancient stars is enriched in heavy elements, new generations of stars formed in spiral arms are likely to be more metal rich than their ancestors. The *Cosmic Connections* figure illustrates this cycle of star birth and death in the disk of our galaxy.

The density-wave model is still under development. One problem is finding a driving mechanism that keeps density waves going in spiral galaxies. After all, density waves expend an enormous amount of energy to compress the interstellar gas and dust. Hence, we would expect that density waves should eventually die away, just as do ripples on a pond. The American astronomers Debra and Bruce Elmegreen have suggested that gravity can supply that needed energy. As mentioned in Section 22-3, the central bulge of our galaxy is elongated into a bar shape, much like the central bulge of the galaxy M83 (see Figure 22-15a and Figure 22-16a). The asymmetric gravitational field of such a bar pulls on the stars and interstellar matter of a galaxy to generate density waves. Another factor that may help to generate and sustain spiral arms is the gravitational interactions *between* galaxies. We will discuss this in Chapter 23.

CONCEPT CHECK 22-11

If our galaxy did not have density waves, how would it be different?

Answer appears at the end of the chapter.

The Self-Propagating Star-Formation Model

Spiral density waves may not be the whole story behind spiral arms in our galaxy and other galaxies. The reason is that spiral density waves should produce very well-defined spiral arms. We do indeed see many so-called **grand-design spiral galaxies** (**Figure 22-26a**), with thin, graceful, and well-defined spiral arms. But in some galaxies, called **flocculent spiral galaxies** (Figure 22-26b), the spiral arms are broad, fuzzy, chaotic, and poorly defined. (*Flocculent* means "resembling wool.")

To explain such flocculent spirals, M. W. Mueller and W. David Arnett in 1976 proposed a theory of **self-propagating star formation**. Imagine that star formation begins in a dense interstellar cloud within the disk of a galaxy that does not yet have spiral arms. As soon as hot, massive stars form, their radiation and stellar winds compress nearby matter, triggering the formation of additional stars in that gas. When massive stars become supernovae, they produce shock waves that further compress the surrounding interstellar medium, thus encouraging still more star formation (see Figure 18-27).

Although all parts of this broad, star-forming region have approximately the same orbital speed about the galaxy's center, the inner regions have a shorter distance to travel to complete one orbit than the outer regions. As a result, the inner edges of the star-forming region move ahead of the outer edges as the galaxy rotates. The bright O and B stars and their nearby glowing nebulae soon become stretched out, forming what appears as a segment of a spiral arm. These spiral arm segments come and go essentially at random across a galaxy. Bits and pieces of spiral arms appear where star formation has recently begun but fade and disappear at other locations where all the massive stars have died off. Self-propagating star formation therefore tends to produce flocculent spiral galaxies that have a chaotic appearance with poorly defined spiral arms, like the galaxy in Figure 22-26b.

The two theories presented here are very different in character. In the density-wave model, star formation is caused by the spiral arms; in the self-propagating star formation model, by contrast, the spiral arms are caused by star formation. A complete and correct description of spiral arms in our galaxy remains a topic of active research.

COSMIC CONNECTIONS

Stars in the Milky Way

Different populations of stars are found in different neighborhoods of our home galaxy. (The galaxy shown here is another spiral galaxy similar to our own.) The variations from one galactic region to another are due to the presence or absence of ongoing star formation.

1. All the stars, gas clouds, and dust particles within the disk orbit the center of the galaxy. They are held in orbit by the combined gravitational effects of the other parts of the galaxy.

Orbital motion

2. Dark dust lanes indicate where spiral density waves cause material to pile up as it orbits the galactic center. Gas is compressed as it passes through these regions, triggering the formation of new stars.

3. Hot, blue, massive O and B stars begin their formation in the dark dust lanes, then appear fully formed some distance "downstream" of the dust lanes. These stars have short main-sequence lifetimes, so they die out before they travel too far around the disk from their birthplace.

6. The density waves do not penetrate to the center of the galaxy, so new stars are not presently forming there. The blue O and B stars have long since died out, so the bulge has a yellowish color. It contains both old Population I stars and even older Population II stars (metal-poor stars from the first generation of stars to form in the galaxy).

5. As a star in the disk evolves, it produces metals. Much of a star's metal-enriched material is returned to the interstellar medium when it dies. When this material is next compressed by a density wave, it is incorporated into a new generation of stars. Thus, the disk contains many metal-rich stars of Population I. Our Sun is one example.

4. Most of the stars produced in the dust lanes are relatively faint, with $1\ M_\odot$ or less. These long-lived stars complete many round trips around the galaxy during a main-sequence lifetime. (Our Sun has made more than 20 round trips in the 4.56 billion years since it formed.)

R I V U X G

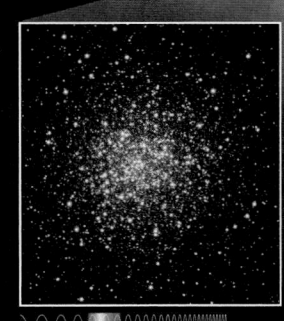

R I V U X G

7. Also orbiting the galaxy are the globular clusters. These star clusters have no mechanism to trigger star formation, so they contain only very old, metal-poor Population II stars. Studies of these clusters suggest that the first stars formed in them about 13.6 billion years ago—a mere 100 million years after the Big Bang.

8. Not visible to any telescope is the galaxy's dark matter, which emits no electromagnetic radiation of any kind. The spherical halo of dark matter that envelops our galaxy has at least 10 times the combined mass of all of the galaxy's stars, planets, gas, and dust. Its fundamental nature remains a mystery.

NASA, ESA; S. Kafka and K. Honeycutt, Indiana University/WIYN/NOAO/NSF

The densest part of this spiral arm (indicated by the presence of dust, shown in red) has not yet moved past Star A . . .

Star A

Rotation of M51

Rotation of M51

Star B

The densest part of this spiral arm (indicated by the presence of dust shown in red) has not yet moved past Star B . . .

(a) An infrared view of M51 shows the locations of dust.

FIGURE 22-25   R **I** V U X G

Star Formation in the Whirlpool Galaxy   The spiral galaxy M51 (called the Whirlpool) is a real-life example of the density-wave model illustrated in Figure 22-24. **(a)** This infrared image shows where dust has piled up as the material within M51 passes through its spiral arms. Radio images of M51 show that hydrogen gas also piles up in the same locations, thus beginning

. . . but some of the recently formed bright blue stars in this spiral arm have already moved past Star A.

Star A

Rotation of M51

Rotation of M51

Star B

. . . but some of the recently formed bright blue stars in this spiral arm have already moved past Star B.

(b) A visible-light view of M51 shows the locations of young stars.

R I **V** U X G

the formation of new stars. **(b)** By the time stars complete their formation process, their motion around the galaxy has swept them "downstream" of the positions of greatest dust density, just as depicted in Figure 22-24. (a: NASA/JPL-Caltech/R. Kennicutt [Univ. of Arizona]; b: NASA/JPL-Caltech)

CONCEPT CHECK 22-12

If astronomers learned that the earliest galaxies in the universe were flocculent spiral galaxies, what would this suggest about which came first: stars or spiral arms?

Answer appears at the end of the chapter.

22-6 There is a supermassive black hole at the center of our galaxy

TUTORIAL 22.2   The innermost region of our galaxy is an active, crowded place. If you lived on a planet near the galactic center, you could see a million stars as bright as Sirius, the brightest single star in our own night sky. The total intensity of starlight from all those nearby stars would be equivalent to 200 of our full moons. In

effect, night would never really fall on a planet near the center of the Milky Way. At the center of this empire of light, however, lies the darkest of all objects in the universe—a black hole millions of times more massive than the Sun.

At the very center of the Milky Way Galaxy is a maelstrom of activity centered on a supermassive black hole.

Sagittarius A*: Heart of Darkness

Because of the severe interstellar absorption at visual wavelengths, most of our information about the galactic center comes from infrared and radio observations. **Figure 22-27** shows three infrared views of the center of our galaxy. Figure 22-27a is a wide-angle view covering a 50° segment of the Milky Way from Sagittarius through Scorpius. (The top photograph that opens this chapter

(a) Grand-design spiral galaxy

(b) Flocculent spiral galaxy

FIGURE 22-26   R I **V** U X G

Variety in Spiral Arms The differences from one spiral galaxy to another suggest that more than one process can create spiral arms. **(a)** NGC 628 is a grand-design spiral galaxy with thin, well-defined spiral arms. **(b)** NGC 4414 is a flocculent spiral galaxy with fuzzy, broken, and poorly defined spiral arms. (a: Gemini Observatory/AURA/NSF; b: Olivier Vallejo [Observatoire de Bordeaux], HST, ESA, NASA)

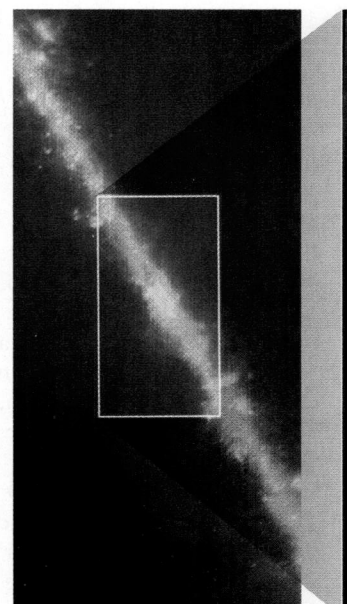

(a) A wide-angle (50°) infrared view.

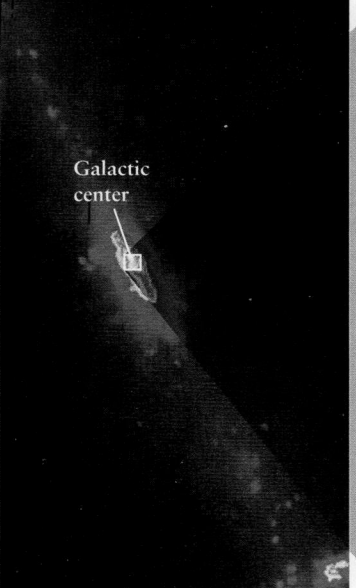

(b) A close-up view shows a more luminous region at the galactic center.

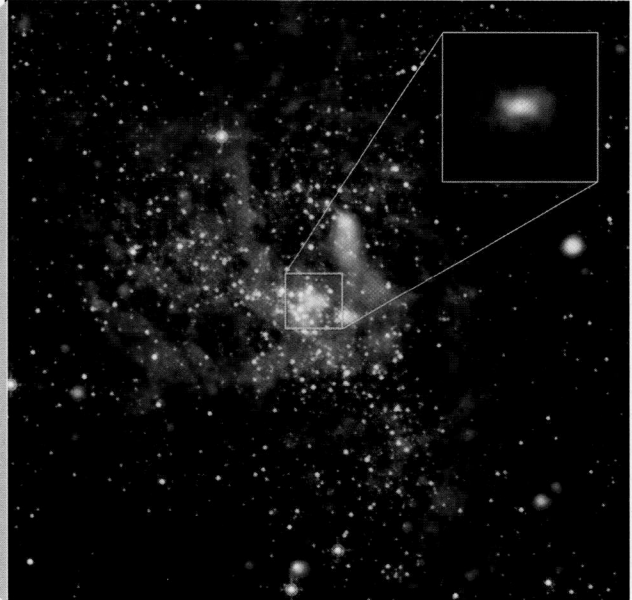

(c) An extreme close-up view centered on Sagittarius A*, a radio and x-ray source at the very center of the Milky Way Galaxy.

FIGURE 22-27   R **I** V U X G

R **I** V U **X** G

The Galactic Center **(a)** In this false-color infrared image, the reddish band is dust in the plane of the galaxy and the fainter bluish blobs are interstellar clouds heated by young O and B stars. **(b)** This close-up infrared view covers the area outlined by the white rectangle in (a). **(c)** An infrared image reveals many stars, displayed in red and yellow. X rays are shown in blue. The inset image only displays x rays, and is half a light-year on each side. (a, b: NASA; c: x ray: NASA/UMass/D. Wang et al., infrared: NASA/STScI/NASA)

0.1"

VIDEO
22.2

FIGURE 22-28 R **I** V U X G

Stars Orbiting Sagittarius A*
The colored dots superimposed on this infrared image show the motion of seven stars in the vicinity of the unseen massive object (denoted by the yellow five-pointed star) at the position of the radio source Sagittarius A*. The orbits were measured over a 21-year period, and for each star, darker shades indicate more recent observations. Analysis of the orbits indicates that the stars are held in orbit by a black hole of 4.1 million $M_\odot$. The blue dots for S0-16 show this star reached 4% the speed of light at its closest approach to the black hole. (Keck/UCLA Galactic Center Group)

S0-2
S0-38
S0-1
S0-3
S0-5
S0-8
S0-16
S0-17
S0-20
S0-102

1995–2016

shows this same region at visible wavelengths, viewed at a different angle.) The prominent reddish band through the center of this false-color infrared image is a layer of dust in the plane of the galaxy. Figure 22-27b is an IRAS view of the galactic center. It is surrounded by numerous streamers of dust (shown in blue). The strongest infrared emission (shown in white) comes from sources that also emit radio waves. One of these sources, **Sagittarius A*** (say "A star"), lies at the very center of the galaxy. (Its position, pinpointed with simultaneous observations by radio telescopes scattered around the world, seems to be very near the gravitational center of the galaxy.) The high-resolution infrared view in Figure 22-27c, made using adaptive optics (see Section 6-3), shows hundreds of stars crowded within 1 ly (0.3 pc) of Sagittarius A*. Compare this region to our region of the galaxy, where the average distance between stars is more than a light-year.

Sagittarius A* itself does not appear in infrared images. Nonetheless, astronomers have used infrared observations to make truly startling discoveries about this object. Since the 1990s, two research groups—one headed by Reinhard Genzel of the Max Planck Institute for Extraterrestrial Physics in Garching, Germany, and another led by Andrea Ghez at the University of California, Los Angeles—have been using infrared detectors to monitor the motions of stars in the immediate vicinity of Sagittarius A*. They

have found a number of stars orbiting Sagittarius A* at speeds in excess of 1500 km/s (**Figure 22-28**). (By comparison, Earth orbits the Sun at a lackadaisical 30 km/s.) In 2000, the UCLA group observed one such star, called S0-16, as its elliptical orbit brought it within a mere 45 au from Sagittarius A* (1½ times the distance from the Sun to Neptune). At its closest approach, S0-16 was traveling at a breathtaking speed of 12,000 km/s, or 4% of the speed of light!

In order to keep stars like S0-16 in such small, rapid orbits, Sagittarius A* must exert a powerful gravitational force and hence must be very massive. By applying Newton's form of Kepler's third law to the motions of these stars around Sagittarius A*, the UCLA group calculated the mass of Sagittarius A* to be a remarkable 4.1 *million* solar masses ($4.1 \times 10^6 M_\odot$). Furthermore, the small separation between S0-16 and Sagittarius A* at closest approach shows that Sagittarius A* can be no more than 45 au in radius. An object this massive and this compact can only be one thing—a supermassive black hole (see Section 22-4). From Box 21-3, a black hole with the mass inferred from the stellar orbits would have a Schwarzschild radius just over 20% the average radius of Mercury's orbit. Just try to imagine it: While we have exactly 1 $M_\odot$ within Mercury's orbit, the black hole at our galactic center has 41 million $M_\odot$ within an even smaller volume.

(a) A radio view of the galactic center   R I V U X G

(b) An x-ray view of the galactic center   R I V U X G

FIGURE 22-29

The Energetic Center of the Galaxy **(a)** The area shown in this radio image has the same angular size as the full moon. Sagittarius A*, at the very center of the galaxy, is one of the brightest radio sources in the sky. Magnetic fields shape nearby interstellar gas into immense, graceful arches. **(b)** This composite of images at x-ray wavelengths from 0.16 to 0.62 nm shows lobes of gas on either side of Sagittarius A*. The character of the x-ray emission shows that the gas temperature is as high as 2×10^7 K. (a: VLA, F. Sadeh et al./NRAO; b: NASA/CXC/MIT/F. K. Baganoff et al.)

CONCEPT CHECK 22-13

How would stars around Sagittarius A* be moving if a supermassive black hole were not located at the galactic center?

Answer appears at the end of the chapter.

X Rays from Around a Supermassive Black Hole

ANIMATION 22.1 Evidence in favor of this picture comes from the Chandra X-ray Observatory, which has observed x-ray flares coming from Sagittarius A*. The flares brighten dramatically over the space of just 10 minutes, which shows that the size of the flare's source can be no larger than the distance that light travels in 10 minutes. (We used a similar argument in Section 21-5 to show that the flickering x-ray source Cygnus X-1 must be very small.) In 10 minutes, light travels a distance of 1.8×10^8 km or 1.2 au, and only a black hole could pack a mass of 4.1×10^6 M$_\odot$ into a volume that size or smaller. The x-ray flares were presumably emitted by blobs of material that were compressed and heated as they fell into the black hole (see Section 21-5).

The x-ray flares from Sagittarius A* are relatively feeble, which suggests that the supermassive black hole is swallowing only relatively small amounts of material. But the region around Sagittarius A* is nonetheless an active and dynamic place. Figure 22-29a is a wide-angle radio image of the galactic center covering an area more than 60 pc (200 ly) across. Huge filaments of gas stretch for 20 pc (65 ly) northward of the galactic center (to the right and upward in Figure 22-29a), then abruptly arch southward (down and to the left in the figure). The orderly arrangement of these filaments is reminiscent of prominences on the Sun (see Section 16-10, especially Figure 16-28). This suggests that, as on the Sun, there is ionized gas at the galactic center that is being controlled by a powerful magnetic field. Indeed, much of the radio emission from the galactic center is synchrotron radiation: As we saw in Section 22-4, such radiation is produced by high-energy electrons spiraling in a magnetic field.

The false-color x-ray image in Figure 22-29b shows the immediate vicinity of Sagittarius A*. The black hole is flanked by lobes of hot, ionized, x-ray–emitting gas that extend for dozens of light-years. These lobes are thought to be the relics of immense explosions that may have taken place over the past several thousand years. Perhaps these past explosions cleared away much of the material around Sagittarius A*, leaving only small amounts to fall into the black hole. This could explain why the x-ray flares from Sagittarius A* are so weak.

There is nonetheless evidence that Sagittarius A* has been more active in the recent past. Between 2002 and 2005, Michael Muno and his colleagues at the California Institute of Technology observed that certain nebulae within about 50 ly of Sagittarius A* suddenly became very bright at x-ray wavelengths. The character of the x-ray light was such that it could not have originated from the nebulae themselves. Muno and colleagues concluded that x rays emitted from Sagittarius A* about 50 years earlier had struck and excited the nebulae, causing the nebulae to glow intensely at x-ray wavelengths. To produce such an intense x-ray glow, an object the size of the planet Mercury must have fallen into the supermassive black hole.

FIGURE 22-30

Galactic Gamma-Ray Bubbles This is an artist's illustration of giant gamma-ray "bubbles" seen above and below our galactic plane, possibly due to the black hole at the galactic center. Each of the bubbles extends about 25,000 ly away from the plane. The bubbles are estimated to be a few million years old. The gamma-ray data also contain hints of linear jetlike features, which are sometimes found around black holes, but the existence of the jets is far from clear. If the jets are real, producing them might have required the black hole to consume a giant hydrogen cloud of about 10,000 $M_\odot$, with only a small portion escaping into the jets. (David A. Aguilar, Harvard-Smithsonian Center for Astrophysics)

CONCEPT CHECK 22-14

Why would the sudden brightening of a nebula near the galactic center suggest the presence of a supermassive black hole?

Answer appears at the end of the chapter.

Gamma Rays Offer Clues to Possible Black Hole Activity

In 2010, astronomers with the Fermi Gamma-ray Space Telescope (FGST) reported seeing enormous gamma-ray "bubbles" above and below the plane of our galaxy (Figure 22-30). The gamma rays are produced when high-speed electrons collide with infrared or visible photons and bump the photon energies up to gamma-ray levels. What remains unclear is what created the high-speed electrons. One idea involves a hot "wind" of electrons and other particles blown away from an accretion disk of matter (see Figure 21-21) around the black hole at our galaxy's center.

It is too early to say if the gamma-ray bubbles are caused by a galactic black hole. However, if these interpretations hold up, they certainly point to interesting behavior by the black hole at the center of our galaxy. As intriguing as it is, the supermassive black hole at the center of our galaxy is not unique. Observations show that such titanic black holes are a feature of most large galaxies. In Chapter 24, we will see how black holes of this kind power *quasars,* the most luminous sustained light sources in the cosmos.

KEY WORDS

Word preceded by an asterisk (*) is discussed in Box 22-1.

central bulge (of a galaxy), p. 700
dark matter, p. 709
density wave, p. 713
disk (of a galaxy), p. 700
far-infrared wavelength, p. 699
flocculent spiral galaxy, p. 714
galactic nucleus, p. 698
globular cluster, p. 697

grand-design spiral galaxy, p. 714
H I, p. 702
halo (of a galaxy), p. 700
interstellar extinction, p. 697
Local Bubble, p. 703
*magnetic resonance imaging (MRI), p. 704
massive compact halo object (MACHO), p. 710

microlensing, p. 710
Milky Way Galaxy, p. 696
near-infrared wavelength, p. 699
rotation curve, p. 708
RR Lyrae variable, p. 698
Sagittarius A*, p. 718
self-propagating star formation, p. 714

spin (of a particle), p. 702
spin-flip transition, p. 703
spiral arm, p. 702
21-cm radio emission, p. 703
weakly interacting massive particle (WIMP), p. 710
winding dilemma, p. 712

KEY IDEAS

The Shape and Size of the Galaxy: Our galaxy has a disk about 50 kpc (160,000 ly) in diameter and about 600 pc (2000 ly) thick, with a high concentration of interstellar dust and gas in the disk (Section 22-2).

- The galactic center is surrounded by a large distribution of stars called the central bulge. This bulge is not perfectly symmetrical, but may have a bar shape.

- The disk of the galaxy is surrounded by a spherical distribution of globular clusters and old stars, called the galactic halo.

- There are about 200 billion (2×10^{11}) stars in the galaxy's disk, central bulge, and halo.

IV Galaxies and Cosmology

This Hubble Ultra Deep Field image shows the distant background galaxies of our universe with more sensitivity than any other image. To accomplish this, the Hubble Space Telescope was pointed at a patch of space with fewer stars in our own galaxy or obscuring nearby galaxies. While there are some stars in this image, every object that doesn't look perfectly round is a galaxy. Because it takes so long for the light to reach us from these distant galaxies, looking far away is also looking back in time. While our universe is 13.7 billion years old, some of these galaxies are only about half a billion years old. As an analogy, if a 75-year-old person looked back that far in time, he would be seeing himself before his 3rd birthday. (R. Windhorst [ASU], H. Yan [SSC, Caltech], et al., ESA, NASA)

R I V U X G

The Whirlpool Galaxy has well-defined spiral arms and a small companion galaxy on its right. These galaxies are about 20 million light-years away and are bright enough and large enough to be viewed with binoculars. The small red patches are typically a few hundred light-years across: Ultraviolet light from hotter stars kicks electrons off from hydrogen atoms, and as the electrons recombine with the hydrogen, the atoms emit red light. Computer modeling, carried out to account for the speed and distance of the companion relative to the larger galaxy, suggests that the companion may have passed right through the spiral plane of the Whirlpool. Rather than disturb the spiral arms, gravitational interactions with the companion may have given the arms their distinct structure. (S. Beckwith [STScI], Hubble Heritage Team [STScI/AURA], ESA, NASA)

R I V U X G

The Sun's Location in the Galaxy: Our Sun lies within the galactic disk, some 8000 pc (26,000 ly) from the center of the galaxy (Section 22-1).

- Interstellar dust obscures our view at visible wavelengths along lines of sight that lie in the plane of the galactic disk. As a result, the Sun's location in the galaxy was unknown for many years. This dilemma was resolved by observing parts of the galaxy outside the disk.

- The Sun orbits the center of the galaxy at a speed of about 790,000 km/h. It takes about 220 million years to complete one orbit.

The Galaxy's Spiral Structure: OB associations, H II regions, and molecular clouds in the galactic disk outline huge spiral arms (Section 22-3).

- Spiral arms can be traced from the positions of clouds of atomic hydrogen. These can be detected throughout the galactic disk by the 21-cm radio waves emitted by the spin-flip transition in hydrogen. These emissions easily penetrate the intervening interstellar dust.

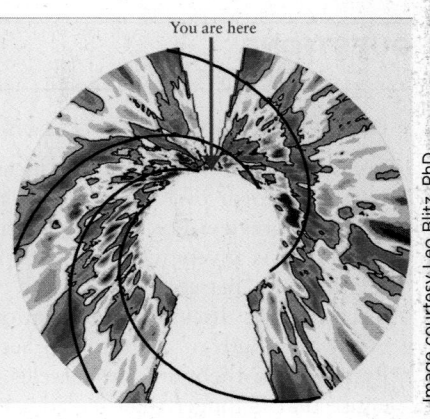

You are here

Image courtesy Leo Blitz, PhD

The Rotation of the Galaxy and Dark Matter: From studies of the rotation of the galaxy, astronomers estimate that the total mass of the galaxy is about 10^{12} $M_\odot$. Only about 10% of this mass is in the form of visible stars, gas, and dust. The remaining 90% is in some nonvisible form, called dark matter, that extends beyond the edge of the luminous material in the galaxy (Section 22-4).

- Our galaxy's dark matter contains only a small amount of MACHOs (dim, star-sized objects) and is hypothesized to consist mostly of WIMPs (relatively massive subatomic particles).

(a) The orbital speed of stars and gas around the galactic center is nearly uniform throughout most of our galaxy.

Stars in smaller orbits gradually overtake those in larger orbits.

Sun's orbit

Galactic center

Theories of Spiral Structure: There are two leading theories of spiral structure in galaxies (Section 22-5).

- According to the density-wave theory, spiral arms are created by density waves that sweep around the galaxy. The gravitational field of this spiral pattern compresses the interstellar clouds through which it passes, thereby triggering the formation of the OB associations and H II regions that illuminate the spiral arms.

- According to the theory of self-propagating star formation, spiral arms are caused by the birth of stars over an extended region in a galaxy. Differential rotation of the galaxy stretches the star-forming region into an elongated arch of stars and nebulae.

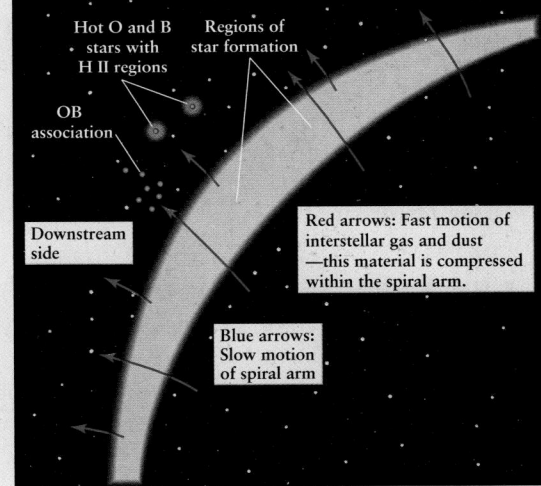

Hot O and B stars with H II regions

Regions of star formation

OB association

Downstream side

Red arrows: Fast motion of interstellar gas and dust —this material is compressed within the spiral arm.

Blue arrows: Slow motion of spiral arm

The Galactic Nucleus: The innermost part of the galaxy, or galactic nucleus, has been studied through its radio, infrared, x-ray, and gamma-ray emissions, which are able to pass through interstellar dust (Section 22-6).

- A strong radio source called Sagittarius A* is located at the galactic center. This marks the position of a supermassive black hole with a mass of about 4.1×10^6 $M_\odot$.

Keck/UCLA Galactic Center Group

QUESTIONS

> **Problem-solving tips and tools**
>
> Several of the following questions make extensive use of Newton's form of Kepler's third law, and you might find it helpful to review Box 4-4. Another useful version of Kepler's third law is given in Section 17-9. Box 17-2 discusses the relationship between luminosity, apparent brightness, and distance. We discussed the relationship between the energy and wavelength of a photon in Section 5-5. According to the Pythagorean theorem, an isosceles right triangle has a hypotenuse that is longer than its sides by a factor of $\sqrt{2} = 1.414$. The formula for the Schwarzschild radius of a black hole is given in Box 21-3. You will find the small-angle formula in Box 1-1 useful. It is also helpful to remember that an object 1 au across viewed at a distance of 1 pc has an angular size of 1 arcsec. Remember, too, that the volume of a cylinder is equal to its height multiplied by the area of its base, the area of a circle of radius r is πr^2, and the volume of a sphere of radius r is $4\pi r^3/3$. You can find other geometrical formulae in Appendix 8.

Section 22-1

1. Why do the stars of the galaxy appear to form a bright band that extends around the sky?

2. **TUTORIAL 22.1** How did interstellar extinction mislead astronomers into believing that we are at the center of our galaxy?

3. How did observations of globular clusters help astronomers determine our location in the galaxy?

4. What are RR Lyrae stars? Why are they useful for determining the distance from our solar system to the center of the galaxy?

5. Discuss how the Milky Way would appear to us if the Sun were relocated to (**a**) the edge of the galaxy; (**b**) the galactic halo; (**c**) the galactic bulge.

6. Explain why globular clusters spend most of their time in the galactic halo, even though their eccentric orbits take them close to the galactic center.

Section 22-2

7. Why are infrared telescopes useful for exploring the structure of the galaxy? Why is it important to make observations at both near-infrared and far-infrared wavelengths?

8. The galactic halo is dominated by Population II stars, whereas the galactic disk contains predominantly Population I stars. In which of these parts of the galaxy has star formation taken place recently? Explain your answer.

9. O or B main-sequence stars are found in the galactic disk but not in globular clusters. Why is this so?

Section 22-3

10. What must happen within a hydrogen atom for it to emit a photon of wavelength 21 cm?

11. Most interstellar hydrogen atoms emit only radio waves at a wavelength of 21 cm, but some hydrogen clouds emit profuse amounts of visible light (see, for example, Figure 18-1b and Figure 18-2). What causes this difference?

12. How do astronomers determine the distances to H I (neutral hydrogen) clouds?

13. The radio map in Figure 22-14 has a large gap on the side of the galaxy opposite to ours. Why is this?

14. What kinds of objects (other than H I clouds) do astronomers observe to map out the galaxy's spiral structure? What is special about these objects? Which of these can be observed at great distances?

15. Why don't astronomers detect 21-cm radiation from the hydrogen in giant molecular clouds?

16. A typical hydrogen atom in interstellar space undergoes a spin-flip transition only once every 10^7 years. How, then, is it at all possible to detect the 21-cm radio emission from interstellar hydrogen?

17. Calculate the energy of the photon emitted when a hydrogen atom undergoes a spin-flip transition. How many such photons would it take to equal the energy of a single H_α photon of wavelength 656.3 nm?

18. Suppose you were to use a radio telescope to measure the Doppler shift of 21-cm radiation in the plane of the galaxy. (**a**) If you observe 21-cm radiation from clouds of atomic hydrogen at an angle of 45° from the galactic center, you will see the highest Doppler shift from a cloud that is as far from the galactic center as it is from the Sun. Explain this statement using a diagram. (**b**) Find the distance from the Sun to the particular cloud mentioned in (a).

Section 22-4

19. In what way are the orbits of stars in the galactic disk different from the orbits of planets in our solar system? What does this difference imply about the way that matter is distributed in the galaxy?

20. How do astronomers determine how fast the Sun moves in its orbit around the galaxy? How does this speed tell us about the amount of mass inside the Sun's orbit? Does this speed tell us about the amount of mass outside the Sun's orbit?

21. How do astronomers conclude that vast quantities of dark matter surround our galaxy? How is this dark matter distributed in space?

22. What is the difference between dark matter and dark nebulae?

23. What proposals have been made to explain the nature of dark matter? What experiments or observations have been made to investigate these proposals? What are the results of this research?

24. Sketch the rotation curve you would obtain if the galaxy were rotating like a rigid body.

25. The mass of our galaxy inside the Sun's orbit is calculated from the radius of the Sun's orbit and its orbital speed. By how much would this estimate be in error if the calculated distance to the galactic center were off by 10%? By how much would this estimate be in error if the calculated orbital velocity were off by 10%? Explain your reasoning.

26. A gas cloud located in the spiral arm of a distant galaxy is observed to have an orbital velocity of 400 km/s. If the cloud is 20,000 pc from the center of the galaxy and is moving in a circular orbit, find (**a**) the orbital period of the cloud and (**b**) the mass of the galaxy contained within the cloud's orbit.

27. According to the galaxy's rotation curve in Figure 22-18, a star 16 kpc from the galactic center has an orbital speed of about 270 km/s. Calculate the mass within that star's orbit.

28. Speculate on the reasons for the rapid rise in the galaxy's rotation curve (see Figure 22-18) at distances close to the galactic center.

29. **Box 22-2** Show that the form of Kepler's third law stated in Box 22-2, $P^2 = 4\pi^2 a^3/G\,(M + M_\odot)$, is equivalent to $M = rv^2/G$, provided the orbit is a circle. (*Hint:* The mass of the Sun [$M_\odot$] is much less than the mass of the galaxy inside the Sun's orbit [M].)

Section 22-5

30. Many old black-and-white photographs of spiral galaxies were taken using film that was most sensitive to blue light. Explain why the spiral arms were particularly prominent in such photographs.

31. What is the winding dilemma? What does it tell us about the nature of spiral arms?

32. Do density waves form a stationary pattern in a galaxy? If not, do they move more rapidly, less rapidly, or at the same speed as stars in the disk?

33. In our galaxy, why are stars of spectral classes O and B found only in or near the spiral arms? Is the same true for stars of other spectral classes? Explain why or why not.

34. Compare the kinds of spiral arms produced by density waves with those produced by self-propagating star formation. By examining Figure 22-16, cite evidence that both processes might occur in our galaxy.

35. The accompanying figure shows infrared images of two spiral galaxies. Explain which of these is a grand-design spiral galaxy and which is a flocculent spiral galaxy. Explain your reasoning.

(a)

(b)
R **I** V U X G

NASA/JPL-Caltech/R. Kennicutt (University of Arizona) and the SINGS Team

36. From what you know about stellar evolution, the interstellar medium, and the density-wave theory, explain the appearance and structure of the spiral arms of grand-design spiral galaxies.

Section 22-6

37. **TUTORIAL 22.2** What is the evidence that there is a supermassive black hole at the center of our galaxy? How is it possible to determine the mass of this black hole?

38. What is the evidence that material has been falling into the supermassive black hole at the galactic center?

39. **Box 21-2** (a) Calculate the Schwarzschild radius of a supermassive black hole of mass $4.1 \times 10^6 \, M_\odot$, the estimated mass of the black hole at the galactic center. Give your answer in both kilometers and astronomical units. (b) What is the angular diameter of such a black hole as seen at a distance of 8 kpc, the distance from Earth to the galactic center? Give your answer in arcseconds. Observing an object with such a small angular size will be a challenge indeed! (c) What is the angular diameter of such a black hole as seen from a distance of 45 au, the closest that the star S0-16 comes to Sagittarius A*? Again, give your answer in arcseconds. Would it be discernible to the naked eye at that distance? (A normal human eye can see details as small as about 60 arcsec.)

40. **Box 21-2** (a) The scale bar in Figure 22-28 shows that at the distance of Sagittarius A*, a length of 1600 au has an angular size of 0.2 arcsec. Use this information to calculate the distance to Sagittarius A*. (b) The star S0-16 moves around Sagittarius A* in an elliptical orbit with semimajor axis 1680 au. Use this and the information given in the caption to Figure 22-28 to find the orbital period of S0-16. (c) Given the period and the semimajor axis of the star's orbit, is it possible to calculate the mass of S0-16 itself? If it is, explain how this could be done; if not, explain why not.

41. **Box 21-2** The stars S0-2 and S0-19 orbit Sagittarius A* with orbital periods of 14.5 and 37.3 years, respectively. (a) Assuming that the supermassive black hole in Sagittarius A* has a mass of $4.1 \times 10^6 \, M_\odot$, determine the semimajor axes of the orbits of these two stars. Give your answers in astronomical units (au). (b) Calculate the angular size of each orbit's semimajor axis as seen from Earth. (See Section 22-1 for the distance from Earth to the center of the galaxy.) Explain why extremely high-resolution infrared images are required to observe the motions of these stars.

42. Consider a star that orbits around Sagittarius A* in a circular orbit of radius 530 au. (a) If the star's orbital speed is 2500 km/s, what is its orbital period? Give your answer in years. (b) Determine the sum of the masses of Sagittarius A* and the star. Give your answer in solar masses. (Your answer is an estimate of the mass of Sagittarius A*, because the mass of a single star is negligibly small by comparison.)

43. Discuss how a supermassive black hole could have formed at the center of our galaxy.

General Problems

44. In a spiral galaxy, are stars in general concentrated in the spiral arms? Why are spiral arms so prominent in visible-light images of spiral galaxies?

45. Another student tells you that the Milky Way Galaxy is made up "mostly of stars." Is this statement accurate? Why or why not?

46. The disk of the galaxy is about 50 kpc in diameter and 600 pc thick. (a) Find the volume of the disk in cubic parsecs. (b) Find the volume (in cubic parsecs) of a sphere 300 pc in radius centered on the Sun. (c) If supernovae occur randomly throughout the volume of the galaxy, what is the probability that a given supernova will occur within 300 pc of the Sun? If there are about three supernovae each century in our galaxy, how often, on average, should we expect to see one within 300 pc of the Sun?

47. **Box 19-2** An RR Lyrae star whose peak luminosity is $100 \, L_\odot$ is in a globular cluster. At its peak luminosity, this star appears from Earth to be only 1.47×10^{-18} as bright as the Sun. Determine the distance to this globular cluster (a) in astronomical units (au) and (b) in parsecs.

48. The accompanying image shows the spiral galaxy M74, located about 55 million ly from Earth in the constellation Pisces (the Fish). It is actually a superposition of two false-color images. The red portion is an optical image taken at visible wavelengths, while the blue portion is an ultraviolet image made by NASA's Ultraviolet Imaging Telescope, which was carried into orbit by the space shuttle *Columbia* during the Astro-1 mission in 1990. Compare the visible and ultraviolet images and, from what you know about stellar evolution and spiral structure, explain the differences you see.

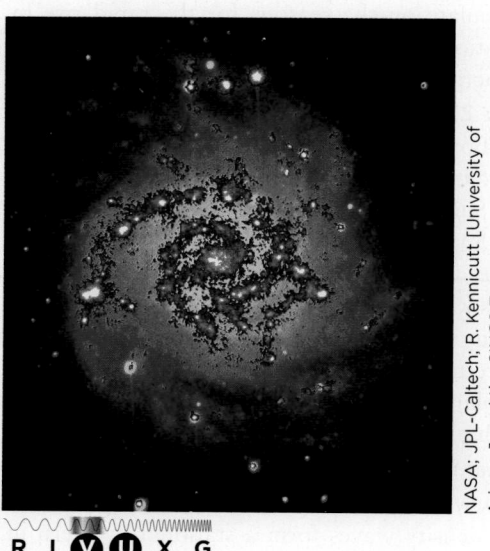

NASA; JPL-Caltech; R. Kennicutt [University of Arizona]; and the SINGS Team

R I V U X G

49. What observations would you make to determine the nature of the dark matter in our galaxy's halo?

50. Describe how the appearance of the night sky might change if dark matter were visible to our eyes.

WEB/E-BOOK QUESTIONS

51. Some scientists have suggested that the rotation curve of the galaxy can be explained by modifying the laws of physics rather than by positing the existence of dark matter. Search the internet for information about these proposed modifications, called MOND (for *MO*dified *N*ewtonian *D*ynamics). Under what circumstances do MOND and conventional physics predict different behavior? What evidence is there that MOND might be correct?

52. **Fast-Moving Stars at the Galactic Center.** Access and and view the video "Stars Orbiting Sagittarius A*" in Chapter 22 of the *Universe* website or e-book. Explain how you can tell which of the stars in the video are actually close to Sagittarius A* and which just happen to lie along our line of sight to the center of the galaxy.

ACTIVITIES

Starry Night™ Explorations

53. Use *Starry Night™* to explore the alignment of the Milky Way with respect to our Earth-based coordinate system and then examine the appearance of the galaxy at different electromagnetic wavelengths. Select **Favorites > Explorations > Milky Way** to display a view of our spiral galaxy seen edge on from Earth looking toward the central region of this galaxy.

 (a) Select **View > Galactic Guides > Equator** from the menu to display the plane of the galaxy and then select **View > Celestial Guides > Equator** to show the projection of the Earth's equator onto the celestial sphere. Change the cursor to the angular separation tool and use this to draw a line from the point of intersection of these two equators along the blue galactic equator toward the left (east) and note the position angle of this line. Repeat this measurement of position angle for the red celestial equator. What is the angle between the planes defined by these two equators?

 (b) Select **File > Revert** to reset the view. Use the view controls to examine this visible-light image of the complete Milky Way to see its significant structure, with many dark regions. These dark regions do not show the absence of material in these directions but the presence of dense dust and gas clouds obscuring the distant stars. Visible light is scattered and absorbed by these dense clouds. Now examine this galactic plane at other electromagnetic wavelengths. Select **Options > Stars > Milky Way…** from the menu to open a window that allows you to display the image of the galactic plane at various wavelengths. Explore the image of the galaxy in the view at each of the offered wavelengths. At which wavelengths does the galactic center show up most prominently? Why do you think this region shows up brighter at these wavelengths?

54. Use *Starry Night™* to examine the appearance and dimensions of our galaxy, gain a sense of our position in the galaxy, and explore the position of the Milky Way among its neighboring galaxies. Select **Favorites > Explorations > Milky Way Galaxy** to display a simulation of a face-on view of the Milky Way, with the position of the Sun indicated. Note the spiral arm structure of the galaxy. (Recent research has revealed that our galaxy appears to have a bar structure across the galactic center.)

 (a) Use the HUD to obtain initial data about the galaxy, including its diameter and thickness. From these dimensions, what do you think is the overall shape of our galaxy? Use the location scroller to confirm your answer.

 (b) Select **File > Revert** to reset the view and then change the cursor to the angular separation tool. Use this tool to measure the distance from the center of the galaxy to the Sun. How far, in light years, is the Sun from the center of the galaxy?

 (c) Most of the bright objects surrounding our galaxy from this view are other galaxies that are members of the Local Group of galaxies. Use the elevation control to move to about 0.5 Mly from our galaxy to examine the distribution of these galaxies. Place the cursor over the brighter central galaxy of the group to the bottom left of the view to identify this galaxy. What is the name of this large companion galaxy?

 (d) Use the angular separation tool to measure the separation between the Milky Way Galaxy and this other dominant galaxy. How far are we from this galaxy? How does this distance compare to the distance between the Sun and the center of the Milky Way?

 (e) Switch back to the location scroller tool and move around this group of galaxies to examine their three-dimensional distribution. How are these galaxies grouped together within this overall collection of galaxies?

55. There are two types of stellar clusters in galaxies. Globular clusters in our galaxy consist of old, Population II stars while open clusters consist of younger, Population I stars. The youngest open clusters are still surrounded by the dust and gas clouds from which these cluster stars were formed and in which active star formation continues. Use the distribution of these two types of star clusters to survey the distribution of Population I and II stars in the Milky Way Galaxy. Open **Favorites > Explorations > Population I and II** to see a view oriented along the plane of the galaxy toward its center from the position of the Earth. The galactic equator and the position of the galactic poles are shown on the celestial sphere.

 (a) Open the **Options** panel and turn on the **Stars > Star Clusters > Globular Clusters** option to display the distribution of these clusters of Population II stars in the galaxy as bold orange circles. Use the view controls to study the distribution of these clusters in the galaxy. Return to the **Options** panel and turn on the **Stars > Star Clusters > Open Clusters** to indicate the locations of these clusters of Population I stars in the galaxy with dotted or faint solid orange outlines. Use the view controls to compare the distribution of these two types of star clusters and, by extension, the distribution of Population I and II stars in the galaxy. What do you notice about the distribution of the globular clusters compared to the open clusters?

 (b) Use the hand tool to move the galactic center to the top of the screen and note the two small concentrations of star clusters southwest of the galactic center. Select **Labels > Bright NGC Objects** from the menu. How do you explain these concentrations of star clusters?

(c) Examine the appearance of several of the clusters in detail. Select **File > Revert** to reset the view and then open the **Observing Lists** panel and expand the two lists: **Globular Star Clusters** and **Open Star Clusters**. Select **Magnify** from the context menu icon for the clusters in each list to see a telescopic image of these objects. What are the notable differences in the nature and structure of open versus globular star clusters?

Collaborative Exercises

56. Student book bags often contain a wide collection of odd-shaped objects. Each person in your group should rummage through their own book bags and find one object that is most similar to the Milky Way Galaxy in shape. List the items from each group member's belongings and describe what about the items is similar to the shape of our galaxy and what about the items is not similar; then indicate which of the items is the closest match.

57. One strategy for identifying a central location is called *triangulation*. In triangulation, a central position can be pinpointed by knowing the distance from each of three different places. First, on a piece of paper, create a rough map showing where each person in your group lives. Second, create a circle around each person's home that has a radius equal to the distance that each home is from your classroom. Label the place where the circles intersect as your classroom. Why can't you identify the position of the classroom with only two people's circles?

58. Figure 22-13 shows how emission spectra from hydrogen clouds would be shifted due to their motion around the galaxy. Create a similar sketch showing an oval automobile racetrack with four cars moving on the track and a stationary observer outside the track at one end. Position and label the four moving cars all sounding their horns: (1) one that would have its horn sound shifted to longer wavelengths, (2) one that would have its horn sound shifted to shorter wavelengths, and (3) two cars moving in opposite directions so that their horn sounds would have no Doppler shift at all.

ANSWERS

ConceptChecks

ConceptCheck 22-1: Looking at other galaxies, astronomers have observed that globular clusters tend to form a spherical distribution centered on the center of galaxies. If our own globular clusters appeared in every direction equally, then astronomers would assume that we were near the center of the globular clusters and, subsequently, near the center of our galaxy.

ConceptCheck 22-2: Far-infrared wavelengths best reveal the presence of warmed dust grains, and weak emission from stars is not easily detected with dust around. Near-infrared wavelengths, which can pass through dust, are emitted more strongly from cool stars than are far-infrared wavelengths, making near-infrared wavelengths a better choice.

ConceptCheck 22-3: As the view between the galactic center and the Sun is obscured by intervening dust, the better vantage point would be from outside the disk in the halo, where the path of light is largely unobstructed by dust.

ConceptCheck 22-4: There is virtually no star formation occurring in the surrounding halo, so the youngest stars would be found actively forming in the rich dusty regions along the galaxy's disk.

ConceptCheck 22-5: Because human eyes are only sensitive in the visible range, we would observe virtually no difference in the appearance of the Milky Way.

ConceptCheck 22-6: Neutral hydrogen gas emits 21-cm photons, and this is the wavelength observed when there is no relative motion between the gas and the observer. However, if the gas is moving away from the observer, its emitted light will be Doppler shifted and detected at longer wavelengths.

ConceptCheck 22-7: The Perseus arm (see Figure 22-16).

ConceptCheck 22-8: The most important factor for how fast a star moves around the galaxy is how much mass is closer to the galactic center than to the star.

ConceptCheck 22-9: In the absence of dark matter, the stars most distant from the galactic center should be moving much slower than those stars orbiting closer to the center.

ConceptCheck 22-10: Astronomers detected only a small amount of matter in the form of MACHOs, yet they would have detected more if it were there. Therefore, the source of dark matter still remains to be discovered. Until WIMPs are detected, or ruled out, they remain a good candidate.

ConceptCheck 22-11: Without density waves causing areas of the galaxy to bunch up here and there, the galaxy would have no discernible arms.

ConceptCheck 22-12: The arms of flocculent spiral galaxies are likely the result of star formation; this suggests that stars came before spiral arms in the early universe.

ConceptCheck 22-13: The gravitational effects of a supermassive black hole cause surrounding stars to move at very high speeds; if there were no supermassive black hole, then nearby stars would be moving much more slowly.

ConceptCheck 22-14: The gases in the nebula would need to be energized by an external energy source, and it is proposed that a sudden intense release of x rays from heated material moving into a supermassive black hole could have externally energized the nebula.

CalculationChecks

CalculationCheck 22-1: No matter how many hydrogen atoms are undergoing spin-flip transitions, only 21-cm photons are released from each atom. Fortunately, greater numbers of hydrogen atoms results in more photons and a stronger signal received at radio telescopes.

CalculationCheck 22-2: Remembering that it takes 2.2×10^8 years for the Sun to make one orbit around the galaxy, 4.5×10^9 years $\times$ (1 orbit/2.2×10^8 years) = 20 orbits.

23 | Galaxies

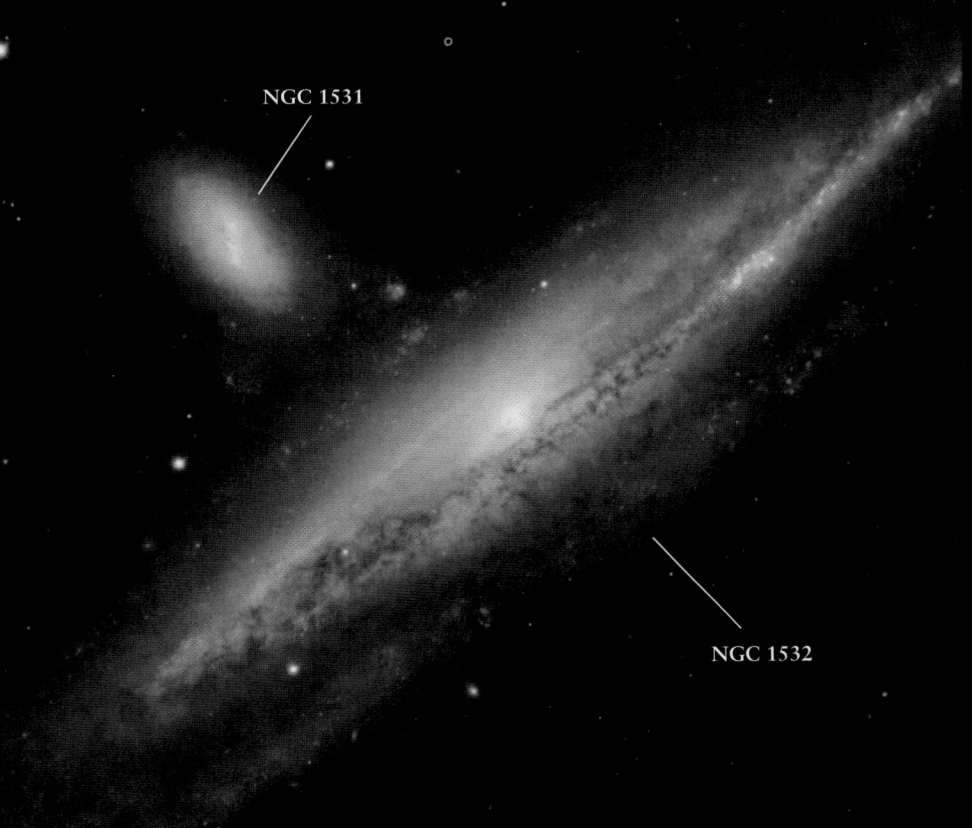

NGC 1531

NGC 1532

LEARNING GOALS

A century ago, most astronomers thought that the entire universe was only a few thousand light-years across and that nothing lay beyond our Milky Way Galaxy. One of the most important discoveries of the twentieth century was that this conception is utterly wrong. We now understand that the Milky Way is just one of billions of galaxies strewn across billions of light-years. The opening image shows two of them, denoted by rather mundane catalog numbers (NGC 1531 and NGC 1532) that give no hint as to the magnificence of these galaxies.

Some galaxies are spirals like NGC 1532 or the Milky Way, with arching spiral arms that are active sites of star formation. (The bright pink bands in NGC 1532 are H II regions, clouds of excited hydrogen that are set aglow by ultraviolet radiation from freshly formed massive stars.) Others, like NGC 1531, are featureless, ellipse-shaped agglomerations of stars, virtually devoid of interstellar gas and dust. Some galaxies are only one one-hundredth the size and one ten-thousandth the mass of the Milky Way. Others are giants, with 5 times the size and 50 times the mass of the Milky Way. Only about 10% of a typical galaxy's mass emits radiation of any kind; the remainder is made up of the mysterious dark matter.

The two galaxies NGC 1531 and NGC 1532 are so close together that they exert strong gravitational forces on each other. Both galaxies are about 17 million pc (55 million ly) from us in the constellation Eridanus. (Gemini Observatory/AURA/NSF)

R I **V** U X G

Just as most stars are found within galaxies, most galaxies are located in groups and clusters. These clusters of galaxies extend throughout the universe, forming huge, lacy patterns. Remarkably, remote clusters of galaxies are receding from us; the greater their distance, the more rapidly they are moving away. This relationship between distance and recessional velocity, called the **Hubble law**, reveals that our immense universe is expanding. In Chapters 25 and 26, we will learn what this implies about the distant past and remote future of the universe.

23-1 When galaxies were first discovered, it was not clear that they lie far beyond the Milky Way

As early as 1755, the German philosopher Immanuel Kant suggested that vast collections of stars lie outside the confines of the Milky Way. Less than a century later, an Irish astronomer observed the structure of some of the "island universes" that Kant proposed.

William Parsons, the third Earl of Rosse, was a wealthy amateur astronomer who used his fortune to build immense telescopes. His largest telescope, completed in February 1845, had an objective mirror 1.8 m (6 ft) in diameter (**Figure 23-1**). The mirror was mounted at one end of a 60-ft tube controlled by cables, straps, pulleys, and cranes (Figure 23-1a). For many years, this triumph of nineteenth-century engineering was the largest telescope in the world.

> As late as the 1920s, it was unclear whether spiral nebulae were very remote "island universes" or simply nearby parts of our galaxy.

With this new telescope, Lord Rosse examined many of the nebulae that had been discovered and cataloged by William Herschel. He observed that some of these nebulae have a distinct spiral structure. One of the best examples is M51, also called NGC 5194. (The "M" designations of galaxies and nebulae come from a catalog compiled by the French astronomer Charles Messier between 1758 and 1782. The "NGC" designations come from the much more extensive "New General Catalogue" of galaxies and nebulae published in 1888 by J. L. E. Dreyer, a Danish astronomer who lived and worked in Ireland.)

Lacking photographic equipment, Lord Rosse had to make drawings of what he saw. Figure 23-1b shows a drawing he made of M51, which compares favorably with modern photographs, as shown in **Figure 23-2**. Views such as this inspired Lord Rosse to echo Kant's proposal that such nebulae are actually "island universes."

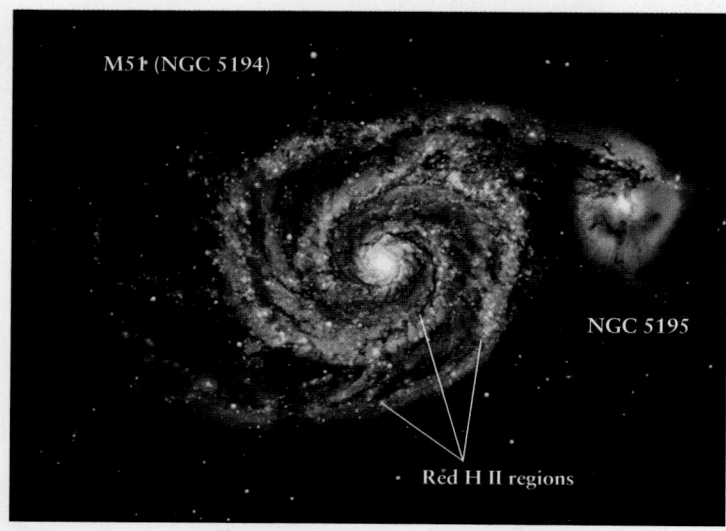

M51 (NGC 5194)

NGC 5195

Red H II regions

FIGURE 23-2 R I V U X G

A Modern View of the Spiral Galaxy M51 This galaxy, also called NGC 5194, has spiral arms that are outlined by glowing H II regions. These regions reveal the sites of star formation (see Section 18-2). One spiral arm extends toward the companion galaxy NGC 5195. (© Jean-Charles Cuillandre, Hawaiian Starlight, CFHT)

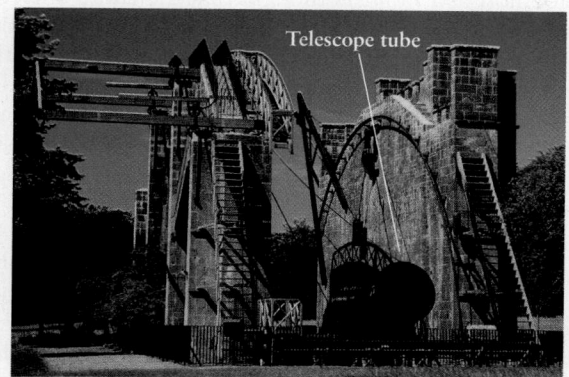

(a) Lord Rosse's "Leviathan of Parsonstown"

Telescope tube

(b) M51 as viewed through the "Leviathan"

FIGURE 23-1

A Pioneering View of Another Galaxy **(a)** Built in 1845, this structure housed the largest telescope of its day. The telescope itself (the black cylinder pointing at a 45° angle above the horizontal) was restored to its original state during 1996–1998. **(b)** Using his telescope, Lord Rosse made this sketch of spiral structure in M51. This galaxy, whose angular size is 8 × 11 arcmin (about one-third the angular size of the full moon), is today called the Whirlpool Galaxy because of its distinctive appearance. (a: Richard T. Nowitz/ Getty Images)

Many astronomers of the nineteenth century disagreed with this notion of island universes, however. A considerable number of nebulae are in fact scattered throughout the Milky Way (see Figures 18-2 and 18-4 for some examples). The safest assumption was that "spiral nebulae," even though they are very different in shape from other sorts of nebulae, could also be components of our galaxy.

The astronomical community became increasingly divided over the nature of the spiral nebulae. In April 1920, two opposing ideas were presented before the National Academy of Sciences in Washington, D.C. On one side was Harlow Shapley from the Mount Wilson Observatory, renowned for his recent determination of the size of the Milky Way Galaxy (see Section 22-1). Shapley thought the spiral nebulae were relatively small, nearby objects scattered around our galaxy like the globular clusters he had studied. Opposing Shapley was Heber D. Curtis of the University of California's Lick Observatory. Curtis championed the island universe theory, arguing that each of these spiral nebulae is a rotating system of stars much like our own galaxy.

The Shapley–Curtis "debate" generated much heat but little light. Nothing was decided, because no one could present conclusive evidence to demonstrate exactly how far away the spiral nebulae are. Astronomy desperately needed a definitive determination of the distance to a spiral nebula. Such a measurement became the first great achievement of a young man who studied astronomy at the Yerkes Observatory, near Chicago. His name was Edwin Hubble.

CONCEPT CHECK 23-1

If you were observing one of the "spiral nebulae" and determined that it was closer than some of the stars of our galaxy, would you be providing evidence in support of Shapley's argument or Curtis's argument?

Answer appears at the end of the chapter.

23-2 Hubble proved that the spiral nebulae are far beyond the Milky Way

After completing his studies, Edwin Hubble joined the staff of the Mount Wilson Observatory in Pasadena, California. On October 6, 1923, he took a historic photograph of the Andromeda "Nebula," one of the spiral nebulae around which controversy raged. (Figure 23-3 is a modern photograph of this object.)

Hubble carefully examined his photographic plate and discovered a *new* bright spot—what he at first thought to be a nova. Referring to previous plates of that region, he soon realized that the object was actually a Cepheid variable star. Further scrutiny of additional plates over the next several months revealed several more Cepheids. Figure 23-4 shows modern observations of a Cepheid in another "spiral nebula."

As we saw in Section 19-6, Cepheid variables help astronomers determine distances. An astronomer begins by carefully

Just as RR Lyrae variable stars demonstrated the extent of the Milky Way, Cepheid variables revealed the immense distances to other galaxies.

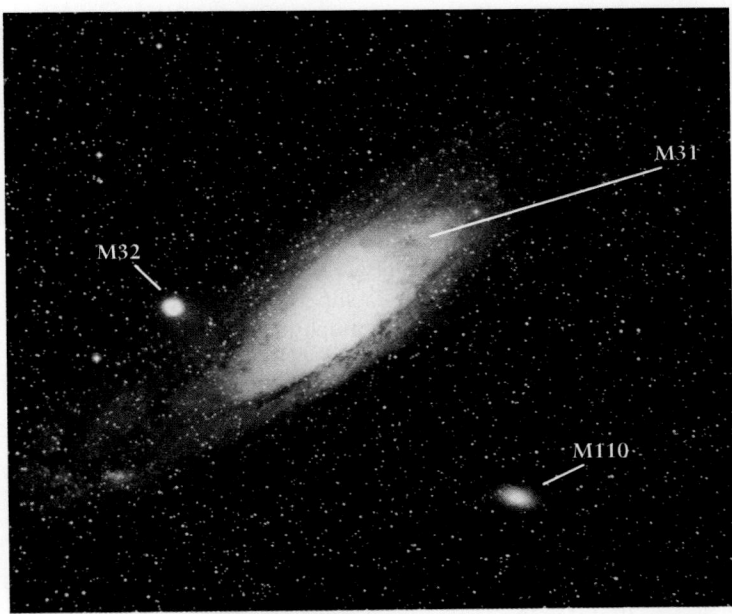

FIGURE 23-3 R I V U X G

The Andromeda Galaxy Also known as M31, the Andromeda Galaxy can be seen with a small telescope, or even the naked eye on a clear night. Edwin Hubble was the first to demonstrate that M31 is actually a galaxy that lies far beyond the Milky Way. M31 is the largest galactic neighbor near our Milky Way. A collision between our Milky Way and the Andromeda Galaxy is expected in about 3.75 billion years. M32 and M110 are two small satellite galaxies that orbit M31. (NASA/Science Source)

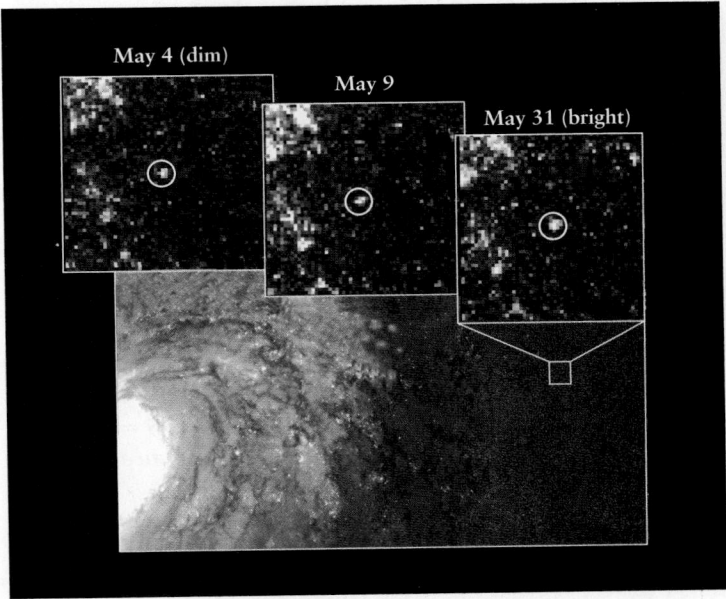

VIDEO 23.1

FIGURE 23-4 R I V U X G

Measuring Galaxy Distances with Cepheid Variables By observing Cepheid variable stars in M100, the galaxy shown here, astronomers have found that it is about 17 Mpc (56 million ly) from Earth. The insets show one of the Cepheids in M100 at different stages in its brightness cycle, which lasts several weeks. The inset containing the variable star is in a region that is still within the galaxy, even though it is not in the brightest region. (NASA, HST, W. Freedman [CIW], R. Kennicutt [U. Arizona], J. Mould [ANU])

BOX 23-1   TOOLS OF THE ASTRONOMER'S TRADE

Cepheids and Supernovae as Indicators of Distance

Because their periods are directly linked to their luminosities, Cepheid variables are one of the most reliable tools astronomers have for determining the distances to galaxies. To this day, astronomers use this link—much as Hubble did back in the 1920s—to measure intergalactic distances. More recently, they have begun to use Type Ia supernovae, which are far more luminous and thus can be seen much farther away, to determine the distances to very remote galaxies.

EXAMPLE: In 1992, a team of astronomers observed Cepheid variables in a galaxy called IC 4182 to deduce this galaxy's distance from Earth. One such Cepheid has a period of 42.0 days and an average apparent magnitude (m) of +22.0. (See Box 17-3 for an explanation of the apparent magnitude scale.) By comparison, the dimmest star you can see with the naked eye has $m = +6$; this Cepheid in IC 4182 appears less than one one-millionth as bright.

According to the period–luminosity relation shown in Figure 19-21, such a Cepheid with a period of 42.0 days has an average luminosity of 33,000 $L_\odot$. An equivalent statement is that this Cepheid has an average absolute magnitude (M) of −6.5. (This compares to $M = +4.8$ for the Sun.) Use this information to determine the distance to IC 4182.

Situation: We are given the apparent magnitude $m = +22.0$ and the absolute magnitude $M = -6.5$ of the Cepheid variable star in IC 4182. Our goal is to calculate the distance to this star, and hence the distance to the galaxy of which it is part.

Tools: We use the relationship between apparent magnitude, absolute magnitude, and distance given in Box 17-3.

Answer: In Box 17-3, we saw that the apparent magnitude of a star is related to its absolute magnitude and distance in parsecs (d) by

$$m - M = 5 \log d - 5$$

This equation can be rewritten as

$$d = 10^{(m - M + 5)/5} \text{ pc}$$

We have $m - M = (+22.0) - (-6.5) = 22.0 + 6.5 = 28.5$. (Recall from Box 17-3 that $m - M$ is called the *distance modulus*.) Hence, our equation becomes

$$d = 10^{(28.5 + 5)/5} \text{ pc} = 10^{6.7} \text{ pc} = 5 \times 10^6 \text{ pc}$$

(A calculator is needed to calculate the quantity $10^{6.7}$.)

Review: Our result tells us that the galaxy is 5 million pc, or 5 Mpc, from Earth (1 Mpc = 10^6 pc). This distance can also be expressed as 16 million ly.

EXAMPLE: Astronomers are interested in IC 4182 because a Type Ia supernova was observed there in 1937. All Type Ia supernovae are exploding white dwarfs that reach nearly the same maximum brightness at the peak of their outburst (see Section 20-9). Once astronomers know the peak absolute magnitude of Type Ia supernovae, they can use these supernovae as distance indicators. Because the distance to IC 4182 is known from its Cepheids, the 1937 observations of the supernova in that galaxy help us calibrate Type Ia supernovae as distance indicators.

At maximum brightness, the 1937 supernova reached an apparent magnitude of $m = +8.6$. What was its absolute magnitude at maximum brightness?

Situation: We are given the supernova's apparent magnitude m, and we know its distance from the previous example. Our goal is to calculate its absolute magnitude M.

Tools: We again use the relationship $m - M = 5 \log d - 5$.

Answer: We could plug in the value of d found in the previous example. But it is simpler to note that the distance modulus $m - M$ has the same value no matter whether it refers to a Cepheid, a supernova, or any other object, just so it is at the same distance d. From the Cepheid example, we have $m - M = 28.5$ for IC 4182, so

$$M = m - (m - M) = 8.6 - (28.5) = -19.9$$

This absolute magnitude corresponds to a remarkable peak luminosity of $10^{10} L_\odot$.

Review: Whenever astronomers find a Type Ia supernova in a remote galaxy, they can combine this absolute magnitude with the observed maximum apparent magnitude to get the galaxy's distance modulus, from which the galaxy's distance can be easily calculated (just as we did above for the Cepheids in IC 4182). This technique has been used to determine the distances to galaxies hundreds of millions of parsecs away (see Section 25-7).

measuring the variations in apparent brightness of a Cepheid variable, then recording the results in the form of a plot of brightness versus time, or *light curve* (see Figure 19-19a). This graph gives the variable star's period and average brightness. Given the star's period, the astronomer then uses the period–luminosity relation shown in Figure 19-21 to find the Cepheid's average luminosity. Knowing both the apparent brightness and luminosity of the Cepheid, the astronomer can then use the inverse-square law to calculate the distance to the star (see Box 17-2). Box 23-1 presents an example of this calculation. This procedure is very similar to that used by Harlow Shapley to measure the distances to the Milky Way's globular clusters using RR Lyrae variable stars (see Section 22-1).

Cepheid variables are intrinsically quite luminous, with average luminosities that can exceed 10^4 L$_\odot$. Hubble realized that for these luminous stars to appear as dim as they were on his photographs of the Andromeda "Nebula," they must be extremely far away. Straightforward calculations using modern data reveal that M31 is some 750 kpc (2.5 million ly) from Earth. Based on its angular size, M31 has a diameter of 70 kpc—larger than the diameter of our own Milky Way Galaxy!

These results prove that the Andromeda "Nebula" is actually an enormous stellar system, far beyond the confines of the Milky Way. Today, this system is properly called the Andromeda Galaxy. (Under good observing conditions, you can actually see this galaxy's central bulge with the naked eye. If you could see the entire Andromeda Galaxy, it would cover an area of the sky roughly 5 times as large as the full moon.) Galaxies are so far away that their distances from us are usually given in millions of parsecs, or *megaparsecs* (Mpc): 1 Mpc = 10^6 pc. For example, the distance to the galaxies in the chapter-opening image is 17 Mpc.

Hubble's results, which were presented at a meeting of the American Astronomical Society on December 30, 1924, settled the Shapley–Curtis "debate" once and for all. The universe was recognized to be far larger and populated with far bigger objects than anyone had seriously imagined. Hubble had discovered the realm of the galaxies.

CAUTION!

In everyday language, many people erroneously use the words *galaxy* and *universe* interchangeably. It is true that before Hubble's discoveries, our Milky Way Galaxy was thought to constitute essentially the entire universe. But we now know that the universe contains literally billions of galaxies. A single galaxy, vast though it may be, is just a tiny part of the entire observable universe.

CONCEPT CHECK 23-2

If a nearby galaxy were discovered, why would astronomers immediately look for Cepheids?

Answer appears at the end of the chapter.

23-3 Galaxies are classified according to their appearance

Millions of galaxies are visible across every unobscured part of the sky. Although all galaxies are made up of large numbers of stars, they come in a variety of shapes and sizes.

TUTORIAL 23.1 Hubble classified galaxies into four broad categories based on their appearance. These categories form the basis for the **Hubble classification**, a scheme that is still used today. The four classes of galaxies are: spirals, classified S; barred spirals, or SB; ellipticals, E; and irregulars, Irr. Table 23-1 summarizes some key properties of each class. These various types of galaxies differ not only in their shapes, but also in the kinds of processes taking place within them.

Spiral Galaxies: Stellar Birthplaces

VIDEO 23.2 M51 (Figure 23-2), M31 (Figure 23-3), and M100 (Figure 23-4) are examples of **spiral galaxies**. Figure 23-5 shows that spiral galaxies are characterized by arched lanes of stars, just as is our own Milky Way Galaxy (see Section 22-3). The spiral arms contain young, hot, blue stars and their associated H II regions, indicating ongoing star formation.

Thermonuclear reactions within stars create *metals*, that is, elements heavier than hydrogen or helium (see Section 17-5). These metals are dispersed into space as the stars evolve and die. So, if new stars are being formed from the interstellar matter in spiral galaxies, they will incorporate these metals and be Population I stars (see Section 19-5 and Section 22-2). Indeed, the visible-light spectrum of the disk of a spiral galaxy has strong metal absorption lines. Such a spectrum is a composite of the spectra of many stars and shows that the stars in the disk are principally of Population I. By contrast, there is relatively little star formation in the central bulges of spiral galaxies, and these regions are dominated by old Population II stars that have a low metal content. The lack of star formation also explains why the central bulges of spiral galaxies have a yellowish or reddish color; as a population of stars ages, the massive, luminous blue stars die off first, leaving only the longer-lived, low-mass red stars.

Hubble further classified spiral galaxies according to the texture of the spiral arms and the relative size of the central bulge. Spirals with smooth, broad spiral arms and a fat central bulge are called Sa galaxies, for spiral type *a* (Figure 23-5a); those galaxies

TABLE 23-1 Some Properties of Galaxies

	Spiral (S) and barred spiral (SB) galaxies	Elliptical galaxies (E)	Irregular galaxies (Irr)
Mass (M$_\odot$)	10^9 to 4×10^{11}	10^5 to 10^{13}	10^8 to 3×10^{10}
Luminosity (L$_\odot$)	10^8 to 2×10^{10}	3×10^5 to 10^{11}	10^7 to 10^9
Diameter (kpc)	5 to 250	1 to 200	1 to 10
Stellar populations	Spiral arms: young Population I Nucleus and throughout disk: Population II and old Population I	Population II and old Population I	Mostly Population I
Percentage of observed galaxies	77%	20%*	3%

*This percentage does not include dwarf elliptical galaxies that are as yet too dim and distant to detect. Hence, the actual percentage of galaxies that are ellipticals may be higher than shown here.

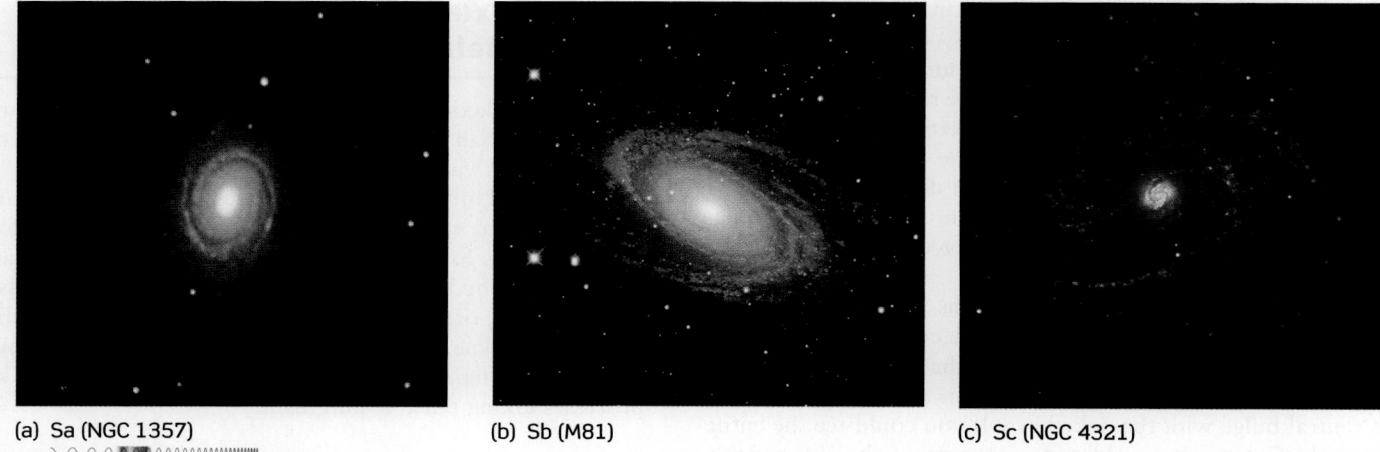

(a) Sa (NGC 1357)   (b) Sb (M81)   (c) Sc (NGC 4321)

FIGURE 23-5 R I V U X G

Spiral Galaxies Edwin Hubble classified spiral galaxies according to the texture of their spiral arms and the relative size of their central bulges. Sa galaxies have smooth, broad spiral arms and the largest central bulges, whereas Sc galaxies have narrow, well-defined arms and the smallest central bulges. (a: Adam Block/Steve Mandel/Jim Rada and Students/NOAO/AURA/NSF; b: Robert Gendler/Stocktrek Images/Science Source; c: ESO)

with moderately well-defined spiral arms and a moderate-sized central bulge, like M31 and M51, are Sb galaxies (Figure 23-5b); and galaxies with narrow, well-defined spiral arms and a tiny central bulge are Sc galaxies (Figure 23-5c).

The differences between Sa, Sb, and Sc galaxies may be related to the relative amounts of gas and dust they contain. Observations with infrared telescopes (which detect the emission from interstellar dust) and radio telescopes (which detect radiation from interstellar gases such as hydrogen and carbon monoxide) show that about 4% of the mass of an Sa galaxy is in the form of gas and dust. This percentage is 8% for Sb galaxies and 25% for Sc galaxies.

Interstellar gas and dust is the material from which new stars are formed, so an Sc galaxy has a greater proportion of its mass involved in star formation than an Sb or Sa galaxy. Hence, an Sc galaxy has a large disk (where star formation occurs) and a small central bulge (where there is little or no star formation). By comparison, an Sa galaxy, which has relatively little gas and dust, and thus less material from which to form stars, has a large central bulge and only a small star-forming disk.

The central bulge contains more than just stars: Each of these features is also thought to contain a supermassive black hole at its center. In each case, the mass of the black hole is about 1/1000 the mass of the central bulge, corresponding to a range from millions to hundreds of millions of solar masses. (The mass of the black hole at the center of our Milky Way is just over 4 million $M_\odot$.) Chapter 24 discusses how these black holes are the "central engines" of quasars and active galaxies.

CONCEPT CHECK 23-3

When making a sketch of a spiral galaxy with considerable star formation, which areas would you label with *more* star formation? Which areas would have *less*?
Answer appears at the end of the chapter.

Barred Spiral Galaxies: Spirals with an Extra Twist

In **barred spiral galaxies**, such as those shown in Figure 23-6, the spiral arms originate at the ends of a bar-shaped region running

(a) SBa (NGC 1291)   (b) SBb (M83)   (c) SBc (NGC 1365)

FIGURE 23-6 R I V U X G

Barred Spiral Galaxies As with spiral galaxies, Hubble classified barred spirals according to the texture of their spiral arms (which correlates to the sizes of their central bulges). SBa galaxies have the smoothest spiral arms and the largest central bulges, while SBc galaxies have narrow, well-defined arms and the smallest central bulges. (a: NASA/JPL-Caltech/CTIO; b, c: ESO) through the galaxy's nucleus rather than from the nucleus itself. As with ordinary spirals, Hubble subdivided barred spirals according to the relative size of the central bulge and the character of the spiral arms. An SBa galaxy has a large central bulge and thin, tightly wound spiral arms (Figure 23-6a). Likewise, an SBb galaxy is a barred spiral with a moderate central bulge and moderately wound spiral arms (Figure 23-6b), while an SBc galaxy has lumpy, loosely wound spiral arms and a tiny central bulge (Figure 23-6c). As for ordinary spiral galaxies, the difference between SBa, SBb, and SBc galaxies may be related to the amount of gas and dust in the galaxy.

Bars appear to form naturally in many spiral galaxies. This conclusion comes from computer simulations of galaxies, which set hundreds of thousands of simulated "stars" into orbit around a common center. As the "stars" orbit and exert gravitational forces on one another, a bar structure forms in most cases. Indeed, barred spiral galaxies outnumber ordinary spirals by about 2 to 1. (As we saw in Section 22-2, there is evidence that the Milky Way Galaxy is a barred spiral.)

Why don't all spiral galaxies have bars? According to calculations by Jeremiah Ostriker and P. J. E. Peebles of Princeton University, a bar will not develop if a galaxy is surrounded by a sufficiently massive halo of nonluminous *dark matter*. (In Section 22-4, we saw evidence that our Milky Way Galaxy is surrounded by such a dark matter halo.) The difference between barred spirals and ordinary spirals may thus lie in the amount of dark matter the galaxy possesses. In Section 23-8, we will see compelling evidence for the existence of dark matter in spiral galaxies.

CONCEPT CHECK 23-4

How are galaxies with loosely wrapped arms categorized using Hubble's scheme?

Answer appears at the end of the chapter.

Elliptical Galaxies: From Giants to Dwarfs

Elliptical galaxies, so named because of their distinctly elliptical shapes, have no spiral arms. Hubble subdivided these galaxies according to how round or flattened they look. The roundest elliptical galaxies are called E0 galaxies and the flattest are called E7 galaxies. Elliptical galaxies with intermediate amounts of flattening are given designations between these extremes (Figure 23-7).

CAUTION!

Unlike the designations for spirals and barred spirals, the classifications E0 through E7 may not reflect the true shape of elliptical galaxies. A flat, disk-shaped galaxy might appear very round to us if we happen to be viewing it face on, earning it an E1 or E2 designation. But if instead it happens to be oriented nearly edge on, then we would view it as a thin, cigar-shaped E7 galaxy. The Hubble scheme classifies galaxies entirely by how they *appear* to us on Earth.

Elliptical galaxies look far less dramatic than their spiral and barred spiral cousins. The reason, as shown by radio and infrared observations, is that ellipticals are virtually devoid of interstellar gas and dust. Consequently, there is little material from which stars could have recently formed, and indeed there is no evidence of young stars in most elliptical galaxies. For the most part, star formation in elliptical galaxies ended long ago. Hence, these galaxies are composed of old, red, Population II stars with only small amounts of metals.

Elliptical galaxies come in a wide range of sizes and masses. Both the largest and the smallest galaxies in the known universe are elliptical. Figure 23-8 shows two **giant elliptical galaxies** that are about 20 times larger than an average galaxy. These giant ellipticals are located near the middle of a large cluster of galaxies in the constellation Virgo. (We will discuss this and other clusters of galaxies in Section 23-6.)

Giant ellipticals are rather rare, but **dwarf elliptical galaxies** are quite common. Dwarf ellipticals are only a fraction the size of their normal counterparts and contain so few stars—only a few million, compared to more than 100 billion (10^{11}) stars in our Milky Way Galaxy—that these galaxies are completely transparent. You can actually see straight through the center of a dwarf galaxy and out the other side, as Figure 23-9 shows.

(a) E0 (M105)

(b) E3 (NGC 4406)

(c) E6 (NGC 3377)

FIGURE 23-7 R I V U X G

Elliptical Galaxies Hubble classified elliptical galaxies according to how round or flattened they look. A galaxy that appears round is labeled E0, and the flattest-appearing elliptical galaxies are designated E7. False coloring is used in these images to bring out fainter features. (a: Karl Gebhardt [University of Michigan], Tod Lauer [NOAO], and NASA; b: Jean-Charles Cuillandre, Hawaiian Starlight, CFHT; c: Karl Gebhardt [University of Michigan], Tod Lauer [NOAO], and NASA)

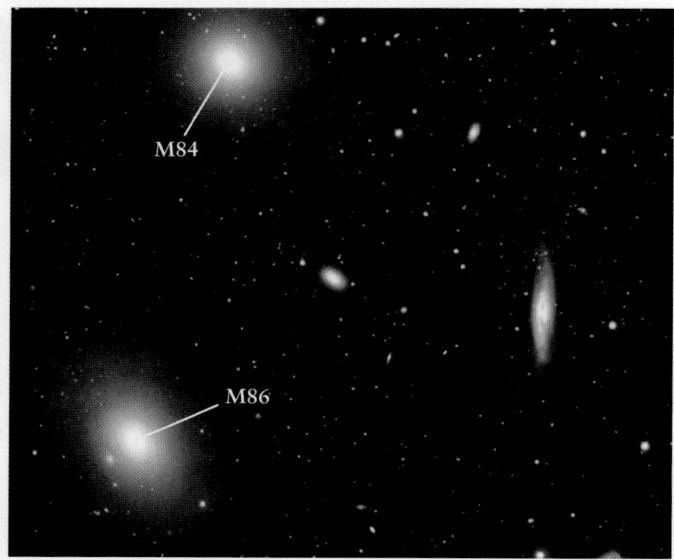

FIGURE 23-8 R I V U X G

Giant Elliptical Galaxies The Virgo Cluster is a rich, sprawling collection of more than 2000 galaxies about 17 Mpc (56 million ly) from Earth. Only the center of this huge cluster appears in this photograph. The two largest members of this cluster are the giant elliptical galaxies M84 and M86. These galaxies have angular sizes of 5 to 7 arcmin. (© Jean-Charles Cuillandre, Hawaiian Starlight, CFHT)

The visible light from a galaxy is emitted by its stars, so the visible-light spectrum of a galaxy has absorption lines. But because a galaxy's stars are in motion, with some approaching us and others moving away, the Doppler effect smears out and broadens the absorption lines. The average motions of stars in a galaxy can be deduced from the details of this broadening.

FIGURE 23-9 R I V U X G

A Dwarf Elliptical Galaxy This diffuse cloud of stars is a nearby E4 dwarf elliptical called Leo I. It actually orbits the Milky Way at a distance of about 180 kpc (600,000 ly). Leo I is about 1 kpc (3000 ly) in diameter but contains so few stars that you can see through the galaxy's center. (NASA and The Hubble Heritage Team [STScI/AURA])

FIGURE 23-10 R I V U X G

A Lenticular Galaxy NGC 2787 is classified as a lenticular galaxy because it has a disk but no discernible spiral arms. Its nucleus displays a faint bar (not apparent in this image), so NGC 2787 is denoted as an SB0 galaxy. It lies about 7.4 Mpc (24 million ly) from Earth in the constellation Ursa Major. (NASA and The Hubble Heritage Team, STScI/AURA)

For elliptical galaxies, studies of this kind show that star motions are quite random. In a truly round (E0) elliptical galaxy, this randomness is **isotropic,** meaning "equal in all directions." Because the stars are whizzing around equally in all directions, the galaxy is genuinely spherical. In a flattened (E7) elliptical galaxy, the randomness of the stellar motions is **anisotropic,** which means that the range of star speeds is different in different directions.

Hubble also identified galaxies that are midway in appearance between ellipticals and the two kinds of spirals. These are denoted as S0 and SB0 galaxies, also called **lenticular galaxies.** Although they look somewhat elliptical, lenticular ("lens-shaped") galaxies have both a central bulge and a disk like spiral galaxies, but no discernible spiral arms. They are therefore sometimes referred to as "armless spirals." Figure 23-10 shows an example of an SB0 lenticular galaxy.

Edwin Hubble summarized his classification scheme for spiral, barred spiral, and elliptical galaxies in a diagram, now called the **tuning fork diagram** for its shape (Figure 23-11).

CAUTION!

When Hubble first drew his tuning fork diagram, he had the idea that it represented an evolutionary sequence. He thought that galaxies evolved over time from the left to the right of the diagram, beginning as ellipticals and eventually becoming either spiral or barred spiral galaxies. We now understand that this evolution of galaxies is not the case at all! For one thing, elliptical galaxies have little or no overall rotation, while spiral and barred spiral galaxies have a substantial amount of overall rotation. There is no way that an elliptical galaxy could suddenly start rotating, which means that it could not evolve into a spiral galaxy.

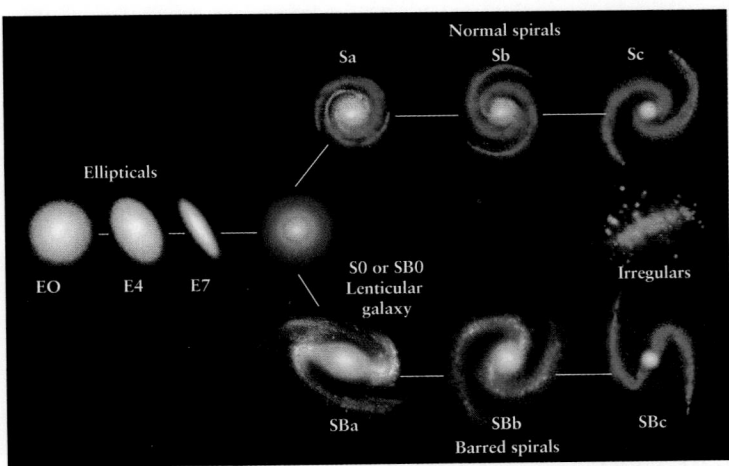

FIGURE 23-11

Hubble's Tuning Fork Diagram Edwin Hubble's classification of regular galaxies is shown in his tuning fork diagram. An elliptical galaxy is classified by how flattened it appears. A spiral or barred spiral galaxy is classified by the texture of its spiral arms and the size of its central bulge. A lenticular galaxy is intermediate between ellipticals and spirals. Irregular galaxies do not fit into this simple classification scheme. The ordering of galaxies does not represent an evolutionary sequence.

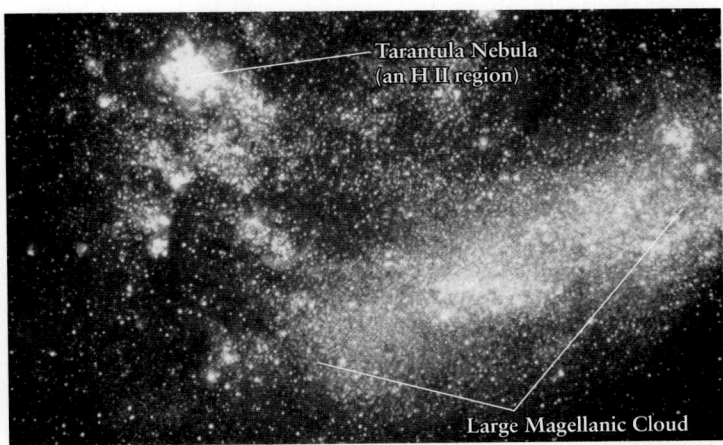

FIGURE 23-12   R I V U X G

The Large Magellanic Cloud (LMC) At a distance of only 55 kpc (179,000 ly), this Irr I galaxy is the third closest known companion of our Milky Way Galaxy. About 19 kpc (62,000 ly) across, the LMC spans 22° across the sky, or about 50 times the size of the full moon. One sign that star formation is ongoing in the LMC is the Tarantula Nebula, whose diameter of 250 pc (800 ly) and mass of 5×10^6 M$_\odot$ make it the largest known H II region. (NOAO/AURA/NSF)

A more modern interpretation of the Hubble tuning fork diagram views it as an arrangement of galaxies according to their overall rotation. A rapidly rotating collection of matter in space tends to form a disk, while a slowly rotating collection does not. Thus, the elliptical galaxies at the far left of the tuning fork diagram have little internal rotation and hence no disk. Sa and SBa galaxies have enough overall rotation to form a disk, though their central bulges are still dominant. The galaxies with the greatest amount of overall rotation are Sc and SBc galaxies, in which the central bulges are small and most of the gas, dust, and stars are in the disk.

Irregular Galaxies: Deformed and Dynamic

Galaxies that do not fit into the scheme of spirals, barred spirals, and ellipticals are usually referred to as **irregular galaxies**. They are generally rich in interstellar gas and dust, and have both young and old stars. For lack of any better scheme, the irregular galaxies are sometimes placed between the ends of the tines of the Hubble tuning fork diagram, as in Figure 23-11.

Hubble defined two types of irregulars. Irr I galaxies have only hints of organized structure, and they have many OB associations and H II regions. The best-known examples of Irr I galaxies are the Large Magellanic Cloud (LMC; see Figure 23-12) and the Small Magellanic Cloud (SMC). Both are nearby companions of our Milky Way and can be seen with the naked eye from southern latitudes. Both these galaxies contain substantial amounts of interstellar gas. Tidal forces exerted on these irregular galaxies by the Milky Way help to compress the gas, which is why both of the Magellanic clouds are sites of active star formation.

The other type of irregulars, called Irr II galaxies, have asymmetrical, distorted shapes that seem to have been caused by collisions with other galaxies or by violent activity in their nuclei. M82,

shown in Figure 21-26 and in the image that opens Chapter 6, is an example of an Irr II galaxy.

CONCEPT CHECK 23-5

Which type of galaxy has almost no ongoing active star formation?

Answer appears at the end of the chapter.

23-4 Astronomers use various techniques to determine the distances to remote galaxies

A key question that astronomers ask about galaxies is "How far away are they?" Knowing the distances to galaxies is essential for learning the structure and history of the universe. Unfortunately, many of the techniques that are used to measure distances within our Milky Way Galaxy cannot be used for the far greater distances to other galaxies. The extremely accurate parallax method that we described in Section 17-1 can be used only for stars within about 500 pc. Beyond that distance, parallax angles become too small to measure. Spectroscopic parallax, in which the distance to a star or star cluster is found with the help of the H-R diagram (see Section 17-8), is accurate out to roughly only 10 kpc from Earth; more distant stars or clusters are too dim to give reliable results.

> The various methods of distance determination are interrelated because one is used to calibrate another.

Standard Candles: Variable Stars and Type Ia Supernovae

To determine the distance to a remote galaxy, astronomers look instead for a **standard candle**—an object, such as a star, that lies within that galaxy and for which we know the luminosity (or, equivalently, the absolute magnitude, described in Section 17-3). By measuring how bright the standard candle appears, astronomers can calculate its distance—and hence the distance to the galaxy of which it is part—using the inverse-square law. (To reflect the fact that there is still progress to be made in determining the object's luminosity as well as we'd like to, some researchers call these objects *standardizable candles*.)

The challenge is to find standard candles that are luminous enough to be seen across the tremendous distances to galaxies. To be useful, standard candles should have four properties:

1. They should be luminous enough that we can see them out to great distances.
2. We should be fairly certain about their luminosities so that we can be equally certain of any distance calculated from a standard candle's apparent brightness and luminosity.
3. They should be easily identifiable—for example, by the shape of the light curve of a variable star.
4. They should be relatively common so that astronomers can use them to determine the distances to many different galaxies.

For nearby galaxies, Cepheid variable stars make reliable standard candles. These variables can be seen out to about 30 Mpc (100 million ly) using the Hubble Space Telescope, and their luminosities can be determined from their periods through the period–luminosity relation depicted in Figure 19-21. Box 23-1 gives an example of using Cepheid variables to determine distances. RR Lyrae stars, which are Population II variable stars often found in globular clusters, can be used as standard candles in a similar way. (We saw in Section 22-1 how RR Lyrae variables helped determine the size of our galaxy.) Because they are less luminous than Cepheids, RR Lyrae variables can be seen only out to 100 kpc (300,000 ly).

Beyond about 30 Mpc, even the brightest Cepheid variables, which have luminosities of about $2 \times 10^4 \, L_\odot$, fade from view. Astronomers have tried to use even more luminous stars such as blue supergiants to serve as standard candles. However, this idea is based on the assumption that there is a fixed upper limit on the luminosities of stars, which may not be the case. Hence, these

standard candles are not very "standard," and distances measured in this way are somewhat uncertain.

One class of standard candles that astronomers have used beyond 30 Mpc is Type Ia supernovae. As we described in Section 20-9, these supernovae occur when a white dwarf in a close binary system accretes enough matter from its companion to blow itself apart in a thermonuclear conflagration. A Type Ia supernova can reach a maximum luminosity of about $3 \times 10^9 \, L_\odot$ (Figure 23-13). If a Type Ia supernova is seen in a distant galaxy and its maximum apparent brightness is measured, the inverse-square law can be used to find the galaxy's distance (see Box 23-1).

One complication is that not all Type Ia supernovae are equally luminous. Fortunately, there is a simple relationship between the peak luminosity of a Type Ia supernova and the rate at which the luminosity decreases after the peak: The more slowly the brightness decreases, the more luminous the supernova. Using this relationship, astronomers have measured distances to supernovae more than 1000 Mpc (3 billion ly) from Earth.

Unfortunately, this technique can be used only for galaxies in which we happen to observe a rare Type Ia supernova. But since telescopic surveys now encompass thousands of galaxies, many dozens of these supernovae are identified every year, and the number of galaxies whose distances can be measured in this way is continually increasing.

CONCEPT CHECK 23-6

Imagine that a thin cloud of intergalactic dust is reducing the observed brightness of a Type Ia supernova found in a distant galaxy. Would astronomers who do not know the dust is there mistakenly assume that the galaxy is farther or closer than it actually is?

Answer appears at the end of the chapter.

Distance Determination without Standard Candles

Other methods for determining the distances to galaxies do not make use of standard candles. One was discovered in the 1970s by the astronomers Brent Tully and Richard Fisher. They found that the width of the hydrogen 21-cm emission line of a spiral galaxy (see Section 22-3) is related to the galaxy's luminosity. This correlation is the **Tully–Fisher relation**—the broader the line, the more luminous the galaxy.

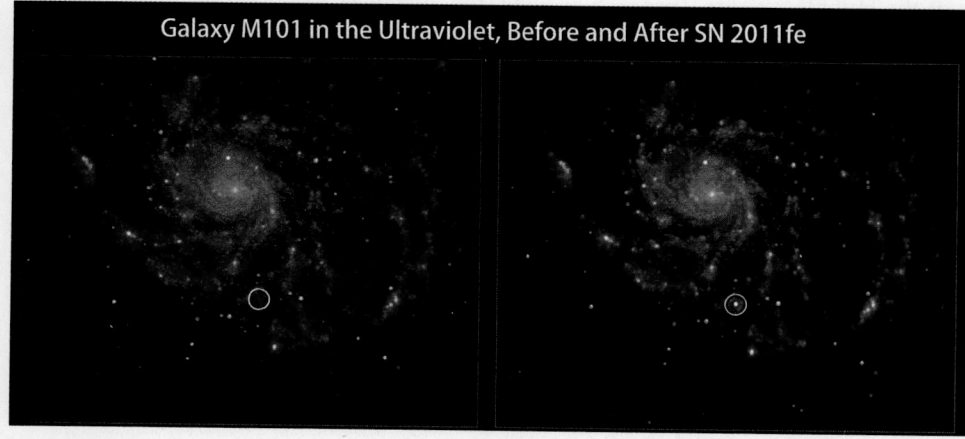

Galaxy M101 in the Ultraviolet, Before and After SN 2011fe

(a) March/April 2007   (b) August through November 2011

FIGURE 23-13 R I V U X G

A Supernova in a Spiral Galaxy These images show the spiral galaxy M101 **(a)** before and **(b)** after a Type Ia supernova exploded within the galaxy in 2011. Such luminous supernovae, which can be seen at extreme distances, are important standard candles used to determine the distances to faraway galaxies. (Swift/Peter Brown, Univ. of Utah/NASA)

Such a relationship exists because radiation from the approaching side of a rotating galaxy is blueshifted while that from the galaxy's receding side is redshifted. Thus, the 21-cm line is Doppler broadened by an amount directly related to how fast a galaxy is rotating. Rotation speed is related to the galaxy's mass by Newton's form of Kepler's third law (see Box 4-4). The more massive a galaxy, the more stars it contains and thus the more luminous it is. Consequently, the width of a galaxy's 21-cm line is directly related to its luminosity.

Because line widths can be measured quite accurately, astronomers can use the Tully–Fisher relation to determine the luminosity of a distant spiral galaxy. By combining this information with measurements of apparent brightness, they can calculate the distance to the galaxy. This technique can be used to measure distances of 100 Mpc or more.

Elliptical galaxies do not rotate, so the Tully–Fisher relation cannot be used to determine their distances. But in 1987, the American astronomers Marc Davis and George Djorgovski pointed out a correlation between the size of an elliptical galaxy, the average motions of its stars, and how the galaxy's brightness appears distributed over its surface. In geometry, three points define a plane, so the relationship among size, motion, and brightness is called the **fundamental plane**. By measuring the last two quantities, an astronomer can use the fundamental plane relationship to determine a galaxy's actual size. And by comparing this to the galaxy's apparent size, the astronomer can calculate the distance to the galaxy using the small-angle formula (see Box 1-1). Ellipticals can be substantially larger and more luminous than spirals, so the fundamental plane can be used at somewhat greater distances than can the Tully–Fisher relation.

CONCEPT CHECK 23-7

Is a galaxy that appears to be quite bright, yet has a narrow width for its emitted hydrogen light, relatively close or very distant from our own galaxy?

Answer appears at the end of the chapter.

The Distance Ladder

Figure 23-14 shows the ranges of applicability of several important means of determining astronomical distances. Because these ranges overlap, one technique can be used to calibrate another. As an example, astronomers have studied Cepheids in nearby galaxies that have also been host to Type Ia supernovae. The Cepheids provide the distances to these nearby galaxies, making it possible to determine the peak luminosity of each supernova using its maximum apparent brightness and the inverse-square law. Once the peak luminosity is known, it can be used to determine the distance to Type Ia supernovae in more distant galaxies. Because one measuring technique leads us to the next one like rungs on a ladder, the techniques shown in Figure 23-14 (along with others) are referred to collectively as the **distance ladder**.

ANALOGY

If you slightly shake the bottom of a tall ladder, the top can wobble back and forth alarmingly. A change in distance-measuring techniques used for nearby objects can also have substantial effects on the distances to remote galaxies. For instance, if astronomers discovered that the distances to nearby Cepheids were in error, distance measurements using any technique that is calibrated by Cepheids would be affected as well. (As an example, the distance to the galaxy M100 shown in Figure 23-4 is determined using Cepheids. A Type Ia supernova has been seen in M101, and its luminosity is determined using the Cepheid-derived distance to M101. Any change in the calculated distance to M101 would change the calculated luminosity of the Type Ia supernova, and so it would have an effect on all distances derived from observations of how bright these supernovae appear in other galaxies.) For this reason, astronomers go to great lengths to check the accuracy and reliability of their standard candles.

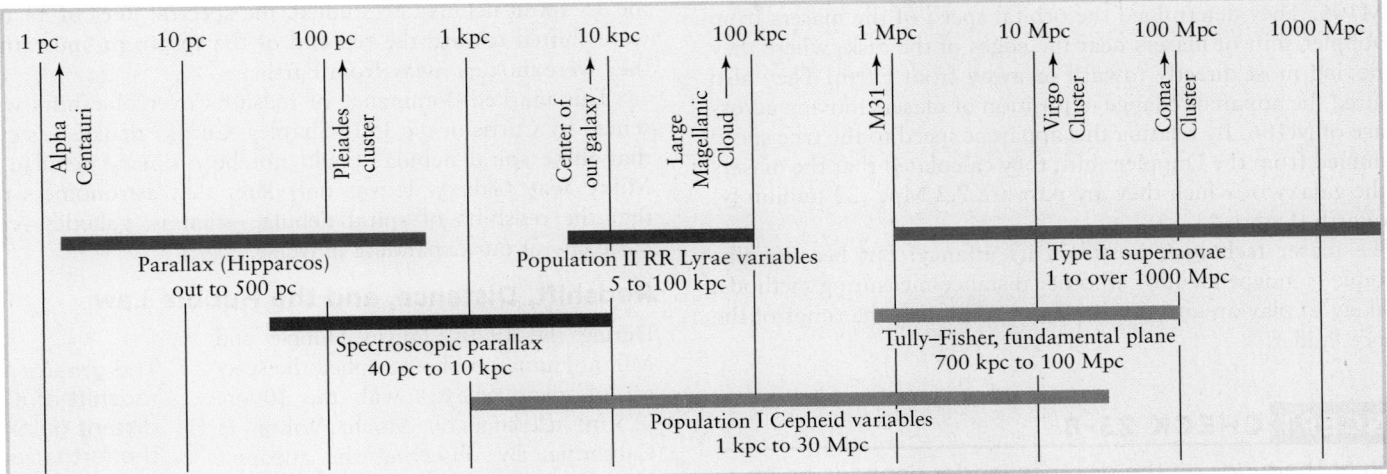

FIGURE 23-14

The Distance Ladder Astronomers employ a variety of techniques for determining the distances to objects beyond the solar system. Because their ranges of applicability overlap, one technique can be used to calibrate another. The arrows indicate distances to several important objects. Note that each division on the scale indicates a tenfold increase in distance, such as from 1 to 10 Mpc.

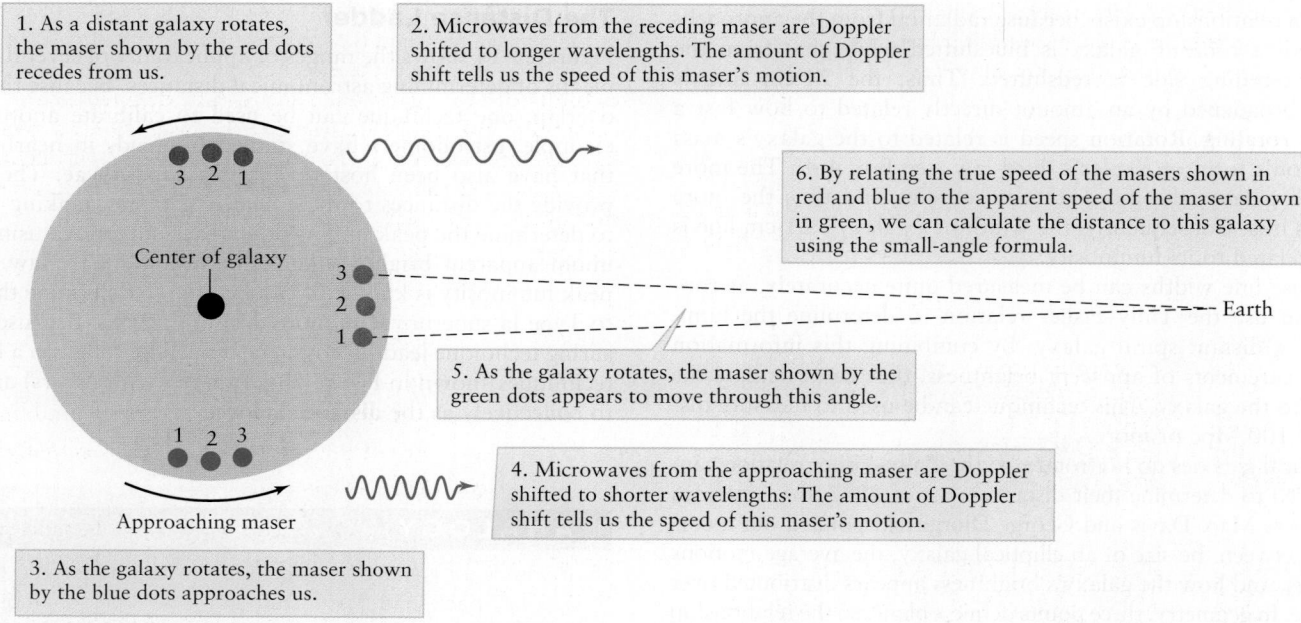

1. As a distant galaxy rotates, the maser shown by the red dots recedes from us.

2. Microwaves from the receding maser are Doppler shifted to longer wavelengths. The amount of Doppler shift tells us the speed of this maser's motion.

6. By relating the true speed of the masers shown in red and blue to the apparent speed of the maser shown in green, we can calculate the distance to this galaxy using the small-angle formula.

Center of galaxy

Earth

5. As the galaxy rotates, the maser shown by the green dots appears to move through this angle.

4. Microwaves from the approaching maser are Doppler shifted to shorter wavelengths: The amount of Doppler shift tells us the speed of this maser's motion.

Approaching maser

3. As the galaxy rotates, the maser shown by the blue dots approaches us.

FIGURE 23-15

Measuring the Distance to a Galaxy Using Masers This drawing shows interstellar clouds called masers (the colored dots) moving from position 1 to 2 to 3 as they orbit the center of a galaxy. The redshift and blueshift of microwaves from the masers shown in red and blue tell us their orbital speed. By relating this to the angle through which the masers shown in green appear to move in a certain amount of time, we can calculate the distance to the galaxy.

One distance-measuring technique that has broken free of the distance ladder uses observations of molecular clouds called **masers**. (*Maser* is an acronym for *m*icrowave *a*mplification by *s*timulated *e*mission of *r*adiation.) Just as an electric current stimulates a laser to emit an intense beam of visible light, nearby luminous stars can stimulate water molecules in a maser to emit intensely at microwave wavelengths. This radiation is so intense that masers can be detected millions of parsecs away.

During the 1990s, Jim Herrnstein and his collaborators used the Very Long Baseline Array (see Section 6-6) to observe a number of masers orbiting in a disk around the center of the spiral galaxy M106. They determined the orbital speed of the masers from the Doppler shift of masers near the edges of the disk, where they are moving most directly toward or away from Earth. They also measured the apparent change in position of masers moving across the face of M106. By relating this apparent speed to the true speed determined from the Doppler shift, they calculated that the masers and the galaxy of which they are part are 7.2 Mpc (23 million ly) from Earth (Figure 23-15).

The maser technique is still in its infancy. But because this technique is independent of all other distance-measuring methods, it is likely to play an important role in calibrating the rungs of the distance ladder.

CONCEPT CHECK 23-8

Which of the rungs on the distance ladder depends on an accurate measurement of parallax?

Answer appears at the end of the chapter.

23-5 The Hubble law relates the redshifts of remote galaxies to their distances from Earth

Whenever an astronomer finds an object in the sky that can be seen or photographed, the natural inclination is to attach a spectrograph to a telescope and record the spectrum. As long ago as 1914, Vesto M. Slipher, working at the Lowell Observatory in Arizona, began taking spectra of "spiral nebulae"—a name used before they were known to be galaxies. He was surprised to discover that, of the 15 spiral nebulae he studied, the spectral lines of 11 of them were shifted toward the red end of the spectrum, indicating that they were moving away from Earth.

This marked dominance of redshifts over blueshifts was presented by Curtis in the 1920 Shapley–Curtis "debate" as evidence that these spiral nebulae could not be ordinary nebulae in our Milky Way Galaxy. It was only later that astronomers realized that the redshifts of spiral nebulae—that is, galaxies—reveal a basic law of our expanding universe.

Redshift, Distance, and the Hubble Law

During the 1920s, Edwin Hubble and Milton Humason photographed the spectra of many galaxies with the 100-in. (2.5-m) telescope on Mount Wilson in California. By observing the apparent brightnesses and pulsation periods of Cepheid variables in these galaxies, they were also able to measure the distance to

The greater the redshift of a distant galaxy, the farther away it is.

each galaxy (see Section 23-2). Hubble and Humason found that most galaxies show a redshift in their spectrum. They also found a direct correlation between the distance to a galaxy and its redshift:

The more distant a galaxy, the greater its redshift and the more rapidly it is receding from us.

In other words, nearby galaxies are moving away from us slowly, and more distant galaxies are rushing away from us much more rapidly. **Figure 23-16** shows this relationship for five representative

elliptical galaxies. This universal recessional movement is referred to as the **Hubble flow** and is described by what is called *Hubble's law*.

Hubble estimated the distances to a number of galaxies and the redshifts of those galaxies. The **redshift**, denoted by the symbol z, is found by taking the wavelength (λ) observed for a given spectral line, subtracting from it the ordinary, unshifted wavelength of that line (λ_0) to get the wavelength difference ($\Delta\lambda$), and then dividing that difference by λ_0:

Redshift of a receding object

$$z = \frac{\lambda - \lambda_0}{\lambda_0} = \frac{\Delta\lambda}{\lambda_0}$$

z = redshift of an object

λ_0 = ordinary, unshifted wavelength of a spectral line

λ = wavelength of that spectral line that is actually observed from the object

From the redshifts, Hubble used the Doppler formula to calculate the speed at which these galaxies are receding from us. **Box 23-2** describes this calculation. Plotting the data on a graph of distance versus speed, Hubble found that the points lie near a straight line. **Figure 23-17** is a modern version of Hubble's graph based on recent data.

This relationship between the distances to galaxies and their redshifts was one of the most important astronomical discoveries of the twentieth century. As we will see in Chapter 26, this relationship tells us that we are living in an expanding universe. In 1929, Hubble published this discovery, which is now known as the Hubble law. The Hubble law is most easily stated as a formula:

The Hubble law

$$v = H_0 d$$

v = recessional velocity of a galaxy

H_0 = Hubble constant

d = distance to the galaxy

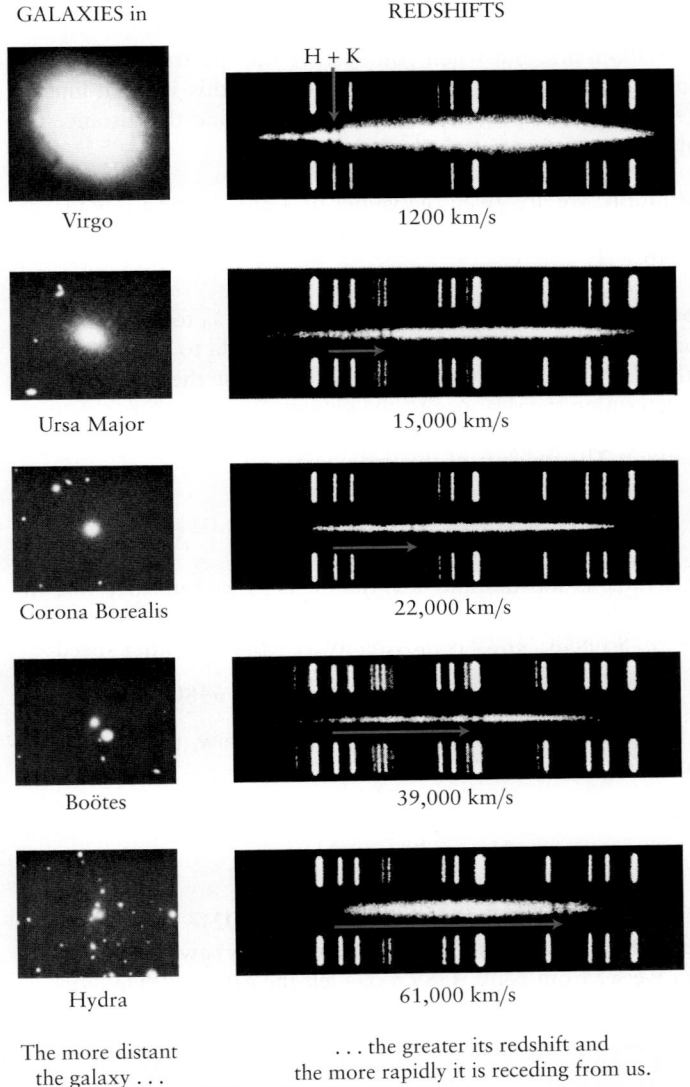

GALAXIES in   REDSHIFTS

H + K

Virgo   1200 km/s

Ursa Major   15,000 km/s

Corona Borealis   22,000 km/s

Boötes   39,000 km/s

Hydra   61,000 km/s

The more distant   . . . the greater its redshift and
the galaxy . . .   the more rapidly it is receding from us.

FIGURE 23-16   R I V U X G

Relating the Distances and Redshifts of Galaxies These are several of the original images Hubble and Humason used to establish the Hubble law. The five galaxies are arranged, from top to bottom, in order of increasing distance from us. All are shown at the same magnification. Each galaxy's spectrum is a bright band with dark absorption lines; the bright lines above and below it are a reference emission spectrum of a light source at the observatory on Earth. The horizontal red arrows show how much the H and K lines of singly ionized calcium are redshifted in each galaxy's spectrum. Below each spectrum is the recessional velocity calculated from the redshift. The more distant a galaxy is, the greater its redshift. Since these galaxies are arranged in order of increasing distance from top to bottom, the red arrows indicating redshift are increasing (left to right) as well. (Carnegie Observatories)

FIGURE 23-17

The Hubble Law This graph plots the distances and recessional velocities of a sample of galaxies. The straight line is the best fit to the data. This linear relationship between distance and recessional velocity is called the Hubble law.

BOX 23-2   TOOLS OF THE ASTRONOMER'S TRADE

The Hubble Law, Redshifts, and Recessional Velocities

Suppose that you aim a telescope at a distant galaxy. The galaxy will be moving away from you in the Hubble flow. You take a spectrum of the galaxy and find that the spectral lines are shifted toward the red end of the spectrum. For example, a particular spectral line whose normal wavelength is λ_0 appears in the galaxy's spectrum at a longer wavelength λ. The spectral line has thus been shifted by an amount $\Delta\lambda = \lambda - \lambda_0$. The redshift of the galaxy, z, is given by

$$z = \frac{\lambda - \lambda_0}{\lambda_0} = \frac{\Delta\lambda}{\lambda_0}$$

The redshift means that the galaxy is receding from us. According to the Hubble law, the recessional velocity v of a galaxy is related to its distance d from Earth by

$$v = H_0 d$$

where H_0 is the Hubble constant. We can rewrite this equation as

$$d = \frac{v}{H_0}$$

Given the value of H_0, we can find the distance d to the galaxy if we know how to determine the recessional velocity v from the redshift z.

If the redshift is not too great (so that the redshift z is much less than 1), we can use the following Doppler shift equation to find the recessional speed:

$$z = \frac{v}{c} \text{ (valid for low speeds only)}$$

where c is the speed of light. For example, a 5% shift in wavelength ($z = 0.05$) corresponds to a recessional velocity of 5% of the speed of light ($v = 0.05c$).

For larger redshifts (about $z = 0.1$ or greater), a more complicated equation is needed to determine the recessional velocity. A small redshift can be explained by the Doppler shift as we have used here. However, as we will see in Chapter 25, larger redshifts arise from a large *expansion of space* and cannot be described as a simple Doppler shift. In fact, for galaxies at redshifts greater than about $z > 1.5$, recessional speeds exceed the speed of light.

Due to the unique attributes of expanding space, faster-than-light recession is not a violation of Einstein's laws. Nothing can move through space faster than light, but the expansion of space can carry galaxies with it at speeds greater than light.

EXAMPLE: When measured in a laboratory on Earth, the so-called K line of singly ionized calcium has a wavelength $\lambda_0 = 393.3$ nm. But when you observe the spectrum of the giant elliptical galaxy NGC 4889, you find this spectral line at $\lambda = 401.8$ nm. Using $H_0 = 73$ km/s/Mpc, find the distance to this galaxy.

Situation: We are given the values of λ and λ_0 for a line in this galaxy's spectrum. Our goal is to determine the galaxy's distance d.

Tools: We use the relationship $z = (\lambda - \lambda_0)/\lambda_0$ to determine the redshift. We then use the appropriate formula to determine the galaxy's recessional velocity v, and finally use the Hubble law to determine the distance to the galaxy.

Answer: The redshift of this galaxy is

$$z = \frac{401.8 \text{ nm} - 393.3 \text{ nm}}{393.3 \text{ nm}} = 0.0216$$

This value is substantially less than 0.1, so we can safely use the low-redshift relationship between recessional speed and redshift: $v = zc$. So NGC 4889 is moving away from us with a speed

$$v = zc = (0.0216)(3 \times 10^5 \text{ km/s}) = 6480 \text{ km/s}$$

Using $H_0 = 73$ km/s/Mpc in the Hubble law, the distance to this galaxy is

$$d = \frac{v}{H_0} = \frac{6480 \text{ km/s}}{73 \text{ km/s/Mpc}} = 89 \text{ Mpc}$$

Review: This galaxy is receding from us at 0.0216 (2.16%) of the speed of light, and it is 89 Mpc (290 million ly) away. Thus, the light we see from NGC 4889 today left the galaxy 290 million years ago, even before the first dinosaurs appeared on Earth.

This formula is the equation for the straight line displayed in Figure 23-17, and the **Hubble constant** H_0 is the slope of this straight line. From the data plotted on this graph, we find that $H_0 = 73$ km/s/Mpc. In other words, for each million parsecs to a galaxy, the galaxy's speed away from us increases by 73 km/s. For example, a galaxy located 100 million pc from Earth should be rushing away from us with a speed of 7300 km/s. (In other books, you may see the units of the Hubble constant written with exponents: 73 km s^{-1} Mpc^{-1}.) In addition, this value is still being investigated; for example, while we use $H_0 = 73$ km/s/Mpc here, estimates using data from the Planck telescope put the value at 68 km/s/Mpc.

CAUTION!

A common misconception about the Hubble law is that *all* galaxies are moving away from the Milky Way. If the Hubble flow were the only thing causing galaxy motion, then this would be true. However, galaxies also have their own neighbor-induced motions relative to one another, due to their mutual gravitational attraction of neighboring galaxies. Astronomers call this neighbor-induced motion *intrinsic velocity*. For nearby galaxies, the speed of the Hubble flow is small compared to these intrinsic velocities. Hence, some of the nearest galaxies, including the Andromeda Galaxy (shown in Figure 23-3), are actually approaching us and have blueshifts rather than redshifts. But for distant galaxies, the Hubble speed $v = H_0 d$ is much greater than any intrinsic motion that the galaxies might have. Even if the intrinsic velocity of such a distant galaxy is toward the Milky Way, the fast-moving Hubble flow sweeps that galaxy away from us.

CONCEPT CHECK 23-9

In Figure 23-17, which follows Hubble's law, consider galaxies at 100 Mpc and 200 Mpc. Which galaxy has the larger redshift?

Answer appears at the end of the chapter.

Pinning Down the Hubble Constant

A precise value of the Hubble constant has been a topic of heated debate among astronomers for several decades. The problem is that while redshifts are relatively easy to measure in a reliable way, distances to galaxies (especially remote galaxies) are not, as we saw in Section 23-4. Hence, astronomers who use different methods of determining galactic distances have obtained different values of H_0. To see why different values are measured, it is helpful to rewrite the Hubble law as

$$H_0 = \frac{v}{d}$$

This equation shows that if galaxies of a given recessional velocity (v) are far away (so d is large), the Hubble constant H_0 must be relatively small. But if these galaxies are relatively close (so d is small), then H_0 must have a larger value.

In the past, astronomers who used Type Ia supernovae for determining galactic distances found galaxies to be farther away than did their colleagues who employed the Tully–Fisher relation. Therefore, the supernova adherents found values of H_0 in the range from about 40 to 65 km/s/Mpc, whereas the values from the Tully–Fisher relation ranged from about 80 to 100 km/s/Mpc.

However, more precise measurements of Cepheid variables have been made by the Hubble Space Telescope (HST) in galaxies as far away as 30 Mpc (100 million ly). These observations and others suggest a value of the Hubble constant of about 73 km/s/Mpc, with an uncertainty of no more than 10%. At the same time, reanalysis of the supernova and Tully–Fisher results have brought the values of H_0 from these techniques closer to the HST Cepheid value. We will adopt the value $H_0 = 73$ km/s/Mpc in this book. One of the reasons the HST was named after Hubble was that one if its primary early missions was to observe these Cepheid variables, allowing a more precise determination of the Hubble constant.

Determining the value of H_0 has been an important task of astronomers for a very simple reason: The Hubble constant is one of the most important numbers in all of astronomy. It expresses the rate at which the universe is expanding and, as we will see in Chapter 25, even helps give the age of the universe. Furthermore, the Hubble law can be used to determine the distances to extremely remote galaxies. If the redshift of a galaxy is known, the Hubble law can be used to determine its distance from Earth. Thus, the value of the Hubble constant helps determine the distances of the most remote objects in the universe that astronomers can observe.

Because the value of H_0 remains somewhat uncertain, astronomers often express the distance to a remote galaxy simply in terms of its redshift z (which can be measured very accurately). Given the redshift, the distance to this galaxy can be calculated from the Hubble law, but the distance obtained in this way will depend on the particular value of H_0 adopted. Rather than going through these calculations, an astronomer might simply say that a certain galaxy is "at $z = 0.128$." From the Hubble law relating redshift and distance, this redshift makes it clear that the galaxy in question is more distant than one at $z = 0.120$ but not as distant as one at $z = 0.130$. When astronomers use redshift to describe distance, they are making use of the following general rule:

The greater the redshift of a distant galaxy, the greater its distance.

CONCEPT CHECK 23-10

Although nearly all distant galaxies have measurable redshifts, the relatively nearby Andromeda Galaxy exhibits an overall blueshift. What does this mean about the Andromeda Galaxy's movement?

CALCULATION CHECK 23-1

What is the redshift z-value for a galaxy that has a galaxy spectral line shifted to 725.6 nm when a stationary object would emit the line at 656.3 nm?

CALCULATION CHECK 23-2

What is the distance to a galaxy that is observed to have a recessional velocity of 10,000 km/s?

Answers appear at the end of the chapter.

23-6 Galaxies are grouped into clusters and superclusters

Galaxies are not scattered randomly throughout the universe but are found in **clusters**. Figure 23-18 shows one such cluster. Like stars within a star cluster, the members of a cluster of galaxies are

FIGURE 23-18 **R I V U X G**

The Hercules Cluster This irregular cluster of galaxies is about 200 Mpc (650 million ly) from Earth. The Hercules Cluster contains many spiral galaxies, often associated in pairs and small groups. (NOAO/AURA/NSF)

in continual motion around one another. They appear to be at rest only because they are so distant from us.

Clusters of Galaxies: Poor and Rich, Regular and Irregular

A cluster is said to be either **poor** or **rich,** depending on how many galaxies it contains. Poor clusters, which far outnumber rich ones, are often called **groups.** For example, the Milky Way Galaxy, the Andromeda Galaxy (M31), and the Large and Small Magellanic Clouds belong to a poor cluster familiarly known as the **Local Group.** The Local Group contains more than 40 galaxies, most of which are dwarf ellipticals (see Figure 23-9). **Figure 23-19** is a map of a portion of the Local Group.

In recent years, astronomers have discovered several previously unknown dwarf elliptical galaxies in the Local Group. Two recently discovered dwarf elliptical galaxies are Leo T and the Canis Major Dwarf. Tidal forces exerted by the Milky Way on the Canis Major Dwarf are causing this dwarf galaxy to gradually disintegrate and leave a trail of debris behind it (Figure 23-20). This galaxy is about 13 kpc (42,000 ly) from the center of the Milky Way Galaxy and a mere 8 kpc (25,000 ly) from Earth (about the same as the distance from Earth to the center of the Milky Way).

We may never know the total number of galaxies in the Local Group, because dust in the plane of the Milky Way obscures our view over a considerable region of the sky. Nevertheless, we can be certain that no additional large spiral galaxies are hidden by the Milky Way. Radio astronomers would have detected 21-cm radiation from them, even though their visible light would be completely absorbed by interstellar dust.

The nearest fairly rich cluster is the Virgo Cluster, a collection of more than 2000 galaxies covering a $10° \times 12°$ area of the sky. Figure 23-8 shows a portion of this cluster. One member of this cluster not shown in Figure 23-8 is the spiral galaxy M100; measurements of Cepheid variables in M100 (see Figure 23-4) give a distance of about 17 Mpc (56 million ly). The Tully–Fisher relation and observations of Type Ia supernovae (see Figure 23-13) give similar distances to this cluster. The overall diameter of the cluster is about 3 Mpc (9 million ly).

The center of the Virgo Cluster is dominated by three giant elliptical galaxies. You can see two of these, M84 and M86, in Figure 23-8. The diameter of each of these enormous galaxies is comparable to the 750-kpc distance between the Milky Way and the Andromeda Galaxy. In other words, one giant elliptical is approximately the same size as the entire Local Group!

Astronomers also categorize clusters of galaxies as regular or irregular, depending on the overall shape of the cluster. The Virgo Cluster, for example, is called an **irregular cluster,** because its galaxies are scattered throughout a sprawling region of the sky. Our

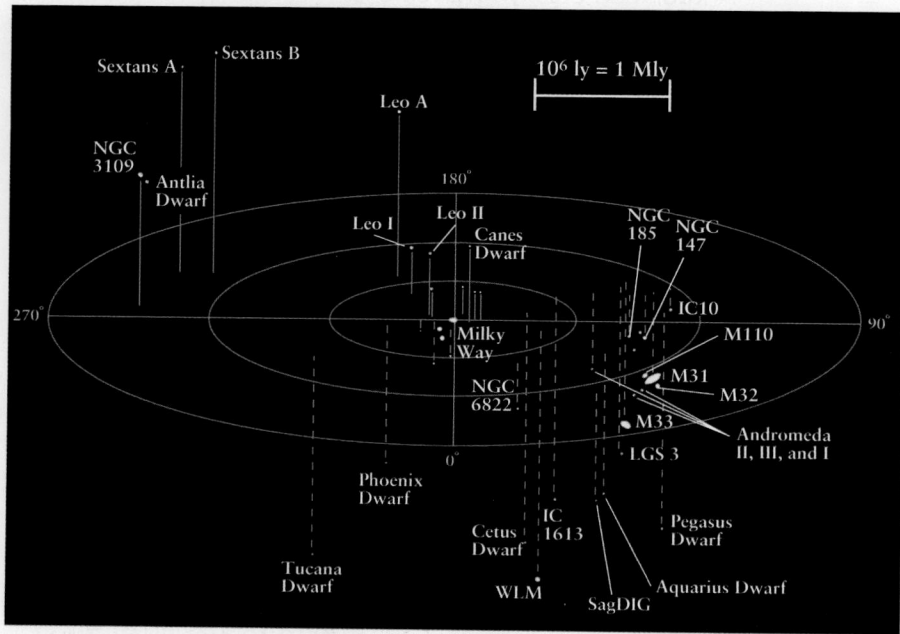

FIGURE 23-19

The Local Group This illustration shows the relative positions of the galaxies that comprise the Local Group, a poor, irregular cluster of which our galaxy is part. (The yellow rings represent the plane of the Milky Way's disk; 0° is the direction from Earth toward the Milky Way's center. Solid and dashed lines point to galaxies above and below the plane, respectively.) The largest and most massive galaxy in the Local Group is M31, the Andromeda Galaxy; in second place is the Milky Way, followed by the spiral galaxy M33. Both the Milky Way and M31 are surrounded by a number of small satellite galaxies. (Data from Richard Powell, www.atlasoftheuniverse.com)

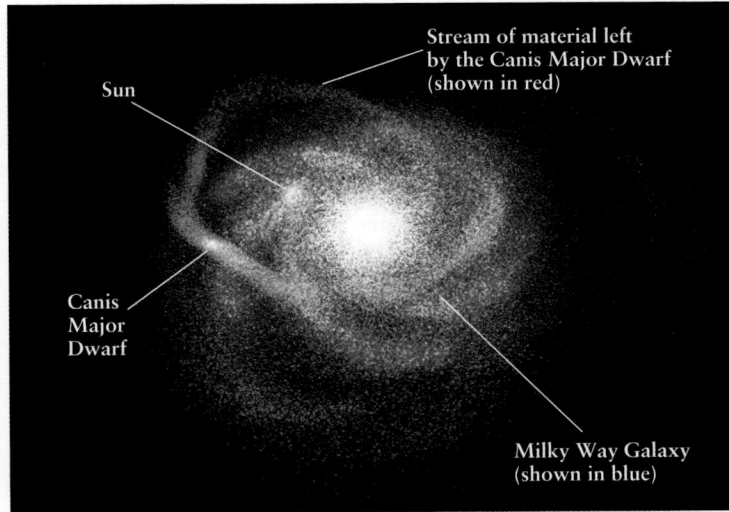

ANIMATION **FIGURE 23-20** R I ⓥ U X G
23.1 ➤

The Canis Major Dwarf Discovered in 2003, this dwarf elliptical galaxy is actually slightly closer to Earth than is the center of the Milky Way Galaxy. This illustration shows the stream of material left behind by the Canis Major Dwarf as it orbits the Milky Way. This material is torn away by the Milky Way's tidal forces (see Section 4-8). (R. Ibata [Strasbourg Observatory, ULP] et al., 2MASS, NASA)

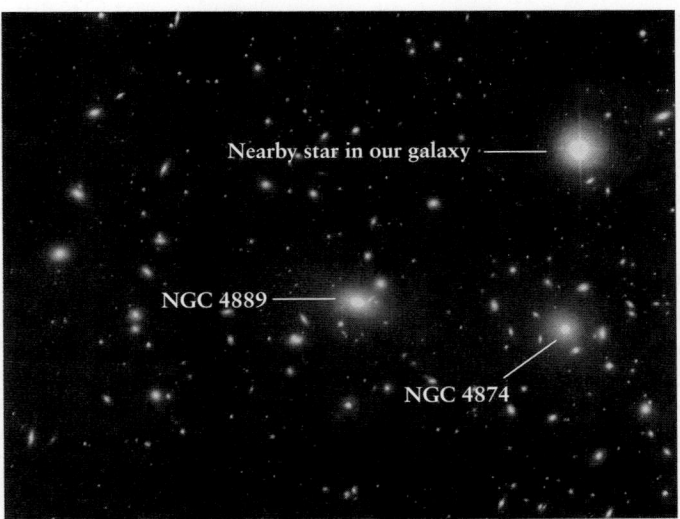

FIGURE 23-21 R I ⓥ U X G

The Coma Cluster This rich, regular cluster is about 90 Mpc (300 million ly) from Earth. Almost all of the spots of light in this image are individual galaxies of the cluster. Two giant elliptical galaxies, NGC 4889 and NGC 4874, dominate the center of the cluster. The bright star at the upper right is within our own Milky Way Galaxy, a million times closer than any of the galaxies shown here. (Omar Lopez-Cruz & Ian Shelton/NOAO/AURA/NSF)

own Local Group is also an irregular cluster. In contrast, a **regular cluster** has a distinctly spherical appearance, with a marked concentration of galaxies at its center.

The nearest example of a rich, regular cluster is the Coma Cluster, located about 90 Mpc (300 million ly) from us toward the constellation Coma Berenices (Berenice's Hair), shown in **Figure 23-21**. Despite its great distance, telescopic images of this cluster show more than 1000 galaxies. The Coma Cluster almost certainly contains many thousands of dwarf ellipticals, so the total membership of the cluster may be as many as 10,000 galaxies. The core of the Coma Cluster is dominated by two giant ellipticals surrounded by many normal-sized galaxies.

The overall shape of a cluster is related to the dominant types of galaxies it contains. Rich, regular clusters contain mostly elliptical and lenticular galaxies. For example, about 80% of the brightest galaxies in the Coma Cluster (see Figure 23-21) are ellipticals; only a few spiral galaxies are scattered around the cluster's outer regions. Irregular clusters, such as the Virgo Cluster and the Hercules Cluster shown in Figure 23-18, have a more even mixture of galaxy types.

CONCEPT CHECK 23-11

How would a photograph of a rich cluster's appearance compare to that of a poor cluster?

Answer appears at the end of the chapter.

Superclusters: Clusters of Clusters of Galaxies and the Great Attractor

Clusters of galaxies are themselves grouped together in huge associations called **superclusters**. A typical supercluster contains dozens of individual clusters spread over a region of space up to 45 Mpc (150 million ly) across. **Figure 23-22** shows the distribution

of clusters in our part of the universe. The nearer ones out to the Virgo Cluster, including our own Local Group, are members of the **Local Supercluster**. The other clusters shown in Figure 23-22 belong to other superclusters. The most massive cluster in the local universe is called the **Great Attractor**; its gravity is so great that the

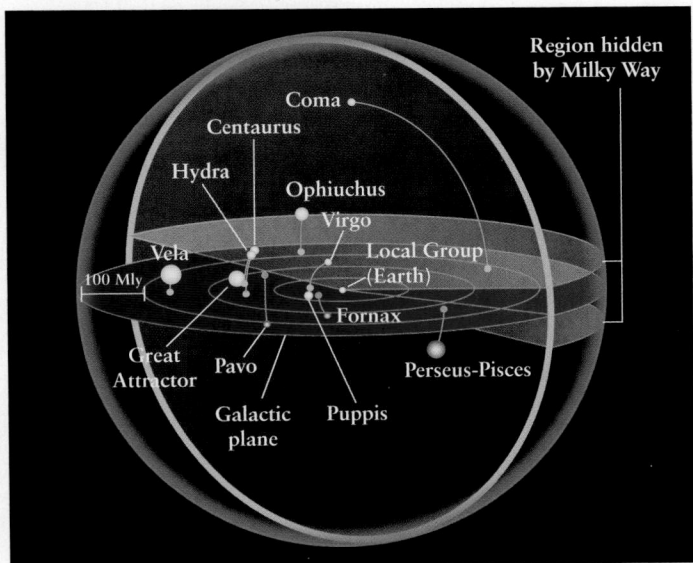

FIGURE 23-22

Nearby Clusters of Galaxies This illustration shows a sphere of space 800 million ly (250 Mpc) in diameter centered on Earth in the Local Group. Each spherical dot represents a cluster of galaxies. To better see the three-dimensionality of this figure, colored arcs are drawn from each cluster to the green plane, which is an extension of the plane of the Milky Way outward into the universe. Note that clusters of galaxies are unevenly distributed here, as they are elsewhere in the universe.

Milky Way, as well as the rest of the Local Supercluster, is moving toward it at speeds of several hundred kilometers per second.

Observations indicate that unlike clusters, superclusters are not bound together by gravity. That is, most clusters in each supercluster are drifting away from most of the other clusters in that same supercluster. Furthermore, the superclusters are all moving away from each other due to the Hubble flow.

Cosmic Voids and Sheets: The Distribution of Superclusters

Since the 1980s, astronomers have been working to understand how superclusters are distributed in space. Some of this structure is revealed by maps such as that shown in Figure 23-23, which displays the positions on the sky of 1.6 million galaxies. Such maps reveal that superclusters are not randomly distributed, but seem to lie along filaments. But to comprehend more fully the distribution of superclusters, it is necessary to map their positions in three dimensions. Their coordinates are mapped by measuring both the position of a galaxy on the sky as well as the galaxy's redshift. Using the Hubble law (see Section 23-5), astronomers can use each galaxy's redshift to estimate its distance from Earth and thus its position in three-dimensional space.

> Superclusters of galaxies are not spread uniformly across the universe, but are found in vast sheets separated by truly immense voids.

The first three-dimensional maps of this kind were made in the 1970s and included measurements of a few hundred galaxies. Collecting the data for such maps required many months or years of telescopic observations. Technology for astronomy has advanced tremendously since then, and it is now possible to measure the redshifts of 400 galaxies in a single hour!

Two extensive galaxy maps are from the Sloan Digital Sky Survey and from the Two Degree Field Galactic Redshift Survey (2dF). (The name refers to the 2° field of view of the telescope used for the observations, which is unusually wide for a research telescope.) Figure 23-24a shows a map made from 2dF measurements of more than 60,000 galaxies. This particular map encompasses two wedge-shaped slices of the universe, one on either side of the plane of the Milky Way (Figure 23-24b). Earth (in the Milky Way) is at the apex of the wedge-shaped map; each dot represents a galaxy. The measurements used to create this map included galaxies with redshifts as large as $z = 0.25$, corresponding to a recessional velocity of 66,000 km/s. Using a Hubble constant of 73 km/s/Mpc, we can determine that the map in Figure 23-24a extends out to a distance of nearly 1000 Mpc, or 3 *billion* ly from Earth.

Maps such as that shown in Figure 23-24a reveal enormous **voids** where exceptionally few galaxies are found. (These voids were first discovered in 1978 in a pioneering study by Stephen Gregory and Laird Thompson at the Kitt Peak National Observatory.) These voids are roughly spherical and measure 30 to 120 Mpc (100 million to 400 million ly) in diameter. They are not entirely empty, however. There is evidence for hydrogen clouds in some voids, while others may be subdivided by strings of dim galaxies.

FIGURE 23-23   R I V U X G

Structure in the Nearby Universe This composite infrared image from the 2MASS (Two Micron All Sky Survey) project shows the light from 1.6 million galaxies. In this image, the entire sky is projected onto an oval; the blue band running vertically across the center of the image is light from the plane of the Milky Way. Note that galaxies form a lacy, filamentary structure. Note also that there are large regions with fewer galaxies that appear as darker patches in the map. These are called *voids*, and they contain fewer galaxies. (2MASS, T. H. Jarrett, J. Carpenter, & R. Hurt)

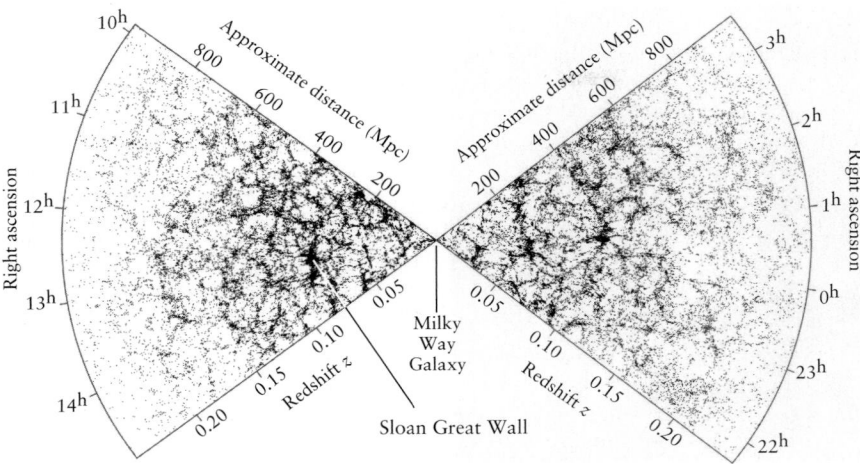

(a) The 2dF galaxy survey

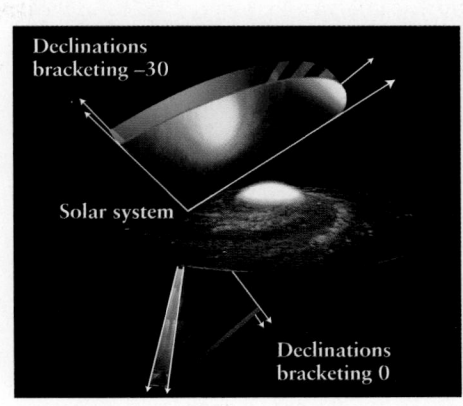

(b) Fields of view in the 2dF survey

ANIMATION **FIGURE 23-24**

23.2

The Large-Scale Distribution of Galaxies (a) This map shows the distribution of 62,559 galaxies in two wedges extending out to redshift z = 0.25. (For an explanation of right ascension and declination, see Box 2-1.) Note the prominent voids surrounded by thin regions full of galaxies. **(b)** The two wedges shown in (a) lie roughly perpendicular to the plane of the Milky Way. These were chosen to avoid the obscuring dust that lies in our galaxy's plane. (Courtesy of the 2dF Galaxy Redshift Survey Team/Australian Astronomical Observatory)

Figure 23-24a shows that most galaxies are concentrated in sheets on the surfaces between voids. These sheets can be more than 100 Mpc long and several megaparsecs thick. This pattern is similar to the structure of a sponge or soap bubbles, with sheets of soap film (analogous to galaxies) surrounding air bubbles (analogous to voids) (Figure 23-25). These titanic sheets of galaxies are the largest structures known in the universe: On scales much larger than 100 Mpc, the distribution of galaxies in the universe appears to be roughly uniform. As we will see in Chapter 26, this pattern of sheets and voids contains important clues about how clusters of galaxies formed in the early universe.

There is an unusually large and dense sheet in Figure 23-24a called the Sloan Great Wall; this is the largest known structure in the universe! This structure is about 1.4 billion ly in length (which is almost 2% of the visible universe). The Sloan Great Wall contains several superclusters, and astronomers still have not figured out how this large structure formed.

CONCEPT CHECK 23-12

Are superclusters evenly distributed throughout the universe?

CONCEPT CHECK 23-13

What is the most common type of galaxy in our Local Group?

Answers appear at the end of the chapter.

23-7 Colliding galaxies produce starbursts, spiral arms, and other spectacular phenomena

Occasionally, two galaxies within a cluster or from adjacent clusters can collide with each other. Past collisions have hurled vast numbers of stars into intergalactic space. In some cases, we can even observe a collision in progress, a cosmic catastrophe that gives birth to new stars. And astronomers can predict collisions that will not take place for billions of years, such as the collision that is fated to occur between the Andromeda Galaxy and our own Milky Way Galaxy.

FIGURE 23-25

Foamy Structure of the Universe A sponge that recreates the distribution of bright clusters of galaxies throughout the universe. The empty spaces in the foam are analogous to the voids found throughout the universe. The spongy regions are analogous to the locations of most of the superclusters of galaxies. (Svetlana Zhukova/Shutterstock.com)

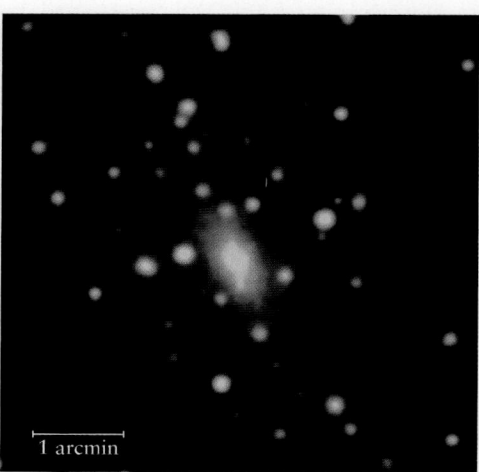

(a) An x-ray image of Abell 2029 shows emission from hot gas.

R I V U **X** G

(b) A visible-light image of Abell 2029 shows the cluster's galaxies.

R I **V** U X G

FIGURE 23-26

X-ray Emission from a Cluster of Galaxies
(a) An x-ray image of this cluster of galaxies shows emission from hot gas between the galaxies. The gas was heated by collisions between galaxies within the cluster. **(b)** The galaxies themselves are too dim at x-ray wavelengths to be seen in (a), but they are apparent at visible wavelengths. This cluster, one of many cataloged by the UCLA astronomer George O. Abell, is about 300 Mpc (1 billion ly) from Earth in the constellation Serpens. (a: NASA/CXC/UCI/A. Lewis et al.; b: Palomar Observatory DSS)

High-Speed Galaxy Collisions: Shredding Gas and Dust

When two galaxies collide at high speed, the huge clouds of interstellar gas and dust in the galaxies slam into each other—like two cars locking bumpers in a collision. In this way, two colliding galaxies can leave behind their interstellar gas and dust as the stars in each galaxy pass through the impact site.

The best evidence that such collisions take place is that many rich clusters of galaxies are strong sources of x rays (Figure 23-26). This emission reveals the presence of substantial amounts of hot **intracluster gas** (that is, gas within the cluster) at temperatures between 10^7 and 10^8 K. The only way that such large amounts of gas could be heated to such extremely high temperatures is in violent collisions between galaxies.

CAUTION!

Although galaxies can and do collide, it is highly unlikely that the *stars* from two colliding galaxies actually run into each other. The reason is that the stars within a galaxy are very widely separated from one another, with a tremendous amount of space between them.

Gentle Galactic Collisions and Starbursts

In a less violent collision or a near miss between two galaxies, the compressed interstellar gas may have more time to cool, allowing many protostars to form. Such collisions may account for **starburst galaxies** such as M82 (**Figure 23-27**), which blaze

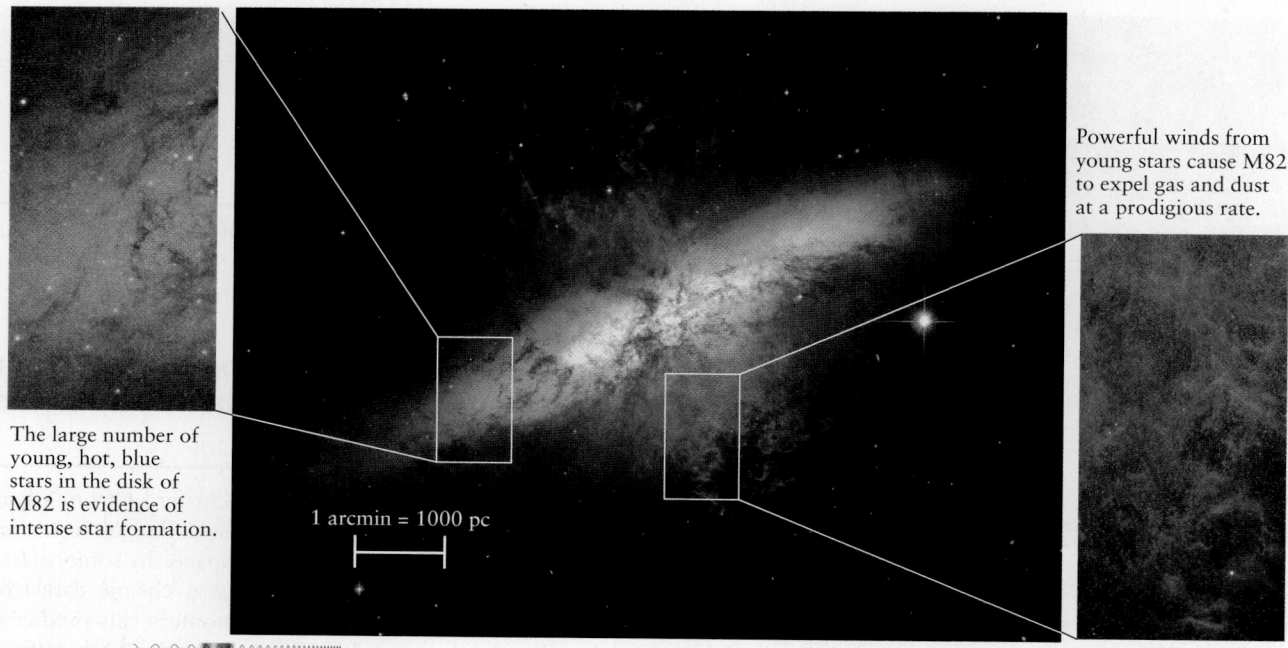

The large number of young, hot, blue stars in the disk of M82 is evidence of intense star formation.

Powerful winds from young stars cause M82 to expel gas and dust at a prodigious rate.

1 arcmin = 1000 pc

VIDEO 23.3

FIGURE 23-27 R **I** V U X G

A Starburst Galaxy Prolific star formation is occurring at the center of the irregular galaxy M82, which lies about 3.6 Mpc (12 million ly) from Earth in the constellation Ursa Major. M82 also contains an unusual x-ray source, shown in Figure 21-26. (NASA, ESA and the Hubble Heritage Team [STScI/AURA])

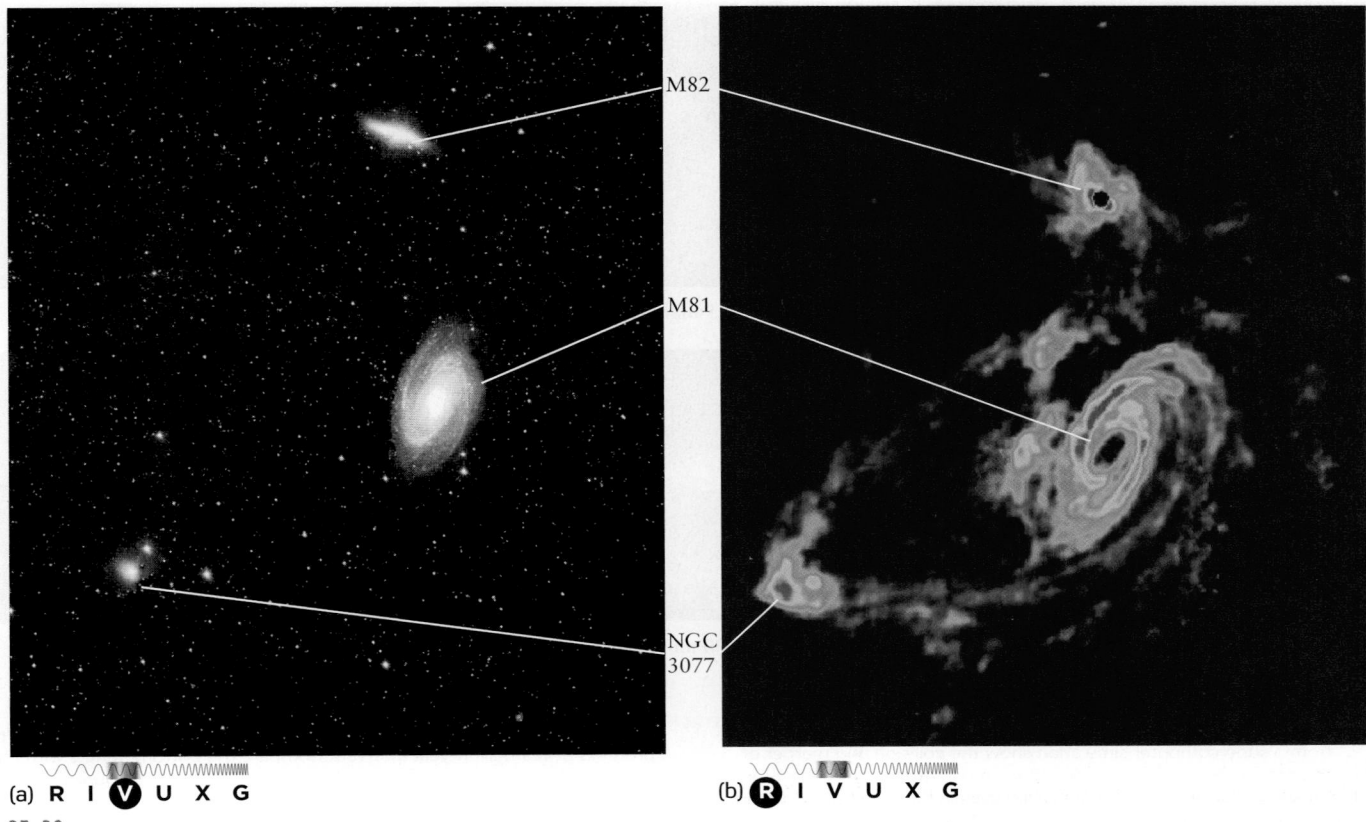

M82

M81

NGC
3077

(a) R I V U X G

(b) R I V U X G

FIGURE 23-28

The M81 Group (a) The starburst galaxy M82 is part of a cluster of about a dozen galaxies. This wide-angle visible-light photograph shows the three brightest galaxies of the cluster. The area shown is about 1° across. **(b)** This false-color radio image of the same region, created from data taken by the Very Large Array, shows streamers of hydrogen gas that connect the three bright galaxies as well as several dim ones. (a: Palomar Sky Survey; b: NRAO/AUI)

with the light of numerous newborn stars. These galaxies have bright centers surrounded by clouds of warm interstellar dust, indicating recent, vigorous star birth. Their warm dust is so abundant that starburst galaxies are among the most luminous objects in the universe at infrared wavelengths. (The right-hand image at the opening of Chapter 6 shows the infrared emission from M82's warm dust.)

The starburst galaxy M82 shown in Figure 23-27 also shows the effects of strong winds from young, luminous stars. It also contains a number of luminous globular clusters. Unlike the globular clusters in our galaxy, whose stars are about 12.5 billion years old, those in M82 are no more than 600 million years old. These young globular clusters are another sign of recent star formation.

M82 is one member of a nearby cluster of galaxies that also includes the beautiful spiral galaxy M81 and a fainter elliptical companion called NGC 3077 (Figure 23-28a). Radio surveys of that region of the sky reveal enormous streams of hydrogen gas connecting the three galaxies (Figure 23-28b). The loops and twists in these streamers suggest that the three galaxies have had several close encounters over the ages. A similar stream of hydrogen gas connects our galaxy with its second nearest neighbor, the Large Magellanic Cloud (see Figure 23-12), suggesting a history of close encounters between our galaxy and the LMC.

CONCEPT CHECK 23-14

What energy source accounts for intergalactic gas between galaxies that is often quite hot?

Answer appears at the end of the chapter.

Tidal Forces and Galaxy Mergers

Tidal forces between colliding galaxies can deform the galaxies from their original shapes, just as the tidal forces of the Moon on Earth deform the oceans and help give rise to the tides (see Section 4-8, especially Figure 4-29). The galactic deformation is so great that thousands of stars can be hurled into intergalactic space along huge, arching streams. (This same effect has stripped material away from the Canis Major Dwarf Galaxy as it orbits the Milky Way, as shown in Figure 23-20.) Supercomputer simulations of such collisions show that while some of the stars are flung far and wide, other stars slow down and the galaxies may merge.

Figure 23-29 shows one such simulation of stars (but no gas or dust). As the two galaxies pass through each other, they are severely distorted by gravitational interactions and throw out a pair of extended tails. The interaction also prevents the galaxies in the simulation from

> Galaxies need not actually collide to exert strong forces on each other.

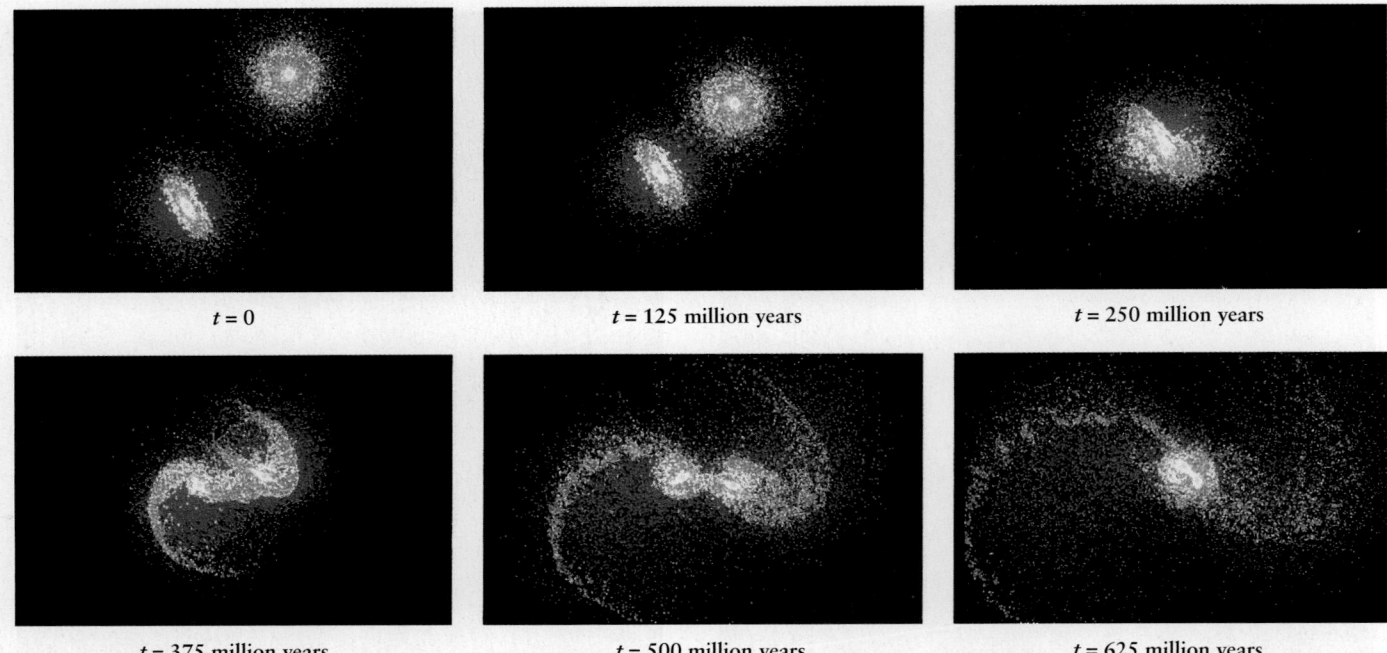

$t = 0$ $t = 125$ million years $t = 250$ million years

$t = 375$ million years $t = 500$ million years $t = 625$ million years

ANIMATION FIGURE 23-29
23.3
A Simulated Collision Between Two Galaxies These frames from a supercomputer simulation show the collision and merger of two galaxies accompanied by an ejection of stars into intergalactic space. Stars in the disk of each galaxy are colored blue, while stars in their central bulges are yellow-white. Red indicates dark matter that surrounds each galaxy. The frames progress at 125-million-year intervals. Compare the bottom frames with the image of the Antennae galaxies in the *Cosmic Connections* figure. (Joshua Edward Barnes, Institute for Astronomy, University of Hawaii)

continuing on their original paths. Instead, they fall back together for a second encounter (at 625 million years). The simulated galaxies merge soon thereafter, leaving a single object. The *Cosmic Connections* figure on the next page explores a real-life example of two galaxies that are colliding in just this way.

Our own Milky Way Galaxy is expected to eventually undergo a galactic collision like that shown in Figure 23-28. The Milky Way and the Andromeda Galaxy, shown in Figure 23-3, are actually approaching each other fast enough to travel the Earth–Moon distance in 1 hour and should collide in 3.75 billion years or so (Figure 23-30). (Recall that our Sun will not become a red giant for more than 5 billion years.) During the collision, the sky will light up with a plethora of newly formed stars, followed in rapid succession by a string of supernovae.

The complete merger between our Milky Way and the Andromeda Galaxy is expected to last about 2 billion years. Recall that stars are very far apart compared to their size: A galaxy scaled down in size to the United States would have stars the size of human cells, each about a football field apart. In the galactic collision, stars are not expected to hit each other and our solar system should remain intact.

ANIMATION When two galaxies merge, the result is a bigger
23.4
galaxy. If this new galaxy is located in a rich cluster, it may capture and devour additional galaxies, growing to enormous dimensions by **galactic cannibalism**. Cannibalism differs from mergers in that the galaxy that does the devouring is bigger than its "meal," whereas merging galaxies are about the same size.

Many astronomers suspect that galactic cannibalism is the reason that giant ellipticals are so huge. As we have seen, giant galaxies typically occupy the centers of rich clusters. In many cases, smaller galaxies are located around these giants (see Figure 23-8 and Figure 23-21). As they pass through the extended halo of a giant elliptical, these smaller galaxies slow down and are eventually devoured by the larger galaxy.

Galaxy Interactions and Spiral Arms

ANIMATION Close encounters between galaxies provide a third way
23.5
of forming spiral arms (in addition to density waves and self-propagating star formation, discussed in Section 22-5). Computer simulations clearly demonstrate that spiral arms can be created during a collision, either by drawing out long streamers of stars or by compressing clouds of interstellar gas. For example, the spiral arms of M51 (examine Figure 23-2) may have been produced by a close encounter with a second galaxy. The disruptive galaxy, NGC 5195, is now located at the end of one of the spiral arms created by the collision. The two galaxies shown in the chapter-opening image are thought to be interacting in the same way.

The very fact of our existence may be intimately related to interactions between galaxies. Some astronomers argue that the spiral arms of our Milky Way Galaxy were produced by a close encounter with the Large Magellanic Cloud. As we saw in Section 22-5, spiral arms compress the interstellar medium in the Milky Way's disk to form Population I stars like our own Sun, which have enough heavy elements to produce Earthlike planets. Thus, the chain of events that led to the formation of our Sun, our solar system, and life on our planet may have been initiated by a long-ago interaction between two galaxies.

COSMIC CONNECTIONS

When Galaxies Collide

 ANIMATION 23.8 VIDEO 23.4 Although galaxies can collide at very high speeds by Earth standards, they are so vast that a collision can last hundreds of millions of years. Understanding what happens during a galactic collision requires ideas about tidal forces (Chapter 4), star formation (Chapter 18), and stellar evolution (Chapter 19).

1. One example of a galactic collision is the pair of galaxies called the Antennae, which lie 19 Mpc (16 million ly) from Earth in the constellation Corvus (the Crow). They probably began to interact several hundred million years ago.

2. As the gas and dust clouds of the two galaxies collide with each other, they are greatly compressed. This compression causes stars to form in tremendous numbers.

Tidal forces between the galaxies pulled out these long "tidal tails" 200 to 300 million years ago.

R I V U X G

Bob and Bill Twardy/Adam Block/NOAO/AURA/NSF

Brown:
Dense dust clouds
Blue:
Hot, recently formed stars
Red:
H II regions caused by the hot stars

R I V U X G

ESA/Hubble & NASA

4. The globular clusters that orbit our Milky Way Galaxy contain only old stars; all of the short-lived blue stars have long since died. But some of the globular clusters that orbit the Antennae galaxies do have hot blue stars. Hence these clusters must be young. These, too, are a result of the compression of gas and dust that takes place in a collision between galaxies.

NASA/JPL-Caltech/Harvard-Smithsonian CfA/NOAO/AURA

Red:
Infrared emission from warm dust
Green:
Visible light from stars

a. Two galaxies, each with old globular clusters (yellow), begin to interact.

b. The two galaxies swoop past each other before finally settling down as a single merged galaxy. Gas and dust is compressed in both galaxies in the process, creating new star clusters.

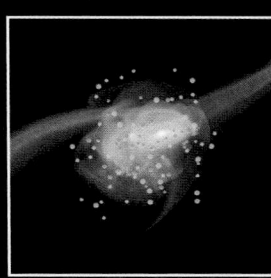

(John Dubinski, University of Toronto)

c. The combined galaxy has the original globular clusters (shown in yellow) as well as new ones (shown in blue).

R I V U X G

3. This composite infrared and visible-light image of the Antennae allows us to see inside the two galaxies and reveals clouds of dust warmed by the light of hot young stars.

FIGURE 23-30

Collision Between Andromeda and the Milky Way Computer
simulations of the merger show what the night sky might look like from
Earth. Two billion years (By) from now, Andromeda will look much bigger, but
the real collision will not begin until about 3.75 billion years from now. In
3.85 billion years from now, the sky will be engulfed in new star formation,
which will have largely ended by 4 billion years, leaving the galaxies to
completely merge. After 7 billion years, our solar system is expected to end
up much farther from the galactic core than it is today. (NASA; ESA; Z. Levay
and R. van der Marel, STScI; T. Hallas, and A. Mellinger)

CONCEPT CHECK 23-15

When the Milky Way Galaxy and the Andromeda
Galaxy finish colliding with each other, what will
be left over?

Answer appears at the end of the chapter.

23-8 Most of the matter in the universe is mysterious dark matter

Galaxies in a cluster typically move at large speeds and must be
held in the cluster by gravity. In other words, there must be enough
matter in the cluster to prevent the galaxies from drifting away.

Nevertheless, careful examination of a rich cluster, like the Coma Cluster shown in Figure 23-21, reveals that the mass of the visually luminous matter (principally the stars in the galaxies) is not at all sufficient to bind the cluster gravitationally. The observed line-of-sight speeds of the galaxies, measured by Doppler shifts, are so large that the cluster should have broken apart long ago. Considerably more mass than is visible is needed to keep the galaxies bound in orbit about the center of the cluster.

We encountered a similar situation in studying our own Milky Way Galaxy in Section 22-4: The total mass of our galaxy is more than the amount of visible mass. As for our galaxy, we conclude that clusters of galaxies must contain significant amounts of nonluminous *dark matter*. If this dark matter were not there, the galaxies would have long ago dispersed in random directions and the cluster would no longer exist today. Analyses demonstrate that the total mass needed to bind a typical rich cluster is about 10 times greater than the mass of material that shows up on visible-light images.

Dark Matter and Rotation Curves

It is possible to investigate how dark matter is distributed in galaxies and clusters of galaxies. It appears that dark matter lies within and immediately surrounding galaxies, with less in the vast spaces between galaxies. The evidence for this dark matter distribution comes principally from observations of the rotation curves of galaxies and of the gravitational bending of light by clusters of galaxies.

As we saw in Section 22-4, a rotation curve is a graph that shows how fast stars in a galaxy are moving at different distances from that galaxy's center. For example, Figure 22-18 is the rotation curve for our galaxy. In the 1970s, Vera Rubin and her colleagues showed that many other spiral galaxies have similar rotation curves that remain remarkably flat out to surprisingly great distances from each galaxy's center (Figure 23-31). In other words, the orbital speed of the stars remains roughly constant out to the visible edges of these galaxies.

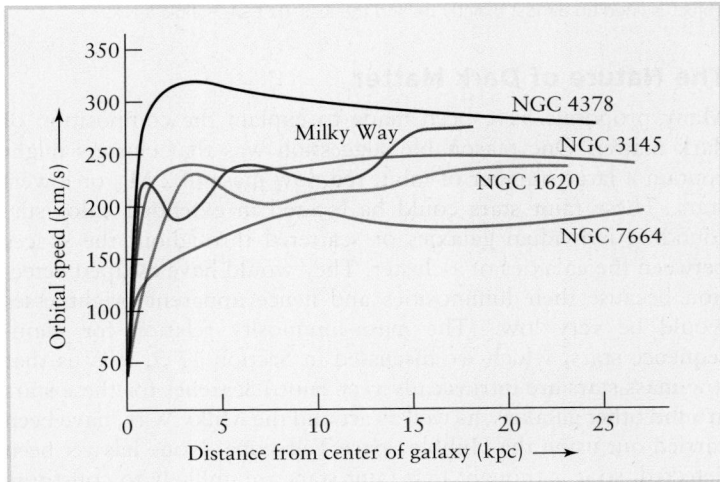

ANIMATION **FIGURE 23-31**
23.6

The Rotation Curves of Four Spiral Galaxies This graph shows how the orbital speed of material in the disks of four spiral galaxies varies with the distance from the center of each galaxy. If most of each galaxy's mass were concentrated near its center, these curves would fall off at large distances. But these and many other galaxies have flat rotation curves that do not fall off, even beyond the bulk of the visible matter in these galaxies. This indicates the presence of extended halos of dark matter. (Data from V. Rubin and K. Ford)

This observation tells us that we still have not detected the *true* edges of these galaxies (and many similar ones). Near the true edge of a galaxy, we should see a decline in orbital speed, in accordance with Kepler's third law (see Figure 22-18). Because this decline has not been observed, astronomers conclude that there must be a considerable amount of dark material that extends well beyond the visible portion of the disk.

"Seeing" Dark Matter with Gravitational Lensing

Further evidence about how dark matter is distributed comes from the gravitational bending of light rays, which we described in Section 21-3. As Figure 21-12 shows, the gravity of a single star like the Sun can deflect light by only a few arcseconds. But a more massive object such as a galaxy can produce much greater deflections, and the amount of this observed deflection can be used in equations from Einstein's general theory of relativity to determine the galaxy's total mass. For example, suppose that Earth, a massive galaxy, and a background light source (such as a more distant galaxy) are in nearly perfect alignment, as sketched in Figure 23-32a. Because of the warped space around the massive galaxy, light from the distant background source curves around the galaxy as it heads toward us. As a result, light rays can travel along two paths from the background source to us here on Earth. Thus, we should see two images of the background source. Furthermore, these observations can be used to determine the total mass of the galaxy.

> Vast assemblages of dark matter reveal their presence by bending passing rays of light.

A powerful source of gravity that distorts background images is called a **gravitational lens.** For gravitational lensing to work, the alignment must be very good between Earth, the massive galaxy acting as a lens, and a remote background light source. As the alignment worsens, the second image of the background galaxy is too faint to be noticeable, leaving a regular view of the background galaxy. There is even a case that produces four images of the same distant object. Beginning in 1979, astronomers have discovered a great number of examples of gravitational lensing.

When the alignment between Earth, a massive gravitational lens, and a distant object becomes nearly perfect, a ring-like image called an *Einstein ring* is formed (Figure 23-32b). As discussed next, beyond the interesting images, gravitational lensing can also be used as a tool to study dark matter.

The observed geometry of a gravitational lens and the resulting distorted image allows astronomers to deduce the mass of the gravitational lensing galaxy. The results agree with studies that use rotation curves: *Most of the galactic matter is dark.* This agreement leads to an important point. Early on, some astronomers proposed that there might not be any dark matter at all, and that the unexpected rotation curves were actually telling us that Newton's law of gravity needed to be refined. However, because gravitational lensing measures the same amount of missing matter without relying on Newton's law of gravity, all of these observations (and more described next) are best explained in terms of dark matter.

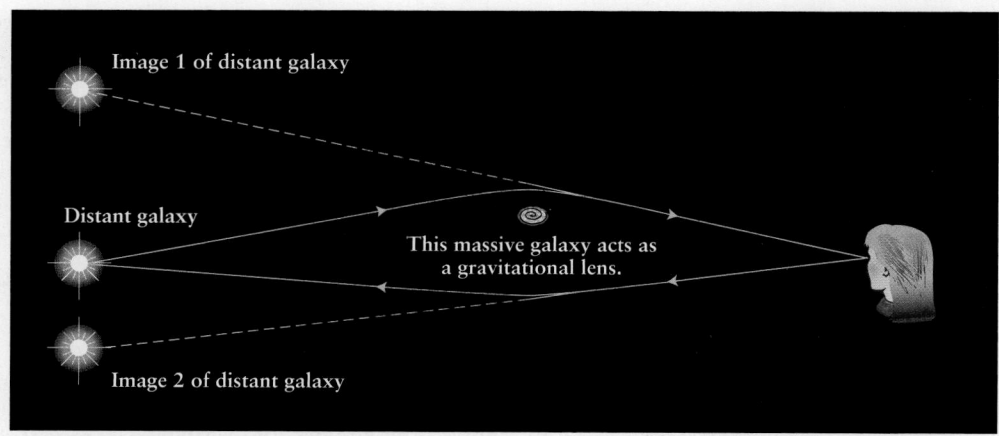

(a) How gravitational lensing happens

(b) An Einstein ring

VIDEO 23.5   **FIGURE 23-32** R I V U X G

Gravitational Lensing **(a)** A massive object such as a galaxy can deflect light rays like a lens so that an observer sees more than one image of a more distant galaxy. **(b)** An Einstein ring is produced when a distant object is lined up almost exactly behind the lens. (b: ESA/Hubble & NASA)

Figure 23-33 shows a situation in which an entire cluster of galaxies acts as a gravitational lens. The image shows an ordinary-looking rich cluster of yellowish elliptical and spiral galaxies, but with a number of curious blue arcs. Reconstruction of the light paths through the cluster shows that all these blue arcs are actually distorted images of a single galaxy that lies billions of light-years beyond the cluster.

By measuring the distortion of the images of such background galaxies, the amount of total matter required to create those distortions can be calculated. Then, by comparing the visible matter from the galaxies, gas, and dust in the cluster to the total inferred mass, the amount of dark matter can be estimated. By analyzing the gravitational lensing from numerous clusters, we have found that about 85% of a cluster's mass is dark matter (whereas about 90% of a galaxy is dark matter). Gravitational lenses are common enough that a map of the distribution of dark matter has been made, as shown in Figure 23-34.

CONCEPT CHECK 23-16

If dark matter cannot be seen, what are the two primary lines of evidence that it exists?

Answer appears at the end of the chapter.

The Nature of Dark Matter

Many proposals have been made to explain the composition of dark matter. One reasonable suggestion was that clusters might contain a large number of faint, red, low-mass ($0.2 \, M_\odot$ or lower) stars. These faint stars could be located in extended halos surrounding individual galaxies or scattered throughout the spaces between the galaxies of a cluster. They would have escaped detection because their luminosities and hence apparent brightnesses would be very low. (The mass–luminosity relation for main-sequence stars, which we discussed in Section 17-9, tells us that low-mass stars are intrinsically very faint.) Searches for these stars around other galaxies, as well as around the Milky Way, have been carried out using the Hubble Space Telescope. None has yet been detected, so it is thought that faint stars are unlikely to constitute the majority of the dark matter in the universe.

As we described in Section 22-4, other dark matter candidates include massive neutrinos, subatomic particles called WIMPs (weakly interacting massive particles), and MACHOs (massive compact halo objects, such as small black holes or brown dwarfs). Although WIMPs have never been detected, they are the favored hypothesis. Interacting very weakly with regular matter, about a billion WIMPs might pass through you each second, and about

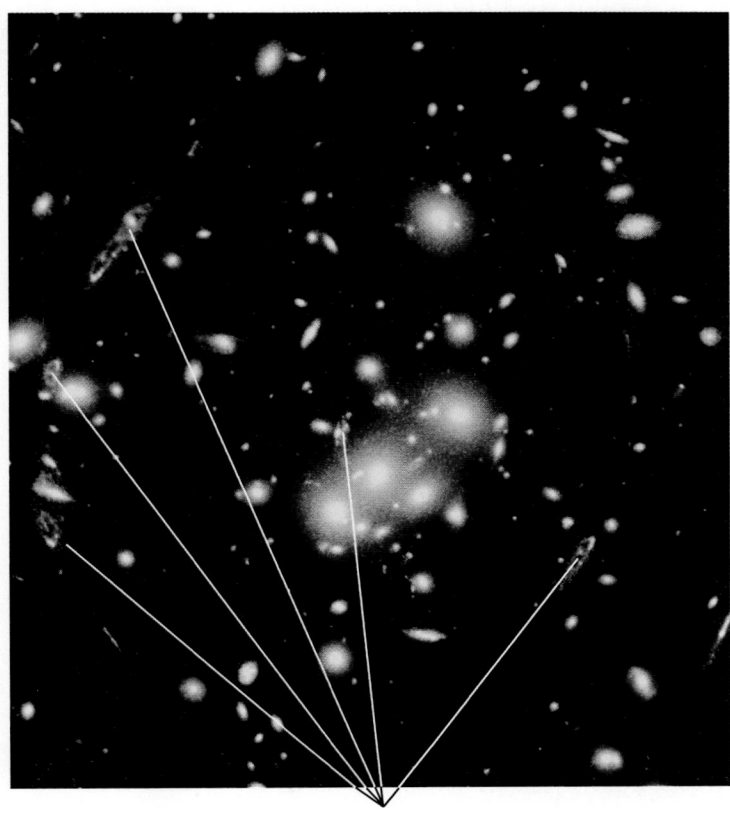

All of these blue arcs are images of the same distant galaxy.

FIGURE 23-33 R I 🅥 U X G

Gravitational Lensing by a Cluster of Galaxies The blue arcs in this image of the rich cluster CL 0024+1654 are distorted multiple images of a single, more distant galaxy. These images are the result of gravitational lensing by the matter in CL 0024+1654. The cluster is about 1600 Mpc (5 billion ly) from Earth; the blue galaxy is about twice as distant. The blue color of the remote galaxy suggests that it is very young and is actively forming stars. (Space Telescope Science Institute/NASA/Science Source)

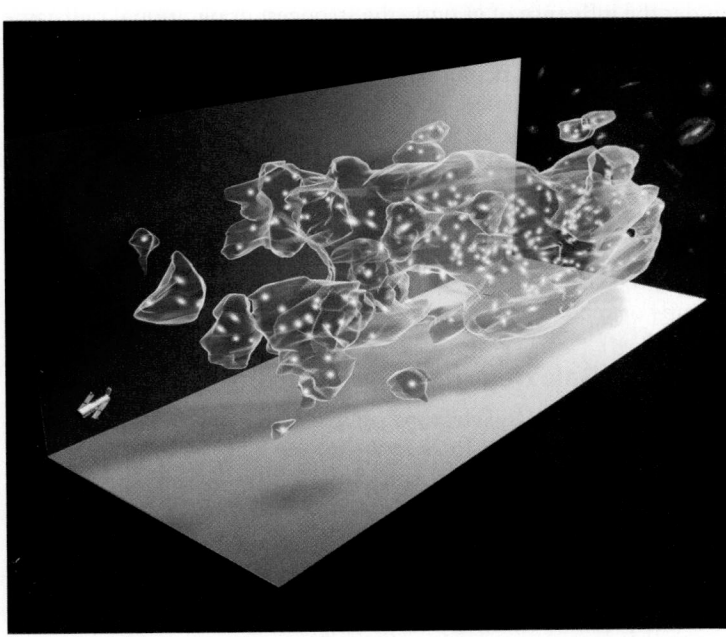

FIGURE 23-34

Three-Dimensional Map of Dark Matter Computers were used to analyze the distortions from gravitational lensing of more than half a million background galaxies in a patch of sky nearly 9 times the diameter of the full moon. This three-dimensional map shows the inferred distribution of dark matter from all of the gravitational lensing. The distorted and distant background galaxies are not shown here, only the closer inferred dark matter lenses as the larger shaded clumps, and the nearby galaxies visible as small bright spots. Further analysis reveals that the dark matter started clumping before the regular matter that makes stars and galaxies. It was the clumping of dark matter that led to the gravitational attraction, which then pulled together regular matter to form galaxies. This observational result is also consistent with computer simulations of how structures formed in the universe from regular and dark matter. (NASA, ESA, and R. Massey [California Institute of Technology])

1 kg of WIMPs might be passing through Earth at any moment. To date, however, the true nature of dark matter remains unknown.

An important clue about the nature of dark matter was discovered in 2006 by examining a rich cluster of galaxies called the Bullet Cluster. Remarkably, the visible matter and dark matter in this cluster do *not* have the same distribution (Figure 23-35a). The best explanation for how this could have come about is that the Bullet Cluster is the result of a collision between two galaxy clusters, one larger than the other (Figure 23-35b). (In Section 23-7, we asked you to visualize collisions between entire galaxies; now you must imagine a vastly more immense collision between entire clusters of galaxies!)

During the collision, the gas from one cluster slams into the gas from the other cluster and slows down due to *fluid resistance*. (You feel the force of fluid resistance pushing against you whenever you try to move through a liquid or gas—for example, when you swim in a lake or put your hand outside the window of a fast-moving car.) Fluid resistance is a consequence of the electric forces between adjacent atoms and molecules in a fluid. But if dark matter is made up of some curious material that responds only to *gravitational* forces, it is unaffected by fluid resistance. As a result, during the collision sketched in Figure 23-35b, the gas is slowed by fluid resistance but the dark matter is not. The agreement of the simulation in Figure 23-35b with the observations in Figure 23-35a strongly reinforces the idea that dark matter, though mysterious, is quite real.

The study of dark matter forces us to reconsider our impression of the universe based on luminous matter alone. As we will see in Chapter 25, we now know that when the whole universe is considered, beyond galaxies and clusters, there is about 5 times as much dark matter than there is visible matter. Because dark matter is so dominant, "ordinary" visible matter—including this book, the air you breathe, all the planets and stars, and your own body—is in fact relatively rare. Thus, the vast majority of mass in the universe is of completely unknown composition.

Exotic Dark Matter

The mystery of dark matter is even greater for another reason. If dark matter consisted of unseen but otherwise regular objects made of atoms from the periodic table of the elements (objects such as brown dwarfs), dark matter would still be a familiar form of matter. However, there is very strong evidence that dark matter is not made up of this "ordinary" matter consisting of protons and neutrons.

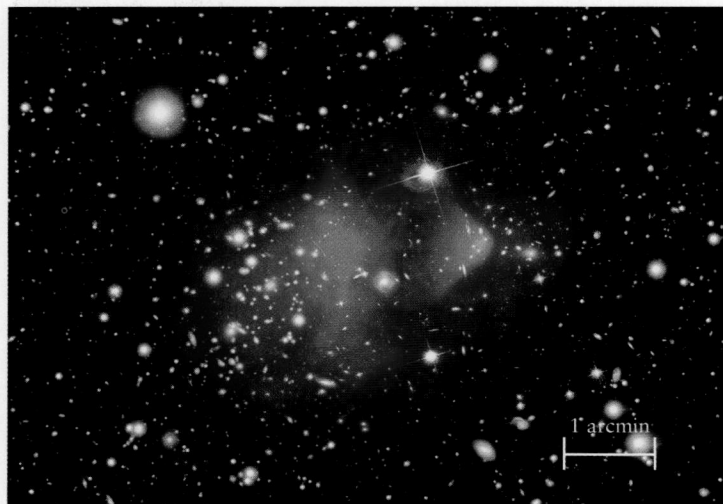

(a) Composite image of galaxy cluster 1E0657-56 showing visible galaxies, x-ray–emitting gas (red) and dark matter (blue)

1. Two galaxy clusters approach each other.

Cluster 1

Cluster 2

2. The two clusters begin to collide.

Cluster 1

Cluster 2

3. Fluid resistance slows the gas down . . .

Cluster 2 dark matter

Cluster 1 gas

Cluster 2 gas

Cluster 1 dark matter

4. . . . but the dark matter keeps on moving.

Cluster 1 gas

Cluster 2 dark matter

Cluster 2 gas

Cluster 1 dark matter

(b) A model of how the gas and dark matter in 1E0657-56 could have become separated

ANIMATION 23.7   **FIGURE 23-35**   R I **V** U **X** G

Isolated Dark Matter in a Cluster of Galaxies (a) This visible-light image of the Bullet Cluster shows more than 1000 galaxies. The superimposed image in red shows the distribution of the cluster's hot, x-ray–emitting gas, and the blue image shows the distribution of dark matter as determined by gravitational lensing of numerous background galaxies (see Figure 23-32). **(b)** We can understand the separation of dark matter and gas in this cluster if we assume that dark matter does not feel any force of fluid resistance. This is what we would expect if dark matter responds to gravitational forces only. (a: NASA/STScI; Magellan/U. Arizona/D. Clowe et al.; b: NASA/CXC/M. Weiss)

As we will see in Chapter 25 (especially Section 25-4, which describes the cosmic microwave background radiation), the amount of matter in the form of protons and neutrons has been measured. Because the amount of dark matter is much greater than the amount of atomic matter, most of the dark matter is in a truly unknown form. The WIMPs, if ever discovered, would be an example of a completely new form of matter. We use the term **exotic dark matter** for these hypothesized new forms of matter that are not made of protons and neutrons. We have invited Gregory Bothun, a dark matter researcher, to discuss his broad perspective on the topic in an essay at the end of this chapter, *The Unseen Universe: Dark Matter.*

CONCEPT CHECK 23-17

If some dark matter consists of neutron stars, would this be "ordinary" matter or exotic dark matter?
Answer appears at the end of the chapter.

23-9 Initiated by dark matter, galaxies formed from the merger of smaller objects

How do galaxies form and how do they evolve? Astronomers can gain important clues about galactic evolution simply by looking deep into space. The more distant a galaxy is, the longer its light takes to reach us. As we examine galaxies that are at increasing distances from Earth, we are actually looking further and further back in time. By looking into the past, we can see galaxies in their earliest stages.

One of the first steps to galaxy formation begins with dark matter. In fact, gravity leads to the clumping of dark matter before the regular matter starts being drawn together. Therefore, the hydrogen and helium gas that makes up most of the regular matter of the early universe was spread out quite uniformly, whereas the dark matter had already begun to form more dense concentrations.

In Chapter 26, we'll see why dark matter was able to clump first, as well as what happened later to allow regular matter to start clumping. But, once regular matter could clump together under the influence of gravity, the strongest gravitational pull came from the already-forming clumps of dark matter. What this means is that wherever dark matter formed clumps, it attracted regular matter, which ultimately became the galaxies and galaxy clusters we see. With 90% of galactic mass typically in the form of dark matter, and dark matter clumps providing the landscape on which galaxies form, our beautiful galaxies are small players in the structure of the universe.

Dark matter only starts the process of galaxy formation. Having been drawn in by dark matter, we next see how smaller clumps of regular matter came together to form galaxies.

Building Galaxies from the "Bottom Up"

The Hubble Space Telescope images in **Figure 23-36** provide a glimpse of galaxy formation in the early universe. Figure 23-36 shows a number of galaxylike objects some 13 billion ly (3988 Mpc) away, which are thus seen as they were 13 billion years ago. These objects are between one-tenth and one-half the size of our Milky Way Galaxy and have unusual, irregular shapes. Furthermore, computer simulations of these objects show that they will likely collide to form a large elliptical galaxy.

FIGURE 23-36 R I V U X G

The Building Blocks of Galaxies In this Hubble Space Telescope image, the objects outlined by circles are about 13 billion ly from Earth and only a fraction of the size of our Milky Way. They are the building blocks that merge to form larger galaxies and clusters. While barely visible, these are the brightest objects at this great distance and early time, and numerous objects even smaller and dimmer are expected. (NASA, ESA, M. Trenti [University of Colorado, Boulder, and Institute of Astronomy, University of Cambridge, UK], L. Bradley [STScI], and the BoRG team)

Images of the young universe such as those in Figure 23-36 lead astronomers to conclude that galaxies formed "from the bottom up"—that is, by the merger of smaller objects to form full-size galaxies. (These same images provide evidence against an older idea that galaxies formed "from the top down"—that is, fragmenting or breaking apart from immense, cluster-sized clouds of material.) We will see evidence in Chapter 26 that the clumps that formed early on evolved into objects like those shown in Figure 23-36, which in turn merged to form the population of galaxies that we see today.

CONCEPT CHECK 23-18

Did today's galaxies form from combining smaller galaxies or from the separation of larger galaxies?

Answer appears at the end of the chapter.

Forming Spirals, Lenticulars, and Ellipticals

Once a number of subgalactic units combine, they make an object called a *protogalaxy*. The rate at which stars form within a protogalaxy may determine whether this protogalaxy becomes a spiral or an elliptical. If stars form relatively slowly, the gas surrounding them has enough time to settle by collisions to form a flattened disk, much as happened on a much smaller scale in the solar nebula (see Section 8-4). Star formation continues because the disk contains an ample amount of hydrogen from which to make new stars. The result is a spiral or lenticular galaxy (**Figure 23-37a**). But if stars initially form in the protogalaxy at a rapid rate, virtually all of the available gas is used up to make stars before a disk can form. In this case, what results is an elliptical galaxy (Figure 23-37b).

(a) Formation of a spiral galaxy

(b) Formation of an elliptical galaxy

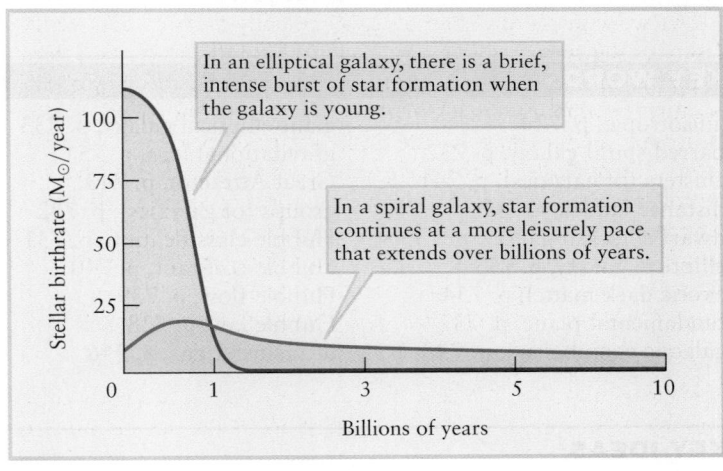

(c) The stellar birthrate in galaxies

FIGURE 23-37

The Formation of Spiral and Elliptical Galaxies **(a)** If the initial star formation rate in a protogalaxy is low, it can evolve into a spiral galaxy with a disk. **(b)** If the initial star formation rate is rapid, no gas is left to form a disk. The result is an elliptical galaxy. **(c)** This graph shows how the rate of star birth (in solar masses per year) varies with age in spiral and elliptical galaxies.

Figure 23-37c compares the stellar birthrate in the two types of galaxies. This graph helps us understand some of the differences between spiral and elliptical galaxies that we described in Section 23-3. Protogalaxies are thought to have been composed almost exclusively of hydrogen and helium gas, so the first stars were Population II stars with hardly any metals (that is, heavy elements).

As stars die and form planetary nebulae or supernovae, they eject into the interstellar medium gases enriched in metals. In a spiral galaxy, there is ongoing star formation in the disk, so these metals are incorporated into new generations of stars, making relatively metal-rich Population I stars like the Sun. By contrast, an elliptical galaxy has a single flurry of star formation when it is young, after which star formation ceases. Elliptical galaxies therefore contain only metal-poor Population II stars.

Figure 23-37c shows that both elliptical and spiral galaxies form stars most rapidly when they are young. This idea is borne out by the observation that very distant galaxies tend to be blue, which means that galaxies were bluer in the distant past than they are today. (Note the very blue colors of the distant, gravitationally lensed galaxies shown in Figure 23-32 and Figure 23-33.) Spectroscopic studies of such galaxies demonstrate that most owe their blue color to vigorous star formation, often occurring in intense, episodic bursts. The hot, luminous, and short-lived O and B stars produced in these bursts of star formation give blue galaxies their characteristic color.

CONCEPT CHECK 23-19

Why do elliptical galaxies contain only Population II stars?
Answer appears at the end of the chapter.

An Evolving Universe of Galaxies

In addition to changes in galaxy colors, the character of the galactic population has also changed over the past several billion years.

In nearby rich clusters, only about 5% of the galaxies are spirals. But observations of rich clusters at a redshift of $z = 0.4$—which corresponds to looking about 4 billion years into the past—show that about 30% of their galaxies were spirals.

Why were spiral galaxies more common in rich clusters in the distant past? Galactic collisions and mergers are probably responsible. During a collision, interstellar gas in the colliding galaxies is vigorously compressed, triggering a burst of star formation (see the *Cosmic Connections* figure in Section 23-7, on page 749). A succession of collisions produces a series of star-forming episodes that create numerous bright, hot O and B stars that become dispersed along arching spiral arms by the galaxy's rotation. Eventually, however, the gas is used up; star formation then ceases and the spiral arms become less visible.

As mentioned earlier, a full description of galaxy formation must include the effects of dark matter, and we discuss this in Chapter 26. As we have seen, only about 10% of the mass of a galaxy—its stars, gas, and dust—emits electromagnetic radiation or light of any kind. As yet, we have no idea what the remaining 90% looks like or what it is made of. The dilemma of dark matter is one of the most challenging problems facing astronomers today.

CONCEPT CHECK 23-20

What happens to spiral galaxies during galaxy mergers?
Answer appears at the end of the chapter.

KEY WORDS

anisotropic, p. 734	giant elliptical galaxy, p. 733	irregular cluster, p. 742	regular cluster, p. 743
barred spiral galaxy, p. 732	gravitational lens, p. 751	irregular galaxy, p. 735	rich cluster, p. 742
clusters (of galaxies), p. 741	Great Attractor, p. 743	isotropic, p. 734	spiral galaxy, p. 731
distance ladder, p. 737	groups (of galaxies), p. 742	lenticular galaxy, p. 734	standard candle, p. 736
dwarf elliptical galaxy, p. 733	Hubble classification, p. 731	Local Group, p. 742	starburst galaxy, p. 746
elliptical galaxy, p. 733	Hubble constant, p. 740	Local Supercluster, p. 743	supercluster, p. 743
exotic dark matter, p. 754	Hubble flow, p. 739	maser, p. 738	Tully–Fisher relation, p. 736
fundamental plane, p. 737	Hubble law, p. 728	poor cluster, p. 742	tuning fork diagram, p. 734
galactic cannibalism, p. 748	intracluster gas, p. 746	redshift, p. 739	void, p. 744

KEY IDEAS

The Hubble Classification: Galaxies can be grouped into four major categories: spirals, barred spirals, ellipticals, and irregulars (Section 23-3).

- The disks of spiral and barred spiral galaxies are sites of active star formation.
- Elliptical galaxies are nearly devoid of interstellar gas and dust, so star formation is severely limited.
- Lenticular galaxies are intermediate between spiral and elliptical galaxies.
- Irregular galaxies have ill-defined, asymmetrical shapes. They are often found associated with other galaxies.

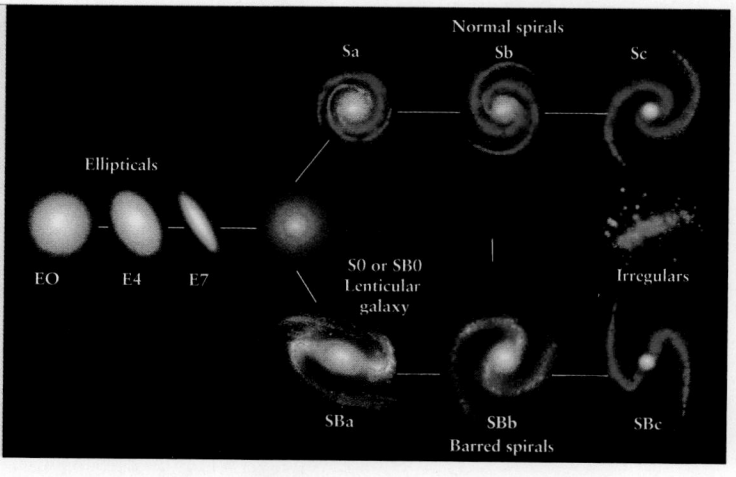

Distance to Galaxies: Standard candles—such as Cepheid variables and the most luminous supergiants, globular clusters, H II regions, and supernovae in a galaxy—are used in estimating intergalactic distances (Section 23-4).

- The Tully–Fisher relation, which correlates the width of the 21-cm line of hydrogen in a spiral galaxy with its luminosity, can also be used for determining distance. A method that can be used for elliptical galaxies is the fundamental plane, which relates the galaxy's size to its surface brightness distribution and to the motions of its stars.

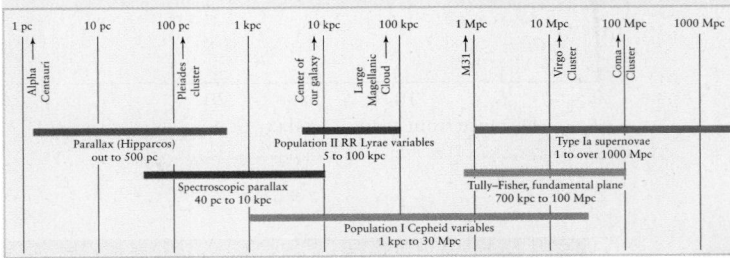

The Hubble Law: There is a simple linear relationship between the distance from Earth to a remote galaxy and the redshift of that galaxy (which is a measure of the speed at which it is receding from us). This relationship is the Hubble law, $v = H_0 d$ (Section 23-5).

- The value of the Hubble constant, H_0, is not known with certainty but is close to 73 km/s/Mpc.

Clusters and Superclusters: Galaxies are grouped into clusters rather than being scattered randomly throughout the universe (Section 23-6).

- A rich cluster contains hundreds or even thousands of galaxies; a poor cluster, often called a group, may contain only a few dozen.
- A regular cluster has a nearly spherical shape with a central concentration of galaxies; in an irregular cluster, galaxies are distributed asymmetrically.
- Our galaxy is a member of a poor, irregular cluster called the Local Group.
- Rich, regular clusters contain mostly elliptical and lenticular galaxies; irregular clusters contain spiral, barred spiral, and irregular galaxies along with ellipticals.
- Giant elliptical galaxies are often found near the centers of rich clusters.

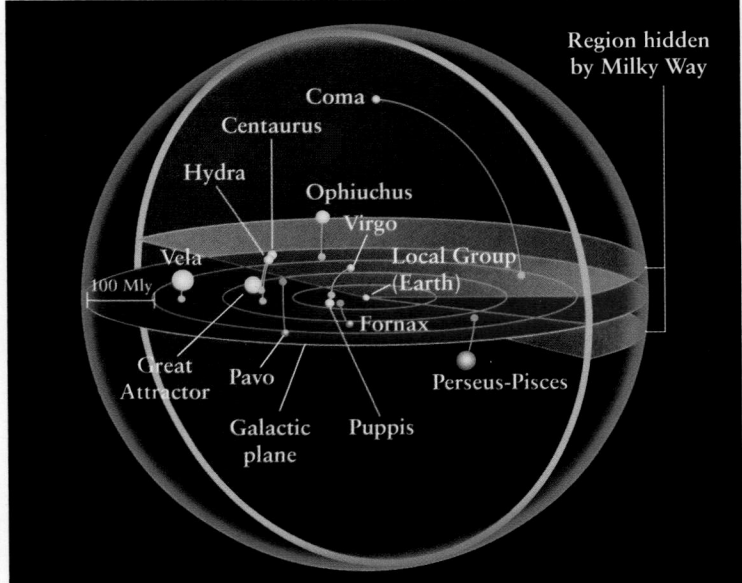

Galactic Collisions and Mergers: When two galaxies collide, their stars pass each other, but their interstellar media collide violently, either stripping the gas and dust from the galaxies or triggering prolific star formation (Section 23-7).

- The gravitational effects during a galactic collision can throw stars out of their galaxies into intergalactic space.
- Galactic mergers may occur; a large galaxy in a rich cluster may tend to grow steadily through galactic cannibalism, perhaps producing in the process a giant elliptical galaxy.

a. Two galaxies, each with old globular clusters (yellow), begin to interact.

b. The two galaxies swoop past each other before finally settling down as a single merged galaxy. Gas and dust is compressed in both galaxies in the process, creating new star clusters.

c. The combined galaxy has the original globular clusters (shown in yellow) as well as new ones (shown in blue).

(John Dubinski, University of Toronto)

Dark Matter: The regular luminous mass of a cluster of galaxies is not large enough to account for the observed motions of its galaxies; a large amount of unobserved mass must also be present. This is one of the indicators that there is matter we don't see because it neither emits, absorbs, nor reflects light. We call this dark matter (Section 23-8).

- Rotation curves for spiral galaxies do not show a drop off in orbital speed even when it appears, based on visible light, that the bulk of the galaxy's mass is well inside the orbiting star. This indicates that there is some unseen mass, or dark matter, that extends well beyond the visible mass of a galaxy.
- Gravitational lensing of remote galaxies by a foreground cluster enables astronomers to glean information about the distribution of dark matter in the foreground cluster.

Formation and Evolution of Galaxies: Observations indicate that galaxies arose from mergers of smaller collections of stars (Section 23-9).

- Clumping of dark matter created the "seeds" of gravitational attraction that later pulled in regular matter to form galaxies and clusters of galaxies.
- Whether a protogalaxy evolves into a spiral galaxy or an elliptical galaxy depends on its initial rate of star formation.

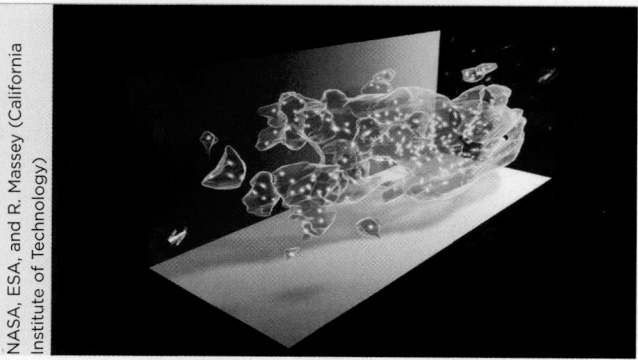

NASA, ESA, and R. Massey (California Institute of Technology)

QUESTIONS

Problem-solving tips and tools

Box 1-1 explains the small-angle formula, and Box 17-3 discusses the relationship among apparent magnitude, absolute magnitude, and distance. As Box 22-2 explains, a useful form of Kepler's third law is $M = rv^2/G$, where M is the mass within an orbit of radius r, v is the orbital speed, and G is the gravitational constant. Another form of Kepler's third law, particularly useful for two stars or two galaxies orbiting each other, is given in Section 17-9. The volume of a sphere of radius r is $4\pi r^3/3$. The mass of a hydrogen atom (^{1}H) is given in Appendix 7.

Section 23-1

1. Why did many nineteenth-century astronomers think that the "spiral nebulae" are part of the Milky Way?
2. What was the Shapley–Curtis "debate" all about?

Section 23-2

3. How did Edwin Hubble prove that the Andromeda "Nebula" is not a nebula within our Milky Way Galaxy?

4. Are any galaxies besides our own visible with the naked eye from Earth? If so, which one(s)?
5. Who won the Shapley–Curtis debate, and how?
6. Hubble made his observations of Cepheids in the Andromeda Galaxy (M31) using the 100-in. (2.5-m) telescope on Mount Wilson. Completed in 1917, this was the largest telescope in the world when Hubble carried out his observations in 1923. Why was it helpful to use such a large telescope?
7. **Box 23-1** Astronomers often state the distance to a remote galaxy in terms of its distance modulus, which is the difference between the apparent magnitude m and the absolute magnitude M (see Box 17-3). (a) By measuring the brightness of supernova 1994I in the galaxy M51 (see Figure 23-2), the distance modulus for this galaxy was determined to be $m - M = 29.2$. Find the distance to M51 in megaparsecs (Mpc). (b) A separate distance determination, which involved measuring the brightnesses of planetary nebulae in M51, found $m - M = 29.6$. What is the distance to M51 that you calculate from this information? (c) What is the difference between your answers to parts (a) and (b)? Compare this difference with the 750-kpc distance from Earth to M31, the Andromeda Galaxy. The difference between your answers illustrates the uncertainties involved in determining the distances to galaxies!

8. **Box 23-1** Suppose you discover a Type Ia supernova in a distant galaxy. At maximum brilliance, the supernova reaches an apparent magnitude of +10. How far away is the galaxy? (*Hint:* See Box 23-1.)

Section 23-3

9. **TUTORIAL 23.1** What is the Hubble classification scheme? Which category includes the largest galaxies? Which includes the smallest? Which category of galaxy is the most common?

10. Which is more likely to have a blue color, a spiral galaxy or an elliptical galaxy? Explain why.

11. Which types of galaxies are most likely to have new stars forming? Describe the observational evidence that supports your answer.

12. Explain why the apparent shape of an elliptical galaxy may be quite different from its real shape.

13. Why do astronomers suspect that the Hubble tuning fork diagram does not depict the evolutionary sequence of galaxies?

14. The following image shows the Small Magellanic Cloud (SMC), an irregular galaxy that orbits the Milky Way. The SMC is 63 kpc (200,000 ly) from Earth and 8 kpc (26,000 ly) across, and it can be seen with the naked eye from southern latitudes. What features of this image indicate that there has been recent star formation in the SMC? Explain your answer.

R I V U X G
© Australian Astronomical Observatory/David Malin Images

15. Earth is composed principally of heavy elements, such as silicon, nickel, and iron. Would you be likely to find such planets orbiting stars in the disk of a spiral galaxy? In the nucleus of a spiral galaxy? In an elliptical galaxy? In an irregular galaxy? Explain your answers.

16. Discuss what observations you might make to determine whether or not the various Hubble types of galaxies represent some sort of evolutionary sequence.

Section 23-4

17. Why are Cepheid variable stars useful for finding the distances to galaxies? Are there any limitations on their use for this purpose?

18. Why are Type Ia supernovae useful for finding the distances to very remote galaxies? Can they be used to find the distance to any galaxy you might choose? Explain your answers.

19. What is the Tully–Fisher relation? How is it used for measuring distances? Can it be used for galaxies of all kinds? Why or why not?

20. What are masers? How can they be used to measure the distance to a galaxy?

21. The masers that orbit the center of the spiral galaxy M106 travel at an orbital speed of about 1000 km/s. Astronomers observed these masers at intervals of 4 months. (a) What distance does a single maser move during a 4-month period? Give your answer in kilometers and in astronomical units. (b) During this period, a maser moving across the line of sight (like the maser shown in green in Figure 23-15) appeared to move through an angle of only 10^{-5} arcsec. Calculate the distance to the galaxy.

22. Discuss the advantages and disadvantages of using the various standard candle distance indicators to obtain extragalactic distances.

Section 23-5

23. What is the Hubble law? How can it be used to determine distances?

24. How did the discovery of the Hubble Law reinforce the idea that the spiral "nebulae" could not be part of the Milky Way?

25. Why do you suppose it has been so difficult to determine the value of H_0?

26. Some galaxies in the Local Group exhibit blueshifted spectral lines. Why aren't these blueshifts violations of the Hubble law?

27. The average radial velocity of galaxies in the Hercules Cluster pictured in Figure 23-18 is 10,800 km/s. (a) Using $H_0 = 73$ km/s/Mpc, find the distance to this cluster. Give your answer in megaparsecs and in light-years. (b) How would your answer to (a) differ if the Hubble constant had a smaller value? A larger value? Explain your answers.

28. A certain galaxy is observed to be receding from the Sun at a rate of 7500 km/s. The distance to this galaxy is measured independently and found to be 1.4×10^8 pc. Using these data, what is the value of the Hubble constant?

29. **Box 23-2** In the spectrum of the galaxy NGC 4839, the K line of singly ionized calcium has a wavelength 403.2 nm. (a) What is the redshift of this galaxy? (*Hint:* See Box 23-2.) (b) Determine the distance to this galaxy using the Hubble law with $H_0 = 73$ km/s/Mpc.

30. **Box 23-2** The galaxy RD1 has a redshift of $z = 5.34$. (a) Determine its recessional velocity v in km/s and as a fraction of the speed of light. (b) What recessional velocity would you have calculated if you had erroneously used the low-speed formula relating z and v? Would using this formula have been a small or large error? (c) According to the Hubble law, what is the distance from Earth to RD1? Use $H_0 = 73$ km/s/Mpc for the Hubble constant, and give your answer in both megaparsecs and light-years.

31. How might you determine what part of a galaxy's redshift is caused by the galaxy's orbital motion about the center of mass of its cluster?

Section 23-6

32. What are the differences between regular and irregular clusters?

33. What is the difference between a cluster and a supercluster? Are both clusters and superclusters held together by their gravity?

34. What measurements do astronomers make to construct three-dimensional maps of the positions of galaxies in space?

35. What is the Sloan Great Wall, and why is it interesting to astronomers?

36. Describe what voids are and what they tell us about the large-scale structure of the universe.

37. Why do giant elliptical galaxies dominate rich clusters but not poor clusters?

38. Describe what sorts of observations you might make to search for as-yet-undiscovered galaxies in our Local Group. How is it possible that such galaxies might still remain to be discovered? In what part of the sky would these galaxies be located? What sorts of observations might reveal these galaxies?

Section 23-7

39. Why is the intracluster gas in galaxy clusters at such high temperatures?

40. What are starburst galaxies? How can they be produced by collisions between galaxies?

41. The accompanying images show the unusual elliptical galaxy NGC 5128 in visible and infrared wavelengths. Explain how the properties of this galaxy seen in the infrared image can be explained if NGC 5128 is the result of a merger of an elliptical galaxy and a spiral galaxy.

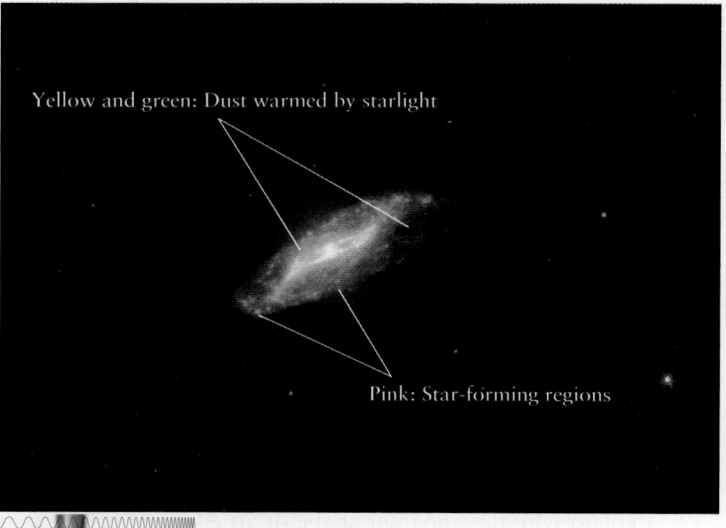

Yellow and green: Dust warmed by starlight

Pink: Star-forming regions

R I V U X G

NASA/JPL-Caltech/J. Keene (SSC/Caltech)

NGC 5128 is an elliptical galaxy . . . .

. . . . with a dark dust lane.

Eric Peng, Herzberg Institute of Astrophysics, and NOAO/AURA/NSF

R I V U X G

Section 23-8

42. What evidence is there for the existence of dark matter in clusters of galaxies?

43. What is gravitational lensing? Why don't we notice the gravitational lensing of light by ordinary objects on Earth?

44. How do observations of the Bullet Cluster help constrain the nature of dark matter?

45. What is the favored hypothesis for what dark matter is made of? What properties do these particles have?

46. Would the existence of WIMPs be a form of exotic dark matter or "ordinary" matter? What is exotic dark matter?

47. Figure 23-31 shows the rotation curve of the Sa galaxy NGC 4378. Using data from that graph, calculate the orbital period of stars 20 kpc from the galaxy's center. How much mass lies within 20 kpc of the center of NGC 4378?

48. Explain why the dark matter in galaxy clusters could not be neutral hydrogen.

49. What would rotation curves look like if there was no dark matter?

50. Explain how the distribution of dark matter was determined for Figure 23-34.

Section 23-9

51. What observations suggest that present-day galaxies formed from smaller assemblages of matter?

52. On what grounds do astronomers think that in the past, spiral galaxies were more numerous in rich clusters than they are today? What could account for this excess of spiral galaxies in the past?

53. According to Figure 23-37c, elliptical galaxies continue to form stars for about a billion years after they form. Give an argument why we might expect to find some Population I

stars in an elliptical galaxy. (*Hint:* Table 19-1 gives the main-sequence lifetimes for stars of different masses.)

General Problems

54. An educational publication for children included the following statement: "The Sun is in fact the only star in our galaxy. All of the other stars in the sky are located in other galaxies." How would you correct this statement?

55. It is estimated that the Coma Cluster (see Figure 23-21) contains about 10^{13} $M_\odot$ of intracluster gas. (**a**) Assuming that this gas is made of hydrogen atoms, calculate the total number of intracluster gas atoms in the Coma Cluster. (**b**) The Coma Cluster is roughly spherical in shape, with a radius of about 3 Mpc. Calculate the number of intracluster gas atoms per cubic centimeter in the Coma Cluster. Assume that the gas fills the cluster uniformly. (**c**) Compare the intracluster gas in the Coma Cluster with the gas in our atmosphere (3×10^{19} molecules per cubic centimeter, temperature 300 K); a typical gas cloud within our own galaxy (a few hundred molecules per cubic centimeter, temperature 50 K or less); and the corona of the Sun (10^5 atoms per cubic centimeter, temperature 10^6 K).

56. Two galaxies separated by 600 kpc are orbiting each other with a period of 40 billion years. What is the total mass of the two galaxies?

57. How would you distinguish star images from unresolved images of remote galaxies on a CCD?

WEB/E-BOOK QUESTIONS

58. When galaxies pass close to one another, as should happen frequently in a rich cluster, tidal forces between the galaxies can strip away their outlying stars. The result should be a loosely dispersed sea of "intergalactic stars" populating the space between galaxies in a cluster. Search the internet for information about intergalactic stars. Have they been observed? If so, where are they found? What would our nighttime sky look like if our Sun were an intergalactic star?

59. The Hubble Space Telescope (HST) has made extensive observations of very distant galaxies. Visit the HST website to learn about these investigations. How far back in time has HST been able to look? What sorts of early galaxies are observed? What is the current thinking about how galaxies formed and evolved?

60. ANIMATION **Radiation from a Rotating Galaxy.** Access and view
23.6 the animation "Radiation from a Rotating Galaxy" in Chapter 23 of the *Universe* website or e-book. Describe how the animation would have to be changed for a spiral galaxy of the same size but (**a**) greater mass and (**b**) smaller mass.

ACTIVITIES

Observing Projects

61. Using a telescope with an aperture of at least 30 cm (12 in.), observe as many of the spiral galaxies listed in the following table as you can. If you have a copy of the *Starry Night*™ program, use it to help determine when these galaxies can best be viewed. Many of these galaxies are members of the Virgo Cluster, which can best be seen from March through June. Because all galaxies are quite faint, be sure to schedule your observations for a moonless night. The best view is obtained when a galaxy is near the meridian. While at the eyepiece, make a sketch of what you see. Can you distinguish any spiral structure? After completing your observations, compare your sketches with photographs found in an online catalog of Messier objects (the "M" in the galaxy designations stands for Messier).

Spiral galaxy	Right ascension	Declination	Hubble type
M31 (NGC 224)	0^h 42.7^m	+41° 16′	Sb
M58 (NGC 4579)	12^h 37.7^m	+11° 49′	Sb
M61 (NGC 4303)	12^h 21.9^m	+4° 28′	Sc
M63 (NGC 5055)	13^h 15.8^m	+42° 02′	Sb
M64 (NGC 4826)	12^h 56.7^m	+21° 41′	Sb
M74 (NGC 628)	1^h 36.7^m	+15° 47′	Sc
M83 (NGC 5236)	13^h 37.0^m	−29° 52′	Sc
M88 (NGC 4501)	12^h 32.0^m	+14° 25′	Sb
M90 (NGC 4569)	12^h 36.8^m	+13° 10′	Sb
M91 (NGC 4548)	12^h 35.4^m	+14° 30′	SBb
M94 (NGC 4736)	12^h 50.9^m	+41° 07′	Sb
M98 (NGC 4192)	12^h 13.8^m	+14° 54′	Sb
M99 (NGC 4254)	12^h 18.8^m	+14° 25′	Sc
M100 (NGC 4321)	12^h 22.9^m	+15° 49′	Sc
M101 (NGC 5457)	14^h 03.2^m	+54° 21′	Sc
M104 (NGC 4594)	12^h 40.0^m	−11° 37′	Sa
M108 (NGC 3556)	11^h 11.5^m	+55° 40′	Sc

Note: The right ascensions and declinations are given for epoch 2000.

62. Using a telescope with an aperture of at least 30 cm (12 in.), observe as many of the following elliptical galaxies as you can. Six of these galaxies are in the Virgo cluster, which is conveniently located in the evening sky from March through June. If you have a copy of the *Starry Night*™ program, use it to help determine when these galaxies can best be viewed. As in the previous exercise, be sure to schedule your observations for a moonless night, when the galaxies you wish to observe will be near the meridian. Do these elliptical galaxies differ in appearance from spiral galaxies?

Elliptical galaxy	Right ascension	Declination	Hubble type
M49 (NGC 4472)	12^h 29.8^m	+8° 009′	E4
M59 (NGC 4621)	12^h 42.0^m	+11° 39′	E3
M60 (NGC 4649)	12^h 43.7^m	+11° 33′	E1
M84 (NGC 4374)	12^h 25.1^m	+12° 53′	E1
M86 (NGC 4406)	12^h 26.2^m	+12° 57′	E3
M89 (NGC 4552)	12^h 35.7^m	+12° 33′	E0
M110 (NGC 205)	00^h 40.4^m	+41° 41′	E6

Note: The right ascensions and declinations are given for epoch 2000.

63. Using a telescope with an aperture of at least 30 cm (12 in.), observe as many of the following interacting galaxies as you can. If you have a copy of the *Starry Night™* program, use it to help determine when these galaxies can best be viewed. As in the previous exercises, be sure to schedule your observations for a moonless night, when the galaxies you wish to observe will be near the meridian. While at the eyepiece, make a sketch of each galaxy. Can you distinguish hints of interplay among the galaxies? After completing your observations, compare your sketches with photographs found in an online catalog of Messier objects (the "M" in the galaxy designations stands for Messier).

Interacting galaxies	Right ascension	Declination
M51 (NGC 5194)	13^h 29.9^m	+47° 12′
NGC 5195	13^h 30.0^m	+47° 16′
M65 (NGC 3623)	11^h 18.9^m	+13° 05′
M66 (NGC 3627)	11^h 20.2^m	+12° 59′
NGC 3628	11^h 20.3^m	+13° 36′
M81 (NGC 3031)	9^h 55.6^m	+69° 04′
M82 (NGC 3034)	9^h 55.8^m	+69° 41′
M95 (NGC 3351)	10^h 44.0^m	+11° 42′
M96 (NGC 3368)	10^h 46.8^m	+11° 49′
M105 (NGC 3379)	10^h 47.8^m	+12° 35′

Note: The right ascensions and declinations are given for epoch 2000.

Starry Night™ Explorations

64. Use *Starry Night™* to examine the appearances of a variety of galaxies: spiral, barred spiral, elliptical, lenticular, and irregular. Select **Favorites > Explorations > Galaxy Classification** from the menu. Click the **Quick Search Items** icon in the toolbar to open the **Observing Lists . . .** panel. Locate and expand the **Galaxy Classification** observing list in this panel.

(a) Click the menu icon beside the name of each of the galaxies in the observing list and select **Magnify** to study the appearance of the galaxy in the view, adjusting the zoom as necessary to allow you to classify the galaxy as a spiral (S),

barred spiral (SB), elliptical (E), or irregular (Irr), and the sub-classification of each galaxy (e.g., Sa, E5).

(b) Select **File > Revert** to reset the view. Click the name in the label for the galaxy M51 in the view and select **Magnify** to study the appearance of this galaxy. What is the classification of this galaxy? Comment on anything unusual about its structure and/or its neighboring objects.

65. In *Starry Night™*, open **Favorites > Explorations > Virgo Cluster** to see this cluster of galaxies (highlighted in yellow) centered in the view. This view shows the distribution of galaxies in the region of the universe that contains the Milky Way, which is labeled.

(a) Use the location scroller to study the Virgo Cluster. What is the overall shape of the Virgo Cluster? Can you see any general pattern in the distribution of the galaxies that surround the Virgo Cluster?

(b) Use the location scroller to find other clusters of galaxies in this view. Select a sample of these clusters and right-click (Windows) or control-click (Mac) on a galaxy within the group to open the contextual menu and select **Highlight**. Reopen the contextual menu over the highlighted cluster and select **Center** before using the location scroller to study the appearance of this cluster. How does the Virgo Cluster compare to some of the other galaxy clusters you observe? Now try to identify and highlight the walls of galaxies surrounding large voids in space that link these concentrated clusters of galaxies.

66. Clusters of galaxies contain different numbers and distributions of galaxies and harbor significant amounts of dark matter. Use *Starry Night™* to see the gravitational effect of this mysterious dark matter. Open **Favorites > Explorations > Gravitational Lens** to see a Hubble Space Telescope image that shows a rich cluster of ordinary-looking yellowish galaxies surrounded by blue arcs of light. These blue arcs are multiple images of a very distant galaxy, located well beyond the cluster. The images are formed when the light from the distant galaxy is focused and distorted by the gravitational lens effect of the dark matter that pervades the cluster of galaxies. The distribution of this mysterious substance within clusters of galaxies can be inferred from the analysis of such images of distant galaxies.

(a) What does the predominantly blue color of the light from this very distant galaxy indicate about the average age of the stars in this galaxy?

(b) If the cluster of galaxies is 5 billion ly from Earth, what can you say about how long the light in the blue arcs has been traveling from the source galaxy compared to the age of the universe?

Collaborative Exercises

67. In the early twentieth century, there was considerable debate about the nature of spiral nebulae and their distance from us, but the debate was resolved by improvements in technology. As a group, list three issues that we, as a culture, did not understand in the past but understand today, and explain why we now have that understanding.

68. Even though there are billions of galaxies, there are not billions of different kinds. In fact, galaxies are classified according to their appearance. As a group, dig into your book bags and put all of the writing implements (pens, pencils, highlighters, and so on) you have in a central pile. Remember which ones are yours! Determine a classification scheme that sorts the writing implements into at least three to six piles. Write down the scheme and the number of items in each pile. Ask the group next to you to use your scheme and sort your materials. Correct any ambiguities before submitting your classification scheme.

69. Imagine your company, Astronomical Artistry, has been contracted by the local marching band to create a football halftime show about spiral galaxies. How exactly would you design the positions of the band members on the field to represent the different spiral galaxies of classes Sa, Sb, and Sc? Create two columns on your paper by drawing a line from top to bottom, drawing sketches in the left-hand column and writing a description of each sketch in the right-hand column. Also include what the band's opening formation and final formation should be.

ANSWERS

ConceptChecks

ConceptCheck 23-1: If spiral nebulae are closer than some of the stars of our galaxy, then the evidence presented supports Shapley's argument that spiral nebulae are within our galaxy and are not themselves large galaxies very far away.

ConceptCheck 23-2: Cepheid variable stars have well-known luminosities, and can even be identified in other galaxies. Comparing a Cepheid's luminosity to apparent brightness reveals a precise distance to the galaxy.

ConceptCheck 23-3: Sc galaxies have the most active star formation, so a sketch would have a smaller central bulge and a relatively large star-forming disk with considerable gas and dust.

ConceptCheck 23-4: In both spiral and barred spirals, the designation a is used for tightly wound arms and c for loosely wrapped arms.

ConceptCheck 23-5: Elliptical galaxies have almost no gas or dust available for the formation of stars.

ConceptCheck 23-6: If the Type Ia supernova appeared to be dimmer because of intervening dust, then astronomers would mistakenly believe that the supernova was farther away and, subsequently, that the galaxy was farther away than it really is.

ConceptCheck 23-7: A galaxy has a narrow width to its emitted hydrogen light if it rotates slowly. This is because the width arises from the Doppler shift of emitted light from approaching and receding sides of the rotating galaxy. Smaller galaxies rotate more slowly than larger galaxies, which means that galaxies with narrower emission line widths are also smaller. If, as in this question, the galaxy appears bright, then it must be close to our own galaxy.

ConceptCheck 23-8: Because the base of the distance ladder depends on parallax, all rungs of the ladder are completely dependent on an accurate understanding of brightness as determined by parallax.

ConceptCheck 23-9: From Figure 23-17, the farther away the galaxy, the larger its recessional velocity in the Hubble flow. Furthermore, a larger recessional velocity produces a larger redshift, so the farther galaxy has the largest redshift.

ConceptCheck 23-10: The spectral shift toward the shorter blue wavelengths means that the Andromeda Galaxy is moving toward our Milky Way Galaxy.

ConceptCheck 23-11: Rich clusters are observed to contain significantly more galaxies within them than poor clusters.

ConceptCheck 23-12: No, although on the largest scales, galaxies appear in every direction. Observations of superclusters show they are clumped into uneven groups and into long filaments.

ConceptCheck 23-13: The Local Group, dominated by the Milky Way Galaxy and the Andromeda Galaxy, contains about 40 dwarf ellipticals.

ConceptCheck 23-14: Collisions between gas and dust in colliding galaxies will warm gas, causing it to glow in x rays.

ConceptCheck 23-15: Simulations suggest that a single elliptical galaxy will form with many newly formed stars.

ConceptCheck 23-16: The best evidence is that the rotation curves in galaxies cannot be accounted for by the observed mass and the ability of galaxies and galaxy clusters to gravitationally lens more distant sources of light.

ConceptCheck 23-17: All matter consisting of neutrons and protons is called "ordinary" matter, so a neutron star is still made of ordinary matter, even though a neutron star is not an ordinary object like a planet or star.

ConceptCheck 23-18: Earlier galaxies appear to have been much smaller, suggesting that today's large galaxies formed by collisions and combinations of smaller galaxies.

ConceptCheck 23-19: Elliptical galaxies create most of their stars during the initial formation of the galaxy. Metal-rich, Population I stars can only form from the remains of Population II stars, which can only occur in later generations of star formation, which do not occur in elliptical galaxies.

ConceptCheck 23-20: The arms become distorted during collisions, resulting in astronomers observing far fewer spiral galaxies today than in the past.

CalculationChecks

CalculationCheck 23-1: Because z = the change in wavelength $(\lambda - \lambda_0)$ divided by the original λ_0, $z = (725.6 \text{ nm} - 656.3 \text{ nm}) \div 656.3 \text{ nm} = 0.11$.

CalculationCheck 23-2: Hubble's law, $v = H_0 d$, can be rearranged as $d = v \div H_0$. Using $H_0 = 73$ km/s/Mpc, $d = v \div H_0 = 10{,}000$ km/s $\div 73$ km/s/Mpc $= 134$ Mpc (million parsecs).

The Unseen Universe: Dark Matter

by Gregory Bothun

Objects that emit light, whether they are cigars, light-bulbs, cows, stars, or galaxies, can be characterized by their emitted energy per unit mass. This is parameterized as the **mass** to **luminosity** ratio, or *M/L*. For regular matter like the Sun, we know how much mass corresponds to how much light is emitted. *While dark matter (DM) is truly dark, it is typically found in regions that also contain light-emitting matter; therefore, observing higher-than-expected M/L values indicates the presence of DM.* In general, the evidence for DM is a result of analyzing the motions of test particles (for example, stars or galaxies) orbiting a region of mass and determining if there is enough mass in the region, which can be accounted for by the light-emitting objects within that region, to produce the observed motion. If there is not enough mass, as calculated from the observed light, then some form of DM must be providing the "missing" mass.

Historically, the planet Neptune was the first detected (in 1846) DM object as it represented a hitherto unseen mass that was perturbing the orbit of Uranus. During the 1970s, it became apparent that galaxies must contain DM to explain the orbital velocities of their constituent stars. Plotting rotational velocity against distance from the center of a galaxy yields a *rotation curve*. If light in a galaxy implies mass, then the mass should be centrally concentrated and the rotation curve would be like that of the planets orbiting the Sun—rotation velocity would decrease with increasing distance from the Sun (dashed line in Figure 1). For real galaxies, however, the rotation curve does not decrease but reaches a constant (flat) velocity. Flat rotation curves imply the kind of galaxy structure shown in Figure 2,

Gregory Bothun

FIGURE 2

where the small-scale disk of regular visible matter (orange) is embedded in a large-scale DM halo (blue).

On a scale larger than galaxies, it is has been known for 35 years that clusters of galaxies must also contain a large amount of DM to (a) bind the cluster together and (b) produce the kind of gravitational lens images easily observed in the first few years of the Hubble Space Telescope (Figure 3). So one question now emerges: Is the DM presence in the halos of galaxies the same stuff as the DM in clusters of galaxies? In fact, is there also DM between clusters of galaxies located in intergalactic space? If there is none, then DM is strictly confined to galaxies and clusters of galaxies, and its existence may be related to the formation of that structure. But on the grander scale—the space between clusters of galaxies—we do have an embarrassing problem. Measurements indicate that only about 4% of the total content of the universe is in the form of the ordinary matter that makes up stars and galaxies. The remainder is mostly in the newly discovered dark energy (DE) as well as the DM sector. The fact that 96% of the universe is composed of mysterious quantities means we know next to nothing about the fundamental nature of the universe, and so there is still a lot more to learn.

But what is the nature of this DM? There are basically two families of candidates: ordinary matter with high M/L or exotic particles. Normal DM is composed of baryons and is referred to as baryonic DM, but baryonic DM might not always have been dark. A stellar mass black hole is the prime example, as it once was a star radiating energy. Exotic dark matter would be

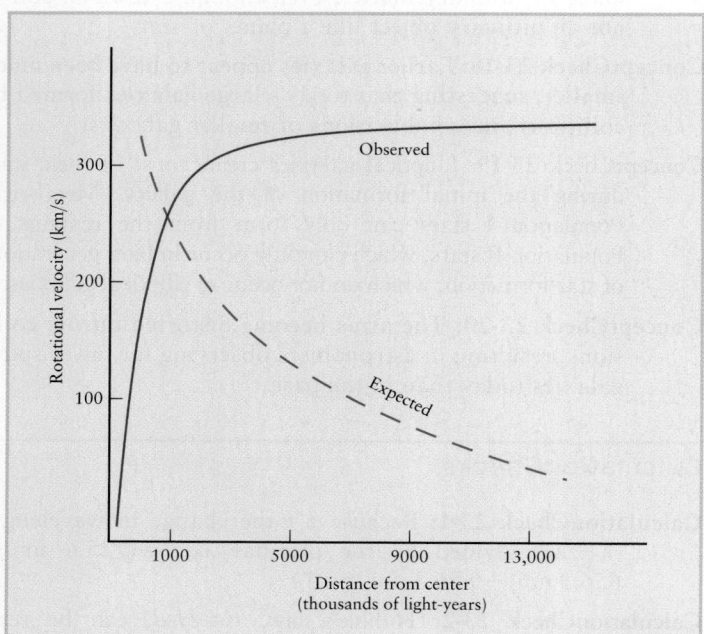

FIGURE 1

(continued on the next page)

The Unseen Universe: Dark Matter *(continued)*

Gravitational lens in Abell 2218
PF95-14 · ST ScI OPO · April 5, 1995 · W. Couch (UNSW), NASA

HST · WFPC2

W. Couch (University of New South Wales), R. Ellis (Cambridge University)/NASA

FIGURE 3

R I V U X G

nonbaryonic in nature and is most probably in the form of some strange particle, hitherto undetected.

Let's look at some candidate forms of dark matter and assess whether they'd fit the observations.

BEER CANS: While aluminum cans have very high **M/L**, they are not a possible form of DM because they are made of heavy elements and would have therefore required a lot of (luminous) massive stars to exist previously to make these heavy elements. This is not observed, so the universe is not filled with beer cans.

SMALL BALLS OF HYDROGEN: A Jupiter-mass ball of hydrogen would have high **M/L**, as would a brown dwarf star. These objects are examples of potential MACHOs, and their existence as the dominant source of DM in our galaxy has been convincingly ruled out.

STELLAR REMNANTS: In a few billion years, when star formation has ceased in our galaxy, its stars will fade into their stellar remnants (black holes, neutron stars, or white dwarfs)—these are all normal matter objects with high M/L in visible light. But these objects produce some radiation at wavelengths other than the visible, and their presence as this form of DM has been ruled out for some time. So, 0 for 3 so far.

BIG STELLAR REMNANTS: Globular clusters are the oldest known stars in our galaxy and may have been the first objects to form in the universe. However, globular clusters do show signs of chemical enrichment. To account for this enrichment, a prior

population of very massive stars (10,000 $M_\odot$ and higher) might have formed and rapidly evolved to large black hole masses (which might also be the source of black hole mergers recently detected as the source of gravitational waves). If these very massive stars are distributed in intergalactic space and were never part of forming galaxies, then it is possible that some of the DM component comes from them.

QUANTUM BLACK HOLES: An intersection between general relativity and the precepts of quantum mechanics allows for the existence of a very unusual particle—a mini black hole. A mini black hole has a mass equivalent to that of a large terrestrial mountain, about 10^{15} g and a radius of 10^{-13} cm. Such an object could only be created by tremendous compressional forces, which might have been present in the very early universe.

NONBARYONIC DARK MATTER: This form of DM would be a particle that must be distributed more evenly than the light we see from galaxies, which is concentrated toward the center. It would also need to be distributed between galaxies as well as within them. Nonbaryonic DM is considered to come in two basic forms: hot and cold. Hot dark matter (HDM) refers to relatively light particles that would have moved at nearly the speed of light when they were initially created in the early universe. Cold dark matter (CDM) consists of weakly interacting massive particles (WIMPs) that are able to cool and clump together, even at relatively high temperatures. Thanks to this gravitational clumping, CDM may be the necessary precursor to galaxy formation, and therefore it is associated with structure and not likely to be evenly distributed throughout interstellar space compared to HDM.

So that is the list of viable candidates, but what does the data say? What is this DM? Well, to date, despite 40 years of experimental searches for particle-based DM, there have been no secure detections, so the exact nature of DM continues to elude us. While this might be an unfortunate state of affairs, perhaps the real universe is not conforming to our desire to fully understand it. Perhaps the presumed presence of DM instead is a manifestation that we don't yet fully understand how gravity works on a large scale and that alternate forms of gravity are the likely solution. Indeed, this possibility is gaining some traction due to our continued failure to produce any direct detection of DM. In this way, DM might be a kind of necessary operational device to make our cosmological model work, the same way as Ptolemy's original epicycle did. Whatever is the case, DM is a solid reminder that, in most respects, the universe remains a mysterious place.

Gregory Bothun is a scientist at the University of Oregon who is known on the internet as Dr. DarkMatter. He was the original discoverer of low-surface-brightness galaxies, which have high M/L, and he continues to remain perplexed about galaxy formation and evolution to this day. In addition to observational cosmology, he does research into regional climate change as well as energy-sustainability issues.

24 Quasars and Active Galaxies

An ordinary star emits radiation primarily at ultraviolet, visible, and infrared wavelengths, in varying proportions that are governed by the star's surface temperature. Composed of many stars, ordinary galaxies, too, emit most strongly in these wavelength regions. But the object illustrated here, called a quasar, is outrageously different: It emits strongly over an immense range of wavelengths from radio to x ray.

The name *quasar* refers to one of the most fantastic types of objects in all of astronomy. At a quasar's core is a supermassive black hole that can be a *billion* times more massive than our Sun! Hot gas around this black hole can give off a tremendous amount of visible light—enough to shine with a luminosity *10,000* times more than a typical galaxy.

As in this illustration, some quasars even have strong jets. Unlike young stars that form smaller and slower-speed jets, matter in quasar jets travels near the speed

This artist's conception of a quasar shows a thick doughnut-shaped torus surrounding a small, bright inner region of hot swirling gas. The inner region is about the size of our solar system and is called an accretion disk. At the very center is a supermassive black hole that steadily consumes, or accretes, the surrounding gas. As illustrated here, some quasars have strong jets beaming away perpendicularly to the disk of swirling matter. While the outer torus can sometimes block much of the light emitted by the interior, when the interior light is visible, it can be 100,000 times as bright as our Milky Way Galaxy! (NASA/CXC/M. Weiss) of light and can shoot well beyond the galaxy that hosts the quasar. Galaxies that hold a quasar—always at the galaxy's center—are called *active galaxies*. Untangling the mysterious properties of quasars has been in process for more than half a century and is still a major topic of research today. Physicist George Gamow captured the intriguing nature of quasars in his 1964 revised lullaby:

Twinkle, twinkle quasi-star
Biggest puzzle from afar
How unlike the other ones
Brighter than a billion suns.
Twinkle, twinkle quasi-star
How I wonder what you are.

24-1 Quasars are the ultraluminous centers of distant galaxies

The discovery of quasars begins in the Illinois backyard of Grote Reber, a radio engineer who built the first true radio telescope in 1936. One of the objects he discovered is called Cygnus A; a modern radio and optical image of this quasar is shown in **Figure 24-1**. The object's thin jets shoot matter well beyond its host galaxy; to appreciate the scale, consider that about four Milky Way galaxies could fit between the ends of the two lobes of Cygnus A.

Examining the spectra of quasars revealed that they are immensely distant.

In a visible-wavelength image, a distant quasar often appears as a point source of light, like a star, because the host galaxy containing the quasar is so far away that the galaxy's light is not easily

observed (**Figure 24-2**). Along with intense radio emission, these properties led to the name **quasar**, which means "quasi-stellar radio source."

While the visible light from a quasar might appear pointlike, the way a star does, its spectrum tells a very different story. Stars, and the galaxies they make up, show absorption lines in their spectra. Quasars, on the other hand, show strong *emission* lines.

Clues from Emission Lines

Although quasars were clearly oddballs, many astronomers thought they were just strange stars in our own galaxy. A breakthrough occurred in 1963, when Maarten Schmidt at Caltech took another look at the emission lines of the quasar 3C 273 (**Figure 24-3**). Emission lines are caused by excited atoms, which emit radiation at specific wavelengths unique to each atom (see Figure 6-21). However, the lines of quasar 3C 273 did not match the known emission line wavelengths of hydrogen (or any other known atoms or molecules), so astronomers could not identify the composition of quasars.

Schmidt realized that four of 3C 273's brightest emission lines are positioned relative to one another in precisely the same way as four common emission lines of hydrogen (these are the Balmer lines; see Figure 5-24). However, these emission lines from 3C 273 were all shifted to *much longer wavelengths* than had ever been seen before in the Balmer lines. Were these lines really from hydrogen, but Doppler shifted from an object receding away at great speed? Schmidt determined that 3C 273 has a redshift of $z = 0.158$, corresponding to a recessional velocity equal to 16% of the speed of light (44,000 km/s). No star could be moving this fast and remain within our galaxy for very long. Thus, Schmidt concluded that 3C 273 could not be a nearby star, but must lie outside the Milky Way.

(a) Radio image of Cygnus A
R I V U X G

(b) Visible-light close-up of the central galaxy
R I V U X G

FIGURE 24-1

Cygnus A (a) This false-color radio image from the Very Large Array shows that most of the emission from Cygnus A comes from luminous radio lobes located on either side of a peculiar galaxy. (Red indicates the strongest radio emission, while blue indicates the faintest.) Each lobe extends about 230,000 ly from the galaxy. **(b)** The galaxy at the heart of Cygnus A has a substantial redshift, so it must be extremely far from Earth (it is about 740 million ly). Only the very brightest central region of the galaxy is visible here. To be so distant and yet be one of the brightest radio sources in the sky, Cygnus A must have an enormous energy output. (a: NRAO/AUI; b: Hubble Space Telescope/NASA/ESA)

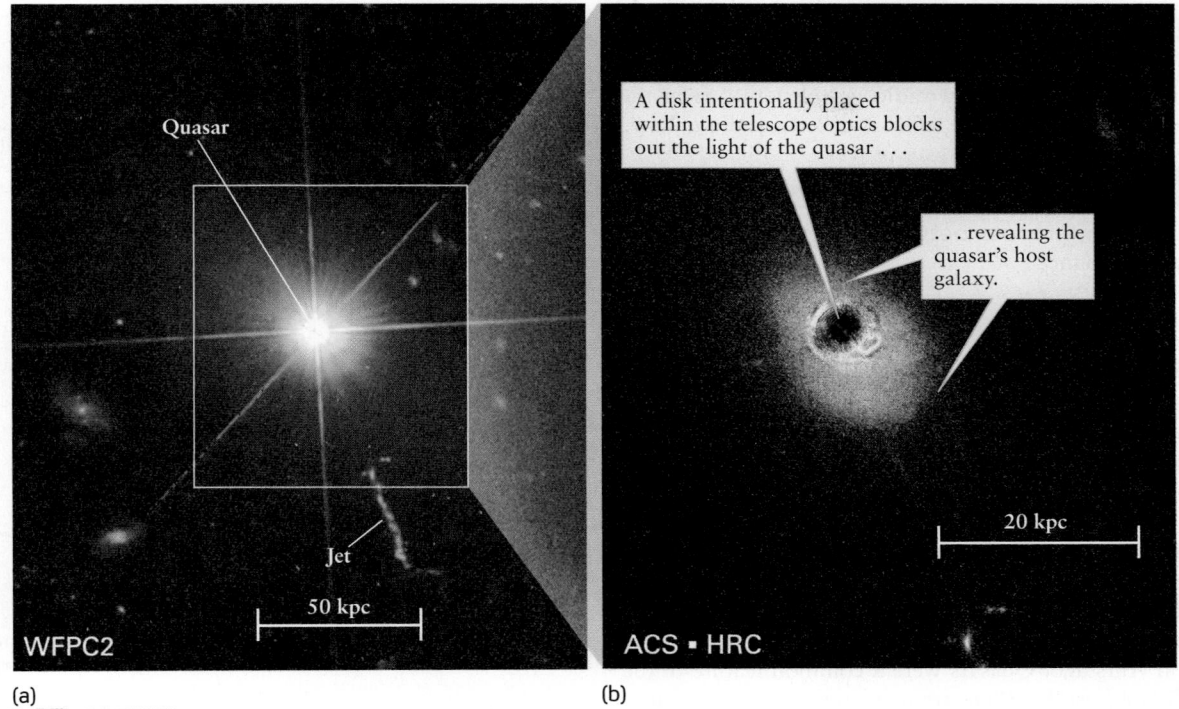

(a)

(b)

FIGURE 24-2 R I (V) U X G

A Quasar and Its Host Galaxy (a) Appearing pointlike, a quasar can look like a star in images. In this HST image, the bright glare of a quasar named 3C 273 hides its fainter host galaxy. Quasar 3C 273 also has a jet seen in visible light, although quasar jets are usually seen at radio wavelengths.

(b) Using a special technique, the quasar is blocked out to reveal the host galaxy. (a: NASA, HST [WFPC2], J. Bahcall; b: NASA, A. Martel [JHU], the ACS Science Team, J. Bahcall [IAS], and ESA)

According to the Hubble law, the recessional velocity of 3C 273 implies that its present distance from us is 2 billion ly—more than 800 times farther away than the Andromeda Galaxy! To be detected at such distances, 3C 273 must be an extraordinarily powerful source of both visible light and radio emission.

A quasar's luminosity can be calculated from its apparent brightness and distance using the inverse-square law (see Section 17-2). For example, 3C 273 has a luminosity of about 10^{40} watts, which is equivalent to 2.5×10^{13} (25 trillion) Suns. (Recall that the Sun's luminosity is $L_\odot = 3.90 \times 10^{26}$ watts.) Generally, quasar luminosities range from about 10^{38} watts up to nearly 10^{42} watts. For comparison, a typical large galaxy, like our own Milky Way, shines with a luminosity of 10^{37} watts, which equals 2.5×10^{10} (25 billion) Suns. Thus, a bright quasar can be many thousands of times more luminous than our entire Milky Way Galaxy!

> The most luminous quasars emit 100,000 times more radiation than the entire Milky Way Galaxy.

Quasars are members of a larger class of objects called **active galactic nuclei**, or **AGN** (the singular form, active galactic nucleus, is also called an AGN). AGN include a range of similar objects that, technically speaking, are not all luminous enough to be called quasars. Nonetheless, all AGN are powered by hot gas accreting around

FIGURE 24-3

The Spectrum of a Quasar The visible-light and infrared spectrum of quasar 3C 273 is dominated by four bright emission lines of hydrogen (see Section 5-8). The redshift is $z = 0.158$, so the wavelength of each line is 15.8% greater than for a sample of hydrogen on Earth. For example, the wavelength of H_β is shifted from 486 nm (a blue-green wavelength) to $1.158 \times (486 \text{ nm}) = 563 \text{ nm}$ (a yellow wavelength).

a supermassive black hole, and astronomers often use the terms *quasar* and *AGN* interchangeably. If the host galaxy that contains an AGN can be easily observed, the object is called an **active galaxy**. If the active galaxy emits strongly at radio wavelengths, as in the case of Cygnus A in Figure 24-1, the AGN is called a **radio galaxy**.

CONCEPT CHECK 24-1

What are two main differences between the spectrum of a star and that of a quasar?

Answer appears at the end of the chapter.

Quasars: High Redshifts, Extreme Distances

More than 200,000 quasars have been discovered. Most quasars have redshifts of 0.3 or more, which implies that they are more than 3 billion ly from Earth. Because there are no quasars with small redshifts, it follows that *there are no nearby quasars*. The nearest one is some 800 million ly from Earth. Furthermore, light takes time to travel across space, so when we observe a very remote object with a large redshift, we are seeing an image propagating from the remote past. Hence, the absence of nearby quasars means that *the era of quasars ended long ago*. Indeed, the number of quasars began to decline precipitously roughly 10 billion years ago. Quasars were a common feature of the universe in the distant past, but there are none in the present-day universe (Figure 24-4). Quasars not only occurred early in our universe, but whenever they occurred, their brief lifetimes were only about 1% as long as the lives of their host galaxies.

CONCEPT CHECK 24-2

From Figure 24-4, about how old was the universe when quasars were most abundant? What is the approximate redshift of objects from this time?

CONCEPT CHECK 24-3

Why can we say that any quasar observed from Earth today must have been very luminous when its light was emitted?

Answers appear at the end of the chapter.

24-2 Supermassive black holes are the "central engines" that power active galactic nuclei

What produces the intense emission from quasars? The basic model of quasar emission involves **accretion**—the gravitational accumulation of matter—around a supermassive black hole. As the matter (mostly gas) works its way closer to the black hole, it heats up and gives off the visible light we see from a quasar. In this section, we look at the quasar behaviors that pointed to a central role for black holes, and how to estimate the black hole masses.

Size of the Light Source

One characteristic that is common to *all* types of active galactic nuclei is variability. For example, Figure 24-5 shows brightness fluctuations of the quasar 3C 273 as determined from 29 years of observations. The brightness of 3C 273 increased by 60% from the

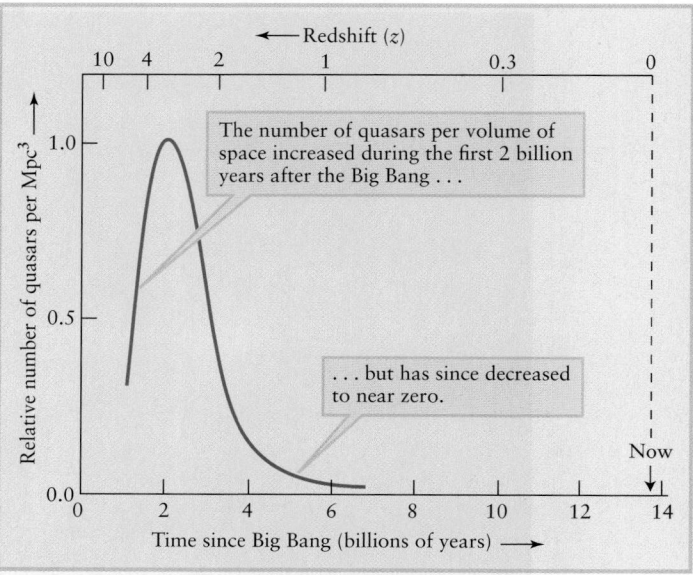

FIGURE 24-4

Quasars Are Extinct The greater the redshift of a quasar, the farther it is from Earth and the farther back in time we are seeing it. By observing the number of quasars found at different redshifts, astronomers can calculate how the density of quasars in the universe has changed over the history of the universe. The peak of quasar activity occurred more than 10 billion years ago, and there is no significant quasar activity today. The history of quasars is reminiscent of the history of the dinosaurs, which once populated the entire Earth but today are extinct. (Data from Peter Shaver, European Southern Observatory)

beginning to the end of 1982, then declined to the starting value in just 5 months. Other AGN undergo even greater fluctuations in brightness (by a factor of 25 or more) that occur even more rapidly (x-ray observations reveal that some AGN vary in brightness over time intervals as short as 3 hours).

The crucial aspect of these fluctuations in brightness is that they allow astronomers to place fundamental limits on the maximum size of a light source. This strict limit arises because *an object cannot vary in brightness faster than light can travel across that object*. For example, an object that is 1 ly in diameter cannot vary significantly in brightness over a period of less than 1 year.

To understand this limitation, imagine an object that measures 1 ly across, as in Figure 24-6. Suppose the entire object suddenly brightens, emitting a brief flash of light. Photons from the part of the object nearest Earth arrive at our telescopes first. Photons from the middle of the object arrive at Earth 6 months later. Finally, light from the far side of the object arrives a year after the first photons. Although the object emitted a sudden flash of light, we observe only a gradual increase in brightness that lasts a full year. In other words, the flash is stretched out over an interval equal to the difference in the light travel time between the nearest and farthest observable regions of the object.

The rapid flickering exhibited by AGN means that they emit their energy from a small volume—in some cases, less than half a light-day across (which is about the diameter of Neptune's orbit). Thus, from quasar variability, astronomers discovered that *a region about the size of our solar system can emit more energy per second than*

The rapid variability of active galactic nuclei tells us that they must be very small.

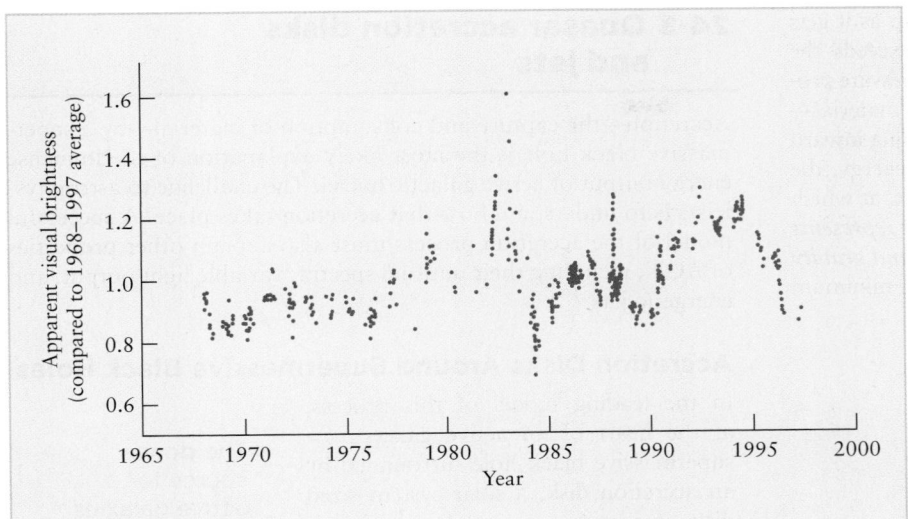

FIGURE 24-5

Brightness Variations of an AGN   This graph shows variations over a 29-year period in the apparent brightness of the quasar 3C 273 (see Figure 24-2a, b). Note the large outburst in 1982–1983 and the somewhat smaller ones in 1988 and 1992. (Data from M. Türler, S. Paltani, and T. J.-L. Courvoisier)

1000 galaxies! More than anything else, this size constraint points to supermassive black holes, as no other known object can power the release of so much energy from such a small volume.

CAUTION!

Quasars are intensely luminous, but their central black holes do not emit this light. Instead, the light is emitted by hot gas swirling around the black hole. The black hole is still considered the "central engine" of the quasar because it is the black hole's gravity that pulls in the surrounding gas and heats it up.

While the light-emitting region in a quasar might be about the size of our solar system, the black hole is smaller still. Just how big and massive are the black holes in quasars? Surprisingly, the size

and mass of the black hole can be estimated from great distances where nothing is seen but the light from its hot accreting gas.

The Eddington Limit and Black Hole Sizes

There is a relationship between the size of a black hole and the maximum luminosity that can be emitted by the hot gas falling into it. Even if there is plenty of gas around to act as "fuel," there is a natural limit to the luminosity that can be radiated by gas that is pulled into a black hole. This limit is called the **Eddington limit**, after the British astrophysicist Sir Arthur Eddington who described it. The Eddington limit applies to any object held together by its own gravity and applies to stars as well as quasars.

The accretion (or gravitational accumulation) of fuel by a black hole is thought to be a smooth and steady process, but let's imagine what might happen if a quasar were to suddenly brighten. Keep in

1. An object 1 ly across emits a sudden flash of light.

⟵ 1 ly ⟶

2. The first light that we receive comes from *A* (the part of the object nearest to Earth).

A

B   To Earth ⟶

C

3. The light from *B* (the center of the object) has to travel an additional ½ ly to reach Earth, so we see this light half a year later than the light from *A*.

4. We see the light from *C* (the far side of the object) half a year later than the light from *B* and 1 year later than the light from *A*. Hence, we see the sudden flash of light spread over a full year.

FIGURE 24-6

A Limit on the Speed of Variations in Brightness
The rapidity with which the brightness of an object can vary significantly is limited by the time it takes light to travel across the object. If an object 1 ly in size emits a sudden flash of light, the flash will be observed from Earth to last 1 full year. If the object is 2 ly in size, brightness variations will last at least 2 years as seen from Earth, and so on.

mind that a quasar gets its luminosity from gas heating up as it gets closer and closer to the black hole. If the luminosity exceeds the Eddington limit, there is so much *radiation pressure*—the pressure produced by photons streaming outward from the hot infalling material— that the surrounding gas is pushed outward rather than falling inward onto the black hole. Without a source of gas to provide energy, the luminosity naturally decreases to below the Eddington limit, at which point gas can again fall inward. Thus, *the Eddington limit represents the balance between radiation pushing gas-fuel outward and gravity pulling the gas inward.* This limit allows us to calculate the minimum mass of the black hole in a quasar.

Numerically, the Eddington limit is:

The Eddington limit

$$L_{Edd} = 30,000\left(\frac{M}{M_\odot}\right)L_\odot$$

L_{Edd} = maximum luminosity that can be radiated by accretion around a black hole

M = mass of the black hole

$M_\odot$ = mass of the Sun

$L_\odot$ = luminosity of the Sun

The tremendous luminosity of a quasar must be less than or equal to its Eddington limit (or it would blow away its accreting gas), so this limit must be very high indeed. Furthermore, quasars are very luminous, so the mass of the black hole must also be quite large to keep pulling in fuel. For example, consider the quasar 3C 273, which has a luminosity of about $3 \times 10^{13}\,L_\odot$. To calculate the minimum mass of a black hole that could continue to attract gas to power the quasar, assume that the quasar's luminosity equals the Eddington limit. Inserting $L_{Edd} = 3 \times 10^{13}\,L_\odot$ into the above equation, we find that $M = 10^9\,M_\odot$. Therefore, if a black hole is responsible for the energy output of 3C 273, its mass must be greater than a billion Suns!

Astronomers have indeed found evidence for such **supermassive black holes** at the centers of many nearby normal galaxies (see Section 22-5); these are thought to be the remnants of AGN. As we saw in Section 22-6, at the center of our own Milky Way Galaxy lies what is almost certainly a black hole of about 4.1×10^6 solar masses—supermassive in comparison to a star, but less than 1% the mass of the behemoth black hole at the center of 3C 273.

> Most or all large galaxies have supermassive black holes at their centers.

CONCEPT CHECK 24-4

Suppose a quasar's luminosity is initially right at the Eddington limit. What would happen if the quasar were somehow able to temporarily become even brighter without a significant change in its mass?

CALCULATION CHECK 24-1

What is the maximum luminosity that a 1-$M_\odot$ black hole could have, if it derived its luminosity (outside its event horizon) by heating gas during accretion? Is this larger or smaller than our Sun's luminosity?

Answers appear at the end of the chapter.

24-3 Quasar accretion disks and jets

Accretion—the capture and consumption of material—by a supermassive black hole is the most likely explanation of the immense energy output of active galactic nuclei. The challenge to astrophysicists is to understand how that accretion takes place. A successful model of the accretion process must also explain other properties of AGN, including their unusual spectra, variable light output, and energetic jets.

Accretion Disks Around Supermassive Black Holes

In the leading model of this process, at the heart of an active galaxy is a supermassive black hole surrounded by an **accretion disk**, a solar system–sized disk of matter captured by the hole's gravity and spiraling into it. We saw in Section 18-5 that accretion disks are found around stars in the process of formation. The accretion disks that astrophysicists envision around supermassive black holes are similar but far larger and far more dynamic. The *Cosmic Connections* figure depicts the physics of such accretion disks.

> The power source for active galaxies is gravitational energy released by material falling toward a central black hole.

Imagine a 1-billion-$M_\odot$ black hole sitting at the center of a galaxy, surrounded by a rotating accretion disk. According to Kepler's third law, the inner regions of this accretion disk would orbit the hole more rapidly than would the outer parts. Thus, the rapidly spinning inner regions would constantly rub against the slower moving gases in the outer regions. This friction, aided by magnetic forces within the disk, would cause the gases to lose energy and spiral inward toward the black hole.

As the gases move inward within the accretion disk, they are compressed and heated to very high temperatures. This causes the accretion disk to glow, thus producing the brilliant luminosity of an active galactic nucleus. The temperature of material in the accretion disk reaches 100,000 K or more, which emits thermal radiation primarily at UV and visible wavelengths. (By comparison, the Sun is about 5800 K and the hottest blue stars are about 50,000 K.)

In this model, the fundamental source of the energy output by an active galactic nucleus is *gravitational* energy converted into, and released as, thermal radiation by infalling material in the accretion disk. We saw in Section 20-6 that gravitational energy is also what powers the immense light output of a core-collapse supernova. The difference between such a supernova and an active galactic nucleus is that the infall of a supernova's outer layers is a one-time event, whereas gas in an AGN's accretion disk spirals inward continuously.

This model of accretion works well for smaller disks, such as those around neutron stars or stellar mass black holes. However, as much as it is the favored model, detailed accretion disk calculations for AGN are a poor fit to actual observations. Whether quasars derive their intense luminosity through accretion disks, a more complicated accretion scenario, or some other process is still unknown. Until the emission mechanism for quasars is understood, it is likely to remain on the frontier of modern research.

COSMIC CONNECTIONS
Accretion Disks

ANIMATION
24.3
Quasars and active galaxies are thought to be powered by gravitational energy released when gas falls toward a central supermassive black hole. Due to conservation of angular momentum, the gas does not fall straight into the black hole. Instead, it must negotiate its way through an accretion disk.

How an accretion disk forms around a black hole

Slowly rotating cloud

1. Gravity causes a cloud of gas to coalesce around a black hole. Initially, the cloud has some slight rotation.

Contraction is slower perpendicular to the rotation axis

Contraction is more rapid parallel to the rotation axis

The cloud rotates faster.

2. Due to conservation of angular momentum, gas moving parallel to the rotation axis contracts more quickly than gas moving perpendicular to the rotation axis. This leads to a flattening of the gas cloud.

All of the gas has flattened into a plane perpendicular to the rotation axis.

The rotational motion of the disk holds it up against gravity, and it is stable.

3. The gas cloud becomes thinner and thinner. Eventually, the gas forms a flattened disk.

Central supermassive black hole

Orbital motion of gas in disk

4. Like planets orbiting around the Sun, the closer a blob of gas is to the center of the disk, the faster its orbital speed.

Why gas in an accretion disk spirals in and radiates

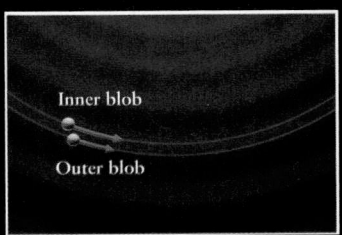

Inner blob

Outer blob

5. Consider two adjacent blobs of gas in the accretion disk. The inner blob moves a bit faster than the outer blob (see 4).

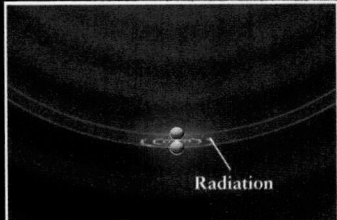

Radiation

6. When the two blobs rub past each other, energy is transferred from the inner blob to the outer blob. The heated gas emits radiation.

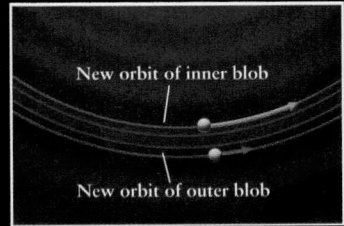

New orbit of inner blob

New orbit of outer blob

7. Having lost energy, the inner blob moves inward toward the center of the disk and speeds up. The outer blob moves outward and slows down.

8. Conservation of angular momentum slows the inward motion of the blobs of gas.

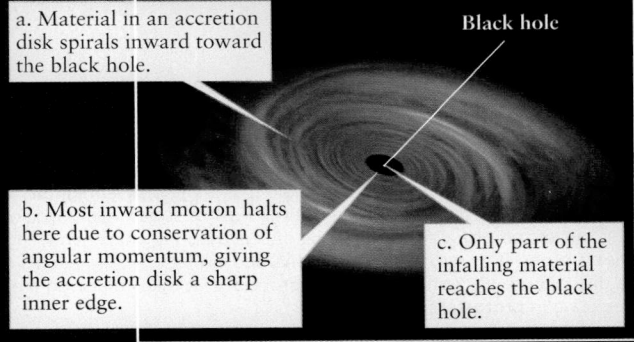

a. Material in an accretion disk spirals inward toward the black hole.

Black hole

b. Most inward motion halts here due to conservation of angular momentum, giving the accretion disk a sharp inner edge.

c. Only part of the infalling material reaches the black hole.

Michael Owen and John Blondin (North Carolina State University)

Jets of Charged Particles

Because of the constant inward crowding of hot gas in the plane of the accretion disk, pressures climb rapidly near the center. These pressures expel matter at extremely high speeds. This ejected material escapes at right angles to the plane of the accretion disk and the material forms into two jets (**Figure 24-7**). However, the active galaxy M87 only shows a single jet. Similar to a flashlight beam, we can clearly see the jet pointed toward us, but the jet pointing away is not visible (**Figure 24-8**) For AGN, the matter in the jets travels at nearly the speed of light, and this leads to beamed emission that is only visible when pointed somewhat toward the observer. For the jet we can see, the beam is pointed toward us; for the jet we don't see, another beam is pointed away.

ANIMATION 24.1 ➤ Magnetic forces play a crucial role in steering these fast-moving particles. These forces arise because the hot gases in the accretion disk are ionized, forming a plasma, and the motions of this plasma generate a magnetic field (see Section 16-9). As the plasma in the disk rotates around the black hole, it pulls the magnetic field along with it. But because the disk rotates faster in its inner regions than at its outer rim,

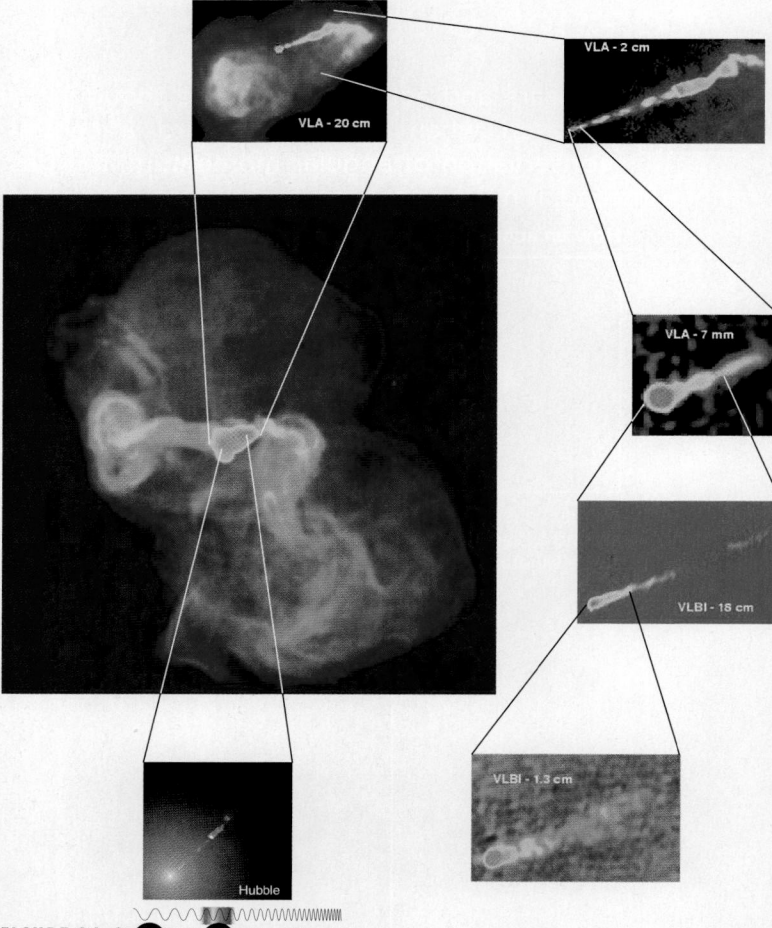

FIGURE 24-8 Ⓡ I Ⓥ U X G

Giant Elliptical Galaxy M87   M87 is located near the center of the sprawling, rich Virgo Cluster, which is about 50 million ly from Earth. Embedded in this radio image of gas is the galaxy M87 from which the gas has been ejected (bottom inset). The bottom inset is a visible light image from the Hubble Space Telescope, and the rest are at radio wavelengths. As the images zoom into smaller scales, they reveal a variety of details about the structure of the jets of gas. M87's extraordinarily bright nucleus and the gas jets result from a 3-billion-$M_\odot$ black hole, whose gravity causes huge amounts of gas and an enormous number of stars to crowd around it. (NASA and The Hubble Heritage Team [STScI/AURA]; inset: F. N. Owen [NRAO] et al., VLA, NRAO, AUI)

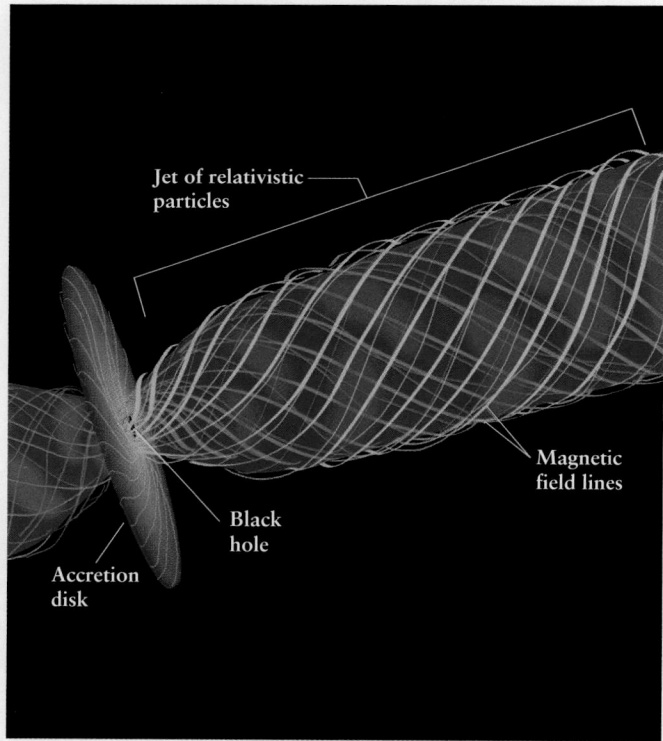

FIGURE 24-7

Jets from a Supermassive Black Hole   The rotation of the accretion disk surrounding a supermassive black hole twists the disk's magnetic field lines into a helix. The field then channels the flow of subatomic particles pouring out of the disk. Over a distance of less than 1 ly, this channeling changes a broad flow into a pair of tightly focused jets, one on each side of the disk. Figure 18-17 shows a similar process that takes place on a much smaller scale in protostars. (NASA)

the magnetic field becomes severely twisted. This twisted field forms two helix shapes, one on either side of the plane of the disk.

Charged particles flowing outward from the accretion disk tend to follow these magnetic field lines. The result is that the outflowing beams of particles are focused into two jets oriented perpendicular to the plane of the accretion disk. The chapter-opening figure is an artist's conception of the accretion disk and jets. The highly collimated jets shoot well beyond the host galaxy, and very diffuse gas in the space between galaxies spreads the jets into radio lobes, as in Figure 24-1. Jets are primarily observed at radio wavelengths, but some jets can also be seen at visible wavelengths and even x rays (**Figure 24-9**). It is not known if the charged particles in

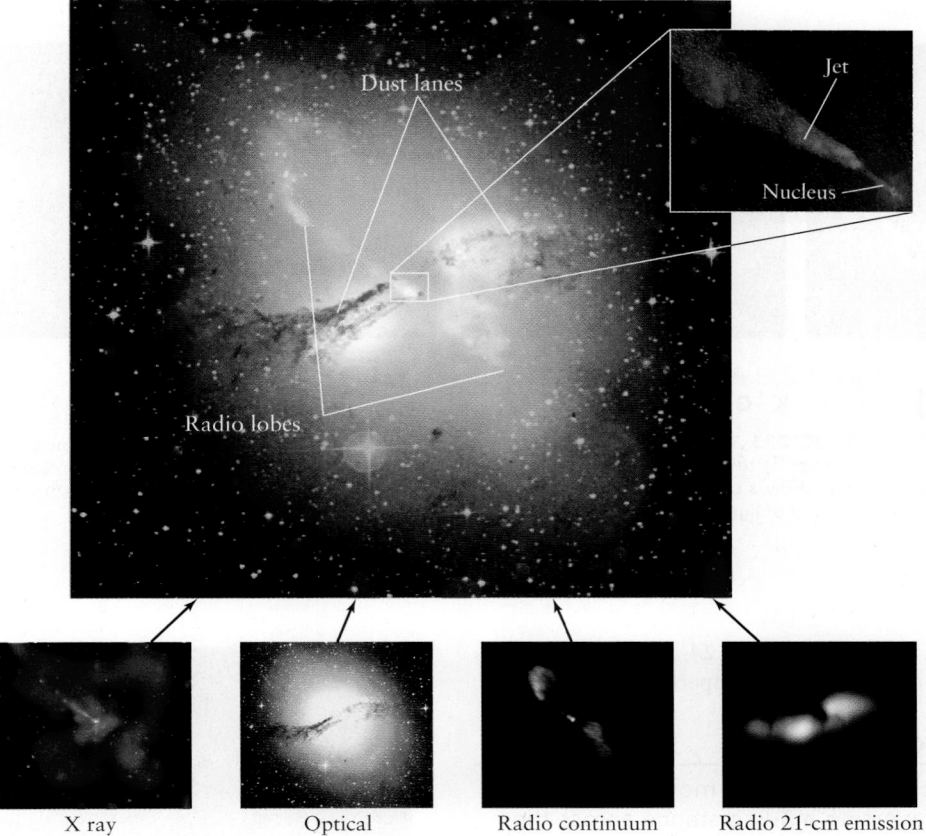

FIGURE 24-9 R I V U X G

Peculiar Galaxy NGC 5128 (Centaurus A) This extraordinary radio galaxy is located in the constellation Centaurus, 11 million ly from Earth. At visible wavelengths, a dust lane crosses the face of the galaxy. Superimposed on this visible image is a false-color radio image (green) showing that vast quantities of radio emission pour from matter ejected from the galaxy perpendicular to the dust lane, along with radio emission (rose-colored) along the dust lane, and x-ray emission (blue) detected by NASA's Chandra X-ray Observatory. The x rays may be from material ejected by the black hole or from the collision of Centaurus A with a smaller galaxy. Upper right inset: This x-ray image from the Einstein Observatory shows that NGC 5128 has a bright x-ray nucleus. An x-ray jet protrudes from the nucleus along a direction perpendicular to the galaxy's dust lane. (X ray: NASA/CXC/SAO/M.Karovska et al.; optical: NASA/Digital Sky Survey/U.K. Schmidt Image/STScI; radio continuum: NRAO/AUI/NSF/J. Condon et al.; radio 21-cm: NRAO/VLA/J. Van Gorkom/ Schminovich et al.; inset: X ray: NASA/CXC/Bristol U./M. Hardcastle et al.; radio: NRAO/AUI/NSF/Bristol U./M. Hardcastle; x ray [NASA/CXC/M. Karovska et al.])

jets are composed of electrons and protons, or electrons and their antimatter partners, positrons.

CONCEPT CHECK 24-5

In the accretion disk model, what happens when an inner blob of gas passes an outer blob of gas during accretion (see the *Cosmic Connections* figure)? How is light produced in an accretion disk?

Answer appears at the end of the chapter.

The Strange Case of Superluminal Motion

Nothing travels faster than the speed of light, but if something did, it would be called **superluminal motion**. Initially quite a mystery, some quasars seemed to show superluminal motion when looking very closely at "blobs" of radio-emitting particles near the very base of their radio jets. For example, Figure 24-10 shows four high-resolution images of the quasar 3C 273 spanning 3 years. During this interval, a blob moved away from the quasar at a rate of almost 0.001 arcsec per year. Taking into account the distance to the source, this rate of angular separation corresponds to a speed 10 times that of light!

Was Albert Einstein wrong that nothing travels faster than the speed of light? No. The blobs in the jets move *near* the speed of light—these are called relativistic jets—but never faster than light. Astronomers soon realized that apparent super-luminal motion can be explained by movement slower than light—once we take into account the *angle* at which we view the relativistic jet.

In order to see material that appears to move faster than the speed of light, two conditions must be met: The beam of material must be aimed close to your line of sight (within about 5° or so), and material in the jet must travel near the speed of light. Given the number of quasars with radio jets, it is not surprising that a

July 1977   March 1978   July 1979   July 1980

VIDEO 24.1

FIGURE 24-10 Ⓡ I V U X G

Superluminal Motion in 3C 273 These four images are false-color, high-resolution radio maps of the quasar 3C 273 (shown in the image that opens this chapter). They show a blob that seems to move away from the quasar at 10 times the speed of light. In fact, a beam of relativistic particles from 3C 273 is aimed almost directly at Earth, giving the illusion of faster-than-light motion. Each image is about 7 milliarcsec (0.007 arcsec) across. (DR S C UNWIN/Science Source)

few just happen to point toward us. **Figure 24-11** shows how jets and their geometry can produce apparent superluminal motion.

CONCEPT CHECK 24-6

In Figure 24-11, a bright blob appears to move faster than the speed of light by traversing a 3-ly distance across the sky in only 2 ly. How far did the blob actually travel, and how long did it take?

Answer appears at the end of the chapter.

24-4 The unified model explains much of the diversity of AGN

Untangling the mystery of active galactic nuclei was extraordinarily difficult because AGN can appear very different from each other. In fact, it took several decades before researchers determined that the diverse group of AGN could all be "unified" into one physical model. **Box 24-1** describes some different properties of AGN and their naming conventions. For example, one AGN, such as 3C 273, might have a visible spectrum with unusually strong and broad hydrogen emission lines (Figure 24-3) and show strong variability

FIGURE 24-11

An Explanation of Superluminal Motion (a) If a blob of material ejected from a quasar moves at 5/6 of the speed of light, it covers the 5 ly from point A to point B in 6 years. In the case shown here, it moves 4 ly toward Earth and 3 ly in a sideways, transverse direction. The light emitted by the blob at A reaches us in 2014. The light emitted by the blob at B reaches us in 2016. The light left the blob at B 6 years later than the light from A but had 4 fewer light-years to travel to reach us. (*Note:* The superluminal effect is only significant when the jet is pointing within about 5° of our line of sight, but the drawing here shows a larger angle in order to make room for labels.) **(b)** From Earth, we can see only the blob's sideways, transverse motion across the sky, as in Figure 24-10. It appears that the blob has traveled 3 ly in just 2 years, so its apparent speed is 3/2 of the speed of light, or 1.5*c*.

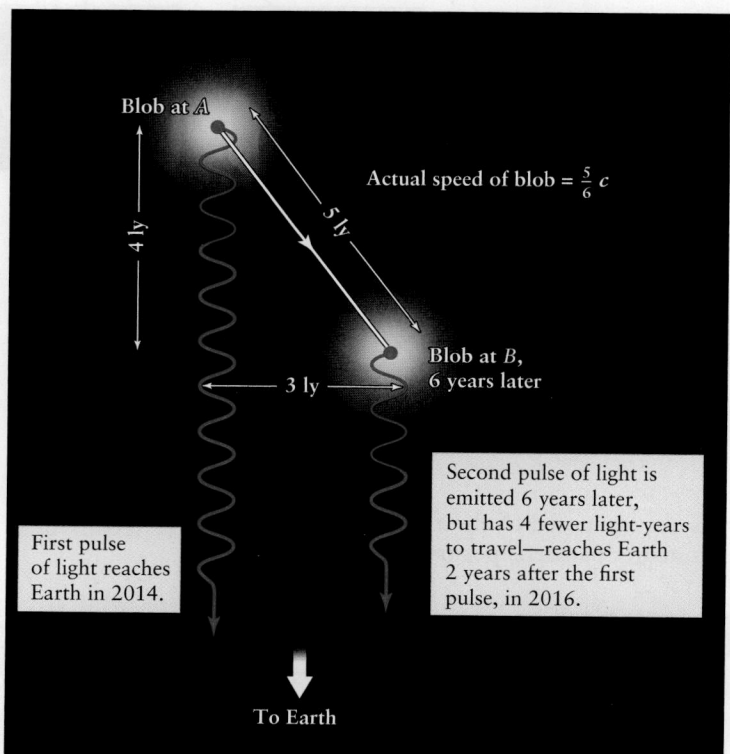

Blob at *A*

Actual speed of blob = $\frac{5}{6}c$

4 ly

5 ly

3 ly

Blob at *B*, 6 years later

First pulse of light reaches Earth in 2014.

Second pulse of light is emitted 6 years later, but has 4 fewer light-years to travel—reaches Earth 2 years after the first pulse, in 2016.

To Earth

(a) View from above

2014   2016

3 ly

Apparent speed of blob = 1.5*c*

(b) View from Earth

BOX 24-1   TOOLS OF THE ASTRONOMER'S TRADE

The Diversity of AGN

Active galactic nuclei, or AGN, come in two unique varieties: those with radio jets and lobes (called *radio loud*), and those without (called *radio quiet*). Radio galaxies, like those in Figures 24-1 and 24-8, are radio-loud AGN. About 10% of quasars are radio loud, and, when observable, their host galaxies are elliptical. The remaining quasars are radio quiet and, when observable, their hosts are spiral galaxies. (It is not understood why radio-loud and radio-quiet AGN are each found in ellipticals and spirals, respectively—this remains a mystery.)

TUTORIAL 24.1 The light emitted from jets is focused into a strong beam because the jets' emitting plasma travels near the speed of light. In cases where the jets are pointed almost directly toward Earth, the jets' beamed emission swamps the usual AGN emission, and these objects have a very different spectrum (often without any detected emission lines). These objects are called **blazars**, and while originally thought to be distinct from quasars due to their different appearance, they are now understood as special cases of jet alignment with our line of sight from Earth.

Quite similar to quasars, **Seyfert galaxies** also show bright, compact nuclei with emission lines. Eventually, it was realized that both quasars and Seyfert galaxies contain the same type of object, and both are examples of AGN. Historically, bright and distant AGN, where the host galaxies were not initially observed, are called quasars; nearby cases where the host galaxies can be easily seen are called Seyfert galaxies.

Seyfert galaxies and quasars also come in two further varieties: those with strong broad hydrogen emission lines (called Type 1) and those with weak narrow hydrogen emission lines (called Type 2). Furthermore, the continuum emission in the Type 1 AGN can vary significantly in brightness, whereas the continuum in the Type 2 AGN does not vary.

Explaining Type 1 and Type 2 AGN in terms of the *same central engine* was a major advance in understanding quasars, and the explanation is called the unified model. Ultimately, appearing as Type 1 or Type 2 depends on the line of sight, as shown in Figure 24-13. The table below summarizes some properties of AGN; the modern distinction of Type 1 or Type 2 is in parenthesis after the object's name.

Properties of Active Galactic Nuclei

Object (Type)	Found in which type of galaxy	Strength of radio emission	Type of emission lines in spectrum	Intrinsic luminosity (watts)	(Milky Way Galaxy = 1)
Blazar (1)	Elliptical	Strong	None	10^{38} to 10^{42}	10 to 10^5
Radio-loud quasar (1)	Elliptical	Strong	Broad	10^{38} to 10^{42}	10 to 10^5
Radio galaxy (2)	Elliptical	Strong	Narrow	10^{36} to 10^{38}	0.1 to 10
Radio-quiet quasar (1)	Spiral or elliptical	Weak	Broad	10^{38} to 10^{42}	10 to 10^5
Type 1 Seyfert	Spiral	Weak	Broad	10^{36} to 10^{38}	0.1 to 10
Type 2 Seyfert	Spiral	Weak	Narrow	10^{36} to 10^{38}	0.1 to 10

(see Figure 24-5). Another AGN might have very weak and narrow hydrogen emission lines with little variability. Given the difficulty of explaining how one type of "central engine" powering their luminosity could explain such diverse behavior, these observations suggested—incorrectly—that these two AGN contain different emission mechanisms. In fact, the **unified model** explains how different AGN, at their core, are fundamentally the same.

The astronomer Robert Antonucci (now at University of California, Santa Barbara) put together the main picture of the unified model and uncovered its key ingredient: a thick dusty doughnut-shaped torus around the central engine's accretion disk. The presence of such a torus seems to be a natural result of the overall accretion process. One such dusty torus is shown in Figure 24-12. Rather than emitting visible light, the dusty torus can *block* light from passing through it from the central engine. In

contrast to the torus, the accretion disk within the torus is much too small to be seen—its emission appears pointlike—so no structural features of an accretion disk have ever been imaged.

If there were no dusty torus around an accreting supermassive black hole, an observer could view such a black hole from any angle and see the intense radiation from the accretion disk. But from certain angles, the torus blocks the view of the accretion disk. This idea offers a single explanation for several types of seemingly different active galaxies.

Figure 24-13 illustrates the unified model for a luminous active galactic nucleus with radio jets. If an imaginary observer looks straight down the axis of the jet (#1), the observed radiation is dominated by the jet and has a unique spectrum with emission lines often undetected. (When viewed from this angle, an AGN is a blazar, as described in Box 24-1.)

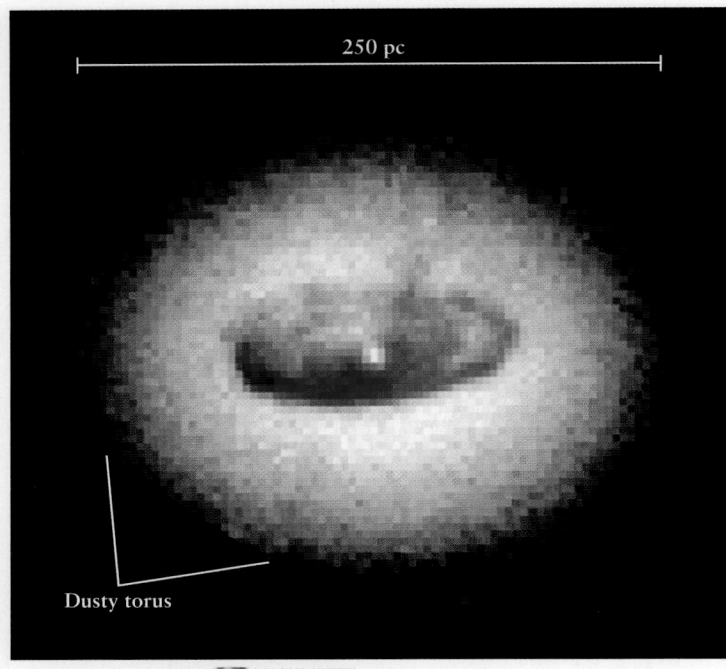

250 pc

Dusty torus

FIGURE 24-12   R I V U X G

A Dusty Torus Around a Supermassive Black Hole This immense doughnut of dust and gas orbits the black hole at the center of the active galaxy NGC 4261. Radio jets emerge perpendicular to the plane of this torus (see Figure 21-26). (H. Ford, L. Ferrarese [JHU]/NASA)

At an intermediate angle (#2), the observer gets a clear view of the luminous accretion disk and the turbulent region around the black hole. Because gases move at many different velocities in this turbulent region, the observer sees hydrogen spectral lines (and other lines) that have been broadened by the Doppler effect. The observer also sees intense thermal radiation from the hot accretion disk. This line of sight, and the resulting features of the emission, account for the properties of quasar 3C 273 and the majority of quasars. (The Type 1 objects referred to in Box 24-1 are viewed from these intermediate angles.)

If the observer looks nearly edge on at the torus (#3), the accretion disk is completely hidden. Some of the light reaching the observer comes from hot gas flowing out of the accretion disk, and this light has an emission-line spectrum. But given the edge-on orientation, this gas is not moving rapidly toward or away from the observer, so there is little Doppler shift and the hydrogen emission lines are narrow. (From this edge-on view, the AGN would be a Type 2 object as described in Box 24-1.) No radio jets are visible at this angle, but diffuse radio lobes are visible.

Unlike our imaginary observer, we cannot move the vast distances through space that would be needed to view a given active galaxy from different angles. Instead, we may see a given active galaxy differently, depending on how the accretion disk and torus are oriented to our line of sight. In other words, "What you see depends on the line of sight you get!"

CAUTION!

Note that we are really making use of *two* different but complementary models here. The unified model says that different types of active galactic nuclei are really different views of the same type of "central engine," while the accretion-disk model explains how the "central engine" works.

CONCEPT CHECK 24-7

Is the large object in Figure 24-12 an accretion disk? Does it emit intense visible light?

CONCEPT CHECK 24-8

From which viewing angle in Figure 24-13 would an AGN appear bright in visible light and show broad emission lines?

Answers appear at the end of the chapter.

Supermassive black hole
with accretion disk and jets

Receding jet

Approaching jet

Torus

1. This observer sees emission dominated by the jet.

2. This observer sees emission from the accretion disk.

3. The torus blocks this observer's view of the accretion disk, but diffuse radio lobes are still visible.

ANIMATION **FIGURE 24-13**
24.2
A Unified Model of Active Galaxies The AGN model consists of a supermassive black hole surrounded by an intensely luminous accretion disk. Farther out is a dusty doughnut-shaped torus. A pair of jets is also shown on this object, although not every AGN has jets. While this basic structure is at the heart of the randomly oriented AGN in space, each AGN can appear different depending on the viewing angle.

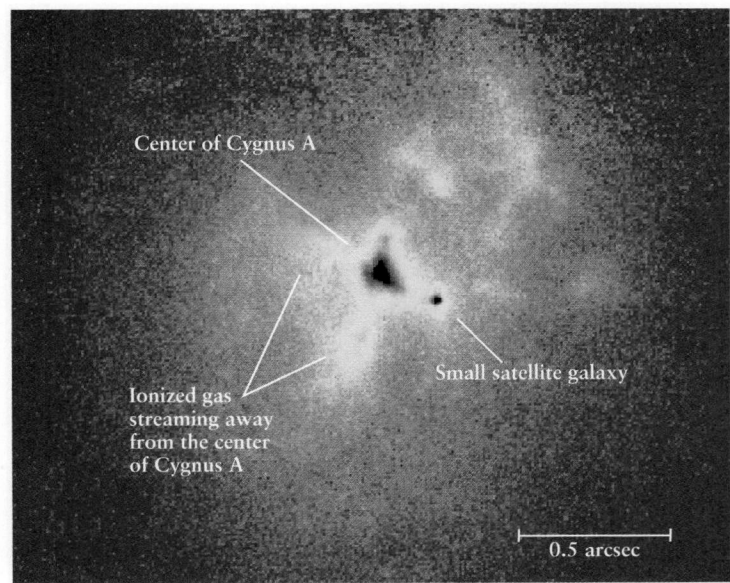

FIGURE 24-14 R I **V** U X G

Providing Fresh "Fuel" for a Supermassive Black Hole A small satellite galaxy has fallen into the central regions of the radio galaxy Cygnus A (see Figure 24-1). Its material may eventually accrete onto the supermassive black hole at Cygnus A's very center. This false-color extreme close-up was made using adaptive optics, described in Section 6-3. (NASA/ESA Hubble Space Telescope: Gabriela Canalizo, Claire Max, David Whysong, Robert Antonucci, Scott E. Dahm)

24-5 The evolution of AGN leads to very different behavior over time

The accretion-disk idea can explain what happens when active galaxies collide or merge with other galaxies. Such collisions and mergers can transfer gas and dust from one galaxy to another, providing more "fuel" for a supermassive black hole. For example, the image in **Figure 24-14** shows a close-up view of a supermassive

black hole about to take in more "fuel" from a merger. Thus, it is not surprising that many of the most luminous active galaxies are also those that have recently undergone collisions, such as the radio galaxy Centaurus A shown in **Figure 24-15**.

The accretion-disk idea also helps to explain why there are no nearby quasars (see Figure 24-4). Over time, much of the available gas and dust surrounding a quasar passes into the accretion disk and then enters the black hole. Eventually, the central engine has less and less infalling matter to act as fuel, and the quasar becomes inactive. Since quasars started shining in the early universe (about 12 billion years ago), and they consumed their substantial fuel rather quickly (in a few hundred million years), the only quasars visible today in our expanding universe are very far away at large redshifts.

On the other hand, the smaller amounts of fuel available today do produce some lower-luminosity AGN observed nearby at lower redshifts. While the unified model brings together many objects that only appear to be different, there is a real range in the intrinsic luminosity of AGN as well.

Dead Quasars and Flares

When a quasar runs out of accreting fuel altogether and becomes inactive, some astronomers call the leftover supermassive black hole a **dead quasar**. The vast majority of galaxies appear to have supermassive black holes at their centers that are not presently active. Therefore, dead quasars seem to be in almost all luminous galaxies, so most galaxies once led much more dramatic lives as quasars when they were young.

Even our own Milky Way Galaxy was probably an active galaxy in the past. We have a black hole with a mass of about $4.1 \times 10^6 \, M_\odot$, which is small compared to the black holes in the brightest AGN. Therefore, our active phase probably wasn't very bright, and once the fuel was consumed by the black hole, we stopped hosting an AGN. Interestingly, in about 4 billion more years, our merger with the Andromeda Galaxy (see Figure 23-30) might yield more fuel and another phase of accretion and emission as an active galaxy.

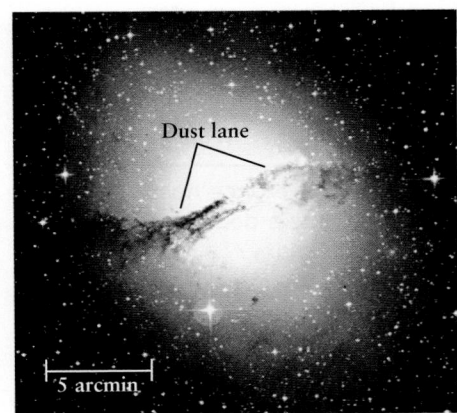

(a) Centaurus A: light from stars

R I **V** U X G

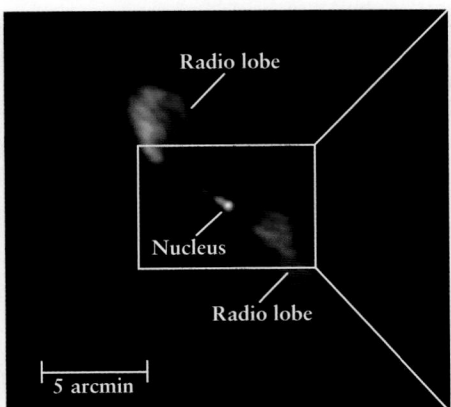

(b) Centaurus A: radio lobes

R I V U X G

(c) An x-ray–emitting jet emanates from the nucleus

R I V U **X** G

FIGURE 24-15

The Radio Galaxy Centaurus A **(a)** This elliptical galaxy, called NGC 5128, lies about 4 Mpc (13 million ly) from Earth at the location of the radio source Centaurus A. The dust lane is evidence of a collision with another galaxy. **(b)** At radio wavelengths, we see synchrotron radiation from

two radio lobes centered on the galaxy's nucleus. **(c)** A luminous jet extends from the nucleus directly toward one of the lobes. (a: NASA/Digital Sky Survey/U.K. Schmidt Image/STScI; b: NRAO/AUI/NSF/J. Condon et al.; c: R. Kraft et al./SAO/NASA)

1. A Sun-like star on an eccentric orbit plunges toward the supermassive black hole in the heart of a distant galaxy.

2. Strong tidal forces near the black hole increasingly distort the star. If the star passes too close, it is ripped apart.

3. The part of the star facing the black hole streams toward it and forms an accretion disk. The remainder of the star just expands into space.

4. Near the black hole, magnetic fields power a narrow jet of particles moving near the speed of light. Viewed head on, the jet is a brilliant x-ray and radio source.

FIGURE 24-16

Dead Quasar Flare A star wandering too close to a supermassive black hole can provide temporary fuel for the dead quasar. Modeling indicates that this event in 2011, called Swift J1644+57, occurred around a black hole about twice the mass of the 4-million-$M_\odot$ black hole at the center of our Milky Way Galaxy. The x-ray emission from this flare lost most of its initial intensity over a period of several months. (NASA/Goddard Space Flight Center/Swift)

Beyond our own galaxy, is there a way to test the idea that most galaxies harbor a supermassive black hole? Yes. Fortunately, there is a chance that a dead quasar can light up once again, even if only briefly. If a star wanders too close to the supermassive black hole, tidal forces (which act to stretch the star) can tear the star apart and pull in some of its matter. This newly acquired stellar gas acts as temporary accretion fuel for the otherwise dormant black hole. Models of this process suggest that it might occur about once every 10,000 years in a typical galaxy. As the star's matter is accreted and consumed by the black hole over a period of a year or so, it can shine brightly and is called a **dead quasar flare**. Contrast this approximate 1-year timescale for bright emission with the minimum age for quasars of about 10,000 years; that's why the shorter event is called a flare.

Very few events provide convincing evidence for a flare. After all, there are other objects that can give a burst of light. One case for a flare came in March 2011, when a burst of x rays (named Swift J1644+57) was observed at the center of a galaxy. This flare lasted only 2 days, but was observed from beginning to end. The event was much brighter and shorter than expected for a typical flare, and astronomers think the emission came from temporarily formed jets, as opposed to an accretion disk (**Figure 24-16**).

Much is still unknown about quasars. How do they form? Does their visible light really come from an accretion disk? Why do some AGN have strong radio emission, while others do not? What happens if a star wanders too close to a dead quasar—does its captured gas form an accretion disk or jets? All of these questions are part of ongoing research in the field of active galaxies.

CONCEPT CHECK 24-9

Could our own Milky Way Galaxy have once hosted a quasar? If so, are we likely to witness a dead quasar flare from the center of our Milky Way Galaxy?

Answer appears at the end of the chapter.

KEY WORDS

Words preceded by an asterisk (*) are discussed in Box 24-1.

accretion, p. 770
accretion disk, p. 772
active galactic nucleus
  (AGN), p. 769

active galaxy, p. 770
*blazar, p. 777
dead quasar, p. 779
dead quasar flare, p. 780

Eddington limit, p. 771
quasar, p. 768
radio galaxy, p. 770
*Seyfert galaxy, p. 777

superluminal motion, p. 775
supermassive black hole,
  p. 772
unified model, p. 777

KEY IDEAS

Quasars or active galactic nuclei (AGN): A quasar looks like a star but has a huge redshift. According to the Hubble law, these redshifts show that quasars are typically billions of light-years away from Earth (Section 24-1).

- To be seen at such large distances, quasars must be very luminous, typically about 1000 times brighter than an ordinary galaxy.

- The peak of quasar activity took place when the universe was just over 2 billion years old; the era of quasar activity has long since passed.

Supermassive black holes power quasars: At the center of a quasar is a black hole with millions or even billions of solar masses (Section 24-2).

- Rapid fluctuations in the brightness of quasars indicate that the region that emits radiation is quite small; the only known way to produce so much energy in such a small space is with a black hole.

- The Eddington limit relates the mass of a black hole to the maximum luminosity it can emit from its surroundings; if more luminous, radiation pressure would push away accreting gas and shut off the black hole's fuel source.

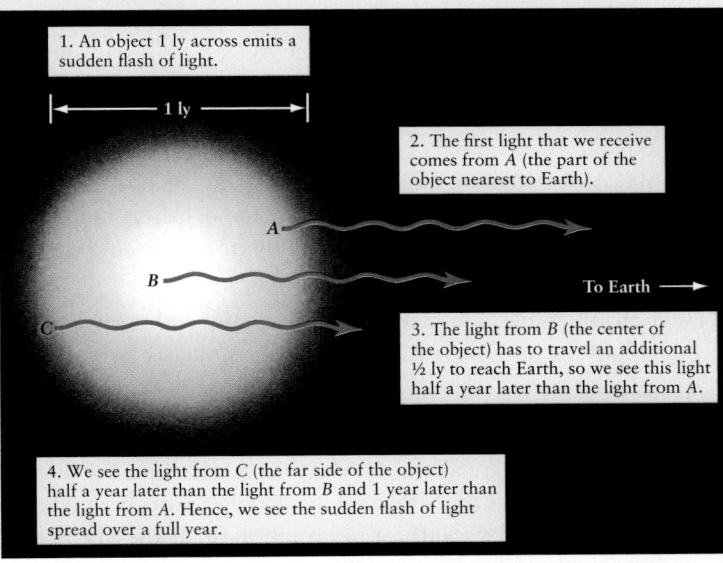

Accretion and jets around black holes: Hot gas accreting around the black hole is thought to give a quasar its intense luminosity. About 10% of AGN also have strong radio emission coming from jets, as well as lobes of radio emission as the jets collide with surrounding gas on scales much larger than the host galaxy (Section 24-3).

- Models for a thin accretion disk describe heat and emission produced as portions of gas on different orbits rub past each other, although no specific model of accretion has been verified.

- As gases spiral in toward the supermassive black hole, some of the gas may be redirected to form two jets of high-speed particles that are aligned perpendicularly to the accretion disk. When a jet is directed toward us, we can observe apparent superluminal motion.

NASA/CXC/M. Weiss

Unified model of AGN: A thick dusty doughnut-shaped torus surrounds a quasar that, when viewed edge on, blocks a direct view of the luminous accreting gas (Section 24-4).

- A quasar is visible only if our viewing angle allows us to see the region in the center of the torus.

- When viewing the torus somewhat edge on, the most luminous part of a quasar is blocked. Some emission is still observed, and the object appears as an active galaxy with a different emission spectrum.

- While the viewing angle determines what we observe, the central engine—an accreting black hole—is fundamentally the same. This insight unifies a diverse group of AGN.

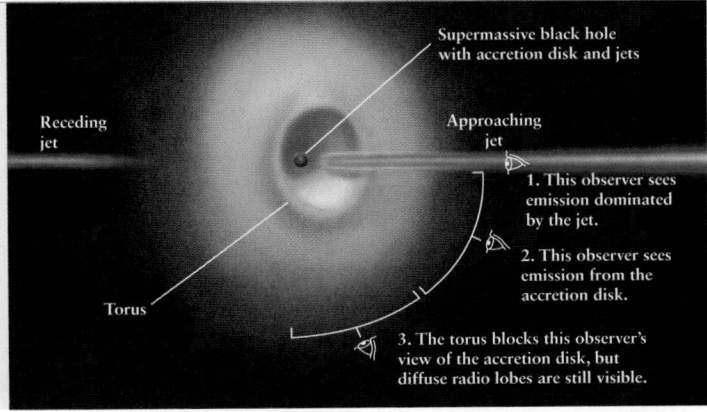

QUESTIONS

> **Problem-solving tips and tools**
>
> Box 21-1 describes time dilation in the special theory of relativity. You may find it useful to recall from Section 1-7 that 1 ly = 63,240 au. A speed of 1 km/s is the same as 0.211 au/yr. You can find Newton's form of Kepler's third law in Box 4-4 and the formula for the Schwarzschild radius in Box 21-3.

Section 24-1

1. When quasi-stellar radio sources were first discovered and named, why were they called *quasi-stellar*?
2. How were quasars first discovered? How was it known that they are very distant objects?
3. Suppose you saw an object in the sky that you suspected might be a quasar. What sort of observations might you perform to test your suspicion?
4. How does the spectrum of a quasar differ from that of an ordinary galaxy? How do spectral lines help astronomers determine the distances to quasars?
5. Since quasars lie at the centers of galaxies, why don't we see strong absorption lines from a galaxy's stars when we look at the spectrum of a quasar (like that shown in Figure 24-3)?
6. It was suggested in the 1960s that quasars might be compact objects ejected at high speeds from the centers of nearby ordinary galaxies. Explain why the absence of blueshifted quasars disproves this hypothesis. (We see only quasars with redshifted lines.)
7. How do astronomers know that quasars are located in galaxies? Are the host galaxies easy to observe for distant quasars?
8. What is a radio galaxy?
9. When we observe a quasar with redshift $z = 1.0$, about how far into the past are we looking (see Figure 24-4)? If we could see that object as it really is right now (that is, if the light from the object could somehow reach us instantaneously), would it still look like a quasar? Explain why or why not.

Section 24-2

10. What do the brightness fluctuations of a particular active galaxy tell us about the size of the energy-emitting region within that galaxy?
11. How could a supermassive black hole, from which nothing—not even light—can escape, be responsible for the extraordinary luminosity of a quasar?
12. What is the Eddington limit? Explain how it can be used to set a limit on the mass of a supermassive black hole, and explain why this limit represents a minimum mass for the black hole.
13. (a) Calculate the maximum luminosity that could be generated by accretion onto a black hole of 3.7×10^6 M$_\odot$. (This is the size of the black hole found at the center of the Milky Way.) Compare this to the total luminosity of the Milky Way, about 2.5×10^{10} L$_\odot$. (b) Speculate on what we might see if the center of our galaxy became an active galactic nucleus with the luminosity you calculated in (a).

Section 24-3

14. Some quasars appear to be ejecting material at speeds faster than light. Is the material really moving that fast? If so, how is this possible? If not, why does the material appear to be traveling so fast?
15. How does matter falling inward toward a central black hole find itself being ejected outward in a high-speed jet?
16. Suppose the distance from point A to point B in Figure 24-11a is 26 ly and that the blob moves at $^{13}/_{15}$ of the speed of light. As the blob moves from A to B, it moves 24 ly toward Earth and 10 ly in a sideways, transverse direction. (a) How long does it take the blob to travel from A to B? (b) If the light from the blob at A reaches Earth in 2020, in what year does the light from B reach Earth? (c) As seen from Earth, at what speed does the blob appear to move across the sky?

Section 24-4

17. **Box 24-1** What is the main difference between a Type 1 and a Type 2 AGN?
18. In the unified model, why do some AGN have broad spectral lines while others do not?
19. **Box 24-1** How does the unified model of active galaxies explain that quasars and blazars are the same kind of object viewed from different angles?
20. **Box 24-1** According to the unified model, is a radio-loud quasar an object that can be seen only from a specific viewing angle or from a range of viewing angles?
21. **TUTORIAL 24.1** What is a blazar? What is unique about its spectrum? How is it related to other active galaxies?
22. What is a Type 1 Seyfert galaxy? Does the active nucleus of a Type 1 Seyfert galaxy differ from that of a quasar?
23. Suppose there are aliens out there with telescopes looking at the same distant AGN that we are. Furthermore, pretend that Earthlings observe diffuse radio lobes. (a) If the aliens also live in our Milky Way Galaxy, can they see emission dominated by a jet? (b) If the aliens live in a different galaxy, could they see emission dominated by a jet?

Section 24-5

24. What is it about the fueling of quasars that explains why there are no quasars relatively near to our galaxy?
25. Figure 24-14 shows ionized gas streaming away from the "central engine" of the radio galaxy Cygnus A. Instead of spreading outward equally in all directions, the gas appears to be funneled into two oppositely directed cones. Discuss how this could be caused by a dusty torus surrounding a supermassive black hole at the center of Cygnus A.

General Problems

26. Suppose a blazar at $z = 1.00$ goes through a fluctuation in brightness that lasts 1 week (168 hours) as seen from Earth. (**a**) At what speed does the blazar seem to be moving away from us? (**b**) Using the idea of time dilation, determine how long this fluctuation lasted as measured by an astronomer within the blazar's host galaxy. (**c**) What is the maximum size (in au) of the region from which this blazar emits energy?

27. Observations of a certain galaxy show that stars at a distance of 16 pc from the center of the galaxy orbit the center at a speed of 200 km/s. Use Newton's form of Kepler's third law to determine the mass of the central black hole.

28. Calculate the Schwarzschild radius of a 10^9-$M_\odot$ black hole. How does your answer compare with the size of our solar system (given by the diameter of Pluto's orbit)?

29. Figure 24-15 shows the double radio source Centaurus A. Is it possible that somewhere in the universe there is an alien astronomer who observes this same object as a blazar? Explain your answer with a drawing showing the relative positions of Earth, the alien astronomer, and Centaurus A.

30. The accompanying image from the Very Large Array (VLA) shows the radio galaxy 3C 75 in the constellation Cetus. This galaxy has several radio-emitting jets. High-resolution optical photographs reveal that the galaxy has two nuclei, which are the two red spots near the center of the VLA image. Propose a scenario that might explain the appearance of 3C 75.

R I V U X G

31. Some quasars show several sets of absorption lines whose redshifts are less than the redshifts of their emission lines. For example, the quasar PKS 0237-23 has five sets of absorption lines with redshifts in the range from 1.364 to 2.202, whereas the quasar's emission lines have a redshift of 2.223. Propose an explanation for these sets of absorption lines.

32. The Milky Way Galaxy is in the process of absorbing the satellite galaxy called the Canis Major Dwarf (see Section 23-6). Discuss whether this process could cause the Milky Way to someday become an active galaxy.

WEB/E-BOOK QUESTIONS

33. Search the internet for information about "microquasars." These are objects that are found *within* the Milky Way Galaxy. How are they detected? What are the similarities and differences between these objects and true quasars? Are they long-lasting or short-lived?

34. The Lockman hole is a region in the constellation Ursa Minor where the Milky Way's interstellar hydrogen is the thinnest. By observing in this part of the sky, astronomers get the clearest possible view of distant galaxies and quasars. Search the internet for information about observations of the Lockman hole. What has been learned through x-ray observations made by spacecraft such as ROSAT and XMM-Newton?

ACTIVITIES

Observing Projects

35. Use a telescope with an aperture of at least 20 cm (8 in.) to observe the Seyfert Galaxy NGC 1068 (also known as M77). Located in the constellation Cetus (the Whale), this galaxy is most easily seen from September through January. The epoch 2000 coordinates are R.A. = $2^h\, 2.7^m$ and Decl. = $-0°\, 01'$. Sketch what you see. Is the galaxy's nucleus diffuse or starlike? How does this compare with other galaxies you have observed?

36. Use a telescope with an aperture of at least 20 cm (8 in.) to observe the two companions of the Andromeda Galaxy, M32 and M110. Both are small elliptical galaxies, but only M32 is suspected of harboring a supermassive black hole. They are located on opposite sides of the Andromeda Galaxy and are most easily seen from September through January. The epoch 2000 coordinates are

Galaxy	Right ascension	Declination
M32 (NGC 221)	$0^h\, 42.7^m$	$+40°\, 52'$
M110 (NGC 205)	$0^h\, 40.4^m$	$+41°\, 41'$

Make a sketch of each galaxy. Can you see any obvious difference in the appearance of these galaxies? How does this difference correlate with what you know about these galaxies?

37. If you have access to a telescope with an aperture of at least 30 cm (12 in.), you should definitely observe the Sombrero Galaxy, M104, which is suspected to have a supermassive black hole at its center. It is located in Virgo and can most easily be seen from March through July. The epoch 2000 coordinates are R.A. = $12^h\, 40.0^m$ and Decl. $-11°\, 37'$. Make a sketch of the galaxy. Can you see the dust lane? How does the nucleus of M104 compare with the centers of other galaxies you have observed?

38. If you have access to a telescope with an aperture of at least 30 cm (12 in.), observe M87 and compare it with the other two giant elliptical galaxies, M84 and M86, that dominate the central regions of the Virgo Cluster. If you have access to the *Starry Night*™ program, use it to help you plan your observations. The epoch 2000 coordinates are:

Galaxy	Right ascension	Declination
M84 (NGC 4374)	12^h 25.1^m	+12° 53′
M86 (NGC 4406)	12^h 26.2^m	+12° 57′
M87 (NGC 4486)	12^h 30.8^m	+12° 24′

39. If you have access to a telescope with an aperture of at least 40 cm (16 in.), you might try to observe the brightest-appearing quasar, 3C 273, which has an apparent magnitude of nearly +13. It is located in Virgo at coordinates R.A. = 12^h 29^m 07^s and Decl. +2° 3′ 8″.

Starry Night™ Explorations

40. Most quasars are star-like in appearance, but their spectra show very large redshifts. If these redshifts are cosmological (i.e., caused by the expansion of the universe), and quasars obey Hubble's law, then their large redshifts show that they must be very far away. Since they appear bright in our sky, they must be very luminous, with outputs equivalent to that of billions and even trillions of Suns. Quasars are part of a group of objects known as Active Galactic Nuclei (AGN). The source of energy output of these very luminous objects is probably a supermassive black hole at the object's center.

    (a) To explore these objects, select **Favorites > Explorations > Atlas** and open the **Observing Lists** panel. Locate and expand the **Quasars** observing list and double-click the entry for **Quasar 2345** in this list. Zoom in to a field of view about 15′ wide. Using the HUD, how far is this pair of quasars from Earth?

    (b) Using a Hubble constant of 73 km/s/Mpc in the Hubble equation, calculate the expected recessional velocity of this quasar pair (1 pc = 3.26 ly and 1 Mpc = 10^6 pc). What is the ratio of the recessional velocity of these quasars to the speed of light?

    (c) Quasar pair 2345 have a redshift $z = 2.15$. Recall that z is the ratio of the shift in the wavelength of the light to the emitted wavelength, $z = \Delta\lambda/\lambda_0 = (\lambda - \lambda_0)/\lambda_0$, where λ_0 is the emitted or unshifted wavelength, and λ is the wavelength received at Earth. During the time that the light takes to travel from the source to Earth, the wavelength of the light increases by the same proportion as the amount of expansion of the universe (that is, if the universe doubles in size during this time, then the wavelength of the light also doubles [$\lambda/\lambda_0 = 2$]). Rearranging the earlier equation, $\lambda/\lambda_0 = 1 + z$. Thus, for distant galaxies, $1 + z$ is a measure of the amount by which the universe has expanded during the travel time of the light from the galaxy to the Earth. Zoom back out to a field of 150° and use the "K" keyboard shortcut to display the constellations on the celestial sphere. Use the remaining entries in the Quasars observing list to find their redshifts and locations in the sky. For each of the other quasars in this small sample of all known quasars, double-click the name in the list and then click the info icon in its label in the view to obtain its redshift. Which of these other galaxies is farther from Earth than quasar pair 2345? Which of this small sample of quasars is farthest from Earth and which is nearest to Earth?

41. Energetic jets of material moving at relativistic speeds are observed to be blasting outward in opposite directions from the centers of many active galaxies. These are thought to be projected at right angles to the accretion disks of supermassive black holes at the centers of these galaxies. Use *Starry Night*™ to examine an example of one of these energetic jets in the vicinity of the galaxy M87. Select **Favorites > Explorations > M87** to see the sky from Earth's north pole, looking toward this galaxy.

    (a) According to the HUD, what is the distance of this galaxy from Earth?

    (b) Zoom in to a field of view of 30 arcmin. What type of galaxy is M87?

    (c) Select **View > Deep Space > Hubble Images** from the menu to display a high-resolution Hubble Space Telescope image of M87 to the upper right from the galaxy in the view. Center this image in the view and zoom in to view one of the jets emanating from the galaxy. (If our galaxy were to produce a jet like this, it would extend across more than 13° in our sky.) Use the angular separation cursor to measure the angular extent of this jet in the image in the view. What is the projected length, in light-years, of this jet into space? (*Hint:* Use the small-angle formula.)

Collaborative Exercises

42. Make a labeled sketch clearly showing how the spectrum of 3C 273, shown in Figure 24-3, would be different if the object were moving toward us at the same velocity. Compare your sketch to that of another group and resolve any inconsistencies.

43. Two dramatic images of M87 are shown in Figure 24-8. Invent an imaginary scenario that would be analogous to a local dance club where two pictures would show different aspects of the event and write a description of each. Be sure to include a description of how this is analogous to the images in Figure 24-8.

ANSWERS

ConceptChecks

ConceptCheck 24-1: Here are three differences: A quasar has emission lines, whereas a star has absorption lines. A quasar also emits more strongly than a star in the radio and x-ray wavelengths. The Doppler shift of spectral lines (whether emission or absorption) is much larger for quasars than for the stars in our galaxy, indicating that quasars recede at much higher speeds than do stars in our galaxy.

ConceptCheck 24-2: The universe was about 2 billion years old when quasar activity peaked, and that time corresponds to a redshift z of around 3.

ConceptCheck 24-3: The farther away an object is observed, the lower its apparent brightness. The redshifts in Figure 24-5 all correspond to very distant objects that can only be observed because they were very luminous at the early times and remote distances when they shined bright.

ConceptCheck 24-4: Radiation pressure would push outward on potentially accreting gas-fuel more than gravity could pull it in. This would slow the inward flow of gas (slowing accretion), and since accretion of gas fuels the quasar's luminosity, the quasar would soon become less luminous. This reduction in luminosity would actually allow accretion to then rise to its original Eddington limit.

ConceptCheck 24-5: Orbits closer to the central mass have higher speeds so that an inner blob rubs against an outer blob in the disk, causing the inner blob to lose energy and spiral inward. Also, as the blobs rub past each other, they produce heat. Thermal radiation from this heat produces much of a quasar's visible light.

ConceptCheck 24-6: The blob travels 5 ly (even though we see only the 3 ly of its sideways, transverse motion). The blob takes 6 years to travel this distance (and it travels slower than the speed of light).

ConceptCheck 24-7: The answer is "no" to both questions. The object is a dim dusty torus that can block light, depending on the viewing angle. The intense visible light from a quasar comes from the much smaller accretion disk, which appears pointlike in images.

ConceptCheck 24-8: Viewing angle #2. At these intermediate angles, visible light can be seen directly from the accretion disk without getting blocked by the torus. This line of sight also receives Doppler-shifted light from fast-moving clouds in a turbulent region near the black hole; these clouds produce the broad emission lines.

ConceptCheck 24-9: Yes, the supermassive black hole at our galaxy's center is probably a dead quasar. However, since a supermassive black hole is expected to capture a star and produce a dead quasar flare only about once every 10,000 years, we are unlikely to witness this event.

CalculationCheck

CalculationCheck 24-1: From the formula for the Eddington limit, if $M = 1\text{-}M_\odot$, then $L_{Edd} = 30,000\ L_\odot$. Therefore, L_{Edd} is much, much greater (30,000 times greater) than the Sun's luminosity.

The Most Majestic Objects in the Universe

by Robert Antonucci

When I went off to college (1963), quasars had been discovered only eight years prior. They are mysterious and awesome objects—to me, they are still the most majestic entities in the universe. We knew that they must be very distant, based on their redshifts, and the distances and apparent brightnesses led (via the inverse square law modified by general relativity) to the realization that a single quasar can be as luminous as thousands of giant galaxies. Yet their rapid variability required that their "central engines" be not much bigger than our solar system! As the years went by, we discovered that quasars are closely related to radio galaxies, as well as to weaker objects that we see more locally, called Seyfert galaxies. Superluminal motion was discovered! The mere fact that most quasars are many billions of light-years away tells us that they were much more common in the distant past than they are today, so that the universe must be *evolving* over time. Thus, the universe can't be both eternal and unchanging. That was very much up for debate at the time.

Now, having spent a lifetime studying quasars and related objects, I have mixed emotions. On the one hand, some subsequent discoveries have pointed very strongly to copious accretion flows around supermassive black holes as the basic power source. That's huge progress. But true understanding requires that we have theories with *predictive power*, and success in that effort has been disappointing to me. I developed a controversial opinion as to the reason for this circumstance. And I have some anxiety that we may not have good predictive theories of how they work before I leave this world!

What are our successes, and why do we think we are on the right track regarding the ultimate power source? Again, most of the powerful quasars are at distances of about 10 billion ly. Therefore, they were shining when the age of the universe was about 4 billion years, compared with today's 14-billion-year-old universe. All evidence indicates that the only differences between the distant reaches of space and our nearby neighborhood are due to the difference in the ages of the contents. Consider this thought experiment. Suppose we lived on a planet orbiting a star in the host galaxy of one of these quasars today. Our magnificent quasar would very likely have gone quiescent and dark. Suppose we looked across the universe to the Milky Way and its neighbors in the Local Group of galaxies. We would look back at our Milky Way neighborhood and see it as it was 10 billion years ago. The patch of space near the Milky Way would be seen in its youth, so our region would be rich in luminous quasars as seen from that great distance. The situation is exactly symmetric!

This was pointed out by astronomer Donald Lynden-Bell in 1969. Why don't we see quasars in our own neighborhood today? The supermassive black holes powering them when the universe was young must still be around, but they are now almost entirely starved of fuel. They must have eaten all the available matter in their vicinities, which is actually quite reasonable, considering the universe has greatly thinned out since then and the galaxies have settled down into mature quiescent forms. But the black holes must still be here! So the fundamental profound prediction of the supermassive black hole hypothesis for distant quasars is that there must be similar black holes nearby, though they would be mostly dark. These predicted supermassive black holes have been found, at the centers of almost all big galaxies in the Milky Way neighborhood, including in our Milky Way itself!

I'll tell you about two super-fun discoveries from my early career. I was in graduate school at UC Santa Cruz from 1975 until 1982. My thesis advisor Joseph Miller had an idea for getting information on active galactic nuclei which few people had exploited, namely the simultaneous measurement of the *polarization* along with the brightness of the light at each wavelength of the spectrum. He built the best instrument in the world to do this (called a *spectropolarimeter*), and we used it at the University of California's Lick Observatory, a couple hours' drive from Santa Cruz.

For my thesis project, we observed radio galaxy nuclei. We had little idea of what we'd find, and it took us quite a while to figure out what the data were telling us. What we discovered was that there is a trace of polarized light from the nuclear regions of radio galaxies; that the polarization was caused by reflection; and that the reflection was due to ambient gas near the black holes, acting as a natural periscopic mirror, allowing us effectively to see the radio galaxies from totally different directions in space. What we saw through these periscopic mirrors were the spectra of hidden quasars! This led to the Unified Model for quasars and radio galaxies, as well as for the two types of AGN called Seyfert galaxies and radio-quiet quasars.

At about the same time, radio astronomers were discovering superluminal motion, posing a great dilemma about whether there was a real violation of the theory of relativity. One alternative considered was that perhaps Hubble's law was incorrect and it was giving us the wrong distances to the quasars; if they were really much closer than the law indicated, then the observed speeds would come out to be less than the speed of light. But another alternative was the Beam Model, proposed in its unifying form by R. Blandford and M. Rees in 1978. It's explained in the text because it now seems very likely to be the correct explanation.

In this theory, superluminal sources were just like ordinary, giant double-jet radio galaxies and quasars, except their axes happened to lie very close to (but not exactly along) our line

(continued on the next page)

of sight. In that case, the effects of changing light travel time would cause an illusion of faster-than-light motion; furthermore, the apparent brightnesses of the tiny radio cores would be greatly boosted by an effect of special relativity. If so, then the giant radio lobes should be detectable as roughly round halos of diffuse radio emission, surrounding the superluminal cores. Following preliminary work by James Ulvestad, Ian Browne, and others, Ulvestad and I set out to look for this diffuse emission in a survey of all the blazars known at that time. Diffuse radio emission hadn't been detected in most cases, and our hope was that the new Very Large Array radio telescope, together with powerful new mathematical algorithms for making radio maps, would reveal it. We did detect the predicted emission, greatly strengthening the case for the Beam Model.

Not everything is well in this field, however. We don't really understand how energy is generated around the black holes, or the

Robert Antonucci, Professor of Physics at the University of California, Santa Barbara, has worked with most of the major ground- and space-based observatories, covering the full range of the electromagnetic spectrum. He did some of the earliest work on several topics in the field of active galactic nuclei and quasars, including the response of the emission lines to continuum changes; their dust and molecular emission in the millimeter band; and the nature of their radio, infrared, optical, ultraviolet, and x-ray radiation. He is also known for several comprehensive review papers and critiques of theoretical models.

production of the immense luminosity in the optical and ultraviolet bands. The chapter represents the majority view that this light is emitted by opaque hot gases spiraling toward the black hole in an accretion disk. I think that may be true, but that our *specific models* of the process aren't correct, and lack predictive power.

This has been quite frustrating to me, and I've controversially argued that we astronomers are somewhat at fault for the lack of more progress. My contention is that the community has the intelligence, the perseverance, and the facilities needed to solve the problem, but that we have been held back in part because of a lack of sufficiently critical thinking. I believe that the AGN community has continued to work on certain models long after those models have been disproven by observation! In particular, many workers have continued to rely on a simple "quasi-static" (slowly changing) accretion-disk model, whereas the true situation is more complicated and messy than that. The simple picture of gas heating up and emitting blackbody radiation as it spirals into the center of attraction *has* been shown to apply roughly to accretion by stellar-mass black holes (the fuel being drawn from their companion stars). But in several important ways, quasars don't behave in the same fashion. The luminosities vary too quickly; as output of a quasar varies, its spectrum doesn't change in a manner consistent with the blackbody laws; and the size of the emission region, though tiny compared with a galaxy, is much too big for a blackbody disk.

I think we should have abandoned this simple picture long ago. Today, we have some hope of modeling the accretion flows with supercomputers, although in my opinion so far no model has shown any significant predictive power. I dearly hope this will change while I'm still here to enjoy it!

25

Cosmology: The Origin and Evolution of the Universe

Supernova HST05Str

$z = 1.010$

Lookback time = 7.7 billion years

Supernova HST05Lan

$z = 1.230$

Lookback time = 8.5 billion years

Supernova HST05Sas

$z = 1.390$

Lookback time = 9.0 billion years

LEARNING GOALS

By reading the sections of this chapter, you will learn

25-1 Why the darkness of the night sky presents a mystery

25-2 What it means to say that the universe is expanding

25-3 How to estimate the age of the universe from its expansion rate

25-4 How astronomers detect the afterglow of the Big Bang

25-5 What the universe was like during its first 380,000 years

25-6 How the curvature of the universe reveals its matter and energy content

25-7 What distant supernovae tell us about the expansion history of the universe

25-8 How cosmic sound waves reveal details of our universe

So far in this book, we have cataloged the contents of the universe. Our scope has ranged from subatomic objects to superclusters of galaxies hundreds of millions of light-years across. In between, we have studied planets, moons, and stars.

But now we turn our focus beyond the objects we find in the universe to the nature of the universe itself—the subject of the science called *cosmology*. How large is the universe? What is its structure? How long has it existed, and how has it changed over time?

In this chapter, we will see that the universe is expanding. This expansion began with an event at the beginning of time called the *Big Bang*. We will see direct evidence of the Big Bang in the form of microwave radiation from space. This radiation is the faint afterglow of a primordial fireball that filled all space shortly after the beginning of the universe.

Will the universe continue to expand forever, or will it eventually collapse back on itself? We will find that to predict the future of the universe, we must first understand what happened in the remote past. To this end, astronomers study luminous

Very distant supernovae—which we see as they were billions of years ago—help us understand the evolution of the universe. (NASA, ESA, and A. Riess [STScI])

R I V U X G

supernovae like the example shown in the opening images. These can be seen across billions of light-years and can therefore tell us about conditions in the universe billions of years ago. We will see how recent results from such supernovae, as well as from studies of the Big Bang's afterglow, have revolutionized our understanding of cosmology and given us new insights into our place in the cosmos.

25-1 The darkness of the night sky tells us about the nature of the universe

Cosmology is the science concerned with the structure and evolution of the universe as a whole. One of the most profound and basic questions in cosmology may at first seem foolish: Why is the sky dark at night? This question, which haunted Johannes Kepler as long ago as 1610, was brought to public attention in the early 1800s by the German amateur astronomer Heinrich Olbers.

Olbers's Paradox and Newton's Static Universe

Olbers and his contemporaries pictured a universe of stars scattered more or less randomly throughout infinite space. Isaac Newton himself thought that no other model made sense. The gravitational forces between any *finite* number of stars, he argued, would in time cause them all to fall together, and the universe would soon be a compact blob.

Obviously, this has not happened. Newton concluded that we must be living amid a static, infinite expanse of stars. In this model, the universe is infinitely old, and it will exist forever without major changes in its structure. Olbers noticed, however, that a static, infinite universe presents a major puzzle.

If space goes on forever, with stars scattered throughout it, then any line of sight must eventually hit a star. In this case, no matter where you look in the night sky, you should ultimately see a star. The entire sky should be as bright as an average star, so, even at night, the sky should blaze like the surface of the Sun. **Olbers's paradox** is that the night sky is actually *dark* (**Figure 25-1**).

Olbers's paradox suggests that something is wrong with Newton's infinite, static universe. According to the classical, Newtonian picture of reality, space is like a gigantic flat sheet of inflexible, rectangular graph paper. (Space is actually three-dimensional, but it is easier to visualize just two of its three dimensions with graph paper.)

This rigid, flat, Newtonian space extends unchanged, on and on, totally independent of stars or galaxies or anything else. The same is true of time in Newton's view of the universe; a Newtonian clock ticks steadily and monotonously forever, never slowing down or speeding up. Furthermore, Newtonian space and time are unrelated, in that a clock runs at the same rate no matter where in the universe it is located.

Einstein's Revolution and His "Greatest Blunder"

Albert Einstein overturned this view of space and time. His special theory of relativity (recall Section 21-2 and Box 21-2) shows that measurements with clocks and rulers depend on the motion of the observer. What's more, Einstein's general theory of relativity

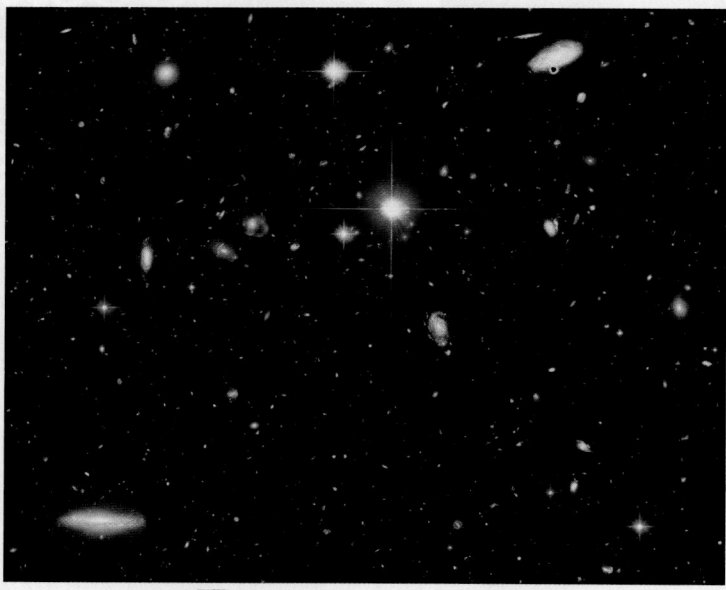

FIGURE 25-1 R I V U X G

The Dark Night Sky If the universe were infinitely old and filled uniformly with stars that were fixed in place, the night sky would be ablaze with light. In fact, the night sky is dark, punctuated only by the light from isolated stars and galaxies, Hence, this simple picture of an infinite, static universe cannot be correct. (NASA, ESA, and The Hubble Heritage Team [STScI/AURA])

(Section 21-3) tells us that gravity curves the fabric of space. As a result, just as one massive object can bend the space around it, the matter that occupies the universe influences the overall shape of space throughout the universe.

If we represent the universe as a sheet of graph paper, the sheet is not perfectly flat but has a dip wherever there is a concentration of mass, such as a person, a planet, or a star (see Figure 21-11). Also, because of gravitational effects, clocks run at different rates depending on whether they are close to or far from a massive object, as Figure 21-14a shows.

What does the general theory of relativity, with its many differences from the Newtonian picture, have to say about the structure of the universe as a whole? Einstein attacked this problem shortly after formulating his general theory in 1915. At that time, the prevailing view was that the universe was static, just as Newton had thought.

> Einstein narrowly missed predicting that our universe is not static.

Einstein was therefore dismayed to find that his calculations could not produce a truly static universe. According to general relativity, the universe must be either expanding or contracting. In a desperate move to force his theory to predict a static universe, he added to the equations of general relativity a term called the **cosmological constant** (denoted by Λ, the capital Greek letter lambda). The cosmological constant was intended to represent a pressure that tends to make the universe expand as a whole. Einstein's idea was that this pressure would just exactly balance gravitational attraction, so that the universe would be static and not collapse.

away from us with speeds proportional to their distances. Now let us see how Hubble's law reveals an expanding universe.

The Hubble Law and the Expanding Universe

The Hubble law can be stated as a very simple equation that relates the recessional velocity v of a galaxy to its distance d from Earth:

$$v = H_0 d$$

where H_0 is the Hubble constant. Because remote galaxies are getting farther and farther apart as time goes on, astronomers say that the universe is expanding. We say these galaxies are in the Hubble flow.

ANALOGY

What does it actually mean to say that the universe is expanding? According to general relativity, space itself is not rigid. The amount of space between widely separated locations increases over time. A good analogy is that of baking a chocolate chip cake, as in **Figure 25-2**. As the cake expands during baking, the chocolate chips do not move through the cake but the amount of space between the chocolate chips gets larger and larger. In the same way, as the universe expands, galaxies in the Hubble flow do not move through space, but the amount of space between widely separated galaxies increases. The expansion of the universe *is* the expansion of space.

CAUTION!

It is important to realize that the expansion of the universe occurs primarily in the vast spaces that separate clusters of galaxies. Just as the individual chocolate chips in Figure 25-2 do not expand as the cake expands during baking, galaxies themselves do not expand. Einstein and others have established that an object that is held together by its own gravity, such as a galaxy or a cluster of galaxies, creates a region of nonexpanding space. Likewise, molecular forces, which are much stronger than gravity, prevent expansion. For example, Earth and your body, which are held together by molecular forces, are not getting any bigger. Only the distance between widely separated galaxies increases with time. This expansion actually creates space. The **Cosmic Connections** figure has more to say about several misconceptions concerning the expanding universe.

ANALOGY

Einstein's cosmological constant is analogous to the pressure of gas inside a balloon once the balloon has already been inflated. This pressure exactly balances the inward force exerted by the stretched rubber of the balloon itself, so the balloon maintains the same size. (*Note:* This analogy is limited because while there is an inside and outside to a balloon, there is no "outside" of the universe.)

Unlike other aspects of Einstein's theories, the cosmological constant did not have a firm basis in physics. He just added it to make the general theory of relativity agree with the prejudice that the universe is static.

Because Einstein doubted his original equations, he missed an incredible opportunity: He could have postulated that we live in an expanding universe. Einstein has been quoted as saying in his later years that the cosmological constant was "the greatest blunder of my life." (In fact, the cosmological constant plays an important role in modern cosmology, although a very different one from what Einstein proposed. We will explore this in Section 25-7.)

Instead, the first hint that we live in an expanding universe came more than a decade later from the observations of Edwin Hubble. As we will see in Section 25-3, Hubble's discovery provides the resolution of Olbers's paradox.

CONCEPT CHECK 25-1

In an infinite, static universe uniformly filled with stars, would space be bright or dark when viewed from the Moon (looking away from the Sun)?

Answer appears at the end of the chapter.

25-2 The universe is expanding

TUTORIAL 25.1 Hubble is usually credited with discovering that our universe is expanding. He found a simple linear relationship between the distances to remote galaxies and the redshifts of the spectral lines of those galaxies (review Section 23-5, especially Figures 23-16 and 23-17). This relationship, now called the *Hubble law*, states that the greater the distance to a galaxy, the greater is the galaxy's redshift. Thus, remote galaxies are moving

Six chocolate chips are evenly spaced within an unbaked cake.

1 2 3 4 5 6

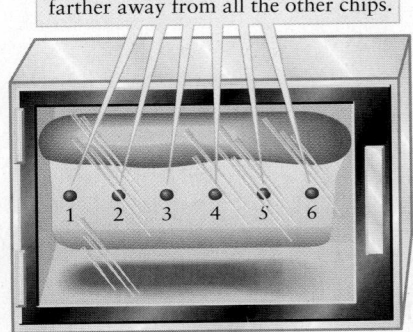

Each chocolate chip has moved farther away from all the other chips.

1 2 3 4 5 6

FIGURE 25-2

The Expanding Chocolate Chip Cake Analogy The expanding universe can be compared to what happens inside a chocolate chip cake as the cake expands during baking. All of the chocolate chips in the cake recede from one another as the cake expands, just as all the galaxies recede from one another as the universe expands.

COSMIC CONNECTIONS

"Urban Legends" about the Expanding Universe

ANIMATION
25.6

Incorrect Urban Legend #1:
The expansion of the universe means that as time goes by, galaxies move away from each other through empty space. In this picture, space is simply a background upon which the galaxies act out their parts.

Time ⟶

Reality:
The expansion of the universe means that as time goes by, space itself expands (shown here by the expanding grid). As it expands, it carries the galaxies along with it.

Time ⟶

Incorrect Urban Legend #2:
The redshift of light from distant galaxies is a Doppler shift produced by galaxies moving away from us through an unchanging space. Once the light is emitted with its Doppler shift, the wavelength remains the same during its journey to us.

Time

Time

Reality:
As light travels through intergalactic space, its wavelength gradually expands as the space through which it is traveling expands. This is called a cosmological redshift.

Incorrect Urban Legend #3:
As the universe expands, so do objects within the universe. Hence galaxies within a cluster are now more spread out than they were billions of years ago.

A cluster of galaxies

Time ⟶

In this picture, the cluster expands as the universe expands.

Cluster has stabilized, and its size no longer changes.

Reality:
At first, the expansion of the universe tends to pull the galaxies of a cluster away from each other. But the force of gravitational attraction binds the members of the clusters together, so the cluster stabilizes at a certain size.

A cluster forms

Time ⟶

792

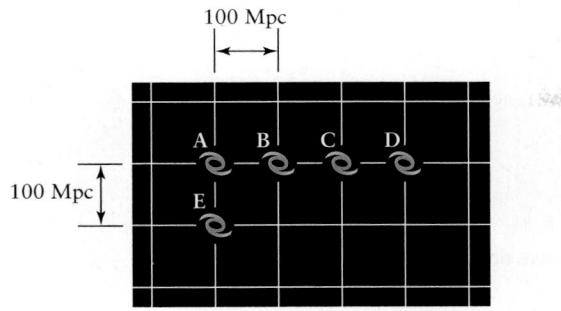

(a) Five galaxies spaced 100 Mpc apart.

(b) The expansion of the universe spreads the galaxies apart.

	Original distance from A (Mpc)	Later distance from A (Mpc)	Change in distance from A (Mpc)
A ↔ B	100	150	50
A ↔ C	200	300	100
A ↔ D	300	450	150
A ↔ E	100	150	50

ANIMATION 25.1

FIGURE 25-3

The Expanding Universe and the Hubble Law
(a) Imagine five galaxies labeled A, B, C, D, and E. At the time shown here, adjacent galaxies are 100 Mpc apart. **(b)** As the universe expands, by some later time the spacing between adjacent galaxies has increased to 150 Mpc. The table shows that the greater the original distance between galaxies, such as galaxy A and another, the greater the amount that distance has increased. Galaxy A is used only as an example in the table; this behavior applies to any two galaxies, as you can see when comparing (a) and (b)—the greater the original distance, the greater it has increased after further expansion. This agrees with the Hubble law.

The Hubble law is a direct proportionality—that is, a galaxy twice as far away is receding from us twice as fast. This property is just what we would expect in an expanding universe. To see why this property results from uniform expansion, imagine a grid of parallel lines (as on a piece of graph paper) crisscrossing the universe. Figure 25-3a shows a series of such gridlines 100 Mpc apart, along with five galaxies labeled A, B, C, D, and E that happen to lie where gridlines cross. As the universe expands in all directions, the gridlines and the attached galaxies spread apart. (This is just what would happen if the universe were a two-dimensional rubber

sheet that was being pulled equally on all sides. Alternatively, you can imagine that Figure 25-3a depicts a very small portion of the chocolate chip cake in Figure 25-2, with galaxies taking the place of chocolate chips.)

Figure 25-3b shows the universe at a later time, when the gridlines are 50% farther apart (150 Mpc) and all the distances between galaxies are 50% greater than in Figure 25-3a. Imagine that A represents our galaxy, the Milky Way. The table accompanying Figure 25-3 shows how far each of the other galaxies has moved away from us during the expansion: Galaxies A and B and galaxies A and E were originally 100 Mpc apart and have moved away from each other by an additional 50 Mpc; A and C, which were originally 200 Mpc apart, have increased their separation by an additional 100 Mpc; and the distance between A and D, originally 300 Mpc apart, has increased by an additional 150 Mpc. In other words, the increase in distance between any pair of galaxies is in direct proportion to the original distance; if the original distance is twice as great, the increase in distance is also twice as great.

To see what these distances tell us about the recessional velocities of galaxies, remember that velocity is equal to the distance moved divided by the elapsed time. (For example, if you traveled in a straight line for 360 km in 4 hours, your velocity was $(360 \text{ km})/(4 \text{ h}) = 90 \text{ km/h}$.) Because the distance that each galaxy moves away from A during the expansion is directly proportional to its original distance from A, it follows that the velocity v at which each galaxy moves away from A is also directly proportional to the original distance d. This result is just the Hubble law, $v = H_0 d$.

CAUTION!

It may seem that if the universe is expanding, and if we see all the distant galaxies rushing away from us, then we must be in a special position at the very center of the universe. In fact, the expansion of the universe looks the same from the vantage point of *any* galaxy. For example, as seen from galaxy D in Figure 25-3, the initial distances to galaxies A, B, and C are 300 Mpc, 200 Mpc, and 100 Mpc, respectively. Between parts (a) and (b) of the figure, these distances increase by 150 Mpc, 100 Mpc, and 50 Mpc, respectively. So, as seen from D as well, the recessional velocity increases in direct proportion with the distance, and in the same proportion as seen from A. In other words, no matter which galaxy you call home, you will see all the other galaxies receding from you in accordance with the same Hubble law (and the same Hubble constant) that we observe from Earth.

Figure 25-2 also shows that the expansion of the universe looks the same from one galaxy as from any other. An insect sitting on any one of the chocolate chips would see all the other chips moving away. If the cake were infinitely long, it would not actually have a center; as seen from any chocolate chip within such a cake, the cake would extend off to infinity to the left and to the right, and the expansion of the cake would appear to be centered on that chip. Likewise, because *every* point in the universe appears to be at the center of the expansion, it follows that our universe has no center at all.

"If the universe is expanding, what is it expanding into?" This commonly asked question arises only if we take our chocolate chip cake analogy too literally. In Figure 25-2, the cake (representing the universe) expands in three-dimensional space into the surrounding air. But the actual universe includes *all* space; there is no extra or additional space for the universe to expand into. Asking "What lies beyond the universe?" is like asking "Where on Earth is north of the North Pole?"—there just isn't a place on Earth that this point could be.

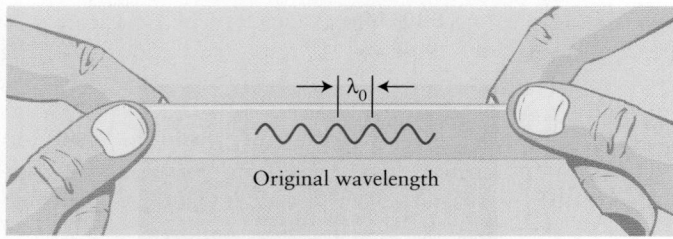

(a) A wave drawn on a rubber band . . .

(b) . . . increases in wavelength as the rubber band is stretched.

FIGURE 25-4

Cosmological Redshift A wave drawn on a rubber band stretches along with the rubber band. In an analogous way, a light wave traveling through an expanding universe "stretches"; that is, its wavelength increases.

The ongoing expansion of space explains why the light from remote galaxies is redshifted. Imagine light or a photon coming toward us from a distant galaxy. As the photon travels through space, the space is expanding, so the photon's wavelength becomes stretched. When the photon reaches our eyes, we see an increased wavelength: The photon has been redshifted. The longer the photon's journey, the more its wavelength will have been stretched along the way. Thus, photons from distant galaxies have larger redshifts than those of photons from nearby galaxies, as expressed by the Hubble law.

> The redshifts of distant galaxies are not Doppler shifts; they are caused by the expansion of space itself.

A redshift caused by the expansion of the universe is properly called a **cosmological redshift**. It is *not* the same as a Doppler shift. Doppler shifts are caused by an object's *motion through space*, whereas a cosmological redshift is caused by the *expansion of space* (Figure 25-4).

In Hubble's law, the original interpretation of redshifts was that they were due to Doppler shifts by receding galaxies; this works because for small redshifts, the Doppler shift can account for the recession of galaxies in expanding space. However, the full properties of expanding space, described by Einstein's general theory of relativity, must be taken into account for larger redshifts. In fact, at redshifts (z) greater than about 1.5, galaxies are observed with recessional velocities greater than the speed of light. While this would violate the laws of physics in a static universe, recessional speeds can exceed the speed of light in an expanding universe.

CONCEPT CHECK 25-2

Although nearly all distant galaxies have measurable redshifts, the relatively nearby Andromeda Galaxy exhibits an overall blueshift. What does this mean about the Andromeda Galaxy's movement?

Answer appears at the end of the chapter.

Cosmological Redshift and Lookback Time

We can calculate the factor by which the universe has expanded since some ancient time from the redshift of light emitted by objects at that time. As we saw in Section 23-5, redshift (z) is defined as

$$z = \frac{\lambda - \lambda_0}{\lambda_0}$$

In this equation, λ_0 is the unshifted wavelength of a photon and λ is the wavelength we observe. For example, λ_0 could be the wavelength of a particular emission line in the spectrum of light leaving a remote galaxy. As the galaxy's light travels through space, its wavelength is stretched by the expansion of the universe. Thus, at our telescopes we observe the spectral line to have a wavelength λ. The ratio λ/λ_0 is a measure of the amount of stretching. By rearranging terms in the preceding equation, we can solve for this ratio to obtain

$$\frac{\lambda}{\lambda_0} = 1 + z$$

For example, consider a galaxy with a redshift $z = 3$. Since the time that light left that galaxy, the universe has expanded by a factor of $1 + z = 1 + 3 = 4$. In other words, when the light left that galaxy, representative distances between widely separated galaxies were only one-quarter as large as they are today. A representative volume of space, which is proportional to the cube of its dimensions, was only $(1/4)^3 = 1/64$ as large as it is today. Thus, the density of matter in such a volume was 64 times greater than it is today.

If you know the redshift z of a distant object such as a remote galaxy, you can calculate that object's recessional velocity v (see Box 23-2 for this calculation). Then, using the Hubble law, you can determine the distance d to that object if you know the value of the Hubble constant H_0. This distance also tells you the **lookback time** of that object, that is, how far into the past you are looking when you see that object. For example, if the lookback time for a distant galaxy is 1 billion years, that means the light from the galaxy took 1 billion years to reach us, so we are seeing it as it was 1 billion years ago. The chapter-opening images show supernovae in distant galaxies with lookback times from 7.7 to 9.0 billion years.

Distances and lookback times determined in this way are somewhat uncertain, because there is some uncertainty in the value of the Hubble constant. Furthermore, as we will see in Section 25-8, the universe has not always expanded at the same rate. This variable rate of expansion means that the value of the Hubble constant H_0 was different in the distant past. (In other words, the Hubble "constant" is not actually constant in time, but at a given time it is still constant throughout space.) Furthermore, the correct distance d to use in the Hubble law is not the distance at which we *see* the object, but rather the distance between us and the object *now*. This latter distance is larger because during the time that it takes light to reach us from a distant object, that object has moved farther away due to the expansion of the universe.

To avoid dealing with these uncertainties and complications, astronomers commonly refer to times in the distant past in terms of redshift rather than years. For example, instead of asking how common quasars were 5 billion years ago, an astronomer might ask, "How common were quasars at $z = 1.0$?" In this question, "at $z = 1.0$" is a shorthand way of saying "at the lookback time that corresponds to objects at $z = 1.0$." We will use this terminology later in this chapter. Remember that the greater the redshift, the greater the lookback time, and hence the farther back into time we are peering.

The Cosmological Principle

In cosmology, there is only one universe that we can observe. Unlike other sciences, we cannot carry out controlled experiments or even make comparisons between two similar systems, when the system is the universe itself. But then a question arises: Are we observing the universe from a special location, resulting in a unique, nonrepresentative appearance? Or would observers anywhere see a universe with the same properties on large scales? The answer to this question determines how we interpret Hubble's law, because receding galaxies could be interpreted to mean that we are at the center of the universe. The idea that we are at the center of the universe was rejected, however, because it violates a cosmological extension of Copernicus's argument that we do not occupy a special location in space. An expanding universe, on the other hand, has no center; at all locations in space, observers would see galaxies recede according to Hubble's law.

When Einstein began applying his general theory of relativity to cosmology, he made a daring assumption: Over very large distances, the universe is **homogeneous**, meaning that every region is the same as every other region, and **isotropic**, meaning that the universe looks the same in every direction. In other words, if you could stand back and look at a very large region of space, any one part of the universe would look basically the same as any other part, with the same kinds of galaxies distributed through space in the same way. The assumption that the universe is homogeneous and isotropic constitutes the **cosmological principle**. It gives precise meaning to the idea that we do not occupy a special location in space.

CAUTION!

On small scales, the universe is not homogeneous or isotropic at all. For example, a single galaxy is a region of much more mass than average, and the same goes for a cluster of galaxies. Roughly speaking, the average distance between galaxies is about 1 Mpc, and the universe is only homogeneous and isotropic (approximately) on size scales above 100 Mpc.

Models of the universe based on the cosmological principle have proven remarkably successful in describing the structure and evolution of the universe and in interpreting observational data. More recently, direct observations reveal a homogeneous and isotropic universe on the largest scales—a validation of the cosmological principle. A homogenous and isotropic universe is revealed in maps made of large-scale structure, as shown in Figures 23-23 and 23-24. As we will see in Section 25-5, radiation left over from the Big Bang produces a similar result for the early universe.

CONCEPT CHECK 25-3

If we notice in the night sky that there are more stars along the Milky Way than in other regions of the sky, is this consistent or inconsistent with the cosmological principle?

CALCULATION CHECK 25-1

What is the redshift z-value for a galaxy that has a galaxy spectral line shifted to 725.6 nm when a stationary object would emit the line from 656.3 nm?

CALCULATION CHECK 25-2

What is the distance to a galaxy that is observed to have a recessional velocity of 10,000 km/s?

Answers appear at the end of the chapter.

25-3 The expanding universe emerged from a cataclysmic event called the Big Bang

The Hubble flow shows that the universe has been expanding for billions of years. Therefore, we can conclude that in the past, the matter in the universe must have been closer together and therefore denser than it is today. Indeed, as we look farther into the past, we clearly see the density of the universe increasing. There is even strong evidence of increasing density back to the first moments of time. Therefore, some sort of tremendous event caused high-density matter to begin the expansion that continues to the present day. This event, called the **Big Bang**, marks the creation of the universe.

CAUTION!

It is not correct to think of the Big Bang as an explosion. When a bomb explodes, pieces of debris fly off *into space* from a central location. If you could trace all the pieces back to their origin, you could find out exactly where the bomb had been. This process is not possible with the universe, however, because the universe itself always has and always will consist of all space, with no center. There is no single or central location in space where the Big Bang occurred, because the Big Bang is the expansion of all space (see the discussion of the expanding chocolate chip cake in Section 25-2).

Estimating the Age of the Universe

How long ago did the Big Bang take place? To estimate an answer, imagine two galaxies that today are separated by a distance d and receding from each other (that is, the space between them is expanding) with a velocity v. Keep in mind that the galaxies are not moving through space; they are receding as space expands. A movie of these galaxies would show them flying apart. If you run the movie backward, you would see the two galaxies approaching each other as time runs in reverse. We can calculate the time T_0 it will take for the reverse-run galaxies to collide by using the equation

$$T_0 = \frac{d}{v}$$

This expression says that the time to travel a distance d at velocity v is equal to the ratio d/v. [As an example, to travel 360 km at a velocity of 90 km/h takes (360 km)/(90 km/h) = 4 hours.] If we use the Hubble law $v = H_0 d$ to replace the velocity v in this equation, we get

$$T_0 = \frac{d}{H_0 d} = \frac{1}{H_0}$$

Note that the distance of separation, d, has canceled out and does not appear in the final expression. Distance not appearing in the expression means that T_0 is the same for *all* galaxies. This period is the time in the past when all the matter in galaxies was crushed together, the time back to the Big Bang. In other words, the reciprocal (or inverse) of the Hubble constant H_0 gives us an estimate of the age of the universe called the **Hubble time**; this is one reason why H_0 is such an important quantity in cosmology.

Using $H_0 = 73$ km/s/Mpc, our estimate for the age of the universe is

$$T_0 = \frac{1}{73 \text{ km/s/Mpc}}$$

To convert this into units of time, we simply need to remember that 1 Mpc equals 3.09×10^{19} km and 1 year equals 3.156×10^7 s. Using the technique we discussed in Box 1-3 for converting units, we get

$$T_0 = \frac{1}{73} \frac{\text{Mpc s}}{\text{km}} \times \frac{3.09 \times 10^{19} \text{ km}}{1 \text{ Mpc}} \times \frac{1 \text{ year}}{3.156 \times 10^7 \text{ s}}$$
$$= 1.34 \times 10^{10} \text{ years} = 13.4 \text{ billion years}$$

By comparison, the age of our solar system is only 4.56 billion years, or about one-third the age of the universe. Thus, the formation of our home planet is a relatively recent event in the history of the cosmos.

The value of H_0 has an uncertainty of about 5%, so our simple estimate of the age of the universe is likewise uncertain by at least 5%. Furthermore, the formula $T_0 = 1/H_0$ is at best an approximation, because in deriving it we assumed that the universe expands at a constant rate. In Section 25-8, we will discuss how the expansion rate of the universe has changed over its history; it's still constant throughout space, but not constant in time. When these factors are taken into consideration, we find that the age of the universe is 13.7 billion years, with an uncertainty of about 0.2 billion years. This time period is remarkably close to our simple estimate.

Whatever the true age of the universe, it must be at least as old as the oldest stars. The oldest stars that we can observe readily lie in the Milky Way's globular clusters (see Section 19-4 and Section 22-1). The most recent observations, combined with calculations based on the theory of stellar evolution, indicate that these stars are about 13.4 billion years old (a time period with an uncertainty of about 6%). This time period is less than the calculated age of the universe: As required, the oldest stars in our universe are younger than the universe itself.

What was our universe like 15 billion years ago? The passage of time (as we understand time) in the only universe we know of did not exist prior to the Big Bang. Nor did space (as we understand space) exist in the universe until expansion from the Big Bang. Time began to pass and space began to exist at the Big Bang. Since the universe, as we understand it, is only 13.7 billion years old, there does not appear to be physical meaning to the question of what things were like in the universe 15 billion years ago.

CONCEPT CHECK 25-4

If the Hubble constant were smaller than it actually is, would the universe be younger or older than 13.7 billion years?

Answer appears at the end of the chapter.

Our Observable Universe and the Dark Night Sky

The Big Bang helps resolve Olbers's paradox, which we discussed in Section 25-1. We know that the universe had a definite beginning, and thus its age is finite (as opposed to infinite). If the universe is 13.7 billion years old, then the most distant objects that we can see are those whose light has traveled 13.7 billion years to reach us. (Due to the expansion of the universe, these objects are now more than 13.7 billion ly away.) As a result, we can see objects that lie only within an immense sphere centered on Earth (Figure 25-5). This is true even if the universe is infinite, with galaxies scattered throughout its limitless expanse.

The surface of the sphere depicted in Figure 25-5 is called our **cosmic particle horizon**. Our entire **observable universe** is located inside this sphere. We cannot see anything beyond our cosmic particle horizon, because the time required for light to reach us from these incredibly remote distances is greater than the present age of the universe. As time goes by, light from more distant parts of the universe reaches us for the first time, and the size of the cosmic particle horizon—the size of our observable universe—increases. The finite

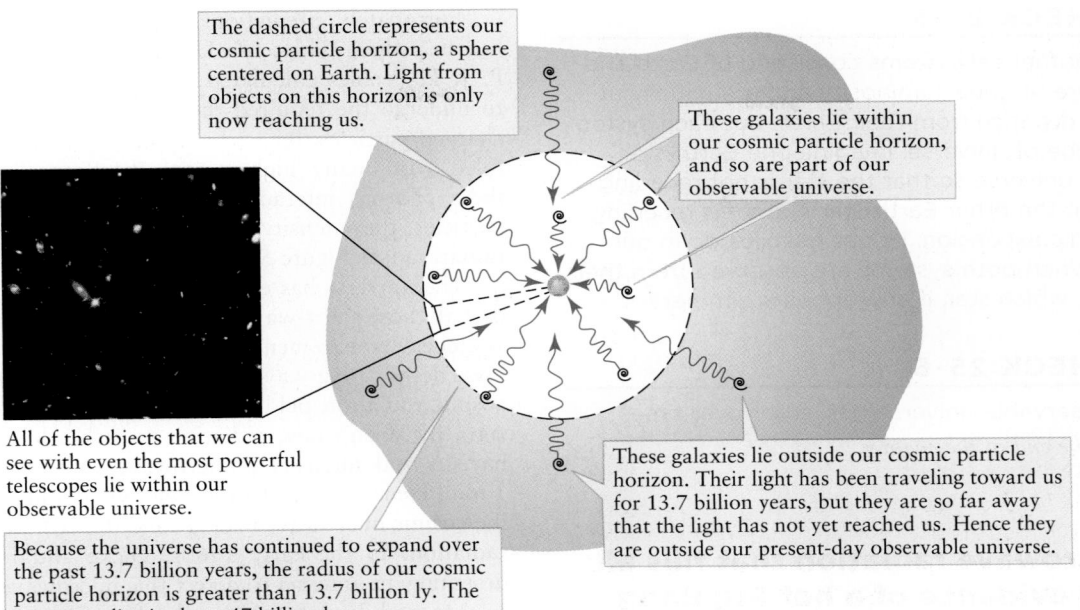

The dashed circle represents our cosmic particle horizon, a sphere centered on Earth. Light from objects on this horizon is only now reaching us.

These galaxies lie within our cosmic particle horizon, and so are part of our observable universe.

All of the objects that we can see with even the most powerful telescopes lie within our observable universe.

Because the universe has continued to expand over the past 13.7 billion years, the radius of our cosmic particle horizon is greater than 13.7 billion ly. The present radius is about 47 billion ly.

These galaxies lie outside our cosmic particle horizon. Their light has been traveling toward us for 13.7 billion years, but they are so far away that the light has not yet reached us. Hence they are outside our present-day observable universe.

FIGURE 25-5 R I **V** U X G

Our Observable Universe The part of the universe that we can observe lies within our cosmic particle horizon. The galaxies that we can just barely make out with our most powerful telescopes lie inside our cosmic particle horizon; we see them as they were less than 1 billion years after the Big Bang.

We cannot see objects beyond our cosmic particle horizon, because in the 13.7 billion years since the Big Bang, their light has not had enough time to reach us. (Inset: NASA/JPL/STScI Hubble Deep Field Team)

size of the observable universe, with a finite number of stars and galaxies, also helps to resolve Olbers's paradox: Galaxies are distributed sparsely enough in our observable universe that there are no stars along most of our lines of sight. This sparse distribution of visible stars is one reason why the night sky is dark.

Besides the finite age of the universe, a second effect also contributes significantly to the darkness of the night sky—the redshift. According to the Hubble law, the greater the distance to a galaxy, the greater the redshift. When a photon is redshifted, its wavelength becomes longer, and its energy—which is inversely proportional to its wavelength (see Section 5-5)—decreases.

> Measuring the recessional velocities of galaxies allows us to estimate the age of the universe.

Consequently, even though there are many galaxies far from Earth, they have large redshifts and their light does not carry much energy. To understand this effect, let's consider an imaginary galaxy nearly at the cosmic particle horizon (even though galaxies didn't exist that early in the universe). A galaxy nearly at the cosmic particle horizon has a nearly infinite redshift, meaning that the light we receive from that galaxy carries practically no energy at all. For real galaxies that are not quite as old or distant, this decrease in photon energy, because of the expansion of the universe, decreases the brilliance of these remote galaxies. This effect helps to make the night sky dark.

The concept of a Big Bang origin for the universe is a straightforward, logical consequence of an expanding universe. In Section 25-4, we will see direct evidence of the primordial fireball associated with the Big Bang and other confirmations of the Big Bang

back to the first few moments. But how far back in time can our laws describe the universe? If you can just imagine far enough back into the past, you can arrive at a time 13.7 billion years ago, when the density throughout the universe was nearly infinite. As a result, throughout the universe, space and time were jumbled up with nearly infinite curvature. A full description of this earliest instant requires (as do aspects of black holes) a valid mathematical theory of quantum gravity, which is a work in progress. But when in the past did the known laws of physics begin to apply?

A very short time after the Big Bang, space and time began to behave in the way we think of them today. This short time interval, called the **Planck time** (t_P), is given by the following expression:

The Planck time

$$t_P = \sqrt{\frac{Gh}{c^5}} = 1.35 \times 10^{-43}\ \text{s}$$

t_P = the Planck time

G = universal constant of gravitation

h = Planck's constant

c = speed of light

We do not yet understand how space, time, and matter behaved in that brief but important interval from the beginning of the Big Bang to the Planck time, about 10^{-43} seconds later. (Indeed, the laws of physics suggest that it might be impossible ever to know what happened during this extremely short time interval.) Hence, the Planck time represents a limit to our knowledge of conditions at the very beginning of the universe.

CONCEPT CHECK 25-5

Consider two hypothetical systems consisting of Earth and a star. The stars are of equal luminosity and are initially at the same remote distance from their Earth, but each system is in a different type of universe. Imagine one Earth-star system in a static universe so that the star is not receding away. Imagine that the other Earth-star system is receding away due to cosmic expansion, just as galaxies do in our actual universe. When both systems are observed from their respective Earths, which star, if any, appears dimmer?

CONCEPT CHECK 25-6

Why does our observable universe get larger over time?

Answers appear at the end of the chapter.

25-4 The microwave radiation that fills all space is evidence of a hot Big Bang

TUTORIAL 25.2 One of the major advances in twentieth-century astronomy was the discovery of the origin of the heavy elements. We know today that essentially all the heavy elements are created by thermonuclear reactions at the centers of stars, in supernovae, and in neutron star mergers (see Chapters 20 and 21). The starting point of all these reactions is the fusion of hydrogen into helium, which we described in Section 16-1. But as astronomers began to understand the details of thermonuclear synthesis in the 1960s, they were faced with a dilemma: There is far more helium in the universe than could have been created by hydrogen fusion in stars.

For example, the Sun consists of about 74% hydrogen and 25% helium by mass, and the universe as a whole is also about 25% helium. Some helium was produced by the thermonuclear fusion of hydrogen within stars, but calculations show that the amount of helium produced in this way is not nearly enough to account for the large fraction of helium observed in the universe. Because it was thought that the universe originally contained only hydrogen—the simplest of all the chemical elements—the presence of so much helium posed a major dilemma.

A Hot Big Bang and the Cosmic Microwave Background

Shortly after World War II, Ralph Alpher, George Gamow, and Robert Hermann proposed that the universe immediately following the Big Bang must have been so incredibly hot that thermonuclear reactions occurred everywhere throughout space. Quite impressively, calculations for the amount of helium (and some other elements) produced by thermonuclear reactions in the first few minutes after the Big Bang are in extraordinary agreement with the amount of helium we observe. This astounding prediction of the Big Bang—that about 25% of mass in the universe is helium—was confirmed by detailed observations, which means that strong evidence for the Big Bang goes all the way back to the first few minutes of the universe. This topic is discussed further in the next chapter (Section 26-5).

Fortunately, predictions coming from the Big Bang theory did not stop there. Princeton University physicists Robert Dicke and P. J. E. Peebles calculated that in order for the entire universe to undergo the required thermonuclear reactions, in addition to charged particles, the early universe must have been very hot and filled with many high-energy, short-wavelength photons. With these photons interacting strongly with the "soup" of charged particles, the intensity distribution of photons formed blackbody radiation (see Figure 5-12).

The universe has expanded so much since those ancient times that all those short-wavelength photons have had their wavelengths stretched by a tremendous factor. As a result, they have become low-energy, long-wavelength photons. The temperature of this cosmic radiation field is now only a few degrees above absolute zero. By Wien's law, radiation at such a low temperature should have its peak intensity at microwave wavelengths of approximately 1 mm. Hence, this radiation field, which fills all of space, is called the **cosmic microwave background** (CMB), or **cosmic background radiation**. In the early 1960s, Dicke and his colleagues began designing an antenna to detect this microwave radiation.

Meanwhile, just a few miles from Princeton University, Arno Penzias and Robert Wilson of Bell Telephone Laboratories in New Jersey were working on a new microwave horn antenna designed to relay telephone calls to Earth-orbiting communications satellites (**Figure 25-6**). Penzias and Wilson were deeply puzzled when, no matter where in the sky they pointed their antenna, they detected faint background noise. Thanks to a colleague, they learned about the work of Dicke and Peebles and realized that they had discovered the cooled-down cosmic background radiation left over from the hot Big Bang. Penzias and Wilson shared the 1978 Nobel Prize in Physics for their discovery.

FIGURE 25-6

The Bell Labs Horn Antenna Using this microwave horn antenna, originally built for communications purposes, Arno Penzias and Robert Wilson detected a signal that seemed to come from all parts of the sky. After carefully removing all potential sources of electronic "noise" (including bird droppings inside the antenna) that could create a false signal, Penzias and Wilson realized that they were actually detecting radiation from space. This radiation is the afterglow of the Big Bang. (Bell Labs/NASA)

(a) The COBE spacecraft

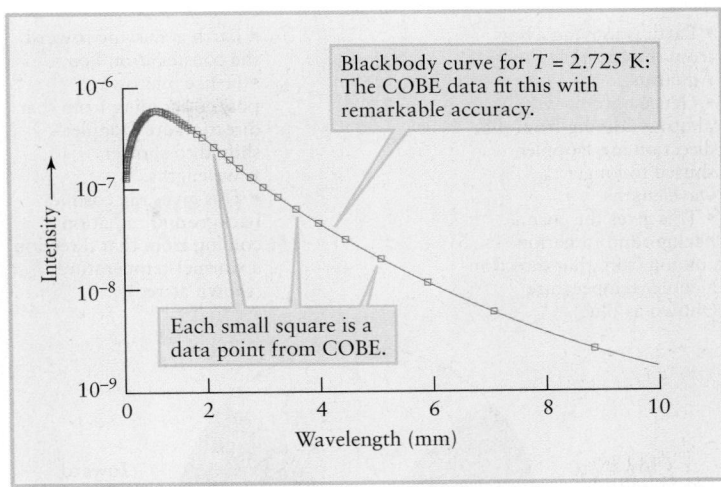

(b) The spectrum of the cosmic microwave background

FIGURE 25-7

COBE and the Spectrum of the Cosmic Microwave Background
(a) The Cosmic Background Explorer (COBE), launched in 1989, measured
the spectrum and angular distribution of the cosmic microwave background

over a wavelength range from 1 μm to 1 cm. **(b)** A blackbody curve gives an
excellent match to the COBE data. (a: J. Mather/NASA)

A TV using an antenna for signal
reception (as opposed to cable or a
satellite dish) can actually detect cosmic
background radiation. This radiation
is responsible for about 1% of the random noise or "hash" that appears on
the screen when you tune a television
to a station that is off the air. Using
far more sophisticated detectors than
TV sets, scientists have made many
measurements of the intensity of the
background radiation at a variety of
wavelengths. Unfortunately, Earth's atmosphere is almost totally
opaque to wavelengths between about 10 μm and 1 cm (see Figure
6-25), which is just the wavelength range in which the background
radiation is most intense. As a result, scientists have had to place
detectors either on high-altitude balloons (which can fly above the
majority of the obscuring atmosphere) or, even better, on board
orbiting spacecraft.

> The afterglow
> of the Big
> Bang was first
> discovered
> by a happy
> coincidence—
> and can be
> detected with
> an ordinary
> television.

A Detailed Look at the Cosmic Microwave Background

The first high-precision measurements of the cosmic microwave
background came from the Cosmic Background Explorer (COBE,
pronounced "coe-bee") satellite, which was placed in Earth orbit
in 1989 (**Figure 25-7a**). Data from COBE's spectrometer, shown in
Figure 25-7b, demonstrate that this ancient radiation has the spectrum of a blackbody with a temperature of 2.725 K. In recognition
of this discovery, as well as others that we will discuss in Section
25-5, COBE team leaders John Mather of NASA and George
Smoot of the University of California, Berkeley were awarded the
2006 Nobel Prize in Physics.

An important feature of the microwave background is that
its intensity is almost perfectly isotropic, that is, the same in all

directions. In other words, we detect nearly the same background
intensity from all parts of the sky. This is a striking confirmation
of Einstein's assumption that the universe is isotropic (see Section
25-2). However, extremely accurate measurements, first made
from high-flying airplanes and later from high-altitude balloons
and from COBE, reveal a very slight variation in temperature
across the sky. The microwave background appears slightly
warmer than average toward the constellation of Leo and slightly
cooler than average in the opposite direction toward Aquarius.
Between the warm spot in Leo and the cool spot in Aquarius,
the background temperature declines smoothly across the sky.
Figure 25-8 is a map of the microwave sky showing this variation.

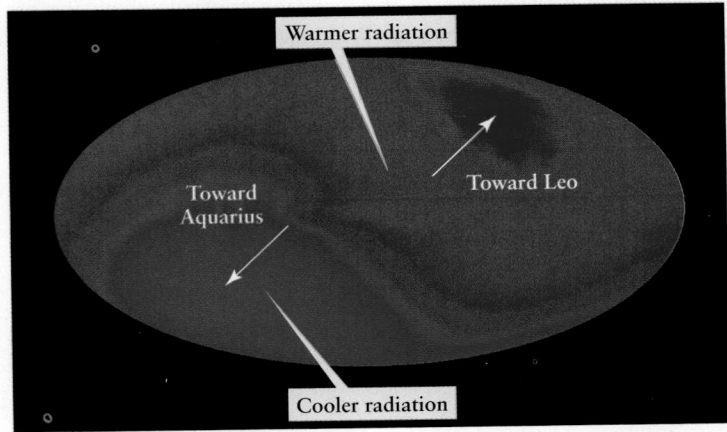

FIGURE 25-8 R I V U X G

The Microwave Sky   In this map of the entire sky made from COBE data,
the plane of the Milky Way runs horizontally across the map, with the galactic
center in the middle. Color indicates temperature—red is warm and blue is
cool. The small temperature variation across the sky—only 0.0033 K above
or below the average radiation temperature of 2.725 K—is caused by Earth's
motion through the microwave background. (NASA)

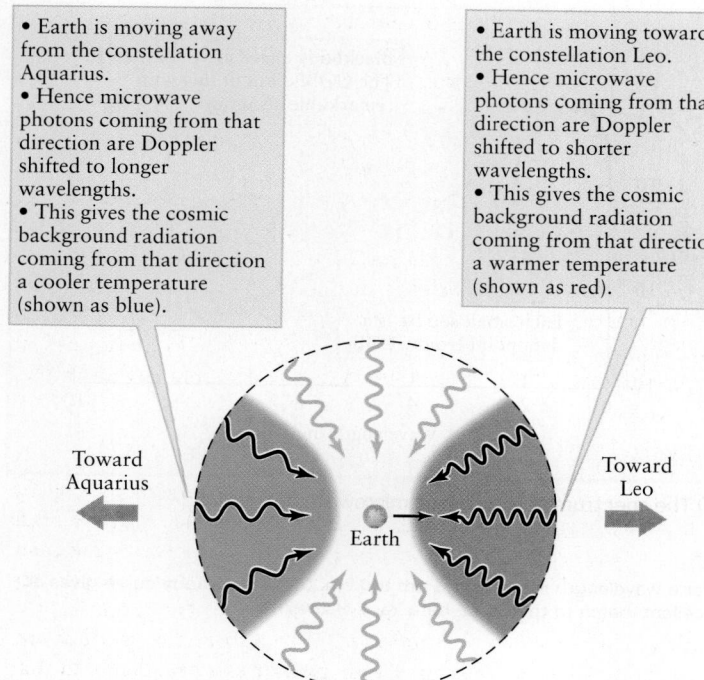

- Earth is moving away from the constellation Aquarius.
- Hence microwave photons coming from that direction are Doppler shifted to longer wavelengths.
- This gives the cosmic background radiation coming from that direction a cooler temperature (shown as blue).

- Earth is moving toward the constellation Leo.
- Hence microwave photons coming from that direction are Doppler shifted to shorter wavelengths.
- This gives the cosmic background radiation coming from that direction a warmer temperature (shown as red).

Toward Aquarius

Earth

Toward Leo

FIGURE 25-9

Our Motion Through the Microwave Background Because of the Doppler effect, we detect shorter wavelengths in the microwave background and a higher temperature of radiation in that part of the sky toward which we are moving. This part of the sky is the area shown in red in Figure 25-8. In the opposite part of the sky, shown in blue in Figure 25-8, the microwave radiation has longer wavelengths and a cooler temperature.

This apparent variation in temperature is caused by Earth's motion through the cosmos. If we were at rest with respect to the microwave background, the radiation would be even more nearly isotropic. Because we are moving through this radiation field, however, we see a Doppler shift. Specifically, we see shorter-than-average wavelengths in the direction toward which we are moving, as drawn in **Figure 25-9**. A decrease in wavelength corresponds to an increase in photon energy and thus an increase in the temperature of the radiation. The slight temperature excess observed, about 0.00337 K, corresponds to a speed of 371 km/s. Conversely, we see longer-than-average wavelengths in that part of the sky from which we are receding. An increase in wavelength corresponds to a decline in photon energy and, hence, a decline in radiation temperature.

CAUTION!

Due to the Doppler shift, light from a source that is moving away from an observer experiences an increase in wavelength. We call this a redshift. However, in the context of the CMB, longer wavelengths also correspond to cooler temperatures. Since it is common for people in general to represent cold with the color blue, and hot as red, a patch represented by the color blue in the CMB corresponds to redshifted light. Likewise, a red patch in the CMB corresponds to blueshifted light. Both cases can be seen in Figure 25-9.

Our solar system is thus traveling away from Aquarius and toward Leo at a speed of 371 km/s. Taking into account the known velocity of the Sun around the center of our galaxy, we find that the entire Local Group of galaxies, including our Milky Way Galaxy, is moving at about 620 km/s toward the Hydra–Centaurus Supercluster. Observations show that thousands of other galaxies are being carried in this direction, as is the Hydra–Centaurus Supercluster itself. This tremendous flow of matter is thought to be due to the gravitational pull of an enormous collection of visible galaxies and dark matter lying in that direction. This immense object, dubbed the *Great Attractor*, lies about 50 Mpc (150 million ly) from Earth (see Figure 23-22).

The existence of such concentrations of mass, as well as the existence of superclusters of galaxies (see Section 23-6), shows that the universe is rather "lumpy" on scales of 100 Mpc or smaller. It is only on larger scales that the universe is homogeneous and isotropic.

CONCEPT CHECK 25-7

If the early universe was filled with high-energy, short-wavelength photons, why are these observed today to be low-energy, long-wavelength microwave photons?

Answer appears at the end of the chapter.

25-5 The universe was a hot, opaque plasma during its first 380,000 years

Energy in the universe falls into one of two categories—radiation or matter. (We will encounter another type of energy in Section 25-7.) Photons are massless particles of light and are a form of radiation. There are many photons of starlight traveling across space, but the vast majority of photons in the universe are not visible and belong to the cosmic microwave background. The matter in the universe is contained in such luminous objects as stars, planets, and galaxies, as well as in nonluminous dark matter. A natural question to ask is which plays a more important role in the universe: radiation or matter? As we will see, the answer to this question for the early universe differs from the answer for our universe today.

Radiation and Matter in the Universe

To make a comparison between radiation and matter, recall Einstein's famous equation $E = mc^2$ (see Section 16-1). We can think of the photon energy in the universe (E) as being equal to a quantity of mass (m) multiplied by the square of the speed of light (c). The amount of this equivalent mass in a volume V, divided by that volume, is the **mass density of radiation** (ρ_{rad}; say "rho sub rad").

We can combine $E = mc^2$ with the Stefan–Boltzmann law (see Section 5-4) to give the following formula:

Mass density of radiation

$$\rho_{rad} = \frac{4\sigma T^4}{c^3}$$

ρ_{rad} = mass density of radiation

T = temperature of radiation

σ = Stefan–Boltzmann constant = 5.67×10^{-8} W m^{-2} K^{-4}

c = speed of light = 3.00×10^8 m/s

For the present-day temperature of the cosmic background radiation, $T = 2.725$ K, this equation yields

$$\rho_{rad} = 4.6 \times 10^{-31} \text{ kg/m}^3$$

The **average density of matter** (ρ_m; say "rho sub em") in the universe is harder to determine. To find this density, we look at a large volume (V) of space, determine the total mass (M) of all the stars, galaxies, and dark matter in that volume, and divide the volume into the mass: $\rho_m = M/V$. (We emphasize that this quantity is the *average* density of matter. It would be the actual density if all the matter in the universe were spread out uniformly rather than being clumped into galaxies and clusters of galaxies.) Determining how much mass is in a large volume of space is a challenging task. A major part of the challenge involves dark matter, which neither emits nor reflects electromagnetic radiation (or light), and can be detected only by its gravitational influence (see Section 22-4 and Section 23-8).

One method that appears to deal successfully with this challenge of measuring both visible and invisible (dark) matter is to observe clusters of galaxies. Galaxy clusters contain most of the luminous (regular, non-dark) mass in the universe and, crucially, the contribution from dark matter can also be estimated. Here's how: Many clusters are surrounded by halos of hot, x-ray–emitting gas, typically at temperatures of 10^7 to 10^8 K (see Figure 23-26). Such a halo should be in hydrostatic equilibrium, so that it neither expands nor contracts but remains the same size (see Section 16-2, especially Figure 16-2). The outward gas pressure associated with the halo's high temperature would then be balanced by the inward gravitational pull due to the total mass of the cluster. Thus, by measuring the temperature of the halo—which can be determined from the properties of the halo's x-ray emission—astronomers can infer the cluster's mass, including the contribution from dark matter!

From galaxy clusters and other measurements, the present-day average density of matter in the universe is estimated to be

$$\rho_m = 2.4 \times 10^{-27} \text{ kg/m}^3$$

with an uncertainty of about 15%. The mass of a single hydrogen atom is 1.67×10^{-27} kg. Hence, if the mass of the universe were spread uniformly over space, there would be the equivalent of 1.5 hydrogen atoms per cubic meter of space. By contrast, there are 5×10^{25} atoms in a cubic meter of the air you breathe! The very small value of ρ_m shows that our universe has a very low average density. There are also other methods (including using the CMB) for estimating the density of different forms of matter in the universe, and each gives a slightly different numerical value.

Furthermore, by counting galaxies and other measurements, astronomers determine that the average density of "ordinary," non-dark matter throughout the universe is about 4.2×10^{-28} kg/m^3. This density is only about 17% of the average density of matter of all forms. Thus, dark matter is actually the predominant form of matter in our universe. The ordinary non-dark matter (mostly protons and neutrons) of which the stars, gas, the planets, and ourselves are made is only about one-fifth of the total matter! While stars are the easiest form of ordinary matter to observe, about 90% of ordinary matter is simply in the form of diffuse hydrogen gas in the vast spaces between the galaxies.

The fact that there is more dark matter than regular matter does not make it any easier to detect. By interacting with light, regular matter can cool and condense into stars and planets. However, dark matter is more evenly and sparsely distributed. The actual amount of dark matter within Earth's volume at any given time is only about 1 kg. It is thought that dark matter particles stream right through Earth with only rare interactions. Due to its elusive nature, yet dominance over regular matter, the detection of dark matter particles is considered one of the "Holy Grails" of physics and astronomy and would surely result in a Nobel Prize.

When Radiation Held Sway Over Matter

Although the average density of matter in the universe is tiny by Earth standards, it is thousands of times larger than ρ_{rad}, the mass density of radiation. However, this ratio was not always the case. Matter prevails over radiation today only because the energy now carried by microwave photons is so small. Nevertheless, the number of photons in the microwave background is astounding. Using the physics of blackbody radiation, it can be calculated that today there are 410 million (4.1×10^8) photons in every cubic meter of space. In other words, the photons in space outnumber atoms by roughly a billion (10^9) to one. In terms of total number of particles, the universe thus consists almost entirely of microwave photons. This radiation field no longer has much effect on the universe, however, because its photons have been redshifted to long wavelengths and low energies after 13.7 billion years of being stretched by the expansion of the universe.

In contrast, think back toward the Big Bang. The universe becomes increasingly compressed, and thus the density of matter increases as we go back in time. The photons in the background radiation also become more crowded together as we go back in time. But, in addition, the photons become less redshifted and thus have shorter wavelengths and higher energy than they do today. Because of this added energy, the mass density of radiation (ρ_{rad}) increases more quickly as we go back in time than does the average density of matter (ρ_m). In fact, as Figure 25-10 shows, there was a time in the ancient past when ρ_{rad} equaled ρ_m. Before this time, ρ_{rad} was greater than ρ_m, and radiation thus held sway over matter. Astronomers call this state a **radiation-dominated universe**. After ρ_m became greater than ρ_{rad}, so that matter prevailed over radiation, our universe became a **matter-dominated universe**.

This transition from a radiation-dominated universe to a matter-dominated universe occurred about 24,000 years after the Big Bang, at a time that corresponds to a redshift of about $z = 5200$. Since that time, the wavelengths of photons have been stretched by a factor of $1 + z$, or about 5200. Today, these microwave photons typically have wavelengths of about 1 mm. But when the universe was about 24,000 years old, they had wavelengths of about 190 nm in the ultraviolet part of the spectrum.

> The temperature of the background radiation has declined over the eons thanks to the expansion of the universe.

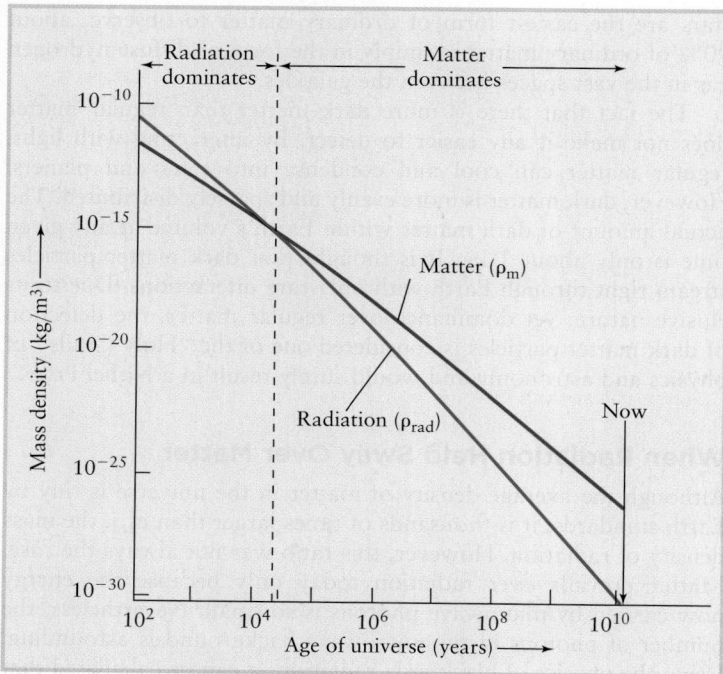

FIGURE 25-10

The Evolution of Density For approximately 24,000 years after the Big Bang, the mass density of radiation (ρ_{rad}, shown in red) exceeded the matter density (ρ_m, shown in blue), and the universe was radiation-dominated. Later, however, continued expansion of the universe caused ρ_{rad} to become less than ρ_m, at which point the universe became matter-dominated. (Graph courtesy of Clem Pryke, University of Minnesota)

To calculate the temperature of the cosmic background radiation at the time of this transition from a radiation-dominated universe to a matter-dominated one, we use Wien's law (see Section 5-4). This law says that the wavelength of maximum emission (λ_{max}) of a blackbody is inversely proportional to its temperature (T): A *decrease* of λ_{max} by a factor of 2 corresponds to an *increase* of T by a factor of 2.

The present-day peak wavelength of the cosmic background radiation corresponds to a blackbody temperature of 2.725 K. Hence, a peak wavelength 5200 times smaller corresponds to a temperature 5200 times greater: $T = 5200 \times 2.725$ K = 14,000 K. In other words, the radiation temperature at redshift z was greater than the present-day radiation temperature by a factor of $1 + z$. Therefore, the temperature of the radiation background was once much greater and has been declining over the ages, as **Figure 25-11** shows.

CONCEPT CHECK 25-8

At the time Earth formed, was the universe dominated by radiation or by matter?

Answer appears at the end of the chapter.

When the First Atoms Formed

The nature of the universe changed again in a fundamental way about 380,000 years after the Big Bang, when z was roughly 1100 and the temperature of the radiation background was about

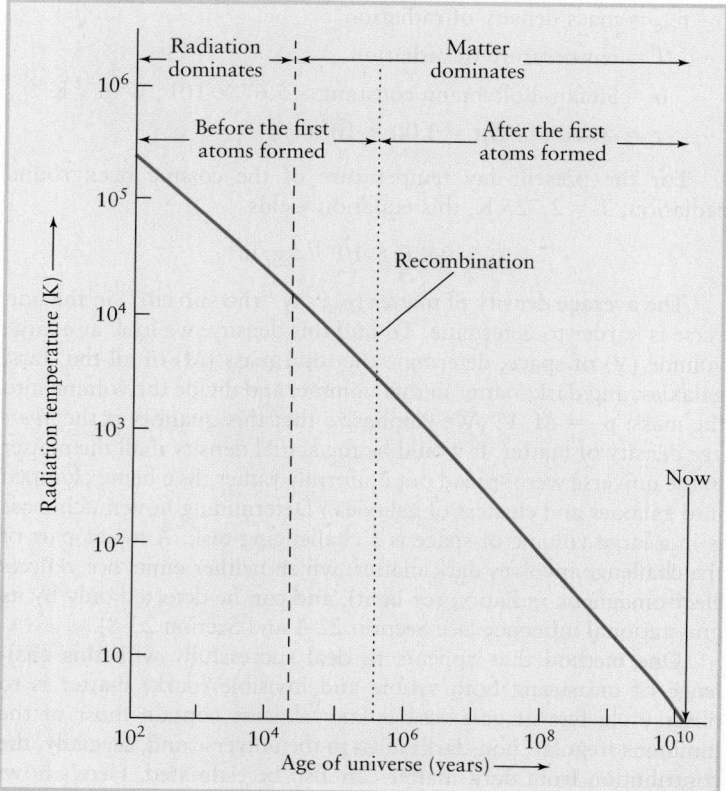

FIGURE 25-11

The Evolution of Radiation Temperature As the universe expanded, the photons in the radiation background became increasingly redshifted and the temperature of the radiation fell. Approximately 380,000 years after the Big Bang, when the temperature fell below 3000 K, hydrogen atoms formed and the radiation field "decoupled" from the matter in the universe. After that point, the temperature of matter in the universe was not the same as the temperature of radiation. The time when the first atoms formed is called the era of recombination (see Figure 25-12). (Graph courtesy of Clem Pryke, University of Minnesota)

1100×2.725 K = 3000 K. (At a temperature of 3000 K, the entire universe was filled with visible orange light.) To see the significance of this moment in cosmic history, recall that hydrogen is by far the most abundant element in the universe—hydrogen atoms outnumber helium atoms by about 12 to 1. A hydrogen atom consists of a single proton orbited by a single electron, and it takes relatively little energy to knock the electron completely out of its orbit around the proton. In fact, ultraviolet radiation warmer than about 3000 K easily ionizes hydrogen. Thus, neutral hydrogen atoms could not survive in the universe that existed before $z = 1100$. That is, in the first 380,000 years after the Big Bang, the background photons had energies great enough to prevent electrons and protons from binding to form hydrogen atoms (**Figure 25-12a**). Only since $z = 1100$ (that is, since $t = 380,000$ years) have the energies of these photons been low enough to permit hydrogen atoms to exist (Figure 25-12b).

The epoch when atoms first formed at $t = 380,000$ years is called the **era of recombination**. This name refers to electrons "recombining" to form atoms. (The name is a bit misleading, because the electrons and protons had never before combined into atoms.)

FIGURE 25-12

The Era of Recombination
(a) Before recombination, the energy of photons in the cosmic background was high enough to prevent protons and electrons from forming hydrogen atoms. **(b)** Some 380,000 years after the Big Bang, the energy of the cosmic background radiation became low enough that hydrogen atoms could survive.

(a) Before recombination:
- Temperatures were so high that electrons and protons could not combine to form hydrogen atoms.
- The universe was opaque: Photons underwent frequent collisions with electrons.
- Matter and radiation were at the same temperature.

(b) After recombination:
- Temperatures became low enough for hydrogen atoms to form.
- The universe became transparent: Collisions between photons and atoms became infrequent.
- Matter and radiation were no longer at the same temperature.

Prior to $t = 380,000$ years, the universe was completely filled with a shimmering expanse of high-energy photons colliding vigorously with protons and electrons. This state of matter, called a **plasma**, is opaque, which means that light cannot pass through it without being strongly scattered and absorbed. The surface and interior of the Sun are also a hot, glowing, opaque plasma (see Section 16-9). P. J. E. Peebles coined the term **primordial fireball** to describe the universe during its first 380,000 years of existence.

After $t = 380,000$ years, the photons no longer had enough energy to keep the protons and electrons apart. As soon as the temperature of the radiation field fell below about 3000 K, protons and electrons began combining to form hydrogen atoms. These atoms do not absorb low-energy photons, so space became transparent! All the photons that heretofore had been vigorously colliding with charged particles could now stream unimpeded across space. Today, these same photons constitute the microwave background.

The significance of the era of recombination becomes clear by considering how it affects an astronomer's ability to form images of the early universe. After recombination, the photons from a large clump of matter can stream to us while having very little interaction with the matter they pass through along their journey. These undisturbed photons can carry images of different clumps of matter in the early universe. On the other hand, photons emitted prior to recombination are scattered so strongly that any image from that early time is completely blurred out before its photons reach us. Therefore, *the era of recombination marks the earliest time and farthest distance that astronomers can collect images of the universe.*

Before recombination, matter and the radiation field had the same temperature because photons, electrons, and protons were all in continuous interaction with one another. After recombination, photons and atoms hardly interacted at all, and thus the temperature of matter in the universe was no longer the same as the temperature of the background radiation. Thus, $T = 2.725$ K is the temperature of the present-day background radiation field, *not* the temperature of the matter in the universe. Note that while the temperature of the background radiation is very uniform, the temperature of matter in the universe is anything but: It ranges

from hundreds of millions of kelvins in the interiors of giant stars to a few tens of kelvins in the interstellar medium.

ANALOGY

A good analogy is the behavior of a glass of cold water. If you hold the glass in your hand, the water will get warmer and your hand will get colder until both the water and your hand are at the same temperature. But if you set the glass down and do not touch it, so that the glass and your hand do not interact, their temperatures are decoupled: The water will stay cold and your hand will stay warm for much longer.

Because the universe was opaque prior to $t = 380,000$ years, we cannot see any farther into the past than the era of recombination. In particular, we cannot see back to the era when the universe was radiation-dominated. The microwave background, whose photons have suffered a redshift of $z = 1100$, contains the most ancient photons we will ever be able to observe.

Note that a detailed description of the universe in the time around recombination is beyond the scope of this textbook. For example, the expansion of the universe also acts to "decouple" the photons from electrons. This is because expansion spreads the electrons out, making them less and less likely to interact with photons. Recombination aids in this decoupling by drastically lowering the number of free electrons for photons to scatter off of.

CONCEPT CHECK 25-9

If the universe had cooled more slowly, would the first atoms have appeared more quickly or more slowly?

Answer appears at the end of the chapter.

Nonuniformities in the Early Universe and the Origin of Galaxies

The COBE data show that the universe is full of blackbody radiation with a temperature of 2.725 K. (This is the cosmic microwave background radiation, or CMB, discussed in Section 25-4.)

–200   –100   0   +100   +200

Temperature difference from average (μK)

ANIMATION 25.2   ANIMATION 25.3

FIGURE 25-13 ⓡ I V U X G

Temperature Variations in the Cosmic Microwave Background This map from Wilkinson Microwave Anisotropy Probe (WMAP) data shows small variations in the temperature of the cosmic background radiation across the entire sky. (The variations due to Earth's motions through space, shown in Figure 25-8, have been factored out.) Lower-temperature regions (shown in blue) show where the early universe was slightly denser than average; warmer regions (shown in red) correspond to regions where the density of matter was less dense than average. Note that the temperature variations in this figure are no more than 200 μK, or 2×10^{-4} K; these are tiny fluctuations around the blackbody radiation measured by COBE in Figure 25-7. (NASA/WMAP Science Team)

However, more sensitive observations show that there are slight temperature variations spread across the sky (Figure 25-13). What do these temperature variations tell us?

The distribution of matter in the early universe was not perfectly uniform, and these temperature variations reveal the nonuniformities—the clumps and voids—in the mass distribution. It is important to understand that the temperature fluctuations do not arise from the temperature of the matter. Instead, clumps in the distribution of matter produce a gravitational redshift (see Figure 21-14b) on the CMB photons as they leave a denser region. Redshifted CMB photons not only have longer wavelengths and lower energy, but they also correspond to lower CMB temperatures. Thus, in Figure 25-13, the cooler bluer spots correspond to denser regions.

Astronomers place great importance on studying temperature and density variations in the cosmic background radiation. The reason is that clumps of matter seen at early times are the "seeds" of structures that continued to "grow" into today's superclusters of galaxies through gravitational contraction. Within these immense concentrations formed the galaxies, stars, and planets. Thus, by studying these nonuniformities, we are really studying the origins of our present-day structure.

Temperature variations in the cosmic background radiation do more than show us the origins of large-scale structure in the universe. As we will see in the next two sections, these temperature variations actually reveal the shape of the universe as a whole, including many of its most fundamental properties.

CONCEPT CHECK 25-10

Consider a blue patch in Figure 25-13. If astronomers looked at everything they could see in the same region with a visible-light telescope, what are they likely to see? In other words, what has probably become of a blue region in the 13.7 billion years since the CMB photons began streaming toward us?

Answer appears at the end of the chapter.

25-6 The shape of the universe indicates its matter and energy content

We have seen that by following the mass densities of radiation (ρ_{rad}) and of matter (ρ_m), we can learn about the evolution of the universe. But it is equally important to know the combined mass density of *all* forms of matter and energy. (In an analogous way, an accountant needs to know the overall financial status of a company, not just individual profits or losses.) Remarkably, we can do this by investigating the overall shape of the universe.

The Curvature of the Universe

Einstein's general theory of relativity explains that gravity curves the fabric of space. Furthermore, the equivalence between matter and energy, expressed by Einstein's equation $E = mc^2$, tells us that either matter or energy produces gravity. Thus, the matter and energy scattered across space should give the universe an overall curvature. The degree of curvature depends on the **combined average mass density** of all forms of matter and energy. (Again, because of Einstein's equivalence between matter and energy, some also refer to this as the combined average energy density.) This quantity, which we call ρ_0 (say "rho sub zero"), is the sum of the average mass densities of matter, radiation, and any other form of energy. Thus, by measuring the curvature of space, we should be able to determine the value of ρ_0 and, hence, learn about the content of the universe as a whole.

To see what astronomers mean by the curvature of the universe, imagine shining two powerful laser beams out into space so that they are perfectly parallel as they leave Earth. Furthermore, suppose that nothing gets in the way of these two beams, so we can follow them for billions of light-years as they travel across the universe and across the space whose curvature we wish to detect. Figure 25-14 illustrates the only three possibilities:

1. We might find that our two beams of light remain perfectly parallel, even after traversing billions of light-years. In this case, space would not be curved: The universe would have **zero curvature**, and space would be **flat**.

2. Alternatively, we might find that our two beams of light gradually converge. In such a case, space would not be flat. Recall that lines of longitude on Earth's surface are parallel at the equator but intersect at the poles. Thus, in this case, the three-dimensional geometry of the universe would be analogous to the two-dimensional geometry of a spherical surface. We would then say that space is **spherical** and that the universe

(a) Flat space, $\rho_0 = \rho_c$, $\Omega_0 = 1$:
The sum of angles in a triangle is 180°.

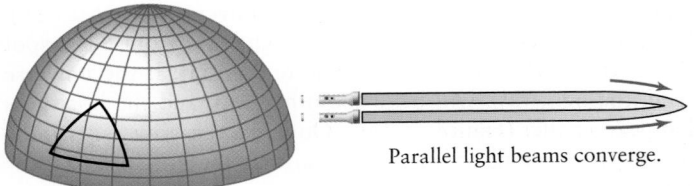

(b) Spherical (closed) space, $\rho_0 > \rho_c$, $\Omega_0 > 1$:
The sum of angles in a triangle is greater than 180°.

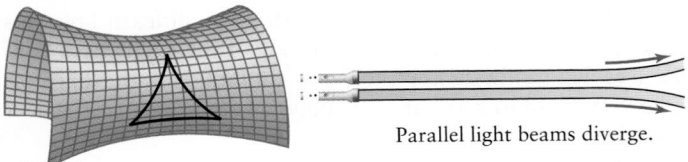

(c) Hyperbolic (open) space, $\rho_0 < \rho_c$, $\Omega_0 < 1$:
The sum of angles in a triangle is less than 180°.

FIGURE 25-14

The Geometry of the Universe The curvature of the universe is either
(a) zero (for flat space), **(b)** positive (for a spherical space), or **(c)** negative
(for a hyperbolic space). The curvature depends on whether the combined
mass density is greater than, equal to, or less than the critical density, or,
equivalently, on whether the density parameter Ω_0 ("omega sub zero")
is greater than, equal to, or less than 1. In theory, the curvature could be
determined by seeing whether two laser beams initially parallel to each
other would converge, remain parallel, or spread apart. Another difference
between these spaces is the sum of the angles in a triangle—the sum is only
equal to 180° in a flat space. While the shapes shown here are drawn as if
they are in a three-dimensional space, the actual space represented in these
universes is only the two-dimensional surface. An actual universe can have
three-dimensional versions of each of these cases (without requiring a fourth
spatial dimension), but these spaces can't be easily illustrated.

has **positive curvature**. Such a universe is also called **closed**,
because if you travel in a straight line in any direction in such
a universe, you will eventually return to your starting point.

3. Finally, we might find that the two initially parallel beams
of light would gradually diverge, becoming farther and farther apart as they move across the universe. In this case, the
universe would still have to be curved, but in the opposite
sense from the spherical model. We would then say that the
universe has **negative curvature**. In the same way that a sphere
is a positively curved two-dimensional surface, a saddle is a
good example of a negatively curved two-dimensional surface.
Parallel lines drawn on a sphere always converge, but parallel
lines drawn on a saddle always diverge. Mathematicians say
that saddle-shaped surfaces are hyperbolic. Thus, in a negatively curved universe, we would describe space as **hyperbolic**.

Such a universe is also called **open** because if you were to travel
in a straight line in any direction, you would never return to
your starting point.

Figure 25-14 summarizes the three cases of zero curvature,
positive curvature, and negative curvature. Real space is three-
dimensional, but we have drawn the three cases as analogous,
more easily visualized two-dimensional surfaces. Therefore, as
you examine the drawings in Figure 25-14, remember that the real
universe has one more dimension. For example, if the universe is in
fact hyperbolic, then the geometry of space must be the (difficult to
visualize) three-dimensional analog of the two-dimensional surface
of a saddle.

Note that in accordance with the cosmological principle, none
of these models of the universe has an "edge" or a "center." You
can begin to picture this lack of a center in Figure 25-14 for both
the flat and hyperbolic universes, because these spaces would be
infinite if drawn in full and would extend forever in all directions.
A spherical universe is finite, but it also lacks a center and an edge.
To understand this, remember that in Figure 25-14, the space is
only the two-dimensional surface of the sphere; no center or edge
on a spherical surface. You could walk forever around the surface
of a sphere (like the surface of Earth) without finding a center or
an edge. While Figure 25-14 shows only two-dimensional versions
of these different spaces, the actual three-dimensional spaces that
would make a universe also lack a center or edge.

CONCEPT CHECK 25-11

If two lasers are aligned to be perfectly parallel out in a
region of space far from any galaxies, could the geometry
of space ever cause their light beams to cross?
Answer appears at the end of the chapter.

Density Determines Curvature

The curvature of the universe is determined by the actual value of
the combined mass density ρ_0 and a reference density called the
critical density ρ_c (say "rho sub cee"). If ρ_0 is greater than the critical density ρ_c, the universe has positive curvature and is closed. If
ρ_0 is less than ρ_c, the universe has negative curvature and is open.
In the special case that ρ_0 is exactly equal to ρ_c, the universe is flat.
Clearly, the critical density plays a crucial role in determining the
geometry of the universe. It is given by the expression:

Critical density of the universe

$$\rho_c = \frac{3H_0^2}{8\pi G}$$

ρ_c = critical density of the universe

H_0 = Hubble constant

G = universal constant of gravitation

Using a Hubble constant $H_0 = 73$ km/s/Mpc, we get

$$\rho_c = 1.0 \times 10^{-26} \text{ kg/m}^3$$

This critical density amounts to only about five hydrogen atoms per cubic meter! Even with the vast empty spaces between the stars, this density is about 1 million times less than the density of matter from stars and gas in the disk in our Milky Way Galaxy.

Many astronomers prefer to characterize the combined average mass density of the universe in terms of the **density parameter** Ω_0 (say "omega sub zero"). This parameter is just the ratio of the combined average mass density to the critical density:

Density parameter

$$\Omega_0 = \frac{\rho_0}{\rho_c}$$

Ω_0 = density parameter

ρ_0 = combined average mass density

ρ_c = critical density

An open universe (negative curvature) has a density parameter Ω_0 between 0 and 1, and a closed universe (positive curvature) has Ω_0 greater than 1. In a flat universe, Ω_0 is equal to 1. Thus, we can use the value of Ω_0 as a measure of the curvature of the universe (Table 25-1). It is important to understand that every form of energy contributes to Ω_0. One way to think of this is that, according to $E = mc^2$, energy itself can be equated to mass. As we will see shortly, something called dark energy is the largest contributor to Ω_0.

CALCULATION CHECK 25-3

What is the combined average mass density (in kg/m³) in a flat universe?

Answer appears at the end of the chapter.

Measuring the Cosmic Curvature

How can we determine the curvature of space across the universe? In theory, if you drew an enormous triangle whose sides were each 1 billion ly long (see Figure 25-14), you could determine the curvature of space by measuring the three angles of the triangle. If their sum equaled 180°, space would be flat. If the sum was greater than 180°, space would be spherical. And if the sum of the three angles was less than 180°, space would be hyperbolic. Unfortunately, this direct method for measuring the curvature of space is not practical.

A way to determine the curvature of the universe that is both practical and precise is to see if light rays bend toward or away from each other, as shown in Figure 25-14. The greater the distance a pair of light rays has traveled, and hence, the longer the time the light has been in flight, the more pronounced any such

bending should be. Therefore, astronomers test for the presence of such bending by examining the oldest radiation in the universe—the cosmic microwave background.

If the cosmic microwave background were truly isotropic, so that equal amounts of radiation reached us from all directions in the sky, it would be impossible to tell whether individual light rays have been bent. However, as we saw in Section 25-5, there are localized "hot spots" in the cosmic microwave background due to density variations in the early universe. The apparent size of these hot spots depends on the curvature of the universe (Figure 25-15). If the universe is closed, the bending of light rays from a hot spot will make the spot appear larger (Figure 25-15a); if the universe is open, the light rays will bend the other way and the hot spots will appear smaller (Figure 25-15c). Only in a flat universe will the light rays travel along straight lines, so that the hot spots appear with their true size (Figure 25-15b).

By calculating what conditions were like in the primordial fireball, astrophysicists find that in a flat universe, the dominant "hot spots" in the cosmic background radiation should have an angular size of about 1°. (In Section 25-9, we will learn how this is deduced.) This is just what observations of the CMB have confirmed (see Figure 25-13). Hence, the curvature of the universe must be very close to zero, and the universe must be either flat or very nearly so.

As Table 25-1 shows, once we know the curvature of the universe, we can determine the density parameter Ω_0 and hence the combined average mass density ρ_0. By analyzing the data shown in Figure 25-13, astrophysicists find that $\Omega_0 = 1.0$ with an uncertainty of about 2%. In other words, ρ_0 is within 2% of the critical density ρ_c.

The flatness of the universe poses a major dilemma: Even if you include dark matter, there is not enough matter to make the universe flat. We saw in Section 25-5 that the average mass density of matter in the universe, ρ_m, is 2.4×10^{-27} kg/m³. This density is only 0.24 of the critical density ρ_c. We can express this ratio in terms of the **matter density parameter** Ω_m (say "omega sub em"), equal to the ratio of ρ_m to the critical density:

$$\Omega_m = \frac{\rho_m}{\rho_c} = 0.24$$

If matter and radiation were all there is in the universe, the combined average mass density ρ_0 would be equal to ρ_m (plus a tiny contribution from radiation, which we can neglect because the average mass density of radiation is only about 0.02% that of matter). Then the density parameter Ω_0 would be equal to Ω_m—that is, equal to 0.24—and the universe would be open. But the temperature variations in the cosmic microwave background clearly show that the universe is either flat or very nearly so.

TABLE 25-1 The Geometry and Average Density of the Universe

Geometry of space	Curvature of space	Type of universe	Combined average mass density (ρ_0)	Density parameter (Ω_0)
Spherical	Positive	Closed	$\rho_0 > \rho_c$	$\Omega_0 > 1$
Flat	Zero	Flat	$\rho_0 = \rho_c$	$\Omega_0 = 1$
Hyperbolic	Negative	Open	$\rho_0 < \rho_c$	$\Omega_0 < 1$

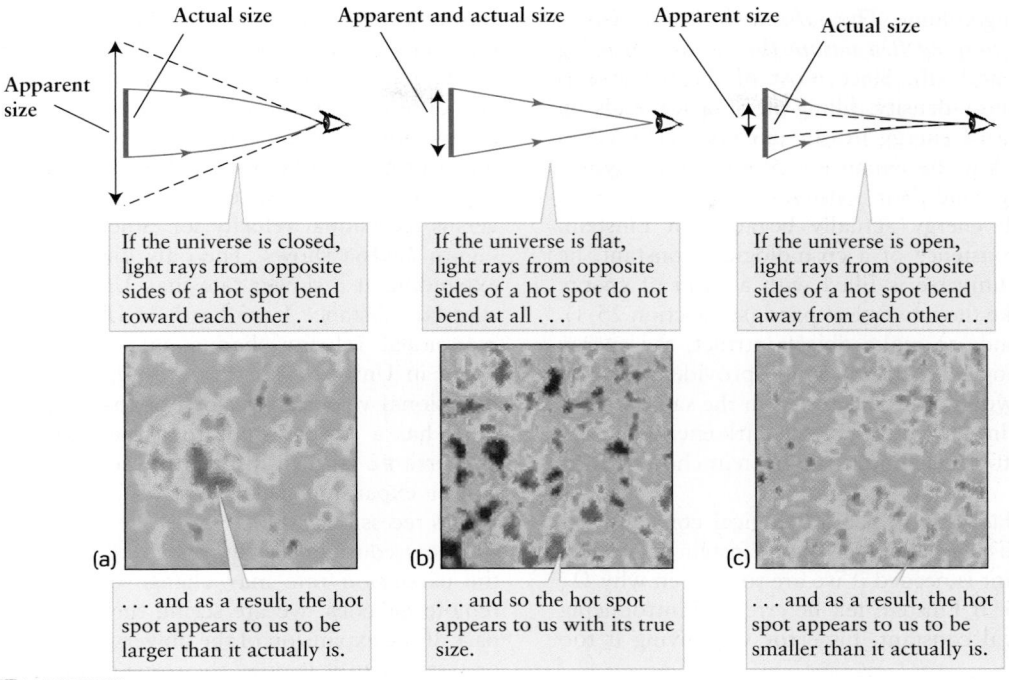

Actual size

Apparent size

Apparent and actual size

Apparent size   Actual size

If the universe is closed, light rays from opposite sides of a hot spot bend toward each other . . .

If the universe is flat, light rays from opposite sides of a hot spot do not bend at all . . .

If the universe is open, light rays from opposite sides of a hot spot bend away from each other . . .

(a)

(b)

(c)

. . . and as a result, the hot spot appears to us to be larger than it actually is.

. . . and so the hot spot appears to us with its true size.

. . . and as a result, the hot spot appears to us to be smaller than it actually is.

FIGURE 25-15 Ⓡ I V U X G

The Cosmic Microwave Background and the Curvature of Space Temperature variations in the early universe appear as "hot spots" in the cosmic microwave background. The apparent size of these spots depends on the curvature of space. (The Boomerang Collaboration)

These variations also show that the density parameter Ω_0, which includes the effects of *all* kinds of matter and energy, is equal to 1.0. In other words, radiation and matter, including dark matter, together account for only 24% of the total density of the universe! The dilemma is to account for the rest of the density.

CONCEPT CHECK 25-12

In a closed universe, a 1-m stick is moved from Earth to a specific location in the distant cosmos. Assuming it could be observed from Earth, would the stick appear larger, smaller, or the same in size than in a flat universe at that same distance?

Answer appears at the end of the chapter.

Dark Energy

The source of the missing energy density must be some form of energy that we cannot detect through the various gravitational effects that astronomers use to detect dark matter. The missing energy density also does not appear to emit detectable radiation of any kind, so we cannot directly detect it with light. With these properties in mind, we refer to this mysterious energy as **dark energy.**

The geometry of space reveals that the universe is filled with dark energy.

Just as we express the average density of matter and radiation by the matter density parameter Ω_m, we can express the average density of dark energy in terms of the **dark energy density**

parameter Ω_Λ (say "omega sub lambda"). This parameter is equal to the average mass density of dark energy, ρ_Λ, divided by the critical density ρ_c:

$$\Omega_\Lambda = \frac{\rho_\Lambda}{\rho_c}$$

We can determine the value of Ω_Λ by noting that the combined average mass density ρ_0 must be the sum of the average mass densities of matter, radiation, and dark energy. As we have seen, the contribution of radiation is so small that we can ignore it, so we have

$$\rho_0 = \rho_m + \rho_\Lambda$$

If we divide this through by the critical density ρ_c, we obtain

$$\Omega_0 = \Omega_m + \Omega_\Lambda$$

That is, the density parameter Ω_0 is the sum of the matter density parameter Ω_m and the dark energy density parameter Ω_Λ. Solving for Ω_Λ, we find

$$\Omega_\Lambda = \Omega_0 - \Omega_m$$

Since Ω_0 is close to 1.0 and Ω_m is 0.24, we conclude that Ω_Λ must be $1.0 - 0.24 = 0.76$. Thus, whatever dark energy is, it accounts for 76% (about three-quarters) of the energy content of the universe!

To put the dark energy density into context, consider the kinetic energy of a jumping flea. The flea's kinetic energy is very small (about the same amount of electrical energy used in turning on a single LED light for one ten-millionth of a second). Since an energy density is the amount of energy contained within a certain volume, to put the dark energy density into context, we also need to indicate a corresponding volume. Thus, *the dark energy density is equivalent to a single jumping flea within the volume of a big football stadium* (dome and all). Since most of the universe is devoid of matter, this energy density, filling all of space, ends up being the dominant form of energy in our universe. (The CMB contains many photons, but the expansion of space has significantly lowered their energy and their density.)

The concept of dark energy actually began with Einstein. When he proposed the existence of a cosmological constant, he was suggesting that the universe is filled with a form of energy that by itself tends to make the universe expand (see Section 25-1). Unlike gravity, which tends to make objects attract, the energy associated with a cosmological constant would provide a form of "antigravity." Hence, it would not be detected in the same way as matter. (The subscript Λ in the symbol for the dark energy density parameter pays homage to the symbol that Einstein chose for the cosmological constant.)

If dark energy is in fact due to a cosmological constant, the value of this constant must be far larger than Einstein suggested. This change in the constant is needed if we are to explain why Ω_Λ has a large value of 0.76. If Einstein felt he erred by introducing the idea of a cosmological constant, his error was giving it too small a value!

These ideas concerning dark energy are extraordinary, and extraordinary claims require extraordinary evidence to confirm them. As we will see in the next section, another way to measure dark energy is to examine how the rate of expansion of the universe has evolved over the eons.

CONCEPT CHECK 25-13

If we do not know what makes up dark energy, how can we estimate its energy density?

Answer appears at the end of the chapter.

25-7 Observations of distant supernovae reveal that we live in an accelerating universe

We have seen that the universe is expanding. But does the rate of expansion stay the same? Because there is matter in the universe, and because gravity tends to pull the bits of matter in the universe toward one another, we would expect the expansion to slow down with time. (In the same way, a cannonball shot upward from the surface of Earth will slow down as it ascends because of Earth's gravitational pull.) If there is a cosmological constant, however, its associated dark energy will exert an outward pressure that tends to accelerate the expansion. Which of these effects is more important?

Modeling the Expansion History of the Universe

To determine whether the expansion of the universe is slowing down or speeding up, astronomers study the relationship between redshift and distance for extremely remote galaxies. We see these galaxies as they were billions of years ago. If the rate of expansion was the same in the distant past as it is now, the same Hubble law should apply to distant galaxies as to nearby ones. But if the

rate of expansion has either increased or decreased, we will find important deviations from the Hubble law.

To see how astronomers approach this problem, first imagine two different parallel universes. Both Universe #1 and Universe #2 are expanding at constant rates, so for both universes there is a direct proportion between recessional velocity v and distance d as expressed by the Hubble law $v = H_0 d$. Hence, a graph of distance versus recessional velocity for either universe is a straight line, as Figure 25-16a shows. The only difference is that Universe #1 is expanding at a slower rate than Universe #2. Hence, a galaxy at a certain distance from Earth in Universe #1 will have a slower recessional velocity than a galaxy at the same distance from Earth in Universe #2. As a result, the graph of distance versus recessional velocity for slowly expanding Universe #1 (shown in blue) has a steeper slope than the graph for rapidly expanding Universe #2 (shown in green). Keep this observation in mind: A slower expansion means a steeper slope on a graph of distance versus recessional velocity.

Now consider *our* universe and allow for the possibility that the expansion rate may change over time. If we observe very remote galaxies, we are seeing them as they were in the remote past. If the expansion of the universe in the remote past was slower or faster than it is now, the slope of the graph of distance versus recessional velocity will be different for those remote galaxies. If the expansion was slower, then the slope will be steeper for distant galaxies (shown in blue in Figure 25-16b); if the expansion was faster, the slope will be shallower for distant galaxies (shown in green in Figure 25-16b). In either case, there will be a deviation from the straight-line Hubble law (shown in red in Figure 25-16b).

Measuring Ancient Expansion with Type Ia Supernovae

Which of the possibilities shown in Figure 25-16b represents the actual history of the expansion of our universe? In Section 23-5, we looked at the observed relationship between distance and recessional velocity for galaxies. Figure 23-17 is a plot of some representative data. The data points appear to lie along a straight line, suggesting that the rate of cosmological expansion has not changed. (Figure 23-17 is actually a graph of recessional velocity versus distance, not the other way around. But a straight line on one kind of graph will be a straight line on the other, because in either case there is a direct proportion between the two quantities being graphed.) However, the graph in Figure 23-17 was based on measurements of galaxies no farther than 400 Mpc (1.3 billion ly) from Earth, which means we are looking only 1.3 billion years into the past. The straightness of the line in Figure 23-16 means only that the expansion of the universe has been relatively constant over the past 1.3 billion years—only 10% of the age of the universe and a relatively brief interval on the cosmic scale.

Now suppose that you were to measure the redshifts and distances of galaxies *several* billion light-years from Earth. The light from these galaxies has taken billions of years to arrive at your telescope, so your measurements will reveal how fast the universe was expanding billions of years ago. To determine the expansion, we need a technique that will allow us to find the distances to these very remote galaxies. We saw in Section 23-4 that one way to determine distances is to identify Type Ia supernovae in such

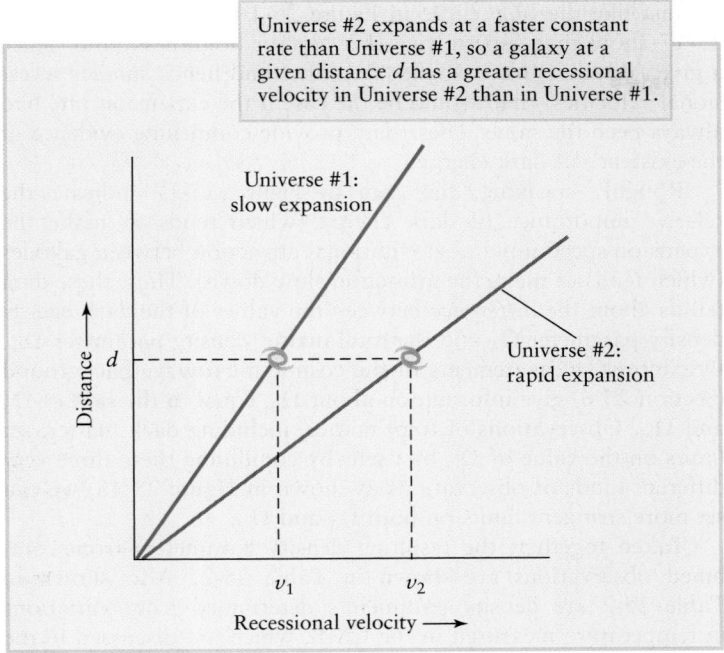

(a) Two universes with different expansion rates

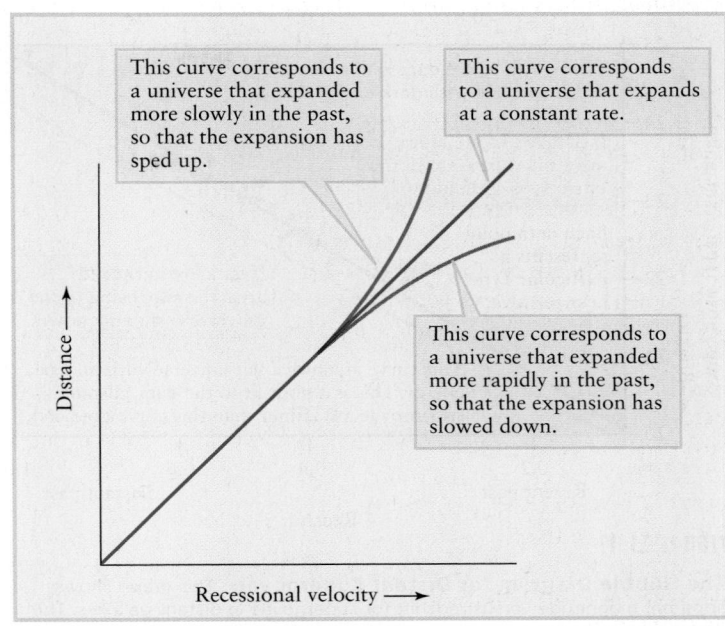

(b) Possible expansion histories of the universe

FIGURE 25-16

Varying Rates of Cosmic Expansion **(a)** Imagine two universes, #1 and #2. Each expands at its own constant rate. For a galaxy at a given distance, the recessional velocity will be greater in the more rapidly expanding universe. Hence, the graph of distance d versus recessional velocity v will have a shallower slope for the rapidly expanding universe and a steeper slope for the slowly expanding one. **(b)** If the rate of expansion of our universe was more rapid in the distant past, corresponding to remote distances, the graph of d versus v will have a shallower slope for large distances (green curve). If the expansion rate was slower in the distant past, the graph will have a steeper slope for large distances (blue curve).

galaxies. These supernovae are among the most luminous objects in the universe, and hence can be detected even at extremely large distances (see Figure 23-14). The maximum brightness of a supernova tells astronomers its distance through the inverse-square law for light, and the redshift of the supernova's spectrum tells them its recessional velocity. As an example, the chapter-opening image shows Type Ia supernovae with redshifts $z = 1.010$, 1.230, and 1.390, corresponding to recessional velocities of 60%, 67%, and 70% of the speed of light. We see these supernovae as they were 7.7 to 9.0 billion years ago, when the universe was less than half of its present age.

In 1998, two large research groups—the Supernova Cosmology Project, led by Saul Perlmutter of Lawrence Berkeley National Laboratory, and the High-Z Supernova Search Team, led by Brian Schmidt of the Mount Stromlo and Siding Springs Observatories in Australia—reported their results from a survey of Type Ia supernovae in galaxies at redshifts of 0.2 or greater, corresponding to distances beyond 750 Mpc (2.4 billion ly). Figure 25-17 shows some of their data, along with more recent observations, on a graph of apparent magnitude versus redshift. Recall that a greater apparent magnitude corresponds to a dimmer supernova (see Section 17-3), which means that the supernova is more distant. A greater redshift implies a greater recessional velocity. Hence, this graph is basically the same as those in Figure 25-16 (distance versus recessional velocity).

To interpret these results, we need guidance from **relativistic cosmology**. This field provides a theoretical description of the universe and its expansion, based on Einstein's general theory of relativity, and was developed in the 1920s by Alexander Friedmann in Russia, Georges Lemaître in France, and Willem de Sitter in the Netherlands. Given values of the mass density parameter Ω_m and the dark energy density parameter Ω_Λ, cosmologists can use the equations of relativistic cosmology to predict how the expansion rate of the universe should change over time. Such predictions can be expressed as curves on a graph of distance versus redshift, as demonstrated in Figure 25-17.

The lower, gray curve in Figure 25-17 shows what would be expected in a flat universe with $\Omega_0 = 1.00$ but with no dark energy, so that $\Omega_\Lambda = 0$ and $\Omega_m = \Omega_0$ (that is, the density parameter is due to matter and radiation alone). In this model, and in fact in any model whose curve lies in the red area in Figure 25-17, the absence of dark energy means that gravitational attraction between galaxies would cause the expansion of the universe to slow down with time. Hence, the expansion rate would have been greater in the past (compare with the green curve in Figure 25-16b).

In fact, the data points in Figure 25-17 are almost all in the *blue* region of the graph, and agree very well with the curve shown in black. This curve also assumes a flat universe, but with an amount of dark energy consistent with the results from the cosmic microwave background ($\Omega_m = 0.24$, $\Omega_\Lambda = 0.76$, $\Omega_0 = \Omega_m + \Omega_\Lambda = 1.00$). In this model, and indeed in any model whose curve lies in the blue region of Figure 25-17, dark energy has made the expansion of the universe speed up over time. Hence, the expansion of the universe was slower in the distant past, which means that we live in an *accelerating* universe.

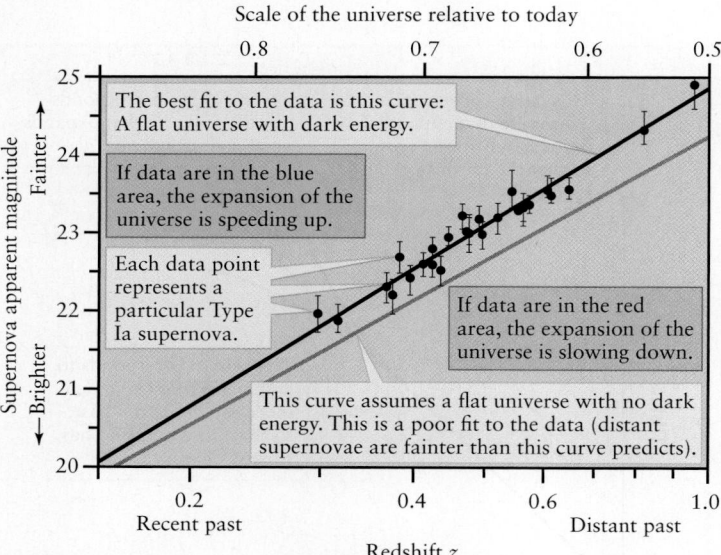

FIGURE 25-17

The Hubble Diagram for Distant Supernovae This graph shows apparent magnitude versus redshift for supernovae in distant galaxies. The greater the apparent magnitude, the dimmer the supernova and the greater the distance to it and its host galaxy. If the expansion of the universe is speeding up, the data will lie in the blue area; if it is slowing down, the data will lie in the red area. The data show that the expansion is in fact speeding up. (The Supernova Cosmology Project/S. Perlmutter)

Just like the blue curve in Figure 25-16b, the data in Figure 25-17 show that supernovae of a certain brightness (and hence a given distance) have smaller redshifts (and hence smaller recessional velocities) than would be the case if the expansion rate had always been the same. These data provide compelling evidence of the existence of dark energy.

Roughly speaking, the data in Figure 25-17 indicate the relative importance of dark energy (which tends to make the expansion speed up) and gravitational attraction between galaxies (which tends to make the attraction slow down). Thus, these data tell us about the *difference* between the values of the dark energy density parameter Ω_Λ and the total matter density parameter Ω_m. By contrast, measurements of the cosmic microwave background (Section 25-6) give information about Ω_0, equal to the *sum* of Ω_Λ and Ω_m. Observations of total matter, including dark matter, set limits on the value of Ω_m by itself. By combining these three very different kinds of observations as shown in Figure 25-18, we can set more stringent limits on both Ω_Λ and Ω_m.

Taken together, the resulting density parameters from combined observations are shown in Table 25-2. Also shown in Table 25-2 are density parameters determined from variations in temperature measured in the CMB, which are discussed in the next section. The values in this table also depend on the value of the Hubble constant. Recall from Section 23-5 that while we use $H_0 = 73$ km/s/Mpc in this text, the value from the Planck CMB observations is $H_0 = 68$ km/s/Mpc—this is what is assumed in

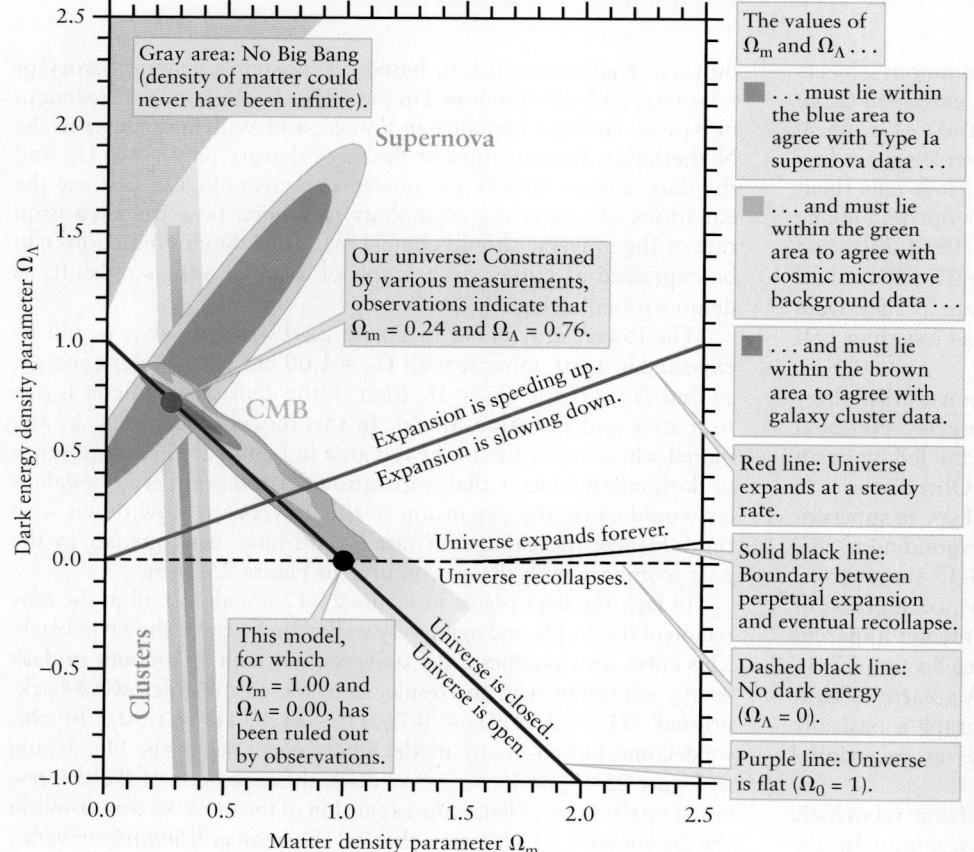

FIGURE 25-18

Limits on the Nature of the Universe The three regions on this graph show values of the mass density parameter Ω_m and the dark energy density parameter Ω_Λ that are consistent with various types of observations. Galaxy cluster measurements (in brown) set limits on Ω_m. Observations of the cosmic microwave background (in green) set limits on the sum of Ω_m and Ω_Λ: A larger value of Ω_m (to the right in the graph) implies a smaller value of Ω_Λ (downward in the graph) to keep the sum the same, which is why this band slopes downward. Observations of Type Ia supernovae (in blue) set limits on the difference between Ω_m and Ω_Λ; this band slopes upward since a larger value of Ω_m implies a larger value of Ω_Λ to keep the difference the same. The best agreement to all these observations is where all three regions overlap (the red dot). (The Supernova Cosmology Project/S. Perlmutter)

TABLE 25-2

Density parameters	Combined observations	Planck CMB variations
Ω_m	0.24	0.31
Ω_Λ	0.76	0.69
$\Omega_0 = \Omega_m + \Omega_\Lambda$	1.00	1.00

determining the corresponding Planck density parameters in the table. As yet, there is no consensus on the precise value of the Hubble constant.

This collection of numbers points to a radically different model of the universe from what was suspected just a few years ago. In the 1980s, there was no compelling evidence for an accelerating expansion of the universe, so it was widely assumed that $\Omega_\Lambda = 0$. Evidence from distant galaxies suggested a flat universe, so it was presumed that $\Omega_m = 1$. Figure 25-18 shows that modern data rule out this model.

> Dark energy became the dominant form of energy in the universe at about the same time that our solar system formed.

The model we are left with is one in which the universe is suffused with a curious dark energy due to a cosmological constant. Unlike matter or radiation, whose average densities decrease as the universe expands and thins out, the average density of this dark energy remains constant throughout the history of the universe (Figure 25-19). The dark energy was relatively unimportant over most of the early history of the universe. Today, however, the density of dark energy is greater than that of matter (Ω_Λ is greater than Ω_m). In other words, we live in a **dark-energy–dominated universe**.

CONCEPT CHECK 25-14

What assumption must one make about Type Ia supernovae in order to use them to measure distances to galaxies?

Answer appears at the end of the chapter.

Cosmic Expansion: From Slowing Down to Speeding Up

ANIMATION 25.4 ANIMATION 25.5

As Figure 25-19 shows, the dominance of dark energy is a relatively recent development in the history of the universe. Prior to about 5 billion years ago, the density of matter should have been greater than that of dark energy. Hence, we would expect that up until about 5 billion years ago, the expansion of the universe should have been slowing down rather than speeding up. Recently, astronomers have found evidence of this picture by using the Hubble Space Telescope to observe extremely distant Type Ia supernovae with redshifts z greater than 1. (The chapter-opening image shows three of these supernovae.) When astronomers compare the distance to these supernovae (determined from their brightness) to their redshifts, they find that the redshifts are *greater* than would be the case if the expansion of the universe had always been at the same rate or had always been speeding up (see Figure 25-16).

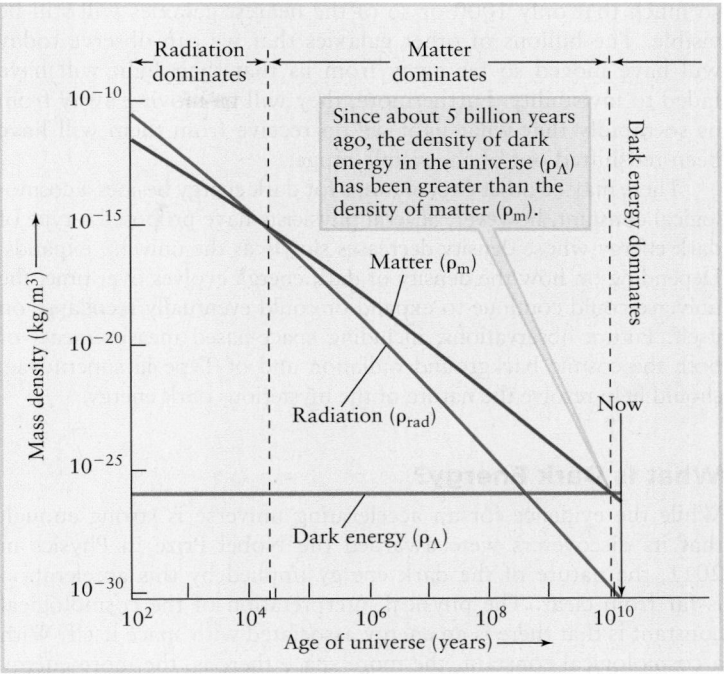

FIGURE 25-19

The Evolution of Density, Revisited The average mass density of matter, ρ_m, and the average mass density of radiation, ρ_{rad}, both decrease as the universe expands and becomes more tenuous. But if the dark energy is due to a cosmological constant, its average mass density ρ_Λ remains constant. In this model, our universe became dominated by dark energy about 5 billion years ago.

This is just what would be expected if the expansion was slowing down in the very early universe. After about 5 billion years ago, the effects of dark energy became dominant and the expansion began to speed up.

ANALOGY

If you see a red light up ahead while driving, you would probably apply the brakes to make the car slow down. But if the light then turns green before your car comes to a stop, and the road ahead is clear, you would step on the gas to make the car speed up again. The expansion of the universe has had a similar history. The mutual gravitational attraction of all the matter in the universe means that "the brakes were on" for about the first 9 billion years after the Big Bang, so that the expansion slowed down. But for about the past 5 billion years, dark energy has "had its foot on the gas," and the expansion has been speeding up.

So far, observations of distant supernovae support dark energy that is described by a cosmological constant. If dark energy truly is a cosmological constant, the density of dark energy will continue to remain constant, as shown in Figure 25-19. Due to the effects of this dark energy, the universe will keep on expanding forever, and the rate of expansion will continue to accelerate. Eventually, some 30 billion years from now, the universe will have expanded so much that only 1000 or so of the nearest galaxies will still be visible. The billions of other galaxies that we can observe today will have moved so far away from us that their light will have faded to invisibility. Furthermore, they will be moving away from us so rapidly that what light we do receive from them will have been redshifted out of the visible range.

There may be other explanations for dark energy besides a cosmological constant, however. Several physicists have proposed a type of dark energy whose density decreases slowly as the universe expands. Depending on how the density of dark energy evolves over time, the universe could continue to expand or could eventually recollapse on itself. Future observations, including space-based measurements of both the cosmic background radiation and of Type Ia supernovae, should help resolve the nature of the mysterious dark energy.

What Is Dark Energy?

While the evidence for an accelerating universe is strong enough that its discoverers were awarded the Nobel Prize in Physics in 2011, the nature of the dark energy implied by this acceleration is far from clear. The physical interpretation of the cosmological constant is that there is an energy associated with space itself. With a cosmological constant, the more space there is, the more energy there is. In other words, as the universe expands, the additional space leads to more energy. Just as the energy content in the universe determines its shape (Section 25-6), the energy content can also cause the universe to accelerate its expansion. This is what dark energy is doing—but how does energy arise from seemingly empty space?

The answer might come from **virtual particle pairs**—imperceptible particles made of matter and antimatter. As mentioned in Section 21-11, quantum physics reveals that seemingly empty space is not so empty after all. Everywhere, at all times, virtual particle pairs pop up from nothing and then disappear again on very short timescales. While the virtual particles themselves are not directly observable, the phenomenon is real and secondary effects of the virtual particles have been observed. In fact, virtual particle effects between microscopic nanotechnology devices are strong enough that this phenomenon can produce practical problems—forces that cause some devices to clump together. In Chapter 26, we will also discuss virtual particles for their possible role in the early universe.

Because virtual particles appear in the otherwise empty "vacuum" of space, they are called *vacuum fluctuations*. Overall, the effect of virtual particles, or vacuum fluctuations, is to produce an energy associated with space itself. This seems to be just what is needed to produce a cosmological constant. However, there is a big problem with this explanation for dark energy. When the effects of virtual particles are calculated, they predict a cosmological constant that is 100 orders of magnitude larger than what is actually observed; that number has a 1 with a *hundred zeroes* after it! With such a large numerical discrepancy between theory and observation, our current understanding of virtual particles and vacuum fluctuations cannot explain dark energy. Perhaps a modification of the vacuum fluctuation model can explain dark energy, but each theoretical modification also introduces new problems.

There are other theories offered to explain dark energy. Some propose that Einstein's general theory of relativity needs modifying on very large scales. Others propose that we are in a special part of the universe that only gives the appearance of accelerated expansion, so that there is actually no dark energy at all. However, each theory has its own difficulties and no explanation has gained wide acceptance. We explore these theories in more detail in the essay at the end of this chapter. While the observations march forward, they are difficult, and dark energy is likely to remain one of the universe's greatest mysteries for some time to come.

25-8 Primordial sound waves help reveal the character of the universe

We have seen how studying the "hot spots" in the cosmic background radiation reveals that we live in a flat universe. In fact, temperature variations reveal more: They give us a window into conditions in the early universe and actually help us pin down the values of other important quantities, such as the Hubble constant and the density of matter in the universe. The key to extracting this additional information from the cosmic background radiation is recognizing that the hot and cold spots in a map such as Figure 25-13 actually result from sound waves.

Sound Waves in the Early Universe

Sound waves can travel in gases and fluids of all kinds. Sound waves in air are used in human speech and hearing, while whales communicate using high-frequency underwater clicks and whistles. If you could take a snapshot of a sound wave, you would see that at any moment there are some regions, called **compressions**, where the gas or fluid is squeezed together, and other regions, called **rarefactions**, where the gas or fluid is thinned out or rarefied (**Figure 25-20**).

> Immense sound waves in the early universe left their imprint as variations in the cosmic microwave background.

There would also have been sound waves in the early universe before recombination. During the first 380,000 years after the Big Bang, the universe was filled with a fluidlike medium composed primarily of photons, electrons, and protons, with a density more than 10^9 times greater than that of our present-day universe. Just as water molecules in a glass of water collide with each other, photons and particles collided frequently with each other in this primordial fluid, triggering random sound waves with compressions and rarefactions.

Before discussing the sound waves in more depth, it is worth revisiting the relationship (discussed in Section 25-5) between temperature and density variations. Because there is more mass in a compression than in a rarefaction, photons emerging from a compression experience a greater gravitational redshift than do photons emerging from a rarefaction (see Section 21-3). As a result, the light from a compression is shifted to slightly longer wavelengths. We saw in Sections 5-3 and 5-4 that a blackbody spectrum dominated by longer wavelengths corresponds to a lower blackbody temperature (see Figure 5-12). Hence, we see compressions as the blue cold spots in Figure 25-13, and we see rarefactions as the red hot spots. The overall pattern of cold and hot spots is thus a record of the density variations due to sound waves that were present just as the universe became transparent, some 380,000 years after the Big Bang. From the denser compressions arose our present-day population of galaxies.

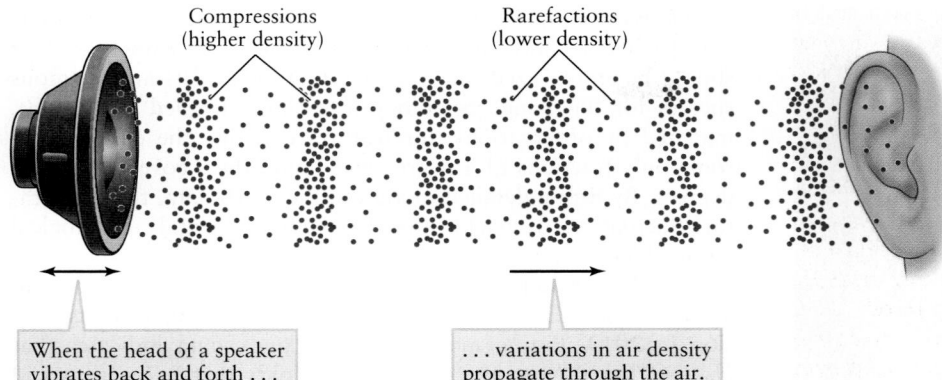

FIGURE 25-20

Snapshot of a Sound Wave When the head of an audio speaker oscillates back and forth, higher-density compressions are created that propagate through the air. The sound from clapping your hands is produced in a similar way, but for a shorter length of time. When a sound wave enters a human ear, the air next to the eardrum is alternately compressed and rarefied, which makes the eardrum flex back and forth. This flexing is translated into an electrical signal that is sent to the brain. Sound waves produced shortly after the Big Bang (inaudible to the human ear) also produced variations in the density of matter throughout the universe.

Compressions (higher density)

Rarefactions (lower density)

When the head of a speaker vibrates back and forth . . .

. . . variations in air density propagate through the air.

The nature of a sound wave depends on the material through which it passes. For example, sound waves travel faster in helium than they do in air (because helium is less dense) and faster still in water (which, while denser than air, is much more resistant to compression). So, by studying the primordial sound waves recorded in Figure 25-13, we can learn about the properties of the fluid that made up the early universe. These properties include the average densities of matter and dark energy in the fluid, as well as the value of the Hubble constant (which helps determine how rapidly the fluid was expanding and thinning out as the universe expanded). We can also determine the age of the universe at the time that the cosmic background radiation was emitted, since this determines the maximum size to which a hot spot (rarefaction) or cold spot (compression) could have grown since the Big Bang.

Figure 25-21 shows an important way in which astronomers systematize their data about hot and cold spots in the cosmic

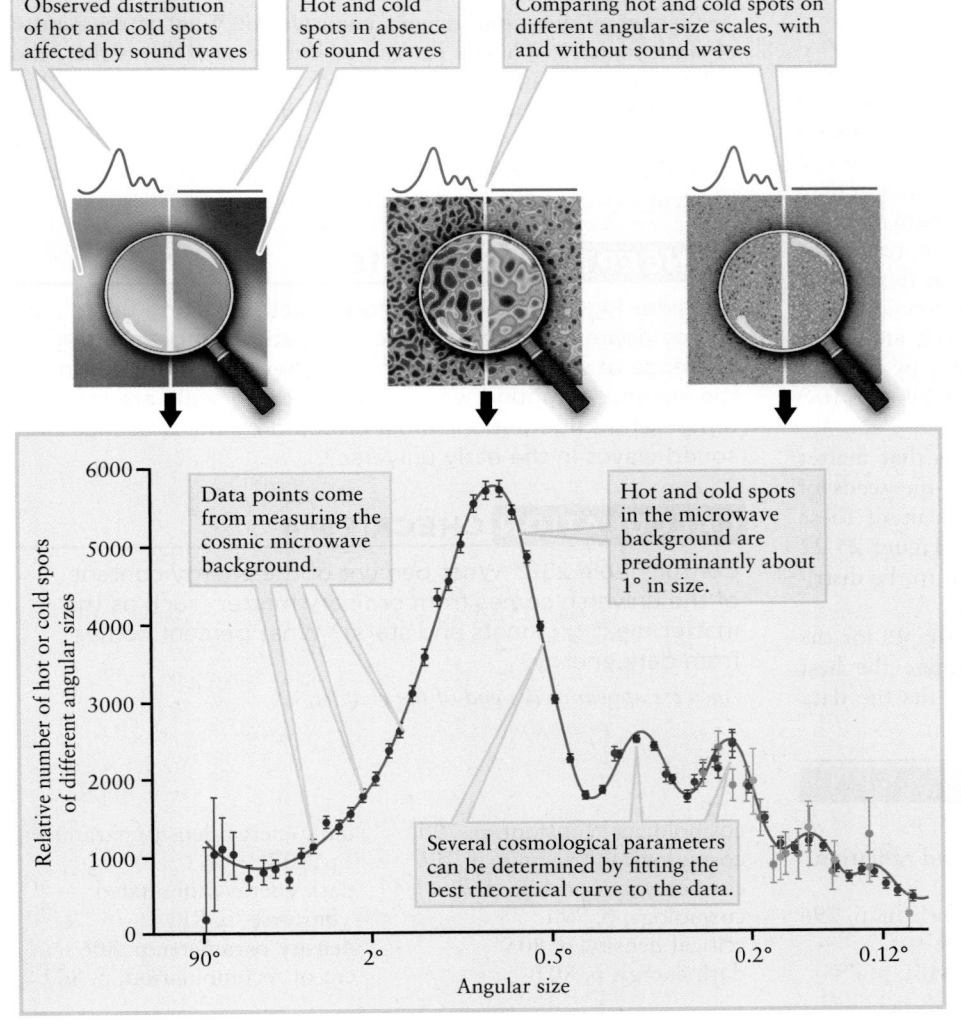

Observed distribution of hot and cold spots affected by sound waves

Hot and cold spots in absence of sound waves

Comparing hot and cold spots on different angular-size scales, with and without sound waves

Data points come from measuring the cosmic microwave background.

Hot and cold spots in the microwave background are predominantly about 1° in size.

Several cosmological parameters can be determined by fitting the best theoretical curve to the data.

FIGURE 25-21

Sound Waves in the Early Universe Top panel: At three different angular sizes (90°, 1°, and 0.25°), a comparison is made between the observed temperature variations in the cosmic microwave background resulting from sound waves and the temperature variations in a hypothetical universe with no sound waves. Sound waves affect the distribution of temperature fluctuations but do not entirely create them. Bottom panel: Observations of the cosmic background radiation show that hot and cold spots of certain angular sizes are more common than others. Matching a model that describes these observations (the red curve) helps to determine the values of important cosmological parameters. Most of the data shown here are from WMAP; the data for the smallest angles (at the right of the graph) come from the CBI detector in the Chilean Andes and the ACBAR and BOOMERANG detectors at the south pole. (Top panel: NASA/WMAP Science Team)

FIGURE 25-22

Structure Evolution from Sound Waves This illustration shows the evolution of early density variations seen in the CMB to the present-day distribution of galaxies. The size scales of the resulting variations are much larger than galaxies or clusters, and the variations show up as filaments and voids traced out by the distribution of superclusters of galaxies. The Baryon Oscillation Spectroscopic Survey (BOSS) has observed these variations at their predicted size scale of about 150 Mpc. (Sam Moorfield and Chris Blake [Swinburne University])

background radiation. This graph shows the number of observed hot or cold spots of different angular sizes, with larger spots on the left and smaller spots on the right. The presence of peaks in the graph shows that spots of certain sizes are more common than others. The largest peak tells us that the predominant angular size is about 1°, which corresponds to a region of compression or rarefaction that was about 1 million (10^6) ly across at the time of recombination at $z = 1100$. (By contrast, the compressions and rarefactions in the sound waves most used in speech are a few meters across.) Since then, the universe has expanded by a factor of about 1100, so that same region is now about 1 billion (10^9) ly across.

The peaks in Figure 25-21 tell us the size scales that matter formed into clumps; these overdense regions were the seeds of structure formation. As predicted, we see the imprint of these early sound waves in today's distribution of mass. Figure 25-22 illustrates this evolution from variations in the CMB to the distribution of galaxies.

Different cosmological models predict different shapes for the curve shown in Figure 25-21. Astronomers determine the best model by seeing which one gives a curve that best fits the data

points. For example, the peak of the curve at an angular size of 1° is just what would be expected for a flat universe with $\Omega_0 = 1$. It should be emphasized that Figure 25-21 represents an enormous success for Big Bang cosmology in making detailed predictions, matched by observations, for features covering the entire sky—some only fractions of a degree wide and others covering tens of degrees. By fitting detailed models to the CMB data, the temperature fluctuations can yield values for the fundamental cosmological parameters.

The best results using the CMB come from the European Space Agency (ESA)'s Planck satellite, which completed its mission in 2013. Fitting models to the Planck data indicates that $H_0 = 68$ km/s/Mpc, the density of ordinary matter (protons and neutrons) is $\Omega_b = 0.05$, dark matter is $\Omega_{DM} = 0.26$, dark energy is $\Omega_\Lambda = 0.69$, and the universe is flat with $\Omega_0 = 1.00$.

Our understanding of the universe as a whole has increased tremendously over the past several years. We have found compelling evidence that dark energy exists and that it is the dominant form of energy in the universe. Studies of supernovae, galaxy clusters, and the cosmic background radiation have provided us with so much high-quality data that we can now express the key parameters of the universe with very high accuracy. When we look back to the situation in the 1980s, when the value of the Hubble constant was uncertain by at least 50%, it is no exaggeration to say that we have entered an age of *precision* cosmology.

Yet many questions remain unanswered. What is the nature of dark matter? What is leading to dark energy? Together, the unknown nature of dark matter and dark energy indicate that we do not understand about 95% of the energy content in our universe. Can either of these mysterious entities be detected and studied in the laboratory? These and other questions will continue to occupy cosmologists for many years to come.

CONCEPT CHECK 25-15

Consider Figure 25-21. The colorful pictures along the top compare how temperature fluctuations appear in the presence of sound waves with how they would appear in the absence of sound waves. On what size scale are the temperature fluctuations most affected by the presence of sound waves in the early universe?

CALCULATION CHECK 25-4

Consult Table 25-2. What percent of the energy content of the universe comes from ordinary matter (such as the matter making planets and stars)? What percent comes from dark energy?

Answers appear at the end of the chapter.

KEY WORDS

KEY IDEAS

The Expansion of the Universe: The Hubble law describes the continuing expansion of space. On large scales, galaxies spread out in this cosmic flow, also called the Hubble flow (Section 25-2).

- In the Hubble flow, galaxies are not actually moving through space, but the space between them is increasing as space expands.

- The redshifts that we see from distant galaxies are caused by this cosmic expansion and are called cosmological redshifts. These redshifts develop as photons travel through expanding space, getting stretched along their journey.

- The redshift of a distant galaxy is a measure of the relative size and age of the universe at the time the galaxy emitted its light.

- There is no center or edge of the known universe.

Original wavelength

(a) A wave drawn on a rubber band . . .

Stretched (redshifted) wavelength

(b) . . . increases in wavelength as the rubber band is stretched.

The Cosmological Principle: Cosmological theories are based on the idea that on large scales, the universe looks roughly the same at all locations and in every direction (Section 25-2).

The Big Bang: The universe began with nearly infinite density and began its expansion in the event called the Big Bang, which can be described as the beginning of time and space (Section 25-3).

- The observable universe extends about 14 billion ly in every direction from Earth. We cannot see objects beyond this distance because light from these objects has not had enough time to reach us.

- During the first 10^{-43} second after the Big Bang, the universe was too dense to be described by the known laws of physics.

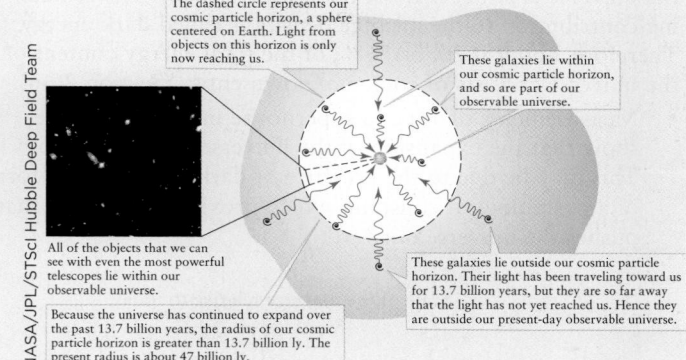

The dashed circle represents our cosmic particle horizon, a sphere centered on Earth. Light from objects on this horizon is only now reaching us.

These galaxies lie within our cosmic particle horizon, and so are part of our observable universe.

All of the objects that we can see with even the most powerful telescopes lie within our observable universe.

These galaxies lie outside our cosmic particle horizon. Their light has been traveling toward us for 13.7 billion years, but they are so far away that the light has not yet reached us. Hence they are outside our present-day observable universe.

Because the universe has continued to expand over the past 13.7 billion years, the radius of our cosmic particle horizon is greater than 13.7 billion ly. The present radius is about 47 billion ly.

NASA/JPL/STScI Hubble Deep Field Team

Cosmic Background Radiation and the Evolution of the Universe: The cosmic microwave background radiation, corresponding to radiation from a blackbody at a temperature of nearly 3 K, is the greatly redshifted remnant of the hot universe as it existed about 380,000 years after the Big Bang (Section 25-4).

- The background radiation was hotter and more intense in the past. During the first 380,000 years of the universe, radiation and matter formed an opaque plasma called the primordial fireball.

- Light from this early era gets scattered before it can carry images to us, but then there is an abrupt change. When the temperature of the radiation fell below 3000 K, protons and electrons could combine to form hydrogen atoms and the universe became transparent. We can look all the way back to this transition, which is what we see in images of the cosmic microwave background radiation.

- The universal abundance of helium is much more than stars can produce. Most helium was produced by thermonuclear reactions occurring throughout the universe during its first few minutes. It was the high temperatures required for these thermonuclear reactions that led to the prediction of the cosmic microwave background radiation, which was much hotter in the early universe.

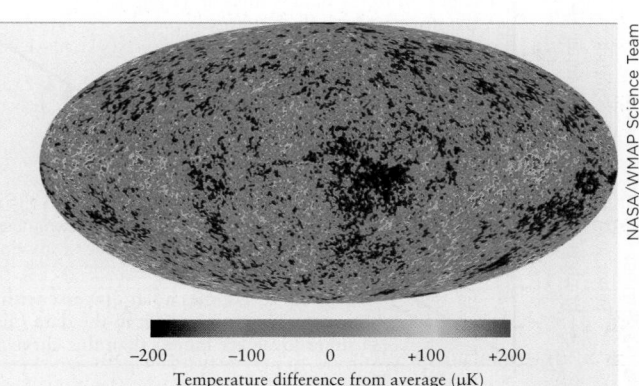

−200   −100   0   +100   +200
Temperature difference from average (µK)

NASA/WMAP Science Team

The Geometry of the Universe: The curvature of the universe as a whole depends on how the combined average mass density ρ_0 compares to a critical density ρ_c. Due to the equivalence of mass and energy, ρ_0 also includes contributions from dark energy (Section 25-6).

- If ρ_0 is greater than ρ_c, the density parameter Ω_0 has a value greater than 1, the universe is closed, and space is spherical (with positive curvature).

- If ρ_0 is less than ρ_c, the density parameter Ω_0 has a value less than 1, the universe is open, and space is hyperbolic (with negative curvature).

- If ρ_0 is equal to ρ_c, the density parameter Ω_0 is equal to 1 and space is flat (with zero curvature).

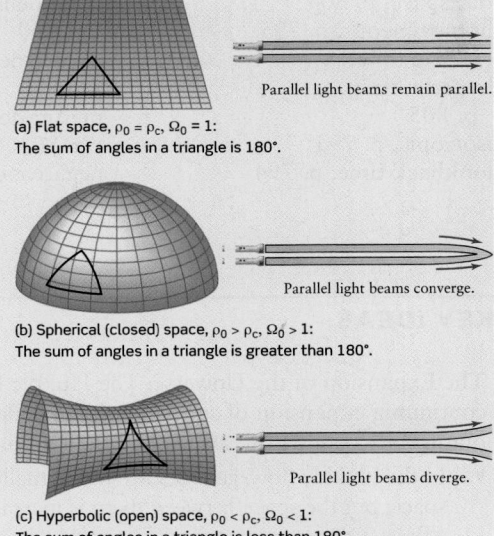

Parallel light beams remain parallel.

(a) Flat space, $\rho_0 = \rho_c$, $\Omega_0 = 1$:
The sum of angles in a triangle is 180°.

Parallel light beams converge.

(b) Spherical (closed) space, $\rho_0 > \rho_c$, $\Omega_0 > 1$:
The sum of angles in a triangle is greater than 180°.

Parallel light beams diverge.

(c) Hyperbolic (open) space, $\rho_0 < \rho_c$, $\Omega_0 < 1$:
The sum of angles in a triangle is less than 180°.

Cosmological Parameters and Dark Energy: Observations of temperature variations in the cosmic microwave background indicate that the universe is flat or nearly so, with a combined average mass density equal to the critical density. Observations of galaxy clusters suggest that the average density of matter in the universe is about 0.24 of the critical density. The remaining contribution to the average density is called dark energy. Therefore, about 0.76, or 76%, of the total energy content of the universe consists of some unknown entity (Section 25-6).

- Measurements of Type Ia supernovae in distant galaxies show that the expansion of the universe is speeding up. This may be due to the presence of dark energy in the form of a cosmological constant, which provides a pressure that pushes the universe outward.

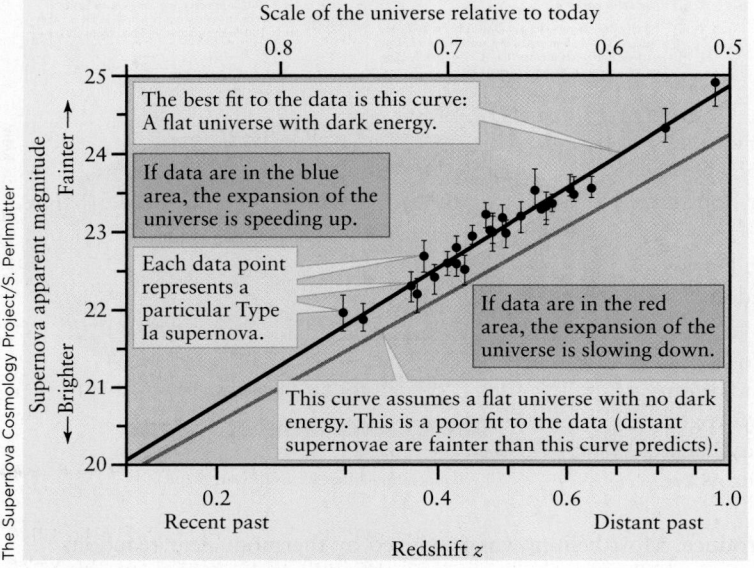

Scale of the universe relative to today

The best fit to the data is this curve: A flat universe with dark energy.

If data are in the blue area, the expansion of the universe is speeding up.

Each data point represents a particular Type Ia supernova.

If data are in the red area, the expansion of the universe is slowing down.

This curve assumes a flat universe with no dark energy. This is a poor fit to the data (distant supernovae are fainter than this curve predicts).

Recent past   Distant past

Redshift z

The Supernova Cosmology Project/S. Perlmutter

Cosmological Parameters and Primordial Sound Waves: Temperature variations in the cosmic background radiation are a record of sound waves in the early universe. Studying the character of these sound waves helps to determine that the universe is flat and other fundamental properties of the universe (Section 25-8).

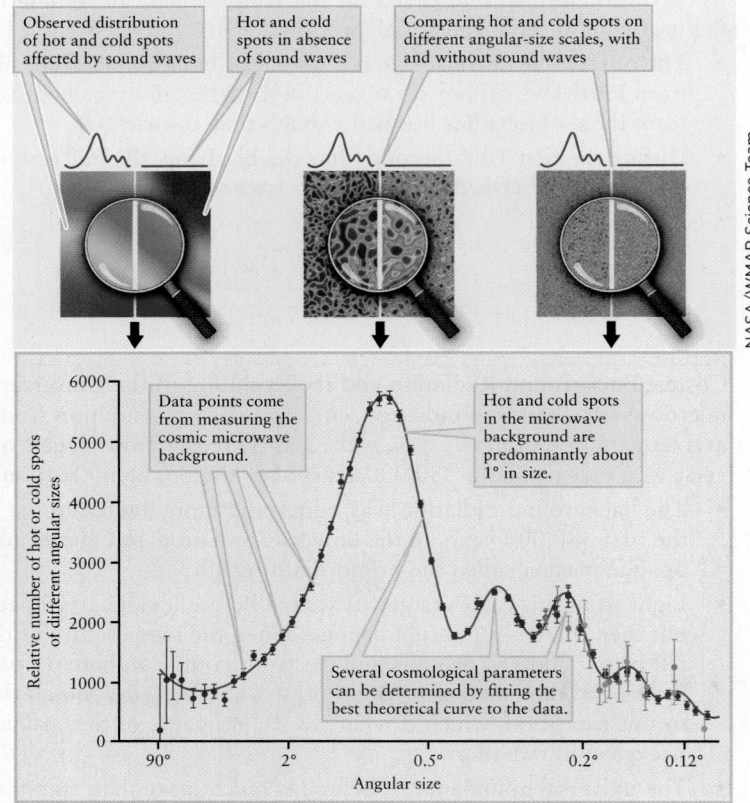

Observed distribution of hot and cold spots affected by sound waves

Hot and cold spots in absence of sound waves

Comparing hot and cold spots on different angular-size scales, with and without sound waves

NASA/WMAP Science Team

Data points come from measuring the cosmic microwave background.

Hot and cold spots in the microwave background are predominantly about 1° in size.

Relative number of hot or cold spots of different angular sizes

Several cosmological parameters can be determined by fitting the best theoretical curve to the data.

Angular size

QUESTIONS

> **Problem-solving tips and tools**
>
> We discussed Wien's law in Section 5-4. You may find it useful to recall that 1 pc equals 3.26 ly, that 1 Mpc equals 3.09×10^{19} km, and that a year contains 3.16×10^7 s.

Section 25-1

1. Why did Isaac Newton conclude that the universe was static? Was he correct?

2. What is Olbers's paradox? How can it be resolved?

3. What is a cosmological constant? Why did Einstein introduce it into cosmology?

Section 25-2

4. **TUTORIAL** What does it mean when astronomers say that we **25.1** live in an expanding universe? What is actually expanding?

5. Describe how the expansion of the universe explains Hubble's law.

6. Would it be correct to say that due to the expansion of the universe, Earth is larger today than it was 4.56 billion years ago? Why or why not?

7. Using a diagram, explain why the expansion of the universe as seen from a distant galaxy would look the same as seen from our galaxy.

8. Despite your intuition from common experience, explain why an infinite expanding universe doesn't have either a center or an edge.

9. Explain the difference between a Doppler shift and a cosmological redshift.

10. Explain how redshift can be used as a measure of lookback time. In what ways is it superior to time measured in years?

11. By what factor has the universe expanded since $z = 1$? Explain your reasoning.

12. What does it mean to say that the universe is homogeneous? That it is isotropic?

13. What is the cosmological principle? How is it justified? Is it a law of nature?

14. **(a)** For what value of the redshift z were representative distances between galaxies only 20% as large as they are now? **(b)** Compared to representative distances between galaxies in the present-day universe, how large were such distances at $z = 8$? Compared to the density of matter in the present-day universe, what was the density of matter at $z = 8$? **(c)** If dark energy is in the form of a cosmological constant, how does its present-day density compare to the density of dark energy at $z = 2$? At $z = 5$? Explain your answers.

15. The host galaxy of the supernova HST04Sas (see the chapter-opening image) has a redshift $z = 1.390$. The light from this galaxy includes the Lyman-alpha (L_α) spectral line of hydrogen, with an unshifted wavelength of 121.6 nm. Calculate the wavelength at which we detect the Lyman-alpha photons from this galaxy. In what part of the electromagnetic spectrum does this wavelength lie?

Section 25-3

16. How was the Big Bang different from an ordinary explosion? Where in the universe did it occur?

17. Some people refer to the Hubble constant as "the Hubble variable." What is constant about the Hubble constant, and what isn't?

18. What is meant by "the observable universe"?

19. **(a)** Explain why the radius of the observable universe is continually increasing. **(b)** Although the universe is 13.7 billion years old, the observable universe includes objects that are more than 13.7 billion ly away from Earth. Explain why.

20. Imagine an astronomer living in a galaxy 1 billion ly away. Is the observable universe for that astronomer the same as for an astronomer on Earth? Why or why not?

21. Estimate the age of the universe for a Hubble constant of **(a)** 50 km/s/Mpc, **(b)** 75 km/s/Mpc, and **(c)** 100 km/s/Mpc. On the basis of your answers, explain how the ages of globular clusters could be used to place a limit on the maximum value of the Hubble constant.

22. Some people claim that the universe came into being about 6000 years ago. Find the value of the Hubble constant for such a universe. Is this a reasonable value for H_0? Explain your answer.

23. The quasar HS 1946+7658 has a redshift $z = 3.02$. At the time when the light we see from HS 1946+7658 left the quasar, how many times more dense was the matter in the universe than it is today?

Section 25-4

24. **TUTORIAL** How did the abundance of helium in the universe **25.2** suggest the existence of the cosmic background radiation?

25. Can you see the cosmic background radiation with the naked eye? With a visible-light telescope? Explain why or why not.

26. If the universe continues to expand forever, what will eventually become of the cosmic background radiation?

27. If the mass density of radiation in the universe were 625 times larger than it is now, what would the background temperature be?

28. How can astronomers be certain that the cosmic microwave background fills the entire cosmos, not just the vicinity of Earth?

Section 25-5

29. What does it mean to say that the universe was once radiation-dominated? What happened when the universe changed from being radiation-dominated to being matter-dominated? When did this happen?

30. What was the era of recombination? What significant events occurred in the universe during this era? Was the universe matter-dominated or radiation-dominated during this era?

31. (a) Was there ever an era when the universe was radiation-dominated and matter and radiation were at the same temperature? If so, approximately when was this, and were there atoms during that era? If not, explain why not. (b) Was there ever an era when the universe was radiation-dominated and matter and radiation were *not* at the same temperature? If so, approximately when was this, and were there atoms during that era? If not, explain why not.

32. Describe two different cases in which the cosmic microwave background is not isotropic.

33. Suppose that the present-day temperature of the cosmic background radiation were somehow increased by a factor of 100, from 2.725 K to 272.5 K. (a) Calculate ρ_{rad} in this situation. (b) If the average density of matter (ρ_m) remained unchanged, would it be more accurate to describe our universe as matter-dominated or radiation-dominated? Explain your answer.

34. Calculate the mass density of radiation (ρ_{rad}) in each of the following situations, and explain whether each situation is matter-dominated or radiation-dominated: (a) the photosphere of the Sun ($T = 5800$ K, $\rho_m = 3 \times 10^{-4}$ kg/m^3); (b) the center of the Sun ($T = 1.55 \times 10^7$ K, $\rho_m = 1.6 \times 10^5$ kg/m^3); (c) the solar corona ($T = 2 \times 10^6$ K, $\rho_m = 5 \times 10^{-13}$ kg/m^3).

35. If a photon from the cosmic microwave background had wavelength λ_0 when it was emitted at redshift z, its wavelength today is $\lambda = \lambda_0(1 + z)$. (a) Let T be the symbol for the temperature of the cosmic microwave background today. Explain why the radiation temperature was $T_0 = T(1 + z)$ at redshift z. (b) What was the radiation temperature at $z = 1$? (c) At what redshift was the radiation temperature equal to 293 K (a typical room temperature)?

36. Suppose we were living in a radiation-dominated universe. Discuss how such a universe would be different from the universe we now observe.

Section 25-6

37. What is meant by the critical density of the universe? Why is this quantity important to cosmologists?

38. Describe how astronomers use the cosmic background radiation to determine the geometry of the universe.

39. What is dark energy? Describe two ways that we can infer its presence.

40. What does it mean to say that the universe is dark energy–dominated? What happened when the universe changed from being matter-dominated to being dark-energy–dominated?

41. What would be the critical density of matter in the universe (ρ_c) if the value of the Hubble constant were (a) 50 km/s/Mpc? (b) 100 km/s/Mpc?

42. Consider the quasar HS 1946+7658 (see Question 23), which has $z = 3.02$. (a) Suppose that in the present-day universe, two clusters of galaxies are 500 Mpc apart. At the time that the light was emitted from HS 1946+7658 to produce an image on Earth tonight, how far apart were those two clusters? (b) What was the average density of matter (ρ_m) at that time? Assume that in today's universe, $\rho_m = 2.4 \times 10^{-27}$ kg/m^3. (c) What were the temperatures of the cosmic background radiation and the mass density of radiation (ρ_{rad}) at that time? (d) At this time in the remote past, was the universe matter-dominated, radiation-dominated, or dark-energy–dominated? Explain your answer.

Section 25-7

43. Explain why it is important to measure how the expansion rate of the universe has changed over time. How is this rate measured?

44. Does dark energy have the same effect on the universe as the gravitational pull between masses?

45. For a galaxy with a given recessional velocity, would it be closer to or farther from us compared to the distance in a universe that expands at a constant rate? (*Hint:* See Figure 25-16.)

46. What form of energy is the dominant contributor to the total energy in the universe today? Was this form of energy always dominant? Explain your answer.

47. How might virtual particle pairs explain dark energy? Are there problems with this hypothesis?

Section 25-8

48. How can we detect the presence of sound waves in the early universe? What do these sound waves tell us?

49. Explain how compressions and rarefactions result in colder and warmer spots in the CMB. Which effect leads to which temperature variation?

50. What size scale is most likely to have different CMB temperatures?

General Problems

51. Use Wien's law (Section 5-4) to calculate the wavelength at which the cosmic microwave background ($T = 2.725$ K) is most intense.

52. Whether the expansion of the universe is speeding up or slowing down, it can be expressed in terms of a quantity called the *deceleration parameter*, denoted by q_0. The expansion is slowing down if q_0 is positive and speeding up if q_0 is negative; if $q_0 = 0$, the expansion proceeds at a constant rate. If we assume that the dark energy is due to a cosmological constant, the deceleration parameter can be calculated using the formula

$$q_0 = \frac{1}{2}\Omega_0 - \frac{3}{2}\Omega_\Lambda$$

(Recall that the density parameter Ω_0 is equal to $\Omega_m + \Omega_\Lambda$.) **(a)** Show that if there is no cosmological constant, the expansion of the universe must slow down. **(b)** Using the values of Ω_m and Ω_Λ from combined observations given in Table 25-2, find the value of the deceleration parameter for our present-day universe. Based on this, is the expansion of the universe speeding up or slowing down? **(c)** Imagine a universe that has the same value of Ω_m as our universe but in which the expansion of the universe is neither speeding up nor slowing down. What would be the value of Ω_Λ in such a universe? Which would be dominant in such a universe, matter or dark energy? Explain your answer.

53. In general, the deceleration parameter (see Question 52) is not constant but varies with time. For a flat universe, the deceleration parameter at a redshift z is given by the formula

$$q_z = \frac{1}{2} - \frac{3}{2}\left[\frac{\Omega_\Lambda}{\Omega_\Lambda + (1 - \Omega_\Lambda)(1 + z)^3}\right]$$

where Ω_Λ is the dark energy density parameter. Using the value of Ω_Λ given in Table 25-2, find the value of q_z for **(a)** $z = 0.5$ and **(b)** $z = 1.0$. **(c)** Explain how your results show that the expansion of the universe was actually decelerating at $z = 1.0$, but changed from deceleration to acceleration between $z = 1.0$ and $z = 0.5$.

54. The dark energy density parameter Ω_Λ is related to the value of the cosmological constant Λ by the formula

$$\Omega_\Lambda = \frac{\Lambda c^2}{3H_0^2}$$

where $c = 3.00 \times 10^8$ m/s is the speed of light. Determine the value of Λ using the combined observation values in Table 25-2. (*Hint:* You will need to convert units to eliminate kilometers and megaparsecs.)

55. Do you think there can be "other universes," regions of space and time that are not connected to our universe? The term for such a collection of universes is the *multiverse*. Should astronomers be concerned with such possibilities? Why or why not?

WEB/E-BOOK QUESTIONS

56. Before the discovery of the cosmic microwave background, it seemed possible that we might be living in a "steady-state universe" with overall properties that do not change with time. The steady-state model, like the Big Bang model, assumes an expanding universe, but it does not assume a "creation event" such as the Big Bang. Instead, matter is assumed to be created continuously everywhere in space to ensure that the average density of the universe remains constant. Search the internet for information about the steady-state theory. Explain why the existence of the cosmic microwave background was a fatal blow to the steady-state theory.

ACTIVITIES

Starry Night™ Explorations

57. In an attempt to collect light from the far reaches of the universe, the Hubble Space Telescope (HST) took long-exposure images of very dark regions of space that appear to contain no bright stars or galaxies. These images, known as the Hubble Deep Field and Hubble Ultra Deep Field images, reveal very rich fields of faint and very distant galaxies. The light now arriving at Earth from some of these galaxies has traveled for over 13 billion years and was collected by the HST at a rate of a few photons per minute! This light was emitted very early in the life of the universe, only a few hundred million years after the Big Bang. You can examine and measure these two images. In *Starry Night™*, open **Favorites > Explorations > Hubble Deep Field** to see a view from a position in orbit around Earth alongside the HST—centered therein is the region of sky containing the first of these remarkable deep field images. Slowly zoom in to a field of view about 20′ wide and note that the region still appears to be devoid of objects. Then select **View > Deep Space > Hubble Images** from the menu to see the HST image of this seemingly blank region of Earth's sky. Zoom in again until the image fills the view. The two bright objects with spikes radiating from them are stars in our own galaxy, where the spikes are caused by diffraction by the supports for Hubble's secondary mirror. All the other objects that appear in this long-exposure image are galaxies containing millions of stars.

   (a) Examine this image carefully and attempt to identify some examples of each kind of galaxy (spiral, barred spiral, elliptical, and irregular) in this field. Choose 5 or 6 of the largest galaxies in this field, record their shapes and galaxy types, and use the angular separation tool to measure carefully and record their angular dimensions.

   (b) Open **Favorites > Explorations > Hubble Ultra Deep Field** to center the view onto this region of the sky. Zoom in and see that, even at a field of view as small as 15′ wide, no objects can be seen in the position of this long-exposure image. Select **View > Deep Space > Hubble Images** from the menu to reveal what this amazing telescope saw in this "empty" region of space. Carefully examine this rich field of faint and very distant galaxies and attempt to identify each kind of galaxy (spiral, barred spiral, elliptical, and irregular) in this field. Again, select 5 or 6 of the largest galaxies in this field, record their shapes and galaxy types, and use the Angular Separation tool to measure their dimensions.

   (c) Consider the mix of different kinds of galaxies and assess whether the proportions of different kinds are the same in these two images. Compare the angular sizes of the largest galaxies in these two images.

Collaborative Exercises

58. As a group, create a four- to six-panel cartoon strip showing a discussion between two individuals describing why the sky is dark at night.

59. Imagine your firm, Creative Cosmologists Coalition, has been hired to create a three-panel, folded brochure describing the principal observations that astronomers use to infer the existence of a Big Bang. Create this brochure on an $8\frac{1}{2}'' \times 11''$ piece of paper. Be sure each member of your group supervises the development of a different portion of the brochure and that the small print acknowledges who in your group was primarily responsible for which portion.

60. The three potential geometries of the universe are shown in Figure 25-14. To demonstrate this, ask one member of your group to hold a piece of paper in one of the positions while another member draws two parallel lines that never change in one geometry, eventually cross in another geometry, and eventually diverge in another.

ANSWERS

ConceptChecks

ConceptCheck 25-1: Bright. Even though more distant stars would individually appear dimmer, that loss of intensity is made up for by there being a greater number of distant stars in any given patch of the sky. Taken together, these two effects would tend to make the universe appear bright, instead of dark as it actually does. This is Olbers's paradox.

ConceptCheck 25-2: The spectral shift toward the shorter blue wavelengths means that the Andromeda Galaxy is moving toward our Milky Way Galaxy.

ConceptCheck 25-3: The cosmological principle applies over very large regions of outer space; it does not apply to the distribution of stars in objects as small as a galaxy.

ConceptCheck 25-4: Older. A smaller Hubble constant corresponds to slower expansion. That means it would have taken longer for galaxies to reach their present separation distances.

ConceptCheck 25-5: The receding star appears dimmer. As its light travels through expanding space, the wavelength of its light is stretched. This lowers the light's energy and the star appears dimmer.

ConceptCheck 25-6: The observable universe becomes larger as more light from never-before-seen distant galaxies has sufficient time to reach us.

ConceptCheck 25-7: The expansion of the universe over the past 13.7 billion years has stretched and redshifted these photons to such a great degree that they are now low-energy, long-wavelength microwave photons.

ConceptCheck 25-8: Earth formed about 4.5 billion years ago, when the universe was about 9 billion years old, and the universe has been dominated by matter for most of its history.

ConceptCheck 25-9: The universe had to be sufficiently cool for atoms to form, so if the early universe was hotter for longer, the first atoms would have appeared more slowly.

ConceptCheck 25-10: Cooler blue regions correspond to denser concentrations of mass in the early universe. These are often the seeds of structures that form later in that same region. A visible-light telescope would probably find a supercluster of galaxies in that same region.

ConceptCheck 25-11: Yes, as in Figure 25-14b. In a universe with the geometry of spherical space, parallel light beams will eventually cross. In this case, the universe is considered a *closed* space and has positive curvature.

ConceptCheck 25-12: It would appear larger in size. As illustrated in Figure 25-15a, since the light rays are bent inward as they travel from the object toward an observer on Earth, the light rays appear to be coming from ends of the stick that are farther apart than as would be observed in a flat universe (Figure 25-15b). This makes the stick appear larger in the closed universe.

ConceptCheck 25-13: The total energy density of the universe determines whether the universe is open, closed, or flat. The cosmic microwave background reveals that the universe is nearly flat with Ω_0 close to 1.0. Since matter can only account for about 0.24 of this energy density (and radiation is negligible), the remaining dark energy density is about 0.76.

ConceptCheck 25-14: Astronomers must make the assumption that all Type Ia supernovae have identical luminosities, regardless of the galaxy in which they are found.

ConceptCheck 25-15: The comparison in the middle, corresponding to size scales of about 1°, shows the greatest difference between the actual fluctuations (which include sound waves) and the variations that would be observed without sound waves. This is consistent with the peak in the lower part of the figure, showing that the effects of sound waves are strongest on scales of 1°.

CalculationChecks

CalculationCheck 25-1: Because z = the change in wavelength $(\lambda - \lambda_0)$ divided by the original λ_0, $z = (725.6 \text{ nm} - 656.3 \text{ nm}) \div 656.3 \text{ nm} = 0.11$.

CalculationCheck 25-2: Hubble's law, $v = H_0 d$ can be rearranged as $d = v \div H_0$. Using $H_0 = 73$ km/s/Mpc, $d = v \div H_0 = 10{,}000$ km/s $\div 73$ km/s/Mpc $= 137$ Mpc.

CalculationCheck 25-3: In a flat universe, Ω_0 is equal to 1. That means $\rho_0 =$ (the combined average mass density) equals $\rho_c =$ (the critical density). Therefore, $\rho_0 = 1.0 \times 10^{-26}$ kg/m^3.

CalculationCheck 25-4: The combined total density parameter for both mass and energy has a value of nearly $\Omega_0 = 1$. The density parameter for ordinary matter is $\Omega_b = 0.05$, so ordinary matter contributes only 5% to the total energy content of the universe. For dark energy, $\Omega_\Lambda = 0.76$, so dark energy makes up about 76% of the universe's energy content.

What Is Making the Expansion of the Universe Accelerate?

by Matthew R. Francis

Dark energy is the name we give to cosmic acceleration: Instead of slowing down in its expansion, the expansion of the universe is speeding up. However, that doesn't tell us what is *causing* the acceleration. Since the discovery of dark energy in 1998, many researchers have tried and tested ideas to explain dark energy, but none of those ideas are entirely satisfactory.

First of all, physicist Michael Turner picked the name to remind us of dark matter, but the two substances behave in opposite ways. Dark matter's gravity pulls things together to make galaxies, whereas dark energy pushes things apart, driving the expansion of the universe to faster and faster rates. (Turner's original name was "funny energy," but his colleagues didn't like it.)

THE ENERGY OF NOTHING

The simplest explanation for dark energy comes from an entirely different branch of physics. According to quantum theory—the same theory that tells us how atoms absorb and emit light, and how stars fuse atomic nuclei together—empty space is a seething soup of "virtual particles," which pop in and out of existence.

These virtual particles, which have been experimentally verified, affect how the more normal particles like electrons behave; but more importantly for us, they carry energy. Since it's the energy of empty space, we call it "vacuum energy." The larger the amount of empty space, the more vacuum energy is contained in that volume.

So, the idea goes like this: As the universe expands, there's more empty space for vacuum energy to occupy. That energy feeds into the expansion of spacetime. More expansion, more vacuum energy, more expansion, and so forth at ever-accelerating rates. That looks like . . . dark energy.

The problem is, if you calculate the amount of vacuum energy, you get *way* too much of it. As in, more than 10^{100} times too much. That's not a typo: 10^{100} is a 1 followed by 100 zeroes. This is a *huge* mismatch, and one that can't just be explained away.

However, because vacuum energy acts like dark energy in its general behavior, physicists looked at ways to fix the problem. One of those possibilities comes from string theory, a branch of research that attempts to unify quantum physics with gravity by replacing particles like electrons or photons with vibrations of strings that are coiled up much smaller than the nucleus of an atom.

String theory calculates the energy of empty space in a different way than straight-up quantum theory. Each possible way of curling strings up corresponds to a different amount of vacuum energy, and there are a *lot* of those possibilities, which are called the "string theory landscape." Most of the curlings don't look anything like the universe we inhabit—they produce universes that collapse right after the Big Bang, or expand too fast, or too

slow, or any number of other scenarios. But a few of them look like our universe.

Unfortunately, string theory doesn't give us a reason why the universe picked our specific amount of dark energy out of all the possible curlings in the string theory landscape. Maybe all the possibilities exist in some huge multiverse; we happen to exist in one of the universes that allows galaxies and planets to form. That's unsatisfactory to a lot of scientists, though, because there's no way to test that hypothesis. It might be *true*, but without a way to check it directly using experiments or observations, we may never know.

DEFYING GRAVITY

When existing theories don't have an easy answer, physicists sometimes look to new ideas. Maybe dark energy is a sign that our current theory of gravity, Einstein's general theory of relativity (GR), breaks down at distances much larger than galaxies. Since the best tests of GR are on smaller scales—the solar system, colliding black holes, and binary neutron star systems—it's not *too* far-fetched to think gravity might have some surprises on the cosmological scale.

However, it's surprisingly hard to come up with a new theory of gravity that both explains dark energy *and* describes all the material general relativity already handles so well. Even though gravitation is by far the weakest force in the universe, its effects are well understood, from pendulums up to clusters of galaxies.

One idea researchers have considered is that maybe gravity itself has mass. More specifically, maybe the graviton, which is the quantum particle carrying the gravitational force, has a very tiny mass. If it does, the effect is to make gravity much *weaker* than the inverse-square law at large distances, cutting the force that binds the cosmos together. The result: The more spacetime expands, the less gravity acts to hold it together. Boom: dark energy. In other words, dark energy in this scheme is the *absence* of large-scale gravity.

We can't measure a graviton's mass the way we can measure particles of matter, such as electrons. However, a massive graviton wouldn't travel at the speed of light, so gravitational waves and other phenomena would be a little slower, and we might learn something about a graviton's mass from future Laser Interferometer Gravitational-Wave Observatory (LIGO) detections. We can also measure the "cut-off" scale for gravity, using the deflection of light, or an unexpected lack of deflection, from galaxies due to gravitational lensing.

As a result, we know that if the graviton has mass, it must be about 10^{-67} kg or less. That's *tiny*—remember an electron's mass is 9×10^{-31} kg. We don't have any concept of a mass that

(continued on the next page)

small, which is only part of the reason why the massive graviton hypothesis is hard to test. To make matters worse, the theory describing massive gravitons is a lot more complicated than general relativity, to ensure compatibility with the universe we observe. However, with better data from lensing, gravitational waves, and the cosmic microwave background, we'll continue to either place limits on the maximum mass of the graviton down below, where it would be responsible for dark energy, or—just maybe—detect and measure that gravity *does* have mass.

Meanwhile, scientists have come up with many other ways to modify the theory of gravity to produce dark energy, which is

generally a sign that they don't know the right way to proceed. (The other possibility, of course, is that general relativity is right, and we need another approach like the string theory landscape after all.) Most of these alternative ideas haven't worked out. Either they've failed to reproduce the existing successes of general relativity, or they've predicted consequences that haven't showed up with better data.

However, some of these modified gravity theories have survived the carnage, and the search for the solution to the dark energy mystery goes on. We're pushing the boundaries of knowledge in modern cosmology, trying to grasp the structure, history, and evolution of the entire universe, so maybe it shouldn't be surprising to find that cosmic acceleration isn't easy to figure out. That's part of what makes this study exciting, though: understanding the limits of our current knowledge, striving to do better through more sophisticated telescopes, more minds applied to the problem, and sheer human ingenuity.

Matthew R. Francis is a science writer, physicist, and frequent wearer of jaunty hats. Before embarking on a life of journalism, he taught astronomy and physics, and wrote research articles on gravitation. His website is BowlerHatScience.org.

26 | Exploring the Early Universe

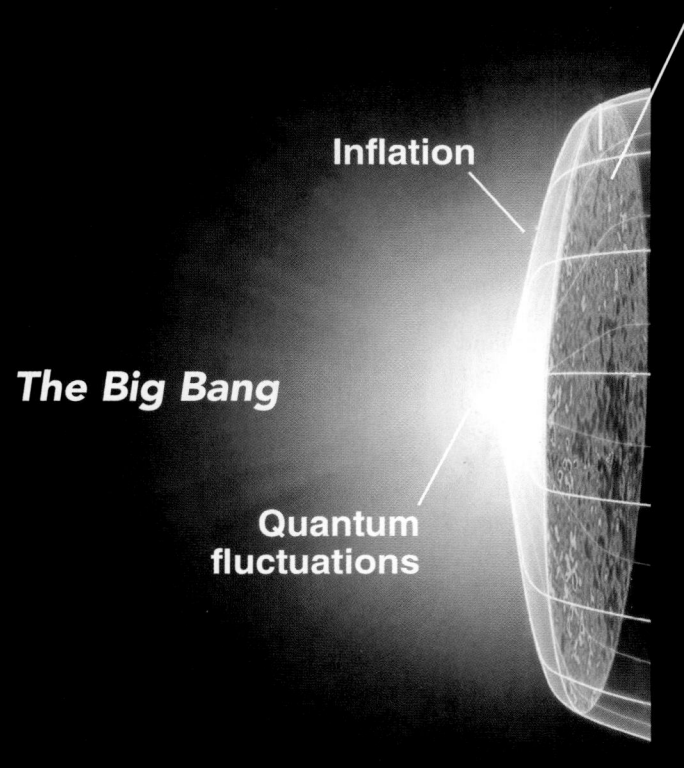

Cosmic microwave
background radition

Inflation

The Big Bang

Quantum
fluctuations

Time →

LEARNING GOALS

By reading the sections of this chapter, you will learn

26-1 How the very young universe expanded enormously in a brief instant of time

26-2 How the fundamental forces of nature and the properties of empty space changed during the first second after the Big Bang

26-3 How the physics of subatomic particles affected the evolution of the early universe

26-4 How, as the early universe expanded and cooled, most of the matter and antimatter annihilated each other

26-5 Which chemical elements in today's universe are remnants of the primordial fireball

26-6 How the first stars and galaxies formed in the early universe

26-7 What steps scientists are taking in the quest toward an all-encompassing "theory of everything"

There is a lot more to the Big Bang than the expansion of space. The extremely high energies in the early universe also produced some very exotic physics. While quantum physics plays out on only small scales in today's universe, it dominated the early universe. As depicted in the opening figure, the early universe is thought to have begun with quantum fluctuations and then to have undergone a brief burst of rapid inflation. Some day soon, we might even be able to detect an imprint of these quantum fluctuations in the cosmic microwave background—and learn if they acted as "seeds" for the formation of galaxies and other structures in today's universe.

Clearly, there is tremendous physics we would like to see from the early universe. However, no telescope observing light can ever hope to directly see events that unfolded during the first 380,000 years after the Big Bang, when the universe was so opaque that it blocked the free passage of light. The earliest light we can see is the cosmic microwave background radiation. We can nonetheless reconstruct some of the events of that hidden epoch, because many aspects of today's universe are relics of the earliest events in the cosmos.

A simple diagram of the early universe from the Big Bang to the cosmic microwave background radiation. (NASA/JPL-Caltech)

In this chapter, we will see evidence that during the first minuscule fraction of a second after the Big Bang, the universe inflated in size by a stupendously large factor of about 10^{50}. During the next 15 minutes, after inflation came to an end, the universe was so dense and hot that particles were constantly colliding at high speeds. As we will see, the events of those 15 violent minutes set the stage for all that came afterward—from the formation of atoms 380,000 years later, to the appearance of the first stars and galaxies some 400 million years after the Big Bang, down to the diverse present-day universe of which we are part.

26-1 The newborn universe underwent a brief period of vigorous expansion

With the discovery of the cosmic microwave background, astronomers had direct evidence that the universe began with a hot Big Bang (see Section 25-4). Remarkably, the microwave background is incredibly uniform, or *isotropic,* across the sky. If we subtract the effects of our own motion through the microwave background (see Figure 25-9), we find that the temperature of the microwave background is the same in all parts of the sky, with deviations typically less than 0.01%. These small deviations, or nonuniformities, in the microwave background are also remarkable, in part because their angular sizes help indicate that our universe has a flat geometry (see Section 25-6, especially Figure 25-15). As striking as these observations are, they pose substantial challenges to the standard theory of the expansion of the universe.

The Isotropy Problem: Why Is the Microwave Background So Uniform?

To appreciate why the uniformity of the microwave background poses a problem, think about two opposite parts of the sky, labeled A and B in **Figure 26-1**. Both of these points lie on a sphere centered on Earth; our *observable universe* lies within the sphere, and the surface is called our *cosmic particle horizon* (see Section 25-3, especially Figure 25-5). At present, the radius of our observable universe is about 46.5 billion ly; due to expansion, this is much larger than the age of the universe (13.7 billion years) times the speed of light. We can see light coming from any object on or inside our cosmic particle horizon, but we cannot yet see objects beyond this horizon. Even traveling at the speed of light—the ultimate speed limit—no information of any kind from objects beyond our cosmic particle horizon has had time to reach us over the entire 13.7-billion-year history of the universe. Thus, the cosmic particle horizon defines the limits of our observable universe.

Points A and B in Figure 26-1 lie near the distant edge of the observable universe, just inside our cosmic particle horizon; when we look at these points, we are seeing about as far back into the past as possible—that is, the light we see from these points is the cosmic background radiation. In order for the radiation that reaches us from A and B to be nearly the same, the material of the early universe at A must have had the same temperature as at B. But for two objects to be at the same temperature, they should have been in contact (in the young universe) and been able to exchange heat with each other. (A hot cup of coffee and a cold spoonful of

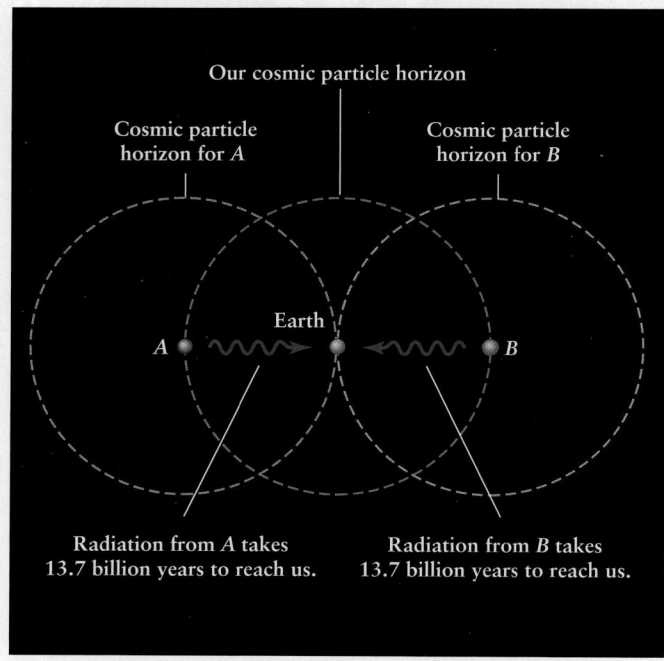

FIGURE 26-1

The Isotropy Problem Regions A and B, both of which lie on our cosmic particle horizon, are so far apart that they seem never to have been in communication over the lifetime of the universe. Yet the cosmic background radiation from A and B and from all other parts of the sky shows that these disconnected regions have almost exactly the same temperature. The dilemma of why this should be is called the isotropy problem.

cream reach the same temperature only after you pour the cream into the coffee.)

The problem is that the widely separated regions at A and B have absolutely no connection with each other. As Figure 26-1 shows, point A lies outside the cosmic particle horizon for point B, and point B lies outside the cosmic particle horizon for point A. (Put another way, point A lies outside the observable universe of point B and vice versa.) Hence, no information has had time to travel between points A and B over the entire history of the universe. How is it possible, then, that these unrelated parts of the universe have almost exactly the same temperature? This dilemma is called the **isotropy problem** or the **horizon problem**. (*Isotropy* means "uniformity in all directions.")

Even when we consider that the contents of the universe were closer together at earlier times in the universe, the isotropy problem does not go away. While it takes less time for closer regions to come into contact with each other, there is also less time for this to occur because the universe is not as old. When both of these effects are taken into account, there does not seem to be a way for the most distant regions to have reached the same temperature.

The Flatness Problem: Why Is $\Omega_0 = 1$?

The flatness of our universe presents us with a second enigma. Recall that the geometry of our universe depends on the density parameter Ω_0, which is the ratio of the combined average mass density in the universe (ρ_0) to the critical density (ρ_c) (this includes dark energy; see Section 25-6). Observations of temperature

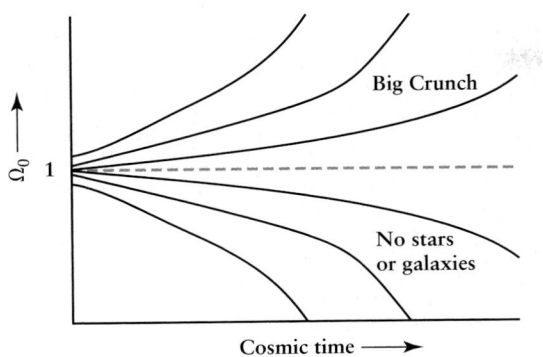

FIGURE 26-2

The Flatness Problem We observe that Ω_0 has a value close to $\Omega_0 = 1$ today. This presents a problem, in the sense that it seems to be very unlikely. If the universe started with $\Omega_0 > 1$, then Ω_0 would not stay constant, but rather would grow exponentially large. This would lead to a Big Crunch, as gravitational attraction would quickly overpower the effect of cosmic expansion. And, if the universe started with $\Omega_0 < 1$, then Ω_0 would not stay constant, but rather would decrease exponentially. This would lead to a rapidly expanding universe with no structure such as stars and galaxies, as the effect of expansion would quickly overcome gravitational attraction. In order to "fine tune" Ω_0 so that we observe nearly $\Omega_0 = 1$ today, at the Big Bang Ω_0 would have had to be 1.000 . . . 01 with 50 zeroes! That's a very tight requirement on the Big Bang model that requires an explanation.

variations in the cosmic microwave background indicate that Ω_0 is very close to 1, which corresponds to a flat universe (Figure 26-2).

For the density parameter Ω_0 to be close to 1 today, it must have been *extremely* close to 1 during the Big Bang. In other words, the density of the early universe was almost exactly equal to the critical density. (The density was much higher than it is today, but the value of the critical density was also much higher. In a flat universe, the average mass density and the critical density decrease together as the universe expands, so that they remain equal at all times.)

The equations for an expanding universe show that any deviation from exact equality would have mushroomed within a fraction of a second. Had the average mass density been slightly less than ρ_c, the universe would have expanded so rapidly that matter could never have clumped together to form galaxies. Without galaxies, there would be no stars or planets, and humans would never have evolved. If, on the other hand, the density had been slightly greater than ρ_c, the universe would soon have become tightly packed with matter. Had this been the case, the gravitational attraction of this matter would long ago have collapsed the entire cosmos in a reversed Big Bang or "Big Crunch," and again humans would never have evolved (see Figure 26-2).

In other words, immediately after the Big Bang, the fate of the universe hung in the balance. The tiniest deviation from the precise equality $\rho_0 = \rho_c$ would have rapidly propelled the universe away from the special case of $\Omega_0 = 1$. Had there been such a deviation, we would not be here to contemplate the nature of the universe.

Happily, our universe is one in which galaxies, stars, planets, and humans do exist, a Big Crunch has not taken place, and Ω_0 is very close to 1. These conditions in the universe today tell us that the density of the universe immediately after the Big Bang must have been equal to the critical density to an incredibly high order of precision. In order for space to be as flat as it is today, the value of ρ_0 right after the Big Bang must have been equal to ρ_c to more

than 50 decimal places! Unless there was a physical mechanism to bring these two parameters into such close agreement, this extraordinary match is an unappealing characteristic for a scientific theory because it seems to depend on a lucky coincidence.

What could have happened during the first few moments of the universe to ensure that $\rho_0 = \rho_c$ to such an astounding degree of accuracy? Because $\rho_0 = \rho_c$ means that space is flat, this enigma is called the **flatness problem**.

Solving the Problems: The Inflationary Model

In the early 1980s, a remarkable solution to both the isotropy problem and the flatness problem was proposed. Several physicists suggested that the universe might have experienced a brief period of **inflation**, or extremely rapid expansion, shortly after the Planck time. (As we saw in Section 25-3, before the Planck time the normal laws of physics do not properly describe the behavior of space, time, and matter.) During this **inflationary epoch**, the universe expanded outward in all directions by a factor of about 10^{50}. This epoch of dramatic expansion may have lasted only about 10^{-32} of a second (Figure 26-3).

> The inflationary model explains not only why the cosmic microwave background is so uniform, but also how stars, galaxies, and humans can exist.

Inflation accounts for the isotropy of the microwave background. During the inflationary epoch, much of the material that was originally near our location was moved out to tremendous distances. Over the past 13.7 billion years, our cosmic particle horizon has expanded so that we can see radiation from these distant regions. Hence, when we examine microwaves from opposite parts of the sky, we are seeing radiation from parts of the universe that were originally in intimate contact with one another. This contact allowed for the exchange of heat and is the reason why all parts of the sky have almost exactly the same temperature.

ANALOGY

An inflationary epoch can also account for the flatness of the universe. To see why, think about a small portion of Earth's spherical surface, such as a small lake. For all practical purposes, it is impossible to detect Earth's curvature over such a small area, and a small lake looks flat. Similarly, the observable universe is such a tiny fraction of the entire inflated universe that any overall curvature in it is virtually undetectable (**Figure 26-4**). Like a small lake, the segment of space we can observe looks flat.

Not only does tremendous inflation produce a nearly flat universe, but it does so for just about *any* original degree of curvature prior to inflation. Let us consider why this is very appealing for a scientific theory. Recall that the flatness of our present universe could be explained (in an unsatisfactory way) if, by extraordinary coincidence alone, the value of ρ_0 right after the Big Bang was equal to ρ_c for the first 50 decimal places. Also recall that curvature is determined by ρ_0 (Section 25-6 and Table 25-1), where $\rho_0 = \rho_c$ is flat. Since Figure 26-4 shows that inflation flattens out just about any original degree of curvature, inflation also *drives*

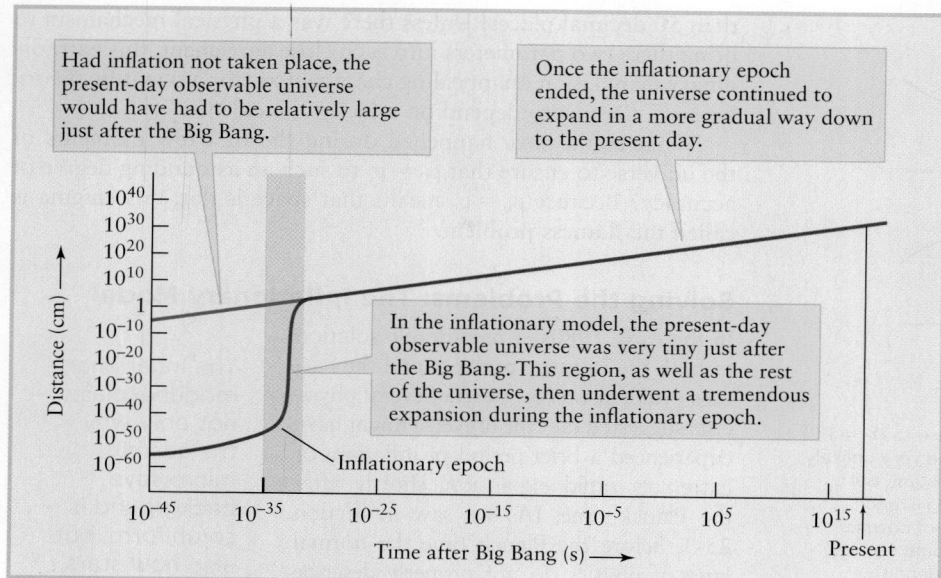

FIGURE 26-3

The Observable Universe With and Without Inflation   According to the inflationary model (shown in purple), the universe expanded by a factor of about 10^{50} shortly after the Big Bang. This growth in the size of the present-day observable universe—the portion of the universe that lies within our present cosmic particle horizon—occurred during a very brief interval, as indicated by the vertical shaded area on the graph. The blue line shows the projected size of the present-day observable universe soon after the Big Bang if inflation had not taken place. (Data from A. Guth)

Within the figure 26-3:

Had inflation not taken place, the present-day observable universe would have had to be relatively large just after the Big Bang.

Once the inflationary epoch ended, the universe continued to expand in a more gradual way down to the present day.

In the inflationary model, the present-day observable universe was very tiny just after the Big Bang. This region, as well as the rest of the universe, then underwent a tremendous expansion during the inflationary epoch.

Inflationary epoch

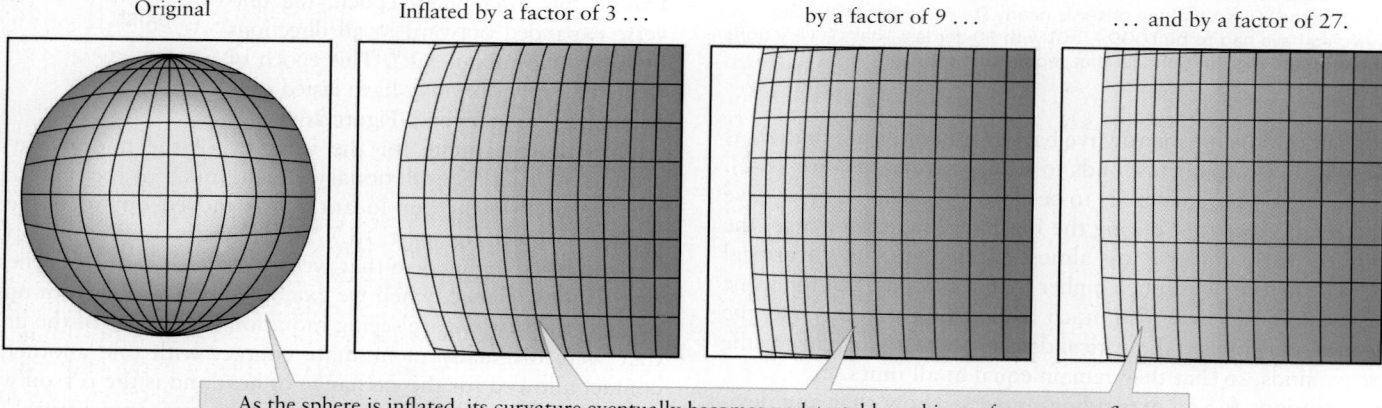

Original   Inflated by a factor of 3 . . .   by a factor of 9 . . .   . . . and by a factor of 27.

As the sphere is inflated, its curvature eventually becomes undetectable and its surface appears flat.

FIGURE 26-4

Inflation Solves the Flatness Problem   This sequence of drawings shows how inflation can produce a locally flat geometry. In each successive frame, the sphere is inflated by a factor of 3 (and the number of grid lines on the sphere is increased by the same factor). Note how the curvature of the surface quickly becomes undetectable on the scale of the illustration. Inflation models expand the universe by about a factor 10^{50}, so nearly any original degree of curvature (indicated by the radius of the sphere on the left), will end up nearly flat after inflation. (Data from A. Guth and P. Steinhardt)

any initial value of ρ_0 toward ρ_c and removes the need for an unlikely coincidence alone to account for a flat universe.

CAUTION!

It is important to note that the concept of inflation does not violate Einstein's law that nothing can travel faster than the speed of light. Remember that the expansion of the universe is the expansion of space itself, not the motion of objects through space. During the inflationary epoch, the distances between particles increased faster than the speed of light, but this was entirely the result of a sudden vigorous *expansion of space*. Particles did not move through space faster than the speed of light; space itself inflated to increase the distance between the particles. Even today, galaxies at redshifts greater than about $z = 1.5$ are receding faster than the speed of light.

Inflation and Polarization of the Cosmic Microwave Background

As we saw in Section 25-8, the detailed properties of temperature fluctuations in the cosmic microwave background, or CMB, can reveal a number of the fundamental parameters that describe our universe. This impressive list includes the Hubble constant, the density parameters for the total energy and for ordinary matter, and also for dark energy. In the near future, even the inflation model might gain observational support by measuring the *polarization* of light in the CMB.

As illustrated in **Figure 26-5**, light can be polarized. Ordinary light from a lightbulb or from the Sun is *unpolarized*, which means that the electric fields of many light waves are oriented in random directions. But when light collides with and bounces off an object, it tends to become *polarized*, with its electric fields oriented in a specific direction. In a similar way, the cosmic background

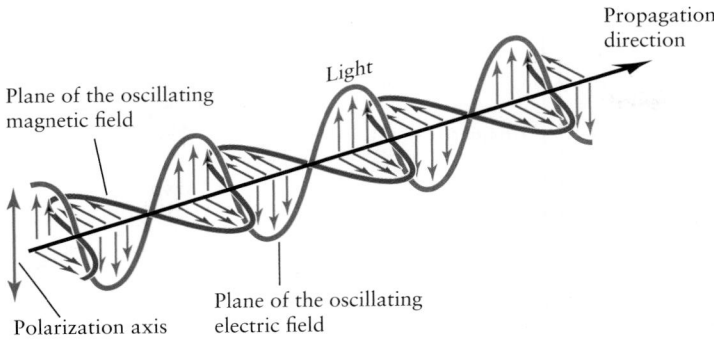

FIGURE 26-5

Polarized Light A light wave consists of oscillating electric and magnetic fields, and the electric field determines the polarization axis. In this illustration, the electric field oscillates up and down in a vertical plane, so the polarization axis is also vertical. Polarized glasses take advantage of the fact that much of the reflective glare from sunlight has its polarizing axis parallel to the ground (perpendicular to the example in this figure). By wearing glasses that block this horizontally polarized light, much of the Sun's glaring reflections are eliminated.

radiation acquires a polarization when it scatters from material in a dense cold spot (Figure 26-6).

The effects of inflation are also expected to produce an additional and uniquely polarized CMB signature, although measurements from the Wilkinson Microwave Anisotropy Probe (WMAP) (Figure 26-6) were unable to detect these weak signals. However, there are several experiments that hope to detect CMB polarization produced by inflation (these include the BICEP3, POLARBEAR, CMB-S4, and others). These are crucial experiments, as inflation is one of the most profound predictions about our early universe, and CMB polarization appears to be our best hope of finding evidence to support the inflation model.

CONCEPT CHECK 26-1

In Figure 26-1, points A and B appear too far apart to reach the same temperature by exchanging energy with each other. Nonetheless, distant locations like this in the universe do show nearly the same 2.7 K temperatures. How does the inflation model solve this problem?

CONCEPT CHECK 26-2

Suppose the curvature of the universe before inflation was represented by the surface of a sphere that was 1 billion times smaller than the sphere on the left in Figure 26-4. With this high degree of initial curvature, could inflation have produced a universe that appears nearly flat today?

Answers appear at the end of the chapter.

26-2 Inflation extends the principles that govern the fundamental forces of nature

If the universe went through an episode of extreme inflation, what could have triggered it? Our understanding is that inflation was one of a sequence of remarkable events during the first 10^{-12} seconds after the Big Bang. In each of these events, there was a fundamental transformation of the basic physical properties of the universe. To understand what happened during that brief moment of time, when the universe was a hot, dense sea of fast-moving particles and energetic photons colliding with each other, we must first understand how particles interact at very high energies.

The Fundamental Forces of Nature and the Standard Model

Just *four* fundamental forces—*gravitation, electromagnetism,* and the *strong* and *weak forces*—explain the interactions of everything in the universe. Of these forces, gravitation is the most familiar (Figure 26-7a). It is a long-range force that dominates the universe over astronomical distances. Electromagnetism, which accounts for the electric and magnetic forces, creates a long-range force, but it is intrinsically much stronger than the gravitational force. For example, the electric force between an electron and a proton is about 10^{39} times stronger than the gravitational force between those two particles. Because the electromagnetic force is stronger by a factor of 10^{39}, electromagnetism, not the gravitation, holds electrons in orbit about the nuclei in atoms.

We do not generally observe longer-range effects of the electromagnetic force because there is usually a negative electric

FIGURE 26-6 R I V U X G

Polarization of the Cosmic Microwave Background The polarization axes of CMB light are indicated by the orientation of the small black lines. In the two spots on the left, as CMB radiation scatters off of denser regions in the early universe, a unique ring-structured polarization pattern is created. These are the regions that show up as cooler blue spots in images of the CMB temperature fluctuations (such as Figure 25-13).

In the two spots on the right, a unique pattern is also created when the CMB scatters off of warmer regions of matter. While these polarization measurements agree with predictions, they are not sensitive enough to test the inflation model, and more sensitive experiments have already begun. (NASA/WMAP Science Team)

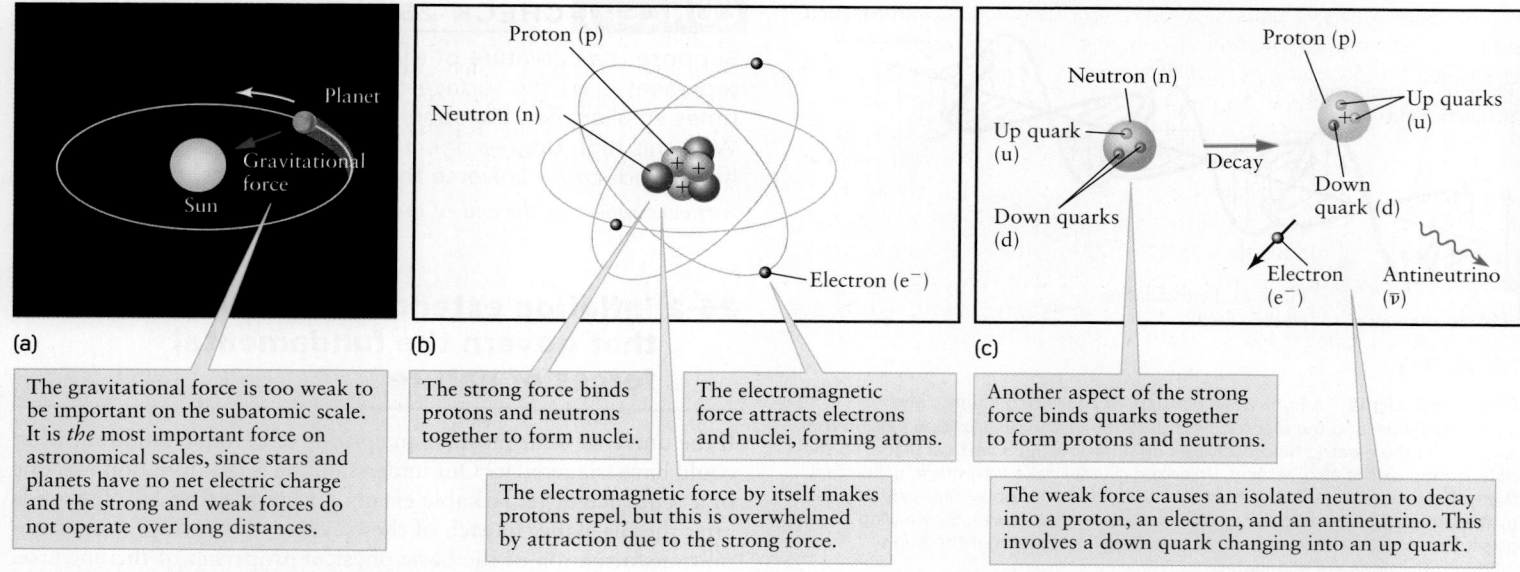

The gravitational force is too weak to be important on the subatomic scale. It is *the* most important force on astronomical scales, since stars and planets have no net electric charge and the strong and weak forces do not operate over long distances.

The strong force binds protons and neutrons together to form nuclei.

The electromagnetic force attracts electrons and nuclei, forming atoms.

The electromagnetic force by itself makes protons repel, but this is overwhelmed by attraction due to the strong force.

Another aspect of the strong force binds quarks together to form protons and neutrons.

The weak force causes an isolated neutron to decay into a proton, an electron, and an antineutrino. This involves a down quark changing into an up quark.

Force	Relative strength	Particles exchanged	Particles on which the force can act	Range	Example
Strong	1	Gluons	Quarks	10^{-15} m	Holding protons, neutrons, and nuclei together
Electromagnetic	1/137	Photons	Charged particles	Infinite	Holding atoms together
Weak	10^{-4}	Intermediate vector bosons	Quarks, electrons, neutrinos	10^{-16} m	Radioactive decay
Gravitational	6×10^{-39}	Gravitons	Everything	Infinite	Holding the solar system together

The Standard Model of the Elementary Particles

FIGURE 26-7

The Four Forces and the Standard Model **(a)** Gravitation is dominant on the scales of planets, star systems, and galaxies, while **(b)**, **(c)** the strong, electromagnetic, and weak forces hold sway on the scale of atoms and nuclei. **(d)** The Standard Model describes the particles involved in the electromagnetic, weak, and strong forces. Bosons transmit the different forces; the photon transmits the electromagnetic force between charged particles (holding electrons in atoms); the gluon transmits the force between quarks (holding together atomic nuclei); and the intermediate vector bosons (Z, W) transmit the weak force (resulting in certain radioactive decays). These are all the known fundamental particles found in nature.

The **strong force** holds protons and neutrons together to form the nuclei of atoms (Figure 26-7b). It is said to be a *short-range force*: Its influence extends only over distances less than about 10^{-15} m, about the diameter of a proton. Without the strong force, nuclei would disintegrate because of the electric repulsion of the positively charged protons. In fact, the strong force overpowers the electric forces inside nuclei. While there are technical differences, the strong force is also referred to as the **nuclear force**.

The **weak force**, which is also a short-range force, is at work in certain kinds of radioactive decay (Figure 26-7c). An example is the transformation of a neutron (n) into a proton (p), in which an electron (e^-) is released, along with a nearly massless particle called an antineutrino ($\bar{\nu}$):

$$n \rightarrow p + e^- + \bar{\nu}$$

charge for every positive charge and a south magnetic pole for every north magnetic pole. Thus, over great volumes of space, the effects of electromagnetism effectively cancel out. No similar canceling occurs with gravity because there is no equivalent "negative mass." Because the effects of gravity do not cancel out, the force that holds Earth in orbit around the Sun is gravitational, not electromagnetic.

Numerous experiments in nuclear physics demonstrate that protons and neutrons are themselves composed of even more basic particles called **quarks**, the most common varieties being "up" (u) quarks and "down" (d) quarks. A proton is composed of 2 up quarks and 1 down quark; a neutron is composed of 2 down quarks and 1 up quark.

In the 1970s, the concept of quarks led to a breakthrough in our understanding of the strong and weak forces. The strong force holds quarks together, while the weak force is at work whenever a quark changes from one variety to another. For example, when a neutron decays into a proton, one of the neutron's down quarks changes into an up quark, as shown in Figure 26-7c. Thus, the weak force is responsible for transformations such as

$$d \to u + e^- + \bar{\nu}$$

In the 1940s, physicists Richard P. Feynman and Julian S. Schwinger (working independently in the United States) and Sin-Itiro Tomonaga (in Japan) succeeded in developing a basic description of what we mean by force. Focusing their attention on the electromagnetic force, they tried to describe exactly what happens when two charged particles interact. According to their theory, now called **quantum electrodynamics**, charged particles interact by exchanging *virtual* photons that cannot be directly observed. (These virtual photons will be explored further in Section 26-3.)

Quantum electrodynamics has proved the most successful theory in modern physics. It describes with remarkable accuracy many details of electromagnetic forces and interactions between charged particles. Inspired by these successes, physicists developed similar theories for the other three forces. A key feature of a microscopic description of force is that two particles exert a force on each other by exchanging a third and different particle (Figure 26-8). For example, the

> Experimental and theoretical research into the basic forces of nature helps us understand the evolution of the cosmos.

weak force occurs when particles exchange particles called **intermediate vector bosons**, and quarks stick together by exchanging elementary particles called **gluons**. These theories have been verified by experiments, and our overall understanding of the known particles is called the **Standard Model** (Figure 26-7d). The table in Figure 26-7 summarizes these features of the four fundamental forces.

CONCEPT CHECK 26-3

A neutron consists of 1 up quark (u) and 2 down quarks (d) as in Figure 26-7c. Are there any other particles in the Standard Model that you would expect to be involved in holding a neutron together?

Answer appears at the end of the chapter.

Experimentally Verified Unification Theories and the Higgs Particle

Physicists made important progress in understanding the weak force during the 1970s. Steven Weinberg, Sheldon Glashow, and Abdus Salam proposed a theory with three types of intermediate vector bosons, which are exchanged in various manifestations of the weak force. These three particles were actually discovered in experiments in the 1980s, providing strong support for the theory.

A startling prediction of the Weinberg–Glashow–Salam theory is that the weak force and the electromagnetic force should be identical to each other for particles with energies greater than 100 GeV (1 GeV equals 10^9, or 1 billion, electron volts; see Section 5-5). In other words, if particles are slammed together with a total energy greater than 100 GeV, then electromagnetic interactions become indistinguishable from weak interactions. We say that above 100 GeV, the electromagnetic force and the weak force are "unified" into a single **electroweak force**. This is our first example of the "unification" of forces, and the only one to be experimentally verified.

This unification occurs because the three types of intermediate vector bosons behave just like photons above 100 GeV. At such high energies, these three particles actually lose their mass, and the weak force becomes a long-range force with the same intrinsic strength as electromagnetism. Physicists describe this similarity by saying that "symmetry is restored" above 100 GeV.

In the world around us, however, the typical energies with which particles interact are much lower, on the order of 1 eV or less. Below 100 GeV, intermediate vector bosons behave like massive particles, but photons have no mass. Because intermediate vector bosons and photons are not similar at low energies, we say that "symmetry is broken" below 100 GeV, which is why the electromagnetic and the weak forces behave so differently in the lower-energy world around us. In the language of physics, the electroweak force experiences a **spontaneous symmetry breaking**; above 100 GeV, the electroweak force represents a form of symmetry, and below 100 GeV, the symmetry is broken and two distinct forces emerge: the electromagnetic and weak forces.

FIGURE 26-8

Forces Due to Particle Exchange Imagine each skateboarder as a particle that can move freely. By tossing a basketball, the right skateboarder is pushed backward, and by catching the ball, the left skateboarder is pushed away as well. This example of particle exchange illustrates a repulsive force; in quantum physics, particle exchange can produce both repulsive and attractive forces.

FIGURE 26-9 R I V U X G

Spontaneous Symmetry Breaking of Water   As liquid water freezes, it transitions from a highly uniform and symmetric state to ice crystals that break this uniformity. Since lower temperatures correspond to lower average energy for the water's particles, we say that water experiences a spontaneous symmetry breaking at lower energy. By analogy, the high temperature symmetric water is like the electroweak force, and the ice crystals are like the electromagnetic and weak forces. (Frank Siteman/Aflo Co., Ltd./Alamy Stock Photo)

ANALOGY

As abstract as spontaneous symmetry breaking sounds, it is a familiar phenomenon when describing water. At room temperature, water is liquid. If you were to take a vantage point from within liquid water, it would look the same in all directions. In this sense, liquid water is uniform and symmetric. At lower temperatures, however, water freezes into ice crystals and no longer appears the same in every direction (**Figure 26-9**). While commonly referred to as a "phase transition," this change can also be described as the spontaneous symmetry breaking of water when it freezes.

As strange as these ideas sound, they have been tested to great precision in particle accelerators that smash particles together at extremely high energies. The world's largest particle accelerator is CERN (the European Organization for Nuclear Research) in Europe (**Figure 26-10**), which enabled physicists to make a monumental discovery in 2012. For more than 40 years, there was one major prediction of the Standard Model that had not been observed—the Higgs particle (see Figure 26-7d). This key particle was predicted as part of the electroweak symmetry breaking that gives the Standard Model some of the properties we observe. In fact, one of these properties is mass: The Higgs particle "gives mass" to the other particles (except for the massless photons and gluons) by interacting with them. Since you and I, the planets, and the stars are all made of these particles, we owe our heft to the Higgs particle.

Long before the Higgs particle was observed, the Standard Model's overall success at unifying the electromagnetic force with the weak force led to attempts at unifying the remaining two forces: the strong force and gravity.

ANALOGY

How does the Higgs particle give other particles mass? First of all, in the realm of quantum physics, every particle is associated with an extended entity called a *field*. Thus, the Higgs particle is associated with a Higgs field, and this field is present even at low energy when the Higgs particle itself is absent. To acquire the property of mass, particles in the Standard Model interact with this Higgs field, which produces a force on them. This force is somewhat similar to what a crumb experiences when it's surrounded by syrup; analogously, the Higgs field is like a syrup that pervades all of space. The interactions of particles with this "syrup" make it harder to set particles into motion, giving them a property of mass. (Interestingly, it is not known why different particles interact more strongly to have different masses; this is an unsolved problem in physics.)

CONCEPT CHECK 26-4

Are the photons (the W and the Z bosons of the Standard Model in Figure 26-7d), "symmetric," meaning that they all have the same physical properties?

CONCEPT CHECK 26-5

Symmetry breaking, which refers to the fact that below 100 GeV, the electroweak force appears as two distinct forces (electromagnetic and weak), is said to be "spontaneous." What is spontaneous about symmetry breaking?

Answers appear at the end of the chapter.

Proposed Unification Theories of all the Forces

In the 1970s, several physicists proposed **grand unified theories** (or **GUTs**), which predict that the strong force becomes unified with the weak and electromagnetic forces (but not gravity) at energies above 10^{14} GeV. In other words, if particles were to collide at energies greater than 10^{14} GeV, the strong, weak, and electromagnetic interactions would all be long-range forces and would be indistinguishable from each other.

Many physicists suspect that all four forces, including gravitation, may be unified at energies greater than 10^{19} GeV (**Figure 26-11**). That is, if particles were to collide at these colossal energies, there would be no difference between the gravitational, electromagnetic, and nuclear forces. However, no one has yet succeeded in working out the details of such a **supergrand unified theory**, which is sometimes called a **theory of everything** (or **TOE**). The favored TOE to describe all of the forces is called *string theory*, which we will discuss further in Section 26-7.

A supergrand unified theory or TOE that includes gravitation would also be a theory of *quantum gravity* (which would describe the effects of strong gravity on microscopic scales). We have already seen with black holes and the Big Bang that developing the answers to some questions requires a theory of quantum gravity (such as what happens to the information about material after it falls into a black hole). Similar to the particle exchange mechanism between the other forces, the gravitational force in a quantum

(a)

(b)

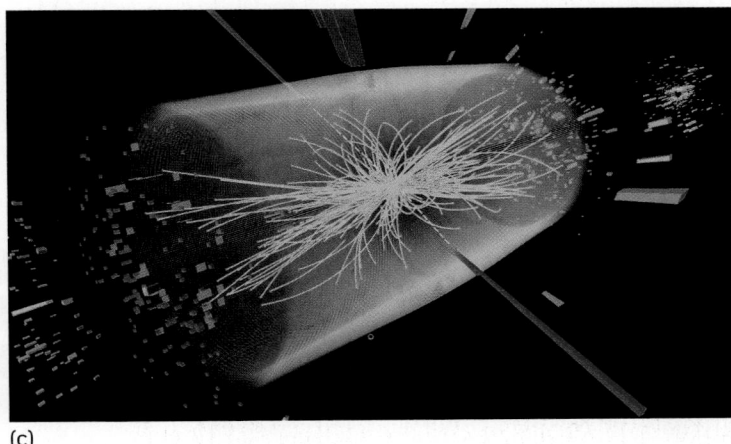

(c)

FIGURE 26-10 R I (V) U X G

Discovery of the Higgs Particle **(a)** The CERN particle collider is 27 km (16.8 mi) in diameter and straddles the border between Switzerland and France. Protons are accelerated to speeds very close to the speed of light in a tube (shown in red) that is about 100 m (330 ft) below ground. **(b)** When these protons collide, their energy produces numerous additional particles that are tracked and detected in enormous particle detectors, as shown here (under construction). This is a direct result of converting energy to mass via $E = mc^2$. To get a sense of scale, note that there is a person standing up on the right side of the yellow platform near the center. **(c)** By analyzing the particle tracks (in yellow), the mass, charge, and various other properties of the particles can be determined. This is how new particles are discovered, including the Higgs particle. (a, b: © CERN; c: CERN/NOTIMEX/Newscom)

gravity theory is predicted to arise when particles exchange **gravitons**, although gravitons themselves have not been observed.

The Fundamental Forces in the Early Universe

Figure 26-11 shows how the various forces are thought to have changed during the first fraction of a second after the Big Bang. Before the Planck time (from $t = 0$ second to $t = 10^{-43}$ second), particles collided with energies greater than 10^{19} GeV, and all four forces were unified. Because we do not yet have a TOE that properly describes the behavior of gravity, we remain ignorant of what was going on during the first 10^{-43} second of the universe's existence. We know, however, that by the end of the Planck time, the expansion and cooling of the universe had caused the energy of

particles to fall to 10^{19} GeV. At energies below this level, gravity is not unified with the other three forces.

In the language of physics, at $t = 10^{-43}$ second, there was a spontaneous symmetry breaking in which gravity was "frozen out" of the otherwise unified hot soup that filled all space. In such a "soup," the typical energy of a particle (E) is related to temperature (T) by $E = kT$, where k is the Boltzmann constant (about 10^{-4} eV/K, or 10^{-13} GeV/K). Thus, the temperature of the universe was 10^{32} K when gravity emerged as a separate force.

As the universe expanded, its temperature decreased and the energy of particles decreased as well. (We discussed this property of gases in Box 19-1.) At $t = 10^{-35}$ second, the energy of particles in the universe had fallen to 10^{14} GeV, equivalent to a temperature of 10^{27} K. Below these energies and temperatures, the strong force is no longer unified with the electromagnetic and weak forces. Thus, at $t = 10^{-35}$ second, there was a second spontaneous symmetry breaking, at which time the strong force "froze out."

The inflationary epoch is thought to have begun at some point after the symmetry breaking between the strong force and the other forces. At the start of inflation, physicists hypothesize that the universe was in an unstable state called a **false vacuum**. In this state, physicists hypothesize that the energy associated with a quantity called the *inflaton field* (or inflation field) had a nonzero value of energy (**Figure 26-12a**). The inflaton field is a hypothesized field that contains most of the energy in the universe by the end of inflation. (Just as the space around a magnet is permeated

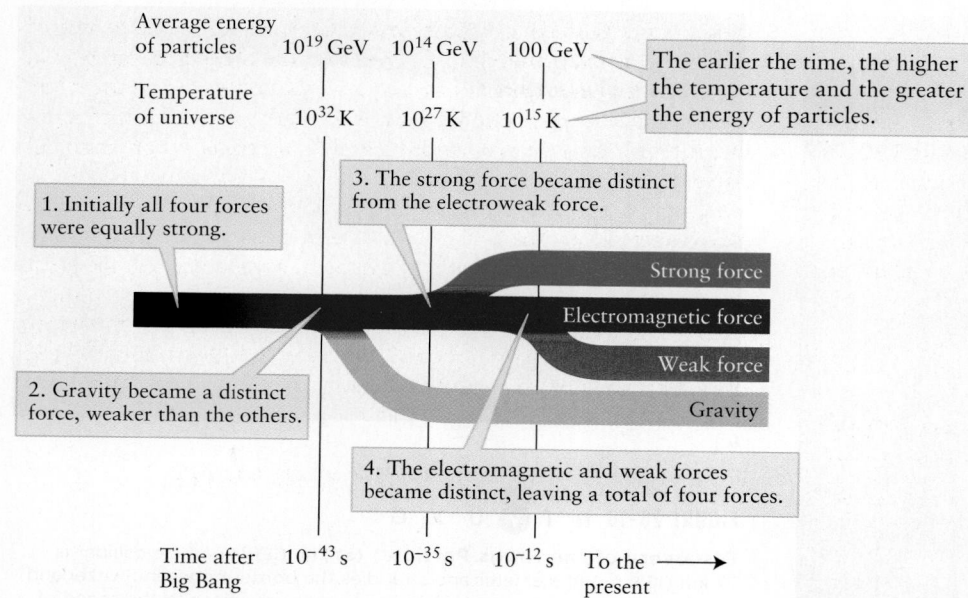

Average energy of particles 10^{19} GeV 10^{14} GeV   100 GeV

Temperature of universe 10^{32} K 10^{27} K 10^{15} K

The earlier the time, the higher the temperature and the greater the energy of particles.

1. Initially all four forces were equally strong.

3. The strong force became distinct from the electroweak force.

Strong force

Electromagnetic force

2. Gravity became a distinct force, weaker than the others.

Weak force

Gravity

4. The electromagnetic and weak forces became distinct, leaving a total of four forces.

Time after Big Bang 10^{-43} s 10^{-35} s 10^{-12} s   To the present →

FIGURE 26-11

Unification of the Four Forces   The strength of the four fundamental forces depends on the energy of the particles that interact. As shown in this schematic diagram, the higher the energy, the more the forces resemble each other. Also included here are the temperature of the universe and the time after the Big Bang when the strengths of the forces are thought to have been equal. At energies between 10^{14} and 10^{19} GeV, the forces might be described by a grand unified theory (GUT), and at energies above 10^{19} GeV, the forces might all appear similar in a supergrand unified theory (or theory of everything).

Energy associated with the inflaton field →

Initially, the universe (represented by the black dot) was in a high-energy "false-vacuum" state.

True vacuum

The universe began a "slow roll" toward the lower-energy true-vacuum state, triggering a tremendous expansion of the universe.

The universe "rolled" back and forth around the true-vacuum state, converting its energy into mass, and eventually settling down in the state of minimum energy.

—Value of the inflaton field →

(a) The initial false vacuum   (b) Inflationary expansion   (c) Reheating

FIGURE 26-12

Inflation: Transitioning from the False Vacuum to the True Vacuum   (a) The energy of the vacuum is thought to be determined by the value of a quantity called the inflaton field. As shown by the red curve in this diagram, this energy is at a minimum for a certain value of this field. The state of lowest energy is called the true vacuum. (b) Inflation is associated with the universe making a transition from its initial false-vacuum state to a true-vacuum state. This expansion took place over a period of about 10^{-32} second, during which the universe cooled to a temperature of about 3 K. (c) As the universe "settled in" to the true-vacuum state, energy was released that reheated the universe to a temperature of 10^{27} K.

by a magnetic field, like that shown in Figure 7-14a, the entire universe is thought to be permeated by the inflaton field.) This state was unstable in the same sense as a ball perched atop a cone with a pointed top: The ball will stay there if left undisturbed, but it will roll downhill if even slightly disturbed. In an analogous way, it is thought that the inflaton field of the universe "rolled downhill" to the *true vacuum,* a state of lower energy. It was the high energy in the false vacuum that produced tremendous inflationary expansion, and inflation comes to an end as the universe transitions to its true vacuum with minimum energy (Figure 26-12b). By the time the inflationary epoch had ended, about 10^{-32} second after the Big Bang, the universe had increased in scale by a factor of roughly 10^{50}. That expansion is so large that a region initially the size of an atom would be inflated to a size larger than today's observable universe!

The rapid expansion of the universe also gave rise to a rapid cooling. At the end of the inflationary epoch, the temperature of the universe may have dropped to about 3 K, about the same as the

temperature of the cosmic background radiation today. But as the universe finally settled into the vacuum-energy state, an additional amount of energy was released from the inflaton field that went into *reheating* the universe to a temperature of 10^{27} K—about the same as it had before inflation began (Figure 26-12c). Thus, inflation caused the universe to expand tremendously while having no net effect on its temperature. We'll look at this period again in the next section because it leads to the origin of the matter that makes up today's universe.

After the end of the inflationary epoch at $t = 10^{-32}$ second, the universe continued to expand and cool at a more sedate rate. At $t = 10^{-12}$ second, the temperature of the universe had dropped to 10^{15} K, the energy of the particles had fallen to 100 GeV, and there was a final spontaneous symmetry breaking as described by the Standard Model. That last symmetry breaking separated the electromagnetic force from the weak force, and from that moment on, all four forces have interacted with particles essentially as they do today.

The electroweak symmetry breaking around 100 GeV has been verified, including the predicted Higgs particle. While the two symmetry breaking events predicted to occur at higher energies (and earlier times) have not been observed, they follow a similar mechanism for what has been observed at lower energies with the electroweak force. This lower-energy success certainly does not prove that these higher energy theories are correct, but it provides motivation for formulating and testing them. For a "report card" on cosmology theories, including inflation, see the essay at the end of this chapter.

26-3 Following inflation, energy is converted to matter and antimatter

If the ideas of inflation are correct, it was a brief but stupendous expansion of the universe soon after the beginning of time. Physicists now realize that inflation helps explain where all the matter and radiation in the universe came from. To see how a violent expansion of space could create particles, we must first understand what quantum mechanics tells us about the creation of particles.

Quantum Mechanics and the Heisenberg Uncertainty Principle

Quantum mechanics is the branch of physics that explains the behavior of nature on the atomic scale and smaller. For example, quantum mechanics tells us how to calculate the structure of atoms and the interactions between atomic nuclei. **Elementary particle physics** is the branch of quantum mechanics that deals with individual subatomic particles and their interactions, including the strong, weak, and electromagnetic forces that we discussed in Section 26-2.

The microscopic world of quantum mechanics is significantly different from the ordinary world around us. In the ordinary world, we have no trouble knowing where things are. You know where your house is, you know where your car is, you know where this book is. In the subatomic world of electrons and nuclei, however, you can no longer speak with this same confidence. A certain amount of fuzziness, or uncertainty, enters into the description of reality at the incredibly small dimensions of the quantum world.

To appreciate the reasons for this uncertainty, imagine trying to measure the position of a single electron. To find out where it is located, you must observe it. And to observe it, you must shine a light on it. However, the electron is so tiny and has such a small mass that the photons in your beam of light possess enough energy to give the electron a mighty kick. As soon as a photon strikes the electron, the electron recoils in some unpredictable direction. Consequently, no matter how carefully you try to measure the precise location of an electron, you necessarily introduce some uncertainty into the speed of that electron.

These ideas are at the heart of the **Heisenberg uncertainty principle**, first formulated in 1927 by the German physicist Werner Heisenberg, one of the founders of quantum mechanics. This principle states that there is a reciprocal uncertainty between position and momentum (momentum is equal to the mass of a particle multiplied by its velocity). The more precisely you try to

measure the position of a particle, the more unsure you become of how the particle is moving. Conversely, the more accurately you determine the momentum of a particle, the less sure you are of its location. These restrictions are not a result of errors in making measurements; they are fundamental limitations inherent to all natural phenomena.

There is an analogous uncertainty involving energy and time. You cannot know the energy of a system with infinite precision at every moment in time. Over short time intervals, there can be great uncertainty about the amounts of energy in a subatomic system. Specifically, let ΔE be the smallest possible uncertainty in energy measured over a short interval of time Δt. (Astronomers and physicists often use the capital Greek letter delta, Δ, as a prefix to denote a small quantity or a small change in a quantity.)

Heisenberg uncertainty principle for energy and time

$$\Delta E \times \Delta t = \frac{h}{2\pi}$$

ΔE = uncertainty in energy

Δt = time interval over which energy is measured

h = Planck's constant = 6.625×10^{-34} J s

This equation says that the shorter the time interval Δt, the greater the energy uncertainty ΔE must be in order to ensure that the product of ΔE and Δt is equal to $h/2\pi$.

We might look upon the Heisenberg uncertainty principle as merely an unfortunate limitation on our ability to know everything with infinite precision. But, in fact, this principle provides startling insights into the nature of the universe.

Spontaneously Created Matter and Antimatter

We have seen that one of the important conclusions of Einstein's special theory of relativity is the equivalence of mass and energy: $E = mc^2$ (see Section 16-1). There is nothing uncertain about the speed of light (c), which is an absolute constant. Therefore, any uncertainty in the energy of a physical system can be attributed to an uncertainty Δm in the mass. Thus,

$$\Delta E = \Delta m \times c^2$$

Combining this expression with the previous equation, we obtain:

Heisenberg uncertainty principle for mass and time

$$\Delta m \times \Delta t = \frac{h}{2\pi c^2}$$

Δm = uncertainty in mass

Δt = time interval over which mass is measured

h = Planck's constant = 6.625×10^{-34} J s

c = speed of light = 3.00×10^8 m/s

This result is astonishing. It means that over a very brief interval Δt of time, we cannot be sure how much matter there is in a particular location, even in "empty space." During this brief moment, matter can spontaneously appear and then disappear. The greater the amount of matter (Δm) that appears spontaneously, the shorter the time interval (Δt) during which it can exist before disappearing into nothingness. This bizarre state of affairs is a natural consequence of quantum mechanics.

No particle can appear spontaneously by itself, however. For each particle created, so is a second, almost identical **antiparticle**; particles are made of matter, and antiparticles are made of antimatter. In other words, equal amounts of matter and **antimatter** come into existence and then disappear. (These are the virtual particle pairs discussed in Section 25-7, and in more depth later on.)

Despite its exotic name, there is actually nothing terribly mysterious about antimatter. A particle and an antiparticle are identical in almost every respect; their main distinction is that they carry opposite electric charges. For example, an ordinary electron (e^-) carries a negative charge; the corresponding antiparticle has the same mass but a positive charge, which is why it is called a **positron** (e^+, and also called an antielectron). Protons (with positive charge) also have an antimatter partner called an **antiproton** (with negative charge). Particles that have no electric charge can also have corresponding antiparticles. An example is the neutrino (ν); we met its antiparticle, the antineutrino ($\overline{\nu}$), in Section 26-2. The antineutrino is also electrically neutral, but it differs from the neutrino in having opposite values of other, more subtle physical properties.

A spontaneously created particle–antiparticle pair lasts for only an incredibly brief time. For example, consider an electron and a positron, each with a mass of 9.11×10^{-31} kg, If we rewrite the Heisenberg uncertainty principle for mass and time to solve for Δt and then substitute the combined mass of $2 \times 9.11 \times 10^{-31}$ kg, we find that a spontaneously created electron–positron pair can last for a time

$$\Delta t = \frac{1}{\Delta m}\frac{h}{2\pi c^2} = \frac{1}{2 \times 9.11 \times 10^{-31}} \times \frac{6.625 \times 10^{-34}}{2\pi(3.00 \times 10^8)^2}$$

$$= 6.43 \times 10^{-22}\ \text{s}$$

In other words, an electron and a positron can spontaneously appear and then disappear without violating any laws of physics—but they cannot remain in existence for much longer than 6.43×10^{-22} second. In fact, the laws of quantum physics *require* that electrons and positrons constantly pop in and out of existence. The same behavior applies to all massive particles, but the more massive the particle, the shorter the time interval it can exist. Next, we look at this process in more detail.

Particle–Antiparticle Pairs and the Origin of Matter in the Universe

Spontaneous creation can and does happen absolutely anywhere and at any time, not just under the unusual conditions of the early universe. (It is happening right now in the space between this book and your eyes.) Quantum mechanics tells us that if a process is not strictly forbidden, then it must occur. Pairs of every conceivable

> Quantum mechanics reveals that "empty" space is not empty, but is seething with particles and antiparticles that appear and then annihilate.

particle and antiparticle are constantly being created and destroyed at every location across the universe. We have no way of observing these pairs directly, but we can observe their effects on regular particles. For this reason, they are called **virtual pairs**. They do not "really" exist in the same sense as ordinary particles; but they "virtually" exist and have been experimentally verified. The particles that are exchanged in the four fundamental forces (see Section 26-2) are also virtual particles.

Although virtual pairs of particles and antiparticles cannot be observed directly, their effects have nonetheless been predicted and detected to very high precision. Imagine, for example, an electron in orbit about the nucleus of an atom, such as a hydrogen atom. Ideally, the electron should follow its orbit in a smooth, unhampered fashion. However, because of the constant brief appearance and disappearance of pairs of particles and antiparticles, minuscule electric fields exist for extremely short intervals of time. These tiny, fleeting electric fields cause the electron to jiggle slightly in its orbit. This jiggling produces slight changes in the energies of different electron orbits in the hydrogen atom, which manifest themselves as a minuscule shift in the wavelengths of the hydrogen spectral lines. (We discussed the connection between the energies of electron orbits in hydrogen and the hydrogen spectrum in Section 5-8.)

This shift was first detected in 1947 and today is known as the **Lamb shift**. The Lamb shift and more recent experiments provide powerful evidence that every point in space, all across the universe, is seething with virtual pairs of particles and antiparticles. In this sense, "empty space" is actually not empty at all. **Figure 26-13** sketches the constant appearance and disappearance of virtual particles and antiparticles.

Recall that, in considering what could be the source of dark energy in Section 25-7, one idea is that it comes from virtual particle pairs. This source of energy has the perfect property required for dark energy: it is proportional to how much space there is. Unlike regular particles that just get farther apart as the universe expands, if the size of the universe doubles, the energy from virtual particle pairs also doubles. In this sense, virtual particle pairs result in a true cosmological constant, for a constant energy associated with each volume of space. However, as we mentioned,

A particle–antiparticle pair can appear anywhere in space . . .   . . . but must disappear after a very short time interval.

FIGURE 26-13

Virtual Pairs Pairs of particles and antiparticles can appear and then disappear anywhere in space provided that each pair exists only for a very short time interval, as dictated by the uncertainty principle. In this sketch, electrons are shown in green and positrons are shown in red.

FIGURE 26-14

Pair Production and Annihilation **(a)** Pairs of virtual particles can be converted into real particles by high-energy gamma-ray photons. In this illustration, an electron (shown in green) and a positron (in red) are produced. This process can take place only if the combined energy of the two photons is no less than Mc^2, where M is the total mass of the electron and positron. **(b)** Conversely, a particle and an antiparticle can annihilate each other and be transformed into energy in the form of gamma rays.

the numerical value of the cosmological constant derived from virtual particle pairs is 100 orders of magnitude larger than what is observed. So, while the ideas of virtual particles may play a role in dark energy, there is no clear understanding yet of how these ideas all come together.

Not all pairs of particles and antiparticles are virtual, however. In a phenomenon called **pair production**, pairs of highly energetic gamma rays (photons) can convert their energy into pairs of real particles and antiparticles (Figure 26-14a). Quite simply, the gamma-ray photons disappear upon colliding into each other and convert their energy into a particle and an antiparticle. As real particles, they are directly observable and do not have to annihilate. While Figure 26-14a shows this process of pair production, Figure 26-14b shows the inverse process of **annihilation**, in which a particle and antiparticle collide with each other and are converted into high-energy gamma rays.

Pair production is routinely observed in high-energy particle accelerators. Indeed, it is one of the ways in which physicists manufacture exotic species of particles and antiparticles. The only requirement is that nature's balance sheet be satisfied. To create a particle and an antiparticle having a total mass M, the incoming gamma-ray photons must possess an amount of energy E that is greater than or equal to Mc^2. If the photons carry too little energy (less than Mc^2), pair production will not occur. Likewise, the more energetic the photons, the more massive the particles and antiparticles that can be manufactured.

Furthermore, the energy that gets converted into pairs of particles and antiparticles doesn't need to come from photons. For example, the particle accelerator at CERN that discovered the Higgs particle collided protons together. While protons do have mass, the Higgs particle they created had more than 50 times the regular mass of the colliding protons; the large mass of the Higgs largely came from the *energy* of the collisions. In order to have a large enough kinetic energy to produce the Higgs, protons were accelerated at 0.99999999% the speed of light! This speed is only 3.1 m/s less than the speed of light—a mere difference of 7 mi/h!

We have seen that the conversion of energy into mass described by $E = mc^2$ can come from a variety of energy sources (gamma-ray photons, high-speed protons, etc.). *On the grandest scale of all, converting energy into mass is thought to be the origin of mass in our universe.* Unlike the chicken-and-egg question of which came first, we think energy came first. The energy could have been in the inflaton field—the field posited to give rise to inflation—as illustrated in Figure 26-12. The universe starts out in an energy state determined from the black dot in Figure 26-12a, where the initial energy is indicated along the vertical axis. After the universe expands, and as the energy of the inflaton field decreases (Figure 26-12c), it is thought that the field energy is converted into the mass of many particles.

Inflationary expansion of the universe led to its rapid cooling, but these new particles zooming around brought the temperature of the universe back up. That is why this process is referred to as reheating.

The creation of mass from pure energy produces equal amounts of matter and antimatter. Thus, the early universe was awash in particles and antiparticles. This is the first step for the origin of matter in the universe, but it is not the last.

CONCEPT CHECK 26-6

Protons are about 2000 times more massive than electrons. If a virtual proton and antiproton spontaneously appear out of otherwise empty space, would this virtual pair last longer or shorter than a pair of virtual electrons before annihilating?

CONCEPT CHECK 26-7

Suppose two gamma rays collide and produce an electron and its antiparticle. Does the appearance of this pair violate the physical law that the total charge of an isolated system remains constant? (This law is also referred to as *conservation of charge*.)

Answers appear at the end of the chapter.

26-4 The origin of the matter and radiation that we see today

As soon as the flood of matter and antimatter appeared in the universe, collisions between particles and antiparticles began to produce numerous high-energy gamma rays. As these gamma rays collided, they promptly turned back into the particles and antiparticles from which they came. As a result, the rate of pair production soon equaled the rate of annihilation. For example, for every electron and positron that annihilated each other to create gamma rays (Figure 26-14b), two gamma rays collided elsewhere to produce an electron and a positron (Figure 26-14a). In other words, annihilation and pair production reactions proceeded with equal vigor, and as many particles and antiparticles were being created as were being destroyed.

As the universe continued to expand, all the gamma-ray photons became increasingly redshifted. As a result, the temperature of the radiation field fell. Due to their frequent interaction, radiation and particles of all kinds were in **thermal equilibrium:** All particle species, including photons, were at the same temperature. Hence, as the radiation temperature decreased, the temperature of particles of different types decreased as well. As we'll see next, a decreasing temperature has a different effect on the production of particle pairs compared to their annihilation.

From Quark Confinement to Particle–Antiparticle Annihilation

The first change in the population of particles and antiparticles occurred at $t = 10^{-6}$ second, when the temperature was 10^{13} K and particles were colliding with energies of roughly 1 GeV. Prior to this moment, particles collided so violently that individual protons and neutrons could not exist, being constantly fragmented into quarks. After this time, appropriately called the period of **quark confinement,** quarks were finally able to stick together and became confined within individual protons and neutrons.

LOOKING DEEPER 26.1 As the universe continued to expand, temperatures eventually became so low that the gamma rays, according to $E = mc^2$, no longer had enough energy to create particular kinds of particles and antiparticles. We say that the temperature dropped below the particular particle's **threshold temperature.** Collisions between these types of particles and antiparticles continued to add photons to the cosmic radiation background, but collisions between photons with decreasing energy could no longer convert into mass and replenish the supply of particles and antiparticles.

> The cosmic background radiation we see today was spawned from a vast sea of particles and antiparticles in the early universe.

At the same time that quark confinement became possible so that protons and neutrons appeared, the universe also became cooler than the 10^{13}-K threshold temperatures of both protons and neutrons. No new protons or neutrons were formed by pair production, but the annihilation of protons by antiprotons and of neutrons by antineutrons continued vigorously everywhere throughout space. This wholesale annihilation dramatically lowered the matter content (particles and antiparticles) of the universe, while simultaneously increasing the radiation (photon) content.

A little later, when the universe was about 1 second old, its temperature fell below 6×10^9 K, the threshold temperature for electrons and positrons. As with protons and neutrons, a similar annihilation of pairs of electrons and positrons further decreased the matter content of the universe while raising its radiation content. This radiation field, which fills all space, is the *primordial fireball* discussed in Section 25-5. Therefore, this fireball, which later became the CMB, filled the universe for the next 380,000 years, deriving much of its energy from the annihilation of particles and antiparticles during the first second after the Big Bang.

Now we have a dilemma. If there had been perfect symmetry between particles and antiparticles, then for every proton there should have been an antiproton. For every electron, there should likewise have been a positron. Consequently, by the time the universe was 1 second old, every particle would have been annihilated by an antiparticle, leaving no matter at all in the universe.

Obviously, a total annihilation of all matter and antimatter never happened. The planets, stars, and galaxies we see in the sky are made of matter, not antimatter. If there were still substantial amounts of antimatter in the universe, it would eventually collide with ordinary matter. We would then see copious amounts of gamma rays being emitted from the entire sky. While we do observe gamma-ray photons from various locations in the universe, they are neither numerous enough nor of the right energy to indicate the presence of much antimatter. Thus, there must have been an excess of matter over antimatter immediately after the Big Bang so that the particles outnumbered the antiparticles.

Quite remarkably, we can estimate the extent of this initial asymmetry between matter and antimatter. In other words, while we essentially see a universe consisting of matter today, we can still infer how much antimatter there was before the antimatter portion of the universe annihilated with most of the matter portion. The key is light: Each annihilation of two particles produces exactly two photons (Figure 26-14b), and after being stretched to longer wavelengths, these are the photons we observe in the cosmic microwave background. As noted in Section 25-5, there are roughly 10^9 (1 billion) photons today in the microwave background for each proton and neutron in the universe. Thus, for every 10^9 antiprotons, there must have been 10^9 plus 1 ordinary protons, leaving one surviving proton after annihilation. Similarly, for every 10^9 positrons, there must have been 10^9 plus 1 ordinary electrons.

While we can infer that the initial matter–antimatter asymmetry was equal to about one additional particle made of regular matter per billion matter–antimatter pairs, we do not understand what process led to this initial asymmetry. There are several ideas for explaining the asymmetry but no hard evidence or agreement on a likely solution, and this initial asymmetry remains one of the great unsolved problems in physics and astronomy.

CONCEPT CHECK 26-8

Why does the following hypothesis seem unlikely? There is a perfect symmetry between matter and antimatter, but the antimatter is "somewhere else" and avoids annihilating with matter.

Answer appears at the end of the chapter.

26-5 A background of neutrinos and most of the helium in the universe are relics of the primordial fireball

The early universe must have been populated with vast numbers of neutrinos (ν) and their antiparticles, the antineutrinos ($\bar{\nu}$). These particles have a very small mass, so their threshold temperature is quite low. These particles take part in the nuclear reactions that transform neutrons into protons and vice versa. For example, a neutron can decay into a proton by emitting an electron and an antineutrino:

$$n \rightarrow p + e^- + \bar{\nu}$$

This radioactive decay happens quickly (its half-life is about 10.5 min), which is why we do not find free neutrons floating around in the universe today. In the first 2 seconds after the Big Bang, however, neutrons were also created by collisions between protons and electrons:

$$p + e^- \rightarrow n + \nu$$

This reaction kept the number of neutrons approximately equal to the number of protons—a balance that was maintained only as long as collisions between protons and electrons were frequent. But the number of electrons decreased precipitously as the temperature of the universe fell below 6×10^9 K, and electrons and positrons annihilated each other without being replenished (Section 26-4). By the time the universe was about 2 seconds old, no new neutrons were being formed, the natural tendency for neutrons to decay into protons took over, and the number of neutrons began to decline.

Spawning the First Nuclei

Before many of the neutrons could decay into protons, they began to combine with protons to form nuclei. Nuclei of helium, the first element more massive than hydrogen, consist of either two protons and two neutrons (^{4}He) or two protons and a single neutron (^{3}He). It is exceedingly improbable that two protons and one or two neutrons should all simultaneously collide with one another to form a helium nucleus. Instead, helium nuclei are built in a series of steps. The first step is to have a single proton and a single neutron combine to form deuterium (^{2}H), sometimes called "heavy hydrogen." A photon (γ) is emitted in this process, so we write this reaction as

> The first atomic nuclei formed within a quarter-hour after the Big Bang.

$$p + n \rightarrow {}^2H + \gamma$$

Forming deuterium, however, does not immediately lead to the formation of helium. The problem is that deuterium nuclei are easily destroyed, because a proton and a neutron do not stick together very well. Indeed, in the early universe, high-energy gamma rays easily broke deuterium nuclei back down into independent protons and neutrons. As a result, the synthesis of helium could not get beyond the first step. This block to the creation of helium is called the **deuterium bottleneck.**

When the universe was about 3 minutes old, the background radiation had cooled enough that its photons no longer had enough energy to break up the deuterium. By this time, most of the neutrons had decayed into protons, and protons outnumbered neutrons by about 6 to 1. Because deuterium nuclei could now survive, the remaining neutrons combined with protons and rapidly produced helium. (The *Cosmic Connections* figure in Chapter 16 depicts a similar sequence of reactions that take place in the core of the present-day Sun.)

The result was what we find in the universe today—about 1 helium atom for every 10 hydrogen atoms. In addition to helium, nuclei of lithium (Li, which has 3 protons) and beryllium (Be, which has 4) were also produced in small numbers. The process of building up nuclei such as deuterium and helium from protons and neutrons is called **nucleosynthesis** (**Figure 26-15**).

FIGURE 26-15

Nucleosynthesis in the Early Universe   This graph shows how nuclei were produced between 10 seconds and 10 hours after the Big Bang. The vertical axis shows the fraction of the total mass that was in each type of particle or nucleus (p = proton, n = neutron, ^{2}H = deuterium, ^{3}He and ^{4}He = helium, ^{6}Li and ^{7}Li = lithium, ^{7}Be = beryllium). Very few nuclei were formed before the universe was 10 seconds old, due to the phenomenon of the deuterium bottleneck, which occurred at times earlier than those shown here. By about 10^3 seconds (roughly 15 min) after the Big Bang, the temperature had dropped below 4×10^8 K, no further nucleosynthesis was possible, and the relative amounts of different nuclei stabilized. The number of free neutrons declined rapidly as these particles decayed into protons, electrons, and antineutrinos. Agreement between these predicted mass fractions with what is observed today provides strong evidence for nuclear reactions in the early universe. (Data from R. V. Wagoner)

Because nuclei have positive electric charges, bringing them together to form more massive nuclei requires that they overcome their mutual electric repulsion. They are unable to do so if they are moving too slowly, which will be the case if the temperature is too low. As a result, by about 15 minutes after the Big Bang, the universe was no longer hot enough for nucleosynthesis to take place. Only the four lightest elements (hydrogen, helium, lithium, and beryllium) were present in appreciable numbers. The heavier elements would be formed only much later, once stars had formed and nuclear reactions within those stars could manufacture carbon, nitrogen, oxygen, and all the other elements.

Nucleosynthesis was one of the great early successes of the Big Bang theory. At lower temperatures (to the right in Figure 26-15), the amounts of different nuclei leveled off and remain unchanged. That allows us to compare the amount of each predicted atomic element to the amount observed today, and the agreement is excellent. Furthermore, these nucleosynthesis calculations also predicted the density of ordinary matter in the universe, and this prediction has been verified by the CMB fluctuations, as in Section 25-8 and Figure 25-22. Taken together, the success of nucleosynthesis provides very strong evidence for a hot Big Bang, and that during its first few minutes the entire universe was one big nuclear reactor.

The Neutrino–Antineutrino Background

While nuclei were being formed in the early universe, what happened to all those primordial neutrinos and antineutrinos that had interacted so vigorously with the protons and neutrons before the universe was 2 seconds old? The answer is that by $t = 2$ seconds, matter was sufficiently spread out so that the universe became transparent to neutrinos and antineutrinos. From that time on, neutrinos and antineutrinos could travel across the universe unimpeded. Their interaction with matter is so weak that about 1 trillion neutrinos pass right through our bodies every second. Even Earth itself is virtually transparent to neutrinos from the Sun (Section 16-4).

The neutrinos and antineutrinos that were liberated at $t = 2$ seconds should now fill the universe much as the cosmic microwave background does. Indeed, these ancient neutrinos and antineutrinos may be about as populous today as the photons in the microwave background (of which there are 4.1×10^8 per cubic meter). The *neutrino–antineutrino background* should be slightly cooler than the photon background, which received extra energy from electron–positron annihilations. Physicists estimate that the current temperature of the neutrino–antineutrino background is about 2 K, as opposed to 2.725 K for the microwave background. Unfortunately, because neutrinos and antineutrinos are so difficult to detect, we do not yet have direct evidence of the neutrino–antineutrino background.

CONCEPT CHECK 26-9

How does the changing temperature of the universe confine the production of the elements helium, lithium, and beryllium to a period ranging from about 3 minutes to 15 minutes after the Big Bang?

Answer appears at the end of the chapter.

26-6 Galaxies and the first stars formed from density fluctuations in the early universe

The distribution of matter in the universe today is quite lumpy. Stars are grouped together in galaxies, galaxies into clusters, and clusters into superclusters that stretch across 50 Mpc (150 million ly) or more (see Section 23-6). Furthermore, galaxies seem to be concentrated along enormous sheets, which in turn surround voids measuring 30 to 120 Mpc (100 million to 400 million ly) across. Figures 23-23 and 23-24 show these features, which characterize the large-scale structure of the universe. How did this large-scale structure arise from the chaos of the primordial fireball? When did stars first appear in the universe? And when and how did galaxies first form?

Density Fluctuations and the Jeans Length

At first glance, the origin of large-scale structure seems puzzling, because the early universe must have been exceedingly smooth. To see why, think back to the era of recombination that occurred 380,000 years after the Big Bang (see Section 25-5). Before recombination, high-energy photons were constantly and vigorously colliding with charged particles throughout all space. After recombination, the universe became transparent, and these photons stopped interacting with the matter in the universe. Astronomers say that matter "decoupled" from radiation during the era of recombination. Because the cosmic microwave background is extremely isotropic, we can conclude that the matter with which these photons once collided so frequently must also have been spread smoothly across space.

The distribution of matter during the early universe could not have been *perfectly* uniform, however. If it had been, it would still have to be absolutely uniform today; there would now be only a few atoms per cubic meter of space, with no stars and no galaxies. Consequently, there must have been slight lumpiness, or **density fluctuations**, in the distribution of matter in the early universe. These fluctuations are thought to have been produced by inflation. According to the Heisenberg uncertainty principle (see Section 26-3), even the inflaton field itself undergoes quantum fluctuations. Some fluctuations result in a region containing slightly more energy in the inflaton field, and some regions contain slightly less energy. After the energy in the inflaton field converts into particles with mass, there is a higher density of particles in regions where the energy was slightly greater. Likewise, regions with lower energy yielded a lower density of particles.

It was over time, through the action of gravity, that these density fluctuations from inflation initiated the clumping that eventually grew to become the galaxies and clusters of galaxies that we see today throughout the universe. We call these early density fluctuations the "seeds" of structure formation. As we saw in Section 25-5 and Section 25-8, the pattern of density fluctuations became imprinted on the cosmic background radiation during the era of recombination. Figure 25-13 shows a map of these fluctuations obtained from the WMAP microwave background spacecraft.

Our understanding of how gravity can amplify density fluctuations dates back to 1902, when the British physicist James Jeans solved a problem first proposed by Isaac Newton. Suppose that you have a gas with only very tiny fluctuations in density, as shown in Figure 26-16a. These regions of higher density will then gravitationally attract nearby material and thus gain mass. As the regions become more massive, however, the pressure of the gas inside these regions will also increase, which can make these regions expand and disperse. Jeans analyzed these opposing effects and asked the question: Under what conditions does gravity overwhelm gas pressure so that a permanent object can form?

Jeans proved that an object will grow due to a density fluctuation, provided that the fluctuation extends over a distance that exceeds the so-called **Jeans length** (L_J):

Jeans length for density fluctuations

$$L_J = \sqrt{\frac{\pi k T}{m G \rho_m}}$$

L_J = Jeans length

k = Boltzmann constant = 1.38×10^{-23} J/K

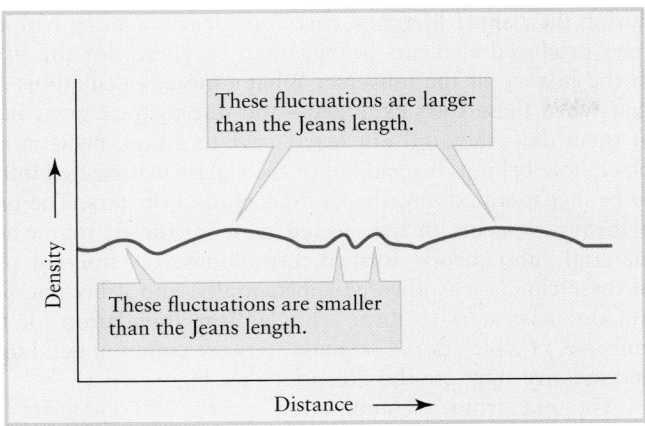

(a) At an early time

(b) At a later time

FIGURE 26-16

The Growth of Density Fluctuations **(a)** This conceptual illustration shows small density fluctuations in the distribution of matter shortly after the era of recombination. **(b)** If the size of a fluctuation is greater than the Jeans length (L_J), it becomes gravitationally unstable and can grow in amplitude.

T = temperature of the gas (in kelvins)

m = mass of a single particle in the gas (in kilograms)

G = universal constant of gravitation
   = $6.67 \times 10^{-11}\ \mathrm{N \cdot m^2/kg^2}$

ρ_m = average density of matter in the gas

Density fluctuations that extend across a distance larger than the Jeans length tend to grow, whereas fluctuations smaller than L_J tend to disappear (Figure 26-16b).

We can apply the Jeans formula to the conditions that prevailed during the era of recombination, when $T = 3000\ \mathrm{K}$ and $\rho_m = 10^{-18}\ \mathrm{kg/m^3}$. Taking m to be the mass of the hydrogen atom ($m = 1.67 \times 10^{-27}\ \mathrm{kg}$), we find that $L_J = 100$ ly, the diameter of a typical globular cluster (Figure 26-17). Furthermore, the mass contained in a cube whose sides are 1 Jeans length in size (equal to the product of the density, ρ_m, and the volume of the cube, L_J^3) is about $5 \times 10^5\ \mathrm{M_\odot}$, equal to the mass of a typical globular cluster. These calculations suggest that structures the size of globular clusters were among the first objects to form after recombination.

CONCEPT CHECK 26-10

Consider a density fluctuation around the era of recombination when the universe had a temperature of $T = 3000\ \mathrm{K}$. If the density fluctuation is about 10 ly across, will the matter within that fluctuation begin clumping together? What causes or prevents clumping?

Answer appears at the end of the chapter.

Population III Stars: The "Zeroth" Generation

We saw in Section 19-4 that globular clusters contain the most ancient stars we can find in the present-day universe. These are Population II stars with a low percentage of metals (elements heaver than hydrogen and helium), and they are of an earlier stellar generation than metal-rich Population I stars like the Sun (see Section 19-5). However, the Population II stars in globular clusters

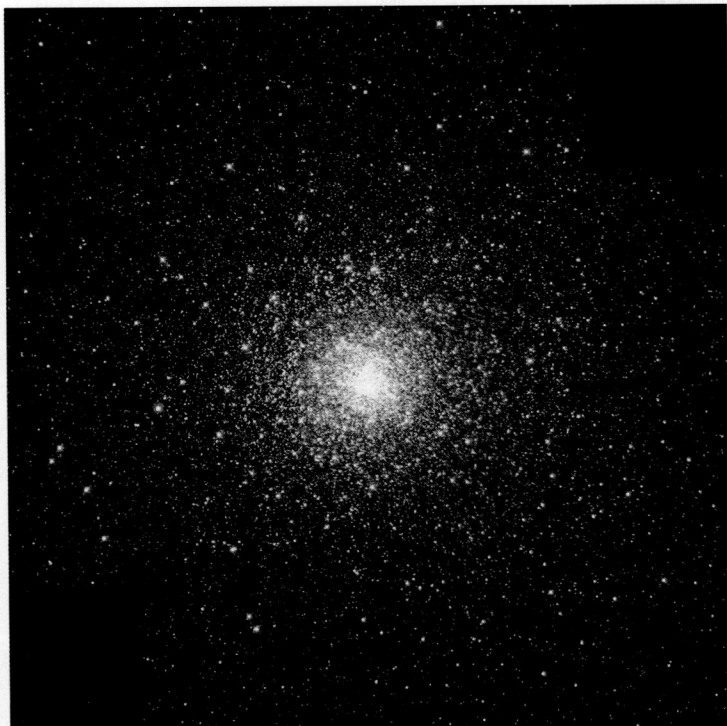

FIGURE 26-17 R I **V** U X G

A Globular Cluster A typical globular cluster contains 10^5 to 10^6 stars, each with an average mass of about $1\ \mathrm{M_\odot}$, so the total mass of a typical cluster is 10^5–$10^6\ \mathrm{M_\odot}$. Cluster diameters range from about 6 to 120 pc (20 to 400 ly). Because these masses and diameters are comparable to the Jeans length (L_J) during the era of recombination, astronomers suspect that structures the size of globular clusters were among the first objects to form in the universe. (Hubble Heritage Team/AURA/SIScI/NASA)

cannot be the very first stars to have formed after the Big Bang. Those first stars could have contained only hydrogen, helium, and tiny amounts of lithium and beryllium; as Figure 26-15 shows, these were the only elements whose nuclei formed in the early universe. Hence, these original stars would have contained an even smaller percentage of metals than the Population II stars found in globular clusters. Such "zeroth-generation" stars are called *Population III stars* (see Section 19-5).

ANIMATION
26.1 ➤ Like stars in the present-day universe, Population III stars would have formed from clouds of gas. These ancient gas clouds were composed almost exclusively of hydrogen and helium atoms, and such clouds have higher internal pressures than do metal-rich clouds of the same temperature. A star can form only when the mutual gravitation of the various parts of a cloud (which tends to make the cloud collapse) overcomes the internal pressure of the cloud (which tends to prevent collapse). Hence, Population III stars could form only if their mass (and hence their mutual gravitation) was rather large. Calculations suggest that these stars had masses from 30 to 1000 $M_\odot$, compared to the range of 0.4 to roughly 120 $M_\odot$ for modern stars. Even the smallest Population III star would rank among the largest stars observed today.

> The "zeroth generation" of stars were much more massive and luminous than stars today.

Since Population III stars were all very massive, their lifetimes were short and none could have survived to the present day. But during their short lifetimes, thermonuclear reactions within these stars produced elements heavier than beryllium for the first time in the history of the universe. What's more, calculations suggest that when these stars exploded—and due to their great mass, all of them did—they did not leave a white dwarf, neutron star, or black hole behind. Instead, all of their mass was ejected into space to be incorporated into the next generation of stars. The presence of heavy elements in the ejected material meant that when this material subsequently formed into clouds, the internal pressure of these clouds was lowered substantially, and it became possible for low-mass stars to form. This laid the foundation for today's universe, in which dim, low-mass stars are common and luminous, and massive stars are the exception (see Figure 17-5).

The era from recombination at $t = 380{,}000$ years to the first stars at about $t = 700$ million years is called the **dark ages**. The only photons present at that time were those that make up today's cosmic background radiation. The dark ages ended when the universe was filled for the first time with the light from stars (**Figure 26-18**).

Forming Large-Scale Structure

Once clumps the size of globular clusters had formed in the universe, how did they form into galaxies, clusters of galaxies, and

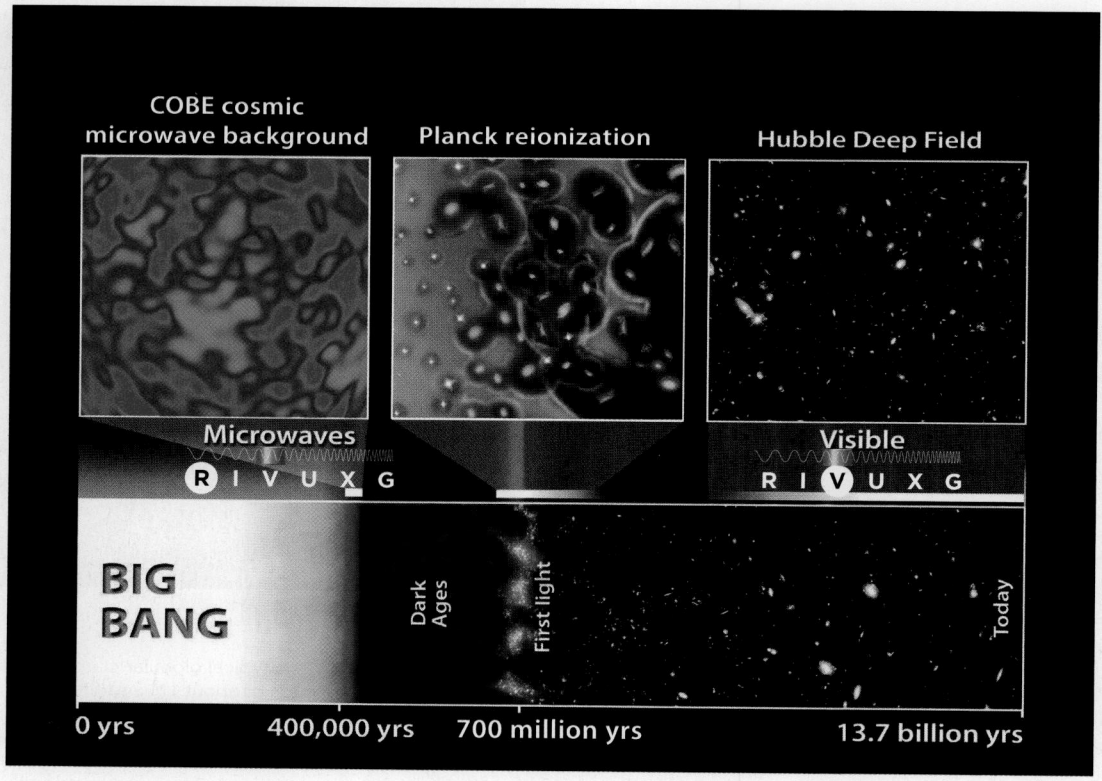

FIGURE 26-18

A Timeline of Light in the Universe   The oldest light that we can see today is the cosmic background radiation, which comes from a time 380,000 years after the Big Bang when the universe first became transparent. This light has a redshift of about $z = 1100$ and now appears in the microwave spectrum. Some 700 million years later, at a redshift of about $z = 7.5$, the first stars appeared. Ultraviolet light from these first stars (first appearing on the left of the middle illustration) ionized the hydrogen gas that had been neutral since recombination (depicted as yellow). This process is called reionization and was about 50% complete after 700 million years. Much of the light from the first stars is now redshifted to infrared wavelengths. Galaxies formed more recently and can be seen at visible wavelengths. (Top left: NASA; top center: ESA/C. Carreau; top right: NASA/JPL/STScI Hubble Deep Field Team) larger structures? One issue that complicates this matter is the presence of dark energy, which acts to accelerate the expansion of the universe (see Section 25-6). This accelerated expansion pulls clumps of material away from each other and makes it more difficult for them to coalesce into larger structures. Another complication is that about 85% of the mass in the universe is in the form of dark matter, whose nature is not known (see Section 23-8). Researchers have hypothesized different types of dark matter in the hope of explaining the large-scale structure that we see. Neutrinos are an example of **hot dark matter**, so named because it consists of lightweight particles traveling at high speeds. **Cold dark matter**, on the other hand, consists of massive particles traveling at slow speeds. Examples include WIMPs (weakly interacting massive particles, which we discussed in Section 22-4) as well as other exotic, speculative particles.

Scientists use supercomputer simulations to see how different types of dark matter would influence the development of large-scale structure. **Figure 26-19** shows the results of such a simulation for a flat universe with dark energy and *cold* dark matter. The simulation follows the motions of 2 million particles of cold dark matter in a box that expands as the universe expands. The box at the lower right of the figure, representing the present time (redshift $z = 0$) is 43 Mpc (160 million ly) a side. At earlier times, the box

represents a volume whose side is smaller by a factor of $1/(1 + z)$. For example, each side of the box for $z = 0.99$ is actually $1/1.99$ as long as the box for $z = 0$.

The simulation begins 120 million years after the Big Bang with an almost perfectly uniform distribution of particles, mimicking the tiny density fluctuations that must have been present just after inflation. A supercomputer then calculates how these particles move, based on Newton's laws in an expanding universe. As time goes on, the fluctuations grow into small, bright clumps whose sizes and masses are similar to those of galaxies. A large filament also forms, spanning the entire box from left to right. The simulation shows that few additional structures formed after the universe was about 6 billion years old, corresponding to redshift $z = 1$. The explanation is that after this time, the accelerating expansion of the universe becomes more important than gravitational attraction. The final frame of the simulation strongly resembles actual maps of galaxies in our present-day universe (see Figure 23-24).

Simulations similar to those in Figure 26-19 have also been carried out using *hot* dark matter in the form of neutrinos. A massless neutrino would always travel at the speed of light, just as a photon does. However, experiments show that neutrinos do have a small mass. (This nonzero mass is what allows one type of neutrino to

| z = 27.36   Universe 120 million years old | z = 9.83   Universe 490 million years old | z = 4.97   Universe 1.2 billion years old |

| z = 2.97   Universe 2.2 billion years old | z = 0.99   Universe 6.0 billion years old | z = 0.00   Universe 13.7 billion years old |

ANIMATION 26.2 ANIMATION 26.3 FIGURE 26-19

A Cold Dark Matter Simulation with Dark Energy These six views show the evolution of dark matter particles in a large, box-shaped volume of space. The box actually expands with time to follow the expansion of the universe and holds the same number of particles. In this figure, the boxes have been rescaled so they all appear at the same size. Small fluctuations in density are put into the simulation at the beginning (at upper left); these evolve over time to form structures that resemble those actually observed in our present-day universe (z = 0.00, shown at the lower right). (Applications by Andrey Kravtsov [The University of Chicago] and Anatoly Klypin [New Mexico State University], visualizations by Andrey Kravtsov)

FIGURE 26-20

"Bottom-Up" Galaxy Formation: Simulation   This image is taken from a cold dark matter simulation like that shown in Figure 26-15. A portion of the universe is shown at a time 2.2 billion years after the Big Bang, corresponding to redshift $z = 3.04$. The colors indicate the density of gas: Yellow is highest, red is medium, and blue is the lowest density. Over time, the gas tends to pile up at points where filaments intersect, forming galaxies and clusters of galaxies. (ESO)

transform into another. We saw in Section 16-4 that such transformations provided the explanation to the long-standing solar neutrino problem.) Hence, neutrinos travel slower than light, and they slow down as the universe expands and cools. Slow-moving neutrinos would accumulate over time within density fluctuations, and the gravitational pull of these neutrinos on surrounding matter could eventually lead to the formation of clusters of galaxies.

A primary difference between simulations based on cold and hot dark matter is the way in which galaxies form. In calculations based on cold dark matter, the formation of galaxies takes place from the "bottom up." In these simulations, the densest gas undergoes collapse early in the history of the universe and stars begin to form. The regions of star formation stream along the filaments (Figure 26-20). When they meet at the intersections between filaments, they merge and group together into galaxies, then clusters of galaxies, then superclusters. But in calculations based on hot dark matter, galaxies form from the "top down." Huge supercluster-sized sheets of matter form first and then fragment into galaxies. Observations of remote galaxies show that galaxies actually formed from the "bottom-up" scenario. One piece of evidence for this is the image in Figure 26-21a, which shows a handful of "galaxy building blocks" at $z = 3.04$ (when the universe was 2.2 billion years old). These building blocks, which have not yet coalesced into galaxies, lie within a long filament (see Figure 26-21b) that resembles those shown in the simulation of Figure 26-20. Figure 26-22 shows a collection of building blocks at a later stage in the process of merging into a galaxy. These observations strongly suggest that the dominant form of dark matter is cold, not hot.

What Large-Scale Structure Reveals

How might the universe have evolved if it had contained different amounts of cold dark matter and dark energy? Figure 26-23 shows some simulations designed to explore these possibilities. If the density of matter in the universe is kept constant for various simulations, the simulations predict approximately the same structure for different values of the dark energy density parameter Ω_Λ defined in Section 25-6 (see Figure 26-23a and Figure 26-23b). But if too large a matter density is used in the simulation, the voids between galaxies are smaller

(a) High-redshift objects that lie within a filament

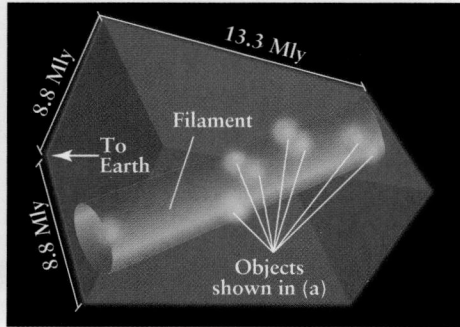

(b) Illustration of the filament

FIGURE 26-21 R I U X G

"Bottom-Up" Galaxy Formation: Observation   (a) The hexagons in this image from the Very Large Telescope show the positions of a number of sub-galaxy-sized objects at a redshift $z = 3.04$, the same as in the simulation shown in Figure 26-16. Excited hydrogen atoms in these objects emit ultraviolet photons, which are redshifted to visible wavelengths. This

redshifted emission gives these objects a characteristic green color. (The object at lower left actually lies in front of a much brighter quasar.) (b) The objects in (a) all lie within an immense filament. The purple box shows the volume of space studied in this observing program. The dimensions are given in millions of light-years (Mly). (ESO)

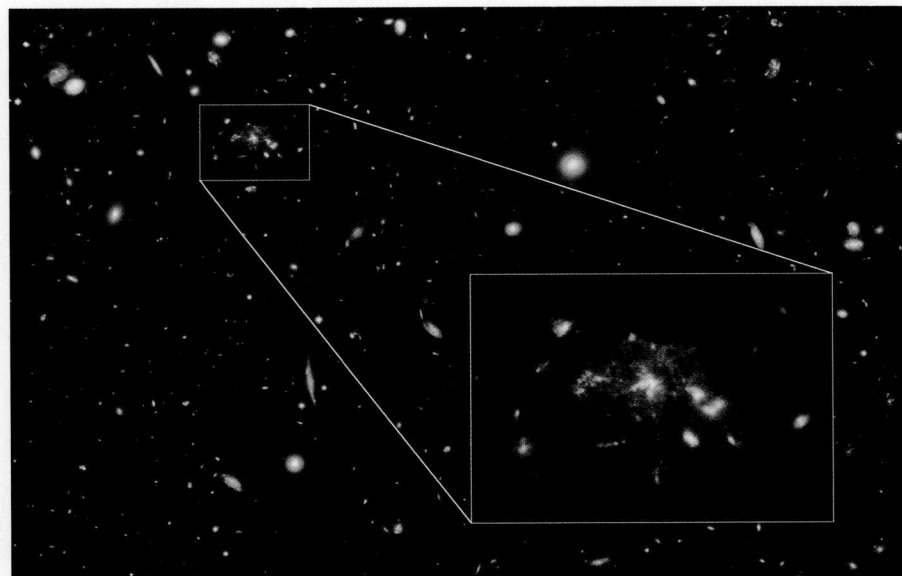

FIGURE 26-22 R I V U X G

A Galaxy Under Construction   This Hubble Space Telescope image shows dozens of small galaxies in the process of merging into a single large galaxy. We see this galaxy at a redshift $z = 2.2$, corresponding to a time 3.1 billion years after the Big Bang. (NASA, ESA, G. Miley and R. Overzier [Leiden Observatory], and the ACS Science Team)

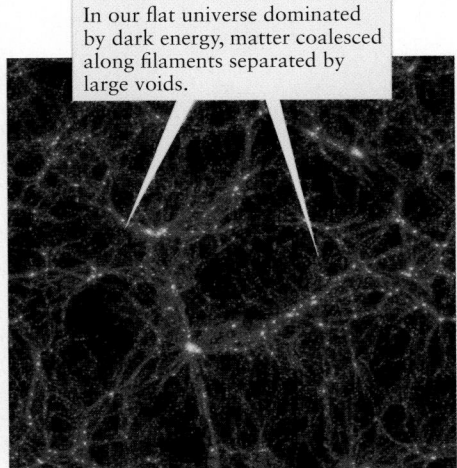

In our flat universe dominated by dark energy, matter coalesced along filaments separated by large voids.

(a) A flat universe with dark energy: $\Omega_m = 0.3$, $\Omega_\Lambda = 0.7$

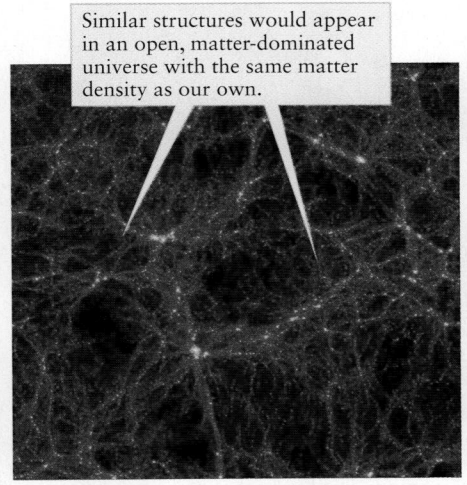

Similar structures would appear in an open, matter-dominated universe with the same matter density as our own.

(b) An open universe without dark energy: $\Omega_m = 0.3$, $\Omega_\Lambda = 0$

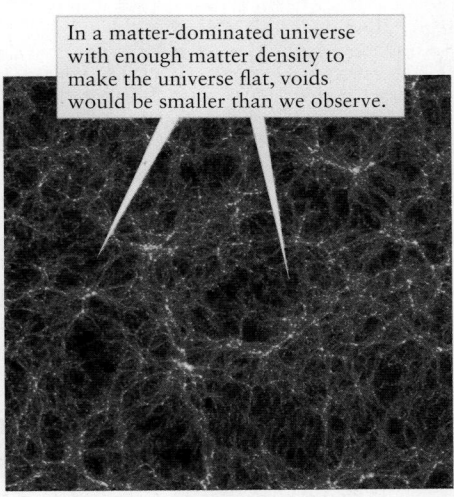

In a matter-dominated universe with enough matter density to make the universe flat, voids would be smaller than we observe.

(c) A flat universe without dark energy: $\Omega_m = 1.0$, $\Omega_\Lambda = 0$

FIGURE 26-23

Using Simulations to Constrain the Matter Density of the Universe   Cold dark matter simulations like those in Figures 26-15 and 26-16 help astronomers determine the value of the matter density parameter Ω_m. These three simulations show a portion of the universe at $z = 0$. **(a)** A simulation with $\Omega_m = 0.3$ and $\Omega_\Lambda = 0.7$, close to the values for our universe, gives a good match to the observed distribution of filaments and voids. **(b)** Nearly as good a match is obtained if we keep $\Omega_m = 0.3$, but eliminate dark energy so that $\Omega_\Lambda = 0$. **(c)** If we use a larger value of Ω_m, the distribution of matter in the simulation is a poor match to our universe. (Max Planck Institute for Solar System Research)

than what we actually observe in our universe (Figure 26-23c). Hence, observations of galaxy clustering coupled with supercomputer simulations of galaxy formation help determine the matter density of our universe. (We made use of this idea in Section 25-7. The brown band in Figure 25-18 shows the constraints on cosmological parameters from these observations and simulations of galaxies.)

The best match to the observed distribution of galaxy clusters and to the cosmic background radiation data (see Figure 25-22) is a model like that shown in Figure 26-19, with dark energy and cold dark matter in the proportions listed in Section 25-8.

The *Cosmic Connections* figure summarizes the past history of our universe down to the present day. As we discussed in Section 25-7, the *future* of our universe is less certain and depends on the detailed character of dark energy. More detailed data about galaxy clusters and the cosmic background radiation will be needed to pin down the future evolution of our universe.

OSMIC CONNECTIONS

e History of the Universe

Research in astronomy, elementary particle physics, and nuclear physics has allowed scientists to piece together the grand sweep of events over the first billion years of cosmic history. The greatest drama occurred in the first few minutes, and these graphs have been drawn to emphasize those earliest moments.

...ing the inflationary epoch, the ...ma of photons and charged ...icles expanded by an immense ...or.

...ing the 15 minutes after the inflationary ...ch, most matter and antimatter annihilated, ...ing a sea of photons and little matter. ...ons, neutrons, and the lightest nuclei formed.

During the recombination period about 380,000 years later, the first atoms formed and the cosmic microwave background (CMB) radiation was emitted.

After another 700 million years, radiation from the first stars reionized most of the hydrogen and helium. Galaxies later formed from building blocks that were initially about the size of a globular cluster.

(Above: from "The Cosmic Symphony" by W. Hu and M. White, *Scientific American*, February 2004, pp. 44–53, http://background.uchicago.edu/~whu/Papers/HuWhi04.pdf)

This graph shows the temperature history of the universe.
- As the universe cooled, the four forces "froze out" of their unified state as a result of spontaneous symmetry breaking.
- Neutrons and protons froze out of the hot "quark soup" during the quark confinement stage, which occurred 10^{-6} second after the Big Bang.

CONCEPT CHECK 26-11

How does structure formation in the early universe provide clues to the nature of dark matter?

Answer appears at the end of the chapter.

26-7 String theory attempts to unify the fundamental forces and predicts that the universe may have 11 dimensions

While we have a growing understanding of the early universe, there remains a veil obscuring the first 10^{-43} seconds after the Big Bang. These very *first* moments in the history of the universe, whose duration was the Planck time, set the stage for what would come after. To understand this brief interval—where the universe is compact and energetic enough to require quantum physics, and dense enough to require an understanding of gravity—we need to construct a quantum-mechanical theory that is unified with gravity: a theory of **quantum gravity**. While unification remains an unfinished task, remarkable progress has been made in recent years. One major development is that physicists have abandoned the idea that there are only three dimensions of space.

Beyond Four Dimensions

In his special and general theories of relativity, Einstein combined time with the three known dimensions of ordinary space, resulting in a four-dimensional combination called *spacetime* (see Section 21-2). In 1919, the Polish physicist Theodor Kaluza proposed the existence of a *fifth* dimension. Kaluza hoped to describe both gravity and electromagnetism in terms of the curvature of five-dimensional spacetime, just as Einstein had explained gravity by

itself in terms of the curvature of four-dimensional spacetime (see Section 21-3).

In Kaluza's theory, a particle always follows the straightest possible path in the four space dimensions he proposed. But in the three dimensions of ordinary space, the path appears curved. Hence, it appears to us that the particle has been deflected—by force—and this is Kaluza's mechanism to describe gravitational and electromagnetic forces. Kaluza's hypothetical fifth dimension exists at every point in ordinary space but is curled up so tightly, like a very tiny sphere, that it is not directly observable (Figure 26-24).

In 1926, the Swedish physicist Oskar Klein attempted to make Kaluza's five-dimensional theory compatible with quantum mechanics. While he was not successful, Klein discovered that particles of different masses could be identified with different vibrations occurring in the tiny sphere of Kaluza's fifth dimension.

String Theory and Speculative Models of the Universe

When Kaluza and Klein developed their theories, gravity and electromagnetism were the only known forces of nature. Today we know of four fundamental forces, which suggests that modern unification theories might require even more than five dimensions. Edward Witten at Princeton University has argued that a geometric theory for describing all four forces would work best with 11 dimensions, 10 of space and 1 of time. These models that attempt to unify all the forces of nature, including a quantum mechanical description of gravity, are called **string theory**. (There are also related names, such as superstring theory and **M-theory**.)

> Subatomic particles may actually be multidimensional membranes.

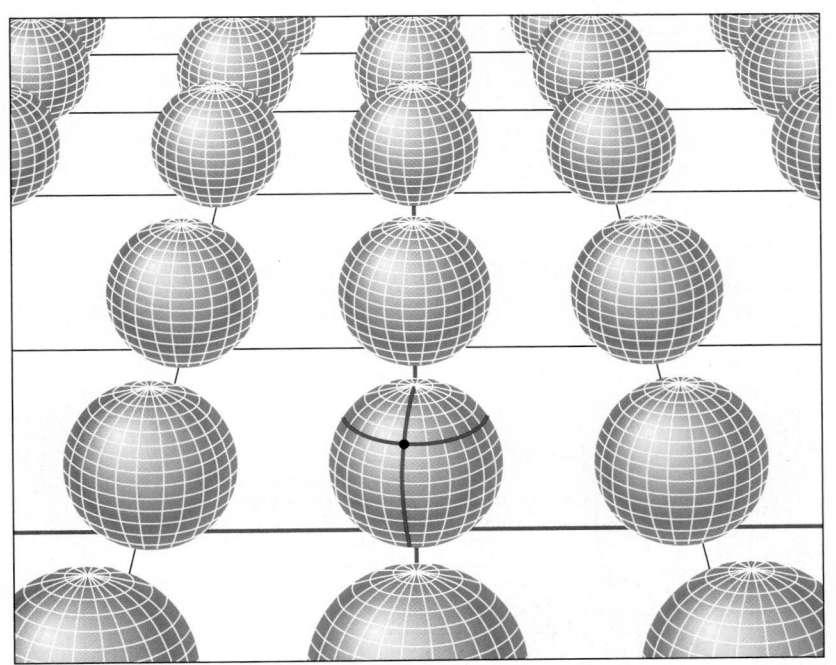

FIGURE 26-24

Hidden Dimensions of Space   Hidden dimensions of space might exist, provided they are curled up so tightly that we cannot observe them. This drawing shows how an ordinary two-dimensional plane might contain two additional dimensions. At every point on the plane, there is a very tiny sphere so small that it cannot be seen. To pinpoint a particular location, you need to give not only a position on the plane but also a position on the sphere. Thus, if at every point, space really has additional curled-up dimensions represented by these spheres, then space has more than the three regular dimensions of space we are familiar with. Like a hamster running endlessly on its wheel, moving in a curled-up dimension is not the same as traveling through regular space. (Ian Warpole)

We see only the regular three dimensions of space, and one dimension of time. As with Kaluza's theory, particles traveling straight in 11-dimensional space would appear to take curved paths in ordinary three-dimensional space. From our perspective, these curved paths would be the result of the four known forces acting on the particles.

Theoretical physicists have shown that if there are indeed 11 dimensions, there must also exist very massive particles that have not yet been discovered. The more massive the particle, the more energy is required in particle accelerators to create and observe it, and particle accelerators might not ever be able to reach these very high energies. Some of these speculative particles may be the dark matter that pervades the universe. Even more bizarre, the new theories no longer regard fundamental particles, such as electrons and quarks, as tiny points of mass. Instead, these particles may actually be multidimensional strings or membranes, wrapped so tightly around the extra dimensions of space that they appear to us as points. Just as a guitar can vibrate in different ways to make different sounds, vibrations of these fundamental strings correspond to different particles and the forces between them.

One criticism of string theory is that its predictions are too far beyond the energies that can foreseeably be tested. However, when the extremely large energies of the Big Bang are considered, proponents of string theory hope that the early universe itself will have acted as a powerful cosmic particle accelerator that left behind telling evidence. In this manner, astronomy might be used to gather evidence for strings whose vibrations lead to all the known forces in nature.

CONCEPT CHECK 26-12

If, as suggested by string theory, there really are 11 dimensions, why don't we notice them? How many dimensions appear to be hidden?

Answer appears at the end of the chapter.

We shall not cease from exploration
And the end of all our exploring
Will be to arrive where we started
And know the place for the first time.

T. S. Eliot, *Four Quartets*

KEY WORDS

annihilation, p. 835
antimatter, p. 834
antiparticle, p. 834
antiproton, p. 834
cold dark matter, p. 841
dark ages, p. 840
density fluctuation, p. 838
deuterium bottleneck, p. 837
electroweak force, p. 829
elementary particle physics, p. 833
false vacuum, p. 831
flatness problem, p. 825

gluon, p. 829
grand unified theory (GUT), p. 830
graviton, p. 831
Heisenberg uncertainty principle, p. 833
hot dark matter, p. 841
inflation, p. 825
inflationary epoch, p. 825
intermediate vector boson, p. 829
isotropy problem (horizon problem), p. 824

Jeans length, p. 838
Lamb shift, p. 834
M-theory, p. 845
nuclear force, p. 828
nucleosynthesis, p. 837
pair production, p. 835
positron, p. 834
quantum electrodynamics, p. 829
quantum gravity, p. 845
quantum mechanics, p. 833
quark, p. 829
quark confinement, p. 836

spontaneous symmetry breaking, p. 829
Standard Model, p. 829
string theory, p. 845
strong force, p. 828
supergrand unified theory, p. 830
theory of everything (TOE), p. 830
thermal equilibrium, p. 836
threshold temperature, p. 836
virtual pairs, p. 834
weak force, p. 828

KEY IDEAS

Cosmic Inflation: A brief period of rapid expansion, called inflation, is thought to have occurred immediately after the Big Bang. During a tiny fraction of a second, the universe expanded to a size many times larger than it would have reached through its normal expansion rate (Section 26-1).

• Inflation explains why the universe is nearly flat and the 2.725-K microwave background is almost perfectly isotropic.

Had inflation not taken place, the present-day observable universe would have had to be relatively large just after the Big Bang.

Once the inflationary epoch ended, the universe continued to expand in a more gradual way down to the present day.

In the inflationary model, the present-day observable universe was very tiny just after the Big Bang. This region, as well as the rest of the universe, then underwent a tremendous expansion during the inflationary epoch.

Inflationary epoch

The Four Forces and Their Unification: Four basic forces—gravity, electromagnetism, the strong force, and the weak force—explain all the interactions observed in the universe (Section 26-2).

- The Standard Model accurately describes all the known particles in nature and their observed interactions (except for gravity).
- The weak force and electromagnetic force become unified into a single force called the electroweak force at higher energies than those typically found in today's universe. This unification has been observed in high-energy particle accelerators.
- Grand unified theories (GUTs) are attempts to explain three of the forces (strong force, weak force, and electromagnetic force) in terms of a single force. This has not been observed, and particle accelerators fall far short of having the energy to directly probe the high energy where this unification is predicted to occur.
- A supergrand unified theory would explain all four forces (including gravity) at extremely high energies as a single force acting similarly on all the particles in nature. String theory attempts to make this unification, and it would describe the quantum nature of gravity. Supergrand unification is hypothesized to occur before the Planck time ($t = 10^{-43}$ seconds after the Big Bang).

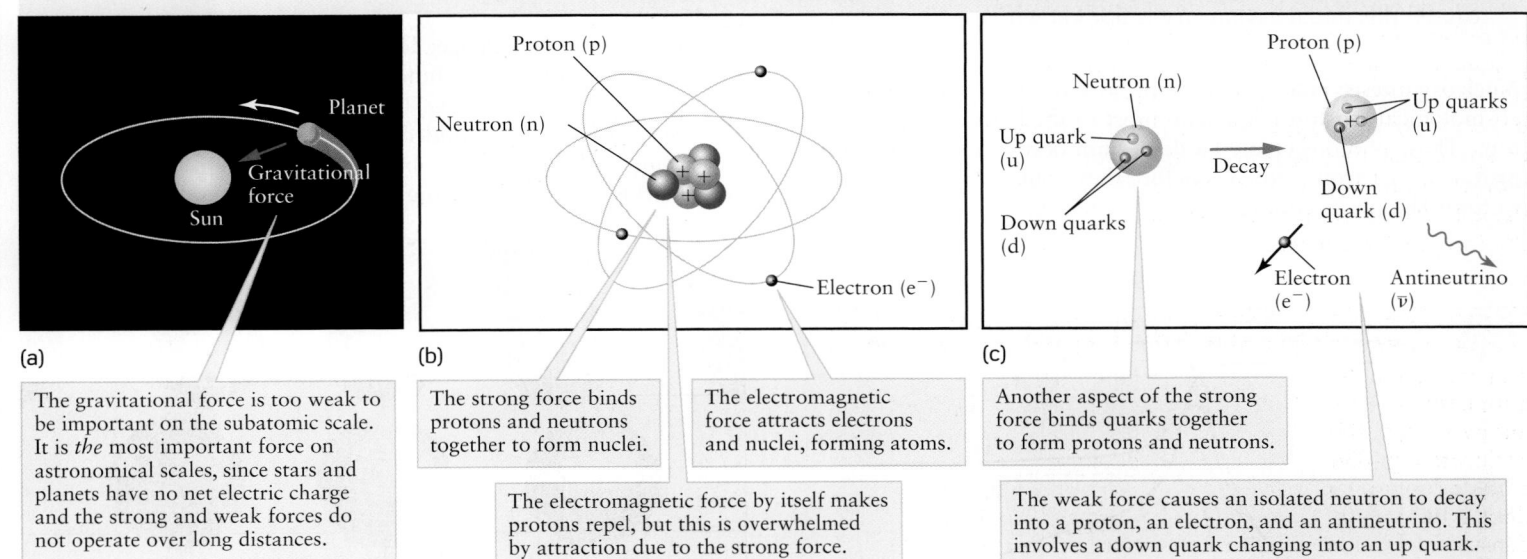

(a) The gravitational force is too weak to be important on the subatomic scale. It is *the* most important force on astronomical scales, since stars and planets have no net electric charge and the strong and weak forces do not operate over long distances.

(b) The strong force binds protons and neutrons together to form nuclei.

The electromagnetic force attracts electrons and nuclei, forming atoms.

The electromagnetic force by itself makes protons repel, but this is overwhelmed by attraction due to the strong force.

(c) Another aspect of the strong force binds quarks together to form protons and neutrons.

The weak force causes an isolated neutron to decay into a proton, an electron, and an antineutrino. This involves a down quark changing into an up quark.

Spontaneous Symmetry Breaking: As the universe expands and cools, the unified forces break into separate forces. Starting around the Planck time, gravity became a distinct force through a spontaneous symmetry breaking. During a second spontaneous symmetry breaking, the strong nuclear force became a distinct force. A final spontaneous symmetry breaking separated the electromagnetic force from the weak nuclear force; from that moment on, the universe behaved as it does today (Section 26-2).

Frank Siteman/Aflo Co., Ltd./Alamy Stock Photo

(a) Pair production   (b) Annihilation

Particles and Antiparticles: Heisenberg's uncertainty principle states that the amount of uncertainty in the mass of a subatomic particle increases as it is observed for shorter and shorter time periods (Section 26-3).

- Because of the uncertainty principle, particle–antiparticle pairs can spontaneously form and disappear within a fraction of a second. These pairs, whose presence can be detected only indirectly, are called virtual pairs.
- The collision of two high-energy photons can produce a real particle–antiparticle pair. In this process, called pair production, the photons disappear, and their energy is transformed into the masses of the particle–antiparticle pair. In the process of annihilation, a colliding particle–antiparticle pair disappears and two high-energy photons appear.

The Origin of Matter: Just after the inflationary epoch, the universe was filled with particles and antiparticles formed from numerous high-energy photons. The particles also annihilated to produce a state of thermal equilibrium between the particles and the photons (Section 26-4).

- As the universe expanded, its temperature decreased. When the temperature fell below the threshold temperature required to produce each kind of particle, annihilation of that kind of particle began to dominate over production.

- Matter is much more prevalent than antimatter in the present-day universe, because particles and antiparticles were not created or maintained in exactly equal numbers. What caused this initial asymmetry is not known.

Nucleosynthesis: Helium could not have been produced until the cosmological redshift eliminated most of the high-energy photons. These photons created a deuterium bottleneck by breaking protons apart from neutrons before they could combine further to form helium (Section 26-5).

Density Fluctuations and the Origin of Stars and Galaxies: The large-scale structure of the universe arose from primordial density fluctuations (Section 26-6).

- The first stars were much more massive and luminous than stars in the present-day universe. The material that they ejected into space seeded the cosmos for all later generations of stars.

- Galaxies are generally located on the surfaces of roughly spherical voids. Models based on dark energy and cold dark matter give good agreement with details of this large-scale structure.

The Frontier of Knowledge: The search for a theory that unifies gravity with the other fundamental forces suggests that the universe might actually have 11 dimensions (10 of space and 1 of time), 7 of which are folded on themselves so that we cannot see them. The fundamental objects of our universe may be very small strings, rather than pointlike particles (Section 26-7).

Ian Warpole

QUESTIONS

Problem-solving tips and tools

We described Wien's law for blackbody radiation in Section 5-4. If light with wavelength λ_0 is emitted by an object at redshift z, the wavelength that we measure is $\lambda = \lambda_0 (1 + z)$. As explained in Section 26-6, the critical density ρ_c is equal to $3H_0^2/8\pi G$; that is, ρ_c is proportional to the square of the Hubble constant H_0. If $H_0 = 73$ km/s/Mpc, the critical density is equal to about 1.0×10^{-26} kg/m³. It is also useful to know that the mass of a proton is 1.67×10^{-27} kg, the mass of an electron is 9.11×10^{-31} kg, the mass of the Sun is 1.99×10^{30} kg, that 1 m³ = 10^6 cm³, that 1 ly = 9.46×10^{15} m, and that 1 GeV = 10^3 MeV = 10^9 eV.

Section 26-1

1. **TUTORIAL 26.1** What is the horizon problem? What is the flatness problem? How can these problems be resolved by the idea of inflation?

2. The inflationary epoch lasted a mere 10^{-32} second. Why, then, is it worthy of so much attention by scientists?

3. In what ways is inflation similar to the present-day expansion of the universe? In what ways is it different?

Section 26-2

4. Explain why the inflationary model does not violate the principle that the speed of light in a vacuum represents the ultimate speed limit.

5. Describe an example of each of the four basic types of interactions in the physical universe. Do you think it possible that a fifth force might be discovered someday? Explain your answer.

6. If gravity is intrinsically so weak compared to the strong force, why do we say that gravity rather than the strong force keeps the planets in orbit around the Sun? (*Hint:* Forces have a strength and a range.)

7. How many times stronger than the weak force is the electromagnetic force? How many times stronger than the electromagnetic force is the strong force? Use this information to suggest one reason why the electromagnetic and weak forces can become unified at a lower energy than do the electroweak and strong forces.

8. If you hold an iron rod next to a strong magnet, the rod will become magnetized; one end will be a north pole and the other a south pole. But if you heat the iron rod to 1043 K (770°C = 1418°F) or higher, it will lose its magnetization and there will be no preferred magnetic direction in the rod. This demagnetization is an example of *restoring* a broken symmetry. Explain why.

9. What is the difference between a true vacuum and a false vacuum? What role do these vacuums play in inflation? (*Hint:* Consult Figure 26-12.)

Section 26-3

10. What is the Heisenberg uncertainty principle? How does it lead to the understanding that all space is filled with virtual particle–antiparticle pairs?

11. What is the difference between an electron and a positron?

12. What happens when an electron and a positron annihilate?

13. Is it possible for photons to create a particle pair of matter and antimatter? If so, what kind of photons?

14. Is it possible for a single hydrogen atom, with a positively charged proton and a negatively charged electron, to be created as a virtual pair? Why or why not?

15. Which can exist for a longer time, a virtual electron–positron pair or a virtual proton–antiproton pair? Explain your reasoning.

16. An electron has a lifetime of 1.0×10^{-8} seconds in a given energy state before it makes a transition to a lower state. What is the uncertainty in the energy of the photon emitted in this process?

17. How long can a proton–antiproton pair exist without violating the principle of the conservation of mass?

18. What is the difference between a pair of virtual particles and a pair of real particles?

19. Explain the origin of the mass we see in today's universe.

Section 26-4

20. Explain why antimatter was present in copious amounts in the early universe but is very rare today.

21. What is meant by the threshold temperature of a particle?

22. Explain the connection between the fact that humans exist and the imbalance between matter and antimatter in the early universe.

23. Explain the connection between particles and antiparticles in the early universe and the cosmic microwave background that we observe today.

24. The mass of the intermediate vector boson W^+ (and of its antiparticle, the W^-) is 85.6 times the mass of the proton. The weak nuclear force involves the exchange of the W^+ and the W^-. (**a**) Find the rest energy of the W. Give your answer in GeV. (**b**) Find the threshold temperature for the W^+ and W^-. (**c**) From Figure 26-11, how long after the Big Bang did W^+ and W^- particles begin to disappear from the universe? Explain your answer.

Section 26-5

25. What is the deuterium bottleneck? Why was it important during the formation of nuclei in the early universe?

26. How does nucleosynthesis provide strong evidence for a hot Big Bang?

27. Why were only the four lightest chemical elements produced in the early universe?

28. The first generation of stars in the universe are thought to have appeared hundreds of millions of years after the Big Bang. Once these stars formed, thermonuclear fusion reactions began in their interiors. Explain why these were the first fusion reactions to occur since the universe was 15 minutes old.

29. Why is it reasonable to suppose that all space is filled with a neutrino background analogous to the cosmic microwave background?

Section 26-6

30. Explain the connection between quantum fluctuations and the seeds of structure formation.

31. What is the Jeans length? Why is it significant for the formation of structure in the universe?

32. Why do astronomers suspect that structures the size of globular clusters were among the first objects to form in the history of the universe? Why not something larger and more massive?

33. What are Population III stars? How do they differ from stars found in the present-day universe? Why are they so difficult to detect directly?

34. Describe the large-scale structure of the universe as revealed by the distribution of clusters and superclusters of galaxies.

35. What is the difference between hot and cold dark matter? How do astronomers decide which was more important in the formation of large-scale structures such as clusters of galaxies?

36. How did the presence of dark energy help to "turn off" the process of structure formation in the universe?

37. Using the physical conditions present in the universe during the era of recombination ($T = 3000$ K and $\rho_m = 10^{-18}$ kg/m^3), show by calculation that the Jeans length for the universe at that time was about 100 ly and that the total mass contained in a sphere with this diameter was about 4×10^5 M$_\odot$.

38. A typical dark nebula (see Figure 18-4) has a temperature of 30 K and a density of about 10^{-12} kg/m^3. (a) Calculate the Jeans length for such a dark nebula, assuming that the nebula is composed mostly of hydrogen. Express your answer in meters and in light-years. (b) A typical dark nebula is several light-years across. Is it likely that density fluctuations within such nebulae will grow with time? (c) Explain how your answer to (b) relates to the idea that protostars form within dark nebulae (see Section 18-3).

39. At the time labeled $z = 4.97$ in Figure 26-19, how large was the length of each side of the box used in the simulation compared to its size in the present day ($z = 0$)? How much greater was the density at $z = 4.97$ than the present-day density?

Section 26-7

40. What is meant by the term quantum gravity? What situation requires this kind of theory? Has a theory of quantum gravity been experimentally verified?

41. Could there possibly be more than the three spatial dimensions we currently recognize? Explain how the existence of a fourth spatial dimension might not be so obvious.

42. Is string theory an attempt to formulate a theory of quantum gravity? How many dimensions of space and of time might exist if string theory is correct?

43. If there are more than three spatial dimensions, do objects necessarily travel in straight lines when viewed from our three-dimensional perspective? How do extra dimensions relate to forces?

44. Can particle accelerators generate the necessary energies required to test string theory? Is there any other system that might provide a test of string theory?

General Problems

45. Describe the observational evidence, if any, for (a) the Big Bang, (b) the inflationary epoch, (c) the confinement of quarks, and (d) the era of recombination.

46. (a) If the Hubble constant is 73 km/s/Mpc, the critical density ρ_c is 1.0×10^{-26} kg/m^3. The average density of dark matter is known to be about 0.20 times the critical density. Suppose that massive neutrinos constitute this dark matter, and the average density of neutrinos throughout space is 100 neutrinos per cubic centimeter. (In fact, the density of neutrinos is far less than this.) Under these assumptions, what must be the mass of the neutrino? Give your answers in kilograms and as a fraction of the mass of the electron. (b) Why do astronomers think that massive neutrinos are *not* the dominant type of dark matter in the universe?

47. Some GUTs predict that the proton is unstable, although with a half-life far longer than the present age of the universe. What would it be like to live at a time when protons were decaying in large numbers?

WEB/E-BOOK QUESTIONS

48. Search the internet for information about the top quark. What kind of particle is it? How does it compare with the up and down quarks found in protons and neutrons? Why did physicists work so hard to try to find it?

49. Search the internet for information about primordial deuterium (that is, deuterium that was formed in the very early universe). Why are astronomers interested in knowing how abundant primordial deuterium is in the universe? What techniques do they use to detect it?

50. Search the internet for information about the South Pole Telescope. What is the purpose of this telescope? Why is it sited at the South Pole? How will it help us understand the early universe?

ACTIVITIES

Starry Night™ Explorations

51. Use *Starry Night™* to observe globular star clusters in the Milky Way and see how they are distributed in the galaxy. Select **Favorites > Explorations > Globular Clusters** to see a view looking from Earth along the plane of the galaxy (indicated by the blue line) toward the center of our galaxy, indicated by the label. Each of the numerous green dots scattered in the view represents a globular star cluster in the Milky Way. A sample of the brighter globular star clusters observable from Earth with binoculars or small telescopes is indicated by the labels. Use the hand tool to explore the distribution of globular clusters around the galactic center. The distribution of globular clusters in most galaxies is similar to that in our galaxy. The age of these clusters is determined by examining the Hertzsprung–Russell (H-R) diagrams of their stars and by determining the relative amounts of heavy elements compared to those of the lighter elements, hydrogen and helium. Heavy elements build up over time in the interstellar medium, so the lower the heavy-element composition of a star, the earlier the star must have formed from the interstellar medium and therefore the older the star. Such investigations have shown that globular clusters contain the oldest stars that we see in our sky.

    (a) What is the distribution of globular clusters around our galaxy? Are they (i) concentrated along the entire plane of the galaxy (not just near the center), and much less numerous above and below this plane; (ii) concentrated only near the center of the galaxy, and much less numerous everywhere away from the center; or, (iii) uniformly distributed everywhere in and around the galaxy?

(b) Click on the names of some of the labeled clusters in the view and select **Magnify** from the menu that appears, to see a telescopic view of the cluster. Use the HUD or click the info icon in the cluster's label to obtain more information about it, such as its distance from Earth and angular size in Earth's sky. How would you describe the characteristic appearance of a typical globular cluster? Do you see evidence of a preponderance of redder stars in these clusters?

52. Use *Starry Night*™ to compare the distances of the galaxies in the *Starry Night 3D Galaxy Database* with the radius of the limit of our observable universe, the cosmic particle horizon. As you will find, the most distant galaxies in this database are a long way away from Earth, yet these distances are only a small fraction of the distances from which we can see light in our universe. Select **Favorites > Explorations > Database** to see a view centered on the Milky Way from a location 500 million ly from the Sun. Each dot represents the 3D position in space of the more than 2 million galaxies contained within the database. Use the location scroller to explore the distribution of galaxies surrounding the Milky Way, noting the wall-like structures surrounding large voids that contain few galaxies. Click on **File > Revert** to return to the original view and switch to the angular separation cursor.

(a) Use the angular separation cursor to measure the distances from the Milky Way to a selection of galaxies at the perimeter of this database. How do the distances to the most remote galaxies in the database compare with the cosmic particle horizon?

(b) What fraction of the radius of the observable universe is covered by this database of galactic positions?

Collaborative Exercises

53. The four fundamental forces of nature are the strong force, the weak force, the gravitational force, and the electromagnetic force. List four things at your school that rely on one of these fundamental forces, and explain how each thing is dependent on one of the fundamental forces.

54. Consider the following hypothetical scenario adapted from a daytime cable television talk show. Chris states that Pat borrowed Chris's telescope without permission. Tyler purchased balloons and a new telescope eyepiece without telling Chris. Sean borrowed star maps from the library, with the library's permission, but without telling Pat. Eventually, when the four met on Sunday evening, Chris was crying and speechless. Can you create a "grand unified theory" that explains this entire situation?

55. The *Cosmic Connections* figure shows the history of the universe in the form of a graph of the temperature versus the time after the Big Bang. Create a similar history of your class, starting with estimated outside temperature on the vertical axis and number of days since the beginning of the academic term on the horizontal axis. Include dates for major exams and assignments up through today. In different color ink,

show your predictions for temperatures, days, and events from today until the end of the course.

ANSWERS

ConceptChecks

ConceptCheck 26-1: Early expansion is different in the inflation model compared to the gradual Hubble expansion. As shown in Figure 26-3, the inflationary model predicts an even smaller universe at early times than that described by gradual expansion. In this more compact condition, much more material would be in contact and reach the same temperature. Then, during inflation, this matter would get spread out so much that today this matter is at opposite ends of our observable universe (such as points *A* and *B*).

ConceptCheck 26-2: Yes. Inflation expands the universe by about a factor of 10^{50}. With the curved space of the universe represented on the surface of a sphere, after the radius of the sphere increases during inflation by a factor of 10^{50}, the surface (and universe) would appear flat. Due to tremendous inflation, just about any initial curvature ends up nearly flat after inflation.

ConceptCheck 26-3: Yes. The up and down quarks are held together by the strong force. Since the strong force is transmitted by gluons, these gluons are also involved in holding a neutron together.

ConceptCheck 26-4: No. Photons are massless, whereas W and Z particles have mass. The particles of the Standard Model appear *after* spontaneous symmetry breaking has changed the electroweak force into two distinct forces (electromagnetic and weak), which are transmitted by distinct particles (photons and intermediate vector bosons).

ConceptCheck 26-5: As the universe expands, it cools down, and the average energy of colliding particles decreases. Therefore, as the universe expands, there is a spontaneous symmetry breaking when particles naturally begin to collide with energies below 100 GeV.

ConceptCheck 26-6: Shorter. The greater the mass of the virtual particles, the shorter they last. A virtual pair of protons lasts about 1/2000th as long as a virtual pair of electrons.

ConceptCheck 26-7: No. Electrically charged particle pairs come with one positively charged particle and one negative particle, so the pair's total charge is zero.

ConceptCheck 26-8: First, because the amount of antimatter would be equal to the amount of matter in our universe, this hypothesis would imply a lot of "hidden" antimatter. Second, near the boundary or boundaries of the hidden cache of antimatter, matter and antimatter would annihilate to produce gamma rays. And third, the gamma rays produced by these annihilations would be very easy to recognize because they all occur at the same energy determined by $E = mc^2$. Taken together, these effects would make it very hard to hide so much antimatter.

ConceptCheck 26-9: The temperature of the universe steadily decreases with time as the universe expands. When the universe was less than 3 minutes old, hot radiation filling the universe was energetic enough to break apart protons and neutrons before they could begin to build atomic nuclei. However, the temperature also determines the average speed at which atomic nuclei collide into each other. After 15 minutes, the temperature was not high enough for colliding nuclei, with their charged protons, to overcome their electric repulsion and build larger nuclei.

ConceptCheck 26-10: A clump of matter will not form. When the universe reaches $T = 3000$ K, the Jeans length is $L_J = 100$ ly. Fluctuations shorter in length will not contract into clumps because there is not enough gravitational attraction to overwhelm the increased pressure that clumping would produce.

ConceptCheck 26-11: Cold dark matter particles (which are heavier and move at lower speeds) lead to "bottom-up" formation, which is what we observe. Hot dark matter, such as neutrinos, leads to "top-down" formation, which is not supported by observations.

ConceptCheck 26-12: We only notice four dimensions—three for space and one for time. That would leave seven hidden dimensions of space. If these dimensions are curled up tightly, as in Figure 26-24, we would not notice them until experiments could probe the tiny microscopic sizes scales in which they are curled.

BRAVE NEW COSMOS

Confused by all those theories? Good.

Making Sense of Modern Cosmology

BY P. JAMES E. PEEBLES
(From P. James E. Peebles,
"Making Sense of Modern Cosmology,"
Scientific American, October 2002, 2–3)

This is an exciting time for cosmologists: Findings are pouring in, ideas are bubbling up, and research to test those ideas is simmering away. But it is also a confusing time. All the ideas under discussion cannot possibly be right; they are not even consistent with one another. How is one to judge the progress? Here is how I go about it.

For all the talk of overturned theories, cosmologists have firmly established the foundations of our field. Over the past 70 years, we have gathered abundant evidence that our universe is expanding and cooling. First, the light from distant galaxies is shifted toward the red, as it should be if space is expanding and galaxies are pulled away from one another. Second, a sea of thermal radiation fills space, as it should if space used to be denser and hotter. Third, the universe contains large amounts of deuterium and helium, as it should if temperatures were once much higher. Fourth, distant galaxies, seen as they were in the past because of light's travel time, look distinctly younger, as they should if they are closer to the time when no galaxies existed. Finally, the curvature of spacetime seems to be related to the material content of the universe, as it should be if the universe is expanding according to the predictions of Einstein's gravity theory, the general theory of relativity.

That the universe is expanding and cooling is the essence of the Big Bang theory. You will notice I have said nothing about an "explosion"—the Big Bang theory describes how our universe is evolving, not how it began.

I compare the process of establishing such compelling results, in cosmology or any other science, to the assembly of a framework. We seek to reinforce each piece of evidence by adding cross bracing from diverse measurements. Our framework for the expansion of the universe is braced tightly enough to be solid. The Big Bang theory is no longer seriously questioned; it fits together too well. Even the most radical alternative—the latest incarnation of the steady state theory—does not dispute that the universe is expanding and cooling. You still hear differences of opinion in cosmology, to be sure, but they concern additions to the solid part.

For example, we do not know what the universe was doing before it was expanding. A leading theory, inflation, is an attractive addition to the framework, but it lacks cross bracing. That is precisely what cosmologists are now seeking. If measurements in progress agree with the unique signatures of inflation, then we will count them as a persuasive argument for this theory. But until that time, I would not settle any bets on whether inflation really happened. I am not criticizing the theory; I simply mean that this is brave, pioneering work still to be tested.

More solid is the evidence that most of the mass of the universe consists of dark matter clumped around the outer parts of galaxies. We also have a reasonable case for Einstein's infamous cosmological constant or something similar; it would be the agent of the acceleration that the universe now seems to be undergoing. A decade ago, cosmologists generally welcomed dark matter as an elegant way to account for the motions of stars and gas within galaxies. Most researchers, however, had a real distaste for the cosmological constant. Now the majority accept it, or its allied concept, quintessence. Particle physicists have come to welcome the challenge that the cosmological constant poses for quantum theory. This shift in opinion is not a reflection of some inherent weakness; rather, it shows the subject in a healthy state of chaos around a slowly growing fixed framework. We students of nature adjust our concepts as the lessons continue.

The lessons, in this case, include the signs that cosmic expansion is accelerating: the brightness of supernovae near and far; the ages of the oldest stars; the bending of light around distant masses; and the fluctuations of the temperature of the thermal radiation across the sky. The evidence is impressive, but I am still skeptical about details of the case for the cosmological constant, including possible contradictions with the evolution of galaxies and their spatial distribution. The theory of the accelerating universe is a work in progress. I admire the architecture, but I would not want to move in just yet.

How might one judge reports in the media on the progress of cosmology? I feel uneasy about articles based on an interview with just one person. Research is a complex and messy business. Even the most experienced scientist finds it hard to keep everything in perspective. How do I know that this individual has managed it well? An entire community of scientists can head off in the wrong direction, too, but it happens less often. That is why I feel better when I can see that the journalist has consulted a cross section of the community and has found agreement that a certain result is worth considering. The result becomes more interesting when others reproduce it. It starts to become convincing when independent lines of evidence point to the same conclusion. To my mind, the best media reports on science describe not only the latest discoveries and ideas but also the essential, if sometimes tedious, process of testing and installing the cross bracing.

(continued on the next page)

Making Sense of Modern Cosmology *(continued)*

Report Card for Major Theories

Concept	Grade	Comments
The universe evolved from a hotter, denser state.	A+	Compelling evidence drawn from many corners of astronomy and physics.
The universe expands as the general theory of relativity predicts.	A−	Passes the tests so far, but few of the tests have been tight.
Dark matter made of exotic particles dominates galaxies.	B+	Many lines of indirect evidence, but the particles have yet to be found and alternative theories have yet to be ruled out.
Most of the mass of the universe is smoothly distributed; it acts like Einstein's cosmological constant, causing the expansion to accelerate.	B−	Encouraging fit from recent measurements, but more must be done to improve the evidence and resolve the theoretical conundrums.
The universe grew out of inflation.	Inc	Elegant, but lacks direct evidence and requires huge extrapolation of the laws of physics.

Over time, inflation, quintessence and other concepts now under debate either will be solidly integrated into the central framework or will be abandoned and replaced by something better. In a sense, we are working ourselves out of a job. But the universe is a complicated place, to put it mildly, and it is silly to think we will run out of productive lines of research anytime soon. Confusion is a sign that we are doing something right: It is the fertile commotion of a construction site.

P. JAMES E. PEEBLES is one of the world's most distinguished cosmologists, a key player in the early analysis of the cosmic microwave background radiation and the bulk composition of the universe. He has received some of the highest awards in astronomy, including the 1982 Heineman Prize, the 1993 Henry Norris Russell Lectureship of the American Astronomical Society, and the 1995 Bruce Medal of the Astronomical Society of the Pacific. He is emeritus professor at Princeton University.

FURTHER INFORMATION

"The Evolution of the Universe." P. J. E. Peebles, D. N. Schramm, E. L. Turner, and R. G. Kron. *Scientific American* 271 (October 1994): 52–57.

A. H. Guth. *The Inflationary Universe: The Quest for a New Theory of Cosmic Origins.* Perseus Press, 1997.

M. Rees. *Before the Beginning: Our Universe and Others.* Perseus Press, 1998.

M. Livio and A. Sandage. *The Accelerating Universe: Infinite Expansion, the Cosmological Constant, and the Beauty of the Cosmos.* John Wiley & Sons, 2000.

LEARNING GOALS

By reading the sections of this chapter, you will learn

27-1 How comets and meteorites could have aided the origin of life on Earth

27-2 On which other worlds of our solar system life might have evolved

27-3 About the controversy over proposed fossilized life from Mars

27-4 How astronomers estimate the number of other civilizations in the galaxy

27-5 Why radio telescopes may be a useful tool for contacting alien civilizations

27-6 How astronomers hope to detect Earthlike planets orbiting other stars

One of the most compelling questions in science is also one of the simplest: Are we alone? That is, does life exist beyond Earth? As yet, we have no definitive answer to this question. None of our spacecraft has found life elsewhere in the solar system, and radio telescopes have yet to detect signals of intelligent origin coming from space. Claims of aliens visiting our planet and abducting humans make compelling science fiction, but none of these stories has ever been verified.

Yet there are reasons to suspect that life might indeed exist beyond Earth. One is that biologists find living organisms in some of the most "unearthly" environments on our planet. An example (shown on this page) is at the bottom of the Gulf of Mexico, where the crushing pressure and low temperature cause methane—normally a gas—to form solid, yellowish mounds. Amazingly, these mounds teem with colonies of pink, eyeless, alien-looking worms the size of your thumb. These worms feed off of bacteria that feed off the methane, and methane is found elsewhere in the solar system (recall Titan's methane lakes in Chapter 13). If life can

An "alien" life-form on Earth: pink, eyeless worms in an underwater mound of solid methane clathrate, a compound in which methane molecules are trapped in the crystal structure of water ice. (Image courtesy of the NOAA Office of Ocean Exploration and Research, Gulf of Mexico 2012)

R I **V** U X G

flourish in extreme environments on Earth, might it not also exist in the seemingly hostile conditions found on other worlds?

There are even bigger reasons to suspect that life might exist beyond our Earth. We now have estimates that about 1 in 5 stars in our galaxy has an Earth-sized planet orbiting within a habitable zone that could allow for liquid water (Chapter 8). With about 200 billion stars in our galaxy, that's about 40 billion places for life to start with conditions that could be similar to Earth! The odds only improve when you consider that there are about 100 billion galaxies in our observable universe.

In this chapter, we will look for places in our solar system where life may once have originated, and where it may exist today. We will see how scientists estimate the chances of finding life beyond our solar system, and how they search for signals from other intelligent species. And we will learn how a new generation of telescopes may make it possible to detect the presence of even single-celled organisms on worlds many light-years away.

27-1 The chemical building blocks of life are found in space

Suppose you were the first visitor to a new and alien planet. If you saw a three-headed lizard running by, you would be sure it was an alien life-form. But how can you distinguish an alien microbe—or even just a fossil-like remnant of a microbe—from a dust grain? What might alien life look like? Questions such as these are central to **astrobiology**, the study of life in the universe. Most astrobiologists suspect that if we find living organisms on other worlds, they will be "life as we know it"—that is, their biochemistry will be based on the unique properties of the carbon atom, as is the case for all life on Earth.

$$Z-X-X-X-X-Y$$

(a)   Linear molecule

(b)   Glucose

FIGURE 27-1

Complex Molecules and Carbon **(a)** Atoms that can bond to only two other atoms, like the atoms denoted X shown here, can form a chain of atoms called a linear molecule. The chain stops where we introduce an atom, such as those labeled Y and Z, that can bond to only one other atom. **(b)** A carbon atom (denoted C) can bond with up to four other atoms. Hence, carbon atoms can form more complex, nonlinear molecules like glucose. Most of the molecules that are found in living organisms are made predominantly of carbon atoms.

Organic Molecules in the Universe

Why carbon? The reason is that carbon has the most versatile chemistry of any element. Carbon atoms can form chemical bonds to create especially long and complex molecules (**Figure 27-1**). These carbon-based compounds, called **organic molecules**, are a central part of life's molecular machinery. Silicon has some chemical similarities to carbon, and it can also form complex molecules. For this reason, astrobiologists keep an open mind about silicon-based life forms. But as **Figure 27-2** shows, complex silicon molecules do not appear to have the right properties to make living organisms.

(a)

(b)

FIGURE 27-2 R I V U X G

Why Silicon Appears Less Suitable than Carbon for Making Living Organisms Like carbon, silicon atoms can bond with up to four other atoms. However, silicon typically reacts too much or too little to be suitable for use in living organisms. Perhaps this is why, despite there being about 1000 times as much silicon in Earth's crust than carbon, we find no silicon-based life forms. **(a)** Silicone has a backbone of silicon and oxygen atoms. The silicon-oxygen bonds within them, however, react too slowly to undergo the rapid chemical changes required of molecules in organisms. **(b)** Other molecules can also be made with silicon, but the results (like this quartz crystal) are too rigid for use in organisms. (a: © 2004 Richard Megna—Fundamental Photographs; b: Ispace/Shutterstock)

Organic carbon-based molecules can be linked together to form elaborate structures, such as chains, lattices, and fibers. Some of these structures are capable of complex, self-regulating chemical reactions. Furthermore, the primary constituents of organic molecules—carbon, hydrogen, nitrogen, oxygen, sulfur, and phosphorus—are among the most abundant elements in the universe. The versatility and abundance of carbon suggest that extraterrestrial life, like life on Earth, is likely to be based on organic chemistry.

If life is based on organic molecules, then these molecules must initially be present on a planet in order for life to arise from non-living matter. We now understand that many simple carbon-based molecules originate from nonbiological processes in interstellar space. One such molecule is carbon monoxide (CO), which is made when a carbon atom and an oxygen atom collide and bond together. Carbon monoxide is found in abundance within giant interstellar clouds that lie along the spiral arms of our Milky Way Galaxy (see Figure 1-7) as well as in other galaxies (see Figure 1-9). Carbon atoms have also combined with other elements to produce an impressive array of small organic molecules in the space between the stars, including ethyl alcohol (CH_3CH_2OH), formaldehyde (H_2CO), methylcyanoacetylene (CH_3C_3N), and acetaldehyde (CH_3CHO). Radio astronomers have detected these molecules by looking for the telltale microwave emission lines of these carbon-based chemicals in interstellar clouds.

The planets of our solar system formed out of interstellar material (see Section 8-5), and some of the organic molecules in that material must have ended up on the planets' surfaces. Evidence for this comes from meteorites called **carbonaceous chondrites**, like the one shown in Figure 27-3. These meteorites are ancient, date from the formation of the solar system, and are often found to contain a variety of carbon-based molecules. The Murchison meteorite, which fell on Australia in 1969, contains more than 70 amino acids, and these organic molecules are some of the building blocks of life.

The spectra of comets (see Sections 7-3 and 7-5)—which are also among the most primitive (least changed since their origins) objects in the solar system—show that they, too, contain an assortment of organic compounds. In 2006, NASA's Stardust mission returned samples from the atmosphere (or coma) of Comet Wild 2. The samples were very limited, but one type of amino acid was found. In 2014, the Philae probe from the European Space Agency landed on a comet's surface for direct samples of its nucleus. However, the probe bounced a few times off the surface before coming to rest in a mostly unusable state. It will be up to future missions to directly sample the organic compounds of a comet's nucleus.

Comets and meteoroids were much more numerous in the early solar system than they are today, and they were correspondingly more likely to collide with a planet. These collisions would have seeded the planets with organic compounds from the very beginning of our solar system's history. Organic compounds are also found in interplanetary dust particles (see Figure 8-9 and Section 15-5), which continually rain down on the planets. Once meteorites, comets, and interplanetary dust particles bring simple organic chemicals to a planet's surface, additional chemical reactions can produce an even wider range of the complex organic compounds needed for life. Similar processes are thought to take place in other planetary systems, which are thought to form in basically the same way as did our own (see Figure 8-14 and the image that opens Chapter 8).

CONCEPT CHECK 27-1

Why would life-forms throughout the cosmos likely be based on carbon?

CONCEPT CHECK 27-2

How can organic molecules end up on the surfaces of planets?

Answers appear at the end of the chapter.

The Miller–Urey Experiment

Comets and meteorites would not have been the only sources of organic material on the young planets of our solar system. In 1952, the American chemists Stanley Miller and Harold Urey demonstrated

FIGURE 27-3 R I V U X G

A Carbonaceous Chondrite Carbonaceous chondrites are primitive meteorites that date back to the very beginning of the solar system. This sample is a piece of the Allende meteorite, a large carbonaceous chondrite that fell in Mexico in 1969. Chemical analyses of newly fallen specimens disclose that they are rich in organic molecules, many of which are the chemical building blocks of life. (Detlev van Ravenswaay/Science Source) that under conditions that are thought to have prevailed on the young Earth, simple chemicals can combine to form the chemical building blocks of life. In a closed container, they prepared a sample of "atmosphere": a mixture of hydrogen (H_2), ammonia (NH_3), methane (CH_4), and water vapor (H_2O), the most common molecules in the solar system. Miller and Urey then exposed this mixture of gases to an electric arc (to simulate lightning) for a week. At the end of this period, the inside of the container had become coated with a reddish-brown "tar" rich in amino acids and other compounds that form the basis of life's chemistry.

Since Miller and Urey's original experiment, most scientists have come to the conclusion that Earth's primordial atmosphere was composed of carbon dioxide (CO_2), nitrogen (N_2), and water vapor outgassed from volcanoes, along with some hydrogen. Modern versions of the Miller–Urey experiment (Figure 27-4) using these gases, however, have also succeeded in synthesizing a wide variety of organic compounds. The combination of comets and meteorites falling from space and chemical synthesis in the atmosphere could have made the chemical building blocks of life available in substantial quantities on the young Earth.

> Organic molecules do not have to come from living organisms—they can also be synthesized in nature from simple chemicals.

CAUTION!

It is important to emphasize that scientists have *not* (yet) created life in a test tube. While organic molecules may have been available on the ancient Earth, biologists have yet to figure out how these molecules first gathered themselves into cells and developed systems for self-replication. Nevertheless, because so many chemical components of life are so easily synthesized under conditions that simulate the primordial Earth, it is reasonable to suppose that life could have originated as the result of such chemical processes. Furthermore, because the molecules that combine to form these compounds are rather common, it seems equally reasonable that life could have originated in the same way on other planets.

Organic building blocks are commonplace throughout the universe, but their abundance does not guarantee that life is equally commonplace. If a planet's environment is hostile, life may never get started or may quickly be extinguished. But we now have evidence that numerous Earth-sized planets orbit other stars (see Section 8-7). It seems increasingly likely that Earthlike planets will be found orbiting other stars, and that conditions on some of these worlds may be suitable for life as we know it. The moons of some planets, in our solar system or associated with other stars, might also have environments suitable for life.

CONCEPT CHECK 27-3

What did Miller and Urey create when they passed electricity through their sample of "atmosphere" containing a mixture of hydrogen (H_2), ammonia (NH_3), methane (CH_4), and water vapor (H_2O)?

Answer appears at the end of the chapter.

FIGURE 27-4

An Updated Miller–Urey Experiment Modern versions of this classic experiment prove that numerous organic compounds important to life can be synthesized from gases that were present in Earth's primordial atmosphere. This experiment supports the hypothesis that life on Earth arose as a result of ordinary chemical reactions.

27-2 Water and the potential for life

If life evolved on Earth from nonliving organic molecules, might the same process have taken place elsewhere in our solar system? Scientists are carefully scrutinizing the planets and their moons in an attempt to answer this question.

The Importance of Liquid Water

One major constraint is that liquid water is essential for the survival of life as we know it. However, the water need not be pleasant by human standards—terrestrial organisms have been found in water that is boiling hot, fiercely acidic, or ice cold (see the chapter-opening image)—but it must be liquid. Organisms living in such extreme conditions are called **extremophiles**, such as the thermophiles that thrive at high temperatures (Figure 27-5). Other extremophiles include microorganisms that are resistant to high levels of nuclear radiation, and those that live *within* the tiny

Living organisms called hyperthermophiles

(a) Grand prismatic spring, Yellowstone National Park

(b) Tube worms at ocean vents

FIGURE 27-5 R I **V** U X G

Extremophiles Can Take the Heat **(a)** These microscopic thermophiles (heat-loving organisms) live in water that is between 80°C and 100°C (85°F–140°F). **(b)** Tube worms (light-green tubes) with hemoglobin-rich red plumes reside around black smokers—vents in the ocean bottom that are in the same temperature range as the hot springs shown in (a). The tube worms don't feed directly on the rising black mineral-rich water, but on the bacteria that do. These vents are more than 3 km (2 mi) under water. (a: Jim Peaco; July 2001 Yellowstone National Park Image by NPS Photo; inset: NASA; b: Verena Tunnicliffe, University of Victoria, School of Earth & Ocean Sciences, Department of Biology)

porous spaces inside of rocks. All of these life-forms, however, still require at least some liquid water; even the extremophile microbes living within ice contain a form of antifreeze to keep their internal water from solidifying.

In order for water on a planet's surface to remain liquid, the temperature cannot be too hot or too cold. Furthermore, there must be a relatively thick atmosphere to provide enough pressure to keep liquid water from evaporating under the ambient temperature. Of all the worlds in our present-day solar system, only Earth has the right conditions for water to remain liquid on its surface for long periods of time.

The Oceans of Europa and Enceladus

Of course, water need not necessarily be on the surface; there is now compelling evidence that Europa, one of the large satellites of Jupiter (see Table 7-2 and Figure 7-4), has an ocean of water *beneath* its icy crust. As it orbits Jupiter, Europa is caught in a tug-of-war between gravitational forces from Jupiter and Jupiter's other large satellites. These forces flex the interior of Europa, which in turn generates enough heat to keep subsurface water from freezing. Images indicate that chunks of ice have floated around on the surface, and magnetic field measurements reveal that Europa has a subsurface ocean (see Section 13-6).

No one knows whether life exists in Europa's ocean. But interest in this exotic little world is great, and scientists have proposed several missions to explore the moon in more detail. Unfortunately, the ice layer is probably too thick (many kilometers) for a robotic probe to penetrate the ice and sample the underlying ocean anytime soon. However, just as microorganisms on Earth can survive within ice, perhaps life on Europa has spread beyond its ocean toward the surface?

A mission to search Europa's surface ice is still very challenging, as high levels of radiation from Jupiter impinge on the surface of the atmosphere-free moon and would likely destroy any organisms within the first few meters of ice. A thermal probe designed to simply melt its way down a few meters also faces challenges because the fine rocky dust (regolith) that is expected to exist near the surface likely does not melt. To search for life about 10 m (33 ft) down, scientists are testing a number of ideas, including a thermal probe with a drill tip to get through any patches of dust.

Saturn's moon Enceladus has also revealed its subsurface ocean through salty water spewing out in plumes from ice volcanoes (Figure 27-6 and Figure 13-30). The salts, which are composed of a variety of dissolved minerals, form when the water makes extensive contact with the rocky interior. This evidence points strongly toward a large interior ocean. Enceladus's ocean of liquid water with dissolved minerals, along with some organic material detected in its plumes, suggests that its ocean could make a suitable environment for life.

While Enceladus's ocean could be kilometers beneath the surface and difficult to explore, its plume vents offer a location where microorganisms seeping up from the ocean might be ejected from the surface. Without the challenges of drilling or melting through ice, as on Europa, a probe can pass right through the plumes of Enceladus's ice volcanoes and test samples for life.

Even if there is no life in the Enceladean ocean, the plume vents themselves might harbor life. These ice volcanoes contain liquid

FIGURE 27-6 R I (V) U X G

Subsurface Ocean on Enceladus Plumes containing water vapor and dissolved minerals indicate a subsurface ocean on Saturn's moon Enceladus. Some very simple organic compounds, in concentrations 20 times higher than expected, were also detected in the plumes. As on Earth, liquid water with dissolved minerals and organic compounds could be a suitable setting for life. (NASA/JPL/SSI)

water, organic compounds, and are about 200 K warmer than the surrounding terrain. Within these vents, the presence of liquid water and energy could make a hospitable environment for life.

CONCEPT CHECK 27-4

Why do dissolved salts (or minerals) ejected from Enceladus suggest a large ocean beneath its surface?

Answer appears at the end of the chapter.

Life in Similar Environments on Earth

Here on Earth, there are environments somewhat similar to those expected in the subsurface oceans of Europa and Enceladus. By investigating these environments, we learn if life might be possible in such extremes. One such place is Lake Vida in Antarctica. Its liquid water is covered by 20 m (65 ft) of frozen ice. At that depth, virtually no sunlight reaches the water. The water is also 6 times saltier than seawater, which is too salty for most organisms. However, the high concentration of salt also prevents the water from freezing, where it drops to temperatures of −13.5°C (7.7°F). Incredibly, in 2012, researchers found that bacteria are thriving in these subfreezing waters.

The most extreme liquid water environment known on Earth is Lake Vostok in Antarctica (**Figure 27-7**). This enormous subsurface lake is completely buried by a whopping 2.2 mi (3.3 km) of glacial ice. With this lake so cut off from its surroundings, its water has probably been isolated for more than 15 million years! In 2012, Russian scientists very carefully drilled more than 2 mi into Lake Vostok, piercing its icy depths to sample the waters beneath. When water was finally reached, it rushed up and gushed out of the borehole. In 2013, pristine samples of the lake water were collected and more than 3500 different microbial species were found. This discovery certainly adds to the plausibility of life in the subsurface oceans of Europa, Enceladus, and other moons. As we discuss in Section 27-3, the possible *transfer* of microbes from Earth to these moons makes life on them even more possible.

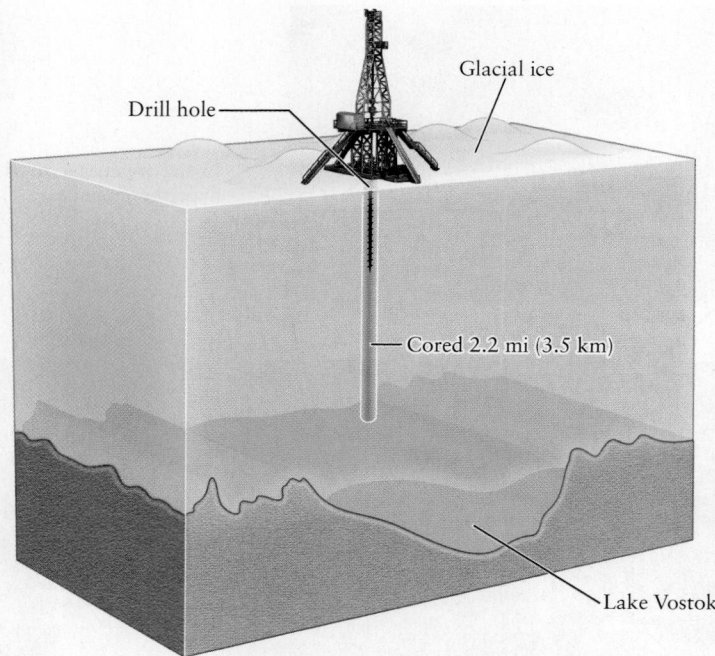

FIGURE 27-7

Exploring Lake Vostok Russian scientists search for life in an environment that might be similar to the oceans of Europa or Enceladus. The coldest temperature recorded on Earth was above Lake Vostok, at −89°C (−128°F). This is similar to typical Martian surface temperatures. However, the lake's liquid water below is much warmer at only about −3°C (26.6°F). This is encouraging because thick ice can act as insulation to keep in a planet or moon's internal heat.

Searching for Life on Mars

Another possibility for the existence of life in our solar system is Mars. The present-day Martian atmosphere is so thin that water can exist only as ice or as a vapor. However, images captured from Martian orbit show dried-up rivers, and data from the Curiosity rover found that one riverbed very likely held water for thousands to millions of years (see Section 11-8 and the *Popular Mechanics* article excerpt "NASA's Next Great Mars Rover" at the end of Chapter 11). These features are evidence that the Martian atmosphere was once thicker and that liquid water once coursed over the planet's surface. Could life have evolved on Mars during its "warm, wet period"? If so, could life have survived in the form of extremophile microorganisms as the Martian atmosphere thinned and the surface water either froze or evaporated? While Curiosity is designed to find suitable habitats where life may have existed, the rover is not equipped to test for the presence of life itself. Nonetheless, some tests for Martian life have already been carried out, and future tests for life will likely involve similar experiments.

In 1976, the Viking Lander 1 and Viking Lander 2 landed on different parts of Mars. Each spacecraft carried a scoop at the end of a mechanical arm to retrieve soil samples (**Figure 27-8**). These samples were deposited into a compact on-board biological laboratory that carried out three different tests aimed at detecting living Martian microorganisms.

> Some areas on Mars were covered with liquid water for extended periods.

Because the present-day surface of Mars is bone dry, these chemicals had nothing to react with until they were placed inside the moist interior of the Viking Lander laboratory.

At best, the results from the Viking Lander biological experiments were inconclusive. Perhaps life never existed on Mars at all. Or perhaps it did originate there, but failed to survive the thinning of the Martian atmosphere, the unstable chemistry of the planet's surface, and exposure to ultraviolet radiation from the Sun. (Unlike Earth, Mars has no ozone layer to block ultraviolet rays.) Another possibility is that Martian microorganisms have survived only in certain locations that the Viking Landers did not sample, such as isolated spots on the surface or beneath the ground. Yet another option is that there is life on Mars, but the experimental apparatus on board the Viking Lander spacecraft was not sophisticated enough to detect it.

The Power of Measuring Carbon Isotopes

An entirely different type of biological experiment can look for evidence of life having existed in the distant past. To test for past life on Mars, a rover takes a sample of an ancient lakebed region potentially containing the carbon-rich molecules of organisms—*dead or alive*. Then, signs of past or present life might be found by measuring how many of the organic molecules contain the isotope ^{12}C, which appears preferentially in biological molecules, and how many contain ^{13}C, which does not. (See Box 5-5 for a description of isotopes.) Thus, not only can ^{12}C isotopes indicate the existence of life, but this test works even when the organisms have died long ago.

The main mission of the Curiosity rover is to find sites on Mars that were once habitable; these will be the best places to measure the ^{12}C and ^{13}C isotopes in future missions. However, a number of other measurements must be taken along with ^{12}C and ^{13}C in order to draw strong conclusions regarding life. For this reason, such an analysis is more suitable back on Earth. The Mars 2020 rover is expected to drill sample cores in the most promising areas, for later pickup with a sample-return mission back to Earth.

Methane from Microbes?

Methane spikes have been detected in the Martian atmosphere. Microorganisms on Earth can gain energy by converting carbon dioxide to methane, and if Martian microorganisms do the same, it could explain the unexpected presence of Martian methane. Once it appears, methane rapidly decomposes in the Martian atmosphere, so we know that Mars presently has a source of methane. We also know that the spikes are not a simple result of seasonal changes. What we do not know is if this source of methane is geochemical or biological.

CONCEPT CHECK 27-5

Could any of the Viking Lander measurements detect signs of life on Mars that died out long ago?

Answer appears at the end of the chapter.

The Martian Civilization That Never Was

In 1976, while the Viking Landers were carrying out their biological experiments on the Martian surface, the companion

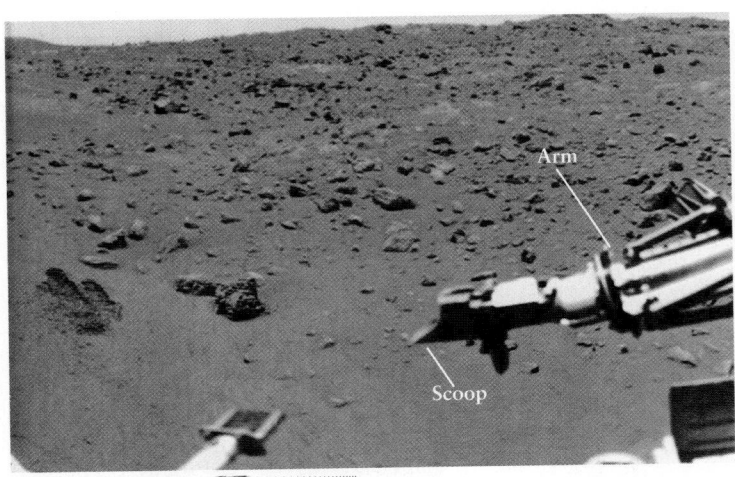

FIGURE 27-8 R I V U X G

Digging in the Martian Surface This view from the Viking Lander 1 spacecraft shows the mechanical arm with its small scoop against the backdrop of the Martian terrain. The scoop was able to dig about 30 cm (12 in.) beneath the surface. (NASA)

1. The *gas-exchange experiment* was designed to detect gases released due to chemical processes involving nutrients. For example, in the process of extracting energy from food, many organisms absorb oxygen and produce carbon dioxide, which is why we exhale this gas in our breath. Plants, conversely, take in carbon dioxide and release oxygen. To detect the exchange of carbon dioxide and oxygen, this experiment involved a soil sample that was placed in a sealed container along with nutrient-rich water. The gases in the container were then monitored to check for any microbial *exhalation*.

2. The *labeled-release experiment* was designed to detect metabolic processes involving oxygen, like those that we humans employ. A soil sample was moistened with nutrients containing low levels of radioactive carbon atoms. If any organisms in the sample consumed the nutrients, their waste carbon dioxide should have included the easily detectable radioactive carbon.

3. The *pyrolytic-release experiment* was designed to detect photosynthesis, the biological process by which terrestrial plants use solar energy to help synthesize organic compounds from carbon dioxide. A surface sample was placed in a container along with slightly radioactive carbon dioxide and exposed to artificial sunlight. If plantlike photosynthesis occurred, microorganisms in the sample would have taken in some of the radioactive carbon from the gas and brought it into their cells in the soil. The soil was later heated to drive off any of this "fixed" carbon, which could then be detected with a radiation-detecting Geiger counter.

The first data returned from these experiments caused great excitement, for in almost every case, rapid and extensive changes were detected inside the sealed containers. Further analysis of the data, however, led to the conclusion that these changes were more likely due to nonbiological chemical processes. It appears that the Martian surface is rich in unstable chemicals that react with water to release oxygen, or with nutrients to release carbon dioxide.

(a)

(b)

(c)

FIGURE 27-9   R I V U X G

A "Face" on Mars? **(a)** This 1976 image from Viking Orbiter 1 shows a Martian surface feature that resembles a human face. Some suggested that this feature might have been made by intelligent beings. **(b)** This 1998 Mars Global Surveyor (MGS) image, made under different lighting conditions with a far superior camera, reveals the "face" to be just an eroded hill. **(c)** This MGS image shows features of natural origin within a 215-km (134-mi) wide crater on Mars. Can you see this "face"? (a: NSSDC/NASA and Dr. Michael H. Carr; b: NASA/JPL/Malin Space Science Systems; c: Malin Space Science Systems, MGS, NASA)

Viking Orbiter spacecraft photographed some surface features that at first glance seemed to have been crafted by *intelligent* life on Mars. The Viking Orbiter 1 image in Figure 27-9a shows what appears to be a humanlike face, perhaps the product of an advanced and artistic civilization. However, when the more advanced Mars Global Surveyor spacecraft viewed the surface in 1998 using a superior camera (Figure 27-9b), it sadly found no evidence for the facial features, which were simply shadows on windblown hills. Microscopic life may once have existed on Mars, and may yet exist today, but there is no evidence that the red planet has ever been the home of intelligent beings.

27-3 Asteroid impacts might have spread Earth's microbes to other planets and moons

Asteroids might have led to the spread of life between bodies in the solar system. This could be, for example, the spread of life from Earth to Europa or Mars, and it could be that even life on Earth may have come from Mars.

For some of the largest asteroid impacts, rocks from the resulting crater can get ejected into space with enough speed to escape Earth and orbit the solar system (recall the escape speed in Section 4-7). Furthermore, rocks often have tiny cracks that harbor microbial life. If some of the microbes in these rocks can survive the journey elsewhere—and it appears that they could—then they can seed life on other bodies in the solar system.

Larger impacts were much more frequent in the past, and planets and their moons were often warmer and wetter as well. A few billion years ago, the ice crust on Europa could have been much thinner, or there may have been no ice crust at all; both scenarios would have increased the chance that microbes transferred from Earth could reach Europa's ocean.

As for Mars and Earth, if microbes were transferred to Mars from Earth shortly after life appeared on Earth (about 3.8 billion years ago), the microbes may have seeded Martian oceans of water. And most interesting of all, perhaps, is that if life started on Mars, it could have transferred to Earth—so there's a chance that we are Martians.

Surviving the Journey

Could microbes survive such a journey? There's no proof that they could, but evidence suggests that they might. Simulations of asteroid impact debris show that most transfers between solar system bodies are completed within 10 million years. Yet bacteria have been revived after lying dormant as "spores" for 250 million years! So the duration of the trip doesn't appear to be a problem.

What about shocks during the violent ejection? Simulations estimate that acceleration during ejection could be as high as 300,000 times the Earth's acceleration of gravity. These accelerations were created in a spinning centrifuge machine, and it took 65 hours at this enormous acceleration to kill 90% of the bacteria; much more survived for the shorter times that would be experienced during an actual ejection.

Would the microbes get cooked in a fiery trip through Earth's atmosphere? The microbes in deeper cracks are probably safe because only the outer few millimeters get hot. This is because rock is a very good thermal insulator—which slows down the flow of heat—and it only takes a few seconds for the rock to pass through Earth's atmosphere. Analysis of the interior of one meteorite on Earth that came from Mars (ALH84001) shows that its interior did not exceed a temperature of 104°F (40°C), which is an unpleasant fever for a human, but nothing threatening for most bacteria.

Another threat is exposure to radiation during the long voyage, which is normally blocked by our atmosphere. (This is also the top threat astronauts are expected to face in a mission to Mars.) While there is a strain of bacteria found near nuclear reactors that is highly resistant to radiation, most bacteria isn't. For this reason, scientists who calculate how many rocks make the journey from one object to another limit their analysis to rocks that are at least 3 m (10 ft) in radius. Rocks of this size can adequately shield most bacteria for a 10-million-year journey.

Overall, there is no reason to rule out the possibility of microbes surviving a trip from one celestial body to another.

Is Rock Transfer Likely?

Earlier work on the likelihood of Earth rocks making the journey to other objects suggested that the chances were very small. However, more recent and powerful simulations (for example, by Rachel Worth and others) have shown that impact debris kicked up by large asteroids can reach the required speeds to escape Earth.

For example, the dinosaur-killing asteroid that created the Chicxulub crater (see Section 15-4) would have ejected about 700 billion kg of rocky debris into space. This is about 70 times the mass of the largest Egyptian pyramid! However, there is only a 60% chance that this event resulted in a rock more than 3 m in radius reaching Europa. The odds improve when all impacts well after the end of the Late Heavy Bombardment (Section 8-6) are simulated, which limits the period to the last 3.5 billion years. The result is that we expect about six rocks capable of transferring microbes to have spread from Earth to Europa. Enceladus is farther out in the solar system, and only about three suitable rocks are expected to have made the journey from Earth.

Earth and Mars, on the other hand, have engaged in a heavy exchange of debris over the last 3.5 billion years. About 360,000 rocks suitable for microbial transfer are predicted to have gone from Earth to Mars. Ejection from Mars is even easier due to its lower gravity, and about 21 million suitable rocks are expected to have made the journey from Mars to Earth! Of course, it remains to be seen if any of these rocks seeded life.

CALCULATION CHECK 27-1

If we assume rocks have a density of 3000 kg/m³, how many kilograms is the smallest rock that is expected to shield wayward microbes from radiation?

Answer appears at the end of the chapter.

Scrutinizing Meteorites from Mars for Life-Forms

Aside from the chance that microbes could survive a journey to Earth, astrobiologists are deeply interested in analyzing rocks from other celestial bodies for signs of life. Such a monumental discovery would be the first evidence that life has existed beyond Earth. Rocks from other bodies can be analyzed by visiting rovers, brought back to Earth as envisioned for a "Mars sample-return" mission, or by simply finding meteorites on Earth that came from

Mars or elsewhere. Next, we discuss rocks from Mars for any signs that Mars may have once harbored life.

More than 100 meteorites that appear to have formed on Mars have been found at a variety of locations on Earth. The feature that identifies meteorites as having come from Mars is the chemical composition of trace amounts of gas trapped within them. This composition is very different from that of Earth's atmosphere, but it is a nearly perfect match to the composition of the Martian atmosphere found by the Viking Landers.

How could a rock have traveled from Mars to Earth? When an asteroid collides with a planet's surface and forms an impact crater, most of the material thrown upward by the impact falls back onto the planet's surface. But some extraordinarily powerful impacts have produced large craters on Mars—roughly 100 km in diameter or larger. These tremendous impacts can eject some rocks with such speed that they escape the planet's gravitational attraction and fly off into space.

In addition to simulations predicting large impacts, there are numerous large craters on Mars, so Martian rocks have probably been blasted into space over the planet's history. Some of these ejected rocks then go into elliptical orbits around the Sun; about 21 million of them, measuring more than 3 m in radius, are expected to hit Earth. These, and smaller ones, are the rocks that scientists find as meteorites from Mars. In fact, estimates are that several tons (about 1 cubic meter) of Martian rocks land on Earth *each year*.

Using the radioisotope age-dating technique (see Section 8-3), scientists find that most meteorites from Mars are between 200 million and 1.3 billion years old, much younger than the 4.56-billion-year age of the solar system. Indeed, this was the first hint that these rocks came from Mars; most meteors are thought to come from the asteroids, which are so small that their geologic activity would have ceased many billions of years earlier. But one meteorite from Mars, denoted by the serial number ALH 84001 and found in Antarctica in 1984, was found to be 4.5 billion years old (Figure 27-10a). Thus, ALH 84001 is a truly ancient piece of Mars. More detailed analysis of its radioisotopes suggests that ALH 84001 was ejected from Mars by an impact about 16 million years ago and landed in Antarctica a mere 13,000 years ago.

ALH 84001 was on Mars during the era when liquid water existed on the planet's surface. Scientists have therefore investigated this rock carefully for clues about Martian water and possible life. One such clue is the presence of rounded grains of minerals called *carbonates*. Analysis of these carbonates indicates that they formed in liquid water at a comfortable temperature of about 64°F, or about 18°C.

In 1996, David McKay and Everett Gibson of the NASA Johnson Space Center, along with several collaborators, began reporting the results from studies of the carbonate grains in ALH 84001. They provided several pieces of evidence that life may have once existed within the rock's cracks while it was still on Mars:

1. There are large numbers of elongated, tubelike structures around the carbonate grains that could be fossilized microorganisms (Figure 27-10b).

2. Microscopic cracks contain organic molecules—consistent with what is expected from the decay of microorganisms.

3. The carbonate grains contain small magnetic crystals called magnetite, which are similar to magnetite produced by bacteria on Earth.

(a)

FIGURE 27-10 R I V U X G

A Meteorite from Mars **(a)** This 1.9-kg meteorite, known as ALH 84001, formed on Mars some 4.5 billion years ago. About 16 million years ago, a massive impact blasted it into space, where it drifted in orbit around the Sun until landing in Antarctica 13,000 years ago. The small cube at lower right is 1 cm (0.4 in.) across. **(b)** This electron microscope image, magnified

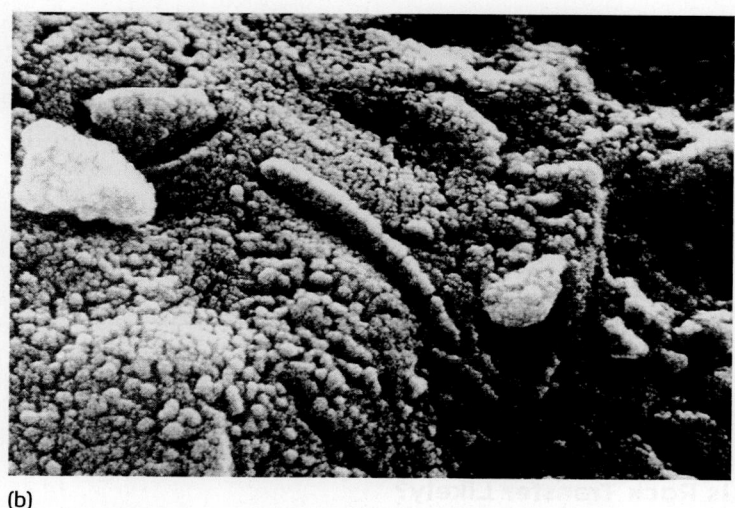

(b)

some 100,000 times, shows tubular structures about 100 nm (10^{-7} m) in length found within the Martian meteorite ALH 84001. One controversial interpretation is that these are the fossils of microorganisms that lived on Mars billions of years ago. (a: NASA Johnson Space Center; b: NASA)

Are McKay and Gibson's conclusions correct? Their claims of ancient life on Mars are extraordinary, and they require extraordinary proof. Further studies found that the organic compounds detected are not a very good match for decayed organisms. Also, it was later found that the tubelike structures and magnetite crystals found in ALH 84001, while consistent with life, could have been formed through geological processes on Mars. Thus, while some of the evidence is consistent with Martian life, it is far from compelling, and scientific consensus does not back up this claim.

> Claims that scientists have found Martian microorganisms are intriguing but very controversial.

CONCEPT CHECK 27-6

If bacteria on Earth are known to create magnetite crystals like those from Mars, why isn't this proof of life on Mars?

Answer appears at the end of the chapter.

Possible Microbial Tunnels

The most recent indications of possible Martian microbial remnants involve microscopic tunnels found in meteorites from Mars. In 2006, Martin Fisk from Oregon State University led a team that discovered small tunnels in a Martian meteorite named Nakhla. These tunnels are similar to structures found in rocks on Earth that are thought to be produced by rock-eating microbes. The Nakhla meteorite is not the only one with tunnels; tunnels are also found in ALH 84001, and a 2013 study at the Johnson Space Center led by Lauren White finds them in the Martian meteorite Y000593 (Figure 27-11).

Could the tunnels have been formed after the meteorites hit Earth? That is possible for ALH 84001, and Y000593, which sat

Microtunnels

5 μm

FIGURE 27-11 R I V U X G

Microscopic Tunnels in Martian Meteorite This figure shows microtunnels in the Martian meteorite Y000593. Although not common, very similar microscopic tunnels are found in rocks on Earth that contain DNA and are thought to be produced by rock-eating microbes. Even though Y000593 sat on Earth for thousands of years before being collected, the tunnels themselves were likely created on Mars: the tunnels in Y000593 are filled with Martian clay, as determined by an analysis of the clay's composition. The microtunnels emanate from larger fractures in the rock and are about 1–5 μm long. (Lauren White)

on Earth for thousands of years before being collected. But the Nakhla meteorite was observed falling on June 28, 1911, and its debris was collected too quickly for microbes on Earth to have created its tunnels. Are the microtunnels produced by Martian microbes? As in the case of the tubelike structures and magnetite crystals from ALH 84001, features that *appear* biological in origin might simply occur naturally through Martian geology. Thus, the Martian tunnels lead us back to Earth, where we must determine if these tunnels can instead be created through geological processes. For now, the origin of these tunnels is uncertain. Though they provide only weak evidence at this early stage of analysis, these microscopic tunnels open up a whole new line of inquiry in the search for Martian life.

CONCEPT CHECK 27-7

Why are scientists convinced that these meteorites actually came from Mars and not from our Moon?

Answer appears at the end of the chapter.

27-4 The Drake equation helps scientists estimate how many civilizations may inhabit our galaxy

 TUTORIAL 27.1 We have seen that, at best, only a few locations in our solar system are currently suitable to support life. But what about planets and moons around other stars? The existence of life on Earth seems to suggest that extraterrestrial life, possibly including intelligent species, might evolve on terrestrial planets around other stars, given sufficient time and hospitable conditions. How can we learn whether such worlds exist, given the tremendous distances that separate us from them? This is the great challenge facing the **search for extraterrestrial intelligence, or SETI.**

> Are we alone, or does the galaxy teem with intelligent life? Or is the truth somewhere in between?

Close Encounters Versus Remote Communication

A common belief is that alien civilizations do exist and that their spacecraft have visited Earth. Indeed, surveys show that between one-third and one-half of all Americans believe in unidentified flying objects (UFOs) of alien origin. A somewhat smaller percentage believe that aliens have landed on Earth. But, in fact, there is absolutely *no* scientifically verifiable evidence of alien visitations. As an example, many UFO proponents believe that the U.S. government is hiding evidence of an alien spacecraft that crashed near Roswell, New Mexico, in 1947. However, the bits of "spacecraft wreckage" found near Roswell turned out to be nothing more than remnants of a research balloon. Atomic isotopes are one way to tell if a material comes from Earth or not, because we know that the abundances of many isotopes vary in our solar system and beyond, yet no alien spacecraft debris with non-Earth isotope signatures has been found.

While there is no evidence of alien visits to Earth, alien civilizations could exist around some of the billions of stars in our galaxy. To find real evidence for intelligent civilizations on other worlds, we must look to the stars.

With our present technology, sending even a small robotic spacecraft to another star requires a flight time of tens of thousands of years. However, some astrophysicists are exploring the idea of ground-based lasers accelerating a small electronics-filled

chip up to about 20% the speed of light, which would then take about 20 years to get to the nearest star. Nonetheless, to search for life among many stars, astronomers search for radio transmissions from technologically advanced civilizations. Radio waves are a logical choice for interstellar communication because they can travel immense distances without being significantly degraded by the interstellar medium, the thin gas and dust found between the stars (see Section 8-1).

Over the past several decades, astronomers have proposed various ways to search for alien radio transmissions, and several searches have been undertaken. In 1960, Frank Drake first used a radio telescope at the National Radio Astronomy Observatory in West Virginia to monitor two Sunlike stars, Tau Ceti and Epsilon Eridani, without success. Since then, many SETI searches have taken place using radio telescopes around the world. Occasionally, a search has detected an unusual or powerful signal. But none has ever repeated, as a signal of intelligent origin might be expected to do. To date, we have no evidence of radio transmissions from other worlds.

CONCEPT CHECK 27-8

Why might the use of radio waves for exploration for life in the galaxy be more fruitful than using robotic interstellar spaceships?

Answer appears at the end of the chapter.

Is There Anybody Out There?

Should we be discouraged by this failure to make contact? What are the chances that a radio astronomer might someday detect radio signals from an extraterrestrial civilization? The first person to tackle this issue was Frank Drake, who proposed that the number of technologically advanced civilizations in the galaxy could be estimated by a simple equation. This is now called the **Drake equation:**

Drake equation

$$N = R_* f_p n_e f_l f_i f_c L$$

N = number of technologically advanced civilizations in the galaxy whose messages we might be able to detect

R_* = the rate at which solar-type stars form in the galaxy

f_p = the fraction of stars that have planets

n_e = the number of planets per solar system that are Earthlike (that is, suitable for life)

f_l = the fraction of those Earthlike planets on which life actually arises

f_i = the fraction of those life-forms that evolve into intelligent species

f_c = the fraction of those species that develop adequate technology and then choose to send messages into space

L = the lifetime of a technologically advanced civilization

The Drake equation is enlightening because it expresses the number of extraterrestrial civilizations that we might ever encounter as the product of a simple series of terms. We can estimate some of these terms from what we know about stars, stellar evolution, and planetary orbits. The *Cosmic Connections* figure depicts some of

COSMIC CONNECTIONS

Habitable Zones for Life

ANIMATION 27.1 Intelligent civilizations in our Milky Way Galaxy can evolve only in a certain region called the galactic habitable zone. In that zone, a suitable planet must lie within the planetary habitable zone of its parent star.

Galactic habitable zone

Too close to the center of the Milky Way Galaxy:
• The distances between stars are small, so there can be close encounters between stars that would disrupt a planetary system.
• There are also frequent outbursts of potentially lethal radiation from supernovae and from the supermassive black hole at the very center of the galaxy.

Too far from the center of the Milky Way Galaxy:
• Stars are deficient in elements heavier than hydrogen and helium, so they lack both the materials needed to form Earthlike planets and the chemical substances required for life as we know it.

Planetary habitable zone

Inside the galactic habitable zone:
• If the star's mass is too large, it will use up its hydrogen fuel so rapidly that it will burn out before life can evolve on any of its planets.
• Low-mass (class M) stars can have planets with conditions that support life.

The planet:
• Must be a terrestrial planet with a solid surface.
• Must have enough mass to provide the gravity needed to retain an atmosphere and oceans.
• Must be at a comfortable distance from the star so that the temperatures are neither too high nor too low.
• Must be in a stable, nearly circular orbit. (A highly elliptical orbit would cause excessively large temperature swings as the planet moved toward and away from the star.)

The neighborhood:
• There needs to be one or more large jovian planets whose gravitational forces will clear away comets and meteors.

these considerations. Keep in mind that some of these parameters are uncertain, and different groups come up with different values.

For example, the first two factors, R_* and f_p, can be determined by observation. In estimating R_*, some astronomers exclude stars with masses greater than about 1.5 times that of the Sun. These more massive stars use up the hydrogen in their cores in 3 billion (3×10^9) years or less. On Earth, by contrast, human intelligence developed only within the last million years or so, some 4.56 billion years after the formation of the solar system. If that is typical of the time needed to evolve higher life-forms, then a star of 1.5 $M_\odot$ or more probably fades away or explodes into a supernova before intelligent creatures like us can evolve on any of that star's planets.

Habitable zones (see the *Cosmic Connections* figure) are considered a possibility around some red dwarfs, but the lower range of mass is uncertain. Based on statistical studies of star formation in the Milky Way, a safe estimate is that roughly one suitable star forms in the galaxy each year in the galactic habitable zone. This result sets R_* at 1 per year.

As we saw in Section 8-7, planets are commonplace around stars. Astronomers estimate that most stars have planets, so they give f_p a value of 1.

Unfortunately, the rest of the terms in the Drake equation are very uncertain. Let's play with some hypothetical values. The chances that a planetary system has an Earthlike world suitable for life are not known. While exoplanet studies so far allow for estimates on the number of Earth-sized planets in habitable zones within our solar system, there is still guesswork in estimating how many of these are Earthlike in the sense that they are suitable for life. Were we to consider our own solar system as representative, we would put n_e at 1, but there's no evidence to support that. Let's be more conservative, however, and suppose that 1 in 10 suitable stars is orbited by a habitable planet, making $n_e = 0.1$.

What about the fraction of Earthlike planets that develop life? Given the existence of life on Earth, we might assume that, given appropriate conditions, the development of life is quite likely, and we could set $f_l = 1$. However, it is also possible that life on Earthlike planets is rare. After all, we do not know how it started here, so we cannot say whether its start was a likely event or super unlikely. With only one Earthlike planet (our own Earth) known to develop life, there is simply no way to deduce the probability of Earthlike planets in general to form life. Furthermore, biochemists do not understand the steps of life's origins and therefore cannot predict the probability for life to form. This is a topic of intense interest and much debate among astrobiologists, but we will be optimistic and assume here that $f_l = 1$.

For the sake of argument, we might also assume that evolution might naturally lead to the development of intelligence (a conjecture that is hotly debated—after all, this required perhaps 3 billion years to occur on our planet) and also make $f_i = 1$. It is anyone's guess as to whether these intelligent extraterrestrial beings would attempt communication with other civilizations in the galaxy, but were we to assume that they would, f_c would also be valued at 1. This number might be optimistic, considering that none of Earth's technologically advanced civilization have chosen to send messages for the purpose of being discovered. Furthermore, our own emissions for radio and TV (even before cables replaced many antenna transmissions) would not likely be detectable beyond a small region of the galaxy around our Sun.

The last variable, L, involving the longevity of an advanced civilization, might be the most uncertain of all. Looking at our own example, we see a planet whose atmosphere and oceans are increasingly polluted by creatures that possess nuclear weapons. If we are typical, perhaps L is as short as 100 years. Putting all these numbers together, we arrive at

$$N = 1/\text{year} \times 1 \times 0.1 \times 1 \times 1 \times 1 \times 100 \text{ years} = 10$$

In other words, even if we make many highly optimistic predictions about the numbers in the Drake equation, we would estimate that, out of the hundreds of billions of stars in the galaxy, only 10 host technologically advanced civilizations from which we might receive communications.

A wide range of values has been proposed for the terms in the Drake equation, and these various guesses produce vastly different estimates of N. Some scientists argue that there is likely only one advanced civilization in the galaxy and that we alone are it. Others speculate that there may be hundreds or thousands of planets inhabited by intelligent creatures. While the numerical result of the Drake equation is uncertain, the impact has been large. As we'll discuss next, the search is on to detect alien civilizations.

CONCEPT CHECK 27-9

What makes the longevity of an advanced civilization, L, so difficult to estimate?

Answer appears at the end of the chapter.

27-5 Radio searches for alien civilizations are underway

Even if only a few alien civilizations are scattered across the galaxy, we have the technology to detect radio transmissions from them. However, there are challenges! On the one hand, the signals would be too weak to detect if they were similar in strength to the radio waves we use for communication and radio stations here on Earth. Such waves are like ripples on a pond that diminish as they spread outward. On the other hand, if an advanced civilization wanted to make its presence known, it *might* sweep the skies with a focused beam of radio waves, which is much easier to detect. Even if we could not decode such a signal, it should be easy to tell that it is the product of an advanced technology.

But if other civilizations are trying to communicate with us using radio waves, what frequency are they using? This is an important question, because if we fail to tune our radio telescopes to the right frequency, we would not detect them in the first place.

A reasonable choice would be a frequency that is fairly free of interference from extraneous sources. SETI pioneer Bernard Oliver was the first to draw attention to a range of relatively noise-free frequencies in the neighborhood of the microwave emission lines of hydrogen (H) and hydroxide (OH) (Figure 27-12). This region of the spectrum can be called microwave or radio emission and is called the **water hole**, because H and OH together make H_2O.

In 1989, NASA began work on the High Resolution Microwave Survey (HRMS), an ambitious project to scan the entire sky at frequencies spanning the water hole from 10^3 to 10^4 MHz. HRMS

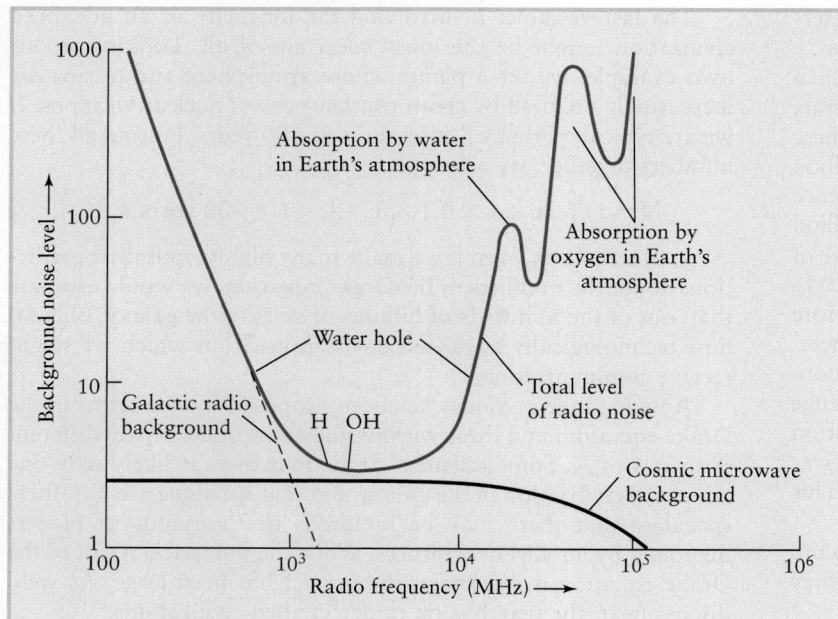

FIGURE 27-12

The Water Hole This graph shows the background noise level from the sky at various radio and microwave frequencies. The so-called water hole is a range of radio frequencies from about 10^3 to 10^4 megahertz (MHz), in which there is little noise and little absorption by Earth's atmosphere. Some scientists suggest that this noise-free region would be well suited for interstellar communication. Within the water hole itself, the principal source of noise is the afterglow of the Big Bang, called the cosmic microwave background. To put this graph in perspective, a frequency of 100 MHz corresponds to "100" on an FM radio, and 10^3 MHz is a frequency used for various types of radar. (Data from C. Sagan and F. Drake)

would have observed more than 800 nearby solar-type stars, but, unfortunately, its funding was cut just one year after becoming operational.

Even though NASA no longer funds searches for extraterrestrial intelligence, several teams of scientists remain actively involved in SETI programs. Funding for these projects has come from nongovernmental organizations such as the Planetary Society and from private individuals. In 1995, the SETI Institute in California began Project Phoenix, the direct successor to HRMS. In 2004, after careful observation of the nearest 800 solar-type stars, scientists for Project Phoenix announced that they had not found any promising signals.

A major challenge facing SETI is the tremendous amount of computer time needed to analyze the mountains of data returned by radio searches. To this end, scientists at the University of California, Berkeley, have recruited more than 5 million personal computer users to participate in a project called SETI@home. Each user receives actual data from a detector called SERENDIP IV (*Search for Extraterrestrial Radio Emissions from Nearby, Developed, Intelligent Populations*) and a data analysis program that also acts as a screensaver. When the computer's screensaver is on, the program runs, the data are analyzed, and the results are reported via the internet to the researchers at Berkeley. The program then downloads new data to be analyzed. Since 1999, SETI@home users have provided the equivalent of more than 3 million years of computing time!

In 2007, the SETI Institute began operation of a large radio telescope dedicated to the search for intelligent signals. This telescope, called the Allen Telescope Array (ATA), is actually hundreds of relatively small and inexpensive radio dishes working together. Designed for rapid and sensitive searches, the ATA hopes to search the nearest 100,000 solar-type stars. If astronomers knew which stars had Earthlike planets, SETI's telescopes could focus

> Millions of personal computers have helped scan for alien radio transmissions.

their attention on these objects, and the effort to find these stars is underway.

CONCEPT CHECK 27-10

Why should the range of frequencies in the water hole of Figure 27-12 be an attractive range of frequencies for an alien civilization to send out messages?

Answer appears at the end of the chapter.

27-6 Telescopes have begun searching for Earthlike planets

Although no longer involved in SETI, NASA is carrying out a major effort to search for Earthlike planets suitable for the evolution of an advanced civilization. Such a search poses a major challenge. One problem is that Earth-sized planets are too dim to be seen in visible light against the glare of their parent star. Using a different approach, astronomers have discovered many Jupiter-sized planets by detecting the "wobble" that these planets produce in their parent star (see Section 8-7). But a planet the size of Earth exerts only a weak gravitational force on its parent star. Given these challenges, finding large numbers of Earth-sized planets requires another method: transits.

Searching for Planetary Transits

An alternative technique has been used by an orbiting telescope called Kepler, which began detecting planets in 2009 and has discovered hundreds of Earth-sized planets. Recall from Section 8-7 that if a star is orbited by a planet whose orbital plane is oriented edge on to our line of sight, once per orbit the planet will pass in front of the star in an event called a *transit* (Figure 8-20). By blocking some of the star's light, this transit causes a temporary dimming of the light we see from that star.

FIGURE 27-13

Biosignature in the Infrared Spectrum of a Simulated Planet The image on the left is a simulation of what a space-based infrared telescope might see once one is launched. The white dot at the center is a nearby Sunlike star, and the smaller dots around it are planets orbiting the star. On the right is the simulated infrared spectrum of one of the planets. The red curve is the smooth continuous emission expected in the absence of biosignatures. The black dots and the blue line are simulated data showing broad absorption lines of water vapor (H_2O), ozone (O_3), and carbon dioxide (CO_2). Measuring the abundances of these gases can tell us about the planet's conditions, as well as whether it might be harboring oxygen-producing life. (Jet Propulsion Library/NASA)

Once a transiting planet is detected, astronomers determine the transiting planet's size (inferred from how much dimming takes place), as well as the size of its orbit. Recall from Johannes Kepler's third law that the size of the orbit can be calculated using the orbital period of the planet, which is the same as the time interval between successive transits. Given the distance from the planet to its parent star and the star's luminosity, astronomers will even be able to estimate the planet's average temperature. However, this estimated temperature would not include the warming effects of an atmosphere without additional observations.

To determine a planet's atmosphere, detailed spectra are required—and these observations are only just beginning. It is important to keep in mind that just because a planet is Earth-sized and at a certain distance from its sun, it does not mean it is necessarily Earthlike. Finding a truly Earthlike planet with liquid water is difficult but could be discovered with the current generation of telescopes, or the next generation designed specifically for this goal.

CONCEPT CHECK 27-11

What exactly was the Kepler telescope watching for as it searched for planets around stars?

CONCEPT CHECK 27-12

Which planets were the Kepler telescope not able to find, even if quite nearby?

Answers appear at the end of the chapter.

Biosignatures: Signs of Life

This next generation of telescopes might be able to measure a "biosignature" of life in a planet's atmosphere. By analyzing the infrared spectra of potentially Earthlike planets, the telescopes could reveal the atmospheric composition from the unique absorption features of gases such as ozone, carbon dioxide, and water vapor (**Figure 27-13**). Ozone (O_3) is made from ultraviolet sunlight acting on oxygen (O_2), which could indicate the presence

of life. This is significant not because oxygen *sustains* many life-forms on Earth (although it does), but because Earth's oxygen-rich atmosphere was *produced* by living organisms, and does not seem attainable through nonbiological means. Therefore, oxygen is a key atmospheric biosignature that will be the target of future studies.

If a planet with a thick oxygen atmosphere is close enough to us, its biosignature could be detected by the James Webb Space Telescope (JWST, with an expected 2021 launch). If we aren't that lucky, the next generation of infrared space telescopes would have a better chance of detecting an oxygen biosignature.

Sometime during the twenty-first century, we will probably answer the question "Are there worlds like Earth orbiting other stars?" If the answer is yes, then searches for life around other stars will gain even more impetus.

> Future telescopes may be able to detect levels of atmospheric oxygen that arise from life.

The potential rewards from such searches are great. Detecting a message from an alien civilization could dramatically change the course of our own civilization, through the sharing of scientific information with another species or an awakening of social or humanistic enlightenment. In only a few years, our technology, industry, and social structure might advance the equivalent of centuries into the future. Such changes would touch every person on Earth. Mindful of these profound implications, scientists push ahead with the search for extraterrestrial intelligence.

CONCEPT CHECK 27-13

How does ozone act as a biosignature for life?

CONCEPT CHECK 27-14

Why not look for oxygen gas (O_2) directly? Consult Figure 27-13.

Answers appear at the end of the chapter.

KEY WORDS

astrobiology, p. 856
carbonaceous chondrite, p. 857

Drake equation, p. 865
extremophile, p. 858
habitable zone, p. 867

organic molecules, p. 856
search for extraterrestrial intelligence (SETI), p. 865

water hole, p. 867

KEY IDEAS

Organic Molecules in the Universe: All life on Earth, and presumably on other worlds, depends on organic (carbon-based) molecules. These molecules occur naturally throughout interstellar space (Section 27-1).

- The organic molecules needed for life to originate were probably brought to the young Earth by comets, meteorites, and interplanetary dust particles. Another likely source for organic molecules is chemical reactions in Earth's primitive atmosphere. Similar processes may occur on other worlds.

Detlev van Ravenswaay/Science Source

Radio Searches for Extraterrestrial Intelligence: Astronomers have carried out a number of searches for radio signals from other stars. No signs of intelligent life have yet been detected, but searches are continuing and use increasingly sophisticated techniques (Sections 27-4 and 27-5).

- The Drake equation is a tool for estimating the number of intelligent, communicative civilizations in our galaxy.

Microtunnels

5 μm

Lauren White

Life in the Solar System: Besides Earth, other worlds in our solar system—the planet Mars, Jupiter's satellite Europa, and Saturn's satellite Enceladus—may have had the right conditions for the origin of life (Sections 27-2 and 27-3).

- Europa appears to have extensive liquid water beneath its icy surface. Future missions may search for the presence of life there. Enceladus also appears to contain a large subsurface ocean, and some of this material is ejected through ice volcanoes for easier sampling.

- Mars once had liquid water on its surface, though it has none today. Life may have originated on Mars during the liquid water era.

- The Viking Lander spacecraft searched for microorganisms on the Martian surface, but found no conclusive sign of their presence. The Mars Science Laboratory is designed to assess the past and present suitability of Mars for microbial life.

- An ancient Martian rock (ALH 84001) that came to Earth as a meteorite shows features that were suggested as remnants of Martian life. Most of these features can be explained by geologic processes. However, microscopic tunnels in other Martian meteorites are suggestive of microbes on Mars if the tunnels can't be produced through geological processes.

NASA/JPL/SSI

Telescope Searches for Earthlike Planets: A future generation of orbiting telescopes may be able to detect numerous terrestrial planets around nearby stars. If such planets are found, their infrared spectra may reveal the presence or absence of life (Section 27-6).

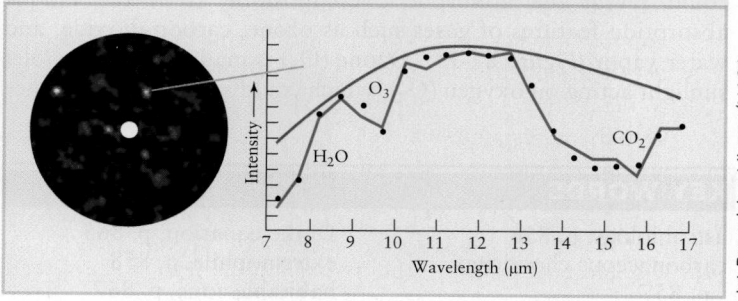

Jet Propulsion Library/NASA

QUESTIONS

> **Problem-solving tips and tools**
>
> The small-angle formula, discussed in Box 1-1, will be useful. Section 5-2 gives the relationship between wavelength and frequency, while Section 5-9 and Box 5-6 discuss the Doppler effect. Section 6-3 gives the relationship between the angular resolution of a telescope, the telescope diameter, and the wavelength used. You will find useful data about the planets in Appendix 1.

Section 27-1

1. What is meant by "life as we know it"? Why do astrobiologists suspect that extraterrestrial life is likely to be of this form?

2. How have astronomers discovered organic molecules in interstellar space? Does this discovery mean that life of some sort exists in the space between the stars?

3. Science-fiction television shows and movies often depict aliens as looking very much like humans. Discuss the likelihood that intelligent creatures from another world would have (a) a biochemistry similar to our own, (b) two legs and two arms, and (c) about the same dimensions as a human.

Section 27-2

4. Why do astronomers think Enceladus could support life? What would be the easiest way to search for this life?

5. Why are scientists interested in searching for life in Lake Vostok, Antarctica?

6. Many science-fiction stories and movies—including *The War of the Worlds, Invaders from Mars, Mars Attacks!,* and *Martians, Go Home*—involve invasions of Earth by intelligent beings from Mars. Why Mars rather than any of the other planets?

7. Summarize the three tests performed by the Viking Landers to search for Martian microorganisms.

8. What arguments can you give against the idea that the "face" on Mars (Figure 27-9) is of intelligent origin? What arguments can you give in favor of this idea?

9. Suppose the Curiosity rover measured an enhanced abundance of the carbon isotope ^{12}C in an ancient riverbed. What might this tell us?

10. Suppose someone told you that the Viking Landers failed to detect life on Mars simply because the tests were designed to detect terrestrial life-forms, not Martian life-forms. How would you respond?

11. Why is methane on Mars a possible sign of microorganisms?

Section 27-3

12. Explain how large asteroid impacts could lead to the spread of life to planets and moons in the solar system.

13. What evidence is there that microbes in the cracks of rocks could survive a trip to another body in the solar system? Consider the length of the trip, accelerations, temperatures while passing through an atmosphere, and radiation exposure.

14. Suppose someone brought you a rock that he claimed was a Martian meteorite. What scientific tests would you recommend be done to test this claim?

15. The Martian meteorite ALH 84001 contains magnetite, which can be produced by microorganisms on Earth. Why isn't this good evidence for Martian life?

16. Why are astronomers interested in the microtunnels within several Martian meteorites? Why don't these prove the existence of Martian life?

17. Consider the debris estimated to be ejected into space from the dinosaur-killing asteroid that created the Chicxulub crater. If all of the debris consisted of rocks 3 m in radius with a density of 3000 kg/m^3, how many rocks would have been ejected?

18. Why would it be easier for rocky debris to spread from Mars to Earth than from Earth to Mars?

19. Are more rocks expected to have transferred between celestial objects in the early period of the solar system or more recently? If microbes from Earth were to land elsewhere, would the conditions required for their survival be more likely in a young solar system or more recent? Explain your answer.

Section 27-4

20. **TUTORIAL 27.1** Assume that all the terms in the Drake equation have the values given in the text, except for N and L. (a) If there are 1000 civilizations in the galaxy today, what must be the average lifetime of a technological civilization? (b) What if there are 1 million such civilizations?

21. (a) Of the visually brightest stars in the sky listed in Appendix 5, which might be candidates for having Earthlike planets on which intelligent civilizations have evolved? Explain your selection criteria. (b) Repeat part (a) for the nearest stars, listed in Appendix 4.

22. How do you think our society would respond to the discovery of intelligent messages coming from a civilization on a planet orbiting another star? Explain your reasoning.

23. What do you think will set the limit on the lifetime of our technological civilization? This factor is L in the Drake equation. Explain your reasoning.

Section 27-5

24. Why are most searches for extraterrestrial intelligence made using radio telescopes? Why are most of these carried out at frequencies between 10^3 MHz and 10^4 MHz?

25. It has been suggested that extraterrestrial civilizations would choose to communicate at a wavelength of 21 cm. Hydrogen atoms in interstellar space naturally emit at this wavelength, so astronomers studying the distribution of hydrogen around

the galaxy would already have their radio telescopes tuned to receive extraterrestrial signals. (**a**) Calculate the frequency of this radiation in megahertz. Is this inside or outside the water hole? (**b**) Discuss the merits of this suggestion.

26. Imagine that a civilization in another planetary system is sending a radio signal toward Earth. As our planet moves in its orbit around the Sun, the wavelength of the signal we receive will change due to the Doppler effect. This gives SETI scientists a way to distinguish stray signals of terrestrial origin (which will not show this kind of wavelength change) from interstellar signals. (**a**) Use the data in Appendix 1 to calculate the speed of Earth in its orbit. For simplicity, assume the orbit is circular. (**b**) If the alien civilization is transmitting at a frequency of 3000 MHz, what wavelength (in meters) would we receive if Earth were moving neither toward nor away from their planet? (**c**) The maximum Doppler shift occurs if Earth's orbital motion takes it directly toward or directly away from the alien planet. How large is that maximum wavelength shift? Express your answer both in meters and as a percentage of the unshifted wavelength you found in (b). (**d**) Discuss why it is important that SETI radio receivers be able to measure frequency and wavelength to very high precision.

Section 27-6

27. Explain why telescopes that would look for the infrared spectra of other planets need to be placed in space. (*Hint:* See spectrum in Figure 27-13.)

28. Astronomers have proposed using interferometry to make an extremely high-resolution telescope. This proposal involves placing a number of infrared telescopes in space, separating them by thousands of kilometers and combining the light from the individual telescopes. One design of this kind has an effective diameter of 6000 km and uses infrared radiation with a wavelength of 10 mm. If it is used to observe an Earthlike planet orbiting the star Epsilon Eridani, 3.22 pc (10.5 ly) from Earth, what is the size of the smallest detail that this system will be able to resolve on the face of that planet? Give your answer in kilometers. (*Hint:* Use the diffraction-limited angular resolution from Section 6-3.)

General Problems

29. Why are extreme life-forms on Earth, such as those shown in the chapter-opening image, of interest to astrobiologists?

30. Mercury, Venus, and the Moon are all considered unlikely places to find life. Suggest why this should be.

31. In 1802, when it seemed likely to many scholars that there was life on Mars, the German mathematician Karl Friedrich Gauss proposed that we signal the Martian inhabitants by drawing huge geometric patterns in the snows of Siberia. His plan was never carried out. (**a**) Suppose patterns had been drawn that were 1000 km across. What minimum diameter would be needed for the objective of a Martian telescope in

order to resolve these patterns? Assume that the observations are made at a wavelength of 550 nm, and assume that Earth and Mars are at their minimum separation. (*Hint:* Use the diffraction-limited angular resolution from Section 6-3.) (**b**) Ideally, the patterns used would be ones that could not be mistaken for natural formations. They should also indicate that they were created by an advanced civilization. What sort of patterns would you have chosen?

32. The late, great science-fiction editor John W. Campbell exhorted his authors to write stories about organisms that think as well as humans but not *like* humans. Discuss the possibility that an intelligent being from another world might be so alien in its thought processes that we could not communicate with it.

33. The first of all Earth spacecraft to venture into interstellar space were Pioneer 10 and Pioneer 11, which were launched in 1972 and 1973, respectively. Their missions took them past Jupiter and Saturn and eventually beyond the solar system. Both spacecraft carry a metal plaque with artwork (reproduced below) that shows where the spacecraft is from and what sort of creatures designed it. If an alien civilization were someday to find one of these spacecraft, which of the features on the plaque do you think would be easily understandable to them? Explain your reasoning.

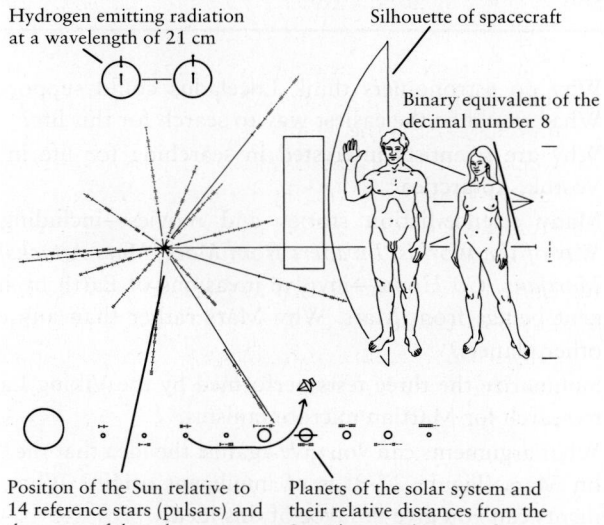

Hydrogen emitting radiation at a wavelength of 21 cm

Silhouette of spacecraft

Binary equivalent of the decimal number 8

Position of the Sun relative to 14 reference stars (pulsars) and the center of the galaxy

Planets of the solar system and their relative distances from the Sun (shown in binary code)

NASA

WEB/E-BOOK QUESTIONS

34. Any living creatures in the subsurface ocean of Europa would have to survive without sunlight. Instead, they might obtain energy from Europa's inner heat. Search the internet for information about "black smokers," which are associated with high-temperature vents at the bottom of Earth's oceans. What kind of life is found around black smokers? How do these life-forms differ from the more familiar organisms found in the upper levels of the ocean?

35. Search the internet for information about the *Mars Express* orbiter and the *Spirit* and *Opportunity* rovers. What discoveries have these missions made about water on Mars? Have they found any evidence that liquid water has existed on Mars in the recent past? Describe the evidence, if any.

36. Like other popular media, the internet is full of claims of the existence of "extraterrestrial intelligence," namely, UFO sightings and alien abductions. (a) Choose a website of this kind and analyze its content using the idea of Occam's razor, the principle that if there is more than one viable explanation for a phenomenon, one should choose the simplest explanation that fits all the observed facts. (b) Read what a skeptical website has to say about UFO sightings. A good example is the website of the Committee for Skeptical Inquiry, or CSI. After considering what you have read on both sides of the UFO debate, discuss your opinions about whether intelligent aliens really have landed on Earth.

ACTIVITIES

Starry Night™ Explorations

37. Use *Starry Night™* to view Earth as seen from space. Select **Favorites > Explorations > Earth** from the menu to see Earth from a distance of 12,000 km in the direction toward the Sun. Use the location scroller, time flow, and zoom controls to examine the surface of our home planet.
    (a) Can you see any evidence of human-made objects or other features suggesting the presence of intelligent life on the planet in the view?
    (b) Click the name in the label for Earth and select **Surface Image/Model > EarthNight** from the menu. Does this view change your answer to the preceding question? If so, why?
    (c) Use the elevation buttons to increase the distance from Earth to about 380,000 km (the approximate distance of the Moon from Earth). Without changing the zoom, can you detect evidence of intelligence on Earth in this view of the planet? Now zoom in to simulate the use of a telescope to view Earth from the Moon, at this distance. Based on what you observe, if you were to design a spacecraft for obtaining visual clues to the presence of intelligent life on another planet or moon in the solar system, what features would be important when designing the spacecraft's trajectory and imaging instruments?

38. In *Starry Night™*, open **Favorites > Explorations > Mars** and use the location scroller and zoom and time controls to explore the many features on the surface of this planet. In particular, conduct a search for suitable landing sites for a spacecraft designed to look for signs of the presence of life on the Martian surface. List several of your chosen sites, providing rationale for your choices, and discuss the relative potential hazards to landing at these sites.

39. Use *Starry Night™* to examine Jupiter's moon Europa. Select **Favorites > Explorations > Europa** and use the location scroller, zoom, and time flow controls to examine the surface of this enigmatic moon. Are there any surface features that suggest this moon might harbor life? Is there any other evidence, perhaps not visible in this simulation, which suggests the possibility that life might exist here?

40. Use *Starry Night™* to explore how developments in the search for planets around other stars affect the Drake equation, which was formulated to estimate the number of civilizations in the universe that might be sufficiently advanced technologically to be able to signal their presence to humans on Earth. The complete equation ($N = R_* \times f_p \times n_e \times f_l \times f_i \times f_c \times L$) includes the factor f_p—the fraction of Sun-like stars having orbiting planets—and the factor n_e—the number of planets per system that are suitable for life. Space exploration was in its infancy when astronomer Frank Drake proposed this equation in 1961. Since then, and particularly in recent years with the Kepler telescope mission, the search for exoplanets has finally produced reliable evidence for the existence of many thousands of stars with one or more planets in our near neighborhood. Indeed, it is confidently predicted that every star in our galaxy has at least one planet in orbit around it!
    (a) Select **Favorites > Explorations > Drake fp** to see a view of the stars of the Milky Way in the vicinity of Earth. Each star marked with a blue circle in this view is known to have at least one orbiting planet. Use the hand tool to look around the sky and locate one particular region of the sky with a very dense pattern of marked stars. Sample the names of the marked stars in this region with the HUD. What explanation can you provide for the density and pattern of this region? What does this map of stars with exoplanets suggest for the most likely value of the factor f_p in the Drake equation?
    (b) If our own solar system is typical, then a reasonable proportion of all stars is likely to have at least one planet whose surface conditions are suitable for the development of life. This implies that a reasonable value for the factor accounting for the number of habitable planets at any star is $n_e \approx 1$. Simplify the Drake equation to reflect these likely values.
    (c) Select **Favorites > Explorations > Drake Overview** to see a view of the Milky Way showing stars with known exoplanets once again marked in blue. Use the location scroller to gain an appreciation for the proportion of our galaxy that scientists have so far surveyed for exoplanets. Then use the elevation controls to move at least 400 Mly from Earth. Each glowing dot on the screen represents a galaxy in our region of the universe, each containing a large number of stars. From your observations, what do you deduce about the likelihood that life, including intelligent life, exists somewhere else in the universe than on Earth? (Of course, it is not obvious that such a form of

life actually exists on Earth!) Why do you suppose contact with intelligent civilizations elsewhere in the universe has not yet been made?

Collaborative Exercise

41. Imagine that astronomers have discovered intelligent life in a nearby star system. Your group is submitting a proposal for who on Earth should speak for the planet and what 50-word message should be conveyed. Prepare a maximum one-page proposal that states (**a**) who should speak for Earth and why; (**b**) what this person should say in 50 words; and (**c**) why this message is the most important compared to other things that could be said. Only serious responses receive full credit.

ANSWERS

ConceptChecks

ConceptCheck 27-1: Carbon atoms can combine with other elements to form an impressively wide array of different molecules, and this complexity is needed for even the smallest living organisms. Furthermore, carbon is produced in stars and is widely available.

ConceptCheck 27-2: Organic molecules are commonly found in meteorites and comets, which frequently crashed onto planet surfaces during the early formation of the solar system. Also, interplanetary dust grains containing organic compounds steadily fall to a planet's surface.

ConceptCheck 27-3: They created amino acids and other compounds that living organisms need—they did not create living entities.

ConceptCheck 27-4: The salts dissolve in water when the water is in contact with rock. Rather than a small pocket of water within ice, the salts suggest a much larger ocean is in contact with a rocky mantle.

ConceptCheck 27-5: No. The three experiments were sensitive to only living organisms.

ConceptCheck 27-6: Geological processes (involving no biology) can also produce this type of magnetite, so there is no compelling reason to assume that the magnetite was produced by Martian microbes. (*Note:* The claims that this magnetite was of biological origin came *before* it was figured out that this magnetite can be produced through geology.)

ConceptCheck 27-7: The Martian meteorites have trace amounts of gas trapped within them that are nearly identical to the unique Martian atmosphere.

ConceptCheck 27-8: At the speed of light, radio waves travel at more than 100,000 km every second, whereas our fastest spacecraft can travel a maximum of only tens of thousands of kilometers every hour. Therefore, the vast distances between the stars make it seemingly impossible for spacecraft to carry out exploration of the stars.

ConceptCheck 27-9: L is particularly difficult to estimate because at about the same time the technology to communicate beyond one's own planet is developed, the technology of weapons of mass destruction can develop, which could end the civilization just as it starts (alternatively, it can learn to live without war and exist for a very long time).

ConceptCheck 27-10: At other frequencies, there is more background noise, and more energy would need to be used to reliably transmit the same message across the galaxy. Any advanced civilization could figure this out and would see the benefits of sending messages in this narrow range.

ConceptCheck 27-11: The Kepler telescope was looking for an ever-so-slight dimming of a star as an orbiting planet passes in front of the star, partially blocking the star from our view. This is called a transit.

ConceptCheck 27-12: If the orbital plane of a planet around its star is tilted up or down from our perspective, then the planet will not block out any of the star's light during its orbit. Since Kepler looked for the dimming of stars as they were blocked by planets, Kepler would not have detected a planet with this kind of orbit.

ConceptCheck 27-13: Ozone comes from oxygen gas, O_2, which is produced by living organisms on Earth. Scientists do not know of any plausible nonbiological way to create a dense oxygen atmosphere.

ConceptCheck 27-14: In the infrared where a planet's spectrum can be more easily measured, an absorption line from oxygen gas is not present, but absorption from ozone (O_3) is.

CalculationCheck

CalculationCheck 27-1: The smallest rocks considered are 3 m in radius. Mass = (volume) × (density): $(4/3\pi(3 \text{ m})^3) \times (3000 \text{ kg/m}^3) = 340,000$ kg.

A Biologist's View of Astrobiology

by Kevin W. Plaxco

In the early 1960s, flush with the excitement of Sputnik and the first crewed space missions, the Nobel Prize–winning biologist Joshua Lederberg coined the word *exobiology* to describe the scientific study of extraterrestrial life. But after a brief flurry of popularity in the 1960s and 1970s (due partly to the pioneering research and public outreach work of Carl Sagan), exobiology fell out of fashion among scientists; it was, after all, the only field of scientific study without any actual subject material to research.

In the last decade, however, the field of astrobiology has filled the void left by exobiology's demise by being simultaneously more encompassing and more practical. Astrobiology removes exobiology's distinction between life on Earth and life "out there" and focuses on broader, but more tractable, questions regarding the relationship between life (any life, anywhere) and the universe. In a nutshell, astrobiology uses our significant (if incomplete) knowledge of life on Earth to address three broad questions about life in the universe:

- Which of the physical attributes of our universe were necessary in order for it to support the origins and further evolution of life?

- How did life arise and evolve on Earth, and how might it have arisen and evolved elsewhere?

- Where else might life have arisen in our universe, and how do we best search for it?

Much of the worth and appeal of astrobiology lie in the fact that these questions, which are among the most fundamental posed by science, address the most profound issues regarding our origins and our nature.

Perhaps not surprisingly, some aspects of astrobiology are far better understood than others. We understand, for example, the broad details of how the almost metal-free hydrogen and helium left over after the Big Bang was enriched in the metallic elements by fusion reactions occurring in the more massive stars and how these elements—absolutely critical for biology because only they support the complex chemistry required for life—were then returned to the cosmos in supernova explosions. Similarly, we understand in detail how rocky planets—which we see as potentially habitable—form out of the dust and gas of prestellar nebulae. In contrast, however, much of the *biology* in astrobiology remains speculative at best. There is, for example, nothing even remotely resembling consensus regarding how life first arose on Earth, much less how—or if—it might have arisen elsewhere in the cosmos.

Perhaps equally unsurprisingly, some branches of the astrobiology community are more "optimistic" than others. Astronomers generally view astrobiology in terms of the processes by which potentially habitable planets form. As these processes are reasonably well understood, astronomers tend to be rather upbeat regarding the events surrounding life's origins. The radio astronomer Frank Drake, who pioneered the search for extraterrestrial intelligence (SETI) in the 1960s, is a relatively extreme example of this hopefulness: When trying to estimate the number of intelligent, "communicating" civilizations in the galaxy, he assumed that rocky, water-soaked planets almost invariably give rise to life. Biochemists, in contrast, tend to view the key issues in astrobiology as relating to the first steps in the origins of life—that is, the detailed processes by which inanimate chemicals first combine to form a living, evolving, self-replicating system.

To date, about a half dozen theories have been put forth to explain how this might have occurred on Earth, but all are fraught with enormous scientific difficulties. Specifically, every single theory of the origins of life postulated to date either utterly lacks experimental evidence to back it up or requires events that are so improbable they make even the "astronomical" number of planets in the universe look small by comparison. When faced with what is, in effect, a complete unknown, biochemists tend to be much more cautious about estimating the probability of life arising and the chances of our being alone in the cosmos. To quote Francis Crick, codiscoverer of the structure of DNA, "We cannot decide whether the origin of life on Earth was an extremely unlikely event or almost a certainty, or any possibility in between these two extremes."

Irrespective of our degree of personal optimism or pessimism, though, no one would be more thrilled than us biologists by the detection of life that had arisen independently of life on Earth. Such a find would revolutionize our field by providing the first, tangible evidence of whether there are viable routes to the evolution of life other than the DNA/proteins/cells path that it took on Earth. But even barring such a momentous discovery as the detection of extraterrestrial life, astrobiology still provides a significant service to biology: By placing life into a broader context, astrobiology provides the perspective that we need in order to truly understand who we are and where we come from.

Prior to becoming a professor of biochemistry at the University of California at Santa Barbara, **Kevin Plaxco** earned his BS in chemistry from the University of California at Riverside and his PhD in molecular biology from the California Institute of Technology. While his research focuses predominantly on the study of proteins, the "macromolecular machines" that perform the lion's share of the action in our cells, he has also performed NASA-funded research into methods for the detection of extraterrestrial life. In addition, he teaches a course in astrobiology and has published a popular science book on the subject.

Appendices

APPENDIX 1 The Planets: Orbital Data

Planet	Semimajor axis (10^6 km)	Semimajor axis (au)	Sidereal period (years)	Sidereal period (days)	Synodic period (days)	Average orbital speed (km/s)	Orbital eccentricity	Inclination of orbit to ecliptic (°)
Mercury	57.9	0.387	0.241	87.969	115.88	47.9	0.206	7.00
Venus	108.2	0.723	0.615	224.70	583.92	35.0	0.007	3.39
Earth	149.6	1.000	1.000	365.256	—	29.79	0.017	0.00
Mars	227.9	1.524	1.88	686.98	779.94	24.1	0.093	1.85
Jupiter	778.3	5.203	11.86		398.9	13.1	0.048	1.30
Saturn	1429	9.554	29.46		378.1	9.64	0.053	2.48
Uranus	2871	19.194	84.10		369.7	6.83	0.043	0.77
Neptune	4498	30.066	164.86		367.5	5.5	0.010	1.77

APPENDIX 2 The Planets: Physical Data

Planet	Equatorial diameter (km)	Equatorial diameter (Earth = 1)	Mass (kg)	Mass (Earth = 1)	Average density (kg/m³)	Rotation period* (solar days)	Inclination of equator to orbit (°)	Surface gravity Earth = 1	Albedo	Escape speed (km/s)
Mercury	4880	0.383	3.302×10^{23}	0.0553	5430	58.646	0.5	0.38	0.12	4.3
Venus	12,104	0.949	4.868×10^{24}	0.8149	5243	243.01^R	177.4	0.91	0.59	10.4
Earth	12,756	1.000	5.974×10^{24}	1.000	5515	0.997	23.45	1.000	0.39	11.2
Mars	6794	0.533	6.418×10^{23}	0.107	3934	1.026	25.19	0.38	0.15	5.0
Jupiter	142,984	11.209	1.899×10^{27}	317.8	1326	0.414	3.12	2.36	0.44	59.5
Saturn	120,536	9.449	5.685×10^{26}	95.16	687	0.444	26.73	1.1	0.47	35.5
Uranus	51,118	4.007	8.682×10^{25}	14.53	1318	0.718^R	97.86	0.92	0.56	21.3
Neptune	49,528	3.883	1.024×10^{26}	17.15	1638	0.671	29.56	1.1	0.51	23.5

*For Jupiter, Saturn, Uranus, and Neptune, the internal rotation period is given.
A superscript R means that the rotation is retrograde (opposite the planet's orbital motion).

APPENDIX 3 Major Satellites of the Planets by Mass

Planet	Satellite	Year of discovery	Average distance from center of planet (km)	Orbital (sidereal) period (days)*	Orbital eccentricity	Size of satellite (km)**	Mass (kg)
EARTH	Moon	—	384,400	27.322	0.0549	3476	7.349×10^{22}
MARS	Phobos	1877	9378	0.319	0.01	$28 \times 23 \times 20$	1.1×10^{22}
	Deimos	1877	23,460	1.263	0.00	$16 \times 12 \times 10$	1.1×10^{22}
JUPITER	Ganymede	1610	1,070,000	7.155	0.0011	5268	1.482×10^{23}
	Callisto	1610	1,883,000	16.689	0.0074	4800	1.077×10^{23}
	Io	1610	421,600	1.769	0.0041	3642	8.932×10^{22}
	Europa	1610	670,900	3.551	0.0094	3120	4.791×10^{22}
	Himalia	1904	11,461,000	250.56	0.1623	184	6.7×10^{18}
	Amalthea	1892	181,400	0.498	0.0031	$270 \times 200 \times 155$	2.1×10^{18}
	Elara	1905	11,741,000	259.64	0.2174	78	8.7×10^{17}
	Thebe	1979	221,900	0.675	0.0177	98	1.5×10^{18}
	Pasiphaë	1908	23,624,000	743.63R	0.4090	58	3.0×10^{17}
	Carme	1938	23,404,000	734.17R	0.2533	46	1.3×10^{17}
SATURN	Titan	1655	1,221,870	15.95	0.0288	5150	1.34×10^{23}
	Rhea	1672	527,070	4.518	0.001	1528	2.3×10^{21}
	Iapetus	1671	3,560,840	79.33	0.0283	1472	2.0×10^{21}
	Dione	1684	377,420	2.737	0.0022	1123	1.1×10^{21}
	Tethys	1684	294,670	1.888	0.0001	1066	6.2×10^{20}
	Enceladus	1789	238,040	1.37	0.0047	504	1.1×10^{20}
	Mimas	1789	185,540	0.942	0.0196	397	3.8×10^{19}
	Phoebe	1898	12,947,780	550.31R	0.1635	$230 \times 220 \times 210$	8.3×10^{18}
	Hyperion	1848	1,500,880	21.28	0.0274	$360 \times 280 \times 225$	5.7×10^{18}
	Janus	1980	151,460	0.695	0.0068	$193 \times 173 \times 137$	1.9×10^{18}
URANUS	Titania	1787	436,300	8.706	0.0011	1578	3.53×10^{21}
	Oberon	1787	583,500	13.46	0.0014	1522	3.01×10^{21}
	Ariel	1851	190,900	2.52	0.0012	1158	1.35×10^{21}
	Umbriel	1851	266,000	4.144	0.0039	1169	1.2×10^{21}
	Miranda	1948	129,900	1.413	0.0013	471	6.59×10^{19}
	Puck	1985	86,000	0.762	0.0001	162	2.9×10^{18}
	Sycorax	1997	12,179,000	1288.3R	0.5224	150	5.4×10^{18}
	Portia	1986	66,100	0.513	0.0001	156×126	1.7×10^{18}
	Juliet	1986	64,400	0.493	0.0007	150×74	5.6×10^{17}
	Belinda	1986	75,300	0.624	0.0001	128×64	3.6×10^{17}

APPENDIX 3 Major Satellites of the Planets by Mass   *(continued)*

Planet	Satellite	Year of discovery	Average distance from center of planet (km)	Orbital (sidereal) period (days)*	Orbital eccentricity	Size of satellite (km)**	Mass (kg)
NEPTUNE	Triton	1846	354,800	5.877^R	0	2706	2.15×10^{22}
	Proteus	1989	117,647	1.122	0.0005	$440 \times 416 \times 404$	4.4×10^{19}
	Nereid	1949	5,513,400	360.14	0.7512	340	3.1×10^{19}
	Larissa	1989	73,548	0.555	0.0014	$216 \times 204 \times 164$	4.2×10^{18}
	Galatea	1989	61,953	0.429	0	$204 \times 184 \times 144$	2.1×10^{18}
	Despina	1989	52,526	0.335	0.0002	$180 \times 150 \times 130$	2.1×10^{18}
	Thalassa	1989	50,075	0.311	0.0002	$108 \times 100 \times 52$	3.5×10^{17}
	Naiad	1989	48,227	0.294	0.0004	$96 \times 60 \times 52$	1.9×10^{17}
	Halimede	2002	15,728,000	1879.71^R	0.5711	62	1.8×10^{17}
	Neso	2002	22,422,000	2914.07	0.2931	44	6.3×10^{16}

This table was compiled from data provided by the Jet Propulsion Laboratory.
*A superscript R means that the satellite orbits in a retrograde direction (opposite to the planet's rotation).

**The size of a spherical satellite is equal to its diameter.

APPENDIX 4 The Nearest Stars

Name*	Parallax (arcsec)	Distance (pc)	Distance (ly)	Spectral type	Proper motion (arcsec/yr)	Apparent visual magnitude	Absolute visual magnitude	Mass (Sun = 1)
Proxima Centauri	0.772	1.30	4.22	M5.5 V	3.853	+11.09	+15.53	0.107
Alpha Centauri A	0.747	1.34	4.36	G2 V	3.710	−0.01	+4.36	1.144
Alpha Centauri B	0.747	1.34	4.36	K0 V	3.724	+1.34	+5.71	0.916
Barnard's Star	0.547	1.83	5.96	M4.0 V	10.358	+9.53	+13.22	0.166
Wolf 359	0.419	2.39	7.78	M6.0 V	4.696	+13.44	+16.55	0.092
Lalande 21185	0.393	2.54	8.29	M2.0 V	4.802	+7.47	+10.44	0.464
Sirius A	0.380	2.63	8.58	A1 V	1.339	−1.43	+1.47	1.991
Sirius B	0.380	2.63	8.58	white dwarf	1.339	+8.44	+11.34	0.500
UV Ceti	0.374	2.68	8.73	M5.5 V	3.368	+12.54	+15.40	0.109
BL Ceti	0.374	2.68	8.73	M6.0 V	3.368	+12.99	+15.85	0.102
Ross 154	0.337	2.97	9.68	M3.5 V	0.666	+10.43	+13.07	0.171
Ross 248	0.316	3.16	10.32	M5.5 V	1.617	+12.29	+14.79	0.121
Epsilon Eridani	0.310	3.23	10.52	K2 V	0.977	+3.73	+6.19	0.850
Lacaille 9352	0.304	3.29	10.74	M1.5 V	6.896	+7.34	+9.75	0.529
Ross 128	0.299	3.35	10.92	M4.0 V	1.361	+11.13	+13.51	0.156
EZ Aquarii A	0.290	3.45	11.27	M5.0 V	3.254	+13.33	+15.64	0.105
EZ Aquarii B	0.290	3.45	11.27	—	3.254	+13.27	+15.58	0.106
EZ Aquarii C	0.290	3.45	11.27	—	3.254	+14.03	+16.34	0.095
Procyon A	0.286	3.50	11.40	F5 IV-V	1.259	+0.38	+2.66	1.569
Procyon B	0.286	3.50	11.40	white dwarf	1.259	+10.70	+12.98	0.500
61 Cygni A	0.286	3.50	11.40	K5.0 V	5.281	+5.21	+7.49	0.703
61 Cygni B	0.286	3.50	11.40	K7.0 V	5.172	+6.03	+8.31	0.630
GJ725 A	0.283	3.53	11.53	M3.0 V	2.238	+8.90	+11.16	0.351
GJ725 B	0.283	3.53	11.53	M3.5 V	2.313	+9.69	+11.95	0.259
GX Andromedae	0.281	3.56	11.62	M1.5 V	2.918	+8.08	+10.32	0.486
GQ Andromedae	0.281	3.56	11.62	M3.5 V	2.918	+11.06	+13.30	0.163
Epsilon Indi A	0.276	3.63	11.82	K5 V	4.704	+4.69	+6.89	0.766
Epsilon Indi B	0.276	3.63	11.82	T1.0	4.823			0.044
Epsilon Indi C	0.276	3.63	11.82	T6.0	4.823			0.028
DX Cancri	0.276	3.63	11.83	M6.5 V	1.290	+14.78	+16.98	0.087
Tau Ceti	0.274	3.64	11.89	G8 V	1.922	+3.49	+5.68	0.921
RECONS 1	0.272	3.68	11.99	M5.5 V	0.814	+13.03	+15.21	0.113
YZ Ceti	0.269	3.72	12.13	M4.5 V	1.372	+12.02	+14.17	0.136
Luyten's Star	0.264	3.79	12.37	M3.5 V	3.738	+9.86	+11.97	0.257
Kapteyn's Star	0.255	3.92	12.78	M1.5 V	8.670	+8.84	+10.87	0.393
AX Microscopium	0.253	3.95	12.87	M0.0 V	3.455	+6.67	+8.69	0.600

This table, compiled from data reported by the Research Consortium on Nearby Stars, lists all known stars within 4.00 pc (13.05 ly).

Stars that are components of multiple star systems are labeled A, B, and C.

APPENDIX 5 The Visually Brightest Stars

Name	Designation	Distance (pc)	Distance (ly)	Spectral type	Radial velocity (km/s)*	Proper motion (arcsec/yr)	Apparent visual magnitude	Apparent visual brightness (Sirius = 1)**	Mass (Sun = 1)
Sirius A	α CMa A	2.63	8.58	A1 V	−7.6	1.34	−1.43	1.000	+1.46
Canopus	α Car	95.9	313	F0 II	+20.5	0.03	−0.72	0.520	−5.63
Arcturus	α Boo	11.3	36.7	K1.5 III	−5.2	2.28	−0.04	0.278	−0.30
Alpha Centauri A	α Cen A	1.34	4.36	G2 V	−25	3.71	−0.01	0.270	+4.36
Vega	α Lyr	7.76	25.3	A0 V	−13.9	0.35	+0.03	0.261	+0.58
Capella	α Aur	12.9	42.2	G5 III	+30.2	0.43	+0.08	0.249	−0.48
Rigel	α Ori A	237	773	B8 Ia	+20.7	0.002	+0.12	0.240	−6.75
Procyon	α CMi A	3.50	11.4	F5 IV-V	−3.2	1.26	+0.34	0.196	+2.62
Achernar	α Eri	44.1	144	B3 V	+16	0.10	+0.50	0.169	−2.72
Betelgeuse	α Ori	131	427	M1 Iab	+21	0.03	+0.58	0.157	−5.01
Hadar	β Cen	161	525	B1 III	+5.9	0.04	+0.60	0.154	−5.43
Altair	α Aql	51.4	168	A7 V	−26.1	0.66	+0.77	0.132	−2.79
Aldebaran	α Tau A	20	65.1	K5 III	+54.3	0.20	+0.85	0.122	−0.65
Spica	α Vir	80.4	262	B1 III-IV	+1	0.05	+1.04	0.103	−3.49
Antares	α Sco A	185	604	M1.5 Iab	−3.4	0.03	+1.09	0.098	−5.25
Pollux	β Gem	10.3	33.7	K0 IIIb	+3.3	0.63	+1.15	0.093	+1.08
Fomalhaut	α PsA	7.69	25.1	A3 V	+6.5	0.37	+1.16	0.092	+1.73
Deneb	α Cyg	990	3230	A2 Ia	−4.5	0.002	+1.25	0.085	−8.73
Mimosa	β Cru	108	353	B0.5 IV	+15.6	0.05	+1.297	0.081	−3.87
Regulus	α Leo A	23.8	77.5	B7 V	+5.9	0.25	+1.35	0.077	−0.53

Data in this table were compiled from SIMBAD database operated at the Centre de Données Astronomiques de Strasbourg, France.

*A positive radial velocity means that the star is receding; a negative radial velocity means that the star is approaching.

**This is a ratio of the star's apparent brightness to that of Sirius, the brightest star in the night sky.

Note: Acrux, or α Cru (the brightest star in Crux, the Southern Cross) appears to the naked eye as a star of apparent magnitude +0.87, the same as Aldebaran. However, it does not appear in this table because Acrux is actually a binary star system. The blue-white component stars of this binary system have apparent magnitudes of +1.4 and +1.9, and so they are dimmer than any of the stars listed here.

APPENDIX 6 Some Important Astronomical Quantities

Astronomical unit	1 au = 1.4960×10^{11} m = 1.4960×10^{8} km
Light-year	1 pc = 3.2616 ly = 63,240 au
Parsec	1 pc = 3.2616 ly = 3.0857×10^{16} m = 206,265 au
Year	1 y = 365.2564 days = 3.156×10^{7} s
Solar mass	1 $M_\odot$ = 1.989×10^{30} kg
Solar radius	1 $R_\odot$ = 6.9599×10^{8} m
Solar luminosity	1 $L_\odot$ = 3.90×10^{26} W

APPENDIX 8 Some Useful Mathematics

Area of a rectangle of sides a and b	$A = ab$
Volume of a rectangular solid of sides a, b, and c	$V = abc$
Hypotenuse of a right triangle whose other sides are a and b	$c = \sqrt{a^2 + b^2}$
Circumference of a circle of radius r	$C = 2\pi r$
Area of a circle of radius r	$A = \pi r^2$
Surface areas of a sphere of radius r	$A = 4\pi r^2$
Volume of a sphere of radius r	$V = 4\pi r^3/3$
Value of π	$\pi = 3.1415926536$

APPENDIX 7 Some Important Physical Constants

Speed of light	$c = 2.9979 \times 10^{8}$ m/s
Gravitational constant	$G = 6.673 \times 10^{-11}$ N m^2/kg^2
Planck's constant	$h = 6.6261 \times 10^{-34}$ J s = 4.1357×10^{-15} eV s
Boltzmann constant	$k = 1.3807 \times 10^{-23}$ J/K = 8.6173×10^{-5} eV/K
Stefan–Boltzmann constant	$\sigma = 5.6704 \times 10^{-8}$ W m^{-2} K^{-4}
Mass of electron	$m_e = 9.1094 \times 10^{-31}$ kg
Mass of proton	$m_p = 1.6726 \times 10^{-27}$ kg
Mass of neutron	$m_n = 1.6749 \times 10^{-27}$ kg
Mass of hydrogen atom	$m_H = 1.6735 \times 10^{-27}$ kg
Rydberg constant	$R = 1.0973 \times 10^{-7}$ kg m^{-1}
Electron volt	1 eV = 1.6022×10^{-19} J

Glossary

You can find more information about each term in the indicated chapter or chapters. See the Key Words section in each chapter for the specific page on which the meaning of a given term is described.

A ring One of three prominent rings encircling Saturn. (Chapter 12)

absolute magnitude The apparent magnitude that a star would have if it were at a distance of 10 pc. (Chapter 17)

absolute zero A temperature of −273°C (or 0 K), at which all molecular motion stops; the lowest possible temperature. (Chapter 5)

absorption line spectrum Dark lines superimposed on a continuous spectrum. (Chapter 5)

acceleration The rate at which an object's velocity changes due to a change in speed, a change in direction, or both. (Chapter 4)

accretion The gradual accumulation of matter in one location, typically due to the action of gravity. (Chapters 8, 18, 24)

accretion disk A disk of gas orbiting a star or black hole. (Chapter 24)

active galactic nucleus (AGN) The center of an active galaxy. (Chapter 24)

active galaxy A galaxy that is emitting exceptionally large amounts of energy; a Seyfert galaxy, radio galaxy, or quasar. (Chapter 24)

active optics A technique for improving a telescopic image by altering the telescope's optics to compensate for variations in air temperature or flexing of the telescope mount. (Chapter 6)

adaptive optics A technique for improving a telescopic image by altering the telescope's optics in a way that compensates for distortion caused by Earth's atmosphere. (Chapter 6)

aerosol Tiny droplets of liquid dispersed in a gas. (Chapter 13)

AGB star See *asymptotic giant branch star.*

AGN See *active galactic nucleus.*

air drag A force that opposes the motion of an object through air; a portion of the object's energy is transferred to the air molecules. (Chapter 4)

albedo The fraction of sunlight that a planet, asteroid, or satellite reflects. (Chapter 9)

alpha particle The nucleus of a helium atom, consisting of two protons and two neutrons. (Chapter 19)

amino acids The chemical building blocks of proteins. (Chapter 15)

angle The opening between two lines that meet at a point. (Chapter 1)

angular diameter The angle subtended by the diameter of an object. (Chapter 1)

angular distance The angle between two points in the sky. (Chapter 1)

angular momentum See *conservation of angular momentum.*

angular resolution The angular size of the smallest feature that can be distinguished with a telescope. (Chapter 6)

angular size See *angular diameter.*

anisotropic Possessing unequal properties in different directions; not isotropic. (Chapter 23)

annihilation The process by which the masses of a particle and antiparticle are converted into energy. (Chapter 26)

annular eclipse An eclipse of the Sun in which the Moon is too distant to cover the Sun completely; a ring of sunlight is seen around the Moon at mid-eclipse. (Chapter 3)

anorthosite Rock commonly found in ancient, cratered highlands on the Moon. (Chapter 10)

Antarctic Circle A circle of latitude 23½° north of Earth's south pole. (Chapter 2)

antimatter Matter consisting of antiparticles such as antiprotons, antineutrons, and positrons. (Chapter 26)

antiparticle A particle with the same mass as an ordinary particle but with other properties, such as electric charge, reversed. (Chapters 21, 26)

antipode The point on a planet exactly opposite a specific location. (Chapter 11)

antiproton A particle with the same mass as a proton but with a negative electric charge. (Chapter 26)

aphelion The point in its orbit where a planet is farthest from the Sun. (Chapter 4)

apogee The point in Earth's orbit where a satellite or the Moon is farthest away. (Chapter 3)

apparent brightness The flux of a star's light arriving at Earth. (Chapter 17)

apparent magnitude A measure of the brightness of light from a star or other object as measured from Earth. (Chapter 17)

apparent solar day The interval between two successive transits of the Sun's center across the local meridian. (Chapter 2)

apparent solar time Time reckoned by the position of the Sun in the sky. (Chapter 2)

arcminute One-sixtieth (1/60) of a degree, designated by the symbol ′. (Chapter 1)

arcsecond One-sixtieth (1/60) of an arcminute, or 1/3600 of a degree, designated by the symbol ″. (Chapter 1)

Arctic Circle A circle of latitude 23½° south of Earth's north pole. (Chapter 2)

asteroid One of tens of thousands of small, rocky, planetlike objects in orbit about the Sun. Also called *minor planet*. (Chapters 7, 15)

asteroid belt A region between the orbits of Mars and Jupiter that encompasses the orbits of many asteroids. (Chapters 7, 15)

asthenosphere A warm, plastic layer of the mantle beneath the lithosphere of Earth. (Chapter 9)

astrobiology The study of life in the universe. (Chapter 27)

astrometric method A technique for detecting extrasolar planets by looking for stars that "wobble" periodically. (Chapter 8)

astronomical unit (au) The semimajor axis of Earth's orbit; the average distance between Earth and the Sun. (Chapter 1)

asymptotic giant branch A region on the Hertzsprung–Russell (H-R) diagram occupied by stars in their second and final red giant phase. (Chapter 20)

asymptotic giant branch star (AGB star) A red giant star that has ascended the giant branch for the second and final time. (Chapter 20)

atmosphere (atm) A unit of atmospheric pressure. (Chapter 9)

atmospheric pressure The force per unit area exerted by a planet's atmosphere. (Chapter 9)

atom The smallest particle of an element that has the properties characterizing that element. (Chapter 5)

atomic number The number of protons in the nucleus of an atom of a particular element. (Chapters 5, 8)

au See *astronomical unit*.

aurora (*plural* **aurorae**) Light radiated by atoms and ions in Earth's upper atmosphere, mostly in the polar regions. (Chapter 9)

aurora australis Aurorae seen from southern latitudes; the southern lights. (Chapter 9)

aurora borealis Aurorae seen from northern latitudes; the northern lights. (Chapter 9)

autumnal equinox The intersection of the ecliptic and the celestial equator, at which the Sun crosses the equator from north to south. Also used to refer to the date on which the Sun passes through this intersection. (Chapter 2)

average density The mass of an object divided by its volume. (Chapter 7)

average density of matter In cosmology, the average density of all the material substance in the universe. (Chapter 25)

B ring One of three prominent rings encircling Saturn. (Chapter 12)

Balmer line An emission or absorption line in the spectrum of hydrogen caused by an electron transition between the second and higher energy levels. (Chapter 5)

Balmer series The entire pattern of Balmer lines. (Chapter 5)

Barnard object One of a class of dark nebulae discovered by E. E. Barnard. (Chapter 18)

barred spiral galaxy A spiral galaxy in which the spiral arms begin from the ends of a "bar" running through the nucleus rather than from the nucleus itself. (Chapter 23)

baseline In interferometry, the distance between two telescopes whose signals are combined to give a higher-resolution image. (Chapter 6)

belt A dark band in Jupiter's atmosphere. (Chapter 12)

Big Bang An explosion of all space that took place roughly 13.7 billion years ago and marks the beginning of the universe. (Chapters 1, 25)

binary star (also **binary star system** or **binary**) Two stars orbiting each other. (Chapter 17)

biosphere The layer of soil, water, and air surrounding Earth in which living organisms thrive. (Chapter 9)

bipolar outflow Oppositely directed jets of gas expelled from a young star. (Chapter 18)

black hole An object whose gravity is so strong that the escape speed exceeds the speed of light. (Chapters 1, 25)

black hole evaporation The process by which black holes emit particles. (Chapter 21)

blackbody A hypothetical perfect radiator that absorbs and re-emits all radiation falling upon it. (Chapter 5)

blackbody curve The intensity of radiation emitted by a blackbody plotted as a function of wavelength or frequency. (Chapter 5)

blackbody radiation The radiation emitted by a perfect blackbody. (Chapter 5)

blazar A type of active galaxy whose nucleus has a featureless spectrum without prominent spectral lines. (Chapter 24)

blueshift A decrease in the wavelength of photons emitted by an approaching source of light. (Chapter 5)

Bohr orbits In the model of the atom described by Niels Bohr, the only orbits in which electrons are allowed to move about the nucleus. (Chapter 5)

Bok globule A small, roundish, dark nebula. (Chapter 18)

bow shock A region where charged particles in the solar wind abruptly slow down as they encounter a planet's magnetosphere. (Chapters 9, 12)

bright terrain (on Ganymede) Young, reflective, relatively crater-free terrain on the surface of Ganymede. (Chapter 13)

brightness See *apparent brightness*.

brown dwarf A starlike object that is not massive enough to sustain hydrogen fusion in its core. (Chapter 17)

brown ovals Elongated, brownish features usually seen in Jupiter's northern hemisphere. (Chapter 12)

C ring One of three prominent rings encircling Saturn. (Chapter 12)

capture theory The hypothesis that the Moon was gravitationally captured by Earth. (Chapter 10)

carbon fusion The thermonuclear fusion of carbon to produce heavier nuclei. (Chapter 20)

carbon star A peculiar red giant star whose spectrum shows strong absorption lines of carbon and carbon compounds. (Chapter 20)

carbonaceous chondrite A type of meteorite that has a high abundance of carbon and volatile compounds. (Chapters 15, 27)

Cassegrain focus An optical arrangement in a reflecting telescope in which light rays are reflected by a secondary mirror to a focus behind the primary mirror. (Chapter 6)

Cassini division An apparent gap between Saturn's A and B rings. (Chapter 12)

CCD See *charge-coupled device.*

celestial equator A great circle on the celestial sphere 90° from the celestial poles. (Chapter 2)

celestial sphere An imaginary sphere of very large radius centered on an observer; the apparent sphere of the sky. (Chapter 2)

center of mass The point between a star and a planet, or between two stars, around which both objects orbit. (Chapters 8, 10, 17)

central bulge (of a galaxy) A spherical distribution of stars around the nucleus of a spiral galaxy. (Chapter 22)

Cepheid variable A type of yellow, supergiant, pulsating star. (Chapter 19)

Cerenkov radiation The radiation emitted by a particle traveling through a substance at a speed greater than the speed of light in that substance. (Chapter 20)

Chandrasekhar limit The maximum mass (1.4 $M_\odot$) of a white dwarf. (Chapter 20)

charge-coupled device (CCD) A type of solid-state device designed to detect photons. (Chapter 6)

chemical composition A description of which chemical substances make up a given object. (Chapter 7)

chemical differentiation The process by which the heavier elements in a planet sink toward its center while lighter elements rise toward its surface. (Chapter 8)

chemical element See *element.*

chondrule A glassy, roughly spherical blob found within meteorites. (Chapter 8)

chromatic aberration An optical defect whereby different colors of light passing through a lens are focused to different locations, resulting in rainbow artifacts. (Chapter 6)

chromosphere A layer in the atmosphere of the Sun between the photosphere and the corona. (Chapter 16)

circumpolar A term describing a star that neither rises nor sets but appears to rotate around one of the celestial poles. (Chapter 2)

circumstellar accretion disk An accretion disk that surrounds a protostar. (Chapter 18)

close binary A binary star system in which the stars are separated by a distance roughly comparable to their diameters. (Chapter 19)

closed universe A universe with positive curvature, so that its geometry is analogous to that of the surface of a sphere. (Chapter 25)

cluster (of galaxies) A collection of galaxies containing a few to several thousand member galaxies. (Chapter 23)

cluster (of stars) A group of stars that formed together and have remained together because of their mutual gravitational attraction. (Chapter 18)

CNO cycle A series of nuclear reactions in which carbon is used as a catalyst to transform hydrogen into helium. (Chapter 16)

cocoon nebula The nebulosity surrounding a protostar. (Chapter 18)

co-creation theory The hypothesis that Earth and the Moon formed at the same time from the same material. (Chapter 10)

cold dark matter Slowly moving, weakly interacting particles presumed to constitute the bulk of matter in the universe. (Chapter 26)

collapsar A proposed type of supernova in which a black hole forms at the center of a dying star before the outer layers of the star have time to collapse. (Chapter 21)

collisional ejection theory The hypothesis that the Moon formed from material ejected from Earth by the impact of a large asteroid. (Chapter 10)

color ratio The ratio of the apparent brightness of a star measured in one spectral region to its brightness measured in a different region. (Chapter 17)

color-magnitude diagram A plot of the apparent magnitudes (that is, apparent brightnesses) of stars in a cluster versus their color indices (a measure of their surface temperatures). (Chapter 19)

coma (of a comet) The diffuse gaseous component of the head of a comet. (Chapter 15)

coma (optical) The distortion of off-axis images formed by a parabolic mirror. (Chapter 6)

combined average mass density The average density in the observable universe of all forms of matter and energy, measured in units of mass per volume. (Chapter 25)

comet A small body of ice and dust in orbit about the Sun. While passing near the Sun, a comet's vaporized ices give rise to a coma and tail. (Chapters 7, 15)

compound A substance consisting of two or more chemical elements in a definite proportion. (Chapter 5)

compression A region in a sound wave where the medium carrying the wave is compressed. (Chapter 25)

condensation temperature The temperature at which a particular substance in a low-pressure gas condenses into a solid. (Chapter 8)

conduction The transfer of heat by directly passing energy from atom to atom. (Chapter 16)

conic section The curve of intersection between a circular cone and a plane; this curve can be a circle, ellipse, parabola, or hyperbola. (Chapter 4)

conjunction The geometric arrangement of a planet in the same part of the sky as the Sun, so that the planet is at an elongation of 0°. (Chapter 4)

conservation of angular momentum A law of physics stating that in an isolated system, the total amount of angular momentum—a measure of the amount of rotation—remains constant. (Chapter 8)

conservation of energy A law of physics stating that energy can change forms and be transferred from one object to another but cannot be created or destroyed. (Chapter 4)

constellation A configuration of stars in the same region of the sky. (Chapter 2)

contact binary A binary star system in which both members fill their Roche lobes. (Chapter 19)

continuous spectrum A spectrum of light over a range of wavelengths without any spectral lines. (Chapter 5)

convection The transfer of energy by moving currents of fluid or gas containing that energy. (Chapters 9, 16)

convection cell A circulating loop of gas or liquid that transports heat from a warm region to a cool region. (Chapter 9)

convection current The pattern of motion in a gas or liquid in which convection is taking place. (Chapter 9)

convective zone The region in a star where convection is the dominant means of energy transport. (Chapter 16)

core (of Earth) The iron-rich inner region of Earth's interior. (Chapter 9)

core accretion model The hypothesis that each of the jovian planets formed by accretion of gas onto a rocky core. (Chapter 8)

core helium fusion The thermonuclear fusion of helium at the center of a star. (Chapters 19, 20)

core hydrogen fusion The thermonuclear fusion of hydrogen at the center of a star. (Chapter 19)

core-collapse supernova A supernova that occurs at the end of a massive star's lifetime, when the star's core collapses to high density and the star's outer layers are expelled into space. (See *Type Ib supernova, Type Ic supernova,* and *Type II supernova*) (Chapter 20)

corona (of the Sun) The Sun's outer atmosphere, which has a high temperature and a low density. (Chapter 16)

coronal heating problem The unsolved scientific question surrounding the physical mechanism responsible for heating the Sun's corona to multi-million-degree temperatures. (Chapter 16)

coronal hole A region in the Sun's corona that is deficient in hot gases. (Chapter 16)

coronal mass ejection An event in which billions of tons of gas from the Sun's corona are suddenly blasted into space at high speed. (Chapter 16)

cosmic background radiation See *cosmic microwave background (CMB)*.

cosmic microwave background (CMB) An isotropic radiation field with a blackbody temperature of about 2.725 K that permeates the entire universe. (Chapter 25)

cosmic particle horizon An imaginary sphere, centered on Earth, whose radius equals the distance light has traveled since the Big Bang. (Chapters 25, 26)

cosmic rays Fast-moving subatomic particles that enter our solar system from interstellar space. (Chapter 11)

cosmological constant (Λ) In the equations of general relativity, a quantity signifying a pressure throughout all space that helps the universe expand; a type of dark energy. (Chapter 25)

cosmological principle The assumption that the universe is homogeneous and isotropic on the largest scale. (Chapter 25)

cosmological redshift A redshift that is caused by the expansion of the universe. (Chapter 25)

cosmology The study of the structure and evolution of the universe. (Chapter 25)

coudé focus An optical arrangement with a reflecting telescope; a series of mirrors is used to direct light to a remote focus away from the moving parts of the telescope. (Chapter 6)

crater See *impact crater*.

critical density The value of the combined average mass density throughout the universe for which space would be flat. (Chapter 25)

crust (of a planet) The surface layer of a terrestrial planet. (Chapter 9)

crustal dichotomy (on Mars) The contrast between the young northern lowlands and older southern highlands on Mars. (Chapter 11)

cryovolcano An ice volcano that erupts or oozes volatiles, such as water, ammonia, or methane, instead of lava. (Chapter 13)

crystal A material in which atoms are arranged in orderly rows. (Chapter 9)

D ring One of several faint rings encircling Saturn. (Chapter 12)

dark ages The era between the formation of the first atoms (when the universe was about 380,000 years old) and the formation of the first stars (when the universe was about 400 million years old). (Chapter 26)

dark energy A form of energy that appears to pervade the universe and causes the expansion of the universe to accelerate but has no discernible gravitational effect. (Chapter 25)

dark energy density parameter (V_L) The ratio of the average mass density of dark energy in the universe to the critical density. (Chapter 25)

dark matter Nonluminous matter that is the dominant form of matter in galaxies and throughout the universe. (Chapter 22)

dark nebula A cloud of interstellar gas and dust that obscures the light of more distant stars. (Chapter 18)

dark terrain (on Ganymede) Older, heavily cratered, dark-colored terrain on the surface of Ganymede. (Chapter 13)

dark-energy–dominated universe A universe in which the mass density of dark energy exceeds both the average density of matter and the mass density of radiation. (Chapter 25)

dead quasar A supermassive black hole that has already consumed the gas and dust that previously provided fuel for quasar activity. Any supermassive black hole in the center of a galaxy is presumed to have gone through a brief period when it exhibited quasar activity. (Chapter 24)

dead quasar flare The flare that results if a star wanders too close to a supermassive black hole in the center of a galaxy (which is presumably a dead quasar); gravitational tidal forces can tear some matter away from the star and provide a small source of fuel to briefly "light up" the dead quasar with an approximately year-long flare of light. (Chapter 24)

declination Angular distance of a celestial object north or south of the celestial equator. (Chapter 2)

deferent A stationary circle in the Ptolemaic system along which another circle (an epicycle) moves, carrying a planet, the Sun, or the Moon. (Chapter 4)

degeneracy The phenomenon, due to quantum mechanical effects, whereby the pressure exerted by a gas does not depend on its temperature. (Chapter 19)

degenerate-electron (-neutron) pressure The pressure exerted by degenerate electrons (neutrons). (Chapters 19, 20)

degree A basic unit of angular measure, designated by the symbol °. (Chapter 1)

degree Celsius A basic unit of temperature, designated by the symbol °C and used on a scale where water freezes at 0° and boils at 100°. (Chapter 5)

degree Fahrenheit A basic unit of temperature, designated by the symbol °F and used on a scale where water freezes at 32° and boils at 212°. (Chapter 5)

density See *average density*.

density fluctuation A variation of density from one place to another. (Chapter 26)

density parameter (V_0) The ratio of the average density of matter and energy in the universe to the critical density. (Chapter 25)

density wave In a spiral galaxy, a localized region in which matter piles up as it orbits the center of the galaxy; this is a proposed explanation of spiral structure. (Chapter 22)

detached binary A binary star system in which neither star fills its Roche lobe. (Chapter 19)

deuterium bottleneck The situation during the first 3 minutes after the Big Bang when deuterium (an isotope of hydrogen whose nucleus contains one proton and one neutron) inhibited the formation of heavier elements. (Chapter 26)

differential rotation The rotation of a nonrigid object in which parts adjacent to each other at a given time do not always stay close together. (Chapters 12, 16)

differentiated asteroid An asteroid in which chemical differentiation has taken place so that denser material is toward the asteroid's center. (Chapter 15)

differentiation See *chemical differentiation*.

diffraction The spreading out of light passing through an aperture or opening in an opaque object. (Chapter 6)

diffraction grating An optical device consisting of thousands of closely spaced lines etched in glass or metal that disperses light into a spectrum. (Chapter 6)

direct motion The apparent eastward movement of a planet seen against the background stars. (Chapter 4)

disk (of a galaxy) The disk-shaped distribution of Population I stars that dominates the appearance of a spiral galaxy. (Chapter 22)

distance ladder The sequence of techniques used to determine the distances to very remote galaxies. (Chapter 23)

distance modulus The difference between the apparent and absolute magnitudes of an object. (Chapter 17)

diurnal motion Any apparent motion in the sky that repeats on a daily basis, such as the rising and setting of stars. (Chapter 2)

Doppler effect The apparent change in the wavelength of radiation due to relative motion between the source and the observer along the line of sight. (Chapter 5)

double star A pair of stars located at nearly the same position in the night sky. Some, but not all, double stars are binary stars. (Chapter 17)

Drake equation An equation used to estimate the number of intelligent civilizations in the galaxy with whom we might communicate. (Chapter 27)

dredge-up The transporting of matter from deep within a star to its surface by convection. (Chapter 20)

dust devil Whirlwind found in dry or desert areas on both Earth and Mars. (Chapter 11)

dust grain A microscopic bit of solid matter found in interplanetary or interstellar space. (Chapter 18)

dust tail The tail of a comet that is composed primarily of dust grains. (Chapter 15)

dwarf elliptical galaxy A low-mass elliptical galaxy that contains only a few million stars. (Chapter 23)

dwarf planet A solar system body that is large enough to be spherical in shape and have a circular orbit around the Sun, but not large enough to clear its own path of other bodies. This term is used for Ceres, Pluto, and Eris. (Chapter 14)

dynamo The mechanism whereby electric currents within an astronomical body generate a magnetic field. Also referred to as a *magnetic dynamo*. (Chapters 7, 8)

E ring A very broad, faint ring encircling Saturn. (Chapter 12)

earthquake A sudden vibratory motion of Earth's surface. (Chapter 9)

eccentricity A number between 0 and 1 that describes the shape of an ellipse. (Chapter 4)

eclipse The cutting off of part or all the light from one celestial object by another. (Chapter 3)

eclipse path The track of the tip of the Moon's shadow along Earth's surface during a total or annular solar eclipse. (Chapter 3)

eclipse year The interval between successive passages of the Sun through the same node of the Moon's orbit. (Chapter 3)

eclipsing binary A binary star system in which, as seen from Earth, stars periodically pass in front of each other. (Chapter 17)

ecliptic The apparent annual path of the Sun on the celestial sphere. (Chapter 2)

ecliptic plane The plane of Earth's orbit around the Sun. (Chapter 2)

Eddington limit A constraint on the rate at which a black hole can accrete matter. (Chapter 24)

electromagnetic radiation Radiation consisting of oscillating electric and magnetic fields. Examples include gamma rays, x rays, visible light, ultraviolet light, infrared radiation, radio waves, and microwaves. (Chapter 5)

electromagnetic spectrum The entire array of electromagnetic radiation. (Chapter 5)

electromagnetism Electric and magnetic phenomena, including electromagnetic radiation. (Chapter 5)

electron A subatomic particle with a negative charge and a small mass, usually found in orbits about the nuclei of atoms. (Chapter 5)

electron volt (eV) The energy acquired by an electron accelerated through an electric potential of 1 volt. (Chapter 5)

electroweak force A unification of the electromagnetic and weak forces that occurs at high energies. (Chapter 26)

element A chemical that cannot be broken down into more basic chemicals. (Chapter 5)

elementary particle physics The branch of quantum mechanics that deals with individual subatomic particles and their interactions. (Chapter 26)

ellipse A conic section obtained by cutting completely through a circular cone with a plane. (Chapter 4)

elliptical galaxy A galaxy with an elliptical shape and no conspicuous interstellar material. (Chapter 23)

elongation The angular distance between a planet and the Sun as viewed from Earth. (Chapter 4)

emission line spectrum A spectrum that contains bright emission lines. (Chapter 5)

emission nebula A glowing gaseous nebula whose spectrum has bright emission lines. (Chapter 18)

Encke gap A narrow gap in Saturn's A ring. (Chapter 12)

energy flux The rate of energy flow, usually measured in joules per square meter per second. (Chapter 5)

energy level In an atom, a particular amount of energy possessed by an atom above the atom's least energetic state. (Chapter 5)

energy-level diagram A diagram showing the arrangement of an atom's energy levels. (Chapter 5)

epicenter The location on Earth's surface directly over the focus of an earthquake. (Chapter 9)

epicycle A moving circle in the Ptolemaic system about which a planet revolves. (Chapter 4)

epoch The date used to define the coordinate system for objects on the sky. (Chapter 2)

equal areas, law of See *Kepler's second law.*

equinox One of the intersections of the ecliptic and the celestial equator. Also used to refer to the date on which the Sun passes through such an intersection. (Chapter 2) See also *autumnal equinox* and *vernal equinox.*

equivalence principle In the general theory of relativity, the principle that in a small volume of space, the downward pull of gravity can be accurately and completely duplicated by an upward acceleration of the observer. (Chapter 21)

era of recombination The moment, approximately 380,000 years after the Big Bang, when the universe had cooled sufficiently to permit the formation of hydrogen atoms. (Chapter 25)

ergoregion The region of space immediately outside the event horizon of a rotating black hole where it is impossible to remain at rest. (Chapter 21)

escape speed The speed needed by an object (such as a spaceship) to leave a second object (such as a planet or star) permanently and to escape into interplanetary space. (Chapter 7)

eV See *electron volt.*

event horizon The location around a black hole where the escape speed equals the speed of light; the "surface" of a black hole. (Chapter 21)

evolutionary track The path across an H-R diagram that a star follows as it evolves. (Chapter 18)

excited state A state of an atom, ion, or molecule with a higher energy than the ground state. (Chapter 5)

exoplanet (extrasolar planet) A planet orbiting a star other than our Sun. (Chapter 8)

exotic dark matter Dark matter that is not made up of regular matter. Regular matter is made of protons, neutrons, and electrons, including all atoms and molecules. There is evidence that the vast majority of matter is not made of regular matter, but little else is known about what it might be. (Chapter 23)

exponent A number placed above and after another number to denote the power to which the latter is to be raised, as n in 10^n. (Chapter 1)

extinction (interstellar) See *interstellar extinction.*

extrasolar planet See *exoplanet.*

extremophile Microorganisms that live in extreme conditions such as hot or cold temperatures, high acidity, or high radioactivity. (Chapter 27)

eyepiece lens A magnifying lens used to view the image produced at the focus of a telescope. (Chapter 6)

F ring A thin, faint ring encircling Saturn just beyond the A ring. (Chapter 12)

false color In astronomical images, color used to denote different values of intensity, temperature, or other quantities. (Chapter 6)

false vacuum Empty space with an abnormally high energy density that is thought to have filled the universe immediately after the Big Bang. (Chapter 26)

far side (of the Moon) The side of the Moon that faces perpetually away from Earth. (Chapter 10)

far-infrared wavelengths The part of the infrared spectrum most different in wavelength from visible light. (Chapter 22)

filament A portion of the Sun's chromosphere that arches to high altitudes. (Chapter 16)

first quarter moon The phase of the Moon that occurs when the Moon is 90° east of the Sun. (Chapter 3)

fission theory The hypothesis that the Moon was pulled out of a rapidly rotating proto-Earth. (Chapter 10)

flare See *solar flare.*

flat space Space that is not curved; space with zero curvature. (Chapter 25)

flatness problem The dilemma posed by the fact that the combined average mass density of the universe is very nearly equal to the critical density. (Chapter 26)

flocculent spiral galaxy A spiral galaxy with fuzzy, poorly defined spiral arms. (Chapter 22)

fluorescence A process in which high-energy ultraviolet photons are absorbed and the absorbed energy is radiated as lower-energy photons of visible light. (Chapter 18)

focal length The distance from a lens or mirror to the point where converging light rays meet. (Chapter 6)

focal plane The plane in which a lens or mirror forms an image of a distant object. (Chapter 6)

focal point The point at which a lens or mirror forms an image of a distant point of light. (Chapter 6)

focus (of a lens or mirror) The point to which light rays converge after passing through a lens or being reflected from a mirror. (Chapter 6)

focus (of an ellipse) (*plural* foci) One of two points inside an ellipse such that the combined distance from the two foci to any point on the ellipse is a constant. (Chapter 4)

force A push or pull that acts on an object. (Chapter 4)

frequency The number of crests or troughs of a wave that cross a given point per unit time; also, the number of vibrations per unit time. (Chapter 5)

full moon A phase of the Moon during which its fully lit hemisphere can be seen from Earth. (Chapter 3)

fundamental plane A relationship among the size of an elliptical galaxy, the average motions of its stars, and how the galaxy's brightness appears distributed over its surface. (Chapter 23)

fusion crust The coating on a stony meteorite caused by the heating of the meteorite as it descended through Earth's atmosphere. (Chapter 15)

G ring A thin, faint ring encircling Saturn. (Chapter 12)

galactic cannibalism A collision between two galaxies of unequal mass and size in which the smaller galaxy seems to be absorbed by the larger galaxy. (Chapter 23)

galactic cluster See *open cluster.*

galactic nucleus The center of a galaxy. (Chapter 22)

galaxy A large assemblage of stars, nebulae, and interstellar gas and dust. (Chapters 1, 22)

Galilean satellites The four large moons of Jupiter. (Chapter 13)

gamma rays The most energetic form of electromagnetic radiation. (Chapter 5)

gamma-ray bursters Objects found in all parts of the sky that emit a one-time intense burst of high-energy radiation. (Chapter 21)

general theory of relativity A description of gravity formulated by Albert Einstein. It states that gravity affects the geometry of space and the flow of time. (Chapter 21)

geocentric model An Earth-centered theory of the universe. (Chapter 4)

giant (star) A star whose diameter is typically 10 to 100 times that of the Sun and whose luminosity is roughly that of 100 Suns. (Chapter 17)

giant elliptical galaxy A large, massive, elliptical galaxy containing many billions of stars. (Chapter 23)

giant molecular cloud A large cloud of interstellar gas and dust in which temperatures are low enough and densities high enough for atoms to form into molecules. (Chapter 18)

global warming The upward trend of Earth's average temperature caused by increased amounts of greenhouse gases in the atmosphere. (Chapter 9)

globular cluster A large spherical cluster of stars, typically found in the outlying regions of a galaxy. (Chapters 19, 22)

gluon A particle that is exchanged between quarks. (Chapter 26)

Grand Tack model A model in which the jovian planets migrate toward the Sun in the first few million years of the solar system and then move back outward. This model can help explain Mars's small mass and icy objects in the asteroid belt. (Chapter 8)

grand unified theory (GUT) A theory that describes and explains the four physical forces. (Chapter 26)

grand-design spiral galaxy A galaxy with well-defined spiral arms. (Chapter 22)

granulation The rice grain-like structure found in the solar photosphere. (Chapter 16)

granule A convective cell in the solar photosphere. (Chapter 16)

grating See *diffraction grating.*

gravitational force See *gravity.*

gravitational lens A massive object that deflects light rays from a remote source, forming an image much as an ordinary lens does. (Chapter 23)

gravitational potential energy Energy associated with Earth's pull on an object toward Earth's center. This energy increases with an object's distance from the center of Earth. (Chapter 4)

gravitational radiation (gravitational waves) Oscillations of space produced by changes in the distribution of matter. (Chapter 21)

gravitational redshift The increase in the wavelength of a photon as it climbs upward in a gravitational field. (Chapter 21)

graviton The particle that is theorized to be responsible for the gravitational force. (Chapter 26)

gravity The force with which all matter attracts all other matter. (Chapter 4)

Great Attractor An apparent region of significant gravitational attraction in intergalactic space at the center of the Local Supercluster. (Chapter 23)

Great Red Spot A prominent high-pressure system in Jupiter's southern hemisphere. (Chapter 12)

greatest eastern elongation The configuration of an inferior planet at its greatest angular distance east of the Sun. (Chapters 4, 11)

greatest western elongation The configuration of an inferior planet at its greatest angular distance west of the Sun. (Chapters 4, 11)

greenhouse effect The trapping of infrared radiation near a planet's surface by the planet's atmosphere. (Chapter 9)

greenhouse gas A substance whose presence in a planet's atmosphere enhances the greenhouse effect. (Chapter 9)

ground state The state of an atom, ion, or molecule with the least possible energy. (Chapter 5)

group (of galaxies) A poor cluster of galaxies. (Chapter 23)

GUT See *grand unified theory.*

H I Neutral, unionized hydrogen. (Chapter 22)

H II region A region of ionized hydrogen in interstellar space. (Chapter 18)

habitable zone Regions of a galaxy or around a star in which conditions may be suitable for life to have developed. (Chapters 8, 27)

half-life The time required for one-half of a quantity of a radioactive substance to decay. (Chapter 8)

halo (of a galaxy) A spherical distribution of globular clusters and Population II stars that surround a spiral galaxy. (Chapter 22)

Hawking radiation Light and particles made of matter and antimatter that emanate away from a region just outside a black hole's event horizon. The radiation occurs when one member of a virtual particle pair is captured by the black hole; the other member of the pair becomes a real particle and carries away some of the black hole's energy (and therefore, some of its mass). This loss of energy leads to black hole evaporation. (Chapter 21)

Heisenberg uncertainty principle A principle of quantum mechanics that places limits on the precision of simultaneous measurements. (Chapters 21, 26)

heliocentric model A Sun-centered theory of the universe. (Chapter 4)

helioseismology The study of the vibrations of the Sun as a whole. (Chapter 16)

helium flash The nearly explosive beginning of helium fusion in the dense core of a red giant star. (Chapter 19)

helium fusion The thermonuclear fusion of helium to form carbon and oxygen. (Chapter 19)

helium shell flash A brief thermal runaway that occurs in the helium-fusing shell of a red supergiant. (Chapter 20)

Herbig–Haro object A small, luminous nebula associated with the end point of a jet emanating from a young star. (Chapter 18)

Hertzsprung–Russell (H-R) diagram A plot of the luminosity (or absolute magnitude) of stars against their surface temperature (or spectral type). (Chapter 17)

highlands (on Mars) See *southern highlands*.

highlands (on the Moon) See *lunar highlands*.

Hirayama family A group of asteroids that have nearly identical orbits about the Sun. (Chapter 15)

homogeneous Having the same property in one region as in every other region. (Chapter 25)

horizon problem See *isotropy problem*.

horizontal branch A group of stars on the color-magnitude diagram of a typical globular cluster that lies near the main sequence and has roughly constant luminosity. (Chapter 20)

horizontal-branch star A low-mass, post–helium-flash star on the horizontal branch. (Chapter 19)

hot dark matter Dark matter consisting of low-mass particles moving at high speeds. (Chapter 26)

hot Jupiter A Jupiter-sized planet orbiting a star other than our Sun with an orbital distance similar to Mercury's, and thus a very hot planet. (Chapter 8)

hot-spot volcanism Volcanic activity that occurs over a hot region buried deep within a planet. (Chapter 11)

H-R diagram See *Hertzsprung–Russell (H-R) diagram*.

Hubble classification A method of classifying galaxies as spirals, barred spirals, ellipticals, or irregulars according to their appearance. (Chapter 23)

Hubble constant (H_0) In the Hubble law, the constant of proportionality between the recessional velocities of remote galaxies and their distances. (Chapter 23)

Hubble flow The recessional motions of remote galaxies caused by the expansion of the universe. (Chapter 23)

Hubble law The empirical relationship stating that the redshifts of remote galaxies are directly proportional to their distances from Earth. (Chapter 23)

Hubble time A fairly accurate estimate of the age of the universe. It is equal to the inverse of the Hubble constant. This age estimate makes a simplifying assumption that the expansion of the universe has occurred at precisely the same rate since the Big Bang, but the expansion rate was slightly different in the past. (Chapter 25)

hydrocarbon Any one of a variety of chemical compounds composed of hydrogen and carbon. (Chapter 13)

hydrogen envelope A huge, tenuous sphere of gas surrounding the head of a comet. (Chapter 15)

hydrogen fusion The thermonuclear conversion of hydrogen into helium. (Chapter 16)

hydrostatic equilibrium A balance between the weight of a layer in a star and the pressure that supports it. (Chapter 16)

hyperbola A conic section formed by cutting a circular cone with a plane at an angle steeper than the sides of the cone. (Chapter 4)

hyperbolic space Space with negative curvature. (Chapter 25)

hypothesis An idea or collection of ideas that seems to explain a specified phenomenon; a conjecture. (Chapter 1)

ice rafts (on Europa) Segments of Europa's icy crust that have been moved by tectonic disturbances. (Chapter 13)

ices Solid materials with low condensation temperatures, including ices of water, methane, and ammonia. (Chapter 7)

ideal gas A gas in which the pressure is directly proportional to both the density and the temperature of the gas; an idealization of a real gas. (Chapter 19)

igneous rock A rock that formed from the solidification of molten lava or magma. (Chapter 9)

imaging The process of recording the image made by a telescope of a distant object. (Chapter 6)

impact breccia A type of rock formed from other rocks that were broken apart, mixed, and fused together by a series of meteoritic impacts. (Chapter 10)

impact crater A circular depression on a planet or satellite caused by the impact of a meteoroid. (Chapters 7, 10)

induced electric current Electric current resulting from a magnetic field sweeping over an electrically conductive material. (Chapter 13)

induced magnetic field A magnetic field resulting from an induced electric current. (Chapter 13)

inertia, law of See *Newton's first law of motion*.

inferior conjunction The configuration when an inferior planet is between the Sun and Earth. (Chapter 4)

inferior planet A planet that is closer to the Sun than Earth is. (Chapter 4)

inflation The expansion of space. (Chapter 26)

inflationary epoch A brief period shortly after the Big Bang during which the scale of the universe increased very rapidly. (Chapter 26)

infrared radiation Electromagnetic radiation of wavelength longer than visible light but shorter than radio waves. (Chapter 5)

inner core (of Earth) The solid innermost portion of Earth's iron-rich core. (Chapter 9)

inner Lagrangian point The point between the two stars constituting a binary system where their Roche lobes touch; the point across which mass transfer can occur. (Chapter 19)

instability strip A region of the H-R diagram occupied by pulsating stars. (Chapter 19)

interferometry A technique that combines the observations of two or more telescopes to produce images better than one telescope alone could make. (Chapter 6)

intermediate vector boson The particle that is responsible for the weak force. (Chapter 26)

intermediate-mass black hole See *mid-mass black hole*.

interstellar extinction The dimming of starlight as it passes through the interstellar medium. (Chapters 18, 22)

interstellar matter Diffuse gas and dust spread out in the galaxy between the stars. Interstellar matter is mostly hydrogen. (Chapter 22)

interstellar medium Gas and dust in interstellar space. (Chapters 8, 18)

interstellar reddening The reddening of starlight passing through the interstellar medium as a result of blue light being scattered more than red. (Chapter 18)

intracluster gas Gas found between the galaxies within a cluster of galaxies. (Chapter 23)

inverse-square law The statement that the apparent brightness of a light source varies inversely with the square of the distance from the source. (Chapter 17)

Io torus A doughnut-shaped ring of gas circling Jupiter at the distance of Io's orbit. (Chapter 13)

ion tail The relatively straight tail of a comet produced by the solar wind acting on ions. (Chapter 15)

ionization The process by which a neutral atom becomes an electrically charged ion through the loss or gain of electrons. (Chapter 5)

iron meteorite A meteorite composed primarily of iron. (Chapter 15)

irregular cluster (of galaxies) A sprawling collection of galaxies whose overall distribution in space does not exhibit any noticeable spherical symmetry. (Chapter 23)

irregular galaxy An asymmetrical galaxy having neither spiral arms nor an elliptical shape. (Chapter 23)

isotope Any of several forms for the same chemical element whose nuclei all have the same number of protons but different numbers of neutrons. (Chapter 5)

isotropic Having the same property in all directions. (Chapters 23, 25)

isotropy problem The dilemma posed by the fact that the cosmic microwave background is isotropic; also called the *horizon problem*. (Chapter 26)

Jeans length The smallest scale over which a density fluctuation in a medium will contract to form a gravitationally bound object. (Chapter 26)

joule (J) A unit of energy. (Chapter 5)

jovian planet Low-density planet composed primarily of hydrogen and helium; the jovian planets include Jupiter, Saturn, Uranus, and Neptune. (Chapter 7)

Jupiter-family comet A comet with an orbital period of less than 20 years. (Chapter 15)

kelvin (K) A unit of temperature on the Kelvin temperature scale, equivalent to 1°C. (Chapter 5)

Kelvin–Helmholtz contraction The contraction of a gaseous body, such as a star or nebula, during which gravitational energy is transformed into thermal energy. (Chapter 8)

Kepler's first law The statement that each planet moves around the Sun in an elliptical orbit with the Sun at one focus of the ellipse. (Chapter 4)

Kepler's second law The statement that a planet sweeps out equal areas in equal times as it orbits the Sun; also called the *law of equal areas*. (Chapter 4)

Kepler's third law A relationship between the period of an orbiting object and the semimajor axis of its elliptical orbit. (Chapter 4)

kiloparsec (kpc) One thousand parsecs (pc); about 3260 light-years (ly). (Chapter 1)

kinetic energy The energy possessed by an object because of its motion. (Chapters 4, 7)

Kirchhoff's laws Three statements about circumstances that produce absorption lines, emission lines, and continuous spectra. (Chapter 5)

Kirkwood gaps Gaps in the spacing of asteroid orbits, discovered by Daniel Kirkwood. (Chapter 15)

Kuiper belt A region that extends from around the orbit of Pluto to about 500 au from the Sun where many icy objects orbit the Sun. (Chapters 7, 14, 15)

Lamb shift A tiny shift in the spectral lines of hydrogen caused by the presence of virtual particles. (Chapter 26)

Late Heavy Bombardment A brief, intense period of large impacts on all bodies of the solar system theorized to have happened about 600 million years after formation of the solar system. (Chapter 8)

lava Molten rock flowing on the surface of a planet. (Chapter 9)

law of equal areas See *Kepler's second law.*

law of inertia See *Newton's first law of motion.*

law of universal gravitation A formula deduced by Isaac Newton that expresses the strength of the force of gravity that two masses exert on each other. (Chapter 4)

laws of physics A set of physical principles with which we can understand natural phenomena and the nature of the universe. (Chapter 1)

length contraction In the special theory of relativity, the shrinking of an object's length along its direction of motion. (Chapter 21)

lens A piece of transparent material (usually glass) that can bend light and bring it to a focus. (Chapter 6)

lenticular galaxy A galaxy with a central bulge and a disk but no spiral structure; an S0 galaxy. (Chapter 23)

libration An apparent rocking of the Moon whereby an Earth-based observer can, over time, see slightly more than one-half the Moon's surface. (Chapter 10)

light curve A graph that displays how the brightness of a star or other astronomical object varies over time. (Chapter 17)

light pollution Light from cities and towns that degrades telescope images. (Chapter 6)

light scattering The process by which light bounces off particles in its path. (Chapters 5, 12)

light-gathering power A measure of the amount of radiation brought to a focus by a telescope. (Chapter 6)

light-year (ly) The distance light travels in a vacuum in 1 year. (Chapter 1)

limb darkening The phenomenon whereby the Sun looks darker near its apparent edge, or limb, than near the center of its disk. (Chapter 16)

line of nodes The line where the plane of Earth's orbit intersects the plane of the Moon's orbit. (Chapter 3)

liquid metallic hydrogen Hydrogen compressed to such a density that it behaves like a liquid metal. (Chapters 7, 12)

lithosphere The solid, upper layer of Earth; essentially Earth's crust. (Chapter 9)

Local Bubble A large cavity in the interstellar medium within which the Sun and nearby stars are located. (Chapter 22)

Local Group The cluster of galaxies of which our galaxy is a member. (Chapter 23)

local meridian See *meridian.*

Local Supercluster A large group of galaxies that contains the Local Group, within which lies our Milky Way Galaxy. (Chapter 23)

long-period comet A comet that takes hundreds of thousands of years or more to complete one orbit of the Sun. (Chapter 15)

long-period variable A variable star with a period longer than about 100 days. (Chapter 19)

lookback time How far into the past we are looking when we see a particular object. (Chapter 25)

Lorentz transformations Equations that relate the measurements of different observers who are moving relative to each other at high speeds. (Chapter 21)

lower meridian The half of the meridian that lies below the horizon. (Chapter 2)

lowlands (on Mars) See *northern lowlands*.

luminosity The rate at which electromagnetic radiation is emitted from a star or other object. (Chapters 5, 16, 17)

luminosity class A classification of a star of a given spectral type according to its luminosity. (Chapter 17)

luminosity function The numbers of stars of differing brightness per cubic parsec. (Chapter 17)

lunar eclipse An eclipse of the Moon by Earth; a passage of the Moon through Earth's shadow. (Chapter 3)

lunar highlands Ancient, high-elevation, heavily cratered terrain on the Moon. (Chapter 10)

lunar month See *synodic month*.

lunar phase The appearance of the illuminated area of the Moon as seen from Earth. (Chapter 3)

Lyman series A series of spectral lines of hydrogen produced by electron transitions to and from the lowest energy state of the hydrogen atom. (Chapter 5)

MACHO See *massive compact halo object*.

magma Molten rock beneath a planet's surface. (Chapter 9)

magnetic axis A line connecting the north and south magnetic poles of a planet or star possessing a magnetic field. (Chapter 14)

magnetic dynamo The mechanism whereby electric currents within an astronomical body generate a magnetic field. Sometimes simply referred to as a *dynamo*. (Chapter 9)

magnetic reconnection An event where two oppositely directed magnetic fields approach and cancel, thus releasing energy. (Chapter 16)

magnetic resonance imaging (MRI) A technique for viewing the interior of the human body that makes use of the way protons respond to magnetic fields. (Chapter 22)

magnetic-dynamo model A theory that explains the solar cycle as a result of the Sun's differential rotation acting on the Sun's magnetic field. (Chapter 16)

magnetogram An image of the Sun that shows regions of different magnetic polarity. (Chapter 16)

magnetometer A device for measuring magnetic fields. (Chapter 7)

magnetopause That region of a planet's magnetosphere where the magnetic field counterbalances the pressure from the solar wind. (Chapter 9)

magnetosphere The region around a planet occupied by its magnetic field. (Chapter 9)

magnification The factor by which the apparent angular size of an object is increased when viewed through a telescope. (Chapter 6)

magnifying power See *magnification*.

magnitude scale A system for denoting the brightnesses of astronomical objects. (Chapter 17)

main sequence A grouping of stars on the Hertzsprung–Russell (H-R) diagram extending diagonally across the graph from hot, luminous stars to cool, dim stars. (Chapter 17)

main-sequence lifetime The total time that a star spends fusing hydrogen in its core, and hence the total time that it will spend as a main-sequence star. (Chapter 19)

main-sequence star A star whose luminosity and surface temperature place it on the main sequence on an H-R diagram; a star that derives its energy from core hydrogen fusion. (Chapter 17)

major axis (of an ellipse) The longest diameter of an ellipse. (Chapter 4)

mantle (of a planet) That portion of a terrestrial planet located between its crust and core. (Chapter 9)

mare (*plural* maria) Latin for "sea"; a large, relatively crater-free plain on the Moon. (Chapter 10)

mare basalt A type of lunar rock commonly found in the mare basins. (Chapter 10)

maser An interstellar cloud in which water molecules emit intensely at microwave wavelengths. (Chapter 23)

mass A measure of the total amount of material in an object. (Chapter 4)

mass density of radiation The energy possessed by a radiation field per unit volume divided by the square of the speed of light. (Chapter 25)

mass loss A process by which a star gently loses matter. (Chapter 19)

mass transfer The flow of gases from one star in a binary system to the other. (Chapter 19)

massive compact halo object (MACHO) A dim star or low-mass black hole that may constitute part of the unseen dark matter. (Chapter 22)

mass–luminosity relation A relationship between the masses and luminosities of main-sequence stars. (Chapter 17)

mass–radius relation A relationship between the masses and radii of white dwarf stars. (Chapter 20)

matter density parameter (V_m) The ratio of the average density of matter in the universe to the critical density. (Chapter 25)

matter-dominated universe A universe in which the average density of matter exceeds both the mass density of radiation and the mass density of dark energy. (Chapter 25)

mean solar day The interval between successive meridian passages of the mean Sun; the average length of a solar day. (Chapter 2)

mean sun A fictitious object that moves eastward at a constant speed along the celestial equator, completing one circuit of the sky with respect to the vernal equinox in one tropical year. (Chapter 2)

medium (*plural* media) A material through which light travels. (Chapter 6)

medium, interstellar See *interstellar medium*.

megaparsec (Mpc) One million parsecs. (Chapter 1)

melting point The temperature at which a substance changes from solid to liquid. (Chapter 9)

meridian (or **local meridian**) The great circle on the celestial sphere that passes through an observer's zenith and the north and south celestial poles. (Chapter 2)

meridian transit The crossing of the meridian by any astronomical object. (Chapter 2)

mesosphere A layer in Earth's atmosphere above the stratosphere. (Chapter 9)

metal In reference to stars and galaxies, any element other than hydrogen and helium; in reference to planets, a material such as iron, silver, and aluminum that is a good conductor of electricity and heat. (Chapter 17)

metal-poor star A star that, compared to the Sun, is deficient in elements heavier than helium; also called a *Population II star*. (Chapter 19)

metal-rich star A star whose abundance of heavy elements is roughly comparable to that of the Sun; also called a *Population I star*. (Chapter 19)

metamorphic rock A rock whose properties and appearance have been transformed by the action of pressure and heat beneath Earth's surface. (Chapter 9)

meteor The luminous phenomenon seen when a meteoroid enters Earth's atmosphere; a "shooting star." (Chapter 15)

meteor shower An event wherein many meteors seem to radiate from a common point in the sky. (Chapter 15)

meteorite A fragment of a meteoroid that has survived passage through Earth's atmosphere. (Chapters 1, 8, 15)

meteoritic swarm A collection of meteoroids moving together along an orbit about the Sun. (Chapter 15)

meteoroid A small rock in interplanetary space. (Chapters 7, 15)

microlensing A phenomenon in which a compact object such as a MACHO acts as a gravitational lens, focusing the light from a distant star. (Chapter 22)

microwaves Short-wavelength radio waves. (Chapter 5)

mid-mass black hole A black hole with a mass of hundreds of Suns; also referred to as an *intermediate-mass black hole*. (Chapter 21)

migration Changes in a planet's average orbital distance from the Sun. (Chapter 8)

Milky Way Galaxy Our galaxy; the band of faint stars seen from Earth in the plane of our galaxy's disk. (Chapter 22)

mineral A naturally occurring solid composed of a single element or chemical combination of elements, often in the form of crystals. (Chapter 9)

minor planet See *asteroid*.

minute of arc See *arcminute*.

model (1) A hypothesis that has withstood experimental or observational tests. (2) The results of a theoretical calculation that gives the values of temperature, pressure, density, and so forth throughout the interior of an object such as a planet or star. (Chapter 1)

molecule A combination of two or more atoms. (Chapter 5)

moonquake Sudden, vibratory motion of the Moon's surface. (Chapter 10)

MRI See *magnetic resonance imaging*.

M-theory A theory that attempts to describe fundamental particles as 11-dimensional membranes. (Chapter 26)

nanometer (nm) One billionth of a meter: $1 \text{ nm} = 10^{-9} \text{ m} = 10^{-6} \text{ mm} = 10^{-3} \text{ } \mu\text{m}$. (Chapter 5)

neap tide An ocean tide that occurs when the Moon is near first-quarter or third-quarter phase. (Chapter 4)

near-Earth object (NEO) An asteroid whose orbit lies wholly or partly within the orbit of Mars. (Chapter 15)

near-infrared wavelengths The part of the infrared spectrum closest in wavelength to visible light. (Chapter 22)

nebula (*plural* **nebulae**) A cloud of interstellar gas and dust. (Chapters 1, 18)

nebular hypothesis The idea that the Sun and the rest of the solar system formed from a cloud of interstellar material. (Chapter 8)

nebulosity See *nebula*.

negative curvature The curvature of a surface or space in which parallel lines diverge and the sum of the angles of a triangle is less than 180°. (Chapter 25)

negative hydrogen ion A hydrogen atom that has acquired a second electron. (Chapter 16)

NEO See *near-Earth object*.

neon fusion The thermonuclear fusion of neon to produce heavier nuclei. (Chapter 20)

neutrino A subatomic particle with no electric charge and very little mass, yet one that is important in many nuclear reactions. (Chapter 16)

neutrino oscillation The spontaneous transformation of one type of neutrino into another type. (Chapter 16)

neutron A subatomic particle with no electric charge and with a mass nearly equal to that of the proton. (Chapter 5)

neutron capture The buildup of neutrons inside nuclei. (Chapter 20)

neutron star A very compact, dense stellar remnant composed almost entirely of neutrons. (Chapter 21)

new moon The phase of the Moon when the dark hemisphere of the Moon faces Earth. (Chapter 3)

Newton's first law of motion The statement that a body remains at rest, or moves in a straight line at a constant speed, unless acted upon by a net outside force; the law of inertia. (Chapter 4)

Newton's form of Kepler's third law A relationship among the period of two objects orbiting each other, the semimajor axis of their orbit, and the masses of the objects. (Chapter 4)

Newton's second law of motion A relationship among the acceleration of an object, the object's mass, and the net outside force acting on the mass. (Chapter 4)

Newton's third law of motion The statement that whenever one body exerts a force on a second body, the second body exerts an equal and opposite force on the first body. (Chapter 4)

Newtonian mechanics The branch of physics based on Newton's laws of motion. (Chapters 1, 4)

Newtonian reflector A reflecting telescope that uses a small mirror to deflect the image to one side of the telescope tube. (Chapter 6)

Nice model A model in which the jovian planets began in a more compact configuration in the early solar system and then migrated outward. This model can explain aspects of the Kuiper belt, the Oort cloud, and the Late Heavy Bombardment. (Chapter 8)

noble gas An element whose atoms do not combine into molecules. (Chapter 12)

node See *line of nodes*.

no-hair theorem A statement of the simplicity of black holes. (Chapter 21)

north celestial pole The point directly above Earth's north pole where Earth's axis of rotation, if extended, would intersect the celestial sphere. (Chapter 2)

northern lights See *aurora borealis*.

northern lowlands (on Mars) Relatively young and crater-free terrain in the Martian northern hemisphere. (Chapter 11)

nova (*plural* **novae**) A star that experiences a sudden outburst of radiant energy, temporarily increasing its luminosity roughly a thousandfold. (Chapter 20)

nuclear density The density of matter in an atomic nucleus; about 4×10^{17} kg/m^3. (Chapter 20)

nuclear force The force that holds protons and neutrons together in an atom. It arises from the strong force and is strong enough to hold protons closely together despite their electrical repulsion. (Chapter 26)

nucleosynthesis The process of building up nuclei such as deuterium and helium from protons and neutrons. (Chapter 26)

nucleus (of an atom) The massive part of an atom, composed of protons and neutrons, about which electrons revolve. (Chapter 5)

nucleus (of a comet) A collection of ices and dust that constitute the solid part of a comet. (Chapter 15)

nucleus (of a galaxy) See *galactic nucleus*.

OB association A grouping of hot, young, massive stars, predominantly of spectral types O and B. (Chapter 18)

OBAFGKM The temperature sequence (from hot to cold) of spectral classes. (Chapter 17)

objective lens The principal lens of a refracting telescope. (Chapter 6)

objective mirror The principal mirror of a reflecting telescope. (Chapter 6)

oblate Flattened at the poles. (Chapter 12)

oblateness A measure of how much a flattened sphere (or spheroid) differs from a perfect sphere. (Chapter 12)

observable universe The portion of the universe inside our cosmic particle horizon. (Chapters 25, 26)

Occam's razor The notion that a straightforward explanation of a phenomenon is more likely to be correct than a convoluted one. (Chapter 4)

occultation The eclipsing of an astronomical object by the Moon or a planet. (Chapters 13, 14)

oceanic rift A crack in the ocean floor that exudes lava. (Chapter 9)

Olbers's paradox The dilemma associated with the fact that the night sky is dark despite the multitude of stars emitting light in the universe. (Chapter 25)

1-to-1 spin-orbit coupling See *synchronous rotation*.

Oort cloud A presumed accumulation of comets and cometary material surrounding the Sun at distances of roughly 50,000 au. (Chapters 7, 15)

opaque When light cannot travel through an object without being scattered many times or absorbed. Unlike glass or our atmosphere, an opaque material is not transparent. (Chapter 5)

open cluster A loose association of young stars in the disk of our galaxy; a galactic cluster. (Chapters 18, 19)

open universe A universe with negative curvature, so that its geometry is analogous to a saddle-shaped hyperbolic surface. (Chapter 25)

opposition The configuration of a planet when it is at an elongation of 180° and thus appears opposite the Sun in the sky. (Chapter 4)

optical double star Two stars that lie along nearly the same line of sight but are actually at very different distances from us. (Chapter 17)

optical telescope A telescope designed to detect visible light. (Chapter 6)

optical window The range of visible wavelengths to which Earth's atmosphere is transparent. (Chapter 6)

orbit The trajectory followed by an object revolving around another object under the influence of gravity. (Chapter 8)

orbital energy The net energy of an orbiting object due to its kinetic and gravitational potential energy. The farther a planet orbits from the Sun, the greater its orbital energy. (Chapter 4)

orbital resonance The resonance that occurs whenever the orbital periods of two objects are related by a ratio of small integers—for example, when one moon has twice the orbital period of another. The rhythmic gravitational interactions in an orbital resonance can have either a stabilizing or a destabilizing effect. (Chapters 12, 13, 15)

organic molecules Molecules containing carbon, some of which are the molecules of which living organisms are made. (Chapter 27)

outer core (of Earth) The outer, molten portion of Earth's iron-rich core. (Chapter 9)

outgassing The release of gases into a planet's atmosphere by volcanic activity. (Chapter 9)

overcontact binary A close binary system in which the two stars share a common atmosphere. (Chapter 19)

oxygen fusion The thermonuclear fusion of oxygen to produce heavier nuclei. (Chapter 20)

ozone A type of oxygen whose molecules contain three oxygen atoms. (Chapter 9)

ozone hole A region of Earth's atmosphere over Antarctica, where the concentration of ozone is abnormally low. (Chapter 9)

ozone layer A layer in Earth's upper atmosphere, where the concentration of ozone is high enough to prevent much ultraviolet light from reaching the surface. (Chapter 9)

P wave One of three kinds of seismic waves produced by an earthquake; a primary, or pressure, wave. (Chapter 9)

pair production The creation of a particle and its antiparticle from energy. (Chapter 26)

parabola A conic section formed by cutting a circular cone at an angle parallel to one of the sides of the cone. (Chapter 4)

parallax The apparent displacement of an object due to the motion of the observer. (Chapters 4, 17)

parsec (pc) A unit of distance; 3.26 light-years (ly). (Chapters 1, 17)

partial lunar eclipse A lunar eclipse in which the Moon does not appear completely covered. (Chapter 3)

partial solar eclipse A solar eclipse in which the Sun does not appear completely covered. (Chapter 3)

Paschen series A series of spectral lines of hydrogen produced by electron transitions between the third and higher energy levels. (Chapter 5)

Pauli exclusion principle A principle of quantum mechanics stating that no two electrons can have the same position and momentum. (Chapter 19)

penumbra (of a shadow) (*plural* **penumbrae**) The portion of a shadow in which only part of the light source is covered by an opaque body. (Chapter 3)

penumbral eclipse A lunar eclipse in which the Moon passes only through Earth's penumbra. (Chapter 3)

perigee The point in its orbit where a satellite or the Moon is nearest Earth. (Chapter 3)

perihelion The point in its orbit where a planet or comet is nearest the Sun. (Chapter 4)

period (of a planet) The interval of time between successive geometric arrangements of a planet and an astronomical object, such as the Sun. (Chapter 4)

periodic table A listing of the chemical elements according to their properties, invented by Dmitri Mendeleev. (Chapter 5)

period–luminosity relation A relationship between the period and average density of a pulsating star. (Chapter 18)

photodisintegration The breakup of nuclei by high-energy gamma rays. (Chapter 20)

photometry The measurement of light intensities. (Chapters 6, 17)

photon A discrete unit of electromagnetic energy. (Chapter 5)

photosphere The region in the solar atmosphere from which most of the visible light escapes into space. (Chapter 16)

photosynthesis A biochemical process in which solar energy is converted into chemical energy, carbon dioxide and water are absorbed, and oxygen is released. (Chapter 9)

physics, laws of See *laws of physics.*

pixel A picture element in a camera or digital display. (Chapter 6)

plage A bright region in the solar atmosphere as observed in the monochromatic light of a spectral line. (Chapter 16)

Planck time A fundamental interval of time defined by basic physical constants; the shortest possible time interval that can be measured, equal to 1.35×10^{-43} s. (Chapter 25)

Planck's law The relationship between the energy of a photon and its wavelength or frequency; $E = hc/\lambda = h\nu$. (Chapter 5)

planetary nebula A luminous shell of gas ejected from an old, low-mass star. (Chapter 20)

planetesimal One of many small bodies of primordial dust and ice that combined to form the planets. (Chapter 8)

plasma A hot ionized gas. (Chapters 13, 16, 25)

plastic The attribute of being nearly solid yet able to flow. (Chapter 9)

plate A large section of Earth's lithosphere that moves as a single unit. (Chapter 9)

plate tectonics The motions of large segments (plates) of Earth's surface over the underlying mantle. (Chapters 7, 9)

plutino One of about 100 objects in the Kuiper belt that orbit the Sun with nearly the same semimajor axis as Pluto. (Chapter 14)

poor cluster (of galaxies) A cluster of galaxies with very few members; a group of galaxies. (Chapter 23)

Population I star A star whose spectrum exhibits spectral lines of many elements heavier than helium; a metal-rich star. (Chapter 19)

Population II star A star whose spectrum exhibits comparatively few spectral lines of elements heavier than helium; a metal-poor star. (Chapter 19)

Population III star One of the first stars to form after the Big Bang, composed of only hydrogen, helium, and tiny amounts of lithium and beryllium; a "zeroth-generation" star, yet to be observed directly. (Chapter 26)

positional astronomy The study of the apparent positions of the planets and stars and how those positions change. (Chapter 2)

positive curvature The curvature of a surface or space in which parallel lines converge and the sum of the angles of a triangle is greater than 180°. (Chapter 25)

positron An electron with a positive rather than negative electric charge; the antiparticle of the electron. (Chapters 16, 26)

power of ten The exponent n in 10^n. (Chapter 1)

powers-of-ten notation A shorthand method of writing numbers, involving 10 followed by an exponent. (Chapter 1)

precession (of Earth) A slow, conical motion of Earth's axis of rotation caused by the gravitational pull of the Moon and Sun on Earth's equatorial bulge. (Chapter 2)

precession of the equinoxes The slow westward motion of the equinoxes along the ecliptic due to precession of Earth. (Chapter 2)

primary mirror See *objective mirror.*

prime focus The point in a telescope where the objective focuses light. (Chapter 6)

primitive asteroid See *undifferentiated asteroid.*

primordial black hole A type of black hole that may have formed in the very early universe. (Chapter 21)

primordial fireball The extremely hot gas that filled the universe immediately following the Big Bang. (Chapter 25)

principle of equivalence See *equivalence principle.*

progenitor star A star that later explodes into a supernova. (Chapter 20)

prograde orbit An orbit of a satellite around a planet that is in the same direction as the rotation of the planet. (Chapter 13)

prograde rotation A situation in which an object (such as a planet) rotates in the same direction that it orbits around another object (such as the Sun). (Chapter 11)

projection The projection of a point on Earth is made by an imaginary line from the point, extending perpendicular to the surface of Earth, until the line intersects the imaginary celestial sphere. (Chapter 2)

prominence Flamelike protrusions seen near the limb of the Sun and extending into the solar corona. (Chapter 16)

proper length (proper distance) A length measured by a ruler at rest with respect to an observer. (Chapter 21)

proper motion The angular rate of change in the location of a star on the celestial sphere, usually expressed in arcseconds per year. (Chapter 17)

proper time A time interval measured with a clock at rest with respect to an observer. (Chapter 21)

proplyd See *protoplanetary disk.*

proton A heavy, positively charged subatomic particle that is one of two principal constituents of atomic nuclei. (Chapter 5)

proton–proton chain A sequence of thermonuclear reactions by which hydrogen nuclei are built up into helium nuclei. (Chapter 16)

protoplanet A Moon-sized object formed by the coalescence of planetesimals. (Chapter 8)

protoplanetary disk (proplyd) A disk of material encircling a protostar or a newborn star. (Chapters 8, 18)

protostar A star in its earliest stages of formation. (Chapter 18)

protosun The part of the solar nebula that eventually developed into the Sun. (Chapter 8)

Ptolemaic system The definitive version of the geocentric cosmogony of ancient Greece. (Chapter 4)

pulsar A pulsating radio source thought to be associated with a rapidly rotating neutron star. (Chapters 1, 20)

pulsating variable star A star that pulsates in size and luminosity. (Chapter 19)

quantum electrodynamics A theory that describes details of how charged particles interact by exchanging photons. (Chapter 26)

quantum gravity A field of theoretical physics that seeks to describe situations that require both quantum mechanics and gravity. At present, there is no accepted theory of quantum gravity. (Chapter 26)

quantum mechanics The branch of physics dealing with the structure and behavior of atoms and their constituents as well as their interaction with light. (Chapters 5, 26)

quark One of several particles thought to be the internal constituents of certain heavy subatomic particles such as protons and neutrons. (Chapter 26)

quark confinement The permanent bonding of quarks to form particles like protons and neutrons. (Chapter 26)

quasar A very luminous object with a very large redshift and a starlike appearance. (Chapters 1, 24)

radial velocity That portion of an object's velocity parallel to the line of sight. (Chapters 5, 17)

radial velocity curve A plot showing the variation of radial velocity with time for a binary star or variable star. (Chapter 17)

radial velocity method A technique used to detect extrasolar planets by observing Doppler shifts in the spectrum of the planet's star. (Chapter 8)

radiation darkening The darkening of methane ice by electron impacts. (Chapter 14)

radiation pressure Pressure exerted on an object by radiation falling on the object. (Chapter 15)

radiation-dominated universe A universe in which the mass density of radiation exceeds both the average density of matter and the mass density of dark energy. (Chapter 25)

radiative diffusion The random migration of photons from a star's center toward its surface. (Chapter 16)

radiative zone A region within a star where radiative diffusion is the dominant mode of energy transport. (Chapter 16)

radio galaxy A galaxy that emits an unusually large amount of radio waves. (Chapter 24)

radio telescope A telescope designed to detect radio waves. (Chapter 6)

radio waves The longest-wavelength electromagnetic radiation. (Chapter 5)

radio window The range of radio wavelengths to which Earth's atmosphere is transparent. (Chapter 6)

radioactive dating A technique for determining the age of a rock sample by measuring the radioactive elements and their decay products in the sample. (Chapter 8)

radioactive decay The process whereby certain atomic nuclei spontaneously transform into other nuclei. (Chapter 8)

rarefaction A region in a sound wave where the medium carrying the wave is spread out or rarefied. (Chapter 25)

recombination The process in which an electron combines with a positively charged ion. (Chapter 18)

red dwarf A main-sequence star with a mass between about 0.08 M$_\odot$ and 0.4 M$_\odot$ that has a fully convective interior and that never goes through a red giant stage. (Chapter 19)

red giant A large, cool star of high luminosity. (Chapters 17, 19)

reddening (interstellar) See *interstellar reddening.*

red giant branch The region of an H-R diagram occupied by stars in their first red giant phase. (Chapter 20)

redshift The shifting to longer wavelengths of the light from remote galaxies and quasars; the Doppler shift of light from a receding source. (Chapters 5, 23)

reflecting telescope A telescope in which the principal optical component is a concave mirror. (Chapter 6)

reflection The return of light rays by a surface. (Chapter 6)

reflection nebula A comparatively dense cloud of dust in interstellar space that is illuminated by a star. (Chapter 18)

reflector A reflecting telescope. (Chapter 6)

refracting telescope A telescope in which the principal optical component is a lens. (Chapter 6)

refraction The bending of light rays when they pass from one transparent medium to another. (Chapter 6)

refractor A refracting telescope. (Chapter 6)

regolith The layer of rock fragments covering the surface of the Moon. (Chapter 10)

regular cluster (of galaxies) A spherical cluster of galaxies. (Chapter 23)

relativistic cosmology A cosmology based on the general theory of relativity. (Chapter 25)

residual polar cap An ice-covered polar region on Mars that does not completely evaporate during the Martian summer. (Chapter 11)

respiration A biological process that produces energy by consuming oxygen and releasing carbon dioxide. (Chapter 9)

retrograde motion The apparent westward motion of a planet with respect to background stars. (Chapter 4)

retrograde orbit An orbit of a satellite around a planet that is in the direction opposite to which the planet rotates. (Chapter 13)

retrograde rotation A situation in which an object (such as a planet) rotates in the direction opposite to which it orbits around another object (such as the Sun). (Chapter 11)

rich cluster (of galaxies) A cluster of galaxies containing many members. (Chapter 23)

ridge push A force on the oceanic plate that occurs when molten rock emerging from an oceanic rift is pulled by gravity down the ridge's slopes. This segment of ocean seafloor on the slope pushes on the larger seafloor plate and is one of the main forces driving plate tectonics. (Chapter 9)

right ascension A coordinate for measuring the east–west positions of objects on the celestial sphere. (Chapter 2)

ring particles Small particles that constitute a planetary ring. (Chapter 12)

ringlet One of many narrow bands of particles of which Saturn's ring system is composed. (Chapter 12)

Roche limit The smallest distance from a planet or other object at which a second object can orbit and still be held together by purely gravitational forces. (Chapter 12)

Roche lobe A teardrop-shaped volume surrounding a star in a binary system inside which gases are gravitationally bound to that star. (Chapter 19)

rock A mineral or combination of minerals. (Chapter 9)

rotation curve A plot of the orbital speeds of stars and nebulae in a galaxy versus distance from the center of the galaxy. (Chapter 22)

RR Lyrae variable A type of pulsating star with a period of less than 1 day. (Chapters 19, 22)

runaway greenhouse effect A greenhouse effect in which the temperature continues to increase due to a positive feedback cycle. (Chapter 11)

runaway icehouse effect A situation in which a decrease in atmospheric temperature causes a further decrease in temperature. (Chapter 11)

S wave One of three kinds of seismic waves produced by an earthquake; a secondary, or shear, wave. (Chapter 9)

Sagittarius A* A powerful radio source at the center of our galaxy. (Chapter 22)

saros A particular cycle of similar eclipses that recur about every 18 years. (Chapter 3)

scarp A line of cliffs formed by the faulting or fracturing of a planet's surface. (Chapter 11)

scattering of light See *light scattering.*

Schwarzschild radius (R_{Sch}) The distance from the singularity to the event horizon in a nonrotating black hole. (Chapter 21)

scientific method The basic procedure used by scientists to investigate phenomena. (Chapter 1)

seafloor spreading The separation of plates under the ocean due to lava emerging in an oceanic rift. (Chapter 9)

search for extraterrestrial intelligence (SETI) The scientific search for evidence of intelligent life on other planets. (Chapter 27)

second of arc See *arcsecond.*

sedimentary rock A rock that is formed from material deposited on land by rain or winds, or on the ocean floor. (Chapter 9)

seeing disk The angular diameter of a star's image. (Chapter 6)

seismic wave A vibration traveling through a terrestrial planet, usually associated with earthquake-like phenomena. (Chapter 9)

seismograph A device used to record and measure seismic waves, such as those produced by earthquakes. (Chapter 9)

selection effect The bias that occurs when an observational technique is more sensitive to one type of system than another. For example, the radial velocity method for detecting exoplanets can more easily detect large planets orbiting closely to their stars. This method is unlikely to detect many planets with Earth's characteristics, but it does not mean they are not there. (Chapter 8)

self-propagating star formation The process by which the formation of stars in one location in a galaxy stimulates the formation of stars in a neighboring location. (Chapter 22)

semidetached binary A binary star system in which one star fills its Roche lobe and the other does not. (Chapter 19)

semimajor axis One-half of the major axis of an ellipse. For an orbiting object, the semimajor axis is also equal to the average orbital distance. (Chapter 4)

SETI See *search for extraterrestrial intelligence.*

Seyfert galaxy A spiral galaxy with a bright nucleus whose spectrum exhibits emission lines. (Chapter 24)

shell helium fusion The thermonuclear fusion of helium in a shell surrounding a star's core. (Chapter 20)

shell hydrogen fusion The thermonuclear fusion of hydrogen in a shell surrounding a star's core. (Chapter 19)

shepherd satellite A satellite whose gravity restricts the motions of particles in a planetary ring, preventing them from dispersing. (Chapter 12)

shield volcano A volcano with long, gently sloping sides. (Chapter 11)

shock wave An abrupt, localized region of compressed gas caused by an object traveling through the gas at a speed greater than the speed of sound. (Chapter 10)

short-period comet A comet that originated in the Kuiper belt and has a period of less than 200 years. (Chapter 15)

SI units The International System of Units, based on the meter (m), the second (s), and the kilogram (kg). (Chapter 1)

sidereal clock A clock that measures sidereal time. (Chapter 2)

sidereal day The interval between successive meridian passages of the vernal equinox. (Chapter 2)

sidereal month The period of the Moon's revolution about Earth with respect to the stars. (Chapter 3)

sidereal period The orbital period of one object about another as measured with respect to the stars. (Chapter 4)

sidereal time Time reckoned by the location of the vernal equinox. (Chapter 2)

sidereal year The orbital period of Earth about the Sun with respect to the stars. (Chapter 2)

silicon fusion The thermonuclear fusion of silicon to produce heavier elements, especially iron. (Chapter 20)

singularity A place of infinite spacetime curvature; the center of a black hole. (Chapter 21)

slab pull A force on seafloor plates that occurs when a portion (or slab) of the seafloor plate sinks down into the mantle at a subduction zone. Gravity pulls on this higher-density slab more than on the surrounding mantle material. The pull on this slab results in a pull on the entire seafloor plate, which is a main force driving plate tectonics. (Chapter 9)

small-angle formula A relationship between the angular and linear sizes of a distant object. (Chapter 1)

snow line The distance from the Sun at which temperatures are low enough for water vapor to solidify into ice or frost. Formation of this solid material beyond the snow line helped to build the large jovian planets. (Chapter 8)

solar constant The average amount of energy received from the Sun per square meter per second, measured just above Earth's atmosphere. (Chapters 5, 17)

solar corona Hot, faintly glowing gases seen around the Sun during a total solar eclipse; the uppermost regions of the solar atmosphere. (Chapter 3)

solar cycle See *22-year solar cycle.*

solar eclipse An eclipse of the Sun by the Moon; a passage of Earth through the Moon's shadow. (Chapter 3)

solar flare A sudden, temporary outburst of light from an extended region of the solar surface. (Chapter 16)

solar nebula The cloud of gas and dust from which the Sun and solar system formed. (Chapter 8)

solar neutrino A neutrino emitted from the core of the Sun. (Chapter 16)

solar neutrino problem The discrepancy between the predicted and observed numbers of solar neutrinos. (Chapter 16)

solar system The Sun, planets, and their satellites, asteroids, comets, and related objects that orbit the Sun. (Chapter 1)

solar wind An outward flow of particles (mostly electrons and protons) from the Sun. (Chapters 9, 16)

south celestial pole The point directly above Earth's south pole where Earth's axis of rotation, if extended, would intersect the celestial sphere. (Chapter 2)

southern highlands (on Mars) Older, cratered terrain in the Martian southern hemisphere. (Chapter 11)

southern lights See *aurora australis.*

space velocity The speed and direction in which a star moves through space. (Chapter 17)

space weather Variations in the solar wind and magnetic field, which can affect satellites and astronauts. (Chapter 16)

spacetime A four-dimensional combination of time and the three dimensions of space. (Chapter 21)

special theory of relativity A description of mechanics and electromagnetic theory formulated by Albert Einstein, which explains that measurements of distance, time, and mass are affected by the observer's motion. (Chapter 21)

spectral analysis The identification of chemical substances from the patterns of lines in their spectra. (Chapter 5)

spectral class A classification of stars according to the appearance of their spectra. (Chapter 17)

spectral line In a spectrum, an absorption or emission feature that is at a particular wavelength. (Chapter 5)

spectral type A subdivision of a spectral class. (Chapter 17)

spectrograph An instrument for photographing a spectrum. (Chapter 6)

spectroscopic binary A binary star system whose binary nature is deduced from the periodic Doppler shifting of lines in its spectrum. (Chapter 17)

spectroscopic parallax The distance to a star derived by comparing its apparent brightness to a luminosity inferred from the star's spectrum. (Chapter 17)

spectroscopy The study of spectra and spectral lines. (Chapters 5, 6, 7)

spectrum (*plural* spectra) The result of dispersing a beam of electromagnetic radiation so that components with different wavelengths are separated in space. (Chapter 5)

spectrum binary A binary star whose binary nature is deduced from the presence of two sets of incongruous spectral lines. (Chapter 17)

speed Distance traveled divided by the time elapsed to cover that distance. (Chapter 4)

spherical aberration The distortion of an image formed by a telescope due to differing focal lengths of the optical system. (Chapter 6)

spherical space Space with positive curvature. (Chapter 25)

spicule A narrow jet of rising gas in the solar chromosphere. (Chapter 16)

spin (of a particle) The intrinsic angular momentum possessed by certain particles. (Chapter 22)

spin-flip transition A transition in the ground state of the hydrogen atom that occurs when the orientation of the electron's spin changes. (Chapter 22)

spin-orbit coupling See *1-to-1 spin-orbit coupling* and *3-to-2 spin-orbit coupling.*

spiral arms Lanes of interstellar gas, dust, and young stars that wind outward in a plane from the central regions of a galaxy. (Chapter 22)

spiral galaxy A flattened, rotating galaxy with pinwheel-like spiral arms winding outward from the galaxy's nucleus. (Chapter 23)

spontaneous symmetry breaking A process by which certain symmetries in the mathematics of particle physics are suddenly altered to produce new particles and forces. (Chapter 26)

spring tide An ocean tide that occurs at new moon and full moon phases. (Chapter 4)

stable Lagrange points (of Jupiter) Locations along Jupiter's orbit where the combined gravitational effects of the Sun and Jupiter cause asteroids to collect. (Chapter 15)

standard candle An astronomical object of known intrinsic brightness that can be used to determine the distances to other galaxies. (Chapter 23)

Standard Model The detailed theory, verified by experiments, that explains all of the known forces between all of the known particles (not including the force of gravity). Phenomena such as dark matter and dark energy may or may not be related to the Standard Model. (Chapter 26)

star cluster See *cluster (of stars).*

starburst galaxy A galaxy that is experiencing an exceptionally high rate of star formation. (Chapter 23)

stationary absorption line An absorption line in the spectrum of a binary star that does not show the same Doppler shift as other lines, indicating that it originates in the interstellar medium. (Chapter 18)

Stefan–Boltzmann law A relationship between the temperature of a blackbody and the rate at which it radiates energy. (Chapter 5)

stellar association A loose grouping of young stars. (Chapter 18)

stellar evolution The changes in size, luminosity, temperature, and so forth that occur as a star ages. (Chapter 18)

stellar parallax The apparent displacement of a star due to Earth's motion around the Sun. (Chapter 17)

stellar-mass black hole A black hole with a mass comparable to that of a star. (Chapter 21)

stony iron meteorite A meteorite composed of both rock and iron. (Chapter 15)

stony meteorite A meteorite composed of rock. (Chapter 15)

stratosphere A layer in Earth's atmosphere directly above the troposphere. (Chapter 9)

string theory The currently favored theory for combining all the known forces in nature. (Chapter 26)

strong force The force that binds protons and neutrons together in nuclei. (Chapter 26)

subduction zone A location where colliding tectonic plates cause Earth's crust to be pulled down into the mantle. (Chapter 9)

subtend To extend over an angle. (Chapter 1)

summer solstice The point on the ecliptic where the Sun is farthest north of the celestial equator. Also used to refer to the date on which the Sun passes through this point. (Chapter 2)

sunspot A temporary cool region in the solar photosphere. (Chapter 16)

sunspot cycle The semiregular 11-year period with which the number of sunspots fluctuates. (Chapter 16)

sunspot maximum/minimum That time during the sunspot cycle when the number of sunspots is highest/lowest. (Chapter 16)

supercluster A collection of clusters of galaxies. (Chapter 23)

supergiant A very large, extremely luminous star of luminosity class I. (Chapters 17, 20)

supergrand unified theory A complete description of all forces and particles, as well as the structure of space and time; a *theory of everything* (TOE). (Chapter 26)

supergranule A large convective feature in the solar atmosphere, usually outlined by spicules. (Chapter 16)

superior conjunction The configuration of a planet being behind the Sun as viewed from Earth. (Chapter 4)

superior planet A planet that is more distant from the Sun than is Earth. (Chapter 4)

superluminal motion Motion that appears to involve speeds greater than the speed of light. (Chapter 24)

supermassive black hole A black hole with a mass of a million or more Suns. (Chapters 21, 24)

supernova (*plural* **supernovae**) A stellar outburst during which a star suddenly increases its brightness roughly a millionfold. (Chapters 1, 15, 20)

supernova remnant The gases ejected by a supernova. (Chapters 18, 20)

supersonic Faster than the speed of sound. (Chapter 18)

surface wave A type of seismic wave that travels over Earth's surface, rather than through its interior. (Chapter 9)

synchronous rotation The rotation of a body with a period equal to its orbital period; also called *1-to-1 spin-orbit coupling.* (Chapters 3, 10)

synodic month The period of revolution of the Moon with respect to the Sun; the length of one cycle of lunar phases. Also called the *lunar month.* (Chapter 3)

synodic period The interval between successive occurrences of the same configuration of a planet. (Chapter 4)

T Tauri stars Young variable stars associated with interstellar matter that show erratic changes in luminosity. (Chapter 18)

tail (of a comet) Gas and dust particles from a comet's nucleus that have been swept away from the comet's head by the radiation pressure of sunlight and the solar wind. (Chapter 15)

tangential velocity The portion of an object's velocity perpendicular to the line of sight. (Chapter 17)

temperature See *degree Celsius, degree Fahrenheit,* and *kelvin.*

terminator The line dividing day and night on the surface of the Moon or a planet; the line of sunset or sunrise. (Chapter 10)

terrae Cratered lunar highlands. (Chapter 10)

terrestrial planet A high-density world with a solid surface, including Mercury, Venus, Earth, and Mars. (Chapter 7)

theory A hypothesis that has withstood experimental or observational tests. (Chapter 1)

theory of everything (TOE) See *supergrand unified theory.*

thermal energy An energy that arises from the kinetic energy of the individual atoms and molecules in a substance. The hotter an object, the greater its thermal energy. (Chapter 5)

thermal equilibrium A balance between the input and outflow of heat in a system. (Chapters 16, 26)

thermal pulse A brief burst in energy output from the helium-fusing shell of an aging low-mass star. (Chapter 20)

thermonuclear Describes nuclear reactions that fuse atoms together as a result of very high temperatures. (Chapter 20)

thermonuclear fusion The combining of nuclei under conditions of high temperature in a process that releases substantial energy. (Chapter 16)

thermonuclear supernova A supernova that occurs when a white dwarf in a close binary system accretes so much matter from its companion star that the white dwarf explodes in a blast of thermonuclear fusion; a Type Ia supernova. (Chapter 20)

thermosphere A region in Earth's atmosphere between the mesosphere and the exosphere. (Chapter 9)

third quarter moon The phase of the Moon that occurs when the Moon is 90° west of the Sun. (Chapter 3)

3-to-2 spin-orbit coupling The rotation of Mercury, which makes three complete rotations on its axis for every two complete orbits around the Sun. (Chapter 11)

threshold temperature The temperature above which photons spontaneously produce particles and antiparticles of a particular type. (Chapter 26)

tidal force A gravitational force whose strength and/or direction varies over a body and thus tends to deform the body. (Chapters 4, 12)

tidal heating The heating of the interior of a satellite by continually varying tidal stresses. (Chapter 13)

time dilation The slowing of time due to relativistic motion. (Chapter 21)

time zone A region on Earth where, by agreement, all clocks have the same time. (Chapter 2)

TNO See *Trans-Neptunian object (TNO)*.

TOE See *supergrand unified theory*.

total lunar eclipse A lunar eclipse during which the Moon is completely immersed in Earth's umbra. (Chapter 3)

total solar eclipse A solar eclipse during which the Sun is completely hidden by the Moon. (Chapter 3)

totality (of lunar eclipse) The period during a total lunar eclipse when the Moon is entirely within Earth's umbra. (Chapter 3)

totality (of solar eclipse) The period during a total solar eclipse when the disk of the Sun is completely hidden. (Chapter 3)

transient lunar phenomena A flash of light that can occur when meteorites as small as 10 cm hit the lunar surface. Enough heat is generated in the impact to produce light visible from Earth by the naked eye. (Chapter 10)

transit An event in which an astronomical body moves in front of another. See also *meridian transit*. (Chapter 8)

transit method A method for detecting extrasolar planets that come between us and their parent star, dimming the star's light. (Chapter 8)

Trans-Neptunian object (TNO) Any small body of rock and ice that orbits the Sun within the solar system, but beyond the orbit of Neptune. (Chapters 7, 14)

triple alpha process A sequence of two thermonuclear reactions in which three helium nuclei combine to form one carbon nucleus. (Chapter 19)

Trojan asteroid One of several asteroids that share Jupiter's orbit about the Sun. (Chapter 15)

Tropic of Cancer A circle of latitude 23½° north of Earth's equator. (Chapter 2)

Tropic of Capricorn A circle of latitude 23½° south of Earth's equator. (Chapter 2)

tropical year The period of revolution of Earth about the Sun with respect to the vernal equinox. (Chapter 2)

troposphere The lowest level in Earth's atmosphere. (Chapter 9)

Tully–Fisher relation A correlation between the width of the 21-cm line of a spiral galaxy and the total luminosity of that galaxy. (Chapter 23)

tuning fork diagram A diagram that summarizes Edwin Hubble's classification scheme for spiral, barred spiral, and elliptical galaxies. (Chapter 23)

turnoff point The point on an H-R diagram where the stars in a cluster are leaving the main sequence. (Chapter 19)

21-cm radio emission Radio radiation emitted by neutral hydrogen atoms in interstellar space. (Chapter 22)

22-year solar cycle The semiregular 22-year interval between successive appearances of sunspots at the same latitude and with the same magnetic polarity. (Chapter 16)

Type I supernova A supernova whose spectrum lacks hydrogen lines. Type I supernovae are further classified as Type Ia, Ib, or Ic. (Chapter 20)

Type Ia supernova A supernova whose spectrum lacks hydrogen lines but has a strong absorption line of ionized silicon. (Chapter 20)

Type Ib supernova A supernova whose spectrum lacks hydrogen and silicon lines but has a strong helium absorption line. (Chapter 20)

Type Ic supernova A supernova whose spectrum is almost devoid of emission or absorption lines. (Chapter 20)

Type II supernova A supernova with hydrogen emission lines in its spectrum, caused by the explosion of a massive star. (Chapter 20)

UBV photometry A system for determining the surface temperature of a star by measuring the star's brightness in the ultraviolet (U), blue (B), and visible (V) spectral regions. (Chapter 17)

ultraviolet radiation Electromagnetic radiation of wavelengths shorter than those of visible light but longer than those of x rays. (Chapter 5)

umbra (of a shadow) (*plural* **umbrae**) The central, completely dark portion of a shadow. (Chapter 3)

undifferentiated asteroid An asteroid within which chemical differentiation did not occur. (Chapter 15)

unified model Describes a physical system at the core of active galactic nuclei (AGN) that appears very different, depending on the line of sight from which it is viewed from Earth. The basic model involves a bright accretion disk that may or may not be obscured by a thick dusty torus. (Chapter 24)

universal constant of gravitation (G) The constant of proportionality in Newton's law of gravitation. (Chapter 4)

upper meridian The half of the meridian that lies above the horizon. (Chapter 2)

Van Allen belts Two doughnut-shaped regions around Earth where many charged particles (protons and electrons) are trapped by Earth's magnetic field. (Chapter 9)

velocity The speed and direction of an object's motion. (Chapter 4)

vernal equinox The point on the ecliptic where the Sun crosses the celestial equator from south to north. Also used to refer to the date on which the Sun passes through this intersection. (Chapter 2)

very-long-baseline interferometry (VLBI) A method of connecting widely separated radio telescopes to make very high-resolution observations. (Chapter 6)

virtual pair (also **virtual particle pair**) A particle and antiparticle that exist for such a brief interval that they cannot be observed. (Chapters 21, 25, 26)

virtual particle pair See *virtual pair.*

visible light Electromagnetic radiation detectable by the human eye. (Chapter 5)

visual binary A binary star in which the two components can be resolved through a telescope. (Chapter 17)

VLBI See *very-long-baseline interferometry.*

void A large volume of space, typically 30 to 120 Mpc (100 to 400 million ly) in diameter, that contains very few galaxies. (Chapter 23)

volatile element An element with low melting and boiling points. (Chapters 10, 11)

waning crescent moon The phase of the Moon that occurs between third quarter and new moon. (Chapter 3)

waning gibbous moon The phase of the Moon that occurs between full moon and third quarter. (Chapter 3)

water hole A range of frequencies in the microwave spectrum suitable for interstellar radio communication. (Chapter 27)

watt A unit of power equal to 1 joule of energy per second. (Chapter 5)

wavelength The distance between two successive wave crests. (Chapter 5)

wavelength of maximum emission The wavelength at which a heated object emits the greatest intensity of radiation. (Chapter 5)

waxing crescent moon The phase of the Moon that occurs between new moon and first quarter. (Chapter 3)

waxing gibbous moon The phase of the Moon that occurs between first quarter and full moon. (Chapter 3)

weak force The short-range force that is responsible for transforming certain particles into other particles, such as the decay of a neutron into a proton. (Chapter 26)

weakly interacting massive particle (WIMP) A hypothetical massive particle that may make up part of the unseen dark matter. (Chapter 22)

weight The force with which gravity acts on a body. (Chapter 4)

white dwarf A low-mass star that has exhausted all its thermonuclear fuel and contracted to a size roughly equal to that of Earth. (Chapters 17, 20)

white ovals Round, whitish features usually seen in Jupiter's southern hemisphere. (Chapter 12)

Widmanstätten patterns Crystalline structure seen in certain types of meteorites. (Chapter 15)

Wien's law A relationship between the temperature of a blackbody and the wavelength at which it emits the greatest intensity of radiation. (Chapter 5)

WIMP See *weakly interacting massive particle.*

winding dilemma The problem that the spiral arms of a galaxy like the Milky Way do not disappear. (Chapter 22)

winter solstice The point on the ecliptic where the Sun reaches its greatest distance south of the celestial equator. Also used to refer to the date on which the Sun passes through this point. (Chapter 2)

x-ray burster A nonperiodic x-ray source that emits powerful bursts of x rays. (Chapter 21)

x rays Electromagnetic radiation whose wavelength is between that of ultraviolet light and gamma rays. (Chapter 5)

ZAMS See *zero-age main sequence.*

Zeeman effect A splitting or broadening of spectral lines due to a magnetic field. (Chapter 16)

zenith The point on the celestial sphere directly overhead an observer. (Chapter 2)

zero curvature The curvature of a surface or space in which parallel lines remain parallel and the sum of the angles of a triangle is exactly 180°. (Chapter 25)

zero-age main sequence The main sequence of young stars that have just begun to burn hydrogen at their cores. (Chapter 19)

zero-age main-sequence star A newly formed star that has just arrived on the main sequence. (Chapter 19)

zodiac A band of 12 constellations around the sky centered on the ecliptic. (Chapter 2)

zonal winds Eastward and westward winds, such as those found in the atmospheres of Jupiter and Saturn. (Chapter 12)

zone A light-colored band in Jupiter's atmosphere. (Chapter 12)

Answers to Selected Questions

Chapter 1

16. (a) 1.5 m (b) 89 m (c) 5.4×10^3 m **17.** 6.9 m **18.** 3.4×10^3 km
19. 3.7 km **20.** 0320 arcmin **25.** 8.5×10^3 km **26.** 2.8×10^7 Suns
27. about 3×10^{36} times larger **28.** 8.94×10^{56} hydrogen atoms
29. (a) 1.581×10^{-5} ly (b) 4.848×10^{-6} pc **30.** 4.99×10^2 s
31. 4.3×10^9 km **32.** (a) 1.59×10^{14} km (b) 16.8 years
33. 4.32×10^{17} s

Chapter 2

8. around 8:02 P.M. **13.** (a) about 9 hours **42.** 50° **43.** 3:50 A.M. local time **55.** (a) 6:00 P.M. (b) September 21 **60.** October 25, 1917

Chapter 3

10. (a) full (b) last quarter (c) new (d) first quarter **11.** (a) full at the vernal equinox (b) last quarter at the summer solstice (c) new at the autumnal equinox (d) first quarter at the winter solstice **18.** (a) 0.91 hour (b) 11° **19.** 49 arcsec **21.** about one more sidereal month than synodic month in each calendar year **23.** shorter **31.** west **41.** (a) more common (b) The Moon would not be large enough to cover the Sun's disk.

Chapter 4

14. 87.97 days **23.** semimajor axis = 0.25 au, period = 0.125 year
24. average = 25 au, farthest = 50 au **25.** (a) 4 au (b) 8 years
26. (a) 16.0 au (b) 0.5 au **27.** (a) 1.26 au (b) 268 days **32.** 6 newtons, 4 m/s² **35.** Earth exerts a force of 1.98×10^{20} newtons on the Moon.
36. Forces are approximately the same; Earth's acceleration is 100 times greater. **37.** ¼ as much as on Earth **38.** 62 newtons, 0.13 **51.** 0.5 year
52. (a) 24 hours (b) 43,200 km **55.** (a) 5 years (b) 2.92 au
58. (a) 3.43×10^{-5} newton (b) 3.21×10^{-5} newton (c) 2.2×10^{-6} newton

Chapter 5

2. 7.5 times **3.** 500 s **7.** 0.340 m **8.** 2.61×10^{14} Hz **15.** about 10 μm
19. 135 nm **20.** 3400 K **21.** 9400 nm **22.** 2.9 nm **23.** 3.9×10^{26} W
24. 190 times more **25.** (a) 4890 nm = 4.89 μm (b) 540 times more
26. (a) 5.75×10^8 W/m (b) 10,000 K **30.** 2.43×10^{-3} nm
38. (a) 1005 nm **41.** possible transitions: energy = 1 eV, λ = 1240 nm; energy = 2 eV, λ = 620 nm; energy = 3 eV, λ = 414 nm
44. 8.6×10^4 km/s **45.** coming toward us at 13.0 km/s

Chapter 6

17. 1/25 = 0.04 **23.** (a) 222× (b) 100× (c) 36× (d) 0.75 arcsec
38. 300 km (Hubble Space Telescope, Jupiter's moons); 110 km (human eye, our Moon) **39.** (a) 34 ly (b) 37 km **41.** (a) 5.39×10^{-4} m (c) 0.56 m

Chapter 7

6. mass = 6.4×10^{23} kg, average density = 3900 kg/m³ **11.** (a) 1.0×10^{13} kg (b) 1.2 m/s **12.** (a) 3.3×10^{21} J (b) equivalent to 3.9×10^7 Hiroshima-type weapons **13.** (a) 2.02 km/s (b) 19.4 km/s **14.** 1.2 km/s
15. (a) 618 km/s **22.** 1.76 years **24.** (a) 1000 years (b) 17 days
31. In 100 years, probability is 8.3×10^{-8} (1 chance in 12 million); in 10^6 years, probability is 8.3×10^{-4} (1 chance in 1200).

Chapter 8

11. 0.40 kg after 1.3 billion years; 0.20 kg after 2.6 billion years; 0.10 kg after 3.9 billion years **12.** 2.6 billion years **19.** (a) about 180 au
20. (a) about 600 au (b) about 5×10^{40} m³ (c) about 10^{55} atoms
(d) about 3×10^{14} atoms per m³ **52.** 2.9×10^7 au = 140 pc = 460 ly

Chapter 9

8. (a) 6.8×10^{16} W (b) 1.07×10^{17} W (c) 209 W/m² (d) 246 K = −27°C
15. 4 km **16.** core: 17%; mantle: 82%; crust: 1% **17.** 0.020 (2% of the total mass) **18.** (b) about 15,000 kg/m³

Chapter 10

11. 2.56 s **21.** (a) 4671 km (b) 1707 km below the surface (c) 449 km
22. (a) 5.78×10^{-5} newton (b) 4.15×10^{-5} newton (c) 1.39
23. 10^6 times greater **41.** 130 newtons on the Moon, 780 newtons on Earth

Chapter 11

5. 13.0 arcsec **12.** (a) 3.03 m/s (b) 1.26 nm **22.** Phobos: about 16 arcmin; Deimos: about 2.7 arcmin **27.** For T = 460°C, $\lambda_{max} = 4.0$ μm
39. (a) 3770 km (b) 370 km **40.** Difference between the round-trip times is 2.0×10^{-4} s. **49.** 920 m **65.** (a) radius = 20,400 km, altitude = 17,000 km
69. (a) 0.16 au **71.** 0.615 au

Chapter 12

6. 12.7 km/s **12.** roughly 600 km/h **13.** 127 K **40.** (a) 14.4 hours for inner edge of A ring, 7.9 hours for inner edge of B ring **52.** 8.5×10^{53} hydrogen atoms, 7.1×10^{52} helium atoms **54.** 59.5 km/s **55.** 8300 newtons

Chapter 13

6. 760 km **7.** about 8.1 min **24.** 2.8×10^{11} (280 billion) years **25.** about 4×10^{-11} (one part in 3×10^{10}) **48.** 0.09 arcsec **55.** (a) 65.7 hours (b) 27.6 arcsec **66.** escape speed = 2.6 km/s; mass of molecule = 2.0×10^{-26} kg, corresponding to a molecular weight of 12

Chapter 14

3. Sun–Uranus force = 1.39×10^{11} newtons, Neptune–Uranus force = 2.24×10^{17} newtons; Neptune reduces the Sunward gravitational pull on Uranus by 1.61×10^4, or 0.0161%. 16. (a) 2900 kg/m^3 41. about 7200 km 50. (a) 8.4 hours 53. (a) 1.13×10^{-3} as bright as on Earth (b) 4.10×10^{-4} as bright as on Earth (c) 2.77 times as bright 55. Without the boost, a one-way trip would take 30.5 years. 63. 0.95 arcsec

Chapter 15

12. (a) 3.6×10^{12} kg (b) 0.83 m/s 20. about 4% 21. 3.6×10^7 km = 0.24 au 43. (a) 10^{15} kg (b) 10^{-16} kg/m^3 50. 2.82 au 57. 30 km 58. (a) period = 350 years, lifetime = 3.5×10^4 years (b) period = 1.1×10^4 years, lifetime = 1.1×10^6 years (c) period = 3.5×10^5 years, lifetime = 3.5×10^7 years (d) period = 1.1×10^7 years, lifetime = 1.1×10^9 years

Chapter 16

8. (a) 1.8×10^{-9} J (b) 9.0×10^{16} J (c) 5.4×10^{41} J 9. (a) 4.6×10^{-36} s (b) 2.3×10^{10} s (c) 1.4×10^{15} s = 4.4×10^7 years 10. 0.048 (4.8%) of the Sun's mass will be converted from hydrogen to helium; chemical composition of the Sun (by mass) will be 69% hydrogen, 30% helium. 11. (a) 8.8×10^{29} kg of hydrogen consumed, 6.2×10^{26} kg lost 13. (a) 1.64×10^{-13} J (b) 2.43×10^{-3} nm 14. 1.4×10^{13} kg/s 22. 98,600 kg/m^3 24. 1.9×10^{-7} nm 33. (a) 1700 nm 43. for the umbra, 670 nm; for the penumbra, 580 nm 44. (a) (flux from patch of penumbra)/(flux from patch of photosphere) = 0.55 (b) (flux from patch of penumbra)/(flux from patch of umbra) = 1.8 56. for the photosphere, 500 nm; for the chromosphere, 58 nm; for the corona, 1.9 nm

Chapter 17

4. (a) 9.7 pc (b) 0.10 arcsec 6. 6.54 pc 7. (a) 161 km/s (b) 294 km/s 8. Distance = 105 pc; tangential velocity would have to be about 5200 km/s. 9. 110 km/s 10. (a) +59.4 km/s (c) 486.23 nm 12. about 2500 pc, about 125 times greater volume 22. 37.0 au 23. 6.1 L$_\odot$ 24. 0.38 pc = 7.9×10^4 au 25. (a) +13.8 (b) (luminosity of HIP 72509)/(luminosity of Sun) = 2.4×10^{-4} 26. +17 27. 6300 pc 28. (a) Brightest star has $M = -2.37$; dimmest star has $M = +4.32$. (b) $M = +0.79$ 33. (b) $m_B - m_V = -0.23$ (Bellatrix), +0.68 (Sun), +1.86 (Betelgeuse) 47. 99 R$_\odot$ 49. The radius of star X is 17 times larger than the radius of star Y. 50. Radius increases by a factor of 2, luminosity increases by a factor of 64. 52. $T = 10,000$ K, $R = 3.8$ R$_\odot$ 53. $L = 35$ L$_\odot$, so distance = 14 pc 54. (b) 0.26 R$_\odot$ = 1.8×10^5 km 67. 9700 years 68. (a) 5.0 pc (b) 22.5 au (c) 1.5 M$_\odot$ 69. (a) 40 M$_\odot$

Chapter 18

12. 0.34% 18. 3.4×10^4 atoms per cm^3 28. 250 au = 3.7×10^{10} km 38. 1.3×10^{31} m^3 42. 2.2×10^3 km/s, or 7.5×10^{-3} (0.75%) of the speed of light 46. 100 R$_\odot$ = 7.4×10^8 km = 4.9 au

Chapter 19

7. (a) 4.9×10^7 years (b) 3.8×10^{11} years 34. about 650 pc 35. about 370 pc 42. (a) 1.65×10^6 km 45. 657,000 km 46. (a) 618 km/s (b) 61.8 km/s 47. (a) 12.0 km/s (b) 9.3 km/s 48. 2.3×10^{29} kg, or 0.15 (15%) of the original mass of hydrogen 49. about 1900 K (1600°C, or 2900°F) 50. 3.4×10^7 years

Chapter 20

17. We see the nebula about 5900 to 8200 years after the central star shed its outer layers. 27. 0.11 R$_\odot$ 28. (a) 97 nm 30. (a) 1.8×10^9 kg/m^3

(b) 6.5×10^3 km/s 31. about 4×10^6 kg 40. (a) 1.7×10^{30} kg = 0.84 M$_\odot$ (b) 1.1×10^{12} newtons (c) 1.5×10^8 m/s, or 0.5 (50%) of the speed of light 45. (a) 6.4×10^5 m/s 53. (maximum luminosity of SN 1993J)/(maximum luminosity of SN 1987A) = 4.5 58. 3.2×10^9 km = 22 au 62. (a) 7×10^{-7} b$_\odot$ (b) It would be about 700 times brighter than Venus. 63. 1.3×10^8 pc = 130 Mpc

Chapter 21

27. 0.6 of the speed of light 28. 8.3×10^{-8} s 29. 0.8 of the speed of light 30. (a) 25 years (b) 20 ly as measured by an Earth observer, 12 ly as measured by the astronaut 36. 2.8 M$_\odot$ 37. 5.7×10^8 years 38. (a) 2.01×10^6 km 56. 0.32 year 59. 0.84 m 60. (a) $R_{Sch} = 8.9$ mm, density = 2.0×10^{30} kg/m^3 (b) $R_{Sch} = 3.0$ km, density = 1.8×10^{19} kg/m^3 (c) $R_{Sch} = 3.5 \times 10^9$ km = 24 au, density 13 kg/m^3 61. 7.4×10^{30} kg 62. 2.9×10^{17} kg/m^3 63. 2.7×10^{38} kg = 1.4×10^8 M$_\odot$

Chapter 22

17. 9.5×10^{-25} J; it takes 3.2×10^5 such photons to equal the energy of one H$_\alpha$ photon. 18. (b) 5700 pc 25. A 10% error in radius results in a 10% error in mass; a 10% error in velocity results in a 20% error in mass. 26. (a) 3.1×10^8 years (b) 7.4×10^{11} M$_\odot$ 27. 2.7×10^{11} M$_\odot$ 42. (a) 6.3 years (b) 3.7×10^6 M$_\odot$ 46. (a) 1.2×10^{12} cubic parsecs (b) 1.1×10^8 cubic parsecs (c) Probability = 9.6×10^{-5}; we can expect to see a supernova within 300 pc once every 350,000 years. 47. (a) 8.3×10^9 au (b) 4.0×10^4 pc

Chapter 23

7. (a) 6.9 Mpc (b) 8.3 Mpc (c) 1.4 Mpc 8. 9.5 Mpc 21. (a) 1.1×10^{10} km = 70 au (b) 7.0 Mpc 27. (a) 152 Mpc = 4.96×10^8 ly 28. 54 km/s/Mpc 29. (a) $z = 0.0252$ (b) 106 Mpc 30. (a) 2.85×10^5 km/s, or 0.951 (95.1%) of the speed of light (b) 1.60×10^6 km/s (c) 4030 Mpc = 1.31×10^{10} ly 47. period = 4.4×10^8 years; mass = 3.6×10^{11} M$_\odot$ 55. (a) 1.22×10^{70} atoms (b) 3.6×10^{-6} atom per cm^3 56. 1.2×10^{12} M$_\odot$

Chapter 24

13. (a) 1.2×10^{11} L$_\odot$ 16. (a) 30 years (b) 2014 (c) $^5/_3$ of the speed of light 26. (a) 1.80×10^5 km/s, or 0.600 (60%) of the speed of light (b) 134 hours (c) 970 au 27. 1.5×10^8 M$_\odot$ 28. 2.9×10^9 km = 20 au

Chapter 25

14. (a) $z = 4$ (b) Distances were $^1/_9$ as great and the density of matter was 729 times greater. 21. (a) 20 billion years (b) 13 billion years (c) 9.7 billion years 22. 1.6×10^8 km/s/Mpc 23. 65.0 times denser 27. 13.6 K 33. (a) 4.6×10^{-23} kg/m^3 34. (a) 9.5×10^{-18} kg/m^3 (b) 4.8×10^{-18} kg/m^3 (c) 1.3×10^{-7} kg/m^3 41. (a) 4.7×10^{-27} kg/m^3 (b) 1.9×10^{-26} kg/m^3 51. 1.06×10^{-3} m = 1.06 mm 52. (b) $q_0 = -0.595$ (b) $\Omega_\Lambda = 0.135$ 53. (a) -0.17 (b) $+0.12$ 54. 1.3×10^{-53} m^{-2}

Chapter 26

16. 1.1×10^{-26} J 17. 3.5×10^{-25} s 24. (a) 80.5 GeV (b) 9.34×10^{14} K 38. (a) 1.1×10^4 m = 0.011 ly 39. Length was 0.168 of its present-day value; density was 213 times the present-day value.

Chapter 27

20. (a) 10,000 years (b) 10 million years 25. (a) 1430 MHz 26. (a) 29.8 km/s (b) 0.10 m (c) 9.9×10^{-6} m, or 9.9×10^{-3}% of the unshifted wavelength 28. 200 km 31. (a) 3.7 cm

Index

Page numbers in *italics* indicate figures and captions.

Star Charts

The following set of star charts, one for each month of the year, are from *Griffith Observer* magazine. They are useful in the northern hemisphere only. For a set of star charts suitable for use in the southern hemisphere, see the *Universe* website.

To use these charts, first select the chart that best corresponds to the date and time of your observations. Hold the chart vertically as shown in the illustration below and turn it so that the direction you are facing is shown at the bottom.

THE NIGHT SKY IN JANUARY

Chart time (Local Standard Time):

10 P.M. . . . First of January
9 P.M. . . . Middle of January
8 P.M. . . . Last of January

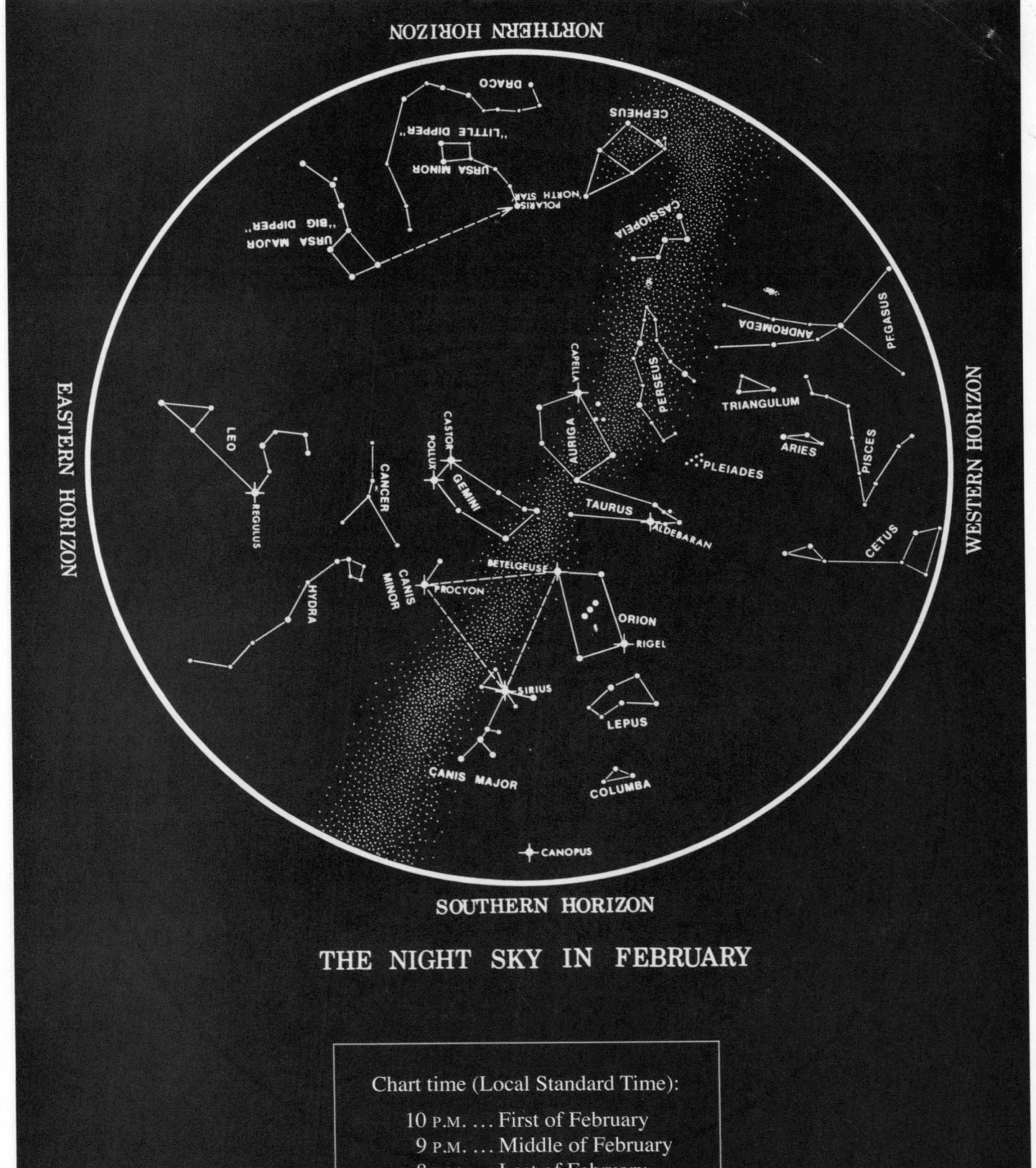

THE NIGHT SKY IN FEBRUARY

Chart time (Local Standard Time):

10 P.M. ... First of February
9 P.M. ... Middle of February
8 P.M. ... Last of February

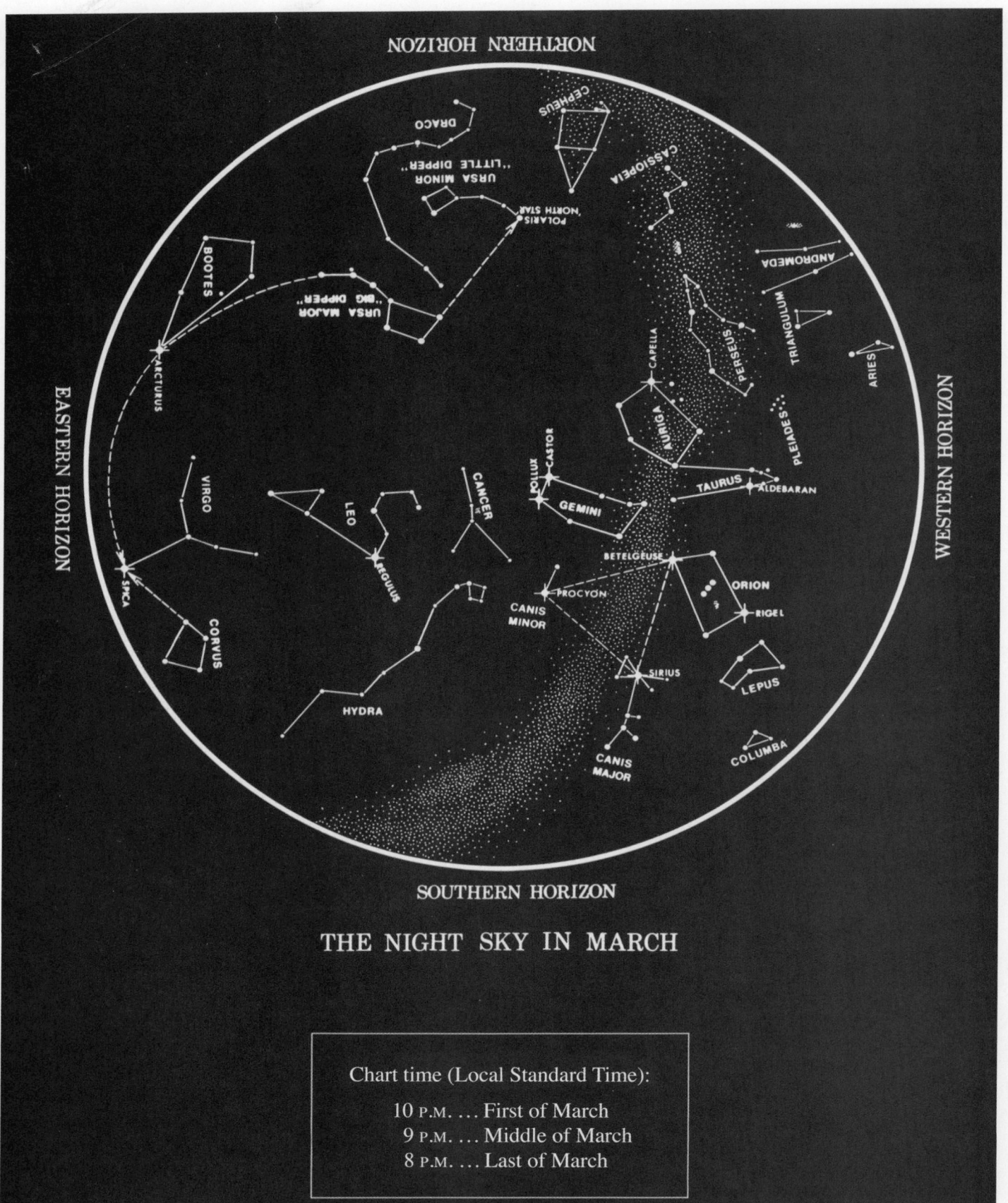

THE NIGHT SKY IN MARCH

Chart time (Local Standard Time):

10 P.M. ... First of March
9 P.M. ... Middle of March
8 P.M. ... Last of March

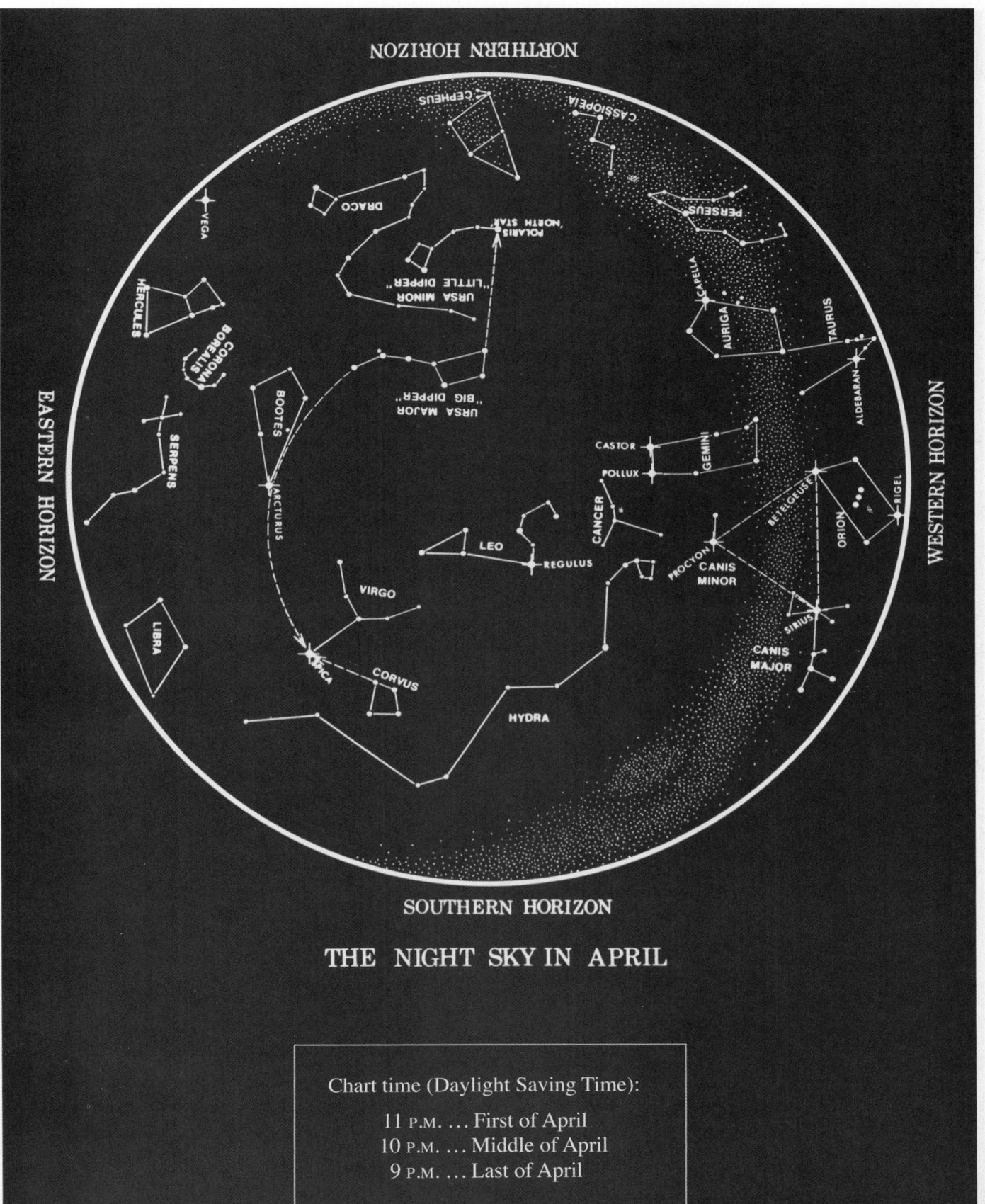

THE NIGHT SKY IN APRIL

Chart time (Daylight Saving Time):

11 P.M. ... First of April
10 P.M. ... Middle of April
9 P.M. ... Last of April

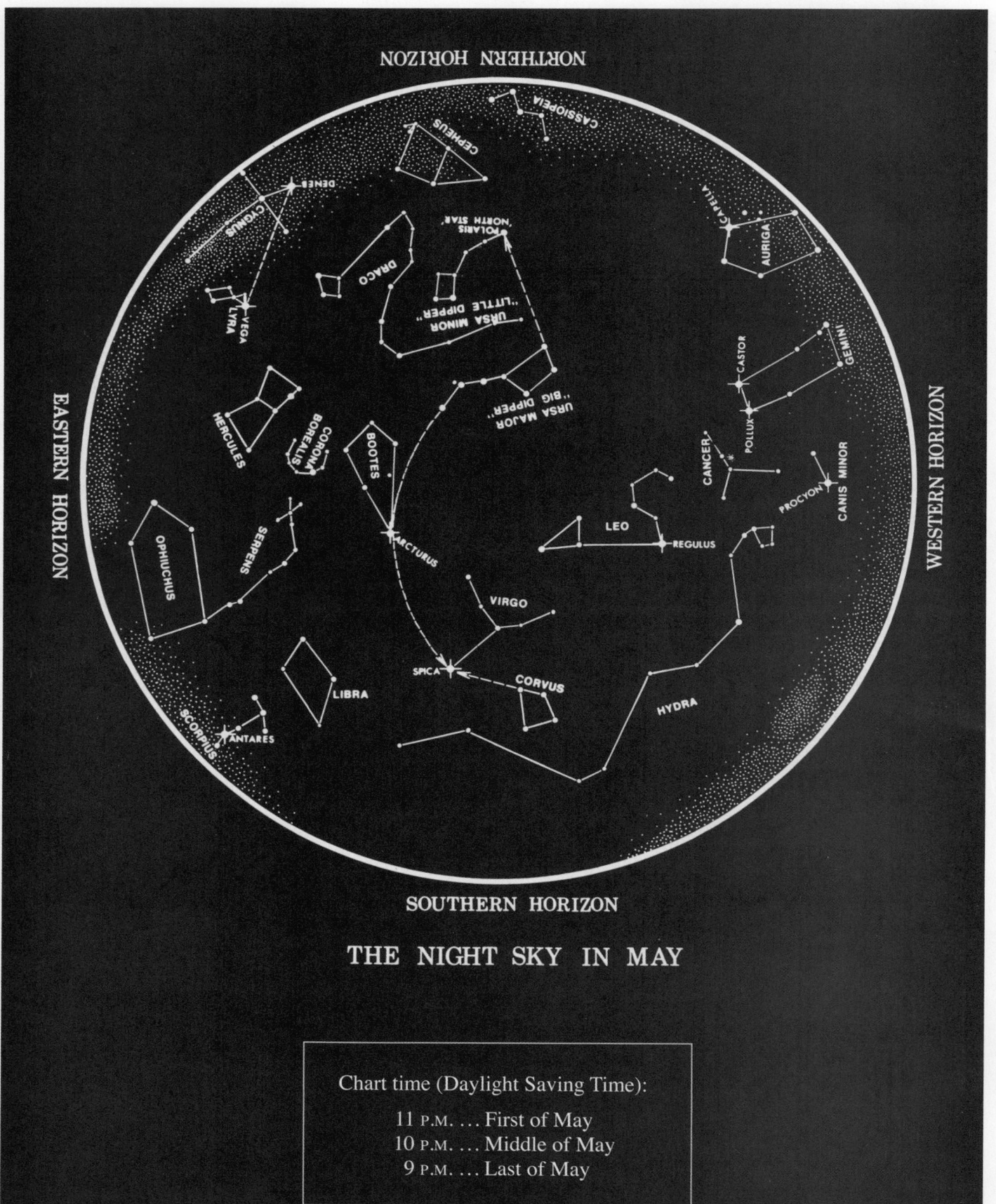

THE NIGHT SKY IN MAY

Chart time (Daylight Saving Time):

11 P.M. ... First of May
10 P.M. ... Middle of May
9 P.M. ... Last of May

THE NIGHT SKY IN JUNE

Chart time (Daylight Saving Time):

11 P.M. ... First of June
10 P.M. ... Middle of June
9 P.M. ... Last of June

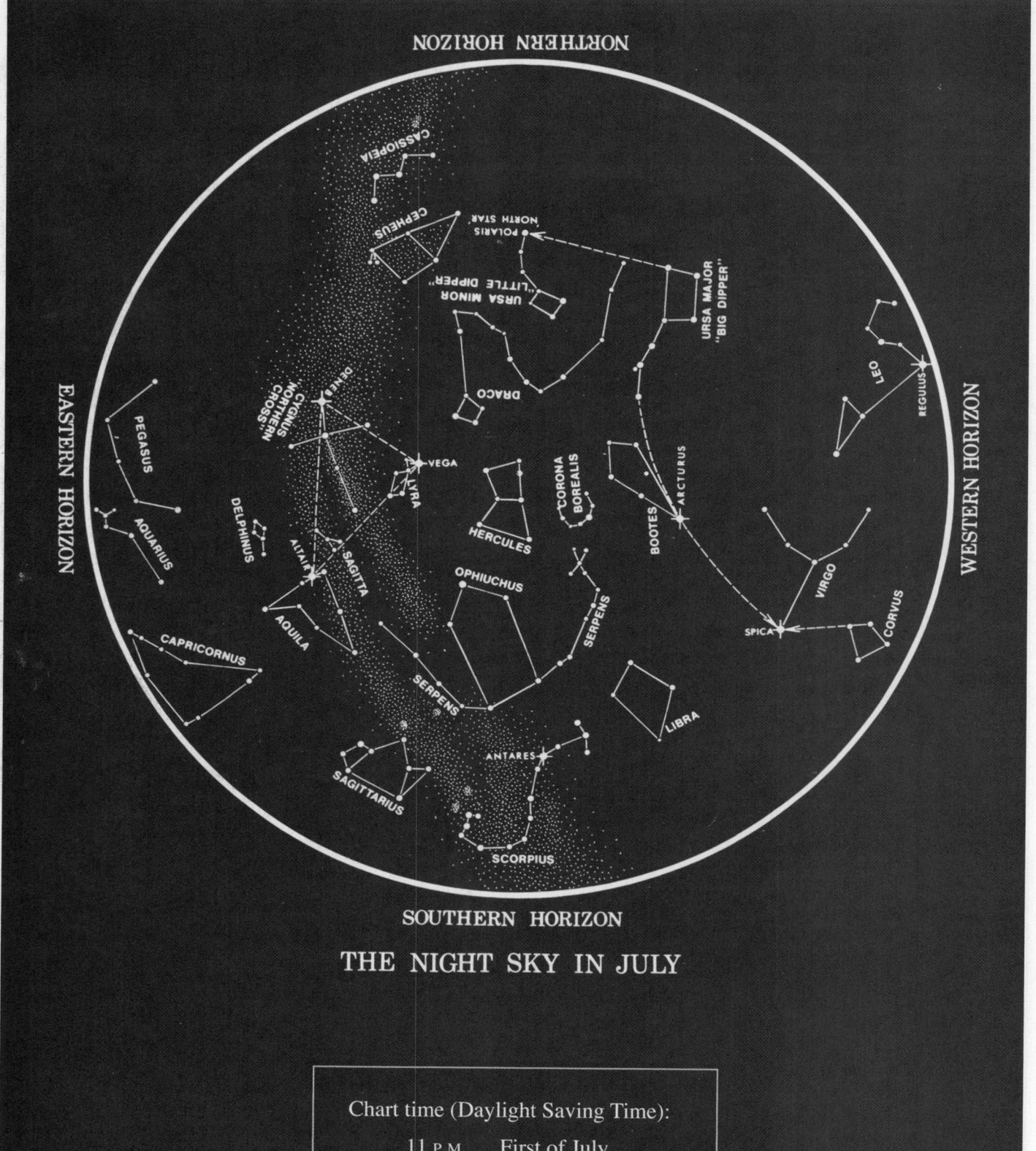

THE NIGHT SKY IN JULY

Chart time (Daylight Saving Time):

11 P.M. ... First of July
10 P.M. ... Middle of July
9 P.M. ... Last of July

SOUTHERN HORIZON

THE NIGHT SKY IN AUGUST

Chart time (Daylight Saving Time):

11 P.M. ... First of August
10 P.M. ... Middle of August
9 P.M. ... Last of August

THE NIGHT SKY IN SEPTEMBER

Chart time (Daylight Saving Time):

11 P.M. ... First of September
10 P.M. ... Middle of September
9 P.M. ... Last of September

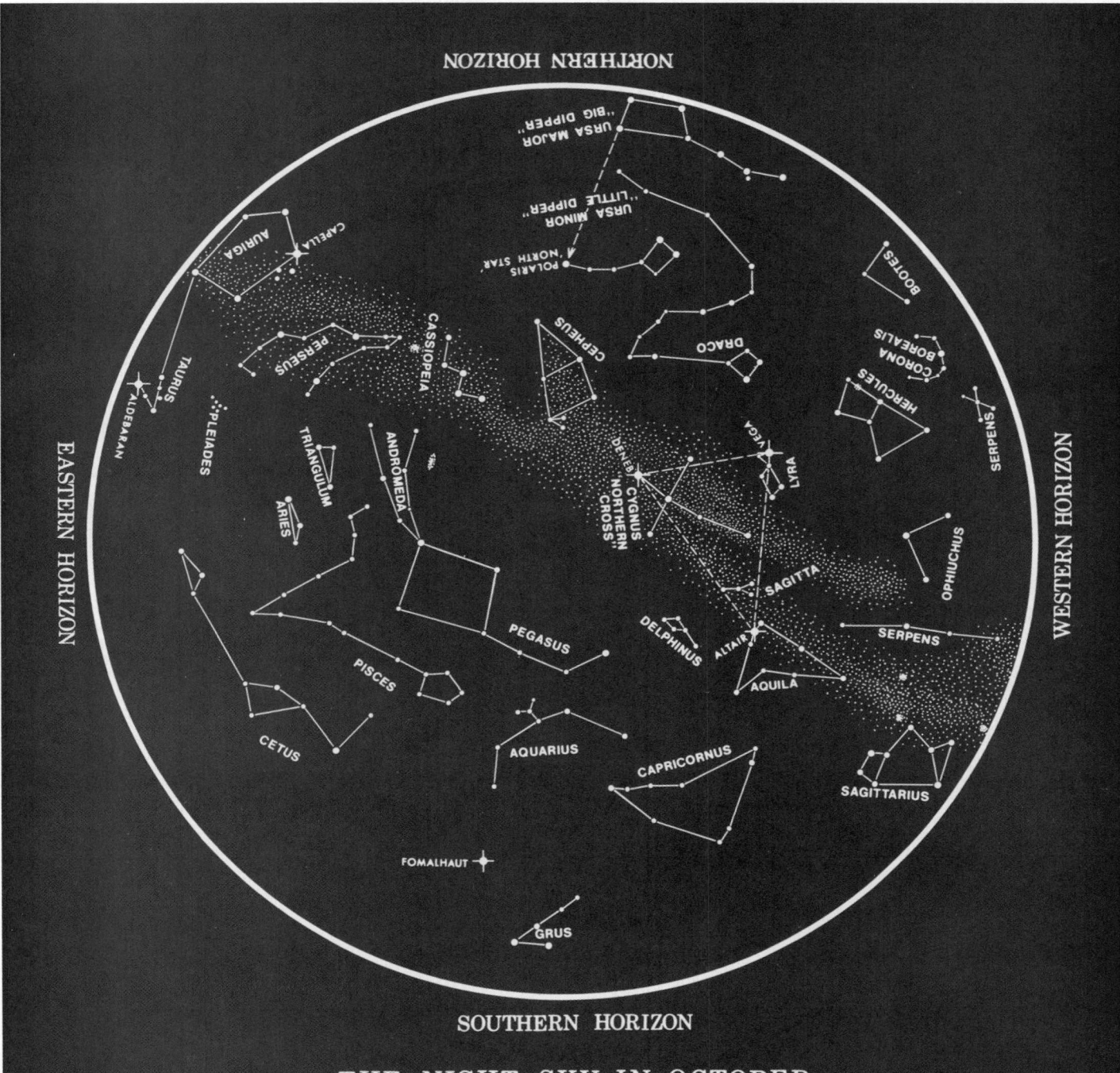

THE NIGHT SKY IN OCTOBER

Chart time (Daylight Saving Time):

11 P.M. ... First of October
10 P.M. ... Middle of October
9 P.M. ... Last of October

THE NIGHT SKY IN NOVEMBER

Chart time (Local Standard Time):

10 P.M. ... First of November
9 P.M. ... Middle of November
8 P.M. ... Last of November

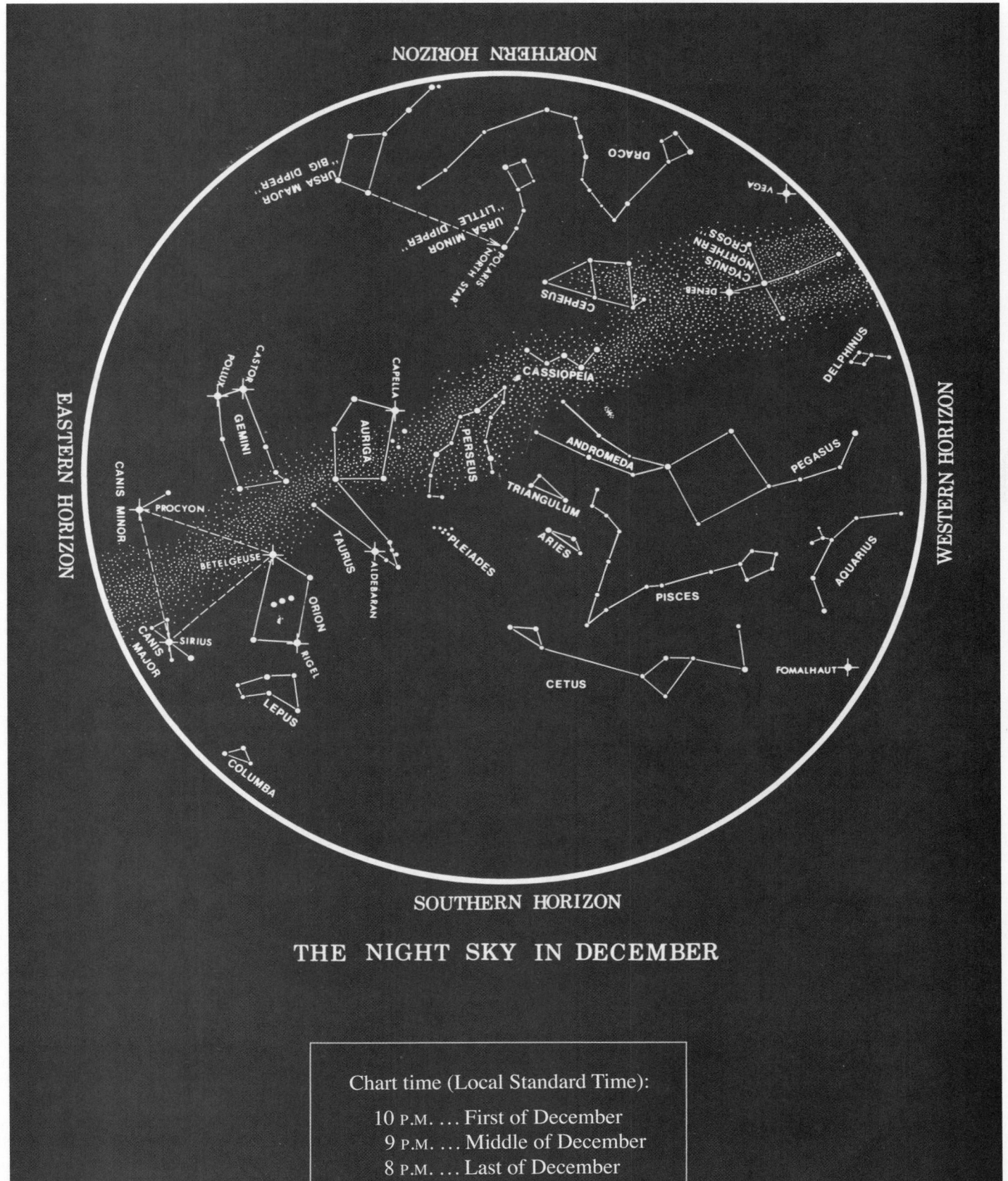

THE NIGHT SKY IN DECEMBER

Chart time (Local Standard Time):

10 P.M. ... First of December
9 P.M. ... Middle of December
8 P.M. ... Last of December